MONOGRAPH OF THE UROSTYLOIDEA (CILIOPHORA, HYPOTRICHA)

MONOGRAPHIAE BIOLOGICAE

VOLUME 85

Series Editor

H. J. Dumont

Aims and Scope

The Monographiae Biologicae provide a forum for top-level, rounded-off monographs dealing with the biogeography of continents or major parts of continents, and the ecology of well individualized ecosystems such as islands, island groups, mountains or mountain chains. Aquatic ecosystems may include marine environments such as coastal ecosystems (mangroves, coral reefs) but also pelagic, abyssal and benthic ecosystems, and freshwater environments such as major river basins, lakes, and groups of lakes. In-depth, state-of-the-art taxonomic treatments of major groups of animals (including protists), plants and fungi are also elegible for publication, as well as studies on the comparative ecology of major biomes. Volumes in the series may include single-author monographs, but also multi-author, edited volumes.

The titles published in this series are listed at the end of this volume.

Monograph of the Urostyloidea (Ciliophora, Hypotricha)

by

HELMUT BERGER
Consulting Engineering Office for Ecology
Salzburg, Austria

and

University of Salzburg
Department of Organismal Biology
Salzburg, Austria

Springer

A C.I.P. Catalogue record for this book is available from the Library of Congress

ISBN-10 1-4020-5272-3 (HB)
ISBN-13 978-1-4020-5272-9 (HB)
ISBN-10 1-4020-5273-1 (e-book)
ISBN-13 978-1-4020-5273-6 (e-book)

Published by Springer,
P.O. Box 17, 3300 AA Dordrecht, The Netherlands.

www.springer.com

Printed on acid-free paper

Caudiholosticha sylvatica, illustration by Berger H. and Foissner W. (1989): Morphology and biometry of some soil hypotrichs (Protozoa, Ciliophora) from Europe and Japan. – Bull. Br. Mus. Nat. Hist. (Zool.) 55: 19-46.

Dedication

For my wife, Elisabeth, and my daughters, Magdalena, Eva, and Helena

Contents

Preface

The present book is a monograph about a relatively large group of hypotrichous ciliates. It is the second of several volumes, which review the Hypotricha, one of the three major parts of the spirotrichs. The first volume treats the Oxytrichidae, also a large group, most species of which have 18 highly characteristically arranged frontal-ventral-transverse cirri and, more importantly, a comparatively complex dorsal ciliature due to fragmentation of dorsal kineties during cell division (Berger 1999).

The present volume treats the Urostyloidea, which are characterised by a zigzag-arrangement of the ventral cirri. Although this pattern is often very impressive, it is a rather simple feature originating by a more or less distinct increase of the number of frontal-ventral-transverse cirral anlagen to produce cirral pairs, which are serially arranged in non-dividing specimens. Some users will be astonished that the monograph does not include *Uroleptus*, a group of tailed species, which also have a distinct zigzagging cirral pattern. However, molecular and morphological data indicate that the zigzag pattern of *Uroleptus* evolved independently, that is, convergently to that of the urostyloids. Thus, it was excluded from the present review.

Urostyloids are common in all major habitats, that is, freshwater, sea, and soil. The last detailed illustrated guide to this group of hypotrichs was provided by Kahl (1932). Of course, Kahl's book – which comprises all hypotrichs and the euplotids – is outdated in many respects, for example, synonymy and faunistics. Moreover, in Kahl's review the urostyloids are not treated as a group because he did not accept the "Urostylinae Bütschli, 1889". Borror & Wicklow (1983) briefly reviewed the urostyloids and provided a valuable introduction, a partly illustrated key to 48 species, and a synonymy, which is, however, not very detailed. Thus it was not too early for a monographic treatment of this group, which comprises 154 species at the present state of knowledge. As in the first volume, almost all available data (morphology, ontogenesis, ecology, faunistics) of each species have been included. For each species, a detailed list of synonyms is provided, followed by a nomenclature section. In the remarks, all important data concerning taxonomy, synonymy, phylogeny, and similar taxa are treated. The morphology section contains a thorough description of the species, following the same sequence in every species. If the data on various populations or synonyms do not agree very well, then the morphology data are kept separate so that even workers who do not agree with the synonymy proposed can use the revision. For several species, cell division data are available. They are also included because ontogenetic data are often important to understand the interphasic cirral pattern correctly. The occurrence and ecology section contains a description of the type locality and all other localities where a species was recorded. In addition, almost all illustrations published so far have been included. Thus, the present book is not only a "field guide" like Kahl's paper, but also a reference book so that the general microscopist need not refer back to the widely scattered original literature. Specialists, however, should always check both the present treatise *and* the original description or authoritative redescription when redescribing a known species.

The most prominent and productive workers on urostyloid taxonomy are, in chronological order, Ehrenberg, Stein, Claparède & Lachmann, Stokes, Kahl, Jerka-Dziadosz, Borror, Foissner, Hemberger, Song, and Hu. However, many others wrote important papers on the alpha-taxonomy of the urostyloid hypotrichs. In total they described about 260 species in urostyloid genera from 1758 to 2005. 154 species are considered as valid in the present revision, 67 are synonyms, that is, the synonymy rate is about 45%, which is very similar to that of the Oxytrichidae (48%). 29 species (= 20%) have one or more synonyms; the record holder is *Holosticha pullaster* with 16 synonyms! 17 species are species indeterminata, two are nomina nuda, and 29 species belong to non-urostyloid taxa. *Anteholosticha*, which is likely not monophyletic, comprises 37 species, two genera comprise 10 species each, and three genera include 9 species each. Thus, these six genera include about 57% of the valid species. 10 genera are monotypic, that is, comprise only the type species.

Recently I started on the next volumes of the series, which treat the remaining groups, for example, the Amphisiellidae and the Kahliellidae. Fortunately, the Austrian Academy of Sciences is sponsoring the last part of the series so that the monographic treatment of the hypotrichs can be completed in a few years. I hope that many ciliate-lovers gain from the series.

Salzburg, May 2006 Helmut Berger

Acknowledgements and Permissions

I wish to express my appreciation to those who helped me prepare this book, in particular, to Wilhelm Foissner (University of Salzburg) for supplying original micrographs, faunistic data, and fruitful discussions; to Martin Schlegel, Stefanie Schmidt, and Detlef Bernhard (University of Leipzig, Germany) for providing molecular biological data; to Erna Aescht (Upper Austrian Museum in Linz), Hannes Augustin (Naturschutzbund, Salzburg), Bruno Ganner (Consulting Engineering Office for Ecology, Salzburg), Xiaozhang Hu and Weibo Song (Ocean University Qingdao, P. R. China), Horst Hemberger (Germany), Maria Jerka-Dziadosz (Polish Academy of Sciences), Wolfgang Petz (University of Salzburg), Ute Seiler (Hydrologische Untersuchungsstelle Salzburg), Tadao Takahashi (Nishikyusyu University, Japan), Norbert Wilbert (University of Bonn, Germany), Ryozo Yagiu (Japan), and Wei-Jen Chang (Princeton University, USA) for supplying information, data, slides, samples, and literature. Thanks to the staff at the Salzburg University Library, for their patient assistance in locating the vast literature on hypotrichs. Many thanks to Eric Strobl (Salzburg) for improving the English – I take full responsibility for any mistakes that remain. I also wish to acknowledge the generosity of the Springer Publisher, especially Tamara Welschot, Senior Publishing Editor Paleo-Environmental Sciences, and the series editor Henri J. Dumont (The State University of Ghent, Belgium) for printing this book. The work was generously supported by a three-year research grant from the Austrian Science Fund FWF (Vienna; Project P14778-B06), based on three independent reviews. Many thanks to the reviewers and to Rudolf Novak, Jörg Ott, and their colleagues from the Science foundation.

I thank my wife, Elisabeth, and my daughters, Magdalena, Eva, and Helena, for their understanding of this time-consuming job.

The figures are either originals or reproductions from the vast literature of the past 220 years. My sincere thanks to the following publishers and organisations who largely freely granted permission to use published drawings and photographs:

Acta Biochimica Polonica, Warszawa (http://www.actabp.pl): *Acta Biochimica Polonica.*

Ilham Alekperov: *An atlas of free-living ciliates* - Publishing House Borcali, Baku.

American Geophysical Union, Washington (http://www.agu.org): *Antarctic Research Series.*

Asociación Latinoamericana de Microbiología, Cuernavaca (http://www.medigraphic.com/espanol/e-htms/e-lamicro/em-mi.htm): *Revista Latinoamericana Microbiología.*

Hungarian Academy of Sciences, Balaton Limnological Research Institute, Tihany (http://tres.blki.hu/BLRI.htm): *Annales Instituti Biologici (Tihany) Hungariae Academiae Scientiarum.*

Bayerisches Landesamt für Wasserwirtschaft, München (http://www.bayern.de/LWF/): *Informationsberichte des Bayerischen Landesamtes für Wasserwirtschaft.*

Blackwell Publishing, Oxford (http://www.blackwellpublishing.com): *Transactions of the American Microscopical Society*.

Cambridge University Press, Cambridge (http://www.cambridge.org): *Biological Reviews*; *Bulletin of the British Museum of Natural History*; *Journal of the Marine Biological Association of the U.K.*

Chinese Academy of Sciences, Beijing (http://zss.ioz.ac.cn): *Acta Zoologica Sinica*; *Acta Zootaxonomica Sinica*.

CNRS Editions, Paris (http://www.cnrseditions.fr): *Annales de Speleologie*; *Protistologica*.

Duncker & Humblot GmbH, Berlin (http://www.duncker-humblot.de): *Zoologische Beiträge*.

Editura Academiei Romane, Bucaresti (http://www.ear.ro): *Studii şi cercetări de biologie, Seria "biologie animală"*.

Elsevier, Amsterdam (http://www.elsevier.com): *Annales des Sciences Naturelles - Zoologie et Biologie Animale*; *Archiv für Protistenkunde*; *European Journal of Protistology*; *Zoologischer Anzeiger*; *Zoologische Jahrbücher Anatomie*; *Zoologische Jahrbücher Systematik*.

Erik Mauch Verlag, Dinkelscherben (http://www.lauterbornia.de): *Lauterbornia*.

Finnish Zoological and Botanical Publishing Board, Helsinki (www.sekj.org/Acta Zool.html): *Acta Zoologica Fennica*.

Instituto de Biología, UNAM, Coyoacán (http://www.ibiologia.unam.mx): *Anales del Instituto de Biología UNAM, Series Botánica y Zoología*; *Cuadernos del Instituto de Biología*.

International Society of Protistologists, Lawrence (http://www.uga.edu/protozoa/index.html): *The Journal of Eukaryotic Microbiology*; *The Journal of Protozoology*.

Les Naturalistes Verviétois asbl, Verviers (http://betula.br.fgov.be/BIODIV/instit.html; Association): *Revue Verviétoise d'Histoire Naturelle*.

Magnolia Press, St. Lukas, Auckland (http://www.mapress.com): *Zootaxa*.

Muzeul National de Istorie Naturală "Grigore Antipa", Bucharest (http://www.antipa.ro/index.html): *Travaux du Muséum National d'Histoire Naturelle "Grigore Antipa"*.

Nencki Institute of Experimental Zoology, Polish Academy of Sciences, Warszawa (www.nencki.gov.pl/ap.htm): *Acta Protozoologica*.

Northern Michigan University Press, Marquette (http://www.nmu.edu/nmupress): *Monographic Series of the Northern Michigan College Press*.

Oberösterreichisches Landesmuseum Biologiezentrum, Linz (http://www.biologie zentrum.at): *Beiträge zur Naturkunde Oberösterreichs*; *Denisia*; *Stapfia*.

Ocean University of China, Qingdao (http://www.ouc.edu.cn): *Journal of the Ocean University of Qingdao*.

Scientific and Technical Research Council of Turkey (http://journals.tubitak.gov.tr/zoology/index.php): *Turkish Journal of Zoology*.

Springer Science and Business Media, Berlin (http://www.springer.com): *Biodiversity and Conservation*; *Hydrobiologia*; *Carey, Marine interstitial ciliates*.

Station Biologique de Roscoff, Roscoff (http://www.sb-roscoff.fr/CBM): *Cahiers de Biologie Marine*.

Taylor & Francis Group Ltd, Oxford (http://www.tandf.co.uk/journals): *Journal of Natural History*; *Sarsia*.

Universität Bonn, Institut für Landwirtschaftliche Zoologie und Bienenkunde, Bonn (http://www.zoobee.uni-bonn.de): *Arbeiten aus dem Institut für landwirtschaftliche Zoologie und Bienenkunde*.

I also ask understanding from those publishers whose permission was not obtained due to my oversight.

Specific acknowledgements are made in the list of synonyms and the figure legends where the authors of the papers and the journals/books, in which the illustrations originally appeared, are named. All sources are cited in the reference section.

A General Section

1 Morphology, Biology, and Terminology

In the following chapters the general external and internal morphology of the Urostyloidea (= urostyloids)[1] and terms specific to this group are described and explained (Fig. 1a–g). For explanation of other terms, see Corliss (1979), Corliss & Lom (1985, 2002), Lynn & Corliss (1991), Hausmann & Hülsmann (1996), and Hausmann et al. (2003). Moreover, some other topics (e.g., parasitism, ecology and distribution) are briefly discussed. For discussion of the ground pattern of the Urostyloidea see the systematic section.

1.1 Size and Shape

The body length of urostyloid ciliates ranges from about 50 µm (e.g., small specimens of *Holosticha pullaster*) to ca. 850 µm (*Urostyla gigas*); the majority is between 100 µm and 300 µm long. The body length:width ratio ranges from about 3:1 or less (e.g., some *Urostyla* species) to about 10:1 in some *Anteholosticha* species, for example, *Anteholosticha fasciola*. Consequently, the body outline of urostyloids is basically either broadly elliptical, elongate elliptical, or almost vermiform. The ventral side is, as in most other hypotrichs, usually flat, the dorsal side more or less distinctly vaulted (Fig. 1d, f). The dorsoventral flattening given in the descriptions is the (usually roughly) estimated ratio of body width to body height (Fig. 1d). For example, a specimen with a body width of 30 µm and a body height of 10 µm is flattened 3:1 dorsoventrally.

Urostyloid hypotrichs are flexible (supple), and almost acontractile to distinctly (up to about 30%) contractile. So far no urostyloid with a rigid body is reliably described. A rigid body/cortex in the Hypotricha is only known from the Stylonychinae (Fig. 14a). Very likely this conspicuous feature evolved convergently in the euplotids and stylonychines (Berger 1999). The adoral zone of membranelles ("oral apparatus") is, as is usual, in the left anterior portion of the cell, and usually less than 40% of body length, in most species around 30%. Hypotricha-species with a longer adoral zone (more than 40%) are either immature postdividers or, if their body is inflexible, stylonychines for which a relative length of 40% or more is characteristic. Moreover, some stylonychines, for example, *Pattersoniella vitiphila* (for review see Berger 1999, p. 766), have a cirral pattern very similar to the midventral pattern of the urostyloids. The biomass of urostyloids ranges from about 12 mg (e.g., *Holosticha pullaster*) to about 8000 mg for the huge *Urostyla gigas* which is nearly 1 mm long.

[1] For names of higher taxa used in the present book see Figs. 13a, 14a and Table 1.

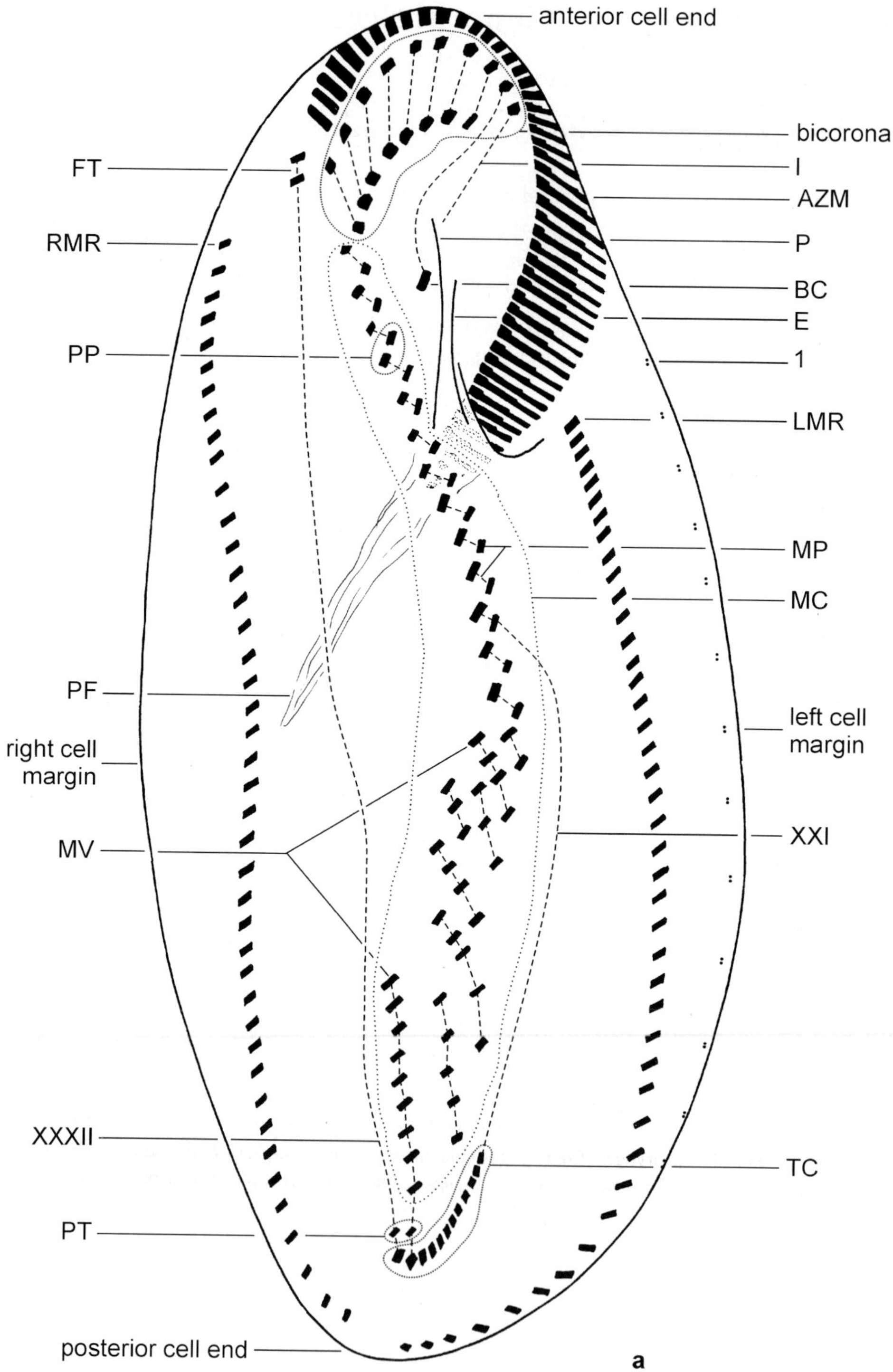
anterior cell end
bicorona
I
AZM
P
BC
E
1
LMR
MP
MC
left cell margin
XXI
TC
FT
RMR
PP
PF
right cell margin
MV
XXXII
PT
posterior cell end
a

1.2 Nuclear Apparatus

Urostyloids have – like most ciliates – a homomerous, polyploid macronucleus (for reviews see Raikov 1969, 1982 and Prescott 1994, 1998). It is composed either of two relatively large nodules (e.g., *Holosticha pullaster*; Fig. 28a); several, more or less scattered nodules (e.g., *Caudiholosticha islandica*; Fig. 48d); more or less moniliformly arranged nodules (e.g., *Anteholosticha monilata*; Fig. 57c); or very many scattered, rather small nodules (e.g., *Urostyla grandis*; Fig. 208h). Species with two macronuclear nodules have either one or more micronuclei attached to each nodule (e.g., *Caudiholosticha stueberi*; Fig. 44e), or a single micronucleus between the two nodules (e.g., *Caudiholosticha navicularum*; Fig. 51a). Fragmentation of the two macronuclear nodules into more than two pieces probably occurred several times independently within the urostyloids. In contrast to other hypotrich taxa, the urostyloids contain a very high number of species with many nodules, whereas the Oxytrichidae are dominated by species with only two macronuclear nodules (for review, see Berger 1999). As in other ciliates, the nuclear apparatus pattern is a very important feature for identification.

Suganuma & Inaba (1966, 1967) and Inaba & Suganuma (1966) studied the fine structure of the macronucleus of *Urostyla grandis*. The chromatin-material in the macronucleus forms a large, irregular network composed of threads of up to 500 nm across. The fundamental components of these threads are pairs of fine fibrils, each 10 nm thick. The nucleoli are about 1–2 µm across. They appear to be composed partly of fine fibrils about 10 µm across, and partly of granular appearing material. The nuclear envelope is double, about 21 nm thick, and there are many discontinuities (50 nm across), which may represent the pores. The micronuclei of *U. grandis* are bounded by a double, porous envelope similar to that of the macronuclear nodules. Chromatic material in the micronucleus forms a small network of comparatively thin threads, 60–80 µm thick. The macronuclear nodules of *Pseudokeronopsis carnea* appear moderately dense and homogeneous with few nucleoli, or with a single, central endosome, whereas other

← **Fig. 1a** Terminology of urostyloid ciliates (from Berger 2004b, supplemented). Infraciliature (after protargol impregnation) of ventral side of a species with a bicorona. Frontal-midventral-transverse cirri which originate from the same anlage are connected by a broken line (for the sake of clarity only the leftmost transverse cirrus, and the two rightmost transverse cirri and pretransverse ventral cirri are connected with the corresponding midventral pair, respectively, midventral rows). AZM = adoral zone of membranelles, BC = buccal cirrus, E = endoral (endoral plus paroral are the undulating membranes), FT = frontoterminal cirri (= migratory cirri), LMR = anterior end of left marginal row, MC = midventral complex (= midventral pairs plus midventral rows), MP = midventral pairs, MV = midventral rows, P = paroral (paroral plus endoral are the undulating membranes), PF = pharyngeal fibres (= cytopharynx), PP = pseudo-pair (composed of rear [= left] cirrus of an anlage and front [= right] cirrus of next anlage, that is, the cirri of a pseudopair do not originate from the same anlage), PT = pretransverse ventral cirri (= cirri ahead the two rightmost transverse cirri; = accessory transverse cirri according to Wicklow 1981), RMR = anterior end of right marginal row, I = first (= leftmost) frontal-(midventral-transverse) cirral anlage (forms always the leftmost frontal cirrus and the undulating membranes), TC = transverse cirri (form a pseudorow), XXI = 21. frontal-midventral-transverse cirral anlage (forms the leftmost [anteriormost] transverse cirrus in this specimen), XXXII = last (= rightmost; = 32. from left, respectively, from the front) frontal-midventral-transverse cirral anlage (number of anlagen varies among species and often within species), 1 = dorsal kinety 1 (= leftmost kinety).

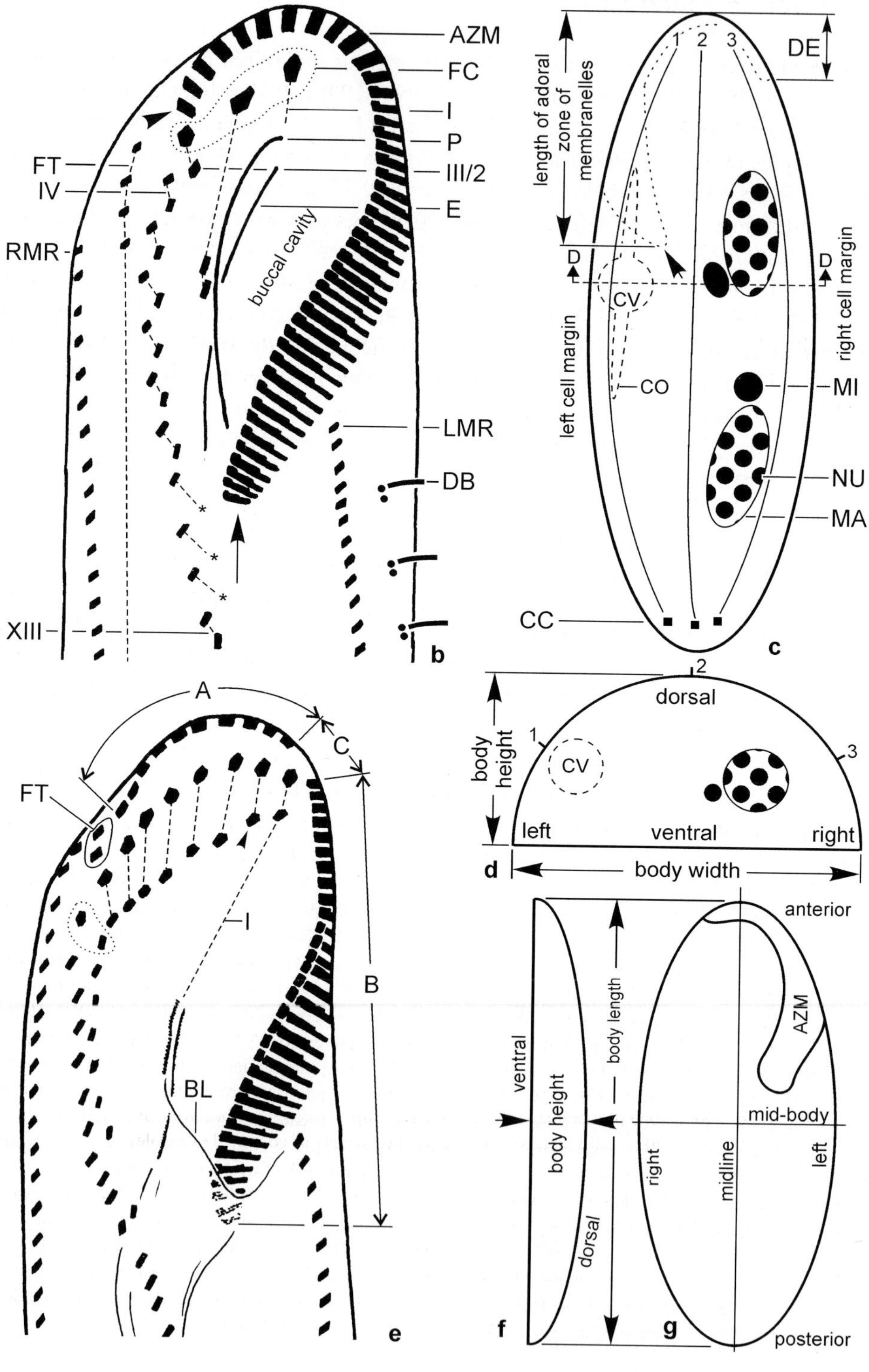
AZM
FC
I
P
III/2
E
FT
IV
RMR
buccal cavity
LMR
DB
XIII
b
length of adoral zone of membranelles
DE
1 2 3
D
D
CV
CO
MI
NU
MA
left cell margin
right cell margin
CC
c
A
C
FT
I
B
BL
e
2
dorsal
body height
1
CV
3
left
ventral
right
d
body width
anterior
AZM
body length
ventral
body height
mid-body
right
midline
left
dorsal
f
g
posterior

nodules may be crammed with condensed chromatin (Wirnsberger & Hausmann 1988b). Micronuclei are spherical and very densely stained.

The macronuclear nodules of urostyloids posses a replication (or reorganisation) band, a feature which evolved in the stem-line of the spirotrichs (Fig. 2a–g). In this band, which is a clear disc that gradually moves through the whole macronuclear nodule, DNA is replicated (for reviews see Raikov 1982 and Prescott 1994). The replication band of *Urostyla grandis* is, like that of other hypotrichs, divided into two zones: the forward zone consists of fine, twisted fibrils 30 nm thick formed by pairs of parallel fibrils, each about 10 nm thick. The rear zone is composed of small chromatin bodies and thin threads, both about 80 nm across, consisting of fine fibrils 10 nm thick. Reorganised chromatin threads appear to be thin spiral threads 80–130 nm thick in striking contrast to the thick (100–500 nm in diameter) threads composing the large, irregular network observed in early interphase macronucleus (Suganuma & Ibana 1966, 1967, Ibana & Suganuma 1966; for review see Olins & Olins 1994, p. 150).

The development of the urostyloid nuclear apparatus during cell division is obviously the same as in the other hypotrichs, that is, the individual macronuclear nodules fuse to a single mass and divide again in the species-specific number. In *Urostyla grandis* the many macronucleus-nodules are small, about the size of micronuclei, and scattered throughout the cell. The fusion of the individual nodules of *U. grandis* into a single macronucleus within a matter of minutes is a rather impressive event (Fig. 3a–l). As cytokinesis begins, the composite macronucleus in various species quickly undergoes one or more rapid, successive amitotic divisions to produce the appropriate number of daughter macronuclei in each filial product (Prescott 1994). Nothing is known about the triggering of macronuclear fusion at the beginning of ontogenesis, or its mo-

---

← **Fig. 1b–g** Terminology of urostyloid ciliates (b–d, from Berger 2004b, supplemented; e–g, originals). Frontal-midventral cirri which originate from the same anlage are connected by a broken line. **b:** Infraciliature (after protargol impregnation) of a species with three frontal cirri. Arrow marks proximal end, arrowhead distal end of adoral zone of membranelles. Asterisks mark anlagen, which eventually produce only a single midventral cirrus, that is, one cirrus (in the present case the left one) of a pair is resorbed in late dividers. **c:** Schematic illustration of infraciliature of dorsal side, nuclear apparatus, and contractile vacuole. **d:** Schematic cross section (about at level D-D of Fig. 1c) showing, inter alia, dorsoventral flattening and contractile vacuole. Arrow marks proximal end of adoral zone of membranelles. **e:** Infraciliature (after protargol impregnation) of a species with a gap in the adoral zone. First midventral pair encircled by dotted line. "Buccal cirrus" marked by arrowhead. **f, g:** Left lateral view and ventral view showing some terms used in the species descriptions. A = distal (= frontal) portion of adoral zone of membranelles, AZM = adoral zone of membranelles, B = proximal (= ventral) portion of adoral zone of membranelles, BL = buccal lip, C = gap in adoral zone of membranelles, CC = caudal cirri (at rear end of dorsal kineties), CO = collecting canal (of contractile vacuole), CV = contractile vacuole, DB = anteriormost dorsal bristle of kinety 1 (= leftmost kinety), DE = distance between anterior body end and distal end of adoral zone of membranelles (for DE-value see chapter 1.8), E = endoral, FC = frontal cirri (left = cirrus I/1; middle = homologous to cirrus II/3 of the 18-cirri oxytrichids; right = homologous to cirrus III/3), FT = frontoterminal cirri, LMR = left marginal row, MA = posterior macronuclear nodule, MI = micronucleus, NU = nucleolus, P = paroral, PC (= III/2) = parabuccal cirrus(i) (= cirrus behind right frontal cirrus), RMR = right marginal row, I, IV, XIII = cirri which originated from the first, fourth, and thirteenth frontal-midventral-transverse cirral anlage, III/2 (= PC) = cirrus behind right frontal cirrus (also designated parabuccal cirrus/cirri), 1–3 = dorsal kineties (kinety 1 is the leftmost kinety).

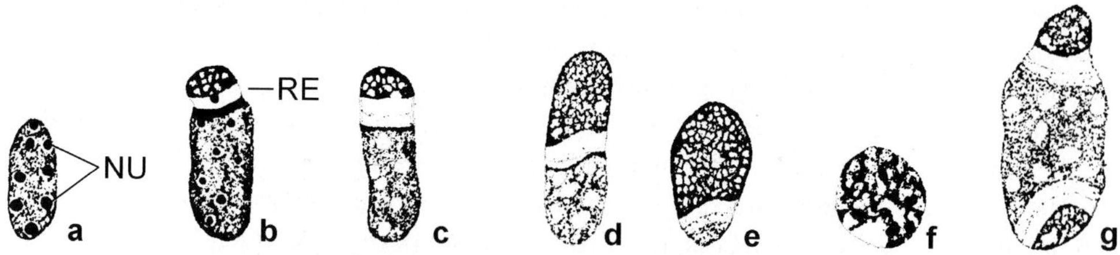

Fig. 2a–g Macronuclear nodules of *Urostyla grandis* during first stages of cell division (from Tittler 1935. a, c, fixed with Schaudinn's fluid; b, fixed with Flemming's fluid; d–g, fixed with Gilson Carnoy's fluid; a, b, stained with Heidenhain's iron haematoxylin; c, Feulgen stain; d–g, Mayer's haemalaun stain) **a:** Interphasic nodule, about 8 μm long (all other nodules drawn to same scale). **b–f:** Passing of replication band. **g:** Macronucleus nodules with two replication bands occur very rarely. NU = nucleoli, RE = replication (reorganisation) band.

lecular mechanism or what controls and accomplishes amiotic division at the end of cell division. Perhaps the cytoskeleton mediates these events (Prescott 1994).

The multiple micronuclei in a single cell are all genetically identical: they are all derived by mitosis from one original micronucleus formed by fertilisation at cell mating. Micronuclei divide mitotically during vegetative growth, but the form of mitosis is different from that of plant and animal cells (for details see Prescott 1994). Mitosis occurs intranuclearly, that is, without breakdown of the nuclear envelope, and individual chromosomes are not distinguishable. Rather, the mitotic micronucleus contains long strands of chromatin that distribute to produce two genetically equivalent daughter micronuclei (Fig. 4a–h). Details of the process are poorly understood (Prescott 1994).

Only in the pseudokeronopsines, which have many macronuclear nodules, does the macronuclear development differ from that in other hypotrichs, a feature reviewed by Raikov (1982, p. 348). For example, the macronuclear anlage of *Pseudokeronopsis rubra* contains paired filamentous chromosomes (possibly in a state of somatic conjugation), but they are not clearly polytenic (Ruthmann 1972). Neither the transverse fragmentation of chromosomes nor the "achromatic" phase in macronuclear development have been found. Ruthmann (1972) has shown by electron microscopy that the chromatin of the macronuclear anlage gradually condenses into compact bodies that are separated from the anlage into the cytoplasm and become small definitive macronuclei, each of which contains a paradiploid amount of DNA. Ruthmann (1972) believed that the chromatin bodies preformed in the anlage are diploid subnuclei that later become individual macronuclei. These numerous macronuclei divide without prior fusion to a single mass. Often, they divide into parts with an unequal DNA content. This means that the genome of the *Pseudokeronopsis* macronuclei fragments into subunits smaller than even the haploid genome (Ruthmann 1972). Consequently, the macronucleus apparatus of *Pseudokeronopsis* has features of both the subnuclear and the chromomeric types. This type of macronuclear division was already discovered by Gruber (1884a).

The just mentioned mode of macronucleus-development also occurs in *Uroleptopsis* (Mihailowitsch & Wilbert 1990, Berger 2004b) and therefore has to be considered as

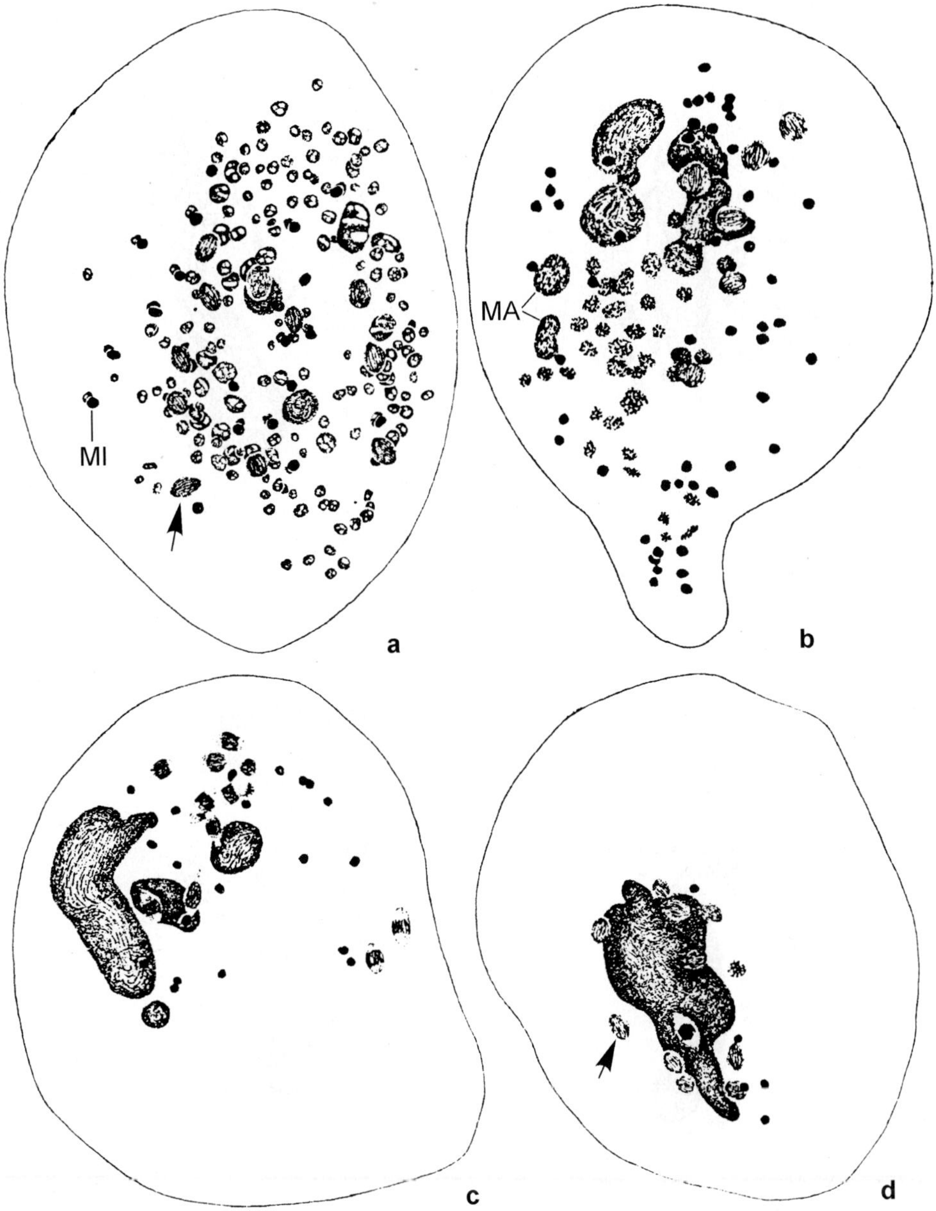

Fig. 3a–d Division of macronucleus in *Urostyla grandis* (from Tittler 1935. a, b, d, fixed with Gilson-Carnoy's fluid; c, fixed with Schaudinn's fluid; a–c, Mayer's haemalaun stain; d, Heidenhain's iron haematoxylin stain). The many individual macronuclear nodules present in specimen (a) fuse to a single mass (d). a = 200 µm long, b = 162 µm; c = 160 µm, d = 152 µm. Arrows in (a, d) mark dividing micronuclei. Following stages, see Fig. 3e–h. For details, see text. MA = non-dividing macronuclear nodules, MI = non-dividing micronucleus.

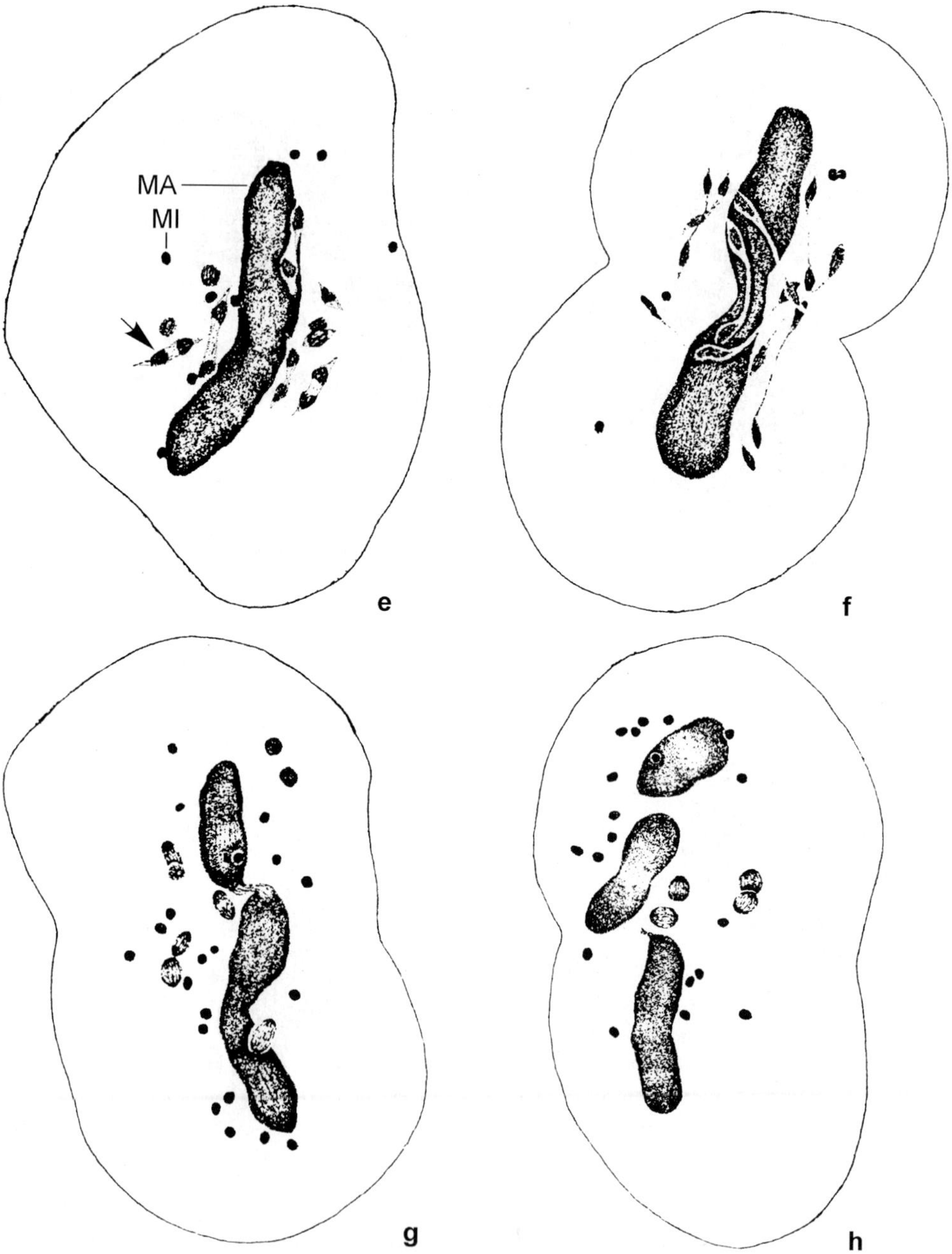

Fig. 3e–h Division of macronucleus in *Urostyla grandis* (from Tittler 1935. a, b, fixed with Schaudinn's fluid; c, fixed with Gilson-Carnoy's fluid; d, fixed with Bouin's fluid; a, c, d, Mayer's haemalaun stain; b, Feulgen stain). The fused macronucleus (e) begins to divide (f–h). Arrow in (e) marks dividing micronucleus. e = 170 µm long, f = 185 µm, g = 180 µm, h = 195 µm. For details, see text. MA = fused macronucleus, MI = non-dividing micronucleus.

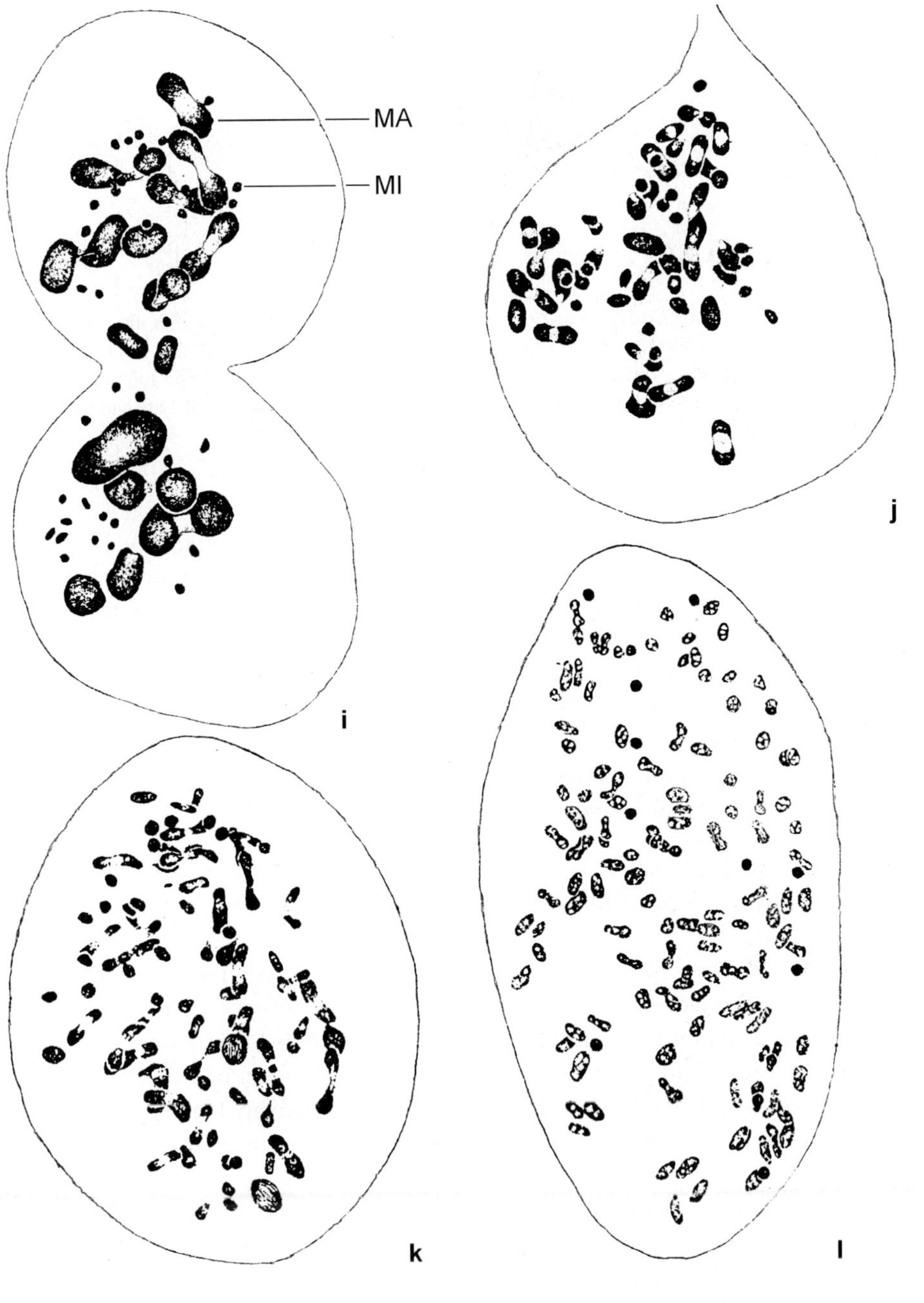

Fig. 3i–l Division of macronucleus in *Urostyla grandis* (from Tittler 1935. i–l, fixed with Schaudinn's fluid; i, Heidenhain's iron haematoxylin stain; j, k, Mayer's haemalaun stain; l, Feulgen stain). **i:** Late divider, 230 μm. **j:** Very late divider (proter not illustrated), 138 μm. **k:** Post-divider, 128 μm. **l:** Specimen with normal nucleus-apparatus, 188 μm. MA = dividing macronucleus, MI = non-dividing micronucleus.

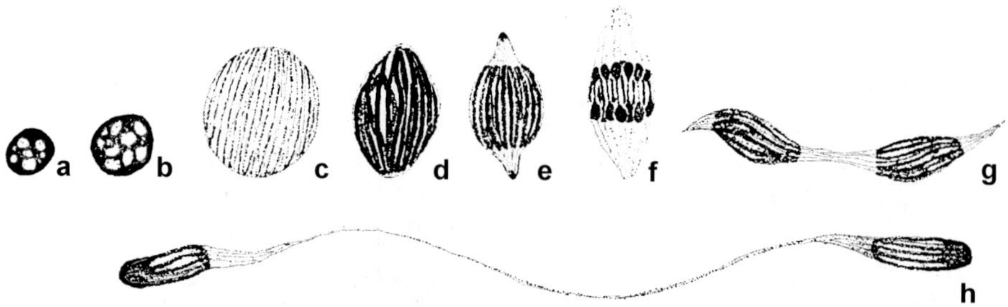

Fig. 4a–h Micronucleus of *Urostyla grandis* during division (from Tittler 1935. a, d–h, fixed with Schaudinn's fluid; b, fixed with Gilson Carnoy's fluid; c, fixed with Bouin's fluid; a, d, f–h, Feulgen stain; b, c, Mayer's haemalaun stain; e, Haindenhain's iron haematoxylin stain). **a:** Interphasic micronucleus, about 3.5 µm across (all other micronuclei drawn to same scale). **b:** Early prophase. **c, d:** Late and very late prophase. **e:** Metaphase. **f:** Early anaphase. **g:** Late anaphase. **h:** Telophase.

autapomorphy of the Pseudokeronopsinae (Fig. 167a, autapomorphy 3). *Thigmokeronopsis antarctica* and *T. crystallis* also have many macronuclear nodules, which do not fuse to a single mass, but to several parts (Petz 1995). This state can be interpreted as a transitional state between the total fusion, for example, in *Urostyla grandis*, and the specific mode described for the Pseudokeronopsinae. Berger (2004b) considered the *Thigmokeronopsis* type of macronuclear division as autapomorphy of the Pseudokeronopsidae (Fig. 167a, autapomorphy 1).

Maula et al. (1993) found prokaryotic endosymbionts in the macronucleus, but not in the micronuclei of *Pseudokeronopsis* sp.

1.3 Contractile Vacuole and Cytopyge

Many urostyloids have, like most other Hypotricha, a single contractile vacuole near the left cell margin about or slightly behind the level of the cytostome (Fig. 1c, d). The very common *Holosticha pullaster* has this organelle distinctly behind mid-body so that it is very easy to identify (Fig. 28f–i). Few species have more than one contractile vacuole (e.g., *Pseudokeronopsis sepetibensis*; Fig. 186a). For a relatively high number of marine species no contractile vacuole is described, possibly because it is lacking. In some marine species a vacuole is present, but contracts in rather long intervals. Very little is known about the excretory pore in the urostyloids. Probably it is, as in other Hypotricha, on the dorsal surface.

The cytopyge of the urostyloids is a little-known organelle which is usually – as in other hypotrichs (Berger 1999) – located in the posterior portion of the cell (Fig. 135b, 181c, 226n).

1.4 Cytoplasm, Cortex, and Colouring

The cytoplasm of many urostyloids is more or less colourless. Some species have, however, a yellow (e.g., *Anteholosticha xanthichroma*, *Uroleptopsis citrina*) or reddish (e.g., *Diaxonella pseudorubra*) coloured cytoplasm. Few species (e.g., *Caudiholosticha viridis*, *Urostyla viridis*) are green due to symbiotic algae. Symbiotic algae must not be confused with ingested algae, which are enclosed in usually distinctly recognisable food vacuoles. The size of the food vacuoles depends mainly on the size of the species and the size of the ingested diet. By contrast, symbiotic green algae usually occur in high numbers, are of same size (usually 4–6 µm across) and morphology, have a distinct membrane, but are not in vacuoles, and have a dark central or acentral globule (Foissner et al. 1999). In many (all?) pseudokeronopsids blood-cell-shaped structures occur underneath the cortex (see next chapter).

In *Pseudokeronopsis carnea* a typical plasma membrane covers flat alveoli that are frequently insignificant or invisible in main parts of the somatic region, but very distinct in the buccal area (Wirnsberger & Hausmann 1988b). Below the somatic pellicle is a single layer of longitudinal subpellicular microtubules. By contrast, the rigid stylonychines have several layers of subpellicular microtubules, which are arranged crosswise in *Stylonychia* (Calvo et al. 1986, Puytorac et al. 1976).

The plasma membrane of some (all?) Hypotricha (e.g., *Urostyla grandis*, *Pseudokeronopsis rubra*, *Pseudourostyla cristata*, *Uroleptus caudatus*, *Paraurostyla weissei*, *Oxytricha fallax*, *Stylonychia mytilus*, *Urosoma* sp.) and oligotrichs (e.g., *Strombidium*) is covered by an additional layer called perilemma (Bardele 1981, Grimes 1972, Laval 1971, Laval-Peuto 1975, Wasik & Mokolajczyk 1992). This outer coating also covers cilia, membranelles, and cirri. The perilemma is lacking in *Halteria* and the euplotids. Bardele (1981) assumed that the perilemma in hypotrichs is a temporary structure which is discarded quite often because numerous layers of the perilemma are usually seen in the buccal cavity. Unfortunately, nothing is known as to how the perilemma is derived or replenished. Lynn & Corliss (1991) supposed that it may be a special kind of fixation artefact of the glycocalyx, that is, the protein and glycoprotein layer of the plasma membrane.

The endoplasm of *Pseudokeronopsis carnea* is characterised by many mitochondria, reserve organelles such as paraglycogen granules and lithosomes, and the nuclear apparatus (Wirnsberger & Hausmann 1988b).

1.5 Cortical Granules

These organelles have various names in the urostyloid literature, for example, Öltröpfchen (= oil droplets; e.g., Stein 1859), protrichocysts (Kahl 1932), Perlen (pearls; e.g., Kahl 1932), subpellicular granules (e.g., Berger & Foissner 1987), or pigmentocysts (Wirnsberger & Hausmann 1988b). Cortical granules occur in many species of the Urostyloidea (e.g., Fig. 208r), but also in many species of the Oxytrichidae (for review see Berger 1999) and other ciliate groups (e.g., Foissner et al. 2002), that is, these organelles are an old feature (homology of these organelles assumed). Their absence in the Stylonychinae is successfully used to characterise this monophylum (Berger & Foissner 1997). Likely most of the granules in the Urostyloidea belong to the mucocyst type. Their colour, size, shape, and arrangement are very important features, which cannot usually be seen after protargol impregnation. Live observation is therefore absolutely necessary for reliable identification of urostyloids (e.g., Stein 1859, Kahl 1932, Berger & Foissner 1987a, Borror & Wicklow 1983).

Wirnsberger & Hausmann (1988b) studied the fine structure of *Pseudokeronopsis carnea*. The striking orange-red colour of this species is caused by two types of pigment structures, the pigment vacuoles and the pigmentocysts. The pigment vacuoles are not extrusive and are confined to a characteristic ectoplasmic zone, about 1.5–3 µm thick, where they form 2–5, but usually three layers. Only a few mitochondria can be found in this area. The pigment vacuoles show a loose, fluffy periphery and a central, more intensely stained part, which is elliptic with a sometimes lamellar appearance. The pigmentocysts are narrowly arranged around the ciliary organelles on the ventral and dorsal side of the cell. A few also occur in the endoplasm and between the ciliary organelles. Under the light microscope, the pigmentocysts appear darker red than the pigment vacuoles. They are globular to oviform and about 0.5–1.0 µm long. A short, electron-dense channel is oriented, to and connected with, the pellicular membranes. However, Wirnsberger & Hausmann (1988b) never observed the discharge of the pigmentocyst content.

Several pseudokeronopsid species have a distinct layer of curious organelles underneath the cell surface (e.g., Fig. 180c, 185l–n, 192e, f, i, j, 193b, c). They have the shape of the erythrocytes of mammals, are colourless and therefore sometimes difficult to recognise although about 2.0 µm across. However, these structures are also described for some non-pseudokeronopsid species, for example, *Anteholosticha warreni*. Their function is unknown, ultrastructure data are lacking.

1.6 Movement

There exist only very few detailed studies about the movement of urostyloids. Most urostyloids are, like the majority of the hypotrichs, thigmotactic, that is, they adhere more or less strongly to the substrate whenever the opportunity arises. They creep on their flattened ventral side by means of the cirri; usually the specimens move hastily to and fro, but sometimes they remain immobile for more or less long periods during feeding. Verworn (in Pütter 1900, p. 284) found that *Urostyla grandis* bends both left and right when unimpededly swimming. When it makes spontaneous reverse movements or when it is bumped, for example, due to shaking the slide, then it always changes direction rightwards.

All urostyloids have a flexible body which bends to varying degrees. Thus, if you see a rigid, freely motile hypotrich you can exclude that it is a urostyloid.

1.7 Somatic Ciliature and Ultrastructure

The somatic ciliature of the Urostyloidea is composed of rows and localised groups of cirri on the flattened ventral side, and several (about 3–10) rows of more or less widely spaced, usually short (2–5 μm) and stiff cilia (bristles) on the vaulted dorsal side (Fig. 1a–e).

The Urostyloidea are characterised by the midventral complex, which is usually composed of ventral cirral pairs forming a more or less distinct zigzag pattern (Fig. 1a). Although this pattern very likely evolved convergently in, for example, *Uroleptus*, *Territricha*, *Pattersoniella* (see the phylogeny chapter and Fig. 16a–o) the "midventral"-species treated in the present book form very likely a monophylum. The only exceptions are *Neokeronopsis spectabilis* and *Urostyloides sinensis*, which have a pronounced midventral pattern, but also dorsomarginal kineties and a fragmenting dorsal kinety (Fig. 242a–h, 243a–m). The dorsal ciliature features assign them unequivocally to the Oxytrichidae (Fig. 14a).

The arrangement of the cirri is a very important feature for urostyloid/hypotrich systematics. Therefore, an unambiguous terminology is necessary. Fig. 1a–g show many important features necessary for the understanding of the urostyloid morphology and phylogeny. In the following paragraphs the individual cirri and, respectively, cirral groups are discussed. Note that many cirri of the various taxa of the Hypotricha (e.g., Urostyloidea, Oxytrichidae) can be homologised and therefore some of them have, of course, the same designation in these taxa (for discussion of the confusing terminology of some cirri see Berger 1999). As in volume I of the revision of Hypotricha (Berger 1999), I use the well-established numbering system introduced by Wallengren (1900a). Note that cirral groups/rows can be true rows (e.g., marginal rows) or pseudorows (e.g., transverse cirri). The cirri of a true row originate from the same anlage, whereas the cirri of a pseudorow originate from different anlagen. For details on the homology see the phylogeny chapter. The oral apparatus is described in the next chapter.

Frontal cirri (FC). These are the cirri near the anterior end of the cell (Fig. 1b). Many urostyloid species have, likely most oxytrichids (Berger 1999), three more or less distinctly enlarged frontal cirri. They are homologous in all groups (Fig. 16). The left frontal cirrus (= cirrus I/1) is usually ahead of the anterior end of the paroral. During cell division it originates from the same anlage (= anlage I) as the undulating membranes. The middle frontal cirrus is homologous to cirrus II/3 of the 18-cirri oxytrichids. During morphogenesis it originates, like the buccal cirrus, from anlage II. The right frontal cirrus is homologous to cirrus III/3 of the 18-cirri oxytrichids. Usually this cirrus is arranged close to the distal end of the adoral zone of membranelles. *Biholosticha* obviously has only two frontal cirri, many other urostyloid taxa, however, have an increased number of frontal cirri. The increase is due to the insertion of additional cirral anlagen, which produce – like those of the midventral complex – usually only two cirri. This results in the formation of a so-called bicorona (Fig. 1a). Usually, anlage I produces only the leftmost frontal cirrus. However, in *Uroleptopsis citrina* it forms two cirri (Fig. 192v–x). If more than two cirri per anlage are produced, then a tricorona (*Tricoronella*; Fig. 147f, h) or a (more or less regular) multicorona (e.g., *Uro-*

styla grandis, *Epiclintes*, *Eschaneustyla*) is formed. Live, it is rather easy to recognise whether a specimen has three enlarged frontal cirri or, for example, a bicorona.

Buccal cirrus (BC). This term is used for the cirrus immediately right of the paroral (Fig. 1a). It is homologous with the buccal cirrus (= cirrus II/2) of the other hypoptrichs (e.g., Berger 1999). For a discussion of the confusing terminology, see Berger (1999). Borror & Wicklow (1983) introduced the term malar cirrus in their revision on the urostyloids. Most urostyloid species have, like the majority of the Hypotricha, a single buccal cirrus, which is often slightly to distinctly behind the anterior end of the paroral (Fig. 1a). Some taxa, for example, *Paragastrostyla* have lost the buccal cirrus, some species have two (Fig. 1b) or more such cirri (e.g., *Anteholosticha adami*; Fig. 74b, i). In *Uroleptopsis citrina* the buccal cirrus is not right of the paroral, but forms a part of the bicorona (Fig. 1e, arrowhead). In life, the buccal cirrus is sometimes difficult to recognise because it is often fine and therefore easily misinterpreted as paroral cilia.

Parabuccal cirrus (PC). This is cirrus III/2 according to Wallengren's (1900) terminology. Most species with three frontal cirri have a single parabuccal cirrus (Fig. 1b). A taxon with more than one such cirrus is *Bakuella*.

Frontoterminal cirri (FT). This term was introduced by Hemberger (1982, p. 11). Most species have two such cirri which are homologous to the frontoventral cirri VI/3 and VI/4 of the 18-cirri oxytrichids (Fig. 16b; for review see Berger 1999). Borror & Wicklow (1983) introduced the term migratory cirri because of the conspicuous migration from posterior to near the distal end of the adoral zone of membranelles during late stages of cell division (Fig. 1a). They always originate from the rightmost (= rearmost) frontal-midventral-transverse cirral anlage. In some species they possibly occur from the two rightmost anlagen (e.g., *Bakuella edaphoni*); however, these data should be checked again. As already mentioned, most species have, like many Oxytrichidae, two frontoterminal cirri. Unfortunately, the frontoterminal cirri are very difficult to recognise in life, and even in protargol preparations they are sometimes hardly recognisable. In some cases ontogenetic stages are needed to be certain whether or not this cirral group is present. Some taxa, for example, *Holostichides* and *Keronella*, have more than two frontoterminal cirri (Fig. 1b, 201l–s, 202a). Few taxa, for example, *Urostyla grandis* and *Australothrix* and *Parabirojimia* lack frontoterminal cirri.

Midventral complex (MC). The terminology for the autapomorphy of the urostyloids, the midventral cirri, was rather confusing since the term midventral row has not been used uniformly. Thus I introduced the term "midventral complex" (Berger 2004b; Fig. 1a). The expression midventral cirri was introduced by Borror (1972) as follows: "Between the right and left marginal cirri in members of the Holostichidae is a double row of cirri that often is arranged in a zigzag position. The midventral cirri arise from a longitudinal series of transverse streaks in *Urostyla cristata*, ...". However, this term was not used in all subsequent papers on urostyloid hypotrichs. For example, Buitkamp (1977) designated the two rows formed by the zigzagging cirri as ventral rows (note that these two rows are pseudorows!). Hemberger (1982) and Foissner (1982) basically accepted Borror's expression and designated the two pseudorows as right and left midventral row. In several urostyloid taxa (e.g., *Bakuella*, *Keronella*) not only cirral pairs, but also more or less long rows are formed by the midventral anlagen. Wiackowski (1985) summarised both the cirral pairs and the cirral rows under the term midventral cirri. By contrast, Song et al. (1992) confined the expression midventral row to the zigzagging cirral pairs and designated the cirral rows in the posterior body portion as ventral rows. In 1994, Eigner introduced two terms for these cirral rows in the posterior body portion of some taxa, namely (i) short midventral row composed of 3–4 cirri, and (ii) long midventral row composed of more than four cirri. According to Eigner's terminology, for example, a *Bakuella* species has (i) a "midventral row" (composed of zigzagging cirral pairs), (ii) one or more "short midventral rows", and (iii) one or more "long midventral rows". Since the midventral row mentioned under (i) can also be either short or long, the terms introduced by Eigner are somewhat misleading. In addition, the left cirrus of several cirral pairs is lacking in non-dividers of *Uroleptopsis citrina* further complicating the terminology (see below). To overcome these terminological problems, the various structures are designated as shown in Figs. 1a, b, e. The generic term is "midventral complex", which can be composed of various structures. For example, in *Holosticha* species the midventral complex consists of midventral pairs only, whereas in *Bakuella* it is composed of midventral pairs and midventral rows. In *Epiclintes* and *Eschaneustyla* the midventral complex is composed of midventral rows only, that is, midventral pairs and therefore the characteristic urostyloid zigzag pattern is lacking. In species with three enlarged frontal cirri, the distinction between the frontal cirri and the midventral complex is straightforward (Fig. 1b). In taxa with a bicorona – for example, *Kerononella* and *Uroleptopsis* – it is sometimes difficult to define the beginning of the midventral complex (Fig. 1a, e). However, usually the cirri of the anterior corona and even those of the posterior are slightly to distinctly larger than the midventral cirri and often at least slightly set off from them.

The right cirrus of a midventral pair is often larger than the left cirrus. Likely this is due to the fact that the right cirri are homonomous to the anterior cirri of a bicorona (Fig. 1a), respectively, the enlarged frontal cirri (Fig. 1b), which are more or less distinctly larger then the other cirri (e.g., buccal cirrus, cirrus III/2) of the same anlage. However, the enlargement is sometimes indistinct and in many cases such details are neither mentioned nor illustrated in the individual descriptions. Only in *Uroleptus*, which is very likely not a urostyloid, is the difference usually very distinct (Fig. 16j).

Pretransverse ventral cirri (PT). This term was introduced by Berger & Foissner (1997) for two, often inconspicuous cirri immediately ahead of the transverse cirri (Fig. 1a). Unfortunately, we overlooked the older term accessory transverse cirri introduced by Wicklow (1981, p. 348). According to Wallengren's (1900) numbering system they have the designation V/2 and VI/2, that is, they originate from the two rightmost frontal-ventral-transverse cirral anlagen. Interestingly, these two cirri are also present in some urostyloids, for example, *Anteholosticha australis* and *A. mancoidea*. Of course, in these species they do not originate from the anlagen V and VI, but from the two rightmost anlagen, which, however, are homologous with the anlagen V and VI of the 18-cirri oxytrichids and the amphisiellids (Berger 2004a). In many urostyloid species pretransverse cirri are lacking, in others they have likely been subsumed under the term transverse cirri.

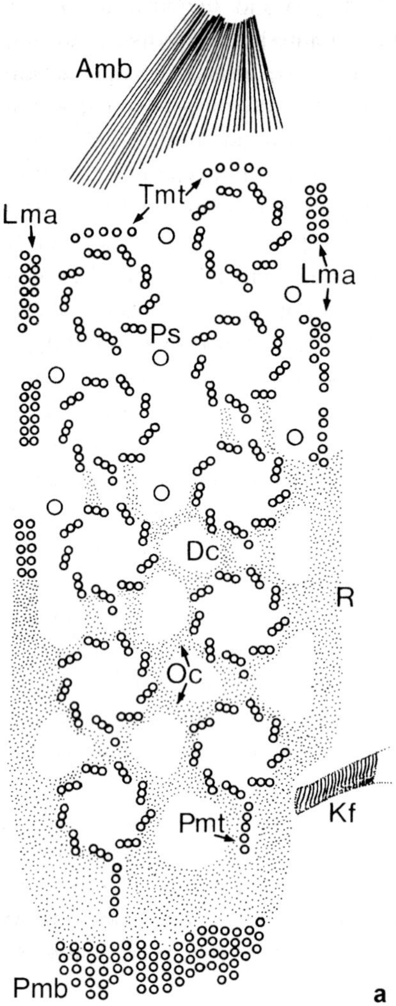

Fig. 5a *Pseudokeronopsis carnea* (from Wirnsberger & Hausmann 1988b). Scheme of the fine structure of a left marginal cirrus illustrating the positions of the various fibres. The top part of the figure represents a distal level of section, whereas the lower parts correspond to proximal sections. Amb = anterior microtubular bundles, DC = double connections, Kf = kinetodesmal fibre, Lma = linear microtubular arrays, Oc = oblique connections, Pmb = posterior microtubular bundles, Pmt = postciliary microtubules, Ps = parasomal sacs, R = rampart, Tmt = transverse microtubulus.

Transverse cirri (TC). This cirral group, which is a pseudorow, is usually in the posterior quarter of the cell (Fig. 1a). Transverse cirri are present in most hypotrichs. A transverse cirrus is, per definition, the rearmost cirrus produced by a frontal-(mid)ventral-transverse cirral anlage. It forms – usually together with other rearmost cirri – a "transverse" pseudorow (typically it is more or less obliquely arranged). In the Urostyloidea the transverse cirri are usually not or only slightly larger than, for example, the midventral cirri. By contrast, in many Stylonychinae they are very large and therefore prominent (Berger 1999). In most urostyloids only the rearmost cirral anlagen produce a transverse cirrus. Only in few taxa, for example, *Holosticha* and *Pseudoamphisiella*, does each anlage (except the anteriormost anlagen) produce a transverse cirrus resulting in a rather uncommon cirral pattern. Other taxa (e.g., *Holostichides*) lack this cirral group.

Marginal cirri (LMR, RMR). These cirri run along the left and right body margin. Many urostyloids have one left and one right marginal row (Fig. 1a, 5a). Some taxa, for example, *Pseudourostyla*, *Urostyla*, or *Diaxonella* have more than two marginal rows. However, the increase in number certainly occurred several times independently, as indicated by the rather different morphogenetic pattern (see the cell division chapter). Usually the marginal rows are more or less distinctly separated posteriorly. However, the gap is often difficult to recognise because it is seemingly occupied by the caudal cirri, which, however, insert on the dorsal surface (Fig. 1a, c).

Dorsal cilia (DB; 1, 2, 3, ...). The dorsal side of all hypotrichs and euplotids is covered with a more or less high number of kineties, which are therefore named dorsal kineties or dorsal bristle rows (e.g., Fig. 101g). Many urostyloids have three kineties, but species with up to 10 bristle rows are known. The kineties of the urostyloids are basically bipolar, that is, they extend from near the anterior end to the rear body end. Dorsomarginal kineties (originating from/near the right marginal primordium; Fig. 243j, l) and fragmenting kineties (one, usually kinety 3, or more kineties fragments into an anterior and posterior portion; Fig. 243k, m) are lacking. The dorsomarginal kineties are the morphological apomorphy of the Dorsomarginalia, the fragmentation the apomorphy of the Oxytrichidae (Fig. 14a). Thus, *Neokeronopsis* and *Urostyloides* belong to the Oxytrichidae (p. 1190, 1205).

As in other hypotrichs, the dorsal kineties of the urostyloids consist of basal body pairs. The bristle originates from the anterior basal body, is usually short (about 2–5 µm), and more or less stiff. The number of bristle rows is difficult to recognise in life, that is, usually protargol preparations are needed to know the number. However, in species with distinct (usually coloured) cortical granules the number of kineties corresponds with the number of stripes formed by the cortical granules. The fine structure is likely identical to that of the Oxytrichidae (see Berger 1999 for review). The function of the dorsal bristles is not known. Likely they are remnants of the ciliature of an early ancestor. The dorsal kinety which is closest to the left marginal row is designated as kinety 1 (Fig. 1a–d).

Caudal cirri (CC). These cirri originate at the rear end of the bipolar dorsal kineties (Fig. 1c). Dorsomarginal kineties and the anterior portion of a fragmenting kinety are never associated with a caudal cirrus (Berger 1999). Usually they are inserted at the rear tip of the cell, often above the gap formed by the rear end of the marginal rows. Thus, study your slides (in vivo and protargol!) carefully and do not misinterprete caudal cirri as marginal cirri! Some species produce more than one caudal cirrus per dorsal kinety. On the other hand, several urostyloids which lack these cirri exist. Very likely, the loss occurred several times independently. In one case this feature is characteristic for a relatively large group, which is therefore named Acaudalia (Fig. 144a). The caudal cirri of the urostyloids are usually inconspicuous, that is, neither very long and/or strong. By contrast, the caudal cirri of some oxytrichids (e.g., *Stylonychia*) are rather long and therefore very conspicuous (Berger 1999).

Fine structure of cirri and membranelles. There exist only few data on the fine structure of urostyloids (Yasuzumi et al. 1972, Wicklow 1981, Carey & Tatchell 1983, Wirnsberger & Hausmann 1988b, Wicklow & Borror 1990). In *Pseudokeronopsis carnea* the marginal, frontoterminal, and midventral cirri have the same microtubular and microfibrillar associates (Fig. 5a). The anterior and posterior microtubular bundles of several cirri overlap and accompany the single layer of subpellicular microtubules. The pairs of midventral cirri are very closely set; thus, each kinetodesmal fibre of the left midventral cirrus is in contact with the margin of the right midventral cirrus (Wirnsberger & Hausmann 1988b).

Linear microtubular arrays which characteristically comprise two rows of 5–7 serially arranged microtubules border the longer sides of the cirral bases and extend toward the pellicle, probably contributing to the single layer of subpellicular microtubules. Likewise, they occur to the right and to the left of each adoral membranelle as well as to the right of the paroral and to the left of the endoral. The left microtubular arrays of the membranelles may contribute to the postmembranellar fibre, and the right ones probably line the buccal cavity and build the highly ordered structure to the left of the endoral. Between the groups of three sheets of 7–8 microtubules, there are prominent vacuoles and alveoli beneath the pellicle. Thus, the whole system resembles oral ribs (Wirnsberger & Hausmann 1988b).

Wirnsberger & Hausmann (1988b) discussed some fine structural features that may unify urostyloid hypotrichs based on the data about *Thigmokeronopsis jahodai* and *Pseudokeronopsis carnea*. Unfortunately, Wicklow (1981) gave sparse details concerning the fine structure of the buccal region; therefore, Wirnsberger & Hausmann's findings are difficult to compare and no definite taxonomic conclusion could be derived. They found that both taxa share some ultrastructural characters that are perhaps restricted to urostyloids. (i) At present, the additional linear microtubular arrays bordering the cirri are unique for urostyloid taxa. This microtubular system reminds one of that found in the heterotrich ciliate, *Plagiotoma lumbrici* (Wicklow 1981). (ii) In *Pseudokeronopsis carnea* these linear microtubular arrays are also present beside the left and the right buccal organelles. Although Wicklow (1981) described them to the right of the membranelles only, they are also visible in the paroral of *T. jahodai*. (iii) Wicklow (1981) considered the urostyloid midventral cirral pairs to be linked in a ladder-like array owing to the anterior microtubular bundles joining them in *Thigmokeronopsis*. This is obviously not the case in *Pseudokeronopsis*, but the pairs of midventral cirri are in fact very closely set and seem to be "linked" by the kinetodesmal fibre of the left midventral cirrus. (iv) In contrast to oxytrichid taxa, the anterior frontal cirri have not been found to be linked with the frontal adoral membranelles in urostyloids. For details on the fine structure of *Epiclintes auricularis*, see species description.

1.8 Oral Apparatus

The oral apparatus of the Urostyloidea is composed, as in the remaining Hypotricha, of an adoral zone of membranelles, two undulating membranes (paroral and endoral), the buccal cavity (buccal field, oral field), and associated fibres including the cytopharynx (e.g., Fig. 1a, b, e, 151g, 208j–l, n, o). For details, see Foissner & AL-Rasheid (2006).

The adoral zone of membranelles, the most prominent part of the oral apparatus, extends from the anterior body end along the left anterior body margin to near midline of the cell and usually terminates at about 25–35% of body length. Usually, it is roughly the shape of a question mark. In some taxa the distal (= frontal) portion extends far onto the right body margin. This feature was used by Wicklow (1981) to characterise the Keronopsidae (now Pseudokeronopsidae). Wiackowski (1988) quantified this character in that he divided the distance between the anterior body end and the distal end of adoral zone by the distance between the anterior body end and the proximal end of the adoral zone (Fig. 1c). He distinguished four ranges: less than 0.11 (designated as plesiomorph by Wiackowski); 0.11–0.20; 0.21–27; and 0.28 or more (most derived). Whether a low or high value is apomorph is not yet certain because we do not know the state in the last common ancestor of the Hypotricha. Preliminarily, I accept Wiackowski's assumption that high values are derived. For the sake of simplicity this quotient introduced by Wiackowski (1988) is named "DE-value" (for *D*istal *E*nd of adoral zone).

High DE-values occur not only in the pseudokeronopsids, but also in the pseudourostylids, the Epiclintidae, *Pseudoamphisiella*, and some taxa outside the Urostyloidea (e.g., *Amphisiella namibiensis*, *Pseudouroleptus caudatus*; Foissner et al. 2002), indicating that this feature occurred, like many others, convergently.

Some species, for example, *Holosticha* spp., *Uroleptopsis citrina* or *Afrothrix* spp. have a more or less distinct gap (break) in the zone (Fig. 31b, 104a, f, 105a, f, 192k). The proximal (= ventral) portion is sometimes distinctly spoon-shaped. In some *Holosticha*-species the proximalmost membranelles are slightly to distinctly wider than the remaining membranelles (Fig. 29b, 34b, f).

Species of the Hypotricha are characterised by two undulating membranes, the paroral and the endoral (Fig. 1a, b, e, 61t, 151g). For a detailed discussion of the rather bewildering terminology of these structures, see Berger & Foissner (1997) and Berger (1999). In general, the paroral extends between two usually inconspicuous cytoplasmic lips at the right outer margin of the buccal cavity, that is, on the cell surface, while the endoral is on the bottom and right wall of the cavity. This means that the membranes extend at different levels. However, if the cell is viewed from the ventral side, they appear to lie side by side (e.g., *Pseudokeronopsis*, *Uroleptopsis*; Fig. 192k) or to intersect (e.g., *Urostyla*; 208h, j, o), depending on their shape and arrangement.

In the Oxytrichidae, the shape and arrangement of the adoral zone and especially the undulating membranes is often used to recognise subgroups (for reviews see Kahl 1932, Berger & Foissner 1997, and Berger 1999). This is also possibly within the Urostyloidea, however, to a distinctly smaller extent. Especially the undulating membranes show a lower diversity than in the Oxytrichidae, where rather curious patterns occur (e.g., *Steinia* pattern with a fragmented endoral; Berger & Foissner 1997). However, there exist urostyloid groups with a characteristic undulating membrane pattern, for example, the pseudokeronopsids where membranes are rather short and arranged more or less parallel. Very likely some further patterns can be recognised (distuingished) when more detailed data become available.

The buccal cavity is also different in shape and size and usually described by the terms flat or deep and wide and narrow. Flat means that the cavity is only slightly hollowed, whereas a deep cavity extends to near the dorsal side of the cell, making the oral field conspicuously bright. In species with a wide cavity, the right margin of the cavity is in the midline of the cell, whereas in a narrow cavity it is arranged close to the right margin of the adoral zone (further details see Berger 1999 and Foissner & Al-Rasheid 2006).

Wirnsberger & Hausmann (1988b) studied the fine structure of the oral apparatus of *Pseudokeronopsis carnea*. The basic features resemble those of other hypotrichs. Therefore they described only details that are peculiar to the urostyloid *P. carnea*. The cilia of the endoral are connected by a microfibrillar material (Fig. 6a). These peculiar connections have not been found between the cilia of the paroral. Proximally, the basal bodies of the endoral are linked by amorphous connectives, which join the triplets 7, 8 of the anterior basal body with the triplets 2, 3 of the posterior one. About five transverse microtubules are associated with the triplets 9, 1, and 2; two postciliary microtubules are oriented to the right, and parasomal sacs are situated to the left of the endoral. The two

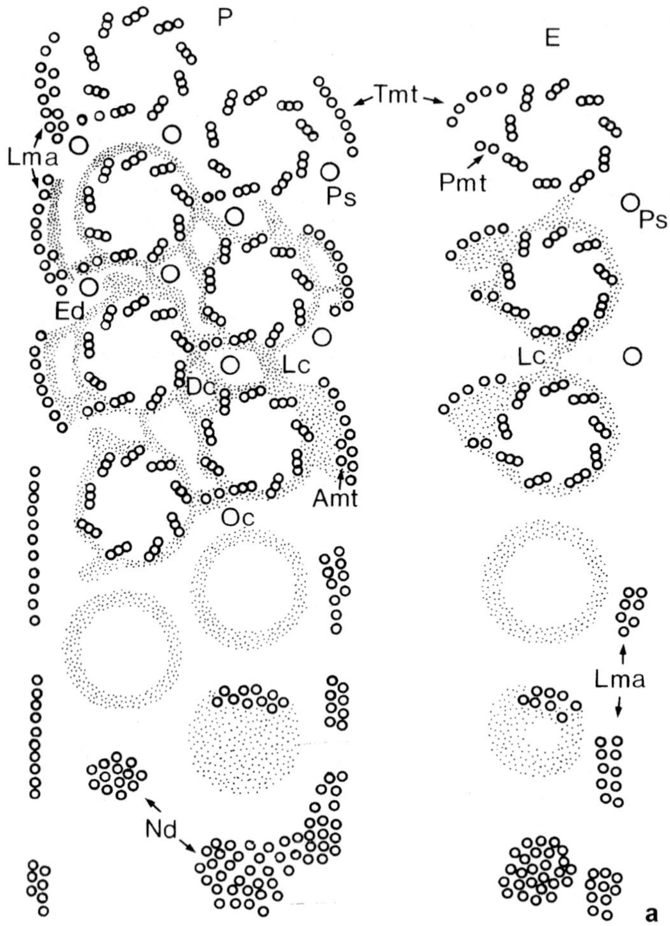

Fig. 6a *Pseudokeronopsis carnea* (from Wirnsberger & Hausmann 1988b). Scheme of the fine structure of the paroral (P) and endoral (E). The top part of the figures represent a distal level of section, whereas the lower parts correspond to proximal sections. The basal bodies are embedded in electron-dense material (Ed) which forms longitudinal (Lc), oblique (Oc), and double connections (Dc). Two postciliary microtubules (Pmt) are associated with basal bodies. The transverse microtubules (Tmt) of both membranes are differently oriented, the paroral ones to the left and the endoral ones to the right. Sometimes additional microtubules (Amt) occur in a second row of the paroral transverse microtubules. The characteristic linear microtubular arrays (Lma) are associated with the right of the paroral and with the left endoral. Ps = parasomal sac, Nd = nematodesmata.

rows of paroral basal bodies are proximally linked by electron-dense material at four different locations: two of them connect both neighbouring basal bodies, which are reminiscent of dikinetids; in addition, longitudinal linkages join the basal bodies within one row and oblique ones connect the neighbouring "pairs". About 7–9 transverse microtubules originate beside the left row of paroral basal bodies and a second sheet of microtubules may appear at a more proximal level, possibly corresponding to nematodesmata. Two postciliary microtubules appear at the right of each paroral basal body, and parasomal sacs are situated to the right and to the left. Nematodesmata emerge from the proximal part of all endoral and paroral basal bodies, contributing to the pharyngeal basket (Wirnsberger & Hausmann 1988b).

1.9 Silverline System

The silverline system of urostyloids is, like that of the Oxytrichidae (for review see Berger 1999), composed of small (1–2 μm) polygonal meshes (Fig. 103b; Foissner 1980a, 1982). It has no systematic value in the hypotrichs, whereas it is successfully used to characterise euplotids (e.g., Borror & Hill 1995).

1.10 Life Cycle

The Urostyloidea have, like most other Hypotricha, a normal life cycle, that is, the theronts feed, become trophonts and divide, encyst, or conjugate. There is much less specific literature available about these topics than for the Oxytrichidae (for review of this group see Berger 1999).

1.10.1 Cell Division

Urostyloid ciliates divide by isotomic transverse fission, like many other ciliates (Foissner 1996c; e.g., Fig. 7a–t). The anterior filial product is the proter, the posterior the opisthe. Early in division, a replication (= reorganisation) band traverses each macronuclear nodule. In species with many tiny nodules, this feature is often difficult to recognise (Fig. 2a–g). The two to very many nodules fuse to a single mass during early and middle stages of cell division. The macronucleus divides amitotically just before cytokinesis (Fig. 3a–l). By contrast, the micronuclei(eus) divide(s) mitotically (Fig. 4a–h). Only the pseudokeronopsids show a deviating pattern in that the many macronuclear nodule divide individually (e.g., Fig. 192r).

The changes of the ciliature during cell division are known from a relatively low number of species. Morphogenetic data allowed homologising the individual cirri of the urostyloid with those of the 18-cirri oxytrichids. For example, the frontoterminal cirri of the urostyloid are certainly homologous with the frontoventral cirri VI/3 and VI/4 of the 18-cirri oxytrichids, because in both cases these are the two anteriormost cirri (of a total of four) of the rightmost (= rearmost) frontal-(mid)ventral-transverse cirral anlage (e.g., Hemberger 1982, Wirnsberger 1987, Berger 1999). Moreover, in both groups these two cirri migrate anteriorly in the area between the distal end of the adoral zone and the anterior end of the right marginal row (Fig. 1a, 192w, x).

As in other hypotrichs, the parental ventral and dorsal somatic ciliature of the urostyloids is completely renewed during cell division. The parental oral apparatus is either retained after a more or less distinct reorganisation, or it is completely renewed as, for example, in the pseudokeronopsids.

In the urostyloids, the ventral somatic ciliature develops from more than six, more or less obliquely arranged frontal-midventral-transverse cirral anlagen (Fig. 7a–e, k–o). Usually these anlagen are numbered from I to n (I = leftmost anlage forming cirrus I/1 and undulating membranes; anlage n = rightmost anlage usually forming the frontoter-

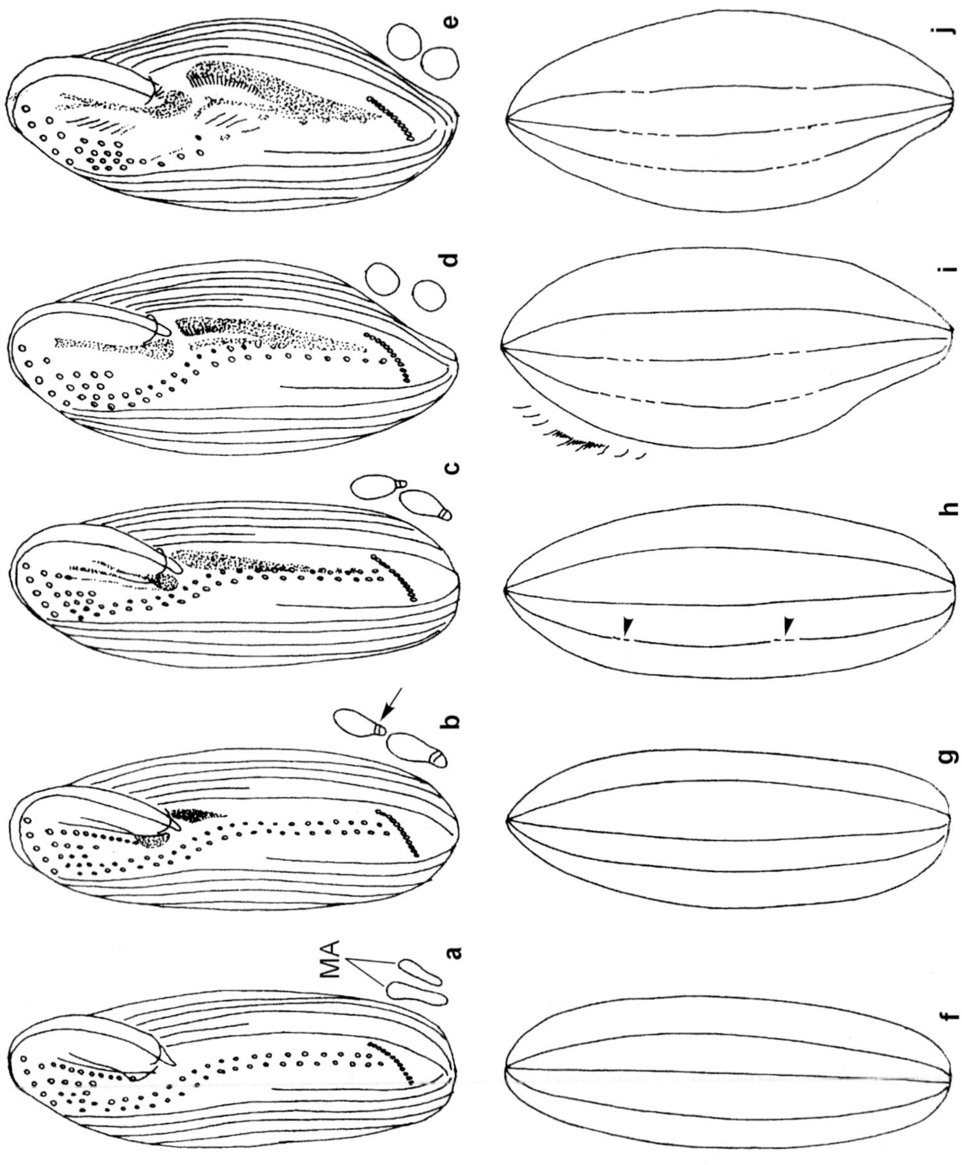

Fig. 7a–j *Urostyla grandis* (from Jerka-Dziadosz 1972. After protargol impregnation). Schematic illustrations of an interphasic specimen (a, f) and early to middle stages of cell division in ventral (b–e) and dorsal (g–j) view. Arrowheads in (h) mark the beginning of the intrakinetal formation of the dorsal kinety primordia. Arrow in (b) marks replication band. Note that in *U. grandis*, many other urostyloid species, and all remaining hypotrichs the macronuclear nodules fuse to a single mass prior to cell division. For details see text. MA = two of many macronuclear nodules.

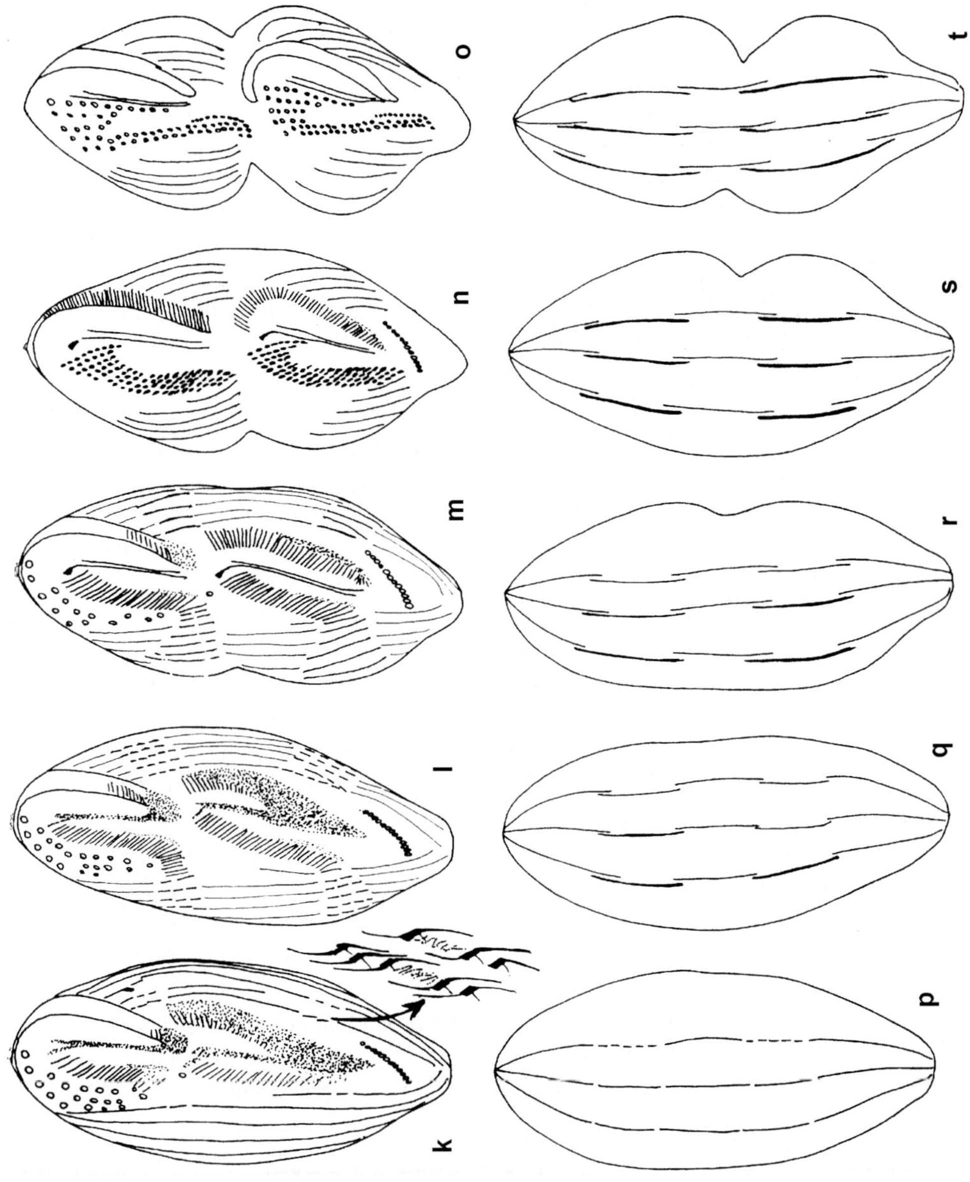

Fig. 7k–t *Urostyla grandis* (from Jerka-Dziadosz 1972. After protargol impregnation). Schematic illustrations of late to very late stages of cell division in ventral (k–o) and dorsal (p–t) view. Note that the frontal-midventral-transverse cirral pattern of urostyloid hypotrichs originates from (usually) many obliquely arranged cirral anlagen. The many marginal rows of *Urostyla grandis* divide, like the dorsal kineties, individually. By contrast, in *Pseudourostyla* the marginal rows of each side are formed from a common anlage. For details see text.

minal cirri, the right pretransverse ventral cirrus, and the rightmost transverse cirrus; Fig. 1a)[1]. The cirral pattern of the urostyloids is therefore much more variable than that of the 18-cirri oxytrichids, which usually have, as indicated by the designation, 18 frontal-ventral-transverse cirri. In the urostyloids, the number of frontal-midventral-transverse cirri varies not only among the species, but also within them because the number of anlagen forming the midventral pattern is usually more or less variable. The supposed relationships of the hypotrichs with the euplotids require the assumption that the urostyloids evolved from an ancestor which had six anlagen (Fig. 12, 13a, 14a). This hypothesis can also explain the fact that the so-called midventral pattern, that is, the zigzag pattern formed by the ventral cirri evolved several times independently in rather different groups of hypotrichs (see chapter phylogeny).

The frontal-midventral-transverse cirral anlagen of the urostyloids arise (i) from parental ciliature, (ii) new, and (iii) from the oral primordium. However, it is often rather difficult to recognise how an anlage originates. The conspicuous zigzag cirri-pattern of the urostyloid ciliates is formed from anlagen, which produce only two cirri. Only the posterior anlagen form a pair plus a transverse cirrus. In a relatively high number of species not only cirral pairs but also more or less long rows are formed per midventral anlage (Fig. 1a). However, this feature conflicts with the frontal ciliature.

Berger & Foissner (1997) and Berger (1999) used several morphogenetic peculiarities of the 18-cirri oxytrichids to elucidate the phylogeny of this group. This is not yet possible in such a big way for the urostyloids because their cirral pattern is much more variable, and much less relevant data are available. In spite of this, cell division data are useful markers to elucidate the evolutionary relationships among the Urostyloidea. For example, in *Holosticha* species the midventral cirral anlagen originate largely right of the parental midventral complex, a feature well supporting other morphological traits characterising seven species as a monophyletic group (Berger 2003). Ontogenetic data are also useful to understand deviating cirral patterns. *Uroleptopsis citrina* has a curious midventral complex, that is, the zigzag pattern is lacking in the central portion of the complex. Cell division data showed that this is due to the resorption of the left cirrus of some pairs (Fig. 192v, w; Berger 2004b).

Dorsal morphogenesis proceeds simply in the urostyloids because each dorsal kinety forms one anlage each in the proter and the opisthe by intrakinetal proliferation of basal bodies (Fig. 7f–j, p–t; Foissner & Adam 1983). Dorsomarginal kineties and fragmentation of dorsal kineties – characteristic features for most non-urostyloid hypotrichs, respectively, the Oxytrichidae (for review see Berger 1999) – are lacking in the urostyloids. Only in very few urostyloids (e.g., *Holosticha bradburyae*) did a more complex dorsal morphogenesis evolve. Caudal cirri originate at the rear end of a dorsal kinety anlage. As stated above, proliferation of basal bodies begins at two levels within parental dorsal kineties (Fig. 7h–j). These two regions correspond to the same levels within which the marginal cirri proliferate on the ventral surface (Fig. 7l, q). Several urostyloid

[1] This numbering system has the disadvantage that the two rightmost (= rearmost) anlagen, which produce, inter alia, the two pretransverse ventral cirri, do not have the same Roman numbers in the urostyloids and oxytrichids.

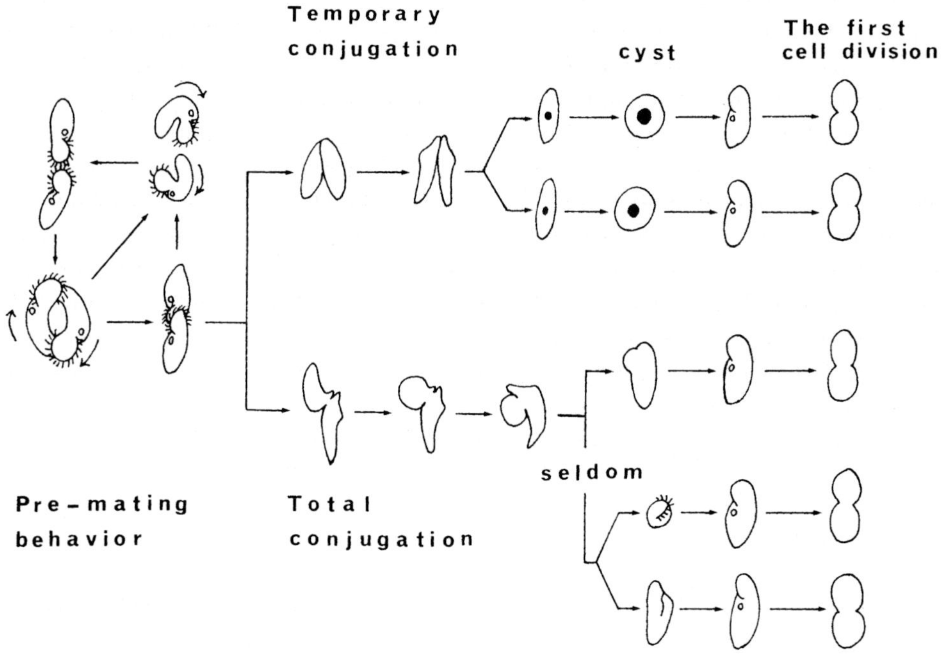

Fig. 8a Schematic illustration of temporary and total conjugation in *Pseudourostyla levis* (from Takahashi 1973). For details see text.

taxa with more than two marginal rows are known. In *Urostyla grandis* the marginal rows divide individually (Fig. 7l). By contrast, in *Pseudourostyla cristata* all marginal rows of a side originate from a common anlage (Fig. 149f–h).

1.10.2 Conjugation

Relatively little is known about this part of the urostyloid life cycle. Takahashi (1973) found two types of conjugation in *Pseudourostyla levis*, namely temporary conjugation and total conjugation (Fig. 8a). In a mixture of two opposite mating types, cells gave no sign of mating during at least five or more hours (a refractory period), but then showed characteristic pre-mating behaviour before forming conjugating pairs. In pre-mating behaviour, two specimens came closer by creeping on the bottom of the container and came in contact. A cell attached with its anterior end to the posterior part of the other specimen. The contacted cells revolved clockwise for several minutes, and then united with their mouths in straight fashion through further complicated behaviour. During this process, the united cells were suddenly separated by interference of another cell, but the disjoined cells again made contact with each other. The reunited head-to-head pair remained as it was for about 20 min, and then changed into a typical pair with side-to-side contact (Fig. 8a).

The conjugating pair underwent meiosis of one micronucleus, which lay near the posterior end of the cytostome, exchanged migratory pronuclei, and formed a synkaryon in each conjugant within 20 h of the onset of conjugation. The old macronuclei of the pair decreased gradually in number as a result of absorption into cytoplasm during this period. Each exconjugant derived from the pair contained a macronuclear primordium, two new micronuclei, and several old macronuclei. Thereafter the exconjugant fell into encystment without cell division two or three days after the separation. The cyst formed was about 50–100 µm across, had no cyst wall, and the macronuclear primordium developed into a new elongate macronucleus 5–8 days after the encystment. The cyst excysted and produced a swimming cell, which bore a new nodulated macronucleus, two new micronuclei, and several old macronuclear nodules. The swimming specimen underwent the first cell division within 48 h after excystment. The nuclear processes during conjugation are described in detail by Takahashi (1974).

For a brief description of the cortical reorganisation during conjugation see Fig. 149s–z.

1.10.3 Cyst

Knowledge about this stage of the life cycle is modest compared to the Oxytrichidae (for review of this group see Berger 1999). Resting cysts are described only for a few urostyloids (e.g., 44f–i, 206c, d, 208d). Reproductive cysts are not known in this group. Factors that induce encystment are, inter alia, desiccation (especially in soil species) and deficiency of food (e.g., Corliss & Esser 1974, Gutiérrez & Martin-González 2002).

The classification of resting cysts is based on a system proposed by Walker & Maugel (1980), who designated the cysts of the euplotids as NKR (non-kinetosome-resorbing) and those of the oxytrichids as KR (kinetosome-resorbing; for review on oxytrichids cyst literature see Berger 1999). Matsusaka et al. (1989) studied the resting

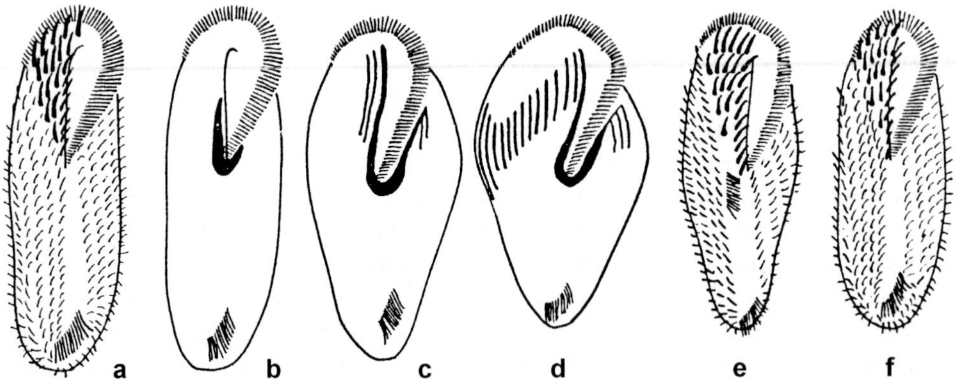

Fig. 9a–f Schematic illustrations of physiological reorganisation in *Urostyla grandis* (from Jerka-Dziadosz 1963). For details see text.

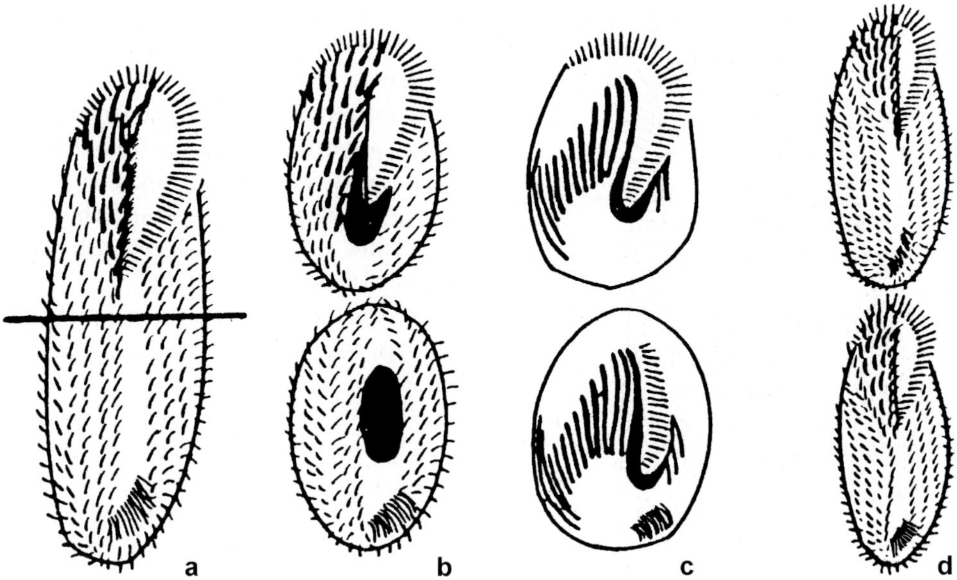

Fig. 10a–d Schematic illustrations of posttraumatic regeneration in *Urostyla grandis* (from Jerka-Dziadosz 1963). For details see text.

cysts of *Anteholosticha adami*, *Pseudourostyla levis*, and *Gonostomum affine* and found that they produce an intermediate type. They therefore distinguished an *Oxytricha*-type cyst (= KR-type), a *Urostyla*-type cyst, and a *Euplotes*-type cyst (NKR-type). Martin-González et al. (1991, 1992) proposed a modification of Walker & Maugel's system in that they named the *Urostyla*-type PKR-cysts (partial-kinetosome-resorbing).

The review by Gutiérrez et al. (2003) indicates that oxytrichids have four cyst wall layers (including the granular layer), whereas the urostyloids have only three. More data on the fine structure (number of cyst wall layers; state of macronuclear nodules, that is, fused or not fused; degree of ciliature resorption) of resting cysts will very likely increase our insights into the phylogeny of the hypotrichs.

1.10.4 Reorganisation, Regeneration, Doublets

Like other hypotrichs, the Urostyloidea produce ciliature not only during cell division or other normal parts of the life cycle (conjugation, excystment), but also during physiological reorganisation and post-traumatic regeneration.

Physiological reorganisation. This part of the life cycle is defined as morphogenesis which re-establishes a complete set of ciliary structures in an intact morphostatic (non-dividing) cell (Grimes & Adler 1978). Usually, this process is a response to an altered nutritional status induced by unfavourable culture conditions (e.g., starvation) or other more subtle changes in the environment. For example, Wirnsberger (1987)

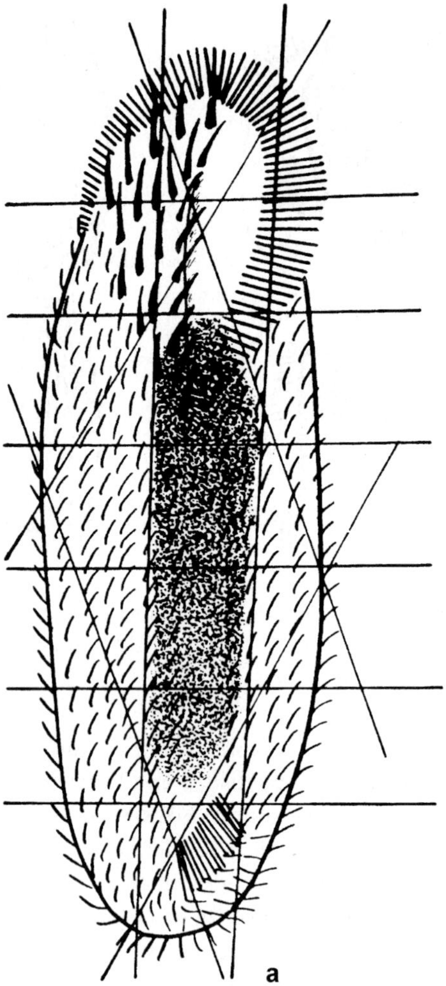

Fig. 11a Presumptive organisation area (dotted) in *Urostyla grandis* (from Jerka-Dziadosz 1964). For details see text.

studied reorganisational morphogenesis in *Pseudokeronopsis rubra* (Fig. 179q; see species description for details). As in other hypotrichs, the morphogenetic processes occurring during physiological reorganisation are rather similar to those during cell division (Fig. 9a–f). Thus it is sometimes difficult to distinguish early stages of these two processes.

Regeneration. Fauré-Fremiet (1910b; 1948, p. 46) made merotomy experiments on cell division by cutting middle dividers at various sites. Jerka-Dziadosz (1967), who also made microsurgical experiments, observed that an individual continued to divide normally when the section line ran somewhat farther from the division furrow. Jerka-Dziadosz (1963, 1964, 1965) found that in *Urostyla grandis* the postoral region is the morphogenetically most active region and thus named it the presumptive organisation area (Fig. 10a–d, 11a). This area is able to develop the primordia of ciliature in division and regeneration. Jerka-Dziadosz (1974) studied, inter alia, the formation of primordia in right fragments. In the right fragments obtained after longitudinal section along the central meridian of the ventral side, in which the wound repair occurs in situ, the primordium of the adoral zone is formed near the wounded margin. The primordium first appears as a small group of basal bodies located near the postoral part of the ventral cirri. In later stages of regeneration it can be seen that such fragments are able to form all of the kinds of ventral and dorsal primordia. For review on this topic, see Frankel (1974; 1989, p. 119).

Doublets. There is little information available about urostyloid doublets as compared to the oxytrichids (for review see Berger 1999). Altmann & Ruthmann (1979) studied doublet formation in *Urostyla grandis*. Accordingly, the formation of homopolar doublets can be induced by the action of antibiotics, which inhibit the growth of cytoplasmic bacterial symbionts whose cell cycle appears to be controlled by the host. At 50 $\mu g\ ml^{-1}$ the rate of doublet formation, expressed as a percentage of the total number of cells, was 0.43%, at 100 μg ml^{-1} 0.61%, and at 200 μg ml^{-1} 1.3%. The symbionts multiply during the macronuclear S-phase of the ciliate, and are enclosed in vesicles and largely destroyed just before cell division is completed. Since doublet formation is due to incomplete cell division, and because experimental disturbances at the cell cortex of dividing ciliates also led to doublets, the symbionts are thought to contribute some factor which is essential for normal cytokinesis of *Urostyla grandis* (Altmann & Ruthmann 1979). Homopolar doublets of *Urostyla trichogaster*, a synonym of *Urostyla grandis*, were analysed by Fauré-Fremiet (1945a, b; 1948, p. 49). The illustrations in Fauré-Fremiet (1967, p. 264) do not show *Urostyla trichogaster*, as mistakenly indicated in the legend, but *Paraurostyla weissei* (see Fig. 3, 5, 6 in Fauré-Fremiet 1945b).

2 Phylogeny

2.1 Notes on the Spirotricha Bütschli, 1889 [1]

The urostyloids are part of the spirotrichs, a large group likely comprising 2000 or more extant species. The main (sole?) "morphological" apomorphy of the Spirotricha is the replication band where DNA is replicated locally and sequentially along the macronucleus (Raikov 1982).[2] Whether the more or less well-developed adoral zone of membranelles is a further apomorphy or a plesiomorphy is not yet clear. The perilemma, considered as still doubtfull apomorphy of the spirotrichs by Petz & Foissner (1992), is possibly lacking in the euplotids (Bardele 1981, Agatha 2004).

All authorities agree that the oligotrichs, the euplotids, and the hypotrichs are the three major components of the spirotrichs. Moreover, the monophyly of each of these three groups is widely accepted[3]. There are three ways to arrange these taxa (Fig. 12a–c),

---

[1] For names of higher taxa (see Fig. 13a, 14a and Table 1).

[2] Very likely *Phacodinium* does not have a replication band (Lynn & Small 2002, p. 420, 421). If we assume that this feature is primarily lacking in *Phacodinium*, then it does not belong to the spirotrichs. If the replication band was lost during evolution in *Phacodinium*, then it has to be included in the Spirotricha (if we use the replication band to limit the group). Molecular data about *Phacodinium* are rather contradictory. According to Shin et al. (2000), it clusters between the oligotrichs and the euplotids. Bernhard et al. (2001), Petroni et al. (2002), and Johnson et al. (2004) found that it is the sistergroup of the unit formed by the three major taxa of the spirotrichs, and according to Strüder-Kypke & Lynn (2003) it is the sistergroup of the euplotids. By contrast, *Protocruzia* with its highly interesting nuclear apparatus (Ammermann 1968) is the adelphotaxon to the unit formed by all taxa mentioned above in all studies using small subunit rRNA gene sequences (Shin et al. 2000, Bernhard et al. 2001, Petroni et al. 2002). However, phylogeny derived from histone H4 analyses shows a quite different position for *Protocruzia* (Bernhard & Schlegel 1998).

[3] The fact that the halterids are assigned either to the oligotrichs (morphologists; e.g., Foissner et al. 1999, Lynn & Small 2002) or to the hypotrichs near *Oxytricha* (molecular biologists; e.g., Strüder-Kypke & Lynn 2003, Dalby & Prescott 2004, Adl et al. 2005) has no influence on the monophyly of the oligotrichs. In the first case the halterids are a subgroup of the oligotrichs, in the second the halterids are a subgroup of the oxytrichids. For a discussion of the "*Halteria*-problem" see Foissner et al. (2004a).

Interestingly, some molecular trees indicate that the euplotids (e.g., *Euplotes*, *Uronychia*, *Diophrys*) do not

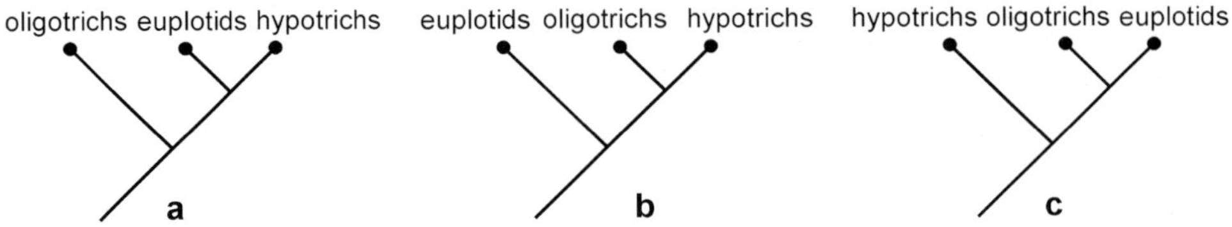

Fig. 12a–c The three possibilities to arrange the three major taxa of the spirotrichs (original). *Protocruzia* and *Phacodinium*, each composed of only one or very few species, are not considered. For explanation see text.

and if the assumptions just mentioned are correct, only one of the trees reflects the truth. However, each of these three trees has potential, depending on the features used. The arrangement shown in Fig. 12a is mainly suggested by morphologists, with the presence of cirri as main apomorphy for the unit euplotids + hypotrichs (e.g., Petz & Foissner 1992, Agatha 2004). By contrast, many molecular studies indicate that the oligotrichs and the hypotrichs are adelphotaxa (e.g., Fleury et al. 1995, Shin et al. 2000, Bernhard et al. 2001, Petroni et al. 2002, Strüder-Kypke & Lynn 2003, Agatha et al. 2004, Johnson et al. 2004; Fig. 12b). Less often, molecular data suggest a common ancestor for euplotids and oligotrichs (e.g., Snoeyenbos-West et al. 2004, Foissner et al. 2004a; Fig. 12c, 15). To decide which of the three hypothesis is correct, further data (e.g., fate of the somatic ciliature in cysts, different molecular markers) on more species are likely needed.

Although it is rather a nomenclatural than a taxonomic problem, the naming of the spirotrich taxa has to be discussed. For a long time (1859–1985) the name Hypotricha (or its derived forms Hypotrichea, Hypotrichia, Hypotrichida, Hypotrichina, depending on the category assigned; for review see Berger 2001) was used in a rather uniform way, that is, for a group comprising the euplotids, the oxytrichids, the urostyloids, etc. Based on features of the dorsal somatic kinetids, Small & Lynn (1985) suggested that the hypotrichs be divided between the Postciliodesmatophora and the Cyrtophorea. Their monotypic and therefore redundant subclass Hypotrichia contained the order Euplotida. The (non-euplotid) hypotrichs were assigned to the likewise monotypic (and therefore redundant) subclass Stichotrichia Small & Lynn, 1985 with the order Stichotrichina Fauré-Fremiet, 1961 (Table 1).

Lynn & Sogin (1988) analysed the 16S-like ribosomal RNA and found that the classification of euplotids and stichotrichids in different higher taxa proposed by Small & Lynn (1985) was incorrect (for review see Lynn 1991). However, they retained Small & Lynn's subclasses Hypotrichia and Stichotrichia and suggested the introduction of the class Hypotrichea to include the Hypotrichia (in their sense), the Stichotrichia, and the Oligotrichia. However, for this group the name Spirotricha (originally incorrectly

form a monophyletic group (e.g., Baroin-Tourancheau et al. 1992, Chen & Song 2002, Lynn 2003). Moreover, *Prodiscocephalus* and related taxa very likely do not belong to the Euplota as suggested by Lynn & Small (2002), but to the hypotrichs, as, inter alia, indicated by the presence of two undulating membranes (Lin et al. 2004).

Table 1 Comparison of names of some higher spirotrich taxa used in four recent books and in the present review

Present book [a]	Corliss 1979	Small & Lynn 1985	Tuffrau & Fleury 1994	Lynn & Small 2002
Spirotricha Bütschli, 1889 (spirotrichs)	Polyhymenophora Jankowski, 1967 [b]	Spirotrichea Bütschli, 1889	Spirotricha Bütschli, 1889	Spirotrichea Bütschli, 1889
Euplota Ehrenberg, 1830 (euplotids)	Euplotidae Ehrenberg, 1838; Aspidiscidae Ehrenberg, 1838; Gastrocirrhidae Fauré-Fremiet, 1961	Hypotrichia Stein, 1859 [d]	Euplotia author? [f]	Hypotrichia Stein, 1859
Oligotricha Bütschli, 1889 (oligotrichs)	Oligotrichida Bütschli, 1889	Choreotrichia n. subclass.	Oligotrichea Bütschli, 1887 [g]	Choreotrichia Small & Lynn, 1985; Oligotrichia Bütschli, 1889
Hypotricha Stein, 1859 (hypotrichs)	Hypotrichida Stein, 1859 [c]	Stichotrichia n. subclass. [e]	Oxytrichia author? [f]	Stichotrichia Small & Lynn, 1985
Urostyloidea Bütschli, 1889 (urostyloids)	Urostylidae Bütschli, 1889; Holostichidae Fauré-Fremiet, 1961	Urostylina Jankowski, 1979	Urostylida Jankowski, 1979	Urostylida Jankowski, 1979

[a] Note that in the present book the names are usually used as originally introduced (see Berger 2001). Vernacular names in brackets.

[b] With the single subgroup Spirotricha Bütschli, 1889 (for a brief discussion of the redundancy of the name Polyhymenophora, see Berger & Foissner 1992).

[c] Also includes the Euplotidae, the Aspidiscidae, and the Gastrocirrhidae.

[d] With the single subgroup Euplotida n. ord.

[e] With the single subgroup Stichotrichida Fauré-Fremiet, 1961.

[f] According to Berger (2001) the authors of this subclass are Tuffrau & Fleury (1994).

[g] Incorrect year.

also including the heterotrichs and peritrichs) was introduced by Bütschli (1889, p. 1719), a name now generally used for the monophylum formed by the oligotrichs, the euplotids, the hypotrichs, *Phacodinium*, and *Protocruzia* (e.g., Foissner et al. 1999, Lynn & Small 2002, Hausmann et al. 2003, Strüder-Kypke & Lynn 2003; Fig. 13a). By contrast, the name Hypotricha (or one of its derived forms) is not used uniformly since the publication of Small & Lynn's (1985) paper, that is, either only for the euplotids (mainly by molecular biologists) or non-euplotid hypotrichs (mainly morphologists; e.g., Lemullois et al. 2004).

The ordername Stichotrichina was introduced by Fauré-Fremiet (1961) for hypotrichs with a frontoventral ciliature mainly composed of distinct cirral rows (e.g., *Urostyla*, *Holosticha*, *Strongylidium*). Simultaneously, Faurè-Fremiet (1961) introduced the

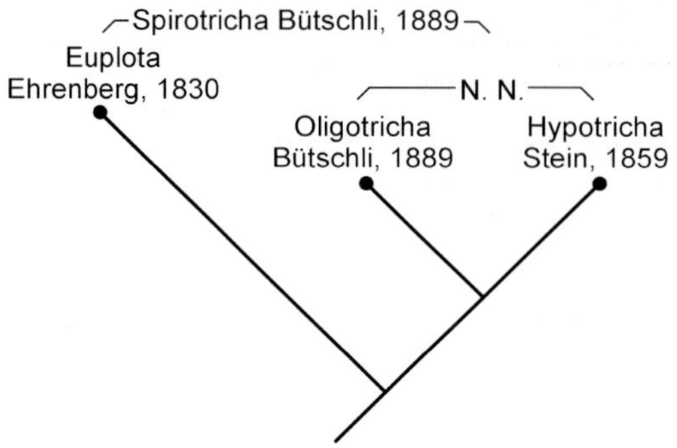

Well-known representatives

Hypotricha (hypotrichs): *Oxytricha, Stylonychia, Sterkiella, Amphisiella, Uroleptus, Urostyla, Holosticha*

Oligotricha (oligotrichs): *Strobilidium, Lohmanniella, Strombidium, Tontonia, Limnostrombidium, Tintinnidium, Codonella*

Euplota (euplotids): *Euplotes, Aspidisca, Uronychia, Diophrys*

Fig. 13a Names of the three major taxa of the Spirotricha as used in the present book (original). Note that (i) I usually use the same spellings as in the original descriptions, and (ii) I do not use categories (e.g., family, order), simply because they do not exist in nature. N. N. = Nomen nominandum. This term was introduced by Ax (1999, p. 18) for a supposed monophylum, which is not yet well founded. In the present case several morphological features indicate that the euplotids, and not the oligotrichs, are the sister group of the hypotrichs (Fig. 12a). Further details see text.

suborder Sporadotrichina for taxa with "sporadically" distributed frontoventral cirri originating from six anlagen (e.g., *Steinia*, *Gastrostyla*, *Euplotes*). Thus, the name Stichotrichia – although established as new category by Small & Lynn (1985) – for all non-euplotid hypotrichs is somewhat misleading. I do not use the name Stichotrichina (or one of its derived forms) because there are enough older names available to handle the situation (Fig. 14a). However, a discussion about this topic is difficult because the nomenclature above the "familylevel" is not regulated by the ICZN.[1]

2.2 The Hypotricha Stein, 1859

The name Hypotricha was introduced by Stein (1859, p. 72, 73) to include the "Oxytrichinen Ehrenberg" (e.g., *Oxytricha*, *Stylonychia*, *Urostyla*, *Holosticha*, *Uroleptus*), the "Euplotinen Ehrenberg" (e.g., *Euplotes*), the "Aspidiscinen Ehrenberg" (e.g., *Aspidisca*), and the "Chlamydodonten Stein" (e.g., *Chlamydodon*, *Trochilia*). The misclassification of the chlamydodonts in the hypotrichs was recognised very early and therefore the Hypotricha consisted of the oxytrichids, the euplotids, and the aspidiscids over a long period (e.g., Kahl 1932, Corliss 1961). In the present book I confine the name Hypotricha to the non-euplotid hypotrichs because there is some evidence, especially from molecular data, that not the euplotids, but the oligotrichs are the sistergroup of the hy-

[1] The examples hypotrichs and stichotrichs show very impressively how different the authorship of higher taxa is handled. Stein (1859) established the Hypotricha as order, and no author dared to add his own name when he lowered or raised the rank (see, for example, Small & Lynn 1985, Lynn & Small 2002, and Tuffrau & Fleury 1994 for classifications including authorships). In the case of the suborder Stichotrichina Fauré-Fremiet, 1961 the situation is different. Small & Lynn (1985) introduced the monotypic subclass Stichotrichia, which is now generally assigned to Small & Lynn (1985).

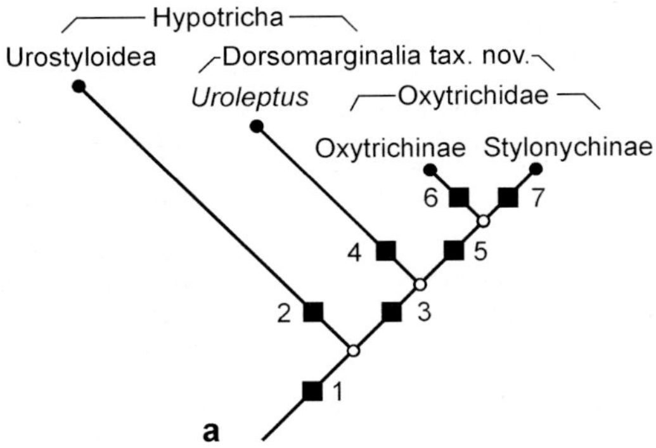

Fig. 14a Diagram of phylogenetic relationships within the Hypotricha (original). This arrangement is roughly supported by a tree based on rDNA (Hewitt et al. 2003), but also supports the CEUU-hypothesis proposed by Foissner et al. (2004a). For characterisation of the taxon Dorsomarginalia see chapter 2.4. For details see chapter 2 of the general section and ground pattern of the Urostyloidea (systematic section). Autapomorphies (black squares 1–7): **1** – oral primordium on cell surface; two macronuclear nodules; three dorsal kineties; 18 frontal-ventral-transverse cirri; somatic ciliature new; somatic ciliature largely lost in cyst (PKR-cyst; see chapter 1.10.3); endoral present (that is, two undulating membranes); one left and one right marginal row; high agreement in SSU rRNA gene sequences. **2** – more than six frontal-ventral-transverse cirral anlagen produce a distinct zigzag pattern of ventral cirri (that is, midventral complex composed of pairs only); more than five transverse cirri. **3** – dorsomarginal kineties present; micronuclear DNA polymerase alpha genes scrambled. **4** – more than 6 frontal-ventral-transverse cirri anlagen; body slender and tailed; actin I gene scrambling type *Uroleptus*. **5** – fragmentation of dorsal kinety 3; 4-layered cyst wall; actin I gene scrambling type *Oxytricha*; 2 or more macronuclear molecules encode histone H4. **6** – no apomorphy known, therefore likely a paraphyletic group. **7** – body rigid; adoral zone of membranelles ≥40% of body length; cortical granules lacking.

potrichs (Fig. 12b, 13a). Likely it would be more fair to use the older name Oxytrichina Ehrenberg, 1830 (or one of its derivatives; see Berger 2001) for this group instead of Hypotricha. However, at present it seems wise to retain the name Hypotricha and to use the name Oxytrichina, respectively its derived form Oxytrichidae, for a subgroup of the Hypotricha.

As just mentioned, the Hypotricha comprise all non-euplotid hypotrichs (Fig. 14a). Most morphologists agree that oxytrichids and urostyloids are two monophyletic lineages (e.g., Borror 1972, Corliss 1979, Borror & Wicklow 1983, Lynn & Small 1997, 2002, Shi et al. 1999). In contrast, Eigner (1997) proposed a non-monophyly of the oxytrichids, that is, he assumed that the characteristic "18 frontal-ventral-transverse cirral pattern" of this group and the specific ontogenetic processes producing this pattern evolved several times. However, the pattern and the ontogenesis are too complex to assume a convergent evolution. On the other hand, Eigner (2001) also supports a monophyly of the urostyloids. For further details on the this taxon see the next chapter.

The monophyly of the Oxytrichidae is, from the morphological point of view, mainly based on the fragmentation of dorsal kinety 3 (Fig. 14a; apomorphies 5).[1] The

[1] A fragmentation is also known from the "urostyloid" *Neokeronopsis spectabilis* (Fig. 243k, m). This indicates that *Neokeronopsis* is not a urostyloid, but an oxytrichid which convergently produced a zigzagging cirral pattern feigning a urostyloid origin.

18 frontal-ventral-transverse cirri, proposed as apomorphy by Berger & Foissner (1997) and Berger (1999), are likely a plesiomorphy at this level. Probably this pattern occurred for the first time in the last common ancestor of the Hypotricha (Fig. 14a, apomorphies 1). The dorsal kinety fragmentation is rather curious and therefore a convergent evolution is very unlikely. Of course, this pattern was transformed in various ways, for example, from simple to multiple fragmentation in *Pattersoniella* (for review see Berger 1999). Berger & Foissner (1997) proposed a rather distinct separation within the Oxytrichidae in the Oxytrichinae and the Stylonychinae. The Stylonychinae are characterised by at least three apomorphies, namely a rigid body, a long (more than 40% of body length) adoral zone of membranelles, and the lack of cortical granules (Fig. 14a, apomorphies 7). This group is supported by almost all molecular studies (e.g., Bernhard et al. 2001, Chen & Song 2002, Agatha et al. 2004, Foissner et al. 2004a), strongly indicating that the Stylonychinae are indeed a monophylum. Berger (1999) recognised from morphological data that *Pattersoniella*, *Laurentiella*, and *Onychodromus* are also members of the Stylonychinae. This was confirmed by molecular data some years later (Bernhard et al. 2001, Foissner et al. 2004a).

In contrast to the Stylonychinae, the Oxytrichinae are less well defined, both from the morphological and molecular point of view, strongly indicating the Oxytrichinae sensu Berger & Foissner (1997) and Berger (1999) are paraphyletic (Schmidt et al. 2004a, b). Various molecular markers indicate that almost all hypotrich species which do not belong to the urostyloids, to *Uroleptus*, or to the Stylonychinae, cluster somewhere inside the oxytrichids. This means that taxa like *Amphisiella* (dorsomarginal kineties and fragmenting dorsal kinety lacking; Wicklow 1982, Berger 2004a), *Kahliella* and *Parakahliella* (dorsomarginal kineties present, kinety fragmentation lacking; Berger et al. 1985, Berger & Foissner 1989b, Eigner 1995), *Engelmanniella* (dorsomarginal kineties and kinety fragmentation lacking; Wirnsberger-Aescht et al. 1989), *Paraurostyla* (dorsomarginal kineties and kinety fragmentation present; Wirnsberger et al. 1985, for review see Berger 1999) are more or less strongly modified 18-cirri oxytrichids. For example, the cirral pattern of *Amphisiella* species can be rather easily derived from the pattern of 18-cirri oxytrichids by a more or less distinct increase of cirri produced per anlage (Berger 2004a). However, since we do not know whether the lack of dorsomarginal kineties in *Amphisiella* is primary or secondary, we cannot estimate its position in the phylogenetic system; that is, we do not know whether or not it belongs to the Dorsomarginalia (Fig. 14a). Unfortunately, the position of other "difficult" taxa (e.g., *Engelmanniella*, *Gonostomum*) is rather different in various molecular trees.

2.3 The Urostyloidea Bütschli, 1889

The separation of the urostyloids from the oxytrichids dates back to Bütschli (1889), who established it as subfamily of the Oxytrichina family. Originally, the Urostylinae included *Trichogaster*, *Urostyla*, *Kerona*, *Epiclintes*, *Stichotricha*, *Strongylidium*, *Holosticha*, *Amphisia*, *Uroleptus*, and *Sparotricha*. Simultaneously, Bütschli established the

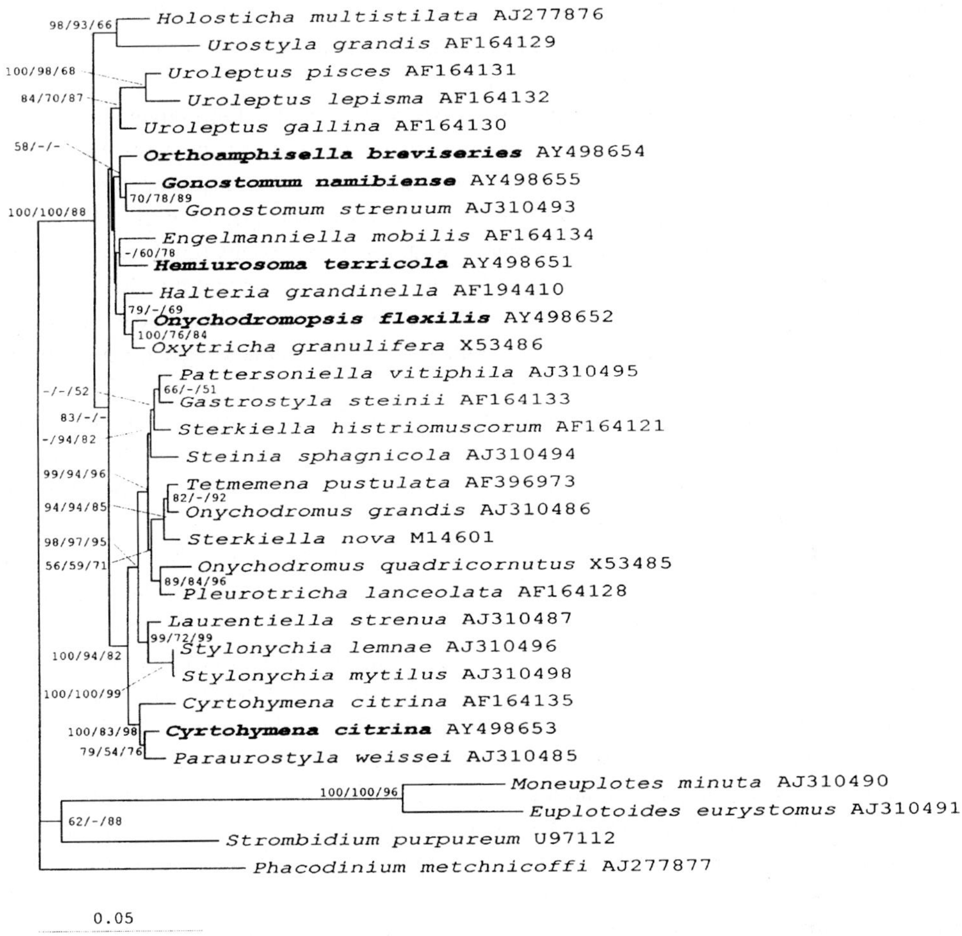

Fig. 15a Neighbour-joining tree of spirotrichs (mainly urostyloid and oxytrichid hypotrichs) based in 18S rRNA gene sequences (from Foissner et al. 2004a). The codes following the names are the GenBank Accession Numbers. The numbers at nodes represent the neighbour-joining and maximum-parsimony bootstrap percentages from 100 replicates (values below 50% not shown) and the quartet puzzle support values obtained with 10000 puzzling steps, respectively. The scale bar corresponds to a distance of 5 substitutions per 100 nucleotide positions. For details see text. *Holosticha multistilata* = *Anteholosticha intermedia* in present book.

subfamilies Pleurotrichina (e.g., *Pleurotricha*, *Oxytricha*, *Histrio*) and Psilotrichina (*Psilotricha* and *Balladina*).

Kahl (1932), the next revisor, ignored Bütschli's Urostylinae and again divided the hypotrichs in the Oxytrichidae, the Euplotidae, and the Aspidiscidae, that is, he included all non-euplotid hypotrichs in the Oxytrichidae without further division. Corliss (1961) accepted Kahl's scheme. By contrast, Faurè-Fremiet (1961) divided the hypotrichs into two new taxa, the Stichotrichina and the Sporadotrichina. The first group comprised the Urostylidae, which he incorrectly assigned to Calkins, the Keronidae, the

Holostichidae, and the Strongylidae. The Sporadotrichina contained the Pleurotrichidae, the Euplotidae, the Gastrocirrhidae, and the Aspidiscidae. We now know that Fauré-Fremiet's classification was rather artificial.

Based on the observations on *Pseudourostyla cristata* by Jerka-Dziadosz (1964) and own data, Borror (1972) refined terminology by introducing the term midventral cirri (see chapter 1.7 for details), which are paired and form a highly characteristic zig-zag pattern (Fig. 1a). He also recognised the importance of the special origin of this pattern from rather many oblique cirral anlagen. Within the hypotrichs (euplotids and non-euplotids) he distinguished six families including the Urostylidae and the Holostichidae (Table 3). Some years later, he came to the conclusion that the Holostichidae are a junior synonym of the Urostylidae because both groups are characterised by midventral cirri (Borror 1978, 1979).

Borror & Wicklow (1983) made the last revision of the urostyloids. They provided a key to all species and list of synonyms, but neither detailed discussions nor descriptions. They introduced the Pseudokeronopsidae comprising the two subgroups Pseudokeronopsinae and Thigmokeronopsinae (Table 8).

The Urostyloidea are a large group of hypotrichs; however, their phylogenetic position in the hypotrichs is not yet certain. Molecular data about the origin of the urostyloids are rather contradictory. In some trees they are classified within the 18-cirri oxytrichids (e.g., Shin et al. 2000, Bernhard et al. 2001, Snoeyenbos-West et al. 2002 [their Fig. 2], Strüder-Kypke & Lynn 2003), in others they cluster outside the oxytrichids (e.g., Lozupone et al. 2001, Snoeyenbos-West et al. 2002 [their Fig. 1], Hewitt et al. 2003, Croft et al. 2003, Foissner et al. 2004a, Dalby & Prescott 2004, Coleman 2005; Fig. 15a).

The neighbour-joining and parsimony analyses by Foissner et al. (2004a) support the morphological and morphogenetical data on the monophyly of the *Urostyla/Anteholosticha* clade, which is the sister group of all other hypotrichs (Fig. 15a). However, the three *Uroleptus* species do not cluster with the urostyloids, but with the oxytrichids, although they look like typical urostyloids differing from *Anteholosticha* only in body shape (tailed vs. untailed) and the presence/absence of dorsomarginal kineties (Borror 1972, Martin et al. 1981, Hemberger 1982, Borror & Wicklow 1983, Eigner 2001). A similar result was obtained by Dalby & Prescott (2004) using *Urostyla grandis* and *Holosticha polystylata* (= *Diaxonella pseudorubra* in the present book) as representatives of the urostyloids. This suggests that *Uroleptus*-species are not urostyloids, a hypothesis supported by the rather similar trees by Snoeyenbos-West et al. (2002) and Hewitt et al. (2003). In contrast, the trees by Shin et al. (2000), Bernhard et al. (2001), and Chen & Song (2002) cluster "*Holosticha*" (= *Anteholosticha* in present book) very near to *Oxytricha granulifera*. Likely, this is because they did not include *Urostyla*. This indicates that such profound differences in treestructures are caused by insufficient taxa sampling. Sequences are available from only about 40 of the more than 300 known oxytrichids (Berger 1999) and urostyloids (present book). Further, more than 50% of the ciliate species are probably undescribed (for review see Foissner et al. 2002). Accordingly, molecular trees contain less than 5% of the hypotrich species that probably exist. Nonetheless, differences in alignment, outgroup, phylogenetic algorithm, and clustering method may also contribute to the differences in the molecular trees available. Lastly, the low sequence divergence among the hypotrichs hampers phylogenetic analysis and makes SSU-rRNA gene analyses very sensitive to undersampling (Foissner et al. 2004a). However, the morphological situation is not much better because the many conflicting features prevent the construction of a convincing tree.

For some comments on the phylogenetic relationships within the Urostyloidea see the systematic section.

2.4 Is *Uroleptus* a Subgroup of the Urostyloidea?

Uroleptus is generally assigned to the urostylids because it has – like, for example, *Holosticha* or *Urostyla* – zigzagging ventral cirri (e.g., Bütschli 1889, Borror 1972, Borror & Wicklow 1983, Tables 2–8, 10). Only rarely is it assigned to other higher taxa, for example, the Kahliellidae (Tuffrau & Fleury 1994; p. 137).

However, several molecular studies suggest that the inclusion of *Uroleptus* in the urostyloids is incorrect (Snoeyenbos-West et al. 2002, Hewitt et al. 2003, Croft et al. 2003; Foissner et al. 2004a, Fig. 15a; Dalby & Prescott 2004). According to these molecular data, *Uroleptus* is more closely related to the oxytrichids than to the urostyloids. This requires the assumption that the conspicuous zigzag-pattern formed by the ventral cirri evolved convergently in the Urostyloidea and in *Uroleptus* (if we assume that the last common ancestor of the Hypotricha had six frontal-ventral-transverse cirral anlagen; see below). Thus, we proposed the CEUU (*C*onvergent *E*volution of *U*rostylids and *U*roleptids) hypothesis, which tries to combine morphological and molecular data (Foissner et al. 2004a). Traditionally, 18-cirri[1] oxytrichids and euplotids – that is, spirotrichs with relatively few and "sporadically" arranged cirri – are regarded as derived from a *Urostyla*-like ancestor which had many cirral rows (e.g., Kahl 1932, Borror 1972, Wirnsberger 1987). The CEUU hypothesis, however, tries to explain the opposite; that is, an euplotid-like ancestor because the euplotids have an "*Oxytricha*-like" cirral pattern originating from six anlagen (or vice versa; Wallengren 1900) and are the adelphotaxon of the group formed by the oligotrichs and the hypotrichs in many molecular trees (Fig. 12b; e.g., Eisler et al. 1995, Shin et al. 2000, Bernhard et al. 2001, Petroni et al. 2002, Modeo et al. 2003, Strüder-Kypke & Lynn 2003). Moreover, the hypothesis proposes that cirri- and anlagen-multiplication are not necessarily correlated with the production of a midventral complex and occurred several times in the Oxytrichidae (e.g., *Paraurostyla*, *Laurentiella*, *Styxophrya*), including midventral complex-like arrangements as, for example, in *Pattersoniella* and *Territricha* (for review see Berger 1999). Indeed, *Pattersoniella* and *Territricha* have been united in the Pattersoniellidae and included in the urostylids by Shi et al. (1999; Table 10). However, detailed morphological and molecular studies clearly show that *Pattersoniella* and *Territricha* belong to the Oxytrichidae (see above; Berger 1999, Bernhard et al. 2001, Foissner et al. 2004a).

[1] The term "18-cirri oxytrichids" is an abbreviation for oxytrichids with 18 frontal-ventral-transverse cirri arranged in the highly characteristic pattern discussed in detail by Berger (1999).

Fig. 16 shows the proposed homology of the euplotid, urostyloid, and oxytrichid cirri. The homology of the euplotid and oxytrichid cirral pattern was already explained in detail by Wallengren (1900). If we assume that the last common ancestor of the euplotids and hypotrichs (Fig. 12a), respectively, the spirotrichs (Fig. 12b), had six cirral anlagen then we have to assume that the urostyloid midventral cirral pattern originated by inserting additional anlagen, each producing a pair of cirri, among the basic six anlagen (I–VI). If the tree shown in Fig. 14a is basically correct then this process must have occurred at least twice. The first separation process caused the common ancestor to split in a urostyloid and oxytrichid (including *Uroleptus*) lineage. Later, a similar zigzag pattern evolved again either outside (Fig. 14a) or within the Oxytrichidae (Foissner et al. 2004a) to form the *Uroleptus* lineage. Although cirri and anlagen multiplication are obviously not correlated with the generation of a midventral pattern, the scenario above is quite likely, considering the many cirral patterns found in the oxytrichids (for review see Berger 1999). Possibly, the second separation event was driven by ecological constraints because all *Uroleptus* species are, per definition, more or less distinctly tailed (Foissner et al. 2004a).

Although the CEUU hypothesis is reasonable and in accordance with several molecular trees as well as with the high diversity of the oxytrichid cirral pattern in general, Foissner et al. (2004a) did not have a specific morphological proof. The tree proposed in Fig. 14a is basically in accordance with the molecular tree presented by Hewitt et al. (2003)[1], especially in that *Uroleptus* clusters outside the oxytrichids. For this hypothesis the morphology can provide a very good apomorphy for the group *Uroleptus* + Oxytrichidae (Fig. 14a, apomorphies 3), namely, the presence of dorsomarginal kineties. These kineties, which are never associated with caudal cirri, originate from/very near the right marginal row primordium and occur only in *Uroleptus* species (e.g., Martin et al. 1981, Foissner et al. 1991, Eigner 2001), very many oxytrichids (for review see Berger 1999; Fig. 243j, l, m), and some other taxa, for example, *Kahliella*, *Parakahliella, Nudiamphisiella* (Berger et al. 1985, Eigner 1995, Foissner et al. 2002). This feature is rather conspicuous and therefore has to be interpreted as synapomorphy implying that *Uroleptus* is more closely related to the oxytrichids than to the urostyloids.[2] On the other hand, fragmentation of dorsal kineties is lacking in *Uroleptus*, which is a distinct hint that it splits off outside the Oxytrichidae. The close relationship of *Uroleptus* and the Oxytrichidae is not only indicated by the dorsomarginal row, but also by molecular features. Thus, the monophylum composed of *Uroleptus* and Oxytrichidae is rather certain and therefore named Dorsomarginalia taxon novum[3]: Hypotricha with dorsomarginal kineties and scrambled micronuclear DNA polymerase alpha genes (Fig. 14a, apomorphies 3). For a discussion of the features, see the ground pattern chapter of the Urostyloidea. *Amphisiella* and some other taxa lack a midventral pattern and dorsomarginal

[1] Unfortunately, the three *Uroleptus* species used in this paper do not form a monophylum.

[2] Hemberger (1982, p. 89) described, but did not illustrate (!), the de novo origin of a dorsal kinety beside the right marginal row in *Holosticha pullaster*. Whether or not this feature is homologous to the dorsomarginal kineties is not known.

[3] The name refers to the main (sole?) morphological autapomorphy of this group, the dorsomarginal kineties, which are formed at/near the right marginal row primordium.

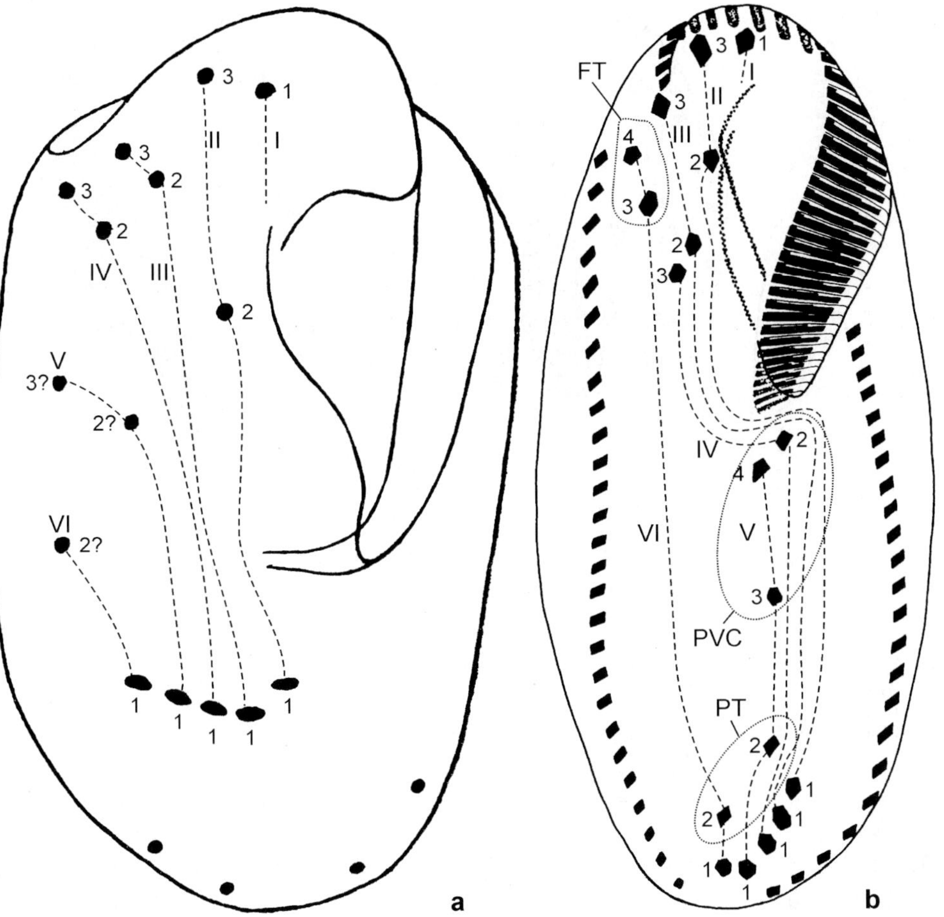

Fig. 16a, b Homology of cirri in a euplotid (a, *Euplotes harpa*; from Wallengren 1900) and an 18-cirri oxytrichid (b, *Sterkiella histriomuscorum*; from Augustin & Foissner 1992), as representative of the hypotrichs (further hypotrichs see Fig. 16c–o). Numbering of frontal-ventral-transverse cirral anlagen (I–VI) and cirri (1–4) according to Wallengren (1900). Cirri originating from the same anlage are connected by a broken line. The rearmost cirrus (1) of each anlage is the so-called transverse cirrus; these cirri form the transverse cirral row which is a pseudorow. The main difference between *Euplotes* and *Sterkiella* is the different number of cirri formed by the anlagen V and VI: *Euplotes* forms only three, respectively, two cirri, whereas *Sterkiella* produces each four cirri. Unfortunately, we are unable to say which cirri are lacking in *Euplotes*, respectively, supernumerary in *Sterkiella*; thus the question marks in *Euplotes*. Anyhow, the similarities between the cirral patterns and their origin in these two representatives are too high to be explained by chance, that is, we have to assume that the last common ancestor of the euplotids and hypotrichs produced a relatively low number of distinct cirri from six (I–VI) anlagen. Thus, six frontal-ventral-transverse cirri anlagen is obviously a plesiomorphy in the stem-lineage of the Hypotricha (Fig. 14a). FT = frontoterminal cirri, PT = pretransverse ventral cirri, PVC = postoral ventral cirri, I–VI = frontal-midventral-transverse cirral anlagen (primordia, streaks), 1–4 = cirri formed within an anlage (the rearmost cirrus has the number 1).

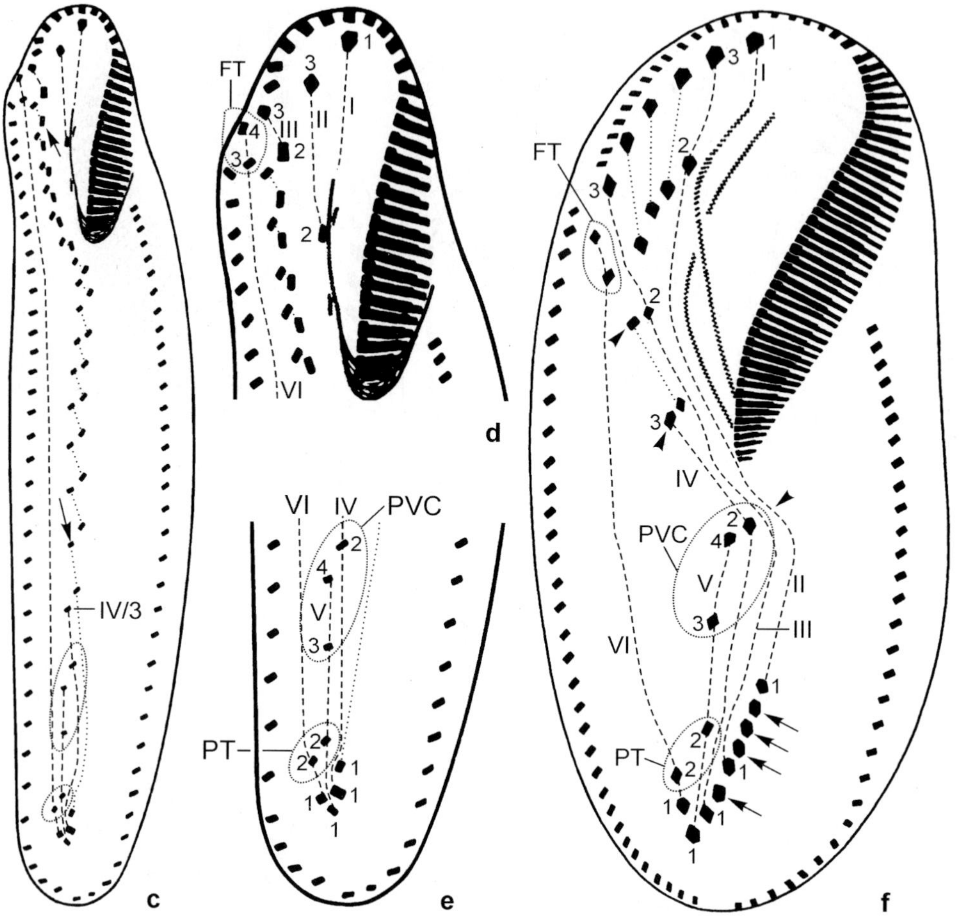

Fig. 16c–f Homology of cirri in a urostyloid (c–e; *Anteholosticha australis*) and a stylonychine (f; *Pattersoniella vitiphila*, from Foissner 1987b) hypotrich (after Foissner et al. 2004a, supplemented; see also Fig. 16a, b, g–o). Urostyloid hypotrichs likely evolved from an 18-cirri ancestor by inserting additional anlagen generating cirral pairs, which produce the highly characteristic zigzagging midventral pattern (first and last additional cirral pair marked by an arrow each in (c). Cirri of each additional pair connected by dotted line; for zigzag-pattern see Fig. 16l). Numbering of frontal-ventral-transverse cirral anlagen (I–VI) and cirri (1–4), which are homologous to those in euplotids and 18-cirri oxytrichids (Fig. 16a, b) according to Wallengren (1900). Note that the insertion of additional anlagen did not occur only in the urostyloids, but also in oxytrichids as indicated in (f) which shows *Pattersoniella*, a stylonychine oxytrichid "feigning" a bicorona and a urostyloid midventral pattern (pseudopairs marked by arrowheads). Arrows denote the additional cirral anlagen. FT = frontoterminal cirri, PT = pretransverse ventral cirri, PVC = postoral ventral cirri (in urostyloids distinctly dislocated posteriorly), I–VI = frontal-midventral-transverse cirral anlagen (primordia, streaks), 1–4 = cirri formed within an anlage (the rearmost cirrus has the number 1).

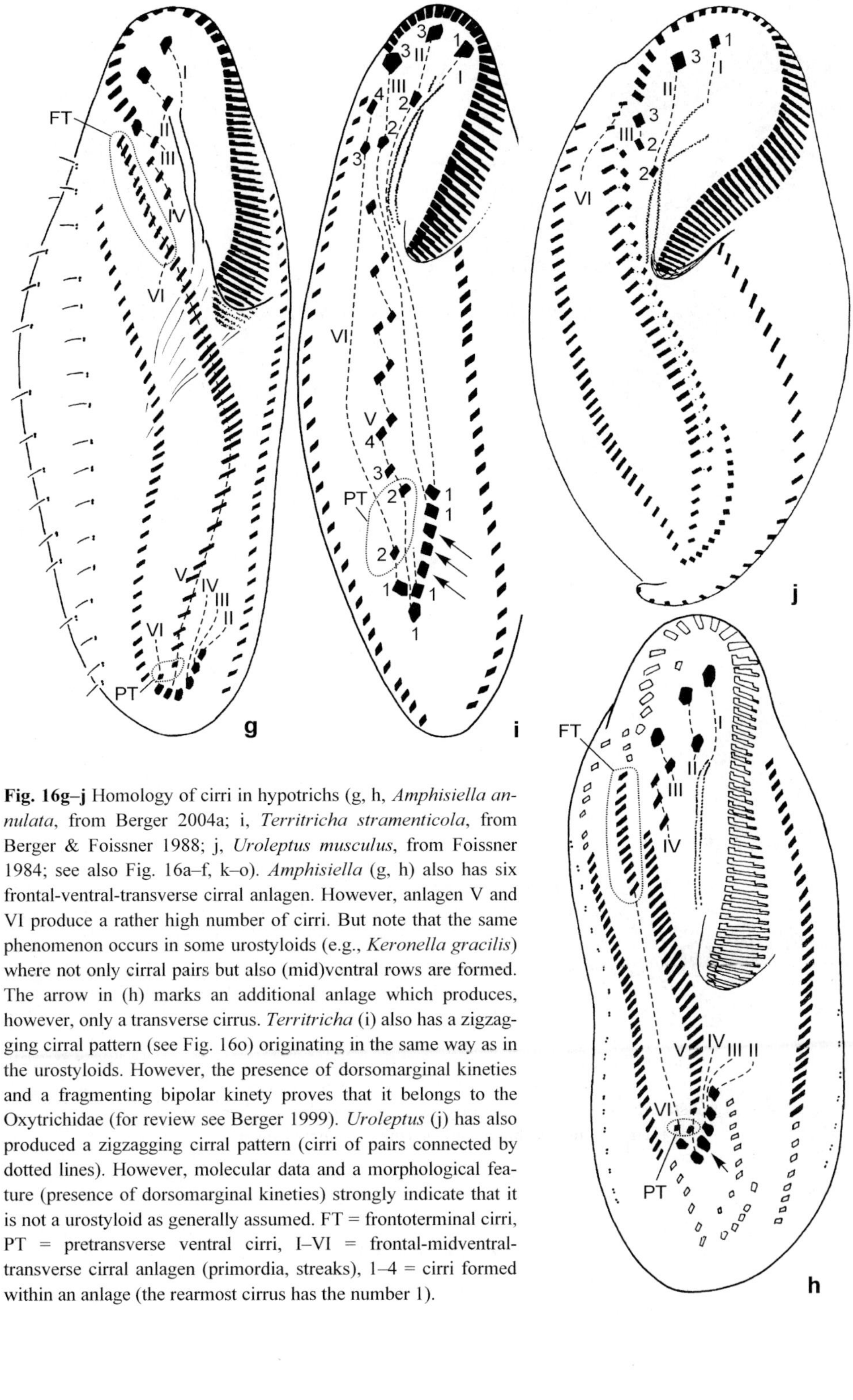

Fig. 16g–j Homology of cirri in hypotrichs (g, h, *Amphisiella annulata*, from Berger 2004a; i, *Territricha stramenticola*, from Berger & Foissner 1988; j, *Uroleptus musculus*, from Foissner 1984; see also Fig. 16a–f, k–o). *Amphisiella* (g, h) also has six frontal-ventral-transverse cirral anlagen. However, anlagen V and VI produce a rather high number of cirri. But note that the same phenomenon occurs in some urostyloids (e.g., *Keronella gracilis*) where not only cirral pairs but also (mid)ventral rows are formed. The arrow in (h) marks an additional anlage which produces, however, only a transverse cirrus. *Territricha* (i) also has a zigzagging cirral pattern (see Fig. 16o) originating in the same way as in the urostyloids. However, the presence of dorsomarginal kineties and a fragmenting bipolar kinety proves that it belongs to the Oxytrichidae (for review see Berger 1999). *Uroleptus* (j) has also produced a zigzagging cirral pattern (cirri of pairs connected by dotted lines). However, molecular data and a morphological feature (presence of dorsomarginal kineties) strongly indicate that it is not a urostyloid as generally assumed. FT = frontoterminal cirri, PT = pretransverse ventral cirri, I–VI = frontal-midventral-transverse cirral anlagen (primordia, streaks), 1–4 = cirri formed within an anlage (the rearmost cirrus has the number 1).

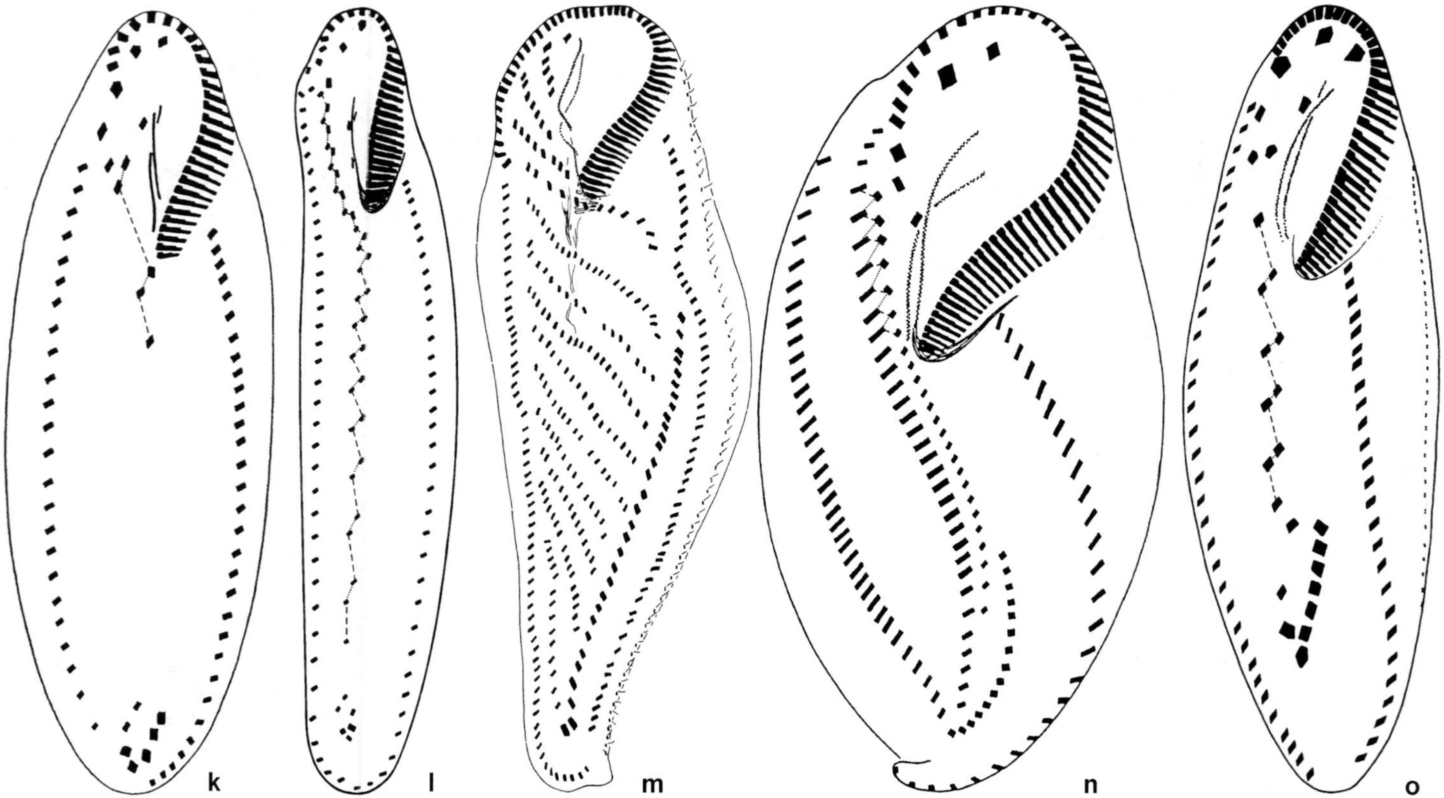

Fig. 16k–o "Zigzag" pattern in *Oxytricha lanceolata* (k; from Foissner 1996a), *Anteholosticha australis* (l), *Uroleptus musculus* (n, from Foissner 1984), and *Territricha stramenticola* (o, from Berger & Foissner 1988; see also Fig. 16a–j). Cirri of a true pair are connected by a broken line, cirri of a pseudopair (see Fig. 1a) are connected by a dotted line. The higher the number of cirral pairs, the better becomes recognisable the zigzag pattern. *Epiclintes auricularis* (m), a urostyloid, has lost the characteristic zigzagging midventral pattern because the anlagen do not produce cirral pairs, but cirral rows.

kineties (Wicklow 1982, Berger 2004a). Molecular data are needed to understand whether dorsomarginal kineties are primarily or secondarily lacking in these groups.

Kelminson et al. (2002) found that in *Uroleptus* sp. only one macronuclear molecule encodes histone H4, whereas in oxytrichids (e.g., *Sterkiella*, *Stylonychia*, *Pleurotricha*, *Oxytricha*) two or more molecules encode histone H4. Harper & Jahn (1989) detected only a single histone H4 gene in *Moneuplotes crassus*, indicating that this is the plesiomorphic state within the spirotrichs. This would suggest that (i) *Uroleptus* split off outside the Oxytrichidae (Fig. 14a), and not inside as suggested by Foissner et al. (2004a), and (ii) the increased number of histone H4 encoding macronuclear molecules are an apomorphy of the Oxytrichidae. Possibly, the four-layered cyst wall is a further apomorphy of the Oxytrichidae (for review on cyst wall data, see Gutiérrez et al. 2003).

Anyhow, at the present state of knowledge there is strong evidence that *Uroleptus* is a distinct group more closely related to the Oxytrichidae than to the Urostyloidea. Thus, it is not treated in the present review.

3 Previous Classifications and Revisions

Several classifications of urostyloids exist. The original classification by Bütschli (1889) and some modern classification schemes are shown in Tables 2–11. I did not change the original presentation, for example, original spelling of names; moreover, authors are partially not included in my reference list. Kahl (1932) included all non-euplotid hypotrichs in the Oxytrichidae without further subdivision. Thus, his classification is not presented.

Kahl (1932) provided the last detailed revision of urostyloid taxa; that is, his paper included a key to all species and a description and illustration of each species. Later reviews contained either only a list of genera and species (e.g., Borror 1972), or descriptions (Hemberger 1982), or a key and list of synonyms (Borror & Wicklow 1983). Thus, it was not too early to review the data on the Urostyloidea thoroughly.

For a brief discussion of the various schemes presented below, see the systematic section. Several taxa (e.g., *Amphisiella*, *Gonostomum*, *Pattersoniella*, *Uroleptoides*) are not considered in the present book, although classified by some authors in the urostyloids or holostichids. For an explanation of the exclusion, see the chapter "Taxa not considered" at the end of the book.

Table 2 Classification of urostyloid ciliates according to Bütschli (1889)

Family Oxytrichina (Ehrbg) Stein, 1859
  Subfamily Urostylinae Bütschli
    Trichogaster Sterki, 1878
    Urostyla Ehrenberg, 1830
    Kerona Müller, 1786
    Epiclintes Stein, 1862
    Stichotricha Perty, 1849
    Strongylidium Sterki, 1878

Table 2 Continued

Holosticha Wrzesniowski, 1877
Amphisia Sterki, 1876
Uroleptus Ehrenberg, 1831
Sparotricha Entz, 1879

Table 3 Classification of urostyloid ciliates according to Borror (1972)

Family Urostylidae Bütschli, 1889
Urostyla Ehrenberg, 1830
Amphisiella Gourret & Roeser, 1887
Balladyna Kowalewski, 1882
Banyulsella Dragesco, 1953
Epiclintes Stein, 1862
Kahliella Corliss, 1960
Kerona Ehrenberg, 1835
Lacazea Dragesco, 1960
Paraurostyla n. g.
Family Holostichidae Fauré-Fremiet, 1961
Holosticha Wrzesniowski, 1877
Keronopsis Penard, 1922
Paraholosticha Kahl, 1932
Pseudourostyla n. g.
Trichotaxis Stokes, 1891
Uroleptopsis Kahl, 1932
Uroleptus Ehrenberg, 1831

Table 4 Classification of urostyloid ciliates according to Borror (1979)

Family Urostylidae Bütschli, 1889
Urostyla Ehrenberg, 1830
Bakuella Agamaliev & Alekperov, 1976
Holosticha Wrzesniowski, 1877
Keronopsis Penard, 1922
Pseudourostyla Borror, 1972
Uroleptus Ehrenberg, 1831

Table 5 Classification of urostyloid ciliates according to Corliss (1979)

Family Urostylidae Bütschli, 1889
Banyulsella Dragesco, 1953
Hemicycliostyla Stokes, 1886
Isosticha Kiesselbach, 1936
Paraholosticha Kahl, 1932
Paraurostyla Borror, 1972
Urostyla Ehrenberg, 1838
Family Holostichidae Fauré-Fremiet, 1961
Amphisiella Gourret & Roeser, 1888
Bakuella Agamaliev & Alekperov, 1976
Balladyna Kowalewski, 1882
Balladynella Stiller, 1974

Table 5 Continued

Gonostomum Sterki, 1878
Holosticha Wrzesniowski, 1877
Keronopsis Penard, 1922
Lamtostyla Buitkamp, 1977
Laurentiella Dragesco & Njiné, 1971
Paruroleptus Kahl, 1932
Parurosoma von Gelei, 1954
Psammomitra Borror, 1972
Pseudourostyla Borror, 1972
Trachelochaeta Sramek-Husek, 1954
Trachelostyla Kahl, 1932
Trichotaxis Stokes, 1891
Uncinata Bullington, 1940
Uroleptoides Wenzel, 1953
Uroleptus Ehrenberg, 1831
Wallackia Foissner, 1977

Table 6 Classification of urostyloid ciliates according to Wicklow (1981)

Suborder Urostylina Jankowski, 1979
- Superfamily Urostyloidea Bütschli, 1889
  - Family Urostylidae Bütschli, 1889
    - Subfamily Holostichinae (n. subfam.)
      - *Holosticha*
      - *Bakuella*
      - *Uroleptus*
    - Subfamily Urostylinae (n. subfam.)
      - *Urostyla*
  - Family Keronopsidae Jankowski, 1979
    - Subfamily Keronopsinae (n. subfam.)
      - *Keronopsis*
    - Subfamily Thigmokeronopsinae (n. subfam.)
      - *Thigmokeronopsis*
- Superfamily Pseudourostyloidea (n. superfam.)
  - Family Pseudourostylidae Jankowski, 1979
    - *Pseudourostyla*

Table 7 Classification of urostyloid ciliates according to Hemberger (1982)

Family Urostylidae Bütschli, 1889
- *Urostyla* Ehrenberg, 1838
- *Bakuella* Agamaliev & Alekperov, 1976
- *Holosticha* Wrzesniowski, 1877
- *Periholosticha* n. gen.
- *Trichototaxis* Stokes, 1891
- *Uroleptopsis* Kahl, 1932
- *Uroleptus* Ehrenberg, 1831

Table 8 Classification of urostyloid ciliates according to Borror & Wicklow (1983)

Urostylina Jankowski, 1979
- 1. Urostyloidea Bütschli, 1889
  - 1. Urostylidae Bütschli, 1889
    - 1. Urostylinae Bütschli, 1889
      - 1. *Urostyla* Ehrenberg, 1838
    - 2. Holostichinae Fauré-Fremiet, 1961
      - 1. *Holosticha* Wrzesniowski, 1877
      - 2. *Bakuella* Agamaliev & Alekperov, 1976
      - 3. *Uroleptus* Ehrenberg, 1831
  - 2. Pseudokeronopsidae fam. nov.
    - 1. Pseudokeronopsinae subfam. nov.
      - 1. *Pseudokeronopsis* gen. nov.
    - 2. Thigmokeronopsinae Wicklow, 1981
      - 1. *Thigmokeronopsis* Wicklow, 1981
- 2. Pseudourostyloidea Jankowski, 1979
  - 1. Pseudourostylidae Jankowski, 1979
    - 1. *Pseudourostyla* Borror, 1972

Table 9 Classification of urostyloid ciliates according to Tuffrau & Fleury (1994)

Order Urostylida Jankowski, 1979
- Family Urostylidae Bütschli, 1889
  - *Bakuella* Agamaliev & Alekperov, 1976
  - *Isosticha* Kiesselbach, 1936
  - *Urostyla* Ehrenberg, 1830
- Family Pseudourostylidae Jankowski, 1979
  - *Pseudourostyla* Borror, 1972
- Family Holostichidae Fauré-Fremiet, 1961
  - *Holosticha* Wrzesniowski, 1877
  - *Paruroleptus* Kahl, 1932
  - *Periholosticha* Hemberger, 1982
  - *Trichotaxis* Stokes, 1891
- Family Pseudokeronopsidae Borror & Wicklow, 1983
  - *Keronella* Wiackowski, 1985
  - *Pseudokeronopsis* Borror & Wicklow, 1983
  - *Thigmokeronopsis* Wicklow, 1981
  - *Uroleptopsis* Kahl, 1932

Table 10 Classification of urostyloid ciliates according to Shi et al. (1999)

Suborder Urostylina Jankowski, 1979
- Family Urostylidae Bütschli, 1889
  - *Urostyla* Ehrenberg, 1830
  - *Metabakuella* Alekperov, 1989
  - *Pseudourostyla* Borror, 1972
  - *Metaurostylopsis* Song & Petz, in press
  - *Australothrix* Blatterer & Foissner, 1988
  - *Birojimia* Berger & Foissner, 1989
  - *Thigmokeronopsis* Wicklow, 1981
- Family Holostichidae Fauré-Fremiet, 1961
  - *Keronella* Wiackowski, 1985

Table 10 Continued

Bakuella Agamaliev & Alekperov, 1976
Parabakuella Song & Wilbert, 1987
Pseudokeronopsis Borror & Wicklow, 1983
Tricoronella Blatterer & Foissner, 1988
Holosticha Wrzesniowski, 1877
Uroleptus Ehrenberg, 1831
Periholosticha Hemberger, 1985
Notocephalus Petz et al., 1995
Family Pseudoamphisiellidae Song et al., 1997
Pseudoamphisiella Song, 1996
Family Pattersoniellidae Foissner, 1987
Pattersoniella Foissner, 1987
Territricha Berger & Foissner, 1988

Table 11 Classification of urostyloid ciliates according to Lynn & Small (2002)[a]

Order Urostylida Jankowski, 1979
Family Urostylidae Bütschli, 1889
Notocephalus Petz, Song & Wilbert, 1995
Eschaneustyla Stokes, 1886
Birojima Berger & Foissner, 1989
*Urostyla*Ehrenberg, 1830
Australothrix Blatterer & Foissner, 1988
Paruroleptus Kahl, 1932
Parabakuella Song & Wilbert, 1988
Bakuella Agamaliev & Alekperov, 1976
Holosticha Wrzesniowski, 1877
Territricha Berger & Foissner, 1988
Family Pseudourostylidae Jankowski, 1979
Pseudourostyla Borror, 1972
Family Pseudokeronopsidae Borror & Wicklow, 1983
Thigmokeronopsis Wicklow, 1981
Keronella Wiackowski, 1985
Tricoronella Blatterer & Foissner, 1988
Bicoronella Foissner, 1995
Pseudokeronopsis Borror & Wicklow, 1983

[a] Lynn & Small (2002) listed only representative genera.

For the classification used in the present book see table of content.

4 Parasitism

Urostyla grandis is sometimes attacked or infected by the suctorian *Podophrya urostylae* (Maupas, 1881) Jankowski, 1963 (basionym *Sphaerophrya urostylae*)[1]. The first record of this parasite was likely provided by Cohn (1851, p. 277, Tafel VII, Fig. 11, 12), who found *Urostyla*-cells packed with black-grey globules. Cohn, Lachmann (1856, p.

[1] According to Dovgal (2002, p. 245) the original combination, *Sphaerophrya urostylae*, is correct.

386), and Stein (1859) mistakenly interpreted the globules as embryos (embryonal hypothesis) of *Urostyla grandis*. But even Balbiani (1858, 1860), Engelmann (1876), and Bütschli (1876) recognised the parasitic nature of the suctorians, whose life cycle was described in detail by Stein (1859, Fig. 17a–y) and Jankowski (1963).

Adult specimens of the suctorian ciliate are about 35 μm across (Matthes 1988, p. 165). The swarmer has seven ciliary wreathes and tentacles with which it adheres to the host. It loses the cilia and causes an invagination at the host, which it penetrates and starts to suck with the tentacles. The invagination produced by the suctorian ciliate is not closed during the development to a globular, adult suctorian which has one or two contractile vacuoles, a spherical macronucleus, and one micronucleus. The swarmer is formed by external budding (Fig. 18a). Adult specimens can also be found outside the host. They are stalked or unstalked and have tentacles of ordinary length (Fig. 18b). *Podophrya urostylae* forms stalked resting cysts, which deviate distinctly from the *Podophrya*-type (Fig. 18c).

5 Ecology, Occurrence, and Geographic Distribution

Urostyloids live, throughout the year, in almost all biotopes, for example, freshwater (brooks, rivers, lakes, ponds), brackish water, sea, semiterrestrial habitats, and soil (e.g., Borror & Wicklow 1983, Foissner 1987a, 1998, Foissner et al. 1995, 1995a, Kahl 1932, Patterson et al. 1989, Petz & Leitner 2003). No symbiotic or parasitic species is known.

Very likely all limnetic and marine species are, as in most other hypotrichs, bottom-dwellers creeping on, for example, detritus, stones, or macrophytes. No species is obligatorily pelagic, however, several species can be occasionally found in the plankton community of large rivers, lakes, ponds, and the sea (for review see Foissner et al. 1999 and Petz 1999).

Many species are confined to one of the three major habitats, freshwater, sea, or soil. Only few species are reliably recorded from two habitats. For example, *Holosticha pullaster* is very common in limnetic and marine habitats, and *Anteholosticha intermedia* (= *Holosticha multistilata* of earlier papers) is present both in soil and freshwater. However, *Holosticha pullaster* was never reliably recorded from terrestrial habitats, and the large limnetic *Urostyla grandis* obviously does not occur in the sea or the soil. Possibly, the populations of a species inhabiting different habitats (e.g., freshwater and sea) are sibling species because gene flow among these populations is hampered (not existent?). Interestingly, a rather high percentage of urostyloid species occurs in marine habitats. Some groups are confined to the sea (*Thigmokeronopsis*), or at least most included species occur exclusively in this habitat (*Pseudokeronopsis*, *Holosticha*).

Fig. 17a–c *Urostyla grandis* parasitised by *Podophrya urostylae* (from Stein 1859). Stein and some other workers misinterpreted the parasitisation as embryonic reproduction. The suctors are usually scattered throughout the cytoplasm. Three suctors in the *Urostyla* specimen shown in (a) divide; some adult suctors have formed swarmers, which leave the host (b, c). Further details see text and Stein's (1859) exhaustive description. →

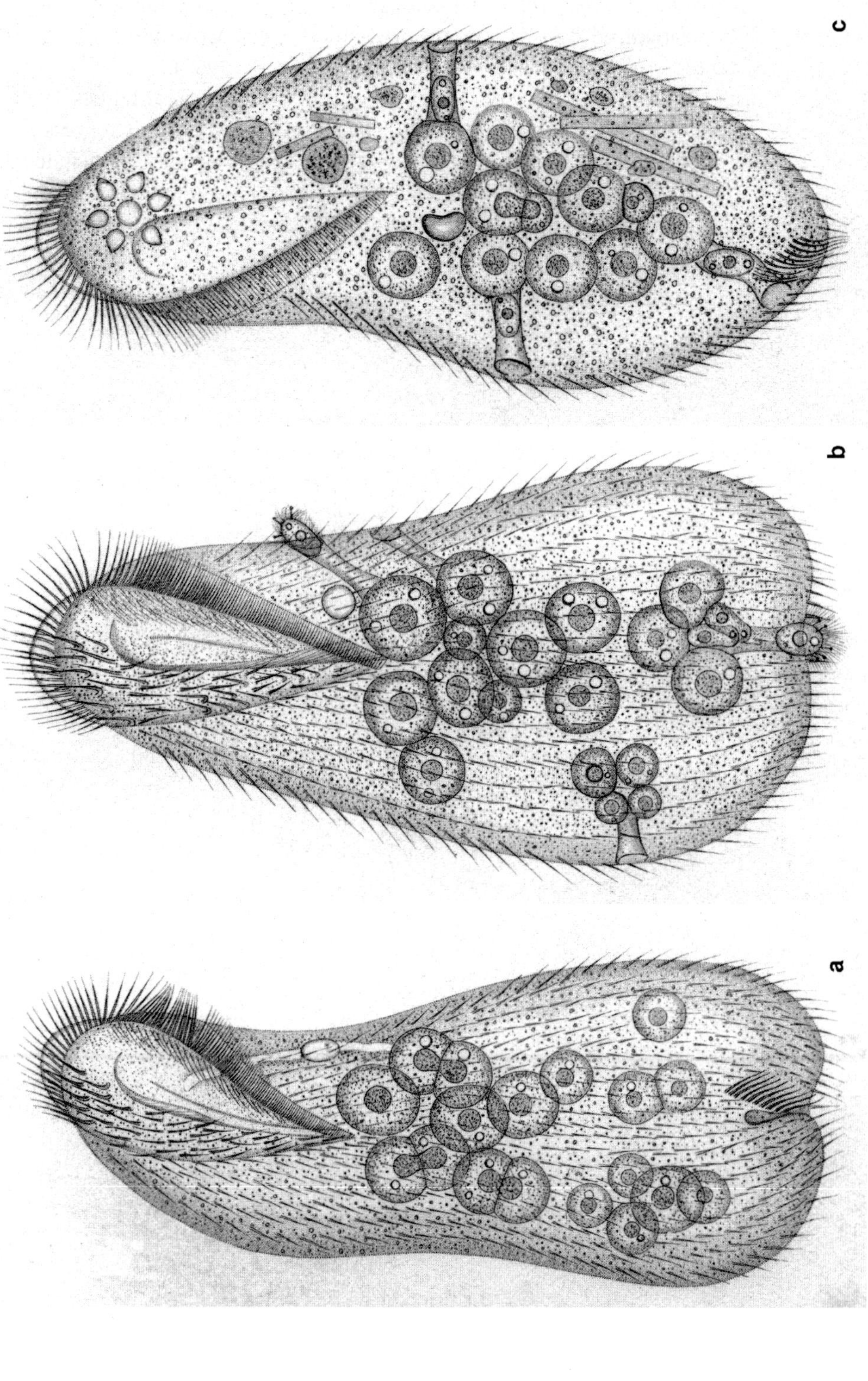
c
b
a

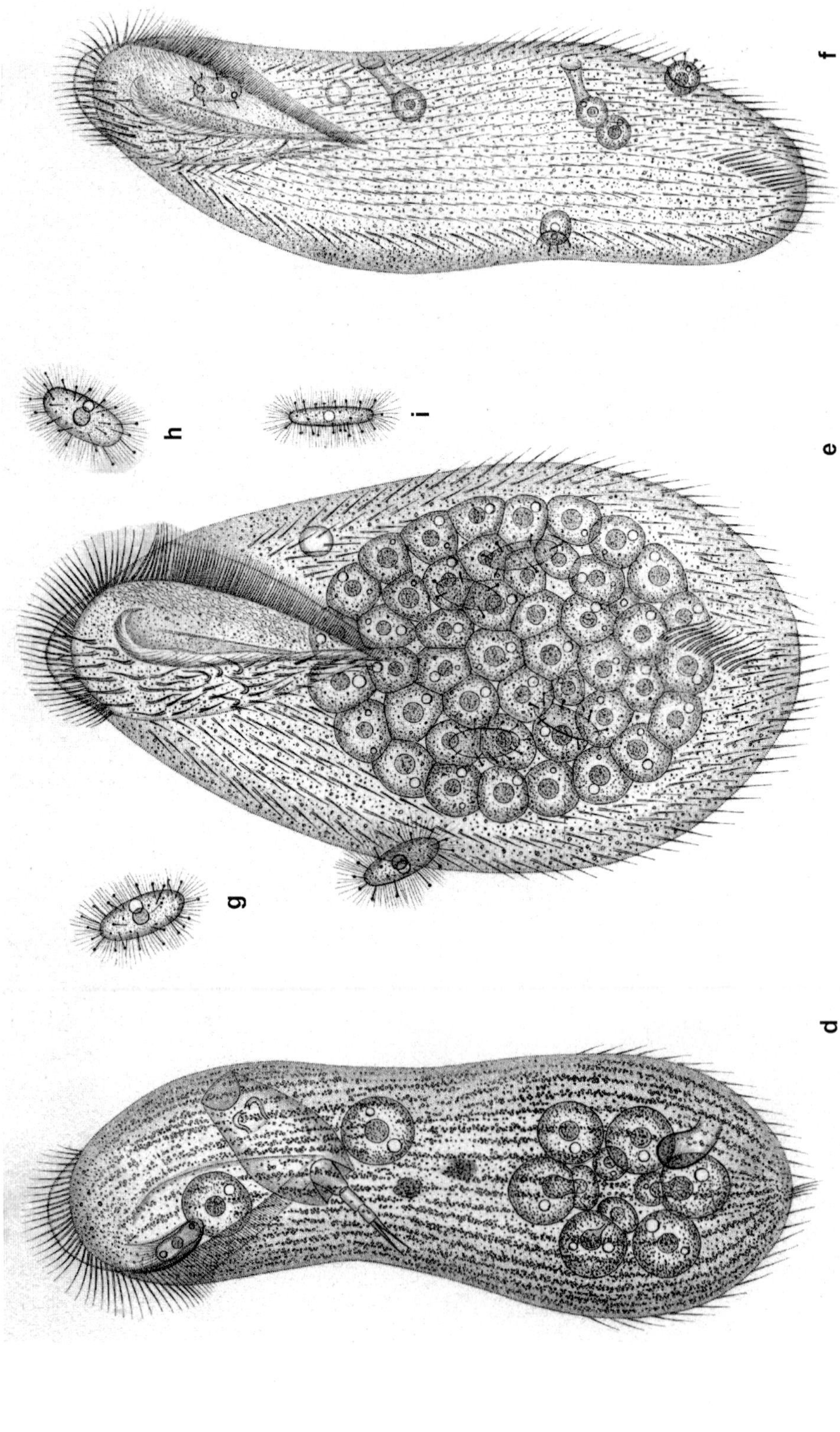
f
h
i
e
g
d

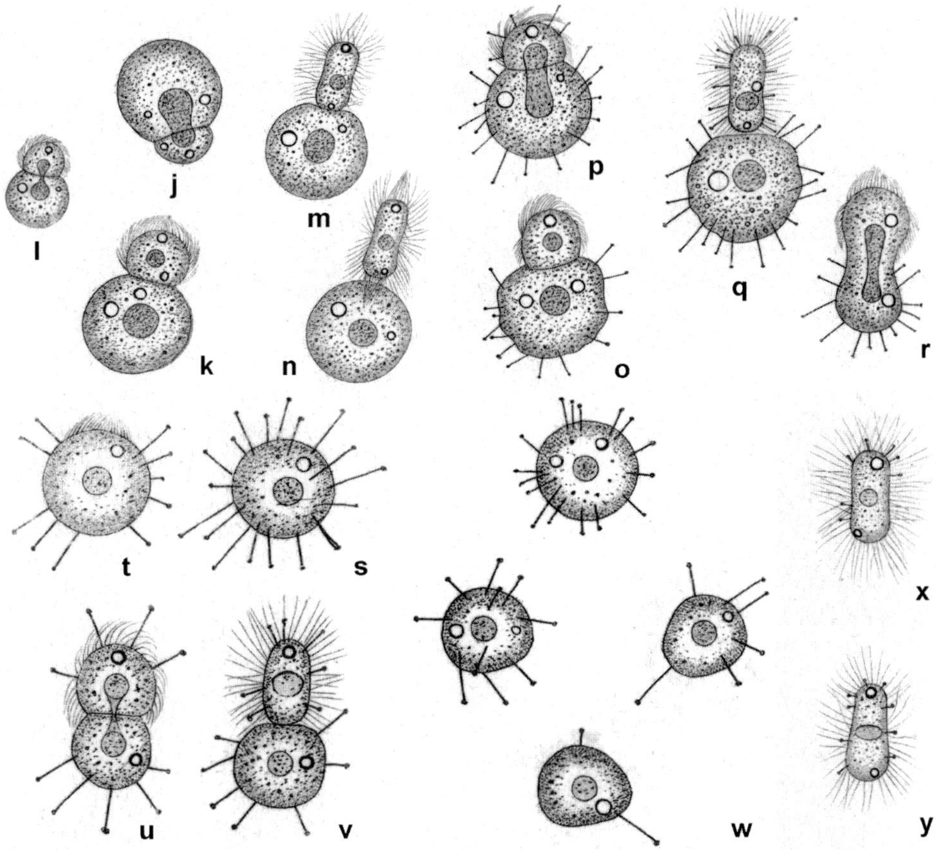

Fig. 17j–y *Podophrya urostylae*, a suctorian parasite of *Urostyla grandis* (from Stein 1859; see also Figs. 17a–i). The swarmers of the parasite are formed by external budding. Further details see text and Stein's (1859) detailed description.

Urostyloid hypotrichs have about the same food spectrum as other hypotrichs (Berger 1999), that is, they feed on bacteria, cyanobacteria, algae (including diatoms), hyphae and spores of fungi, auto- and heterotrophic flagellates, other ciliates, and small metazoans, for example, rotifers (Fig. 138a, 206b). The spectrum of the individual species is of course usually much narrower.

Very little is known about the geographic distribution of urostyloids. The group as such is likely distributed world-wide. However, we are unable to say whether individual

← **Fig. 17d–i** *Urostyla grandis* parasitised by *Podophrya urostylae* (from Stein 1859). Various stages of parasitisation. The host shown in (e) contains about 50 parasites. The body outline of the swarmers varies from slender to broad elliptical (g–i). Further details see text and Stein's (1859) exhaustive description.

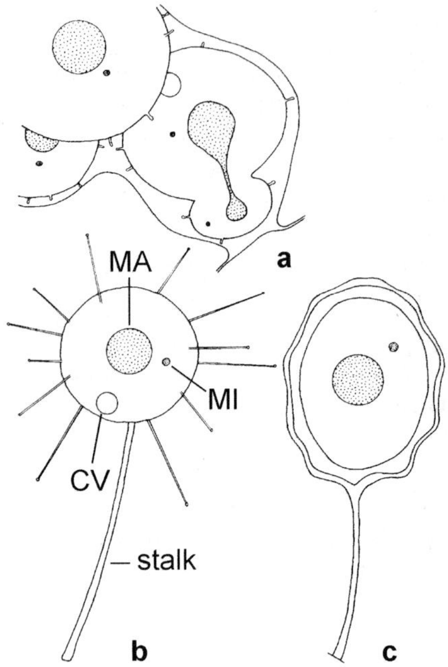

Fig. 18a–c *Podophrya urostylae*, a suctor parasitising *Urostyla grandis* (after Jankowski 1963 from Matthes 1988). **a:** Parasitic stage with beginning external budding in *Urostyla*. **b:** Stalked, free-living (adult) specimen. **c:** Resting cyst. CV = contractile vacuole, MA = macronucleus, MI = micronucleus.

species or subgroups are confined to certain biogeographic regions or not, simply because too few reliable data are available. In the descriptions of the individual species I mention all published records I know from all over the world. There is no doubt that several determinations are incorrect. Thus, records which are not substantiated by *serious* morphological data should be used with caution (or better not at all) for biogeographical interpretations. Certainly, many more urostyloid species than reviewed in the present book exist because little is known about ciliates outside Europe. Moreover, the sea likely harbours a considerable number of not yet known species (e.g., Wanick & Silva-Neto 2004).

Six urostyloids are used as indicators of water quality (Table 12). By contrast, 24 oxytrichid species, four *Uroleptus* species, and seven euplotids are included in relevant lists (Foissner et al. 1991, Berger 1999, Berger & Foissner 2003). A detailed description of the morphology and ecology of these hypotrichs, euplotids, and other species is given in our "ciliate atlas" (Foissner et al. 1991, 1992b, 1994, 1995). Keys and revised lists with the saprobic classification can also be found in Foissner et al. (1995a), Foissner & Berger (1996), Berger et al. (1997), and Berger & Foissner (2003). Note that *Holosticha gibba* is marine and *Urostyla viridis* is little known. *Holosticha pullaster* is very common in freshwater, but unfortunately euryoecious and therefore has the lowest indicative weight. *Anteholosticha intermedia*, *A. monilata*, and *Urostyla grandis* occur regularly in running waters, but usually with low abundance.

Only few urostyloids (*Anteholosticha monilata*, *Diaxonella pseudorubra*, *Pseudourostyla cristata*) are reliably recorded from activated sludge plants (Augustin & Foissner 1992, Oberschmidleitner & Aescht 1996). *Holosticha pullaster*, although rather common in stagnant and running waters, was never reliably recorded from activated sludge.

Species found in the marine interstitial are summarised by Carey (1992) and Patterson et al. (1989). Likely no species is obligatorily anaerobic, although *Anteholosticha fasciola* can be maintained in anaerobic cultures (Fenchel & Finlay 1991, 1995; identification uncertain).

Table 12 Saprobic classification of urostyloid ciliates (from Foissner et al. 1991)[a]

Species[b]	S	Valency				I	SI	Page
		o	b	a	p			
Anteholosticha intermedia (Bergh, 1889) comb. nov.[c]	a–b	-	4	5	1	2	2.7	317
Anteholosticha monilata (Kahl, 1928) Berger, 2003	a–b	-	3	6	1	3	2.8	297
Holosticha gibba (Müller, 1786) Wrześniowski, 1877	a–b	-	4	5	1	2	2.7	99
Holosticha pullaster (Müller, 1773) Foissner, Blatterer, Berger & Kohmann, 1991	b–a	1	4	4	1	1	2.5	128
Urostyla grandis Ehrenberg, 1830	a	-	3	7	-	4	2.7	1048
Urostyla viridis Stein, 1859	b–a	-	5	5	-	3	2.5	1106

[a] S = indication of saprobity by simple letter, o = oligosaprobic, b = betamesosaprobic, a = alphamesosaprobic, p = polysaprobic, I = indicative weight (1, 2, 3, 4, or 5) of species, SI = saprobic index (ranging from 1 to 4 in the limnosaprobic area).

[b] Some species have a different name in the present book and in Foissner et al. (1991): *Anteholosticha intermedia* = *Holosticha multistilata* in Foissner et al. (1991); *Anteholosticha monilata* = *Holosticha monilata*; *Holosticha gibba* = *Holosticha kessleri*; *Urostyla viridis* = *Paraurostyla viridis*.

[c] See description for combination.

6 Collecting, Culturing, Observing, and Staining of Urostyloid Ciliates

A detailed description of these topics for all ciliates is given by Foissner (1991, 1993) and Foissner et al. (1991, 1999, 2002).

6.1 Collecting and Culturing

Urostyloids occur in terrestrial (litters, humic and mineral soil horizons), semiterrestrial (e.g., astatic puddles, mosses, flood plains), limnetic (e.g., ponds, lakes, brooks, rivers, sewage treatment plants), and brackish water biotops. Furthermore, a rather high number of marine species is described.

There are two principle techniques available for collecting protozoans from waters: either direct sampling of natural substrates, or artificial substrate sampling. Urostyloids – and hypotrichs in general – can be sampled from natural substrates by collecting algae masses, mud, debris, macrophytes, small stones, and leaves, and by brushing off the aufwuchs from stones, twigs, and vegetation (e.g., Berger et al. 1997, Blatterer 1995, Foissner et al. 1991, 1992a, Heuss 1976, Liebmann 1962). Plankton samples (mesh size ≤10 μm) should be fixed with saturated, aqueous mercuric chloride (formalin destroys all [?] urostyloids) or studied in life (Foissner et al. 1999). For a detailed description of foam sampling, see Cairns & Henebry (1982) and Pratt & Kepner (1992). Samples should be collected in at least 0.5–1.0 l wide-necked bottles and transported to the laboratory in a cooler. The investigation should be done within 24 h after collecting because the ciliate biocoenosis changes very rapidly. Occasionally some urostyloids thrive in older, slightly fouling cultures.

Water samples can be studied by the so-called cover glass method, a simple but very effective technique (for review see Berger & Foissner 2003). The bottles containing the collected material are opened in the laboratory. Place a cover glass on the water surface of the bottle and remove it with a cover glass forceps after 30 min or more. Put the cover glass on an ordinary microscope slide and look for the numbers and kinds of ciliates present. Ciliates accumulate on the cover glass due to oxygen depletion in the deeper zones of the bottle and because of their life style, that is, many are aufwuchs inhabitants and therefore attach to solid surfaces, that is, the cover glass. The ciliate community obtained in this way is very clean and rich. Do not distribute the material collected in a large Petri dish! This would slow down oxygen depletion and ciliate attachment to the cover glass. It was just this mistake why Krieg (in Tümpling & Friedrich 1999) did not succeed with the cover glass method. Finally, take some drops from the bottle's sediment surface and investigate it for bottom-dwellers, which usually do not, or with low abundance, attach to the cover glass. For detailed water quality assessment follow the method described by Blatterer (1995; see also Berger et al. 1997 and Moog et al. 1999), which is now even available as Austrian Standard (ÖNORM M6118).

Activated sludge samples can be analysed as follows (for review see Berger & Foissner 2003): use fresh sludge, which is taken from the plant with a trowel, put into a 500 ml bottle, and transported to the laboratory under cool conditions. Take care for anaerobic zones, which must be sampled and assessed separately. For investigation, shake the bottle, take a small drop (about 0.1 ml) with an ordinary pipette, put it on a microscope slide, and cover the preparation with a cover glass. Three replications should be investigated to obtain reliable data on the species present. Usually, semiquantitative investigation with a rating scale will be sufficient. However, quantitative investigation is also possible and easily performed with the method described by Augustin et al. (1989) and Augustin & Foissner (1992a). Sludge quality can be assessed with the sludge biotic index (SBI) of Madoni (1994) or the method by Großmann et al. (1999).

The most effective means for collecting and culturing urostyloids and other ciliates from soils and mosses is the non-flooded petri dish method as described by Foissner (1987a; see also Foissner 1993 and Foissner et al. 2002). Here, 10–200 g of fresh or air-dried soil or litter sample are placed in a petri dish (10–20 cm across) and saturated, but not flooded, with distilled water. A ciliate, flagellate, and naked amoeba fauna, often very rich, develops within a few days. Inspection of the cultures on days 2, 4, 6, 10, 14, and 20 usually suffices. Subsequent inspections reveal only few species due to the effects of ciliatostasis (Lüftenegger et al. 1987). *Paraholosticha* and *Keronopsis* species usually occur after few hours, the very common *Gonostomum affine* can be found also in old cultures. Several conditions influence the outcome of the method: (i) air-dried soil often yields more individuals and species than fresh soil, perhaps due to reduced microbiostasis; (ii) the sample should contain ample litter and plant debris and must be spread over the bottom of the petri dish in an at least 1 cm thick layer; (iii) the soil may not be flooded. Water should be added to the sample until 5–20 ml drains off when the petri dish is tilted and the soil is gently pressed with a finger. This run-off contains the protozoa and can be used for further preparations such as silver staining.

The methods for culturing hypotrichous ciliates are treated only briefly here as detailed culturing methods – if available – are provided in the species descriptions. Furthermore, the general procedures as described, for instance by Dragesco & Dragesco-Kernéis (1986), Finlay et al. (1988), Foissner et al. (1991, 2002), Galtsoff et al. (1959), Lee et al. (1985), Mayer (1981), and Provasoli et al. (1958) apply also to the hypotrichs.

Some of the bacteriovorous urostyloids thrive on various media (e.g., diluted lettuce and/or hay extracts, table waters [e.g., Volvic], tap water) enriched with a little dried yolk, rice grains, or crushed wheat grains to promote bacterial growth. Some predatory species grow well with small ciliates (e.g., species of the *Tetrahymena pyriformis* complex, *Glaucoma scintillans*) as food. For marine species, artificial sea water (e.g., the supersoluble seasalt Biosal by Aqualine Buschke, Berg, Germany) can be used.

6.2 Observing Living Hypotrichs

Many physical and chemical methods have been described for retarding the movement of ciliates in order to observe structural details (for literature see Foissner 1991). Chemical immobilisation – for example, by nickel sulfate – or physical slowing down by increasing the viscosity of the medium (e.g., methyl cellulose) are rarely helpful. These procedures often change the shape of the cell or cause pre-mortal alterations of various cell structures. The following simple method is therefore preferable: place about 0.5 ml of the raw sample on a slide and pick out (collect) the desired specimens with a micropipette under a compound microscope with low magnification (for example, objective 4:1, ocular 10 ×). If specimens are large enough, they can be picked out from a petri dish under a dissecting microscope. Working with micro-pipettes, the diameter of which must be adjusted to the size of the specimens, requires some training. Transfer the collected specimens, which are now in a very small drop of fluid, onto a slide. Apply small dabs of Vaseline (Petroleum jelly) to each of the four corners of a small cover glass (Fig. 19a; the four dabs can be also applied to the slide); it is useful to apply the jelly by an ordinary syringe with a thick needle. Place the cover glass on the droplet containing the ciliates. Press on the vaselined corners with a mounted needle until ciliates are held firmly between slide and cover glass (Fig. 19b–d). As the pressure is increased the ciliates gradually become less mobile and more transparent. Hence, first the location of the main cell organelles (e.g., nuclear and oral apparatus, contractile vacuole) and then details (e.g., cortical granules, micronucleus) can easily be observed under low (100–400 ×) and high (1000 ×; oil immersion objective) magnification. The colour of the cortical granules and/or the cytoplasm must be studied with well-adjusted bright field.

The shape of the cells is of course altered by this procedure. Therefore, specimens taken directly from the raw culture with a large-bore (opening about 1 mm) Pasteur pi-

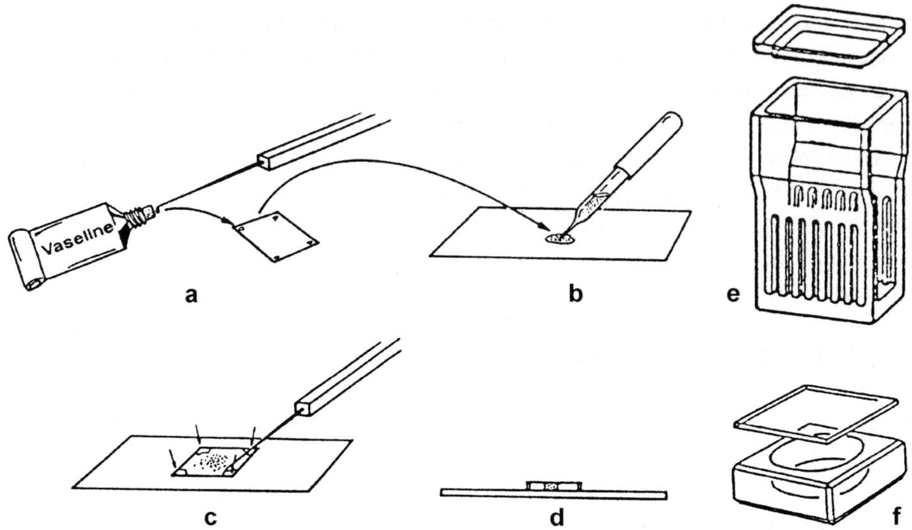

Fig. 19a–f Live observation and staining of urostyloid ciliates (from Foissner 1991). **a–d:** Preparation of slides for observing living ciliates. **e:** Staining jar for 8 and 16 (back to back) slides, respectively. **f:** Watch-glass for protargol procedure according to Wilbert.

pette must first be investigated under low magnification (100–400 ×), that is, without cover glass. Some species are too fragile to withstand handling with micro-pipette and cover glass trapping without deterioration. Investigation with low magnification also requires some experience, but it guarantees that the outline of undamaged cells are recorded. Video-microscopy is very useful at this point of investigation, especially for the registration of the swimming behaviour.

A compound microscope equipped with Normarski differential interference contrast optics is best for discerning the arrangement of the cirri and dorsal cilia in living hypotrichs. If not available, use bright-field. The nuclear apparatus is well-recognisable with differential interference contrast or phase-contrast when specimens are strongly squeezed.

Species that were not observed in life often cannot be identified after silver impregnation with certainty because important characters (e.g., shape, colour of cortical granules, colour of cytoplasm) are not known. As already mentioned above, the correct colour can only be seen with a well-adjusted bright field illumination.

6.3 Staining Procedures

There are many methods for staining ciliates, but only protargol silver impregnation yields (usually) good results in urostyloid hypotrichs. Thus, familiarity with this method is an absolute prerequisite for the description of urostyloids. It is thus described in detail. Simple, namely molecular, formulae are given for the chemicals used, since usually only these are found in the catalogues of the suppliers (e.g., Merck). Other silver impregnation methods (dry silver nitrate method, wet silver nitrate method, silver carbonate method), detailed literature, and some general instructions are to be found in the reviews by Foissner (1991, 1993) and Foissner et al. (1991, 1999).

Apart from silver impregnation, some other staining techniques are useful for taxonomic work with ciliates, especially the Feulgen nuclear reaction and supravital staining with methyl green-pyronin in order to reveal the nuclear apparatus and, respectively, the extrusomes.

6.3.1 Feulgen Nuclear Reaction

Descriptions of this method are to be found, for example, in Dragesco & Dragesco-Kernéis (1986) and Lee et al. (1985). The Feulgen reaction reveals the nuclear apparatus very distinctively, but, because these organelles usually stain well with protargol, it is seldom used for hypotrichs.

6.3.2 Supravital Staining with Methyl Green-Pyronin

This simple method was described by Foissner (1979a). It is an excellent technique for revealing the mucocysts of most ciliates. Mucocysts are stained deeply and very distinctively blue or red, and can be observed in various stages of explosion because the cells are not killed instantly. The nuclear apparatus is also stained.

Procedure (after Foissner 1991)

1. Pick out desired ciliates with a micro-pipette and place the small drop of fluid in the centre of a slide.

2. Add an equally sized drop of methyl green-pyronin and mix the two drops gently by swivelling the slide.
Remarks: If ciliates were already mounted under the coverslip, add a drop of the dye at one edge of the coverslip and pass it through the preparation with a piece of filter paper placed at the other end of the coverslip.

3. Place a coverslip with vaselined corners on the preparation.
Remarks: Observe immediately. Cells die in the stain within 2 min. Mucocysts stain very quickly and many can be observed at various stages of explosion. To reveal the nuclear apparatus, cells should be fairly strongly squashed (= flattened). The preparation is temporary. After 5–10 min the cytoplasm often becomes heavily stained and obscures other details.

Reagents

1 g methyl green-pyronin (Chroma-Gesellschaft, Schmid GmbH and Co., Köngen/N., Germany)
add 100 ml distilled water
This solution is stable and can be used for years.

6.3.3 Protargol Methods

Protargol methods are indispensable for descriptive research on urostyloids and hypotrichs in general. The first procedures were provided by Kirby (1945), Moskowitz (1950), Dragesco (1962), and Tuffrau (1964, 1967), and many more modifications were subsequently proposed (see Foissner 1991 for references). Here, two variations which produce good results are described. These procedures work well with most ciliate species, but require at least 20 specimens. A single specimen cannot usually be handled successfully. Depending on the procedure used, protargol can reveal many cortical and internal structures, such as basal bodies, fibrillary systems, nuclear apparatus. The silverlines (which have no systematic value in the urostyloids), however, never impregnate. The shape of the cells is usually well preserved in permanent slides, which is an advantage for the investigation, but makes photographic documentation more difficult. However, pictures as clear as those taken from wet silver carbonate impregnations can be obtained with the Wilbert modification (procedure B) if the cells are photographed prior to embedding in the albumen-glycerol.

Procedure A (after Foissner 1991)

The quality of the slides is usually adequate but frequently not as good as with the Wilbert modification. The latter demands more material and experience; inexperienced workers may easily lose all the material. As in all protargol methods, the procedure is rather time-consuming and complicated because subject to many factors. A centrifuge may be used for step 2; staining jars (Fig. 19e) are necessary for steps 6–16.

1. Fix organisms in Bouin's or Stieve's fluid for 10–30 min.
Remarks: The fixation time has little influence on the quality of the preparation within the limits given. Ratio fixative to sample fluid should be at least 2:1. Pour ciliates into fixative using a wide-necked flask in order to bring organisms in contact with the fixative as quickly as possible. Both fixatives work well but may provide different results with certain organisms. Stieve's fluid may be supplemented with some drops of 2 % osmium tetroxide for better fixation of very fragile hypotrichs, for example, *Pseudoamphisiella*. This increases the stability of the cells but usually reduces their impregnability.

2. Concentrate by centrifugation and wash organisms 3 to 4 times in distilled water.
Remarks: There are now 2 choices: either to continue with step 3, or to transfer the material through 30–50–70 % alcohol into 70 % alcohol (ethanol), where it remains stable for several years. Transfer preserved material back through the graded alcohol series into distilled water prior to continuing with the next step. Impregnability of preserved material may be slightly different.

3. Clean 8 ordinary slides (or less if material is very scarce) per sample. The slides must be grease-free (clean with alcohol and flame).
Remarks: Insufficiently cleaned slides may cause the albumen to detach. Mark slides on back if several samples are prepared together. Alternatively you can use SuperFrost slides, which are ready to use. In addition, these slides have a field enabling simple labelling with a pencil. Use staining jars with 8 sections so that you can work with 16 slides simultaneously by putting them back to back (Fig. 19e).

4. Put 1 drop each of albumen-glycerol and concentrated organisms in the centre of a slide. Mix drops with a mounted needle and spread over the middle third.
Remarks: Use about equally sized drops of albumen-glycerol and suspended (in distilled water) organisms to facilitate spreading. The size of the drops should be adjusted so that the middle third of the slide is covered after spreading. Now remove sand, grains, etc. The thickness of the albumen layer should be equal to that of the organisms. Some thicker and thinner slides should, however, also be prepared because the thickness of the albumen layer greatly influences the quality of the preparation. Cells may dry out and/or shrink if the albumen layer is too thin; if it is too thick, it may detach, or the cells may become impossible to study with the oil immersion objective.

5. Allow slides to dry for at least 2 h at room temperature.
Remarks: We usually dry slides overnight, that is, for about 12 h. However, slides may be allowed to dry for up 24 h, but no longer if quality is to be maintained. Oven-dried (2 h at 60 °C) slides are usually also of poorer quality.

6. Place slides in a staining jar (Fig. 19e) filled with 95 % alcohol (ethanol) for 20 to 30 min. Place a staining jar with protargol solution into an oven (60° C).
Remarks: Slides should not be transferred through an alcohol series into concentrated alcohol as this causes the albumen layer to detach! Decrease hardening time to 20 min if albumen is already rather old and/or not very sticky.

7. Rehydrate slides through 70 % alcohol and 2 distilled water steps for 5 min each.

8. Place slides in 0.2 % potassium permanganate solution. Remove first slide (or pair of slides) after 60 s and the rest at 15 s intervals. Collect slides in a staining jar filled with distilled water.
Remarks: Bleaching is by permanganate and oxalic acid (step 9). The procedure described above is necessary because each species has its optimum bleaching time. The sequence in which slides are treated should be recorded because the immersion time in

oxalic acid must be proportional to that in the permanganate solution. The albumen layer containing the organisms should swell slightly in the permanganate solution and the surface should become uneven. If it remains smooth, the albumen is too sticky and this could decrease the quality of the impregnation. If the albumen swells strongly, it is possibly too weak (old) and liable to detach. Use fresh $KMnO_4$ solution for each series.

9. Quickly transfer slides to 2.5 % oxalic acid. Remove first slide (or pair of slides) after 160 s, the others at 20 s intervals. Collect slides in a staining jar filled with distilled water.
Remarks: Same as for step 8! Albumen layer becomes smooth in oxalic acid.

10. Wash slides three times in distilled water for 3 min each.

11. Place slides in warm (60° C) protargol solution and impregnate for 10–15 min at 60° C.
Remarks: Protargol solution can be used only once.

12. Remove staining jar with the slides from the oven and allow to cool for 10 min at room temperature.
Remarks: In the meantime organise six staining jars for developing the slides: distilled water – distilled water – fixative (sodium thiosulfate) – distilled water – 70 % alcohol – 100 % alcohol (ethanol).

13. Remove the first slide from the protargol solution and drop some developer on the albumen layer. Move slide gently to spread developer evenly. As soon as the albumen turns yellowish, pour off the developer, dip slide in the first 2 distilled water steps for about 2 s each and stop development by submerging the slide in the fixative (sodium thiosulfate), where it can be left for 1–5 min.
Remarks: Now control impregnation with the compound microscope. The impregnation intensity is sufficient if the infraciliature is just recognisable. The permanent slide will be too dark if the infraciliature is distinct at this stage of the procedure! The intensity of the impregnation can be controlled by the concentration of the developer and the time of development. 5–10 s usually suffice for the diluted developer! Development time increases with bleaching time. Therefore commence development with those slides, which were in the bleaching solutions for 60 and 120 s, respectively. The thinner the albumen layer, the quicker the development.

14. Collect slides in the fixative (sodium thiosulfate) and transfer to distilled water for up to 5 min.
Remarks: Do not wash too long; the albumen layer is very fragile and detaches rather easily!

15. Transfer slides to 70 % – 100 % – 100 % alcohol for 5 min each.

16. Clear by two 10 min transfers through xylene.

17. Mount in synthetic neutral mounting medium.
Remarks: Do not dry slides between steps 16 and 17! Mounting medium should be rather viscous to avoid air-bubbles being formed when solvent evaporates during drying. If air-bubbles develop in the mounted and hardened slide, re-immerse in xylene for some days until the coverslip drops off. Remount using a more viscous medium and remove possible sand grains protruding from the gelatine. Usually, some air-bubbles are found immediately after mounting; these can be pushed to the edge of the coverslip with a finger or mounted needle. The preparation is stable.

Reagents

a) Bouin's fluid (prepare immediately before use; components can be stored)
15 parts saturated, aqueous picric acid ($C_6H_3N_3O_7$; preparation: add an excess of picric crystals to, for example, 1 litre of distilled water; shake solution several times within a week; some undissolved crystals should remain; filter before use)
5 parts formalin (HCHO; commercial concentration, about 37 %)
1 part glacial acetic acid (= concentrated acetic acid; $C_2H_4O_2$)

b) Stieve's fluid (slightly modified; prepare immediately before use; components can be stored)
38 ml saturated, aqueous mercuric chloride (dissolve 60 g $HgCl_2$ in 1 litre of boiling distilled water)
10 ml formalin (HCHO; commercial concentration, about 37 %)
3 ml glacial acetic acid (= concentrated acetic acid; $C_2H_4O_2$)

c) Albumen-glycerol (2–4 month stability)
15 ml egg albumen
15 ml concentrated (98–100 %) glycerol ($C_3H_8O_3$)

Pre-treatment of the egg albumen and preparation of the albumen-glycerol: separate the white carefully from the yolk and embryo of three eggs (free range eggs are preferable to those from battery chickens, whose egg white is less stable and sticky). Shake the white by hand (do not use a mixer!) for some minutes in a narrow-mouthed 250 ml Erlenmeyer flask until a stiff white foam is formed. Allow the flask to stand for about 1 min. Then pour the viscous rest of the egg white in a second Erlenmeyer flask and shake again until a stiff foam is formed. Repeat until most of the egg white is either stiff or becomes watery; usually 4–6 Erlenmeyer flasks of foam are obtained. Leave all flasks undisturbed for about 10 min and discard the watery albumen from the last flask. During this time a glycerol-like fluid percolates from the foam. This fluid is collected and used. Add an equal volume of concentrated glycerol and a small thymol crystal ($C_{10}H_{14}O$) for preservation to the mixture. Mix by shaking gently and pour mixture into a small flask. Leave undisturbed for two weeks. A whitish slime settles at the bottom of the flask. Decant the clear portion, discard slime and thymol crystal. A "good" albumen-glycerol drags a short thread when touched with a needle. The albumen is too thin (not sticky enough) or too old if this thread is not formed. Fresh albumen which is too thin may be concentrated by leaving it open for some weeks so that water can evaporate. If the albumen is too sticky, which may cause only one side of the organisms to impregnate well, it is diluted with distilled water or old, less sticky albumen to the appropriate consistency. The preparation of the albumen-glycerol must be undertaken with great care because much depends on its quality. Unfortunately, all commercial products which have been tried detach during impregnation. A somewhat simpler method to produce the albumen-glycerol is described by Adam & Czihak (1964, p. 274): the white of one or two fresh chicken egg(s) and the same amount of concentrated glycerol are well stirred to a homogenous, thick fluid (a magnetic stirrer can be used). Then filter through cotton wool. Add a small thymol crystal to the filtrate. The albumen-glycerol can be used right away.

d) 0.2 % potassium permanganate solution (stable for about 1 d)
0.2 g potassium permanganate ($KMnO_4$) are dissolved in 100 ml distilled water

e) 2.5 % oxalic acid solution (stable for about 1 d)
2.5 g oxalic acid ($C_2H_2O_4 \cdot 2H_2O$) are dissolved in 100 ml distilled water

f) 0.4–0.8 % protargol solution (stable for about 1 d)
100 ml distilled water
add 0.4–0.8 g protargol
Remarks: Use light-brown "protargol for microscopy" (for example, Merck's "Albumosesilber für die Mikroskopie" or "Proteinate d'Argent", Roques, Paris, France). Some dark-brown, cheaper products do not work! Sprinkle powder on the surface of the water in the staining jar and allow to dissolve without stirring.

g) Developer (mix in sequence indicated; sodium sulphite must be dissolved before hydroquinone is added)
95 ml distilled water
5 g sodium sulphite (Na_2SO_3)
1 g hydroquinone ($C_6H_6O_2$)
Remarks: This recipe yields the stock solution which is stable for some weeks and should be used undiluted for certain ciliates (step 13). Usually, however, it must be diluted with tap water in a ratio of 1:20 to 1:50 to avoid too rapid development and one-sided impregnation of the organisms. Freshly prepared developer is usually inadequate (the albumen turns greenish instead of brownish). The developer should thus be prepared from equal parts of fresh and old (brown) stock solutions. Take great care with the developer as its quality contributes highly to that of the slides. If the developer has lost its activity (which is not always indicated by a brown colour!) the silver is not or only insufficiently reduced and the organisms stain too faintly. A fresh developer should therefore be prepared for each "impregnation week", and some old developer

kept. Fresh developer can be artificially aged by adding some sodium carbonate (Na_2CO_3). However, better results are obtained with air-aged solutions, that is, by a developer which has been kept uncovered for some days in a wide-mouthed bottle. It first turns yellowish, then light brown (most effective), and later dark brown and viscous (at this stage the developer has lost most of its activity, but is still suitable for artificial ageing of fresh developer = 1:1 mixture mentioned above).
During the last years, we obtained very good slides with the low-speed developer used by Fryd-Versavel (pers. comm. to W. Foissner). It is composed of 7 g boric acid, 1.5 g hydroquinone, 10 g sodium sulphite, and 75 ml acetone, all solved, one by one, in 420 ml distilled water. This developer is stable for some weeks and should be used only once. Pour developer into a staining jar and immerse slides, one by one, controlling impregnation intensity after 30–60 s. Usually, developing is finished within 1–5 min (if not, double protargol concentration because slides should not be too long in the developer, as the albumen may detach). The further procedure is as described above (steps 14–17).
In many cases commercial paper developers (for example, Ilford Multigrade) yield very good results.

h) Fixative for impregnation (stable for several years)
25 g sodium thiosulfate ($Na_2S_2O_3 \cdot 5H_2O$) are dissolved in 1000 ml distilled water

Procedure B (after Wilbert 1975 and Foissner 1991)

This modification produces excellent results but demands much experience. Manipulate large cells with micropipettes in a watch-glass, whereas the centrifuge is used for steps 1–4, 7, 8 if cells are smaller than about 150 μm. The watch-glass method is used when there are only few specimens of larger cells; thus an attempt is worthwhile even if only 20 cells are available. The organisms are very soft after development and fixation, and are thus easily compressed between slide and coverslip, which greatly facilitates photographic documentation.

1. Fix organisms as described in protargol procedure A.

2. Wash and, if so desired, preserve organisms as described in protargol procedure A.
Remarks: Wash cells either in the centrifuge (small species) or in a watch-glass (Fig. 19f). To change fluids allow cells to settle on bottom of watch-glass and remove supernatant with a micro-pipette under the dissecting microscope; concentrate cells in the centre of the watch-glass by gentle swirling.

3. Leave organisms in a small amount of distilled water and add, drop by drop, diluted sodium hypochlorite (NaClO) and bleach for about 1–3 min under the dissecting microscope (for small specimens, various concentrations of NaClO can be applied in centrifuge tubes, keeping the reaction time constant, for example, 1 min).

Remarks: This is the critical step in this modification. If bleaching is too strong or too weak all is lost: cells either dissolve or do not impregnate well. Systematic investigations showed that not the bleaching time but the amount of active chloride in the sodium hypochlorite and the pre-treatment of the cells (fixation method, fresh or preserved material) are decisive for the quality of the preparation. Different species need different concentrations. Unfortunately, the concentration of active chloride in the commercial products varies (10–13 %) and is dependent on the age of the fluid. It is thus impossible to provide more than only a few guidelines: 100 ml distilled water + 0.2–0.4 ml NaClO (if product is fresh and cells were not stored in alcohol) or 100 ml distilled water + 0.5–1.6 ml NaClO (if product is older and cells were stored in alcohol). The transparency of the cells under the dissecting microscope may serve as a further indicator: fixed, unbleached cells appear dark and opaque, whereas accurately bleached cells are almost colourless and rather transparent (depends, however, also on size and thickness of the cell). Thus, increase the concentration of sodium hypochlorite stepwise if cells appear too dark with the recommended concentrations. We routinely start with 3 different hypochlorite concentrations if enough material is available.

4. Wash organisms at least 3 times with distilled water and finally once in the protargol solution.
Remarks: Wash thoroughly, especially when fluids are changed with micro-pipettes, because even the slightest traces of the sodium hypochlorite disturb impregnation.

5. Transfer to 1 % protargol solution and impregnate for 10–20 min at 60° C.
Remarks: This and the next step should be carried out in a watch-glass even for material which is otherwise manipulated with the centrifuge. The impregnation time depends on the kind of material and the degree of bleaching. Check the progress of impregnation every 3–4 min under the compound microscope by picking out a few cells with the micro-pipette under the dissecting microscope; add these to 1 drop of developer. Dilute developer and/or interrupt development of adding a little fixative (sodium thiosulfate) if impregnation is strong enough.

6. Remove most of the protargol solution with a micro-pipette and add some drops of developer to the remainder containing the organisms.
Remarks: Fresh, undiluted developer is usually used (but see step 5). Control development in compound or dissecting microscope. As soon as the infraciliature becomes faintly visible, development must be stopped by adding a few drops of sodium thiosulfate. Judging the right moment is a question of experience; the permanent slide will be too dark if the infraciliature is very distinct at this stage of the procedure! Generally: if bleaching was correct, specimens cannot be over-impregnated.

7. Stabilise the impregnation by 2, approximately 5-minute transfers through sodium thiosulfate.

Remarks: The developer need not be removed before fixation. For small species this and the next step can be carried out in a centrifuge. Larger species must be manipulated with micro-pipettes because cells become very fragile and would be damaged in a centrifuge. Cells are very soft at this stage and can thus be easily compressed and photographed. Transfer some of the more darkly impregnated specimens with a very small amount of the fixative onto a clean slide using a micro-pipette and cover with a coverslip. Organisms are usually flattened by the weight of the coverslip; excess fluid my be removed from the edge of the coverslip with a piece of filter paper. Frequently, even better micrographs are obtained if specimens are flattened before fixed with sodium thiosulphate; that is, together with some developer.

8. Wash very thoroughly in distilled water (3 times with the centrifuge; 7–10 times in watch-glass with micro-pipettes). Finally remove as much of the water as possible.
Remarks: Even the slightest traces of the fixative destroy the impregnation within a few days or weeks.

9. Smear a moderately thick layer of albumen-glycerol on a clean slide with a finger. Drop impregnated, washed cells on the albumised slide with a large-bore pipette (opening about 1 mm) and dry preparation for at least 2 h.
Remarks: The cells are too fragile to be spread with a needle. With much care it is possible to orientate cells using a mounted eyelash. Commercial albumen-glycerol can be used.

10. Harden albumen by two 10-minute transfers through concentrated alcohol (ethanol).
Remarks: This and the next step are best carried out in staining jars. The albumen layer turns milky and opaque.

11. Clear by two 5-minute transfers through xylene.
Remarks: The albumen layer turns transparent.

12. Mount in synthetic neutral mounting medium.
Remarks: Do not dry slides between steps 11 and 12! Mounting medium should be rather viscous to avoid air-bubbles being formed when solvent evaporates during drying. If air-bubbles develop in the mounted and hardened slide, re-immerse in xylene for some days until the coverslip drops off. Remount using a more viscous medium and remove possible sand grains protruding from the albumen. Usually, some air-bubbles are found immediately after mounting; these can be pushed to the edge of the coverslip with a finger or mounted needle. The preparation is stable.

Reagents

If not stated otherwise, the same reagents as in the protargol procedure A are to be used (see above).

6.4 Preparation for Scanning Electron Microscopy

Hypotrichs, and especially urostylids, cannot usually be identified solely by scanning electron microscopy because only a limited number of characters is revealed. However, this method is useful in that it allows a three-dimensional view of the object, as well as for documenting details which are difficult to reveal with other methods. For a detailed instruction of preparation for scanning electron microscopy, see Foissner (1991, 1993), Foissner et al. (1991, 1999), and other textbooks.

7 Species Concept and Nomenclature

7.1 Species Concept

The species/subspecies concept used in the present book is the same as described by Foissner et al. (2002). Briefly, we usually apply the "morphospecies" concept as basically defined by Nixon & Wheeler (1990): "A species is the smallest aggregation of populations (sexual) or lineages (asexual) diagnosable by a unique combination of character states in comparable individuals (semaphoronts)." That is, I consider two populations as belonging to two different species if they differ from each other in at least one "important" morphological feature (e.g., number of macronuclear nodules; presence/absence of cortical granules). Of course, there is no strict consensus about the importance of various features and, unfortunately, for many species several features (e.g., presence/absence of cortical granules or caudal cirri; number of dorsal kineties; length of dorsal bristles; exact arrangement of cirri) are not known, making revisions rather difficult. Often it is a matter of taste whether or not two species are synonymised or not. To overcome these difficulties I have kept the descriptions (and the ecological data) of synonyms separate, especially when the descriptions did not fit in all important details.

The presence/absence of a certain cirral group (e.g., frontoterminal cirri, caudal cirri, transverse cirri) is generally considered as diagnostic character, that is, such features are usually used to characterise supraspecific taxa. However, features of the cirral pattern are certainly not the sole source to elucidate the phylogenetic relationships. In the Oxytrichidae the consistence of the cell (flexible vs. rigid), the presence/absence of cortical granules, and the relative length (i.e., a quantitative feature!) of the adoral zone have been successfully used to characterise the Stylonychinae (Berger & Foissner 1997, Berger 1999). Moreover, molecular markers will significantly increase our knowledge on the phylogeny of hypotrichs (Fig. 14a).

For a discussion of the advantages and disadvantages of various species concepts see textbooks on evolution (e.g., Ax 1984, Wägele 2001) and references cited by Foissner et al. (2002, p. 35).

7.2 Notes on Nomenclature

In the case of nomenclatural problems the ICZN (1964, 1985, 1999) have been consulted, depending on the date when the paper was published. For explanation of nomenclatural terms (e.g., nomen nudum, holotype) see the glossary of the ICZN (1999) or various textbooks (e.g., Lincoln et al. 1985, CBE 1996, Winston 1999).

I tried to explain the meaning and origin of the scientific names using, inter alia, the ICZN (1985, 1999), Werner (1972), Hentschel & Wagner (1996), and Latin/German dictionaries. Only in few cases (likely less than 5%) does the original description contain an etymology section. The gender of ciliate genus-group names can be found in the valuable catalogue by Aescht (2001). I did not consult a Latin/Greek linguist; thus, improprieties cannot be excluded.

Note that Kahl (1932, 1933) divided *Holosticha* into several subgenera, a fact very often overlooked. Consequently, many species names including the combining authorities have been written incorrectly in many post-Kahlian papers.

For authorship and date of non-urostyloid hypotrichs see Berger (1999, 2001). A permanently updated version of the "Catalogue of Ciliate Names. 1. Hypotrichs" is available at http://protozoology.com.

As in the first volume of the revision of hypotrichs (Berger 1999), higher taxa are not provided with categories (e.g., family, order), simply because categories do not contain information and cannot be defined objectively (e.g., Ax 1995, Westheide & Rieger 1996, Wägele 2001). For example, the taxon Hypotricha was established as order by Stein (1859). Since then it also attained the categories suborder, subclass, and even class (for review see Berger 2001). However, to avoid inflation of names I use those which are available. Therefore the "defined" endings (ICZN 1999, Article 29.2; e.g., -idae, -inae) have **no meaning** in the present book.

7.3 Summary of New Taxa and Nomenclatural Acts

Within the framework of the revision of the Urostyloidea, three books (Berger 2001, Berger & Foissner 2003, present book), five papers (Berger 2003, 2004a, b, Berger et al. 2001, Foissner et al. 2004a), and six abstracts (Berger 2001a, 2003a, b, Berger et al. 2004, Schmidt et al. 2004a, b) have been published. In these publications the nomenclatural acts listed below have been undertaken.

New species: *Anteholosticha antecirrata* (present book, p. 370).

New combinations: *Anteholosticha adami* (Foissner, 1982) Berger, 2003 (p. 377; basionym: *Holosticha adami*); *Anteholosticha alpestris* (Kahl, 1932) comb. nov. (present book, p. 403; basionym: *Holosticha (Keronopsis) alpestris*); *Anteholosticha arenicola* (Kahl, 1932) Berger, 2003 (p. 377; basionym: *Holosticha arenicola*); *Anteholosticha australis* (Blatterer & Foissner, 1988) Berger, 2003 (p. 377; basionym: *Holosticha australis*); *Anteholosticha azerbaijanica* (Alekperov & Asadullayeva, 1999) comb. nov.

(present book, p. 454; basionym: *Holosticha azerbaijanica*); *Anteholosticha bergeri* (Foissner, 1987) Berger, 2003 (p. 377; basionym *Holosticha bergeri*); *Anteholosticha brachysticha* (Foissner, Agatha & Berger, 2002) Berger, 2003 (p. 377; basionym: *Holosticha brachysticha*); *Anteholosticha brevis* (Kahl, 1932) Berger, 2003 (p. 377; basionym: *Holosticha brevis*); *Anteholosticha camerounensis* (Dragesco, 1970) Berger, 2003 (p. 377; basionym: *Holosticha camerounensis*); *Anteholosticha distyla* (Buitkamp, 1977) Berger, 2003 (p. 377; basionym: *Holosticha distyla*); *Anteholosticha estuarii* (Borror & Wicklow, 1983) Berger, 2003 (p. 377; basionym: *Holosticha estuarii*); *Anteholosticha extensa* (Kahl, 1932) Berger, 2003 (p. 377; basionym: *Holosticha extensa*); *Anteholosticha fasciola* (Kahl, 1932) Berger, 2003 (p. 377; basionym: *Holosticha fasciola*); *Anteholosticha gracilis* (Kahl, 1932) Berger, 2003 (p. 377; basionym: *Holosticha gracilis*); *Anteholosticha* grisea (Kahl, 1932) Berger, 2003 (p. 377; basionym: *Holosticha grisea*); *Anteholosticha intermedia* (Bergh, 1889) comb. nov. (present book, p. 317; basionym: *Urostyla intermedia*); *Anteholosticha longissima* (Dragesco & Dragesco-Kernéis, 1986) comb. nov. (present book, p. 437; basionym: *Keronopsis longissima*); *Anteholosticha macrostoma* (Dragesco, 1970) comb. nov. (present book, p. 365; basionym: *Pleurotricha macrostoma*); *Anteholosticha manca* (Kahl, 1932) Berger, 2003 (p. 377; basionym: *Holosticha manca*); *Anteholosticha mancoidea* (Hemberger, 1985) Berger, 2003 (p. 377; basionym: *Holosticha mancoidea*); *Anteholosticha monilata* (Kahl, 1928) Berger, 2003 (p. 377; basionym: *Holosticha monilata*); *Anteholosticha multistilata* (Kahl, 1928) Berger, 2003 (p. 377; basionym: *Holosticha multistilata*); *Anteholosticha muscicola* (Gellért, 1956) Berger, 2003 (p. 377; basionym: *Holosticha muscicola*); *Anteholosticha muscorum* (Kahl, 1932) Berger, 2003 (p. 377; basionym: *Holosticha muscorum*); *Anteholosticha oculata* (Mereschkowsky, 1877) Berger, 2003 (p. 377; basionym: *Oxytricha oculata*); *Anteholosticha plurinucleata* (Gellért, 1956) comb. nov. (present book, p. 399; basionym: *Holosticha manca plurinucleata*); *Anteholosticha pulchra* (Kahl, 1932) Berger, 2003 (p. 377; basionym: *Holosticha pulchra*); *Anteholosticha randani* (Grolière, 1975) Berger, 2003 (p. 377; basionym: *Holosticha randani*); *Anteholosticha scutellum* (Cohn, 1866) Berger, 2003 (p. 377; basionym: *Oxytricha scutellum*); *Anteholosticha sigmoidea* (Foissner, 1982) Berger, 2003 (p. 377; basionym: *Holosticha sigmoidea*); *Anteholosticha sphagni* (Grolière, 1975) Berger, 2003 (p. 377; basionym: *Holosticha sphagni*); *Anteholosticha thononensis* (Dragesco, 1966) Berger, 2003 (p. 377; basionym: *Keronopsis thononensis*); *Anteholosticha violacea* (Kahl, 1932) Berger, 2003 (p. 377; basionym: *Holosticha violacea*); *Anteholosticha vuxgracilis* (Berger, 2005) comb. nov. (present book, p. 369; basionym: *Holosticha vuxgracilis* nom. nov., present book, p. 369); *Anteholosticha warreni* (Song & Wilbert, 1997) Berger, 2003 (p. 377; basionym: *Holosticha warreni*); *Anteholosticha xanthichroma* (Wirnsberger & Foissner, 1987) Berger, 2003 (p. 377; basionym: *Holosticha xanthichroma*); *Apoamphisiella vernalis* (Stokes, 1887) comb. nov. (present book, p. 98; basionym: *Holosticha vernalis*); *Biholosticha discocephalus* (Kahl, 1932) Berger, 2003 (p. 378; basionym: *Holosticha discocephalus*); *Biholosticha arenicola* (Dragesco, 1963) Berger, 2003 (p. 378; basionym: *Keronopsis arenicola*); *Caudiholosticha algivora* (Kahl, 1932) Berger, 2003 (p. 377; basionym: *Holosticha algivora*); *Caudiholosticha gracilis* (Foissner, 1982) comb.

nov. (present book, p. 266; basionym: *Perisincirra gracilis*); *Caudiholosticha interrupta* (Dragesco, 1966) Berger, 2003 (p. 377; basionym: *Holosticha interrupta*); *Caudiholosticha islandica* (Berger & Foissner, 1989) Berger, 2003 (p. 377, 378; basionym: *Holosticha islandica*); *Caudiholosticha multicaudicirrus* (Song & Wilbert, 1989) Berger, 2003 (p. 378; basionym: *Holosticha multicaudicirrus*); *Caudiholosticha navicularum* (Kahl, 1932) Berger, 2003 (p. 378; basionym: *Holosticha navicularum*); *Caudiholosticha notabilis* (Foissner, 1982) comb. nov. (present book, p. 260; basionym: *Paruroleptus notabilis*); *Caudiholosticha paranotabilis* (Foissner, Agatha & Berger, 2002) comb. nov. (present book, p. 254; *Uroleptus paranotabilis*); *Caudiholosticha setifera* (Kahl, 1932) Berger, 2003 (p. 378; basionym: *Holosticha setifera*); *Caudiholosticha stueberi* (Foissner, 1987) Berger, 2003 (p. 378; basionym: *Holosticha stueberi*); *Caudiholosticha sylvatica* (Foissner, 1982) Berger, 2003 (p. 378; basionym: *Holosticha sylvatica*); *Caudiholosticha tetracirrata* (Buitkamp & Wilbert, 1974) Berger, 2003 (p. 377; basionym: *Holosticha tetracirrata*); *Caudiholosticha viridis* (Kahl, 1932) Berger, 2003 (p. 378; basionym: *Holosticha viridis*); *Diaxonella pseudorubra* (Kaltenbach, 1960) comb. nov. (present book, p. 463; basionym: *Keronopsis pseudorubra*); *Diaxonella pseudorubra polystylata* (Borror & Wicklow, 1983) comb. nov. (present book, p. 479; basionym: *Holosticha polystylata*); *Diaxonella pseudorubra pulchra* (Borror, 1972) comb. nov. (present book, p. 483; basionym: *Trichotaxis pulchra*); *Hemisincirra gellerti* (Foissner, 1982) Foissner in Berger, 2001 (p. 71: basionym: *Perisincirra gellerti*); *Hemisincirra gracilis* (Foissner, 1982) Foissner in Berger, 2001 (p. 71; basionym: *Perisincirra gracilis*); *Hemisincirra interrupta* (Foissner; 1982) Foissner in Berger, 2001 (p. 72; basionym: *Perisincirra interrupta*); *Paragastrostyla terricola* (Foissner, 1988) comb. nov. (present book, p. 631; basionym: *Holostichides terricola*); *Paraholosticha ovata* (Horváth, 1933) Berger, 2001 (p. 68; basionym: *Paraholosticha ovata*); *Paraholosticha vitrea* (Vörösváry, 1950) Berger, 2001 (p. 68; basionym: *Paraholosticha vitrea*); *Pseudourostyla magna* (Alekperov, 1984) comb. nov. (present book, p. 809; basionym: *Metaurostyla magna*); *Pseudourostyla raikovi* (Alekperov, 1984) comb. nov. (present book, p. 807; basionym: *Metaurostyla raikovi*); *Tetmemena bifaria* (Stokes, 1887) Berger, 2001 (p. 52, 53; basionym: *Oxytricha bifaria*); *Thigmokeronopsis crassa* (Claparède & Lachmann, 1858) comb. nov. (present book, p. 873; basionym: *Oxytricha crassa*); *Trichototaxis multinucleatus* (Burkovsky, 1970) Berger, 2001 (p. 95, 96; basionym: *Trichotaxis multinucleatus*); *Uroleptopsis (Plesiouroleptopsis) ignea* (Mihailowitsch & Wilbert, 1990) Foissner, 1995 in Berger (2004b, p. 115); *Uroleptopsis (Uroleptopsis) citrina* Kahl, 1932 in Berger (2004b, p. 114); *Uroleptopsis (Uroleptopsis) roscoviana* (Maupas, 1883) Kahl, 1932 in Berger (2004b, p. 114); *Uroleptopsis tannaensis* (Shigematsu, 1953) Berger, 2004b and *Uroleptopsis (Uroleptopsis) tannaensis* (Shigematsu, 1953) Berger, 2004b (p. 111, 114; basionym: *Keronopsis tannaensis*); *Uroleptopsis (Uroleptopsis) viridis* (Pereyaslawzewa, 1886) Kahl, 1932 in Berger (2004b, p. 114); *Urostyla variabilis* (Borror & Wicklow, 1983) comb. nov. (present book, p. 1104; basionym: *Bakuella variabilis*).

New subgenus: *Uroleptopsis (Plesiouroleptopsis)* Berger, 2004b (p. 114).

New genera: *Anteholosticha* Berger, 2003 (p. 377); *Biholosticha* Berger, 2003 (p. 378); *Caudiholosticha* Berger, 2003 (p. 377); *Styxophrya* Foissner et al., 2004a (p. 279).

New higher taxa: Acaudalia (present book, p. 749); Dorsomarginalia (present book, p. 38); Retroextendia (present book, p. 732).

New ranks: *Anteholosticha plurinucleata* (Gellért, 1956) (species rank; present book, p. 399); *Bakuella (Pseudobakuella)* Alekperov, 1992 (subgenus rank; present book, p. 576); *Diaxonella pseudorubra polystylata* (Borror & Wicklow, 1983) (subspecies rank; present book, p. 479); *Diaxonella pseudorubra pseudorubra* (Kaltenbach, 1960) (subspecies rank; present book, p. 468); *Diaxonella pseudorubra pulchra* (Borror, 1972) (subspecies rank; present book, p. 483); *Uroleptopsis (Uroleptopsis)* Kahl, 1932 (subgenus rank in Berger 2004b, p. 114).

New names: *Holosticha holomilnei* Berger, 2001 (p. 35) for *Holosticha (Holosticha) milnei* Kahl, 1932; *Holosticha vuxgracilis* (present book, p. 369, for *Holosticha gracilis* Vuxanovici 1963).

Corrected names: Urostylididae Dallas, 1851 (Insecta, Heteroptera) in Berger et al. (2001, p. 301).

Neotypifications: *Amphisiella annulata* (Kahl, 1932) Borror, 1972 in Berger (2004a, p. 13); *Anteholosticha intermedia* (Bergh, 1889) comb. nov. (present book, p. 317); *Pseudourostyla levis* Takahashi, 1973 (present book, p. 778); *Uroleptopsis citrina* Kahl, 1932 in Berger (2004a, p. 109).

New synonyms (including supposed synonyms): *Bakuella kreuzkampii* Song, Wilbert & Berger, 1992 is synonymous with *Bakuella agamalievi* Borror & Wicklow, 1983 (present book, p. 541); *Bakuella muensterlandii* Alekperov, 1992 is synonymous with *Bakuella agamalievi* Borror & Wicklow, 1983 (present book, p. 541); *Diaxonella trimarginata* Jankowski, 1979 is synonymous with *Diaxonella pseudorubra* (Kaltenbach, 1960) comb. nov. (present book, p. 463); *Holosticha corlissi* Fernandez-Galiano & Calvo, 1992 is synonymous with *Anteholosticha monilata* (Kahl, 1928) Berger, 2003 (present book, p. 314); *Holosticha (Keronopsis) muscorum* Kahl, 1932 is synonymous with *Anteholosticha intermedia* (Berger, 1889) comb. nov. (present book, p. 318); *Holosticha manca mononucleata* Gellért, 1956 is synonymous with *Anteholosticha plurinucleata* (Gellèrt, 1956) comb. nov. (present book, p. 400); *Holosticha nagasakiensis* Hu & Sudzuki, 2004 is synonymous with *Anteholosticha gracilis* (Kahl, 1932) Berger, 2003 (present book, p. 426); *Keronopsis macrostoma* Reuter, 1963 is synonymous with *Anteholosticha intermedia* (Berger, 1889) comb. nov. (present book, p. 317); *Keronopsis multiplex* Ozaki & Yagiu, 1943 is synonymous with *Uroleptopsis roscoviana* (Maupas, 1883) Kahl, 1932 (Berger 2004b, p. 111); *Oxytricha kessleri* Wrzesniowski, 1877 (and its synonyms) is synonymous with *Holosticha gibba* (Müller, 1786)

Wrzesniowski, 1877 (Berger 2003b, p. 376); *Oxytricha pernix* Wrzesniowski, 1877 is synonymous with *Holosticha pullaster* (Müller, 1773) Foissner, Blatterer, Berger & Kohmann, 1991 (present book, p. 146); *Periholosticha wilberti* Song, 1990 is synonymous with *Paragastrostyla lanceolata* Hemberger, 1985 (present book, p. 618); *Pseudokeronopsis trinesestra* Dragesco & Dragesco-Kernéis, 1991 is synonymous with *Diaxonella pseudorubra* (Kaltenbach, 1960) comb. nov. (present book, p. 463); *Trichototaxis rubra* Plückebaum, Winkelhaus & Hauser, 1997 is synonymous with *Diaxonella pseudorubra* (Kaltenbach, 1960) comb. nov. (present book, p. 463); *Urostyla algivora* Gellért & Tamás, 1958 is synonymous with *Pseudourostyla urostyla* (Claparède & Lachmann, 1858) Borror, 1972 (present book, p. 806); *Urostyla chlorelligera* Foissner, 1980 is synonymous with *Urostyla grandis* Ehrenberg, 1830 (present book, p. 1086); *Urostyla pseudomuscorum* Wang, 1940 is synonymous with *Pseudourostyla urostyla* (Claparède & Lachmann, 1858) Borror, 1972 (present book, p. 804).

7.4 Deposition of Slides

If mentioned in the individual papers, the site where the slide(s) (holotype; paratype; neotype; voucher) has (have) been deposited, is given in the corresponding entry of the list of synonyms. For a detailed list of type specimens deposited in the collection "diverse invertebrates" (except insects) of the Biology Centre Linz (Upper Austria), see Aescht (2003; also available at http://www.biologiezentrum.at). Slides used for original observations are also deposited in this collection.

B Systematic Section

Urostyloidea Bütschli, 1889

1889 **Urostylinae Bütschli**[1] – Bütschli, Protozoa, p. 1741 (original description). Type genus: *Urostyla* Ehrenberg, 1830.
1926 **Urostylidae** – Calkins, Protozoa, p. 390 (brief review).
1932 **Oxytrichidae Ehrenberg, 1838** – Kahl, Tierwelt Dtl., 25: 537, pro parte (last detailed revision).
1972 **Urostylidae Bütschli, 1889**[2] – Borror, J. Protozool., 19: 8 (revision of hypotrichs and euplotids).
1979 **Urostyloidea Bütschli, 1889, superfam. n.** – Jankowski, Trudy zool. Inst., Leningr., 86: 73 (revision).
1979 **Urostylina subordo. n.** – Jankowski, Trudy zool. Inst., 86: 84 (original description). Type genus: *Urostyla* Ehrenberg, 1830.
1979 **Holostichina subordo n.** – Jankowski, Trudy zool. Inst., 86: 84 (original description). Type genus: *Holosticha* Wrzesniowski, 1877.
1979 **Urostylidae Bütschli, 1889**[3] – Corliss, Ciliated Protozoa, p. 309 (revision).
1979 **Urostylidae Bütschli, 1889**[4] – Borror, J. Protozool., 26: 548 (redefinition).
1981 **Urostylina Jankowski, 1979**[5] – Wicklow, Protistologica, 17: 348 (revision).
1981 **Urostyloidea Butschli, 1889**[6] – Wicklow, Protistologica, 17: 348 (revision).
1982 **Urostylidae Bütschli, 1889**[7] – Hemberger, Dissertation, p. 75 (revision of hypotrichs).
1983 **Urostylidae** – Curds, Gates & Roberts, Synopses of the British Fauna, 23: 390 (guide to genera).
1983 **Urostylina Jankowski, 1979** – Borror & Wicklow, Acta Protozool., 22: 120 (revision of urostylines).
1985 **Urostylina** – Small & Lynn, Phylum Ciliophora, p. 450 (guide to representative genera).

---

[1] The diagnosis by Bütschli (1889) is as follows: Stets eine grössere oder geringere Zahl, zum mindesten zwei ununterbrochene Bauchreihen, wozu sich noch zwei ununterbrochene Randreihen gesellen. Differenzierung von Stirn- und Aftercirren meist deutlich, selten die eine Sorte, oder beide undeutlich. Hinter dem Mund fast nie grössere Bauchcirren im Verlauf der Bauchreihen differenziert.

[2] The diagnosis by Borror (1972) is as follows: Cirri in 3–12 ventral rows. No anatomically or morphogenetically distinct marginal cirri, frontoventral cirri, or midventral cirri.

[3] Corliss (1979) provided the following characterisation: Ventral cirri in straight rows (of variable number), generally with only transverse cirri (near posterior end) morphologically distinct and conspicuous; body elongate-elliptical in outline, but quite broad, and often of large size (up to 800 μm).

[4] Borror (1979) provided the following diagnosis: Somatic ciliature including row of dorsal cilia, and one or more rows of right and left marginal cirri; frontal ciliature with variously arranged frontal, midventral, and transverse cirri (sometimes reduced, inconspicuous, or absent) that differentiate during prefission morphogenesis from longitudinal field of oblique ciliary streaks.

[5] Wicklow (1981) provided the following diagnosis: Frontal ciliature includes midventral cirri that develop during division morphogenesis from a longitudinal series of oblique streaks; somatic ciliature includes dorsal bristle rows and marginal cirral rows (Marginal cirri are replaced in one group by longitudinal ventral cirral rows that differentiate from ventral primordia).

[6] Wicklow (1981) provided the following diagnosis: In addition to midventral cirri, 5 other frontal derivatives *may* be present: malar, migratory, transverse, accessory transverse, and thigmotactic cirri. All non-midventral, longitudinal cirral rows arise by somatic (within row) development and are considered marginal cirral rows.

[7] Hemberger (1982) provided the following diagnosis: Hypotrichida mit mindestens je 1 rechten und linken Marginalreihe; Cirren der beiden Ventralreihen (= Midventral-Reihen) in einer typischen Zick-Zack-Anordnung (= "Diaxoneme" nach Jankowski, pers. Mitt. an Buitkamp); diese differenzieren sich aus (meist zahlreichen) schrägen Anlagen; die Frontalcirren entwickeln sich aus den vorderen Teilen der frontalen Anlagen oder den vorderen Anlagen insgesamt; die Transversalcirren entwickeln sich aus den hinteren Enden der caudalen Anlagen; bei der Gattung *Bakuella* Agamaliev & Alekperov differenzieren sich aus den Anlagen mehrere Cirren, so daß in der Infraciliatur kurze Schrägreihen von Cirren vorhanden sind.

1987 **Urostylidae Butschli, 1889** – Tuffrau, Annls Sci. nat., 8: 115 (brief revision).
1994 **Urostylida Jankowski, 1979**[1] – Tuffrau & Fleury, Traite de Zoologie, 2: 126 (revision of hypotrichs and euplotids).
1999 **Urostylina Jankowski, 1979** – Shi, Song & Shi, Progress in protozoology, p. 110 (revision of hypotrichs).
2001 **Urostyloidea Bütschli, 1889** – Berger, Catalogue of ciliate names 1. Hypotrichs, p. 114 (nomenclator containing all basionyms, combinations, and higher taxa of hypotrichs).
2001 **Urostylidae Bütschli, 1889**[2] – Eigner, J. Euk. Microbiol., 48: 78 (redefinition).
2002 **Urostylida Jankowski, 1979**[3] – Lynn & Small, Phylum Ciliophora, p. 441 (guide to representative genera).

Nomenclature: The name Urostylinae and its derived forms (e.g., Urostyloidea) are based on the genus-group name *Urostyla*. Originally established as subfamily, later categorised as family (Calkins 1926, Borror 1972, 1979, Corliss 1979, Hemberger 1982, Curds et al. 1983, Tuffrau 1987, Eigner 2001), superfamily (Jankowski 1979), suborder (Jankowski 1979, Wicklow 1981, Borror & Wicklow 1983, Small & Lynn 1985, Shi et al. 1999), and order (Tuffrau & Fleury 1994, Lynn & Small 2002). I do not categorise the supraspecific taxa. However, to avoid inflation of names I use those which are available.

Since the content of a supraspecific taxon (e.g., Urostyloidea) is different in almost each paper, it would be appropriate to write "pro parte" in each entry of the list of synonyms. However, since this is already obvious I confine it to the present note.

Characterisation (Fig. 14a, autapomorphies 2): Hypotricha with paired ventral cirri producing a conspicuous zigzag pattern (= midventral complex) and with more than five transverse cirri. Frontal-midventral-transverse cirri formed by more than six anlagen.[4]

The ground pattern of the Urostyloidea: The Urostyloidea are very likely the sistergroup to the remaining Hypotricha for which the name Dorsomarginalia is suggested (Fig. 14a). In the present chapter I discuss the ground pattern of the Urostyloidea. Briefly, the ground pattern of a monophylum (evolutionary unit) is the combination of features of the stem-species from which the monophylum evolved, that is, it is a summary of apomorphies and plesiomorphies (Ax 1995). However, usually only more or less young plesiomorphies are included. Old plesiomorphies – for example, the presence

[1] Tuffrau & Fleury (1994) provided the following diagnosis: Ciliature ventrale composée de cirres fronto-ventraux et éventuellement transversaux qui se différencient au cours de la morphognèse à partir d'un champ longitudinal de nombreux alignements obliques. Ciliature marginale souvent développée. Macronoyau souvent composé de nombreux lobes (2 à 100).

[2] Eigner (2001) provided the following updated diagnosis: Hypotrichida that develop zigzag midventral cirri. Each cirral pattern for proter and opisthe including the two rightmost ventral anlagen develops independently during divisional morphogenesis (i.e. "long primary primordia" do not develop). More than three "within-row" anlagen may develop for dorsal kineties in the proter and opisthe. "Split dorsal kineties" and "fragmented dorsal kineties" are absent.

[3] The diagnosis by Lynn & Small (2002) is as follows: Frontoventral cirri as 2 or more zig-zag files almost full length of ventral surface; these zig-zag files may range from "single" file of zig-zag or offset cirri to multiple and short files of cirri that are offset (= developed zig-zag) at their anterior and posterior ends (e.g., *Eschaneustyla*).

[4] Note that this characterisation reflects the situation in the last common ancestor of the Urostyloidea. The characterisation does not exclude taxa with more or less distinct deviations (e.g., transverse cirri lacking).

of a micronucleus and a macronucleus in the Urostyloidea (an apomorphy of the Ciliophora) – are usually not included the ground pattern.

Apomorphies of the Urostyloidea: The following two features are morphological apomorphies of the Urostyloidea. The first is generally accepted, the second is new.

More than six frontal-midventral-transverse cirral anlagen produce a distinct zigzag-cirral pattern (= midventral pattern). This well-known pattern, which is often rather conspicuous, is – at the present state of knowledge – the most important morphological apomorphy of the group (chapter 1.7, Fig. 1a). Although it looks rather impressive, especially in species with many zigzagging pairs, we can be almost certain that such a pattern evolved several times within the Hypotricha by simple insertion of additional anlagen between the ordinary six cirral anlagen, which each produce primarily a cirral pair, except anlage I which usually produces only the left frontal cirrus and the undulating membranes (Fig. 1a, 16a–0; see chapter 2).

Number of transverse cirri distinctly increased. The frontal-ventral-transverse cirri pattern of euplotids, oxytrichids, and several (basically all?) other groups originates from six anlagen, strongly indicating that this number occurred for the first time in the last common ancestor of the euplotids and hypotrichs (Fig. 12a, 16a, b), respectively, spirotrichs (Fig. 12b). Basically each of the anlagen II–VI produces a transverse cirrus. The CEUU hypothesis (see chapter 2.4) assumes that the midventral complex of the Urostyloidea originated by the insertion/production of additional anlagen. Thus, it is very likely that the additional anlagen did not produce only a midventral pair, but also (as the anlagen II–VI of the euplotids and hypotrichs; Fig. 16a, b) a transverse cirrus. Such a pattern is present only in a low number of urostyloid taxa, for example, *Holosticha* and *Pseudoamphisiella* (Fig. 20a, b). Therefore, the low number of transverse cirri (compared to the number of midventral pairs) in most genera has to be interpreted as apomorphy. But very likely the reduction of the anterior (= left) transverse cirri occurred several times independently within the urostyloids. However, this is not a great problem because the reduction of structures is a rather simple feature. The same phenomenon occurred within the Oxytrichidae. In this group we have several genera with the ordinary frontal-ventral cirral pattern, but a reduced number (i.e., less than five) of transverse cirri (e.g., *Urosomoida*). In other oxytrichid taxa the number of frontal-ventral-transverse cirral anlagen increased so that they have not only a "midventral pattern", but also an increased number of transverse cirri (e.g., *Pattersoniella*, *Territricha*; 16f, i; for review see Berger 1999). In *Pattersoniella* a bicorona has even been formed! Interestingly there is no(?) hypotrich known which has reduced the rightmost (rearmost) transverse cirri. I assume that in the last common ancestor of the Urostyloidea the number of anlagen was only slightly higher than six, that is, a high number of streaks, for example, in *Pseudokeronopsis*, is derived. However, the opposite cannot be excluded.

Plesiomorphies of the Urostyloidea: In the following paragraphs the most important more or less young plesiomorphies of the Urostyloidea are discussed.

Body length about 100–200 μm. There is no evidence that the last common ancestor of the Urostyloidea was very large or very small. Many species are, like many non-urostyloid hypotrichs, between 100 μm and 200 μm long, indicating that the last common ancestor of the Hypotricha had a length in this range. Within the urostylines some groups (e.g., *Urostyla*) are characterised by a rather large body.

Body elongate elliptical and distinctly dorsoventrally flattened. The body outline of many urostyloids and non-urostyloid hypotrichs is elongate elliptical (body length:width ratio roughly around 2–3:1; the cell shown in Fig. 1c has a length:width ratio of 2.9:1), indicating that this is an apomorphy of the Hypotricha. Euplotids have a lower ratio. By contrast, the distinct dorsoventral flattening (Fig. 1d) is an older feature because it also occurs in the euplotids and even in *Phacodinium*.

Body flexible. All urostyloids and most Dorsomarginalia have a flexible body when freely motile (if cells are squeezed, then even the body of rigid species becomes more or less flexible). This indicates that the last common ancestor of the Hypotricha had a flexible body. Within the hypotrichs only the Stylonychinae – a comparatively small subgroup of the Oxytrichidae – have a rigid body (for review see Berger 1999).

Two macronuclear nodules. Several urostyloids have, like many oxytrichids and *Uroleptus* species, two macronuclear nodules, indicating that this pattern evolved in the last common ancestor of the Hypotricha. Many euplotids, oligotrichs, and *Phacodinium* have only one macronucleus, indicating that the stem-species of the spirotrichs (and ciliates) had a single macronucleus. The division into two or more parts obviously evolved convergently in the euplotids, the olgiotrichs, and the hypotrichs. Borror & Wicklow (1983) supposed that a high number of macronucleus-nodules is the "primitive" (= plesiomorphic) condition. However, this assumption is very likely incorrect.

Contractile vacuole near left cell margin about at level of cytostome. This is the most common position both in the urostyloids and the non-urostyloid hypotrichs, indicating that the vacuole was at this site in the stem-lineage of the Hypotricha. Euplotids (e.g., *Euplotes*, *Aspidisca*, *Cytharoides*, *Uronychia*) have the contractile vacuole subterminally near the right body margin (e.g., Petz et al. 1995). In the oligotrichs the situation is rather diverse because the vacuole is either lacking, terminal, subterminal, or – as in the hypotrichs – at the level of the oral apparatus (halterids, tintinnids; Foissner et al. 1999). In the halterids (e.g., *Meseres corlissi*; Petz & Foissner 1992) the vacuole empties via a permanent excretion pore between the second and third somatic kinety, which is reminiscent, like some other morphological features and the molecular data, of the hypotrichs (for review see Foissner et al. 2004a). Unfortunately I did not find data about the excretion pore in tintinnids.

Cortical granules present. These organelles are present in many urostyloid and non-urostyloid hypotrichs. Thus, we have to assume that they were already present in the last common ancestor of the hypotrichs. The lack of the granules in various species or genera evolved very likely convergently both in the urostyloids and the remaining hypotrichs. Only the Stylonychinae are a relatively large group having the lack of cortical granules as apomorphy (Berger & Foissner 1997, Berger 1999).

Adoral zone of membranelles "short" and continuous. In most species of the urostyloids and in many species of the remaining hypotrichs, the adoral zone occupies only about 25–35% of body length, indicating that this relative size is an autapomorphy of the Hypotricha. Moreover it lacks a distinct gap (Fig. 1a–c, e). The last common ancestor of the Stylonychinae evolved a large adoral zone usually occupying more than 40% of body length (Berger & Foissner 1997). In euplotids and *Phacodinium* the relative size of the adoral zone is much larger.

Distal end of adoral zone of membranelles does not extend far posteriorly. This feature was quantified by Wiackowski (1988). For explanation see chapter 1.8.

Endoral membrane: The presence of a second undulating membrane (paroral and endoral) is certainly an autapomorphy of the Hypotricha because the euplotids and the oligotrichs lack an endoral. *Prodiscocephalus* and some related taxa also have a paroral and an endoral (Lin et al. 2004). Therefore these taxa, which are usually assigned to the euplotids (e.g., Lynn & Small 2002), belong to the Hypotricha. For a discussion of the two undulating membranes of *Halteria* see Foissner et al. (2004a).

Undulating membranes long and curved. This is a common pattern in urostyloids, but also occurs in the oxytrichids (e.g., *Oxytricha*, *Sterkiella*, *Histriculus*) and *Uroleptus*. In the oxytrichids, this plesiomorphic pattern was modified several times (e.g., *Cyrtohymena* pattern, *Notohymena* pattern, *Steinia* pattern; for review see Berger & Foissner 1997 and Berger 1999). By contrast, the diversity of these structures within the urostyloids is comparatively low. Several taxa, for example, pseudokeronopsids and *Holosticha*, have rather short, straight, and parallel undulating membranes.

Three frontal cirri. This feature is likely a comparatively old plesiomorphy because there is strong evidence that it evolved in the last common ancestor of the spirotrichs (if Fig. 12b is true), respectively, the group euplotids + hypotrichs (if Fig. 12a is true).

One buccal cirrus. This cirrus, which is usually located immediately right of the paroral, is certainly homologous with the buccal cirrus of the oxytrichids and the euplotids (Fig. 16a, d), indicating that it evolved in the last common ancestor of the group euplotids + hypotrichs. For the complicated terminology of this cirrus see Berger (1999). Most urostyloids retained the plesiomorphic single cirrus; some taxa lack such a cirrus (e.g., *Paragastrostyla*, *Periholosticha*), some have two or more. In *Uroleptopsis citrina* it is not right of the paroral but migrated anteriorly and is therefore a part of the bicorona. Basically the buccal cirrus is homonomous to the left cirrus of a midventral pair.

Two frontoterminal cirri. The two frontoterminal cirri present in many urostyloids are undoubtedly homologous with the cirri VI/3 and VI/4 of the Oxytrichidae (Fig. 1a, 16b, d; e.g., Wirnsberger 1987, Berger 1999, Foissner et al. 2004a). This can be concluded (i) from the same position during interphase (near the distal end of the adoral zone of membranelles); (ii) from the origin from the rightmost frontal-ventral-transverse cirral anlage; and (iii) by a conspicuous migration of these cirri to the anterior body portion. This implies that these two cirri occurred in the last common ancestor of the Hypotricha and not in the stem-lineage of the Urostyloidea. Within the urostyloids this feature changed in two ways: (i) the number of migrating cirri increased (e.g., *Keronella*); (ii) frontoterminal cirri are lost (e.g., *Australothrix*). In *Bakuella edaphoni* and few other hypotrichs the frontoterminal cirri possibly originate from the two rightmost anlagen. Such observations should be checked. The euplotids obviously lack frontoterminal cirri (Fig. 16a).

Two pretransverse ventral cirri. These two cirri are not very common in urostyloids. In several descriptions they are incorrectly considered as transverse cirri. They can be homologised with the pretransverse ventral cirri V/2 and VI/2 of the oxytrichids or *Amphisiella* without problems (Fig. 16b, c, e, g, h). However, it is much more difficult to decide whether or not these cirri are homologous with the cirri V/2 and VI/2 of *Euplotes* species with 10 frontal-ventral cirri because in *Euplotes* they are not very near to the corresponding transverse cirri (Fig. 16a). Anyhow, in the stem-lineage of the Urostyloidea the presence of pretransverse ventral cirri is a plesiomorphy.

One left and one right marginal row. Marginal rows are basically lacking in euplotids (16a; note that *Prodiscoephalus* and related taxa are very likely not euplotids, but hypotrichs because they have two undulating membranes; Lin et al. 2004). One left and one right marginal row obviously evolved in the stem-lineage of the Hypotricha and therefore two marginal rows are a plesiomorphy for the Urostyloidea. In some taxa (e.g., *Urostyla*, *Pseudourostyla*, *Diaxonella*) the number of rows increased. Different morphogenetic patterns indicate that these species with more than two rows evolved convergently.

Three dorsal kineties. Very likely this is an apomorphy of the Hypotricha (Fig. 1c; 14a, apomorphies 1). This number is obviously retained in the stem-lineage of the Urostyloidea because many species of this group also have three bipolar dorsal kineties. In some groups of the urostyloids the number of bipolar kineties increased more or less distinctly (e.g., *Pseudokeronopsis*). By contrast, in the lineage to the *Uroleptus* + Oxytrichidae group (Fig. 14a) two other methods evolved to increase the number of kineties, namely (i) the formation of dorsomarginal kineties (Fig. 14a; apomorphies 3; Fig. 243j), and (ii) the fragmentation of bipolar kineties (Fig. 14a; apomorphies 5). Within the Oxytrichidae further modifications, for example, multiple fragmentation (Fig. 243k, m) or retention of parental kineties, occurred (for review see Berger 1999). *Gonostomum* and *Wallackia* also have only three dorsal kineties (Berger 1999, Foissner et al. 2002). Berger & Foissner (1997) and Berger (1999) supposed that this "*Gonostomum* pattern" evolved – via the *Urosomoida* pattern – from the *Oxytricha* pattern by the loss of both the fragmentation of dorsal kinety 3 and the dorsomarginal kineties. However, it also cannot be excluded that the ancestor of *Gonostomum* (and *Wallackia*?) split off before the dorsomarginal kineties evolved. This hypothesis is supported by the fact that *Gonostomum* has only three cyst wall layers and not four like the Oxytrichidae (for review see Gutiérrez et al. 2003).

Three caudal cirri. Caudal cirri originate at the end of the bipolar dorsal kineties (Fig. 1c). Dorsomarginal kineties and the anterior fragments of splitting dorsal kineties are never associated with such cirri. Many urostyloids and non-urostyloid hypotrichs have one caudal cirrus at the rear end of each bipolar kinety. This strongly suggests that one caudal cirrus at the end of dorsal kineties 1, 2, and 3 (see previous paragraph) is a novelty for the Hypotricha and therefore a relatively young plesiomorphy for the Urostyloidea. However, caudal cirri in general are older because *Euplotes* also forms such cirri at the end of at least two dorsal kineties (e.g., Voss 1989).

Resting cyst of PKR-type with three cyst wall layers. Very likely this resting cyst type occurred for the first time in the stem-lineage of the Hypotricha (see chapter 1.10.3 of general section). The urostyloids likely retained this type whereas in the stem-lineage of the oxytrichids a cyst of the KR-type with four cyst wall layers evolved.

Oral primordium originates on cell surface. In all Hypotricha the oral primordium of the opisthe originates more or less on the cell surface, that is, epiapokinetal. By contrast, in the euplotids and the oligotrichs the oral primordium develops hypoapokinetally (e.g., Foissner 1996c, Agatha 2004). The development in a pouch is considered as derived state within the spirotrichs (Agatha 2004) and the development on the cell surface in the urostyloids therefore a rather old plesiomorphy.

Parental adoral zone of membranelles not or only proximally reorganised during cell division. In *Holosticha* and several other urostyloid taxa the parental adoral zone is not or only proximally reorganised. This pattern also occurs in the Dorsomarginalia (= *Uroleptus* + Oxytrichidae; e.g., Berger 1999, Eigner 2001), but also, for example, in *Euplotes*, *Diophrys* (e.g., Voss 1989, Song & Wilbert 1994) and oligotrichs (e.g., Agatha 2003), indicating that this state is a plesiomorphy in the stem-lineage of the Urostyloidea. Only in several urostyloids is the parental adoral zone completely replaced (e.g., pseudokeronopsids). The formation of a new oral apparatus for the proter in *Uronychia transfuga* (e.g., Hill 1990) is very likely a convergence.

Frontal-(mid)ventral-transverse cirral anlagen of proter and opisthe develop independently. This feature is (very likely) present in *Euplotes* (e.g., Wise 1965, Voss 1989) and the Urostyloidea and many Dorsomarginalia. By contrast, in several oxytrichids (e.g., *Urosoma*), but also in some euplotids (e.g., *Uronychia*, Hill 1990), some anlagen of the proter and the opisthe have a common origin; that is, are formed from so-called primary primordia[1] (for review see Berger 1999). Unfortunately, we cannot decide unequivocally which of the two methods was present in the last common ancestor of the euplotids and hypotrichs. Anyhow, the independent development is no apomorphy for the Urostyloidea.

Marginal rows and dorsal kineties divide intrakinetally. This mode of marginal row and dorsal kinety formation is a very old plesiomorphy because even present in, for example, the spathidiids (Berger et al. 1983). Only in few urostyloids (e.g., *Thigmokeronopsis*), do these structures originate de novo, which has to be interpreted as apomorphy.

Marginal rows and dorsal kineties originate independently for proter and opisthe. This mode was at least present in the last common ancestor of the Hypotricha because it is the sole mode in this group. Consequently, it is a plesiomorphy in the stem-lineage of the Urostyloidea.

Parental somatic ciliature completely replaced during cell division. Both in the urostyloids and in the Dorsomarginalia, no part of the parental somatic ciliature (frontal-ventral-transverse cirri, marginal rows, dorsal kineties, caudal cirri) is retained in the postdividers. By contrast, in the euplotids and the oligotrichs, and in ciliates in general, at least parts of the parental somatic ciliature are present in the post-dividers (e.g., Foissner 1996c, Agatha 2004). Thus, the complete replacement is likely an apomorphy of the Hypotricha and therefore a young plesiomorphy for the Urostyloidea. This state is present in all urostyloids. However, in some Oxytrichidae parental structures (e.g.,

---

[1] *Anteholosticha warreni* also forms "primary primordia" (Fig. 85k, l). Whether or not this pattern is homologous to that of, for example, *Urosoma*, is unknown.

parts of the marginal rows in *Coniculostomum*, or parts of the dorsal ciliature in *Parakahliella*; Kamra & Sapra 1990, Berger & Foissner 1989b, Berger et al. 1985) are retained in the post-dividers. These specific features have to be interpreted as apomorphies of the individual groups.

Silverline system. The silverline system of the Urostyloidea and the Dorsomarginalia is fine-meshed. That is, we have to assume that this pattern was present at least in the last common ancestor of the Hypotricha.

Macronuclear nodules fuse to a single mass during cell division. In ciliate species with two or more macronuclear nodules, these pieces fuse to a single mass during cell division (e.g., Raikov 1982, Petz & Foissner 1993, Song & Wilbert 1994, Berger 1999, Foissner et al. 2002). Consequently this is a rather old plesiomorphy in the stem-lineage of the Urostyloidea. The individual division of the many nodules in the Pseudokeronopsinae is an apomorphy for this group (Fig. 167; Berger 2004b).

Micronuclear DNA polymerase alpha genes not scrambled. Chang et al. (2003) found that the micronuclear DNA polymerase alpha genes of the urostyloids *Urostyla grandis* and *Holosticha kessleri* (= *H. gibba* in present paper) are not scrambled. On the other hand they found that in *Uroleptus* sp. and *Paraurostyla weissei* these genes are scrambled with a configuration similar to that in the oxytrichids *Stylonychia lemnae* and *Oxytricha*. Thus, they suggested that the DNA polymerase alpha gene became scrambled in a species diverging later than *Holosticha* (that is, urostyloids), but earlier than *Uroleptus*. This molecular feature therefore supports the Dorsomarginalia (Fig. 14a, apomorphies 3). Obviously the unscrambled configuration of the urostyloids is the plesiomorphic state.

Micronuclear actin I gene MDSs (macronuclear destined segments) not scrambled. The micronuclear actin I gene of *Urostyla grandis* and *Engelmanniella mobilis* consists of three and four nonscrambled MDSs, respectively (Hogan et al. 2001). Dalby & Prescott (2004) concluded that the nonscrambling is the plesiomorphic state. By contrast, the actin I gene became scrambled both in the *Uroleptus* and the Oxytrichidae branch, but in two different, completely unrelated patterns (Dalby & Prescott 2004). Thus, the "actin I gene scrambling type *Uroleptus*" is obviously an apomorphy for the *Uroleptus* group (Fig. 14a, apomorphies 4) and the "actin I gene scrambling type Oxytrichidae", beside the fragmentation of dorsal kinety 3, a further apomorphy of the Oxytrichidae cluster (Fig. 14a, apomorphies 5).

Remarks: The characterisation of the Urostyloidea includes only the apomorphic features in the last common ancestor of this group. I hope that molecular studies reveal a further apomorphy of this group. The ground pattern of the Urostyloidea contains a more detailed analyses of the group. For a discussion of the history see chapter 2.3 in the general section. Unfortunately, the apomorphies of the Urostyloidea are not unique; that is, a zigzagging ventral cirral pattern originating from more than six anlagen evolved convergently several times (Fig. 16c, f, i, n). This phenomenon can be explained by the CEUU-hypothesis discussed in detail in chapter 2.4 of the general section. However, there are some features, which show whether a species with zigzagging arranged ventral cirri is a urostyloid or belongs to a different group (see next chapter).

It is impossible to compare all systems of the urostyloids established so far (Tables 2–11) with the classification proposed in the present paper (see content). Thus only some examples can be discussed. Kahl (1932) classified all non-euplotid hypotrichs in the Oxytrichidae, that is, he did not accept Bütschli's taxon. Moreover, he divided *Holosticha* into five subgenera (*Paruroleptus*, *Keronopsis*, *Holosticha*, *Trichotaxis*, *Amphisiella*).

Borror (1972, Table 3) did not summarise the urostylids and the holostichids in a single taxon because the cirral pattern and the morphogenesis of *Urostyla grandis* were not yet described in detail at that time. Thus, he assumed that the urostylids lack midventral cirri (see footnote of corresponding entry in list of synonyms).

Wicklow (1981, Table 6) and Borror & Wicklow (1983, Table 8) provided a rather detailed system. They distinguished two major groups, namely the Pseudourostyloidea with the single genus *Pseudourostyla* and the Urostyloidea with the remaining taxa. I agree with these authors that *Pseudokeronopsis* and *Thigmokeronopsis* are closely related.

Eigner & Foissner (1992) re-evaluated the classification of urostyloid hypotrichs. They recognised that evolutionary relationships within the urostyloids are little known. In spite of this they presented an argumented tree including only taxa, which might be representatives of higher taxa to stimulate the discussion. Briefly, the tree has the following structure: ((((*Pseudokeronopsis*, *Thigmokeronopsis*) *Holosticha*) *Pseudourostyla*) (*Urostyla*, *Australothrix*)). Only few points should be discussed: (i) Eigner & Foissner considered, like Borror & Wicklow (1983), *Pseudokeronopsis* and *Thigmokeronopsis* as sistergroups, however, without providing an apomorphy. But obviously they incorrectly applied their character 4 (many frontal cirri forming bicorona [apomorphy] vs. 3–4 frontal cirri [plesiomorphy]) because they used this feature to characterise *Holosticha*, which, however, has only three frontal cirri. Also in error, they used the same feature as apomorphy for *Australothrix*. (ii) In the tree they assumed that many marginal rows (e.g., *Urostyla*) is the plesiomorphic state. In the present book I assume the opposite (see ground pattern above). (iii) They united the (*Pseudourostyla* ((*Pseudokeronopsis*, *Thigmokeronopsis*) *Holosticha*)) group by the presence of (two) frontoterminal cirri. By contrast I suppose that the presence of frontoterminal cirri is a plesiomorphy in the stem-lineage of the urostyloids. On the other hand they found no autapomorphy for their *Urostyla* + *Australothrix* group, both of which lack frontoterminal cirri. However, they differ significantly in the frontal ciliature (many cirri vs. three). In spite of these differences to my review, Eigner & Foissner (1992) correctly concluded that phylogeny within the urostylids is little known since character states are uncertain and morphogenetic data are still too sparse or inaccurate.

I also failed to create a usable diagram of the phylogenetic relationships within the Urostyloidea using both Hennig86 and the head and hand method. As just mentioned, this is mainly due to the lack of some important data (e.g., on cell division) and the fact that at least one of two important urostyloid features (bicorona or midventral rows) must have evolved convergently in the Urostyloidea. I assigned most species to four more or less large (monophyletic?) groups using the frontal ciliature and the midventral complex as main features.

• Holostichidae: three more or less enlarged frontal cirri[1], midventral complex composed of cirral pairs only
• Bakuellidae: three more or less enlarged frontal cirri, midventral complex composed of cirral pairs and midventral rows
• Urostylidae: many frontal cirri form a bicorona, tricorona, or multicorona, midventral complex composed of cirral pairs only or cirral pairs and midventral rows
• Epiclintidae: many frontal cirri arranged in oblique rows, midventral complex composed of midventral rows only

Only few taxa cannot be assigned to one of the four groups (see key below). *Neokeronopsis* and *Urostyloides*, so far assigned to the urostyloids because of the zigzagging cirral pattern, belong to the Oxytrichidae because they have dorsomarginal kineties and fragmenting bipolar dorsal kineties. They are reviewed at the end of the book.

How to recognise an urostyloid hypotrich in practice

Before using the key below you have to know whether the hypotrich in question belongs to the Urostyloidea at all. If one of the two following points applies then you can be rather certain that your population belongs to an urostyloid species.

- The specimens have (i) zigzagging ventral cirri; that is, a midventral complex composed of midventral pairs (Fig. 1a). In several taxa the complex is not only composed of pairs (anterior portion), but also of midventral rows (posterior portion). (ii) The dorsal ciliature is composed of bipolar kineties only (Fig. 1c), that is, your species lacks dorsomarginal kineties (Fig. 243j) and/or fragmented kineties (Fig. 243m). Several species, for example, *Pattersoniella* (Fig. 16f), *Territricha* (Fig. 16i; for review see Berger 1999), *Neokeronopsis* (p. 1190), and *Urostyloides* (p. 1205) feign a urostyloid relationship. However, their complex dorsal ciliature reveals that they belong to the Oxytrichidae (note that *Holosticha bradburyae* [Fig. 36e–l] also has a rather complex dorsal morphogenesis, which proceeds, however, differently from that of the Oxytrichidae). *Uroleptus* species, which have a characteristic tailed and contractile body (Fig. 16j), posses a dorsomarginal kinety and therefore belong to the Dorsomarginalia (Fig. 14a). (iii) The body is flexible when freely motile. If your specimens are rigid like, for example, *Stylonychia mytilus*, then your population does not belong to the Urostyloidea; it belongs to the Stylonychinae (for review see Berger 1999) or to the euplotids.

- Your specimens belong to the Epiclintidae (Fig. 225a–c). In this group the midventral complex is composed of more or less oblique midventral rows only. In other taxa with ventral cirral rows (e.g., *Amphisiella*; Fig. 16g), these rows are usually longitudinally arranged or the body, and therefore also the cirral rows, is distinctly

[1] Note that some workers designate cirrus III/2 (= parabuccal cirrus; Fig. 1b) also as frontal cirrus. Then this is the fourth "frontal cirrus".

twisted. The Epiclintidae also lack dorsomarginal kineties and fragmented bipolar kineties and have a flexible body.

Key to the subgroups of the Urostyloidea

The Urostyloidea are divided into four higher taxa (Holostichidae, Bakuellidae, Urostylidae, Epiclintidae)[1] and four (largely monotypic) genera, which cannot be assigned to any of these groups. As usual in the Hypotricha, the separation is mainly based on the cirral pattern. Thus, silver preparations or at least very detailed live observations (interference contrast) are needed to use the following key and the subsequent keys successfully. Moreover you must be familiar with the terms specific to the Urostyloidea (Fig. 1a–g). *Paramitrella* must not be confused with *Psammomitra* (Fig. 42, 43) and *Epiclintes* (Fig. 226–228). Please note that, as in other ciliate groups, only a limited number of the extant species is known, that is, it can be possible that your population belongs to a not yet described species.

Please read the previous chapter (How to recognise ...) before using the following key. The basic cirral patterns of the urostyloid genera treated in the present book are summarised on several plates (Fig. 20a–h, 112a–i, 145a–e, 168a–f, 200a–c, 237e, 238a, 240a, 241a).

1 Body outline very slender and tripartite, that is, with more or less distinct head, widened central body portion (trunk), and tail (Fig. 42j, 43a; 228t; 240a; 241a) **7**

\- Body outline not as above **2**

2 Body outline roughly metopid, that is, with oblique head (Fig. 237c) . . . . *Notocephalus* (p. 1169)

\- Body outline not as above **3**

3 Two enlarged frontal cirri (Fig. 238a, 239a) . . . . *Biholosticha* (p. 1176)

\- Three more or less distinctly enlarged frontal cirri (Fig. 1b) or many frontal cirri (e.g., Fig. 1a) **4**

4 Three enlarged frontal cirri[2]; midventral complex composed of cirral pairs only (e.g., Fig. 20a–h) . . . . Holostichidae (p. 84)

\- Cirral pattern not as above **5**

5 Three enlarged frontal cirri[3]; midventral complex composed of midventral pairs and at least one midventral row (Fig. 112a–i) . . . . Bakuellidae (p. 527)

\- Cirral pattern not as above **6**

6 Many frontal cirri arranged in a bicorona, tricorona, or multicorona; midventral complex composed of cirral pairs only or cirral pairs and midventral rows (Fig. 145a–e, 200a–c) . . . . Urostylidae (p. 731)

---

[1] I am not convinced that all these taxa are monophyletic in my review. For details see individual groups.

[2] *Urostyla* contains several little known species, which also have only three frontal cirri. However, they have either midventral rows and/or more than two marginal rows. *Trichototaxis aeruginosa* has a difficult to interpret ventral cirral pattern (Fig. 103a–d).

[3] *Urostyla* contains several little-known species, which also have only three frontal cirri. Thus, check the *Urostyla* key if you cannot identify your population with the Bakuellidae key and its subsequent keys.

- Many frontal cirri arranged in oblique rows; midventral complex composed of midventral rows only (Fig. 225a–c) .... Epiclintidae (p. 1113)

7 (1) Midventral complex composed of cirral pairs (Fig. 20c, 43a, 240a, 241a) .... **8**[1]

- Midventral complex composed of midventral rows only (Fig. 225a–c) .... Epiclintidae (p. 1113)

8 Anterior body end usually narrow (Fig. 42j, 43a) and midventral complex terminate about in middle of trunk (Fig. 20c) .... *Psammomitra* (p. 221)

- Anterior body end distinctly wider and midventral complex extends in tail-region (Fig. 240a, 241a) **9**

9 (1) Body length 264–1100 µm (average 618 µm); anterior body end with beak-like, leftwards curved projection; posterior body portion moderately thin (Fig. 241a–c) .. .. *Uncinata* (p. 1186)

- Body length 260–440 µm (average 354 µm); anterior body end round; posterior body portion very thin (Fig. 240a, b) .... *Paramitrella* (p. 1183)

Holostichidae Fauré-Fremiet, 1961

1961 **Holostichidae n. nom.** – Fauré-Fremiet, C. r. hebd. Seanc. Acad. Sci. Paris, 252: 3517 (original description). Type genus: *Holosticha* Wrzesniowski, 1877.

1972 **Holostichidae Fauré-Fremiet, 1961**[2] – Borror, J. Protozool., 19: 10 (revision).

1979 **Holostichoidea superfam. n.** – Jankowski, Trudy zool. Inst., 86: 74 (revision).

1979 **Holostchina subordo. n.** – Jankowski, Trudy zool. Inst., 86: 84 (original description; incorrect original spelling). Type genus: *Holosticha* Wrzesniowski, 1877.

1979 **Holostichidae Fauré-Fremiet, 1961**[3] – Corliss, Ciliated Protozoa, p. 309 (revision).

1981 **Holostichinae (n. subfam.)** – Wicklow, Protistologica, 17: 348 (original description). Type genus: *Holosticha* Wrzesniowski, 1877.

1983 **Holostichinae Fauré-Fremiet, 1961** – Borror & Wicklow, Acta Protozool., 19: 121 (revision).

2001 **Holostichidae Fauré-Fremiet, 1961** – Berger, Catalogue of ciliate names 1. Hypotrichs, p. 108 (nomenclator containing all basionyms, combinations, and higher taxa of hypotrichs).

Nomenclature: The name Holostichidae and its derived forms for the various categories are based on the genus-group name *Holosticha*. Originally established as family, later categorised as superfamily (Jankowski 1979), suborder (correct spelling is Holostichina; Jankowski 1979), and subfamily (Borror & Wicklow 1983). Jankowski (1979) and Wicklow (1981) obviously overlooked the principle of coordination (ICZN 1964 and 1999, Article 36). I use the name as introduced by Fauré-Fremiet (1961), however, without category (see chapter 7.2 of general section).

---

[1] If you are uncertain check all leads of couplets 7–9.

[2] The diagnosis by Borror (1972) is as follows: Right marginal cirri, left marginal cirri, and midventral cirri present. Midventral cirri usually present in zigzag series that arise (at least in *Holosticha*, *Keronopsis* and *Uroleptus*) from alignment of transverse streaks of cilia during cell division.

[3] Corliss (1979) provided the following characterisation: Right and left marginal cirri present, with variable number of rows of other ventral cirri; transverse and frontal cirri often differentiated; body elongate; macronuclei two to many in number.

Characterisation: Urostyloidea with three frontal cirri and a midventral complex composed of cirral pairs only.

Remarks: The list of synonyms contains only the original description and the papers where the category (and therefore the spelling) was changed. The characterisation above is, due to the lack of apomorphies, only a combination of the most important plesiomorphies, indicating that the group is non-monophyletic. However, the genera included can be arranged in three (artificial?) groups: (i) in *Holosticha*, *Pseudoamphisiella*, and *Psammomitra* the number of transverse cirri (roughly) equals the number of frontal-midventral-transverse cirral anlagen. However, in *Holosticha* and *Pseudoamphisiella* the number is comparatively high, whereas *Psammomitra* has only about eight cirral anlagen. Transverse cirri often prominent. (ii) In *Caudiholosticha*, *Anteholosticha*, and *Diaxonella* the number of transverse cirri is distinctly lower than the number of cirral anlagen, respectively, midventral pairs; that is, only the rightmost streaks produce transverse cirri. Moreover, the transverse cirri are usually not very prominent. *Diaxonella* is the sole species in the Holostichidae with more than two marginal rows. (iii) *Periholosticha* and *Afrothrix* have a rather short midventral complex and a distinct gap in the adoral zone.

Jankowski (1979, p. 84) established the Psammomitrinae (type genus: *Psammomitra*) as subgroup of the Oxytrichidae. Originally it contained *Balladynella*, *Onychodromus*, *Stylonethes*, *Tetrastyla*, *Epiclintes*, *Pescozoon*, *Uroleptoides*, *Perisincirra*, *Plagiostyla*, *Gastrosticha*, and *Banyulsella*. This seems a rather artificial assemblage.

Song et al. (1997, p. 266) established the Pseudoamphisiellidae[1] with the sole genus *Pseudoamphisiella* (see also Shi et al. 1999, p. 117). Since the group is monotypic (at the present state of knowledge) it is redundant and thus not used in the present review.

Key to the genera of the Holostichidae

This group contains genera/species, which have three (more or less) distinctly enlarged frontal cirri. Distinction between *Caudiholosticha* and *Anteholosticha* is difficult because the sole difference is the presence/absence of caudal cirri. Thus I recommend checking both alternatives. *Psammomitra* must not be confused with *Paramitrella* (Fig. 240a) and *Epiclintes* (Fig. 226–228), which have a similar body shape. Moreover most of them are marine and likely psammophilous.

Urostyla contains several little-known species which also have only three frontal cirri. However, they have either midventral rows and/or more than two marginal rows.

1 Body outline club-shaped (Fig. 20c, 43a) . . . . *Psammomitra* (p. 221)
- Body outline not as above **2**
2 Two or more left marginal rows (e.g., Fig. 20h) **9**

[1] Song et al. (1997) provided the following diagnosis: Hypotrichida (Discocephalina? Urostylina?) with differentiated frontal and highly developed transverse cirri; two widely separated midventral rows, which originate from a series of oblique FVT-streaks during morphogenesis; without frontoterminal cirri; right marginal row derived from the rightmost streak of FVT-anlagen.

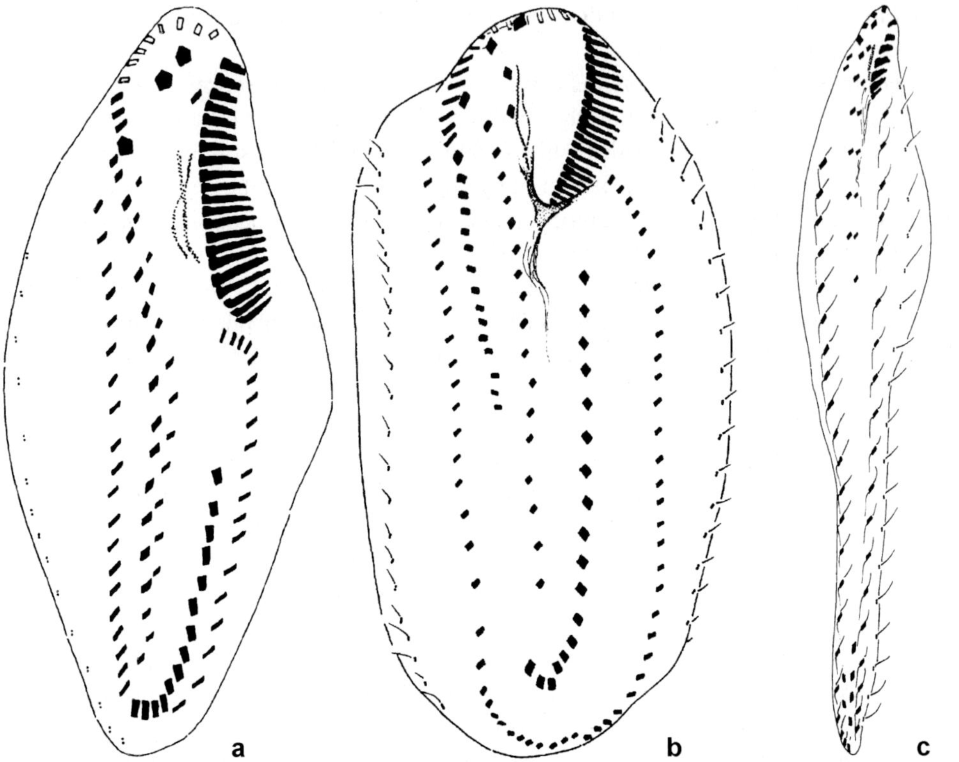

Fig. 20a–c Ventral cirral pattern in members of the Holostichidae (part 1). **a:** *Holosticha heterofoissneri*. **b:** *Pseudoamphisiella lacazei*. **c:** *Psammomitra retractilis*. Sources of illustrations see individual descriptions. Abbreviations used in short characterisations of infraciliature: AZM = adoral zone of membranelles, A//ZM = bipartite adoral zone of membranelles, BC = buccal cirrus, CC = caudal cirri, DK = dorsal kinety, FC = frontal cirri, FT = frontoterminal cirri, LMR = left marginal row, MC(MP) = midventral complex composed of cirral pairs only, RMR = right marginal row, TC = transverse cirri. Explanation of supplemental signs and figures (explained on examples): A//ZM = bipartite adoral zone of membranelles (that is, with gap), 3FC = three frontal cirri, CC- = caudal cirri lacking, TC+ = transverse cirri present (+ = number of transverse cirri less than number of frontal-midventral-transverse cirral anlagen; ++ = same number as anlagen; +++ = more than ordinary anlagen, e.g., *Epiclintes*, *Parabirojimia*), >2FT = more than two frontoterminal cirri, >1(1)BC = usually more than one buccal cirrus.

- One left marginal row (Fig. 20a, b, d–g; do not interpret the long transverse cirral row of *Pseudoamphisiella* or *Holosticha* as [second] left marginal row) **3**

3 Two or more right marginal rows (Fig. 134) . . . . *Birojimia terricola* (p. 678)

- One right marginal row **4**

4 Midventral complex short and inconspicuous (Fig. 20f, g) **5**

- Midventral complex extends at least to near mid-body, usually to near transverse cirri (e.g., Fig. 20a, b, d, e) **6**

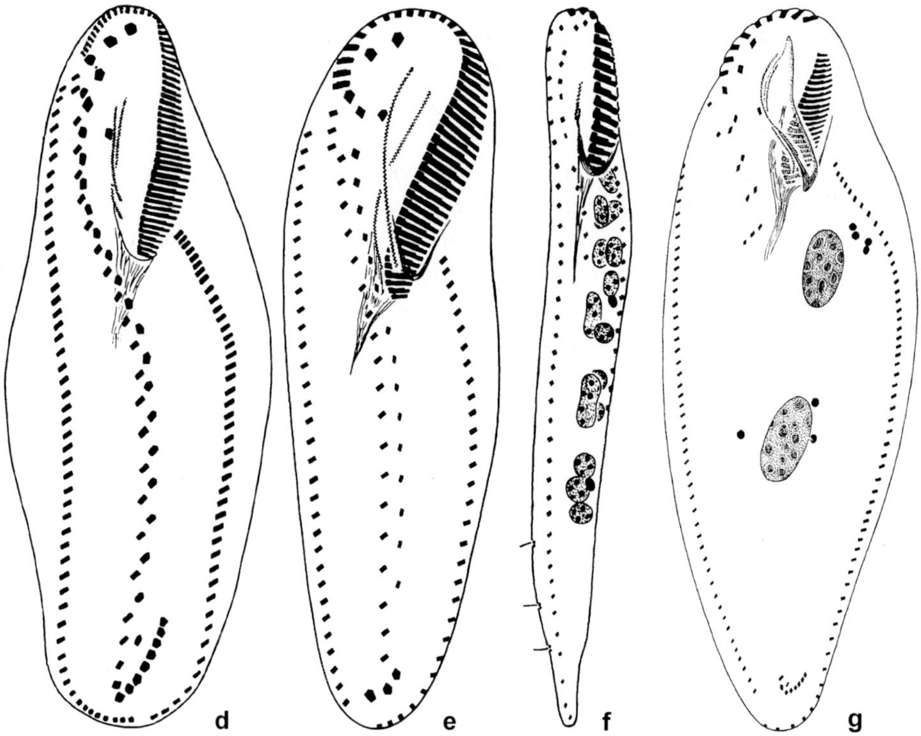

d	e	f	g
Anteholosticha	*Caudiholosticha*	*Periholosticha*	*Afrothrix*
AZM, 3FC, BC+, FT+, MC(MP), TC+, CC-, 1LMR, 1RMR, ≥3DK	AZM, 3FC, BC+, FT+, MC(MP), TC+, CC+, 1LMR, 1RMR, <3–>3DK	A//ZM, 3FC, BC-, FT-+, MC(MP), TC-+, CC+, 1LMR, 1RMR, 2–3DK	A//ZM, 3FC, BC+, FT+, MC(MP), TC+, CC-, 1LMR, 1RMR, 2–3DK

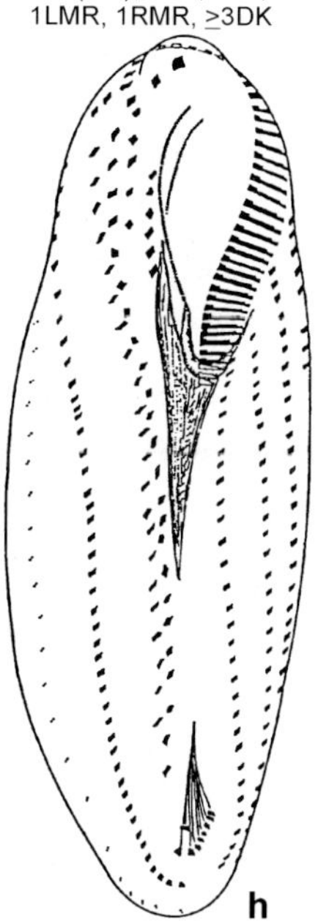

h

Diaxonella
AZM, 3FC, BC+, FT+,
MC(MP), TC+, CC-,
>1LMR, 1RMR, 3DK

Fig. 20d–h Ventral cirral pattern in members of the Holostichidae (part 2). **d:** *Anteholosticha monilata.* **e:** *Caudiholosticha stueberi.* **f:** *Periholosticha lanceolata.* **g:** *Afrothrix darbyshirei.* **h:** *Diaxonella pseudorubra.* Sources of illustrations see individual descriptions. Abbreviations used in short characterisations of infraciliature: AZM = adoral zone of membranelles, A//ZM = bipartite adoral zone of membranelles, BC = buccal cirrus, CC = caudal cirri, DK = dorsal kinety, FC = frontal cirri, FT = frontoterminal cirri, LMR = left marginal row, MC(MP) = midventral complex composed of cirral pairs only, RMR = right marginal row, TC = transverse cirri. Explanation of supplemental signs and figures see legend to Fig. 20a–c.

5 Buccal cirrus present (Fig. 20g) .... *Afrothrix* (p. 486)
- Buccal cirrus absent (Fig. 20f .... *Periholosticha* (p. 498)
6 (3) Many (about as much as midventral pairs) prominent transverse cirri form distinct, roughly longitudinal row (e.g., Fig. 20a, b) **7**
- Few (distinctly less than midventral pairs) not distinctly enlarged transverse cirri form short oblique row (e.g., Fig. 20d, e) **8**
7 Adoral zone with gap; anterior end of left marginal row curved rightwards; cortical seam lacking; midventral complex distinctly zigzagging (Fig. 20a) *Holosticha* (p. 88)
- Adoral zone without gap; anterior end of left marginal row not curved rightwards; cortical seam present; zigzag-pattern of midventral complex inconspicuous (lacking) because cirri of each pair distinctly separated so that two ventral rows are feigned (Fig. 20b) .... *Pseudoamphisiella* (p. 191)
8 (5) Caudal cirri present (Fig. 1c, 20e) .... *Caudiholosticha* (p. 232)
- Caudal cirri lacking (Fig. 20d) .... *Anteholosticha* (p. 292)
9 One right marginal row (Fig. 20h) .... *Diaxonella* (p. 461)
- Two or more right marginal rows (Fig. 133.2g) ... *Metaurostylopsis songi* (p. 672)

Holosticha Wrześniowski, 1877

1877 ***Holosticha*** – Wrześniowski, Z. wiss. Zool., 29: 278 (original description; no formal diagnosis provided). Type species (by subsequent designation by Borror 1972, p. 10): *Oxytricha kessleri* Wrześniowski, 1877.
1878 ***Amphisia* n. gen.**[1] – Sterki, Z. wiss. Zool., 31: 57 (original description of subjective synonym). Type species (by original designation on p. 57): *Trichoda gibba* Müller, 1786.
1882 ***Amphisia*, Sterki** – Kent, Manual Infusoria II, p. 767 (revision).
1882 ***Holosticha*, Wrz.** – Kent, Manual Infusoria II, p. 769 (revision).
1889 ***Holosticha* (Wrzesniowski 1877) emend. Entz 1884** – Bütschli, Protozoa, p. 1744 (revision).
1889 ***Amphisia* Sterki 1878** – Bütschli, Protozoa, p. 1745 (revision).
1932 ***Holosticha* Wrzesniowski, 1877 emend.**[2] – Kahl, Tierwelt Dtl., 25: 570 (revision of hypotrichs).
1932 ***Holosticha* subgen. n.** – Kahl, Tierwelt Dtl., 25: 578 (revision of hypotrichs; see nomenclature).
1933 ***Holosticha* (s. lat.) Wrzesniowski, 1884** – Kahl, Tierwelt N.- u. Ostsee, 23: 108 (guide to marine ciliates; incorrect year).
1933 ***Holosticha* (s. str.) Kahl 1932** – Kahl, Tierwelt N.- u. Ostsee, 23: 109 (guide to marine ciliates; classified as subgenus).
1965 ***Holosticha*** – Lepsi, Protozoologie, p. 970 (revision).
1972 ***Holosticha* Wrzesniowski, 1877**[3] – Borror, J. Protozool., 19: 10 (revision of hypotrichs; fixation of type species by subsequent designation).
1974 ***Holosticha* Wrzesniowski** – Stiller, Fauna hung., 115: 68 (revision).

---

[1] Sterki (1878) did not provide a formal diagnosis. However, to document the somewhat cryptic designation of the type species his text is repeated: "5. O. gibba St., die freilich am besten mit dem Wortlaute der Stein'schen Diagnose für O. übereinstimmt, indem sie durch » drei griffelförmige Stirnwimpern « und zwei mediane Reihen borstenförmiger Bauchwimpern charakterisiert ist, welch letztern die Randreihen sehr genähert sind. **Gerade dies zeigt die Nothwendigkeit einer anderen Bestimmung**. Es fand sich nun eine neue ..."
[2] The diagnosis by Kahl (1932) is as follows: Oxytrichidae mit Transversalreihe und 1–3 geschlossenen Ventralreihen.
[3] Borror (1972) provided the following diagnosis: One row each of right and left marginal cirri. Transverse cirri present. Three frontal cirri differentiated from midventral cirri. Usually 2 macronuclei.

1979 ***Holosticha***[1] – Borror, J. Protozool., 26: 547, 549 (systematics of Urostylidae).
1979 ***Amphisia* Sterki, 1878** – Jankowski, Trudy zool. Inst., Leningr., 86: 50 (generic catalogue of hypotrichs).
1979 ***Holosticha* Wrześniowski, 1877** – Jankowski, Trudy zool. Inst., Leningr., 86: 55 (generic catalogue of hypotrichs).
1979 ***Holosticha* Wrzesniowski, 1877** – Corliss, Ciliated protozoa, p. 309 (generic revision).
1982 ***Holosticha* Wrzesniowski, 1877**[2] – Hemberger, Dissertation, p. 82 (revision of non-euplotid hypotrichs).
1983 ***Holosticha* Wrześniowski, 1877** – Borror & Wicklow, Acta Protozool., 22: 121 (revision of urostylids).
1983 ***Holosticha* Wrzesniowski, 1877** – Curds, Gates & Roberts, British and other freshwater ciliated protozoa, p. 406 (guide to ciliate genera).
1985 ***Holosticha*** – Small & Lynn, Phylum Ciliophora, p. 451 (guide to ciliate genera).
1992 ***Holosticha* Wrzesniowski, 1877** – Carey, Marine interstitial ciliates, p. 180 (guide).
1999 ***Holosticha* Wrzesniowski, 1877** – Shi, Song & Shi, Progress in Protozoology, p. 116 (revision of hypotrichs).
2001 ***Holosticha* Wrześniowski 1877** – Aescht, Denisia, 1: 82 (catalogue of generic names of ciliates).
2001 ***Holosticha* Wrzesniowski, 1877** – Berger, Catalogue of ciliate names 1. Hypotrichs, p. 32 (nomenclator containing all basionyms, combinations, and higher taxa of hypotrichs).
2002 ***Holosticha* Wrzesniowski, 1877** – Lynn & Small, Phylum Ciliophora, p. 444 (guide to ciliate genera).
2003 ***Holosticha* Wrzesniowski, 1877** – Berger, Europ. J. Protistol., 39: 374 (redefinition).

Nomenclature: No derivation of the names *Holosticha* and *Amphisia* are given in the original descriptions. *Holosticha* is a composite of the Greek adjective *hol-* (whole, complete), the thematic vowel ·*o*- (at the end of the first root when the second begins with a consonant; Werner 1972, p. 37), the Greek substantive *stich-* (row, line), and the inflectional ending ·*a*. The name refers to the midventral pairs, which usually form two long, continuous, and narrowly spaced ventral rows in *Holosticha* species (in contrast to the "sporadically" arranged cirri of the oxytrichids, especially the 18-cirri oxytrichids). Feminine gender. Incorrect subsequent spellings: *Holisticha* (Merriman 1937, p. 428); *Holostica* (El-Serehy 1993, p. 132; on page 138 he/she wrote the incorrect year 1887); *Holostricha* (Dragesco & Njiné 1971, p. 125); *Holostycha* sp. (Gracia et al. 1985; Gracia & Igual 1987, p. 3).

Amphisia is a composite of the Greek *amphis-* (on both sides, all the way around) and the suffix -*i*·*a* (provided with). I do not know exactly to which feature this name alludes; perhaps, as in *Holosticha*, it refers to the continuous ventral rows. Feminine gender. *Amphisca kessleri* in Chatton & Seguela (1940, p. 353) is an incorrect subsequent spelling.

Wrześniowski (1877) assigned seven *Oxytricha* species – including two new ones – to *Holosticha*, but without combining their species-group names formally with *Holosticha*. His new species were *Oxytricha pernix* – provisionally classified as supposed

[1] The diagnosis by Borror (1979) is as follows: One row of right marginal cirri; one or more rows of left marginal cirri; transverse cirri present; midventral cirri in typical zigzag series, 2 cirri per original oblique ciliary streak, except that 3 or more frontal cirri differentiate from midventral cirri.

[2] The diagnosis by Hemberger (1982) is as follows: Je 1 linke und rechte Marginalreihe; 2 familientypische Midventral-Reihen; mindestens 3 differenzierte Frontalcirren; Transversalcirren vorhanden (meist zahlreich und auffallend).

synonym of *H. pullaster* in the present book – and *Oxytricha kessleri*. However, more importantly, he definitely did not fix one of these seven species as type. This deficit was not eliminated by Kent (1882), Entz (1884), and Bütschli (1889). Kahl (1932, p. 570) discussed that "Kent transferred the typical species Wrześniowski's, *Holosticha kessleri* to *Amphisia*". I do not exactly know whether or not this can be considered as subsequent designation of a type species because according to Article 67.5 of the ICZN (1999) the term designation must be rigidly construed. In 1972, Borror declared that *Holosticha kessleri* is the type species of *Holosticha* by monotypy. Although the statement "by monotypy" is certainly incorrect, Borror (1972) should be considered as author who subsequently fixed the type species of *Holosticha*. In contrast, Aescht (2001) wrote that Jankowski (1979) subsequently designated *Oxytricha kessleri* as type species, but I suppose that Jankowski (1979) simply followed Kahl and Borror.

Sterki (1878) unequivocally established *Amphisia* for "*Oxytricha gibba* Stein" (the correct basionym is *Trichoda gibba* Müller, 1786), although Sterki's text is formulated rather cryptically (see list of synonyms and corresponding footnote). Jankowski (1979) was of the same conviction. Kahl (1932) and Aescht (2001) incorrectly assumed that *Amphisia multiseta* Sterki, 1878 is the type species.

Kahl (1932) divided *Holosticha* into subgenera (see below). The term "subgen. n." for *Holosticha (Holosticha)* is incorrect because the nominotypical subgenus has the same authority (and type species) as the genus, that is, Wrześniowski (1877) (ICZN 1999, Article 43.1). The correct term would have been "status novus".

Of course each entry in the list of synonyms would need the additional remark "pro parte" because in each paper *Holosticha* is more extensive than in Berger (2003) and the present review.

Characterisation (A = supposed apomorphies): Body anteriorly and posteriorly usually more or less distinctly narrowed. Adoral zone of membranelles bipartite in proximal and distal portion by more or less distinct gap (A). Rear membranelles of proximal portion slightly to distinctly wider than remaining (A). Undulating membranes short and in parallel. 3 enlarged frontal cirri. Buccal cirrus distinctly ahead of paroral (A). 2 frontoterminal cirri. Midventral complex composed of midventral pairs only. 2 pretransverse ventral cirri. Number of transverse cirri equal to or only slightly lower than number of midventral pairs. 1 left and 1 right marginal row. Anterior end of left marginal row composed of narrowly spaced cirri and distinctly curved rightwards (A). Caudal cirri lacking. Nuclear apparatus right of or in midline or scattered (A). Frontal-midventral-transverse cirral anlagen originate mainly from right midventral cirri and thus basically occur right of the parental midventral complex (A). Parental adoral zone remains more or less unchanged for proter or proximalmost portion reorganised (*H. bradburyae*). Left marginal row anlage for proter originates de novo (A). More than 3 dorsal kineties.

Remarks: Wrześniowski (1877) recognised that *Oxytricha* sensu Stein (1859) is heterogeneous. Thus, he suggested splitting it into two groups: (i) species with interrupted ventral cirral rows which could keep the name *Oxytricha* (basically, these are the flexible 18-cirri oxytrichids); and (ii) species with continuous ventral cirral rows for which he proposed the name *Holosticha* (he did not mention a number of ventral rows!). It is noteworthy that Wrześniowski did not include the frontal ciliature into the definition of *Holosticha*. As mentioned in the nomenclature section, he assigned seven species to his new genus: *Oxytricha gibba* Müller, 1786; *O. mystacea* Stein, 1859; *O. crassa* Claparède & Lachmann, 1858; *O. micans* Engelmann, 1862; *O. velox* Quennerstedt, 1869; *O. pernix* Wrześniowski, 1877; and *O. kessleri* Wrześniowski, 1877. Five of these species, namely *O. gibba*, *O. micans* and *O. pernix* (synonyms of *Holosticha pullaster*), and *O. kessleri* and *O. velox* (synonyms of *O. gibba*) are still in *Holosticha. Oxytricha mystacea* (incorrectly spelled *O. mystacina* by Wrześniowski 1877) is now the type species of *Gastrostyla (Spetastyla)* Foissner, Agatha & Berger, 2002; its complete actual name is therefore *Gastrostyla (Spetastyla) mystacea* (Stein, 1859) Sterki, 1878. *Oxytricha crassa* lacks three enlarged frontal cirri and has three ventral rows; it is now classified in *Thigmokeronopsis*.

Just one year after the description of *Holosticha*, Sterki (1878) established *Amphisia*. His characterisation contained – beside the feature two continuous ventral rows – the presence of three frontal cirri. Further, Sterki definitely established *Amphisia* for *Oxytricha gibba*, which is therefore the type species (see nomenclature). In addition, he assigned his own species, *Amphisia multiseta*, and Engelmann's *Oxytricha micans*; both are now junior synonyms of *Holosticha pullaster*.

Kent (1882), the first revisor, as well as Kowalewskiego (1882; see also Kowalewski 1883), Bütschli (1889), and Hamburger & Buddenbrock (1929) accepted both *Holosticha* and *Amphisia*; by contrast, Entz (1884) synonymised *Amphisia* with *Holosticha* because he considered the inwardly shifted marginal rows of Sterki's *Amphisia* as insufficient difference to *Holosticha*, where the marginal cirri distinctly project beyond the body margin.

Kahl (1932) followed Entz's proposal to merge *Amphisia* into *Holosticha*. He mentioned – beside Entz's argument – that the typical species of *Amphisia*, namely *A. multiseta* Sterki, 1878 is (in his opinion) insufficiently diagnosed. Since then, *Amphisia* did not occur in the literature (except in catalogues), and nothing should be changed in this situation to keep the nomenclature stable. In spite of this, the somewhat incorrect actions by Entz and Kahl should be discussed briefly. Entz and Kahl focused too strongly on the closely spaced ventral and marginal rows mentioned by Sterki. They neglected that Sterki had characterised the present group more precisely than Wrześniowski in that he included the three enlarged frontal cirri in his definition. Interestingly, Entz (1884, p. 359, 360, footnote 1) already recognised the inhomogeneity of *Holosticha*. He distinguished two groups, namely those species which have three enlarged frontal cirri, and those which lack such distinct cirri. The best solution would have been if he had used the name *Amphisia* for the first group (with three cirri) and *Holosticha* for the species without three enlarged frontal cirri (that is, for species with a bicorona) as already proposed by Kowalewskiego (1882, p. 411). The latter species are now in *Pseudokeronopsis*. As already stated above, Sterki established *Amphisia* for *Oxytricha gibba* and not for *O. multiseta*, as incorrectly assumed by Kahl (1932) and Aescht (2001). And the redescription of *O. gibba* by Stein (1859) has at least the same quality as the original description of *O. kessleri*, which was subsequently fixed as type species of *Holosticha*.

Kahl (1932) classified 50% or more of the urostyloid hypotrichs in *Holosticha* (most other species belonged to *Urostyla* and *Uroleptopsis*). He divided *Holosticha* into five subgenera, namely, *H. (Paruroleptus)* Kahl, 1932; *H. (Keronopsis)* Penard, 1922; *H. (Holosticha)* Wrześniowski, 1877; *H. (Trichototaxis)* Stokes, 1891; and *H. (Amphisiella)* Gourret & Roeser, 1888. Most species assigned to *H. (Paruroleptus)* are now in *Uroleptus*. The species assigned to *H. (Keronopsis)* are now in different genera, inter alia, in *Pseudokeronopsis*. Most of the 26 species which Kahl classified in *H. (Holosticha)* have been transferred to other taxa (mainly *Anteholosticha* and *Caudiholosticha*) because they lack the numerous apomorphies of *Holosticha* (Berger 2003). *Trichototaxis* is likely a valid, but little known urostyloid group, and *Amphisiella* is not a urostyloid (e.g., Berger 2004a).

Borror (1972) listed 23 valid *Holosticha* species, Borror & Wicklow (1983) only 22. Only few of these assignments agree with my estimation because of the more precise definition of *Holosticha*. According to Borror's (1979) diagnosis, *Holosticha* comprises species with one or more left marginal rows.

Foissner (1982, p. 46) wrote that all species assigned to the genus *Keronopsis* by Kahl (1932) have to be transferred to *Holosticha* because *Keronopsis* Penard, 1922 must be confined to species without midventral cirri (Hemberger & Wilbert 1982). Obviously Foissner has overlooked that Kahl had classified *Keronopsis* only as subgenus of *Holosticha*. One year later, Borror & Wicklow (1983) assigned species with a bicorona and midventral cirri to a new genus (*Pseudokeronopsis*), a distinction proposed already 100 years ago (see above).

The characterisation above excludes most species classified in *Holosticha* so far. The seven species now assigned to *Holosticha* share so many good synapomorphies that it would have been unwise to include other species (Berger 2003; see addenda for the 8th species). In the following paragraphs these supposed apomorphies are discussed.

Gap in adoral zone of membranelles. A distinct gap (break) in the adoral zone is not very common in hypotrichs. Within the urostylids such a bipartite adoral zone is described for *Afrothrix*, *Parabirojimia*, *Uroleptopsis*, and some (all?) *Periholosticha* species. *Uroleptopsis* has a double row of frontal cirri (bicorona) and a special nuclear apparatus indicating that it is not closely related to *Holosticha*. *Parabirojimia* has, inter alia, very few midventral pairs (against several to many in *Holosticha*) and two or more right marginal rows (against one), but lacks frontoterminal cirri (against present). *Afrothrix* has basically more or less the same cirral components as *Holosticha*, that is, three frontal cirri, a buccal cirrus, a midventral complex composed of cirral pairs only, transverse cirri, two frontoterminal cirri, no caudal cirri. However, there are many differences between *Holosticha* and *Afrothrix* in the arrangement of these cirral groups, for example, midventral complex (long against short), transverse cirri (prominent cirri forming long row against inconspicuous cirri forming short row), left marginal row (anterior end curved rightwards against not curved), buccal cirrus (ahead of undulating membranes against right of undulating membranes), undulating membranes (short and parallel against rather long and curved), nuclear apparatus (macronuclear nodules basically right of midline against left of midline), habitat (marine and freshwater against terrestrial). Thus, synonymy of these two taxa can be excluded. By contrast, a close phylogenetic relationship could be possible although the overall similarity is rather low. If they are in fact sister groups, the gap in the adoral zone would of course be a symplesiomorphy for *Holosticha*.

Proximalmost adoral membranelles widened. The widening is rather different within *Holosticha*, for example, rather inconspicuous in *H. diademata* (Fig. 24n) to very pronounced in *H. spindleri* (Fig. 34b, e, f) and *H. bradburyae* (Fig. 35h). I do not know another group where such a curious pattern occurs. In some other taxa – for example, the pseudokeronopsids – the proximal portion of the adoral zone is distinctly spoon-shaped.

Buccal cirrus ahead of undulating membranes. Usually this cirrus is right of the paroral; by contrast, in all *Holosticha* species it is distinctly ahead of the undulating membranes. In *Uroleptopsis citrina* the buccal cirrus is also distinctly ahead of the paroral, and, since this species has a bicorona, it is part of the rear bow of frontal cirri (Fig. 192k). However, *Holosticha* and *Uroleptopsis* are very likely not really closely related so that a convergent evolution of this feature has to be assumed.

Anterior end of left marginal row composed of narrowly spaced cirri and curved rightwards. This is likely the best apomorphy of *Holosticha* because it is distinct in all species (e.g., Fig. 24p) and does not occur in any other group of hypotrichs. Possibly this curve is the reason why the anlage for the left marginal row of the proter originates basically de novo (for details on *H. bradburyae* see there) left of the proximal portion of the adoral zone. This is exactly the region where, in species with a normal, that is, straight left marginal row, the anlage occurs within the parental row.

Anlage for left marginal row of proter originates de novo. See previous feature.

Frontal-midventral-transverse cirral anlagen originate basically right of the midventral complex. At least in *H. diademata* (Fig. 25b, c), *H. pullaster* (Fig. 28b, c), *H. heterofoissneri* (Fig. 33c, d), and *H. bradburyae* (Fig. 36d, e) the cirral anlagen for the midventral complex originate mainly right of the parental midventral complex. Further, mainly the right cirri of the pairs are included in primordia formation. In all other taxa these cirral anlagen occur left of the midventral complex and usually the left midventral cirri contribute to primordia formation.

Nuclear apparatus in body midline or right of it. Usually the macronuclear nodules are arranged left of midline. Surprisingly, in all *Holosticha* species, which have two or several nodules, they are dislocated distinctly rightwards or at least in midline (Fig. 22a, p, 24e, 29, 31a, 32f, 34a). I assume that the exact location was often not quite correctly recorded so that deviating data must not be over-interpreted.

There are two further features which could be apomorphies of *Holosticha*, namely, (i) the contractile vacuole is usually in mid-body or behind it, whereas in most other urostyloids, oxytrichids, etc. this organelle is usually slightly behind the proximal end of the adoral zone, that is, ahead of mid-body; and (ii) the undulating membranes are more or less straight and roughly in parallel. All eight species assigned to *Holosticha* are marine, except for *H. pullaster*, which occurs both in marine and limnetic habitats.

The proximal portion of the adoral zone of membranelles is obviously not reorganised in *H. pullaster* and *H. heterofoissner* during cell division. In *H. diademata*, the proximal portion shows distinct signs of reorganisation, whereas in *H. bradburyae* this part of the adoral zone is completely replaced by a new array of membranelles.

The characterisation above is basically according to Berger (2003). I included the feature "2 pretransverse ventral cirri" because these cirri are demonstrable in the modern original descriptions and redescriptions. I am convinced that these cirri are also present in the type species, which, however, awaits detailed redescription. The two pretransverse ventral cirri are a plesiomorphic feature (see ground pattern of the Urostyloidea). In contrast, the lack of caudal cirri is a much younger character. However, I do not know at which level it is an autapomorphy.

Some species (*H. heterofoissneri*, *H. spindleri*, *H. bradburyae*) have conspicuous extrusomes with a fine thread when they are exploded. Possibly this is an apomorphy unifying these three species. At least *H. bradburyae* has blood-cell shaped structures, which are reminiscent of those of many pseudokeronopsines.

Species included in *Holosticha* (alphabetically arranged basionyms are given): (1) *Amphisia diademata* Rees, 1884; (2) *Holosticha bradburyae* Gong, Song, Hu, Ma & Zhu, 2001; (3) *Holosticha foissneri* Petz, Song & Wilbert, 1995; (4) *Holosticha hamulata* Lei, Xu & Choi, 2005 (see addenda for brief description); (5) *Holosticha heterofoissneri* Hu & Song, 2001; (6) *Holosticha spindleri* Petz, Song & Wilbert, 1995; (7) *Trichoda gibba* Müller, 1786; (8) *Trichoda pullaster* Müller, 1773.

Species misplaced in *Holosticha*: So far *Holosticha* has been a melting pot for all urostyloids with three frontal cirri, one left and one right marginal row, a midventral complex composed of cirral pairs only, transverse cirri, and with or without caudal cirri. More than 100 species were originally assigned to *Holosticha* (Berger 2001). Before Berger (2003) redefined *Holosticha*, it contained about 49 species. However, a detailed analysis shows that only the eight species mentioned above can be assigned to *Holosticha* with *Trichoda gibba*, or its synonym *Oxytricha kessleri*, as type species.

The following species – originally classified in *Holosticha* – are now assigned to other genera within the urostyloids, or they do not belong to the urostyloids at all. Synonyms of true *Holosticha* species are not included in the following list. Names are listed alphabetically according to species-group names, that is, subgenera are not considered. If you do not find a certain name in the list below, please refer to the index.

Most *Holosticha* species which lack caudal cirri have been assigned to *Anteholosticha*, whereas the majority of the species with caudal cirri has been transferred to *Caudiholosticha* (Berger 2003). However, very likely this is only a preliminary assignment because new data will probably provide more proper groupings for some of these species.

Species indeterminata, nomina nuda, and insufficient redescriptions can be found after the description of the last *Holosticha* species.

Holosticha adami Foissner, 1982 (now *Anteholosticha adami*)

Holosticha (Holosticha) algivora Kahl, 1932 (now *Caudiholosticha algivora*)

Holosticha (Keronopsis) alpestris Kahl, 1932 (now *Anteholosticha alpestris*)

Holosticha alpestris Foissner, 1981 (nomen nudum)

Holosticha (Holosticha) alveolata Kahl, 1932 (now *Pseudoamphisiella alveolata*)

Holosticha annulata Kahl, 1928, Arch. Hydrobiol., 19: 212, Abb. 44f. Remarks: This is, according to own observations (Berger 2004a), a very well defined and rather easy to identify *Amphisiella* species, *Amphisiella annulata* (Kahl, 1932) Borror, 1972.

Holosticha aquarumdulcium Bürger, 1905 (species indeterminata)

Holosticha (Holosticha) arenicola Kahl, 1932 (now *Anteholosticha arenicola*)

Holosticha australis Blatterer & Foissner, 1988 (now *Anteholosticha australis*)

Holosticha azerbaijanica Alekperov & Asadullayeva, 1999 (now *Anteholosticha azerbaijanica*).

Holosticha begoniensis Fernandez-Leborans, 1990 (species indeterminata)

Holosticha bergeri Foissner, 1987b (now *Anteholosticha bergeri*)

Holosticha binucleata Foissner, Peer & Adam, 1985 (nomen nudum)

Holosticha brachysticha Foissner, Agatha & Berger, 2002 (now *Anteholosticha brachysticha*)

Holosticha (Holosticha) brevis Kahl, 1932 (now *Anteholosticha brevis*)

Holosticha camerounensis Dragesco, 1970 (now *Anteholosticha camerounensis*)

Holosticha caudata Stokes, 1886 (now *Uroleptus caudatus*)

Holosticha contractilis Dragesco, 1970 (junior synonym of *Uroleptus musculus*)

Holosticha corlissi Fernandez-Galiano & Calvo, 1992 (supposed synonym of *Anteholosticha monilata*)

Holosticha coronata Gourret & Roeser, 1888, Archs Biol., 8: 182, Planche XV, Fig. 1 (Fig. 100d). Remarks: *Holosticha coronata* Gourret & Roeser is the primary senior homonym of *H. coronota* Vuxanovici, 1963. Hamburger & Buddenbrock (1929, p. 93) transferred it, although with doubt, to *Gastrostyla* (*Gastrostyla coronata* (Gourret & Roeser, 1888) Hamburger & Buddenbrock, 1929; combination overlooked by Berger 2001), whereas it was classified as *Keronopsis coronoata*[1] by Kahl (1932, p. 576, Fig. 101_{21}) in the subgenus *Holosticha (Keronopsis)*. Thus, the correct name in Kahl's revision is *Holosticha (Keronopsis) coronata* Gourret & Roeser, 1888. Kahl (1932) considered *H. coronata* as valid species although he had recognised some misobservations, for example, the contractile vacuole in the posterior body portion; by contrast, Borror (1972, p. 14) synonymised it with *Gastrostyla pulchra* (Pereyaslawzewa, 1886) Kahl, 1932. Unfortunately, I overlooked this synonymy (with which I agree) in the review on oxytrichids (Berger 1999, p. 818).

Holosticha decolor Wallengren, 1900 (now *Pseudokeronopsis decolor*)

Holosticha (Holosticha) discocephalus Kahl, 1932 (now *Biholosticha discocephalus*)

Holosticha distyla Buitkamp, 1977 (now *Anteholosticha distyla*)

Holosticha dragescoi Borror & Wicklow, 1983 (now *Biholosticha arenicola*)

Holosticha (Parurosoma) dubium Gelei, 1954 (now *Parurosoma dubium* (Gelei, 1954) Berger, 1999; for review, see Berger 1999, p. 492)

Holosticha estuarii Borror & Wicklow, 1983 (now *Anteholosticha estuarii*)

Holosticha (Holosticha) extensa Kahl, 1932 (now *Anteholosticha extensa*)

Holosticha (Holosticha) fasciola Kahl, 1932 (now *Anteholosticha fasciola*)

[1] Actually, Kahl's heading was "*Keronopsis (Holosticha) coronata* (Gourret u. R., 1888)" because he omitted the genus name *Holosticha* and put the original generic classification (also *Holosticha*!) in parentheses between the subgenus-group name (*Keronopsis*) and the species-group name (*coronata*).

Holosticha (Keronopsis) flavicans Kahl, 1932 (now *Pseudokeronopsis flavicans*)

Holosticha flavorubra Entz, 1884 (synonym of *Pseudokeronopsis rubra*)

Holosticha fontinalis Lepsi, 1926 (species indeterminata)

Holosticha (Trichototaxis) fossicola Kahl, 1932 (supposed synonym of *Paraurostyla granulifera* Berger & Foissner, 1989a; for review, see Berger 1999, p. 875)

Holosticha geleii Wilbert, 1986 (now *Oxytricha geleii* (Wilbert, 1986) Berger, 1999; for review, see Berger 1999, p. 232)

Holosticha (Keronopsis) globulifera Kahl, 1932 (now synonym of *Pseudokeronopsis decolor*)

Holosticha (Keronopsis) gracilis Kahl, 1932 (now *Anteholosticha gracilis*)

Holosticha gracilis Vuxanovici, 1963 (now *Anteholosticha vuxgracilis*)

Holosticha (Holosticha) grisea Kahl, 1932 (now *Anteholosticha grisea*)

Holosticha hymenophora Stokes, 1886 (now *Apoamphisiella hymenophora* (Stokes, 1886) Berger, 1999; for review, see Berger 1999, p. 786)

Holosticha interrupta Dragesco, 1966 (now *Caudiholosticha interrupta*)

Holosticha islandica Berger & Foissner, 1989 (now *Caudiholosticha islandica*)

Holosticha lacazei Maupas, 1883 (now *Pseudoamphisiella lacazei*)

Holosticha (Paruroleptus) lacteus Kahl, 1932 (now *Uroleptus lacteus*)

Holosticha (Paruroleptus) lepisma Wenzel, 1953 (now *Uroleptus lepisma*)

Holosticha longiseta Lepsi, 1951 (species indeterminata)

Holosticha macronucleata in Gelei et al. (1954, p. 367). Remarks: According to Berger (2001, p. 37) this is a nomen nudum. Now I am convinced that it is simply an incorrect subsequent spelling of *Holosticha mononucleata* Gelei, 1954. This species is very likely a synonym of *Parurosoma dubium* (Gelei, 1954) Berger, 1999 (for details, see Berger 1999, p. 492)

Holosticha (Paruroleptus) magnificus Kahl, 1932 (now *Uroleptus magnificus*)

Holosticha (Holosticha) manca Kahl, 1932 (now *Anteholosticha manca*)

Holosticha manca var. *mononucleata* Gellért, 1956 (synonym of next species)

Holosticha manca var. *plurinucleata* Gellért, 1956 (now *Anteholosticha plurinucleata*)

Holosticha mancoidea Hemberger, 1985 (now *Anteholosticha mancoidea*)

Holosticha (Holosticha) milnei Kahl, 1932 (now junior synonym of *Anteholosticha oculata*)

Holosticha (Amphisiella) milnei Kahl, 1932 (now *Amphisiella milnei* (Kahl, 1932) ?Horvath, 1950)

Holosticha monilata Kahl, 1928 (now *Anteholosticha monilata*)

Holosticha mononucleata Gelei, 1954 (synonym of *Parurosoma dubium* (Gelei, 1954) Berger, 1999; for review, see Berger 1999, p. 492).

Holosticha multicaudicirrus Song & Wilbert, 1989 (now *Caudiholosticha multicaudicirrus*)

Holosticha multinucleata Maupas, 1883 (now *Pseudokeronopsis multinucleata*)

Holosticha (Keronopsis) multiplex Ozaki & Yagiu, 1943 (now junior synonym of *Uroleptopsis roscoviana*)

Holosticha multistilata Kahl, 1928 (now *Anteholosticha multistilata*)

Holosticha muscicola Gellért, 1956 (now *Anteholosticha muscicola*)

Holosticha (Keronopsis) muscorum Kahl, 1932 (now *Anteholosticha intermedia*)

Holosticha (Paruroleptus) musculus Kahl, 1932 (now *Uroleptus musculus*)

Holosticha (Paruroleptus) musculus var. *simplex* Kahl, 1932 (now *Uroleptus musculus*)

Holosticha nagasakiensis Hu & Suzuki, 2004 (now *Anteholosticha gracilis*)

Holosticha (Holosticha) navicularum Kahl, 1932 (now *Caudiholosticha navicularum*)

Holosticha obliqua Kahl, 1928 (species indeterminata)

Holosticha (Keronopsis) ovalis Kahl, 1932 (now *Pseudokeronopsis ovalis*)

Holosticha (Keronopsis) ovalis f. *arenivora* Kahl, 1932 (now *Pseudokeronopsis ovalis*)

Holosticha oxytrichoidea Németh in Gelei, 1950 (species indeterminata)

Holosticha polystylata Borror & Wicklow, 1983 (now *Diaxonella pseudorubra*)

Holosticha (Keronopsis) pulchra Kahl, 1932 (now *Anteholosticha pulchra*)

Holosticha randani Grolière, 1975 (now *Anteholosticha randani*)

Holosticha (Keronopsis) rubra forma *heptasticha* Kahl, 1932 (now *Pseudokeronopsis rubra*).

Holosticha (Keronopsis) rubra forma *pentasticha* Kahl, 1932 (now *Pseudokeronopsis rubra*)

Holosticha salina Fernandez-Leborans & Novillo, 1993 (species indeterminata)

Holosticha (Holosticha) setifera Kahl, 1932 (now *Caudiholosticha setifera*)

Holosticha setigera in Conn (1905) (species indeterminata)

Holosticha sigmoidea Foissner, 1982 (now *Anteholosticha sigmoidea*)

Holosticha similis Stokes, 1886 (now *Pseudokeronopsis similis*)

Holosticha sp. in Wilbert & Song 2005, J. nat. Hist., 39: 958, Fig. 10A–C (Fig. 37.1a–c). Remarks: Wilbert & Song (2005) found only a small number of protargol-impregnated specimens, that is, live data are lacking. They assigned it to *Holosticha* although it lacks some important features, namely the rightwards curved anterior end of the left marginal row and the break in the adoral zone of membranelles. Moreover, the buccal cirrus is right of the anterior portion of the undulating membranes and not ahead of the membranes as in the other *Holosticha* species. *Holosticha* sp. sensu Wilbert & Song (2005) is very likely an *Anteholosticha* species and possible identical with *A. scutellum* (Fig. 94a–k). Body size of live specimens of *Holosticha* sp. lacking; after protargol impregnation about 65 × 40 μm. Body shape of prepared specimens broadly oval. About 30 macronuclear nodules scattered throughout cell; individual nodules ovoid, about 4 μm long. Four micronuclei, globular, about 3 μm across. On dorsal side always several argentophilic extrusome-like structures (cortical granules?) along dorsal kineties; individual structures about 3 μm long, vase-shaped with rounded posterior end. Adoral zone occupies about 40% of body length, composed of about 17 membranelles; distal end extends distinctly onto right cell margin; bases of largest membranelles about 8–10 μm wide. Paroral and endoral short, roughly straight, and almost in parallel. Pharyngeal fibres more than 20 μm long. Cirral pattern as shown in Fig. 37.1a. Two frontal cirri and one cirrus left behind right frontal cirrus (Wilbert & Song interpreted

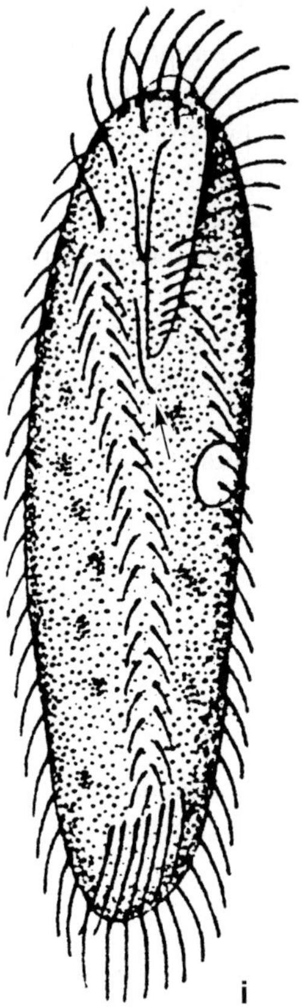

Fig. 20i *Apoamphisiella vernalis* (from Stokes 1887). Ventral view, 180 µm. Arrow marks postoral ventral cirrus, indicating that this species is not a urostyloid. Page 98.

them as three frontal cirri); buccal cirrus right of anterior portion of undulating membranes; two frontoterminal cirri more or less in common position; midventral complex composed of 3–4 pairs and a short row composed of 3–5 cirri, complex terminates slightly behind mid-body (more detailed data, including ontogenetic one, are needed for a correct interpretation of the cirral pattern!). Often two pretransverse ventral cirri. 6–7 distinctly enlarged transverse cirri arranged in J-shaped pattern. Right marginal row composed of about 17 cirri; left row consists of about 14 cirri, anterior portion straight, that is, not curved rightwards as in *Holosticha*; marginal rows widely separated posteriorly. Dorsal cilia about 5 µm long, arranged in four kineties with 6–15 cilia each; caudal cirri lacking.

Holosticha (Keronopsis) spectabilis Kahl, 1932 (now *Neokeronopsis spectabilis*)

Holosticha stueberi Foissner, 1987e (now *Caudiholosticha stueberi*)

Holosticha sylvatica Foissner, 1982 (now *Caudiholosticha sylvatica*)

Holosticha tenuiformis Vuxanovici, 1963 (species indeterminata)

Holosticha tetracirrata Buitkamp & Wilbert, 1974 (now *Caudiholosticha tetracirrata*)

Holosticha vernalis Stokes, 1887, Ann. Mag. nat. Hist., 20: 108, Plate III, fig. 5 (Fig. 20i). Remarks: Kahl (1932, p. 585) and Borror & Wicklow (1983, p. 122) accepted Stokes' classification in *Holosticha*. Borror & Wicklow synonymised it with *Keronopsis thononensis* Dragesco, 1966, which indeed has a rather similar cirral pattern. However, there is an important difference between the two illustrations, namely, *Holosticha vernalis* has a distinct postoral ventral cirrus (Fig. 20i, arrow; cirrus unfortunately not mentioned in original description), which is lacking in *K. thononensis* (Fig. 63a). Since misobservations are unlikely (Stokes was a good observer and this cirrus is rather easily recognisable in life; Dragesco had protargol preparations), Stokes' species is very probably not a urostyloid, but belongs to the oxytrichid genus *Apoamphisiella* Foissner, 1997. Species of this group have two ventral rows (not a midventral complex!) and a postoral ventral cirrus (for review, see Berger 1999, p. 781). I thus transfer *Holosticha vernalis* Stokes, 1887 to *Apoamphisiella*: *Apoamphisiella vernalis* (Stokes, 1887) comb. nov. Stokes did not observe the nuclear apparatus, strongly indicating that several to many macronuclear nodules are present. This feature separates it from *Apoamphisiella tihanyiensis* (Gellért & Tamás, 1958) Foissner, 1997 and *Apoamphisiella hymenophora*

(Stokes, 1886) Berger, 1999, both of which have only two nodules (note that Stokes 1886 described two macronuclear nodules for *A. hymenophora*).

Holosticha vesiculata Vuxanovici, 1963 (species indeterminata)
Holosticha violacea Kahl, 1928 (now *Anteholosticha violacea*)
Holosticha (Holosticha) viridis Kahl, 1932 (now *Caudiholosticha viridis*)
Holosticha warreni Song & Wilbert, 1997 (now *Anteholosticha warreni*)
Holosticha wrzesniowskii var. *punctata* Rees, 1884 (species indeterminata)
Holosticha xanthichroma Wirnsberger & Foissner, 1987 (now *Anteholosticha x.*)

Key to *Holosticha* species

If you know that your specimen/population is a *Holosticha* species, identification is relatively simple. Main features are the nuclear apparatus, contractile vacuole, body size, and number of transverse cirri. If your material is from freshwater, it is certainly *H. pullaster*. See addenda for the eighth species (*H. hamulata*), which has a rather long, narrowed posterior body portion and the contractile vacuole ahead of mid-body.

1 Two macronuclear nodules **2**
\- More than 2 macronuclear nodules **4**
2 Contractile vacuole distinctly behind mid-body; freshwater and saltwater (Fig. 28f–i) *Holosticha pullaster* (p. 128)
\- Contractile vacuole in mid-body; saltwater **3**
3 Body usually 120–160 μm long; 12–20 transverse cirri (Fig. 22i, p) .... *Holosticha gibba* (p. 99)
\- Body usually 70–140 μm long; 6–11 transverse cirri (Fig. 24b) .... *Holosticha diademata* (p. 115)
4 (1) Usually 4 macronuclear nodules (Fig. 34a) ...... *Holosticha spindleri* (p. 163)
\- Usually more than 4 macronuclear nodules **5**
5 5–11, usually 8 macronuclear nodules (Fig. 31a) .... *Holosticha foissneri* (p. 149)
\- Usually more than 13 macronuclear nodules **6**
6 14–24, usually 16–17 macronuclear nodules; body length 110–150 μm (Fig. 32g, h) .... *Holosticha heterofoissneri* (p. 152)
\- 28–33, usually about 30 macronuclear nodules; body length 150–320 μm (Fig. 35a, j) .... *Holosticha bradburyae* (p. 167)

Holosticha gibba (Müller, 1786) Wrześniowski, 1877
(Fig. 21a–i, 22a–r, 23a–f, Table 12)

1786 ***Trichoda gibba***[1] – Müller, Animalcula Infusoria, p. 179, Tab. XXV, Fig. 16–20 (Fig. 21a–c; original description; no type material available).

[1] The diagnosis by Müller (1786) is as follows: *Trichoda* oblonga, dorso gibbera, ventre excavata, antice ciliata; extremitatibus obtusis.

1786 ***Trichoda foeta***[1] – Müller, Animalcula Infusoria, p. 180, Tab. XXV, Fig. 11–15 (Fig. 21d; original description of synonym; no type material available).

1824 ***Oxitricha gibbosa*** – Bory de Saint-Vincent in Lamouroux, Bory de Saint-Vincent & Deslongchamps, Encyclopédie methodique, p. 596 (revision; combination with *Oxitricha*, the original spelling of *Oxytricha*).

1838 ***Oxytricha gibba*** – Ehrenberg, Infusionsthierchen, p. 365, pro parte (revision).

1841 ***Oxytricha gibba*** – Dujardin, Histoire Naturelle zoophytes, p. 418, Planche XI, fig. 12 (Fig. 21g; redescription).

1859 ***Oxytricha gibba*. Stein** – Stein, Organismus der Infusionsthiere, p. 184, Tafel XI, Fig. 9, 10 (Fig. 21h, i; redescription).

1869 ***Oxytricha velox*. n. sp.** – Quennerstedt, Acta Univ. lund., 6: 20, Fig. 20, 21 (Fig. 23d; original description of synonym; no type material available and no formal diagnosis provided).

1877 ***Oxytricha kessleri*, nov. sp.**[2] – Wrześniowski, Z. wiss. Zool., 29: 275, Tafel XIX, Fig. 12–15 (Fig. 22a–d; original description of synonym; no type material available).

1877 ***Holosticha gibba*** – Wrześniowski, Z. wiss. Zool., 29: 278 (combination with *Holosticha*; see nomenclature).

1877 ***Holosticha velox*** – Wrześniowski, Z. wiss. Zool., 29: 278 (combination with *Holosticha*; see nomenclature).

1877 ***Holosticha kessleri*** – Wrześniowski, Z. wiss. Zool., 29: 278 (combination with *Holosticha*; see nomenclature).

1877 ***Oxytricha wrzesniowskii* nova species** – Mereschkowsky, Trudy imp. S-peterb. Obshch. Estest., 8: 231, Plate II, Fig. 6 (Fig. 23a; original description of synonym; see remarks; no type material available and no formal diagnosis provided).

1878 ***Amphisia gibba*** – Sterki, Z. wiss. Zool., 31: 57 (designation as type species of *Amphisia* and thus combination with *Amphisia*).

1879 ***Oxytricha wrzesniowskii*, n. sp.** – Mereschkowsky, Arch. mikrosk. Anat. EntwMech., 16: 162, Tafel X, Fig. 35 (Fig. 23a; German translation of Russian original description from 1877; no formal diagnosis provided).

1882 ***Amphisia gibba*, Müll. sp.** – Kent, Manual infusoria II, p. 767 (revision).

1882 ***Amphisia kessleri*, Wrz. sp.** – Kent, Manual infusoria II, p. 768 (revision; combination with *Amphisia*).

1882 ***Amphisia velox*, Quenn. sp.** – Kent, Manual infusoria II, p. 768 (revision; combination with *Amphisia*).

1882 ***Holosticha wrzesniowskii*, Mereschk. sp.** – Kent, Manual infusoria II, p. 771 (revision; combination with *Holosticha*).

1902 ***Amphisia kessleri* Wrzes. '77** – Calkins, Bull. U. S. Fish Commn, 21: 454, Fig. 51 (Fig. 22f; redescription).

1929 ***Amphisia gibba* O. F. M.** – Hamburger & Buddenbrock, Nord. Plankt., 7: 89, Fig. 108 (redrawing of Fig. 21h; guide to marine ciliates).

1929 ***Amphisia kessleri* Wrzesn.** Hamburger & Buddenbrock, Nord. Plankt., 7: 90, Fig. 109 (redrawing of Fig. 22a, d; guide to marine ciliates).

1929 ***Amphisia wrzesniowskii* Mereschk.** – Hamburger & Buddenbrock, Nord. Plankt., 7: 91, Fig. 112 (redrawing of Fig. 23a; combination with *Amphisia*; guide to marine ciliates).

1932 ***Amphisia kessleri* (Wrzesniowsky 1877)** – Wang & Nie, Contr. biol. Lab. Sci. Soc. China, Zoological Series, 8: 356, Fig. 65 (Fig. 22h; redescription, see remarks).

1932 ***Holosticha (Oxytricha) kessleri* (Wrzesniowski, 1877)** – Kahl, Tierwelt Dtl., 25: 581, Fig. $106_{1,\ 13}$ (Fig. 22g, i; revision; authoritative redescription).

1932 ***Holosticha gibba* (Müller, 1786) Stein, 1859** – Kahl, Tierwelt Dtl., 25: 583, Fig. 106_{17} (Fig. 21e; revision; incorrect combining author).

---

[1] The diagnosis by Müller (1786) is as follows: *Trichoda* oblonga, dorso protuberante, antice ciliata, extemitatibus obtusis.

[2] The diagnosis by Wrześniowski (1877) is as follows: Körper in hohem Grade retractil und flexil, gestreckt, flachgedrückt, vorn und hinten verschmälert; die Oberlippe doppelt; vier hakenförmige Stirnwimpern; zwei continuirliche Bauchwimperreihen; zwölf bis fünfzehn Afterwimpern.

1932 ***Holosticha (Oxytricha) wrzesniowskii* (Mereschkowsky, 1877)** – Kahl, Tierwelt Dtl., 25: 583, Fig. 106_{15} (Fig. 23b; revision).
1932 ***Trichotaxis (Oxytricha) velox* Quennerstedt, 1869** – Kahl, Tierwelt Dtl., 25: 589, Fig. 106_{19} (Fig. 23c; revision).
1933 ***Holosticha kessleri* (Wrzesniowski 1877)** – Kahl, Tierwelt N.- u. Ostsee, 23: 111, Fig. 17.7 (Fig. 22p; guide to marine ciliates).
1933 ***Holosticha gibba* Stein 1859** – Kahl, Tierwelt N.- u. Ostsee, 23: 111, Fig. 17.15 (Fig. 21e; guide to marine ciliates; incorrect author).
1933 ***Trichotaxis velox* (Quennerstedt 1869)** – Kahl, Tierwelt N.- u. Ostsee, 23: 111, Fig. 17.16 (Fig. 23c; guide to marine ciliates).
1933 ***Holosticha wrzesniowskii* (Mereschkowsky 1877)** – Kahl, Tierwelt N.- u. Ostsee, 23: 111, Fig. 17.13 (Fig. 23b; guide to marine ciliates).
1963 ***Holosticha kessleri* Wrzesniowski, 1877** – Biernacka, Polskie Archwm Hydrobiol., 11: 49, Abb. 94 (Fig. 22e; possibly a redrawing from Kahl 1932; see also next entry).
1967 ***Holosticha kessleri* Wrzesniowski, 1877** – Biernacka, Wiss. Z. Ernst. Mortz Arndt-Univ. Greifswald, 16: 242, Abb. 16 (Fig. 22e; possibly a redrawing from Kahl 1932).
1972 ***Holosticha kessleri* Wrzesniowski, 1877** – Borror, J. Protozool., 19: 10 (revision of hypotrichs).
1972 ***Holosticha gibba* (Stein, 1859) Kahl, 1932** – Borror, J. Protozool., 19: 11 (revision; incorrect author and combining author).
1972 ***Trichotaxis velox* (Quennerstedt, 1869) Kahl, 1932** – Borror, J. Protozool., 19: 11 (revision; combination with *Trichotaxis*, see nomenclature).
1972 ***Paraurostyla gibba* (Müller, 1786) n. comb.** – Borror, J. Protozool., 19: 10 (pro parte, see remarks; combination with *Paraurostyla*).
1972 ***Holosticha* sp.** – Fenchel & Lee, Arch. Protistenk., 114: 233, Fig. 2 (Fig. 23f; illustrated record from Antarctica).
1974 ***Holosticha kessleri* Wrzesniowski** – Stiller, Fauna hung., 115: 76, Fig. 45B (Fig. 22i; revision).
1979 ***Holosticha kessleri*** – Borror, J. Protozool., 26: 547, Fig. 4 (Fig. 22j, k; illustrated record).
1982 ***Holosticha kessleri* Wrzesniowski, 1887** – Ricci, Santangelo & Luporini, Monitore zool. ital., Suppl. 17: 143, Fig. 35A, B (Fig. 22l, m; redescription from life; incorrect date).
1983 ***Holosticha kessleri* Wrzesnionwski, 1877** – Borror & Wicklow, Acta Protozool., 22: 121, Fig. 17 (Fig. 22o; revision of urostylids).
1983 ***Holosticha gibba* (Stein, 1859) Kahl, 1932** – Borror & Wicklow, Acta Protozool., 22: 121 (revision of urostylids; incorrect author and combining author).
1983 ***Holosticha velox* (Quennerstedt, 1869) nov. comb.** – Borror & Wicklow, Acta Protozool., 22: 122 (revision of urostylids; combination with *Holosticha*).
1985 ***Holosticha kessleri* Wrzesnioski, 1877** – Aldro Lubel, An. Inst. Biol. Univ. Méx., Ser. Zoologia, 55: 26, Lámina 12, Fig. 6 (Fig. 22q; brief redescription; incorrect spelling of author's name).
1990 ***Holosticha kessleri* (Wrzesniowski, 1877)** – Aladro Lubel, Martínez Murillo & Mayén Estrada, Manual de Ciliados, p. 131, Figure on p. 131 (Fig. 22r; review).
1991 ***Holosticha kessleri* (Wrześniowski, 1877) Wrześniowski, 1877** – Foissner, Blatterer, Berger & Kohmann, Informationsberichte des Bayer. Landesamtes für Wasserwirtschaft, 1/91: 228, Abb. 1–5 (Fig. 22a, h, i, j, 23a; guide to ciliates of the saprobic system).
1992 ***Holosticha kessleri* (Wrzesniowski, 1877) Kahl, 1930–5** – Carey, Marine interstitial ciliates, p. 182, Fig. 717 (Fig. 22n; incorrect combining author; guide).
1992 ***Trichotaxis velox* (Quennerstedt, 1869) Kahl, 1935** – Carey, Marine interstitial ciliates, p. 187, Fig. 743 (Fig. 23e; guide).
2001 ***Holosticha gibba* (Müller, 1786) Wrzesniowski, 1877** – Berger, Catalogue of ciliate names 1. Hypotrichs, p. 92 (nomenclator containing all basionyms, combinations, and higher taxa of hypotrichs).
2003 ***Holosticha gibba* (Müller, 1786) Wrzesniowski, 1877** – Berger, Europ. J. Protistol., 39: 376, Fig. 8 (Fig. 22i; brief review).

Nomenclature: The species-group name *gibb·us -a -um* (Latin adjective; vaulted, humpbacked, convex) refers to the vaulted dorsal side. The species-group name *velox* (Latin adjective; fast, swiftly, quickly) refers to the fast movement. The species *Oxytricha kessleri* was dedicated to Professor Kessler, Petersburg, Russia (Wrześniowski 1877, p. 276). The species *Oxytricha wrzesniowskii* was dedicated to August Wrześniowski, Warsaw (Mereschkowsky 1879, p. 63).

Wrześniowski (1877, p. 278) established *Holosticha* and included seven species without combining them formally with *Holosticha*. In spite of this, he should be considered as combining author as indicated in the headline above and already proposed by Foissner et al. (1991).

Incorrect subsequent spellings: *Amphisia kessleria* (Wrzesniowsky) (Wang & Nie 1934, p. 4210); *Amphisia (wrzesmovskii?*) (Bervoets 1940, p. 124); *Amphysia gibba* O. F. Müll. (Dagajeva 1930, p. 35); *Amphysia kessleri* Wrzsn. (Dagajeva 1930, p. 35); *Holosticha kesleri* (Wrz.) (Detcheva 1986, p. 63); *Holostricha kessleri* Wrzesnioski (Guillén et al. 2003, p. 180)

Kent (1882), who accepted both *Holosticha* and *Amphisia* (see genus section), assigned *Oxytricha kessleri* to *Amphisia*. Obviously simultaneously, Kowalewskiego (1882, p. 411; see Kowalewski 1883, p. 243 for German text) had the same idea and also assigned the present species to *Amphisia* Sterki, however, without combining the species-group name with *Amphisia* formally.

Kahl's (1932) confusing spellings mean that the present species and its synonyms were originally described in *Oxytricha*. Since he classified three species in the subgenus *Holosticha*, the correct names in Kahl (1932, 1933) are *Holosticha (Holosticha) kessleri* (Wrześniowski, 1877) Wrześniowski, 1877; *Holosticha (Holosticha) gibba* (Müller, 1786) Wrześniowski, 1877; and *Holosticha (Holosticha) wrzesniowskii* (Mereschkowsky, 1877) Kent, 1882. The correct name of *Oxytricha velox* in Kahl (1932, 1933) is *Holosticha (Trichototaxis) velox* (Quennerstedt, 1869) Wrzesniowski, 1877 because he classified *Trichotaxis*, an incorrect subsequent spelling of *Trichototaxis*, also as subgenus of *Holosticha*.

Berger (2001) mentioned Rees (1884) as author for the combination *Holosticha wrzesniowskii*. However, this is incorrect because this act was already done by Kent (1882).

Borror (1972) and Borror & Wicklow (1983) did not realise that Wrześniowski's species was established in *Oxytricha* and not in *Holosticha*. Further, they assumed – likely par lapsus – that *H. kessleri* is the type species of *Holosticha* by monotypy (for details, see the nomenclature chapter of the genus section). Borror (1972) incorrectly assumed that Kahl (1932) transferred *Oxytricha velox* to *Trichotaxis*. However, Kahl (1932) classified *Trichotaxis* only as subgenus of *Holosticha* and therefore he cannot be the author for this combination. Borror & Wicklow (1983) transferred it to *Holosticha*, an act already done by Wrześniowski (1877).

Oxytricha kessleri is type species of *Holosticha* by subsequent designation by Borror (1972), and *Trichoda gibba* is type species of *Amphisia* by original designation.

Remarks: The systematics of the present species is rather complicated and therefore has to be discussed in detail. The original description of *Trichoda gibba* by Müller

(1786) is insufficient so that, as in many other cases by this author, an identification will always be arbitrary. However, the general appearance of *T. gibba* sensu Müller is not in contradiction with the authoritative redescription provided by Kahl (1932; see below). Müller (1786) found *T. gibba* "In aqua littorali passim". I do not know if this refers to a marine or limnetic habitat, but Stein (1859, p. 184) writes firmly that Müller frequently found *T. gibba* in seawater.

The synonymy of *Trichoda gibba* and *T. foeta* was proposed by Ehrenberg (1838, p. 366), although with doubt. I accept this proposal simply because it is not really possibly to check this decision. Ehrenberg reviewed the present species and provided own data from a freshwater population which is now designated as *Australothrix gibba* (see there for details).

Dujardin (1841) redescribed Müller's species (Fig. 21g). However, the data and the illustration did not provide usable news.

Stein (1859) described a *Trichoda gibba* population from a sea port. His population is basically characterised by the marine habitat, enlarged frontal cirri, about five enlarged transverse cirri, and two macronuclear nodules (Fig. 21h, i). He found this species several times indicating that it was common. However, a thorough review of the literature yielded that *O. gibba* was never reliably redescribed after Stein (1859). This hints that Stein's description is not quite correct, so that most later workers where uncertain about the identity of *O. gibba* and *O. kessleri* Wrześniowski, a species described about 20 years after Stein's important contribution to the biology of hypotrichs. However, there is no doubt that Stein's description could also be correct, which would prevent a synonymy with *H. kessleri*.

Wrześniowski (1877) originally described his new species in *Oxytricha* and simultaneously assigned it to his newly established genus *Holosticha* (see nomenclature). *Holosticha kessleri* has a rather long row of transverse cirri largely extending very close and parallel to the left marginal row so that it is difficult to recognise as a distinct row in life. The anterior portion of the left marginal row is not curved rightwards in his illustrations (Fig. 22a, b). Thus, the sole significant difference between Stein's *O. gibba* (Fig. 21h, i) and Wrzesniowski's *O. kessleri* (Fig. 22a–d) is in the number of transverse cirri. However, as just mentioned, this feature is rather difficult to recognise in life so that we can assume that Stein overlooked the longitudinally arranged portion of the transverse cirral row. In addition, this portion of the row is composed of cirri, which are distinctly finer than those of the posterior, transversely arranged part (Wrzesniowski 1877). Stein was not the sole worker who overlooked this special feature; Wang & Nie (1932) also did not recognise the cirral pattern in detail (see below; Fig. 22h).

Wrześniowski compared his species in detail with *Oxytricha velox* Quennerstedt and *O. micans* Engelmann, but not with *O. gibba* sensu Stein. He stated that *O. velox* has five cirral rows including the marginal rows. Kahl (1932) classified *O. velox* in *Holosticha (Trichotaxis)*, but simultaneously stated that it is likely a synonym of *O. kessleri*. He assumed, and I agree with him, that Quennerstedt very likely misinterpreted the transverse cirral row as ventral row. Further, the general morphology of *Trichoda gibba*, *O. velox*, and *O. kessleri* agree rather well so that conspecificity of these three taxa is very likely. Unfortunately, the long description of *O. velox* is in Swedish and thus not readable for me. *Oxytricha micans* is a junior synonym of *H. pullaster*.

Simultaneously with Wrześniowski (1877), Mereschkowsky (1877, 1879) described another new species, *Oxytricha wrzesniowskii*. Synonymy of this species and *Trichoda gibba* was first proposed by Borror (1972). There is some confusion about the size of this species. In the Russian original description, Mereschkowsky (1877, p. 232) wrote 0,06''', which means 0.06 Linija. According to Hellwig (1988, p. 148) the size of this old linear measure depends on the country. For Russia, where Mereschkowsky lived and worked, the value is 2.54 mm, that is, a tenth of an inch. Thus, 0.06 Linija corresponds to 152.4 μm. In the German translation of the original description the length is given as 0.01 Linija (= 25 μm), which must be interpreted as a printer's error; by contrast, Kahl (1932) mentioned 200 μm and therefore wrote that *O. wrzesniowskii* is an imposing species. The illustration of *O. wrzesniowskii* shows several enlarged frontal cirri. However, Mereschkowsky (1879) mentioned that he did not count them, so that it cannot be excluded that he overestimated the number. Very likely, Mereschkowsky did not recognise, like several other workers, the transverse cirral pattern correctly. He found that the left marginal row curves distinctly rightwards posteriorly, where the cirri are strong and elongated, indicating that he misinterpreted the posteriormost transverse cirri as marginal cirri.

Kent (1882), who did not provide own data on the present species, did not synonymise *Amphisia* with *Holosticha* and thus assigned Wrześniowski's species to Sterki's *Amphisia*, which is, inter alia, defined by the three enlarged frontal cirri; by contrast, *Holosticha* originally comprised not only species with three enlarged frontal cirri (e.g., *O. kessleri*), but also species without enlarged frontal cirri (e.g., *Oxytricha pernix*). Due to the increased number of frontal cirri in *O. wrzesniowskii*, Kent assigned it to *Holosticha*. Thus, Kent's decision was anticipatory in many ways, but unfortunately neglected by later workers. Kent (1882) considered *Oxytricha crassa* Claparède & Lachmann (= *Thigmokeronopsis crassa* in present book) as a variety of the present species.

Calkins (1902) redescribed *O. kessleri* in some detail. He even recognised the break in the adoral zone. However, he interpreted the anteriorly directed portion of the transverse cirral row as half ventral row and counted three frontal cirri, obviously including the buccal cirrus. Likely, he misinterpreted the right frontal cirrus as distal end of the adoral zone of membranelles.

Kahl (1932, 1933) provided a small, but elegant illustration of *H. kessleri* showing most (all?) important features (Fig. 22i, p). The cirral pattern agrees very well with that described recently for other marine *Holosticha* species. Kahl (1932, p. 582, 583) was the first to suppose synonymy of *Trichoda gibba* and *Oxytricha kessleri*. His text on *H. gibba* is somewhat confusing because, on the one hand, he obviously made own observations on this species, but, on the other hand, he stated that the marginal and transverse ciliature have to be checked. Anyhow, since Kahl's time no new data became available for *T. gibba*. Thus, I consider Kahl's description of *H. kessleri* as authoritative redescription of *Holosticha gibba*. Surprisingly, Kahl (1932) did not mention the very common *Holosticha pullaster* or one of its synonyms, the limnetic, distinctly smaller counterpart to *H. gibba*. Likely due to this deficiency of Kahl's revision, many post-1932 workers identified *H. pullaster* as *H. kessleri* (sensu Kahl). Consequently, all limnetic records of *H. kessleri* are assigned to *H. pullaster*.

Hamburger & Buddenbrock (1929) and Wang & Nie (1932) were the last workers who accepted both *Amphisia* and *Holosticha*. Wang & Nie (1932) described and illustrated only five or six transverse cirri (Fig. 22h). In addition, the figure does not show the characteristic bend of the anterior end of the left marginal row. However, both features are difficult to recognise in life and therefore these differences must not be over-interpreted. Since size, shape, nuclear apparatus, and the other details of the ciliature agree with the original description of *H. kessleri* and the data by Kahl (1932) and Borror (1979), I accept Wang & Nie's identification.

The illustrations provided by Biernacka (1963, 1967) are very similar to Kahl's figure (compare Fig. 22e, i), indicating that Biernacka's sketch is a redrawing. The illustration by Petran (1963) looks like Stein's figure (Fig. 21h); thus, I assume that Pertran's illustration is also a redrawing.

Borror (1972) accepted both *H. kessleri* and *H. gibba*. For the latter species he proposed *Oxytricha wrzesniowskii* and *Holosticha wrzesniowskii punctata* Rees as synonyms. However, I agree with Kahl (1932) that Rees' taxon should not be synonymised with the present species (see species indeterminata). Borror's (1979) data agree rather well with the redescription by Kahl (1932), although he did not mention or illustrate the two macronuclear nodules and the gap in the adoral zone. Borror & Wicklow (1983, p. 121) added four further species to the list of synonyms of *Holosticha gibba*, namely, *Holosticha arenicola* Kahl, *Holosticha viridis* Kahl, *Holosticha algivora* Kahl, and *H. rhomboedrica* Vuxanovici; by contrast, I consider them as distinct species or synonyms of other species. Borror (1972) mentioned *Trichoda gibba* Müller, 1786 and *Oxytricha gibba* Claparède & Lachmann, 1858 under the heading *Paraurostyla gibba* (Müller). In the present revision, Müller's species is now the type of *Holosticha*, whereas Claparède & Lachmann's species was transferred to *Australothrix* (lacks transverse cirri) by Blatterer & Foissner (1988).

Holosticha sp. sensu Fenchel & Lee (1972) is certainly identical with *H. gibba* as already supposed by the authors themselves. Especially the large size (140–150 µm) indicates that it is *H. gibba*, and not *H. diademata* or *H. pullaster*, which are distinctly smaller.

Ricci et al. (1982) redescribed this species from the Somalian coast. Although the frontal portion of the ciliature is not described, the identification is beyond reasonable doubt. Aladro Lubel (1985) and Aladro Lubel et al. (1990) also reported only five transverse cirri. However, some other features, for example, the nuclear apparatus and the anteriorly curved left marginal row, indicate that the identification is correct. Carey (1992) wrote that *Holosticha gibba* is indistinguishable from *H. kessleri* and thus must be considered a synonym. However, since *H. gibba* is older, it would have been consistent to apply the name of the senior synonym. Al-Rasheid (1996a) provided a very brief description and a micrograph which, however, does not show any details. He mentioned a body length of 155 µm and two macronuclear nodules indicating that the identification could be correct although he found it in a so-called freshwater lake in an oasis, which, however, had a salinity of 13‰.

Holosticha gibba sensu Hofker (1931, Fig. 21f) is likely identical with *H. foissneri* as indicated by the nuclear apparatus and the large gap in the adoral zone of membranelles. *Holosticha kessleri aquae-dulcis* Buchar is a synonym of *H. pullaster* (Fig. 26i). *Holosticha kessleri* sensu Sarmiento & Guerra (1960) is insufficiently redescribed (Fig. 191l). Gellért (1956a, p. 345) described a population from a terrestrial habitat, strongly indicating that the identification is incorrect.

The synonymy of *Trichoda gibba*, *Oxytricha velox*, *Oxytricha wrzesniowskii*, and *Oxytricha kessleri* has been discussed for a long time. Now I summarise these four species under the oldest name. This is not in contradiction with the stabilisation of nomenclature because none of these names is in extensive use. I keep the data separate so that the individual species can be reactivated when new data become available. Detailed redescription necessary.

Oxytricha crassa Claparède & Lachmann, 1858 is, according to Hemberger (1982, p. 117), a further junior synonym of *Holosticha gibba*. However, in the present book it is classified as valid species in *Thigmokeronopsis* (see there for details). *Amphisia kessleri* sensu Schewiakoff (193) is very likely a *Holosticha pullaster*.

The generic characterisation above is rather detailed. Unfortunately, the type species still awaits neotypification based on a thorough redescription. Thus, one cannot exclude that the definition of *Holosticha* has to be modified when new data on the type species become available. However, I am convinced that the changes will be minimal.

I do not know this species. But very likely it is easy to separate from most other species listed in the key above because it is distinctly larger than *Holosticha diademata* (marine) and *H. pullaster* (marine and limnetic), which also has the contractile vacuole distinctly dislocated posteriorly.

Morphology: As mentioned above, the sparse morphological data for the four species synonymised are kept separate. I provide the description of *Holosticha kessleri* sensu Kahl (1932) first because his illustrations very likely shows the morphology and cirral pattern best. Some reliable, supplementary data from other sources are added. The descriptions of *H. gibba*, *O. velox*, and *H. wrzesniowskii*, which I keep brief, are based on Stein (1859), Quennerstedt (1869), respectively, Mereschkowsky (1879).

Description based on *H. kessleri* data (unless otherwise indicated from Kahl 1932; Fig. 22i, p): Body size 120–160 × 30–40 μm (width estimated via body length:width ratio of 4:1 of specimen shown in Fig. 22i); body length:width ratio according to Wrześniowski 3:1. Further measurements: body length up to 150 μm (Wrześniowski 1877);

Fig. 21a–e, g–i *Holosticha gibba* from life (a–d, from Müller 1786; e, after Stein 1859 from Kahl 1932; g, from Dujardin 1841; h, i, from Stein 1859). **a, b, d:** Ventral and/or dorsal views, size? **c:** Lateral view. **e, h, i:** Ventral views showing body outline, cirral pattern (presumably not quite correct), and nuclear apparatus. **g:** Cell seen from dorsal. TC = transverse cirri. Page 99. →

Fig. 21f *Holosticha gibba* (from Hofker 1931. Heidenhain stain). Part of ciliature and nuclear apparatus as seen from dorsal side, 110 μm? The wide gap in the adoral zone (arrow) and the nuclear apparatus are reminiscent of *H. foissneri* (see Fig. 31a–d).

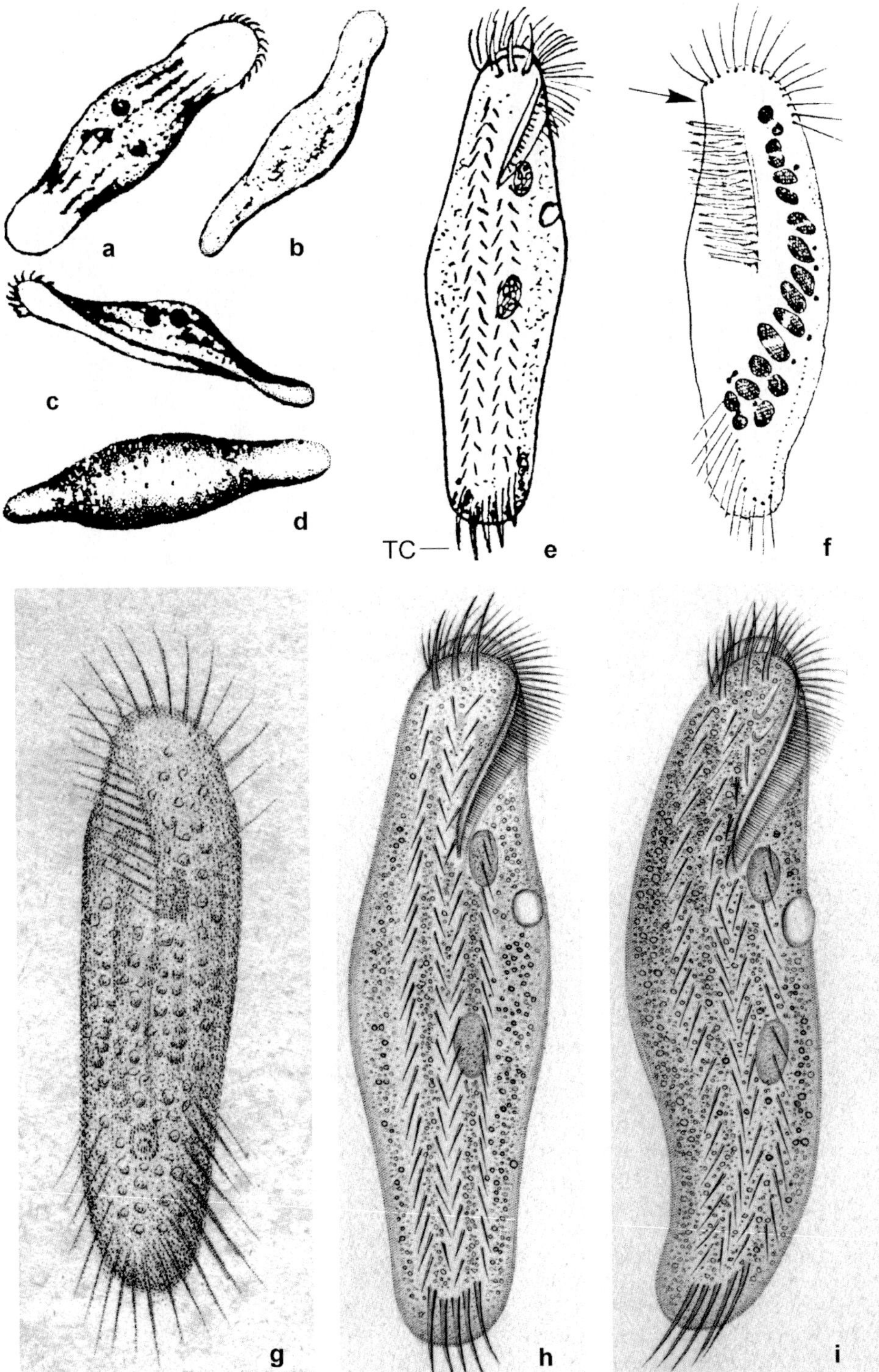
a
b
c
d
TC
e
f
g
h
i

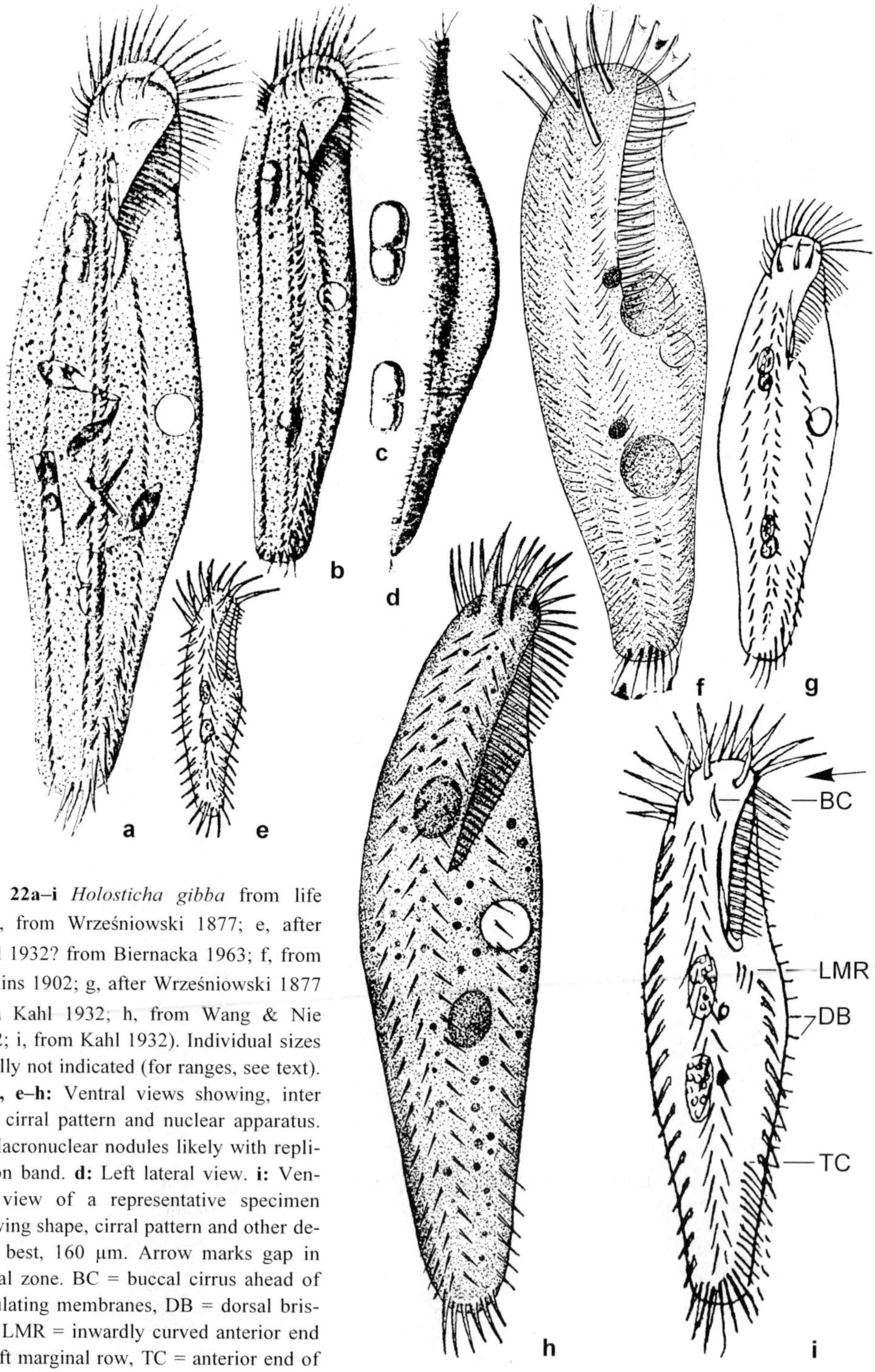

Fig. 22a–i *Holosticha gibba* from life (a–d, from Wrześniowski 1877; e, after Kahl 1932? from Biernacka 1963; f, from Calkins 1902; g, after Wrześniowski 1877 from Kahl 1932; h, from Wang & Nie 1932; i, from Kahl 1932). Individual sizes usually not indicated (for ranges, see text). **a, b, e–h:** Ventral views showing, inter alia, cirral pattern and nuclear apparatus. **c:** Macronuclear nodules likely with replication band. **d:** Left lateral view. **i:** Ventral view of a representative specimen showing shape, cirral pattern and other details best, 160 μm. Arrow marks gap in adoral zone. BC = buccal cirrus ahead of undulating membranes, DB = dorsal bristles, LMR = inwardly curved anterior end of left marginal row, TC = anterior end of transverse cirral row. Page 99.

70–235 µm (Florentin 1899; the wide range indicates that he mixed up two or more species); 155 µm (Al-Rasheid 1996a; see remarks); about 140 µm long (Fenchel & Lee 1972); body size 135 × 40 µm (Calkins 1902); 150–170 × 34 µm (Wang & Nie 1932); 135 × 41 µm (Mayén Estrada 1987); 100–110 × 30–35 µm (Ricci et al. 1982; rather small! other species?); 91–105 × 28–35 µm (Aladro Lubel 1984; rather small); 91–135 × 28–41 µm (Aladro Lubel et al. 1990). Body outline roughly spindle-shaped, that is, widest in mid-body and distinctly narrowed anteriorly and posteriorly, anterior end often slightly curved leftwards. Body slightly contractile; according to Wrześniowski highly contractile and flexible. Two ellipsoidal macronuclear nodules in or slightly right of midline; each nodule with a micronucleus. Wrześniowski (1877) observed specimens prior to division because nodules had replication bands; anterior nodule right of peristome, rear one behind cytopyge. Contractile vacuole not illustrated by Kahl (1932, Fig. 22i), left of anterior end of left marginal row in Kahl's (1933, Fig. 22p) illustration. According to Wrześniowski, contractile vacuole about in mid-body near left margin; contracts rarely and surrounded by many other vacuoles so that it is difficult to recognise. Cytopyge slit-like, dorsal at 66% of body length (Wrześniowski 1877). Cortical granules very likely lacking because never mentioned in the descriptions. Cytoplasm with black (likely refractive) granules (Wrześniowski 1877), according to Wang & Nie (1932) with yellowish tint of unknown origin. Movement lively, twitching (Kahl); never resting, moves to and fro, always contracting, extending, and bending in all directions, posterior body portion very mobile (Wrześniowski 1877).

Adoral zone occupies about 33% of body length, extends distinctly onto right body margin, likely bipartite. 14–15 long, thick membranelles in proximal portion of adoral zone and 4–5 short membranelles in anterior portion (Wrześniowski 1877); in total composed of 24 (Ricci et al. 1982) to about 40 (Fenchel & Lee 1972) membranelles. Buccal area narrow. Three enlarged frontal cirri; buccal cirrus ahead of undulating membranes. Frontal cirri about 14 µm long (Aladro Lubel 1984). Midventral complex extends from near distal end of adoral zone close to rear portion of transverse cirral row; left cirrus of each midventral pair stronger than right one and hooking anteriorly (Wrześniowski 1877); total complex composed of about 30 cirri (Ricci et al. 1982). 12–20 transverse cirri in J-shaped pattern; rearmost five or six cirri enlarged and distinctly projecting beyond rear body end, remaining, longitudinally arranged cirri finer and very close and parallel to posterior portion of left marginal row. These data agree very well with Wrześniowski's (1877) observations: 12–15 transverse cirri; the rearmost three parallel to rear body end, remaining longitudinally arranged; length increases from rightmost cirrus to fifth from right, length of others decreases in anteriad direction; rightmost five cirri strong and long and thus distinctly protruding beyond rear end, remaining transverse cirri fine and not projecting beyond body margin. According to Ricci et al. (1982) 12 transverse cirri (erroneously designated as caudal cirri); Wang & Nie (1932) and Fenchel & Lee (1972) counted only 5–6, indicating that they overlooked the anteriorly extending portion of the transverse cirral row (see remarks); Calkins

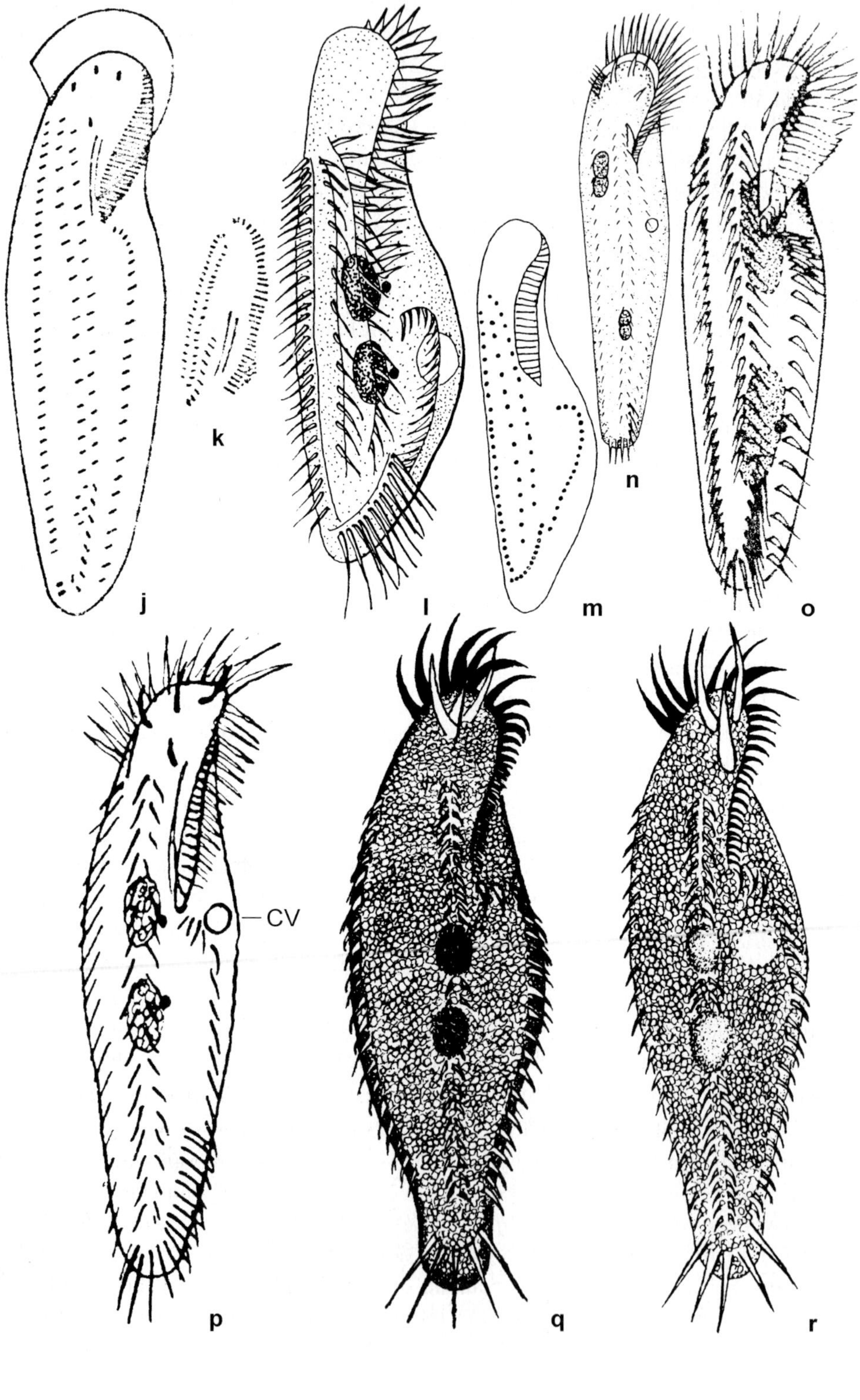
j
k
l
m
n
o
p
CV
q
r

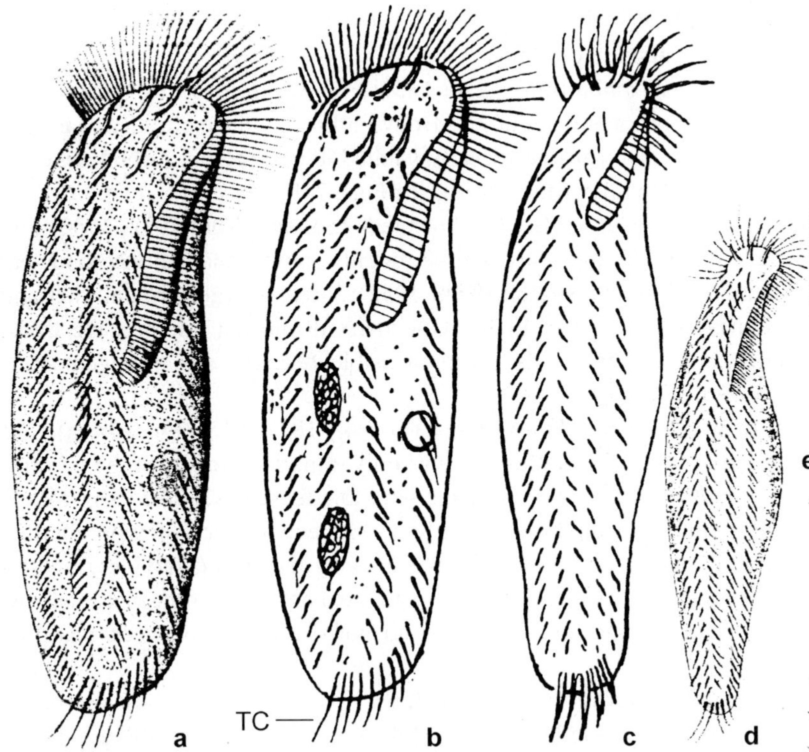

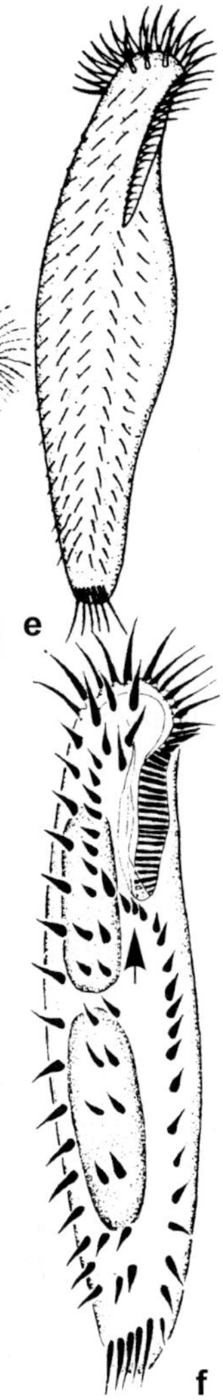

Fig. 23a–f *Holosticha gibba* from life (a, from Mereschkowsky 1877; b, after Mereschkowsky 1877 from Kahl 1932; c, after Quennerstedt 1869 from Kahl 1932; d, from Quennerstedt 1869; e, after Quennerstedt 1869 from Carey 1992; f, from Fenchel & Lee 1972). **a, b:** Ventral views, 152 µm. **c–e:** The body outline and the shape indicates that Quennerstedt's *Oxytricha velox* is in fact identical with *Holosticha gibba*. **f:** Ventral view, 150 µm. Fenchel & Lee did not identify this population, however, they supposed that it could be *H. kessleri*, a synonym of *H. gibba*. I agree with this proposal especially because of the large size. Arrow marks the rightwardly curved anterior end of left marginal row. The transverse cirral pattern was obviously not exactly recognised. TC = transverse cirri. Page 99.

(1902) also counted only five transverse cirri and misinterpreted the anterior, longitudinally arranged portion as third ventral row. Right marginal row begins near distal end of adoral zone, ends subterminally. Left marginal row distinctly curved rightwards anteriorly (Kahl); composed of 26–30 cirri (Ricci et al. 1982); each marginal row composed of about 18 cirri according to Fenchel & Lee (1972). Marginal rows distinctly

← **Fig. 22j–r** *Holosticha gibba* (j, k, from Borror 1979; l, m, from Ricci et al. 1982; n, after Wrześniowski 1877? from Carey 1992; o, from Borror & Wicklow 1983; p, from Kahl 1933; q, from Aladro Lubel 1985; r, from Aladro Lubel et al. 1990. j, k, o, protargol impregnation?; l–n, p–r, from life). **j, o:** Cirral pattern, j = 148 µm, o = 102 µm. **k:** Anlagen of a middle divider. **l, m:** Ventral view, 91 µm. **n, p–r:** Ventral views, n = 150 µm, p = 120–160 µm, q = 91 µm, r = 88 µm. CV = contractile vacuole. Page 99.

dislocated inwards so that cirri do not project beyond lateral body margins (Wrześniowski 1877). Dorsal cilia about 3 μm long.

Description of *Oxytricha gibba* (after Stein 1859; Fig. 21e, h, i): Body length up to 182 μm[1]; ratio of body length:width about 3:1. Body outline elongate elliptical, bulbous widened about in mid-body. Dorsal side strongly vaulted in mid-region. Adoral zone occupies about 33% of body length. Cirral pattern, see Figs. 21h, i.

Description of *Oxytricha velox* (after Quennerstedt 1869 and Kahl 1932, 1933; Fig. 23c–e): Body length about 125 μm; body outline roughly spindle-shaped; macronuclear nodules neither mentioned nor illustrated (Kahl 1932); three enlarged frontal cirri and about five large transverse cirri (very likely, Quennerstedt did not interpret the transverse cirral pattern correctly, which is comprehensible because this arrangement is rather difficult to recognise in life).

Description of *O. wrzesniowskii* (after Mereschkowsky 1877, 1879; Fig. 23a, b): The reader is mainly referred to the illustration because only important details of the description are provided. Length 152 μm (for problems with this feature, see remarks). Likely two macronuclear nodules. Cytoplasm colourless. Movement very slowly. Outline elongate-ovoid. Adoral zone occupies almost 50% of body length (postdivider?). Number of frontal cirri not counted; Mereschkowsky estimated about six. Left marginal row J-shaped curving posteriorly, indicating that he did not distinguish between marginal and transverse cirri.

Cell division (Fig. 22k): Borror (1979) provided a small illustration showing the arrangement of the anlagen in a middle divider. However, no further details are given. Dembowska (1926, p. 486, 498) studied the regeneration of the ciliature of *Amphisia kessleri*, however, without distinguishing between the results on the present species and an *Actinotricha* species.

Molecular data: Chang et al. (2004) discovered a three-gene macronuclear chromosome in a *Holosticha* sp., which was similar to *H. kessleri* (= *H. gibba* in present book) according to Mann-Kyoon Shin. However, since this population was isolated from lawn mosses (Plainsboro, New Jersey), that is, a terrestrial habitat, it was likely an *Anteholosticha* species because *Holosticha* species (e.g., *H. gibba, H. pullaster*) do not occur in soil.

Occurrence and ecology: *Holosticha gibba* is obviously common in marine habitats, but absent in fresh water! Usually benthic (upper sediment layer, aufwuchs), but also in the neuston (Webb 1956). It was likely often confused with the limnetic counterpart *H. pullaster* and thus classified as holo-euryhalin by Albrecht (1984, p. 145). In contrast, Riedel-Lorjé (1981) considered it as characteristic for brackish and marine habitats. Type locality of *H. gibba* not given in detail; very likely Müller (1786) collected it from the Danish coast of the Baltic Sea. Stein (1859) found it in the harbour of Travemünd (Baltic Sea), a village north-east of the city of Hamburg, Germany. He also found it in a sample from the Adriatic Sea collected in the harbour of the city of Trieste, Italy.

---

[1] Length according to Stein (1859) up to 1/12 ''' (= Linie, an old linear measure, which was 2.18 mm in Prussia; Hellwig 1988). Kahl (1932) mentioned a length of 170 μm.

The type locality of the synonym *O. velox* is likely the Baltic Sea at the city of Visby, Gotlandslan, Sweden (Quennerstedt 1869).

The type locality of *O. kessleri* is the eastern coast of Rügen, a German island in the Baltic Sea. Wrzesniowski (1877) found it together with *Oxytricha pernix* (in the present book considered as supposed synonym of *Holosticha pullaster*) among algae washed ashore. The type locality of *Oxytricha wrzesniowskii* is the White Sea, where Mereschkowsky (1877, 1879) discovered it in the Kloster-Bay at the Solowetzky Islands (Russia); it was highly abundant among algae in "not completely fresh water".

Records of *H. gibba* largely not substantiated by morphological data: Mediterranean Sea (Dujardin 1841; Fig. 21g); Bay of Naples, Mediterranean Sea (Entz 1884, p. 294); harbour of Trieste, Adriatic Sea (Gruber 1884b, p. 482); Liguarian Sea at the Italian city of Rapallo (Gunea 1891, p. 146); Brazomar Beach, Castro Urdiales, Bay of Biscay, Spain (Fernandez-Leborans et al. 1999, p. 742); Black Sea (Mereschkovsky 1880, p. 29; Bacescu et al. 1967, p. 7; Pavlovskaya 1969; Petran 1963, p. 193; 1967, p. 174; 1971, p. 154); Caspian Sea (Agamaliev 1971, p. 383); Baltic Sea (Eichwald 1847, p. 334; Quennerstedt 1869, p. 2, 32; Kahl 1932); together with "*A. kessleri*" in saline lakes in Ukraine (Butchinsky 1895, p. 145; Dagajeva 1930); saline lakes in Romania (Entz 1904a, p. 113); saline lake in Azerbaijan (Aliev 1982, p. 87); salt-water basins in Ukraine (Gayewskaya 1924).

Records of the synonym *Oxytricha velox* not substantiated by morphological data: Bay of Kiel, Germany (Bock 1952, p. 83; Hartwig 1974, p. 17); groundwater of coast region of the Hiddensee Island, Germany (Münch 1956, p. 434); Gulf of Mexico (Borror 1962, p. 342).

Records of the synonym *H. kessleri* substantiated by morphological data and/or illustrations: Elbe estuary (Northern Sea) and harbour of the city of Kiel (Baltic Sea), Germany (Kahl 1932); littoral of Hiddensee, a German island in the Baltic Sea (Biernacka 1967); Bay of Danzig, Baltic Sea, Poland (Biernacka 1962; p. 75, 78; 1963); among algae from Bay of Amoy, Yellow Sea?, China (Wang & Nie 1932); lake (Ash Shu'bah) with 13‰ salinity from Al-Hasa Oasis, Eastern Region of Saudi Arabia (Al-Rasheid 1996a, p. 198); Woods Hole area, Massachusetts, USA (Calkins 1902); intertidal sand and gravel near the Jackson Estuarine Laboratory, Adams Point, New Hampshire, USA, at a salinity of 30‰ and other site in the USA (Borror 1979, Borror & Wicklow 1983); sediment (0–2 cm) of a marine grass community of *Thalassia testudinum* from Enmedio Island, Veracruz, Mexico (Aladro Lubel 1985; Aladro Lubel et al. 1990); sediment samples from Somalian coast near Chisimaio (Ricci et al. 1982); interstitial sea ice from the Antarctic pack ice (Fenchel & Lee 1972).

Marine and brackish water records of the synonym *H. kessleri* not substantiated by morphological data: Nova Bay, Denmark (Fenchel 1968); epizoic on European oysters (*Ostrea edulis*) from Conway, North Wales, England and St. Andrews, New Brunswick, Canada (Laird 1961, p. 457); throughout the year (sometimes abundantly) in the Dee estuary at Parkgate, Cheshire, England and other sites in Great Britain (Webb 1956, p. 152; Carey & Maeda 1985, p. 568); saline pools near Nancy, France (Florentin 1899, p. 249; see also Hammer 1986, p. 371); coast of Mediterranean sea near Marseille, France (Vacelet 1961, p. 3; 1961a, p. 15); sparsely to abundantly on sand with little to much debris,

Hamburg Harbour, Germany (Bartsch & Hartwig 1984, p. 556); in September 12 ind. cm^{-2} on exposed slides in the Elbe estuary between Cuxhaven and Brunsbüttel, Germany (Knüpling 1979, p. 277, 281; for further records from the Elbe estuary upstream and downstream of the city of Hamburg, see Grimm 1968, p. 365; Riedel-Lorjé 1981, p. 164); Bay of Kiel, Germany (Bock 1952, p. 83); groundwater of coastal area of the Hiddensee Island, Germany (Münch 1956, p. 434); Königshafen near List, Sylt, Germany (Küsters 1974, p. 174); Venice Lagoon, Italy (Coppellotti & Matarazzo 2000, p. 426); Netherlands (Verschaffelt 1929, p. 53); sites polluted by pulp-mill effluents on the west coast of Scotland (Wyatt & Pearson 1982; p. 298, 301); marine sediments from the Scolt Head Island (52°59'15"N 0°41'39"E), Scotland (Barnes et al. 1976, p. 507); Caspian Sea (Agamaliyev 1974, p. 21); Kandalaksha Bay, White Sea, Russia (Burkovsky 1970a, p. 189; 1970b, p. 11; 1970c, p. 56); Barents Sea near Novaya Zemlya, Russia (Azovsky 1996, p. 6); 3 ind. cm^{-2} in the sandy bottom (0–1 cm) of the Odessa Bay, Black Sea, Ukraine (Dzhurtubayev 1978, p. 65); in sulfur and non-sulfur patches from aquaria filled with material from Narragansett area, Rhode Island, USA (Lackey 1961, p. 276): in debris from the Woods Hole area, Massachusetts, USA (Lackey 1936, p. 269; 1938, p. 510); up to 400 ind. cm^{-2} in Friday Harbor Waters, San Juan Island, Washington, USA (Eddy 1925, p. 104); Gulf of Mexico at the coast of Wakulla and Franklin Counties, Florida, USA (Borror 1962, p. 342); Laguna La Mancha, Veracruz, Mexico (Mayén Estrada 1987, p. 74; for review, see Aladro-Lubel et al. 1988, p. 437); polluted Almendares River estuary in Havana City, Cuba (Diaz Pérez & Montoto Lima 1989, p. 92; as *Amphisia kessleri*); Sepetiba Bay, Rio de Janeiro, Brazil (Wanick & Silva-Neto 2004, p. 5); benthic in Great Bitter Lake, the central and most important water body of the Suez Canal, Egypt, at 40–45‰ salinity, pH 8.1–8.4 (El-Serehy 1993, p. 138).

Records of *O. kessleri* from inland saltwater: salt polluted running waters of the Weser River Basin, Germany (Albrecht 1983, p. 100; 1986, p. 193); saline lake near Sevastopol, Ukraine (Dagajeva 1930, p. 35).

Papers which list both *O. kessleri* and *H. pullaster* from freshwater habitats where the records of *O. kessleri* must be interpreted as misidentification: pelagial of an Upper Austrian lake (Nauwerck 1996, p. 156; ciliates identified by R. Xu); Stirone stream, northern Italy (Madoni & Bassanini 1999, p. 394); Covolo della Guerra, a cave in Berici Hills, Vicenza, Italy (Coppellotti & Guidolin 1999, p. 75; Guidolin & Coppellotti Krupa 1999, p. 76); Slovakia (Tirjaková 1992, p. 293); Karasu River, Gevas River, and Engilsu River, Turkey (Senler et al. 1996, p. 186; 1998, p. 41; Senler & Yildiz 1998, p. 5); ponds of the Panatanos de Villa, Chorrillos, Lima, Peru (Guillén et al. 2003, p. 180). The terrestrial record of *H. kessleri* from Slovakia by Tirjaková (1988, p. 499) is certainly a misidentification.

Records of *O. wrzesniowskii* not substantiated by morphological data: among algae from the Bay of Concarneau, French Atlantic coast (Fabre-Domergue 1885, p. 568); Belgium (Bervoets 1940, p. 124; according to the associated species it is a freshwater habitat).

Fernandez-Leborans & Novillo (1994, p. 203) found *H. kessleri* in the control, but not in the treatment with 1 mg l^{-1} lead; the sediment sample was collected in Castro Urdiales, Spain, that is, the Atlantic Ocean.

The synonym *Holosticha kessleri* feeds on diatoms (Kahl 1932) and purple sulphur bacteria (for review, see Fenchel 1968, p. 116; 1969, p. 25); according to Webb (1956, p. 169) on bacteria, small diatoms, and detritus. Fenchel & Lee's (1972) population fed on 50–100 μm long diatoms.

Holosticha kessleri, a junior synonym of *H. gibba*, is classified as indicator for beta- to alphamesosaprobic waters (a–b; b = 4, a = 5, p = 1, I = 2, SI = 2.7; Table 12; Foissner et al. 1991, 1995, Foissner & Berger 1996, Sládeček & Sládečková 1997, Berger & Foissner 2003). However, note that this species is confined to marine or brackish habitats.

***Holosticha diademata* (Rees, 1884) Kahl, 1932**
(Fig. 24a–w, 25a–o, Table 13)

1884 ***Amphisia diademata*, mihi** – Rees, Tijdschr. ned. dierk. Vereen, Supplement Deel I: 650, 651, Planche XVI, Fig. 21; Fig. 23, as indicated in the heading of the original description, is incorrect (Fig. 24a; original description; no type material available and no formal diagnosis provided).

1928 ***Holosticha thiophaga*** – Kahl, Arch. Hydrobiol., 19: 212, Abb. 44g (Fig. 24r; original description of synonym; no type material available and no formal diagnosis provided).

1929 ***Amphisia diademata* Rees** – Hamburger & Buddenbrock, Nord. Plankt., 7: 90, Fig. 110 (redrawing of Fig. 24a; guide to marine ciliates).

1932 ***Holosticha (Amphisia) diademata* (Rees, 1884)** – Kahl, Tierwelt Dtl., 25: 582, Fig. $106_{2,10}$ (Fig. 24b, e; revision; combination with *Holosticha*).

1932 ***Amphisiella (Holosticha) thiophaga* Kahl, 1928** – Kahl, Tierwelt Dtl., 25: 591, Fig. 112_2 (Fig. 24s; revision).

1933 ***Holosticha diademata* (Rees 1884)** – Kahl, Tierwelt N.- u. Ostsee, 23: 111, Fig. 17.8 (Fig. 24c; guide to marine ciliates).

1933 ***Amphisiella thiophaga* Kahl 1928** – Kahl, Tierwelt N.- u. Ostsee, 23: 112, Fig. 17.27 (Fig. 24t; guide to marine ciliates).

1962 ***Holosticha teredorum* n. sp.** – Tucolesco, Arch. Protistenk., 106: 31, Fig. 50 (Fig. 24v; original description of synonym; very likely no type material available and no formal diagnosis provided).

1963 ***Holosticha diademata* (Rees, 1884)** – Borror, Arch. Protistenk., 106: 510, Fig. 117 (Fig. 24d; redescription).

1972 ***Holosticha diademata* (Rees, 1883) Kahl, 1932** – Borror, J. Protozool., 19: 11, Fig. 21 (Fig. 24w; revision of hypotrichs).

1983 ***Holosticha diademata* (Rees, 1883) Kahl, 1932** – Borror & Wicklow, Acta Protozool., 22: 121, Fig. 18 (Fig. 24f; revision of urostyloids).

1985 ***Holosticha diademata* (Rees, 1883)** – Aladro Lubel, An. Inst. Biol. Univ. Méx., Ser. Zoologia, 55: 26, Lámina 12, Fig. 7 (Fig. 24g; brief redescription).

1986 ***Holosticha diademata* (Rees) Kahl** – Wilbert, Symposia Biologica Hungarica, 33: 253, Fig. 5 (Fig. 24j; brief redescription).

1990 ***Holosticha diademata* (Rees, 1883)** – Aladro Lubel, Martínez Murillo & Mayén Estrada, Manual de Ciliados, p. 129, Figure on p. 129 (Fig. 24h; review).

1992 ***Holosticha diademata* (Rees, 1884) Kahl, 1930–5** – Carey, Marine interstitial ciliates, p. 182, Fig. 715 (Fig. 24i; guide).

1999 ***Holosticha diademata* (Rees, 1883) Kahl, 1932** – Hu & Song, J. Ocean Univ. Qingdao, 29: 469, Fig. 1a–e, Tables 1, 2 (Fig. 24k–o; redescription).

1999 ***Holosticha diademata* (Rees, 1883) Kahl, 1932** – Petz, Ciliophora, p. 299, Fig. 8.66 (Fig. 24j; review).
2000 ***Holosticha diademata* (Kahl, 1932)** – Hu, Wang & Song, J. Zibo Univ., 2: 78, Fig. 1a–f, 2a–f, 3a–c (Fig. 25a–o; cell division; incorrect author).
2001 ***Holosticha diademata* (Rees, 1884) Kahl, 1932** – Berger, Catalogue of ciliate names 1. Hypotrichs, p. 7 (nomenclator containing all basionyms, combinations, and higher taxa of hypotrichs).
2002 ***Holosticha diademata* (Rees, 1884)** – Song & Wilbert, Acta Protozool., 41: 53, Fig. 13A, B, Table 5 (Fig. 24p, q; redescription; voucher slides are deposited in the Laboratory of Protozoology, College of Fisheries, Ocean University of Qingdao, China).
2003 ***Holosticha diademata* (Rees, 1884) Kahl, 1932** – Berger, Europ. J. Protistol., 39: 375, 376, Fig. 6 (Fig. 24b; brief review).

Nomenclature: The species-group name *diademátus -a -um* (Greek adjective; decorated, having a diadem) refers to the anterior portion of the adoral zone of membranelles which looks like a diadem (Rees 1884). The species-group name *thiophaga* (sulphur feeding) obviously alludes to the sulphur bacterium *Thiovolum* on which *H. thiophaga* fed. The species-group name *teredorum* obviously refers to the fact that this species was discovered on *Teredo nivalis* (shipworm; Mollusca).

Kahl (1932, 1933) divided *Holosticha* into several subgenera. The correct names in his papers are thus *Holosticha (Holosticha) diademata* (Rees, 1884) Kahl, 1932 and *Holosticha (Amphisiella) thiophaga* Kahl, 1928. The somewhat confusing spellings in Kahl (1932; see list of synonyms) should indicate that the species were originally classified in *Amphisia*, respectively, *Holosticha*. Tucolesco (1962c) also described his species under the heading "subgenus *Holosticha*". Thus, the correct basionym of his species is *Holosticha (Holosticha) teredorum* Tucolesco, 1962.

Lopez-Ochoterena et al. (1976) mentioned Kahl's species under the heading *Amphisiella tiophaga* Kahl, 1928, which is not only an incorrect subsequent spelling of the species-group name (see below), but also a new combination with *Amphisiella* because this species was never before in the genus *Amphisiella* (an exception is the species list by Agamaliev 1971; see faunistic records). Very likely the Mexican authors assumed that this species was transferred to *Amphisiella* by Kahl (1932), who, however, classified *Amphisiella* only as subgenus of *Holosticha*.

Incorrect subsequent spellings: *Amphisiella tiophaga* Kahl, 1928 (Lopez-Ochoterena et al. 1976, p. 217); *Amphista diademata* Rees, 1884 (Faria et al. 1922, p. 113, 197); *Holosticha diademata* Pees (incorrect spelling of author; Burkovsky 1971a, p. 1774). According to the Zoological Record, Rees' paper was published in 1884 indicating that 1883, for example used by Borror (1972), is incorrect.

Remarks: There has been great confusion about this species for many years. Unfortunately, I cannot clear up the situation completely because I do not know *H. diademata* from my own experience.

Fig. 24a–h *Holosticha diademata* (a, from Rees 1884; b, from Kahl 1932; c, after Kahl 1932 from Kahl 1933; d, from Borror 1963; e, after Rees 1884 from Kahl 1932; f, from Borror & Wicklow 1983; g, from Aladro Lubel 1985; h, from Aladro Lubel et al. 1990. a–e, g, h, from life; f, protargol impregnation?). Ventral views, a, e = 100 µm, b, c = 70–100 µm, d = 80 µm, f = 40 µm, g = 53 µm, h = 54 µm. Some specimens are rather small (f) and the contractile vacuole is not shown; thus, one cannot exclude that these small individuals belong to *H. pullaster*. Arrow in (b) marks gap in adoral zone which is also recognisable in the type population (a). CV = contractile vacuole, LMR = rightwards curved anterior end of left marginal row. Page 115. →

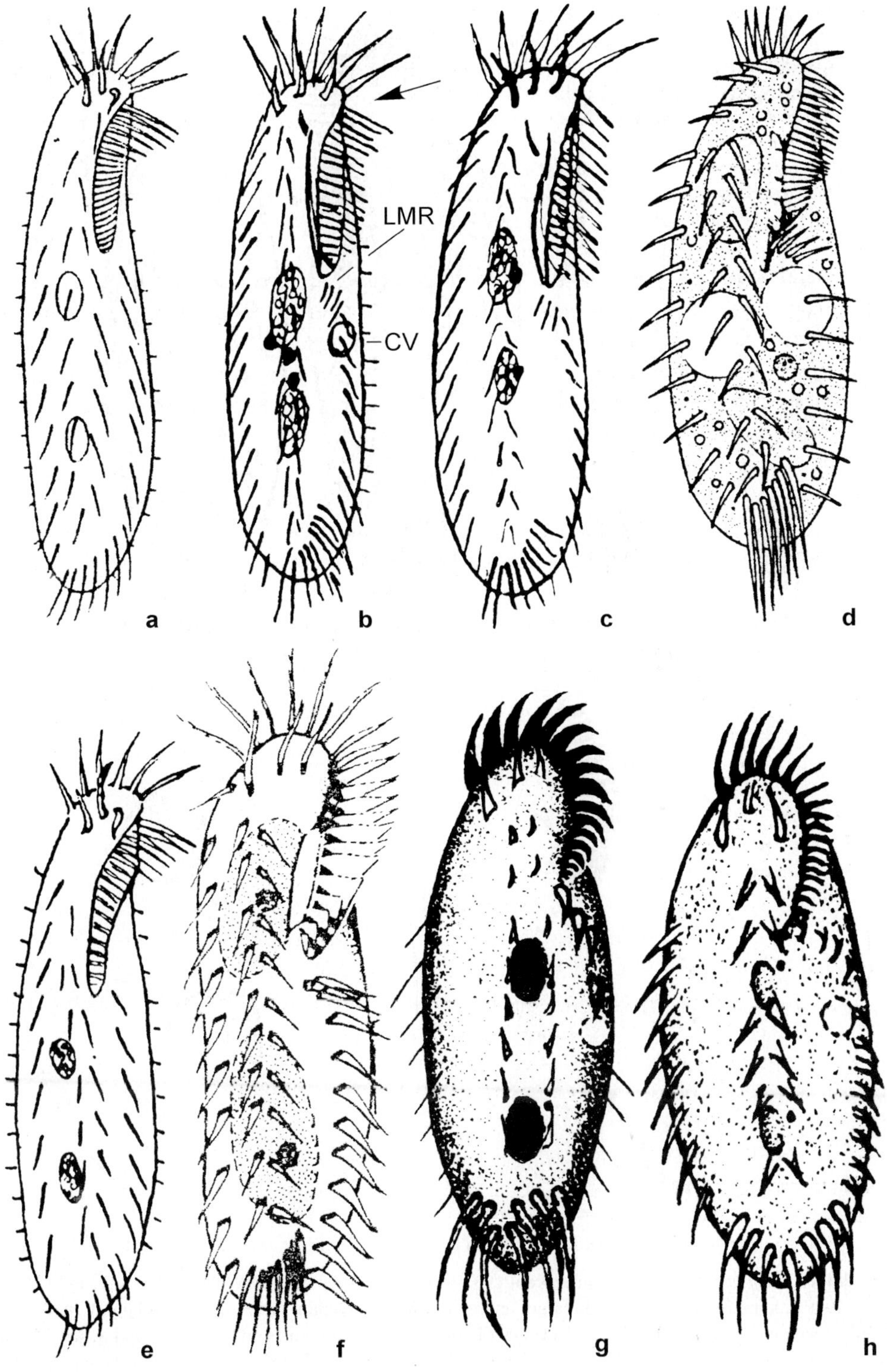
LMR
CV
a
b
c
d
e
f
g
h

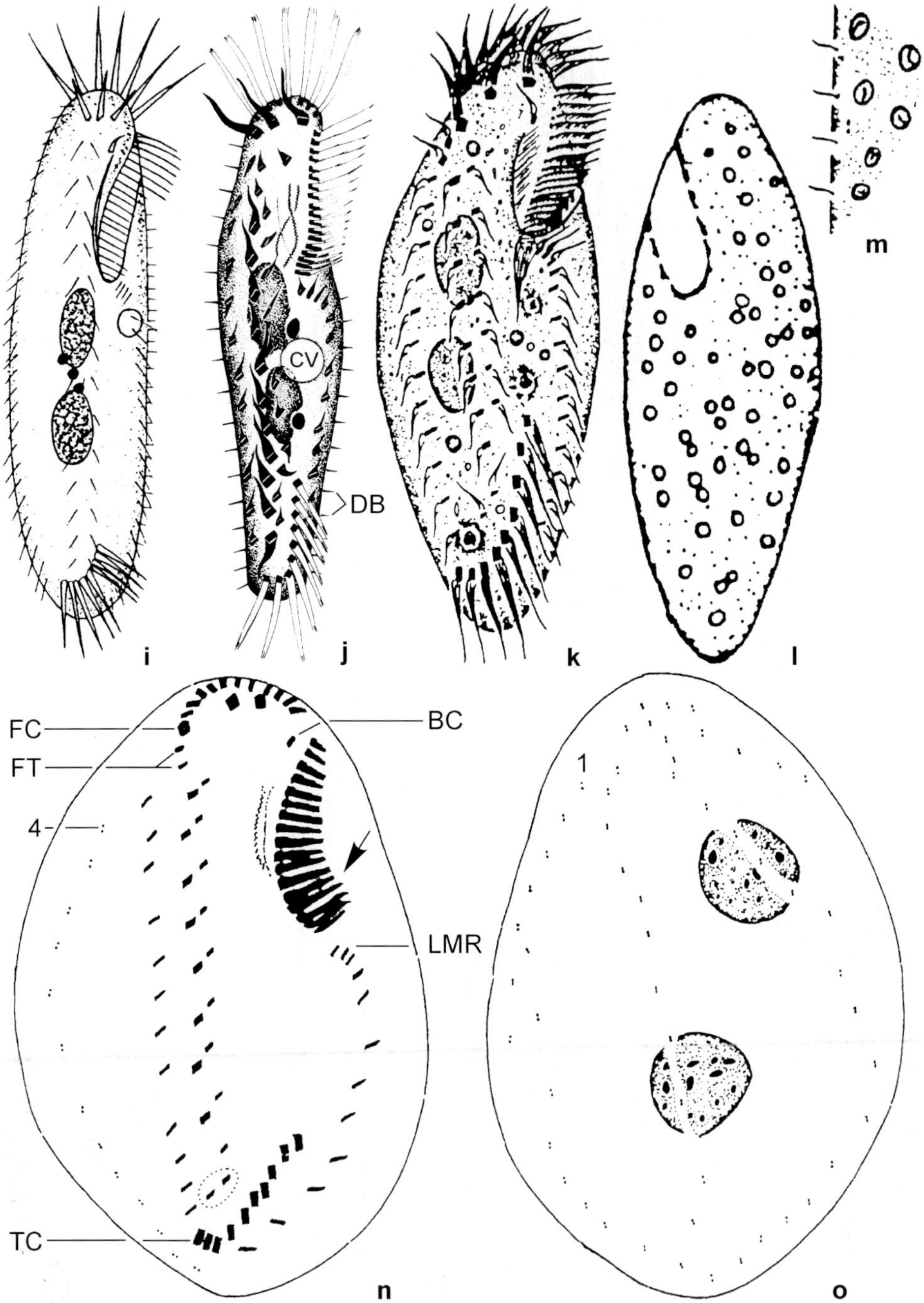

Fig. 24i–o *Holosticha diademata* (i, after Kahl 1932 from Carey 1992; j, from Wilbert 1986; k–o, from Hu & Song 1999. i, k–m, from life; j, n, o, protargol impregnation). **i, k:** Ventral views, a = 70–100 µm, k = 107 µm. **j:** Infraciliature of ventral side, 69 µm. **l, m:** Distribution of cortical granules on dorsal side and detail, 74 µm. **n, o:** Infraciliature of ventral and dorsal side of same (?) specimen, 67 µm. Arrow marks widened membranelles

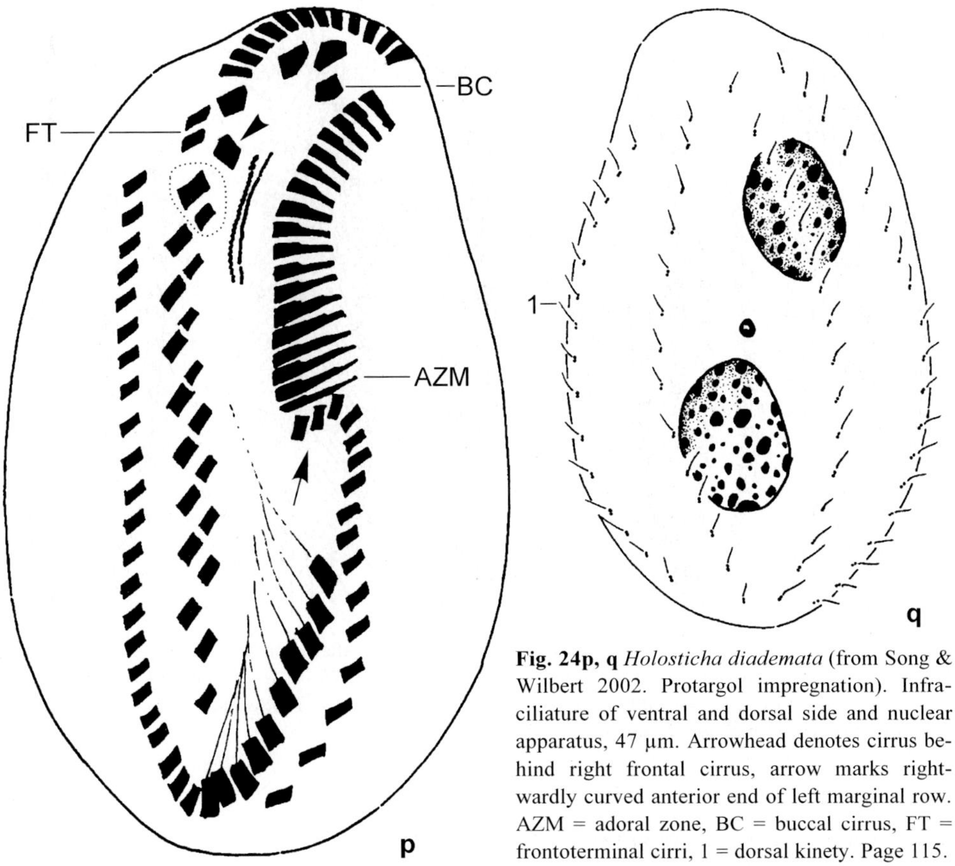

Fig. 24p, q *Holosticha diademata* (from Song & Wilbert 2002. Protargol impregnation). Infraciliature of ventral and dorsal side and nuclear apparatus, 47 µm. Arrowhead denotes cirrus behind right frontal cirrus, arrow marks rightwardly curved anterior end of left marginal row. AZM = adoral zone, BC = buccal cirrus, FT = frontoterminal cirri, 1 = dorsal kinety. Page 115.

Holosticha diademata was established in *Amphisia* by Rees (1884) because it has three enlarged frontal cirri (see genus section). The general morphology and the ciliature agree with *Holosticha gibba*, type of genus, indicating that the classification in *Holosticha* is correct (see previous species). Rees (1884) gave a range of body length from 85–140 µm. In addition, he could not observe a contractile vacuole (Fig. 24a). Kahl (1932) transferred it to *Holosticha (Holosticha)* and provided his own illustration and a reliable redescription which shows that the contractile vacuole is about in mid-body (Fig. 24b). Kahl separated it from *H. gibba* by the smaller size, the shorter adoral zone of membranelles, and the lower number of transverse cirri. Surprisingly, Kahl (1932) did not mention *H. pullaster* or one of its synonyms as valid species. *Holosticha pullaster* is one of the most common freshwater hypotrichs and is very easily recognisable,

---

← of posterior portion of adoral zone. Pretransverse ventral cirri encircled. BC = buccal cirrus, CV = contractile vacuole about in mid-body, DB = dorsal bristles of kinety 1, FC = rightmost frontal cirrus, FT = frontoterminal cirri, LMR = left marginal row, TC = transverse cirri, 1, 4 = dorsal kineties. Page 115.

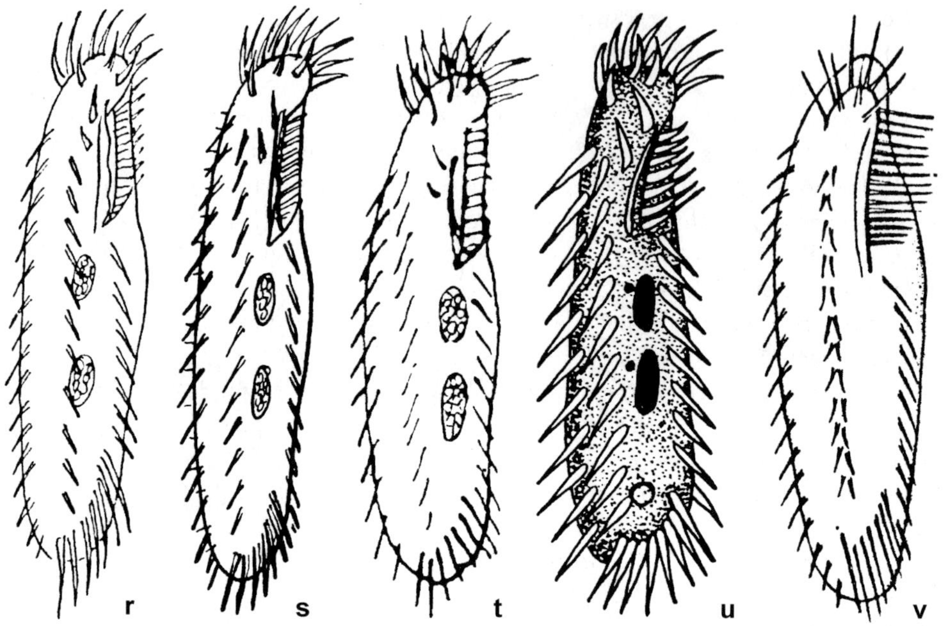

Fig. 24r–v *Holosticha diademata* from life (r, from Kahl 1928; s, after Kahl 1928 from Kahl 1932; t, after Kahl 1932? from Kahl 1933; u, from Lopez-Ochoterena et al. 1976; v, from Tucolesco 1962). Ventral views, r = 50–70 μm, s, t = 70–100 μm, u = 103 μm, v = 100 μm, (r–s) show the synonym *H. thiophaga*, (v) shows the synonym *H. teredorum*. Page 115.

even at lowest magnification (40×), by its posteriorly dislocated contractile vacuole. I suppose that due to this deficiency of Kahl's revision, many post-1932 workers sometimes erroneously identified the small freshwater *Holosticha* as *H. diademata*. There exist few further redescriptions of *H. diademata* showing the contractile vacuole in mid-body (Fig. 24d, g, j). Recently, Song's group redescribed *H. diademata* several times (Hu & Song 1999, Hu et al. 2000, Song & Wilbert 2002). Unfortunately, they never clearly described and illustrated the position of the contractile vacuole in their populations. Only Song & Wilbert (2002) mentioned that it is in a post-equatorial position, which contradicts the data by Kahl who illustrated the vacuole exactly in mid-body (Fig. 24b). The problem is complicated due to the fact that *H. pullaster*, that is, the species with the posteriorly dislocated contractile vacuole is reliably redescribed also from marine habitats (Petz et al. 1995), indicating that *Holosticha pullaster* is euryhaline. On the other hand, *Holosticha diademata* is also reliably recorded from inland saltwater, but never from freshwater. Recently, Song & Wilbert (2002) confined *H. pullaster* to freshwater populations, which lack cortical granules, while *H. diademata* lives in saltwater and has cortical granules on the dorsal side (Fig. 24m). As a consequence of these uncertainties, all limnetic records of *H. diademata* are assigned to *H. pullaster*, and the faunistic data are kept separate.

Borror (1972) synonymised *Holosticha thiophaga* Kahl (Fig. 24r–t) and *H. simplicis* Wang & Nie (Fig. 29d) with *H. diademata*, however, without foundation. I accept only *H. thiophaga* as synonym of the present species and put *H. simplicis* into the synonymy of *H. pullaster* because its contractile vacuole is distinctly dislocated posteriorly. *Holosticha thiophaga* is a very little known species, discovered by Kahl (1928) in inland saltwater. Kahl (1932), who classified it in the subgenus *Holosticha (Amphisiella)*, did not provide new data although he mentioned a greater length than in the original description (70–100 µm vs. 50–70 µm). He stated that he cannot exclude that this species has two narrowly spaced ventral rows; that is, a midventral complex composed of midventral pairs instead of a single cirral row as is characteristic for amphisiellids. Further, the adoral zone is bipartite by a wide gap and the 7–8 transverse cirri are distinctly J-shaped as in other *Holosticha* species. In addition, amphisiellids usually produce their frontal-ventral-transverse ciliature from five or six anlagen only (Eigner & Foissner 1994; Berger 2004a), which is a further indicator that *H. thiophaga* is not an amphisiellid (*H. diademata* has 6–11 transverse cirri, that is, forms at least 7 anlagen). However, Lopez-Ochoterena et al. (1976) basically confirmed Kahl's (1928) observations (Fig. 24u). Thus, one cannot exclude that such a species with a linear (not zigzagging) ventral row really exists, although the observations by Lopez-Ochoterena et al. (1976) must not be over-interpreted.

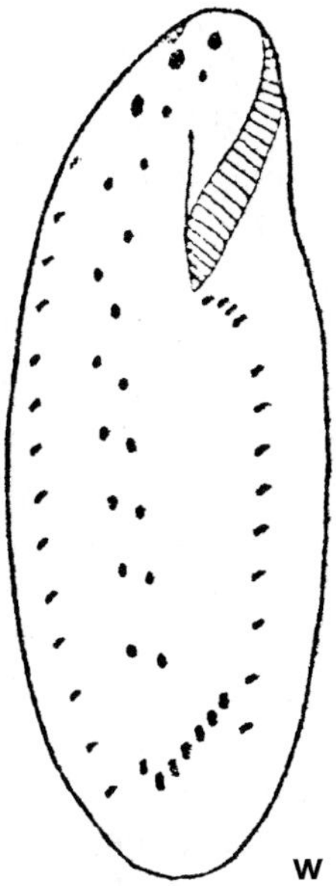

Fig 24w *Holosticha diademata* (from Borror 1972. Protargol impregnation). Infraciliature of ventral side, 95 µm. Page 115.

Borror & Wicklow (1983) synonymised, beside the two species discussed in the previous paragraph, three further species with *H. diademtata*, namely *H. milnei* Kahl, *H. coronata* Vuxanovici, and *H. teredorum* Tucolesco. *Holosticha milnei* is a junior synonym of *Anteholosticha oculata*. *Holosticha coronata* is obviously a junior synonym of *H. pullaster*. But *Holosticha teredorum* indeed resembles the present species, especially as concerns the adoral zone (obviously with gap), the size (100 µm), and the transverse cirri (7 arranged in J-shape; Fig. 24v). I thus accept the decision by Borror & Wicklow to put Tucolesco's species into the synonymy of *H. diademata*. Borror (1972, p. 19) has classified *H. teredorum* as invalid species. The redescription and illustration of *H. diademata* provided by Alzamora (1929) is insufficient (Fig. 191k).

Summarising the data, we can distinguish *H. gibba*, *H. diademata*, and *H. pullaster* by the combination of features used in the key above. If somebody considers the differences among these three species as insufficient, then *H. diademata* is either the junior synonym of *H. gibba* or *H. pullaster*.

Morphology: Unfortunately, there exists no more or less complete description of this species, except for that by Hu & Song (1999), which is, however, in Chinese and does not show the location of the contractile vacuole in the illustrations. The following description is thus a combination of the original description and the redescriptions mentioned above. The data of the synonyms (*Holosticha thiophaga*, *H. teredorum*) are kept separate.

Body length 85–140 μm (Rees), 70–140 μm (Kahl), degenerating specimens according to Kahl only 50–60 μm long (possibly confused with *H. pullaster*); however, Aladro Lubel's (1985) specimen is also only 53 × 21 μm; 67–80 × 22–30 μm (Borror 1963). Body outline elliptical with anterior body portion slightly narrowed and curved leftwards (Rees, Kahl); according to Hu & Song spindle-shaped. Two ellipsoidal macronuclear nodules, front one – as in other *Holosticha*-species – in or slightly right of median, rear one somewhat dislocated leftwards (Fig. 24a, j, o, q); individual nodules about 18 μm long (Borror 1963). Contractile vacuole not seen by Rees (1884), according to Figs. 24b, d, g, h, i, j near left margin about in mid-body; obviously this is a very important difference to *Holosticha pullaster*, which has the contractile vacuole invariably distinctly behind mid-body (according to Song & Wilbert 2002, the contractile vacuole of *H. diademata* is also post-equatorial, although they did not illustrate it). Disc-shaped cortical granules on dorsal side (Fig. 24l, m). Moves rapidly to and fro (Rees, Kahl).

Adoral zone of membranelles occupies about 34–38% of body length in life (Fig. 24a, b, j, k), bipartite with seven (Fig. 24a, b, j; Wilbert 1981), about 11 (Fig. 24n), or 8–12 (Fig. 24p, Song & Wilbert) membranelles in distal portion, and 13–23 in posterior portion of population described by Song & Wilbert; Borror's population in total with around 28 membranelles, Wilbert's (1981) population with a total number of 22–25 membranelles. Membranelles in proximal portion become wider posteriad (Fig. 24n, p). Undulating membranes of equal length (20 μm with about 4 μm long cilia; Borror 1963), more or less straight, arranged in parallel, and right of mid-portion of proximal part of adoral zone (Fig. 24n, p).

Cirral pattern exactly *Holosticha*-like (Fig. 24a, b, j, n, p; Table 13) and thus not described in detail. Additional morphometric data: 10–12 midventral pairs (Wilbert 1981). Number of transverse cirri ranging from 6–8 (Kahl), 6–10 (Hu & Song), 7 (Borror 1963), 7 (rarely 8; Wilbert 1981), to 7–11 (Song & Wilbert); cirri arranged in J-shape and bases slightly to distinctly enlarged. Wilbert's (1981) population with 12–16 right and 9–13 left marginal cirri with anteriormost three, as usual, transversely arranged; Borror's (1963) specimen with about 10 right marginal cirri. Cirri about 8–9 μm long (Borror 1963). Dorsal cilia short, that is, around 2–4 μm (Fig. 24a, b, j, m, q), arranged in usually four more or less bipolar kineties (Fig. 24o, q; Table 13). Caudal cirri lacking.

Synonym *H. thiophaga* according to Kahl (1928) 50–70 μm in life, according to Kahl (1932, 1933) 70–100 μm. Adoral zone bipartite, proximal portion extending at left cell margin (Fig. 24r–t). Cytoplasm without large, ring-shaped structures, as they occur, for example, in *Amphisiella annulata*. Seven transverse cirri arranged in J-shape (Fig. 24s). Population described by Lopez-Ochoterena et al. (1976) 103 × 28 μm in life(?); cirral pattern amphisiellid (however, I am uncertain about the quality of the observation; Fig. 24u).

Synonym *H. teredorum* 100 µm long, body outline lanceolate. Adoral zone bipartite, occupies about 40% of body length. Three enlarged frontal cirri, seven transverse cirri (Fig. 24v).

Cell division: Hill (1980) briefly reported some data on the morphogenesis of *H. diademata*, *Anteholosticha multistilata*, and *A. scutellum*. However, he did not report the results on these three species separately. Hu et al. (2000) studied the ontogenesis of *H. diademata* in detail. Unfortunately, the illustrations are very small and the description in Chinese so that the reader is mainly referred to the illustrations (Fig. 25a–o).

The data can be summarised as follows: (i) The proter gets the proximal portion of the parental adoral zone of membranelles which shows distinct signs of reorganisation; the undulating membranes are reorganised. (ii) The oral primordium of the opisthe forms the adoral zone of membranelles and the undulating membranes and frontal cirrus I/1. (iii) Right(!) of the parental midventral complex (possibly some midventral cirri are involved in primordia formation? Fig. 25b, c) 8–10 streaks originate de novo each in the anterior and posterior body portion to form the frontal-midventral-transverse cirral anlagen. These anlagen produce the middle and right frontal cirrus, about 11–15 midventral cirri, 7–9 transverse cirri, and two frontoterminal cirri. (iv) Division of marginal rows and dorsal kineties proceeds in ordinary manner, that is, two anlagen each occur; Fig. 25g indicates that the anlage for the left marginal row of the proter originates de novo, as in congeners. (v) The two macronuclear nodules fuse during division. For a further discussion of some ontogenetic details see the genus section.

Occurrence and ecology: Marine, but also in inland saltwaters. The type locality of *H. diademata* is in the Oosterschelde, a large bay in the southern Netherlands (Rees 1884; see also Verschaffelt 1930, p. 53). Kahl (1928) discovered *H. thiophaga* in the Brennermoor, a saline (25‰), silt peat bog near the north German village of Bad Oldesloe (see also Kahl 1928a). The type locality of the synonym *H. teredorum* is the Romanian coast of the Black Sea at the village of Eforia, where Tucolesco (1962c) discovered it in the littoral on the shipworm *Teredo navalis* in November 1957. No further records of *H. teredorum* published.

In freshwater *H. diademata* was confused with *H. pullaster*. This mistake falsely indicated a very wide ecological range, especially as concerns salinity (Albrecht 1983, p. 99; 1984, p. 145; Hammer 1986, p. 371). Limnetic records of *H. diademata* are assigned to *H. pullaster* without exception.

Borror (1963a) found *H. diademata* in the benthal region of Alligator Harbour, Florida. It was uncommon in sand and diatom detritus and was recorded from eight stations, often flourishing in cultures developing a surface scum. Usually it occurred in low numbers, but became locally abundant in regions of decaying organic material. According to Borror (1972a) this common species is one of the first bacteriovores of early successional stages.

Further records of *H. diademata* substantiated by morphological data and/or illustrations: among detritus in the Baltic Sea, North Sea, and saline waters near the German village of Bad Oldesloe (Kahl 1932); littoral region of the German islands Sylt and Helgoland, North Sea (Hartwig 1973, p. 451); at 80–180‰ salinity in Solar Lake on the Sinai east coast (Wilbert & Kahan 1981; p. 85); mollusc cultures off the Chinese city of

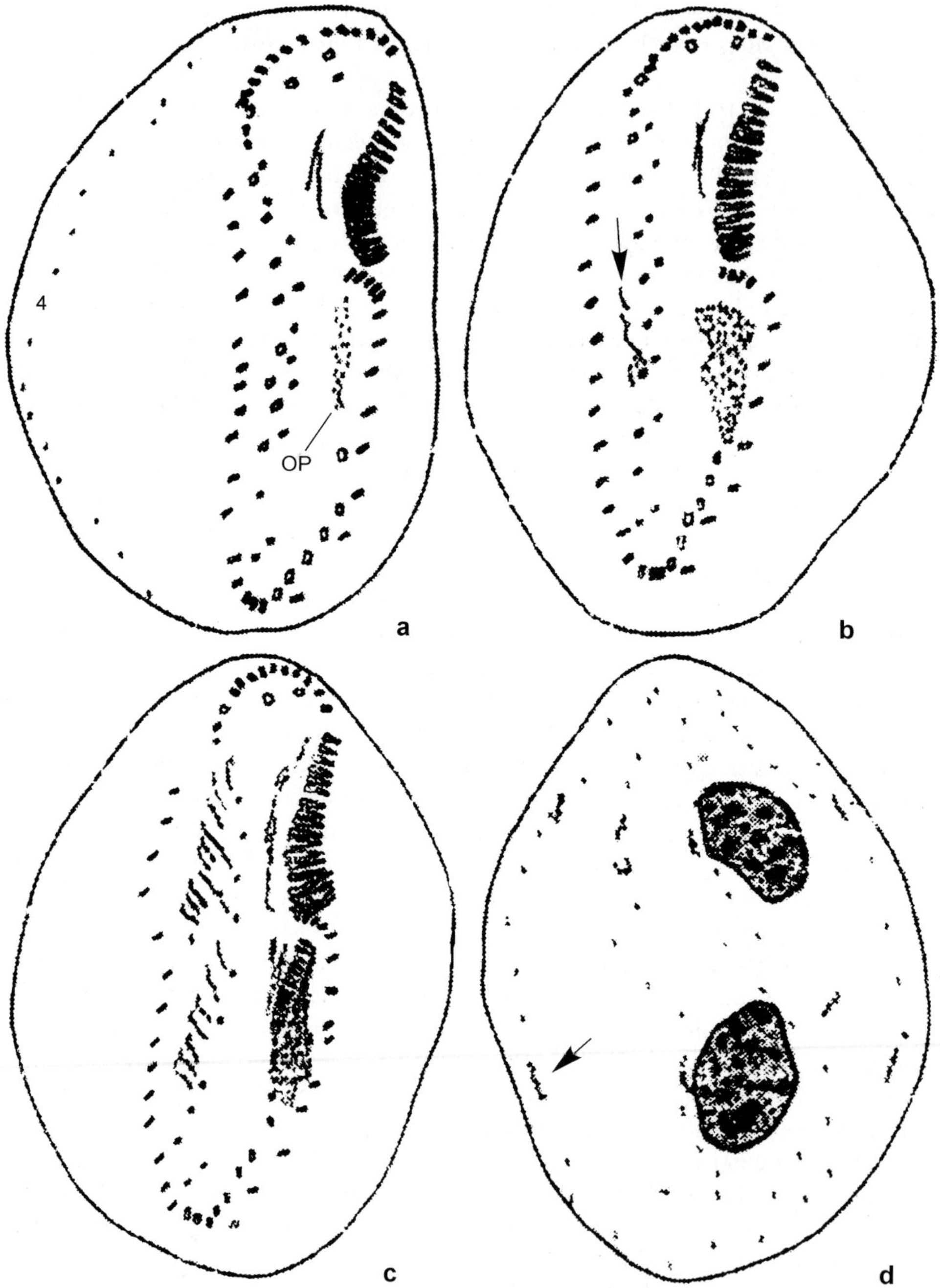

Fig. 25a–d *Holosticha diademata* (from Hu et al. 2000. Protargol impregnation). Unfortunately, the printing quality in the original paper is rather low and the illustrations rather small; thus, the graphical quality of the figures is limited. Early to middle dividers, a = 83 µm, b = 62 µm, c, d = 67 µm. Arrow in (b) denotes formation of frontal-midventral-transverse cirral anlagen for opisthe right (!) of midventral complex. Arrow in (d) denotes anlage for new dorsal kinety 1 of opisthe. OP = oral primordium. Page 115.

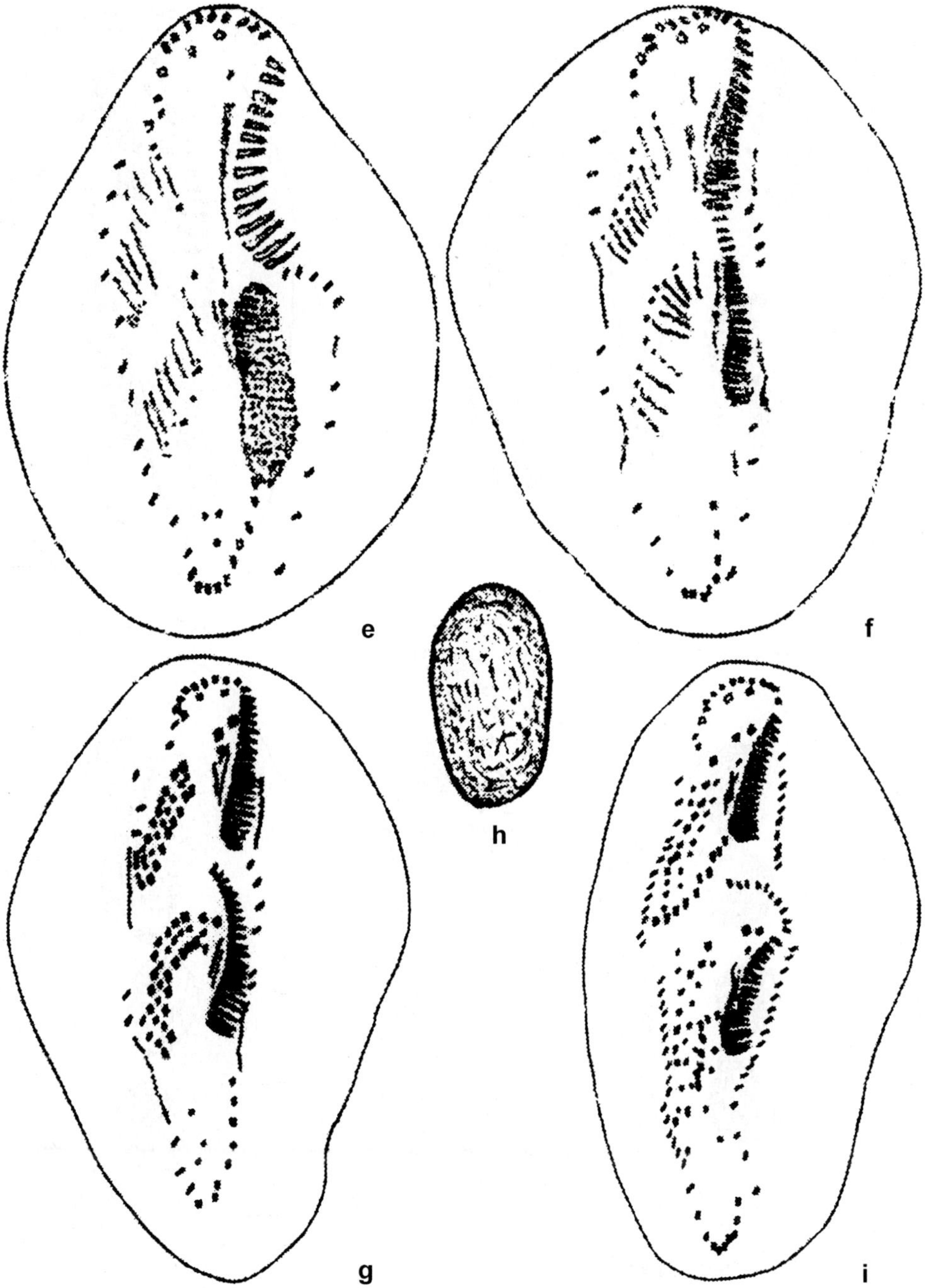

Fig. 25e–i *Holosticha diademata* (from Hu et al. 2000. Protargol impregnation). For general comment, see Fig. 25a–d. Infraciliature of ventral side of middle to late dividers, e = 72 μm, f = 67 μm, g = 104 μm, i = 88 μm. The fused macronucleus likely belongs to (i). Note that the proximal portion of the parental adoral zone shows distinct signs of reorganisation. Page 115.

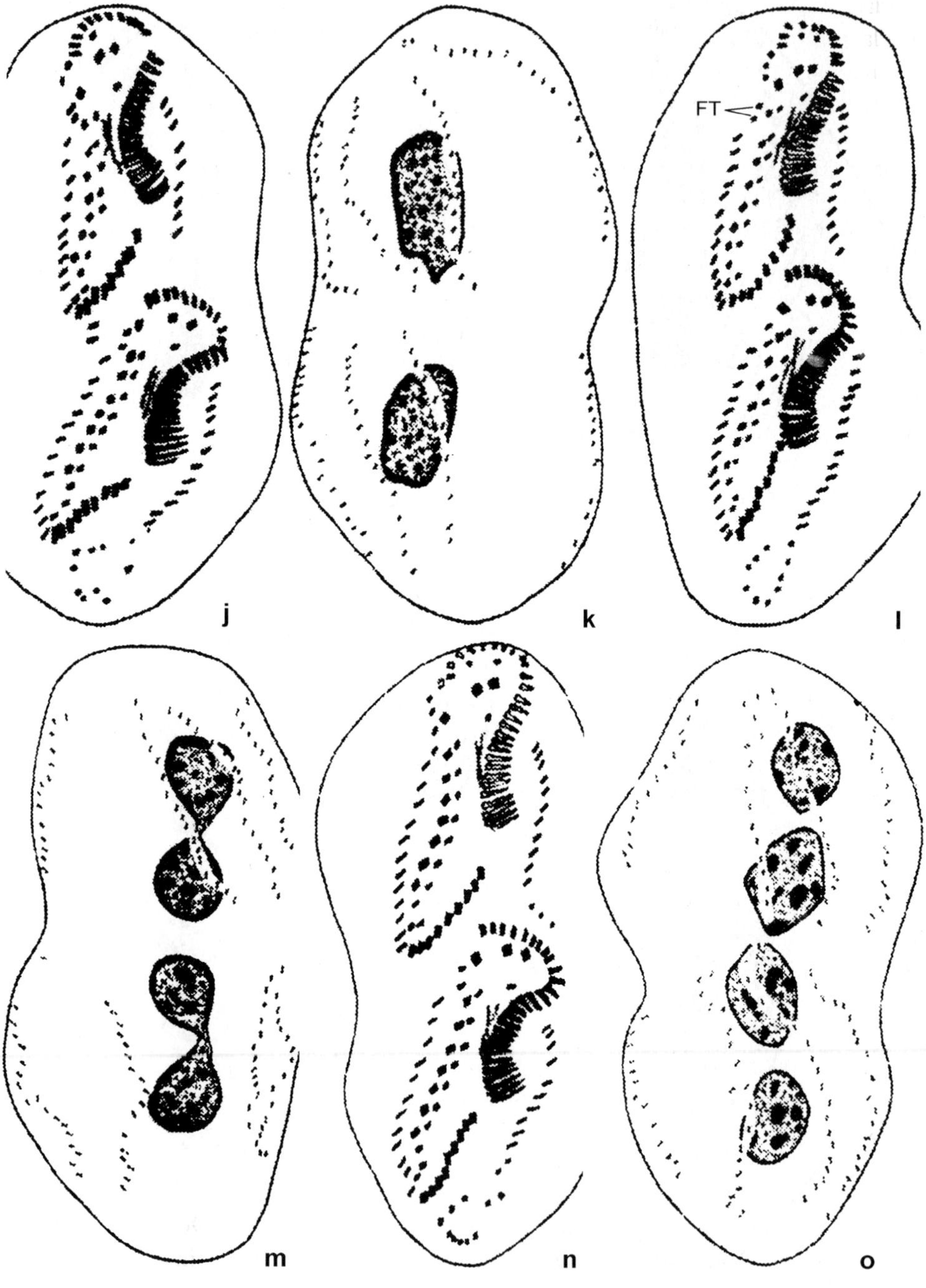

Fig. 25j–o *Holosticha diademata* (from Hu et al. 2000. Protargol impregnation). Infraciliature of ventral and dorsal side and nuclear apparatus of late and very late dividers, j, k = 87 µm, l, m = 110 µm, n, o = 126 µm. FT = new frontoterminal cirri of proter. Page 115.

Qingdao, Yellow Sea (Hu & Song 1999; Hu et al. 2000); saline lake in Saskatchevan, Canada (Wilbert 1986); USA? (Borror & Wicklow 1983); sediment (0–2 cm) of a marine grass community of *Thalassia testudinum* from Enmedio Island, Veracruz, Mexico (Aladro Lubel 1985; Aladro Lubel et al. 1986, p. 240; 1987, p. 437; 1990; Mayén Estrada et al. 1987, p. 74); planktonic in South Atlantic (Petz 1999); rock pool and littoral of Potter Cove, King George Island, Antarctica (Song & Wilbert 2002).

Records of *H. diademata* from marine habitats not substantiated by morphological data: harbour of Ostend, Belgium (Persoone 1968, p. 187); marine sediments polluted by effluents of pulp and paper mill on the west coast of Scotland (Wyatt & Pearson 1982, p. 301); seaport (Königshafen) near List on the German island of Sylt, North Sea (Küsters 1974, p. 174); periphyton of the brackish-water region of the Elbe estuary, Germany (Knüpling 1979, p. 277); Schlei, a brackish water near the north German city of Kiel (Bock 1960, p. 63; Jaeckel 1962, p. 13); Italy (Dini et al. 1995, p. 70); at sites with reduced salinity (16–24‰) and with normal salinity in the White Sea, Russia (Burkovsky 1970a, p. 190; 1970b, p. 11; 1970c, p. 56; 1971, p. 1570; 1971a, p. 1774; 1976, p. 288); polluted and unpolluted areas of the White Sea, Russia (Azovsky et al. 1996, p. 30); Barents Sea (Azovsky 1996, p. 6); periphyton of Yellow Sea, China (Song & Wang 1993, p. 43); Gulf of Mexico and New Hampshire tidal marshes (Borror 1962, p. 342; 1972a, p. 63); abundant in samples collected in the Bay of Rio de Janeiro, Brazil (Faria et al. 1922, p. 113).

Records of *H. diademata* from saline inland waters not substantiated by morphological data (confusion with *H. pullaster* cannot be excluded): salt polluted drainage ditch system west of the Bad Waldliesborn district of the German city of Lippstadt (Mihailowitsch 1989, p. 165); salt-loaded (up to 6.6 g l^{-1} Cl^{-}) running waters in Germany (Albrecht 1986; p. 203); in saline habitats near the Black Sea in Bulgaria at about 24°C and 17.5‰ salinity (Detcheva 1980, p. 34; further records from similar habitats: Detcheva 1982, p. 249; 1983, p. 72); saline lake in Romania (Tucolesco 1962a, p. 813; 1965, p. 160); brackish marsh on the north shore of Lake Pontchartrain, near Sidell, Louisiana, USA (Elliott & Bamforth 1975, p. 516); Lake Qarun, a salt lake (24.8‰) in the Fayum Oasis, Egypt (Wilbert 1995, p. 282); saline lakes in Australia (Wilbert 1995, p. 283).

Records of *H. thiophaga* largely not substantiated by morphological data: Laguna de Términos, Campeche, Gulf of Mexico (Lopez-Ochoterena et al. 1976; Mayén Estrada et al. 1987, p. 74; Aladro-Lubel et al. 1988, p. 437); sediment cores with *Thiovolum* and other bacteria from Nivå Bay, Øresund, Denmark (Bernard & Fenchel 1995, p. 176; as *Amphisiella thiophaga*); west coast of Caspian Sea (Agamaliev 1971, p. 383).

The record of *H. diademata* by Tirjaková (1988, p. 499; Matis et al. 1996, p. 12) from agricultural soils is certainly a misidentification.

Feeds on bacteria and other microflora (Borror 1963a, Fenchel 1968, p. 116). *Holosticha diademata* itself is the main (exclusive?) food of the suctorid ciliate *Lecanophrya drosera* Kahl, 1934 (p. 199). The synonym *H. thiophaga* ingested exclusively the sulphur bacterium *Thiovolum* (Kahl 1928).

Holosticha pullaster (Müller, 1773) Foissner, Blatterer, Berger & Kohmann, 1991

(Fig. 26a–z, 27a, 28a–i, 29a–d, Tables 12–14)

1773 ***Trichoda pullaster***[1] – Müller, Vermium Terrestrium et Fluviatilium, p. 81 (original description with Latin diagnosis, but without illustration; no type material available).

1776 ***Trichoda pullaster*** – Müller, Zoologiœ Danicœ, p. 208 (list of animals found in Denmark; no illustration).

1786 ***Kerona pullaster***[2] – Müller, Animalcula Infusoria, p. 241, Tab. XXXIII, Fig. 21–23 (Fig. 26a–c; combination with *Kerona* Müller, 1786; redescription with illustration).

1790 ***Trichoda pullaster*** – Gmelin, Systema Naturae, p. 3888 (catalogue).

1824 ***Oxitricha pullaster*** – Bory de Saint-Vincent in Lamouroux, Bory de Saint-Vincent & Deslongchamps, p. 595 (combination with *Oxitricha* Bory de Saint-Vincent in Lamouroux, Bory de Saint-Vincent & Deslongchamps, 1824; no illustration).

1838 ***Oxytricha pullaster***[3] – Ehrenberg, Infusionsthierchen, p. 366, Tafel XLI, Fig. III (Fig. 26d–g; redescription).

1850 ***Oxytricha pullaster* Bory** – Diesing, Systema Helminthum, p. 159 (review; incorrect author).

1862 ***Oxytricha micans*** – Engelmann, Z. wiss. Zool., 11: 387 (original description of synonym; no illustration and no formal diagnosis provided; no type material available).

1876 ***Oxytricha alba*** – Fromentel, Microzoaires, p. 268, Planche XIII, Fig. 16 (Fig. 26h; original description of synonym; no type material available and no formal diagnosis provided; in the legend to the figure erroneously named *Oxytricha leucoa*).

1877 ***Holosticha micans*** – Wrześniowski, Z. wiss. Zool., 29: 278 (combination of synonym with *Holosticha*).

1878 ***Amphisia multiseta*** – Sterki, Z. wiss. Zool., 31: 57 (original description of synonym; no illustration and type material available and no formal diagnosis provided).

1878 ***Amphisia micans*** – Sterki, Z. wiss. Zool., 31: 57 (combination of synonym with *Amphisia*; see remarks).

1905 ***Holosticha* sp. (?)** – Conn, Bull. Conn. St. geol. nat. Hist. Surv., 2: 60, Fig. 242 (Fig. 27a; illustrated record).

1906 ***Amphisia* Núm. 2** – Izquierdo, Protozoos, p. 188, Lám. XII, Fig. 476–483 (Fig. 37e, f; see remarks).

1926 ***Holosticha* sp.** – Lepsi, Arch. Hydrobiol., 17: 755 (record from Romania).

1932 ***Holosticha simplicis* sp. nov.** – Wang & Nie, Contr. biol. Lab. Sci. Soc. China, 8: 352, Fig. 62 (Fig. 29d; original description of synonym; no type material available and no formal diagnosis provided).

1957 ***Holosticha kessleri* Wrzesniowski var. *aquae-dulcis* var. n.** – Buchar, Cas. národ. Mus., 126: 139, Fig. 2D (Fig. 26i; original description of synonym; no type material available and no formal diagnosis provided).

1958 ***Keronopsis litoralis* n. sp.** – Gellért & Tamás, Annls Inst. biol. Tihany, 25: 228, Fig. 6 (Fig. 26j; original description of synonym; likely no type material available and no formal diagnosis provided).

1960 ***Holosticha danubialis* sp. n.**[4] – Kaltenbach, Wass. Abwass. Wien, 1960: 167, Abb. 3d (Fig. 26k; original description of synonym; likely no type material available).

1962 ***Holosticha retrovacuolata* n. sp.** – Tucolesco, Annls Spéléol., 17: 105, Fig. 32 (Fig. 26l; original description of synonym; likely no type material available and no formal diagnosis provided).

---

[1] The diagnosis by Müller (1773) is as follows: *Trichoda* ovata, antice sinuata, fronte cristita, basi crinita.

[2] The diagnosis by Müller (1786) is as follows: *Kerona* subovata, antice sinuata, fronte corniculata, basi crinita.

[3] The characterisation by Ehrenberg (1838) is as follows: *Oxytricha* corpore albicante, lanceolato, utrinque obtuso, ventre medio nudo, capite aliqunatum dicreto caudaque hirtis, oris rima angusta.

[4] The diagnosis by Kaltenbach (1960) is as follows: Kleinere Form mit verhältnismäßig stark entwickeltem Peristomfeld, 1/3 bis 2/5 körperlang. Frontalmembranellen nicht dicht stehend, adorale Zone kurz bewimpert. 2 Kernsegmente. Kontraktile Vakuole dem Hinterende genähert. 5 Transversalcirren von der Ventralreihe getrennt.

1963 ***Holosticha rhomboedrica* n. sp.** – Vuxanovici, Studii Cerc. Biol., 15: 206, Plansa III, Fig. 15 (Fig. 26m–s; original description of synonym; likely no type material available and no formal diagnosis provided; distinguished two forms, see next entries).

1963 ***Holosticha rhomboedrica* forma *eliptica*** – Vuxanovici, Studii Cerc. Biol., 15: 207, Plansa III, Fig. 15A, 15a (Fig. 26n, o; original description of form; likely no type material available and no formal diagnosis provided).

1963 ***Holosticha rhomboedrica* forma *lata*** – Vuxanovici, Studii Cerc. Biol., 15: 207, Plansa III, Fig. 15b (Fig. 26p; original description of form; likely no type material available and no formal diagnosis provided).

1963 ***Holosticha coronata* n. sp.** – Vuxanovici, Studii Cerc. Biol., 15: 205, Plansa II, Fig. 11 (Fig. 26z; original description of synonym; likely no type material available and no formal diagnosis provided).

1963 ***Holosticha minima* n. sp.** – Vuxanovici, Studii Cerc. Biol., 15: 205, Plansa II, Fig. 13 (Fig. 26t; original description of synonym; likely no type material available and no formal diagnosis provided).

1963 ***Holosticha rostrata* n. sp.** – Vuxanovici, Studii Cerc. Biol., 15: 205, Plansa III, Fig. 14, 14A (Fig. 26u, v; original description of synonym; likely no type material available and no formal diagnosis provided).

1963 ***Holosticha rostrata* forma *pitica*** – Vuxanovici, Studii Cerc. Biol., 15: 205, Plansa III, Fig. 14A (Fig. 26v; original description of synonym; likely no type material available and no formal diagnosis provided).

1972 ***Keronopsis retrovacuolata* (Tucolesco, 1952) n. comb.** – Borror, J. Protozool., 19: 11 (combination of synonym with *Keronopsis*; incorrect year).

1974 ***Holosticha diademata* (Rees) Kahl** – Pätsch, Arb. Inst. landw. Zool. Bienenk., 1: 56, Abb. 45 (Fig. 26w; misidentification).

1974 ***Holosticha rostrata* Vuxanovici, 1963 var. *mononucleata* n. n.** – Stiller, Annls hist.-nat. Mus. natn. hung., 66: 132 (new name for the monomacronucleate form of *H. rostrata*, Fig. 26v; see nomenclature).

1980 ***Holosticha retrovacuolata* Tucolesco, 1962** – Foissner, Ber. Nat.-Med. Ver. Salzburg, 5: 104, Fig. 24a, b, 56 (Fig. 26x, y; redescription of synonym from life).

1982 ***Holosticha diademata* (Rees 1884) Kahl 1930** – Bernerth, Cour. Forsch.-Inst. Senckenberg, 57: 193, Abb. 100 (misidentification; micrograph-documented record from cooling system of power station).

1982 ***Holosticha diademata* (Rees, 1883) Kahl, 1932** – Hemberger, Dissertation, p. 85, Abb. 12a–e (Fig. 28a–e; misidentification; cell division).

1983 ***Pseudokeronopsis retrovacuolata* (Tucolesco, 1962) nov. comb.** – Borror & Wicklow, Acta Protozool., 22: 124 (combination of synonym with *Pseudokeronopsis*).

1987 ***Holosticha danubialis* Kaltenbach, 1960** – Foissner, Arch. Protistenk., 133: 229, Abb. 1–5 (reactivation of the forgotten species *H. danubialis* and discussion of synonymy).

1991 ***Holosticha pullaster* (Mueller, 1773) nov. comb.** – Foissner, Blatterer, Berger & Kohmann, Informationsberichte des Bayer. Landesamtes für Wasserwirtschaft, 1/91: 240, Tabelle on p. 242, Abb. 1–14 (Fig. 28f–i; taxonomic and ecological monograph; combination with *Holosticha*).

1995 ***Holosticha pullaster* (Mueller, 1773) Foissner et al., 1991** – Petz, Song & Wilbert, Stapfia, 40: 166, Fig. 49a–c, Table 24 (Fig. 29a–c; detailed redescription of marine population; at least 1 voucher slide is deposited in the Oberösterreichische Landesmuseum in Linz [LI], Upper Austria).

2001 ***Holosticha pullaster* (Müller, 1773) Foissner, Blatterer, Berger and Kohmann, 1991** – Berger, Catalogue of ciliate names 1. Hypotrichs, p. 94 (nomenclator containing all basionyms, combinations, and higher taxa of hypotrichs).

2001 ***Holosticha diademata*** – Eigner, J. Euk. Microbiol., 48: 777, Fig. 32 (Fig. 28a modified; brief review on Urostylidae).

2003 ***Holosticha pullaster* (Müller, 1773) Foissner et al., 1991** – Berger, Europ. J. Protistol., 39: 375, 376, Fig. 10 (Fig. 29a; brief review).

2005 ***Holosticha pullaster* (Müller) Foissner, Blatterer, Berger & Kohmann (1991)** – Petz, Ciliates, p. 395, Fig. 14.87a–c (Fig. 29a–c; guide to Antarctic marine ciliates).

Nomenclature: In most cases no derivation of the species-group name is given in the original description. The numerous names are treated chronologically. I do not know the origin and meaning of the species-group name *pullaster*. The species-group name *micans* is likely a composite of the Latin verb *micare* (twitch) and the suffix *~ans* (meaning an activity) and likely refers to fact that this species has a contractile, flexible body making a twitching movement (Engelmann 1862). The species-group name *alb·us -a -um* (Latin adjective; white, bright) likely refers to the white appearance of this species; the name *leucoa* (Latin) used by Fromentel (1876) in the figure legend, has the same meaning. The species-group name *multiseta* (Latin adjective; having many bristles) is a composite of the Latin indefinite numeral *mult-* (many, numerous), the thematic vowel *·i-*, and the Latin noun *seta* (bristle), and likely refers to the many (10) transverse cirri. The species-group name *simplicis* (Latin adjective; simple) alludes to the "simple" organisation of the body (Wang & Nie 1932). The variety name *aquaedulcis* (freshwater) is a composite of the Latin noun *aquae* (water in the sense of a habitat) and the Latin adjective *dulcis* (sweet, delightful) and refers to the habitat (freshwater) where the population was discovered. The species-group name *litorál·is -is -e* (Latin adjective; belonging to the shore) obviously alludes to fact that Gellért & Tamas (1958) discovered this species in the littoral. The species-group name *danubial·is -is -e* (Latin adjective; occurring in the Danube River region) refers to the locus typicus, namely the Danube River. The species-group name *retrovacuolata* is a composite of the Latin prefix *retro+* (dislocated posteriorly) and the Latin *vacuolata* (having a vacuole) and refers to the posteriorly displaced contractile vacuole. The species-group name *rhomboedrica* (Latin; having the shape of a rhombus) refers to the rhomboidal body outline (Vuxanovici 1963; Fig. 26m); the form-name *eliptica*, obviously an incorrect spelling of *elliptic·us -a -um* (Greek adjective; elliptical), refers to the elliptical body outline (Fig. 26n); the form-name *lat·us -a -um* (Latin adjective; wide, extended) also refers to the body outline (Fig. 26p). The species-group name *coronát·us -a -um* (Latin; having a crown) possibly refers to the adoral zone of membranelles. The species-group name *minim·us -a -um* (Latin adjective; smallest; superlative of *párvus*) alludes to the small size (35–40 µm) of this species. I do not know to which feature the species-group name *rostrat·us -a -um* (Latin; having a bill, rostratum) refers. Further, I do not know the origin of the variety name *pitica*.

Tucolesco (1962b) classified *H. retrovacuolata* in the subgenus *Holosticha (Holosticha)*; thus the correct name in the original description is *Holosticha (Holosticha) retrovacuolata*. Borror & Wicklow (1983, p. 124) wrote "*Keronopsis retrovacuolata* Tucolesco, 1962" as basionym, which is incorrect. Matis et al. (1996, p. 12) incorrectly mentioned "Berger (1992)" as combining author of *Holosticha pullaster*.

Holosticha coronata Vuxanovici, 1963 is the junior primary homonym of *H. coronata* Gourret & Roeser, 1888. Since Vuxanovici's species has synonyms, the oldest of these becomes the valid name of the taxon (ICZN 1999, Article 60.2); that is, the name *H. coronata* Vuxanovici must not be replaced by a new name.

Incorrect subsequent spellings: *Holosticha clanabialis* (Song et al. 1993, p. 101); *Holosticha pulaster* (Mueller, 1773) (Senler et al. 1996, p. 186); *Oxytricha pulaster* (Dumas 1929, Legend to Planche XXVII).

Remarks: This is the sole limnetic *Holosticha* species, but simultaneously one of the most common hypotrichs in freshwater and the sea. Likely for the latter reason it has been described 12 times as new species (and at least 5 additional names for varieties and forms have been created!), although it has a very characteristic shape and, more important, an extraordinary position of the easy-to-recognise contractile vacuole, namely in the posterior body portion which makes this species – together with its two macronuclear nodules – unmistakable.

When we wrote the first volume of our guide to the ciliates of the saprobic system (Foissner et al. 1991), we also included this species which was lacking in such important reviews like those by Kahl (1932) and Sládeček (1973). Only some relatively young synonyms, for example, *Holosticha danubialis* and *H. retrovacuolata*, have sometimes been mentioned in the literature. The other freshwater records are usually misidentifications as *H. kessleri* and *H. diademata*, which are confined to marine, respectively, saltwater habitats. Since the present species is so common, we thought that it must have been known to the first protozoologists like O. F. Müller and C. G. Ehrenberg. And indeed we found that this species was already recognisably described by Ehrenberg as a species discovered by Müller (1773), although neither Ehrenberg nor Müller mentioned the most important feature, namely the posteriorly dislocated contractile vacuole. However, the shape, the small size, and the 10 transverse cirri are unmistakable signs that "*Trichoda pullaster*" must be this common hypotrich. Since this species occurs in many freshwater samples, at least in Europe, its further history and its synonymy are discussed in detail.

Dujardin (1841, p. 421) and Claparède & Lachmann (1858, p. 149) mentioned *H. pullaster*, but did not provide new data. Surprisingly, Stein did not describe this common species in his 1859-monograph.

The first subjective synonym of *H. pullaster* is *Oxytricha micans* Engelmann, which is only briefly described in a footnote, but not illustrated (Engelmann 1862). However, the features provided (often in community with *Tachysoma pellionellum*; 8–10 transverse cirri with rearmost strongest; body very flexible and contractile) clearly indicate synonymy with *H. pullaster*, as already suggested by Foissner et al. (1991). Kahl (1932) and other workers – for example, Borror (1972) – did not mention this synonym. Sterki (1878) transferred it, although not formally, to *Amphisia*.

Oxytricha alba Fromentel is rather easy to assign to *H. pullaster* because its description is the first one where the posteriorly dislocated contractile vacuole is mentioned and illustrated (Fig. 26h; Foissner et al. 1991). Fromentel (1876) simultaneously redescribed *H. pullaster* (Fig. 191g, i, n). However, this redescription is too superficial to accept the identification (see below).

Amphisia multiseta Sterki is not illustrated, but the important features (common; similarity of cirral pattern with that of *H. gibba*; 10 transverse cirri; posteriorly dislocated contractile vacuole) are clearly described by Sterki (1878). Kahl (1932, p. 570), who considered it as nomen nudum, erroneously assumed that this is the type species of *Amphisia* Sterki, the junior synonym of *Holosticha*. Synonymy of Sterki's species and *H. pullaster* was already proposed in the Ciliate Atlas (Foissner et al. 1991). *Amphisia multiseta* sensu Milne (1886) is classified as synonym of *Anteholosticha oculata* (Fig. 95d).

Schewiakoff (1893, p. 67) briefly redescribed "*Amphisia kessleri*". The specimens of his limnetic population had the contractile vacuole in the posterior body portion, proving that he had observed *H. pullaster*.

Holosticha sp. sensu Conn (1905, Fig. 27a) is certainly *H. pullaster* as indicated by body size (62 µm), body shape, and ciliature. *Amphisia* Núm. 2 in Izquierdo (1906) is, at least partially (Fig. 37e, f), identical with *H. pullaster* as indicated by the posteriorly dislocated contractile vacuole, the body size (80 × 32 µm), the body shape, and the cirral pattern. The other figures of Izquierdo (1906; Fig. 37g–l) very likely show another (indeterminable) species. Lepsi (1926a) designated the present species as *Holosticha* sp. without giving an illustration. However, since he mentioned the most important feature, that is, the post-equatorial contractile vacuole, the identification as *H. pullaster* is beyond reasonable doubt.

Kahl (1932) supposed that *H. simplicis* is a synonym of *H. diademata*. I agree with Kahl (1932) that Wang & Nie (1932) overlooked the enlarged frontal cirri. However, because of the posteriorly dislocated contractile vacuole, synonymy with *H. pullaster* is much more likely. Further, we have to assume that Wang & Nie (1932) overlooked the gap in the adoral zone, a feature which is rather difficult to recognise in this small species. As already mentioned above, Kahl (1932) did not mention *H. pullaster* with its unmistakable contractile vacuole in his review, which was the main guide to ciliates for many decades. Probably the reason so many synonyms were created in the second half of the twentieth century.

The first post-Kahlian synonym is that produced by Buchar (1957), who classified it as variety of *Holosticha kessleri*. It was synonymised with *H. pullaster* by Foissner et al. (1991).

The synonymy of *Keronopsis litoralis* Gellért & Tamás and *H. pullaster* was also already proposed by Foissner et al. (1991). Although some important features (contractile vacuole, frontal cirri) are lacking, synonymy is indicated by the small size (70 µm) and the prominent transverse cirri (Fig. 26j).

Holosticha danubialis was described by Kaltenbach (1960) as "similar to the marine *H. kessleri*, but smaller and more stocky" (Fig. 26k). Although the illustration is not quite perfect, synonymy with *H. pullaster* is beyond reasonable doubt.

Holosticha retrovacuolata Tucolesco is rather well described from life and shows all important features very well (Fig. 26l). Borror (1972) transferred it to *Keronopsis* and Borror & Wicklow (1983) to *Pseudokeronopsis*. Both combinations are hardly comprehensible because Tucolesco (1962) wrote that the cirral pattern is *Holosticha (Holosticha)*-like, that is, three distinctly enlarged frontal cirri are present (Kahl 1932). Foissner (1980a) redescribed *H. retrovacuolata* from life and considered three species described by Vuxanovici (1963), namely *H. coronata*, *H. rostrata*, and *H. rhomboedrica*, as synonyms (see below). Later he synonymised *H. retrovacuolata* and *H. rhomboedrica* with *H. danubialis* (Foissner 1987d). In 1991, we put all these species into the synonymy of *H. pullaster* (Foissner et al. 1991), and there is no reasonable doubt.

Holosticha coronata Vuxanovici is a rather small species (45–50 µm) with a posteriorly dislocated contractile vacuole and a long adoral zone indicating that it is a post-divider of *Holosticha pullaster*. Synonymy of *H. coronata* and *H. pullaster* (actually *H.*

danubialis) was first proposed by Foissner (1987d) and this act is also beyond reasonable doubt.

Holosticha minima Vuxanovici was synonymised with *H. diademata* by Hemberger (1982) who, however, assumed that in *H. diademata* the contractile vacuole is dislocated posteriorly (see below). Size, position of contractile vacuole, and transverse cirri strongly indicate that *H. minima* is indeed not a distinct species, but a tiny specimen of *H. pullaster*.

Holosticha rostrata Vuxanovici was synonymised with *H. retrovacuolata* by Foissner (1980a, p. 105). I agree with this proposal, that is, synonymy with *H. pullaster*, because size, position of contractile vacuole, and prominent transverse cirri agree very well with the situation in *H. pullaster*. For the mono-macronucleate form, Vuxanovici (1963) proposed the name *H. rostrata* forma *pitica*. Obviously, this was overlooked by Stiller (1974a), who introduced the name *H. rostrata* var. *mononucleata* for this specimen/population. I assume that this is a postconjugate, postdivider, or malformed specimens because of the single macronucleus.

Pätsch (1974) provided the first illustration of a protargol-impregnated specimen of the present species (misidentified as *H. diademata*). It shows all features occurring also in the other *Holosticha* species, namely, gap in adoral zone, proximal-most adoral membranelles widest, anterior end of left marginal row curved rightwards, J-shaped transverse cirral row. Surprisingly, the total number of adoral membranelles is rather high in Pätsch's population, namely 28, whereas other freshwater populations and the marine population described by Petz et al. (1995) usually have less than 20 membranelles. Perhaps Pätsch did not use a camera lucida and overestimated the number of membranelles.

Hemberger (1982) also misidentified the present species as *H. diademata*. He did not consider (i) that Rees (1884) did not describe the position of the contractile vacuole and (ii) that Kahl (1932) illustrated this organelle in mid-body in his figure of *H. diademata* (Fig. 24b). Consequently, Hemberger confused the synonyms of *H. diademata* and *H. pullaster*.

In 1991 we reactivated *H. pullaster*, which was distinctly underrepresented in faunal lists. Since then the number of records of this common species has increased significantly, showing clearly that it is worth making a good guide (Foissner et al. 1991).

Al-Rasheid (1996a, p. 198) provided a small micrograph (his Fig. 4i), which, however, does not show any details. The outline is rather wide and no contractile vacuole is recognisable. Thus, it remains uncertain whether or not he found *H. pullaster* or another species.

The redescriptions of *H. pullaster* by Fromentel (1876) and Dumas (1929) and the redescription of *Oxytricha alba* by Dumas (1929) are insufficient.

Since the synonymy above is beyond reasonable doubt I summarise the morphological data in a single description. I only keep the descriptions of marine and freshwater populations separate because Petz et al. (1995) stated that limnetic and marine populations differ physiologically (transfer from saltwater to freshwater is seemingly impossible) although they are morphologically inseparable (see ecology). Since gene transfer is thus prevented between marine and limnetic populations (except via populations in es-

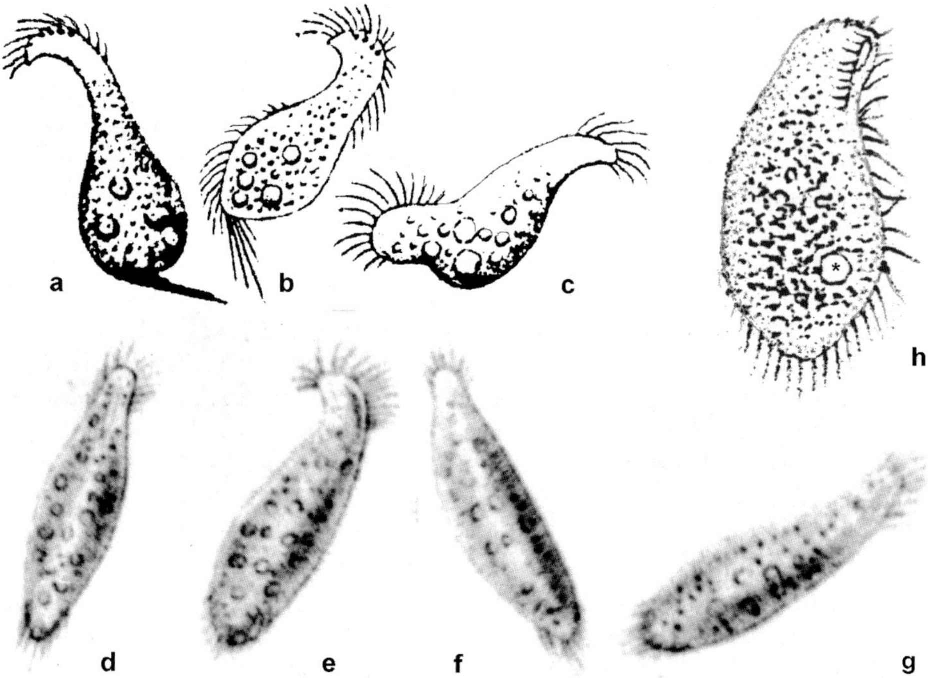

Fig. 26a–h *Holosticha pullaster* (a–c, from Müller 1786; d–g, from Ehrenberg 1838; h, from Fromentel 1876. a–h, from life). Dorsal and ventral views, d–g = 57 μm, h = about 70 μm. Asterisk in (h) marks posteriorly dislocated contractile vacuole in the synonym *Oxytricha alba*. Page 128.

tuaries) we have to assume a species separation process or the presence of sibling species.

In freshwater habitats *Holosticha pullaster* is unmistakable because of the posteriorly dislocated contractile vacuole. The 18-cirri oxytrichids *Tachysoma pellionellum* and species of the *Oxytricha setigera* group, which have a similar shape and size, have the contractile vacuole in mid-body, long (>8 μm) dorsal bristles, and a single micronucleus between the two macronuclear nodules, which are in the left body portion (for review, see Berger 1999). The posteriorly dislocated contractile vacuole is also the most important feature for the separation of marine *H. pullaster* populations from *H. diademata*, which has this organelle in mid-body. *Holosticha foissneri* an *H. heterofoissneri*, which have the contractile vacuole also distinctly behind mid-body, have 5–11, respectively, 14–24 macronuclear nodules (Fig. 31a, 32g–i).

Morphology: Several freshwater populations have been described. Thus many features show a rather high variability. Body size ranges from 30–100 × 20–50 μm, usually, however, *H. pullaster* is only 50–70 μm long; individual values provided are: body length 30–35 μm (Vuxanovici 1963, for *H. rostrata pitica*); 35–40 μm (Vuxanovici 1963, for *H. minima*); 45–50 μm (Vuxanovici 1963, for *H. coronata*); 45–55 μm (Vuxanovici 1963, for *H. rostrata*); 57 μm (Ehrenberg 1838); 60–80 μm (Buchar 1957);

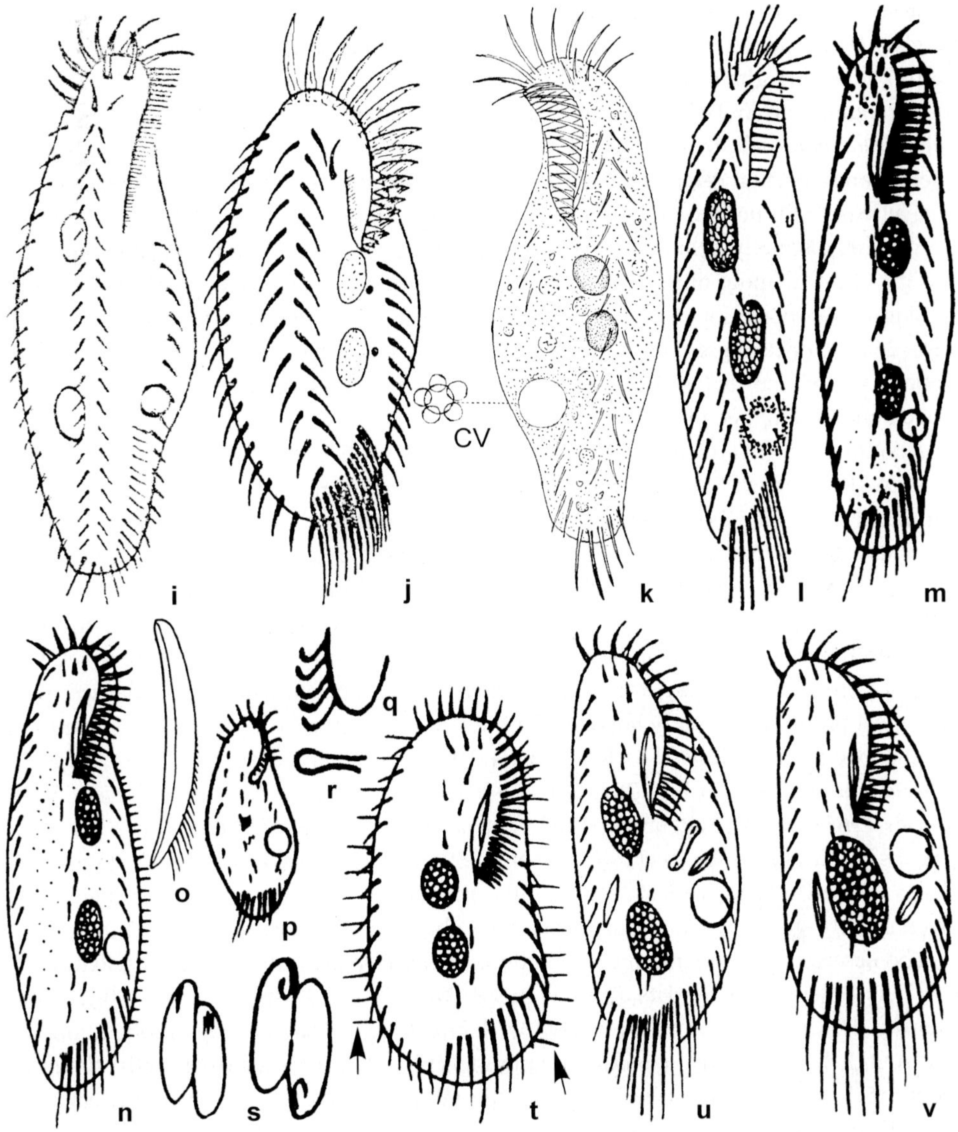

Fig. 26i–v *Holosticha pullaster* (i, from Buchar 1957; j, from Gellért & Tamás 1958; k, from Kaltenbach 1960; l, from Tucolesco 1962; m–v, from Vuxanovici 1963. i, k–v, from life; j, Bresslau-opalblue preparation). *Holosticha pullaster* has a huge number of synonyms. **i:** *Holosticha kessleri aquadulcis*, 80 μm. **j:** *Keronopsis litoralis*, 70 μm. **k:** *Holosticha danubialis*, 80 μm. Ventral cirral pattern as seen from dorsal. The contractile vacuole is formed by several small vesicles during diastole. **l:** *Holosticha retrovacuolata*, 90 μm. **m–s:** *Holosticha rhomboedrica*, m = 75 μm, n = 30 μm (forma *eliptica*), p = size not indicated (forma *lata*). (o) is a lateral view of forma *eliptica*. (q, r) show transverse cirri in lateral and ventral view? (s) shows conjugation. **t:** *Holosticha minima*, 35 μm. The dorsal bristles (arrows) appear rather long; however, they have a length of only about 3 μm (estimated via body length). **u, v:** *Holosticha rostrata*, u = 52 μm, v (forma *pitica*) = 30 μm. The monomacronucleate specimen is either a postdivider or postconjugate. CV = contractile vacuole. Page 128.

about 70 µm (Fromentel 1876, Gellért & Tamás 1958); 50–80 µm, usually 70 µm (for *H. rhomboedrica*, Vuxanovici 1963); 70–90 µm, contracted about 60 µm (Foissner 1980a); 80 µm (Kaltenbach 1960); 80–90 µm (Heuss 1976); 90 µm (Tucolesco 1962b); 70 × 20 µm (Pätsch 1974); 70–100 × 20–30 µm (Bernerth 1982). Body outline typically *Holosticha*-like, that is, narrow to wide spindle-shaped; widest about in mid-body, anterior body portion often slightly curved leftwards. Invariably two macronuclear nodules, at least front nodule, but usually both nodules, right of midline (Fig. 26i). Contractile vacuole distinctly behind mid-body (Fig. 26h–n, p, t–x, z, 29a), in specimens shown in Figs. 28f–i at about 60% of body length, according to Tucolesco (1962b) at the beginning of the posterior cell quarter (Fig. 26l); vacuole originates from small vesicles (Fig. 26k). Cortical granules lacking. Cytoplasm colourless, with moderately many, globular, yellowish inclusions about 5 µm across and food vacuoles (Foissner 1980a). Movement without peculiarities.

Adoral zone of membranelles occupies about 25–33% (Tucolesco 1962b) to 33–40% (Kaltenbach 1960) of body length; for details, see marine populations below; composed of about 17 (Gellért & Tamás 1958) to 20 (Fig. 28a) membranelles; population illustrated by Pätsch (1974, Fig. 26w) with 9 distal and 19 proximal membranelles, that is, in total 28, which is a rather high number. Cirral pattern as described below; midventral complex composed of about 8–10 cirral pairs (Fig. 26w, 28a), right cirrus of midventral pairs usually composed of three, left one of only two basal body rows, that is, right cirrus distinctly larger than left. Transverse cirri prominent, bases enlarged, posteriormost protrude distinctly beyond rear body end; number varies from 5 to 12 with individual values as follows: 5 (Kaltenbach 1960, likely underestimated); 6–7 (for *H. coronata*, 6 in text, 7 in Fig. 26z; Vuxanovici 1963); 7–8 (for *H. rhomboedrica*; Vuxanovici 1963); 8–10 (Engelmann 1862; Vuxanovici 1963, for *H. minima*); 10 (Ehrenberg 1838, Sterki 1878); 9–11 (Foissner 1980a); 9–12 (Pätsch 1974); 12 (Gellért & Tamás 1958). Anterior end of left marginal row (three cirri) transversely arranged and cirri narrowly spaced, each marginal row with about 15 cirri (Foissner 1980a).

The description of the marine populations is based mainly on the paper by Petz et al. (1995), supplemented by data from Wang & Nie (1932). In life 50–70 × 20–26 µm; synonym *H. simplicis* 68 × 22 µm (Wang & Nie 1932). Body outline elliptical, anteriorly rounded, posteriorly sometimes tapering (Fig. 29a, d); dorso-ventrally flattened about 1.5:1; somewhat flexible and retractile (Wang & Nie 1932). Macronuclear nodules ellipsoidal, about 10 × 5 µm, in centre or in right body half, with several globular nucleoli up to 4 µm across. Micronuclei spherical, 2 µm in diameter, each one in indentation of a maconuclear nodule; usually not impregnated with protargol. Contractile vacuole near left margin, behind mid-body (at 68% of body length in specimens shown in Fig. 29a, d), with inconspicuous collecting canals, pulsation interval about 15 min (in marine population)! Cytoplasm hyaline, contains small, greasily shining globules and food vacuoles, which contain, inter alia, unidentified greenish material. Movement slowly crawling on substrate, rests for long periods, sometimes jerking back and forth short distances; thigmotactic, that is, not easily removable from solid surface with pipette.

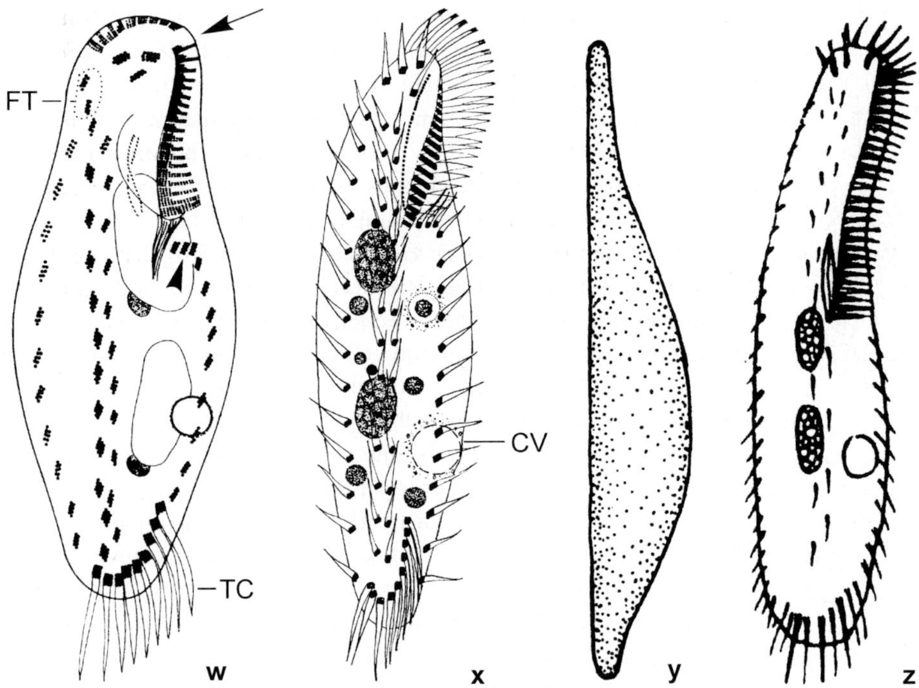

Fig. 26w–z *Holosticha pullaster* (w, from Pätsch 1974; x, y, from Foissner 1980a; z, from Vuxanovici 1963. w, protargol impregnation; x–z, from life). **w:** Infraciliature of ventral side and nuclear apparatus, 78 µm. Arrow marks gap in adoral zone, arrowhead denotes rightwardly curved anterior end of left marginal row. Very likely the right frontal cirrus is lacking (or overlooked) in this specimen. **x, y:** Ventral view and left lateral view, 88 µm. **z:** Ventral view of synonym *Holosticha coronata*, 50 µm. CV = contractile vacuole, FT = frontoterminal cirri, TC = transverse cirri. Page 128.

Adoral zone occupies about 38% of body length, bipartite, with 3–8 (usually 6) membranelles in distal portion and 7–16 membranelles in proximal portion; gap between portions 3–5 µm wide; bases of membranelles 3–7 µm wide, gradually lengthened posteriad, cilia 11–13 µm long. Buccal area rather narrow (Wang & Nie 1932). Undulating membranes equally short (only about 6 µm in Fig. 29b), paroral optically slightly crosses endoral. Pharyngeal fibres 8–13 µm long.

Cirral pattern and number of cirri of usual variability (Table 13). Invariably three slightly enlarged frontal cirri and a single buccal cirrus distinctly ahead of undulating membranes. Frontoterminal cirri, as is usual, between distal end of adoral zone and anterior end of right marginal row. Midventral complex composed of around five pseudopairs (Fig. 29b). Obviously two pretransverse ventral cirri present (for explanation of cirral pattern, see Fig. 29b). Transverse cirri arranged in J-shaped row, 10–17 µm long, project by about half of their length beyond body margin; synonym *H. simplicis* with 5–6 transverse cirri (number likely underestimated). Marginal rows separated posteriorly, cirri 8–12 µm long, right row commences at level of undulating membranes, terminates

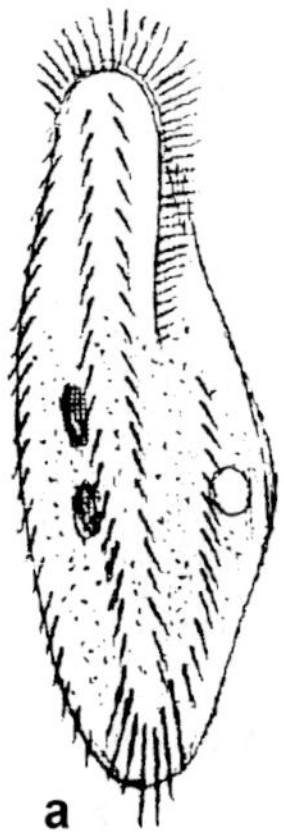

Fig. 27a *Holosticha pullaster* from life (a, from Conn 1905). Ventral view (62 µm) showing ciliature, nuclear apparatus right of midline, and contractile vacuole behind mid-body. Page 128.

near rightmost transverse cirrus. Left row begins *Holosticha*-like, that is, with anterior end (anteriormost three cirri) distinctly curved rightwards, ends slightly subterminally (Fig. 29a, b).

Dorsal cilia 2–3 µm long, arranged in usually four, rarely five kineties (Fig. 29c, Table 13). Kinety 1 usually slightly shortened anteriorly, others more or less bipolar, kineties 2 and 3 composed of 9–11, others of 6–9 dikinetids. Caudal cirri lacking (Fig. 29c).

Cell division (Fig. 28a–e): Hemberger (1982) provided five stages, which show that morphogenesis proceeds basically as in congeners. It commences with the formation of an oral primordium left of the midventral complex about in mid-body (Fig. 28a). Somewhat later the right cirri of the middle midventral pairs differentiate to cirral anlagen (Fig. 28b). In the posterior portion of the right marginal row the anlage of the opisthe's right marginal row begins to form. Ahead of the undulating membranes, a small anlage is recognisable.

The next stage shows an distinctly enlarged oral primordium from which the primordium for the undulating membranes begins to split off (Fig. 28c). The right cirri and obviously some left cirri of midventral pairs right of the parental adoral zone and right of the oral primordium have modified to frontal-midventral-transverse cirral anlagen. The parental undulating membranes (only paroral?) have modified to anlage I of the proter. Left of the oral primordium one or two left marginal cirri have formed the anlage for the left marginal row of the opisthe.

In a late stage all anlagen for the frontal-midventral-transverse cirral anlagen are recognisable (Fig. 28d). Usually 9–10 anlagen (obviously without anlagen I and II) each are formed for both filial products. All these anlagen form three cirri because each anlage produces a transverse cirrus and a midventral pair. Anlage II likely formed de novo, although Hemberger could not clarify its origin completely; it forms, as is usual, the middle frontal cirrus and the buccal cirrus (= cirrus II/2), which is, however, not right of the undulating membranes in non-dividers, but distinctly displaced anteriorly. The parental adoral zone of *H. pullaster* does not change and thus forms the adoral zone of the proter.

A very late stage shows the migration of the cirri to their final position (Fig. 28e). The most interesting feature is that the two frontoterminal cirri originate from the two rightmost anlagen, obviously as in *Uroleptus caudatus* (Eigner 2001) and *Bakuella edaphoni* (Song et al. 1992). Usually – for example, in *H. diademata* and *H. heterofoissneri* – the frontoterminal cirri are the two anteriormost cirri of the rightmost anlage. Thus, this feature has to be re-examined in these species.

The formation of the marginal rows shows one peculiarity, namely the anlage for the left row of the proter originates de novo (Fig. 28d). This feature agrees with the conditions described for *H. heterofoissneri* (Fig. 33e) and *H. diademata* (Fig. 25f), indicating that this is an apomorphy of *Holosticha*. According to Hemberger (1982) a dorsomarginal kinety is formed; this is, however, not substantiated by an illustration. Division of the nuclear apparatus proceeds in ordinary manner (Fig. 28a–e).

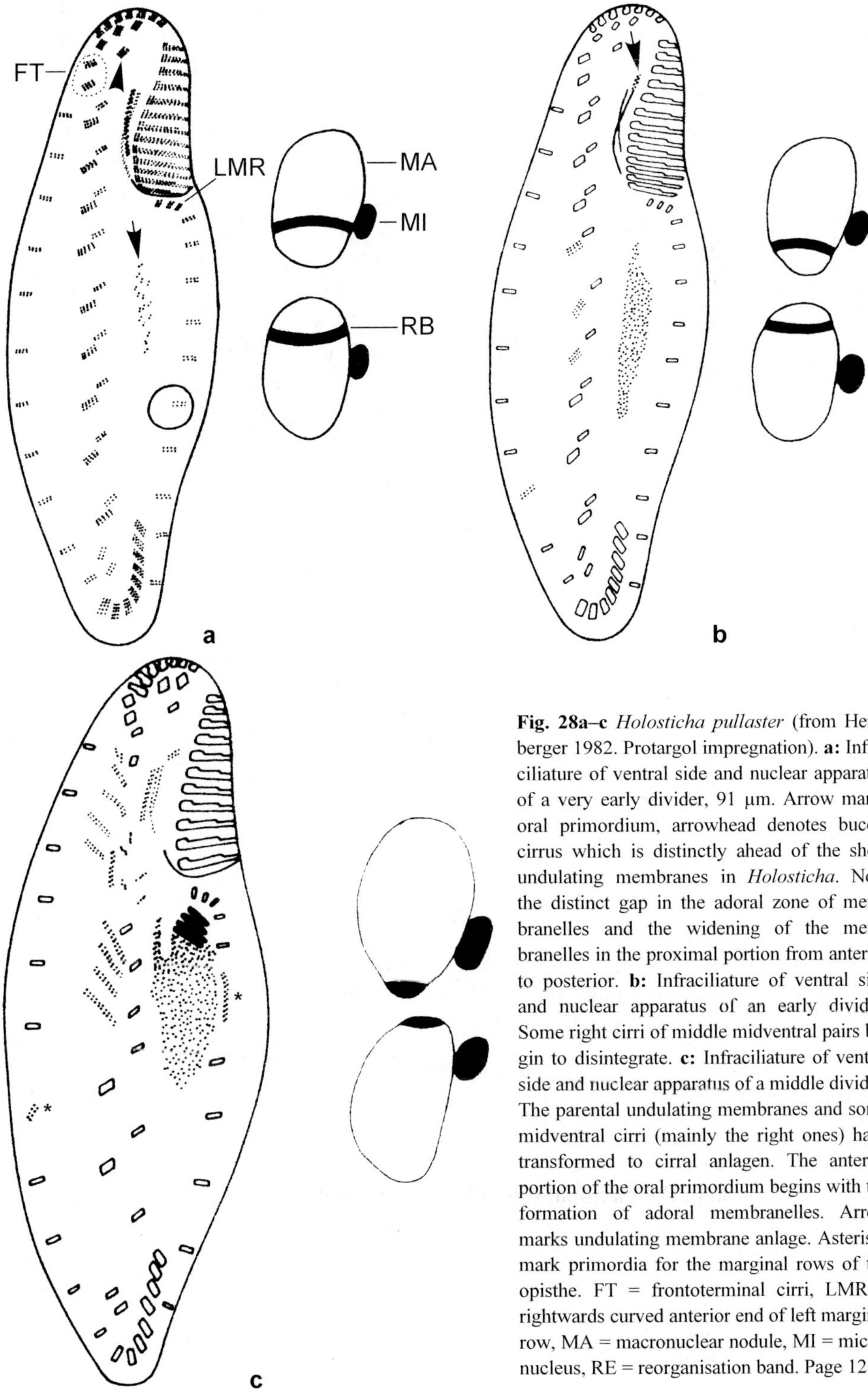

Fig. 28a–c *Holosticha pullaster* (from Hemberger 1982. Protargol impregnation). **a:** Infraciliature of ventral side and nuclear apparatus of a very early divider, 91 µm. Arrow marks oral primordium, arrowhead denotes buccal cirrus which is distinctly ahead of the short undulating membranes in *Holosticha.* Note the distinct gap in the adoral zone of membranelles and the widening of the membranelles in the proximal portion from anterior to posterior. **b:** Infraciliature of ventral side and nuclear apparatus of an early divider. Some right cirri of middle midventral pairs begin to disintegrate. **c:** Infraciliature of ventral side and nuclear apparatus of a middle divider. The parental undulating membranes and some midventral cirri (mainly the right ones) have transformed to cirral anlagen. The anterior portion of the oral primordium begins with the formation of adoral membranelles. Arrow marks undulating membrane anlage. Asterisks mark primordia for the marginal rows of the opisthe. FT = frontoterminal cirri, LMR = rightwards curved anterior end of left marginal row, MA = macronuclear nodule, MI = micronucleus, RE = reorganisation band. Page 128.

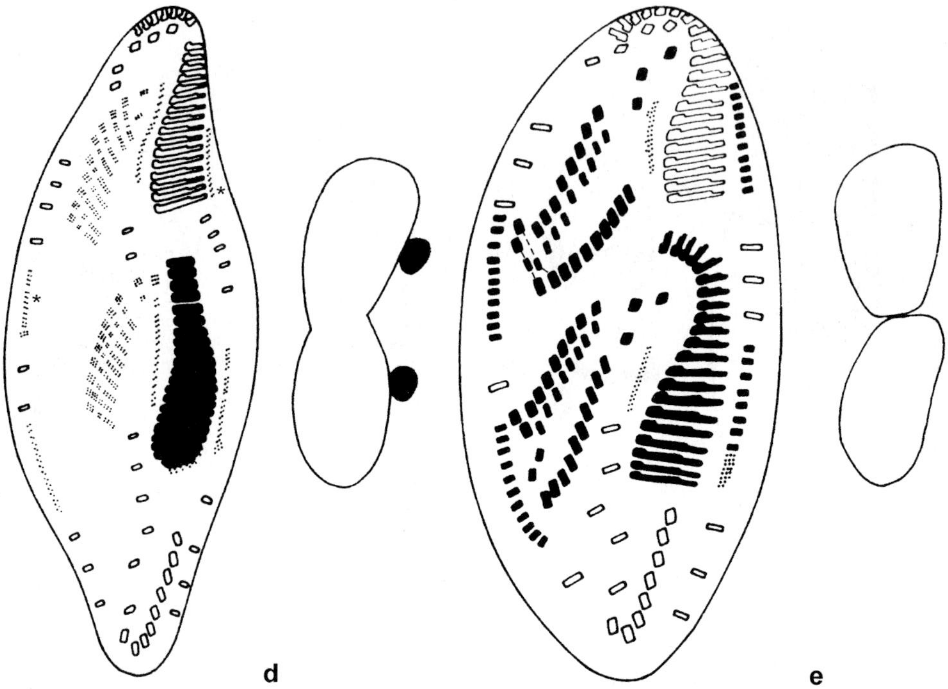

Fig. 28d, e *Holosticha pullaster* (from Hemberger 1982. Protargol impregnation). **d:** Infraciliature of ventral side and nuclear apparatus of a late divider, 95 μm. Asterisks mark marginal row anlagen for proter (anlage for left marginal row originates likely de novo). Parental cirri white, new dotted. Note that the parental adoral zone is more or less completely retained for the proter. **e:** Infraciliature of ventral side and nuclear apparatus of a very late divider, size not indicated. Almost all new cirri (black) are formed. According to Hemberger, frontal-midventral-transverse cirri anlagen III to n produce three cirri each, namely a midventral pair and one transverse cirrus. The frontoterminal cirri are formed by the anteriormost cirri of the two rightmost anlagen, whose cirri are connected by broken lines in the proter (usually both frontoterminal cirri originate from the rightmost anlage). Page 128.

Occurrence and ecology: *Holosticha pullaster* is one of the most common freshwater hypotrichs and occurs more or less throughout the year in all habitats, often rather abundant (own observations). No published record from a sewage treatment plant, although occurrence cannot be excluded. Does not occur in terrestrial habitats! The type locality is a freshwater habitat in the city of Copenhagen, Denmark, where Müller (1773, 1786) discovered *Holosticha pullaster* below *Lemna* sp.

Type localities of synonyms: *Oxytricha micans* was discovered in freshwater habitats in the surroundings of the German city of Leipzig, where it usually occurred in association with *Tachysoma pellionellum* (Engelmann 1862); type locality of *Oxytricha alba* not known, likely a freshwater habitat in France (Fromentel 1876); *Amphisia multiseta* was discovered in Switzerland, where it was very common (Sterki 1878); *Holosticha simplicis* was discovered in the Bay of Amoy, where Wang & Nie (1932) found it together with *Pseudokeronopsis rubra* (according to Wang & Nie 1934, p. 4210

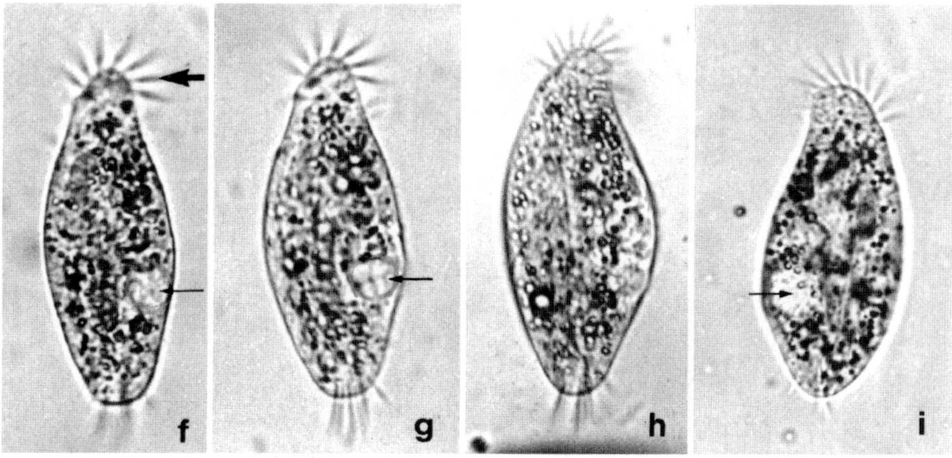

Fig. 28f–i *Holosticha pullaster* (from Foissner et al. 1991. Bright field micrographs). Ventral (f–h) views and dorsal view (i). Small arrows mark posteriorly dislocated contractile vacuole which is the most important feature of this species. Further, it makes this common freshwater species very easily determinable and unmistakable. Large arrow denotes the anterior portion of the adoral zone of membranelles. Page 128.

it appeared mainly in old cultures); *Holosticha kessleri aquaedulcis* was discovered in a brook (Botic) in/near the Czech capital, Prague, where it was abundant from April to October (Buchar 1957); *Keronopsis litoralis* was discovered in the littoral area from the eastern shore of the peninsula of Tihany, Lake Balaton, Hungary (Gellért & Tamás 1958; further records from same lake, see Gellért & Tamás 1959, p. 122; 1960, p. 60; Tamás & Gellért 1959, p. 239; 1960, p. 68); *Holosticha danubialis* was discovered in mosses and *Cladophora* aufwuchs in the littoral region of the Danube River in the city of Vienna (Austria) at the sites Nußdorf, Haslau, and Deutsch-Altenburg from November to August (Kaltenbach 1960; for review of Danube river ciliates, see Enăceanu & Brezeanu 1970, p. 234); *Holosticha retrovacuolata* was discovered in the waters of the Ialomicioara cave, which is near the villages of Baia de Fier and Polovragi in the southern Carpathian Mountains, Romania (Tucolesco 1962b; for review, see Gittleson & Hoover 1969, p. 750); *Holosticha rhomboedrica* was discovered in Lacul Tei, likely a lake in Bucharest, Romania, where it occurred rather abundantly under meso- to polysaprobic conditions (Vuxanovici 1963); *Holosticha coronata* was discovered in Lake Floreasca, Bucharest, Romania, where Vuxanovici (1963) found few specimens in the sapropel; type locality of *H. minima* is Lake Herastrau, Bucharest, Romania (Vuxanovici 1963).

Holosticha pullaster is also reliably recorded from marine habitats, for example, by Petz et al. (1995) from the Wedell Sea, where it occurred common in the endopagial, mainly in the brown layer, of multiyear and pancake sea ice between latitude 68°38'S to 70°21'S and longitude 06°05'W to 11°00'W. Up to 12.405 active ind. l^{-1} melted ice were found (biomass 0.16 mg l^{-1}), comprising up to 41% of the total ciliate community; only in multiyear sea ice on average 5819 ind. l^{-1} (n = 5) occurred. Occurs together with a

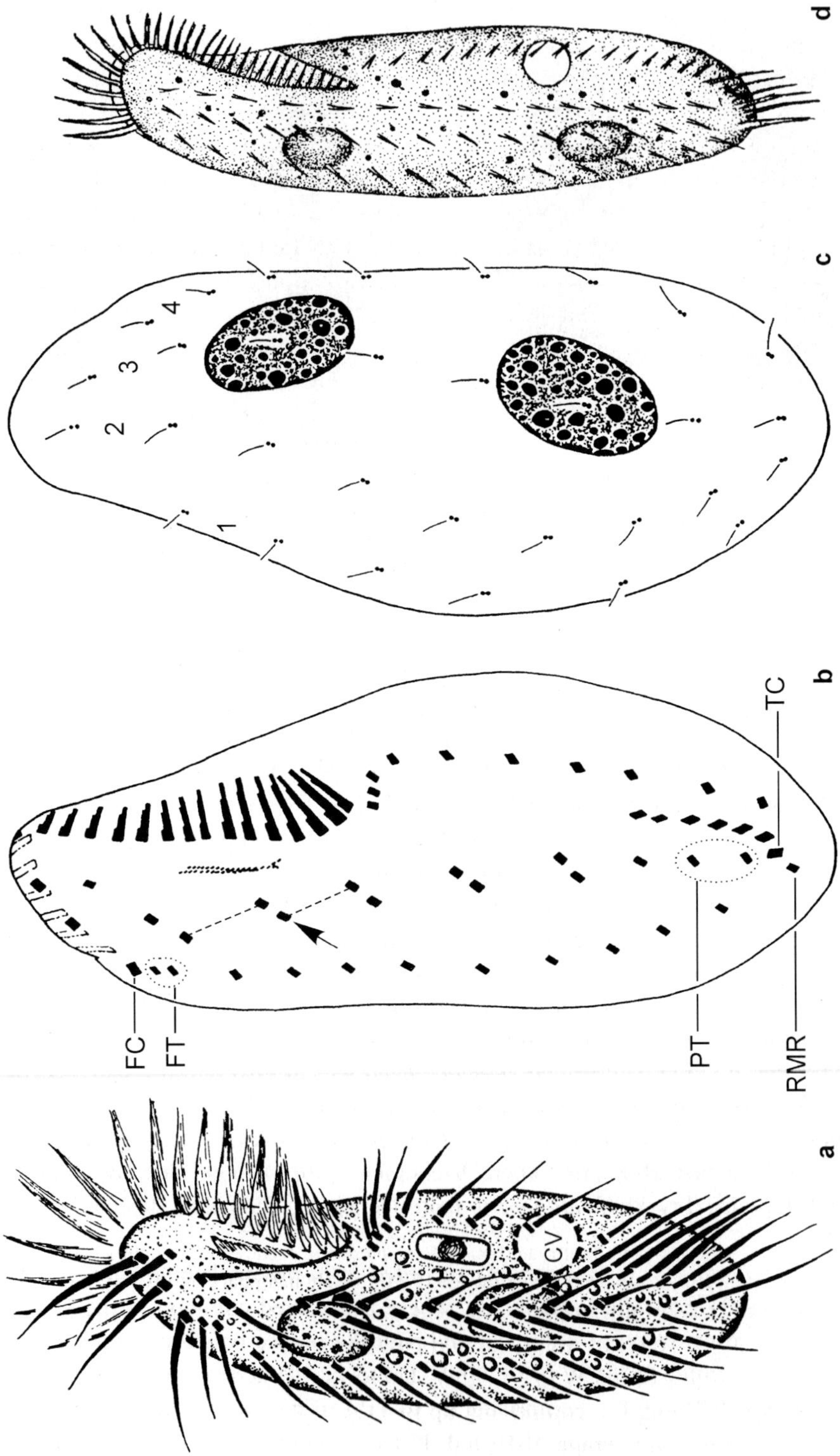
d
c
4
3
2
1
b
TC
FC
FT
PT
RMR
a
CV

variety of other organisms, for example, diatoms, flagellates, and ciliates. Petz et al. (1995) provided the following environmental parameters in brine: -3.4 to -3.0°C, 51.8–59.1‰ salinity, 7.2 μmol l^{-1} NO_3, 17.2 μmol l^{-1} Si; in melted ice: 2.8 μmol l^{-1} PO_4, 3.5 μmol l^{-1} NH_4, 4.3–80.1 μg l^{-1} chlorophyll *a*. In raw cultures also at +1°C and at a salinity of 21.3 ‰. Does not burst at room temperature. According to Petz et al. (1995), freshwater and marine populations cannot be separated morphologically. However, there seem to be physiological differences. In cultures, a population from the Antarctic endopagial could not be adapted to freshwater and a limnetic population could not be transferred to saltwater (Andermahr & Wilbert unpublished in Petz et al. 1995). Petz (2005) recorded *H. pullaster* from sea-ice of the Ross Sea region.

Further records of *H. pullaster* from freshwater habitats substantiated by morphological data: not very common/abundant? in water bottles and plant infusions from Berlin, Germany (Ehrenberg 1838); rheocrene spring in south-eastern Romanian Dobrudscha (Lepsi 1926a; as *Holosticha* sp.).

Freshwater records of *H. pullaster* not substantiated by morphological data (post 1991 records are based on Foissner et al. 1991 and are thus reliable): Klosterneuburg, a small town in Austria (Riess 1840, p. 38); beta- to alpha-mesosaprobic sites in the Traun river, Upper Austria (Foissner & Moog 1992, p. 101); clean to distinctly polluted rivers in Upper Austria (for example, AOÖLR 1993a, p. 78; 1993b, p. 36); pelagial of the meromictic lake Höllerersee, Upper Austria (Nauwerck 1996, p. 156); karstic waters (Plitvice Lakes) in Croatia (Primc-Habdija et al. 2000a, p. 2603); 14 ind. cm^{-2} at a current velocity of 20–50 cm s^{-1}, 41 at 50–100, 58 at 100–130, and 24 at more than 130 in bryophyte covered substrate from travertine barriers in Plitvice Lakes, Croatia (Primc-Habdija et al. 2000, p. 283); karstic river in Croatia (Primc-Habdija et al. 2001, p. 92); mesosaprobic rivers (Amper, Illach) and brooks in Bavaria (Foissner 1997a, p. 184; Foissner et al. 1992, p. 49; 1992a, p. 101); in periphyton and sediment of an unpolluted stream (Breitenbach) in Germany (Packroff & Zwick 1996, p. 258); detached biofilm in backwash water from a rapid gravity filter using tertiary groundwater from Munich region, Germany (Foissner 1996b[1], p. 16; see also Gierig 1993, p. 34, as *H. danubialis*); infusion of tree mosses from Germany? (Ehrenberg 1849, p. 97; possibly a misidentification); hyporheic interstitial of a German brook (Cleven 2004, p. 77); water from the carst cave Covollo della Guerra in Vizenza, Italy (Coppellotti & Guidolin

← **Fig. 29a–d** *Holosticha pullaster* (a–c, from Petz et al. 1995; d, from Wang & Nie 1932. a, d, from life; b, c, protargol impregnation). Marine populations. **a–c:** Ventral view of a representative specimen (a; 66 μm) and infraciliature of ventral and dorsal side and nuclear apparatus of same specimen (54 μm). Arrow marks a pseudopair (corresponding midventral pairs connected by broken lines). **d:** Ventral view, 63 μm. Very likely, Wang & Nie overlooked the frontal cirri. CV = contractile vacuole, FC = right frontal cirrus, FT = frontoterminal cirri, PT = pretransverse ventral cirri, RMR = right marginal row, TC = rearmost transverse cirrus, 1–4 = dorsal kineties. Page 128.

[1] *Tachysoma pellionellum* sensu Foissner (1996b, Fig. 9) is likely also a *Holosticha pullaster* as indicated by the body shape and the increased number of transverse cirri.

1999, p. 75; Guidolin & Coppellotti Krupa 1999, p. 76); geothermal sulphur spring in province Piacenza, Northern Italy (Madoni & Uluhogian 1997, p. 165); polluted running water (saprobic index 2.2 to 3.4) in northern Italy (Madoni & Bassanini 1999, p. 394); in the biofilm of sand filters and 17 ind. ml^{-1} in granular activated carbon filters of a drinking water biofilter in Northern Italy (Madoni et al. 2001, p. 459); St. Petersburg, Russia (Eichwald 1844, p. 583; Weisse 1845, p. 22); dominant in the polluted Manzanares River near the village of La Pedriza, Spain (Fernandez-Leborans & Novillo 1996, p. 315); ponds with spring water in Switzerland (Perty 1852, p. 154); eutrophic pond, Karasu river, and other Turkish running waters (Senler et al. 1996, p. 186, 187; 1998, p. 40, 41; Senler & Yildiz 1998, p. 5; 2004, p. 248).

Record of the synonym *Oxytricha micans* not substantiated by morphological data: Warsaw, Poland (Wrzesniowskiego 1866, p. 18).

Freshwater records of *H. kessleri* (for explanation, see remarks): Tundza River, Bulgaria (Detcheva 1986, p. 63); Schwalm River, Germany (Heuss et al. 1972, p. 93); small ditches in Germany and throughout the year with maxima during fall and winter in alphamesosaprobic running waters near the German city of Krefeld (Heuss 1975, p. 151; 1976, p. 146, Table 14); in March about 20 ind. cm^{-2} on slides exposed in the Poppelsdorfer Weiher, a eutrophic pond in the German city of Bonn (Wilbert 1969, p. 491); dominant at a rate of flow of 0.3–0.5 m s^{-1} in the beta- to alphamesosaprobic river Rhine near the German city of Bonn (Schmitz 1985, p. 2295); Lake Lough Neagh, the largest freshwater lake in Northern Ireland (Xu & Wood 1999, p. 105); Stirone stream, northern Italy (Madoni 2000a, b); River Lielupe, Latvia (Liepa 1973, p. 33); Turiec river, Slovakia (Tirjaková 1993, p. 133); Zemplínska Sírava reservoir in eastern Slovakia (Matis 1977, p. 30; as variety *H. kessleri aquadulcis*); dead arm of Danube River in Slovakia (Matis & Tirjaková 1994, p. 51); submerged and wet mosses in Slovakia (Tirjaková & Matis 1987a, p. 8); brook and thermal lakes in Bojnice spa, Slovakia (Matis & Straková-Striešková 1991, p. 114); Hanjiang River, China (Shen et al. 1994, p. 207); Yellow River in Lanzhou, China (Ma 1994, p. 95); river and lake in China (Ma & Dang 1994, p. 453); freshwater sites from the Yuelushan area, China (Yang 1989, p. 157; further record from China: Su et al. 1988, p. 3); Chaohu Lake, China (Xu et al. 2005, p. 188); during April in the Savannah river, USA (Patrick et al. 1967, p. 320); in July in the lake of the extinct volcano Tantalus, Oahu, Sandwich Osland (Schewiakoff 1893, p. 67).

Freshwater records of *Holosticha diademata* (for explanation, see remarks): England (Craigie 1921, p. 119); dominant (up to 12 ind. ml^{-1}) in the slightly to distinctly polluted, eutrophic (about 0.6 mg l^{-1} PO_4^{3-}-P) La Dore River, France (Grolière et al. 1990, p. 387; Sparagano & Grolière 1991, p. 53); brook (Duvebach) in the Rhineland, Germany (Pätsch 1974; Fig. 26w); alpine brook polluted by waste water, south Germany (Bauer 1987, p. 19); mesosaprobic brooks and river Rhine near the German city of Bonn at a streaming velocity of 0.30–0.49 m s^{-1} (Jutrzenki 1982, p. 108; Schmitz 1983, p. 4); common and often abundant in the aufwuchs, only once in the plankton community, of a cooling system of a conventional power station in Germany (Bernerth 1982, Table 14); brook in Hungary (Vörösváry 1950, p. 376); spring in Italy (Stella

1947, p. 25); at a depth of 10 m on exposed substrate in Douglas Lake, USA (Jones et al. 1976, p. 4).

Records of the synonym *H. simplicis*: two sites (with a salinity of 4–10 ‰) from the Al-Hassa Oasis in Saudi Arabia (Al-Rasheid 1996a, p. 198; see remarks for identification); Kandalakshskii Bay, White Sea (Burkovsky 1970a, p. 190; 1970b, p. 11; 1970c, p. 56, species incorrectly assigned to Kahl); pelagic in Szolnok region of the Tisza river, Hungary (Jósa 1974, p. 56).

Records of the synonym *Holosticha danubialis*: pond at Salzburg University, Austria (Blatterer 1989, p. 9); Oichten River, Austria (Augustin et al. 1987a, p. 75); running waters in the Upper Austrian city of Linz (Augustin et al. 1987b, p. 216); Danube River, Austria (Humpesch & Moog 1994, p. 91); abundant throughout the year, especially during summer and November in the Poppelsdorfer Weiher, a eutrophic pond in the German city of Bonn (Song & Wilbert 1989, p. 159); aufwuchs in two Eifel maar lakes, Germany (Packroff 1992, p. 211; Packroff & Wilbert 1991, p. 123); Morava River system in Slovakia (Matis & Tirjaková 1994a, p. 57); Danube River, side branch of Danube River, and Turiec River in Slovakia (Szentivány & Tirjaková 1994, p. 93; Tirjaková 1992, p. 293; 1992a, p. 77; 1993, p. 133; Matis et al. 1996, p. 12); periphyton of Lake Xiaoxihu in the Qingdao region, China (Song et al. 1993, p. 101).

Records of the synonym *H. retrovacuolata* partially substantiated by morphological data: pasture pond, ponds with melting snow, mosses from small running waters, and sprayed rocks from the Großglockner area, Austrian Alps (Foissner 1980a, 1980b, p. 107).

The record of the synonym *H. retrovacuolata* by Tirjaková (1988; see Matis et al. 1996, p. 12) from an agricultural soil in Czechoslovakia is likely a misidentification, although one cannot exclude that *H. pullaster* occurs in smallest water bodies of fields.

Holosticha pullaster likely avoids low oxygen concentrations (Bernerth 1982). Possibly for that reason it does not occur in sewage treatment plants (for example, Ganner et al. 2002) where the O_2-content is usually below 2 mg l^{-1}. Lethal concentration (24-h LC_{50}) of nickel is 1.1 mg l^{-1} (Madoni 2000a; see also Madoni 2000b; misidentified as *H. kessleri*).

Feeds on bacteria and flagellates (Pätsch 1974), bacteria and green algae (Foissner 1980a), pennate diatoms 6–12 µm long (Petz et al. 1995; Vuxanovici 1963), and small, green globular algae (Gellért & Tamás 1958, Tamás & Gellért 1958). Biomass of 10^6 individuals about 12 mg (Foissner et al. 1991).

As mentioned above, *Holosticha pullaster* – including its numerous synonyms – was never classified saprobiologically. Albrecht (1984) considered a saprobiological classification as meaningless because it is euryoecious. In the Ciliate Atlas we reported on a distribution from water quality I–II to III–IV at saprobic indices from 1.8 to 3.1. Abundant and mass occurrence was observed at saprobic indices from 2.1 to 3.1. However, the majority of the records is in the betameso- to alphamesosaprobic region. Thus, we proposed the following classification: b–a; o = 1, b = 4, a = 4, p = 1, I = 1, SI = 2.5 (Table 12; Foissner et al. 1991, 1995; see also Foissner & Berger 1996, Sládeček & Sládečková 1997, Berger & Foissner 2003).

Supposed synonym of *Holosticha pullaster*

Oxytricha pernix Wrześniowski, 1877 (Fig. 30a–e)

1877 ***Oxytricha pernix*, nov. sp.**[1] – Wrześniowski, Z. wiss. Zool., 29: 273, Tafel XIX, Fig. 10, 11 (Fig. 30a, b; original description; no type material available).

1877 ***Holosticha pernix*, mihi** – Wrześniowski, Z. wiss. Zool., 29: 278 (combination with *Holosticha*; see nomenclature).

1882 ***Amphisia pernix*, Wrz sp.** – Kent, Manual infusoria II, p. 768, Plate XLIII, Fig. 12 (Fig. 30c; revision and combination with *Amphisia*).

1929 ***Holosticha pernix* Wrzesn.** – Hamburger & Buddenbrock, Nord. Plankt., 7: 86, Fig. 103 (Fig. 30a, b; guide to marine ciliates).

1932 ***Keronopsis (Oxytricha) pernix* (Wrzesniowski, 1877)** – Kahl, Tierwelt Dtl., 25: 575, Fig. 101 $_{26}$ (Fig. 30d; revision; see nomenclature).

1933 ***Keronopsis pernix* (Wrzesniowski 1877)** – Kahl, Tierwelt N.- u. Ostsee, 23: 109, Fig. 16.34 (Fig. 30e; guide to marine ciliates).

1972 ***Keronopsis pernix* (Wrzeniowski, 1877) Kahl, 1932** – Borror, J. Protozool., 19: 11 (revision of hypotrichs; combination with *Keronopsis*; incorrect spelling of Wrześniowski).

1979 ***Holosticha pernix* (Wrześniowski, 1877) comb. n.** – Jankowski, Trudy zool. Inst., Leningr., 86: 57 (combination with *Holosticha*; see nomenclature).

1983 ***Pseudokeronopsis pernix* (Wrześniowski, 1877) nov. comb.** – Borror & Wicklow, Acta Protozool., 22: 124 (revision of urostyloids; combination with *Pseudokeronopsis*).

1992 ***Keronopsis pernix* (Wrzesniosky, 1877) Kahl, 1930-5** – Carey, Marine interstitial ciliates, p. 184, Fig. 727 (guide; incorrect spelling of Wrześniowski).

2001 ***Pseudokeronopsis pernix* (Wrzesniowski, 1877) Borror and Wicklow, 1983** – Berger, Catalogue of ciliate names 1. Hypotrichs, p. 61 (nomenclator containing all basionyms, combinations, and higher taxa of hypotrichs).

Nomenclature: The species-group name *pernix* (Latin adjective; fast, quick, rapid) refers to the rapid movement of this species (Wrześniowski 1877, p. 273). Wrześniowski (1877, p. 278) introduced the genus *Holosticha* and suggested that several species, inter alia, *Oxytricha pernix*, should be assigned to this genus. Although Wrześniowski did not use the binomen *Holosticha pernix*, he is the combining author for *Holosticha*. Thus, when classified in *Holosticha*, the correct name is *Holosticha pernix* (Wrześniowski, 1877) Wrześniowski, 1877. In my catalogue I assumed that Kahl (1932) is the combining author for *Holosticha* because I overlooked the previous act. Jankowski (1979) also overlooked that this species had already been transferred from *Oxytricha* to *Holosticha* by Wrześniowski (1877).

Kahl (1932, 1933) classified *Keronopsis* as subgenus of *Holosticha*; consequently, the correct name in his papers is *Holosticha (Keronopsis) pernix* (Wrześniowski, 1877)

[1] The diagnosis by Wrześniowski (1877) is as follows: Körper extensil, höchst beugsam, lancetförmig, verdickt; keine Stirnwimpern; Bauchwimpern in zwei continuirlichen Reihen; Randwimpernreihen weit nach innen gerückt; fünf borstenförmige Afterwimpern.

Wrześniowski, 1877. The misleading spelling *Keronopsis (Oxytricha) pernix* in Kahl (1932) should simply indicate that this species was previously (originally) classified in *Oxytricha*. By contrast, Borror (1972) incorrectly assumed that Kahl (1932) had transferred the present species to the genus *Keronopsis* and mentioned Kahl as combining author for *Keronopsis*. Later, he listed himself as combining author for *Keronopsis* (see list of synonyms in Borror & Wicklow 1983). Šrámek-Hušek (1957) also used the binomen *Keronopsis pernix* so that this author could be responsible for the now outdated transfer from *Holosticha* to *Keronopsis*.

Remarks: Wrześniowski (1877) provided a rather detailed description from life. He wrote that enlarged frontal cirri are lacking, respectively, the frontal cirri do not differ from the (first) ventral cirri (= midventral complex). Although this contradicts a classification in *Holosticha*, I suppose that it is member of this genus as indicated by the two rightwards displaced macronuclear nodules and the transverse cirri, which are obviously indistinctly set off from the left marginal row. The posteriorly located contractile vacuole leads me to classify *O. pernix* a supposed synonym of *H. pullaster*. The classification in *Pseudokeronopsis* proposed by Borror & Wicklow (1983) is unlikely because these species are usually rather slender and have many macronuclear nodules, which do not fuse prior to division. I do not expect that the two macronucleus-nodules of *O. pernix* divide individually. The anterior portion of the midventral complex is not distinctly curved leftwards in *O. pernix* (Fig. 30a). In this respect, it resembles *Pseudokeronopsis multinucleata* (Fig. 187a). These two independent observations indicate that such a pattern could exist; however, I suppose that both observations are not quite correct.

The illustrated record of *O. pernix* by Šrámek-Hušek (1957; Fig. 143k) from freshwater is considered a misidentification. Perhaps it is a true *Keronopsis* as indicated by the widely separated ventral cirral rows. The description by Wailes (1943) is insufficient and thus the identification cannot be accepted (Fig. 143j).

Morphology: Size around 108 × 36 µm; width estimated via the body length:width ratio (3:1) of the specimen shown in Fig. 30a; size according to Bullington (1925) 104 × 34–52 µm. Body lancet-shaped, anterior and posterior portion narrowed and rounded, sometimes anterior region more narrowed than posterior; ventral side plane, dorsal side strongly vaulted and humped in mid-body (Fig. 30b). Lateral regions thick and rounded. Body flexible and contractile (extent of contractility not mentioned). Frontal scutum narrow, thick, crescent-shaped, extends onto ventral surface on right side. Two ellipsoidal macronuclear nodules, anterior one right of proximal end of adoral zone, rear one right of contractile vacuole. Contractile vacuole behind mid-body, at 66% of body length in specimen illustrated (Fig. 30a). Cytopyge behind contractile vacuole. Cytoplasm colourless, transparent, with few small granules. Movement restless, rotates leftwards (Bullington 1925).

Adoral zone occupies about 33% of body length, of usual shape; membranelles of moderate dimension. Buccal area narrow. Paroral distinct, rather short. Buccal cirrus neither mentioned nor illustrated (this must not be over-interpreted because this cirrus is often difficult to recognise in life). Enlarged frontal cirri lacking, respectively, not dis-

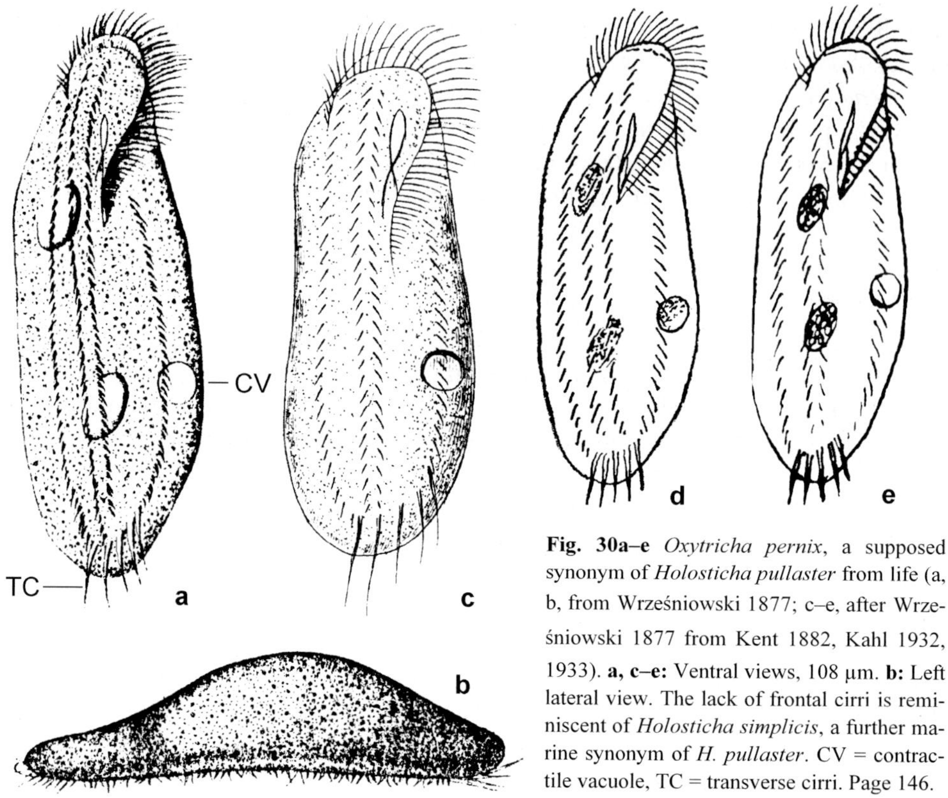

Fig. 30a–e *Oxytricha pernix*, a supposed synonym of *Holosticha pullaster* from life (a, b, from Wrześniowski 1877; c–e, after Wrześniowski 1877 from Kent 1882, Kahl 1932, 1933). **a, c–e:** Ventral views, 108 µm. **b:** Left lateral view. The lack of frontal cirri is reminiscent of *Holosticha simplicis*, a further marine synonym of *H. pullaster*. CV = contractile vacuole, TC = transverse cirri. Page 146.

tinguishable from fine, bristle-like cirri of two narrowed ventral rows (= midventral complex obviously composed of many cirral pairs; number likely distinctly overestimated); both cirral rows extend from anterior body end to near transverse cirri. Five slightly enlarged transverse cirri arranged in short oblique row near rear body end, thus projecting for nearly half their length beyond posterior border. Marginal rows terminate near outermost transverse cirri, distinctly displaced inwards, cirri short. No dorsal bristles observed indicating that they are short, that is, around 3 µm.

Occurrence and ecology: Marine. The type locality of *Oxytricha pernix* is the Baltic Sea where Wrześniowski (1877) discovered it on the eastern coast of the island of Rügen. It occurred with high abundance among algae washed ashore. Ax & Ax (1960, p. 12; without morphological data) write that *O. pernix* is, according to literature data, characteristic (confined to) for brackish water. Ax & Ax (1960, p. 15) themselves recorded *O. pernix* from 0.8–4% salt content.

Records not substantiated by illustrations and/or morphological data: Black Sea, inter alia, 5 ind. cm^{-2} at 24°C in the 0–1 cm layer of the sandy bottom in the Odessa Bay (Jeliaskova-Paspalewa 1933, p. 22; Dzhurtubayev 1978, p. 65); littoral of islands in the Caspian Sea (Agamaliev 1972, p. 7); Gulf of Kola, Russia (Gassovsky 1916, p. 142); attached to debris in the Woods Hole Area, USA (Lackey 1936, p. 269). The records from the river Lyna in Poland by Hul (1986, p. 154; 1987, p. 208) have to be treated as misidentifications.

Holosticha foissneri Petz, Song & Wilbert, 1995 (Fig. 21f, 31a–d, Table 13)

1931 ***Amphisia gibba* O. F. Müller** – Hofker, Arch. Protistenk., 75: 394, Fig. 89 (Fig. 21f; misidentification).

1995 ***Holosticha foissneri* nov. spec.**[1] – Petz, Song & Wilbert, Stapfia, 40: 159, Fig. 47a–d, Table 24 (Fig. 31a–d; original description; 1 holotype slide [2001/129] and 1 paratype slide [2001/18] of protargol-impregnated specimens are deposited in the Oberösterreichische Landesmuseum in Linz [LI], Upper Austria).

2001 ***Holosticha foissneri* Petz, Song and Wilbert, 1995** – Berger, Catalogue of ciliate names 1. Hypotrichs, p. 34 (nomenclator containing all basionyms, combinations, and higher taxa of hypotrichs).

2003 ***Holosticha foissneri* Petz et al., 1995** – Berger, Europ. J. Protistol., 39: 375, 376, Fig. 7 (Fig. 31a; brief review).

2005 ***Holosticha foissneri* Petz, Song & Wilbert (1995)** – Petz, Ciliates, p. 395, Fig. 14.86a–c (Fig. 31a–c; guide to Antarctic marine ciliates).

Nomenclature: This species was dedicated to Wilhelm Foissner, University of Salzburg, Austria (Petz et al. 1995). "*Holosticha foissnseri*" in Song et al. (2002, p. 155) is an incorrect subsequent spelling.

Remarks: *Amphisia gibba* sensu Hofker (1931, Fig. 21f) is likely identical with *H. foissneri* as indicated by the nuclear apparatus and the large gap in the adoral zone.

The present species unequivocally belongs to *Holosticha* as indicated by the gap in the adoral zone, the short and parallel undulating membranes, the widening of the proximal adoral membranelles, the anteriorly displaced buccal cirrus, the rightwards curved anterior end of the left marginal row, the long transverse cirral row, and the nuclear apparatus in the right body portion. Especially the nuclear apparatus, composed of eight serially arranged macronuclear nodules, makes the species easily determinable (see key).

According to the scale bar, the specimen shown in Fig. 31a is 125 µm long, which is slightly below the (approximate) minimum value provided in the diagnosis.

Morphology: Size about 130–170 × 40 µm in life, body length:width ratio 3.3:1 on average in protargol preparations (Table 13); body size according to Petz (2005) 120–190 × 35–60 µm. Body outline elongate to slightly fusiform, that is, left and right margin convex, anterior and posterior end narrowly rounded (Fig. 31a). Dorsoventrally flattened about 2:1 (Fig. 31d). Macronuclear nodules arranged in series, extending from second to fourth fifth of body right of midline; individual nodules globular, about 10 µm across in life, usually with one large nucleolus (5–7 µm across) each (Fig. 31a, c). Rarely specimens with more than eight (once; reorganiser or divider?) or fewer macro-

[1] Petz et al. (1995) provided the following diagnosis: In vivo about 130–170 × 40 µm, elongate. Contractile vacuole subequatorial. Left marginal row with usually 4 transversely arranged cirri at anterior end. 2 frontoterminal cirri, 1 buccal and 9–17 transverse cirri. Midventral row extending to transverse cirri. 4 dorsal kineties. Adoral zone bipartite, consists of 26–36 membranelles. Usually 8 macronuclear nodules in right half of cell. Marine.

nuclear nodules (apparently fused, that is, one large nodule with two large nucleoli or a replication band). Micronuclei lenticular to globular, in indentation of macronuclear nodules; lightly stained with protargol. Contractile vacuole behind mid-body (at 62% of body length in specimen shown in Fig. 31a) near left body margin; pulsation, however, not observed. No cortical granules mentioned. Cytoplasm hyaline, contains many colourless globules 3–5 µm across. Food vacuoles about 13 µm in diameter. Slowly crawling on substrate, sometimes jerking back and fourth; thigmotactic.

Adoral zone occupies about 35–39% of body length, bipartite by 3–7 µm wide gap in left anterior region, extends relatively far onto right body margin; 7–13 membranelles in anterior portion, 17–24 membranelles of ordinary fine structure in roughly spoon-shaped posterior portion, that is, width of membranelles increases from anterior to posterior; largest bases about 8 µm wide, cilia ca. 17 µm long. Undulating membranes each about 14 µm long in Fig. 31b, almost in parallel so that double-rowed paroral optically intersects single-rowed endoral only inconspicuously. Pharyngeal fibres about 30 µm long (Fig. 31b).

Number of cirri of usual variability, except for number of transverse cirri which varies rather strongly (Table 13). Three slightly enlarged frontal cirri, all obviously right of midline. Single buccal cirrus also slightly enlarged, distinctly ahead of undulating membranes about at level of anterior end of proximal portion of adoral zone (Fig. 31b). Frontoterminal cirri between right frontal cirrus and anteriormost midventral pair. Midventral complex composed of about 18 midventral pairs, extends to near anteriormost transverse cirri; both cirri of each pair of about same size. Likely two pretransverse ventral cirri present (encircled in Fig. 31b; interpreted as rearmost midventral pair by Petz et al. 1995). 9–17 transverse cirri arranged in J-shape, of same size as midventral and marginal cirri, project distinctly beyond rear and left posterior body margin (Fig. 31a, b). Right marginal row commences obviously distinctly behind anterior body end (at 22% of body length in Fig. 31b), terminates near right (= posterior) end of transverse cirral row. 3–4 anteriormost and narrowly spaced cirri of left marginal row transversely arranged; row ends somewhat subterminally, thus marginal rows slightly separated posteriorly (Fig. 31b); marginal cirri 16–18 µm long, bases composed of two basal body rows.

Dorsal cilia 2.5–3.5 µm long, arranged in four kineties with 13–20 cilia each; kinety 1 distinctly shortened anteriorly. Caudal cirri lacking (Fig. 31c).

Occurrence and ecology: Marine. Type locality is the sea ice of the Weddell Sea, Antarctica (69°46'S 11°00'W). Petz et al. (1995) found it frequently in endopagial of pancake and, more often, multiyear sea ice (brown layer) between latitude 69°02'–71°00'S and longitude 08°02'–11°80'W. Up to 4000 active ind. l^{-1} melted ice were found (biomass 0.3 mg l^{-1}), comprising up to 6% of total ciliate community. Environmental parameters in brine: -3.4° to -3.0°C, salinity 52–59‰; in melted ice: 2.8 µmol l^{-1} PO_4, 6.1 µmol l^{-1} NO_3, 3.5 µmol l^{-1} NH_4, 14.8 µmol l^{-1} Si, 11.1–80.1 µg l^{-1} chlorophyll *a*. In raw cultures also at a salinity of 16–21‰ and a temperature of +1°C; does not burst at room temperature. Petz (2005) found it in sea-ice and plankton of the Ross Sea region. *Holosticha foissneri* feeds on small pennate diatoms and, very likely, autotrophic flagellates as indicated by green debris in food vacuoles. Biomass of 10^6 individuals 80 mg.

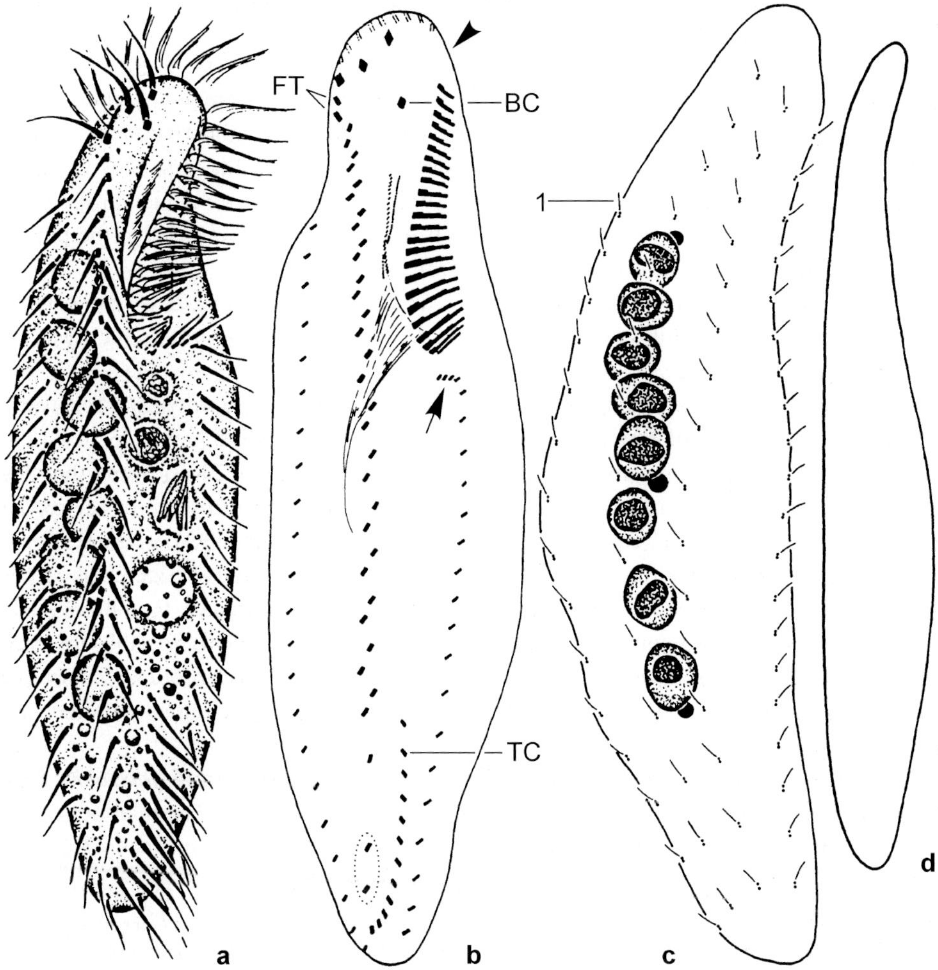

Fig. 31a–d *Holosticha foissneri* (from Petz et al. 1995. a, d, from life; b, c, protargol impregnation). **a:** Ventral view of a representative specimen, 125 µm. **b:** Ventral view of infraciliature, 134 µm. Arrowhead marks huge gap in adoral zone, arrow denotes rightwards curved anterior end of left marginal row. The two cirri encircled by a doted line are very likely pretransverse ventral cirri and not the last midventral pair as described by Petz et al. (1995). Note the spoon-shape of the postoral portion of the adoral zone. **c:** Infraciliature of dorsal side and nuclear apparatus, 135 µm. Note the large nucleoli. **d:** Left lateral view. BC = buccal cirrus, FT = frontoterminal cirri, TC = transverse cirri, 1 = dorsal kinety 1. Page 149.

Holosticha heterofoissneri Hu & Song, 2001 (Fig. 32a–p, 33a–p, Table 13, Addenda)

2001 ***Holosticha heterofoissneri* nov. spec.**[1] – Hu & Song, Hydrobiologia, 448: 172, Fig. 1a–i, 2a–f, 3a–g, 4–15, Table 1 (Fig. 32a–f, 33a–p; original description; 1 holotype and 1 paratype slide of protargol-impregnated specimens are deposited in the Laboratory of Protozoology, College of Fisheries, Ocean University of Qingdao, China).

2002 ***Holosticha heterofoissneri* Hu & Song, 2001**[2] – Song, Wilbert & Warren, Acta Protozool., 41: 159, Fig. 21–29, 46–51, Tables 1, 4 (Fig. 32g–p; redescription).

2003 ***Holosticha heterofoissneri* Hu and Song, 2001** – Berger, Europ. J. Protistol., 39: 375, 376, Fig. 9 (Fig. 32g; brief review).

Nomenclature: No derivation of the name is given in the original description. The species-group name *heterofoissneri* is a composite of the Greek adjective *heter-* (different), the thematic vowel ·*o*- (in compound words at the end of the first root when the second root begins with a consonant; Werner 1972, p. 37), and the species-group name *foissneri* (a species named after W. Foissner). It likely refers to the fact that the present species is different from the closely related *H. foissneri*.

"*Holosticha foissneri sinensis* nov. spec." in the legend to Fig. 2 of the original description is likely a nomen nudum. Possibly, *sinensis* was a draft title which was not deleted.

Remarks: The present species unequivocally belongs to *Holosticha* because it shows all apomorphies of this group, as discussed in detail in the genus section (e.g., rightwards curved anterior end of left marginal row). Song et al. (2002) found that dorsal kinety 1 shows a kind of fagmentation, a feature which is reminiscent of the oxytrichids where kinety 3 splits (for review of this group, see Berger 1999). The question is whether or not these two fragmentation processes are homologous? Since this is certainly a difficult task, and because this feature is of great importance for phylogenetic analysis, further populations of *H. heterofoissneri* should be studied. Possibly the fragmentation in *H. heterofoissneri* is more closely related to the curious dorsal kinety formation described for *H. bradburyae*, which is very likely not homologous with the fragmentation widely distributed in the oxytrichids.

The posterior portion of the adoral zone has the same shape as in *H. foissneri*. From this species it differs, inter alia, by the higher number of adoral membranelles (32 against 42–45; Table 13), the smaller gap in the adoral zone, the presence of cortical

[1] The diagnosis by Hu & Song (2001a) is as follows: Marine *Holosticha* in vivo 110–150 × 30–50 µm with elongate to fusiform body shape; adoral zone slightly bipartite consisting of 33–47 membranelles; 23–31 midventral and 11–18 transverse cirri; 18–23 left and 18–33 right marginal cirri; constantly 5 dorsal kineties; 14–16 macronuclei.

[2] The improved diagnosis by Song et al. (2002) is as follows: Marine *Holosticha*, about 110–150 × 30–60 µm in vivo; adoral zone of membranelles comprising ca 50 membranelles and with a distinct gap between anterior and posterior parts of AZM; one anteriorly positioned buccal cirrus; 4 frontal, 2 frontoterminal and 12–17 transverse cirri; midventral rows comprising about 13 pairs of cirri, which extend almost to posterior end of cell; 5 dorsal kineties; cortical granules small and sparsely distributed on dorsal side; one post-equatorially located contractile vacuole. 14–21 macronuclear segments connected to each other by thread-like structure (or funiculus) and forming an elongated U-shape.

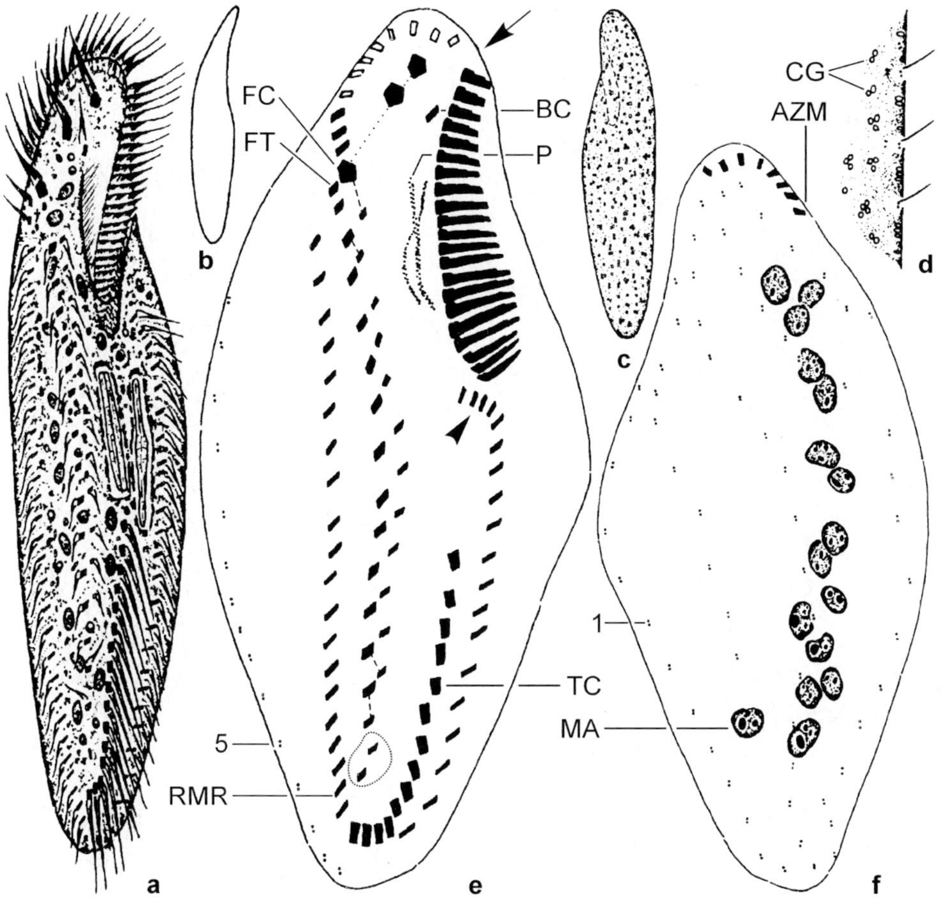

Fig. 32a–f *Holosticha heterofoissneri* (from Hu & Song 2001. a–d, from life; e, f, protargol impregnation). **a:** Ventral view of representative specimen, 141 μm. **b:** Lateral view, according to scale bar only 92 μm long. **c, d:** Dorsal view and detail of cortex showing distribution of cortical granules. **e, f:** Infraciliature of ventral and dorsal side and nuclear apparatus of same specimen, 125 μm. Arrow marks gap in adoral zone of membranelles; arrowhead denotes transversely arranged anterior end of left marginal row. Some cirral pairs are connected by broken lines; pretransverse ventral cirri are encircled. AZM = adoral zone of membranelles, BC = buccal cirrus, CG = cortical granules, FC = frontal cirri (connected by dotted line), FT = frontoterminal cirri, MA = macronuclear nodule, P = paroral, RMR = right marginal row, TC = transverse cirri, 1, 5 = dorsal kineties. Page 152.

granules, and the nuclear apparatus (5–11 macronuclear nodules vs. 14–24). For separation from other *Holosticha* species, see key.

Morphology: The population described by Song et al. (2002) agrees very well with the type material. Thus, the descriptions are combined.

Body size 110–150 × 30–50 μm, usually 115–135 × 32–45 μm in life (Hu & Song 2001a), according to Song et al. (2002) up to 60 μm wide and usually around 140 ×

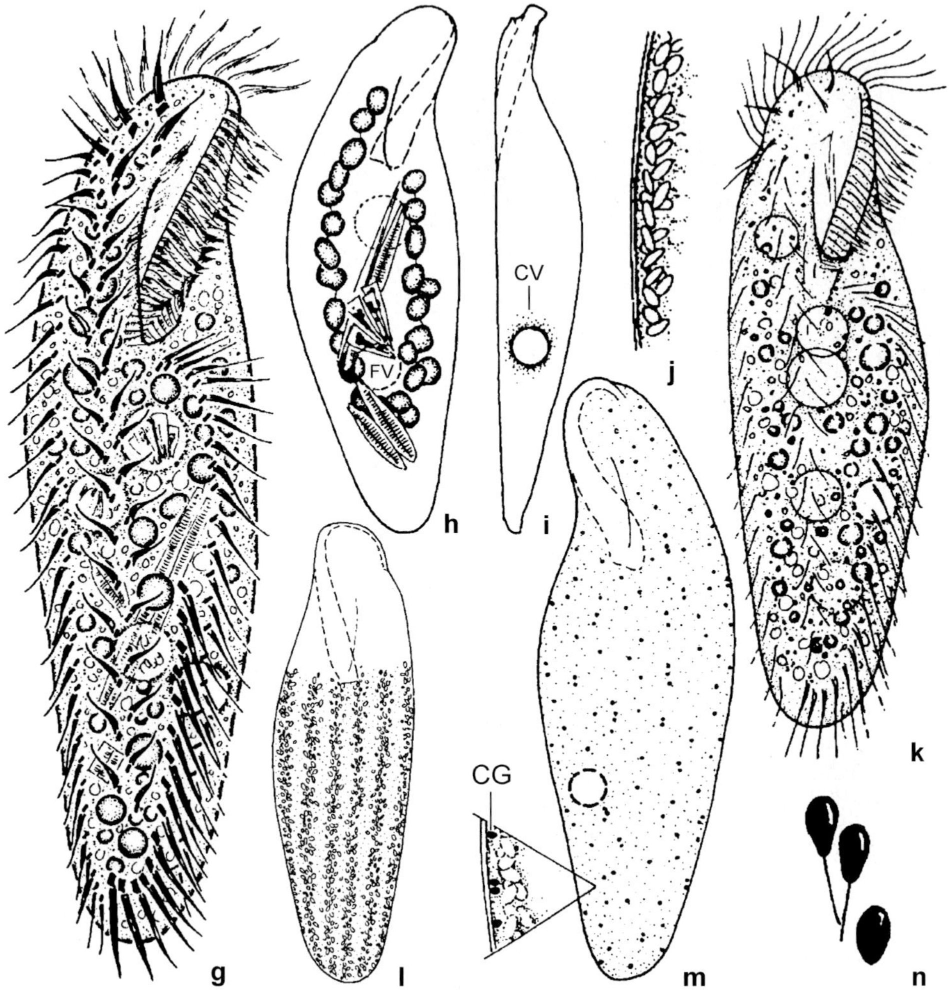

Fig. 32g–n *Holosticha heterofoissneri* from life (from Song et al. 2002). **g:** Ventral view of representative specimen, 141 µm. **h:** Ventral view of a contracted specimen showing U-shaped nuclear apparatus and ingested diatoms. **i:** Left lateral view showing contractile vacuole. **j, l:** Mitochondria in optical section and their distribution in longitudinal rows on dorsal side. **k:** Ventral view of a broader specimen. **m:** Dorsal view showing distribution of cortical granules (detail in inset). **n:** Cortical granules of disturbed specimen. Possibly, these granules are some kind of extrusomes. CG = cortical granules, CV = contractile vacuole, FV = food vacuole. Page 152.

50 µm. Body outline long elliptical to fusiform, that is, anterior and posterior body portion with converging margins; anterior portion sometimes slightly cephalised and curved leftwards; when contracted, right margin more convex than left (Fig. 32h). Body flexible, dorsoventrally flattened 2:1 (Fig. 32b, i). Pellicle thin. Macronuclear nodules form an irregular series in right body-half or a roughly U-shaped pattern (Fig. 32f, h, p); individual nodules about 9 × 7 µm, globular to ovoid, with several large nucleoli,

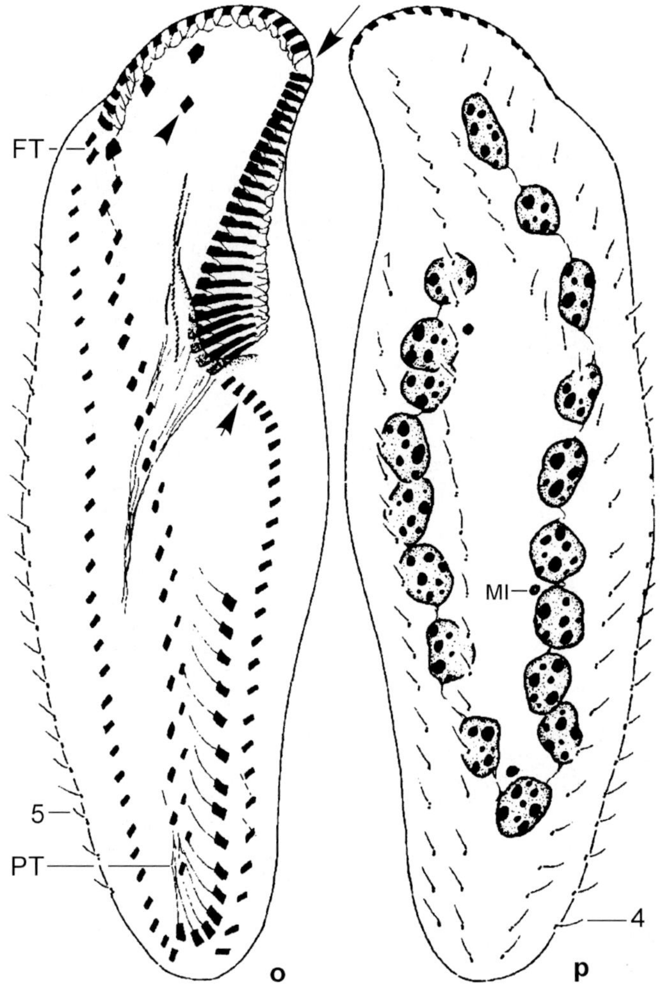

Fig. 32o, p *Holosticha heterofoissneri* after protargol impregnation (from Song et al. 2002). Infraciliature of ventral and dorsal side and nuclear apparatus of same specimen, 131 µm. Long arrow marks gap in adoral zone separating it in an anterior (distal) portion and a spoon-shaped posterior (proximal) portion. Arrowhead denotes the buccal cirrus, which is distinctly ahead of the anterior end of the undulating membranes. The macronuclear nodules are arranged in U-shape and connected by a thin thread. The bases of the transverse cirri are slightly enlarged and form a J-shaped figure which begins at about 60% of body length. The anterior end of the left marginal row is distinctly curved inwards (short arrow). FT = frontoterminal cirri, MI = micronucleus, PT = pretransverse ventral cirrus, 1, 4, 5 = dorsal kineties. Page 152.

connected by an inconspicuous thread-like funiculus. Micronuclei not observed in type population, 3–5 present in population described by Song et al. (2002); individual micronuclei ovoid, about 3 µm across, adjacent to macronuclear nodules (Fig. 32p). Contractile vacuole not found in type population (Fig. 32a), distinctly behind mid-body (at 70% of body length in Fig. 32g) in population described by Song et al. (2002; Fig. 32g, i, k, m). Cortical granules sparsely distributed on dorsal side, form no or small groups composed of 2–3 granules (Fig. 32c, d, m); individual granules spherical to ellipsoidal, 0.5–0.8 µm across, colourless to slightly greenish; granules of disrupted specimens become pear-shaped with an about 1–2 µm long thread (Fig. 32n), do not stain with protargol. Underneath cortex a dense layer of mitochondria forms several longitudinal rows on dorsal side (Fig. 32j, l); individual mitochondria ellipsoidal, about 1.5 µm long, well recognisable at high magnification. Cytoplasm colourless to slightly greyish, with

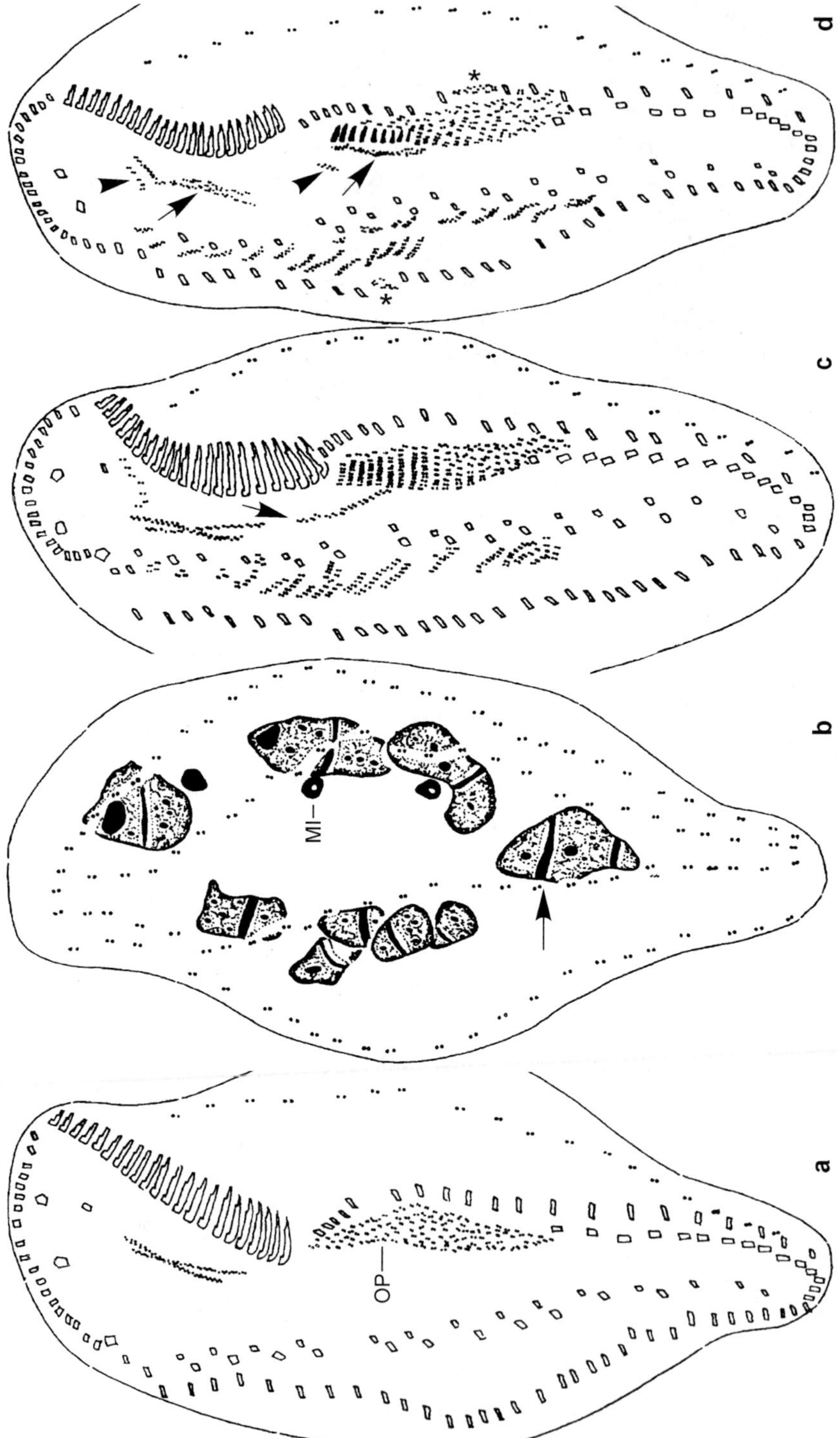
d
c
b
MI
a
OP

numerous granular inclusions 3–5 µm across and several to many colourless globules 5–10 µm in diameter. Movement inconspicuous, that is, more or less slowly crawling, occasionally jerking back and forth, reacting to disturbance by contracting and remaining motionless for a short while.

Adoral zone occupies about 41% of body length on average (Table 13), bipartite by an about one membranelle-long gap (about 3 µm), extends far onto right side; 11–20 membranelles with up to 20 µm long cilia in anterior portion, 20–27 membranelles in posterior portion. Membranelles in posterior portion widen distinctly posteriad (Fig. 32e, o), according to Song et al. (2002) width ranges from 8–15 µm. Buccal field narrow (Fig. 32a, g). Undulating membranes almost equally long, slightly curved and inconspicuously optically crossing, in type population terminating distinctly ahead of proximal end of adoral zone, in population described by Song et al. (2002) paroral extending to second membranelle (Fig. 32e, o).

Cirral pattern and number of cirri basically of usual variability (Fig. 32e, o, Table 13). Frontal cirri distinctly enlarged and 15–20 µm long, right one at distal end of adoral zone. Remaining cirri about 12–15 µm long. Cirrus behind right frontal cirrus of same size as midventral cirri (Fig. 32e, o). Buccal cirrus slightly enlarged, distinctly ahead of anterior end of undulating membranes. Frontoterminal cirri behind or right of right frontal cirrus. Midventral complex composed of cirral pairs only, extends to about 80–85% of body length (Fig. 32e, o), that is, terminates close ahead of pretransverse ventral cirri. Transverse cirri arranged in J-shaped pattern which commences at about 60% of body length; cirral bases distinctly enlarged; cirri about 15 µm long, only posterior ones distinctly projecting beyond body margin (Fig. 32a, e, g, k, o). Right marginal row commences about at level of anterior end of midventral complex, terminates about at rear end of transverse cirral row, separated from left marginal row by distinct gap sometimes occupied by rearmost transverse cirri; anterior end of left marginal row *Holosticha*-specific curved rightwards (Fig. 32e, o); marginal cirri 7–8 µm long, composed of two short kineties.

Dorsal cilia 3–4 µm long, arranged in five kineties; kinety 1 distinctly, kinety 2 only in type population somewhat shortened anteriorly (Fig. 32f, p). Caudal cirri lacking.

Cell division (Fig. 33a–p): This process is described in the original description, where the illustrations are rather small (Hu & Song 2001). Most stages are documented by micrographs, which, however, do note show details clearly. Especially the development of the dorsal ciliature proceeds rather curiously. Unfortunately, the description of this is rather brief so that some uncertainties remain. Thus, I recommend a reinvestigation of the ontogenesis of *H. heterofoissneri*.

Stomatogenesis commences with the formation of a longish oral primordium right of anterior and middle portion of left marginal row (Fig. 33a). Somewhat later, a narrow

← **Fig. 33a–d** *Holosticha heterofoissneri* (from Hu & Song 2001. Protargol impregnation). **a, b:** Very early divider, 126 µm. Arrow marks a replication band. **c:** Early divider with several frontal-midventral-transverse cirral streaks right of the parental midventral complex. Arrow denotes undulating membrane anlage of the opisthe. **d:** Early to middle divider, 149 µm. Arrows mark undulating membrane anlage (= anlage I), arrowheads denote anlage II, and asterisks mark anterior right and posterior left marginal row anlage. MI = micronucleus, OP = oral primordium. Page 152.

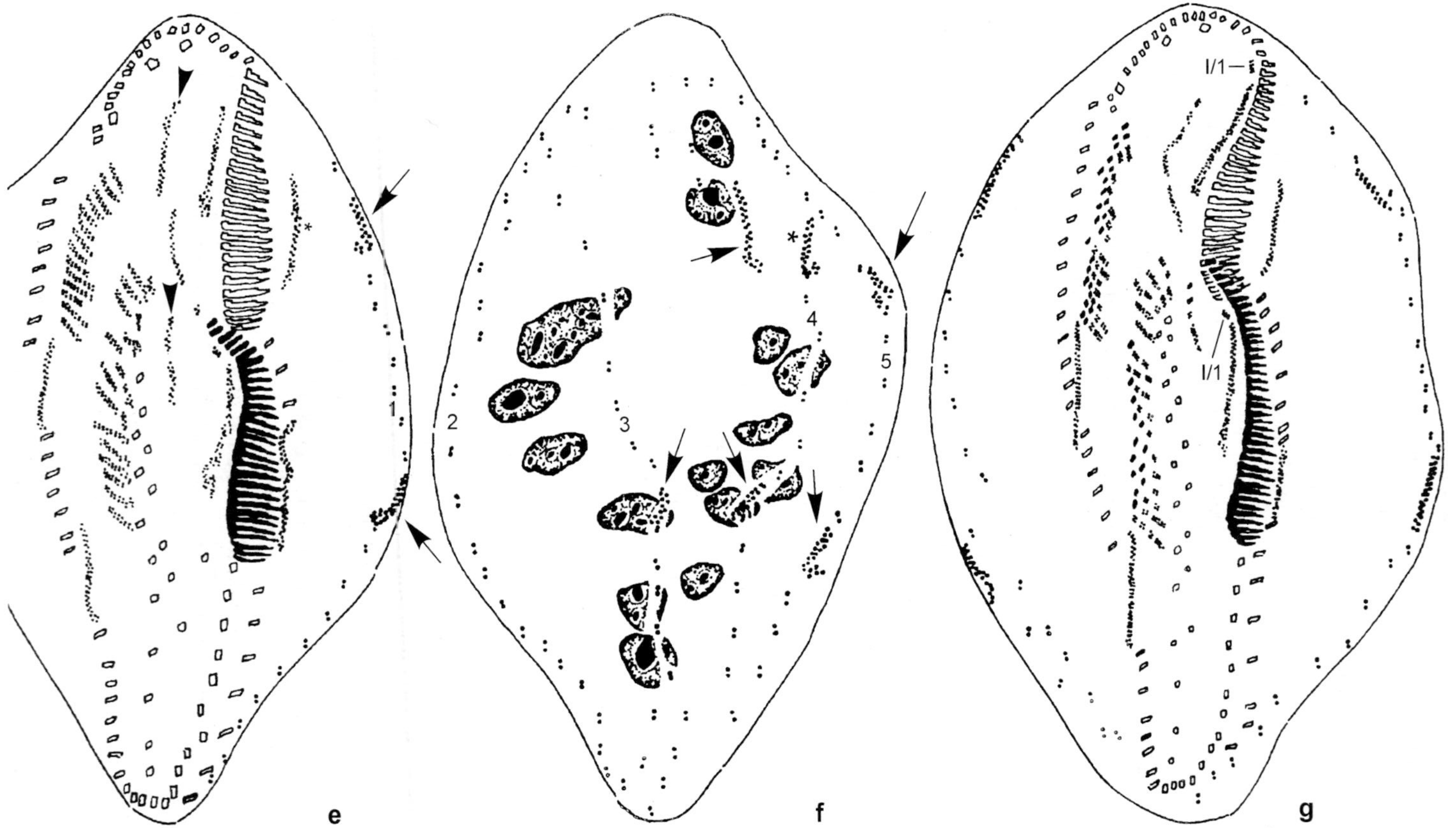

Fig. 33e–g *Holosticha heterofoissneri* (from Hu & Song 2001. Protargol impregnation). **e, f:** Middle divider, 146 µm. Arrowheads denote anlage II, arrows mark anlagen within dorsal kineties. Note that within kinety 2 no anlage is formed. Likewise, no anlage is produced in the anterior portion of kinety 3. Instead, a primordium (asterisk in f) occurs between the anterior anlagen in kinety 4 and kinety 5. Asterisk in (e) marks de novo anlage of left marginal row for proter. **g:** Middle divider. I/1 = leftmost frontal cirrus, 1–5 = parental dorsal kineties. Page 152.

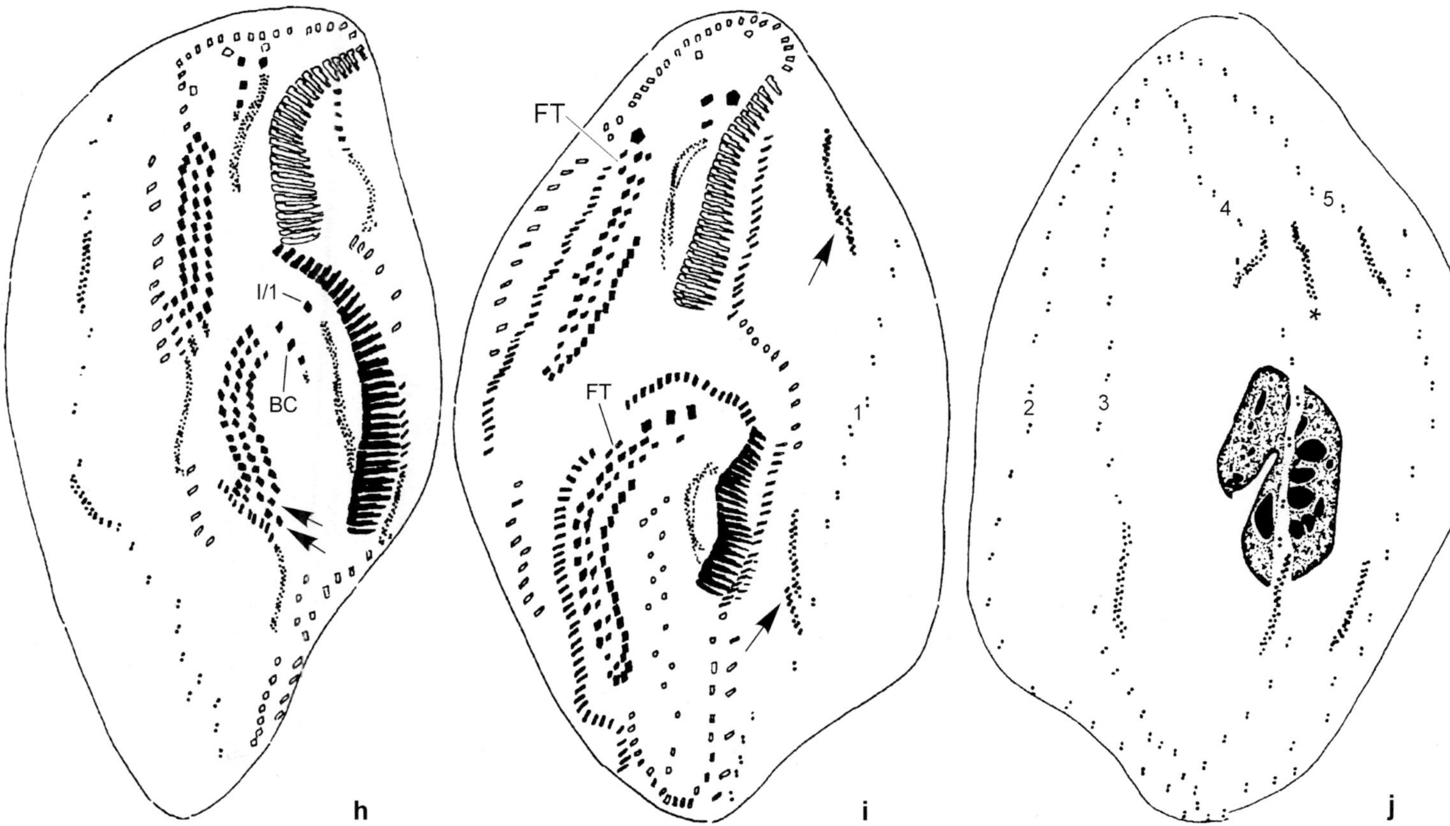

Fig. 33h–j *Holosticha heterofoissneri* (from Hu & Song 2001. Protargol impregnation). Parental infraciliature white, new black. **h:** Late divider, 110 µm. Arrows mark the two rearmost frontal-midventral-transverse cirral anlagen. **i, j:** Late divider, 106 µm. Arrows mark fragmentation of dorsal kinety anlagen in parental kinety 1; asterisk denotes the anlage between the anterior anlagen in kineties 4 and 5. I/1 = leftmost frontal cirrus, BC = buccal cirrus (= cirrus II/2), FT = new frontoterminal cirri, 1–5 = parental dorsal kineties. Page 152.

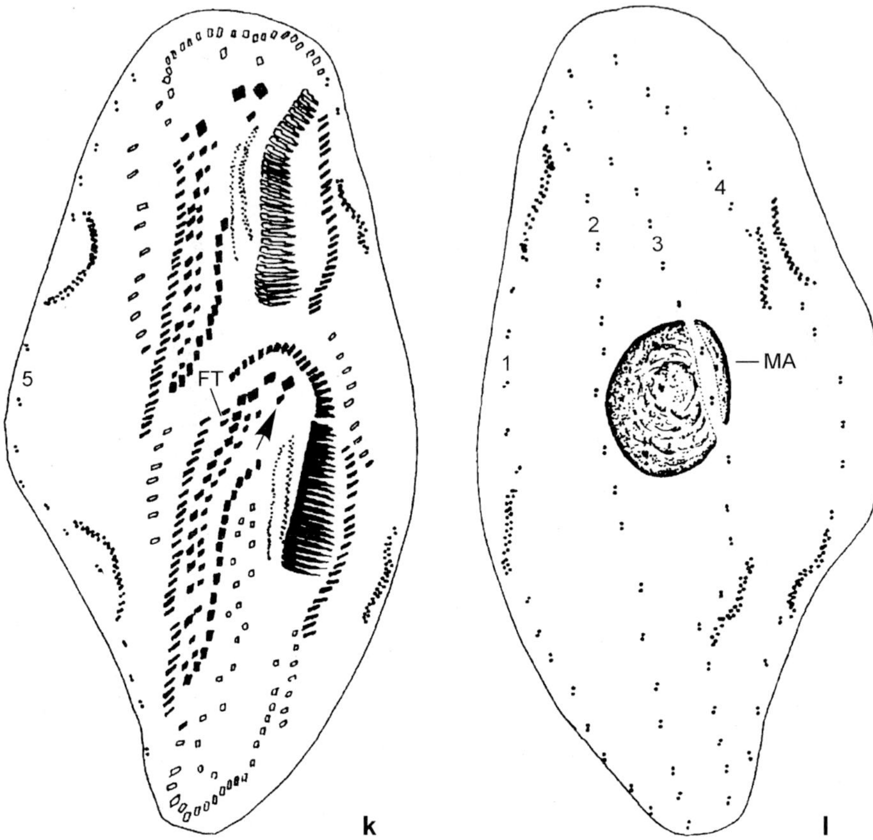

Fig. 33k, l *Holosticha heterofoissneri* (from Hu & Song 2001. Protargol impregnation). Parental infraciliature white, new black. Late divider. Arrow marks buccal cirrus of opisthe, FT = new frontoterminal cirri of opisthe, MA = fused macronucleus, 1–5 = parental dorsal kineties. Page 152.

anlage is formed right of the oral primordium (Fig. 33c, arrow). This anlage later modifies to the undulating membranes of the opisthe. The parental adoral zone is not observably disorganised and is thus completely retained for the proter. The parental undulating membranes disorganise to later become the undulating membranes of the proter again (Fig. 33d, e, g–i). As is usual, at the anterior end of this anlage the leftmost frontal cirrus (= cirrus I/1) is formed (Fig. 33e, g, h).

While in the oral primordium the first membranelles are formed, oblique streaks occur right of the middle portion of the midventral complex (Fig. 33c). Obviously, most midventral pairs remain intact. According to Hu & Song (2001a), each of these streaks breaks into two so that two groups of cirral anlagen, each consisting of 14–17 short streaks, are formed (Fig. 33d, e). By contrast, I suppose that just the middle streaks divide (transversely?) whereas the anterior and posterior streaks do not divide. Meanwhile, a further anlage (called "extra anlage" in the original description) occurs be-

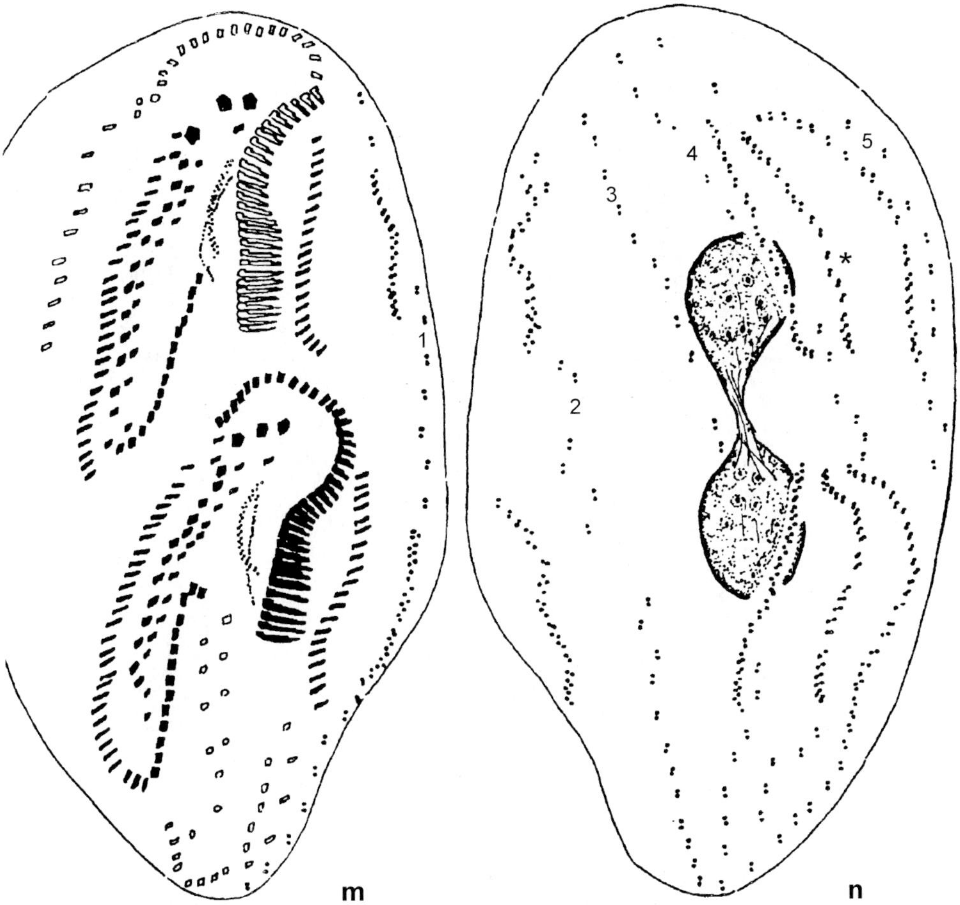

Fig. 33m, n *Holosticha heterofoissneri* (from Hu & Song 2001. Protargol impregnation). Parental infraciliature white, new black. Very late divider. Note that the anterior portion of parental dorsal kinety 3 is almost unchanged, very likely because it does not produce an anlage. Asterisk marks the dorsal kinety anlage which originates between the anterior anlagen originating in parental kineties 4 and 5. 1–5 = parental dorsal kineties. Page 152.

tween undulating membrane anlage and the other frontal-midventral anlagen (arrowheads in Fig. 33d, e). Later, this streak forms the middle frontal cirrus and the buccal cirrus, and even a transverse cirrus (Fig. 33h, i). Hu & Song (2001a) do not write anything about the origin of this anlage. According to Figs. 33c, d, it is the disintegrated parental buccal cirrus, and in the opisthe it very likely originates from the undulating membrane anlage.

During middle stages, all of the frontal-midventral-transverse cirri anlagen form three cirri each, namely a midventral pair (with a slightly enlarged right cirrus) and a transverse cirrus. The rearmost two anlagen form four cirri each. The penultimate streak produces the posteriormost midventral pair, a pretransverse ventral cirrus, and a transverse

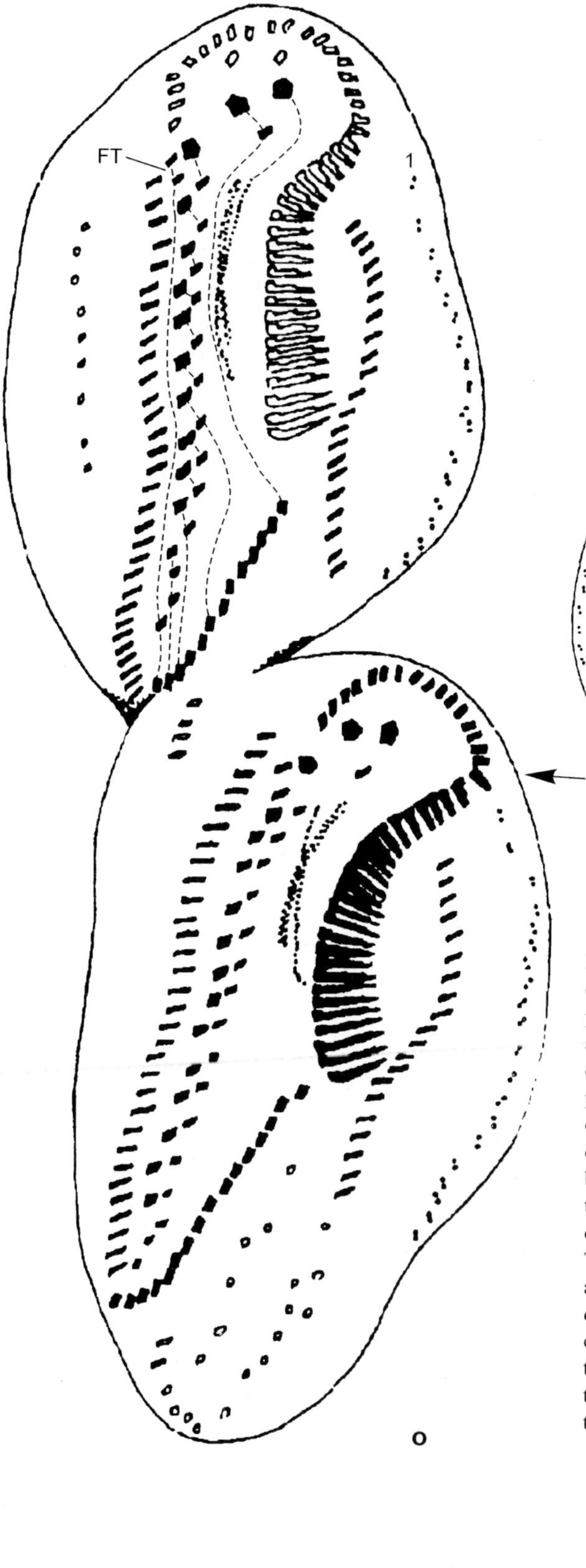

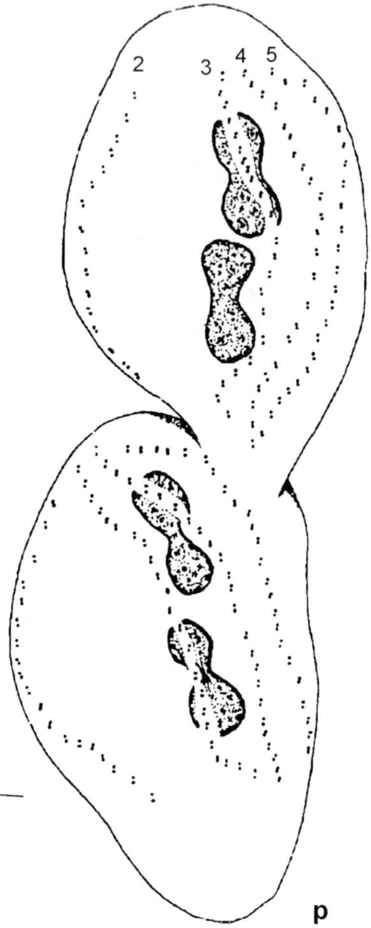

Fig. 33o, p *Holosticha heterofoissneri* (from Hu & Song 2001. Protargol impregnation). Very late divider immediately prior to separation of proter and opisthe. Figure (o) is a very large depiction so that the broken lines, which connect frontal-midventral-transverse cirri originating from same anlage, can be clearly marked in the proter. Note that only 5 of the 13 transverse cirri are connected to their corresponding streak. The anteriormost cirri of the rightmost anlage migrate anteriorly and are thus designated migratory or frontoterminal cirri. Arrow marks site where the gap in the adoral zone will occur. FT = frontoterminal cirri, 1–5 = new dorsal kineties of proter. Page 152.

cirrus. The last anlage produces the two frontoterminal cirri, which migrate anteriorly, a pretransverse ventral cirrus, and the last transverse cirrus (Fig. 33g–i, k, m, o).

According to Hu & Song (2001a), marginal cirral anlagen develop in ordinary manner within the parental marginal rows each at two levels (Fig. 33d, e, g–i, k, k, o) and the anlagen for the proter might be formed somewhat later than those of the opisthe (Fig. 33d, asterisks). However, very likely they mixed up the anterior with the posterior anlage in the right marginal row as clearly recognisable by the next stage (Fig. 33e). Further, the anlage for the left marginal row of the proter obviously originates de novo (Fig. 33e) as in *H. pullaster* (Fig. 28d) and likely also in *H. diademata* (Fig. 25f), strongly indicating that this is an apomorphy of *Holosticha*.

The formation of the five new dorsal kineties commences with the formation of four anlagen each for the proter and the opisthe. Hu & Song (2001a) write that these anlagen are formed within kineties 1, 3, 4, and 5 and refer to Figs. 33e, f. However, this divider and later stages show that two anlagen occur only within kineties 1, 4, and 5 (Fig. 33e–l). Kinety 2 does in fact not contribute to primordia formation. However, Figs. 33f, j, l show that no primordium is formed in the anterior portion of kinety 3. Instead, an anlage occurs between the anterior anlagen of kineties 4 and 5. Supposing that these illustrations are correct this would be a very curious type of dorsal kinety formation, in that the anlagen within kinety 1 fragment to make two anlagen each (Fig. 33i). Such a fragmentation is reminiscent of the oxytrichids, where usually kinety 3 fragments to form kineties 3 and 4. Because of the difference between the description and the illustrations provided by Hu & Song (2001a), I suggest reinvestigating ontogenesis.

Division of the nuclear apparatus proceeds more or less ordinarily, that is, the macronuclear nodules fuse to a single mass which then divides again (Fig. 33b, f, j, l, n, p).

Occurrence and ecology: As yet found only at the marine type locality. Hu & Song (2001a) discovered *H. heterofoissneri* in samples from a mollusc culture in the Yellow Sea near Qingdao (36°08'N 120°43'E), China. Salinity was about 29‰, water temperature 5–13° C, and pH about 8.0. Specimens were cultured in boiled sea water to which squeezed rice grains were added. Song et al. (2002, p. 146) found two populations in the same area in an open scallop (*Chlamys farreri*) farming pond (31–31‰ salinity) on 31 March 1995 and 28 October 1997. I found it in the northern Adriatic Sea near the village of Caorle, Italy, in May 2005. *Holosticha heterofoissneri* feeds, inter alia, on bacteria, diatoms, and flagellates (Hu & Song 2001a, Song et al. 2002).

Holosticha spindleri Petz, Song & Wilbert, 1995 (Fig. 34a–f, Table 13)

1995 ***Holosticha spindleri*** **nov. spec.**[1] – Petz, Song & Wilbert, Stapfia, 40: 164, Fig. 48a–d, 59, 60, Table 24 (Fig. 34a–f; original description; 1 holotype slide [2001/138] and 1 paratype slide [2001/53] of

[1] Petz et al. (1995) provided the following diagnosis: In vivo about 100–115 × 45 µm, elongate. Contractile vacuole equatorial. Left marginal row with 4–8 transversely arranged cirri anteriorly. 2 frontoterminal cirri, 1 buccal and 7–14 transverse cirri. Midventral row long. Usually 4 dorsal kineties. Adoral zone bipartite, consists of 22–43 membranelles, posteriormost membranelle distinctly elongate. 4 macronuclear nod-

protargol-impregnated specimens are deposited in the Oberösterreichische Landesmuseum in Linz [LI], Upper Austria).

2001 ***Holosticha spindleri* Petz, Song and Wilbert, 1995** – Berger, Catalogue of ciliate names 1. Hypotrichs, p. 39 (nomenclator containing all basionyms, combinations, and higher taxa of hypotrichs).

2003 ***Holosticha spindleri* Petz et al., 1995** – Berger, Europ. J. Protistol., 39: 375, 376, Fig. 11 (Fig. 34a; brief review).

2005 ***Holosticha spindleri* Petz, Song & Wilbert (1995)** – Petz, Ciliates, p. 395, Fig. 14.88a–c, 14.175, 14.176 (Fig. 34a–c, e, f; guide to Antarctic marine ciliates).

Nomenclature: This species was dedicated to Michael Spindler, University of Kiel, Germany. "*Holosticha spindeleri*" in Song et al. (2002, p. 155) is an incorrect subsequent spelling.

Remarks: According to the diagnosis, this species is up to 115 μm long in life. By contrast, the maximum value in protargol preparations is 208 μm (Table 13). Although it is known that specimens inflate distinctly by Wilbert's modification, this value (181% of maximum live value!) seems unrealistic if the live data are correct. Thus, likely either the maximum live value (115 μm) is too low, or the maximum value from Table 13 (208 μm) is too high.

Petz et al. (1995) did not mention cortical granules. However, in protargol preparations they found enigmatic, globular structures along the dorsal kineties (Fig. 34d, e). The same structures have been described for *H. bradburyae* (Fig. 35h) as exploded cortical granules (extrusomes), which are blood-cell shaped in live specimens. Possibly, Petz et al. (1995) overlooked these curious structures (mitochondria?), which usually occur in pseudokeronopsids, during live observation of *H. spindleri*. Similar structures are known from *Holosticha heterofoissneri* (Fig. 32n).

I suppose that the cirri circled in Fig. 34b are pretransverse ventral cirri and not the rearmost midventral pair, as suggested by Petz et al. (1995). For general discussion of this detail, see genus section.

For a separation from its congeners, see key. The four macronuclear nodules right of mid-line, the contractile vacuole at or behind mid-body, and the bipartite adoral zone allow a rather simple identification of this little known species even in life.

Morphology: Size 100–115 × 45 μm in life (problems with size, see remarks), body length:width ratio 3.0:1 on average in protargol preparations (Table 13); body size according to Petz (2005) 100–200 × 45–75 μm. Body outline elongate elliptical to spindle-shaped, that is, right and left margin convex, anteriorly rounded, tapering posteriorly (Fig. 34a). Body dorsoventrally flattened about 1.5:1; rather fragile, that is, bursts easily. Macronuclear nodules narrowly spaced, arranged in a longitudinal row in right body portion; individual nodules slightly ellipsoidal, with few large nucleoli 4–7 μm across (Fig. 34a, c, e). Usually four globular, rarely lenticular micronuclei, each attached to a macronuclear nodule (Fig. 34c). Contractile vacuole in or slightly behind mid-body, near left body margin. No cortical granules described (however, see dorsal ciliature below). Cytoplasm contains many slightly greenish shining globules 2–3 μm across and food vacuoles. Movement without peculiarities, that is, crawling on substrate.

---

ules in right body half. Marine.

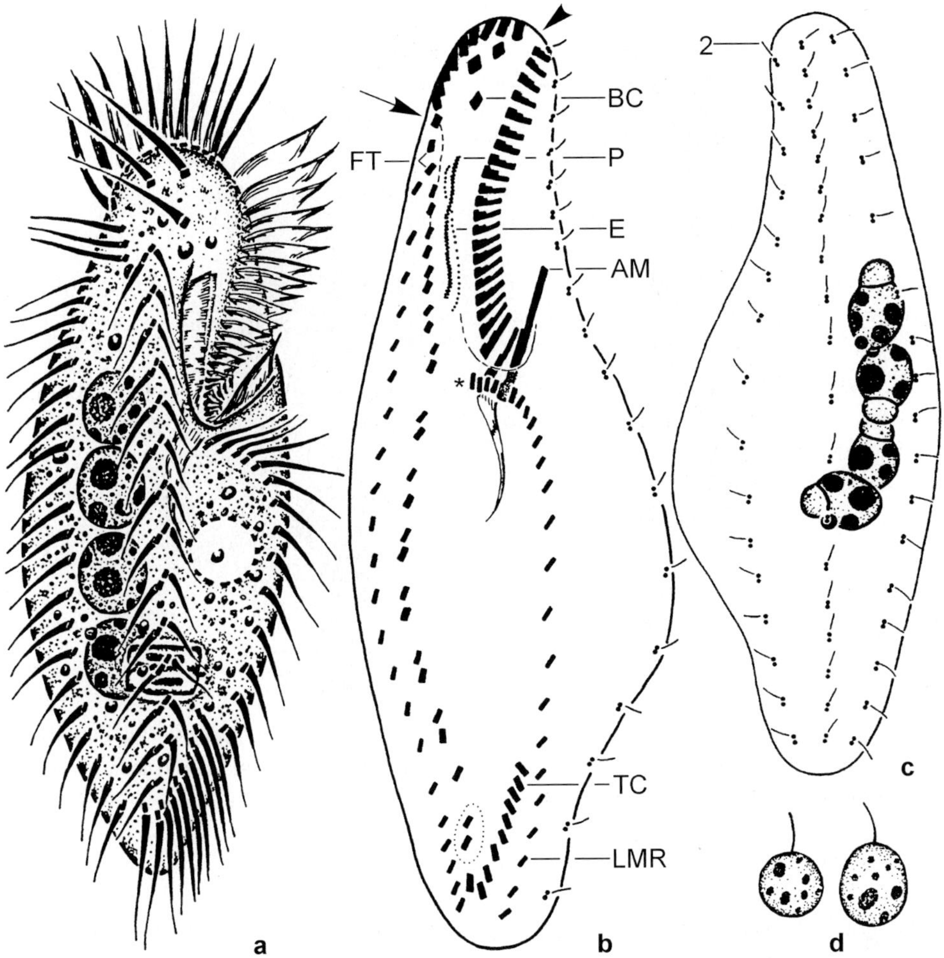

Fig. 34a–d *Holosticha spindleri* (from Petz et al. 1995. a, from life; b–d, protargol impregnation). **a:** Ventral view of a representative specimen, 117 µm. **b, c:** Infraciliature of ventral and dorsal side of same specimen, 131 µm. Arrow marks right frontal cirrus, arrowhead denotes gap in adoral zone. Asterisk marks anterior end of left marginal row which is more or less rectangularly curved rightwards in *Holosticha*. **d:** Enigmatic structures along dorsal kineties (see text). AM = proximalmost adoral membranelle which is distinctly elongated, BC = buccal cirrus, E = endoral, FT = frontoterminal cirri, LMR = left marginal row, P = paroral, TC = transverse cirri, 2 = dorsal kinety 2. Page 163.

Adoral zone occupies about 31–35% of body length, bipartite by about 3 µm wide gap, extends far onto right side; 4–18, on average 14 membranelles in anterior portion, 17–26 membranelles in posterior portion, which is slightly C-shaped in protargol preparations. Proximal-most membranelle about twice as wide as next membranelle; anterior-most membranelle of posterior portion only about 3 µm wide; membranelles composed of four basal body rows, cilia 13–21 µm long. Undulating membranes short, about of same length, arranged almost in parallel; 2-rowed paroral optically intersecting single-

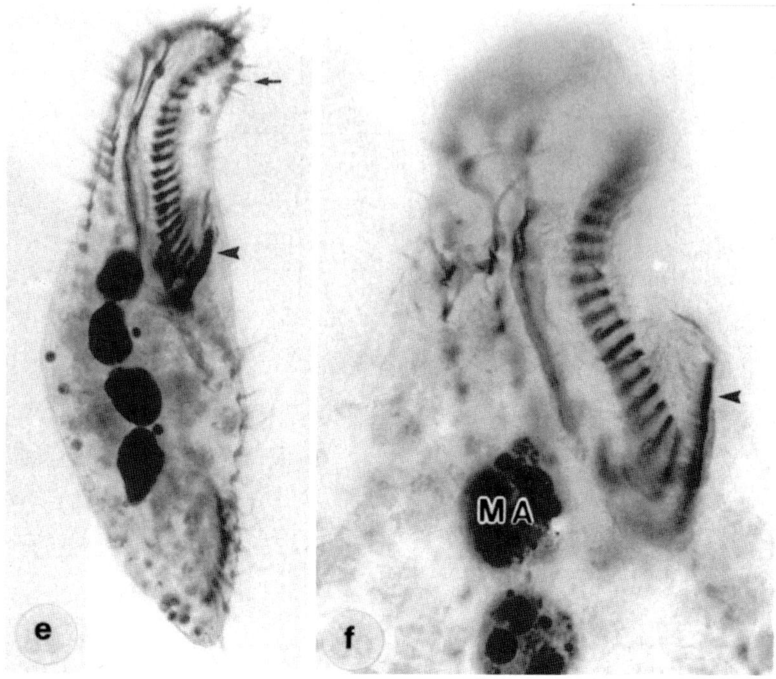

Fig. 34e, f *Holosticha spindleri* after protargol impregnation (from Petz et al. 1995). Ventral view and detail of oral apparatus. Arrowhead in (e) denotes elongated proximalmost adoral membranelle, short arrow marks enigmatic globules along dorsal kineties. Arrowhead in (f) marks elongated adoral membranelle. MA = macronuclear nodule. Page 163.

rowed endoral, cilia about 7 µm long. Buccal field narrow. Pharyngeal fibres about 15–30 µm long (Fig. 34b).

Number of cirri of usual variability (Table 13). Three enlarged frontal cirri, right one very close to distal end of adoral zone of membranelles. Single buccal cirrus also slightly enlarged, arranged immediately behind middle frontal cirrus, that is, distinctly ahead of anterior end of paroral; buccal cirrus and frontal cirri about 17 µm long, bases 4- or 5-rowed. Frontoterminal cirri between right frontal cirrus and anterior end of midventral complex, bases 2- or 3-rowed, cirri about 17 µm long; difficult to distinguish from right marginal row or midventral cirri. Midventral complex composed of cirral pairs only, terminates right of anteriormost transverse cirri; midventral cirri 2- or 3-rowed, about 13 µm long. Very likely two pretransverse ventral cirri present (encircled in Fig. 34b). Transverse cirri not distinctly enlarged, arranged in J-shape, protrude by about half their length beyond rear body end (Fig. 34a, b; length of transverse cirri not mentioned). Marginal rows almost confluent posteriorly, bases consist of two rows of kinetids, cirri 13–20 µm long; anteriormost cirri of left row narrowly spaced and almost transversely arranged immediately behind proximal end of adoral zone.

Dorsal cilia 2.5–6.0 μm long, usually arranged in four, rarely five bipolar kineties composed of about 16–20 cilia each. Caudal cirri lacking (Fig. 34c). In protargol-stained specimens often numerous small globules 2.0–3.5 μm across with a hair-like process 2–3 μm long, positioned along dorsal kineties, nature unknown (extrusomes, parasites?; Fig. 34d, e).

Occurrence and ecology: Marine. As yet found only in the Antarctic region. Type locality is sea ice (core number AN 103107b) of the Weddell Sea, Antarctica (70°21'S 08°53'W) and in areas nearby. Petz et al. (1995) found it together with *H. foissneri*, but less frequently between latitude 69°26'–70°21'S and longitude 07°19'–11°00'W. Occurred together with diatoms, autotrophic and heterotrophic flagellates, and ciliates. Environmental parameters in brine (1 measurement): -3.4°C; salinity 5.9%; in melted ice: chlorophyll *a* 49.3 μg l^{-1}; in raw cultures it was observed at +1°C and a salinity of 2.1%. Petz (2005) recorded it from sea-ice and (rarely) plankton of the Ross Sea area. Feeds on pennate and centric diatoms and green flagellates. Biomass of 10^6 specimens 75 mg.

Holosticha bradburyae Gong, Song, Hu, Ma & Zhu, 2001 (Fig. 35a–k, 36a–q, Table 13, Addenda)

2001 ***Holosticha bradburyae* nov. spec.**[1] – Gong, Song, Hu, Ma & Zhu, Hydrobiologia, 464: 65, Fig. 1–18, Tables 1, 2 (Fig. 35a–k; original description; 1 holotype and 1 paratype slide of protargol-impregnated specimens are deposited in the Laboratory of Protozoology, College of Fisheries, Ocean University of Qingdao, China).

2003 ***Holosticha bradburyae*** – Hu, Song & Suzuki, Europ. J. Protistol., 39: 173, Fig. 1A–H, 2A–G, 3A–K, 4A–J, Table 1 (Fig. 36a–q; morphogenesis).

2003 ***Holosticha bradburyae* Gong et al., 2001** – Berger, Europ. J. Protistol., 39: 375, 376, Fig. 5 (Fig. 35a; brief review).

Nomenclature: This species was dedicated to Phyllis C. Bradbury, North Carolina State University, USA (Gong et al. 2001).

Remarks: This is the largest species of *Holosticha*. As in *H. spindleri* the posteriormost adoral membranelles are very conspicuously widened. This feature and the curious structures (extrusomes?; Fig. 34d, 35b–d, h) in the dorsal side are possibly synapomorphies of these two species. The huge size, the high number of transverse cirri, the brown colour, and the snout-like protrusion at the left anterior corner of the cell make this species distinctive (see key).

Morphology: The following paragraphs are based on the type population. For a brief characterisation of the population studied by Hu et al. (2003), see Table 13.

[1] The diagnosis by Gong et al. (2001) is as follows: Brown-coloured marine *Holosticha* in vivo about 150–320 × 25–75 μm, ca. 53 adoral membranelles and 1 anteriorly positioned buccal cirrus; 3 frontal, 2 frontoterminal and 20–26 transverse cirri; midventral rows comprising 27–32 pairs of cirri; one conspicuous gap always present between anterior and posterior parts of AZM, and 2–5 distinctly elongated membranelles are always present at the posteriormost end; cortical granules conspicuous, round and flattened with central depression, arranged in about 10 lines on dorsal side; 28–33 macronuclear nodules irregularly arranged; 9–11 complete dorsal kineties; caudal cirri absent.

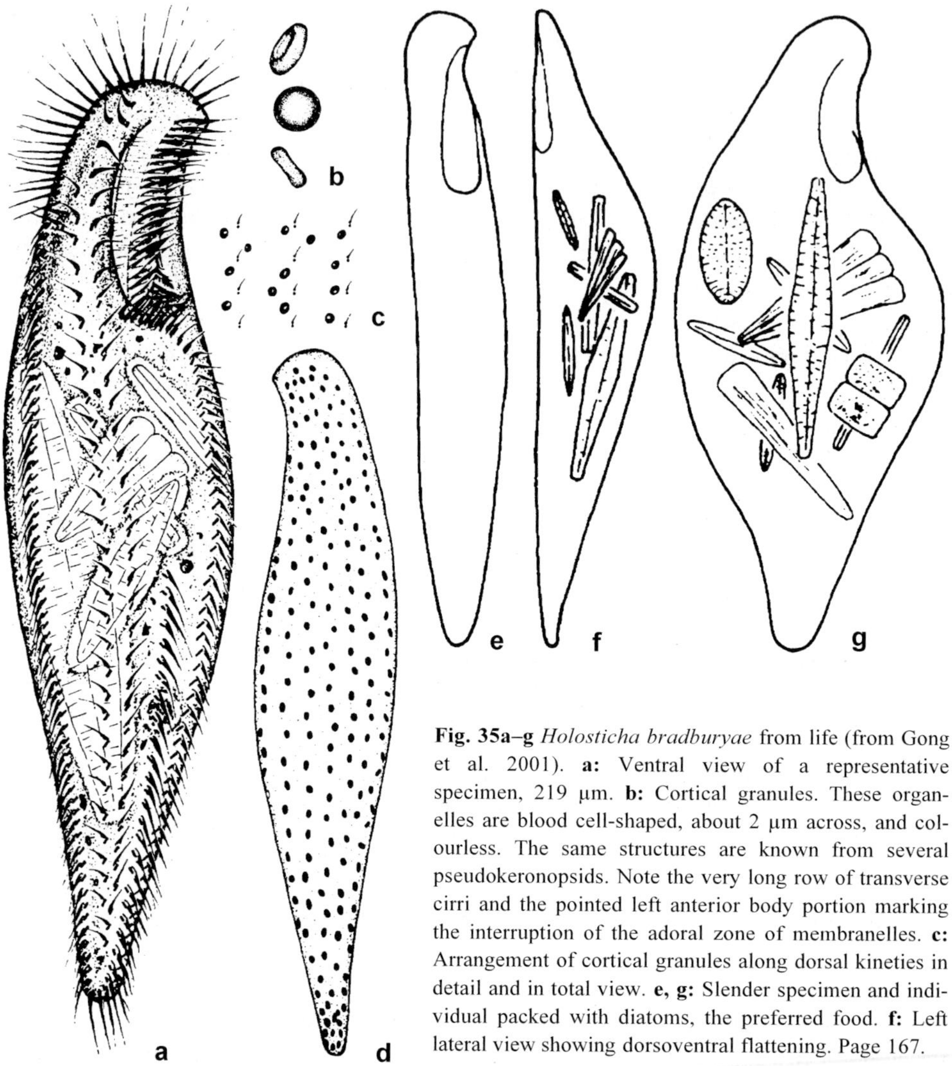

Fig. 35a–g *Holosticha bradburyae* from life (from Gong et al. 2001). **a:** Ventral view of a representative specimen, 219 μm. **b:** Cortical granules. These organelles are blood cell-shaped, about 2 μm across, and colourless. The same structures are known from several pseudokeronopsids. Note the very long row of transverse cirri and the pointed left anterior body portion marking the interruption of the adoral zone of membranelles. **c:** Arrangement of cortical granules along dorsal kineties in detail and in total view. **e, g:** Slender specimen and individual packed with diatoms, the preferred food. **f:** Left lateral view showing dorsoventral flattening. Page 167.

Body size 150–320 × 25–75 μm in life, ratio of body length:width 2.4:1 on average in protargol preparations (Table 13). Body outline elongate fusiform, largely depending on feeding status; starved specimens almost band-shaped, that is, with more or less parallel margins; well fed individuals very wide spindle-shaped and flattened; anterior body portion often slightly bent leftwards with anterior-left region protruding snout-like; margins of posterior body portion usually distinctly converging (Fig. 35a, e, g). Body flexible, dorsoventrally flattened about 2:1 (Fig. 35f). On average 30 macronuclear nodules scattered throughout cytoplasm, individual nodules about 11 × 5 μm (Fig. 35j). Contractile vacuole not observed. Cortical granules about 2 μm across, colourless, discoid with central depression and thus resembling erythrocytes of mammals

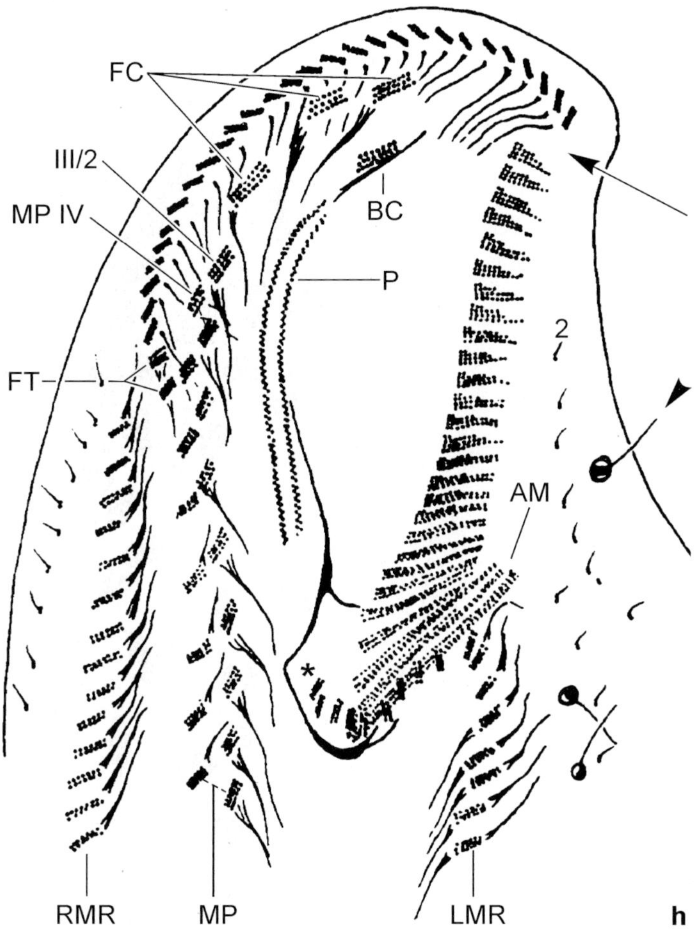

Fig. 35h *Holosticha bradburyae* after protargol impregnation (from Gong et al. 2001). Infraciliature of ventral side of anterior body portion. Arrow marks gap in adoral zone, which is thus divided in an anterior (distal) portion and a posterior (proximal) portion; arrowhead indicates (exploded) cortical granule. The asterisk is at the anterior end of the left marginal row, which is conspicuously curved rightwards anteriorly. AM = elongated proximal adoral membranelles, BC = buccal cirrus, FC = frontal cirri, FT = frontoterminal cirri, III/2 = cirrus III/2, LMR = left marginal row, MP = midventral pair, MP IV = anteriormost midventral pair (originates from anlage IV), P = paroral, RMR = right marginal row, 2 = dorsal kinety 2. Page 167.

(Fig. 35b), longitudinally arranged in about 10 rows on dorsal side more or less along dorsal kineties (Fig. 35c, d); according to Gong et al. (2001) these granules might be a kind of extrusomes because in protargol-impregnated specimens they are often ejected with a hair-like process (Fig. 35h, arrowhead). Cytoplasm brownish to dark brown, especially at low magnification, seemingly due to "dissolved" pigments (the colour is definitely not from the cortical granules or from food inclusions). Food vacuoles large. Movement moderately rapid, crawling on substrate, and always jerking back and forth.

Adoral zone occupies about 33% of body length, bipartite by 4–6 μm wide gap at left anterior body corner, extends moderately far onto right body margin (14% of body length in Fig. 35i); 24–29 membranelles in anterior portion, 24–40 membranelles of ordinary fine structure in posterior portion; cilia of membranelles in anterior portion (according to text) 12–16 μm longer than cilia of posterior membranelles. Bases of posteriormost 2–5 membranelles distinctly wider than remaining membranelles (Fig. 35h, i;

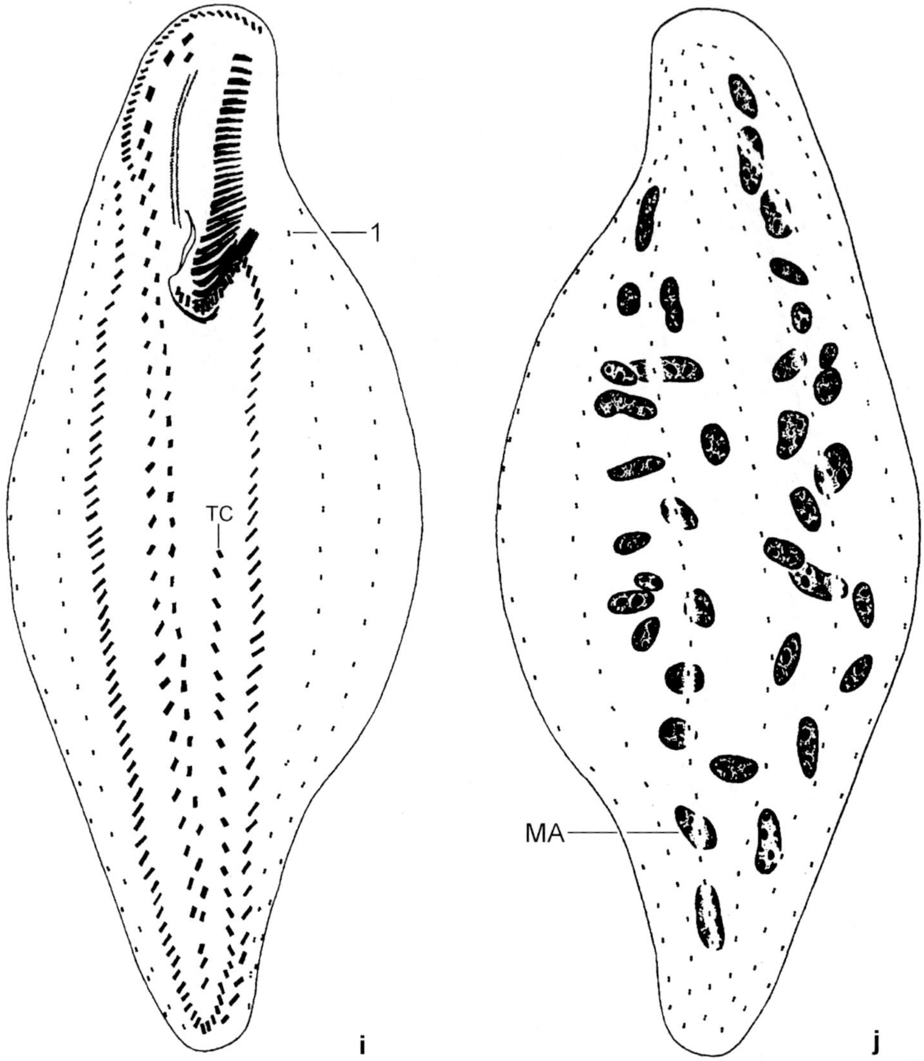

Fig. 35i, j *Holosticha bradburyae* after protargol impregnation (from Gong et al. 2001). Infraciliature of ventral and dorsal side and nuclear apparatus of same specimen, 194 μm. For detailed labelling, see Fig. 35h. MA = macronuclear nodule, TC = transverse cirri, 1 = dorsal kinety 1. Page 167.

Table 13). Undulating membranes of about same length, slightly curved, arranged in parallel (Fig. 35h, i).

Number of cirri of usual variability (Table 13). Three frontal cirri distinctly enlarged, very obliquely arranged. Buccal cirrus between left frontal cirrus and anterior end of undulating membranes, distinctly enlarged (Fig. 35h, i). Two fine frontoterminal cirri very close to distal end of adoral zone and thus difficult to recognise even in protargol preparations. Cirrus II/2 almost continuous with slightly curved row formed by

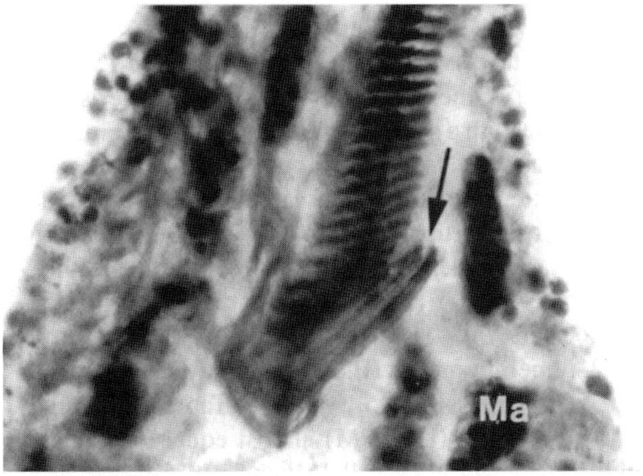

Fig. 35k *Holosticha bradburyae* after protargol impregnation (from Gong et al. 2001). Proximal portion of adoral zone showing the rearmost, strongly widened adoral membranelles (arrow). Ma = macronuclear nodule. Page 167.

frontal cirri, slightly enlarged. Midventral complex composed of cirral pairs only, forms characteristic zigzag pattern, extends far posteriorly and terminates ahead of some (three in Fig. 35i) single cirri, two of which are, as usual, very likely pretransverse ventral cirri; both cirri of each midventral pair of about same size. On average 23 transverse cirri arranged in longitudinal row extending from about mid-body to very close to rear end of right marginal row; bases of transverse cirri not distinctly enlarged. Right marginal row commences near distal end of adoral zone, terminates at rear body end. Anterior end (7–10 cirri) of left marginal row curved rightwards, often underlies[1] proximalmost adoral membranelles (Fig. 35h, i); row terminates at rear end, separated from right row by rear end of transverse cirral row.

Dorsal cilia 3–4 µm long, arranged in 9–11 kineties; central kineties more or less bipolar, marginal kineties partially distinctly shortened anteriorly. Cilia narrowly spaced within rows. Caudal cirri lacking (Fig. 35h, j).

Cell division (Fig. 36a–q): Hu et al. (2003) studied the ontogenesis of *H. bradburyae*. The illustrations are rather small so that some features are very difficult to recognise. Important details are documented in 21 micrographs not shown in the present review. Morphogenesis proceeds basically as in the other *Holosticha* species investigated so far. Thus, the reader is mainly referred to the illustrations and legends because only interesting and deviating features are briefly described below.

The oral primordium of the opisthe originates left of the anteriormost transverse cirri. Apparently, no transverse cirri contribute to the formation of this anlage (Fig. 36c). As in other *Holosticha* species, the frontal-midventral-transverse cirral anlagen originate right of the parental midventral complex. Most parental midventral cirri remain intact (Fig. 36d–f, h). Anlage I produces the leftmost frontal cirrus (= cirrus I/1); anlage II produces the middle frontal cirrus (II/3), the buccal cirrus (II/2), which is distinctly ahead of the undulating membranes, and the leftmost (= anteriormost transverse

[1] Gong et al. (2001, p 68) wrote "overlie" which is incorrect because the marginal row is on the ventral surface and thus on the underside of the specimen; the proximalmost membranelles are more dorsal, that is, closer to the top side, and thus the membranelles overlie the anterior end of the marginal row (and not vice versa!)

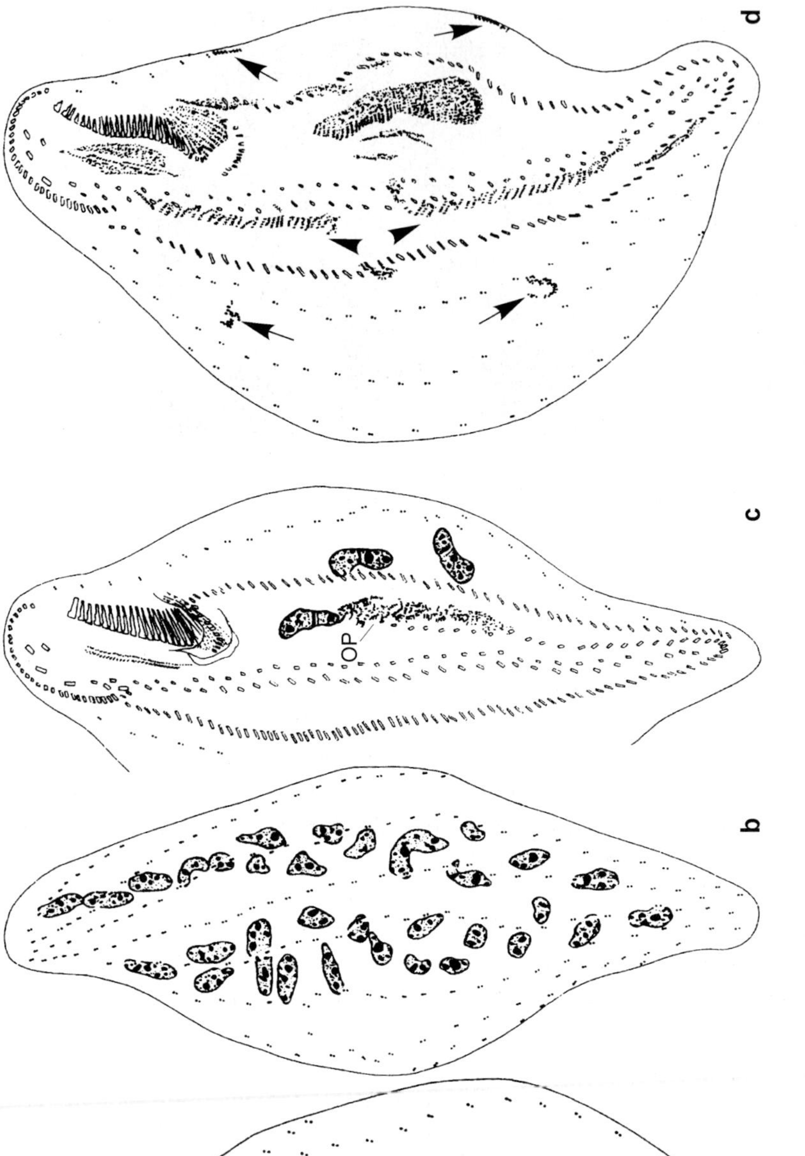

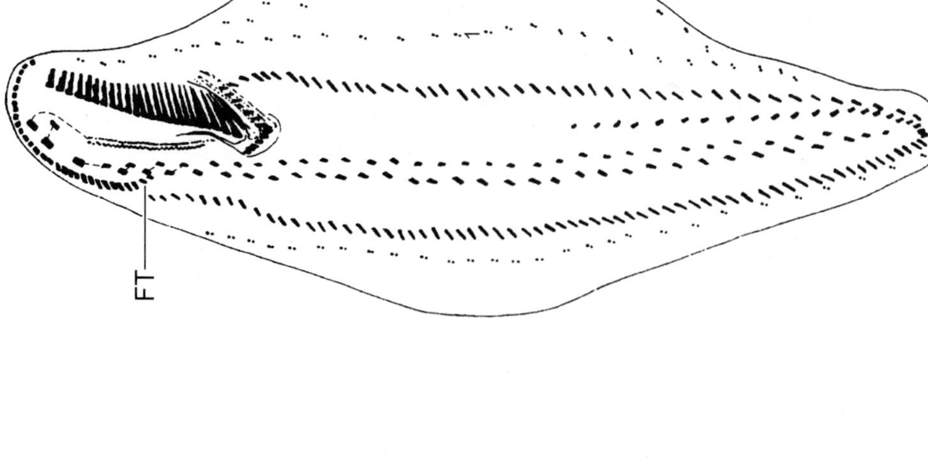

Fig. 36a–d *Holosticha bradburyae* (from Hu et al. 2003. Protargol impregnation). **a, b:** Infraciliature of ventral and dorsal side and nuclear apparatus of non-dividing specimen, 225 µm. **c:** Infraciliature of ventral side and nuclear apparatus of a very early divider, 234 µm. **d:** Infraciliature of ventral side of early to middle divider, size not indicated. Arrows mark anlagen for right and left dorsal kineties, arrowheads denote frontal-midventral-transversal cirral anlagen which originate, like in other *Holosticha* species, right of the parental midventral complex. FT = frontoterminal cirri, OP = oral primordium of opisthe. Page 167.

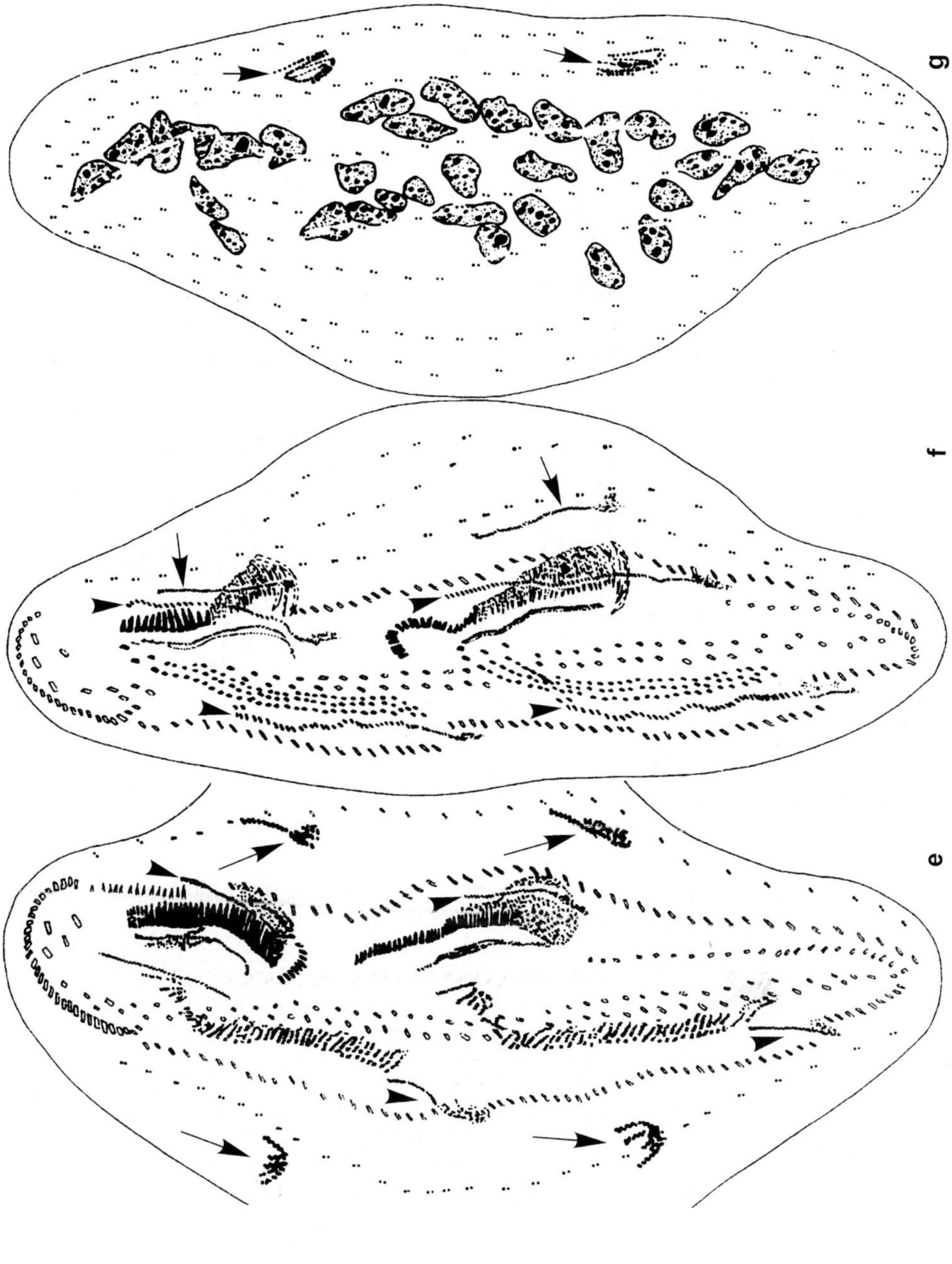

Fig. 36e–g *Holosticha bradburyae* (from Hu et al. 2003. Protargol impregnation). **e:** Infraciliature of ventral side of a middle divider, 270 µm. Arrows mark right and left dorsal kinety anlagen. Arrowheads denote the anlagen for the new right marginal rows. **f, g:** Infraciliature of ventral and dorsal side and nuclear apparatus of a middle divider, size not indicated. Arrows in (f) mark left dorsal kinety anlagen, arrowheads denote new marginal rows (note that the parental marginal rows are hardly involved in anlagen formation). Arrows in (g) denote the right dorsal kineties anlagen. Page 167.

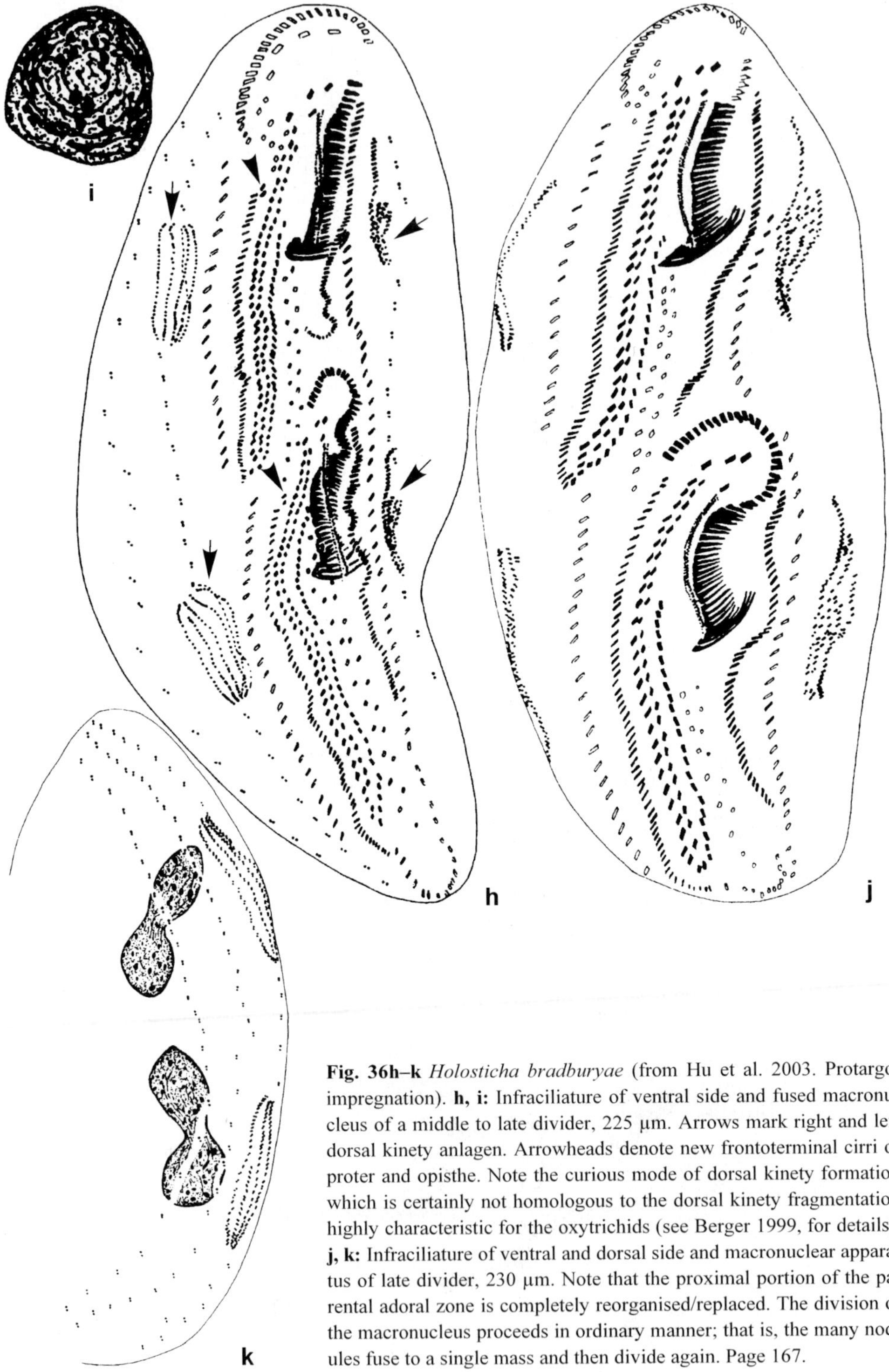

Fig. 36h–k *Holosticha bradburyae* (from Hu et al. 2003. Protargol impregnation). **h, i:** Infraciliature of ventral side and fused macronucleus of a middle to late divider, 225 µm. Arrows mark right and left dorsal kinety anlagen. Arrowheads denote new frontoterminal cirri of proter and opisthe. Note the curious mode of dorsal kinety formation which is certainly not homologous to the dorsal kinety fragmentation highly characteristic for the oxytrichids (see Berger 1999, for details). **j, k:** Infraciliature of ventral and dorsal side and macronuclear apparatus of late divider, 230 µm. Note that the proximal portion of the parental adoral zone is completely reorganised/replaced. The division of the macronucleus proceeds in ordinary manner; that is, the many nodules fuse to a single mass and then divide again. Page 167.

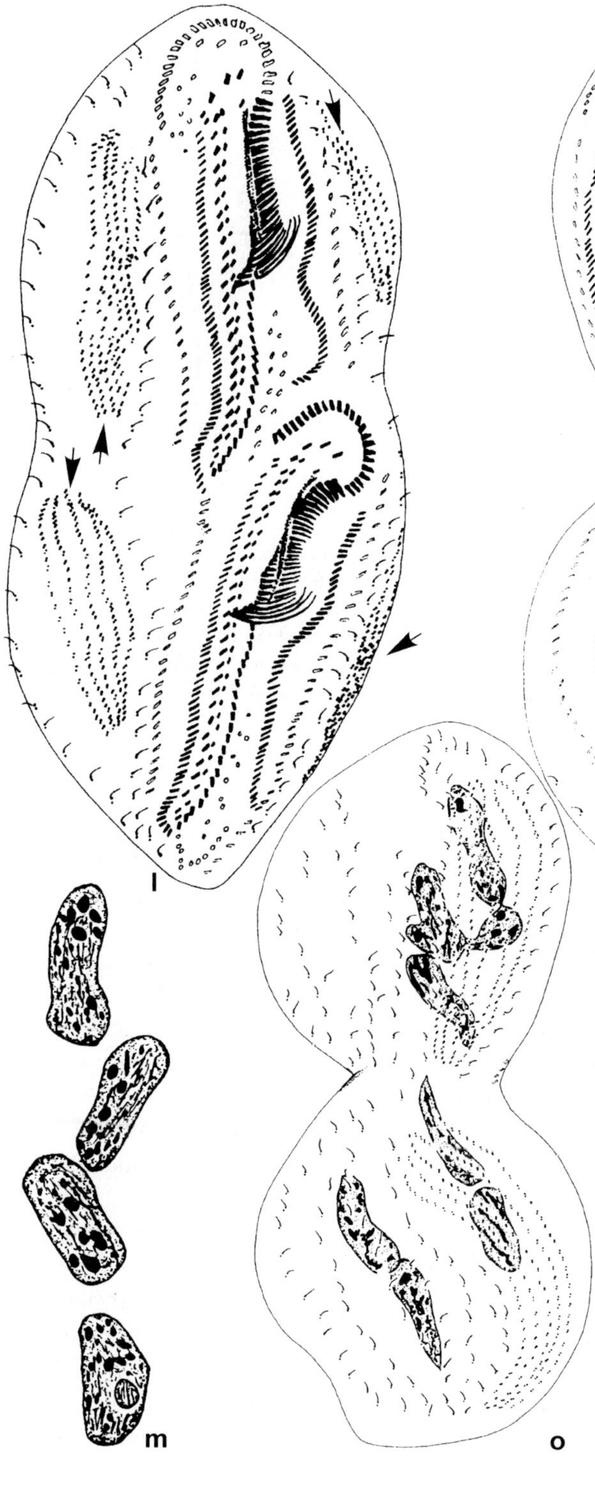

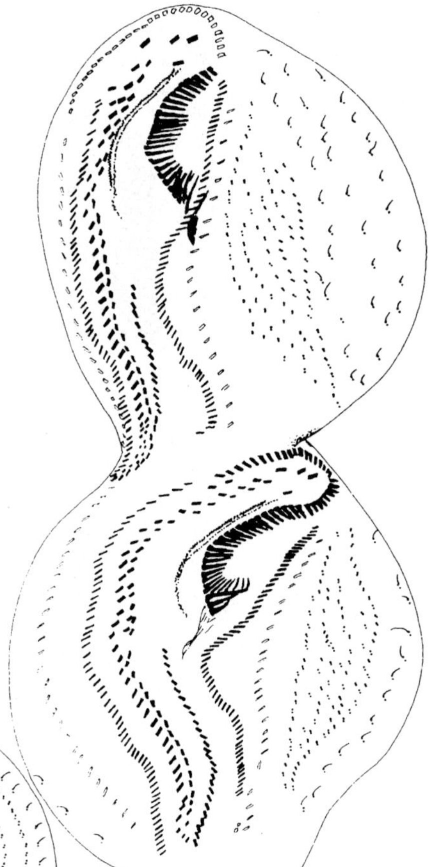

Fig. 36l–o *Holosticha bradburyae* (from Hu et al. 2003. Protargol impregnation). **l, m:** Infraciliature of ventral side and nuclear apparatus of a late divider, size not indicated. Arrows mark the right and left dorsal kinety anlagen. **n, o:** Infraciliature of ventral and dorsal side and nuclear apparatus of a very late divider, 297 µm. Note that the about 10 dorsal kineties of each filial product originate from two anlagen fields, both of which produce about five kineties. The proximal portion of the parental adoral zone, which is separated from the distal portion by a distinct gap, is completely reorganised, respectively, replaced. Page 167.

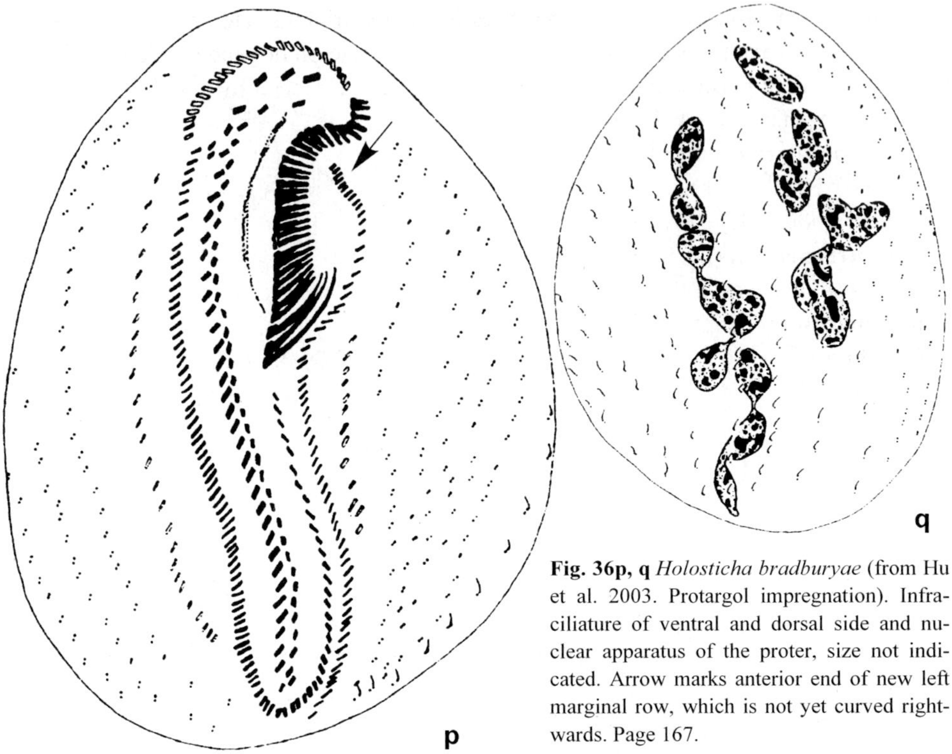

p q

Fig. 36p, q *Holosticha bradburyae* (from Hu et al. 2003. Protargol impregnation). Infraciliature of ventral and dorsal side and nuclear apparatus of the proter, size not indicated. Arrow marks anterior end of new left marginal row, which is not yet curved rightwards. Page 167.

cirrus, Fig. 36f, opisthe; possibly this anlage does not always produce a transverse cirrus); anlage III generates the rightmost frontal cirrus (III/3), the cirrus behind the right frontal cirrus (III/2), and a transverse cirrus; the anlagen IV to n–2 produce a midventral pair each and a transverse cirrus; anlage n–1 generates the rearmost midventral pair, the left pretransverse ventral cirrus, and the penultimate (second from right) transverse cirrus; anlage n generates the two frontoterminal cirri, the right pretransverse ventral cirrus, and the rightmost transverse cirrus (Fig. 36f, h, j, l, n, p).

The development of the marginal rows shows some peculiarities: (i) only very few parental cirri are involved in primordia formation; (ii) the new right rows extend left of the parental row, the new left rows extend right of the parental row; (iii) the left marginal row for the proter very likely originates de novo (as in congeners), although this is not clearly recognisable from the illustrations (Fig. 36d, e).

The formation of the dorsal kineties proceeds very interestingly. Each two anlagen originate de novo very close to the leftmost kinety (= dorsal kinety 1) and the rightmost kinety (Fig. 36d). These anlagen elongate due to the proliferation of basal bodies. In middle dividers, each left anlage is composed of a long row and several short rows, while the right anlagen comprise 5–6 equally long kineties (Fig. 36e–h, j–l). Subsequently, these anlagen migrate to their final positions and replace the parental kineties (Fig. 36n–q). At first glance this mode of dorsal kinety formation is somewhat reminiscent of the fragmentation of dorsal kineties in the oxytrichids. However, I suppose that the processes are not homologous, that is, do not indicate phylogenetic relationship.

The proximal portion of the parental adoral zone is replaced by a new one, inter alia, due to reorganisation of the parental portion. The distal membranelles of the parental adoral zone are retained for the proter (Fig. 36d–f, h, j, l, n, p).

The division of the nuclear apparatus shows no peculiarities; that is, the many macronuclear nodules fuse to a single mass prior to the division of the cell.

Occurrence and ecology: As yet found only at the type locality. *Holosticha bradburyae* was discovered from an open scallop (*Chlamys farreri*) farm in the Yellow Sea near Qingdao on 22 December 2000. Gong et al. (2001) provided the following data: 8°C, salinity 29–31‰, pH 7.4, 8.5 mg l^{-1} O_2. Hu et al. (2003) found it in the same area (36°08'N; 120°43'E) in December 2001. They used boiled seawater with crushed rice grains to support microbial growth.

Species indeterminata

***Holosticha aquarumdulcium* Bürger, 1905**, An. Univ. Chile, 117: 437, Lamina VIII, Fig. 1 (Fig. 164i, j). Remarks: The species-group name refers to the habitat (freshwater) where the species was discovered. The original description lacks some important details of the cirral pattern so that it will very likely never be possibly to identify this species with certainty. Kahl (1932, p. 589, Fig. 101_{28}) mentioned it under the somewhat confusing name *Trichotaxis (Holosticha) aquarum dulcium*; since he classified *Trichototaxis* as subgenus of *Holosticha*, the correct name in his paper is *Holosticha (Trichototaxis) aquarumdulcium* Bürger, 1905. Borror (1972, p. 11) synonymised it with *T. stagnatilis*, type of *Trichototaxis*. Likely for that reason it was not mentioned in the review by Borror & Wicklow (1983, p. 119), who eliminated *Trichototaxis*. *Trichotaxis aquarumdulcium* (Bürger, 1905) in Stiller (1974b, p. 53) is a new combination because she considered Bürger's species as valid. Body 320 × 80 µm in size, very flexible and contractile. Roughly elliptical, right margin straight, left distinctly curved, both ends rounded. Ventral side plane, dorsal vaulted. Many macronuclear nodules dispersed throughout cell. Contractile vacuole about in mid-body near left cell margin. Presence/absence of cortical granules unknown. Movement serpentine. Adoral zone occupies about 40% of body length. Buccal field moderately wide. Distalmost 16 adoral membranelles larger and stronger than remaining. Cirral pattern rather fragmentarily described so that identification will never be possible. Type of frontal ciliature (three enlarged frontal cirri, bicorona, or other pattern) neither described nor illustrated. Presence/absence of midventral complex unknown. Four enlarged transverse cirri protrude distinctly beyond rear body end. In total likely five cirral rows, that is, one each near cell margins; two further rows extend in left body portion, that is, these are likely two additional left marginal rows; and one row commences near distal end of adoral zone and terminates about in mid-body (note that Bürger confused left and right). Dorsal ciliature unknown. Type locality is Santiago de Chile, where Bürger (1905) discovered it in a freshwater habitat (briefly also mentioned by Bürger 1908, p. 188).

Table 13 Morphometric data on *Holosticha bradburyae* (br1, from Gong et al. 2001; br2, from Hu et al. 2003), *Holosticha diademata* (di1, from Hu & Song 1999; di2, from Song & Wilbert 2002), *Holosticha foissneri* (foi, from Petz et al. 1995), *Holosticha heterofoissneri* (he1, from Hu & Song 2001; he2, from Song et al. 2002), *Holosticha pullaster* (pu1, from Petz et al. 1995), and *Holosticha spindleri* (spi, from Petz et al. 1995)

Characteristics[a]	Species	mean	M	SD	SE	CV	Min	Max	n
Body, length	br1	208.6	–	20.6	4.4	9.9	148.0	236.0	22
	br2	211.1	–	15.0	3.7	7.1	186.0	232.0	16
	di1	55.9	–	14.55	1.0	26.0	51.0	65.0	15
	di2	66.7	–	9.2	3.5	13.7	52.0	77.0	12
	foi	134.1	130.0	22.5	4.7	16.8	93.0	174.0	23
	he1	107.6	–	14.8	3.7	13.8	86.0	130.0	16
	he2	116.5	–	18.0	6.4	15.5	90.0	133.0	8
	pu1	48.2	48.0	5.9	1.2	12.3	36.0	59.0	23
	spi	139.9	139.0	27.5	5.0	19.6	86.0	208.0	30
Body, width	br1	85.8	–	10.4	2.6	12.1	64.0	96.0	16
	br2	85.1	–	8.4	2.4	9.9	68.0	96.0	12
	di1	42.6	–	3.0	0.8	7.0	38.0	48.0	15
	di2	42.0	–	6.5	1.9	15.5	21.0	54.0	12
	foi	40.4	39.0	8.7	2.0	21.6	29.0	55.0	23
	he1	53.8	–	10.0	2.5	18.6	30.0	70.0	16
	he2	44.9	–	10.7	4.0	23.8	40.0	56.0	7
	pu1	21.8	22.0	3.6	0.8	16.5	14.0	28.0	23
	spi	45.9	42.0	15.3	3.1	33.3	25.0	84.0	30
Anterior body end to proximal end of adoral zone, length	br1	67.4	–	5.3	1.2	7.9	58.0	76.0	21
	br2	68.3	–	5.5	1.4	8.0	56.0	75.0	16
	di1	24.3	–	3.1	0.8	12.7	20.0	29.0	15
	di2	28.6	–	6.4	1.8	22.4	20.0	38.0	12
	foi	48.4	48.0	7.8	1.9	16.1	36.0	61.0	23
	he2	48.7	–	2.7	1.0	5.5	44.0	52.0	7
	he1	45.7	–	6.1	1.5	13.2	35.0	55.0	16
	pu1	18.4	17.5	2.4	0.5	13.3	14.0	25.0	23
	spi	51.9	54.0	9.4	2.2	18.1	30.0	64.0	30
Adoral membranelle 2[j], width	foi	7.5	7.0	1.1	0.3	14.4	6.0	8.0	23
	spi	9.5	10.0	1.3	0.3	13.2	8.0	12.0	30
Adoral membranelle 1[j], width	br1	28.5	–	4.3	0.9	15.2	20.0	36.0	23
	foi	6.8	7.0	0.8	0.3	11.6	6.0	8.0	23
	spi	19.2	20.0	4.1	1.0	21.2	11.0	28.0	30
Adoral membranelles, total number	br1	52.8	–	1.6	0.4	3.1	50.0	56.0	21
	br2	54.2	–	2.5	0.6	4.6	51.0	60.0	15
	di1	27.0	–	1.9	0.5	6.9	24.0	30.0	15
	di2	28.1	–	3.8	1.2	13.6	21.0	33.0	10
	foi	32.1	32.0	2.9	0.6	9.0	26.0	36.0	23
	he1	42.6	–	3.5[c]	0.9	8.2	33.0	47.0	14
	he2	44.6	–	3.3	1.1	7.3	41.0	49.0	9
	pu1	18.4	19.0	2.5	0.6	13.6	10.0	21.0	23
	spi	35.8	37.0	4.0	0.7	11.2	22.0	43.0	30
Elongate membranelles, number	br1	3.2	–	0.6	0.1	18.3	2.0	5.0	23
	br2	2.9	–	0.3	0.1	8.5	2.0	3.0	16
Undulating membranes, length	foi	22.0	23.0	4.5	1.2	20.2	14.0	28.0	23
	pu1	6.5	6.5	0.8	0.2	11.7	5.0	8.0	23
	spi	21.8	22.0	5.6	1.3	25.6	11.0	35.0	30

Table 13 Continued

Characteristics[a]	Species	mean	M	SD	SE	CV	Min	Max	n
Macronuclear nodule, length	br2	12.9	–	1.5	0.4	11.8	11.0	16.0	16
	di1	11.5	–	0.9	0.2	8.0	11.0	14.0	15
	foi	10.7	11.0	3.3	0.6	31.0	6.5	20.0	23
	he1	6.8	–	1.3	0.3	18.7	4.0	9.0	16
	he2	8.8	–	2.1	0.7	23.8	6.0	12.0	16
	pu1[g]	11.3	11.0	2.6	0.6	23.2	7.0	18.0	23
	pu1[h]	10.1	10.5	1.9	0.4	18.8	6.0	14.0	23
	spi	16.8	16.5	3.9	0.7	23.2	11.0	26.0	30
Macronuclear nodule, width	br2	6.5	–	0.7	0.2	11.2	6.0	8.0	16
	di1	9.2	–	0.8	0.2	8.2	8.0	10.0	15
	foi	8.7	8.0	2.2	0.4	25.9	6.0	13.0	23
	he1	4.9	–	1.0	0.3	20.2	3.0	7.0	16
	he2	7.0	–	2.3	0.7	32.9	4.0	10.0	16
	pu1[g]	6.6	6.0	1.4	0.3	20.8	5.0	11.0	23
	pu2[h]	6.8	6.5	1.4	0.3	21.2	5.0	12.0	23
	spi	12.3	12.0	2.5	0.5	20.5	8.0	17.0	30
Macronuclear nodules, number	br1	29.9	–	2.0	0.4	6.6	28.0	33.0	21
	br2	32.9	–	2.6	0.6	7.8	27.0	39.0	16
	di1	2.0	–	0.0	0.0	0.0	2.0	2.0	15
	foi	7.5	8.0	1.1	0.2	15.2	5.0	11.0	23
	he1	15.7	–	0.6	0.2	3.8	14.0	16.0	16
	he2[k]	17.3	–	2.0	0.6	11.4	14.0	21.0	12
	pu1	2.0	2.0	0.0	0.0	0.0	2.0	2.0	23
	spi	3.7	4.0	0.7	0.1	18.5	1.0	4.0	30
Micronucleus, length	foi	2.7	3.0	0.4	0.1	15.0	2.0	3.0	23
	pu1	–	–	–	–	–	2.0	2.0	?
	spi	2.4	2.5	0.3	0.1	14.0	2.0	3.0	30
Micronucleus, width	foi	2.4	2.3	0.5	0.1	18.9	2.0	3.0	23
	pu1	–	–	–	–	–	2.0	2.0	?
	spi	2.4	2.5	0.3	0.1	14.0	2.0	3.0	30
Micronuclei, number	foi	3.2	3.0	1.0	0.3	32.3	2.0	6.0	23
	he1	15.7	–	0.6	0.2	3.8	14.0	16.0	16
	pu1	–	–	–	–	–	2.0	2.0	?
	spi	3.3	4.0	0.9	0.2	27.8	2.0	4.0	30
Frontal cirri, number	br1	3.0	–	0.0	0.0	0.0	3.0	3.0	23
	br2	3.0	–	0.0	0.0	0.0	3.0	3.0	16
	di1	3.0	–	0.0	0.0	0.0	3.0	3.0	15
	di2[f]	4.0	–	0.0	0.0	0.0	4.0	4.0	12
	he1	3.0	–	0.0	0.0	0.0	3.0	3.0	15
	he2[f]	4.0	–	0.0	0.0	0.0	4.0	4.0	12
Frontoterminal cirri, number	br1	2.0	–	0.0	0.0	0.0	2.0	2.0	23
	br2	2.0	–	0.0	0.0	0.0	2.0	2.0	16
	di1	2.0	–	0.0	0.0	0.0	2.0	2.0	15
	di2	2.0	–	0.0	0.0	0.0	2.0	2.0	13
	foi	2.1	2.0	0.3	0.1	13.4	2.0	3.0	23
	he1	2.0	–	0.0	0.0	0.0	2.0	2.0	15
	he2	2.0	–	0.0	0.0	0.0	2.0	2.0	12
	pu1	2.0	2.0	0.0	0.0	0.0	2.0	2.0	23
	spi	2.0	2.0	0.0	0.0	0.0	2.0	2.0	30

Table 13 Continued

Characteristics[a]	Species	mean	M	SD	SE	CV	Min	Max	n
Buccal cirri, number	br2	1.0	–	0.0	0.0	0.0	1.0	1.0	16
	di1	1.0	–	0.0	0.0	0.0	1.0	1.0	15
	di2	1.0	–	0.0	0.0	0.0	1.0	1.0	12
	foi [e]	1.0	–	0.0	0.0	0.0	1.0	1.0	23
	he1	1.0	–	0.0	0.0	0.0	1.0	1.0	15
	he2	1.0	–	0.0	0.0	0.0	1.0	1.0	12
Midventral cirral pairs, number	br1	29.0	–	1.8	0.6	6.1	27.0	32.0	10
	br2 [d]	61.2	–	4.9	1.2	8.0	53.0	69.0	16
	di1 [d]	14.7	–	1.5	0.4	10.1	13.0	17.0	15
	di2	8.7	–	1.0	0.3	12.4	7.0	10.0	12
	foi [b]	17.7	17.0	1.3	0.4	7.4	16.0	20.0	23
	he1 [d]	26.5	–	6.7	1.6	24.8	23.0	31.0	18
	he2	13.4	–	1.3	0.5	9.5	12.0	15.0	7
	pu1 [i]	10.1	10.0	0.4	0.1	3.5	10.0	11.0	23
	spi [b]	13.5	14.0	2.1	0.5	15.8	9.0	17.0	30
Pretransverse ventral cirri, number	di1	2.0	–	0.0	0.0	0.0	2.0	2.0	15
	he1	2.0	–	0.0	0.0	0.0	2.0	2.0	15
Transverse cirri, number	br1	22.7	–	2.4	0.7	10.6	20.0	26.0	11
	br2	25.0	–	1.9	0.5	7.6	21.0	27.0	15
	di1	8.9	–	1.1	0.3	12.0	6.0	10.0	15
	di2	9.8	–	1.2	0.4	12.5	7.0	11.0	12
	foi	11.5	11.0	2.5	0.7	22.0	9.0	17.0	23
	he1	15.4	–	17	0.4	10.8	11.0	18.0	17
	he2	14.3	–	1.5	0.5	10.4	12.0	17.0	8
	pu1	6.7	6.5	0.8	0.2	11.7	6.0	8.0	23
	spi	10.8	11.0	1.7	0.3	16.0	7.0	14.0	30
Left marginal row, number of cirri	br1	55.6	–	3.3	0.9	6.0	49.0	60.0	14
	br2	54.8	–	2.8	0.8	5.1	50.0	59.0	13
	di1	10.7	–	1.2	0.3	10.8	8.0	12.0	15
	di2	14.3	–	1.5	0.4	10.4	12.0	16.0	12
	foi	21.6	21.0	2.0	0.7	9.3	20.0	26.0	23
	he1	20.6	–	1.4	0.3	6.6	18.0	23.0	16
	he2	21.8	–	2.25	0.8	10.4	19.0	25.0	10
	pu1	9.5	9.0	1.1	0.3	11.2	9.0	12.0	23
	spi	22.2	22.0	2.1	0.4	9.4	19.0	26.0	30
Right marginal row, number of cirri	br1	58.6	–	3.0	0.8	5.1	53.0	63.0	16
	br2	59.3	–	3.7	0.9	6.2	55.0	67.0	16
	di1	11.8	–	1.1	0.3	9.2	10.0	14.0	15
	di2	17.7	–	2.0	0.6	11.0	14.0	21.0	11
	foi	23.4	23.0	2.0	0.7	8.5	21.0	27.0	23
	he1	27.7	–	3.7	1.0	13.3	18.0	33.0	15
	he2	30.4	–	2.0	0.7	6.6	27.0	33.0	10
	pu1	10.0	10.0	0.7	0.2	7.1	9.0	11.0	23
	spi	21.7	22.0	3.0	0.6	13.8	11.0	26.0	30
Dorsal kineties, number	br1	10.0	–	0.7	0.2	7.4	9.0	11.0	12
	br2	10.1	–	1.0	0.3	10.1	8.0	11.0	16
	di1	4.0	–	0.0	0.0	0.0	4.0	4.0	8
	di2	4.3	–	0.5	0.1	10.6	4.0	5.0	12
	foi	4.0	4.0	0.0	0.0	0.0	4.0	4.0	23
	he1	5.0	–	0.0	0.0	0.0	5.0	5.0	16

Table 13 Continued

Characteristics[a]	Species	mean	M	SD	SE	CV	Min	Max	n
Dorsal kineties, number	he2	5.0	–	0.0	0.0	0.0	5.0	5.0	16
	pu1	4.2	4.0	0.4	0.1	9.3	4.0	5.0	23
	spi	4.1	4.0	0.3	0.1	7.3	4.0	5.0	30

[a] All measurements in µm. ? = number of individuals investigated (= sample size) not indicated; if only one value is known it is listed in the mean column; if two values are available they are listed as Min and Max. CV = coefficient of variation in %, M = median, Max = maximum value, mean = arithmetic mean, Min = minimum value, n = number of individuals investigated, SD = standard deviation, SE = standard error of arithmetic mean. Data based on protargol-impregnated specimens.

[b] The last two cirri (encircled in Fig. 31b, 34b) are not a midventral pair, but the pretransverse ventral cirri. Very likely, pseudo-pairs have been counted (see Fig. 1a for terminology).

[c] The value 34.8 mentioned in the original description is likely incorrect. The value presented above was calculated from the standard error of arithmetic mean.

[d] The number of midventral cirri is given.

[e] From Hu & Song (2001).

[f] Cirrus behind right frontal cirrus (= cirrus III/2) included.

[g] Anterior macronuclear nodule.

[h] Posterior macronuclear nodule.

[i] Obviously the number of midventral cirri is given (pretransverse cirri likely not included).

[j] Adoral membranelle 1 is the proximalmost one, membranelle 2 the next one.

[k] Specimen shown in Fig. 32h has 24 macronuclear nodules.

Table 14 Autecological data of *Holosticha pullaster*. References: column **1**, from Heuss (1976; 163 analyses from German running waters); **2**, **3**, from Bernerth (1982; 2 = total range, 3 = optimum range; many analyses from the river Main near the inlet of the cooling water of a power station); **4**, **5**, from Foissner et al. (1991; 4 = total occurrence in various Austrian running waters; 5 = abundant and very abundant occurrence only; number of analyses in brackets)

Parameter[a]	Reference				
	1	2	3	4	5
Frequency (%)	–	–	–	32	15
pH	6.2–9.6	6.6–8.1	–	7.1–8.6 (44)	7.2–8.4 (20)
Temperature (°C)	0–21	2–30	–	3–13 (44)	3.5–11 (11)
O_2 (mg l^{-1})	2.4–27.0	0.2–12	>5	8–14 (44)	8.1–13.7 (20)
O_2 (% saturation)	26–240	–	–	68–128 (44)	70–114 (20)
BOD_5 (mg l^{-1})	1.3–44	–	–	0.4–>8.6 (44)	0.9–6.8 (20)
DOC (mg l^{-1})	–	6.4–27	–	1.7–2.7 (6)	–
Total hardness (°dH)	–	–	–	4.1–13.7 (33)	4.1–12.8 (17)
$KMnO_4$-consumption (mg l^{-1})	6.3–170	–	–	6–51 (37)	8–39 (18)
NH_4^+-N (mg l^{-1})	0.08–13.5	–	–	<0.01–2.4 (44)	<0.01–0.8 (20)
NO_3^--N (mg l^{-1})	–	–	–	0.4–15 (41)	0.4–5.2 (18)
NO_2^--N (mg l^{-1})	–	–	–	0–0.06 (33)	0.002–0.02 (14)
PO_4^{3-}-P (mg l^{-1})	0.07–4.0	–	–	0.006–0.8 (40)	0.006–0.25 (16)
Bacteria (ind. ml^{-1}; plate method; × 10^6)	–	0.005–0.2	0.14–0.2	0.0001–1.3 (40)	0.0003–1.3 (20)

[a] The figure "0" denotes only that the parameter in question could not be detected with the analytical method used.

***Holosticha begoniensis* Fernandez-Leborans, 1990**, Arch. Protistenk., 138: 51, Fig. 1–8 (Fig. 99e). Remarks: I do not know to which feature the species-group name refers. This species is undoubtedly misclassified in *Holosticha*. Fig. 3 in the original description clearly shows a bicorona, strongly indicating that Fernandez-Leborans observed a *Pseudokeronopsis* species. Since the adoral zone of membranelles lacks a gap and since transverse cirri are present it can be excluded that the population belongs to *Uroleptopsis*. Unfortunately, the original description does not contain live data and the illustration is rather misleading. Since live observations (colour of cytoplasm; cortical granulation) are indispensable for a reliable identification of *Pseudokeronopsis* species, Fernandez-Leborans' population is likely indeterminable. I provide only the illustration and some important morphometric data (arithmetic mean, and minimum and maximum value; n = 80!): body length = 193.9 µm, 188–199 µm; body width = 55.1 µm, 48.2–67.5 µm; number of macronuclear nodules = 57.8, 55–64; length of macronuclear nodule = 6.7 µm, 5.4–8.4 µm; length of PF1 (paroral?) = 27.7 µm, 25.8–30.6 µm; length of adoral zone = 80.7 µm, 72.0–85.8 µm; number of left ventral cirri (is about number of midventral pairs) = 25.1; 24–26; number of adoral membranelles = 49–55 (the mean with 27.1 is obviously a wrong value); number of right marginal cirri = 42.8, 39–45; number of left marginal cirri = 47.8, 46–50; number of frontal cirri (cirri forming the bicorona) = 15.9; 12–20; number of pretransverse ventral cirri = 2.1, 2–3. Coast of Pedreña (Cantabria, Spain), Cantabrian Sea.

***Holosticha fontinalis* Lepsi, 1926**, Arch. Hydrobiol., 17: 755 (Fig. 100a, 191m). Remarks: The species-group name (Latin adjective; *fontinal·is -is -e*; living in clear water) refers to the habitat (spring) where the species was discovered. The original description is rather detailed, however, without illustration. Figures were provided by Lepsi (1926b, p. 85, Tafel XIII, Fig. 450) and Lepsi (1927, p. 120, Fig. 60). I agree with Kahl (1932, p. 577, Fig. 101_{22}) that it is very likely a mutilated specimen. Kahl (1932) wrote that the size is not given, indicating that he did not have the paper by Lepsi (1926a). I provide a brief translation of the original description by Lepsi (1926a) and the illustrations given in Lepsi (1926b, 1927). The paper by Lepsi (1927) is in Romanian and thus not considered further. Size in life 126 × 45 µm. Body anteriorly slightly wider than posteriorly, body ends widely rounded; flexible, colourless. Eight elliptical macronuclear nodules, rosary arrangement (according to this feature it could be a synonym of *Anteholosticha monilata*). Contractile vacuole near left margin at beginning of second body third, contracts every 20 seconds at 15 °C; one collecting canal. Cytoplasm finely granulated, ectoplasm of greater viscosity and with radial, rod-shaped structures underneath pellicle (likely cortical granules around dorsal bristles and/or cirri). Feeds on bacteria and diatoms. Creeps for a long time in the same place and often moves backwards very rapidly. Frontal scutum lacking. Adoral zone 27 µm long, 9.5 µm wide, composed of about 15 well-developed membranelles with frontmost 13 µm long and 9 µm wide. Adoral zone conspicuous because extending in midline of cell. Undulating membrane (paroral?) widest at beginning of last third. Frontal cirri indistinctly set off from ventral rows which extend over the frontal area (Stirnfeld). Near the right peristome margin some slightly stronger cirri, which are, however, not distinctly set off from the ventral rows. Transverse cirri numerous, but not distinctly set off from the two ventral rows, indicating that a midventral complex was present. Caudal cirri lacking. Discovered in a rheocrene spring near the city of Cavarna, south-eastern Dobrudscha, Romania/Bulgaria (Lepsi 1926a; see also Detscheva 1972, p. 77).

***Holosticha longiseta* Lepsi, 1951**, Buletin sti. Acad. Repub. pop. rom., 3: 519, Fig. 11 (Fig. 100b, c). Remarks: The species-group name is a composite of the Latin adjective *longus* (long) and the Latin substantive *seta* (bristle) and obviously refers to the long cirri (dorsal bristles). This species is also briefly described by Lepsi (1965, p. 218, Fig. 97k). The original description is in Romanian and thus not translated by me. However, the illustration is to superficial to allow an identification. There are only five cirri on the frontal area and some behind the adoral zone of membranelles; possibly this is the oral primordium of an early divider. Body length about 80 μm. Discovered in an agricultural soil in Romania.

***Holosticha obliqua* Kahl, 1928**, Arch. Hydrobiol., 19: 211, Abb. 44c (Fig. 100g). Remarks: The species-group name *obliqu·us -a -um* (Latin adjective; oblique) possibly refers to the oblique body outline. The description contains most data necessary for an identification of hypotrichs (see below). However, Kahl (1932, p. 582, Fig. 106 $_{26}$; Fig. 100h in present book) stated that *H. obliqua* seems to be a teratological (pathological) form of another species, *Holosticha setifera* Kahl, 1932 (now *Caudiholosticha setifera*). Borror (1972, p. 11) considered the name *H. obliqua* as valid and put *H. setifera* in its synonymy; a procedure approved by Hemberger (1982, p. 91). According to Borror & Wicklow (1983, p. 118), *Holosticha setifera* (Fig. 54a) has some oxytrichid features (caudal cirri, un-urostyline-like arrangement of cirri) and thus supposed that it is related to *Gastrostyla*. I do not understand this proposal because *H. setifera* has a distinct midventral cirral pattern and the caudal cirri indicate that it belongs to *Caudiholosticha*. According to the ICZN (1964, Article 1; see also ICZN 1999, Article 1.3.2), names given to teratological specimens are excluded from zoological nomenclature and are thus not available. Of course, nobody knows whether or not Figure 100g indeed shows a teratological specimen. However, I accept Kahl's (1932) statement and thus mention *Holosticha obliqua* under the heading species indeterminata (see also *Caudiholosticha setifera*, for discussion). Most features provided in Kahl's (1928) description are shown in the illustration. Characters not shown: body length 90 μm; dorsal side strongly vaulted; cytoplasm with ring-shaped structures. Collected in a slightly saline (2.9‰) water containing many rhodobacteria near the German village of Alt-Fresenburg. The redescriptions of *H. obliqua* by Aladro Lubel et al. (1986, 1990) are insufficient (see below).

***Holosticha oxytrichoidea* Németh in Gelei, 1950**, Acta biol. hung., 1: 80, Abb. 8d (see Berger 1999, Fig. 222a). Remarks: The species-group name refers to the similarity with an *Oxytricha*. The reference "Nemeth" in Gelei (1950) is given without source, indicating that it is an unpublished paper/observation. A description in Gelei (1950) is lacking.

The cirral pattern (18-frontal-ventral-transverse cirri) indicates that it is an 18-cirri oxytrichid, possibly a *Sterkiella* species. *Holostischa oxytrichoidea* in the legend is an incorrect subsequent spelling of *Holosticha*. Likely found in Hungary.

Holosticha salina **Fernandez-Leborans & Novillo, 1993**[1], Zool. Jb. Anat., 123: 207, Fig. 1, Table 1 (Fig. 37a). Remarks: No derivation of the name is given in the original description. The species-group name *salin·us -a -um* (Latin adjective; salty) likely refers to the habitat (sea) where the species was discovered. Locality where type slides are deposited not mentioned.

The description and the illustration of this species show many shortcomings, which likely make an identification impossible. I list only some of the deficiencies: (i) according to the table and the text of the original description the present species has 38–40 adoral membranelles, whereas the specimen illustrated has only 23 (!) membranelles (Fig. 37a); (ii) according to the table, 68–70 right marginal cirri against only 59 cirri in specimen illustrated; (iii) according to the table, 57–62 left marginal cirri against 52 in Fig. 37a (in addition, the variability for these features is unbelievable low for n = 80!); (iv) no transverse cirri are mentioned, but the rearmost four cirri of the right marginal row are slightly set off from the remaining cirri (Fig. 37a), indicating that these could be transverse cirri; in addition, such a strong overlapping of the marginal rows is unlikely; (v) arrangement of cirri in frontal region likely not correctly shown (probably due to the silver carbonate impregnation technique which inflates the hypotrichs and thus disfigures the cirral pattern); (vi) presence/absence of cortical granules unknown. Because of these discrepancies this population has to be classified as species indeterminate. I omit most morphometric data presented in Table 1 of the original description and simple show the specimen illustrated (diagnosis see footnote). Body size 210–241 × 36–48 µm after silver carbonate impregnation; 50–62 macronuclear nodules; 4–6 dorsal kineties; usually four, rarely three frontoterminal cirri. Collected at Castrourdiales beach (Santander, Spain), Biscay Bay, Atlantic Ocean.

Holosticha setigera **(?)** – Conn, 1905, Bull. Conn. St. geol. nat. Hist. Surv., 2: 60, Fig. 265, 265a (Fig. 100i, j). Remarks: In the review on ciliate names (Berger 2001, p. 39) I wrote that I do not know the original description of this species. Now I am certain that Conn (1905) is the original description because in cases of "redescriptions" Conn always added the name of the author to the binomen. In the present case he did it not, but the question mark indicates that he was not quite sure about the validity of his species. Since Conn provided no description, but only two tiny figures, which just show that this species indeed has a midventral complex, I classify it as species indeterminata.

---

[1] The diagnosis provided by Fernandez-Leborans & Novillo (1993) is as follows: Elongated ciliates with the anterior end rounded and the posterior end slightly more narrow than the rest. The body, dorsoventrally compressed, is 200 to 240 µm long and 36 to 48 µm wide. 50 to 62 macronuclear nodules and 4 to 8 spheric micronuclei. Oral area [(4 to 57 µm) × (16 to 19 µm)] in V-shape composed of 38 to 40 adoral organelles and a paroral formation as a stichomonad. 5 frontal cirri. 4 (3) frontoterminal cirri. Ventral cirri (50 to 54) in zig-zag with the greater of them located on the left. 57 to 62 left marginal cirri. 68 to 70 right marginal cirri that extend posteriorly towards to left border. Marine habitat.

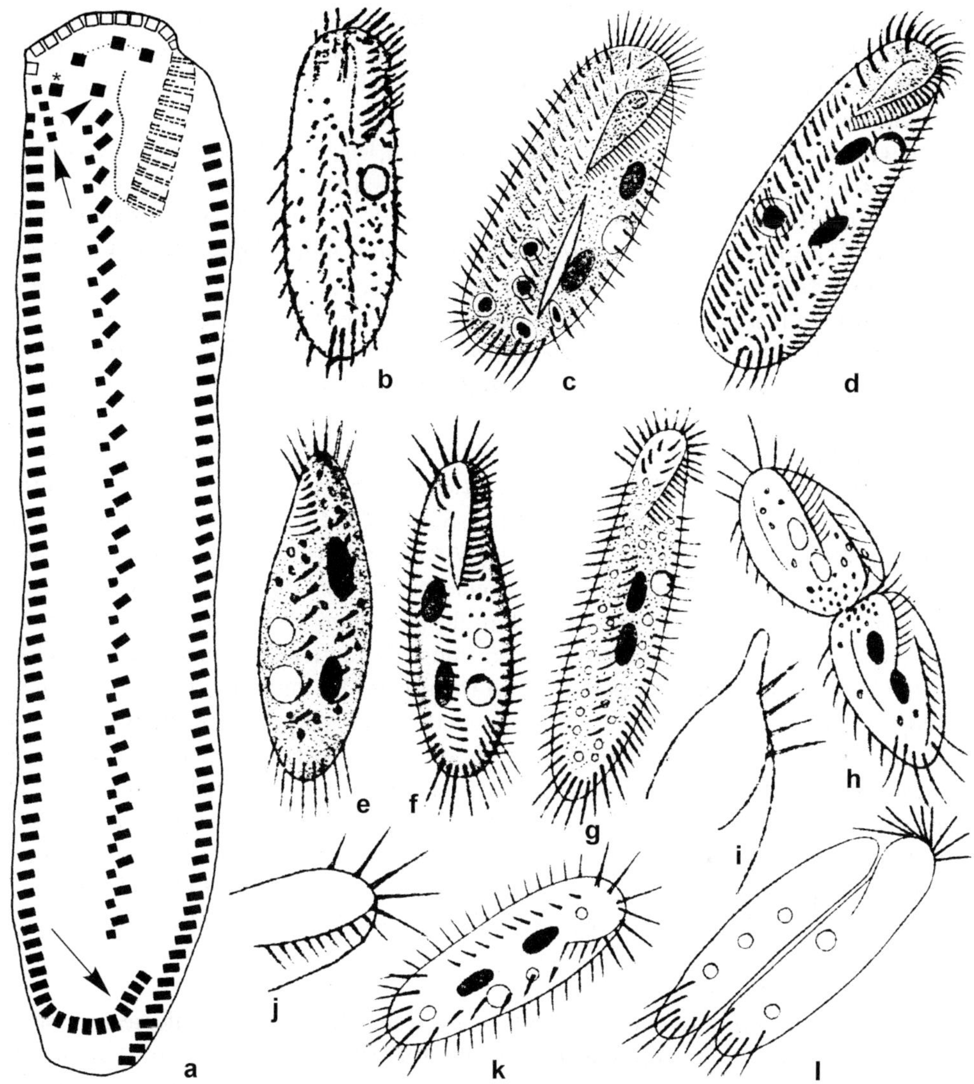

Fig. 37a–d, g–l Species indeterminata and non-identified populations (ventral view from life unless otherwise indicated). **a:** *Holosticha salina* (from Fernandez-Leborans & Novillo 1993. Silver carbonate impregnation). Infraciliature of ventral side, size not indicated. Frontal cirri connected by dotted line. Arrowhead likely marks buccal cirrus; asterisk likely denotes cirrus III/2; short arrow denotes likely frontoterminal cirri; and long arrow marks possibly the rightmost transverse cirrus; p. 184. **b:** *Amphisia* sp., 103 μm (from Lepsi 1932); p. 188. **c:** *Amphisia* Núm 1, 180 μm (from Izquierdo 1906); p. 187. **d:** *Amphisia* Núm 3, 80–100 μm (from Izquierdo 1906); p. 187. **g–l:** *Amphisia* Núm 2, 80 μm (from Izquierdo 1906). Two illustrations (Fig. 37e, f) of *Amphisia* Num. 2 show *Holosticha pullaster*; however, the present figures do not allow an unequivocal identification (g, k = ventral view; h = later divider; i = rear end?; j = adoral zone; l = conjugation); p. 132.

Fig. 37e, f: *Holosticha pullaster* (from Izquierdo 1906. From life). Cirral pattern as seen from dorsal (e) and ventral (f) side, 80 μm. Identification is based on the posteriorly located contractile vacuole, the rightwards dislocated macronuclear nodules, the body shape, and the arrangement and high number of transverse cirri. Page 128.

No type material available. No derivation of the name is given in the original description. The species-group name *setigera* (*saetigera*; Latin adjective; bristle bearing) obviously alludes to the long marginal cirr. The following brief description is based on the two illustrations (Fig. 100i, j) only. Body size about 36 × 15 µm, that is, body length:width ratio about 2.3:1; body outline elliptical; slightly flattened dorso-ventrally (about 1.4:1); one vacuole each in anterior and posterior body portion (Fig. 100i; reminiscent of *Anteholosticha oculata*, Fig. 95a–e); adoral zone occupies only about 22% of body length, distinctly curved inwards proximally; midventral complex obviously present, extends from near anterior to posterior end; marginal rows with relatively long cirri; number of macronuclear nodules, presence/absence of cortical granules and basic cirral pattern (frontal ciliature, transverse cirri) not known. Discovered in a freshwater habitat in Connecticut, USA. No further records available.

Holosticha tenuiformis **Vuxanovici, 1963**, Studii Cerc. Biol., Seria "biologie animala", 15: 203, Plansa II, Fig. 10, 10a (Fig. 100k). Remarks: The species-group name is a composite of the Latin adjective *tenuis -is -e* (thin, fine) and the Latin *-formes* (-shaped) and obviously refers to the slender body shape. Vuxanovici (1963) described this species from a single and likely teratological specimen. I did not find it in the revision by Borror (1972). Borror & Wicklow (1983) put it into the synonymy of *Anteholosticha intermedia*. According to my catalogue it was never transferred to another genus (Berger 2001, p. 39). I agree with Hemberger (1982) that it is a species indeterminata because the morphology is described too superficially to allow a reliable identification. Body size 80 × 16 µm (width calculated from figure). Outline roughly rectangular. Several scattered macronuclear nodules. Cytoplasm transparent, with dark inclusions. Large (contractile?) vacuole in rear body end (pathological feature?). The cirral pattern – three frontal cirri, a midventral complex extending to near transverse cirri, 7–8 transversely arranged transverse cirri, one left and one right marginal row – is *Anteholosticha*-like. Dorsal cilia about 5 µm long, very narrowly spaced within kineties. Caudal cirri obviously lacking. Freshwater in Colentina, Bucharest, Romania.

Holosticha vesiculata **Vuxanovici, 1963**, Studii Cerc. Biol., Seria "biologie animala", 15: 206, Plansa III, Fig. 16 (Fig. 100l). Remarks: The species-group name is a composite of the Latin noun *vesicula* (small vesicle) and the suffix *-atus* (provided with something) and likely refers to the acontractile vesicle in the posterior body portion. Borror & Wicklow (1983, p. 121) synonymised this species with *Urostyla intermedia* (= *Anteholosticha intermedia* in present book). However, I agree with Hemberger (1982, p. 277), who classified it as species indeterminata because neither the description nor the illustration allow a reliable determination. Vuxanovici (1963) himself was uncertain about the generic classification. Body 70–100 µm long, pyriform, flexible. Many small macronuclear nodules. Contractile vacuole obviously with strongly dilated anterior collecting canal (likely Vuxanovici illustrated a squeezed specimen); in some specimens an acontractile vacuole in posterior body portion. Cytoplasm cloudy, with ingested algae, and small brown granules. Adoral zone occupies about 25% of body length. Three frontal cirri, two cirri behind right frontal cirrus; midventral complex possibly composed of cirral pairs only; 5(?) transverse cirri; one left and one right marginal row. Many specimens among decomposing plants in Lake Floreasca, Bucharest, Romania during October 1961.

Holosticha wrzesniowskii punctata **Rees, 1884**, Tijdschr. ned. dierk. Vereen, Suppl. I: 644, 645, 672, 673, Plaat XVI, Fig. 18 (Fig. 99h). Remarks: The name *punctatus -a -um* (Latin adjective; dotted) possibly refers to the cortical granules. In my opinion the description and illustration are too superficial for a reliable identification; by contrast, Kahl (1932, p. 583, Fig. 106 $_{16}$) raised it to species rank (as *Holosticha (Holosticha) punctata* Rees, 1884). Borror (1972, p. 11) and Borror & Wicklow (1983, p. 121) synonymised it with *Holosticha gibba*. I suppose that it is a postdivider or a malformed specimen of a pseudokeronopsid species, as indicated by the far posteriorly extending distal end of the adoral zone, the frontal ciliature, and the cortical granulation. Body length about 80 µm; the conspicuous dots at the cell margin are likely cortical granules (protrichocysts according to Kahl 1932). Marine habitat (Oosterschelde) in Belgium. Bamforth (1963, p. 133) found it rare in a limnetic habitat in Louisiana, USA.

Insufficient redescriptions

Amphisia diademata **Rees** – Alzamora, 1929, Notas Resúm. Inst. esp. Oceanogr., 2: 13, Fig. 29 (Fig. 191k). Remarks: Both the description and the illustration are too superficial so that the identification cannot be accepted. Body length 80 µm; outline variable. Three frontal and seven transverse cirri. Mediterranean Sea, Bay of Palma de Mallorca, Spain.

Amphisia **Núm. 1** – Izquierdo, 1906, Protozoos, p. 188, Lám. XII, Fig. 475 (Fig. 37c). Remarks: The widely separated ventral rows indicate that it is not a urostyloid; possibly it is a *Paraurostyla* species. Body length 180 µm; two macronuclear nodules; rear body portion with food vacuoles containing diatoms; eight transverse cirri. In a pool (pond?) in Santiago de Chile, Chile.

Amphisia **Núm. 3** – Izquierdo, 1906, Protozoos, p. 189, Lám. XII, Fig. 484 (Fig. 37d). The description and the illustration are too superficial for an identification. Body length 80–100 µm. Two macronuclear nodules. Ventral cirral rows widely separated, indicating that it is not a urostyloid. Cytoplasm with small granules (symbiotic algae? cortical granules?) so that cells appear green. Brook in Chile.

Amphisia pernix – Wailes, 1943, Can. Pacif. Fauna 1: 28, Fig. 81 (Fig. 143j). Remarks: The data are too inaccurate to accept the identification. Body size 55 × 22–30 µm; body ovoid; two macronuclear nodules; contractile vacuoles located posteriorly; adoral zone of membranelles about one third of body length; five transverse cirri. Departure Bay among algae, Canadian Pacific Coast.

***Amphisia* sp.** – Lepsi, 1926a, Arch. Hydrobiol., 17: 756. Remarks: The description is too superficial for an identification. Single ventral row, indicating that it is an amphisiellid, two marginal rows. Frontal and transverse cirri distinct. Adoral zone inverse sigmoidal. Very rare in a spring near the city of Cavarna, Romania.

***Amphisia* sp.** – Lepsi, 1927, Studii Cerc. Acad, RPR, 12: 120. Remarks: Lepsi (1927) provided no illustration and from the description (in Romanian) alone an identification is impossible. Body size about 90 × 31 µm; body flexible, fusiform. Adoral zone about 1/3 of body length. Two macronuclear nodules. Contractile vacuole in ordinary position. Cytoplasm colourless. Cirral pattern (not translated in detail) composed of two cirral rows, 10 transverse cirri, and frontal cirri; caudal cirri lacking? Likely found in a rheocrene spring near the city of Cavarna, south-eastern Dobrudscha, Romania/Bulgaria.

***Amphisia* sp.** – Lepsi, 1932, Publnile Soc. Nat. Rom., 10: 30, Fig. 57 (Fig. 37b). Remarks: The tiny illustration and the description are too superficial for a reliable identification. Body length 103 µm; body margins parallel; contractile vacuole in ordinary position; adoral zone about 1/3 of body length; infraciliature, see Figure. Limnetic habitat in Romania.

Holosticha – Gruber, 1888, Ber. naturf. Ges. Freiburg i. B., 3: 61, Tafel VII, Fig. 5, 6. Remarks: In his paper on *Pseudokeronopsis* species, Gruber mentioned and illustrated a brown *Holosticha* species (his Fig. 5) and a bright, long *Holosticha* species (his Fig. 6). However, the data provided are too sparse for a reliable identification. Sea at Genoa?, Italy.

***Holosticha kessleri* Wrzesniowski** – Gellért, 1956, Acta Biol. Hung., 6: 345. Remarks: No illustration, but a short description, which is somewhat confusing, is provided. In addition, the habitat – humus underneath moss on rock in Hungary – clearly shows that the identification is incorrect. Length 110 µm; right portion of midventral row terminates about at 25% of body length, left one at 50%; beside the (buccal?) lip 10 cirri; two pretransverse ventral cirri; 11 transverse cirri; contractile vacuole empties every 17 seconds; feeds mainly on testate amoebae.

***Holosticha kessleri* Wrzesniowski** – Sarmiento & Guerra, 1960, Publnes Mus. Hist. nat., Lima, 19: 18, Fig. 29 (Fig. 191l). Remarks: The description and illustration are too superficial so that the identification cannot be accepted. Size 77–132 × 22–30 µm; one macronuclear nodule; five transverse cirri. Found in the laguna de Villa, Peru. Unfortunately I was unable to find this location in the map so that I do not know whether it is a freshwater or marine habitat. The other species described in this paper indicate that is freshwater.

***Holosticha obliqua* Kahl, 1928** – Aladro Lubel, Martínez-Morillo & Mayén Estrada, 1986, An. Inst. Biol. Univ., 57: 9, Lámina 5, figure 1 (Fig. 164c) and Aladro Lubel,

Martínez Murillo & Mayén Estrada, 1990, Manual de ciliados, p. 132, Figure on same page (Fig. 164d). Remarks: The illustrations and the description are too superficial for a reliable identification. Body size 130–140 × 32–36 µm; macronuclear nodules about 8 × 3 µm; adoral zone about 45 µm long; transverse cirri about 25 µm long; remaining features, see figures. Laguna de Mandinga, Veracruz, Mexico (see also Aladro-Lubel et al. 1988, p. 437).

***Holosticha* sp.** – Chardez, 1981, Revue verviét. Hist. nat., 38: 53, Fig. 3 (Fig. 164b). Remarks: Chardez did not provide a description, but only the illustration which is, however, too superficial for a reliable identification. Possibly it is an *Anteholosticha* or *Caudiholosticha* species (however, the specimen illustrated has two left marginal rows!). Found in mosses in Belgium.

***Holosticha* sp.** – Lepsi, 1932, Publnile Soc. Nat. Rom., 10: 31, Fig. 58 (Fig. 164a). Remarks: The tiny illustration and the description are too superficial for a reliable identification. Possibly an amphisiellid because of the ventral cirral row. Size not indicated; about eight macronuclear nodules possibly mainly right of midline; infraciliature, see Figure. Limnetic habitat in Romania.

***Keronopsis pernix* (Wrzesniowski 1887)** – Šrámek-Hušek, 1957, Vest. csl. zool. Spol., 21: 12, obr. 15 (Fig. 143k). Remarks: *Oxytricha pernix* was originally found in the Baltic Sea. Marine species usually do not occur in freshwater, indicating that this record from a betamesosaprobic area of the Morava River at the village of Zimrovicemi, Czechoslovakia, is wrong. Possibly, the illustration shows a true *Keronopsis* species. No morphological details, except illustration, given.

***Oxytricha alba*, de Fromentel** – Dumas, 1929, Les Microzoaires, p. 73, Planche XXVII, Fig. 16a (Fig. 191j). Remarks: The figure numbers on the plate are very difficult to read (it can be that I confused 14 and 16; however, this does not matter because all illustrations on hypotrichs are not usable). The illustration does not show any resemblance to *H. pullaster*, the senior synonym of *O. alba*. France.

Oxytricha pullaster – Fromentel, 1876, Microzoaires, p. 267, Planche XIII, Fig. 12, 12a, 12b (Fig. 191g, i, n). Remarks: The organisms illustrated do not show a resemblance to *H. pullaster*. Possibly a fragment of a hypotrich or even a *Lacrymaria*? Note that Fromentel described *H. pullaster* as *Oxytricha alba* (Fig. 26h), which is very likely the first paper where the posteriorly located contractile vacuole is described.

***Oxytricha pullaster*, Müller** – Dumas, 1929, Les Microzoaires, p. 70, Planche XXVII, Fig. 7 (Fig. 191h). Remarks: Possibly a fragment of another hypotrich or perhaps a *Lacrymaria*. France.

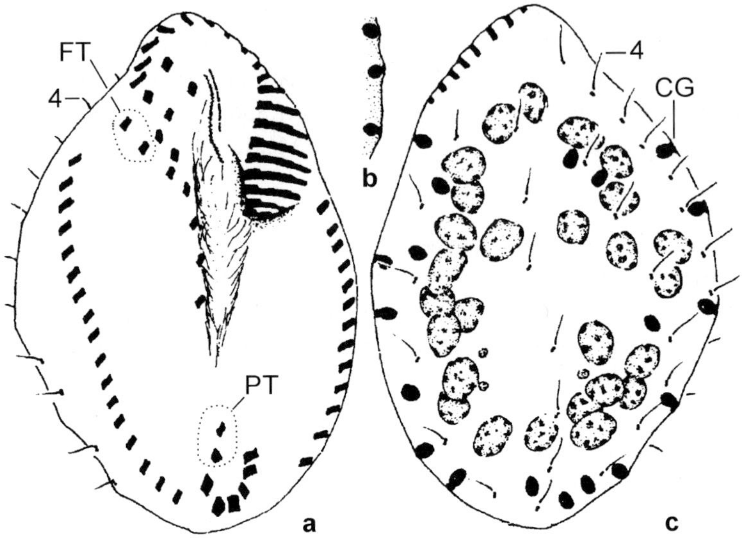

Fig. 37.1a–c *Holosticha* sp. (from Wilbert & Song 2005. Protargol impregnation). **a, c:** Infraciliature of ventral and dorsal side and nuclear apparatus, 62 µm. Morphogenetic data are needed for a correct interpretation of the cirral pattern. **b:** Argentophilic, extrusome-like structures (cortical granules?) on dorsal side, about 3 µm. CG = cortical granules?, FT = frontoterminal cirri?, PT = pretransverse ventral cirri, 4 = dorsal kinety 4 (= rightmost kinety). Page 97.

Nomina nuda

***Holosticha alpestris* Foissner, 1981**, Veröff. Österr. MaB-Programms, 4: 18. Remarks: Foissner (1982, p. 50) identified this population with *H. multistilata* (Foissner & Foissner 1988, p. 84, 105; Berger 2001, p. 33); now this population is assigned to *Anteholosticha intermedia* (Table 18).

***Holosticha binucleata* Foissner, Peer & Adam, 1985**, Mitt. öst. bodenk. Ges., 30: 108. Remarks: Later identified with *Keronopsis algivora* (Gellért, 1942) Foissner, 1987a (Foissner & Foissner 1988, p. 105; Berger 2001, p. 34).

Pseudoamphisiella Song, 1996

1996 ***Pseudoamphisiella* nov. gen.**[1] – Song, Oceanologia Limnol. sin., 27: 18, 21 (original description). Type species (by original designation on p. 18, 21): *Holosticha lacazei* Maupas, 1883.

1997 ***Pseudoamphisiella* Song, 1996** [2] – Song, Warren & Hu, Arch. Protistenk., 147: 266 (improved diagnosis, and redescription and ontogenesis of type species).

1999 ***Pseudoamphisiella* Song, 1996** – Shi, Acta Zootax. sinica, 24: 367 (revision of hypotrichs).

1999 ***Pseudoamphisiella* Song, 1996** – Shi, Song & Shi, Progress in Protozoology, p. 117 (revision of the Hypotrichida).

2001 ***Pseudoamphisiella* Song 1996** – Aescht, Denisia, 1: 135 (catalogue of generic names of ciliates).

2001 ***Pseudoamphisiella* Song, 1996** – Berger, Catalogue of ciliate names 1. Hypotrichs, p. 76 (nomenclator containing all basionyms, combinations, and higher taxa of hypotrichs).

Nomenclature: No derivation of the name is given in the original description. *Pseudoamphisiella* is a composite of the Greek adjective *pseudo-* (wrong, lying) and the name of the hypotrichous genus *Amphisiella* Gourret & Roeser, 1888 which is a composite of *Amphisia* Sterki, 1878 and the diminutive suffix *-ella*. For a derivation of the name *Amphisia*, see *Holosticha*. Feminine gender because ending with the suffix *-ella* (ICZN 1999, Article 30.1.3).

Characterisation (A = supposed apomorphies): Adoral zone of membranelles continuous, extends far onto right body margin (A?). 3 frontal cirri. Buccal cirri(us) present. Cirrus III/2 behind buccal cirrus (A). Frontoterminal cirri lacking (A). Midventral complex composed of midventral pairs only, zigzagging pattern, however, not recognisable because cirri of each pair widely separated (A). Many transverse cirri because almost all frontal-midventral-transverse cirri anlagen form a transverse cirrus. 1 right and 1 left marginal row. Caudal cirri present. Number of dorsal kineties relatively high (around 10). Extrusomes rod-shaped, form distinct seam because perpendicularly arranged to pellicle (A). Adoral zone of proter originates behind parental adoral zone (A). Anterior cirrus of rightmost midventral anlagen resorbed in late dividers so that pseudorow formed by left cirri of midventral pairs is longer than right row in non-dividers (A). New right marginal row originates (likely de novo) left of parental right marginal row.

Remarks: Besides the features mentioned above, the two species assigned have the following characters in common: body large, that is, length around 200 µm, broad elliptical, flexible; contractile vacuole likely lacking; adoral zone of membranelles short rather than long, proximal end curves distinctly towards mid-line; frontal cirri form oblique row; pseudorow formed by right midventral cirri runs, possibly in a more or less distinct groove; dorsal cilia short (<5 µm); rather high number of caudal cirri, form a more or less continuous row with the marginal rows; marine.

---

[1] The diagnosis by Song (1996) is as follows: long ellipsoid Amphisiellidae (?) with one left and one right marginal row each, which are connected posteriorly; frontoterminal, transverse and ventral rows derived from "numerous" FVT-anlagen; frontal and buccal cirri present, transverse cirri extremely developed, not typical mid-ventral rows.

[2] The improved diagnosis by Song et al. (1997) is as follows: Pseudoamphisiellidae with one left and one right marginal row of cirri; three enlarged frontal cirri and two long ventral (midventral) rows; caudal cirri completely aligned with marginal rows.

Maupas (1883) assumed that *Holosticha lacazei* has only one "abdominal" cirral row because he considered the right part of the midventral complex as narrow groove (gouttière étroite) and not as cirral row. At first, this feature inspired him to create a new genus. But finally, he placed it in *Holosticha*, which was established by Wrzesniowski some years ago. Kahl (1932), who redescribed *H. lacazei* from life, confirmed Maupas' classification in that he put it into *Holosticha (Holosticha)*. The assignment to *Holosticha* was also accepted by Borror (1972) and Borror & Wicklow (1983), who already recognised the special mode of right marginal row formation (see below). Song (1996) reinvestigated this species and established *Pseudoamphisiella*. However, his diagnosis was not very convincing because he assigned it to the amphisiellids, admittedly with doubt, although he knew that, like in urostyloids, numerous cirral anlagen are formed. Later, when he had studied the ontogenesis, he made the diagnosis more precise and, in addition, established the monotypic family Pseudoamphisiellidae[1] (Song et al. 1997, p. 266). Shi (1999a, p. 367) and Shi et al. (1999, p. 117) considered this family, beside the Urostylidae, the Holostichidae, and the Pattersoniellidae, as member of the Urostylina.

Song & Warren (2000) reinvestigated *Holosticha alveolata* Kahl and found that this species is, as already suggested by Kahl (1932) himself, closely related to *Pseudoamphisiella lacazei*. Thus, they transferred it to *Pseudoamphisiella* which comprises two species now.

Song & Warren (2000) found also that *Keronopsis macrostoma* Dragesco, 1963 is the junior synonym of *Pseudoamphisiella lacazei* Maupas, 1883. Yankovskij (1978) fixed *K. macrostoma* as type species of *Erionella*. Thus, *Erionella* is the senior synonym of *Pseudoamphisiella*, a problem not discussed by Song & Warren. Following Song & Warren (2000) strictly, *Pseudoamphisiella* had to be replaced by *Erionella*, a taxon classified in the discocephalines (Small & Lynn 1985, p. 516; Tuffrau & Fleury 1994, p. 133; Lynn & Small 2002, p. 423). Since I am not completely convinced that *Holosticha lacazei* Maupas and *Keronopsis macrostoma* Dragesco are the same species, I preliminarily do not replace *Pseudoamphisiella* by *Erionella* (see *Keronopsis macrostoma* for details).

Holosticha (Holosticha) arenicola Kahl, 1932 also has an alveolar seam, but lacks extrusomes and the widely spaced midventral cirri (Fig. 96a). Thus, this species is assigned to *Anteholosticha*.

The characterisation above is rather comprehensive, inter alia, because *Pseudoamphisiella* shows some curious features. The adoral zone of membranelles forms almost a circle due to two peculiarities: (i) The distal end of adoral zone extends far posteriorly; that is, the DE-value (see Fig. 1c for explanation) is very high. Such high values are characteristic for the Retroextendia, a subgroup of the Urostylidae. However, the Retroextendia have coronally arranged frontal cirri so that a close relationship with *Pseudoamphisiella* is unlikely; that is, a far posteriorly extending distal end has evolved at

[1] The diagnosis by Song et al. (1997) is as follows: Hypotrichida (Discocephalina? Urostylina?) with differentiated frontal and highly developed transverse cirri; two widely separated midventral rows, which originate from a series of oblique FVT-streaks during morphogenesis; without frontoterminal cirri; right marginal row derived from the rightmost streak of FVT-anlagen.

least twice. (ii) The proximal end of the adoral zone extends to near the cell midline (e.g., Fig. 41k).

In *Pseudoamphisiella* the zigzag pattern of the midventral complex is very indistinct (basically lacking) because the cirri of each pair are widely separated. This is somewhat reminiscent of *Thigmokeronopsis*. Moreover, the anterior cirri of the rearmost midventral anlagen are resorbed in late dividers so that the pseudorow formed by the right cirri is usually distinctly shorter than the left pseudorow (Fig. 39d). Thus the urostyloid origin of *Pseudoamphisiella* is difficult to recognise. A similar phenomenon (resportion of midventral cirri) is known from *Uroleptopsis citrina* where, however, the left cirri are resorbed (Fig. 192v, w).

Pseudoamphisiella is classified in the Holostichidae because it has three frontal cirri and the midventral complex is composed of cirral pairs only. The high number of transverse cirri (that is, each anlage forms a transverse cirrus) indicates that it is closely(?) related to *Holosticha*.

Species included in *Pseudoamphisiella* (alphabetically arranged basionyms are given): (1) *Holosticha (Holosticha) alveolata* Kahl, 1932; (2) *Holosticha lacazei* Maupas, 1883.

Key to *Pseudoamphisiella* species

The two species assigned can be easily distinguished by the number of macronuclear nodules. Thus, simple live observation is sufficient for reliable identification, provided that one knows that it is a *Pseudoamphisiella*.

1 Two macronuclear nodules (Fig. 41a, r) .... *Pseudoamphisiella alveolata* (p. 211)

\- Many (24–60) macronuclear nodules (Fig. 38b, e, o) .... *Pseudoamphisiella lacazei* (p. 193)

Pseudoamphisiella lacazei (Maupas, 1883) Song, 1996 (Fig. 38a–o, 39a–p, Table 15)

1883 ***Holosticha lacazei* (nov. sp.)** – Maupas, Archs Zool. exp. gén., 1: 556, Planche XXIII, Fig. 5–8 (Fig. 38a–d; original description; no type material available and no formal diagnosis provided).

1929 ***Amphisia (Holosticha) lacazei*** – Hamburger & Buddenbrock, Nord. Plankt., 7: 91 (combination with *Amphisia*).

1932 ***Holosticha lacazei* Maupas, 1888** – Kahl, Tierwelt Dtl., 25: 579, Fig. 106_4, 111_2 (Fig. 38e, f; redescription and revision; incorrect year).

1933 ***Holosticha lacazei* Maupas 1888** – Kahl, Tierwelt N.- u. Ostsee, 23: 109, Fig. 17.5 (Fig. 38g; guide to marine ciliates; incorrect year).

1970 ***Holosticha lacazei* Maupas, 1888** – Kattar, Zoologia e biologia marinha N. S., 27: 191, Fig. 35 (Fig. 38h; redescription after protargol impregnation; incorrect year).

1972 ***Holosticha lacazei* Maupas, 1883** – Borror, J. Protozool., 19: 11 (revision of hypotrichs).

1982 ***Holosticha lacazei* Maupas, 1888** – Hemberger, Dissertation, p. 94 (revision of hypotrichs).

1983 ***Holosticha lacazei* Maupas, 1883** – Borror & Wicklow, Acta Protozool., 22: 109, 114, 121, Fig. 12 (Fig. 38i; revision; brief note on ontogenesis).

1992 ***Holosticha lacazei* Maupas, 1888** – Carey, Marine interstitial ciliates, p. 182, Fig. 716 (redrawing of Fig. 38a; guide; incorrect year).

1996 ***Pseudoamphisiella lacazei* (Maupas, 1883) nov. comb.**[1] – Song, Oceanologia Limnol. sin., 27: 18, 21, Fig. 1a–d, 2e, f (Fig. 38j–o; redescription and combination with *Pseudoamphisiella*).

1997 ***Pseudoamphisiella lacazei* (Maupas, 1883) Song, 1996**[2] – Song, Warren & Hu, Arch. Protistenk., 147: 266, Fig. 1–25, Table 1 (Fig. 39a–p; improved diagnosis, redescription, and ontogenesis; protargol slides are deposited in the Laboratory of Protozoology, College of Fisheries, Ocean University of Qingdao, China and two voucher slides are deposited at the Natural History Museum, London, registration numbers 1996:5:31:1 and 1996:5:31:2).

2001 ***Pseudoamphisiella lacazei* (Maupas, 1883) Song, 1996** – Berger, Catalogue of ciliate names 1. Hypotrichs, p. 37 (nomenclator containing all basionyms, combinations, and higher taxa of hypotrichs).

2001 ***Pseudoamphisiella lacazei*** – Eigner, J. Euk. Microbiol., 48: 77, Fig. 30 (Fig. 39d modified; brief review of the Urostylidae).

Nomenclature: This species was named in honour of M. de Lacaze-Duthiers (Maupas 1883, p. 562). Kahl (1932, p. 570ff; 1933) divided *Holosticha* into five subgenera; thus, the correct name in his papers is *Holosticha (Holosticha) lacazei* Maupas, 1883, a name overlooked by Berger (2001). Song (1996) fixed *Holosticha lacazei* as type species of *Pseudoamphisiella* by original designation.

Remarks: Maupas (1883) described *Holosticha lacazei* in detail, including the cortical granules, which form a distinct seam (Fig. 38a–d). Some misobservations, respectively, misinterpretations, for example the lack of the right part of the midventral complex, could be clarified by Kahl (1932), who studied this species from life. Kattar (1970) was the first who made protargol preparations of *H. lacazei*. However, he found two left marginal rows, which is a distinct difference to the other populations, which have only one. Borror & Wicklow (1983, Fig. 38i) provided a further illustrated record and they found another significant difference to all other populations, namely, in the number of buccal cirri. While Borror & Wicklow found only one such cirrus, all other authors observed two cirri along the buccal lip, a feature supported by the detailed redescriptions by Song (1996) and Song et al. (1997); however, according to ontogenesis the rear buccal cirrus is homologous to the cirrus (III/2) behind the right frontal cirrus, that is, originates from anlage III and not from anlage II. Surprisingly, the second species assigned to *Pseudoamphisiella* has two buccal cirri too (Fig. 41a, k). Whether or not the deviations described by Kattar (1970) and Borror & Wicklow (1983) indicate the presence of geographical races can be clarified only after the investigation of further populations from these areas. Since the data by Song et al. (1997) agree very well with the original description and because they also studied ontogenesis, I declare this paper as authoritative redescription. If taxonomic problems arise, the population described by

---

[1] The improved diagnosis by Song (1996, p. 22) is as follows: Large marine form in vivo 200–300 × 70–100 μm with numerous short-bar-shaped subpellicular extrusomes. 50–60 macronucleus segments; about 11 dorsal kineties and 66–73 marginal cirri; ca. 20 transverse cirri which extend onto the anterior 1/2 of the body. One frontoterminal and one ventral row, each with 14–16 and 16–21 cirri separately. Each of the obliquely arranged ventral cirri oriented (characteristically!) in an upper-right-to-down-left direction.

[2] The improved diagnosis by Song et al. (1997) is as follows: Large, marine *Pseudoamphisiella*, in vivo 120–300 μm × 40–80 μm with numerous cortical granules. On average 50 adoral membranelles and 2 buccal cirri, 50 macronuclear segments, 20 transverse cirri extending to the anterior 1/2 of the body; two ventral rows, with 14–16 (left ventral row) and 16–21 (right ventral row) cirri respectively, 8–11 dorsal kineties and 8–11 caudal cirri; 21–34 and 20–31 cirri in left and right marginal rows respectively.

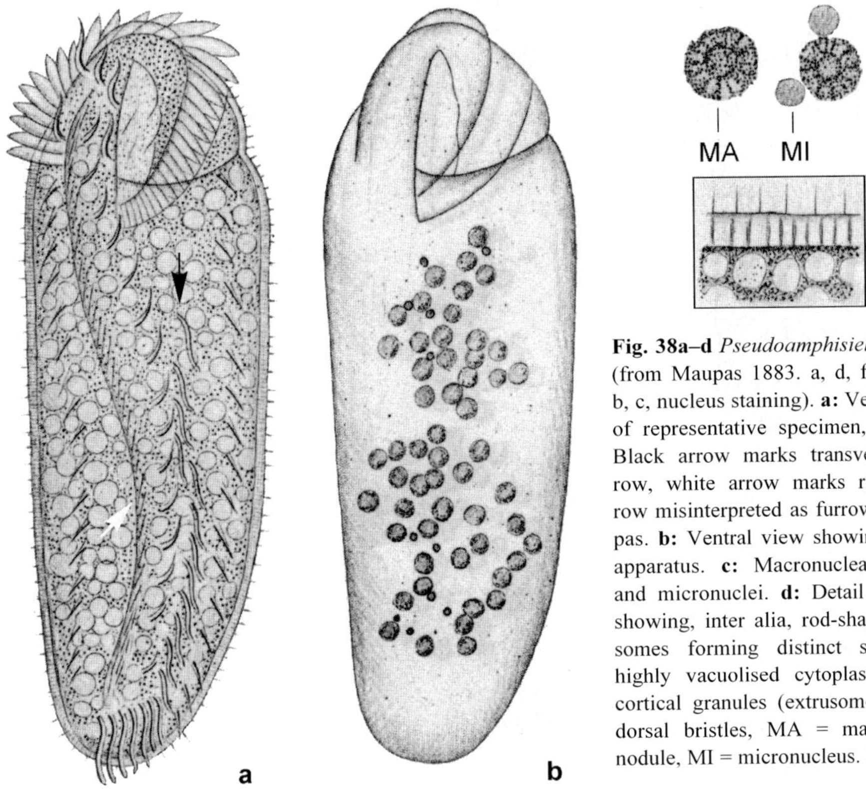

Fig. 38a–d *Pseudoamphisiella lacazei* (from Maupas 1883. a, d, from life?; b, c, nucleus staining). **a:** Ventral view of representative specimen, 195 µm. Black arrow marks transverse cirral row, white arrow marks right cirral row misinterpreted as furrow by Maupas. **b:** Ventral view showing nuclear apparatus. **c:** Macronuclear nodules and micronuclei. **d:** Detail of cortex showing, inter alia, rod-shaped extrusomes forming distinct seam, and highly vacuolised cytoplasm. CG = cortical granules (extrusomes), DB = dorsal bristles, MA = macronuclear nodule, MI = micronucleus. Page 193.

Song et al. (1997) should be designated as neotype. Carey (1992) did not provide own data or a significant discussion.

According to Song (1996) and Song et al. (1997, p. 272), Gelei (1937) mentioned the present species. I checked Gelei's paper, but could not find any hint to *Holosticha lacazei*. Shi et al. (1999, p. 118) provided two small illustrations of *P. lacazei*; these are schematic redrawings of Song's (1996) illustrations and are thus not given in the present paper.

According to Borror (1972), *Anteholosticha intermedia* (Bergh, 1889) is the junior synonym of *H. lacazei*. However, this is incorrect since Bergh's species lacks the long transverse cirral row. Song & Warren (2000, p. 456) synonymised the marine *Keronopsis macrostoma* Dragesco, 1963 with *Pseudoamphisiella lacazei* because the species agree basically in the ventral ciliature and the nuclear apparatus. I am not sure that this synonymisation is correct although there is no doubt that the overall similarity is rather high. I keep *Keronopsis macrostoma* as supposed synonym of *P. lacazei* (see below for details). For the synonymy of *Erionella* and *Pseudoamphisiella*, see remarks on *Pseudoamphisiella*.

Pseudoamphisiella lacazei is rather easily identified by the long transverse cirral row, the size (around 200 µm), and the seam formed by the rod-shaped cortical granules

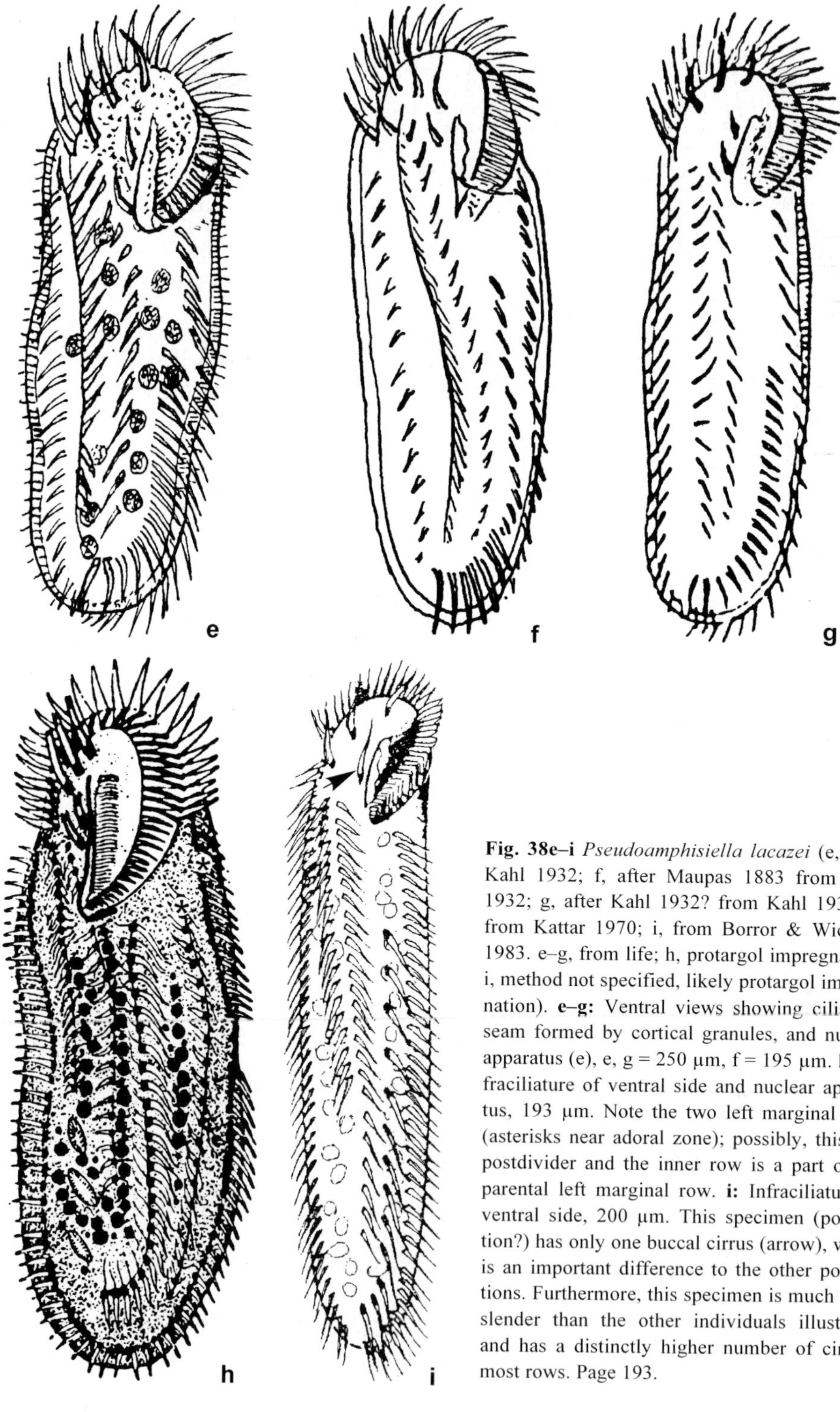

Fig. 38e–i *Pseudoamphisiella lacazei* (e, from Kahl 1932; f, after Maupas 1883 from Kahl 1932; g, after Kahl 1932? from Kahl 1933; h, from Kattar 1970; i, from Borror & Wicklow 1983. e–g, from life; h, protargol impregnation; i, method not specified, likely protargol impregnation). **e–g:** Ventral views showing ciliature, seam formed by cortical granules, and nuclear apparatus (e), e, g = 250 μm, f = 195 μm. **h:** Infraciliature of ventral side and nuclear apparatus, 193 μm. Note the two left marginal rows (asterisks near adoral zone); possibly, this is a postdivider and the inner row is a part of the parental left marginal row. **i:** Infraciliature of ventral side, 200 μm. This specimen (population?) has only one buccal cirrus (arrow), which is an important difference to the other populations. Furthermore, this specimen is much more slender than the other individuals illustrated and has a distinctly higher number of cirri in most rows. Page 193.

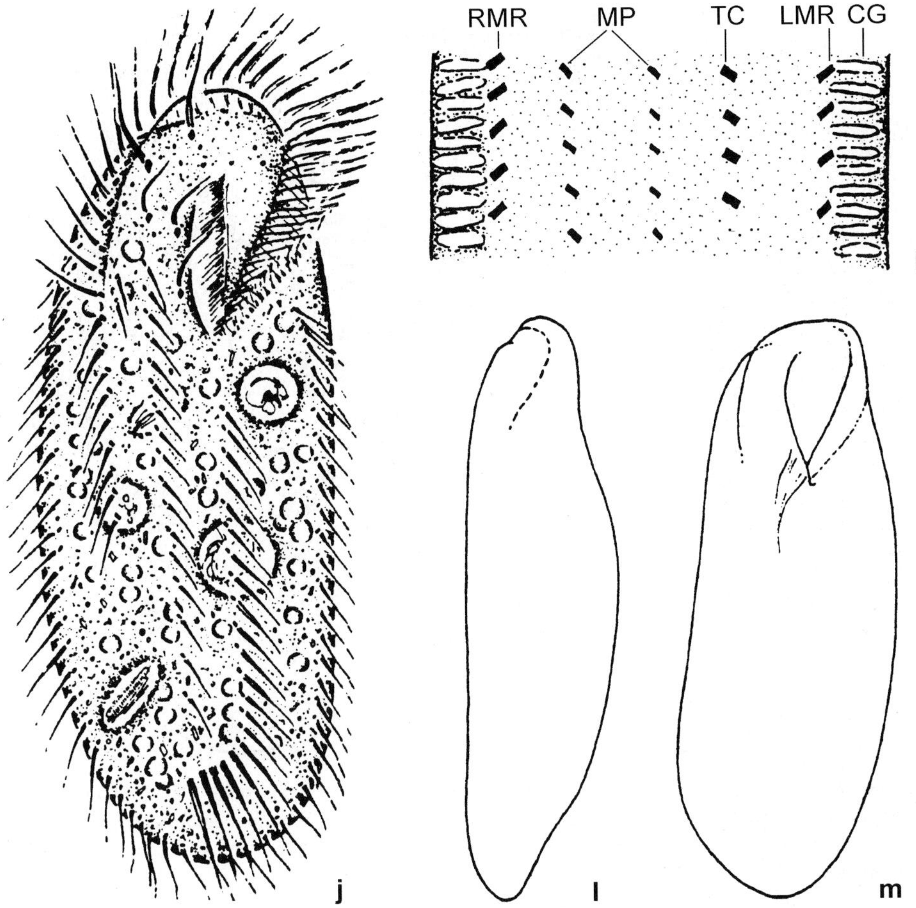

Fig. 38j–m *Pseudoamphisiella lacazei* (from Song 1996. j, l, m, from life; k, protargol impregnation). **j, l, m:** Ventral views (j, m) and left lateral view, j = 244 µm. **k:** Part of the central body portion showing the five cirral rows and the seam formed by the cortical granules. CG = rod-shaped cortical granules, LMR = left marginal row, MP = midventral pairs forming a right and a left cirral row without distinct zigzagging pattern, RMR = right marginal row, TC = transverse cirral row. Page 193.

already recognisable at middle magnification. It differs from *P. alveolata* not only in the number of macronuclear nodules (many vs. two), but also in a higher number of cirri in most rows (see Table 15). *Holosticha* species also have many transverse cirri arranged in J-shape. However, they have, inter alia, a distinct zigzagging midventral pattern, the buccal cirrus ahead of the undulating membranes, a distinct gap in the adoral zone, and the anterior end of the left marginal row is curved rightwards.

Neokeronopsis spectabilis (Fig. 242–244), which also has a very prominent row of transverse cirri, lives in freshwater and has a distinct bicorona (against 3 frontal cirri). Some *Thigmokeronopsis* species, which have a similar general appearance mainly due

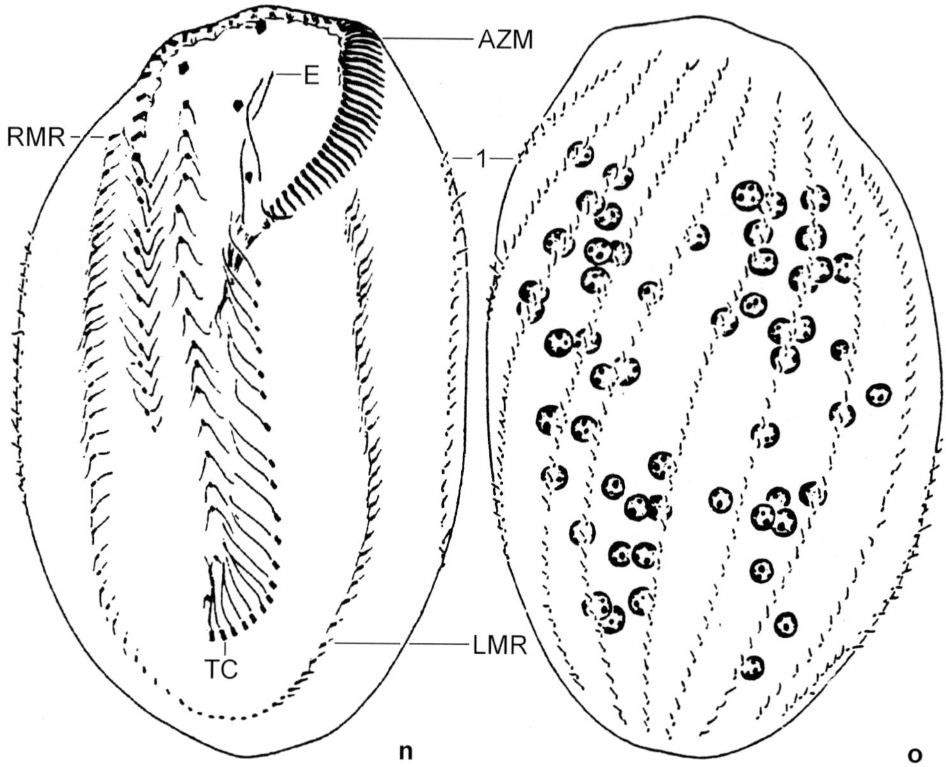

Fig. 38n, o *Pseudoamphisiella lacazei* after protargol impregnation (from Song 1996). Infraciliature of ventral and dorsal side and nucleus apparatus, 218 µm. AZM = adoral zone of membranelles, E = endoral, LMR = left marginal row, RMR = anterior end of right marginal row, TC = transverse cirri, 1 = dorsal kinety (= leftmost kinety). Page 193.

to the prominent transverse ciliature, have a distinct bicorona and thus cannot be confused with *Pseudoamphisiella*. For a more detailed comparison of these two genera, see remarks at *Thigmokeronopsis*. *Holosticha lacazei* sensu Pätsch (1974; Fig. 55i) is an *Anteholosticha monilata* population.

Morphology: Several populations have been described in more or less detail, namely by Maupas (1883), Kahl (1932), Kattar (1970), Borror & Wicklow (1983), Song (1996), Song et al. (1997). Some of them (Maupas, Kahl, Song, Song et al.) agree rather well, so that the conspecificity is beyond reasonable doubt. The populations found by Kattar and Borror & Wicklow in South, respectively, North America deviate in some features so that identity with the other populations is not quite certain; possibly they are subspecies. The authoritative redescription by Song et al. is given first, supplemented by some data by Song (1996). This is followed by important deviating and additional observations from the other populations.

Description of Chinese populations (from Song et al. 1997; Fig. 39a–d, Table 15; corresponding data by Song 1996 [Fig. 38j–o] in brackets!): Size in life 150–250 ×

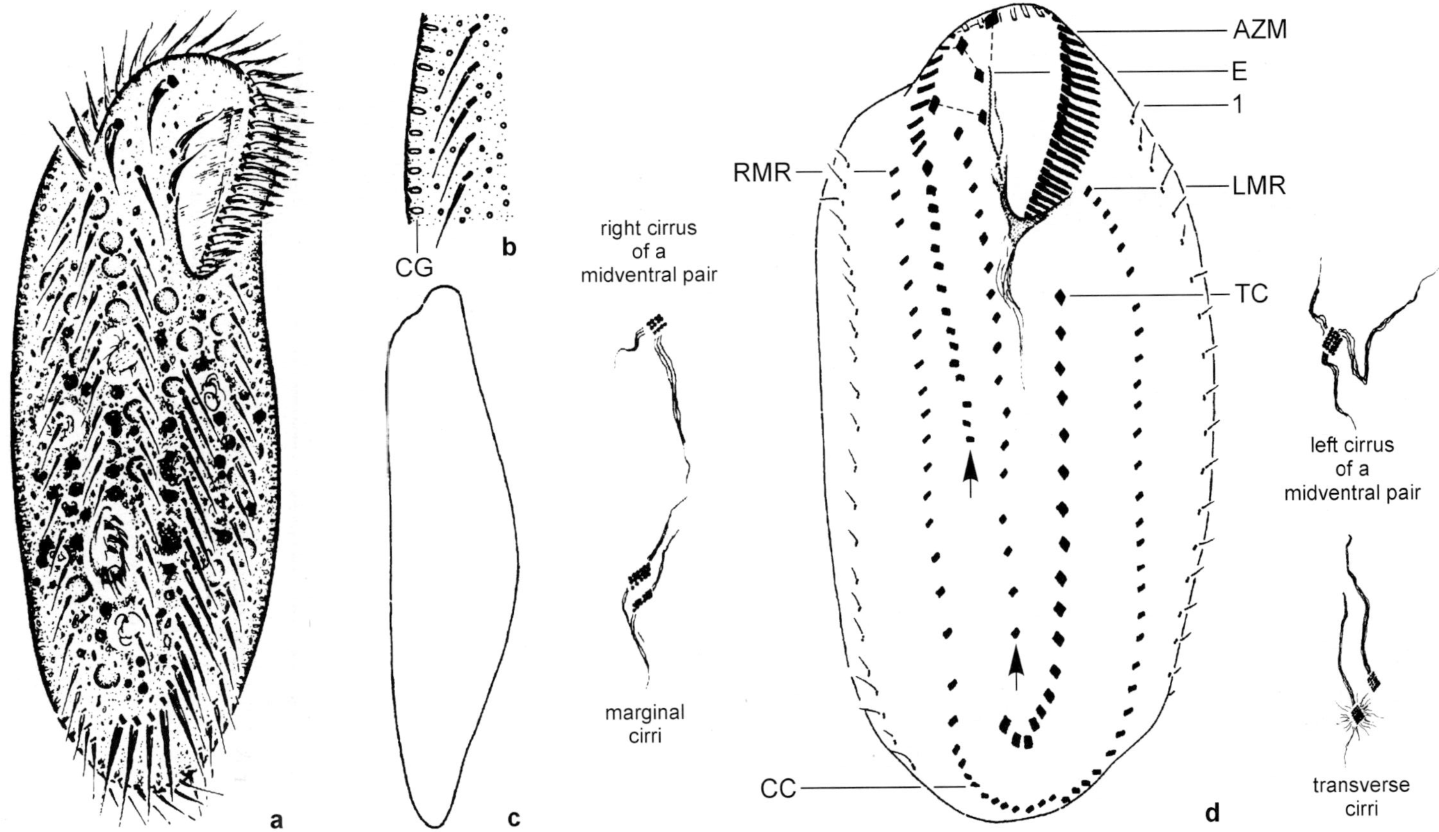

Fig. 39a–d *Pseudoamphisiella lacazei* (from Song et al. 1997. a–c, from life; d, protargol impregnation). **a, c:** Ventral and left lateral view of representative specimens, a = 253 µm. **b:** Cortex. **d:** Infraciliature of ventral side, 162 µm. Arrows mark right and left cirral row formed by midventral pairs. Dotted lines connect frontal and buccal cirrus originating from same anlage. Insets show some cirri with associated fibrils. AZM = adoral zone of membranelles, CC = caudal cirri, CG = cortical granules, E = endoral, LMR = left marginal row, RMR = right marginal row, TC = leftmost (= anteriormost) transverse cirrus, 1 = dorsal kinety 1. Page 193.

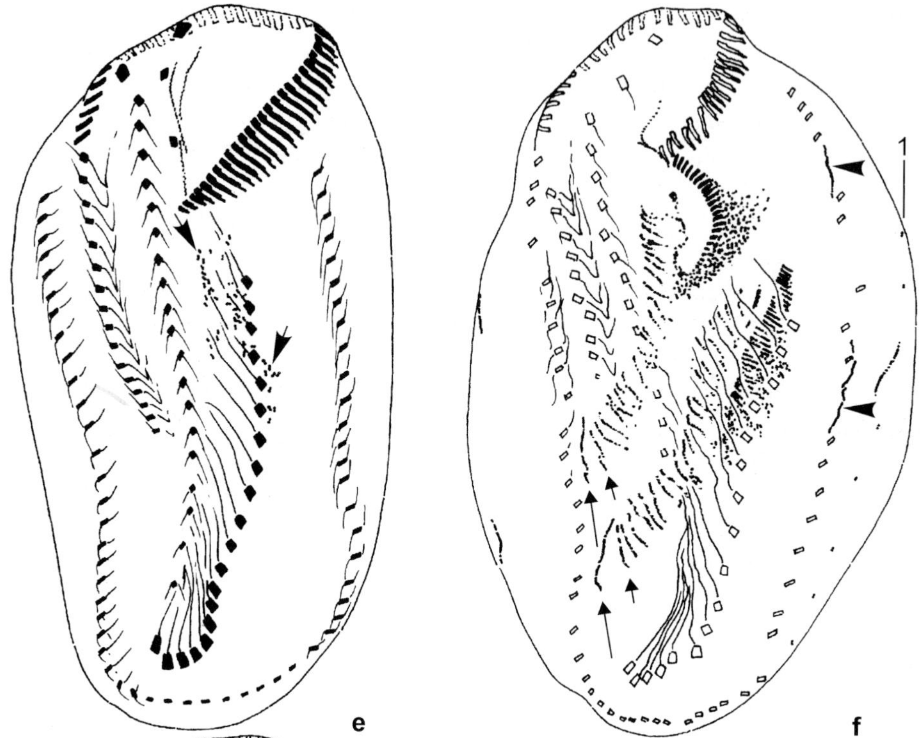

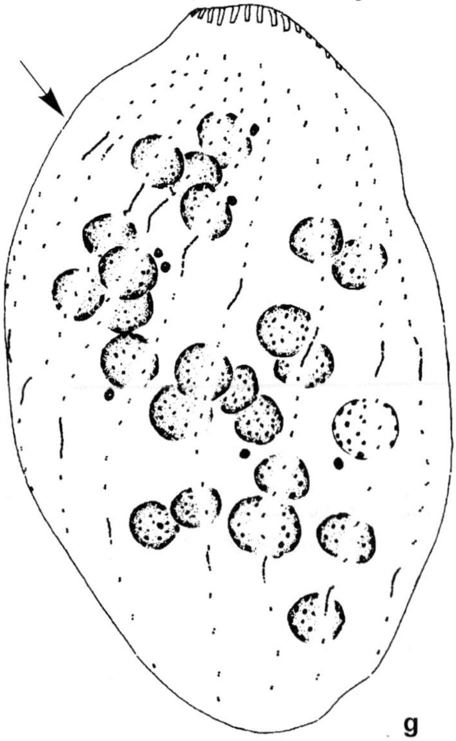

Fig. 39e–g *Pseudoamphisiella lacazei* (from Song et al. 1997. Protargol impregnation). **e:** Infraciliature of ventral side of a very early divider, 153 µm. Arrows mark basal bodies forming the oral primordia. **f, g:** Infraciliature of ventral and dorsal side and nuclear apparatus of an early divider, 164 µm. Arrowheads mark left marginal primordia, long arrows in (f) denote the primordia for the right marginal rows, which do not originate within the parental row, but left of it. The exact origin, either de novo or as rightmost frontal-midventral-transverse cirral anlage, is not yet definitely known. Short arrows in (f) mark the rightmost frontal-midventral-transverse cirral anlage. Note that the anlagen of the opisthe are likely beneath the cortex because the parental cirri and their fibrils are still clearly recognisable. The arrow in (g) denotes the dorsal kinety anlagen for the proter. 1 = dorsal kinety 1 (= leftmost kinety). Page 193.

45–70 µm (200–300 × 70–100 µm), length:width ratio 2.5–4.0:1, after protargol impregnation distinctly stouter because ratio only 2.0:1 on average (Table 15). Body shape variable, usually elongate elliptical, both ends broadly rounded with left anterior margin often distinctly vaulted so that left margin appears slightly sigmoidal; right body margin more or less straight; distal end of adoral zone of membranelles forms marked furrow so that anterior body portion appears slightly cephalised (Fig. 39a); rear body region only slightly narrowed; starved cells with inconspicuous fold (site of fold neither described nor labelled); body rather fragile. Flexibility of body not mentioned, but likely, as in other urostyloids, rather high. On average 37 macronuclear nodules scattered almost throughout body, about 5 µm (4–6 µm) across (in life?). Micronuclei difficult to recognise. Contractile vacuole not mentioned and illustration indicates that it is lacking. Cortical granules inconspicuous, rod-shaped, about 2 µm long, arranged perpendicularly to cell surface and thus forming distinct seam; colour not mentioned, indicating that they are colourless. Cytoplasm greyish, usually with many granular inclusions 1–5 µm across, giving cells a dark, opaque appearance. Usually slowly crawling on bottom of Petri dish, but moving more quickly when disturbed.

Adoral zone occupies up to one third of body length (29% on average in protargol-impregnated specimens; Table 15), extends far onto right side (19% of body length in specimen shown in Fig. 39d; DE-value about 0.69!), composed of an average of 44 membranelles of usual shape and structure (Fig. 39a, d); largest membranelles about 10 µm wide with cilia up to 25 µm long. Buccal cavity obviously of ordinary size. Undulating membranes slightly curved to straight, optically crossing in posterior region; endoral long, paroral distinctly shorter. Pharyngeal fibres distinct after protargol impregnation, extending posteriorly.

Cirral pattern and number of cirri of usual variability (Fig. 39d, Table 15). Three enlarged, 25–30 µm long frontal cirri. All other cirri, except transverse cirri, about 20 µm long. Two buccal cirri, that is, cirri close to right side of paroral, namely the ordinary buccal cirrus (cirrus II/2) near anterior end of paroral and cirrus III/2 (usually located behind right frontal cirrus) about in mid-region of paroral (Fig. 39d). Frontoterminal cirri lacking. Midventral complex composed of cirral pairs only; however, pairs do not form a zigzag-pattern in non-dividing specimens, but two conspicuous longitudinal pseudorows! Left row (labelled ventral row 1 or VR1 by Song et al. 1997, but VR2 by Song 1996) commences between distal end of adoral zone and rear buccal cirrus and terminates near last fifth of body length (at 77% of body length in Fig. 39d); right row (labelled VR2 by Song et al. but VR1 by Song 1996) commences immediately behind distal end of adoral zone, terminates in mid-body (at 54% of body length in Fig. 39d; possibly Song et al. 1997 interchanged the cirral values for the two rows, see Table 15 and corresponding footnotes). Transverse cirri conspicuous because of the high number (up to 23!), distinctly enlarged, 25–30 µm long in life, form distinct J-shaped pattern with anteriormost cirrus near proximal end of adoral zone; rearmost transverse cirrus inserted at 90% of body length in specimen shown in Fig. 39d. Right marginal row begins at level of distal end of adoral zone, terminates at about level of rearmost transverse cirri; caudal cirri not distinctly set off from right and left marginal row, which thus form a U-shaped figure (Fig. 39d). All cirri associated with conspicuous fibrils (Fig. 39d, insets).

Dorsal cilia 3–5 μm (2–3 μm) long, usually arranged in about nine almost bipolar kineties (Table 15; according to Song and co-workers' text, p. 272, usually 11 kineties are present); one or two bristle rows extend laterally (Fig. 38o, 39d). Each kinety with single caudal cirrus; however, they are distinctly set off from kineties to form an almost continuous row with left and right marginal cirri.

Additional or deviating observations from other populations (Fig. 38a–i, Table 15): Population described by Maupas (1883) in life about 200 × 65 μm (width estimated from Fig. 38a). Body widest at level of proximal end of adoral zone, with distinct notch at this region on left margin due to conspicuous transverse structure (buccal lip? structure not described or illustrated by other authors) covering the proximal portion of the adoral zone of membranelles. Body acontractile, but flexible. Ventral side flat, dorsal side curved, thickness about one third of body length. Macronuclear nodules about 6 μm across, likely with central nucleolus. Micronuclei about 2.5 μm across (Fig. 38c). Contractile vacuole very likely absent! Cytoplasm colourless, highly vacuolised. Cortical granules rod-shaped, form 2.5 μm thick seam already recognisable at middle magnification (Fig. 38a, d). Movement conspicuous because specimens sometimes totally immobile for a very long time, then suddenly restlessly creep very quickly when searching for food. Adoral zone occupies about 28% of body length, extends far onto right side forming almost a circle; buccal area rather prominent, undulating membranes (paroral) form distinct velum (Fig. 38a). Maupas basically recognised the cirral pattern, especially the three enlarged frontal cirri, the two buccal cirri, the cirral row formed by the left cirri of the midventral pairs, the long row of transverse cirri, and the marginal rows; the only serious misobservation is that he obviously interpreted the right cirral row (Fig. 38a, white arrow) as furrow. Dorsal cilia short (Fig. 38a, d).

Kahl's (1932, 1933) population in life 200–250 μm long, stout to slender ellipsoidal (body length:width ratio 3–5:1), sometimes even ovoid (Fig. 38e–g); body thick, flexible and contractile. Seam formed by rod-shaped cortical granules about 3 μm thick. Macronuclear nodules scattered, several homogenous micronuclei. No contractile vacuole illustrated. Usually slowly rooting in the detritus, but can also make jerky movements; often stands totally still for a long time. Adoral zone about 29% of body length (Fig. 38e), composed of strong, but short membranelles. Buccal lip on right side of buccal cavity soft, thick, and granulated, makes jerky movements; Kahl did not observe the large undulating membrane (paroral?) described by Maupas, but found only a row of short cilia on the margin of the lip. Three frontal cirri, usually two buccal cirri. Cirral row formed by right cirri of midventral pairs run in furrow; in resting specimens these cirri are directed right-forwards and thus can be misinterpreted as fibrils (Kahl 1932).

Fig. 39h–k *Pseudoamphisiella lacazei* (from Song et al. 1997. Protargol impregnation). **h, i:** Infraciliature of ventral and dorsal side and nuclear apparatus of a middle divider, 167 μm. Arrows mark the anlagen for the right marginal rows, which originate left of the parental row. **j, k:** Infraciliature of ventral and dorsal side and nuclear apparatus of a late divider, 174 μm. Parental structures white, new black. CC = caudal cirri, MA = fused macronucleus, MI = micronucleus, RMR = new right marginal rows. Page 193. →

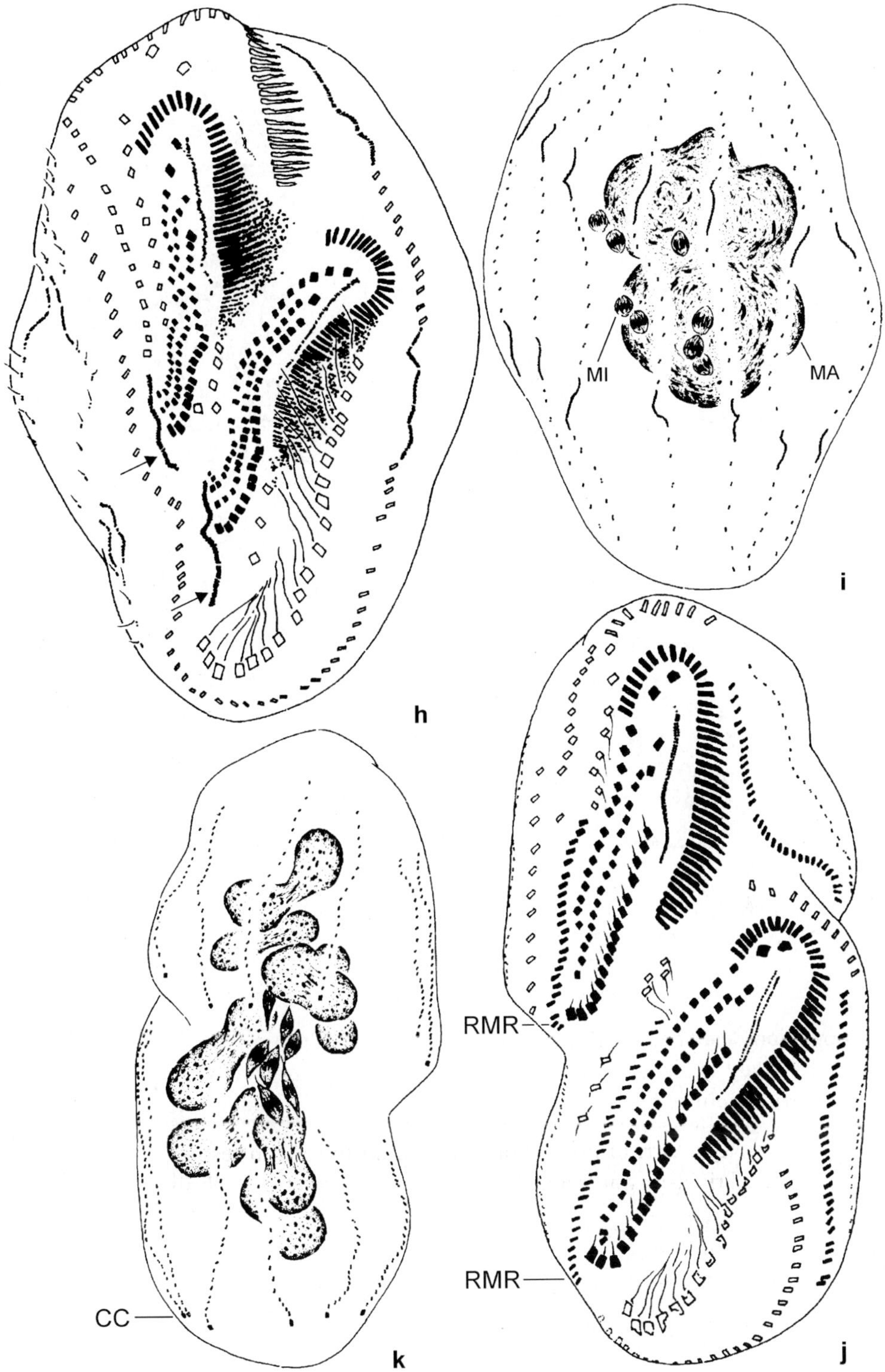
MI
MA
i
h
RMR
RMR
CC
k
j

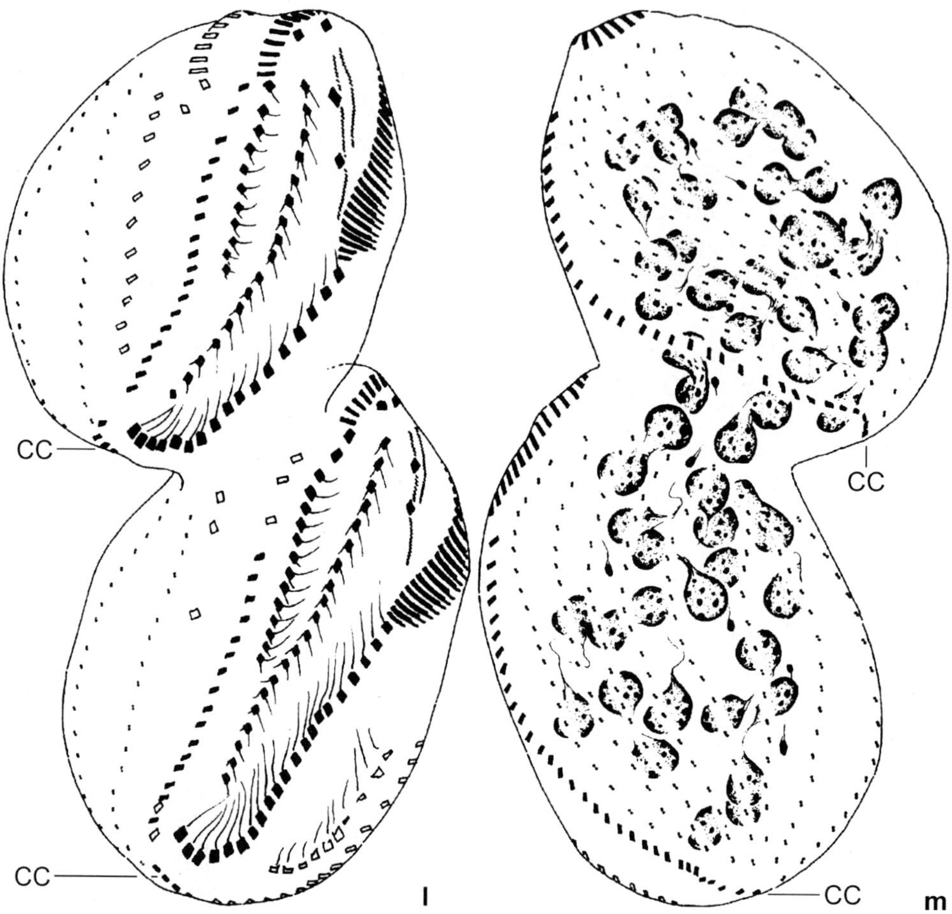

Fig. 39l, m *Pseudoamphisiella lacazei* (from Song et al. 1997. Protargol impregnation). Infraciliature of ventral and dorsal side and nuclear apparatus of a very late divider, 180 µm. Parental structures white, new black. CC = new caudal cirri. Page 193.

Population described by Kattar (1970) in life 200–220 µm long, sometimes shorter (Fig. 38h). Body ellipsoidal, flexible, but fragile. Adoral zone about one third of body length. Cirri pattern basically as in type population and populations described by Song et al. (1997), except for the second left marginal row so that in total six cirral rows are present (from right): (i) right marginal row; (ii) right and (iii) left cirral row formed by midventral pairs; (iv) transverse cirral row (with 26 cirri according to Fig. 38h); (v) inner left marginal row; and (vi) outer left marginal row which is continuous (likely via the caudal cirri) with the right marginal row. In addition, Kattar counted only six dorsal kineties (against 9–11 in Chinese populations). These two deviations (second left marginal row, only six dorsal kineties) indicate that it could be a different species.

However, the data are too scanty at the moment; that is, further populations from this region have to be studied.

Borror & Wicklow (1983) provided an illustration (Fig. 38i, Table 15), but no description. Size of specimen illustrated 200 × 40 µm, that is, body length:width ratio about 5:1. Macronuclear nodules scattered. Cortical granules (seam) not illustrated. Adoral zone of membranelles only 20% of body length. Number of transverse cirri distinctly below, values of most other features distinctly above values of other populations (Table 15). One of the most important deviations is the single buccal cirrus (against two such cirri in other populations). Since all these deviations are quantitative and not qualitative and since the data provided are scanty, a separation at species level seems unwise at the present state of knowledge; possibly, the American population is a geographical race.

Cell division (Fig. 39e–m): This process is described in detail by Song et al. (1997). Some of these data (specific origin of right marginal row, lack of frontoterminal cirri) caused them to establish the Pseudoamphisiellidae family. The following description contains only the main events and is thus rather short; for a more detailed description of the whole process, see Song et al. (1997).

Ontogenesis commences with the formation of basal bodies around the anterior end of the transverse cirri row (Fig. 39e, arrows). The next stage found by Song et al. is distinctly later and shows the oral primordia and frontal-midventral-transverse cirri anlagen for the proter and the opisthe (Fig. 39f). The oral primordium of the proter develops behind the parental adoral zone and replaces the old zone completely. The most important event in this stage is the formation of the right marginal primordia. They do not originate, as in most species, within the parental row, but left of it, exactly in line with the frontal-midventral-transverse cirri streaks. According to Song et al. the two right marginal row anlagen (Fig. 39f, long arrows) are each the rightmost anlage of the frontal-midventral-transverse cirri streak complex. However, they do not show or know a stage where the connection between this marginal primordium and the next streak (Fig. 39f, short arrows) is more clearly recognisable. At the present state of knowledge it thus cannot be excluded that the right marginal primordia originate de novo left of the parental row at the same level as the frontal-midventral-transverse cirral streaks. The formation of the right marginal primordia at this same level is also known from other taxa, for example, *Metaurostylopsis marina* (Fig. 39l–n), and de novo marginal row formation is also reported for *Thigmokeronopsis*, where, however, the primordia originate on the right side of the right marginal row (Fig. 174f). By contrast, left marginal row formation proceeds in ordinary manner in *P. lacazei* (Fig. 39f, h, j), whereas in *Holosticha* the left marginal row of the proter is formed de novo. Each of the two frontal-midventral-transverse cirri anlagen complexes of *P. lacazei* comprises 20–25 streaks.

Next, the anlagen of the proter, especially the adoral zone, migrate anteriad. The anlagen of the opisthe are obviously still beneath the cortex because the parental cirri and their fibrils are not yet resorbed (Fig. 39h). A cirrus originates at the anterior end of the undulating membranes anlagen; as is usual, it becomes the left frontal cirrus (Song et al., obviously erroneously, wrote that this cirrus becomes the rightmost frontal cirrus).

Each frontal-midventral-transverse streak, except the leftmost ones, separates into three cirri to form the midventral complex and the long transverse cirri row. In late stages, the anterior cirrus of the rightmost anlage(n) is usually resorbed so that the right row (Fig. 39d, right arrow) is then shorter than the left.

There is also a peculiarity in the ontogenesis of the buccal ciliature. Usually, *Pseudoamphisiella lacazei* has two so-called buccal cirri (Fig. 38a, e, j, 39a, Table 15). In most species with two or more buccal cirri, they originate from the same anlage, that is, from anlage II, which forms the middle frontal cirrus. In *P. lacazei*, only the anterior of the two buccal cirri is formed from anlage II, while the rear one is the second cirrus (counted from anterior) of anlage III. Thus, this rear buccal cirrus is homologous with cirrus III/2, which is usually located behind the right frontal cirrus (Fig. 39h, j, l, o).

Pseudoamphisiella lacazei lacks frontoterminal cirri. This is confirmed by the ontogenetic data showing that the rightmost frontal-midventral-transverse cirri streak does not form cirri which migrate anteriorly (Fig. 39h, j, l). Another important feature of this species is, as already described above, the location of the proter's oral primordium behind the parental adoral zone, which is completely replaced. Furthermore, no part of the parental ciliature, except of the left marginal row and the dorsal kineties, is involved in primordia formation.

Dorsal ontogenesis shows no peculiarities; each parental kinety produces, as is usual, two anlagen. At the rear end of each new kinety a single caudal cirrus, composed of about six basal bodies, is formed (Fig. 39g, i, k, m). During interphase these numerous caudal cirri "connect" the rear end of the two marginal rows (Fig. 38n, 39d).

Nucleus division proceeds basically in ordinary manner, that is, the macronuclear nodules fuse to a single mass which divides in late stages of ontogenesis. Song et al. could not observe replication bands (Fig. 39g, i, k, m). Division of the nuclear apparatus was already described by Torch (1960). In this population, the number of macronuclear nodules varied from 32 to 71, with most specimens having 40–50 (4–8 μm across, with central chromatin mass surrounded by numerous small, achromatic spheres); 5–13 micronuclei, about 2 μm across, were present. Torch's observation agree very well with the data by Song et al. (1997). Surprisingly, he also could not observe macronuclear reorganisation, indicating that the lack of this feature is an autapomorphy of *Pseudoamphisiella lacazei.*

Physiological reorganisation (Fig. 39n–p): This process is described briefly by Song et al. (1997). Obviously, it proceeds very similarly to divisional morphogenesis. They mention the following main results: (i) the parental adoral zone is replaced; (ii) the new right marginal row originates left of the parental one (obviously de novo, as in dividers); (iii) the whole parental ciliature is replaced.

Occurrence and ecology: Marine; not very common. The type locality of *Pseudoamphisiella lacazei* is the Mediterranean Sea at the beach of Bab el Qued (36°47'51"N 02°02'34"E), near Algiers, Algeria (Maupas 1883, p. 561). Kahl (1932, 1933) found it mainly in the oligo- to mesosaprobic detritus of marine aquaria from Helgoland and Kiel, Germany. Song et al. (1997; see also Song & Wang 1999, p. 73) isolated *P. lacazei* in May 1995 in Taipingjia, Qingdao (36°08'N 120°43'E), from a eutrophic pond used for storing marine molluscs (salinity about 32‰, 15–16°C, pH 8.3).

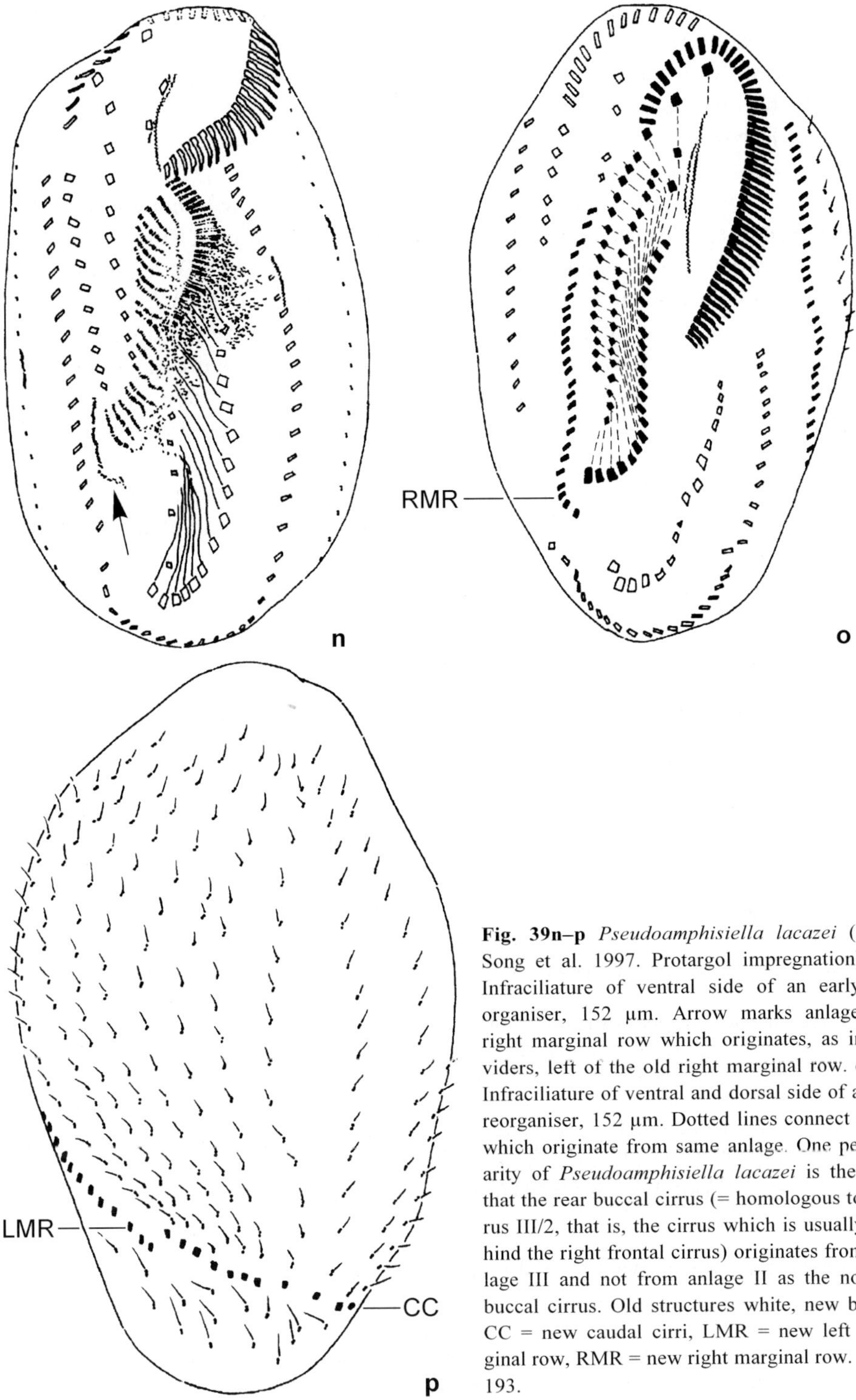

Fig. 39n–p *Pseudoamphisiella lacazei* (from Song et al. 1997. Protargol impregnation). **n:** Infraciliature of ventral side of an early reorganiser, 152 µm. Arrow marks anlage for right marginal row which originates, as in dividers, left of the old right marginal row. **o, p:** Infraciliature of ventral and dorsal side of a late reorganiser, 152 µm. Dotted lines connect cirri, which originate from same anlage. One peculiarity of *Pseudoamphisiella lacazei* is the fact that the rear buccal cirrus (= homologous to cirrus III/2, that is, the cirrus which is usually behind the right frontal cirrus) originates from anlage III and not from anlage II as the normal buccal cirrus. Old structures white, new black. CC = new caudal cirri, LMR = new left marginal row, RMR = new right marginal row. Page 193.

Specimens were cultured in boiled sea water supplemented with squeezed wheat grains. Song (1996) likely found it in the same locality (shellfish culture pond in Qingdao), but obviously two years earlier, that is, in 1993 (see Song et al. 1997, p. 265). Song et al. (1997, p. 268) morphometrically characterised a second population collected in 1991; it is thus unclear whether or not this is the same population as described by Song (1996), or a third one from the Chinese sea.

Further records substantiated by illustrations: common on the (oligosaprobic?) beach of Urca (Kattar's coordinates 22°53'08"S 43°10'00"W), a district of Rio de Janeiro, Brazil (Kattar 1970); marine habitat in USA (Borror & Wicklow 1983; no details given). Torch (1960) collected the present species at Woods Hole, USA on June 23, and cultured it for two month to study nucleus events during division (see cell division).

The unsubstantiated records from a brook near Bonn (Germany) by Jutrczenki (1982, p. 108) and from the river Rhine (Germany) by Schmitz (1986, p. 66) are likely misidentifications because this species is not reliable recorded from freshwater (possibly confused with *Neokeronopsis spectabilis*, see above).

Pseudoamphisiella lacazei feeds on (heterotrophic) flagellates and small ciliates in nature, but can be maintained in culture with bacteria alone (Song et al. 1997). Kattar (1970) reported diatoms and peridinia as main food.

Supposed synonym of *Pseudoamphisiella lacazei*

Keronopsis macrostoma Dragesco, 1963
(Fig. 40a)

1963 ***Keronopsis macrostoma* n. sp.** – Dragesco, Cah. Biol. mar., 4: 262, Fig. 9 (Fig. 40a; original description; site of deposition of type slides not indicated; no formal diagnosis provided).

1978 ***Keronopsis macrostoma* Dragesco** – Yankovskij[1], Tezisy Dokl. zool. Inst. Akad. Nauk SSSR, 1978: 40 (original description of *Erionella* with *K. macrostoma* as type species, see nomenclature).

1979 ***Keronopsis macrostoma* Dragesco, 1963** – Jankowski, Trudy zool. Inst., Leningr., 86: 53 (mentioned as type species of *Erionella*).

1979 ***Holosticha macrostoma* (Dragesco, 1963)** – Jankowski, Trudy zool. Inst., Leningr., 86: 56 (combination with *Holosticha*).

1982 ***Erionella macrostomum* (Dragesco, 1963) Jankowski, 1978** – Wicklow, Protistologica, 18: 328 (establishment of the family Erionellidae).

1985 ***Erionella* (formerly *Keronopsis*) *macrostoma*** – Small & Lynn, Phylum Ciliophora, p. 517, Fig. 30A (Fig. 40a; guide to ciliate genera).

2001 ***Keronopsis macrostoma* Dragesco, 1963** – Berger, Catalogue of ciliate names 1. Hypotrichs, p. 44 (nomenclator containing all basionyms, combinations, and higher taxa of hypotrichs).

2002 ***Erionella* (formerly *Keronopsis*) *macrostoma*** – Lynn & Small, Phylum Ciliophora, p. 423, Fig. 4A (Fig. 40a; guide to ciliate genera).

Nomenclature: No derivation of the name is given in the original description. The species-group name *macrostoma* is a composite of the Greek adjective *macr-* (large,

[1] This spelling is from the Zoological Record, vol. 16, 1979, p. 144 because the paper is completely written in Cyrillic. Usually, the name of this Russian author is spelled Jankowski.

long), the thematic vowel ·*o*-, and the Greek substantive *stoma* (mouth, opening) and obviously refers to the large, triangular buccal field of this species.

Yankovskij (1978) established *Erionella* (see below for derivation of name) with *Keronopsis macrostoma* Dragesco as type species. Thus, he is the combining author for *Erionella* although he did not combine the species with the new genus formally. Thus, the correct citation, when accepting Yankovkij's decision, would be *Erionella macrostoma* (Dragesco, 1963) Yankovskij, 1978.

Both *Keronopsis* and *Erionella* are feminine (ICZN 1999, Articles 30.1.2 and 30.1.3). Thus, the change of *K. macrostoma* to *E. macrostomum* as proposed by Wicklow (1982) is incorrect.

Keronopsis macrostoma Dragesco, 1963 is a primary homonym of *Keronopsis macrostoma* Reuter, 1963 (see Berger 2001). Since both species are described in the same year it is difficult to find out the junior homonym because one has to search for the exact publication date of both papers. Dragesco's species was transferred to *Erionella* by Yankovskij (1978) and to *Holosticha* by Jankowski (1979); Song & Warren (2000) consider it as synonym of *Pseudoamphisiella lacazei*. Reuter's species was never transferred to another genus (Berger 2001); I classify it in *Anteholosticha*. Since Dragesco's species certainly does not belong to *Anteholosticha* these two species are not congeneric and according to the ICZN (1999, Article 23.9.5) the junior primary homonym (which I do not know) must not be replaced automatically.

No derivation of *Erionella* is given in the original description (Yankovskij 1978). I cannot explain the name; possibly the first part stems from the Greek substantive *eri*- (wool); because the name ends with *-ella* (Latin diminutive suffix), it has feminine gender (ICZN 1999, Article 30.1.3). Yankovskij (1978) included only *Keronopsis macrostoma* Dragesco in *Erionella*.

Remarks: Dragesco (1963) discovered *Keronopsis macrostoma* in the mesopsammon of the Atlantic Ocean near Roscoff (France) and described it after protargol preparations. Borror (1972) obviously overlooked this species in his revision. Yankovskij (1978) thought that it requires a new genus, which is basically correct when we consider that about 20 years later, Song established *Pseudoamphisiella* for its supposed senior synonym *Holosticha lacazei*. However, one year later Jankowski obviously was no longer convinced of the necessity of *Erionella* and thus transferred it to *Holosticha* (Jankowski 1979). In contrast, Wicklow (1982) accepted *Erionella* and even established the family Erionellidae within the suborder Discocephalina for this single species. The Erionellidae and its classification in the discocephalines was accepted by Small & Lynn (1985, p. 516), Tuffrau & Fleury (1994, p. 133), and Lynn & Small (2002, p. 423).

Small & Lynn (1985, p. 517) and Lynn & Small (2002, p. 423) classified *Keronopsis arenicola* Dragesco, 1963 in *Erionella*. Very likely, Small & Lynn are the authors for this combination, which was done, however, without foundation. *Keronopsis arenicola*, which is somewhat superficially described, was transferred to *Biholosticha* by Berger (2003).

Just recently, Song & Warren (2000) synonymised *Keronopsis macrostoma* with *P. lacazei*, the type of *Pseudoamphisiella*. However, they overlooked that this synonymy

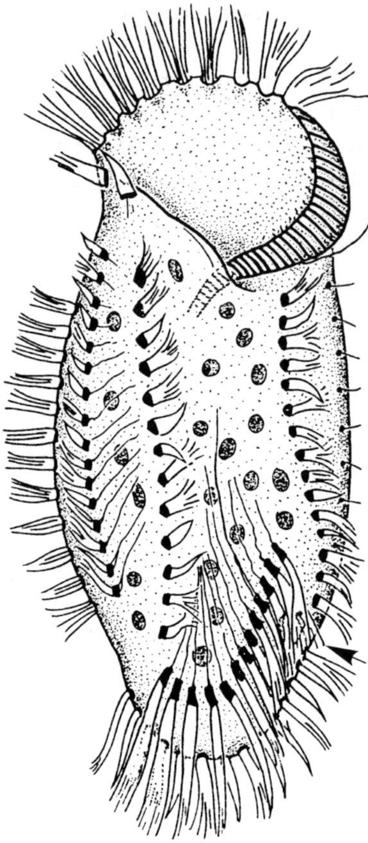

Fig. 40a *Keronopsis macrostoma* (from Dragesco 1963. Protargol impregnation). Infraciliature of ventral side, 60 µm. Possibly this species, which is the type species of *Erionella*, is the junior synonym of *Pseudoamphisiella lacazei*. The arrow marks some small cirri between the transverse cirral row and left marginal row (for details see text). Page 208.

eliminates Song's (1996) genus *Pseudoamphisiella* because *K. macrostoma* is the type species of *Erionella*, which was already established by Yankowskij in 1978. In fact, the general appearance of these two species is rather similar, especially as concerns the nuclear apparatus and the ventral ciliature, namely, the two ventral cirral rows, which do not show a zigzagging pattern, the increased number of transverse cirri, and the short adoral zone (Fig. 39a, 40a). However, there are some differences, partially also discussed by Song & Warren, which indicate that the synonymisation is not quite sure: body length (60 µm vs. 140 µm on average in *P. lacazei*; Table 15), number of transverse cirri (13–14 vs. 16 or more in Song's populations of *P. lacazei*; however, 13 cirri are also reported by Kattar for *P. lacazei*), number of dorsal kineties (6 vs. 8–11; however, 6 kineties are reported by Kattar), buccal field (large, triangular vs. of ordinary size and outline), and frontal cirri (lacking vs. three). Further, Dragesco (1963) did not mention the extrusome seam, possibly because he did not make live observation, which is necessary to see this feature. Because of these differences and uncertainties I classify *Keronopsis macrostoma* only as supposed synonym of *Pseudoamphisiella lacazei*. Thus, I do not finally synonymise *Pseudoamphisiella* Song, 1996 (type species *Holosticha lacazei* Maupas) with *Erionella* Yankovskij, 1978 (type species *Keronopsis macrostoma* Dragesco).

Morphology: Since no live observations were made, some important data (e.g., size, body shape, cortical granules) are not provided. Body length about 60 µm after protargol impregnation. 22–25 macronucleus nodules. Tuffrau & Fleury (1994, p. 133) obviously erroneously wrote that *Erionella macrostoma* has two macronuclear nodules. Oral apparatus conspicuous due to large triangular, funnel-shaped buccal field (Fig. 40a). Unfortunately it is unclear whether or not this is the true, live condition of the oral apparatus, or if this is an artificial, strongly inflated condition due to the preparation procedure. If this triangular buccal field is the normal condition, then identity with *Pseudoamphisiella lacazei* can be almost certainly excluded. Undulating membranes not observed. Two cirri at anterior margin of buccal lip (possibly these are two buccal cirri); no frontal cirri described or recognisable in illustration. For arrangement of cirri, see Fig. 40a. 16 cirri in right pseudorow formed by midventral pairs, 13 cirri in left pseudorow. 13–14 prominent transverse cirri in J-shaped pattern. According to the text, two or three small cirri between transverse cirri and left marginal row; I do not see them clearly in the illustration (Fig. 40a, arrow). 12–13 right marginal cirri; 30–32 left marginal cirri, possibly including caudal cirri. Dorsal cilia short, according to Fig. 40a (60 μm body length assumed) about 2–3 μm, arranged in six kineties, one of which near left marginal row.

Ecology: Marine. The type locality of *Keronopsis macrostoma* is the mesopsammon of the English Channel (Atlantic Ocean) near Roscoff (France) where Dragesco found it only in one sample with low abundance (for review, see Bocquet 1971, p. 388). Song & Wang (1993, p. 43) found it in the periphyton of marine culture water bodies in the Bohai Bay and the Yellow Sea (China).

Pseudoamphisiella alveolata (Kahl, 1932) Song & Warren, 2000 (Fig. 41a–r, Table 15)

1932 ***Holosticha alveolata* spec. n.** – Kahl, Tierwelt Dtl., 25: 581, Fig. 107 (Fig. 41a; original description and revision; no type material available and no formal diagnosis provided).

1933 ***Holosticha alveolata* Kahl 1932** – Kahl, Tierwelt N.- u. Ostsee, 23: 109, Fig. 17.9 (Fig. 41b; guide to marine ciliates).

1963 ***Holosticha alveolata* Kahl, 1932** – Borror, Arch. Protistenk., 106: 510, Fig. 119, 120 (Fig. 41c, d; redescription).

1972 ***Holosticha alveolata* Kahl, 1932** – Borror, J. Protozool., 19: 10 (revision of hypotrichs).

1982 ***Holosticha alveolata* Kahl, 1932** – Hemberger, Dissertation, p. 83 (revision of hypotrichs).

1983 ***Holosticha alveolata* Kahl, 1932** – Borror & Wicklow, Acta Protozool., 22: 121 (revision of urostyloids).

1992 ***Holosticha alviolata* Kahl, 1930–5** – Carey, Marine interstitial ciliates, p. 181, Fig. 710 (redrawing of Fig. 41a; guide; incorrect subsequent spelling).

2000 ***Pseudoamphisiella alveolata* (Kahl, 1932) nov. comb.**[1] – Song & Warren, Europ. J. Protistol., 36: 452, Fig. 1–17, Table 1 (Fig. 41e–r; fixation of neotype, redescription, and combination with *Pseudoamphisiella*. One neotype slide of protargol-impregnated specimens has been deposited in the Natural History Museum, London, U. K.; slide reference number: 1999:12:7:1).

2001 ***Holosticha (Holosticha) alveolata* Kahl, 1932** – Berger, Catalogue of ciliate names 1. Hypotrichs, p. 35 (nomenclator containing all basionyms, combinations, and higher taxa of hypotrichs).

Nomenclature: No derivation of the name is given in the original description. The species-group name *alveolata* (adjective) is a composite of the Latin noun *alvéolus* (small cave, small hollow, vesicle; diminutive of *álveus*) and the suffix *~āt·a* (provided with something; having something) and refers to the alveolar seam of the species. Kahl

[1] The improved diagnosis by Song & Warren (2000, p. 452) is as follows: Slightly contractile marine *Pseudoamphisiella*, in vivo 120–240 × 50–80 μm with 2 macronuclei and conspicuous extrusomes within outer alveolar layer. On average, 50 adoral membranelles; 2 buccal cirri; 12–16 transverse cirri extending about 2/5 of cell length; two genus-typical, widely separated ventral rows, each with about 12 cirri; 14–20 and 12 –14 cirri in left and right marginal rows respectively; 11–16 caudal cirri; 10–12 dorsal kineties.

(1932, 1933) divided *Holosticha* into several subgenera; thus, the correct name in his papers is *Holosticha (Holosticha) alveolata* Kahl, 1932.

Song & Warren (2000) designated a neotype for *Holosticha alveolata* although there was no exceptional need because the species was described and illustrated with sufficient accuracy by Kahl (1932) and synonymy with another species was never supposed. Thus, the taxonomic status was beyond reasonable doubt because identity with *H. lacazei* could be excluded due to the different nuclear apparatus (two against many macronuclear nodules), a feature generally accepted as species character. Borror (1963a) basically confirmed Kahl's data and did not mention a complex problem with *H. alveolata*. Song & Warren (2000) could not have checked whether or not a prior neotype was present (ICZN 1999, Article 75.3.4) because they overlooked[1] the redescription by Borror (1963a). Further, the type locality (Yellow Sea) is not very near to the original type locality (Baltic). However, there is no doubt that the ocean is a rather homogenous habitat, at least compared to various terrestrial or limnetic habitats. Although Song & Warren did not publish all particulars demanded by the code (ICZN 1999, Article 75.3), the neotypification by these authors should be accepted because it is likely that further *Pseudoamphisiella* species exist so that it will be an advantage if the known species are defined via permanent preparations.

Remarks: Kahl (1932) found this species regularly when his paper was in press. He wrote that the frontal cirri are schematically illustrated because he was unable to immobilise or fix the specimens due to their fragility. Kahl recognised the relationship with *Holosticha lacazei*, now type species of *Pseudoamphisiella*, because both have a similar, short, curved adoral zone and a seam formed by longish extrusomes (Fig. 41a). The resemblance of these two species was confirmed by Borror (1963a) and Song & Warren (2000), who thus transferred *H. alveolata* to *Pseudoamphisiella*.

Kahl (1933, p. 109) provided a simple key and wrote that the species with the letters *a* to *h* have many macronuclear nodules whereas the species *i* to *r* have only two nodules. Thus, the designation of *H. alveolata* with the letter *e* is incorrect although Kahl correctly wrote that it has only two macronuclear nodules.

Borror's (1963a) specimens differ from Kahl's description in the larger size (145–220 µm against 80–150 µm) and the somewhat higher number of transverse cirri (14 against 8–9). The other features, including the short adoral zone and the alveolar layer, fit well so that conspecificity is beyond reasonable doubt. In contrast, body size, number of transverse cirri, and the conspicuous caudal cirri of Borror's population agree very well with the neotype population described by Song & Warren (2000). The main difference between these populations is in the number of adoral membranelles (80 against 47–59); however, Borror (likely) did not impregnate his specimens and thus an overestimation of these difficult-to-count organelles cannot be excluded. Interestingly, both Kahl and Borror describe a rather short adoral zone (below 25% of body length), whereas in the neotype population it is 30–40% long.

Borror (1972) synonymised *Keronopsis arenicola* Dragesco, 1963 (= *Biholosticha arenicola* in present review; Fig. 239a) with the present species. However, Dragesco's

[1] However, on page 452 they cite Hemberger (1982) who definitely mentioned the redescription by Borror (1963a).

species has a distinct zigzagging midventral pattern, many more midventral and marginal cirri, and lacks an alveolar layer. I thus agree with Song & Warren (2000, p. 456), who did not accept this synonymy, but considered *K. arenicola* as valid species.

Obviously, Song & Warren (2000) overlooked Borror's redescription (however, see footnote from nomenclature section) and overlooked the other papers concerning the present species because on page 451 they wrote that *H. alveolata* has never been redescribed or even recorded from the time it was discovered by Kahl (1932). The data provided by Biernacka (1963, Fig. 143h) and Chardez (1986, Fig. 143i) are superficial and thus mentioned under insufficient redescriptions.

Pseudoamphisiella alveolata is easily distinguished from *P. lacazei* by the two macronuclear nodules (against many). *Neokeronopsis spectabilis* (Fig. 242–244), which also has a long row of transverse cirri and two macronuclear nodules, lives in limnetic habitats (against marine), is more than 250 µm long (against distinctly below 250 µm), has a bicorona (against 3 frontal cirri), and yellow cortical granules, which look like those of *Urostyla grandis* (against alveolar seam with rods). *Thigmokeronopsis* species, which have a similar general appearance especially as concerns the prominent transverse ciliature, have a distinct bicorona and many macronuclear nodules and thus cannot be confused with *P. alveolata*. *Holosticha* species have, inter alia, distinctly zigzagging midventral cirri, a gap in the adoral zone, and the anterior end of the left marginal row is strongly curved to the right.

Morphology: The description of the neotype population provided by Song & Warren (2000) is presented first, followed by Kahl's (1932) and Borror's (1963a) data. As already mentioned above, conspecificity of these three populations is beyond reasonable doubt.

Neotype population (from Song & Warren 2000; Fig. 41e–r, Table 15): Body size usually 120–200 × 50–70 µm in life, but occasionally up to 240 × 80 µm-sized specimens occur. Body length:width ratio 2.5–4.0:1; postdividers may be less than 100 µm long and relatively wider (Fig. 41a). Body outline variable, that is, slightly fusiform, slender to broad oval, or elliptical. Left anterior body margin usually slightly vaulted, such cells thus indistinctly cephalised. Body shape seemingly more influenced by physiological state, that is, stage in life cycle, than by the nutritional situation. Cells slightly to conspicuously contractile when stimulated (Fig. 41j) and fragile, readily bursting when manipulated. Dorsoventrally flattened about 2:1. Often two distinct furrows on ventral side along cell margin within which the marginal rows, and possibly the caudal cirri, are located (Fig. 41h, i). Invariably two macronuclear nodules in or left of cell midline. In total 2–5 micronuclei about 3 µm across attached to macronuclear nodules at various position (Fig. 41r). No contractile vacuole found, possibly lacking. Cortex with conspicuous alveolar layer forming an about 3 µm thick seam visible even under low magnification (Fig. 41e); irregular polygonal structure of alveolar layer best seen in dorsal view; within layer a relatively low number of rod-shaped, 2–3 µm long extrusomes perpendicularly arranged to cell surface; obviously, they are mainly located in the corners of the polygons (Fig. 41f, g). Cytoplasm usually dark grey, typically packed with granular inclusions 5–10 µm across giving specimens a dark, opaque appearance, especially at low magnification (Fig. 41h, i). Food vacuoles difficult to recog-

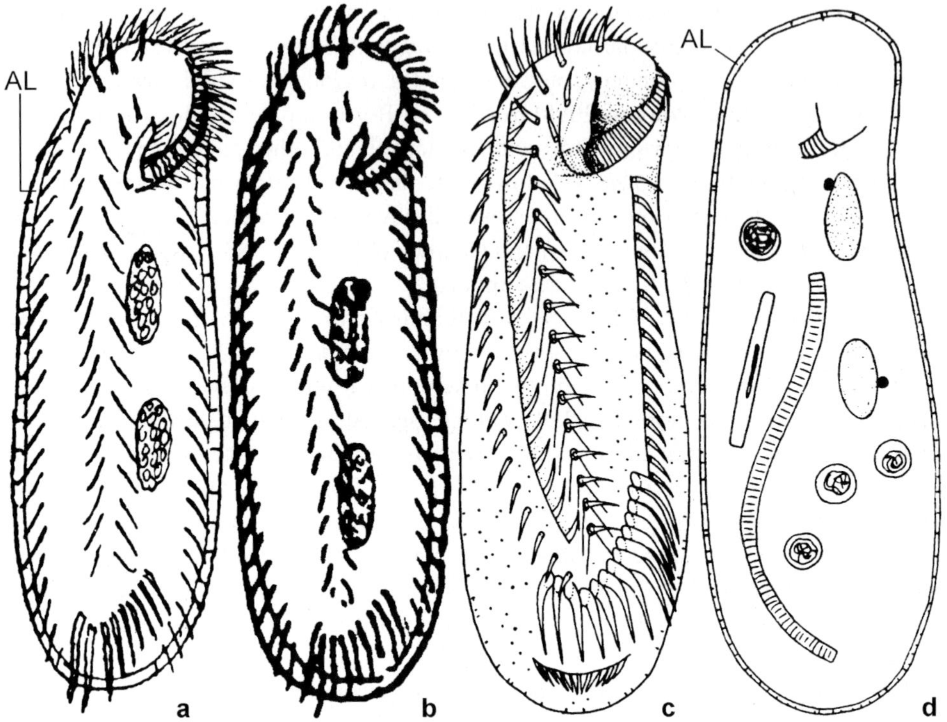

Fig. 41a–d *Pseudoamphisiella alveolata* from life (a, from Kahl 1932; b, after Kahl 1932 from Kahl 1933; c, d, from Borror 1963a). **a–c:** Ventral views, a, b = no specific size indicated, c = 200 µm. **d:** Outline with alveolar layer, food vacuoles with a diatom, cyanobacteria, nuclear apparatus, and proximal portion of adoral zone, 200 µm. AL = alveolar layer. Page 211.

nise. Continuously crawl on substrate or on bottom of Petri dish, reacting quickly when disturbed by contracting and remaining motionless for a short while.

Adoral zone of membranelles occupies around one third of body length in life[1], 36% on average after protargol impregnation (Table 15), extends far onto right side (34% of body length in specimen shown in Fig. 41k, DE-value about 0.79!), composed of an average of 51 membranelles of usual shape and fine structure. Bases of membranelles 10–15 µm wide, cilia up to 25 µm long. Buccal cavity seemingly narrow because right wall (buccal lip) strongly protruding. Paroral long, parallel to endoral. Pharyngeal fibres extend rightwards and posteriorly, obviously without peculiarities.

Cirral pattern and number of cirri of usual variability (Fig. 41k, Table 15). Three enlarged frontal cirri 30–40 µm long, form rather oblique (almost longitudinal) row; rightmost cirrus at distal end of adoral zone. All other cirri, except transverse cirri, 20–30 µm long. Two buccal cirri right of mid-region of paroral; probably they show the same,

[1] Song & Warren (p. 452) write 2/5 (40%) of cell length, which seems too high because most of their illustrations show distinctly lower values: 33% (Fig. 41e), 31% (Fig. 41h), 31% (Fig. 41i), 33% (Fig. 41j, left), 32% (Fig. 9 in Song & Warren 2000).

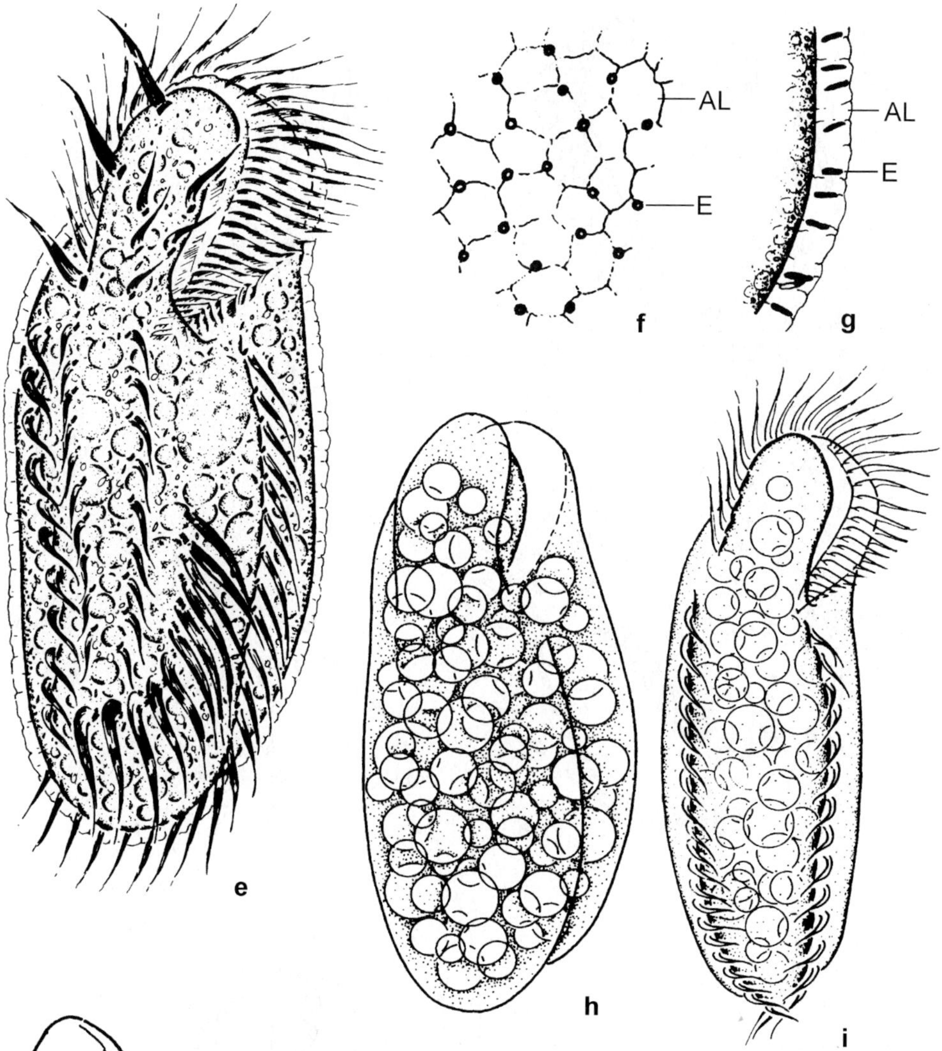

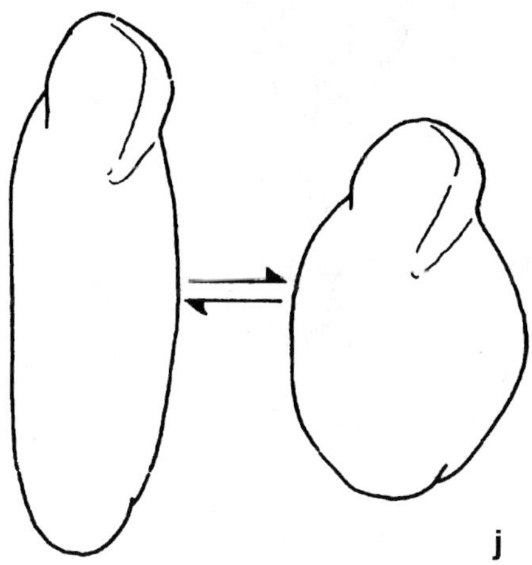

Fig. 41e–j *Pseudoamphisiella alveolata* (neotype population from Song & Warren 2000. e–j, from life). **e:** Ventral view of a representative specimen, 196 µm. Note the alveolar layer (species name), the prominent transverse cirri, and the adoral zone which extends far onto the right side. **f, g:** Top and lateral view of seam formed by extrusomes (2–3 µm long) mainly arranged in corners of polygonal alveole. **h, i:** Wide and slender specimen both packed with large globular inclusions. **j:** *Pseudoamphisiella alveolata* contracts distinctly when it is stimulated. AL = alveolar layer, E = extrusomes. Page 211.

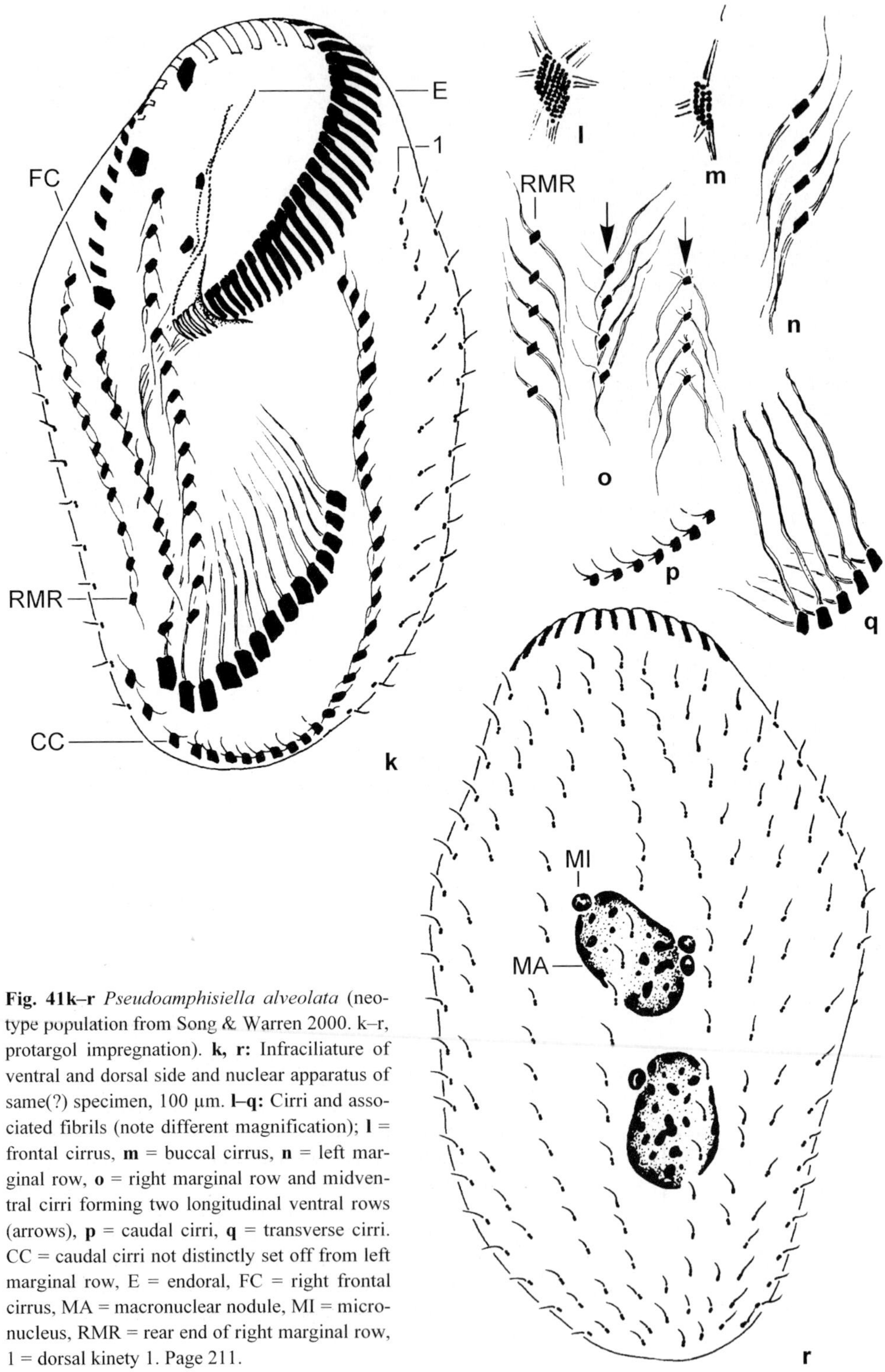

Fig. 41k–r *Pseudoamphisiella alveolata* (neotype population from Song & Warren 2000. k–r, protargol impregnation). **k, r:** Infraciliature of ventral and dorsal side and nuclear apparatus of same(?) specimen, 100 µm. **l–q:** Cirri and associated fibrils (note different magnification); **l** = frontal cirrus, **m** = buccal cirrus, **n** = left marginal row, **o** = right marginal row and midventral cirri forming two longitudinal ventral rows (arrows), **p** = caudal cirri, **q** = transverse cirri. CC = caudal cirri not distinctly set off from left marginal row, E = endoral, FC = right frontal cirrus, MA = macronuclear nodule, MI = micronucleus, RMR = rear end of right marginal row, 1 = dorsal kinety 1. Page 211.

somewhat curious ontogenesis as in *P. lacazei* (see there, for details). Frontoterminal cirri lacking. Midventral complex composed of cirral pairs only; however, pairs do not form a zigzagging pattern in non-dividing specimens, but two longitudinal, slightly curved rows (Fig. 41e, k); left row commences near middle frontal cirrus, terminates, like right row, slightly ahead of right portion of transverse cirral row; right row begins close behind right frontal cirrus. Transverse cirri not or indistinctly protruding posteriorly although 30–40 µm long, rather strong and thus forming prominent, close-set J-shaped figure (Fig. 41e, k); according to the morphometric analysis, 12–16 transverse cirri are present, whereas Fig. 41e shows only 11 cirri. Right marginal row commences right of distal end of adoral zone, terminates at, or slightly ahead of, level of right transverse cirrus; left marginal cirri obviously continuous with caudal cirri and thus forming J-shaped row; marginal cirri usually bending towards cell mid-line or lying along the furrows, thus usually not protruding beyond the cell margin, except in posterior region. Fibrils, which are highly developed and group-dependent in terms of length and structure, characteristically associated with various cirral groups (Fig. 41l–q).

Dorsal cilia about 3–5 µm long, seem to be fixed in deep layer of pellicle or at base of alveolar layer; arranged in 10–12 bipolar rows. Caudal cirri narrowly spaced and continuous with rear end of left marginal row; likely some dorsal kineties produce more than one caudal cirrus because the numbers do not agree (10–12 vs. 11–16; Table 15).

Kahl's specimens 80–150 µm long in life, Borror's 145–220 × 58–90 µm; body length:width ratio 3.1:1 in specimen shown in Fig. 41a, about 2:1 flattened dorsoventrally (Borror). Body outline ellipsoidal with posterior end broader rounded than anterior one. Cell greyish granulated, very soft and flexible, but fragile, slightly contractile, and very agile (Kahl). Invariably two macronuclear nodules (Kahl, Borror), 20 × 10 µm (Borror), each nodule with a single micronucleus 4 µm across (Fig. 41d). Contractile vacuole neither mentioned nor illustrated by Kahl and Borror indicating that it is lacking. Alveolar seam distinct, subdivided by few rod-shaped extrusomes perpendicularly arranged to pellicle. Adoral zone of membranelles short (distinct difference to neotype population), that is, 20% (Fig. 41c) to 24% (Fig. 41a), proximal end distinctly curved and extending rightwards. Paroral covered by buccal lip (Borror). Three frontal cirri, which are, however, only schematically illustrated by Kahl. According to Borror six frontal cirri present; obviously, this number includes two buccal cirri, and a cirrus behind a frontal cirrus or a cirrus of the anteriormost midventral pair (Fig. 41c). Midventral complex extends from near distal end of adoral zone to near transverse cirri (Kahl, Borror). "Right ventral cirrus row" (= pseudorow formed by right cirri of midventral pairs) in a groove (Borror). According to Kahl (1932) 8–9 strong, almost not projecting transverse cirri (Fig. 41a), according to Borror 14, which are somewhat dislocated anteriorly (Fig. 41c). Number of dorsal kineties and length of dorsal cilia neither mentioned nor illustrated by Kahl and Borror.

Cell division: Hu & Suzuki (2005) studied this part of the life cycle in a Japanese population, which corresponds very well with the neotype population. They summarised the results as follows: (i) the oral primordium and the cirral anlagen of the opisthe are derived from an anarchic field originating independently between the left ventral row

Table 15 Morphometric data on *Pseudoamphisiella alveolata* (al1, from Song & Warren 2000; al2, from Borror 1963a) and *Pseudoamphisiella lacazei* (la1, population 1 from Song et al. 1997; la2, population 2 from Song et al. 1997; la3, from Maupas 1883; la4, from Kattar 1970; la5, from Borror & Wicklow 1983; la6, from Song 1996)

Characteristics [a]	Population	mean	M	SD	SE	CV	Min	Max	n
Body, length	al1	96.6	–	24.5	5.8	25.3	75.0	159.0	18
	la1	140.2	–	18.0	3.5	13.3	110.0	178.0	26
	la2	–	–	–	–	–	156.0	204.0	4
Body, width	al1	47.3	–	15.9	3.8	33.7	34.0	86.0	18
	la1	67.7	–	8.9	1.9	12.2	46.0	81.0	23
	la2	–	–	–	–	–	59.0	77.0	4
Anterior body end to proximal end of adoral zone, distance	al1	35.1	–	5.6	1.3	16.1	24.0	42.0	18
	la1	40.9	–	4.4	0.9	11.0	33.0	51.0	25
	la2	–	–	–	–	–	49.0	60.0	4
Macronucleus nodule, length	al1	15.5	–	3.0	0.7	19.1	11.0	20.0	16
	la1	6.3	–	1.5	0.3	24.5	4.0	10.0	29
	la2	6.2	–	0.9	0.2	14.9	5.0	8.0	16
Macronucleus nodule, width	al1	9.8	–	1.8	0.4	18.2	7.0	12.0	16
Macronucleus nodules, number	al1	2.0	–	0.0	0.0	0.0	2.0	2.0	16
	la1	36.8	–	8.1	1.8	22.0	24.0	57.0	20
	la2			about 50					
	la3	–	–	–	–	–	43.0	54.0	2
	la4	60.0	–	–	–	–	–	–	?
Micronucleus, length	la1	2.2	–	0.2	0.0	9.4	2.0	3.0	9
Micronuclei, number	al1	3.5	–	1.2	0.4	35.1	2.0	5.0	11
	la1	–	–	–	–	–	7.0	10.0	6
	la3	11.0	–	–	–	–	–	–	1
Adoral membranelles, number	al1	50.9	–	3.7	1.2	7.3	47.0	59.0	9
	al2	80.0	–	–	–	–	–	–	1
	la1	44.1	–	3.3	0.7	7.3	39.0	49.0	25
	la2	–	–	–	–	–	46.0	52.0	4
	la4	45.0	–	–	–	–	–	–	?
	la5	40.0	–	–	–	–	–	–	1
Buccal cirri, number	al1	2.0	–	0.0	0.0	0.0	2.0	2.0	18
	la1	2.0	–	0.0	0.0	0.0	2.0	2.0	25
	la2	–	–	–	–	–	2.0	2.0	4
	la4	2.0	–	–	–	–	–	–	?
	la5	1.0	–	–	–	–	–	–	1
Frontal cirri, number	al1	3.0	–	0.0	0.0	0.0	3.0	3.0	18
	la1	3.0	–	0.0	0.0	0.0	3.0	3.0	25
	la2	3.0	–	–	–	–	3.0	3.0	4
	la4	3.0	–	–	–	–	–	–	?
	la5	3.0	–	–	–	–	–	–	1
Midventral pairs, number of left cirri	al1	12.2	–	1.5	0.4	12.1	10.0	14.0	11
	la1 [b]	13.2	–	1.2	0.2	9.4	11.0	15.0	26
	la2 [b]	–	–	–	–	–	14.0	16.0	4
	la5	26.0	–	–	–	–	–	–	1
	la6 [b]	–	–	–	–	–	16.0	21.0	?
Midventral pairs, number of right cirri	al1	12.5	–	1.6	0.5	12.5	11.0	15.0	11
	la1 [c]	19.1	–	1.8	0.4	9.0	16.0	23.0	26
	la2 [c]	–	–	–	–	–	16.0	21.0	4
	la5	21.0	–	–	–	–	–	–	1

Table 15 Continued

Characteristics [a]	Population	mean	M	SD	SE	CV	Min	Max	n
Midventral pairs, number of cirri	la6 [c]	–	–	–	–	–	14.0	16.0	?
Transverse cirri, number	al1	14.2	–	1.6	0.4	11.0	12.0	16.0	16
	al2	14.0	–	–	–	–	–	–	1
	la1	19.2	–	1.7	0.3	9.2	16.0	23.0	26
	la2	–	–	–	–	–	17.0	20.0	4
	la5	13.0	–	–	–	–	–	–	1
Left marginal cirri, number	al1	17.0	–	1.8	0.6	10.9	14.0	20.0	11
	al2	17.0	–	–	–	–	–	–	1
	la1	26.1	–	3.4	0.9	13.1	21.0	31.0	16
	la2			about 34					
	la5	42.0	–	–	–	–	–	–	1
Right marginal cirri, number	al1	13.1	–	0.8	0.3	6.3	12.0	14.0	11
	al2	19.0	–	–	–	–	–	–	1
	la1	24.0	–	2.9	0.7	12.3	20.0	29.0	16
	la2			about 31					
	la5 [d]	45.0	–	–	–	–	–	–	1
Dorsal kineties, number	al1	11.2	–	0.8	0.2	6.7	10.0	12.0	16
	la1	9.1	–	0.7	0.1	7.3	8.0	11.0	25
	la2	–	–	–	–	–	10.0	11.0	4
	la4	6.0	–	–	–	–	–	–	?
	la6	–	–	–	–	–	9.0	11.0	?
Caudal cirri, number	al1	13.9	–	1.9	0.6	13.7	11.0	16.0	9
	al2	12.0	–	—		–	–	–	1
	la1	–	–	–	–	–	9.0	10.0	5
	la2			about 10					

[a] All measurements in µm. Data by Song (1996), Song & Warren (2000), Song et al. (1997), and Kattar (1970) are based on protargol-impregnated specimens; data by Borror (1963a) likely from life; Borror & Wicklow (1983) did not specify the method. Data by Borror & Wicklow (1983) from Fig. 38i. Possibly, the data from "la2" and "la6" belong to the same population (not clearly stated by Song et al. 1997). CV = coefficient of variation in %, M = median, Max = maximum value, mean = arithmetic mean, Min = minimum value, n = number of individuals investigated (? = number not given. If only one value is available, then it is listed under the column mean; if two values are given, then they are listed under Min and Max), SD = standard deviation, SE = standard error of arithmetic mean.

[b] This is the ventral row 1 (VR1) according to Song et al. (1997). The specimen shown in Fig. 39d has 17 cirri in this row, whereas the maximum value given in the table is 15; possibly, Song et al. confused the values for VR1 and VR2 (in Song 1996 these two rows are designated just opposite!). I assume that the lower values apply to the right row, and the higher values to the left row.

[c] This is the ventral row 2 (VR2) according to Song et al. (1997). The specimen shown in Fig. 39d has 15 cirri in this row whereas the minimum value in the table is 16; possibly, Song et al. confused the values for VR1 and VR2 (in Song 1996 these two rows are designated just opposite!). I assume that the lower values apply to the right row, and the higher values to the left row.

[d] The anteriormost two cirri of the right marginal row are slightly set off; possibly, these cirri are frontoterminal cirri, which are, however, not present in the Chinese population described by Song et al. (1997).

and the anterior transverse cirri; (ii) numerous oblique frontal-ventral-transverse cirral anlagen are formed in the proter and the opisthe by the division of primary primordia; (iii) the parental adoral zone is partly renewed; (iv) the anlagen for the right marginal rows originate between the parental right marginal row and the rightmost frontal-ventral-transverse cirral anlage; (v) the left marginal rows and dorsal kineties originate by intrakinetal anlagen formation; (vi) caudal cirri are formed at the end of each dorsal kinety anlage; (vii) several cirri near the rightmost transverse cirri, previously known as caudal cirri, are marginal cirri because they originate from a small anlage from the posterior end of the right marginal row.

The data by Hu & Suzuki (2005) show two distinct differences to *P. lacazei*, namely, (i) in the formation of the adoral zone of the proter (partly reorganised vs. completely new), and (ii) in the formation of some additional right marginal cirri from a small anlage (vs. lacking in *P. lacazei*).

Occurrence and ecology: Marine; not very common. According to Article 76.3 of the ICZN (1999), the place of origin of the neotype becomes the type locality of the nominal species. Song & Warren (2000) found the neotype population in coastal waters (salinity 32‰, 24°C, pH 8.0) off Qingdao (China) at the co-ordinates 36°08'N 120°43'E. In contrast, Kahl (1932) discovered *P. alveolata* on/in various sandy sediments of the Baltic sea near the city of Kiel, Germany. Borror (1962, 1963a) found it in the benthal of Alligator Harbor, Franklin County (Florida), on the northern edge of the Gulf of Mexico; it was uncommon in sand and was not recorded from diatom detritus although it appeared in three stations all rich in diatoms and cyanobacteria. Borror's population has survived in week-old cultures, becoming numerous in illuminated sand cultures containing blooms of algae; it may be a denizen of detritus and algae at the sand surface. The population studied by Hu & Suzuki (2005) was isolated from the coastal waters off Nagasaki, Japan. The limnetic record, which is not substantiated by morphological data, from a brook in Bonn (Germany) by Jutrczenki (1982, p. 108) must be interpreted as misidentification because *P. alveolata* is likely confined to marine habitats.

Pseudoamphisiella alveolata feeds on diatoms (Kahl 1932, Borror 1963a) and up to 100 μm long cyanobacteria (Borror 1963a; for review, see Fenchel 1968). Song & Warren (2000) did not mention food vacuole content; they cultured it in boiled seawater to which squeezed wheat grains were added, indicating that it feeds on bacteria.

Insufficient redescriptions

***Holosticha alveolata* Kahl 1932** – Biernacka, 1963, Polskie Archwm Hydrobiol., 11: 49, Fig. 93 (Fig. 143h). Remarks: This population from brackish water in the Danzig Bay, Baltic Sea (Poland) is rather insufficiently documented and differs, according to Biernacka, from Kahl's description in two of three features, namely, colour yellowish against grey and movement slowly against quickly. Biernacka (1962, 1963) found this population at 6.5–7.5‰ salt content and 11–15°C water temperature.

***Holosticha alveolata* Kahl 1932** – Chardez, 1986, Revue verviét. Hist. nat., 43: 21, Fig. 7 (Fig. 143i). Remarks: Chardez found this population in a limnetic habitat near Liège, Belgium, strongly indicating that this is a misidentification because *Pseudoamphisiella alveolata* is very likely confined to marine habitats. The illustration is too superficial for a reliable identification and no further details have been provided.

Psammomitra Borror, 1972

1866 ***Claparedia*** – Diesing, Sber. Akad. Wiss. Wien, 52: 519 (original description; see nomenclature). Type species (by subsequent designation by Jankowski 1979, p. 51): *Oxytricha retractilis* Claparède & Lachmann, 1858.

1867 ***Mitra radiosa*** – Quennerstedt, Acta Univ. lund., 4: 41 (original description of *Mitra*; no formal diagnosis provided). Type species (by monotypy): *Mitra radiosa* Quennerstedt, 1867.

1933 ***Micromitra* nom. n.** – Kahl, Tierwelt N.- u. Ostsee, 23: 112 (replacement name for *Mitra* Quennerstedt, 1867; see nomenclature). Type species (same as for *Mitra*): *Mitra radiosa* Quennerstedt, 1867.

1935 ***Micromitra* Kahl, 1933** – Kahl, Tierwelt Dtl., 30: 842 (revision; see remarks at single species).

1972 ***Psammomitra* n. nom.**[1] – Borror, J. Protozool., 19: 8, 15 (replacement name for *Micromitra* Kahl, 1933; see nomenclature). Type species (same as for *Mitra* and *Micromitra*): *Mitra radiosa* Quennerstedt, 1867.

1979 ***Psammomitra* Borror, 1972** – Corliss, Ciliated Protozoa, p. 309 (revision).

1979 ***Psammomitra* Borror, 1972** – Jankowski, Trudy zool. Inst., Leningr., 86: 63 (catalogue of generic names of hypotrichs).

1982 ***Psammomitra* Borror, 1972**[2] – Hemberger, Dissertation, p. 226 (revision of non-euplotid hypotrichs).

1985 ***Psammomitra*** – Small & Lynn, Phylum Ciliophora, p. 461 (guide to ciliate genera).

1987 ***Psammomitra* Borror, 1972** – Tuffrau, Annls Sci. nat. (Zool.), 8: 115 (brief revision of hypotrichs).

1992 ***Psammomitra* Borror, 1972** – Carey, Marine interstitial ciliates, p. 185 (guide).

1994 ***Psammomitra* Borror, 1972** – Tuffrau & Fleury, Traite de Zoologie, 2: 141 (revision of hypotrichs).

2001 ***Psammomitra* Borror 1972** – Aescht, Denisia, 1: 135 (catalogue of generic names of ciliates).

2001 ***Psammomitra* Borror, 1972** – Berger, Catalogue of ciliate names 1. Hypotrichs, p. 76 (nomenclator containing all basionyms, combinations, and higher taxa of hypotrichs).

2002 ***Psammomitra* Borror, 1972** – Lynn & Small, Phylum Ciliophora, p. 451 (guide to ciliate genera).

Nomenclature: *Claparedia* was named according to Édouard Claparède who described, together with Johannes Lachmann, the single species of the present genus. However, the same name was used just one year before by another author for a different genus so that *Claparedia* Diesing, 1866 is a junior homonym (details, see Jankowski 1979, p. 52). Moreover, Diesing (1866a, b) did not fix a type species, a shortcoming corrected by Jankowski (1979). Simultaneously, Corliss (1979. p. 208) put *Claparedia* into the list of nomina oblita.

Mitra (Latin noun; mitre, feminine) likely refers to the outline of the anterior body portion. Kent (1882, p. 775) recognised that *Mitra* Quennerstedt, 1867 is preoccupied by a mollusc-genus[3], but did not introduce a new name because he classified both *Oxytricha retractilis* and *Mitra radiosa* in *Epiclintes*. Kahl (1933) did not accept the classification of *O. retractilis* in *Epiclintes* and therefore introduced the replacement name *Micromitra*, which is a composite of the Greek adjective *micro-* (small) and the genus-group name *Mitra* and obviously refers to the small size of the ciliate-*Mitra* (as compared to the mollusc *Mitra*). Unfortunately, Kahl's replacement name was also preoccupied, namely by *Micromitra* Meek, 1873 (brachipod) and *Micromitra* Bellardi, 1889

[1] The diagnosis by Borror (1972) is as follows: Elongate, highly contractile forms with narrowed anterior end, psammobiotic.

[2] The diagnosis by Hemberger (1982) is as follows: Je 1 linke und recht Marginalreihe; Angeblich keine Ventral- und Transversalcirren; Körper langgestreckt, mit schwanzartigem Hinterende; sehr kontraktil; psammobiont.

[3] Kent did not mention the author of the senior homonym. According to Jankowski (1979, p. 58), *Mitra* was used **seven** times (including Quennerstedt 1867) as genus-group name.

(mollusc; papers by Meek and Bellardi not mentioned in reference section). Thus, Borror (1972) replaced *Micromitra* Kahl, 1933 by *Psammomitra*, which is a composite of the Greek noun *he psammos* (the sand) and *Mitra*; it refers to the fact the ciliate species lives in the psammal. Interestingly, Corliss (1960) overlooked this homonymy in his review. According to the International Code of Zoological Nomenclature a genus introduced as a replacement name for a prior name must have, respectively, has the same type species as the original genus (ICZN 1964, Article 67i; ICZN 1999, Article 67.8). Thus, Borror's (1972) statement that *Psammomitra retractilis* (Claparède & Lachmann, 1858) is the type species of *Psammomitra* is not quite correct because the type species of *Mitra* and *Micromitra* is *Mitra radiosa* Quennerstedt, 1867. Aescht (2001, p. 101) even assumed that *Micromitra* Kahl, 1933 is a nomen nudum because Kahl (1933) did not fix a type species. She therefore assigned *Micromitra* to Jankowski (1979) who mentioned *Oxytricha retractilis* as type.

Incorrect subsequent spelling: *Psammonitra* sp. (Mahachanchawalit et al. 1986, p. 12).

Characterisation (A = supposed apomorphies): Body tripartite in (narrrow) head, trunk, and tail (A). Adoral zone of membranelles continuous. Undulating membranes short and in parallel (A?). Buccal cirrus, frontoterminal cirri, and transverse cirri present. Midventral complex composed of midventral pairs only. 1 right marginal row and 1 left marginal row. Caudal cirri lacking.

Remarks: See same chapter at single species.

Species included in *Psammomitra*: (1) *Oxytricha retractilis* Claparède & Lachmann, 1858.

Single species

Psammomitra retractilis (Claparède & Lachmann, 1858) Borror, 1972 (Fig. 42a–v, 43a–k, Table 16, Addenda)

1858 ***Oxytricha retractilis***[1] – Claparède & Lachmann, Mém. Inst. natn. génev., 5: 148, Planche V, Fig. 3, 4 (Fig. 42a, c; original description; no type material available).

1862 ***Oxytricha longi-caudata*, n. sp.** – Strethill Wright, Q. Jl microsc. Sci., 2: 220, Plate IX, Fig. 7, 8 (Fig. 42b, d; original description of synonym; no formal diagnosis provided and no type material available).

1864 ***Epiclintes retractilis*** – Stein, Sber. K. böhm. Ges. Wiss., 1864: 46 (combination with *Epiclintes*).

1866 ***Claparedia retractilis* Diesing** – Diesing, Sber. Akad. Wiss. Wien, 53: 98 (combination with *Claparedia*).

1866 ***Claparedia longicaudata* Diesing** – Diesing, Sber. Akad. Wiss. Wien, 53: 98 (combination of junior synonym with *Claparedia*).

1867 ***Mitra radiosa*** – Quennerstedt, Acta Univ. lund., 4: 41, Plate II, Fig. 12, 13 (Fig. 42e, f; original description of synonym; no formal diagnosis provided and no type material available).

1882 ***Epiclintes retractilis*, C. & L. sp.** – Kent, Manual Infusoria II, p. 774, Plate XLIII, Fig. 23, 24 (redrawing of Fig. 42a, c; revision).

---

[1] The diagnosis by Claparède & Lachmann (1858) is as follows: Oxtrique à partie antérieure trés-étroite; une queue rétractile.

1882 ***Epiclintes radiosa*, Quenn. sp.** – Kent, Manual Infusoria II, p. 774, Plate XLIII, Figs. 31, 32 (redrawing of Fig. 42e, f; revision; combination of synonym with *Epiclintes*).

1889 ***Oxytricha retractilis* Clap. und L.** – Bütschli, Protozoa, p. 1744, Tafel 70, Fig. 13 (redrawing of Fig. 42a, c; revision).

1902 ***Epiclintes radiosa* Quenn.** – Calkins, Bull. U. S. Fish Comm., 21: 453, Fig. 50 (Fig. 42g, h; redescription).

1929 **? *Epiclintes retractilis* Cl. u. L.** – Hamburger & Buddenbrock, Nord. Plankt., 7: 83, Fig. 100 (redrawing of Fig. 42a, c; guide to marine plankton).

1932 ***Mitra (Oxytricha) retractilis* (Clap. u. L., 1858)** – Kahl, Tierwelt Dtl., 25: 570, Fig. 97.$_{18}$ (Fig. 42i; revision; combination with *Mitra*).

1933 ***Micromitra retractilis* Clap. & L.** – Kahl, Tierwelt N.- u. Ostsee, 23: 113, Fig. 16.23, 17a.1–3 (Fig. 42j–l; combination with *Micromitra*; redescription from life).

1933 ***Micromitra brevicauda* spec. n.**[1] – Kahl, Tierwelt N.- u. Ostsee, 23: 113, Fig. 17a.4–6 (Fig. 42m–o; original description of junior synonym; no type material available).

1935 ***Micromitra retractilis* Clap. u. L. 1858** – Kahl, Tierwelt Dtl., 30: 842, Fig. 155.$_{9,\,9a}$ (Fig. 42p, q; revision; see nomenclature and remarks).

1935 ***Micromitra brevicauda* Kahl, 1933** – Kahl, Tierwelt Dtl., 30: 842, Fig. 155.10, 10a (Fig. 42r, s; revision; combination with *Oxytricha*, see nomenclature and remarks).

1938 ***Micromitra retractilis* Clap. et Lachm.** – Delphy, Bull. Stn. biol. Arcachon, 35: 68, Fig. 21 (Fig. 42t; illustrated record).

1943 ***Epiclintes radiosa* (Quennerstedt)** – Wailes, Can. Pacif. Fauna, 1: 28, Fig. 80 (Fig. 42u; redescription).

1972 ***Psammomitra retractilis* (Claparède & Lachmann, 1858) n. comb.** – Borror, J. Protozool., 19: 15, Fig. 44 (Fig. 42v; combination with *Psammomitra*; revision of hypotrichs).

1972 ***Psammomitra brevicauda* (Kahl, 1933) n. comb.** – Borror, J. Protozool., 19: 15 (combination with *Psammomitra*; revision of hypotrichs).

1983 ***Epiclintes radiosa* Calkins, 1902** – Carey & Tatchell, Bull. Br. Mus. nat. Hist. (Zool.), 45: 53, Fig. 11–15 (redrawings from various authors; revision; incorrect author and year, see nomenclature).

1985 ***Psammomitra retractilis*** – Small & Lynn, Phylum Ciliophora, p. 461, Fig. 43 (Fig. 42v; guide to ciliate genera).

1992 ***Psammomitra brevicaudata* (Kahl, 1933) Borror, 1972** – Carey, Marine interstitial ciliates, p. 185, Fig. 735 (redrawing of Fig. 42r; guide; incorrect subsequent spelling).

1994 ***Psammomitra retractilis*** – Tuffrau & Fleury, Traite de Zoologie, 2: 142, Fig. 53b (redrawing of Fig. 42p; revision).

1996 ***Uroleptus retractilis* comb. n.** – Song & Warren, Acta Protozool., 35: 228, Fig. 1–11, Table 1 (Fig. 43a–k; detailed redescription from life and after protargol impregnation; combination with *Uroleptus*; voucher slides are deposited in the Laboratory of Protozoology, Ocean University of Qingdao, China).

2001 ***Psammomitra retractilis* (Claparède & Lachmann, 1858) Borror, 1972** – Berger, Catalogue of ciliate names 1. Hypotrichs, p. 62 (nomenclator containing all basionyms, combinations, and higher taxa of hypotrichs).

2002 ***Psammomitra retractilis*** – Lynn & Small, Phylum Ciliophora, p. 451, Fig. 31 (Fig. 42v; guide to ciliate genera).

Nomenclature: The species-group name *retractilis* is a composite of *retracto* (Latin verb; draw back) and the Latin suffix *~il·is* (passive or active ability or possibility; Werner 1972) and refers to the fact that the present species can retract the tail. The species-group name *longicaudatus -a -um* (Latin adjective; having a long tail; Hentschel & Wagner 1996) refers to the very long tail. The species-group name *radiosus -a -um* (Latin adjective; radiating) alludes to the radiating appearance of the anterior portion

[1] Kahl (1933) provided the following brief description: Größe (im gedehnten Zustand) 80 bis 110 µm; Schwanz höchstens halb so lang wie der übrige Körper; stets nur 4 Frontalmembranellen; 2 Kernteile (nur einmal durch Färbung festgestellt); Dorsalborsten 5 µm hoch.

Table 16 Morphometric data on *Psammomitra retractilis* (from Song & Warren 1996)

Characteristics [a]	mean	M	SD	SE	CV	Min	Max	n
Body, length	41.5	–	13.8	4.4	33.3	30.0	71.0	10
Body, width	23.8	–	2.9	0.9	12.2	19.0	28.0	10
Anterior body end to proximal end of adoral zone, distance	10.0	–	1.3	0.4	12.5	8.0	12.0	10
Macronuclear nodules, length	8.5	–	1.1	0.4	12.6	7.0	10.0	8
Macronuclear nodules, width	5.2	–	1.3	0.4	24.9	4.0	8.0	9
Macronuclear nodules, number	2.0	–	0.0	0.0	0.0	2.0	2.0	16
Adoral membranelles, number	11.1	–	0.7	0.2	6.3	10.0	12.0	9
Frontal cirri, number	–	–	–	–	–	1.0	3.0	5
Frontoterminal cirri, number	2.1	–	0.3	0.1	13.9	2.0	3.0	12
Midventral complex, number of pairs [b]	5.6	–	0.5	0.2	9.6	5.0	6.0	7
Transverse cirri, number	6.7	–	1.0	0.3	14.2	5.0	8.0	10
Left marginal cirri, number	20.0	–	1.8	0.6	9.1	18.0	23.0	10
Right marginal cirri, number	20.5	–	2.0	0.6	9.6	18.0	24.0	10
Dorsal kineties, number	4.0	–	0.0	0.0	0.0	4.0	4.0	11

[a] All measurements in µm. Data are based on protargol-impregnated specimens. CV = coefficient of variation in %, M = median, Max = maximum value, mean = arithmetic mean, Min = minimum value, n = number of individuals investigated, SD = standard deviation, SE = standard error of arithmetic mean.

[b] Very likely Song & Warren (1996) counted the pseudopairs.

of the adoral zone of membranelles. The species-group name *brevicauda* (short-tail) is a composite of the Latin adjective *brevis -is -e* (short, small, low) and the Latin noun *cauda* (tail, feminine) and refers to the short tail (not more than half as long as the remaining body).

Kahl (1935) wrote that *Micromitra* belongs, according to the scattered ventral cirri, to the genus *Oxytricha* and therefore forms a further subgenus. Consequently, the correct names in Kahl (1935) are *Oxytricha (Micromitra)* Kahl, 1933, *Oxytricha (Micromitra) retractilis* Claparède & Lachmann, 1858, and *Oxytricha (Micromitra) brevicauda* (Kahl, 1933) Kahl, 1935. These names were overlooked by Berger (2001).

Carey & Tatchell (1983) revised *Epiclintes* and considered the present species as congeneric with *E. auricularis*, type of *Epiclintes*. Interestingly, they overlooked the papers by Kahl (1933) and Borror (1972) and therefore did not discuss the replacement names *Micromitra* and *Psammomitra* and the species *Micromitra brevicauda* Kahl. Moreover, they mistakenly assigned the name *Epiclintes radiosa* to Calkins (see list of synonyms); the correct designation of this combination would be *Epiclintes radiosa* (Quennerstedt, 1867) Kent, 1882. Last but not least, they did not use the oldest species-group name *retractilis*, although they mentioned it in their list of synonyms.

Incorrect subsequent spellings: *Epiclintes retracta* (Lackey 1936, p. 269); *Psammomitra brevicaudata* (Kahl, 1933) and *Micromitra brevicaudata* Kahl, 1933 (Dini et al. 1995, p. 71; Fauré-Fremiet 1950, p. 74; Agamaliyev 1974, p. 21). In my previous books (Berger 1999, 2001) I incorrectly wrote "Wright T. S." instead of "Strethill Wright T." as author of the synonym *Oxytricha longicaudata*.

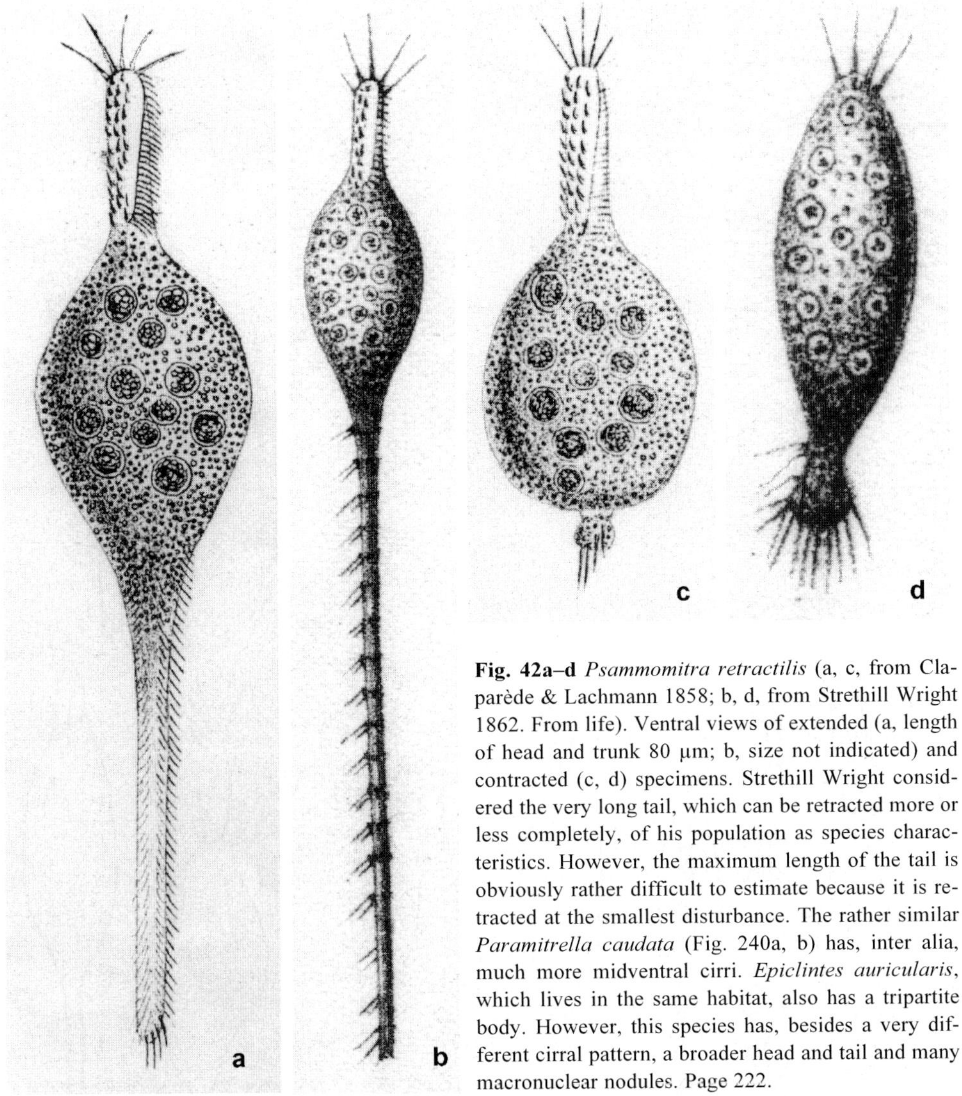

Fig. 42a–d *Psammomitra retractilis* (a, c, from Claparède & Lachmann 1858; b, d, from Strethill Wright 1862. From life). Ventral views of extended (a, length of head and trunk 80 μm; b, size not indicated) and contracted (c, d) specimens. Strethill Wright considered the very long tail, which can be retracted more or less completely, of his population as species characteristics. However, the maximum length of the tail is obviously rather difficult to estimate because it is retracted at the smallest disturbance. The rather similar *Paramitrella caudata* (Fig. 240a, b) has, inter alia, much more midventral cirri. *Epiclintes auricularis*, which lives in the same habitat, also has a tripartite body. However, this species has, besides a very different cirral pattern, a broader head and tail and many macronuclear nodules. Page 222.

Remarks: This species was described three times within the period 1858 to 1867 (see list of synonyms). Both Strethill Wright (1862) and Quennerstedt (1867) knew the original description of the oldest synonym *O. retractilis*, but both authors found differences to their populations which justified, according to their opinion, the establishment of species. Strethill Wright (1862) emphasised the very long tail of his species, whereas *Mitra radiosa* is obviously characterised by the conspicuously radiating distal adoral membranelles. Synonymy of *O. retractilis* and *O. longicaudata* was first proposed by Stein (1867, p. 151). About seventy years later, Kahl (1933) described *Micromitra*

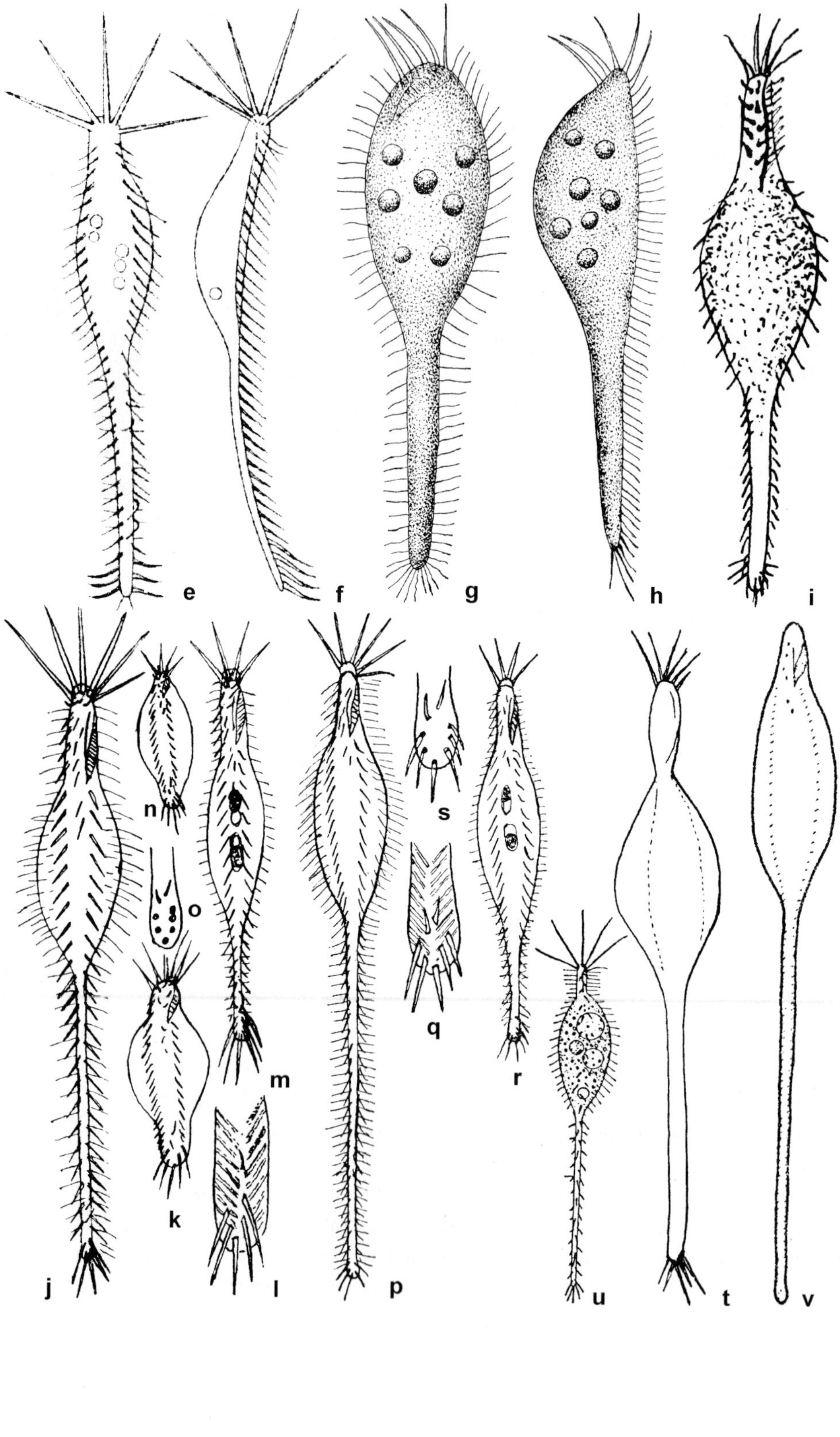
e
f
g
h
i
j
n
o
k
l
m
p
s
q
r
u
t
v

brevicauda which has a very short tail. Synonymy of *Oxytricha retractilis*, *Oxytricha longicaudata*, and *Mitra radiosa* was proposed by Bütschli (1889). However, he was uncertain about the generic assignment of *O. retractilis*; he supposed a close relationship to *Epiclintes* or *Stichotricha*. Bütschli's synonymy was generally accepted, for example, by Kahl (1933), Borror (1972), Hemberger (1982), and Carey & Tatchell (1983). Only few earlier authors (e.g., Kent 1882, Calkins 1902) considered them as two distinct species. Kahl (1932) mentioned the present species within the genus *Epiclintes*, but simultaneously transferred it to *Mitra* Quennerstedt, an act overlooked by Berger (2001). Kahl (1935) made a note that *Micromitra retractilis* and *M. brevicauda* have only few scattered ventral cirri and therefore considered *Mitra* as further subgenus of *Oxytricha*.

Micromitra brevicauda was accepted by Borror (1972), but Hemberger (1982) doubted that it is a distinct species and proposed that it can also be considered as a form of *P. retractilis*. By contrast, Carey (1992) considered Kahl's species as the sole member of *Psammomitra*; however, this is impossible because it is not the type species.

Song & Warren (1996) provided the first detailed description of this species. They found that the cirral pattern is not oxytrichid as supposed by Kahl (1935), but urostyloid, that is, composed of few midventral pairs, two frontoterminal cirri, and some transverse cirri (Fig. 43). Since the body shape is reminiscent of *Uroleptus*, they transferred *P. retractilis* to *Uroleptus* and simultaneously synonymised *Psammomitra* with *Uroleptus*. By contrast, I preliminarily accept *Psammomitra* (although it is monotypic), because the present species lacks caudal cirri and dorsomarginal kineties, whereas both structures are present in most (all?) *Uroleptus* species (for brief review, see Foissner et al. 1991; Eigner 2001). Moreover, *Uroleptus* is likely confined to freshwater, whereas the present species is marine.

Psammomitra is classified in various higher taxa, for example, the Holostichidae (Corliss 1977, 1979, Tuffrau 1979, 1987, Detcheva 1992), the Trachelostylidae (Small & Lynn 1985, Dini et al. 1995), the Amphisiellidae (Tuffrau & Fleury 1994, Lynn & Small 2002), and the Oxytrichidae (Kahl 1932, 1935, Borror 1972). Jankowski (1979) even established an own group, the Psammomitrinae. Wicklow & Borror (1990, p. 192) correctly excluded it from *Epiclintes* and considered it a member incertae sedis. The redescription by Song & Warren (1996) strongly indicates that the present species is a urostyloid. However, I am uncertain about the sistergroup of *Psammomitra*. Possibly it is related to *Paramitrella* which has a similar body shape and also lives in the sea. Kahl's assumption that the present species is an oxytrichid "sensu stricto" cannot be entirely excluded. Perhaps it formed one, two, or three additional cirral anlagen

---

← **Fig. 42e–v** *Psammomitra retractilis* (e, f, from Quennerstedt 1867; g, h, from Calkins 1902; i, from Kahl 1932; j–o, from Kahl 1933; p–s, from Kahl 1935; t, from Delphy 1938; u, from Wailes; v, after Kahl 1933 from Borror 1972. From life). **e–h:** Ventral and right lateral views (for sizes, see text). Note that the illustrations by Calkins are rather superficial, but the habitus indicates that the identification is correct. **i:** Inaccurately observed specimen, 150 μm. **j–l, p, q:** *Psammomitra retractilis* extended (j, p), contracted (k), and tip of tail (l, q), j, p = 160 μm. Kahl was the first worker who found that *P. retratilis* has two macronuclear nodules. **m–o, r, s:** The junior synonym *Micromitra brevicauda* extended (m, r), contracted (n), and tip of tail (o, s), 110 μm. **t–v:** Ventral views (for sizes, see text). Page 222.

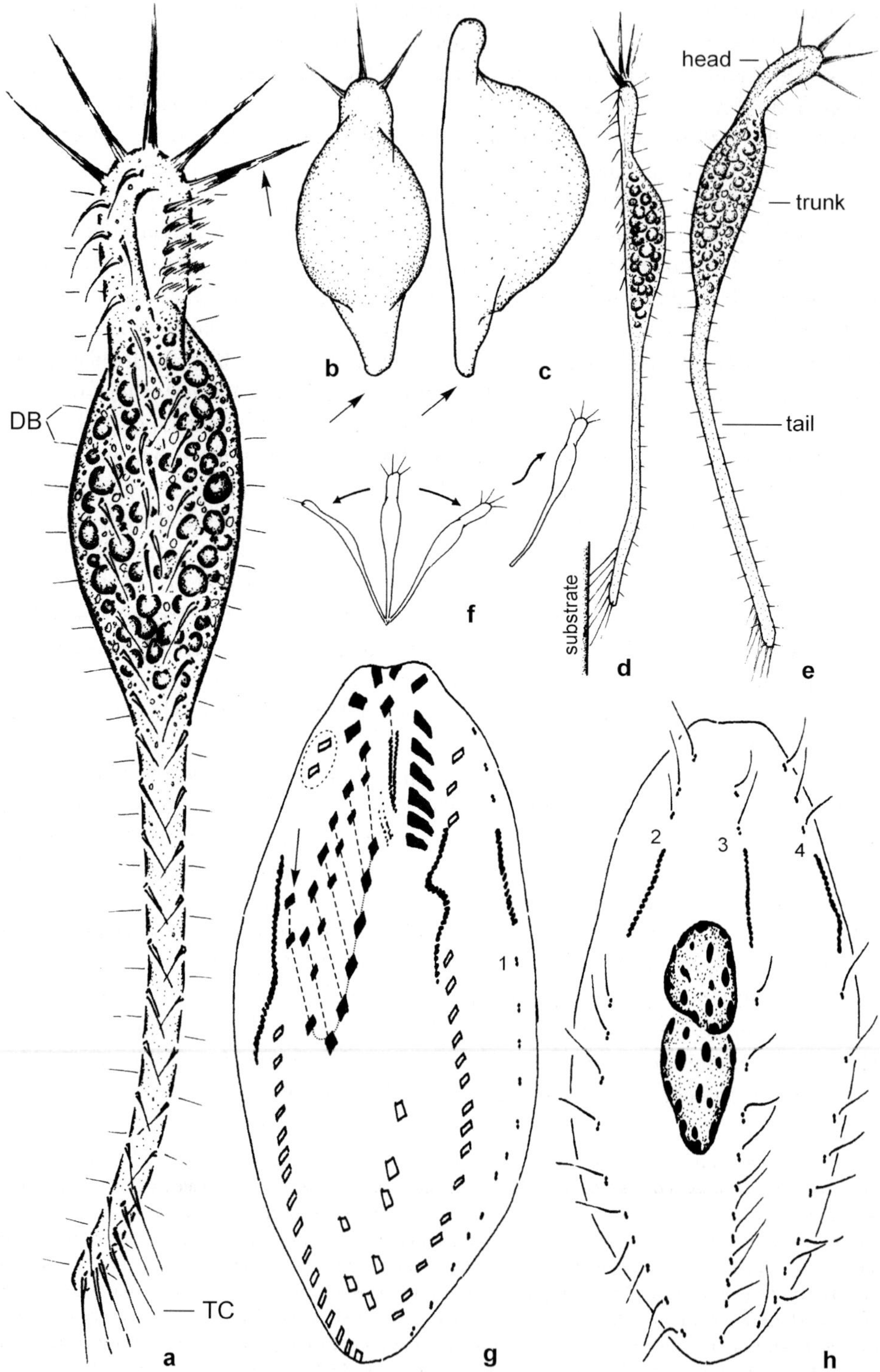
DB
TC
a
b
c
f
head
trunk
tail
substrate
d
e
1
2
3
4
g
h

(that is, in total about 7–9), a process which likely occurred several times independently in the 18-cirri oxytrichids, for example, in *Pattersoniella* and *Onychodromus* (for review, see Berger 1999). However, the lack of dorsomarginal kineties (Fig. 43g) strongly indicates that it is a urostyloid. It is classified in the Holostichidae because it has three frontal cirri and a midventral complex composed of cirral pairs only. Possibly, molecular data can provide a further hint to elucidate the phylogenetic relationships of *Psammomitra*.

Morphology: This chapter is based on the very detailed redescription by Song & Warren (1996; Fig. 43a–k). Additional and/or deviating data from other populations are added below. Body size 140–200 × 20–30 µm, but specimens at maximum extension can be up to 300 µm long. Body slender, fragile, and highly contractile; tripartite in head, trunk, and tail. Head region about 12–20% of extended cell length, narrow, colourless, and dorsoventrally flattened. Trunk elliptical in outline, 20–33% of cell length; ventral side flat and straight, dorsal side slightly (Fig. 43d) to highly (Fig. 43c) convex. Tail stalk-like, that is, very thin and long; hyaline and highly contractile (can be almost completely retracted; Fig. 43b) and, when fully extended, about twice as long as in normal state (Fig. 43a, d, e). Invariable two distinctly separated macronuclear nodules in central portion of trunk; individual nodules about 8 × 5 µm (Fig. 43j). Micronuclei not observed. Presence/absence of contractile vacuole not mentioned, indicating that this organelle is lacking. Pellicle thin. Cortical granules lacking. Cytoplasm colourless, within the trunk portion always with many large (3–8 µm), colourless globules masking the nuclear apparatus in life (Fig. 43a, d, e).

Psammomitra retractilis has a rather conspicuous behaviour: it usually attaches firmly to the substratum by its transverse cirri on the tail, which is usually extended to the maximum length in this case (Fig. 43d, e). Sometimes the cell bends from side to side (Fig. 43f). Undisturbed specimens may remain motionless for 1–3 min. After stirring, they usually contract and spring rapidly forwards or backwards over a short distance. When swimming, locomotion is usually very slow, the cell being suspended in the water with cirri and membranelles almost motionless.

Adoral zone, although rather short, very conspicuous because distalmost 3–5 (usually 4 or 5) membranelles about 20 µm long and forming "crown" at anterior end of cell; proximal membranelles only about 10 µm long. Length of adoral zone of contracted, protargol-impregnated specimens 24% of body length on average (Table 16); in more or less extended specimens only about 12% (Fig. 43i). Distalmost 3–4 membranelles distinctly separated from remaining membranelles. Buccal field comparatively

← **Fig. 43a–h** *Psammomitra retractilis* (from Song & Warren 1996. a–f, from life; g, h, protargol impregnation). **a:** Ventral view of an extended specimen, 158 µm. Arrow marks the leftmost elongated adoral membranelle, which forms, together with the other distal membranelles, a conspicuous crown. **b, c:** Dorsal and left lateral view of contracted specimen. Note the retracted tail (arrows) and the strong vaulting of the dorsal side of the trunk. **d, e:** Left lateral and dorsal view of extended specimens (190 µm) attached to the substrate by the transverse cirri. **f:** Moving specimen (for details, see text). **g, h:** Infraciliature of ventral and dorsal side of a reorganiser, size not indicated. Parental frontoterminal cirri circled, new marked by arrow. Cirri which very likely originate from the same anlage are connected by broken lines. New transverse cirri connected by dotted line. Parental cirri white, new black. DB = dorsal bristles, TC = transverse cirri, 1–4 = dorsal kineties. Page 222.

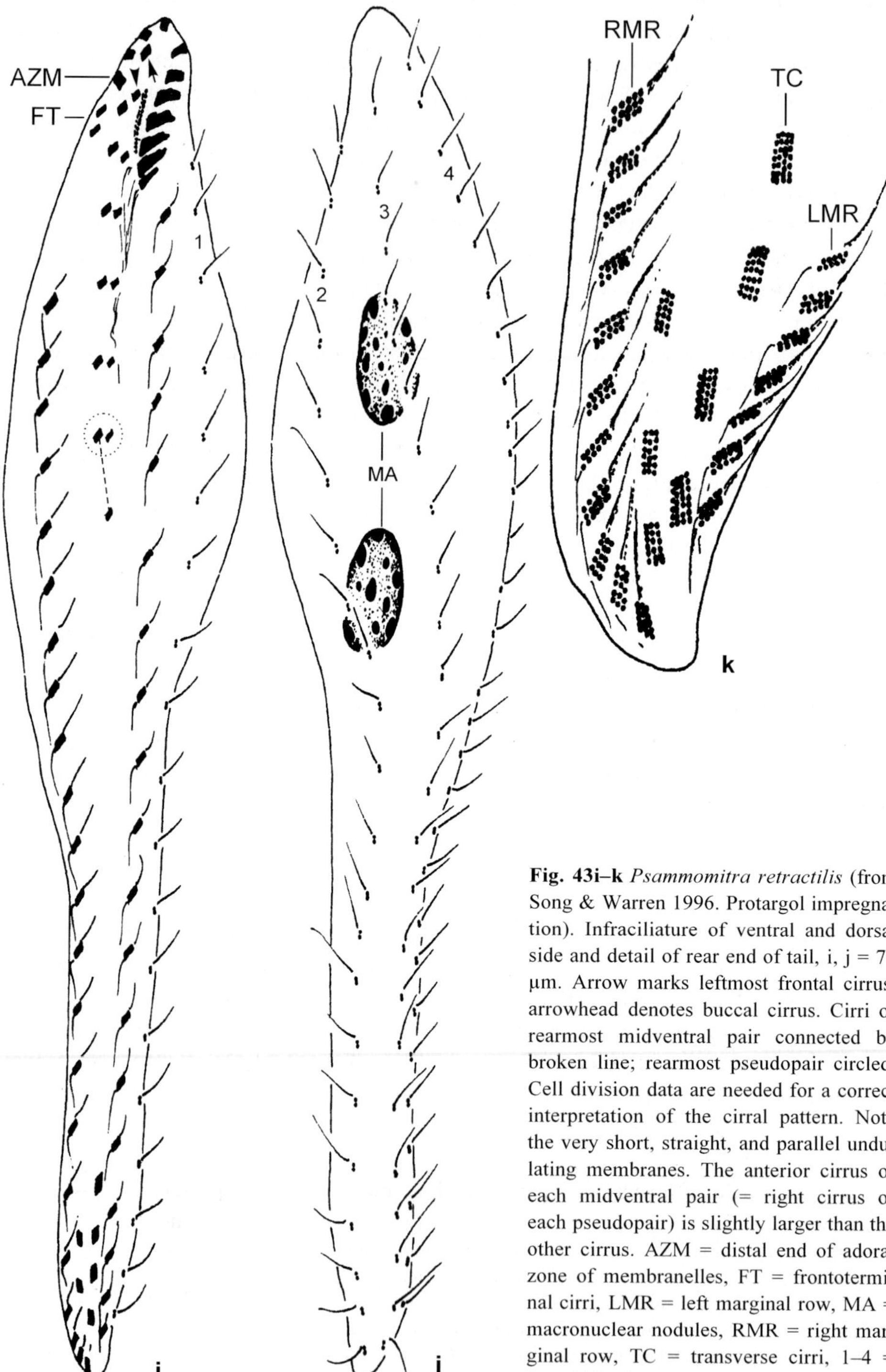

Fig. 43i–k *Psammomitra retractilis* (from Song & Warren 1996. Protargol impregnation). Infraciliature of ventral and dorsal side and detail of rear end of tail, i, j = 70 µm. Arrow marks leftmost frontal cirrus, arrowhead denotes buccal cirrus. Cirri of rearmost midventral pair connected by broken line; rearmost pseudopair circled. Cell division data are needed for a correct interpretation of the cirral pattern. Note the very short, straight, and parallel undulating membranes. The anterior cirrus of each midventral pair (= right cirrus of each pseudopair) is slightly larger than the other cirrus. AZM = distal end of adoral zone of membranelles, FT = frontoterminal cirri, LMR = left marginal row, MA = macronuclear nodules, RMR = right marginal row, TC = transverse cirri, 1–4 = dorsal kineties, Page 222.

wide, forms shallow depression. Endoral and paroral short, straight, parallel (Fig. 43i). Pharyngeal fibres pronounced after protargol impregnation.

Cirral pattern at first glance rather simple because composed of comparatively few cirri. However, because of the slender body shape and the movement it is difficult to recognise in life. Even in protargol preparations it is difficult to designate the cirri correctly because they are rather narrowly spaced and details on cell division are lacking (Fig. 43i). Leftmost frontal cirrus (from anlage I) very close to anterior end of cell. Close behind distal end of adoral zone of membranelles two cirri, the left one is very likely the buccal cirrus, the right likely the middle frontal cirrus (cirrus II/3). Right of proximal end of adoral zone two further cirri which are difficult to interpret (right frontal cirrus III/3 and cirrus III/2 or first pseudopair of midventral complex?). Two frontoterminal cirri near distal end of adoral zone. Midventral complex composed of cirral pairs, which form five distinct pseudopairs in specimen shown in Fig. 43i; complex extends about half the length of the trunk. Transverse cirri slightly enlarged (Fig. 43k), about 20 µm long, project distinctly beyond rear body end, form hook-shaped pattern near rear end of tail and are used to attach the cell to the substrate (Fig. 43d). Whether or not pretransverse ventral cirri are present can be decided only after analysis of cell division. Left marginal row commences near (slightly behind) proximal end of adoral zone, extends to near rear body end. Right marginal row begins distinctly behind level of proximal end of adoral zone (at 20% of body length in specimen shown in Fig. 43i), terminates at tip of tail. Marginal cirri about 10 µm long.

Dorsal bristles conspicuous because 5–8 µm long and rather stiff, invariably arranged in four more or less bipolar kineties. Caudal cirri lacking.

Additional and/or deviating observations from other authors: length of head and trunk about 80 µm (Claparède & Lachmann 1858), about 100 µm (Quennerstedt 1867); tail about twice as long as that of the population described by Claparède & Lachmann (Strethill Wright 1862); total? length 45 µm (Calkins 1902); total body length of extended specimens 80–110 µm (*brevicauda*), respectively, 160 µm (*retractilis*; Kahl 1933); total length 125 µm (Delphy 1938); 150 × 22 µm (Kiesselbach 1936a); 150 µm total length, tail 80 µm long, body width 26 µm (Wailes 1943). Anteriormost five (Claparède & Lachmann 1858), respectively, 5–6 (Kahl 1933) adoral membranelles distinctly elongated; length of frontal adoral membranelles 30 µm (Wailes 1943); dorsal bristles 5 µm (*brevicauda*), respectively, 10 µm (*retractilis*) long (Kahl 1935). Wailes (1943) described a single macronucleus (7 µm across) and two contractile vacuoles indicating misobservations.

Occurrence and ecology: Common in/on marine sediments (e.g., Hartwig 1974, Patterson et al. 1989, p. 210). *Psammomitra retractilis* was discovered in the fjord of Bergen, North Sea, Norway (Claparède & Lachmann 1858). Type locality of the synonym *Oxytricha longicaudata* is the sea at Largo, Fifeshire, Great Britain, where Strethill Wright (1862) found it in great abundance. *Mitra radiosa* was discovered in the Kattegat near Varberg, Sweden (Quennerstedt 1867, 1869). Type locality of the synonym *Micromitra brevicauda* is marine sand from near the "Helgoländer Düne" on the island Helgoland, North Sea (Germany), where Kahl (1933) found it abundantly in an agar-culture.

Records of *P. retractilis* substantiated by morphological data: sandy Bay at the island of Brioni, Adriatic Sea, Croatia (Kiesselbach 1936a, p. 20; figure after Kahl); Arcachon, Bay of Biscay, France (Delphy 1938); eutrophic pond (32‰ salinity, 12–15°C, pH 8.3) used for storing marine shellfish in Taipingjiao, Qingdao, China (Song & Warren 1996); Woods Hole area, Massachusetts, USA (Calkins 1902; Lackey 1936 [without morphological data] recorded both *Epiclintes retractilis* and *E. radiosa*); on *Zostera* in Departure Bay, Canadian Pacific Coast (Wailes 1943).

Records not substantiated by morphological data: Bay of Marseille, Mediterranean Sea, France (Vacelet 1960, p. 54); Gulf of Naples, Italy (Nobili 1957, Tab. I; for review see Dini et al. 1995, p. 71); during summer and winter in the Dee estuary at Parkgate, Cheshire, England (Webb 1956, p. 153); infrequent in mesopsammon (21° C, 18‰ salinity) of Bulgarian coast of Black Sea (Jeliaskowa-Paspalewa 1933, p. 22; Detcheva 1982, p. 249; 1983, p. 72; further records from Bulgaria, see Detcheva 1992, p. 103); White Sea, inter alia, Gulf of Kola and Kandalaksha Bay (Mereschkowsky 1877, p. 231; 1879, p. 216; Gassovsky 1916, p. 143; Burkovsky 1971a, p. 1774); gulf of Finland, inter alia, among *Cordylophora*, and brackish water bays near Helsingfors, Finland (Levander 1894, p. 212; 1894a, p. 93; 1901a, p. 11; 1901b, p. 8).

Records of the synonym *Micromitra brevicauda*: Bay of Kiel, Baltic Sea, Germany (Bock 1952, p. 83); Gulf of Naples, Italy (Nobili 1957, Tab. I); western coast of Caspian Sea (Agamaliev 1971, p. 383; 1983, p. 36; Agamaliyev 1974, p. 21).

Records from freshwater habitats (Hausman 1917, p. 166; Webb 1961, p. 141, with doubt) are possibly based on misidentifications of *Uroleptus* species or *Ancystropodium maupasi* which has a similar habitus (for review see Berger 1999, p. 777).

Unidentified *Psammomitra* species were recorded, inter alia, from the following sites: Ligurian Sea near Pisa, Italy (Santangelo & Lucchesi 1995, p. 51); in the 25 foot deep one million gallon ocean of the Biosphere II project (Spoon & Alling 1993); Aburatsubo Inlet, Kanagawa Prefecture, Japan (Mahachanchawalit et al. 1986, p. 12).

Psammomitra retractilis feeds on diatoms and other small algae so that the trunk is yellow- or green brown. However, in laboratory cultures, specimens were observed to ingest bacteria (Song & Warren 1996).

Caudiholosticha Berger, 2003

2003 ***Caudiholosticha* nov. gen.** – Berger, Europ. J. Protistol., 39: 377 (original description). Type species (by original designation on p. 377): *Holosticha stueberi* Foissner, 1987.

Nomenclature: Composite of the Latin noun *caud·a* (tail), the thematic vowel *·i-*, and the genus-group name *Holosticha*, indicating that the species included have caudal cirri and were previously classified in *Holosticha* (Berger 2003). Like *Holosticha*, feminine gender.

Characterisation: Adoral zone continuous. Rearmost membranelles not wider than remaining membranelles of proximal portion. 3 enlarged frontal cirri. Buccal cirrus/cirri right of paroral. Frontoterminal cirri present. Midventral complex composed of midventral pairs only. Pretransverse ventral cirri present or absent. Number of transverse cirri usually distinctly lower than number of midventral pairs. 1 left and 1 right marginal row. Anterior end of left marginal row more or less straight, commences left of adoral zone. Caudal cirri present. Nuclear apparatus left of midline or scattered.

Remarks: The characterisation is basically according to the diagnosis of the original description, except for the pretransverse ventral cirri, which are present or absent in *Caudiholosticha*. It contains mainly those species previously classified in *Holosticha* which lack the apomorphies (e.g., anterior end of left marginal row curved rightwards; further features and details, see *Holosticha*) of *Holosticha* as defined by Berger (2003) and which have caudal cirri. The presence/absence of this or another cirral group is generally used to define genera, for example, *Tachysoma* (for review see Berger 1999, p. 431) or *Paragonostomum* (Foissner et al. 2002).

I fixed *C. stueberi* as type because this species has distinct caudal cirri and is very well defined morphologically (Foissner 1987e, Berger 2003)[1]. The presence of caudal cirri is certainly not a derived feature for *Caudiholosticha* because such cirri occur even in the euplotids and have been retained in many groups of hypotrichs (see ground pattern of the Urostyloidea). Thus, the loss of caudal cirri is the apomorphic state. However, the loss of a (simple) structure is not a complex feature which explains the fact that caudal cirri were certainly lost several times independently.

The characterisation above is, due to the lack of apomorphies, only a combination of more or less young plesiomorphies, indicating that *Caudiholosticha* is – like *Anteholosticha* – heterogeneous. Non-monophyly is also indicated by the rather different shape and arrangement of the paroral and endoral (cf. Fig. 44d, 48c). The various, highly characteristic types of the paroral and endoral have been successfully used to characterise oxytrichid genera (Berger & Foissner 1997, Berger 1999). Another candidate for a possible apomorphy of a subgroup is the origin of caudal cirri from the rightmost dorsal kinety only, as is the case in *C. sylvatica*, but likely also in *C. islandica* and *C. tetracirrata*. But of course ontogenetic data are needed to check the real situation. As in *Anteholosticha*, splitting of *Caudiholosticha* should be done only when monophyletic groups can be reliably extracted.

Besides the species originally assigned to *Caudiholosticha*, *Uroleptus paranotabilis*, *Paruroleptus notabilis*, and *Perisincirra gracilis* are preliminary classified in this genus because they have basically the same infraciliature as the other *Caudiholosticha* species (deviations of individual species, see descriptions below). However, their body is rather slender and narrowly rounded posteriorly, whereas the other *Caudiholosticha* species have a more or less wide and broadly rounded body. The slender body could be used to establish a separate taxon. However, I avoid such a step at the present state of knowledge because there are some features (cytopharynx, number of dorsal kineties, shape of undulating membranes), indicating that these three slender species do not form a monophyletic group.

[1] Now I am no longer convinced that *Caudiholosticha stueberi* is a urostyloid because of the dorsal infraciliature (see *C. stueberi*).

Interestingly, *Caudiholosticha* does not contain a real marine species, whereas *Holosticha* comprises exclusively marine species, except for *H. pullaster*, which is common in freshwater and seawater.

Species included in *Caudiholosticha* (alphabetically arranged according to basionym; subgenus name ignored): (1) *Holosticha (Holosticha) algivora* Kahl, 1932; (2) *Holosticha interrupta* Dragesco, 1966; (3) *Holosticha islandica* Berger & Foissner, 1989; (4) *Holosticha multicaudicirrus* Song & Wilbert, 1989; (5) *Holosticha (Holosticha) navicularum* Kahl, 1932; (6) *Holosticha (Holosticha) setifera* Kahl, 1932; (7) *Holosticha stueberi* Foissner, 1987; (8) *Holosticha sylvatica* Foissner, 1982; (9) *Holosticha tetracirrata* Buitkamp & Wilbert, 1974; (10) *Holosticha (Holosticha) viridis* Kahl, 1932; (11) *Paruroleptus notabilis* Foissner, 1982; (12) *Perisincirra gracilis* Foissner, 1982; (13) *Uroleptus paranotabilis* Foissner, Agatha & Berger, 2002.

The species are arranged according to the habitat, namely soil (*C. stueberi*, *C. sylvatica*, *C. tetracirrata*, *C. islandica*, *C paranotabilis*, *C. notabilis*, *C. gracilis*), freshwater (*C. algivora*, *C. viridis*, *C. navicularum*, *C. multicaudicirrus*, *C. interrupta*), and saline waters (*C. setifera*).

Key to *Caudiholosticha* species

If you know that your specimen/population is a *Caudiholosticha* species, identification is still rather difficult. Main features are, inter alia, habitat, nuclear apparatus, body size, and details of the ventral and dorsal infraciliature, that is, protargol preparation is recommended (essential for inexperienced workers) for reliable identification. As usual, use only healthy organisms for determination.

1 Two macronuclear nodules **2**
\- 10 or more macronuclear nodules **6**
2 Single micronucleus between macronuclear nodules (Fig. 51a, 54a) **3**
\- Usually 2 or more micronuclei, and if only 1 micronucleus present than not exactly in between the macronuclear nodules (e.g., Fig. 44e) **4**
3 6–8 transverse cirri inserted near end of middle body third (Fig. 51a) .... *Caudiholosticha navicularum* (p. 274)
\- 5 transverse cirri near rear body end (Fig. 54a) .. *Caudiholosticha setifera* (p. 281)
4 (2) Body length 200–300 μm; terrestrial (Fig. 44a–e) .... *Caudiholosticha stueberi* (p. 235)
\- Body length below 150 μm; limnetic **5**
5 Symbiotic algae present; cortical granules lacking (Fig. 50a) .... *Caudiholosticha viridis* (p. 272)
\- Symbiotic algae lacking (note: possibly green due to ingested euglenids/algae); cortical granules (colourless, arranged in rows) present (Fig. 49a) .... *Caudiholosticha algivora* (p. 270)
6 (1) Behind left frontal cirrus a short row of 2–5 cirri (Fig. 45b, h) .... *Caudiholosticha sylvatica* (p. 239)

- Behind left frontal cirrus no such row **7**
7 Terrestrial **8**
- Limnetic or marine **9**
8 Rear body end broadly rounded (e.g., Fig. 46b) **8a**
- Rear body end narrowly rounded, that is, rear body portion more or less tailed (e.g., Fig. 48.4a) **11**
8a Body about 80–100 μm long; about 14–17 macronuclear nodules; cortical granules (yellowish; globular, <0.5 μm; irregularly and loosely arranged) present; 3 dorsal kineties (Fig. 48a–d) .... *Caudiholosticha islandica* (p. 252)
- Body 100–160 μm long; 25–60 macronuclear nodules; cortical granules lacking; 4 dorsal kineties (Fig. 46a–f) .... *Caudiholosticha tetracirrata* (p. 246)
9 (7) Marine; body very slender (length:width ratio about 10:1; Fig. 93a, b; presence/absence of caudal cirri uncertain) .... *Anteholosticha fasciola* (p. 441)
- Limnetic; less slender (e.g., Fig. 52a) **10**
10 Body length around 200 μm; midventral complex terminates in mid-body; about 14–16 caudal(?) cirri between end of marginal rows (Fig. 53a) .... *Caudiholosticha interrupta* (p. 279)
- Body length about 80–120 μm; midventral complex extends to near transverse cirri; usually 6 caudal cirri (Fig. 52a–e) ..... *Caudiholosticha multicaudicirrus* (p. 276)
11 25–70 macronuclear nodules; undulating membranes long and curved (Fig. 48.3f) . ... *Caudiholosticha notabilis* (p. 260)
- Usually 10–20, rarely up to 30 macronuclear nodules; undulating membranes short and almost straight (Fig. 48.1a, h, j, 48.4a–g) **12**
12 Cortical granules lacking; 13–19, on average 16 adoral membranelles; 2 dorsal kineties (Fig. 48.4a–g) .... *Caudiholosticha gracilis* (p. 266)
- Cortical granules present; 19–26, on average 22–24 adoral membranelles; 3–4 dorsal kineties (Fig. 48.1a–m) .... *Caudiholosticha paranotabilis* (p. 254)

Caudiholosticha stueberi (Foissner, 1987) Berger, 2003 (Fig. 44a–i, Table 17)

1987 ***Holosticha stueberi* nov. spec.**[1] – Foissner, Jber. Haus Nat. Salzburg, 10: 59, Abb. 1a–g, 2, 3, Tabelle 3 (Fig. 44a–i; original description; 1 holotype slide [registration number 1988/110] and 3 paratype slides [1988/111–113] are deposited in the Oberösterreichische Landesmuseum in Linz [LI], Upper Austria).

2001 ***Holosticha stueberi* Foissner, 1987** – Berger, Catalogue of ciliate names 1. Hypotrichs, p. 39 (nomenclator containing all basionyms, combinations, and higher taxa of hypotrichs).

2003 ***Caudiholosticha stueberi* (Foissner, 1987) n. comb.** – Berger, Europ. J. Protistol., 39: 377 (combination with *Caudiholosticha*).

[1] The diagnosis by Foissner (1987e) is as follows: In vivo etwa 200–300 × 60–100 μm große, farblose *Holosticha* mit großem Buccalfeld, weit voneinander entfernten, bis zu den Transversalcirren reichenden Midventralreihen und 2 Makronucleus-Teilen. Durchschnittlich 45 adorale Mebranellen und 6 Dorsalkineten. 3 Transversalcirren nahe dem hinteren Körperrand, 3 Caudalcirren. Cysten kugelig, außen mit zahlreichen pfeilerartigen Fortsätzen.

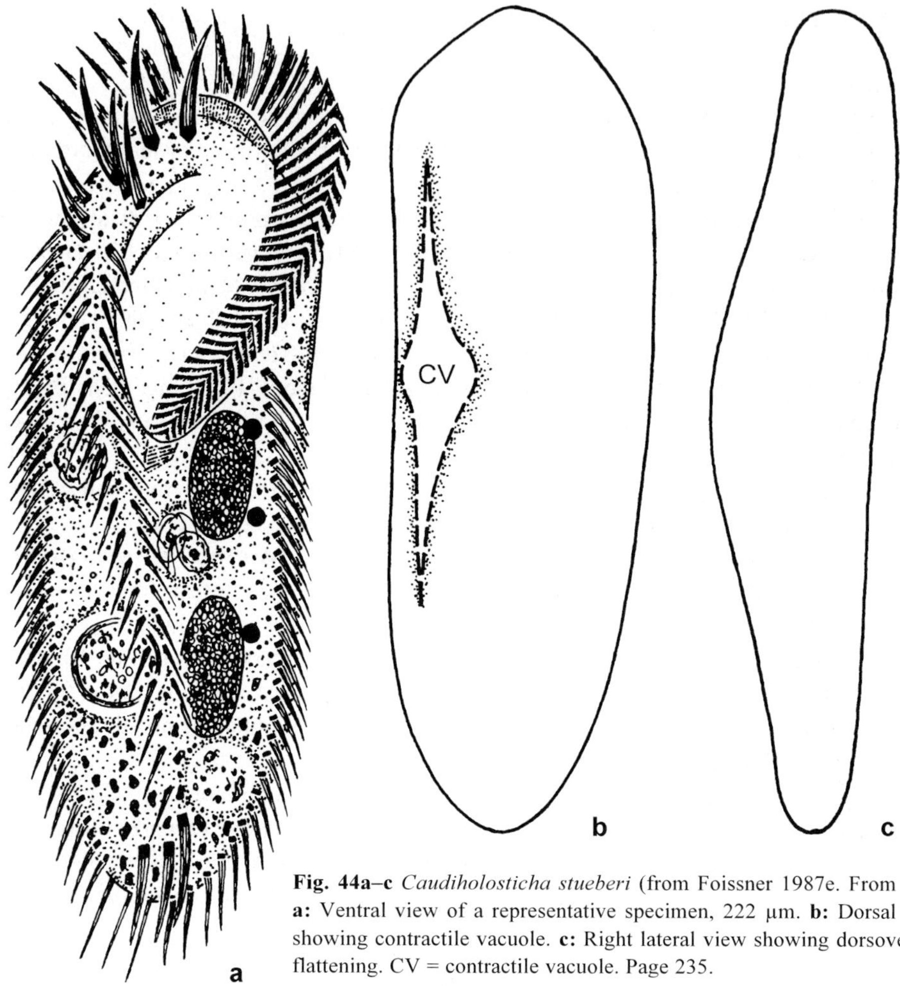

Fig. 44a–c *Caudiholosticha stueberi* (from Foissner 1987e. From life). **a:** Ventral view of a representative specimen, 222 µm. **b:** Dorsal view showing contractile vacuole. **c:** Right lateral view showing dorsoventral flattening. CV = contractile vacuole. Page 235.

Nomenclature: Foissner (1987e) dedicated this species to Eberhard Stüber, head of the Natural History Museum (Haus der Natur) in Salzburg, Austria. *Holosticha stüberi* Foissner, 1987 in Blatterer & Foissner (1988, p. 67) is an incorrect subsequent spelling. *Caudiholosticha stueberi* was fixed as type species of *Caudiholosticha* by original designation.

Remarks: This species is described in great detail from life and after protargol impregnation (Foissner 1987e). Since it has rather distinct caudal cirri, Berger (2003) designated it as type species of *Caudiholosticha* (for further details, see the genus section). At superficial observation it resembles *Australocirrus oscitans* Blatterer & Foissner, 1988, which, however, is an 18-cirri oxytrichid (for review, see Berger 1999, p. 470). The large buccal field and the body shape are reminiscent of *Cyrtohymena* species (for review see Berger 1999, p. 279).

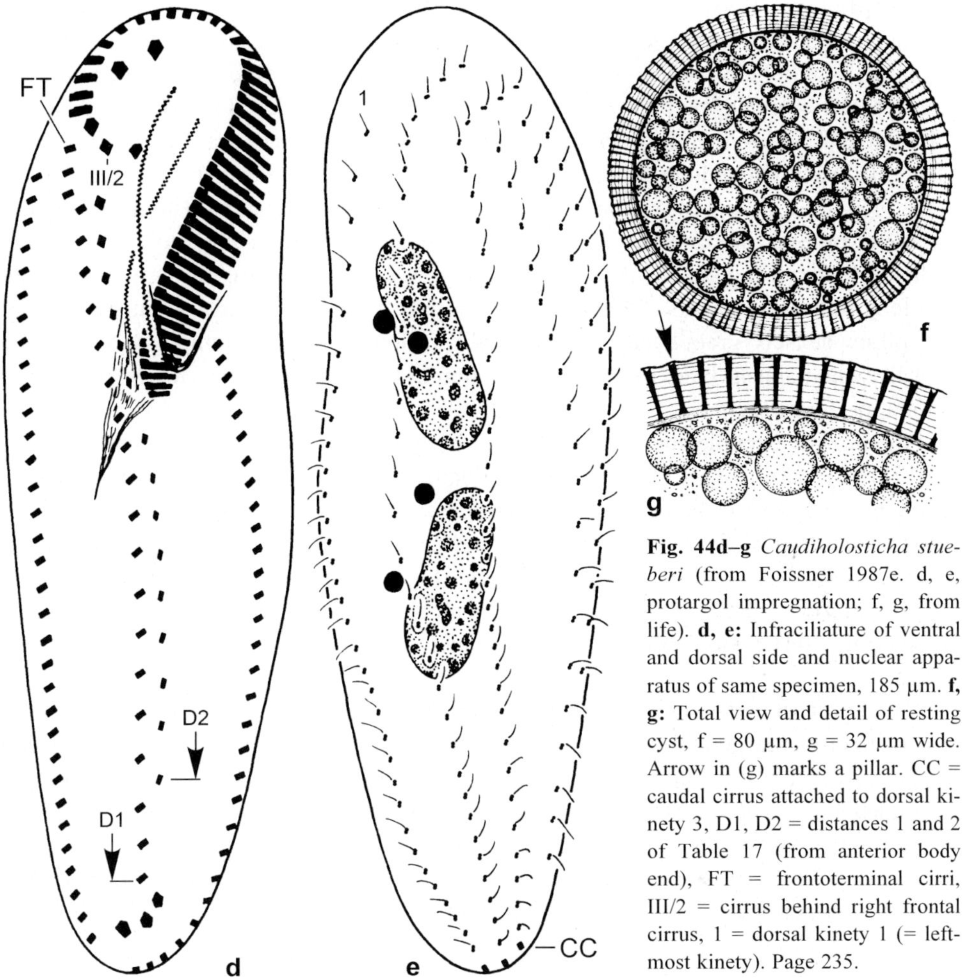

Fig. 44d–g *Caudiholosticha stueberi* (from Foissner 1987e. d, e, protargol impregnation; f, g, from life). **d, e:** Infraciliature of ventral and dorsal side and nuclear apparatus of same specimen, 185 µm. **f, g:** Total view and detail of resting cyst, f = 80 µm, g = 32 µm wide. Arrow in (g) marks a pillar. CC = caudal cirrus attached to dorsal kinety 3, D1, D2 = distances 1 and 2 of Table 17 (from anterior body end), FT = frontoterminal cirri, III/2 = cirrus behind right frontal cirrus, 1 = dorsal kinety 1 (= leftmost kinety). Page 235.

Caudiholosticha stueberi has a very interesting dorsal ciliature because kineties 1–3 are more or less bipolar and are associated with caudal cirri, whereas kineties 4–7 are posteriorly successively shortened, usually indicating that these are dorsomarginal rows (Fig. 44e). However, according to Fig. 14a, dorsomarginal kineties are a novelty for the non-urostyloid hypotrichs. Thus if *C. stueberi* indeed has dorsomarginal kineties (at least) two possibilities exist: (i) *Caudiholosticha stueberi* is not a urostyloid, but, for example, a *Uroleptus* with a broadly rounded, not narrowed posterior body end; or (ii) the hypothesis proposed in Fig. 14a is incorrect; that is, dorsomarginal kineties evolved at least twice, or evolved earlier and were lost in almost all urostyloids, but not in *C. stueberi*. Ontogenetic and molecular data are needed for a more detailed discussion.

Morphology: Body size 200–300 × 60–100 µm, body length:width ratio 3.1:1 on average in protargol preparations (Table 17). Body outline elongate, left margin slightly sigmoidal, right one moderately strongly convex, anterior end widely rounded, rear margins distinctly converging posteriorly, end thus narrowly rounded (Fig. 44a). Body about 2:1 flattened dorso-ventrally, ventral side plane, dorsal distinctly vaulted in mid-body. Macronuclear nodules long-ellipsoidal (2.8:1 on average in protargol preparations), arranged near mid-body left of midline; rarely (1 out of 12) specimens with three nodules (must not be over-interpreted). Micronuclei globular, closely attached to macronuclear nodules. Contractile vacuole about in mid-body at left cell margin, during diastole with two long collecting canals. Pellicle strong, slightly flexible. Cortical granules lacking. Cytoplasm densely granulated, posterior cell portion with many lumpy, 3–5 µm long crystals which appear blackish due to refraction at low magnification. Movement rapid, without peculiarities.

Adoral zone occupies on average 37% of body length in protargol preparations (Table 17; in life almost half of body length), distinctly ?-shaped, on average composed of 45 membranelles of ordinary fine structure; bases of largest membranelles in life about 14 µm wide. Buccal field in life very wide and deep, in protargol preparations, however, often less conspicuous likely due to shrinkage (Fig. 44a, d). Paroral and endoral moderately to strongly curved and optically intersecting, both composed of zigzagging basal bodies (Fig. 44d). Cytopharynx extends obliquely backwards, obviously without peculiarities.

Cirral pattern and number of cirri of usual variability (Fig. 44d; Table 17). Frontal cirri distinctly enlarged, about 30 µm long. Buccal cirrus distinctly behind anterior end of paroral about at level of cirrus III/2. 2–3 frontoterminal cirri in ordinary position, that is, immediately behind distal end of adoral zone. Midventral complex composed of cirral pairs only, extends to near transverse cirri, left cirri of pairs postorally finer than right cirri; at rear end slightly more right cirri than left (ontogenetic data are needed to show the correct designation [for example, pretransverse ventral cirri] of these cirri). Invariably (n = 12) three distinctly enlarged, about 30 µm long transverse cirri, which project distinctly beyond rear body end. Right marginal row commences at level of frontoterminal cirri, terminates about at level of transverse cirri; left row begins distinctly ahead of level of buccal vertex, terminates in mid-body at rear end of cell; marginal cirri about 20 µm long.

Dorsal cilia arranged in 5–7, usually six kineties; length of cilia not mentioned, but according to Fig. 44e about 3 µm long. Kineties 1–3 more or less bipolar, kineties 4–7 posteriorly successively shortened (see remarks). One very fine caudal cirrus each at rear end of dorsal kineties 1–3.

Resting cysts colourless, spherical, in life (with pillars) 70–90 µm (mean = 80.2 µm; S = 6.0; n = 12) across. Cyst wall with very many 5–6 µm long, distally slightly widened and notched pillars (pillars in top view roughly triangular) embedded in a 5–6 µm thick, very finely striated cover distally marked by slightly more refractive membrane (Fig. 44f–i). Endoplasm with very many greasily shining globules 1–5 µm across.

Occurrence and ecology: Terrestrial in Holarctis, Palaeotropics, and Australis (Foissner 1998). Type locality is the soil (0–5 cm) of a meadow (1270 m above sea-

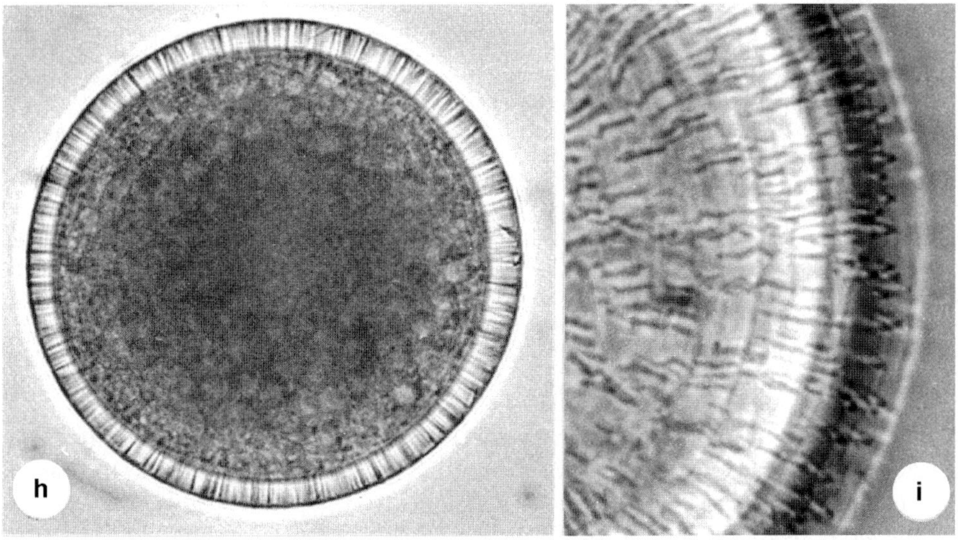

Fig. 44h, i *Caudiholosticha stueberi* (from Foissner 1987e. From life). Total view and detail of resting cyst (see also Fig. 44f, g). Page 235.

level; pH 7.5) with orchids opposite of the Pfiffmoos (a bog), respectively, about 1 km south of the Schupfer Grund-Alm left of the river Fuscher Ache, Fuscher Tal, Salzburg, Austria. I found it in a terrestrial moss from the bank of the Garstner Bach, a small brook in Upper Austria. In Namibia it occurred only in two of 73 samples (Foissner et al. 2002, p. 60). Foissner et al. (2005) found it in a flood plain soil (*Pruno-Fraxinetum*) in the Müllerboden region, Austria.

Caudiholosticha stueberi feeds on naked amoebas, heterotrophic flagellates, ciliates (*Cyclidium* sp.), and fungi (Foissner 1987e); African specimens ingested diatoms and fungal spores (Foissner, pers. comm.). Biomass of 10^6 specimens about 525 mg (Foissner 1998, p. 204).

Caudiholosticha sylvatica (Foissner, 1982) Berger, 2003
(Fig. 45a–m, Table 17)

1982 ***Holosticha sylvatica* nov. spec.**[1] – Foissner, Arch. Protistenk., 126: 58, Abb. 11a–e, 51, Tabelle 12 (Fig. 45a–e; original description; the holotype slide [registration number 1981/95] is deposited in the Oberösterreichische Landesmuseum in Linz [LI], Upper Austria).

[1] The diagnosis by Foissner (1982) is as follows: In vivo etwa 150–210 × 45–60 µm große, lang ellipsoide *Holosticha* mit auffallend langen Pharynxfibrillen und 1–2 Cirren dicht unterhalb des linken Frontalcirrus. Midventralreihen deutlich verstärkt. 2 Caudalcirren leicht rechts der Medianen. Durchschnittlich 35 adorale Membranellen und 7 Transversalcirren, in deren Nähe 2 Ventralcirren inserieren. 5 Dorsalkineten.

1983 ***Holosticha sylvatica* Foissner, 1982** – Borror & Wicklow, Acta Protozool., 22: 122, Fig. 5 (Fig. 45j; revision of urostylids).
1989 ***Holosticha sylvatica* Foissner, 1982** – Berger & Foissner, Bull. Br. Mus. nat. Hist. (Zool.), 55: 21, Fig. 5–8, Table 2 (Fig. 45f–i; description of Japanese population; a voucher slide [registration number 1988:2:1:14] is deposited in the Natural History Museum in London, see remarks).
1992 ***Holosticha sylvatica* Foissner, 1982** – Shen, Liu, Song & Gu, Protozoa, p. 153, Fig. 2-24Aa, Ab (slightly modified re-drawings of Fig. 45f, h; review on fauna of subtropical China).
1993 ***Holosticha sylvatica* Foissner, 1982** – Shin & Kim, Korean J. syst. Zool., 1: 254, Fig. 2A–C, Table 1 (Fig. 45k–m; description of Korean population).
1994 ***Holosticha sylvatica* Foissner, 1982** – Shin, Dissertation, p. 73, Fig. 9A–C, Table 8 (Fig. 45k–m; data already published by Shin & Kim 1993, see previously entry).
2001 ***Holosticha sylvatica* Foissner, 1982** – Berger, Catalogue of ciliate names 1. Hypotrichs, p. 39 (nomenclator containing all basionyms, combinations, and higher taxa of hypotrichs).
2003 ***Caudiholosticha sylvatica* (Foissner, 1982) n. comb.** – Berger, Europ. J. Protistol., 39: 378 (combination with *Caudiholosticha*).

Nomenclature: No derivation of the name is given in the original description. The species-group name *sylvatic·us -a -um* (Latin; living in the forest) refers to the habitat (soil of a mixed forest) where the species was discovered.

Berger & Foissner (1989, p. 19) wrote "One slide of holotype specimens and 1 slide of paratype specimens of the new species and 1 neotype-slide of each other species described have been deposited in the British Museum (Natural History) in London". Unfortunately, this sentence is incorrect as concerns the neotypes because it was not our intention to fix a neotype for each species redescribed, like, for example *C. sylvatica*, whose original type slides are deposited in Linz, Upper Austria. In this and some other cases voucher slide would have been the correct term.

Remarks: This species lacks most (all?) apomorphies of *Holosticha* and has distinct caudal cirri. Thus, it was transferred from *Holosticha* to *Caudiholosticha* by Berger (2003). For a detailed foundation of the transfer of many species from *Holosticha* to *Caudiholosticha*, see the genus section.

Caudiholosticha sylvatica is very conspicuous due to the more or less long row of cirri behind the left frontal cirrus. *Bicoronella costaricana* has a very similar (homologous?) row. Probably all these cirri originate from anlage I, which usually forms only the undulating membranes and the left frontal cirrus (= cirrus I/1). Only rarely, for example in *Uroleptopsis*, more than one cirrus is formed from this anlage. However, *Uroleptopsis* is a pseuokeronopsid so that the increased number of cirri in these two taxa must be interpreted as convergence. In addition, ontogenetic data are needed to show how these cirri originate in *C. sylvatica*.

Foissner (1982) did not consider the cortical granules in the diagnosis. Although they are rather small, they are an important and stable feature, as indicated by the redescription of a Japanese population, which also showed these organelles (Berger & Foissner 1989). Borror & Wicklow (1983) and Shin & Kim (1993) did not mention a cortical granulation. I suppose that they overlooked or ignored these organelles in the present case. In spite of this problem and the rather high inter-population variability in some features (e.g., number of transverse cirri and cirri behind left frontal cirrus, number of adoral membranelles and macronuclear nodules; see Table 17 for details), I do not doubt the conspecificity of all populations described so far.

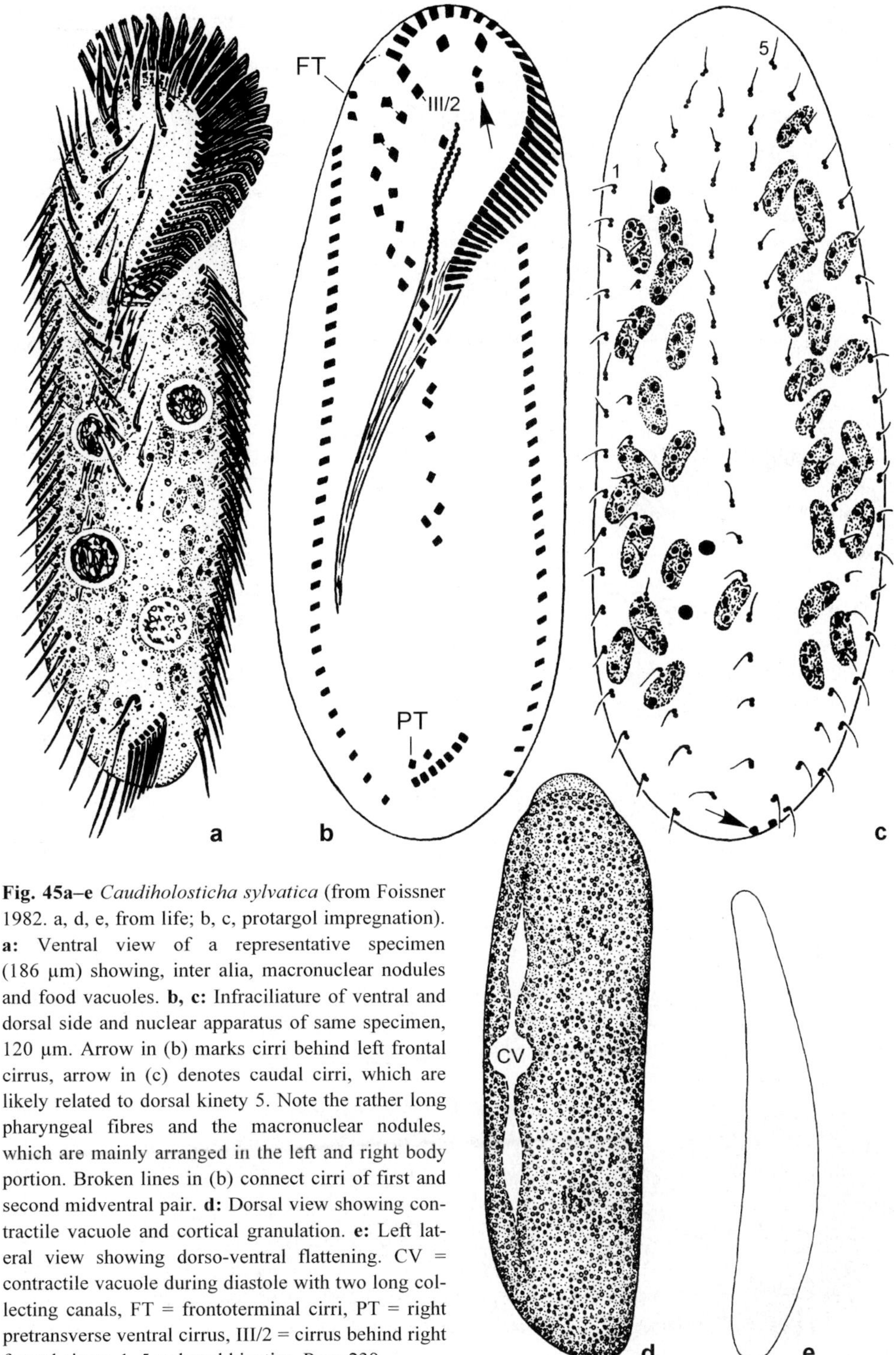

Fig. 45a–e *Caudiholosticha sylvatica* (from Foissner 1982. a, d, e, from life; b, c, protargol impregnation). **a:** Ventral view of a representative specimen (186 µm) showing, inter alia, macronuclear nodules and food vacuoles. **b, c:** Infraciliature of ventral and dorsal side and nuclear apparatus of same specimen, 120 µm. Arrow in (b) marks cirri behind left frontal cirrus, arrow in (c) denotes caudal cirri, which are likely related to dorsal kinety 5. Note the rather long pharyngeal fibres and the macronuclear nodules, which are mainly arranged in the left and right body portion. Broken lines in (b) connect cirri of first and second midventral pair. **d:** Dorsal view showing contractile vacuole and cortical granulation. **e:** Left lateral view showing dorso-ventral flattening. CV = contractile vacuole during diastole with two long collecting canals, FT = frontoterminal cirri, PT = right pretransverse ventral cirrus, III/2 = cirrus behind right frontal cirrus, 1, 5 = dorsal kineties. Page 239.

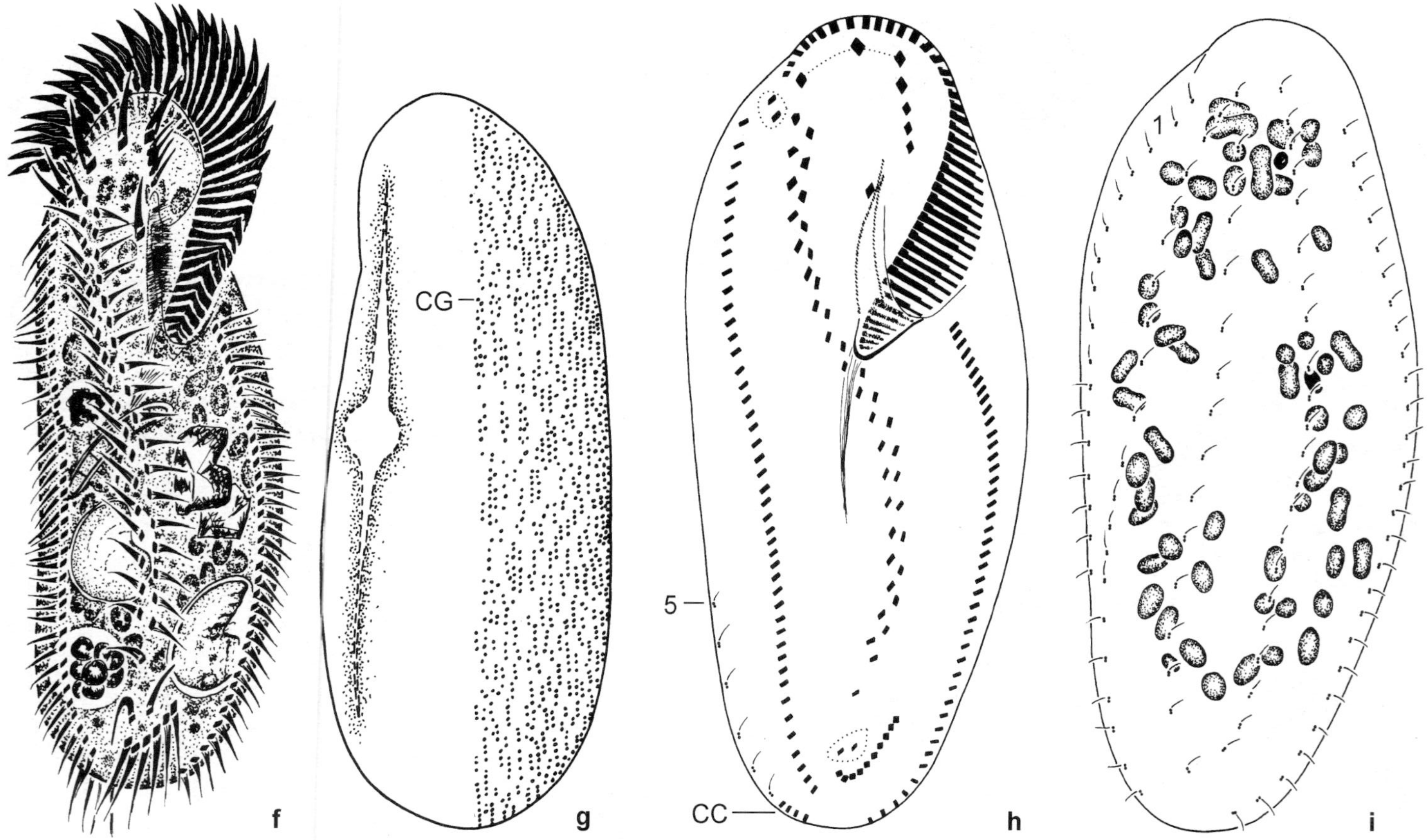

Fig. 45f–i *Caudiholosticha sylvatica* (from Berger & Foissner 1989. f, g, from life; h, i, protargol impregnation). **f, g:** Ventral and dorsal view of representative specimens, f = 175 µm. **h, i:** Infraciliature of ventral and dorsal side and nuclear apparatus of same specimen, 114 µm. Frontoterminal cirri and pretransverse ventral cirri circled by dotted line; three enlarged frontal cirri connected by dotted line. CC = caudal cirri attached to dorsal kinety 5, CG = cortical granules, 1, 5 = dorsal kineties. Page 239.

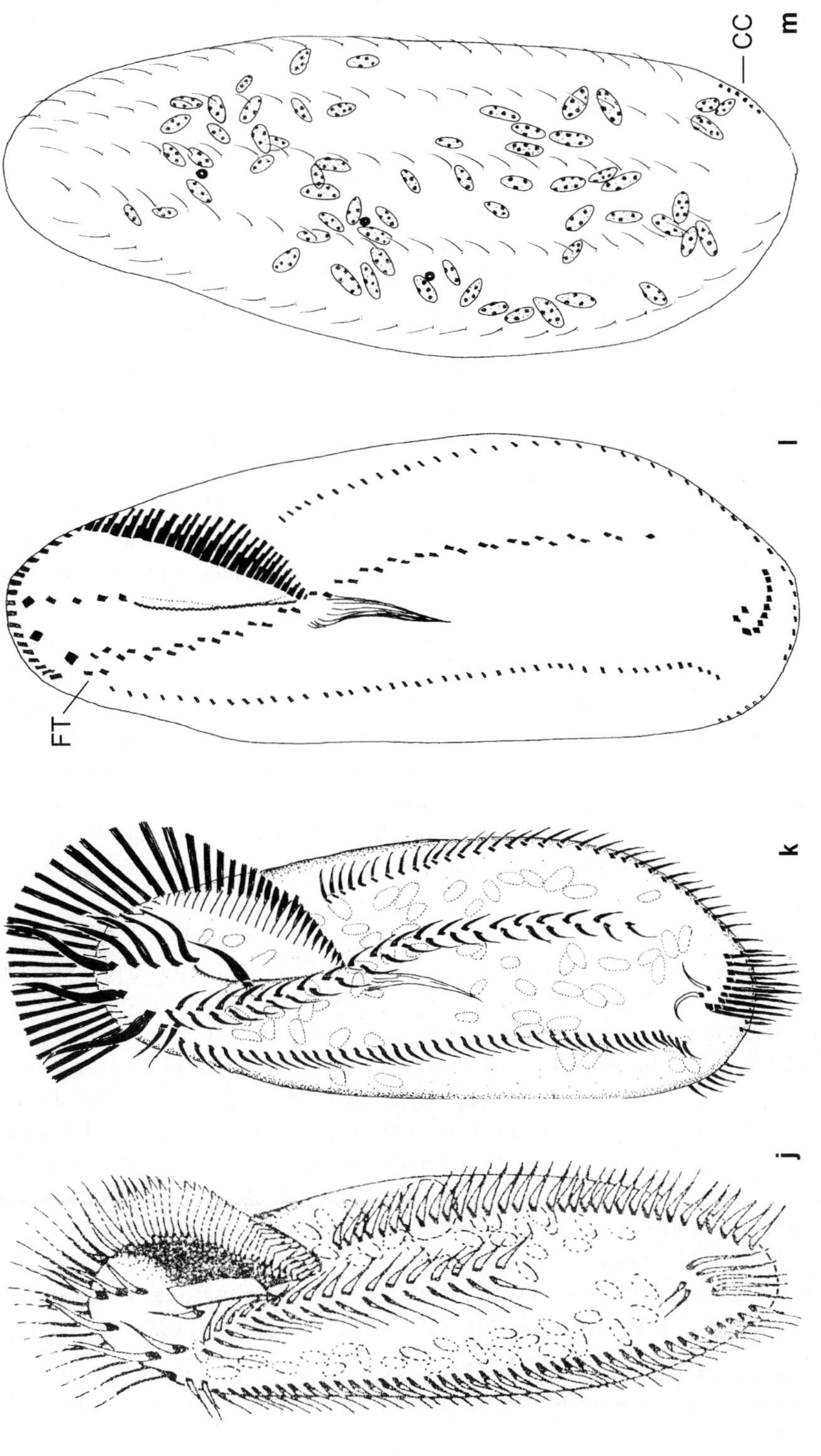

Fig. 45j–m *Caudiholosticha sylvatica* (j, from Borror & Wicklow 1983; k–m, from Shin & Kim 1993. j, protargol impregnation?; k, from life?; l, m, protargol impregnation). **j, k:** Ventral views, j = 144 µm, k = 188 µm. **l, m:** Infraciliature of ventral and dorsal side and nuclear apparatus of same(?) specimen, 185 µm. Note that the specimen illustrated lacks, obviously par lapsus, a buccal cirrus (cp. with Fig. 45k). CC = caudal cirri attached to dorsal kinety 5, FT = frontoterminal cirri. Page 239.

Morphology: The following description is based, unless otherwise indicated, on the original description, supplemented with data by Berger & Foissner (1989). For some supplementary and deviating data on Borror & Wicklow's (1983) and Shin & Kim's (1993) populations, see last paragraph of the present chapter and Table 17 and Figs. 45j–m.

Type population 150–210 × 45–60 µm in life. Body outline elongate elliptical, right margin often slightly concave, left convex, both ends broadly rounded; usually slightly sigmoidal and margins inconspicuously converging posteriorly (Fig. 45a). Body flattened about 2:1 dorso-ventrally, ventral side concave, dorsal convex. Macronuclear nodules mainly scattered along cell margins; individual nodules about 5 × 2.5 µm in life. Several micronuclei, each about 2.5 µm across. Contractile vacuole in mid-body close to left cell margin, during diastole with two long collecting canals (Fig. 45a). Pellicle colourless, flexible. Cortical granules colourless, about 1 µm across (Fig. 45d); according to Berger & Foissner (1989) the colourless granules are less than 0.5 µm in diameter and arranged in short longitudinal rows (Fig. 45g). In a population from Hawaii, the granules were slightly ellipsoidal, 0.3–0.5 µm long, and weakly yellowish (W. Foissner, pers. comm.).[1] Cytoplasm colourless, without conspicuous inclusions. Japanese specimens filled with 2–5 µm large, greasy globules, many food vacuoles, and some pieces of quartz; there was always an accumulation of greasy globules in the frontal area (Berger & Foissner 1989). Cytopyge left, near rear end of cell (Berger & Foissner 1989). Movement moderately rapid, nestling on soil particles.

Adoral zone occupies 30% of body length on average (in Japanese population about 35%; Table 17), composed of about 35–44 membranelles of ordinary fine structure; bases of largest membranelles in life about 8 µm wide (Berger & Foissner 1989). Buccal field of ordinary size, but rather shallow. Undulating membranes more short than long, almost straight, inconspicuously intersecting optically in posterior portion. Pharyngeal fibres extending obliquely backwards to near last quarter of cell in type population (Fig. 45b).

Three enlarged frontal cirri; behind left cirrus 1–2 smaller cirri (Fig. 45b), Japanese population invariably with four such cirri (Fig. 45h). Cirrus behind right frontal cirrus (= III/2) slightly enlarged. Buccal cirrus near anterior end of paroral. Frontoterminal cirri in ordinary position. Midventral complex composed of cirral pairs only, extends slightly sigmoidally from near right frontal cirrus to about 57% of body length on average (Table 17); midventral cirri about 10 µm long, right cirrus of each pair slightly stronger than left, generally cirri become smaller from anterior to posterior; midventral complex of Japanese population terminates at 70% of body length on average (Table 17). Two pretransverse ventral cirri ahead of right portion of oblique, slightly curved transverse cirral row; transverse cirri about 15 µm long, not distinctly enlarged, right ones project slightly beyond rear body end. Right marginal row commences near frontoterminal cirri, ends subterminally. Left row begins left of proximal portion of adoral zone, inconspicuously (Fig. 45b) to rather distinctly J-shaped (Fig. 45h); marginal cirri

[1] Recently I found *Caudiholosticha sylvatica* in the Lammertal area (see occurrence and ecology). The cortical granules were very small (around 0.5 µm across) and more or less colourless and thus they are rather difficult to recognise.

10 µm (Fig. 45a) to 13 µm (Fig. 45f) long in life, become slightly finer from anterior to posterior, distance between individual cirri of about same size.

Dorsal cilia about 4 µm long in life, arranged in five kineties. Kineties 1, 3, and 4 slightly to distinctly shortened anteriorly. Two fine caudal cirri somewhat right of midline, likely associated with kinety 5; Japanese population with 3–5 cirri on kinety 5 (Fig. 45c, i).

For data on Borror & Wicklow's and Shin & Kim's populations from the USA, respectively, Korea, see mainly Fig. 45j–m and Table 17. Supplementary and deviating data of Korean population: body size 150–240 × 50–100 µm; body soft and flexible; contractile vacuole spindle-shaped; cortical granules neither mentioned nor illustrated (see remarks); adoral zone about 36% of body length; buccal field deep; dorsal kinety 3 with 20–30 bristles; dorsal bristles about 5 µm long.

Occurrence and ecology: Terrestrial in all biogeographical regions, except Archinotis (Foissner 1998). Type locality is a mixed forest (*Fagus*, *Pinus*, *Betula*, *Quercus*; about 260 m above sea-level; brown soil with tendency to leaching) near the village of Baumgarten, Lower Austria, where Foissner (1982) discovered *C. sylvatica* in the upper soil layer. For a detailed description of this site, see Foissner et al. (1985, p. 89, Profil 7). Berger & Foissner (1989) found it in yellow soil with poorly decomposed needles (pH 3.8; 10–20 m above sea-level) from Mea-Shima, Amakusa, Kumamoto Prefecture, Japan. Borror & Wicklow (1983) likely collected their samples somewhere in the USA. Kin & Shim (1993) isolated *C. sylvatica* from moss-covered soils in the campus of Seoul National University, Korea. Shen et al. (1992) found it in subtropical regions of China.

Further reliable records: alpine soil in the Gastein area, Salzburg, Austria (Foissner & Peer 1985, p. 41; includes a chemical analysis of the soil); litter of natural forest stands in Austria (Foissner et al. 2005); beech litter from the "Lammertaler Urwald", a forest with rather large trees near the village of Lungötz, Salzburg (original observation); soil sample from Santa Rosa National Park, Costa Rica, about 5 km east of the ranch house "La Casona", near a path to the Pacific Ocean (Foissner 1995, p. 39); five soil samples from near Manaus, Amazonia (Foissner 1997, p. 322); four soil samples from Ullung Island, Korea (Shin & Kim 1995, p. 160); upper soil layer (pH 4.5; about 100 m above sea-level) with much litter from the bush in the Royal National Park south of Sydney, and bark grown with lichen and moss from a pine forest between Cairns and Innisfall, Australia (Blatterer & Foissner 1988, p. 8); soil samples from Hawaii and Kenya (Foissner 1999, p. 323; W. Foissner, pers. comm.). Not recorded in an extensive study on soil ciliates from Namibia (Foissner et al. 2002).

Feeds on cyanobacteria, fungal spores, and ciliates like *Colpoda inflata* (Berger & Foissner 1989). Kin & Shim (1993) maintained laboratory cultures in commercial mineral water provided with boiled wheat grains and shrimp meats to support fungal and bacterial growth. Biomass of 10^6 specimens about 181 mg (Foissner 1987a, p. 124; 1998, p. 204).

Caudiholosticha tetracirrata (Buitkamp & Wilbert, 1974) Berger, 2003 (Fig. 46a–i, Table 17)

1974 ***Holosticha tetracirrata* n. sp.** – Buitkamp & Wilbert, Acta Protozool., 13: 206, Abb. 5 (Fig. 46a; original description; site where type material deposited not mentioned, likely in the Zoological Institute of the University of Bonn; no formal diagnosis provided).

1976 ***Holosticha tetracirrata* Buitkamp & Wilbert** – Bick & Buitkamp, Trans. Am. microsc. Soc., 95: 491, Fig. 1 (Fig. 46a; brief comment).

1982 ***Holosticha tetracirrata* Buitkamp & Wilbert, 1974** – Hemberger, Dissertation, p. 110 (revision of non-euplotid hypotrichs).

1982 ***Holosticha tetracirrata* Buitkamp und Wilbert, 1974** – Foissner, Arch. Protistenk., 126: 55, Abb. 10a–e, 52, Tabelle 11 (Fig. 46b–f; detailed description of Austrian populations; a voucher slide [registration number 1981/99] is deposited in the Oberösterreichische Landesmuseum in Linz [LI], Upper Austria; see nomenclature).

1988 ***Holosticha tetracirrata* Buitkamp & Wilbert, 1974** – Blatterer & Foissner, Stapfia, 17: 48, Abb. 17a–c, Tabelle 10 (Fig. 46g–i; description of an Australian population; at least 1 voucher slide is deposited in the Oberösterreichische Landesmuseum in Linz [LI], Upper Austria).

2001 ***Holosticha tetracirrata* Buitkamp and Wilbert, 1974** – Berger, Catalogue of ciliate names 1. Hypotrichs, p. 39 (nomenclator containing all basionyms, combinations, and higher taxa of hypotrichs).

2003 ***Caudiholosticha tetracirrata* (Buitkamp and Wilbert, 1974) n. comb.** – Berger, Europ. J. Protistol., 39: 378 (combination with *Caudiholosticha*).

Nomenclature: The species-group name *tetracirrata* (having four cirri) is a composite of the Greek numeral *tetra-* (four), the Latin noun *cirrus* (cirrus), and the suffix *-at·us ·a* (having something, provided with [a feature]), and refers to the four transverse cirri (Buitkamp & Wilbert 1974, p. 207). According to Aescht (2003, p. 398), Foissner (1982) designated a neotype. This is incorrect, inasmuch as the holotype slide is likely deposited in the University of Bonn, Germany.

Remarks: The original description is based on live observation and data from protargol-impregnated specimens. Unfortunately, it does not contain data about the dorsal infraciliature, especially number of kineties and presence/absence of caudal cirri. Foissner's (1982) Austrian populations agree rather well with the type material from Canada. He described two very fine caudal cirri, which are the reason for the classification in *Caudiholosticha*. Ontogenetic data are needed to clarify whether or not these cirri are indeed (vestigial?) caudal cirri. The population described by Blatterer & Foissner (1988) has, on average, more transverse cirri than the type population (5.9 against usually 4) with which it, however, agrees rather well in all other features. In spite of this, they provided a complete description of the Australian population. Blatterer & Foissner (1988) briefly mentioned that a population from Denmark also had an increased number of transverse cirri (mean = 5.6; range 4–6; CV = 20.1%) and 0–2 (mean = 0.5) caudal cirri.

Borror & Wicklow (1983, p. 122) synonymised *H. tetracirrata* with *Anteholosticha scutellum*. However, this species, which is not yet redescribed in detail, lives in marine habitats and has likely more adoral membranelles and transverse cirri. *Anteholosticha manca* is also marine, lacks caudal cirri, and has cortical granules and only three dorsal kineties. *Anteholosticha plurinucleata* is smaller (around 60 µm) and therefore has only about 14 adoral membranelles (Fig. 81a). For separation of *C. tetracirrata* from congeners, see key.

Holosticha manca sensu Hemberger (1982, p. 94; Fig. 47a, b) is possibly identical with *C. tetracirrata* because it has caudal cirri and four dorsal kineties; by contrast, *Anteholosticha manca* lacks caudal cirri and has only three dorsal kineties (Fig. 87f, g). However, the three caudal cirri of Hemberger's specimens are widely spaced, which is a conspicuous difference to Foissner's (1982) population. To complete the picture, the illustration and a brief characterisation of this population are provided (see last paragraph of morphology section).

Morphology: The description contains data from the original description, supplemented with data from Foissner's (1982) population (for detailed morphometric characterisation, see Table 17). The description of the Australian population is kept separate because one cannot exclude that it is a distinct (sub)species, as indicated by the increased number of transverse cirri and the presence of a pretransverse ventral cirrus.

Body length of type population 120–160 µm, usually 130 µm in life(?); body length:width ratio 4:1; Austrian specimens 120–150 × 30–50 µm in life (Foissner 1982). Body outline slender elliptical, usually curved rightwards. Austrian specimens orthogonal, both ends widely rounded, anterior portion sometimes slightly set off from remaining body; during swimming slightly curved leftwards, during creeping usually slightly sigmoidally curved (Foissner 1982). Body about 2:1 flattened dorso-ventrally (Foissner 1982). Around 30 ellipsoidal, granulated macronuclear nodules scattered throughout cytoplasm; individual nodules of Austrian population in life about 9 × 5 µm, ellipsoidal to slightly bean-shaped, during reorganisation drumstick-shaped, like micronuclei scattered throughout cytoplasm (Foissner 1982). Contractile vacuole ahead of mid-body (about 40% of body length) near left cell margin (Buitkamp & Wilbert 1974, Foissner 1982), during diastole with two long collecting canals (Fig. 46d). Pellicle very flexible (Buitkamp & Wilbert 1974, Foissner 1982), colourless. Cortical granules lacking because neither mentioned nor illustrated by either Buitkamp & Wilbert (1974) or by Foissner (1982). Cytoplasm cloudily granulated; Austrian specimens with moderately many, slightly yellowish, globules about 3 µm across and many fine granules making cells brownish at low magnification. Movement slow, nestling on soil particles (Foissner 1982).

Adoral zone occupies 25% of body length, composed of 24 membranelles on average (Fig. 46a, Table 17). Buccal field moderately wide; in Austrian specimens small, shallow, anteriorly bordered by hook-shaped lip (Foissner 1982). Undulating membranes curved, paroral about 18 µm long, composed of 5 µm long cilia. Pharynx rather long, extends obliquely backwards beyond mid-body.

Three slightly enlarged frontal cirri. Buccal cirrus right of anterior end of paroral. Two frontoterminal cirri ahead of midventral complex (not mentioned by Buitkamp & Wilbert 1974; Fig. 46a); specimens with only one frontoterminal cirrus rarely occur (Foissner 1982). Midventral complex composed of cirral pairs only, extends from near level of buccal cirrus to about last quarter of cell; specimen shown in Fig. 46a with about 18 midventral cirri (cirrus III/2 likely included); both cirri of each pair of about same size. Midventral complex in Austrian populations on average 66% (Glockner

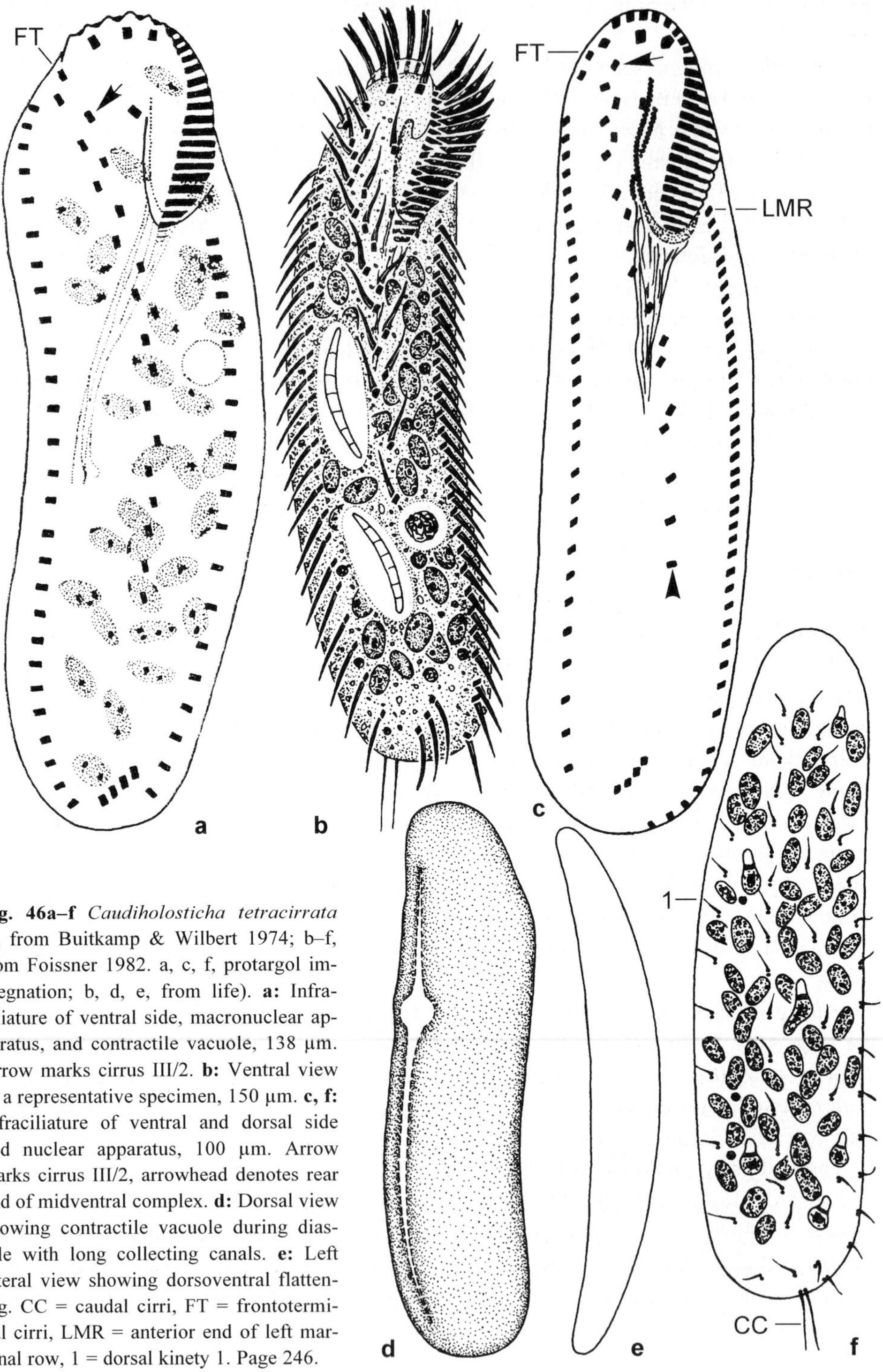

Fig. 46a–f *Caudiholosticha tetracirrata* (a, from Buitkamp & Wilbert 1974; b–f, from Foissner 1982. a, c, f, protargol impregnation; b, d, e, from life). **a:** Infraciliature of ventral side, macronuclear apparatus, and contractile vacuole, 138 µm. Arrow marks cirrus III/2. **b:** Ventral view of a representative specimen, 150 µm. **c, f:** Infraciliature of ventral and dorsal side and nuclear apparatus, 100 µm. Arrow marks cirrus III/2, arrowhead denotes rear end of midventral complex. **d:** Dorsal view showing contractile vacuole during diastole with long collecting canals. **e:** Left lateral view showing dorsoventral flattening. CC = caudal cirri, FT = frontoterminal cirri, LMR = anterior end of left marginal row, 1 = dorsal kinety 1. Page 246.

area) and 58% (Tullnerfeld) of body length (Foissner 1982, Table 17). Distinct pretransverse ventral cirri lacking. Transverse cirri in oblique row near rear body end, about 10 µm long; rarely specimens with three cirri. Right marginal row commences about at level of buccal cirrus, ends in type population, like left row, slightly subterminally; left row commences left of proximal end of adoral zone; marginal cirri about 10 µm long. In Austrian populations left marginal row extends to near midline (Fig. 46c).

Dorsal cilia 3 µm long (number of kineties and presence/absence of caudal cirri not mentioned in original description). Austrian populations with four dorsal kineties (Fig. 46f); kineties 1 and 3 distinctly, kineties 2 and 4 only slightly shortened anteriorly; two very fine caudal cirri somewhat right of midline (Foissner 1982); ontogenetic data are needed to know to which dorsal kinety(ies) the caudal cirri are attached.

Australian population (Fig. 46g–i) described by Blatterer & Foissner (1988) in life about 100–150 × 30–40 µm. Body outline elongate, slightly sigmoidal, both ends moderately widely rounded. Body very flexible. Macronuclear nodules scattered throughout cytoplasm, globular to ellipsoidal, with globular nucleoli. Micronuclei globular to ellipsoidal, near macronuclear nodules. Contractile vacuole near mid-body, during diastole with two collecting canals. Cortical granules lacking. Cytoplasm colourless, with few greasily shining globules about 3 µm across. Food vacuoles 4–6 µm in diameter. Movement slow.

Adoral zone occupies about 31% of body length on average in protargol preparations (Table 17), composed of about 26 membranelles of ordinary fine structure; largest membranelles about 7 µm wide. Pharyngeal fibres strong and long. Buccal field very shallow, which is, according to Blatterer & Foissner, a characteristic feature of this species. Undulating membranes slightly optically crossing anteriorly, each likely composed of two rows of basal bodies.

Marginal and midventral cirri 8–10 µm long, frontal and transverse cirri about 14 µm. Frontal cirri distinctly enlarged. Buccal cirrus near anterior end of paroral. Cirrus III/2 enlarged. Midventral complex usually terminates slightly behind mid-body, rarely (e.g., as in specimen illustrated, Fig. 46h) it extends almost to near transverse cirri. Ahead of rightmost transverse cirrus usually a single pretransverse ventral cirrus. Transverse cirri arranged in J-shaped pattern, project distinctly beyond rear body end. Right marginal row commences near frontoterminal cirri, ends subterminally, left row commences slightly left ahead of buccal vertex, terminates almost in midline at rear end.

Dorsal cilia about 3 µm long, arranged in four kineties. Kineties 1 and 3 shortened anteriorly. Behind last bristle of the fourth kinety 0–2 caudal cirri, which often differ only in their length from dorsal bristles (the illustration obviously shows a specimen without caudal cirri, Fig. 46i).

Holosticha manca sensu Hemberger (1982; Fig. 47a, b) in life(?) about 100–120 × 30–40 µm; length:width ratio 3–4:1, that is, body slender; flexible; 16–25 macronuclear nodules, form longitudinal figure, likely left of midline; 3–4 micronuclei; contractile vacuole slightly ahead of mid-body; adoral zone occupies 20–25% of body length, composed of about 21 membranelles; 3 distinctly enlarged frontal cirri; 1 buccal cirrus;

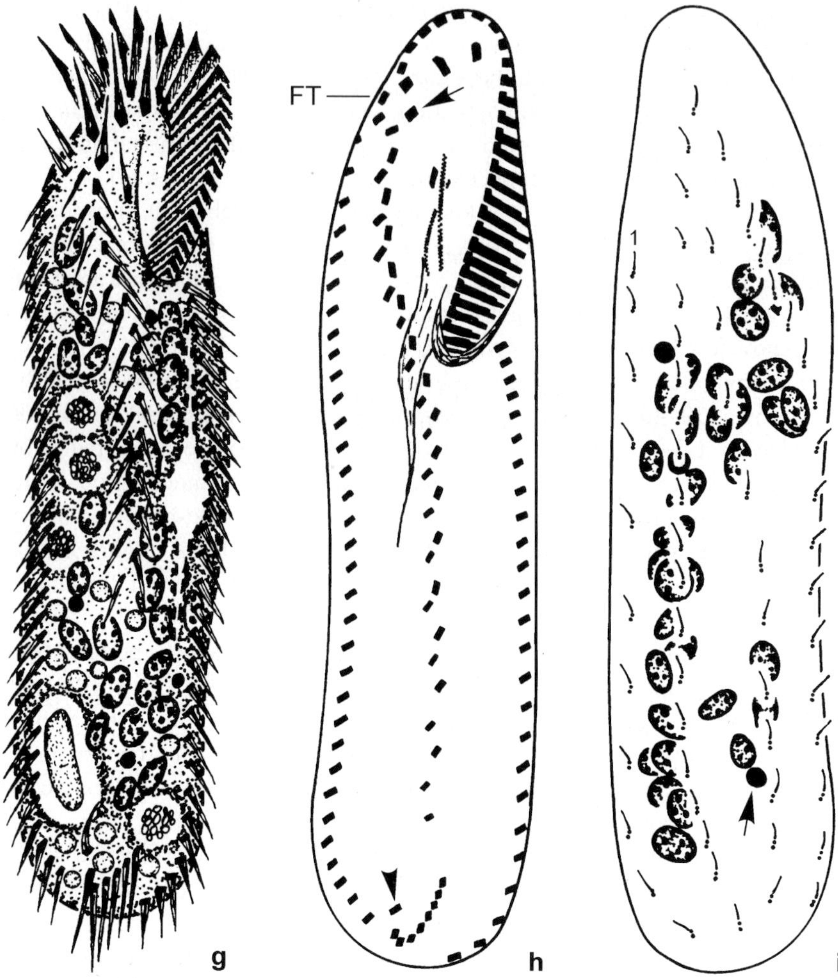

Fig. 46g–i *Caudiholosticha tetracirrata* (from Blatterer & Foissner 1988. g, from life; h, i, protargol impregnation). **g:** Ventral view of a representative specimen of an Australian population, 130 µm. **h, i:** Infraciliature of ventral and dorsal side and nuclear apparatus, 97 µm. Arrow in (h) marks cirrus III/2, arrowhead denotes pretransverse ventral cirrus. Arrow in (i) marks rearmost micronucleus. FT = frontoterminal cirri, 1 = dorsal kinety 1. Page 246.

2 frontoterminal cirri; midventral complex extends distinctly beyond mid-body, composed of 17–24 cirri; 5 transverse cirri; 26–35 right marginal cirri; 21–27 left marginal cirri; dorsal cilia 3–6 µm long, arranged in 4 kineties; 3 widely spaced caudal cirri. Hemberger found this population in a forest soil sample. Unfortunately, he did not mention from which country (Peru or Germany) this sample was.

Cell division: The macronuclear nodules fuse to a single mass prior to division (Buitkamp & Wilbert 1974).

Occurrence and ecology: Terrestrial; common throughout the world (Foissner 1998). Type locality is a grassland (*Sceptre* association) near the village of Matador (50°48'07"N 107°56'55"W), South Saskatchewan, Canada. Samples (0–10 cm) from this natural grassland (chernozemic brown soils) were collected on 16.06.1971 (Buitkamp & Wilbert 1974). Foissner (1982) found *C. tetracirrata* in an alpine soil sample (*Caricetum curvulae*; alpine brown soil with mica slate; 2310 m above sea-level; details, see Foissner 1981, p. 18) from near the Wallackhaus, a hotel at the Glocknerstraße, Austria, and in various soil samples from the Tullnerfeld, Lower Austria (details, see Foissner et al. 1985, p. 108). Blatterer & Foissner (1988) recorded it in five soil samples from various sites in Australia, and in a soil sample from Århus, Denmark.

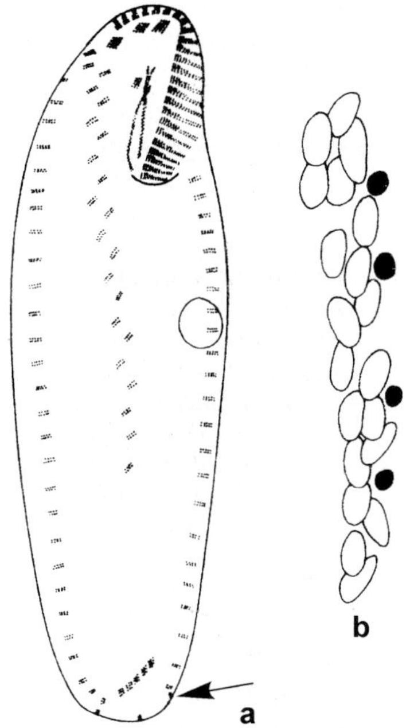

Fig. 47a, b *Holosticha manca* sensu Hemberger (1982; protargol impregnation). Infraciliature of ventral side and nuclear apparatus, 100–120 μm. Arrow marks left caudal cirrus. This is possibly a synonym of *Caudiholosticha tetracirrata* (see text). Page 246.

Records not substantiated by morphological data: soil from subalpine grassland field trial in Aiglern near the village of Aigen, Styria, Austria (Foissner et al. 1990, p. 18); upper soil layer (0–5 cm; about 1240 m above sea-level; pH 7.5) with lichens and calciphil plants on rocks near the Vögerl-Alm, left of the Fuscher Ache, a river in Salzburg, Austria (Foissner 1987e, p. 57); litter of natural forest stands in Austria (Foissner et al. 2005); alkaline soil (salt content 0.18%) of the Hortobágy National Park, Hungary (Szabó 1999b, p. 229; 2000a, p. 8); forest soil in Shimba Hills Nature Reserve in Kenya (Foissner 1999, p. 323); soil from fields in Ostrov (48°38'N 17°46'E), southwest Slovakia (Tirjaková 1988, p. 499; Matis et al. 1996, p. 12); rain forests near Manaus, Amazonia (Foissner 1997, p. 322); two out of 73 soil samples from Namibia (Foissner et al. 2002, p. 60); fern bush (*Histiopteris incisa*) peat litter mixed with some soil and moss (pH 4.4; 30 m above sea level) from Transvaal Bay, Gough Island (40°21'S 09°53'E) and fern bush (*Blechnum pennmarina*) material mixed with some soil (pH 5.2; 30 m above sea-level) near meteorological station on Marion Island (46°52'S 37°51'E; Foissner 1996, p. 284); terrestrial moss (*Sarconeurum glaciale*) from Robertskollen, a group of nunataks fringing an ice-rise in the northern Ahlmannryggen, western Dronning Maud Land, Antarctica (Ryan et al. 1989, p. 18); soil from Antarctica (Sudzuki 1979, p. 123).

Feeds on algae, diatoms, and bacteria (Buitkamp & Wilbert 1974); according to Foissner (1982) and Blatterer & Foissner (1988) it ingests bacteria, ciliates (*Cyclidium glaucoma*), and large fungal spores. *Caudiholosticha tetracirrata* grew well in cultures (boiled stinging nettle infusion with rice grain and dried yolk); sometimes, malformed specimens occurred (Buitkamp & Wilbert 1974). Biomass of 10^6 specimens about 39 mg (Foissner 1987a, p. 124; 1998, p. 204).

Caudiholosticha islandica (Berger & Foissner, 1989) Berger, 2003
(Fig. 48a–d, Table 17)

1989 ***Holosticha islandica* nov. spec.**[1] – Berger & Foissner, Bull. Br. Mus. nat. Hist., 55: 19, Fig. 1–4, Table 2 (Fig. 48a–d; original description; 1 slide of holotype specimens [reference number 1988:2:1:10] and 1 slide of paratype specimens [1988:2:1:13] have been deposited in the British Museum of Natural History in London; see nomenclature).

2001 ***Holosticha islandica* Berger and Foissner, 1989** – Berger, Catalogue of ciliate names 1. Hypotrichs, p. 36 (nomenclator containing all basionyms, combinations, and higher taxa of hypotrichs).

2003 ***Caudiholosticha islandica* (Berger and Foissner, 1989) n. comb.** – Berger, Europ. J. Protistol., 39: 377, 378 (combination with *Caudiholosticha*).

Nomenclature: No derivation of the name is given in the original description. The species-group name *islandic·us -a -um* (Latin; living in Iceland) refers to the site (Iceland) where the species was discovered. In Table 1 of the original description only the reference number of the paratype slide is given under the heading *Holosticha islandica*. The holotype slide is that mentioned under the heading *Hemisincirra inquieta* because these two species occurred in the same sample.

Remarks: The present species lacks most (all?) apomorphies of *Holosticha* and was therefore transferred to *Caudiholosticha* by Berger (2003). For a detailed discussion and foundation of the transfer of many species from *Holosticha* to *Caudiholosticha*, see the genus section. The caudal cirri of *C. islandica* are not very conspicuous (sometimes even lacking, Table 17) and thus can be overlooked even in protargol preparations. Consequently, look carefully for this feature.

Caudiholosticha islandica closely resembles *C. tetracirrata*, especially in that both species have two fine caudal cirri. However, *Caudiholosticha tetracirrata* lacks cortical granules and has, inter alia, more dorsal kineties (4 against 3 in *C. islandica*), adoral membranelles (24 against 17 on average), and macronuclear nodules (30 or more against 16 on average).

Morphology: Body size 80–100 × 25 µm, body length:width ratio 3.6:1 on average in protargol preparations (Table 17). Body outline roughly long elliptical, both margins straight, both ends rounded. Body about flattened 2:1 dorsoventrally (Fig. 48b). Macronuclear nodules in life 6–8 × 3–4 µm; most nodules arranged left of midline, some always right of midline at about level of cytostome. Contractile vacuole about in mid-body near left cell margin, during diastole with inconspicuous collecting canals. Cortical granules irregularly and loosely arranged (Fig. 48b), individual granules yellowish, globular (<0.5 µm across). Cytoplasm colourless with some 1–4 µm large, yellowish grease globules and 5–8 µm large food vacuoles.

[1] The diagnosis by Berger & Foissner (1989a) is as follows: In vivo about 80–100 × 25 µm, long ellipsoid. Subpellicular granules yellowish, spherical (<0.5 µm). Usually 6–7 midventral pairs and 3–4 transverse cirri. 17 adoral membranelles and 16 macronuclear segments on average. 3 dorsal kineties.

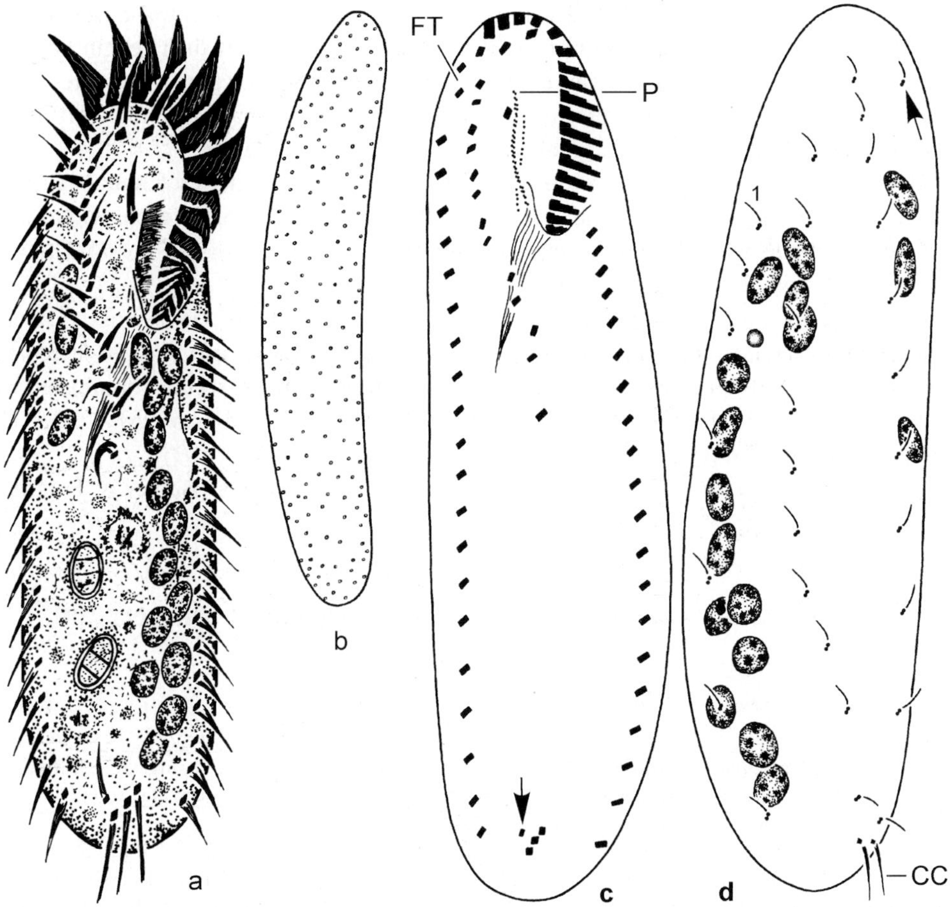

Fig. 48a–d *Caudiholosticha islandica* (from Berger & Foissner 1989a. a, b, from life; c, d, protargol impregnation). **a:** Ventral view of a representative specimen, 100 µm. **b:** Right lateral view (95 µm) showing dorsoventral flattening and arrangement of the tiny (<0.5 µm), yellowish cortical granules. **c, d:** Infraciliature of ventral and dorsal side and nuclear apparatus of same specimen, 80 µm. Arrow in (c) denotes the pretransverse ventral cirrus, arrow in (d) marks the single dorsal bristle right of anterior end of dorsal kinety 3. CC = caudal cirri, FT = frontoterminal cirri, P = paroral, 1 = dorsal kinety 1 (= leftmost kinety). Page 252.

Adoral zone occupies 28% of body length on average, composed of about 17 membranelles of ordinary fine structure. Bases of largest membranelles in life about 5 µm wide. Buccal field flat and narrow. Undulating membranes of about equal length, not distinctly curved and overlapping (Fig. 48c).

Cirral pattern and number of cirri of ordinary variability (Fig. 48c, Table 17). Frontal cirri only slightly enlarged. Buccal cirrus near anterior end of paroral. Frontoterminal cirri in ordinary position, that is, right of anterior end of midventral complex. Midventral complex consists of cirral pairs only, short, that is, terminates about at mid-body (at 46% of body length in specimen illustrated); zigzagging pattern indistinct because cirri arranged almost in line. Usually a small cirrus (pretransverse ventral cirrus?) near the about 14 μm long transverse cirri, which project by about half their length beyond rear body end. Right marginal row commences distinctly subapically, terminates, like left row, about at level of transverse cirri; left marginal row commences about at level of proximal end of adoral zone; marginal cirri in life about 10 μm long.

Dorsal cilia about 3 μm long, arranged in three kineties. Kineties 1 and 2 slightly shortened anteriorly, kineties 2 and 3 usually with a small caudal cirrus each. Right (and not left, as erroneously written in original description) of anterior end of dorsal kinety 3 invariably a single basal body pair (Fig. 48d).

Occurrence and ecology: Rare, terrestrial. Type locality is Gooa Foss (120 m above sea-level), Bardårdalur, North-Iceland, where we found *C. islandica* in litter and the dark (volcanic) upper soil layer (pH 5.5) of a heath with dwarf shrubs (dominated by *Betula nana*, *Empetrum nigrum*, *Vaccinium uliginosum*, *Arctostaphylos uva-ursi*). The sample was collected on 08.08.1985.

Caudiholosticha islandica feeds on bacteria and fungal spores (Berger & Foissner 1989a). Biomass of 10^6 specimens about 30 μg (Foissner 1998, p. 204).

Caudiholosticha paranotabilis (Foissner, Agatha & Berger, 2002) comb. nov.

(Fig. 48.1a–m, 48.2a–h, Table 17)

1996 ***Paruroleptus notabilis* Foissner, 1982** – Foissner, Acta Protozool., 35: 117, Fig. 57–63, Table 8 (Fig. 48.2a–h; misidentification; see remarks; at least one voucher slide is deposited in the Oberösterreichische Landesmuseum in Linz [LI], Upper Austria).

2002 ***Uroleptus paranotabilis* nov. spec.**[1] – Foissner, Agatha & Berger, Denisia, 5: 566, Fig. 130a–i, 381e, h, Table 112 (Fig. 48.1a–m; original description; the holotype slide [accession number 2002/383] and 2 paratype slides [2002/384, 385] and 3 voucher slides [2002/386–388] of a second population are deposited in the Oberösterreichische Landesmuseum in Linz [LI], Upper Austria).

Nomenclature: The species-group name *paranotabil·is -is -e* is a composite of the Greek word *para* (beside) and the Latin adverb *notabilis* (Latin; curious), meaning a ciliate similar to *Uroleptus notabilis* (= *Caudiholosticha notabilis* in present book) (Foissner et al. 2002).

Remarks: The original generic assignment of the present species was uncertain because of nomenclatural and taxonomic problems in the similar genera *Holosticha*, *Paruroleptus*, and *Uroleptus* (Foissner et al. 2002). We preliminary assigned it to the oldest genus, that is, *Uroleptus*. For a foundation of the transfer from *Uroleptus* to *Caudiholosticha*, see remarks at *C. notabilis*.

[1] Foissner et al. (2002) provided the following diagnosis: Size about 130 × 20 μm; very slenderly lanceolate. Usually 16 macronuclear nodules forming ± distinct strand left of midline. Cortical granules yellowish to citrine, ≤ 0.5 μm across, scattered throughout cortex and in clusters around cirri and dorsal bristles. Midventral row terminates above mid-body, composed of about 14 cirri. On average 22–24 adoral membranelles, about 30 cirri each in right and left marginal row, 1 buccal cirrus, 1 cirrus behind right frontal cirrus, 2 frontoterminal cirri, 2 transverse cirri very near to posterior body end, 4 dorsal kineties, and 3 caudal cirri.

The overall appearance of the present species is very similar to *Caudiholosticha notabilis* of which four populations have been described, namely from Austria, Germany, Australia, and Antarctica. However, the populations differ distinctly in the oral apparatus. Therefore, the Antarctic population of *C. notabilis* was assigned to the present species by Foissner et al. (2002, p. 569). The buccal field is large and deep in *C. notabilis* against narrow and flat in *C. paranotabilis*; the undulating membranes are long and distinctly curved against short and almost straight; the distance between the anterior end of the undulating membranes and the buccal cirrus is considerable larger in *C. notabilis* than in *C. paranotabilis*; and the pharynx is surrounded by conspicuous, rod-shaped organelles only in *C. notabilis*. Furthermore, the cortical granules are usually larger (about 2 × 1 µm) in *C. notabilis* than in *C. paranotabilis* (< 1 µm across), and the macronuclear nodules are more numerous (25–70 vs. up to 30) and scattered (vs. in more or less distinct series left of midline) in the former than the latter. For a separation of *C. paranotabilis* from *C. gracilis*, see *C. gracilis*. In *Caudiholosticha islandica* and *C. tetracirrata* the posterior body end is broadly rounded (vs. narrowly in the present species).

The Antarctic population of *C. paranotabilis* (Fig. 48.2a–h) differs rather distinctly from the Namibian specimens (Fig. 48.1a), indicating that it is a distinct subspecies: three dorsal kineties vs. four (Fig. 48.1g, 48.2f); cortical granules colourless vs. yellowish; buccal cirrus at posterior vs. anterior half of paroral (Fig. 48.1j, 48.2e); six vs. two transverse cirri including pretransverse ventral cirri (Fig. 48.1f, 48.2e). However, further populations should be studied before *C. paranotabilis* is divided into subspecies. Similar differences occur in *C. notabilis* populations (Fig. 48.3a–o).

In life, *Caudiholosticha paranotabilis* is not only easily confused with *C. notabilis* (look at the buccal field!), but also with some other, slender soil hypotrichs, such as *Periholosticha* species (buccal cirrus lacking, adoral zone bipartite; see present book), or *Hemisincirra* and *Hemiurosoma* species (both taxa lack a midventral complex; see Foissner et al. 2002). Thus, both live observation (cortical granules) and protargol impregnation are necessary for a reliable identification. However, if a specimen has the following combination of features in vivo, then it is likely *C. paranotabilis*: size about 130 × 20 µm, that is, slenderly lanceolate; buccal field moderately narrow and flat; about 20 macronuclear nodules in indistinct series left of midline; midventral complex slightly to distinctly longer than adoral zone and composed of rather indistinct cirral pairs; cortical granules colourless to yellowish, minute (< 1 µm) and mainly around cirri and dorsal bristles, do not form distinct rows; 3–4 dorsal kineties; 1 buccal cirrus and 1–6 transverse cirri.

Morphology: The two Namibian populations studied by Foissner et al. (2002) agree very well in all features so that conspecificity is beyond reasonable doubt. The diagnosis (see corresponding footnote) contains data from both populations while the morphometric data are kept separate (Table 17). The Antarctic population investigated by Foissner (1996a) is described separately because it cannot be excluded that it is a distinct subspecies.

Body size of Namibian populations 110–170 × 15–30 µm, on average about 130 × 20 µm. Body length:width ratio 5–8:1 in life, 4.7–9.1:1, on average 6.6:1 in protargol

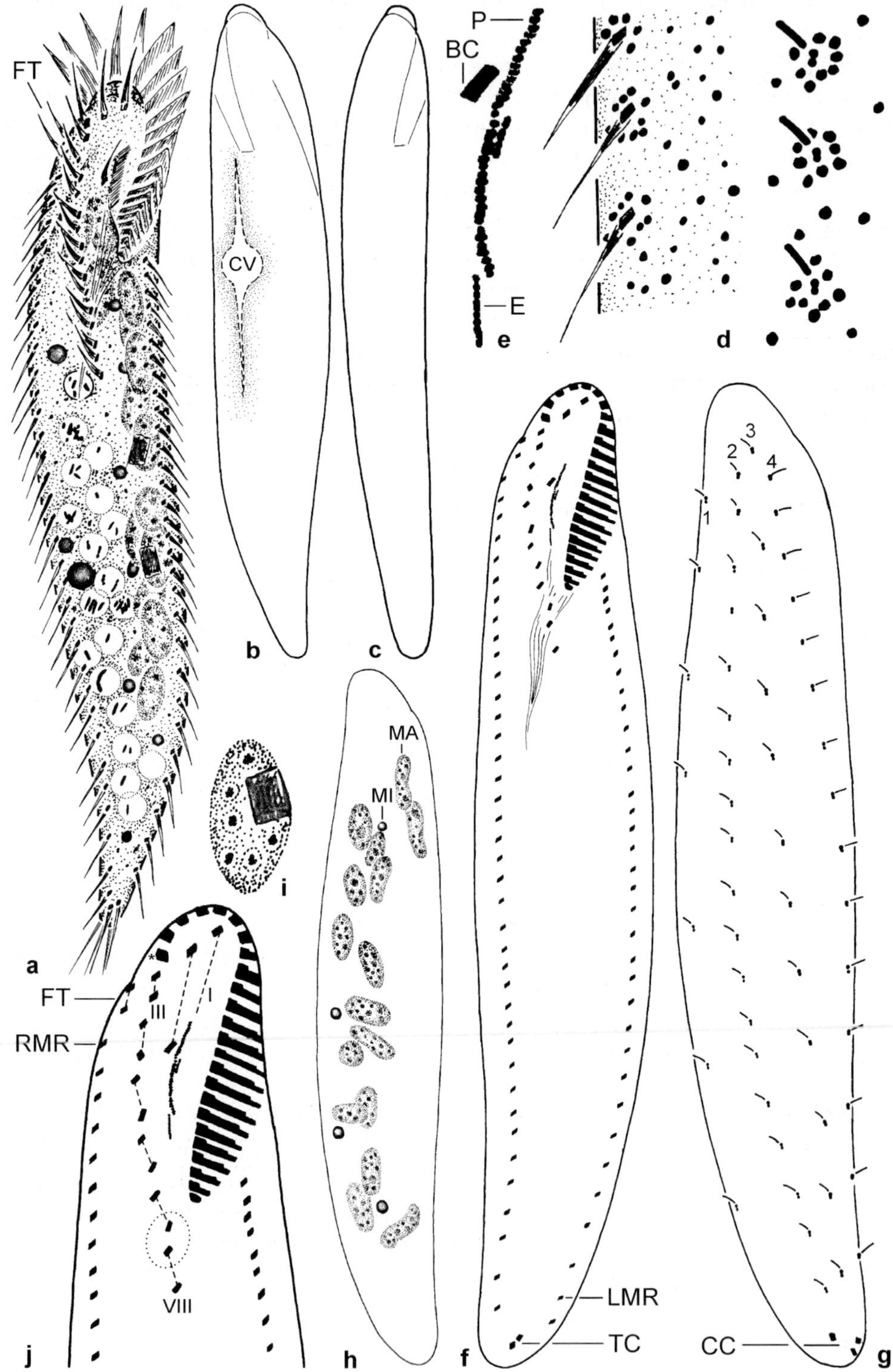
FT
a
CV
b
c
P
BC
E
e
d
i
MA
MI
h
LMR
TC
f
1
2
3
4
CC
g
FT
RMR
III
I
VIII
j

preparations (Table 17). Body outline slenderly lanceolata and often sigmoidal with posterior portion narrowed and curved rightwards; anterior region usually with distinct furrow along dorsal kinety 4; body very flexible but acontractile, frequently slightly to distinctly twisted about main body axis. Body up to 2:1 flattened dorsoventrally. Macronuclear nodules mainly left of midline in a ventral and dorsal strand one upon the other in about 50% of specimens, while slightly scattered in 15% and distinctly so in 35% of cells; anteriormost nodules usually dislocated rightwards; individual nodules ellipsoidal to elongate ellipsoidal, rarely globular to reniform, contain small to medium-sized nucleoli and sometimes a cubic protein crystal (Fig. 48.1a, j, l). On average 2–3 micronuclei within macronuclear strand; individual micronuclei globular to ellipsoidal, up to 5.0 × 2.5 µm in life. Contractile vacuole with inconspicuous collecting canals slightly above mid-body at left cell margin. Cortical granules clustered around cirri and dorsal bristles and scattered throughout cortex, although only less than 0.5 µm in size well recognisable because brilliant citrine in type population and yellowish in Namibian site (4) specimens (Fig. 48.1d); impregnate slightly to heavily with the protargol method used. Cytoplasm colourless, contains some greasily shining globules 1–5 µm across and many, almost empty food vacuoles 4–6 µm across. Movement without peculiarities, that is, swims and glides rather rapidly on microscope slide and debris showing great flexibility.

Adoral zone occupies 17–26%, on average 21% of body length, of usual shape and structure; composed of an average of 23 membranelles, bases of largest membranelles up to 7 µm wide in life (Fig. 48.1j, Table 17). Buccal cavity flat and narrow, right margin forms inconspicuous lip partially covering proximal portion of adoral zone and bearing paroral composed of an about 9 µm long, slightly curved, zigzagging series of 5 µm long cilia; endoral about as long and curved as paroral, but shifted slightly backwards. Pharynx without peculiarities in live and protargol preparations, that is, lacks the rod-shaped structures described in all *C. notabilis* populations.

Cirral pattern and number of cirri of usual variability, except for the rather strongly varying number of midventral cirri (Fig. 48.1f, Table 17). Most cirri 8–10 µm long in life and of similar size. Frontal cirri slightly enlarged, form oblique row with right cirrus, as is usual, behind distal end of adoral zone of membranelles; invariably one cirrus

---

← **Fig. 48.1a–j** *Caudiholosticha paranotabilis* (from Foissner et al. 2002. a, b, d–j, type population; c, flood plain population. a–d, i, from life; e–h, j, protargol impregnation). **a:** Ventral view of a representative specimen (130 µm) with macronuclear nodules arranged in two distinct strands one upon the other. **b:** Dorsal view showing contractile vacuole with conspicuous collecting canals and dorsal furrow. **c:** Ventral view of slender shape variant. **d:** Minute (0.3–0.5 µm), yellow to brilliant citrine cortical granules are scattered throughout the cortex and clustered around dorsal bristles and cirri. **e:** Undulating membranes and buccal cirrus. **f–h, j:** Infraciliature of ventral and dorsal side and nuclear apparatus of holotype specimen, 143 µm. Broken lines in (j) connect cirri which originate (very likely) from the same anlage. Asterisk marks distal end of adoral zone of membranelles. Rearmost pseudopair of midventral complex encircled by dotted line. **i:** Macronuclear nodule with a cubic, about 3 µm large protein crystal. BC = buccal cirrus, CC = caudal cirri, CV = contractile vacuole, E = endoral, FT = frontoterminal cirri, LMR = left marginal row, MA = macronuclear nodule, MI = micronucleus, P = paroral, RMR = right marginal row, TC = transverse cirri, I, III, VIII = cirral anlagen, 1–4 = dorsal kineties. Page 254.

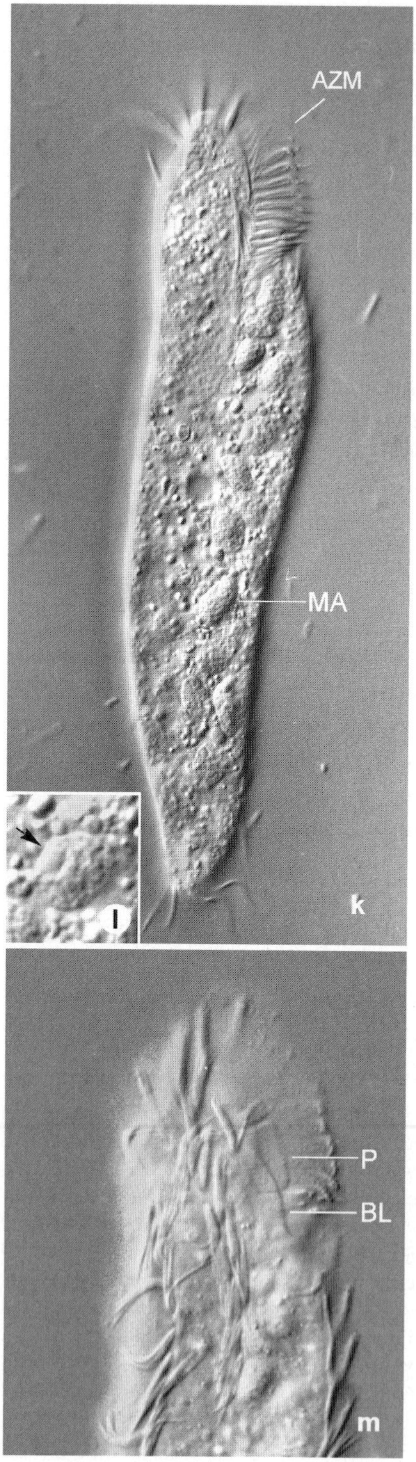

(= cirrus III/2) behind right frontal cirrus. Buccal cirrus slightly behind anterior end of paroral and near level of anterior end of endoral. Two frontoterminal cirri slightly ahead of anterior end of right marginal row. Midventral cirri only indistinctly zigzagging, midventral pattern thus difficult to recognise in life and in prepared specimens; midventral complex terminates between 26% and 46%, on average at 33% of body length. Transverse cirri terminal and thus distinctly projecting, although only inconspicuously longer than marginal cirri. Marginal rows end subterminally; right row commences near level of buccal cirrus, left begins at level of proximal end of adoral zone.

Dorsal bristles about 3 μm long in life, arranged in four rows usually easily recognisable both in live and protargol preparations due to the cortical granules clustered around the bristles (Fig. 48.1g). Kinety 1 composed of only few (about 5–7), widely spaced bristles and likely without caudal cirrus indicating that it is a vestige from the previous generation. Kineties 2–4 almost bipolar and associated with one caudal cirrus each. Caudal cirri not longer than marginal and transverse cirri.

The Antarctic population described by Foissner (1996a, Fig. 48.2a–g, Table 17) differs in some features (see remarks) from the Namibian populations so that one cannot exclude that it is a distinct subspecies or even species. Thus, the descriptions are kept separate. Specimens of Antarctic population in life 90–140 × 15–20 μm. Body outline slenderly elliptical to slightly sigmoidal and/or pisciform,

Fig. 48.1k–m *Caudiholosticha paranotabilis* (from Foissner et al. 2002. Specimen of type population under interference contrast). **k:** Total view, showing, inter alia, the adoral zone of membranelles, the flat and narrow buccal cavity, and the nuclear apparatus. **l:** Macronuclear nodule with protein crystal (arrow). **m:** Oral region showing, inter alia, buccal lip, paroral, and frontal ciliature. AZM = adoral zone of membranelles, BL = buccal lip, MA = macronuclear nodule, P = paroral. Page 254.

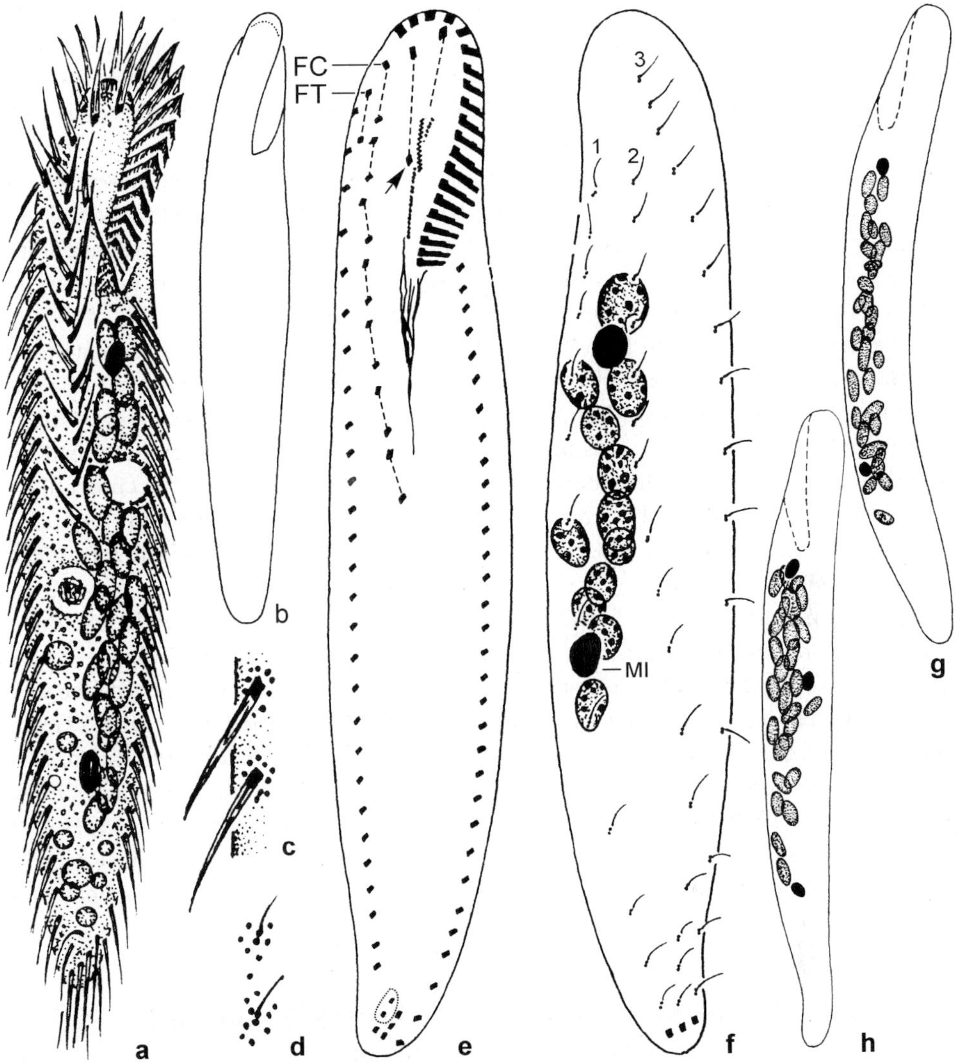

Fig. 48.2a–h *Caudiholosticha paranotabilis* (Antarctic population from Foissner 1996a. a–d, from life; e–h, protargol impregnation). **a, b:** Ventral views of typical specimens, a = 125 µm. **c, d:** Cortical granules around cirri and dorsal bristles. **e, f:** Infraciliature of ventral and dorsal side of a specimen with few macronuclear nodules, 100 µm. Arrow marks buccal cirrus; broken lines connect cirri which originate (very likely) from same anlage. Pretransverse ventral cirri encircled. **g, h:** Variability of nuclear apparatus. FC = rightmost frontal cirrus, FT = frontoterminal cirri, MI = micronucleus, 1–3 = dorsal kineties. Page 254.

that is, rather distinctly narrowed posteriorly. Body very flexible and dorsoventrally inconspicuously to distinctly (up to 2:1) flattened. Macronuclear nodules globular to ellipsoidal, arranged in 2–3 loose rows mainly in left half of cell (Fig. 48.2f–h). Micronuclei conspicuous because compact and large, in life up to 7 × 3 µm, usually one each near end of macronuclear chain, stain weakly with protargol. Contractile vacuole slightly ahead of mid-body. Cortex flexible, contains tiny (<0.5 μm across), colourless granules mainly around cirral bases and dorsal bristles. Cytoplasm colourless, without crystalline inclusions, posterior portion usually filled with 2–5 μm sized, colourless fat globules. Glides and swims rather slowly.

Adoral zone continuous, occupies 24% of body length on average, composed of about 21 membranelles of ordinary structure; bases of largest membranelles about 6 μm wide in life. Buccal cavity narrow and flat, almost entirely covered by very hyaline lip. Paroral only about half as long as endoral. Cytopharynx of ordinary structure, that is, without rod-shaped structures as described in *C. notabilis*.

Cirral pattern as shown in Fig. 48.2e. Frontal cirri about of same size as remaining cirri. Buccal cirrus right of posterior portion of paroral. Frontoterminal cirri left of anterior end of right marginal row. Midventral complex of specimen illustrated composed of six pseudopairs; the isolated cirrus at the end of the midventral complex described by Foissner (1996a) is the rear (left) cirrus of the rearmost midventral pair. Specimen illustrated with four transverse cirri arranged in hook-shape and two rather small pretransverse ventral cirri ahead (Fig. 48.2e). Likely invariably three dorsal kineties (n = 3), each with a single caudal cirrus (Fig. 48.2f).

Occurrence and ecology: Terrestrial. Type locality of *C. paranotabilis* are dung balls formed by a large *Scarabaeus* at the Bambatsi Guest Farm, Mopane forest (20°10'S 15°25'E) in Namibia (Foissner et al. 2002). The abundance at the type locality was high, whereas it was low in a sample (litter sieved off a dry sand bank; pH 6.2) from a floodplain of the Bukaos River about 80 km north of the town Keetmanshoop, Namibia. *Caudiholosticha paranotabilis* is well adapted to soil life with its slender body shape. Foissner (1996a) found it in a *Drepanocladus uncinatus* moss sample (pH 6.9) from Byers Peninsula (Sealer Hill), Livingstone Island, South Shetland Islands. Recently we found it in some Austrian natural forest stands (Foissner et al. 2005). Food not known.

Caudiholosticha notabilis (Foissner, 1982) comb. nov.
(Fig. 48.3a–o, Table 17)

1982 ***Paruroleptus notabilis* nov. spec.**[1] – Foissner, Arch. Protistenk., 126: 64, Fig. 13a–f, Table 13 (Fig. 48.3a–g; original description; type slides not mentioned by Aescht 2003 in her list on slides deposited in the Oberösterreichische Landesmuseum in Linz [LI], Upper Austria).

1987 ***Paruroleptus notabilis* Foissner, 1982** – Berger & Foissner, Zool. Jb. Syst., 114: 206, Fig. 30–35, Table 6 (Fig. 48.3h–m; description of a German population; a voucher slide is deposited in the Oberösterreichische Landesmuseum in Linz [LI], Upper Austria).

1988 ***Paruroleptus notabilis* Foissner, 1982** – Blatterer & Foissner, Stapfia, 17: 51, Abb. 14a, b, Tabelle 11 (Fig. 48.3n, o; description of an Australian population; a voucher slide is deposited in the Oberösterreichische Landesmuseum in Linz [LI], Upper Austria).

---

[1] The diagnosis by Foissner (1982) is as follows: In vivo etwa 180 × 25 μm großer, sehr schlank keilförmiger, häufig leicht S-formig gebogener *Paruroleptus* mit ungefähr 1/3 körperlangen Midventralreihen. Etwa 70 ellipsoide Makronucleus-Teile. 3 Dorsalkineten.

1997 ***Uroleptus notabilis* (Foissner 1982)** – Foissner, Biol. Fertil. Soils, 25: 324 (combination with *Uroleptus*; see nomenclature).
1998 ***Uroleptus notabilis* (Foissner, 1982) nov. comb.** – Foissner, Europ. J. Protistol., 34: 210 (combination with *Uroleptus*; see nomenclature).
2001 ***Uroleptus notabilis* (Foissner, 1982) Foissner, 1998** – Berger, Catalogue of ciliate names 1. Hypotrichs, p. 70 (nomenclator containing all basionyms, combinations, and higher taxa of hypotrichs).
2002 ***Uroleptus notabilis*** – Foissner, Agatha & Berger, Denisia, 5: 568, Fig. 130j–o (Fig. 48.3f, g, l–o; comparison with *C. paranotabilis*).

Nomenclature: No derivation of the name is given in the original description. The species-group name *notabil·is -is -e* (Latin adverb; conspicuous, remarkable, curious) possibly refers to the cytopharynx, which has a conspicuous structure (see below). *Paruroleptus notabilis* Foissner in Foissner (1981, p. 19) is a nomen nudum.

Foissner (1998) transferred *Paruroleptus notabilis* to *Uroleptus* (see list of synonyms), likely because Foissner et al. (1991, p. 248) considered *Holosticha (Paruroleptus)* Kahl, 1932[1] as junior synonym of *Uroleptus* Ehrenberg, 1831. However, the binomen *Uroleptus notabilis* appeared for the first time in Foissner (1997, p. 324). Since neither Foissner (1997) nor Foissner (1998) provided an explanation of the combination with *Uroleptus* it seems wise to consider Foissner (1997) as author of this combination.

Remarks: The present species was originally assigned to *Paruroleptus* likely because of the slender, tailed body shape and the presence of a midventral complex. Later it was transferred to *Uroleptus*, which is generally considered as senior synonym of *Paruroleptus* (e.g., Borror & Wicklow 1983, Foissner et al. 1991; for nomenclatural problems with *Paruroleptus* see previous chapter). However, classical *Uroleptus*-species have a dorsomarginal kinety, usually many transverse cirri, and only two macronuclear nodules. By contrast, the present species very likely lacks a dorsomarginal kinety, has only three transverse cirri, but many macronuclear nodules, strongly indicating that the classification of the present species in *Uroleptus* is incorrect. Molecular studies suggest that classical *Uroleptus*-species do not belong to the urostyloids, although they have a distinct zigzagging midventral complex (see CEUU-hypothesis discussed in chapter 2.4). Since the cirral pattern of the present species, of *Uroleptus paranotabilis* (Fig. 48.1a–m, 48.2a–h), and of *Hemisincirra gracilis* (Fig. 48.4a–i) basically agrees with that of many *Caudiholosticha* species these three tailed species are preliminary assigned to *Caudiholosticha*, which is, however, very likely only a melting pot for holostichids without a conspicuous feature (see genus section). Additional morphological data and ontogenetic and molecular studies are needed to extract monophyletic groups.

The three *C. notabilis* populations described so far differ distinctly in some features. Especially the population from Germany described by Berger & Foissner (1987) is smaller and has therefore a distinctly lower number of adoral membranelles and midventral and marginal cirri (Table 17). In addition, it lacks pretransverse ventral cirri, which are present in the other populations. Berger & Foissner therefore discussed the

[1] Kahl (1932, p. 586) established *Paruroleptus* as subgenus of *Holosticha*, a fact overlooked for a long time. Moreover, he did not fix a type species so that the subgenus is invalid. Later established by Wenzel (1953, p. 109; details see Berger 2001, p. 33).

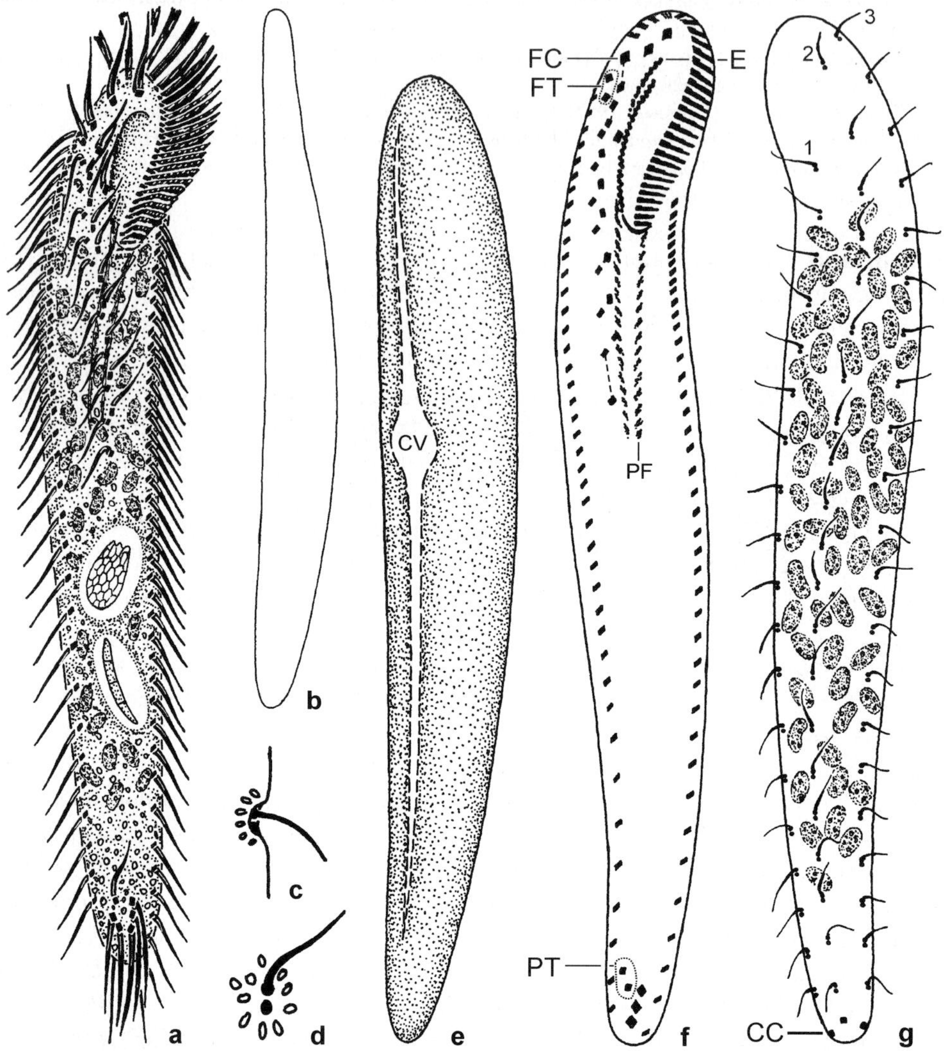

Fig. 48.3a–g *Caudiholosticha notabilis* (from Foissner 1982. a–e, from life; f, g, protargol impregnation). **a:** Ventral view of representative specimen, 162 µm. **b:** Left lateral view. **c, d:** Cortical granules (size not indicated) around dorsal bristles in lateral and top view. **e:** Dorsal view showing contractile vacuole with collecting canals. **f, g:** Infraciliature of ventral and dorsal side and nuclear apparatus of same specimen, 150 µm. Right frontal cirrus and cirrus III/2 and cirri of first and last midventral pair connected by broken lines. CC = caudal cirri, CV = contractile vacuole, E = endoral, FC = right frontal cirrus, FT = frontoterminal cirri, PF = pharyngeal fibres with oblique, rod-shaped structures, PT = pretransverse ventral cirri, 1–3 = dorsal kineties. Page 260.

Fig. 48.3h–m *Caudiholosticha notabilis* (from Berger & Foissner 1987. h–k, from life; l, m, protargol impregnation). **h:** Ventral view of a representative specimen, 108 µm. **i:** Right lateral view of posterior body portion. **j:** Cortical granules (0.5 µm across) around cirri. **k:** Dorsal view. **l, m:** Infraciliature of ventral and dorsal side of same specimen, 97 µm. Arrow marks posterior end of midventral complex. Cirri which (very likely) originate from the same anlage are connected by broken lines. CC = caudal cirri, FC = right frontal cirrus, FT = frontoterminal cirri, MA = macronuclear nodule, PF = pharyngeal fibres with oblique, rod-shaped structures, TC = transverse cirri, 1–4 = dorsal kineties. Page 260. →

possibility that their population could be a distinct species. By contrast, Blatterer & Foissner (1988) assumed a high interpopulation variability. Anyhow, I keep the data separate because the description of *C. paranotabilis* shows that similar species exist.

The cytopharynx of *C. notabilis* is strongly reminiscent of that of *Birojimia* species because of the oblique rod-shaped structures (Fig. 48.3f, l, n, 134b, 135c, h). However, *Birojimia* species have, inter alia, more than two marginal rows. Ontogenetic and molecular data are likely needed to elucidate the correct position of *C. notabilis*. *Paruroleptus notabilis* sensu Foissner (1996a) was identified with *Uroleptus paranotabilis* by Foissner et al. (2002; details see previous species).

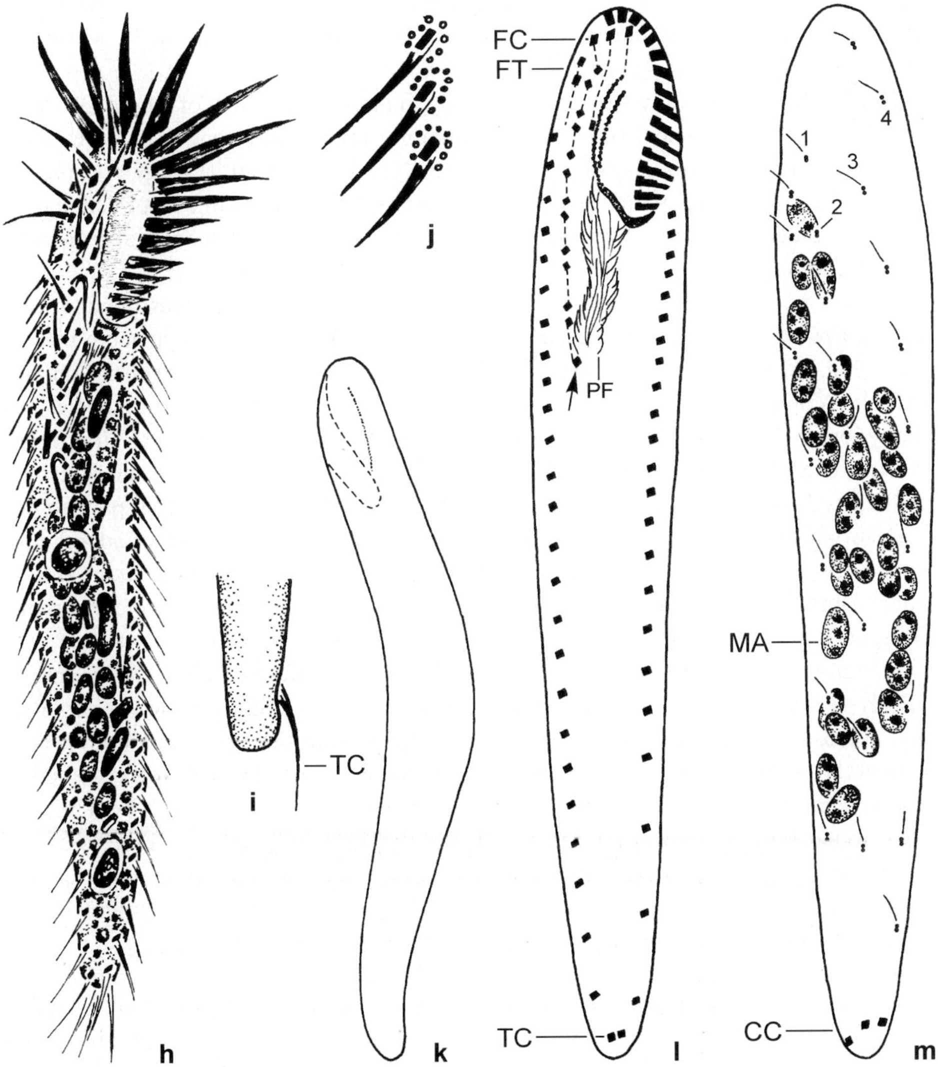

Morphology: The three populations described so far differ distinctly in some features. Blatterer & Foissner (1988) therefore assumed a high interpopulation variability.

Body size of type population 180 × 25 µm in life. Body outline slender wedge-shaped, often slightly sigmoidal. Anterior body end slightly wider rounded than posterior end. Anterior portion about 3:1, middle and posterior body portion about 2:1 flattened dorsoventrally (Fig. 48.3b). Macronuclear nodules scattered throughout cytoplasm; individual nodules about 5 × 2 µm, containing about five nucleoli. Contractile vacuole near left cell margin slightly ahead of mid-body, during diastole with long collecting canals. Pellicle flexible, colourless. Many cortical granules arranged in rows (not illustrated!); they also form rosettes around bases of cirri and dorsal bristles (Fig. 48.3c, d); individual cortical granules cylindrical, size, however, not indicated. Cytoplasm colourless; anterior two body portions with many food vacuoles, rear portion with many slightly yellowish granules, which appear dark at low magnification. Movement slow, adheres closely to soil particles.

Adoral zone occupies about 20% of body length, composed of about 30 membranelles (Table 17). Buccal field deep. Undulating membranes long and more or less distinctly curved, optically slightly intersecting. Cytopharynx tubular, very conspicuous because of beating cilia, which are either endoral cilia or emerging from rod-shaped, protargol-affine organelles (short kineties?) in the wall of the cytopharynx; at rear end of tubular section of pharynx some fibres extend to near rear body end.

Frontal cirri slightly enlarged, arranged in oblique row. Buccal cirrus about at mid-portion of paroral. Frontoterminal cirri not described, but clearly recognisable in illustration (Fig. 48.3f); likely Foissner (1982) included them in the midventral complex. Midventral complex composed of eight pairs in specimen illustrated, terminates at 38% of body length (Fig. 48.3f). Usually one or two small pretransverse ventral cirri. Transverse cirri slightly enlarged, project distinctly beyond rear body end. Right marginal row commences dorsolaterally about at level of frontoterminal cirri; distinctly separated posteriorly from left marginal row; left row commences left of proximal end of adoral zone; distance among individual cirri in posterior portion about twice as large as in anterior portion. Dorsal bristles arranged in small dimples, about 5 µm long in protargol preparations, arranged in three kineties; row 1 slightly shortened anteriorly (Fig. 48.3g).

German population described by Berger & Foissner in life 80–110 × 15–25 µm (n = 4). Body vermicular, very flexible, tapered posteriorly, only slightly flattened dorsoventrally. Macronuclear nodules in life about 5 × 3 µm. Two or three (n = 2) kidney-shaped micronuclei, in life about 7 × 2 µm. Contractile vacuole near left cell margin about in mid-body, during diastole with channels. Cortical granules colourless, about 0.5 µm across, found only around cirral bases and dorsal bristles. Cytoplasm colourless, with few yellowish crystals and small plates of mice and some food vacuoles. Posterior body portion filled with many 1–3 µm large yellowish globules, making specimens dark at low magnification. Movement winding and fast gliding.

Adoral zone occupies about 20–25% of body length, cilia of distal membranelles about 15 µm long. Buccal area deep, anteriorly distinctly bent to the left, but undulating membranes only slightly curved. Pharyngeal fibres conspicuous in life and after protargol impregnation (Fig. 48.3l).

Cirral pattern and number of cirri of usual variability (Fig. 48.3h–m, Table 17). Frontal cirri slightly enlarged, buccal cirrus about at mid-portion of paroral. Frontoterminal cirri distinctly set off from anterior end of right marginal row. Midventral complex commences at level of frontoterminal cirri, terminates about at one third of body length. Transverse cirri in life about 12 µm long, distinctly projecting beyond rear body end (Fig. 48.3i, l; whether these cirri are transverse cirri or the rearmost marginal cirri has to be checked by ontogenetic data). Marginal cirri in life about 10 µm long, composed of four cilia only; going backwards distances among cirri become distinctly wider. Dorsal ciliature composed of four kineties; rows 1–3 slightly shortened anteriorly, row 4 is probably the continuation of the right marginal row, as in some other hypotrichs.

For a characterisation of the Australian population see Fig. 48.3n, o and Table 17. Cortical granules 2.5 × 1.0 µm, colourless.

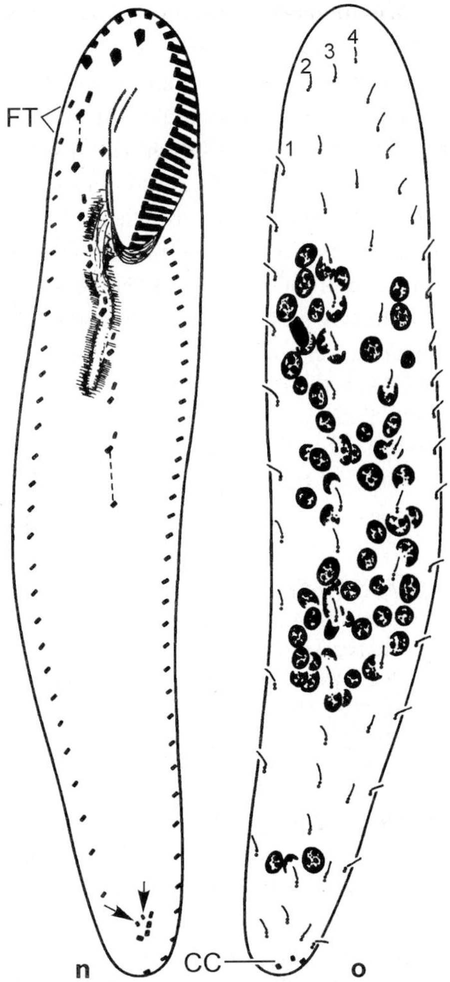

Fig. 48.3n, o *Caudiholosticha notabilis* (from Blatterer & Foissner 1988. Protargol impregnation). Infraciliature of ventral and dorsal side and nuclear apparatus, 135 µm. Cirri of first and last midventral pair connected by broken lines. Arrows in (n) mark pretransverse ventral cirri. CC = caudal cirri, FT = frontoterminal cirri, 1–4 = dorsal kineties. Page 260.

Occurrence and ecology: *Caudiholosticha notabilis* is probably a cosmopolitan and likely confined to terrestrial habitats (Foissner 1987a, p. 126; 1998, p. 210). Type locality is the soil of a highly eutrophic alpine pasture of the Hochmais-Alm (1850 m above sea-level) near the Großglockner-Hochalpenstraße (an alpine road), Salzburg, Austria (Foissner 1982); Foissner (1981, p. 19) found it in the same area in a eutrophic soil sample from near the hotel Wallackhaus. Berger & Foissner (1987) isolated *C. notabilis* from the upper soil layer (0–5 cm) of a spruce forest near the city of Ulm, Germany. Blatterer & Foissner (1988) found it in two soil samples (upper soil layer of mouldy spruce litter and roots; pH 5.1) from a secondary pine forest near the beach of Adelaide, Australia. Foissner et al. (2002, p. 63a, b) recorded it in various soil samples from Namibia, including the Etosha region.

Records not substantiated by morphological data: field with crop rotation (wheat, maize) in the Tullnerfeld area, Lower Austria (Foissner et al. 1985, p. 108); soil of spruce forests in the northern Upper Austria (Petz et al. 1988, p. 83); soil of spruce forest in the Upper Austrian part of the Bohemian Forest treated with organically enriched magnesite fertilisers (Aescht & Foissner 1993, p. 328); natural forest stands in eastern Austria (Foissner et al. 2005); soil samples from the experimental site at the Macaulay Land Use Research Station, near Kelso, Southern Scotland (Finlay et al. 2001, p. 363); soil samples from a floodplain primary(?) rain forest and a blackwater inundation primary(?) rain forest in the Manaus region, Brazil (Foissner 1997).

Biomass of 10^6 specimens about 32 mg (Foissner 1987a, p. 126; 1998). Feeds on fungal spores and *Euglypha laevis*; Berger & Foissner's (1987) population ingested flagellates and fungal spores.

Caudiholosticha gracilis (Foissner, 1982) comb. nov.
(Fig. 48.4a–i, Table 17)

1982 ***Perisincirra gracilis* nov. spec.**[1] – Foissner, Arch. Protistenk., 126: 95, Abb. 25a–g, 67, Tabelle 22 (Fig. 48.4a–i; original description; the holotype slide [accession number 1981/87] is deposited in the Oberösterreichische Landesmuseum in Linz [LI], Upper Austria).

1984 ***Hemisincirra gracilis* (Foissner, 1982) nov. comb.** – Foissner, Stapfia, 12: 119 (combination with *Hemisincirra*; see nomenclature).

2001 ***Hemisincirra gracilis* (Foissner, 1982) comb. nov.** – Berger, Catalogue of ciliate names 1 Hypotrichs, p 71 (combination with *Hemisincirra*, see nomenclature).

Nomenclature: No derivation of the name is given in the original description. The species-group name *gracil·is -is -e* (Latin adjective; slender, fine, delicate) likely refers to the slender body shape. Foissner (1984) made the new combination before *Hemisincirra* Hemberger, 1985 was available (further details see remarks). *Perisincira gracilis* in Chaouite et al. (1990) is an incorrect subsequent spelling of *Perisincirra*.

Remarks: This species was originally assigned to *Perisincirra* Jankowski, 1979 (type species *Uroleptus kahli* Groliere, 1975). Hemberger (1985) recognised that all species so far assigned to *Perisincirra* by Hemberger (1982) and Foissner (1982) belong to a new genus for which he proposed the name *Hemisincirra* (type species *Uroleptus kahli* Buitkamp, 1977). Foissner (1984) knew Hemberger's not yet published paper and therefore transferred his own species to *Hemisincirra*, which, however, became available only in 1985. Since it is impossible to transfer a species in a not yet available genus, I again transferred *Perisincirra gracilis* to *Hemisincirra* (Berger 2001, p. 71). However, I recommended citing this species, when classified in *Hemisincirra*, as *H. gracilis* (Foissner, 1982) Foissner in Berger, 2001.

---

[1] Foissner (1982) provided the following diagnosis: In vivo etwa 90–180 × 13–25 µm große, schmal lanzettförmige *Perisincirra* mit 2 Dorsalkineten, durchschnittlich 14 ellipsoiden Makronucleus-Teilen und 16 adoralen Membranellen, von denen die vorderen 3–4 durch eine schmale Lücke von den hinteren getrennt sind. Frontalreihe ungefähr so lang oder etwas länger als die adorale Membranellenzone.

Now I am convinced that *P. gracilis* is more closely related to *Caudiholosticha paranotabilis* than to the type species of *Hemisincirra*, *Uroleptus kahli* Buitkamp, 1977 (= *Perisincirra buitkampi* Jankowski, 1979 because of primary homonymy with *U. kahli* Groliere, 1975). *Perisincirra gracilis* and *U. kahli* Buitkamp differ, inter alia, in the caudal cirri (likely present vs. absent) and very likely also in the number of the frontal-ventral-transverse cirral anlagen (about 8 vs. 6). The cirral pattern of the present species is very similar to that of *Caudiholosticha* species. The slender body shape and the length of the indistinctly zigzagging midventral complex is reminiscent of *C. paranotabilis*, which, however, has, inter alia, a continuous adoral zone of membranelles (vs. bipartite in present species), four dorsal kineties (vs. two), and cortical granules (vs. lacking). Thus, I transfer *Perisincirra gracilis* to *Caudiholosticha*. However, cell division and molecular data are needed to show whether my assumption is correct that it is indeed a urostyloid. Moreover, only morphogenetic studies will show the correct cirral pattern (presence/absence of transverse/caudal cirri) of the very narrow and thus difficult to observe posterior cell end. *Periholosticha* species lack both frontoterminal cirri and a buccal cirrus (both present in *C. gracilis*).

Morphology: Foissner (1982) described two populations from Austria, which show a rather high intra- and interpopulational variability in some features (Table 17). However, the main characteristics are very similar so that conspecificity is beyond reasonable doubt. Thus, Foissner's description is a mixture of at least two populations.

Body size of one population 150–180 × 13–20 µm, of a second population about 90–150 × 12–25 µm in life. Body slightly contractile; outline always more or less distinctly pisciform, anterior end narrowly rounded, middle portion slightly widened, rarely almost parallel-sided, posterior portion distinctly converging and rear end more or less distinctly pointed (Fig. 48.4a, c); body sometimes slightly twisted about main body axis. Only middle third of body slightly to distinctly flattened dorsoventrally; cells rapidly inflate and become circular in cross-section under coverglass (Fig. 48.4b). Macronuclear nodules in two more or less distinct strands in middle cell portion near left cell margin; individual nodules 6.0 × 3.5 µm in life, with many nucleoli. Several compact micronuclei about 3.5 × 2.0 µm in size; usually they do not stain with the protargol method used (Fig. 48.4e). Contractile vacuole slightly ahead of mid-body near left cell margin, during diastole with short, inconspicuous collecting canals. Cytopyge subterminal, faecal balls fluffy (Fig. 48.4c). Pellicle colourless, very fine and flexible. Distinct cortical granules lacking. Cytoplasm translucent, colourless, posterior cell portion sometimes packed with yellowish, about 2 µm-sized crystals, which are sometimes enclosed in vacuoles about 5 µm across. Food vacuoles about 12 µm in diameter. Movement moderately rapid, swims often hastily to and fro, adheres to soil particles.

Adoral zone occupies 19% (Tullnerfeld population) to 22% (type population) of body length on average, composed of about 16 membranelles of ordinary fine structure (Table 17); zone bipartite, that is, distalmost 3–4 membranelles distinctly set off from proximal portion, which extends to near cell midline. Bases of largest membranelles about 3.5 µm wide. Pharyngeal fibres distinct in life, form a thick, short bundle. Buccal field very small, flat. Paroral about 7 µm long, slightly curved, anterior portion of type population distinctly two-rowed, in Tullnerfeld population likely composed of single

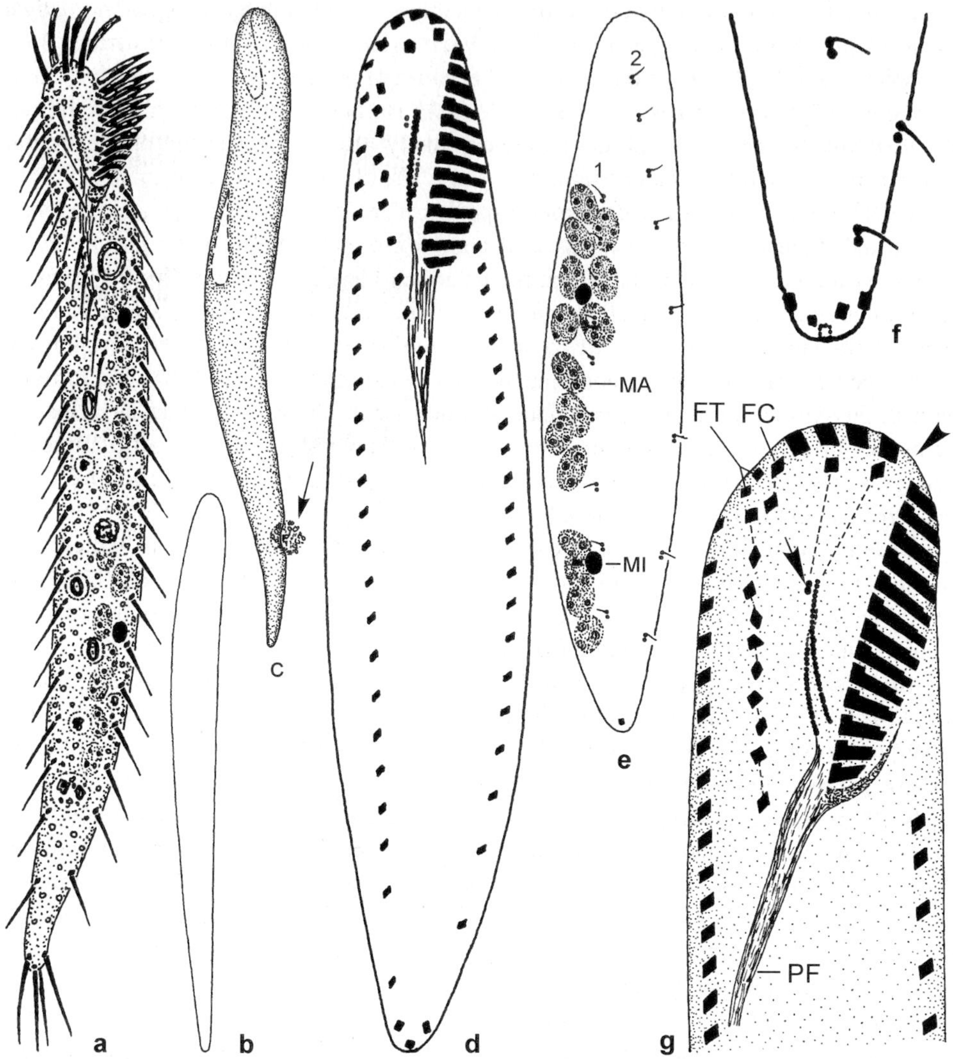

Fig. 48.4a–g *Caudiholosticha gracilis* (from Foissner 1982. Type population from the Schloßalm area [a–f] and Tullnerfeld population [g]. a–c, from life; d–g, protargol impregnation). **a:** Ventral view of a representative specimen, 164 µm. **b, c:** Right lateral view and dorsal view of a sigmoidal specimen showing, inter alia, contractile vacuole and defecation (arrow). **d, e:** Infraciliature of ventral and dorsal side and nuclear apparatus, 62 µm. **f:** Infraciliature of dorsal side in posterior body region. Whether the cirri are caudal cirri or transverse cirri is not known (morphogenetic data are needed). **g:** Infraciliature of ventral side in oral region. Arrowhead marks gap in adoral zone, arrow denotes buccal cirrus. Frontal-midventral cirri, which very likely originate from the same anlage, are connected by broken lines. FC = rightmost frontal cirrus, FT = frontoterminal cirri, MA = macronuclear nodule, MI = micronucleus, PF = pharyngeal fibres, 1, 2 = dorsal kineties. Page 260.

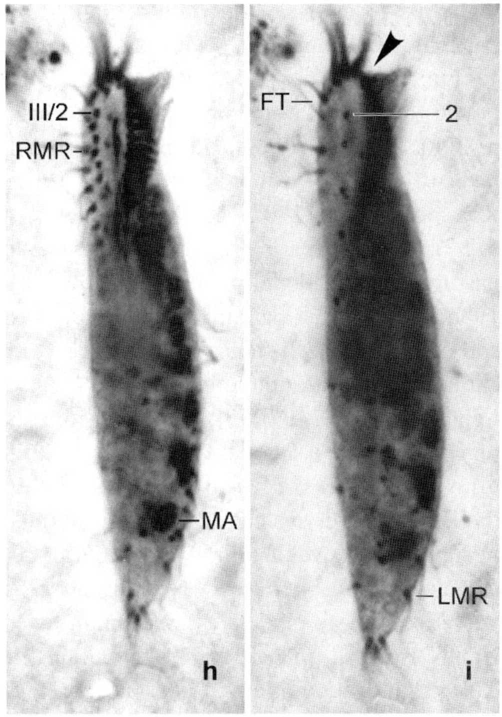

Fig. 48.4h, i *Caudiholosticha gracilis* (type population from Foissner 1982. Protargol impregnation). Infraciliature of oral region and dorsal side. Arrowhead in (i) denotes gap in adoral zone. FT = frontoterminal cirri, LMR = left marginal row, MA = rearmost macronuclear nodule, RMR = anterior end of right marginal row, III/2 = cirrus behind right frontal cirrus, 2 = right dorsal kinety. Page 266.

row of basal bodies. Endoral likely composed of monokinetids, almost in parallel with paroral (Fig. 48.4a, d, g).

Cirral pattern and number cirri of usual variability (Fig. 48.4d, g, Table 17). All cirri very fine and about 10 μm long. Behind distal portion of adoral zone three slightly enlarged frontal cirri in almost transverse row. Buccal cirrus right of anterior end of paroral, composed of 2–3 basal body pairs. One cirrus (= cirrus III/2) behind right frontal cirrus. Two frontoterminal cirri indistinctly set off from anterior end of right marginal row (Foissner therefore included it in the number of right marginal cirri). Midventral complex likely composed of cirral pairs only, that is, (distinct) cirral rows obviously lacking; cirri in anterior portion indistinctly zigzagging (Fig. 48.4d); cirri of rear portion arranged in line so that one cannot exclude that it is a midventral row; midventral complex of type population extends distinctly behind level of proximal end of adoral zone, that is, terminates at 33% of body length in specimen illustrated (Fig. 48.4d). Midventral complex of second population terminates at about same level as adoral zone of membranelles (Fig. 48.4g). Posterior body end with several about 15 μm long cirri, likely transverse and/or caudal cirri (ontogenetic data are needed for correct interpretation), which are not distinctly set off from marginal cirri. Right marginal row commences at 14% of body length in specimen illustrated; left row begins at level of proximal end of adoral zone; distance between marginal cirri in posterior portion about twice (right) to three-times (left) as large as in anterior portion (Fig. 48.4d).

Dorsal bristles about 3 μm long in life, arranged in two slightly curved rows; kinety 1 distinctly shortened anteriorly and posteriorly. Presence/absence of caudal cirri not known; cell division data needed for final decision (Fig. 48.4e, f).

Occurrence and ecology: *Caudiholosticha gracilis* is likely confined to terrestrial habitats (Foissner 1987a). Type locality is the soil of the Schloßalm area (1950 m above sea level) near the village of Bad Hofgastein, Salzburg, Austria. Also common in alluvial soils from the Tullnerfeld region, Lower Austria (Foissner et al. 1985, p. 108). Autecological data on the Gastein area populations see Foissner & Peer (1985, p. 42).

Further records: soil of spruce forest in the Upper Austrian part of the Bohemian Forest treated with organically enriched magnesite fertilisers (Aescht & Foissner 1993, p. 328); subalpine grassland in Styria, Austria (Foissner et al. 1990, p. 18); upper soil layer (0–5 cm) of a floodplain forest (*Alnetum incanae*) in the Fuscher valley near the Rotmoos (about 1280 m above sea level, pH 7.2) in Salzburg, Austria (Foissner 1987e, p. 57); rare in experiments on soil compaction and on effects of fertilisers on an alpine pasture in the Schloßalm area, that is, near the type locality (Berger et al. 1985a, p. 107; 1986, p. 268); thermal and mineral waters in Auvergne, France (Chaouite et al. 1990); soil of spruce beech stands in the Black Forest and the Ulm area, Germany, including a site fertilised with ammonia-sulphate (Lehle 1989, p. 141; 1993, p. 17; 1994, p. 115; Lehle et al. 1992, p. 279); anthropogenic stressed soils from an abandoned textile factory in Nordhorn, Germany (Niebuhr 1989, p. 81; identified by W. Foissner); soil from primary dunes of the Dutch island Terschelling (Verhoeven 1999, p. 64); soils from near the Danube river system in Slovakia (Tirjaková 1992a, p. 77); soil samples from fields in southwest Slovakia (Tirjaková 1988, p. 500); four sites (forest around picnic site; forest surrounding Sheldrick waterfalls; grassland downhill path the Sheldrick waterfalls; young secondary pine forest near main gate to Nature Reserve) from the Shimba Hills Nature Reserve, that is, in Kenya about 40 km south of Mombassa (Foissner 1999, p. 323); sand-litter mixture from the Corel Pink Sand Dunes in Utah, USA (Foissner 1994a, p. 164); soil sample (about 5 km east of the ranch house La Casona) from a tropical dry forest in the Santa Rosa National Park, Costa Rica, Central America (Foissner 1995, p. 39).

Caudiholosticha gracilis feeds on small fungal spores and likely also on bacteria (Foissner 1982). Biomass of 10^6 specimens 11 mg (Foissner 1987a, p. 124).

Caudiholosticha algivora (Kahl, 1932) Berger, 2003 (Fig. 49a, b)

1932 ***Holosticha algivora* spec. n.** – Kahl, Tierwelt Dtl., 25: 586, Fig. 110 $_5$ (Fig. 49a; original description; no type material available and no formal diagnosis provided).

1972 ***Holosticha algivora* Kahl** – Bick, Ciliated protozoa, p. 19, Fig. 8C (Fig. 49b; guide to indicator species; see remarks).

2001 ***Holosticha (Holosticha) algivora* Kahl, 1932** – Berger, Catalogue of ciliate names 1. Hypotrichs, p. 35 (nomenclator containing all basionyms, combinations, and higher taxa of hypotrichs).

2003 ***Caudiholosticha algivora* (Kahl, 1932) n. comb.** – Berger, Europ. J. Protistol., 39: 377 (combination with *Caudiholosticha*).

Nomenclature: No derivation of the name is given in the original description. The species-group name *algivora* (Latin; algae-feeding) obviously alludes to the preferred food, namely green algae. Kahl (1932) divided *Holosticha* into subgenera. Thus, the correct name in his paper was *Holosticha (Holosticha) algivora* Kahl, 1932. *Holosticha algirora* in Han & Hao (1995, p. 4) is an incorrect subsequent spelling.

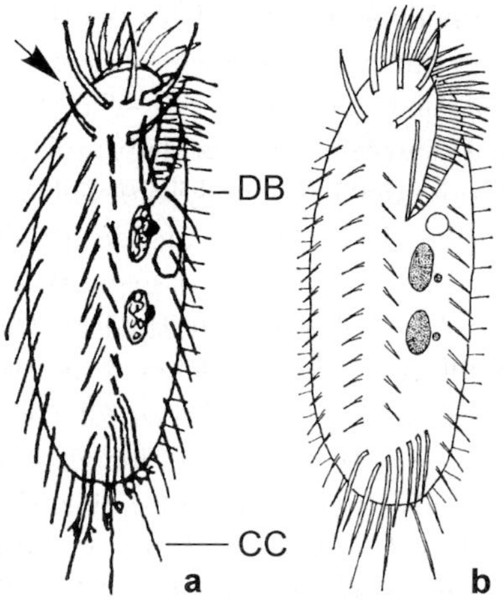

Fig. 49a, b *Caudiholosticha algivora* (a, from Kahl 1932; b, modified after Kahl 1932 from Bick 1972. From life). Ventral view, 75 µm. Arrow marks cirrus III/2 behind right frontal cirrus. CC = caudal cirri, DB = dorsal bristle. Page 270.

Remarks: The present species lacks most (all?) apomorphies of *Holosticha* and was thus transferred to *Caudiholosticha* (Berger 2003). For a detailed discussion and foundation of the transfer of many species from *Holosticha* to *Caudiholosticha*, see the genus section.

The original description of *C. algivora* is rather short, but together with the illustration it should be possible to identify it (Fig. 49a). Kahl (1932) himself was uncertain about the validity of *C. viridis* (Fig. 50a), *C. algivora* (Fig. 49a), and *Anteholosticha brevis* (Fig. 69a). Likely for that reason Borror (1972, p. 11) and Hemberger (1982, p. 92) put the present species into the synonym of *C. viridis*. Indeed, these two species have a very similar habitus according to the original descriptions. However, *Caudiholosticha viridis* has symbiotic algae, whereas *C. algivora* was obviously green due to ingested euglenids. In addition, *C. algivora* has cortical granules, whereas such organelles are not described for *C. viridis*. Because of these differences I consider them as distinct species, both of which, however, have to be redescribed in detail. Redescription should include neotypification because no type material is available, the type locality is not known, and the validity was doubted.

Buck (1961) provided a very tiny illustration which lacks most details. I am thus uncertain about the identification and classify it as insufficient redescription (Fig. 99a). The same is true of Chardez's (1981) illustrated record (Fig. 99b). Bick (1972) mentioned *C. algivora* in his guide to indicator species without, however, giving details. His illustration is a "modified redrawing" from Kahl (1932) and thus difficult to interpret; that is, we do not know whether or not he made his own observations (Fig. 49b).

Borror & Wicklow (1983, p. 121) synonymised both *Caudiholosticha algivora* and *C. viridis* with *Holosticha gibba*, however, without explanation.

Morphology: The following description is based solely on Kahl's text and illustration (Fig. 49a). Body length about 75 µm, body length:width ratio about 3:1. Body outline slender oval. Two ellipsoidal macronuclear nodules in mid-body left of midline, each with a micronucleus. Contractile vacuole in ordinary position, that is, slightly ahead of mid-body near left cell margin. Cortical granules colourless and arranged in rows (size and shape not mentioned). Adoral zone of membranelles almost 33% of body length, buccal field likely of ordinary structure. Five conspicuously long frontal cirri (Kahl 1932), that is, three frontal cirri, one cirrus (III/2) behind right frontal cirrus, and one buccal cirrus. Midventral complex obviously composed of cirral pairs only (about 8 in specimen illustrated), extends from near frontal cirri to near transverse cirri. Seven transverse cirri, fringed at rear end, long and thus distinctly projecting beyond rear body end. Marginal rows obviously not overlapping at rear body end, gap between rows, however, occupied by caudal cirri. Dorsal cilia almost 5 µm long. Three distinctly elongated caudal cirri (Fig. 49a).

Occurrence and ecology: Limnetic. Kahl (1932) did not mention the sample site (I suppose that he found it somewhere in northern Germany). Kahl observed it "not rarely" in a freshwater habitat on detritus covered with algae. Bick (1972) likely did not know this species from his own experience.

Records not substantiated by morphological data: pond near Ceska Lipa, northern Bohemia (Hassdenteufelová-Moravcová 1955, p. 215; body size 74–84 × 24–29 µm); benthic in the dammed river Elbe upstream from the city of Hamburg, Germany (Grimm 1968, p. 365); sometimes abundant in ponds in Greece (Stephanides 1948, p. 146, average body length 75 µm); freshwater in Hengshui, China (Han & Hao 1995, p. 4); river in China (Shen & Jiang 1979, p. 163). *Caudiholosticha algivora* feeds on green algae, mainly on euglenids (Kahl 1932).

Caudiholosticha viridis (Kahl, 1932) Berger, 2003
(Fig. 50a–f)

1932 ***Holosticha viridis* spec. n.** – Kahl, Tierwelt Dtl., 25: 586, Fig. 110 $_6$ (Fig. 50a; original description; no type material available and no formal diagnosis provided).

1972 ***Holosticha viridis* Kahl, 1932** – Borror, J. Protozool., 19: 11 (revision of hypotrichs).

1983 ***Holosticha viridis* Kahl, 1932** – Shen, Protozoa, p. 204, Plate XLII, Fig. 359 (Fig. 50f; illustrated record).

2001 ***Holosticha (Holosticha) viridis* Kahl, 1932** – Berger, Catalogue of ciliate names 1. Hypotrichs, p. 35 (nomenclator containing all basionyms, combinations, and higher taxa of hypotrichs).

2003 ***Caudiholosticha viridis* (Kahl, 1932) n. comb.** – Berger, Europ. J. Protistol., 39: 378 (combination with *Caudiholosticha*).

Nomenclature: No derivation of the name is given in the original description. The species-group name *viridis -is -e* (Latin; green, emerald-green) obviously refers to the presence of zoochlorellae. Kahl (1932) divided *Holosticha* into several subgenera. Thus, the correct, complete name in the original description is *Holosticha (Holosticha) viridis* Kahl, 1932.

Remarks: This species lacks most (all?) apomorphies of *Holosticha* and has distinct caudal cirri. Thus, it was transferred from *Holosticha* to *Caudiholosticha* by Berger (2003). For a detailed foundation of the transfer of many species from *Holosticha* to *Caudiholosticha*, see the genus section.

Borror (1972) and Hemberger (1982, p. 92) considered the present species as valid, with *H. algivora* Kahl, 1932 as synonym (for discussion, see *C. algivora*). Later, Borror & Wicklow (1983, p. 121) put *C. viridis* – however, without foundation and together with five other species – into the synonymy of *Holosticha gibba*, type of *Holosticha*. The paper by Shen (1983) is in Chinese; thus, only the illustration, which is possibly a modified redrawing of Fig. 50a, is presented (Fig. 50f). The redescription of a marine population by Aladro-Lubel et al. (1986; Fig. 99f, g) is too inspecific to accept the iden-

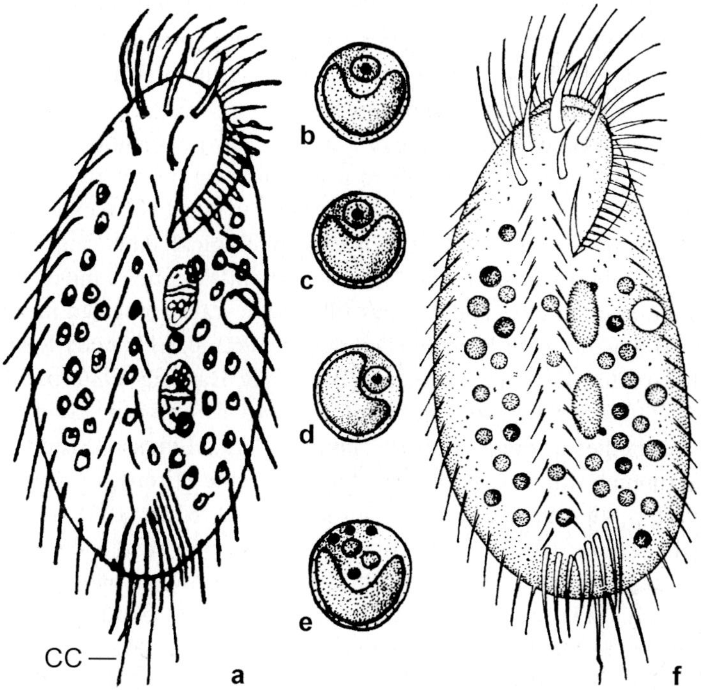

Fig. 50a–f *Caudiholosticha viridis* (a, from Kahl 1932; b–e, from Sud 1969; f, from Shen 1983. From life). **a, f:** Ventral views, a = 100 µm, f = size not indicated. **b–e:** *Chlorella miniata* (typical variety) from *Caudiholosticha viridis*. (b–d) show resting cells, (e) is a cell with divided nucleus. CC = caudal cirri. Page 272.

tification; it is thus classified as insufficient. Detailed redescription, including neotypification, of *Caudiholosticha viridis* needed.

Morphology: The description is based solely on the original data by Kahl (1932). Body length 100–110 µm; length:width ratio about 2:1 (Kahl), specimen illustrated about 2.3:1 (Fig. 50a). Body outline broad elliptical to oval. Two ellipsoidal macronuclear nodules, each with a large micronucleus. Contractile vacuole slightly ahead of mid-body near left cell margin. Cortical granules neither mentioned nor illustrated, indicating that they are lacking. Symbiotic algae (zoochlorellea) present (see below). Adoral zone occupies about 33% of body length (Fig. 50a). Three enlarged frontal cirri. Buccal cirrus near anterior end of undulating membranes. One cirrus behind right frontal cirrus; according to Kahl 5–6 frontal cirri present, indicating that a further, enlarged cirrus is sometimes present. Midventral complex obviously distinct, extends from near anterior end to transverse cirri. Eight transverse cirri in rather oblique row, at least rightmost ones project distinctly beyond rear body end (Fig. 50a). Marginal cirri distinctly protruding beyond rear body margin. Dorsal bristles neither mentioned nor illustrated, indicating that they are of ordinary (about 3 µm) length. Three long caudal cirri (Fig. 50a; according to Kahl these cirri are part of the left marginal row).

Sud (1969, p. 435) described the symbiotic green algae in detail. Accordingly, *Caudiholosticha viridis* harbours *Chlorella miniata* Oltmanns (Fig. 50b–e). Algae distributed fairly evenly throughout the cell; spherical with firm cell wall; pyrenoid absent; chloroplast parietal and cup-shaped, fills much of algae cell and has hollow space in middle where the nucleus lies, does not reveal reticulate structure; in some cells the chloroplast appears to be a thick, massive structure with a little lumen inside (Fig. 50c, d); during division, chloroplast begins to shrink and the nucleus divides into two (Fig. 50e), resulting in two aplanospores.

Occurrence and ecology: Limnetic. Type locality not mentioned; likely Kahl (1932) discovered it somewhere near Hamburg, Germany, where he lived and worked. He found *C. viridis*, sometimes abundant, on detritus covered with algae.

Records not substantiated by morphological data (marine and terrestrial records questionable): mossy ponds near Clermont-Ferrand, France (Grolière 1978, p. 300); common in hollows and furrows filled with water, irrigated meadows, and soil infusions at pH 6.6–7.4 from Bavaria (Dingfelder 1962, p. 617; body length 91–110 µm); benthic in Hamburg Harbour, Germany (Bartsch & Hartwig 1984, p. 556); during April (at pH 6.7–7.9; 5.1–7.5 °C, 6.4–8.5 mg l^{-1} O_2, 340–387 µS cm^{-1} conductivity) in small, alkaline water bodies of the Hortobágy National Park, Hungary (Szabó 1999a, p. 229; 2000a, p. 8); running waters (e.g., River Po and Torrente Stirone) in Northern Italy (Madoni 1979; 1980, p. 49; 1983, p. 89; Madoni & Ghetti 1981, p. 147); soil (macchia) in Italy (Luzzatti 1938, p. 101; body length only 65 µm); Slovakia (Matis et al. 1996, p. 12); benthic in lakes in Azerbaijan (Aliev 1982, p. 87); freshwater habitats of Tibetan Plateau (Wang 1977, p. 145); various freshwater habitats (lakes, ponds, rivers) in China (Han & Hao 1995, p. 4; Ma 1994, p. 95; Ning et al. 1993, p. 2; Shen & Jiang 1979, p. 163; Song et al. 1993, p. 101); fertilised, nitrogen rich, spring-fed, artificial farm pond (Ferrier's Pond) in Mountain Lake Region, Giles County, Virginia, USA (Bovee 1960, p. 357); limnetic habitats near Madison, Wisconsin, USA (Sud 1969).

Caudiholosticha navicularum (Kahl, 1932) Berger, 2003 (Fig. 51a, b)

1932 ***Holosticha navicularum* spec. n.** – Kahl, Tierwelt Dtl., 25: 586, Fig. 110 $_{14}$ (Fig. 51a; original description; no type material available and no formal diagnosis provided).

1968 ***Holosticha navicularum* Kahl, 1932** – Chorik, Free-living ciliates, p. 130, Fig. 120 (Fig. 51b; redescription).

1972 ***Holosticha navicularum* Kahl, 1932** – Borror, J. Protozool., 19: 11 (revision of hypotrichs).

1982 ***Holosticha navicularum* Kahl, 1932** – Hemberger, Dissertation, p. 91 (revision of non-euplotid hypotrichs).

1983 ***Holosticha navicularum* Kahl, 1932** – Borror & Wicklow, Acta Protozool., 22: 122 (revision of urostylids).

2001 ***Holosticha (Holosticha) navicularum* Kahl, 1932** – Berger, Catalogue of ciliate names 1. Hypotrichs, p. 35 (nomenclator containing all basionyms, combinations, and higher taxa of hypotrichs).

2003 ***Caudiholosticha navicularum* (Kahl, 1932) n. comb.** – Berger, Europ. J. Protistol., 39: 378 (combination with *Caudiholosticha*).

Nomenclature: No derivation of the name is given in the original description. The species-group name *navicularum* (ingesting diatoms, obviously *Navicula*) alludes to the preferred food, namely diatoms. Kahl (1932) divided *Holosticha* into several subgenera. Thus, the correct name in his paper is *Holosticha (Holosticha) navicularum*

Kahl, 1932. Incorrect subsequent spellings: *Holosticha naviculatum* Kahl (Berecky & Nosek 1995, p. 126); *Holosticha naviculorum* (Gellért & Tamás 1958, p. 234).

Remarks: The present species lacks most (all?) apomorphies of *Holosticha* and was therefore transferred to *Caudiholosticha* (Berger 2003). For a detailed discussion and foundation of the transfer of many species from *Holosticha* to *Caudiholosticha*, see the genus section. The original illustration shows three slightly elongated cirri at the posterior body end (Fig. 51a). According to Kahl (1932) these are posteriorly protruding marginal cirri; in contrast, I suppose that these are caudal cirri and thus preliminarily classify the present species in *Caudiholosticha*. However, only a detailed redescription of this obviously very easily identified species will show if my assumption and thus classification is correct. The validity of *Caudiholosticha navicularum* was never doubted, not even by the lumper Borror.

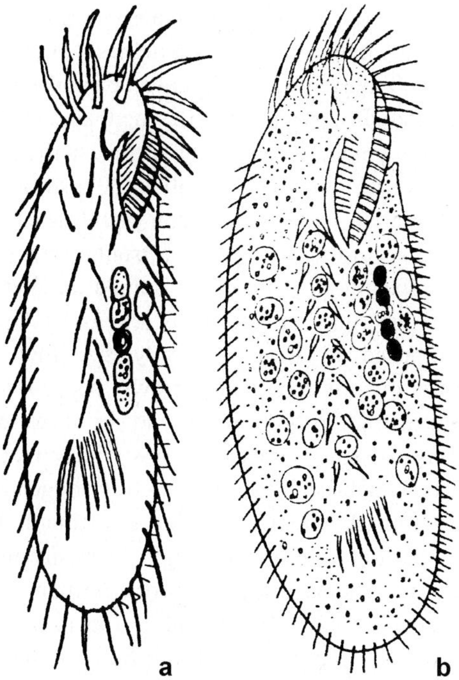

Fig. 51a, b *Caudiholosticha navicularum* (a, from Kahl 1932; b, from Chorik 1968. From life). **a:** Ventral view, 200 μm. Note the conspicuous nuclear apparatus (two macronuclear nodules with single micronucleus in between) and the far anteriorly displaced transverse cirri. The three rearmost "marginal" cirri are slightly elongated, strongly indicating that they are caudal cirri; however, this has to be confirmed by detailed redescription. **b:** Ventral view of a Moldavian specimen, 188 μm. Page 274.

Especially the nuclear apparatus (two macronuclear nodules with a single micronucleus in between) and the far anteriorly displaced transverse cirri separate this species from other *Caudiholosticha* and *Anteholosticha* species. *Territricha stramenticola* Berger & Foissner, 1988 has a similar habitus. However, this species lives in beech litter, has a different nuclear apparatus (each of the two macronuclear nodules have one or more micronuclei attached), the transverse cirri terminate at the posterior body end, and the number of cirral pairs is slightly lower. Further, this species has large extrusomes, which are, however, not very distinct. Since dorsal kinety 3 shows fragmentation it is classified in the oxytrichids (for review, see Berger 1999, p. 884).

The redescription by Chorik (1968) is in Russian. The illustration basically agrees with Kahl's data except for the nuclear apparatus, which does not show the micronucleus. In spite of this, the identification is accepted (Fig. 51b).

Morphology: The following description is solely based on the original description. Body length around 200 μm, length:width ratio of specimen illustrated 4.1:1 (Fig. 51a). Body outline long elliptical, posteriorly slightly converging. Body very flat and thus translucent. Two elongate macronuclear nodules in mid-body left of midline; in between a single, rather large micronucleus. Contractile vacuole in ordinary position, that is, near left cell margin distinctly ahead of mid-body, that is, at 42% of body length in specimen illustrated. Cortical granules neither mentioned nor illustrated, strongly indicating that such organelles are lacking. Cells usually packed with large diatoms.

Adoral zone occupies about 30% of body length in specimen illustrated. Buccal field moderately wide. Undulating membranes likely slightly curved. Three enlarged frontal cirri; according to Kahl 4–5 frontal cirri (likely this number includes the buccal cirrus near the anterior end of the undulating membranes and cirrus III/2, that is, the cirrus behind the right frontal cirrus; cirrus III/2 not shown in Fig. 51a). Midventral complex composed of cirral pairs only, specimen illustrated with six, rather loosely arranged pairs; cirri of each pair narrowly spaced, long, and thin. 6–8 transverse cirri inserted at end of middle body third, end far ahead of (about 1/6 of body length) rear body end. Marginal rows protruding posteriorly (Kahl 1932); I suppose that the rearmost three cirri, which he illustrated slightly longer than the other marginal cirri, are caudal cirri (see remarks). Dorsal bristles according to Fig. 51a about 4 μm long.

Occurrence and ecology: In freshwater and slightly saline waters. Type locality is the Bad Oldesloe area in north Germany near the city of Hamburg, where Kahl (1932) discovered it in slightly saline (up to 3%) waters. It was very common on dead leaf-litter after the winter-rotting; he classified *C. navicularum* as mesosaprobic to almost katharobic (see also Mauch 1976, p. 422). Chorik (1968) found it in Moldavian waters.

Records not substantiated by morphological data: eastern bank of peninsula Tihany, Lake Balaton, Hungary (Gellért & Tamás 1958, p. 234); Szigetkösz (Hungary) side-arm system of the Danube river (Bereczky & Nosek 1995, p. 126); freshwater lagoon (Little Kysylagach Bay) of the Caspian Sea (Agamaliev 1986, p. 207); cooling plant of a Moldavian power station (Chorik & Vikol 1973, p. 69); dominant in the Terterchay reservoir and other Azerbaidzhanian reservoirs (Alekperov 1981, p. 57; 1982, p. 45; 1982a, p. 87; 1983, p. 22). The record from the leaf litter of a Danish beech forest by Brunberg-Nielsen (1968, p. 85) is very likely a misidentification; I suppose that she observed *Territricha stramenticola* (see remarks).

Caudiholosticha navicularum feeds on diatoms (Kahl 1932), likely by naviculaceans as indicated by the species-group name.

Caudiholosticha multicaudicirrus (Song & Wilbert, 1989) Berger, 2003 (Fig. 52a–e, Table 17)

1989 ***Holosticha multicaudicirrus* nov. spec.**[1] – Song & Wilbert, Lauterbornia, 3: 160, Abb. 89a–e, Tabelle 30 (Fig. 52a–e; original description; type slides are deposited in the College of Fisheries, Ocean University of Qingdao, China).

2001 ***Holosticha multicaudicirrus* Song and Wilbert, 1989** – Berger, Catalogue of ciliate names 1. Hypotrichs, p. 37 (nomenclator containing all basionyms, combinations, and higher taxa of hypotrichs).

---

[1] The diagnosis by Song & Wilbert (1989 is as follows: In vivo 80–120 × 20–40 μm große, farblose bis leicht gelbliche *Holosticha* mit ca. 30 Makronucleus-Teilen; 5–7 Caudalcirren, 7 Midventralcirrenpaaren und 6 Dorsalkineten; rechte Marginalreihe auffallend vor den Transversalcirren verkürzt.

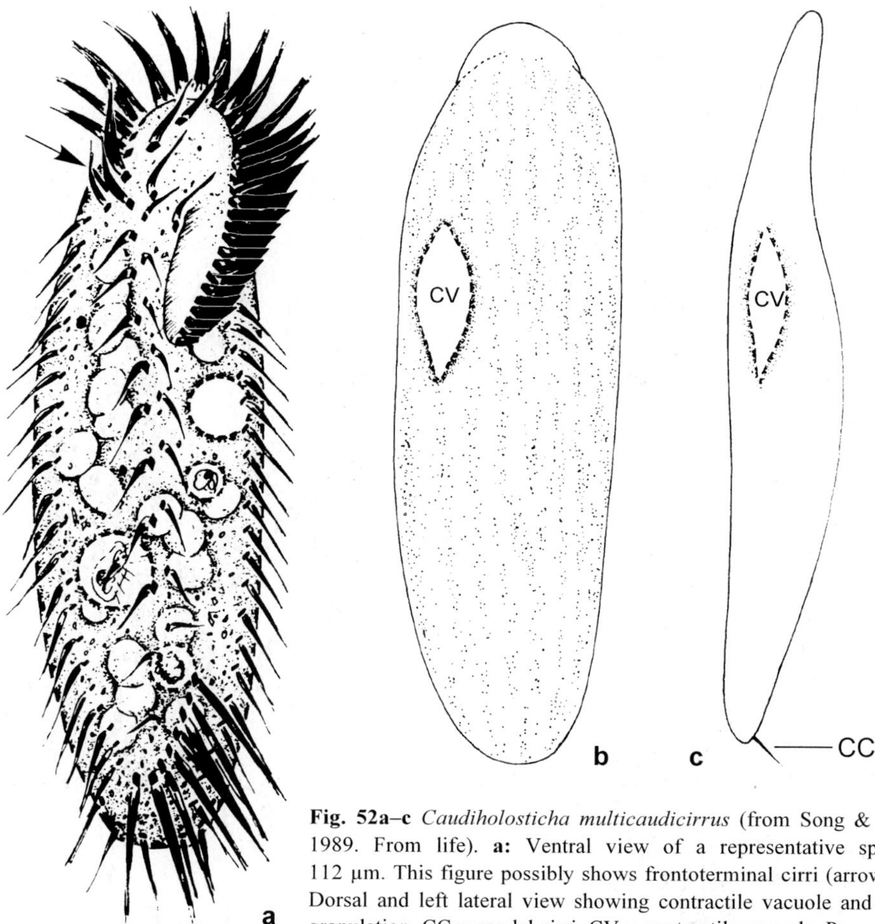

Fig. 52a–c *Caudiholosticha multicaudicirrus* (from Song & Wilbert 1989. From life). **a:** Ventral view of a representative specimen, 112 µm. This figure possibly shows frontoterminal cirri (arrow). **b, c:** Dorsal and left lateral view showing contractile vacuole and cortical granulation. CC = caudal cirri, CV = contractile vacuole. Page 276.

2003 ***Caudiholosticha multicaudicirrus* (Song and Wilbert, 1989) n. comb.** – Berger, Europ. J. Protistol., 39: 378 (combination with *Caudiholosticha*).

Nomenclature: No derivation of the name is given in the original description. The species-group name *multicaudicirrus* (having many caudal cirri) is a composite of the Latin indefinite numeral *multus -a -um* (many, numerous), the Latin noun *cauda* (tail), and the Latin noun *cirrus* (bundle, cirrus) and refers to the many (around 6) caudal cirri.

Remarks: The more or less equal number of transverse cirri and midventral pairs indicates that each frontal-midventral-transverse anlage forms – as in *Holosticha* and *Pseudoamphisiella* – a transverse cirrus. However, the present species lacks most other apomorphies of *Holosticha* (e.g., anterior end of left marginal cirral row not curved rightwards) and has distinct caudal cirri. Thus it was transferred from *Holosticha* to *Caudiholosticha* (Berger 2003).

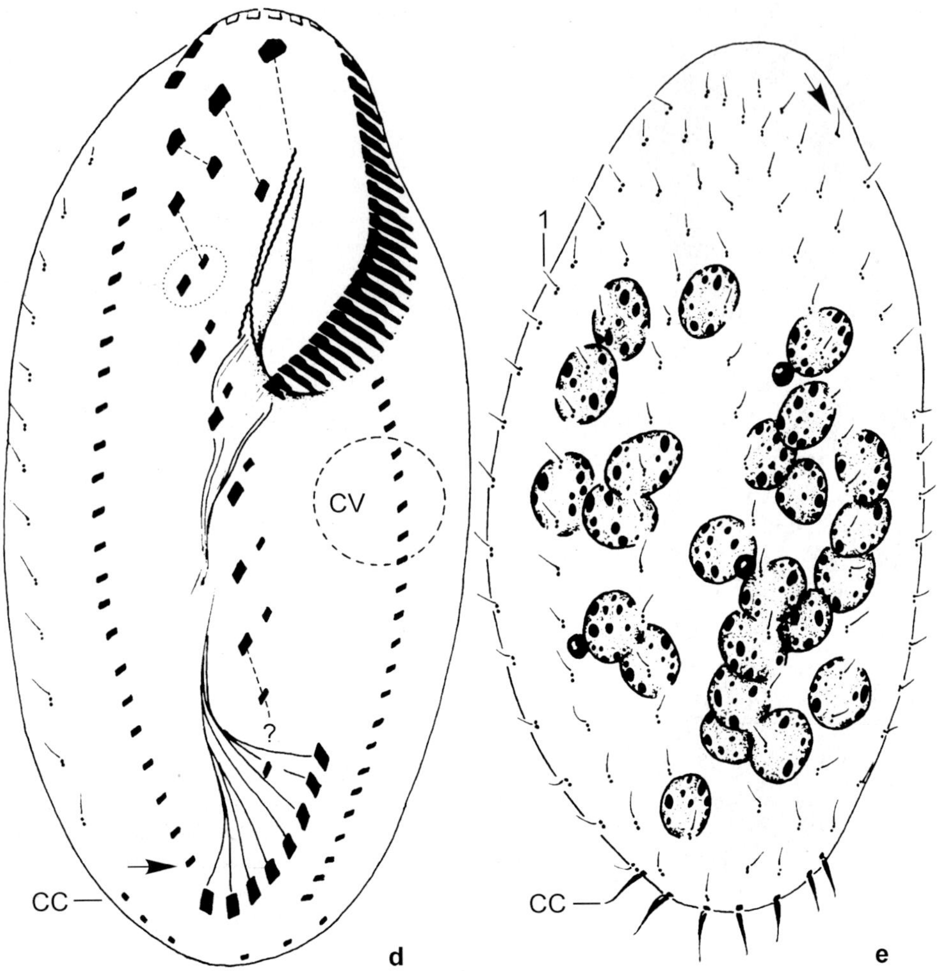

Fig. 52d, e *Caudiholosticha multicaudicirrus* (from Song & Wilbert 1989. Protargol impregnation). Infraciliature of ventral and dorsal side and nuclear apparatus of two specimens, d = 101 µm. Broken lines connect cirri (anlagen I–IV and X [forming last midventral pair]) which very likely originate from the same anlage. Dotted line encircles anteriormost pseudopair. The right marginal row terminates ahead of the right transverse cirrus (arrow in d). Arrow in (e) marks short (seventh) dorsal kinety. CC = caudal cirri, CV = contractile vacuole, 1 = dorsal kinety 1. Page 276.

I do not know whether or not *C. multicaudicirrus* has frontoterminal cirri because in Fig. 52a (arrow) two inconspicuous cirri ahead of the anterior end of the right marginal row are illustrated; by contrast, no such cirri are shown in Fig. 52d and they are neither mentioned in the text nor in the table.

Morphology: Body size 80–120 × 20–40 µm, body length:width ratio 2.3:1 in protargol preparations (Table 17). Body outline elliptical, both ends broadly rounded (Fig.

52a–c). Body flexible. Around 30 ellipsoidal macronuclear nodules scattered in central portion of cell, easily recognisable in life. Contractile vacuole in ordinary position, that is, near left cell margin about in mid-body. Pellicle fragile, very finely granulated (obviously this species has very small cortical granules roughly arranged along dorsal kineties; Fig. 52b; however, size, shape, colour, and arrangement of granules not described). Cells colourless to slightly yellowish (due to cortical granules?). Rear cell portion often with many dark granules. Movement moderately rapid.

Adoral zone occupies 39% of body length on average (Table 17), composed of about 24 membranelles (obviously of ordinary fine structure). Buccal field moderately wide. Paroral slightly curved, endoral more or less straight and intersecting paroral optically in rear portion (Fig. 52a, d).

Cirral pattern and number of cirri of usual variability (Fig. 52d; Table 17). Three enlarged frontal cirri obliquely arranged. Buccal cirrus near anterior end of paroral. Frontoterminal cirri neither mentioned in text and table nor shown in Fig. 52d, but possibly illustrated in Fig. 52a (arrow; see remarks). Cirrus III/2 left behind right frontal cirrus. Midventral complex composed of about seven cirral pairs only, extends to near level of anteriormost transverse cirri; cirri of individual pairs rather widely spaced thus forming distinct pseudopairs; right cirrus of each pair distinctly larger than left. 7–10 enlarged transverse cirri forming roughly J-shaped pattern, project distinctly beyond rear body end. Right marginal row commences about at level of buccal cirrus, terminates slightly ahead of right transverse cirrus. Left marginal row begins left of proximal portion of adoral zone, ends subterminally.

Dorsal cilia arranged in six more or less bipolar kineties; rarely a seventh very short (usually two bristles) kinety right of anterior end of kinety 6. Length of dorsal cilia not mentioned, according to Fig. 52e about 2–3 μm long. At end of each kinety a single, about 10 μm long, relatively weak caudal cirrus (Fig. 52a, d, e).

Occurrence and ecology: Limnetic. Type locality is the Poppelsdorfer Weiher, a eutrophic pond in the botanical gardens in Bonn, Germany, where it occurred moderately abundantly during June (see also Song & Chen 1999, p. 344). No further records published. Food not described, possibly small ciliates (Fig. 52a).

Caudiholosticha interrupta (Dragesco, 1966) Berger, 2003 (Fig. 53a)

1966 ***Holosticha interrupta*** **n. sp.** – Dragesco, Protistologica, 2: 87, Fig. 26A (Fig. 53a; original description; site where type material deposited not mentioned, likely in Dragesco's private collection; no formal diagnosis provided).

1982 ***Holosticha interrupta*** **Dragesco, 1966** – Hemberger, Dissertation, p. 93 (revision of hypotrichs).

2001 ***Holosticha interrupta*** **Dragesco, 1966** – Berger, Catalogue of ciliate names 1. Hypotrichs, p. 36 (nomenclator containing all basionyms, combinations, and higher taxa of hypotrichs).

2003 ***Caudiholosticha interrupta*** **(Dragesco, 1966) n. comb.** – Berger, Europ. J. Protistol., 39: 377 (combination with *Caudiholosticha*).

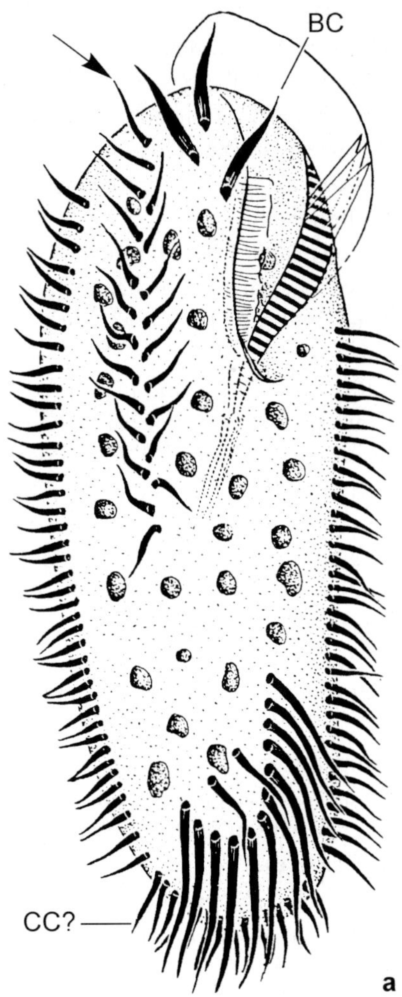

Fig. 53a *Caudiholosticha interrupta* (from Dragesco 1966. Protargol impregnation). Infraciliature of ventral side and nuclear apparatus, 195 µm. Arrow marks a cirrus which is possibly the third (right) frontal cirrus. BC = buccal cirrus?, CC? = caudal cirri? Page 279.

Nomenclature: No derivation of the name is given in the original description. The species-group name *interrupt·us -a -um* (Latin adjective; interrupted) obviously alludes to the short ("interrupted") midventral complex.

Remarks: The present species lacks most (all?) apomorphies of *Holosticha* and was therefore transferred to *Caudiholosticha* by Berger (2003). For a detailed discussion and foundation of the transfer of many species from *Holosticha* to *Caudiholosticha*, see the genus section.

The present species is only described after protargol preparations. Thus detailed redescription is needed. I agree with Hemberger (1982) that *C. interrupta* is a valid species because it shows a rather unique combination of features, namely, body length is about 200 µm, the midventral complex is rather short, the number of transverse cirri is rather high, and between the rear end of the marginal rows is a conspicuous row of cirri. Hemberger supposed that this conspicuous cirral row is part of a marginal row. However, he also correctly stated that the origin (from marginal row or dorsal kineties) can be recognised only after investigation of the ontogenesis.

Borror (1972) obviously overlooked this species, and Borror & Wicklow (1983, p. 122) synonymised it with *Anteholosticha scutellum*. However, this species is marine (vs. limnetic), smaller (around 100 µm vs. about 200 µm), and has less transverse cirri (7–8 vs. 11–12), strongly indicating that synonymy of *A. scutellum* and *C. interrupta* is incorrect. *Anteholosticha mancoidea* is smaller (about 120 µm long), lacks caudal cirri, and has less macronuclear nodules (usually 8 vs. about 30), transverse cirri (5 vs. 11–12), and dorsal kineties (3 vs. 6).

Morphology: Body length about 200 µm (in protargol preparations?). Body outline of prepared (?) specimens cylindrical with both ends rounded (data must not be overinterpreted). Consistency of body not mentioned, likely rather flexible like congeners. About 30 macronuclear nodules scattered throughout cell; individual nodules about

6.5 µm across. Number of micronuclei not known. Contractile vacuole neither mentioned nor illustrated. Presence/absence of cortical granules not known.

Adoral zone occupies about 36% of body length (value from Fig. 53a), composed of 30–31 membranelles. Buccal field likely of ordinary size, undulating membranes not described in detail (Fig. 53a). Three enlarged cirri in frontal area (I agree with Hemberger 1982 that their arrangement is not correctly observed); the rearmost cirrus is possibly a buccal cirrus (I suppose that this species has an ordinary frontal cirral pattern). Midventral complex consists of cirral pairs only, in total composed of 19–23 cirri (that is, about 9–11 pairs), terminates at 52% of body length in specimen illustrated. Two pretransverse ventral cirri. 11–12 slightly enlarged transverse cirri arranged in J-shaped pattern (the high number indicates that each frontal-midventral-transverse cirral anlage forms a transverse cirrus!). Right marginal row composed of 30–40 cirri, left row of 37–41 cirri; both rows end distinctly subterminally; between ends of marginal rows, a curved row of 14–16 cirri whose origin (from marginal row or from dorsal kineties) and thus correct designation (marginal cirri or caudal cirri) remains unknown. Six dorsal kineties; length of bristles not mentioned, but likely short (probably around 3 µm).

Occurrence and ecology: Limnetic. Type locality is a small pond near Thonon-les-Bains at the southern bank of lake Geneva, France. No further records known.

Caudiholosticha setifera (Kahl, 1932) Berger, 2003 (Fig. 54a, b)

1932 ***Holosticha setifera* spec. n.** – Kahl, Tierwelt Dtl., 25: 582, Fig. 106 $_9$ (Fig. 54a; original description; no type material available and no formal diagnosis provided).

1933 ***Holosticha setifera* Kahl 1932** – Kahl, Tierwelt N.- u. Ostsee, 23: 111, Fig. 17.36 (Fig. 54b; guide to marine ciliates).

2001 ***Holosticha (Holosticha) setifera* Kahl, 1932** – Berger, Catalogue of ciliate names 1. Hypotrichs, p. 35 (nomenclator containing all basionyms, combinations, and higher taxa of hypotrichs).

2003 ***Caudiholosticha setifera* (Kahl, 1932) n. comb.** – Berger, Europ. J. Protistol., 39: 378 (combination with *Caudiholosticha*).

Nomenclature: No derivation of the name is given in the original description. The species-group name *setifera* (bearing bristles) is a composite of the Latin noun *seta* (bristle, hair) and the Latin verb *fero, ferre* (to bear, to carry) and obviously refers to the three distinct caudal cirri. Kahl (1932) divided *Holosticha* into several subgenera. Thus, the correct name in the original description is *Holosticha (Holosticha) setifera* Kahl, 1932. Chardez (1987) incorrectly mentioned 1930 as year of publication.

Borror & Wicklow (1983, p. 118) excluded this species from the urostyloids and supposed that it belongs to *Gastrostyla*, however, without combining it formally. If this combination is accepted, the name of Kahl's species would be the junior secondary homonym of *Gastrostyla setifera* (Engelmann, 1862) Kent, 1882 (for review, see Berger 1999, p. 816).

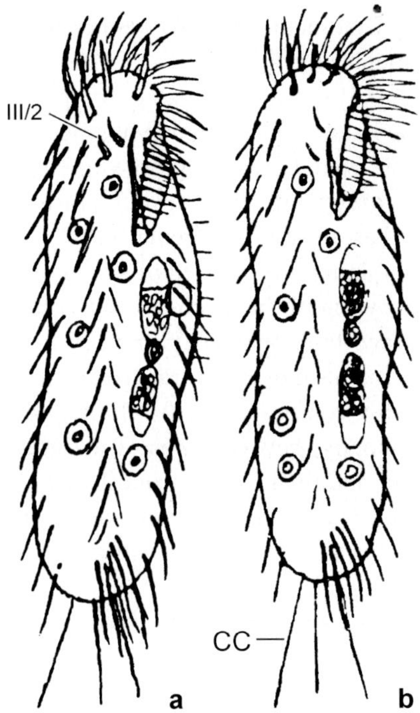

Fig. 54a, b *Caudiholosticha setifera* (a, from Kahl 1932; b, after [?] Kahl 1932 from Kahl 1933. From life). Ventral view, 140 µm. CC = caudal cirri, III/2 = cirrus behind right frontal cirrus? Page 281.

Remarks: This species lacks most apomorphies of *Holosticha* and has distinct caudal cirri. Thus, it was transferred from *Holosticha* to *Caudiholosticha* (Berger 2003). For a detailed discussion and foundation of the transfer of many species from *Holosticha* to *Caudiholosticha*, see the genus section.

Borror (1972) and Hemberger (1982) put the present species into the synonymy of *Holosticha obliqua* Kahl, 1928, which I classify as species indeterminata (for discussion, see p. 183). Borror & Wicklow (1983, p. 118) suspected that it belongs to *Gastrostyla* (for nomenclatural problems, see previous chapter). However, the ventral cirri of the present species are unequivocally arranged in a zigzag-pattern, strongly indicating that Kahl's original classification in *Holosticha* was correct. Further, it lacks a postoral ventral cirrus which is easily recognisable even in life and which is characteristic for *Gastrostyla* species (for review on the oxytrichid *Gastrostyla*, see Berger 1999, p. 789).

Kahl (1932) found two populations, namely one with ring-shaped structures (Fig. 54a; occurred in water of high salinity), and the other (not figured) without ring-shaped structures (lived in low salinity waters). Kahl was uncertain whether these populations are conspecific or belong to very closely related species. Thus, detailed redescription should consider this problem.

The illustration by Kahl (1933) deviates in some details from the figure of the original description (for example, lack of contractile vacuole and dorsal bristles; cirral pattern). I do not know whether this is a new illustration based on original data or if these deviations are due to inaccurate redrawing.

Caudiholosticha setifera has the same nuclear apparatus as *C. navicularum*, namely two macronuclear nodules with a single micronucleus in between (Fig. 51a, 54a). However, the species can be easily distinguished by the position of the transverse cirri, namely, near rear body end in *C. setifera* against at the end of second body third in *C. navicularum*. Consequently, synonymy can be excluded.

Morphology: Body length 120–150 µm, body length:width ratio of specimen illustrate 3.5:1 (Fig. 54a). Body outline elongate elliptical, left margin usually slightly vaulted at level of contractile vacuole. Two macronuclear nodules with single micronucleus in between, about in mid-body left of midline. Cortical granules neither mentioned nor illustrated, indicating that such organelles are lacking. Cytoplasm with several ring-

Table 17 Morphometric data on *Caudiholosticha gracilis* (gr1, gr2, type population from the Schloßalm area and population from the Tullnerfeld region from Foissner 1982), *Caudiholosticha islandica* (isl, from Berger & Foissner 1989a), *Caudiholosticha multicaudicirrus* (mul, from Song & Wilbert 1989), *Caudiholosticha notabilis* (no1, type population from Foissner 1982; no2, from Berger 1987; no3, from Blatterer & Foissner 1988), *Caudiholosticha paranotabilis* (pa1, pa2, type population and flood plain population from Foissner et al. 2002; pa3, from Foissner 1996a), *Caudiholosticha stueberi* (stu, from Foissner 1987e), *Caudiholosticha sylvatica* (sy1, from Foissner 1982; sy2, from Berger & Foissner 1989; sy3 to sy7, five populations from Borror & Wicklow 1983; sy8, from Shin & Kim 1993), *Caudiholosticha tetracirrata* (te1, from Buitkamp & Wilbert 1974; te2, population from Glockner-area from Foissner 1982; te3, population from Tullnerfeld from Foissner 1982; te4, from Blatterer & Foissner 1988)

Characteristics [a]	Species	mean	M	SD	SE	CV	Min	Max	n
Body, length	gr1	68.7	68.0	5.7	1.7	8.3	60.0	80.0	11
	gr2	78.0	78.0	9.4	3.0	12.0	66.0	93.0	10
	isl	74.5	75.0	6.0	1.7	8.1	66.0	86.0	13
	mul	73.6	–	10.8	3.1	14.7	61.0	95.0	12
	no1	163.3	–	–	–	–	150.0	180.0	3
	no2	93.6	94.0	11.2	3.4	11.9	74.0	110.0	11
	no3	158.8	153.0	21.8	6.6	13.7	136.0	198.0	11
	pa1	118.1	113.0	15.6	3.2	13.2	99.0	164.0	23
	pa2	128.0	128.0	10.8	3.8	8.5	113.0	146.0	8
	stu	225.2	230.0	31.2	–	13.9	178.0	270.0	12
	sy1	117.3	121.0	8.1	2.5	6.9	106.0	126.0	10
	sy2	127.0	126.0	9.0	2.7	7.1	112.0	140.0	11
	sy8	184.4	183.0	22.7	5.2	12.3	140.0	227.0	19
	te2	79.4	78.0	4.8	1.6	6.0	73.0	92.0	9
	te3	79.5	79.0	10.1	3.2	12.6	60.0	95.0	10
	te4	102.8	102.5	11.8	3.4	11.5	79.0	120.0	12
Body, width	gr1	11.6	12.0	1.1	0.3	9.2	9.0	13.0	11
	gr2	10.2	10.0	1.5	0.5	14.4	8.0	13.0	10
	isl	20.5	21.0	2.7	0.7	13.0	16.0	25.0	13
	mul	31.6	–	5.4	1.7	17.2	27.0	45.0	10
	no1	17.3	–	–	–	–	13.0	20.0	3
	no2	16.5	15.0	3.9	1.2	23.3	13.0	25.0	11
	no3	36.5	38.0	6.9	2.1	18.8	23.0	46.0	11
	pa1	18.0	18.0	2.6	0.5	14.5	14.0	25.0	23
	pa2	19.0	18.5	1.6	0.6	8.4	17.0	22.0	8
	stu	71.9	72.5	11.6	–	16.1	55.0	95.0	12
	sy1	44.0	41.5	5.9	1.9	13.4	40.0	60.0	10
	sy2	52.0	49.0	8.3	2.5	16.0	43.0	70.0	11
	sy8	79.3	85.0	14.3	3.3	18.1	55.0	100.0	19
	te2	21.8	21.0	3.8	1.3	17.4	17.0	29.0	9
	te3	24.0	24.0	3.2	1.0	13.2	19.0	29.0	10
	te4	24.6	26.0	2.6	0.7	10.5	19.0	27.0	12
Body length:width, ratio	pa1	6.6	6.7	1.0	0.2	15.4	4.7	9.1	23
	pa2	6.8	6.8	0.8	0.3	11.6	5.6	8.1	8
	sy8	2.4	2.4	0.3	0.1	10.8	2.0	2.8	19
Distance 1 [d]	gr1	21.8	21.0	2.4	0.7	11.2	19.0	27	11
	gr2	15.5	14.0	3.3	1.0	21.4	11.0	21.0	10
	no1	74.0	–	–	–	–	67.0	81.0	2
	no2	32.5	32.0	4.3	1.3	13.3	25.0	39.0	11
	no3	80.2	83.0	9.3	2.8	11.6	67.0	95.0	11
	pa1	39.6	38.0	5.6	1.2	14.0	33.0	50.0	22
	pa2	48.0	47.0	4.2	1.5	8.8	43.0	56.0	8

Table 17 Continued

Characteristics [a]	Species	mean	M	SD	SE	CV	Min	Max	n
Distance 1 [d]	stu	208.3	213.5	31.5	–	15.1	165.0	252.0	12
	sy1	67.1	66.5	10.3	3.3	15.3	53.0	90.0	10
	sy2	88.9	88.0	5.3	1.6	6.0	81.0	98.0	11
	te2	53.0	53.0	5.3	1.8	9.9	45.0	63.0	9
	te3	46.6	46.5	4.9	1.6	10.5	40.0	57.0	10
	te4	67.9	69.5	10.4	3.0	15.3	51.0	86.0	12
Distance 2 [d]	no1	65.5	–	–	–	–	57.0	74.0	2
	stu	188.8	194.5	30.3	–	16.0	145.0	234.0	12
Distance between end of midventral complex and transverse cirri	te4	26.2	27.0	10.6	3.1	40.6	6.0	45.0	12
Adoral zone of membranelles, length	gr1	15.1	15.0	0.8	0.2	5.3	13.0	16.0	11
	gr2	14.8	14.5	2.0	0.6	14.0	12.0	19.0	10
	isl	20.6	21.0	0.7	0.2	3.2	19.0	21.0	13
	mul	28.9	–	1.9	0.6	6.6	27.0	33.0	12
	no1	32.7	–	–	–	–	32.0	33.0	3
	no2	21.2	21.0	2.3	0.7	10.9	18.0	27.0	11
	no3	40.4	41.0	4.1	1.2	10.1	35.0	47.0	11
	pa1	25.7	26.0	1.9	0.4	7.4	20.0	29.0	23
	pa2	30.0	30.5	2.2	0.8	7.3	26.0	32.0	8
	stu	82.7	84.0	12.1	–	14.7	60.0	98.0	12
	sy1	35.6	35.0	2.6	0.8	7.3	32.0	40.0	10
	sy2	44.5	43.0	2.9	0.9	6.6	40.0	49.0	11
	sy8	65.9	61.0	9.8	2.2	14.8	52.0	86.0	19
	te2	23.2	24.0	1.9	0.6	8.1	20.0	25.0	9
	te3	24.0	24.0	2.3	0.7	9.7	21.0	28.0	10
	te4	31.4	31.5	3.1	0.9	9.8	27.0	38.0	12
Body length:length adoral zone, ratio	pa1	4.6	4.5	0.5	0.1	11.1	3.8	5.9	23
	pa2	4.3	4.2	0.5	0.2	11.6	3.5	5.0	8
	sy8	2.8	2.8	0.2	0.0	6.4	2.5	3.2	19
Body length:length of midventral row, ratio	pa1	3.0	3.1	0.5	0.1	16.6	2.2	3.9	22
	pa2	2.7	2.7	0.4	0.1	13.4	2.0	3.2	8
Anterior body end to buccal cirrus, distance	pa1	12.7	13.0	1.5	0.3	11.5	10.0	16.0	23
	pa2	13.8	14.0	1.8	0.6	12.7	10.0	16.0	8
Anterior body end to paroral, distance	pa1	9.9	10.0	1.0	0.2	10.3	8.0	13.0	23
	pa2	11.6	12.0	1.6	0.6	13.7	8.0	13.0	8
Anterior body end to endoral, distance	pa1	13.2	13.0	1.4	0.3	10.5	11.0	16.0	20
	pa2	14.6	15.0	1.7	0.6	11.8	11.0	16.0	7
Adoral membranelles, number	gr1	16.8	17.0	0.7	0.2	4.3	16.0	18.0	11
	gr2	15.8	15.0	2.1	0.7	13.5	13.0	19.0	10
	isl	17.5	17.0	0.8	0.2	4.4	16.0	19.0	13
	mul	24.0	–	1.7	0.5	6.9	21.0	26.0	12
	no1	29.7	–	–	–	–	28.0	35.0	3
	no2	16.7	17.0	0.7	0.2	3.9	16.0	18.0	11
	no3	31.6	31.0	3.2	1.0	10.0	26.0	36.0	11
	pa1	21.9	22.0	1.5	0.3	6.9	19.0	25.0	23
	pa2	23.9	24.5	2.2	0.8	9.1	19.0	26.0	8
	stu	45.4	44.0	3.4	–	7.4	43.0	54.0	12
	sy1	34.8	34.5	1.5	0.5	4.4	33.0	38.0	10
	sy2	44.1	44.0	3.7	1.1	8.3	39.0	49.0	11
	sy8	46.7	46.0	4.8	1.1	10.4	38.0	55.0	19

Table 17 Continued

Characteristics[a]	Species	mean	M	SD	SE	CV	Min	Max	n
Adoral membranelles, number	te1	24.0	–	–	–	–	–	–	?
	te2	23.6	23.0	0.7	0.2	2.9	23.0	25.0	9
	te3	23.6	23.5	0.7	0.2	2.8	23.0	25.0	10
	te4	26.1	26.5	3.4	1.0	13.0	20.0	31.0	12
Undulating membrane (paroral?), length	sy8	40.5	40.0	4.8	1.1	11.9	32.0	49.0	19
Paroral, length	pa1	8.7	8.0	–	–	–	8.0	10.0	18
	pa2	9.6	10.0	–	–	–	9.0	10.0	5
Endoral, length	pa1	8.8	8.5	1.0	0.2	11.8	7.0	10.0	18
	pa2	9.0	10.0	1.7	1.0	19.2	7.0	10.0	3
Anterior body end to first fronto-terminal cirrus, distance	pa1	7.0	7.0	1.0	0.2	14.9	6.0	9.0	23
	pa2	8.3	9.0	1.4	0.5	16.8	6.0	10.0	8
Anterior body end to second fronto-terminal cirrus, distance	pa1	10.7	10.5	1.1	0.2	10.0	9.0	13.0	22
	pa2	12.5	12.5	1.2	0.4	9.6	11.0	14.0	8
Anterior body end to right marginal row, distance	pa1	12.2	13.0	2.5	0.6	20.8	6.0	17.0	21
	pa2	14.1	14.0	1.6	0.6	11.6	11.0	16.0	8
Posterior body end to posteriormost transverse cirrus, distance	pa1	1.2	1.0	–	–	–	0.0	3.0	18
	pa2	1.1	1.0	–	–	–	1.0	2.0	8
Anterior body end to first macronuclear nodule, distance	pa1	20.4	21.0	3.6	0.7	17.6	12.0	28.0	23
	pa2	27.5	28.5	3.1	1.1	11.2	23.0	31.0	8
Length paroral:length adoral zone, ratio	sy8	0.6	0.6	0.0	0.0	4.6	0.6	0.7	19
Pharyngeal fibres, length	sy8	31.3	31.0	2.6	0.8	8.4	27.0	37.0	12
Macronuclear nodules, number	gr1	13.7	15.0	1.9	0.6	13.9	10.0	16.0	11
	gr2	15.5	15.0	3.3	1.0	21.2	12.0	22.0	10
	is1	16.0	16.0	0.8	0.2	5.1	14.0	17.0	13
	mu1[f]	30.6	–	0.9	0.3	3.0	29.0	32.0	14
	no1			about 70					3
	no2	30.6	31.0	3.1	0.9	10.2	25.0	35.0	11
	no3	55.5	55.0	6.9	2.1	12.5	47.0	69.0	11
	pa1	15.4	16.0	1.0	0.2	6.7	13.0	18.0	23
	pa2	15.8	16.0	–	–	–	15.0	16.0	8
	stu	2.0	2.0	–	–	–	2.0	3.0	12
	sy1	32.3	32.5	2.8	0.9	8.5	28.0	36.0	10
	sy2	56.0	56.0	4.1	1.2	7.3	50.0	61.0	11
	sy8	61.6	60.0	8.4	2.0	13.6	52.0	88.0	17
	te2	35.9	31.0	10.2	3.4	28.5	25.0	60.0	9
	te3	31.1	31.5	2.7	0.9	8.7	27.0	35.0	10
	te4	35.6	32.0	8.5	2.4	23.8	30.0	58.0	12
Nuclear figure, length	pa1	74.8	73.0	11.7	2.4	15.7	56.0	99.0	23
	pa2	74.6	73.5	8.0	2.8	10.7	64.0	86.0	8
Anteriormost macronuclear nodule, length	pa1	7.7	8.0	1.8	0.4	23.0	4.0	11.0	23
	pa2	7.3	7.5	1.2	0.4	16.1	6.0	9.0	8
Anteriormost macronuclear nodule, width	pa1	3.4	3.0	0.6	0.1	18.8	2.5	5.0	23
	pa2	4.4	4.8	0.7	0.3	16.4	3.0	5.0	8
Posterior macronuclear nodule, length	gr1[e]	4.4	4.0	0.9	0.3	21.0	3.3	6.6	11
	gr2[e]	5.6	5.3	0.8	0.2	14.3	4.0	6.6	10
	is1	5.6	6.0	1.1	0.3	20.0	4.0	7.0	13
	no1[e]	5.0	–	–	–	–	3.9	6.0	3
	no2	4.9	5.0	1.0	0.3	20.4	3.0	6.0	10
	no3[e]	5.0	4.6	1.9	0.6	37.8	3.0	7.6	11
	pa1	7.9	8.0	2.0	0.4	25.2	5.0	12.0	22
	pa2	8.4	8.5	1.3	0.5	15.6	7.0	10.0	8

Table 17 Continued

Characteristics [a]	Species	mean	M	SD	SE	CV	Min	Max	n
Posterior macronuclear nodule, length	stu [e]	39.4	36.5	7.4	–	18.7	28.0	50.0	12
	sy1 [e]	7.9	7.5	1.3	0.4	16.8	6.6	10.6	10
	sy2	6.1	7.0	1.5	0.4	23.7	4.0	8.0	11
	sy8 [e]	8.5	8.0	0.8	0.2	9.9	7.0	11.0	19
	te2 [e]	5.7	5.3	0.8	0.3	14.5	4.0	6.6	9
	te3 [e]	5.7	6.0	1.0	0.3	17.9	4.0	6.6	10
	te4 [e]	4.9	4.7	1.8	0.5	36.5	2.3	7.6	12
Posterior macronuclear nodule, width	gr1 [e]	2.6	2.7	0.2	0.1	7.2	2.2	2.8	11
	gr2 [e]	2.5	2.5	0.6	0.2	22.2	2.0	3.9	10
	isl	3.2	3.0	0.6	0.2	18.5	3.0	5.0	13
	no1 [e]	2.2	–	–	–	–	1.7	2.7	3
	no2	2.7	2.9	0.4	0.1	14.6	2.0	3.0	10
	no3 [e]	3.1	3.0	0.2	0.1	6.5	2.8	3.5	11
	pa1	3.8	4.0	0.8	0.2	20.2	2.5	5.0	22
	pa2	4.1	4.0	0.6	0.2	15.5	3.0	5.0	8
	stu [e]	14.1	14.0	2.5	–	17.5	10.0	18.0	12
	sy1 [e]	3.9	4.0	0.8	0.3	21.5	2.7	5.3	10
	sy2	4.1	4.0	0.7	0.2	17.1	3.0	5.0	11
	sy8 [e]	3.9	4.0	0.2	0.1	5.4	3.5	4.0	19
	te2 [e]	3.0	3.1	0.6	0.2	21.5	1.8	4.0	9
	te3 [e]	2.8	2.7	0.5	0.2	18.3	2.0	4.0	10
	te4 [e]	2.9	3.0	0.8	0.2	28.6	2.0	4.5	12
Micronuclei, number	isl	2.2	2.0	1.6	0.5	75.5	1.0	7.0	13
	mul	2.1	–	0.3	0.1	14.4	2.0	3.0	12
	no3	2.2	2.0	0.8	0.2	34.4	1.0	4.0	11
	pa1	3.0	3.0	0.9	0.2	28.4	2.0	5.0	23
	pa2	2.1	2.0	1.2	0.4	58.7	1.0	4.0	8
	stu	3.7	4.0	1.1	–	30.4	2.0	6.0	12
	sy2	2.2	2.0	0.4	0.1	18.5	2.0	3.0	11
	sy8	3.3	3.0	0.8	0.2	24.2	2.0	5.0	15
	te2	3.6	3.0	1.4	0.5	40.0	2.0	6.0	9
	te3	2.5	2.0	0.7	0.2	28.3	2.0	4.0	8
	te4	3.8	3.5	1.3	0.4	34.9	2.0	6.0	12
Anterior micronucleus, length	pa1	2.6	2.5	–	–	–	1.5	3.0	23
	pa2	2.5	2.5	–	–	–	2.0	3.0	8
Anterior micronucleus, width	pa1	2.2	2.0	–	–	–	1.5	3.0	23
	pa2	2.4	2.5	–	–	–	2.0	3.0	8
Posterior micronucleus, length	isl	1.6	1.6	0.1	0.0	8.5	1.4	1.8	13
	no3 [e]	4.0	3.9	0.9	0.3	23.0	2.8	6.0	10
	stu [e]	4.6	4.5	0.3	–	7.5	4.2	5.3	12
	sy1 [e]	3.1	2.7	0.6	0.2	19.3	2.7	4.0	10
	sy2	3.0	3.0	0.0	0.0	0.0	3.0	3.0	11
	sy8 [e]	2.9	3.0	0.2	0.1	7.2	2.5	3.0	19
	te2 [e]	1.5	1.5	0.1	0.0	6.3	1.3	1.6	9
	te3 [e]	1.7	1.6	0.2	0.1	14.5	1.3	2.0	9
	te4 [e]	1.6	1.6	0.1	0.0	9.1	1.5	2.0	12
Posterior micronucleus, width	isl	1.5	1.5	0.1	0.0	5.1	1.4	1.6	13
	no3 [e]	2.4	2.3	0.6	0.2	27.2	1.5	3.2	10
	sy1	2.3	2.4	0.2	0.1	10.5	2.0	2.7	10
	sy2	3.0	3.0	0.0	0.0	0.0	3.0	3.0	11
	te4 [e]	1.6	1.6	0.1	0.0	9.1	1.5	2.0	12

Table 17 Continued

Characteristics[a]	Species	mean	M	SD	SE	CV	Min	Max	n
Frontal cirri, number	gr1	3.0	3.0	0.0	0.0	0.0	3.0	3.0	11
	gr2	3.0	3.0	0.0	0.0	0.0	3.0	3.0	10
	isl	3.0	3.0	0.0	0.0	0.0	3.0	3.0	13
	mul[c]	4.3	–	0.6	0.2	11.1	4.0	5.0	14
	no1	3.0	–	–	–	–	3.0	3.0	3
	no2	3.0	3.0	0.0	0.0	0.0	3.0	3.0	11
	no3	3.0	–	–	–	–	2.0	3.0	11
	pa1	3.0	3.0	0.0	0.0	0.0	3.0	3.0	21
	pa2	3.0	3.0	0.0	0.0	0.0	3.0	3.0	8
	stu[c]	4.0	4.0	0.0	–	0.0	4.0	4.0	12
	sy1[c]	5.7	6.0	0.5	0.1	8.0	5.0	6.0	10
	sy2[c]	7.0	7.0	0.0	0.0	0.0	7.0	7.0	11
	sy3[c]	–	–	–	–	–	3.0	5.0	30
	sy4[c]	–	–	–	–	–	4.0	6.0	30
	sy5[c]	–	–	–	–	–	4.0	4.0	4
	sy6[c]	–	–	–	–	–	4.0	5.0	10
	sy7[c]	–	–	–	–	–	3.0	6.0	19
	sy8[c]	7.4	7.0	0.6	0.1	8.2	6.0	8.0	19
	te1	3.0	–	–	–	–	–	–	?
	te2	3.0	3.0	0.0	0.0	0.0	3.0	3.0	9
	te3	3.2	3.0	0.6	0.2	18.8	3.0	5.0	10
	te4	2.9	3.0	0.3	0.1	9.9	2.0	3.0	12
Buccal cirri, number	gr1	1.0	1.0	0.0	0.0	0.0	1.0	1.0	11
	isl	1.0	1.0	0.0	0.0	0.0	1.0	1.0	13
	mul	1.0	–	0.0	0.0	0.0	1.0	1.0	14
	no1	1.0	–	–	–	–	1.0	1.0	3
	no2	1.0	1.0	0.0	0.0	0.0	1.0	1.0	11
	no3	1.0	1.0	0.0	0.0	0.0	1.0	1.0	11
	pa1	1.0	1.0	0.0	0.0	0.0	1.0	1.0	23
	pa2	1.0	1.0	0.0	0.0	0.0	1.0	1.0	8
	stu	1.0	1.0	0.0	–	0.0	1.0	1.0	12
	sy1	1.0	1.0	0.0	0.0	0.0	1.0	1.0	10
	sy2	1.0	1.0	0.0	0.0	0.0	1.0	1.0	11
	sy8	1.0	1.0	0.0	0.0	0.0	1.0	1.0	16
	te2	1.0	1.0	0.0	0.0	0.0	1.0	1.0	9
	te3	1.0	1.0	0.0	0.0	0.0	1.0	1.0	10
	te4	1.1	1.0	0.3	0.1	26.6	1.0	2.0	12
Frontoterminal cirri, number	isl	2.0	2.0	0.0	0.0	0.0	2.0	2.0	13
	no2	2.0	2.0	0.0	0.0	0.0	2.0	2.0	11
	no3	2.0	2.0	0.0	0.0	0.0	2.0	2.0	11
	pa1	2.0	2.0	0.0	0.0	0.0	2.0	2.0	23
	pa2	2.0	2.0	0.0	0.0	0.0	2.0	2.0	8
	stu	2.2	2.0	0.5	–	20.1	2.0	3.0	12
	sy2	2.0	2.0	0.0	0.0	0.0	2.0	2.0	11
	sy8	2.0	2.0	0.0	0.0	0.0	2.0	2.0	18
	te4	1.8	2.0	0.6	0.2	31.5	0.0	2.0	12
Cirri behind right frontal cirrus, number	pa1	1.0	1.0	0.0	0.0	0.0	1.0	1.0	22
	pa2	1.0	1.0	0.0	0.0	0.0	1.0	1.0	8
Midventral complex, number of cirral pairs	mul	7.2	–	0.7	0.2	9.7	5.0	8.0	14
	pa1	6.6	6.0	1.4	0.3	20.5	5.0	9.0	21

Table 17 Continued

Characteristics[a]	Species	mean	M	SD	SE	CV	Min	Max	n
Midventral complex, number of cirral pairs	pa2	7.3	7.0	0.8	0.3	10.4	6.0	8.0	7
	sy2	16.7	16.5	1.5	0.4	8.8	14.5	19.5	11
	sy3	–	–	–	–	–	14.0	19.0	30
	sy4	–	–	–	–	–	13.0	19.0	30
	sy5	–	–	–	–	–	15.0	17.0	4
	sy6	–	–	–	–	–	17.0	19.0	10
	sy7	–	–	–	–	–	12.0	19.0	19
Midventral cirri, number[g]	gr1	11.1	11.0	1.2	0.4	11.2	9.0	13.0	11
	gr2	8.5	7.0	2.3	0.7	27.5	7.0	13.0	10
	no2	12.5	13.0	1.9	0.6	14.8	9.0	15.0	11
	no3	19.7	18.0	3.1	0.9	19.7	15.0	27.0	11
	pa1	13.4	12.0	3.0	0.7	22.3	10.0	19.0	21
	pa2	15.4	16.0	0.8	0.3	5.1	14.0	16.0	7
Midventral complex, number of left cirri	is1	5.5	6.0	0.5	0.1	9.4	5.0	6.0	13
	no1	9.5	–	–	–	–	9.0	10.0	2
	stu	19.2	17.0	7.1	–	36.7	15.0	41.0	12
	sy1	13.1	13.5	1.9	0.6	14.3	10.0	15.0	10
	sy8	16.6	16.0	2.3	0.6	13.8	14.0	21.0	15
	te2	8.4	9.0	0.7	0.2	8.1	7.0	9.0	9
	te3	9.6	9.5	1.4	0.4	14.1	8.0	12.0	10
	te4	11.9	12.0	2.6	0.8	22.2	8.0	17.0	12
Midventral complex, number of right cirri	is1	6.7	7.0	1.0	0.3	14.2	5.0	8.0	13
	no1	11.0	–	–	–	–	10.0	12.0	2
	stu	21.8	20.5	4.7	–	21.5	19.0	36.0	12
	sy1	13.1	13.5	1.9	0.6	14.3	10.0	15.0	10
	sy8	17.7	17.0	2.1	0.6	11.8	15.0	21.0	14
	te2	10.9	11.0	1.4	0.5	12.6	9.0	13.0	9
	te3	11.3	11.0	1.3	0.4	11.9	10.0	14.0	10
	te4	12.3	12.0	2.5	0.7	20.0	9.0	16.0	12
Pretransverse ventral cirri, number	no3	1.8	2.0	0.5	0.1	25.8	1.0	2.0	12
	sy1	2.0	2.0	0.0	0.0	0.0	2.0	2.0	10
	sy2	2.0	2.0	0.0	0.0	0.0	2.0	2.0	11
	sy8	1.9	2.0	0.5	0.1	25.0	1.0	3.0	18
	te2	0.0	0.0	0.0	0.0	0.0	0.0	0.0	9
	te3	0.0	0.0	0.0	0.0	0.0	0.0	0.0	10
	te4	0.8	1.0	0.4	0.1	57.4	0.0	1.0	29
Transverse cirri, number	is1[b]	3.6	4.0	0.7	0.2	18.0	2.0	4.0	13
	mu1	8.6	–	0.9	0.3	10.7	7.0	10.0	14
	no1	4.2	–	–	–	–	2.0	6.0	5
	no2	1.7	2.0	0.7	0.2	37.4	0.0	2.0	11
	no3	2.9	3.0	0.9	0.2	29.5	1.0	4.0	13
	pa1	2.2	2.0	–	–	–	2.0	3.0	20
	pa2	1.9	2.0	–	–	–	0.0	3.0	8
	stu	3.0	3.0	0.0	–	0.0	3.0	3.0	12
	sy1	7.1	7.0	0.9	0.3	13.3	6.0	9.0	10
	sy2	8.3	8.0	1.1	0.3	13.3	6.0	10.0	11
	sy3	–	–	–	–	–	6.0	9.0	30
	sy4	–	–	–	–	–	6.0	8.0	30
	sy5	–	–	–	–	–	7.0	8.0	4
	sy6	–	–	–	–	–	9.0	11.0	10

Table 17 Continued

Characteristics [a]	Species	mean	M	SD	SE	CV	Min	Max	n
Transverse cirri, number	sy7	–	–	–	–	–	9.0	11.0	19
	sy8	8.5	8.5	1.4	0.4	16.6	6.0	11.0	16
	te1	4.0	–	–	–	–	3.0	4.0	?
	te2	2.7	3.0	1.0	0.3	39.5	1.0	5.0	9
	te3	3.4	3.5	1.5	0.5	44.0	0.0	5.0	10
	te4	5.9	6.0	1.3	0.2	21.6	3.0	8.0	29
Left marginal row, number of cirri	gr1 [h]	20.7	21.0	1.4	0.4	6.8	19.0	23.0	11
	gr2 [h]	25.1	24.5	2.9	0.9	11.5	22.0	32.0	10
	isl	17.9	18.0	1.6	0.4	8.8	14.0	20.0	13
	mul	22.2	–	6.3	1.8	29.6	20.0	26.0	12
	no1	39.5	–	–	–	–	32.0	45.0	3
	no2	24.0	24.0	1.3	0.4	5.3	22.0	27.0	11
	no3	39.5	39.0	6.4	1.9	16.3	32.0	55.0	11
	pa1	31.4	32.5	3.9	0.8	12.4	25.0	38.0	22
	pa2	28.0	27.0	3.7	1.4	13.0	23.0	34.0	7
	stu	32.1	33.0	4.3	–	13.5	23.0	38.0	12
	sy1	30.8	30.5	2.9	0.9	9.4	25.0	35.0	10
	sy2	37.2	39.0	4.3	1.3	11.5	31.0	43.0	11
	sy8	40.3	40.0	6.2	1.7	15.4	33.0	51.0	13
	te1	20.0	–	–	–	–	–	–	?
	te2	28.1	27.0	3.1	1.0	10.9	25.0	36.0	9
	te3	24.6	25.0	2.3	0.7	9.3	20.0	28.0	10
	te4	29.3	27.5	5.2	1.5	17.9	24.0	40.0	12
Right marginal row, number of cirri	gr1 [i]	25.1	25.0	1.5	0.5	6.0	22.0	28.0	11
	gr2 [i]	28.3	27.0	5.3	1.7	18.8	22.0	40.0	10
	isl	17.9	18.0	1.9	0.5	10.8	15.0	21.0	13
	mul	23.9	–	1.6	0.5	6.5	22.0	27.0	12
	no1	36.3	–	–	–	–	32.0	42.0	3
	no2	22.5	23.0	2.2	0.7	9.6	18.0	25.0	11
	no3	33.5	33.0	3.7	1.1	11.0	28.0	41.0	11
	pa1	33.6	34.0	4.2	0.9	12.6	26.0	42.0	23
	pa2	33.4	33.5	1.8	0.7	5.5	31.0	37.0	8
	stu	32.4	32.0	5.2	–	15.9	20.0	42.0	12
	sy1	32.1	32.0	3.4	1.1	10.7	28.0	40.0	10
	sy2	40.9	42.0	4.7	1.4	11.5	30.0	47.0	11
	sy8	45.8	47.0	7.8	2.2	17.1	34.0	60.0	13
	te1	27.0	–	–	–	–	–	–	?
	te2	25.6	25.0	1.4	0.5	5.6	24.0	29.0	9
	te3	26.6	26.0	2.2	0.7	8.3	23.0	30.0	10
	te4	30.2	29.5	4.1	1.2	13.6	25.0	39.0	12
Dorsal kineties, number	gr1	2.0	2.0	0.0	0.0	0.0	2.0	2.0	11
	gr2	2.0	2.0	0.0	0.0	0.0	2.0	2.0	10
	isl	3.0	3.0	0.0	0.0	0.0	3.0	3.0	13
	mul	6.1	–	0.3	0.1	4.4	6.0	7.0	14
	no1	3.0	–	–	–	–	3.0	3.0	3
	no2 [j]	4.0	4.0	0.0	0.0	0.0	4.0	4.0	11
	no3 [k]	4.0	4.0	0.0	0.0	0.0	4.0	4.0	10
	pa1	4.0	4.0	0.0	0.0	0.0	4.0	4.0	18
	pa2	4.0	4.0	0.0	0.0	0.0	4.0	4.0	5
	stu	6.0	6.0	0.4	–	6.8	5.0	7.0	13

Table 17 Continued

Characteristics[a]	Species	mean	M	SD	SE	CV	Min	Max	n
Dorsal kineties, number	sy1	5.0	5.0	0.0	0.0	0.0	5.0	5.0	10
	sy2	5.0	5.0	0.0	0.0	0.0	5.0	5.0	11
	sy8	5.0	5.0	0.0	0.0	0.0	5.0	5.0	14
	te2	4.0	4.0	0.0	0.0	0.0	4.0	4.0	9
	te3	4.0	4.0	0.0	0.0	0.0	4.0	4.0	10
	te4	4.0	4.0	0.0	0.0	0.0	4.0	4.0	12
Caudal cirri, number	isl	1.5	2.0	0.7	0.2	42.9	0.0	2.0	13
	mul	6.0	–	0.7	0.2	11.3	5.0	7.0	14
	no1	3.0	–	–	–	–	2.0	4.0	3
	no2	2.3	2.0	0.7	0.2	28.5	1.0	3.0	11
	pa1	3.0	3.0	0.0	0.0	0.0	3.0	3.0	14
	pa2	3.0	3.0	0.0	0.0	0.0	3.0	3.0	7
	stu	3.0	3.0	0.0	–	0.0	3.0	3.0	12
	sy2	4.0	4.0	0.6	0.2	15.8	3.0	5.0	11
	sy8	5.3	5.0	1.7	0.5	31.1	3.0	8.0	13
	te4	0.8	1.0	0.6	0.1	76.0	0.0	2.0	25

[a] All measurements in µm. ? = number of individuals investigated (= sample size) not indicated; if only one value is known it is listed in the mean column; if two values are available they are listed as Min and Max. CV = coefficient of variation in %, M = median, Max = maximum value, mean = arithmetic mean, Min = minimum value, n = number of individuals investigated, SD = standard deviation, SE = standard error of arithmetic mean. Data based on protargol-impregnated specimens.

[b] Likely sometimes at least one pretransverse cirrus is present (e.g., Fig. 48c, arrow).

[c] *Caudiholosticha multicaudicirrus*: cirrus III/2 (= cirrus behind right frontal cirrus) and likely a further cirrus (right cirrus of anteriormost midventral pair?) included. *Caudiholosticha stueberi*: cirrus III/2 included. *Caudiholosticha sylvatica* (sy1, sy2, and sy8): 3 frontal cirri plus cirri behind left frontal cirrus; *Caudiholosticha sylvatica* (sy3 to sy7): only left frontal cirrus and cirri behind.

[d] Distances see Fig. 44d. In other species distance 1 is equivalent to the distance between anterior body end and rear end of midventral complex.

[e] Nodule/micronucleus (anterior or posterior) not mentioned. For *C. sylvatica* sy6 the diameter is given.

[f] Specimen illustrated has only 24 nodules (Fig. 52e)!

[g] Cirrus behind right frontal cirrus not included.

[h] Transverse and caudal cirri, if present at all, included.

[i] Frontoterminal cirri included.

[j] The short dorsal kinety 4 is very probably the continuation of the right marginal row.

[k] From the original notes kindly supplied by H. Blatterer and W. Foissner.

shaped structures (Fig. 54a), which, however, can also be lacking (see remarks); rings of specimen illustrated about 8 µm across.

Adoral zone occupies 33% of body length in specimen illustrated (Fig. 54a), obviously without peculiarities. Buccal field likely narrow. Frontal cirri as in *Oxytricha* (Kahl 1932); the illustration shows three enlarged frontal cirri, one buccal cirrus near the anterior end of the undulating membranes, and one cirrus (cirrus III/2) behind right frontal cirrus. Ahead of the (zigzagging) midventral complex three cirri arranged in line (possibly this is a special feature of this species, or Kahl did not illustrate the pattern correctly). Midventral complex commences about at level of buccal vertex, extends to very near transverse cirri, composed of widely spaced pairs (specimen illustrated with 8 pairs; value must not be over-interpreted); cirri of pairs narrowly spaced; right cirrus of each pair larger than left. Five transverse cirri, which protrude by about half their length beyond rear body end. Marginal rows obviously without peculiarities, distinctly separated posteriorly. Dorsal bristles about 4 μm long. Three long caudal cirri (Fig. 54a).

Occurrence and ecology: Type locality is the Bad Oldesloe area in north Germany near the city of Hamburg, where Kahl (1932) discovered *C. setifera* in waters with high salinity (percentage not given); a further population without ring-shaped structures was found in slightly saline (up to 3‰) waters of the same region. Records not substantiated by morphological data: Belgium (Chardez 1987, p. 13); benthic in the Krasnovodsk Bay, eastern shore of the Caspian Sea (Agamaliev 1973, p. 1598).

Feeds on *Euglena* (Kahl 1932; type population with ring-shaped structures), respectively, small diatoms (population without rings).

Insufficient redescriptions

***Holosticha algivora* Kahl** – Bick, 1961, Jh. Ver. vaterl. Naturk. Württ., 116: 211, Abb. 29 (Fig. 99a). Remarks: According to Bick (1961) this species is very common at all sites (about 150 sites) in running waters of Nordwürttemberg, a province in Germany. Thus, I suppose that he confused it with *Holosticha pullaster*, which has a similar habitus and is very common in all limnetic habitats. The illustration is very tiny and too general to allow a reliable identification. Once he found it in large quantities in a trickling filter where it mainly ingested bacteria. Alphamesosaprobic.

***Holosticha algivora* Kahl 1930** – Chardez, 1981, Revue verviét. Hist. nat., 38: 53, Fig. 16 (Fig. 99b). Remarks: Chardez (1981) provided only an inspecific illustration. It differs from the original description in the lower number of transverse cirri (5 vs. 7) and the lack of caudal cirri. Further, the cortical granules mentioned by Kahl (1932) are neither mentioned nor illustrated by Chardez (1981). Thus it would be unwise to accept the identification. Body length 75–85 μm. Roadside ditch (pH 5.80), Belgium (see also Chardez 1987, p. 13).

***Holosticha viridis* Kahl, 1932** – Aladro-Lubel, Martínez-Murillo, Mayén-Estrada, Hernández & Sánchez-Calderon, 1986, Rev. Lat-amer. Microbiol., 28: 239, Lámina III, Figura 6 (Fig. 99f). Remarks: This redescription basically does not contradict the original description (for review, see Aladro Lubel et al. 1990, p. 132; Fig. 99g). However, the description and the illustration are rather inspecific so that the identification is not certain. The main difference is that Aladro-Lubel et al. (1986) found this species in the sea, whereas Kahl (1932) discovered it in a limnetic habitat. Body length 73 μm. Adoral zone about one third of body length; two macronuclear nodules; for cirral pattern, see Fig. 99f; cytoplasm with green algae; Mexican beach Boca del Rio, Veracruz (Gulf of Mexico) during July (see also Aladro-Lubel et al. 1988, p. 437).

Anteholosticha Berger, 2003

2003 ***Anteholosticha* nov. gen.** – Berger, Europ. J. Protistol., 39: 377 (original description). Type species (by original designation on p. 377): *Holosticha monilata* Kahl, 1928.

Nomenclature: *Anteholosticha* is a composite of the Latin prefix *ante+* (temporal: before) and the genus-group name *Holosticha*, indicating that the species included were previously classified in *Holosticha* (Berger 2003). Like *Holosticha*, feminine gender.

Characterisation: Adoral zone of membranelles continuous. Rearmost membranelles not wider than remaining membranelles of proximal portion. 3 enlarged frontal cirri. Buccal cirrus/cirri right of paroral. Frontoterminal cirri present. Midventral complex composed of midventral pairs only. Pretransverse ventral cirri present or absent. Number of transverse cirri usually distinctly lower than number of midventral pairs. 1 left and 1 right marginal row. Anterior end of left marginal row ± straight, commences left of adoral zone. Caudal cirri lacking. Nuclear apparatus left of midline or scattered.

Remarks: The characterisation above is basically in accordance with the diagnosis of the original description, except for the pretransverse ventral cirri included in the present review. *Anteholosticha* comprises mainly those species previously classified in *Holosticha* which lack the apomorphies (e.g., anterior end of left marginal row curved rightwards, proximalmost membranelles widened; details see *Holosticha*) of this genus and which also lack caudal cirri. The presence/absence of this or another cirral group is generally used to define genera, for example, *Tachysoma* (for review see Berger 1999, p. 431).

The characterisation is, due to the lack of derived features, only a combination of plesiomorphies. This indicates that *Anteholosticha* is heterogeneous like *Caudiholosticha*, a further group extracted from *Holosticha* by Berger (2003). *Caudiholosticha* contains those "*Holosticha*" species which have more or less distinct caudal cirri. Since the presence of this cirral group is certainly the plesiomorphic state within the hypotrichs (see ground pattern of the Urostyloidea), the absence of these organelles in *Anteholosticha* could be interpreted as autapomorphy. On the other hand, it might also be a synapomorphy with another group, for example, *Holosticha*, which also lacks caudal cirri.

The species assigned to *Anteholosticha* have an unspectacular cirral pattern. Very likely morphogenetic studies are needed to extract monophyletic groups. Hemberger (1982) found that *A. monilata* and *A. multistilata* differ significantly in the formation of proter's adoral zone of membranelles. In *A. monilata* only the proximal portion of the parental zone is replaced (Fig. 56a–h), whereas in *A. multistilata* the parental zone is completely resorbed (Fig. 84c–j). *Anteholosticha warreni* also replaces the parental adoral zone completely, indicating that it is more closely related to *A. multistilata* than *A. monilata* (Fig. 85p, r, t, v). The replacement of only the proximal portion is reminiscent of some *Holosticha* species (e.g., *H. bradburyae*; Fig. 36l, n, p).

The descriptions of many *Anteholosticha* species lack details, for example, presence or absence of cortical granules, exact arrangement of cirri and dorsal bristles. Thus, the differences between some of these species are indistinct, making the construction of a good key rather difficult. The species are arranged according to the habitats (freshwater, soil, salt water); within these groups species with many macronuclear nodules are described first. However, note that some species occur, for example, both in freshwater and soil. Please also consider that few common species, e.g., the *Holosticha multistilata* in Foissner et al. (1991), have changed the name; Table 18 gives a brief overview about the new situation, which will be hopefully stable for many years.

The insufficient data caused previous workers to synonymise several of these species (Borror 1972, Hemberger 1982, Borror & Wicklow 1983), although some differences between the species (e.g., habitat, specific food) strongly indicate that they are indeed distinct taxa. Consequently, I accept some species, which were put into the synonymy of other species by these authorities.

Species included in *Anteholosticha* (alphabetically arranged basionyms are given; subgenera ignored): (1) *Anteholosticha antecirrata* nov. spec.; (2) *Holosticha adami* Foissner, 1982; (3) *Holosticha (Keronopsis) alpestris* Kahl, 1932; (4) *Holosticha (Holosticha) arenicola* Kahl, 1932; (5) *Holosticha australis* Blatterer & Foissner, 1988; (6) *Holosticha azerbaijanica* Alekperov & Asadaullayeva, 1999; (7) *Holosticha bergeri* Foissner, 1987; (8) *Holosticha brachysticha* Foissner, Agatha & Berger, 2002; (9) *Holosticha (Holosticha) brevis* Kahl, 1932; (10) *Holosticha camerounensis* Dragesco, 1970; (11) *Holosticha distyla* Buitkamp, 1977; (12) *Holosticha estuarii* Borror & Wicklow, 1983; (13) *Holosticha (Holosticha) extensa* Kahl, 1932; (14) *Holosticha (Holosticha) fasciola* Kahl, 1932; (15) *Holosticha (Keronopsis) gracilis* Kahl, 1932; (16) *Holosticha (Holosticha) grisea* Kahl, 1932; (17) *Holosticha (Holosticha) manca* Kahl, 1932; (18) *Holosticha manca plurinucleata* Gellért, 1956; (19) *Holosticha mancoidea* Hemberger, 1985; (20) *Holosticha monilata* Kahl, 1928; (21) *Holosticha multistilata* Kahl, 1928; (22) *Holosticha muscicola* Gellért, 1956; (23) *Holosticha (Keronopsis) pulchra* Kahl, 1932; (24) *Holosticha randani* Grolière, 1975; (25) *Holosticha sigmoidea* Foissner, 1982; (26) *Holosticha (Holosticha) violacea* Kahl, 1928; (27) *Holosticha vuxgracilis* nom. nov.; (28) *Holosticha warreni* Song & Wilbert, 1997; (29) *Holosticha xanthichroma* Wirnsberger & Foissner, 1987; (30) *Keronopsis longissima* Dragesco & Dragesco-Kernéis, 1986; (31) *Keronopsis sphagni* Grolière, 1975; (32) *Keronopsis thononensis* Dragesco, 1966; (33) *Oxytricha oculata* Mereschkowsky, 1877; (34) *Oxytricha scutellum* Cohn, 1866; (35) *Pleurotricha macrostoma* Dragesco, 1970; (36) *Urostyla intermedia* Bergh, 1889.

At the end of the *Anteholosticha*-section is an unidentified *Tachysoma* species, which possibly belongs to the present genus.

Key to *Anteholosticha* species

If you know that your specimen/population belongs to *Anteholosticha*, identification is still rather difficult because many species are not described in detail. Key features are the nuclear apparatus, body size, cortical granulation, ventral and dorsal ciliature, and habitat. Thus, very detailed life observation and protargol impregnation is needed for a reliable identification of many species. Check also the *Caudiholosticha* key if you are uncertain about the presence or absence of caudal cirri.

1 Two macronuclear nodules **2**
- More than two macronuclear nodules **11**
2 Single micronucleus in between macronuclear nodules (Fig. 51a, 69a, 82a, 83a) . **3**
- Different nucleus pattern (e.g., Fig. 96a) **6**
3 Transverse cirri displaced very far anteriorly, that is, inserted at end of second body third (Fig. 51a; caudal cirri likely present) . . *Caudiholosticha navicularum* (p. 274)
- Transverse cirri not so far displaced anteriorly **4**
4 About 10 enlarged frontal cirri (Fig. 83a) . . . . *Anteholosticha alpestris* (p. 403)
- Three enlarged frontal cirri (Fig. 69a, 82a) **5**
5 Body length:width ratio about 2:1; midventral complex extends to near transverse cirri; limnetic (Fig. 69a) . . . . *Anteholosticha brevis* (p. 360)
- Body length:width ratio about 3:1; midventral complex terminates slightly behind mid-body; terrestrial (Fig. 82a) . . . . *Anteholosticha muscicola* (p. 401)
6 (2) One conspicuous ring-shaped structure each in anterior and posterior body portion; marine (Fig. 95a, e) . . . . *Anteholosticha oculata* (p. 448)
- No such structures in this arrangement; marine or limnetic **7**
7 Ectoplasm with alveolar seam; marine (Fig. 96a) *Anteholosticha arenciola* (p. 452)
- Alveolar seam lacking; limnetic or slightly saline waters **8**
8 Body length below 100 µm (Fig. 72a) . . . . *Anteholosticha vuxgracilis* (p. 369)
- Body length more than 100 µm **9**
9 Body length:width ratio around 6–8:1; adoral zone occupies about 30% of body length (Fig. 67a) . . . . *Anteholosticha violacea* (p. 342)
- Body length:width ratio around 2–3:1; adoral zone occupies about 40% or more of body length (Fig. 70a, 71a) **10**
10 About 11 midventral cirral pairs; 9–15 transverse cirri (Fig. 70a) . . . . *Anteholosticha camerounensis* (p. 361)
- About 5 (very loosely arranged) midventral cirral pairs; 4–6 transverse cirri (Fig. 71a) . . . . *Anteholosticha macrostoma* (p. 365)
11 (1) Four (rarely 3) longitudinally arranged pairs of macronuclear nodules, each with a single micronucleus in between; dorsal bristles about 10 µm long (Fig. 92a) . . . . *Anteholosticha extensa* (p. 438)
- No such nuclear pattern; dorsal bristles distinctly shorter (in *A. violacea* bristles are about 8 µm long) **12**
12 Many small macronuclear nodules more or less scattered throughout cytoplasm (e.g., Fig. 61j) **22**
- About 4 to about 20 macronuclear nodules, often in more or less distinct row in left body portion (e.g., Fig. 57c) **13**
13 Cytoplasm diffuse yellow (Fig. 68a) . . . . . . *Anteholosticha xanthichroma* (p. 345)
- Cytoplasm not distinctly yellow **14**

14 Cortical granules (e.g., Fig. 78b) or extrusomes (e.g., Fig. 55q, r) present **15**
- Cortical granules or extrusomes lacking **19**

15 Cortical granules/extrusomes colourless, rod-shaped (about 2–3 × 1–1.5 μm), form (indistinct) seam, ejected when methyl-green pyronin is added (Fig. 55w, x, 75b–e) **16**
- Cortical granules more or less globular to slightly ellipsoidal, colourless or coloured (e.g., Fig. 77h, i, m) **17**

16 Seven to about 12 transverse cirri; usually 6 dorsal kineties; mainly limnetic (Fig. 55b, t, v) .... *Anteholosticha monilata* (p. 297)
- 3–6 transverse cirri; 4 dorsal kineties; terrestrial (Fig. 75a–g) .... *Anteholosticha australis* (p. 382)

17 (15) Midventral complex extends beyond mid-body; cortical granules in longitudinal rows, colourless; 4 dorsal kineties; on average 20 or more adoral membranelles (Fig. 77a–m) .... *Anteholosticha sigmoidea* (p. 387)
- Midventral complex terminates ahead of mid-body; cortical granules around dorsal bristles, yellowish, orange, or pink; 3 dorsal kineties; on average about 15 adoral membranelles (e.g., Fig. 78a–e) **18**

18 Cortical granules pink or orange, globular; right marginal row commences about at level of proximal end of adoral zone (Fig. 78a–j) .. *Anteholosticha bergeri* (p. 393)
- Cortical granules yellowish, ellipsoidal; right marginal rows commences about at level of buccal cirrus (number of macronuclear nodules 28–38; Fig. 80a–g) .... *Anteholosticha brachysticha* (p. 397)

19 (14) Three dorsal kineties (Fig. 64a–c) .... *Anteholosticha mancoidea* (p. 336)
- 4 dorsal kineties (Fig. 65a, 66a, 76a) **20**

20 Four distal adoral membranelles distinctly set off from remaining part of adoral zone (Fig. 65a) .... *Anteholosticha randani* (p. 338)
- Adoral zone of membranelles continuous (Fig. 66a, 76a; separation of following 2 species difficult) **21**

21 About 16 macronuclear nodules; 30–33 adoral membranelles; midventral complex terminates at about 80% of body length; usually two enlarged transverse cirri; terrestrial (Fig. 76a) .... *Anteholosticha distyla* (p. 385)
- 8–11 macronuclear nodules; 24–25 adoral membranelles; midventral complex terminates at about 60% of body length; usually 5 transverse cirri; limnetic (Fig. 66a) .... *Anteholosticha sphagni* (p. 340)

22 (12) Cortical granules red (dark-red, yellow-reddish) **23**
- Cortical granules not red or lacking **24**

23 Body length 200–300 μm; dark-red cortical granules around cirri and dorsal bristles (Fig. 90a, b) .... *Anteholosticha pulchra* (p. 433)
- Body length around 120 μm; yellowish-red granules arranged in rows (Fig. 88a) .... *Anteholosticha gracilis* (p. 426)

24 (22) Cytoplasm diffuse yellow (colour is not due to yellow cortical granules!) (e.g., Fig. 68a) **24a**
- Cytoplasm not *distinctly* yellow (yellow cortical granules are present in some species!) **25**

24a Limnetic; buccal cirri present; body not distinctly tailed (Fig. 68a) .... *Anteholosticha xanthichroma* (p. 345)
- Saltwater; buccal cirri lacking; posterior body end tail-like narrowed (Fig. 97a) .... *Anteholosticha azerbaijanica* (p. 454)

25 (24) More than 1 buccal cirrus present (e.g., Fig. 74a, b) **26**
- Buccal cirrus present or lacking (e.g., Fig. 85a, g) **30**

26 More than 3 frontal cirri (Fig. 84a–c) .... *Anteholosticha multistilata* (p. 405)
- 3 enlarged frontal cirri (e.g., Fig. 61a, f, i) **27**

27 Marine (salt water) (Fig. 86a) .... *Anteholosticha estuarii* (p. 420)
- Limnetic or terrestrial **28**

28 Body length above 200 μm in life; transverse cirri hardly project beyond rear body end (Fig. 73a–h) .... *Anteholosticha antecirrata* (p. 370)
- Body length below 200 μm in life; transverse cirri project distinctly beyond rear body end (e.g., Fig. 61f) **29**

29 Marginal rows overlapping posteriorly (Fig. 61i, l, n) .... *Anteholosticha intermedia* (p. 317)
- Marginal rows distinctly separated posteriorly (Fig. 74b, g, i) .... *Anteholosticha adami* (p. 376)

30 (25) Buccal cirrus lacking (Fig. 97a) .... *Anteholosticha azerbaijanica* (p. 454)
- Buccal cirrus present **30a**

30a Terrestrial **31**
- Limnetic or marine (saltwater) **32**

31 Cortical granules around dorsal bristles; about 28–38 macronuclear nodules (Fig. 80a–g) .... *Anteholosticha brachysticha* (p. 397)
- Cortical granules lacking (Fig. 81a) .... *Anteholosticha plurinucleata* (p. 399)

32 (30a) Limnetic (see also marine, if your sample site is slightly brackish) **33**
- Marine (saltwater; see also limnetic, if the sample site is brackish) **36**

33 Four distalmost adoral membranelles distinctly set off from remaining membranelles; 16–20 macronuclear nodules (Fig. 65a) .... *Anteholosticha randani* (p. 338)
- Adoral zone continuous; more than 20 macronuclear nodules **34**

34 Body length 140–250 μm; body rather slender (Fig. 62a, 67b) **35**
- Body length 100–140 μm; body elliptical (Fig. 63a) .... *Anteholosticha thononensis* (p. 334)

35 Body 140–180 μm long; dorsal bristles about 3 μm long (Fig. 62a) .... *Anteholosticha grisea* (p. 332)
- Body 180–250 μm long; dorsal bristles about 8 μm long (Fig. 67b) .... *Anteholosticha violacea* (p. 342)

36 (32) Body length 200–300 μm; body very slender (length:width ratio about 10:1; Fig. 93a, b) .... *Anteholosticha fasciola* (p. 441)
- Body length below 200 μm; body length:width ratio distinctly below 10:1 ..... **37**

37 Cortical granules yellow-greenish or yellow-red (Fig. 88a, e) .... *Anteholosticha gracilis* (p. 426)
- Cortical granules(?) colourless or lacking (note: for *A. longissima* the presence/absence of cortical granules is not known) **38**

38 Cortical granules shaped like mammalian erythrocytes (Fig. 85a–g) .... *Anteholosticha warreni* (p. 412)
- Cortical granules of other shape **39**

39 Midventral complex extends to near transverse cirri (Fig 91a) .... *Anteholosticha longissima* (p. 437)
- Midventral complex terminates at about 2/3 of body length (e.g., Fig. 87a, 94f) . **40**

40 Body length about 100–130 µm; about 5 transverse cirri (Fig. 87a–g) .... *Anteholosticha manca* (p. 422)
- Body length below 100 µm (around 50–80 µm); about 7–8 transverse cirri (Fig. 94f) .... *Anteholosticha scutellum* (p. 443)

Anteholosticha monilata (Kahl, 1928) Berger, 2003
(Fig. 55a–z, 56a–i, 57a–d, Tables 12, 19, Addenda)

1928 ***Holosticha monilata*** – Kahl, Arch. Hydrobiol., 19: 212, Abb. 44e (Fig. 55a; original description. No type material available and no formal diagnosis provided).

1932 ***Keronopsis (Holosticha) monilata* (Kahl, 1928)** – Kahl, Tierwelt Dtl., 25: 577, Fig. 104 $_1$, not Fig. 104 $_3$, 110 $_4$ (Fig. 55b; revision of hypotrichs; see remarks).

1945 ***Keronopsis monilata* Kahl** – Šrámek-Hušek, Veda prír., 23: 247, Fig. 2, not Fig. 1 (Fig. 55c; illustrated record).

1963 ***Keronopsis monilata* (Kahl, 1928)** – Reuter, Sarsia, 10: 5, Fig. 9 (Fig. 55d; illustrated record).

1966 ***Keronopsis monilata* (Kahl)** – Dragesco, Protistologica, 2: 84, Fig. 24 (Fig. 55f; redescription after protargol impregnation).

1968 ***Keronopsis monilata* (Kahl, 1928)** – Chorik, Free-living ciliates; p. 128, Fig. 117 (Fig. 55e; redescription).

1970 ***Keronopsis monilata* (Kahl, 1928)** – Dragesco, Annls Fac. Sci. Univ. féd. Cameroun (Numero Hors-série): 107, Fig. 79A, B (Fig. 55g, h; see remarks for the somewhat deviating frontal ciliature of specimen shown in Fig. 55h).

1972 ***Keronopsis monilata* (Kahl, 1928) Kahl, 1932** – Borror, J. Protozool., 19: 11 (revision of hypotrichs; see nomenclature).

1974 ***Holosticha lacazei* Maupas** – Pätsch, Arb. Inst. landw. Zool. Bienenkd., 1: 57, Abb. 46 (Fig. 55i; illustrated record; misidentification).

1974 ***Keronopsis similis* (Stokes, 1886) Kahl, 1932** – Jones, Univ. South Alabama Monogr., 1: 40, Plate XXVIII, Fig. 7 (Fig. 55j; misidentification, see remarks).

1975 ***Keronopsis monilata* Kahl, 1932** – Grolière, Protistologica, 11: 484, Fig. 4 (Fig. 55k; description of a French population; see remarks).

1981 ***Holosticha similis* Stokes, 1886** – Foissner & Didier, Annls Stn limnol. Besse, 15: 260, Abb. 5a–g, Tabelle 4 (Fig. 55m–s; misidentification; detailed redescription).

1982 ***Holosticha similis* Stokes, 1886** – Hemberger, Dissertation, p. 105, Abb. 16a–i (Fig. 56a–h; revision of non-euplotid hypotrichs; cell division).

1983 ***Holosticha intermedia* (Bergh, 1889) Kahl, 1972** – Borror & Wicklow, Acta Protozool., 22: 121, Fig. 15 (Fig. 55l; misidentification; see remarks).

1986 ***Keronopsis monilata* (Kahl, 1928) Kahl, 1932** – Dragesco & Dragesco-Kernéis, Faune tropicale, 26: 440, Planche 129B–F (Fig. 56i; see nomenclature and remarks).

1989 ***Holosticha similis* Stokes, 1886** – Song & Wilbert, Lauterbornia, 3: 159, Tabelle 30 (misidentification).

1991 ***Holosticha monilata* Kahl, 1928** – Foissner, Blatterer, Berger & Kohmann, Informationsberichte des Bayer. Landesamtes für Wasserwirtschaft, 1/91: 231, Fig. 1–14, 16, 17 (Fig. 57a, c, d; guide to ciliates of saprobic system; discussion of synonymy).

1992 ***Holosticha monilata* Kahl, 1928** – Augustin & Foissner, Arch. Protistenk., 141: 274, Fig. 12a–i, Tabelle 9 (Fig. 55t–z, 57b; description of a population from activated sludge; a voucher slide is deposited in the Oberösterreichische Landesmuseum in Linz [LI], Austria).

2000 ***Pseudokeronopsis monilata*** – Liu & Jin, Acta Scientiarum Naturalium Universitatis Sunyatseni, 39: 82, Fig. 1-1–6, 2-1–11 (combination with *Pseudokeronopsis*; morphogenesis; see remarks).

2001 ***Holosticha monilata* Kahl, 1928** – Berger, Catalogue of ciliate names 1. Hypotrichs, p. 37 (nomenclator containing all basionyms, combinations, and higher taxa of hypotrichs).

2003 ***Anteholosticha monilata* (Kahl, 1928) nov. comb.** – Berger, Europ. J. Protistol., 39: 377 (combination with *Anteholosticha*).

Nomenclature: No derivation of the name is given in the original description. The species-group name *monilata* is a composite of the Latin noun *monile* (necklace) and the suffix *~at·us* (having a feature or organ) and obviously alludes to the nuclear apparatus whose macronucleus-nodules form a chain. Kahl (1928) establish this species in the genus *Holosticha*. In 1932, he classified it in the subgenus *Holosticha (Keronopsis)*. Thus, the correct name in Kahl (1932) is *Holosticha (Keronopsis) monilata* Kahl, 1928, although he wrote on page 577 the somewhat confusing heading *Keronopsis (Holosticha) monilata* (Kahl, 1928). Consequently, Kahl (1932) did not change the generic combination, as incorrectly assumed by several authors (e.g., Borror 1972, Dragesco & Dragesco-Kernéis 1986). *Keronopsis monolita* Kahl, 1932 in Chardez (1987, p. 13) is an incorrect subsequent spelling. *Holosticha monilata* was fixed as type species of *Anteholosticha* by original designation (Berger 2003).

Berger (2001) supposed that Šrámek-Hušek (1945, p. 247) had transferred the present species from *Holosticha* to *Keronopsis* (mentioned as *Keronopsis monilata* (Kahl, 1928) ?Šrámek-Hušek, 1945).

Remarks: The systematics of *Anteholosticha monilata* is rather complicated and confusing and thus has to be discussed in detail. Kahl (1928) described the present species very briefly and provided a somewhat ambiguous illustration (Fig. 55a). On page 211 he wrote that he would like to investigate it again, indicating that he was uncertain about some details. In his 1932 revision, Kahl gave three illustrations, two of which show a distinct bicorona (Fig. 190d, e) like Stokes' *Pseudokeronopsis similis* (Fig. 190a–c), with which these two populations are therefore synonymised. The third illustration shows a specimen which has three distinctly enlarged frontal cirri (Fig. 55b). This figure agrees rather well with the original illustration (Fig. 55a). Kahl (1932) explained the differences among his populations by intraspecific variation depending on sampling site. His description is somewhat confusing because in the text he wrote that the three anteriormost cirri are usually set off; by contrast, in the legend to Fig. 190e he wrote "typical form" although this specimen shows a distinct bicorona.

Šrámek-Hušek (1945) described two populations of *H. monilata*. However, one is certainly *Pseudokeronopsis similis* because it has a distinct bicorona (Fig. 190f). The other could indeed be identical with the present species although the frontal ciliature is still reminiscent of a (weak) bicorona (Fig. 55c). The populations described by Reuter (1963), Dragesco (1966), and Chorik (1968) agree rather well with Kahl's (1932) population, as indicated by the three enlarged frontal cirri, the rather long midventral complex, and the nuclear apparatus (Fig. 55d–f).

Dragesco (1970) provided two illustrations of *A. monilata*. The specimen shown in Fig. 55g obviously has three distinct frontal cirri and a further enlarged cirrus which is likely the cirrus behind the right frontal cirrus; by contrast, the specimen shown in Fig. 55h and those described by Dragesco & Dragesco-Kernéis (1986, Fig. 56i) and Dragesco (2003, Fig. 59a–e, Table 19) have around six enlarged frontal cirri. Thus, I doubt that these populations belong to *A. monilata*. Dragesco (2003) identified his population as "*Holosticha similis*". However, this species (= *Pseudokeronopsis similis* in present book) has a distinct bicorona. This suggests that the specimens with the "monocorona" (e.g., Fig. 56i) are a different, not yet described species. More detailed data are needed for a proper classification of Dragesco's populations.

Borror (1972) accepted the present species. However, very likely because of the bicorona shown in Figs. 190d, e, he classified it in *Keronopsis* (now most species of this genus are assigned to *Pseudokeronopsis*).

Hartwig (1973) described, but did not illustrate the present species from a brackish water pond. The cells were very large (160–350 μm long) and the number of macronuclear nodules was rather high so that it cannot be excluded that it is a distinct species. Thus, his data are kept separate.

Hemberger (1982) recognised that Pätsch (1974) misidentified the present species as *Holosticha lacazei*, which is now the type species of *Pseudoamphisiella*. Jones (1974) redescribed both *Pseudokeronopsis similis* and *A. monilata*. However, I have no doubt that he confused them, as indicated by body size and cirral pattern. Thus, I assign his *Keronopsis similis* to the present species and his *Keronopsis monilata* to *Pseudokeronopsis similis*.

Keronopsis monilata sensu Grolière (1975) is very large and has very many midventral pairs (about 50 in specimen shown in Fig. 55k) and a rather high number of macronuclear nodules (15–18). Identity with *Pseudokeronopsis similis* can be excluded because it has three enlarged frontal cirri. Since it cannot be excluded that it is a distinct species I keep the data separate.

Hemberger (1982) synonymised the present species – together with *Holosticha vernalis*, *Urostyla intermedia*, *Holosticha globulifera*, and *Keronopsis clavata* – with *Holosticha similis* (now *Pseudokeronopsis similis*). I do not agree with these synonymies because there are distinct differences in the ciliature and/or habitat (the classification of these species in the present book can be found via the index).

Foissner & Didier (1981) accepted Hemberger's synonymy and simultaneously provided a very detailed redescription of a true *Anteholosticha monilata* population. They wrote that this is the first description where the rod-shaped extrusomes are described. This is obviously not quite correct because Dragesco (1970, p. 109) already mentioned large protrichosts. These extrusomes are rather difficult to recognise in life because they are colourless and obviously have almost the same refractive index as the cytoplasm. In spite of this, conspecificity with Kahl's populations (Fig. 55a, b), for which no extrusomes are described, is beyond reasonable doubt.

Borror & Wicklow (1983) put *A. monilata*, together with some other species, into the synonymy of *Anteholosticha intermedia*, a species with many macronuclear nodules (vs. moniliform) and globular, brightly shining cortical granules (vs. ellipsoidal and colourless).

Song & Wilbert (1985) provided a morphometric analysis of a population which "agrees both in life and in protargol preparations with the data by Foissner & Didier (1981)".

Foissner et al. (1991) solved the complicated synonymy. They recognised that Kahl's (1932) redescription is heterogeneous and confined *Anteholosticha monilata* to species with a moniliform macronucleus and three distinctly enlarged frontal cirri. Augustin & Foissner (1992) provided detailed data on a population from an activated sludge plant, but neither Foissner & Didier (1981) and Foissner et al. (1991) nor Augustin & Foissner (1992) fixed a neotype. However, these descriptions should be considered as authoritative

In 1992, Fernandez-Galiano & Calvo described *Holosticha corlissi* (Fig. 58a–d). The description is rather detailed, but unfortunately they did not compare their population with Foissner & Didier's redescription of *A. monilata*. I do not find a distinct difference, except the end of the marginal rows, and thus classify *H. corlissi* preliminary as supposed synonym of *A. monilata* (for details see below).

Fernandez-Leborans (1985, p. 368) described the undulating membranes of the present species. However, the identification is not substantiated by other morphological data; thus this paper is not mentioned in the list of synonyms. The populations described by Jirovec et al. (1953), Vuxanovici (1963, p. 203), and Chardez (1986) are classified as *Pseudokeronopsis similis*.

Foissner & Didier (1981) were uncertain about the presence or absence of caudal cirri because 2–3 cirri at the end of the right marginal row are sometimes slightly set off from the remaining cirri (Fig. 55s). Song & Wilbert (1989) also found 0–3, on average 0.5 caudal cirri, whereas according to Foissner et al. (1991; Fig. 57c, d) and Augustin & Foissner (1992; Fig. 55u, v; their Table 9) caudal cirri are lacking. Hemberger (1982), who studied cell division, also clearly stated that no caudal cirri are produced. This strongly indicates that Foissner & Didier, respectively, Song & Wilbert slightly misinterpreted their slides in this respect. There are several other misobservations/misinterpretations (e.g., Reuter 1963 did not illustrate a buccal cirrus), which must not be overinterpreted.

Liu & Jin (2000) studied the morphogenesis of *A. monilata*. They designated it as *Pseudokeronopsis monilata*, which has to be interpreted as new combination because this species was never before transferred to this genus. Since the paper is written in Chinese (with English summary), I could not check whether or not the authors transferred the species formally to *Pseudokeronopsis*. The identification is likely correct, that is,

---

Fig. 55a–h *Anteholosticha monilata* (a, from Kahl 1928; b, from Kahl 1932; c, from Šrámek-Hušek 1945; d, from Reuter 1963; e, from Chorik 1968; f, from Dragesco 1966; g, h, from Dragesco 1970. a–e, from life; f–h, protargol impregnation). **a–e:** Ventral views showing cirral pattern and nuclear apparatus, a = 150–200 µm, b = 150 µm, c = 100–200 µm, d = 130–150 µm, e = 194 µm. **f–h:** Cirral pattern and nuclear apparatus, f = 160 µm, g = 176 µm, h = size not indicated. Note that the frontal cirral pattern of specimen shown in (h) is deviating from the standard (3 frontal cirri) pattern, indicating that it belongs to a different species. Frontoterminal cirri circled by dotted line in Fig. 55h. Page 297. →

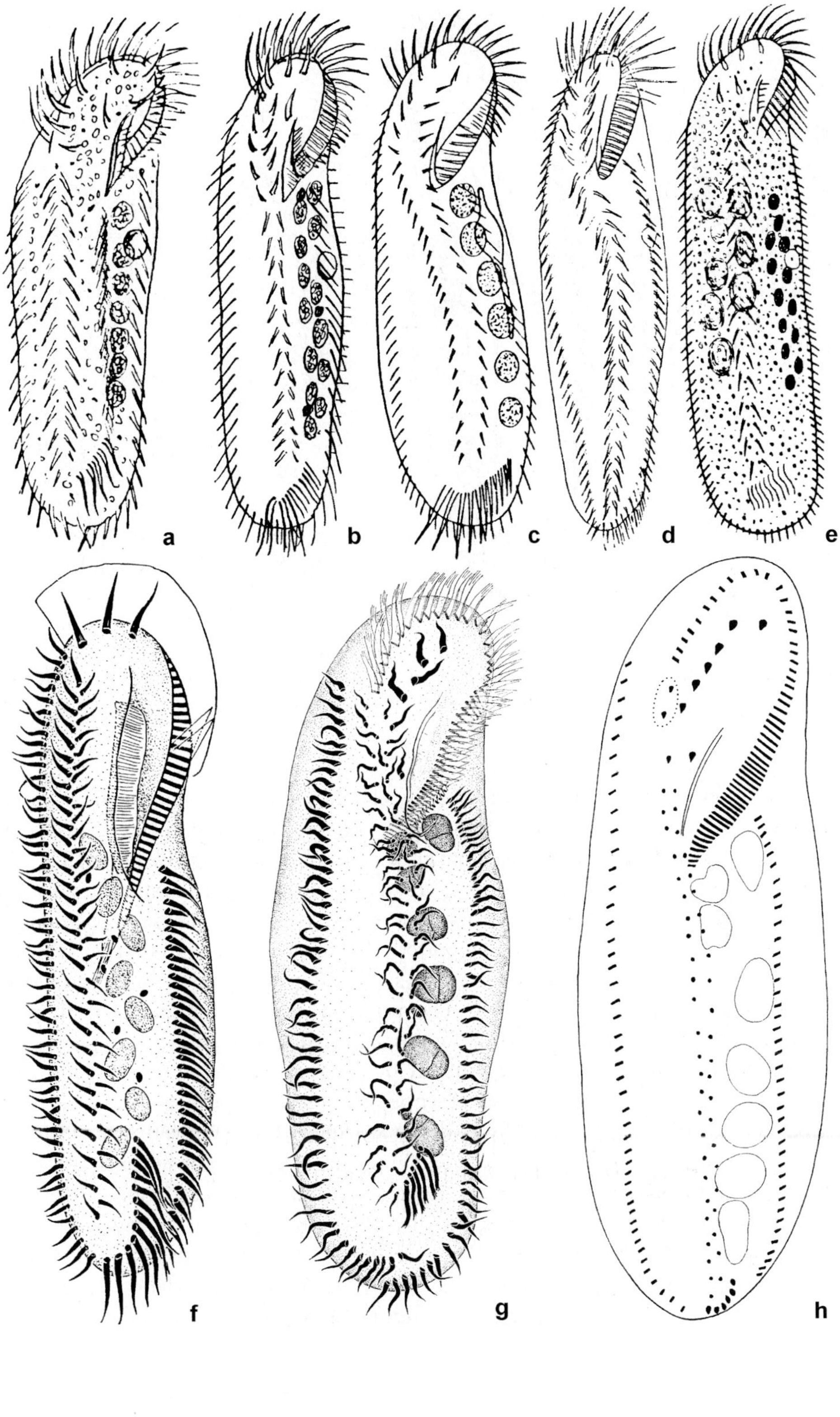
a
b
c
d
e
f
g
h

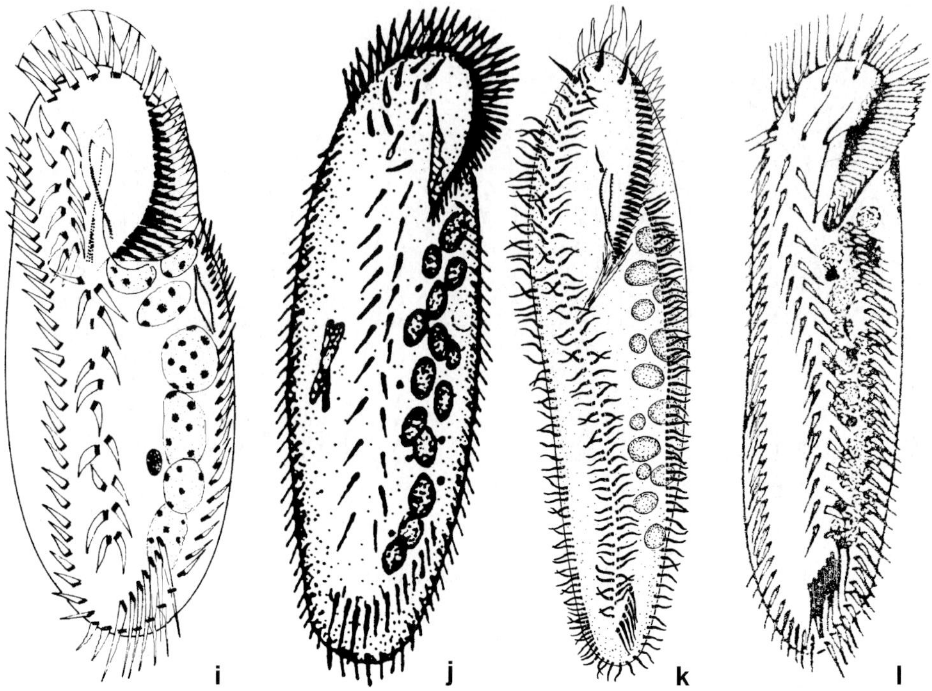

Fig. 55i–l *Anteholosticha monilata* (i, from Pätsch 1974; j, from Jones 1974; k, from Grolière 1975; l, from Borror & Wicklow 1983. i, k, l, protargol impregnation; j, from life). Infraciliature of ventral side and nuclear apparatus, i = 116 μm, j = 105 μm, k = 240 μm, l = 120 μm. Page 297.

their population has three frontal cirri and the characteristic macronucleus of *A. monilata,* whose individual nodules fuse to a single mass during division (see their Figure 1-4). Thus, I do not understand the transfer to *Pseudokeronopsis*. Unfortunately, the printing quality of the illustrations and micrographs of this paper is rather low so that they cannot be shown in the present book. Interestingly, in this Chinese population the marginal rows are distinctly overlapping posteriorly (see their Figures 1-1, 1-2), which is strongly reminiscent of *Holosticha corlissi*, a supposed synonym of *A. monilata*. Possibly, this feature is, as in *A. adami* and *A. intermedia*, relevant for species distinction. However, more detailed data are needed for a final decision.

Morphology: The detailed redescription by Foissner & Didier is provided first. Supplementary and deviating data from other populations are kept separate.

Body size of specimens studied by Foissner & Didier (1981) 90–160 × 30–45 μm in life; body length:width ratio 3.5:1 on average in protargol preparations (Table 19). Body outline long orthogonal, left side slightly, right side distinctly convex, both ends moderately widely rounded; anterior portion usually distinctly narrowed on left side. Body very flexible and slightly contractile, flattened about 2:1 dorso-ventrally; ventral side slightly concave, dorsal side distinctly convex (Fig. 55m, n, q). Macronuclear nodules globular to ellipsoidal, always left of midline largely between oral apparatus and

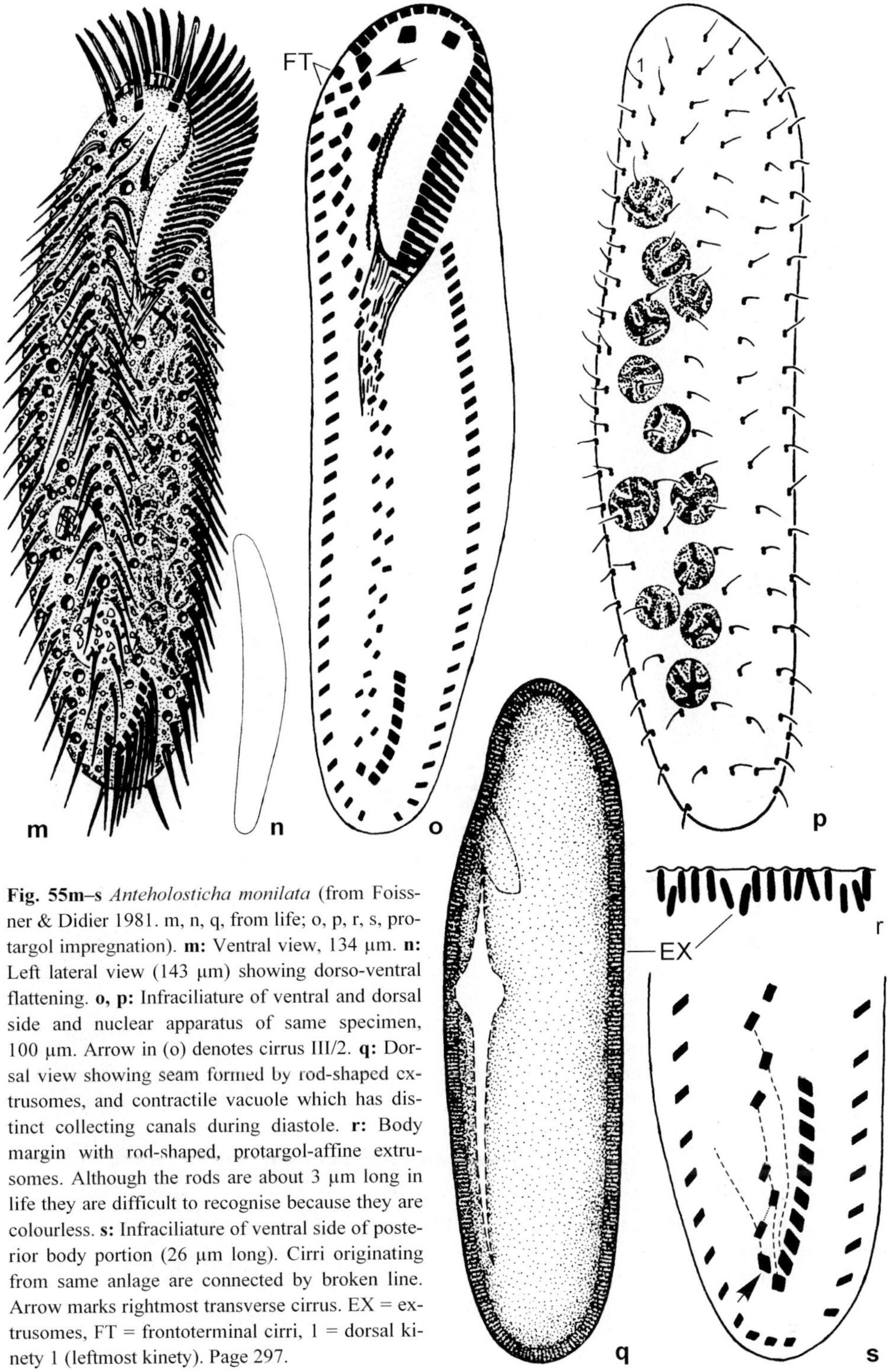

Fig. 55m–s *Anteholosticha monilata* (from Foissner & Didier 1981. m, n, q, from life; o, p, r, s, protargol impregnation). **m:** Ventral view, 134 µm. **n:** Left lateral view (143 µm) showing dorso-ventral flattening. **o, p:** Infraciliature of ventral and dorsal side and nuclear apparatus of same specimen, 100 µm. Arrow in (o) denotes cirrus III/2. **q:** Dorsal view showing seam formed by rod-shaped extrusomes, and contractile vacuole which has distinct collecting canals during diastole. **r:** Body margin with rod-shaped, protargol-affine extrusomes. Although the rods are about 3 µm long in life they are difficult to recognise because they are colourless. **s:** Infraciliature of ventral side of posterior body portion (26 µm long). Cirri originating from same anlage are connected by broken line. Arrow marks rightmost transverse cirrus. EX = extrusomes, FT = frontoterminal cirri, 1 = dorsal kinety 1 (leftmost kinety). Page 297.

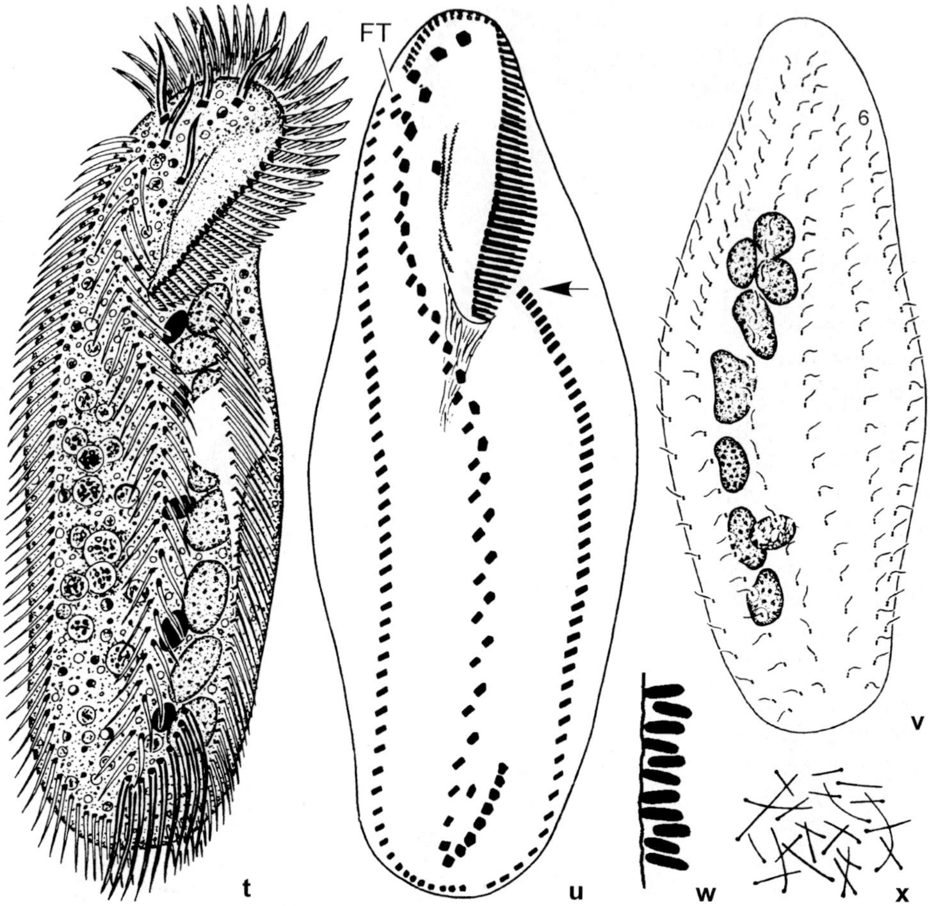

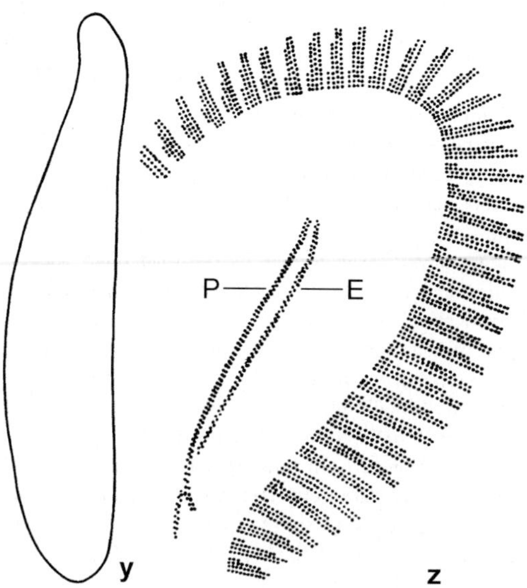

Fig. 55t–z *Anteholosticha monilata* (from Augustin & Foissner 1992. t, y, from life; u–w, protargol impregnation; x, methyl-green pyronin stain; z, silver carbonate impregnation). **t:** Ventral view, 155 µm. **u, v:** Infraciliature of ventral and dorsal side and nuclear apparatus of same specimen, 132 µm. Arrow marks anterior end of left marginal row which shows the ordinary pattern, that is, it is straight and not curved rightwards behind the proximal end of the adoral zone of membranelles as in *Holosticha*. **w, x:** Resting and ejected extrusomes. **y:** Right lateral view showing dorsoventral flattening. **z:** Fine structure of adoral zone of membranelles and undulating membranes. This detail shows that most membranelles have the ordinary fine structure. E = endoral, FT = frontoterminal cirri, P = paroral, 6 = dorsal kinety 6. Page 297.

transverse cirri; nucleoli reticular to clod-shaped (Fig. 55p). Several ellipsoidal micronuclei. Contractile vacuole slightly ahead of mid-body near left cell margin, during diastole with two long collecting canals (Fig. 55q). Pellicle colourless, underneath about 30 rows of short, colourless rods (protrichocysts?), which can be ejected and stain heavily with protargol (Fig. 55q, p). Cytoplasm colourless, usually packed with yellowish-shining globules 2–5 µm across and food vacuoles. Ahead of transverse cirri usually a large defecation vacuole with loose, granular content (Fig. 55m). Movement rapid, able to nestle against debris.

Adoral zone occupies about 33% of body length (Table 19), composed of 37 membranelles on average; without peculiarities. Buccal field small, rather flat (Fig. 55m). Undulating membranes slightly curved and almost parallel, but in spite of this optically intersecting about in mid-portion; paroral composed of two rows of basal bodies, endoral likely of only one.

Cirral pattern and number of cirri of usual variability (Fig. 55o, Table 19). 3–4 enlarged frontal cirri (including cirrus III/2; Fig. 55o). Buccal cirrus distinctly behind anterior end of paroral (Fig. 55o), relatively often lacking in this population (that is, in 4 out of 11 specimens). Frontoterminal cirri in ordinary position, that is, between distal end of adoral zone and anterior end of right marginal row (as is usual, difficult to recognise in life). Midventral complex composed of cirral pairs only, extends from near frontal cirri to transverse cirri (according to Foissner & Didier, the complex is right of midline, a feature which likely must not be over-interpreted). Right cirrus of each pair slightly larger than left cirrus, and cirri of anterior pairs larger than those of posterior pairs. Two pretransverse ventral cirri (Fig. 55o). Transverse cirri distinctly enlarged, arranged in J-shape; rearmost cirri project distinctly beyond rear body end (Fig. 55m, o). Right marginal row begins about at level of frontoterminal cirri, extends to near midline at rear end; 2–3 rearmost cirri sometimes slightly set off feigning caudal cirri (see remarks). Left marginal row commences slightly ahead of proximal end of adoral zone, extends to near midline and thus only slightly separated from rear end of right row; marginal cirri about 10 µm long, bases wide, composed of two rows of basal bodies, those of right row equidistant, those of left row rather narrowly spaced in anterior portion.

Dorsal bristles about 4 µm long in life, arranged in six more or less bipolar rows; rows 1 and 6 slightly shortened anteriorly. Distance between individual bristles increases slightly from anterior to posterior (Fig. 55p). Caudal cirri lacking (see remarks).

Supplementary and deviating data from other populations (see also figures and Table 19): body size 100–170 × 30–50 µm (Augustin & Foissner 1992); body length 150 to 200 µm (Kahl 1928), 100–150 µm (Šrámek-Hušek 1945), 130–170 µm (Dragesco 1966), 250–280 µm (Chorik 1968), 170–180 µm (Dragesco 1970), 90–100 µm (Jones 1974). Anterior body portion rather thin, distinctly curved leftwards, dorsal side strongly vaulted (Kahl 1928). Number of macronuclear nodules: 6–8 (Kahl 1928), 8–10 (Dragesco 1966), 7–8 (7–13 µm long; Dragesco 1970), 14–16 (7–9 µm long; Jones 1974). Number of micronuclei: 4–6 (Dragesco 1966), 6–8 (about 2 µm across; Jones 1974). Extrusomes very difficult to recognise in life, 2.5 µm long, stain intensely with protargol, when ejected of course clearly recognisable in scanning electron micrographs

(Fig. 57a, b); ejected after addition of methyl-green pyronin (Fig. 55x) and are then 4–15 µm long, needle-shaped, straight or curved, and knobbed at one end (Augustin & Foissner 1992); length of extrusomes about 3 µm (Song & Wilbert 1989). Fine structure of adoral zone, see Fig. 55z. Number of adoral membranelles: 33–47 (Dragesco 1966), 53–57 (Dragesco 1970). Number of frontal cirri: 8 (Kahl 1928; this value obviously includes the 3 frontal cirri, the buccal cirrus, cirrus III/2, frontoterminal cirri, and possibly the anterior end of the midventral complex), 4–6 (Dragesco 1970). Borror & Wicklow (1983) studied one population (n = 4) with a single buccal cirrus and three populations with 2–5 (n = 17), 2–3 (n = 19), and 3 (n = 4) buccal cirri, strongly indicating that they confused *A. monilata* with other species. Number of midventral pairs: 19–21 (n = 4), 14–21 (n = 17), 9–26 (n = 19), 25 (n = 4; 4 populations described by Borror & Wicklow 1983). Five transverse cirri, ahead of left cirrus a longitudinal furrow (Kahl 1928); specimen illustrated by Kahl (1932, Fig. 55b) with 12 transverse cirri; further values: 10–13 (Šrámek-Hušek 1945); 6–7 (Dragesco 1970); 12–13 (Jones 1974); 10–12 (n = 4), 8–11 (n = 17), 6–11 (n = 19), 10–12 (n = 4; 4 populations described by Borror & Wicklow 1983). Number of right marginal cirri: 50–59 (Dragesco 1966), 45–55 (Dragesco 1970). Number of left marginal cirri: 37–41 (Dragesco 1966), 40–53 (Dragesco 1970).

Population described by Hartwig (1973) 160–350 µm long; body margins in parallel, both ends rounded. 8–36 macronuclear nodules in left body portion. Cytoplasm hyaline. Eight or more transverse cirri, project beyond rear body end. Three frontal cirri, one (enlarged) cirrus (obviously cirrus III/2) ahead of midventral complex. One buccal cirrus. Marginal rows converging posteriorly. Feeds on diatoms.

The description provided by Grolière (1975) obviously contains several wrong values (see below). Body size 240–270 × 40–70 µm. 15–18 macronuclear nodules (specimen illustrated, however, with only 14 nodules). Adoral zone about 70 µm long. Three enlarged frontal cirri, one buccal cirrus at anterior end of paroral. 60–65 midventral cirral pairs (specimen illustrated, however, with only about 50 pairs). 7–8 (specimen illustrated with only 6) transverse cirri. In total 180–200 marginal cirri, which is likely distinctly over-estimated (specimen illustrated with less than 140 marginal cirri!). Feeds on diatoms, flagellates, and small ciliates.

Fernandez-Leborans (1985) studied the fine structure of the undulating membranes of *A. monilata*. Unfortunately, no general view of a specimens is provided so that identification cannot be checked.

Cell division (Fig. 56a–h): This process is described by Hemberger (1982) and Liu & Jin (2000). As already mentioned above, the paper by the latter authors is in Chinese and of rather low printing quality. Thus, the data presented below are exclusively from

---

Fig. 56a–d *Anteholosticha monilata* (from Hemberger 1982. Protargol impregnation). Infraciliature of ventral side and nuclear apparatus of very early to middle morphogenetic stages, a = 230 µm. Arrow in (a) denotes the oral primordium which originates left of the anteriormost transverse cirri. Arrow in (d) marks the anlage of the left marginal row of the proter which does not originate de novo as in *Holosticha*, but from the anterior end of the parental left marginal row. Note that the anterior end of the midventral complex does not show the characteristic zigzag pattern; possibly some (left?) cirri are resorbed in this population. FT = frontoterminal cirri, OP = oral primordium. Page 297. →

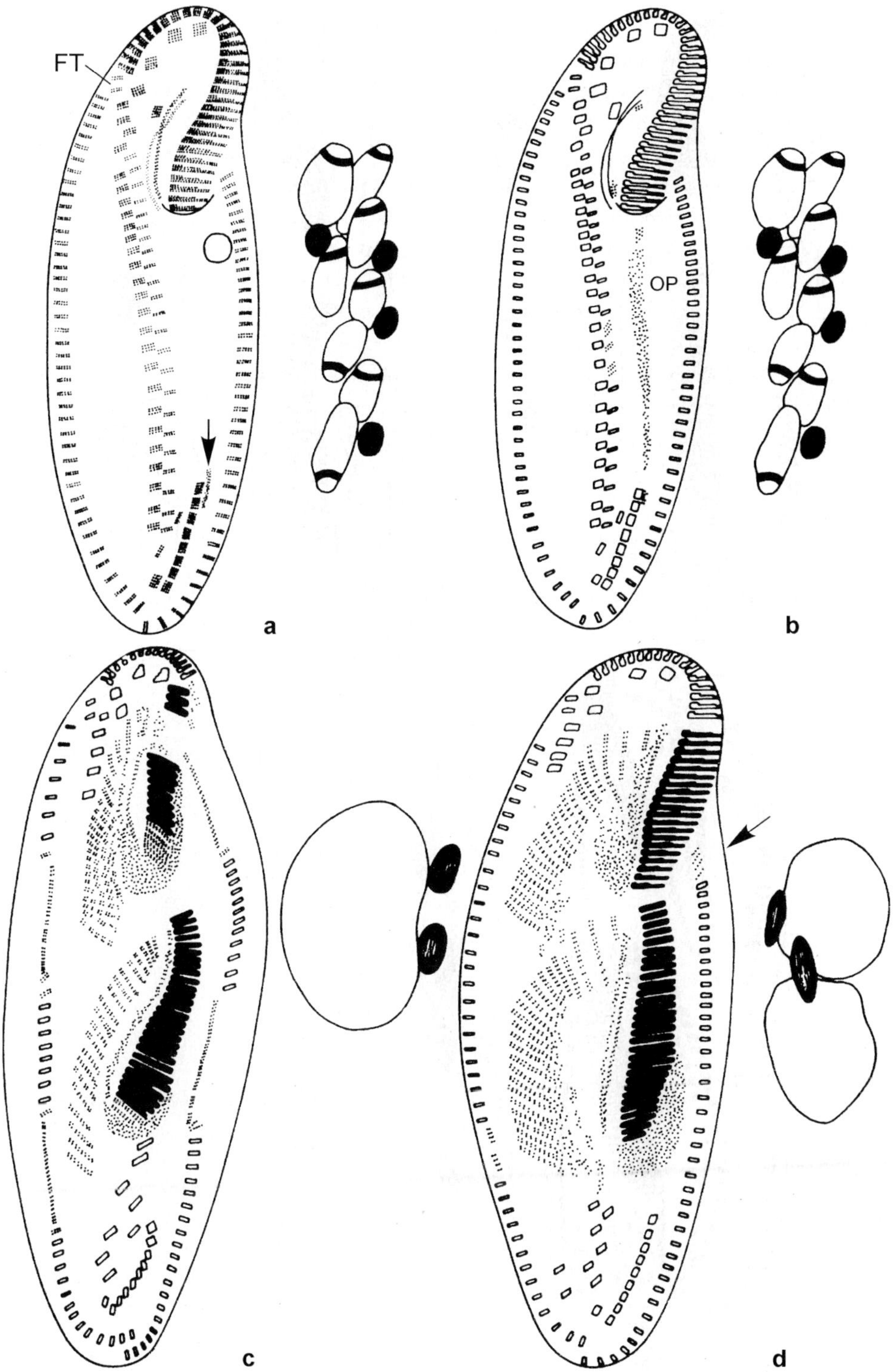
FT
OP
a
b
c
d

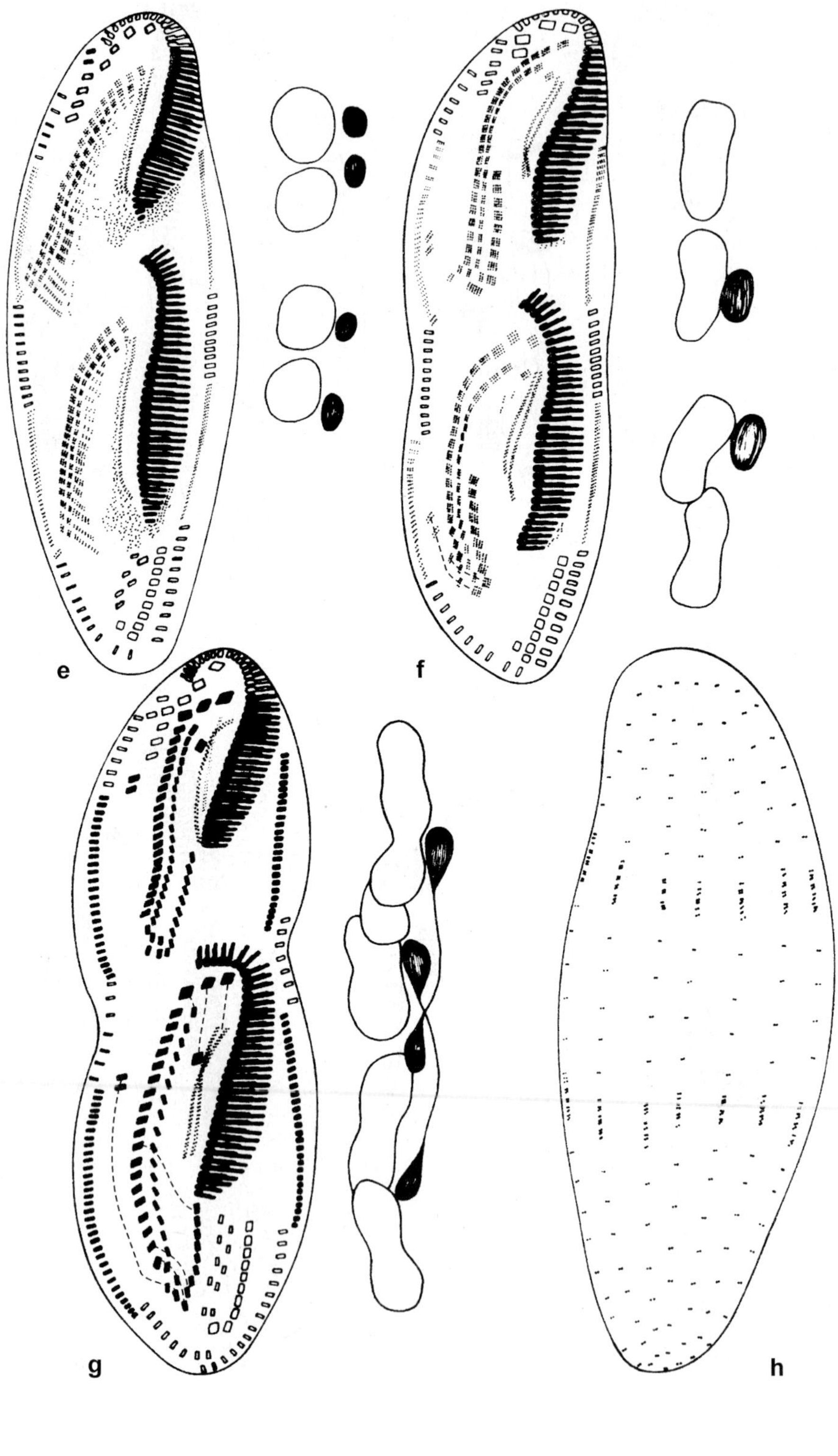
e
f
g
h

Hemberger's dissertation. However, the differences between the two populations seem to be inconspicuous.

The anterior portion of the midventral complex of his population lacks the characteristic zigzag-pattern which is due to the resorption of the anteriormost two left cirri (see below). In addition, his specimens invariably (n = ?) have seven dorsal kineties, whereas other populations usually have six (Table 19). The other data (for example, number of transverse cirri, adoral membranelles, and macronuclear nodules) agree rather well with those of the other populations. Hemberger (1982) did not provide information about the sample site so that one cannot exclude that his population is from an unusual locality. Consequently, I am not quite certain that Hemberger's population is conspecific with, for example, Foissner & Didier's population.

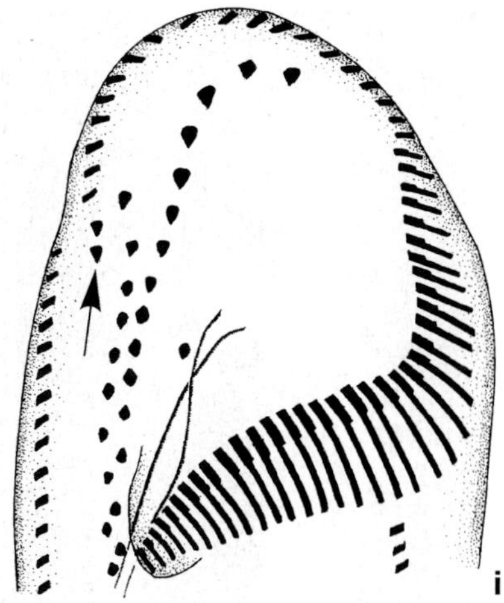

i

Fig. 56i *Anteholosticha monilata* (from Dragesco & Dragesco-Kernéis 1986. Protargol impregnation). The frontal ciliature of this specimen/population differs distinctly from the ordinary *Anteholosticha* pattern indicating that the identification is incorrect. Arrow marks (likely) frontoterminal cirri. Page 297.

Division commences with the formation of an oral primordium near the left transverse cirri (Fig. 56a). This primordium extends anteriad and some modifications occur at the parental undulating membranes. In addition, some left cirri of the midventral complex modify to primordia (Fig. 56b). The next figure shows a rather late stage so that the origin of the various frontal-midventral-transverse primordia remains unknown (Fig. 56c). The proximal portion of the parental adoral zone is resorbed and replaced by new membranelles. The buccal cirrus and several midventral cirri are resorbed, respectively, modified to primordia.

In middle and late dividers the numerous oblique frontal-midventral-transverse cirral anlagen are clearly recognisable (Fig. 56e–g). Basically no peculiarities occur; only in a very late stage are the left cirri of the anteriormost two midventral pairs resorbed (Fig. 56g), producing a somewhat curious cirral pattern in interphasic specimens (Fig. 56a). The rearmost (rightmost) anlage produces, as is usual, four cirri, namely the rightmost transverse cirrus, the corresponding pretransverse ventral cirrus, and two frontoterminal cirri which migrate anteriorly. The second anlage from the right also forms four cirri, namely a transverse cirrus (the posteriormost in interphasic specimens), the corresponding pretransverse ventral cirrus, and the rearmost midventral pair (Fig. 56g).

← **Fig. 56e–h** *Anteholosticha monilata* (from Hemberger 1982. Protargol impregnation). Infraciliature of ventral (e–g) and dorsal (h) side and nuclear apparatus of late to very late morphogenetic stages. Old structures white, new black. Broken lines in (f, g) connect cirri which originate from the same anlage (only cirri of some anlagen connected). Further details, see text. Page 297.

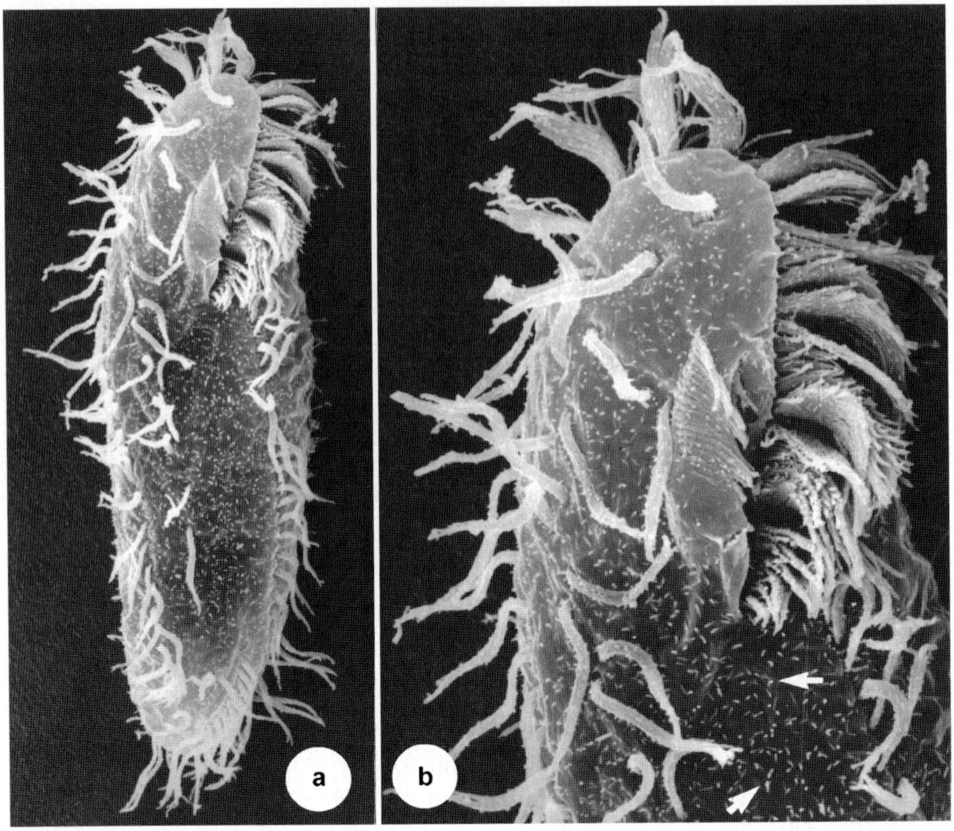

Fig. 57a, b *Anteholosticha monilata* (a, from Foissner et al. 1991; b, from Augustin & Foissner 1992. Scanning electron micrographs). Arrows mark ejected extrusomes. The midventral complex of this population is rather short. Page 297.

Formation of marginal primordia and dorsal kineties proceeds in ordinary manner; that is, each two primordia originate within the parental marginal rows and dorsal kineties (Fig. 56c–h). No caudal cirri are formed at the end of the dorsal kineties. Division of the nuclear apparatus also does not show peculiarities; the macronuclear nodules fuse to a single mass and later divide into the species-specific number of nodules (Fig. 56b–g).

Occurrence and ecology: Common in freshwater habitats, but usually not abundant (Foissner et al. 1991; own observations). Rare in terrestrial habitats, marine records not substantiated by morphological data (see below). Type locality is a slightly saline (2.9‰) ditch with many rhodobacteria near the village of Alt-Fresenburg (Bad Oldesloe region, north Germany), where *Anteholosticha monilata* occurred for a while (Kahl 1928, 1928a).

Records substantiated by morphological data: activated sludge plant from Zellhof, Salzburg, Austria, during late September (Augustin & Foissner 1992); Austrian (beta-

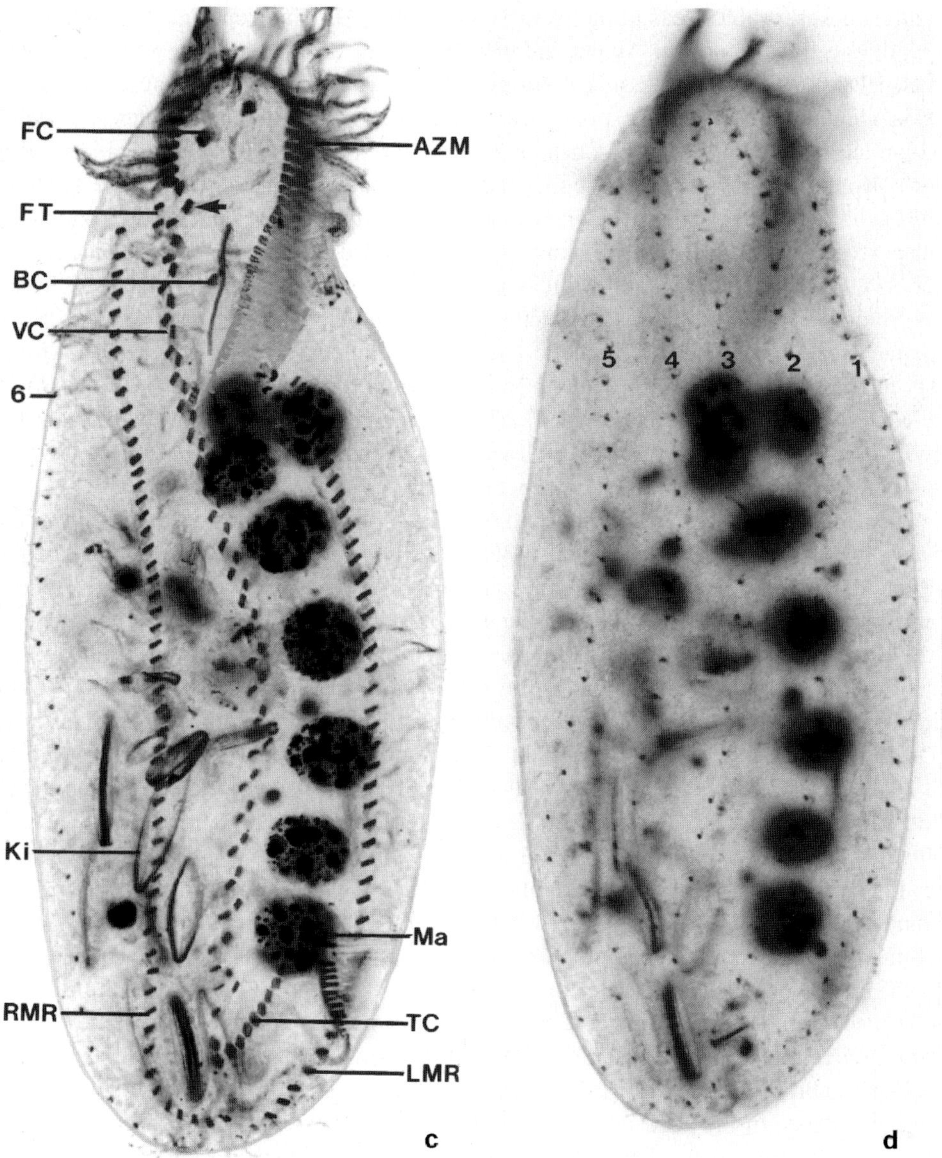

Fig. 57c, d *Anteholosticha monilata* (from Foissner et al. 1991. Protargol impregnation). Infraciliature of ventral and dorsal (as seen from ventral) side and nuclear apparatus of same specimen. Explanation of original labelling: AZM = adoral zone of membranelles, BC = buccal cirrus, FC = middle frontal cirrus, FT = frontoterminal cirri, Ki = ingested diatom, LMR = left marginal row, Ma = rearmost macronuclear nodule, RMR = right marginal row, TC = transverse cirri, VC = midventral complex, 1–6 = dorsal kineties. Page 297.

to alphamesosaprobic areas of the river Traun, Upper Austria in spring and autumn; Fig. 57c, d) and Bavarian (river Amper and tributaries; Fig. 57a, b) running waters (Foissner et al. 1991; 1992, p. 49; Foissner & Moog 1992, p. 101); limnetic habitat in Czechoslovakia (Šrámek-Hušek 1945); pond near Thonon-les-Bains, France (Dragesco 1966); sphagnum pond near Besse-en-Chandesse, France (Grolière 1975; identification uncertain); running waters near Besse-en-Chandesse, France, during May (Foissner & Didier 1981); limnetic habitats in Rhineland, Germany (Pätsch 1974); eutrophic pond (Poppelsdorfer Weiher) in Bonn, Germany, frequently during fall and spring (Song & Wilbert 1989); limnetic habitats in Moldavia (Chorik 1968); small brook near the Biological Station Espegrend, Norway (Reuter 1963); limnetic habitats near Madrid, Spain (Fernandez-Leborans 1985); limnetic habitats in Cameroon (Dragesco 1970); Mobile Bay at Choctaw Point (salinity about 2‰), Alabama, USA (Jones 1974); USA (Borror & Wicklow 1983; no details provided). The population from Ruanda (Butaré) described by Dragesco (2003, Fig. 59a–e) very likely belongs neither to *A. monilata* nor to *Pseudokeronopsis similis*; it is likely a distinct species.

Records not substantiated by morphological data and/or illustrations: various running waters in Austria (own observations; Blatterer 1994; AOÖLR 1992, 1993b, c, 1994, 1995a–d, 1996a–c, 1997a, c); littoral of the Danube River in Vienna (Austria) during January, that is, when the water temperature is only slightly above 0° C (Kaltenbach 1960, p. 170); Lake Traunsee, Upper Austria (Griebler et al. 2002, p. 49); in an experiment on the degradation of artificial sewage (Weninger 1967, p. 318); Belgium (Chardez 1987, p. 13); river Iskar, a tributary of the Danube River and other rivers in Bulgaria (Detcheva 1979, p. 364; 1979a; 1993, p. 34); in moss (20.2 ind. cm^{-2} at current velocity of 20–50 cm^{-1}; 2.8 at 50–100; 0.9 at 100–130; 3.6 at >130) from lotic habitats on travertine barriers of the cascade hydrosystem Plitvice Lakes, Croatia (Primc-Habdija et al. 2000, p. 283); up to 28 ind. ml^{-1} in the river La Dore, a French running water (Grolière et al. 1990, p. 386; Sparagano & Grolière 1991, p. 53, 54); peat-bog moss biotopes in France (Grolière & Njine 1973, p. 14; Grolière 1977, p. 338; 1978); periphyton and sediment of an unpolluted foothill stream in Hesse, Germany (Packroff & Zwick 1996, p. 258); Mettma, a running water in the Southern Black Forest, Germany (Bauer 1987, p. 16); sometimes very abundant (especially during winter) in the aufwuchs and on sand with moderate debris in the Hamburg Harbour, Germany, a polluted and likely slightly brackish habitat (Tent 1981, p. 11; Bartsch & Hartwig 1984, p. 556); with middle to high abundance in the aufwuchs in an alphamesosaprobic reservoir in Germany (Nusch 1970, p. 300); freshwater flats of "Fährmannssand", Elbe estuary, Germany (Pfannkuche et al. 1975, p. 482); littoral of river Elbe near Geesthacht, Germany (Grimm 1968, p. 365); common in a polluted but salt-free groundwater well on the island of Hiddensee, Germany (Münch 1956, p. 434); two clean small rivers (Illach, Eger) in Bavaria, Germany (Foissner 1997a, p. 184); mesosaprobic rivers in Germany (Foissner et al. 1992, p. 101); mesotrophic lake (Heiliges Meer) near the city of Münster, Germany (Mücke 1979, p. 266); backwash sludge of a German waterwork (Foissner 1996b, p. 16); hyporheic interstitial of a German brook (Cleven 2004, p. 77); Covolo della Guerra, a karst cave in the Berici Hills, Vicenza, Italy (Coppellotti & Guidolin 1999, p. 75); brook with geothermal sulphur water near Bibbio in Northern Italy (Madoni & Uluhogian 1997, p.

165); mesosaprobic sites of Italian rivers (Madoni & Bassanini 1999, p. 394; Madoni 2005, p. 60; Madoni & Zangrossi 2005, p. 23); river Lielupe and small rivers in Latvia (Liepa 1973, p. 33; 1983, p. 137; Veylande & Liyepa 1985, p. 82); freshwater (cooling plant) in Moldavia (Chorik & Vikol 1973, p. 68); in bottom of rearing fishponds in the Golysz complex and other fishponds in Poland (Czapik 1959, p. 190; Grabacka 1971, p. 12; Sieminska & Sieminska 1967, p. 56); dystrophic lakes in the Wigry National Park (Poland) in June and September (Czapik & Fyda 1995, p. 67); rare in the benthos of Karlova Ves, a branch of the Danube River in Bratislava, and other running waters in Slovakia (Szentivány & Tirjaková 1994, p. 93; Tirjaková 1992, p. 293; 2003, p. 36; Matis et al. 1996, p. 12); freshwater habitats near Barcelona, Spain (Margalef López 1945, p. 376); benthos of the Terterchay Reservoir and other reservoirs in Azerbaijan (Alekperov 1981, p. 57; 1982, p. 46; 1982a, p. 87; 1983, p. 22); reservoir in Ukraine (Kravchenko 1969, p. 73); freshwater in the Suoxiyu Nature Reserve in Hunan Province, China (Shen & Gong 1989, p. 83); Hanjiang river, China (Shen et al. 1994, p. 207); freshwaters in the Yyuelushan area, China (Yang 1989, p. 157); Yellow river in Lanzhou, China (Ma 1994, p. 95); freshwater habitats in the Chongqing area, China (Su et al. 1988, p. 3); at 19.5° C in littoral area of the Mountain Lake (1179 m above sea level), Giles County, Virginia, USA (Bovee 1960, p. 357); rare in a lake about 80 km north of Montreal, Canada (Puytorac et al. 1972, p. 435).

Hartwig (1973) found *A. monilata* in a brackish water pond on Jordsand, a small island near Sylt (Germany, North Sea; see remarks). The records from a brackish pond in the Camarque (France) by Dragesco (1960, p. 312), salt lakes in Azerbaijan by Aliev (1982, p. 87), a salt lake in Romania (Tucolesco 1965, p. 160), and the littoral region of the Black Sea by Tucolesco (1962a, p. 813) are not substantiated by morphological data.

Records from terrestrial habitats: dry mosses from Bavaria (Wenzel 1953, p. 111); in the lichen *Usnea filipendula* on *Picea abies* from the Haitzingalm (about 1700 m above sea level), Gastein area, Salzburg, Austria (Foissner 1986, p. 44; as *Holosticha similis*); soil samples from near Kelso, Southern Scotland, UK (Finlay et al. 2001, p. 363); straw from rice stubble and soil of drained rice fields in Japan (Takahashi & Suhama 1991b, p. 106); soil in Antarctica (Sudzuki 1979, p. 123).

Feeds on ciliates and diatoms (Kahl 1932, Jones 1974), according to Foissner & Didier (1981) on diatoms and bacteria. Biomass of 10^6 specimens about 51 mg (Nesterenko & Kovalchuk 1991, Foissner et al. 1991). *Anteholosticha monilata* is rather resistant against heavy metal pollution. According to Fernandez-Leborans & Antonio-Garcia (1986, p. 209), it survived nearly until the end of an experiment in fractions with 500 μg l^{-1} lead and 200 μg l^{-1} zinc.

According to Kahl (1932), *Anteholosticha monilata* occurs in katharobic and mesosaprobic waters. Foissner et al. (1991) classified it as indicator of alpha- to betamesosaprobic water quality (a–b; b = 3, a = 6, p = 1, I = 3, SI = 2.8; Table 12); see also Sládeček (1973), Sládeček et al. (1981), Wegl (1983), Foissner (1988a), Foissner et al. (1995, 1995a), Sládeček & Sládečková (1997), Foissner & Berger (1996), and Berger & Foissner (2003). Sládeček (1988) proposed a slightly different classification (b = 1, a = 8, p = 1, I = 4, SI = 3.0).

Supposed synonym of *Anteholosticha monilata*

Holosticha corlissi Fernandez-Galiano & Calvo, 1992 (Fig. 58a–d, Table 20)

1992 ***Holosticha corlissi* n. sp.** – Fernandez-Galiano & Calvo, J. Protozool., 39: 600, 603, Fig. 1–9, 10B, Table 1 (Fig. 58a–d; original description; the holotype slide is deposited in the Non Insect Invertebrate Collection, National Museum of Natural Sciences of Madrid, Spain, accession number MNCN-39.03/1).

2001 ***Holosticha corlissi* Fernandez-Galiano and Calvo, 1992** – Berger, Catalogue of ciliate names 1. Hypotrichs, p. 34 (nomenclator containing all basionyms, combinations, and higher taxa of hypotrichs).

Nomenclature: This species was dedicated to John O. Corliss on the occasion of his 70th birthday (Fernandez-Galiano & Calvo 1992, p. 600).

Remarks: This species is well characterised by Fernandenz-Galiano & Calvo (1992). They compared it with *A. xanthichroma*, *A. sigmoidea*, and *A. monilata*, and supplemented Borror & Wicklow's (1983) key using only the end of the marginal rows (overlapping in *H. corlissi*; non-overlapping in *A. xanthichroma*) as discriminating feature. However, there are further characteristics which show that these two species are clearly separated, namely number of dorsal kineties (6 vs. 4), macronuclear nodules (on average 14 vs. 28; ranges slightly overlapping), and midventral cirral pairs (on average 25 vs. 41; ranges slightly overlapping). Unfortunately they overlooked the redescription of *A. monilata* by Foissner & Didier (1981). Both populations have, besides the more or less same cirral pattern, the same number of macronuclear nodules (15 on average) and dorsal kineties (6). In addition, Fernandez-Galiano & Calvo (1992) described subpellicular granules of mucocyst type, however, without mentioning details (size, colour, arrangement). Thus, the sole difference between *A. monilata* and *Holosticha corlissi* is in the end of the marginal rows, namely not overlapping (Fig. 55o, s, u) against overlapping (Fig. 58b, d). Since the same feature is used to separate *A. adami* (Fig. 74b, g, i) and *A. intermedia* (Fig. 61i, l), I do not finally synonymise *H. corlissi* with *A. monilata*. Further populations of *H. corlissi* should be studied to show whether or not the main species feature (overlapping marginal rows) is stable. If the feature can be confirmed, then Fernandez-Galianos & Calvo's species has to be transferred to *Anteholosticha*. The description below is kept rather short and the reader is mainly referred to the illustrations and to Table 20. The original description contains five very good micrographs of silver carbonate-impregnated specimens documenting most features.

Morphology: Body size about 160–220 × 40–70 μm in life. Body outline elliptical with more or less parallel margins (slightly narrowed at level of adoral zone), anterior end somewhat more widely rounded than posterior. On average 14 macronuclear nodules in left body portion; individual nodules with small spherical nucleoli. Micronuclei arranged near macronuclear nodules. "Endoplasm contains numerous subpellicular granules that may be mucocysts" (unfortunately no details about the cortical granulation are provided). Movement moderately rapid. Adoral zone occupies about 33% of body length (about 72 μm), composed of 44 membranelles on average of ordinary fine structure (2 long rows of basal bodies, 1 row with about 15 basal bodies, shortest row with

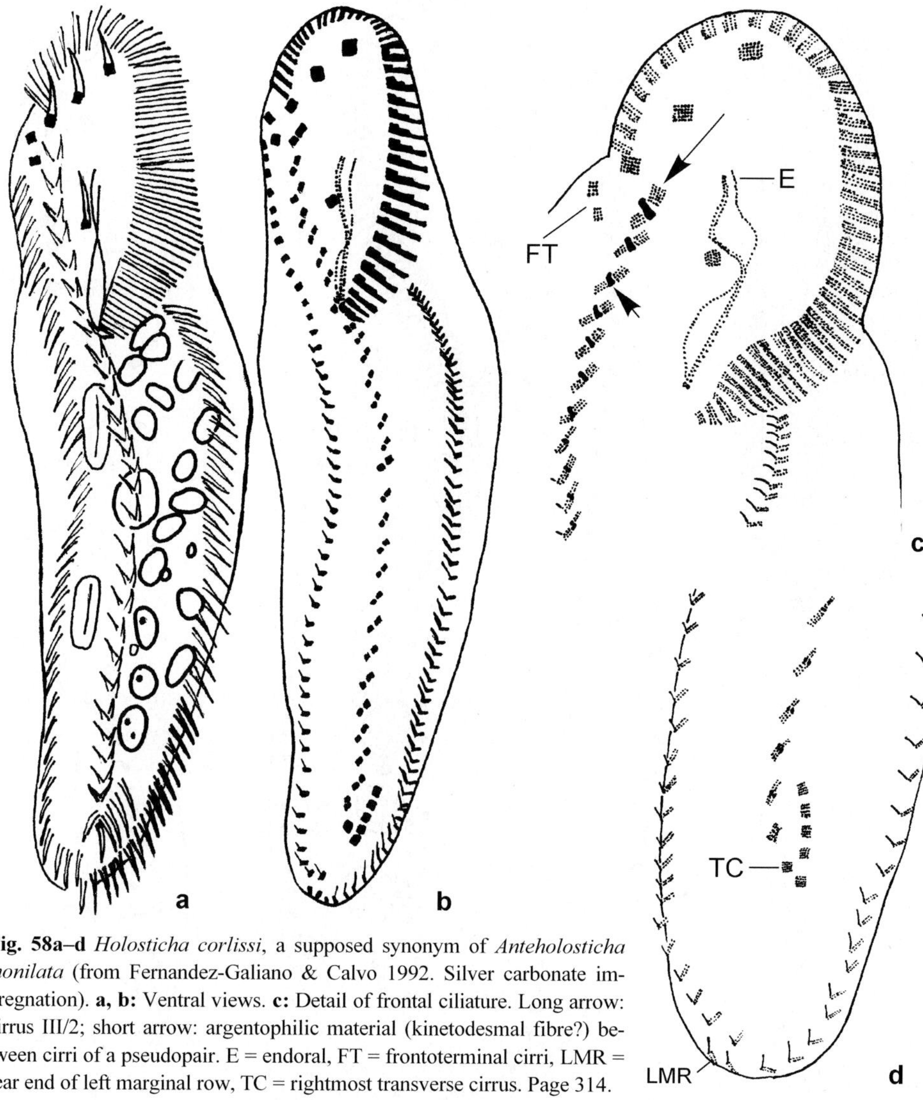

Fig. 58a–d *Holosticha corlissi*, a supposed synonym of *Anteholosticha monilata* (from Fernandez-Galiano & Calvo 1992. Silver carbonate impregnation). **a, b:** Ventral views. **c:** Detail of frontal ciliature. Long arrow: cirrus III/2; short arrow: argentophilic material (kinetodesmal fibre?) between cirri of a pseudopair. E = endoral, FT = frontoterminal cirri, LMR = rear end of left marginal row, TC = rightmost transverse cirrus. Page 314.

3–6 basal bodies). Paroral composed of two rows of basal bodies, endoral of a single row. Fine structure of cirri, see Figs. 58c, d. Single buccal cirrus behind anterior end of paroral. Midventral complex composed of about 25 cirral pairs; between the two cirri of each pseudopair argentophilic material (possibly a kinetodesmal fibre; Fig. 58c). Usually six (range 5–8; in the text, they erroneously wrote 7 as maximum value) transverse cirri. Left marginal row extends far onto right body margin posteriorly so that it overlaps distinctly with right marginal row (Fig. 58b, d). Invariably six dorsal kineties. Caudal cirri obviously lacking because neither mentioned nor illustrated.

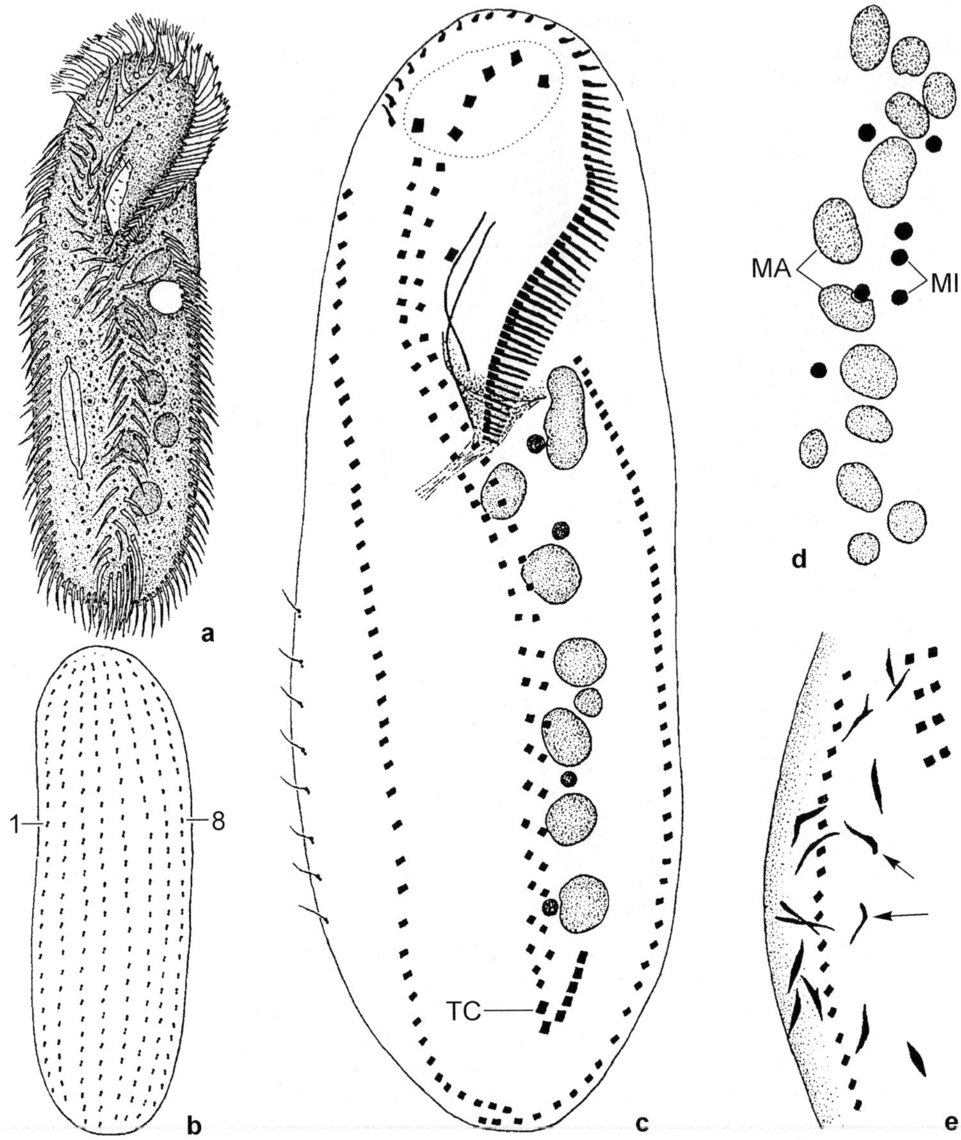

Fig. 59a–e *Holosticha similis* sensu Dragesco (2003; a from life; b–e, protargol impregnation). The nuclear apparatus of this population is reminiscent of *Anteholosticha monilata* which, however, has only three frontal cirri, whereas the specimens shown on this plate has about six frontal cirri (dotted line in c; details see remarks on *A. monilata*). Arrows in (e) mark exploded extrusomes. MA = macronuclear nodules, MI = micronuclei, TC = rightmost transverse cirrus, 1, 8 = dorsal kineties. Page 299.

Occurrence and ecology: As yet found only at the type location, which is the beech wood of Montejo de la Sierra in/near the Spanish capitol Madrid. Fernandez-Galiano & Calvo (1992) discovered it in a sample of the moss *Calliergonella cuspidata*, where *H.*

corlissi occurred abundantly together with other hypotrichs and heterotrichs. Feeds on green algae and diatoms. The ciliates were cultivated in mineral water with *Chlorogonium elongatum* as food.

Anteholosticha intermedia (Bergh, 1889) comb. nov. (Fig. 60a–d, 61a–u, Tables 12, 19)

1889 ***Urostyla intermedia* n. sp.** – Bergh, Archs Biol., 9: 497, 512, Planche XXXV, Fig. 11–18 (Fig. 60a, b; original description; no formal diagnosis provided and no type material available).

1932 ***Keronopsis (Holosticha) multistilata* Kahl, 1928** – Kahl, Tierwelt Dtl., 25: 574, Fig. 104_2 (Fig. 61b; misidentification; revision).

1932 ***Keronopsis muscorum* spec. n.** – Kahl, Tierwelt Dtl., 25: 576, Fig. 101_{24} (Fig. 61a; original description of new synonym; no type material available and no formal diagnosis provided).

1932 ***Holosticha (Urostyla) intermedia* (Bergh, 1889)** – Kahl, Tierwelt Dtl., 25: 585, Fig. 106_{29} (Fig. 60c; revision; combination with *Holosticha*, see nomenclature).

1933 ***Keronopsis multistilata* Kahl 1928** – Kahl, Tierwelt N.- u. Ostsee, 23: 109, Fig. 16.31 (Fig. 61c; revision of marine ciliates).

1963 ***Keronopsis macrostoma* sp. n.** – Reuter, Sarsia, 10: 6, Fig. 7, 12 (Fig. 61d; original description of new synonym; no type material available and no formal diagnosis provided).

1972 ***Holosticha multistilata* Kahl, 1928** – Borror, J. Protozool., 19: 11 (revision of hypotrichs).

1977 ***Holosticha multistilata* Kahl** – Buitkamp, Acta Protozool., 16: 269, Abb. 13 (Fig. 61n; redescription).

1982 ***Holosticha multistilata* (Kahl, 1932)** – Foissner, Arch. Protistenk., 126: 50, Abb. 7a–d, Tabelle 9 (Fig. 61f, h–j; incorrect date; a voucher slide [1981/90] is deposited in the Oberösterreichische Landesmuseum in Linz [LI], Austria; now this slide/population is the neotype slide/population of *A. intermedia*, see remarks).

1983 ***Holosticha multistylata* Kahl, 1928** – Borror & Wicklow, Acta Protozool., 22: 121, Fig. 16, 24, 25 (Fig. 61e; revision of urostylids; see remarks; incorrect subsequent spelling).

1986 ***Holosticha multistilata* Kahl, 1928** – Dragesco & Dragesco-Kernéis, Faune tropicale, 26: 448, Planche 131E (Fig. 61n; review).

1991 ***Holosticha multistilata* Kahl, 1928** – Foissner, Blatterer, Berger & Kohmann, Informationsberichte des Bayer. Landesamtes für Wasserwirtschaft, 1/91: 236, Abb. 1–5, 7, 8, 10–14 not Abb. 6, 9 (Fig. 61g, o–s, u; review and guide to freshwater ciliates).

1992 ***Holosticha multistilata* Kahl, 1928** – Shen, Liu, Song & Gu, Protozoa, p. 154, Fig. 2-24Ca, Cb (redrawings of Fig. 61f, i; record from China).

1993 ***Holosticha multistylata* Kahl, 1928** – Shin & Kim, Korean J. Syst. Zool., 1: 252, Fig. 1A–C, Table 1 (Fig. 61k–n; redescription; incorrect subsequent spelling).

1994 ***Holosticha multistylata* Kahl, 1928** – Shin, Dissertation, p. 66, Fig. 8, Table 7 (Fig. 61k–n; data already published in previous entry; incorrect subsequent spelling).

2000 ***Holosticha multistylata*** – Shin, Hwang, Kim, Wright, Krawoczyk & Lynn, Europ. J. Protistol., 36: 295 (sequence of ssrRNA genes; incorrect subsequent spelling).

2001 ***Urostyla intermedia* Bergh, 1889** – Berger, Catalogue of ciliate names 1. Hypotrichs, p. 101 (nomenclator containing all basionyms, combinations, and higher taxa of hypotrichs).

2003 ***Anteholosticha muscorum* (Kahl, 1932) n. comb.** – Berger, Europ. J. Protistol., 39: 377 (combination of synonym with *Anteholosticha*).

Nomenclature: Bergh (1889) selected the species-group name *intermédi·us -a -um* (Latin; lying in between) likely because he assumed that the present species is "between *Urostyla grandis* and *Urostyla weissei* (now *Paraurostyla weissei*)" as concerns the ciliature and the nuclear apparatus. The species-group name *muscorum* (Latin; Genitive Plural) refers to the occurrence in moss-habitats. The species-group name *macrostoma* is a composite of the Greek adjective *makrós* (long, large) and the Latin noun *stóma* (mouth, opening) and refers to the large oral apparatus (more than 33% of body length). For derivation of the name *multistilata*, see *Anteholosticha multistilata*.

When I redefined *Holosticha* (Berger 2003), I was uncertain about the status of *Urostyla intermedia* (often also classified in *Holosticha*) and therefore did not include it in the list of species. Thus, it has to be transferred to *Anteholosticha* in the present book.

Kahl (1932) classified *Keronopsis* as subgenus of *Holosticha*. Thus, the correct names in Kahl (1932) are *Holosticha (Holosticha) intermedia* (Bergh, 1889) Kahl, 1932, *Holosticha (Keronopsis) multistilata* Kahl, 1928, and *Holosticha (Keronopsis) muscorum* Kahl, 1932. *Ceronopsis muscorum* Kahl in Bajomi (1969, p. 236) is an incorrect subsequent spelling of the genus-group name *Keronopsis*.

Dingfelder (1962, p. 616) recognised that Kahl (1932) has classified *Keronopsis* and other groups (e.g., *Paruroleptus*) as subgenera of *Holosticha*. He did not agree with this classification and treated Kahl's subgenera – "according to the wide use in the literature" – as genera. Thus, Dingfelder can be considered as author who formally transferred *H. muscorum* to *Keronopsis*. In my catalogue (Berger 2001), I supposed that Gellèrt (1956) had done this act.

Remarks: The original description of *Urostyla intermedia* Bergh, 1889 is rather detailed. Bergh compared his species with *Urostyla grandis*, but he also discussed that it could belong to *Holosticha* according to the cirral pattern. Finally he classified it in *Urostyla* because the nuclear apparatus and the cortical granulation closely resembled that of *U. grandis*. Kahl (1932) recognised that Bergh's species does not belong to *Urostyla*, but to *Holosticha* and thus transferred it into the subgenus *Holosticha (Holosticha)*. Simultaneously, he described *H. muscorum* (Fig. 61a), but did not recognise that this species and his *H. multistilata* (Fig. 61b) agree very well with *H. intermedia*. Borror (1972, p. 11) synonymised *H. intermedia* with *Pseudoamphisiella lacazei*, which, however, is rather different. By contrast, Borror & Wicklow (1983) accepted *H. intermedia*, but synonymised it with *Anteholosticha monilata* Kahl, 1928, which has a rather different nuclear apparatus and cortical granulation.

The illustration and the description provided by Bergh (1889, Fig. 60a) agree rather well with the data on *Holosticha muscorum* Kahl, 1932 (Fig. 61a), and *H. multistilata* sensu Kahl (1932; Fig. 61b) and sensu Foissner (1982, Fig. 68f–j). Especially the size, the cirral pattern including the three frontal cirri, the nuclear apparatus (many macronuclear nodules), and – very important – the cortical granules ("oil droplets arranged in longitudinal rows"; Fig. 60a) strongly indicate that these populations are conspecific. Bergh (1889) did not draw buccal cirri, but illustrated the left cirri of each midventral pair of the anterior midventral complex portion very close to the paroral. This indicates that he confused these parts of the cirral pattern, a fact which must not be overinterpreted because these details are rather difficult to recognise without staining. In addition, it is very unlikely that this species, which is rather common, but usually not abundant in terrestrial and limnetic habitats, was discovered just 70 years ago.

For further details on the complicated systematics of this species, see also this chapter at *A. multistilata* and Table 18. *Anteholosticha intermedia*, respectively its most important synonym *Holosticha muscorum*, is the species with the three enlarged frontal cirri and a single enlarged cirrus (= cirrus III/2) behind the right frontal cirrus. The populations described after protargol impregnation differ in the number of dorsal kineties (3 in Eurasian populations, 4 in African population; Table 19), indicating a speciation process in these populations from different biogeographic regions.

The general morphology and the cirral pattern of *Keronopsis macrostoma* Reuter, 1963 agree rather well with *A. intermedia*. Hemberger (1982) and Borror & Wicklow (1983) synonymised it with *Holosticha multistilata*. However, in their papers *H. multistilata* also included *H. muscorum*, which is classified as junior synonym of *A. intermedia* in the present review.

As discussed above, there are great systematic problems with the present species; that is, there is an exceptional need to define *A. intermedia* (Bergh, 1889) objectively. Since no type material of *A. intermedia* or its synonyms is available, and since Bergh (1889) did not fix the type locality in detail it seems appropriate to designate a neotype (ICZN 1999, Article 75).

Aescht (2003) erroneously assumed that most/all(?) voucher slides deposited in the Upper Austrian Museum in Linz are automatically neotypes. For example, she designated the voucher slide 1981/84 of "*Histriculus muscorum*" deposited by Foissner (1982) as neotype slide of *Sterkiella histriomuscorum* (Aescht 2003, p. 391). However, neotypification of *S. histriomuscorum* was just done by Foissner & Berger (1999, p. 217), who deposited two neotype slides (accession numbers 1999/109, 110). Aescht's (2003) "neotypifications" fall under Article 75.2 of the ICZN (1999; curatorial routine) and are therefore invalid. Consequently, the voucher slide 1981/90 of *Holosticha multistilata* – deposited by Foissner (1982), and designated as neotype slide of "*H. multistilata* Kahl 1928" by Aescht (2003) – is still (until now) a voucher slide and not a neotype slide. In the present book I fix this slide (1981/90) as neotype slide/population of *Anteholosticha intermedia*.

To make the neotypification valid, I publish the particulars according to Article 75.3 of the ICZN (1999):

(i) See above for details on the systematics of *A. intermedia*, and read also the remarks on *Anteholosticha multistilata* for a discussion of the tricky history of *Holosticha muscorum*, the junior synonym of *A. intermedia*. In addition, Bergh (1889) did not fix the type locality in detail. The designation of Foissner's (1982) slide (accession number in Linz: 1981/90) as neotype of "*Holosticha multistilata*" by Aescht (2003, p. 391) is invalid (see above).

(ii) *Anteholosticha intermedia* agrees very well with *Holosticha muscorum* so that conspecificity is beyond reasonable doubt. *Holosticha muscorum* Kahl, 1932 was classified as synonym of *A. multistilata* from 1972 (Borror 1972) until now. A thorough check of the available data, including the original descriptions of *A. intermedia* and its synonym *H. muscorum*, yielded the following main differentiating feature between *A. intermedia* and *A. multistilata*: three enlarged frontal cirri and one cirrus (= cirrus III/2) behind right frontal cirrus against 8–10 enlarged frontal cirri in *A. multistilata*. This difference is not only evident from the original descriptions (Fig. 60a, 84a), but also from redescriptions after protargol impregnation (e.g., Fig. 61i, 84c).

(iii) The most detailed redescription of the present species is that by Foissner (1982; designated as *Holosticha multistilata*). Thus, the specimen illustrated in Fig. 61i, j (Abb. 7b, c in Foissner 1982) is designated as neotype. For a description of this population, see below.

(iv) There is no indication that Bergh (1889) made slides which are still available. It is also generally known, that no type material is available from the species described by Kahl (1932). Reuter (1963), the author of the synonym *Keronopsis macrostoma*, wrote that "No preparates have been made because ..".

(v) The population described by Foissner (1982; as *Holosticha multistilata*) agrees basically with the original descriptions of *Anteholosticha intermedia* and *H. muscorum* and with the redescription of *H. multistilata* by Buitkamp (1977). It should be stated expressly that the description by Foissner (1982) does not agree with the original description of *Holosticha multistilata* by Kahl (1928; Fig. 84a). Kahl's (1928) paper is not listed by Foissner (1982) and other workers, indicating that they did not consult the original description, which differs from the redescription by Kahl (1932). Bergh's (1889) paper is listed by Foissner (1982), however, only mentioned in the context with *Holosticha sylvatica* and not with *H. multistilata*.

(vi) Bergh (1889) discovered *A. intermedia* in an infusion of beech litter which he collected from a pond. Probably this pond was in/near Copenhagen (Denmark), where he likely lived and worked; however, the type locality is not explicitly mentioned so that we cannot be certain that Copenhagen is indeed the locus classicus. The ecological and faunistic data show that *A. intermedia* is one of few species which are rather common both in freshwater and soil (Foissner et al. 1991). Thus, it seems justifiable to select Foissner's (1982) population, which is from a field in Lower Austria, as neotype material.

(vii) The protargol slide (accession number 1981/90) of the neotype population is deposited in the Oberösterreichische Landesmuseum in Linz (LI), Upper Austria. In Aescht (2003, p. 391) this slide is erroneously designated as neotype slide of *Holosticha multistilata* (see above for explanation).

Borror & Wicklow (1983) provided one illustration (Fig. 61e) and three scanning electron micrographs (not shown in present book). I am uncertain whether or not the SEM micrographs in fact show the present species because the number of cirri in the frontal area is rather high. Furthermore, one (teratological?) specimen has several partly shortened left marginal rows and supplementary midventral cirri. For a discussion of *Keronopsis muscorum* sensu Gellèrt (1956), see insufficient redescriptions. *Holosticha intermedia* sensu Wiackowski (1988) lacks cortical granules (mucocysts; his character 26) and therefore cannot be identical with the present species. The illustration by Haneda (1971) is rather simple; however, since it basically agrees with the significant descriptions it can be accepted (Fig. 60d).

Anteholosticha intermedia differs from *A. multistilata* by the lower number of frontal cirri (4 if cirrus III/2 is included vs. about 10). Unfortunately we do not have detailed live data from *A. multistilata*. Thus, it cannot be excluded that the two species differ in other features too, for example, the cortical granulation. *Anteholosticha estuarii*, which is also rather similar, has up to three cirri behind the right frontal cirrus

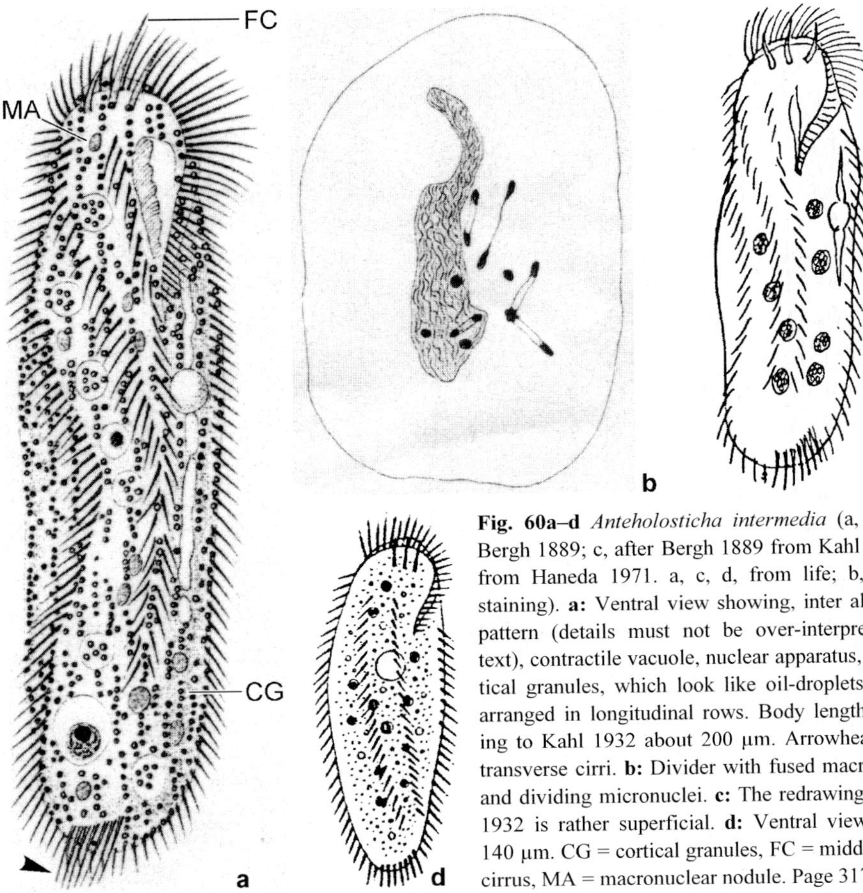

Fig. 60a–d *Anteholosticha intermedia* (a, b, from Bergh 1889; c, after Bergh 1889 from Kahl 1932; d, from Haneda 1971. a, c, d, from life; b, nuclear staining). **a:** Ventral view showing, inter alia, cirral pattern (details must not be over-interpreted; see text), contractile vacuole, nuclear apparatus, and cortical granules, which look like oil-droplets and are arranged in longitudinal rows. Body length according to Kahl 1932 about 200 µm. Arrowhead marks transverse cirri. **b:** Divider with fused macronucleus and dividing micronuclei. **c:** The redrawing by Kahl 1932 is rather superficial. **d:** Ventral view, 100 to 140 µm. CG = cortical granules, FC = middle frontal cirrus, MA = macronuclear nodule. Page 317.

(identity with the present species cannot be excluded). In *Anteholosticha adami* the marginal rows are distinctly separated posteriorly (Fig. 74b, g, i).

Morphology: This species has been redescribed several times (see list of synonyms). The neotype population (= *Holosticha multistilata* sensu Foissner 1982) is described first. This is followed by data from the original descriptions of *U. intermedia* and *H. muscorum* and supplementary and deviating data from other sources.

Description of neotype population (Fig. 61f–j, Table 19): Size 130–170 × 35–45 µm, ratio of body length:width about 3.7:1 in life (Fig. 61f), 3.1:1 on average after protargol impregnation (Table 19). Body outline elliptical, margins usually distinctly converging posteriorly (Fig. 61f, h); body distinctly flattened dorsoventrally (Fig. 61g). About 100 macronuclear nodules scattered throughout cytoplasm (Fig. 61j). Contractile vacuole slightly ahead of mid-body near left body margin, during diastole with two short collecting canals (Fig. 61h). Pellicle delicate, flexible. Cortical granules arranged in rows, about 0.5 µm across and yellow-green, so that cells appear slightly yellow-green at low magnification (Fig. 61h). Cytoplasm colourless, rather densely filled with

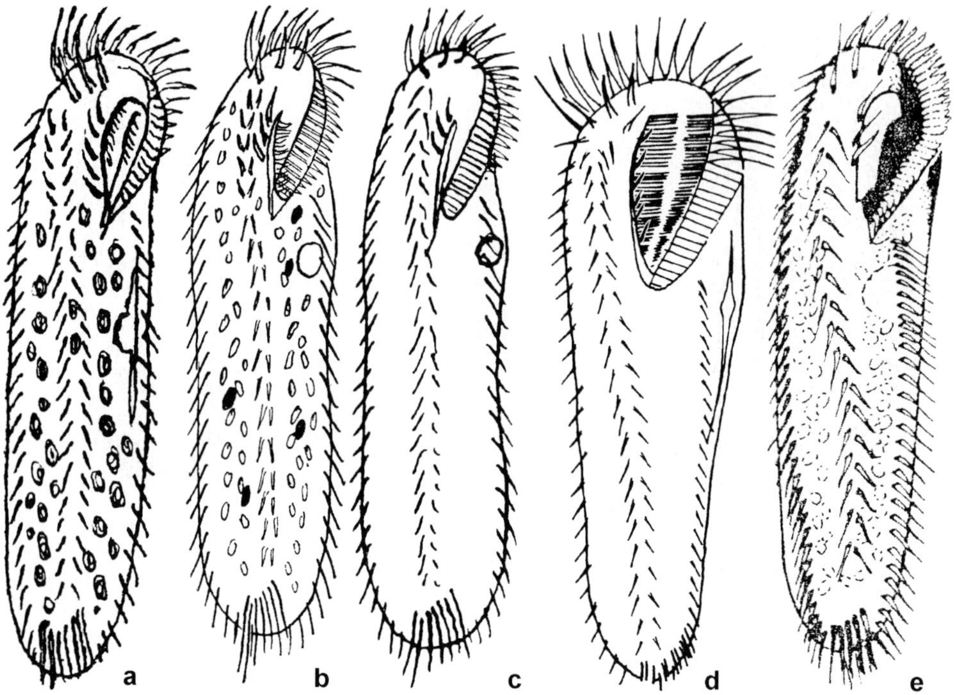

Fig. 61a–e *Anteholosticha intermedia* (a, *Holosticha muscorum* from Kahl 1932; b, c, *Holosticha multistilata* from Kahl 1932, 1933; d, from Reuter 1963; e, from Borror & Wicklow 1983. a–d, from life; e, from life?). Ventral views, a = 250 μm, b = 250 μm, c, d = individual size not indicated, e = 104 μm. In all specimens illustrated the midventral complex extends very close to transverse cirri, whereas in Foissner's (1982) population the complex terminates distinctly ahead of transverse cirri. Page 317.

yellow-green globules 1–3 μm across. Food vacuoles up to 20 μm in diameter. Movement rapid, gliding, adheres closely to soil particles.

Adoral zone occupies 35% of body length on average (Table 19), of ordinary shape, composed of 30 membranelles of usual fine structure (Fig. 61i). Buccal area deep, at the base (roof) a delicate, obliquely striated argyrophilic structure (Fig. 61i). Paroral and endoral composed of basal body pairs, distinctly curved and optically intersecting about at mid-portion; paroral commences almost near left frontal cirrus and slightly ahead of endoral, which is, however, somewhat longer (Fig. 61i); cilia of undulating membranes about 25 μm long.

Cirral pattern and number of cirri of usual variability (Fig. 61i, Table 19). Three frontal cirri distinctly, buccal cirri and cirrus (= III/2) behind right frontal cirrus slightly enlarged, all about 20 μm long. Two frontoterminal cirri in ordinary position. Midventral complex basically composed of midventral pairs, cirri of each pair about of same size, however, cirri become smaller from anterior to posterior, rather irregularly arranged; usually more right cirri than left; midventral cirri about 15 μm long (ontogenetic data are needed to clarify midventral pattern). 1–2 pretransverse ventral cirri,

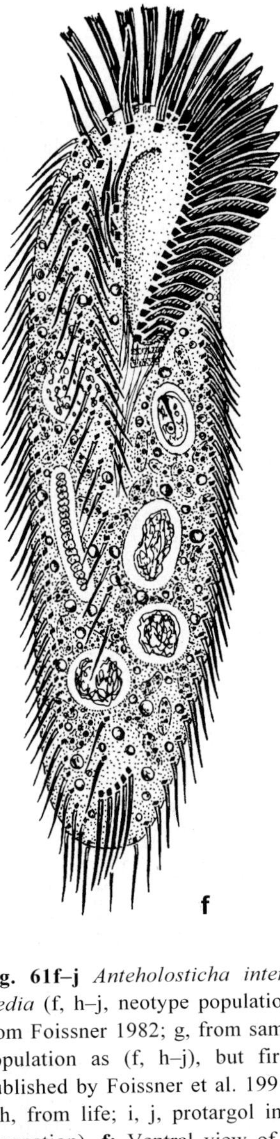

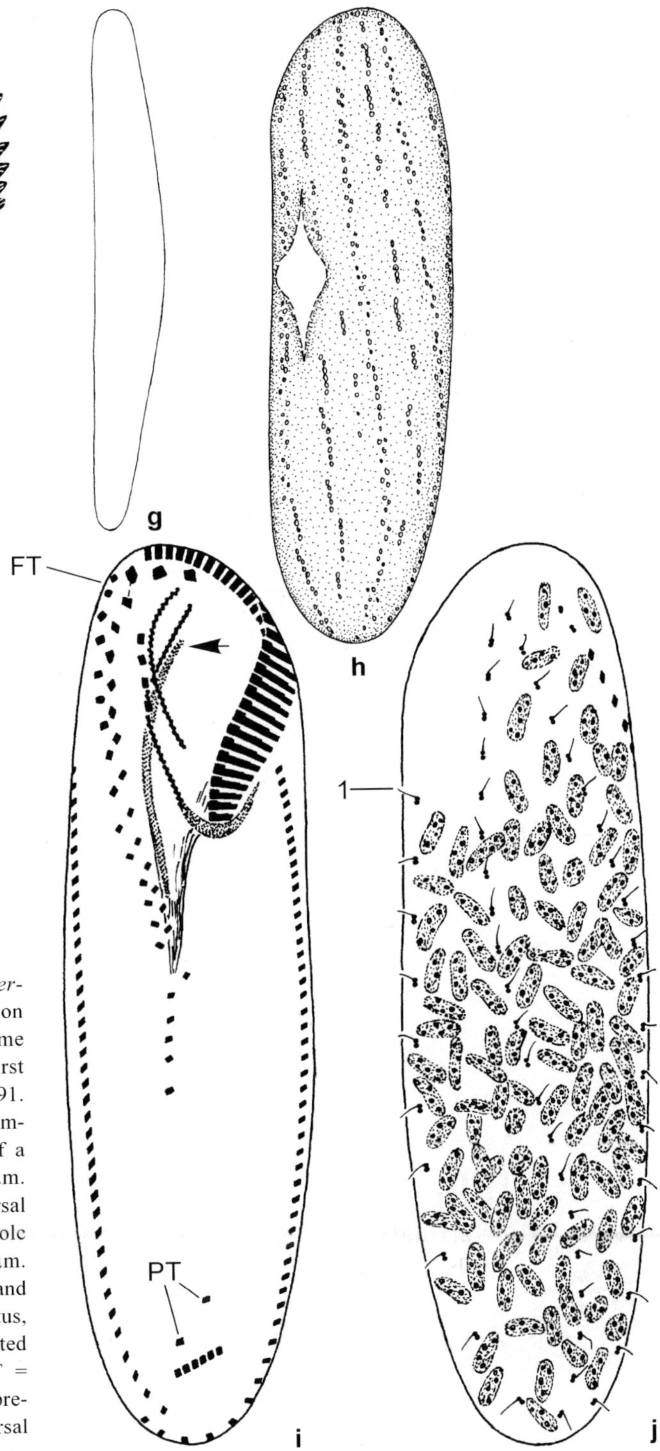

Fig. 61f–j *Anteholosticha intermedia* (f, h–j, neotype population from Foissner 1982; g, from same population as (f, h–j), but first published by Foissner et al. 1991. f–h, from life; i, j, protargol impregnation). **f:** Ventral view of a representative specimen, 165 µm. **g:** Left lateral view. **h:** Dorsal view showing contractile vacuole and cortical granulation, 147 µm. **i, j:** Infraciliature of ventral and dorsal side and nuclear apparatus, 114 µm. Arrow marks striated structure in buccal cavity. FT = frontoterminal cirri, PT = pretransverse ventral cirri, 1 = dorsal kinety 1. Page 317.

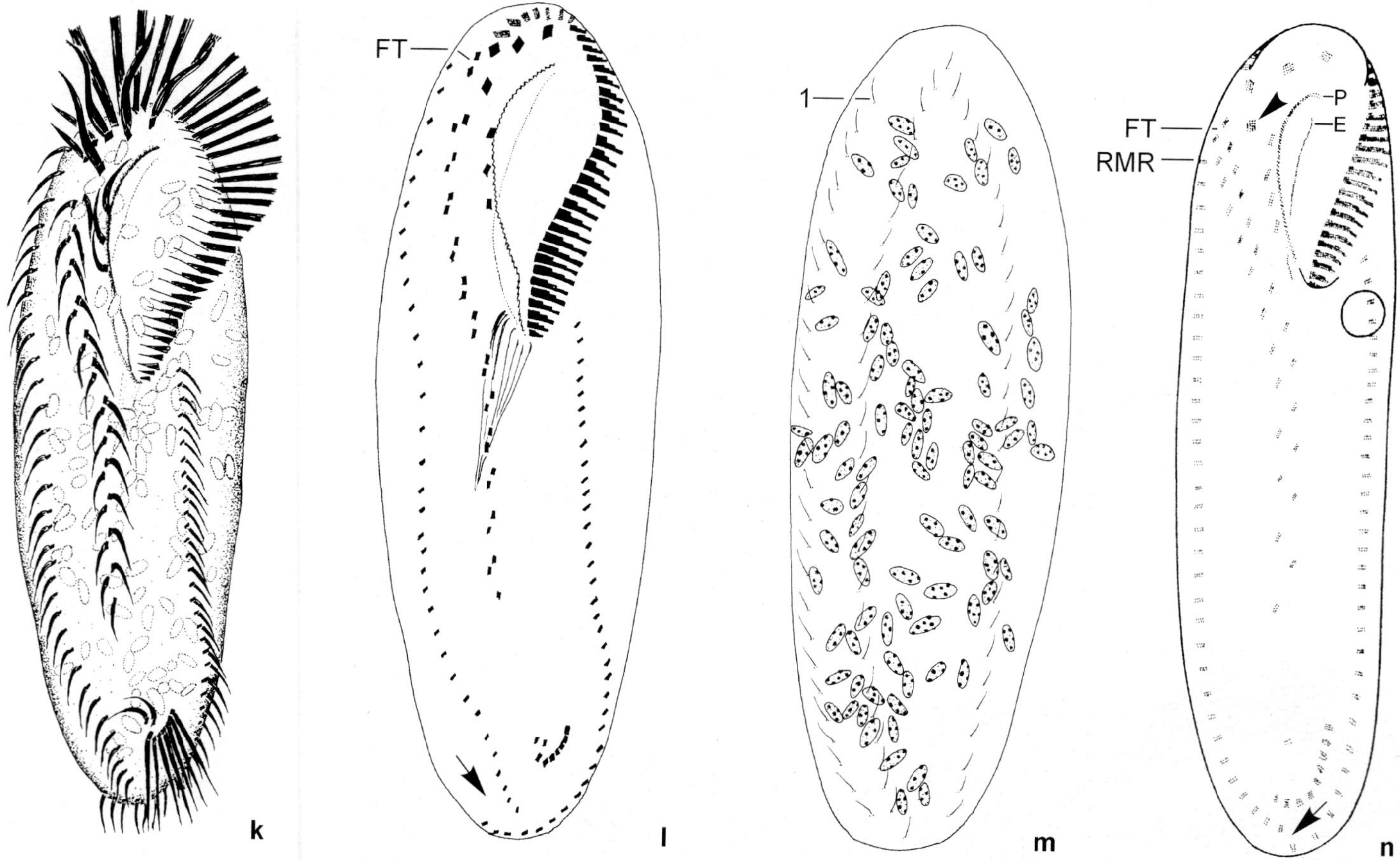

Fig. 61k–n *Anteholosticha intermedia* (k–m, from Shin & Kim 1993; n, from Buitkamp 1977. k, from life; l–n, protargol impregnation). **k:** Ventral view, 134 µm. **l–n:** Infraciliature of ventral and dorsal side and nuclear apparatus, l, m = 133 µm, n = 146 µm. Arrows mark confluent to overlapping marginal rows. Arrowhead denotes cirrus III/2. E = endoral, FT = frontoterminal cirri, P = paroral, RMR = right marginal row, 1 = dorsal kinety 1. Page 317.

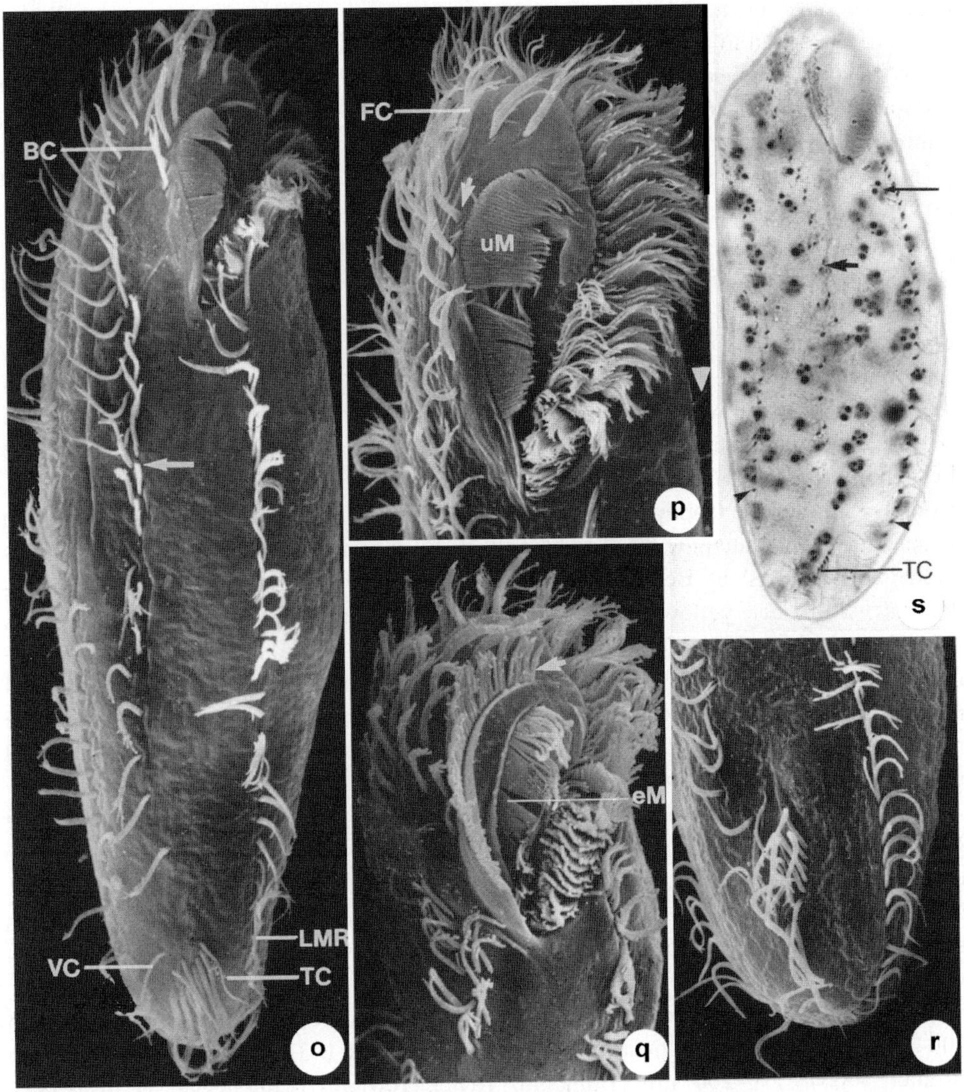

Fig. 61o–s *Anteholosticha intermedia* (from Foissner et al. 1991. o–r, scanning electron micrographs; s, protargol impregnation). **o:** Ventral view showing cirral pattern. Arrow marks rear portion of midventral complex. **p, q:** Oral region showing "closed" (p) and "open" (q) paroral. Arrow in (p) marks anterior buccal cirrus, arrowhead denotes a dorsal bristle. Arrow in (q) marks anterior end of paroral. **r:** Ventral view of rear body end showing marginal and transverse cirri and rear portion of midventral complex. **s:** Ventral view showing cirral pattern and nuclear apparatus. Small arrow marks a macronuclear nodule, large arrow denotes a pseudopair. Arrowheads marks the rear portion of the marginal rows. Explanation of original labelling: BC = buccal cirral row, eM = endoral, FC = right frontal cirrus, LMR = left marginal row, TC = transverse cirri, uM = paroral, VC = last midventral(?) cirrus. Page 317.

sometimes only indistinctly set off from rearmost cirri of midventral complex. Transverse cirri arranged in oblique, indistinctly J-shaped row; about 20 μm long, project distinctly beyond rear body end. Marginal rows overlapping posteriorly by 1–3 cirri with left row behind right (Fig. 61i). Right marginal row extends dorsolaterally anteriorly, commences with 1–3 bristle-like cilia; distance between individual cirri of right row almost equal, about twice as long posteriorly as anteriorly in left marginal row; marginal cirri about 15 μm long.

Dorsal cilia about 4 μm long in life, arranged in three slightly curved longitudinal rows; kineties 2 and 3 almost bipolar, kinety 1 distinctly shortened anteriorly. Caudal cirri lacking (Fig. 61j).

Description of *Urostyla intermedia* population of Bergh (1889; Fig. 60a, b): body size obviously not mentioned, Kahl (1932) estimated from the illustration a length of 200 μm; body length:width ratio about 4:1 (estimated from Fig. 60a); both ends broadly rounded. Many macronuclear nodules (slightly less than *Urostyla grandis*) scattered throughout cell, individual nodules about 5 μm long; 3–16, usually 7–8 micronuclei, about 1.5 μm across. Contractile vacuole near left margin at about 40% of body length in specimen illustrated, with distinct collecting canals (Fig. 60a). Cortical granules (called "oil droplets" by Bergh) as in *Urostyla grandis*, respectively, *Paraurostyla weissei*, that is, distinct, yellowish/greenish, and arranged in longitudinal rows (Fig. 60a). Adoral zone (about 30% of body length in specimen illustrated) and buccal field of ordinary size. Three enlarged frontal cirri; midventral complex extends to about 75% of body length (buccal cirri likely misinterpreted as left portion of midventral complex, see remarks); 7–8 transverse cirri, which project distinctly beyond rear body end; marginal rows without peculiarities, confluent posteriorly.

Description of the *Holosticha muscorum* population by Kahl (1932; Fig. 61a, Table 19): body length 200–300 μm, body length:width ratio 4.4:1 (Fig. 61a); outline slender elliptical; soft and flexible; cortical granules brownish, arranged in longitudinal rows. Many scattered macronuclear nodules. Peristomial lip usually distinctly curved leftwards anteriorly. Frontal cirri distinctly enlarged; only right transverse cirri project slightly beyond rear body end.

Supplementary and deviating data from other sources (Table 19): Body length 200–350 μm in life (*Holosticha multistilata* sensu Kahl 1932; Fig. 61b), 200–250 μm (Reuter 1963), 170–200 μm (Buitkamp 1977), 100 μm (Lüftenegger et al. 1988), 232–262 μm (Dingfelder 1962); size 130–190 × 35–70 μm (Shin & Kim 1993). Body parallel-sided (Kahl 1932, Fig. 61a–c), elongate elliptical, or distinctly converging posteriorly (Reuter 1963, Fig. 61d). Ventral side flattened to slightly concave, dorsal surface convex (Shin & Kim 1993). Body soft and flexible (Shin & Kim 1993), contractile (Reuter 1963). Many macronuclear nodules (Reuter 1963, Buitkamp 1977). Cortical granules light-yellow, arranged in short (each with 3–5 granules), longitudinal rows (Kahl 1932, mentioned at *H. multistilata*; Fig. 61u). Movement rapid, changing direction frequently (Shin & Kim 1993). Oral apparatus rather large, more than one third of body length (Reuter 1963); buccal area deep (Shin & Kim 1993; Fig. 61t); paroral composed of two rows, endoral of a single row of narrowly spaced cilia (Buitkamp 1977). Frontal cirri enlarged (Reuter 1963); about 15 μm long (Buitkamp 1977); midventral

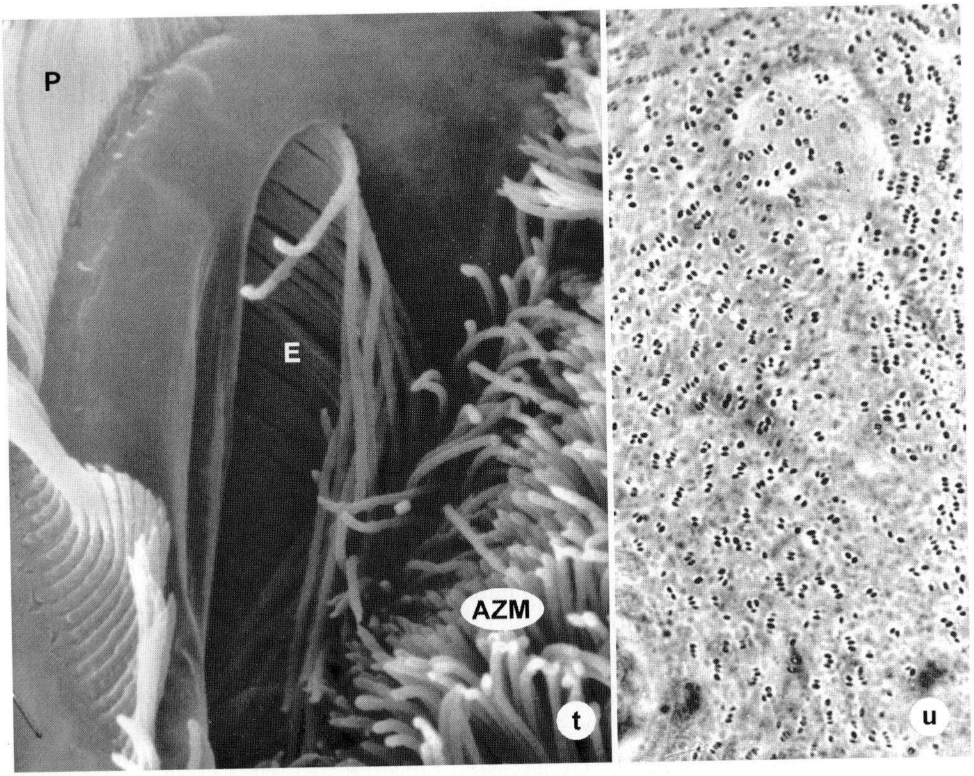

Fig. 61t, u *Anteholosticha intermedia* (t, original kindly supplied by W. Foissner; u, from Foissner et al. 1991. t, scanning electron micrograph; u, from life). **t:** The buccal cavity contains the endoral (E) and is bordered by the paroral (P) and the adoral zone of membranelles (AZM). Population from Stampfltal near Vienna, Austria. **u:** Cortical granules (dark dots) in the frontal region. Note that the specimen is squeezed and therefore the arrangement of the granules is changed. Page 317.

cirri about 10 μm long (Buitkamp 1977); transverse cirri slightly enlarged, J-shaped arranged (Buitkamp 1977). Dorsal cilia about 4 μm long (Buitkamp 1977).

Molecular data: Shin et al. (2000) sequenced the small subunit rRNA genes of *Anteholosticha intermedia* (designated as *Holosticha multistilata*; identified according to Shin & Kim 1993).

The complete sequences of *A. intermedia* are 1778 nulceotides long (GenBank accession No. AJ277876; named "*Holosticha multistylata*"). In both the distance matrix tree and the maximum likelihood tree reconstruction based on 10,000 puzzling quartets, *Anteholosticha intermedia* is in a clade with *Oxytricha granulifera* and *Halteria grandinella*. These three species are the "sistergroups" of three stylonychines (*Tetmemena pustulata*, *Sterkiella nova*, *Onychodromus quadricornutus* (now *Styxophrya quadricornuta*). A similar position was obtained in a tree by Strüder-Kypke & Lynn (2003, p. 92). By contrast, in a maximum parsimony tree by Shin et al. (2000), *Anteholosticha intermedia* clustered outside the 18-cirri oxytrichids, a position also estimated by Modeo et al. (2003), Affa'a et al. (2004), Foissner et al. (2004a; Fig. 15a), and Song et al. (2004).

Cell division: Only few data are available about the division of *A. intermedia*. Bergh (1889) studied the division of the nuclear apparatus (Fig. 60b). It shows the plesiomorphic state; that is, the many macronuclear nodules fuse to a single mass which divides again. No data are available about the infraciliature. However, from the interphasic cirral pattern one can conclude that the cirral anlagen III–VI produce only two cirri each, and not up to three as in *A. multistilata* (Fig. 84i, j). If the parental adoral zone is not completely replaced as in *A. multistilata* (Fig. 84g–j), then a close relationship of *A. multistilata* and *A. intermedia* is unlikely.

Occurrence and ecology: Due to the complicated systematics of *A. intermedia*, its synonym *H. muscorum*, and *A. multistilata*, it is difficult to assign the faunistic records and ecological data correctly. Generally, *Anteholosticha intermedia* is a common, but usually not abundant species occurring in freshwater (e.g., Foissner & Moog 1992), terrestrial habitats (e.g., Buitkamp 1977, Foissner 1982, Shin & Kim 1993, Foissner et al. 2005), and saline inland waters (Kahl 1932). Marine records are available, but doubtful (see below). Kahl (1932) briefly mentioned but did not illustrate a form similar to his "*H. multistilata*" (Fig. 61b). This population was slightly smaller (100–150 μm) and lived in the Baltic Sea near the city of Kiel; however, he supposed that this marine population belongs to a distinct species.

Bergh (1889) discovered *Anteholosticha intermedia* in a infusion of beech litter collected from a pond likely in/near the city of Copenhagen, Denmark. Due to the designation of Foissner's (1982) population as neotype material, its sample site ("field B") becomes the (new) type locality of *A. intermedia*: aperiodically flooded field (sandy-silty clay; about 189 m above sea-level; 15°46'35"E 48°23'37"N) near the village of Grafenwörth, Lower Austria. This field with crop rotation (wheat, maize, potato) was treated with inorganic fertiliser and pesticids (Foissner 1982, Foissner et al. 1985, p. 89). Foissner (1982) found the present species (as *H. multistilata*) in all sites from the Tullnerfeld region (Lower Austria), including forests, flood plains, fields, and dry meadows (for details on the sample sites, including chemical parameters, see Foissner et al. 1985).

Only few records are available where the present species was designated with the species-group name *intermedia*. By contrast, there exist a considerable number of records of *Holosticha muscorum*. Identifications of *H. muscorum* based on the descriptions by Grolière (1975; Fig. 73f–h) or Foissner (1982; Fig. 73a–e) are assigned to *A. antecirrata*. Some other post 1975/1982 records are also listed under *A. antecirrata*, especially when one can assume that the worker knew these two descriptions (e.g., Bamforth 1995). All other records concerning *H. muscorum* are listed here. Further, the present chapter contains all records from *A. multistilata*, except for those by Kahl (1928), Jutrczenki (1982), and Hemberger (1982). However, I keep the data separate so that workers who do not agree with the present systematics can use the records too.

Records of *A. intermedia* largely not substantiated by morphological data: during March in a draw-well in Italy (Grispini 1938, p. 153; 131 μm long); soil from the macchia in Italy (Luzzatti 1938, p. 101); terrestrial(?) moss and freshwater from Antarctica (Hada 1966, p. 212; Haneda 1971, Fig. 60d; Sudzuki 1979, p. 123).

No type locality was fixed for *H. muscorum* by Kahl (1932; Fig. 61a). According to him it is common in mosses from "all areas", and he even found it in America. Once he observed *H. muscorum* among the limnetic macrophyte *Hottonia*.

The type locality of *Keronopsis macrostoma* Reuter is near the Biological Station in Espegrend, Blomsterdalen, Western Norway where Reuter (1963) discovered it in a rockpool with a salinity of 0.3‰. Kahl (1932) found "*H. multistilata*" (Fig. 61b) in the saltwater sites near Oldesloe, especially in the area of Fresenburg, about 30 km north east of the city of Hamburg, Germany. Shen et al. (1992) recorded it from subtropical soils in China.

Further records of *A. intermedia* substantiated by morphological data and/or illustrations (designated as *Holosticha multistilata* in these papers): moss-covered soils from the campus of the Seoul National University, Korea (Shin & Kim 1993; the mosses studied by Shin et al. 2000 are likely from the same location); USA (Borror & Wicklow 1983; no details provided); soil samples from a riverine forest in Ivory Coast (Buitkamp 1977; 1979, p. 225).

Records of "*Holosticha muscorum*" (or *Keronopsis muscorum*) not substantiated by morphological data and/or illustrations (possibly some of the post-1975 records refer to *A. antecirrata*): various Danish beech wood sites (Brunberg-Nielsen 1968, p. 85; Stout 1968, p. 394); in three soil types (mor, acid mull, calcareous mull) in England (Stout 1963, p. 285); among *Sphagnum* in a small lake (Heiliges Meer) near Munster, Germany (Mücke 1979, p. 273); dry mosses, leaf and coniferous litter, heather, and *Sphagnum* from near the city of Erlangen, Germany (Wenzel 1953, p. 111); at pH 7.2 in field-furrows and sod-infusions from near Erlangen, Germany (Dingfelder 1962, p. 616); litter of deciduous forests (all, inter alia, with *Fagus* sp.) in Hungary (Varga 1959, p. 458); Danube River in Hungary (Bereczky 1969, p. 96; Nosek & Bereczky 1981, p. 179); cave in Hungary (Bajomi 1969, p. 236); terrestrial moss from the Pisa-region, Italy (Verni & Rosati 2000, p. 68); at 16° C and pH 6.9 in a lake in Slovakia (Matis & Straková-Striesková 1991, p. 114); various moss-species (e.g., *Brachythecium plumosum*) in Japan (Sudzuki 1964a, p. 168, 171; 1964b, p. 249; 1965, p. 138); soil from Scoresby Land, East Greenland (Stout 1970, p. 20); desert sites in Northern Utah, USA (Bamforth & Bennett 1985, p. 424); in litters, soils, and cryptogamic crusts in deserts and semi-arid woodlands in Arizona, USA (Bamforth 1984, p. 135); prairie soil horizons (L, F–H), forest soils, and deltaic soils in Louisiana, USA (Bamforth 1967, p. 15; 1968, p. 14; 1969, 73); grass and shrub samples from a Cedar glade in Rutherford County, Tennessee, USA (Martin & Sharp 1983, p. 34); lake near Montreal, Canada (Puytorac et al. 1972, p. 435); common in various tussock-grassland soils, pastures, and burnt sites in New Zealand (Stout 1958, p. 977; 1960, p. 240; 1961, p. 745; 1962, p. 318; 1984, p. 123); moss and soil from the Antarctic region (Sudzuki 1979, p. 123).

Foissner and some others found "*Holosticha multistilata*" sensu Kahl (1932) and sensu Foissner (1982) in various limnetic and terrestrial sites almost throughout the world: mesosaprobic river Traun in Upper Austria (Foissner & Moog 1992, p. 101); rivers in Upper Austria (Blatterer 1994, p. 158); soils in Austria (Lüftenegger et al. 1988, p. 96); soil from spruce forests of the northern region of Upper Austria (Aescht & Foissner 1993, p. 358; Petz et al. 1988, p. 82); litter of various natural forest stands in

Austria (Foissner et al. 2005; Fig. 61t); soil from a humus forest near Fontainebleau, France (Palka 1991, p. 127); litter of beech forests near the cities of Göttingen and Kassel, Germany (Bonkowski 1996, p. 35); soil samples from forests in Germany, inter alia, from an acid spruce forest, from a limed and fertilised spruce forest, and from a beech forest in the Ulm region, Germany (Funke 1986, p. 72; Lehle 1989, p. 141); soil from a mixed forest near Bonn, Germany (Buitkamp 1977a, p. 116; 1979, p. 225); clean rivers (Illach, Zinnbach) in Bavaria, Germany (Foissner 1997a, p. 184); mesosaprobic river system (SI = 2.5–2.7) near Munich, Germany (Foissner et al. 1992, p. 49; 1992a, p. 101); soil (0–5 cm) from an area used to store chemicals near the German village of Nordhorn (Niebuhr 1989, p. 81; identifications checked by W. Foissner); soil samples from the Macaulay Land Use Research Institute's Sourhope Research Station near Kelso in Southern Scotland, UK (Finlay et al. 2001, p. 362); at saprobic values between 3.0 and 2.0 in four sites of the River Stirone, Northern Italy (Madoni & Bassanini 1999, p. 394); stream formed by water discharged from a geothermal sulfur spring in Bobbio, Northern Italy (Madoni & Uluhogian 1997, p. 165); dystrophic lakes in Wigry National Park, north east Poland (Czapik & Fyda 1995, p. 67); moss on calcareous rock in Poland (Wiackowski 1988, p. 4); agricultural soils from near the village of Ostrov, Slovakia (Tirjaková 1988, p. 499; Matis et al. 1996, p. 12); submerged, wet, moist, and dry mosses from the area of Slovensky raj and Bratislava, Slovakia (Tirjaková & Matis 1987a, p. 8; 1987b, p. 22); soil from a tropical dry forest in the Santa Rosa National Park, Costa Rica (Foissner 1995, p. 39); soil of non-flooded areas of primary and secondary rain forests near Manaus, Brazil (Foissner 1997, p. 322); soil from the Antarctic region (Sudzuki 1979, p. 123); in *Deschampsia antarctica* grass swards from the Antarctic region (Foissner 1996a, p. 100).

Records of "*H. multistilata*" sensu Kahl (1932) from salt waters (possibly these records refer to *A. estuarii*): Schlei, a brackish water near the city of Kiel, Germany (Bock 1960, p. 63; Jaeckel 1962, p. 13); groundwater (7.9‰) of the coast region of the Island Hiddensee, Germany (Münch 1956, p. 434).

As already mentioned above, "*Holosticha multistilata*" sensu Kahl (1932) was also recorded from true marine habitats. However, these records are not substantiated by morphological data and/or illustrations (I suppose that these are misidentifications; possibly they refer to *A. estuarii*): Plymouth area, England (Lackey & Lackey 1963, p. 802); mainly from August to January in the Atlantic Ocean with a biomass of up to 15 mg C m^{-3} in the Bay of Biscay at Brazomar Beach, Castro Urdiales, Spain (Fernandez-Leborans et al. 1999, p. 730; see also Fernandez-Leborans 2001, p. 740); Caspian Sea (Agamaliev 1971, p. 383). Fernandez-Leborans & Novillo (1994, p. 203) found it in laboratory marine microecosystems in the control, but not in the treatment with 1 mg l^{-1} lead.

Anteholosticha intermedia feeds on heterotrophic flagellates and filamentous cyanobacteria (Foissner 1982), on testate amoebae and ciliates (Buitkamp 1977), for example, *Chilodonella uncinata* (Wiackowski 1988). Shin & Kim (1993) cultured it in commercial mineral water enriched with boiled wheat grain and shrimp meats to support bacterial growth. Later, they established and maintained clonal cultures in neutral Pringsheim solution or autoclaved commercial mineral water with appropriate prey (Shin et

Table 18 Assignment of various *Holosticha*- and *Keronopsis*-populations to *Anteholosticha adami*, *A. antecirrata*, *A. multistilata*, and *A. intermedia*. Names are written as in the individual papers

Designation in present book	Designation in								
	Kahl (1928)	Kahl (1932)	Borror (1972)	Grolière (1975)	Buitkamp (1977)	Hemberger (1982)	Jutrczenki (1982)	Foissner (1982)	Foissner et al. (1991)
***Anteholosticha adami* (Foissner, 1982) Berger, 2003**	–	–	–	–	–	*Holosticha multistilata* Kahl, Variante aus Mullrendzina (Abb. 15i)	–	*Holosticha adami* nov. spec.	–
Anteholosticha antecirrata	–	–	–	*Keronopsis muscorum* Kahl, 1932	–	–	–	*Holosticha muscorum* (Kahl, 1932)	–
***Anteholosticha multistilata* (Kahl, 1928) Berger, 2003**	*Holosticha multistilata*	–	–	–	–	*Holosticha multistilata* Kahl, 1928 (Abb. 15a–h)	*Holosticha multistilata* Kahl	–	*Holosticha multistilata* Kahl, 1928 (Abb. 6, 9)
***Anteholosticha intermedia* (Bergh, 1889) comb. nov.**	–	*Keronopsis muscorum* nov. spec. and *Keronopsis (Holosticha) multistilata* Kahl, 1928	*Holosticha multistilata* Kahl, 1928	–	*Holosticha multistilata* Kahl	–	–	*Holosticha multistilata* (Kahl, 1932)	*Holosticha multistilata* Kahl, 1928 (Abb. 1–5, 7, 8, 10–14)

al. 2000). Biomass of 10^6 specimens about 109–130 mg (Foissner 1987a, p. 124; Buitkamp 1979). Buitkamp (1979, p. 228) made experiments on the effect of temperature (5–40° C). At 15° C he found 66 specimens g^{-1} dry soil (corresponding a biomass of 6864 g ha^{-1}), at 20° C 89 specimens (9256 g), and at 25° C again 66 specimens.

We classified the present species (under the name *Holosticha multistilata*) as alpha-mesosaprobic to betamesosaprobic indicator of water quality (a–b; b = 4, a = 5, p = 1, I = 2, SI = 2.7; Foissner et al. 1991, p. 237; 1995, p. 97; 1995a; Foissner & Berger 1996, p. 380, 414; Sládeček & Sládečková 1997, p. 138; Table 12).

Anteholosticha grisea (Kahl, 1932) Berger, 2003
(Fig. 62a, b)

1932 ***Holosticha grisea* spec. n.** – Kahl, Tierwelt Dt., 25: 585, Fig. 110 II (Fig. 62a; original description; no type material available and no formal diagnosis provided).

1968 ***Holosticha grissea* Kahl, 1932** – Chorik, Free-living ciliates, p. 128, Fig. 118 (Fig. 62b; redescription; incorrect subsequent spelling).

2001 ***Holosticha (Holosticha) grisea* Kahl, 1932** – Berger, Catalogue of ciliate names 1. Hypotrichs, p. 35 (nomenclator containing all basionyms, combinations, and higher taxa of hypotrichs).

2003 ***Anteholosticha grisea* (Kahl, 1932) n. comb.** – Berger, Europ. J. Protistol., 39: 377 (combination with *Anteholosticha*).

Nomenclature: No derivation of the name is given in the original description. The species-group name *grise·us -a -um* (Latin adjective; ash-coloured, ash-grey; grey) likely alludes to the blackish food reserves with which the species is packed. Incorrect subsequent spelling: *Holosticha grissea* Kahl (Agamaliev 1973, p. 61; Liepa 1983, p. 137).

Remarks: The present species very likely lacks caudal cirri and all(?) apomorphies of *Holosticha* and was thus transferred to *Anteholosticha* (Berger 2003). For a detailed discussion and foundation of the transfer of many species from *Holosticha* to *Anteholosticha* see the genus section.

Borror (1972, p. 11) and Borror & Wicklow (1983, p. 122) synonymised *H. grisea* with *H. violacea* Kahl, 1932 which is indeed rather similar, as already stated by Kahl (1932) himself. However, Kahl (1932) was a very good observer and thus I accept both species by considering the following differences as sufficient, respectively, interesting: body length 140–180 μm against 180–250 μm; dorsal bristles obviously of ordinary length (about 3 μm) against rather long (about 8 μm according to Fig. 67a); colour (invariably blackish due to food vacuoles against purple due to ingested rhodobacteria); body very flat against moderately flat. There is no doubt that further populations have to be redescribed in detail to show finally if they are indeed distinct species.

According to Borror (1972) *A. grisea* is also identical with *A. fasciola*. However, this species is likely confined to saltwater, distinctly longer (200–300 μm) and more slender (10:1), and has conspicuous cortical granules along the marginal rows (Fig. 93a, b). In contrast, Borror & Wicklow (1983) synonymised *A. grisea* not with *A. fasciola*, but with *A. vuxgracilis*. But this species has, inter alia, only two macronuclear nodules so that we can exclude synonymy.

The redescription by Chorik (1968) is in Russian. However, the illustration presented (if it is indeed an original), indicates that the identification is correct (Fig. 62b).

Morphology: The following description is based on Kahl's data unless otherwise indicated. Body length 140–180 μm, body length:width ratio of specimen illustrated 7.4:1 (Fig. 62a); specimens of Chorik's population 150–160 μm long. Body short worm- to band-shaped, margins of anterior portion slightly converging anteriorly, rear body end broadly rounded; moderately flattened dorso-ventrally. Many macronuclear nodules; specimen illustrated with about 24 macronuclear nodules and micronuclei(?) each about half left and right of midline (Chorik's specimen with 21 nuclei; Fig. 62b). Contractile vacuole at about one third of body length at left cell margin (Fig. 62a). Cortical granules neither mentioned nor illustrated. Cells always packed with blackish inclusions ("Nahrungsreserven", food reserve). Movement not described.

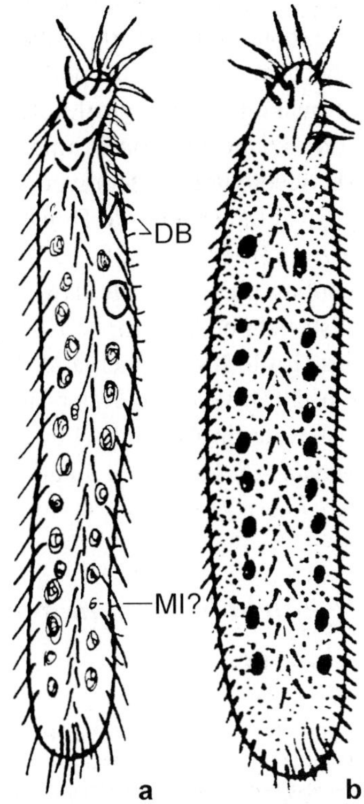

Fig. 62a, b *Anteholosticha grisea* (a, from Kahl 1932; b, from Chorik 1968. From life). Ventral views showing, inter alia, body outline, contractile vacuole, nuclear apparatus, and basic cirral pattern, a = 160 μm, b = 150 μm. DB = dorsal bristles, MI? = micronucleus? Page 332.

The following data about the basic cirral are not described by Kahl but from his illustration and thus must not be over-interpreted in each detail. Adoral zone occupies about 23% of body length, proximal portion possibly slightly sigmoidal. Three (slightly enlarged?) frontal cirri; behind frontal cirri and ahead of midventral complex six also slightly enlarged cirri (possibly including a buccal cirrus); midventral complex composed of cirral pairs (16 fine pairs are illustrated), extends to near transverse cirri; five transverse cirri, which protrude by about half their length beyond rear body end. Marginal rows obviously without peculiarities. Dorsal bristles about 3 μm long (Fig. 62a). Caudal cirri likely lacking because neither illustrated nor mentioned; however, note that these cirri are often very fine and difficult to distinguish from marginal cirri, even in protargol preparations!

Occurrence and ecology: Freshwater, likely confined to the sapropel. Type locality not mentioned by Kahl (1932); he found it, although never abundant, in several sites invariably in the sapropel, likely somewhere in (North?) Germany. Chorik (1968) found *A. grisea* in Moldavian water basins.

I found it in the sapropelic littoral area of the Salzach River in the village of Kaltenhausen (about 10 km south of the city of Salzburg, Austria) in late October 1989; at this time the river was heavily polluted by the waste water of a paper mill (unfortunately I did not make morphological notes). Further records not substantiated by morphological data: Novozámecký rybnik, a pond in northern Bohemia near Ceská Lipa, Czechoslovakia (Hassdenteufelová-Moravcová 1955, p. 215; Matis et al. 1996, p. 12); pond near the village of Randan (Puy-de-Dôme), France (Grolière & Njine 1973, p. 14); Latvian running waters (Liepa 1983, p. 137; 1986, p. 229; 1990, p. 67; Veylande & Liyepa 1985, p. 82); cooling plant of a Moldawian power station (Chorik & Vikol 1973, p. 69); benthic in Polish fishponds (Czapik 1959, p. 190); microbenthos and periphyton of freshwater lagoons (Little Kysylagach and Agrakhansky) of Caspian Sea (Agamaliev 1973, p. 61; 1986, p. 207); volcanic crater-lake (about 15 km west of Warrnambool, Victoria) with brackish water in Australia (Finlay et al. 1999, p. 140; Esteban et al. 2000, p. 163).

Anteholosticha grisea likely feeds exclusively on heterotrophic flagellates (Kahl 1932).

Anteholosticha thononensis (Dragesco, 1966) Berger, 2003 (Fig. 63a)

1963 ***Keronopsis thononensis* n. sp.** – Dragesco, Protistologica, 2: 83, Fig. 23 (Fig. 63a; original description; site where type material deposited not mentioned; no formal diagnosis provided).

1979 ***Holosticha thononensis* (Dragesco, 1966)** – Jankowski, Trudy zool. Inst., 86: 56 (combination with *Holosticha*).

2001 ***Holosticha thononensis* (Dragesco, 1966) Jankowski, 1979** – Berger, Catalogue of ciliate names 1. Hypotrichs, p. 44, 45 (nomenclator containing all basionyms, combinations, and higher taxa of hypotrichs).

2003 ***Anteholosticha thononensis* (Dragesco, 1966) n. comb.** – Berger, Europ. J. Protistol., 39: 377 (combination with *Anteholosticha*).

Nomenclature: No derivation of the name is given in the original description. The species-group name *thononensis* is a composite of the name of the village Thonon-les-Bains at lake Geneva and the suffix *~ens·is* (habitat in geographical sense), indicating that this species was discovered in this area.

Remarks: The present species likely lacks caudal cirri and all(?) apomorphies of *Holosticha* and was thus transferred to *Anteholosticha* (Berger 2003). For a detailed discussion and foundation of the transfer of many species from *Holosticha* to *Anteholosticha* see the genus section.

Borror (1972, p. 11) classified *K. thononensis* as junior synonym of *Anteholosticha gracilis*. However, this marine species has yellow-red cortical granules (lacking in present species) and more transverse cirri (9 vs. 6). Since *A. gracilis* is likely confined to marine habitats, all freshwater records of this species are listed here in a separate paragraph of the occurrence and ecology chapter.

Jankowski (1979) transferred the present species to *Holosticha*, which was basically correct because of the frontal ciliature. Admittedly, the midventral cirri do not form a distinct zigzag-pattern so that the classification as urostyloid hypotrich is not quite certain. However, I assume that the cell is inflated due to the preparation procedure, which possibly slightly disturbed the midventral pattern. Further, the many macronuclear nodules also indicate that it is a urostyloid.

Borror & Wicklow (1983) synonymised *K. thononensis* with *Holosticha vernalis* Stokes. However, Stokes' species obviously has a single postoral ventral cirrus and thus belongs to *Apoamphisiella* Foissner, 1997 (for details see *Holosticha*).

Anteholosticha thononensis is basically described only after protargol preparations. Thus, redescription – including detailed life observations – is recommended.

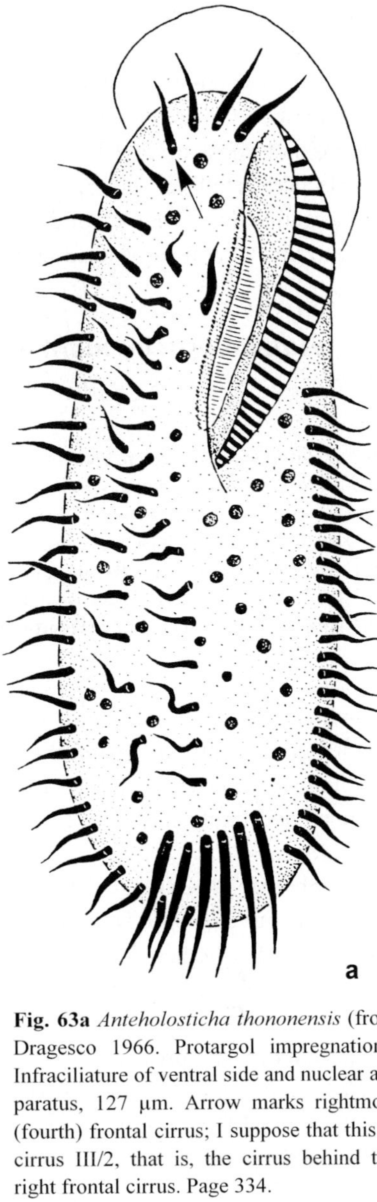

Fig. 63a *Anteholosticha thononensis* (from Dragesco 1966. Protargol impregnation). Infraciliature of ventral side and nuclear apparatus, 127 μm. Arrow marks rightmost (fourth) frontal cirrus; I suppose that this is cirrus III/2, that is, the cirrus behind the right frontal cirrus. Page 334.

Morphology: Body length 100–140 μm. Body outline elongate elliptical (the outline of the prepared specimen shown is likely not very meaningful). Many macronuclear nodules; individual nodules 2–4 μm across. Micronuclei small. Cortical granulation (presence or absence) not described. Adoral zone composed of 34–40 membranelles (relative length 44%! in specimen illustrated). Buccal field likely narrow. Paroral roughly straight. Four slightly enlarged frontal cirri (likely three frontal cirri and cirrus III/2). Buccal cirrus distinctly behind anterior end of paroral. Midventral complex extending to near transverse cirri; (pseudo)row formed by right cirri of pairs in line with frontal cirri, composed of 14–21 cirri; left row composed of 13–18 cirri. Six moderately enlarged and likely slightly subterminally arranged transverse cirri. Right marginal row commences somewhat behind distal end of adoral zone, terminates at rear end in midline, composed of 26–32 cirri. Left marginal row terminates at level of transverse cirri, composed of 25–30 cirri. Dorsal cilia arranged in four kineties, length not mentioned, but likely, as in most species, short, that is, around 3 μm. Caudal cirri probably lacking because neither mentioned nor illustrated.

Occurrence and ecology: *Anteholosticha thononensis* was discovered in freshwater habitats near Thonon-les-Bains at the southern bank of lake of Geneva, France (exact sample site not mentioned). No further records published.

Records of "*Keronopsis gracilis*" (now *Anteholosticha gracilis*) from freshwater habitats not substantiated by morphological data (for explanation see remarks): mesotrophic lake (Heiliges Meer) near the German city of Münster (Mücke 1979, p. 266); submerged and wet mosses of Slovenský raj, Slovakia (Tirjaková & Matis 1987a, p. 9); Guadarrama river near Madrid, Spain (Fernandez-Leborans et al. 1990, p. 512).

Anteholosticha mancoidea (Hemberger, 1985) Berger, 2003 (Fig. 64a–c)

1982 ***Holosticha mancoidea* n. spec.**[1] – Hemberger, Dissertation, p. 96, Fig. 14a, b (Fig. 64a, b).

1985 ***Holosticha mancoidea* n. spec.**[2] – Hemberger, Arch. Protistenk., 130: 403, Abb. 7 (Fig. 64a, b; original description; type material likely deposited in the Institut für landwirtschaftliche Zoologie und Bienekunde, Universität Bonn, Germany).

2001 ***Holosticha mancoidea* Hemberger, 1985** – Berger, Catalogue of ciliate names 1. Hypotrichs, p. 37 (nomenclator containing all basionyms, combinations, and higher taxa of hypotrichs).

2003 ***Anteholosticha mancoidea* (Hemberger, 1985) n. comb.** – Europ. J. Protistol., 39: 377 (combination with *Anteholosticha*).

Nomenclature: No derivation of the name is given in the original description. The species-group name *mancoidea* is a composite of the species-group name *manc·a* (see *Anteholosticha manca* for derivation), the thematic vowel *·o-*, and the Latin suffix *~idea* (similar) and likely refers to the fact that the present species has a similar cirral pattern as *A. manca*.

Hemberger (1985) stated for every new species that the type slides are deposited in the University of Bonn. However, for the present species this statement was not made. I did not check whether or not the type slide(s) of this species are available or not at the Bonn University. I assume that he simply forgot to add the standard sentence in the present case.

Remarks: The present species very likely lacks caudal cirri and most apomorphies of *Holosticha* and was thus transferred to *Anteholosticha* (Berger 2003). For a detailed discussion and foundation of the transfer of many species from *Holosticha* to *Anteholosticha* see the genus section.

Anteholosticha mancoidea is described mainly after protargol preparations. Unfortunately, nothing is known about the presence or absence of cortical granules. Thus, a redescription is necessary.

The cirral pattern of *A. manca* is rather similar, especially as concerns the posteriorly distinctly shortened midventral complex. However, this species has more macronu-

[1] This name is disclaimed for nomenclatural purposes (ICZN 1999, Article 8.3); see footnotes at genus *Periholosticha*.

[2] The diagnosis provided by Hemberger (1985) is as follows: L: 120 μm; B: 20–30 μm; L:B = 4–5:1; PerL:L = 1:4. Körper: Langgestreckt, fast parallelseitig, sehr flexibel, meist frontal etwas verjüngt.

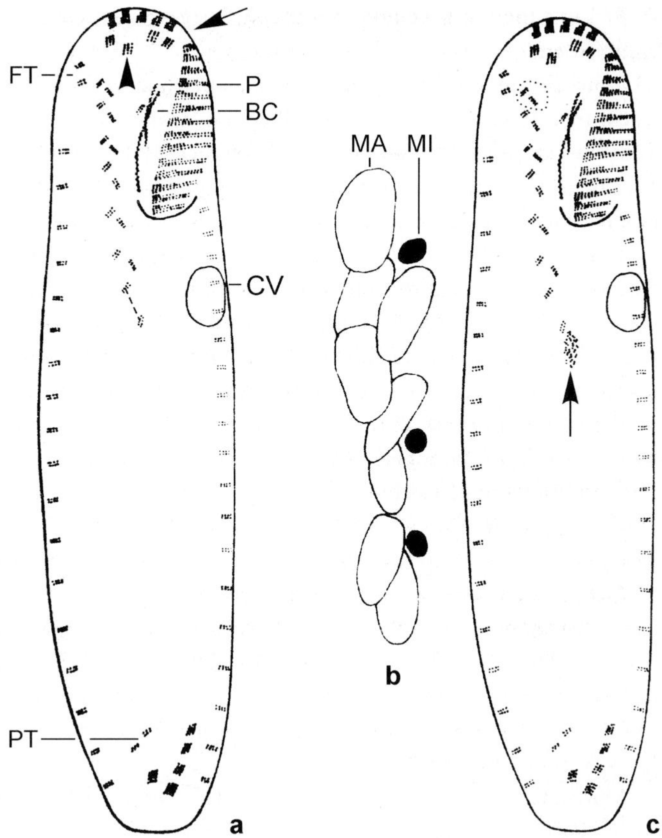

Fig. 64a–c *Anteholosticha mancoidea* (from Hemberger 1982. Protargol impregnation). **a, b:** Infraciliature of ventral side and nuclear apparatus, 120 µm. Arrow marks gap in adoral zone, arrowhead denotes (likely) cirrus III/2. Rearmost midventral cirral pair connected by broken line. **c:** Very early divider with oral primordium (arrow). Anteriormost midventral cirral pair circled by dotted line. There is no difference in the cirral pattern and nuclear apparatus (not shown) to the specimen illustrated in (a) indicating that Hemberger (1982, 1985) draw in (c) or deleted (a) the oral primordium. BC = buccal cirrus, CV = contractile vacuole, FT = frontoterminal cirri, MA = anteriormost macronuclear nodule, MI = anteriormost micronucleus, P = paroral?, PT = pretransverse ventral cirri. Page 336.

clear nodules (about 50–70 vs. 6–12, usually 8 in *A. mancoidea*) and lacks pretransverse ventral cirri. Hemberger (1985) compared his species with *A. sphagni* (Grolière), which is indeed very similar. At first I thought these two species could be synonyms. However, a closer inspection revealed that there are some differences, which indicate that Hemberger's and Grolière's population are not conspecific. The most conspicuous differences concern the number of dorsal kineties (3 in *A. mancoidea* vs. 4) and the structure of the adoral zone of membranelles (gap present vs. lacking). The other morphometric differences are inconspicuous and do not clearly separate *A. mancoidea* and *A. sphagni*.

Morphology: Body size 120 × 20–30 µm (in life?), length:width ratio 4–5:1. Body outline elongate, margins almost parallel, anterior portion usually slightly narrowed. Body very flexible. Usually eight macronuclear nodules forming slightly irregular, longitudinal row likely somewhat left of midline. 2–3 globular micronuclei. Contractile vacuole in ordinary position, that is, slightly behind proximal end of adoral zone near left body margin. Presence/absence of cortical granules not known.

Adoral zone occupies about 25% of body length, composed of 17–20 membranelles, distalmost 4–5 membranelles set off from proximal portion by a gap (Fig. 64a, arrow).

Undulating membranes commence about at same level, paroral(?) only about half of length of endoral. Buccal field narrow.

Four slightly enlarged frontal cirri; likely one of these is cirrus III/2 (Fig. 64a, arrowhead), that is, the cirrus "behind" the right frontal cirrus. One buccal cirrus slightly behind anterior end of undulating membranes. Two frontoterminal cirri slightly behind level of frontal cirri. Midventral complex composed of about seven cirral pairs forming distinct zigzag-pattern; terminates at 38% of body length in specimen illustrated. Two pretransverse ventral cirri ahead of right portion of transverse cirral group, which is composed of five slightly enlarged cirri arranged in hook-shape (transverse cirri likely distinctly projecting beyond rear body end in life specimens). Right marginal row obviously distinctly shortened anteriorly (commences at 17% of body length in specimen illustrated), composed of 16–20 cirri, terminates, like left row, subterminally. Left marginal row commences at level of buccal vertex, composed of 17–19 cirri. Dorsal cilia about 3 µm long, arranged in three kineties. Caudal cirri obviously lacking because neither mentioned nor illustrated.

Cell division (Fig. 64c): Hemberger (1982, 1985) found only one very early divider showing that the oral primordium originates near the posteriormost midventral cirrus.

Occurrence and ecology: Freshwater. Type locality is a eutrophic pond (Heinrichsweiher) at the Institut für landwirtschaftliche Zoologie und Bienenkunde, University of Bonn, Germany. Hemberger found it there in sediment samples during winter. Later, he found *A. mancoidea* in the sediment of a brook (Forstbach, Gemarkung Bornich) in the same region also during winter; these specimens were slightly longer (150 µm) than the type population and anteriorly narrowed.

Martin Cereceda et al. (2002) recorded *A. mancoidea* from a rotating biological contactor (RBC) placed at a wastewater treatment plant in Madrid (Spain). It was related – together with some other species – to high dissolved oxygen content, lowest levels of BOD_5 and ammonium, and highest levels of nitrates, and was therefore associated with the best RBC performance.

Anteholosticha randani (Grolière, 1975) Berger, 2003 (Fig. 65a)

1975 ***Holosticha randani* n. sp.** – Grolière, Protistologica, 11: 486, Fig. 7, 11 (Fig. 65a; original description; locality where type slide is deposited not mentioned; no formal diagnosis provided).

1982 ***Holosticha randani* Grolière, 1975** – Hemberger, Dissertation, p. 102 (revision of non-euplotid hypotrichs).

2001 ***Holosticha randani* Grolière, 1975** – Berger, Catalogue of ciliate names 1. Hypotrichs, p. 38 (nomenclator containing all basionyms, combinations, and higher taxa of hypotrichs).

2003 ***Anteholosticha randani* (Grolière, 1975) n. comb.** – Berger, Europ. J. Protistol., 39: 377 (combination with *Anteholosticha*).

Nomenclature: The species-group name *randani* refers to the region (Randan, France) where the species was discovered.

Remarks: The present species lacks caudal cirri and most apomorphies of *Holosticha* and was thus transferred to *Anteholosticha* (Berger 2003). For a detailed discussion and foundation of the transfer of many species from *Holosticha* to *Anteholosticha* see the genus section.

Anteholosticha randani is mainly described after protargol preparations. Grolière (1975) likely did not look at cortical granules; thus, it is not known whether or not such organelles are present. Hemberger (1982) considered the present species as valid; by contrast, Borror & Wicklow (1983, p. 121) synonymised it with *A. intermedia* (Bergh), which has indeed a similar general appearance. However, there are also some distinct differences preventing a synonymy of *A. randani* and *A. intermedia* (different size, that is, about 110 μm long against about 200 μm; gap in the adoral zone present against lacking).

I agree with Hemberger (1982), that Grolière (1975) misinterpreted the four set off distal adoral membranelles as frontal cirri, although Grolière also discussed the possibility that these could be adoral membranelles. However, he could not imagine that the adoral zone is interrupted. In this feature, the present species agrees with Hemberger's *Anteholosticha mancoidea*, which, however, has, inter alia, less macronuclear nodules (16–20 vs. usually 8), but more midventral cirral pairs (20–27 vs. 7) and dorsal kineties (4 vs. 3).

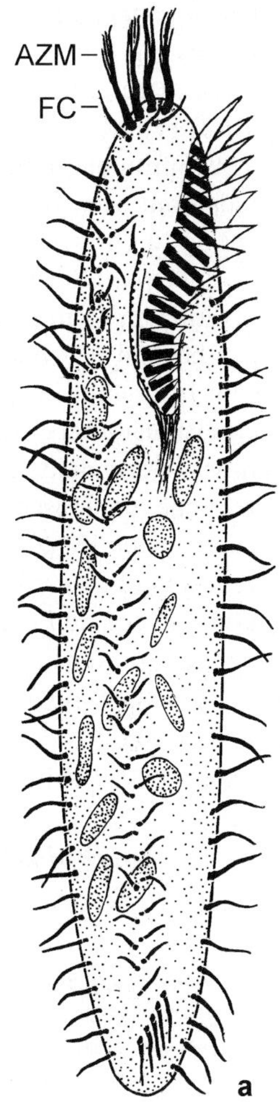

Fig. 65a *Anteholosticha randani* (from Grolière 1975. Protargol impregnation). Infraciliature of ventral side and macronuclear apparatus, 111 μm. AZM = distal portion of adoral zone, FC = rightmost frontal cirrus. Page 338.

Morphology: Body size 95–125 × 15–30 μm, on average 110 × 22 μm in life?; body length:width ratio about 5:1. Body outline elongate elliptical. Body very flexible. 16–20 elongate to globular macronuclear nodules irregularly arranged mainly in right body portion; micronuclei small. Contractile vacuole near left margin underneath proximal end of adoral zone. Presence/absence of cortical granules not mentioned. Movement rapid, glides quickly among leaves.

Adoral zone occupies 34% of body length in specimen illustrated, bipartite by distinct gap (about 4 μm in Fig. 65a). Posterior membranelles of proximal portion distinctly displaced inwards; proximal portion thus sigmoidally curved, composed of 18–29 membranelles (range rather high for this feature!); distal portion of adoral zone composed of four membranelles (misinterpreted as frontal cirri in original description, see remarks). Buccal field very narrow, right margin bordered by relatively short (12–15 μm) paroral; endoral not described, indicating that it is difficult to recognise even in protargol preparations.

Cirral pattern likely without peculiarities. Grolière (1975) provided a good micrograph (his Fig. 11) of the anterior body half showing some details, which he did not illustrate and describe. Three not distinctly enlarged frontal cirri. Likely one cirrus (= cirrus III/2) behind right frontal cirrus. One buccal cirrus at anterior end of paroral. Very likely two frontoterminal cirri in ordinary position (see his Fig. 11). Midventral complex extends from near frontal cirri to transverse cirri, composed of about 20–27 cirral pairs; according to Grolière, "left row" always with one additional cirrus (Grolière obviously counted the pseudopairs). Usually five, rarely four or only three transverse cirri arranged in subterminal, oblique row; cirri obviously illustrated much too short so that they do not project beyond rear body end. Right marginal row likely commences distinctly behind anterior body end (I suppose that Grolière misinterpreted the frontoterminal cirri as anteriormost marginal cirri), composed of 28–41, on average 34 cirri, ends – like left one – slightly subterminally; left row commences distinctly ahead of level of proximal end of adoral zone, composed of 24–35, on average 29 cirri. Four dorsal kineties, each composed of about 18–20 dorsal cilia, which are short, that is, around 3 µm. Caudal cirri likely lacking because neither mentioned nor illustrated.

Occurrence and ecology: Type locality of *A. randani* are two pools in a forest between the two French villages of Randan and Saint-Yorre. Grolière (1975) discovered it there, sometimes highly abundant, in the marginal *Sphagnum* girdle mainly during spring and autumn. A detailed mathematical analysis of the ciliate cenosis of this habitat and other sites investigated by the French author was provided by Grolière (1978).

Anteholosticha sphagni (Grolière, 1975) Berger, 2003 (Fig. 66a, b)

1975 ***Keronopsis sphagni* n. sp.** – Grolière, Protistologica, 11: 484, Fig. 3 (Fig. 66a; original description; locality where type slide is deposited not mentioned; no formal diagnosis provided).

1983 ***Holosticha sphagni* (Grolière, 1975) nov. comb.** – Borror & Wicklow, Acta Protozool., 22: 108, 122, Fig. 11 (Fig. 66b; brief, illustrated redescription and combination with *Holosticha*; revision of urostylids; see nomenclature).

2001 ***Holosticha sphagni* (Grolière, 1975) Borror and Wicklow, 1983** – Berger, Catalogue of ciliate names 1. Hypotrichs, p. 44 (nomenclator containing all basionyms, combinations, and higher taxa of hypotrichs).

2003 ***Anteholosticha sphagni* (Grolière, 1975) n. comb.** – Berger, Europ. J. Protistol., 39: 377 (combination with *Anteholosticha*).

Nomenclature: No derivation of the name is given in the original description. The species-group name *sphagni* obviously refers to the habitat (*Sphagnum* ponds) where the species was discovered. "*Holosticha sphangi* (Groliére, 1975) nov. comb." in Borror & Wicklow (1983, p. 120) is an incorrect subsequent spelling.

Hemberger (1982, p. 110) transferred this species to *Holosticha* one year before Borror & Wicklow (1983). However, generally dissertations do not constitute publications in the sense of zoological nomenclature although in the ICZN (1964, Article 9) – which was relevant for Hemberger (1982) – theses are not definitely excluded (possibly it falls under point 6 of this article: "mere deposit of a document in a library").

Remarks: The present species very likely lacks caudal cirri and most apomorphies of *Holosticha* and was thus transferred to *Anteholosticha* (Berger 2003). For a detailed discussion and foundation of the transfer of many species from *Holosticha* to *Anteholosticha* see the genus section.

The two populations described by Borror & Wicklow (1983) agree very well with the type material (Fig. 66b). *Anteholosticha sphagni* is mainly described after protargol preparations. The ventral and dorsal infraciliature is more or less identical to that of *A. sigmoidea* (for discussion and comparison see this chapter at *A. sigmoidea*). The main difference is in the cortical granulation, which is lacking in *A. sphagni* (pers. comm. of C.-A. Grolière to Foissner 1982), but pronounced in *A. sigmoidea*. Thus, it would not be wise to merge both species. The lack of cortical granules separates it also from *A. monilata*, which has a similar nuclear apparatus, but more dorsal kineties (4 vs. usually 6) and midventral pairs (about 10 vs. on average more than 20). Redescription, especially "documentation of lacking cortical granules", recommended.

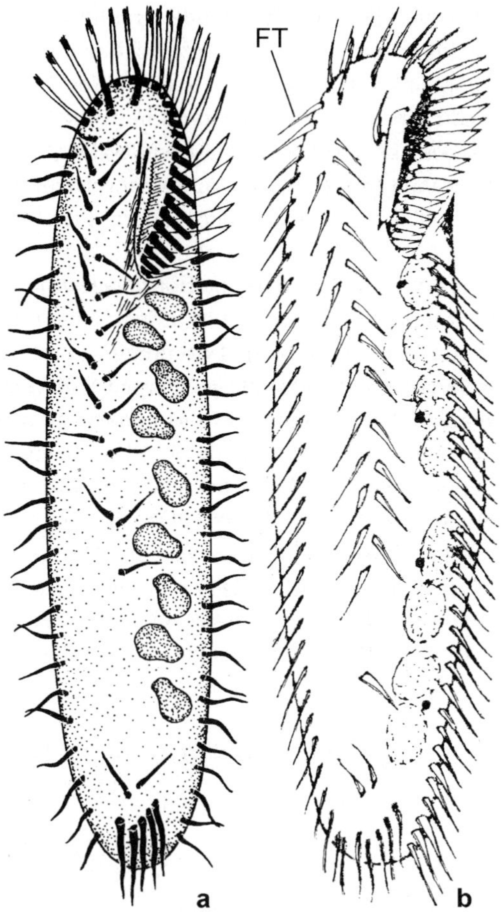

Fig. 66a, b *Anteholosticha sphagni* (a, from Grolière 1975; b, from Borror & Wicklow 1983. a, protargol impregnation; b, protargol impregnation?). Infraciliature of ventral side and nuclear apparatus, a = 83 µm, b = 84 µm. Differs from *A. sigmoidea* mainly by the lack of cortical granules. Brief characterisation of Borror & Wicklow's populations: 11–20, respectively, 7–10 midventral pairs; 4–5, respectively, 2–7 transverse cirri; 1–2 buccal cirri in both populations (n = 8, respectively, 12). Page 340.

Morphology: The description below is from the type population. For a brief characterisation of Borror & Wicklow's specimens, see figure legend.

Body size 60–90 × 16–20 µm (78 × 14 µm on average) in life?, body length:width ratio of specimen illustrated about 5.0:1. Body outline elongate elliptical. 8–11 macronuclear nodules serially arranged in left body portion behind adoral zone (Fig. 66a). Contractile vacuole neither mentioned nor illustrated, indicating that life observation was not very detailed (lack of this organelle is very unlikely in this freshwater species). According to a personal communication by A.-C. Grolière to Foissner (1982), the present species lacks cortical granules. Adoral zone about 17–20 µm long, in specimen illustrated it occupies 26% of body length (Fig. 66a); composed of about 24–25 membranelles. Buccal field narrow, paroral straight, composed of zigzagging basal bodies. Cirral pattern more or less as in *A. sigmoidea*. Only two enlarged frontal cirri (3 × 3 basal bodies; other cirri 2 × 3 basal bodies) illustrated; if this observation is constant and correct then this would be an apomorphy of this species; however, misinterpretation of right frontal cirrus as distalmost adoral membranelle cannot be excluded. One buccal cirrus right of anterior portion of paroral. Frontoterminal cirri not mentioned, indicating that they are misinterpreted as anteriormost (two) cirri of right marginal row. Midventral complex composed of about 10 cirral pairs and terminating at 63% of body length in specimen illustrated (Fig. 66a). Two pretransverse ventral cirri ahead of five more or less terminally arranged and slightly enlarged transverse cirri (Hemberger 1982 designated these two pretransverse ventral cirri as midventral cirri and thus counted in total 23 midventral cirri). Right marginal row composed of 23–26 cirri, commences near distal end of adoral zone, terminates, like left row, about at level of rearmost transverse cirrus. Left marginal row composed of 24–27 cirri. Dorsal cilia about 5 µm long, arranged in four kineties. Caudal cirri likely lacking because neither mentioned nor illustrated.

Occurrence and ecology: The type location of *Anteholosticha sphagni* are bogs (Bargeresse; la Landie) near Besse-en-Chandesse (Puy-de-Dôme; France). Borror & Wicklow (1983) did not provide details; likely they found it somewhere in the USA.

Anteholosticha violacea (Kahl, 1928) Berger, 2003
(Fig. 67a–c)

1928 ***Holosticha violacea* spec. n.** – Kahl, Arch. Hydrobiol., 19: 210, Abb. 39b (Fig. 67a; original description; no type material available and no formal diagnosis provided).

1932 ***Holosticha violacea* Kahl, 1928** – Kahl, Tierwelt Dtl., 25: 585, Fig. 106 22 (Fig. 67b; revision of hypotrichs; see remarks).

1972 ***Holosticha violacea* Kahl, 1928** – Borror, J. Protozool., 19: 11 (revision of hypotrichs).

1983 ***Holosticha violacea* Kahl, 1928** – Borror & Wicklow, Acta Protozool., 22: 122, Fig. 14 (Fig. 67c; revision of urostylids; see remarks).

2001 ***Holosticha violacea* Kahl, 1928** – Berger, Catalogue of ciliate names 1. Hypotrichs, p. 39 (nomenclator containing all basionyms, combinations, and higher taxa of hypotrichs).

2003 ***Anteholosticha violacea* (Kahl, 1928) n. comb.** – Berger, Europ. J. Protistol., 39: 377 (combination with *Anteholosticha*).

Nomenclature: No derivation of the name is given in the original description. The species-group name *violáce·us -a -um* (violet, violet-like) obviously refers to the violet colour of the species which is due to violet vacuoles filled with rhodo-bacteria. Kahl (1932) divided *Holosticha* into subgenera. Thus, the correct name in his revision is *Holosticha (Holosticha) violacea* Kahl, 1928.

Remarks: The present species probably lacks caudal cirri and most apomorphies of *Holosticha* and was thus transferred to *Anteholosticha* (Berger 2003). For a detailed discussion and foundation of the transfer of many species from *Holosticha* to *Anteholosticha* see the genus section.

The taxonomy of *A. violacea* is rather complicated. Kahl (1928) was uncertain whether or not his population was a new species or identical with Stein's (1859) *Uroleptus violaceus*. As main difference he mentioned the lack of transverse cirri in *U. violaceus*. Although Stein (1859) did not describe transverse cirri he mentioned that the rearmost marginal cirri are stronger, longer, and of different shape than the other marginal cirri, indicating that Stein misinterpreted (inconspicuous?) transverse cirri as marginal cirri.

Kahl (1928) provided a more or less good illustration (Fig. 67a) and a rather detailed description. Accordingly, his species has, like *U. violaceus*, two macronuclear nodules. In Kahl's 1932 revision, however, he mentioned a multinodular macronucleus (see his key on page 585 "29 (38) Kern in zahlreiche Brocken zerlegt"), and his new illustration also shows several macronuclear nodules (Fig. 67b). Possibly for that reason Kahl (1932) wrote that his species is hardly identical with Stein's species. Kahl (1928) described and illustrated 6–7 frontal cirri ("Stirnzirren"; Fig. 67a likely shows 3 frontal cirri and 3 buccal cirri), whereas the illustration from the 1932 revision shows an ordinary pattern of three enlarged frontal cirri, cirrus III/2, and one buccal cirrus (Fig. 67b).

Unfortunately, Kahl (1928) did not discuss whether all data on *H. violacea* are original observations, or, whether some of them – for example, number of macronuclear nodules – were taken over from Stein's description of *Uroleptus violaceus*. In his revision, he also forgot to discuss the higher number of macronuclear nodules (Kahl 1932). Usually, such a difference in the number of macronuclear nodules (2 against many) is characteristic for different species. However, since *Anteholosticha violacea* is not known in detail, it would be unwise to establish a new species for the multi-macronuclear population described by Kahl (1932).

I simply keep Kahl's data separate because it cannot be excluded that he mixed two species. Borror & Wicklow (1983) provided no description and only one illustration, which, however, fits Kahl's (1932) data rather well. Al-Rasheid (1996a) redescribed *A. violacea* from a saline (18‰) pond. The micrograph and the description (vermiform to cylindrical; 200–230 μm long; adoral zone one fourth of body length; right and left marginal cirri; two ovoid macronuclear nodules) are too inaccurate for a reliable identification.

According to Borror (1972) and Hemberger (1982, p. 111), *Anteholosticha grisea* and *A. fasciola* are junior synonyms of the present species. By contrast, I consider all of them as valid although there is no doubt that such populations have to be redescribed in detail to show whether or not my decision is correct (for comparison, see this chapter at *A. grisea*). In addition, Hemberger (1982) synonymised *Keronopsis rubra* sensu Borror (1963) with the present species; by contrast, I assign Borror's population to *Anteholosticha pulchra*. Borror & Wicklow (1982) considered *A. grisea* and *A. vuxgracilis* as synonyms of *A. violacea*. However, the binucleate *A. vuxgracilis* is distinctly smaller (80 μm vs. around 200 μm).

Anteholosticha extensa, which has a similar habitus, is marine and has a very conspicuous nuclear apparatus, namely, usually four pairs of macronuclear nodules in a line and each pair with a single micronucleus in between.

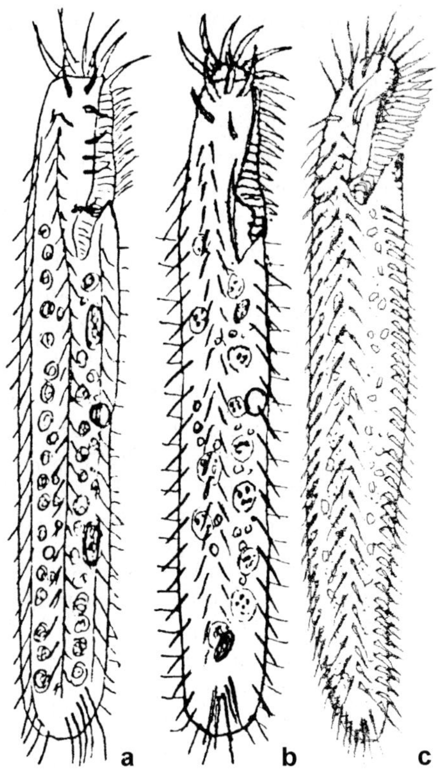

Fig. 67a–c *Anteholosticha violacea* (a, from Kahl 1928; b, from Kahl 1932; c, from Borror & Wicklow 1983. a, b, from life; c, protargol impregnation?). **a:** Ventral view, individual size not indicated (length 180–250 µm). Note that the type population has only 2 macronuclear nodules and some buccal cirri (?). **b:** Ventral view, 220 µm. This population has several macronuclear nodules and only 1 buccal cirrus. **c:** Infraciliature of ventral side and nuclear apparatus, 99 µm. Page 342.

Morphology: Specimens of Kahl's (1928) population 180–250 µm long, body length:width ratio of specimen illustrated about 6.7:1 (Fig. 67a). Body outline rectangular, sometimes slightly converging posteriorly; anterior end transversely truncated, with distinct frontal scutum. One macronuclear nodule each in anterior and posterior body portion left of midline. Contractile vacuole near left cell margin about in mid-body. Cytoplasm packed with violet vacuoles containing rhodobacteria. Adoral zone about 28% of body length (Fig. 67a), roughly as in *Gonostomum*, that is, middle portion extending along cell margin, proximal part distinctly curving inwards. In total 6–7 frontal cirri ("Stirnzirren"); according to Fig. 67a three frontal cirri and three buccal cirri are present. Midventral complex likely composed of cirral pairs only, extends from near anterior end to transverse cirri. Specimen illustrated with five inconspicuous transverse cirri. One right and one left marginal row. Dorsal cilia fine (Fig. 67a); length not mentioned, indicating that they are not very long. Caudal cirri likely lacking because neither mentioned nor illustrated.

Kahl (1932) mentioned the same body length as in the original description. Body length:width ratio 7–8:1. Body outline slender band-shaped, adoral zone portion narrowed slightly trapezoidally, posterior portion not distinctly narrowed. Adoral zone roughly as in *Gonostomum*. Buccal area narrow, with closely attached lip. Three enlarged frontal cirri, one cirrus (III/2) close behind right frontal cirrus, buccal cirrus distinctly ahead of small buccal field. Remaining cirral pattern basically as in type population. Five fine transverse cirri which project only slightly beyond rear body end. Dorsal bristles long (in specimen illustrated at least about 8 µm; Fig. 67b). Caudal cirri likely lacking because neither mentioned nor illustrated.

Individual illustrated by Borror & Wicklow (1983) with about 28 adoral membranelles (adoral zone about 23% of body length), three frontal cirri, one cirrus behind right frontal cirrus, one buccal cirrus right of anterior portion of undulating membrane, about 25 midventral pairs, about four (?) transverse cirri, and 38 right and 41 left marginal cirri.

Occurrence and ecology: Type locality of *A. violacea* is a slightly saline (2.9‰) ditch beside a footpath near the village of Alt-Fresenburg, Germany (Kahl 1928, 1928a). Kahl (1932) found it rather constantly in some ditches with very low salinity in the Bad Oldesloe area near the north German city of Hamburg. Very likely a freshwater species (Kahl 1932). Borror & Wicklow (1983) did not provide detailed data on the individual sample sites; possibly they found it somewhere in New Hampshire (USA) where they worked.

Records from freshwater habitats not substantiated by detailed morphological data and/or illustrations: Lake Balaton, Hungary (Gellért & Tamás 1958, p. 234); alkaline ponds in the Hortobágy National Park, Hungary (Szabó 1999a, p. 229); 30 ind. cm^{-2} in river Stirone (northern Italy) at alphamesosaprobic conditions (Madoni 1979, p. 306); cooling plant of Moldavian power station (Chorik & Vikol 1973, p. 69). Saltwater records: marine sediments polluted by effluents of pulp and paper mill on the west coast of Scotland (Wyatt & Pearson 1982, p. 301); Krasnovodsk Bay and Bol'shoy Kyzylachag Bay at eastern, respectively, western shore of the Caspian Sea (Agamaliev 1973, p. 1599; Agamaliyev 1974, p. 21); pond (Al 'Uyun) with 18‰ salinity in Al-Hassa Oasis, Saudi Arabia (Al-Rasheid 1996a, p. 198). Szabó (1993, 1995, 1999b, 2000) recorded *A. violacea* from crusty meadow solonetz soils in the Hortobágy National Park (Hungary), respectively, a chernozem soil from the centre of the Great Hungarian Plain.

Feeds on purple sulphur bacteria (Kahl 1928, 1932; Fenchel 1968, p. 116). Cytokinesis takes 40–50 min (Zaika 1988).

Anteholosticha xanthichroma (Wirnsberger & Foissner, 1987) Berger, 2003 (Fig. 68a–i, Table 20)

1987 ***Holosticha xanthichroma* sp. n.**[1] – Wirnsberger & Foissner, Acta Protozool., 26: 3, Abb. 1–8, Tabelle 1 (Fig. 68a–i; original description; the holotype slide and a paratype slide are deposited in the Oberösterreichische Landesmuseum in Linz [LI], Austria).

2001 ***Holosticha xanthichroma* Wirnsberger and Foissner, 1987** – Berger, Catalogue of ciliate names 1. Hypotrichs, p. 39 (nomenclator containing all basionyms, combinations, and higher taxa of hypotrichs).

2003 ***Anteholosticha xanthichroma* (Wirnsberger and Foissner, 1987) n. comb.** – Berger, Europ. J. Protistol., 39: 377 (combination with *Anteholosticha*).

Nomenclature: No derivation of the name is given in the original description. The species-group name *xanthichroma* is a composite of the Greek adjective *xanth-* (yellow, brownish, blond), the thematic vowel *·i-*, and the Greek substantive *to chróma* (colour); it refers to the yellow colour of the cytoplasm. The present species is not listed by Aescht (2003).

[1] Wirnsberger & Foissner (1987) provided the following diagnosis: In vivo etwa 110–220 × 26–45 µm große, leicht kontraktile, linealische *Holosticha*, deren Plasma gelb gefärbt ist. Midventralreihen unverkürzt. Vier Dorsalkineten. Durchschnittlich 28 ellipsoide Makronucleus-Teile.

Remarks: The present species very likely lacks caudal cirri (see below) and all apomorphies of *Holosticha* and was thus transferred to *Anteholosticha* (Berger 2003). For a detailed discussion and foundation of the transfer of many species from *Holosticha* to *Anteholosticha* see the genus section.

This species has a rather ordinary ciliature. In spite of this, it can be easily distinguished from the other *Anteholosticha* species by the diffuse yellow cytoplasm. For example, *Anteholosticha grisea* (Fig. 62a) and *A. violacea* (Fig. 67a) resemble the present species in size, shape, and cirral pattern. However, *Anteholosticha grisea* is usually packed with black food inclusions, and *A. violacea* is purple due to ingested rhodobacteria. Further studies must show whether or not these differences are stable and the sole ones.

Anteholosticha xanthichroma possibly has caudal cirri (see end of morphology section). Thus, the assignment to *Anteholosticha* has to be checked by ontogenetic data.

Morphology: Size about 110–220 × 26–45 µm in life, body length:width ratio around 5:1 in life (Fig. 68a) and 4.1:1 on average after protargol impregnation (Table 20). Body outline usually slightly sigmoidal, rear end transversely truncated, anterior end rounded (Fig. 68a, b). Body slightly contractile, about 1.5–2.0:1 flattened dorsoventrally (Fig. 68c). Macronuclear nodules basically arranged along midline with first nodules near proximal end of adoral zone; individual nodules 4–13 × 3–4 µm in life, with few large to many small nucleoli (Fig. 68a, f, h). Usually three slightly argyrophilic micronuclei 4–6 µm across in life (Fig. 68f). Contractile vacuole at about 45% of body length, that is, slightly ahead of mid-body and, as is usual, near left cell margin; during diastole with distinct collecting canals extending to near body ends (Fig. 68b). Pellicle soft and flexible, crenelated along cirral rows. Cytoplasm diffuse yellow (appears brownish at low magnification); possibly this diffuse colour is due to finest pigment particles, which are not recognisable with ordinary light microscopic methods; sometimes moderately many 2 µm–sized dark inclusions in rear body portion. Cortical granules lacking. Always moving, gliding, or creeping worm-like.

Adoral zone occupies 27% of body length on average and is of usual shape, composed of about 37 membranelles of ordinary fine structure (Fig. 68a, e, g, i, Table 20). Buccal area small, not conspicuously deepened. Paroral and endoral short, slightly curved (Fig. 68i). Pharyngeal fibres fine and long.

Cirral pattern without peculiarities, except for the rather high coefficient of variability in some features (Table 20). Three slightly enlarged frontal cirri along anterior portion of adoral zone, invariably (n = 26) one cirrus (= cirrus III/2) behind right frontal cirrus. Buccal cirrus fine, slightly behind anterior end of paroral. Two frontoterminal cirri in ordinary position, that is, between anterior end of right marginal row and midventral complex (Fig. 68i). Midventral complex genus-specifically composed of cirral pairs only, extends slightly sigmoidally to near pretransverse ventral cirri, respectively, transverse cirri. Transverse cirri short, do not project beyond body end, arranged in almost longitudinal row; in 22 out of 79 specimens (28%!) transverse cirri were lacking (Fig. 68d); possibly in protargol preparations they are sometimes difficult to distinguish from midventral cirri. Marginal rows open posteriorly; right one distinctly curved at rear end and terminating almost in cell midline; left one not distinctly curved

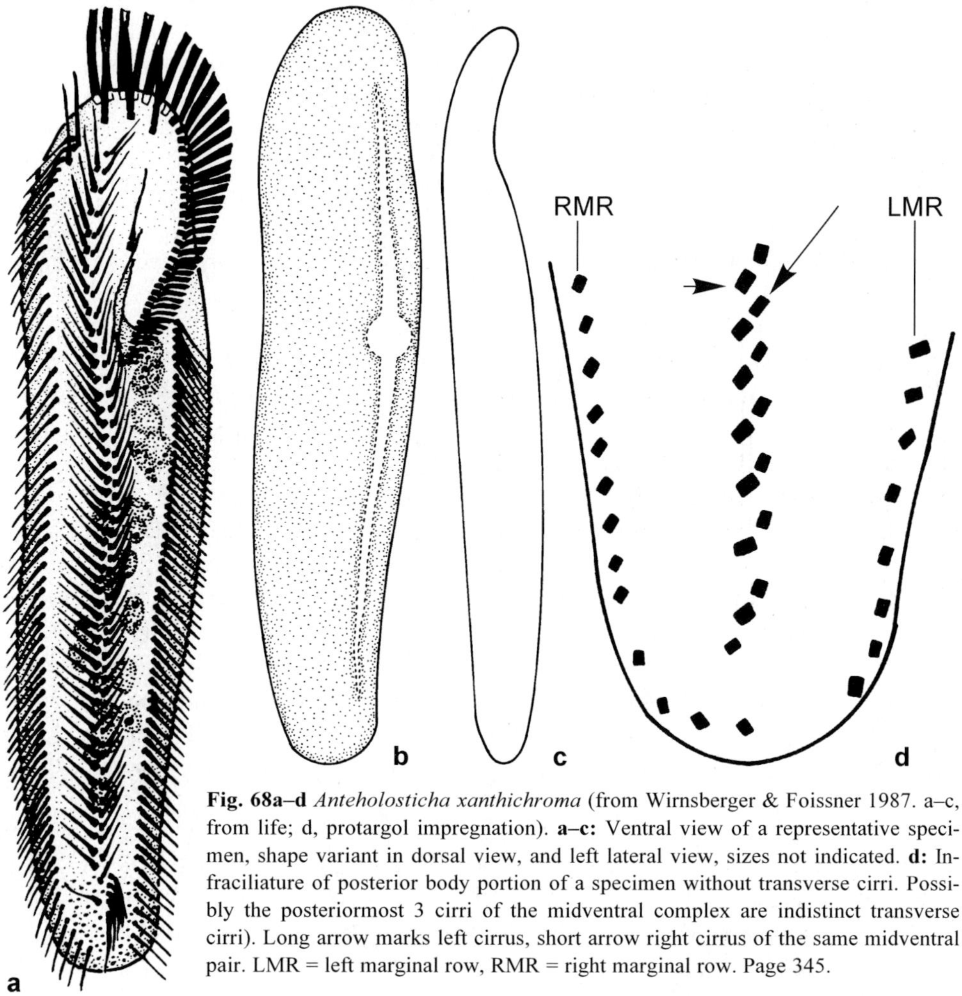

Fig. 68a–d *Anteholosticha xanthichroma* (from Wirnsberger & Foissner 1987. a–c, from life; d, protargol impregnation). **a–c:** Ventral view of a representative specimen, shape variant in dorsal view, and left lateral view, sizes not indicated. **d:** Infraciliature of posterior body portion of a specimen without transverse cirri. Possibly the posteriormost 3 cirri of the midventral complex are indistinct transverse cirri). Long arrow marks left cirrus, short arrow right cirrus of the same midventral pair. LMR = left marginal row, RMR = right marginal row. Page 345.

posteriorly, ends slightly subterminally (Fig. 68d, e, g). Invariably four dorsal kineties, one or two of them anteriorly or posteriorly more or less distinctly shortened (Fig. 68f, h); length of dorsal cilia not mentioned, indicating that they are of ordinary length, that is, about 3–4 µm. Perhaps 1–3 caudal cirri, which are, however, if present all, composed of only few cilia (morphogenetic data are needed to check presence of caudal cirri).

Occurrence and ecology: As yet found only at the type locality, which is in the Schloßalm area – an alpine pasture region used for skiing – near the village of Bad Hofgastein, Salzburg, Austria. Wirnsberger & Foissner (1987) discovered *A. xanthichroma* in the littoral of a perennial, slightly polluted pasture pond about 1950 m above sea level. Food vacuoles contained diatoms and undefined material.

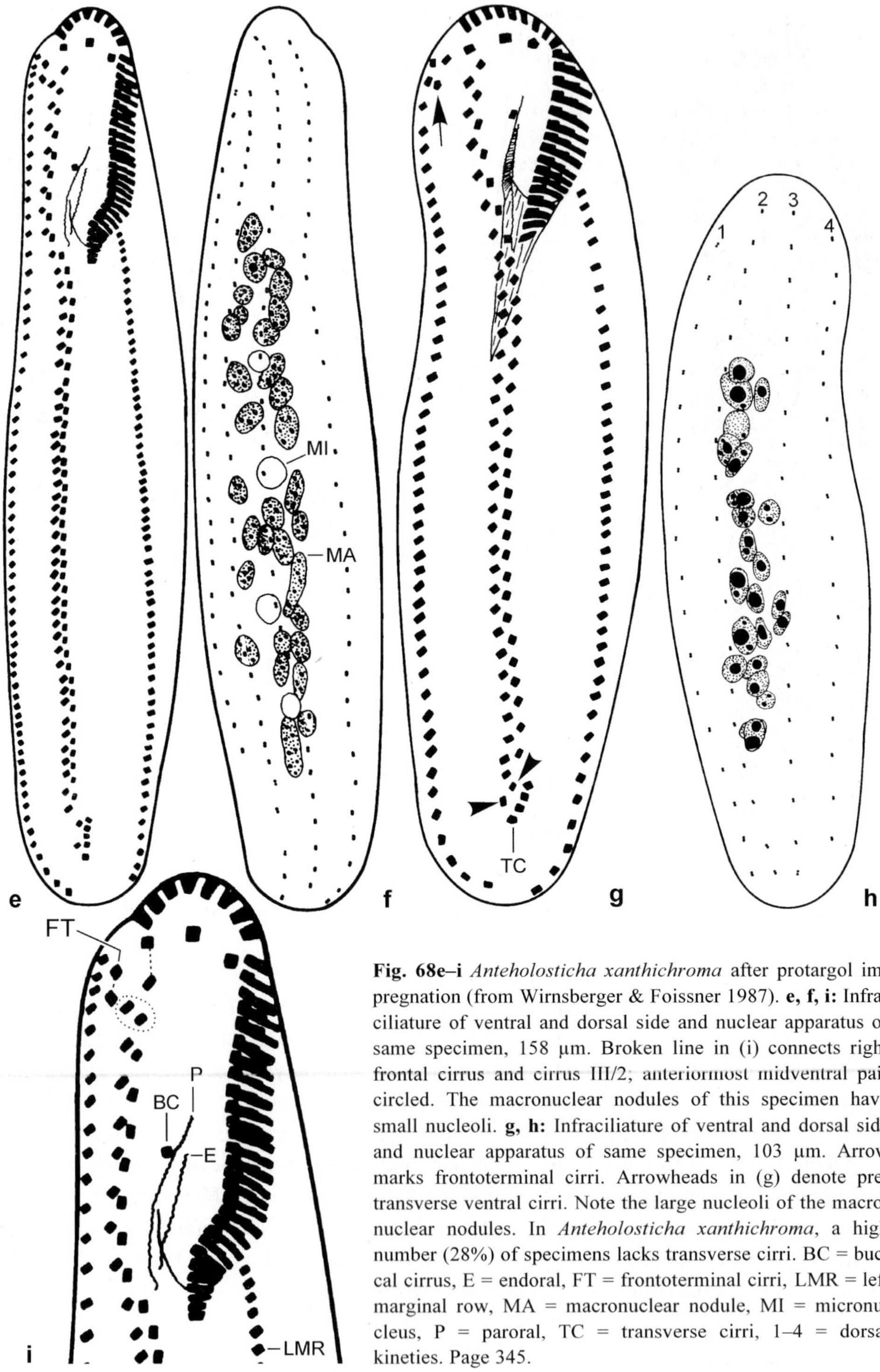

Fig. 68e–i *Anteholosticha xanthichroma* after protargol impregnation (from Wirnsberger & Foissner 1987). **e, f, i:** Infraciliature of ventral and dorsal side and nuclear apparatus of same specimen, 158 μm. Broken line in (i) connects right frontal cirrus and cirrus III/2; anteriormost midventral pair circled. The macronuclear nodules of this specimen have small nucleoli. **g, h:** Infraciliature of ventral and dorsal side and nuclear apparatus of same specimen, 103 μm. Arrow marks frontoterminal cirri. Arrowheads in (g) denote pretransverse ventral cirri. Note the large nucleoli of the macronuclear nodules. In *Anteholosticha xanthichroma*, a high number (28%) of specimens lacks transverse cirri. BC = buccal cirrus, E = endoral, FT = frontoterminal cirri, LMR = left marginal row, MA = macronuclear nodule, MI = micronucleus, P = paroral, TC = transverse cirri, 1–4 = dorsal kineties. Page 345.

Table 19 Morphometric data on *Anteholosticha adami* (ad1, from Foissner 1982; ad2, from Hemberger 1982 [the values are either from Fig. 74g, then n = 1, or from Tabelle 1 in Hemberger 1982 with unknown number of specimens investigated]; ad3, original data from population found by Berger et al. 1985a), *Anteholosticha antecirrata* (an1, from Foissner 1982; an2, from Grolière 1975), *Anteholosticha australis* (aus, from Blatterer & Foissner 1988), *Anteholosticha bergeri* (be1, from Foissner 1987b; be2, Australian population from Blatterer & Foissner 1988; be3, Antarctic population from Blatterer & Foissner 1988), *Anteholosticha brachysticha* (bra, from Foissner et al. 2002), *Anteholosticha gracilis* (gr1, gr2, two populations [gr1 = type population] from Hu & Sudzuki 2004), *Anteholosticha intermedia* (in1, neotype population from Foissner 1982; in2, *Anteholosticha intermedia* from Kahl 1932; in3, from Shin & Kim 1993; in4, from Reuter 1963; in5, from Buitkamp 1977), *Anteholosticha manca* (man, from Song & Wilbert 1997a), *Anteholosticha monilata* (mo1, from Foissner & Didier 1981; mo2, from Song & Wilbert 1989; mo3, from Augustin & Foissner 1992; mo4, from Hemberger 1982; mo5, "*Holosticha similis*" sensu Dragesco 2003), *Anteholosticha multistilata* (ml1, from Jutrczenki 1982, Fig. 84b; ml2, from Hemberger 1982, Fig. 84c), *Anteholosticha sigmoidea* (si1, population from the Glockner area; si2, population from the Schloßalm area; si3, population from Tulnerfeld; si1–si3, from Foissner 1982; si4, alpine population from Foissner 1984), *Anteholosticha warreni* (war, from Song & Wilbert 1997)

Characteristics [a]	Species	mean	M	SD	SE	CV	Min	Max	n
Body, length	ad1	102.6	105.5	16.6	5.9	16.1	82.0	132.0	8
	ad2	–	–	–	–	–	170.0	200.0	?
	ad3	112.8	116.0	14.1	3.6	12.5	83.0	136.0	15
	an1	104.9	106.5	16.9	6.0	10.2	135.0	190.0	8
	aus	131.0	129.5	16.1	5.1	12.3	105.0	163.0	10
	be1	59.2	56.0	7.3	–	12.4	49.0	70.0	11
	be2	73.1	76.0	8.0	2.0	11.0	60.0	82.0	16
	be3	65.8	67.0	5.3	2.7	8.1	59.0	70.0	4
	bra	78.8	82.0	9.5	2.6	12.0	57.0	90.0	13
	gr1	119.4	–	16.9	4.2	14.1	80.0	140.0	16
	gr2	128.1	–	14.4	3.6	11.3	94.0	158.0	16
	in1	108.9	110.0	14.2	4.3	13.1	84.0	132.0	8
	in3	132.1	124.0	25.3	6.7	19.2	108.0	181.0	14
	man [i]	76.3	–	19.0	4.8	24.9	42.0	106.0	16
	ml1	198.0	–	–	–	–	–	–	1
	ml2	203.0	–	–	–	–	–	–	1
	mo1	97.3	97.0	13.4	4.0	13.8	73.0	126.0	11
	mo2	103.9	–	31.2	9.4	30.6	72.0	135.0	11
	mo3	117.3	117.0	10.3	–	8.8	102.0	135.0	13
	mo5	191.0	190.0	20.0	5.4	10.0	158.0	228.0	15
	si1	79.4	84.0	8.9	3.4	11.2	61.0	90.0	7
	si2	70.9	69.0	5.3	1.6	7.5	65.0	83.0	11
	si3	73.7	73.0	–	–	–	70.0	79.0	4
	si4	82.6	84.0	11.4	2.9	13.8	50.0	98.0	15
	war	87.1	–	10.2	2.5	11.7	69.0	104	16
Body, width	ad1	29.7	29.0	4.8	1.7	16.0	22.0	37.0	8
	ad2	–	–	–	–	–	45.0	60.0	?
	ad3	37.3	38.0	8.2	2.1	22.1	24.0	52.0	15
	an1	56.7	53.0	10.6	3.7	18.6	44.0	75.0	8
	aus	25.2	24.5	3.2	1.0	12.7	20.0	31.0	10
	be1	15.7	15.0	2.0	–	12.4	14.0	20.0	11
	be2	15.4	15.0	1.7	0.5	11.1	11.0	18.0	13
	be3	14.5	14.0	1.9	1.0	13.2	13.0	17.0	4
	bra	15.2	14.0	2.1	0.6	14.0	12.0	20.0	13
	gr1	41.3	–	8.5	2.5	20.7	29.0	55.0	12

Table 19 Continued

Characteristics[a]	Species	mean	M	SD	SE	CV	Min	Max	n
Body, width	gr2	35.9	–	5.9	1.5	16.3	26.0	52.0	16
	in1	35.4	35.0	3.8	1.1	10.8	29.0	45.0	11
	in3	55.9	57.0	12.3	3.3	21.9	33.0	78.0	14
	man	26.0	–	6.6	1.6	25.2	15.0	42.0	16
	ml1	64.0	–	–	–	–	–	–	1
	ml2	68.0	–	–	–	–	–	–	1
	mo1	28.2	28.0	2.9	0.9	10.3	23.0	32.0	11
	mo2	33.9	–	4.3	1.3	12.6	29.0	46.0	11
	mo3	35.4	34.0	5.4	–	15.4	30.0	45.0	13
	mo5	72.0	70.0	10.3	2.7	14.0	55.0	90.0	15
	si1	15.9	16.0	2.5	1.0	16.0	12.0	19.0	7
	si2	16.4	16.0	1.9	0.6	11.7	13.0	20.0	11
	si3	18.7	19.0	–	–	–	17.0	20.0	4
	si4	25.7	25.0	2.5	0.7	9.8	21.0	30.0	15
	war	37.4	–	3.7	0.9	9.9	30.0	44.0	16
Body length:width, ratio	ad3	3.1	3.2	0.6	0.2	19.0	2.4	4.9	15
	bra	5.3	5.3	1.0	0.3	18.7	3.8	6.7	13
	in3	2.4	2.4	0.4	0.1	18.1	1.8	3.3	14
	ml1	3.1	–	–	–	–	–	–	1
	ml2	3.0	–	–	–	–	–	–	1
Anterior body end to proximal end of adoral zone, distance	ad1	34.4	34.5	4.1	1.4	11.8	27.0	40.0	8
	an1	66.0	67.5	7.2	2.5	10.9	48.0	73.0	8
	ad3	36.3	36.0	5.1	1.3	14.0	26.0	46.0	15
	aus	35.1	34.0	3.3	1.0	9.4	30.0	41.0	10
	be1	15.3	15.0	1.2	–	7.8	14.0	17.0	11
	be2	18.3	18.0	1.3	0.3	7.4	16.0	21.0	16
	be3	16.0	16.0	1.2	0.6	7.2	15.0	17.0	4
	gr1	41.1	–	4.3	1.1	10.4	30.0	48.0	16
	gr2	42.5	–	3.4	0.8	7.9	38.0	49.0	16
	in1	37.7	37.0	4.1	1.2	11.0	30.0	46.0	11
	in3	56.0	56.0	8.7	2.3	15.5	43.0	71.0	14
	bra	17.6	18.0	1.6	0.4	9.1	14.0	20.0	13
	man	25.6	–	5.6	1.4	22.1	17.0	35.0	16
	ml1	74.0	–	–	–	–	–	–	1
	ml2	68.0	–	–	–	–	–	–	1
	mo1	32.7	33.0	4.0	1.2	12.2	26.0	39.0	11
	mo2	39.5	–	5.2	1.6	13.1	33.0	54.0	11
	mo3	43.3	43.0	3.9	–	9.1	37.0	50.0	13
	mo5	80.0	79.0	9.2	2.4	11.0	65.0	93.0	16
	si1	18.6	19.0	1.0	0.4	5.6	17.0	20.0	7
	si2	17.0	17.0	0.9	0.3	5.0	16.0	19.0	11
	si3	21.7	21.5	–	–	–	20.0	24.0	4
	si4	25.4	25.0	2.2	0.6	8.7	22.0	29.0	15
	war	32.2	–	2.4	0.6	7.5	29.0	36.0	16
Body length:length adoral zone, ratio	ad3	3.1	3.0	0.2	0.1	7.5	2.9	3.6	15
	bra	4.5	4.4	0.4	0.1	9.2	3.7	5.0	13
	in3	2.4	2.4	0.3	0.1	11.9	2.0	2.8	14
Adoral membranelles, number	ad1	29.7	30.0	3.5	1.2	11.8	24.0	36.0	8
	ad2	–	–	–	–	–	26.0	35.0	?
	ad3	30.4	30.0	3.6	0.9	11.7	24.0	37.0	15

Table 19 Continued

Characteristics [a]	Species	mean	M	SD	SE	CV	Min	Max	n
Adoral membranelles, number	an1	47.9	48.0	6.0	2.1	12.6	38.0	55.0	8
	an2	50.0	–	–	–	–	–	–	?
	aus	29.5	30.0	1.8	0.6	6.2	27.0	33.0	10
	be1	13.3	14.0	0.9	–	6.8	12.0	14.0	11
	be2	15.8	16.0	0.8	0.2	5.1	14.0	17.0	14
	be3	14.8	15.0	0.5	0.3	3.4	14.0	15.0	4
	bra	15.8	16.0	1.2	0.3	7.5	13.0	17.0	12
	gr1	27.4	–	1.7	0.4	6.2	24.0	30.0	16
	gr2	28.4	–	1.9	0.5	6.5	25.0	32.0	16
	in1	30.6	30.0	2.8	0.8	9′.2	28.0	38.0	11
	in3	32.4	32.0	3.6	1.0	11.1	26.0	38.0	14
	in5	26.0	–	–	–	–	–	–	?
	man	24.3	–	2.3	0.7	9.4	21.0	27.0	11
	ml1	25.0	–	–	–	–	–	–	1
	ml2	29.0	–	–	–	–	–	–	1
	mo1	37.1	36.0	4.4	1.3	11.9	30.0	44.0	11
	mo2	36.5	–	2.9	1.0	7.9	33.0	41.0	8
	mo3	43.2	42.0	4.5	–	10.4	35.0	50.0	13
	mo4	–	–	–	–	–	42.0	46.0	?
	mo5	59.0	59.0	4.0	1.0	7.0	50.0	60.0	16
	si1	19.9	20.0	2.6	1.0	13.0	16.0	24.0	7
	si2	19.5	20.0	1.2	0.3	5.9	17.0	21.0	11
	si3	21.0	21.0	–	–	–	20.0	22.0	4
	si4	26.3	27.0	1.1	0.3	4.23	25.0	28.0	15
	war	27.8	–	1.7	0.6	6.2	26.0	31.0	16
Paroral, length	bra	7.8	8.0	1.3	0.6	16.7	6.0	9.0	5
	in3	46.9	47.5	7.0	1.9	15.0	35.0	60.0	14
Paroral length:length adoral zone, ratio	in3	0.8	0.0	0.0	0.0	4.5	0.0	0.0	14
Endoral, length	bra	8.0	8.0	0.0	0.0	0.0	8.0	8.0	3
Pharyngeal fibres, length	in3	25.9	25.0	3.2	1.1	12.4	22.0	30.0	9
Distance 1 [e]	ad3	7.4	7.0	1.5	0.4	20.9	6.0	10.0	15
Distance 2 [e]	ad3	5.6	6.0	2.0	0.5	36.2	2.0	9.0	15
	bra	2.5	2.3	1.3	0.4	54.0	1.0	5.0	10
Distance 3 [e]	ad3	6.4	6.0	2.4	0.7	38.3	3.0	12.0	14
Distance 4 [e]	bra	5.0	5.0	0.9	0.3	17.9	3.0	6.0	11
Distance 5 [e]	bra	8.5	8.0	1.4	0.4	16.2	6.0	11.0	11
Distance 6 [e]	bra	13.4	13.5	2.3	0.7	17.2	11.0	18.0	12
Distance 7 [e]	bra	13.5	14.0	3.2	0.9	23.7	6.0	19.0	13
Anterior body end to paroral, distance	bra	6.2	6.0	1.0	0.3	16.3	5.0	8.0	13
Anterior body end to endoral, distance	bra	7.9	8.0	1.3	0.4	16.1	6.0	10.0	9
Transverse cirri, relative position [g]	ad3	95.1	95.2	1.7	0.4	1.8	92.2	97.6	15
Nuclear figure, length	bra	55.0	58.0	7.5	2.1	13.6	41.0	65.0	13
Macronuclear nodule, length	ad1	5.0	5.3	0.8	0.3	16.2	3.8	6.6	8
	ad3	5.4	5.0	1.2	0.3	22.7	4.0	8.0	14
	an1	7.9	6.6	2.9	1.0	37.1	5.3	15.0	9
	aus	8.7	7.8	3.0	0.9	34.1	5.5	13.0	10
	be1	3.8	3.5	1.0	–	25.8	2.5	5.6	11
	be2	4.4	4.5	1.2	0.3	27.9	2.8	6.2	15
	be3	4.4	4.2	0.9	0.4	20.1	3.5	5.6	4
	bra [h]	4.7	5.0	0.9	0.3	20.2	3.0	6.0	13

Table 19 Continued

Characteristics[a]	Species	mean	M	SD	SE	CV	Min	Max	n
Macronuclear nodule, length	gr1	5.1	–	1.9	0.5	37.6	2.0	8.0	15
	gr2	4.6	–	1.2	0.3	25.7	3.0	7.0	15
	in1	5.2	5.3	1.0	0.3	19.1	4.0	6.6	11
	in3	5.5	5.0	0.7	0.2	12.3	5.0	6.5	13
	mo1	7.5	8.0	1.6	0.5	21.3	5.3	11.0	11
	mo3	11.2	11.0	2.9	–	25.5	7.0	18.0	13
	mo5	11.0	10.0	3.4	0.6	31.0	5.2	18.0	30
	si1	6.6	6.6	1.0	0.4	15.4	5.3	8.0	7
	si2	5.8	6.6	1.0	0.3	16.5	4.0	7.0	11
	si3	6.7	6.6	–	–	–	6.5	7.0	3
	si4	8.8	8.4	2.1	0.5	23.2	7.0	13.0	15
Macronuclear nodule, width	ad1	2.0	1.9	0.5	0.2	25.0	1.4	2.6	8
	ad3	2.5	2.3	0.8	0.2	32.3	1.5	4.0	14
	an1	2.2	2.4	0.4	0.1	18.7	1.5	2.7	9
	aus	3.8	3.7	0.8	0.3	22.1	2.8	5.0	10
	be1	2.7	3.0	0.4	–	16.6	1.8	3.0	11
	be2	2.2	2.0	0.6	0.2	27.9	1.5	3.2	15
	be3	2.5	2.8	0.6	0.3	24.0	1.6	2.8	4
	bra[h]	2.3	2.5	–	–	–	1.5	3.0	13
	gr1	3.7	–	1.2	0.3	33.2	2.0	6.0	15
	gr2	3.0	–	0.7	0.2	21.7	2.0	4.0	15
	in1	2.1	2.1	0.4	0.1	18.5	1.4	2.6	11
	in3	2.0	2.0	0.1	0.0	6.8	2.0	2.5	13
	mo1	4.9	5.3	0.6	0.2	11.5	4.5	3.0	11
	mo3	6.3	6.0	1.2	–	18.7	^4.0	9.0	13
	si1	3.2	2.8	0.7	0.3	22.5	2.3	4.0	7
	si2	3.6	3.6	0.5	0.2	14.9	2.7	4.5	11
	si3	3.6	3.8	–	–	–	3.0	4.0	3
	si4	5.7	5.6	0.9	0.24	16.4	4.0	7.0	15
Nuclear apparatus, length	si4	52.9	50.0	7.3	1.9	13.7	42.0	63.0	15
Macronuclear nodules, number	aus	12.5	12.0	2.1	0.7	16.5	10.0	16.0	10
	be1	15.0	15.0	1.0	–	6.7	13.0	16.0	11
	be2	28.6	28.0	2.8	0.8	9.8	25.0	34.0	13
	be3	15.8	15.5	1.7	0.9	10.8	14.0	18.0	4
	bra	31.7	31.0	2.9	0.8	9.1	28.0	38.0	13
	gr1	41.5	–	11.7	3.0	28.1	23.0	62.0	15
	gr2	48.2	–	11.5	3.0	23.9	26.0	63.0	15
	in3	102.7	106.0	18.8	5.7	18.3	78.0	135.0	11
	mo1	15.0	15.0	3.3	1.0	21.8	9.0	23.0	11
	mo2	9.1	–	3.6	0.9	39.9	4.0	15.0	17
	mo3	8.4	8.0	0.9	–	10.6	7.0	10.0	16
	mo4	–	–	–	–	–	8.0	11.0	?
	mo5	9.0	8.0	2.4	0.6	27.0	6.0	16.0	15
	si1	7.7	8.0	0.7	0.3	9.1	6.0	8.0	7
	si2	8.1	8.0	0.3	0.1	3.6	8.0	9.0	11
	si3	11.7	12.0	–	–	–	11.0	12.0	4
	si4	7.9	8.0	1.1	0.3	14.3	5.0	9.0	15
Micronucleus, length	aus	3.9	4.0	0.6	0.2	16.8	3.0	4.5	10
	be1	1.6	1.4	0.3	–	17.7	1.4	2.0	7
	be2	2.8	3.0	0.4	0.1	12.5	2.2	3.3	12
	bra[h]	1.8	1.5	–	–	–	1.5	3.0	7

Table 19 Continued

Characteristics[a]	Species	mean	M	SD	SE	CV	Min	Max	n
Micronucleus, length	bra[h]	1.8	1.5	–	–	–	1.5	3.0	7
	mo3	4.9	5.0	0.9	–	18.9	4.0	7.0	13
	mo5	4.0	4.0	0.4	0.1	9.0	3.5	4.8	11
	si1	2.3	2.3	0.2	0.1	8.1	2.0	2.5	7
	si4	2.5	2.7	0.4	0.1	17.4	1.6	3.0	15
Micronucleus, width	aus	2.3	2.2	0.5	0.1	20.2	1.8	3.0	10
	be1	1.4	1.4	0.1	–	8.3	1.2	1.6	7
	be2	2.0	2.0	0.5	0.1	23.2	1.5	2.8	12
	bra[h]	1.6	1.5	–	–	–	1.0	2.4	7
	mo3	3.3	3.0	0.6	–	18.3	2.5	4.0	13
	mo5	4.0	4.0	0.4	0.1	9.0	3.5	4.8	11
	si1	2.2	2.2	0.1	0.1	5.7	2.0	2.4	7
	si4	1.8	1.6	0.5	0.1	27.8	1.4	3.0	15
Micronuclei, number	aus	2.8	3.0	1.0	0.3	36.9	1.0	4.0	10
	be2	2.3	2.0	1.0	0.3	42.9	1.0	4.0	12
	bra	2.0	2.0	1.1	0.3	52.7	1.0	4.0	10
	gr1	3.4	–	1.7	0.4	48.5	1.0	7.0	14
	gr2	3.9	–	1.6	0.5	41.0	2.0	6.0	10
	mo2	4.6	–	1.4	0.5	31.3	2.0	6.0	9
	mo3	3.8	4.0	1.0	–	25.7	2.0	5.0	13
	mo5	5.0	5.0	1.4	0.3	28.0	3.0	7.0	16
	si1	2.2	2.0	0.4	0.1	17.2	2.0	3.0	6
Frontal cirri, number	ad1[b]	4.0	4.0	0.0	0.0	0.0	4.0	4.0	8
	ad3	3.0	3.0	0.0	0.0	0.0	3.0	3.0	15
	an1	4.5	5.0	0.7	0.2	14.9	3.0	5.0	10
	an2[c]	5.0	–	–	–	–	–	–	?
	aus	3.0	3.0	0.0	0.0	0.0	3.0	3.0	10
	be1	3.0	3.0	0.0	–	0.0	3.0	3.0	11
	bra	2.9	3.0	–	–	–	2.0	3.0	13
	gr1[b]	4.0	–	0.0	0.0	0.0	4.0	4.0	16
	gr2[b]	4.0	–	0.0	0.0	0.0	4.0	4.0	15
	in1[b]	4.0	4.0	0.0	0.0	0.0	4.0	4.0	11
	in2	3.0	–	–	–	–	–	–	?
	in3[b]	4.0	4.0	0.0	0.0	0.0	4.0	4.0	14
	in4	3.0	–	–	–	–	–	–	?
	in5	3.0	–	–	–	–	–	–	?
	man	4.0	–	0.0	0.0	0.0	4.0	4.0	16
	ml1	8.0	–	–	–	–	–	–	1
	ml2	9.0	–	–	–	–	–	–	1
	mo1[b]	3.7	4.0	0.4	0.1	11.9	3.0	4.0	11
	mo2[b]	4.0	–	0.0	0.0	0.0	4.0	4.0	14
	mo3[b]	4.0	4.0	0.0	–	0.0	4.0	4.0	13
	mo4[b]	4.0	–	–	–	–	–	–	?
	mo5	7.0	7.0	0.4	0.1	6.0	7.0	8.0	16
	si1	3.0	3.0	0.0	0.0	0.0	3.0	3.0	7
	si2	2.9	3.0	0.3	0.1	9.9	2.0	3.0	11
	si3	3.0	3.0	–	–	–	3.0	3.0	4
	si4	1.0	1.0	0.0	0.0	0.0	1.0	1.0	15
	war	3.0	–	0.0	0.0	0.0	3.0	3.0	30

Table 19 Continued

Characteristics[a]	Species	mean	M	SD	SE	CV	Min	Max	n
Buccal cirri, number	ad1	3.4	3.5	0.7	0.2	20.6	2.0	4.0	8
	ad2	–	–	–	–	–	4.0	5.0	?
	ad3	3.6	3.0	0.7	0.2	20.5	3.0	5.0	15
	an1	6.7	6.5	1.2	0.4	17.8	5.0	9.0	8
	an2	–	–	–	–	–	4.0	6.0	?
	aus	1.0	1.0	0.0	0.0	0.0	1.0	1.0	10
	be1	1.0	1.0	0.0	–	0.0	1.0	1.0	11
	bra	1.0	1.0	0.0	0.0	0.0	1.0	1.0	13
	gr1	1.0	–	0.0	0.0	0.0	1.0	1.0	16
	gr2	1.0	–	0.0	0.0	0.0	1.0	1.0	16
	in1	3.7	4.0	0.6	0.2	16.5	2.0	4.0	11
	in2	–	–	–	–	–	1.0	5.0	?
	in3	2.9	3.0	0.3	0.1	11.5	2.0	3.0	9
	in4	5.0	–	–	–	–	–	–	?
	in5	–	–	–	–	–	3.0	4.0	?
	man	1.0	–	0.0	0.0	0.0	1.0	1.0	16
	ml1	4.0	–	–	–	–	–	–	1
	ml2	4.0	–	–	–	–	–	–	1
	mo1	0.6	1.0	0.5	0.1	75.6	0.0	1.0	11
	mo2	1.0	–	0.0	0.0	0.0	1.0	1.0	14
	mo3	1.0	1.0	0.0	–	0.0	1.0	1.0	13
	mo4	1.0	–	–	–	–	–	–	?
	si1	1.0	1.0	0.0	0.0	0.0	1.0	1.0	6
	si2	1.0	1.0	0.0	0.0	0.0	1.0	1.0	11
	si3	1.0	1.0	–	–	–	1.0	1.0	4
	war	1.0	–	0.0	0.0	0.0	1.0	1.0	30
Cirri behind right frontal cirrus, number	ad3	2.1	2.0	0.3	0.1	12.5	2.0	3.0	15
	in4	1.0	–	–	–	–	–	–	?
	in5	1.0	–	–	–	–	–	–	
Frontoterminal cirri, number	ad3	2.0	2.0	0.0	0.0	0.0	2.0	2.0	15
	aus	2.0	2.0	0.0	0.0	0.0	2.0	2.0	9
	be1	2.0	2.0	0.0	–	0.0	2.0	2.0	11
	bra	2.0	2.0	0.0	0.0	0.0	2.0	2.0	10
	gr1	2.0	–	0.0	0.0	0.0	2.0	2.0	16
	gr2	2.0	–	0.0	0.0	0.0	2.0	2.0	16
	in3	2.0	2.0	0.0	0.0	0.0	2.0	2.0	12
	man	2.8	–	0.7	0.2	24.8	2.0	4.0	16
	ml2	2.0	–	–	–	–	–	–	1
	mo3	2.0	2.0	0.0	–	0.0	2.0	2.0	9
	war	2.0	–	0.0	0.0	0.0	2.0	2.0	30
Anterior body end to buccal cirrus, distance	bra	8.6	9.0	1.3	0.3	14.6	6.0	10.0	13
Anterior body end to rearmost left midventral cirrus, distance	ad1	54.4	53.0	10.0	3.8	18.3	40.0	70.0	8
	ad3	68.7	66.0	11.8	3.1	17.2	51.0	92.0	15
	an1	144.5	144.0	17.9	6.3	12.4	110.0	170.0	8
	aus	111.2	108.0	13.0	4.1	11.7	91.0	138.0	10
	be1	24.0	25.0	2.1	–	8.7	18.0	27.0	11
	bra	26.7	26.0	2.8	0.8	10.5	22.0	32.0	13
	in1	55.7	53.0	8.0	2.4	14.3	44.0	73.0	11
	ml1	153.0	–	–	–	–	–	–	1
	ml2	156.0	–	–	–	–	–	–	1

Table 19 Continued

Characteristics[a]	Species	mean	M	SD	SE	CV	Min	Max	n
Anterior body end to rearmost left midventral cirrus, distance	si1	42.0	41.5	4.0	1.6	9.5	37.0	47.0	6
	si2	35.1	35.0	2.4	0.7	6.9	30.0	40.0	11
	si3	43.0	43.0	–	–	–	40.0	46.0	4
Anterior body end to rearmost right midventral cirrus, distance	be1	26.4	27.0	2.3	–	8.8	21.0	29.0	11
	be2	31.2	30.0	3.3	0.8	10.4	28.0	39.0	16
	be3	27.8	27.5	–	–	–	26.0	30.0	4
Midventral complex, relative length[f]	ad3	60.8	61.1	6.1	1.6	10.1	50.8	74.6	15
Body length:length of midventral complex, ratio	bra	3.0	3.1	0.4	0.1	14.4	2.2	3.5	13
Midventral complex, number of cirral pairs	bra	4.7	4.5	0.0	–	–	4.0	6.0	12
	man	12.4	–	2.5	0.8	20.6	9.0	16.0	11
	mo1	23.0	24.0	2.6	0.8	11.3	19.0	27.0	11
	mo2	19.6	–	1.6	0.4	8.1	18.0	23.0	14
	mo3	22.1	22.0	1.7	–	7.5	20.0	25.0	13
	war	7.6	–	0.7	0.2	8.8	7.0	9.0	11
Midventral complex, number of cirri	be2	13.7	13.0	0.9	0.2	6.4	13.0	15.0	15
	be3	11.0	11.0	–	–	–	11.0	11.0	4
	bra	10.3	10.5	1.4	0.4	13.3	9.0	13.0	12
	gr1	39.6	–	4.1	1.1	10.2	34.0	48.0	15
	gr2	42.0	–	2.5	0.7	5.9	36.0	44.0	15
	mo4	–	–	–	–	–	40.0	47.0	?
	mo5	58.0	58.0	4.7	1.2	8.0	50.0	64.0	16
Midventral complex, number of left cirri	ad1	10.6	10.0	1.5	0.6	14.2	8.0	13.0	7
	ad2	14.0	–	–	–	–	–	–	1
	ad3[d]	12.8	12.0	2.0	0.5	15.5	10.0	16.0	15
	an1	26.6	28.0	4.5	1.6	17.0	19.0	31.0	8
	aus	19.0	19.0	2.9	1.0	15.3	15.0	23.0	9
	be1	5.3	5.0	0.5	–	8.8	5.0	6.0	11
	in1	12.1	12.0	1.4	0.4	11.9	10.0	15.0	11
	in3	10.8	11.0	0.4	0.2	4.1	10.0	11.0	9
	in5[d]	13.0	–	–	–	–	–	–	?
	ml1	21.0	–	–	–	–	–	–	1
	ml2	17.0	–	–	–	–	–	–	
	si1	9.2	9.5	0.9	0.4	9.8	8.0	10.0	6
	si2	9.3	9.0	1.5	0.5	16.6	7.0	12.0	11
	si3	11.0	10.5	–	–	–	10.0	13.0	4
	si4	15.1	16.0	1.6	0.4	10.9	12.0	18.0	15
Midventral complex, number of right cirri	ad1	12.9	12.0	1.8	0.7	14.0	10.0	15.0	7
	ad2	14.0	–	–	–	–	–	–	1
	an1	27.5	28.0	3.4	1.2	12.5	22.0	31.0	8
	aus	19.8	20.0	2.8	0.9	14.0	15.0	23.0	10
	be1	6.3	6.0	0.5	–	7.4	6.0	7.0	11
	in1	18.5	18.0	1.8	0.5	9.8	15.0	21.0	11
	in3	12.6	12.0	0.7	0.2	5.8	12.0	14.0	9
	ml1	20.0	–	–	–	–	–	–	1
	ml2	18.0	–	–	–	–	–	–	1
	si1	10.2	10.5	0.9	0.4	8.8	9.0	11.0	6
	si2	9.5	9.0	1.3	0.4	13.7	8.0	12.0	11
	si3	12.0	12.0	–	–	–	11.0	14.0	4
	si4	15.1	16.0	1.6	0.4	10.9	12.0	18.0	15

Table 19 Continued

Characteristics[a]	Species	mean	M	SD	SE	CV	Min	Max	n
Pretransverse ventral cirri, number	ad1	1.1	1.0	0.3	0.1	33.1	1.0	2.0	7
	ad2	2.0	–	–	–	–	–	–	1
	ad3	1.2	1.0	0.6	0.1	46.7	0.0	2.0	15
	aus	1.7	2.0	0.5	0.2	30.0	1.0	2.0	9
	be1	2.0	2.0	0.0	–	0.0	2.0	2.0	11
	be2	1.0	1.0	0.0	0.0	0.0	1.0	1.0	12
	be3	2.0	2.0	–	–	–	2.0	2.0	4
	bra	1.0	1.0	0.0	0.0	0.0	1.0	1.0	10
	in1	1.6	2.0	0.5	0.1	29.4	1.0	2.0	11
	in3	1.8	2.0	0.4	0.1	21.2	1.0	2.0	12
	in5	2.0	–	–	–	–	–	–	?
	ml1	2.0	–	–	–	–	–	–	1
	ml2	2.0	–	–	–	–	–	–	1
	si1	2.0	2.0	0.0	0.0	0.0	2.0	2.0	7
	si2	1.9	2.0	0.3	0.1	15.1	1.0	2.0	11
	si3	2.0	2.0	–	–	–	2.0	2.0	4
Transverse cirri, number	ad1	6.0	6.0	0.9	0.3	14.4	5.0	7.0	8
	ad2	–	–	–	–	–	4.0	6.0	?
	ad3	5.3	5.0	1.0	0.3	19.6	4.0	7.0	15
	an1	12.0	12.5	2.2	0.8	17.7	8.0	15.0	8
	an2	–	14.0	–	–	–	11.0	14.0	?
	aus	4.6	5.0	0.8	0.3	18.3	3.0	6.0	10
	be1	3.9	4.0	–	–	–	3.0	4.0	11
	be2	4.8	5.0	0.4	0.1	7.7	4.0	5.0	13
	be3	4.3	4.0	0.5	0.3	11.8	4.0	5.0	4
	bra	2.1	2.0	–	–	–	2.0	3.0	10
	gr1	8.4	–	1.0	0.3	11.8	7.0	10.0	15
	gr2	8.9	–	0.8	0.2	8.7	8.0	10.0	16
	in1	5.5	5.0	0.8	0.2	14.3	4.0	7.0	11
	in2	–	–	–	–	–	6.0	9.0	?
	in3	6.9	7.0	0.8	0.2	11.5	6.0	8.0	12
	in4	10.0	–	–	–	–	–	–	?
	in5	–	–	–	–	–	6.0	11.0	?
	man	4.9	–	0.6	0.1	11.6	4.0	6.0	16
	ml1	8.0	–	–	–	–	–	–	1
	ml2	8.0	–	–	–	–	–	–	1
	mo1	9.0	9.0	1.0	0.3	10.6	7.0	10.0	11
	mo2	9.4	–	1.4	0.4	15.3	7.0	11.0	11
	mo3	10.0	10.0	1.1	–	10.8	8.0	12.0	13
	mo4	–	–	–	–	–	8.0	10.0	?
	mo5	7.4	7.5	0.6	0.1	8.0	7.0	8.0	16
	si1	4.4	4.0	0.9	0.3	20.4	3.0	6.0	7
	si2	4.0	4.0	0.0	0.0	0.0	4.0	4.0	11
	si3	5.0	5.0	–	–	–	5.0	5.0	4
	war	10.9	–	0.5	0.1	4.6	10.0	12.0	16
Transverse cirri and pretransverse ventral cirri, number	bra	3.1	3.0	–	–	–	3.0	4.0	10
	si4	5.1	5.0	0.7	0.2	14.5	4.0	6.0	15
Left marginal row, number of cirri	ad1	37.5	37.5	4.7	1.7	12.6	30.0	45.0	8
	ad2	34.0	–	–	–	–	–	–	1
	ad3	41.7	39.0	5.9	1.5	14.2	34.0	54.0	15

Table 19 Continued

Characteristics [a]	Species	mean	M	SD	SE	CV	Min	Max	n
Left marginal row, number of cirri	an1	57.1	57.5	12.6	4.4	22.0	28.0	72.0	8
	an2	60.0	–	–	–	–	–	–	?
	aus	38.5	38.0	4.5	1.4	11.6	33.0	48.0	10
	be1	14.4	14.0	1.4	–	10.0	13.0	16.0	11
	be2	21.4	21.0	3.3	0.8	15.6	16.0	28.0	16
	be3	16.3	16.5	1.0	0.5	5.9	15.0	17.0	4
	bra	23.1	23.0	2.1	0.6	9.3	19.0	27.0	13
	gr1	32.8	–	4.2	1.1	12.9	28.0	41.0	16
	gr2	37.6	–	2.6	0.7	6.9	33.0	42.0	16
	in1	42.2	41.0	4.8	1.5	11.5	36.0	51.0	11
	in3	36.1	36.0	4.0	1.3	11.0	31.0	41.0	9
	in5	–	–	–	–	–	30.0	52.0	?
	man	26.1	–	2.3	0.3	8.9	22.0	29.0	11
	ml1	28.0	–	–	–	–	–	–	1
	ml2	40.0	–	–	–	–	–	–	1
	mo1	37.1	39.0	5.6	1.7	15.2	25.0	45.0	11
	mo2	41.7	–	2.8	1.1	6.7	39.0	46.0	7
	mo3	43.0	42.0	3.7	–	8.4	40.0	50.0	9
	mo4	–	–	–	–	–	47.0	52.0	?
	mo5	54.0	53.0	5.6	1.4	10.0	46.0	65.0	16
	si1	19.8	19.0	3.1	1.3	15.8	16.0	24.0	6
	si2	26.4	27.0	4.1	1.2	15.9	19.0	31.0	11
	si3	27.0	27.0	–	–	–	23.0	31.0	4
	si4	29.0	29.0	2.8	0.7	9.5	24.0	35.0	15
	war	23.4	–	2.1	0.5	8.9	22.0	27.0	16
Right marginal row, number of cirri	ad1	34.7	35.0	4.3	1.5	12.5	29.0	42.0	8
	ad2	34.0	–	–	–	–	–	–	1
	ad3	42.7	44.0	6.5	1.7	15.3	32.0	55.0	15
	an1	67.7	69.5	7.4	2.6	10.9	54.0	80.0	8
	an2	60.0	–	–	–	–	–	–	?
	aus	37.5	36.5	2.9	0.9	7.8	34.0	43.0	10
	be1	12.5	13.0	1.8	–	14.0	10.0	15.0	11
	be2	20.9	21.0	2.7	0.7	13.0	16.0	27.0	15
	be3	15.5	15.5	0.6	0.3	3.7	15.0	16.0	4
	bra	22.6	22.5	2.5	0.7	11.1	19.0	26.0	12
	gr1	30.4	–	3.2	0.8	10.4	25.0	35.0	16
	gr2	32.6	–	2.2	0.6	6.8	28.0	38.0	16
	in1	41.9	40.0	4.2	1.3	10.0	37.0	52.0	11
	in3	33.6	35.0	2.9	0.9	8.6	30.0	38.0	10
	in5	–	–	–	–	–	35.0	55.0	?
	man	29.5	–	3.0	0.9	10.0	6.0	34.0	11
	ml1	35.0	–	–	–	–	–	–	1
	ml2	46.0	–	–	–	–	–	–	1
	mo1	38.8	42.0	7.8	2.3	20.0	21.0	46.0	11
	mo2	45.3	–	3.8	1.4	8.3	41.0	50.0	7
	mo3	49.0	48.0	2.7	–	5.6	46.0	54.0	9
	mo4	–	–	–	–	–	48.0	53.0	?
	mo5	55.0	57.0	5.7	1.5	10.0	44.0	62.0	16
	si1	18.3	19.0	2.6	1.0	14.2	15.0	23.0	7
	si2	24.4	24.0	2.1	0.6	8.8	21.0	28.0	11

Table 19 Continued

Characteristics[a]	Species	mean	M	SD	SE	CV	Min	Max	n
Right marginal row, number of cirri	si3	27.7	27.5	–	–	–	24.0	32.0	4
	si4	28.7	29.0	2.6	0.7	9.2	22.0	33.0	15
	war	22.8	–	1.9	0.5	8.5	21.0	26.0	16
Number 1 [e]	ad3	1.8	2.0	0.6	0.2	35.5	0.0	2.0	12
Dorsal kineties, number	ad1	3.0	3.0	0.0	0.0	0.0	3.0	3.0	8
	ad2	–	–	–	–	–	3.0	3.0	?
	ad3	3.1	3.0	0.3	0.1	8.7	3.0	4.0	14
	an1	3.0	3.0	0.0	0.0	0.0	3.0	3.0	4
	an2	4.0	–	–	–	–	–	–	?
	aus	4.0	4.0	0.0	0.0	0.0	4.0	4.0	5
	be1	3.0	3.0	0.0	–	0.0	3.0	3.0	11
	bra	3.0	3.0	0.0	0.0	0.0	3.0	3.0	11
	gr1	3.0	3.0	0.0	0.0	0.0	3.0	3.0	12
	gr2	3.0	3.0	0.0	0.0	0.0	3.0	3.0	15
	in3	3.0	3.0	0.0	0.0	0.0	3.0	3.0	11
	in5	4.0	–	–	–	–	–	–	?
	man	3.0	–	0.0	0.0	0.0	3.0	3.0	16
	ml2	3.0	–	–	–	–	–	–	?
	ms1	3.0	3.0	0.0	0.0	0.0	3.0	3.0	11
	mo1	6.0	6.0	0.0	0.0	0.0	6.0	6.0	11
	mo2	5.9	–	0.3	0.1	5.1	5.0	6.0	11
	mo3	6.2	6.0	0.4	–	6.6	6.0	7.0	6
	mo4	7.0	–	–	–	–	–	–	?
	mo5	8.0	8.0	0.6	0.1	8.0	7.0	9.0	14
	si1	4.0	4.0	0.0	0.0	0.0	4.0	4.0	7
	si2	4.0	4.0	0.0	0.0	0.0	4.0	4.0	11
	si4	4.0	4.0	0.0	0.0	0.0	4.0	4.0	15
	war	3.0	–	0.0	0.0	0.0	3.0	3.0	16

[a] All measurements in µm. ? = number of individuals investigated (= sample size) not indicated; if only one value is known it is listed in the mean column; if two values are available they are listed as Min and Max. CV = coefficient of variation in %, M = median, Max = maximum value, mean = arithmetic mean, Min = minimum value, n = number of individuals investigated, SD = standard deviation, SE = standard error of arithmetic mean. Data are based on protargol-impregnated specimens.

[b] Cirrus (= cirrus III/2) behind right frontal cirrus included.

[c] Two enlarged cirri behind right frontal cirrus included (Fig. 73g).

[d] This is the number of midventral pairs.

[e] Distance 1: distance between anterior body end and cirrus III/2 (marked with an asterisk in Fig. 74i). Distance 2: distance between posterior body end and rearmost transverse cirrus. Distance 3: distance between rear end of left and right marginal row. Distance 4: distance between anterior body end and first frontoterminal cirrus. Distance 5: distance between anterior body end and second frontoterminal cirrus. Distance 6: distance between anterior body end and right marginal row. Distance 7: distance between anterior body end and first macronuclear nodule. Number 1: number of basal body pairs ahead of right marginal row.

[f] Percentage of body length at which the rearmost cirrus of the midventral complex is arranged.

[g] Percentage of body length at which the rearmost transverse cirrus is arranged.

[h] Anteriormost macronuclear nodule, respectively, micronucleus.

[i] Specimen shown in Fig. 87f is 111 µm long according to the scale bar indicating that this individual is not included in the morphometrics which shows 106 µm as maximum value.

Table 20 Morphometric data on *Holosticha corlissi* (cor, from Fernandez-Galiano & Calvo 1992; this species is classified as supposed synonym of *Anteholosticha monilata*) and *Anteholosticha xanthichroma* (xan, from Wirnsberger & Foissner 1987)

Characteristics[a]	Species	mean	M	SD	SE	CV	Min	Max	n
Body, length	cor	213.0	–	9.4	1.8	4.4	183.0	246.0	25
	xan	126.2	122	24.7	4.9	19.5	90.0	173.0	26
Body, width	cor	70.2	–	9.7	1.9	13.9	76.0	110.0	25
	xan	30.9	30.0	4.6	0.9	15.0	25.0	45.0	26
Body length:width, ratio	xan	4.1	4.4	0.7	0.1	16.5	3.0	6.0	26
Anterior body end to proximal end of adoral zone, distance	cor	72.1	–	7.1	1.4	9.8	66.6	78.7	25
	xan	34.5	33.5	6.9	1.4	20.0	22.0	50.0	26
Adoral membranelles, number	cor	44.3	–	5.8	1.1	13.1	38.0	55.0	25
	xan	37.0	38.5	6.9	1.4	18.8	21.0	46.0	26
Endoral, length[b]	cor	40.7	–	3.1	0.6	7.4	29.9	50.0	25
Body length:length adoral zone, ratio	xan	3.7	3.7	0.5	0.1	13.4	3.0	5.0	26
Macronuclear nodule, length	xan	5.5	5.0	2.5	0.3	45.1	2.0	10	52
Macronuclear nodule, width	xan	2.9	3.0	0.9	0.1	32.1	2.0	5.0	52
Macronuclear nodules, number	cor	14.0	–	4.4	0.8	31.9	8.0	16.0	25
	xan	28.4	28.5	7.7	1.5	27.1	11.0	42.0	26
Micronuclei, number	cor	5.0	–	1.5	0.3	30.4	2.0	8.0	25
Frontal cirri, number	xan	3.0	3.0	0.0	0.0	0.0	3.0	3.0	26
Buccal cirri, number	xan	1.0	1.0	0.0	0.0	0.0	1.0	1.0	26
Frontoterminal cirri, number	xan	2.0	2.0	0.3	0.1	14.0	1.0	3.0	26
Midventral complex, total number of cirri	cor	54.0	–	2.5	0.5	4.6	52.0	55.0	25
Midventral complex, number of left cirri	xan	41.7	41.0	10.4	2.0	24.9	22.0	65.0	26
Midventral complex, number of right cirri	xan	41.2	40.0	10.4	2.0	25.3	20.0	62.0	26
Transverse cirri, number	cor	6.0	–	1.5	0.3	25.0	5.0	8.0	25
	xan	4.0	4.0	1.9	0.4	48.2	0.0	6.0	26
Left marginal row, number of cirri	cor	45.0	–	3.0	0.6	6.6	42.0	48.0	25
	xan	53.5	56.0	12.7	2.5	23.8	32.0	74.0	26
Right marginal row, number of cirri	cor	42.0	–	3.6	0.7	8.5	40.0	45.0	25
	xan	49.4	49.5	12.3	2.4	25.0	26.0	72.0	26
Dorsal kineties, number	xan	4.0	4.0	0.0	0.0	0.0	4.0	4.0	26

[a] All measurements in µm. CV = coefficient of variation in %, M = median, Max = maximum value, mean = arithmetic mean, Min = minimum value, n = number of individuals investigated, SD = standard deviation, SE = standard error of arithmetic mean. Data from *Holosticha corlissi* are based on silver carbonate-impregnated specimens, data from *A. xanthichroma* are based on mounted, protargol-impregnated (Foissner's method), and randomly selected specimens.

[b] Fernandez-Galiano & Calvo (1992) measured and listed the length of the paroral. However, they confused the two undulating membranes (see text). Thus, I changed the parameter from paroral to endoral.

Anteholosticha brevis (Kahl, 1932) Berger, 2003
(Fig. 69a)

1932 ***Holosticha brevis* spec. n.**[1] – Kahl, Tierwelt Dtl., 25: 586, Fig. 106_8 (Fig. 69a; original description; no type material available).

1972 ***Holosticha brevis* Kahl, 1932** – Borror, J. Protozool., 19: 10 (revision of hypotrichs).

1983 ***Holosticha brevis* Kahl, 1932** – Borror & Wicklow, Acta Protozool., 22: 121 (revision of urostylids).

2001 ***Holosticha (Holosticha) brevis* Kahl, 1932** – Berger, Catalogue of ciliate names 1. Hypotrichs, p. 35 (nomenclator containing all basionyms, combinations, and higher taxa of hypotrichs).

2003 ***Anteholosticha brevis* (Kahl, 1932) n. comb.** – Berger, Europ. J. Protistol., 39: 377 (combination with *Anteholosticha*).

Nomenclature: No derivation of the name is given in the original description. The species-group name *brev·is -is -e* (Latin adjective; short, small, low) likely alludes to the short ovoid body outline (Fig. 69a). Kahl (1932) divided *Holosticha* into subgenera. Thus, the correct name in the original description is *Holosticha (Holosticha) brevis* Kahl, 1932.

Remarks: The present species very likely lacks caudal cirri and all apomorphies of *Holosticha* and was thus transferred to *Anteholosticha* (Berger 2003). For a detailed discussion and foundation of the transfer of many species from *Holosticha* to *Anteholosticha* see the genus section.

The original description is brief, but the data provided indicate that it is a distinct species which, of course, needs detailed redescription. *Caudiholosticha algivora* has a rather similar general appearance, but possesses caudal cirri and each macronuclear nodule has a micronucleus attached (Fig. 49a).

Kahl (1932) wrote that "in the remaining features this species is identical with the previous species", which is *Caudiholosticha algivora*. I do not exactly know to which features this statement refers; possibly he meant the rows of colourless cortical granules and likely not the three fine caudal(?) cirri which he illustrated in Fig. 49a.

Borror (1972) and Borror & Wicklow (1983) synonymised *Keronopsis longicirrata* Gelei & Szabados, 1950 with the present species. However, this species, whose nuclear apparatus is not described, is an 18-cirri oxytrichid with very long cirri. Thus, it was synonymised with *Oxytricha longicirrata* Kahl, 1932 by Berger (1999, p. 163). According to Borror & Wicklow (1983), *Holosticha rostrata* (Fig. 26u) is a further junior synonym of the present species. However, in *H. rostrata* the contractile vacuole is distinctly behind mid-body, indicating that it is a synonym of *H. pullaster*. Further, the micronuclear situation of *H. rostrata* is unknown; thus, it would be unwise to synonymise it with *A. brevis*, which has a rather conspicuous nuclear apparatus.

Anteholosticha alpestris has the same nuclear apparatus, but more frontal cirri and the transverse cirri distinctly displaced anteriad (Fig. 83a); identity of these two species can therefore be excluded.

[1] Kahl (1932) provided the following brief description: Größe 100 µm. Rechte Seite gestreckt, l. stark konvex. 2 Kernteile und stets nur 1 Mi. dazwischen. 5 Trv.cirren, die links der Mediane inserieren und stark nach links hinten überragen. Im übrigen der vorigen Art gleich. Sapropel.

Morphology: Body size about 100 × 50 µm in life, ratio of body length:width 2:1 (Kahl 1932). Body outline wide elliptical, right margin more or less straight, left strongly convex. In left body portion two narrowly spaced macronuclear nodules with single micronucleus in between. Contractile vacuole about in mid-body near left margin (Fig. 68a). Possibly rows of colourless cortical granules present (see remarks). Shape of adoral zone obviously without peculiarities, but occupies about 43% of body length in specimen illustrated. Three enlarged frontal cirri; one buccal cirrus; midventral complex composed of few (about six in specimen illustrated; value must not be over-interpreted) cirral pairs; five transverse cirri left of midline and distinctly extending beyond left posterior body margin. Dorsal bristles obviously short (Fig. 69a). Caudal cirri likely lacking.

Fig. 69a *Anteholosticha brevis* from life (from Kahl 1932). Ventral view, 100 µm. Note the single, large micronucleus in between the two narrowly spaced macronuclear nodules. Page 360.

Occurrence and ecology: Limnetic. Kahl (1932) discovered this species in the sapropel. Unfortunately, he did not mention the type locality, but likely he found it in a freshwater habitat in Germany.

Records not substantiated by morphological data: freshwater in La Vesdre, Belgium (Chardez 1984, p. 20); Potomac River, USA (Patrick 1961, p. 243); rare in limnetic habitats in southeastern Louisiana, USA (Bamforth 1963, p. 133). Feeds on ciliates (Kahl 1932), according to Chardez (1984) it is omnivorous.

Anteholosticha camerounensis (Dragesco, 1970) Berger, 2003 (Fig. 70a–e)

1970 ***Holosticha camerounensis* n. sp.** – Dragesco, Annls Fac. Sci. Univ. féd. Cameroun (Numero Hors-série): 111, Fig. 81 (Fig. 70a; original description; site where type material deposited not known, likely in private collection of J. Dragesco; no formal diagnosis provided).

1972 ***Holosticha camerounensis* Dragesco, 1970** – Borror, J. Protozool., 19: 10 (revision of hypotrichs).

1983 ***Holosticha camerounensis* Drageco, 1970** – Borror & Wicklow, Acta Protozool., 22: 121, Fig. 19 (Fig. 70e; revision of urostylids; see remarks).

1986 ***Holosticha camerounensis* Dragesco, 1970** – Dragesco & Dragesco-Kernéis, Faune tropicale, 26: 447, Planche 131 A–D (Fig. 70b–d; review on African ciliates; redescription; Figure A is a redrawing of Fig. 70a and thus not shown in the present book).

2001 ***Holosticha camerounensis* Dragesco, 1970** – Berger, Catalogue of ciliate names 1. Hypotrichs, p. 34 (nomenclator containing all basionyms, combinations, and higher taxa of hypotrichs).

2003 ***Anteholosticha camerounensis* (Dragesco, 1970) n. comb.** – Berger, Europ. J. Protistol., 39: 377 (combination with *Anteholosticha*).

Nomenclature: The species-group name *camerounensis* (occurring in Cameroon) refers to the country where the species was discovered.

Remarks: The present species lacks caudal cirri and all apomorphies of *Holosticha* and was thus transferred to *Anteholosticha* (Berger 2003). For a detailed discussion and

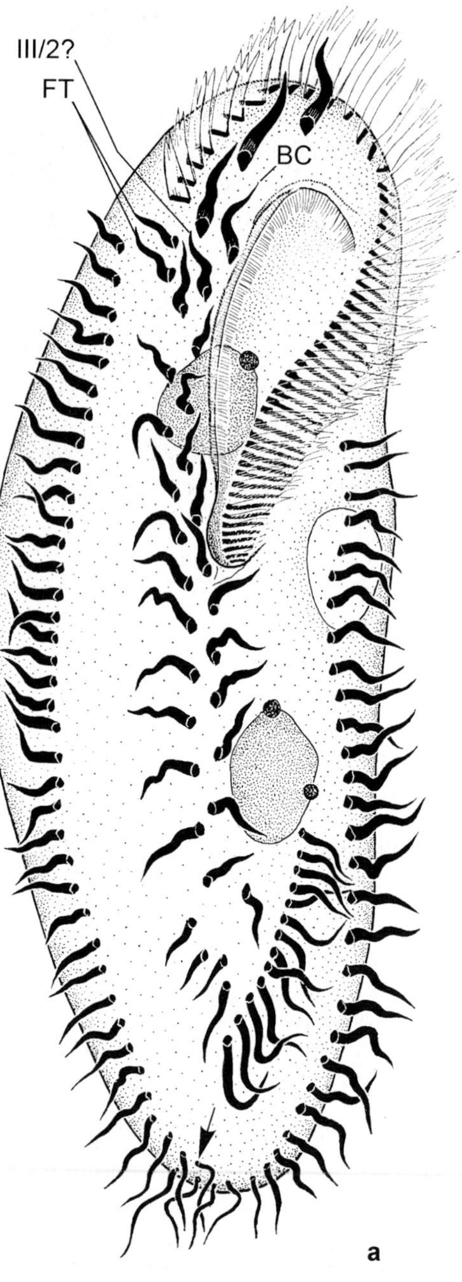

Fig. 70a *Anteholosticha camerounensis* (from Dragesco 1970. Protargol impregnation). Infraciliature of ventral side and nuclear apparatus of a specimen of the type population, 150 µm. Arrow denotes last cirrus of right marginal row. BC = buccal cirrus, FT = frontoterminal cirri?, LMR = last cirrus of left marginal row, III/2? = cirrus behind right frontal cirrus? Page 361.

foundation of the transfer of many species from *Holosticha* to *Anteholosticha* see the genus section.

The population described by Borror & Wicklow (1983, Fig. 70e) differs from Dragesco's populations from Africa in the number of transverse cirri (6 vs. 9–15). Thus, identity is doubtful. Borror & Wicklow (1983) synonymised *A. camerounensis* with *Holosticha contractilis* Dragesco, 1970. However, this highly contractile species, which differs from *A. camerounensis* in many morphometric features, is very likely a junior synonym of *Uroleptus musculus*.

A population from Benin was studied by Dragesco & Dragesco-Kernéis (1986). This second African population agrees very well with the type material so that conspecificity is beyond reasonable doubt. Thus, I summarise both descriptions. Very likely, some details are not quite correctly recognised, respectively, drawn, especially in the anterior body portion. Further, no data are available about cortical granulation (present or absent) and the dorsal infraciliature (number of dorsal kineties; presence or absence of caudal cirri). Thus, redescription from life and after protargol impregnation recommended.

Morphology: The following chapter is based solely on Dragesco's data. For a brief morphometric characterisation of the specimen illustrated by Borror & Wicklow (1983), see legend of Fig. 70e.

Body size 130–160 × 43–58 µm after fixation with Bouin. Body length:width ratio of specimens shown in Figs. 70a, b about 2.8:1. Body outline of live specimens not described, according to

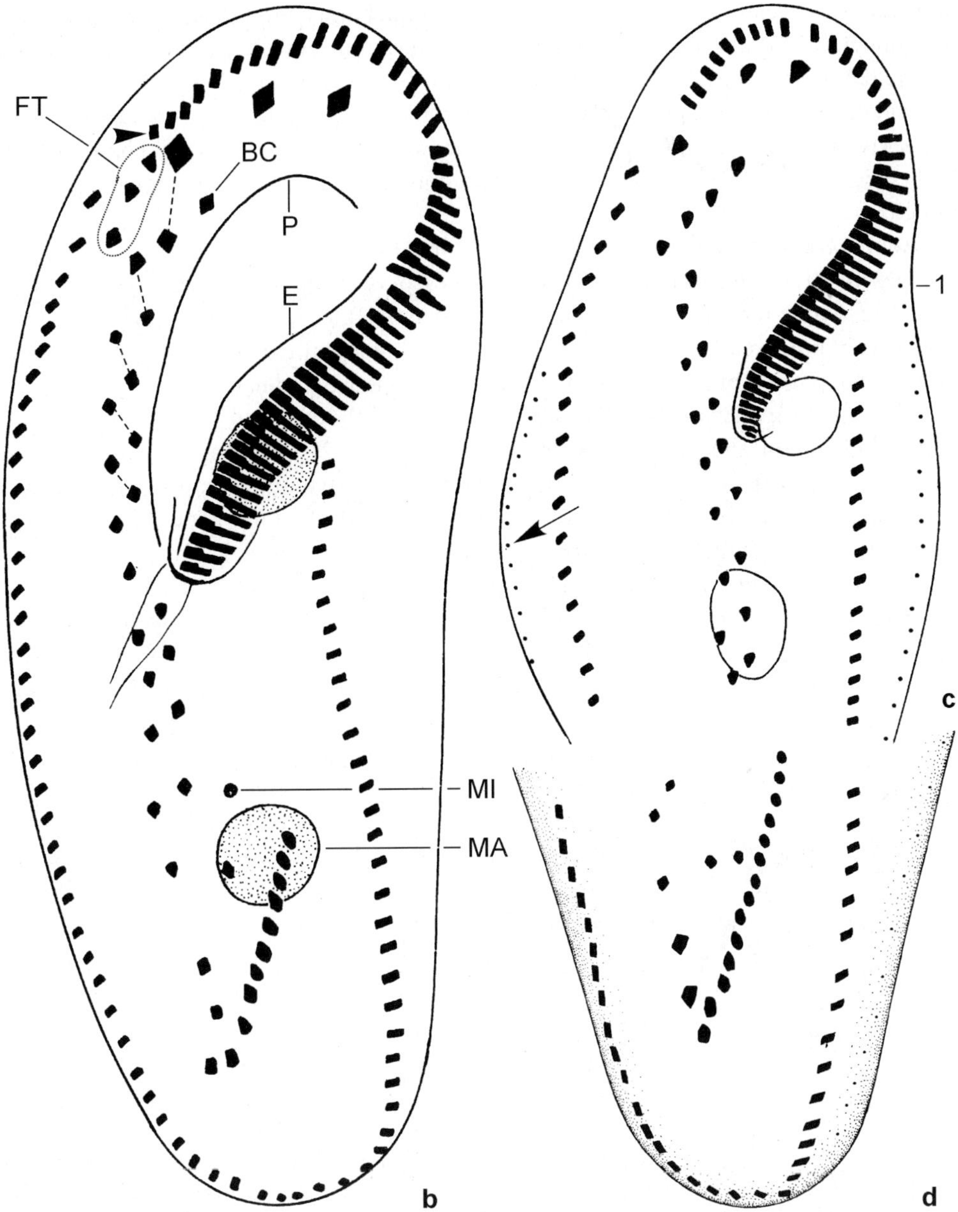

Fig. 70b–d *Anteholosticha camerounensis* (from Dragesco & Dragesco-Kernéis 1986. Protargol impregnation). Infraciliature of ventral side and nuclear apparatus of specimens of the population from Benin, b = 143 µm, c = 103 µm, d = 51 µm. Arrowhead in (b) denotes distal end of adoral zone. Right frontal cirrus and cirrus III/2 as well as anteriormost four midventral pairs connected by broken lines. Arrow in (c) marks rightmost dorsal kinety. BC = buccal cirrus, E = endoral, FT = frontoterminal cirri (possibly the anteriormost cirrus is the distalmost adoral membranelle?), MA = posterior macronuclear nodule, MI = micronucleus, P = paroral, 1 = dorsal kinety 1. Page 361.

prepared specimens likely elongate elliptical with both ends broadly rounded. Two macronuclear nodules, one each in anterior and posterior body half; individual nodules ellipsoidal to globular, 15–18 µm (Dragesco 1970), respectively, 11–19 µm (Dragesco & Dragesco-Kernéis 1986) across. 2–3 micronuclei about 2.0–2.5 µm across (Fig. 70a–c). Contractile vacuole in ordinary position, that is, near left margin about in mid-body (Fig. 70a). Presence/absence of cortical granules and kind of movement not mentioned.

Adoral zone occupies about 44% of body length (Fig. 70a, b), composed of 52–55 (Dragesco 1970), respectively, 52–58 (Dragesco & Dragesco-Kernéis 1986) membranelles, which are likely of ordinary fine structure. Buccal area large, right and anterior margin bordered by *Cyrtohymena*-like paroral; endoral extends along right margin of proximal portion of adoral zone. Cytopharynx obviously ordinary.

Cirral pattern without peculiarities. Three distinctly enlarged frontal cirri in ordinary arrangement and likely a single cirrus (cirrus III/2) behind right frontal cirrus. One buccal cirrus about at level of beginning of midventral complex. Likely two frontoterminal cirri in ordinary position near distal end of adoral zone (Fig. 70a; possibly specimens with 3 frontoterminal cirri occur, Fig. 70b). Midventral complex composed of around 11 cirral pairs (Fig. 70a–c); arrangement of cirri in posterior portion of complex possibly not quite correctly interpreted; according to Dragesco (1970) 28–30, according to Dragesco & Dragesco-Kernéis (1986) 20–30 cirri present (except frontal cirri, buccal cirrus, and transverse cirri) forming distinct zigzag-pattern in central body portion. 12–13 (Dragesco 1970), respectively, 9–15 (average 12, n = 13; Dragesco & Dragesco-Kernéis 1986) transverse cirri obviously distinctly displaced anteriad. Right marginal row commences slightly behind distal end of adoral zone, terminates in midline of posterior body end, composed of 36–42 (Dragesco 1970), respectively, 33–44 (Dragesco & Dragesco-Kernéis 1986) cirri. Left marginal row begins distinctly ahead of level of proximal end of adoral zone of membranelles, J-shaped and thus terminating almost on right body margin behind right marginal row, with which it slightly overlaps; left row composed of 24–32 cirri (Dragesco 1970; Dragesco & Dragesco-Kernéis 1986; obviously Dragesco & Dragesco-Kernéis 1986 confused left and right). Dorsal infraciliature (dorsal kineties, caudal

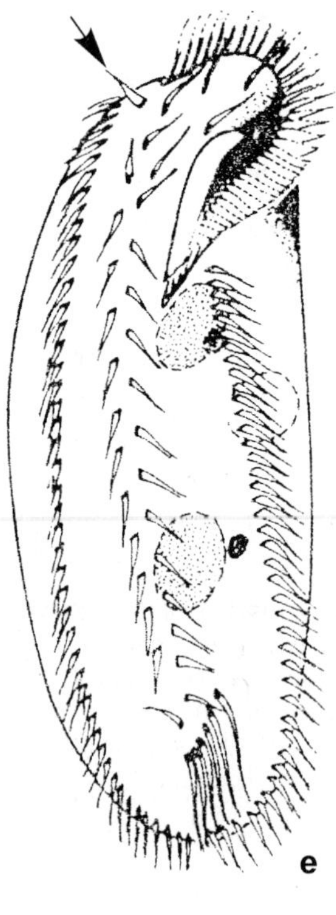

e

Fig. 70e *Anteholosticha camerounensis* (from Borror & Wicklow 1983. Protargol impregnation?). Infraciliature of ventral side, nuclear apparatus, and contractile vacuole, 108 µm. This specimen has 3 frontal cirri, 1 buccal cirrus, only 1 (?) frontoterminal cirrus (arrow), about 13 midventral pairs, 2 pretransverse ventral cirri, 6 transverse cirri, 46 right marginal cirri, 41 left marginal cirri, and about 40 adoral membranelles. Mainly because of the low number of transverse cirri (6 vs. 9–15 in Dragesco's populations) identity is uncertain. Page 361.

cirri) neither mentioned nor illustrated. According to Fig. 70c at least two dorsal kineties present. Caudal cirri likely lacking although confusion with marginal cirri cannot be excluded (Fig. 70a, b, d).

Occurrence and ecology: Reliable records confined to freshwater in Africa. Type locality is the city of Yaoundé, Cameroon, where Dragesco (1970) discovered it in saprobic pools (Dragesco 1970). The second population is from Benin, unfortunately no details are given (Dragesco & Dragesco-Kernéis 1986). Njine (1977, p. 24) found *Anteholosticha camerounensis* with a frequency of 21% in temporary pools in Cameroon at 23–27°C. Borror & Wicklow (1983; identification uncertain, see remarks) found their population likely somewhere in North America. Food unknown.

Anteholosticha macrostoma (Dragesco, 1970) comb. nov. (Fig. 71a–d)

1970 ***Pleurotricha macrostoma* n. sp.** – Dragesco, Annls Fac. Sci. Univ. féd. Cameroun (Numero Hors-série): 123, Fig. 88 (Fig. 71a; original description; site where type material deposited not known, likely in private collection of J. Dragesco; no formal diagnosis provided).

1971 ***Pleurotricha macrostoma*, Dragesco** – Dragesco & Njine, Annls Fac. Sci. Univ. féd. Cameroun, 7–8: 132, Fig. 29 (Fig. 71b; redescription).

1972 ***Holosticha macrostoma* (Dragesco, 1970) n. comb.** – Borror, J. Protozool., 19: 11 (revision of hypotrichs; combination with *Holosticha*).

1980 ***Pleurotricha macrostoma* Dragesco, 1970** – Dragesco, Office Rech. Sci. Tech. Outre, 44: 183, Planche VII, Fig. 68 (Fig. 71b; redrawing of Fig. 71a; review on African ciliates).

1986 ***Holosticha macrostoma* (Dragesco, 1970) Borror, 1972** – Dragesco & Dragesco-Kernéis, Faune tropicale, 26: 449, Planche 133 A, B (Fig. 71d; Fig. A is a slightly modified redrawing of Fig. 71a and thus not shown in the present paper).

2001 ***Holosticha macrostoma* (Dragesco, 1970) Borror, 1972** – Berger, Catalogue of ciliate names 1. Hypotrichs, p. 74 (nomenclator containing all basionyms, combinations, and higher taxa of hypotrichs).

Nomenclature: No derivation of the name is given in the original description. The species-group name *macrostoma* is a composite of the Greek adjective *makrós* (long, large) and the Latin noun *stóma* (mouth, opening) and refers to the large oral apparatus (about 55% of body length). After the transfer from *Pleurotricha* to *Holosticha*, the present species was a junior secondary homonym of *Holosticha macrostoma* (Dragesco, 1963) Jankowski, 1979. However, this species is now classified as supposed synonym of *Pseudoamphisiella lacazei* so that this secondary homonymy no longer exists.

Remarks: The present species is difficult to classify because some important features (consistency of body; presence/absence of cortical granules, caudal cirri, and fragmented dorsal kineties) are not described. The original assignment to *Pleurotricha* is not well founded since this group is characterised by at least two distinct right marginal cirral rows and an 18-frontal-ventral-transverse cirral pattern (Berger 1999). By contrast, *Anteholosticha macrostoma* has only one right[1] and one left marginal row and

[1] Of course one cannot exclude that the cirral pairs are additional right marginal rows as obviously supposed by Dragesco (1970). However, this is very unlikely because in this case the species would not have ventral cirri, preventing a classification in *Pleurotricha* (Fig. 71a).

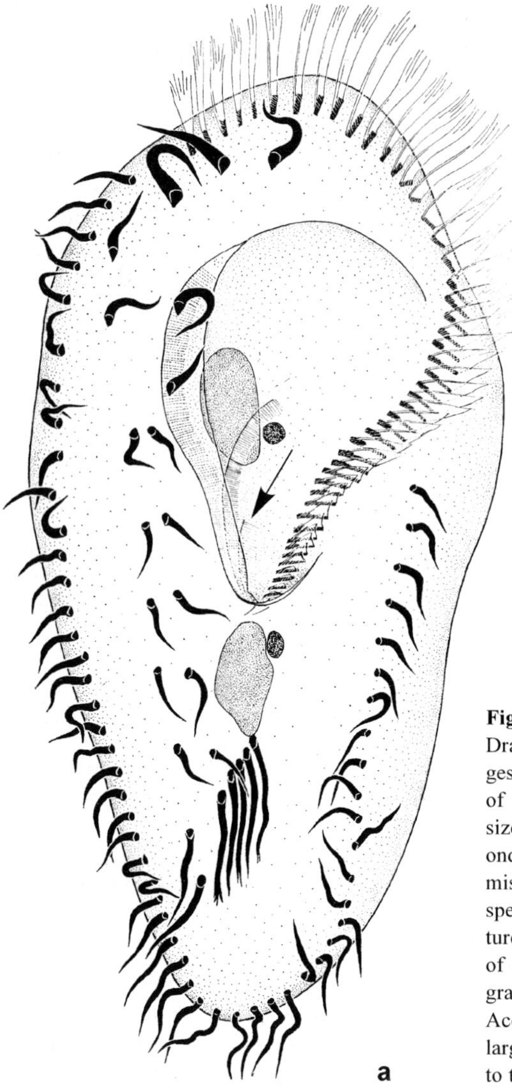

a

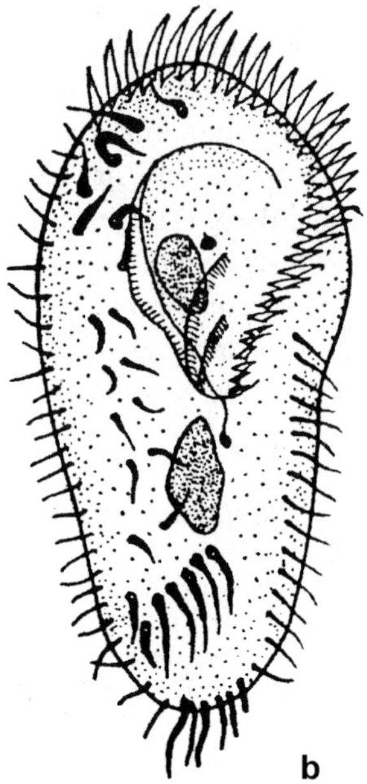

b

Fig. 71a, b *Anteholosticha macrostoma* (a, from Dragesco 1970; b, after Dragesco 1970 from Dragesco 1980. Protargol impregnation). Infraciliature of ventral side and nuclear apparatus, individual size not indicated. The arrow in (a) denotes a second, very short endoral which is very likely a misobservation. Note that the classification of this species is very uncertain since some important features are not known, for example, the consistency of the body, the presence or absence of cortical granules, the exact cirral and dorsal kinety pattern. According to the general appearance (wide body, large adoral zone of membranelles) it could belong to the oxytrichids. Page 365.

more ventral cirri, which indeed show a more or less distinct midventral pattern. Obviously these cirral pairs induced Borror (1972) to transfer it from *Pleurotricha* to *Holosticha*. Borror & Wicklow (1983) did not mention this species in their list on valid urostylids; they again considered it as *Pleurotricha* species (Borror & Wicklow 1983, p. 118). Dragesco & Dragesco-Kernéis (1986) accepted the assignment to *Holosticha*. We compared it with *Territricha stramenticola* Berger & Foissner, 1988 and assumed a close relationship of these two species, however, without changing anything (Berger & Foissner 1988). In my revision on the oxytrichids I did not discuss all these problems, but simply accepted Borror's transfer to *Holosticha* (Berger 1999, p. 699).

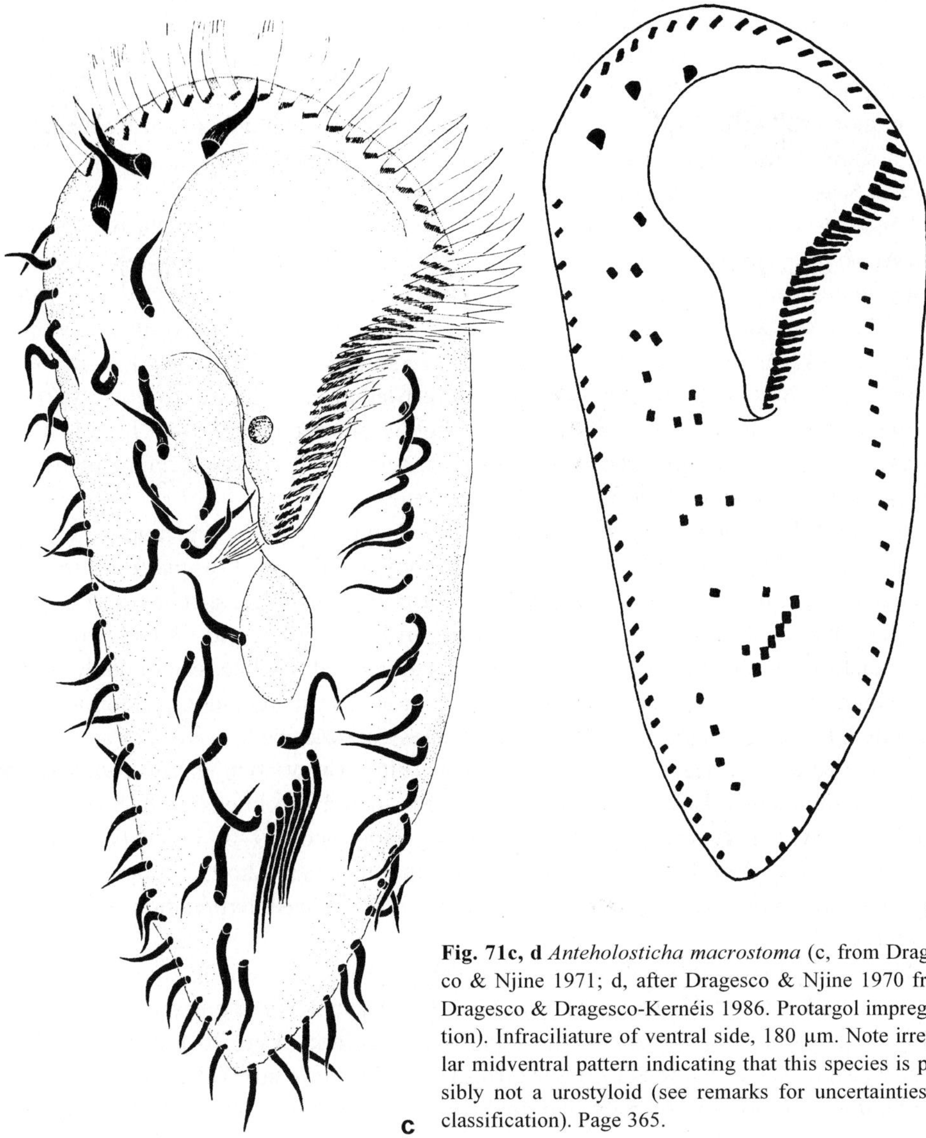

c

d

Fig. 71c, d *Anteholosticha macrostoma* (c, from Dragesco & Njine 1971; d, after Dragesco & Njine 1970 from Dragesco & Dragesco-Kernéis 1986. Protargol impregnation). Infraciliature of ventral side, 180 µm. Note irregular midventral pattern indicating that this species is possibly not a urostyloid (see remarks for uncertainties of classification). Page 365.

One of several features which contradicts the classification in *Holosticha*, respectively, *Anteholosticha* is the very large relative size (around 50% of body length) of the adoral zone. Such a high percentage is usually characteristic for stylonychines, that is, rigid 18-cirri oxytrichids. However, as mentioned above the lack of data prevents a well founded classification of the present species in this group of oxytrichids. Thus, I simply transfer it from *Holosticha* to *Anteholosticha* because the classification in *Holosticha* would make this well defined group unnecessarily heterogeneous (see genus section of

Holosticha for autapomorphies). The assignment to *Anteholosticha* is – as the classification in *Pleurotricha* (Dragesco 1970, Borror & Wicklow 1983) or *Holosticha* (Borror 1972, Dragesco & Dragesco-Kernéis 1986) – certainly only provisional until a detailed redescription is available.

Dragesco & Njine (1971) described and illustrated a second population from Cameroon. It shows a less regular pattern of cirral pairs, but conspecificity with the type population is beyond reasonable doubt.

For separation from other *Anteholosticha* species, see key. It must not be confused with *Neokeronopsis spectabilis*, which also lives in freshwater, has only two macronuclear nodules and a large buccal field (Fig. 242–244). However, this oxytrichid species is very large (usually more than 250 μm long vs. below 200 μm) and has distinctly more cirral pairs (12 or more vs. about 5) and transverse cirri (15–22 vs. 4–6). *Territricha stramenticola* Berger & Foissner, 1988 is a terrestrial species mainly (exclusively?) occurring in beech litter. It has, inter alia, a slightly shorter adoral zone (ca. 40% vs. around 50%) and a complicated dorsal kinety pattern due to multiple fragmentation of dorsal kinety 3 and about four dorsomarginal kineties.

Morphology: Unless otherwise indicated, the data are from the type population. Unfortunately, no live observations are available. Body length 130–170 μm after protargol impregnation, length:width ratio of specimen illustrated about 2:1; specimens of population described by Dragesco & Njine 150–190 μm long and around 68 μm wide. Two macronuclear nodules (25 μm long on average) along cell midline, each nodule with a single micronucleus 4 μm across on average. Contractile vacuole neither mentioned nor illustrated. Presence or absence of cortical granules not described.

Adoral zone occupies 47% (Fig. 71c) to 55% (Fig. 71a) of body length; composed of 44–50 membranelles. Buccal field very large in prepared specimens, right margin bordered by long and curved paroral; surprisingly, Dragesco (1970) found two endoral membranes, one about half as long as paroral, the other very short; I suppose that the very short membrane is a misobservation, respectively, a misinterpretation of fibrillary structures.

The cirral pattern, although studied in protargol preparations, is likely not described exactly. Three enlarged frontal cirri. Two cirri (two buccal cirri or one buccal cirrus and cirrus III/2?) along middle and anterior portion of paroral (Fig. 71a), specimen illustrated by Dragesco & Njine with single cirrus (Fig. 71c). Two cirri (likely frontoterminal cirri) left of anterior end of right marginal row. Behind these "seven frontal cirri" 4–5 cirral pairs. 4–6, usually five transverse cirri in short, oblique, and subterminal (anteriormost cirrus at 70% of body length!) row; cirri thus not projecting beyond rear body end, not distinctly enlarged. Right behind transverse cirri 2–3 cirri; possibly these are also transverse cirri, which are set off, as in some 18-cirri oxytrichids, from the other cirri. Right marginal row composed of 27–39 cirri, commences about at level of right frontal cirrus, posteriorly not separated from left marginal row, which commences in ordinary position and is composed of 20–26, on average 23 cirri; the continuous marginal rows indicate that Dragesco misinterpreted caudal cirri as marginal cirri. Dorsal cilia short (that is, likely around 3 μm), arranged in five kineties.

Ventral cirral pattern of population described by Dragesco & Njine less regular than that of type population (cp. Fig. 71a with 71c, d). They were uncertain about the correct designation of the ventral cirri and interpreted them as marginal rows: 29 cirri in outermost right marginal row and 6, 8, and 5 cirri in inner rows. Further morphometrics of this population: 10–19 ventral cirri (or inner right marginal cirri); 5–6 transverse cirri; 27–32 right marginal cirri (in outermost row); 21 left marginal cirri; 45 adoral membranelles (value from Fig. 71c).

Occurrence and ecology: Likely confined to freshwater. Type locality are freshwater habitats in the surroundings of the city of Yaoundé, Cameroon (Dragesco 1970). Dragesco & Njine (1971) found it in the same region but did not provide any details about the individual sample site. No further records published. Alekperov (1986, p. 56) mentioned the present species in a review.

Anteholosticha vuxgracilis (Berger, 2005) comb. nov. (Fig. 72a)

1963 ***Holosticha gracilis* n. sp.** – Vuxanovici, Studii Cerc. Biol., 15: 205, Plansa II, Fig. 12 (Fig. 72a; original description; likely no type material available; no formal diagnosis provided).

2001 ***Holosticha gracilis* Vuxanovici, 1963** – Berger, Catalogue of ciliate names 1. Hypotrichs, p. 35 (nomenclator containing all basionyms, combinations, and higher taxa of hypotrichs).

2005 ***Holosticha vuxgracilis* nom. nov.** – Original nomenclatural act (see nomenclature).

Nomenclature: No derivation of the name is given in the original description. The species-group name *grácil·is -is -e* (Latin adjective; thin, soft, slender, delicate) likely alludes to the delicate general appearance of this species.

Holosticha gracilis Vuxanovici, 1963 is the junior primary homonym of *Holosticha (Keronopsis) gracilis* Kahl, 1932, and the junior has to be substituted when no synonym is available (ICZN 1999, Article 60.1); note that the presence of the subgeneric name *Keronopsis* is irrelevant to the homonymy between the names (ICZN 1999, Article 57.4). Since the homonymy exists in *Holosticha*, the substitution of the name has to be done in *Holosticha* (see list of synonyms for this act). The species-group name *vuxgracilis* is a composite of *vux-* (the first three letters of the surname of the original author Vuxanovici) and *gracilis*, the original species-group name. Thus, this species is easily recognisable as Vuxanovici's *Holosticha gracilis* even with the new name. Since the species very likely does not belong to *Holosticha* it has to be transferred to *Anteholosticha*, an act done in the heading.

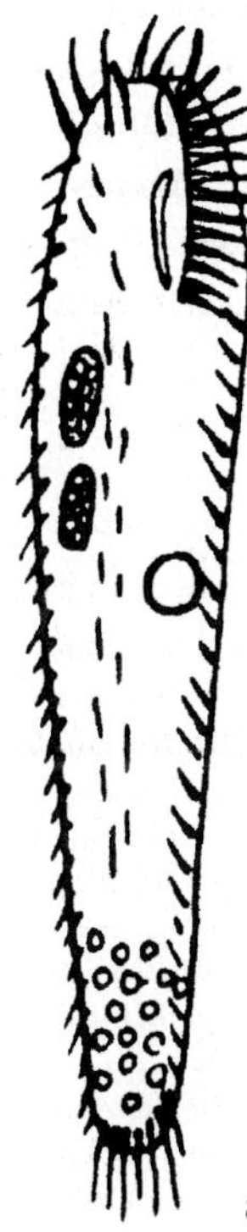

a

Fig. 72a *Anteholosticha vuxgracilis* (from Vuxanovici 1963. From life). Ventral view (80 µm) showing, inter alia, body outline, basic cirral pattern, macronuclear nodules, contractile vacuole, and refractive globules in posterior body portion. Page 369.

Remarks: Although the species is described and illustrated in little detail, the quality of the data is not so bad that it has to be classified as species indeterminata. It obviously lacks most (all?) apomorphies of *Holosticha* and since there is no evidence on caudal cirri it is transferred to *Anteholosticha*. For a detailed discussion and foundation of the transfer of many species from *Holosticha* to *Anteholosticha* see the genus section. Detailed redescription necessary.

Borror (1972) obviously overlooked this species and Borror & Wicklow (1983, p. 122) synonymised it with *Anteholosticha violacea* Kahl, 1928. However, Kahl's species is distinctly larger (180–250 µm vs. about 80 µm) and has, inter alia, many macronuclear nodules (vs. two; however, see description of *A. violacea* for problems with the macronucleus). Hemberger (1982, p. 277) mentioned it in a list containing species indeterminata and species of uncertain systematic position.

Morphology: Note that data about the cirral pattern must not be over-interpreted. Body length 80 µm, body length:width ratio 6:1, according to Fig. 72a about 4.6:1. Body outline wedge-shaped, that is, anterior portion wide and margin continuously converging posteriorly. Body soft, anterior portion mobile. Two narrowly spaced macronuclear nodules right of midline (usually characteristic for true *Holosticha* species) slightly ahead of mid-body. Contractile vacuole near left cell margin about in mid-body; at 18°C diastole takes 4 sec, systole about 10 sec. Cytoplasm translucent and colourless, rear body portion with refractive inclusions (likely lipid globules). Adoral zone of membranelles about 20% of body length. Three frontal cirri and three further cirri behind (including a buccal cirrus?). Midventral complex according to Fig. 72a extending from proximal end of adoral zone to about 70% of body length (Fig. 72a). About 6–7 (Fig. 72a) transverse cirri near rear body end. One left and one right marginal row. Dorsal infraciliature not known.

Occurrence and ecology: Limnetic. Type locality is the botanical garden in the Romanian capital Bucharest, where Vuxanovici (1963) found some specimens in a culture (collected in October) containing decomposing leaves. No further records known.

Anteholosticha antecirrata nov. spec. (Fig. 73a–h, Tables 18, 19)

1975 ***Keronopsis muscorum* Kahl, 1932** – Grolière, Protistologica, 11: 486, Fig. 6a–c (Fig. 73f–h; misidentification).

1982 ***Holosticha muscorum* (Kahl, 1932)**[1] – Foissner, Arch. Protistenk., 126: 51, Fig. 8a–e, 43, 44, 46, Tabelle 9 (Fig. 73a–e; misidentification, now type population of *A. antecirrata*; slides, which are now the type slides, are [likely] deposited in the Oberösterreichische Landesmuseum in Linz [LI], Austria).

[1] Foissner (1982) provided the following new diagnosis: In vivo etwa 220–330 × 70–90 µm große, durch subpelliculäre Granula gelbgrün gefärbte *Holosticha*, deren deutlich getrennte Midventralreihen bis zu den Transversalcirren reichen, die eine schräge bis leicht J-förmige Reihe bilden und den hinteren Körperrand nicht oder sehr wenig überragen. Durchschnittlich 48 adorale Membranellen, 7 buccale und 12 transversale Cirren. Makronucleus in etwa 200 ellipsoide bis bohnenförmige Fragmente zerteilt. Locus typicus nach Grolière (1975).

1992 ***Holosticha muscorum* (Kahl, 1932) Foissner, 1982** – Shen, Liu, Song & Gu, Protozoa, p. 154, Fig. 2-24Ba, Bb (record from China).

Nomenclature: The name *antecirrata* is a composite of the Latin adverb *ante* (spatial: in front) and the Latin noun *cirrata* ([a specimen] having cirri) and refers to the anteriorly displaced transverse cirri.

Neither Grolière (1975) and Foissner (1982, 1987a, 1998) nor Shen et al. (1992) wrote the authorship correctly because obviously all assumed that Kahl (1932) had established *Holosticha muscorum* in *Keronopsis* and not in *Holosticha*. Unfortunately, this is not the only case where Kahl's excessive usage of subgenera caused great nomenclatural confusion.

Diagnosis: In life about 220–330 × 70–90 µm, with yellow-green cortical granules and 8–15 anteriorly displaced transverse cirri so that they do not or only slightly project beyond rear body end. On average 48 adoral membranelles, seven buccal cirri, and 200 macronuclear nodules. Midventral complex extends to transverse cirri.

Type locality: Aperiodically flooded field (sandy-silty clay; about 190 m above sea-level; 15°46'35"E 48°23'37"N) near the village of Grafenwörth, Lower Austria.

Remarks: In 1975, Grolière described a population which he identified as *Keronopsis muscorum* (for the somewhat confusing nomenclature, see previous chapter) although he stated that there are some differences between the type population described by Kahl (1932, Fig. 61a) and his material (Fig. 73f–h). Foissner (1982) – who found a population closely resembling Grolière's description – confirmed that Grolière's and his own data (Fig. 73a–e) do not agree with those of *Holosticha muscorum* sensu Kahl (1932, Fig. 61a). The most important deviations are in the number of buccal cirri (1–5 in type population *H. muscorum* vs. 4–6 in Grolière's and 5–9 in Foissner's population) and transverse cirri (6–9 vs. 11–14 and 8–15). Thus, Foissner correctly stated that the populations described by Grolière (Fig. 73f–h) and himself (Fig. 73a–e) belong to a new species. Instead of establishing a new species, however, Foissner (1982) proposed keeping (conserving) the name *Holosticha muscorum* Kahl, 1932 and provided a new diagnosis (see corresponding footnote). However, previously Borror (1972) had synonymised *Holosticha muscorum* Kahl, 1932 with *H. multistilata* Kahl, 1928 (for details, see *Anteholosticha multistilata*), and this synonymy was accepted by Buitkamp (1977), Hemberger (1982), and Foissner (1982). The acceptation of this synonymy and the simultaneous proposal of a new diagnosis for *H. muscorum* by Foissner (1982) resulted in a rather tricky situation. The solution is that, as already suggested by Foissner (1982), the populations studied by Grolière (1975) and Foissner (1982) is a new species, *Anteholosticha antecirrata*. This is important, because data by Buitkamp (1977), Foissner (1982), Hemberger (1982), and Jutrczenki (1982) show that *A. intermedia* and its junior synonym *Holosticha muscorum* (Fig. 60a–d, 61a–u), and *A. multistilata* (Fig. 84a–k) are distinct taxa. For a more detailed discussion of this problem, see this chapter at *A. multistilata*.

In the ciliate atlas (Foissner et al. 1991, p. 236) we wrote that Foissner (1982) and "Berger (1992)"[1] do not agree with the synonymy of *Holosticha multistilata* and *H.*

[1] This book did not appear and is thus, of course, not cited in the reference section.

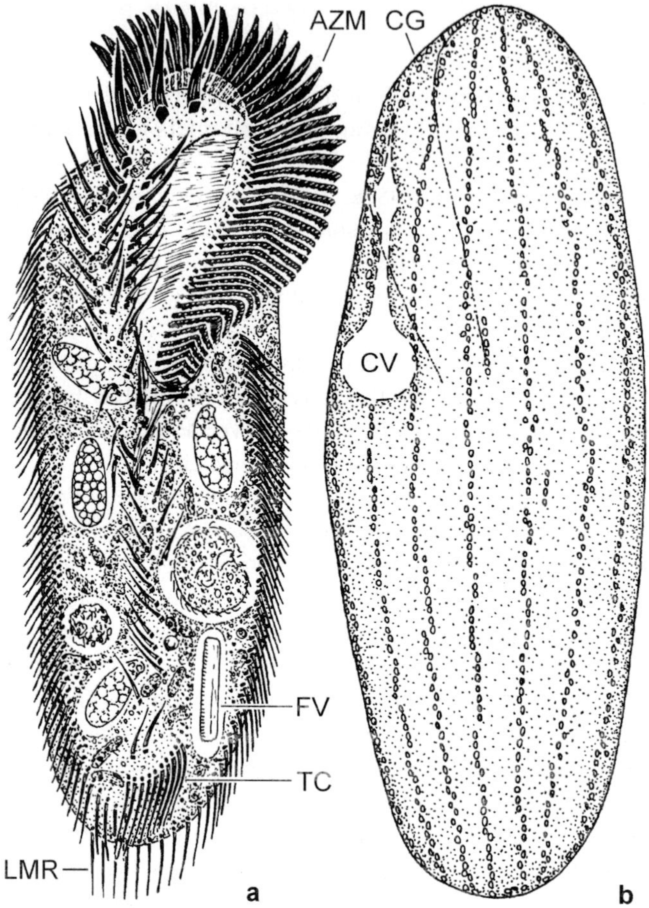

Fig. 73a, b *Anteholosticha antecirrata* from life (from Foissner 1982). **a:** Ventral view of a representative specimen, 292 μm. This specimens has ingested, inter alia, some testate amoebae (*Trinema lineare*), ciliates, and diatoms. **b:** Dorsal view showing contractile vacuole and cortical granules about 1 μm across which are arranged in longitudinal rows. AZM = adoral zone of membranelles, CG = cortical granules, CV = contractile vacuole with anteriorly extending and slightly dilated collecting canal, FV = food vacuole containing a diatom, LMR = left marginal row which terminates right of midline, TC = transverse cirri distinctly displaced anteriad (species-group name!) so that they do not or only slightly project beyond rear body end. Page 370.

muscorum proposed by Borror (1972). However, in the atlas we used Foissner's (1982) "new diagnosis of *H. muscorum*" so that our statement was basically correct and is not in contradiction with the present systematics (Table 18). By contrast, Borror & Wicklow (1983, p. 121) synonymised *Keronopsis muscorum* sensu Grolière (1975) with *Anteholosticha multistilata*, which is likely incorrect because there are distinct differences in several features (Table 19).

The two populations of *Anteholosticha antecirrata* described so far agree very well; however, there is at least one noteworthy difference, namely in the number of dorsal kineties. Grolierè's population has four rows (Fig. 73h), whereas Foissner counted, as in closely related species, only three rows (Fig. 73d; Table 19). Usually, the number of dorsal kineties is rather stable within populations and species. Thus, we cannot exclude that the two populations represent two distinct (sub)species. However, further populations should be studied before the quadrikinetid form is separated at (sub)species level.

The illustrations provided by Shen et al. (1992) are modified redrawings of Figs. 73a, c and are thus not shown in the present book. *Holosticha muscorum* sensu Wiackowski (1988) has 43 or more adoral membranelles and nine or more transverse cirri on average, indicating that he worked with the present species.

For a separation of *A. antecirrata* from the other *Anteholosticha* species with several buccal cirri, see key. This feature allows a rather simple identification of this species, even in life.

Morphology: Since the two populations differ in the number of dorsal kineties – usually an important feature at species level – the descriptions are kept separate because it cannot be excluded that they are distinct (sub)species.

Type population (Foissner 1982) 220–330 × 70–90 μm in life, ratio of body length:width about 3:1 (Fig. 73a, b, Table 19). Body outline wide elliptical, anterior and posterior end broadly rounded, usually widest at level of buccal cirri (Fig. 73a), rarely left margin distinctly vaulted at level of contractile vacuole (Fig. 73b). Body soft and flexible, about 2:1 flattened dorso-ventrally. Many (around 200) macronuclear nodules scattered throughout cytoplasm (Fig. 73d). Contractile vacuole in ordinary position, that is, left of proximal end of adoral zone of membranelles, during diastole with short anteriorly extending collecting canal with indistinct dilatation (Fig. 73b). Cortical granules arranged in many longitudinal rows (Fig. 73b); individual granules yellow-green, about 1 μm across, stain red, but do not swell and are not ejected when methyl green-pyronin is added; specimens yellow-green at low magnification. Cytoplasm densely granulated; food vacuoles up to 50 μm across. Movement rapid.

Adoral zone prominent because rather long, that is, occupies 40% of body length on average (Table 19), composed of 38–55, on average 48 membranelles (Fig. 73e). Bases of largest membranelles about 11 μm wide, composed of two portions (Fig. 73e). Buccal field very large, deep, on right side bordered by almost quadrant-shaped paroral. Endoral slightly longer than paroral, anterior portion straight, posterior portion curved leftwards, optically intersecting with paroral at level of buccal cirri; paroral distinctly, endoral indistinctly, two-rowed (Fig. 73c, e). At base (roof) of buccal field an obliquely striated structure (kineties? fibres?) extending from slightly behind anterior end of undulating membranes into cytopharynx (Fig. 73c, e).

Cirral pattern of usual variability (Fig. 73c, e, Table 19), except for the number of left marginal cirri, which varies rather strongly (however, this must not be over-interpreted). Usually three distinctly enlarged, more or less transversely arranged frontal cirri; cirrus behind right frontal cirrus (= cirrus III/2; arrow in Fig. 73c) also rather large. Buccal cirri along mid-portion of paroral, become slightly smaller from anterior to posterior (Fig. 73c, e). Two frontoterminal cirri near distal end of adoral zone. Midventral complex (very likely) composed of cirral pairs only; according to Foissner (1982), cirri often very irregularly arranged, in posterior portion often forming short rows (ontogenetic data are needed to clarify whether or not true midventral rows are present); midventral complex extends to very near transverse cirri (Fig. 73c). Transverse cirri rather small and narrowly spaced, form almost straight, oblique row, distinctly dislocated anteriad so that they do not or only slightly project beyond rear body end (distance between rearmost transverse cirrus and body end about 30 μm in specimen shown in Fig. 73c, that is, the posteriormost cirrus is at 84% of body length. In *A. adami*, the rearmost transverse cirrus is at 95% on average). Right marginal row

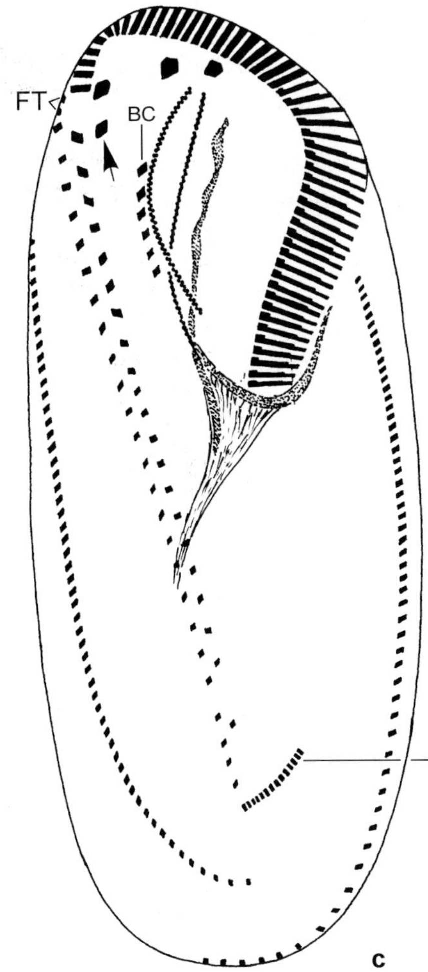

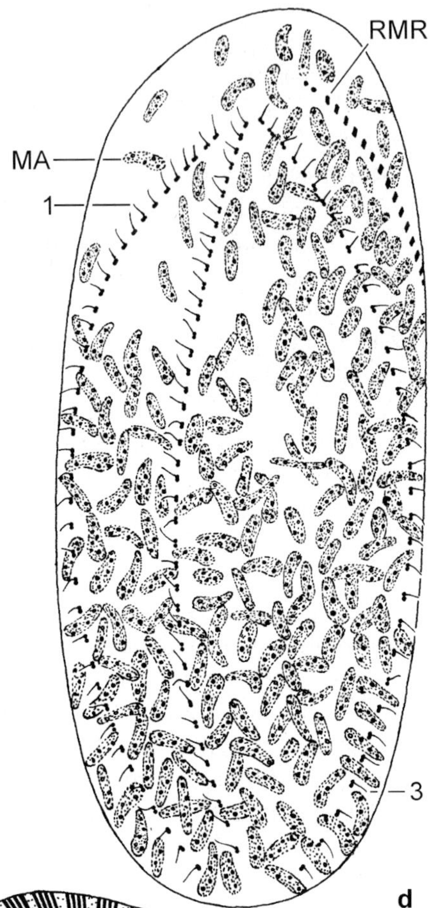

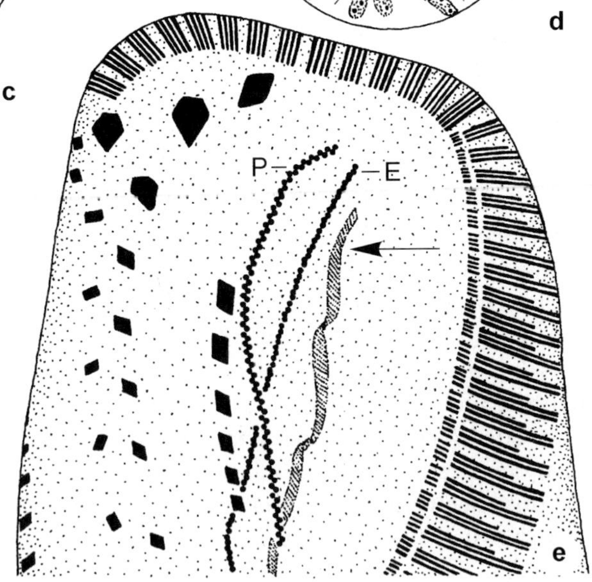

Fig. 73c–e *Anteholosticha antecirrata* after protargol impregnation (from Foissner 1982). **c, d:** Infraciliature of ventral and dorsal side and nuclear apparatus (likely of same specimen), 180 µm. Arrow denotes fourth frontal cirrus (= cirrus III/2). **e:** Infraciliature of anterior body portion, 45 µm. Arrow marks obliquely striated structure in the buccal field. BC = row of buccal cirri, E = endoral, FT = frontoterminal cirri, MA = macronuclear nodule, P = paroral, RMR = right marginal row, TC = transverse cirri, 1, 3 = dorsal kineties. Page 370.

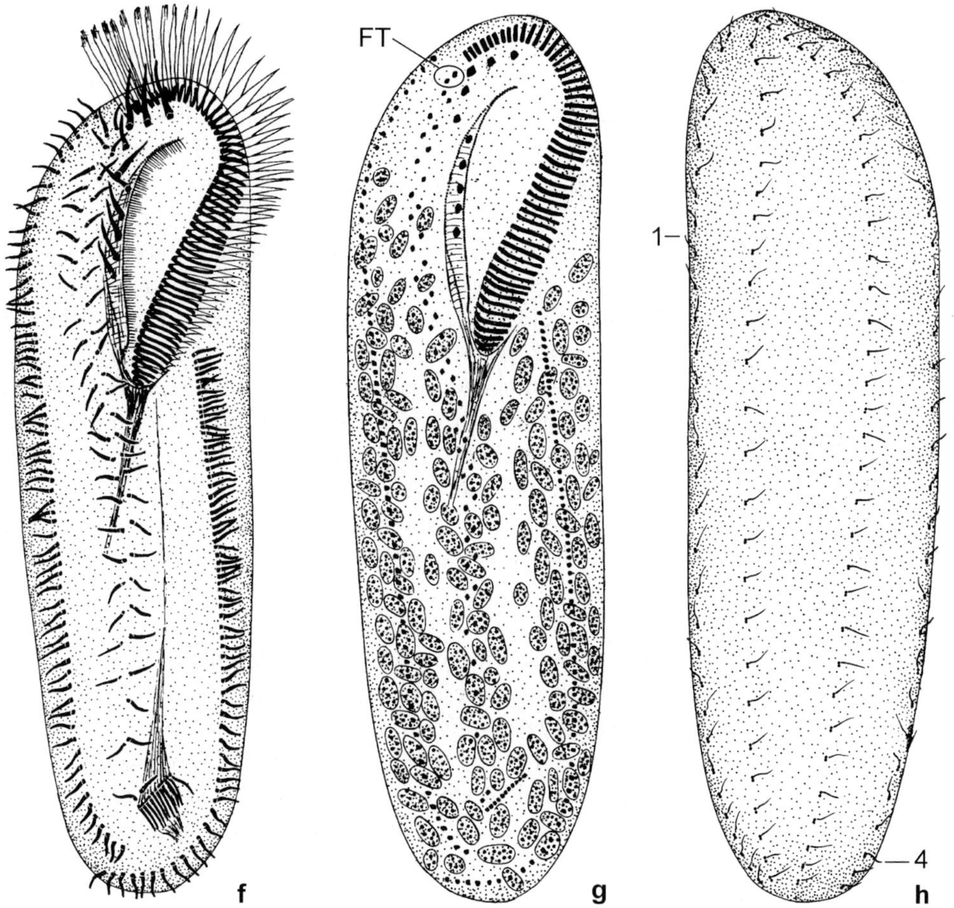

Fig. 73f–h *Anteholosticha antecirrata* (from Grolière 1975. f, general aspect?; g, h, protargol impregnation). **f:** Ventral view showing cirri and oral apparatus, 297 µm (likely same specimen as in Fig. 73g). **g, h:** Infraciliature of ventral and dorsal side and nuclear apparatus (of same specimen?). FT = frontoterminal cirri?, 1, 4 = dorsal kineties. Page 370.

commences dorsoventrally with two or three bristles ("dorsal cilia"), inverted J-shaped; left marginal row begins, as is usual, left of proximal portion of adoral zone, extends along rear body margin above midline; marginal rows distinctly overlapping, but clearly separated in longitudinal direction at least in specimen shown in Fig. 73c.

Dorsal cilia about 4 µm long, invariably arranged in three longitudinal rows. Caudal cirri lacking (Fig. 73d).

French population described by Grolière (1975; only supplementary and deviating data are provided; Fig. 73f–h, Table 19): Body size 210–300 × 48–85 µm; adoral zone about 85–120 µm long (in life?), that is, occupies about one third (text of Grolière 1975) of body length (according to Fig. 73g, however, 41% of body length); usually two cirri behind right frontal cirrus (Fig. 73g), similar to in *A. adami* (Fig. 74i); right cirri of midventral pairs always larger than left; occasionally some cirri form a "third ventral row"; a prominent bundle of fibres extends from transverse cirri to level of cytostome (Fig. 73g); marginal cirri usually composed of 2 × 6 basal bodies.

Occurrence and ecology: This chapter contains the faunistic and ecological data by Grolière (1975) and Foissner (1982) and some records which are obviously based on these two descriptions.

Anteholosticha antecirrata occurs both in limnetic and terrestrial habitats. According to Grolière (1978, p. 309) it is a sphagnophilic species. Foissner found it in three of seven alluvial soils in the "Tullner Feld" region (Lower Austria), namely, in a xerothermic site and two fields (Foissner 1982, p. 24; Foissner et al. 1985, p. 108). The population studied by Foissner (1982), which is here fixed as type population, is from field B with crop rotation (wheat, maize, potato). This field was treated with inorganic fertiliser and pesticids (Foissner 1982, Foissner et al. 1985, p. 89). Reliably recorded throughout the world, except for the Antarctic region (Foissner 1998, p. 204).

Further records: litter of various natural forest stands in Austria (Foissner et al. 2005); pond near the village of Randan, and about 10 ind. l^{-1} in a peat-bog near Besse-en-Chandesse, France during autumn (Grolière 1975; 1977, p. 338; Grolière & Njine 1973, p. 14); soil of spruce forests in Germany (Funke 1986, p. 72; Lehle 1989, p. 141); moss on soil likely collected in Poland (Wiackowski 1988); paddy fields in Japan (Takahashi & Suhama 1991b, p. 106); various terrestrial habitats in the USA, Costa Rica, and Africa (Bamforth 1995, p. 181; no details given); bark of an *Acacia* tree, Santa Rosa National Park, Costa Rica (Foissner 1994, p. 295); litter, soil, and roots (pH 5.1) from flood-plain primary (?) rain forest about 20 km east of the city of Manaus, Brazil (Foissner 1997, p. 322); arid Australian soils (Robinson et al. 2002, p. 452; as *Holoticha muscorum*). According to Bamforth (1995), *Anteholosticha antecirrata* is a strongly edaphic, K-selected species.

Feeds on diatoms, testate amoebae (*Trinema lineara*), and ciliates (*Colpoda cucullus*; Foissner 1982, Fig. 73a); Grolière's specimen were usually packed with vacuoles containing flagellates, diatoms, and ciliates. Wiackowski's (1988) specimens also feed on ciliates, for example, small *Colpoda* species. Biomass of 10^6 specimens about 360 mg (Foissner 1987a, p. 124).

Anteholosticha adami (Foissner, 1982) Berger, 2003
(Fig. 74a–i, Tables 18, 19)

1982 ***Holosticha multistilata* Kahl** – Hemberger, Dissertation, p. 102, Fig. 15i, not Fig. 15a–h (Fig. 74g; misidentification; the slide containing the illustrated specimen is likely deposited in the Institut für landwirtschaftliche Zoologie und Bienenkund, University of Bonn, Germany).

1982 ***Holosticha adami* nov. spec.**[1] – Foissner, Arch. Protistenk., 126: 46, Abb. 6a–e, 53, 56, Tabelle 9 (Fig. 74a–f; original description. The holotype slide [accession number 1981/89; Aescht 2003] is deposited in the Oberösterreichische Landesmuseum in Linz [LI], Austria).

[1] The diagnosis by Foissner (1982) is as follows: In vivo etwa 120–170 × 30–50 µm große, ellipsoide *Holo-*

about 10 μm long. Dorsal cilia about 3 μm long, arranged in four bipolar kineties. Caudal cirri lacking (Fig. 75g).

Blatterer & Foissner (1988) provided supplementary data on two African populations: body about 2:1 flattened dorsoventrally, very flexible, slightly contractile. Cortical granules (in one population colourless to slightly yellowish) easily ejected in squeezed cells, thus feigning lack of these organelles! Granules form loose, not puffy cover. Cytoplasm with many greasy-shining globules 0.5–2.0 μm across and rod-shaped crystals 0.5–2.0 μm in length.

Occurrence and ecology: Likely confined to terrestrial habitats. Reliably recorded from all biogeographic regions, except for the Antarctica (Foissner 1998, p. 204). Type locality of *A. australis* is a secondary pine forest in the suburb of the Australian city of Adelaide, where Blatterer & Foissner (1988) discovered it moderately abundantly in the mouldy litter (site FO 10; 0–5 cm; pH 5.1; about 100 m above sea-level; collected by W. Foissner on 17.02.1987). They found it also in litter with much sand from a bush in the Royal National Park in the south of Sydney (FO 1; pH 4.5; about 50 m above sea-level; collected by H. Blatterer on 18.10.1986) and in the upper soil layer (0–5 cm) with litter and roots of an *Eucalyptus* forest at the South Para Reservoir near the city of Adelaide (FO 11; pH 4.7; 50 m above sea-level; collected by W. Foissner on 17.02.1987). Further records: litter from Oak-hornbeam (Kolmber region) and Austrian pine (Merkenstein region) forests in Austria (Foissner et al. 2005); soil sample from a tropical dry forest in the Santa Rosa National Park, Costa Rica (Foissner 1995, p. 39); soil samples from a flood-plain primary rain forest and in a blackwater inundation primary rain forest near the city of Manaus, Brazil (Foissner 1997, p. 322); forest soils in the Shimba Hills Nature Reserve in Kenya (Foissner 1999, p. 323); two out of 73 soil samples from Namibia (Foissner et al. 2002, p. 60).

Anteholosticha australis feeds on testate amoebae (*Trinema lineare*), ciliates (*Drepanomonas* sp.), bacteria, and heterotrophic flagellates (Blatterer & Foissner 1988). Biomass of 10^6 specimens about 65 mg (Foissner 1998, p. 204).

Anteholosticha distyla (Buitkamp, 1977) Berger, 2003 (Fig. 76a)

1977 ***Holosticha distyla* n. spec.** – Buitkamp, Acta Protozool., 16: 268, Abb. 12 (Fig. 76a; original description; the type slides are deposited in the Institut für landwirtschaftliche Zoologie of the University of Bonn, Germany; no formal diagnosis provided).

1982 ***Holosticha distyla* Buitkamp, 1977** – Hemberger, Dissertation, p. 93 (revision of non-euplotid hypotrichs).

1986 ***Holosticha distyla* Buitkamp, 1977** – Dragesco & Dragesco-Kernéis, Faune tropicale, 26: 451, Planche 132B, C (modified redrawing of Fig. 76a; review of African ciliates).

2001 ***Holosticha distyla* Buitkamp, 1977** – Berger, Catalogue of ciliate names 1. Hypotrichs, p. 34 (nomenclator containing all basionyms, combinations, and higher taxa of hypotrichs).

2003 ***Anteholosticha distyla* (Buitkamp, 1977) n. comb.** – Berger, Europ. J. Protistol., 39: 377 (combination with *Anteholosticha*).

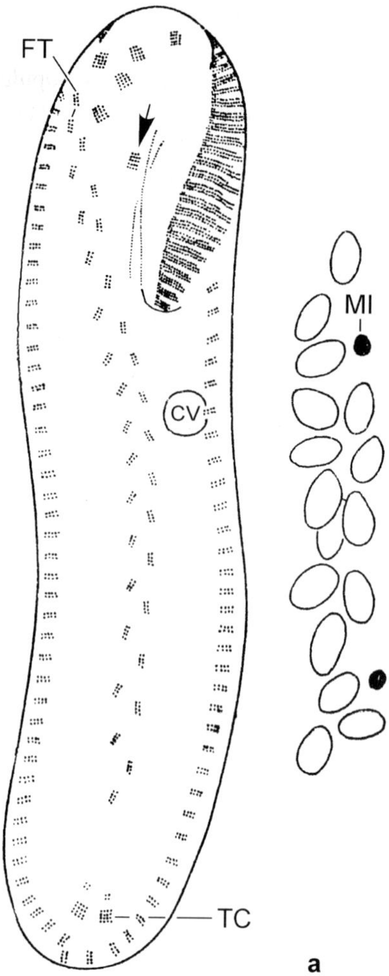

Fig. 76a *Anteholosticha distyla* after protargol impregnation (from Buitkamp 1977). Infraciliature of ventral side and nuclear apparatus, 144 µm. Arrow marks buccal cirrus. CV = contractile vacuole, FT = frontoterminal cirri, MI = micronucleus, TC = transverse cirri. Page 385.

Nomenclature: The species-group name *distyla* is a composite of the Greek numeral *di-* (two) and the Greek substantive *ho stylos* (style, cirrus) and refers to the two enlarged transverse cirri (Buitkamp 1977, p. 268).

Remarks: The present species lacks caudal cirri and all apomorphies of *Holosticha* and was thus transferred to *Anteholosticha* (Berger 2003). For a detailed discussion and foundation of the transfer of many species from *Holosticha* to *Anteholosticha* see the genus section.

Borror & Wicklow (1983, p. 121) synonymised Buitkamp's species with *A. intermedia* Bergh, 1889. However, this species has, inter alia, more (about seven) transverse cirri and macronuclear nodules, strongly indicating that this synonymy is incorrect. The two transverse cirri and the nuclear apparatus separate *A. distyla* from its congeners. However, Hemberger (1982) studied the type slides and found that the number of transverse cirri varies from 2–4; further, the posteriormost three cirri of the left marginal row could be caudal cirri. Consequently, the lack of cortical granules is basically the sole difference to *A. australis*. *Anteholosticha distyla* must therefore be redescribed in detail, especially to show whether or not it has cortical granules and to document the variability of the number of transverse cirri.

Dragesco & Dragesco-Kernéis (1986) mentioned this species in the figure legend. However, I did not find it in their text.

Morphology: Body length 150–180 µm, specimen shown in Fig. 76a about 144 × 31 µm, that is, body length:width ratio 4.6:1. Body ribbon-shaped, that is, margins parallel and ends broadly rounded. According to Hemberger (1982) very flexible. 16 macronuclear nodules extending from second to fourth fifth of body length in left body half; individual nodules ellipsoidal and about 6 µm long; two micronuclei 2–3 µm across. Contractile vacuole near left body margin at about 40% of body length. Presence/absence of cortical granules not known.

Adoral zone occupies about 33% of body length, composed of 30–33 membranelles of ordinary fine structure. Buccal area likely narrow. Paroral and endoral single-rowed, slightly curved and optically intersecting about at level of buccal cirrus.

4–5 enlarged cirri on frontal area (likely the ordinary three frontal cirri and one or two cirri behind the right frontal cirrus; ontogenetic data needed). One buccal cirrus slightly behind anterior end of paroral. Two frontoterminal cirri (presence confirmed by reinvestigation of type slides by Hemberger 1982) in more or less ordinary position. Midventral complex composed of cirral pairs only, commences behind right frontal cirrus, terminates slightly ahead of transverse cirri (in specimen shown in Fig. 76a at 83% of body length); complex composed of 33–34 cirri (frontoterminal cirri included, that is, about 15–16 cirral pairs are present). Two small pretransverse cirri ahead of two distinctly enlarged, about 18 µm long transverse cirri (Hemberger counted 2–4 transverse cirri, see remarks). Right marginal row composed of about 43 cirri, begins about at level of frontoterminal cirri, terminates behind transverse cirri; left marginal row commences left of proximal end of adoral zone, J-shaped, composed of around 34 cirri, terminates behind rear end of right marginal row (rearmost three cirri are possibly caudal cirri, see remarks). Marginal and midventral cirri consist of 2 × 4 basal bodies each bearing about 12 µm long cilia. Dorsal cilia about 5 µm long, arranged in four kineties each composed of about 20 bristles. Caudal cirri "totally absent" (however, see remarks).

Occurrence and ecology: As yet found only in soil samples (0–5 cm) from the Ivory Coast, where Buitkamp (1977; see also Buitkamp 1979) discovered it at two sites, namely, (i) in an unburnt savanna and (ii) in a riverine forest (Gallery Woodland). Both sites are near the Research Station for Tropical Ecology Lamto, about 250 km northwest of the city of Abidjan. *Anteholosticha distyla* feeds on testate amoebae and flagellates. Biomass of 10^6 specimens about 136 mg (Foissner 1987a, p. 124; 1998, p. 204).

Anteholosticha sigmoidea (Foissner, 1982) Berger, 2003 (Fig. 77a–m, Table 19)

1982 ***Holosticha sigmoidea* nov. spec.**[1] – Foissner, Arch. Protistenk., 126: 53, Fig. 9a–f, 48–50, Tabellen 10, 12 (Fig. 77a–f; original description; the holotype slide [accession number 1982/56; Aescht 2003] is deposited in the Oberösterreichische Landesmuseum in Linz [LI], Austria).

1983 ***Holosticha sigmoidea* Foissner, 1982** – Borror & Wicklow, Acta Protozool., 22: 115 (revision of urostylids; see remarks).

1984 ***Holosticha sigmoidea* Foissner, 1982** – Foissner, Stapfia, 12: 107, Abb. 56a–f, Tabelle 27 (Fig. 77g–l; redescription; a voucher slide [1984/98] is deposited in the Oberösterreichische Landesmuseum in Linz [LI], Austria; see remarks).

2001 ***Holosticha sigmoidea* Foissner, 1982** – Berger, Catalogue of ciliate names 1. Hypotrichs, p. 39 (nomenclator containing all basionyms, combinations, and higher taxa of hypotrichs).

2003 ***Anteholosticha sigmoidea* (Foissner, 1982) n. comb.** – Berger, Europ. J. Protistol., 39: 377 (combination with *Anteholosticha*).

[1] The diagnosis provided by Foissner (1982) is as follows: In vivo etwa 90–130 × 20–30 µm große, leicht kontraktile, sehr biegsame *Holosticha* mit vielen Längsreihen farbloser, protargolaffiner, subpelliculärer Granula, die ausgestoßen werden können.

Nomenclature: No derivation of the name is given in the original description. The species-group name *sigmoíde·us -a -um* (Latin adjective; S-shaped, sigmoidal) obviously alludes to the S-shaped body outline. "*Holosticha sigmoidea* Foissner" in Foissner (1981, p. 18) is a nomen nudum because the name is not accompanied by an illustration or description (ICZN 1999, Article 13). Aescht (2003, p. 396) designated the slide deposited by Foissner (1984) as paratype. However, this is incorrect because the specimens on this slide are not from the type series.

Remarks: The present species lacks caudal cirri and most apomorphies of *Holosticha* and was thus transferred to *Anteholosticha* (Berger 2003). For a detailed discussion and foundation of the transfer of many species from *Holosticha* to *Anteholosticha* see the genus section.

Foissner (1982) described three populations from Austria (two alpine sites, one lowland site), which agree rather well. Somewhat later, Foissner (1984) found a further population in the Austrian Alps which differs rather distinctly from the other populations in some features. However, since all these populations agree in the key characteristics, Foissner (1984) designated his population as local variety; I agree with Foissner, but keep the data separate. The analyses of further populations will show whether or not this assumption is correct.

Anteholosticha sigmoidea is obviously the sister species of *A. sphagni*, which basically differ only in the presence (*A. sigmoidea*), respectively, absence (*A. sphagni*) of cortical granules. The close relationship was already recognised by Foissner (1982), who was personally informated by Grolière that *A. sphagni* lacks cortical granules. Thus, I do not merge these two species. The other differences mentioned by Foissner (1982), namely, two frontal cirri in *A. sphagni* (possibly the distalmost adoral membranelle is in fact the right frontal cirrus) against three in *A. sigmoidea* (at least one specimen with only two frontal cirri observed) and different length of marginal rows (both rows terminate behind level of transverse cirri in *A. sphagni* against only left one terminates behind level of transverse cirri) must not be overinterpreted.

Anteholosticha monilata (type species) and *A. australis* are likely closely related to the present species. However, they differ from *A. sigmoidea* in the shape of the cortical granules (ellipsoidal vs. globular). In addition, the type species has more dorsal kineties than *A. australis* and *A. sigmoidea* (6 vs. 4; Table 19).

Anteholosticha mancoidea also has a very similar cirral pattern and nuclear apparatus, but only three dorsal kineties (against invariably 4 in *A. sigmoidea* and *A. sphagni*) and lives, like *A. sphagni*, in freshwater (against *A. sigmoidea* in soil). Possibly there are also differences in the cortical granulation, a feature not yet known for *A. mancoidea*. The marine *Anteholosticha manca* has many more macronuclear nodules than the present species (50–70 vs. 5–12).

Foissner (1982) mentioned a Figure 9g, which is, however, not available. Borror & Wicklow (1983) keyed out *A. sigmoidea* at the same couplet as *A. sphagni*, but did not mention it in the list of species on their page 122.

Morphology: As mentioned above, the data provided by Foissner (1982) and Foissner (1984) are kept separate because conspecificity of populations is not beyond reasonable doubt.

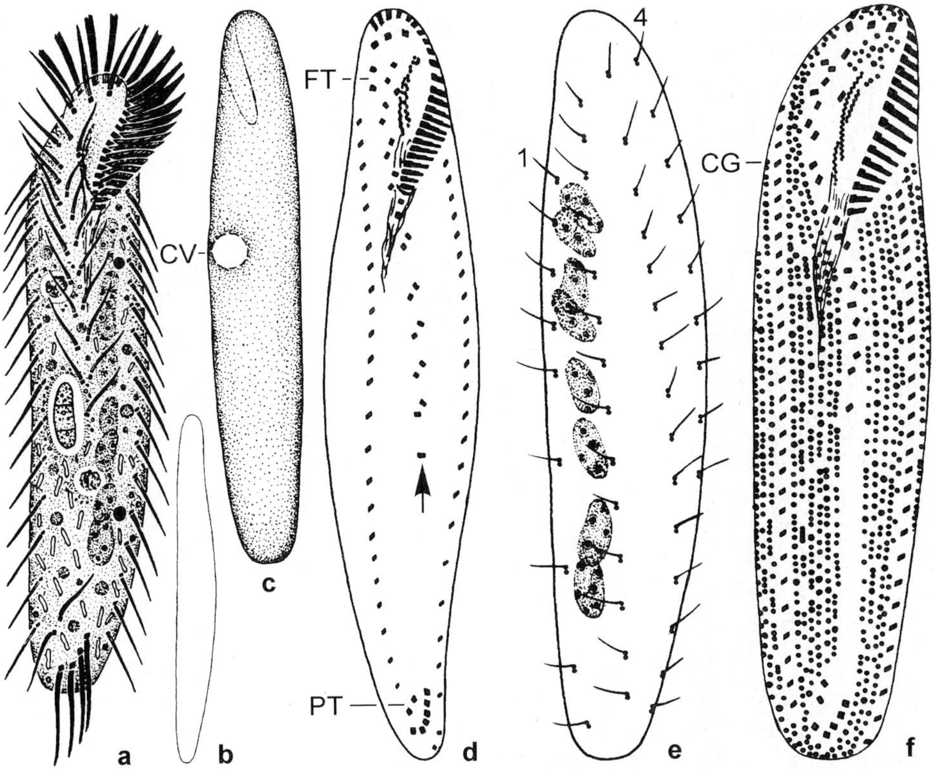

Fig. 77a–f *Anteholosticha sigmoidea* (from Foissner 1982. a–c, from life; d–f, protargol impregnation). **a:** Ventral view of a representative specimen, 115 µm. **b:** Left lateral view, 121 µm. **c:** Dorsal view, 119 µm. **d, e:** Infraciliature of ventral and dorsal side and nuclear apparatus of a specimen from type location, 83 µm. Arrow marks rear end of midventral complex. **f:** Infraciliature of ventral side and cortical granules of specimen from the Tullnerfeld area, 75 µm. CG = cortical granules, CV = contractile vacuole, FT = frontoterminal cirri, PT = pretransverse ventral cirri, 1, 4 = dorsal kineties. Page 387.

Type population about 90–130 × 20–30 µm in life, body length:width ratio 3.9 to 5.0:1 on average in protargol preparations (Table 19). Body outline usually slightly sigmoidal, sometimes roughly orthogonal; rear end mostly truncated transversally. Body slightly contractile, about 2:1 flattened dorsoventrally (Fig. 77a–c). Macronuclear nodules roughly serially arranged in mid-portion of body left of midline; individual nodules ellipsoidal, in life about 7 × 4 µm, contain small nucleoli. Micronuclei compact, shining, in life about 2.6 µm across (Fig. 77a, e). Contractile vacuole slightly ahead of mid-body near left cell margin, without collecting canals. Pellicle colourless, very flexible. Cortical granules in more or less distinct longitudinal rows, dorsally sometimes spirally arranged, lacking along marginal rows, in buccal field, and in median of cell. Individual granules colourless and globular (ca. 0.5–1.0 µm across), after protargol impregnation usually very conspicuous, sometimes increased to 3 µm. Number of granule-rows rather variable in protargol preparations because granules can be ejected in rows

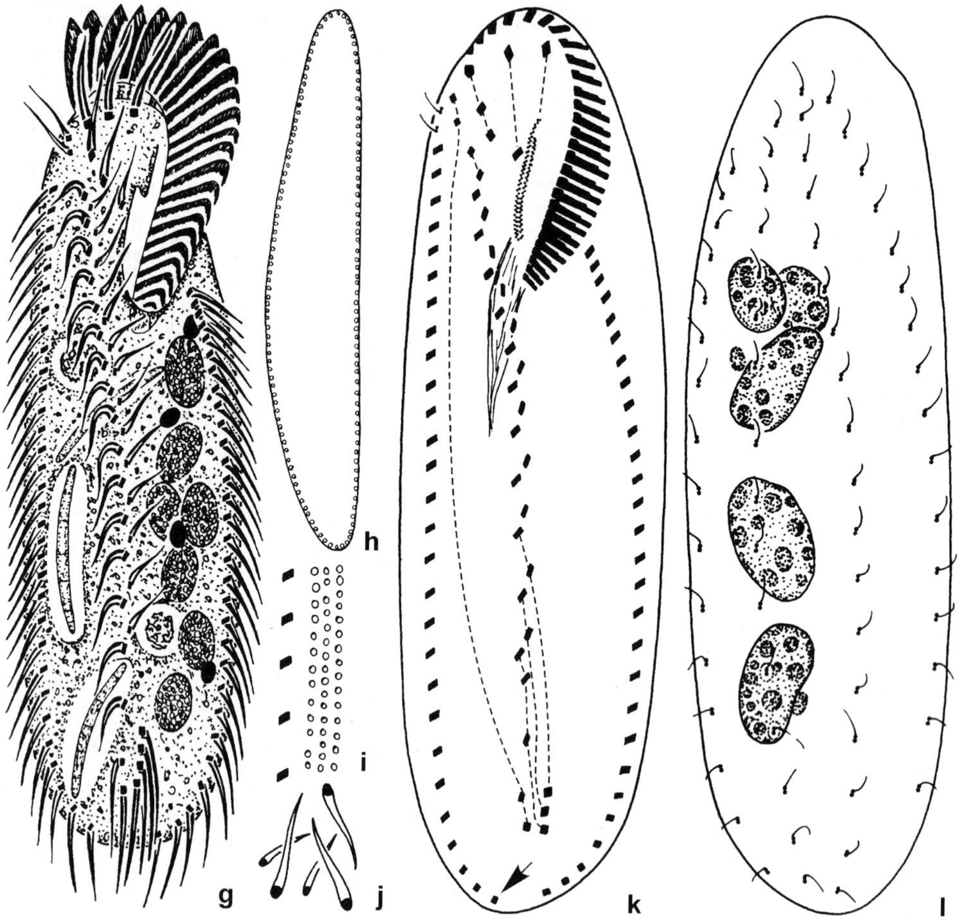

Fig. 77g–l *Anteholosticha sigmoidea* (from Foissner 1984. g–i, from life; j, methyl-green pyronin stain; k, l, protargol impregnation). **g:** Ventral view, 130 μm. **h:** Right lateral view (124 μm) showing seam formed by cortical granules. **i:** Arrangement of cortical granules. **j:** Cortical granules are ejected after methyl-green pyronin stain and get a spike-shaped protrusion. **k, l:** Infraciliature of ventral and dorsal side and nuclear apparatus of same specimen, 84 μm. Broken lines connect some relevant cirri which very likely originate from the same anlagen. Page 387.

(Fig. 77f). In the protargol preparations of the population from the Schlossalm area almost all specimens had ejected the granules; in the slightly brownish pellicle only many bright patches arranged in longitudinal rows were recognisable. Endoplasm colourless, with many, slightly yellow crystals and few to many shining, slightly yellow globules 1–3 μm across. Movement rapid, gliding, attaches closely to soil particles.

Adoral zone of membranelles occupies about 25–33% of body length in life, in protargol preparations 23–25% on average (Table 19); proximal end usually near midline. Buccal field small, almost plane. Paroral likely two-rowed, about 7 μm long in specimen shown in Fig. 77d, terminates distinctly ahead of rear end of adoral zone; endoral of about same length as paroral, commences slightly behind paroral; endoral and paroral arranged almost in parallel or slightly crossing optically (Fig. 77d, f). Cytopharynx without peculiarities.

Cirral pattern and number of cirri of usual variability (Fig. 77a, d, f; Table 19). Three (very rarely only two) slightly enlarged frontal cirri almost transversely arranged. Buccal cirrus fine, slightly to distinctly behind anterior end of paroral. Invariably two frontoterminal cirri in ordinary position, that is, between distal end of adoral zone and anterior end of right marginal row. Midventral complex terminates at 50–58% of body length on average (Table 19), composed of ca. 9–12 cirral pairs; cirri of each pair of about same size. Usually two, rarely only one pretransverse ventral cirrus. Transverse cirri about 17 µm long, only inconspicuously enlarged, terminally roughly arranged in J-shape pattern. Marginal cirri in life about 10 µm long. Right marginal row commences about at level of buccal cirrus, usually terminates ahead of or at level of pretransverse ventral cirri; left row commences, as is usual, slightly ahead of level of proximal end of adoral zone, terminates behind level of transverse cirri. Number of marginal cirri in type population slightly lower than in other populations (Table 19). Distance between individual marginal cirri increases posteriad, especially in type population, where cirri also become finer.

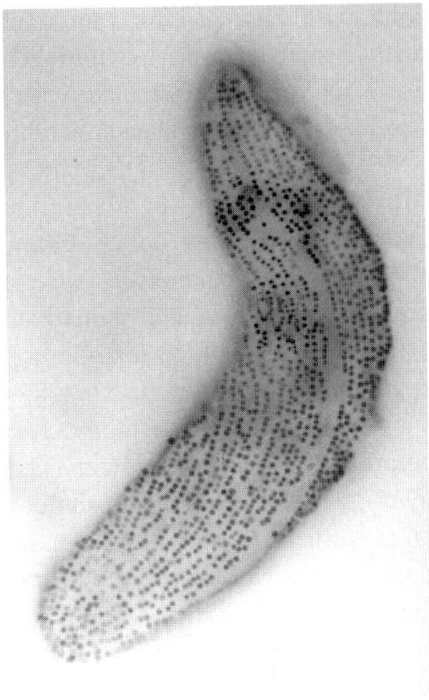

Fig. 77m *Anteholosticha sigmoidea* (original kindly supplied by W. Foissner. Methyl-green pyronin stain). The cortical granules are densely arranged in narrowly spaced longitudinal rows. Page 387.

Dorsal cilia about 3 µm long in life, invariably arranged in four kineties (Fig. 77e). Kinety 1 anteriorly, kinety 4 posteriorly more or less distinctly shortened. Foissner (1982), likely par lapsus, wrote that in one population five kineties are present. However, in his Table 10 both populations have four kineties, and in the population listed in his Table 12 the number of kineties was not recognisable. Caudal cirri lacking.

As mentioned above, the population described by Foissner (1984; Fig. 77g–l) differs from Foissner's (1982) populations in some features, namely, (i) body size, especially width (140 × 40 µm; values of other populations, see above); (ii) number of adoral membranelles and marginal cirri (Table 19; possibly correlated with size); (iii) length of midventral complex (75% of body length); (iv) end of right marginal row (distinctly behind level of transverse cirri); and (v) cirri of anteriormost midventral pair distinctly enlarged (the arrangement of these two cirri indicates that they originate from anlage III which produces the right frontal cirrus, that is, this population has two cirri behind the right frontal cirrus (Fig. 77k); other populations likely have only one, namely cirrus III/2. Foissner (1984) mentioned further supplementary observations:

contractile vacuole with short collecting canals; cortical granules stain red-blue with methyl-green pyronin and form loose cover around cell after ejection, ejected granules 5–10 µm long, spike-shaped, with distinct granule at wider end (Fig. 77i, j); usually (?) two basal body pairs ahead of right marginal row (Fig. 77k).

Foissner (pers. comm.) made additional observations on the cortical granules (Fig. 77m). In very good life preparations granules clearly recognisable with bright field and interference contrast optics. Granules globular, colourless and moderately refractive, about 1.0–1.2 µm across. However, usually specimens are very fragile under cover glass pressure, eject granules immediately leaving a honeycomb pattern forming an about 1 µm thick seam in cortex. Granules stain well with methyl-green pyronin, swell up to 5 µm long rods, which fuse to a loose, unstructured cover.

Occurrence and ecology: Likely confined to terrestrial habitats, where it is rather common. Type locality is very near (about 30 m) the Wallackhaus (2290 m above sea level), a hotel beside the Großglockner Hochalpenstraße, an alpine road in the Glockner area, Austrian Central Alps. Foissner (1982) discovered it there in the soil (alpine pseudogley on mica slate) of an alpine pasture heavily fertilised by the outflow of a septic tank. Further, Foissner (1982) found it in the soil of an alpine pasture and the marginal area of a ski trail in the Schlossalm are (Bad Hofgastein, Austria), and in a forest and a xerothermic site in the Zwentendorf area, Lower Austria. For details on these sample sites and autecological data, see Foissner (1981, p. 18), Foissner & Peer (1985, p. 40), and Foissner et al. (1985, p. 108). Foissner (1984) found his population at the Stubnerkogel, a mountain near the village of Bad Gastein (Salzburg, Austria); it occurred in a soil sample (0–10 cm; alpine pseudogley; moder; 1820 m above sea level) from alder wood (*Alnetum viridis*; for details on site, see Foissner & Peer 1985).

Further reliable records: Hüttschlag, Salzburg, Austria (Foissner, pers. comm., Fig. 77m); in experiments on the compaction and fertilisation of an alpine pasture soil in the Schlossalm area, Austria (Berger et al. 1985, p. 107; 1986, p. 268); soil of spruce forest in the Mühlviertel, the northern region of Upper Austria (Petz et al. 1988, p. 83); litter of various natural forest stands in Austria (Foissner et al. 2005); spruce and beech forests near the city of Ulm and in the Black Forest and Swabian Alb, Germany (Lehle 1989, p. 141; 1993, p. 17; 1994, p. 115; Lehle et al. 1992, p. 279); soil samples near Kelso in southern Scotland (Finlay et al. 2001, p. 363); moss from Casey Station and algal ornithological soil from Shirley Island and Beall Island, Antarctica (Petz & Foissner 1996, p. 259; 1997, p. 309); grass sward and moss (pH 4.8) from Signy Island, Antarctica (Foissner 1996a, p. 100); primary and secondary rain forests near and about 40 km west of Manaus, Brazil (Foissner 1997, p. 322). Likely lacking in the Paleotropis (Foissner 1998, p. 204); also not found in a detailed study on soil ciliates from Namibia (Foissner et al. 2002). The record by Fernandez-Leborans et al. (1990, p. 513) from the La Jarosa Reservoir about 48 km away from Madrid (Spain) is not substantiated by morphological data.

Feeds on bacteria and elongate fungal spores (Foissner 1982; Fig. 77a). According to Foissner (1985, p. 85) possibly an indicator for moder. Biomass of 10^6 specimens 38 mg (Foissner 1987a, p. 124; 1998, p. 204).

Anteholosticha bergeri (Foissner, 1987) Berger, 2003
(Fig. 78a–j, Table 19)

1987 ***Holosticha bergeri* nov. spec.**[1] – Foissner, Zool. Beitr., 31: 197, Abb. 4a–e, Tabelle 3 (Fig. 78a–e; original description; the holotype slide [accession number 1988/138; Aescht 2003] and paratype slides [1988/139, 140] are deposited in the Oberösterreichische Landesmuseum in Linz [LI], Austria).

1988 ***Holosticha bergeri* Foissner, 1987** – Blatterer & Foissner, Stapfia, 17: 50, Abb. 15a–e, Tabelle 11 (Fig. 78f–j; redescription of two populations).

2001 ***Holosticha bergeri* Foissner, 1987** – Berger, Catalogue of ciliate names 1. Hypotrichs, p. 33 (nomenclator containing all basionyms, combinations, and higher taxa of hypotrichs).

2003 ***Anteholosticha bergeri* (Foissner, 1987) n. comb.** – Berger, Europ. J. Protistol., 39: 377 (combination with *Anteholosticha*).

Nomenclature: This species is named after me (Foissner 1987b, p. 198).

Remarks: The present species lacks caudal cirri and all apomorphies of *Holosticha* and was thus transferred to *Anteholosticha* (Berger 2003). For a detailed discussion and foundation of the transfer of many species from *Holosticha* to *Anteholosticha* see genus section.

Blatterer & Foissner (1988) described two populations. The population from Australia (Fig. 78f, g) differs from the type material mainly in the number of macronuclear nodules (on average 28 vs. 15). The specimens from the Antarctic region have orange cortical granules, whereas they are pink in the type population. Blatterer & Foissner thus provided a more or less complete description of these populations because one cannot exclude that they are distinct species. However, at the present state of knowledge separation at species or subspecies level does not seem justified.

Foissner (1982, p. 90) described two populations of *Perisincirra gellerti* (now *Hemisincirra gellerti gellerti*). The cirral pattern of the Tullnerfeld material (Fig. 79a) looks very similar to that of *A. bergeri*. Foissner (1987b, p. 200) therefore assumed a misidentification of the specimen shown in Fig. 79a. However, since no live data are available for the Tullnerfeld population, the situation regarding the cortical granules (colour, arrangement) and thus the identification remains unclear. Foissner (2000, p. 278) described the subspecies *Hemisincirra gellerti verrucosa* (colourless to yellowish cortical granules around the dorsal bristles; usually four dorsal kineties; usually only about 8–9 macronuclear nodules, however, specimens with up to 16 nodules have been observed), which has basically the same ventral cirral pattern as *A. bergeri*, including the inconspicuous zigzagging arrangement of the ventral cirri. I thus suppose that *H. gellerti verrucosa* also belongs to *Anteholosticha*. Possibly, the nominotypical subspecies *Hemisincirra gellerti gellerti* also has a (indistinct) midventral pattern which was not recognised in the original description. However, this analysis is beyond the scope of the present book. *Hemisincirra gellerti gellerti* differs from *A. bergeri*, inter alia, by the cortical granulation (colourless granules narrowly spaced in about 30 longitudinal rows). Anyhow, identification of these small *Anteholosticha* and *Hemisincirra* species

[1] The diagnosis provided by Foissner (1987b) is as follows: In vivo etwa 55–80 × 15–20 µm große *Holosticha* mit leuchtend rosa gefärbten kugelförmigen subpelliculären Granula und stark verkürzten Midventralreihen. Durchschnittlich 13 adorale Membranellen, 4 Transversalcirren und 15 Makronucleus-Teile. 3 Dorsalkineten.

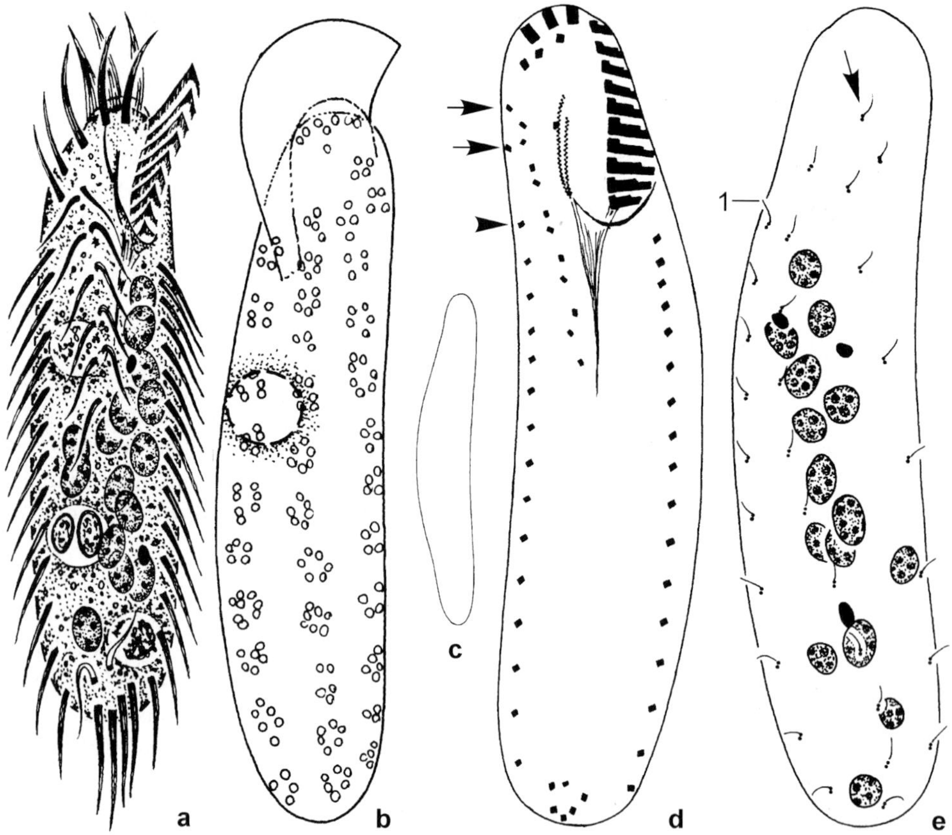

Fig. 78a–e *Anteholosticha bergeri* (from Foissner 1987b. a–c, from life; d, e, protargol impregnation). **a:** Ventral view of a representative specimen, 70 µm. **b:** Dorsal view showing cortical granules and contractile vacuole, 60 µm. **c:** Right lateral view, 65 µm. **d, e:** Infraciliature of ventral and dorsal side and nuclear apparatus of same specimen, 58 µm. Arrows in (d) mark widely spaced frontoterminal cirri, arrowhead in (d) denotes anterior end of right marginal row. Arrow in (e) marks two dorsal bristles right of anterior end of kinety 3. 1 = dorsal kinety 1. Page 393.

needs both detailed live observation and correct interpretation of the cirral pattern after protargol impregnation.

Anteholosticha brachysticha has ellipsoidal, yellowish cortical granules, and the right marginal row begins more anteriorly (about level of buccal cirrus) than in *A. bergeri* (about level of buccal vertex). *Anteholosticha plurinucleata* is a further rather similar species. However, cortical granules are neither mentioned nor illustrated by Gellért (1956) so that synonymy is very unlikely.

Morphology: I do not mix the original description with the data provided by Blatterer & Foissner (1988) because one cannot exclude that this is a sibling species complex.

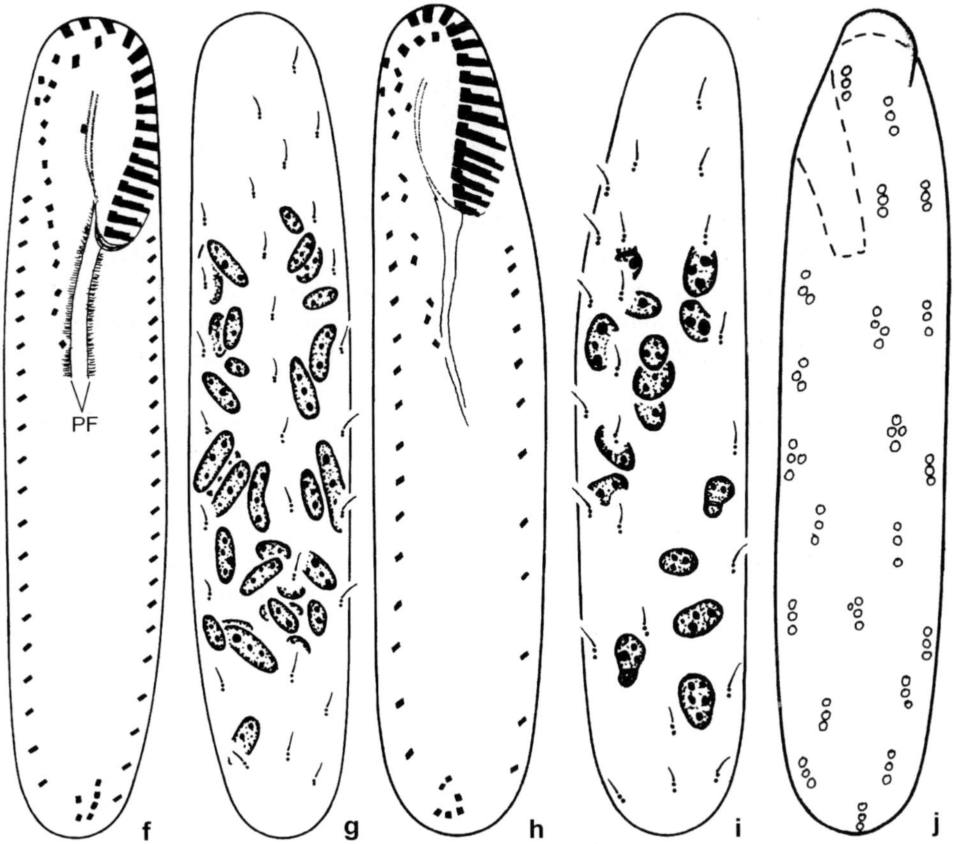

Fig. 78f–j *Anteholosticha bergeri* (from Blatterer & Foissner 1988. f–i, protargol impregnation; j, from life. f, g, Australian population; h–j, Antarctic population). **f–i:** Infraciliature of ventral and dorsal side and nuclear apparatus, f, g = 77 µm, h, i = 64 µm. **j:** Dorsal view showing arrangement of cortical granules, which are orange in this population. PF = pharyngeal fibres with rod-seam. Page 393.

Type population in life about 55–80 × 15–20 µm, body length:width ratio about 3.8:1 in protargol preparations (Table 19). Body outline rather constant, oblong, anterior and posterior end widely rounded. Body slightly flattened, acontractile (Fig. 78c). On average 15 macronuclear nodules, most of them left of midline, individual nodules ellipsoidal, nucleoli of ordinary size. At least two, but likely more globular micronuclei, which usually do not stain with protargol. Contractile vacuole near left body margin slightly ahead of mid-body (Fig. 78a, b, e). Pellicle very flexible and soft, thus cells sensitive to coverglass pressure. Cortical granules in small groups around cirri and dorsal bristles; individual granules globular, about 1 µm across, shining pink making cells slightly rose-coloured at low magnification (Fig. 78b); stain brownish with protargol. Cytoplasm colourless. Movement moderately rapid, without peculiarities.

Adoral zone occupies 26% of body length on average (Table 19), anteriormost three membranelles inconspicuously set off from remaining ones (Fig. 78a, d). Buccal area small, moderately deep. Undulating membranes only slightly curved, not distinctly crossing optically.

Cirral pattern and number of cirri of usual variability (Fig. 78d, Table 19). Most cirri about 15 µm long in life. Three enlarged frontal cirri in straight, slightly oblique row. Buccal cirrus fine, composed of single row (that is, likely of two or three cilia only), arranged near anterior end of paroral. Invariably two, relatively widely separated frontoterminal cirri about at level of buccal cirrus. Midventral complex usually composed of 5–6 cirral pairs, therefore terminates slightly ahead of mid-body (at 44% of body length in specimen shown in Fig. 78d); zigzagging pattern indistinct at least in protargol preparations. Two pretransverse ventral cirri slightly ahead of transverse cirri. Transverse cirri not enlarged, very close to rear body end and thus distinctly projecting beyond rear body end. Both marginal rows commence about at level of proximal end of adoral zone, terminate distinctly ahead of rear body end, marginal rows thus clearly separated posteriorly. Dorsal cilia about 3 µm long in life, arranged in three anteriorly more or less distinctly shortened kineties. Left ahead of kinety 3 two single bristles. Caudal cirri lacking (Fig. 78e, Table 19).

As mentioned above, Blatterer & Foissner (1988) studied two populations for which they provided a common description and a separate morphometric characterisation (I provide only important, supplementary, and/or deviating data; see also Fig. 78f–j and Table 19). Body size about 80–100 × 15–20 µm in life. Body outline oblong, anterior and posterior end moderately widely rounded. Body slightly flattened dorsoventrally, very flexible. Macronuclear nodules ellipsoidal with few globular nucleoli, irregularly distributed mainly in second and third body fourth. Micronuclei do not always impregnate with protargol. Contractile vacuole about in mid-body near left body margin. Cortical granules orange, globular, in small groups around/along cirri and dorsal bristles (Fig. 78j), hardly impregnate with protargol. Cytoplasm colourless, with moderately many greasy-shining globules 1–3 µm across. Movement without peculiarities. Adoral zone occupies 24–25% of body length on average (Table 19). Bases of largest adoral membranelles about 6 µm wide. Undulating membranes composed of 1–2 basal body rows. Pharyngeal fibres well recognisable in life, in most specimens covered by conspicuous seam of rods in protargol preparations (Fig. 78f). Transverse cirri about 15 µm long, protrude far beyond body end. Marginal and midventral cirri about 12 µm long. Antarctic population without the two additional bristles right-ahead of dorsal kinety 3 (Fig. 78i); in the Australian population they are likely continuous with the right marginal row (Fig. 78f, g). Caudal cirri lacking.

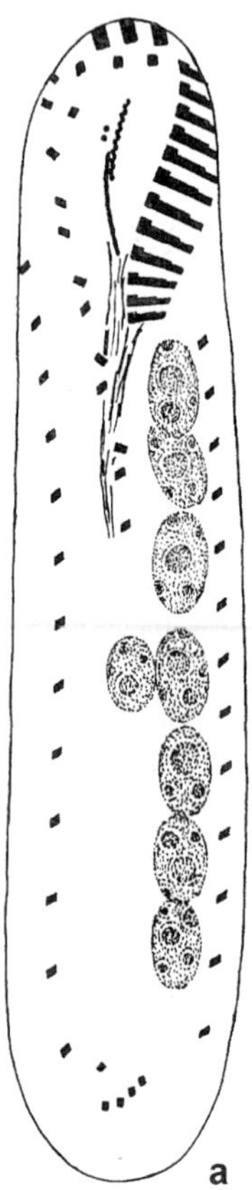

Fig. 79a *Hemisincirra gellerti* (from Foissner 1982. Protargol impregnation). This population from the Tullnerfeld is possibly identical with *Anteholosticha bergeri*, 65 µm. For details, see text. Page 393.

Occurrence and ecology: Likely confined to terrestrial habitats. Type locality of *A. bergeri* is moss from a granite rock near the village of Sandkäs on the northeastern coast of the Danish island Bornholm, Baltic Sea. The sample was collected by F. Wenzel on 26.09.1985. The populations described by Blatterer & Foissner (1988) are from Australia (black litter and sand [pH 4.2] from the bush in Brisbane Water National Park, about 50 km north of Sydney, and litter from sand dunes near a pine forest at the "The Wrecks" on Moreton Island near Brisbane; pH 5.1) and Antarctica (*Deschampsia antarctica* grass sward; Foissner 1996a, p. 100).

Further records: litter from Norway spruce stands (*Picea abies*) in the Bohemian Forest, Upper Austria (Aescht & Foissner 1993, p. 328; see also Petz et al. 1988, p. 82); subalpine grassland in Styria, Austria (Foissner et al. 1990, p. 18); litter bags filled with dried leaves from *Corylus avellana* exposed in a mesosaprobic stream in Bavaria (Foissner et al. 1992a, p. 101; identification uncertain).

Anteholosticha bergeri feeds on coccale cyanobacteria and heterotrophic flagellates (Foissner 1987b), as well as on fungal spores (Blatterer & Foissner 1988). Biomass of 10^6 individuals about 10 mg (Foissner 1998, p. 204).

Anteholosticha brachysticha (Foissner, Agatha & Berger, 2002) Berger, 2003
(Fig. 80a–g, Table 19)

2002 ***Holosticha brachysticha* nov. spec.**[1] – Foissner, Agatha & Berger, Denisia, 5: 579, Fig. 132a–g, Table 114 (Fig. 80a–g; original description; the holotype slide [accession number 2002/754] and 2 paratype slides [2002/754–756] are deposited in the Oberösterreichische Landesmuseum in Linz [LI], Austria).

2003 ***Anteholosticha brachysticha* (Foissner, Agatha and Berger, 2002) n. comb.** – Berger, Europ. J. Protistol., 39: 377 (combination with *Anteholosticha*).

Nomenclature: The species-group name *brachysticha* is a composite of the Greek adjective *brachy* (short) and the Greek noun *sticha* (row) and refers to the short midventral complex (Foissner et al. 2002).

Remarks: The present species lacks caudal cirri and all apomorphies of *Holosticha* and was thus transferred to *Anteholosticha* (Berger 2003). For a detailed discussion and foundation of the transfer of many species from *Holosticha* to *Anteholosticha* see the genus section.

This species certainly belongs to the *Anteholosticha bergeri* group because the ventral and dorsal infraciliature, the nuclear apparatus, and the cortical granulation are very similar. For a detailed comparison with species belonging to the *A. bergeri*-group, see remarks at *A. bergeri* and key.

---

[1] The diagnosis by Foissner et al. (2002) is as follows: Size about 90 × 17 µm in vivo; elongate ellipsoidal. On average 32 scattered macronuclear nodules. Cortical granules around cirri and dorsal bristles, yellowish, about 1.0–1.5 × 0.5–0.8 µm in size. Midventral row terminates at about 1/3 of body length, composed of about 10 cirri. On average 16 adoral membranelles, 23 cirri each in right and left marginal row, 1 buccal cirrus, 2 frontoterminal cirri, 3 transverse cirri, and 3 dorsal kineties.

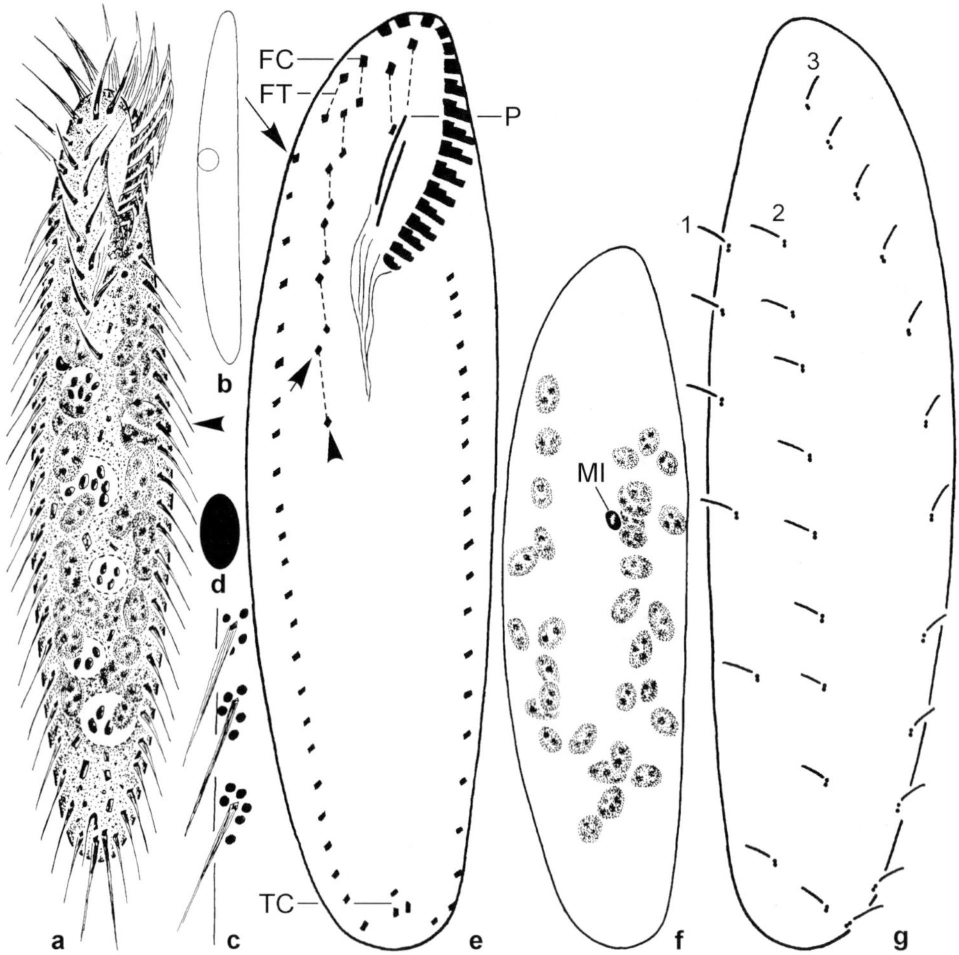

Fig. 80a–g *Anteholosticha brachysticha* (from Foissner et al. 2002. a–d, from life; e–g, protargol impregnation). **a:** Ventral view of a representative specimen, 100 μm. Arrowhead denotes contractile vacuole. **b:** Right lateral view showing dorsoventral flattening and contractile vacuole. **c, d:** Yellowish, compact and thus brilliant, cortical granules 1.0–1.5 × 0.5–0.8 μm in size, occur around cirri and dorsal bristles. **e–g:** Infraciliature of ventral and dorsal side and nuclear apparatus of holotype specimen, 70 μm. Arrowhead in (e) marks last midventral cirrus, short arrow denotes last pseudo pair, long arrow marks beginning of right marginal row. Frontal and midventral cirri very likely originating from same anlage are connected by broken line. FC = right frontal cirrus, FT = frontoterminal cirri, MI = micronucleus, P = paroral, TC = transverse cirri, 1–3 = dorsal kineties. Page 397.

Morphology: Body size 70–110 × 15–20 μm, usually around 90 × 17 μm in life, length:width ratio on average 5.3:1 both in live and protargol preparations (Table 19). Body dorsoventrally flattened up to 2:1, elongate elliptical, very flexible, but acontractile (Fig. 80a, b). Macronuclear nodules usually scattered in U-shaped pattern in central portion of cell; individual nodules usually ellipsoidal, rarely globular, elongate ellipsoidal, or dumb-bell-shaped; nucleoli small. On average two ellipsoidal micronuclei within macronuclear figure (Fig. 80f). Contractile vacuole slightly above mid-body at left cell margin, without distinct collecting canals (Fig. 80a, b). Cortical granules only in clusters around cirri and dorsal bristles, yellowish, conspicuously brilliant because compact, about 1.0–1.5 × 0.5–0.8 μm in size, do not impregnate with the protargol method used (Fig. 80c, d). Cytoplasm colourless, without peculiarities, that is, contains some ordinary, yellowish crystals about 1–2 μm long and some food vacuoles 5–7 μm across. Swims and glides moderately quickly on microscope slide and debris showing great flexibility.

Adoral zone occupies 20–27%, on average 23% of body length, composed of an average of 16 membranelles of usual shape and structure (Fig. 80a, e, Table 19). Buccal cavity of ordinary width, but flat, right margin forms inconspicuous lip partially covering proximal portion of adoral zone of membranelles. Paroral and endoral of about same length and almost in parallel, with endoral on average 2 μm rearward. Pharyngeal fibres distinct after protargol impregnation, extend to near mid-body.

Cirral pattern and number of cirri of usual variability (Fig. 80a, e; Table 19). Marginal and midventral cirri about 10 μm long in life and of similar size, that is, usually composed of 2 × 2 cilia. Frontal cirri slightly enlarged, in transverse, concave line, as in most congeners. Buccal cirrus usually composed of two cilia only, slightly behind anterior end of paroral and thus near level of anterior end of endoral. Two frontoterminal cirri right of anterior end of midventral complex; complex usually composed of 4–5 cirral pairs only, thus terminating at about 34% of body length. Transverse cirri almost terminal, about 15 μm long in vivo and thus distinctly projecting beyond rear body end; anterior cirrus usually smaller than other cirri, indicating that it is a pretransverse ventral cirrus. Right marginal row shortened anteriorly, terminates, like left row, near posterior end of cell.

Dorsal bristles about 3 μm long in life, arranged in three kineties easily recognisable in life due to granule clusters around individual bristles. Rows 1 and 2 distinctly shortened anteriorly, row 3 bipolar. Caudal cirri lacking (Fig. 80g).

Occurrence and ecology: So far only recorded from two sites worldwide. The type locality is a sandy, saline coastal soil (pH 7.6) near Punta Pirikiki, about 54 km south of Limon, Caribbean coast of Costa Rica, Central America (09°40′N 82°40′W). Further, it occurred in Namibian site (53), the Wolfsnes water-hole near the margin of the Etosha Pan (19°S 15°50′E); the sample was from a small salt bush (*Suaeda articulata*) island very near to the pan margin (Foissner et al. 2002, p. 26, 580). Feeds on tuberous bacteria.

Anteholosticha plurinucleata (Gellért, 1956) comb. nov. (Fig. 81a)

1956 ***Holosticha manca*** **var. *plurinucleata*** **n. var.** – Gellért, Acta biol. hung., 6: 95, Abb. 9 (Fig. 81a; original description; no formal diagnosis provided and likely no type material available).

1956 ***Holosticha manca* var. *mononucleata* n. var.** – Gellért, Acta biol. hung., 6: 95 (original description without illustration; no formal diagnosis provided and likely no type material available).
2001 ***Holosticha manca plurinucleata* Gellért, 1956** – Berger, Catalogue of ciliate names 1. Hypotrichs, p. 37 (nomenclator containing all basionyms, combinations, and higher taxa of hypotrichs).

Nomenclature: No derivation of the names is given in the original descriptions. The variety names *plurinucleat·us -a -um* (Latin; having many nuclei) and *mononucleat·us -a -um* (Latin; having one nucleus) refer the number of macronuclear nodules. According to Article 45.6.4 of the ICZN (1999) the names are subspecific, because (i) they are published before 1961; (ii) the author used the term "var."; and (iii) the author did not expressly give infrasubspecific rank. Here I raise *H. manca plurinucleata* from subspecies to species level and then combine it with *Anteholosticha*.

Remarks: Gellért (1956) described both taxa as variety of *Holosticha manca* which is a marine form now also classified in *Anteholosticha*. The subspecies *mononucleata* has only one macronuclear nodule, indicating that Gellért observed an exconjugant with a macronucleus anlage. Since it largely agrees with *H. manca plurinucleata* I synonymise these two subspecies (see nomenclature), raise *plurinucleata* to species rank, and transfer it to *Anteholosticha* because it lacks all(?) apomorphies of *Holosticha* and caudal cirri.

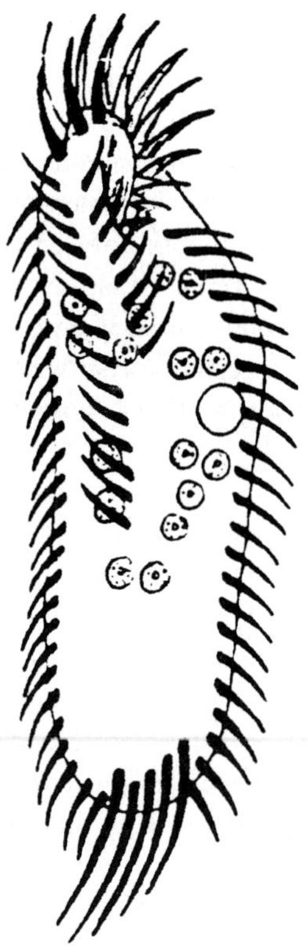

Fig. 81a *Anteholosticha plurinucleata* (from Gellért 1956. sublimat fixation and nigrosin stain). Infraciliature of ventral side, nuclear apparatus, and contractile vacuole, 80 µm. Page 399.

I am not certain that this species is a urostyloid because the zigzagging midventral pattern is very indistinct. Possibly the two ventral cirral rows are true rows (that is, not pseudorows), indicating that the species belongs to the amphisiellids. However, new data are needed for a final classification. If *A. plurinucleata* indeed has a midventral pattern then it is obviously closely related to species such as *A. bergeri* (see there for comparison).

I did not find either taxa (*plurinucleata* and *mononucleata*) in the reviews by Borror (1972) and Hemberger (1982). Borror & Wicklow (1983, p. 122) put *H. manca plurinucleata* into the synonymy of *A. scutellum*, which, however, is likely confined to marine habitats. For separation of *A. plurinucleata* from other congeners, for example, *A. sigmoidea* and *A. mancoidea*, see key.

Morphology: Gellért (1956) described *H. manca mononucleata* in more detail than *H. m. plurinucleate*, where he only mentioned the deviating features to *mononucleata*. The first morphology paragraph thus contains the data of *mononucleata* and the deviations of *plurinucleata* are briefly mentioned in the second paragraph. I suppose that both subspecies were described only from a single specimen each.

Body length 60 µm, length:width ratio 3:1. Right body margin almost straight, anterior third of left margin slightly vaulted. The single macronucleus in cell midline indicates that Gellért observed an exconjugant. Contractile vacuole near left cell margin. Cortical granules neither mentioned nor illustrated. Cytoplasm bright. Movement slowly creeping, sometimes rotating about main body axis. Adoral zone occupies about 1/3 of body length, composed of 14 membranelles. Cytopharynx short. Three enlarged frontal cirri. Single buccal cirrus at anterior end of slightly curved buccal lip. Right ventral row (possibly the species does not have a midventral complex, see remarks) extends to near posterior third of cell, left row ends in mid-body. Four transverse cirri, which project far beyond rear body end. Right marginal row commences at level of buccal lip, ends – like left one – subterminally; left marginal row commences near buccal vertex. Bases of marginal cirri surrounded by a refractive ring. Three "complete" (likely bipolar) dorsal kineties.

According to Gellért (1956), *A. plurinucleata* agrees with the description of its synonym *Holosticha manca mononucleata* (see previous paragraph) except for the following deviations (Fig. 81a): body length 80 µm; 22 macronuclear nodules; buccal lip straight; six transverse cirri; no ring around base of marginal cirri. According to Gellért the length of the ventral cirral rows is also different from that of its synonym; however, the illustration shows that the situation must be very similar.

Occurrence and ecology: Terrestrial. Type locality of *A. plurinucleata* (and its synonym *Holosticha manca mononucleata*) is the southwestern side of a hill (Magoska) near the Hungarian village of Boldogkőváralja, where it occurred in the humus underneath the lichen *Parmelia saxatilis*. Szabó (2000, p. 14) found it in a chernozem soil from the centre of the Great Hungarian Plain. No further records published.

Anteholosticha plurinucleata feeds on hyphae, its synonym on fungal spores and (likely heterotrophic) flagellates (Gellért 1956); Szabó (2000) found bacteria and detritus in the food vacuoles. Biomass of 10^6 specimens according to Foissner (1998, p. 204) 26 mg (*A. plurinucleata*), respectively, 10 mg (for the synonym *H. manca mononucleata*).

Anteholosticha muscicola (Gellért, 1956) Berger, 2003 (Fig. 82a)

1956 ***Holosticha muscicola* n. sp.** – Gellért, Acta biol. hung., 6: 345, Abb. 11 (Fig. 82a; original description; likely no type material available; no formal diagnosis provided).

1972 ***Holosticha muscicola* Gellért, 1956** – Borror, J. Protozool., 19: 11 (revision of hypotrichs).

1982 ***Holosticha muscicola* Gellért, 1956** – Hemberger, Dissertation, p. 90 (revision of non-euplotid hypotrichs).

1983 ***Holosticha muscicola* Gellért, 1956** – Borror & Wicklow, Acta Protozool., 22: 121 (revision of urostylids).

2001 ***Holosticha muscicola* Gellért, 1956** – Berger, Catalogue of ciliate names 1. Hypotrichs, p. 37 (nomenclator containing all basionyms, combinations, and higher taxa of hypotrichs).

2003 ***Anteholosticha muscicola* (Gellért, 1956) n. comb.** – Berger, Europ. J. Protistol., 39: 377 (combination with *Anteholosticha*).

Nomenclature: No derivation of the name is given in the original description. The species-group name *muscicola* (Latin adjective; living in moss; composed of *muscus*

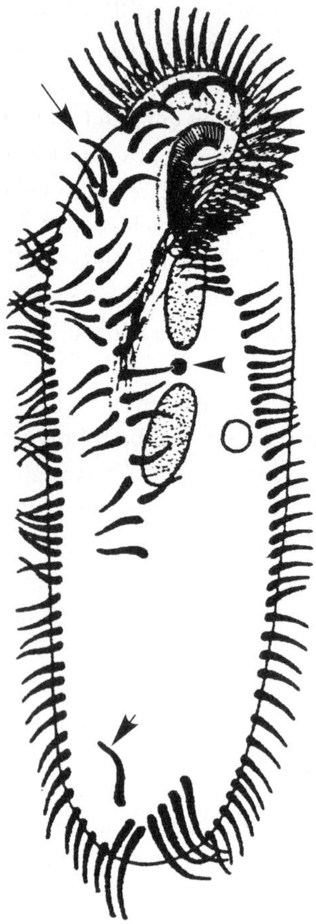

Fig. 82a *Anteholosticha muscicola* (from Gellért 1956a. Sublimate fixation and opalblue staining). Ventral view showing infraciliature, nuclear apparatus, and contractile vacuole, 110 μm. Arrow head marks micronucleus; long arrow denotes the anteriormost of two slightly enlarged cirri which possibly do not belong to the midventral complex; short arrow marks pretransverse ventral cirrus. Page 401.

[moss] and *colére* [inhabit]) obviously refers to the habitat (moss on rock) where the species was discovered. I did not check whether or not permanent preparations are available in Tihany (where Gellért worked) or elsewhere in Hungary.

Remarks: The present species very likely lacks caudal cirri and all(?) apomorphies of *Holosticha* and was thus transferred to *Anteholosticha* (Berger 2003). For a detailed discussion and foundation of the transfer of many species from *Holosticha* to *Anteholosticha* see the genus section.

Gellért (1956a) described *Anteholosticha muscicola* from life and after opalblue staining (Gellért 1956). It's validity was never doubted, not even by the lumper Borror. Some features are unknown, for example, cortical granules (present/absent) or dorsal ciliature. Thus, a detailed redescription is necessary.

The short midventral complex is reminiscent of *A. manca* (Fig. 87a), *A. mancoidea* (Fig. 64a), and *A. sphagni* (Fig. 66a). However, the marine *A. manca* has about 50–70 macronuclear nodules, whereas *A. mancoidea* and *A. sphagni*, which were discovered in limnetic habitats, have about eight nodules. The most interesting feature of the present species is the nuclear apparatus, which is composed of two macronuclear nodules and a single micronucleus in between; usually such a pattern is invariable and thus highly characteristic.

Morphology: Body length 100 μm in life, length:width ratio according to Fig. 82a 3.1:1. Body outline elongate with almost parallel margins, ends broadly rounded. Two ellipsoidal macronuclear nodules with a single globular micronucleus in between. Contractile vacuole near left body margin about in mid-body (according to Fig. 82a slightly ahead); empties every 35–40 sec.

Adoral zone occupies about 25% of body length, composed of about 32 membranelles. Buccal lip strongly curved anteriorly, bears paroral composed of long cilia; endoral below paroral, also composed of long cilia. Cytopharynx with long beating threads (likely cilia of endoral). Beside (left?) the endoral four single "paroral cilia" (no details given). Three enlarged frontal cirri. Behind right frontal cirrus two somewhat smaller cirri[1]; between these cirri and anterior end of midventral complex two further slightly enlarged cirri (possibly these are frontoterminal cirri which are not quite correctly positioned in the illustration). Midventral complex obviously composed of cirral pairs only (number of cirri/pairs not mentioned; specimen illustrated with about 13 pairs with last cirrus at about 60% of body length). One pretransverse ventral cirrus ahead of rightmost transverse cirrus. Six transverse cirri, which project by about half their length beyond rear body end. Right marginal row commences distinctly behind anterior end (at about 16% of body length in specimen illustrated), terminates behind level of transverse cirri, composed of 45 cirri in specimen illustrated. Left marginal row commences near proximal end of adoral zone, composed of 39 cirri (Fig. 82a), ends slightly subterminally; marginal rows thus distinctly separated posteriorly. Dorsal ciliature not described; caudal cirri neither mentioned nor illustrated, indicating that they are lacking.

Occurrence and ecology: Type locality is the south-western region of the hill Magoska north-east of the Hungarian village of Boldogkóváralja, where Gellért (1956) discovered *A. muscicola* in the humus under moss on hypersthenaugitandesit rocks. Biomass of 10^6 specimens about 60 mg (Foissner 1998, p. 204). Fernandez-Leborans & Antonio-García (1988, p. 147, 150) found *A. muscicola* in an experiment on the effects of lead and cadmium. The material was from the Manzanares river in La Pedriza, Madrid, Spain. 11 days after the addition of 1000 µg l^{-1} lead they counted 43 ind. 10 ml^{-1}.

Feeds on debris and bacteria (Gellért 1956a).

Anteholosticha alpestris (Kahl, 1932) comb. nov. (Fig. 83a)

1932 ***Keronopsis alpestris* spec. n.** – Kahl, Tierwelt Dtl., 25: 576, Fig. 110 ı (Fig. 83a; original description; no type material available and no formal diagnosis provided).

1972 ***Paraholosticha alpestris* (Kahl, 1932) n. comb.** – Borror, J. Protozool., 19: 11 (combination with *Paraholosticha*; revision of hypotrichs).

2001 ***Holosticha (Keronopsis) alpestris* Kahl, 1932** – Berger, Catalogue of ciliate names 1. Hypotrichs, p. 36 (nomenclator containing all basionyms, combinations, and higher taxa of hypotrichs).

Nomenclature: No derivation of the name is given in the original description. The species-group name *alpestris* (living in the Alps) refers to the site (Bavarian Alps) where the species was discovered. Kahl (1932) classified *Keronopsis* as subgenus of *Holosticha*. Thus, the correct name in the original description is *Holosticha (Keronopsis) alpestris* Kahl, 1932. Obviously this was overlooked by several authors (e.g., Gellért 1956, p. 94; Borror 1972, p. 11; Foissner 1998, p. 205) who assumed that Kahl (1932) had established this species in the genus *Keronopsis*. *Holosticha alpestris* Foissner, 1981 is a nomen nudum and therefore not a primary homonym of the present species.

[1] Gellért's formulation is somewhat unclear because he wrote "Behind the 3 frontal cirri each two cirri". This would mean that in total 6 cirri are present behind the frontal cirri; however, the illustration shows only two, possibly, four cirri; thus, the following description of the cirral pattern is based mainly on the illustration.

Remarks: The present species has a rather uncommon cirral pattern, namely, a single corona of frontal cirri, a distinct midventral complex composed of cirral pairs only, and transverse cirri. In addition, the nuclear apparatus is very conspicuous (two macronuclear nodules with a large micronucleus in between). Kahl (1932) classified it in the subgenus *Holosticha (Keronopsis)* because of the increased number of frontal cirri.

By contrast, Borror (1972) transferred it to *Paraholosticha* Kahl, 1932, a genus which lacks a midventral complex. It is unlikely that Kahl misinterpreted a *Paraholosticha* cirral pattern (two ventral rows) as midventral pattern because he knew *Paraholosticha* species very well (Kahl 1932, p. 545). The distinct midventral pattern of the present species prevents a classification in *Paraholosticha* (without transverse cirri) or *Keronopsis* (with transverse cirri), although the nuclear apparatus is strongly reminiscent of species of these genera (Berger & Foissner 1987, Dieckmann 1988). In spite of the increased number of frontal cirri, I transfer the present species to the inhomogenous *Anteholosticha* because the assignment to, for example, *Keronopsis*, would make these well-defined genera unnecessarily heterogeneous.

Gellért (1956, p. 94) found a population in the humus formed under the lichen *Parmelia saxialis*. He did not provide a figure but wrote that his specimens basically agree with Kahl's description. However, the following differences are present: (i) anterior end not distinctly curved leftwards; (ii) corona composed of seven cirri (vs. 10; Fig. 83a); (iii) two buccal cirri (vs. one); (iv) right ventral cirral row continuous with corona of frontal cirri (vs. discontinuous); (v) marginal rows very near cell margin (vs. displaced inwards); (vi) transverse cirri project distinctly beyond rear body end (vs. not projecting); (vii) number of adoral membranelles is lower and zone does not extend on right body margin; (ix) three dorsal kineties (number not mentioned by Kahl 1932). Nuclear apparatus and contractile vacuole as in type population. Cytoplasm bright with pale green colour. Movement rapid, crawling. Feeds on plant debris and algae. Common. Since the number of differences is rather high, I doubt that Gellért's population was conspecific with the type population.

Anteholosticha brevis has the same nuclear apparatus. However, this species has only three frontal cirri and the transverse cirri are not distinctly displaced anteriad (Fig. 69a). *Caudiholosticha navicularum* also has the same nuclear apparatus (Fig. 51a). However, it is larger (200 µm vs. about 100 µm), has the transverse cirri displaced much

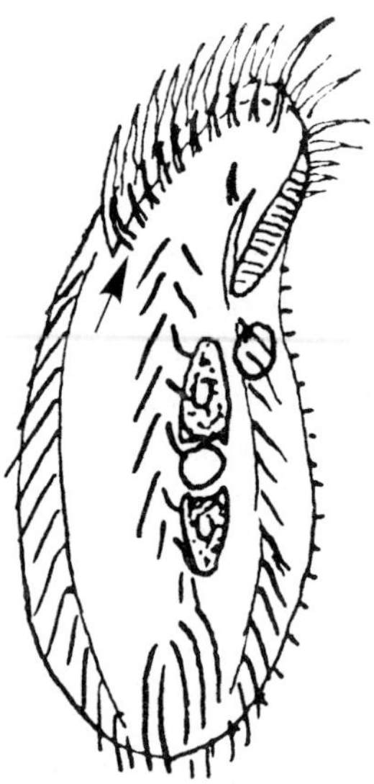

Fig. 83a *Anteholosticha alpestris* from life (from Kahl 1932). Ventral view, 100 µm. This species has a corona of about 10 frontal cirri (rightmost cirrus marked by arrow), a nuclear apparatus composed of two macronuclear nodules with one micronucleus in between, anteriorly displaced transverse cirri, and an adoral zone which extends far onto the right body margin. Because of the increased number of frontal cirri the classification in *Anteholosticha* is of course only preliminary. Page 403.

further anteriorly, has only three enlarged frontal cirri, lives in saline waters, and obviously has distinct caudal cirri.

Morphology: The following description is based solely on Kahl's data. Body length 80–110 µm; body length:width ratio of specimen illustrated 2.2:1 (Fig. 83a). Body outline ovoid; anterior body portion flexible, contractile, and distinctly curved leftwards. Nuclear apparatus very conspicuous because composed of two macronuclear nodules with large micronucleus in between. Contractile vacuole in ordinary position, that is, slightly behind buccal vertex near left cell margin. No cortical granules mentioned, indicating that they are lacking. Adoral zone about 31% of body length in specimen illustrated, extends far onto right body margin. Buccal field likely narrow. About 10 enlarged(?) frontal cirri form distinct corona along anterior body margin. One buccal cirrus near anterior end of undulating membranes. Midventral complex likely composed of cirral pairs only (about 7 in specimen illustrated; value must not be over-interpreted). Five transverse cirri, subterminal and thus hardly reaching rear body margin. Marginal rows slightly displaced inwards (preparation artefact?), otherwise obviously without peculiarities. Dorsal cilia short (likely around 3 µm). Caudal cirri neither mentioned nor illustrated, indicating that they are lacking.

Occurrence and ecology: Likely prefers terrestrial habitats (Foissner 1998, p. 205). Kahl (1932) discovered *A. alpestris* with low abundance in moss cushions from the Bavarian Alps (no details on exact sample site given, that is, type locality not known). Kahl (1932) found a similar species in (terrestrial?) mosses from California. However, he failed to study this population in detail. Gellért (1956) collected his population (see remarks for characterisation) on a hill (Magoska) near the Hungarian village of Boldogköváralja.

I found one further (unsubstantiated) terrestrial record of this interesting species, namely from the litter of a beech forest at Jakëiå Katuni (1800 m above sea-level) in the Durmitor mountains, Montenegro (Varga 1962, p. 154). By contrast, Mücke (1979, p. 266) recorded it in various freshwater habitats near the German city of Münster.

Anteholosticha alpestris feeds on ciliates and algae (Kahl 1932). Biomass of 10^6 specimens about 96 mg (Foissner 1998, p. 205).

Anteholosticha multistilata (Kahl, 1928) Berger, 2003
(Fig. 84a–k, Tables 18, 19)

1928 ***Holosticha multistilata*** – Kahl, Arch. Hydrobiol., 19: 212, Abb. 44d (Fig. 84a; original description; no type material available and no formal diagnosis provided).

1982 ***Holosticha multistilata* Kahl** – Jutrczenki, Decheniana, 135: 109, Abb. 2 (Fig. 84b; redescription after protargol impregnation).

1982 ***Holosticha multistilata* Kahl, 1928** – Hemberger, Dissertation, p. 97, Abb. 15a–h, not Abb. 15i (Fig. 84c–e, g–k; cell division; revision of non-euplotid hypotrichs).

1991 ***Holosticha multistilata* Kahl, 1928** – Foissner, Blatterer, Berger & Kohmann, Informationsberichte des Bayer. Landesamtes für Wasserwirtschaft, 1/91: 236, Abb. 6, 9, not Abb. 1–5, 7, 8, 10–14 (Fig. 84a, b; guide to freshwater ciliates).

1992 ***Holosticha multistilata* Kahl, 1928** – Carey, Marine interstitial ciliates, p. 182, Fig. 725 (Fig. 84f; redrawing from Kahl 1928; guide).

2001 ***Holosticha multistilata*** **Kahl, 1928** – Berger, Catalogue of ciliate names 1. Hypotrichs, p. 37 (nomenclator containing all basionyms, combinations, and higher taxa of hypotrichs).

2003 ***Anteholosticha multistilata*** **(Kahl, 1928) n. comb.** – Berger, Europ. J. Protistol., 39: 377 (combination with *Anteholosticha*).

Nomenclature: No derivation of the name is given in the original description. The species-group name *multistilata* (having many cirri) is a composite of the Latin indefinite numeral adjective *mult-* (many, numerous), the thematic vowel ·*i*-, the Latin noun *stilus* (pointed post, cirrus), and the suffix *~at·us* (having something) and very likely alludes to the increased number of buccal cirri. However, it cannot be excluded that the name refers to the increased number of frontal cirri (about 10 vs. 3–4 in related? species). The name *multistilata* is often incorrectly spelled *multistylata* (e.g., Hill 1980; Bernhard et al. 2001, p. 88, 89; Chen & Song 2002, p. 6; Wiackowski 1988, p. 4). Other incorrect subsequent spelling: *Holosticha multistillata* (Tirjaková 1988, p. 499; Tirjaková & Matis 1987a, p. 8).

Remarks: The present species lacks caudal cirri and all(?) apomorphies of *Holosticha* and was thus transferred to *Anteholosticha* (Berger 2003). For a detailed discussion and foundation of the transfer of many species from *Holosticha* to *Anteholosticha* see the genus section.

The original description of *A. multistilata* is rather brief, but shows the main features (Kahl 1928). The specimen illustrated has a conspicuously high number (about nine) of frontal cirri (Fig. 84a). In his major revision, Kahl (1932) provided new "improved" data and his illustration shows a specimen with only three enlarged frontal cirri (Fig. 61b). Unfortunately, he did not include the original illustration in his revision. Simultaneously, Kahl (1932) described *Holosticha muscorum*, whose description and illustration (Fig. 61a) are more or less identical with those of *H. multistilata* sensu Kahl (1932; Fig. 61b).

Borror (1972) synonymised *H. muscorum* Kahl, 1932 with *H. multistilata* Kahl, 1928, however, without explanation. Very likely, he just compared the illustrations and descriptions provided by Kahl (1932) and did not consider Kahl's (1928) original description of *H. multistilata*.

Buitkamp (1977) described a population with only three frontal cirri and identified it according to Kahl (1932) as *H. multistilata* (Fig. 61n). Like Borror (1972), he put *H. muscorum* into the synonymy of *H. multistilata*. Foissner (1982) found a population which matched the life data of *H. multistilata* sensu Kahl (1932) and the cirral pattern described by Buitkamp (1977). He accepted the synonymy proposed by Borror (1972) and Buitkamp (1977) and correctly stated that *H. muscorum* and *H. multistilata* are inseparable according to Kahl's (1932) diagnoses and illustrations (Foissner 1982, p. 51).

Fig. 84a–f *Anteholosticha multistilata* (a, from Kahl 1928; b, from Jutrczenki 1982; c–e, from Hemberger 1982; f, after Kahl 1928 from Carey 1992. a, f, from life; b–e, protargol impregnation). **a, f:** Ventral view, 180–250 µm. Note the increased number of enlarged frontal cirri. This pattern very closely resembles Hemberger's redescription (c) strongly indicating that these populations are conspecific. **b:** Infraciliature of ventral side, 198 µm. The dotted line circles all (most?) frontal cirri. **c:** Infraciliature of ventral side of a very early divider, 203 µm. The dotted line circles the 10 enlarged frontal cirri. Note the high agreement with the life observations by Kahl (1928, Fig. 84a). **d, e:** Early dividers, sizes not indicated. Details, see text. CV = contractile vacuole, FT = frontoterminal cirri, OP = first sign of oral primordium. Page 405. →

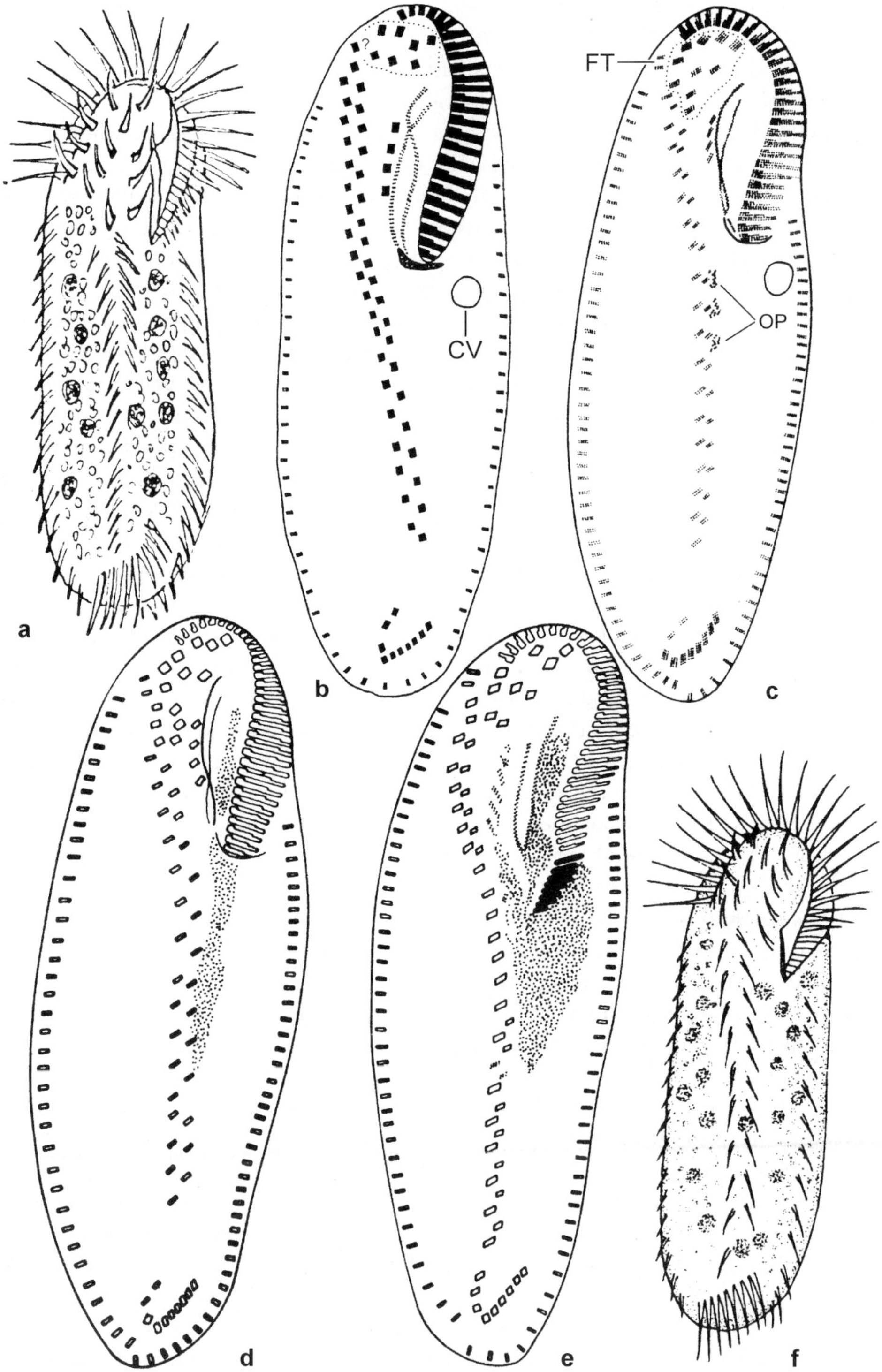
FT
OP
CV
a
b
c
d
e
f

On the same page, however, Foissner (1982) described a population very closely resembling *Keronopsis muscorum* Kahl sensu Grolière (1975). He recognised that these two populations (Fig. 73a–h), which differ significantly from the original description of *H. muscorum* (Fig. 61a), belong to a new species (= *Anteholosticha antecirrata* in the present book). However, instead of describing a new species, he matched the description of *H. muscorum* to his and Grolière's data. His new diagnosis, which is not a neotypification, provided a rather tricky situation which is complicated due to the fact that Jutrczenki (1982; Fig. 84b) and Hemberger (1982; Fig. 84c) described populations, which agree very well with the original description of *H. multistilata* (Fig. 84a): all three populations have a rather high number (8–10) of more or less distinctly enlarged frontal cirri (Fig. 84a–c); by contrast, the *Holosticha multistilata* and *H. muscorum* specimens illustrated by Kahl (1932; Fig. 61a, b) have only three (or four when cirrus III/2 is included) frontal cirri, a pattern confirmed by Buitkamp (1977, Fig. 61n) and Foissner (1982; Fig. 61i). This strongly indicates that both *H. multistilata* as originally illustrated by Kahl (1928) and *Holosticha muscorum* (= *Anteholosticha intermedia* in present book) are valid, that is, the synonymy of *H. multistilata* Kahl, 1928 and *H. muscorum* Kahl, 1932, as suggested by Borror (1972) and Buitkamp (1977), is unlikely.

Admittedly, the frontal ciliature illustrated by Kahl (1928, Fig. 84a) appears incorrect or superficially observed at first glance because most *Anteholosticha* species have only three frontal cirri. However, since the pattern is substantiated almost perfectly by two redescriptions based on protargol preparations, it is beyond reasonable doubt. This means that – as already mentioned above – *Anteholosticha multistilata* (Fig. 84a–c) and *A. intermedia*, the senior synonym of *Holosticha muscorum* (Fig. 61a, i) are very likely not identical as proposed by Borror (1972) and later workers. The ontogenetic data on *A. multistilata* by Hemberger (1982, Fig. 84c–e, g–k) even indicate that they are rather different. Unfortunately, no detailed live observations are available from *A. multistilata* because Kahl's (1928) original description is very brief and both Jutrczenki (1982) and Hemberger (1982) made only protargol preparations. Thus, we do not know, for example, whether or not *A. multistilata* has cortical granules. The *A. multistilata* populations described by Buitkamp (1977; Fig. 61n) and Foissner (1982; Fig. 61i) confirm the cirral pattern of *A. multistilata* sensu Kahl (1932; Fig. 61b) and that of the original description of *H. muscorum* Kahl, 1932 (Fig. 61a). *Anteholosticha intermedia* is thus well defined after live observations and after silver impregnation; however, ontogenetic data are lacking and therefore we do not know whether or not the parental adoral zone is totally replaced as in *A. multistilata*.

Now there are at least three ways to solve the problem: (i) Recognition of both *A. multistilata* (Kahl, 1928) (with many frontal cirri) and *A. intermedia* (Bergh, 1889) with the junior synonym *H. muscorum* Kahl, 1932 (three enlarged frontal cirri); (ii) classification of these two taxa as subspecies of *A. intermedia*; or (iii) keep synonymy of *A. multistilata* and *H. muscorum* (note that in this case the valid name would be *A. intermedia* because identity of *Urostyla intermedia* Bergh, 1889, *Holosticha muscorum* Kahl, 1932, and *H. multistilata* sensu Kahl 1932 is beyond reasonable doubt; see *A. intermedia*, for details). I choose the first possibility because the difference between *A. multistilata* and *A. intermedia* (9–10 vs. 3–4 frontal cirri) is at least as great as the difference between *A. adami* and *H. multistilata* sensu Foissner (1982; = *A. intermedia* in present book; marginal rows not overlapping vs. overlapping). Moreover, the classification as subspecies is somewhat problematic because both taxa have been found in the same region (Europe). However, per definition, subspecies should be clearly separated geographically.

Due to the complicated systematics of the *Anteholosticha multistilata/intermedia*-group, it would be appropriate to fix a neotype for the present species (ICZN 1999, Article 75.1). However, since detailed live data are lacking and since the data by Hemberger (1982) are not published I avoid such an act. However, when further detailed data became available a neotype should certainly be designated.

Keronopsis macrostoma Reuter, 1963 was put into the synonymy of *A. multistilata* by Hemberger (1982) and Borror & Wicklow (1983). The three frontal cirri indicate that it belongs to *A. intermedia*. Redescriptions of further populations are needed to clarify the situation around this species/subspecies group. In our revision on freshwater ciliates we did not distinguish between populations with many and with few frontal cirri (Foissner et al. 1991).

Shin et al. (2000) sequenced the small subunit rRNA genes of "*Holosticha multistylata*" (GenBank Accession No. AJ277876). They identified their material according to Shin & Kim (1993), a population classified as *Anteholosticha intermedia* in the present book (Fig. 61k–m).

Morphology: As mentioned above, the data by Jutrczenki (1982) and Hemberger (1982) confirm the observations by Kahl (1928). Unfortunately, both redescriptions lack life data and a morphometric characterisation. Thus, I extracted at least some data from the illustrations (Fig. 84b, c; Table 19). Body length 180–250 μm in life, length:width ratio about 3.2:1 (Fig. 84a); body regularly flat vaulted, soft and metabolic, that is, flexible (Kahl 1928). Many macronuclear nodules (Kahl 1928). Contractile vacuole obviously in ordinary position (Fig. 84b, c). Very motile (Kahl 1928). Adoral zone occupies about 30% of body length in type population (Fig. 84a), about 35–38% in specimens shown in Fig. 84b, c. Jutrczenki illustrated only 25 membranelles which is a rather low number for such a large species. Paroral high (Kahl 1928). Endoral composed of basal body pairs (Jutrczenki 1982), optically intersecting with paroral. Cirral pattern highly characteristic due to increased number of frontal cirri; Kahl (1928, Fig. 84a) and Hemberger (1982, Fig. 84c) each illustrated 10 enlarged frontal cirri, Jutrczenki's population with "invariably" eight cirri (when cirrus right of question mark is included; Fig. 84b). About 3–4 buccal cirri (Fig. 84a–c). Two frontoterminal cirri in ordinary position, that is, near distal end of adoral zone (Fig. 84c). Midventral complex composed of about 17–20 cirral pairs, extending to about 77% of body length (Fig. 84b, c). Two pretransverse ventral cirri ahead of right portion of transverse cirral row composed of 8–9 slightly enlarged cirri which project somewhat beyond rear body end (Fig. 84a–c). Right marginal row commences ahead of level of anteriormost buccal cirrus, ends slightly subterminally, almost confluent with left marginal row. Three dorsal kineties, caudal cirri lacking (Fig. 84k).

Cell division: Hemberger (1982) studied the ontogenesis of this species in detail (Fig. 84c–e, g–k). In spite of this, some details, for example, a stage between Fig.

84g and 84h are still unknown so that a reinvestigation of the division is recommended.

Ontogenesis commences with the formation of some basal body patches left of the middle portion of the midventral complex (Fig. 84c). A longitudinal oral primordium subsequently extends to the proximal end of the parental adoral zone. In addition, a primordium occurs in the parental buccal field (Fig. 84d). The next stage shows an early divider with already some adoral membranelles in the oral primordium of the opisthe (Fig. 84e). The primordium of the proter covers the parental buccal field. Right of it some oblique anlagen are recognisable. They originated from the parental buccal cirri and likely two or three rear frontal cirri. Two anlagen for the undulating membranes of the proter are present between the oblique streaks and the primordial field covering the parental buccal area. The midventral complex is unchanged in its anterior portion, but in the middle portion the left cirri are very likely dissolved in the large primordial field for the opisthe.

Fig. 84g shows a middle divider with some cirral primordia, 1–2 undulating membranes anlagen, and the primordium for the adoral zone in each filial product. Between this stage and the next stage described (Fig. 84h) several (about 21) frontal-midventral-transverse cirral anlagen have been formed. Hemberger could not clarify the origin of these streaks; he supposed that most midventral cirri are transformed to primordia. Some anterior and posterior parental midventral cirri, most parental frontal cirri, and the frontoterminal cirri are not involved in primordia formation. The resorption of the parental adoral zone commences.

Fig. 84i shows that the anteriormost five cirral anlagen (II–VI), which occurred very early (Fig. 84e, g), form more and distinctly larger cirri than the following anlagen, which each produce, as is usual, only a midventral pair. The posteriormost 6–7 streaks form the transverse cirri, and the posteriormost two anlagen form four cirri each. The penultimate streak forms a midventral pair, one pretransverse ventral cirrus, and one transverse cirrus; the rearmost anlage produces the two frontoterminal cirri and, likewise, one pretransverse ventral cirrus and one transverse cirrus.

In a very late stage the cirri migrate to their final positions (Fig 84j). The parental adoral zone is almost completely resorbed. The marginal rows are formed in an ordinary manner (Fig. 84h–j).

Likewise, the nuclear apparatus divides in ordinary manner; that is, the individual nodules fuse to a single mass in middle dividers (Fig. 84h) and then divide again.

Ontogenesis of dorsal kineties shows no peculiarities; that is, each of the three kineties forms two primordia. No caudal cirri are formed (Fig. 84k).

The division of *Anteholosticha multistilata* shows some interesting details: (i) The increased number of enlarged cirri originating each from the anlagen III–VI. Anlage I produces, as is usual, the left frontal cirrus, and anlage II produces, as in other species of this group, several buccal cirri. In addition, the anlagen II–VI very likely show a somewhat different origin than the remaining streaks (Fig. 84e, g). (ii) A new adoral zone is formed for the proter, that is, the parental adoral zone is totally replaced, as in *A. warreni*. By contrast, in *Holosticha* species at least the distal portion of the parental

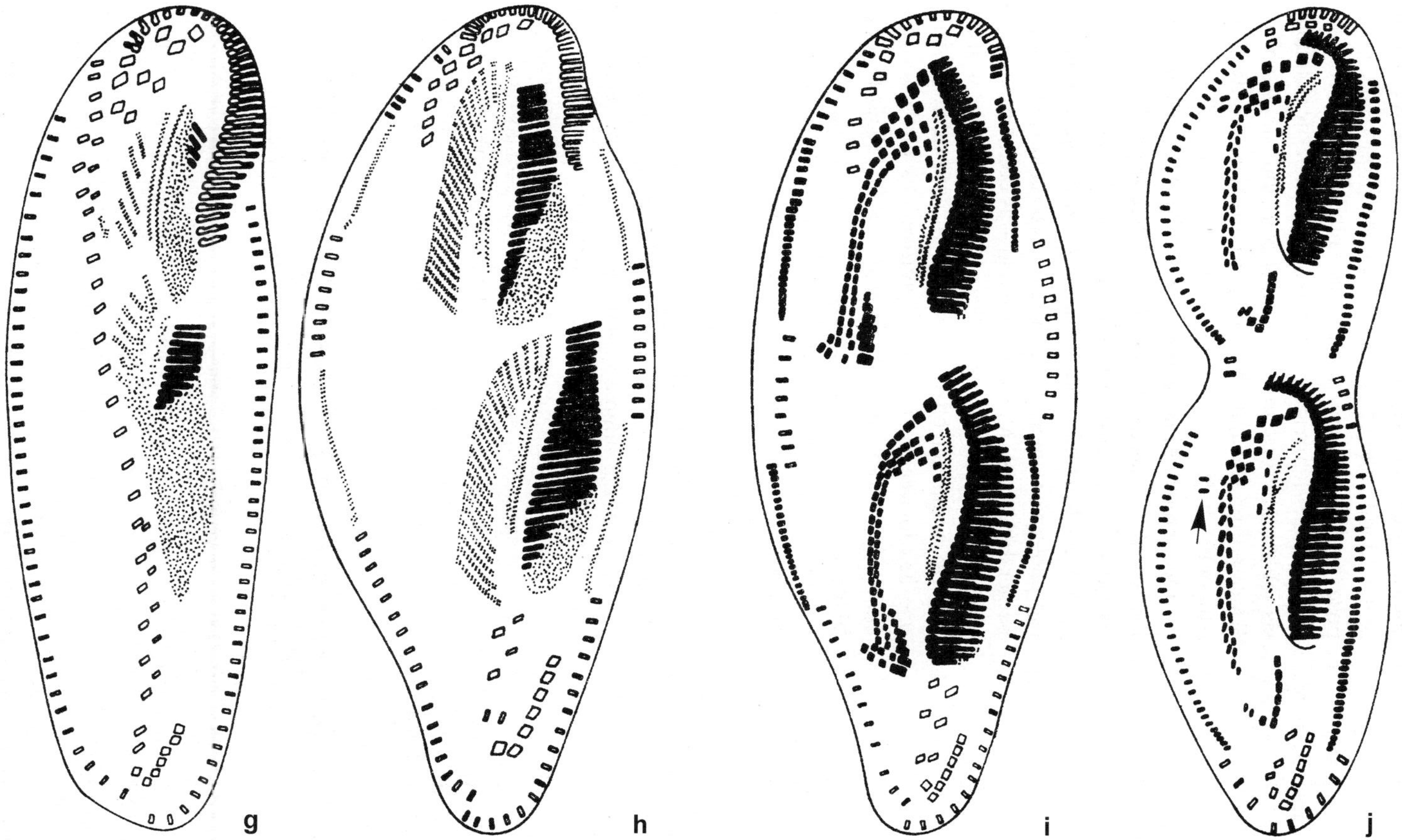

Fig. 84g–j *Anteholosticha multistilata* after protargol impregnation (from Hemberger 1982). Infraciliature of ventral side of middle, late, and very late dividers, sizes not indicated. Arrow in (j) marks anteriorly migrating frontoterminal cirri of opisthe. Note that the parental adoral zone of membranelles is completely replaced by a new adoral zone for the proter. Further details, see text. Page 405.

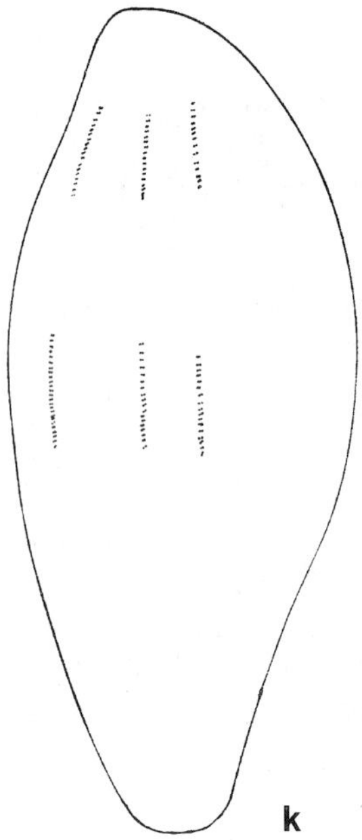
k

Fig. 84k *Anteholosticha multistilata* after protargol impregnation (from Hemberger 1982). Infraciliature of dorsal side of a late divider, size not indicated. Division of dorsal kineties shows no peculiarities. Page 405.

adoral zone is retained. Usually, total replacement of the parental adoral zone occurs in the pseudokeronopsids. I suppose that this is due to convergent evolution.

Hill (1980) provided a brief description of the ontogenesis of *H. multistilata*, *H. diademata*, and *H. scutellum*. However, no illustrations are given, so the identifications cannot be checked.

Occurrence and ecology: Due to the confusing systematics of this species the faunistic and ecological data of *A. multistilata* and *A. intermedia* are somewhat difficult to assign. Here I include only the three records substantiated by morphological data (Kahl 1928, Jutrczenki 1982, Hemberger 1982). No other record refers – as concerns the identification – to one of these three papers, so that it seems wise to assign all other records to *A. intermedia* because one has to assume that the identifications of "*Holosticha multistilata*" were basically done after Kahl (1932; Fig. 61a, b) and Foissner (1982, Fig. 61i).

Anteholosticha multistilata is likely eurytopic because it was recorded from inland saltwater, freshwater, and soil. The type locality is a salt meadow (about 20‰ salinity) near Altfresenburg (53°50'12"N 10°22'03"E), a village about 30 km north-east of the German city of Hamburg (Kahl 1928, 1928a). Kahl (1928) found it occasionally abundant in the sample vessels. In addition, he recorded it rarely in the same area in the Brennermoor, a saline bog with about 25‰ salinity. Jutrczenki (1982) observed the present species in an unpolluted brook (Annaberger Bach) near the city of Bonn, Germany. Hemberger (1982) likely isolated this species from soil samples collected in the Puerto Maldonado area (about 69°12' W 12°36'S) in the Peruvian rain forest. Feeds on diatoms (Kahl 1928, Jutrczenki 1982).

Anteholosticha warreni (Song & Wilbert, 1997) Berger, 2003
(Fig. 85a–w, Table 19, Addenda)

1997 ***Holosticha warreni* nov. spec.**[1] – Song & Wilbert, Europ. J. Protistol., 33: 54, Fig. 16–22, 47–52, Table 3 (Fig. 85a–g; original description. One holotype slide and 2 paratype slides of protargol-impregnated specimens have been deposited in the Laboratory of Protozoology, College of Fisheries, Ocean University of Qingdao, China).

[1] The diagnosis by Song & Wilbert (1997) is as follows: In vivo about 80–120 × 40–55 μm sized, marine *Holosticha* with 1 buccal, 2 frontoterminal, 10–12 transverse cirri and about 28 membranelles. Midventral row with 7–9 cirral pairs extending 2/3 of cell length. Cortical granules conspicuous, ellipsoid and flattened with central depression, arranged in 3 irregular rows. 3 dorsal kineties. About 50 macronuclear segments arranged in a ring-like fashion.

2000 ***Holosticha warreni*** – Hu, Song & Warren, J. mar. biol. Ass., 80: 785, 1A–H, 2A–H, 3A–L (Fig. 85h–w; cell division).

2001 ***Holosticha warreni* Song and Wilbert, 1997** – Berger, Catalogue of ciliate names 1. Hypotrichs, p. 39 (nomenclator containing all basionyms, combinations, and higher taxa of hypotrichs).

2003 ***Anteholosticha warreni* (Song and Wilbert, 1997) n. comb.** – Berger, Europ. J. Protistol., 39: 377 (combination with *Anteholosticha*).

Nomenclature: Song & Wilbert (1997) dedicated this species to Alan Warren of The Natural History Museum in London.

Remarks: The present species lacks caudal cirri and most apomorphies of *Holosticha* and was thus transferred to *Anteholosticha* (Berger 2003). Only the straight and rather short undulating membranes and the high number of transverse cirri, which is equal to that of the frontal-midventral anlagen, are reminiscent of *Holosticha*. However, *Holosticha* species differ from *A. warreni*, inter alia, in the rightwards curved anterior end of the left marginal row (vs. straight) and the gap in the adoral zone (vs. lacking). For a detailed discussion and foundation of the transfer of many species from *Holosticha* to *Anteholosticha* see the genus section. The erythrocyte-like granules (Fig. 85c) occur also in many pseudokeronopsid species and in *Holosticha bradburyae*.

Anteholosticha manca has only about five transverse cirri, which are, in addition, less strong than those of *A. warreni*. *Anteholosticha mancoidea* also has fewer transverse cirri (about 5) and only eight macronuclear nodules, which form a longitudinal figure in the left body portion.

Morphology: Body size 80–120 × 40–55 µm in life; body length:width ratio 2.3:1 on average in protargol preparations (Table 19). Body outline basically as shown in Fig. 85a, that is, widest about at level of buccal vertex and margins more or less distinctly converging posteriorly; left margin sigmoidal at level of buccal vertex. Body fragile, dorsoventrally flattened about 2:1 (Fig. 85d). Nuclear apparatus composed of about 50 macronuclear nodules invariably arranged in an elliptical figure; individual nodules ellipsoidal, 3–5 µm long in life (Fig. 85e, f). Contractile vacuole lacking. Cortical granules conspicuous because about 2 µm across, ellipsoid and flattened with central depression (resembling erythrocytes in shape), arranged dorsally in three loosely organised rows (obviously along dorsal kineties; Fig. 85b, c); granules might be a kind of extrusome because, in protargol-stained specimens, they are often ejected becoming globules with a hair-like process and in total 8–10 µm long (Fig. 85f). Cytoplasm colourless to slightly greyish and somewhat opaque with numerous granules less than 2 µm across.

Adoral zone occupies about 1/3 of body length (in life?), on average 37% in protargol preparations, composed of about 28 membranelles of ordinary fine structure (Fig. 85g, Table 19). Cilia of membranelles up to 15 µm long. Buccal field medium-wide. Undulating membranes straight, of about same length (about 10 µm in specimen shown in Fig. 85f, g), arranged side by side. Cytopharynx without peculiarities; extends longitudinally backwards.

Cirral pattern and number of cirri of usual variability (Fig. 85g; Table 19). Invariably three slightly enlarged, obliquely arranged frontal cirri and one cirrus (III/2) left behind right frontal cirrus. Buccal cirrus right of anterior end of undulating membranes.

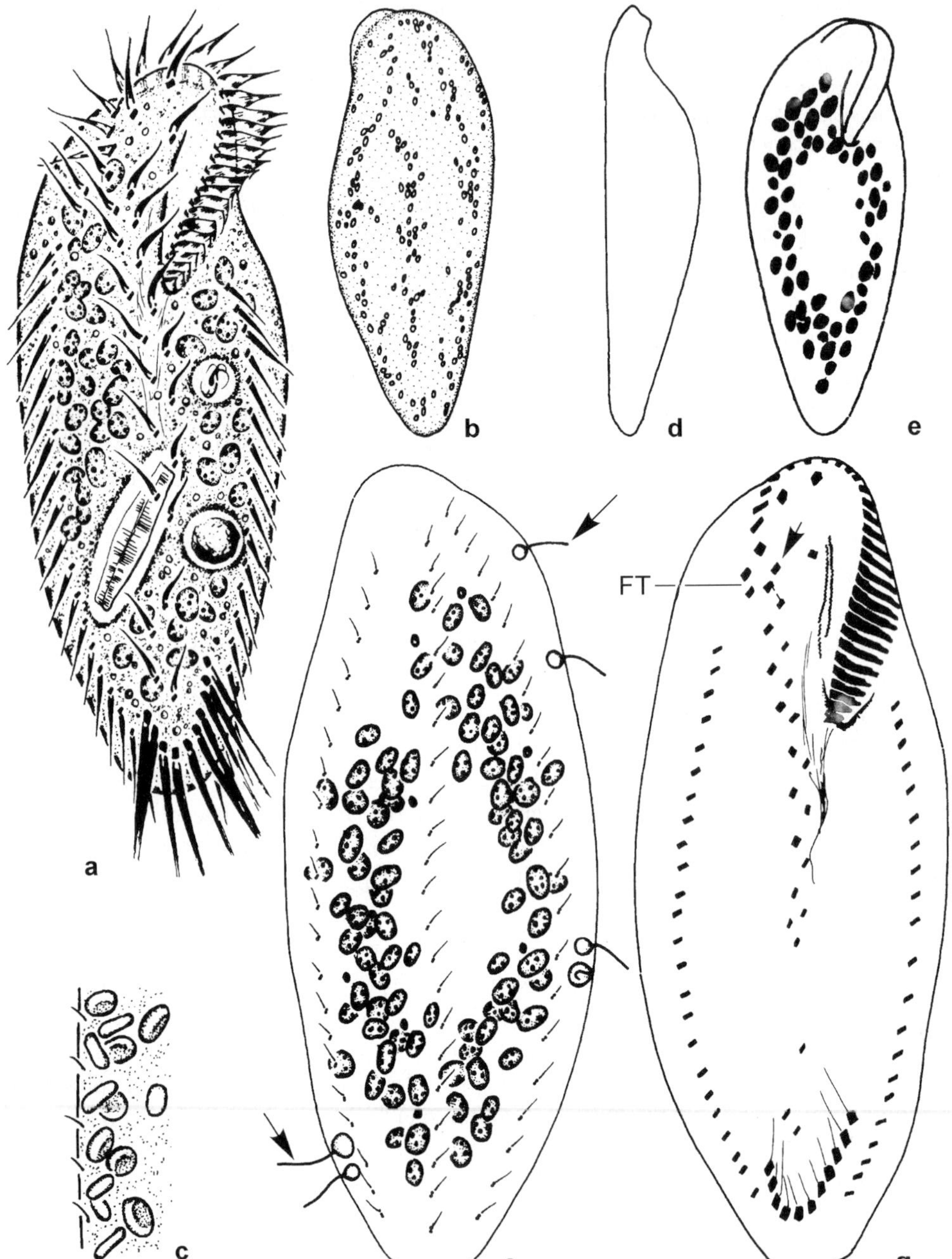

Fig. 85a–g *Anteholosticha warreni* (from Song & Wilbert 1997. a–d, from life; e–g, protargol impregnation). **a:** Ventral view of a representative specimen, 107 µm. **b, c:** Cortical granules (about 2 µm across) are arranged roughly in three longitudinal rows (likely along dorsal kineties) and look like erythrocytes of mammals. **d:** Left lateral view. **e:** Nuclear apparatus. **f, g:** Infraciliature of dorsal and ventral side and nuclear apparatus of same specimen, 74 µm. Arrows in (f) mark ejected granules (extrusomes). Arrow in (g) denotes cirrus III/2. Cirri of first midventral pair connected by broken line. FT = frontoterminal cirri. Page 412.

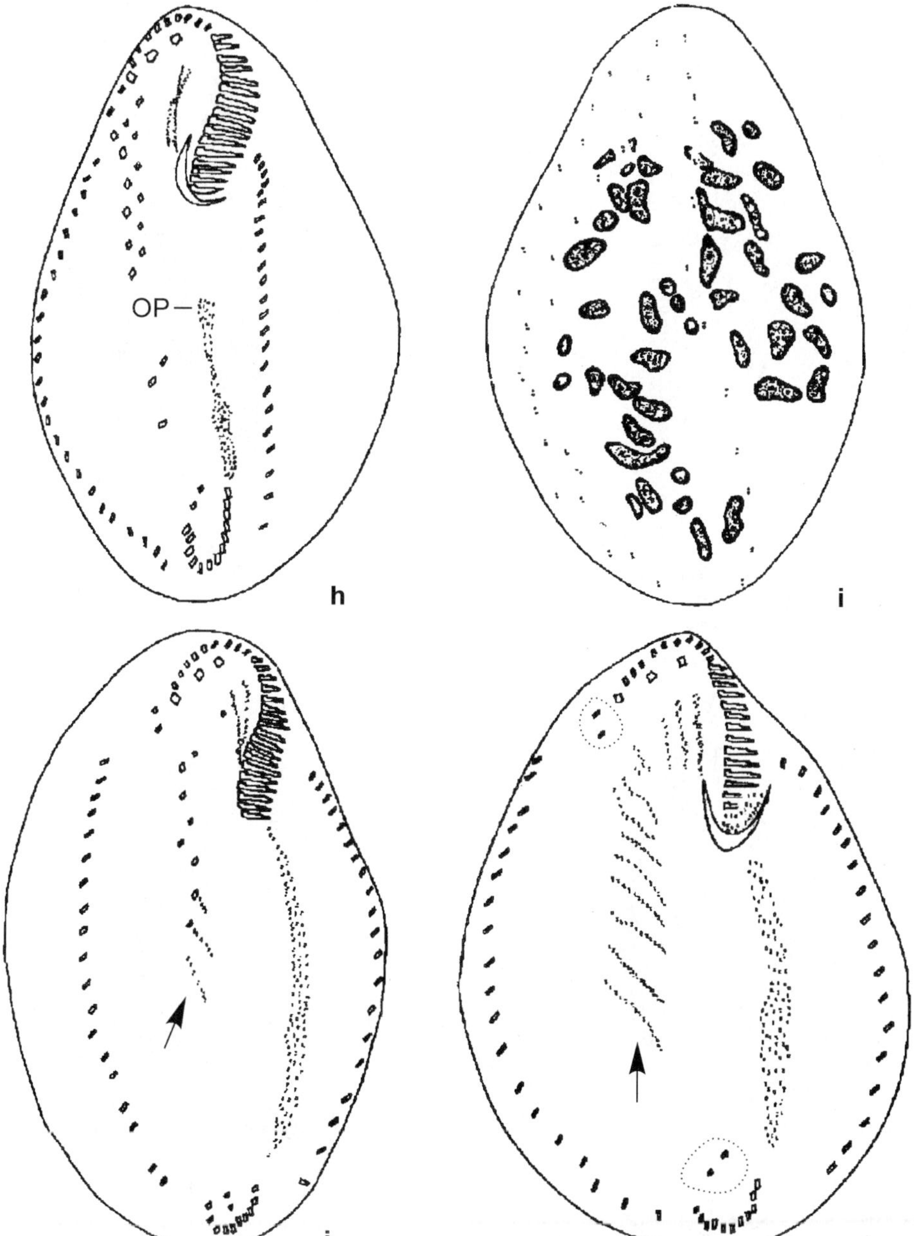

Fig. 85h–k *Anteholosticha warreni* (from Hu et al. 2000a. Protargol impregnation). **h, i:** Infraciliature of ventral and dorsal side and nuclear apparatus of a very early divider, 56 µm. **j:** Early divider (70 µm) with three frontal-midventral-transverse cirral anlagen (arrow). **k:** Early to middle divider, size not indicated. Parental frontoterminal cirri and pretransverse ventral cirri encircled by dotted lines. Arrow marks the frontal-midventral-transverse cirri anlagen (primary primordia) which divide in a later stage (see Figure 85l). OP = oral primordium. Page 412.

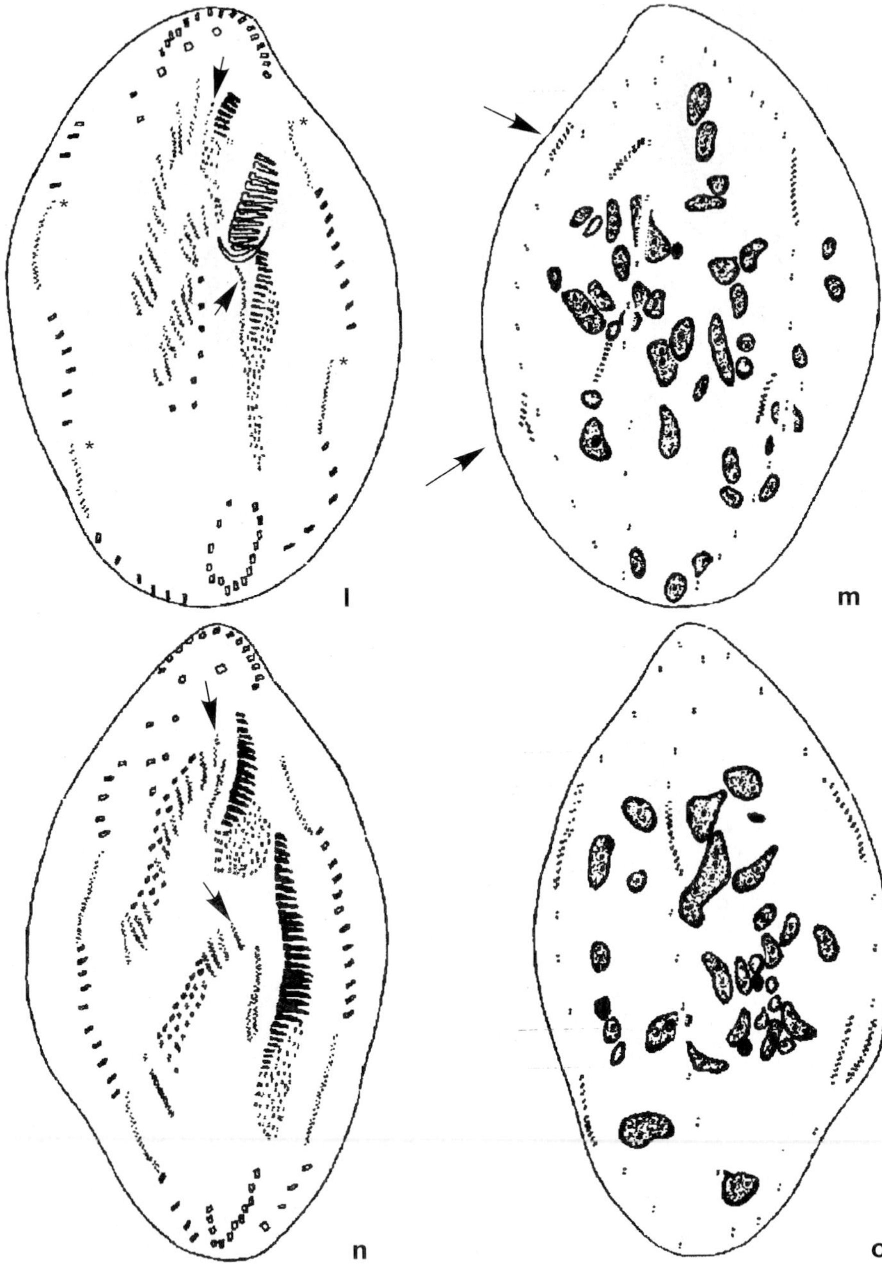

l m n o

Fig. 85l–o *Anteholosticha warreni* (from Hu et al. 2000a. Protargol impregnation). **l, m:** Infraciliature of ventral and dorsal side and nuclear apparatus of same specimen, size not indicated. Arrows in (l) mark undulating membranes anlagen, asterisks denote marginal row anlagen. Arrows in (m) mark the two levels of the dorsal kineties anlagen. **n, o:** Infraciliature of ventral and dorsal side and nuclear apparatus of middle divider. Arrows in (n) denote new left frontal cirri which originate, as usual, from the undulating membrane anlage. The frontal-midventral-transverse cirri anlagen begin with the formation of the individual cirri. Page 412.

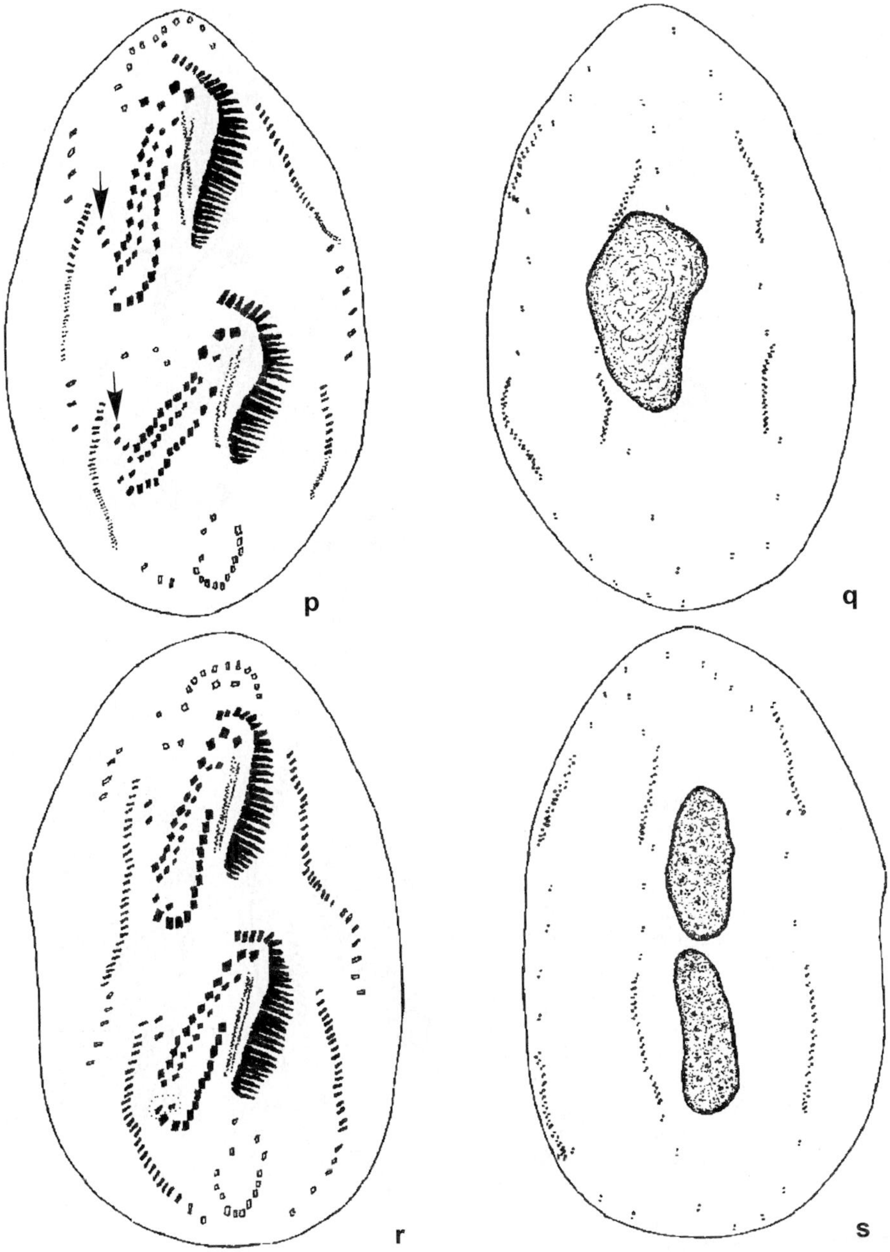

Fig. 85p–s *Anteholosticha warreni* (from Hu et al. 2000a. Protargol impregnation). Parental structures white, new black. **p, q:** Infraciliature of ventral and dorsal side and nuclear apparatus of a late divider, size not indicated. Arrows in (p) denote new frontoterminal cirri. The many macronuclear nodules are, as usual, fused to a single, roundish mass. Each frontal-midventral-transverse cirri anlage forms a transverse cirrus which is reminiscent of *Holosticha*. **r, s:** Infraciliature of ventral and dorsal side and nuclear apparatus of late divider, size not indicated. New pretransverse ventral cirri of opisthe circled by dotted line. Note that a new adoral zone of membranelles is formed both for the proter and the opisthe. Page 412.

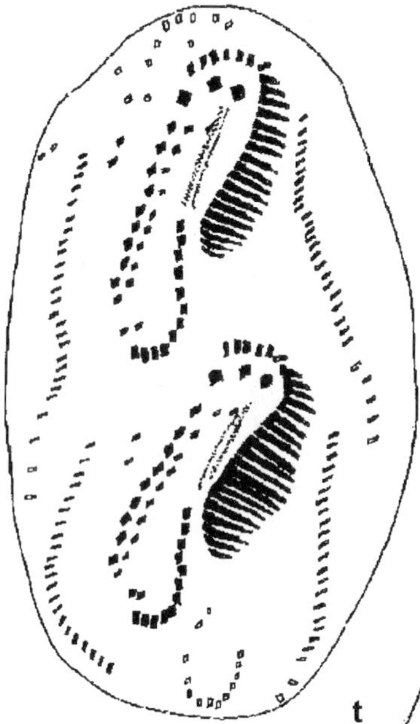

t

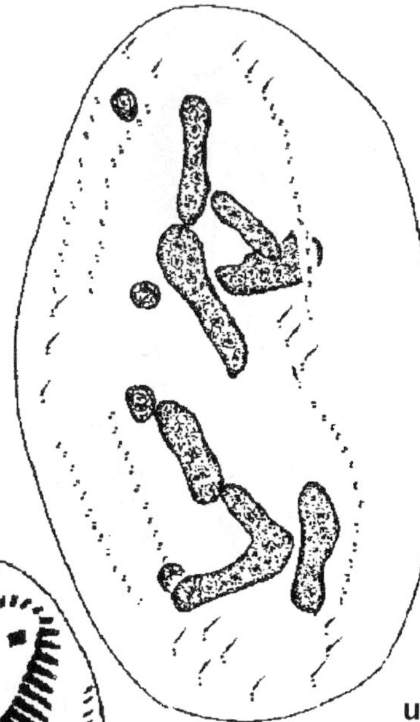

u

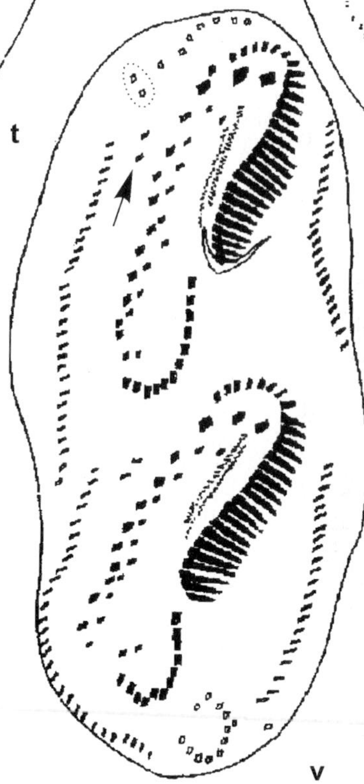

v

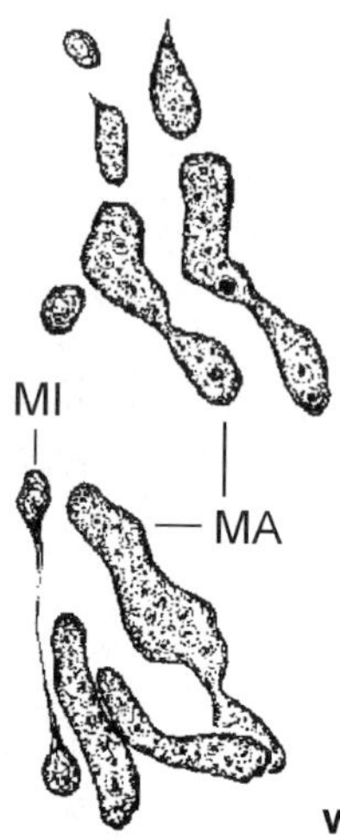

w

Fig. 85t–w *Anteholosticha warreni* (from Hu et al. 2000a. Protargol impregnation). Parental structures white, new black. **t, u:** Infraciliature of ventral and dorsal side and nuclear apparatus of a very late divider, 104 µm. **v, w:** Infraciliature of ventral side and nuclear apparatus of a very late divider, size not indicated. Arrow marks new frontoterminal cirri of proter. Parental frontoterminal cirri encircled by dotted line. MA = macronuclear nodules, MI = dividing micronucleus. Page 412.

Two frontoterminal cirri in ordinary position, that is, slightly right behind right frontal cirrus. Midventral complex composed of cirral pairs only, extends from near right frontal cirrus to slightly behind mid-body (60% in specimen shown in Fig. 85g). 2–3 pretransverse ventral cirri between rear end of midventral complex and transverse cirri. Transverse cirri enlarged, arranged in J-shaped figure, project distinctly beyond rear body end. Right marginal row commences about at level of rear end of undulating membranes, terminates, like left row, slightly ahead of level of rearmost transverse cirri. Left marginal row begins somewhat ahead of level of proximal end of adoral zone.

Dorsal cilia about 3 µm long in life, arranged in three almost bipolar kineties. Caudal cirri lacking (Fig. 85f).

Cell division (Fig. 85h–w): This part of the life cycle is described in detail by Hu et al. (2000a). Unfortunately, the illustrations and especially the micrographs are very small.

Ontogenesis commences with the formation of an oral primordium extending from the leftmost transverse cirrus to above mid-body. Transverse cirri and midventral cirri are obviously intact (Fig. 85h). Somewhat later the rearmost midventral pairs disorganise to frontal-midventral-transverse cirri anlagen (Fig. 85j). Subsequently, all other midventral cirri are modified to anlagen. Per definition these anlagen are primary primordia (definition, see Foissner 1983 and ground pattern of the Urostyloidea) because, somewhat later, they divide into an anterior anlage for the proter and a posterior anlage for the opisthe (Fig. 85l). In addition, the parental undulating membranes and the proximal portion of the adoral zone are disorganised to form the oral primordium of the proter (Fig. 85k).

The further process does not show peculiarities, that is, proceeds as in many other species: (i) each two marginal primordia originate within each parental marginal row (Fig. 85l, asterisks); (ii) the frontal-midventral-transverse cirri anlagen become compact and form the individual cirri; (iii) the undulating membranes anlagen originate from the oral primordia; (iv) the rightmost frontal-midventral-transverse cirri anlage forms, inter alia, the two frontoterminal cirri which migrate anteriorly (Fig. 85l, n, p, r, t, v). However, the two following features are rather interesting: (i) the adoral zone of the proter is newly formed, that is, the parental zone is completely replaced (Fig. 85l, n, p, r, t, v); and (ii) each frontal-midventral-transverse cirri anlage forms a transverse cirrus (Fig. 85p, r, t, v). Feature (i) is reminiscent of *A. multistilata* (Fig. 84g–j), whereas feature (ii) is characteristic for *Holosticha* species.

The ontogenesis of the dorsal ciliature shows no peculiarities. Within each of the three parental kineties two anlagen are formed. Further, no caudal cirri originate at the end of the kineties, which is the main reason for the transfer of the present species from *Holosticha* to *Anteholosticha* and not to *Caudiholosticha* (Fig. 85i, m, o, q, s, u).

The nuclear apparatus divides in ordinary manner; that is, the individual macronuclear nodules fuse to a single, roundish mass, which again divides in late stages of ontogenesis (Fig. 85i, m, o, q, s, u, w).

Occurrence and ecology: Marine. Type locality of *A. warreni* is the eastern China sea (salinity 32‰; pH 8.1–8.3; 5–13° C) near Taipingjiao, where Song & Wilbert (1997) collected it on 12.11.1995. Hu et al. (2000a) did not mention the exact sample site of their material. I suppose that they either used the type population or collected the material from the type locality.

Anteholosticha warreni feeds on diatoms and other algae (Song & Wilbert 1997).

Anteholosticha estuarii (Borror & Wicklow, 1983) Berger, 2003 (Fig. 86a)

1983 ***Holosticha estuarii* sp. n.** – Borror & Wicklow, Acta Protozool., 22: 112, 121, Fig. 3 (Fig. 86a; original description; type slides are deposited in the slide collection of A. C. Borror; no formal diagnosis provided).

2001 ***Holosticha estuarii* Borror and Wicklow, 1983** – Berger, Catalogue of ciliate names 1. Hypotrichs, p. 34 (nomenclator containing all basionyms, combinations, and higher taxa of hypotrichs).

2003 ***Anteholosticha estuarii* (Borror and Wicklow, 1983) n. comb.** – Berger, Europ. J. Protistol., 39: 377 (combination with *Anteholosticha*).

Nomenclature: No derivation of the name is given in the original description. The species-group name *estuarii* (living in an estuary) refers to the habitat where the species was discovered (tidal marshes near Durham).

Borror & Wicklow (1983) described this species on page 112. Only a few pages later, that is, on page 121, they synonymised their own species with *Trichotaxis pulchra* Borror, 1972 (incorrect subsequent spelling of *Trichototaxis*). If Borror & Wicklow had been convinced that these populations are conspecific they could have saved themselves the establishment of *H. estuarii*.

Remarks: Borror & Wicklow (1983) assumed identity of the present species and *Trichototaxis pulchra* (see nomenclature, for explanation of this curiosity). Unfortunately, they did not provide an explanation. I suppose that they synonymised them because *H. estuarii* also sometimes has two left marginal rows, a feature characteristic for *Trichototaxis*; in contrast, I consider *H. estuarii* and *T. pulchra* (= *Diaxonella pseudorubra pulchra* in present book) as distinct species/subspecies for the following reasons: (i) Borror & Wicklow (1983) investigated four specimens (their Table 1 shows four lines each with a sample size of 1). Under the column "Rows of left marginal cirri" they provided the values 1, 1, 1–2, and 1 (I cannot explain the variability 1–2 within only one specimen). Anyhow, *Holosticha estuarii* obviously primarily has only one left marginal row; by contrast *D. pseudorubra pulchra* invariably has two left marginal rows (Borror 1972a, p. 63). (ii) There is a distinct difference in the number of cirral pairs forming the midventral complex, namely 14 pairs in *H. estuarii* (Fig. 86a) against about 43 in *D. pseudorubra pulchra* (Fig. 101b), although the species have the same size. The description of *H. estuarii* includes some discrepancies, namely, (i) in body length which is 140–248 μm according to the text against 130 μm in the specimen illustrated (possibly the larger values are from life while the specimen illustrated is from a preparation), and (ii) in the number of transverse cirri (8–15 according to text against only 5 in specimen illustrated). Detailed redescriptions of further populations are needed to clarify the somewhat confusing situation.

The present species likely lacks caudal cirri and all (?) apomorphies of *Holosticha* and was thus transferred to *Anteholosticha* (Berger 2003). For a detailed discussion and foundation of the transfer of many species from *Holosticha* to *Anteholosticha* see the genus section. *Anteholosticha estuarii* is very similar to *A. adami* (see there for details). For separation from other congeners, see key.

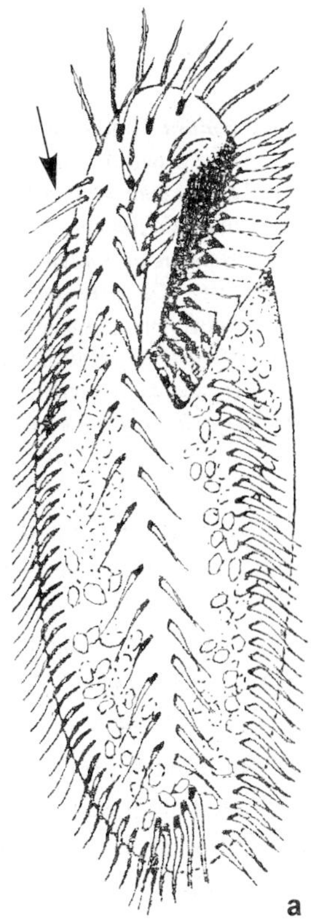

Fig. 86a *Anteholosticha estuarii* (from Borror & Wicklow 1983. Method not indicated). Infraciliature of ventral side and nuclear apparatus, 130 µm. Arrow marks frontoterminal cirri (see text). Page 420.

Morphology: Body length 140–248 µm (from life?), body length:width ratio of specimen illustrated in Fig. 86a about 3.1:1 (for problems with body length, see remarks). Body outline elongate elliptical (Fig. 86a). Numerous macronuclear nodules, obviously scattered mainly left and right of midline (Fig. 86a). Micronuclei and contractile vacuole not mentioned or illustrated. Borror & Wicklow described the cortical granulation as follows: "Cortex heavily granulated with tiny yellow cortical granules generally in areas between ciliary organelles, as well as in groups at bases of cirri. Granules in cortex generally in short rows, except in rosettes about bases of dorsal bristles in some populations". Movement not described.

Adoral zone occupies about 41% of body length in specimen illustrated, composed of about 27 membranelles (Fig. 86a). Buccal field moderately wide, with prominent paroral on lip. Borror & Wicklow (1983) obviously studied four specimens (four lines each with a sample size [= number of specimens investigated] of 1). Three enlarged frontal cirri. 6–8 buccal cirri right along paroral. 0–3 cirri behind right frontal cirrus (individual values: 0, 1, 3, 3). Frontoterminal cirri not mentioned in species description, but likely, as is usual, two such cirri are present (Fig. 86a, arrow). Midventral complex composed of cirral pairs only, no value provided, specimen illustrated with about 14 pairs; extends to near transverse cirri. 8–15 transverse cirri roughly arranged in J-shape, project distinctly (right side) to slightly (left side) beyond rear body end; specimen illustrated with only five transverse cirri. Invariably one right marginal row, commences near distal end of adoral zone, ends slightly subterminally, in specimen illustrated composed of about 45 cirri. Usually one, sometimes two left marginal rows (individual values: 1, 1, 1–2, 1; see remarks), begins near proximal end of adoral zone, terminates slightly subterminally so that marginal rows are distinctly separated posteriorly; specimen illustrated with 31 left marginal cirri. Dorsal infraciliature (length of dorsal bristles, number and arrangement of dorsal kineties) not described; caudal cirri likely lacking because neither mentioned nor illustrated (Fig. 86a).

Occurrence and ecology: Marine. Type locality is a tidal marsh (70°52′W, 43°06′N) at Durham, New Hampshire, USA. Borror & Wicklow (1983) found it also in tidal marshes in Rye, New Hampshire, and Huntington, New York. All populations were isolated from algal mats in panne ponds (I did not find the word panne in any dictionary).

Anteholosticha manca (Kahl, 1932) Berger, 2003
(Fig. 87a–g, Table 19, Addenda)

1932 ***Holosticha manca* spec. n.** – Kahl, Tierwelt Dtl., 25: 579, Fig. 110 $_{17}$ (Fig. 87a; original description; no type material available and no formal diagnosis provided).

1933 ***Holosticha manca* Kahl 1932** – Kahl, Tierwelt N.- u. Ostsee, 23: 109, Fig. 17.6 (Fig. 87b; guide to marine ciliates).

1997 ***Holosticha manca* Kahl, 1932** – Song & Wilbert, Arch. Protistenk., 148: 418, Fig. 4a–e, 14, Table 2 (Fig. 87c–g; redescription and neotypification; 1 neotype slide is deposited in the Protozoological Laboratory, College of Fisheries, Ocean University of Qingdao, China).

2001 ***Holosticha (Holosticha) manca* Kahl, 1932** – Berger, Catalogue of ciliate names 1. Hypotrichs, p. 35 (nomenclator containing all basionyms, combinations, and higher taxa of hypotrichs).

2003 ***Anteholosticha manca* (Kahl, 1932) n. comb.** – Berger, Europ. J. Protistol., 39: 377 (combination with *Anteholosticha*).

Nomenclature: No derivation of the name is given in the original description. The species-group name *mancus* (Latin adjective; incomplete) possibly alludes to the "shortened" midventral complex. Kahl (1932, 1933) divided *Holosticha* into several subgenera. Thus, the correct name in his reviews is *Holosticha (Holosticha) manca* Kahl, 1932. *Holosticha mancha* in Ganapati & Rao (1958, p. 90) is an incorrect subsequent spelling.

Remarks: The present species lacks caudal cirri and all(?) apomorphies of *Holosticha* and was thus transferred to *Anteholosticha* (Berger 2003). For a detailed discussion and foundation of the transfer of many species from *Holosticha* to *Anteholosticha* see genus section.

The original description is rather short, but contains most data necessary for identification. The population described by Ganapati & Rao (1958, Fig. 100f) has only two macronuclear nodules (vs. many; about 50–70 in neotype population) and is distinctly larger (in life 150–190 µm vs. 100–120 µm in Kahl's specimen). By contrast, the cirral pattern is obviously very similar although the Indian authors mentioned that the median ventral rows are widely separated. Since I cannot identify this population with a valid species I mention it under insufficient redescriptions (see end of this genus). *Holosticha manca* sensu Agamaliev (1972) is classified as *Bakuella agamalievi* (see there, for details).

Borror (1972, p. 11), Borror & Wicklow (1983, p. 122), and Carey (1992, p. 183) synonymised the present species with *Anteholosticha scutellum*. However, this species has 7–8 transverse cirri and obviously lacks cortical granules.

Song & Wilbert (1997a) redescribed *Holosticha manca* and simultaneously designated a neotype. The data (size, cortical granulation, cirral pattern, habitat) agree very well so that conspecificity with Kahl's population is beyond reasonable doubt.

Anteholosticha warreni has 11 transverse cirri on average (vs. 5) and a different type of cortical granules (extrusomes shaped like erythrocytes of mammals against globules). *Anteholosticha intermedia* has usually four buccal cirri and the cortical granules are yellow-green. For separation from other *Anteholosticha* species, see key.

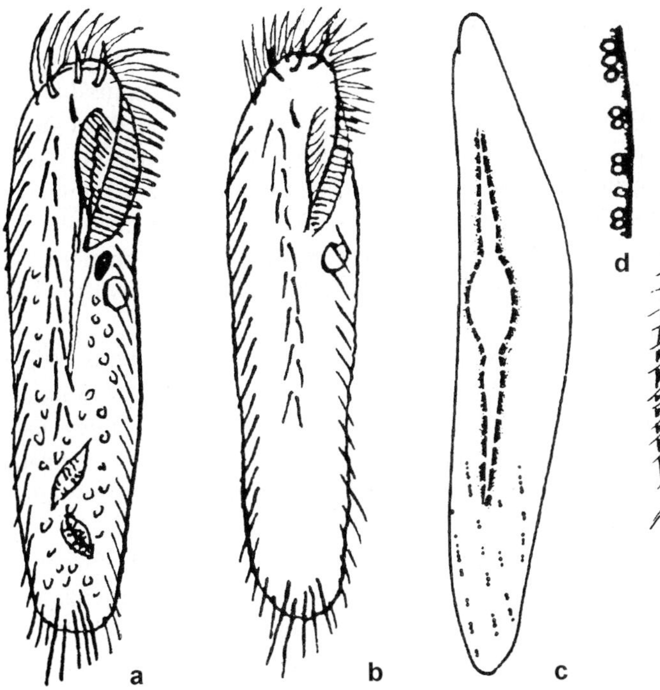

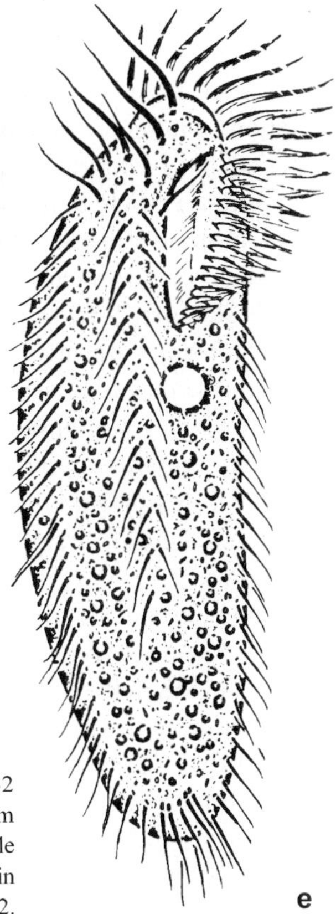

Fig. 87a–e *Anteholosticha manca* (a, from Kahl 1932; b, after Kahl 1932 from Kahl 1933; c–e, neotype population from Song & Wilbert 1997a. From life). **a, b:** Ventral view, 110 μm. **c:** Left lateral view showing contractile vacuole with collecting canals and cortical granules. **d:** Cortical granules in lateral view. **e:** Ventral view of a representative specimen, 127 μm. Page 422.

Morphology: Kahl's population from Oldesloe is first briefly described, followed by the detailed description of the neotype population from the Yellow Sea.

Specimens from Oldesloe 100–120 μm long, body length:width ratio of specimen illustrated 4.4:1 (Fig. 87a, b). Body outline slender, margins slightly converging posteriorly. Cortex (ectoplasm) thin, contains delicate cortical granules arranged in longitudinal rows (colour and size of granules not mentioned). Macronuclear nodules difficult to recognise even after staining; no number mentioned. 1–2 distinct micronuclei. Adoral zone occupies about 1/3 of body length, all membranelles relatively strong; buccal field narrow to moderately wide, buccal lip anteriorly strongly curved. Cytopharynx long. Midventral complex terminates at about 66% of body length, left cirri of pairs very fine. Five transverse cirri project by about half their length beyond rear body end.

Neotype population (Fig. 87c–g; Song & Wilbert 1997): Size of life specimens not mentioned; specimen shown in Fig. 87e about 127 × 38 μm; body length:width ratio of protargol-impregnated specimens 2.9:1 (Table 19). Body outline usually elongate or fusiform, both ends more or less narrowly rounded, left margin almost straight, right distinctly convex (according to the description margins are almost in parallel, however the representative specimen shown in Fig. 87e has a distinctly vaulted right margin).

Body flexible, dorso-ventrally flattened by ca. 2:1 (Fig. 87c). About 50–70 macronuclear nodules scattered throughout cytoplasm; individual nodules 3–5 μm long, containing several large nucleoli. Micronuclei lenticular to globular, only recognisable after protargol staining. Contractile vacuole small, near left body margin and at about 40% of body length; during diastole with two conspicuous collecting canals; pulsation interval up to 5 min. Pellicle thin. Cortical granules about 1 μm across, often several granules attached forming irregular, longitudinally arranged rows (Fig. 87c, d); likely colourless because no colour mentioned. Cytoplasm colourless to greyish, contains numerous refractive globules 2–5 μm across. Movement slowly to moderately fast crawling on substrate, sometimes jerking back and forth.

Adoral zone occupies 33% of body length on average (Table 19; Song & Wilbert wrote, obviously par lapsus, 2/3 of body length), distal end only slightly extending onto right body margin, composed of 24 membranelles on average. Bases of largest membranelles 7–8 μm wide (in life?), cilia of membranelles 15–20 μm long. Undulating membranes almost straight and parallel, sometimes optically intersecting posteriorly (Fig. 87f).

Cirral pattern and number of cirri of usual variability except for number of frontoterminal cirri and midventral pairs (Table 19). Three slightly enlarged frontal cirri about 20–25 μm long, left one obviously somewhat shifted anteriad. Buccal cirrus slightly behind anterior end of paroral. Frontoterminal cirri in ordinary position, that is, between distal end of adoral zone and anterior end of right marginal row, relatively fine, 10–15 μm long. Midventral complex composed of about 12 cirral pairs on average (Table 19), terminates at about 66% of body length; as in Kahl's population, left cirrus of pair obviously smaller than right; at rear end of complex possibly a short midventral row (not described, but possibly present in specimen illustrated, Fig. 87f). Usually five rather fine transverse cirri, only 12–15 μm long, project by about half their length (according to Song & Wilbert only inconspicuously) beyond rear body end (Fig. 87e). Right marginal row commences at level of frontoterminal cirri, terminates, like left one, at level of transverse cirri; marginal rows thus distinctly separated posteriorly; left row begins about at level of buccal vertex. Dorsal bristles about 2–3 μm long, invariably arranged in three almost bipolar kineties. Caudal cirri lacking (Fig. 87g).

Occurrence and ecology: Salt water species. Kahl (1932) discovered *Anteholosticha manca* in a mesosaprobic sample (15‰ salt content) collected in Oldesloe, a region in northern Germany. He found it sometimes abundantly mainly in the algal layer. The type locality, that is, the sample site of the neotype population is a fish- and mollusc-farming pond at the coast of the Yellow Sea at the city of Qingdao, China. Song & Wilbert (1997a) found it there several times, the population described was collected on 21.02.1994.

Records not substantiated by morphological data and/or illustrations (records by Agamaliev are listed under *Bakuella agamalievi*): saline(?) Lake Tekirghiol, Romania (Tucolesco 1962a, p. 813; 1965, p. 160); Bay of Kandalaksha, White Sea (Burkovsky 1970a, p. 190; 1970b, p. 11; 1970c, p. 56). Records from terrestrial habitats are very likely misidentifications (possibly confused with *A. intermedia*): litter from *Robinia*

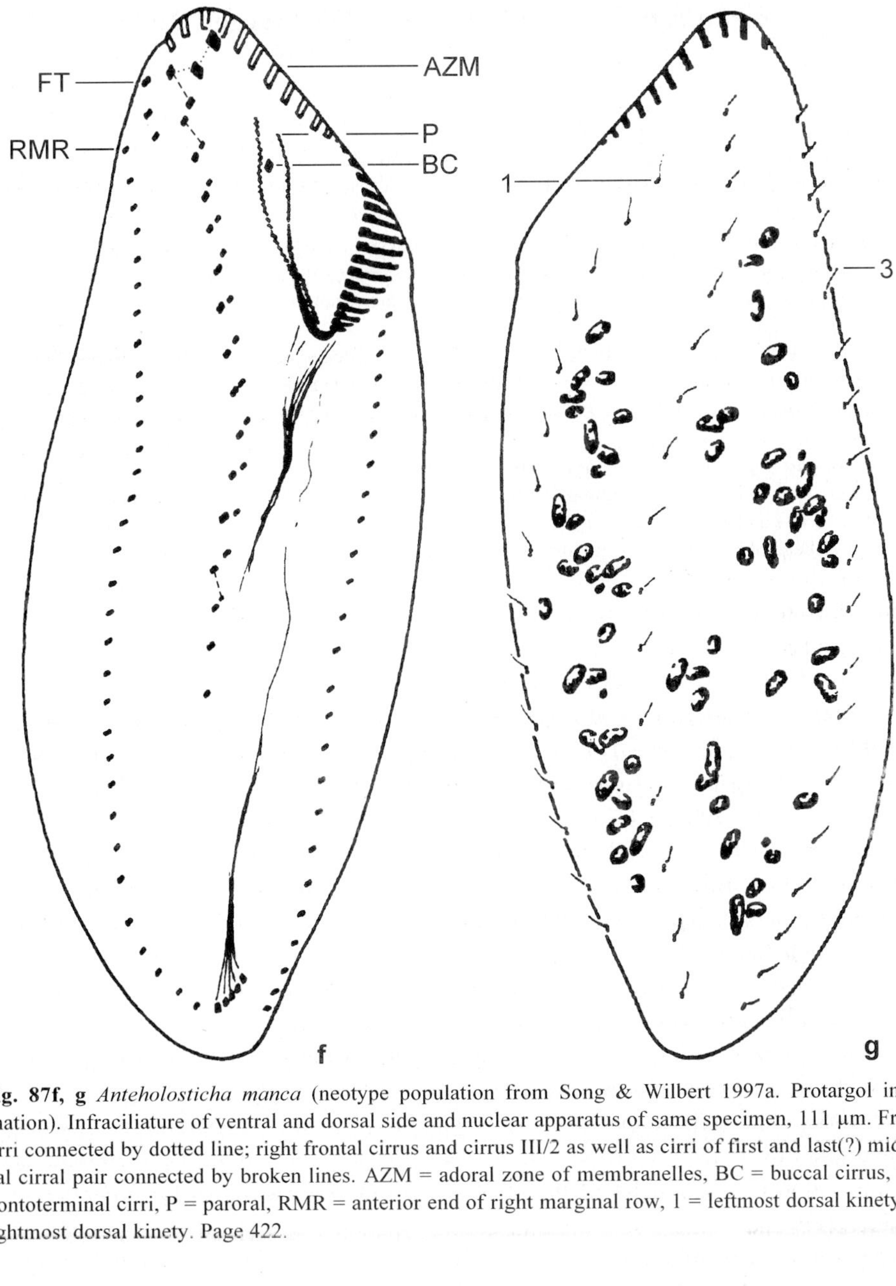

Fig. 87f, g *Anteholosticha manca* (neotype population from Song & Wilbert 1997a. Protargol impregnation). Infraciliature of ventral and dorsal side and nuclear apparatus of same specimen, 111 μm. Frontal cirri connected by dotted line; right frontal cirrus and cirrus III/2 as well as cirri of first and last(?) midventral cirral pair connected by broken lines. AZM = adoral zone of membranelles, BC = buccal cirrus, FT = frontoterminal cirri, P = paroral, RMR = anterior end of right marginal row, 1 = leftmost dorsal kinety, 3 = rightmost dorsal kinety. Page 422.

pseudoacacia and mixed woodland in Hungarian mountains (Varga 1960, p. 221); macchia (coppice forest) in Italy (Luzzatti 1938, p. 101).

Feeds on small diatoms (Kahl 1932).

Anteholosticha gracilis (Kahl, 1932) Berger, 2003
(Fig. 88a, b, d–o, Table 19)

1932 ***Keronopsis gracilis* spec. n.** – Kahl, Tierwelt Dtl., 25: 575, Fig. 110 10 (Fig. 88a; original description; no type material available and no formal diagnosis provided).
1933 ***Keronopsis gracilis* Kahl 1932** – Kahl, Tierwelt N.- u. Ostsee, 23: 109, Fig. 16.35 (Fig. 88b; guide to marine ciliates).
1936 ***Keronopsis gracilis* Kahl, 1932** – Kiesselbach, Thalassia, 2: 20, Abb. 41 (Fig. 88d; illustrated record).
1972 ***Keronopsis gracilis* Kahl, 1932** – Borror, J. Protozool., 19: 11 (revision of hypotrichs).
1990 ***Keronopsis gracilis* Kahl, 1930–5** – Carey, Marine interstitial ciliates, p. 184, Fig. 730 (redrawing from Kahl and thus not shown in present book; guide).
2001 ***Holosticha (Keronopsis) gracilis* Kahl, 1932** – Berger, Catalogue of ciliate names 1. Hypotrichs, p. 36 (nomenclator containing all basionyms, combinations, and higher taxa of hypotrichs).
2003 ***Anteholosticha gracilis* (Kahl, 1932) n. comb.** – Berger, Europ. J. Protistol., 39: 377 (combination with *Anteholosticha*).
2004 ***Holosticha nagasakiensis* sp. nov.**[1] – Hu & Suzuki, J. mar. biol. Ass. U.K., 84: 9, Fig. 1A–I, 2A–J, Tables 1, 2 (Fig. 88e–o; original description of synonym; 1 holotype slide (accession number 2002:12:01) and 1 paratype slide (2002:12:02) have been deposited in the Laboratory of Protozoology, College of Fisheries, Ocean University of China).

Nomenclature: No derivation of the name is given in the original description. The species-group name *grácil·is -is -e* (Latin adjective; thin, soft, slender, delicate) likely alludes to the delicate general appearance of this species. Hu & Sudzuki's species is named after the location (City of Nagasaki, Japan) where the species was discovered.

Kahl (1932, 1933) classified *Keronopsis* as subgenus of *Holosticha*. Thus, the correct name in his reviews is *Holosticha (Keronopsis) gracilis* Kahl, 1932. This was overlooked by all later authors (Vuxanovici 1961, Dragesco 1965, Borror 1972, Borror & Wicklow 1983, Dragesco & Dragesco-Kernéis 1986, Carey 1991), who assumed that Kahl (1932) had described this species in the genus *Keronopsis*. Thus, Vuxanovici (1961) could be considered as combining author for the combination *Keronopsis gracilis*, although neither he nor any other author transferred it to *Keronopsis* formally. However, since the classification in *Keronopsis* is in any case incorrect, a more detailed discussion of this problem is superfluous.

Holosticha gracilis Vuxanovici, 1963 is a junior primary homonym which has to be replaced (see *Anteholosticha vuxgracilis*).

Remarks: The present species lacks caudal cirri and all(?) apomorphies of *Holosticha* and was thus transferred to *Anteholosticha* (Berger 2003). For a detailed discussion and foundation of the transfer of many species from *Holosticha* to *Anteholosticha* see genus section.

The original description of this species is rather short. Kahl (1932) did not recognise the nuclear apparatus, indicating that several to many macronuclear nodules are present.

---

[1] The diagnosis by Hu & Suzuki (2004) is as follows: Medium-sized marine *Holosticha*, in vivo 100–150 × 30–40 μm with elongated body shape and slightly greyish to reddish cell colour; cortical granules arranged in conspicuously longitudinal rows; 24–32 adoral membranelles; ~20 midventral cirral pairs and 7–10 transverse cirri; 18–42 left and 25–38 right marginal cirri; constantly 4 frontal cirri and 3 dorsal kineties; 23–63 macronuclear nodules.

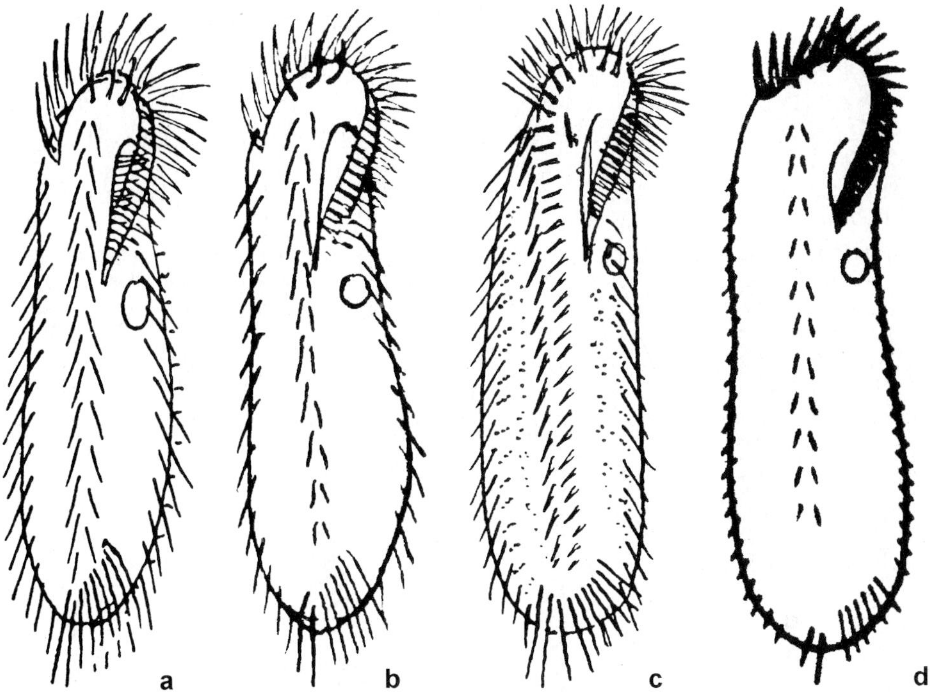

Fig. 88a, b, d *Anteholosticha gracilis* from life (a, from Kahl 1932; b, from Kahl 1933; d, from Kiesselbach 1936a). Ventral views showing, inter alia, body outline, basic cirral pattern, and contractile vacuole, a, b = 120 µm, d = 115 × 35 µm. Page 426.

Fig. 88c *Holosticha* spec. from life (from Kahl 1932). Ventral view, size not indicated. According to Kahl (1932) perhaps identical with *Anteholosticha gracilis*. Main differences are the frontal ciliature (single corona vs. three cirri) and the buccal cirrus (illustrated vs. not illustrated). The small dot patches are likely groups of cortical granules. *Holosticha* spec., which is certainly not a true *Holosticha* species, is a marine sand form likely from Heligoland, Northern Sea. Page 426, 430.

Kahl (1933) provided a new illustration, but no new data, for example, about the nuclear apparatus. Detailed redescription of a saltwater population necessary.

Keronopsis gracilis Kahl sensu Dragesco (1965, p. 395) lacks distinct frontal cirri, has only 5–7 transverse cirri, and the cortical granules are rather blue than yellowish-red, indicating that Dragesco's identification is incorrect (Fig. 89a–d). Unfortunately, I do not have a better idea and since the cirral pattern is not described in detail, I avoid another identification, the establishment of a new species, or the assignment to an already existing or new genus. I simply provide Dragesco's data and illustration (see below). Studies on the ciliate fauna from the sampling site (Lagoon of Abidjan, Ivory Coast) will possibly show that it is a distinct species, as already supposed by Dragesco (1965).

According to Borror (1972) and Hemberger (1982), *Keronopsis thononensis* Dragesco, 1966 is a junior synonym of *A. gracilis*. However, this species lives in freshwater

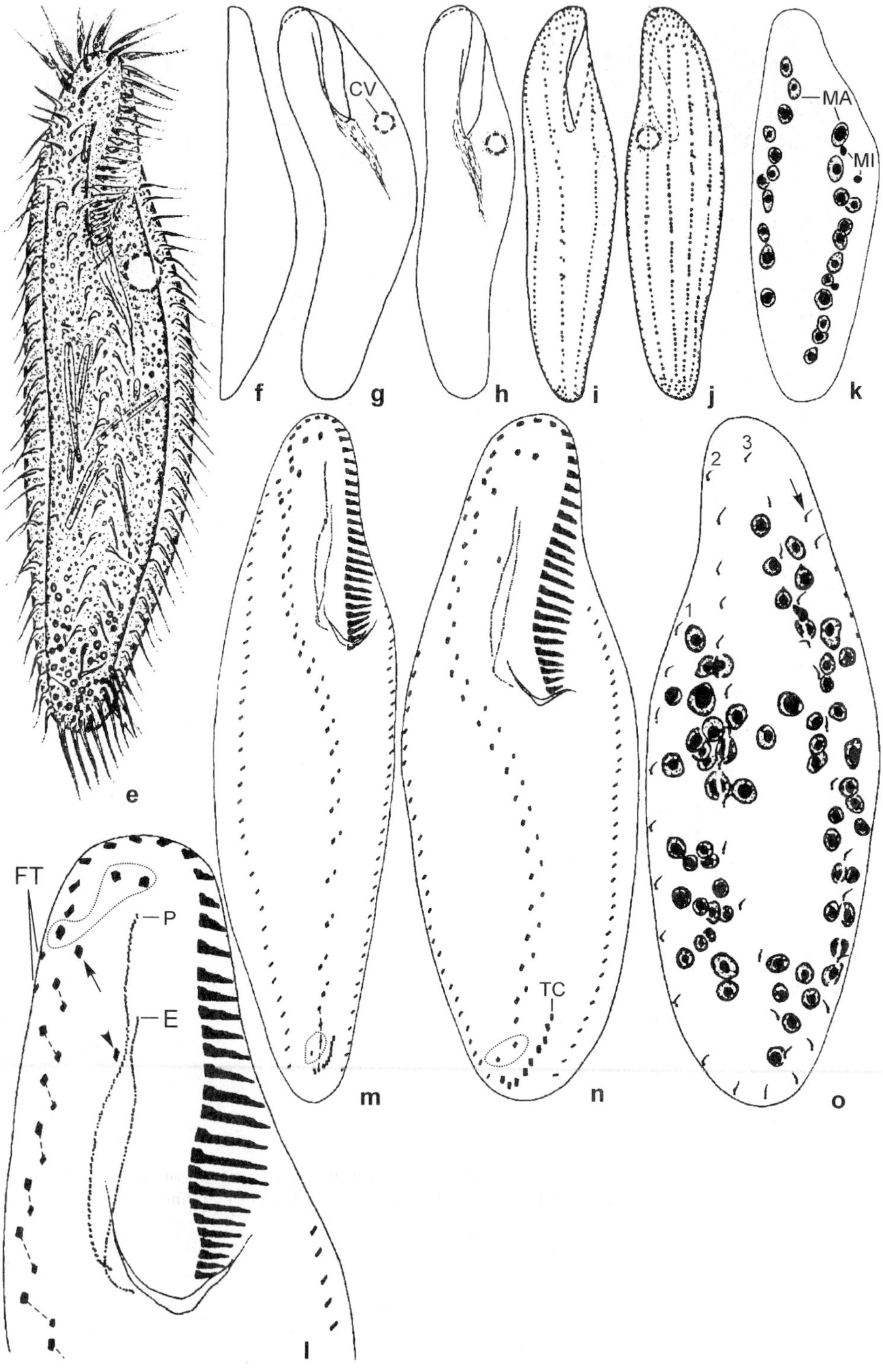
CV
MA
MI
f
g
h
i
j
k
e
1
2
3
FT
P
E
TC
l
m
n
o

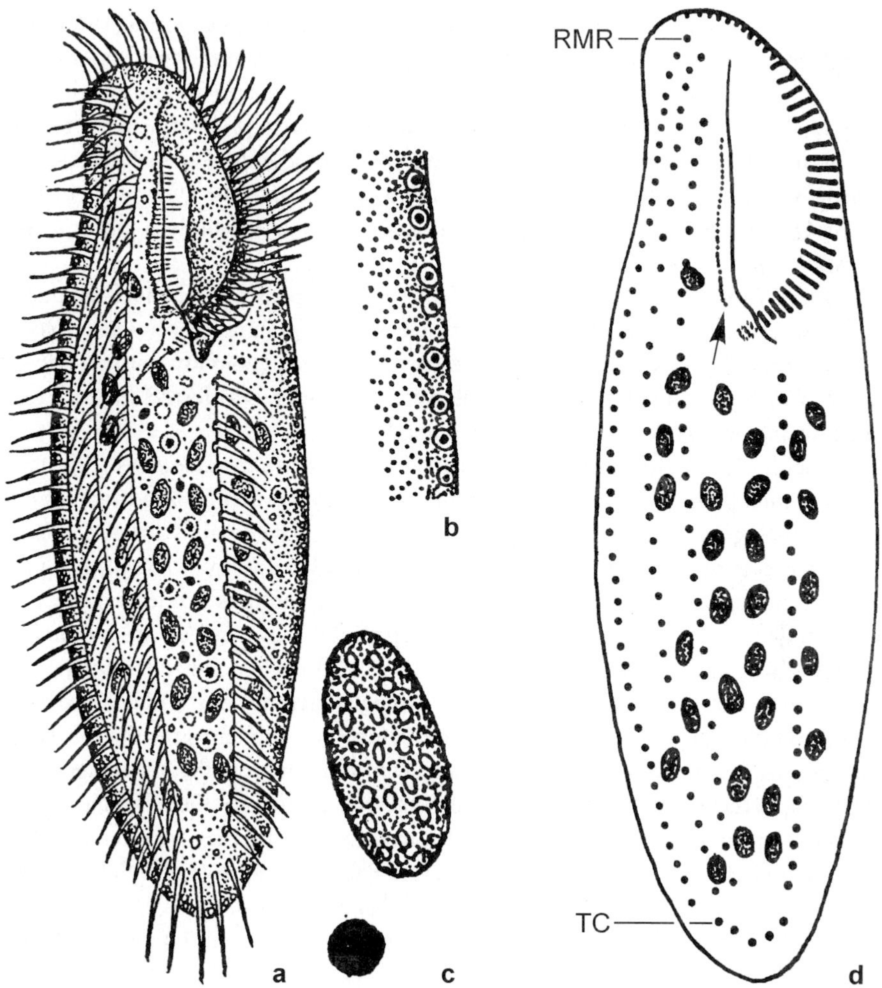

Fig. 89a–d *Keronopsis gracilis* sensu Dragesco (a–c, from Dragesco 1965; d, modified after Dragesco 1965 from Dragesco & Dragesco-Kernéis 1986. a–d, from life). **a, d:** Ventral view (108 µm) showing, inter alia, cirral pattern and nuclear apparatus. Arrow in (d) denotes rear end of paroral. The specimen illustrated has about 40 adoral membranelles, 22 left and 44 right marginal cirri, and 31 cirri in the right and 29 cirri in the left ventral row, that is, about 30 midventral pairs. **b:** This population has small vacuoles (1.5–2.5 µm across) containing tiny, blue granules, obviously a special kind of cortical granulation. **c:** Macronuclear nodule and micronucleus. RMR = anterior end of right marginal row, TC = rightmost transverse cirrus. Page 432.

Fig. 88e–o *Anteholosticha gracilis* (from Hu & Suzuki 2004. e–j, from life; k–o, protargol impregnation). **e–j:** Ventral (e, g–i), left lateral (f), and dorsal (j) views showing inter alia, body outline, contractile vacuole, and cortical granulation, e = 135 µm. **k:** Nuclear apparatus in dorsal view. **l–o:** Infraciliature of ventral and dorsal side and nuclear apparatus, m = 132 µm, n, o = 110 µm. Frontal cirri circled in (l), arrow in (l) marks cirrus III/2, arrowhead denotes buccal cirrus. Midventral pairs connected by broken lines. Pretransverse ventral cirri circled in (m, n). Arrow in (o) marks dorsal bristles ahead of right marginal row. CV = contractile vacuole, E = endoral, FT = frontoterminal cirri, MA = macronuclear nodules, MI = micronuclei, P = paroral, TC = transverse cirri, 1–3 = dorsal kineties. Page 426.

(vs. marine) and has only six transverse cirri (vs. nine). Further, Dragesco (1966) did not mention the yellow-red cortical granules, strongly indicating that the two populations are not conspecific. I consider *K. thononensis* as distinct species and assign, for the sake of simplicity, all limnetic records of *Anteholosticha gracilis* to *A. thononensis*. However, there is no doubt that further populations have to be studied before a final decision.

Borror & Wicklow (1983, p. 122) put Kahl's species into the synonymy of *Oxytricha velox* Quennerstedt. However, *Oxytricha velox* is rather certainly a junior synonym of *Holosticha gibba*.

According to Kahl (1932) *Anteholosticha gracilis* is possibly identical with a *Holosticha* sp. described by himself (Fig. 88c). However, this population obviously has more than three frontal cirri, which form a single corona. Thus, synonymy with an *Anteholosticha* or *Holosticha* species is very unlikely. The reader is referred to the Fig. 88c and the few data available presented in the legend.

Just recently Hu & Suzuki (2004) described *Holosticha nagasakiensis* (Fig. 88e–o). I used the *Anteholosticha* key of the present book and found that *H. nagasakiensis* is basically not distinguishable from *Anteholosticha gracilis*. The "main" difference is in the colour of the cortical granules, which are yellowish-red according to Kahl, but yellow-greenish according to Hu & Suzuki. However, the latter authors wrote that the cortical granules render *H. nagasakiensis* slightly greyish to reddish in colour at low magnification, making this difference irrelevant. Kahl (1932) did not recognise the nuclear apparatus indicating that, as in *H. nagasakiensis*, many macronuclear nodules are present. Moreover, he did not illustrate a buccal cirrus. However, this cirrus is obviously rather delicate so that it is understandable that Kahl, who did not have the advantage of silver preparations, overlooked it. Hu & Suzuki (2004) compared their species with 12 "*Holosticha*" species, but not with *A. gracilis*, explaining why they overlooked the great similarity between their and the present species. The illustration (Fig. 88e) and the micrographs shown in Hu & Suzuki(2004) show that *H. nagasakiensis* is rather delicate, indicating that Kahl (1932) selected the species-group name (*gracilis*) very well (see nomenclature).

Keronopsis gracilis sensu Vuxanovici (1961) from a freshwater lake is rather superficially described. I do not accept the identification, but mention it under the chapter insufficient redescriptions (Fig. 100e). For separation of *A. gracilis* from other *Anteholosticha* species, see key.

Morphology: The following description is bipartite because I do not combine the description of the two synonyms (Kahl 1932, Hu & Suzuki 2004). Kiesselbach (1936a) provided only a small illustration without description.

Body length of Kahl's specimens about 120 µm in life, ratio of body length:width almost 4:1 (Fig. 88a, b). Body outline elongate elliptical. Body very soft, flexible, and slightly contractile. Nuclear apparatus not known ("nucleus not ascertained"; see original figure legend). Contractile vacuole near left body margin slightly behind level of buccal vertex. Cortical granules globular (because termed "pearls"), yellowish-red, arranged in rows. Adoral zone occupies about 38% of body length (Fig. 88a). Buccal field narrow, buccal lip distinctly curved anteriorly. Three enlarged frontal cirri. Buccal cirrus neither mentioned nor illustrated (lacking or overlooked?). Midventral-complex composed of cirral pairs, extends from near frontal cirri to transverse cirri. Specimen illustrated with nine obliquely arranged transverse cirri, right one projects far, left one slightly beyond rear body end. Marginal rows obviously without peculiarities. Dorsal bristles conspicuously widely spaced and short (the illustration, however, shows more or less normally spaced dorsal bristles). Caudal cirri neither mentioned nor illustrated, indicating that they are lacking.

Description of synonym *Holosticha nagasakiensis* (Fig. 88e–o, Table 19): Body size 100–150 × 30–40 µm, body length:width ratio 3–4:1. Body outline usually as shown in Fig. 88e, that is, right margin almost straight, left conspicuous convex, widest usually at mid-body; both ends narrowly rounded. Body dorsoventrally flattened about 3:1 (Fig. 88f) and flexible (Fig. 88g, h). On average about 40–50 macronuclear nodules (Table 19), scattered throughout cell; however, central portion sometimes free of macronuclear nodules (Fig. 88k, o); individual nodules slightly ellipsoidal, usually with only one large nucleolus. 1–7 micronuclei near macronuclear nodules, individual micronuclei globular to ellipsoid, about 1–2 µm long. Contractile vacuole in ordinary position, that is, slightly behind level of buccal vertex at about 1/3 of body length near left cell margin (Fig. 88e, g, h, j). Pellicle soft. Cortical granules conspicuous because about 0.5–1.0 µm across, yellow-greenish at high magnification, arranged in longitudinal row, and densely packed at both cell ends; granules make cells slightly greyish to reddish at low magnification (Fig. 88i, j). Cytoplasm colourless, containing many globules 1–5 µm across. Movement without peculiarities, that is, crawling on debris or substrate.

Adoral zone occupies about 34% of body length in protargol preparations (Table 19), composed of about 28 membranelles of ordinary fine structure (Fig. 88l–n); cilia of membranelles up to 15 µm long. Buccal field moderately wide, right margin bordered by paroral, which is usually distinctly longer than endoral; paroral extends from near left frontal cirrus to level of proximalmost membranelle. Membranes almost straight in anterior portion, distinctly curved leftwards in rear portion.

Cirral pattern of usual variability (Fig. 88l–n, Table 19). Three slightly enlarged frontal cirri, and cirrus III/2 (= parabuccal cirrus) left-behind right frontal cirrus which is very close to distal end of adoral zone of membranelles. Buccal cirrus rather delicate and sometimes very difficult to recognise even in protargol preparations (Fig. 88–n; thus one must not wonder that Kahl overlooked it, see above); arranged slightly behind anterior end of endoral. Invariably two frontoterminal cirri near distal end of adoral zone of membranelles. Midventral complex composed of cirral pairs only, extends from level of frontoterminal cirri to near transverse cirri, slightly sigmoidal, composed of about 20 cirral pairs; right cirrus of each pair slightly larger than left. Likely two pretransverse ventral cirri present. On average 8–9 transverse cirri arranged in roughly J-shaped figure close to rear body end; cirri only inconspicuously stronger than remaining cirri, protrude distinctly beyond rear body end. Right marginal row commences slightly behind anterior end of midventral complex, terminates about at level of rearmost transverse cirri. Left marginal row begins distinctly ahead of level of proximal end of adoral zone, ends about at same level as right row. Marginal cirri composed of two rows of basal bodies. Ahead of right marginal row invariably 2–3 basal body pairs with dorsal bristles (Fig. 88o).

Dorsal cilia about 3 μm long, invariably arranged in three kineties; kinety 1 distinctly shortened anteriorly, rows 2 and 3 bipolar. Caudal cirri lacking (Fig. 88o).

Occurrence and ecology: Likely confined to saltwater habitats. The type locality of *A. gracilis* is a flat, mesosaprobic marine pool at the German island of Helgoland, where Kahl (1932) discovered it with low abundance in the detritus. Type locality of the synonym *Holosticha nagasakiensis* is the Nagasaki New Fishing Port, Nagasaki, Japan (Hu & Suzuki 2004). Kiesselbach (1936, p. 11; 1936a) found it on mud from the island Torcello, Lagoon of Venice, northern Adriatic Sea.

Records from saltwater habitats not substantiated by morphological data: Bay of Marseilles, Mediterranean Sea (Vacelet 1960, p. 54; 1961, p. 3); western coast of Caspian Sea (Agamaliev 1971, p. 383).

Records of "*Keronopsis gracilis*" from limnetic habitats are assigned to *Anteholosticha thononensis*, a freshwater species synonymised with the marine *A. gracilis* by Borror (1972; see remarks). Feeds on bacteria and diatoms (Hu & Suzuki 2004).

Keronopsis gracilis Kahl, 1932 sensu Dragesco (1965) (Fig. 89a–d)

1965 ***Keronopsis gracilis* Kahl (?)** – Dragesco, Cah. Biol. mar., 6: 395, Fig. 31a–c (Fig. 89a–c; redescription).

1968 ***Keronopsis gracilis* Kahl** – Grell, Protozoologie, p. 445, Abb. 407 (very good redrawing after Dragesco 1965; textbook).

1986 ***Keronopsis gracilis* Kahl, 1932** – Dragesco & Dragesco-Kernéis, Faune tropicale, 26: 442, Planche 129G (Fig. 89d; review of African ciliates).

Remarks: See also same chapter at *Anteholosticha gracilis*. *Pseudokeronopsis multinucleata* and *P. decolor* have a very similar cirral pattern. However, *Pseudokeronopsis multinucleata* has brick-red cortical granules and many more (12–13) transverse cirri, and *P. decolor* has only two macronuclear nodules. However, as mentioned above, I avoid the assignment of the present population to a certain genus because the data are not very detailed.

Morphology: Body length 110–125 μm in life(?). Body dorsoventrally flattened, supple, and colourless. 25–31 macronuclear nodules and 3–5 micronuclei scattered in postperistomial body portion (Fig. 89a, c, d). Contractile vacuole not seen, likely lacking (Dragesco & Dragesco-Kernéis 1986). Cortical granules conspicuous because they look like real vacuoles 1.5 to 2.5 μm across containing a small, blue granule (Fig. 89b). Thigmotactic, swims very quickly. Adoral zone of specimen illustrated occupies about 35% of body length and composed of ca. 40 membranelles (Fig. 89d). Buccal field rather wide, right margin bordered by straight paroral (Fig. 89a, d). Enlarged frontal cirri lacking. Buccal cirrus neither mentioned nor illustrated after protargol preparations, strongly indicating that it is lacking. Three cirral rows right of midline; outermost row is the right marginal row which commences unusually near anterior body end and terminates near right end of transverse cirral row. Innermost two rows are very likely a midventral complex composed of cirral pairs only. 5–7 transverse cirri form bow-shaped pseudorow between end of marginal rows. Left marginal row commences slightly behind proximal end of adoral zone, terminates ahead of left transverse cirrus. Dorsal ciliature, namely number of dorsal kineties, length of dorsal bristles, presence/absence of caudal cirri, not known.

Occurrence and ecology: Dragesco (1965) found this species in the Lagoon of Abidjan, Ivory Coast, where it occurred with high abundance. Feeds on diatoms.

Anteholosticha pulchra (Kahl, 1932) Berger, 2003
(Fig. 90a, b, e–g)

1932 ***Keronopsis pulchra* spec. n.** – Kahl, Tierwelt Dtl., 25: 573, Fig. 104 5 (Fig. 90a; original description; no type material available and no formal diagnosis provided).

1933 ***Keronopsis pulchra* Kahl 1932** – Kahl, Tierwelt N.- u. Ostsee, 23: 109, Fig. 16.27 (Fig. 90b; revision of marine ciliates).

1963 ***Holosticha rubra* (Ehrenberg, 1838)** – Borror, Trans. Am. microsc. Soc., 82: 126, Plate I, Fig. 4–6 (Fig. 90e–g; misidentification).

2001 ***Holosticha (Keronopsis) pulchra* Kahl, 1932** – Berger, Catalogue of ciliate names 1. Hypotrichs, p. 36 (nomenclator containing all basionyms, combinations, and higher taxa of hypotrichs).

2003 ***Anteholosticha pulchra* (Kahl, 1932) n. comb.** – Berger, Europ. J. Protistol., 39: 377 (combination with *Anteholosticha*).

Nomenclature: No derivation of the name is given in the original description. The species-group name *pulch·er -ra -um* (Latin adjective; beautiful) possibly alludes to the large, dark-red cortical granules or to the beautiful general appearance of this species. Kahl (1932, 1933) classified *Keronopsis* as subgenus of *Holosticha*. Thus, the correct name in his reviews is *Holosticha (Keronopsis) pulchra* Kahl, 1932, respectively, *Holosticha pulchra* Kahl, 1932.

Remarks: The systematics of this species is somewhat confusing. According to Kahl (1932) it differs from *Pseudokeronopsis rubra* mainly in details of the cortical granulation. In addition, he wrote and illustrated that the three anteriormost frontal cirri are much more pronounced than in *P. rubra*. Further, in the legend to the corresponding figure (his Fig. 5 on his page 577) he wrote, likely par lapsus, *Keronopsis rubra* f. *heptasticha* (Fig. 90a; interestingly, in the original description of *Keronopsis rubra heptasticha*, Kahl made no reference to this illustration!). In his guide to marine ciliates, Kahl (1933) provided a new or slightly modified illustration which unequivocally shows three enlarged frontal cirri so that the frontal cirral pattern is under no circumstances reminiscent of that of *Pseudokeronopsis*. Thus, I do not understand and cannot accept the identifications made by Borror (1972, p. 11; Fig. 90c), Borror (1979, p. 547, his Fig. 4), and Borror & Wicklow (1983, p. 124; Fig. 90d) who transferred *H. pulchra* to *Keronopsis* (although nor formally), respectively, to *Pseudokeronopsis*. Their illustrations obviously show true *Pseudokeronopsis* species with a very pronounced bicorona, which would not have been overlooked by Kahl. Hemberger (1982, p. 103) even synonymised the present species with *Pseudokeronopsis rubra*.

As a consequence of this confusing history and the lack of reliable new data I simply transfer *Holosticha pulchra* to *Anteholosticha*, where it seems best classified, at least according to present state of knowledge. This basically agrees with the opinion of Wirnsberger et al. (1987, p. 85), who also classified "*Keronopsis pulchra*" sensu Kahl (1932) in *Holosticha*. Detailed redescription necessary.

Holosticha rubra sensu Borror (1963; Fig. 90e–g) lacks the distinct bicorona of *Pseudokeronopsis rubra*. Wirnsberger et al. (1987) thus did not accept Borror's identification, but left the population indefinite. I assign it preliminarily to *Anteholosticha pulchra* because the data agree well. Interestingly, both Kahl (1932) and Borror (1963) counted four dorsal kineties and did not illustrate a buccal cirrus. However, the lack of a buccal cirrus in these descriptions must not be over-interpreted because this feature is difficult to recognise in life (Kahl 1932), respectively, after wet silver impregnation (Borror 1963). Differences are present in the length of the midventral complex (almost of body length in Fig. 90a, b vs. distinctly shortened posteriorly in Fig. 90e). Kahl (1933, Fig. 90b) illustrated only three frontal cirri, whereas the specimens shown in Figs. 90a, e have small cirri behind the frontal cirri. Hemberger (1982, p. 111) assigned Borror's population to *Anteholosticha violacea* (Kahl, 1928).

The illustrations provided by Borror (1972) and Borror & Wicklow (1983) are mentioned under insufficient redescriptions since obviously no original life data were provided. Without these data about the cortical granulation and the colour of the cytoplasm, *Pseudokeronopsis* species cannot be identified reliably (Fig. 90c, d).

For separation from other *Anteholosticha* species, see key. *Pseudokeronopsis* species have a distinct bicorona or at least not three distinct (isolated) frontal cirri.

Morphology: The following description is based on the original description by Kahl (1932), respectively, his illustration. Supplementary data of Borror's (1963) population are added below. Body length of type population 200–300 µm, body length:width ratio of specimen illustrated 4.3:1; according to Kahl (1932) body outline very similar to *Pseudokeronopsis rubra*, but usually relatively wider; however, he also saw rather slender specimens. Many macronuclear nodules scattered throughout cytoplasm (not definitely described or illustrated, but this can be concluded from the introduction to the subgenus). Two types of cortical granules: (i) colourless granules (size not mentioned, but rather large because he designated them as coarse) in cortex, make cells, together with cytoplasmic granules, blackish to violet; (ii) dark-red granules of same coarse size as colourless granules at cirral bases and around dorsal bristles. Contractile vacuole neither mentioned nor illustrated, indicating that this organelle is lacking.

Adoral zone occupies about 25% of body length (Fig. 90a). Anteriormost three frontal cirri much stronger than remaining cirri on frontal field. Buccal cirrus neither mentioned nor illustrated, thus possibly lacking because Kahl illustrated this cirrus in similar species. Midventral complex extending from frontal cirri to near transverse cirri; specimen illustrated with about 30 cirral pairs (number must not be over-interpreted). Number of transverse cirri variable, seemingly always more than five (likely 6–9; interestingly, the illustration shows only 5 cirri [Fig. 90a]; later he illustrated 6 cirri [Fig. 90b]); protrude distinctly beyond rear body end. Invariably four dorsal kineties (Kahl

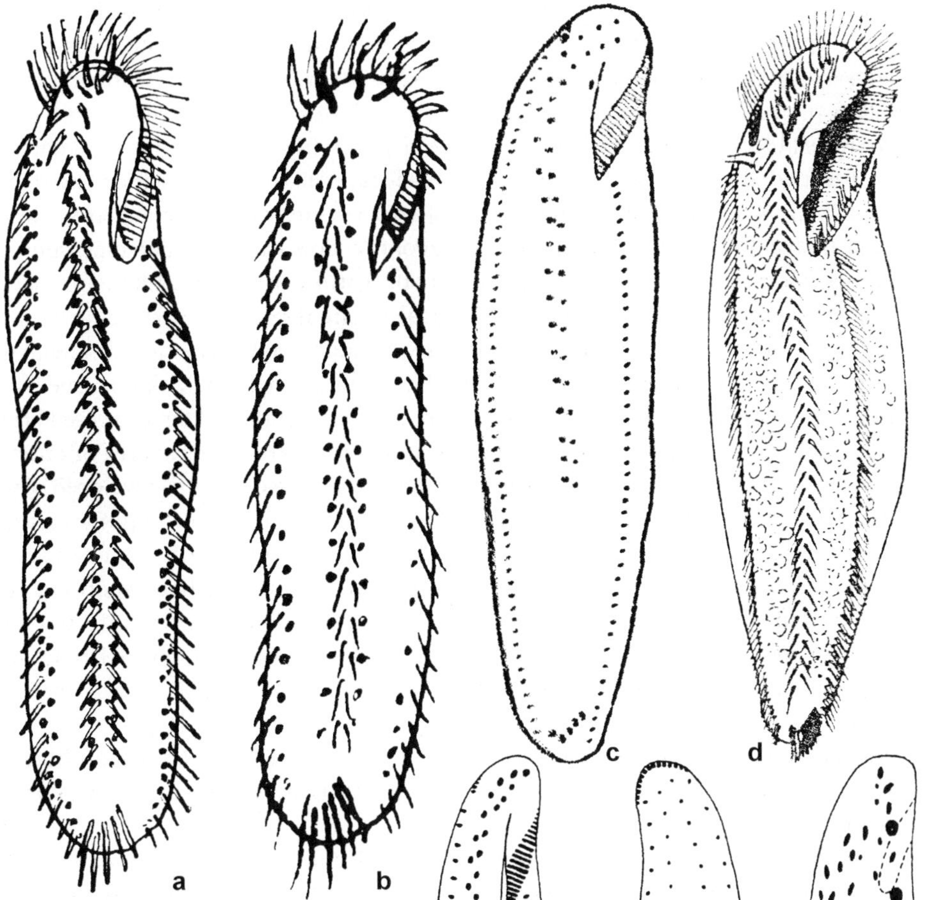

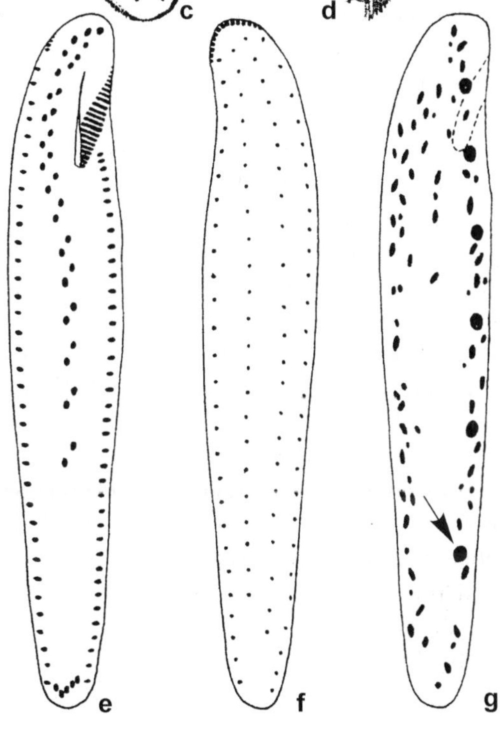

Fig. 90a, b, e–g *Anteholosticha pulchra* (a, from Kahl 1932; b, from Kahl 1933; e–g, from Borror 1963. From life). **a, b:** Ventral view showing, inter alia, cirral pattern and cortical granules near cirri, 200 µm. Note that Kahl illustrated three enlarged frontal cirri, preventing a classification in *Pseudokeronopsis* as suggested by Borror & Wicklow (1983). **e–g:** Infraciliature of ventral and dorsal side and nuclear apparatus, 150 µm. Arrow marks rearmost of 6 micronuclei, which are arranged near left cell margin. Page 433.

Fig. 90c, d *Keronopsis pulchra* and *Pseudokeronopsis pulchra* (c, from Borror 1972; d, from Borror & Wicklow 1983. c, d, protargol impregnation? nigrosin stain?). Infraciliature of ventral side, c = 151 µm. d = 200 µm. The distinct bicorona indicates that these specimens are not conspecific with *A. pulchra* (Fig. 90a, b); I classify them as insufficient redescriptions. Page 460.

saw the wreathes formed by the cortical granules); dorsal bristles, respectively, wreathes narrowly spaced (about 10 µm) within kineties and much more conspicuous than in *Pseudokeronopsis rubra*. Caudal cirri likely lacking because neither mentioned nor illustrated.

Borror's (1963) population 129–330 × 28–51 µm in life?; body worm-like, flattened orally, rounded aborally. Body outline parallel-sided, anterior and posterior end evenly rounded. Macronuclear nodules generally rod-like, 3–5 µm long, very numerous in lateral and midventral strings and scattered in aboral endoplasm. Left dorsolaterally 3–7 dense, spherical or oval micronuclei, each about 3 µm across (Fig. 90g). Contractile vacuole just posterior of buccal cavity. Cortical granules making cells yellow or red, particularly evident at bases of cirri, dorsal cilia, membranelles, and base of undulating membranes. Food vacuoles mostly in central core of endoplasm. Adoral zone 40 to 90 µm long, composed of 40–55 membranelles; each membranelle "three-layered" (possibly a misobservation due to inappropriate staining procedure). Buccal area of ordinary size, buccal lip covering posterior portion of adoral zone; paroral well developed. Cytopharynx 20–30 µm long. Cirral pattern as shown in Fig. 90e. Three slightly enlarged frontal cirri, indistinctly set off from midventral complex composed of about 13 pseudopairs. Behind frontal cirri three somewhat smaller cirri so that indistinct, short bicorona is formed. Buccal cirrus neither mentioned nor illustrated. Five transverse, 16 µm long, overlapping rear end by half their length. 20–43 right marginal cirri, about 10 µm long, 2 µm apart; left marginal row similar. Dorsal bristles 4 µm long, 8–15 µm apart, arranged in four kineties (Fig. 90f).

Walker et al. (1978) studied the macronuclear and micronuclear chromatin fibres by electron microscopy. Unfortunately, the identification is not substantiated by relevant morphological data.

Occurrence and ecology: Marine. According to Kahl *Anteholosticha pulchra* is less widely distributed than *Pseudokeronopsis rubra*. Kahl (1932) did not fix a type locality; he found it (not seldom) in debris collected at the German islands of Helgoland and Sylt (North Sea) and in the Baltic Sea near the city of Kiel. Borror (1963) found his "*Holosticha rubra*" in Alligator Harbor (Florida, USA) where it was uncommon in intertidal sand, but occasionally became abundant in old cultures. It occurred often in aquaria, on egg masses, oyster shells, or on the glass, where it fed on microflora. Borror cultured it by adding a few grains of rice to the culture dish.

Records not substantiated by morphological data: Dnjepr estuary, Black Sea (Kovalchuck 1989); Stoller Grund, a region in the Bay of Kiel, Baltic Sea (Bock 1952, p. 83); Rhode River, a tributary of the Chesapeake Bay in Maryland, USA (Walker et al. 1978).

Fernandez-Leborans (2001, p. 744) obviously found a population resembling *Pseudokeronopsis pulchra* sensu Borror & Wicklow (1983), which is, however, not identical with the present species (see remarks). The limnetic records from the Manzanares River in Madrid (Spain) by Fernandez-Leborans & Antonio-García (1988, p. 147) must be considered as misidentification.

Anteholosticha pulchra feeds on algae and diatoms (Borror 1963).

Anteholosticha longissima (Dragesco & Dragesco-Kernéis, 1986) comb. nov. (Fig. 91a–d)

1986 ***Keronopsis longissima* n. sp.** – Dragesco & Dragesco-Kernéis, Faune tropical, 26: 443, Planche 130, A–D (Fig. 91a–d; original description; no formal diagnosis provided; site where type slides deposited not mentioned, likely in private collection of J. Dragesco).

2001 ***Keronopsis longissima* Dragesco and Dragesco-Kernéis, 1986** – Berger, Catalogue of ciliate names 1. Hypotrichs, p. 44 (nomenclator containing all basionyms, combinations, and higher taxa of hypotrichs).

Nomenclature: No derivation of the name is given in the original description. The species-group name *longissima* is a composite of the Latin adjective *lóng·us -a -um* (long, wide) and the suffix *~issim·us -a -um* (superlative or at least a very high degree of a feature) and obviously refers to the elongate body shape.

Remarks: The distinct midventral pattern shows that this species does not belong to *Keronopsis*, but to a urostyloid genus. Since it lacks a distinct bicorona, midventral rows, and caudal cirri I transfer it to *Anteholosticha*. Dragesco did not provide a detailed description so that some important features (cortical granules present/absent; number of dorsal kineties) are not known. The authors neither mentioned nor illustrated a buccal cirrus, indicating that such a cirrus is lacking. The specific sample site (saline puddle in Benin) indicates that it is a distinct species. Detailed redescription needed.

Morphology: Body size 68–118 × 14–19 μm in protargol preparations. Body outline elongate, margins slightly converging posteriorly. On average 32 macronuclear nodules scattered in central body portion (Fig. 91a); individual nodules ovoid, 4–7 μm long. 4–8 micronuclei about 2.2 μm across. Contractile vacuole and cortical granules neither mentioned nor illustrated. Adoral zone occupies 30% of body length in specimen illustrated (Fig. 91b), composed of 37–48, on average 41 (n = 7) membranelles. Paroral short (Fig. 91a); endoral likely overlooked, possibly because not impregnated. Cirral pattern likely not quite correctly illustrated (Fig. 91a, b). I suppose that it has a more or less ordinary *Anteholosticha* pattern, that is, (i) three enlarged frontal cirri, (ii) one cirrus (= III/2) behind right frontal cirrus, (iii) a midventral complex extending from frontal cirri to near transverse cirri (complex composed of 56–63, on average 60 [n = 4] cirri in total [specimen illustrated in Fig. 91b with only 52 cirri]; zigzag-pattern in anterior portion of complex lacking, possibly due to unusual arrangement of cirri or due to lack [resorption] of the left/right cirrus in each pair), and (iv) three small transverse cirri distinctly projecting beyond rear body end (it is unlikely that these are caudal cirri because they are slightly subterminally on ventral side). A buccal cirrus is neither mentioned nor illustrated, indicating that it is lacking. Frontoterminal cirri also neither mentioned nor illustrated; possibly confused with anterior end of right marginal row as indicated by the far anteriorly extending right row. Marginal rows more or less distinctly separated posteriorly (Fig. 91a, b); 63–69 (mean = 66; n = 5) right marginal cirri (specimen illustrated in Fig. 91b with only 46 cirri!), 40–55 (mean = 51, n = 7) left marginal cirri (specimen illustrated with only 39 cirri); left marginal row commences far ahead of proximal end of adoral zone (Fig. 91b). Length of dorsal bristles and number of kineties not known. Caudal cirri very likely lacking.

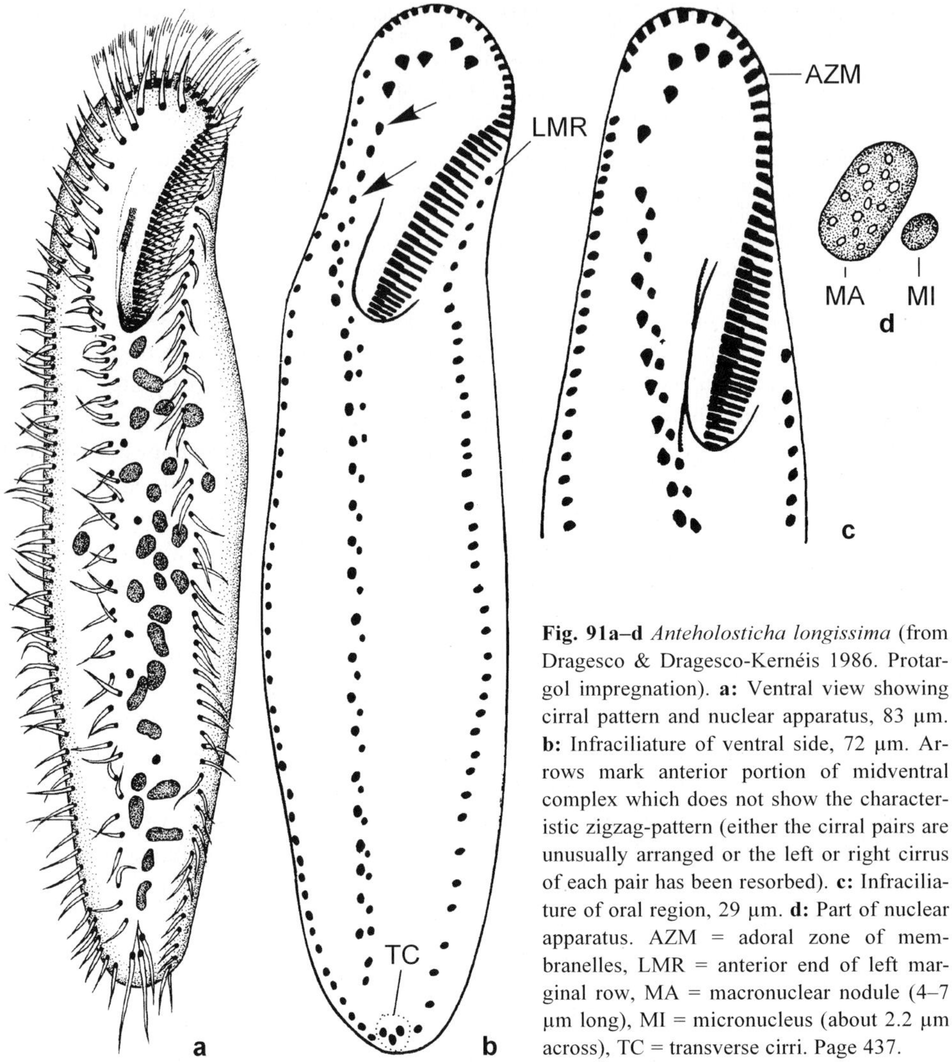

Fig. 91a–d *Anteholosticha longissima* (from Dragesco & Dragesco-Kernéis 1986. Protargol impregnation). **a:** Ventral view showing cirral pattern and nuclear apparatus, 83 μm. **b:** Infraciliature of ventral side, 72 μm. Arrows mark anterior portion of midventral complex which does not show the characteristic zigzag-pattern (either the cirral pairs are unusually arranged or the left or right cirrus of each pair has been resorbed). **c:** Infraciliature of oral region, 29 μm. **d:** Part of nuclear apparatus. AZM = adoral zone of membranelles, LMR = anterior end of left marginal row, MA = macronuclear nodule (4–7 μm long), MI = micronucleus (about 2.2 μm across), TC = transverse cirri. Page 437.

Occurrence and ecology: Type locality of *A. longissima* is a flat, saline puddle near Cotonou Lagoon, Benin (Dragesco & Dragesco-Kernéis 1986).

Anteholosticha extensa (Kahl, 1932) Berger, 2003 (Fig. 92a–d)

1932 ***Holosticha extensa* spec. n.** – Kahl, Tierwelt Dtl., 25: 578, Fig. 105 (Fig. 92a; original description; no type material available and no formal diagnosis provided).

1933 ***Holosticha extensa* Kahl 1932** – Kahl, Tierwelt N.- u. Ostsee, 23: 109, Fig. 17.17 (Fig. 92b; guide to marine ciliates).

1972 ***Holosticha extensa* Kahl, 1932** – Borror, J. Protozool., 19: 11 (revision of hypotrichs).
1985 ***Holosticha extensa* Kahl** – Madrazo-Garibay & López-Ochoterena, Anales Instituto de Ciencas del Mar., 12: 206, Fig. 30 (Fig. 92d; brief description of Mexican population).
1990 ***Holosticha extensa* Kahl, 1932** – Aladro Lubel, Martínez Murillo & Mayén Estrada, Manual de Ciliados, p. 130, 1 figure on same page (Fig. 92c; review of Mexican saltwater ciliates).
2001 ***Holosticha (Holosticha) extensa* Kahl, 1932** – Berger, Catalogue of ciliate names 1. Hypotrichs, p. 35 (nomenclator containing all basionyms, combinations, and higher taxa of hypotrichs).
2003 ***Anteholosticha extensa* (Kahl, 1932) n. comb.** – Berger, Europ. J. Protistol., 39: 377 (combination with *Anteholosticha*).

Nomenclature: No derivation of the name is given in the original description. The species-group name *exténs·us -a -um* (Latin adjective; long, extended) refers to the long, that is, slender body. Kahl (1932, 1933) divided *Holosticha* into several subgenera. Thus, the correct name in his papers is *Holosticha (Holosticha) extensa* Kahl, 1932. *Holosticha extenza* in Azovsky et al. (1996, p. 30) is an incorrect subsequent spelling.

Remarks: The present species lacks caudal cirri and all (?) apomorphies of *Holosticha* and was thus transferred to *Anteholosticha* (Berger 2003). For a detailed discussion and foundation of the transfer of many species from *Holosticha* to *Anteholosticha* see genus section.

This species is not yet redescribed by modern methods. However, it shows a rather unique combination of features (140–240 µm long, that is, rather large and very slender body; conspicuous nuclear apparatus, see below; rather long [about 10 µm] dorsal bristles) so that the validity of this species is beyond reasonable doubt. In spite of this, detailed redescription needed.

Borror & Wicklow (1983, p. 121) put it, together with many other species, into the synonymy of *Anteholosticha intermedia*. However, they differ in several important features, for example, habitat (marine vs. freshwater), nuclear apparatus (about eight macronuclear nodules with four micronuclei between each two nodules in left body portion vs. many scattered macronuclear nodules), and body shape (body length:width ratio about 7:1 vs. 3.8:1).

The descriptions by Madrazo-Garibay & López-Ochoterena (1985) and the review by Aladro Lubel et al. (1990) basically confirm Kahl's data. Since they do not provide new information the reader is mainly referred to their not very detailed illustrations (Fig. 92c, d).

The large ratio of body length to width (7:1), the conspicuous nuclear apparatus, and the long dorsal bristles make this species unmistakable and rather easy to determine. *Anteholosticha fasciola*, which also lives in marine habitats, is longer (200 to 300 µm) and even has a body length:width ratio of 10:1, many scattered macronuclear nodules, and very conspicuous cortical granules, especially along the marginal rows (Fig. 93a).

Morphology: The following description is based solely on Kahl's data: Body length 140–240 µm, body length:width ratio about 7:1, body width thus 20–35 µm. Body outline parallel-sided with anterior (oral) portion slightly (trapezoidally) narrowed, rear body end rounded (Fig. 92a); anterior body portion slightly contractile. Nuclear apparatus conspicuous because 6–8 macronuclear nodules linearly arranged in left body portion; usually four micronuclei, one between each two macronuclear nodules

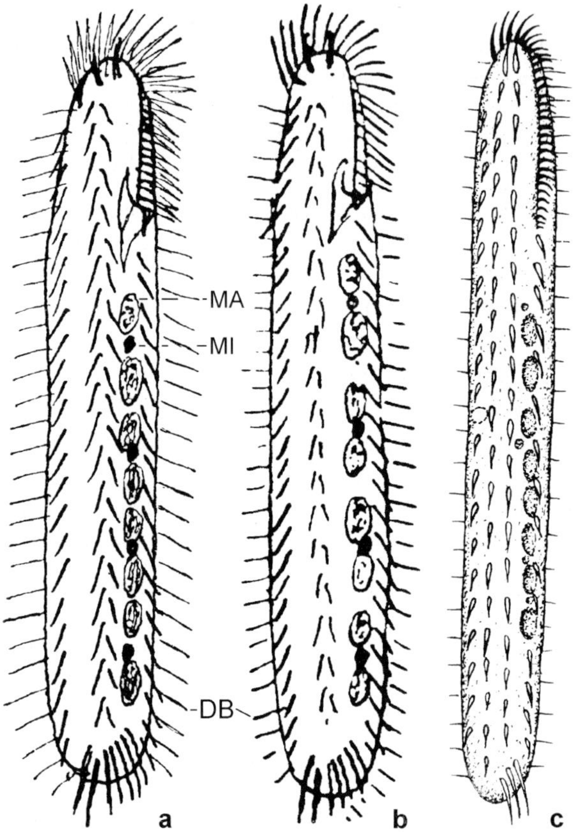

Fig. 92a–d *Anteholosticha extensa* (a, from Kahl 1932; b, after Kahl 1932 (?) from Kahl 1933; c, from Aladro Lubel et al. 1990; d, from Madrazo Garibay & López-Ochoternea 1985. a–d, from life). Ventral views showing, inter alia, cirral pattern and nuclear apparatus, a, b = 140 to 240 µm, c, d = 180 µm. CV = contractile vacuole (location on right body margin is certainly a misobservation), DB = dorsal bristles, MA = macronuclear nodule, MI = micronucleus. Page 438.

(Fig. 92a). Contractile vacuole neither mentioned nor illustrated. Distinct cortical granules likely lacking because not mentioned; however, cells colourless to grey, underneath ectoplasm a layer of longish granules (mitochondria?). Movement not mentioned.

Adoral zone occupies about 20% of body length, roughly in *Gonostomum* pattern; that is, proximal portion extends along left body margin and then abruptly bends towards body centre. Buccal lip short, buccal field tiny and narrow. Two (or three?) enlarged frontal cirri. A single cirrus ahead of the lip (buccal cirrus? obviously not illustrated!). Midventral complex composed of cirral pairs (about 19 pairs present in specimen shown in Fig. 92a; value must not be over-interpreted, likely distinctly too low because body is rather long), extends from near enlarged frontal cirri to near transverse cirri. 6–7 transverse cirri which protrude by about 50% of their length beyond body margin. Dorsal cilia long and fine; if length of specimen illustrated assumed as 200 µm then bristles about 10 µm long (Fig. 92a).

Specimens observed by Madrazo-Garibay & López-Ochoterena (1985) 180 × 20 µm (body length:width ratio 9:1!; according to Fig. 92d, however, only 4.8:1 indicating superficial observations). Contractile vacuole about in mid-body, however, located near right body margin, which is certainly a misobservation.

Occurrence and ecology: Marine. Type locality is a marine sand at Bülk (Baltic Sea) near the German city of Kiel, where the species sometimes occurred abundantly. Madrazo-Garibay & López-Ochoterena (1985) found in the Laguna de Términos, Campeche (19°51'N, 90°32'W), Mexico (for review, see Aladro Lubel et al. 1988, p. 437; 1990).

Records not substantiated by morphological data: Mediterranean Sea in Marseilles, France (Vacelet 1961, p. 3); Kandalaksha Bay, White Sea (Burkovsky 1971a, p. 1774; Azovsky et al. 1996, p. 30).

Anteholosticha fasciola (Kahl, 1932) Berger, 2003 (Fig. 93a, b)

1932 ***Holosticha fasciola* spec. n.** – Kahl, Tierwelt Dtl., 25: 578, Fig. 106 ₃ (Fig. 93a; original description; no type material available and no formal diagnosis provided).

1933 ***Holosticha fasciola* Kahl 1932** – Kahl, Tierwelt N.- u. Ostsee, 23: 109, Fig. 107.2 (Fig. 93b; guide to marine ciliates).

2001 ***Holosticha (Holosticha) fasciola* Kahl, 1932** – Berger, Catalogue of ciliate names 1. Hypotrichs, p. 35 (nomenclator containing all basionyms, combinations, and higher taxa of hypotrichs).

2003 ***Anteholosticha fasciola* (Kahl, 1932) n. comb.** – Berger, Europ. J. Protistol., 39: 377 (combination with *Anteholosticha*).

Nomenclature: No derivation of the name is given in the original description. The species-group name *fasciola* (Latin; the small band) obviously refers to the band-shaped outline of the body (Fig. 93a). Kahl (1932, 1933) divided *Holosticha* into several subgenera. Thus, the correct name in his papers is *Holosticha (Holosticha) fasciola* Kahl, 1932. "*Holostricha fasciola* Kahl" is an incorrect subsequent spelling (Guillén et al. 2003, p. 180).

Remarks: The illustrations provided by Kahl (1932, 1933) are slightly different especially as concerns the ciliature of the posterior body end. In Fig. 93a three very fine, but distinctly elongated (caudal?) cirri are shown which are lacking in Fig. 93b. Kahl (1932) did not mention caudal cirri, indicating that he was uncertain about the presence or absence of this cirral group. Since he omitted these cirri in his 1933 review, I assign *H. fasciola* to *Anteholosticha* and not to *Caudiholosticha*, which is characterised by the presence of caudal cirri. Detailed redescription necessary.

Borror (1972, p. 11) synonymised the present species with *A. violacea* (see next paragraph); by contrast Borror & Wicklow (1983) obviously overlooked it because I did not find it in their lists of species and synonyms.

Anteholosticha extensa, which is slightly broader (about 7:1) and smaller (140 to 240 µm) than *A. fasciola* (10:1; 200–300 µm), has, inter alia, long dorsal bristles (10 µm vs. obviously short, that is, around 3 µm), only 6–8 macronuclear nodules left of body midline with one micronucleus between each two macronuclear nodules (against many elongate nodules), and lacks cortical granules (vs. present). *Anteholosticha violacea* is also slightly wider (7–8:1), but has about 8 µm long dorsal bristles and is usually purple due to ingested rhodobacteria (Fig. 67a). *Anteholosticha grisea* is

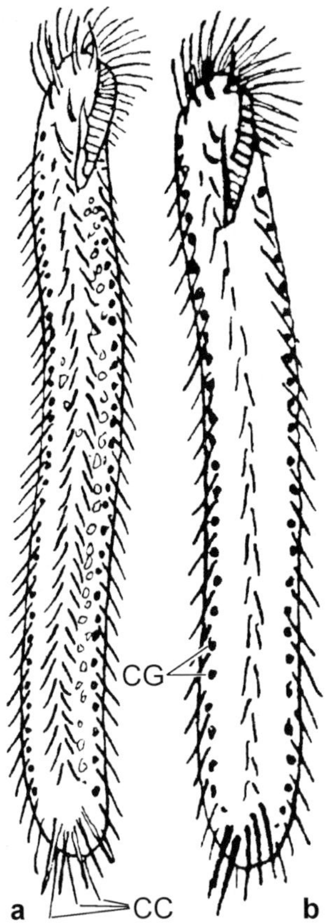

Fig. 93a, b *Anteholosticha fasciola* from life (a, from Kahl 1932; b, from Kahl 1933). Ventral views showing, inter alia, very elongate body shape, basic cirral pattern, and cortical granulation along marginal rows, a = 250 µm, b = size not indicated. CC = caudal cirri? (not shown in b; see remarks), CG = cortical granules. Page 441.

smaller (140–180 µm), lives in the sapropel, and is usually packed with blackish food vacuoles (Fig. 62a, b).

Morphology: Body length 200–300 µm, ratio of body length:width 10:1 (Kahl 1932), length according to Kahl (1933) 100–300 µm. Body outline parallel-sided, both ends rounded, anterior portion not narrowed. Body flat, very flexible, winding, but acontractile. Nuclear apparatus difficult to recognise even in stained condition, composed of many longish macronuclear nodules. Contractile vacuole "not observed" (very likely Kahl meant that he did not find a contractile vacuole). Cell densely, brownish granulated; conspicuously strong cortical granules along marginal cirral rows; individual granules became very distinct after fixation with methyl-green vinegar. Adoral zone occupies about 30 µm, that is, 10–15% of body length. Three enlarged frontal cirri, and two(?) enlarged cirri (buccal cirri?) along undulating membranes. Midventral complex extends to near transverse cirri. Five transverse cirri, which project by about half their length beyond rear body end. Marginal rows according to text distinctly shifted inwards, according to Fig. 93a they are more or less in ordinary position. Dorsal bristles likely of ordinary length (3–4 µm) because neither mentioned nor illustrated. Caudal cirri likely lacking (see remarks).

Occurrence and ecology: Saltwater; records from limnetic habitats (see below) likely based on misidentifications. Type locality is a saltwater site (25% salinity) near the German village of Oldesloe (Hamburg area), where Kahl found it several times. Records from saltwater habitats not substantiated by morphological data: Lower Saxony Wattenmeer National Park, North Sea Coast (Wickham et al. 2000, p. 87); Dnieper-Bug estuary, Ukraine (Kovalchuk 1989, p. 30); Kandalaksha Bay, White Sea (Burkovsky 1970a, p. 190; 1970b, p. 11; 1970c, p. 56); Krasnovodsk Bay, Caspian Sea (Agamaliev 1973, p. 1598); Cape Cod, Massachusets, USA (Fauré-Fremiet 1951, p. 62).

Records from limnetic habitats not substantiated by morphological data (likely confused with similar-shaped species, for example, *Anteholosticha grisea* or *A. violacea*): oligohaline (salinity 1–2%) Massaciuccoli Lake in the Tuscany, Italy (Mori et al. 1998, p. 195); running waters and lakes in Latvia (Liepa 1973, p. 33; 1983, p. 137; Liyepa 1984, p. 13; Veylande & Liyepa 1985, p. 82); Turiec River, Slovakia (Tirjaková 1993, p. 133; Matis et al. 1996, p. 12); ponds of the Pantanos de Villa, Chorrillos, Lima, Peru (Guillén et al. 2003, p. 180).

Fenchel & Finlay (1991, p. 205; 1995, p. 143) maintained the present species in anaerobic cultures for many generations. However, they did not provide morphological data so that the determination is uncertain, especially as Kahl (1932) did not provide any hint that it lives in anaerobic or microaerobic habitats.

Anteholosticha scutellum (Cohn, 1866) Berger, 2003 (Fig. 94a–k)

1866 ***Oxytricha scutellum* n. sp.** – Cohn, Z. wiss. Zool., 16: 287, Tafel XV, Fig. 43–46 (Fig. 94a, b; original description; no type material available and no formal diagnosis provided).

1882 ***Oxytricha scutellum*, Cohn** – Kent, Manual infusoria, II, p. 788, Plate XLIV, Fig. 17, 18 (redrawings; revision).

1884 ***Holosticha scutellum* Cohn sp.** – Entz, Mitt. zool. Stn Neapel, 5: 365, Tafel 22, Fig. 18 (Fig. 94c; redescription and combination with *Holosticha*; see remarks for identification).

1888 ***Holosticha scutellum*, Cohn** – Gruber, Ber. naturf. Ges. Freiburg i. B., 3: 61, Tafel VI, Fig. 12–22 (description of macronuclear division).

1923 ***Holosticha scutellum* Cohn** – Mansfeld, Arch. Protistenk., 46: 125, Fig. 13a, b (Fig. 94j, k; redescription from life).

1929 ***Holosticha scutellum* Cohn** – Hamburger & Buddenbrock, Nord. Plankt., 7: 88, Fig. 106 (Fig. 94c; guide to marine ciliates).

1932 ***Holosticha (Oxytricha) scutellum* Cohn, 1866** – Kahl, Tierwelt Dtl., 25: 579, Fig. 106_7, 110_3 (Fig. 94f, g; revision of hypotrichs; authoritative redescription).

1933 ***Holosticha scutellum* Cohn 1866** – Kahl, Tierwelt N.- u. Ostsee, 23: 111, Fig. 17.3 (Fig. 94e; guide to marine ciliates).

1936 ***Holosticha (Oxytricha) scutellum* Cohn 1866** – Kiesselbach, Thalassia, 2: 20, Abb. 44 (Fig. 94i; illustrated record).

1972 ***Holosticha scutellum* (Cohn, 1866) Kahl, 1932** – Borror, J. Protozool., 19: 11 (revision of hypotrichs).

1982 ***Holosticha scutellum* (Cohn, 1886) Kahl, 1932** – Hemberger, Dissertation, p. 104 (revision of non-euplotid hypotrichs; 1886 is an incorrect year).

1983 ***Holosticha scutellum* (Cohn, 1866) Kahl, 1932** – Borror & Wicklow, Acta Protozool., 22: 122, Fig. 10 (Fig. 94h; revision of urostylids with non-annotated illustration; voucher slides are possibly deposited in the slide collection of A. C. Borror).

1992 ***Holosticha scutellum* Cohn, 1866** – Carey, Marine interstitial ciliates, p. 183, Fig. 723 (redrawing of Fig. 94a; guide).

2001 ***Holosticha (Holosticha) scutellum* (Cohn, 1866) Kahl, 1932** – Berger, Catalogue of ciliate names 1. Hypotrichs, p. 62 (nomenclator containing all basionyms, combinations, and higher taxa of hypotrichs).

2003 ***Anteholosticha scutellum* (Cohn, 1866) n. comb.** – Berger, Europ. J. Protistol., 39: 377 (combination with *Anteholosticha*).

Nomenclature: The species-group name *scutellum* (Latin substantive, diminutive of *scutum*; small shield) refers to the strong dorso-ventral flattening of the body making the cell flat like a shield (Cohn 1866). Kahl (1932, 1933) divided *Holosticha* into several subgenera. Thus, the correct name in his reviews is *Holosticha (Holosticha) scutellum* (Cohn, 1866) Entz, 1884. The confusing spelling *Holosticha (Oxytricha) scutellum* in Kahl (1932) simply indicates that this species was originally classified in *Oxytricha*.

Borror (1972), Hemberger (1982), Borror & Wicklow (1983), Patterson et al. (1989, p. 210), and I myself (Berger 2001) incorrectly assumed that Kahl (1932) had transferred the present species to *Holosticha*. However, this transfer was already done by

Entz (1884) in his well known paper on the ciliates from the Gulf of Naples. Chardez (1987, p. 13) incorrectly assigned the species to Kahl, as "*Holosticha scutellum* Kahl, 1930".

Remarks: The present species very likely lacks caudal cirri and all(?) apomorphies of *Holosticha* – where it was previously classified – and was thus transferred to *Anteholosticha* (Berger 2003). For a detailed discussion and foundation of the transfer of many species from *Holosticha* to *Anteholosticha* see genus section.

The original description, although rather long, does not contain detailed data about the cirral pattern because the outer cell layer ("cuticula") was packed with blackish granules making cells dark-grey and opaque so that Cohn could not see the arrangement of the cirri. It is difficult to decide whether these granules were ordinary cortical granules or highly refractive inclusions in the outer layer of the cytoplasm. Cortical granules are usually not blackish, however Cohn also wrote that the cells were also packed with food vacuoles and fat globules so that a confusion of these blackish granules with fat globules is unlikely.

Kent (1882), who did not provide own observations, considered the classification in *Oxytricha* only as provisional because of the lack of data on the cirral pattern. Entz (1884) made a detailed redescription which basically agrees with the original description and Kahl's data (see below). However, while Cohn did not mention the nuclear apparatus because it was masked by the other cell inclusions mentioned above, Entz described and illustrated two macronuclear nodules (Fig. 94c). In contrast, Gruber (1884a, p. 142, 147, Tafel IX, Fig. 34–37; 1888), Mansfeld (1923), and Kahl (1932, 1933) found many macronuclear nodules. Gruber (1888) thus doubted Entz's identification although Gruber himself wrote that he is not quite certain about his identification. Entz mentioned that the 7–11 transverse cirri are extremely strong and crescent-shaped, indicating – together with the two macronuclear nodules – that he observed a *Holosticha diademata*. However, since we will likely never know which species Entz really observed I simply accept Entz's identification and Kahl's explanation for this discrepancy (see next paragraph). By contrast, Mansfeld (1923, p. 127) established a new species, *Holosticha entziella*, for Entz's bimacronucleate population.

Kahl (1932, 1933) provided a short, but likely significant characterisation of *A. scutellum* (Fig. 94e–g) which is thus considered as authoritative redescription. As already mentioned, he found – like Gruber and Mansfeld – many macronuclear nodules, and two large micronuclei, which are easily recognisable even without staining. Thus, he assumed that Entz had misinterpreted the two micronuclei as macronuclear nodules, an interpretation which cannot be excluded.

Biernacka (1967, p. 248, Abb. 53) provided a meagre illustration and a very brief characterisation which likely does not contain original data. I do not mention it in the list of synonyms, but simple show the illustration (Fig. 94d).

Hemberger (1982) synonymised *Pseudokeronopsis ovalis* with *A. scutellum*. However, this species has more than three frontal cirri so that the synonymisation is very doubtful. According to Borror & Wicklow (1983), *Holosticha manca*, *H. manca plurinucleata*, *H. interrupta*, and *H. tetracirrata* are junior synonyms of the present species. But *Anteholosticha manca*, which also lives in marine habitats, has fewer transverse cirri (about 5) and distinct cortical granules. The other species are likely confined to limnetic or terrestrial habitats apart from the fact that they do not agree in some details of the cirral pattern. The illustration provided by Borror & Wicklow (1983; Fig. 94h) agrees rather well with the authoritative redescription by Kahl (1932), indicating that *A. scutellum* is a valid species.

The discussion shows that *A. scutellum* needs a detailed redescription, which should basically agree with the original description and Kahl's redescription. Since there is a problem with the type locality (see below), the redescription should also include a neotypification.

Possibly, *Holosticha* sp. sensu Wilbert & Song (2005) is identical with the present specie (see *Holosticha*; Fig. 37.1a–c).

Morphology: Because of the uncertainties discussed above the data are kept separate. The authoritative redescription by Kahl (1932) is provided first, followed by supplementary data from the original description (Cohn 1866), the redescriptions by Mansfeld (1923) and Kiesselbach (1936a), and the illustration provided by Borror & Wicklow (1983). Finally, Entz's (1884) binucleate population is briefly described.

Populations studied by Kahl 60–120 µm long, body length:width ratio 2.8:1 (Fig. 94f) to 2.4:1 (Fig. 94g). Body outline rather variable, usually oval, that is, body margins slightly converging posteriorly and rear end more widely rounded than anterior; freely swimming specimens slightly extended, creeping individuals slightly contracted and curved leftwards. Many macronuclear nodules, only visible after staining; two micronuclei, even recognisable in life, that is, without staining (thus, Kahl assumed that Entz had misinterpreted the two micronuclei as macronuclear nodules). Larger specimens usually with heavily granulated cytoplasm, smaller ones colourless. Contractile vacuole and cortical granules neither mentioned nor illustrated; however, specimen shown in Fig. 94f possibly with distinct seam (cause?). Adoral zone occupies about 30% of body length. Buccal field moderately wide to narrow, buccal lip short, anteriorly not turned back. Three enlarged frontal cirri; obviously one cirrus behind right frontal cirrus (= cirrus III/2); one buccal cirrus near anterior end of paroral. Midventral complex composed of cirral pairs only (about 5–8 according to Fig. 94e–g); distinctly shortened posteriorly, ends at 57% of body length in specimen shown in Fig. 94f. Two pretransverse ventral cirri ahead of the 7–8 transverse cirri; especially four rightmost transverse cirri protrude distinctly beyond rear body end. Caudal cirri likely lacking because neither mentioned nor illustrated.

Type material in life about 50–80 × 30 µm (Fig. 94a, b), other populations 50–100 × 28–40 µm, on average 75 × 30 µm (Mansfeld 1923), respectively, 80 × 25 µm (Kiesselbach 1936a). Body highly retractile, that is, either oblong or oval; in hypersaline waters it even takes the shape of an irregularly bent disc. Body strongly flattened dorsoventrally, that is, flat like a shield with ventral side concave. Posterior end broadly rounded, anterior one slightly narrowed and oblique-triangularly truncated. Many globular macronuclear nodules (Mansfeld 1923); for details on the nuclear apparatus of Kiesselbach's (1936a) population see remarks. Outer cell layer packed with blackish granules, so that cells – which are, in addition, often filled with numerous food vacuoles and fat globules – become dark-grey and opaque; granules masking nuclear apparatus. Con-

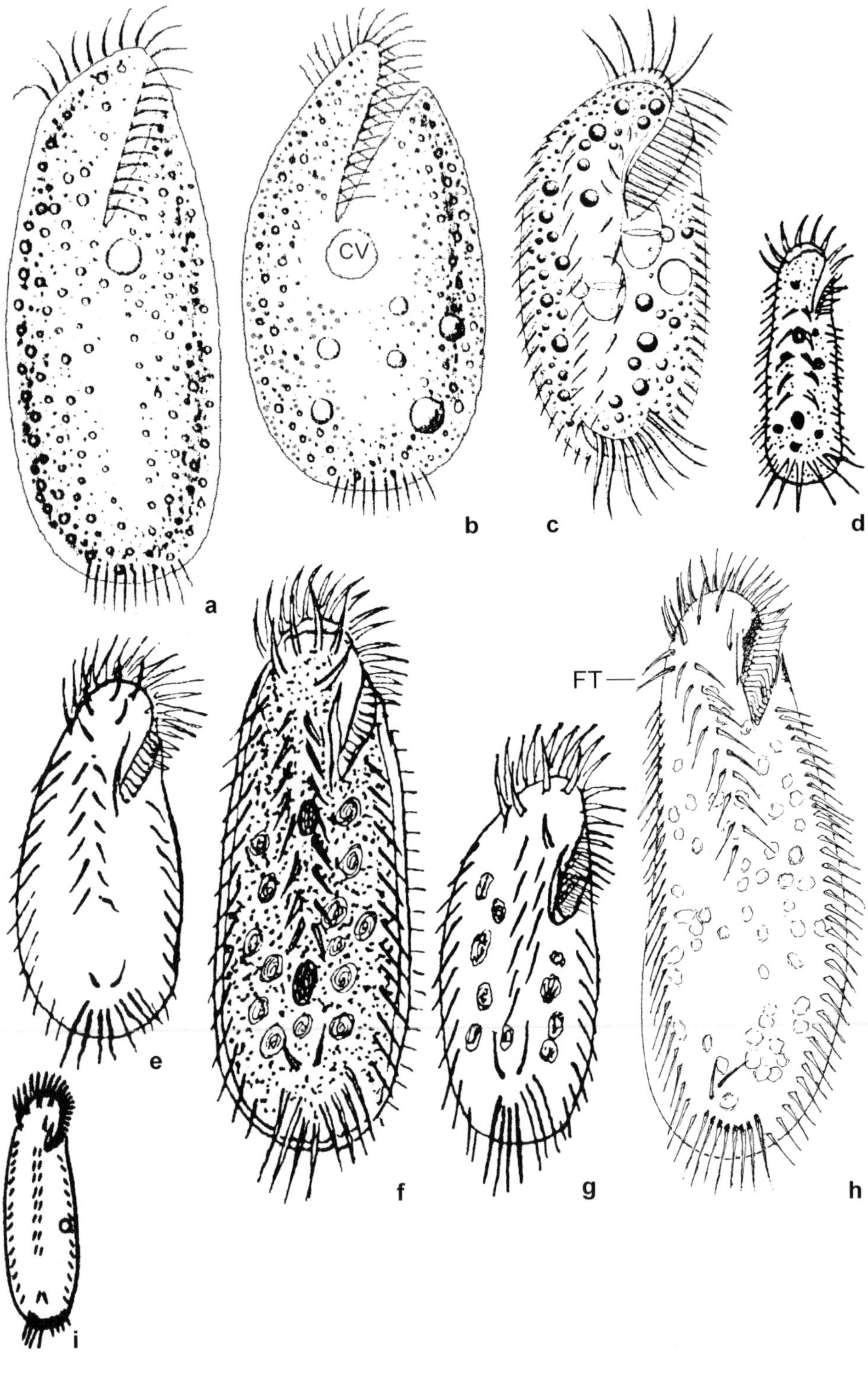
CV
a
b
c
d
FT
e
f
g
h
i

tractile vacuole behind proximal end of adoral zone. Adoral zone occupies about one third of body length. Cirral pattern not recognised due to dark granulation, except for transverse cirri (about 10 illustrated), which protrude beyond rear body end; Mansfeld (1923) counted 8–10, Kiesselbach (1936a) eight transverse cirri. Movement without peculiarities, that is, restlessly swimming to and fro, interrupted by bow-shaped backwards movements.

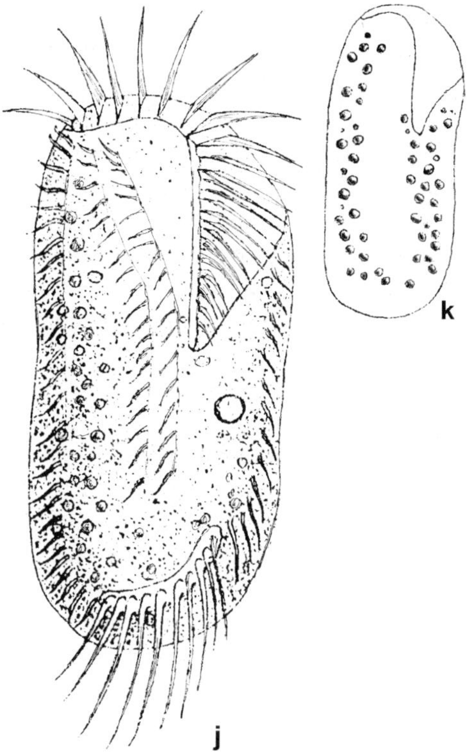

Fig. 94j, k *Anteholosticha scutellum* (from Mansfeld 1923. j, from life; k, nucleus stain). Ciliature of ventral side and nucleus apparatus, average body length 70 μm. Page 443.

For cirral pattern of Borror & Wicklow's population, see Fig. 94h. The specimen illustrated has about 34 adoral membranelles (occupying about 25% of body length); three frontal cirri; one buccal cirrus; about eight midventral pairs with last cirrus at about 45% of body length; two pretransverse ventral cirri; eight almost transversely arranged transverse cirri projecting by half their length beyond rear body end; two frontoterminal cirri near right frontal cirrus; 30 right and 34 left marginal cirri; many macronuclear nodules; caudal cirri obviously lacking.

Brief characterisation of Entz's binucleate population from Bay of Naples (Fig. 94c): Body size 50–70 × 30–40 μm. Two ellipsoidal macronucleus-nodules; anterior one behind proximal end of adoral zone, rear one about in mid-body in or slightly right of midline; the spindle-shaped figure in each nodule is possibly a reorganisation band, indicating that Entz indeed observed a macronuclear nodule and not a micronucleus as supposed by Kahl; each macronuclear nodule with a micronucleus. Contractile vacuole near left cell margin about in mid-body. Cytoplasm colourless with many fatty shining globules of various size. Adoral zone about one third of body length, buccal

← **Fig. 94a–i** *Anteholosticha scutellum* (a, b, from Cohn 1866; c, from Entz 1884; d, from Biernacka 1967; e, from Kahl 1933; f, g, from Kahl 1932; h, from Borror & Wicklow 1983; i, from Kiesselbach 1936a. a–g, i, from life; h, protargol impregnation?). **a, b, d–g, i:** Ventral views showing, inter alia, blackish granules, contractile vacuole, cirral pattern, and nuclear apparatus, a, b = 50–80 μm, d, e = individual size not indicated, f = 90 μm, g = 100 μm, i = 80 μm. The two black globules in the midline of (f) are micronuclei. **c:** Population from Bay of Naples with two macronuclear nodules, a feature contradicting the other descriptions, 50–70 μm. Therefore, Mansfeld (1923) established *Holosticha entziella* for this population. **h:** Infraciliature of ventral side and nuclear apparatus, 87 μm. CV = contractile vacuole, FT = frontoterminal cirri. Page 443.

field deep and narrow. According to Entz, special frontal and caudal cirri lacking (I suppose that he overlooked the frontal cirri). Two widely spaced ventral cirral rows extending from frontal region to near right transverse cirri; cirri of left row stronger and more widely spaced than those of right row. 7–11 strong and sickle-shaped transverse cirri subterminally arranged in curved row; 3–4 rightmost cirri longer than remaining cirri. Marginal rows obviously without peculiarities.

Cell division: Gruber (1884a, 1888) studied the macronuclear division of *A. scutellum*. The division proceeds in plesiomorphic manner, that is, the many macronuclear nodules fuse to a single mass and then divide again. Gruber (1888) assumed that the micronuclei also fuse to a single mass during division, a misinterpretation later discussed by Lewin (1913). Hill (1980) briefly described cell division and reorganisation in three *Holosticha* species, including the present species. However, he did not distinguish between these three species.

Occurrence and ecology: Marine. Type locality is the North Sea at the German island of Helgoland (Cohn 1866, p. 253). Possibly, the aquarium containing the material from Helgoland was contaminated with material from Dorsetshire, a village or region, which I do not find in my atlas (possibly this is a small village in Dorset, South England?). As a consequence of this uncertainty about the type locality, a redescription should also include a neotypification which regulates the type locality. Cohn (1866) found *A. scutellum*, together with *Pseudokeronopsis* species, among rotting substances.

Further records substantiated by morphological data: North Sea (from the islands of Helgoland and Sylt) and Baltic Sea near the German city of Kiel (Kahl 1932, 1933); Baltic Sea near the island of Hiddensee (Biernacka 1967; see remarks); old cultures with decaying algae from the Bay of Naples, Mediterranean Sea (Entz 1884); Gulf of Genoa, Mediterranean Sea (Gruber 1888); seawater aquarium in Berlin, Germany, containing material from the North Sea and the Mediterranean Sea (Mansfeld 1923); sample with *Posidonia* (harbour of Parenzo) and sand sample from Rovigno, northern Adriatic Sea (Kiesselbach 1936a); marine habitat(s) in USA? (Borror & Wicklow 1983).

Records not substantiated by morphological data: benthic in the Krasnovodsk Bay, eastern shore of Caspian Sea (Agamaliev 1973, p. 1598); Belgium (Chardez 1987, p. 13); Bay of Marseille, Mediterranean Sea (Vacelet 1960, p. 54; 1961, p. 4); Black Sea (Pereyaslawzewa 1886, p. 97). The limnetic record from Val Parma, Italy by Madoni & Ghetti (1977, p. 42) is likely a misidentification.

Anteholosticha oculata (Mereschkowsky, 1877) Berger, 2003 (Fig. 95a–e)

1877 ***Oxytricha oculata* nova species** – Mereschkowsky, Trudy imp. S-peterb. Obshch. Estest., 8: 232, Table I, Fig. 9, 10 (Fig. 95a, b; original description in Russian; no type material available and no formal diagnosis provided).

1879 ***Oxytricha oculata*, n. sp.** – Mereschkowsky, Arch. mikrosk. Anat. EntwMech., 16: 163, Tafel X, Fig. 9, 10 (Fig. 95a, b; German translation of original description; see nomenclature).

1886 ***Amphisia multiseta*, Sterki** – Milne, Proc. Manchr Fld Nat. Archaeol. Soc., 18: 54, pro parte, Fig. 12, not Fig. 14, 17 (Fig. 95d; misidentification; see remarks).

1929 ***Amphisia oculata* Mereschk.** – Hamburger & Buddenbrock, Nord. Plankt., 7: 91, Fig. 111 (redrawing of Fig. 95a, b; combination with *Amphisia*; guide to marine ciliates).
1932 ***Holosticha (Oxytricha) oculata* (Mereschkowsky, 1877)** – Kahl, Tierwelt Dtl., 25: 582, Fig. 106 14 (Fig. 95c; revision; combination with *Holosticha*; see nomenclature).
1932 ***Holosticha milnei* spec. n.** – Kahl, Tierwelt Dtl., 25: 582, Fig. 108 (Fig. 95e; original description of synonym; no type material available and no formal diagnosis provided).
1933 ***Holosticha oculata* (Mereschkowsky 1877)** – Kahl, Tierwelt N.- u. Ostsee, 23: 111, Fig. 17.11 (Fig. 95c; guide to marine ciliates).
1933 ***Holosticha milnei* Kahl 1932** – Kahl, Tierwelt N.- u. Ostsee, 23: 111 (guide to marine ciliates).
1992 ***Holosticha oculata* Mereschkowsky, 1879** – Carey, Marine interstitial ciliates, p. 183, Fig. 721 (redrawing of Fig. 95a; guide).
2001 ***Holosticha (Holosticha) oculata* (Mereschkowsky, 1877) Kahl, 1932** – Berger, Catalogue of ciliate names 1. Hypotrichs, p. 59 (nomenclator containing all basionyms, combinations, and higher taxa of hypotrichs).
2001 ***Holosticha holomilnei* nom. nov.** – Berger, Catalogue of ciliate names 1. Hypotrichs, p. 35 (substitute name for primary homonym *Holosticha (Holosticha) milnei*, see nomenclature; nomenclator containing all basionyms, combinations, and higher taxa of hypotrichs).
2003 ***Anteholosticha oculata* (Mereschkowsky, 1877) n. comb.** – Berger, Europ. J. Protistol., 39: 377 (combination with *Anteholosticha*).

Nomenclature: No derivation of the names is given in the original descriptions. The species-group name *ocul̇át·us -a -um* (Latin adjective; having eyes; having dots [like eyes]) obviously alludes to the two ring-shaped (globular?) structures in the anterior and posterior body portion. Kahl (1932) dedicated *Holosticha milnei* to William Milne (England), who misidentified the present species as *Amphisia multiseta* (now *Holosticha pullaster*). The part *holo-* in *holomilnei* indicates that this species was classified in the subgenus *Holosticha*.

The somewhat confusing spelling *Holosticha (Oxytricha) oculata* in Kahl's (1932) heading (see above) should simply indicate that this species was previously classified in *Oxytricha*. Kahl (1932, 1933) divided *Holosticha* into several subgenera. Thus, the correct names of the species listed above in his reviews are *Holosticha (Holosticha) oculata* (Mereschkowsky, 1877) Kahl, 1932 and *Holosticha (Holosticha) milnei* Kahl, 1932. The name of the latter species is thus a (simultaneous, that is, neither a senior nor a junior) primary homonym of *Holosticha (Amphisiella) milnei* Kahl, 1932 (Kahl 1932, p. 590, Fig. 112 3). Kahl (1932, p. 590) suggested synonymy of these two species although the illustrations are rather different (see remarks and cp. Fig. 95e, f). Since subgeneric names are irrelevant to the homonymy (ICZN 1999, Article 57.4) I replaced the species-group name of *H. (Holosticha) milnei* by the name *Holosticha holomilnei* because this species is much less known than *H. (Amphisiella) milnei* (Berger 2001).

Remarks: The present species very likely lacks caudal cirri and all(?) apomorphies of *Holosticha* – the genus where it was previously classified – and was thus transferred to *Anteholosticha* (Berger 2003). For a detailed discussion and foundation of the transfer of many species from *Holosticha* to *Anteholosticha* see genus section.

Mereschkowsky (1877) described *Oxytricha oculata* in the same year as *Holosticha* was established. Thus, he very likely did not know the *Holosticha* paper. Consequently, it is understandable that he classified his species in *Oxytricha*, which was a melting pot for many types of hypotrichs at that time. The present species was first described by

Mereschkowsky (1877) in Russian. Two years later he published a German translation and again designated the present species as new. Thus, strictly speaking, *O. oculata* Mereschkowsky, 1879 is the junior objective synonym and junior primary homonym of *O. oculata* Mereschkowsky, 1877. Mereschkowsky's description lacks some important data, for example, size and nuclear apparatus. However, the two ring-shaped structures in the anterior and posterior body portion in combination with the obviously distinct midventral cirral pattern must be considered as a rather unique combination of features. The same combination was described by Milne (1886), who misidentified his population as *Amphisia multiseta*, a junior synonym of *Holosticha pullaster* (for differences, see below). Milne (1886) obviously mixed two or more species as already recognised by Kahl (1932). I refer only to Fig. 95d.

Kahl (1932) recognised that Mereschkowsky's species is not a true *Oxytricha* because of the zigzagging cirral rows. Thus – and because of the enlarged frontal cirri – he transferred it to *Holosticha* (three years earlier, Hamburger & Buddenbrock had transferred it to *Amphisia*, the junior synonym of *Holosticha*). At first, Kahl (1932) reviewed *Holosticha oculata* and in the next paragraph he established *Holosticha milnei* for *Amphisia multiseta* sensu Milnei. Simultaneously he considered *H. milnei* as supposed synonym of *H. oculata* in that he wrote "... die eine Form Milne's war starr und gelblich; es mag *oculata* gewesen sein.". For a second form roughly mentioned by Milne, Kahl (1932) established *Holosticha (Amphisiella) milnei*, which he knew from his own experience. This species is certainly a true *Amphisiella* (Fig. 95f), whereas *Holosticha (Holosticha) milnei* (Fig. 95d, e) shows a distinct midventral cirral pattern so that conspecificity can be excluded. In spite of this conspicuous difference, Kahl (1932, p. 590) himself suggested synonymy of his two *milnei* species, that is, he actually supposed synonymy of three species, namely *H. oculata*, *H. (Holosticha) milnei*, and *H. (Amphisiella) milnei*.

Borror (1972) obviously mentioned neither *Oxytricha oculata* nor *H. (Holosticha) milnei*. Borror & Wicklow (1983, p. 121) synonymised the latter species with *Holosticha diademata*, however, without foundation and certainly mistakenly; *Oxytricha oculata* was neglected in their review.

Carey (1992), who considered the German translation of Mereschkowsky (1879) as valid original description, considered the two ring-shaped structures in the body ends as macronuclear nodules, which is certainly incorrect.

Although the data about the present species are not very detailed it seems justified to classify it as valid species. Detailed redescription needed.

For separation from other *Anteholosticha* species, see key. *Holosticha pullaster*, which occurs both in fresh and salt water, has the contractile vacuole distinctly behind mid-body (against not described and thus likely lacking in *A. oculata*) and lacks the conspicuous ring-shaped structures (Fig. 95a). *Holosticha diademata*, according to Borror & Wicklow (1983) the senior synonym of *Oxytricha oculata*, also lacks the ring-shaped structures and has the conspicuous apomorphies of *Holosticha* (see there), which are obviously absent in the present species. *Tachysoma pellionellum* (Müller, 1773) Borror, 1972, which very often has similar (identical) ring-shaped structures in the same position, is an 18-cirri oxytrichid which exclusively lives in limnetic habitats

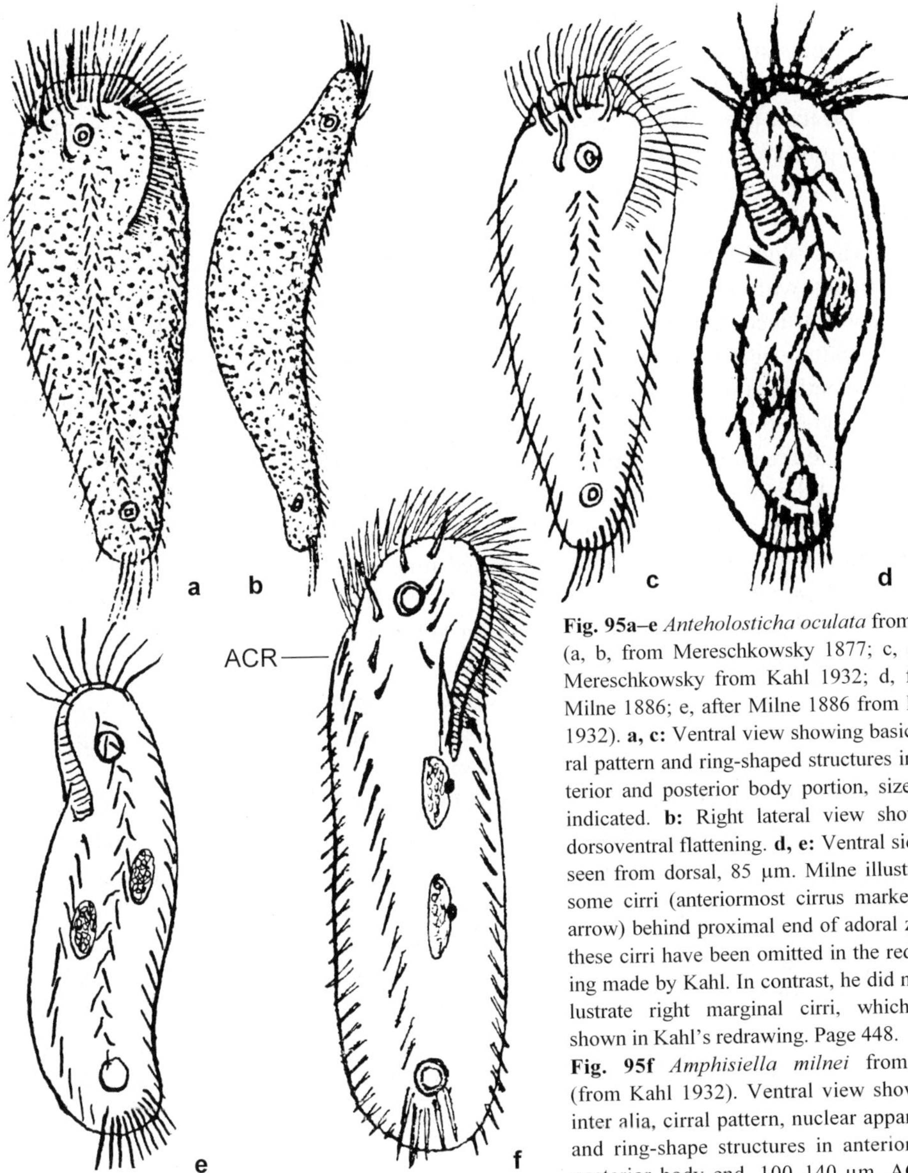

Fig. 95a–e *Anteholosticha oculata* from life (a, b, from Mereschkowsky 1877; c, after Mereschkowsky from Kahl 1932; d, from Milne 1886; e, after Milne 1886 from Kahl 1932). **a, c:** Ventral view showing basic cirral pattern and ring-shaped structures in anterior and posterior body portion, size not indicated. **b:** Right lateral view showing dorsoventral flattening. **d, e:** Ventral side as seen from dorsal, 85 µm. Milne illustrated some cirri (anteriormost cirrus marked by arrow) behind proximal end of adoral zone; these cirri have been omitted in the redrawing made by Kahl. In contrast, he did not illustrate right marginal cirri, which are shown in Kahl's redrawing. Page 448.

Fig. 95f *Amphisiella milnei* from life (from Kahl 1932). Ventral view showing, inter alia, cirral pattern, nuclear apparatus, and ring-shape structures in anterior and posterior body end, 100–140 µm. ACR = amphisiellid median cirral row.

(for review see Berger 1999, p. 433). *Amphisiella milnei* (Kahl, 1932) lacks a midventral pattern and has only five transverse cirri (Fig. 95f).

Morphology: The following description is based on the data by Mereschkowsky (1877[1], 1879) unless otherwise indicated.

---

[1] I did not translate the paper because I assume that his 1879 paper is a more or less exact translation.

Body size not mentioned in original description; size according to Milne about 85 × 34 μm (width estimated from body length:width ratio of 2.5:1 given by Milne; according to Kahl 1933, Milne did not give a size). Body outline rather variable, usually inverted oval or inverted pear-shaped. Body margins converging posteriorly, anterior end broadly rounded, posterior narrowly rounded and according to Merschkowsky even with a short tail; tail sometimes lacking, then outline oval; according to Milne, body outline shoe-shaped, equally rounded at both ends. Body dorsoventrally flattened, always curved in lateral view with distinctly vaulted dorsal side (Fig. 95b). Dorsal surface convex, with three ridges with a double flexure (Milne). Nuclear apparatus not described by Mereschkowsky, according to Milne two large nuclei present (the arrangement [anterior nodule shifted rightwards] is somewhat reminiscent of *Holosticha*). Contractile vacuole neither mentioned nor illustrated (Fig. 95a, b). Anterior and posterior body portion each with a single ring-shaped structure ("Augenkreis"); structures without surrounding pigments; sometimes only one ring-shaped structure present. Cytoplasm packed with granules (size and colour not indicated) and thus cloudy. Movement rapid, never resting.

Adoral zone occupies about one third of body length (Mereschkowsky, Milne), obviously without peculiarities; Milne mentioned nine adoral membranelles indicating that he counted only the distal-most ones. Likely four slightly enlarged frontal cirri. Midventral complex conspicuous, extends from near anterior end to near transverse cirri. About five elongated and slightly rightwards directed transverse cirri (not clearly distinguished from left marginal row by Mereschkowsky); according to Milne, 10 transverse cirri (Fig. 95d shows 9 cirri), which project beyond rear body end. Marginal rows obviously without peculiarities. Dorsal infraciliature unknown.

Occurrence and ecology: Marine. Type locality of *Oxytricha oculata* is the coast of the Kloster-Bucht (cloister bay) at the Solowetzky islands, White Sea, where Mereschkowsky (1877, 1879) discovered it when the abundance was high during June and July 1877. Milne (1886) did not mention a sample site; likely he found it somewhere at the English coast (according to Kahl 1933 at the west coast).

Record of *A. oculata* not substantiated by morphological data: Northern Sea in Sylt, Germany (Küsters 1974, p. 174); harbour of the village of Oostend, Belgium (Persoone 1968, p. 190); western coast of Caspian Sea (Agamaliev 1971, p. 383). The limnetic record from the Iskar river (Bulgaria) by Detcheva (1993, p. 34) is likely a misidentification.

Anteholosticha arenicola (Kahl, 1932) Berger, 2003 (Fig. 96a, b)

1932 ***Holosticha arenicola* spec. n.** – Kahl, Tierwelt Dtl., 25: 583, Fig. 109 (Fig. 96a; original description; no type material available and no formal diagnosis provided).

1933 ***Holosticha arenicola* Kahl 1932** – Kahl, Tierwelt N.- u. Ostsee, 23: 111, Fig. 17.12 (Fig. 96b; guide to marine ciliates).

1972 ***Holosticha arenicola* Kahl, 1932** – Borror, J. Protozool., 19: 10 (revision of hypotrichs).

2001 ***Holosticha (Holosticha) arenicola*** **Kahl, 1932** – Berger, Catalogue of ciliate names 1. Hypotrichs, p. 35 (nomenclator containing all basionyms, combinations, and higher taxa of hypotrichs).

2003 ***Anteholosticha arenicola*** **(Kahl, 1932) n. comb.** – Berger, Europ. J. Protozool., 39: 377 (combination with *Anteholosticha*).

Nomenclature: No derivation of the name is given in the original description. The species-group name *arenicola* (living in the sand) is a composite of the Latin noun *aréna* (sand, sandy site, coast) and the latin verb *cólere* (inhabit) and obviously refers to the habitat (sandy beach) where the species was discovered. Kahl (1932, 1933) divided *Holosticha* into subgenera. Thus, the correct name in his reviews is *Holosticha (Holosticha) arenicola* Kahl, 1932.

Remarks: The present species very likely lacks caudal cirri and most (all?) apomorphies of *Holosticha* and was thus transferred to *Anteholosticha* (Berger 2003). For a detailed discussion and foundation of the transfer of many species from *Holosticha* to *Anteholosticha* see genus section.

Anteholosticha arenicola has a distinct alveolar seam and a rather high number of transverse cirri which is reminiscent of *Pseudoamphisiella* species. However, these species have a very indistinct midventral pattern and the transverse cirral row usually extends more anteriorly. Further, the adoral zone of *Pseudoamphisiella* species extends far onto the right body margin, whereas in *A. arenicola* the zone ends near the anterior body end. A detailed redescription should show which classification is correct.

Borror & Wicklow (1983, p. 121) synonymised *A. arenicola* with *Holosticha gibba*, however, without foundation. The redescriptions of *Holosticha arenicola* by Aladro-Lubel (1985; Fig. 99c) and Aladro Lubel et al. (1990, Fig. 99d) are insufficient.

Morphology: The following description is based solely on Kahl's (1932, 1933) data (Fig. 96a, b). Body length 70–90 µm; body length:width ratio of specimen illustrated about 1.9:1 (Fig. 96a). Body outline usually symmetrical ovoid (according to Kahl 5:2, that is, 2.5:1 which is more slender than the specimen illustrated); some specimens less wide and then right margin almost straight. Body rather flat. Two ellipsoidal macronuclear nodules, specimen illustrated with reorganisation band (Fig. 96a); each macronuclear nodule with one micronucleus. Cytoplasm very soft (flexible?); ectoplasm with honeycomb structure, that is, an alveolar seam which is reminiscent of *Pseudoamphisiella* (see remarks), slightly contractile. Food vacuoles yellowish.

Adoral zone occupies 25–33% of body length (in specimen illustrated about 38%!), distinctly question mark-shaped with distal end only slightly extending onto right body margin. Buccal field very narrow, buccal lip short. Three slightly enlarged frontal cirri. Buccal cirrus neither mentioned nor illustrated (possibly overlooked). 7–10 transverse cirri, right cirri project by about 1/3 of their length beyond rear body margin. Marginal rows distinctly displaced inwards, likely not overlapping posteriorly. Dorsal cilia neither mentioned nor illustrated, indicating that they are short (around 3 µm). Caudal cirri also neither mentioned nor illustrated, strongly indicating that they are lacking.

Occurrence and ecology: Marine. Type locality is the Bay of Kiel (Germany), Baltic Sea, where Kahl (1932) discovered it on polluted sand near the beach; on some sites it was frequent, but obviously not very common (see also Hartwig 1974, p. 17).

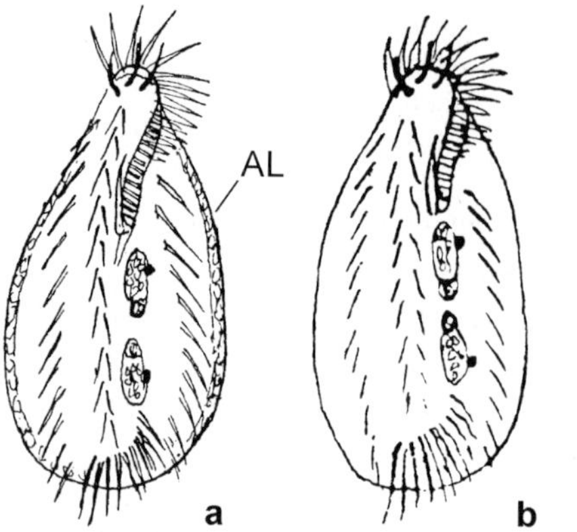

Fig. 96a, b *Anteholosticha arenicola* (a, from Kahl 1932; b, after Kahl 1932 from Kahl 1933). Ventral view, 70–90 µm. AL = alveolar layer. Page 452.

Records not substantiated by morphological data: very abundant in sand from Helsingør Beach (Denmark) during periods when great amounts of seaweed were mixed with sand and in artificial sediments consisting of *Fucus* bits mixed with clean sand (Fenchel 1968, p. 116; see also Fenchel 1967, p. 129); sample from the Atlantic coast at Castro Urdiales, Spain (Fernandez-Leborans & Novillo 1994, p. 203; found only in the control and not in the sample contaminated with 1 mg l^{-1} lead); west coast of Caspian Sea (Agamaliev 1971, p. 383). Also recorded from Italy (Dini et al. 1995, p. 70; no details given, possibly from the limnetic record by Campea & Stella, see next paragraph).

Limnetic records are very likely misidentifications: river Tevere, Italy (Stella & Campea 1948, p. 157); rare in freshwater habitats in southeastern Louisiana, USA (Bamforth 1963, p. 133).

Feeds on heterotrophic (colourless) flagellates (Kahl 1932). Fenchel (1968; 1969, p. 25) cultured *Anteholosticha arenciola* in peptone solutions with a mixed microflora, where it fed on small heterotrophic flagellates and bacteria.

Anteholosticha azerbaijanica (Alekperov & Asadullayeva, 1999) comb. nov.
(Fig. 97a, Addenda)

1999 ***Holosticha azerbaijanica*** **sp. nov. Alekperov & Asadullayeva** – Alekperov & Asadullayeva, Turkish Journal of Zoology, 23: 221, Fig. 9 (Fig. 97a; original description; the holotype slides [23–28] are deposited in the Protistological Laboratory of the Institute of Zoology of the Academy of Sciences of Azerbaijan, Baku).

Nomenclature: No derivation of the name is given in the original description. The species-group name *azerbaijanica* refers to the country (Azerbaijan) where the species was discovered. It is not included in the catalogue by Berger (2001).

Remarks: The present species is definitely not a *Holosticha* as defined by Berger (2003). However, it also does not fit very well in other genera, so that it would be necessary to establish a new genus. Since some important data are unknown, e.g., whether or not midventral rows are present, I preliminarily classify it in *Anteholosticha*, an already heterogeneous group which contains those species previously classified in *Holosticha* which lack caudal cirri and the apomorphies of *Holosticha* sensu Berger (2003). The narrowed posterior body end is reminiscent of *Uroleptus*. However, these species

Fig. 97a *Anteholosticha azerbaijanica* (from Alekperov & Asadullayeva 1999. Wet silver nitrate impregnation?). Infraciliature of ventral side and nuclear apparatus, 178 µm. Note that the classification of this species in *Anteholosticha* is only preliminary. FT = frontoterminal cirri, TC = transverse cirri. Page 454.

a

have (usually) caudal cirri. Because of the increased number of frontoterminal cirri, Alekperov & Asadullayeva (1999) assumed that their species is probably a link between the Holostichidae and the Bakuellidae. Ontogenetic data are needed for a final classification of this interesting species.

Morphology: Body length of life specimens 220–300 µm, after fixation (method not specified) 170–250 µm. Body outline uroleptid, that is, slender elongate with rear portion narrowed tail-like. Body distinctly flattened dorso-ventrally. Many macronuclear nodules scattered throughout cytoplasm; individual nodules and micronuclei spherical or elongated. Cytoplasm transparent, bright yellow. Contractile vacuole in "¼ of the anterior part on the left side" (I suppose that the vacuole is near the proximal end of the adoral zone). Adoral zone occupies about 25% of body length, composed of 110–140 membranelles (number possibly overestimated). Buccal field likely narrow, undulating membranes obviously more or less straight. Three slightly enlarged frontal cirri. Buccal cirrus lacking. About 10 frontoterminal cirri form row between anterior ends of midventral complex and right marginal row. Midventral complex likely composed of midventral pairs only, commences near distal end of adoral zone, terminates near transverse cirri; specimen illustrated with ca. 60 pairs! About five transverse cirri very close to rear body end. Right marginal row composed of about 110 cirri, left of ca. 120 cirri. Four dorsal kineties; arrangement of kineties and length of bristles not mentioned. Caudal cirri lacking.

Occurrence and ecology: As yet found only at the type locality, that is, the south coast of Apsheron, Caspian Sea, Azerbaijan. Alekperov & Asadullayeva (1999) found it there in the psammon and periphyton.

Tachysoma spec. sensu Wilbert & Kahan (1981) (Fig. 98a)

1981 ***Tachysoma* spec.** – Wilbert & Kahan, Arch. Protistenk., 124: 91, Fig. 14 (Fig. 98a; a voucher slide is possibly deposited in the Institut für landwirtschaftliche Zoologie of the University of Bonn, Germany).

Remarks: This population certainly does not belong to *Tachysoma*, a group of 18-cirri oxytrichids. Berger (1999, p. 432) already wrote that the present population is possibly a holostichid because of the zigzagging ventral cirri. The very specific collection site,

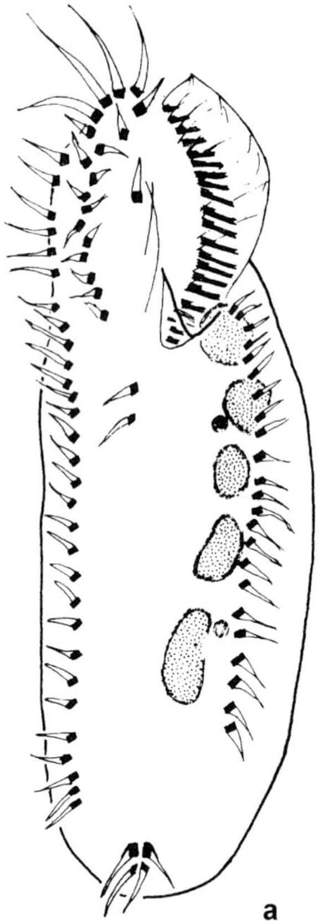

a

Fig. 98a *Tachysoma* spec. (from Wilbert & Kahan 1981. From life and protargol preparations). Ventral view showing, inter alia, cirral pattern and nuclear apparatus, 100 µm. This specimen, respectively, population is only preliminarily assigned to *Anteholosticha*. Page 455.

namely the mesothermal, monomictic Solar Lake (Israel), indicates that it is a distinct species. Because of the zigzagging cirral pattern, which is also mentioned by Wilbert & Kahan, and the lacking caudal cirri I preliminary classify it in *Anteholosticha*. The data are too sparse for a final classification, Consequently, I simply show the illustration and provide the description given by Wilbert & Kahan (1981). The frontal-ventral-transverse cirral pattern originates from six anlagen, which is characteristic for most oxytrichids, as already stated by the authors. A relationship to amphisiellids is unlikely (but not impossible) because Wilbert & Kahan mentioned a zigzagging pattern. Detailed redescription of a population from the original locality, or at least a very similar habitat (highly saline lake), is needed.

Morphology: Body length ca. 100 µm, length:width ratio 3–4:1; body outline elliptical; macronucleus left of midline, ribbon-shaped or divided into 5–8 (usually 5) nodules of different size; two micronuclei; contractile vacuole "on the right in the posterior quarter" (possibly confused with an empty, non-contractile vacuole?); cytoplasm colourless, clear.

Adoral zone of specimen illustrated occupies about 33% of body length, composed of four distal (dorsal) and, separated by a gap, 18–20 proximal (ventral) membranelles; three frontal cirri; 1 buccal cirrus near anterior end of undulating membranes; possibly two frontoterminal cirri present (Fig. 98a); midventral complex (if present at all) composed of about five pairs, ends about at level of buccal vertex (the zigzagging pattern is also mentioned by Wilbert & Kahan); two postoral cirri; four transverse cirri near distal end (likely 2 pretransverse ventral cirri and 2 transverse); right marginal row composed of about 30 cirri, terminates distinctly subterminally, left row composed of about 20 cirri, begins left of proximal portion of adoral zone, terminates at about 80% of body length in specimen illustrated; 3 dorsal kineties, caudal cirri lacking.

Cell division: Wilbert & Kahan provided no illustration of a divider, but a brief description. Briefly, ontogenesis commences with the formation of an oral primordium between the postoral ventral cirri and the buccal vertex. Later, the primordium extends backwards and the frontal and ventral cirri disappear. Their basal bodies form six parallel anlagen in the proter. The membranelles of the adoral zone of the opisthe differentiate in posteriad direction. Right of the new adoral zone six cirral anlagen are formed.

Because of the six anlagen, Wilbert & Kahan classified the population in the oxytrichids, and the lack of caudal cirri caused them to assign it to *Tachysoma*.

Occurrence: Wilbert & Kahan (1981) found this interesting species in spring in the mesothermal, monomictic Solar Lake on the east coast of Sinai, about 20 km south of Elat (Israel). For a detailed description of this shallow (maximum depth 5 m), saline lake, see Wilbert & Kahan (1981).

Insufficient redescriptions

***Holosticha arenicola* Kahl, 1932** – Aladro Lubel, 1985, An. Inst. Biol. Univ. Méx., 55: 26, Lámina 13, Fig. 1 (Fig. 99c) and Aladro Lubel, Martínez Murillo & Mayén Estrada, 1990, Manual de ciliados, p. 128, Figure on same page (Fig. 99d). Remarks: The descriptions and illustrations are too superficial to accept the identification, especially because they do not mention the alveolar layer. Furthermore, the anterior end of the left marginal row curves rightwards, indicating that Aladro Lubel observed a true *Holosticha* species. Body size 70–87.5 × 35–49 µm. For further details, see illustrations. Vera Cruz, Gulf of Mexico. Additional records, see Mayén Estrada & Aladro Lubel (1987, p. 75) and Aladro-Lubel et al. (1988, p. 436).

***Holosticha manca* Kahl, 1932** – Ganapati & Rao, 1958, Andhra Univ. Mem. Oceanogr., 2: 85, Plate I, Figure 9 (Fig. 100f). Remarks: Ganapati & Rao (1958) described and illustrated two macronuclear nodules. However, *Holosticha manca* – now *Anteholosticha manca* – has many (about 50–70) nodules so that conspecificity can be excluded. I cannot identify it with any other species and thus list it under the insufficient redescriptions. Possibly it is a distinct species. Body length 150–190 µm, length:width ratio about 4:1; body elongate, flexible, and contractile; cells light yellow, transparent, or colourless. Two oval macronuclear nodules, each with 1–2 micronuclei. Contractile vacuole somewhat behind buccal vertex near left margin; pulsation frequency very low. Endoplasm often with rows of pearl-like shining globules. Adoral zone about one fourth of body length, cytopharynx rather long. Buccal field moderately wide, buccal lip strongly curved anteriorly. Three enlarged frontal cirri, one buccal cirrus. "Ventral rows" (midventral complex) widely separated, right row extends to 66% of body length, left row terminates at about 50% of body length. Marginal rows not interrupted posteriorly. Feeds on diatoms, especially *Pleurosigma*. Abundant during October and April in backwaters at Visakhapatnam, Indian coast. Few specimens were also collected from the scrapings of pile from harbour jetties.

***Keronopsis gracilis* Kahl, 1935** – Vuxanovici, 1961, Studii Cerc. Biol., 13: 437, Plansa II, Fig. 16 (Fig. 100e). Remarks: Kahl (1932) described this species from marine habitats whereas Vuxanovici's population is limnetic. Furthermore, the illustration is rather superficial so that the identification seems arbitrarily. Possibly it is an *Anteholosticha monilata*. Body length about 120 µm; body outline parallel-sided, slightly sigmoidal, broadly rounded at both ends. Body very soft and flexible, pellicle fragile. Specimen

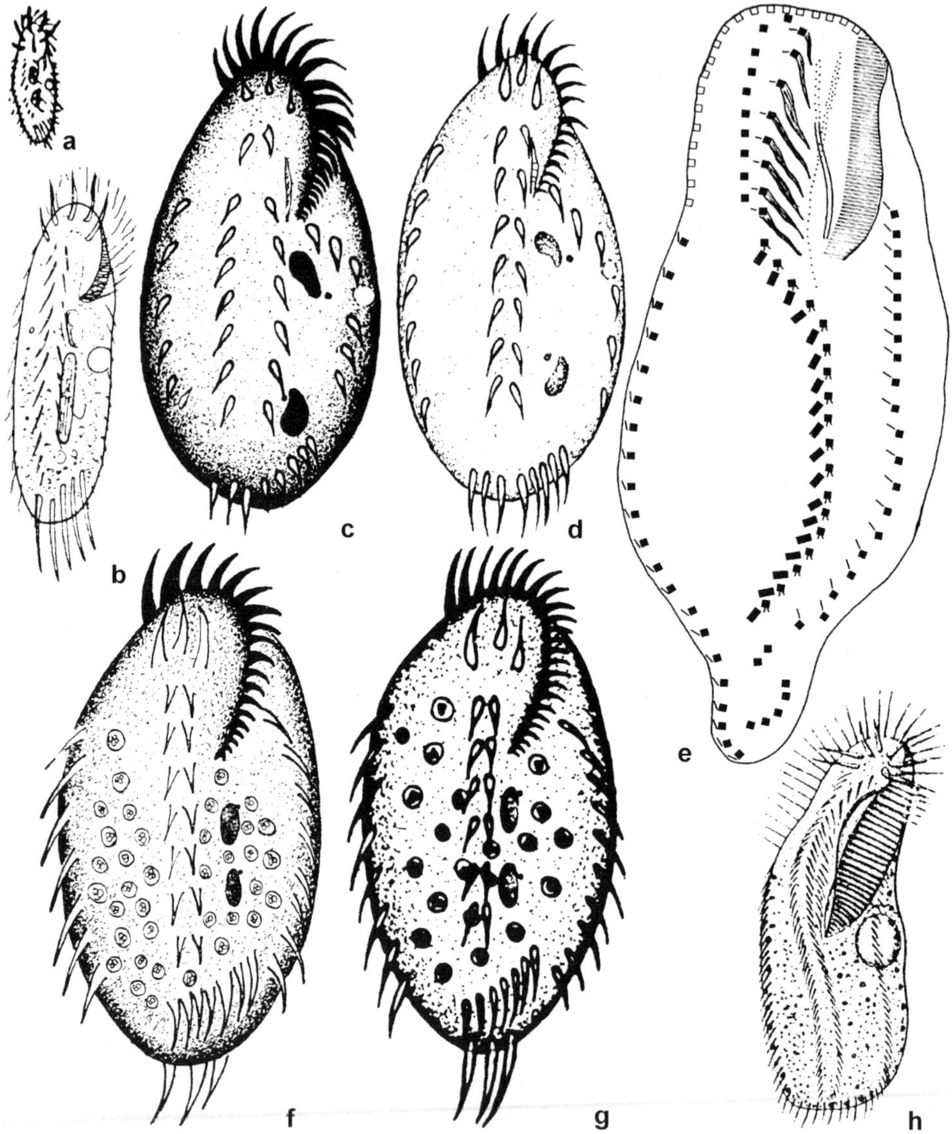

Fig. 99a–h Insufficient redescriptions and species indeterminata (ventral views from life unless otherwise indicated). **a:** *Holosticha algivora* (from Bick 1961), 93 μm; p. 291. **b:** *Holosticha algivora* (from Chardez 1981), 82 μm; p. 291. **c, d:** *Holosticha arenicola* (c, from Aladro Lubel 1985; d, from Aladro Lubel et al. 1990), 71 μm; p. 457. **e:** *Holosticha begoniensis* (from Fernandez-Leborans 1990), infraciliature of ventral side, size not indicated; p. 182. **f, g:** *Holosticha viridis* (f, from Aladro-Lubel et al. 1986; g, from Aladro Lubel et al. 1990), 73 μm; p. 291. **h:** *Holosticha wrzesniowskii punctata* (from Rees 1884), 85 μm; p. 187.

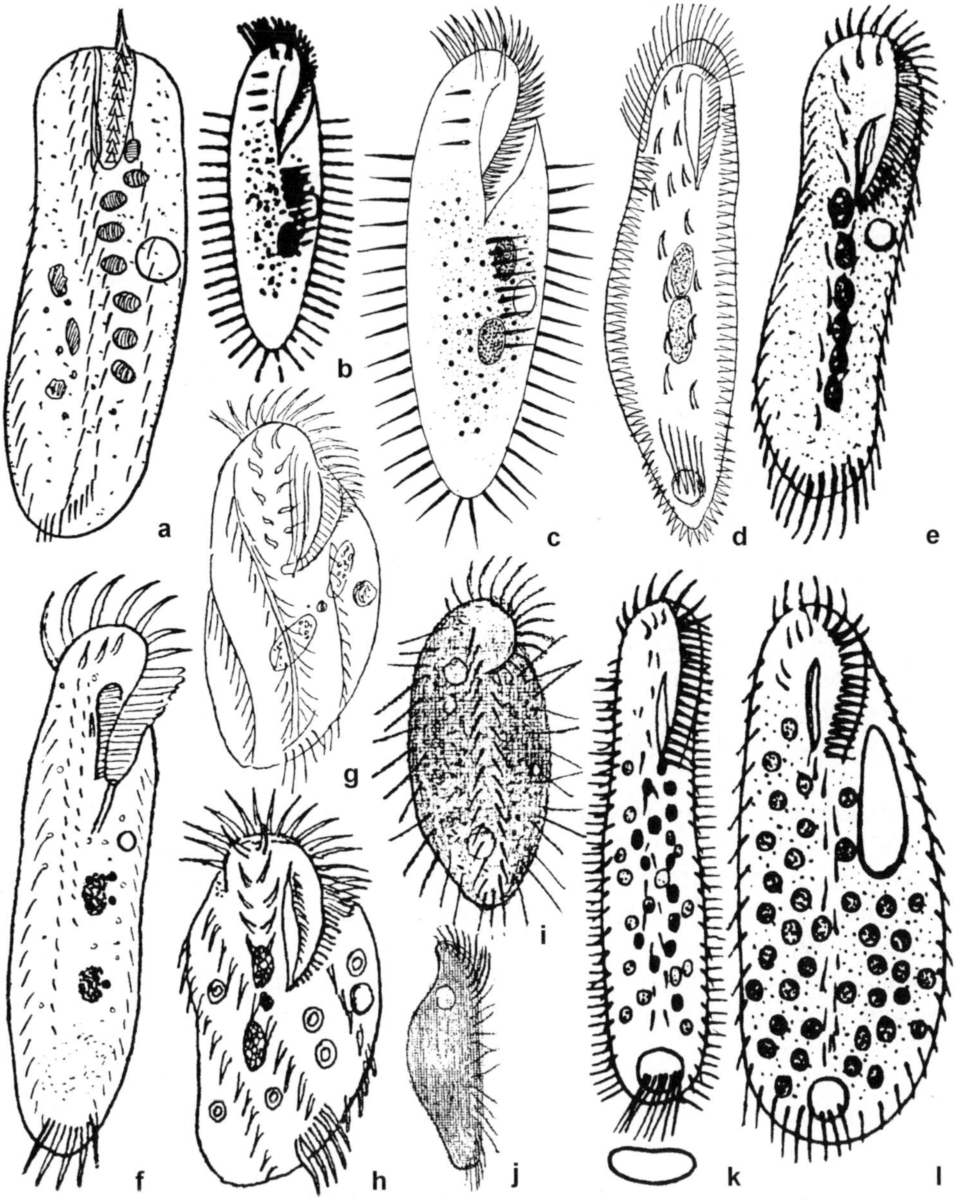

Fig. 100a–l Insufficient redescriptions and species indeterminata (ventral views from life unless otherwise indicated). **a:** *Holosticha fontinalis* (from Lepsi 1927), 126 μm; p. 182. **b, c:** *Holosticha longiseta* (b, from Lepsi 1951; c, from Lepsi 1965), 80 μm; p. 183. **d:** *Holosticha coronata* (after Gourret & Roeser 1888 from Kahl 1932), 250 μm; p. 95. **e:** *Keronopsis gracilis* (from Vuxanovici 1961), 120 μm; p. 457. **f:** *Holosticha manca* (from Ganapati & Rao 1958), 140 μm; p. 457. **g, h:** *Holosticha obliqua* (g, from Kahl 1928; h, after Kahl 1928 from Kahl 1932), 90 μm; p. 183. **i, j:** *Holosticha setigera* (from Conn 1905), ventral and right lateral view, 36 μm; p. 184. **k:** *Holosticha tenuiformis* (from Vuxanovici 1963), ventral view and optical transverse section, 80 μm; p. 186. **l:** *Holosticha vesiculata* (from Vuxanovici 1963), 80 μm; p. 186.

illustrated with six macronuclear nodules arranged in a longitudinal row. Contractile vacuole near left margin slightly behind level of buccal vertex. Cytoplasm translucent, in spite of ingested(?) algae. Cirral pattern not described, obviously three frontal cirri and a midventral complex extending to near transverse cirri, which are only slightly set off from marginal rows. Katharobic to mesosaprobic; in December many specimens in old(?) cultures from a lake (Lacul Fundeni) near/in Bucharest, Romania.

Keronopsis muscorum **Kahl** – Gellért, 1956, Acta biol. hung., 6: 94. Remarks: This specimen/population differs significantly from *H. muscorum* sensu Kahl (1932), for example, in body length (120 µm vs. 200–300 µm), number of buccal cirri (7 vs. 1–5), number of transverse cirri (5 vs. 6–9). The small size and the low number of transverse cirri prevents an identification with *Anteholosticha antecirrata*. Gellért's population had five dorsal kineties and a single globular macronucleus (this indicates that he observed an exconjugant), micronucleus not found. Contractile vacuole (pulsation interval about 10 sec) in ordinary position, with collecting canals. Cortical granules lacking. Movement slowly winding showing great flexibility. Rare. Feeds on hyphae and algae.

Keronopsis pulchra **Kahl, 1932** – Borror, 1972, J. Protozool., 19: 11, Fig. 22 (Fig. 90c). Remarks: The type population of *Holosticha (Keronopsis) pulchra* has three distinctly enlarged frontal cirri (Fig. 90a). In contrast, the illustration provided by Borror (1972) obviously shows a true *Pseudokeronopsis* with a more or less distinct bicorona. Since no further data, especially from life, are provided, Borror's illustration is classified as insufficient redescription (see also next paragraph).

Pseudokeronopsis pulchra **(Kahl, 1932) nov. comb.** – Borror & Wicklow, 1983, Acta Protozool., 22: 124, Fig. 20 (Fig. 90d). Remarks: The specimen illustrated by Borror & Wicklow has a very dense bicorona, whereas Kahl's populations have three enlarged frontal cirri (Fig. 90a). Thus, Borror & Wicklow's identification and transfer to *Pseudokeronopsis* cannot be accepted. Since they did not provide original live data, a reidentification of their *Pseudokeronopsis*-population is impossible. Consequently, the illustration is classified as insufficient redescription.

Diaxonella Jankowski, 1979

1979 ***Diaxonella* gen. n.** – Jankowski, Trudy zool. Inst., Leningr., 86: 83 (original description in Russian). Type species (by original designation on p. 83): *Diaxonella trimarginata* Jankowski, 1979.

1996 ***Diaxonella* Jankowski 1979**[1] – Oberschmidleitner & Aescht, Beitr. Naturk. Oberösterreichs, 4: 21 (improved diagnosis and redescription of type species; see remarks).

2001 ***Diaxonella* Jankowski 1979** – Aescht, Denisia, 1: 58 (catalogue of generic names of ciliates).

2001 ***Diaxonella* Jankowski, 1979** – Berger, Catalogue of ciliate names 1. Hypotrichs, p. 17 (nomenclator containing all basionyms, combinations, and higher taxa of hypotrichs).

Nomenclature: No derivation of the name is given in the original description. *Diaxonella* is obviously a composite of the Greek numeral *di-* (two), the Greek noun *ho áxon* (axis), and the diminutive Latin suffix *-ella*. I do not know whether it refers to the midventral complex whose cirral pairs form two longitudinal axis or to the (usually) two left marginal rows. Feminine gender because ending with *-ella* (ICZN 1999, Article 30.1.3).

Characterisation (A = supposed apomorphy): Adoral zone of membranelles continuous. 3 frontal cirri. Buccal cirrus present. More than 1 parabuccal cirrus. Usually 2 frontoterminal cirri. Midventral complex composed of midventral pairs only. Transverse cirri present. 1 right marginal row and 2 or more left marginal rows, which originate from a common anlage (A?). Caudal cirri absent.

Remarks: The systematics of *Diaxonella* is relatively complicated. Jerka-Dziadosz & Janus (1972) described a *Keronopsis rubra* (now this marine species is named *Pseudokeronopsis rubra*) from a pond near Warsaw. Jankowski (1979) recognised the misidentification and described a new species for this Polish freshwater population. Since it showed a unique combination of features, he also established *Diaxonella*, which differs from *Anteholosticha*, inter alia, by the increased number of left marginal rows and from *Trichototaxis* by the lack of a distinct bicorona. Consequently the validity of *Diaxonella* is beyond reasonable doubt.

It is a matter of taste how to synonymise the populations/species described so far because the differences among them are difficult to interpret. I basically follow Oberschmidleitner & Aescht (1996) who synonymised three species (*D. trimarginata*, *Holosticha polystylata*, *Pseudokeronopsis trisenestra*). During the revision of the urostyloids I found some further species, which very likely belong to *Diaxonella*, for example, *Keronopsis pseudorubra* and *Trichotaxis pulchra*. All these populations can be grouped in three taxa differing in habitat, colour, and midventral complex. One group comprises populations from limnetic habitats whose specimens are red and have a relatively low number of midventral pairs (around 20 or less). The other two groups have a distinctly higher number of midventral pairs (around 30 or more), but differ in habitat and colour (marine coastal waters vs. freshwater; more or less colourless vs. red). Moreover, the populations with the increased number of midventral pairs are so far only known from North America. As already mentioned, the differences are not very

[1] The improved diagnosis by Oberschmidleitner & Aescht (1996) is as follows: Holostichidae mit drei verstärkten Frontalcirren, einer Buccalcirrenreihe, einer Parabuccalcirrenreihe. Eine rechte und mehr als eine linke Marginalreihe. Frontoterminal- und Transversalcirren vorhanden.

pronounced, that is, show transitions. Consequently, I interpret these three groups as subspecies (see key below), which is also indicated by the geographical (Europe, Asia, Africa versus North America), respectively, physiological/ecological (limnetic vs. marine) separation. However, this is only a first trial to gain control of the complicated taxonomy of this group of urostyloids. For routine examinations in, for example, water quality assessment, I recommend the designation *Diaxonella pseudorubra* (Kaltenbach, 1960).

Diaxonella is assigned to the Holostichidae because it has three frontal cirri and the midventral complex is composed of cirral pairs only. Consequently, the formation of the left marginal rows from a common anlage must be interpreted as convergence to *Pseudourostyla* (Urostylidae), which has the same mode.

Trichototaxis aeruginosa is classified as incertae sedis in the present genus (see below). See also *Trichototaxis* for some doubtful species with more than one left marginal row.

Shao et al. (2005) described the species *Trichototaxis hembergeri*. However, according to the ICZN (1999, Article 9), abstracts of lectures and posters, when issued primarily to participants at a congress, do not constitute a published work. Thus, the name is not available from the congress abstract and also not from the present book because I disclaim this name for nomenclatural purposes (ICZN 1999, Article 8.3). In addition, the generic assignment is incorrect because *Trichototaxis* is characterised by a bicorona and not by three frontal cirri (see below). Very likely it belongs to *Diaxonella*. The grass-green cortical granules indicate that it is not identical with *D. pseudorubra* (however, synonymy cannot be excluded!). Body size about 95–185 × 25–50 μm; numerous macronuclear nodules; body flexible, slender; contractile vacuole at 25–33% of body length; body "kermes to rose-colored", likely mainly due to the grass-green and wine-coloured (likely they mean red wine!) cortical granules; about 37 adoral membranelles; 4 frontal cirri (possibly they mean the ordinary three frontal cirri and cirrus III/2, which is also often enlarged); 5–8 buccal cirri; frontal row (likely behind right frontal cirrus) composed of 3–4 cirri; 2 frontoterminal cirri; midventral complex composed of 16–29 pairs; 8–10 transverse cirri; three (plus a fragment) left marginal rows; one right marginal row; 3–4 dorsal kineties; caudal cirri lacking. In addition, Shao et al. (2005) summarised the morphogenesis as follows: (i) the entire parental ciliature, including the oral apparatus, is renewed; (ii) the oral primordium of the proter originates de novo and probably within a pouch beneath the parental buccal cavity; (iii) the right marginal row develops, as is usual, two intrakinetal anlagen; (iv) left marginal rows originate from a single anlage, which is formed to the right of the parental row; (v) no caudal cirri are formed; (vi) the macronuclear nodules fuse to a single mass. Found in a freshwater pond in Tsingtao, China.

Species included in *Diaxonella* (alphabetically arranged according to basionyms): (1) *Keronopsis pseudorubra* Kaltenbach, 1960 (comprising three subspecies, namely *Holosticha polystylata* Borror & Wicklow, 1983; *Keronopsis pseudorubra* Kaltenbach, 1960; and *Trichotaxis pulchra* Borror, 1972a). Incertae sedis: (2) *Trichotaxis aeruginosa* Foissner, 1980.

Single species

Diaxonella pseudorubra (Kaltenbach, 1960) comb. nov. (Fig. 101a–g, 102a–q, Tables 21, 22)

1960 ***Keronopsis pseudorubra* sp. n.** – Kaltenbach, Wass. Abwass. Wien, 1960: 165, Abb. 3c (Fig. 101a; original description; no type material available and no formal diagnosis provided).

1972 ***Keronopsis rubra* (Ehrbg., 1838)** – Jerka-Dziadosz & Janus, Acta Protozool., 10: 249, Fig. 1, 2, Plates I–VI, Fig. 1–33 (Fig. 102a; misidentification, see remarks; morphology and morphogenesis).

1972 ***Trichotaxis pulchra* n. sp.** – Borror, Arch. Protistenk., 10: 63, Fig. 52, Plate I, Fig. 55 (Fig. 101b; original description of subspecies; no formal diagnosis provided; type slides are deposited in the U. S. National Museum; Borror 1972a, p. 30).

1979 ***Diaxonella trimarginata* sp. n.** – Jankowski, Trudy zool. Inst., Leningr., 86: 83 (original description of synonym; established for *Keronopsis rubra* sensu Jerka-Dziadosz & Janus 1972; no formal diagnosis provided).

1982 ***Trichototaxis pulchra* Borror, 1972** – Hemberger, Dissertation, p. 118 (revision of hypotrichs).

1983 ***Holosticha polystylata* sp. n.** – Borror & Wicklow, Acta Protozool., 22: 112, Fig. 4, Table 1 (Fig. 102b; original description of subspecies; the type material is, according to Borror & Wicklow 1983, p. 100, deposited in the slide collection of Arthur C. Borror; no formal diagnosis provided).

1985 ***Keronopsis rubra*** – Zhang, Pang & Gu, Acta zool. sin., 31: 59, Fig. 1, 2, Plate I, II (Fig. 102m–q; misidentification; morphology and morphogenesis of Chinese population).

1987 ***Holosticha pseudorubra* (Kaltenbach, 1960) nov. comb.** – Foissner, Arch. Protistenk., 133: 225 (combination with *Holosticha*).

1987 ***Trichototaxis pulchra* (Borror, 1972) nov. comb.** – Foissner, Arch. Protistenk., 133: 226 (combination with *Trichototaxis*).

1991 ***Pseudokeronopsis trisenestra* nov. spec.**[1] – Dragesco & Dragesco-Kernéis, Europ. J. Protistol., 26: 230, Fig. 10, 11 (Fig. 102c–f; original description of synonym; locality where type material is deposited not mentioned).

1996 ***Diaxonella trimarginata* Jankowski 1979**[2] – Oberschmidleitner & Aescht, Beitr. Naturk. Oberösterreichs, 4: 21, Fig. 23–28, Tabelle 6 (Fig. 102g–l; authoritative redescription and designation of neotype, see remarks).

1997 ***Trichototaxis rubra***[3] – Plückebaum, Winkelhaus & Hauser, J. Euk. Microbiol., 44: 27A, Abstract 103 (original description of synonym; Fig. 101c–g are from Plückebaum's dissertation; see footnote k of Table 21b).

2001 ***Holosticha pseudorubra* (Kaltenbach, 1960) Foissner, 1987** – Berger, Catalogue of ciliate names 1. Hypotrichs, p. 44 (nomenclator containing all basionyms, combinations, and higher taxa of hypotrichs).

Nomenclature: In most cases no derivation of the names is given in the original descriptions. The species-group name *pseudorubra* is a composite of the Greek adjective *pseudo-* (wrong, lying) and the Latin adjective *rub·er -ra -um* (red) and obviously refers

[1] The diagnosis by Dragesco & Dragesco-Kernéis (1991) is as follows: *Pseudokeronopsis trisenestra* is, principally, characterized by its 3 (sometimes 4) left marginal kineties, its numerous frontal cirri and particular disposition of its transverse cirri.

[2] The improved diagnosis by Oberschmidleitner & Aescht (1996) is as follows: In vivo 120–180 × 30–70 µm, mit auffallenden goldgelben cortical Granula und diffus roter Plasmafärbung. 1 Buccal- und 1 Parabuccalcirrenreihe, 1 rechte und 2–4 linke Marginalreihen. Etwa 150 Makronucleus-Teile und 3–4 Dorsalkineten.

[3] The diagnosis by Plückebaum et al. (1997) is as follows: dark red to purple colour, three frontal cirri, two frontoterminal cirri, a distinct midventral row, one right and more than two left marginal rows, 8–9 transversal cirri, about 80 macronuclei and several micronuclei.

to the fact that this species has – like *Pseudokeronopsis rubra* – a red colour. "*Pseudokeronopsis pseudorubra* (Ehrb.)" in Humpesch & Moog (1994, p. 91) is obviously an unintended combination with *Pseudokeronopsis*. In addition, the author is incorrect. The species-group name *trimarginata* is a composite of the Greek numeral *tri-* (three) and the Latin adjective *margināt·us -a -um* (bordered, marginate, lined) and obviously alludes to the increased number (usually three in the population described by Jerka-Dziadosz & Janus 1972) of marginal rows. The species-group name *pulch·er -ra -um* (Latin adjective) means beautiful or excellent and obviously refers to the beauty of the cortical granulation in life and the striking preparations possible with the Nigrosin-mercuric chloride-formalin method (Borror 1972a, p. 64). The species-group name *polystylata* is a composite of the Greek adjective *poly-* (many, large, frequent, common), the Greek noun *styl-* (column, pillar; here the cirri are meant), and the suffix *-ata* (provided or supplied with) meaning a ciliate having many cirri because of the increased number of left marginal rows. *Holosticha polystilata* in Oberschmidleitner & Aescht (1996, p. 24, 27), *Holosticha polystylatu* in Wiackowski (1988, p. 4), and *Holostricha polystylata* and *Holosticha polystalata* in Croft et al. (2003, p. 344, 349; Dalby & Prescott 2004, p. 252) are incorrect subsequent spellings.

I am uncertain about the etymology of the species-group name *trisenestra* because I did not find the word *senestra*; likely it is another spelling of *sinistra* (Latin; left) and refers to the three (*tri-*) left marginal rows usually present. *Pseudokeronopsis trisinestra* in Oberschmidleitner & Aescht (1996, p. 21, 24) is an incorrect subsequent spelling.

Remarks: The taxonomy of this red, limnetic urostyloid is rather complicated because several more or less similar, partially superficially described populations/species are known.

Keronopsis pseudorubra was discovered in the Danube River in Vienna. Kaltenbach's description of this species is not very detailed so that some differences to the authoritative redescription of the junior synonym *Diaxonella trimarginata* by Oberschmidleitner & Aescht (1996) must not be over-interpreted. For example, Kaltenbach neither described nor illustrated a second left marginal row or the frontal cirri. However, he also overlooked the left marginal row in his *Holosticha danubialis*, a junior synonym of *Holosticha pullaster*. On the other hand he correctly illustrated several buccal cirri and recognised the red colour. Kaltenbach (1960) assigned his species to *Keronopsis* (now *Pseudokeronopsis*), a genus characterised by a bicorona. Since the present species lacks such a double row of frontal cirri, Foissner (1987d) transferred it to *Holosticha*. I consider *K. pseudorubra* as senior synonym of *D. trimarginata* and therefore transfer it from *Keronopsis* to the present genus. However, since one cannot exclude that they are distinct species, I keep the description of *K. pseudorubra* separate. Kaltenbach's species was overlooked by Borror (1972) and Borror & Wicklow (1983).

Borror (1972a) likely established his species in *Trichototaxis* because of the two left marginal rows. However, the type species of *Trichototaxis* (*T. stagnatilis*) has a conspicuous bicorona, whereas the present species has three enlarged frontal cirri. Fig. 101b shows, admittedly not very clearly, these three cirri and two smaller (parabuccal) cirri behind. Borror did not mention the frontal cirri explicitly, but he writes "midventral cirri extending entire length of ventral surface, 9 µm long, except anterior most 3–5 that are approximately 12 μm long". The different length and the illustration clearly proves that *T. pulchra* has three enlarged frontal cirri plus some (two) parabuccal cirri. Consequently, the cirral pattern perfectly matches that of *Diaxonella pseudorubra*, type of the genus. The present species is not included in the review by Borror (1972). Hemberger (1982, p. 118) synonymised it with *Keronopsis rubra* sensu Jerka-Dziadosz & Janus (1972), which is the type population of *Diaxonella trimarginata*, a junior synonym of the type species. However, Borror's (1972a) species differs from *D. pseudorubra*, inter alia, in that it lacks the red colour. Thus, it is classified as subspecies.

As already mentioned in the genus section, Jankowski (1979) recognised that *Keronopsis rubra* sensu Jerka-Dziadosz & Janus (1972) is a misidentification which led him to establish a new species and a new genus.

Borror & Wicklow (1983, p. 121) caused great confusion in the taxonomy of the present species in that they put Borror's (1972a) *Trichototaxis pulchra* (Fig. 101b) into the synonymy of a new species *Holosticha estuarii* Borror & Wicklow, 1983 (= *Anteholosticha estuarii* in present book; Fig. 86a). This procedure comprised two incomprehensible steps, namely, (i) it is impossible to describe a new species and, simultaneously, to synonymise it with an older species; and (ii) *Trichototaxis pulchra* and *Holosticha estuarii* are very likely not identical because they have, inter alia, a different number of left marginal rows (two vs. one; note that both populations are well described) and midventral pairs (about 44 vs. 14). Moreover, Borror & Wicklow (1983) described *Holosticha polystylata* (Fig. 102b), which very closely resembles *T. pulchra*. However, according to the descriptions, *H. polystylata* is "pinkish red", whereas in *T. pulchra* the "cytoplasm is generally relatively clear with pale brown inclusions and diatom shells", that is, lacks a red colour. I do not think that "pinkish red" and "pale brown inclusions" describe the same feature and therefore do not synonymise *H. polystylata* and *T. pulchra*, which, moreover, live in different habitats (freshwater against coastal wetlands, that is, marine habitats).

Holosticha polystylata matches *Diaxonella pseudorubra* rather well, except for the number of midventral cirri, which is distinctly higher in *H. polystylata* (29–45 cirral pairs) than in the *D. pseudorubra* populations (on average 17–20, maximum 26). Borror & Wicklow recognised the conspecificity of their species and the Polish population described by Jerka-Dziadosz & Janus (1972). Moreover, they knew Jankowski's paper, but obviously overlooked that Jankowski had already established a new species for the Polish population. Oberschmidleitner & Aescht (1996) therefore synonymised these two species. In my catalogue (Berger 2001) I wrote that *H. polystylata* is an objective synonym of *Diaxonella trimarginata*. However, this is incorrect because the two species do not have the same name-bearing type, that is, are not based on the same population, respectively, specimen.

The papers by Zhang et al. (1985a) and Pang et al. (1986), which were overlooked by all subsequent authors, are in Chinese and are therefore only briefly mentioned. Obviously there is no distinct difference to the Polish and Austrian populations described by Jerka-Dziadosz & Janus (1972) and Oberschmidleitner & Aescht (1996).

Pseudokeronopsis trisenestra, described by Dragesco & Dragesco-Kernéis (1991), is undoubtedly a further synonym of *D. trimarginata* (Oberschmidleitner & Aescht

1996). Dragesco & Dragesco-Kernéis obviously studied Borror & Wicklow's revision, but did not recognise the conspecificity of their species and *Holosticha polystylata*. Dragesco & Dragesco-Kernéis argued that they cannot establish their species in *Holosticha* because this group is characterised by only two marginal rows (one left, one right) and few frontal cirri. Obviously they misinterpreted the frontal ciliature of their species as bicorona and thus included it in *Pseudokeronopsis*, which has such a double row of frontal cirri, but – like *Holosticha* – also only one left and one right marginal row.

Borror & Wicklow (1983) and Dragesco & Dragesco-Kernéis (1991) assigned their species to *Holosticha*, respectively, *Pseudokeronopsis*, which contained only species with just one left and one right marginal row. Thus, Jankowski's decision to establish *Diaxonella* for the present species, which has more than one left marginal row, is comprehensible (Oberschmidleitner & Aescht 1996). The number of marginal rows is also used to characterise some oxytrichids, for example, *Allotricha*, *Coniculostomum*, *Pleurotricha*, which usually have, however, the increased number of marginal rows on the right side (for review see Berger 1999). Surprisingly, *Allotricha* has the same, or at least a very similar type of marginal row formation as *Diaxonella*, namely, from a common anlage which originates from the rightmost right marginal row, respectively, the rightmost left marginal row. At the present state of knowledge I assume that this mode of marginal row formation evolved twice independently; by contrast, marginal row formation proceeds quite differently in *Coniculostomum*.

Within the urostyloids, the marginal row formation from a common anlage occurs also in *Pseudourostyla*. Thus one cannot exclude that *Diaxonella* and *Pseudourostyla* are related with this type of marginal row formation as synapomorphy. Possibly they form a monophyletic group together with *Trichototaxis* (with *T. stagnatilis* as type), which also has an increased number of marginal rows. Unfortunately, nothing is known about the morphogenesis of *T. stagnatilis*. Dragesco & Dragesco-Kernéis (1991) noted that their species perhaps belongs to a new genus which is closely related to *Trichototaxis* and *Pseudokeronopsis*. Hemberger (1982, p. 104) and Wirnsberger et al. (1987, p. 84) even stated that *Diaxonella trimarginata* belongs to *Trichototaxis* because of the increased number of left marginal rows. However, *Trichototaxis stagnatilis* has, like *Pseudourostyla*, a distinct bicorona, whereas *D. trimarginata* has only three frontal cirri (Oberschmidleitner & Aescht 1996). This is the major reason why I classify *Diaxonella* in the Holostichidae and not in the Urostylidae.

Trichototaxis rubra matches the neotype population described by Oberschmidleitner & Aescht (1996) rather perfectly. I recommended that H. Plückebaum identifies his population as *D. trimarginata*. However, he unnecessarily introduced the new name *T. rubra* (Plückebaum et al. 1997).

As already mentioned above, all populations listed in the synonymy agree very well. Only the number of midventral cirral pairs is rather different (Table 21a, b), namely, 14–22 (on average 17) in Plückebaum's population, about 17 in the Polish population, 17–26 in Dragesco & Dragesco-Kernéis' population, 18–27 (on average 24) in Oberschmidleitner & Aescht's population, but about 29–45 in Borror & Wicklow's population, and about 44 in *Trichototaxis pulchra* described by Borror (1972a). Moreover, the latter population is marine and likely colourless indicating that *D. pseudorubra* consists of subspecies. However, further populations, mainly from North America, have to be studied to prove of or disprove this hypothesis.

Oberschmidleitner & Aescht (1996, p. 7), who described an Austrian population of *Diaxonella trimarginata*, deposited one neotype slide in the Oberösterreichische Landesmuseum in Linz. However, they did not discuss the exceptional need of a neotype as demanded by the ICZN (1984, p. 157; 1999, p. 84). Furthermore, some qualifying conditions (ICZN 1984, Article 75) are not published by Oberschmidleitner & Aescht, for instance, the authors' reasons for believing the types to be lost or destroyed, and the steps that had been taken to trace them (Article 75, (d) (3)). To clear up the situation, I supplement the following particulars to Articles 75.3.1 to 75.3.7 of the ICZN (1999):

(i) Oberschmidleitner & Aescht (1996) designated a neotype to characterise *D. trimarginata* (= *D. pseudorubra* in present book) objectively. Furthermore, the neotypification helps to clarify the problem of whether or not *D. pseudorubra* consists of subspecies. The populations studied by Jerka-Dziadosz & Janus (1972), Oberschmidleitner & Aescht (1996), and some other authors have a distinctly lower number of midventral pairs than the American populations (values see above) described by Borror (1972a) and Borror & Wicklow (1983), indicating the presence of allopatric, respectively, physiological/ecological (limnetic vs. marine) subspecies. And finally, Hemberger (1982) suggested the inclusion of *D. trimarginata* in *Trichototaxis*, whose little known type species, however, obviously has a bicorona.

(ii) For a differentiation of *D. pseudorubra* from other taxa, see below.

(iii) For a detailed description of the neotype specimen and population, see Oberschmidleitner & Aescht (1996) and the morphology chapter below.

(iv) Jankowski (1979) introduced the name *Diaxonella trimarginata* for *Keronopsis rubra* sensu Jerka-Dziadosz & Janus (1972). They made protargol preparations which are likely deposited in the laboratory of the senior author. Because of the misidentification, they likely did not realise the importance of the slides with the type material of *Diaxonella trimarginata*. On February 1, 2002, I wrote to Jerka-Dziadosz about the whereabouts of these slides. I did not get an answer, which leads me to believe that the type slides are lost, destroyed, or inadequate, for example, faded. Thus, the designation of a neotype by Oberschmidleitner & Aescht (1996) seems justified. However, there is no doubt that the rediscovery of the Polish slides would set aside the neotype (ICZN 1999, Article 75.8).

(v) The identity of the neotype and the material presented in the original description (Jerka-Dziadosz & Janus 1972, Jankowski 1979) is beyond reasonable doubt and agrees very well with data from other sources (see list of synonyms and description below).

(vi) According to Article 75.3.6 of the ICZN (1999), the neotype should come as nearly as practicable from the original type locality. Considering that *D. pseudorubra*, respectively, its synonym *D. trimarginata* is a widely distributed species (Eurasia, North America, Africa), the distance (about 650 km) between Warsaw in Poland (original type locality) and Linz in Upper Austria (locality of neotype) is relatively short. The site in Poland was a pond, whereas the neotype was found in activated sludge containing several mesosaprobic ciliate species. Considering that the type locality of the senior synonym *Keronopsis pseudorubra* is in the Danube River near Vienna, the distance to the sample site of the neotype population of *D. trimarginata* is only about 200 km.

(vii) The neotype is deposited in the collection of microscopic slides of the Oberösterreichische Landesmuseum in Linz/Donau (LI), Austria since 1996 (Oberschmidleitner & Aescht 1996, p. 7). Interestingly, the slide is not mentioned by Aescht (2003, p. 399).

Diaxonella pseudorubra is rather easily identified because of the conspicuous homogeneously wine-red colour of the cytoplasm, the golden cortical granules, and the increased number of left marginal rows. However, several other red hypotrichs are known. For a separation from the similar limnetic *Trichototaxis aeruginosa*, see below.

The marine *Pseudokeronopsis rubra* has, inter alia, a distinct bicorona and only one left and one right marginal row. The marine *Metaurostylopsis rubra* is longer (150–300 µm) and has, inter alia, more marginal rows (at least six per side; Fig. 133a, h). *Anteholosticha estuarii*, which has a similar size and cirral pattern on the frontal region (several buccal and parabuccal cirri), sometimes has (one of four specimens) more than one left marginal row (Fig. 86a). However, this species from tidal marshes lacks the conspicuous red colour of the cytoplasm. *Rubrioxytricha* species, which have an orange to distinctly pink or brownish cytoplasm, are 18-cirri oxytrichids with one left and one right marginal row and therefore much fewer cirri on the ventral side (for review see Berger 1999, p. 479). Furthermore, they have only two macronuclear nodules against many in *D. pseudorubra* (Foissner & Berger 1996, p. 440). In my experience, *Rubrioxytricha haematoplasma* (Blatterer & Foissner, 1990) Berger, 1999 is more common in running waters than *D. pseudorubra*.

Key to *Diaxonella pseudorubra* subspecies

If you cannot decide on one of the three subspecies briefly characterised in Table 21a below, choose *Diaxonella pseudorubra* (see remarks for details).

Table 21a Brief characterisation of the *Diaxonella pseudorubra* subspecies

Limnetic; cytoplasm red; on average 20 or less midventral pairs (Europe, Asia, Africa)	Limnetic; cytoplasm reddish pink; about 30 or more midventral pairs (North America)	Marine; cytoplasm more or less colourless; about 30 or more midventral pairs (North America)
Diaxonella pseudorubra pseudorubra	*Diaxonella pseudorubra polystylata*	*Diaxonella pseudorubra pulchra*

Diaxonella pseudorubra pseudorubra (Kaltenbach, 1960) comb. nov., stat. nov.
(Fig. 101a, c–g, 102a, c–q, Tables 21a, b, 22)

1960 ***Keronopsis pseudorubra* sp. n.** – Kaltenbach, Wass. Abwass. Wien, 1960: 165, Abb. 3c (Fig. 101a; original description; no type material available and no formal diagnosis provided).

1972 ***Keronopsis rubra* (Ehrbg., 1838)** – Jerka-Dziadosz & Janus, Acta Protozool., 10: 249, Fig. 1, 2, Plates I–VI, Fig. 1–33 (Fig. 102a; misidentification, see remarks; morphology and morphogenesis).
1979 ***Diaxonella trimarginata* sp. n.** – Jankowski, Trudy zool. Inst., Leningr., 86: 83 (original description of synonym; established for *Keronopsis rubra* sensu Jerka-Dziadosz & Janus 1972; no formal diagnosis provided).
1985 ***Keronopsis rubra*** – Zhang, Pang & Gu, Acta zool. sin., 31: 59, Fig. 1, 2, Plate I, II (Fig. 102m–q; misidentification; morphology and morphogenesis of Chinese population).
1987 ***Holosticha pseudorubra* (Kaltenbach, 1960) nov. comb.** – Foissner, Arch. Protistenk., 133: 225 (combination with *Holosticha*).
1991 ***Pseudokeronopsis trisenestra* nov. spec.** – Dragesco & Dragesco-Kernéis, Europ. J. Protistol., 26: 230, Fig. 10, 11 (Fig. 102c–f; original description of synonym; locality where type material is deposited not mentioned).
1996 ***Diaxonella trimarginata* Jankowski 1979** – Oberschmidleitner & Aescht, Beitr. Naturk. Oberösterreichs, 4: 21, Fig. 23–28, Tabelle 6 (Fig. 102g–l; authoritative redescription and designation of neotype, see remarks).
1997 ***Trichototaxis rubra*** – Plückebaum, Winkelhaus & Hauser, J. Euk. Microbiol., 44: 27A, Abstract 103 (original description of synonym; Fig. 101c–g; see footnote k of Table 21b).
2001 ***Holosticha pseudorubra* (Kaltenbach, 1960) Foissner, 1987** – Berger, Catalogue of ciliate names 1. Hypotrichs, p. 44 (nomenclator containing all basionyms, combinations, and higher taxa of hypotrichs).

Nomenclature: This is the nominotypical subspecies. For details see *D. pseudorubra*.

Remarks: See also this chapter at *D. pseudorubra*. As mentioned above, the descriptions and redescriptions mentioned in the list above agree very well. Thus, the following paragraphs are based, unless otherwise indicated, on the detailed redescription of *D. trimarginata* by Oberschmidleitner & Aescht (1996), supplemented by additional or deviating observations made by Jerka-Dziadosz & Janus (1972), Wiackowski (1988), and Dragesco & Dragesco-Kernéis (1991). The sparse and somewhat superficial data by Kaltenbach (1960) are kept separate because one cannot exclude that a species with a single left marginal row exists. Thus the review can be used also by those who do not accept the synonymy of *D. trimarginata* and *K. pseudorubra*.

Morphology: Neotype population described by Oberschmidleitner & Aescht (1996) in life 120–180 × 30–70 μm, body length:width ratio around 4:1 in life (Fig. 102g) and 2.8:1 on average in protargol preparations (Table 21b); length according to Jerka-Dziadosz & Janus 130–170 μm (values likely from protargol preparations), according to Schmitz (1986; identified as *Trichototaxis pulchra*) 140–180 × 60–80 μm; length:width ratio of the synonym *Pseudokeronopsis trisenestra* 3:1 on average after protargol impregnation (Dragesco & Dragesco-Kernéis; Table 21b); Chinese population 100 to 120 μm long (likely after protargol impregnation). Body very flexible; outline elongate rectangular with posterior end somewhat broader rounded than anterior, right margin usually convex, left straight (Fig. 102g); dorsoventrally flattened by about 2:1. Macronuclear nodules scattered, globular to ellipsoidal (length:width ratio 1.6:1 in protargol preparations; Table 21b), with globular nucleoli. Micronuclei scattered, ellipsoidal (Fig. 102l). Contractile vacuole near left body margin slightly ahead of mid-body, with longitudinal collecting canals (Fig. 102g); according to Dragesco & Dragesco-Kernéis, the vacuole is near the adoral zone. Cortical granules between cirral rows; laterally and dorsally in short rows and irregular groups scattered throughout cortex. Individual cortical granules conspicuous because 0.5–1.5μm across and golden; in between, smaller granules

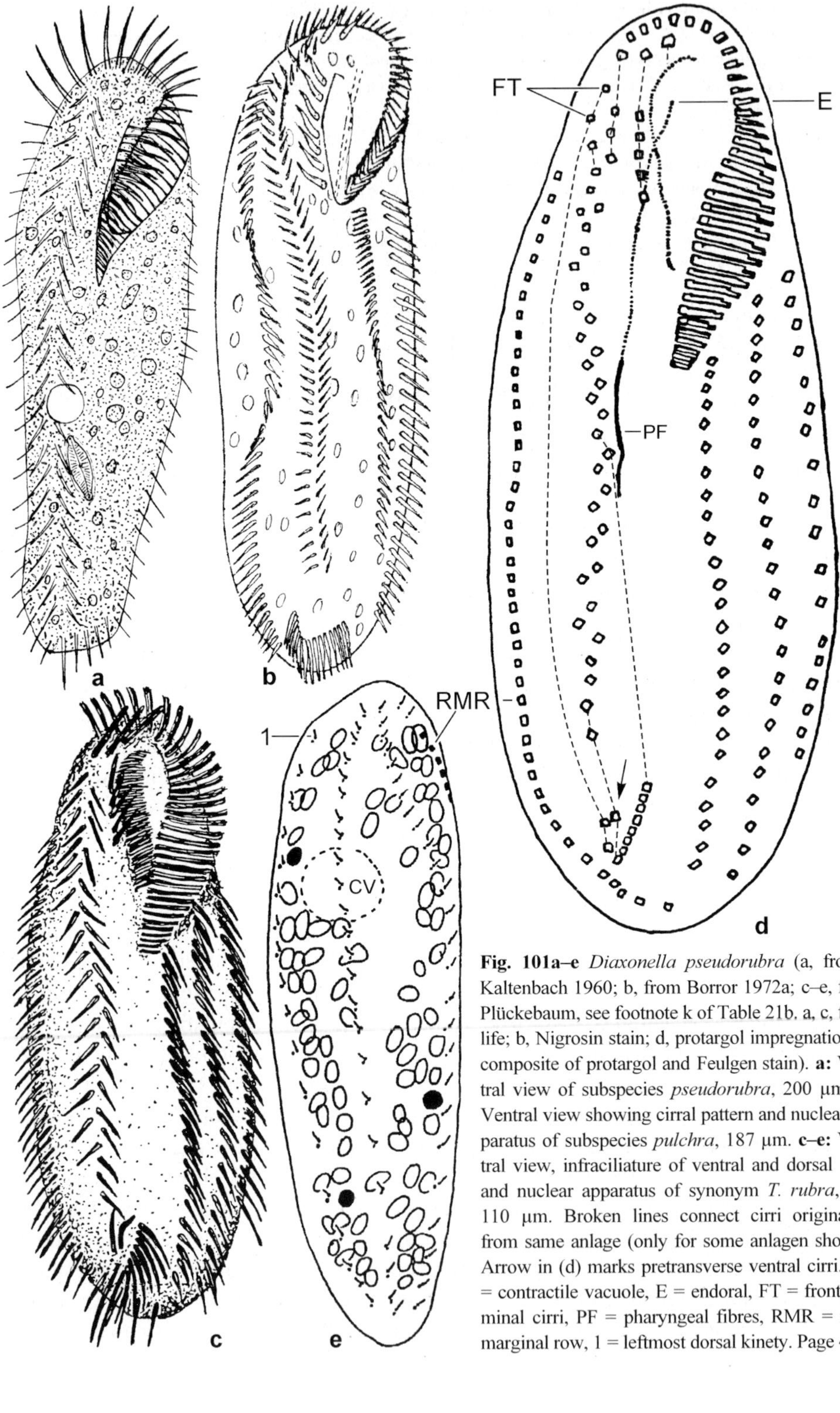

Fig. 101a–e *Diaxonella pseudorubra* (a, from Kaltenbach 1960; b, from Borror 1972a; c–e, from Plückebaum, see footnote k of Table 21b. a, c, from life; b, Nigrosin stain; d, protargol impregnation; e, composite of protargol and Feulgen stain). **a:** Ventral view of subspecies *pseudorubra*, 200 μm. **b:** Ventral view showing cirral pattern and nuclear apparatus of subspecies *pulchra*, 187 μm. **c–e:** Ventral view, infraciliature of ventral and dorsal side, and nuclear apparatus of synonym *T. rubra*, a = 110 μm. Broken lines connect cirri originating from same anlage (only for some anlagen shown). Arrow in (d) marks pretransverse ventral cirri. CV = contractile vacuole, E = endoral, FT = frontoterminal cirri, PF = pharyngeal fibres, RMR = right marginal row, 1 = leftmost dorsal kinety. Page 463.

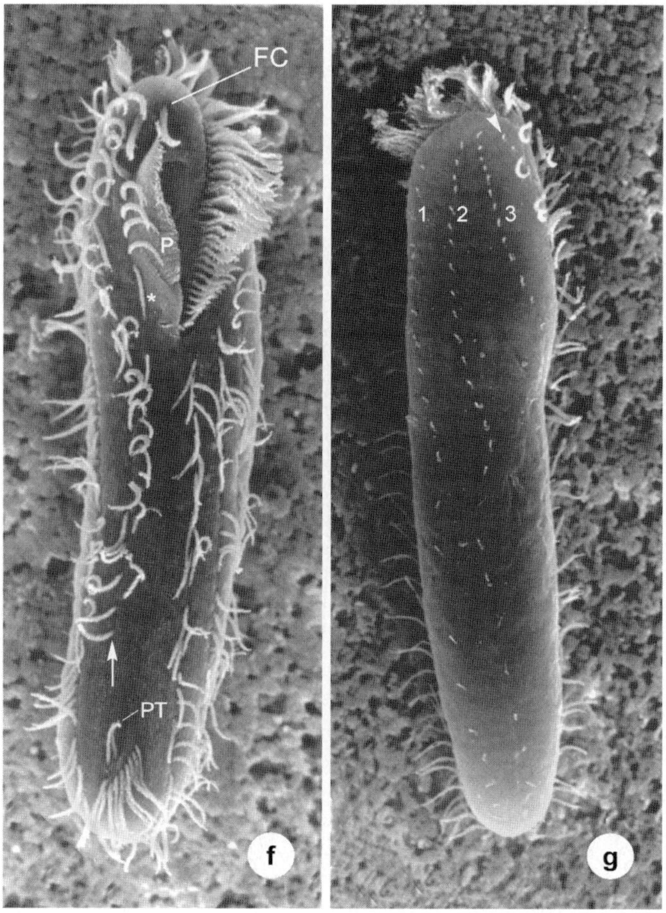

Fig. 101f, g *Diaxonella pseudorubra* (from Plückebaum, see footnote k of Table 21b. Scanning electron micrographs). Ventral and dorsal view showing cirral pattern and arrangement of dorsal kineties. The specimen shown in (f) has three left marginal rows. Arrow in (f) marks rear end of midventral complex, asterisk denotes rearmost buccal cirrus. Arrowhead in (g) marks the two dorsal bristles ahead of the right marginal row. FC = leftmost frontal cirrus, P = paroral, PT = pretransverse ventral cirri, 1–3 = dorsal kineties. Page 463.

which are red in bright field, but yellow in interference contrast. Cortical granules stain intensively crimson, but do not explode when methyl green-pyronin is added; impregnate with protargol (Fig. 102h, i). Cytoplasm conspicuous because of diffuse (homogeneous) red colour, which depends on nutritional condition; well feed specimens with many food vacuoles and greasily shining globules 5–25 μm across are intensively red; starved specimens pale and transparent, cortical granules, however, keep yellow colour (Fig. 102h, i). Polish specimens also dark pink to purple, depending on culture conditions (Jerka-Dziadosz & Janus); African specimens pink in colour, with small red pigment granules in endoplasm (Dragesco & Dragesco-Kernéis). According to Wiackowski (1988), *Holosticha polystylata* lacks mucocysts, that is, cortical granules. However, his *Urostyla grandis* populations also lack cortical granules indicating some misobservations concerning this feature because *U. grandis* always has such organelles. Movement restless, rapidly gliding or swimming while turning around main body axis.

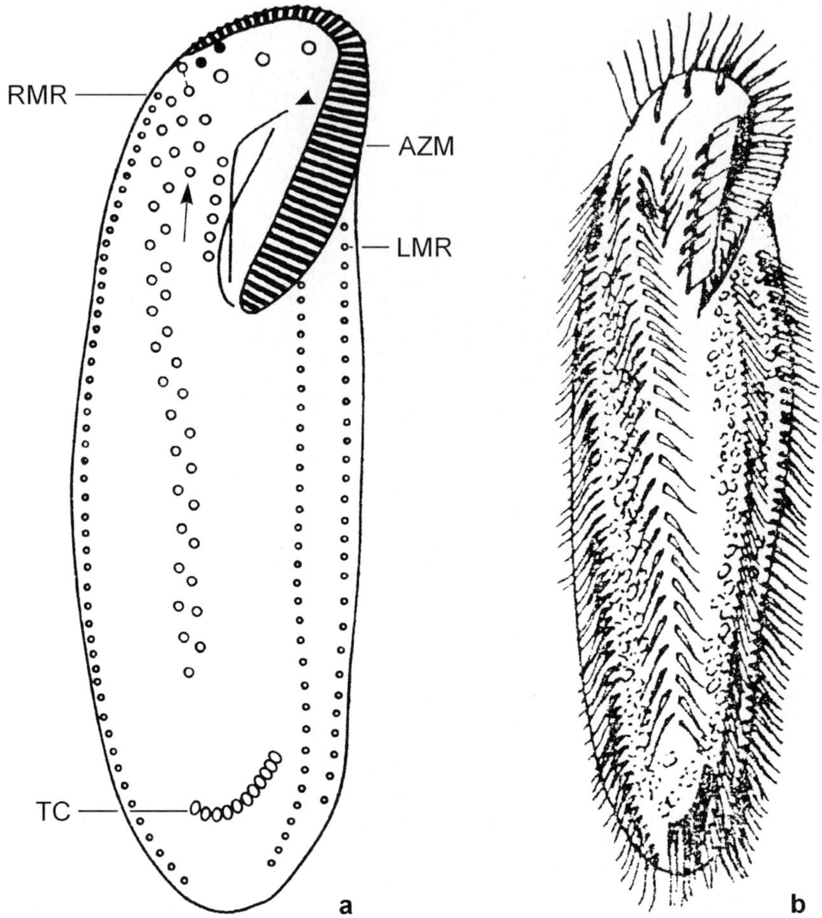

Fig. 102a, b *Diaxonella pseudorubra* (a, from Jerka-Dziadosz & Janus 1972; b, from Borror & Wicklow 1983. a, protargol impregnation; b, protargol impregnation or nigrosin stain). **a:** Infraciliature of ventral side of subspecies *pseudorubra*, slightly schematic; size not indicated. Arrow marks the parabuccal cirral row (originates from same anlage as right frontal cirrus). The two black cirri near the distal end of the adoral zone are likely the frontoterminal cirri, the broken line likely connects the anterior most cirral pair of the midventral complex (however, it cannot be excluded that this specimen has 3 frontoterminal cirri). Jerka-Dziadosz & Janus designated the paroral (arrowhead) as inner undulating membrane (= endoral). **b:** Infraciliature of ventral side and macronuclear nodules of subspecies *polystylata*, 152 µm. Morphometric data of this species: 3 frontal cirri; 7 buccal cirri; 6 parabuccal cirri; frontoterminal cirri not recognisable; 29 midventral pairs; transverse cirri not countable; about 59 right marginal cirri; about 50 cirri in inner left marginal row; about 58 cirri in outer left marginal row; adoral zone occupies about 31% of body length, composed of about 36 membranelles; midventral complex terminates at 80% of body length. AZM = adoral zone of membranelles, LMR = left marginal row 2 (= outer left marginal row), RMR = right marginal row, TC = transverse cirri. Page 463.

Adoral zone occupies about 40% of body length, extends distally near right frontal cirrus, composed of an average of 36 membranelles of usual shape and structure (Fig. 102g, j); cilia up to 15µm long in life. Buccal cavity large and deep and thus bright in

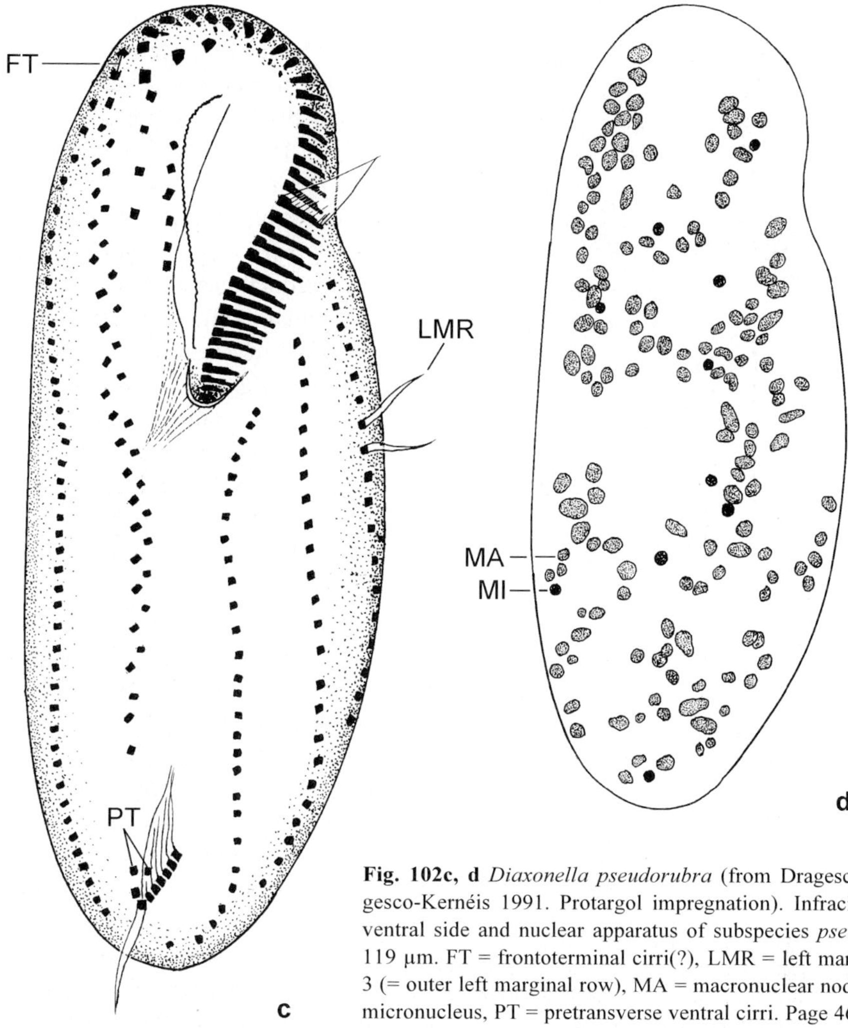

Fig. 102c, d *Diaxonella pseudorubra* (from Dragesco & Dragesco-Kernéis 1991. Protargol impregnation). Infraciliature of ventral side and nuclear apparatus of subspecies *pseudorubra*, 119 μm. FT = frontoterminal cirri(?), LMR = left marginal row 3 (= outer left marginal row), MA = macronuclear nodule, MI = micronucleus, PT = pretransverse ventral cirri. Page 463.

transmitted light, right laterally bordered by wide lip, anterior margin hook-shaped (Fig. 102g). Paroral and endoral composed of two kineties each, both distinctly curved leftwards anteriorly, and optically intersecting in mid-portion. Pharyngeal fibres difficult to recognise in life, extend into posterior body half.

Cirral pattern and number of cirri of usual variability, except for the number of cirri in the fourth left marginal row, which ranges from zero to 13 (Table 21b). Three slightly enlarged, about 12 μm long frontal cirri. Buccal cirral row extends along the right of the mid-portion of paroral. On average five cirri behind right frontal cirrus forming parabuccal cirral row terminating about at level of optical intersection of undulating membranes in neotype specimen illustrated (Fig. 102j). Usually two, occasionally three frontoterminal cirri close to distal end of adoral zone, difficult to recognise in life

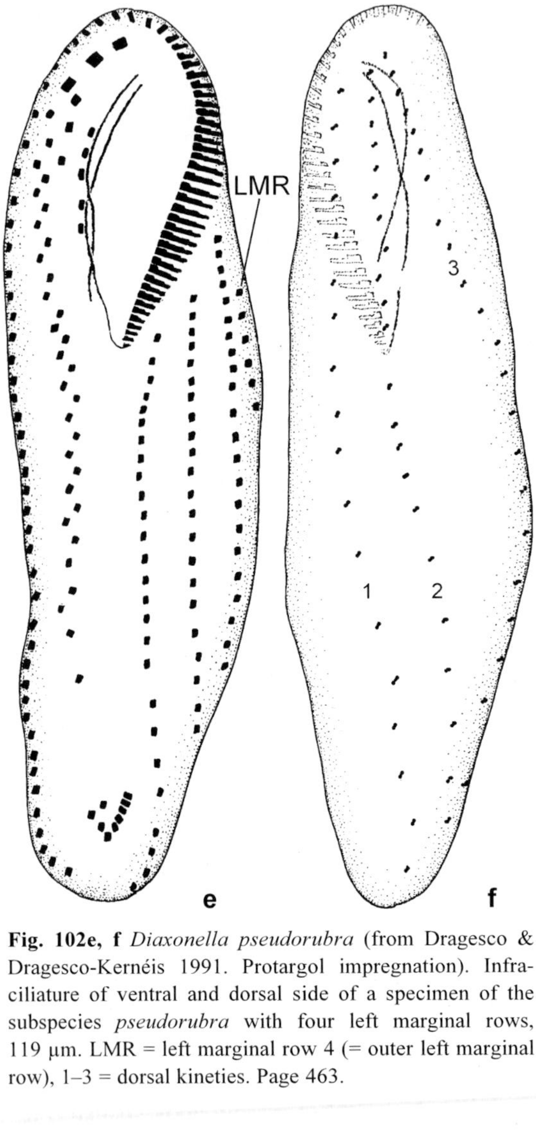

Fig. 102e, f *Diaxonella pseudorubra* (from Dragesco & Dragesco-Kernéis 1991. Protargol impregnation). Infraciliature of ventral and dorsal side of a specimen of the subspecies *pseudorubra* with four left marginal rows, 119 µm. LMR = left marginal row 4 (= outer left marginal row), 1–3 = dorsal kineties. Page 463.

(Fig. 102j; Wiackowski 1988, feature 15, state 1); frontoterminal cirri neither mentioned nor clearly illustrated by Jerka-Dziadosz & Janus[1] (Fig. 102a) and Dragesco & Dragesco-Kernéis (Fig. 102c, e). Midventral complex composed of cirral pairs only, commences near right frontal cirrus, extends slightly sigmoidal somewhat left of cell midline to near (84% of body length in Fig. 102j) transverse cirri; cirri of each pair basically of same size. In specimen shown in Fig. 102a, the midventral complex terminates at 72% of body length (Jerka-Dziadosz & Janus). Three to four "ventral cirri" present; Oberschmidleitner & Aescht, however, neither described nor labelled them so that I do not know which cirri are meant; possibly these ventral cirri are the pretransverse ventral cirri according to my terminology. However, the specimen shown in Fig. 102j obviously has only one such cirrus. African population (Fig. 102c, e; Dragesco & Dragesco-Kernéis) and population described by Plückebaum et al. (1997; Fig. 101d, f) usually with two pretransverse ventral cirri ahead of rightmost transverse cirri. Transverse cirri form J-shaped figure near rear body end, not enlarged, about 15 µm long and projecting beyond rear body end. Right marginal row begins dorsolaterally and terminates behind transverse cirri in cell midline; ahead of right marginal row usually few (according to Plückebaum invariable [n = 15] two) dorsal bristles (Fig. 101j, k). Left marginal rows commence close to left margin of adoral zone, first and second terminate slightly behind or at level of transverse cirri, third row ends subterminally,

[1] Unfortunately, the frontoterminal cirri are also not unequivocally recognisable in the 33 micrographs presented on Plates I–IV by Jerka-Dziadosz & Janus (1972); only in Figure 12 of this paper is the rearmost cirral pair of the proter cirral anlage slightly dislocated anteriad, indicating that these are the migrating frontoterminal cirri.

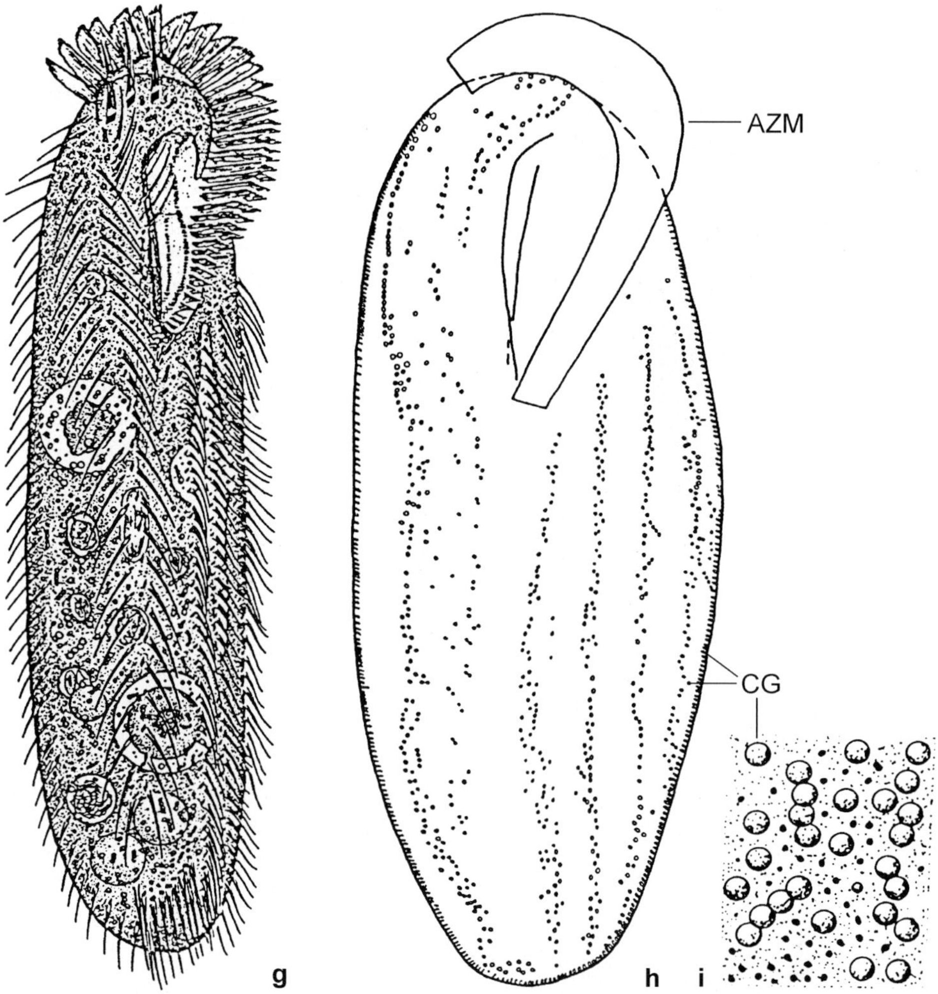

Fig. 102g–i *Diaxonella pseudorubra* (subspecies *pseudorubra*; neotype population of the synonym *D. trimarginata* from Oberschmidleitner & Aescht 1996. g, from life; h, i, protargol impregnation). **g:** Ventral view of a representative specimen showing, inter alia, the contractile vacuole and the deep and thus bright buccal cavity, 153 µm. **h, i:** Ventral view showing arrangement of cortical granules and detail. AZM = adoral zone of membranelles, CG = cortical granules. Page 463.

and fourth row, if present at all, usually rather short (Fig. 102j). Marginal cirri about 9 µm long. Dorsal cilia 3–4 µm long in life, usually arranged in three bipolar kineties. Caudal cirri lacking (Fig. 102j, k). Dorsal kinety 1 composed of 14–21 (average 18.3; n = 14), kinety 2 of 17–27 (average 23.6), and kinety 3 of 13–31 (average 22.6) bristles (Fig. 102k).

Description of *K. pseudorubra* by Kaltenbach (1960): Body length 200 µm in life. Body outline roughly club-shaped, that is, anterior portion wider than posterior, both

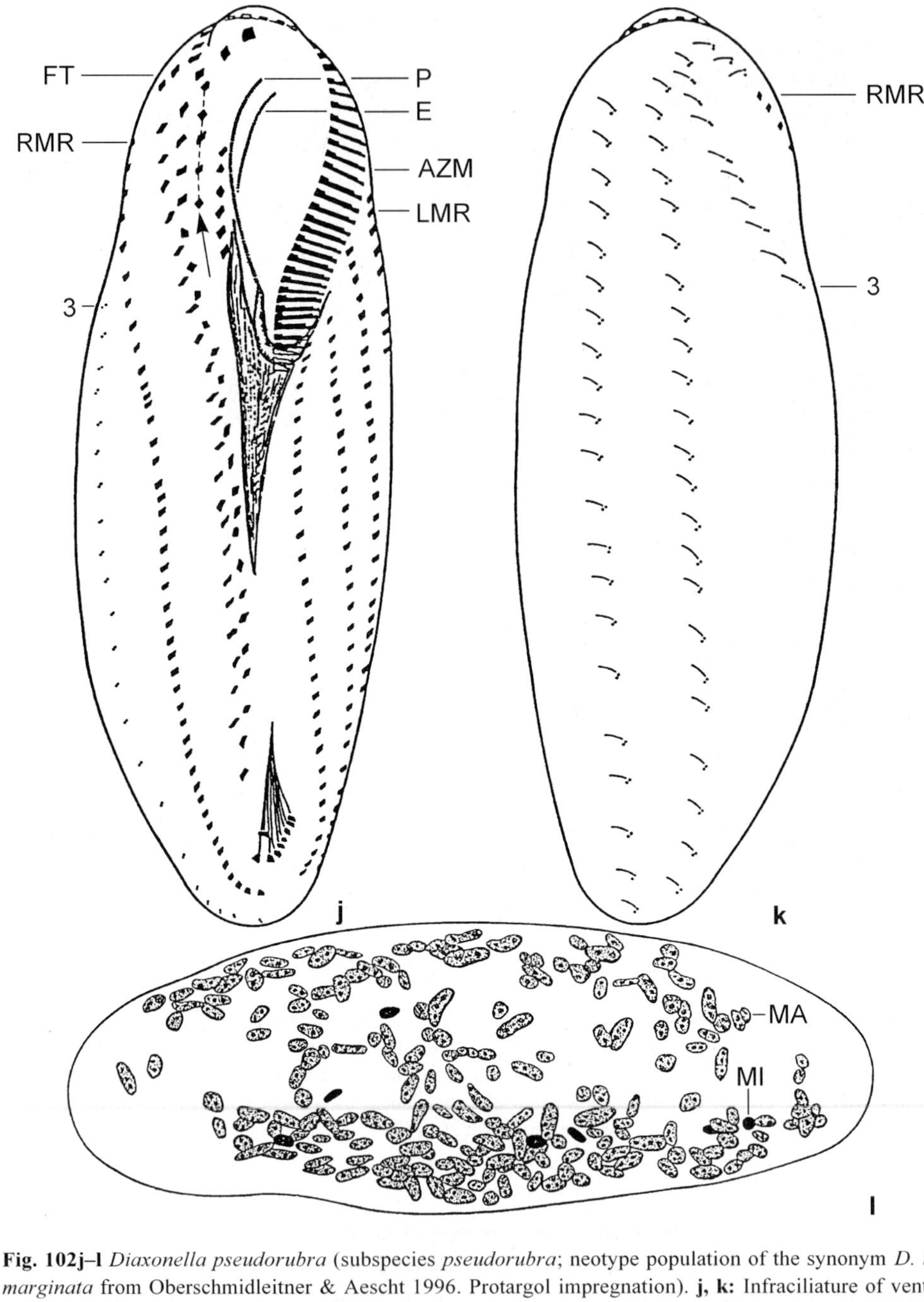

Fig. 102j–l *Diaxonella pseudorubra* (subspecies *pseudorubra*; neotype population of the synonym *D. trimarginata* from Oberschmidleitner & Aescht 1996. Protargol impregnation). **j, k:** Infraciliature of ventral and dorsal side of same specimen, 185 µm. Arrow marks parabuccal cirral row (cirri connected by broken line) behind right frontal cirrus. **l:** Nuclear apparatus, 170 µm. AZM = adoral zone of membranelles, E = endoral, FT = frontoterminal cirri, LMR = left marginal row 4 (= outer left marginal row), MA = macronuclear nodule, MI = micronucleus, P = paroral, RMR = right marginal row, 3 = dorsal kinety 3. Page 463.

ends rounded. Nuclear apparatus masked by cortical granules (I suppose that many macronuclear nodules are present). Contractile vacuole not mentioned; the vacuole illustrated (if this is the contractile vacuole at all) is certainly not in correct position. Cortical granules ("protrichocysts") wine-red, likely make cells dark-red like *Pseudokeronopsis rubra* (size, shape, and arrangement of granules and colour of cytoplasm not mentioned). Adoral zone occupies about 33% of body length. 12–14 distal-most membranelles only indistinctly set off from remaining membranelles of adoral zone. Cirral pattern likely not correctly illustrated in several respects (e.g., frontal cirri not illustrated, second left marginal row overlooked); several (about 7) buccal cirri along paroral (Fig. 101a); midventral complex extends from anterior end of cell to near transverse cirri. Transverse cirri project beyond rear body end. One right and one left marginal row illustrated, that is, second left row obviously overlooked if my synonymy is correct (see also remarks at species and subspecies). Dorsal ciliature (length of bristles, number and arrangement of kineties, presence/absence of caudal cirri) not known.

Cell division: This part of the life cycle is described and documented by several micrographs by Jerka-Dziadosz & Janus (1972). Unfortunately, they did not provide illustrations summarising the main events. Thus, I briefly describe the most important features, first for the proter and then for the opisthe. The paper by Zhang et al. (1985a) is in Chinese. Basically, their data agree rather well with those of the Polish population.

Proter: Stomatogenesis begins with the disaggregation of the inner undulating membrane. On the inner margin of the buccal lip, an extensive proliferation of new kinetosomes takes place. As a result, a longitudinal field of narrowly spaced basal bodies is formed. This field, which is separated from the parental zone, is the primordium of porter's adoral zone of membranelles. Next to and parallel to this primordium, another field of basal bodies, incorporating the parental paroral, is formed; this anlage becomes the primordium of the new undulating membranes. The parental adoral zone is resorbed completely from posterior to anterior. The frontal-midventral-transverse cirri anlagen originate mainly (exclusively?) by the modification of the buccal cirral row. About 24 oblique streaks are formed; no details, for example, migration of frontoterminal cirri, have been observed or described.

Opisthe: The formation of the primordium for the adoral zone commences somewhat earlier than in the proter. First, small groups of basal bodies occur around and to the left of the left cirri of the midventral pairs of the mid-body region. These patches and further basal bodies fuse to a longitudinal oral primordium. This primordium moves leftwards and the undulating membrane anlage is formed similarly to the adoral zone primordium. The frontal-midventral-transverse cirri primordia are formed between the undulating membranes anlagen and the parental midventral complex.

The right marginal row of each filial product is formed in ordinary manner, that is, by proliferation within the parental row. By contrast, both left marginal rows of each daughter cell originate, as in *Pseudourostyla cristata*, from a single anlage within the parental inner row, that is, the outer left marginal row is not involved in primordia formation. This feature was already reported by Wiackowski (1988, p. 6, 7, feature 18) for the synonym *Holosticha polystylata*. Dorsal morphogenesis is not described for the Pol-

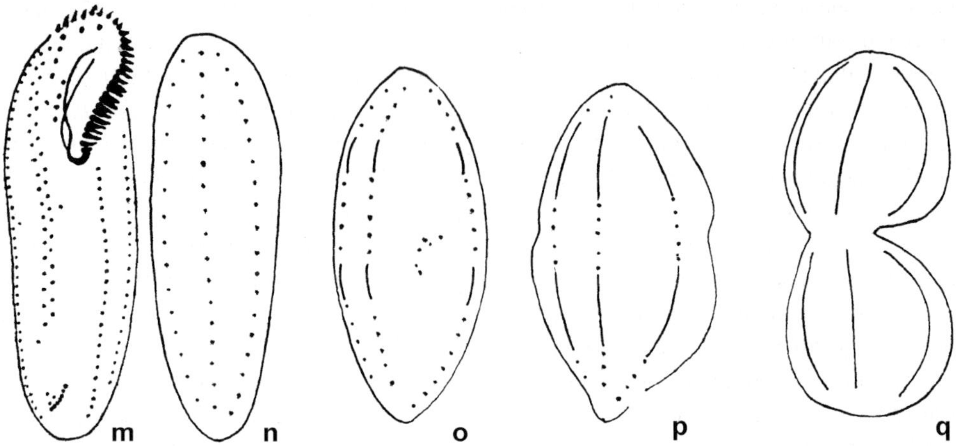

Fig. 102m–q *Diaxonella pseudorubra* (from Zhang et al. 1985a. Protargol impregnation). **m, n:** Infraciliature of ventral and dorsal side of the subspecies *pseudorubra*, 100–120 μm. **o–q:** Morphogenesis of dorsal kineties. Details see text. Page 463.

ish population, but according to Zhang et al. (1985a) it proceeds in an ordinary manner, that is, all kineties are formed by within-row proliferation (Fig. 102n–q).

The division of the nuclear apparatus also proceeds in ordinary manner. During the first stages of morphogenesis a reorganisation band occurs in every macronuclear nodule. When the reorganisation band passes through the nodule it rounds up and then the nodules fuse to a single mass. The division of the macronucleus and the mitosis of the micronuclei occur when the new cirri are well developed.

The formation of the infraciliature during physiological reorganisation proceeds basically as described for cell division (Pang et al. 1986).

Occurrence and ecology: *Diaxonella pseudorubra pseudorubra* lives in freshwater. Type locality of *K. pseudorubra* is the Danube River at Nussdorf, a part of the city of Vienna, Austria. Kaltenbach (1960) found it there in moss (*Fontinalis antipyretica*?) from the river bank during autumn (for review on Danube river fauna see Enãceanu & Brezeanu 1970, p. 234, Humpesch & Moog 1994, p. 91). Feeds on diatoms (Fig. 101a).

Due to the neotype designation for *D. trimarginata* by Oberschmidleitner & Aescht (1996), the type locality of the synonym is the sampling site of the neotype population (ICZN 1999). They found the neotype of *D. trimarginata*, together with *Pseudourostyla cristata*, in an activated sludge sample from the Asten sewage treatment plant near the city of Linz, Upper Austria. The sludge was aerated for three days after sampling and *D. trimarginata* was present for 90 days. On August 28, 1995, when the sample was collected, the activated sludge was characterised as follows: pH 7.5; 182 mg l^{-1} COD; 70 mg l^{-1} BOD_5; 0.7 mg l^{-1} PO_4-P; 4.6 mg l^{-1} NO_3-N; 34.9 mg l^{-1} NH_4-N; 12.6 mg l^{-1} Kjedahl-N. The original type locality of *D. trimarginata* is the Sadyba pond in Warsaw (Poland), where Jerka-Dziadosz & Janus (1972) collected it during summer 1970; in spring 1971, they found it in the Jeziorka River near Warsaw. The type locality of the synonym *Pseudokeronopsis trisenestra* is saprobic sand of temporary ponds fed by the waves of Lake Tanganyika, Burundi (Dragesco & Dragesco-Kernéis 1991). Wiackowski (1988, p. 4) found it in a freshwater aquarium (source of material not mentioned; identified as *Holosticha polystylata*). Zhang et al. (1985a) and likely also Pang et al. (1986) found the present subspecies in Shanghai, China, probably in a freshwater habitat. Schmitz (1986, p. 89) found *D. pseudorubra* (identified as *Trichototaxis pulchra*) pelagic (37 ind. 10 l^{-1}) and benthic (2 ind. cm^{-2}) in the Rhine River, Germany.

I observed *D. pseudorubra* rare, for example, with low abundance in the Glan, a small betamesosaprobic river in Carinthia (Austria) during August. Hubert Blatterer recorded it several times in the aufwuchs, sediment, and mosses during water quality assessment in Upper Austrian running waters (Table 22). The saprobic index ranged from 1.8 to 2.8, the estimated abundance was low in all cases. The average saprobic index (sum of individual index times abundance divided by sum of abundance) of all records is 2.4, indicating that *D. pseudorubra* prefers beta- to alphamesosaprobic conditions. This is supported by the records substantiated by illustrations (ponds; saprobic sand of temporary ponds; activated sludge). Foissner et al. (1995, p. 63) and Foissner & Berger (1996, p. 440) mention *D. trimarginata* in the key, but do not classify it saprobiologically. Oberschmidleitner & Aescht (1996, p. 22), however, write (obviously mistakenly) that we have proposed *D. trimarginata* as indicator of oligosaprobic conditions.

Diaxonella pseudorubra feeds on bacteria, autotrophic flagellates, testate amoeba, (*Arcella* sp.), and ciliates like *Sterkiella histriomuscorum* and *Tetrahymena pyriformis* (Oberschmidleitner & Aescht 1996, Schmitz 1986). Jerka-Dziadosz & Janus (1972) cultured *D. pseudorubra* in Pringsheim solution and added *T. pyriformis* as food. Wiackowski (1988) fed it with *Chlorogonium* sp.

***Diaxonella pseudorubra polystylata* (Borror & Wicklow, 1983) comb. nov., stat. nov.**

(Fig. 102b, Table 21b)

1983 ***Holosticha polystylata* sp. n.** – Borror & Wicklow, Acta Protozool., 22: 112, Fig. 4, Table 1 (Fig. 102b; original description of subspecies; the type material is, according to Borror & Wicklow 1983, p. 100, deposited in the slide collection of Arthur C. Borror; no formal diagnosis provided).

2001 ***Holosticha polystylata* Borror & Wicklow, 1983** – Berger, Catalogue of ciliate names 1. Hypotrichs, p. 38 (nomenclator containing all basionyms, combinations, and higher taxa of hypotrichs).

Nomenclature: See also this chapter at *D. pseudorubra*. Since *H. polystylata* was not yet transferred to *Diaxonella*, a new combination is necessary, even if is classified as subspecies (ICZN 1999, Article 51.3.2).

Remarks: See also this chapter at *D. pseudorubra*. Possibly this taxon is confined to limnetic habitats in North America. Redescription recommended.

Morphology: American population 160–200 μm long after protargol impregnation(?). Body shape as in Fig. 102b, that is, elongate-elliptical with both ends broadly rounded. Ciliates appear light pinkish red in transmitted light. Cortical granules about

Table 21b Morphometric data on *Diaxonella pseudorubra* (ps1, subspecies *pseudorubra* from Jerka-Dziadosz & Janus 1972; ps2, subspecies *pseudorubra* from Dragesco & Dragesco-Kernéis 1991; ps3, subspecies *pseudorubra* from Oberschmidleitner & Aescht 1996; ps4, subspecies *pseudorubra* from Plückebaum et al. 1997[k]; pst, subspecies *polystylata* from Borror & Wicklow 1983; pul, subspecies *pulchra* from Borror 1972a)

Characteristics[a]	Population	mean	M	SD	SE	CV	Min	Max	n
Body, length	ps1	–	–	–	–	–	130.0	170.0	?
	ps2	148.7	–	20.0	3.7	13.4	104.0	180.0	30
	ps3	153.6	150.0	27.0	6.2	17.6	93.0	198.0	19
	ps4	118.6	–	–	–	–	70.0	160.0	104
	pst	–	–	–	–	–	160.0	200.0	10
Body, width	ps2	50.2	–	16.1	3.0	32.1	31.0	108.0	30
	ps3	55.0	56.0	10.9	2.5	19.8	33.0	71.0	19
	ps4	33.7	–	–	–	–	20.0	54.3	104
Macronuclear nodule, length	ps2	1.6	–	1.1	0.2	65.8	0.5	5.5	40
	ps3	4.7	5.0	1.5	0.3	32.0	2.0	9.0	19
Macronuclear nodule, width	ps3	2.9	3.0	0.9	0.2	30.1	2.0	5.0	19
Macronuclear nodules, number	ps2	118.3	–	49.6	12.4	41.9	64.0	243.0	17
	ps3	143.6	150.0	24.6	5.6	17.1	110.0	185.0	19
	ps4	85.7	–	–	–	–	51.0	118	28
	pst	125.0	–	–	–	–	–	–	?
Micronucleus diameter, respectively,	ps2	–	–	–	–	–	2.0	3.0	?
width	ps3	2.5	2.0	0.8	0.2	30.7	2.0	4.0	11
Micronuclei, number	ps2	–	–	–	–	–	2.0	12.0	?
	ps3	5.3	5.0	1.1	0.3	20.0	4.0	7.0	11
Anterior body end to proximal end	ps2	48.3	–	5.8	1.0	12.0	36.0	62.0	30
of adoral zone, distance	ps3	57.3	56.0	8.9	2.0	15.5	40.0	76.0	19
	ps4	43.9	–	–	–	–	34.7	51.3	28
Adoral membranelles, number	ps1	35.0	–	–	–	–	–	–	?
	ps2	36.7	–	2.9	0.5	7.9	31.0	43.0	29
	ps3	35.9	36.0	2.7	0.6	7.4	31.0	40.0	19
	ps4	36.3	–	–	–	–	32.0	40.0	17
	pul	50.0	–	–	–	–	–	–	?
Frontal cirri, number	ps1	3.0	–	–	–	–	–	–	?
	ps2[f]	14.1	–	2.0	0.4	14.5	10.0	17.0	28
	ps3	2.7	3.0	0.5	0.1	17.3	2.0	3.0	19
	ps4	3.0	–	–	–	–	3.0	3.0	21
	pst	3.0	–	–	–	–	–	–	10
Buccal cirri, number	ps1[b]	5.0	–	–	–	–	–	–	?
	ps3	5.8	6.0	0.5	0.1	9.0	5.0	7.0	19
	ps4	4.7	–	–	–	–	4.0	5.0	21
	pst	–	–	–	–	–	6.0	9.0	10
	pul	–	–	–	–	–	6.0	7.0	?
Cirral row 1, number	ps1[c]	–	–	–	–	–	3.0	4.0	?
	ps3	5.1	5.0	0.7	0.2	13.6	4.0	6.0	19
	pst	–	–	–	–	–	5.0	8.0	10
Frontoterminal cirri, number	ps3	2.2	2.0	0.4	0.1	16.9	2.0	3.0	19
Midventral complex, number of cirri	ps1	34.0	–	–	–	–	–	–	?
	ps2	40.5	–	4.6	0.9	11.3	34.0	53.0	28
	ps4	35.2	–	–	–	–	28.0	43.0	23
Midventral pairs, number	ps3	23.1	24.0	2.3	0.5	10.1	18.0	27.0	19
	pst[e]	–	–	–	–	–	30.0	45.0	10
Transverse cirri, number	ps1	10.0	–	–	–	–	–	–	?
	ps2[g]	9.3	–	1.0	0.2	2.4	8.0	12.0	21

Table 21b Continued

Characteristics [a]	Population	mean	M	SD	SE	CV	Min	Max	n
Transverse cirri, number	ps3	7.3	7.0	0.7	0.2	10.0	6.0	9.0	19
	ps4	8.7	–	–	–	–	8.0	9.0	20
	pst	–	–	–	–	–	10.0	11.0	?
	pul	15.0	–	–	–	–	–	–	?
Right marginal row, number of cirri	ps1	–	–	–	–	–	40.0	55.0	?
	ps2	46.2	–	3.2	0.7	7.0	42.0	53.0	24
	ps3	44.8	45.0	5.0	1.1	11.1	34.0	54.0	19
	ps4	44.3	–	–	–	–	36.0	49.0	24
Left marginal rows, number	ps1 [d]	2.0	–	–	–	–	–	–	?
	ps2	–	–	–	–	–	3.0	4.0	?
	ps3	3.8	4.0	0.4	0.1	10.8	3.0	4.0	19
	pst	–	–	–	–	–	1.0	3.0	10
	pul	2.0	–	–	–	–	–	–	?
Left marginal row 1 [j], number of cirri	ps1	35.0	–	–	–	–	–	–	?
	ps2	31.4	–	2.9	0.6	9.2	27.0	37.0	25
	ps3	30.7	31.0	3.2	0.7	10.4	24.0	36.0	19
	ps4	26.7	–	–	–	–	24.0	34.0	20
Left marginal row 2, number of cirri	ps1	30.0	–	–	–	–	–	–	?
	ps2	25.9	–	4.2	0.8	16.1	18.0	33.0	28
	ps3	27.0	27.0	2.1	0.6	7.7	2.4	31.0	12
	ps4	26.6	–	–	–	–	21.0	34.0	15
Left marginal row 3, number of cirri	ps2	19.2	–	4.5	0.9	23.1	11.0	29.0	25
	ps3	21.2	21.0	1.8	0.5	8.6	17.0	24.0 [i]	12
	ps4	20.4	–	–	–	–	15.0	23.0	15
Left marginal row 4, number of cirri	ps2	4.9	–	3.9	0.8	79.0	0.0	14.0	24
	ps3	6.5	6.0	4.7	1.4	72.3	0.0	13.0	12
Ventral cirri, number	ps3 [h]	3.1	3.0	0.3	0.1	9.9	3.0	4.0	19
Dorsal kineties, number	ps1	3.0	–	–	–	–	–	–	?
	ps2	3.0	–	–	–	–	–	–	?
	ps3	3.2	3.0	0.4	0.1	12.2	3.0	4.0	16
	ps4	3.0	–	–	–	–	3.0	3.0	14
	pul			at least 3					

[a] All measurements in µm. Data for populations ps1 to ps4 are based on protargol-impregnated specimens. Borror & Wicklow (1983) did not specify the method (protargol impregnation or nigrosin stain). CV = coefficient of variation in %, M = median, Max = maximum value, mean = arithmetic mean, Min = minimum value, n = number of individuals investigated (? = number not given. If only one value is available, then it is listed under the column mean; if two values are given, then they are listed under Min and Max), SD = standard deviation, SE = standard error of arithmetic mean.

[b] Strain II had 7 or 8 cirri in this row.

[c] Some cells of strain I did not have this cirral row; in strain II, 4 or 5 cirri were present.

[d] Some cells of strain II had 3 left marginal rows.

[e] Fig. 102b shows a specimen with only 29 midventral pairs.

[f] Sum of frontal cirri, buccal cirri, and parabuccal cirri.

[g] Pretransverse ventral cirri (likely usually 2) included.

[h] Not designated in text or illustration.

[i] Specimen shown in Fig. 102j has 25 cirri in this row.

[j] This is the innermost row.

[k] The data and illustrations/micrographs (Fig. 101c–g) are mainly from Plückebaum's dissertation (University Bochum, Germany). However, he did not sent the whole paper and therefore I do not know the title of this work.

Table 22 Records of *Diaxonella pseudorubra pseudorubra* in Upper Austrian running waters [a]

Running water	Date	SI	Abundance	Reference
Mattig	November 1992	2.6	1	AOÖLR 1995a, p. 79
Trattnach	July 1993	2.8	1	AOÖLR 1995b, p. 100
Große Gusen	October 1993	2.7	2	AOÖLR 1996a, p. 104
Gusen	October 1993	2.7	2	AOÖLR 1996a, p. 104
Gusen	October 1993	2.8	1	AOÖLR 1996a, p. 104
Gusen	October 1993	2.8	2	AOÖLR 1996a, p. 104
Waldaist	May 1995	1.8	1	AOÖLR 1996b, p. 101
Waldaist	May 1995	2.2	1	AOÖLR 1996b, p. 101
Feldaist	November 1993	2.5	1	AOÖLR 1996b, p. 101
Aist	November 1993	1.9	2	AOÖLR 1996b, p. 101
Große Mühl	August 1994	2.1	1	AOÖLR 1997c, p. 105
Ranna	August 1994	2.5	1	AOÖLR 1997a, p. 92
Ranna	August 1994	2.6	1	AOÖLR 1997a, p. 92
Pesenbach	July 1994	2.7	1	AOÖLR 1997a, p. 95
Pesenbach	June 1994	2.6	1	AOÖLR 1997a, p. 95
Pesenbach	June 1994	2.3	1	AOÖLR 1997a, p. 95
Große Rodl	August 1994	2.5	2	AOÖLR 1997a, p. 98
Große Rodl	August 1994	2.5	1	AOÖLR 1997a, p. 98
Ager	August 1995	2.6	1	AOÖLR 1997b, p. 51
Ager	August 1995	2.4	2	AOÖLR 1997b, p. 51
Ager	August 1995	1.9	2	AOÖLR 1997b, p. 51
Krems	August 1995	2.5	1	AOÖLR 1997b, p. 70
Traun	November 1995	2.1	1	AOÖLR 1997b, p. 113
Traun	November 1995	2.1	1	AOÖLR 1997b, p. 113
Traun	October 1995	2.6	1	AOÖLR 1997b, p. 113

[a] Identified as *D. trimarginata* by Hubert Blatterer, Linz, Upper Austria. SI = saprobic index of ciliate cenosis according to Blatterer (1995). Abundance 1 = very low, 9 = mass occurrence.

0.75 µm across, golden, occurring as individual granules or in longitudinal groups of two or three between cirral rows and dorsal kineties, but not in obvious rows or in clumps near cirri. Additionally numerous tiny (less than 0.5 µm across) granules just beneath cortex, single or in longitudinal rows of 2–5 granules (colour of this second type of granules not mentioned). Cirral pattern basically as in *D. p. pseudorubra*; details see Table 21b and Fig. 102b. Frontoterminal cirri neither mentioned in description nor illustrated. Specimens with only one left marginal row occur.

Molecular data: Croft et al. (2003) provided some molecular data for *Holosticha polystylata* (GenBank accession number AY044841). Length (in base pairs; excluding telomere sequences) of actin-encoding macronuclear molecule is 1584; length of the 5′leader is 148; length of ORF is 1134; length of 3′trailer is 302; and length of encoded amino acid chain is 377. Hewitt et al. (2003; GenBank accession number AF508760) determined the lengths (bp) of the SSUrDNA (1769), the ITS1 (128), the 5.8S (153), the ITS2 (192), and the LSUrDNA (1356). In all trees illustrated by Hewitt et al. (2003), Kim et al. (2004), Dalby & Prescott (2004), and Coleman (2005) it clustered together with *Urostyla grandis*.

Occurrence and ecology: Type locality of *D. pseudorubra polystylata* are alkaline marl sediments adjacent to Mud Pond, Lloyd-Cornella Reservation, McLean, New York, USA (Borror & Wicklow 1983). They isolated it also from a freshwater aquarium in Durham, New Hampshire. Croft et al. (2003) and Hewitt et al. (2003) isolated it from a pond on the University of Colorado campus, Boulder, USA.

Diaxonella pseudorubra pulchra (Borror, 1972) comb. nov., stat. nov. (Fig. 101b, Table 21b)

1972 ***Trichotaxis pulchra* n. sp.** – Borror, Arch. Protistenk., 10: 63, Fig. 52, Plate I, Fig. 55 (Fig. 101b; original description of subspecies; no formal diagnosis provided; a slide containing paratypes is deposited in the U. S. National Museum; Borror 1972a, p. 64).

1987 ***Trichototaxis pulchra* (Borror, 1972) nov. comb.** – Foissner, Arch. Protistenk., 133: 226 (combination with *Trichototaxis*).

2001 ***Trichototaxis pulchra* (Borror, 1972) Foissner, 1987** – Berger, Catalogue of ciliate names 1. Hypotrichs, p. 96 (nomenclator containing all basionyms, combinations, and higher taxa of hypotrichs).

Nomenclature: See also this chapter at *D. pseudorubra*. Since *T. pulchra* was not yet transferred to *Diaxonella*, a new combination is necessary, even when it is classified as subspecies (ICZN 1999, Article 51.3.2).

Remarks: See same chapter at *D. pseudorubra*. The present taxon is possibly confined to marine habitats in North America. Redescription recommended. The limnetic *Anteholosticha intermedia* has only one left marginal row and fewer midventral pairs. Moreover it has only one parabuccal cirrus (= cirrus III/2).

Morphology: Body size 174–248 × 37–74 μm (from life?); body width usually 50–60 μm. Dorsoventrally flattened by about 2:1. Abnormal specimens shorter, wider, variously with acute or indented posterior end. Body very flexible dorsoventrally as well as laterally. Body outline elongate elliptical. Many macronuclear nodules dispersed throughout cell, difficult to discern in life. Contractile vacuole near left cell margin posterior to buccal cavity. Many short oblique series of 2–6 cortical granules strikingly visible between dorsal kineties. Individual granules about 1 μm "in length". Ventral side likewise granulated as follows: 2–4 oval granules lie between adjacent marginal cirri in all three rows, situated posterior and to the left of each midventral cirrus. Similar granules lie between bases of anterior-most transverse cirri; the posteriormost transverse cirri have practically no granules underlying them. Granules also lie between bases of buccal cirri. Longitudinal groups of closely packed granules, one or two granules wide border each marginal row. Colour of granules not mentioned. Cytoplasm generally relatively clear with pale brown inclusions and diatoms shells (this sentence indicates that *D. p. pulchra* lacks the red colour of the other two subspecies). Worms its way over and among clumps of detritus, crawls forward veering either right or left. Free-swimming occurs in a relatively ineffective counterclockwise helix, whereby the anterior describes a wide arc while the posterior end drags along the axis of the helix.

Adoral zone occupies about one third of body length (specimen illustrated about 28%), composed of circa 50 membranelles; membranelles about 10 μm long. Medial cilia of buccal (= proximal?) membranelles beat separately from remainder of membranelles. Endoral cilia in 6–7 rows(!) on roof of buccal cavity, beat posteriorly. Paroral cilia stiff, vibratile.

3–5 frontal cirri about 12 µm long (likely three frontal and two parabuccal cirri present; possibly the low number of only two parabuccal cirri is a further difference to the other two subspecies). Buccal cirri about 10 µm long, arise from a shallow longitudinal groove right of paroral. Presence/absence of frontoterminal cirri not known. Midventral complex composed of cirral pairs only, commences near anterior end of cell, terminates slightly ahead of transverse cirral row; cirri about 9 µm long. Transverse cirri arranged in inverted L-shaped row near rear end of cell so that they project slightly beyond cell margin. Marginal cirri 8–10 µm long. One right and two left marginal rows. According to text, marginal rows overlapping at posterior end; according to Fig. 101b right and left marginal rows widely separated. Ciliary rootlet fibrils at bases of cirri easily visible in life. Dorsal cilia about 3 µm long, 5 µm apart, erect, vibratile, arranged in at least three kineties.

Occurrence and ecology: Marine (see also Patterson et al. 1989, p. 211). Type locality is very likely a tidal marsh near Adam's Point, Durham, New Hampshire, USA (43°05'22"N 70°52'15"W) where Borror (1972a) discovered it in a collection made in April, 1971. Feeds on diatoms and ciliates.

Incertae sedis in *Diaxonella*

Trichototaxis aeruginosa (Foissner, 1980) Foissner, 1987 (Fig. 103a–d)

1980 ***Trichotaxis aeruginosa* nov. spec.**[1] – Foissner, Ber. Nat.-Med. Ver. Salzburg, 5: Abb. 26a–d (Fig. 103a–d; original description).

1987 ***Trichototaxis aeruginosa* (Foissner, 1980) nov. comb.** – Foissner, Arch. Protistenk., 133: 226 (combination with *Trichototaxis*).

2001 ***Trichototaxis aeruginosa* (Foissner, 1980) Foissner, 1987** – Berger, Catalogue of ciliate names 1. Hypotrichs, p. 95 (nomenclator containing all basionyms, combinations, and higher taxa of hypotrichs).

Nomenclature: No derivation of the name is given in the original description. The species-group name *aeruginos·us -a -um* (Latin adjective; russet) obviously alludes to the colour of the cell. *Trichotaxis* is an incorrect subsequent spelling of *Trichototaxis*. Thus, Foissner (1987d) made a new combination (see also this chapter at *Trichototaxis*).

Remarks: The present species is certainly not a *Trichototaxis* because it lacks a distinct bicorona. The increased number of left marginal rows and the red colour suggest synonymy with *D. pseudorubra*. However, the midventral cirri of *T. aeruginosa* do

[1] The diagnosis provided by Foissner (1980a) is as follows: 120 bis 150 µm große, lang ovale, kontraktile, rostrot gefärbte *Trichotaxis* mit etwa 300, 4–6 × 2–4 µm großen, unregelmäßig geformten Makronuclei. 5 leicht verstärkte Frontalcirren, ca. 8 feine Transversalcirren, 3 Ventralreihen, eine kurze Buccalreihe rechts der paroralen Membran, eine rechte und 2 linke Marginalreihen.

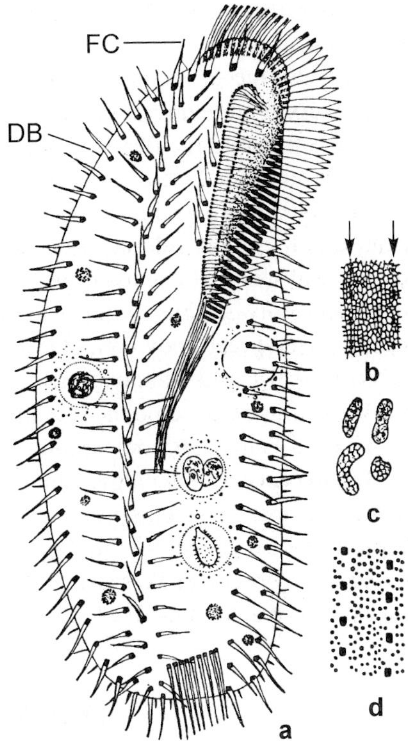

Fig. 103a–d *Trichototaxis aeruginosa* (from Foissner 1980a. a, c, d, from life; b, dry silver nitrate impregnation). This species lacks a bicorona and therefore should not be classified in *Trichototaxis*. The five enlarged frontal cirri (three frontal and two parabuccal?), the increased number of buccal cirri, the two left marginal rows, and the red colour are reminiscent of *Diaxonella pseudorubra*. However, in this species the midventral complex is composed of cirral pairs forming only two pseudorows, whereas Foissner observed three ventral rows. Moreover, *Trichototaxis aeruginosa* has five dorsal kineties (against three in *D. pseudorubra*), indicating that it is indeed a valid species. **a:** Ventral view, 154 μm. **b:** Part of silver-line system of dorsal side showing two kineties (arrows). **c:** Macronuclear nodules. **d:** Cirral rows and russet cortical granules. DB = dorsal bristle of rightmost kinety, FC = rightmost frontal cirrus. Page 484.

not form two pseudorows, but three. Such a pattern is very unusual and difficult to explain.[1] Moreover, *Trichototaxis aeruginosa* seems to have more dorsal kineties than *D. pseudorubra* (five vs. three) so that a synonymy of these two species is not reasonable. Until a redescription is available I classify it as incertae sedis in *Diaxonella*, however, without transferring it formally to this genus.

Foissner (1980a) compared his species with *Trichototaxis rubentis*, which is also red (Fig. 166b). However, since this superficially described species obviously has only two macronuclear nodules, identity can be excluded.

Morphology: Body length 120–150 μm; length:width ratio of specimen illustrated about 2.7:1 (Fig. 103a). Body outline elongate oval, anterior end moderately wide, rear body end broadly rounded; left margin roughly straight, right distinctly convex. Body very flexible, anterior portion distinctly contractile; ventral side plane, dorsal moderately vaulted. About 300 macronuclear nodules dispersed throughout cell; individual nodules 4–6 × 2–4 μm, irregularly shaped, containing several nucleoli causing honeycomb pattern (Fig. 103c). Contractile vacuole about in mid-body. Cortical granules tiny, red, fade rapidly in dying cells (Fig. 103d). Cytoplasm homogeneously reddish,

[1] There are at least three possibilities: (i) each anlage does not produce a midventral pair, but a short midventral row composed of three cirri; (ii) the three rows are not pseudorows, but true rows (then it is not a urostyloid); (iii) the pattern is a misobservation (unlikely).

contains rather many 1–5 µm-sized colourless, greasily shining inclusions and food vacuoles.

Adoral zone of membranelles about 43% of body length (Fig. 103a). Buccal field deep. Paroral and endoral composed of long cilia; bases of both membranes extending almost to frontal cirri. Pharyngeal fibres conspicuous. Five slightly enlarged frontal cirri (likely three frontal cirri and two parabuccal cirri). Specimen illustrated with seven buccal cirri along midportion of paroral. Three(!) narrowly spaced cirral rows extend from near distal end of adoral zone to near transverse cirri (for details see remarks); presence/absence of frontoterminal cirri not known. About eight fine transverse cirri arranged in only slightly oblique terminal row so that cirri project by about half their length beyond rear body end. One almost bipolar right marginal row. Two left marginal rows. Dorsal bristles about 5 µm long, arranged in five bipolar kineties. Caudal cirri likely lacking because neither mentioned nor illustrated. Silverline system between dorsal kineties as in other non-euplotine hypotrichs, that is, narrowly meshed (Fig. 103b).

Occurrence and ecology: Limnetic. Type locality of *T. aeruginosa* is a meander of a brook (Tümpel 58 in Foissner 1980b; about 2000 m altitude) with iron precipitation underneath the so-called "Bretter" near the Großglockner-Hochalpenstrasse, an alpine road in Austria. Moreover, Foissner (1980a; 1980b, p. 111) found it on wet rocks. Feeds on coccale green algae and *Trachelomonas* sp.

Afrothrix Foissner, 1999

1999 ***Afrothrix* nov. gen.**[1] – Foissner, Biodiversity and Conservation, 8: 376 (original description). Type species (by original designation on p. 376): *Afrothrix darbyshirei* Foissner, 1999.

2001 ***Afrothrix* Foissner 1999** – Aescht, Denisia, 1: 19 (Catalogue of generic names of ciliates).

2001 ***Afrothrix* Foissner, 1999** – Berger, Catalogue of ciliate names 1. Hypotrichs, p. 7 (nomenclator containing all basionyms, combinations, and higher taxa of hypotrichs).

Nomenclature: The name *Afrothrix* is a composite of the Latin word *Africa* and the Greek noun *thrix* (hair = ciliate sensu lato), meaning "a ciliate occurring in Africa" (Foissner 1999, p. 376). Feminine gender.

Characterisation (A = supposed apomorphies): Adoral zone of membranelles bipartite (A?). 3 frontal cirri. Buccal cirri present. 2 frontoterminal cirri. Midventral complex composed of midventral pairs only, slightly longer than adoral zone (A?). Transverse cirri present. 1 left and 1 right marginal row. Caudal cirri absent.

Remarks: The two species assigned have – besides the characteristics mentioned above – many other features in common (some of them are possibly further apomorphies of *Afrothrix*), namely body slender and very flexible; contractile vacuole near mid-body; cortical granules about 1 µm across, colourless, only around cirri and dorsal bristles[2], do not stain with Foissner's protargol method; cytoplasm colourless; adoral

[1] The diagnosis by Foissner (1999) is as follows: Hypotrichida with bipartited adoral zone of membranelles, one right and left marginal cirral row, two ventral rows with cirri in midventral pattern, frontoterminal cirri, and transverse cirri.

[2] In *Afrothrix darbyshirei* also on inner surface of buccal cavity.

zone of membranelles less than 25% of body length on average; buccal cavity narrow and rather flat; paroral and endoral distinctly curved, close together, and optically intersecting at level of buccal cirrus, likely composed of closely spaced dikinetids; three slightly enlarged frontal cirri; single buccal cirrus at summit of curve formed by paroral; two frontoterminal cirri with rear cirrus at level of anteriormost midventral pair; midventral complex inconspicuous, composed of a low number (two on average in *A. multinucleata*, four in *A. darbyshirei*) of midventral pairs, respectively, pseudopairs; transverse cirri not enlarged (type species) or only slightly larger than marginal cirri (*A. multinucleata*), form curved row; likely one or two pretransverse ventral cirri present; marginal cirri 10–12 µm long, become gradually thinner posteriorly; right marginal row extends dorsolaterally anteriorly; dorsal bristles 3–4 µm long in life, arranged in a low number (two in *A. multinucleata*, three in *A. darbyshirei*) of almost bipolar kineties; some narrowly spaced and transversely arranged dorsal bristles at rear body end (vestigial caudal cirri?); in sandy soils in Africa.

Foissner (1999) established *Afrothrix* for a single species, *Afrothrix darbyshirei*, mainly because of the curious bipartite adoral zone of membranelles and the midventral pattern of the frontoventral cirri. Only three years later, we found a second species in Africa, *Afrothrix multinucleata*, which matched this combination of features and many other characteristics (see above) very well, indicating that *Afrothrix* is a monophylum whose species are possibly confined to Africa. Foissner (1999) did not classify *Afrothrix* in the urostyloids, but left it unassigned in the Hypotrichida until ontogenetic data becomes available. I preliminarily treat it in this monograph because both species assigned show a – admittedly not very conspicuous – midventral pattern. Since *Afrothrix* has three frontal cirri and a midventral complex which is composed of cirral pairs only it is assigned to the Holostichidae. The low number of frontal-midventral-transverse cirri indicates that only few (about nine in *A. darbyshirei* and about six in *A. multinucleata*) cirral anlagen are formed.

Afrothrix has – like, for example, *Holosticha*, *Erniella* Foissner, 1987b and *Etoschothrix* Foissner, Agatha & Berger, 2002 – a very distinctive oral apparatus. However, this feature probably evolved independently, as indicated by the different cirral pattern and fine structure of the ventral adoral membranelles of the species involved. *Afrothrix* has a more or less distinct midventral pattern and the ventral membranelles show, like those of *Holosticha*, the ordinary fine structure, that is, are composed of two long ciliary rows, one short row, and one very short row (Fig. 104f, 105i). In contrast, *Erniella* has two long ventral cirral rows and *Etoschothrix* has an *Oxytricha*-like frontoventral cirral pattern. Further, both in *Erniella* and *Etoschothrix*, the ventral membranelles are composed of three ciliary rows only, namely two long rows and one very short row. The last mentioned feature indicates that *Etoschothrix* and *Erniella* are possibly closely related because the fine structure of the adoral membranelles is very conservative (Foissner et al. 2002). Interestingly, all species of *Afrothrix*, *Etoschothrix*, and *Erniella* have narrowly spaced, transversely arranged dorsal bristles near the rear body end, possibly vestigial caudal cirri. Likely, ontogenetic data will enable proper classification of *Erniella* and *Etoschothrix*.

Species included in *Afrothrix* (alphabetically arranged according to basionyms): (1) *Afrothrix darbyshirei* Foissner, 1999; (2) *Afrothrix multinucleata* Foissner, Agatha & Berger, 2002.

Key to *Afrothrix* species

1 Body size 230–330 × 40–60 µm in life; 2 macronuclear nodules; 20–25 ventral adoral membranelles (Fig. 104a) .... *Afrothrix darbyshirei* (p. 488)
- Body size 100–200 × 15–30 µm in life; 4–15, usually around 9 macronuclear nodules in strand left of midline; 7–10 ventral adoral membranelles (Fig. 105a, f) .... *Afrothrix multinucleata* (p. 492)

Afrothrix darbyshirei Foissner, 1999
(Fig. 104a–g, Table 23)

1999 ***Afrothrix darbyshirei* nov. spec.**[1] – Foissner, Biodiversity and Conservation, 8: 376, Fig. 13a–g, Table 10 (Fig. 104a–g; original description. 1 holotype slide [registration number 1999/35] and 1 paratype slide [1999/36] with protargol-impregnated specimens are deposited in the Oberösterreichische Landesmuseum in Linz [LI], Austria).

2001 ***Afrothrix darbyshirei* Foissner, 1999** – Berger, Catalogue of ciliate names 1. Hypotrichs, p. 7 (nomenclator containing all basionyms, combinations, and higher taxa of hypotrichs).

Nomenclature: Foissner dedicated this species to John F. Darbyshire, an eminent Scottish soil protozoan ecologist (e.g., Darbyshire 1994). *Afrothrix darbyshirei* was fixed as type species of *Afrothrix* by original designation.

Remarks: For separation from *Afrothrix multinucleata*, see this chapter at this species and key. There are only few species in other genera which are rather similar to *Afrothrix darbyshirei*, namely *Erniella filiformis* Foissner, 1987b, *Etoschothrix terricola* Foissner, Agatha & Berger, 2002, and *Notocephalus parvulus* (Fig. 237a–f). However, *Erniella filiformis* has 31–61 macronuclear nodules (vs. 2), a single dorsal kinety (vs. 3), and two long ventral cirral rows, that is, does not show a midventral pattern. *Etoschothrix terricola* has, like *Erniella filiformis*, many (17–59) macronuclear nodules and only one dorsal kinety, and the frontoventral cirri arranged in an *Oxytricha*-like pattern. Interestingly, all these species have cortical granules only around the cirri and dorsal bristles. The marine *Notocephalus parvulus* has a continuous adoral zone which is about 41% of body length (vs. bipartite and 25%) and 5–9 midventral pairs (vs. four), and lacks a buccal cirrus and frontoterminal cirri (Fig. 237e). In life, *Afrothrix darby-*

[1] The diagnosis by Foissner (1999) is as follows: Size in vivo about 250 × 50 µm, contractile by about 30% of body length. Cylindrical with ends narrowly rounded. Cortical granules colourless, about 1 µm in diameter, mainly around bases of cirri and dorsal bristles. Midventral row usually composed of 4 cirri pairs, thus terminating close underneath adoral zone of membranelles. On average 2 macronuclear nodules, 5 micronuclei, 9 frontal and 23 ventral adoral membranelles, about 52 cirri each in right and left marginal row, 3 frontal cirri, 2 frontoterminal cirri, 1 buccal cirrus, 9 transverse cirri, and 3 dorsal kineties.

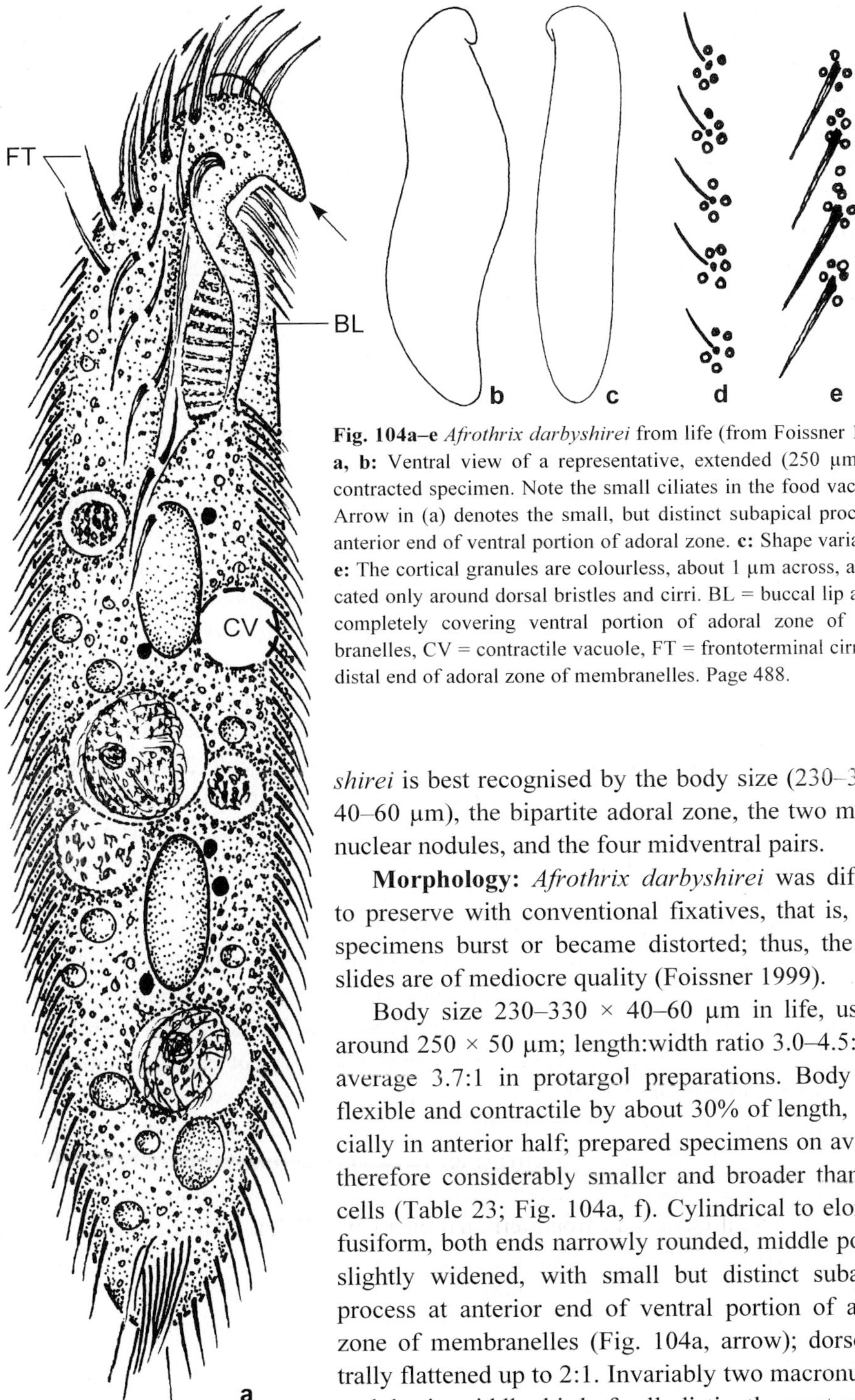

Fig. 104a–e *Afrothrix darbyshirei* from life (from Foissner 1999). **a, b:** Ventral view of a representative, extended (250 µm) and contracted specimen. Note the small ciliates in the food vacuoles. Arrow in (a) denotes the small, but distinct subapical process at anterior end of ventral portion of adoral zone. **c:** Shape variant. **d, e:** The cortical granules are colourless, about 1 µm across, and located only around dorsal bristles and cirri. BL = buccal lip almost completely covering ventral portion of adoral zone of membranelles, CV = contractile vacuole, FT = frontoterminal cirri near distal end of adoral zone of membranelles. Page 488.

shirei is best recognised by the body size (230–330 × 40–60 µm), the bipartite adoral zone, the two macronuclear nodules, and the four midventral pairs.

Morphology: *Afrothrix darbyshirei* was difficult to preserve with conventional fixatives, that is, most specimens burst or became distorted; thus, the type slides are of mediocre quality (Foissner 1999).

Body size 230–330 × 40–60 µm in life, usually around 250 × 50 µm; length:width ratio 3.0–4.5:1, on average 3.7:1 in protargol preparations. Body very flexible and contractile by about 30% of length, especially in anterior half; prepared specimens on average therefore considerably smaller and broader than live cells (Table 23; Fig. 104a, f). Cylindrical to elongate fusiform, both ends narrowly rounded, middle portion slightly widened, with small but distinct subapical process at anterior end of ventral portion of adoral zone of membranelles (Fig. 104a, arrow); dorsoventrally flattened up to 2:1. Invariably two macronuclear nodules in middle third of cell, distinctly apart and ellipsoidal (length:width ratio 2.5:1 on average; Table 23), contain large, globular nucleoli. Micronuclei slightly ellipsoidal, near to or rather distant from macronuclear nodules, compact and thus easy to recognise in life (Fig. 104a, f). Contractile vacuole in mid-body at left cell margin. Cortical granules inconspicuous because only about 1 μm across and colourless, located around cirri and dorsal bristles and on inner surface of buccal cavity; do not stain with Foissner's protargol method (Fig. 104d, e). Cytoplasm colourless, usually packed with yellowish globules up to 13 μm across and food vacuoles containing small ciliates. Swims and glides hastily to and fro, contracting spasmodically and becoming sigmoidal when touching an obstacle (Fig. 104b).

Oral apparatus – though comparatively small, that is, occupying only about 24% of body length on average – conspicuous because of its particular organisation (Fig. 104a, f; Table 23). Adoral zone with distinct break (gap) at left anterior cell margin, dividing zone in a frontal (distal) and ventral (proximal) portion: frontal portion at anterior and right anterior margin of cell, composed of nine short membranelles on average; ventral portion extends obliquely from left anterior cell end, which projects hook-like, to midline of cell, composed of 23 membranelles on average, roughly elongate elliptical because widest (14 μm in life) membranelles in centre of ventral portion. Buccal cavity narrow and rather flat, but with conspicuous, curved roof almost completely covering ventral portion of adoral zone. Undulating membranes about as long as ventral portion of adoral zone, close together and distinctly curved, both very likely composed of closely spaced dikinetids; paroral in distinct cleft on surface of buccal lip, intersects endoral optically in mid of buccal cavity. Pharyngeal fibres distinct in life and protargol preparations, of ordinary length and structure, extend obliquely backwards.

Cirral pattern and number of cirri of usual variability (Table 23). Invariably three frontal cirri, slightly enlarged, in life about 15 μm long, form rather oblique row with right cirrus, as is usual, behind distal adoral membranelle. Buccal cirrus at summit of curve formed by paroral. Invariably two frontoterminal cirri distinctly apart from midventral complex, anterior cirrus right and slightly behind right frontal cirrus, rear cirrus right of first midventral pair (Fig. 104a, f). Midventral complex inconspicuous because composed of an average of four pseudopairs only and thus terminating at 33% of body length on average, that is, slightly behind level of buccal vertex; usually a cirrus near rear end of midventral complex, occasionally forming a distinct triplet with last pseudopair (Fig. 104f, long arrow; Table 23). Transverse cirri not enlarged, in life about 20 μm long, subterminal and thus only slightly projecting beyond rear body margin, form curved, oblique row left of midline; usually two minute (pretransverse?) cirri ahead of two rightmost transverse cirri (Fig. 104a, f). Marginal cirri in life about 12 μm long, become gradually thinner posteriorly. Left marginal row begins left of proximal portion of adoral zone of membranelles, rear end conspicuous because extending around posterior body margin, the last (rightmost) cirri thus easily misinterpreted as caudal cirri; right marginal row commences subapically at level of buccal cirrus and terminates at level of transverse cirri, thus separated from left marginal row by a distinct gap (Fig. 104f).

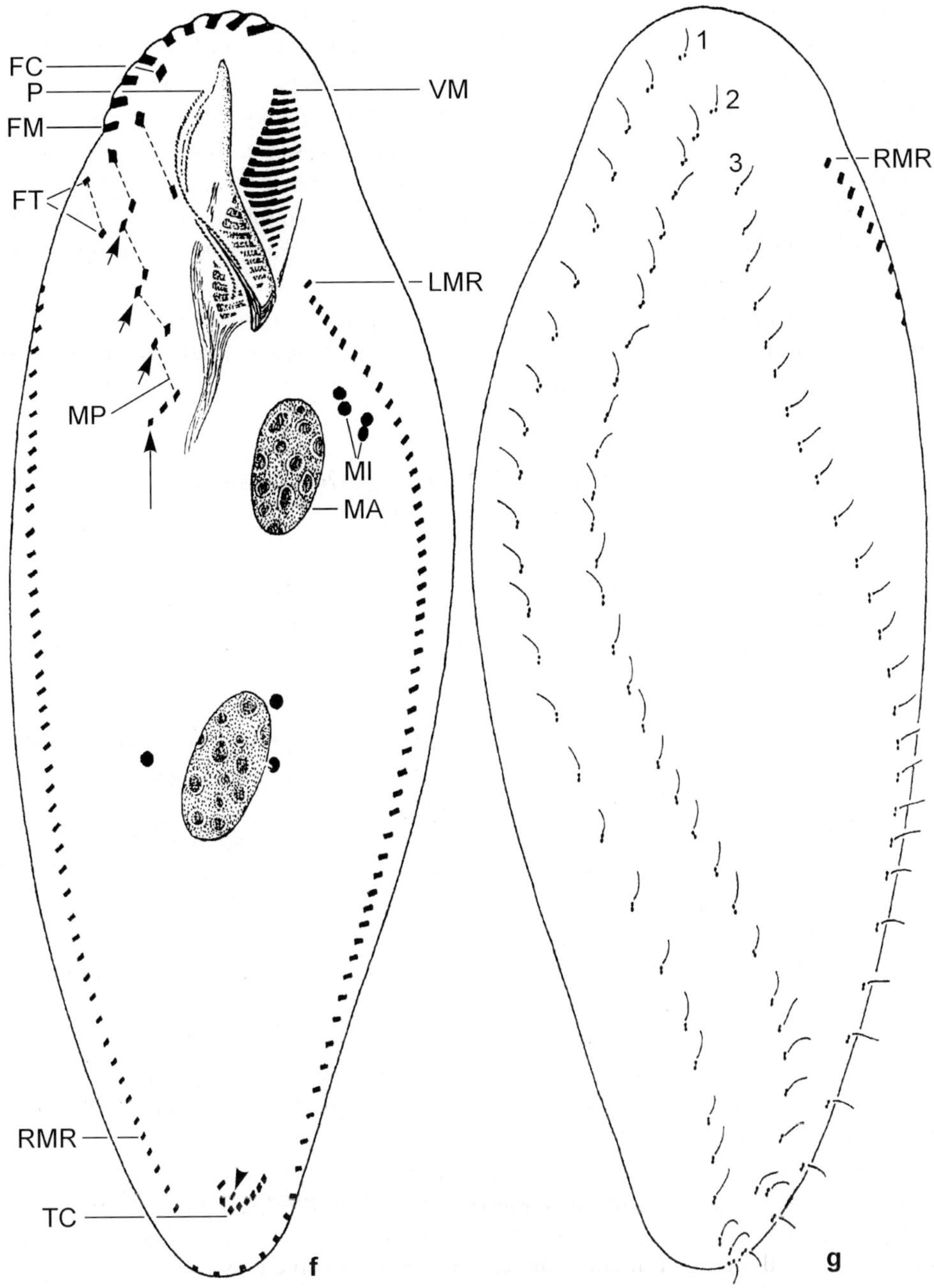

Fig. 104f, g *Afrothrix darbyshirei* (from Foissner 1999. Protargol impregnation). Infraciliature of same specimen, 220 μm. Short arrows mark 3 of 4 pseudopairs; long arrow denotes last cirrus of midventral complex. Arrowhead marks a pretransverse ventral cirrus. Broken lines connect frontoventral cirri, which likely originate from same anlage. FC = left frontal cirrus, FM = frontal membranelle, FT = frontoterminal cirri, LMR = left marginal row, MA = macronuclear nodule, MI = micronuclei, MP = rearmost (?) midventral pair, P = paroral, RMR = right marginal row, TC = transverse cirri, VM = distalmost ventral membranelle, 1–3 = dorsal kineties. Page 488.

Dorsal cilia about 4 µm long in life, arranged in three kineties; row 1 bipolar, rows 2 and 3 slightly shortened anteriorly; invariably two tightly spaced bristles each at posterior end of kineties, possibly vestigial caudal cirri (Fig. 104g).

Occurrence and ecology: As yet found only at type locality, that is, grassland soil near the Sheldrick waterfalls in the Shimba Hills Nature Reserve, Kenya (sample no. 15 in Foissner 1999). This park is about 40 km south of Mombassa and about 20 km west of the Indian Ocean coast (about 04°25'S 39°20'E). The sample is from the downhill path to the waterfalls and composed of the upper 0–5 cm grass sward and very sandy soil layer (pH 6.6). Obviously a rather rare species.

Feeds on small ciliates, like *Cyclidium muscicola*, *Drepanomonas pauciciliata*, *Leptopharynx costatus*, and *Pseudochilodonopsis mutabilis* (Foissner 1999). Biomass of 10^6 specimens about 160 mg (own calculation).

Afrothrix multinucleata Foissner, Agatha & Berger, 2002 (Fig. 105a–l, Table 23)

2002 ***Afrothrix multinucleata* nov. spec.**[1] – Foissner, Agatha & Berger, Denisia, 5: 588, Fig. 134a–l, Table 116 (Fig. 105a–l; original description. 1 holotype slide [registration number 2002/255] and 1 paratype slide [2002/256] with protargol-impregnated specimens are deposited in the Oberösterreichische Landesmuseum in Linz [LI], Austria).

Nomenclature: The name *multinucleata* (having many [macro]nuclei) is a composite of the Latin numeral *mult·us* (many), the thematic vowel *i*, and the Latin adjective *nucleata* (kernel-like), and refers to the numerous macronuclear nodules.

Remarks: The most prominent feature of this species is the bipartite adoral zone of membranelles. We assigned it to *Afrothrix* because the frontoventral cirri form, as in the type species, a midventral-like pattern and the ventral membranelles show the usual fine structure, that is, are composed of two long ciliary rows, one short row, and one very short row (Fig. 105f, i). In contrast, *Erniella* has two long ventral cirral rows and *Etoschothrix* has an *Oxytricha*-like frontoventral cirral pattern. In addition, both in *Erniella* and *Etoschothrix*, the ventral membranelles are composed of three ciliary rows only, namely two long rows and one very short row. *Holosticha* species have, inter alia, a much more prominent transverse cirral row and the anterior end of the left marginal row distinctly curved rightwards.

Afrothrix multinucleata differs from the type species by the smaller size (100–200 × 15–30 µm vs. 230–330 × 40–60 µm), the lower number of midventral pairs (usually 2 vs. usually 4) and dorsal kineties (2 vs. 3), and the higher number of macronuclear nodules (usually 9 vs. 2). *Erniella filiformis* Foissner, 1987b and *Etoschothrix terricola*

[1] The diagnosis by Foissner et al. (2002) is as follows: Size about 150 × 20 µm in vivo. Slender with posterior portion distinctly narrowed. Cortical granules about 1.2 × 0.8 µm, colourless, around cirral bases and dorsal bristles. Midventral row usually composed of 2 cirral pairs, terminates slightly underneath adoral zone of membranelles. On average 10 macronuclear nodules forming strand left of midline, 5 frontal and 8 ventral adoral membranelles, about 44 cirri each in right and left marginal row, 3 frontal cirri, 2 frontoterminal cirri, 1 buccal cirrus, 4 transverse cirri, and 2 dorsal kineties.

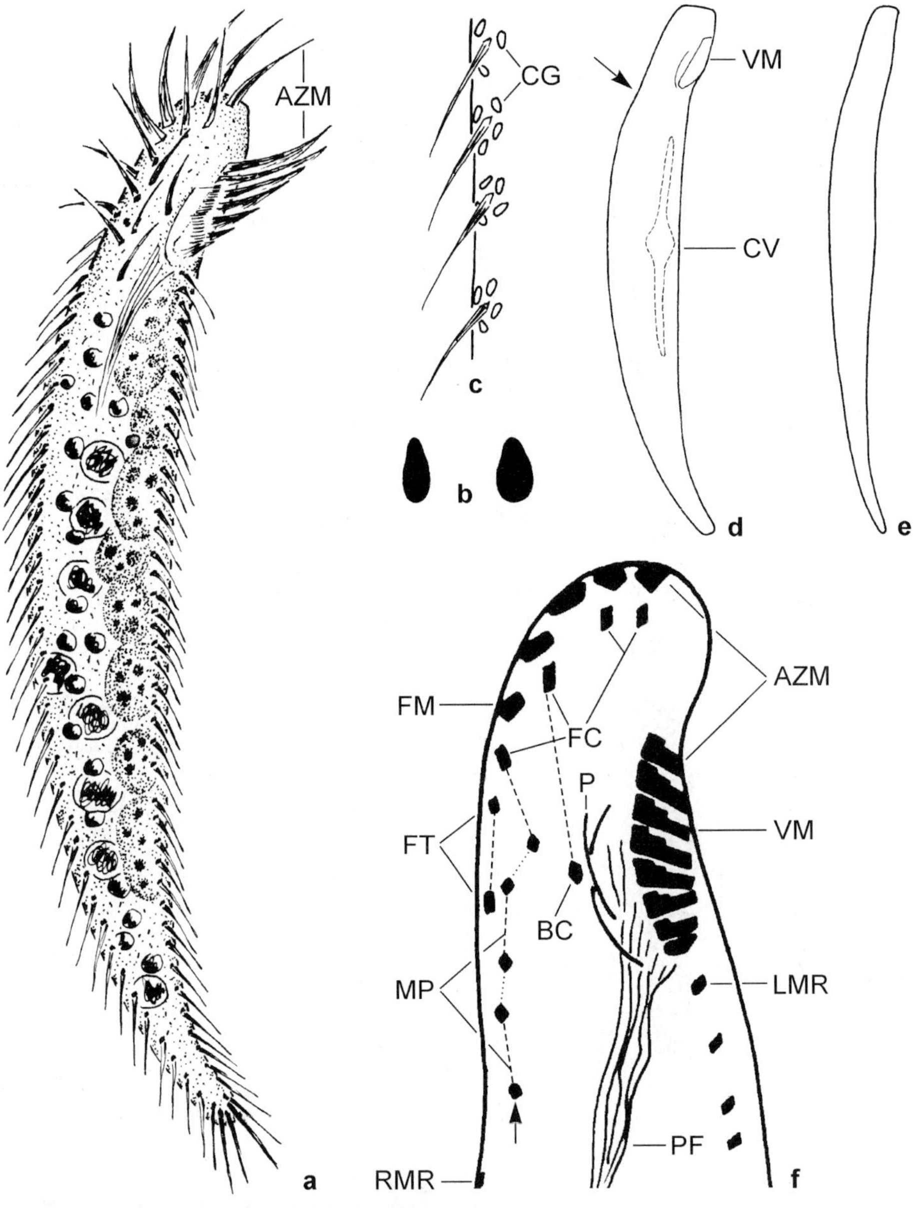

Fig. 105a–f *Afrothrix multinucleata* (from Foissner et al. 2002. a–e, from life; f, protargol impregnation). **a, d:** Ventral view and outline of a representative specimen, 155 μm. Arrow in (d) denotes a slight, but characteristic indentation at level of buccal vertex causing inconspicuous cephalisation. **b, c:** Colourless, about 1.2 × 0.8 μm-sized cortical granules occur around cirri and dorsal bristles. **e:** Lateral view showing dorsoventral flattening. **f:** Infraciliature of oral region of holotype specimen (see also Fig. 105g, h). Arrow marks last cirrus of midventral complex, which is composed of two inconspicuous pseudopairs (dotted lines); broken lines connect cirri likely originating from same anlage. AZM = adoral zone of membranelles, BC = buccal cirrus, CG = cortical granules, CV = contractile vacuole, FC = frontal cirri (left one bipartite in that specimen), FM = frontal adoral membranelles, FT = frontoterminal cirri, LMR = left marginal row, MP = midventral pair, PF = pharyngeal fibres, P = paroral, RMR = right marginal row, VM = ventral adoral membranelles. Page 492.

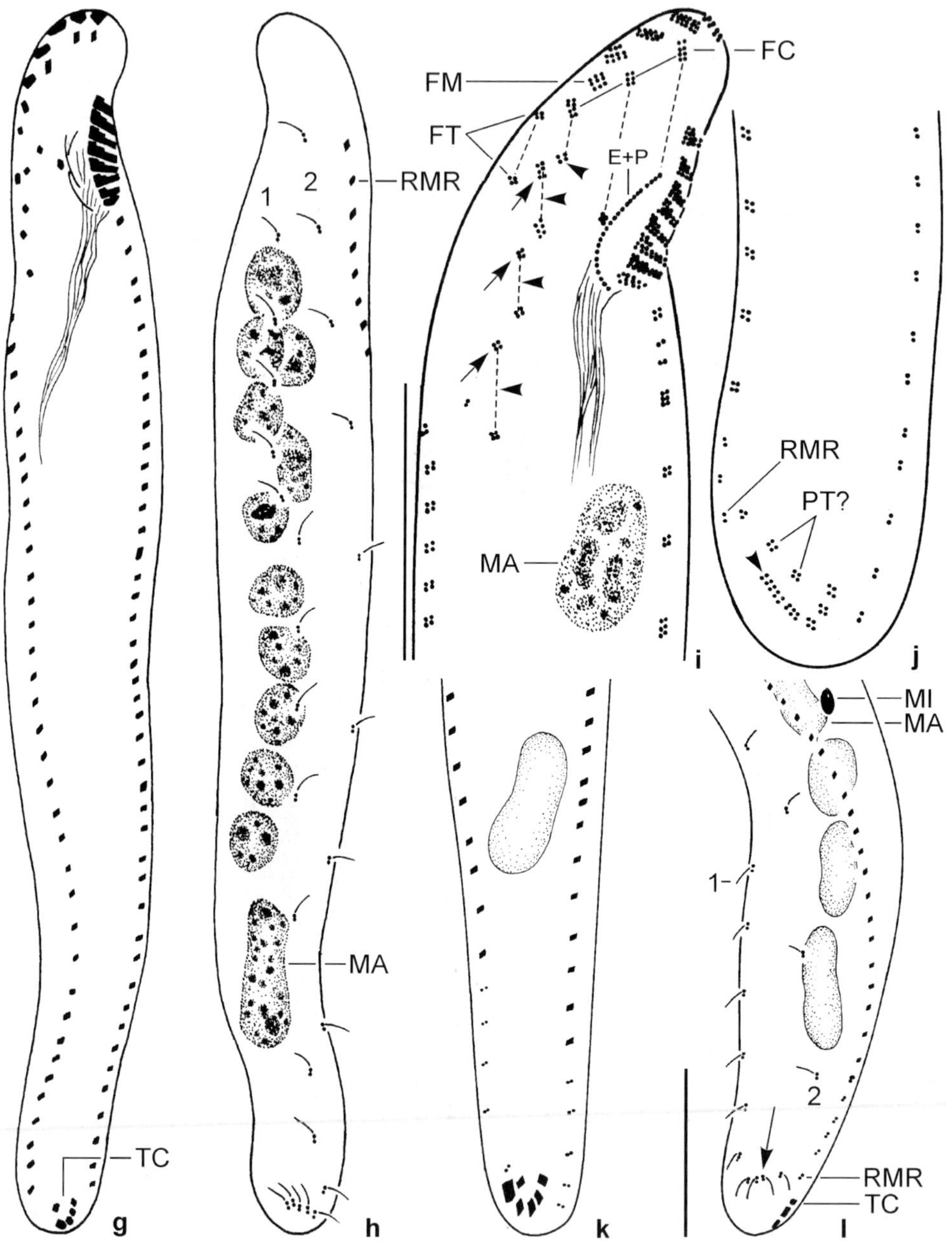

Fig. 105g–l *Afrothrix multinucleata* (from Foissner et al. 2002. Protargol impregnation). **g, h:** Infraciliature and nuclear apparatus of holotype specimen, 120 µm; see also Fig. 105f. **i:** Infraciliature of a specimen with three pseudopairs (arrows), respectively, midventral pairs (horizontal arrowheads). Oblique arrowhead denotes cirrus III/2. Broken lines connect cirri likely originating from same anlage. **j–l:** Details of infraciliature of rear body portion. Arrowhead in (j) denotes hook-shaped arranged transverse cirri. Arrow in (l) denotes narrowly spaced bristles, possibly vestigial caudal cirri. E = endoral, FC = frontal cirri, FM = first frontal membranelle, FT = frontoterminal cirri, MA = macronuclear nodule, MI = micronucleus, P = paroral, PT? = pretransverse (?) ventral cirri, RMR = right marginal row, TC = transverse cirri, 1, 2 = dorsal kineties. Scale bars 20 µm. Page 492.

Table 23 Morphometric data on *Afrothrix darbyshirei* (dar, from Foissner 1999) and *Afrothrix multinucleata* (mul, from Foissner et al. 2002)

Characteristics [a]	Species	mean	M	SD	SE	CV	Min	Max	n
Body, length	dar	217.8	215.0	22.2	6.2	10.2	185.0	280.0	13
	mul	129.3	130.0	24.1	5.0	18.6	84.0	186.0	23
Body, width	dar	60.0	60.0	9.0	2.5	15.0	47.0	75.0	13
	mul	16.6	16.0	2.8	0.6	17.0	13.0	25.0	23
Body length:width, ratio	dar	3.7	3.7	0.5	0.1	13.3	3.0	4.5	13
	mul	7.9	7.7	1.4	0.3	17.2	5.5	11.1	23
Anterior body end to ventral portion of adoral zone, distance	mul	7.1	7.0	1.0	0.2	14.7	5.0	9.0	23
Anterior body end to rear end of ventral portion of adoral zone, distance	dar	52.6	50.0	7.0	1.9	13.3	45.0	70.0	13
	mul	18.5	18.0	1.8	0.4	9.5	14.0	21.0	23
Body length:length of adoral zone, ratio	dar	4.2	4.2	0.4	0.1	10.5	3.2	4.9	13
	mul	7.0	6.9	1.3	0.3	19.2	4.4	10.3	23
Anterior body end to last cirrus of midventral complex, distance [b]	dar	72.5	70.0	13.6	3.8	18.7	50.0	100.0	13
	mul	24.4	24.0	3.4	0.7	13.9	18.0	31.0	23
Midventral complex, relative length (in %) [h]	dar	33.2	32.6	4.2	1.2	12.8	25.0	41.9	13
	mul	19.3	19.7	3.4	0.7	17.6	13.4	24.0	23
Anterior body end to first macronuclear nodule, distance	mul	23.2	23.0	7.2	1.5	30.8	11.0	37.0	23
Anterior macronuclear nodule, length	dar [g]	28.9	27.0	4.8	1.3	17.0	23.0	40.0	13
	mul	11.1	10.0	4.6	1.0	41.5	6.0	26.0	23
Anterior macronuclear nodule, width	dar [g]	11.0	11.0	1.2	0.3	11.1	9.0	13.0	13
	mul	5.3	5.0	0.9	0.2	17.5	4.0	7.0	23
Posterior body end to last macronuclear nodule, distance	mul	28.1	26.0	10.2	2.1	36.2	13.0	53.0	23
Macronuclear nodules, number	dar	2.0	2.0	0.0	0.0	0.0	2.0	2.0	13
	mul [c]	9.9	9.0	2.7	0.6	27.2	4.0	15.0	23
Anterior micronucleus, length	dar [g]	2.8	3.0	–	–	–	2.0	4.0	13
	mul	2.4	2.4	–	–	–	1.6	4.0	13
Anterior micronucleus, width	dar [g]	2.6	2.5	–	–	–	2.0	3.0	13
	mul	1.7	1.6	–	–	–	1.6	3.0	13
Micronuclei, number	dar	5.2	5.0	1.6	0.5	31.5	3.0	9.0	13
	mul [d]	1.1	1.0	–	–	–	0.0	4.0	21
Frontal adoral membranelles, number	dar	8.8	9.0	0.9	0.2	10.2	8.0	10.0	13
	mul	5.0	5.0	–	–	–	4.0	6.0	23
Ventral adoral membranelles, number	dar	22.4	23.0	1.3	0.4	5.9	20.0	25.0	13
	mul	8.1	8.0	0.7	0.2	9.1	7.0	10.0	23
Frontal cirri, number	dar	3.0	3.0	0.0	0.0	0.0	3.0	3.0	10
	mul	3.0	3.0	–	–	–	3.0	4.0	21
Buccal cirri, number	dar	1.0	1.0	0.0	0.0	0.0	1.0	1.0	13
	mul	1.0	1.0	0.0	0.0	0.0	1.0	1.0	21
Frontoterminal cirri, number	dar	2.0	2.0	0.0	0.0	0.0	2.0	2.0	11
	mul	2.2	2.0	–	–	–	2.0	3.0	20
Anterior body end to rear frontoterminal cirrus, distance	mul	14.0	14.0	1.3	0.3	9.3	10.0	16.0	20
Frontoventral cirri, number [e]	dar	11.5	11.0	1.1	0.3	9.8	10.0	13.0	11
	mul	7.5	7.0	1.1	0.2	14.7	5.0	10.0	20
Pseudopairs, number [i]	dar	4.3	4.0	–	–	–	4.0	5.0	13
	mul	2.2	2.0	–	–	–	2.0	3.0	19
Posterior body end to anteriormost transverse cirrus, distance	mul	3.8	4.0	1.1	0.3	29.6	1.0	5.0	17

Table 23 Continued

Characteristics [a]	Species	mean	M	SD	SE	CV	Min	Max	n
Transverse cirri, number [f]	dar	9.1	9.0	1.2	0.3	13.1	7.0	11.0	13
	mul	4.4	5.0	2.1	0.5	47.1	0.0	7.0	18
Anterior body end to first right marginal cirrus, distance	mul	9.6	10.0	1.3	0.3	14.0	7.0	12.0	23
Right marginal cirri, number	dar	53.1	54.0	5.4	1.5	10.2	46.0	66.0	13
	mul	43.2	44.0	5.6	1.2	13.0	32.0	53.0	23
Left marginal cirri, number	dar	50.3	50.0	5.1	1.5	10.1	40.0	59.0	13
	mul	44.4	46.0	7.6	1.6	17.0	23.0	57.0	23
Dorsal kineties, number	dar	3.0	3.0	0.0	0.0	0.0	3.0	3.0	13
	mul	2.0	2.0	–	–	–	1.0	2.0	23
Dorsal kinety 1, number of kinetids	mul	14.3	14.0	1.8	0.4	12.4	11.0	16.0	18
Dorsal kinety 2, number of kinetids	mul	9.3	9.0	3.1	0.7	33.0	5.0	15.0	19

[a] All measurements in µm. In *Afrothrix darbyshirei*, some characteristics not mentioned in the original description have been calculated from the original data. CV = coefficient of variation in %, M = median, Max = maximum value, mean = arithmetic mean, Min = minimum value, n = number of individuals investigated, SD = standard deviation, SE = standard error of arithmetic mean. Data based on mounted, protargol-impregnated (Foissner's method), and randomly selected specimens from non-flooded Petri dish cultures.

[b] See long arrow in Fig. 104f and arrow in Fig. 105f.

[c] The specimen with only four nodules appears otherwise normal.

[d] Difficult to count because faintly impregnated and of similar size as cell inclusions.

[e] All cirri except buccal cirrus and frontal cirri.

[f] Including pretransverse ventral cirri possibly present ahead of transverse cirri.

[g] Macronuclear nodule (anterior or posterior), respectively, micronucleus not specified.

[h] Position (in % of body length) of last cirrus of midventral complex (long arrow in Fig. 104f and arrow in Fig. 105f).

[i] For designation of pseudopairs, see Figs. 104f, 105i.

Foissner, Agatha & Berger, 2002 are easily distinguished from *Afrothrix multinucleata* by the arrangement and higher number of macronuclear nodules (17–61 scattered nodules in *Erniella filiformis* and *Etoschothrix terricola* vs. 4–15 forming a strand). In life, *Afrothrix multinucleata* is therefore easily recognised by the vermiform shape, the bipartite adoral zone, and the macronuclear nodules forming a strand left of midline.

Morphology: Body size 100–200 × 15–30 µm in life, usually around 150 × 20 µm. Body very flexible, but acontractile, length:width ratio thus 7–8:1 both in life and in protargol preparations (Table 23). Slender and often slightly curved, widest near mid-body, anterior portion less distinctly narrowed than posterior, which is often inflated in protargol preparations (cp. Fig. 105a, g, j). Anterior end transverse truncate, without distinct process on ventral side; right body margin with inconspicuous, but rather typical indentation at level of buccal vertex (Fig. 105d, arrow), causing indistinct cephalisation. Dorsoventrally flattened about 2:1. Macronuclear nodules form strand left of midline, rather variable in number and shape, that is, globular, ellipsoidal (2–3:1), or dumbbell-shaped; with small to large nucleoli. Micronuclei globular to ellipsoidal, attached to macronuclear figure in variable positions. Contractile vacuole near mid-body at left cell margin, with two collecting canals extending to body ends. Cortical granules found only around cirri and dorsal bristles, ellipsoidal, inconspicuous because of moderate size (1.2 × 0.8 µm in life) and colourless (Fig. 105b, c); do not stain with methyl green-pyronin and Foissner's protargol method. Cytoplasm colourless, contains many lipid globules about 3 µm across. Glides slowly on microscope slide and soil particles.

Adoral zone of membranelles very short, that is, occupies only 15% of body length on average, with distinct gap in anterior half dividing zone in a distal (frontal) and proximal (ventral) portion (Fig. 105a, g, i): distal portion at anterior and right anterior margin of cell, composed of five (rarely of 4 or 6) minute membranelles; proximal portion begins about 7 µm behind anterior body end and extends along left body margin, composed of eight membranelles on average with ordinary fine structure. Buccal cavity strikingly small and flat, without lip. Endoral slightly, paroral distinctly shorter than proximal portion of adoral zone, both curved and likely composed of closely spaced dikinetids, intersect optically in mid-portion at level of buccal cirrus; cilia of paroral about 8 µm long in life. Pharyngeal fibres prominent in life and in protargol preparations, of ordinary length and structure, extend obliquely backwards.

Cirral pattern and number of cirri of usual variability, except for the number of transverse cirri which varies very strongly. Three slightly enlarged frontal cirri in ordinary position. Buccal cirrus at summit of curves formed by paroral and endoral. 5–10, usually seven frontoventral cirri (including frontoterminal cirri) composed and arranged as shown in Figs. 105f, i: four, rarely six of them form two or three pseudopairs, producing short, inconspicuous midventral complex terminating at 19% of body length on average, that is, slightly behind proximal end of adoral zone. Transverse cirri protrude distinctly beyond rear body end and thus rather conspicuous in vivo, although not or only slightly enlarged and longer than marginal cirri (12 µm vs. 10 µm); form U- or hook-shaped row near rear body end; often one or two cirri ahead of transverse cirri, possibly pretransverse ventral cirri. Marginal cirri about 10 µm long in life, fine, that is, usually composed of four cilia, in posterior body portion often even of two cilia only (Fig. 105i–l); left and right row end slightly subterminally, anterior portion of right row extends onto dorsolateral surface.

Dorsal cilia about 3 µm long in life, arranged in two kineties almost as long as body; bristles in kinety 2 rather widely spaced. Usually some narrowly spaced and transversely arranged bristles near rear end of kineties, possibly vestigial caudal cirri (Fig. 105h, l).

Occurrence and ecology: As yet found only at type locality, that is, the margin of the Namib Desert in litter around *Stipagrostis* roots in a sand dune between the villages of Aus and Helmeringhausen (Namibia, about 26°05'S 16°35'E). *Afrothrix multinucleata* was numerous in the non-flooded Petri dish culture, indicating that many resting cysts were present. The vermiform body indicates that it is a true soil dweller, well adapted to live in the soil pores. Feeds on bacteria and heterotrophic flagellates digested in vacuoles 5–7 µm in diameter (Foissner et al. 2002). Biomass of 10^6 specimens about 20 mg (own calculation).

Periholosticha Hemberger, 1985

1982 ***Periholosticha* n. gen.**[1] – Hemberger, Dissertation[2], p. 76, 111.
1985 ***Periholosticha* n. gen.**[3] – Hemberger, Arch. Protistenk., 130: 403 (original description). Type species (by original designation on p. 403): *Periholosticha lanceolata* Hemberger, 1985.
1987 ***Periholosticha* Hemberger, 1981** – Tuffrau, Annls Sci nat. (Zool.), 8: 115 (classification of hypotrichous ciliates).
1994 ***Periholosticha* Hemberger 1982** – Tuffrau & Fleury, Traite de Zoologie, 2: 128 (revision of hypotrichous ciliates).
1999 ***Periholosticha* Hemberger, 1985** – Shi, Acta Zootax. sinica, 24: 366 (revision of hypotrichous ciliates).
1999 ***Periholosticha* Hemberger, 1985** – Shi, Song & Shi, Progress in Protozoology, p. 117 (revision of hypotrichous ciliates).
2001 ***Periholosticha* Hemberger 1985** – Aescht, Denisia, 1: 123 (catalogue of generic names of ciliates).
2001 ***Periholosticha* Hemberger, 1985** – Berger, Catalogue of ciliate names 1. Hypotrichs, p. 71 (nomenclator containing all basionyms, combinations, and higher taxa of hypotrichs).
2002 ***Periholosticha* Hemberger, 1981** – Lynn & Small, Phylum Ciliophora, p. 453 (guide to ciliate genera).

Nomenclature: No derivation of the name is given in the original description. The name *Periholosticha* is a composite of the Greek prefix *peri+* (around, all around) and the genus name *Holosticha* (see there for derivation). I assume that the name should indicate that the present genus is (closely) related to *Holosticha*. Feminine gender (Aescht 2001, p. 294).

Characterisation (A = supposed apomorphies): Adoral zone bipartite (A?) in a proximal (= ventral) portion and a distal (= frontal) one with 3 or 4 membranelles. 3 frontal cirri. Buccal cirrus absent (A). Frontoterminal cirri lacking (type species) or present (other species). Midventral complex composed of midventral pairs only, rather short (A?). Transverse cirri inconspicuous (low number and very small; *P. paucicirrata*) or lacking (other species). 1 left and 1 right marginal row. 3 (type species and *P. acuminata*) or 2 (other species) dorsal kineties. Caudal cirri present. Parental adoral zone not reorganised during cell division.

Remarks: The species assigned have – beside the characteristics mentioned above – several other features in common, although the difference in two important features

[1] This name is disclaimed for nomenclatural purposes (ICZN 1999, Article 8.3). See next footnote.

[2] According to the ICZN (1964), dissertations are not explicitly mentioned under Article 9, which describes all those acts that do not constitute a publication within the meaning of the Code. Thus, Hemberger's (1982) thesis, for which the ICZN (1964) has to be applied, could possibly also be considered as original description of the present genus and many other taxa first described in this paper. Unfortunately, the situation is rather complicated and almost each thesis would need a detailed analysis whether or not it meets the requirements of publication (P. Tubbs, ICZN, Natural History Museum, London, pers. comm.). Thus, I do not accept Hemberger (1982), but Hemberger (1985) as original description of all the taxa discovered by him. However, I include the thesis in the list of synonyms because it is one of the most important papers on hypotrichs since Kahl (1932). To avoid nomenclatural problems each new name mentioned by Hemberger (1982) is individually disclaimed for nomenclatural purposes (see corresponding footnotes).

[3] The diagnosis by Hemberger (1985) is as follows: Je 1 rechte und linke Marginalreihe; Midventral-Reihen stark verkürzt, nur wenig über den Peristomgrund reichend; Midventral-Anordnung schwer erkennbar; keine oder nur unscheinbare Transversalcirren; einige adorale Membranellen sind frontal von den übrigen abgetrennt; Makronuclei zahlreich; Cirrendifferenzierung während der Morphogenese aus wenigen schrägen Anlagen.

(frontoterminal and transverse cirri absent/present) somewhat weakens the assumed sister-group relationship. Body slender and very flexible; contractile vacuole ahead of mid-body; several (usually between 10 and 20) ellipsoidal macronuclear nodules in strand left of midline; cortical granules present (presence/absence not checked in *P. acuminata*); cytoplasm colourless; adoral zone of membranelles 20–25% of body length on average, composed of 14–19 membranelles of ordinary shape and structure; undulating membranes rather short and slightly curved; buccal cavity narrow and flat (no live data known from *P. acuminata*); frontal cirri not (*lanceolata*) or only slightly (*acuminata*) enlarged; few midventral pairs so that midventral complex terminates at (*acuminata*) or slightly behind (*lanceolata*) buccal vertex; 3 dorsal kineties with short (2–4 µm) bristles; terrestrial.

The taxonomy of *Periholosticha* is complicated and therefore has to be explained in detail. *Periholosticha* was originally classified in the Urostylidae because of the midventral pattern, which is, however, rather indistinct in the type species during interphase (Hemberger 1982, 1985). Two years later, Tuffrau (1987) transferred *Periholosticha* to the Kahliellidae Tuffrau, 1979, but without giving reasons. Later he also classified it in the urostyloids, strictly speaking the Holostichidae (Tuffrau & Fleury 1994; Table 9). Shi (1999a) and Shi et al. (1999; Table 10) classified *Periholosticha* in the Holostichidae too, and simultaneously synonymised *Holostichides* Foissner, 1987 with *Periholosticha* Hemberger, 1985. However, the type species of *Periholosticha* lacks buccal and frontoterminal cirri, whereas these cirral groups are present in *Holostichides*. Moreover, *Periholosticha* has a bipartite adoral zone (vs. continuous in *Holostichides* and *Paragastrostyla*) indicating that Shi's synonymy is incorrect. Recently, Eigner (2001, p. 74) again classified *Periholosticha* in the Urostylidae. Surprisingly, Lynn & Small (2002; Table 11) transferred it to the amphisiellids, likely because of the indistinct zigzag-pattern feigning an amphisiellid median cirral row. However, the ontogenetic data on the type species (Fig. 106a–h) show a midventral pattern, clearly indicating that *Periholosticha* is a urostyloid.

So far, five species have been originally assigned to *Periholosticha* (see sections on species included and species misplaced below). Although the overall morphology is rather homogenous, the four species now included show differences in some features (presence/absence of transverse and frontoterminal cirri), which are usually used to define genera. The *Periholosticha* species are possibly related to *Holostichides* and *Paragastrostyla* species because they lack transverse cirri and have a slender, more or less lanceolate body shape and often a pointed posterior body end. The most important apomorphy of *Periholosticha* is likely the rather distinct gap in the adoral zone of membranelles (Fig. 106a). The absence (primary or secondary?) of a midventral row in *Periholosticha* (see below, for uncertainties in this feature) caused me to assign it (preliminarily) to the Holostichidae. By contrast, *Holostichides* and *Paragastrostyla* are assigned to the Bakuellidae because they have at least one midventral row. I am not certain that these assignments reflect the reality.

On the other hand, one cannot exclude that *Periholosticha* is closely related to *Afrothrix*. Both taxa have a bipartite adoral zone of membranelles and a rather short midventral complex. Thus, the apomorphies marked with a question mark in the characterisations could be synapomorphies of *Afrothrix* + *Periholosticha*. However, further data (cell division, cysts, fine structure, molecular markers) are likely needed to get a better idea about the relationships and members of the Holostichidae.

Differences within *Periholosticha* have been reported for the frontoterminal cirri and the transverse cirri. The lack of frontoterminal cirri in *P. lanceolata* (type species) is confirmed by morphogenetic data (Fig. 106b–h). By contrast, *Periholosticha acuminata* very likely has such a cirral group (Fig. 108a), although this is not confirmed by cell division data. In *P. paucicirrata* and *P. sylvatica* the anteriormost two cirri of the right marginal row are more or less distinctly set off so that one cannot exclude that these are frontoterminal cirri (e.g., Fig. 109c, 110e). However, usually this cirral group is not in line with the right marginal row, but more or less distinctly displaced inwards. Whether or not these cirri are actually frontoterminal cirri can only be decided by the investigation of the cell division. Interestingly, Foissner et al. (2002) found a similar situation in *P. lanceolata* (Fig. 106o, q, r, 107h), which has, as just mentioned, no frontoterminal cirri according to the ontogenetic data of the type specimens (Fig. 106b–h).

All species assigned to *Periholosticha* lack distinct transverse cirri and a distinct midventral row, that is, the midventral complex is composed of cirral pairs only. However, late ontogenetic stages of the type species show that a short midventral row, composed of three cirri only, is formed (Fig. 106g, h). Hemberger (1982) interpreted the rearmost cirrus of this row as transverse cirrus which does not migrate posteriorly. This interpretation seems inappropriate in Song's (1990, p. 230) and my opinion because a transverse cirrus is defined as the rearmost, usually very distinctly set off, and often enlarged cirrus of an anlage. Interestingly, all interphasic specimens of *P. lanceolata* illustrated so far have an even number of midventral cirri, strongly indicating that only midventral pairs remain during interphase (Fig. 106a, o, q, r, 107i). This means, however, that the third cirrus in the posteriormost anlage dissolves during very late ontogenetic stages. A very similar situation, that is, a missing midventral row, occurs also in the other three species. Only in *P. sylvatica* is sometimes a short row present. *Periholosticha paucicirrata* is possibly the sole *Periholosticha* species which has – admittedly inconspicuous – transverse cirri (Fig. 110c, d, j, r). However, whether or not this observation is correct is difficult to ascertain because of the pointed rear body end. Only cell division data will unambiguously show whether or not such cirri are present.

The presence/absence of a cirral group is generally used to define genera. Accordingly *P. acuminata*, which has distinct frontoterminal cirri (vs. lacking in type species) and *P. paucicirrata*, which likely has transverse cirri (vs. lacking in type species), should be transferred to new genera because no matching genera have been described so far. A transfer of *P. acuminata* to *Holostichides* or *Paragastrostyla*, which have frontoterminal cirri, would make these groups unnecessarily inhomogenous, and the establishment of monotypic genera would also not simplify the situation. Consequently, I include all these "deviating" species in *Periholosticha* and characterise the group less rigorously. The description of further species will possibly enable a more proper insight into the phylogeny of this group.

The sistergroup-relationship within *Periholosticha* is difficult to ascertain because of several homoplasies. *Periholosticha lanceolata* and *P. paucicirrata* have three frontal membranelles, whereas *P. acuminata* and *P. sylvatica* have four. "Unfortunately" this grouping differs from that indicated by the number of dorsal kineties because *P. lanceolata* and *P. acuminata* have three (which is likely the plesiomorphic state), whereas *P. paucicirrata* and *P. sylvatica* have only two.

Wiackowski (1988), in his phenetic analysis of urostylids, studied a *Periholosticha* sp., however, without giving an illustration. It lacks buccal and transverse cirri, has three dorsal kineties and caudal cirri, and 13–19 macronuclear nodules. Furthermore, Wiackowski (1988) described two frontoterminal cirri, indicating that he studied a species more closely related to *Periholosticha acuminata* (two frontoterminal cirri) than to *P. lanceolata* (lacks frontoterminal cirri). Probably Wiackowski's population showed a significant difference to *P. acuminata* because otherwise he would have identified it as such.

Species included in *Periholosticha* (alphabetically arranged according to basionyms): (1) *Periholosticha acuminata* Hemberger, 1985; (2) *Periholosticha lanceolata* Hemberger, 1985; (3) *Periholosticha paucicirrata* Foissner, Berger, Xu & Zechmeister-Boltenstern, 2005; (4) *Periholosticha sylvatica* Foissner, Berger, Xu & Zechmeister-Boltenstern, 2005.

Species misplaced in *Periholosticha*: *Periholosticha wilberti* Song, 1990. This species has distinct frontoterminal cirri and a conspicuous midventral row. Since both features are much more pronounced than in the *Periholosticha* species mentioned above and, moreover, confirmed by ontogenetic data, the classification in *Periholosticha* is inappropriate (Fig. 128a–p). Song's (1990) "improved diagnosis"[1] of *Periholosticha* does not contain the bipartite adoral zone and is therefore more or less identical with the diagnosis of *Paragastrostyla*. Eigner (1994) transferred *P. wilberti* to *Holostichides*, whereas in the present book it is synonymised with *Paragastrostyla lanceolata* (see there).

Key to *Periholosticha* species

The species assigned are difficult to distinguish in life. The cortical granules cannot be used in the key because this feature was not checked for *P. acuminata*. Consequently, protargol impregnation is indispensable to recognise the details of the cirral pattern separating the species. See also the *Holostichides* and *Paragastrostyla* key (Bakuellidae) if you have problems identifying your specimen/population with the following key.

1 Three frontal adoral membranelles (e.g., Fig. 106q, FM) **2**
\- Four frontal adoral membranelles (e.g., Fig. 108a) **3**

[1] The "improved diagnosis" for *Periholosticha* by Song (1990) is as follows: Kleine bis mittelgroße Urostylidae mit je einer rechten und linken Marginalreihe; Midventralreihen stark verkürzt, gehen über in eine kurze Ventralreihe; Buccal- und Transversalcirren fehlen; Frontoterminalcirren vorhanden oder fehlend. Caudal schwanzartig, zahlreiche Makronucleus-Teile.

2 Midventral complex terminates distinctly behind level of buccal vertex (Fig. 106a, o, q, r); 3 dorsal kineties (Fig. 106p) .... *Periholosticha lanceolata* (p. 502)
- Midventral complex usually terminates about at level of buccal vertex (Fig. 110c, e, j, r); 2 dorsal kineties (Fig. 110d) .... *Periholosticha paucicirrata* (p. 517)

3 (1) Three dorsal kineties; further morphometric differences (number of marginal cirri and macronuclear nodules) to *P. sylvatica* see Table 24 .... *Periholosticha acuminata* (p. 512)
- Two dorsal kineties (Fig. 109d) .... *Periholosticha sylvatica* (p. 514)

Periholosticha lanceolata Hemberger, 1985
(Fig. 106a–r, 107a–i, Table 24, Addenda)

1982 ***Periholosticha lanceolata* n. spec.**[1] – Hemberger, Dissertation, p. 111, Abb. 17a–h (Fig. 106a–h).

1985 ***Periholosticha lanceolata* n. spec.** – Hemberger, Arch. Protistenk., 130: 403, Abb. 8 (Fig. 106a; original description. The type slide is deposited in the Institut für landwirtschaftliche Zoologie of the University of Bonn, Germany).

2001 ***Periholosticha lanceolata* Hemberger, 1985** – Berger, Catalogue of ciliate names 1. Hypotrichs, p. 71 (nomenclator containing all basionyms, combinations, and higher taxa of hypotrichs).

2002 ***Periholosticha lanceolata* Hemberger, 1985**[2] – Foissner, Agatha & Berger, Denisia, 5: 582, Fig. 133a–l, q–y (Fig. 106i–r, 107a–i; detailed redescription of two populations from life and after protargol impregnation including improved diagnosis. Four voucher slides [registration numbers 2002/456–458, 580] are deposited in the Oberösterreichische Landesmuseum in Linz [LI], Austria).

Nomenclature: No derivation of the species-group name is given in the original description. The Latin adjective *lanceolát·us -a -um* (lanceolate; Hentschel & Wagner 1996, p. 356) likely alludes to the lanceolate body outline. *Periholosticha lanceolata* was fixed as type species of *Periholosticha* by original designation.

Remarks: The type population of *Periholosticha lanceolata* has several specific features (Fig. 106a), all found in the Maldivean and Namibian specimens investigated by Foissner et al. (2002): body slenderly lanceolate; a row of frontoventral cirri forming an indistinct midventral pattern in the frontal body half; lack of buccal, frontoterminal, and transverse cirri; all cirri composed of only 2–4 cilia; three dorsal kineties and caudal cirri. Furthermore, the illustrations and those morphometrics which are independent of the preparation method are highly similar in the type population and the specimens studied by Foissner et al. (2002). Consequently identification is beyond reasonable doubt, although we could not entirely exclude the presence of frontoterminal cirri in our populations (see morphology). There is only one main deviating feature, namely, the

---

[1] This name is disclaimed for nomenclatural purposes (ICZN 1999, Article 8.3). See second footnote at genus section.

[2] The improved diagnosis by Foissner et al. (2002) is as follows (based on original description and data by Foissner et al. 2002): Size about 110 × 12 µm in vivo; slenderly lanceolate with posterior region narrowed, more or less tail-like. On average 16 macronuclear nodules in two strands one upon the other along left postoral body margin. Cortical granules between cirri and around dorsal bristles, colourless, compact and thus refractive, 0.5–1 × 0.4–0.8 µm. Cirri composed of 2–4 cilia. On average 16 adoral membranelles, 3 frontal separated by minute gap; 13 frontoventral cirri forming indistinct midventral pattern in anterior half of row extending to second body third; 30 right and 26 left marginal cirri; 3 dorsal kineties and caudal cirri each.

cortical granules. However, Hemberger studied live cells only superficially, if at all, and thus it is reasonable to assume that he overlooked these colourless and therefore inconspicuous granules.

For separation from other *Periholosticha* species, see key and this chapter at *P. paucirrata* and *P. sylvatica*. *Periholosticha lanceolata* is an inconspicuous species easily confused with many other middle-sized, slender soil hypotrichs, such as *Hemisincirra* spp., *Holostichides* spp., and *Paragastrostyla lanceolata*. Consequently, protargol impregnation is indispensable for reliable identification. In life, the following combination of features indicates *Periholosticha lanceolata*: size about 110 × 20 μm; body lanceolate; about 15 macronuclear nodules in indistinct series left of midline; ellipsoidal, colourless cortical granules 0.5–1.0 μm across around cirri and dorsal bristles; no buccal cirrus; midventral complex extends to near mid-body, that is, longer than adoral zone; buccal cavity flat and narrow; about 16 adoral membranelles; three dorsal kineties.

Since Foissner et al. (2002) could not entirely exclude that the populations from the Maldives and Namibia have frontoterminal cirri, the descriptions are kept separate. The diagnosis provided by Hemberger (1985) is not a diagnosis sensu stricto, but a rather brief description of the species and thus not repeated in a footnote.

Morphology: As mentioned above, descriptions of populations are kept separate and begin with the rather sparse original description, which is based almost entirely on protargol-impregnated material. Moreover, the exact number of specimens investigated for morphometric analysis is not given. Hemberger's "Basis: n = über 100 Ind." is likely not the ordinary sample size.

Description of type population (Fig. 106a, Table 24): Body size in life not indicated, likely of similar size as in protargol preparations (Wilbert's method), that is, around 130 × 20 μm. Body lanceolate, slightly cephalised, posterior end narrowly rounded, very flexible. Macronuclear nodules form strand, individual nodules ellipsoidal to elongate ellipsoidal. Contractile vacuole at about 33% of body length at left cell margin. Adoral zone occupies about 25% of body length, anterior three adoral membranelles set off from other membranelles by more or less distinct gap; undulating membranes almost in line, slightly curved, one (likely the paroral) distinctly shorter than the other. Frontal cirri not enlarged, that is, like other cirri composed of a maximum of four cilia only and about 8–10 μm long. Buccal, frontoterminal, and transverse cirri lacking. Zigzagging midventral pattern very indistinct, midventral complex usually composed of 5–6 pairs with anterior cirrus of each pair often made of four cilia and posterior of two cilia only; midventral complex terminates at 35% of body length in specimen illustrated. Three very indistinct caudal cirri; dorsal cilia 3–4 μm long.

Description of Maldivean population (from Foissner et al. 2002; Fig. 106i–r, Table 24): Body size 70–150 × 10–15 μm in life, usually about 110 × 12 μm; body length:width ratio highly variable, namely, 6.3–11.8:1, on average 8.7:1 in protargol preparations. Body usually only slightly, rarely up to 2:1 flattened dorsoventrally; slenderly lanceolata, posterior region rather abruptly tail-like narrowed, mostly slightly to distinctly twisted about main axis; acontractile and very flexible. Macronuclear nodules in two more or less distinct strands one upon the other along postoral left body margin; individual nodules globular to elongate ellipsoidal, on average 4 × 3 μm in protargol

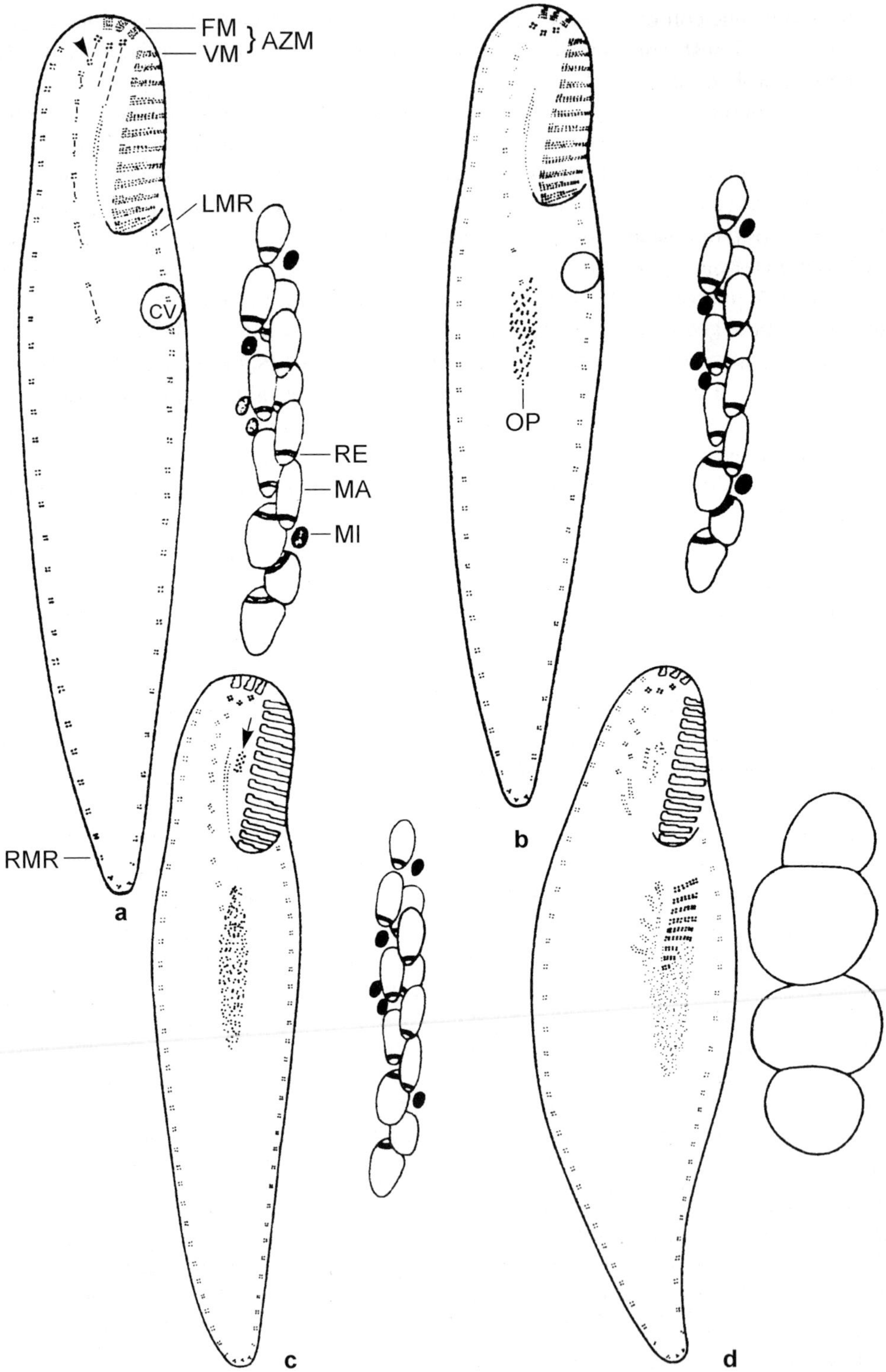
FM
VM
AZM
LMR
CV
RE
MA
MI
RMR
a
OP
b
c
d

Dorsal bristles 2–3 µm long in life and widely spaced, invariably arranged in three rows well recognisable even in live specimens due to cortical granules surrounding individual bristles; rows 1 and 2 slightly shortened anteriorly (Fig. 106p). Likely two or three caudal cirri that are difficult to separate from marginal cirri due to the tail-like body end.

Description of Namibian population (from Foissner et al. 2002; Fig. 107a–i, Table 24): The Namibian specimens are highly similar to the Maldivean ones. Consequently, only deviating or supplementary features are mentioned by Foissner et al.: body size 90–130 × 10–20 µm in life, length:width ratio 7–11:1, on average 9:1 in life, and 5.4–11.0:1, on average 8.8:1 in protargol preparations; micronuclei about 3 × 2 µm; cortical granules smaller (about 0.5 × 0.4 µm) and less compact than in Maldivean specimens; distances between marginal cirri become distinctly wider posteriorly. Similarly to Maldivean specimens, there is often an increased distance between the third and fourth cirrus of the right marginal row (22% of 14 specimens investigated), the second and third (39%), or the second and third plus the third and fourth cirrus (22%); only 17% of the specimens lack this interruption.

Cell division: Hemberger (1982) studied the ontogenesis of *Periholosticha lanceolata* (Fig. 106b–h). Hemberger (1985), who announced the publication of the complete ontogenesis, made some comments about the division, but provided no illustration. Unfortunately, the announced paper never appeared.

Ontogenesis commences with the proliferation of basal bodies very close to the last midventral cirrus (Fig. 106b). This basal body field becomes the elongate oral primordium. Almost simultaneously, an anlage forms within the buccal field (Fig. 106c, arrow), as in *P. paucicirrata* (Fig. 110r, horizontal arrow). The next stage shows the assemblage of adoral membranelles for the opisthe and, at the right side of this primordium, the anlagen for the undulating membranes and the cirral primordia are recognisable (Fig. 106d). In the proter, the parental undulating membranes are modified to primordia. Furthermore, two parental midventral cirri have transformed to cirral primordia. Most other parental cirri, however, do not contribute to anlagen formation.

Fig. 106e shows a middle divider with almost the full number of adoral membranelles in the opisthe oral anlage. Furthermore, in both the proter and the opisthe the anlage for the undulating membranes with the segregated left frontal cirrus (Fig. 106e, arrowheads), and eight oblique cirral anlagen are recognisable. Three of the four marginal cirral anlagen are developed.

Somewhat later, most frontal and midventral cirri are formed (Fig. 106f). From anlage II only one cirrus (= middle frontal cirrus; Fig. 106f, arrowhead) has developed; this explains the lack of a buccal cirrus (= cirrus II/2) in interphasic specimens. The undulating membrane anlage has divided longitudinally in both filial products. Note that in both the proter and the opisthe a more or less high number of irregularly distributed

---

← lines connect cirri originating from same anlage. **r:** Curved and twisted specimen, 70 µm (shortest distance from anterior to posterior body end). Arrow marks cirrus III/2 behind right frontal cirrus. CC = caudal cirri, E = endoral, FC = right frontal cirrus, FM = frontal adoral membranelles, LMR = left marginal row, MA = macronuclear nodule, MI = micronucleus, P = paroral, 1–3 = dorsal kineties. Page 502.

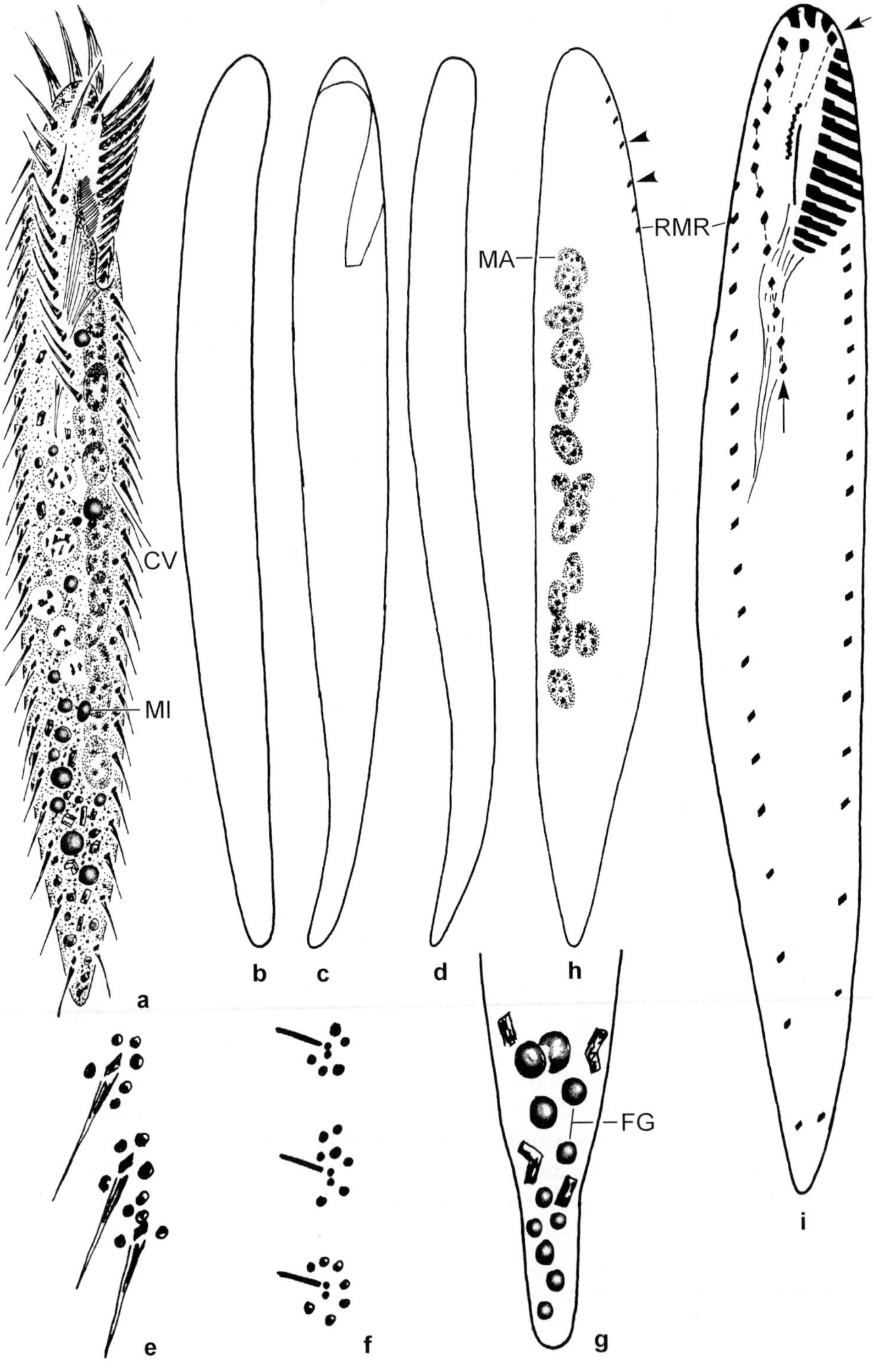
CV
MI
a
b
c
d
h
MA
RMR
i
e
f
FG
g

basal bodies is present near the rear end of the adoral zone. These basal bodies form the rearmost cirral anlage, which does not produce two, but three cirri. The last cirrus of this anlage is designated as transverse cirrus by Hemberger (1982, p. 115). In my and Song's (1990) opinion, this is not a transverse cirrus because it is not distinctly set off posteriorly during interphase, that is, all cirri of the rearmost anlage are immediately behind the buccal vertex in non-dividing specimens. Interestingly, all figured interphasic specimens show an even number of midventral cirri, strongly indicating that the posteriormost cirrus of the last (rightmost) anlage is dissolved in very late ontogenetic stages. Late stages clearly show that no frontoterminal cirri are formed (Fig. 106g, h). The parental adoral zone of membranelles is not reorganised.

The formation of the dorsal infraciliature proceeds in *Gonostomum* pattern, that is, each of the three kineties forms an anterior and posterior anlage. One caudal cirrus originates at the end of each kinety. Fragmentation of dorsal kineties or the formation of dorsomarginal rows does not occur.

The division of the nuclear apparatus proceeds in ordinary manner, that is, the individual macronuclear nodules fuse to a single mass in middle dividers and divide amitotically in later stages. The micronuclei divide mitotically (Fig. 106b–h).

Occurrence and ecology: Likely confined to terrestrial habitats. The type location of *Periholosticha lanceolata* is near Puerto Maldonada (about 12°36'S 69°12'W), Departimento Madre de Dios, Peru, where Hemberger (1985) discovered it in a meadow soil (see also Hemberger 1982, p. 2, for details). Foissner et al. (2002) found it in highly saline and alkaline (pH 8.6) sandy soil and plant litter at the coast of the Maldives (North-Male Atoll, Himmafushi; collected by W. Petz in 1990) and in the floodplain (litter sieved off a dry sand bank; pH 6.2) of the Bukaos River, about 80 km north of the town Keetmanshoop, Namibia. Obviously *Periholosticha lanceolata* has a broad ecological range and possibly cosmopolitan distribution, although Laurasian records are not known.

Periholosticha lanceolata likely feeds on bacteria (Foissner et al. 2002). Biomass of 10^6 specimens about 30 mg (Foissner 1987a, 1998).

← **Fig. 107a–i** *Periholosticha lanceolata* (from Foissner et al. 2002. Namibian specimens from life [a–g] and after protargol impregnation [h, i], Wilbert's method). **a:** Ventral view of a representative specimen, 115 µm. **b–d:** Shape variants in dorsal, ventral, and lateral view. **e, f:** The cortical granules are about 0.5 µm across, colourless, and arranged mainly between cirri and around dorsal bristles. **g:** The narrowed posterior end is usually packed with fat globules and crystals up to 3 µm in size and therefore dark at low magnification. **h, i:** Nuclear apparatus and infraciliature of ventral side of same specimen, 102 µm. Short arrow marks gap separating the adoral zone in a frontal portion with three membranelles and a ventral portion with 11–14 membranelles; the gap is occupied by the left frontal cirrus. The long arrow denotes the end of the midventral complex. Arrowheads mark increased distance between third and fourth cirrus of right marginal row, indicating that frontoterminal cirri might be present. CV = contractile vacuole, FG = fat globules, MA = anteriormost macronuclear nodule, MI = micronucleus, RMR = right marginal row. Page 502.

Periholosticha acuminata Hemberger, 1985 (Fig. 108a, b, Table 24)

1982 ***Periholosticha acuminata* n. spec.**[1] – Hemberger, Dissertation, p. 116, Abb. 18 (Fig. 108a, b; see nomenclature).

1985 ***Periholosticha acuminata* n. spec.** – Hemberger, Arch. Protistenk., 130: 404, Abb. 9 (Fig. 108a, b; original description. The type slide is deposited in the Institut für landwirtschaftliche Zoologie of the University of Bonn, Germany).

2001 ***Periholosticha acuminata* Hemberger, 1985** – Berger, Catalogue of ciliate names 1. Hypotrichs, p. 71 (nomenclator containing all basionyms, combinations, and higher taxa of hypotrichs).

Nomenclature: No derivation of the species-group name is given in the original description. The Latin adjective *acuminat·us -a -um* (pointed; Hentschel & Wagner 1996, p. 62) likely refers to the triangularly pointed posterior body end.

Remarks: *Periholosticha acuminata* is the second species assigned to *Periholosticha* by Hemberger (1985). It differs from *P. lanceolata* (type species) by the possession of frontoterminal cirri so that they are possibly less closely related than indicated by other features (lack of buccal cirrus, transverse cirri, and midventral rows; gap in adoral zone). However, it cannot be excluded that they are in fact very closely related and that the lack of the frontoterminal cirri is an apomorphy of the type species.

Periholosticha acuminata is described from protargol-impregnated specimens only. Thus, redescription from life **and** after protargol impregnation is strongly recommended. Life observation will possibly show that it has, like the other three species, cortical granules. If all other features match rather well, the identification should be beyond reasonable doubt. Hemberger's (1982, 1985) illustration rather clearly shows frontoterminal cirri (Fig. 108a). Unfortunately, he did not mention them in the description although he knew about this cirral group (see Hemberger 1985, p. 11). Thus, a redescription should include a specific statement about this feature.

Periholosticha acuminata is best recognised by the lack of a buccal cirrus and of transverse cirri, the short midventral complex composed of cirral pairs only, the conspicuous nuclear figure, the pointed rear body end, which is likely also present in life specimens, and, most importantly, the four frontal adoral membranelles and the three dorsal kineties. The diagnosis provided by Hemberger (1985) is not a diagnosis sensu stricto, but a rather brief description of the species and thus not repeated in a footnote.

Morphology: Body size in life not indicated, however, likely very similar to that of protargol-impregnated (Wilbert's method) specimens, namely 125–150 × 22–28 μm. Body elongate with margins almost in parallel in second and third quarter; anterior end broadly rounded, posterior quarter triangularly converging with pointed end; very flexible. Macronuclear nodules ellipsoidal, form strand, likely as in congener, left of midline. Micronuclei rather large, attached to macronuclear strand (Fig. 108b). Contractile vacuole at about 28% of body length at left cell margin. Adoral zone occupies 22% of body length in specimen illustrated (Fig. 108a), composed of 14–19, usually 18 membranelles of ordinary fine structure; distalmost four membranelles set off from

[1] This name is disclaimed for nomenclatural purposes (ICZN 1999, Article 8.3). See second footnote of genus section.

rest by small gap. Buccal cavity not described, but likely rather small and flat. Undulating membranes curved, commence and terminate at about same level, optically intersecting. Frontal cirri slightly enlarged, more or less transversely arranged behind frontal adoral membranelles. Buccal and transverse cirri lacking. Two frontoterminal cirri in usual position, that is, between anterior end of right marginal row and right frontal cirrus (although they are rather distinct, their presence should be confirmed by ontogenetic data). Midventral pairs form distinct zigzag about as long as adoral zone of membranelles; right (= anterior) cirrus of each pair slightly larger than left (= posterior). Right marginal row begins near distal end of adoral zone, ends subterminally; left row commences, as is usual, left of buccal vertex and terminates ahead of rear body end; distance between individual marginal cirri increases from anterior to posterior. Dorsal cilia 4 μm long, arranged in three kineties. Three caudal cirri, likely each one per dorsal kinety.

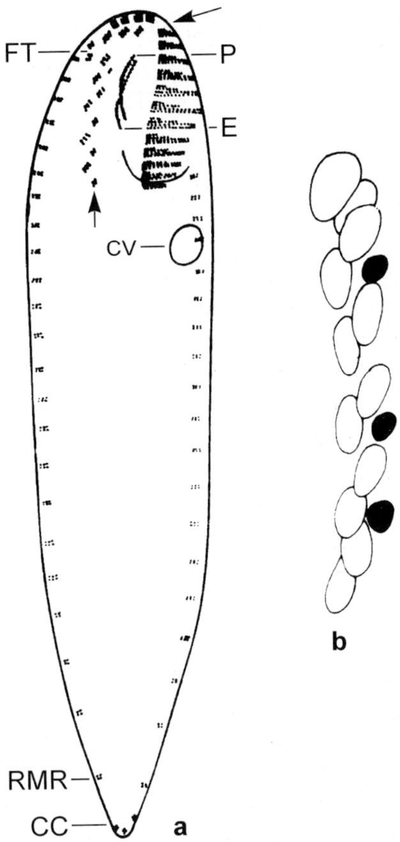

Fig. 108a, b *Periholosticha acuminata* (from Hemberger 1985. Wilbert's protargol impregnation method). **a, b:** Infraciliature of ventral side and nuclear apparatus of representative specimen, 130 μm. Oblique arrow marks gap between frontal and ventral adoral membranelles. Vertical arrow denotes the posteriormost cirrus of the rearmost midventral pair, that is, the posterior end of the midventral complex. In most cirral pairs, the left (= rear) cirrus is smaller (four basal bodies) than the anterior cirrus (six basal bodies). CC = caudal cirri, CV = contractile vacuole, E = endoral, FT = frontoterminal(?) cirri, P = paroral, RMR = right marginal row. Page 512.

Occurrence and ecology: The type location of *Periholosticha acuminata* is not given by Hemberger (1985), who mentions under origin only "suspension of mull rendzina soil". However, this soil sample is not from Peru, the sole country mentioned under materials, but from Germany (see Hemberger 1982, p. 2 and Foissner 2000). According to a personal communication by Norbert Wilbert (University of Bonn, Germany), the site is near the village of Mechernich (50°36'N 6°39'E), Germany. It is a beech forest on the "Kakushöhle Nordhang" and the outcrops are a middle devonian dolomite and a Pleistocene tufaceous limestone.

Biomass of 10^6 specimens about 56 mg (Foissner 1987a, 1998). According to Foissner (1998), *Periholosticha acuminata* is a true soil inhabitant.

Periholosticha sylvatica Foissner, Berger, Xu & Zechmeister-Boltenstern, 2005
(Fig. 109a–k, Table 24)

2005 ***Periholosticha sylvatica* nov. spec.**[1] – Foissner, Berger, Xu & Zechmeister-Boltenstern, Biodiversity and Conservation, 14: 687, Fig. 16a–k, Table 12 (Fig. 109a–k; original description. 1 holotype and 2 paratype slides are deposited in the Oberösterreichische Landesmuseum in Linz [LI], Austria).

Nomenclature: The Latin adjective *sylvatic·us -a -um* (inhabiting the forest) refers to the habitat the species was discovered (Foissner et al. 2005).

Remarks: The present species was classified in *Periholosticha* because the overall morphology and especially the cirral pattern are as in *P. lanceolata* and *P. acuminata*. For general discussion of some features, for example, presence/absence of frontoterminal cirri, see genus section.

Periholosticha acuminata is likely the most similar species because both have four frontal membranelles and a short midventral complex. However, *Periholosticha acuminata* has, like *P. lanceolata*, three dorsal kineties (vs. two in *P. sylvatica*) and is therefore clearly distinguishable, at least in protargol preparations, from *P. sylvatica*, which has only two kineties. Moreover, some further morphometrics are distinctly different: number of right marginal cirri (21–37 in *P. sylvatica* vs. 19–24 in *P. acuminata*), number of left marginal cirri (19–36 vs. 17–22), number of macronuclear nodules (15–25 [average 21] vs. 11–14 [average not known]).

Possibly *Periholosticha sylvatica* was misidentified occasionally as *Paragastrostyla terricola*, which also lacks transverse cirri and a buccal cirrus, and which has only two dorsal kineties too. However, *Paragstrostyla* species have a distinct midventral row (vs. lacking) and therefore the midventral complex extends far behind the buccal vertex (vs. at level of buccal vertex).

Morphology: Body size 90–160 × 15–25 µm in life, usually about 120 × 20 µm; body length:width ratio highly variable, namely, 4.6–8.1:1, on average 5.9:1 in protargol preparations. Body slightly to up to 2:1 flattened dorsoventrally, acontractile, but very flexible. Body outline slender oblanceolate to almost vermiform, frequently slightly sigmoidal and twisted about main body axis; anterior body end moderately broadly rounded, posterior third distinctly narrowed and bluntly pointed (Fig. 109a, e, j, k, Table 24). On average 21 macronuclear nodules in two more or less distinct series one upon the other and/or side by side along postoral left body margin; individual nodules globular to elongate ellipsoidal, on average 7 × 3 µm in protargol preparations; nucleoli scattered, globular, and of ordinary size. Usually a slightly ellipsoidal micronucleus each in anterior and posterior region of nuclear figure. Contractile vacuole slightly ahead of mid-body at left margin of cell. Cortical granules around cirri and dorsal bristles, and loosely scattered throughout cortex, yellowish to citrine and rather refractive,

[1] The diagnosis by Foissner et al. (2005) is as follows: Size about 120 × 20 µm in vivo; narrowly oblanceolate and slightly twisted about main body axis. Cortical granules yellowish, rather loosely arranged, ≤ 1 µm across. On average 21 macronuclear nodules in series left of midline, 19 adoral membranelles, 10 cirri with rather distinct midventral pattern in frontal row, 2 frontoterminal (?) cirri, 30 cirri in right and 27 in left marginal row, 4 caudal cirri, and 2 dorsal kineties.

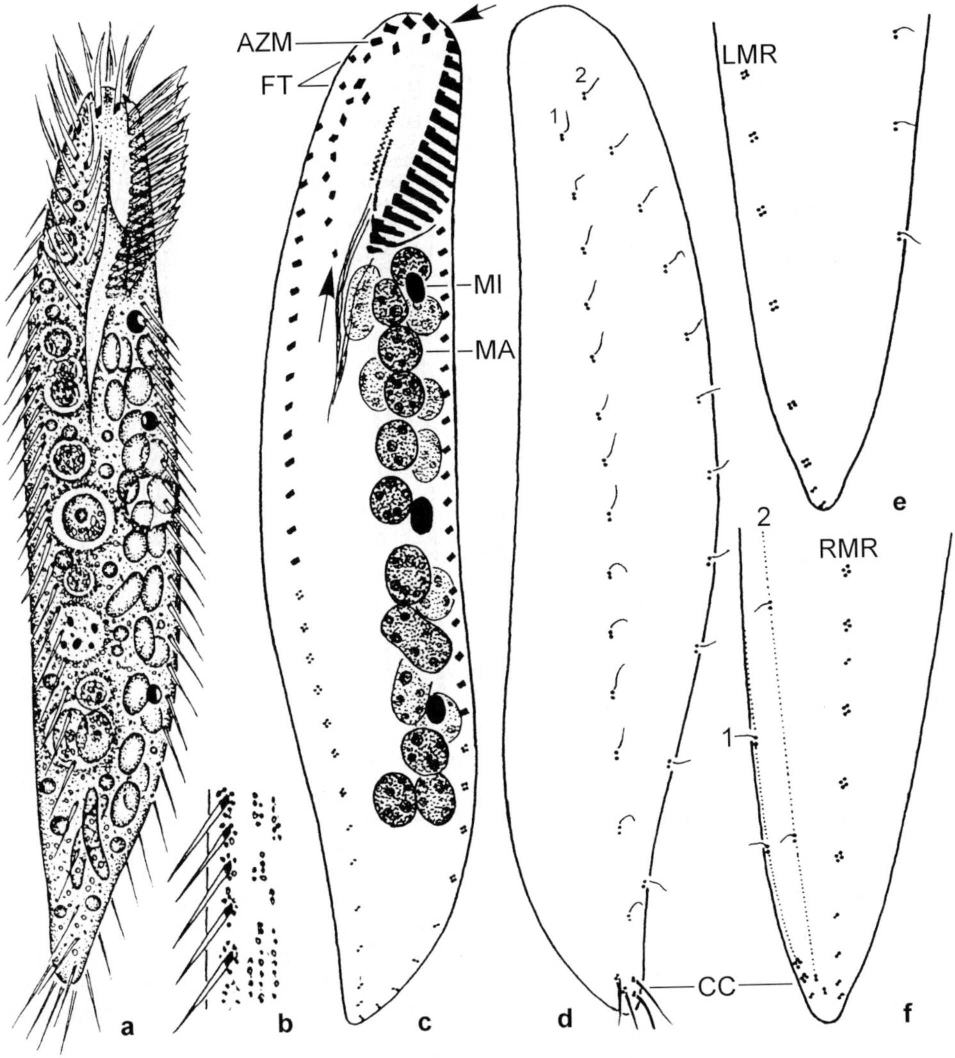

Fig. 109a–f *Periholosticha sylvatica* (from Foissner et al. 2005. a, b, from life; c–f, protargol impregnation). **a:** Ventral view of a representative specimen slightly twisted about main body axis, 120 μm. **b:** Surface view showing citrine cortical granules 0.3–1.0 μm across. **c, d:** Infraciliature of ventral and dorsal side and nuclear apparatus of holotype specimen, 95 μm. The macronuclear nodules form two rough series one upon the other (ventral series dark, dorsal bright). Note distinct size reduction of cirri in posterior portion of marginal rows. Short arrow marks minute gap between frontal (distal) and ventral (proximal) adoral membranelles, long arrow denotes rear end of midventral complex. **e, f:** Lateral views of posterior body region of a specimen with five caudal cirri composed of 2–4 cilia each. Dorsal bristles and their first caudal cirrus connected by dotted lines. CC = caudal cirri, FM = rightmost frontal membranelle (= distal end of adoral zone of membranelles), FT = frontoterminal (?) cirri, LMR = left marginal row, MA = macronuclear nodule, MI = micronucleus, RMR = right marginal row, 1, 2 = dorsal kineties. Page 514.

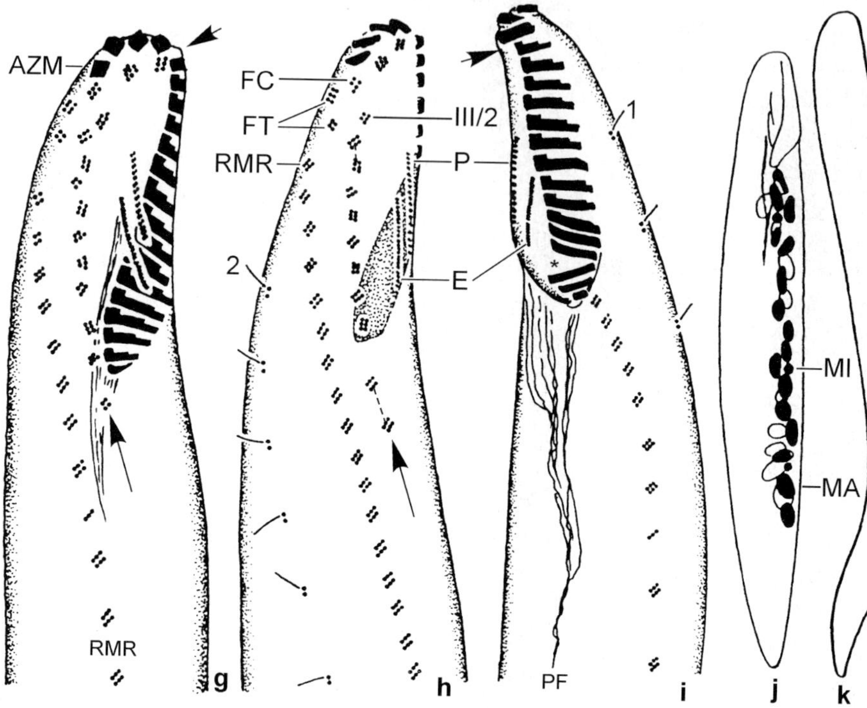

Fig. 109g–k *Periholosticha sylvatica* (from Foissner et al. 2005. Protargol impregnation). **g–i:** Slightly dorsolateral (43 μm), strongly dorsolateral (50 μm), and ventrolateral (52 μm) view of infraciliature in anterior body portion showing, inter alia, the supposed frontoterminal cirri and the variability of the midventral complex. Short arrow in (g, i) marks gap in adoral zone. The ventral (proximal) membranellar ribbon is rather abruptly twisted proximally, as indicated by the increased distance between the membranelles (asterisk). Long arrows mark rear end of midventral complex (cirri of first, respectively, last midventral pair of complex connect by broken line in h). **j, k:** Shape variability of large specimens, 142, μm, 135 μm. The macronuclear nodules are arranged in two rough series one upon the other. AZM = distal end of adoral zone of membranelles, E = endoral, FC = right frontal cirrus, FT = frontoterminal (?) cirri, MA = macronuclear nodule, MI = micronucleus, P = paroral, RMR = right marginal row, III/2 = cirrus behind right frontal cirrus, 1, 2, = dorsal kineties. Page 514.

0.3–1.0 μm in size (Fig. 109b); provide cells with a yellowish shimmer in the bright field microscope. Cytoplasm usually packed with food vacuoles 4–10 μm across and some small lipid droplets. Glides, swims, or winds slowly on microscope slide and between soil particles showing great flexibility.

Adoral zone inconspicuous because occupying only 23% of body length, composed of 19 membranelles on average, distal (frontal) four membranelles invariably set off from ventral membranelles by a small, but distinct gap at left anterior corner of cell; ventral portion of adoral zone twisted along its main axis, as evident from differently oriented last membranelles (Fig. 109c, g–i). Buccal cavity narrow and flat; buccal lip hyaline, projects angularly covering proximal portion of adoral zone. Paroral and endoral appear closely spaced when viewed ventrally, forming slightly curved line along posterior half of adoral zone. Paroral short, slightly curved, composed of distinctly zigzagging basal bodies with about 5 µm long cilia. Endoral almost straight, anterior half appears to overlap posterior half of paroral, composed of very narrowly spaced (di?)kinetids. Pharyngeal fibres of ordinary length and structure (Fig. 109a, c, g–i, Table 24).

Cirral pattern and number of cirri of usual variability, except for the highly variable (40%) number of caudal cirri (Fig. 109a, c–i, Table 24). Cirri 10–12 µm long in life, most composed of six cilia in two rows, except for posterior third marginal cirri, which consist of only 2–4 cilia. Frontal cirri form slightly oblique row subapically, middle cirrus usually slightly enlarged because composed of 8–9 cilia, right cirrus behind distalmost adoral membranelle. Buccal cirrus lacking. Midventral complex composed of cirral pairs forming relatively distinct zigzag, at rear end sometimes a short midventral row composed of three cirri; complex extends right of midline to or slightly behind level of buccal vertex. Two, rarely three cirri right of anterior end of midventral complex and distinctly separate from right marginal cirri, and therefore likely frontoterminal cirri (see genus section for discussion of this feature). Transverse cirri lacking. Marginal rows extend slightly obliquely due to body torsion and abut on bluntly pointed rear body end; right row commences subapically at level of paroral.

Dorsal bristles about 3 µm long in life, arranged in two rows. Bristles more closely spaced in kinety 1 than in kinety 2, both rows slightly shortened anteriorly and posteriorly. Caudal cirri fine, that is, composed of only 2–4 cilia, number highly variable; difficult to distinguish from posteriormost marginal cirri which consist of the same low number of cilia (Fig. 109d–f, h, i).

Cell division: Ontogenesis commences with the production of a long, narrow oral primordium left of midline distinctly behind the rear end of the midventral complex (Foissner et al. 2005). In *P. lanceolata*, which has a longer midventral complex, it begins very close to the last cirrus of the midventral complex (Fig. 106b).

Occurrence and ecology: The type locality of *P. sylvatica* is a *Pinus nigra* forest soil in the Stampfltal (47°53'N 16°02'E) near Vienna (Austria). It was moderately abundant two days after rewetting the sample, suggesting an r-selected strategy, while most hypotrichs are more k- than r-selected (Foissner 1987a). Feeds on bacteria, fungal spores, and heterotrophic flagellates (Foissner et al. 2005).

Periholosticha paucicirrata Foissner, Berger, Xu & Zechmeister-Boltenstern, 2005
(Fig. 110a–s, Table 24)

2005 ***Periholosticha paucicirrata* nov. spec.**[1] – Foissner, Berger, Xu & Zechmeister-Boltenstern, Biodiversity and Conservation, 14: 679, Fig. 15a–s, Table 11 (Fig. 110a–s; original description. 1 holotype and 4

[1] The diagnosis by Foissner et al. (2005) is as follows (averages from four populations): Size about 100 × 13 µm in vivo; very narrowly oblanceolate and slightly twisted about main body axis. Cortical granules mainly around cirri and dorsal bristles, yellowish to yellow-orange, ≤ 0.5 µm across. On average 15–16 macronuclear

paratype slides, and 2–4 voucher slides each from the other populations are deposited in the Oberösterreichische Landesmuseum in Linz [LI], Austria).

Nomenclature: The species-group *paucicirrata* is a composite of the Latin adjectives *pauci* (few) and *cirrat·us -a -um* (having curl of hair), referring to the few cirri composing the midventral complex (Foissner et al. 2005).

Remarks: The assignment of the present species to *Periholosticha* is not quite certain because it has – admittedly inconspicuous – transverse cirri whereas such cirri are lacking in the type species and very likely also in *P. acuminata* and *P. sylvatica*, although this is difficult to ascertain due to the pointed posterior body end.

Periholosticha paucicirrata differs from *P. acuminata* and *P. lanceolata* mainly by the number of dorsal kineties (2 vs. 3) and cirri forming the midventral complex (6–7 vs. 12–15). *Periholosticha sylvatica* is much stouter (body length:width ratio 5.9:1 vs. 8.1:1) and has four (vs. three) frontal membranelles. Moreover, many important morphometrics are significantly different, that is, do not or only slightly overlap, for instance, the number of adoral membranelles (19 vs. 10–15), cirri composing the midventral complex (10 vs. 6–7), and macronuclear nodules (19 vs. 15–16).

In life, the present species highly resembles *Hemisincirra inquieta* Hemberger, 1985, which, however, has three (vs. two) dorsal kineties and a buccal cirrus (vs. none) recognisable in the specimens studied by Foissner et al. (2002; Fig. 110t), and also in a postdivider of the type population (Hemberger 1982). Nonetheless, these two species are easily confused in life, and therefore identifications should be checked in protargol preparations.

Morphology: Foissner et al. (2005) studied four populations of this species. They are so similar that conspecificity is beyond reasonable doubt (Table 24). Consequently, the diagnosis includes all populations, while the in vivo description is based mainly on specimens from Austria (Kolmberg) and Croatia because the other populations were investigated mainly in protargol preparations.

Body size and shape moderately variable and very similar in all populations (Table 24). Size 75–120 × 10–20 µm in life, usually about 100 × 13 µm; body length:width ratio 6.2–10.8:1, on average ca. 8:1 in life and protargol prepartions. Body about 2:1 flattened dorsoventrally, acontractile, but highly flexible, very narrowly oblanceolate and occasionally indistinctly sigmoidal, slightly to distinctly twisted about main body axis; anterior body end narrowly rounded-truncate, posterior gradually narrowed and bluntly pointed to almost tail-like (Fig. 110a–c, j, p–r). An average of 16 macronuclear nodules in two indistinct series one upon the other along postoral left body margin; individual nodules globular to elongate ellipsoidal, on average 5–6 × 3 µm in protargol preparations; nucleoli scattered, globular, small. 2–3 globular to ellipsoidal micronuclei scattered along macronuclear series, rather distinct in life because compact and about 1.5 µm in size. Contractile vacuole in mid-body left of midline, with long collecting canals. Cortical granules located around cirri and dorsal bristles in populations from Croatia

---

nodules in series left of midline, 10–15 adoral membranelles, 6–7 cirri with indistinct midventral pattern in frontal row, 2 frontoterminal (?) cirri, 21–23 cirri in right and 18–21 cirri in left marginal row, 1–2 transverse and 2 caudal cirri, and 2 dorsal kineties.

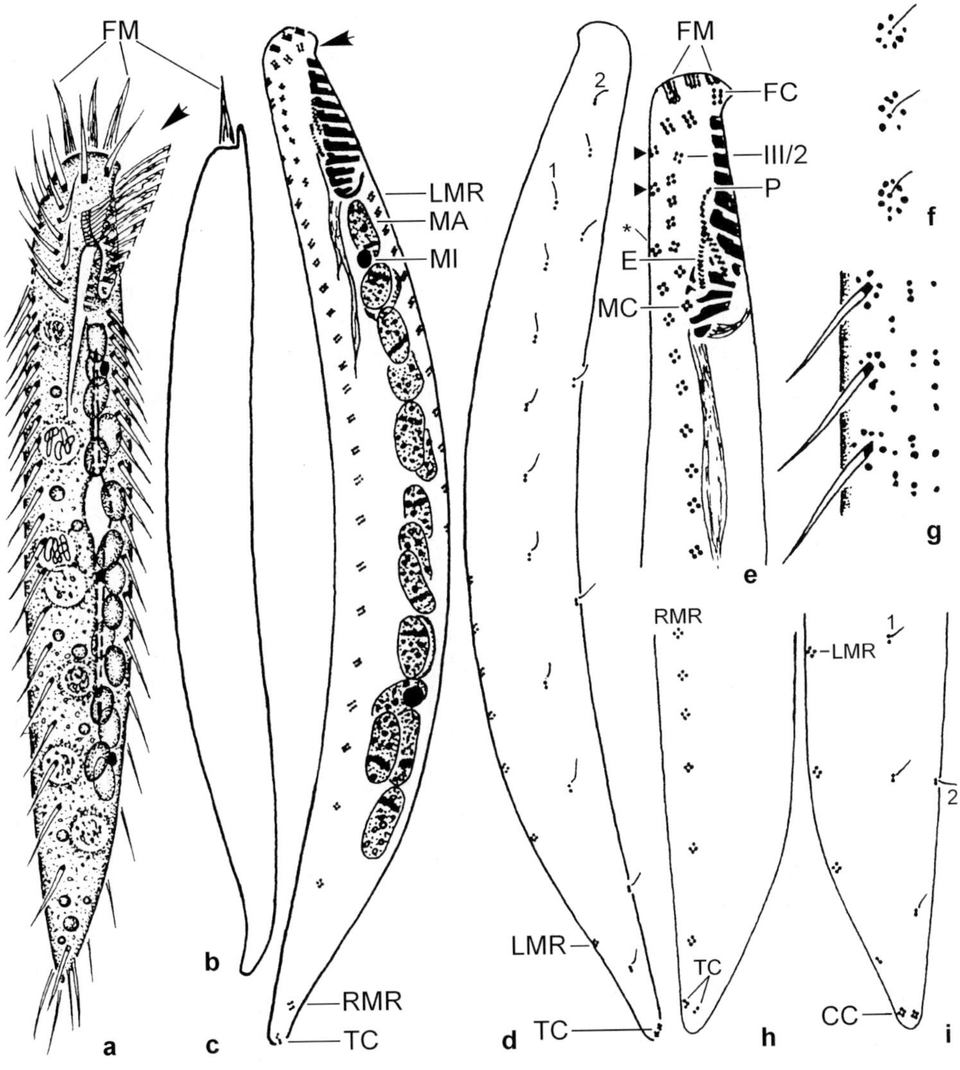

Fig. 110a–i *Periholosticha paucicirrata*, type population from Kolmberg, Austria (from Foissner et al. 2005. a, b, f, g, from life; c–e, h, i, protargol impregnation). **a, b:** Ventral and right lateral view of a representative specimen slightly twisted about main body axis, 98 µm. Arrow denotes gap in adoral zone. **c, d:** Infraciliature of ventral and dorsal side and nuclear apparatus of holotype specimen, 103 µm. Arrow marks broad gap between the three frontal (distal) and the remaining (ventral = proximal) adoral membranelles. **e:** Details in ventral anterior portion. Arrowheads denote supposed frontoterminal cirri, which are distinctly separated from the first right marginal cirrus (asterisk). Note the beak-like projecting left anterior corner bearing the left frontal cirrus and separating frontal and ventral adoral membranelles. **f, g:** The cortex contains loose rows of minute (below 1 µm), yellow granules concentrated around dorsal bristles (f) and cirri (g). **h, i:** Infraciliature of ventral and dorsal side in posterior body portion, where cirri consist of only 2–4 cilia. Two minute transverse cirri and caudal cirri each are recognisable. CC = caudal cirri, E = endoral, FC = left frontal cirrus, FM = frontal adoral membranelles, LMR = left marginal row, MA = macronuclear nodule, MC = rear end of midventral complex, MI = micronucleus, P = paroral, RMR = rear end of right marginal row, TC = transverse cirri, III/2 = cirrus behind right frontal cirrus, 1, 2 = dorsal kineties. Page 517.

and the Müllerboden in Austria, while located around cirri and dorsal bristles and distributed in loose rows throughout cortex in type population; provide cells with a yellowish shimmer in life, never impregnate with the protargol method used. Individual cortical granules distinct, though only 0.2–0.5 µm across, because compact and of a bright citrine to yellow-orange colour (Fig. 110f, g, m). Cytoplasm hyaline, usually contains only few food vacuoles 3–5 µm in diameter and some lipid droplets up to 2 µm across. Glides and swims moderately rapidly on microscope slide and between soil particles showing pronounced flexibility.

Adoral zone inconspicuous because occupying only 20% of body length, composed of an average of 10–15 membranelles, depending on population; frontal (= distal) three membranelles invariably separated from ventral membranelles by a small, but distinct gap at beak-like left anterior corner of cell. Frontal membranelles insert on anterior body end, proximal portion covered by scutum-like projecting ventral surface. Ventral

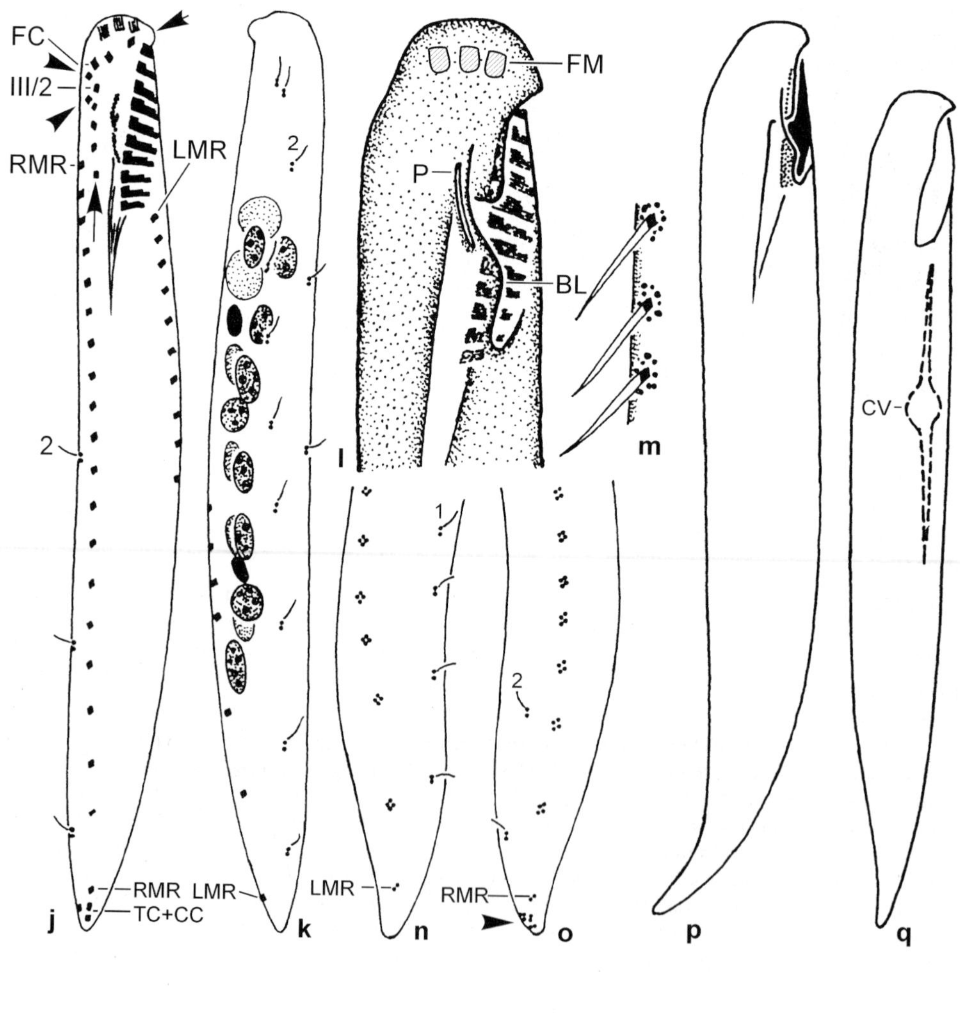

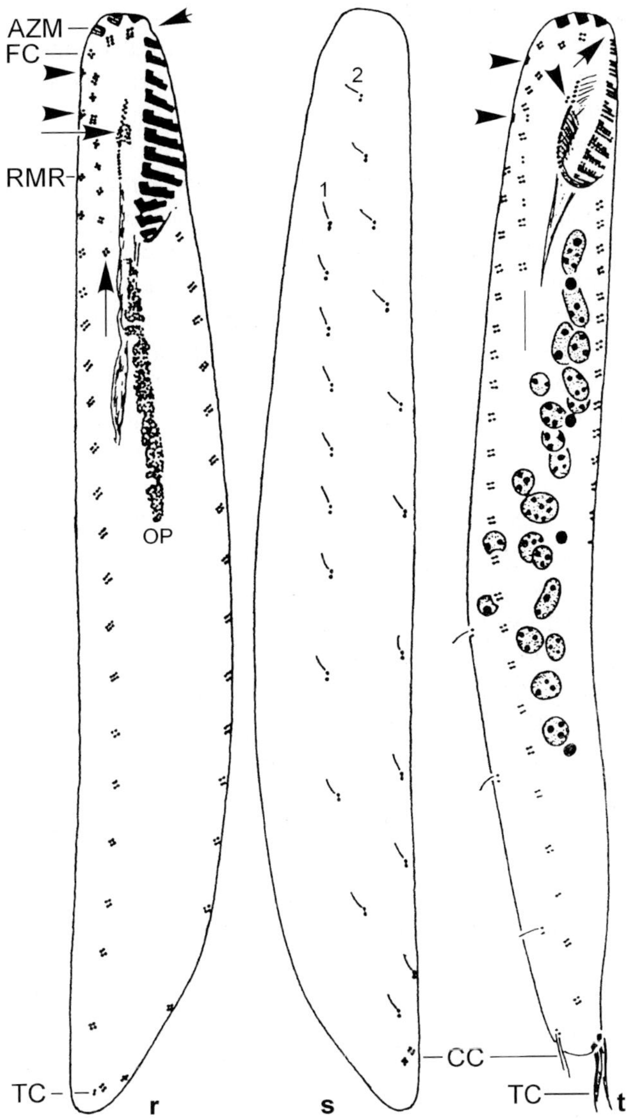

Fig. 110r, s *Periholosticha paucicirrata*, Stampfltal population (from Foissner et al. 2005. Protargol impregnation). Infraciliature of ventral and dorsal side of an early divider, 100 μm. Primordia develop underneath the undulating membranes (long horizontal arrow) and postorally left behind the end (long vertical arrow) of the midventral complex. In the type species of *Periholosticha*, primordia formation commences at the same sites (Fig. 106b). Short arrow marks gap in adoral zone, arrowheads denote supposed frontoterminal cirri. This specimen has two clearly recognisable transverse and caudal cirri. Page 517.

Fig. 110t *Hemisincirra inquieta* Hemberger, 1985 (from Foissner et al. 2002. Protargol impregnation). Infraciliature of ventral side and nuclear apparatus, 102 μm. This species differs from *P. paucicirrata* by having a (indistinct) buccal cirrus (oblique arrowhead) and three dorsal kineties. Arrow marks gap in adoral zone, horizontal arrowheads denote frontoterminal cirri.

CC = caudal cirri, FC = right frontal cirrus, FM = rightmost frontal membranelle (= distal end of adoral zone of membranelles), RMR = right marginal row, TC = transverse cirri, 1, 2 = dorsal kineties.

← **Fig. 110j–q** *Periholosticha paucicirrata*, Croatian population (from Foissner et al. 2005. j, k, n, o, protargol impregnation; l, m, p, q, from life). **j, k:** Infraciliature of ventral and dorsal side of voucher specimen. Note macronuclear nodules arranged in two indistinct series one upon the other, and distinct gap (short arrow) between frontal (distal) and ventral (proximal) adoral membranelles. Long arrow marks rear end of midventral complex composed of two cirral pairs in this specimen. Arrowheads denote supposed frontoterminal cirri, which are almost in line with the right marginal cirri. **l:** Cortex pattern in ventral anterior region. **m:** Minute, yellow-orange granules around cirri and dorsal bristles. **n, o:** Infraciliature of posterior region. Arrowhead marks three minute (transverse and caudal) cirri very near to the posterior end. **p, q:** Shape variants. BL = buccal lip, CC = caudal cirri, CV = contractile vacuole with collecting canals, FC = right frontal cirrus, FM = frontal membranelles of adoral zone, LMR = left marginal row, P = paroral, RMR = right marginal row, TC = transverse cirri, III/2 = cirrus behind right frontal cirrus, 1, 2 = dorsal kineties. Page 517.

portion of adoral zone slightly twisted along main body axis, as described for *P. sylvatica*. Buccal cavity narrow and flat; buccal lip hyaline, projects angularly covering proximal half of adoral zone, anterior half bears paroral in minute cleft. Paroral and endoral appear closely spaced when viewed ventrally, forming a curved line along midportion of adoral zone. Paroral 3–6 µm long, slightly curved, composed of zigzagging basal bodies with 5 µm long cilia; anterior, rarely posterior, third occasionally lacking or incompletely impregnated. Endoral of same shape and length as paroral, anterior half appear to overlap posterior half of paroral, composed of narrowly spaced (di?)kinetids. Pharyngeal fibres of ordinary length and structure (Fig. 110a–c, e, j, l, r).

Cirral pattern and number of cirri of usual variability (Fig. 110c, j, Table 24). Cirri 7–9 µm long in life, most composed of 4–6 cilia, depending on population and specimen, those in posterior body third consist of only 2–4 cilia. Frontal cirri form slightly oblique row behind body end, usually indistinctly enlarged, that is, composed of six cilia. Buccal cirrus lacking. Midventral complex composed of indistinctly zigzagging cirral pairs, extends right of midline, on average slightly shorter than adoral zone of membranelles in all populations; cirri composed of 2–6, usually four cilia. Two (frontoterminal? see genus section) cirri right of anterior end of midventral complex, sometimes indistinctly separated from right marginal row. 2–4, usually 3–4, cirri on pointed rear end; some well-oriented specimens show that these are one or two transverse and two caudal cirri, often difficult to distinguish from posteriormost marginal cirri. Marginal rows extend slightly obliquely due to body torsion from anterior to posterior end; right row commences subapically at level of paroral.

Dorsal bristles 2.5–3.0 µm long in life, invariably arranged in two rows. Bristles more closely spaced in kinety 1 than in kinety 2; kinety 1 distinctly shortened anteriorly, that is, commences at level of buccal vertex; both rows slightly shortened posteriorly. Number of kinetids within rows very similar in all populations. Each kinety with one caudal cirrus (Fig. 110d, i, j, n, o, s, Table 24).

Occurrence and ecology: Type locality of *Periholosticha paucicirrata* is a *Quercus petraea–Carpinus betulus* (oak-hornbeam) forest soil from the Kolmberg (47°58'N 16°41'E) in Lower Austria near Vienna. It occurred in seven out of 12 sites investigated, both in deciduous and coniferous forests, showing that it is a common species with a wide ecological range (Foissner et al. 2005). This is emphasised by populations from Croatia and Greece, which are highly similar to the Austrian specimens (Table 24). Likely, the present species was occasionally misidentified as *Hemisincirra inquieta* (Fig. 110t) in previous studies because they are rather difficult to separate.

In Croatia, *P. paucicirrata* occurred in slightly acidic (pH 6.3 in water) and saline, very sandy coastal soil from Dugi Otok, a small island off the Adriatic Sea coast (sample collected by W. Petz, Austria). In Greece, Foissner et al. (2005) found it in a Peloponnese pine forest between the towns of Katarraklias and Vlasia. The sample (pH 6.3 in water) was a mixture of pine needles, raw humus, and terrestrial mosses. *Periholosticha paucicirrata* was moderately abundant in the non-flooded Petri dish cultures, except for the Greece sample, where it was numerous. It is well adapted to soil life by the slender, flexible body. Feeds on bacteria (Foissner et al. 2005).

Table 24 Morphometric data on *Periholosticha acuminata* (acu, German type population from Hemberger 1985), *Periholosticha lanceolata* (la1, Maldivean population from Foissner et al. 2002; la2, Namibian population from Foissner et al. 2002; la3, Peruvian type population from Hemberger 1985), *Periholosticha paucirrata* (pa1, type population from Kolmberg, Austria; pa2, population from Stampfltal, Austria; pa3, population from Greece; pa4, population from Croatia; all from Foissner et al. 2005), and *Periholosticha sylvatica* (syl, from Foissner et al. 2005)

Characteristics [a]	Species	mean	M	SD	SE	CV	Min	Max	n
Body, length	acu	–	–	–	–	–	125.0	150.0	–
	la1	89.4	87.0	17.4	3.8	19.4	63.0	130.0	21
	la2	97.3	98.0	11.3	3.0	11.6	75.0	110.0	14
	la3	130.0	–	–	–	–	–	–	–
	pa1	85.7	85.0	9.0	2.0	10.5	68.0	100.0	21
	pa2	85.7	84.0	9.2	2.5	10.7	74.0	100.0	13
	pa3	88.9	88.0	7.7	2.0	8.7	75.0	102.0	15
	pa4	87.5	90.0	11.8	3.6	13.5	70.0	103.0	11
	syl	106.3	102.0	15.8	3.4	14.8	83.0	145.0	21
Body, width	acu	–	–	–	–	–	22.0	28.0	–
	la1	10.4	10.0	1.2	0.3	12.0	9.0	14.0	21
	la2	11.2	11.0	1.6	0.4	14.5	10.0	15.0	14
	la3	20.0	–	–	–	–	–	–	–
	pa1	10.7	11.0	1.3	0.3	12.2	8.0	14.0	21
	pa2	11.7	12.0	1.3	0.4	11.3	10.0	15.0	13
	pa3	10.3	10.0	1.1	0.3	10.7	9.0	13.0	15
	pa4	10.6	10.0	1.3	0.4	12.1	9.0	13.0	11
	syl	18.1	18.0	1.8	0.4	10.0	15.0	21.0	21
Body length:width, ratio	acu	–	–	–	–		6.0	7.0	–
	la1	8.7	8.7	1.7	0.4	19.3	6.3	11.8	21
	la2	8.8	9.2	1.4	0.4	16.2	5.4	11.0	14
	la3	–	–	–	–		6.0	7.0	–
	pa1	9.1	7.9	1.2	0.3	14.4	6.3	10.8	21
	pa2	7.4	7.4	0.7	0.2	9.8	6.2	8.4	13
	pa3	8.7	8.8	0.9	0.2	8.9	7.5	10.1	15
	pa4	8.3	8.2	1.3	0.4	15.2	6.5	10.3	11
	syl	5.9	5.8	0.9	0.2	14.8	4.6	8.1	21
Anterior body end to proximal end of adoral zone, distance	la1	18.8	19.0	1.7	0.4	9.2	16.0	22.0	21
	la2	21.9	22.0	1.9	0.5	8.8	18.0	26.0	14
	pa1	16.1	16.0	1.1	0.2	6.7	14.0	18.0	21
	pa2	17.7	18.0	1.6	0.4	9.1	15.0	20.0	13
	pa3	17.3	18.0	1.7	0.4	9.6	14.0	20.0	15
	pa4	15.6	16.0	1.9	0.6	11.9	12.0	20.0	11
	syl	24.3	24.0	2.5	0.5	10.2	19.0	30.0	21
Body length:length of adoral zone, ratio	acu	4.0	–	–	–	–	–	–	–
	la1	4.7	4.6	0.6	0.1	12.2	3.7	5.9	21
	la2	4.5	4.4	0.5	0.1	11.9	3.6	5.5	14
	la3	4.0	–	–	–	–	–	–	–
	pa1	5.4	5.4	0.7	0.2	12.6	4.2	6.3	21
	pa2	4.9	4.7	0.5	0.2	11.2	3.9	6.2	13
	pa3	5.0	5.0	0.4	0.1	7.5	4.6	5.9	15
	pa4	5.7	5.7	0.8	0.3	14.5	4.4	6.9	11
	syl	4.5	4.4	0.6	0.1	13.3	3.5	5.6	21
Anterior body end to last cirrus of midventral complex, distance	la1	30.1	30.0	3.9	0.9	12.9	22.0	40.0	21
	la2	33.5	32.5	4.5	1.2	13.3	29.0	42.0	14
	pa1	15.1	15.0	1.1	0.2	7.2	13.0	18.0	21

Table 24 Continued

Characteristics [a]	Species	mean	M	SD	SE	CV	Min	Max	n
Anterior body end to last cirrus of midventral complex, distance	pa2	17.0	17.0	2.4	0.7	13.8	13.0	22.0	13
	pa3	16.1	16.0	1.8	0.5	11.4	12.0	20.0	15
	pa4	14.6	15.0	2.6	0.8	17.8	10.0	18.0	11
	syl	26.0	26.0	3.3	0.7	12.5	20.0	32.0	21
Anterior body end to first macronuclear nodule, distance	la1	18.0	18.0	2.9	0.6	15.8	10.0	23.0	21
	la2	21.0	21.0	2.4	0.6	11.4	17.0	17.0	14
Nuclear figure, length	la1	49.4	46.0	10.2	2.2	20.7	35.0	78.0	21
	la2	52.1	53.5	6.6	1.8	12.6	38.0	64.0	14
	pa1	50.6	51.0	8.1	1.8	16.0	39.0	65.0	21
	pa2	51.9	51.0	6.7	1.9	12.9	43.0	64.0	13
	pa3	51.5	51.0	6.1	1.6	11.8	42.0	64.0	15
	pa4	50.2	50.0	11.0	3.3	21.9	37.0	68.0	11
	syl	66.4	67.0	10.9	2.4	16.4	47.0	85.0	21
Macronuclear nodules, length	la1	4.6	4.0	1.6	0.3	33.7	2.0	8.0	21
	la2 [c]	5.9	6.0	1.5	0.4	24.9	4.0	10.0	14
	pa1	5.6	6.0	1.2	0.3	22.1	3.0	8.0	21
	pa2	6.2	6.0	1.3	0.4	21.9	5.0	10.0	13
	pa3	4.8	5.0	1.2	0.3	23.9	3.0	7.0	15
	pa4	5.2	5.0	0.9	0.3	16.9	4.0	7.0	11
	syl	6.8	7.0	1.3	0.3	19.2	5.0	9.0	21
Macronuclear nodules, width	la1	2.7	3.0	0.5	0.1	19.7	2.0	4.0	21
	la2 [c]	2.7	3.0	0.6	0.2	23.8	1.5	4.0	14
	pa1	2.6	2.7	0.4	0.1	17.2	2.0	3.0	21
	pa2	2.9	3.0	0.7	0.2	24.7	2.0	4.0	13
	pa3	2.9	3.0	0.6	0.2	19.2	2.0	4.0	15
	pa4	2.9	3.0	0.3	0.1	10.4	2.0	3.0	11
	syl	3.6	3.0	0.8	0.2	21.4	2.5	5.0	21
Macronuclear nodules, number	acu	–	–	–	–	–	11.0	14.0	–
	la1	17.0	16.0	4.0	0.9	23.5	10.0	27.0	21
	la2	15.6	16.0	1.7	0.5	10.9	11.0	18.0	13
	la3	15.0	–	–	–	–	–	–	–
	pa1	16.3	16.0	3.1	0.7	5.4	14.0	29.0	21
	pa2	14.7	16.0	2.4	0.7	16.1	8.0	16.0	13
	pa3	16.3	16.0	1.3	0.3	7.9	14.0	20.0	15
	pa4	15.4	15.0	1.7	0.5	11.0	12.0	18.0	11
	syl	21.1	22.0	3.5	0.8	16.7	12.0	25.0	21
Micronuclei, length	la1	2.3	2.0	–	–	–	2.0	3.0	21
	la2 [c]	2.7	2.5	–	–	–	2.5	3.0	13
	pa1	1.9	2.0	0.3	0.1	13.7	1.5	2.3	21
	pa2	1.9	2.0	–	–	–	1.6	2.0	13
	pa3	2.2	2.2	0.3	0.1	12.7	2.0	3.0	15
	pa4	3.1	3.0	0.4	0.1	12.1	2.5	4.0	11
	syl	3.2	3.2	–	–	–	3.0	4.0	21
Micronuclei, width	la1	1.8	1.8	–	–	–	1.5	2.0	21
	la2 [c]	2.1	2.0	–	–	–	1.5	2.5	13
	pa1	1.7	1.6	0.3	0.1	17.1	1.1	2.2	21
	pa2	1.8	1.8	–	–	–	1.3	2.0	13
	pa3	1.9	2.0	0.2	0.1	10.1	1.6	2.2	15
	pa4	1.9	2.0	0.3	0.1	17.4	1.5	2.5	11
	syl	2.4	2.5	–	–	–	2.0	3.0	21

Table 24 Continued

Characteristics [a]	Species	mean	M	SD	SE	CV	Min	Max	n
Micronuclei, number	acu	–	–	–	–	–	2.0	3.0	–
	la1	2.5	2.0	0.8	0.2	30.3	1.0	4.0	21
	la2	2.2	2.0	0.8	0.2	37.2	1.0	3.0	14
	la3	–	–	–	–	–	4.0	5.0	–
	pa1	3.1	3.0	1.1	0.2	33.8	1.0	5.0	21
	pa2	2.3	2.0	0.5	0.1	20.8	2.0	3.0	13
	pa3	3.4	3.0	0.9	0.2	26.8	2.0	5.0	15
	pa4	2.3	2.0	0.8	0.2	34.6	1.0	4.0	11
	syl	2.5	2.0	0.4	0.1	15.3	2.0	4.0	21
Adoral membranelles, total number	acu	18.0	–	–	–	–	14.0	19.0	–
	la1	15.8	16.0	1.1	0.2	6.8	13.0	18.0	21
	la2	16.1	16.0	0.8	0.2	5.2	14.0	17.0	14
	la3	–	–	–	–	–	15.0	17.0	–
	pa1	13.3	13.0	0.6	0.1	4.3	12.0	14.0	21
	pa2	14.1	14.0	0.6	0.2	4.6	13.0	15.0	13
	pa3	14.9	15.0	1.0	0.3	6.7	12.0	16.0	15
	pa4	10.6	10.0	1.0	0.3	9.8	9.0	13.0	11
	syl	18.9	19.0	1.5	0.3	7.9	14.0	21.0	21
Frontal cirri, number	acu	3.0	–	–	–	–	–	–	–
	la1	3.0	3.0	0.0	0.0	0.0	3.0	3.0	21
	la2	3.0	3.0	0.0	0.0	0.0	3.0	3.0	21
	la3	3.0	–	–	–	–	–	–	–
	pa1	3.0	3.0	0.0	0.0	0.0	3.0	3.0	21
	pa2	3.0	3.0	0.0	0.0	0.0	3.0	3.0	13
	pa3	3.0	3.0	0.0	0.0	0.0	3.0	3.0	15
	pa4	3.0	3.0	0.0	0.0	0.0	3.0	3.0	11
	syl	3.0	3.0	0.0	0.0	0.0	3.0	3.0	21
Frontoterminal cirri, number [e]	pa1	2.0	2.0	0.0	0.0	0.0	2.0	2.0	21
	pa2	2.0	2.0	0.0	0.0	0.0	2.0	2.0	13
	pa3	2.0	2.0	0.0	0.0	0.0	2.0	2.0	15
	pa4	2.0	2.0	0.0	0.0	0.0	2.0	2.0	11
	syl	2.0	2.0	0.0	0.0	0.0	2.0	2.0	21
Midventral cirri, number [d]	acu	–	–	–	–	–	12.0	15.0	–
	la1	12.1	12.0	1.7	0.4	14.0	9.0	15.0	21
	la2	13.6	13.0	1.7	0.5	12.8	11.0	17.0	14
	la3	13.0	–	–	–	–	–	–	–
	pa1	7.1	7.0	0.8	0.2	10.8	5.0	9.0	21
	pa2	6.9	7.0	0.6	0.2	8.1	5.0	7.0	13
	pa3	6.9	7.0	0.9	0.2	13.3	5.0	9.0	15
	pa4	5.6	6.0	1.2	0.4	21.9	4.0	7.0	11
	syl	9.9	10.0	1.3	0.3	12.7	7.0	12.0	21
Left marginal cirri, number	acu	–	–	–	–	–	17.0	22.0	–
	la1	28.0	28.0	3.1	0.7	10.9	23.0	38.0	21
	la2	25.9	26.0	4.9	1.4	18.9	17.0	35.0	13
	la3	25.0	–	–	–	–	–	–	–
	pa1	20.2	20.0	2.0	0.4	9.8	17.0	25.0	21
	pa2	18.5	18.5	2.8	0.8	15.4	12.0	23.0	12
	pa3	19.9	19.0	2.6	0.7	13.1	16.0	26.0	15
	pa4	20.7	21.0	2.7	0.8	13.0	17.0	25.0	11
	syl	27.2	27.0	4.8	1.0	17.6	19.0	36.0	21
Right marginal cirri, number	acu	–	–	–	–	–	19.0	24.0	–

Table 24 Continued

Characteristics [a]	Species	mean	M	SD	SE	CV	Min	Max	n
Right marginal cirri, number	la1 [b]	33.6	33.0	4.3	0.9	12.8	28.0	45.0	21
	la2 [b]	28.5	29.0	4.3	1.2	15.0	20.0	35.0	13
	la3	30.0	–	–	–	–	–	–	–
	pa1	23.0	23.0	2.0	0.4	8.7	20.0	28.0	21
	pa2	20.5	21.0	3.5	1.0	17.0	11.0	26.0	13
	pa3	21.8	22.0	2.6	0.7	11.9	17.0	27.0	15
	pa4	23.1	23.0	3.5	1.1	15.2	18.0	28.0	11
	syl [b]	30.1	30.0	4.9	1.1	16.1	21.0	37.0	21
Caudal cirri, number	pa1 [f]	3.2	3.0	0.7	0.2	21.3	2.0	4.0	21
	pa2 [f]	3.1	3.0	1.2	0.3	38.6	0.0	4.0	13
	pa3 [f]	3.6	4.0	0.6	0.2	17.6	2.0	4.0	15
	pa4 [f]	3.8	4.0	–	–	–	3.0	4.0	11
	syl	4.4	4.0	1.8	0.4	40.1	2.0	10.0	21
Dorsal kineties, number	acu	3.0	–	–	–	–	–	–	–
	la1	3.0	3.0	0.0	0.0	0.0	3.0	3.0	21
	la2	3.0	3.0	0.0	0.0	0.0	3.0	3.0	10
	la3	3.0	–	–	–	–	–	–	–
	pa1	2.0	2.0	0.0	0.0	0.0	2.0	2.0	21
	pa2	2.0	2.0	0.0	0.0	0.0	2.0	2.0	13
	pa3	2.0	2.0	0.0	0.0	0.0	2.0	2.0	15
	pa4	2.0	2.0	0.0	0.0	0.0	2.0	2.0	11
	syl	2.0	2.0	0.0	0.0	0.0	2.0	2.0	21
Dorsal kinety 1, number of bristles	pa1	9.3	9.0	1.0	0.2	10.4	8.0	12.0	21
	pa2	9.5	9.0	1.3	0.4	13.3	7.0	12.0	13
	pa3	9.4	9.0	0.9	0.2	9.7	8.0	11.0	15
	pa4	8.9	9.0	2.1	0.6	23.8	5.0	12.0	11
	syl	15.3	15.0	1.7	0.4	11.2	12.0	19.0	21
Dorsal kinety 2, number of bristles	pa1	6.5	6.0	0.8	0.2	12.6	6.0	9.0	21
	pa2	7.3	7.0	1.6	0.4	21.9	4.0	10.0	13
	pa3	6.5	6.0	0.7	0.2	11.5	5.0	8.0	15
	pa4	6.1	6.0	1.0	0.3	17.2	4.0	8.0	11
	syl	12.9	13.0	1.7	0.4	12.9	8.0	15.0	21

[a] All measurements in µm. Data provided by Foissner et al. (2002, 2005) are based on mounted, protargol-impregnated (Foissner's method), and randomly selected specimens from non-flooded Petri dish cultures. Data provided by Hemberger (1985) are based on specimens impregnated with Wilbert's protargol method. CV = coefficient of variation in %, M = median, Max = maximum value, mean = arithmetic mean, Min = minimum value, n = number of individuals investigated (not known in detail for Hemberger's population; the "basis" for la3 are "more than 100 specimens" and for acu "about 30 specimens"), SD = standard deviation, SE = standard error of arithmetic mean. Note on Hemberger's data: if only one value is given, it is listed as mean; if two values are provided, they are listed as Min and Max.

[b] Inclusive (la1, la2), respectively, exclusive (syl) supposed frontoterminal cirri.

[c] Anteriormost macronuclear nodule, respectively, micronucleus.

[d] Likely in most cases, cirrus III/2 (cirrus behind right frontal cirrus) included.

[e] See genus section for discussion of this feature. Rarely three such cirri occur.

[f] Likely one or two ("on average") of these cirri are transverse cirri (see text). Note that the other *Periholosticha* species have caudal cirri too (see descriptions).

Bakuellidae Jankowski, 1979

1979 **Bakuellidae fam. n.** – Jankowski, Trudy zool. Inst., Leningr., 86: 74 (original description). Type genus: *Bakuella* Agamaliev & Alekperov, 1976.
1988 **Bakuellinae (Jank.) comb. n.** – Alekperov, Zool. Zh., 67: 779, 780 (inclusion of *Keronella*).
1989 **Bakuellidae Jankowski, 1979** – Alekperov, Revision of *Bakuella*, p. 7 (brief revision of the Bakuellidae).
1992 **Bakuellinae Jankowski, 1979**[1] – Song, Wilbert & Berger, Bull. Br. Mus. nat. Hist. (Zool.), 58: 141 (improved diagnosis).
1992 **Bakuellidae Jankowski, 1992** – Alekperov, Zool. Zh., 71: 5 (revision of the Bakuellidae).
1994 **Bakuellinae Jankowski, 1979**[2] – Eigner, Europ. J. Protistol., 30: 473 (revision of the Bakuellinae).
1996 **Bakuellinae Jankowski, 1979** – Franco, Esteban & Téllez, Acta Protozool., 35: 321 (key to the genera of the Bakuellinae).
2001 **Bakuellidae Jankowski, 1979** – Berger, Catalogue of ciliate names 1. Hypotrichs, p. 105 (nomenclator containing all basionyms, combinations, and higher taxa of hypotrichs).

Nomenclature: The name Bakuellidae and its derivatives are based on the genus-group name *Bakuella*. Originally established as family, later categorised as subfamily (Alekperov 1988, Eigner 1994, Franco et al. 1996). I use the name as introduced by Jankowski (1979), however, without category (see chapter 7.2 of general section). The suffix "comb. n." in the Alekperov (1988) entry is incorrect; possibly he meant new status.

Characterisation (Fig. 111a, autapomorphy 1): Urostyloidea with three frontal cirri and a midventral complex composed of midventral pairs in the anterior portion and at least one midventral row in the posterior portion (A).

Remarks: The Bakuellidae have been accepted only by few workers. Wicklow (1981) subsumed *Holosticha* and *Bakuella* in the subfamily Holostichinae and therefore synonymised the bakuellids with the holostichids. Most authors followed this proposal and classified *Bakuella* in the urostylids or holostichids (Tables 4–11).

Alekperov (1989) assigned *Bakuella*, *Metabakuella*, and *Keronella* to the Bakuellidae with a row of frontoterminal cirri and midventral rows as distinctive characters. Song et al. (1992) assigned *Parabakuella* (now a junior synonym of *Holostichides*) to the Bakuellinae. Simultaneously, Alekperov (1992) included *Bakuella*, *Keronella*, *Parabakuella*, *Pseudobakuella*, and *Metabakuella* in the Bakuellidae. In the present paper, *Keronella* and *Metabakuella* are classified in the Urostylidae because they have many frontal cirri forming a bicorona.

Eigner (1994), who studied cell division of *Eschaneustyla brachytona*, included *Bakuella*, *Keronella*, *Holostichides*, and *Eschaneustyla* in the Bakuellinae because he considered the frontal ciliature as less important than the midventral rows. Franco et al. (1996) also briefly reviewed the Bakuellinae and included, besides the nominotypical genus, *Keronella*, *Eschaneustyla*, *Holostichides*, and *Metabakuella*. Like Eigner (1994) they weighted the presence of midventral rows more than the frontal ciliature.

[1] The improved diagnosis by Song et al. (1992) is as follows: Holostichidae with more or less obliquely arranged ventral rows behind the midventral row; 3 more or less distinctly enlarged frontal cirri.

[2] Eigner (1994) provided the following definition: Holostichidae with a midventral row composed of cirral pairs and several short oblique midventral rows in the anterior part and long midventral rows in the posterior part of the ventral surface.

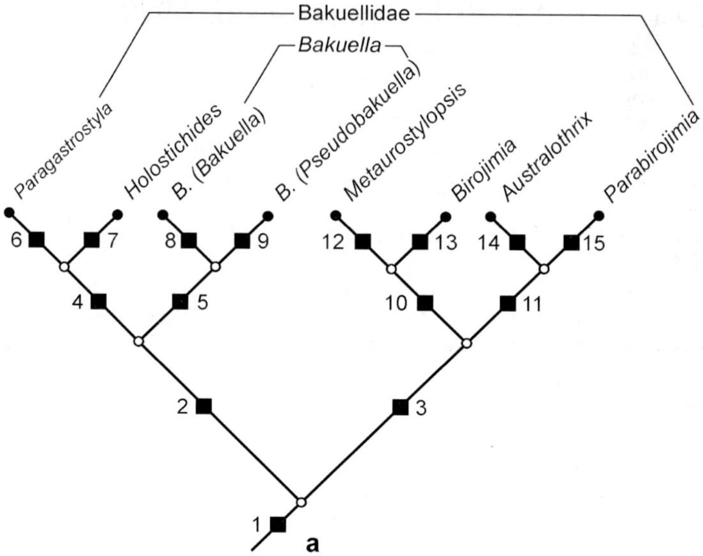

Fig. 111a Diagram of phylogenetic relationships within the Bakuellidae (original). Note that this is only one of several reliable hypothesis; thus, subgroups are not named. Autapomorphies (black squares 1–15). **1** – midventral complex composed, inter alia, of midventral rows. **2** – more than two frontoterminal cirri (convergence to 12; *Bakuella (Pseudobakuella)* species again have two frontoterminal cirri). **3** – more than two marginal rows. **4** – transverse cirri lacking (convergence to 14); body elongate. **5** – caudal cirri lacking (convergence to 12 and 15); more than one buccal cirrus (*B. agamalievi* has again one buccal cirrus). **6** – buccal cirrus lacking; two dorsal kineties. **7** – 4–5 dorsal kineties. **8** – no autapomorphy found. **9** – (again) two frontoterminal cirri. **10** – no autapomorphy found. **11** – frontoterminal cirri lacking. **12** – more than two frontoterminal cirri (convergence to 2); caudal cirri lacking (convergence to 5 and 14). **13** – more than three dorsal kineties; body slender. **14** – transverse cirri lacking (convergence to 4); more than three dorsal kineties. **15** – caudal cirri lacking (convergence to 5 and 12); adoral zone with distinct gap; special mode of transverse cirri formation.

In the present paper, the Bakuellidae comprise all urostyloid taxa which have three more or less enlarged frontal cirri and a midventral complex composed of midventral pairs and at least one midventral row. The three frontal cirri and the anterior portion of the midventral complex composed of cirral pairs are certainly plesiomorphies for the bakuellids whereas the midventral row(s) can be interpreted as novelty for this group. However, midventral rows occur also in the Urostylidae (e.g., *Urostyla grandis*), indicating that such rows evolved convergently. By contrast, summarising all taxa with midventral rows (e.g., *Bakuella* and *Keronella*) requires the assumption of a convergent evolution of a bicorona, for example, in *Keronella* (with midventral rows) and *Pseudokeronopsis* or *Pseudourostyla* (without midventral rows). Both features (midventral rows and bicorona) are not very complex characteristics so that convergent evolution is conceivable for both.

Fig. 111a shows a hypothesis about the relationships within the Bakuellidae. Possibly the stem-lineage of the Bakuellidae split into a group with more than two frontoterminal cirri as apomorphy and a second group with more than two marginal rows as novelty. As expected, the hypothesis includes several convergencies (e.g., loss of caudal

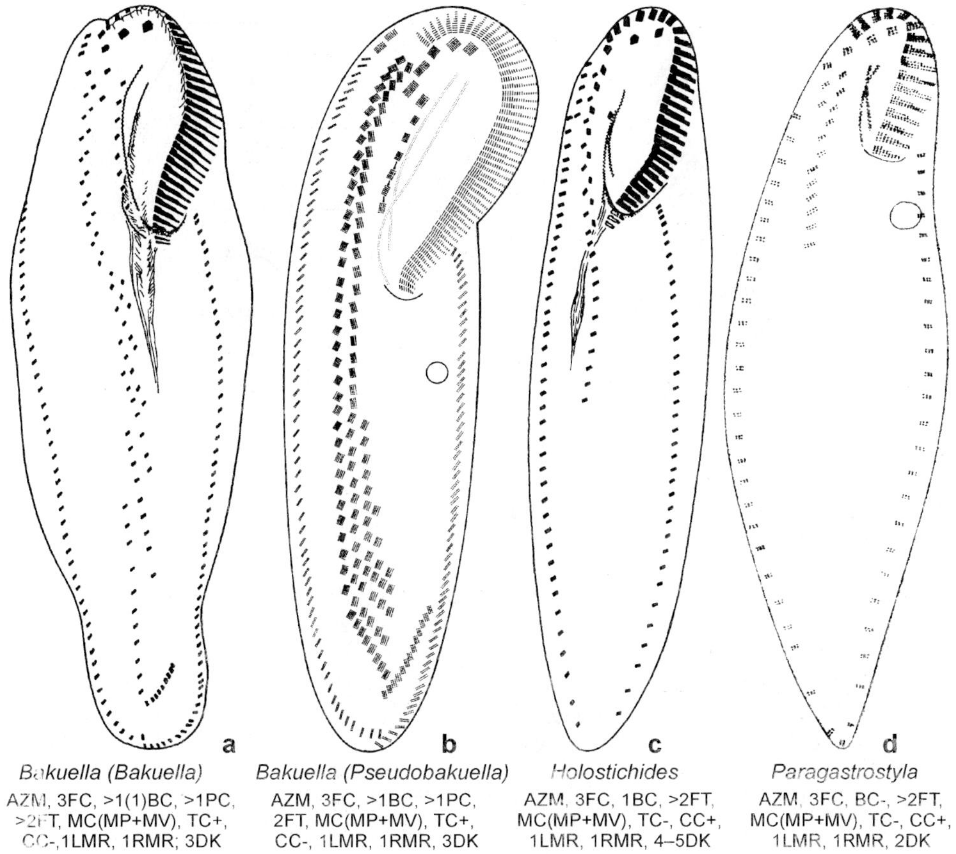

Fig. 112a–d Ventral cirral pattern in members of the Bakuellidae (part 1). **a:** *Bakuella (Bakuella) edaphoni*. **b:** *Bakuella (Pseudobakuella) salinarum*. **c:** *Holostichides chardezi*. **d:** *Paragastrostyla lanceolata*. Sources of illustrations see individual descriptions. Abbreviations used in short characterisation of infraciliature see Fig. 112e–i (explanation of supplemental signs and numbers see legend to Fig. 20a–c).

cirri and transverse cirri; more than two frontoterminal cirri). Further data are needed, including molecular features, to strengthen or weaken the proposal.

Key to the genera of the Bakuellidae

Since the taxa below are mainly distinguished by details of the cirral pattern (certain cirral groups present or not), protargol preparations – or at least very detailed live observations (interference contrast) – are needed for successful identification. Note: *Urostyla* is defined – via the type species *U. grandis* – by many frontal cirri forming a multicorona. However, this genus is a melting pot for several little known urostyloids with more than two marginal rows. It even contains species with only three frontal cirri! Thus, see also the *Urostyla* key if you cannot identify your specimens with the key below.

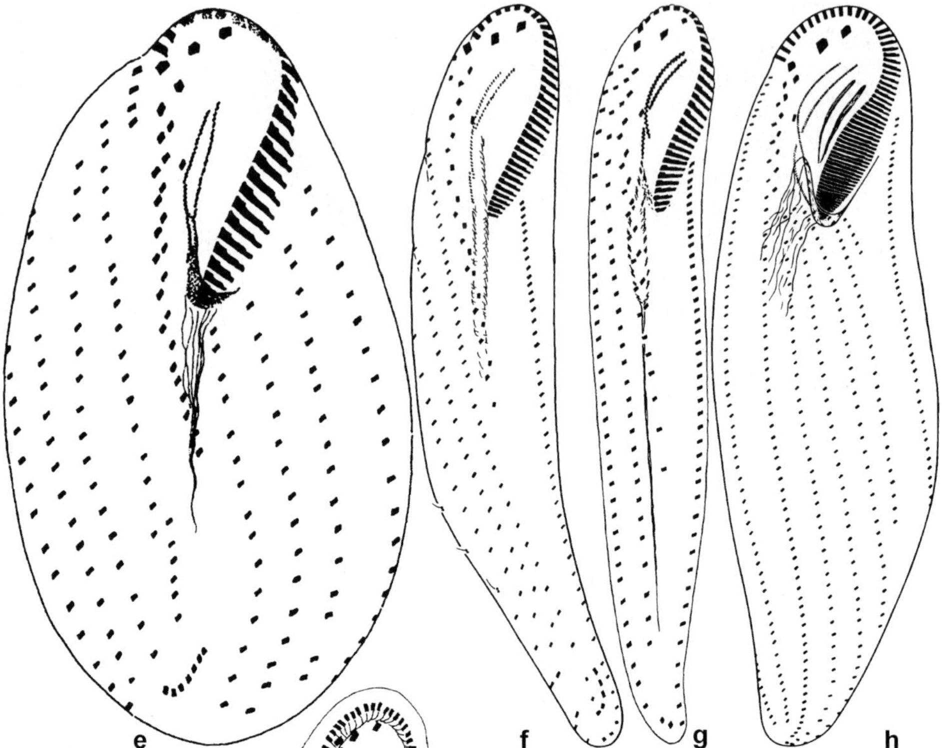

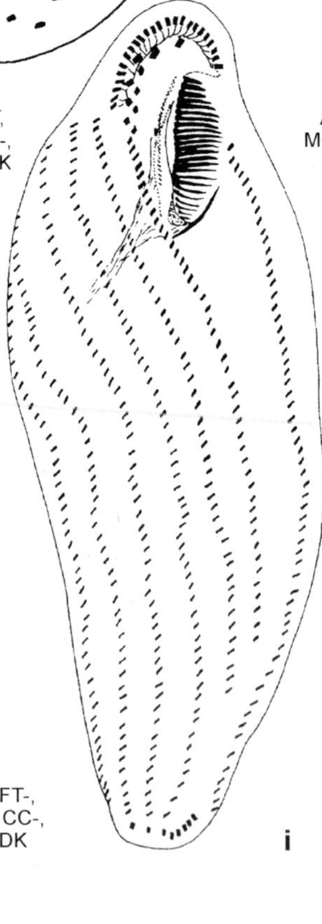

Fig. 112e–i Ventral cirral pattern in members of the Bakuellidae (part 2). **e:** *Metaurostylopsis marina*. **f:** *Birojimia terricola*. **g:** *Birojimia muscorum*. **h:** *Australothrix australis*. **i:** *Parabirojimia similis*. Sources of illustrations see individual descriptions. Abbreviations used in short characterisation of infraciliature (explanation of supplemental signs and numbers see legend to Fig. 20a–c): AZM = adoral zone of membranelles, BC = buccal cirrus, CC = caudal cirri, DK = dorsal kineties, FC = frontal cirri, FT = frontoterminal cirri, LMR = left marginal row, MC(MP+MV) = midventral complex composed of cirral pairs and midventral rows), MR = marginal rows, PC = parabuccal cirrus, RMR = right marginal row, TC = transverse cirri.

1 Transverse cirri absent (Fig. 112c, d, h) **2**
\- Transverse cirri present (Fig. 112a, b, e–g, i) **4**
2 More than 2 marginal rows (Fig. 112h) .... *Australothrix* (p. 703)
\- Two marginal rows (1 left, 1 right; Fig. 112c, d) **3**
3 Buccal cirrus present; 4–5 dorsal kineties (Fig. 112c) ...... *Holostichides* (p. 590)
\- Buccal cirrus lacking; 2 dorsal kineties (Fig. 112d) ..... *Paragastrostyla* (p. 613)
4 (1) Conspicuous snout-like protrusion in gap of adoral zone of membranelles (Fig. 112i, 136a) .... *Parabirojimia* (p. 690)
\- Adoral zone without gap and protrusion (e.g., Fig. 112a) **5**
5 Two (1 left and 1 right) marginal rows (Fig. 112a, b) .... *Bakuella* (p. 531)
\- More than 2 marginal rows (Fig. 112e–g) **6**
6 Caudal cirri absent; marine (Fig. 112e) .... *Metaurostylopsis* (p. 637)
\- Caudal cirri present; terrestrial (Fig. 112f, g) .... *Birojimia* (p. 677)

Bakuella Agamaliev & Alekperov, 1976

1976 ***Bakuella*** **Agamaliev et Alekperov gen. n.** – Agamaliev & Alekperov, Zool. Zh., 55: 128, 131 (original description). Type species (by original designation on p. 129 and p. 131): *Bakuella marina* Agamaliev & Alekperov, 1976.

1977 ***Bakuella*** **Agam. & Alek.** – Corliss, Trans. Am. microsc. Soc., 96: 137 (see nomenclature).

1979 ***Bakuella*** **Agamaliev et Alekperov, 1976** – Jankowski, Trudy zool. Inst., Leningr., 86: 50 (catalogue of generic names of hypotrichs).

1979 ***Bakuella*** **s. str., subgen. n.** – Jankowski, Trudy zool. Inst., Leningr., 86: 50 (catalogue of generic names of hypotrichs). Type species (same as for genus and by original designation): *Bakuella marina* Agamaliev & Alekperov, 1976.

1979 ***Loxocineta*** **subgen. n.** – Jankowski, Trudy zool. Inst., Leningr., 86: 51, 57 (revision). Type species (by original designation): *Bakuella crenata* Agamaliev & Alekperov, 1976.

1979 ***Bakuella*** **Agamaliev & Alekperov, 1976** – Corliss, Ciliated protozoa, p. 309 (establishment of subgenus).

1979 ***Bakuella*** **Agamaliev & Alekperov, 1976** – Tuffrau, Trans Am. microsc. Soc., 98: 526 (classification of hypotrichs).

1979 ***Bakuella*** **Agamaliev & Alekperov, 1976**[1] –Borror, J. Protozool., 26: 549 (redefinition of the Urostylidae).

1981 ***Bakuella*** – Wicklow, Protistologica, 17: 348 (classification of Urostylina).

1983 ***Bakuella*** **Agamaliev et Alekperov, 1976** – Agamaliev, Ciliates of Caspian Sea, p. 104 (review).

1983 ***Bakuella*** **Agamaliev et Alekperov, 1976** – Borror & Wicklow, Acta Protozool., 22: 113, 122 (revision of the Urostylina).

1983 ***Bakuella*** **Agamaliev and Alekperov, 1976** – Curds, Gates & Roberts, British and other freshwater ciliated protozoa, p. 400 (guide to ciliate genera).

1985 ***Bakuella*** – Small & Lynn, Phylum Ciliophora, p. 450 (guide to ciliate genera).

1987 ***Bakuella*** **Agamaliev et Alekperov, 1976** – Tuffrau, Annls Sci. nat. (Zool.), 13: 115 (classification of hypotrichs).

1989 ***Bakuella*** **Agamaliev et Alekperov 1976** – Alekperov, Ecology of marine and freshwater protozoans, p. 7 (short revision).

---

[1] The diagnosis by Borror (1979) is as follows: One row of right marginal cirri; one or more rows of left marginal cirri; some of midventral cirri, especially in the posterior half of the cell, arranged in short oblique rows of more than 2 cirri each.

1990 ***Bakuella* Agamaliev & Alekperov, 1976**[1] – Mihailowitsch & Wilbert, Arch. Protistenk., 138: 208 (improved diagnosis).
1992 ***Bakuella* Agamaliev and Alekperov, 1976** – Carey, Marine interstitial ciliates, p. 179 (guide).
1992 ***Bakuella* Agamaliev & Alekperov, 1976**[2] – Song, Wilbert & Berger, Bull. Br. Mus. nat. Hist. (Zool.), 58: 133, 141 (revision of *Bakuella*).
1992 ***Bakuella* Agamaliev et Alekperov, 1976** – Alekperov, Zool. Zh., 71: 5 (revision of the Bakuellidae).
1994 ***Bakuella* Agamaliev and Alekperov, 1976**[3] – Eigner, Europ. J. Protistol., 30: 473 (revision of the Bakuellinae).
1994 ***Bakuella*** – Corliss, Acta Protozool., 33: 15 (classification of protists; without details about *Bakuella*).
1994 ***Bakuella*, Agamaliev et Alekperov 1976** – Tuffrau & Fleury, Traite de Zoologie, 2: 128 (revision of hypotrichs).
1996 ***Bakuella* Agamaliev and Alekperov, 1976** – Franco, Esteban & Téllez, Acta Protozool., 35: 326, 329 (key to species of Bakuellinae).
1999 ***Bakuella* Agamaliev & Alekperov, 1976** – Shi, Acta Zootax. sinica, 24: 245 (revision of hypotrichs).
1999 ***Bakuella* Agamaliev & Alekerov, 1976** – Shi, Song & Shi, Progress in Protozoology, p. 114 (generic revision of hypotrichs; incorrect spelling of Alekperov).
2001 ***Bakuella* Jankowski 1979** – Aescht, Denisia, 1: 29 (catalogue of generic names of ciliates; see nomenclature).
2001 ***Loxocineta* Jankowski 1979** – Aescht, Denisia, 1: 29 (catalogue of generic names of ciliates; see nomenclature).
2001 ***Bakuella* Agamaliev and Alekperov, 1976** – Berger, Catalogue of ciliate names 1. Hypotrichs, p. 11 (nomenclator containing all basionyms, combinations, and higher taxa of hypotrichs).
2001 ***Bakuella (Bakuella)* Agamaliev and Alekperov, 1976** – Berger, Catalogue of ciliate names 1. Hypotrichs, p. 12 (nomenclator containing all basionyms, combinations, and higher taxa of hypotrichs).
2001 ***Bakuella (Loxocineta)* Jankowski, 1979** – Berger, Catalogue of ciliate names 1. Hypotrichs, p. 12 (nomenclator containing all basionyms, combinations, and higher taxa of hypotrichs).
2002 ***Bakuella* Agamaliev and Alekperov, 1976** – Lynn & Small, Phylum Ciliophora, p. 443 (guide to ciliate genera).

Nomenclature: The name *Bakuella* is a composite of *Baku* (capital of Azerbaijan) and the diminutive suffix *-ella*. Feminine gender (ICZN 1999, Article 30.1.3; Aescht 2001, p. 274). *Loxocineta* is a composite of the Greek adjective *lox-* (laterally curved, lopsided, slanting, oblique), *cin-* (the latinised form of the Greek verb *kin-*; to move, to drive away, to bump into, to shake), the Greek suffix *-et* (doing the activity mentioned by the verb), and the inflectional ending *-a*; it likely refers to the oblique cirral rows forming the posterior portion of the midventral complex. Feminine gender.

Agamaliev & Alekperov (1976) established *Bakuella* with two species, namely *Bakuella marina* and *B. crenata*, and unequivocally designated *Bakuella marina* as type species (see also, e.g., Borror & Wicklow 1983, p. 122). Aescht (2001), probably by mistake, assumed that Agamaliev & Alekperov (1976) did not fix a type species and therefore assigned *Bakuella* to Jankowski (1979) who mentioned, like Agamaliev &

---

[1] The improved diagnosis by Mihailowitsch & Wilbert (1990) is as follows: Je eine rechte und linke Marginalreihe, verstärkte Frontalcirren, familientypische Midventral-Reihen, Cirren vorn in Zick-Zack-Anordnung, Cirren hinten in Schrägreihen, zahlreiche Transversalcirren, mehrere Buccalcirren, zahlreiche Fronto-Midventral-Transversal-Anlagen, aus denen sich die Midventral-Reihe sowie die Transversalcirren differenzieren.

[2] The improved diagnosis by Song et al. (1992) is as follows: Medium sized to large Holostichidae; obliquely arranged ventral rows situated behind midventral row; 3 slightly to distinctly enlarged frontal cirri; 2 or more frontoterminal cirri; transverse cirri present; 1 left and 1 right marginal row; caudal cirri absent.

[3] The improved diagnosis by Eigner (1994) is as follows: More than one buccal cirrus. A midventral row composed of cirral pairs in anterior ventral surface. Transverse cirri present. Caudal cirri absent.

Alekperov (1976), *Bakuella marina* as type species. *Bakyella marina* in Agamaliev (1976, p. 91) is an incorrect subsequent spelling. Corliss (1977, p. 111, 137) erroneously designated *Bakuella* as a nomen nudum.

Characterisation (Fig. 111a, autapomorphies 5): Adoral zone of membranelles continuous. 3 more or less distinctly enlarged frontal cirri. More than 1 buccal cirrus (A; *B. agamalievi* has again only 1 buccal cirrus). 2 or more frontoterminal cirri. Midventral complex composed of midventral pairs and midventral rows. Transverse cirri present. 1 left and 1 right marginal cirral row. Caudal cirri lacking (A). Proximal portion of parental adoral zone reorganised during morphogenesis.

Remarks: Several species are described only after silver impregnation. Consequently, the important feature presence/absence of cortical granules is not known in these taxa. Besides the features mentioned in the characterisation, *Bakuella* species likely have the following characteristics in common: body large, that is, usually more than 150 µm long, flexible, but not distinctly contractile; body outline long elliptical with both ends rounded; movement without peculiarities; adoral zone occupies 30–40% of body length; undulating membranes rather long, curved, and optically intersecting; bases of transverse cirri not distinctly enlarged; 3 dorsal kineties (reinvestigation of *B. crenata* necessary); dorsal cilia short, that is, around 3 µm. *Bakuella* comprises marine, limnetic, and terrestrial species.

Bakuella marina and *B. crenata* were the first urostyloids for which a midventral complex composed of cirral pairs and midventral rows was described in detail. Agamaliev & Alekperov (1976) recognised the urostyloid origin of *Bakuella marina* and *B. crenata*. Since then *Bakuella* has been classified within a multitude of higher taxa, for example, in the Holostichidae (Agamaliev & Alekperov 1976, Corliss 1977, 1979, Tuffrau 1979, Song et al. 1992), the Urostylidae (Hemberger 1982, Small & Lynn 1985, Tuffrau 1987), the Holostichinae (as subfamily of the Urostylidae; Wicklow 1981, Borror & Wicklow 1983, Lynn & Small 2002), the Bakuellidae/Bakuellinae (Wirnsberger 1987, Alekperov 1992, Eigner 1994, Franco et al. 1996), and the Kinetodesmophorida (Shi 1993).

Jankowski (1979) divided *Bakuella* – which contained two species at that time – into two monotypic subgenera, namely, *Bakuella (Bakuella)* with the multimacronucleate *B. marina* as type and *Bakuella (Loxocineta)* with the bimacronucleate *B. crenata*. However, the number of macronuclear nodules is generally not used for supraspecific classification in hypotrichs and thus Jankowski's subgeneric division was not accepted by Song et al. (1992).

In 1983, Borror & Wicklow established *B. agamalievi* for a *Holosticha* species which was somewhat superficially described by Agamaliev (1972). Recently, Song et al. (2002) neotypified this species which is the sole *Bakuella* with only one buccal cirrus. In the same review, Borror & Wicklow described *B. variabilis*, a distinctly deviating species later transferred to *Metabakuella* and now provisionally assigned to *Urostyla*.

Alekperov (1982, 1988) found two further *Bakuella* species in Azerbaijan increasing the total number of species to six. Mihailowitsch & Wilbert (1990) described *B. salinarum* and two unnamed populations. More or less simultaneously, Song et al. (1992)

and Alekperov (1992) established species for these two unnamed populations causing some serious nomenclatural problems. Song et al. (1992) described the first edaphic species (*B. edaphoni*), revised *Bakuella*, and accepted six *Bakuella* species.

Alekperov (1992) established *Pseudobakuella* for *Bakuella* species with only two frontoterminal cirri, in contrast to "true" *Bakuella* species, which have more than two such cirri. Two frontoterminal cirri is likely the plesiomorphic state, which is also present in many other hypotrichs (see ground pattern of the Urostyloidea). Eigner (1994) obviously overlooked Alekperov's (1992) paper and therefore did not discuss *Pseudobakuella*, which includes *Bakuella salinarum* (type) and *Pseudobakuella gracilis* (objective synonym of *Bakuella walibonensis*). Franco et al. (1996) cited Alekperov's revision, but they did not refer to *Pseudobakuella* in their key to the Bakuellinae. I consider *Pseudobakuella* as subgroup of *Bakuella* because it could really be possible that the two species originally included in *Pseudobakuella* form a monophyletic group.

Eigner & Foissner (1992) described the morphology and cell division of *B. pampinaria*. Eigner (1994, p. 473) criticised the *Bakuella* definition by Song et al. (1992) because they used, in Eigner's opionion, an inappropriate term (ventral rows instead of long midventral rows) and in that the feature three enlarged frontal cirri is not stable. However, the latter discrepancy is easily explained because Eigner included the parabuccal cirri (= cirri behind the right frontal cirrus) in the number of frontal cirri, whereas Song et al. (1992) considered only the anteriormost, enlarged, and distinctly set off cirri of the anlagen I–III as frontal cirri. In the same paper, Eigner introduced the terms short midventral row and long midventral row. Eigner (1994) confined *Bakuella* to species with more than one buccal cirrus and therefore excluded *B. kreuzkampii* (now *B. agamalievi*) from *Bakuella*. However, the neotypification of *B. agamalievi* by Song et al. (2002) shows that *Bakuella* species with only one such cirrus exist. Franco et al. (1996, p. 326) accepted Eigner's definition of *Bakuella*. However, in their key to species (supplemented after Song et al. 1992) they also included *B. kreuzkampii* (= junior synonym of *B. agamalievi* in present book), which has, as just mentioned, only one buccal cirrus (Fig. 115a, l).

Shi et al. (1999, p. 114) considered *Pseudobakuella*, which they erroneously assigned to Mihailowitsch & Wilbert (1990), as synonym of *Parabakuella*. However, *Parabakuella* is a junior synonym of *Holostichides*, that is, lacks transverse cirri, whereas this cirral group is present in *Bakuella* and its subgenus *Pseudobakuella*.

The last *Bakuella* species/subspecies was described just recently by Foissner et al. (2002) and Foissner (2004).

Wiackowski (1988, p. 4) mentioned a *Bakuella* sp. He provided no illustration or description and therefore it is indeterminable. It was collected from moss on a tree trunk likely in/near Kraków (Poland) and fed on the cyrtophorid ciliate *Chilodonella uncinata*.

The separation of *Bakuella* from other genera of the Bakuellidae is rather simple so that the reader is referred to the corresponding key.

Species included in *Bakuella* (alphabetically arranged according to basionyms): (1) *Bakuella agamalievi* Borror & Wicklow, 1983; (2) *Bakuella crenata* Agamaliev & Alekperov, 1976; (3) *Bakuella edaphoni* Song, Wilbert & Berger, 1992; (4) *Bakuella* *granulifera* Foissner, Agatha & Berger, 2002; (5) *Bakuella marina* Agamaliev & Alekperov, 1976; (6) *Bakuella pampinaria* Eigner & Foissner, 1992; (7) *Bakuella salinarum* Mihailowitsch & Wilbert, 1990; (8) *Bakuella walibonensis* Song, Wilbert & Berger, 1992; (9) incertae sedis *Paraurostyla pulchra* Buitkamp, 1977.

Key to *Bakuella* species

This key includes the species of both subgenera, which differ only in the number of frontoterminal cirri (many in *Bakuella (Bakuella)* vs. two in *Bakuella (Pseudobakuella)*). Besides the habitat (salt water, freshwater, soil), body size, the nuclear apparatus, the cortical granules, and some details of the cirral pattern are important for identification. Consequently, specimens have to be studied very carefully.

1 Two macronuclear nodules (Fig. 116a) ..... *Bakuella (Bakuella) crenata* (p. 548)
\- Many macronuclear nodules (e.g., Fig. 115f) **2**
2 Soil **3**
\- Salt water **5**
3 Cortical granules lacking[1] (Fig. 117a) **8**
\- Cortical granules present (e.g., Fig. 120a–c) **4**
4 Body length 90–180 μm in life; 2–5 transverse cirri (Fig. 118a) ..... *Bakuella (Bakuella) pampinaria* (p. 559) **9**
\- Body length 270–400 μm; 9–15 transverse cirri (Fig. 119a) ..... *Bakuella (Bakuella) granulifera* (p. 569)
5 (2) One buccal cirrus (Fig. 115a, d, g) .... *Bakuella (Bakuella) agamalievi* (p. 541)
\- More than one buccal cirrus (e.g., Fig. 113a) **6**
6 Two frontoterminal cirri (Fig. 121a, 122a) .. *Bakuella (Pseudobakuella)* (p. 576) **7**
\- 3–11, usually 7–9 frontoterminal cirri (Fig. 113a, k) ..... *Bakuella (Bakuella) marina* (p. 536)
7 Midventral complex terminates near transverse cirri (Fig. 121a) ..... *Bakuella (Pseudobakuella) salinarum* (p. 576)
\- Midventral complex terminates about in mid-body (Fig. 122a) ..... *Bakuella (Pseudobakuella) walibonensis* (p. 581)
8 (3) Body length in life 190–300 μm; 5–9 buccal cirri (Fig. 117a, d) ..... *Bakuella (Bakuella) edaphoni* (p. 551)
\- Body length below 200 μm; about 4 buccal cirri (Fig. 123a) ..... *Paraurostyla pulchra* (p. 588)
9 (4) On average 9 midventral pairs; 2–4, on average 2.6 parabuccal cirri (Fig. 118a, b) ..... *Bakuella (Bakuella) pampinaria pampinaria* (p. 563)
\- On average 4 midventral pairs; 1–2, on average 1.1 parabuccal cirri (Fig. 118p, r) ..... *Bakuella (Bakuella) pampinaria oligocirrata* (p. 566)

[1] For *Paraurostyla pulchra* (Fig. 123a), which closely resembles *Bakuella pampinaria oligocirrata*, it is unknown whether or not cortical granules are present.

Bakuella (Bakuella) Agamaliev & Alekperov, 1976

1976 ***Bakuella* Agamaliev et Alekperov gen. n.** – Agamaliev & Alekperov, Zool. Zh., 55: 128, 131 (original description of genus). Type species (by original designation on p. 129 and p. 131): *Bakuella marina* Agamaliev & Alekperov, 1976.

1979 ***Bakuella* s. str., subgen. n.** – Jankowski, Trudy zool. Inst., Leningr., 86: 50 (catalogue of generic names of hypotrichs). Type species (same as for genus and by original designation): *Bakuella marina* Agamaliev & Alekperov, 1976.

Nomenclature: The nominotypical subgenus *Bakuella (Bakuella)*, which has the same authors and year as the genus, was already created by Jankowski (1979) when he established *Bakuella (Loxocineta)*.

Characterisation (Fig. 111a, no autapomorphy found): *Bakuella* with more than 2 frontoterminal cirri.

Remarks: The list of synonyms contains, for the sake of simplicity, only the original description of *Bakuella* and the establishment of the nominotypical subgenus *Bakuella (Bakuella)*. For further details, see remarks in genus section.

Species included in *Bakuella (Bakuella)* (alphabetically arranged according to basionyms): (1) *Bakuella agamalievi* Borror & Wicklow, 1983; (2) *Bakuella crenata* Agamaliev & Alekperov, 1976; (3) *Bakuella edaphoni* Song, Wilbert & Berger, 1992; (4) *Bakuella granulifera* Foissner, Agatha & Berger, 2002; (5) *Bakuella marina* Agamaliev & Alekperov, 1976; (6) *Bakuella pampinaria* Eigner & Foissner, 1992.

Bakuella marina Agamaliev & Alekperov, 1976
(Fig. 113a–o, 114a, b, Table 25, Addenda)

1976 ***Bakuella marina* Agamaliev et Alekperov sp. n.** – Agamaliev & Alekperov, Zool. Zh., 55: 129, Fig. $1_{1,2}$, 3_1 (Fig. 113a, b; original description, likely no formal diagnosis provided. Type slide(s) likely deposited in the Institute of Zoology, Academy of Sciences of the Azerbaijan SSR, Baku).

1979 ***Bakuella (Bakuella) marina* Agamaliev et Alekperov, 1976** – Jankowski, Trudy zool. Inst., Leningr., 86: 50 (catalogue of generic names of hypotrichous ciliates).

1982 ***Bakuella marina*** – Alekperov, Zool. Zh., 61: 1253, 1254, Fig. $1_{1,2}$ (Fig. 113e, f; review).

1982 ***Bakuella imbricata* Alekperov, sp. n.** – Alekperov, Zool. Zh., 61: 1253, Fig. $1_{5,6}$, 2 (Fig. 114a, b; original description of junior synonym, likely no formal diagnosis provided. Type slide(s) likely deposited in the Institute of Zoology, Academy of Sciences of the Azerbaijan SSR, Baku).

1983 ***Bakuella marina* Agamaliev et Alekperov, 1976** – Agamaliev, Ciliates of Caspian Sea, p. 105, Fig. 55a, b (Fig. 113c, d; review).

1983 ***Bakuella marina* Agamaliev et Alekperov, 1976** – Borror & Wicklow, Acta Protozool., 22: 113, 122 (revision of urostylids).

1986 ***Bakuella marina* Agamaliev** – Wilbert, Symposia Biologica Hungarica, 33: 251, 252, Fig. 3 (Fig. 113g–i; incorrect authorship; illustrated record from a saline lake in Saskatchewan).

1992 ***Bakuella marina* Agamaliev & Alekperov, 1976** – Song, Wilbert & Berger, Bull. Br. Mus. nat. Hist. (Zool.), 58: 142, Fig. 21–35 (Fig. 113j–o; revision of *Bakuella*).

1992 ***Bakuella marina* Agamaliev et Alekperov, 1976** – Alekperov, Zool. Zh., 71: 6, Fig. 1 (redrawing of Fig. 113e; revision of the Bakuellidae).

1994 ***Bakuella marina* Agamaliev and Alekperov, 1976** – Eigner, Europ. J. Protistol., 30: 474 (revision of the Bakuellinae).

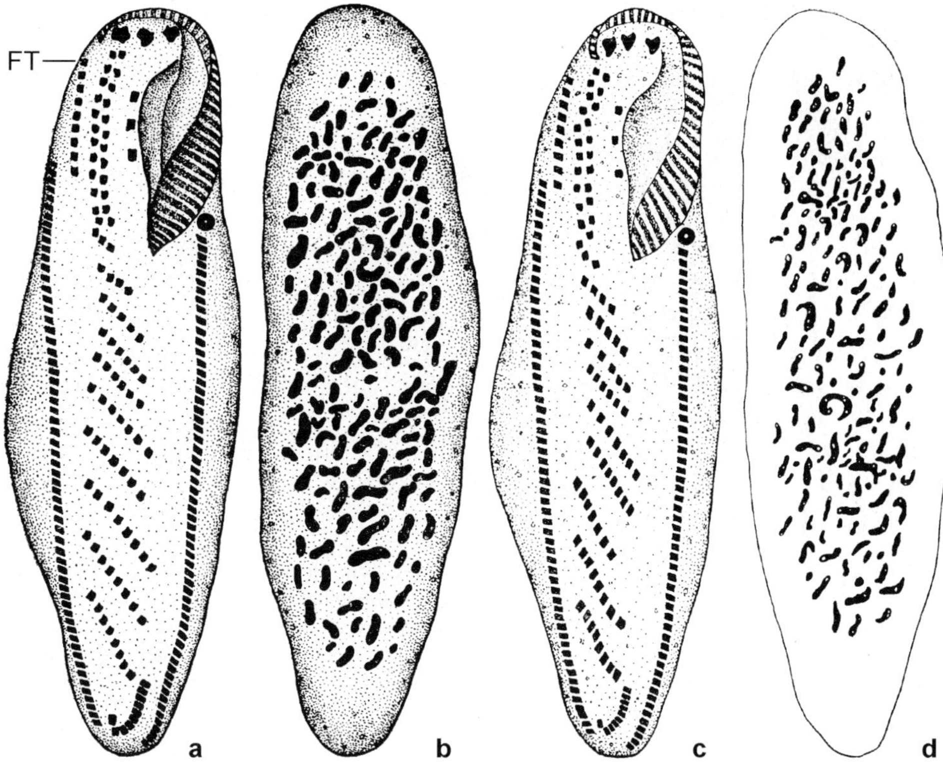

Fig. 113a–d *Bakuella marina* (a, b, from Agamaliev & Alekperov 1976; c, d, from Agamaliev 1983. Wet silver nitrate impregnation). Infraciliature of ventral side (a, c) and nuclear apparatus (b, d), a, b = 130 µm. FT = frontoterminal cirri. Page 536.

1996 ***Bakuella marina* Agamaliev and Alekperov, 1976** – Franco, Esteban & Téllez, Acta Protozool., 35: 329, Fig. 13 (Fig. 113a; key to the species of the Bakuellinae).

2001 ***Bakuella marina* Agamaliev and Alekperov, 1976** – Berger, Catalogue of ciliate names 1. Hypotrichs, p. 12 (nomenclator containing all basionyms, combinations, and higher taxa of hypotrichs).

Nomenclature: The species-group name *marin·us -a -um* (Latin adjective; living in the sea, belonging to the sea), refers to the marine habitat (Caspian Sea) where the species was discovered. The species-group name *imbricata* (Latin *imbrex*, hollow tile; Latin *imbricāre*, to bring rain, to roof with hollow tile) possibly refers to the freshwater habitat where *Bakuella imbricata* was discovered (likely, in contrast to the marine habitat of *B. marina*).

Bakuella marina was fixed as type species of *Bakuella* by original designation. Moreover, it is the type of the nominotypical subgenus *Bakuella (Bakuella)* Agamaliev & Alekperov, 1976 established by Jankowski (1979, p. 50). When *Bakuella marina* is classified in the present subgenus, the correct name is *Bakuella (Bakuella) marina* Agamaliev & Alekperov, 1976 (see Berger 2001, p. 12).

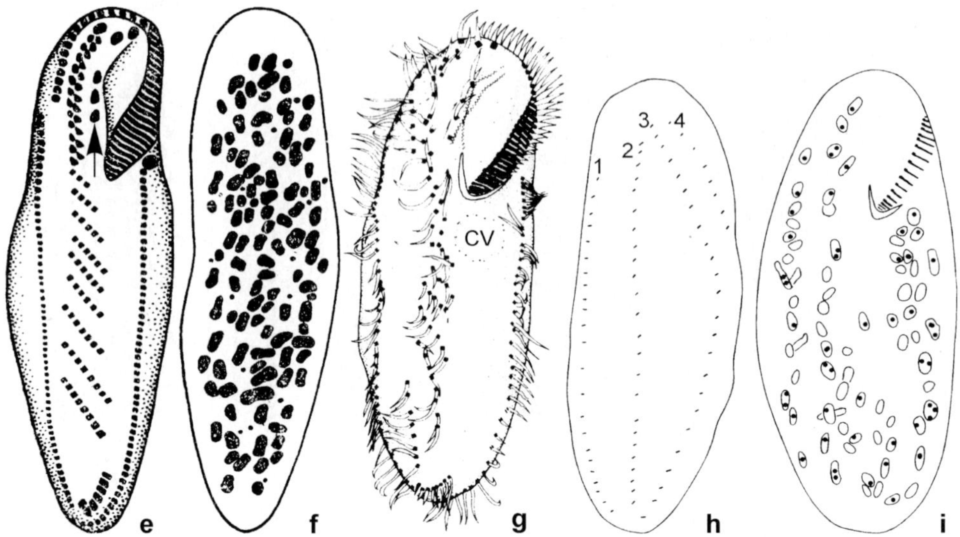

Fig. 113e–i *Bakuella marina* (e, f, from Alekperov 1982b; g–i, from Wilbert 1986. e, wet silver nitrate impregnation; f, Feulgen stain; g–i, protargol impregnation). **e, f:** Infraciliature of ventral side and nuclear apparatus, sizes not indicated. Arrow marks rear end of buccal-cirri row. **g–i:** Infraciliature of ventral and dorsal side and nuclear apparatus, g, h = 210 μm, i = 240 μm. Note that this specimen has an unusual number (4 vs. usually 3) of dorsal kineties. CV = contractile vacuole, 1–4 = dorsal kineties. Page 536.

Remarks: The original description is in Russian and therefore not translated in detail. The illustrations of *B. marina* from Agamaliev & Alekperov (1976; Fig. 113a, b) and Agamaliev (1983; Fig. 113c, d) differ only slightly, whereas Alekperov's (1982b; Fig. 113e, f) figures deviate distinctly from the original illustrations, indicating that Alekperov (1982b) studied a further population.

Bakuella marina and *B. imbricata* were synonymised by Song et al. (1992) because the differences in the features discussed by Alekperov (1982b, p. 1255) – number of midventral rows, transverse and marginal cirri, and adoral membranelles – are rather indistinct (Table 25). Eigner (1994, p. 474) confirmed this synonymy, whereas Franco et al. (1996) considered *B. marina* and *B. imbricata* again as distinct species. In their key, they separated them, inter alia, by the number of midventral rows, namely, more than 10 in *B. marina* and less than 10 in *B. imbricata*. However, Alekperov (1982b; Fig. 113e) himself illustrated a *B. marina* specimen with only nine midventral rows and Song et al. (1992) in their review mentioned a total range of 4–10, indicating that this distinction is unfounded. There is only one interesting difference between *B. marina* and *B. imbricata*, namely the habitat. The former species

---

Fig. 113j–o *Bakuella marina* (from Song et al. 1992. Protargol impregnation). Infraciliature of ventral and dorsal side and nuclear apparatus, j–l = 230–310 μm, m–o = 91–108 μm. Arrow in (k) marks anteriormost midventral row. Very likely Wilbert did not illustrate all macronuclear nodules. E = endoral, FT = frontoterminal cirri, TC = leftmost transverse cirrus. Page 536. →

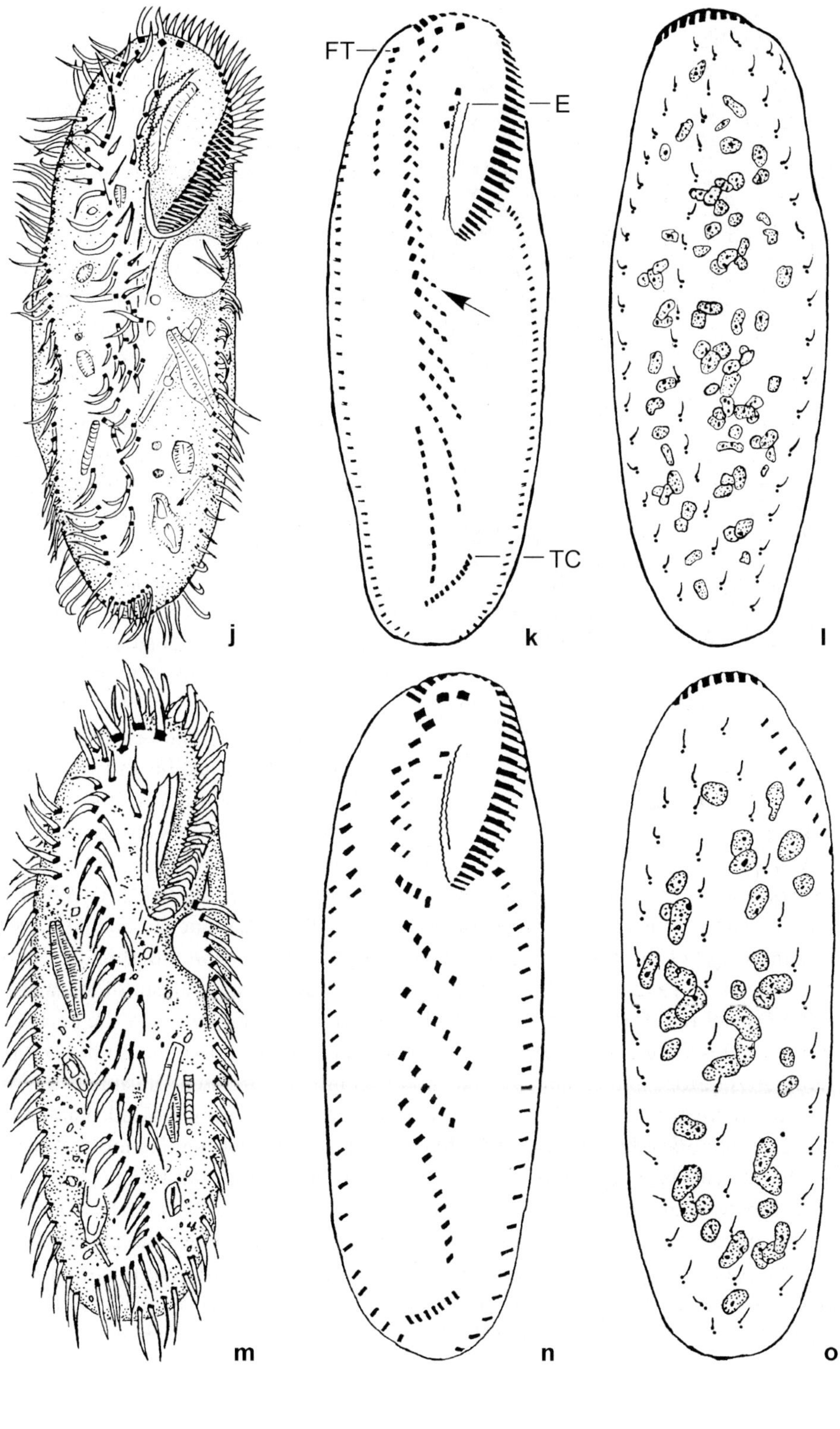
FT
E
TC
j
k
l
m
n
o

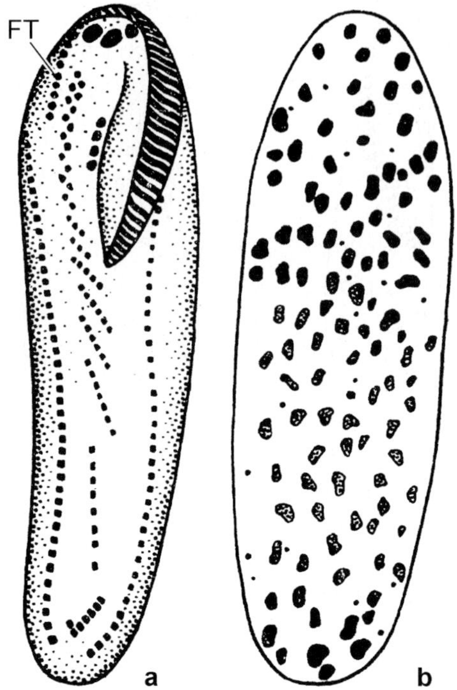

Fig. 114a, b *Bakuella marina* (from Alekperov 1982b. a, wet silver impregnation; b, Feulgen stain). Infraciliature of ventral side and nuclear apparatus of the synonym *Bakuella imbricata*, 110–130 μm. FT = frontoterminal cirri. Page 536.

was isolated from the Caspian Sea, the latter from a freshwater reservoir in the same region. It is very rare within hypotrichs that the same species occurs both in freshwater and sea water; one, possibly the sole example is *Holosticha pullaster*. Unfortunately, no (detailed) live data are available for both *B. marina* and *B. imbricata* so that one cannot exclude that they differ, for example, in the cortical granulation. Consequently, detailed examination of further *B. marina*-like populations from both freshwater and salt water is recommended. Perhaps the populations should be separated at subspecies level to distinguish a marine subspecies (*B. marina marina* Agamaliev & Alekperov, 1976) from a freshwater (*Bakuella marina imbricata* Alekperov, 1982) one.

Bakuella marina sensu Wilbert (1986) is distinctly larger than the specimens of the other populations (Table 25). However, the other morphometrics agree well so that the identification can be accepted for the time being (Fig. 113g–i). Thus, the description below is a combination of the available data.

Morphology: Body length in life around 200 μm (Song et al. 1992). Body outline likely elongate elliptical. Many (probably more than 100) macronuclear nodules of various shape (Fig. 113b, d). Contractile vacuole slightly behind proximal end of adoral zone (Song et al. 1992; Fig. 113g, j, m); contractile vacuole pore ahead of anterior end of left marginal row (Fig. 113a, c, e; I doubt this feature because usually the contractile vacuole empties via the dorsal side). Presence/absence of cortical granules not known.

Adoral zone occupies about 33% of body length in specimen illustrated by Agamaliev & Alekperov (1976; Fig. 113a), composed of 28–51 (range of all populations; Table 25) membranelles. Buccal field likely as in congeners, undulating membranes long, slightly curved, and intersecting (Wilbert 1986, Song et al. 1992; Fig. 113g, k, n).

Cirral pattern and number of cirri of usual variability (Fig. 113a, c, e, k, n, 114a). Three enlarged frontal cirri in almost transverse row. 2–5 buccal cirri along anterior half of paroral. Likely one cirrus behind right frontal cirrus. 5–11 frontoterminal cirri form distinct row commencing slightly behind level of frontal cirri. Midventral complex composed of 4–12 cirral pairs and 4–10 midventral rows. Right cirrus of each pair likely larger than left (Fig. 113k, n). Number of cirri within individual rows increases from anterior (left) to posterior (right); rearmost row extends to near transverse cirral row. 5–11 transverse cirri in oblique, slightly subterminal row; bases of transverse cirri of about same size as those of other cirri. Marginal rows clearly separated posteriorly. Usually three, occasionally four dorsal kineties (Fig. 113h, l, n). Dorsal bristles likely short, that is, probably 2–4 µm. Caudal cirri lacking.

Occurrence and ecology: Marine and limnetic (see remarks). Type locality of *B. marina* is the Caspian Sea (Agamaliev & Alekperov 1976); possibly the eastern part of the Middle Caspian as indicated in Agamaliyev's (1976, p. 91) paper. Wilbert (1986, 1995) and Wilbert in Song et al. (1992; Fig. 113j–o) found it in saline lakes in Saskatchewan, Canada. Type locality of the synonym *B. imbricata* is a freshwater reservoir (Djeiranbatansky) in Azerbaijan (Alekperov 1982b).

Chaouite et al. (1990) found *B. marina* in mineral and hot springs in the Auvergne, France. According to Wilbert & Hammer in Hammer (1986, p. 371), *Bakuella marina* tolerates a salinity by up to 370 ‰! Likely it feeds on diatoms (Fig. 113j, m).

Bakuella agamalievi Borror & Wicklow, 1983 (Fig. 115a–l, Table 25, Addenda)

1972 ***Holosticha manca* Kahl, 1932** – Agamaliev, Acta Protozool., 10: 21, Fig. 11A, B (Fig. 115h, i; misidentification).

1974 ***Keronopsis rubra* (Ehrenberg, 1838)** – Agamaliev, Acta Protozool., 13: 72, Fig. 10A, B, Plate III, Fig. 15 (Fig. 115j, k; misidentification).[1]

1983 ***Keronopsis rubra* (Ehrenberg, 1838)** – Agamaliev, Ciliates of the Caspian Sea, p. 103, Fig. 54a, b (Fig. 115j, k; misidentification).

1983 ***Holosticha manca* Kahl, 1932** – Agamaliev, Ciliates of the Caspian Sea, p. 102, Fig. 53a, b (Fig. 115h, i; misidentification).

1983 ***Bakuella agamalievi* nom. nov.** – Borror & Wicklow, Acta Protozool., 22: 114, 117, 120, 122 (original description; no formal diagnosis provided; see nomenclature).

1989 ***Bakuella* spec. 2** – Mihailowitsch, Dissertation, p. 76, Abb. 8, Tabelle 15 (Fig. 115l; description).

1990 ***Bakuella* spec. 2** – Mihailowitsch & Wilbert, Arch. Protistenk., 138: 213, Abb. 12, Tabelle 4 (Fig. 115l; description after protargol impregnation; slides are likely deposited in the Institut für landwirtschaftliche Zoologie und Bienenkunde, University of Bonn, Germany).

1992 ***Bakuella kreuzkampii* nov. spec.**[2] – Song, Wilbert & Berger, Bull. Br. Mus. nat. Hist. (Zool.), 58: 144, Fig. 52, Table 1 (Fig. 115l; original description of synonym based on previous entry; for details see remarks).

1992 ***Bakuella agamalievi* Borror & Wicklow, 1983** – Song, Wilbert & Berger, Bull. Br. Mus. nat. Hist. (Zool.), 58: 144, Fig. 54, 55, Table 1 (Fig. 115h, i; revision; see remarks).

1992 ***Bakuella muensterlandii* Alekperov, sp. n.** – Alekperov, Zool. Zh., 71: 7, Fig. 5 (Fig. 115l; original description of synonym based on *Bakuella* spec. 2 of Mihailowitsch & Wilbert 1990; see remarks).

2001 ***Bakuella agamalievi* Borror and Wicklow, 1983** – Berger, Catalogue of ciliate names 1. Hypotrichs, p. 12 (nomenclator containing all basionyms, combinations, and higher taxa of hypotrichs).

[1] Note after layout was finished: On page 980 of the present book I mistakenly classified this population as insufficient redescription.

[2] The diagnosis provided by Song et al. (1992) is as follows: After protargol impregnation about 135–175 × 40–50 µm. More than 100 macronuclear segments. 5 transverse cirri, 14 pairs of midventral cirri, and 34 adoral membranelles on average. 3–5 ventral rows and consistently 1 buccal cirrus and 2 frontoterminal cirri.

2002 ***Bakuella agamalievi* Borror & Wicklow, 1983**[1] – Song, Wilbert & Warren, Acta Protozool., 41: 146, Fig. 1–8, 52, 53, Table 1 (Fig. 115a–g; redescription and neotypification; the neotype slide is deposited in the Natural History Museum London, UK, registration number 2001:1z:z8:01).

Nomenclature: Borror & Wicklow (1983) dedicated the species to F. G. Agamaliev (Baku), who provided the first description. They used the expression "nom. nov." which is incorrect because this term is applied only to mark a replacement name (new name in case of homonymy). When *Bakuella agamalievi* is classified in the present subgenus, then the correct name is *Bakuella (Bakuella) agamalievi* Borror & Wicklow, 1983. *Bakuella amagalievi* in Song et al. (2002, p. 148, 151) and *Holosticha manga* in Agamaliyev (1974, p. 21) are incorrect subsequent spellings.

No derivation of the species-group name *kreuzkampii* is given in the original description. Kreuzkamp is the name of a drilling providing saline water for the spa of the city of Lippstadt, Germany. *Bakuella kreuzkampii* was found in a ditch containing the effluent of the spa. The species-group name *muensterlandii* refers to the fact that the sample site (Lippstadt) is in the Münsterland, a region around the city of Münster, Germany. *Bakuella kreuzcampii* in Franco et al. (1996, p. 329) is an incorrect subsequent spelling.

Bakuella kreuzkampii and *B. muensterlandii* are objective synonyms because both species are based on *Bakuella* spec. 2 illustrated and briefly described by Mihailowitsch & Wilbert (1990, Fig. 115l). Unfortunately, both species were described in late 1992. The original description of *B. kreuzkampii* was issued on 26.11.1992 (date printed on reprint!) and *B. muensterlandii* was published in part 12 of volume 71 of the Zoologicheskii Zhurnal. I wrote to Nauka, the publisher of the Russian Journal, about the exact data of publication of part 12. I did not get an answer so that I cannot decide which synonym is the older one. However, since both are very likely junior synonyms of *B. agamalievi* (see next chapter) this question is no longer of great importance.

Remarks: Agamaliev (1972) redescribed *Holosticha manca* Kahl after wet silver impregnation and hemalaun stain (Fig. 115h, i). The cirral pattern agrees in several features (e.g., 3 frontal cirri, shortened midventral complex, 5 transverse cirri) with the original description of *Holosticha manca*, now classified in *Anteholosticha* (Fig. 87a–g). However, there are also some differences, e. g., in the buccal cirrus (lacking vs. present in Kahl's population and in the neotype material of *A. manca*) and the right marginal row which begins at the level of the buccal vertex in Agamaliev's population against distinctly ahead. Kahl (1932) did not mention a concrete number of macronuclear nodules (according to his key it has more than two) and the neotype population has about 50–70 nodules. In contrast, Agamaliev (1972) counted 150–220 nodules which is a distinctly higher value, indicating – together with the other differences – that Agamaliev's specimens do not belong to *A. manca*.

---

[1] The improved diagnosis by Song et al. (2002) is as follows: Marine *Bakuella* with elongated body shape, 100–150 × 30–50 µm in vivo; about 30 adoral membranelles; on average, 33 left and 43 right marginal cirri; 4–7 frontoterminal, 4 frontal, one buccal and 4–7 transverse cirri; 9–13 pairs of midventral cirri distributed mostly in frontal area; 3–6 ventral rows with 3–5 cirri each; consistently 3 dorsal kineties; numerous macronuclei; cortical granules grouped in short rows; one contractile vacuole in anterior 1/3 of cell.

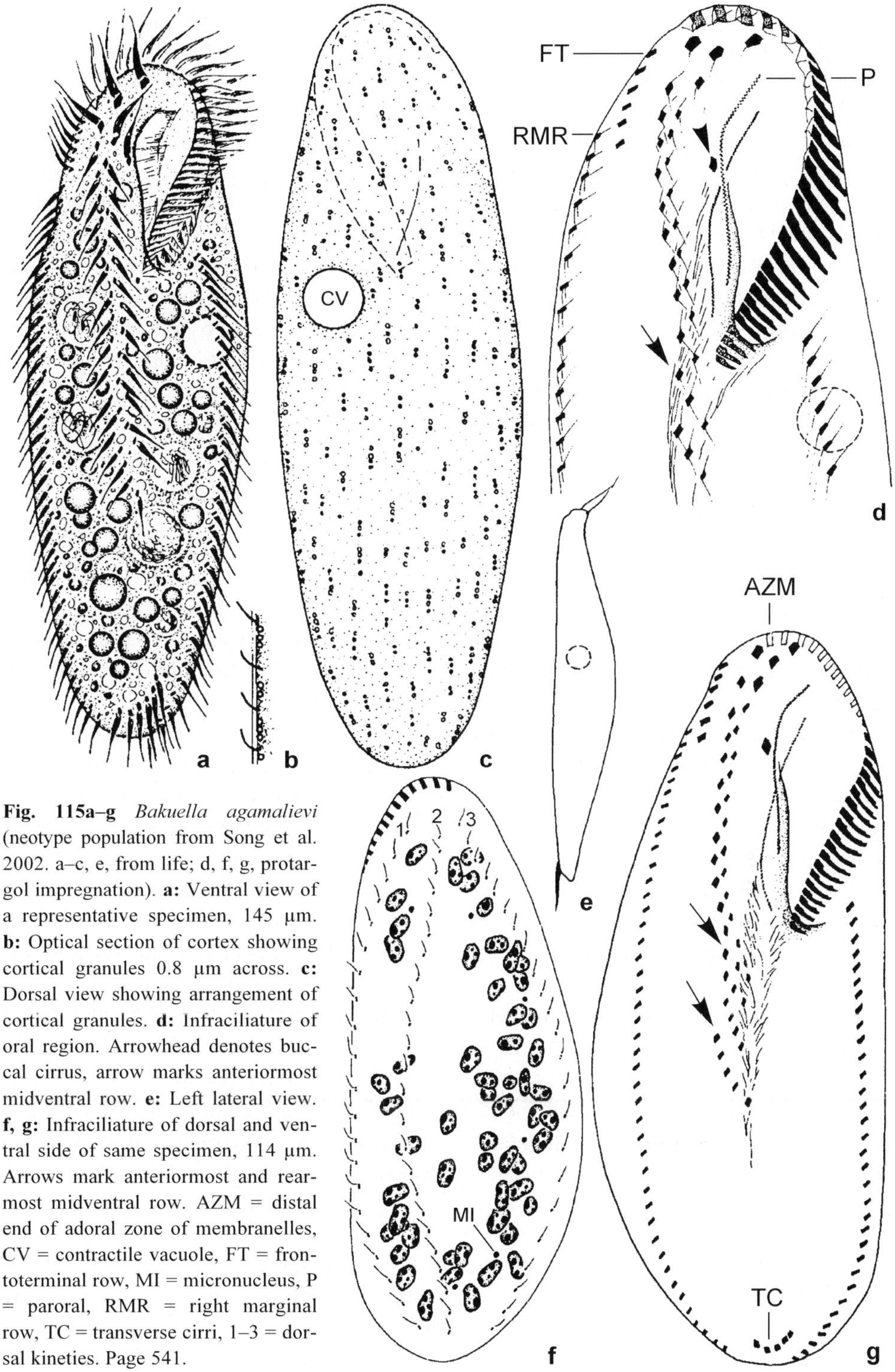

Fig. 115a–g *Bakuella agamalievi* (neotype population from Song et al. 2002. a–c, e, from life; d, f, g, protargol impregnation). **a:** Ventral view of a representative specimen, 145 µm. **b:** Optical section of cortex showing cortical granules 0.8 µm across. **c:** Dorsal view showing arrangement of cortical granules. **d:** Infraciliature of oral region. Arrowhead denotes buccal cirrus, arrow marks anteriormost midventral row. **e:** Left lateral view. **f, g:** Infraciliature of dorsal and ventral side of same specimen, 114 µm. Arrows mark anteriormost and rearmost midventral row. AZM = distal end of adoral zone of membranelles, CV = contractile vacuole, FT = frontoterminal row, MI = micronucleus, P = paroral, RMR = right marginal row, TC = transverse cirri, 1–3 = dorsal kineties. Page 541.

Borror & Wicklow (1983, p. 117) recognised that Agamaliev (1972) made a misidentification. According to these workers irregularities of the cirri of the midventral group suggest that Agamaliev's population (Fig. 115h, i) is a multinucleate marine member of *Bakuella*, and thus established a new species, *Bakuella agamalievi*, for this deviating population. However, the midventral cirral pattern illustrated by Agamaliev (1972) is not very convincing, indicating some misobservations. Especially it is difficult to recognise midventral rows, an important feature of *Bakuella*. Song et al. (1992) therefore excluded Borror & Wicklow's species from *Bakuella* and wrote that it possibly belongs to *Holosticha.* Moreover, we assumed identity with *Keronopsis rubra* sensu Agamaliev (1974; Fig. 115j, k), an opinion which I still hold today because of the largely agreeing cirral patterns. In contrast, Wirnsberger et al. (1987) supposed that the population described by Agamaliev (1974) is a new *Holosticha* species. It is interesting that in both descriptions provided by Agamaliev (1972, 1974) a buccal cirrus is lacking.

Eigner (1994, p. 473) excluded *Bakuella kreuzkampii* from *Bakuella* mainly because it has a single buccal cirrus and an additional row left of the anterior end of the right marginal row. According to him, *Bakuella* species are defined, inter alia, by the possession of more than one buccal cirrus.

Recently, Song et al. (2002) identified a marine *Bakuella* population from China as *B. agamalievi* and fixed it as neotype. However, they did not compare their population with Agamaliev's (1972) population thoroughly, that is, they made no comment on the main differences between the populations, namely the buccal cirrus (present vs. absent) and the number of macronuclear nodules (47–60 vs. 150–220). In addition, Song et al. (2002) did not check whether or not the original slides made by Agamaliev (1972) are still available. I suppose that Agamaliev (1972, Fig. 115h, l) very likely overlooked the buccal cirrus due to the inappropriate preparation procedure, and the distinctly different number of macronuclear nodules must be interpreted as inter-population variability. In addition, Fig. 115h does not show frontoterminal cirri, and the right marginal row commences just at the level of the buccal vertex. This indicates that the specimen illustrated by Agamaliev (1972) was somewhat twisted rightwards anteriorly so that these structures could not be recognised. In spite of the discrepancies between Alekperov's population and the neotype material, the identification by Song et al. (2002) seams correct.

Mihailowitsch (1989, p. 76), respectively, Mihailowitsch & Wilbert (1990) found three *Bakuella* species/populations (*B. salinarum*, *Bakuella* spec. 1, *Bakuella* spec. 2) in salt-loaded waters near the German city of Lippstadt. Song et al. (1992) and Alekperov (1992) independently established two species for the populations spec. 1 and spec. 2 in their revisions on bakuellids (for nomenclatural problems, see above). *Bakuella* spec. 2 has a row composed of about six cirri right of the anterior end of the midventral complex (Fig. 115l). Song et al. (1992) supposed that this is an additional cirral row because they considered, like Mihailowitsch & Wilbert (1990), two cirri ahead of the right marginal row as frontoterminal cirri. Now I suppose – very likely as did Alekperov (1992) in his paper which I did not translate in detail – that this row are the frontoterminal cirri (the anteriormost two cirri of the right marginal row are consequently not frontoterminal cirri, but marginal cirri). However, if one considers this row

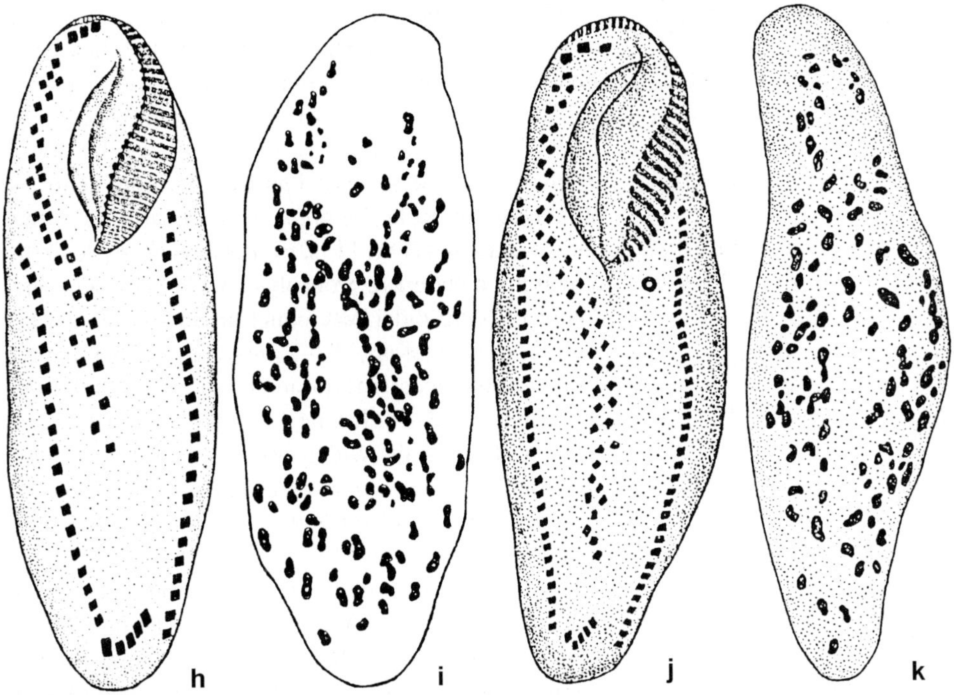

Fig. 115h–k *Bakuella agamalievi* (h, I, from Agamaliev 1972; j, k, from Agamaliev 1974. h, j, wet silver nitrate impregnation; i, k, hemalaun stain). **h, i:** Infraciliature of ventral side and nuclear apparatus of *Holosticha manca* sensu Agamaliev, h = 90 μm. **j, k:** Infraciliature of ventral side and nuclear apparatus of *Keronopsis rubra* sensu Agamaliev, j = 127 μm. Page 541.

as frontoterminal row, one cannot clearly separate *B. kreuzkampii* (respectively, *B. muensterlandii*) from *B. agamalievi* as redescribed by Song et al. (2002), strongly indicating synonymy. The main differences are in the anterior end of the right marginal row (near distal end of adoral zone in *B. kreuzkampii* [Fig. 115l] vs. slightly ahead of level of buccal cirrus in *B. agamalievi* [Fig. 115d, g]) and the distance between the marginal rows at rear body end (lacking vs. present). However, the illustration provided by Mihailowitsch is likely somewhat schematic in this respect so that these differences must not be over-interpreted. The neotype population of *B. agamalievi* has cortical granules (Fig. 115b, c), a feature not described by Mihailowitsch & Wilbert (1990) for *Bakuella* spec. 2. However, according to Mihailowitsch & Wilbert (1990, p. 213) *Bakuella* spec. 1 and 2 differ from *B. salinarum*, which is brownish (p. 208), only in body length indicating that *Bakuella* spec. 1 and 2 are also brownish in life. Thus, one cannot exclude that *Bakuella* spec. 1 and 2 also have a cortical granulation. Furthermore, both Song et al. (2002) and Mihailowitsch & Wilbert (1990) found their populations in waters with a salinity around 10–15‰.

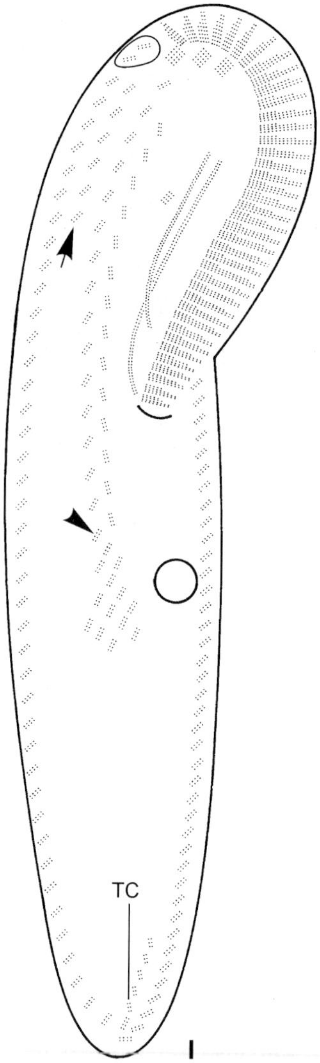

Fig. 115l *Bakuella agamalievi* (from Mihailowitsch & Wilbert 1990. Protargol impregnation). Infraciliature of ventral side, 160 µm. The anteriormost two cirri (encircled) of the right marginal row have been designated as frontoterminal cirri by Mihailowitsch & Wilbert (1990) and Song et al. (1992). Now I assume that the cirral row marked by an arrow are the frontoterminal cirri. Arrowhead denotes anteriormost midventral row. TC = rearmost transverse cirrus. Page 541.

Morphology: The description of the neotype population (Song et al. 2002) is given first, followed by some data of the populations described by Agamaliev (1972, 1974) and Mihailowitsch & Wilbert (1990).

Body size 100–150 × 30–50 µm in life; body length:width ratio about 3:1 in life, but only about 2:1 on average in protargol preparations (Table 25). Body outline elliptical (Fig. 115a, c). Pellicle rigid (flexibility of cell not mentioned, probably flexible like congeners). About 50 ellipsoidal macronuclear nodules; according to the text, nodules about 3–5 µm long (method [in vivo; protargol impregnation] not indicated), according to the table, they are 5–7 µm long (I suppose that Song et al. erroneously used the width instead of the length in the text); each nodule with several large nucleoli. Micronuclei globular, obviously very small, only recognisable after protargol staining (Fig. 115b, c). Contractile vacuole near left body margin at 33–40% of body length (Fig. 115a, c, e); pulsation interval up to 5 min! Cortical granules colourless or slightly greenish at low magnification, about 0.8 µm across, typically grouped together and sparsely arranged in short longitudinal rows both on ventral and dorsal side. Cytoplasm colourless to greyish, usually containing several large food vacuoles and numerous shiny globules 2–5 µm across making cells opaque. Movement moderately fast, crawling on substrate, sometimes jerking back and forth.

Adoral zone occupies about 33–40% of body length, composed of an average of about 30 membranelles of ordinary fine-structure; distal end terminates only slightly right of midline of anterior body end. Largest membranelles 7–8 µm wide, cilia 15–20 µm long. Undulating membranes long and curved, optically crossing about at level of buccal cirrus. Cytopharynx extends posteriorly (Fig. 115a, d, g).

Cirral pattern and number of cirri of usual variability (Fig. 115d, g, Table 25). Three distinctly enlarged, about 20 µm long frontal cirri, more or less transversely arranged. Cirrus behind right frontal cirrus also slightly enlarged. Buccal cirrus about at 1/3 of length of undulating membranes. Usually six relatively fine frontoterminal cirri in region between anterior end of right marginal row and distal end of adoral zone. Midventral complex composed of about 11 cirral pairs extending to about level of cytostome, and about four midventral rows each composed of 3, 4, or 5 cirri. Midventral complex terminates about at end of second body third. Right cirrus of each midventral pair obviously slightly larger then left cirrus (Fig. 115d, g). Transverse cirri terminal, about 15 μm long and thus projecting by about half their length beyond rear body end (Fig. 115a, g). Right marginal row commences slightly ahead of level of buccal cirrus, rear end distinctly separated from left marginal row. Dorsal cilia about 2–3 μm long, arranged in three bipolar kineties. Caudal cirri lacking (Fig. 115f).

Population described by Agamaliev (1972; Fig. 115h, i; description not translated in detail) about 90–100 μm long (120 μm in life?); 150–220 macronuclear nodules (Fig. 115i); adoral zone about 30 μm long, composed of 25–30 membranelles. Specimen illustrated with three enlarged frontal cirri, 38 midventral cirri, five transverse cirri, 21 right marginal cirri, 24 left marginal cirri. *Keronopsis rubra* sensu Agamaliev (1974, Fig. 115j, k; text not translated, thus I do not know whether or not a colour is mentioned) about 130 μm long after silver impregnation; 80–120 macronuclear nodules; 36–40 adoral membranelles; three enlarged frontal cirri (thus, identification as *K. rubra* is certainly incorrect); five transverse cirri.

Bakuella spec. 2 sensu Mihailowitsch & Wilbert (1990; = *B. kreuzkampii* and *B. muensterlandii*) likely brownish in life. More than 100 macronuclear nodules. Contractile vacuole possibly not at exact position in Fig. 115l. Mihailowitsch & Wilbert (1990) interpreted the anteriormost two cirri of the right marginal row as frontoterminal cirri because they are slightly set off from the other marginal cirri. However, now I suppose that these are marginal cirri and the frontoterminal cirri are, as is usual, in between the anterior end of the right marginal row and the midventral complex. Further details, see Fig. 115l and Table 25.

Occurrence and ecology: Likely confined to salt water. Due to the neotypification by Song et al. (2002), the type locality of *B. agamalievi* are mariculture waters near the coast of Qingdao, China, where it was isolated from a semi-closed pond (salinity 10–15‰) used for mollusc culture on 06.05.1997. Agamaliev (1972) found the present species in the benthal of the islands of the Apseronskij and Bakinskij archipelagos of the Caspian Sea. *Keronopsis rubra* sensu Agamaliev (1974) was found in the overgrowth of immobile underwater objects of the eastern coast of the Caspian Sea. For further records of *Holosticha manca* by Agamaliev from various sites and habitats of the Caspian Sea, see Agamaliev (1973, 1974a, 1986), Agamaliyev (1974, 1976), and Agamaliyev & Aliyev (1983).

The type locality of the synonyms *B. kreuzkampii* and *B. muensterlandii* is a salt-loaded ditch in Bad Waldliesborn, city of Lippstadt, Germany (Song et al. 1992, Alekperov 1992; for a more detailed description, see Mihailowitsch 1989).

Bakuella agamalievi feeds on flagellates, small ciliates, and bacteria (Song et al. 2002), according to Mihailowitsch (1989) on bacteria.

Bakuella crenata Agamaliev & Alekperov, 1976 (Fig. 116a–f, Table 25, Addenda)

1976 ***Bakuella crenata*** **Agamaliev et Alekperov sp. n.** – Agamaliev & Alekperov, Zool. Zh., 55: 130, Fig. $2_{1,2}$, $3_{2,3}$ (Fig. 116a; original description, likely no formal diagnosis provided. Type slide(s) likely deposited in the Institute of Zoology, Academy of Sciences of the Azerbaijan SSR, Baku).

1979 ***Bakuella (Loxocineta) crenata*** – Jankowski, Trudy zool. Inst., Leningr., 86: 51 (establishment of subgenus *Loxocineta*).

1982 ***Bakuella crenata*** – Alekperov, Zool. Zh., 61: 1253, 1254, Fig. $1_{3,4}$ (Fig. 116b; comparison with *Bakuella imbricata*).

1983 ***Bakuella crenata*** **Agamaliev et Alekperov, 1976** – Borror & Wicklow, Acta Protozool., 22: 113, 122 (revision of urostylids).

1988 ***Bakuella polycirrata*** **Alekperov, sp. n.** – Alekperov, Zool. Zh., 67: 778, Fig. 2 A–E (Fig. 116c–f; original description of synonym, likely no formal diagnosis provided. Type slide(s) likely deposited in the Institute of Zoology, Academy of Sciences of the Azerbaijan SSR, Baku).

1992 ***Bakuella crenata*** **Agamaliev & Alekperov, 1976** – Song, Wilbert & Berger, Bull. Br. Mus. nat. Hist. (Zool.), 58: 142, Fig. 36–40 (Fig. 116a, c–f; revision of *Bakuella*).

1992 ***Bakuella polycirrata*** – Alekperov, Zool. Zh., 71: 7, Fig. 4 (Fig. 116a; brief revision of the Bakuellinae).

1994 ***Bakuella crenata*** **Agamaliev and Alekperov, 1976** – Eigner, Europ. J. Protistol., 30: 474 (brief revision of the Bakuellinae).

1996 ***Bakuella crenata*** **Agamaliev and Alekperov, 1976** – Franco, Esteban & Téllez, Acta Protozool., 35: 329, Fig. 14 (redrawing of Fig. 116a; key to the genera and species of Bakuellinae).

2001 ***Bakuella crenata*** **Agamaliev and Alekperov, 1976** – Berger, Catalogue of ciliate names 1. Hypotrichs, p. 12 (nomenclator containing all basionyms, combinations, and higher taxa of hypotrichs).

Nomenclature: Likely no derivation of the name is given in the original description. I do not know to which feature the species-group name *crenát·us -a -um* (Latin adjective; notched; having a notch; Hentschel & Wagner 1996, p. 187) refers. The species-group name *polycirrata* is a composite of the Greek adjective *poly-* (many, frequent, large) and the Latin adjective *cirratus* (cirrated) and likely refers to the rather high number of cirri in the posterior portion of the midventral complex.

Bakuella crenata was fixed by original designation as type species of the subgenus *Bakuella (Loxocineta)* Jankowski, 1979, and therefore the correct name of *Bakuella crenata*, if classified in that subgenus, is *Bakuella (Loxocineta) crenata* Agamaliev & Alekperov, 1976. If *Bakuella crenata* is classified in *Bakuella (Bakuella)*, then the correct name is *Bakuella (Bakuella) crenata* Agamaliev & Alekperov, 1976.

Remarks: *Bakuella crenata* and *B. polycirrata* were synonymised by Song et al. (1992) because *B. polycirrata* is almost certainly a postdivider of *B. crenata*. This is indicated by the small body size (50–70 μm), the enormous relative length of the adoral zone of membranelles (more than 50% of body length), and the immature ventral cirral pattern which looks very similar to that of the postdivider of the relative *Bakuella edaphoni* (Fig. 117s). The only significant difference between *B. crenata* and *B. polycirrata* is in the number of dorsal kineties, namely only two[1] in *B. crenata* and five in *B. polycirrata* (Fig. 116d). However, this difference may be due to the wet silver impregnation method employed, which often yields inaccurate data for the dorsal ciliature (e.g., *Stylonychia mytilus* sensu Agamaliev 1978). Another possibility is that in Fig.

[1] This value is from Song et al. (1992). Now I do not know from where we have this value.

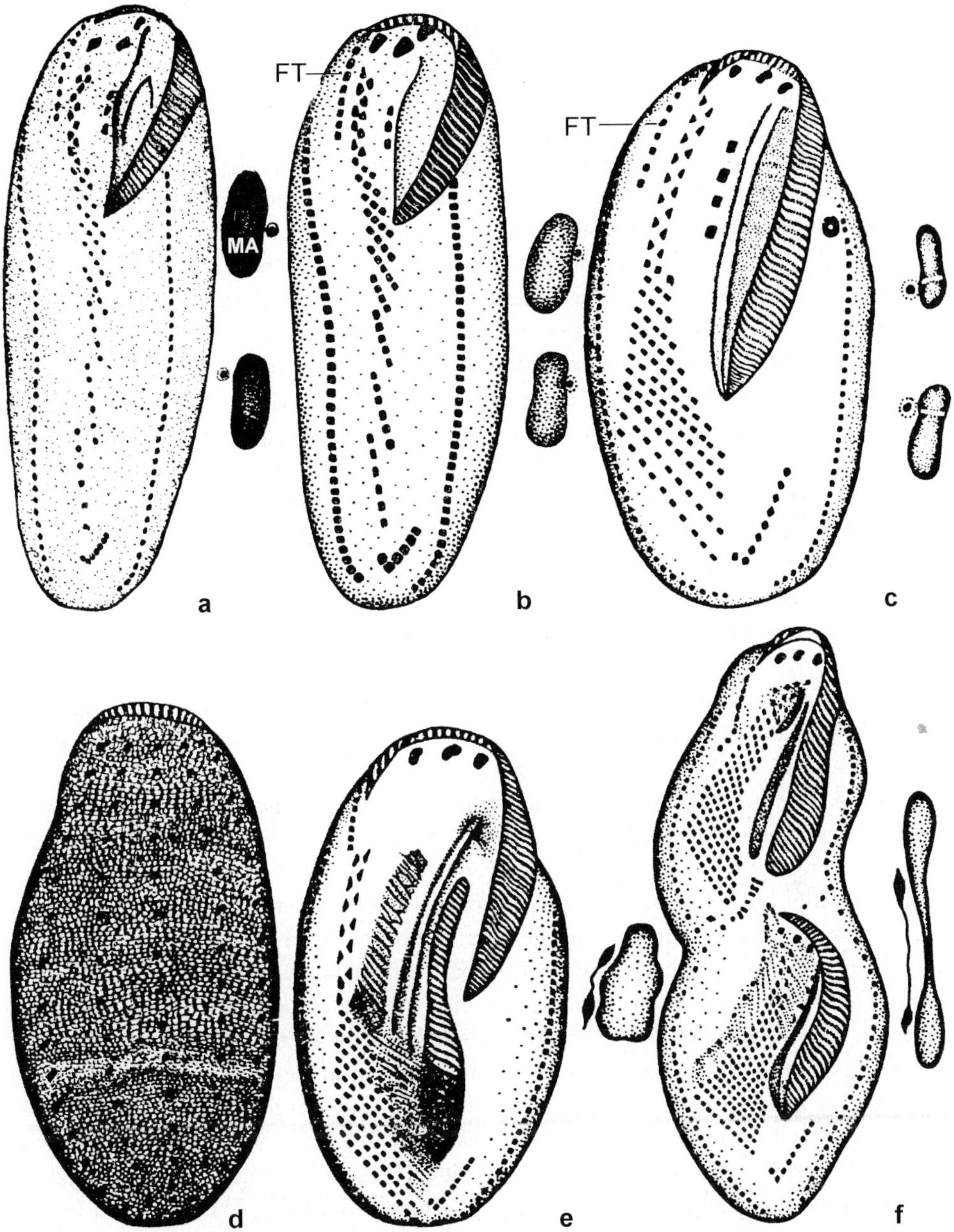

Fig. 116a–f *Bakuella crenata* (a, from Agamaliev & Alekperov 1976; b, after Agamaliev & Alekperov 1976 from Alekperov 1982; c–f, from Alekperov 1988. Wet silver nitrate impregnation; nuclear apparatus after Feulgen stain). **a, b:** Infraciliature of ventral side and nuclear apparatus, 195 µm (according to text, body length after silver impregnation is only 120–150 µm). **c, d:** Infraciliature of ventral and dorsal side and nuclear apparatus of the synonym *B. polycirrata*, 66 µm. Very likely this a postdivider of *B. crenata* (see text for details). (d) shows also the finely meshed silverline system. **e, f:** Infraciliature and nuclear apparatus of reorganiser and late divider. FT = frontoterminal cirri, MA = macronuclear nodule. Page 548.

116d parental dorsal kineties are still preserved. I suppose that *B. crenata* invariably has, like the other *Bakuella* species, three dorsal kineties.

The specimen shown in Fig. 116c indeed has distinctly more cirri than the illustrated specimen of *B. crenata* (Fig. 116a). This is due to a slightly increased number of midventral rows (12 vs. 10) and a higher number of cirri per row (10 vs. 6 in last row). At the present state of knowledge I do not believe that this difference is a species feature. Further populations of bimacronucleate *Bakuella* species should be studied to check all lacking (cortical granulation) and uncertain (e.g., dorsal kineties) features and to decide whether or not the synonymy, which was accepted by Eigner (1994, p. 474) and Franco et al. (1996), is justified. Interestingly both species were discovered in the same reservoir, which is a further hint that they are synonymous.

Bakuella crenata is easily distinguishable from the other bakuellids by the low number of macronuclear nodules (2 vs. several to many). This feature was used by Jankowski (1979) to establish the subgenus *Bakuella (Loxocineta)* with the present species as type.

Morphology: Body length in life about 210 µm, after wet silver nitrate impregnation about 120–150 µm. Body outline in life not known, likely elongate elliptical. Two ellipsoidal macronuclear nodules, micronuclei closely attached (Fig. 116a–c). Contractile vacuole not described; Fig. 116c shows a (excretion?) pore near the anterior end of the left marginal row (usually the pore of the contractile vacuole is on the dorsal side). Presence/absence of cortical granules, cytoplasmic inclusions, food, movement not known.

Adoral zone occupies 35% of body length in specimen illustrated (Fig. 116a), composed of about 28–30 membranelles. Buccal field and undulating membranes too superficially described, likely without peculiarities.

Cirral pattern not yet described in detail because species only studied after wet silver impregnation. Three enlarged frontal cirri, 3–4 buccal cirri along anterior portion of paroral. Frontoterminal cirral row extends from near distal end of adoral zone to about level of middle buccal cirrus. No distinct cirrus behind right frontal cirrus illustrated (lacking? overlooked?). Midventral complex composed of about five midventral pairs with last pair distinctly ahead of level of buccal vertex and about 10 (Fig. 116a, b) to 12 (Fig. 116c) midventral rows; number of cirri per row increases from anterior to posterior; midventral complex terminates slightly ahead of transverse cirri. Six (Fig. 116b) to nine (Fig. 116c) transverse cirri arranged in oblique, hook-shaped row; bases not distinctly enlarged as compared with midventral and marginal cirri. Marginal rows clearly separated posteriorly. *Bakuella crenata* possibly with only two dorsal kineties (Song et al. 1992), synonym *B. polycirrata* with five kineties (Fig. 116d; see remarks); very likely, as in congeners, three bristle rows are present (Table 25). Length of bristles not described, likely, as is usual, 2–4 µm long. Caudal cirri obviously lacking.

Cell division: Alekperov (1988) briefly described the morphogenesis of the synonym *B. polycirrata* (Fig. 116f). This late divider shows no peculiarities, indicating that cell division proceeds basically as in congeners. Fig. 116e obviously shows a reorganiser (Song et al. 1992). According to Alekperov (1988), Fig. 116c shows an interphasic specimen of *B. polycirrata*. However, Song et al. (1992) already recognised that this is very likely a postdivider, as indicated by the small body size (50–70 µm), the large adoral zone (50% of body length), the wide body, and the immature cirral pattern which strongly resembles that of a postdivider of another species (Fig. 116s).

Occurrence and ecology: Likely confined to limnetic habitats. Type locality of *Bakuella crenata* is a freshwater reservoir (Djeiranbatansky) in Azerbaijan. The synonym *B. polycirrata* was obviously discovered in the same reservoir. No further records published.

Bakuella edaphoni Song, Wilbert & Berger, 1992 (Fig. 117a–t, Table 25)

1992 ***Bakuella edaphoni* nov. spec.**[1] – Song, Wilbert & Berger, Bull. Br. Mus. nat. Hist. (Zool.), 58: 142, Fig. 1–20 (Fig. 117a–t; original description and revision of *Bakuella*. The slide with the holotype specimen and one slide of paratype specimens are deposited in the collection of microscopic preparations of the College of Fisheries, Ocean University of Qingdao, China; one paratype slide [reference number 1990:11:19:1] is deposited in the Natural History Museum in London).

1994 ***Bakuella edaphoni* Song, Wilbert and Berger, 1992** – Eigner, Europ. J. Protistol., 30: 474 (revision of the Bakuellinae).

1996 ***Bakuella edaphoni* Song, Wilbert and Berger, 1992** – Franco, Esteban & Téllez, Acta Protozool., 35: 329, Fig. 19 (redrawing of Fig. 117d; key to the genera and species of Bakuellinae).

2001 ***Bakuella edaphoni* Song, Wilbert and Berger, 1992** – Berger, Catalogue of ciliate names 1. Hypotrichs, p. 12 (nomenclator containing all basionyms, combinations, and higher taxa of hypotrichs).

Nomenclature: No derivation of the species-group name *edaphoni* is given in the original description. It is a composite of the Greek substantive *edaph-* (soil, ground, earth), the Greek suffix *~on* (in ecology, the totality of organisms in the same biotope), and the inflectional ending *·i*, and indicates that this species was discovered in soil. When *B. edaphoni* is classified in the present subgenus, then the correct name is *Bakuella (Bakuella) edaphoni* Song, Wilbert & Berger, 1992.

Remarks: *Bakuella edaphoni* differs from the other two terrestrial species, *Bakuella pampinaria* and *Bakuella granulifera*, mainly by the lack of cortical granules. Weibo Song, who investigated the species, informed me that he did not see special cortical granules in life. Since the granules of *Bakuella pampinaria* (Fig. 118d, 120a, b) and *B. granulifera* (Fig. 119f, g, 120c–e) are rather large and therefore very easily recognisable, one must assume that *B. edaphoni* lacks indeed cortical granules. For a brief morphometric comparison of these three soil species, see this chapter at *Bakuella granulifera*.

Morphology: Body size 190–300 × 50–85 µm in life. Body outline long elliptical, posterior quarter usually distinctly narrowed, both ends broadly rounded. Body about 2:1 dorsoventrally flattened, flexible, but not contractile (Fig. 117a, c). Macronuclear nodules scattered throughout cytoplasm; individual nodules usually ellipsoidal, rarely globular. Micronuclei globular, about 3 µm across in protargol preparations (Fig. 117e).

[1] The diagnosis by Song et al. (1992) is as follows: In vivo 190–300 × 50–85 µm. More than 100 macronuclear segments. 7 buccal cirri, 3 frontoterminal cirri, 8 transverse cirri, 9 pairs of midventral cirri, 7 ventral rows, and 39 adoral membranelles on average.

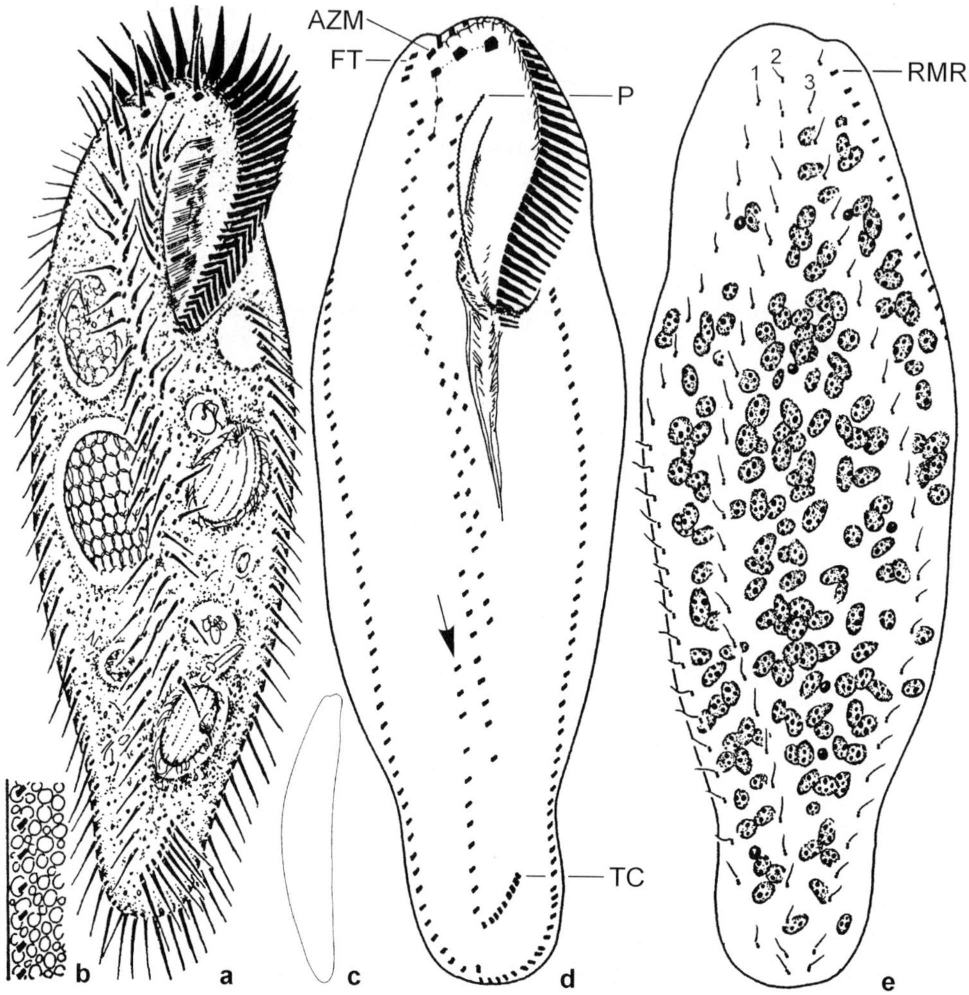

Fig. 117a–e *Bakuella edaphoni* (from Song et al. 1992. a, c, from life; b, d, e, protargol impregnation). **a:** Ventral view of a representative specimen, 225 μm. **b:** Part of pellicle to show the densely packed globules (mitochondria?) up to 5 μm across. **c:** Right lateral view, 215 μm. **d, e:** Infraciliature of ventral and dorsal side and nuclear apparatus of same specimen, 240 μm. Arrow in (d) denotes the rearmost midventral row. Frontal cirri connected by dotted line; cirri originating from anlage III and cirri of rearmost midventral pair connected by broken line. AZM = adoral zone of membranelles, FT = frontoterminal cirri, P = paroral, RMR = right marginal row, TC = transverse cirri, 1–3 = dorsal kineties. Page 551.

Contractile vacuole near left cell margin at level of buccal vertex (Fig. 117a). Cortical granules absent (see remarks). Cytoplasm colourless, often packed with up to 5 μm large, colourless globules (mitochondria?), readily visible in protargol preparations (Fig. 117b). Movement moderately rapid.

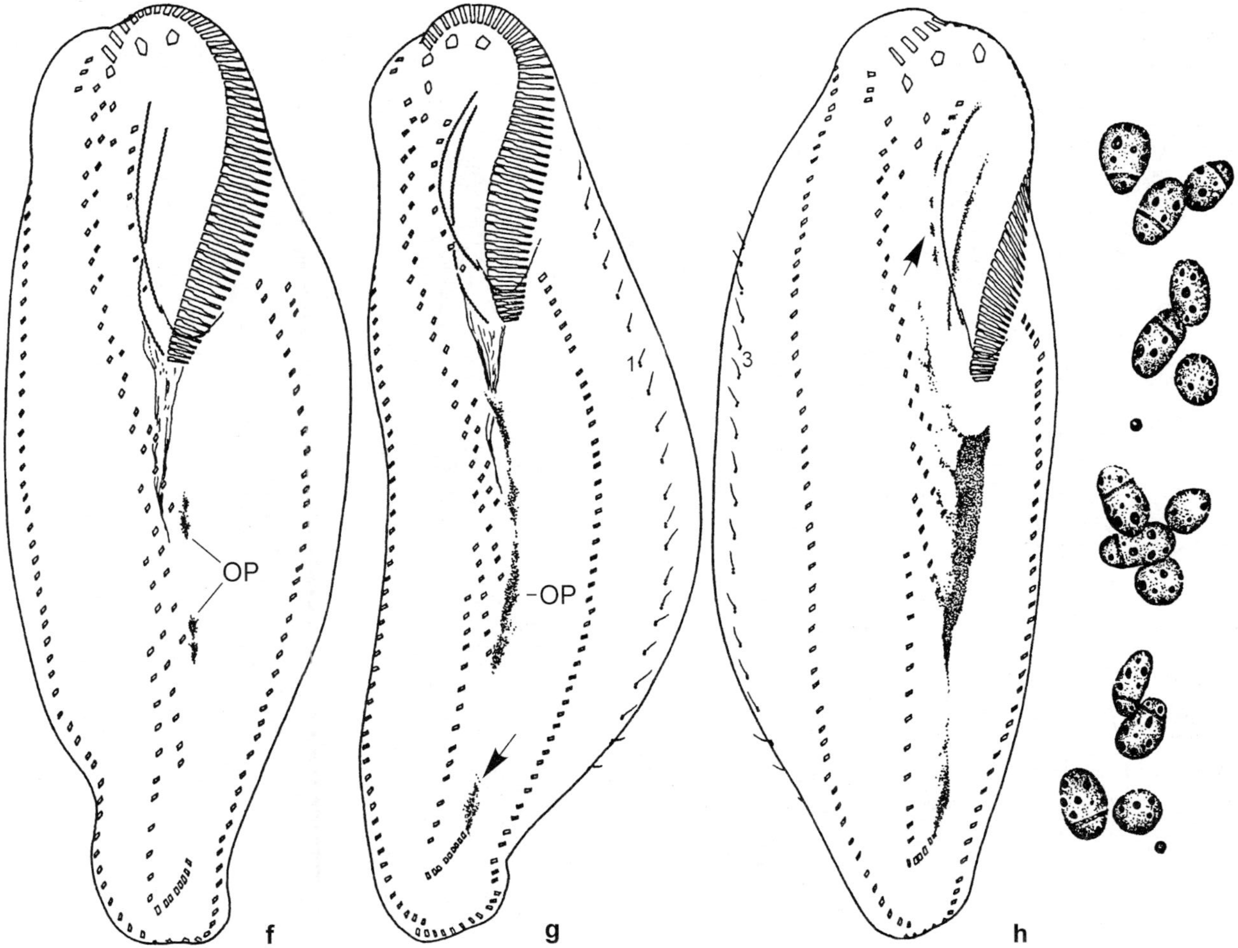

Fig. 117f–h *Bakuella edaphoni* (from Song et al. 1992. Infraciliature of ventral side after protargol impregnation). **f:** Very early divider, 280 µm. **g:** Very early divider, 275 µm. Arrow marks primordium originating close to the transverse cirri. **h:** Early divider, body length 250 µm. Arrow denotes dedifferentiated buccal cirri. Paroral and endoral also begin with dedifferentiation. The many macronuclear nodules are fused to 14 nodules. OP = oral primordium, 1, 3 = dorsal kineties. Page 551.

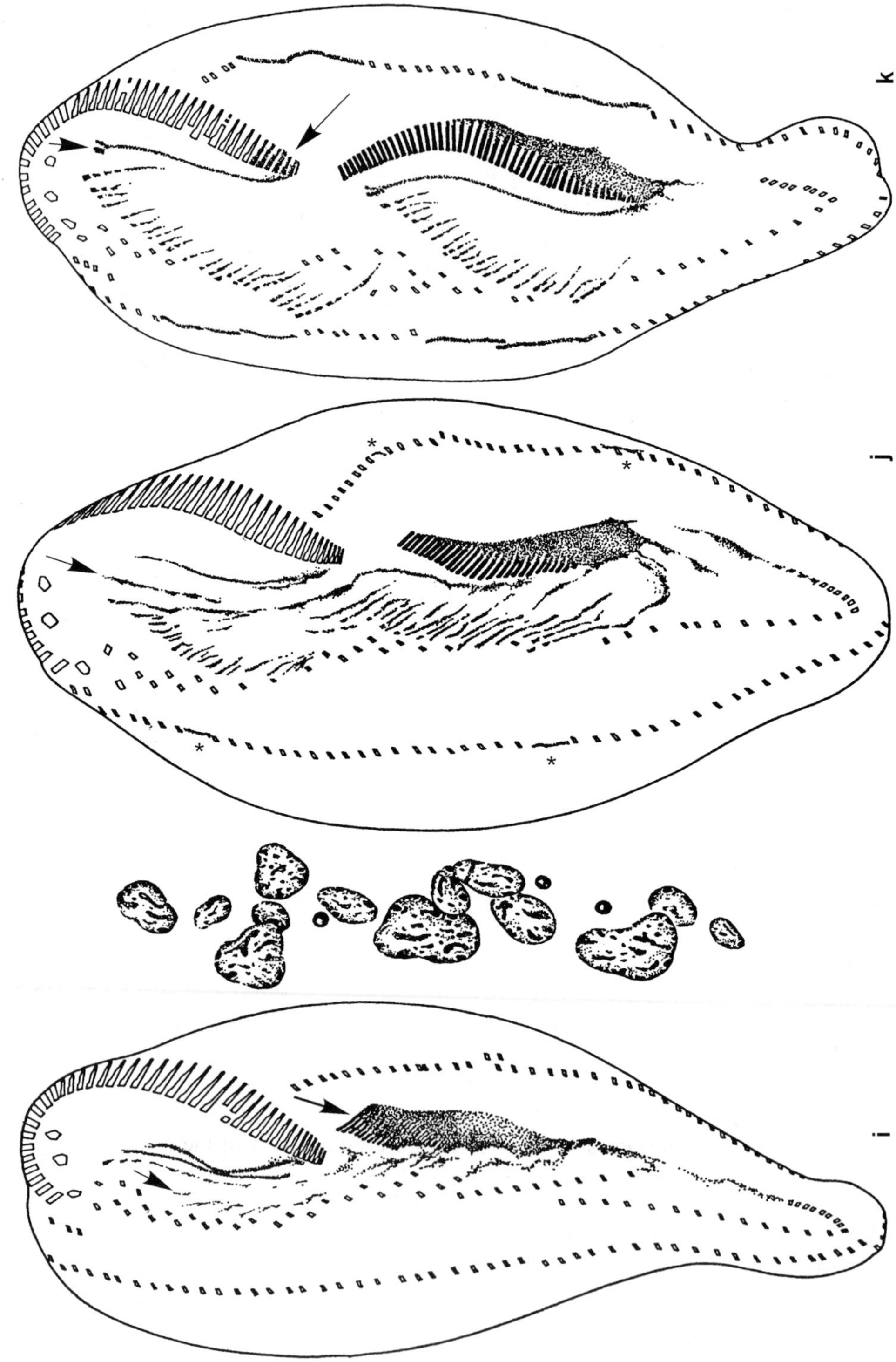
k
j
i

Adoral zone occupies 29% of body length on average, of usual shape and structure, composed of an average of 39 membranelles of ordinary fine structure. Buccal area large. Paroral and endoral long, curved, and optically intersecting. Pharyngeal fibres conspicuous (Fig. 117a, d).

Cirral pattern and number of cirri of usual variability (Fig. 117a, d, Table 25). Three enlarged frontal cirri almost transversely arranged. Buccal cirral row commences somewhat behind anterior end of paroral. 1–7, usually two or three cirri behind right frontal cirrus (Fig. 117d). Frontoterminal row right of distal end of adoral zone of membranelles. Midventral complex composed of about nine cirral pairs with last pair about at level of buccal vertex and about seven midventral rows with rearmost row terminating close to right transverse cirrus; posterior rows almost longitudinally arranged, with 7–14 cirri, anterior rows with 3–5 cirri. Transverse cirri rather small, subterminally arranged and therefore only slightly projecting beyond rear body end. Right marginal row commences with a dorsal bristle on dorsal side near anterior cell end, terminates in cell-midline, and thus confluent with J-shaped left marginal row (Fig. 117d).

Dorsal cilia about 3 μm long, arranged in three bipolar kineties (Fig. 117e). Caudal cirri lacking.

Cell division: This process is described in detail in the original description (Song et al. 1992). The first event is the formation of small groups of basal bodies close to the posterior end of the middle midventral rows (Fig. 117f) and near the left transverse cirrus (Fig. 117g, arrow). The basal bodies increase in number forming a longish field from which some basal bodies migrate forward to the posterior portion of the zigzag portion of the midventral complex (Fig. 117h). The parental endoral and the anterior third of the paroral dedifferentiate. Simultaneously the posterior buccal cirri are modified to the anlage II of the proter (Fig. 117h, arrow). The pharyngeal fibres disappear.

The adoral membranelles of the opisthe develop in a posteriad direction (Fig. 117i, long arrow). At this stage some of the posterior cirri of the middle midventral rows are dedifferentiated and the posterior portion of the paroral becomes disorganised. Between the dedifferentiated buccal row and the unaltered zigzag portion of the midventral complex some small rows of basal bodies occur (Fig. 117i, short arrow).

In the middle stages of morphogenesis most of the cirri of the ventral rows are modified to primordia of the opisthe. The parental buccal cirri are now a narrow band (Fig. 117j, arrow). Some zigzagging cirri are obviously incorporated in the formation of the obliquely arranged primordial streaks.

Somewhat later the proximal parental adoral membranelles begin with reorganisation (Fig. 117k, long arrow). Possibly the posterior end of anlage I of the proter is involved

---

← **Fig. 117i–k** *Bakuella edaphoni* (from Song et al. 1992. Infraciliature of morphogenetic stages in ventral view after protargol impregnation). **i:** Early divider, 250 μm. Long arrow denotes formation of adoral membranelles in oral primordium of opisthe; short arrow marks primordia of the frontal-midventral-transverse cirral anlagen in proter. **j:** Middle divider, 200 μm (dorsal side, see Fig. 117o). Arrow marks anlage II of proter. Asterisks denote primordia of marginal rows. **k:** Middle divider, 270 μm. Long arrow marks reorganising proximal portion of parental adoral zone, short arrow denotes new left frontal cirrus of proter. Page 551.

Fig. 117l–n *Bakuella edaphoni* (from Song et al. 1992. Infraciliature of ventral side of morphogenetic stages after protargol impregnation. New structures black, parental white). **l:** Middle stage, 220 µm (dorsal side, see Fig. 117p). Arrows mark short streaks from which the frontoterminal cirri possibly originate (however, this is uncertain and thus should be checked in other populations; until this feature is reinvestigated it should not be over-interpreted). **m:** Late divider, 205 µm (dorsal side, see Fig. 117q). Arrows mark new frontoterminal cirri. **n:** Very late divider, 220 µm (dorsal side, see Fig. 117r). Arrows mark new frontoterminal cirri which migrate anteriorly to their final position. II = anlage II. Page 551.

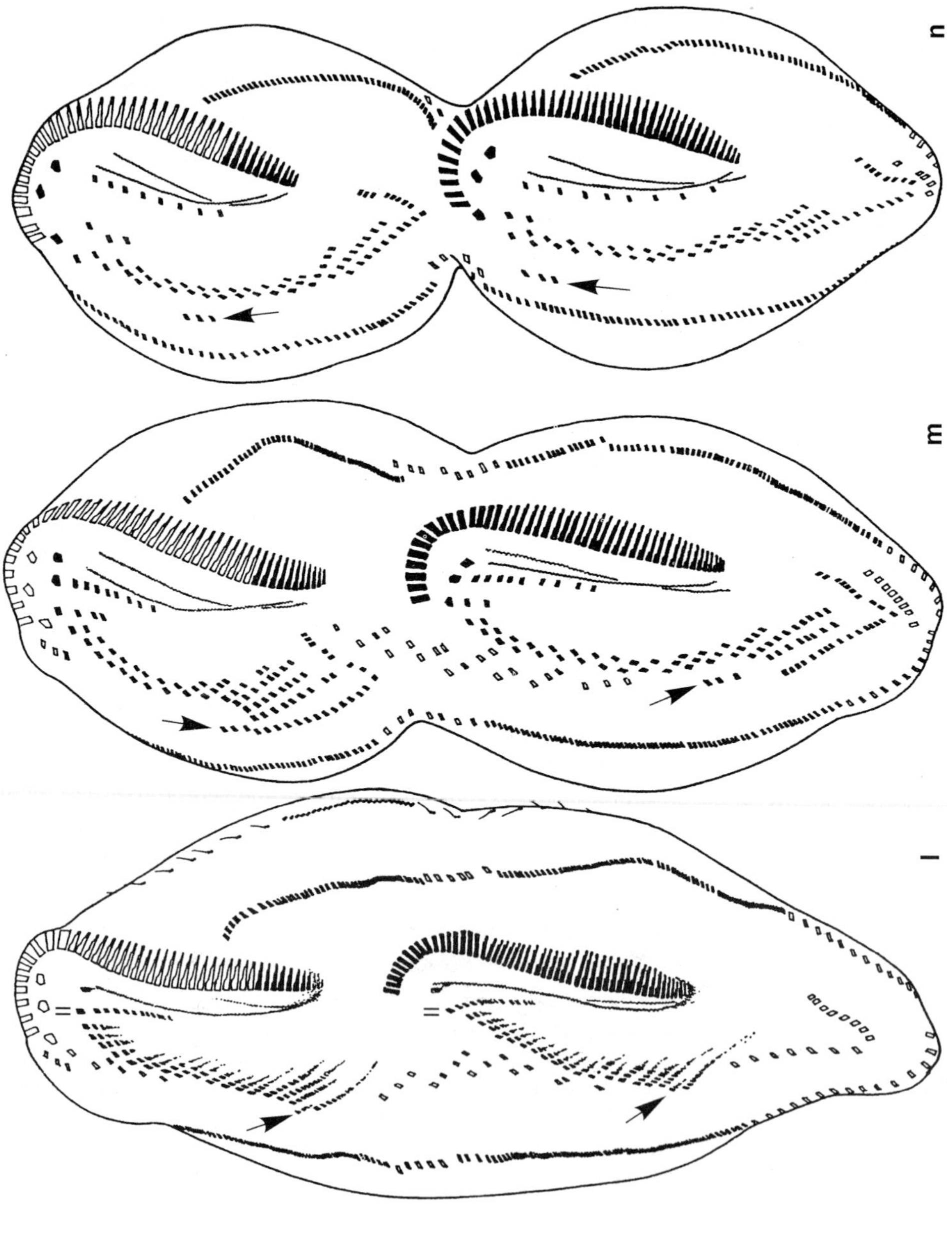

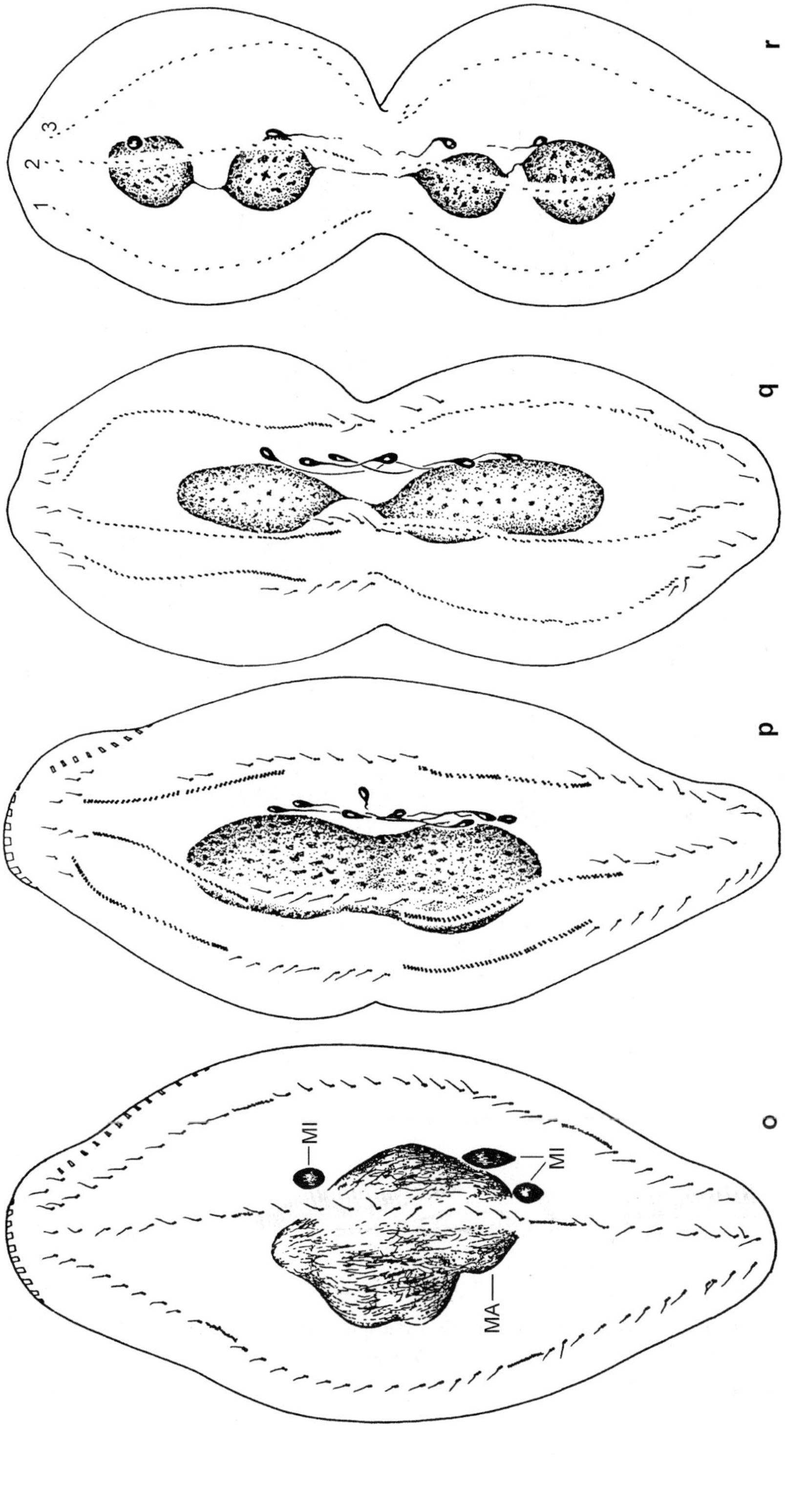

Fig. 117o–r *Bakuella edaphoni* (from Song et al. 1992. Infraciliature of dorsal side and nuclear apparatus of morphogenetic stages after protargol impregnation. Ventral side, see Fig. 117j, l, m, n). The division of the dorsal kineties proceeds in ordinary manner, that is, each two primordia occur within a kinety. The many macronuclear nodules fuse to a single mass in middle dividers, and later divide into the species-specific number of nodules. MA = fused macronucleus, MI = micronuclei, 1–3 = new dorsal kineties of proter. Page 551.

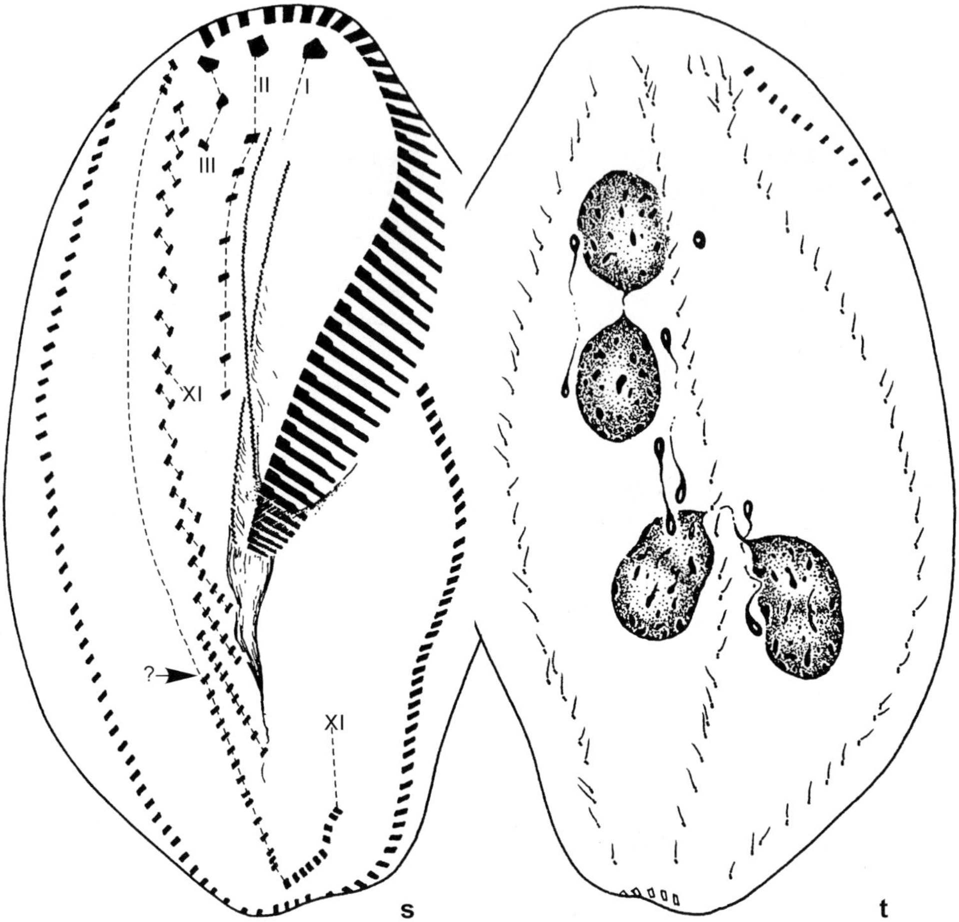

Fig. 117s, t *Bakuella edaphoni* (from Song et al. 1992. Protargol impregnation). Infraciliature of ventral and dorsal side and nuclear apparatus of a postdivider, 120 μm. Cirri originating from same anlage are connected by broken lines. Question mark notes that the origin of the frontoterminal cirri could not be found out unequivocally. Very likely they are, as is usual, the anterior portion of the rightmost cirral streak (see, however, Fig. 117l for other possibility). I, II, III, XI = frontal-midventral-transverse cirral anlagen. Page 551.

in this process. Both in the proter and in the opisthe, about 20 short anlagen and the new left frontal cirrus are recognisable.

In the next stage the oral primordium of the opisthe is already distinctly bent to the right and completely modified to adoral membranelles (Fig. 117l). The reorganisation of the parental adoral zone is restricted to about 10 proximal membranelles. In both the proter and the opisthe the numerous obliquely arranged streaks begin with the differentiation of cirri. The arrows in Fig. 117l mark the penultimate streaks. These small streaks appear to form the frontoterminal cirri although this is unusual because, in general, the frontoterminal cirri originate from the anterior part of the posteriormost (rightmost) anlage (see last paragraph of this chapter).

Fig. 117m shows a late divider. All new frontal-midventral-transverse cirri are differentiated and they begin to migrate to their final positions. Immediately before the separation of the proter and the opisthe, most of the parental cirri and dorsal kineties are resorbed. The postdividers have a wide elliptical outline (Fig. 117s). Now the pharyngeal fibres impregnate again.

The macronuclear nodules fuse to a single mass during division and later divide into the species-specific number (more than 100) of nodules (Fig. 117h, i, o–r, t). Within each marginal row and dorsal kinety two primordia are formed. No caudal cirri are produced at the end of the dorsal kineties (Fig. 117o–r, t).

The frontal-midventral-transverse primordia in the proter have the following origin: anlage I, from the parental undulating membranes, forms the new undulating membranes and the left frontal cirrus. Anlage II, from the parental buccal cirri, forms the middle frontal cirrus and the buccal cirri. Anlage III, probably de novo, forms the right frontal cirrus and one or more cirri behind this cirrus. Anlagen IV to n, probably de novo and by modification of parental midventral cirri, respectively (Fig. 117j, k), form the cirral pairs of the midventral complex, the midventral rows, and the transverse cirri.

The origin of the frontoterminal cirri could not be unequivocally recognised by Song et al. (1992, p. 141). It appears that they originate from a very short streak between the two rightmost anlagen, which is, however, not in accordance with any other morphogenetic pattern of related species. Thus, this stage should be reinvestigated in other populations. In the opisthe, anlagen I–n are derived from the oral primordium and modified midventral rows (Fig. 117s).

Occurrence and ecology: Terrestrial. The type locality of *Bakuella edaphoni* is a hill in the city of Qingdao (36°08′N, 120°43′E), China, where Song et al. (1992) discovered it in the upper soil layer (0–2 cm). Excystment occurred about three days after the incubation of the air-dried soil was begun. It was cultivated in Petri dishes containing Eau de Volvic (France) and few crushed wheat grains to promote bacterial growth. Wilbert (1995, p. 282) found it in the saline soil (0–10 cm) taken from dry lakes in the Coorong National Park near Adelaide, Australia. Feeds on bacteria, diatoms, heterotrophic flagellates, testate amoebae, and small ciliates. Biomass of 10^6 specimens about 600 mg (Foissner 1998, p. 199).

Bakuella pampinaria Eigner & Foissner, 1992 (Fig. 118a–s, 120a, b, Table 25)

1992 ***Bakuella pampinaria* nov. spec.** – Eigner & Foissner, Europ. J. Protistol., 28: 461, Fig. 1–18 (Fig. 118a–o; original description. Two type slides [registration numbers 1993/45; 1993/46] are deposited in the Oberösterreichische Landesmuseum in Linz [LI], Austria).

1996 ***Bakuella pampinaria* Eigner and Foissner, 1992** – Franco, Esteban & Téllez, Acta Protozool., 35: 329, Fig. 20 (redrawing of Fig. 118b; key to the genera and species of Bakuellinae).

2001 ***Bakuella pampinaria* Eigner and Foissner, 1992** – Berger, Catalogue of ciliate names 1. Hypotrichs, p. 12 (nomenclator containing all basionyms, combinations, and higher taxa of hypotrichs).

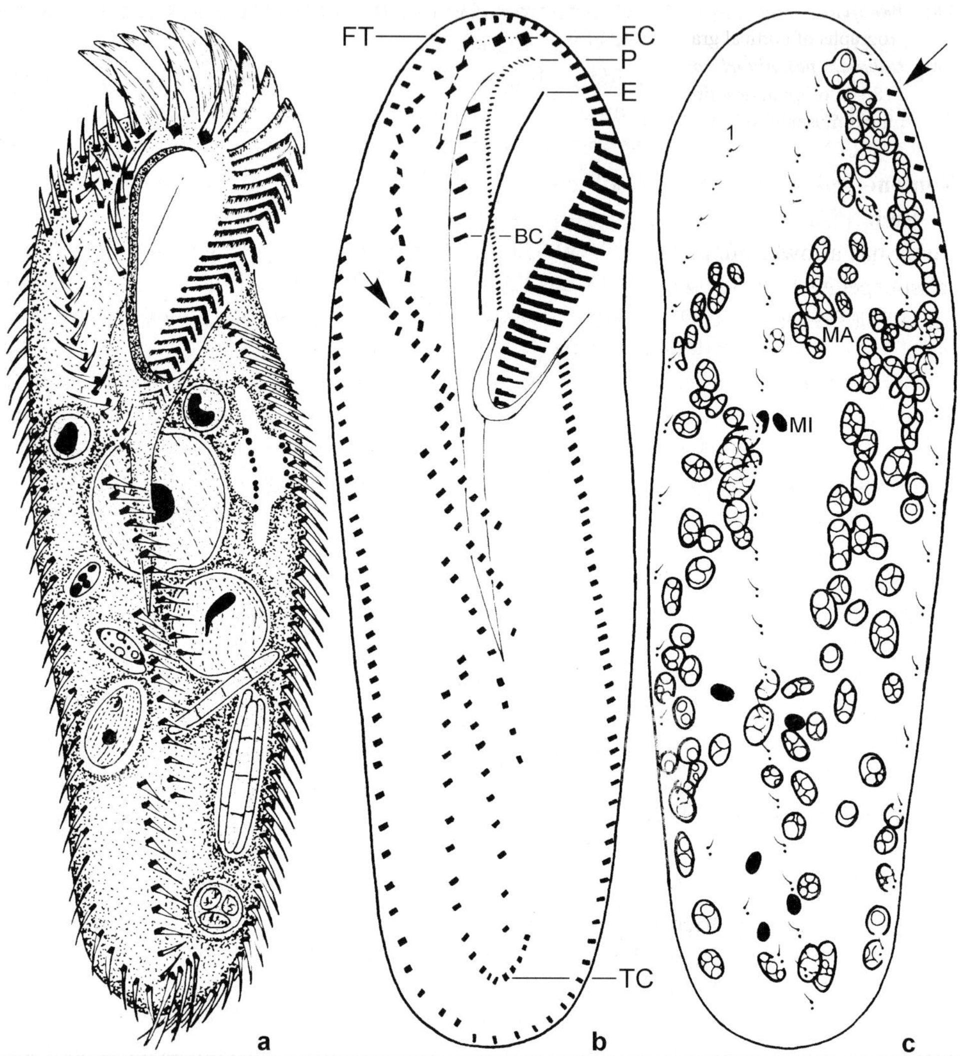

Fig. 118a–c *Bakuella pampinaria pampinaria* (from Eigner & Foissner 1992. a, from life; b, c, protargol impregnation). **a:** Ventral view of a representative specimen, 155 µm. **b, c:** Infraciliature of ventral and dorsal side and nuclear apparatus of same specimen, 170 µm. Arrow in (b) denotes anteriormost midventral row, arrow in (c) marks anterior end of right marginal row. Broken line connects cirri which originate from anlage III (right frontal cirrus and parabuccal cirri [termed ventral row by Eigner & Foissner 1992]). BC = rearmost buccal cirrus, E = endoral, FC = frontal cirri (connected by dotted line), FT = frontoterminal cirri, MA = macronuclear nodules, MI = micronuclei, P = paroral, TC = rearmost (= rightmost) transverse cirrus, 1 = dorsal kinety 1 (= leftmost kinety). Pages 559, 563.

Fig. 118d–h *Bakuella pampinaria pampinaria* (from Eigner & Foissner 1992. d, f–h, protargol impregnation; e, from life). **d:** Infraciliature of ventral side and cortical granulation (black dots), 149 µm. **e:** Left lateral view. **f–h:** Infraciliature of ventral side of very early dividers, f = 134 µm (g is an detail of f), h = 100 µm. In (f) only a small part of the macronuclear nodules is illustrated. Pages 559, 563. →

2002 ***Bakuella pampinaria*** – Foissner, Agatha & Berger, Denisia, 5: 1291, Fig. 380d, e (Fig. 120a, b; micrographs of cortical granules of a Thailand population).

2004 ***Bakuella pampinaria oligocirrata* nov. sspec.** – Foissner, Denisia, 13: 376, Fig. 33–36, Table 2 (Fig. 118p–s; original description; the holotype and a paratype slide are deposited in the Oberösterreichische Landesmuseum in Linz [LI], Austria).

Nomenclature: See subspecies.

Remarks: Foissner (2004) described the subspecies *B. pampinaria oligocirrata* which has a lower number of midventral pairs and parabuccal cirri than the nominotypical subspecies (see *Bakuella* key). I do not provide a description of the species *B. pampinaria*, but the reader is referred to the complete descriptions of the two subspecies.

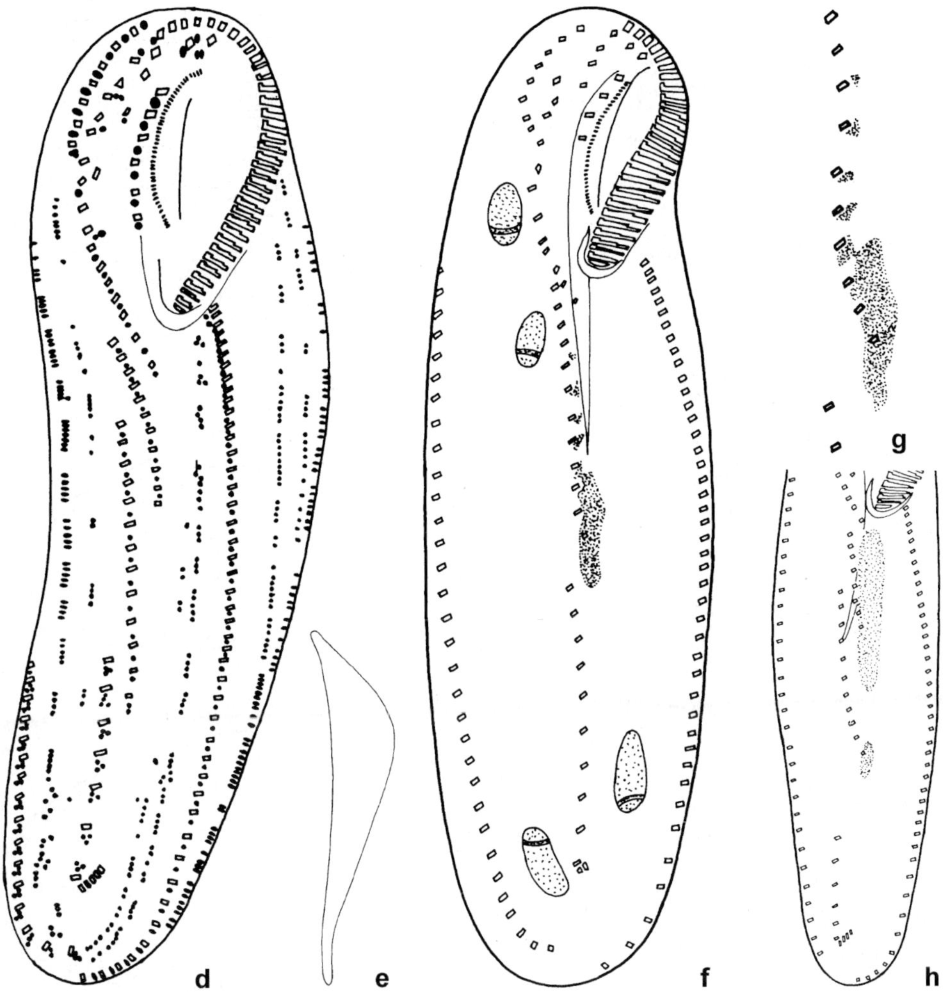

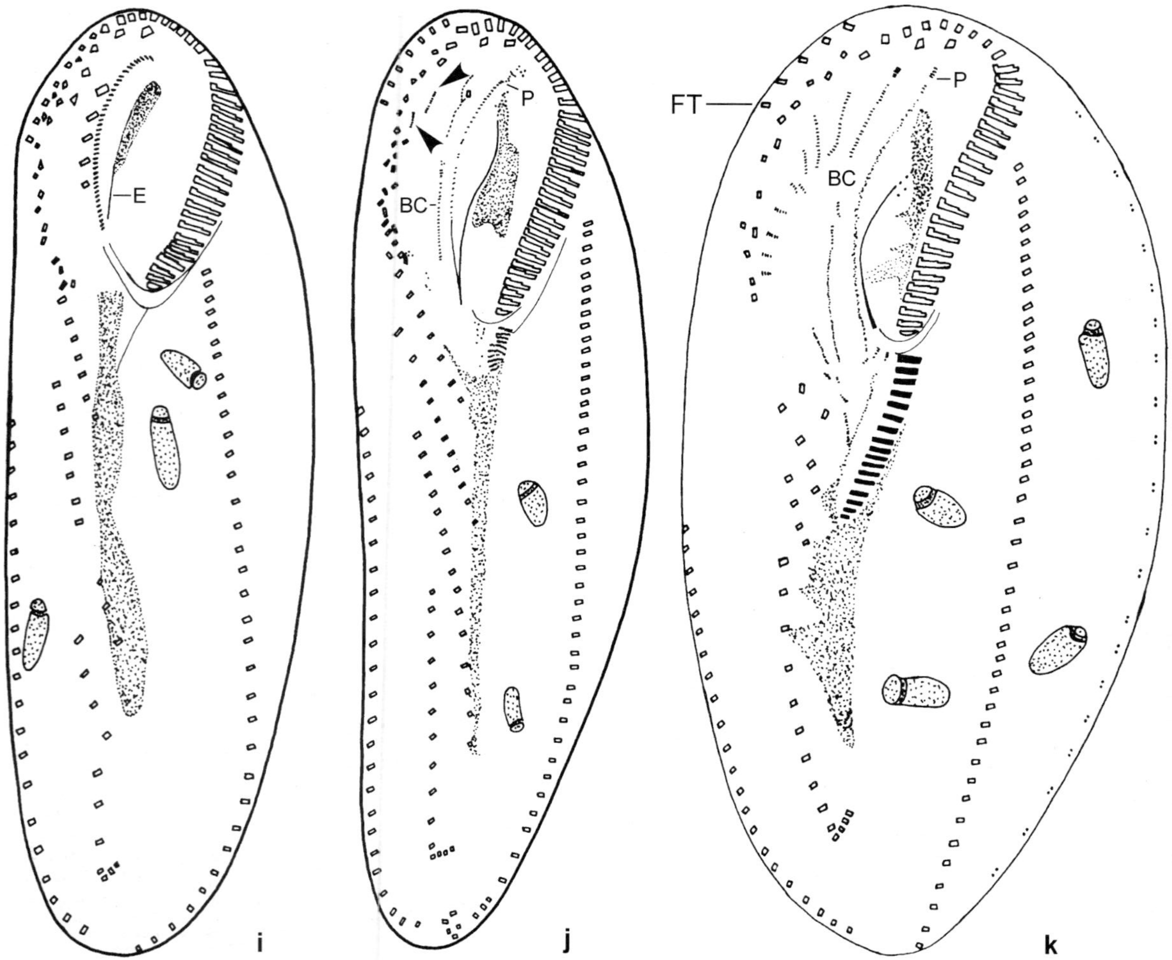

Fig. 118i–k *Bakuella pampinaria pampinaria* (from Eigner & Foissner 1992. Protargol impregnation). Infraciliature of ventral side of morphogenetic stages, i, j = 168 µm, k = 105 µm. Parental structures white, new black. Only few macronuclear nodules illustrated. Arrowheads in (j) mark streaks derived from cirri behind right frontal cirrus. BC = disorganised buccal cirral row, E = endoral, FT = frontoterminal cirri, P = paroral. Pages 559, 563.

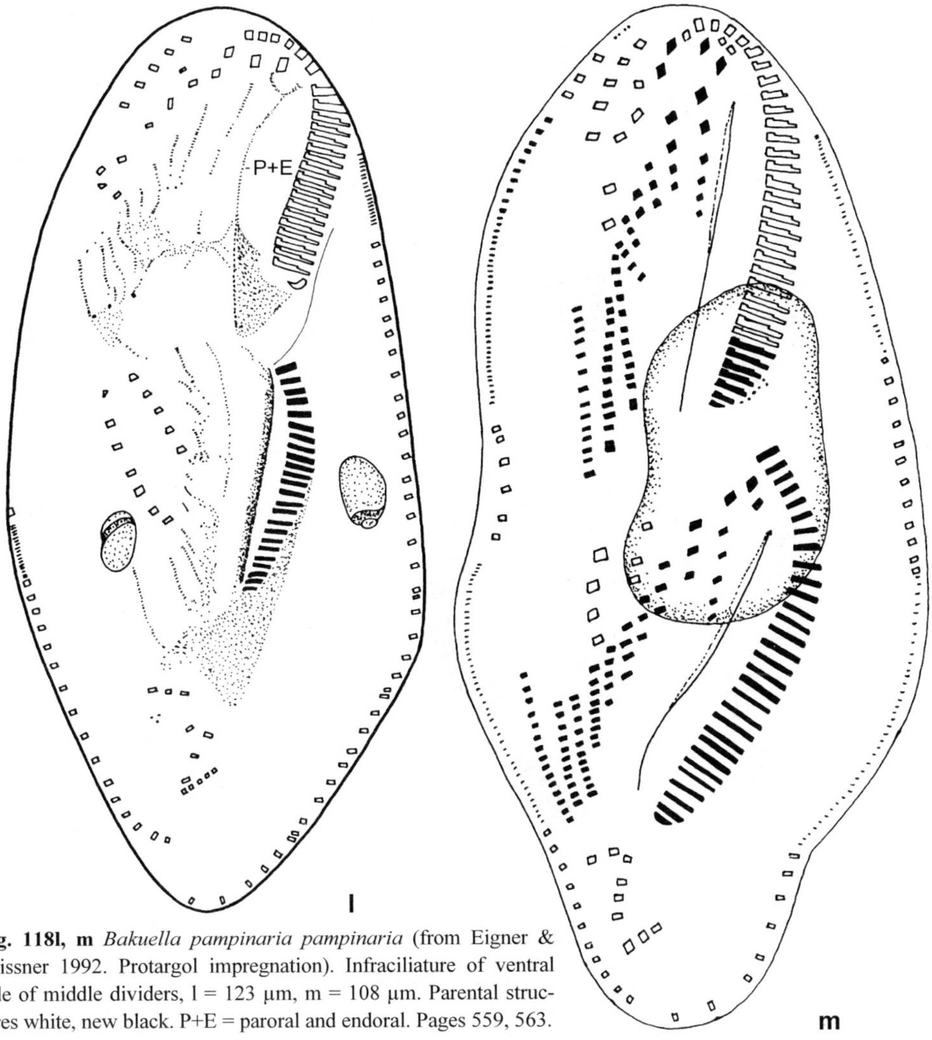

Fig. 118l, m *Bakuella pampinaria pampinaria* (from Eigner & Foissner 1992. Protargol impregnation). Infraciliature of ventral side of middle dividers, l = 123 µm, m = 108 µm. Parental structures white, new black. P+E = paroral and endoral. Pages 559, 563.

Bakuella pampinaria pampinaria Eigner & Foissner, 1992
(Fig. 118a–o, 120a, b, Table 25)

1992 ***Bakuella pampinaria*** **nov. spec.**[1] – Eigner & Foissner, Europ. J. Protistol., 28: 461, Fig. 1–18 (Fig. 118a–o; original description. Two type slides [registration numbers 1993/45; 1993/46] are deposited in the Oberösterreichische Landesmuseum in Linz [LI], Austria).

[1] The diagnosis by Eigner & Foissner (1992) is as follows: Size in vivo 90–180 × 25–60 µm. Distinct rows of yellowish cortical granules. 31 adoral membranelles, 5 buccal cirri, 6 frontoterminal cirri, 9 pairs of midventral cirri, 3 ventral rows, 4 transverse cirri and 100 macronuclear segments on average. Posteriormost ventral row adjacent to right transverse cirrus. Transverse cirri not involved in stomatogenesis.

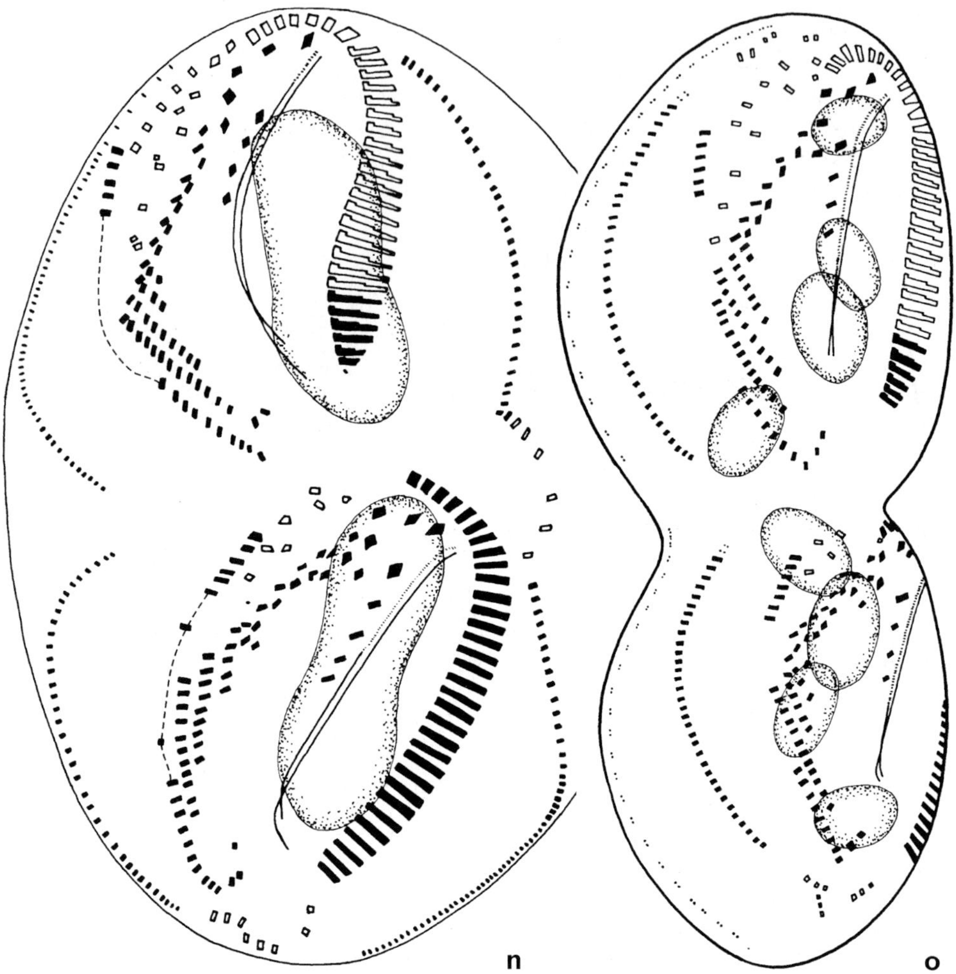

Fig. 118n, o *Bakuella pampinaria pampinaria* (from Eigner & Foissner 1992. Protargol impregnation). Infraciliature of ventral side and nuclear apparatus of very late divider, n = 126 µm, o = 155 µm. Broken lines connect cirri which originate from rightmost anlage of proter and opisthe. Pages 559, 563.

1996 ***Bakuella pampinaria* Eigner and Foissner, 1992** – Franco, Esteban & Téllez, Acta Protozool., 35: 329, Fig. 20 (redrawing of Fig. 118b; key to the genera and species of Bakuellinae).

2001 ***Bakuella pampinaria* Eigner and Foissner, 1992** – Berger, Catalogue of ciliate names 1. Hypotrichs, p. 12 (nomenclator containing all basionyms, combinations, and higher taxa of hypotrichs).

2002 ***Bakuella pampinaria*** – Foissner, Agatha & Berger, Denisia, 5: 1291, Fig. 380d, e (Fig. 120a, b; micrographs of cortical granules of a Thailand population).

Nomenclature: The species-group name *pampinari·us ·a ·um* is a composite of the Latin substantive *pampin·us* (vine leaf), the suffix *~ari* (in species names: habitat; Werner 1972) and the inflectional ending *·a*, indicating that this species was discovered in vine leaf litter. When *B. paminaria* is classified in the subgenus *Bakuella (Bakuella)*, then the correct name is *Bakuella (Bakuella) pampinaria* Eigner & Foissner, 1992.

Foissner (2004) described the subspecies *B. pampinaria oligocirrata*. Thus, he simultaneously established the nominotypical subspecies *B. pampinaria pampinaria* Eigner & Foissner, 1992. If this subspecies is classified in *Bakuella (Bakuella)* then the correct name is *Bakuella (Bakuella) pampinaria pampinaria* Eigner & Foissner, 1992.

Eigner & Foissner (1992, p. 461) wrote that they have deposited a holotype and a paratype slide in the Upper Austrian Museum. By contrast, Aescht (2003, p. 393) designated the slides as syntypes.

Remarks: *Bakuella pampinaria* (both subspecies), although of considerable size (90–180 μm long), is the smallest of the three terrestrial *Bakuella* species. *Bakuella granulifera* is 270–400 μm and *Bakuella edaphoni* 190–300 μm long. The latter species lacks cortical granules, whereas *B. pampinaria* and *B. granulifera* have such organelles. For a detailed comparison of these two species, see this chapter at *B. granulifera*.

The description below is based on the original description, which also includes a detailed analysis of the morphogenesis of cell division (see below).

Morphology: Body size 90–180 × 25–60 μm in life. Body outline long-elliptical, right margin straight to slightly concave, left more or less convex; both ends slightly narrowed and rounded. Body dorsoventrally flattened 2–3:1 (Fig. 118e), highly flexible, but acontractile. Macronuclear nodules scattered, ellipsoidal, in life about 4–6 × 3–4 μm. 3–7 ellipsoidal micronuclei, usually 1–2 near proximal end of adoral zone of membranelles. Contractile vacuole near left cell margin slightly ahead of mid-body, during diastole with inconspicuous collecting canals (Fig. 118a). Cortical granules in distinct rows within and between cirral rows and dorsal kineties, recognisable in life at a magnification of × 100; individual granules yellowish, ellipsoidal, about 1.5–2.0 × 1.0–1.5 μm, impregnate with Foissner's protargol method (Fig. 118d, 120a, b). Cytoplasm brownish at low magnification, contains many food vacuoles up to 35 μm across in well-nourished specimens. Movement rather slow.

Adoral zone occupies 30–40% (average 36%) of body length, of usual shape and structure, composed of about 31 membranelles; cilia about 20 μm long in life (Fig. 118a, b). Buccal cavity deep and large and thus brightly shining in bright field illumination. Paroral conspicuous, anterior portion curved, composed of at least three rows of basal bodies. Endoral crosses buccal field near dorsal inner surface of cell because buccal cavity is very deep; straight in anterior, curved in posterior portion, crosses or parallels paroral optically depending on position of cell. Endoral and paroral terminate at same level near proximal portion of adoral zone. Pharyngeal fibres conspicuous, form curtain-like structure along entire paroral.

Cirral pattern and number of cirri of usual variability (Fig. 118b, d; Table 25). Three slightly enlarged frontal cirri almost transversely arranged. Buccal cirral row begins somewhat behind anterior end of paroral, cirri slightly enlarged as compared to marginal cirri. 2–4 (parabuccal) cirri behind right frontal cirrus. Frontoterminal cirral row begins right of right frontal cirrus, runs along right body margin. Midventral complex composed of midventral pairs and rows; portion with cirral pairs begins about at same level as buccal row, terminates usually ahead of level of posterior end of adoral zone;

distinction between zigzagging anterior portion and posterior portion of midventral complex may be difficult in large specimens; midventral rows more or less obliquely arranged, rearmost row terminates very close to right transverse cirrus. Transverse cirri rather small and subterminal, about 16 µm long, all other cirri 10–12 µm long in life. Marginal rows almost confluent posteriorly, anterior portion of right row extends onto dorsal side commencing with two dorsal bristles.

Dorsal cilia about 3 µm long, arranged in three almost bipolar kineties (Fig. 118c). No caudal cirri.

Cell division: Divisional morphogenesis and reorganisation is described in detail in the original description (Eigner & Foissner 1992). The morphogenesis largely agrees with that described for *B. edaphoni* so that the reader is mainly referred to the illustration (Fig. 118f–n). However, there are three noteworthy differences: (i) the transverse cirri of *B. pampinaria* are not involved in the oral primordium formation (vs. involved in *B. edaphoni* and *B. salinarum*); (ii) left of the parental endoral a rather wide anlagen field originates in *B. pampinaria* and *B. salinarum* (vs. disorganised endoral which forms a narrow band in *B. edaphoni*, Fig. 117h–j); (iii) the frontoterminal cirri of *B. pampinaria* and *B. salinarum* are formed in ordinary manner, that is, from the anterior end of the rightmost midventral row. By contrast, the origin of the frontoterminal cirri could not be unequivocally clarified by Song et al. (1992) in *B. edaphoni* (penultimate anlage or last anlage; Fig. 117l, m, s).

Occurrence and ecology: Likely confined to terrestrial habitats. The type locality is in the village of Schrötten, Styria, Austria (46°47′N 15°49′E; altitude 320 m), where *Bakuella pampinaria* was discovered in fallen leaves of non-grafted vines. Eigner & Foissner (1992) found it also in the litter under a pear tree grown on ecofarmed land in the same area. Specimens were cultivated in a small Petri dish containing local spring water and a crushed wheat grain to support growth of indigenous bacteria and small ciliates which served as food. Foissner et al. (2002, p. 1291) found it in Thailand (Fig. 120a, b) and recently we recorded it in various Austrian forest stands (Foissner et al. 2005). Feeds on small ciliates, heterotrophic flagellates, and fungal spores (Eigner & Foissner 1992). Biomass of 10^6 specimens about 80 mg (Foissner 1998, p. 199).

Bakuella pampinaria oligocirrata Foissner, 2004 (Fig. 118p–s, Table 25)

2004 ***Bakuella pampinaria oligocirrata* nov. sspec.**[1] – Foissner, Denisia, 13: 376, Fig. 33–36, Table 2 (Fig. 118p–s; original description; the holotype and a paratype slide are deposited in the Oberösterreichische Landesmuseum in Linz [LI], Austria).

Nomenclature: The species-group name *oligocirrat·us -a -um* is an adjective composed of the Greek adjective *oligo* (few) and the Latin noun *cirrus* (curl, cilia, cirri), referring to the decreased number (4 in *B. p. oligocirrata* vs. 9 in *B. p. pampinaria*) of

[1] The diagnosis by Foissner (2004) is as follows: Three to seven, usually four pairs of midventral cirri; frontal row composed of a single cirrus on average.

midventral cirral pairs (Foissner 2004). When this subspecies is classified in *Bakuella (Bakuella)* then the correct name is *Bakuella (Bakuella) pampinaria oligocirrata* Foissner, 2004.

Remarks: The present subspecies differs from *B. pampinaria pampinaria* in the number of midventral pairs (on average 4 vs. 9) and the lower number of parabuccal cirri (usually 1 vs. 2). Possibly *Paraurostyla pulchra* (Fig. 123a) is closely related to *B. pampinaria oligocirrata* (see there for details).

Possibly it would be better to raise *B. p. oligocirrata* to species rank because subspecies should be geographically or ecologically separated. Both taxa were discovered in Austrian soils which are, however, rather different (fallen leaves of non-grafted vines [*pampinaria*] vs. floodplain soil [*oligocirrata*]). Likely difficult to separate from *Paraurostyla pulchra* (Fig. 123a; see there).

Morphology: Body size 80–150 × 25–50 µm in life, usually near 115 × 35 µm. Body outline inconspicuous, that is, oblong to elongate elliptical with both ends broadly rounded (Fig. 118p); occasionally slightly curved, that is, left margin concave, right convex. Body dorsoventrally flattened up to 2:1. Nuclear apparatus composed of an average of 90 macronuclear nodules and six globular micronuclei. Macronucleus nodules slightly concentrated along body margins, globular to ellipsoidal (Fig. 118s, Table 25). Contractile vacuole slightly ahead of mid-body, with two collecting canals. Cortex very flexible and yellowish at magnifications of × 100–200 due to rows of citrine cortical granules. Individual granules about 1.0 × 0.7 µm, citrine to yellowish with greenish shimmer, arranged in rather widely spaced, more or less distinct rows (Fig. 118q). Cytoplasm colourless, usually packed with food vacuoles. Movement inconspicuous, that is, glides rather rapidly on microscope slide and soil particles showing great flexibility.

Adoral zone occupies about 34% of body length on average, only slightly curved because commencing in midline of cell; composed of an average of 27 membranelles about 6 µm wide in life (Fig. 118p, r). Buccal cavity wide and deep extending almost to dorsal side, distinctly curved anteriorly; buccal lip inconspicuous. Paroral and endoral membrane slightly curved, optically intersect in mid-buccal cavity, both composed of very narrowly spaced cilia. Pharyngeal fibres of ordinary length and structure, probably mixed with long endoral cilia.

Cirral pattern and number of cirri of usual variability (Fig. 118r, Table 25). Cirri about 10 µm long and of very similar size, except for the slightly enlarged frontal cirri. Three frontal cirri, right one very close to distal end of adoral zone. Usually four buccal cirri along right margin of paroral. Usually only one, rarely two parabuccal cirri. Five frontoterminal cirri in ordinary position. Midventral complex composed of four cirral pairs and four midventral rows on average (Table 25). Cirral pair portion terminates at 26% of body length on average (Table 25). Leftmost midventral row of specimen illustrated with six cirri, next row with nine cirri, next with 11 cirri, and rearmost row with seven cirri extending very close to right transverse cirri. Transverse cirri inconspicuous, project slightly beyond rear body end. Right marginal row commences at level of frontoterminal cirri, ends subterminally. Left marginal row commences left of buccal

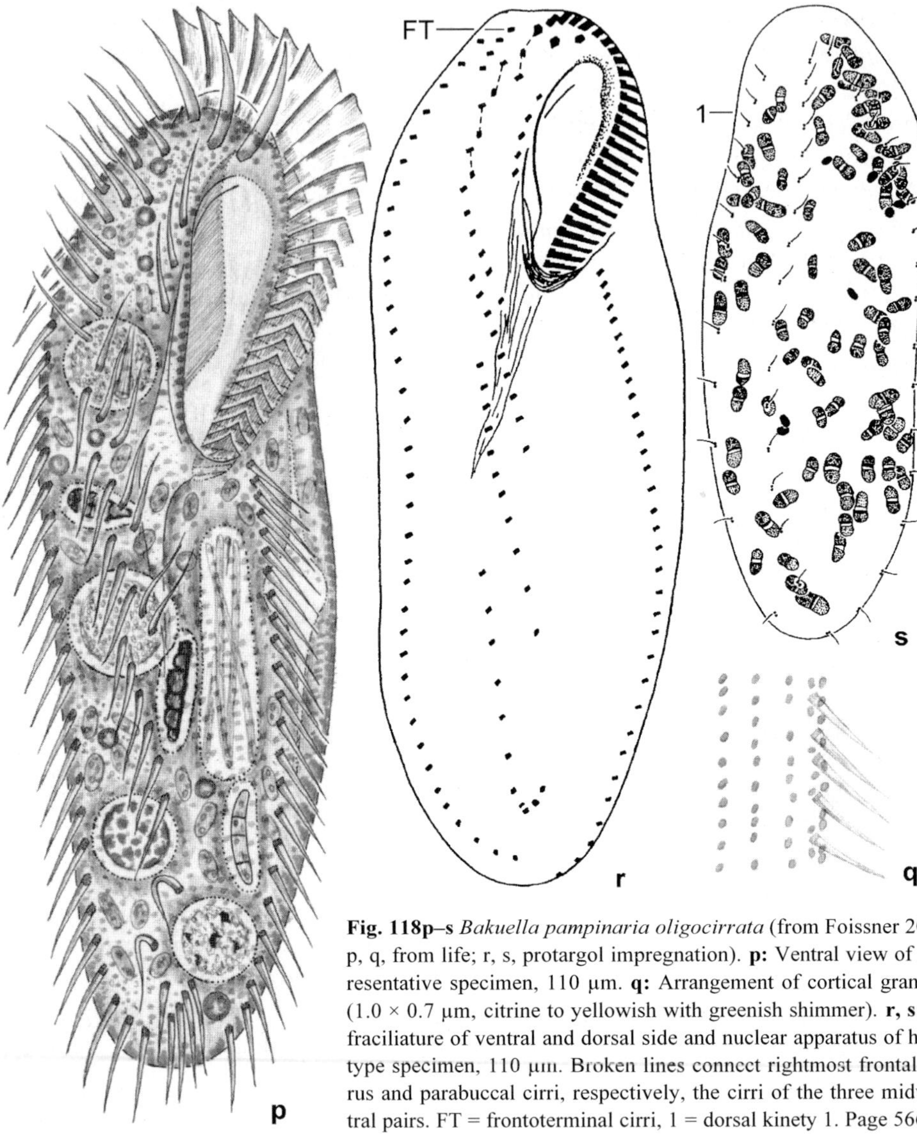

Fig. 118p–s *Bakuella pampinaria oligocirrata* (from Foissner 2004. p, q, from life; r, s, protargol impregnation). **p:** Ventral view of representative specimen, 110 µm. **q:** Arrangement of cortical granules (1.0 × 0.7 µm, citrine to yellowish with greenish shimmer). **r, s:** Infraciliature of ventral and dorsal side and nuclear apparatus of holotype specimen, 110 µm. Broken lines connect rightmost frontal cirrus and parabuccal cirri, respectively, the cirri of the three midventral pairs. FT = frontoterminal cirri, 1 = dorsal kinety 1. Page 566.

vertex, ends terminally, that is, marginal rows usually (>90% of specimens) with small, but distinct gap.

Dorsal bristles about 3 µm long, arranged in three bipolar kineties. Rows 1 and 2 left of midline, row 3 at right body margin, thus, a broad, bare stripe occurs right of midline, as in the other taxa of the *B. edaphoni*-group. Caudal cirri lacking (Fig. 118s).

Cell division: Ontogenesis commences as in nominotypical subspecies, that is, an anarchic field of basal bodies develops in mid-body left of and in connection with the oblique ventral cirral rows (Foissner 2004).

Occurrence and ecology: The type locality of *B. pampinaria oligocirrata* is a floodplain soil (0–5 cm) from the Enns River near the mouth to the Danube River, Upper Austria (48°14′N 14°30′E; altitude 245 m). The floodplains bear an almost original vegetation, mainly composed of *Salix alba*, *Alnus glutinosa*, *Populus alba*, *Fraxinus excelsior*, *Sambucus nigra*, and *Urtica dioica*. The greyish loamy soil was covered by mull humus and a thin litter layer (Foissner 2004).

Omnivorous, that is, feeds on 20–40 µm long bacteria (fungi? found in most specimens), colourless and brown fungal spores, coccal green algae, heterotrophic flagellates, and various ciliates, such as *Colpoda* spp., *Vorticella astyliformis*, *Tetrahymena rostrata*, and *Leptopharynx costatus* (Foissner 2004).

Bakuella granulifera Foissner, Agatha & Berger, 2002
(Fig. 119a–k, 120c–g, Table 25)

2002 ***Bakuella granulifera*** **nov. spec.**[1] – Foissner, Agatha & Berger, Denisia, 5: 561, Fig. 129a–k, 380a–c, f–h, Table 111 (Fig. 119a–k, 120c–g; original description; 1 holotype slide and 6 paratype slides of protargol-impregnated cells are deposited in the Oberösterreichische Landesmuseum in Linz [LI], Austria; accession numbers 2002/459–465).

Nomenclature: The species-group name *granulifer·us -a -um* (bearing granules) is a composite of the Latin noun *granul·um* (small grain, granule), the thematic vowel *·i-*, and the Latin *fero/ferre* (to carry, to keep; carrying) referring to the cortical granules, a main species characteristic. If *B. granulifera* is classified in the present subgenus, then the correct name is *Bakuella (Bakuella) granulifera* Foissner, Agatha & Berger, 2002.

Remarks: *Bakuella granulifera* is very likely closely related to *B. edaphoni*, which is also very large (190–300 × 50–85 µm in vivo) and has an almost identical cirral pattern. The main difference is the conspicuous cortical granulation (Fig. 120c–e), which is lacking in *B. edaphoni* (pers. comm. of Weibo Song, who investigated *B. edaphoni*). However, *Bakuella granulifera* and *B. edaphoni* also differ significantly in several morphometrics, namely, size (319 × 101 µm vs. 219 × 70 µm after protargol impregnation), length of adoral zone (113 µm vs. 63 µm), number of adoral membranelles (54 vs. 39), number of midventral cirral pairs (18 vs. 10), number of midventral rows (4 vs. 7), number of transverse cirri (12 vs. 8), and number of macronuclear nodules (more than 300 vs. likely less than 200).

Bakuella pampinaria, which has, like *B. granulifera*, yellow cortical granules, is distinctly smaller (90–180 × 25–60 µm in vivo), has a lower number of adoral membranelles (31), buccal cirri (5), midventral cirral pairs (9), transverse cirri (4), and macronuclear nodules (about 100), and a slightly higher number of frontoterminal cirri (6 vs. 3–4 in *B. granulifera*). There is no basic morphological difference between these

[1] The diagnosis by Foissner et al. (2002) is as follows: Size about 320 × 100 µm in vivo; oblong. About 350 macronuclear nodules. Cortical granules citrine, in series between cirral and bristle rows. On average 54 adoral membranelles, 8 buccal cirri, 4 frontoterminal cirri, 18 pairs of midventral cirri, 4 short and long midventral rows, and 12 transverse cirri.

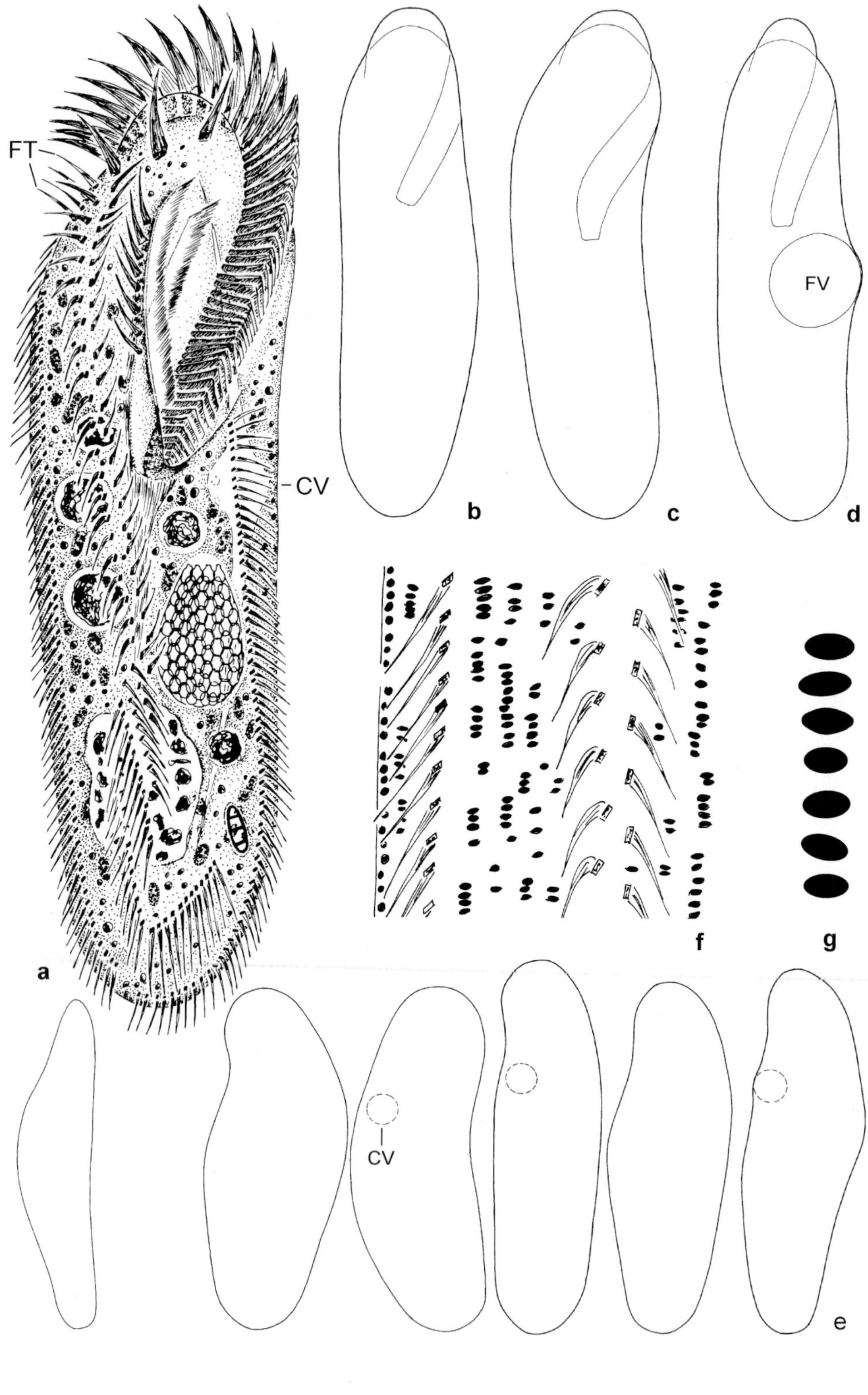

FT
CV
FV
CV
a
b
c
d
e
f
g

two species, and thus subspecies rank would probably be more appropriate. On the other hand, the quantitative differences are conspicuous and not caused by a simple size increase of the cell (which may depend on culture conditions), but concern the entire organism, as evident from, e.g., the doubled number of adoral membranelles, midventral cirral pairs, and macronuclear nodules.

In life, *Bakuella granulifera* is easily confused with *Anteholosticha antecirrata*, which has a similar size (220–330 × 70–90 μm in vivo), cortical granulation, and cirral pattern (Fig. 73a). However, *Anteholosticha* species lack the short and long midventral cirral rows at the posterior end of the midventral complex.

Morphology: Body size 270–400 × 70–120 μm in vivo, usually about 320 × 100 μm, length:width ratio 3.0–4.1:1, on average 3.2:1 in vivo and protargol preparations; Namibian site (60) specimens only 200–300 × 70–120 μm (see occurrence). Outline elongate elliptical, anterior portion curved leftwards during foraging, providing cells with a reniform appearance. Body flattened about 2:1 dorsoventrally, very flexible and contractile by about 10% under mild cover glass pressure (Fig. 119a–e, Table 25). Macronuclear nodules scattered throughout cell, small but numerous and thus difficult to count; type specimen with about 356 nodules, indicating that the range is somewhere between 250 and 400; individual nodules about 7–10 × 3–4 μm in vivo, usually ellipsoidal to elongate ellipsoidal, but also globular, reniform, or dumb-bell-shaped, each with some small nucleoli (Fig. 119j, k, 120f). Micronuclei difficult to recognise in vivo and in protargol preparations, likely several scattered throughout cell, 3–4 μm across after protargol impregnation. Contractile vacuole at left body margin distinctly above midbody at level of buccal vertex, during diastole with two conspicuous collecting canals. Cells yellow at low (×100) magnification due to conspicuous cortical granules in short series mainly between cirral and bristle rows; individual granules 1.3–1.5 × 0.8–1.0 μm in size, that is, ellipsoidal; brilliant citrine, do not impregnate with the protargol method used (Fig. 119f, g, 120c–e). Cytoplasm colourless, contains fat globules 1–3 μm across, 10–20 μm-sized food vacuoles with compact content, and irregular egestion vacuoles up to 50 μm across. Moves quickly on microscope slide and soil particles showing great flexibility.

Adoral zone conspicuous because occupying 30–42%, on average 35% of body length, of usual shape and structure; composed of an average of 54 membranelles, bases of largest membranelles about 20 μm wide in vivo (Fig. 119a, h, 120f, g). Buccal cavity deep and large, at right partially covered by slightly curved, hyaline lip bearing paroral in inconspicuous furrow and covering proximal end of adoral zone of membranelles.

← **Fig. 119a–g** *Bakuella granulifera* (from Foissner et al. 2002. Type population [a–d, f, g] and Namibian site [60] specimens [e] from life). **a:** Ventral view of a representative specimen, 300 μm. Only a small portion of the many macronuclear nodules is shown. **b:** Outline of a slender specimen. **c:** Foraging specimens curve anterior body portion leftwards. **d:** Specimen with large food vacuole vaulting left body margin. **e:** Right lateral view (left figure) and dorsal views, redrawn from video records. **f, g:** The citrine, 1.3–1.5 × 0.8–1.0 μm-sized cortical granules are arranged in short, longitudinal series between the cirral and dorsal bristle rows and make cells yellow at low magnification. The granules are the most important difference to *Bakuella edaphoni*, which has a similar body size and cirral pattern. CV = contractile vacuole, FT = frontoterminal cirri, FV = food vacuole. Page 569.

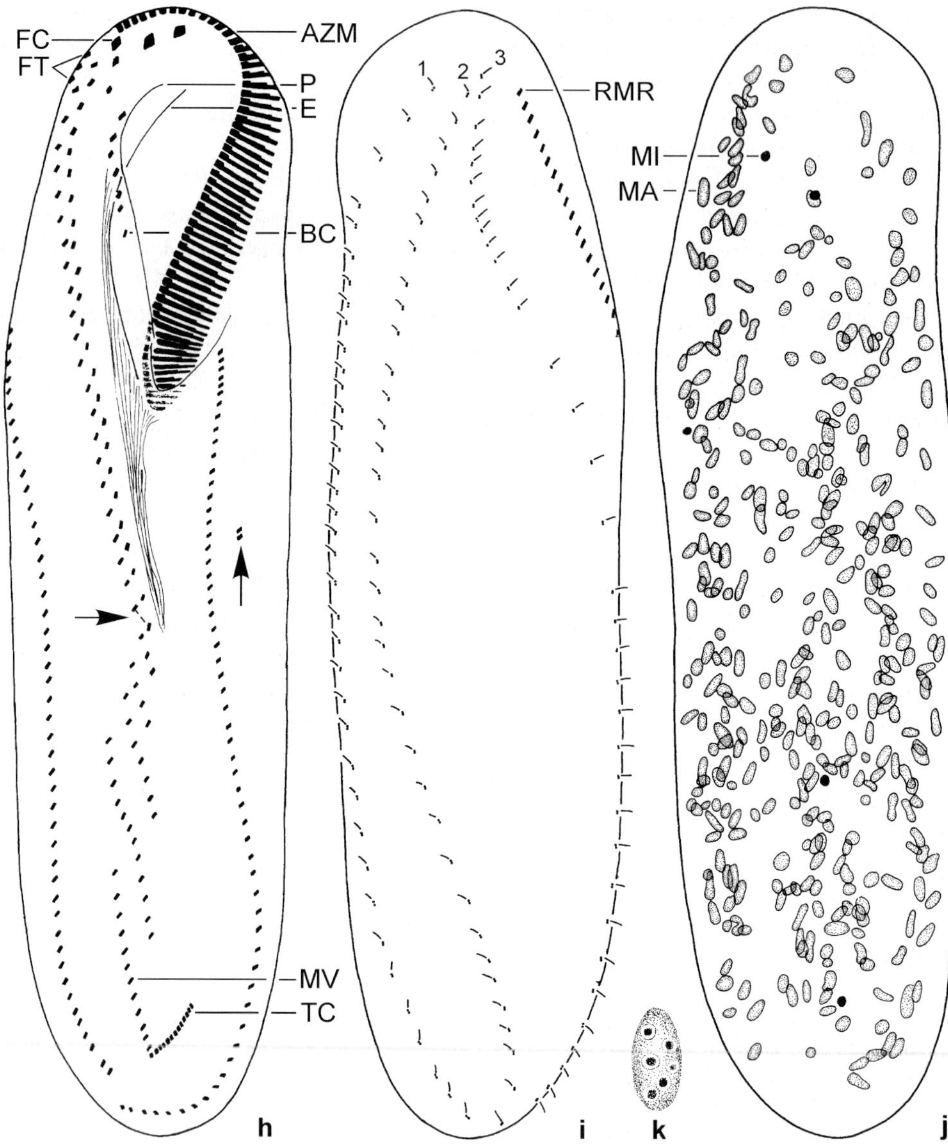

Fig. 119h–k *Bakuella granulifera* (from Foissner et al. 2002. Protargol impregnation). Infraciliature of ventral and dorsal side and nuclear apparatus of holotype specimen, 360 µm. Vertical arrow in (h) marks two supernumerary left marginal cirri, transverse arrow denotes the end of the cirral pair portion of the midventral complex. Figure (k) shows a macronuclear nodule at higher magnification. AZM = adoral zone of membranelles, BC = posteriormost buccal cirrus, E = endoral, FC = right frontal cirrus, FT = frontoterminal cirri, MA = macronuclear nodule, MI = micronucleus, MV = rightmost (rearmost) midventral row, P = paroral, RMR = right marginal row, TC = transverse cirri, 1–3 = dorsal kineties. Page 569.

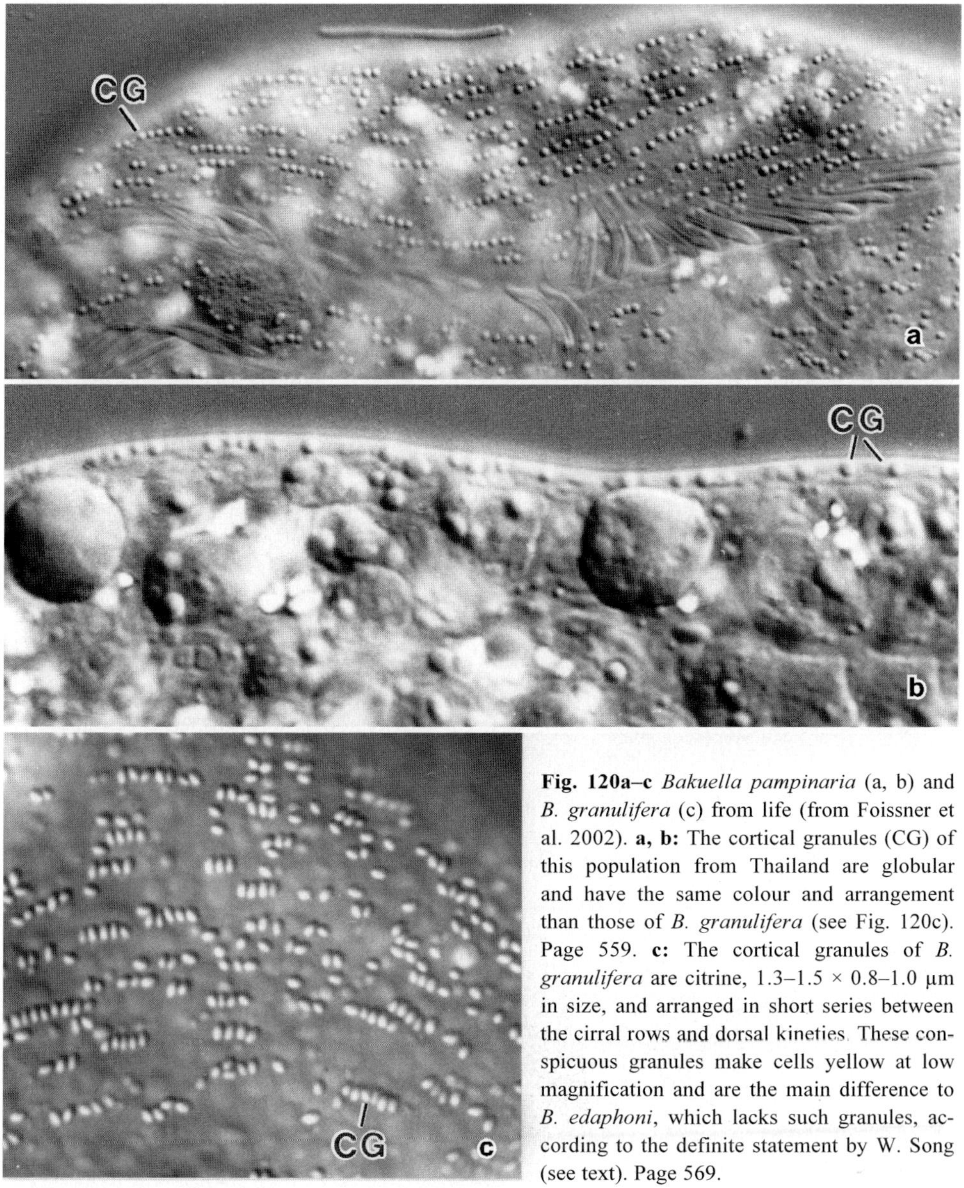

Fig. 120a–c *Bakuella pampinaria* (a, b) and *B. granulifera* (c) from life (from Foissner et al. 2002). **a, b:** The cortical granules (CG) of this population from Thailand are globular and have the same colour and arrangement than those of *B. granulifera* (see Fig. 120c). Page 559. **c:** The cortical granules of *B. granulifera* are citrine, 1.3–1.5 × 0.8–1.0 µm in size, and arranged in short series between the cirral rows and dorsal kineties. These conspicuous granules make cells yellow at low magnification and are the main difference to *B. edaphoni*, which lacks such granules, according to the definite statement by W. Song (see text). Page 569.

Paroral distinctly curved anteriorly, likely composed of zigzagging basal bodies having about 15 µm long cilia, optically intersects with bow-shaped endoral ahead of or near mid of buccal cavity; cilia of endoral anteriorly about 20 µm, posteriorly about 40 µm (!) long extending deeply into the pharynx. Pharyngeal fibres conspicuous in vivo and in protargol preparations, of ordinary length and structure, extend almost longitudinally to posterior half of cell.

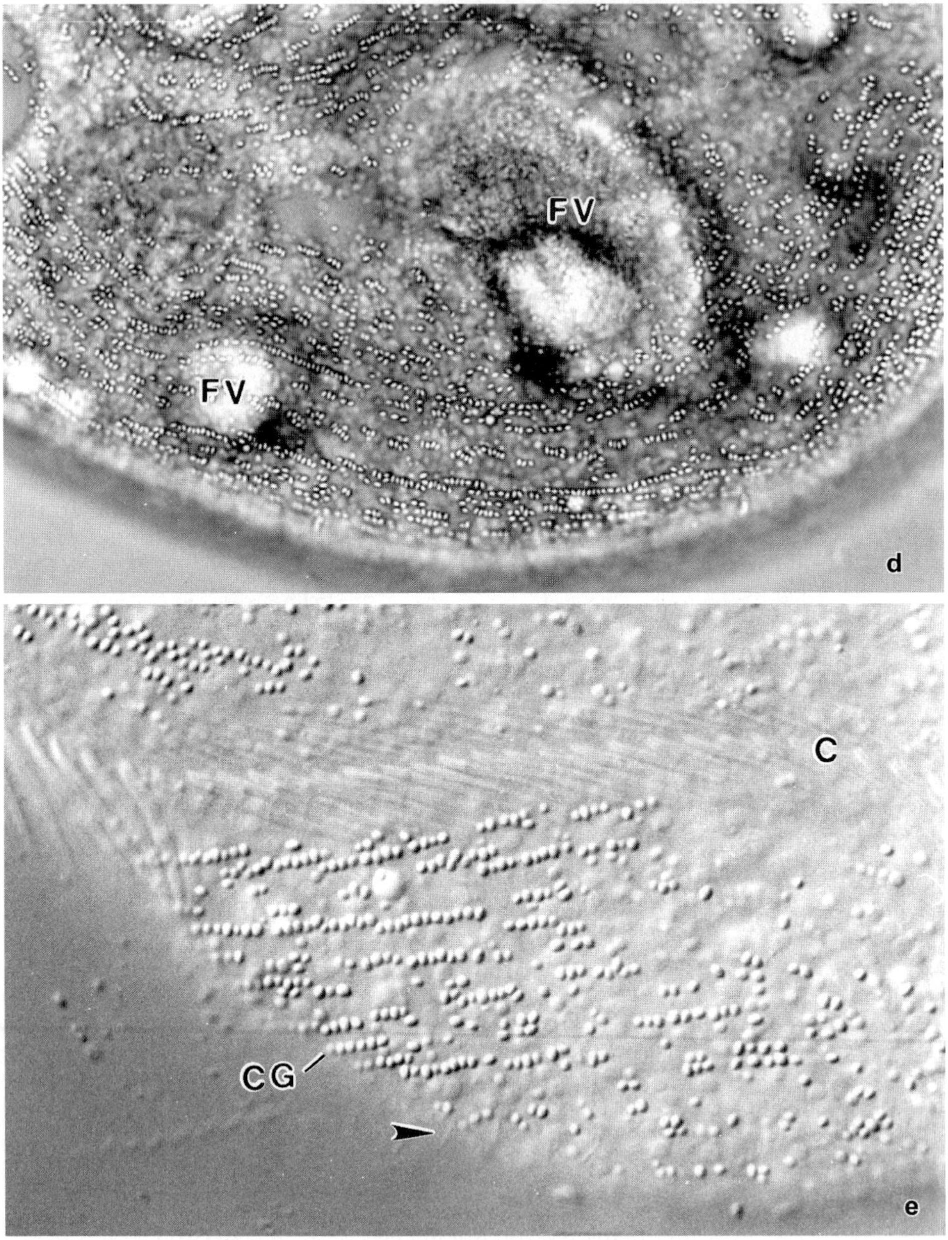

Fig. 120d, e *Bakuella granulifera* (from Foissner et al. 2002. d, bright field illumination; e, interference contrast). **d:** This up to 400 µm long ciliate is a voracious predator, ingesting testate amoebae, small and large ciliates, and even rotifers, which are digested in large food vacuoles often vaulting body margin (Fig. 119d). **e:** The cortical granules are ellipsoidal and arranged in short series. Arrowhead marks dorsal bristle row. C = cirral row, CG = cortical granules, FV = food vacuole. Page 569.

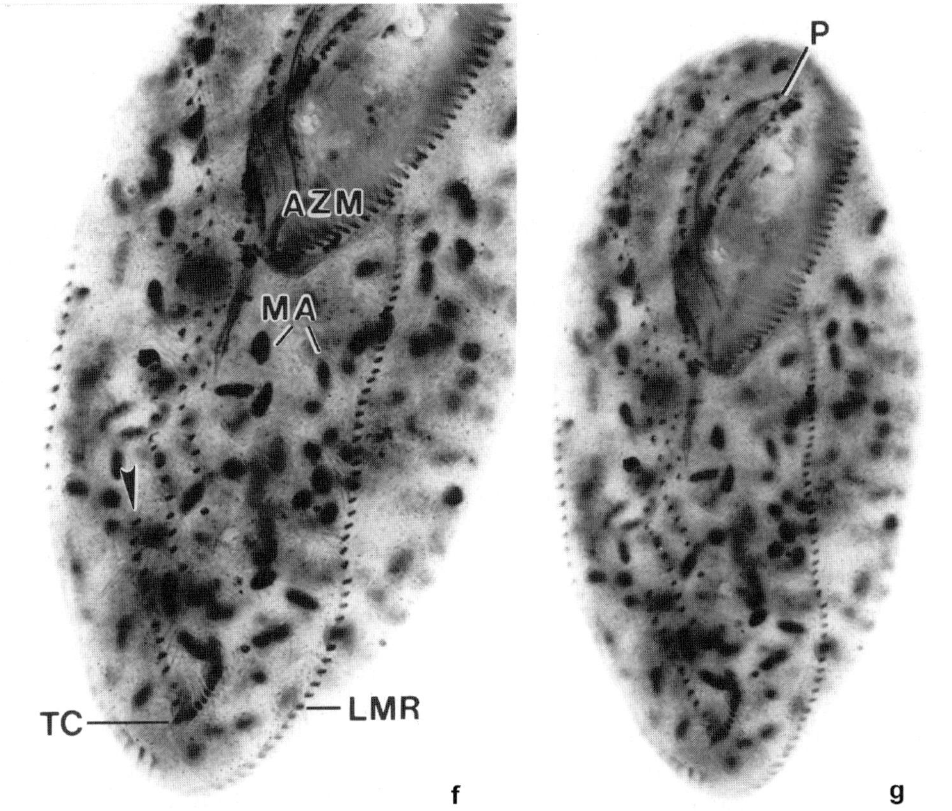

Fig. 120f, g *Bakuella granulifera* (from Foissner et al. 2002. Protargol impregnation). Infraciliature of ventral side. Arrowhead in (f) denotes anterior end of rearmost (rightmost) midventral row. AZM = adoral zone of membranelles, LMR = left marginal row, MA = macronuclear nodules, P = paroral, TC = transverse cirri. Page 569.

Cirral pattern and number of cirri of usual variability (Fig. 119a, h, 120f, g, Table 25). Frontal cirri conspicuously enlarged and about 22 μm long in life, right one, as is usual, at distal end of adoral zone of membranelles; frequently two cirri behind right frontal cirrus. Buccal cirral row right of anterior half of paroral, cirri become slightly smaller from anterior to posterior. Frontoterminal cirral row short, right of anteriormost midventral pair. Midventral complex composed of an anterior portion of about 18 cirral pairs terminating near mid-body and a posterior portion of slightly oblique, short and long midventral rows; rightmost long midventral row terminates at right transverse cirrus, as in some congeners. Transverse cirri about 18 μm long in life, small and closely spaced in oblique row, do not, or only slightly project beyond body end. Marginal, frontoterminal, and midventral cirri 15–17 μm long in life. Right marginal row extends onto dorsolateral surface anteriorly, ends more subterminally than left which is J-shaped, curving along rear body margin.

Dorsal bristles about 4 µm long in life, arranged in, likely invariably, three bipolar rows leaving blank broad, fusiform stripe in midline; rows 1 and 2 extend near left body margin, row 3 runs near right (Fig. 119i). Caudal cirri lacking.

Occurrence and ecology: Type locality is the Bukaos River bank (about S25°40' E18°10'), circa 80 km north of the town of Ketmanshoop (Namibia), where it occurred in sieved litter with moderate abundance. Foissner et al. (2002) found it also in a saline soil from the *Sporobolus* zone around the Etosha Pan, indicating that it is euryhaline (for details see sites 4 [type locality] and 60 in Foissner et al. 2002). Later we found it in an Oak-Hornbeam stand in Kolmberg, Austria (Foissner et al. 2005).

Bakuella granulifera is a voracious predator ingesting fungal spores, diatoms, testate amoebae (*Euglypha* sp., *Trinema lineare*), ciliates (*Colpoda* sp., *Caudiholosticha notabilis*, *Urosoma karinae*), and even rotifers up to 80 µm long (Fig. 119a, d, 120d).

Bakuella (Pseudobakuella) Alekperov, 1992 stat. nov.

1992 ***Pseudobakuella* Alekperov, gen. n.** – Alekperov, Zool. Zh., 71: 8 (original description). Type species (by original designation on p. 8): *Bakuella salinarum* Mihailowitsch & Wilbert, 1990.

2001 ***Pseudobakuella* Alekperov 1992** – Aescht, Denisia, 1: 135 (catalogue of generic names of ciliates).

2001 ***Pseudobakuella* Alekperov, 1992** – Berger, Catalogue of ciliate names 1. Hypotrichs, p. 76 (nomenclator containing all basionyms, combinations, and higher taxa of hypotrichs).

Nomenclature: *Pseudobakuella* is a composite of *pseudo-* (Greek; wrong) and the genus-group name *Bakuella* (for derivation see genus section), likely indicating that the species included are reminiscent or related to *Bakuella* species. The type species for *Bakuella (Pseudobakuella)* is the same as for *Pseudobakuella*, that is, *Bakuella salinarum* Mihailowitsch & Wilbert, 1990.

Characterisation (Fig. 111a, autapomorphy 9): *Bakuella* with 2 frontoterminal cirri (A?).

Remarks: If Fig. 111a reflects reality then we have to interpret the two frontoterminal cirri as atavism. See also same chapter at genus section.

Species included in *Bakuella (Pseudobakuella)* (alphabetically arranged according to basionyms): (1) *Bakuella salinarum* Mihailowitsch & Wilbert, 1990; (2) *Bakuella walibonensis* Song, Wilbert & Berger, 1992.

Bakuella salinarum Mihailowitsch & Wilbert, 1990 (Fig. 121a–l, Table 25)

1989 ***Bakuella salinarum* n. spec.**[1] – Mihailowitsch, Dissertation, p. 62, Abb. 6a–j, Tabelle 13 (Fig. 121a–l; unpublished thesis).

[1] The dissertation by Mihailowitsch (1989) is not a valid publication in the sense of the Code (ICZN 1985, Article 9 (11)). However, to complete the picture I mention this name in the present revision. But to avoid nomenclatural problems, this binomen is disclaimed for nomenclatural purposes (ICZN 1999, Article 8.3).

anterior end of right marginal row. Midventral complex composed of 26 pairs and 16 rows on average; cirral pair portion terminates at 53% of body length in specimen illustrated, rearmost midventral row ends close to right transverse cirri. On average 10 rather small and subterminally arranged transverse cirri. Right marginal row commences about at level of distal end of adoral zone, terminates in mid-body and therefore slightly overlapping with left row which commences somewhat ahead of level of proximal end of adoral zone. Dorsal kineties bipolar; length of dorsal bristles not mentioned, indicating that they are short (2–4 µm). Caudal cirri lacking.

Cell division: Morphogenesis of cell division is described by Mihailowitsch & Wilbert (1990; Fig. 121c–l). Song et al. (1992) discussed two misinterpretations by Mihailowitsch & Wilbert (1990), namely, Fig. 121e does not show a divider, but a reorganiser, and the parental adoral zone does not completely reorganise, but only in the proximal portion (Fig. 121f, g) as in *B. edaphoni* and *B. pampinaria*. However, according to Figs. 121h, k it could be indeed possible that the whole parental adoral zone is reorganised.

The latest stages show that *B. salinarum* forms only two frontoterminal cirri and that it has only the ordinary three frontal cirri (Fig. 121h, k); the other two "frontal cirri" described by Mihailowitsch & Wilbert (1990) originate from the same anlage (III) as the right frontal cirrus and therefore have to be designated as parabuccal cirri.

The macronuclear nodules fuse to a single mass during division (Fig. 121i). The marginal rows and dorsal kineties divide by intrakinetal proliferation at two levels.

Occurrence and ecology: Type locality of *Bakuella salinarum* is a salt-loaded ditch in the village of Bad Waldliesborn, near the town Lippstadt, Germany (Mihailowitsch & Wilbert 1990). No further records published.

Feeds on bacteria. Mihailowitsch & Wilbert (1990) give the following autecological data (n = 28): 5.3–16.3° C; pH 6.99–7.67; 19.1–187.0 mg l^{-1} CO_2; 2.2–9.6 mg l^{-1} O_2; 0.13–7.32 mg l^{-1} NH_4^+-N; 0.02–0.40 mg l^{-1} NO_2^--N; 0.63–10.6 mg l^{-1} NO_3^--N; 300–12763 mg l^{-1} Cl^-; 216–3590 mS m^{-1} spec. conductivity.

Bakuella walibonensis Song, Wilbert & Berger, 1992 (Fig. 122a, Table 25, Addenda)

1989 ***Bakuella* spec. 1** – Mihailowitsch, Dissertation, p. 76, Abb. 7, Tabelle 14 (Fig. 122a; unpublished thesis).

1990 ***Bakuella* spec. 1** – Mihailowitsch & Wilbert, Arch. Protistenk., 138: 213, Abb. 11, Tabelle 3 (Fig. 122a; description; slides are likely deposited in the Institut für landwirtschaftliche Zoologie und Bienenkunde, University of Bonn, Germany).

1992 ***Bakuella walibonensis* nov. spec.**[1] – Song, Wilbert & Berger, Bull. Br. Mus. nat. Hist. (Zool.), 58: 143, Fig. 51, Table 1 (Fig. 122a; original description based on previous entry).

1992 ***Pseudobakuella gracilis* Alekperov, sp. n.** – Alekperov, Zool. Zh., 71: 8, Fig. 11 (Fig. 122a; original description based on *Bakuella* spec. 1 sensu Mihailowitsch & Wilbert; see nomenclature).

1994 ***Bakuella walibonensis* Song, Wilbert and Berger, 1992** – Eigner, Europ. J. Protistol., 30: 474 (brief revision of the Bakuellinae; see remarks).

[1] The diagnosis by Song et al. (1992) is as follows: After protargol impregnation about 180–230 × 60–80 µm. More than 100 macronuclear segments. 6 buccal cirri, 5 transverse cirri, 14 pairs of midventral cirri, and 39 adoral membranelles on average. 2–5 ventral rows and consistently 2 frontoterminal cirri.

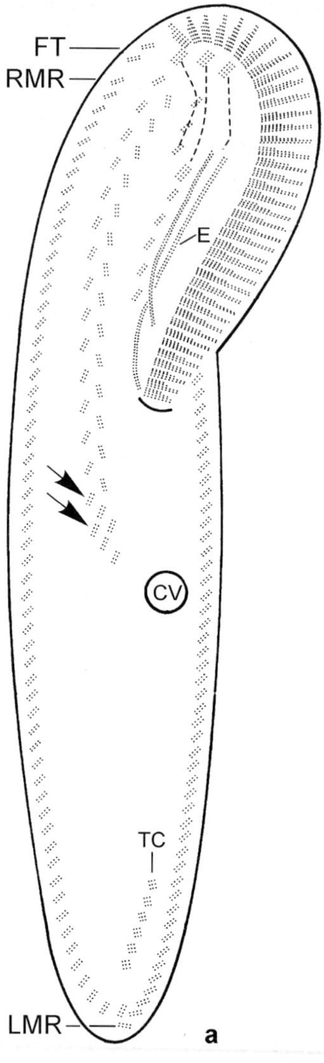

Fig. 122a *Bakuella walibonensis* (from Mihailowitsch 1989. Protargol impregnation). Infraciliature of ventral side, 205 µm (size estimated from Mihailowitsch 1989; size according to bar in Mihailowitsch & Wilbert 1990 [right of their Abb. 12] only 165 µm, which is likely incorrect; see Table 25). Broken lines connect cirri which (very likely) originate from anlagen I–III. Arrows mark midventral rows. CV = contractile vacuole, E = endoral, FT = frontoterminal cirri, RMR = front end of right marginal row, TC = transverse cirri. Page 581.

1996 ***Bakuella walibonensis* Song, Wilbert and Berger, 1992** – Franco, Esteban & Téllez, Acta Protozool., 35: 329, Fig. 17 (Fig. 122a; key to the species of the Bakuellinae).

2001 ***Bakuella walibonensis* Song, Wilbert and Berger, 1992** – Berger, Catalogue of ciliate names 1. Hypotrichs, p. 12 (nomenclator containing all basionyms, combinations, and higher taxa of hypotrichs).

Nomenclature: No derivation of the names are given in the original descriptions. The species-group name *walibonensis* refers to the site (Bad Waldliesborn) where the species was discovered. The species-group name *grácil·is -is -e* (Latin adjective; thin, slender; delicate) possibly alludes to the general appearance of this species.

Bakuella walibonensis and *Pseudobakuella gracilis* are objective synonyms because both species are based on *Bakuella* spec. 1 described by Mihailowitsch (1989) and Mihailowitsch & Wilbert (1990). Unfortunately, both species were established in late 1992 so that it is rather difficult to find out which of the two descriptions is the oldest and therefore valid one. The description of *Bakuella walibonensis* was issued on 26.11.1992 (printed on publication) whereas *P. gracilis* was published in volume 71, Number 12 ("December" number). I contacted Nauka, the publisher of the Zoologicheskii Zhurnal twice, to inform me about the exact date of publication. Unfortunately I did not get an answer and therefore I preliminarily assume that the Bulletin of the British Museum was published earlier (November) than Number 12 (number for December) of the Zoologicheskii Zhurnal. Consequently, *Bakuella walibonensis* is the valid name for the present species. However, there is no doubt that the species name *Pseudobakuella gracilis* becomes valid if its original description was issued before 26.11.1992. However, in that case it has to be transferred to *Bakuella* if the present classification is accepted.

If the subgenus-group name *Pseudobakuella* is included, then the correct name is *Bakuella (Pseudobakuella) walibonensis* Song, Wilbert & Berger, 1992.

Remarks: I did not translate Alekperov's (1992) paper, but since the description is based on the same data (Mihailowitsch & Wilbert 1990) as that of *B. walibonensis*, there cannot be significant differences.

Table 25 Morphometric data on *Bakuella agamalievi* (ag1, from Song et al. 2002; ag2, from Mihailowitsch & Wilbert 1990), *Bakuella crenata* (cr1, from Agamaliev & Alekperov 1976; cr2, synonym *B. polycirrata* from Alekperov 1988), *Bakuella edaphoni* (eda, from Song et al. 1992), *Bakuella granulifera* (gra, from Foissner et al. 2002), *Bakuella marina* (ma1, type population from Agamaliev & Alekperov 1976; ma2, Wilbert 1986 population from Song et al. 1992; ma3, type population of the synonym *B. imbricata* from Alekperov 1982; ma4, from Song et al. 1992), *Bakuella pampinaria oligocirrata* (oli, from Foissner 2004), *Bakuella pampinaria pampinaria* (pam, from Eigner & Foissner 1992), *Bakuella salinarum* (sal, from Mihailowitsch & Wilbert 1990), *Bakuella walibonensis* (wal, from Mihailowitsch & Wilbert 1990), *Paraurostyla pulchra* (pul, from Buitkamp 1977a)

Characteristics [a]	species	mean	M	SD	SE	CV	Min	Max	n
Body, length	ag1	115.7	–	11.4	3.8	9.8	97.0	131.0	19
	ag2	155.8	149.6	16.4	7.3	10.5	136.0	176.0	5
	cr1	–	–	–	–	–	120.0	150.0	?
	cr2	–	–	–	–	–	50.0	70.0	?
	eda	219.6	–	41.5	–	18.9	153.0	283.0	16
	gra	319.0	320.0	33.1	7.6	10.4	274.0	396.0	19
	ma1	–	–	–	–	–	120.0	140.0	?
	ma2	266.7	–	27.6	–	–	230.0	310.0	25
	ma3	–	–	–	–	–	110.0	130.0	?
	ma4	101.4	–	5.7	–	–	91.0	108.0	18
	oli	98.2	95.0	15.7	3.6	16.0	68.0	127.0	19
	pam	113.2	114.0	14.3	–	12.6	84.0	141.0	25
	pul	–	–	–	–	–	150.0	170.0	?
	sal	307.4	319.0	26.6	8.4	8.6	272.0	348.0	10
	wal	202.0	200.0	16.3	5.1	8.0	180.0	229.0	10
Body, width	ag1	58.8	–	7.1	2.3	12.1	48.0	70.0	19
	ag2	46.6	48.4	4.0	1.6	8.6	40.0	51.0	6
	eda	69.4	–	8.1	–	11.7	59.0	82.0	16
	gra	101.2	100.0	14.1	3.2	13.9	77.0	144.0	19
	ma2	85.0	–	14.8	–	–	70.0	110.0	11
	ma4	61.6	–	6.3	–	–	45.0	70.0	18
	oli	31.8	31.0	4.9	1.1	15.3	25.0	43.0	19
	pam	36.1	36.0	5.4	–	15.0	28.0	48.0	25
	sal	99.1	92.8	18.6	5.9	18.8	87.0	145.0	10
	wal	72.3	70.4	8.5	3.2	11.7	62.0	83.0	7
Body length:width, ratio	gra	3.2	3.2	0.5	0.1	15.3	2.3	4.0	19
	oli	3.1	3.0	0.5	0.1	15.5	2.5	4.3	19
AE to proximal end of adoral zone, distance	ag1	43.8	–	5.6	1.7	12.8	37.0	52.0	19
	eda	62.8	–	10.9	–	17.4	47.0	82.0	16
	gra	113.5	113.0	13.0	3.0	11.5	88.0	140.0	19
	oli	33.6	34.0	4.6	1.1	13.6	26.0	45.0	19
	pam	41.1	42.0	5.5	–	13.4	30.0	57.0	25
Body length:length of adoral zone, ratio	gra	2.8	2.8	0.2	0.1	8.3	2.3	3.3	19
	oli	2.9	2.9	0.4	0.1	12.6	2.4	3.7	19
Anterior cell end to paroral, distance	gra	19.9	20.0	3.1	0.9	15.5	14.0	25.0	13
Paroral, length	gra	76.4	76.0	10.2	2.8	13.4	60.0	93.0	13
Anterior cell end to endoral, distance	gra	20.7	21.5	3.5	1.0	17.1	16.0	27.0	12
Endoral, length	gra	86.2	85.0	7.5	2.2	8.7	76.0	102.0	12
AE to first buccal cirrus, distance	gra	30.6	32.0	4.9	1.5	16.1	20.0	37.0	11
AE to last buccal cirrus, distance	gra	75.6	78.0	7.8	2.3	10.3	64.0	86.0	11
AE to first frontoterminal cirrus, distance	gra	16.7	17.0	4.1	1.4	24.6	10.0	24.0	9
AE to last frontoterminal cirrus, distance	gra	29.3	28.0	6.7	2.2	22.9	22.0	44.0	9
AE to rearmost midventral pair cirrus, distance	oli	25.0	24.0	6.1	1.4	24.4	16.0	38.0	19

Table 25 Continued

Characteristics[a]	species	mean	M	SD	SE	CV	Min	Max	n
PE to posteriormost transverse cirrus, distance	gra	28.6	28.0	8.0	1.8	27.9	14.0	48.0	19
	oli	6.3	6.0	1.6	0.4	25.3	4.0	10.0	19
PE to right marginal row, distance	gra	15.8	12.5	9.6	2.3	60.9	4.0	35.0	18
Anteriormost macronuclear nodule, length	ag1[d]	–	–	–	–	–	5.0	7.0	?
	gra	7.4	7.0	1.6	0.6	21.7	6.0	10.0	8
	oli[d]	4.6	4.0	1.4	0.3	30.0	3.0	8.0	19
	pam[d]	5.5	6.2	1.6	–	0.3	2.5	7.4	25
Anteriormost macronuclear nodule, width	ag1[d]	0.0	–	–	–	–	3.0	5.0	?
	gra	4.0	4.0	0.5	0.2	13.4	3.0	5.0	8
	oli[d]	2.8	3.0	0.5	0.1	18.6	2.0	4.0	19
	pam[d]	2.3	2.5	0.4	–	0.2	1.2	2.5	25
Macronuclear nodules, number	ag1	–	–	–	–	–	47.0	60.0	4
	ag2			more than 100					
	cr1	2.0	–	–	–	–	–	–	?
	cr2	2.0	–	–	–	–	–	–	?
	gra			about 300–400					
	oli	91.0	90.0	20.3	4.7	22.3	58.0	125.0	19
	sal			more than 100					
	wal			more than 100					
Anteriormost micronucleus, length	gra	3.3	3.0	–	–	–	3.0	4.0	7
	oli	2.7	3.0	–	–	–	2.0	3.0	19
	pam	2.4	2.5	0.2	–	0.1	2.0	2.5	25
Anteriormost micronucleus, width	gra	3.0	3.0	0.0	0.0	0.0	3.0	3.0	7
	oli	2.7	3.0	–	–	–	2.0	3.0	19
Micronuclei, number	eda	4.6	–	2.0	–	43.0	2.0	7.0	16
	cr1	2.0	–	–	–	–	–	–	?
	oli	5.8	6.0	1.3	0.3	22.3	4.0	8.0	19
Adoral membranelles, number	ag1	30.9	–	4.1	1.6	13.4	26.0	37.0	7
	ag2	33.7	34.0	2.3	0.9	6.8	30.0	37.0	7
	cr1	0.0	–	–	–	–	28.0	30.0	?
	eda	38.9	–	2.8	–	7.3	34.0	45.0	16
	gra	54.6	54.0	4.5	1.0	8.2	44.0	62.0	19
	ma1	–	–	–	–	–	32.0	36.0	?
	ma2	40.4	–	4.2	–	–	34.0	51.0	25
	ma3	–	–	–	–	–	38.0	40.0	?
	ma4	32.4	–	2.4	–	–	28.0	36.0	12
	oli	27.3	27.0	2.9	0.7	10.6	22.0	34.0	19
	pam	31.0	31.0	3.6	–	11.6	22.0	39.0	25
	pul	24.0	–	–	–	–	–	–	?
	sal	55.4	55.5	5.1	1.6	9.2	47.0	63.0	10
	wal	39.2	38.0	4.1	1.0	10.5	33.0	47.0	17
Frontal cirri, number	ag1[e]	4.0	–	0.0	0.0	0.0	4.0	4.0	14
	ag2	3.0	3.0	0.0	0.0	0.0	3.0	3.0	7
	cr1	3.0	–	–	–	–	–	–	?
	cr2[h]	3.0	–	–	–	–	–	–	1
	gra	3.0	3.0	0.0	0.0	0.0	3.0	3.0	19
	oli	3.0	3.0	0.0	0.0	0.0	3.0	3.0	19
	pul	3.0	–	–	–	–	–	–	?
	sal[e]	5.0	5.0	0.0	0.0	0.0	5.0	5.0	10
	wal	3.0	3.0	0.0	0.0	0.0	3.0	3.0	17

Table 25 Continued

Characteristics [a]	species	mean	M	SD	SE	CV	Min	Max	n
Cirri behind right frontal cirrus, number	gra	2.0	2.0	0.0	0.0	0.0	2.0	2.0	7
	oli	1.1	1.0	–	–	–	1.0	2.0	19
	pam	2.6	2.0	0.6	–	25.4	2.0	4.0	25
Buccal cirri, number	ag1	1.0	–	0.0	0.0	0.0	1.0	1.0	14
	ag2	1.0	1.0	0.0	0.0	0.0	1.0	1.0	7
	cr1 [h]	3.0	–	–	–	–	–	–	1
	cr2 [h]	4.0	–	–	–	–	–	–	1
	eda	7.4	–	1.3	–	17.7	5.0	9.0	24
	gra	8.2	8.0	1.1	0.3	13.2	7.0	10.0	11
	ma1	–	–	–	–	–	3.0	4.0	?
	ma2	3.1	–	0.9	–	–	2.0	5.0	34
	ma3	–	–	–	–	–	4.0	5.0	?
	ma4	2.2	–	0.5	–	–	2.0	4.0	9
	oli	3.8	4.0	0.8	0.2	20.8	2.0	5.0	19
	pam	4.9	5.0	0.8	–	16.0	3.0	6.4	25
	pul	4.0	–	–	–	–	–	–	?
	sal	7.1	7.0	0.9	0.3	12.3	6.0	8.0	10
	wal	5.8	6.0	0.4	0.1	17.2	5.0	6.0	17
Frontoterminal cirri, number	ag1	5.6	–	0.9	0.3	15.9	4.0	7.0	10
	cr1 [h]	10.0	–	–	–	–	–	–	1
	cr2 [h]	10.0	–	–	–	–	–	–	1
	eda	3.1	–	0.8	–	24.3	2.0	5.0	22
	gra	3.6	3.5	0.7	0.2	19.4	3.0	5.0	10
	ma1	9.0	–	–	–	–	–	–	?
	ma2	7.6	–	2.0	–	–	5.0	11.0	27
	ma3	9.0	–	–	–	–	–	–	?
	ma4	6.7	–	1.0	–	–	5.0	9.0	14
	oli	4.5	5.0	1.3	0.3	29.2	1.0	7.0	19
	pam	5.8	6.0	1.0	–	17.2	5.0	8.0	25
	sal	2.0	2.0	0.0	0.0	0.0	2.0	2.0	10
	wal	2.0	2.0	0.0	0.0	0.0	2.0	2.0	17
Cirral pairs in midventral complex, number	ag1	10.9	–	1.5	0.5	13.4	9.0	13.0	7
	ag2	14.0	14.5	2.6	1.0	18.6	10.0	18.0	7
	cr1 [h]	5.0?	–	–	–	–	–	–	1
	cr2 [h]	8.0	–	–	–	–	–	–	1
	eda	9.5	–	2.3	–	24.1	5.0	14.0	24
	gra	18.3	19.0	4.0	1.3	21.8	12.0	23.0	9
	ma1	12.0	–	–	–	–	–	–	?
	ma2	9.8	–	4.7	–	–	4.0	12.0	22
	ma3	–	–	–	–	–	4.0	8.0	?
	ma4	4.4	–	0.3	–	–	4.0	5.0	18
	oli	4.4	4.0	1.0	0.2	21.9	3.0	7.0	19
	pam	9.2	9.0	1.6	–	17.7	6.0	13.0	25
	sal	26.0	26.0	7.9	3.0	30.4	22.0	38.0	7
	wal	13.8	14.0	0.4	0.1	3.0	13.0	14.0	15
Number of midventral rows with 3 cirri	eda	2.6	–	1.0	–	37.0	1.0	5.0	22
Number of midventral rows with >3 cirri	eda	4.5	–	0.7	–	16.6	3.0	6.0	22
	ag1	4.0	–	1.2	0.4	28.9	3.0	6.0	10
	ag2	3.6	3.0	1.0	0.4	27.8	3.0	5.0	7
	cr1 [h]	10	–	–	–	–	–	–	1
	cr2 [h]	12	–	–	–	–	–	–	1

Table 25 Continued

Characteristics [a]	species	mean	M	SD	SE	CV	Min	Max	n
Midventral rows, number [b]	eda	7.1	–	1.3	–	18.1	5.0	10.0	22
	gra	4.3	4.0	0.7	0.3	16.6	3.0	5.0	8
	ma1	10.0	–	–	–	–	–	–	?
	ma2	7.0	–	0.8	–	–	5.0	8.0	16
	ma3	–	–	–	–	–	5.0	7.0	?
	ma4	4.3	–	0.5	–	–	4.0	5.0	15
	oli	4.0	4.0	0.6	0.1	14.4	3.0	5.0	19
	pam	3.5	3.0	1.0	–	29.9	2.0	6.0	25
	sal	15.8	14.5	2.9	0.9	18.4	13.0	21.0	10
	wal	2.6	2.0	1.2	0.3	44.6	2.0	5.0	12
First midventral row [f], number of cirri	pam	9.1	9.0	2.2	–	24.0	4.0	14.0	25
Second midventral row, number of cirri	pam	9.6	9.0	3.2	–	33.0	4.0	17.0	25
Third midventral row, number of cirri	pam	9.6	9.0	2.3	–	24.2	5.0	16.0	21
Fourth midventral row, number of cirri	pam	7.1	7.0	1.2	–	25.5	5.0	11.0	11
Fifth midventral row, number of cirri	pam	8.7	9.0	3.4	–	38.9	4.0	12.0	4
Sixth midventral row, number of cirri	pam	13.0	–	–	–	–	–	–	1
Cirri in rightmost midventral row, number [c]	gra	11.4	11.0	1.2	0.4	10.6	9.0	14.0	11
Transverse cirri, number	ag1	5.2	–	1.1	0.4	21.8	4.0	7.0	10
	ag2	5.3	5.0	1.0	0.4	18.9	5.0	7.0	7
	cr1[h]	7.0	–	–	–	–	–	–	1
	cr2[h]	9.0	–	–	–	–	–	–	1
	eda	8.0	–	1.5		18.3	6.0	11.0	16
	gra	11.9	12.0	1.6	0.4	13.6	9.0	15.0	18
	ma1	10.0	–	–	–	–	–	–	?
	ma2	7.0	–	1.6	–	–	5.0	11.0	27
	ma3	7.0	–	–	–	–	–	–	?
	ma4	7.2	–	1.4	–	–	5.0	9.0	20
	oli	4.4	4.0	0.8	0.2	19.0	3.0	7.0	19
	pam	3.8	4.0	0.7	–	19.5	2.0	5.0	25
	pul	6.0	–	–	–	–	–	0	?
	sal	9.5	9.5	2.2	0.7	22.8	7.0	12.0	10
	wal	5.2	5.0	0.7	0.2	13.1	4.0	6.0	15
Left marginal cirri, number	ag1	32.9	–	3.7	1.4	11.2	30.0	38.0	7
	ag2	37.6	38.0	4.5	1.7	12.0	30.0	40.0	7
	cr1	–	–	–	–	–	38.0	40.0	?
	cr2	–	–	–	–	–	36.0	40.0	?
	eda	48.4	–	3.7	–	7.6	44.0	56.0	10
	gra	63.4	64.0	7.0	1.8	11.0	54.0	77.0	15
	ma1	–	–	–	–	–	52.0	56.0	?
	ma2	35.9	–	9.0	–	–	23.0	53.0	23
	ma3	–	–	–	–	–	32.0	35.0	?
	ma4	28.3	–	2.7	–	–	25.0	33.0	9
	oli	31.8	32.0	4.0	0.9	12.7	24.0	42.0	19
	pam	39.5	42.0	5.6	–	14.2	24.0	47.0	25
	pul	35	–	–	–	–	–	–	?
	sal	51.7	50.0	5.2	1.6	10.0	45.0	60.0	10
	wal[g]	51.5	51.0	4.5	1.3	8.8	42.0	58.0	13
Right marginal cirri, number	ag1	43.4	–	3.2	1.4	7.4	40.0	47.0	7
	ag2	41.0	43.0	5.2	2.1	12.7	34.0	46.0	6
	cr1	–	–	–	–	–	40.0	45.0	?
	cr2	–	–	–	–	–	46.0	50.0	?

Table 25 Continued

Characteristics [a]	species	mean	M	SD	SE	CV	Min	Max	n
Right marginal cirri, number	eda	50.6	–	4.1	–	8.1	43.0	55.0	10
	gra	63.2	63.0	5.3	1.5	8.4	54.0	76.0	13
	ma1	–	–	–	–	–	55.0	60.0	?
	ma2	44.4	–	7.6	–	–	34.0	63.0	23
	ma3	–	–	–	–	–	40.0	42.0	?
	ma4	40.7	–	5.6	–	–	34.0	54.0	9
	oli	33.5	34.0	4.3	1.0	12.7	25.0	42.0	19
	pam	39.7	41.0	6.3	–	15.9	23.0	51.0	25
	pul	32.0	–	–	–	–	–	–	?
	sal	62.2	61.5	5.7	1.8	9.2	55.0	66.0	10
	wal	60.5	64.0	6.3	1.9	10.4	51.0	66.0	11
Dorsal kineties, number	ag1	3.0	–	0.0	0.0	0.0	3.0	3.0	14
	ag2	3.0	3.0	0.0	0.0	0.0	3.0	3.0	7
	eda	3.0	–	0.0	–	0.0	3.0	3.0	24
	gra			likely invariably 3					
	ma2	3.0	–	0.0	–	–	3.0	3.0	12
	ma3	3.0	–	–	–	–	–	–	?
	ma4	3.0	–	0.0	–	–	3.0	3.0	25
	oli	3.0	3.0	0.0	0.0	0.0	3.0	3.0	19
	pam	3.0	3.0	0.0	–	0.0	3.0	3.0	25
	pul	3.0	–	–	–	–	–	–	?
	sal	3.0	3.0	0.0	0.0	0.0	3.0	3.0	10
	wal	3.0	3.0	0.0	0.0	0.0	3.0	3.0	17

[a] All data – except for gra, ma1, ma3, oli, and pam – are based on specimens impregnated after Wilbert's protargol protocol. Species gra, oli, and pam are impregnated after Foissner's protargol protocol; ma1 and ma3 are impregnated with the Chatton-Lwoff method. ? = sample size unknown; if only one value is known it is listed in the column mean, if two values are available they are listed as Min and Max. Measurements in µm. AE = anterior end of cell, CV = coefficient of variation in %, M = median, Max = maximum, mean = arithmetic mean, Min = minimum, n = number of individuals investigated, PE = posterior end of cell, SD = standard deviation, SE = standard error of arithmetic mean.

[b] Terminology, see Fig. 1a.

[c] MV in Fig. 119h.

[d] No certain nodule, respectively, micronucleus measured.

[e] Invariably one cirrus (ag1), respectively, two cirri (sal) behind right frontal cirrus included.

[f] This is the anteriormost midventral row.

[g] Mihailowitsch & Wilbert (1990) incorrectly wrote 50 as maximum value.

[h] From Fig. 116a, respectively, Fig. 116c.

According to Eigner (1994) the assignment of the present species to *Bakuella* is uncertain since development of midventral rows is not known. However, the illustration (Fig. 122a) and the morphometric characterisation (Table 25) clearly show that this population has midventral rows. In addition, the differences to other populations are distinct so that I am still certain that this is a valid *Bakuella* species. This decision is supported by the fact that Alekperov, who established *Bakuella*, had the same impression in that he also described it as new species, *Pseudobakuella gracilis* Alekperov, 1992. However, there is no doubt that the species should be reinvestigated, especially as concerns the presence or absence of cortical granules.

Mihailowitsch & Wilbert (1990) used almost the same body outline for *B. walibonensis* (Fig. 122a) and *B. agamalievi* (Fig. 115l). Consequently, the cirral pattern is certainly somewhat schematic and details in the location of some cirri must not be over-interpreted.

Bakuella walibonensis differs from *B. salinarum*, which also has only two frontoterminal cirri, by the shorter midventral complex (terminates near mid-body vs. near transverse cirri). *Bakuella agamalievi* has only one buccal cirrus (vs. several), but more than two frontoterminal cirri.

Morphology: Body size 180–230 × 60–80 µm after protargol impregnation (Wilbert's method), indicating that live specimens are around 200 µm long. Almost no life data available. Contractile vacuole about in mid-body near left cell margin. Adoral zone occupies 38% of body length in specimen illustrated, composed of 39 membranelles on average (Fig. 122a, Table 25). Undulating membranes long, anterior portions straight, posterior curved and optically intersecting; paroral shorter than endoral. Three enlarged frontal cirri. Buccal cirral row commences slightly behind anterior end of paroral, front buccal cirrus slightly enlarged (Fig. 122a). Three (parabuccal) cirri behind frontal cirri (in contrast to *B. salinarum*, these cirri are not included in morphometric characterisation by Mihailowitsch & Wilbert). Two frontoterminal cirri indistinctly set off from anterior end of right marginal row. Midventral complex composed of about 14 midventral pairs and usually only two midventral rows each composed of about three cirri only; midventral complex terminates at 54% of body length in specimen illustrated (Fig. 122a). Transverse cirri not enlarged, arranged in very oblique row slightly subterminally (obviously more anteriorly than in *B. agamalievi*; Fig. 115l). Marginal rows slightly overlapping posteriorly. Three (bipolar?) dorsal kineties; length of bristles not mentioned, indicating that they are short (2–4 µm). Caudal cirri lacking.

Occurrence and ecology: Type locality of *B. walibonensis* is a salt-loaded ditch (sample site P2 in Mihailowitsch 1989; about 51°43′N 8°20′E) in Bad Waldliesborn, near the town of Lippstadt, Germany. At the same site, *B. agamalievi* (Fig. 115l) was found. No further records published. Feeds on bacteria (Mihailowitsch 1989).

Incertae sedis in *Bakuella*

Paraurostyla pulchra Buitkamp, 1977
(Fig. 123a, Table 25)

1977 ***Paraurostyla pulchra* n. spec.** – Buitkamp, Decheniana, 130: 119, Abb. 2 (Fig. 123a; original description; no formal diagnosis provided; site were type slides deposited not mentioned, possibly in the Institut für landwirtschaftliche Zoologie und Bienenkunde, University of Bonn, Germany).

1979 ***Bakuella pulchra* comb. n.** – Jankowski, Trudy zool. Inst., Leningr., 86: 83 (combination with *Bakuella*).

1999 ***Paraurostyla pulchra* Buitkamp, 1977** – Berger, Monographiae biol., 78: 843 (brief comment).

2001 ***Paraurostyla pulchra* Buitkamp, 1977** – Berger, Catalogue of ciliate names 1. Hypotrichs, p. 70 (nomenclator containing all basionyms, combinations, and higher taxa of hypotrichs).

Nomenclature: No derivation of the name is given in the original description. The species-group name *pulch·er -ra -um* (Latin adjective; beautiful) obviously alludes to the general appearance of this species.

Eigner & Foissner (1992, p. 466) incorrectly assumed that Song et al. (1992) have transferred this species to *Bakuella* in that they write "*B. pulchra* (Buitkamp) Song, Wilbert and Berger". However, this was done by Jankowski as already indicated in the list of synonyms by Song et al. (1992).

Remarks: Buitkamp (1977a) assigned this terrestrial species to *Paraurostyla* because of the presence of frontal and transverse cirri and several ventral rows. Jankowski (1979) transferred it to *Bakuella*. Song et al. (1992) did not accept this generic assignment because no zigzagging midventral pattern is recognisable (Fig. 123a). However, they correctly stated that some other features, such as the absence of caudal cirri, the 3 dorsal kineties, the large number of macronuclear nodules, the short cirral row near the anterior end of the right marginal row, and multiple buccal cirri remind one of members of *Bakuella*. In addition, they mentioned that only a detailed redescription, including an investigation of the morphogenesis, will definitely elucidate the correct systematic position of this species. Eigner & Foissner (1992) agreed with Song et al. (1992). Berger (1999) excluded it from *Paraurostyla* because fragmentation of dorsal kineties is obviously lacking in *P. pulchra*.

Although I do not have new data I include *P. pulchra* as incertae sedis in *Bakuella* because the mentioned agreements with other *Bakuella* species are indeed very conspicuous. Moreover, it could be possibly that Buitkamp (1977a) misinterpreted

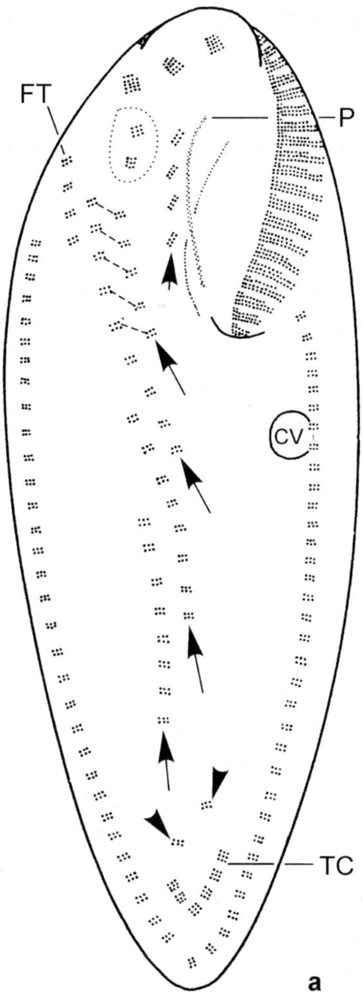

Fig. 123a *Paraurostyla pulchra* (from Buitkamp 1977a. Protargol impregnation). At present the classification of this species is unknown (see text). Infraciliature of ventral side, 158 µm. Cirri behind right frontal cirrus circled; short arrow marks buccal cirri; arrowheads denote pretransverse ventral cirri. Long arrows mark the four ventral cirral rows described by Buitkamp; however, the anterior portions of the two leftmost rows can also be interpreted as zigzag-pattern (note that ontogenetic data are needed for a final explanation of this pattern). CV = contractile vacuole, FT = frontoterminal cirri?, P = paroral, TC = transverse cirri. Page 588.

the zigzag pattern as two longitudinal cirral rows (Fig. 123a, broken lines). If this assumption is correct then it is difficult to separate from *Bakuella pampinaria oligocirrata* (Fig. 118p–s) because it is unknown whether or not *P. pulchra* has cortical granules.

Morphology: Body length 150–170 µm in life? Body outline roughly spindle-shaped with slightly pointed rear end. Many elongated macronuclear nodules; micronucleus not observed. Contractile vacuole ahead of mid-body near left cell margin. Presence/absence of cortical granules not known. Adoral zone occupies about 33% of body length, composed of circa 24 membranelles of ordinary fine structure. Buccal field as in *Bakuella* species, that is, moderately wide. Undulating membranes long, curved, and optically intersecting. Paroral composed of two rows of basal bodies, endoral of single row. Three distinctly enlarged frontal cirri. Four buccal cirri along anterior and middle portion of paroral. Two (parabuccal) cirri behind right frontal cirrus. One short row composed of about four cirri left of anterior end of right marginal row, indicating that these are the frontoterminal cirri. Four cirral rows with five, nine, eight, and eight cirri in specimen illustrated (Fig. 123a); leftmost row and anterior portion of next row are possibly the zigzagging portion of the midventral complex (see remarks). Two pretransverse ventral cirri. Transverse cirri 18 µm long, slightly enlarged, arranged in hook-shape. Marginal cirri 12 µm long, right row commences about at level of rearmost buccal cirrus, ends slightly subterminally; left row begins slightly ahead of level of buccal vertex, terminates at tip of cell. Dorsal bristles about 3 µm long, arranged in three kineties. Caudal cirri lacking.

Occurrence and ecology: Likely confined to terrestrial habitats. Type locality of *P. pulchra* is the upper soil layer (0–5 cm) of a pasture (brown soil mainly grown with *Poa annua*, *Poa pratensis*, *Lolium perenne*, *Taraxacum officinale*, and *Trifolium repens*) on a hill (Venusberg, Melbtal) near the German city of Bonn. Sudzuki (1979, p. 123) found it in Antarctic soil. Feeds on diatoms and ciliates (Buitkamp 1977a). Buitkamp (1979) counted 73 individuals per gram dry soil at a temperature of 30°. Biomass of 10^6 specimens about 300 mg (Foissner 1998, p. 207).

Holostichides Foissner, 1987

1987 ***Holostichides* nov. gen.**[1] – Foissner, Zool. Beitr., 31: 201 (original description). Type species (by original designation on p. 201): *Holostichides chardezi* Foissner, 1987.

1988 ***Parabakuella* nov. gen.**[2] – Song & Wilbert, Arch. Protistenk., 135: 320 (original description of synonym). Type species (by original designation on p. 321): *Parabakuella typica* Song & Wilbert, 1988.

1994 ***Holostichides* Foissner, 1987**[3] – Eigner, Europ. J. Protistol., 30: 473, 474 (synonymy of *Parabakuella* and *Holostichides* and improved diagnosis of *Holostichides*; on p. 462 an incorrect year [1978] is given).

---

[1] The diagnosis by Foissner (1987b) is as follows: Holostichidae mit sehr ungleich langen Midventralreihen und Caudalcirren. Keine Transversalcirren.

[2] The diagnosis by Song & Wilbert (1988) is as follows: Je eine linke und rechte Marginalreihe; verstärkte Frontalcirren; frontal gelegene, familientypische Midventral-Reihe; kurze Schrägreihen von Ventralcirren; eine Frontoterminalreihe; keine Transversalcirren; Caudalcirren vorhanden und zahlreiche Macronucleusteile.

[3] The improved diagnosis by Eigner (1994) is as follows: One or no buccal cirrus. A midventral row composed of cirral pairs in anterior ventral surface. Transverse cirri absent. Caudal cirri present.

1996 ***Holostichides* Foissner, 1987** – Franco, Esteban & Téllez, Acta Protozool., 35: 326, 329, 330 (key to species of subfamily Bakuellinae).
1999 ***Parabakuella* Song & Wilbert, 1988** – Shi, Song & Shi, Progress in Protozoology, p. 114 (revision of hypotrichous ciliates).
1999 ***Parabakuella* Song & Wilbert, 1988** – Shi, Acta Zootax. sinica, 24: 365 (revision of hypotrichous ciliates).
2001 ***Holostichides* Foissner 1987** – Aescht, Denisia, 1: 82 (catalogue of generic names of ciliates).
2001 ***Parabakuella* Song & Wilbert 1988** – Aescht, Denisia, 1: 114 (catalogue of generic names of ciliates).
2001 ***Holostichides* Foissner, 1987** – Berger, Catalogue of ciliate names 1. Hypotrichs, p. 39 (nomenclator containing all basionyms, combinations, and higher taxa of hypotrichs).

Nomenclature: No derivations for the names *Holostichides* and *Parabakuella* are given in the original descriptions. *Holostichides* is a composite of the genus-group name *Holosticha* (see there for derivation) and the suffix *-ides* and means "like *Holosticha*" because the general organisation is as in *Holosticha* (Foissner 1987b). Masculine gender because according to ICZN (1999, Article 30.1.4.4) a compound genus-group name ending with the suffix *-ides* is to be treated as masculine unless its author, when establishing the name, stated that it had another gender or treated it as such by combining it with an adjectival species-group name in another gender form. The name of the synonym *Parabakuella* is a composite of the Greek prefix *para+* (beside, at, along, during) and the urostylid genus-group name *Bakuella* (see there for derivation) and refers to the cirral pattern which is similar in *Bakuella* and *Parabakuella*. *Parabakuella* has, like *Bakuella*, feminine gender.

Characterisation (Fig. 111a, autapomorphy 7): Body slender elongate. Adoral zone of membranelles continuous. 3 frontal cirri. Buccal cirrus(i) present. Usually 3 or more frontoterminal cirri. Midventral complex composed of midventral pairs and one or more midventral rows. Transverse cirri lacking. 1 left and 1 right marginal row. Caudal cirri present. 4–5 dorsal kineties (A).

Remarks: The three species assigned to *Holostichides* have, besides the characteristics mentioned in the paragraph above, some plesiomorphic features in common: body flexible and dorsoventrally flattened; many scattered, ellipsoidal macronuclear nodules; contractile vacuole slightly ahead of mid-body near left cell margin, with distinct, longitudinal collecting canals; soil inhabiting; one buccal cirrus; one cirrus behind right frontal cirrus; marginal rows not overlapping posteriorly; dorsal bristles 3–4 μm long.

Foissner (1987b) established *Holostichides* within the family Holostichidae because the general organisation of the type species resembles *Holosticha* and *Paruroleptus*. Eigner (1994) kept this family classification, but included it in the subfamily Bakuellinae because the midventral complex is composed of pairs and rows. Eigner's classification was taken over by Franco et al. (1996) when they provided a key to the species of the Bakuellinae. Shi et al. (1999) and Shi (1999a) classified the synonym *Parabakuella* also in the Holostichidae.

Foissner (1987b) separated *Holostichides* from *Holosticha* (now most species of this genus are in *Anteholosticha* and *Caudiholosticha*) by the following combination of features: caudal cirri present (vs. lacking); strongly shortened left midventral row, that is, midventral complex composed of low number of midventral pairs and a midventral row in my terminology (vs. long series of midventral pairs present and midventral row lacking); lack of transverse cirri (vs. present); increased number (more than two or three) of frontoterminal cirri (vs. usually two). Foissner (1987b) expected that the midventral row of *H. chardezi* originates from the same anlage as the frontoterminal cirri. However, ontogenetic data on *Holostichides typicus* and the closely related *Paragastrostyla lanceolata* show that the midventral row originates from the penultimate anlage from right, whereas the frontoterminal cirri are produced, as is usual, from the rightmost anlage. The rather high number of frontoterminal cirri in *Holostichides* occurs also in most other bakuellids.

Holostichides was monotypic (*H. chardezi*) when it was established by Foissner (1987b). Only one year later, Foissner (1988) described *Holostichides terricola*, which, however, lacks a buccal cirrus. For that reason, I transfer it to *Paragastrostyla* (see there). Foissner (2000) discovered a third *Holostichides* species, namely *H. dumonti*. Two species have been transferred to *Holostichides* by Eigner (1994), namely *Periholosticha wilberti* Song, 1990 and *Parabakuella typica* Song & Wilbert, 1988. *Periholosticha wilberti* is now the junior synonym of *Paragastrostyla lanceolata* Hemberger, 1985 because it has a midventral row, but lacks a buccal cirrus and cortical granules. The synonymisation of *Parabakuella* and *Holostichides* proposed by Eigner (1994) is accepted because the difference between the type species of *Holostichides* and *Parabakuella* is only quantitative (one midventral row against two or more rows) and not qualitative. However, the synonymy is a matter of taste because a similar feature, namely the number of marginal rows, is generally used as diagnostic feature (e.g., *Metabakuella*). Thus, *Parabakuella* could also be classified as valid taxon with the diagnostic feature "two or more midventral rows" against "one midventral" in *Holostichides*. Then, however, *Holostichides dumonti* has to be transferred to *Parabakuella*. Song & Wilbert (1988) compared *Parabakuella* only with *Bakuella*, from which it differs only by the lack of transverse cirri.

Holostichides is very likely closely related to *Paragastrostyla*, which lacks a buccal cirrus. Ignoring the feature presence/absence of a buccal cirrus at genus level would make *Holostichides* Foissner, 1987b the junior synonym of *Paragastrostyla*, which was established by Hemberger (1985). Foissner (2000), who recognised five *Holostichides* species, also stated that *Holostichides terricola* (= *Paragastrostyla terricola* in present book) and *H. wilberti* (= junior synonym of *Paragastrostyla lanceolata* in present book) form a distinct subgroup within *Holostichides* because both lack a buccal cirrus and are smaller than the other *Holostichides* species.

Shi et al. (1999, p. 117) and Shi (1999a, p. 366) synonymised *Holostichides* with *Periholosticha* Hemberger, 1985. However, *Periholosticha lanceolata*, type species of *Periholosticha*, lacks buccal and frontoterminal cirri, has three dorsal kineties, and the midventral complex is composed of cirral pairs only. Thus, *Periholosticha* is classified in the Holostichidae in the present monograph. By contrast, *Holostichides* has buccal and frontoterminal cirri and at least one midventral row. The proposal by Shi et al. (1999) is likely due to the fact that other species included in *Periholosticha* and *Holostichides* do not show all these differences. For example, *Periholosticha acuminata* has frontoterminal cirri and *Holostichides terricola* lacks a buccal cirrus. Thus, I transfer *H. terricola* to *Paragastrostyla*, which is characterised by the lack of a buccal cirrus. Shi et al. (1999) do not consider *Parabakuella* Song & Wilbert, 1988 as junior synonym of *Holostichides*. They treat it as valid and consider *Pseudobakuella* Alekperov, 1992 as junior synonym of *Parabakuella*. However, *Pseudobakuella*, which they erroneously assigned to Mihailowitsch & Wilbert (1990), has transverse cirri and lacks caudal cirri. Consequently synonymy of *Parabakuella* and *Pseudobakuella* is very unlikely. In the present book, *Pseudobakuella* is treated as subgenus of *Bakuella*.

According to Eigner (1994, p. 472), in *Holostichides* the proximal adoral membranelles are renewed during ontogenesis. Obviously he inferred this information from *Parabakuella typica* and *Periholosticha wilberti*, which he transferred to *Holostichides*. However, neither Song & Wilbert (1988) nor Song (1990) mention such a feature. But Figs. 126f, g, i in fact give the impression that the proximal membranelle(s) is (are) reorganised.

Eigner (1994, p. 472, 473) explains the lack of transverse cirri in *Holostichides* by the fact that all cirri of the rightmost anlage migrate anteriad to form the frontoterminal cirri. However, this explanation is incorrect because transverse cirri are not only formed by the rightmost anlage, but usually by several primordia. Furthermore, per definition only the rearmost, distinctly set off and often enlarged cirrus of an anlage is a transverse cirrus.

Species included in *Holostichides* (alphabetically arranged according to basionym): (1) *Holostichides chardezi* Foissner, 1987; (2) *Holostichides dumonti* Foissner, 2000; (3) *Parabakuella typica* Song & Wilbert, 1988.

Species misplaced in *Holostichides*: *Holostichides terricola* Foissner, 1988 (now *Paragastrostyla terricola*).

Key to *Holostichides* species

The three species assigned can be distinguished by some rather easily recognisable features. Consequently, detailed live observation is sufficient for reliable identification, provided that you know that your specimen/population belongs to *Holostichides*. The body length, although somewhat overlapping, is an important supporting key feature. See also the *Paragastrostyla* key and the *Periholosticha* key if you have problems identifying your specimen/population with the following key.

1 One midventral row present; body length in life 120–180 μm (Fig. 124a, d) .... *Holostichides chardezi* (p. 594)

\- Two or more midventral rows present; body length in life 150–280 μm (Fig. 125c, 126b) **2**

2 Cortical granules (2 × 1 μm, colourless, arranged in longitudinal rows) present; body length in life 170–280 μm (Fig. 125a, d) .... *Holostichides dumonti* (p. 599)

\- Cortical granules lacking; body length in life 150–250 μm (Fig. 126a) .... *Holostichides typicus* (p. 607)

Holostichides chardezi Foissner, 1987
(Fig. 124a–j, Table 26)

1987 ***Holostichides chardezi* nov. spec.**[1] – Foissner, Zool. Beitr., 31: 203, Abb. 6a–g, Tabelle 3 (Fig. 124a–g; original description. The holotype slide [1988/143] and a paratype slide [1988/144] are deposited in the Oberösterreichische Landesmuseum in Linz [LI], Austria).

1994 ***Holostichides chardezi* Foissner, 1987** – Shin, Dissertation, p. 79, Fig. 10A–C, Table 9 (Fig. 124h–j; description of a Korean population).

2001 ***Holostichides chardezi* Foissner, 1987** – Berger, Catalogue of ciliate names 1. Hypotrichs, p. 39 (nomenclator containing all basionyms, combinations, and higher taxa of hypotrichs).

Nomenclature: The species was named in honour of the Belgian protozoologist Didier Chardez, who first used protozoa in criminology. *Holostichides chardezi* was fixed as type species of *Holostichides* by original designation. Song & Wilbert (1989, p. 160) mistakenly assumed that the present species was established in *Holosticha* because they wrote "*Holosticha chardezi* Foissner, 1987".

Remarks: Shin (1994) redescribed this species in his dissertation. The data agree very well with those of the detailed original description, but some information is incorrect, namely (i) the type location of *H. chardezi* is not in Austria (Shin 1994, p. 81), but in Cape Verde Island Santo Antáo; and (ii) Shin (1994, p. 81), wrote that "up to now, only one species has been described in this genus (Foissner 1987d)". However, a second species, *Holostichides terricola*, was described by Foissner (1988). Moreover, Shin's reference "Foissner 1987d" is the same as my reference "Foissner (1987c)", which does not contain the original description of *Holostichides chardezi*.

For separation of *H. chardezi* from congeners, see key and Table 26; also the cortical granules (<1 μm across, yellowish) are different from those of *H. dumonti* (about 2 × 1 μm, colourless). *Holostichides chardezi* should not be confused with *Paragastrostyla* species, which have the same cirral pattern, except for the absent buccal cirrus.

Morphology: Body size 120–160 × 30–40 μm in life, length:width ratio around 4.5:1 in life and 4.1:1 on average in protargol preparations (Fig. 124a, d, Table 26). Body slender sigmoidal, terminally slightly to distinctly narrowed, about 2:1 flattened dorsoventrally, not contractile (Fig. 124c). Macronuclear nodules (36 on average) scattered; individual nodules ellipsoidal (length:width ratio 2:1 on average in protargol preparations) with some nucleoli of ordinary size. 2–6 micronuclei at various positions of macronuclear figure; individual micronuclei ellipsoidal (Fig. 124a, c). Contractile vacuole slightly ahead of mid-body near left cell margin, with two longitudinal collecting canals (Fig. 124b). Pellicle colourless, very flexible. Cortical granules arranged in loose groups forming about 30 longitudinal rows around the cell; individual granules less than 1 μm across, yellowish so that cells are also yellowish at low magnification (Fig. 124b). Cytoplasm colourless, packed with food vacuoles. Movement inconspicuous, glides slowly among soil particles, strongly thigmotactic.

[1] The diagnosis by Foissner (1987b) is as follows: In vivo etwa 120–160 × 30–40 μm große, terminal deutlich verjüngte *Holostichides* mit leicht gelben subpelliculären Granula. Durchschnittlich 36 Makronucleus-Teile, 31 adorale Membranellen und 5 Frontoterminalcirren. Linke Midventralreihe mit durchschnittlich 7, rechte mit 18 Cirren. 4 Dorsalkineten.

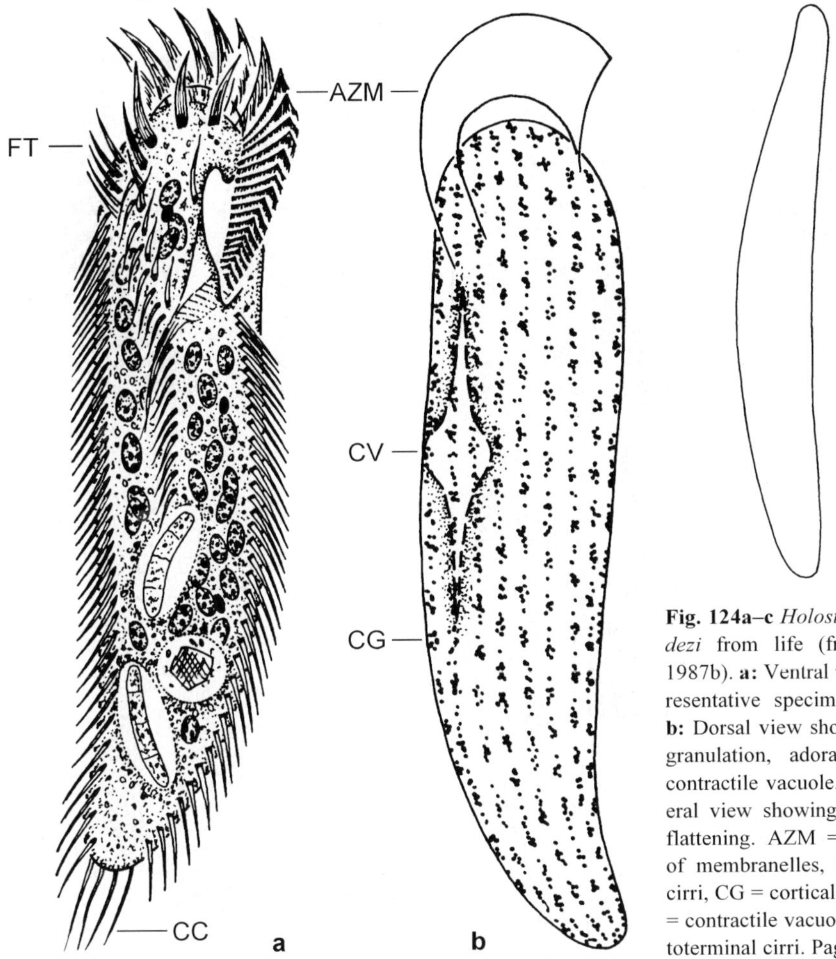

Fig. 124a–c *Holostichides chardezi* from life (from Foissner 1987b). **a:** Ventral view of a representative specimen, 154 μm. **b:** Dorsal view showing cortical granulation, adoral zone, and contractile vacuole. **c:** Right lateral view showing dorsoventral flattening. AZM = adoral zone of membranelles, CC = caudal cirri, CG = cortical granules, CV = contractile vacuole, FT = frontoterminal cirri. Page 594.

Adoral zone occupies 25–33% of body length, arranged roughly in *Gonostomum* pattern (details, see Berger 1999), that is, extends straight along left body margin, performing right bend and slight clockwise rotation to plunge into buccal cavity, composed of 31 membranelles on average. Bases of largest membranelles about 6 μm wide in life. Buccal field flat, narrow, and anterior margin curved like a hook. Buccal lip covers proximal adoral membranelles. Undulating membranes slightly curved, do not intersect optically; paroral begins few micrometres ahead of endoral, which terminates close to proximal adoral membranelles. Pharyngeal fibres conspicuous in life and in protargol preparations, extend obliquely backwards (Fig. 124a, d, f, g).

Cirral pattern and number of cirri of usual variability (Fig. 124d, f, g, Table 26). All cirri about 15 μm long in life and most of similar size. Three somewhat enlarged frontal cirri in slightly oblique row with right cirrus close to distal end of adoral zone of mem-

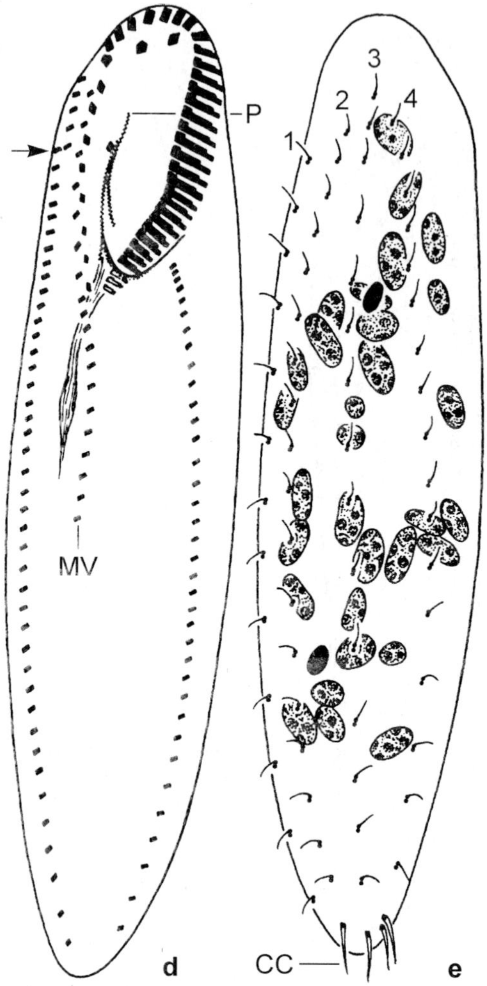

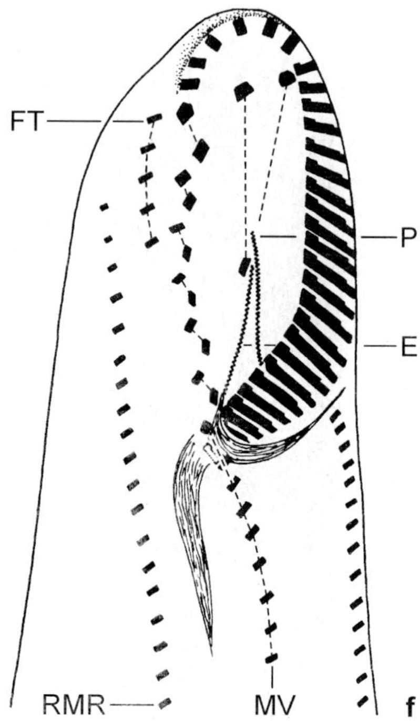

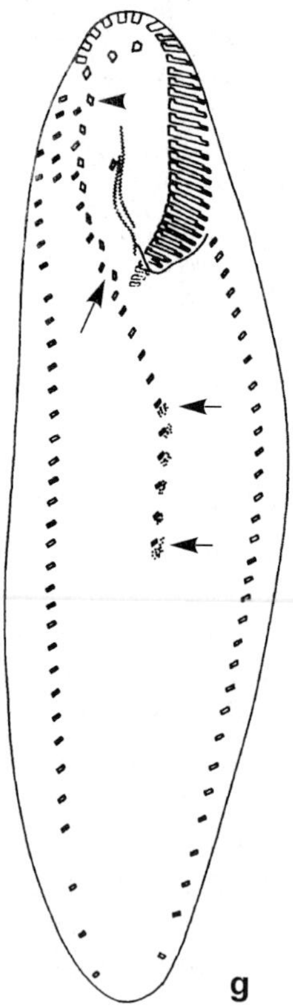

Fig. 124d–g *Holostichides chardezi* after protargol impregnation (from Foissner 1987b). **d, e:** Infraciliature of ventral and dorsal side and nuclear apparatus of representative specimen, 113 µm. This specimen has seven midventral pairs and a midventral row composed of 12 cirri. Arrow marks anterior end of right marginal row. **f:** Detail of anterior region (length of adoral zone 31 µm). Broken lines connect cirri (very likely) originating from same anlage. Probably the rearmost cirri of the midventral row are not shown because Foissner (1987b) mentioned 16 as lowest number of "right midventral cirri" (cirri of midventral row plus right cirri of midventral pairs); here only 14 cirri are shown. **g:** Infraciliature of ventral side of very early divider, 124 µm. Arrowhead denotes cirrus III/2, oblique arrow marks rearmost midventral pair. The oral primordium originates from basal body patches originating on the left and rear side of the rearmost cirri of the midventral row (transverse arrows). CC = caudal cirri, E = endoral, FT = frontoterminal cirri, MV = midventral row, P = paroral, RMR = right marginal row, 1–4 = dorsal kineties. Page 594.

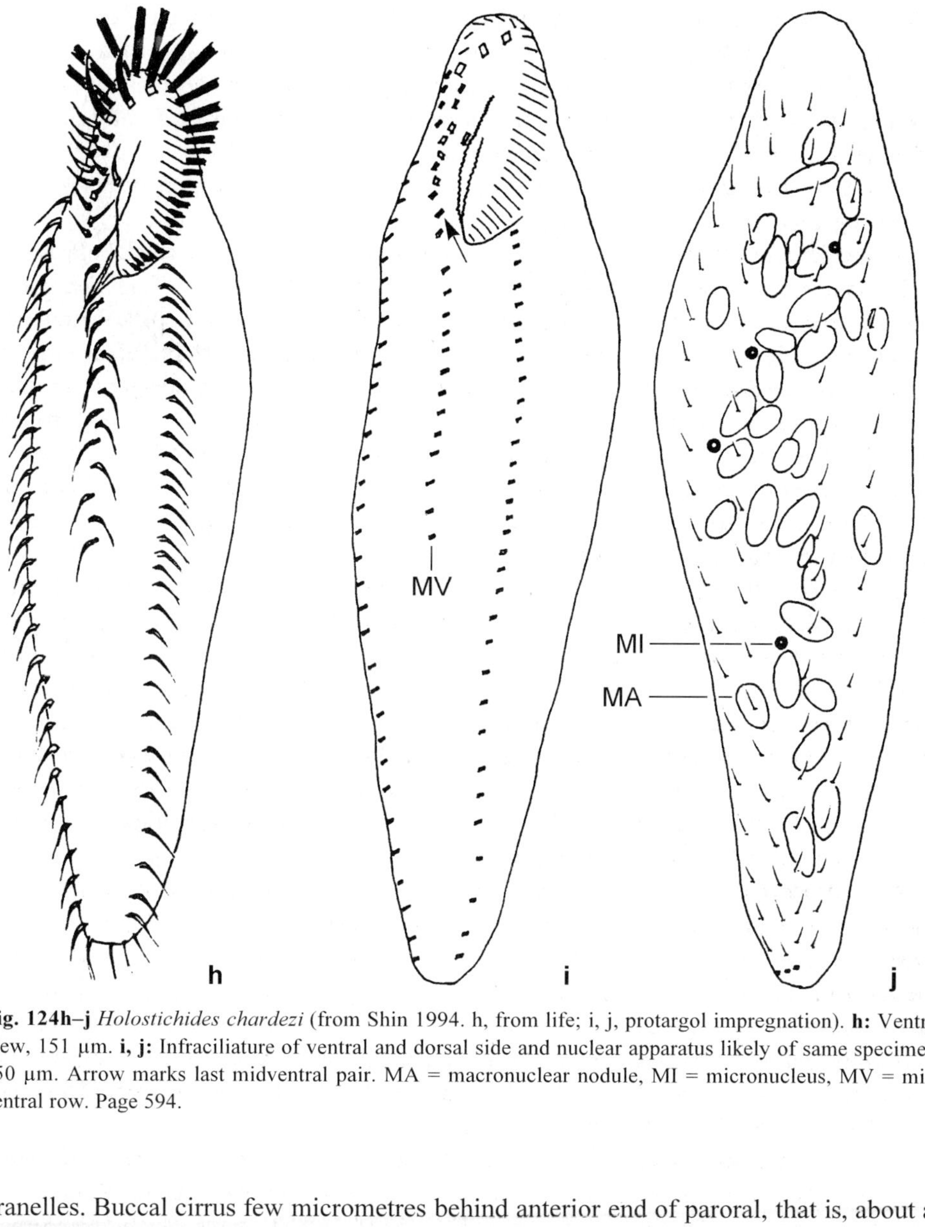

Fig. 124h–j *Holostichides chardezi* (from Shin 1994. h, from life; i, j, protargol impregnation). **h:** Ventral view, 151 μm. **i, j:** Infraciliature of ventral and dorsal side and nuclear apparatus likely of same specimen, 150 μm. Arrow marks last midventral pair. MA = macronuclear nodule, MI = micronucleus, MV = midventral row. Page 594.

branelles. Buccal cirrus few micrometres behind anterior end of paroral, that is, about at level of anterior end of endoral. One cirrus (= cirrus III/2) behind right frontal cirrus. Frontoterminal cirri in usual position, form short row right of anterior end of midventral complex (Fig. 124d, f, g). Midventral complex usually composed of 5–7 midventral pairs and a single midventral row. Both cirri of each midventral pair of about same size, but left (rear) cirrus almost longitudinally arranged whereas right (= front) cirrus oblique to almost transversely oriented; last midventral pair usually near proximal end of adoral zone. Midventral row commences behind last midventral pair, terminates at

54% of body length on average, slightly curved, composed of 11 cirri on average. Transverse cirri lacking. Right marginal row usually commences somewhat ahead of level of buccal cirrus and commonly overlaps slightly with frontoterminal row; ends, like left row, subterminally. In posterior portion marginal cirri become slightly finer and distances among cirri are wider than anteriorly.

Dorsal bristles about 3 µm long in life, invariably (n = 11) arranged in four kineties which begin subapically. Usually each dorsal kinety associated with one caudal cirrus; rarely specimens with only three cirri occur, that is, one kinety without caudal cirrus.

Description of the Korean population (Fig. 124h–j, Table 26): This population agrees rather well with the type population, including the yellowish cortical granules. For example, the average number of adoral membranelles and right marginal cirri are the same as in Foissner's population. Consequently, only important, deviating, or additional observations are given and the reader is mainly referred to the illustrations and Table 26. The only noteworthy differences occur in the "cirral number 1" of the midventral complex (on average 14.4 in Korean population against 18.2; see Table 26 for details) and the number of caudal cirri, which is usually three in the Korean population versus usually four in the type population.

Body size 120–180 × 30–45 µm in life. Body stiff and inflexible; this observation is very surprising because very likely all urostylids are flexible. Only the euplotids, the Stylonychinae, and some taxa of unknown position have a rigid body (Berger 1999); thus, I assume that this is a misobservation. Ventral side slightly concave, anterior body portion more flattened dorsoventrally than posteriorly. Movement usually slow; sometimes specimens change direction frequently. Usually three caudal cirri, that is, one dorsal kinety not associated with a caudal cirrus. Caudal cirri rather fine. Dorsal kineties 2 and 3 with 15–20 bristles each, which is the same as in the type population (Fig. 124e). Dorsal cilia about 4 µm long, some shorter.

Cell division (Fig. 124g): Foissner (1987b) illustrated one very early divider which shows that the oral primordium formation commences, as, for example, in *Paragastrostyla lanceolata* (Fig. 127b, 128e), left of the rear end of the midventral row. Likely, the further process is very similar as in *Holostichides typicus* or *Paragastrostyla lanceolata*. Consequently, one can assume that, for example, the specimen shown in Fig. 124f produced 10 frontal-midventral cirral anlagen which formed, from left to right, the following number of cirri: 1, 2, 2, 2, 2, 2, 2, 2, 9 (probably the rearmost cirri of the midventral row are not illustrated), 5 (frontoterminal cirri).

Occurrence and ecology: The type locality of *Holostichides chardezi* is an eastern slope near the village of Ribeira do Paul on the Cape Verde Island Santo Antáo in the Atlantic Ocean near the Senegalese coast. The sample, which was collected by Herbert Franz (Mödling) about 500 m above sea level on October 10, 1985, was a reddish soil (pH 8.0) under a tussock. *Holostichides chardezi* occurred with low abundance in this sample. Shin (1994) found it at two sites in/near Seoul, South Korea. One sample was collected at Dokchin, Purun-my on, Kanghwa-gun in late March and is described as "ricefields, moss-covered soils and brooklet" indicating that it is possibly semi-terrestrial. The other sample was collected from moss-covered soils and grassland at Namhansansong in Kwangju-gun in early April.

Foissner (1999, p. 323) found *Holostichides chardezi* in a marshy area near the Sheldrick waterfalls, Shimba Hills Nature Reserve (4°25'S 39°20'E), Kenya. Foissner (2000) recorded it from two sites from Germany: (i) A reclaimed, opencast coal mining area near the town of Görlitz (see Dunger 1991 for detailed site description). The sample was collected in late March 1998 about 30 years after reclamation and consisted of the upper 0–3 cm fresh and fermented litter layer (mainly from poplar, beech, acacia), mosses from the soil surface, and the upper (up to 3/4 cm), black moder soil (pH 5.2). (ii) Beech forest near Munich. The sample consisted mainly of fresh and fermented litter and was collected in late August 1987. Recently it was found in the soil of a rice field in Italy (Schwarz & Frenzel 2003, p. 247) and in various Austrian forest stands (Foissner et al. 2005). According to Foissner (1998), *Holostichides chardezi* occurs in the Holarctis, the Palaeotropis, and Australis and shows a strong degree of soil autochthonism.

Feeds on coccale cyanobacteria, fungal spores, and humic particles (Foissner 1987b). Shin (1994) found algae and testate amoebae in the food vacuoles. Biomass of 10^6 specimens 63 mg (Foissner 1998).

Holostichides dumonti Foissner, 2000
(Fig. 125a–h, Table 26)

2000 ***Holostichides dumonti* nov. spec.**[1] – Foissner, Europ. J. Protistol., 36: 273, Fig. 72–78, Tables 6, 7 (Fig. 125a–g; original description. One holotype slide [2000/83] and 3 paratype slides [2000/84–86] with protargol-impregnated specimens are deposited in the Oberösterreichische Landesmuseum in Linz [LI], Austria).

2001 ***Holostichides dumonti* Foissner, 2000** – Berger, Catalogue of ciliate names 1. Hypotrichs, p. 115 (nomenclator containing all basionyms, combinations, and higher taxa of hypotrichs).

Nomenclature: Wilhelm Foissner named this species in honour of Henry J. Dumont, Ghent University, for supporting the publication of my "Monograph of the Oxytrichidae" (Berger 1999).

Remarks: The present species is the third originally assigned to *Holostichides*. It has, like *H. typicus*, more than one midventral row. It differs from *H. chardezi* and *H. typicus* in size (170–280 μm long in life vs. 120–160 μm, respectively, 150–250 μm), and number of macronuclear nodules (122 on average vs. 36, respectively, 30), adoral membranelles (42 on average vs. 31, respectively, 34), and dorsal kineties (5–6 vs. 4); besides *H. chardezi* (type species) has only one midventral row and the cortical granules are smaller (<1 μm vs. 2 × 1 μm) and yellowish (vs. colourless). *Holostichides typicus* lacks cortical granules.

Recently, I found *H. dumonti* in the litter of a blueberry bush in Austria. The specimens of this population agree very well in all features – including cortical granulation – with those of the type population from Finland (Table 26). Thus, I provide only an

[1] The diagnosis by Foissner (2000) is as follows: Size in vivo about 240 × 50 μm; elongate ellipsoidal. Cortical granules in rows, colourless, about 2 × 1 μm. On average, 120 macronuclear nodules, 42 adoral membranelles, 1 buccal cirrus, 4 frontoterminal cirri, 10 midventral pairs, 1 cirral row right of midventral tail, 7 caudal cirri, and 5 dorsal bristle rows.

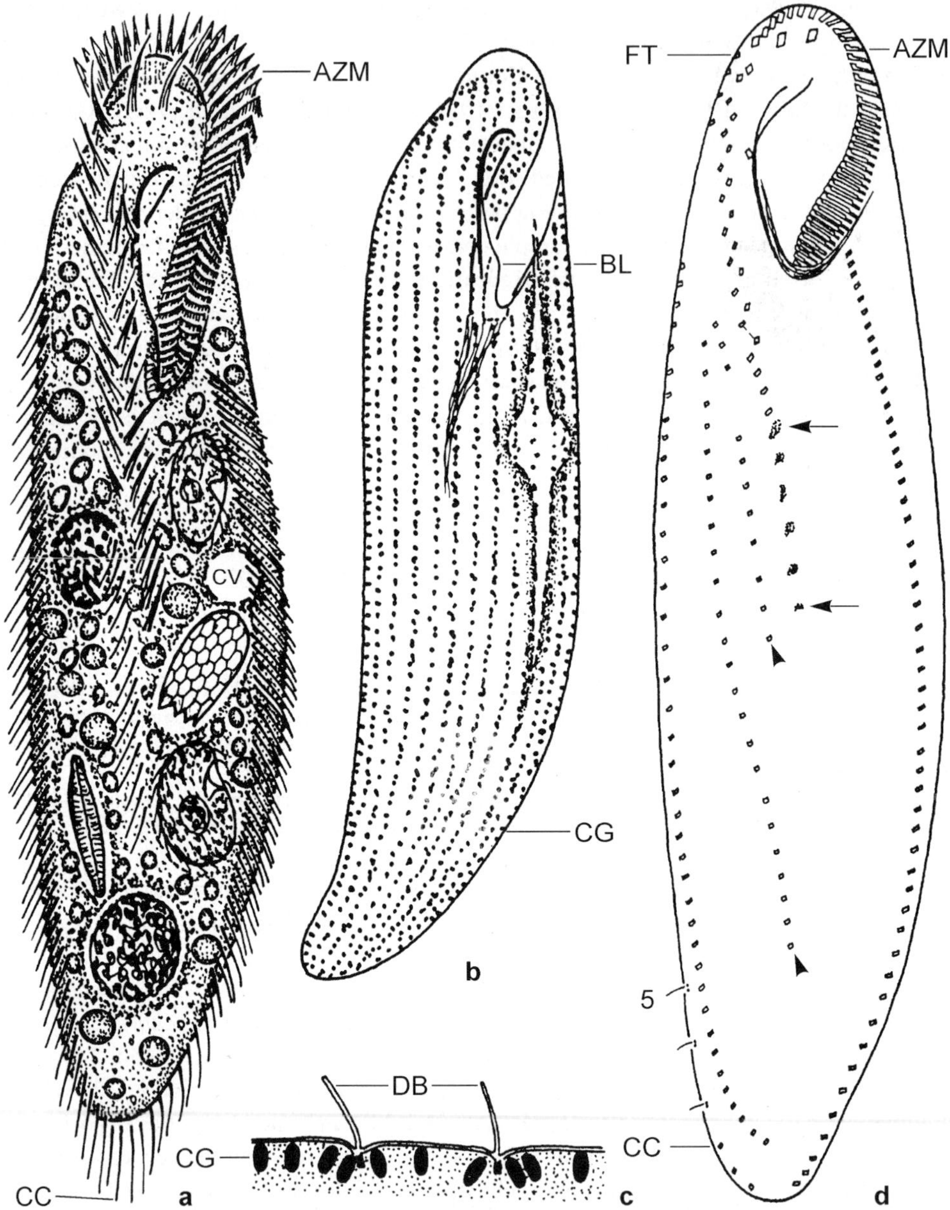

Fig. 125a–d *Holostichides dumonti* (from Foissner 2000. a–c, from life; d, protargol impregnation). **a:** Ventral view of a representative specimen, 228 µm. **b:** Ventral view of a sigmoidal specimen. The cortex contains rows of distinct granules. **c:** Optical section of dorsal surface showing dorsal bristles and compact, colourless cortical granules with a size of about 2 × 1 µm. **d:** Infraciliature of ventral side of a very early divider with some basal body patches close to the rear cirri of the left midventral row (arrows). Broken line connects the two cirri of the rearmost midventral pair. Arrowheads mark middle and right midventral row. AZM = adoral zone of membranelles, BL = buccal lip, CC = caudal cirri, CG = cortical granules, CV = contractile vacuole, DB = dorsal bristles, FT = frontoterminal cirri, 5 = dorsal kinety 5. Page 599.

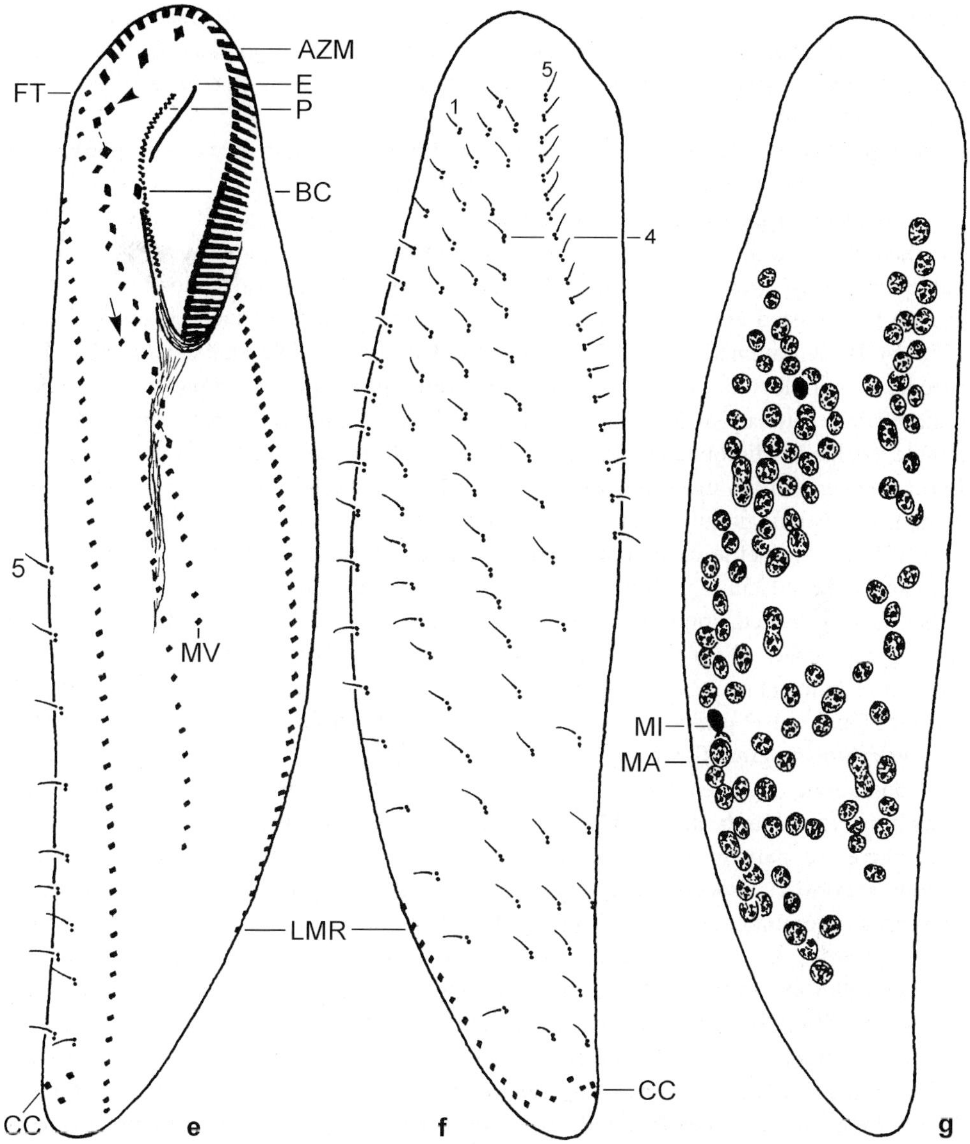

Fig. 125e–g *Holostichides dumonti* after protargol impregnation (from Foissner 2000). Infraciliature of ventral and dorsal side and nuclear apparatus of a representative specimen, 203 µm. The arrowhead marks the cirrus (= cirrus III/2) behind the right frontal cirrus; the cirri of the first and the last midventral pair are connected each other by a broken line. The arrow denotes a solitary cirrus which probably originated from the same anlage as the right midventral row. Note that the anterior and posterior body portion are free of macronuclear nodules and micronuclei. AZM = adoral zone of membranelles, BC = buccal cirrus, CC = caudal cirri, E = endoral, FT = frontoterminal cirri, LMR = left marginal row, MA = macronuclear nodules, MI = micronucleus, MV = left midventral row, P = paroral, 1, 4, 5 = dorsal kineties. Page 599.

illustration of the infraciliature of the ventral side, a morphometric characterisation, and some supplemental data. Few specimens of the Austrian population had only two frontoterminal cirri, whereas usually three or more such cirri are present in other *Holostichides* populations (Table 26).

Morphology: This chapter is divided into the morphology of the type population and some original data from the Austrian population.

Body size of type specimens 190–280 × 40–60 µm, on average 240 × 50 µm in life, length:width ratio 3.8–5.1:1, on average 4.6:1 in protargol preparations (Fig. 125a, e, Table 26). Body outline very similar to *H. chardezi*, that is, usually elongate elliptical and slightly twisted about main body axis, rarely more or less distinctly sigmoidal (Fig. 125a, b). Body flattened about 1.5:1 dorsoventrally, very flexible, but acontractile. Macronuclear nodules in central portion of cell, globular to elongate ellipsoidal (length:width ratio up to 3:1), on average 6 × 3 µm in protargol preparations, contain minute nucleoli. Micronuclei globular to broadly ellipsoidal, all left of cell's midline. Exact number of both macronuclear nodules and micronuclei difficult to count because of high number of nodules and many similar-sized and stained cytoplasmic inclusions (Fig. 125g). Contractile vacuole in or slightly ahead of mid-body, with two long collecting canals (Fig. 125a, b). Cortex thin, fragile, and very flexible. Cortical granules in rather widely spaced, longitudinal rows, colourless, compact and therefore bright and easy to recognise, about 2 × 1 µm, stain red and are extruded when methyl green-pyronin is added (Fig. 125b, c). Cytoplasm colourless, often packed with food vacuoles and fat globules 1–2 µm across. Glides slowly on slide surface and on and between soil particles showing great flexibility.

Adoral zone occupies 28% of body length on average, composed of about 42 membranelles (Fig. 125a, b, d, e, Table 26). Bases of largest membranelles only 7 µm wide; zone therefore narrow as compared with size of cell, proximal portion broadened slightly spoon-like. Buccal cavity rather large and deep, right posterior portion covered by narrow, membranous buccal lip (Fig. 125b). Paroral and endoral distinctly curved, optically cross in mid-buccal cavity; paroral composed of zigzagging dikinetids, emerges from slit in buccal lip, cilia 4 µm long at ends and 10 µm in middle portion. Pharyngeal fibres long, extend posteriorly in midline of cell.

Cirral pattern and number of cirri of usual variability, except for number of midventral rows and caudal cirri which vary rather strong (Fig. 125d–f, Table 26). Cirri about 14 µm long in life, rather thin and short as compared with cell size. Frontal cirri only slightly enlarged and in ordinary arrangement. Buccal cirrus close to paroral near mid-buccal cavity where undulating membranes optically cross. Usually one cirrus (= cirrus III/2) behind right frontal cirrus. Frontoterminal cirri in ordinary position, that is, right of anterior end of midventral complex. Midventral complex composed of 10 midventral pairs and two midventral rows on average (Table 26). Last midventral pair at 37% of body length on average, that is, distinctly behind level of proximal end of adoral zone. Cirri of each midventral pair rather differently arranged (Fig. 125e). Usually two midventral rows; the left row is termed "midventral tail" and the right one "cirral row right of midventral tail" in the original description. About a quarter of specimens (9 out of 35) with three, and one out of 35 specimens with four midventral rows. Left midventral row extends in cell's midline and terminates at 50% of body length on average (Table 26). Right midventral row of specimen shown in Fig. 125e composed of 16 cirri. In this specimen one, and in specimen shown in Fig. 125d three cirri distinctly ahead of right, respectively, middle midventral row. Transverse cirri lacking. Right marginal row commences at level of buccal cirrus, extends slightly obliquely backwards in twisted specimens, usually ends, like left marginal row, subterminally (Fig. 125e, f); sometimes right marginal row terminates more anteriorly than left row (Fig. 125d).

Dorsal bristles 3–4 μm long in life, usually arranged in five rows; rows 1–3, 5 slightly, row 4 distinctly shortened anteriorly. Two out of 20 specimens with six dorsal kineties. On average seven caudal cirri attached to kineties 1–5; specimen shown in Fig. 125e, f with two cirri each associated with dorsal kineties 1–3, 5, and one cirrus attached to kinety 4; maximally four caudal cirri associated with one kinety. One out of 19 specimens without caudal cirri.

Supplemental data from Austrian population (Fig. 125h, Table 26): Size 170–220 × 35–50 μm in life. Body very flexible and very resistant against coverglass pressure. Number of macronuclear nodules not counted in detail, likely between 100 and 150. Food vacuoles about 10 μm across. Cytopyge subterminally on left body margin. Right marginal row commences at level of right frontal cirrus (vs. at level of buccal cirrus in type population).

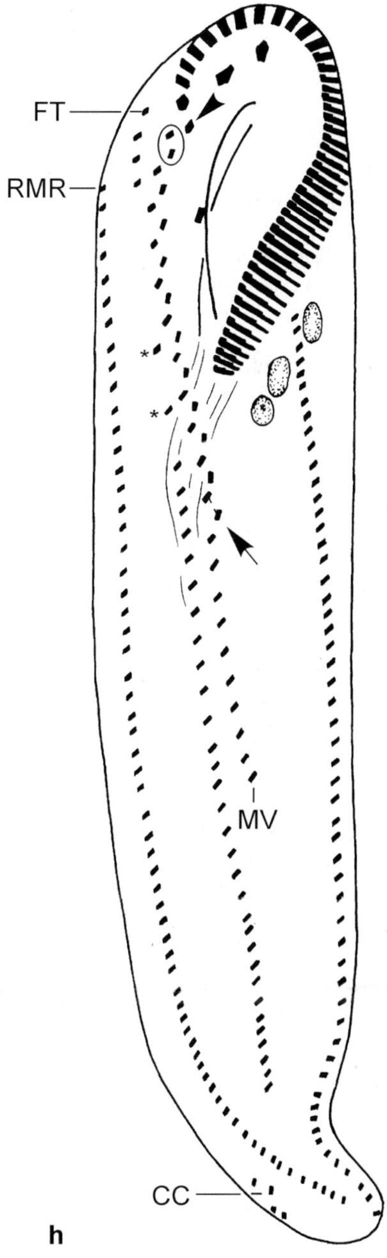

h

Fig. 125h *Holostichides dumonti* after protargol impregnation (original from Austrian population). Infraciliature of ventral side, 170 μm. Arrowhead marks cirrus III/2; ellipse encircles first midventral pair, arrow denotes last midventral pair. Asterisks mark cirri which likely originate from the same anlage as the right midventral row. Note that the right marginal row extends onto the dorsolateral surface up to the level of the right frontal cirrus. Only three of the numerous macronuclear nodules are illustrated. CC = caudal cirri (exact arrangement difficult to recognise because faintly stained), FT = frontoterminal cirri, MV = rear end of left midventral row, RMR = right marginal row. Page 599.

Table 26 Morphometric data on *Holostichides chardezi* (ch1, from Foissner 1987b; ch2, from Shin 1994), *Holostichides dumonti* (du1, from Foissner 2000; du2, original data from Austrian population), and *Holostichides typicus* (typ, from Song & Wilbert 1988)

Characteristics[a]	Species	mean	M	SD	SE	CV	Min	Max	n
Body, length	ch1	111.6	112.0	7.6	–	6.8	100.0	126.0	11
	ch2	148.1	150.0	18.7	7.1	12.6	125.0	178.0	7
	du1	199.0	200.0	20.0	5.5	10.0	168.0	243.0	13
	du2	184.8	183.0	14.9	4.0	8.0	152.0	204.0	14
	typ	133.0	129.0	12.0	3.6	9.0	108.0	153.0	11
Body, width	ch1	27.5	15.0	2.0	–	9.8	24.0	34.0	11
	ch2	37.4	36.0	4.5	1.7	12.1	33.0	45.0	7
	du1	43.5	44.0	3.9	1.1	8.9	35.0	50.0	13
	du2	37.4	40.0	4.1	1.1	11.0	31.0	44.0	13
	typ	41.6	41.0	5.4	1.6	13.0	35.0	54.0	11
Body length:width, ratio	ch2	4.0	3.9	0.1	0.1	3.3	3.8	4.2	7
	du2	5.0	5.0	0.5	0.1	9.5	4.3	6.2	13
Anterior body end to proximal end of adoral zone, distance	ch1	30.6	30.0	2.2	–	7.2	28.0	35.0	11
	ch2	40.6	41.0	2.3	0.9	5.7	38.0	45.0	7
	du1	56.5	58.0	3.2	0.9	5.6	52.0	63.0	13
	du2	49.8	51.5	4.9	1.4	9.7	40.0	56.0	12
Body length:length of adoral zone, ratio	ch2	3.7	3.7	0.4	0.2	11.9	3.0	4.3	7
	du2	3.8	3.7	0.3	0.1	9.1	3.4	4.3	12
Paroral, length	ch2	24.7	25.0	1.0	0.4	3.9	23.0	26.0	7
	du2	32.7	32.0	4.0	1.3	12.2	24.0	40.0	10
Anterior body end to paroral, distance	du2	13.1	12.5	1.5	0.5	11.6	11.0	15.0	10
Paroral length:length of adoral zone, ratio	ch2	0.6	0.6	0.0	0.0	6.8	0.6	0.7	7
Anterior body end to buccal cirrus, distance	du2	28.4	28.0	2.2	0.7	7.8	26.0	32.0	10
Pharyngeal fibres, length	ch2	17.0	17.5	1.4	0.7	8.3	15.0	18.0	4
Anterior body end to last frontoterminal cirrus, distance	du2	22.2	22.0	2.0	0.6	9.0	19.0	26.0	10
Anterior body end to rear end of last midventral pair, distance	ch1	29.9	28.0	2.5	–	8.3	28.0	35.0	11
	du1	74.9	75.0	7.9	2.2	10.5	60.0	93.0	13
	du2	73.4	73.0	9.0	2.9	12.3	52.0	84.0	10
Anterior body end to rear end of (left) midventral row, distance	ch1	60.1	58.0	4.5	–	7.5	55.0	68.0	11
	du1	99.8	100.0	19.5	5.4	19.5	73.0	147.0	13
	du2	116.5	117.5	13.9	4.4	12.0	92.0	133.0	10
Anterior body end to end of midventral complex, distance	du2	150.0	150.0	18.6	5.9	12.4	113.0	186.0	10
Anterior body end to right marginal row, distance	du2	14.5	13.5	3.4	1.1	23.5	10.0	21.0	10
Anterior body end to first macronuclear nodule, distance	du2	32.4	33.0	3.6	1.1	11.2	28.0	39.0	10
Macronuclear nodule, length	ch1	6.7	7.0	0.9	–	13.5	6.0	9.0	11
	ch2	7.3	7.0	1.6	0.6	22.0	5.0	10.0	7
	du1	5.9	6.0	2.1	0.6	36.1	3.0	9.0	13
	du2	4.6	4.0	1.3	0.4	27.5	4.0	8.0	10
Macronuclear nodule, width	ch1	3.4	3.0	0.5	–	14.4	3.0	4.0	11
	ch2	3.6	3.5	0.6	0.2	17.0	3.0	4.5	7
	du1	2.9	3.0	0.5	0.1	16.9	2.0	4.0	13
	du2	3.5	3.0	0.7	0.2	20.2	3.0	5.0	10
Macronuclear nodules, number	ch1	35.8	37.0	5.3	–	14.7	28.0	45.0	11
	ch2	31.9	31.0	4.8	1.8	15.0	28.0	42.0	7
	du1	[b]122.3	120.0	–	–	–	100.0	150.0	13

Table 26 Continued

Characteristics [a]	Species	mean	M	SD	SE	CV	Min	Max	n
Macronuclear nodules, number	typ	29.9	30.0	2.0	0.6	6.6	27.0	34.0	11
Micronucleus, length	ch1	3.7	4.0	0.9	–	23.6	2.0	5.0	11
	ch2 [f]	2.8	3.0	0.3	0.1	9.6	2.5	3.0	7
	du1	4.1	4.0	–	–	–	4.0	5.0	13
	du2	3.8	4.0	–	–	–	3.0	4.0	4
Micronucleus, width	ch1	2.5	2.5	0.5	–	19.2	2.0	3.0	11
	du1	3.0	3.0	–	–	–	3.0	3.0	13
	du2	3.0	3.0	–	–	–	3.0	3.0	4
Micronuclei, number	ch1	4.0	4.0	3.0	–	74.2	2.0	6.0	11
	ch2	3.7	4.0	0.8	0.3	20.3	4.0	5.0	7
	du1 [d]	3.1	3.0	–	–	–	2.0	5.0	13
	typ	3.4	4.0	0.9	0.3	27.1	1.0	4.0	11
Adoral membranelles, number	ch1	30.8	31.0	1.5	–	5.0	28.0	33.0	11
	ch2	29.1	28.0	2.5	1.0	8.7	26.0	34.0	7
	du1	42.4	42.0	1.8	0.5	4.1	40.0	46.0	13
	du2	42.8	43.0	3.4	1.0	8.0	36.0	48.0	12
	typ	33.8	34.0	1.9	0.6	5.7	31.0	37.0	11
Frontal cirri, number	ch1	3.0	3.0	0.0	–	0.0	3.0	3.0	11
	ch2	3.0	3.0	0.0	0.0	0.0	3.0	3.0	7
	du1	3.0	3.0	0.0	0.0	0.0	3.0	3.0	13
	du2	3.0	3.0	0.0	0.0	0.0	3.0	3.0	10
	typ [e]	4.0	4.0	0.0	0.0	0.0	4.0	4.0	11
Buccal cirri, number	ch1	1.0	1.0	0.0	–	0.0	1.0	1.0	11
	ch2	1.0	1.0	0.0	0.0	0.0	1.0	1.0	7
	du1	1.0	1.0	0.0	0.0	0.0	1.0	1.0	13
	du2	1.0	1.0	0.0	0.0	0.0	1.0	1.0	10
	typ	1.0	1.0	0.0	0.0	0.0	1.0	1.0	11
Frontoterminal cirri, number	ch1	4.6	5.0	0.8	–	17.6	3.0	6.0	11
	ch2	4.7	4.0	1.1	0.4	23.6	4.0	7.0	7
	du1	3.8	4.0	0.7	0.2	17.9	3.0	5.0	13
	du2	3.4	3.5	0.7	0.2	20.6	2.0	4.0	10
	typ	8.5	8.0	0.9	0.3	10.9	7.0	10.0	11
Midventral complex, cirral number 1 [c]	ch1	18.2	17.0	1.8	–	10.1	16.0	21.0	11
	ch2	14.5	14.0	2.0	0.8	13.6	13.0	18.0	6
Midventral complex, cirral number 2 [c]	ch1	7.0	7.0	0.6	–	9.0	6.0	8.0	11
	ch2	5.8	6.0	0.8	0.3	12.9	5.0	7.0	6
Midventral pairs, number	du1	10.2	10.0	1.3	0.4	12.7	8.0	12.0	13
	du2	12.0	12.5	2.4	0.7	19.6	7.0	15.0	10
	typ	6.3	6.0	0.5	0.1	7.0	6.0	7.0	11
Midventral rows, number	du1	2.3	2.0	0.5	0.1	22.9	2.0	4.0	35
	du2	2.2	2.0	0.4	0.1	19.2	2.0	3.0	10
	typ	3.1	3.0	1.1	0.3	33.9	2.0	5.0	11
Left marginal cirri, number	ch1	34.2	35.0	4.2	–	12.2	28.0	40.0	11
	ch2	33.7	33.0	3.2	1.2	9.5	29.0	38.0	7
	du1	53.5	53.0	3.6	1.0	6.8	48.0	60.0	13
	du2	53.2	53.5	6.4	2.0	12.1	41.0	63.0	10
	typ	39.6	40.0	3.0	0.9	7.5	35.0	44.0	11
Right marginal cirri, number	ch1	38.5	38.0	4.5	–	11.6	30.0	45.0	11
	ch2	38.4	38.0	4.0	1.5	10.3	33.0	43.0	7
	du1	51.6	51.0	4.7	1.3	9.0	45.0	59.0	13
	du2	55.8	56.0	7.1	2.2	12.7	44.0	66.0	10

Table 26 Continued

Characteristics [a]	Species	mean	M	SD	SE	CV	Min	Max	n
Right marginal cirri, number	typ	47.8	49.0	4.0	1.2	8.3	41.0	53.0	11
Dorsal kineties, number	ch1	4.0	4.0	0.0	–	0.0	4.0	4.0	11
	ch2	4.0	4.0	0.0	0.0	0.0	4.0	4.0	5
	du1	5.1	5.0	–	–	–	5.0	6.0	20
	du2	5.0	5.0	–	–	–	5.0	5.0	4
	typ	4.0	4.0	0.0	0.0	0.0	4.0	4.0	11
Caudal cirri, number (see text for details)	ch1	3.7	4.0	0.5	–	12.6	3.0	4.0	11
	ch2	3.0	3.0	0.6	0.2	19.3	2.0	4.0	7
	du1	6.5	7.0	1.9	0.4	29.5	0.0	9.0	19
	du2 [g]	7.3	7.0	–	–	–	7.0	8.0	4

[a] All measurements in µm. Data provided by Foissner (1987b, 2000) and original data are based on mounted, protargol-impregnated (Foissner's method), and randomly selected specimens from a non-flooded Petri dish culture. Data provided by Song & Wilbert (1988) are based on specimens impregnated with Wilbert's protargol method. Shin (1994) did not mention which method was used in that specific species. CV = coefficient of variation in %, M = median, Max = maximum value, mean = arithmetic mean, Min = minimum value, n = number of individuals investigated, SD = standard deviation, SE = standard error of arithmetic mean.

[b] Rough values because difficult to count.

[c] Cirral number 1 = number of cirri of midventral row plus right (anterior) cirri of midventral pairs. Cirral number 2 = left (posterior) cirri of midventral pairs (cirrus behind right frontal cirrus likely included).

[d] Rough values because sometimes difficult to distinguish from cytoplasmic inclusions and/or macronuclear nodules.

[e] Cirrus behind right frontal cirrus (= cirrus III/2) included.

[f] Diameter of micronucleus.

[g] Difficult to count because only faintly impregnated in most specimens.

Cell division: A very early divider shows that ontogenesis commences left of middle and rear portion of leftmost midventral row (Fig. 125d).

Occurrence and ecology: Likely confined to terrestrial habitats. The type locality of *Holostichides dumonti* is a coniferous forest (mixed with some birch and ash trees) near the town of Savonlinna (61°50'N 29°00'E), Finland. The soil sample, which was collected by Ilse Foissner on June 20, 1987, was composed of litter and some dark moder (pH 4.3). Foissner (2000, p. 254, 276) found it in two sites in Africa, in a soil sample from the Canary Islands, and in a mixed coniferous/deciduous forest on the campus of the University of Kaiserslautern, Germany. I found *Holostichides dumonti* in the litter of a blueberry bush (*Vaccinium myrtillus*) on the southern face of the Kammererköpfl (47°24'14"N 13°10'35"E; about 1300 m altitude; collected on 08.08.2001 by myself), a small mountain near the town of Bischofshofen, Austria.

Feeds on fungal spores, diatoms, heterotrophic flagellates, testate amoebae (*Euglypha*), and small ciliates, like *Drepanomonas pauciciliata* (Foissner 2000); Austrian specimens fed on hyphae and the testate amoebae *Corythion dubium* and *Trinema lineare*. Biomass of 10^6 specimens about 190 mg (own calculation).

Holostichides typicus (Song & Wilbert, 1988) Eigner, 1994 (Fig. 126a–o, Table 26)

1988 ***Parabakuella typica* nov. spec.**[1] – Song & Wilbert, Arch. Protistenk., 135: 321, Fig. 1–12, Tabelle 1 (Fig. 126a–o; original description. The type slide is deposited in the Institute of Zoology of the University of Bonn, Germany).

1994 ***Holostichides typicus* (Song and Wilbert, 1988) nov. comb.** – Eigner, Europ. J. Protistol., 30: 473 (combination with *Holostichides*).

2001 ***Holostichides typicus* (Song and Wilbert, 1988) Eigner, 1994** – Berger, Catalogue of ciliate names 1. Hypotrichs, p. 66 (nomenclator containing all basionyms, combinations, and higher taxa of hypotrichs).

Nomenclature: No derivation of the name is given in the original description. The species-group name *týpic·us -a -um* (Greek, *ho týpos* the picture, the pattern, the examined shape, the type; characteristic, prototypic, normal, original; Hentschel & Wagner 1996, p. 597) should likely indicate that this species is the type of *Parabakuella*. *Parabakuella* is feminine; Eigner (1994) therefore correctly changed the epithet from *typica* to *typicus* when transferring the species to the masculine *Holostichides* (ICZN 1985, Article 30 (b); ICZN 1999, Article 30.1.4.4). *Parabakuella typica* was fixed by original designation as type species of *Parabakuella* Song & Wilbert, 1989 which is now the junior synonym of *Holostichides*.

Remarks: For synonymy of *Parabakuella* and *Holostichides*, see genus section. *Holostichides typicus* has two features in common with *H. dumonti*, namely, (i) two or more midventral rows; and (ii) usually more than one caudal cirrus attached to each dorsal kinety. This combination of features could be used for the definition of *Parabakuella*. On the other hand, the two species differ distinctly in the shape of the undulating membranes. In *Holostichides typicus* (Fig. 126b) they are rather short and straight as in *H. chardezi*, but long and distinctly curved in *H. dumonti* (Fig. 125e, h). Such distinct differences in these structures are not very common in closely related species. Consequently, the description of further species should be awaited before reactivating *Parabakuella*.

Morphology: Body length 150–250 μm in life, specimen shown in Fig. 126a about 190 × 55 μm; the highest value of body length after protargol impregnation is 153 μm (Table 26), indicating that the estimated life size is too high; body length:width ratio according to original description 4:1 in life, according to Fig. 126a about 3.4:1 which is rather similar to the average ratio (3.2:1) in protargol preparations (Table 26). Body outline elongate elliptical, anterior end broadly rounded, posterior portion converging and rear end transversely truncated. Cytoplasm colourless. Presence or absence of cortical granules not mentioned; however, very likely *H. typicus* lacks cortical granules because the authors knew about the importance of this feature (Song & Wilbert 1989). Movement sluggish.

Adoral zone occupies 29% of body length in specimen shown in Fig. 126b, composed of 34 membranelles on average (Table 26). Size and shape of buccal cavity not described, obviously rather small (Fig. 126a, b). Undulating membranes straight to

[1] The diagnosis by Song & Wilbert (1988) is not a diagnosis sensu stricto but a brief description and thus not repeated here.

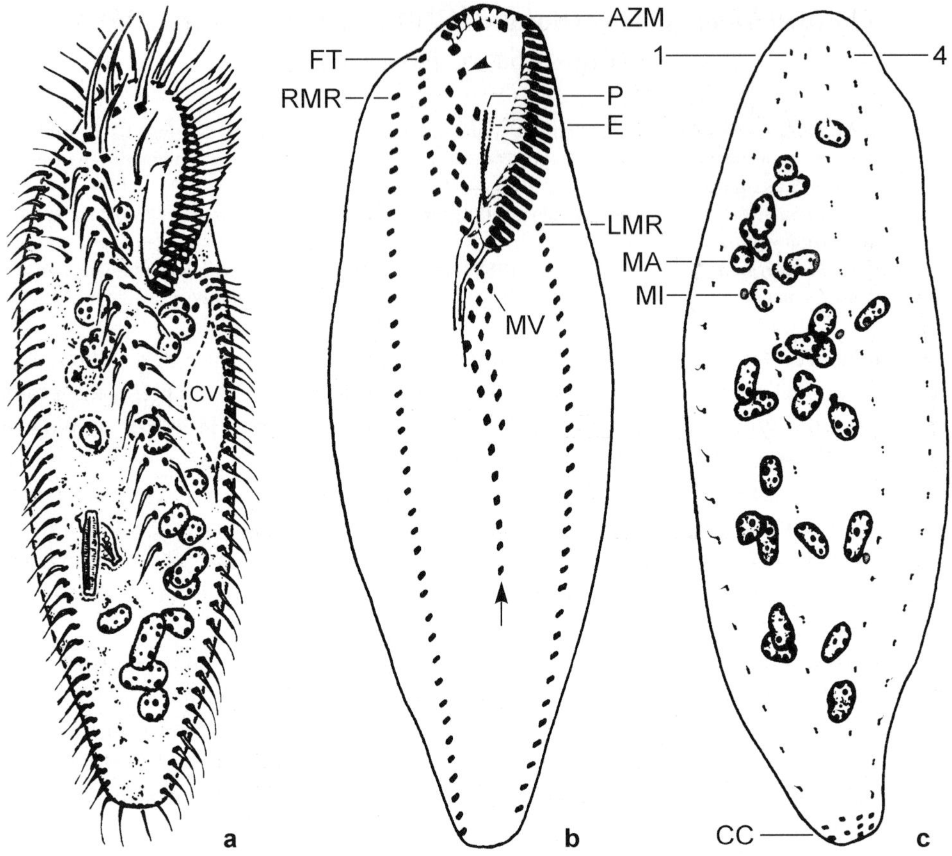

Fig. 126a–c *Holostichides typicus* (from Song & Wilbert 1988. a, from life; b, c, protargol impregnation, Wilbert's method). **a:** Ventral view of a representative specimen, 190 μm. **b, c:** Infraciliature of ventral and dorsal side and nuclear apparatus of same specimen, 135 μm. Arrow marks rightmost midventral row, arrowhead denotes cirrus III/2. This specimen has six midventral pairs and the two anteriormost midventral rows are composed of only three cirri each. AZM = adoral zone of membranelles, CC = caudal cirri, E = endoral, FT = frontoterminal cirri, LMR = left marginal row, MA = macronuclear nodule, MI = micronucleus, MV = leftmost (= anteriormost) midventral row, P = paroral, RMR = right marginal row, 1, 4 = dorsal kineties. Page 607.

slightly curved, optically intersecting about in middle portion. Pharynx obviously without peculiarities.

Cirral pattern and number of cirri of usual variability, except for number of midventral rows which varies from 2–5 (Fig. 126b, c, Table 26). All cirri of about same size except for the three slightly enlarged frontal cirri, which are in usual arrangement. One cirrus (= cirrus III/2) behind right frontal cirrus. Buccal cirrus right of anterior end of paroral. Frontoterminal cirri form rather long row commencing at level of right frontal cirrus and terminating at 23% of body length in specimen shown in Fig. 126b. Midventral complex composed of 6–7 midventral pairs and 2–5 midventral rows, the ante-

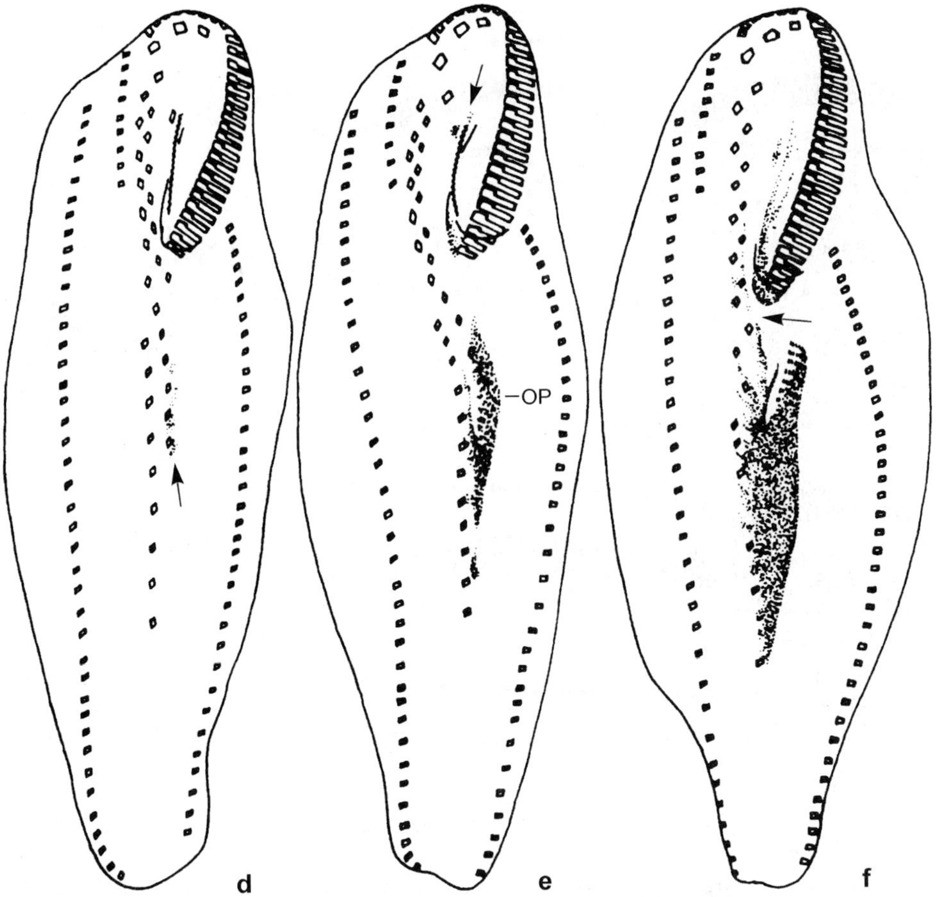

Fig. 126d–f *Holostichides typicus* (from Song & Wilbert 1988. Protargol impregnation, Wilbert's method). Infraciliature of ventral side of very early and early dividers (d = 188 μm; if the scale bar of Fig. 126d applies also to the other specimens, then they have a very similar size). Arrow in (d) denotes basal body patches which become the oral primordium (e). Arrow in (e) marks the beginning disorganisation of the buccal cirrus and the paroral. In early dividers some basal bodies migrate anteriorly (f, arrow) to contribute to anlagen formation of the proter. OP = oral primordium. Page 607.

rior one or two rows often composed of three cirri only (Fig. 126b). Rear (= left) cirrus of last midventral pair at 28% of body length in specimen shown in Fig. 126b, last cirrus of right midventral row at 67%. Transverse cirri lacking. Right marginal row commences about at level of first midventral pair, terminates at rear end; left row often ends subterminally.

Dorsal bristles about 3 μm long, arranged in four bipolar kineties (kinety 2 possibly sometimes slightly shortened posteriorly; Fig. 126c, l). Each dorsal kinety associated with 1–3, usually two caudal cirri (Fig. 126c, h).

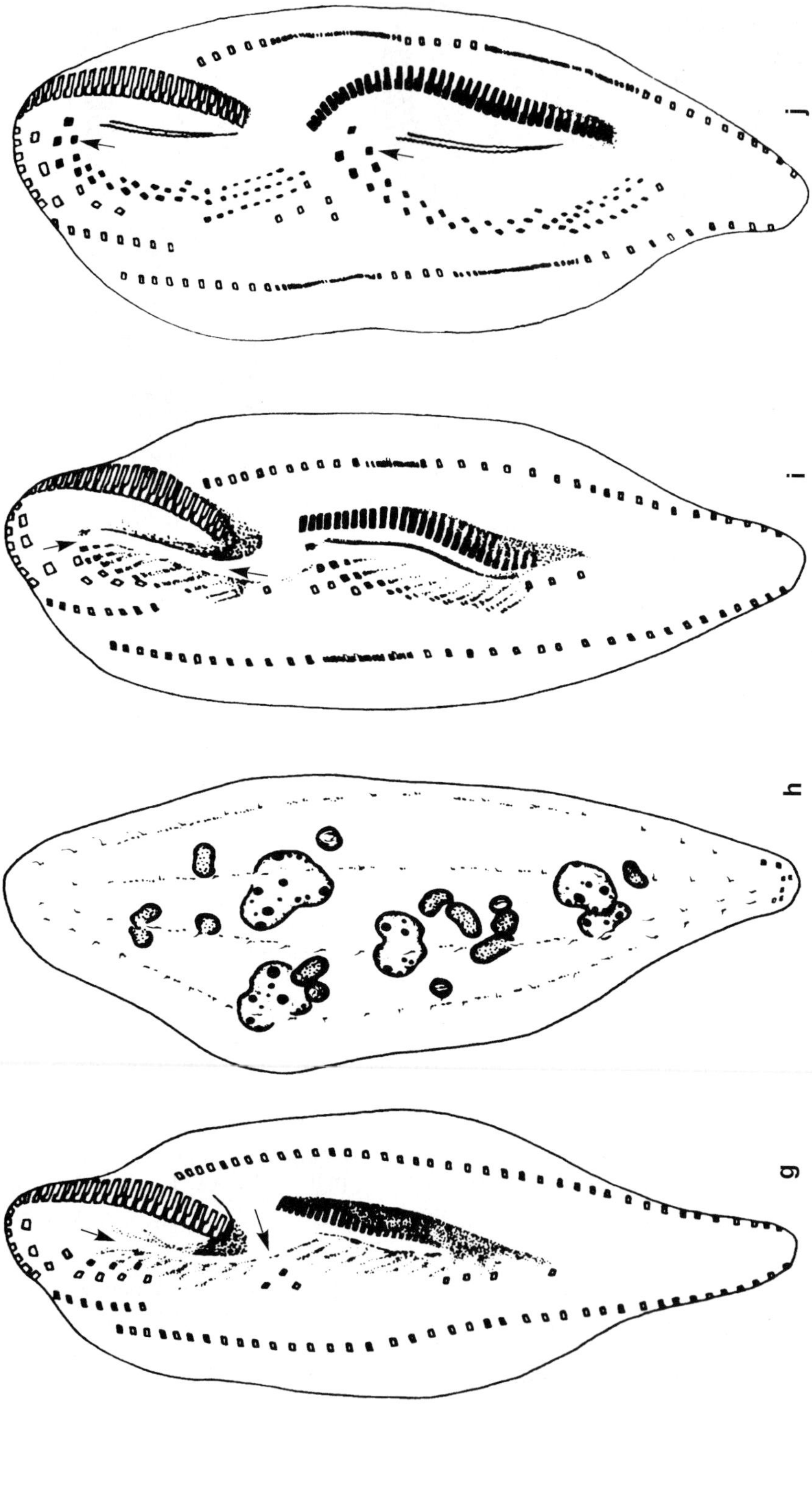

Fig. 126g–j *Holostichides typicus* (from Song & Wilbert 1988. Protargol impregnation, Wilbert's method). Infraciliature of ventral and dorsal side and nuclear apparatus of middle dividers. Long arrow in (g) marks connection of opisthe's and proter's anlagen complex. Short arrow in (g) and arrows in (i) mark separation of proter's anlage I and remaining anlagen. Arrows in (j) denote new buccal cirrus, which originates from anlage II. Parental structures white, new black. Several macronuclear nodules are fused (h). Page 607.

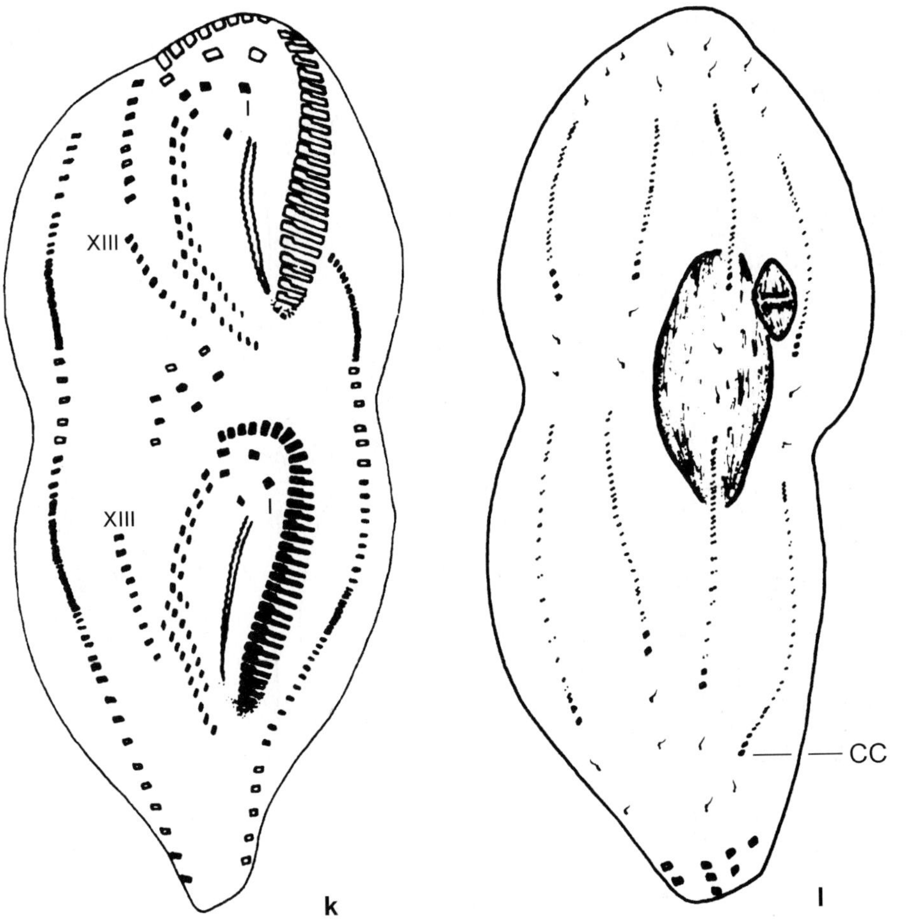

Fig. 126k, l *Holostichides typicus* (from Song & Wilbert 1988. Protargol impregnation, Wilbert's method). Infraciliature of ventral and dorsal side and nuclear apparatus of late divider. Both in proter and opisthe 13 frontal-midventral cirri anlagen (I–XIII) are formed. New structures black, parental white. CC = new caudal cirri on dorsal kinety 4 of opisthe, I, XIII = cirral anlagen. Page 607.

Cell division: Song & Wilbert (1988) described the ontogenesis of *Holostichides typicus* (Fig. 126d–o). The description is rather short and the illustrations very small so that some details, especially of the early events, are not totally clear.

The process commences with the formation of small basal body patches at the last cirri of the leftmost midventral row with more than three cirri (Fig. 126d, arrow). These patches enlarge and fuse to an elongate oral primordium. The buccal cirrus and the anterior end of the paroral begin to disintegrate. At the rear end of the endoral a basal body patch occurs (Fig. 126e). Next, most cirri of the midventral rows disintegrate and likely contribute to the enlargement of the oral primordium and the formation of the frontal-midventral cirri primordia of the opisthe (Fig. 126f, g). Simultaneously, some

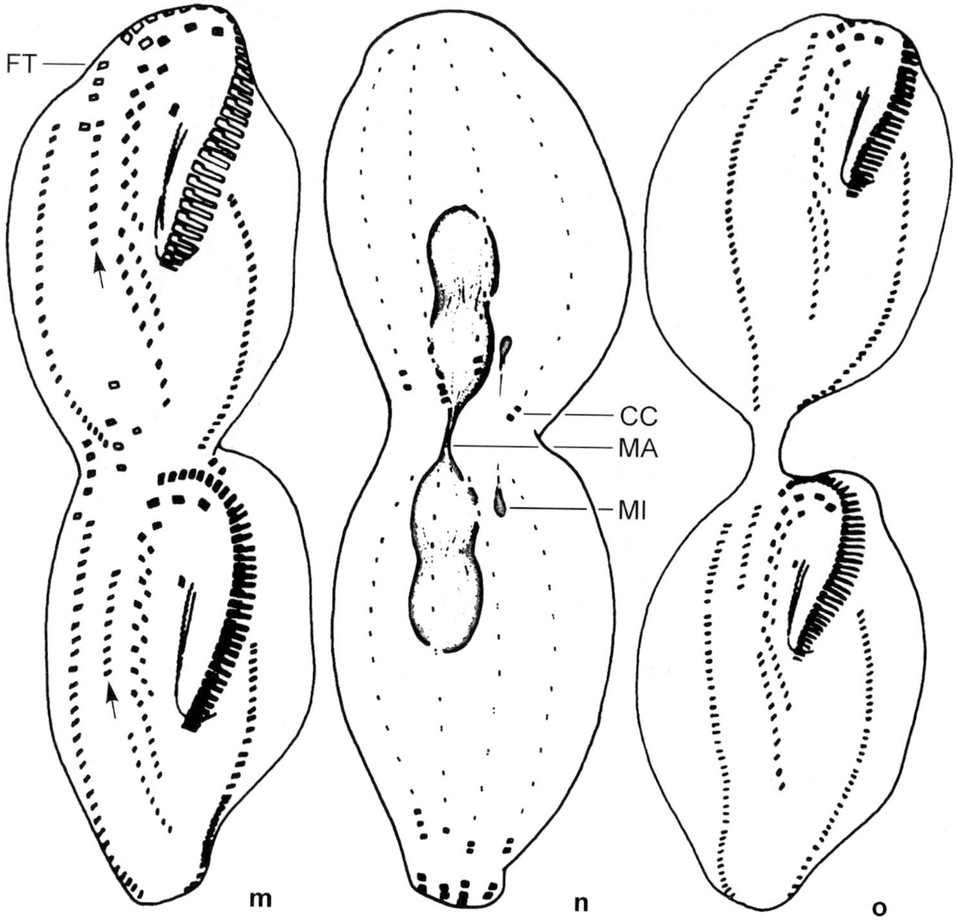

Fig. 126m–o *Holostichides typicus* (from Song & Wilbert 1988. Protargol impregnation, Wilbert's method). Infraciliature of ventral and dorsal side and nuclear apparatus of late (m, n) and very late divider (o). Arrows in (m) denote new frontoterminal cirri of proter and opisthe. CC = caudal cirri of proter, FT = parental frontoterminal cirri, MA = dividing macronucleus, MI = dividing micronucleus. Page 607.

basal bodies, likely mainly from anlage II, migrate anteriad to contribute to the formation of the frontal-midventral cirri complex of the proter. Moreover, some midventral pairs contribute to primordia formation in the proter. The parental undulating membranes and, possibly, the proximal adoral membranelle(s) are disintegrated (the reorganisation of membranelles is not mentioned by Song & Wilbert 1988, but the illustrations indicate that such a process occurs). The frontoterminal cirri, some cirri of the midventral complex, cirrus III/2, and the frontal cirri are not included in primordia formation (Fig. 126g, i–k). Figs. 126g, i show that anlage I of the proter likely originates only from the parental undulating membranes, that is, there is always a clear separation between anlage I and the frontal-midventral cirri anlagen II–n (n = up to XIV).

During middle stages the frontal-midventral cirri anlagen are formed. The anlagen, whose number varies, produce the following number of cirri (e.g., opisthe of Fig. 126k): I = 1 (left frontal cirrus), II = 2 (middle frontal cirrus and buccal cirrus), III = 2 (right frontal cirrus and cirrus III/2), IV–IX = each 2 (each one midventral pair), X = 3 (first or leftmost midventral row), XI = 8 (next midventral row), XII = 10 (last or rightmost midventral row), XIII = 9 (frontoterminal cirri).

The later stages show the migration of the cirri to their final positions (Fig. 126m, o). Ontogenesis clearly shows that no transverse cirri are produced and that the formation of the marginal rows proceeds in ordinary manner.

Dorsal ontogenesis is in *Gonostomum* pattern, that is, all kineties divide intrakinetally. At the end of each kinety 1–2 caudal cirri are formed, rarely three cirral anlagen occur. Each cirrus is composed of two basal pairs only. Division of the nuclear apparatus proceeds in ordinary manner (Fig. 126h, l, n).

According to Song & Wilbert (1988, p. 323), the ontogenesis shows five peculiarities, two of which are noteworthy and have to be briefly discussed: (i) 9–12 frontal-ventral cirral anlagen are formed (anlage I likely not included). However, Fig. 126j (opisthe) shows that up to 14 anlagen, including anlage I, are formed. (ii) The undulating membranes and the buccal cirrus of the proter originate from the same anlage. Very likely, Song & Wilbert confused the buccal cirrus and the left frontal cirrus because their statement does not agree with their illustrations (Fig. 126i, j). As in other hypotrichs, the buccal cirrus (= cirrus immediately right of the paroral) originates from anlage II.

Occurrence and ecology: Likely confined to terrestrial habitats. The type locality of *Holostichides typicus* is the soil of a wooded mountain near Qingdao, a coastal town in the province of Shandong (China), where Song & Wilbert (1988) discovered it in the A-horizon. Food not mentioned; according to Fig. 126a, diatoms have been ingested. Biomass of 10^6 specimens about 170 mg (Foissner 1998).

Paragastrostyla Hemberger, 1985

1982 ***Paragastrostyla* n. gen.**[1] – Hemberger, Dissertation, p. 132, 198.

1985 ***Paragastrostyla* n. gen.**[2] – Hemberger, Arch. Protistenk., 130: 407 (original description). Type species (by original designation on p. 407): *Paragastrostyla lanceolata* Hemberger, 1985.

1987 ***Paragastrostyla* Hemberger, 1981** – Tuffrau, Annls Sci nat. (Zool.), 8: 115 (classification of hypotrichous ciliates; incorrect year).

1994 ***Paragastrostyla* Hemberger, 1981** – Tuffrau & Fleury, Traite de Zoologie, 2: 137 (classification of hypotrichous ciliates; incorrect year).

---

[1] This name is disclaimed for nomenclatural purposes (ICZN 1999, Article 8.3). See second footnote of genus *Periholosticha*.

[2] The diagnosis by Hemberger (1985) is as follows: Je 1 rechte und linke Marginalreihe, terminal getrennt; frontal 3 verstärkte Cirren, dahinter weitere schwächere; 1 kurze Reihe von Cirren (meist 6) in Verlängerung der Frontalcirren auf die Ventralfläche übergreifend; kein Buccalcirrus; keine Transversalcirren; Caudalcirren vorhanden; Morphogenesebeginn mit einer Kinetosomenproliferation am hintersten Cirrus der Cirrenreihe; bei einer Art 5, bei einer anderen 6–7 Fronto-Ventral-Transversalcirren-Anlagen, aus den äußersten rechten Anlagen differenzieren sich zahlreiche Cirren.

1994 ***Paragastrostyla* Hemberger, 1985**[1] – Eigner & Foissner, J. Euk. Microbiol., 41: 258, Fig. 64 (redefinition of the Amphisiellidae).

1996 ***Paragastrostyla* Hemberger, 1985**[2] – Petz & Foissner, Acta Protozool., 35: 277 (improved characterisation of amphisiellid genera).

1999 ***Paragastrostyla* Hemberger, 1985** – Shi, Acta Zootax. sinica, 24: 254 (revision of hypotrichous ciliates).

1999 ***Paragastrostyla* Hemberger, 1985** – Shi, Song & Shi, Progress in Protozoology, p. 102 (revision of hypotrichous ciliates).

2001 ***Paragastrostyla* Hemberger 1985** – Aescht, Denisia, 1: 116 (catalogue of generic names of ciliates).

2001 ***Paragastrostyla* Hemberger, 1985** – Berger, Catalogue of ciliate names 1. Hypotrichs, p. 66 (nomenclator containing all basionyms, combinations, and higher taxa of hypotrichs).

2002 ***Paragastrostyla* Hemberger, 1981** – Lynn & Small, Phylum Ciliophora, p. 453 (guide to ciliate genera).

Nomenclature: No derivation of the name is given in the original description. The name *Paragastrostyla* is a composite of the Greek prefix *para+* (beside, at, along, during) and the name of the oxytrichid genus *Gastrostyla* and refers to the similar ventral ciliature of *Gastrostyla* and *Paragastrostyla*, especially in that both have a long cirral row (Hemberger 1982, p. 132). For *Gastrostyla* also no derivation is given in the original description (Engelmann 1862) and my monograph (Berger 1999). It is a composite of the Greek nouns *he gastér, gastrós* (stomach, abdomen) and *ho stýlos* (pillar, column, style) and possibly refers to the increased number of frontoventral cirri as compared to the 18 cirri of many "ordinary" oxytrichids. Both *Gastrostyla* and *Paragastrostyla* have feminine gender (Aescht 2001, p. 283, 292).

Characterisation (Fig. 111a, autapomorphies 6): Body slender elongate, converging posteriorly. Adoral zone of membranelles continuous. 3 frontal cirri. 3 or more frontoterminal cirri. Buccal cirrus lacking (A). Midventral complex composed of midventral pairs and at least 1 midventral row. Transverse cirri lacking. 1 left and 1 right marginal row. Caudal cirri present. Less than 3 dorsal kineties (A). Parental adoral zone not reorganised during cell division.

Remarks: The classification of *Paragastrostyla* is rather varied. It was established in the Oxytrichidae, but without explanation (Hemberger 1985). However, Hemberger (1982, p. 132) stated that he assigned it to the oxytrichids because the rightmost frontal-ventral-transverse cirral anlagen form several cirri which form, similar as in *Gastrostyla*, a short row during interphase.

Tuffrau (1987) and Tuffrau & Fleury (1994) included *Paragastrostyla*, like many other genera (e.g., *Uroleptus*), in the Kahliellidae Tuffrau, 1979, but without giving detailed reasons. This classification is difficult to comprehend, but very likely it is due to the longitudinal cirral row present during interphase. However, the general cirral pattern of *Paragastrostyla* does not match the kahliellid one.

---

[1] The improved diagnosis by Eigner & Foissner (1994) is as follows: The amphisiellid median cirral row originates from three rightmost anlagen. More than one cirrus left of amphisiellid median cirral row. Caudal cirri present, transverse cirri absent.

[2] The improved characterisation by Petz & Foissner (1996) is as follows: The oral primordium originates in close contact with the amphisiellid median cirral row. The amphisiellid median cirral row commences anlagen formation at its posterior end and originates from three rightmost anlagen. All dorsal kineties develop intrakinetally. More than one cirrus left of amphisiellid median cirral row. Caudal cirri present, transverse cirri absent.

Table 27 Morphometric data on *Paragastrostyla lanceolata* (la1, from Hemberger 1985; la2, the synonym *Periholosticha wilberti* from Song 1990) and *Paragastrostyla terricola* (ter, from Foissner 1988)

Characteristics [a]	Species	mean	M	SD	SE	CV	Min	Max	n
Body, length	la1	–	–	–	–	–	130.0	170.0	–
	la2	99.0	–	25.2	7.3	25.4	59.0	117.0	12
	ter	91.2	92.5	11.4	2.8	12.5	72.0	115.0	16
Body, width	la1	–	–	–	–	–	30.0	40.0	–
	la2	22.6	–	2.0	0.5	8.9	20.0	26.0	12
	ter	16.2	17.0	1.5	0.4	9.1	14.0	19.0	16
Anterior body end to proximal end of adoral zone, distance	la2	24.6	–	2.3	0.7	9.3	21.0	30.0	10
	ter	20.6	20.5	1.1	0.3	5.3	18.0	23.0	16
Distance 1 [b]	ter	35.2	35.5	2.8	0.7	8.1	31.0	39.0	16
Nuclear figure, length	ter	55.5	56.0	7.1	1.8	12.9	42.0	70.0	16
Macronuclear nodule, length	la2	7.4	–	0.7	0.2	10.1	6.3	8.8	12
	ter	6.7	7.0	1.5	0.4	22.6	3.0	10.0	16
Macronuclear nodule, width	la2	4.1	–	0.4	0.1	8.6	3.8	4.7	12
	ter	3.2	3.0	0.6	0.1	17.9	2.0	5.0	16
Macronuclear nodules, number	la1	–	–	–	–	–	15.0	18.0	–
	la2	15.3	–	1.2	0.3	7.6	13.0	17.0	21
	ter	15.7	16.0	1.3	0.3	8.2	14.0	18.0	16
Micronucleus, largest diameter	la2	3.4	–	0.5	0.1	14.4	2.8	4.2	12
	ter	2.3	2.2	0.3	0.1	12.0	2.0	3.0	16
Micronuclei, number	la1	–	–	–	–	–	3.0	4.0	–
	la2	2.1	–	0.9	0.2	41.3	2.0	4.0	34
	ter	3.0	3.0	1.0	0.2	32.2	2.0	5.0	16
Adoral membranelles, number	la1	–	–	–	–	–	16.0	17.0	–
	la2	17.5	–	0.4	0.2	4.2	16.0	21.0	14
	ter	18.9	19.0	1.1	0.3	6.1	17.0	21.0	16
Frontal cirri, number	la1	3.0	–	–	–	–	–	–	–
	la2 [d]	4.2	–	0.4	0.9	9.0	4.0	5.0	19
	ter	3.0	3.0	0.0	0.0	0.0	3.0	3.0	16
Frontoterminal cirri, number	la2	3.5	–	0.5	0.1	14.7	3.0	4.0	13
	ter	3.0	3.0	–	–	–	3.0	4.0	16
Midventral complex, cirral number 1 [c]	ter	9.8	9.5	1.2	0.3	11.9	8.0	12.0	16
Midventral complex, cirral number 2 [c]	ter	3.5	3.0	0.6	0.2	18.1	3.0	5.0	16
Midventral pairs, number of cirri	la2	7.4	–	2.0	0.4	26.6	6.0	10.0	28
Midventral row, number of cirri	la1	6.0	–	–	–	–	–	–	–
	la2	6.8	–	2.3	0.5	33.8	5.0	8.0	28
Left marginal cirri, number	la1	–	–	–	–	–	20.0	25.0	–
	ter	24.4	22.5	4.1	1.0	16.9	20.0	35.0	16
Right marginal cirri, number	la1	–	–	–	–	–	24.0	29.0	–
	ter	24.9	24.5	2.9	0.7	11.8	22.0	31.0	16
Dorsal kineties, number	la1	2.0	–	–	–	–	–	–	–
	la2	2.0	–	0.0	0.0	0.0	2.0	2.0	33
	ter	2.0	2.0	0.0	0.0	0.0	2.0	2.0	16
Caudal cirri, number (see text for details)	la1	2.0	–	–	–	–	–	–	–
	la2	2.0	–	0.0	0.0	0.0	2.0	2.0	13
	ter	3.0	3.0	0.0	0.0	0.0	3.0	3.0	16

[a] All measurements in µm. Data provided by Foissner (1988) are based on mounted, protargol-impregnated (Foissner's method), and randomly selected specimens from a non-flooded Petri dish culture. Data provided by Hemberger (1985) and Song (1990) are based on specimens impregnated with Wilbert's protargol method.

Table 27 Continued

CV = coefficient of variation in %, M = median, Max = maximum value, mean = arithmetic mean, Min = minimum value, n = number of individuals investigated (not known in detail for Hemberger's population; the "basis" for la1 is "more than 100 specimens"), SD = standard deviation, SE = standard error of arithmetic mean. Note on Hemberger's data: if only one value is given, it is listed as mean; if two values are provided, then they are listed as Min and Max.

[b] Anterior body end to rear end of midventral row (Fig. 129e).

[c] Cirral number 1 = number of cirri of midventral row plus right (= anterior) cirri of midventral pairs. Cirral number 2 = left (= posterior) cirri of midventral pairs (cirrus behind right frontal cirrus likely included).

[d] Cirrus III/2 (= cirrus behind right frontal cirrus) and cirrus II/2 (cirrus behind middle frontal cirrus; arrow in Fig. 128e; see text, for details) included.

Eigner & Foissner (1994), Petz & Foissner (1996), Shi (1999), and Shi et al. (1999) classified *Paragastrostyla* in the Amphisiellidae because of the presence of a so-called amphisiellid median cirral row. This row is defined as "a row containing all or most cirri from at least two rightmost anlagen, arranged one behind the other during cytokinesis" (Eigner & Foissner 1994, p. 243). Eigner & Foissner (1994, p. 260) correctly stated that (i) the alignment of the row is less perfect in *P. lanceolata* (in this species the row originates from the 3 rightmost anlagen) than in other amphisiellids; (ii) that an additional anlage may be involved in forming the amphsiellid median cirral row; and (iii) that *Paragastrostyla lanceolata* lacks transverse cirri because all cirri of the rightmost anlage migrate anteriad to form the anterior segment of the amphisiellid median cirral row. Petz & Foissner (1996) added further ontogenetic features (see corresponding footnote) to the definition proposed by Eigner & Foissner (1994). Shi (1999) and Shi et al. (1999) followed Eigner & Foissner (1994) and assigned *Paragastrostyla* to the amphisiellids.

Eigner (1997, p. 555; 1999, p. 46) classified, like Hemberger (1985), *Paragastrostyla lanceolata* in the Oxytrichidae. The oxytrichids sensu Eigner are an assemblage of species united, inter alia, by the neokinetal 3 anlagen formation and the presence of long primary primordia (details see Eigner 1999). In my opinion, this classification is largely artificial because Eigner ignored the most important feature of the oxytrichids, namely, fragmentation of dorsal kineties (for details see Berger 1999). I did not discuss *Paragastrostyla* in the chapter "Taxa not Considered" in my oxytrichid monograph (Berger 1999), simply because I overlooked that Hemberger and Eigner included it in this group.

I classify *Paragastrostyla* in the urostyloids because the type species *P. lanceolata* is very likely synonymous with *Periholosticha wilberti*, respectively, almost indistinguishable from *Holostichides terricola*. The close relationship of the latter two species was recognised by Eigner (1994, p. 473), who transferred *P. wilberti* to *Holostichides*, and by Foissner (2000, p. 277), who suggested synonymy of *H. terricola* and *P. wilberti*. Both *Holostichides terricola* and *Periholosticha wilberti* have been established in the urostyloids, and this classification was never seriously doubted. Thus, it seems reasonably to classify *Paragastrostyla* in the urostyloids as well.

Holostichides is obviously very closely related to, or even synonymous with, *Paragastrostyla*. However, in that I use the presence or lack of the buccal cirrus as generic feature – as it did Hemberger (1985) in the original diagnosis of *Paragastrostyla* (see corresponding footnote) – I consider both *Paragastrostyla* (without buccal cirrus) and *Holostichides* (with buccal cirrus) as valid. But this requires that *Holostichides terricola* and *Holostichides wilberti* (a synonym of *P. lanceolata)* have to be transferred to *Paragastrostyla.* For a detailed discussion of the relationship of the three species, see remarks on *Paragastrostyla lanceolata*.

Paragastrostyla closely resembles *Periholosticha*, which, however, has a bipartite adoral zone of membranelles and lacks a midventral row. For detailed comparison, see *Periholosticha*.

Hemberger (1985) assigned only one species, *Paragastrostyla lanceolata*, to the present genus, although his diagnosis includes a hint to a second species. This is due the fact that he used the diagnosis of his thesis word for word, and in his dissertation he had transferred *Cladotricha variabilis* Ruinen, 1938, to *Paragastrostyla* too. However, this species, which was redescribed by Borror & Evans (1979), has an isolated cirral row behind the buccal vertex, and produces its frontal-ventral ciliature from only six cirral primordia, which is reminiscent on non-urostyloids. Further, *Cladotricha variabilis* has, at least according to the original illustrations, very likely a buccal cirrus and transverse cirri (Ruinen 1938). Therefore it is not treated as member of *Paragastrostyla*. The combination by Hemberger (1982) is considered invalid because it was made in an unpublished thesis. Consequently, it is not listed in my catalogue (Berger 2001). This compilation also shows that no other species was added to *Paragastrostyla* so far.

Pomp & Wilbert (1988, p. 482) and Wilbert (1995, p. 283) mentioned a *Paragastrostyla* sp. However, no description and/or illustration have been provided so far. *Paragastrostyla* spec. sensu Niessen (1984, p. 89, Abb. 22, 22a, b) has a distinct buccal cirrus and a rather different cirral pattern indicating that it is not a *Paragastrostyla*, likely not even a urostyloid. Furthermore, the paper by Niessen is unpublished and thus not considered further.

Species included in *Paragastrostyla* (alphabetically arranged according to basionym): (1) *Holosticha terricola* Foissner, 1988; (2) *Paragastrostyla lanceolata* Hemberger, 1985.

Key to *Paragastrostyla* species

The two species assigned differ only in the presence/absence of cortical granules. Thus, live observation is indispensable for reliable identification. However, it is rather difficult to recognise the *Paragastrostyla* cirral pattern so that genus identification needs protargol impregnation. See also the *Periholosticha* key and the *Holostichides* key if you have problems identifying your specimen/population with the following key.

1 Cortical granules (<1 μm across, yellowish to orange yellow) present (Fig. 129a, d) . . . . *Paragastrostyla terricola* (p. 631)

\- Cortical granules lacking (Fig. 127a, 128a) . . . *Paragastrostyla lanceolata* (p. 618)

Paragastrostyla lanceolata Hemberger, 1985
(Fig. 127a–n, 128a–p, Table 27)

1982 ***Paragastrostyla lanceolata* n. spec.**[1] – Hemberger, Dissertation, p. 198, Abb. 35a–l (Fig. 127a–l).

1985 ***Paragastrostyla lanceolata* n. spec.** – Hemberger, Arch. Protistenk., 130: 408, Abb. 13 (Fig. 127a; original description. The type slide is deposited in the Institut für landwirtschaftliche Zoologie of the University of Bonn, Germany).

1990 ***Periholosticha wilberti* nov. spec.**[2] – Song, Arch. Protistenk., 138: 222, Abb. 1–16, Tabelle 1 (Fig. 128a–p; original description; new synonym. Type slides are deposited in the slide collection of the Zoological Laboratory of the College of Fisheries, Qingdao-University for Marine Science, China).

1994 ***Holostichides wilberti* (Song, 1990) nov. comb.** – Eigner, Europ. J. Protistol., 30: 473 (combination with *Holostichides*).

2001 ***Paragastrostyla lanceolata* Hemberger, 1985** – Berger, Catalogue of ciliate names 1. Hypotrichs, p. 67 (nomenclator containing all basionyms, combinations, and higher taxa of hypotrichs).

2001 ***Periholosticha wilberti* Song, 1990** – Berger, Catalogue of ciliate names 1. Hypotrichs, p. 71 (nomenclator containing all basionyms, combinations, and higher taxa of hypotrichs).

Nomenclature: No derivation of the species-group name *lanceolata* is given in the original description. The Latin adjective *lanceolát·us -a -um* (lanceolate; Hentschel & Wagner 1996, p. 356) obviously alludes to the lancet-shaped body outline. *Paragastrostyla lanceolata* was fixed as type species of *Paragastrostyla* by original designation. The species *Periholosticha wilberti* is named in honour of Norbert Wilbert, University of Bonn, Germany, who taught Weibo Song (Song 1990, p. 230).

Remarks: The taxonomy of the present species is rather complicated because two very similar species have been described. A close relationship of *Holostichides terricola* Foissner, 1988 and *Periholosticha wilberti* Song, 1990 was already recognised by Eigner (1994), who therefore transferred *P. wilberti* to *Holostichides*. As already mentioned in the genus section, Foissner (2000) even suggested synonymy of these two species, although *Holostichides terricola* has cortical granules, whereas they are lacking in *Periholosticha wilberti*[3]. Foissner (1988, p. 112) compared *Holostichides terricola* with *Paragastrostyla lanceolata* and recognised several agreements, for example, the lack of the buccal cirrus and the number of dorsal kineties. As differences, Foissner (1988) mentioned the lack of zigzagging midventral pairs and cortical granules in *Paragastrostyla lanceolata* (Fig. 127a). However, Hemberger studied live specimens only superficially, if at all, and therefore we will never know whether or not his species has cortical granules.

There is no doubt that *Paragastrostyla lanceolata* must be closely related to *Holostichides terricola* and *Periholosticha wilberti*, as indicated by the agreement in most features: (i) body slender; (ii) contractile vacuole ahead of mid-body; (iii) 13–18 ellipsoidal macronuclear nodules form two more or less distinct strands left of midline; (iv)

[1] This name is disclaimed for nomenclatural purposes (ICZN 1999, Article 8.3). See second footnote of genus *Periholosticha*.

[2] The diagnosis by Song (1990) is as follows: In vivo etwa 70–120 × 20–25 μm große lanzettförmige *Periholosticha* mit ca. 15 Makronuclei und 2–4 Mikronuclei. Infraciliatur: 4–5 separate Frontalcirren, 3–4 Frontoterminalcirren, 3–5 Paare Midventralciren, 2 Caudalcirren und 2 Dorsalkineten.

[3] Song (1990, p. 222) wrote "Plasma hell und farblos, keine erkennbaren subpelliculären Granulae oder Kristalle." which means "Cytoplasm colourless and bright, no recognisable cortical granules or crystals".

16–21 adoral membranelles; (v) adoral zone occupies around 22–25% of body length; (vi) lack of buccal cirrus; (vii) three slightly enlarged frontal cirri; (viii) 3–4 frontoterminal cirri; (ix) few midventral pairs (zigzag pattern indistinct in *P. lanceolata*); (x) one midventral row terminating ahead of mid-body; (xi) transverse cirri lacking; (xii) right marginal row commences slightly behind level of last frontoterminal cirrus, composed of 22–31 cirri; (xiii) 20–35 left marginal cirri; (xiv) two almost bipolar dorsal kineties, each usually with one caudal cirrus; (xv) very similar or even identical cell division (investigated for *Paragastrostyla lanceolata* and *Periholosticha wilberti*).

There are only two noteworthy differences between the three taxa, namely in oral apparatus and body size. (i) The adoral zone of membranelles is *Oxytricha*-like in *Paragastrostyla lanceolata*, but more or less *Gonostomum*-like in *Paragastrostyla terricola* and *Periholosticha wilberti*, and the undulating membranes are curved and intersecting in *Paragastrostyla lanceolata*, but straight and in parallel in *Paragastrostyla terricola* and *Periholosticha wilberti*. However, these differences can be explained, at least partially, by the fact that Hemberger (1985, p. 397) used the body outline of live specimens in which the infraciliature was drawn in without the use of a camera lucida! Possibly this drawing method is also the reason why the zigzagging pattern is rather indistinct in *Paragastrostyla lanceolata*. A lateral displacement of plus/minus 1 µm makes a rather distinct difference in the midventral cirral pattern. (ii) The body is distinctly larger in *Paragastrostyla lanceolata* than in the other two species. However, important meristic features, e.g., the number of adoral membranelles and marginal cirri are almost identical, indicating that the different body size must not be over interpreted (Table 27).

Since all three species/populations are rather well investigated, new morphological and ontogenetic data will not increase the knowledge significantly. Thus, I provide a pragmatic solution, namely, I synonymise *Periholosticha wilberti* with *Paragastrostyla lanceolata* and "define" this species by the lack of cortical granules. By contrast, *Paragastrostyla terricola* is characterised by small (less than 1 µm across), yellowish cortical granules. A similar distinction was proposed for *Bakuella edaphoni* (without granules) and *B. granulifera* (with granules) by Foissner et al. (2002). However, I keep all data separate so that every experienced taxonomist can check the situation individually.

For comparison of *Paragastrostyla lanceolata* with similar species see *Paragastrostyla terricola*. The diagnosis of *Paragastrostyla lanceolata* provided by Hemberger (1985) is not a diagnosis sensu stricto, but a rather brief description of the species and therefore not repeated in a footnote.

Morphology: For reasons mentioned above, I keep Hemberger's and Song's data separate.

Type population (Fig. 127a, Table 27): Body size mentioned (130–170 × 30–40 µm) likely from protargol-impregnated (Wilbert's method) specimens; length:width ratio 5–6:1. Body outline rather variable, usually lanceolate (species name), that is, anterior end broadly rounded, widest behind mid-body, posterior margins strongly converging, and rear end narrowly rounded or even slightly pointed. Body flexible and metabolic. Macronuclear nodules form strand left of midline; individual nodules ellipsoidal. Micronuclei globular, attached to macronuclear figure. Contractile vacuole at about 28%

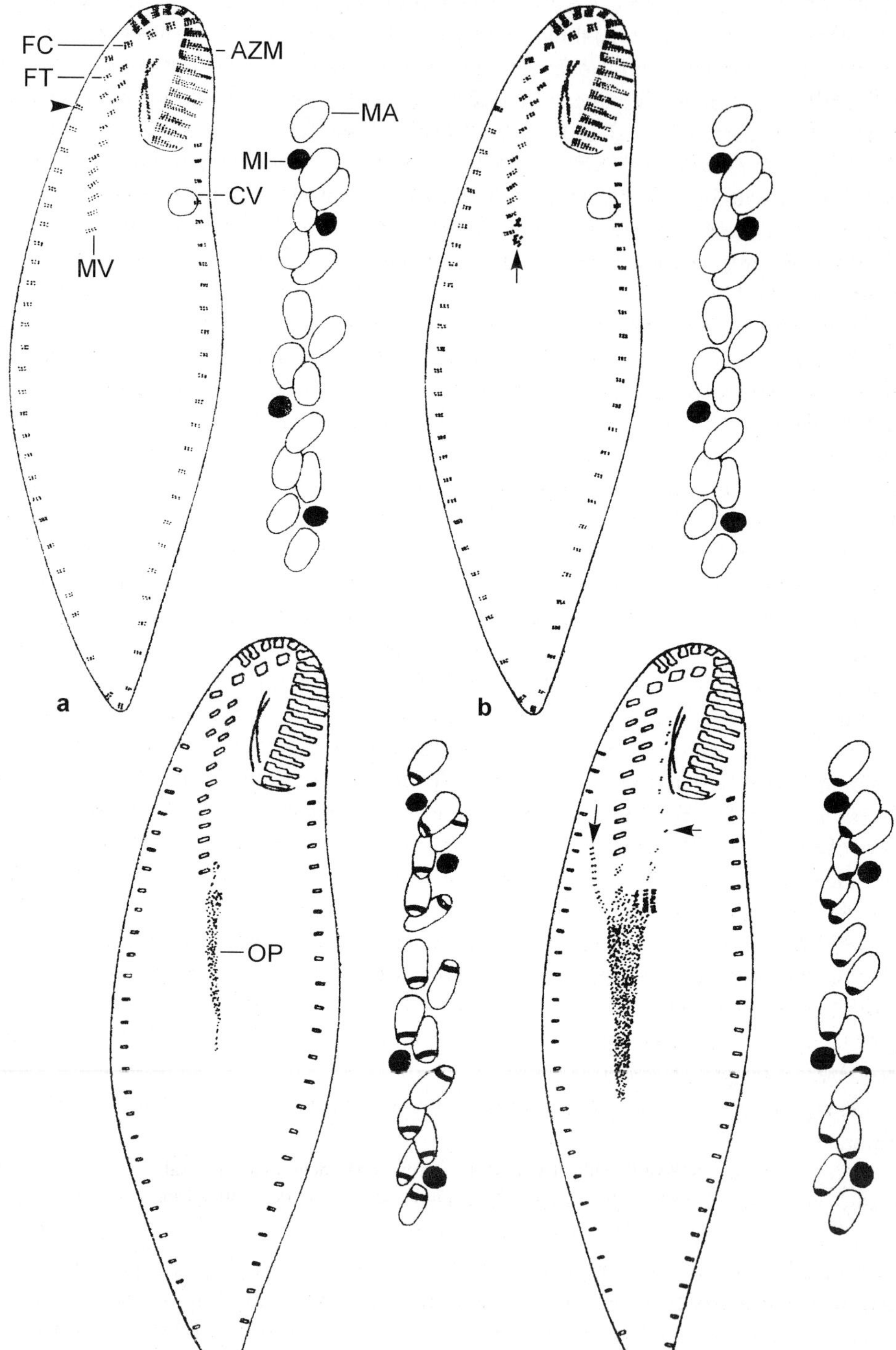
FC
FT
AZM
MA
MI
CV
MV
a
b
OP
c
d

of body length near left cell margin. Cortical granules lacking (see remarks). Adoral zone occupies about 20% of body length, composed of 16–17 membranelles of usual fine structure. Undulating membranes slightly curved, of about equal length, commencing at same level, and optically intersecting in anterior third. Three slightly enlarged frontal cirri (3 × 3 basal bodies) in usual position. Buccal and transverse cirri lacking. Three (rarely four) frontoterminal cirri (2 × 3 basal bodies) forming short row in ordinary position. Behind right frontal cirrus 6–8 cirri (2 × 3 basal bodies) with last cirrus at about level of proximal adoral membranelle (Hemberger 1982, 1985 does not mention a zigzagging pattern which is, indeed, very indistinct; Fig. 127a). Behind this 6–8 cirri a row (this is, in my opinion, a midventral row) which is usually composed of six cirri (2 × 4 basal bodies) and terminates at 31% of body length in specimen illustrated (Fig. 127a). For the interpretation of the midventral pattern, see Figs. 127m, n. Right marginal row commences slightly behind level of last frontoterminal cirrus, terminates, like left row, at/near rear body end; marginal cirri usually composed of 2 × 3 basal bodies bearing about 9 µm long cilia. Dorsal cilia 4–5 µm long, arranged in two kineties each bearing one caudal cirrus composed of four, about 9 µm long cilia. Rarely specimens with long cilia on the rearmost basal body pair occur (Hemberger 1982, p. 203), which is reminiscent of the synonym *Periholosticha wilberti*, where Song also occasionally observed more than one caudal cirrus per dorsal kinety.

Type population of the synonym *Periholosticha wilberti* (Fig. 128a–d, Table 27): Body size about 70–120 × 20–25 µm in life, length:width ratio around 4:1 in life and 4.4:1 on average after protargol preparation (Wilbert's method). Outline hardly variable, slenderly lanceolate, anteriorly rounded, posteriorly almost triangularly pointed. Body distinctly flattened anteriorly and slightly twisted about main axis. Macronuclear nodules form strand left of midline; individual nodules ellipsoidal or globular, contain normal-sized nucleoli. Micronuclei globular, attached to macronuclear figure. Contractile vacuole at about 40% of body length near left cell margin. Cytoplasm bright and colourless. No cortical granules or cytoplasmic crystals. Food vacuoles small. Movement without peculiarities.

Adoral zone occupies 25% of body length on average, roughly arranged in *Gonostomum* pattern, that is, extends straight along left body margin, performing right bend and slight clockwise rotation to plunge into buccal cavity, composed of 18 membranelles

---

← **Fig. 127a–d** *Paragastrostyla lanceolata* (a, from Hemberger 1985; b–c, from Hemberger 1982. Protargol impregnation, Wilbert's method). **a:** Infraciliature of ventral side and nuclear apparatus of representative specimen, 149 µm. Arrowhead marks anterior end of right marginal row. **b–d:** Infraciliature of ventral side and nuclear apparatus of very early and early dividers. Arrow in (b) denotes some basal bodies at the rear end of the midventral row; these patches are the first sign, besides the reorganisation band of the macronuclear nodules, of the beginning cell division. Horizontal arrow in (d) marks a loose row of basal body pairs extending from the left anterior corner of the oral primordium to the right side of the parental undulating membranes. Vertical arrow in (d) denotes anlage which forms the rightmost cirral streak in both the opisthe and the proter; it divides transversely in later stages (between stages shown in f and g) and is thus a primary primordium. AZM = adoral zone of membranelles, CV = contractile vacuole, FC = right frontal cirrus, FT = frontoterminal cirri, MA = macronuclear nodule, MI = micronucleus, MV = midventral row, OP = oral primordium. Page 618.

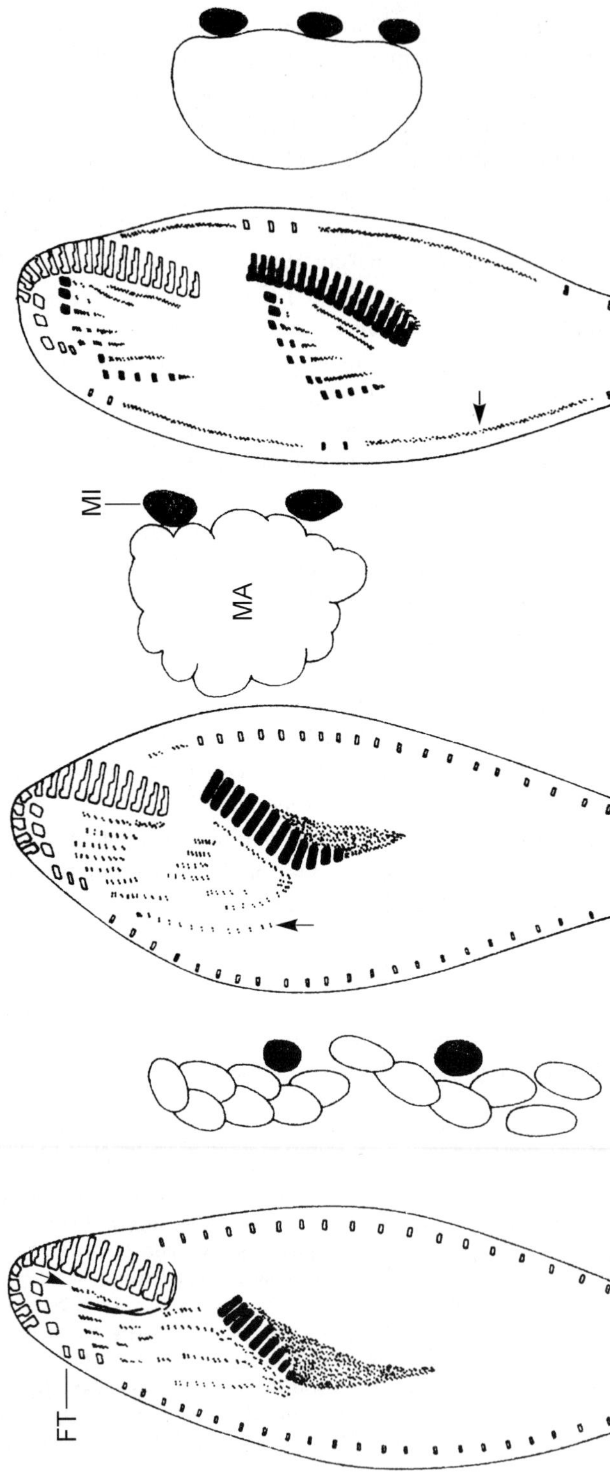

Fig. 127e–g *Paragastrostyla lanceolata* (from Hemberger 1982. Protargol impregnation, Wilbert's method). Infraciliature of ventral side and nuclear apparatus of middle dividers. Parental structures white, news black. Arrow in (e) marks an anlage in the parental buccal field which forms, together with the disorganised parental undulating membranes, the anlage I of the proter. Arrow in (f) denotes the rightmost anlage which is still an undivided primary primordium. Arrow in (g) marks the anlage for the right marginal row of the opisthe. FT = parental frontoterminal cirri, MA = fused macronucleus, MI = micronucleus. Page 618.

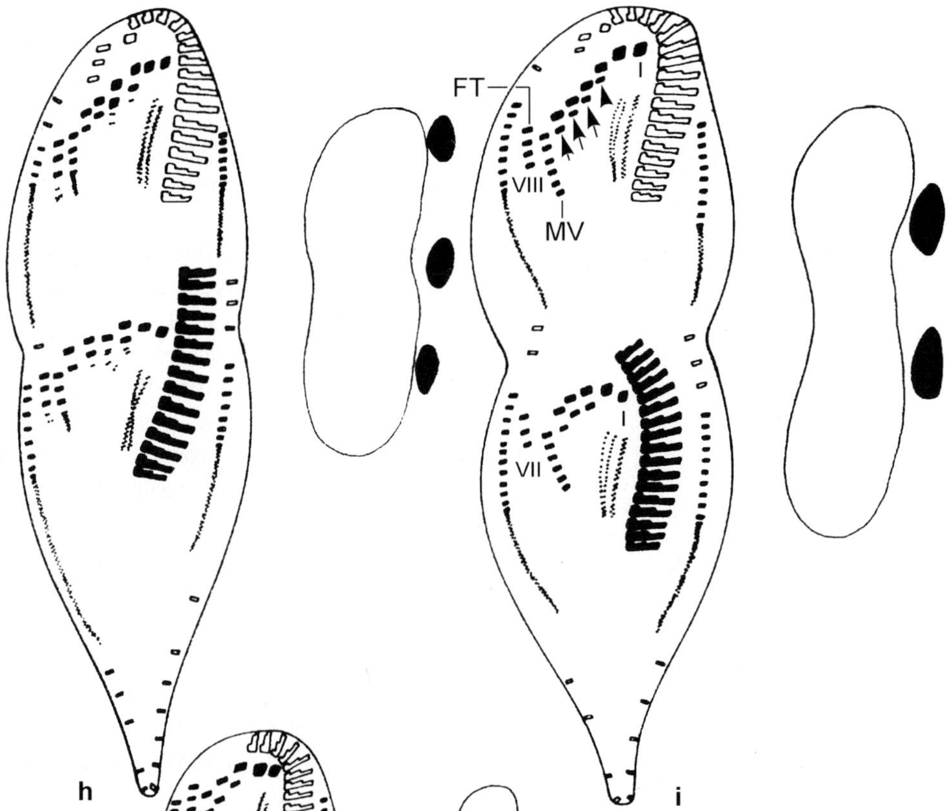

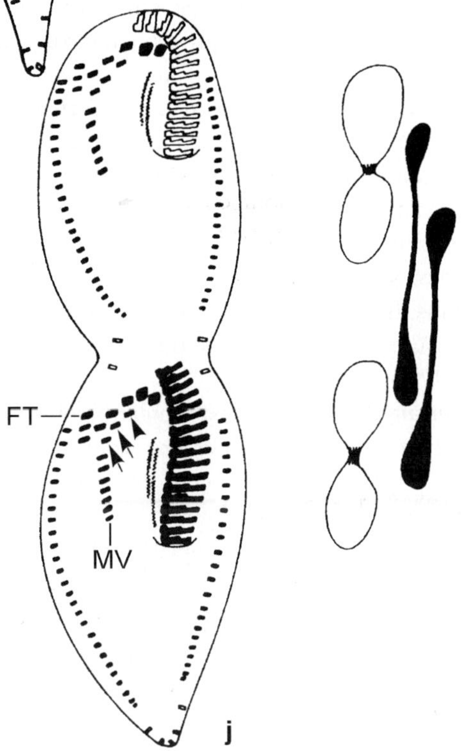

Fig. 127h–j: *Paragastrostyla lanceolata* (from Hemberger 1982. Protargol impregnation, Wilbert's method). Infraciliature of ventral side and nuclear apparatus of late and very late dividers (sizes not indicated). Arrows in (i) and (j) mark new midventral pairs, the arrowhead denotes cirrus III/2, that is, the cirrus behind the right frontal cirrus. Interestingly, in interphasic specimens (Fig. 127a) this cirrus is possibly lacking, indicating that it is resorbed in very late (postdivider) stages. Note the identical cirral formation pattern in the synonym, *Periholosticha wilberti* (Fig. 128m). So far this identity (homology) was not recognised because *P. lanceolata* and *P. wilberti* were classified in different higher taxa. FT = new frontoterminal cirri of proter (i) and opisthe (j), MV = new midventral row of proter (i) and opisthe (j), I, VII, VIII = leftmost and rightmost cirral anlage in a divider with 7 (opisthe) and 8 (proter) cirral anlagen. Page 618.

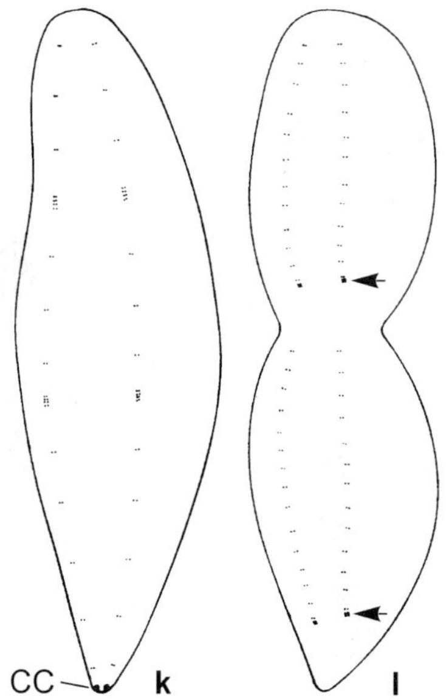

Fig. 127k, l *Paragastrostyla lanceolata* (from Hemberger 1982. Protargol impregnation, Wilbert's method). Infraciliature of dorsal side of a middle and a late divider. Division of dorsal kineties proceeds in *Gonostomum* pattern (see text for explanation). Arrows in (l) mark new caudal cirri of kinety 2. CC = parental caudal cirri. Page 618.

on average; individual membranelles of ordinary fine structure. Buccal field moderately large (Song 1990), according to Figs. 128a, b, however, narrow and likely flat. Undulating membranes of about same length, straight and not intersecting optically. Cytopharynx of usual structure.

Cirral pattern of usual variability, number of frontoterminal cirri and midventral cirri showing high coefficients of variation (Table 27). Cirri 8–10 µm long, for fine structure of individual cirri, see Fig 128d. Invariably three slightly enlarged frontal cirri in usual position. Rarely one cirrus behind the middle frontal cirrus (cirrus II/2); usually this cirrus is the buccal cirrus, however, in specimens where it is present (Fig. 128e, f, h), it is not in the specific position right of the paroral, but only 3 µm behind the frontal cirrus, that is, distinctly away from the paroral. Invariably one cirrus (cirrus III/2) behind the right frontal cirrus. Buccal (see above) and transverse cirri lacking. Frontoterminal cirri form short row commencing right of right frontal cirrus and terminating slightly ahead of anterior end of right marginal row. 6–10 midventral cirri form 3–5 midventral pairs arranged in distinct zigzag-pattern terminating slightly ahead of level of proximal adoral membranelle. Midventral row slightly oblique, composed of 5–8 cirri, terminates at 46% of body length in specimen shown in Fig. 128b. Right marginal row begins, as in Hemberger's population, slightly behind level of last frontoterminal cirrus, ends, like left row, near/at rear body end. Number of marginal cirri not given in original description; according to the illustrations 24–28 right and 19–26 left cirri are present (Fig. 128b, e, f, h, o).

Dorsal bristles about 3 µm long, arranged in two almost bipolar kineties. Caudal cirri fine, almost not set off from marginal cirri and thus difficult to recognise on the pointed posterior body end (Fig. 128b, c). Rarely 2–3 cirri occur on each dorsal anlage; however, they fuse to a single cirrus in non-dividing specimens.

Cell division: Hemberger (1982; Fig. 127b–l) and Song (1990, Fig. 128e–p) investigated the division of the present species, respectively, its synonym *Periholosticha wilberti*. The results are very similar and most details are recognisable from the illustrations to which the reader is mainly referred. Consequently, the description of the process is rather brief. It commences with the formation of a longish oral primordium at/near the left side of the rear end of the midventral row (Fig. 127b, c, 128e).

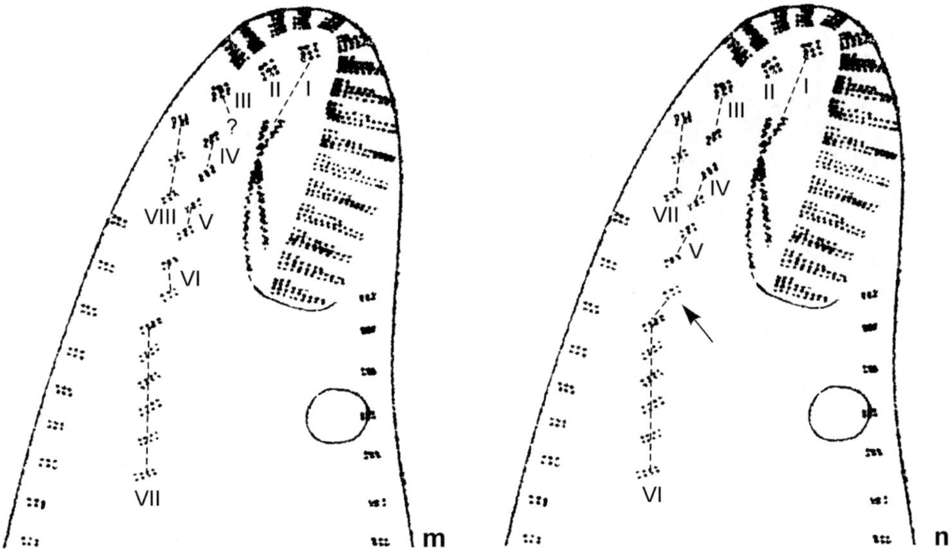

Fig. 127m, n *Paragastrostyla lanceolata* (detail of Fig. 127a). The cirral pattern of the interphasic specimen shown in Fig. 127a is difficult to interpret. Figure m shows one of several possibilities with eight (I–VIII) frontal-midventral cirral anlagen. However, in this case the cirrus (= cirrus III/2) behind the right frontal cirrus is lacking. If the anterior cirrus of streak IV is designated as cirrus III/2 then one of the anlagen IV–VI consists of only one cirrus. However, according to the ontogenetic data none of the anlagen III–VIII produces only one cirrus. Figure n shows the proposal by Eigner (1997) with seven (I–VII) anlagen. However, this assumption also does not seem very reliable because the anteriormost cirrus (arrow) of the midventral row is then not in line with the other cirri of the row, which is very unusual. Possibly, Hemberger overlooked one cirrus when he drew the cirral pattern without a camera lucida. Page 618.

Somewhat later several parental cirri of the midventral row are incorporated into the first cirral anlagen (Fig. 127d, 128f). On the right side of the oral primordium one distinct anlage extends anteriad in both populations (arrow in Fig. 127d, 128h). In Hemberger's population a loose row of basal body pairs extends from the left anterior end of the oral primordium to the right side of the parental undulating membranes (Fig. 127d, horizontal arrow); a feature not so clearly recognisable in Song's population, which, however, does not show the same stage. Subsequently, usually all cirri – except the three frontal cirri and the frontoterminal one – are modified to anlagen (Fig. 127e, f). An additional anlage, which becomes proter's anlage I, is formed in the parental buccal field (Fig. 127e, 128h).

In middle dividers usually seven or eight, rarely nine or ten anlagen are recognisable, including anlage I which forms the undulating membranes. Anlagen I and II form one cirrus each, that is, the left and the middle frontal cirrus. Rarely, anlage II forms a second cirrus which in other hypotrichs migrates to near the paroral and then becomes the buccal cirrus. However, in Song's population it does not migrate into this specific position when it is present, but is rather close behind the middle frontal cirrus (Fig. 128e, f, h, m, o). The next 3–5 (rarely 6) anlagen produce two cirri each, which

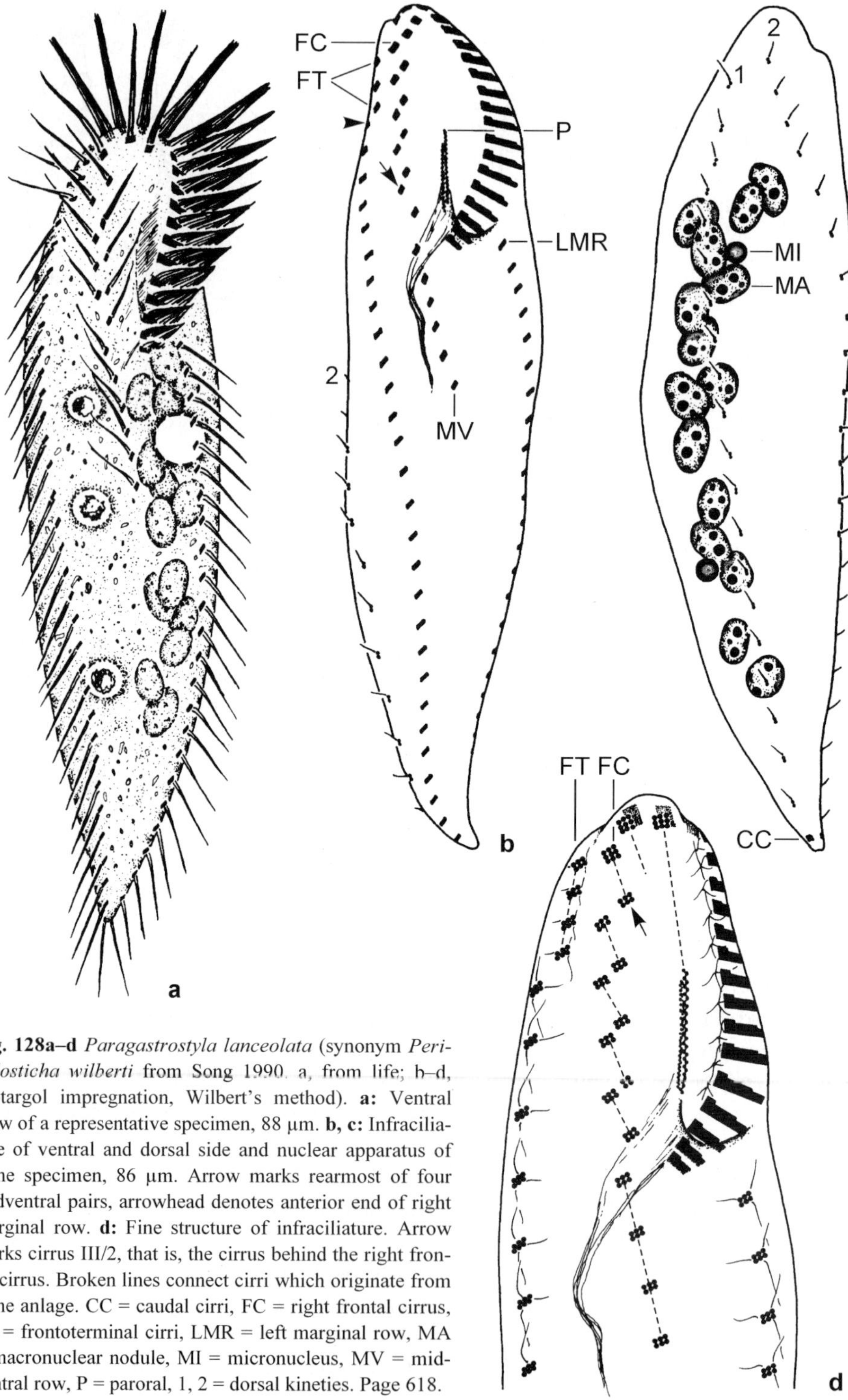

Fig. 128a–d *Paragastrostyla lanceolata* (synonym *Periholosticha wilberti* from Song 1990. a, from life; b–d, protargol impregnation, Wilbert's method). **a:** Ventral view of a representative specimen, 88 µm. **b, c:** Infraciliature of ventral and dorsal side and nuclear apparatus of same specimen, 86 µm. Arrow marks rearmost of four midventral pairs, arrowhead denotes anterior end of right marginal row. **d:** Fine structure of infraciliature. Arrow marks cirrus III/2, that is, the cirrus behind the right frontal cirrus. Broken lines connect cirri which originate from same anlage. CC = caudal cirri, FC = right frontal cirrus, FT = frontoterminal cirri, LMR = left marginal row, MA = macronuclear nodule, MI = micronucleus, MV = midventral row, P = paroral, 1, 2 = dorsal kineties. Page 618.

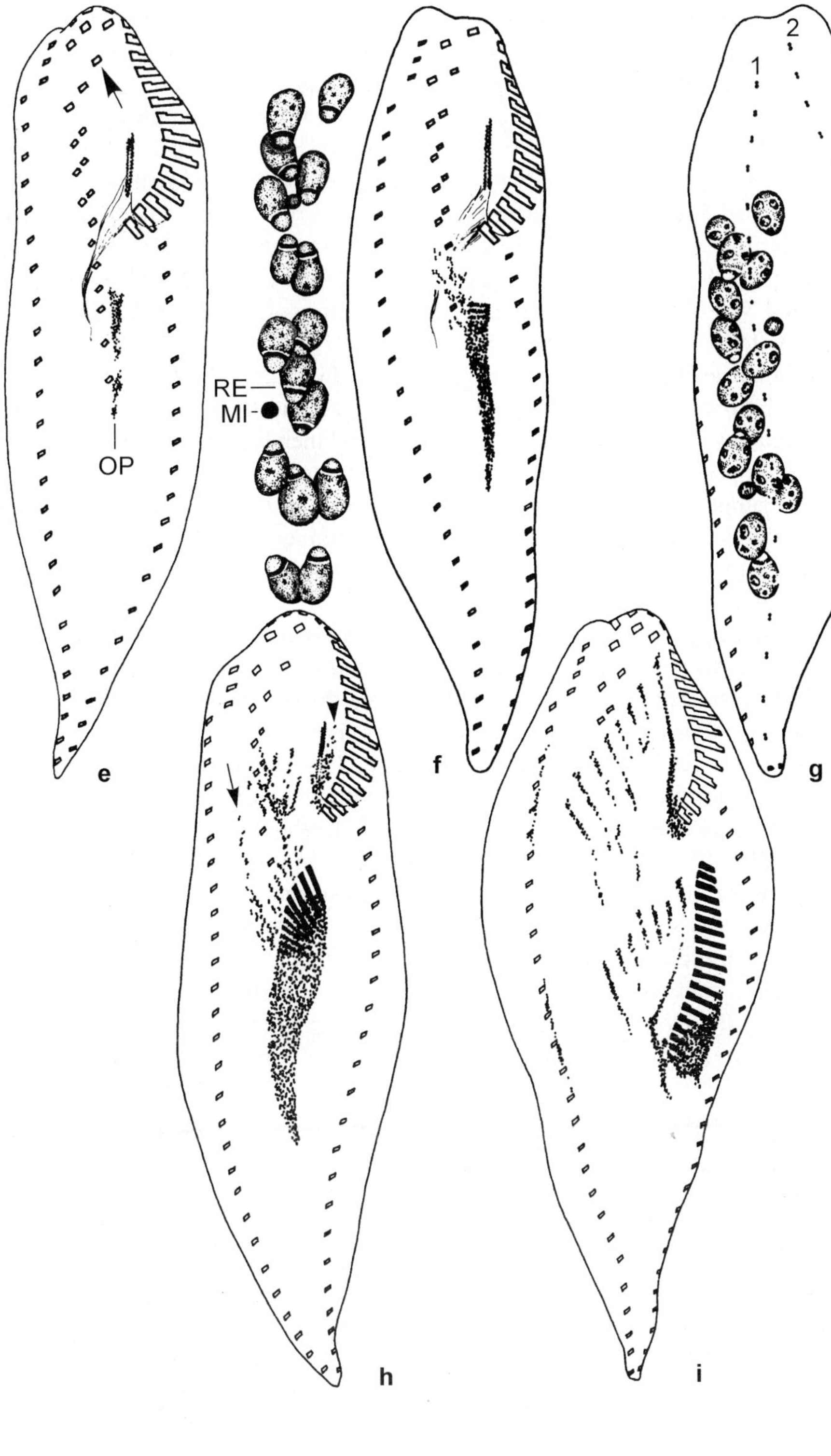
OP
RE
MI
1
2
e
f
g
h
i

form the right frontal cirrus and cirrus III/2 and two to four (rarely five) midventral pairs. The penultimate anlage from right produces 6–7 cirri in Hemberger's population and 5–8 cirri in Song's population; in late dividers these cirri migrate posteriorly and form the midventral row (Fig. 127h, j, 128l, m, o). The rightmost anlage usually produces three cirri, which migrate to near the distal end of the adoral zone of membranelles; consequently, they can be homologised with the frontoterminal cirri of other hypotrichs. No transverse cirri are formed. Summarising, Hemberger (1982) counted the following number of cirri produced within the anlagen I–VIII (in parentheses the designation of the cirri is given): I = 1 cirrus (left frontal cirrus), II = 1 (middle frontal cirrus), III = 2 (right frontal cirrus and cirrus III/2), IV = 2 (first midventral pair), V = 2 (second midventral pair), VI = 2 (third midventral pair; one of the anlagen III–VI is not always present), VII = 6–7 (midventral row), VIII = 3–4 (frontoterminal cirri). Song (1990) provided the following counts: I = 1, II = 1–2, III = 2, IV = 2, V = 2, VI = 2, VII = 2 (rarely a further anlage producing two cirri occurs; Fig. 128l), VIII = 5–8, IX = 3–4. The parental adoral zone of membranelles is retained and therefore forms the adoral zone of the proter (Fig. 127c–j). Possibly the proximal membranelle(s) is (are) disorganised for a short period (Fig. 128i).

The formation of the dorsal infraciliature proceeds in *Gonostomum* pattern and is thus rather simple, that is, both kineties divide by intrakinetal proliferation at two levels. Each kinety forms a caudal cirrus at its rear end. The division of the nuclear apparatus proceeds in ordinary manner (Fig. 127b–j).

Neither Hemberger (1982) nor Song (1990) clearly state or show in their illustrations which frontal-midventral cirral anlagen originate from primary primordia and which are produced independently in proter and opisthe. Only the rightmost anlage, which forms the frontoterminal cirri, almost certainly originates from a primary primordium in both populations (Fig. 127d, f, g, 128h, i). Figs. 127e, f by Hemberger (1982), who does not make any comment about this feature, do not allow a clear decision. According to Eigner (1997, p. 562), anlagen 1–5 (= I–V) possibly develop by long primary primordia. And Song's Fig. 128h, i are also difficult to interpret in this respect. Moreover, Song used the terms primary and secondary primordia mistakenly in that he designated the frontal-ventral cirral streaks of the opisthe as primary primordia and those of the proter as secondary primordia (see his Abb. 9).

Occurrence and ecology: Likely confined to terrestrial habitats. Hemberger (1985, p. 408) discovered *Paragastrostyla lanceolata* in an infusion of garden soil. Unfortunately, he did not explain that this garden soil sample is not from Peru, which

---

← **Fig. 128e–i** *Paragastrostyla lanceolata* (from Song 1990. Protargol impregnation, Wilbert's method). Infraciliature of ventral and dorsal side and nuclear apparatus of early and middle dividers, e = 114 µm, f, g = 106 µm, h = size not indicated, i = 109 µm. Arrow in (e) marks cirrus II/2, which is homologous to the buccal cirrus; usually this cirrus is lacking in *P. lanceolata*, but when it is present it is located immediately behind the middle frontal cirrus and not close to the paroral. Arrow in (h) denotes rightmost anlage which is very probably, like in Hemberger's population (Fig. 127d), a primary primordium. Arrowhead in (h) denotes an anlage in the parental buccal field which becomes, together with the disorganised parental undulating membranes, anlage I of the proter. MI = micronucleus, OP = oral primordium, RE = reorganisation band, 1, 2 = dorsal kineties. Page 618.

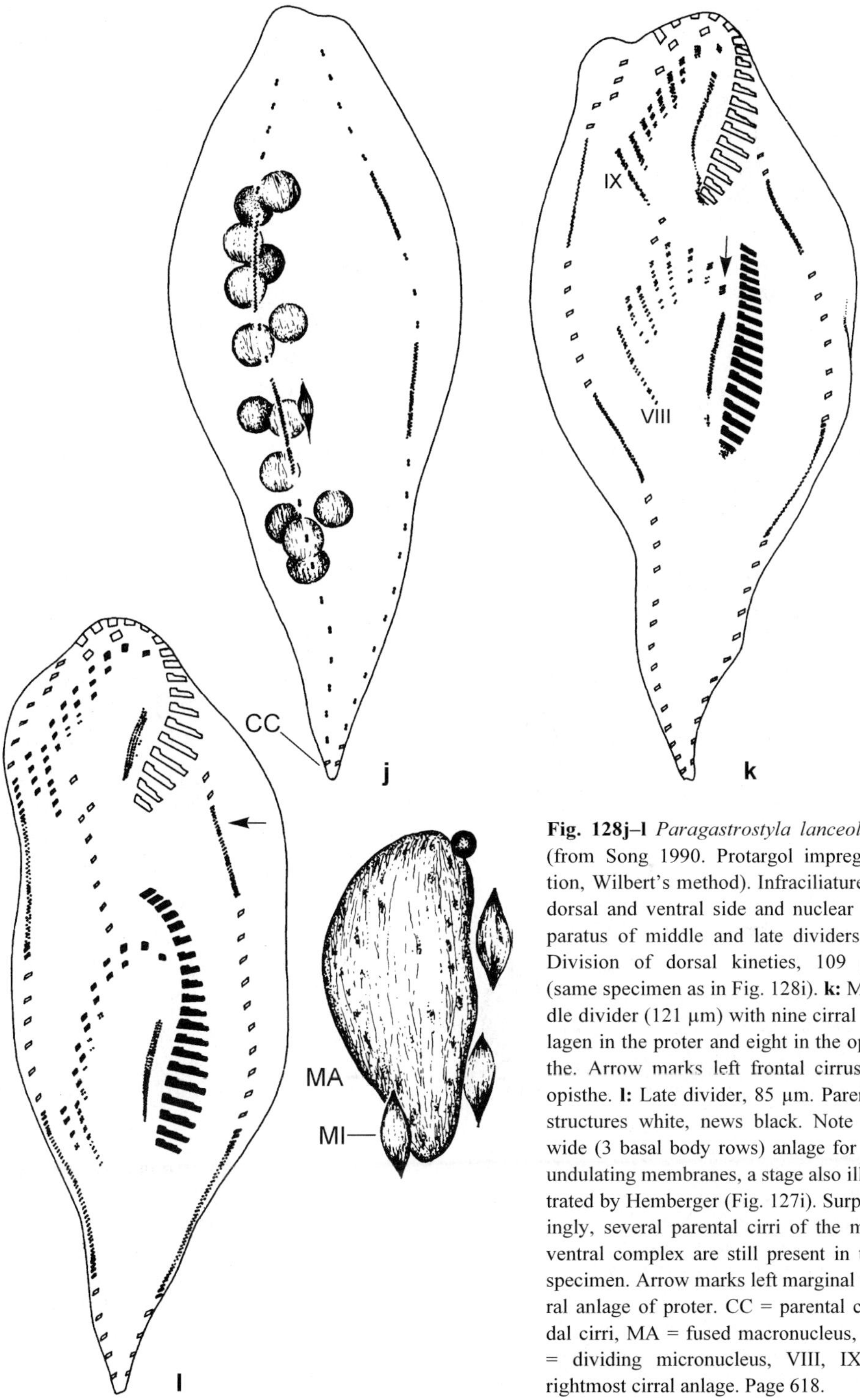

Fig. 128j–l *Paragastrostyla lanceolata* (from Song 1990. Protargol impregnation, Wilbert's method). Infraciliature of dorsal and ventral side and nuclear apparatus of middle and late dividers. **j:** Division of dorsal kineties, 109 μm (same specimen as in Fig. 128i). **k:** Middle divider (121 μm) with nine cirral anlagen in the proter and eight in the opisthe. Arrow marks left frontal cirrus of opisthe. **l:** Late divider, 85 μm. Parental structures white, news black. Note the wide (3 basal body rows) anlage for the undulating membranes, a stage also illustrated by Hemberger (Fig. 127i). Surprisingly, several parental cirri of the midventral complex are still present in this specimen. Arrow marks left marginal cirral anlage of proter. CC = parental caudal cirri, MA = fused macronucleus, MI = dividing micronucleus, VIII, IX = rightmost cirral anlage. Page 618.

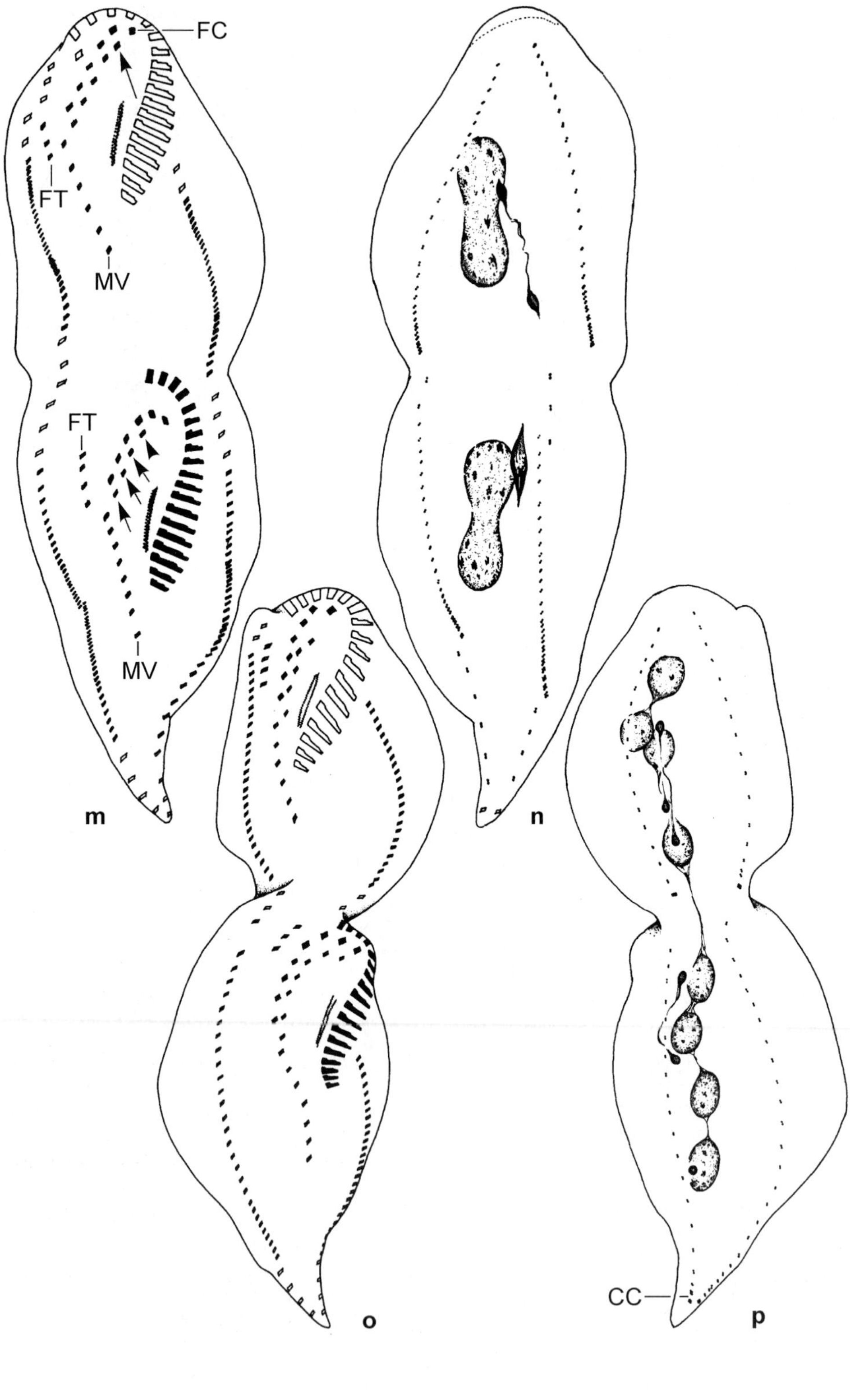
FC
FT
MV
FT
MV
m
n
o
CC
p

is the sole country mentioned in his material section (Hemberger 1985, p. 397). According to his thesis (Hemberger 1982, p. 2), he studied garden soil only from two sites in Germany, namely (i) from Bornich (about 50°10'N 7°48'E), a small village between the towns of Koblenz and Mainz, and (ii) from the campus of the Bonn University, Zoological Institute. H. Hemberger informed me that the type locality is in the garden of the Institut für landwirtschaftliche Zoologie und Bienekunde (University of Bonn) at the building "Gut Melb" (50°43'N 07°50'E). According to Foissner's (1998) compilation, *Paragastrostyla lanceolata* was found in the Neotropis, which is possibly incorrect for reasons mentioned above. However, it is reliable recorded from Namibia (Foissner et al. 2002, p. 61).

The type location of the synonym, *Periholosticha wilberti* is a protected forest plantation area on the campus of the Ocean University of Qingdao (36°8'N 120°43'E), P. R. China, where it occurred in the upper soil layer (0–3 cm). It excysted very rapidly because it occurred 24 h after water saturation, however, with low abundance. Song (1990) established a non-clonal culture with Eau de Volvic (French mineral water) as medium and some squashed wheat grains to support bacterial growth.

Biomass of 10^6 specimens, due to rather different body size (see above), ranging from 27 mg for the synonym *Periholosticha wilberti* to 120 μm for *Paragastrostyla lanceolata* (Foissner 1987a, 1998).

Paragastrostyla terricola (Foissner, 1988) comb. nov. (Fig. 129a–j, Table 27)

1988 ***Holostichides terricola* nov. spec.**[1] – Foissner, Stapfia, 17: 108, Fig. 8a–h, Tabelle 5 (Fig. 129a–h; original description. The holotype slide [1989/18] and 2 paratype slides [1989/19, 20] are deposited in the Oberösterreichische Landesmuseum in Linz [LI], Austria).

2001 ***Holostichides terricola* Foissner, 1988** – Berger, Catalogue of ciliate names 1. Hypotrichs, p. 39 (nomenclator containing all basionyms, combinations, and higher taxa of hypotrichs).

Nomenclature: No derivation of the species-group name *terricola* is given in the original description. It is a composite of the Latin nouns *terra* (soil) and *incola* (inhabitant), and refers to the habitat were the species was discovered, namely the upper soil layer. Usually, species-group names ending with *-cola* are considered as appositive substan-

← **Fig. 128m–p** *Paragastrostyla lanceolata* (from Song 1990. Protargol impregnation, Wilbert's method). Infraciliature of ventral and dorsal side and nuclear apparatus of late and very late dividers, m, n = 116 μm, o, p = 127 μm. Parental structures white, new black. The proter in (m) has a cirrus II/2 (long arrow), that is, a buccal cirrus, which, however, does not migrate to the paroral during interphase. Arrowhead denotes cirrus III/2, that is, the cirrus behind the right frontal cirrus. Short arrows mark the three midventral pairs of the opisthe. CC = new caudal cirri of opisthe, FC = left frontal cirrus of proter, FT = new frontoterminal cirri, MV = new midventral row. Page 618.

[1] The diagnosis by Foissner (1988) is as follows: In vivo etwa 80–140 × 15–30 μm großer, hinten zugespitzter *Holostichides* mit gelben bis orangegelben, kugeligen subpelliculären Granula. Durchschnittlich 16 in 2 Reihen angeordnete Makronucleus-Teile, 19 adorale Membranellen und 3 Frontoterminalcirren. Linke Midventralreihe mit durchschnittlich 3,5, rechte mit 9,8 Cirren. 2 Dorsalkineten.

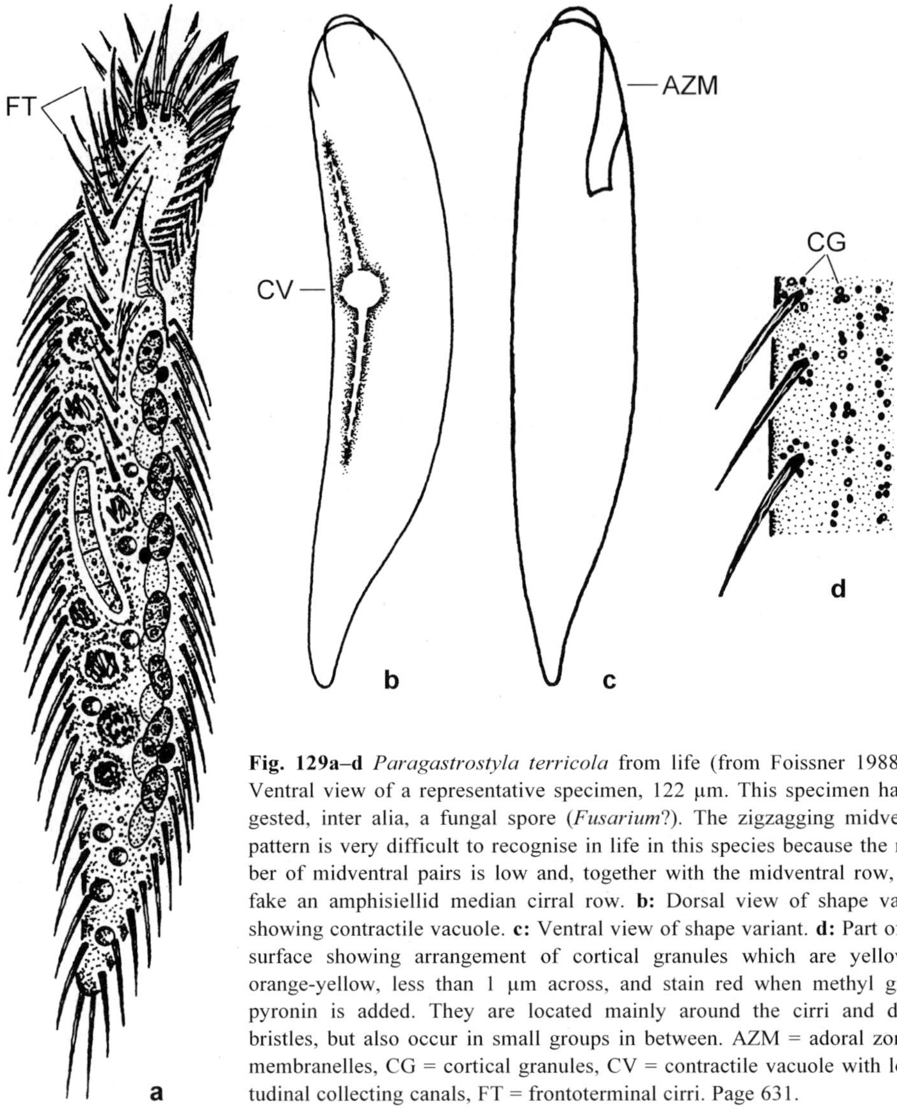

Fig. 129a–d *Paragastrostyla terricola* from life (from Foissner 1988). **a:** Ventral view of a representative specimen, 122 µm. This specimen has ingested, inter alia, a fungal spore (*Fusarium*?). The zigzagging midventral pattern is very difficult to recognise in life in this species because the number of midventral pairs is low and, together with the midventral row, they fake an amphisiellid median cirral row. **b:** Dorsal view of shape variant showing contractile vacuole. **c:** Ventral view of shape variant. **d:** Part of cell surface showing arrangement of cortical granules which are yellow to orange-yellow, less than 1 µm across, and stain red when methyl green-pyronin is added. They are located mainly around the cirri and dorsal bristles, but also occur in small groups in between. AZM = adoral zone of membranelles, CG = cortical granules, CV = contractile vacuole with longitudinal collecting canals, FT = frontoterminal cirri. Page 631.

tives and are therefore unchanged when transferred, like in the present case, to a genus of different gender (Werner 1972, p. 138).

Remarks: The present species was the second established in *Holostichides* (see Berger 2001 for a complete list). Foissner (1988) separated it from *Paragastrostyla lanceolata* by the presence of a distinct zigzagging midventral pattern and the cortical granules. The midventral pattern was not described by Hemberger for *P. lanceolata* and is indeed very indistinctly illustrated (Fig. 127a). However, one must take into account that Hemberger (1985, p. 397) did not use a camera lucida. Thus, such details in the cirral pattern should not be over interpreted. The presence/absence of cortical granules was not checked by Hemberger (1982, 1985) because he studied live specimens, if at all, only very superficially. However, the ontogenetic data by Hemberger (1982) clearly show that *Paragastrostyla lanceolata* (Fig. 127i) must have the same cirral pattern as *Holostichides terricola* and *Periholosticha wilberti* (Fig. 128m). Foissner (2000, p. 277) suggested synonymy of *Holostichides terricola* and *Periholosticha wilberti*, which was previously transferred to *Holostichides* by Eigner (1994). Song (1990), however, clearly stated that *Periholosticha wilberti* lacks cortical granules (see corresponding footnote at previous species). Consequently, I avoid this synonymy and separate *Paragastrostyla terricola* by the presence of cortical granules from *Paragastrostyla lanceolata*, and its new synonym *Periholosticha wilberti*, which lacks such organelles.

Paragastrostyla terricola and *P. lanceolata* differ from *Holostichides* species, inter alia, by the lack of a buccal cirrus and the lower number of macronuclear nodules (16 vs. more than 30 on average), adoral membranelles (19 vs. more than 30 on average), and dorsal kineties (2 vs. 4 or more). Furthermore, they usually have short, straight undulating membranes, whereas they are rather long and curved in *Holostichides*, at least, *Holostichides chardezi* (Fig. 124d) and *H. dumonti* (Fig. 125e, h). *Periholosticha* species, which have a very similar general appearance (size; shape, nuclear apparatus) and which also lack buccal and distinct transverse cirri, have a bipartite adoral zone of membranelles, lack frontoterminal cirri (type species) and (distinct) midventral rows, and have three dorsal kineties. However, these features are difficult to recognise in life so that protargol preparations are indispensable for reliable identification.

Morphology: Body size 80–140 × 15–30 µm in life, length:width ratio 5.6:1 on average in protargol preparations (Table 27; Foissner's method). Body very slender sigmoidal, posteriorly invariably pointed (Fig. 129i), slightly to distinctly (2:1) flattened dorsoventrally, and slightly (up to 10%) contractile. Macronuclear nodules form two strands one upon the other left of midline; individual nodules ellipsoidal (length:width ratio about 2:1). Micronuclei attached to macronuclear figure; if only two micronuclei are present, then they are located at the anterior and posterior end of the macronuclear figure; individual micronuclei globular to slightly ellipsoidal, about 3 × 2 µm in life (Fig. 129a, g). Contractile vacuole slightly ahead of mid-body near left cell margin, with two longitudinal collecting canals. Pellicle colourless, very flexible. Cortical granules mainly along infraciliature, but also in loose rows in between; individual granules small (<1 µm), yellowish to orange-yellow, stain red when methyl green-pyronin is added and make cells yellowish to orange-yellow in life at low magnification, especially at the tail-like posterior end and lateral margins. Number of granules varies within and among populations, but always recognisable when cells observed carefully. Cytoplasm colourless, without crystals, in rear end always some greasily shining globules 1–5 µm across; food vacuoles 3–6 µm in diameter. Movement without peculiarities, that is, slowly gliding showing great flexibility.

Adoral zone occupies about 20–25%, on average 22% of body length, roughly arranged in *Gonostomum* pattern, that is, extends straight along left body margin, performing right bend and slight clockwise rotation to plunge into buccal cavity, composed of 19 membranelles on average (Table 27). Bases of largest membranelles about 6 µm wide in life, anterior 3–4 membranelles, according to original description, visibly set

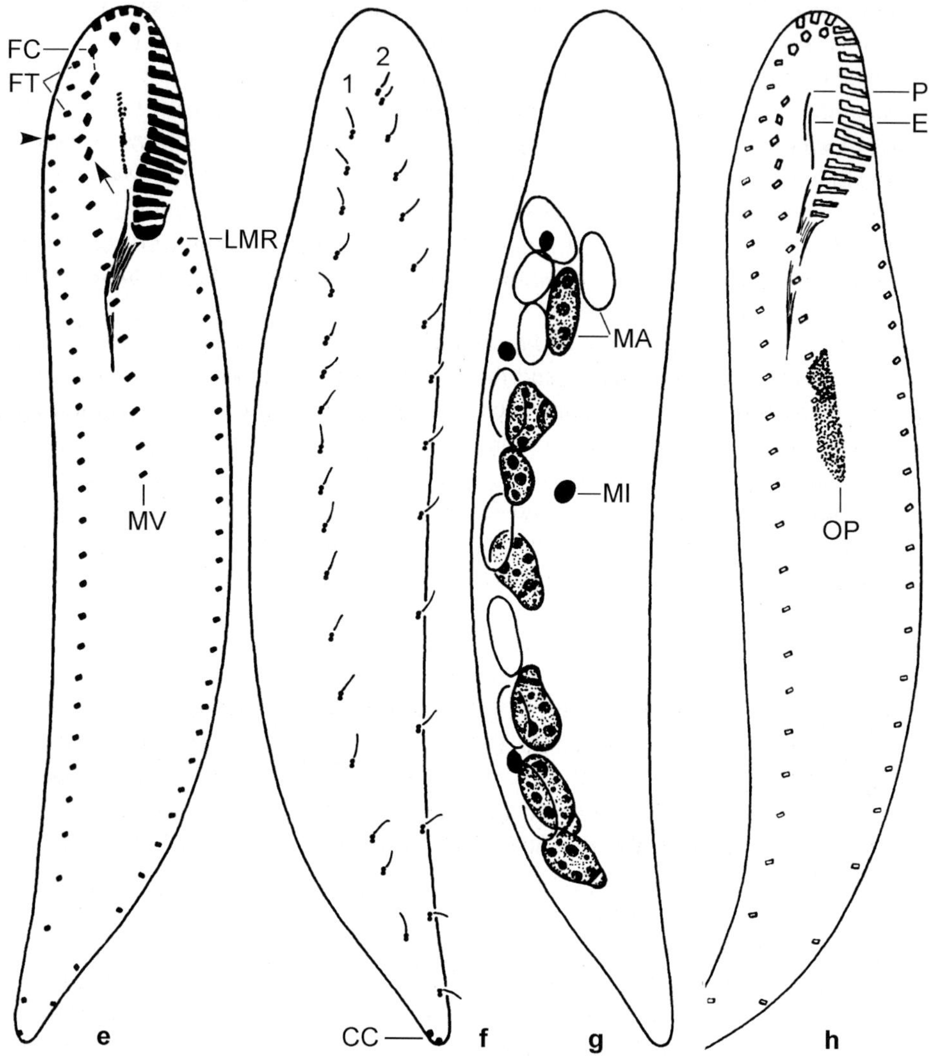

Fig. 129e–h *Paragastrostyla terricola* (from Foissner 1988. Protargol impregnation, Foissner's method). **e–g:** Infraciliature of ventral and dorsal side and nuclear apparatus of same specimen, 98 μm. This specimen has two midventral pairs, the rearmost is marked with an arrow; arrowhead denotes anterior end of right marginal row. Broken line connects right frontal cirrus and cirrus III/2, which originate from the same anlage. **h:** Very early divider. In this specimen cirrus III/2 is likely lacking. CC = caudal cirri, E = endoral, FC = right frontal cirrus, FT = frontoterminal cirri, LMR = left marginal row, MA = macronuclear nodules, MI = micronucleus, MV = rear end of midventral row, OP = oral primordium, P = paroral, 1, 2 = dorsal kineties. Page 631.

Fig. 129i, j *Paragastrostyla terricola* (original scanning electron micrographs kindly supplied by W. Foissner. Australian population). **i:** Ventral view. Arrow marks rear end of midventral complex. **j:** Anterior body portion. Arrow marks cirrus III/2, arrowhead denotes the anterior (= right) cirrus of the anteriormost midventral pair. Note that a buccal cirrus is lacking in *Paragastrostyla* species. FC = right frontal cirrus, LMR = left marginal row, MV = rear end of midventral row, P = paroral, RMR = right marginal row. Page 631. →

off from proximal ones; however, the original illustrations do not show a noteworthy gap (Fig. 129a, e, h; a distinct gap occurs in the closely related group *Periholosticha*). Individual membranelles of ordinary shape and fine structure. Buccal cavity flat and narrow, anteriorly not hook-shaped. Buccal lip rather small, therefore covering proximal portion of adoral zone only partially (Fig. 129j). Undulating membranes of about same length, straight or slightly curved, arranged in parallel or slightly intersecting optically; paroral begins slightly ahead of endoral (Fig. 129e, h). Pharyngeal fibres of ordinary length and structure, clearly recognisable in life and protargol preparations.

Cirral pattern and number of cirri largely of usual variability (Fig. 129e, h–j, Table 27). All cirri about 11 μm long in life and of similar size, except for frontal cirri, which are slightly enlarged. Frontal cirri arranged in oblique row in ordinary position. Usually one cirrus (= cirrus III/2) behind right frontal cirrus. Buccal and transverse cirri lacking. Frontoterminal cirri in usual position, that is, right of anterior end of midventral complex. Midventral complex composed of 2–4, usually only two midventral pairs forming more or less distinct zigzagging pattern and one midventral row terminating at 38% of body length on average (Fig. 129e, i, Table 27). Right marginal row commences behind level of last frontoterminal cirrus, left row near level of proximal end of adoral zone; both rows end subterminally; however, rearmost cirri difficult to distinguish from caudal

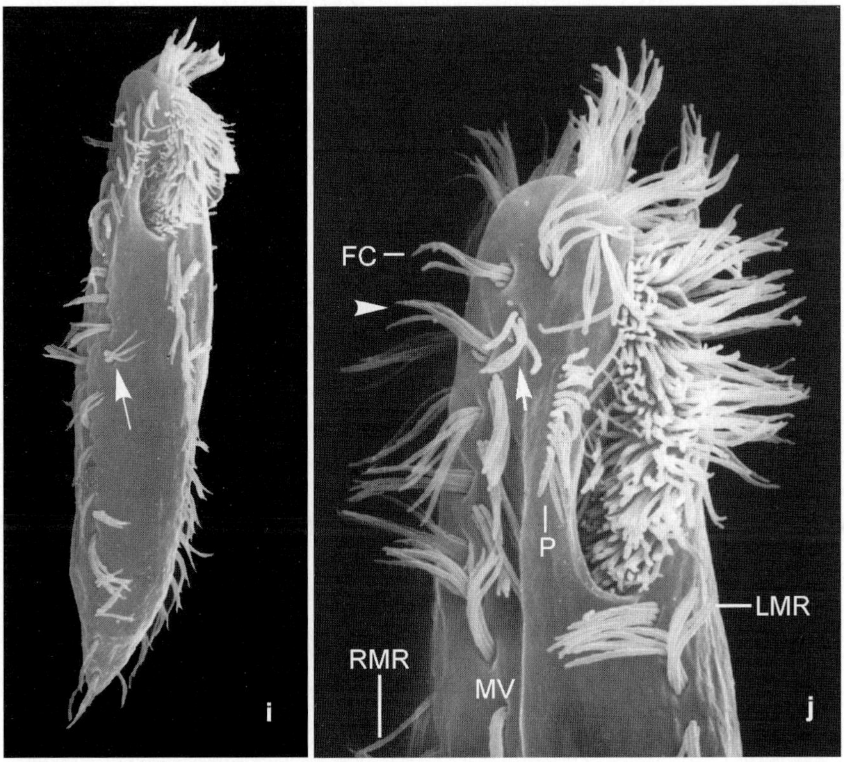

cirri because of pointed body. Individual marginal cirri become somewhat finer posteriorly and distance between them increases from anterior to posterior.

Dorsal bristles about 3 µm long in life, invariably arranged in two almost bipolar rows (Fig. 129f, Table 27). Foissner (1988, Tabelle 5) counted invariably three caudal cirri, that is, one of the two dorsal kineties must be associated with two caudal cirri. However, on page 109 he stated that on the pointed posterior body end, the marginal cirri are not clearly distinguishable from the two or three caudal cirri. A similar situation is described for both populations of *Paragastrostyla lanceolata* (Hemberger 1982, 1985, Song 1990).

Cell division (Fig. 129h): The formation of the oral primordium commences, as in *Paragastrostyla lanceolata* (Fig. 127b, c, 128e), near/at the rearmost cirri of the mid-ventral row where a compact patch of basal bodies occurs. No further details known. However, very likely there is no significant difference to the process described for *Paragastrostyla lanceolata* and its synonym *Periholosticha wilberti*.

Occurrence and ecology: Likely confined to terrestrial habitats. The type locality of *Paragastrostyla terricola* is a savannah in the Samburu National Park (0°37'N 37°32'E), Kenya, where Foissner (1988) discovered it in the upper soil layer (0–5 cm; pH 7.7). Foissner (1999, p. 323) found it also in Kenya, namely near the Sheldrick waterfalls, Shimba Hills Nature Reserve; the sample was composed of the upper 5 cm grass sward and very sandy soil layer (pH 6.6).

Blatterer & Foissner (1988) found *P. terricola* in litter and roots under moss of an autochthonous pine forest (*Callitris* sp.) near Adelaide, Australia. Foissner (1995; cited with incorrect year 1987) recorded it from the soil of a tropical dry forest of Costa Rica. Foissner (1997) found *P. terricola* in sample 23, which he collected about 40 km west of Manaus (Brazil) in Anavilhanas archipelago in the Rio Negro in the vicinity of the Ariau lodge. This site is a blackwater inundation primary (?) rain forest on one of the many small islands of the region, flooded by the Rio Negro during high water periods (further details on soil structure, see Foissner 1997, p. 318). Also rather common in Namibian soils (Foissner et al. 2002, p. 60).

Foissner (2000) reported *Paragastrostyla terricola* from the artificial sand soil from sewage irrigation fields near Berlin, Germany. Recently, Peter Eigner found it in farmland soil, likely from Styria, Austria (pers. comm. to Esteban et al. 2001, p. 141). According to Foissner (1998), *Paragastrostyla terricola* is a true cosmopolitan because it occurs in all continents except Antarctica. Feeds on bacteria, fungal spores, and humus particles (Foissner 1988). Biomass of 10^6 specimens about 33 mg (Foissner 1998).

Metaurostylopsis Song, Petz & Warren, 2001

1999 ***Metaurostylopsis* Song & Petz, in press** – Shi, Acta Zootax. sinica, 24: 245 (revision of hypotrichous ciliates; see nomenclature).
1999 ***Metaurostylopsis* Song & Petz, in press** – Shi, Acta Zootax. sinica, 24: 363 (revision of hypotrichous ciliates; see nomenclature).
1999 ***Metaurostylopsis* Song & Petz, in press** – Song & Wang, Progress in Protozoology, p. 73 (list of marine ciliates from China; see nomenclature).
1999 ***Metaurostylopsis* Song & Petz, in press** – Shi, Song & Shi, Progress in Protozoology, p. 112 (revision of hypotrichous ciliates; see nomenclature).
2001 ***Metaurostylopsis* nov. gen.**[1] – Song, Petz & Warren, Europ. J. Protistol., 37: 64 (original description; see nomenclature). Type species (by original designation on p. 65): *Urostyla marina* Kahl, 1932.
2001 ***Metaurostylopsis*** – Aescht, Denisia, 1: 100 (catalogue of generic names of ciliates).
2001 ***Metaurostylopsis* Song and Petz in Shi, Song and Shi, 1999** – Berger, Catalogue of ciliate names 1. Hypotrichs, p. 47 (nomenclator containing all basionyms, combinations, and higher taxa of hypotrichs).
2002 ***Metaurostylopsis* Song, Petz & Warren, 2001** – Song & Wilbert, Acta Protozool., 41: 52 (brief comment).

Nomenclature: *Metaurostylopsis* is a composite of the Greek prefix *meta+* (in between, together with, whole, after, beside; Werner 1972, p. 70), the urostyloid genus-group name *Urostyla* (see there for derivation), and the Greek suffix *-opsis* (appearance, looking like). Probably the name should indicate the resemblance between the type species of the present genus and a *Urostyla*. Feminine gender because ending with *-opsis* (ICZN 1999, Article 30.1.3).

Unfortunately, the name *Metaurostylopsis* was mentioned in four papers before its "final publication" by Song et al. (2001). In the catalogue of ciliate names (Berger 2001), I wrote that *Metaurostylopsis* should be assigned to "Song and Petz in Shi, Song and Shi" if the Chinese text contains a diagnosis. I did not check this Chinese text and consider now, for the sake of simplicity, the paper by Song et al. (2001) as original description of *Metaurostylopsis*.

Characterisation (Fig. 111a, autapomorphies 12):[2] Adoral zone of membranelles continuous. 3 frontal cirri. Buccal cirrus present. 3 or more frontoterminal cirri (A). Midventral complex composed of midventral pairs and at least 1 midventral row. Transverse cirri present. 2 or more left and 2 or more right marginal rows, which originate from individual anlagen within each parental row. Caudal cirri lacking (A).

Remarks: The four species assigned have, besides the characteristics mentioned above, several features in common, namely, body flexible; many macronuclear nodules; contractile vacuole slightly ahead of mid-body or near mid-body; cortical granules present; adoral zone of membranelles on average 32–40% of body length; adoral zone extends only slightly onto right body margin; buccal cavity moderately large and deep; undulating membranes long and almost straight; buccal cirrus distinctly behind anterior

[1] The diagnosis by Song et al. (2001) is as follows: Urostylidae with frontoterminal cirral row and several clearly differentiated frontal cirri; buccal and transverse cirri present; more than one row of marginal cirri on each side which derive from individual anlagen within each parental row; no caudal cirri; oral primordium for proter probably developing from a "pocket" between pellicle and buccal cavity.

[2] Two features of the characterisation do not support the classification of *M. songi* in *Metaurostylopsis* (details see end of remarks).

Table 28 Morphometric data on *Metaurostylopsis marina* (ma1, from Song et al. 2001; ma2, from Kahl 1932; ma3, from Dragesco 1965; ma4, *Urostyla* sp. from Thompson 1972 [classified as supposed synonym of *M. marina*]; ma5, from Borror 1979; ma6, from Wiackowski 1991; ma7, from Lei et al. 2005), *Metaurostylopsis rubra* (ru1, from Song & Wilbert 2002; ru2, from Wilbert & Song 2005), *Metaurostylopsis salina* (sal, from Lei et al. 2005), and *Metaurostylopsis songi* (son, from Lei et al. 2005)

Characteristics [a]	Species	mean	M	SD	SE	CV	Min	Max	n
Body, length	ma1	107.1	–	9.2	2.1	9.3	86.0	120.0	20
	ma4	124.6	–	–	–	–	103.1	150.0	25
	ma6	89.9	89.9	10.0	–	11.0	69.0	106.1	30
	ma7	–	–	–	–	–	88.0	100.0	5
	ru1	181.1	–	32.5	10.3	18.0	132.0	226.0	10
	sal	52.1	–	8.5	1.8	16.3	38.0	72.0	22
	son	100.9	–	13.5	3.2	13.4	75.0	118.0	18
Body, width	ma1	47.0	–	5.8	1.9	12.2	38.0	60.0	20
	ma4	60.8	–	–	–	–	52.2	74.5	25
	ma6	46.3	46.6	5.6	–	12.0	35.0	60.0	30
	ma7	–	–	–	–	–	45.0	55.0	5
	ru1	83.3	–	11.6	4.8	13.8	72.0	101.0	7
	sal	22.0	–	3.8	0.8	17.2	14.0	28.0	22
	son	21.4	–	2.9	0.7	13.6	16.0	28.0	18
Body length:width, ratio	sal	2.4	–	0.7	0.1	27.1	1.4	4.4	22
	son	4.7	–	0.6	0.1	12.3	3.8	5.8	18
Body, height	sal	22.0	–	4.6	1.9	20.7	17.0	29.0	6
	son	18.3	–	2.7	0.8	15.0	15.0	22.0	12
Anterior body end to proximal end of adoral zone, distance	ma1	40.6	–	2.9	0.7	7.1	35.0	46.0	19
	ma4	47.1	–	–	–	–	–	–	25
	ma6	37.2	38.0	3.0	–	8.0	30.0	41.5	30
	ru1	60.3	–	5.6	2.1	9.3	54.0	72.0	7
	sal	19.8	–	2.6	0.6	13.1	15.0	25.0	22
	son	31.9	–	4.6	1.1	14.3	24.0	40.0	18
Paroral, length	ma6 [e]	28.5	28.8	2.3	–	8.0	24.0	33.0	30
Macronuclear nodules, diameter	ma7	–	–	–	–	–	3.0	6.0	5
Macronuclear nodule, length	sal	2.7	–	0.7	0.2	26.7	2.0	4.0	22
	son	4.2	–	1.3	0.3	30.0	3.0	7.0	18
Macronuclear nodules, number	ma1	50.0	–	–	–	–	–	–	?
	ma3	–	–	–	–	–	28.0	31.0	?
	ma5	32.0	–	–	–	–	–	–	?
	ma6	40.8	41.5	8.2	–	20.0	30.0	54.0	6
	ma7			about 50					?
	ru1			about 100					
	sal	45.4	–	7.8	1.7	17.2	31.0	62.0	22
	son	54.2	–	9.4	2.5	17.4	42.0	70.0	14
Micronucleus, diameter	ma7	–	–	–	–	–	1.0	2.0	5
Micronuclei, number	ma1	–	–	–	–	–	5.0	10.0	?
	ma3	–	–	–	–	–	3.0	6.0	?
	ma5	2.0	–	–	–	–	–	–	?
	ma6	–	–	–	–	–	5.0	8.0	?
	sal	6.0	–	0.9	0.3	15.2	4.0	7.0	13
	son	3.6	–	0.5	0.1	14.8	3.0	6.0	16
Adoral membranelles, number	ma1	28.4	–	1.1	0.3	4.3	27.0	30.0	19
	ma5	–	–	–	–	–	16.0	24.0	?
	ma6	26.0	26.0	1.7	–	7.0	22.0	29.0	30
	ma7			about 27					?

Table 28 Continued

Characteristics[a]	Species	mean	M	SD	SE	CV	Min	Max	n
Adoral membranelles, number	ru1	38.9	–	3.8	1.2	9.7	35.0	46.0	10
	ru2	45.0	–	–	–	–	–	–	?
	sal	20.1	–	1.0	0.2	5.1	18.0	23.0	20
	son	33.9	–	4.4	1.1	13.1	28.0	47.0	15
Buccal cirri, number	ma1	1.0	–	0.0	0.0	0.0	1.0	1.0	25
	ma5	1.0	–	–	–	–	–	–	?
	ma6	1.0	–	–	–	–	–	–	?
	ma7	1.0	–	–	–	–	–	–	?
	ru1	1.0	–	0.0	0.0	0.0	1.0	1.0	20
	ru2	1.0	–	–	–	–	–	–	?
	sal	1.0	–	0.0	0.0	0.0	1.0	1.0	22
	son	1.0	–	0.0	0.0	0.0	1.0	1.0	18
Frontal cirri, number	ma1[i]	4.0	–	0.0	0.0	0.0	4.0	4.0	25
	ma2[b]	3.0	–	–	–	–	–	–	1
	ma5	3.0	–	–	–	–	–	–	?
	ru1[i]	4.0	–	0.0	0.0	0.0	4.0	4.0	14
	ru2[i]	4.0	–	–	–	–	–	–	?
	sal	3.0	–	0.0	0.0	0.0	3.0	3.0	14
	son[j]	5.0	–	0.0	0.0	0.0	5.0	5.0	18
Frontoterminal cirri, number	ma1	4.5	–	1.0	0.2	21.7	3.0	6.0	16
	ma6	4.3	4.0	0.6	–	14.0	3.0	5.0	29
	ma7	–	–	–	–	–	4.0	5.0	?
	ru1	6.2	–	1.0	0.3	15.9	5.0	8.0	11
	ru2			about 10					?
	sal	3.7	–	0.6	0.2	15.9	3.0	5.0	15
	son	2.1	–	0.3	0.1	15.3	2.0	3.0	18
Midventral pairs, number	ma1	9.3	–	1.1	0.3	12.2	7.0	11.0	16
	ma2[b]	6.0	–	–	–	–	–	–	1
	ma6[f]	8.0	–	–	–	–	–	–	1
	ma7	–	–	–	–	–	9.0	10.0	?
	ru1	9.5	–	1.0	0.3	11.0	8.0	11.0	11
	ru2	12.0	–	–	–	–	–	–	?
	sal	4.6	–	0.5	0.1	11.2	4.0	5.0	16
	son	10.6	–	1.0	0.3	9.1	9.0	12.0	10
Midventral row, number of cirri	ma1	–	–	–	–	–	4.0	7.0	5
	ma5	–	–	–	–	–	3.0	4.0	?
	ma6	4.2	4.0	1.0	–	23.0	3.0	7.0	30
	ma7			about 4					
	ru1	9.8	–	1.7	0.5	17.2	8.0	13.0	10
	ru2	13.0	–	–	–	–	–	–	?
	sal	6.4		0.9	0.2	13.3	5.0	7.0	18
	son	1.4	–	0.7	0.2	45.2	1.0	3.0	14
Transverse cirri, number	ma1	6.8	–	1.0	0.2	15.2	5.0	9.0	17
	ma2	–	–	–	–	–	6.0	7.0	?
	ma3[h]	4.0	–	–	–	–	–	–	1
	ma4	–	–	–	–	–	3.0	5.0	25
	ma5	–	–	–	–	–	4.0	5.0	?
	ma6	6.1	6.0	0.7	–	11.0	5.0	8.0	27
	ma7			about 6					
	ru1	4.6	–	0.8	0.2	16.6	4.0	6.0	13
	ru2	5.0	–	–	–	–	–	–	?

Table 28 Continued

Characteristics [a]	Species	mean	M	SD	SE	CV	Min	Max	n
Transverse cirri, number	sal	3.9	–	1.1	0.3	28.5	2.0	5.0	14
	son	6.9	–	0.3	0.1	4.7	6.0	7.0	18
Cirral rows, number [g]	ma3	7.0	–	–	–	–	–	–	?
	ma4	–	–	–	–	–	11.0	13.0	25
Left marginal rows, number	ma1	4.1	–	0.6	0.1	14.1	3.0	5.0	16
	ma2 [b]	4.0	–	–	–	–	–	–	1
	ma5	–	–	–	–	–	2.0	4.0	?
	ma6	3.5	3.0	0.7	–	21.0	3.0	5.0	30
	ma7	4.0	–	–	–	–	–	–	?
	ru1	7.8	–	0.9	0.3	11.2	6.0	9.0	12
	ru2	8.0	–	–	–	–	–	–	?
	sal	3.0	–	0.0	0.0	0.0	3.0	3.0	22
	son	3.0	–	0.0	0.0	0.0	3.0	3.0	18
Left marginal row 1 [c], number of cirri	ma1	15.1	–	2.3	0.8	15.0	11.0	18.0	9
	ma7	–	–	–	–	–	16.0	18.0	?
	sal	13.5	–	2.3	0.6	17.0	10.0	18.0	17
	son	24.5	–	3.0	0.8	12.1	21.0	31.0	14
Left marginal row 2 [c], number of cirri	ma1	17.0	–	3.8	1.3	22.4	9.0	22.0	9
	ma7	–	–	–	–	–	16.0	18.0	?
	sal	14.8	–	2.4	0.6	16.3	10.0	19.0	17
	son	26.7	–	4.8	1.3	17.8	22.0	40.0	14
Left marginal row 3 [c], number of cirri	ma1	18.3	–	2.8	0.9	15.2	14.0	23.0	9
	ma7	–	–	–	–	–	16.0	18.0	?
	sal	13.9	–	2.0	0.5	14.0	11.0	18.0	17
	son	26.4	–	3.4	0.9	12.9	21.0	32.0	14
Left marginal row 4 [c], number of cirri	ma1	19.7	–	2.7	1.1	13.9	16.0	24.0	6
	ma7	–	–	–	–	–	16.0	18.0	?
Left marginal row 5 [c], number of cirri	ma1	–	–	–	–	–	13.0	20.0	3
Right marginal rows, number	ma1	4.1	–	0.4	0.1	10.9	3.0	5.0	16
	ma2 [b]	3.0	–	–	–	–	–	–	1
	ma5	–	–	–	–	–	2.0	4.0	?
	ma6	3.1	3.0	0.3	–	10.0	3.0	4.0	30
	ma7				about 7				
	ru1	6.3	–	0.5	0.1	7.8	6.0	7.0	12
	ru2	6.0	–	–	–	–	–	–	?
	sal	3.0	–	0.0	0.0	0.0	3.0	3.0	22
	son	3.0	–	0.0	0.0	0.0	3.0	3.0	18
Right marginal row 1 [c], number of cirri	ma1	20.1	–	4.4	1.5	21.9	15.0	26.0	9
	ma7	–	–	–	–	–	20.0	25.0	?
	sal	14.2	–	2.8	0.7	20.0	10.0	21.0	17
	son	29.9	–	4.1	1.1	13.6	24.0	36.0	14
Right marginal row 2 [c], number of cirri	ma1	19.8	–	5.8	1.9	29.5	15.0	27.0	9
	ma7	–	–	–	–	–	20.0	25.0	?
	sal	16.7	–	2.9	0.7	17.3	11.0	22.0	17
	son	29.6	–	5.2	1.4	17.5	22.0	41.0	14
Right marginal row 3 [c], number of cirri	ma1	20.6	–	3.2	1.1	15.6	16.0	24.0	9
	ma7	–	–	–	–	–	20.0	25.0	?
	sal	12.9	–	1.4	0.3	10.7	11.0	15.0	17
	son	25.9	–	2.9	0.8	11.2	22.0	32.0	14
Right marginal row 4 [c], number of cirri	ma1	21.9	–	2.7	1.0	12.5	17.0	24.0	7
Right marginal row 5 [c], number of cirri	ma1	–	–	–	–	–	19.0	22.0	2

Table 28 Continued

Characteristics [a]	Species	mean	M	SD	SE	CV	Min	Max	n
Dorsal kineties, number	ma1 [d]	3.2	–	0.4	0.1	12.7	3.0	4.0	16
	ma2	4.0	–	–	–	–	–	–	?
	ma4					about 7			25
	ma5	3.0	–	–	–	–	–	–	?
	ma6	3.3	3.0	0.6	–	20.0	3.0	5.0	30
	ru1 [d]	3.0	–	0.0	0.0	0.0	3.0	3.0	14
	ru2	3.0	–	–	–	–	–	–	?
	sal	3.0	–	0.0	0.0	0.0	3.0	3.0	18
	son	3.0	–	0.0	0.0	0.0	3.0	3.0	18

[a] All measurements in µm. Data provided by Song et al. (2001), Song & Wilbert (2002), and Wiackowski (1991) are based on specimens impregnated with Wilbert's protargol method. Data provided by Kahl (1932) and Dragesco (1965) are based on live observations. Data provided by Thompson (1972) are based on Chatton-Lwoff silver-impregnated specimens, those provided by Borror (1979 on nigrosin-butanol stains. CV = coefficient of variation in %, M = median, Max = maximum value, mean = arithmetic mean, Min = minimum value, n = number of individuals investigated (? = number not given. If only one value is available, then it is listed under the column mean; if two values are given, then they are listed under Min and Max), SD = standard deviation, SE = standard error of arithmetic mean.

[b] Single value from Fig. 130a. The innermost right cirral row (arrow in Fig. 130a) is very likely the midventral row and thus not counted as right marginal row.

[c] Numbered from inside to outside.

[d] Short rows ahead of outermost right marginal rows not included.

[e] Distance between anterior end of paroral and proximal end of adoral zone of membranelles.

[f] Single value from Fig. 130k (last pair marked with MP).

[g] All cirral rows including the midventral complex, which was obviously counted as single row (Fig. 130b, h).

[h] Single value from Fig. 130b.

[i] Cirrus III/2 (= cirrus behind right frontal cirrus) included.

[j] Likely cirrus III/2 and front cirrus of first midventral pair included (ontogenetic data needed for correct interpretation of pattern).

end of paroral; last midventral pair roughly about at level of buccal vertex; only one cirrus behind right frontal cirrus; one midventral row, which terminates slightly behind midbody; transverse cirri inconspicuous; all cirri, except for frontal cirri, of about same size; dorsal cilia about 3 µm long, arranged in three more or less bipolar kineties; some dorsal bristles ahead of outermost right marginal row(s); parental adoral zone is completely replaced during ontogenesis (no data available for *M. salina* and *M. songi*);[1] marine.

Song et al. (2001) established *Metaurostylopsis* to separate *Urostyla marina* Kahl from other *Urostyla*-like species. Song & Wilbert (2002) discovered a second species and Lei et al. (2005) described two further species. Interestingly, all species are very likely confined to marine habitats.

[1] Note that this feature occurs, likely convergently, in the pseudokeronopsids too.

Metaurostylopsis differs from *Urostyla* by several characteristics, namely, the frontal cirri (three vs. distinctly more than three), the frontoterminal cirri (present vs. lacking), and the division mode of the macronucleus. Especially the last feature is very interesting because the individual nodules of *M. marina* do not form, as is usual, a roundish mass during division, but a branched complex (Fig. 131v), which is unique within the hypotrichs. Unfortunately, the division mode of the macronucleus is not known for the other *Metaurostylopsis* species. Thus, we do not know whether or not this feature is an apomorphy for *Metaurostylopsis*, or for *M. marina* only. The branched shape of the macronucleus can also be interpreted as "pre-stage" of the pseudokeronopsid macronuclear apparatus where the numerous nodules do not fuse to a single mass prior to division, but each nodule divides individually (Berger 2004b, p. 116).

Song et al. (2001) included the formation of the proter's oral primordium in the diagnosis. It originates in a pouch beneath the parental buccal cavity. Very likely, *Urostyla grandis* and the pseudokeronopsids have a very similar type of primordia formation. However, electron microscopic observations are needed to show whether or not the two processes are homologous. For separation of *Metaurostylopsis* from other taxa, see Bakuellidae key.

Song et al. (2001, p. 73) also discussed *Metaurostyla* Jankowski, 1979, unfortunately somewhat superficially. At the begin they wrote that "Jankowski (1979) already erected a new genus and species for *Urostyla* sp." of Thompson (1972). This is incorrect because Jankowski did not fix *Metaurostyla thompsoni* (= *Urostyla* sp. sensu Thompson 1972) as type species. Somewhat later they write that Jankowski failed to fix one of the three originally included species as type species, and therefore classified *Metaurostyla* as nomen nudum. However, this is also incorrect because Jankowski applied the expression "gen. et sp. n." only to *Metaurostyla polonica* Jankowski, 1979, which is thus defined as type species[1] (Berger 2001, p. 47). The other two species included and transferred to *Metaurostyla*, *Metaurostyla raikovi* and *M. thompsoni*, by Jankowski (1979) are not accompanied by the expression "gen. n." and therefore cannot be the type species. Aescht (2001, p. 100) also considered *Metaurostyla polonica* as type species and found that *Metaurostyla* is a nomen nudum because established without description or definition. *Metaurostyla polonica* is the *Urostyla grandis* of Jerka-Dziadosz (1972). This population is very likely a true *Urostyla grandis* (see Ganner 1991, p. 123; Foissner et al. 1991, p. 222) so that *Metaurostyla* is not only a nomen nudum, but also a junior synonym of *Urostyla* (for a more detailed discussion, see this chapter in *Urostyla*).

Recently, Lei et al. (2005) described two *Metaurostylopsis* species, namely, *M. salina* (Fig. 133.1a–k) and *M. songi* (133.2a–m). While *M. saline* fits the characterisation above very well, *Metaurostylopsis songi* shows at least two features, which contradict a classification in the present genus. The first feature is the number of frontoterminal cirri. The type species, *M. rubra*, and *M. salina* have three or more such cirri, whereas *M. songi* has only two, which is the plesiomorphic state. The second feature is the midventral complex, which is composed of midventral pairs and one midventral row. In addition,

---

[1] The formula "gen. n., sp. n.", or its exact equivalent is not explicitly forbidden to designate a species as type species after 1930 (ICZN 1964).

the last pair is about at the level of the buccal vertex and the midventral row terminates near mid-body in the type species, *M. rubra*, and *M. salina*. By contrast, *M. songi* lacks a midventral row and the last midventral pair is about in mid-body. Moreover, the frontal ciliature of *M. songi* is somewhat difficult to interpret (see species description). The differences indicate that *M. marina*, *M. rubra*, and *M. salina* form a subgroup; *Metaurostylopsis songi* is perhaps the sister to this subgroup. However, ontogenetic and molecular data are needed for a more detailed discussion.

Species included in *Metaurostylopsis* (alphabetically arranged according to basionym): (1) *Metaurostylopsis rubra* Song & Wilbert, 2002; (2) *Metaurostylopsis salina* Lei, Choi, Xu & Petz, 2005; (3) *Metaurostylopsis songi* Lei, Choi, Xu & Petz, 2005 (generic assignment uncertain); (4) *Urostyla marina* Kahl, 1932.

Key to *Metaurostylopsis* species

1 Body length in life usually below 150 µm; cytoplasm colourless; 2–5 left and 2–5 right marginal rows (e.g., Fig. 130a, i, 131a) **2**

\- Body length in life 150–300 µm; cytoplasm brick-red to dark-red due to fine pigments; 6–9 left and 6–7 right marginal rows (Fig. 133a, h) . . . . *Metaurostylopsis rubra* (p. 663)

2 Body length usually below 80 µm; body outline usually pyriform; cortical granules scattered; 4–5 midventral pairs; 2–5 transverse cirri (e.g., Fig. 133.1a) . . . . *Metaurostylopsis salina* (p. 668)

\- Body length usually above 80 µm; body outline oval to elongate elliptical; cortical granules arranged in rows; 7–12 midventral pairs; 5–9 transverse cirri **3**

3 Cortical granules along dorsal kineties and anterior body margin; distinct midventral row present (Fig. 130k, 131b, d) . . . . *Metaurostylopsis marina* (p. 643)

\- Cortical granules in rows both on dorsal and ventral side; distinct midventral row lacking (Fig. 133.2d, e, g) . . . . *Metaurostylopsis songi* (p. 672)

Metaurostylopsis marina (Kahl, 1932) Song, Petz & Warren, 2001 (Fig. 130a–g, i–l, 131a–x, 132a–g, Table 28)

1932 ***Urostyla marina* spec. n.** – Kahl, Tierwelt Dtl., 25: 567, Fig. 100 (Fig. 130a; original description. No type material available and no formal diagnosis provided).

1933 ***Urostyla marina* Kahl 1932** – Kahl, Tierwelt N.- u. Ostsee, 23: 108, Fig. 16 $_{22}$ (Fig. 130a; guide to marine ciliates from the North Sea and the Baltic).

1963 ***Urostyla marina* Kahl 1930** – Biernacka, Polski Archwm Hydrobiol., 11: 49, Abb. 92 (Fig. 130e; illustrated record from Poland).

1965 ***Urostyla marina* Kahl** – Dragesco, Cah. Biol. mar., 6: 394, Fig. 30 (Fig. 130b–d; description of African population from life).

1972 ***Paraurostyla marina* (Kahl, 1932) n. comb.** – Borror, J. Protozool., 19: 10 (combination with *Paraurostyla* Borror, 1972; revision of hypotrichs).

1979 ***Urostyla marina* Kahl, 1932** – Borror, J. Protozool., 26: 545, Fig. 2, 3 (Fig. 130f, g, 132a–g; description of American population and cell division).

1983 ***Urostyla marina* Kahl, 1932** – Borror & Wicklow, Acta Protozool., 22: 120 (revision of urostylids).

1991 ***Urostyla thompsoni* Jankowski, 1979** – Wiackowski, Acta Protozool., 30: 55, Fig. 1–11, Table 1 (Fig. 131j–l; description of a Polish population; see nomenclature).

1992 ***Urostyla marina* Kahl, 1930** – Carey, Marine interstitial ciliates, p. 178, Fig. 703 (redrawing of Fig. 130a; guide; incorrect year).

1999 ***Metaurostylopsis marina* (Kahl, 1932)** – Song & Wang, Progress in Protozoology, p. 73 (list of marine ciliates from China; see nomenclature).

1999 ***Metaurostylopsis marina* (Kahl, 1932)** – Shi, Song & Shi, Progress in Protozoology, p. 112 (revision of hypotrichous ciliates; see nomenclature).

2001 ***Metaurostylopsis marina* (Kahl, 1932) Song and Petz in Shi, Song and Shi, 1999** – Berger, Catalogue of ciliate names 1. Hypotrichs, p. 102 (nomenclator containing all basionyms, combinations, and higher taxa of hypotrichs).

2001 ***Metaurostylopsis marina* (Kahl, 1932) nov. comb.**[1] – Song, Petz & Warren, Europ. J. Protistol., 37: 65, Fig. 1–33, Tables 1, 2 (Fig. 131a–v; combination with *Metaurostylopsis*; detailed description of Chinese population including ontogenesis and deposition of a neotype slide [registration number 2000:8:1:1] and a voucher slide [2000:8:1:2] in the Natural History Museum in London).

2005 ***Metaurostylopsis marina* (Kahl, 1932) Song, Petz and Warren, 2001** – Lei, Choi, Xu & Petz, J. Euk. Microbiol., 52: 6, Fig. 25, 26, Tables 1, 2 (Fig. 131w, x; brief description of Korean population).

Nomenclature: No derivation of the name was given in the original description. The species-group name *marin·us -a -um* (Latin adjective; living in the sea, belonging to the sea) refers to the habitat where the species was discovered. *Urostyla marina* was fixed as type species of *Metaurostylopsis* by original designation. The citation "*Urostyla thompsoni* Jankowski, 1979" in Wiackowski (1991) is not quite correct because Jankowski established the species in *Metaurostyla* and not in *Urostyla*; thus, the correct original name is "*Metaurostyla thompsoni* Jankowski, 1979" (see below).

The combination *Metaurostylopsis marina* was mentioned twice before its "original" publication by Song et al. (2001). In the catalogue on ciliate names (Berger 2001), I proposed "Song and Petz in Shi, Song and Shi, 1999" as combining authors because Shi et al. (1999) assigned *Metaurostylopsis* to "Song and Petz, in press". In addition, I did not have the paper by Song et al. (2001) when I completed the manuscript of the catalogue. Now I consider – for the sake of simplicity – Song, Petz & Warren as combining authors of *M. marina*.

Remarks: This rather common marine species was described by Kahl (1932) and reinvestigated several times. The last redescription is authoritative because Song et al. (2001) fixed a neotype.

Kahl (1932), who did not have the advantage of silver impregnation, provided a rather detailed description, including the cirral pattern. The sole feature not recognised by him is the frontoterminal cirri. They form, according to the authoritative redescription, a short row, which is only inconspicuously separated from the midventral complex and the inner right marginal row. This explains why these cirri were not recognised by Kahl (1932). Kahl (1933) did not provide new data.

---

[1] The improved diagnosis by Song et al. (2001) is as follows: Medium-sized marine *Metaurostylopsis*, in vivo 80–120 × 50–80 µm with oval to ellipsoidal body shape; conspicuous cortical granules along dorsal kineties and anterior body margin; 3–5 marginal cirral rows on each side of the cell. 27–30 adoral membranelles, 3–4 slightly enlarged frontal, 3–6 frontoterminal, 1 buccal, and 5–9 transverse cirri; 7–11 midventral cirral pairs; usually with 3 complete dorsal kineties; ca. 50 macronuclear nodules and about 5–10 micronuclei.

Biernacka (1963) stated that her specimens agree perfectly with Kahl's description, except for the body length, which is 70–90 μm in the Polish population against 80–120 μm in Kahl's specimens. Although the illustration is rather superficial (Fig. 130e) I accept the identification because this species is, according to my own experience, rather easily identified in life.

The population described by Dragesco (1965) agrees rather well with the original description and the authoritative redescription by Song et al. (2001). The latter authors mention the number of marginal rows (three left and four right[1]) and the colour of the cortical granules (pink or dark red) as differences from their neotype population. However, some specimens of the neotype material also have three left and three right marginal rows (Table 28), and the cortical granules of Dragesco's population are not pink or dark red, but likely colourless. Dragesco wrote that the pink or wine-red colour of the cells is not due to the numerous "protrichocysts", but due to the reddish food vacuoles. Moreover, Song et al. (2001) mention the lack of data on the frontal, frontoterminal and midventral cirri and on the dorsal kineties as the reason for classifying Dragesco's population as unidentifiable. In contrast, I accept Dragesco's identification because it is generally known that the cirral pattern is difficult to recognise in detail in life. In addition, the frontoterminal cirri are also not described by Kahl (1932).

Urostyla sp. sensu Thompson (1972) is possibly the present species. However, since it obviously has more dorsal kineties (about 7) than the other populations (usually 3) it is considered only as supposed synonym and described separately (Fig. 130h).

Borror (1972) transferred *Urostyla marina* to *Paraurostyla* Borror, 1972 mainly because of the clearly differentiated frontal cirri. However, when he later found that *U. marina* has midventral cirri, he discarded this act (Borror 1979, Borror & Wicklow 1983; for review of *Paraurostyla*, see Berger 1999).

Borror (1979) redescribed the present species and provided some ontogenetic data (Fig. 130f, g, 132a–g). All features match Kahl's population very well except for the number of transverse cirri, which is 4–5 against 6–7. Interestingly, frontoterminal cirri were neither mentioned nor drawn by Borror and they are even not recognisable in the dividing specimens (Fig. 132g). Possible this is due to the nigrosin-butanol stain applied. As mentioned above, Kahl (1932) also did not describe or illustrate frontoterminal cirri. Thus, it cannot be excluded that a species without such a cirral group exists. On the other, hand Borror's population agrees rather well with Thompson's (1972) population in the number of transverse cirri (4–5, respectively, 3–5). However, Thompson clearly illustrated frontoterminal cirri. Song et al. (2001), who compared Borror's population thoroughly with Kahl's population and the neotype material, accepted Borror's identification. I agree with them, but keep the data separate.

Borror & Wicklow (1983) mentioned "*Keronopsis gracilis* Dragesco, 1965" as junior synonym of *Urostyla marina* Kahl. However, Dragesco (1965) did not describe a *Keronopsis gracilis*, but redescribed *Holosticha (Keronopsis) gracilis* Kahl. Kahl's *Holosticha gracilis* shows an ordinary *Anteholosticha* cirral pattern, that is, one right marginal row, a long midventral complex composed of cirral pairs only, and one left

[1] If one assumes that the leftmost row of the right set of rows is the midventral complex, then three right marginal rows are present in Dragesco's population (Fig. 130b).

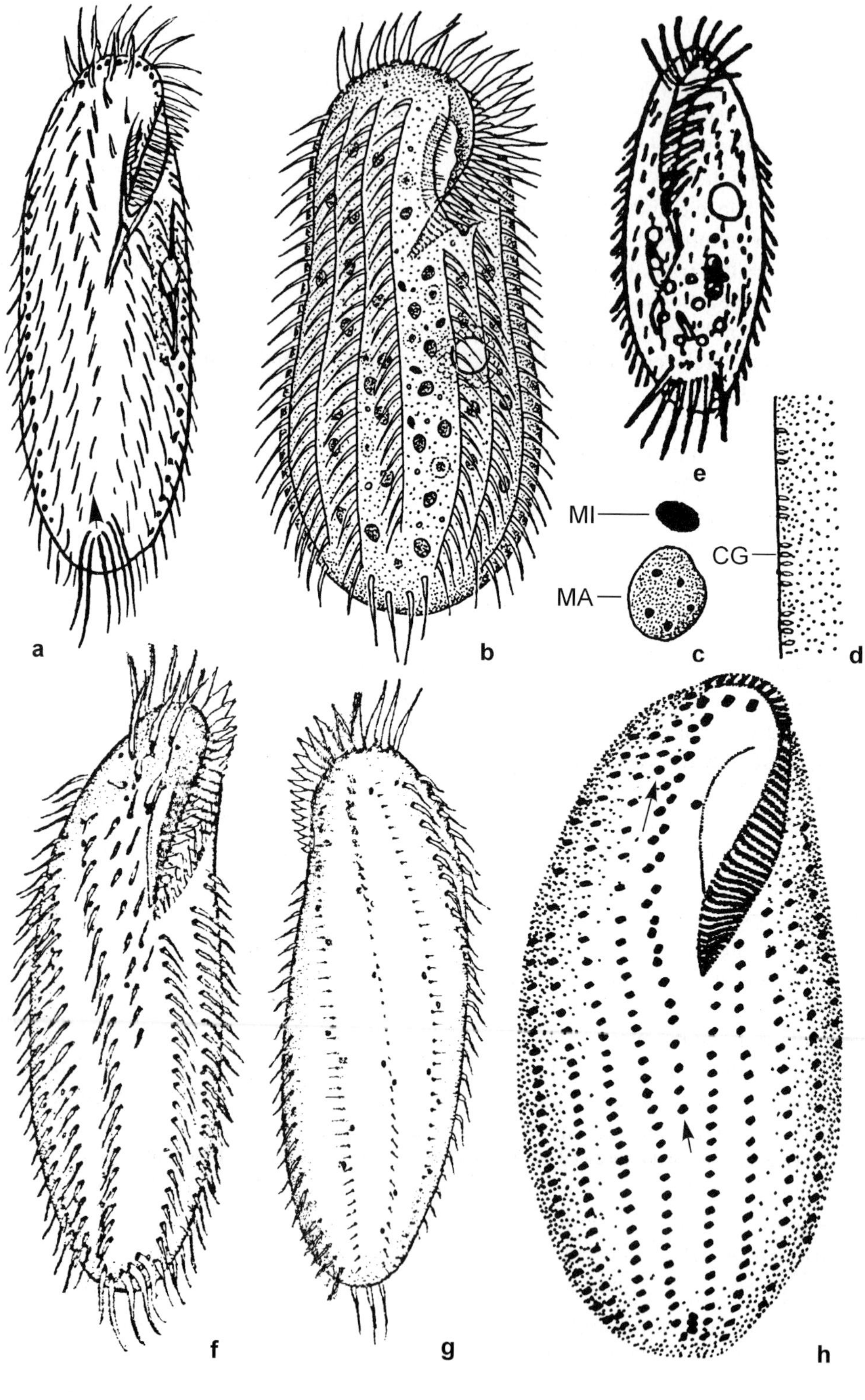
a
b
e
MI
MA
c
CG
d
f
g
h

marginal row. In contrast, Dragesco (1965, p. 395) did not illustrate a distinct zigzagging midventral complex, but three cirral rows right of midline and one left marginal row. Especially the single left marginal row strongly indicates that the synonymy proposed by Borror & Wicklow is not justified because *Metaurostylopsis marina* always has more than one left marginal row.

Wiackowski (1991) identified his Polish population as *Urostyla thompsoni* although it had less dorsal kineties (usually three and rarely up to five vs. about seven), but more transverse cirri (5–8 vs. 3–5) than Thompson's population. Wiackowski, although making live observations, did not describe cortical granules. However, it is known that he was not very familiar with these organelles, because he (Wiackowski 1988) overlooked them also in *Urostyla grandis*, which always has cortical granules. Thus, I agree with Song et al. (2001) and consider Wiackowski's population as synonym of *Metaurostylopsis marina*.

As mentioned above, Song et al. (2001) provided a detailed redescription of *Urostyla marina*, including an analysis of the cell division (Fig. 131h–v). It matches Kahl's population rather well, except for the following details discussed by Song et al. (2001, p. 72, 73): (i) Frontoterminal cirri present against absent. This cirral group is very difficult to recognise in live specimens and even in protargol preparations; hence its absence in Kahl's description must not be over-interpreted. (ii) Body length:width ratio 2.0–2.5:1 (Fig. 131a) against about 3:1 (Fig. 130a). This optically conspicuous, but taxonomically not very important difference is possibly due to the fact that Song et al. (2001) maintained the population for several weeks in the laboratory, where hypotrichs often tend to become stouter. In contrast, Kahl usually studied freshly collected material. (iii) Transverse cirri hardly against distinctly projecting beyond rear body end. This difference is simple due to the fact that Song et al. did not draw the transverse cirri in full length (15 µm).

The neotypification of *Urostyla marina* by Song et al. (2001) seems justified because no fixed or impregnated material of Kahl's population is available. Moreover, the taxonomy of this species is rather complex and a name-bearing type is thus useful and necessary to define the species objectively (ICZN 1999, Article 75). The sole problem is that the neotype locality is not very near the original type locality (China Sea against Baltic Sea). Recently, Lei et al. (2005) found *M. marina* off the Korea coast of the Yellow

← **Fig. 130a–g** *Metaurostylopsis marina* (a, from Kahl 1932; b–d, from Dragesco 1965; e, from Biernacka 1963; f, g, from Borror 1979. a–e, from life; f, g, method not mentioned, possibly the illustration is a composite of live observation and nigrosin-butanol stain). **a:** Ventral view of representative specimen, 80–120 µm. Arrow marks rear end of the midventral complex. **b:** Ventral view of a likely somewhat squeezed specimen, 83 µm. **c:** Macronuclear nodule and micronucleus. **d:** Cortical granules in lateral view. **e:** Superficially drawn specimen from the Danzig Bay, 70–90 µm. **f, g:** Infraciliature of ventral and dorsal side, 100 µm. CG = cortical granules, MA = macronuclear nodule, MI = micronucleus. Page 643.

Fig. 130h *Metaurostyla thompsoni*, a supposed synonym of *Metaurostylopsis marina* (from Thompson 1972. Chatton-Lwoff silver nitrate impregnation). Long arrow marks the frontoterminal cirri; short arrow denotes the rear end of the midventral row. This population has about seven dorsal kineties so that synonymisation with *M. marina* is somewhat doubtful. Page 662.

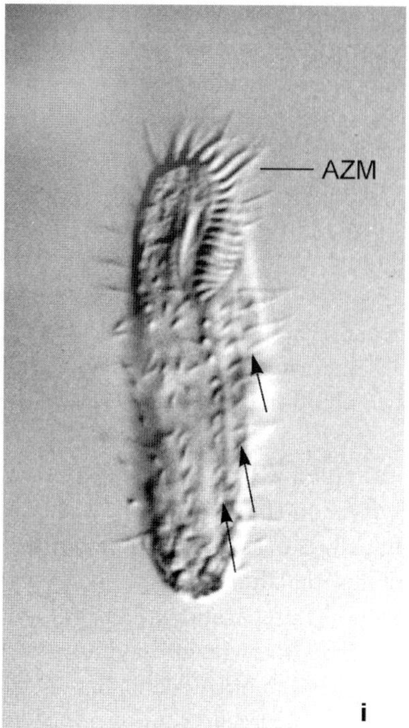
AZM
i

FT
j

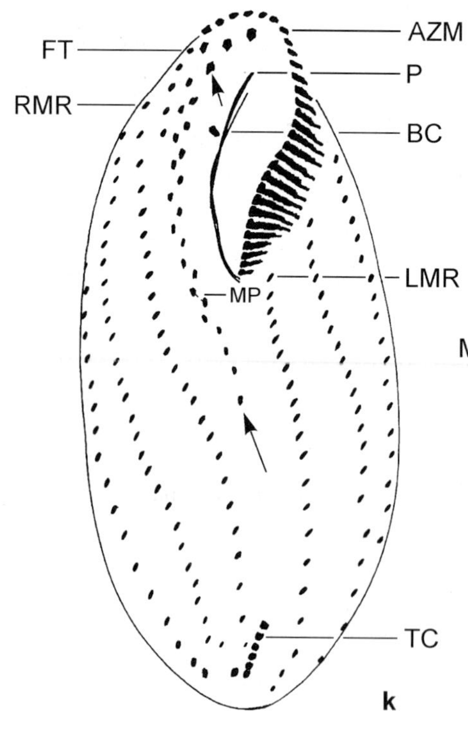
FT
RMR
AZM
P
BC
LMR
MP
TC
k

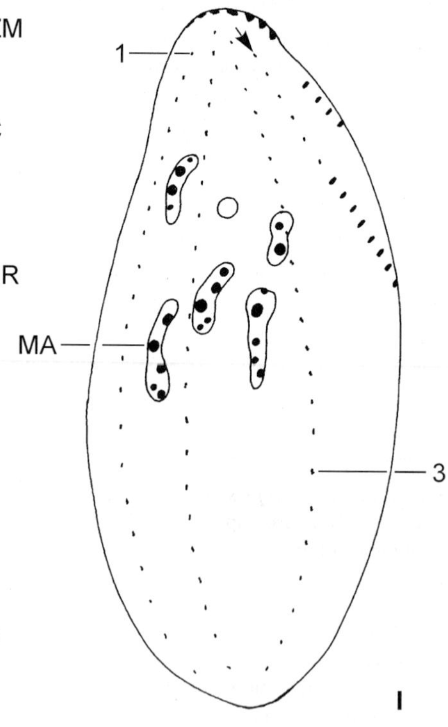
1
MA
3
l

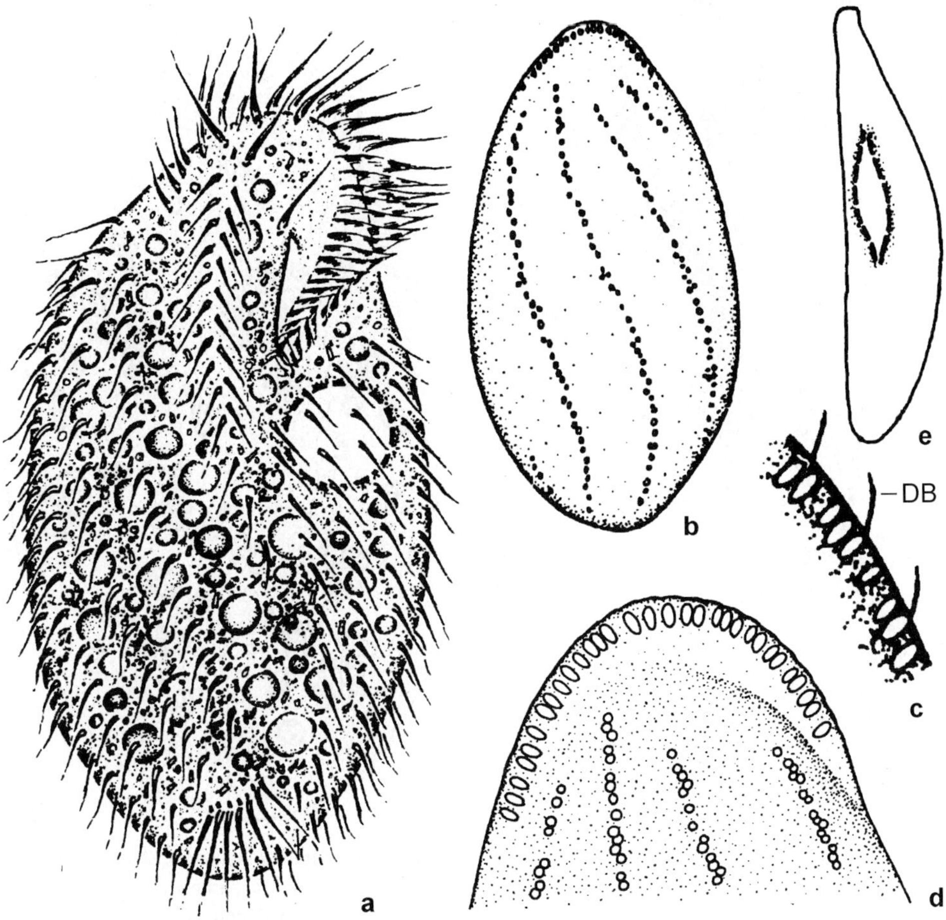

Fig. 131a–e *Metaurostylopsis marina* from life (neotype population from Song et al. 2001). **a:** Ventral view of a rather wide specimen, 93 μm. **b–d:** Dorsal view (b, d) and detail of cortex (c). The cortical granules are more or less colourless, spindle-shaped, 1.5–2.0 μm long, and occur mainly along dorsal kineties and anterior body margin. **e:** Left lateral view showing dorsoventral flattening and contractile vacuole. DB = dorsal bristle. Page 643.

← **Fig. 130i–l** *Metaurostylopsis marina* (i, original; j–l, from Wiackowski 1991. i, Normarski differential interference contrast micrograph; j, from life and after protargol impregnation; k, l, protargol impregnation). **i:** Ventral view of a freely motile specimen showing the oral apparatus with the rather narrow buccal cavity and the cirral rows; the three left marginal rows are marked by arrows. **j–l:** Infraciliature of ventral and dorsal side and nuclear apparatus, j = 90 μm, k, l = 96 μm. Short arrow in (k) marks cirrus (= cirrus III/2) behind right frontal cirrus, long arrow denotes rear end of midventral row. Note that the nuclear apparatus is not completely shown. AZM = adoral zone of membranelles, BC = buccal cirrus, FT = frontoterminal cirri, LMR = inner left marginal row, MA = macronuclear nodule, MP = last midventral pair, P = paroral, RMR = inner right marginal row, TC = transverse cirri, 1, 3 = dorsal kineties. Page 643.

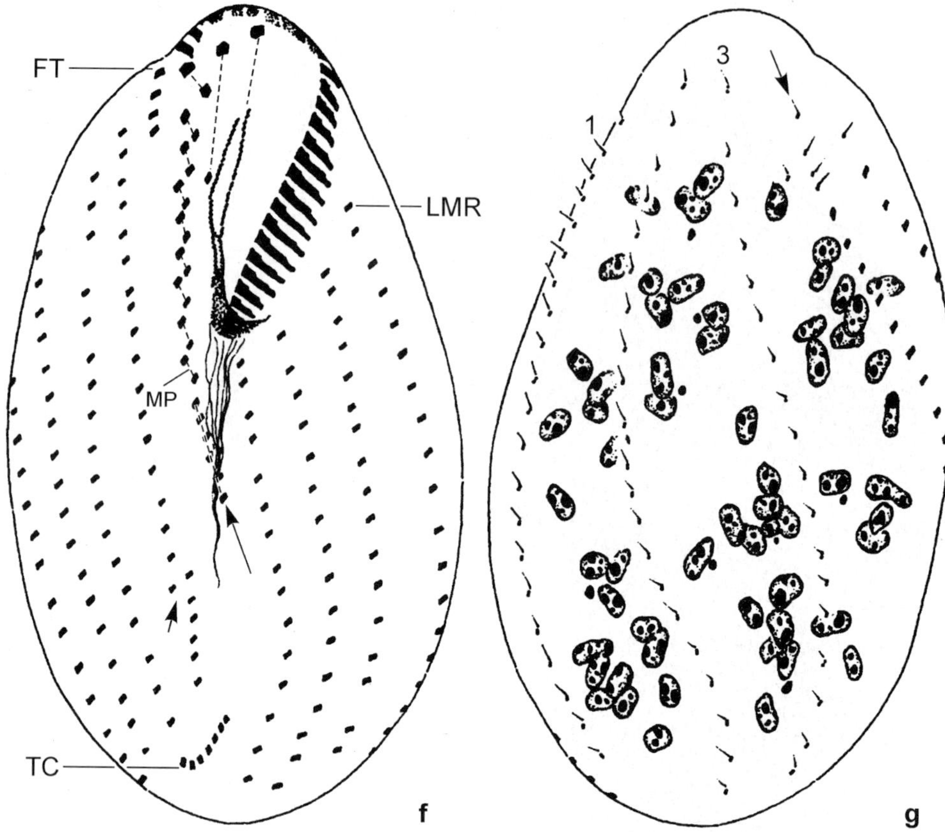

Fig. 131f, g *Metaurostylopsis marina* after protargol impregnation, Wilbert's method (neotype population from Song et al. 2001). Infraciliature of ventral and dorsal side and nuclear apparatus of same specimen, 100 µm. Long arrow in (f) marks rear end of midventral row, short arrow denotes interruption in innermost right marginal row. Frontal-midventral cirri originating from same anlage are connected by broken lines. Arrow in (g) marks some dorsal bristles ahead of the outermost right marginal row. FT = frontoterminal cirri, LMR = outermost left marginal row, MP = last midventral pair, TC = transverse cirri, 1, 3 = dorsal kineties. Page 643.

Sea (Fig. 131w, x). For a brief characterisation of this population from protargol preparations, see figure legend and Table 28.

I identified the present species from a culture provided by Thomas Papenfuss (Rostock University, Germany), who collected it from the Baltic Sea near the island of Rügen. According to my experience, *Metaurostylopsis marina* is easy to identify, mainly due to its rather small size (often less than 100 µm), the high number of cirral rows, and the conspicuous cortical granules.

Urostyla marina sensu Madrazo-Garibay & Lopez-Ochoterena (1985; Fig. 143c) and sensu Aladro Lubel et al. (1990; Fig. 143d) are insufficiently redescribed.

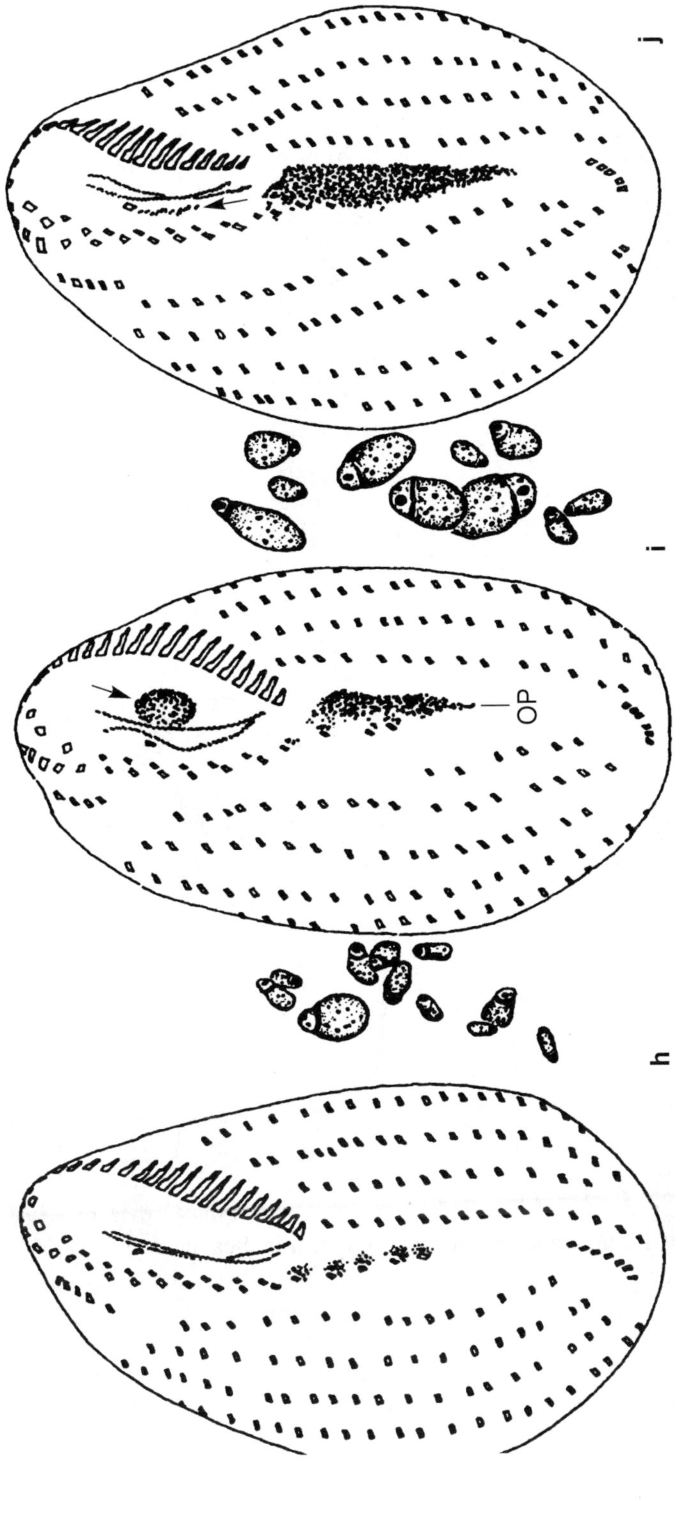

Fig. 131h–j *Metaurostylopsis marina* (neotype population from Song et al. 2001. Protargol impregnation, Wilbert's method). Infraciliature of ventral side and nuclear apparatus (h, i) of very early and early dividers, h = 112 µm, i = 120 µm, j = 128 µm. Ontogenesis commences with the formation of basal body patches close to the cirri of the midventral row (h). Later, they fuse to an elongate oral primordium (i, j). Arrow in (i) denotes the oral primordium of the proter which develops in (beneath) the parental buccal cavity. Arrow in (j) denotes the frontal-midventral-transverse primordium of the proter which originates behind the parental buccal cirrus. The macronuclear nodules fuse to some larger nodules and then form a reticulate structure (Fig. 131v). OP = oral primordium. Page 643.

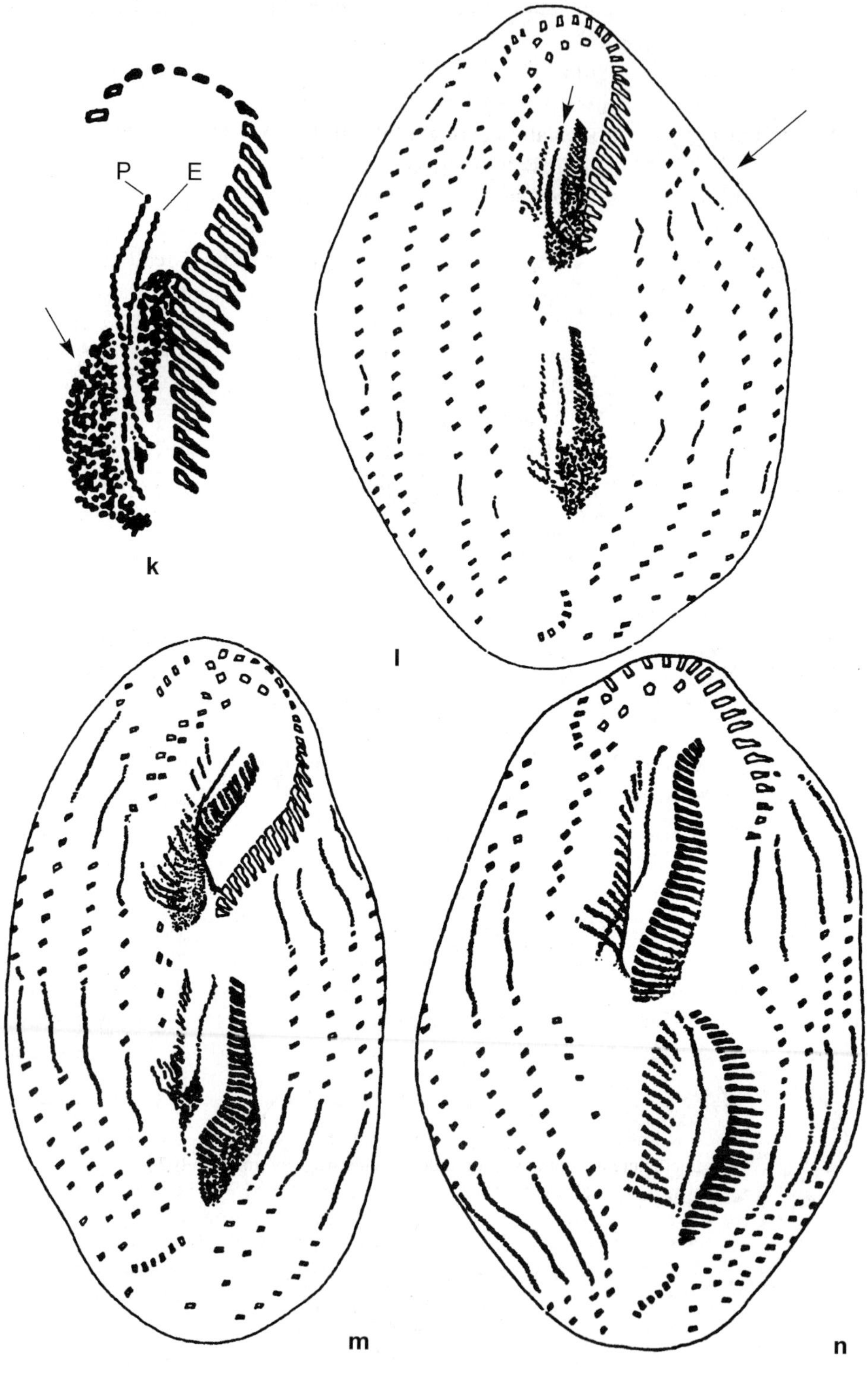
P
E
k
l
m
n

Morphology: Several populations have been described (Kahl 1932, Dragesco 1965, Borror 1979, Wiackowski 1991, Song et al. 2001). The description of the neotype population from China is given first. This is followed by some additional observations from the other descriptions and by original data from live specimens.

Description of neotype population (Song et al. 2001; Fig. 131a–g, Table 28): Body size 80–120 × 50–80 μm in life. Body flexible and slightly contractile, outline thus variable from broadly oval to elongate elliptical (length:width ratio about 2.0–2.5:1[1]), lateral margins evenly convex, anterior end rounded, posterior inconspicuously tapering; body dorsoventrally flattened 2–3:1 (Fig. 131a, e). Macronuclear nodules scattered, difficult to recognise in life; individual nodules oval to elongate ellipsoidal, in life 5–8 μm long, after protargol impregnation only 2–3 μm (last line in Table 1 of Song et al. 2001); likely this discrepancy is due to an error; I assume that the small values refer to the oval micronuclei, which are in life also about 2 μm long, and not to the macronuclear nodules, which have a length of about 6 μm in Fig. 131g. Contractile vacuole large, slightly ahead of mid-body near left cell margin (Fig. 131a, e). Pellicle thin. Cortical granules spindle-shaped and conspicuous because 1.5–2.0 μm long, colourless or slightly greenish, arranged in lines along dorsal kineties and along frontal cell margin (Fig. 131b–d). Cytoplasm colourless to greyish (especially at low magnification), usually containing numerous refractive globules 4–8 μm across and small crystals. Crawls moderately rapidly on debris.

Adoral zone occupies about 40% of body length (Table 28), composed of an average of 28 membranelles of usual shape and structure; bases of largest membranelles 8–10 μm wide, cilia up to about 15 μm long in life. Buccal cavity narrow and rather deep. Undulating membranes long and slightly curved, almost parallel to each other or optically slightly crossing distinctly behind level of buccal cirrus. Pharyngeal fibres conspicuous after protargol impregnation, extend backwards to 71% of body length (Fig. 131f).

Cirral pattern of usually variability while number of some structures shows rather high variation (Table 28), for example, frontoterminal cirri (CV = 22%), dorsal kineties (13%), and number of cirri in marginal rows (up to 30%). All cirri relatively fine and 10–12 μm long, except for frontal and transverse cirri, which are about 15 μm long. Three frontal cirri and one cirrus behind right frontal cirrus in usual position, slightly larger than other cirri. Single buccal cirrus distinctly behind anterior end of paroral. Frontoterminal cirri form short row between distal end of adoral zone and anterior end of innermost right marginal row. Midventral complex composed of 7–11 cirral pairs

← **Fig. 131k–n** *Metaurostylopsis marina* (neotype population from Song et al. 2001. Protargol impregnation, Wilbert's method). Infraciliature of ventral side of middle dividers, l = 99 μm, m = 107 μm, n = 103 μm. **k:** Detail showing proter's oral primordium (arrow) in the parental buccal cavity. **l:** Short arrow marks the undulating membrane anlage, long arrow denotes the primordia for the left marginal rows of the proter. **m, n:** In both the proter and the opisthe a new adoral zone and the oblique streaks for the frontal-midventral-transverse cirri are formed. Note that all marginal rows divide individually which is a difference to *Pseudourostyla* where all marginal rows of one side occur from a single anlage. E = endoral, P = paroral. Page 643.

[1] According to the live measurements, the smallest ratio is at least 120 μm:80 μm, that is, 1.5:1.

and a single midventral row; first midventral pair at level of anterior end of paroral, last pair slightly behind level of proximal adoral membranelle (at 44% of body length in specimen shown in Fig. 131f)[1], midventral row terminates at 61% in specimen shown in Fig. 131f. Size of midventral cirri decreasing towards posterior (not very distinct in Fig. 131f). Transverse cirri form short, oblique, J-shaped row, bases of about same size as those of marginal cirri and thus inconspicuous, subterminal and therefore only slightly projecting beyond rear body end; however, the transverse cirri of the specimen shown in Fig. 131a are not illustrated in full length (15 μm) and thus look shorter than they are. Usually four right and four left marginal rows, occasionally interrupted and/or shortened posteriorly (Fig. 131f, short arrow). Anterior portion of outermost two right marginal rows extends dorsolaterally, composed of basal pairs only, bears a short dorsal bristle (Fig. 131f, g).

Dorsal cilia 3–4 μm long, usually arranged in three almost bipolar rows. 3–5 dorsal bristles ahead of rightmost marginal row and sometimes(?) one bristle ahead of next right marginal row (Fig. 131g). Three out of 16 specimens with a fourth, anteriorly distinctly shortened dorsal kinety either left of kinety 1 or right of kinety 3. Caudal cirri lacking.

Additional observations from other populations (Kahl 1932, Dragesco 1965, Borror 1979, Wiackowski 1991; for data by Lei et al. 2005 see Table 28 and legend to Fig. 131w, x): Body length in life 80–120 μm (Kahl), about 80 μm (Dragesco); according to Borror, size is 68–100 × 24–32 μm (n = 20; in life?), but only 50 μm long specimens are common in older cultures. Body slender oval to almost ellipsoidal (Kahl), according to Wiackowski outline elliptical; Dragesco's specimens stout; soft and flexible, slightly contractile (Kahl), distinctly flattened dorsoventrally with ventral side slightly concave (Wiackowski). Body length:width ratio from about 2:1 (Wiackowski) to about 3:1 (Kahl).

Nuclear apparatus not recognised by Kahl, indicating that many small nodules were present. Individual nodules oval (Borror, Wiackowski), of rather different length, namely, 2–4 μm (Borror), 5–6 μm (Dragesco, Fig. 130b, c), and 9–18 μm (Wiackowski, Fig. 130l). Micronuclei ovoid (Dragesco), about 3 μm across (Wiackowski).

Contractile vacuole slightly ahead of mid-body near left cell margin with longitudinal collecting canals (Kahl, Fig. 130a). According to Borror, the contractile vacuole is oval and located dorsally and slightly left of median. Cytoplasm transparent with a slight yellowish shade (Wiackowski).

Cortical granules conspicuous because strong, yellowish, and oval, arranged in pairs or triplets along dorsal kineties and single granules along the ventral cirral rows; some of the individual granules seem to have a lengthways gap; a wreath of cortical granules seams the frontal scutum (Kahl). According to Borror, the granules are oval too, about 1 μm long, and Borror – like Kahl – observed a longitudinal crease giving the granules the appearance of oat grains; in Borror's specimens the granules occurred in the same areas as in Kahl's specimens, but were lacking near the midventral cirri. Cortical granules ("protrichocysts"; Fig. 130d) of Dragesco's population spindle-shaped and likely colour-

[1] According to the description (Song et al. 2001, p. 68) the last midventral pair is at 50–60% of body length, which is, however, not in accordance with the specimen shown in Fig. 131f.

less and not red as stated by Song et al. (2001; see remarks for details). Cortical granules not mentioned by Wiackowski, according to the character state matrix given by Wiackowski (1988) mucocysts (= cortical granules; feature 26) are lacking.

Adoral zone occupies about 30% of body length in Borror's population (Fig. 130f), on average 41% in Wiackowski's specimens; of usual shape and structure (Kahl, Borror, Wiackowski). Mediad, several cilia of each membranelle in posterior half of adoral zone arise slightly anterior to remaining cilia of that membranelle and beat independently as a ciliary brush along mediad edge of adoral zone, as is typical of most hypotrichs (Borror). Largest membranelles 6–8 µm wide after protargol impregnation (Wiackowski; Wilbert's method). Buccal cavity according to Kahl and Borror rather narrow, according to Wiackowski moderately wide. Paroral and endoral according to Borror each likely composed of a row of single cilia; according to Wiackowski, the paroral is composed of a double row of basal bodies, the endoral of a single row; both rows slightly curved and optically intersecting distinctly behind level of buccal cirrus (Fig. 130k).

Frontal cirri moderately enlarged (Kahl, Wiackowski). One or two buccal cirri (Kahl, Fig. 130a). One cirrus behind right frontal cirrus (Kahl, Borror). First midventral pair about at level of buccal cirrus, last pair about at 46% of body length (Borror; Fig. 130f); midventral cirri distinctly shortened (Wiackowski); midventral complex terminates at about 57% of body length (Borror, Fig. 130f; Wiackowski, Fig. 130k). Transverse cirri protrude distinctly beyond rear body end (Kahl, Borror, Wiackowski), about 15 µm long, frayed at tips, often lacking in small individuals of older cultures (Borror). Innermost left marginal row sometimes shortened posteriorly (Kahl); rightmost two marginal rows extend onto dorsolateral surface anteriorly (Borror); each marginal row composed of about 20, in life circa 8 µm long cirri (Borror).

Dorsal cilia about 4 µm long and about 7 µm apart (Borror). According to Wiackowski usually three dorsal kineties are present, however, three out of 30 specimens had five kineties, and two had four. Caudal cirri lacking (Wiackowski).

I studied some live, cultured specimens from the Baltic Sea and can confirm most observations mentioned above (Fig. 130i): Body size around 80 × 25 µm. Body length:width ratio about 3.0–3.5:1 (Fig. 130i). Body elongate elliptical with margins slightly to distinctly converging posteriorly, both ends broadly rounded; very flexible, but not distinctly contractile, rather resistant against cover glass pressure. Macronuclear nodules difficult to recognise in life. Contractile vacuole at about 45% of body length near left cell margin, sometimes difficult to recognise. Cortical granules strong, almost club-shaped, 1–2 µm long, yellowish, arranged around dorsal bristles and, more rarely, around cirri; some specimens with rather few granules. Cytoplasm colourless with some empty, moderately large vacuoles in posterior body portion. Length of adoral zone 20–25 µm in life, adoral zone of one specimen composed of about 19 membranelles. Buccal cavity rather small and moderately deep. Cirral pattern as described above; three slightly enlarged frontal cirri, one buccal cirrus; frontoterminal row not recognised in life; midventral complex composed of several pairs and one row terminating behind mid-body. Three left and three right marginal rows (n = 1; Fig. 130i); outer left marginal row with 12 cirri, middle with about 13 cirri, and inner with about 11 cirri; five transverse cirri in hook-shaped pattern. Dorsal bristles about 4 µm long in life.

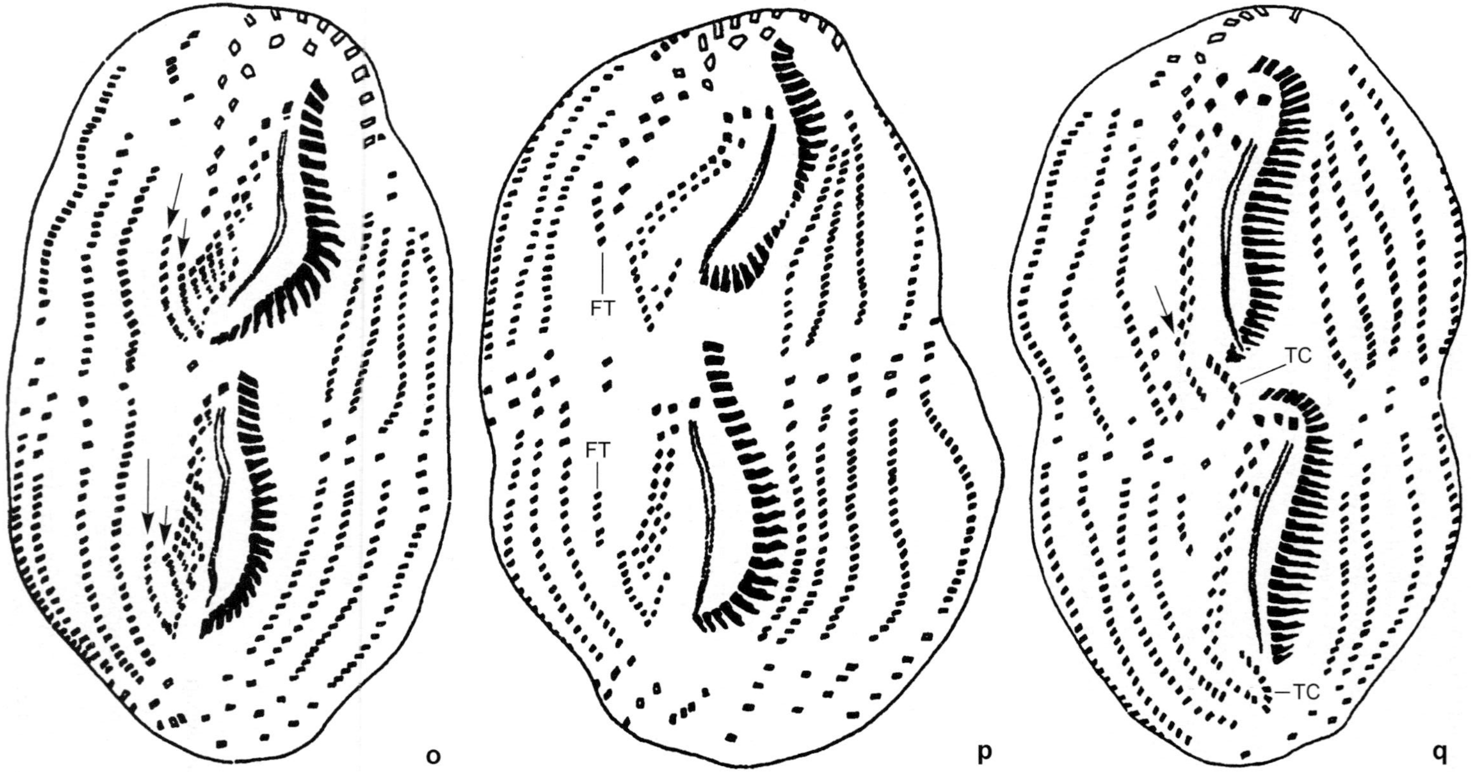

Fig. 131o–q *Metaurostylopsis marina* (neotype population from Song et al. 2001. Protargol impregnation, Wilbert's method). Infraciliature of ventral side of late dividers, o = 102 µm, p = 92 µm, q = 112 µm. New structures black, old white. Long arrows in (o) mark the rightmost cirral streak which forms the frontoterminal cirri and the rightmost transverse cirrus, short arrows in (o) denote the penultimate anlage which forms the midventral row. Note that in (p) several right anlagen form more than three cirri (inclusive the transverse cirrus); however, the surplus will be resorbed in later stages. Arrow in (q) marks the penultimate anlage of the proter which forms the midventral row. FT = new frontoterminal cirri, TC = new transverse cirri. Page 643.

Cell division (Fig. 131h–v, 132a–g): The division process of *Metaurostylopsis marina* was investigated by Borror (1979; Fig. 132a–g), Wiackowski (1991), and, most detailed, by Song et al. (2001; Fig. 131h–v). Unfortunately, the illustrations of the neotype population are very small in the original description. The general pattern is simple in most respects and the reader is therefore referred to the illustrations. Most stages are documented by light micrographs (see Song et al. 2001).

The oral primordium of the opisthe originates from some basal body patches close to the cirri of the midventral row (Fig. 131h). They fuse to a single field, and simultaneously an elliptical patch of basal bodies occurs in the parental buccal cavity at the level of the buccal cirrus (Fig. 131i, arrow). Likely, it originates in a pouch of the buccal cavity wall because its margin is always clearly outlined. This oral primordium of the proter grows rapidly while the parental undulating membranes and adoral zone remain unchanged. A streak of basal bodies occurs behind the parental buccal cirrus (Fig. 131j); later, this becomes the frontal-midventral-transverse cirral anlage of the proter. The parental buccal cirrus dedifferentiates and very likely contributes to this primordium, whose basal bodies begin to align to oblique streaks in the posterior portion of the anlage (Fig. 131l). Further basal bodies occur and streak formation proceeds anteriad until 10–15 oblique streaks have been formed (Fig. 131m); the cirri of the parental midventral complex are not involved.

The parental undulating membranes disorganise and seem to be resorbed, possibly with the exception of the anterior portion of the endoral, which might join the new undulating membrane anlage (Fig. 131l). The proximal part of proter's oral primordium remains in the buccal cavity beneath the frontal-midventral-transverse anlage until the posterior portion of the parental adoral zone is resorbed. The old adoral zone does not contribute to the oral primordium because it is always distinctly separate (Fig. 131m, n).

To the right of opisthe's oral primordium, two longitudinal streaks occur (Fig. 131l). The right streak forms the frontal-midventral-transverse cirral anlagen composed of 10–15 oblique streaks, the left becomes the anlage of the undulating membranes.

Frontal-midventral-transverse cirri differentiate during middle and late stages (Fig. 131n–q). As usual, the left frontal cirrus originates from the anlage of the undulating membranes (Fig. 131n, o). Usually two cirri originate from the following anlagen and they produce, as in other urostylids, the following structures: streak II forms the middle frontal cirrus and the buccal cirrus; streak III forms the right frontal cirrus and the single cirrus (= cirrus III/2) behind; streak IV and on average the next seven streaks form the midventral pairs (the rearmost streaks also produce a transverse cirrus); streak n–1 produces a transverse cirrus and 4–7 cirri, which form the midventral row, that is, the rear portion of the midventral complex (likely the anteriormost two cirri of the midventral row are somewhat separate from the other cirri so that they form a further midventral pair); streak n forms, as is usual, the frontoterminal cirri and the rightmost transverse cirrus. Several streaks (for example, n–2, n–3) often form more than the two cirri of a midventral pair and the transverse cirrus (Fig. 131o, p); however, the surplus cirri will be gradually resorbed during the last stages of ontogenesis. Obviously no distinct pretransverse ventral cirri are formed from the two rightmost anlagen.

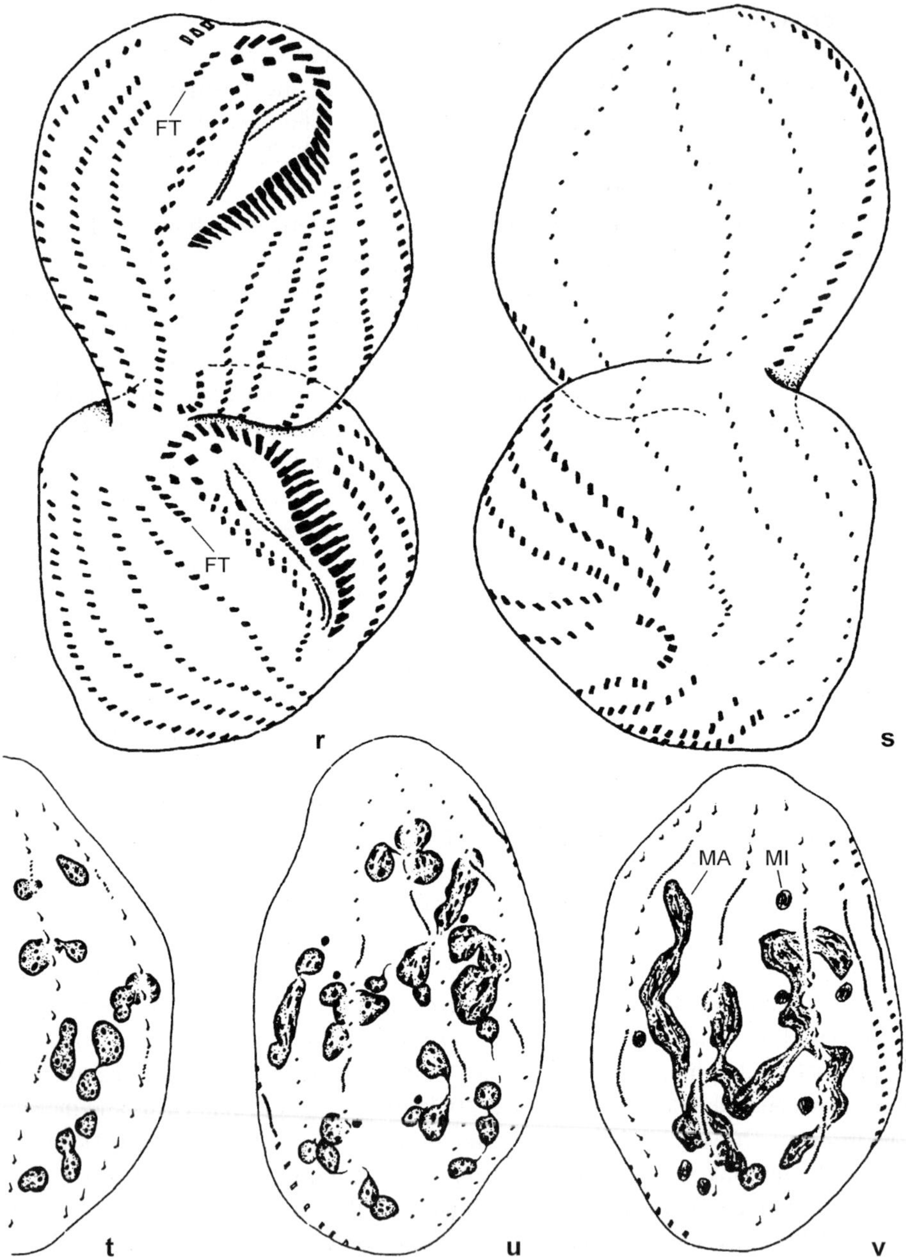

Fig. 131r–v *Metaurostylopsis marina* (neotype population from Song et al. 2001. Protargol impregnation, Wilbert's method). **r, s:** Infraciliature of a very late divider in ventral and dorsal view, 98 µm. **t–v:** Division of dorsal ciliature proceeds in *Gonostomum* pattern, that is, two anlagen develop in each kinety (t shows same specimen as l, u = m and v = n). Note that no caudal cirri are formed. The macronucleus does not form a single roundish mass as in most other hypotrichs, but a branched complex (v). FT = new frontoterminal cirri, MA = branched macronucleus, MI = micronucleus. Page 643.

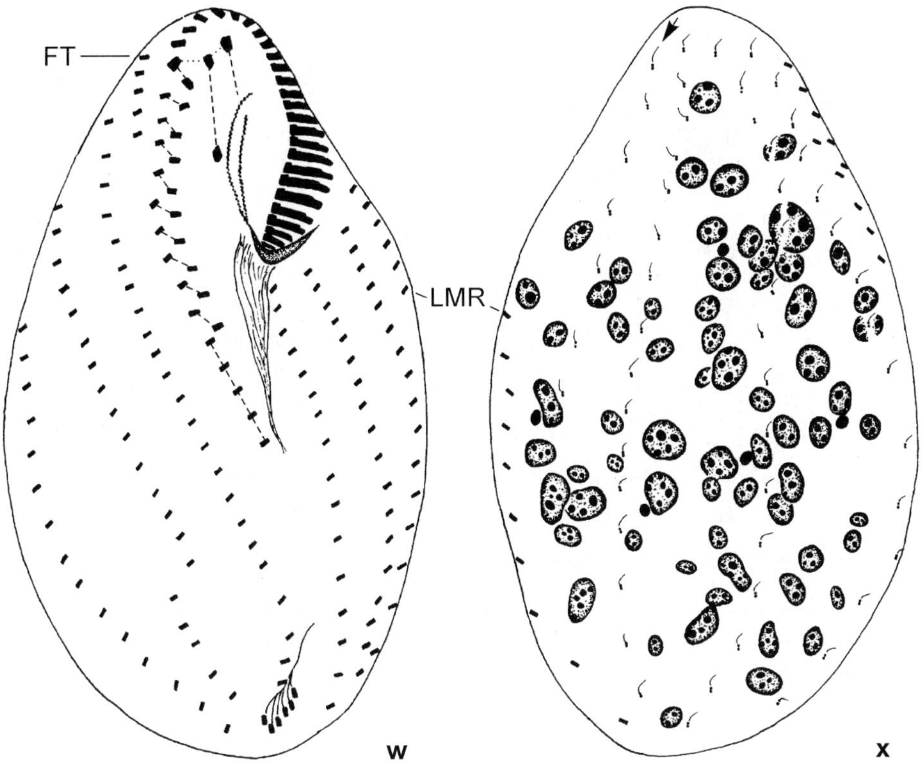

Fig. 131w, x *Metaurostylopsis marina* (from Lei et al. 2005. Protargol impregnation). Infraciliature of ventral and dorsal side and nuclear apparatus of same specimen, 90 µm. Broken lines connect cirri, which originate from the same anlage (corresponding transverse cirri not included). Arrow in (x) marks an additional dorsal kinety on the left side. FT = anterior end of frontoterminal row, LMR = outermost left marginal row. Page 643.

Marginal cirral rows and dorsal kineties divide in the ordinary manner, that is, within each row two anlagen occur (Fig. 131l–r). No caudal cirri are formed at the end of the new dorsal kineties (Fig. 131s–v). Cells with four dorsal kineties are likely post-dividers with a remnant of a parental row. Dorsomarginal kineties are lacking, as in other urostylids.

The nuclear division is rather conspicuous in *Metaurostylopsis marina*. In very early dividers some macronuclear nodules enlarge by up to 3–5 times and almost all have a reorganisation band (Fig. 131h, i, t, u). Subsequently, the nodules fuse and form a single, distinctly branched structure (Fig. 131v). In late dividers, this mass divides into the nodules. The micronuclei remain unchanged for some time, but then enlarge considerably and divide (Fig. 131v).

The data by Borror (1978, 1979) are mainly from nigrosin-butanol stained specimens and thus not very detailed (Fig. 132a–g). Basically the process is very similar to that described by Song et al. (2001). However, Borror did not describe the formation of

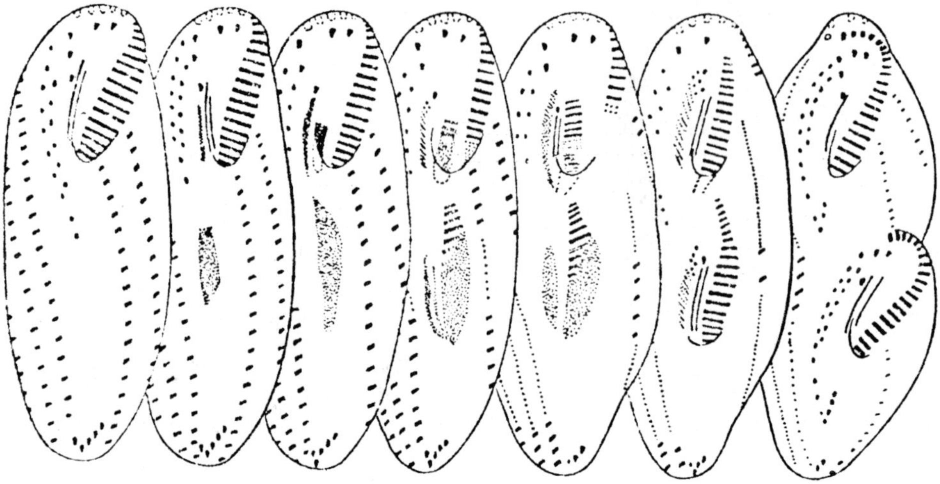

Fig. 132a–g *Metaurostylopsis marina* (from Borror 1979. Nigrosin-butanol stain). **a:** Non-dividing specimen with only two left marginal rows. **b–g:** Infraciliature of ventral side of dividing specimens. Note that Borror did not illustrate frontoterminal cirri although he saw the other structures rather clearly; possibly he overlooked them due to the nigrosin-butanol stain, which generally does not show details as exactly as the protargol method. Page 643.

the frontoterminal or migratory cirri although he observed a stage where the migration of these cirri is usually clearly recognisable (Fig. 132g). Further, the oral primordium of the proter, which occurs in the anterior portion of the buccal cavity in the neotype population (Fig. 131i), is formed near the proximal end of the parental adoral zone (Fig. 132c, d). In both populations, however, the parental adoral zone is completely resorbed and replaced by a new one. A further difference between the populations exists in the division of the macronucleus. In Borror's specimens the nodules condensed into an elongate, slightly coiled single mass while in the neotype population the fused nodules formed a branched complex. Wiackowski (1991) studied some stages of divisional ontogenesis and physiological reorganisation. His data agree with the observations by Song et al. (2001) and are thus not repeated. This indicates that Borror's somewhat deviating data are possibly due to methodological constraints. Although these differences must not be over-interpreted, the investigation of the division process of further populations is recommended to document the variability of the process in more detail.

Occurrence and ecology: Marine. Due to the neotype fixation by Song et al. (2001), the type locality of *Metaurostylopsis marina* is the locality of the neotype population (ICZN 1999, Article 76.3), that is, the Taping Bay (36°08'N 120°43'E), a small bay near Qingdao (Tsingtao), China. The bay is used for farming molluscs and the species was found in the benthal on December 23, 1995 at following conditions: salinity 32‰, water temperature 4–5° C, pH 8.2. Song et al. (2001) maintained *M. marina* in

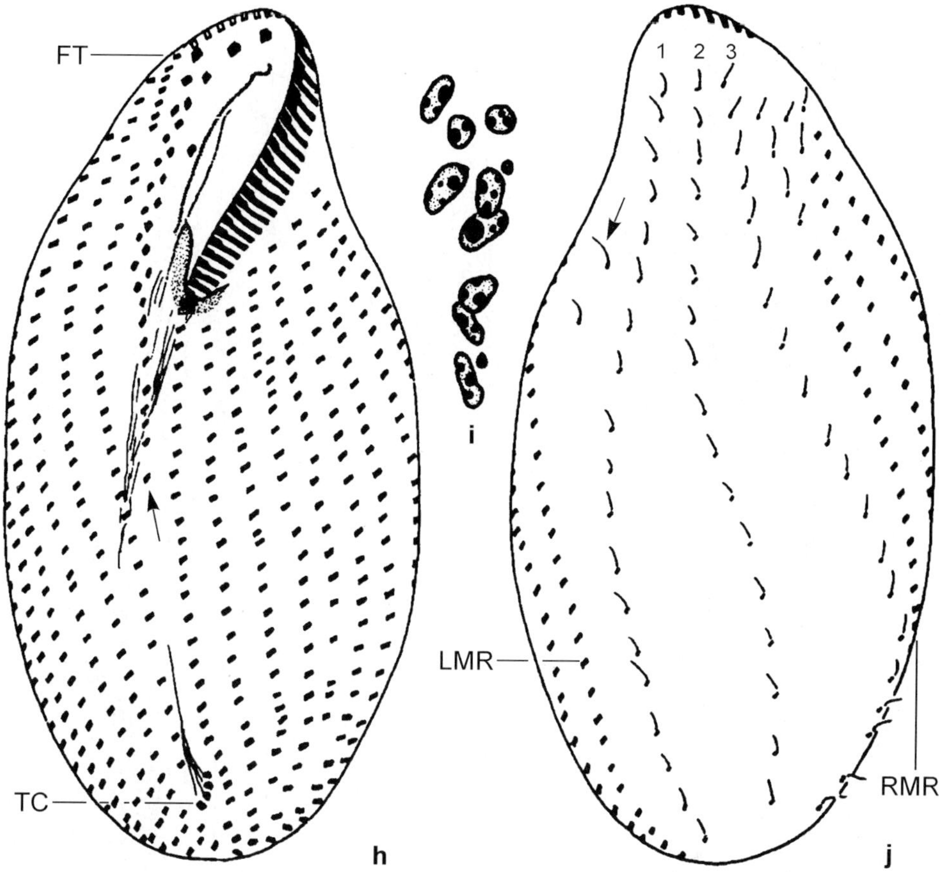

Fig. 133h–j *Metaurostylopsis rubra* (from Song & Wilbert 2002. h–j, protargol impregnation). **h, j:** Infraciliature of ventral and dorsal side of same specimen, 136 µm. Arrow in (h) denotes rear end of midventral row, arrow in (j) marks two supernumerary dorsal bristles between outermost left marginal row and dorsal kinety 1. **i:** Part of nuclear apparatus, length of macronuclear nodules 5–8 µm (Song & Wilbert obviously erroneously wrote "micronuclei" as legend to this figure). FT = frontoterminal cirri, LMR = outermost left marginal row, RMR = outermost right marginal row, TC = transverse cirri, 1–3 = dorsal kineties. Page 663.

Remarks: Song & Wilbert (2002) supposed that *Urostyla marina* sensu Dragesco (1965; Fig. 130b) belongs to *Metaurostylopsis rubra* because Dragesco's specimens are also reddish. However, Dragesco's specimens are reddish due to the food vacuoles, while the red colour of *M. rubra* is caused by pigments in the cytoplasm. Furthermore, *Metaurostylopsis rubra* is distinctly larger and has about twice as many marginal rows as Dragesco's specimens. In contrast to Song & Wilbert, I accept Dragesco's identification.

Metaurostylopsis rubra is up to three times as long as the type species *M. marina* and has a reddish cytoplasm so that they can be easily separated. The marine *Pseudo-*

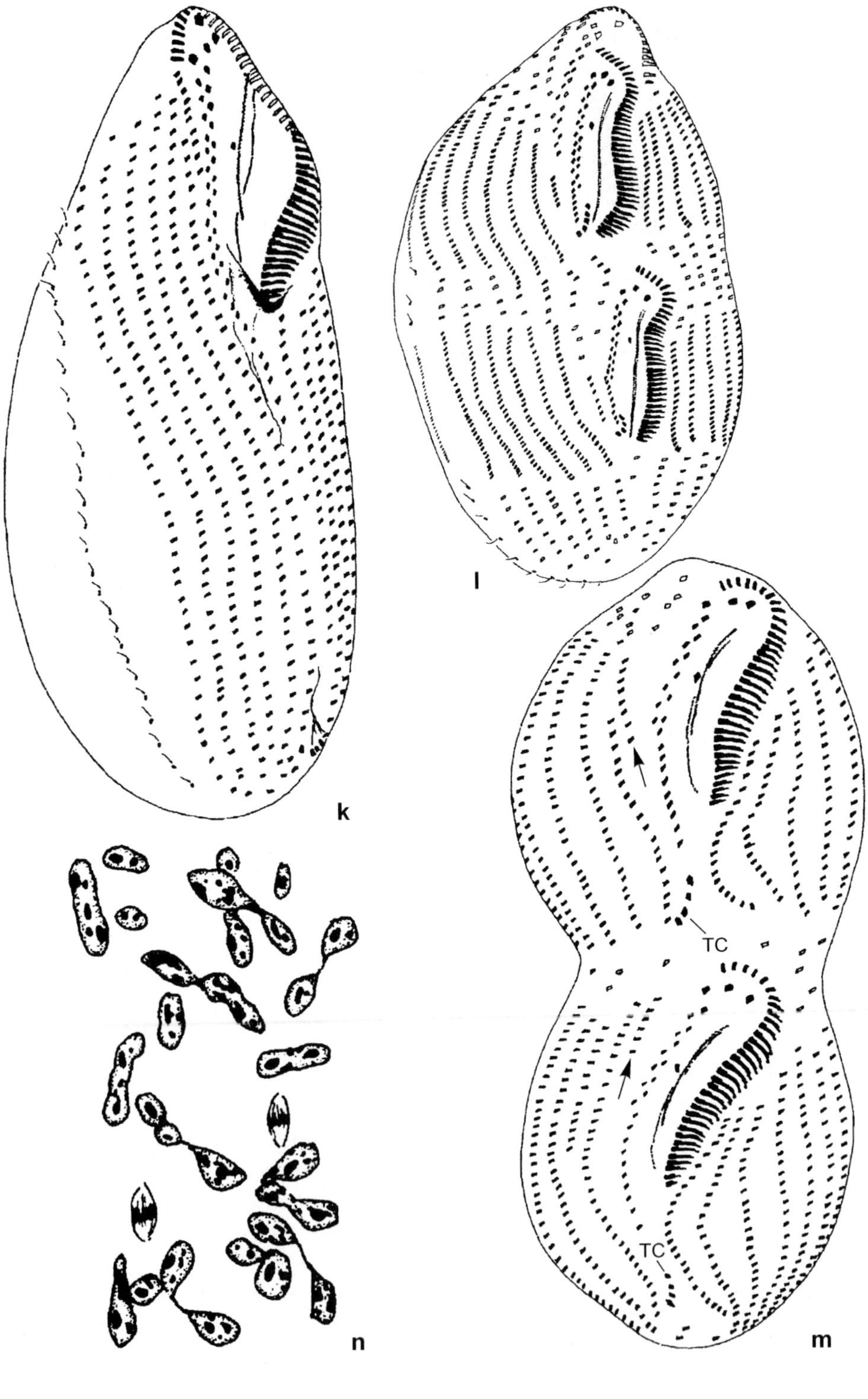
k
l
TC
TC
n
m

keronopsis rubra, which has a similar size and colour, has, inter alia, only one left and one right marginal row (vs. 12 or more in *M. rubra*) and the frontal cirri form a distinct bicorona (vs. three enlarged cirri). The limnetic *Diaxonella pseudorubra* is usually smaller (120–180 × 30–70 μm vs. 150–300 × 50–90 μm) and has, inter alia, 5–9 buccal cirri (vs. one), but only one right marginal row (vs. 6–7).

Morphology: Unless indicated otherwise, the data are from the type population. Body size 150–300 × 50–90 μm in life; specimens described by Wilbert & Song (2005) about 150 × 50 μm in life; ratio of body length:width about 3:1 in life, about 2.2:1 on average in protargol preparations (Table 28). Body flexible and slightly contractile, usually elongate or somewhat sigmoidal, sometimes widest at beginning of posterior third with both ends often narrowly rounded (Fig. 133a–c, f); dorsoventrally flattened about 2:1 (Fig. 133d). Macronuclear nodules (about 100) difficult to recognise in life, scattered throughout cytoplasm; individual nodules oval to elongate, 5–8 μm long after protargol impregnation, each with several large and ordinary sized nucleoli (Fig. 133i). Contractile vacuole large, slightly ahead of mid-body near left cell margin (Fig. 133a, c, d, f). Pellicle thin. Cortical granules conspicuous, packed in irregular lines on dorsal side (Fig. 133b); presence or absence of granules on ventral side, and shape, size, and colour of individual granules not mentioned. Whole cytoplasm brick-red to dark-red due to fine pigments; strongest pigmentation in anterior cell end (Fig. 133e, f). Cytoplasm (endoplasm) with numerous, shining globules 2–5 μm across and food vacuoles. Movement without peculiarities, crawls moderately rapid.

Adoral zone occupies about 33% of body length, composed of an average of 39 membranelles; bases of largest membranelles 12 μm wide, cilia up to 20 μm long; distal end of adoral zone extends only slightly posteriorly on right cell margin. Figs. 133g, h do not show the ordinary 3-step structure of the adoral membranelles, but only a 2-step structure. Since this feature is not mentioned by Song & Wilbert, I assume that this unusual structure is a misobservation. Undulating membranes long and straight to slightly curved; almost parallel. Pharyngeal fibres conspicuous after protargol impregnation, up to 50 μm long, extend obliquely backwards (Fig. 133a, g, h).

Cirral pattern and number of cirri of usual variability (Fig. 133g, h, Table 28). Most cirri relatively fine and about 10–15 μm long. Three enlarged, about 20 μm long frontal cirri in ordinary arrangement; slightly obliquely behind right frontal cirrus a single cirrus (= cirrus III/2). One slightly enlarged buccal cirrus right of mid-paroral. Frontoterminal cirri form short row between anterior end of midventral complex and right marginal rows. Midventral complex composed of 8–11 cirral pairs and a single midventral row, which terminates at 54% of body length in specimen shown in Fig. 133h; first midventral pair commences at level of cirrus III/2, last pair arranged slightly ahead of level of proximal adoral membranelle (Fig. 133g, h); midventral cirri of about same size as marginal cirri. Transverse cirri inconspicuous because of about same length and

---

← **Fig. 133k–n** *Metaurostylopsis rubra* (from Wilbert & Song 2005. Protargol impregnation). **k:** Infraciliature of ventral side, size not indicated. **l, m:** Infraciliature of ventral side of middle and late divider, sizes not indicated. Arrows mark new frontoterminal row of proter and opisthe. **n:** Nuclear apparatus. Unfortunately, the authors did not indicate to which stage this figure belongs. TC = new transverse cirri. Page 663.

size as marginal cirri, subterminally and almost longitudinally arranged and thus only slightly projecting beyond rear body end. Usually eight left and six right marginal rows, which are occasionally shortened or interrupted anteriorly; outermost rows partially extending on dorsolateral surface (Fig. 133h, j).

Dorsal cilia 3–4 µm long, arranged in three almost bipolar kineties. Usually two or three short rows, each composed of 1–4 basal body pairs ahead of outermost right marginal rows; furthermore, some dorsal bristles often occur between ordinary kineties (Fig. 133j). Caudal cirri lacking.

Cell division (Fig. 133l–n): Wilbert & Song (2005) provided two ontogenetic stages showing that cell division proceeds basically as in the type species. They summarised the data as follows (most of the features are rather old plesiomorphies): (i) Parental oral apparatus completely replaced. (ii) Two sets of frontal-ventral-transverse cirral anlagen are produced. (iii) Left frontal cirrus is formed, as is usual, from anlage I. (iv) Anlage II produces the middle frontal cirrus and the buccal cirrus, anlage III forms the right frontal cirrus and the cirrus behind (= cirrus III/2). (v) Anlagen IV to n–2 produce the midventral pairs. (vi) The anlage n–1 (the last but one) produces the midventral row. (vii) The rightmost anlage produces the frontoterminal cirri. (viii) Two anlagen develop within each marginal cirral row, each of which forms a separate marginal row. (ix) Dorsal kineties are formed in common manner, that is, two primordia occur within each parental kinety. (x) No caudal cirri are formed.

Unfortunately, Wilbert & Song (2005) made no comment about the division of the nuclear apparatus, that is, it is unknown whether the macronucleus of *M. rubra* fuses to a roundish mass as in most hypotrichs or is branched as in *M. marina*, type of the genus.

Occurrence and ecology: Likely confined to marine habitats. The type locality of *Metaurostylopsis rubra* is the Potter Cove, King George Island (62°14'S 58°40'W), Antarctica, where Song & Wilbert (2002, p. 24) collected their material from a rock pool and the littoral. Water temperature was about 1°C and salinity about 33‰. No further records published since then. Feeds on flagellates, diatoms, and ciliates.

Metaurostylopsis salina Lei, Choi, Xu & Petz, 2005 (Fig. 133.1a–k, Table 28)

2005 ***Metaurostylopsis salina* n. sp.**[1] – Lei, Choi, Xu & Petz, J. Euk. Microbiol., 52: 4, 8, Fig. 14–24, 27–33, Table 2 (Fig. 133.1a–k; original description. The holotype slide [accession number HS–010614–1] is deposited in the Marine Biological Specimen Depository, Chinese Academy of Sciences, Qingdao, China, while a paratype slide [HS-010614–2] is deposited in the Laboratory of Plankton, Department of Oceanography, Inha University, Inchon, Korea).

[1] The diagnosis by Lei et al. (2005) is as follows: Marine *Metaurostylopsis* in vivo about 60 µm × 25 µm; body pyriform. Colorless cortical granules irregularly arranged. 18–23 adoral membranelles; three frontal, three to five frontoterminal, one buccal, and two to five transverse cirri; four or five midventral cirral pairs followed by five to seven single cirri posteriorly; invariably three marginal cirral rows on each side of cell; three long dorsal kineties. On average 45 scattered macronuclear nodules and six micronuclei.

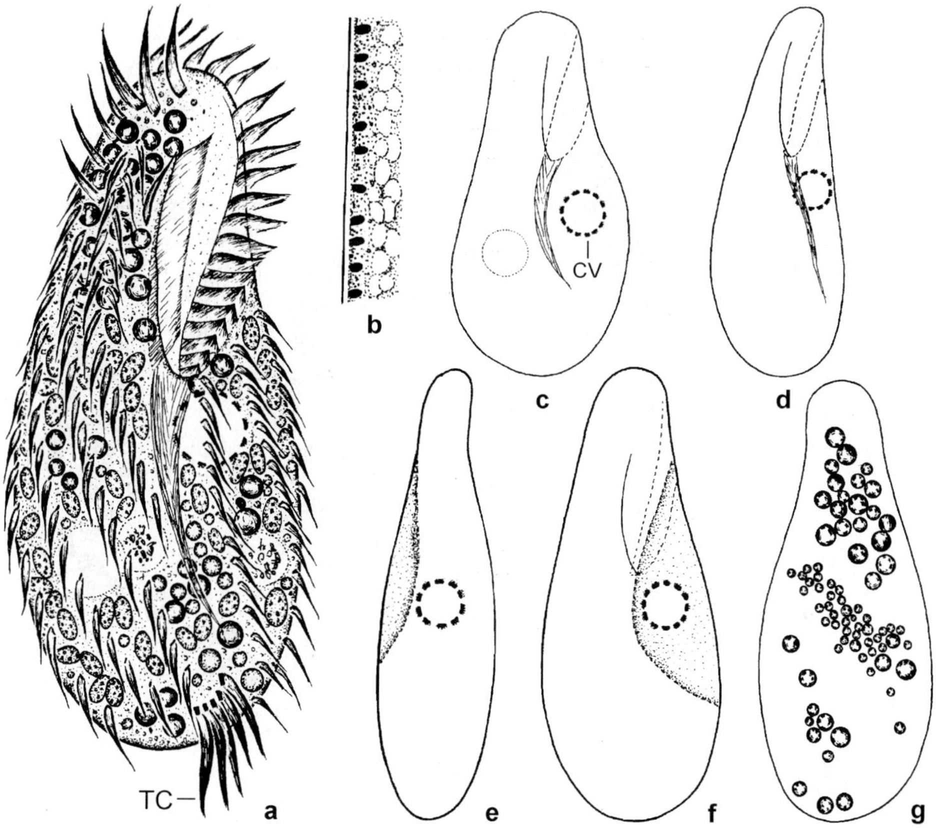

Fig. 133.1a–g *Metaurostylopsis salina* (from Lei et al. 2005. From life). **a:** Ventral view of a representative specimen, 60 μm. **b:** Cortex in cross-section showing, inter alia, scattered colourless cortical granules 0.5–1.0 μm across. **c, d:** Shape variants. **e–g:** Left-lateral, ventral, and dorsal view of same specimen showing a concavity in the middle left portion of the ventral side, the contractile vacuole, and lipid globules. CV = contractile vacuole, TC = transverse cirri. Page 668.

Nomenclature: The Latin adjective *salin·us -a -um* (saline) refers to the habitat where the species was discovered (Lei et al. 2005).

Remarks: *Metaurostylopsis salina* is the smallest of the four species included in *Metaurostylopsis*. The basic cirral pattern is very similar to that of *M. marina* and *M. rubra*, strongly indicating that the assignment to the present genus is correct. The most similar species is *M. marina*, which, however, is larger (80 to about 150 μm vs. 40–80 μm), has an oval or ellipsoidal body shape (vs. usually pyriform), and has scattered cortical granules (vs. arranged in rows). In addition, it differs from the other species in some morphometric data (Table 28).

Morphology: Lei et al. (2005) found several populations from two connected ponds. However, detailed data were available only from a hypersaline population collected in June 2001. The other populations were identified mainly from protargol-impregnated

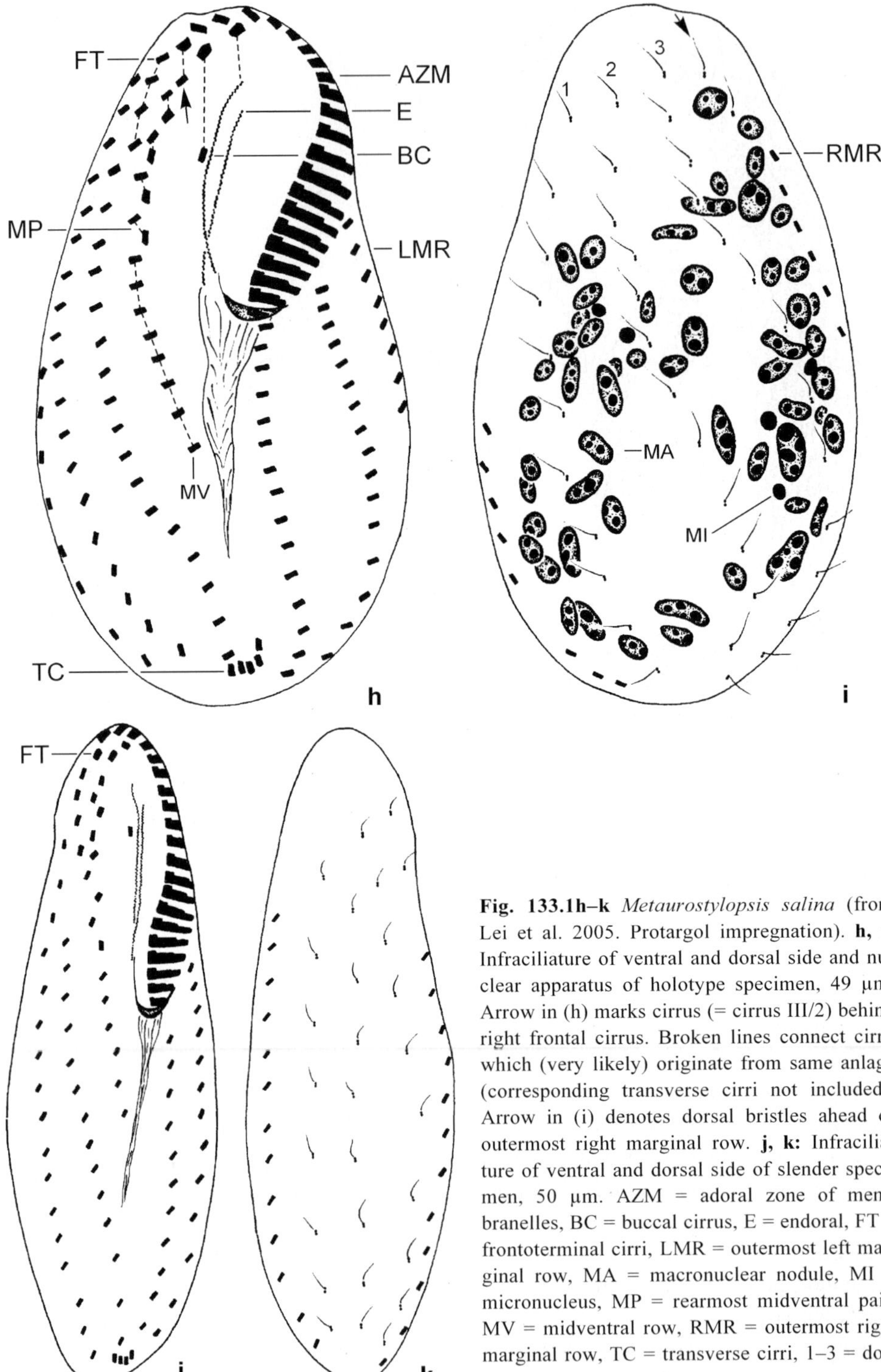

Fig. 133.1h–k *Metaurostylopsis salina* (from Lei et al. 2005. Protargol impregnation). **h, i:** Infraciliature of ventral and dorsal side and nuclear apparatus of holotype specimen, 49 μm. Arrow in (h) marks cirrus (= cirrus III/2) behind right frontal cirrus. Broken lines connect cirri, which (very likely) originate from same anlage (corresponding transverse cirri not included). Arrow in (i) denotes dorsal bristles ahead of outermost right marginal row. **j, k:** Infraciliature of ventral and dorsal side of slender specimen, 50 μm. AZM = adoral zone of membranelles, BC = buccal cirrus, E = endoral, FT = frontoterminal cirri, LMR = outermost left marginal row, MA = macronuclear nodule, MI = micronucleus, MP = rearmost midventral pair, MV = midventral row, RMR = outermost right marginal row, TC = transverse cirri, 1–3 = dorsal kineties. Page 668.

specimens. Thus, the diagnosis (see corresponding footnote) and the description are mainly based on this June population.

Body size 40–80 × 15–30 μm in life, usually about 60 × 25 μm; body length:width ratio highly variable, namely, 1.4–4.4:1 in protargol preparations, on average near 2.4:1 both in life and in protargol slides (Table 28). Body outline usually pyriform, sometimes almost obclavate; frequently a concavity occurs in middle left body portion of ventral side; slightly flattened, flexible but never contractile. Macronucleus consists of 45 scattered nodules on average; nodules globular to ellipsoidal, 2–4 μm across (in protargol preparations?), contain few small nucleoli. On average six globular to ellipsoidal micronuclei scattered between macronuclear nodules, about 1 μm across (Fig. 133.1i, Table 28). Single contractile vacuole located equatorially near left cell margin; frequently an acontractile[1] vacuole up to 8 μm across behind mid-body near right body margin (Fig. 133.1a, c). Cortical granules colourless and scattered, 0.5–1.0 μm across (Fig. 133.1b). Cytoplasm with (i) many refractive lipid globules (1–3 μm across) mainly in anterior body portion; (ii) some reddish-brown inclusions, rendering cell somewhat russet at low magnification; and (iii) food vacuoles. Swims rather slowly by rotation about main body axis.

Adoral zone conspicuous because occupying about 38% of body length, composed of 18–23 membranelles, which become distinctly smaller in anterior half; cilia about 5 μm long in life. Buccal cavity wedge-shaped, usually wide in stout cells. Paroral and endoral optically intersecting in broad specimens when viewed ventrally, while parallel in slender specimens. Pharyngeal fibres conspicuous both in life and protargol preparations because very long and extending to posterior quarter of cell (Fig. 133.1a, h, j; Table 28).

Most cirri fine, 5–8 μm long in life, bases usually composed of six basal bodies. Frontal cirri slightly enlarged, roughly arranged in transverse row; right frontal cirrus, as is usual, behind distal end of adoral zone. Buccal cirrus distinctly behind anterior end of paroral. Cirrus III/2, as is usual, slightly behind right frontal cirrus. 3–5, on average four frontoterminal cirri in area between anterior end of midventral complex and right marginal rows. Midventral complex composed of 4–5 cirral pairs and a single midventral row composed of 5–7 cirri; last midventral pair is at 34% of body length, midventral row terminates at about 65% in holotype (Fig. 133.1h). 2–5, usually four transverse cirri near rear cell end, cirri about 7–8 μm long in life. Invariably three left and three right marginal rows; on average 14 or 15 in left and 13–17 cirri in right marginal rows (Table 28). Leftmost and rightmost marginal rows frequently on dorsomarginal side in slender specimens (Fig. 133.1j, k).

Dorsal cilia 3–4 μm long, basically arranged in three bipolar kineties; frequently a short row composed of two basal body pairs ahead of anterior end of rightmost marginal row (short kinety not included in morphometry). Caudal cirri lacking.

Occurrence and ecology: Marine. Type locality of *M. salina* is a saline pond in Hwasung (37°10'N 126°25'E), Kyeonggi Province, Korea. Lei et al. (2005) found it there frequently in such ponds in June, August, and September 2001 at following conditions:

[1] According to K. Xu (pers. comm.) this vacuole is *acontractile* and not *contractile* as written (due to a printers error) in the original description.

salinity 51–58‰ (June), 31–36‰ (August), 42–53‰ (Septemper); 24–25 °C, 30 °C, 21°C; pH 8.3–8.5, 8.4–8.7, 8.8–8.9; abundance 44–89 cells l^{-1}, 74–197, 26–53. In November 2001 it was also found in intertidal sediments (salinity 31‰; 24 °C; pH 8.1) of Ganghwa Island, Inchon, Korea, where it was very rare.

Metaurostylopsis salina feeds on bacteria and diatoms (Lei et al. 2005).

Metaurostylopsis songi **Lei, Choi, Xu & Petz, 2005**
(Fig. 133.2a–m, Table 28)

2005 ***Metaurostylopsis songi*** **n. sp.**[1] – Lei, Choi, Xu & Petz, J. Euk. Microbiol., 52: 1, 8, Fig. 1–13, 34–38, Table 2 (Fig. 133.2a–m; original description. The holotype slide [accession number KH–010616–01] is deposited in the Marine Biological Specimen Depository, Chinese Academy of Sciences, Qingdao, China, while a paratype slide [KH-010616–02] is deposited in the Laboratory of Plankton, Department of Oceanography, Inha University, Inchon, Korea).

Nomenclature: Lei et al. (2005) dedicated this species to Weibo Song, College of Fisheries, Ocean University of China.

Remarks: *Metaurostylopsis songi* does not fit the characterisation of *Metaurostylopsis* very well (see genus section). The frontal ciliature is difficult to understand from the interphasic pattern alone. Thus, ontogenetic data are needed for a more proper interpretation.

Morphology: Lei et al. (2005) studied several populations collected in 2001, but only the population from June was investigated in detail. Other populations were mainly identified from protargol-impregnated specimens, which match well in all main features. Thus, conspecificity is beyond reasonable doubt. In spite of this, Lei et al. kept the data separate, and the diagnosis (see corresponding footnote) and the description are mainly based on the June (= type) population. Lei et al. (2005) also provided some micrographs, which are, however, not shown in the present book.

Body size 90–150 × 20–35 μm in life, usually about 120 × 25 μm; length:width ratio moderately variable, that is, 3.8–5.8:1, on average 4.7:1 both in life and in protargol preparations (Table 28). Body outline slenderly ellipsoidal, anterior and posterior end narrowly rounded, widest in anterior third of cell; dorsal portion slightly convex. Body about 15–25 μm thick in life, usually near 1.7:1 flattened dorsoventrally in life, flexible but never contractile. Macronuclear nodules scattered, usually globular to ellipsoidal, 3–7 μm long; occasionally ampulliform with a reorganisation band; nucleoli small, mainly in widened portion. Usually four micronuclei scattered among macronuclear nodules, globular or ellipsoidal, about 2 μm across (Fig. 133.2i, k). Contractile vacuole ahead of mid-body near left cell margin. Cortical granules in 8–10 rows on ventral and

[1] The diagnosis by Lei et al. (2005) is as follows: Marine *Metaurostylopsis* in vivo about 120 μm × 25 μm; body slenderly ellipsoidal. Colorless cortical granules in rows on ventral and dorsal side of cell. 28–47 adoral membranelles, five frontal, two or three frontoterminal, one buccal and six or seven transverse cirri; nine to 12 midventral cirral pairs followed by one to three single cirri posteriorly, invariably three marginal cirral rows on each side of cell, usually three long dorsal kineties. On average 54 scattered macronuclear nodules and four micronuclei.

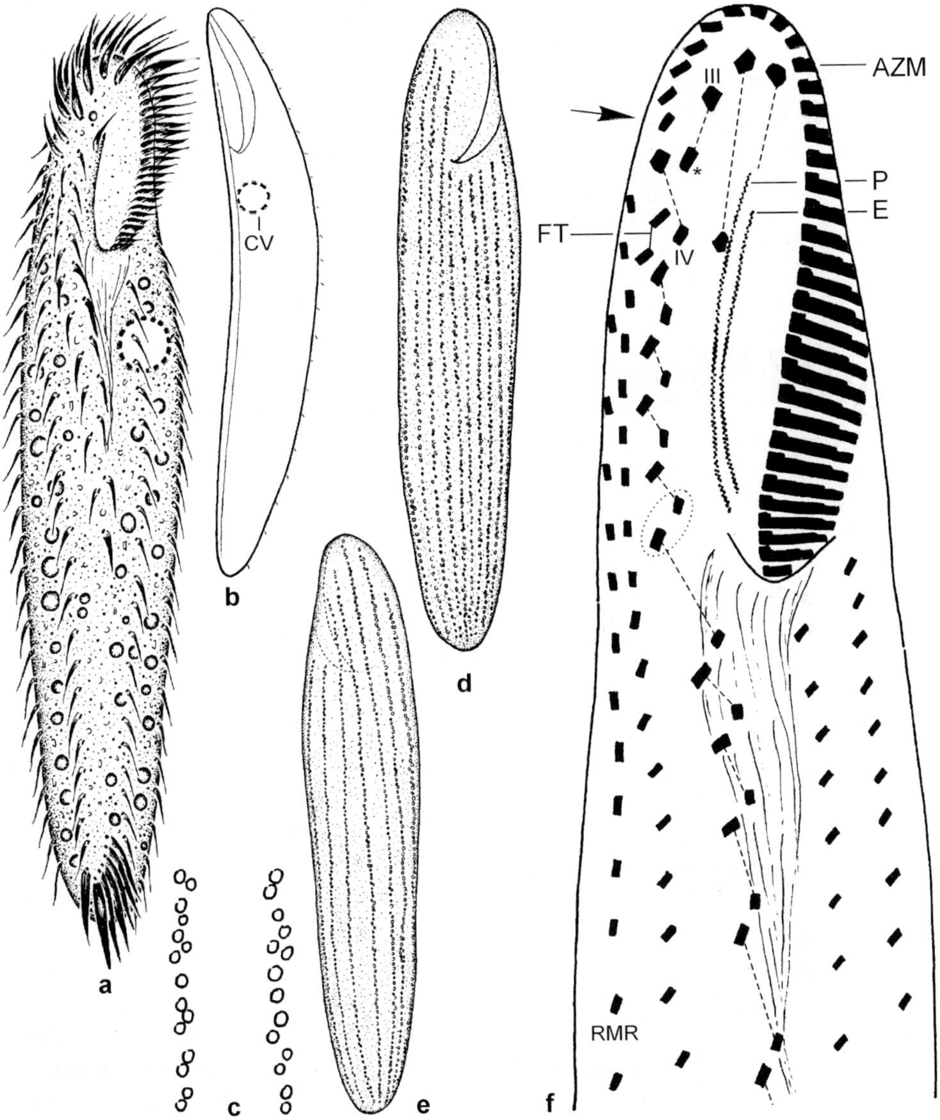

Fig. 133.2a–f *Metaurostylopsis songi* (from Lei et al. 2005. a–e, from life; f, protargol impregnation). **a:** Ventral view of representative specimen, 90–150 μm long (bar not shown in original description). **b:** Left lateral view showing, inter alia, dorsal bristles and contractile vacuole. **c–e:** The cortical granules are about 1 μm across, colourless, and form 8–10 rows on ventral (d) and 6–7 rows on dorsal side (e). **f:** Infraciliature of anterior portion of holotype specimen. Cirri, which very likely originate from the same anlage, are connected by broken lines (has to be checked by cell division data). Arrow marks distal end of adoral zone of membranelles. Asterisks marks (very likely) cirrus III/2. Dotted line circles a pseudopair, that is, a cirral pair whose cirri do not originate from the same anlage but from two neighbouring anlagen. AZM = adoral zone of membranelles, CV = contractile vacuole, E = endoral, FT = frontoterminal cirri, P = paroral, RMR = middle right marginal row, III = cirral anlage III (forms rightmost frontal cirrus and cirrus behind), IV = cirral anlage IV, that is, anteriormost midventral pair. Page 672.

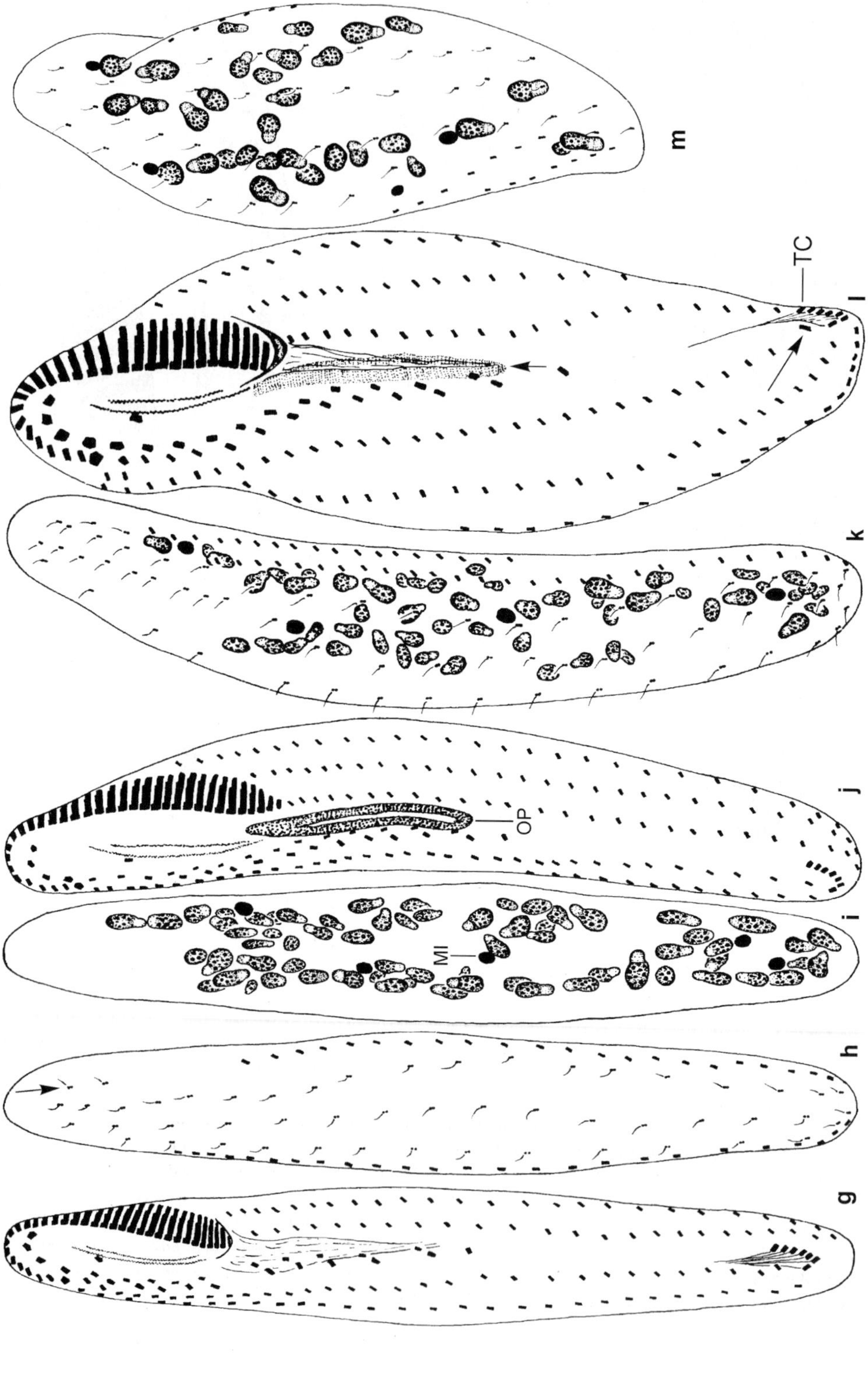
m
TC
l
k
OP
j
MI
i
h
g

6–7 on dorsal side; distinct in life, colourless and about 1 μm across (Fig. 133.2c–e). Cytoplasm colourless, contains many refractive lipid globules usually 3–5 μm across, distributed mostly along left and right body margins, cell rather transparent even at low magnification. Movement crawling.

Adoral zone occupies 25–33% of body length (in life?), on average 31% in protargol preparations (Table 28), composed of 28–47 membranelles, whose bases are distinctly shorter in the anterior portion. Cilia of membranelles about 15 μm long in life. Buccal cavity wide and wedge-shaped. Paroral parallel to endoral, which is slightly shorter than paroral (Fig. 133.2f). Pharyngeal fibres about 25 μm long.

Cirral pattern and number of cirri of usual variability (Table 28). All cirri relatively fine, 10–12 μm long in life, frontal and transverse cirri about 15 μm long. Frontal ciliature difficult to interpret, origin from anlagen possibly as indicated in Fig. 133.2f (however, ontogenetic data are needed to show whether or not the interpretation is correct). Anteriormost five cirri enlarged; rightmost cirrus (likely this is the anterior cirrus of the first midventral pair) behind distal end of adoral zone. Buccal cirrus somewhat behind anterior end of paroral. Usually two, rarely three (in 2 out of 18 specimens analysed) frontoterminal cirri in common position, that is, between anterior end of midventral complex and right marginal row. Midventral complex composed of 9–12 cirral pairs, terminates at 54% of body length in holotype specimen (Fig. 133.2g). Midventral row basically lacking; however, rarely specimens with a short row composed of three cirri. 6–7 transverse cirri near rear cell end, slightly enlarged and distinctly projecting beyond body margin (Fig. 133.2a, g, j). Invariably three left and three right marginal rows (Table 28); outermost left and right row frequently on dorsolateral side.

Dorsal cilia in life about 3 μm long, usually arranged in three bipolar kineties; frequently a short row (not included in morphometry) composed of 2–3 basal body pairs ahead of outermost right marginal row (Fig. 133.2h, k).

Lei et al. (2005) reported the following deviating data for the July-population (from protargol preparations): body size about 90–100 × 30–36 μm; about 35–38 macronuclear nodules; 10 or 11 midventral cirral pairs; one pretransverse ventral cirrus; left marginal rows with 19–23 (innermost row), 23–25, and 20–22 cirri; right marginal rows 27–28, 25–28, and 26–30 cirri; four dorsal kineties including two slightly shorter rows.

← **Fig. 133.2g–m** *Metaurostylopsis songi* (from Lei et al. 2002. Protargol impregnation. g–k, June-population; l, m, July-population). **g–i:** Infraciliature of ventral and dorsal side and nuclear apparatus of holotype specimen, 118 μm. Broken line in (g) connects cirri of rearmost midventral pair (note that Lei et al. 2005 very likely designated the pseudopairs as cirral pairs). Arrow in (h) marks dorsal bristles ahead of outermost right marginal row. Note that some macronuclear nodules (i) are ampulliform and have a reorganisation band. **j, k:** Infraciliature of ventral and dorsal side and nuclear apparatus of an early reorganiser. Lei et al. (2005) likely correctly assumed that it is an reorganiser because in dividers of such a stage the anlagen for both the proter and the opisthe would be recognisable. Note that the oral primordium and the cirral primordia are connected, forming an elongate-elliptical pattern. **l, m:** Infraciliature of ventral and dorsal side and nuclear apparatus of a very early reorganiser. Long arrow marks a pretransverse ventral cirrus, short arrow denotes rear end of oral/cirral primordium. MI = micronucleus, OP = oral primordium, TC = transverse cirri. Page 672.

Physiological reorganisation (Fig. 133.2j, k): Lei et al (2005) observed more than 10 specimens of the type population bearing an oral primordium and cirral primordia posterior to the parental undulating membranes and adoral membranelles. The primordia are connected with each other, forming an elongate-elliptical pattern distinctly underneath the cell surface (Fig. 133.2j). The proter's primordia, however, were not observed. Thus, Lei et al. (2005) concluded that these are reorganisers because in dividers of such a stage the anlagen for both the proter and the opisthe would be recognisable. The macronuclear nodules are ampulliform, some are distinctly larger and bear a reorganisation band, and the nucleoli are mainly arranged in the broad portion (Fig. 133.2k).

One reorganiser was observed in the July-population, which has an oral primordium and cirral primordia, forming an indistinct elongate elliptical pattern (Fig. 133.2l, m).

Occurrence and ecology: Marine. As yet, *Metaurostylopsis songi* is recorded only from the type locality, that is, Ganghwa Island (37°37'N 126°20'E), Inchon, Korea, where Lei et al. (2005) found it in intertidal flat sediments (16–25 °C; salinity about 31 ‰; pH 7.9–8.0). It was usually abundant within 1 cm depth of the mud surface.

Metaurostylopsis songi feeds mainly on diatoms (Lei et al. 2005); in the laboratory, it may survive for weeks consuming bacteria.

Insufficient redescriptions

***Urostyla marina* Kahl** – Madrazo-Garibay & Lopez-Ochoterena, 1985, Anales Instituto de Ciencias del Mar y Limnologia Universidad Nacional Autónomia de México, 12: 206, Fig. 28 (Fig. 143c). Remarks: The description and the illustration are rather superficially. Moreover, the Mexican population has only one large macronuclear nodule so that the identification cannot be accepted (possibly it is an exconjugant with a macronucleusanlage). Some further relevant data: size 120 × 40 µm in life(?); body outline elongate elliptical; body dorsoventrally flattened; eight ventral cirral rows and two marginal rows (obviously only the marginal rows are illustrated!); two micronuclei; contractile vacuole lacking; Laguna de Términos, Campeche, Mexico (see also Aladro-Lubel et al. 1988, p. 436).

***Urostyla marina* Kahl, 1932** – Aladro Lubel, Martínez Murillo & Mayén Estrada, 1990, Manual de Ciliados, p. 124, one illustration on same page (Fig. 143d). Remarks: The description is taken from the previous entry. By contrast, the illustration is different from Fig. 143c, but shows, like this figure, only one macronucleus; possibly, Aladro Lubel et al. (1990) only modified (adapted) the old illustration. Because of the single macronucleus, which is probably a misobservation, the identification is not accepted. No original sample site mentioned.

Birojimia Berger & Foissner, 1989

1989 ***Birojimia* nov. gen.**[1] – Berger & Foissner, Bull. Br. Mus. nat. Hist. (Zool.), 55: 25 (original description). Type species (by original designation on p. 25): *Birojimia terricola* Berger & Foissner, 1989a.

1999 ***Birojimia* Berger & Foissner, 1989** – Shi, Song & Shi, Progress in Protozoology, p. 113 (revision of hypotrichs).[2]

1999 ***Birojimia* Berger & Foissner, 1989** – Shi, Acta Zootax. sinica, 24: 363 (revision of hypotrichs).

2001 ***Birojima* Berger & Foissner 1989** – Aescht, Denisia, 1: 32 (incorrect subsequent spelling; catalogue of generic names of ciliates).

2001 ***Birojimia* Berger and Foissner, 1989** – Berger, Catalogue of ciliate names 1. Hypotrichs, p. 13 (nomenclator containing all basionyms, combinations, and higher taxa of hypotrichs).

2002 ***Birojima* Berger and Foissner, 1989** – Lynn & Small, Phylum Ciliophora, p. 442 (guide to ciliate genera; incorrect subsequent spelling).

Nomenclature: The name *Birojimia* refers to the site (Biro-Jima; Amakusa, Japan) where the type species was discovered. Feminine gender (Aescht 2001, p. 275).

Characterisation (Fig. 111a, autapomorphies 13): Body slender elongate, posteriorly converging (A). Adoral zone of membranelles continuous. 3 frontal cirri. Buccal cirrus present. 2 frontoterminal cirri. Midventral complex composed of midventral pairs only (type species) or of midventral pairs and (usually) 1 midventral row. Transverse and caudal cirri present. 1 left and 2 or more right marginal rows. More than 3 dorsal kineties (A).

Remarks: The two species assigned, *Birojimia terricola* (type) and *B. muscorum*, have, beside the features mentioned in previous paragraph, several other characteristics in common, namely macronuclear nodules scattered mainly in central quarters of cell, many of them near cell margins, individual nodules ellipsoidal; cytoplasm colourless; adoral zone 20–30% of body length on average, of usual shape and structure; buccal cavity deep (and thus bright at bright field illumination) and moderately wide; undulating membranes distinctly curved and rather long, commence at about same level, endoral, however, longer and thus terminating more posteriorly than paroral; pharynx conspicuous in life, in protargol preparations with short, oblique rod-shaped structures; one buccal cirrus right of optical intersection of undulating membranes; all midventral cirri of about same size; transverse and caudal cirri inconspicuous because of about same length and size as marginal cirri; dorsal bristles 3–4 μm long; terrestrial.

Berger & Foissner (1989a) established *Birojimia* to separate two species with two or more right marginal rows from *Uroleptus* and *Paruroleptus* species, which have a similar slender, posteriorly converging body shape and cirral pattern, but only one right marginal row.

The two species assigned to *Birojimia* differ in the composition of the midventral complex. In *Birojimia terricola* (type) it is composed of midventral pairs only, whereas in *Birojimia muscorum* it consists of midventral pairs and one midventral row

[1] The diagnosis by Berger & Foissner (1989a) is as follows: Slender, posteriorly converging Urostylidae with 1 left and 2 or more right marginal rows. 3 slightly to distinctly enlarged frontal cirri. Transverse and caudal cirri present.

[2] Incorrect reference mentioned in literature section (p. 138), namely Berger & Foissner (1989b) instead of Berger & Foissner (1989a).

(Fig. 134b, 135c). In the present book, this would require a classification of *B. terricola* in the Holostichidae (three frontal cirri, midventral complex composed of cirral pairs only). By contrast, *Birojimia muscorum* would belong to the Bakuellidae because this group is defined by three frontal cirri and a midventral complex composed of midventral pairs and midventral rows. Since *B. muscorum* could not be assigned to one of the bakuellid genera, it would be necessary to establish a monotypic genus for this species. To overcome this unsatisfactory situation I preliminarily classify *Birojimia* in the Bakuellidae, but include *B. terricola* also in the Holostichidae key. Possibly the two species are not as closely related as indicated by the similarities in the general cirral pattern. However, one can also not exclude that the definition of the higher taxa (Holostichidae, Bakuellidae) via the frontal ciliature and the midventral complex is incorrect. Ontogenetic data and molecular studies will hopefully allow a more proper classification.

Key to *Birojimia* species

Separation of the two species included can be done by detailed live observation because only the presence/absence of cortical granules and the number of right marginal rows is needed.

1 Six right marginal rows successively shortened anteriorly from left to right; cortical granules lacking (Fig. 134a, b) .... *Birojimia terricola* (p. 678)
\- Two right marginal rows; cortical granules present (Fig. 135a–c) .... *Birojimia muscorum* (p. 683)

Birojimia terricola Berger & Foissner, 1989
(Fig. 134a–c, Table 29)

1989 ***Birojimia terricola* nov. spec.**[1] – Berger & Foissner, Bull. Br. Mus. nat. Hist. (Zool.), 55: 25, Fig. 16–18, Table 4 (Fig. 134a–c; original description. The protargol slides with the holotype specimen [slide number 1988:2:1:5] and with paratype specimens [slide number 1988:2:1:6] are deposited in the British Museum of Natural History in London, United Kingdom).

1992 ***Birojimia terricola* Berger, 1989** – Shen, Liu, Song & Gu, Protozoa, p. 151, Fig. 2-23Ca, Cb (redrawings of illustrations from original description. Authorship ["Berger, 1989"] and source of illustrations ["Berger and Foissner 1987"] incorrect).

2001 ***Birojimia terricola* Berger and Foissner, 1989** – Berger, Catalogue of ciliate names 1. Hypotrichs, p. 13 (nomenclator containing all basionyms, combinations, and higher taxa of hypotrichs).

Nomenclature: No derivation of the species-group name *terricola* is given in the original description. It is a composite of the Latin noun *terr·a* (soil), the thematic vovel *·i-*, and the Latin verb *colere* (to live in) and means living in soil. Usually, species-group

---

[1] The diagnosis by Berger & Foissner (1989a) is as follows: In vivo about 155 × 40 μm. Midventral rows terminate roughly in the middle of the cell. About 6 right marginal rows, successively shortened anteriorly from left to right. 39 adoral membranelles, 5 transverse cirri, and 61 macronuclear segments on average.

names ending with *-cola* are considered as appositive substantives and are thus not changed when transferred to a genus of different gender (Werner 1972, p. 138). In the binomen *Birojima terricola* (Aescht 2001, p. 32, 275; Lynn & Small 2002, p. 442), the genus-group name is incorrectly spelled. *Birojimia terricola* was fixed as type species of *Birojimia* by original designation.

Remarks: *Birojimia terricola* was described both from life and after protargol impregnation. The successive reduction of the number of cirri and the simultaneous increase of the number of bristles in the right marginal rows strongly supports the idea of Borror (1979) and Berger et al. (1985) that marginal rows and dorsal kineties are homonomous structures. It would be interesting to know the division mode of these rows. Very likely they divide like normal (non-fragmenting) dorsal kineties and marginal rows, that is, by simple within-proliferation at two levels.

In life, *Birojimia terricola* is best recognised by the soil habitat, the characteristic ciliature of the right body half, and the scattered macronuclear nodules. *Caudiholosticha paranotabilis* and *C. notabilis* have, inter alia, cortical granules and only one right marginal row (Fig. 48.1a–m, 48.3a–o).

Morphology: Body size in life about 155 × 40 μm according to original description; however, this value is from a single measurement. Specimens impregnated with Foissner's protargol method have a body length between 126 and 215 μm (average 180 μm), indicating that live specimens are 135–235 μm long if a shrinkage of about 10% is assumed. Body length:width ratio around 4.8:1 both in life and in protargol preparations. Body slender, slightly converging posteriorly and somewhat twisted about main axis; both ends rounded (Fig. 134a, b). Macronuclear nodules scattered, except in anterior and posterior body portion where they are usually lacking, more numerous near body margins than in cell centre; individual nodules 6–7 × 4–5 μm in life, that is, globular to ellipsoidal (length:width ratio 1.4:1 on average in protargol preparations). Micronuclei 6 × 4 μm in life and therefore of about same size as macronuclear nodules, do not impregnate with Foissner's protargol method. Contractile vacuole not described in original description, likely – as in most hypotrichs – near mid-body at left cell margin. Cortical granules lacking, cytoplasm colourless. Movement not mentioned in original description; likely without peculiarities.

Adoral zone occupies about 30% of body length, of usual shape and structure, composed of an average of 39 membranelles, bases of largest membranelles 17 μm wide in life; individual membranelles of usual shape and fine structure (Fig. 134b, Table 29). Buccal cavity deep and moderately large. Undulating membranes long and slightly curved, intersect optically about at level of buccal cirrus. Pharyngeal fibres conspicuous after protargol impregnation, with small rod-shaped structures, similar to in *Caudiholosticha notabilis*.

Cirral pattern of usual variability, number of cirri in most cirral groups/rows rather strongly varying (Fig. 134b, c, Table 29). Invariably three distinctly enlarged frontal cirri in an oblique row and a single, non-enlarged cirrus (cirrus III/2) behind right frontal cirrus. Buccal cirrus at summit of paroral. Two frontoterminal cirri close to right frontal cirrus. Midventral complex composed of 11–17 midventral pairs, commences close behind cirrus III/2, extends in cell midline, and terminates at 59% of body length on aver-

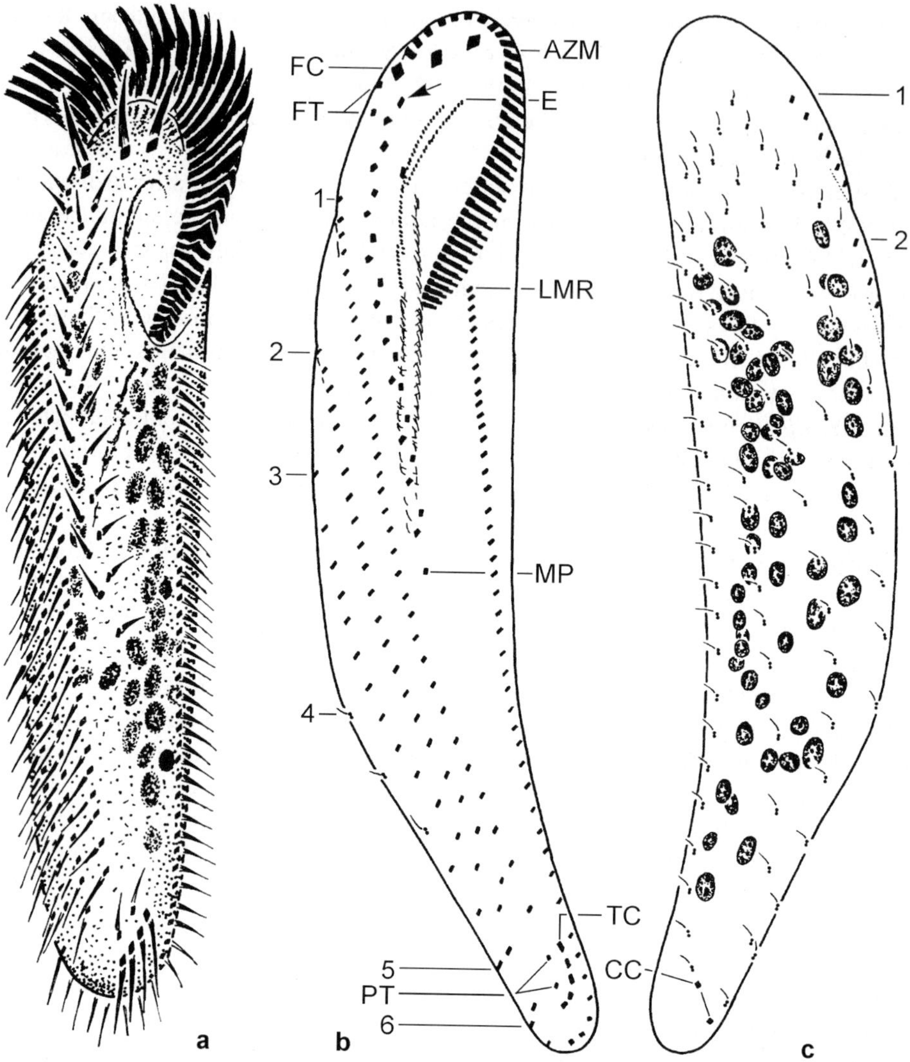

Fig. 134a–c *Birojimia terricola* (from Berger & Foissner 1989a. a, from life; b, c, protargol impregnation). **a:** Ventral view of a representative specimen, 190 μm. **b, c:** Infraciliature of ventral and dorsal side and nuclear apparatus of same specimen, 180 μm. Arrow marks cirrus III/2, that is, the cirrus behind the right frontal cirrus. AZM = adoral zone of membranelles, CC = caudal cirri at end of leftmost bristle row, E = endoral, FC = right frontal cirrus, FT = frontoterminal cirri, LMR = left marginal row, MP = rear end of midventral complex, PT = pretransverse ventral cirri, TC = leftmost (= anteriormost) transverse cirrus, 1 = inner right marginal row (composed of cirri only), 2–6 = (marginal?) rows composed of cirri and dorsal bristles. Page 678.

age; all midventral cirri of about same size and slightly larger than marginal cirri. Midventral row obviously lacking, thus classification in Bakuellidae (as defined in present book) basically incorrect (see remarks at genus section). On average five transverse

Occurrence and ecology: Kahl (1932) found *Birojimia muscorum* common ("nicht selten") in mosses from Upper Bavaria (Germany) and Wisconsin (USA). Unfortunately, he did not fix one of these sites as type locality.

Foissner (1982) found *B. muscorum* at three sites in Lower Austria, namely (i) in a brown soil (with tendency to leaching) of a mixed forest (*Fagus*, *Pinus*, *Betula*, *Quercus*) near the village of Baumgarten (site "SO 16" in Foissner 1982, altitude about 260 m); (ii) in dry grassland on gravel of a xerothermic site without trees near the village of Zwentendorf (site "SO 17", altitude about 181 m); and (iii) in an aperiodically flooded bottom land near the village of Zwentendorf (site "SO 19", altitude about 176 m). For detailed descriptions of these sites, see Foissner et al. (1985), and for autecological data, see Foissner & Peer (1985, p. 42). Foissner (1996, 1997, 1998, 1999 and pers. comm.) and Blatterer & Foissner (1988, p. 6) recorded *Birojimia muscorum* from various terrestrial habitats from all over the world (Austria, Kenya, Namibia, Amazonia, Chile, Australia, Marion Island). Recently, we found it in various forest stands in Austria (Foissner et al. 2005; Fig. 135k, l).

Wiackowski (1988), who provided detailed morphological data, but no illustration, found his population in moss from a tree trunk, likely somewhere in Poland. I found *Biojimia muscorum* in the beech litter beside the path (47°48'39"N 13°05'24"E; about 660 m altitude) from Gnigl, a district of the city of Salzburg (Austria), to the Gaisberg, a mountain nearby. The sketch of the cortical granules (Fig. 135j) is from a population which occurred in a sample of lichen and moss growing on a maple in the Baumschlagreit region near the village of Hinterstoder, Upper Austria.

Records mainly from terrestrial or semiterrestrial habitats not substantiated by morphological data and/or illustrations: upper soil layer (0–5 cm) of fertilised sites of an subalpine grassland field trial in Aiglern near the village of Aigen, Styria, Austria (Foissner et al. 1990, p. 18); litter (80% pine, 20% beech) from a place named Zmijinje Jezero (likely near 43°24'S 18°16'E, about 1500 m altitude) in the Durmitor Mountains, Bosnia Herzegovina (Varga 1962, p. 154); dry mosses, leaf and spruce litter, lichens, and *Sphagnum* samples from Bavaria, Germany (Wenzel 1953, p. 104); in untreated and in fertilised and limed soil from a pine forest, and in soil from a beech forest in and near Ulm, Germany (Lehle 1989, p. 141; see also Funke 1986, p. 72); soil from Vechtedamm, Nordhorn, Germany (Niebuhr 1989, p. 81; identification checked by W. Foissner); humus layer of Hungarian spruce forests (Gellért 1957, p. 15); with a frequency of 3.8% in submerged moss, with 6% in wet moss, with 22.7% in moist moss, and with 56.7% in dry moss [1] in the Slovensky raj in the Stratenská hornatina highlands and with 12% in dry mosses in the surroundings of the city of Bratislava, Slovakia (Tirjaková & Matis 1987a, p. 11, 1987b, p. 23); bogs(?) in Slovakia (Tirjaková 1992, p. 294); with a constancy of about 16% in Slovakian agricultural soils (Tirjaková 1988, p. 500); semitropical forest soils in the Bentley Terrace, Montgomery Terrace, Prairie Terrace, and Deltaic Region in Florida, USA (Bamforth 1967, p. 15, 1968, p. 14); litter and upper 3 cm of topsoil from hardwood forests and grasslands in the Mississippi Delta, USA

[1] The increasing percentage of frequency with decreasing water content is in accordance with the preference of terrestrial habitats.

(Bamforth 1969, p. 73); soil (0–5 cm) from a burnt savannah ("savane brûlée") in the Plateau du Grand Nord, somewhat north of the Station for Tropical Ecology Lamto, Ivory Coast (Buitkamp 1977, p. 252, 1979, p. 225); two out of 73 soil samples from Namibia (Foissner et al. 2002, p. 58); very common in moss and soil, likely from a site named Langhovde (about 69°S 39°30'E) in the Antarctica (Sudzuki 1979, p. 124).

Records from limnetic habitats not substantiated by morphological data and/or illustrations: in two thermal lakes at 16–34°C and pH 5.6–7.4 in the Bojnice spa, Slovakia (Matis & Straková-Striešková 1991, p. 115); freshwater from the Yuelushan area, China (Yang 1989, p. 157).

Birojimia muscorum feeds on fungal spores, ciliates (*Drepanomonas revoluta*) and soil particles (Foissner 1982), on bacteria and small protozoa (Wiackowski 1988), and on heterotrophic flagellates (Gellért 1957). Foissner (1987a, p. 126) estimated a biomass of 40 mg per 10^6 specimens; later, he corrected it to 100 mg, which is a more reliable value (Foissner 1998). According to Foissner (1998), *Birojimia muscorum* has a high degree of soil autochthonism.

Parabirojimia Hu, Song & Warren, 2002

2002 ***Parabirojimia* nov. gen.**[1] – Hu, Song & Warren, Europ. J. Protistol., 38: 352 (original description). Type species (by original designation on p. 352): *Parabirojimia similis* Hu, Song & Warren, 2002.

Nomenclature: The name *Parabirojimia* is a composite of the Latin prefix *para-* (like, equal) and the urostylid generic name *Birojimia* (for derivation, see there) and alludes to the similarity of *Parabirojimia similis* and the *Birojimia* species. Like *Birojimia*, feminine gender.

Characterisation (Fig. 111a, autapomorphies 15): Adoral zone of membranelles bipartite (A). 3 frontal cirri. Buccal cirrus(i) present. Frontoterminal cirri lacking. Midventral complex composed of midventral pairs and one or more midventral rows at rear end of complex. Transverse cirri present. 1 left marginal row and 2 or more right marginal rows which derive from individual anlagen within each parental row. Caudal cirri lacking (A). Transverse cirri originate within frontal-midventral-transverse cirri anlagen and from a part of right marginal cirri anlagen (A).

Remarks: See this chapter at single species.

Species included in *Parabirojimia*: (1) *Parabirojimia similis* Hu, Song & Warren, 2002.

[1] The diagnosis by Hu et al. (2002) is as follows: Urostylidae with short midventral cirral row and clearly differentiated frontal cirri; buccal and transverse cirri present; frontoterminal and caudal cirri absent; one left and two or more right marginal rows which derive from individual anlagen within each parental row; transverse cirri originate within fronto-ventral-transverse cirral anlagen and part of right marginal cirral anlagen.

Single species

Parabirojimia similis Hu, Song & Warren, 2002 (Fig. 136a–t, Table 30, Addenda)

2002 ***Parabirojimia similis* nov. spec.**[1] – Hu, Song & Warren, Europ. J. Protistol., 38: 352, Fig. 1–46, Table 1 (Fig. 136a–t; original description. The protargol slide with the holotype specimens is deposited in the Laboratory of Protozoology, Ocean University of Qingdao, China. One slide with paratype specimens [slide number 2002:5:22:2] is deposited in the British Museum of Natural History in London, UK).

Nomenclature: No derivation of the name is given in the original description. The species-group name *simil·is -is -e* (Latin adjective; similar, identical) likely alludes to the similarity of the present species with *Birojimia* species. *Parabirojimia similis* was fixed as type species of *Parabirojimia* by original designation.

Remarks: The present species has a unique combination of features (see generic characterisation above). Thus, Hu et al. (2002) established *Parabirojimia*. They did not include the bipartite adoral zone in the characterisation of the genus, but used this character to diagnose the sole species, *Parabirojimia similis*. In other cases with such a distinct break (*Erniella* Foissner, 1987b; *Etoschothrix* Foissner, Agatha & Berger, 2002; *Afrothrix*; *Holosticha*; *Uroleptopsis*) this feature was used to define the genera, some of which were also monotypic when they were established. Recently, we found a second *Afrothrix* species which also has a bipartite adoral zone, indicating that this feature is the apomorphy of a supraspecific taxon (*Afrothrix*) and not of a single species. There are many other examples, especially within the 18-cirri oxytrichids, where oral features are successfully used to characterise monophyletic groups (for review see Berger 1999). However, as long as only a single species is known within a supraspecific taxon (including the genus rank) this problem is of course irrelevant because all higher taxa are redundant. But as we know, within the hypotrichs (and the ciliates in general) only a small number of the extant species is described.

The second curious feature of *Parabirojimia* are the transverse cirri. A transverse cirrus is defined as the rearmost cirrus of the frontal-midventral-transverse cirri anlagen (Fig. 1a). Usually these cirri are more or less enlarged and form a rather distinct obliquely (sometimes transversely, thus the name!) arranged pseudorow in non-dividers. In the present species, however, not only the frontal-midventral-transverse cirri anlagen produce transverse cirri, but also most of the several right marginal row anlagen each form a slightly enlarged cirrus at their end. Further, these enlarged marginal cirri are distinctly set off from the ordinary marginal cirri during interphase and form a more or less continuous row with the true transverse cirri (Fig. 136a, f, j, s). In spite of this, we should not change the definition of the term transverse cirrus. I suppose that the

[1] The diagnosis by Hu et al. (2002) is as follows: Marine *Parabirojimia*, in vivo 140–300 × 30–50 μm; elongated elliptical body outline; 5–8 right marginal rows; adoral zone of membranelles bipartite, with a conspicuous snout-like protrusion of the frontal field between anterior and posterior portions; 37–54 adoral membranelles, 3 enlarged frontal, 7–13 midventral, 1 buccal, and 3–11 transverse cirri; usually 1 ventral row composed of 15–56 cirri; usually with 3 complete dorsal kineties; 3–6 macronuclear nodules and about 3–8 micronuclei.

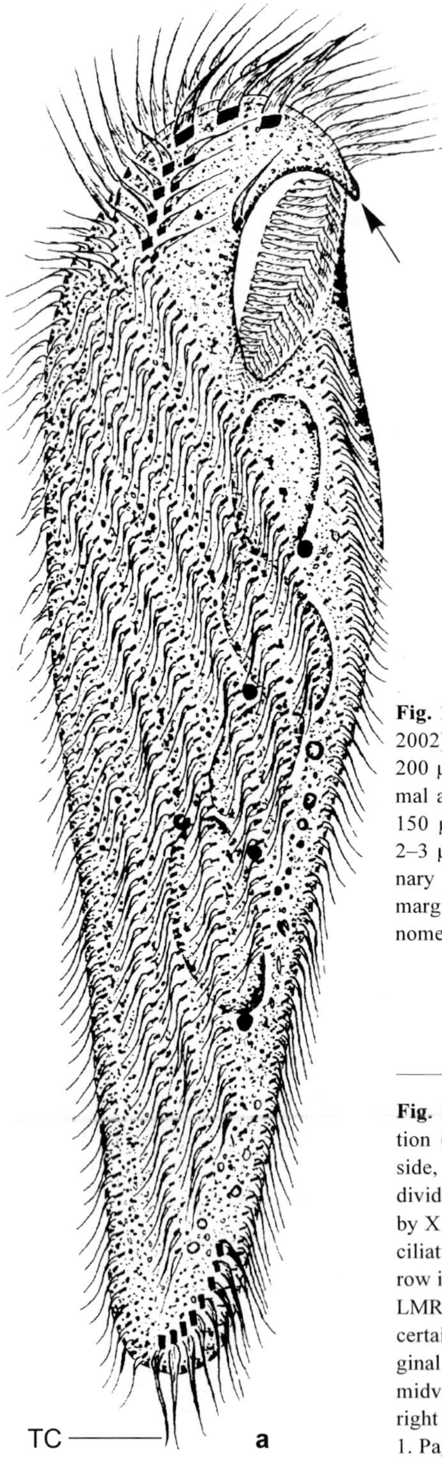

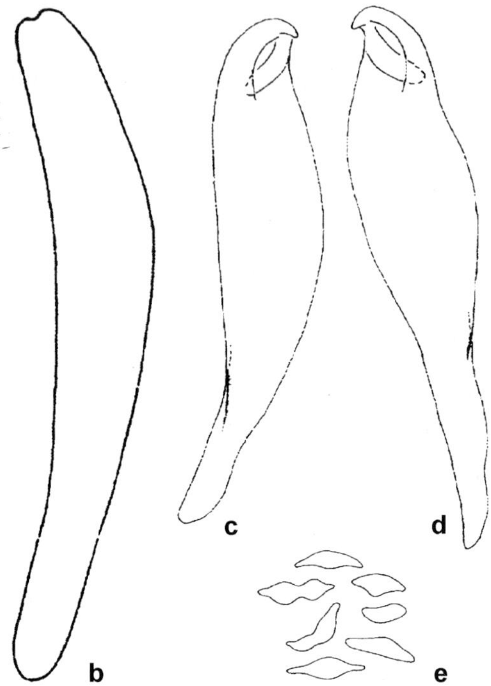

Fig. 136a–e *Parabirojimia similis* from life (from Hu et al. 2002). **a:** Ventral view of a representative specimen, 200 µm. Arrow marks snout-like protrusion between proximal and distal portion of adoral zone. **b:** Left lateral view, 150 µm. **c, d:** Slender specimens. **e:** Cytoplasmic crystals 2–3 µm long. TC = transverse cirri row composed of ordinary transverse cirri and enlarged terminal cirri of right marginal rows (see text for detailed explanation of this phenomenon). Page 691.

Fig. 136f–i *Parabirojimia similis* after protargol impregnation (from Hu et al. 2002). **f, g:** Infraciliature of ventral side, size not indicated. Arrow in (g) indicates gap which divides adoral zone into two. According to a pers. comm. by X. Hu, the inner right marginal row is row 1. **h, i:** Infraciliature of anterior body portion, sizes not indicated. Arrow in (h) denotes buccal cirrus. FC = enlarged frontal cirri, LMR = left marginal row, MV = midventral row (this is certainly a midventral row and not the innermost right marginal row because it commences immediately behind the midventral pairs), RMR1, 6 = innermost and outermost right marginal row, TC = transverse cirri, 1 = dorsal kinety 1. Page 691. →

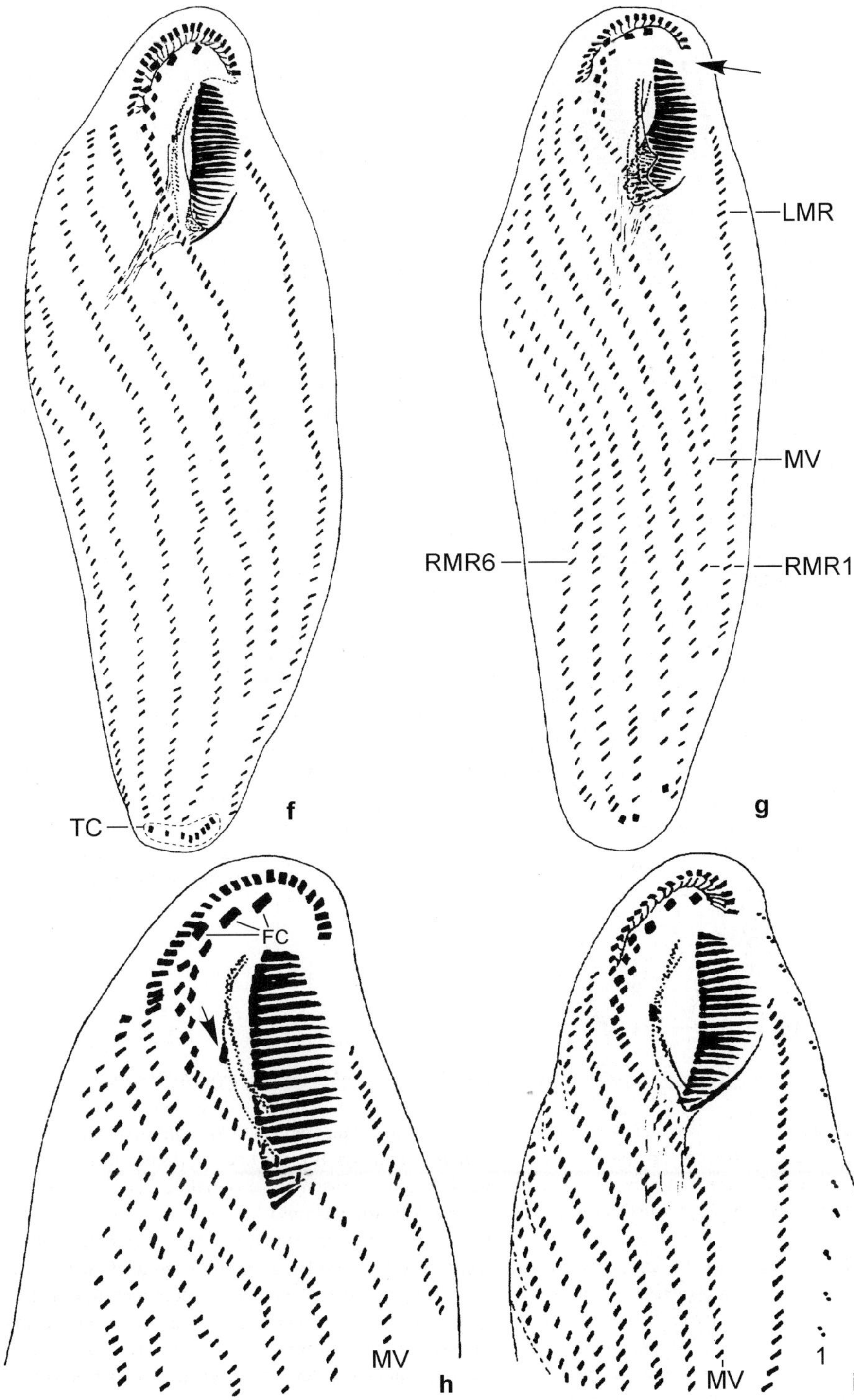
LMR
MV
RMR6
RMR1
TC
f
g
FC
MV
h
1
MV
i

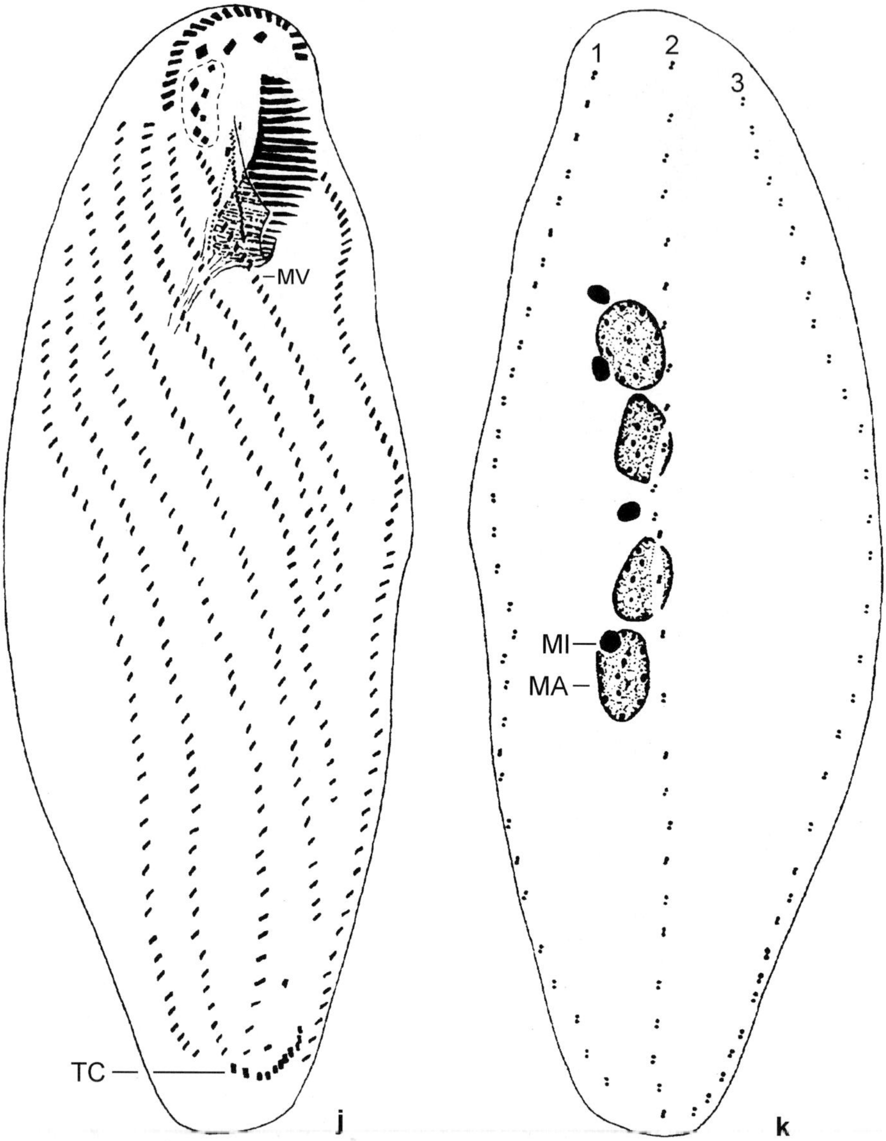

Fig. 136j, k *Parabirojimia similis* after protargol impregnation (from Hu et al. 2002). Infraciliature of ventral and dorsal side and nucleus-apparatus of same specimen, 151 µm. Broken line in (j) encircles the anterior portion of the midventral complex composed of cirri pairs. MA = macronucleus-nodule, MI = micronucleus, MV = midventral row, TC = transverse cirri, 1–3 = dorsal kineties. Page 691.

Fig. 136l–n *Parabirojimia similis* after protargol impregnation (from Hu et al. 2002). **l, m:** Infraciliature of ventral side and nuclear apparatus of early dividers, l = 213 µm, m = size not indicated. Arrow in (l) and short arrow in (m) denote dedifferentiation of buccal cirrus and undulating membranes. Long arrow in (m) marks dedifferentiation of proximal portion of adoral zone; arrowhead denotes cirri anlagen in opisthe. **n:** Infraciliature of ventral side of an early to middle divider (same cell as Fig. 136o), 218 µm. Arrowhead marks anterior primordium in left marginal row, arrows denote anlagen within dorsal kinety 3. OP = oral primordium. Page 691. →

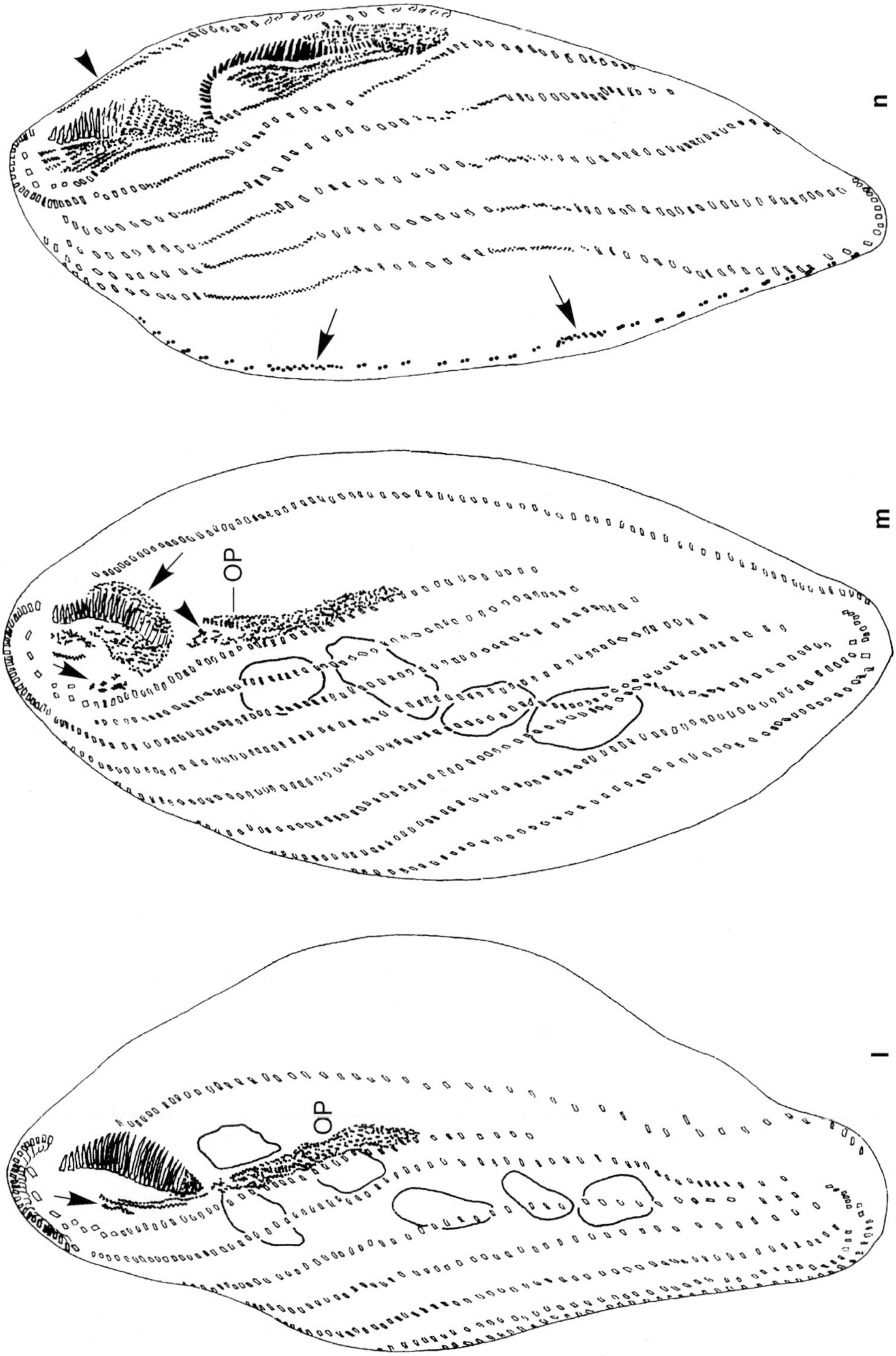
n
m
OP
l
OP

enlarged and set off cirri of the right marginal rows correspond to caudal cirri rather than transverse cirri because we know that dorsal kineties – which usually terminate in caudal cirri – and marginal cirral rows are likely homonomous structures (see Berger et al. 1985, p. 309). A similar phenomenon of a continuous arrangement of different cirral groups is known from *Pseudoamphisiella*, where the caudal cirri are more or less continuous with the marginal rows, feigning a single, U-shaped marginal row. However, in any case the presence of enlarged and set off terminal marginal cirri, which obviously form a functional group with the ordinary transverse cirri, is an autapomorphy of *P. similis*.

The marine *Parabirojimia similis* is easily recognisable by its numerous right marginal rows and the 3–6, on average four macronucleus nodules. The bipartite adoral zone, including the subapical process in the gap, is strongly reminiscent of *Afrothrix darbyshirei* (Fig. 104a–g). However, this species lacks a midventral row (vs. present in *P. similis*), and has frontoterminal cirri (vs. lacking), only one right marginal row (vs. 5–8), two macronuclear nodules (vs. 3–6), and cortical granules (vs. lacking). *Urostyla dispar* Kahl, 1932 has a very similar oral apparatus, but only two macronuclear nodules and very likely a bicorona (Fig. 216a–c). *Etoschothrix terricola* Foissner, Agatha & Berger, 2002 and *Erniella filiformis* Foissner, 1987b, which also have a bipartite adoral zone, lack a midventral complex, that is, do not belong to the urostyloids. *Holosticha* species also have a discontinuous adoral zone (Fig. 20a). However, they have, inter alia, frontoterminal cirri and only one right marginal row.

According to Hu et al. (2002), *Birojimia* is very similar to *P. similis*. However, *Birojimia* species have, inter alia, a continuous adoral zone, frontoterminal cirri, and caudal cirri. In addition, they have much many macronuclear nodules (on average 38 or more; Table 29) and live in terrestrial habitats. *Metaurostylopsis* species have not only two or more right marginal rows, but also two or more left rows. Further, they have a continuous adoral zone of membranelles, frontoterminal cirri, and many macronucleus nodules. *Uroleptopsis* species have, inter alia, a bicorona and many macronuclear nodules. *Parabirojimia* is classified in the Bakuellidae because *P. similis* has three frontal cirri and obviously one midventral row.

Morphology (Fig. 136a–k, Table 30): The paper by Hu et al. (2002) contains, beside the line drawings shown in the present book, 27 micrographs documenting most features.

Body size 140–300 × 30–50 µm in life, on average 180 × 45 µm, body length:width ratio ranges from 4–6:1; ratio in protargol preparations 2.6–2.9:1 on average, indicating that specimens inflate by protargol impregnation (Table 30). Body outline constant,

---

Fig. 136o–q *Parabirojimia similis* after protargol impregnation (from Hu et al. 2002). **o:** Infraciliature of dorsal side and nucleus-apparatus of specimen shown in Fig. 136n. Arrowhead marks posterior primordium in left marginal row, arrows denote anlagen within dorsal kineties 1 and 2. **p:** Infraciliature of ventral side and nuclear apparatus (right above) of a middle divider, size not indicated. Arrows denote the two levels where primordia within right marginal rows occur. The macronucleus-nodules are fused to a single mass. **q:** Infraciliature of ventral side and nuclear apparatus (left) of a middle to late divider, size not indicated. Arrowheads denote short cirri anlagen. Page 691. →

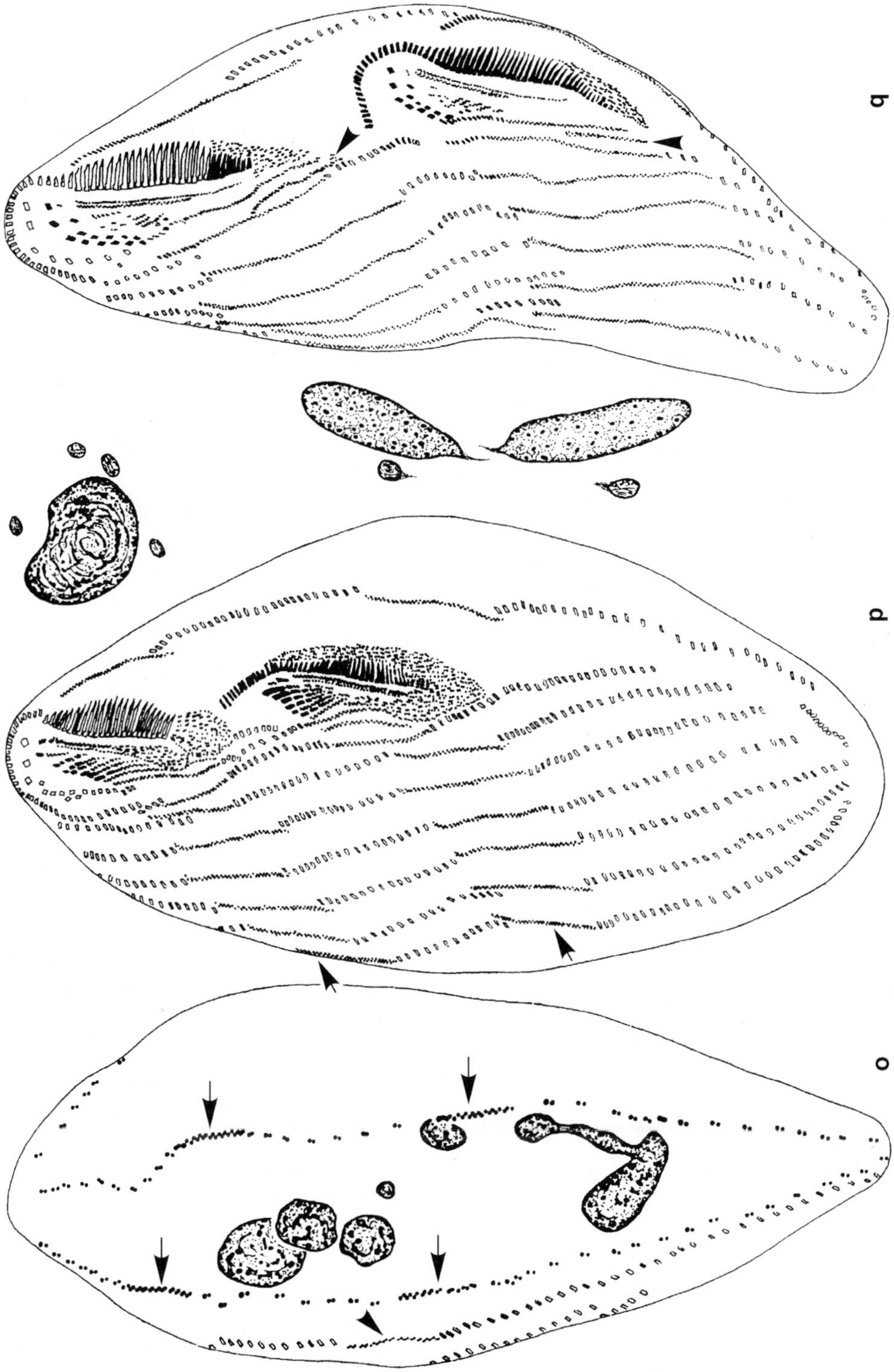

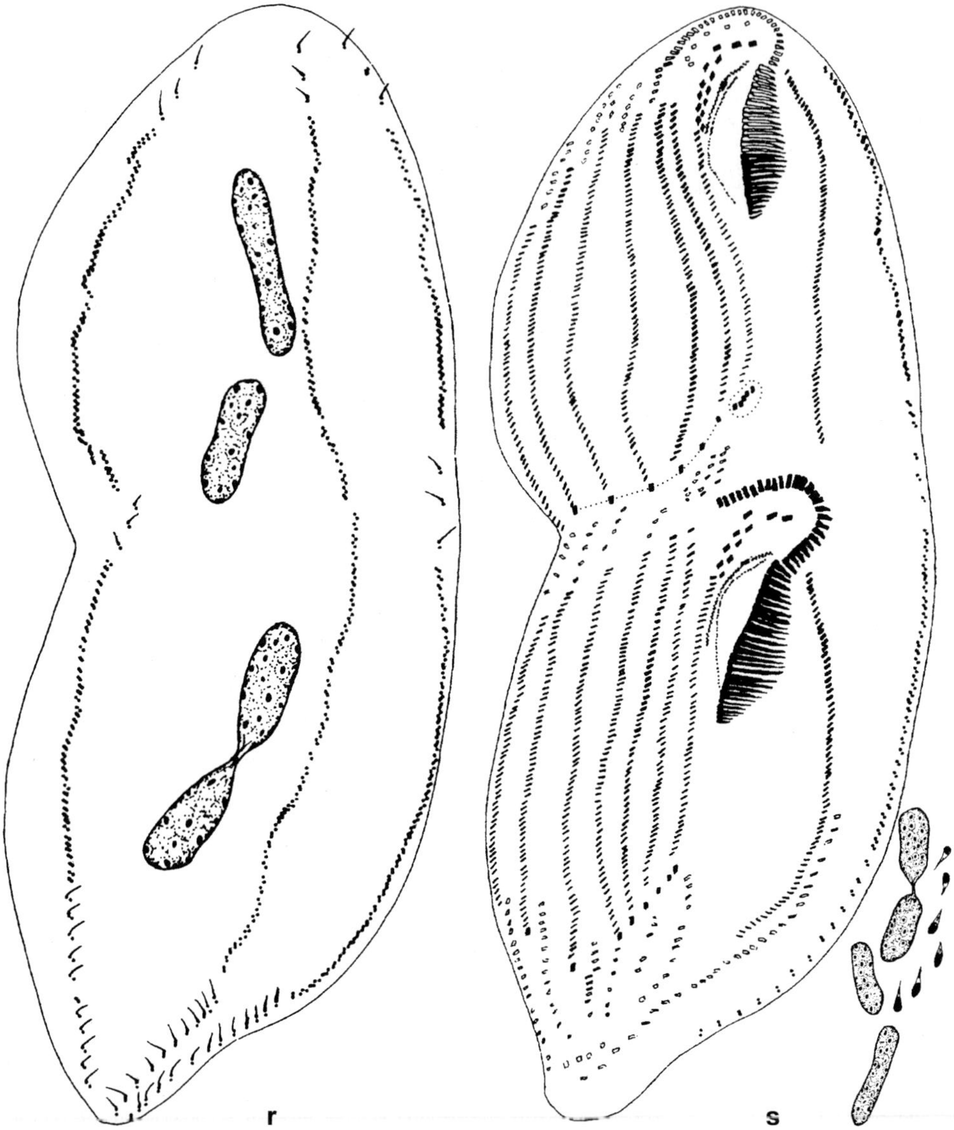

Fig. 136r, s *Parabirojimia similis* after protargol impregnation (from Hu et al. 2002). **r:** Infraciliature of dorsal side and macronuclear nodules of a late diver, 263 µm. **s:** Infraciliature of ventral side and nuclear apparatus of a late divider, size not indicated (nuclear apparatus is reduced to 60% of original size). True transverse cirri (that is, those originating from the frontal-midventral-transverse cirri anlagen) of the proter are circled by a dotted line, the "transverse" cirri formed by the right marginal rows are connected by a dotted line. Page 691.

roughly elongated elliptical, widest in anterior body third, left and right margin converging posteriorly; conspicuous subapical snout-like protrusion in gap between proximal and distal portion of adoral zone (Fig. 136a); after a few days in culture, some slender

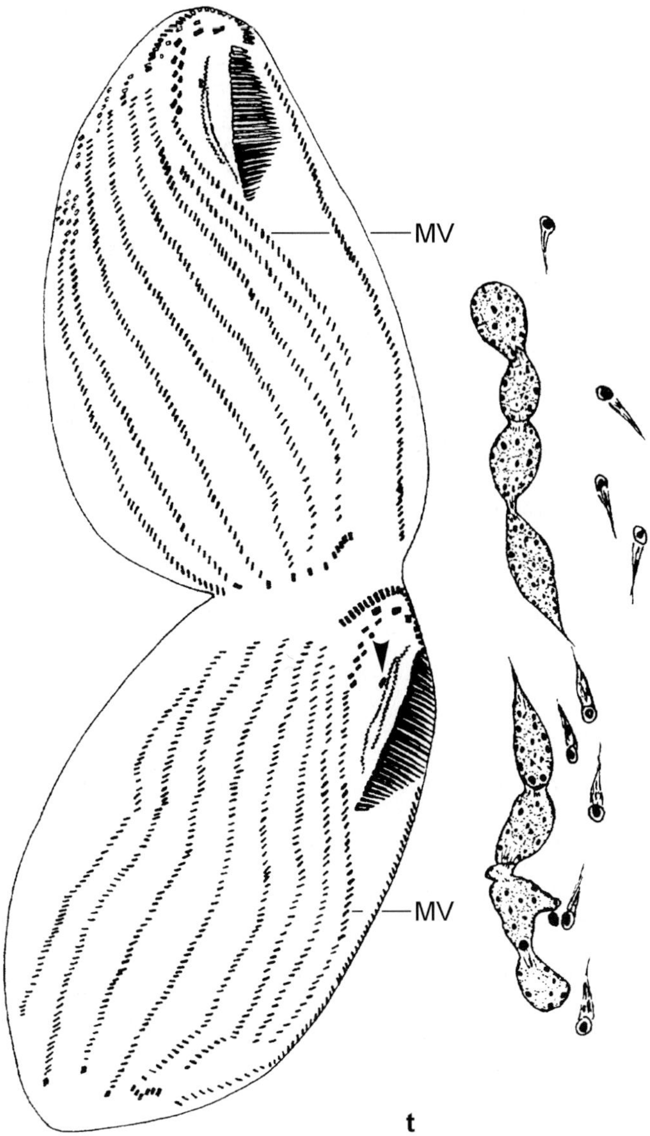

Fig. 136t *Parabirojimia similis* after protargol impregnation (from Hu et al. 2002). Infraciliature of ventral side and nuclear apparatus of a very late divider, size not indicated. Arrow marks buccal cirrus of opisthe. MV = midventral row. Page 691.

specimens were observed with longer tails (Fig. 136c, d). Cells dorsoventrally flattened about 1.5:1 (Fig. 136b). 3–6, on average four macronuclear nodules slightly left of mid-body; individual nodules ovoid to ellipsoidal, in life 15–20 μm long (Fig. 136a). Micronuclei unrecognisable in life, ovoid, about 4 μm large in protargol preparations, adjacent to macronuclear nodules. Contractile vacuole neither mentioned nor illustrated, indicating that this organelle is lacking. No cortical granules observed. Pellicle thin and flexible. Cytoplasm appears greyish to dark at low magnification, but colourless at high magnification, contains numerous refractive globules 2–4 μm across and irregular

Table 30 Morphometric data on *Parabirojimia similis* (si1, si2, two populations from Hu et al. 2002; according to a personal communication by X. Hu, si2 [lower line in original description] is the holotype population)

Characteristics [a]	Species	mean	M	SD	SE	CV	Min	Max	n
Body, length	si1	176.2	–	33.6	6.7	19.1	128.0	272.0	25
	si2	192.5	–	15.3	4.4	8.0	168.0	216.0	12
Body, width	si1	59.8	–	12.7	2.8	21.2	39.0	82.0	20
	si2	73.5	–	11.0	3.3	15.0	60.0	92.0	11
Anterior body end to proximal end of adoral zone, distance	si1	42.2	–	5.3	1.1	12.7	34.0	52.0	23
	si2	43.1	–	3.4	1.0	7.9	36.0	48.0	12
Length of adoral zone/body length, ratio	si1	0.25	–	0.02	0.01	9.2	0.22	0.30	23
	si2	0.22	–	0.03	0.01	12.1	0.18	0.26	12
Macronuclear nodules, number	si1	4.5	–	0.8	0.2	17.3	3.0	6.0	24
	si2	4.0	–	0.0	0.0	0.0	4.0	4.0	12
Macronuclear nodule, length	si1	17.5	–	5.2	1.1	29.5	9.0	30.0	24
	si2	23.1	–	3.4	1.0	14.6	15.0	27.0	12
Macronuclear nodule, width	si1	7.9	–	2.9	0.6	36.4	6.0	18.0	24
	si2	12.3	–	1.9	0.5	15.2	10.0	16.0	12
Micronuclei, number	si1	4.7	–	1.2	0.3	26.5	3.0	8.0	15
Micronuclei, largest diameter	si1	3.5	–	0.9	0.3	26.3	3.0	6.0	14
Adoral membranelles, number [b]	si1	46.7	–	3.3	0.7	7.2	40.0	53.0	23
	si2	47.9	–	4.8	1.4	9.9	37.0	54.0	12
Frontal cirri, number	si1	3.0	–	0.0	0.0	0.0	3.0	3.0	25
	si2	3.0	–	0.0	0.0	0.0	3.0	3.0	12
Buccal cirri, number	si1	1.0	–	0.0	0.0	0.0	1.0	1.0	25
	si2	1.0	–	0.0	0.0	0.0	1.0	1.0	12
Midventral cirri, number [c]	si1	9.0	–	1.3	0.3	14.9	7.0	12.0	21
	si2	9.8	–	1.5	0.4	15.5	8.0	13.0	12
Midventral row, number of cirri	si1	36.5	–	5.1	1.1	13.9	27.0	45.0	21
	si2	41.2	–	9.9	2.9	24.1	15.0	56.0	12
Transverse cirri, number [d]	si1	8.4	–	2.1	0.5	25.3	3.0	11.0	21
	si2	9.7	–	1.1	0.3	11.1	8.0	11.0	12
Left marginal row, number of cirri	si1	61.3	–	6.8	1.5	11.1	50.0	79.0	21
Right marginal rows, number [e]	si1	6.0	–	0.6	0.1	9.1	5.0	7.0	21
	si2	6.8	–	0.8	0.2	11.2	6.0	8.0	12
Right marginal row 1, number of cirri	si1	43.6	–	12.5	2.7	28.8	7.0	64.0	21
	si2	50.8	–	10.7	3.1	21.1	37.0	66.0	12
Right marginal row 2, number of cirri	si1	55.0	–	8.4	1.8	15.3	36.0	68.0	21
	si2	62.2	–	7.2	2.1	11.6	43.0	68.0	12
Right marginal row 3, number of cirri	si1	56.3	–	6.2	1.4	11.1	45.0	67.0	21
	si2	65.6	–	3.7	1.1	5.7	59.0	71.0	12
Right marginal row 4, number of cirri	si1	56.2	–	6.3	1.3	11.2	45.0	66.0	22
	si2	66.8	–	4.7	1.3	7.0	57.0	73.0	12
Right marginal row 5, number of cirri	si1	51.1	–	6.7	1.4	12.9	36.0	64.0	22
	si2	65.8	–	5.9	1.7	39.0	58.0	75.0	12
Right marginal row 6, number of cirri	si1	–	–	–	–	–	39.0	50.0	3
	si2	58.1	–	11.6	3.3	19.9	44.0	74.0	12
Dorsal kineties, number	si1	3.0	–	0.0	0.0	0.0	3.0	3.0	25
	si2	3.0	–	0.0	0.0	0.0	3.0	3.0	12

[a] All measurements in µm. Data based on protargol-impregnated (Wilbert's method) and likely randomly selected specimens. CV = coefficient of variation in %, M = median, Max = maximum value, mean = arithmetic mean, Min = minimum value, n = number of individuals investigated, SD = standard deviation, SE = standard error of arithmetic mean.

Table 30 Continued

[b] This number contains all membranelles because Hu et al. (2002) did not count distal and proximal membranelles separately. For separate values, see text.

[c] This number comprises only the cirri forming the pairs, that is, the cirri of the midventral row are not included. Cirrus III/2, that is, the cirrus behind the right frontal cirrus is likely included.

[d] "Transverse cirri" formed from right marginal rows included (see text).

[e] No counts are available for the right marginal rows 7 and 8. Obviously they are only occasionally present.

yellow-greenish crystals 2–3 µm long (Fig. 136e). Movement without peculiarities, that is, crawls moderately quickly on debris.

Adoral zone occupies about 20–25% of body length, bipartite in a distal, bow-shaped portion along anterior body end and a roughly scalpel-shaped proximal portion distinctly placed inwards, at least in protargol preparations. Fine-structure of distal and proximal membranelles different. Distal portion composed of around 20–26 small membranelles, proximal of 21–25 membranelles of rather distinct length (data from Fig. 136f–j, that is, n = 5). Distal portion extends distinctly posteriorly onto right side making an inconspicuous furrow along anterior-right body margin (Fig. 136a). Cilia of membranelles up to 20 µm long. Paroral and endoral of about same length, slightly curved and optically intersecting. Buccal area narrow, colourless and translucent. Pharyngeal fibres 30–40 µm long, conspicuous after protargol impregnation.

Cirral pattern and number of cirri of usual variability except for number of transverse cirri (see below), number of cirri in midventral row, and number of cirri in some right marginal rows (Fig. 136f–j, Table 30). Invariably three distinctly enlarged, in life about 18 µm long frontal cirri often almost transversely arranged, but nevertheless continuous with midventral complex. Most other cirri relatively fine and 10–12 µm long. Single buccal cirrus near midpoint of paroral. Midventral complex composed of 3.5–6.5 cirral pairs (that is, 7–13 cirri) with rearmost pair about at level of buccal cirrus and basically a single midventral row composed of 15–56 cirri with rearmost cirrus at or distinctly behind mid-body; occasionally a second, short midventral row present (Fig. 136j, q). Usually 7–10 slightly enlarged transverse cirri, in life about 18 µm long, arranged in oblique row near rear body end and thus distinctly (about 70% of their length) projecting beyond cell margin; among the individuals examined (at least 25 or more) Hu et al. (2002) found one with only three transverse cirri (Fig. 136g); transverse row composed of cirri of two different anlagen types, namely, the left, usually narrowly spaced portion stems from the ordinary frontal-midventral-transverse cirri anlagen; the right, usually wider spaced portion originates from the right marginal row anlagen which is unique (Fig. 136f, g, j; for details, see cell division). Invariably one left marginal row, commences above middle of proximal portion of adoral zone, ends slightly to distinctly sub-terminally. 5–8 right marginal rows, obliquely oriented from anterior right to posterior left; rightmost two or three rows anteriorly shortened and turn over slightly to dorsal side; rows terminate successively more posteriorly from left to right (Fig. 136h, j).

Dorsal cilia about 5 µm long, invariably arranged in three bipolar kineties. Caudal cirri lacking (Fig. 136k).

Cell division (Fig. 136l–t): Fortunately, Hu et al. (2002) were able to study the ontogenesis of this curious species so that, inter alia, the origin of the relative many transverse cirri could be clarified. Most features are documented in the original description by micrographs not shown in the present book.

Morphogenesis commences with the apokinetal proliferation of groups of narrowly spaced basal bodies left of the midventral row. These groups subsequently merge by further proliferation of basal bodies forming a single anarchic field, the oral primordium of the opisthe (Fig. 136l). During the formation of the oral primordium parental cirri nearby apparently remain unchanged and do not contribute to this process. Later, some basal body groups evidently appear to the anterior right of the main part of the oral primordium which will develop into opisthe's frontal-midventral-transverse cirri streaks. Meanwhile, the old undulating membranes and the posterior portion of the old adoral zone dedifferentiate (Fig. 136m). In the next stage, the new membranelles begin to develop posteriad in the opisthe, the old adoral zone continues to dedifferentiate anteriad up to half the length of the oral field in the proter, 6–7 frontal-midventral-transverse cirri anlagen and an undulating membrane anlage develop in both dividers (Fig. 136n). Evidently, part of the midventral cirri, the buccal cirrus and the parental undulating membranes join these anlagen. Simultaneously, two anlagen appear within each cirral row and each dorsal kinety (Fig. 136o).

Later, the frontal-midventral-transverse cirri anlagen gradually develop into new cirri and each anlage eventually gives rise to 2–3 cirri; a small basal body patch separates anteriorly from the main part of the undulating membrane anlage and gives rise, as is usual, to a single cirrus (cirrus I/1) in both dividers (Fig. 136p). At the same time, the number of new membranelles continues to increase in the opisthe, and the loosely arranged basal bodies begin to re-organise in the posterior portion of the parental adoral zone. Anlagen within the marginal rows, dorsal kineties, and the midventral row begin to lengthen and finally replace the old structures (Fig. 136q). In one divider, two short midventral cirral anlagen developed to the left of the posterior portion of the normal, long midventral cirri anlage (Fig. 136q). No caudal cirri are formed at the rear end of the dorsal kineties (Fig. 136r). The rearmost cirri derived from the frontal-midventral-transverse cirri anlagen move posteriorly and become the ordinary transverse cirri. In addition, usually each right marginal cirri anlage, except the rightmost one, produce an enlarged terminal cirrus. These enlarged terminal right marginal cirri form a pseudorow with the ordinary transverse cirri (Fig. 136s). The first cirrus from each of the anterior two frontal-midventral-transverse streaks, plus the cirrus derived from the undulating membrane anlage, constitute the ordinary, enlarged frontal cirri (cirri I/1, II/1, III/1). The second cirrus of anlage II becomes, also as usual, the buccal cirrus (cirrus II/2). The remaining cirri of the frontal-midventral-transverse cirri anlagen form the midventral pairs.

Finally, the divider begins to elongate and the new ciliary structures move further apart as they migrate towards their final positions. Almost all parental cirri are now resorbed. The oral apparatus is completed and the daughters begin to separate (Fig. 136t).

Division of the nuclear apparatus proceeds in ordinary manner. At a very early stage, the macronuclear nodules enlarge and almost all have a replication band. Subsequently, the nodules transform and fuse with each other, eventually forming a single, spherical mass (Fig. 136p). In late dividers, the nodules separate again. The micronuclei remain visible and divide during late stages (Fig. 136l, m, o–t).

Occurrence and ecology: Marine. The type location of *Parabirojimia similis* is a marine mollusc culture off the coast of Qingdao (Tsingtao; 36°08'N 120°43'E), China, where Hu et al. (2002) found it in June 2000 and July 2001 at 32–37‰ salinity, 16–24° C water temperature, and pH 8.0–8.2. No further records available.

In nature, *Parabirojimia similis* likely feeds mainly on scuticociliates and diatoms. Hu et al. (2002) maintained it for several weeks in the laboratory. Squeezed wheat grains were added to the medium (likely natural sea water) to provide bacterial growth.

Australothrix Blatterer & Foissner, 1988

1988 ***Australothrix* nov. gen.**[1] – Blatterer & Foissner, Stapfia, 17: 38 (original description). Type species (by original designation on p. 38): *Australothrix australis* Blatterer & Foissner, 1988.

1990 ***Australothrix*** – Foissner & Blatterer, J. Protozool. Suppl., 37: 9A, Abstract 53 (summary of new taxa found in soil samples from Australia and Africa).

1994 ***Australothrix*** – Corliss, Acta Protozool., 33: 15 (classification of protists; list of genera and higher taxa).

1999 ***Australothrix* Blatterer & Foissner, 1988** – Shi, Acta Zootax. sinica, 24: 363 (revision of the Hypotrichida; list of genera and higher taxa).

1999 ***Australothrix* Blatterer & Foissner, 1988** – Shi, Song & Shi, Progress in Protozoology, p. 112 (revision of hypotrichous ciliates).

2001 ***Australothrix* Blatterer & Foissner 1988** – Aescht, Denisia, 1: 28 (catalogue of generic names of ciliates).

2001 ***Australothrix* Blatterer and Foissner, 1988** – Berger, Catalogue of ciliate names 1. Hypotrichs, p. 11 (nomenclator containing all basionyms, combinations, and higher taxa of hypotrichs).

2002 ***Australothrix* Blatterer and Foissner, 1988** – Lynn & Small, Phylum Ciliophora, p. 443 (guide to ciliate genera).

Nomenclature: The name *Australothrix* is a composite of the Latin adjective *australis* (southern) and the Greek noun *trichos* (hair, bristle, cirrus; *thrix* is a form of the nominative only at the end of words; Werner 1972, p. 420), referring to a hypotrichous ciliate occurring in Australia, respectively, southern hemisphere. Feminine gender (Blatterer & Foissner 1988).

Characterisation (Fig. 111a, autapomorphies 14): Adoral zone of membranelles continuous. 3 frontal cirri. Buccal cirrus present. Frontoterminal cirri lacking. Midventral complex composed of midventral pairs and at least 1 midventral row. Transverse cirri lacking (A). More than 2 marginal rows. Caudal cirri present. More than 3 dorsal kineties (A).

Remarks: Blatterer & Foissner (1988) erected *Australothrix* for a group of urostyloids with many ventral and/or marginal rows, few (3–6), enlarged frontal cirri, and

[1] The diagnosis by Blatterer & Foissner (1988) is as follows: Urostylidae mit Caudalcirren, mehreren verstärkten Frontalcirren und mehreren Ventral(Marginal)reihen. Keine Transversalcirren.

Table 31 Morphometric data on *Australothrix australis* (aus, from Blatterer & Foissner 1988 after Wilbert's protargol method), *Australothrix alwinae* (al1, from Blatterer & Foissner 1988 after Wilbert's protargol method; al2, from Blatterer & Foissner 1988 after Foissner's protargol method; alw, al1 and al2 pooled), and *Australothrix steineri* (ste, from Foissner 1995 after Foissner's protargol method)

Characteristics [a]	Species	mean	M	SD	SE	CV	Min	Max	n
Body, length	aus	313.1	312.0	23.6	7.1	7.5	273.0	342.0	11
	al1	281.0	284.5	39.4	13.9	14.0	227.0	340.0	8
	al2	241.0	228.0	38.9	17.4	16.1	196.0	290.0	5
	ste	236.1	240.0	53.3	17.8	22.6	145.0	315.0	9
Body, width	aus	96.7	98.0	14.0	4.2	14.5	73.0	121.0	11
	al1	95.5	90.0	22.2	7.1	21.1	66.0	123.0	8
	al2	72.6	68.0	17.6	7.9	24.2	53.0	101.0	5
	ste	22.7	24.0	4.6	1.5	20.2	16.0	31.0	9
Anterior body end to rear end of adoral zone of membranelles, distance	aus	97.6	98.0	7.0	2.1	7.2	8.6	108.0	11
	al1	87.6	87.0	9.5	3.3	10.8	76.0	101.0	8
	al2	78.8	77.0	12.5	5.6	15.9	61.0	95.0	5
	ste	31.6	32.0	2.2	0.7	7.1	28.0	35.0	9
Macronuclear nodule, length	aus	9.7	10.6	3.5	1.0	35.9	4.5	15.0	11
	alw	7.2	6.0	2.8	0.8	39.1	3.5	12.5	13
	ste	5.3	5.0	0.7	0.2	13.3	4.0	6.0	9
Macronuclear nodule, width	aus	3.9	3.8	1.1	0.3	28.1	2.0	5.3	11
	alw	4.3	4.5	1.2	0.3	27.2	3.0	6.5	13
	ste	2.6	3.0	0.5	0.2	20.6	2.0	3.0	9
Macronuclear nodules, number	aus	184.8	185.0	25.1	7.9	13.6	141.0	225.0	10
	alw	115.8	113.0	19.6	5.7	16.9	93.0	162.0	12
	ste				about 70–150				
Micronucleus, length	aus	4.5	4.5	0.0	0.0	0.0	4.5	4.5	11
	alw	4.8	4.5	1.0	0.3	21.1	3.2	6.5	13
Micronucleus, width	aus	4.5	4.5	0.0	0.0	0.0	4.5	4.5	11
	alw	3.3	3.0	0.9	0.3	27.9	2.5	6.0	13
Micronuclei, number	aus	2.5	3.0	0.7	0.2	28.0	1.0	3.0	11
	alw	5.3	5.0	1.5	0.4	28.1	3.0	8.0	12
Adoral membranelles, number	aus	63.0	64.0	3.6	1.1	5.7	58.0	69.0	11
	alw	53.2	53.5	4.2	1.2	7.8	46.0	61.0	12
	ste	25.2	25.0	1.1	0.4	4.3	24.0	27.0	9
Frontal cirri, number	aus [b]	4.0	4.0	0.0	0.0	0.0	4.0	4.0	11
	alw [g]	–	–	–	–	–	3.0	6.0	15
	ste	3.0	3.0	0.0	0.0	0.0	3.0	3.0	8
Buccal cirri, number	aus	1.3	1.0	–	–	–	1.0	2.0	11
	alw	1.0	1.0	0.0	0.0	0.0	1.0	1.0	15
	ste	1.0	1.0	0.0	0.0	0.0	1.0	1.0	7
Cirri near buccal cirrus, number	alw [h]	0.4	0.0	–	–	–	0.0	1.0	15
Cirral rows, number [c]	aus	8.6	9.0	0.8	0.2	9.4	7.0	10.0	11
	ste	4.8	5.0	0.9	0.3	18.1	4.0	6.0	11
Anterior body end to last midventral pair, distance	ste [f]	24.3	25.0	3.1	1.2	12.7	21.0	30.0	7
Midventral pairs, number	ste [i]	2.3	2.0	–	–	–	2.0	3.0	7
Midventral complex, number of right cirri	aus [d]	50.9	53.5	10.6	3.4	20.9	24.0	62.0	10
	alw	15.8	15.5	3.1	0.9	19.9	11.0	20.0	12
Midventral pairs, number of left cirri	aus	6.2	6.0	0.9	0.3	14.1	5.0	8.0	11
	alw	12.9	13.0	2.2	0.6	16.7	10.0	16.0	12
Right marginal cirri, number [e]	aus	68.0	69.0	4.8	1.5	7.1	59.0	74.0	11

Table 31 Continued

Characteristics [a]	Species	mean	M	SD	SE	CV	Min	Max	n
Right marginal cirri, number [e]	alw	58.1	57.5	7.0	2.2	12.0	50.0	72.0	10
	ste	63.3	75.0	20.7	7.8	32.7	35.0	85.0	7
Left marginal cirri, number [e]	aus	72.2	72.0	6.1	1.8	8.4	64.0	86.0	11
	alw	62.3	60.5	11.1	3.5	17.7	47.0	83.0	10
	ste	58.0	56.5	19.7	7.0	34.0	36.0	82.0	8
Dorsal kineties, number	aus	6.0	6.0	0.5	0.2	9.4	5.0	7.0	11
	alw	5.6	6.0	–	–	–	5.0	6.0	9
	ste	4.0	4.0	0.0	0.0	0.0	4.0	4.0	9
Caudal cirri, number	aus	8.2	8.0	2.0	0.6	24.3	6.0	13.0	11
	alw	11.8	11.5	5.5	2.2	46.1	5.0	19.0	6

[a] All measurements in µm. All data are based on mounted, protargol-impregnated (for method, see table head), and randomly selected specimens. CV = coefficient of variation in %, M = median, Max = maximum value, mean = arithmetic mean, Min = minimum value, n = number of individuals investigated, SD = standard deviation, SE = standard error of arithmetic mean.

[b] Cirrus behind right frontal cirrus included.

[c] Includes marginal and ventral cirral rows.

[d] Includes cirri of third row from right (Fig. 137g, MV) which begins immediately behind the right cirri of the midventral row.

[e] Outermost right, respectively, left cirral row.

[f] Foissner (1995) wrote, due to a slip of the pen, 31 as minimum value.

[g] Very likely almost invariably three distinctly enlarged frontal cirri are present (see Fig. 138g, j).

[h] Likely, this is cirrus III/2, that is, the cirrus behind the right frontal cirrus (for explanation, see Fig. 138g).

[i] Foissner (1995) obviously counted pseudopairs (see Fig. 139f).

lacking transverse cirri. The last feature separates it from many urostyloids, *Urostyla*, *Trichototaxis*, and *Pseudourostyla*. Shi (1999, 1999a), Shi et al. (1999), and Lynn & Small (2002) accepted the classification of *Australothrix* in the Urostylidae. I classify it in the Bakuellidae because it has three frontal cirri and a midventral complex composed of midventral pairs and midventral rows. *Hemicycliostyla*, which also has many marginal rows and lacks transverse cirri, has a frontal ciliature of the bicorona type.

Blatterer & Foissner (1988) discovered two new species in Australian soils. Furthermore, they transferred *Uroleptus zignis* Entz, 1884 and *Oxytricha gibba* Claparède & Lachmann, 1858 to *Australothrix* because these species fit the combination of features mentioned above very well. Later, two further species, *Australothrix simplex* and *A. steineri*, were described, so that *Australothrix* comprises six species now. However, only three of them are known in detail (*A. australis*, *A. alwinae*, *A. steineri*), so that it is impossible to find further characteristics common to all species. Interestingly, the outer left marginal row commences rather far anteriorly in *A. australis* and *A. alwinae* and possibly also in *A. zignis*.

According to Blatterer & Foissner (1988), it cannot be excluded that *Australothrix australis* (type species) and *A. alwinae* are not congeneric because they have a rather differently arranged midventral complex. In the type species it is dislocated to near the right body margin and several long cirral rows are arranged to the left (Fig. 137g). By contrast, the midventral complex extends in cell midline with few cirral rows right of it in *A. alwinae* (Fig. 138g). According to this feature, *Australothrix steineri* is similar to *A. australis*, whereas *A. simplex* resembles *A. alwinae*. No such assignment is possible for *A. zignis* and *A. gibba* because their cirral pattern is not known in detail. However, for a correct interpretation of the pattern ontogenetic data are needed. From interphasic specimens alone it is impossible to decide which rows are midventral and which are marginal ones. The same applies to a further important feature, namely the presence/absence of frontoterminal cirri. It is interesting that in all of the four protargol-impregnated species (*A. australis*, *A. alwinae*, *A. steineri*, *A. simplex*) no such cirral group is recognisable. This strongly suggests that frontoterminals are in fact lacking, as already supposed by Blatterer & Foissner (1988). There are only few further groups, for example, *Parabirojimia*, *Urostyla*, *Notocephalus*, *Periholosticha lanceolata* which lack frontoterminal cirri. Eigner & Foissner (1992) interpreted the lack of this cirral group in the *Australothrix/Urostyla* group as plesiomorphy. However, it is more likely that the loss of the frontoterminal cirri is the apomorphic state because the presence of frontoterminal cirri very likely belongs to the ground pattern of the Urostyloidea. Thus, I unify *Parabirojimia* and *Australothrix* in the Bakuellidae (Fig. 111a).

Species included in *Australothrix* (alphabetically arranged basionyms are given): (1) *Australothrix alwinae* Blatterer & Foissner, 1988; (2) *Australothrix australis* Blatterer & Foissner, 1988; (3) *Australothrix simplex* Liu, Shen, Song & Gu, 1992; (4) *Australothrix steineri* Foissner, 1995; (5) *Oxytricha gibba* Claparède & Lachmann, 1858; (6) *Uroleptus zignis* Entz, 1884.

Key to *Australothrix* species

Identification of *Australothrix* species requires the usual set of features (habitat, size, shape, macronuclear apparatus, cortical granules, cirral pattern).

1 Body length below 150 µm **2**
\- Body length above 150 µm **3**
2 Four cirral rows; many scattered macronuclear nodules (Fig. 140a, b) . . . . *Australothrix simplex* (p. 719)
\- Six cirral rows; two macronuclear nodules (Fig. 142a, b) . . . . *Australothrix gibba* (p. 724)
3 (1) Body length:width ratio 9–12:1 (Fig. 139a, b) . . . *Australothrix steineri* (p. 716)
\- Body length:width ratio 5:1 or below **4**
4 Two macronuclear nodules; cortical granules reddish; marine (Fig. 141a, b) . . . . *Australothrix zignis* (p. 721)
\- Many (more than 30) scattered macronuclear nodules; cortical granules lacking or colourless; terrestrial **5**

5 Body length below 200 μm; 4 cirral rows (Fig. 140a, b) .... *Australothrix simplex* (p. 719)
- Body length above 200 μm; 7–10 cirral rows **6**
6 One left marginal row; midventral complex extends near right body margin; cortical granules around cirri and dorsal bristles (Fig. 137b, c, g) .... *Australothrix australis* (p. 707)
- Two or three left marginal rows; midventral complex extends near cell midline; cortical granules throughout cortex (Fig. 138c, d, g) ... *Australothrix alwinae* (p. 711)

Australothrix australis Blatterer & Foissner, 1988 (Fig. 137a–k, Table 31)

1988 ***Australothrix australis* nov. spec.**[1] – Blatterer & Foissner, Stapfia, 17: 39, Abb. 11a–h, 40, 42, 44, Tab. 9 (Fig. 137a–k; original description. The holotype slide [1989/59] and a paratype slide [1989/60] are deposited in the Oberösterreichische Landesmuseum in Linz [LI], Austria).

1990 ***Australothrix australis*** – Foissner & Blatterer, J. Protozool. Suppl., 37: 9A, Abstract 53 (summary of new taxa found in soil samples from Australia and Africa).

2001 ***Australothrix australis* Blatterer and Foissner, 1988** – Berger, Catalogue of ciliate names 1. Hypotrichs, p. 11 (nomenclator containing all basionyms, combinations, and higher taxa of hypotrichs).

Nomenclature: This species is named after the country/continent (Australia) where it was discovered by Blatterer & Foissner (1988). *Australothrix australis* was fixed as type species of *Australothrix* by original designation.

Remarks: *Australothrix australis* was described in detail both from life and after protargol impregnation by Blatterer and Foissner (1988). However, ontogenetic data are needed for a correct interpretation of the cirral pattern. Among others, the following questions should be answered: (i) Does the species have frontoterminal cirri? (ii) Which cirral rows are marginals and which are midventrals? (iii) Which dorsal kineties are associated with caudal cirri?

For separation from other *Australothrix* species, see key. In life, *Australothrix australis* is best recognised by the soil habitat, the size (250–400 × 60–100 μm), the colourless cortical granules around the cirri and dorsal bristles, the three distinct frontal cirri, the single left marginal row, and the scattered macronuclear nodules.

Morphology: Body size about 250–400 × 60–100 μm in life; body length:width ratio around 4.4:1 in life and 3.2:1 on average in protargol preparations (Table 31). Body pisciform, slightly sigmoidal, narrowed and curved rightwards posteriorly; anterior end broadly, posterior narrowly rounded. Flexible, under coverglass contractile by about 33% of body length, dorsoventrally flattened about 2:1 (Fig. 137a, b, f, Table 31). Macronuclear nodules scattered throughout cytoplasm; individual nodules ellipsoidal

[1] The diagnosis by Blatterer & Foissner (1988) is as follows: In vivo etwa 250–400 × 60–100 μm große *Australothrix* mit durchschnittlich 63 adoralen Membranellen, 9 Ventral(Marginal)reihen, 6 Dorsalkineten und 185 Makronucleus-Teilen. Midventralreihe stark verkürzt, endet in Höhe des posterioren Endes der adoralen Membranellenzone, setzt sich in der dritten rechten Ventral(Marginal)reihe fort. Ellipsoide, etwa 2.5 μm große, farblose subpelliculäre Granula entlang der Infraciliatur.

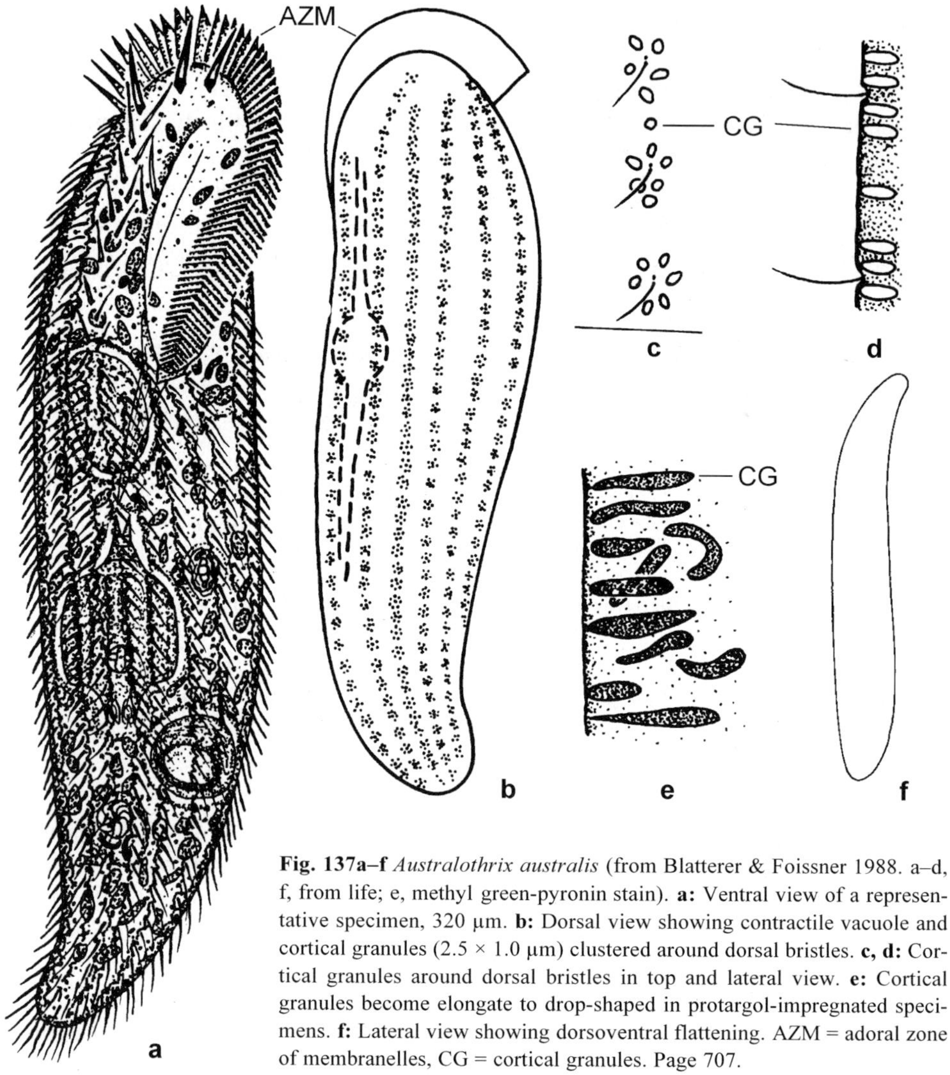

Fig. 137a–f *Australothrix australis* (from Blatterer & Foissner 1988. a–d, f, from life; e, methyl green-pyronin stain). **a:** Ventral view of a representative specimen, 320 µm. **b:** Dorsal view showing contractile vacuole and cortical granules (2.5 × 1.0 µm) clustered around dorsal bristles. **c, d:** Cortical granules around dorsal bristles in top and lateral view. **e:** Cortical granules become elongate to drop-shaped in protargol-impregnated specimens. **f:** Lateral view showing dorsoventral flattening. AZM = adoral zone of membranelles, CG = cortical granules. Page 707.

(length:width ratio 2.5:1 on average in protargol preparations) to dumbbell-shaped, contain many small nucleoli. Micronuclei also scattered, globular, opaque shining. Contractile vacuole slightly ahead of mid-body at left cell margin, with two long collecting canals. Cortical granules around cirri and dorsal bristles (Fig. 137b–e, j, k); individual granules ellipsoidal (about 2.5 × 1.0 µm) and colourless, stain red when methyl green-pyronin is added, and sometimes impregnate with Wilbert's protargol method; stained or impregnated granules elongate to drop-shaped, 3.5–7.5 × 1.0–2.0 µm in size, and partially exploded. Cytoplasm colourless, contains many yellowish, greasily shining globules 1–3 µm across, so that cells appear brownish at low magnification. Glides rapidly on microscope slide and soil particles.

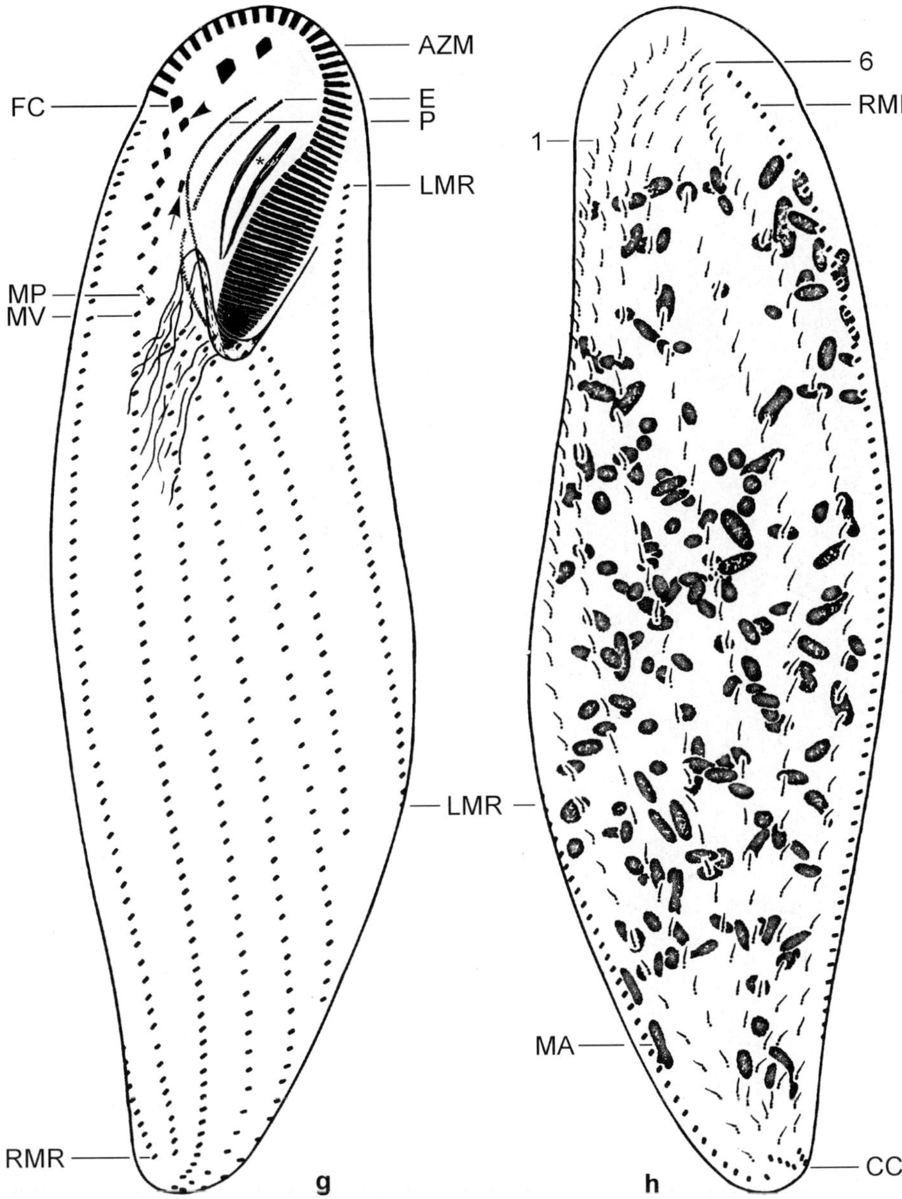

Fig. 137g, h *Australothrix australis* (from Blatterer & Foissner 1988. Protargol impregnation after Wilbert's method). Infraciliature of ventral and dorsal side and nuclear apparatus of same specimen, 333 µm. Arrowhead denotes cirrus (III/2) behind right frontal cirrus, arrow marks buccal cirrus, asterisk is in between two strongly argyrophilic structures in buccal field. The left marginal row does not commence, as in many other species, about at the level of the buccal vertex, but distinctly more anteriorly. AZM = adoral zone of membranelles with spatula-shaped widening of proximal portion, CC = caudal cirri, E = endoral, FC = right frontal cirrus, LMR = left marginal row, MA = macronuclear nodule, MP = posteriormost midventral pair, MV = third cirral row from right (= midventral row?), P = paroral, RMR = right marginal row, 1, 6 = dorsal kineties. Page 707.

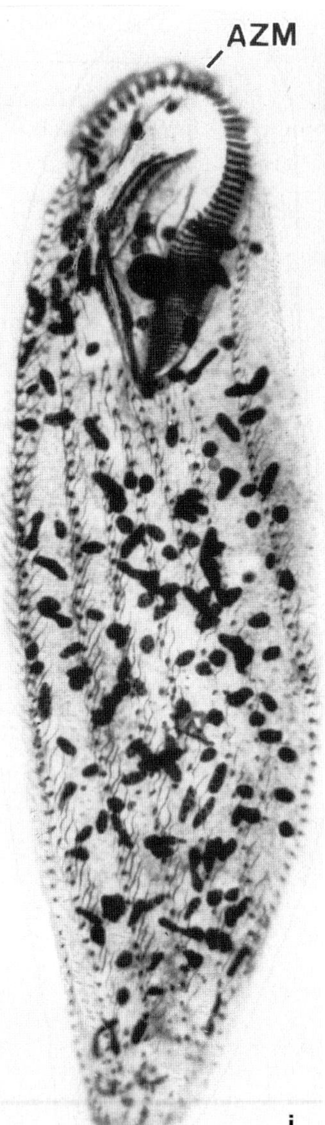

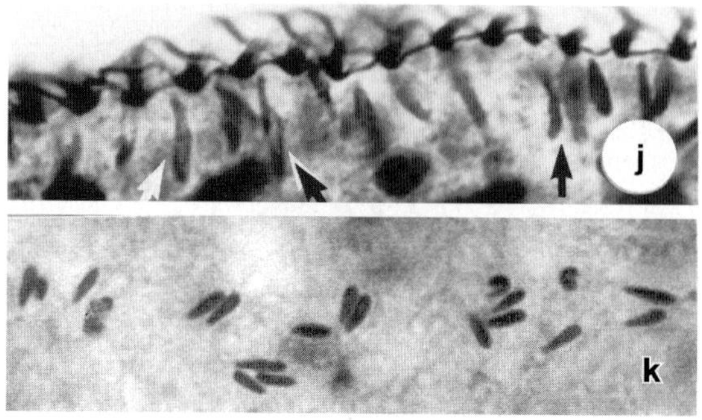

Fig. 137i–k *Australothrix australis* (from Blatterer & Foissner 1988. i, j, protargol impregnation, Wilbert's method; k, methyl green-pyronin stain). **i:** Infraciliature of ventral side and nuclear apparatus of specimen illustrated in Fig. 137g, h; 333 µm. **j, k:** Cortical granules are arranged around cirri and dorsal bristles. Arrows in (j) mark partially exploded cortical granules. AZM = adoral zone of membranelles. Page 707.

Adoral zone occupies about 31% of body length, composed of an average of 63 membranelles, extends rather far onto right body margin, proximal portion conspicuously spatula-shaped, bases of largest membranelles 17 µm wide in life; individual membranelles of usual shape and structure (Fig. 137a, g, i). Buccal cavity large and deep, with two strongly argyrophilic, longish structures (Fig. 137g, i). Undulating membranes strongly curved, anteriorly distinctly separated, intersect optically about at level of buccal cirrus; paroral composed of 2–3 rows of basal bodies, endoral likely composed of monokinetids or dikinetids. Pharyngeal fibres conspicuous in life, extend obliquely backwards.

Cirral pattern and number of cirri of usual variability, except for rather strongly varying number of midventral and caudal cirri (Fig. 137a, g, i, Table 31). Frontal cirri about 20 µm long, others about 15 µm. Invariably three enlarged frontal cirri in an oblique row and a single cirrus (cirrus III/2) behind right frontal cirrus. Usually one, sometimes two buccal cirri at summit of paroral. Frontoterminal cirri probably lacking; however, this has to be checked by ontogenetic data. Midventral complex composed of 4–8 pairs and one or more rows. First midventral pair almost at level of cirrus III/2, last pair about at level of buccal vertex immediately ahead of the third cirral row from right, that is, extends near right cell margin; all midventral cirri of about same size. On average nine (mid)ventral and marginal cirral rows on dorsolateral and ventral side; 2–5 rows begin at level of buccal vertex, left one usually very short, others more or less distinctly shortened posteriorly; one row (Fig. 137g, MV) commences immediately behind last midventral pair (as mentioned above, ontogenetic data are needed to show how this rather curious cirral pattern originates!). Transverse cirri lacking. Marginal rows crenellated, single left row commences distinctly ahead of level of buccal vertex, ends almost terminally. Rightmost two cirral rows commence near distal end of adoral zone of membranelles and end subterminally.

Dorsal cilia about 6 µm long in life, arranged in six more or less bipolar kineties. On average eight caudal cirri; it is unknown from which and from how many dorsal kineties they originate (Fig. 137b, h, Table 31).

Occurrence and ecology: Terrestrial. Type location of *Australothrix australis* is the bark of a tree (presumable *Eukalyptus* sp.) from a rain forest near Cairns (about 16°56'S 145°47'E; Queensland, Australia). This sample ("FO 21" in Blatterer & Foissner 1988) had a pH of 4.2 and was collected by W. Foissner on February 5, 1987. No records published since then. According to Foissner (1998), *Australothrix australis* probably strongly prefers soil habitats.

Polyphagous, that is, ingests heterotrophic flagellates, testate amoebae (*Phryganella acropodia*), ciliates (*Leptopharynx costatus*, *Colpoda* sp.), conidia, mineral particles, and even rotifers (Blatterer & Foissner 1988). Biomass of 10^6 specimens about 1120 mg (Foissner 1998).

Australothrix alwinae Blatterer & Foissner, 1988 (Fig. 138a–j, Table 31)

1988 ***Australothrix alwinae* nov. spec.**[1] – Blatterer & Foissner, Stapfia, 17: 44, Abb. 12a–j, Tab. 9 (Fig. 138a–j; original description. Three protargol slides with the type specimens [accession numbers 1989/62–64] are deposited in the Oberösterreichische Landesmuseum in Linz [LI], Austria; see nomenclature).

1990 ***Australothrix alwinae*** – Foissner & Blatterer, J. Protozool. Suppl., 37: 9A, Abstract 53 (summary of new taxa found in soil samples from Australia and Africa).

2001 ***Australothrix alwinae* Blatterer and Foissner, 1988** – Berger, Catalogue of ciliate names 1. Hypotrichs, p. 11 (nomenclator containing all basionyms, combinations, and higher taxa of hypotrichs).

2002 ***Australothrix alwinae*** – Lynn & Small, Phylum Ciliophora, p. 443, Fig. 6A, B (Fig. 138g, h; guide to ciliate genera).

Nomenclature: This species is named in honour of Hubert Blatterer's mother, Ms. Alwine Blatterer. Blatterer & Foissner (1988, p. 5) wrote that from each new species described in their paper one holotype slide each and 0–2 paratype slides have been depos-

[1] The diagnosis by Blatterer & Foissner (1988) is as follows: In vivo etwa 200–350 × 50–110 µm große *Australothrix* mit median verlagerter Midventralreihe und mehreren rechten (4–6) und linken (2–3) Ventral(Marginal)reihen. Durchschnittlich 53 adorale Membranellen, 115 Makronucleus-Teile und 6 Dorsalkineten. Subpelliculäre Granula farblos, in vivo 3–4 × 1,4 µm groß.

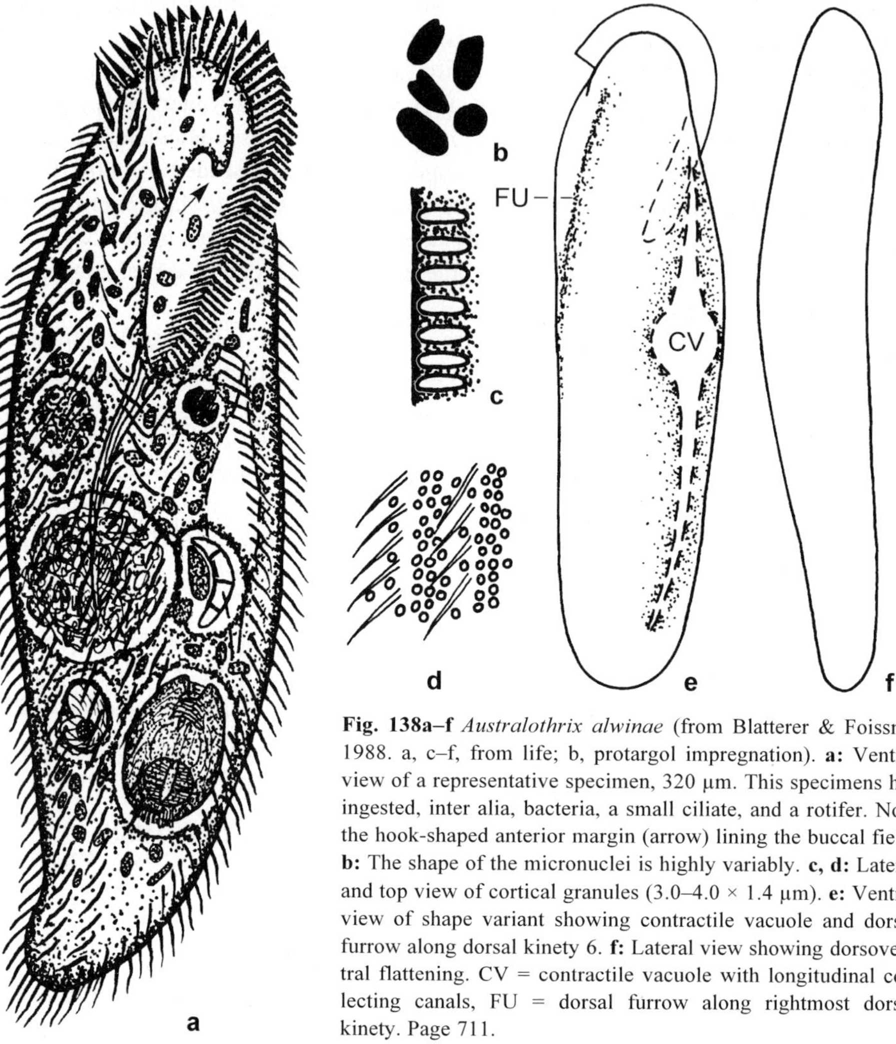

Fig. 138a–f *Australothrix alwinae* (from Blatterer & Foissner 1988. a, c–f, from life; b, protargol impregnation). **a:** Ventral view of a representative specimen, 320 µm. This specimens has ingested, inter alia, bacteria, a small ciliate, and a rotifer. Note the hook-shaped anterior margin (arrow) lining the buccal field. **b:** The shape of the micronuclei is highly variably. **c, d:** Lateral and top view of cortical granules (3.0–4.0 × 1.4 µm). **e:** Ventral view of shape variant showing contractile vacuole and dorsal furrow along dorsal kinety 6. **f:** Lateral view showing dorsoventral flattening. CV = contractile vacuole with longitudinal collecting canals, FU = dorsal furrow along rightmost dorsal kinety. Page 711.

ited in the Landesmuseum in Linz, Upper Austria. Aescht (2003, p. 379) designated these three slides as syntypes.

Remarks: *Australothrix alwinae* was described in detail both from life and after protargol impregnation by Blatterer & Foissner (1988). However, ontogenetic data are needed for a correct interpretation of the cirral pattern, which differs from that of the type species mainly in the position of the midventral complex, namely near cell midline against near right body margin (for details, see remarks in the genus section).

For separation from congeners, see key. In life, *Australothrix alwinae* is best recognised by the soil habitat, the size (200–350 × 50–110 µm), the scattered cortical granules (against around cirri and dorsal bristles in the type species), the three frontal cirri, the two or three left marginal rows, and the scattered macronuclear nodules.

Morphology: Body size about 200–350 × 50–110 μm in life. Body length:width ratio around 4:1 in life and 3.3:1 on average in protargol preparations. Body pisciform, rarely elongate elliptical, narrowed posteriorly, anterior end broadly, posterior narrowly rounded; very flexible, ventral side flat, dorsal convex, dorsoventrally flattened by about 2:1; right anteriorly distinctly vaulted along dorsal kinety causing conspicuous oblique furrow (Fig. 138a, e, f, Table 31). Macronuclear nodules scattered, globular to ellipsoidal (length:width ratio 1.7:1 on average in protargol preparations), contain many small nucleoli. Micronuclei also scattered, globular to ellipsoidal, sometimes notched at one end (Fig. 138b, i). Contractile vacuole near mid-body at left cell margin, with two very long collecting canals. Cortical granules moderately closely spaced throughout cortex, narrowly spaced in oral region; individual granules 3.0–4.0 × 1.4 μm, colourless, stain intensely with methyl green-pyronin (Fig. 138c, d). Cytoplasm colourless, packed with greasily shining globules 0.5–1.0 μm across, so that cells appear brownish at low magnification. Movement without peculiarities, that is, glides moderately rapidly on microscope slide and soil particles.

Adoral zone occupies 32% of body length on average, proximal portion slightly spatula-shaped, composed of an average of 53 membranelles, bases of largest membranelles 11 μm wide in life with cilia up to 15 μm long (Fig. 138a, g, j, Table 31). Buccal field moderately wide and only slightly deepened; anterior margin hook-shaped (Fig. 138a). Undulating membranes long and curved, intersect optically about at level of buccal cirrus, likely composed of two rows of basal bodies. Pharynx extends obliquely backwards, posterior portion with distinct rods (Fig. 138g, j).

Cirral pattern of usual variability, number of cirri in several cirral rows rather strongly varying (Fig. 138a, g, Table 31). Frontal cirri about 20 μm long, others about 15 μm. Three enlarged frontal cirri in slightly oblique row and a single, not distinctly enlarged cirrus (cirrus III/2) behind right frontal cirrus. Invariably one buccal cirrus at optical intersection of undulating membranes. Six of 15 specimens with a cirrus (likely cirrus III/2; see Fig. 138g) close to buccal cirrus. Frontoterminal cirri probably lacking; however, this has to be checked by ontogenetic data. Midventral complex composed of 13 pairs on average and at least one midventral row; midventral pair portion sigmoidal, extends near cell midline from distal end of adoral zone of membranelles to mid-body (terminates at 51% of body length in specimen shown in Fig. 138g); cirri of anterior midventral pairs, buccal cirrus, and cirrus III/2 larger than cirri of rear pairs and (mid)ventral and marginal rows. Behind midventral pairs a short, straight cirral row (Fig. 138g, MV); likely this is the leftmost midventral row in this species. 2–4 shortened cirral rows right of leftmost midventral row (Fig. 138g); as mentioned above, ontogenetic data are needed for a correct interpretation of the cirral rows. Transverse cirri lacking. Two or three right marginal rows commencing near distal end of adoral zone and ending almost terminally. Two or three left marginal rows, distal one commences rather far anteriorly, proximal begins near buccal vertex; all left rows terminate at rear end in cell midline, right and left marginal rows thus not overlapping posteriorly.

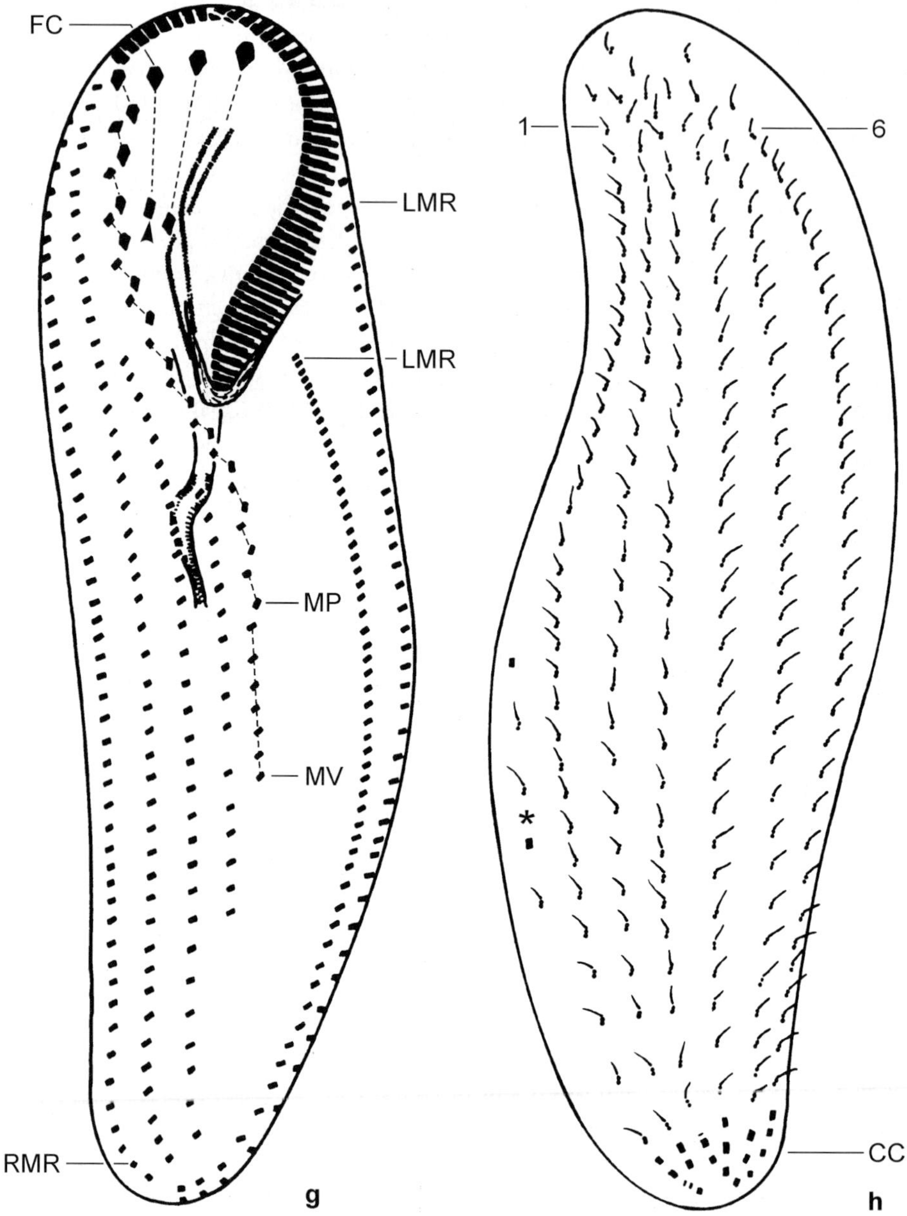

Fig. 138g, h *Australothrix alwinae* (from Blatterer & Foissner 1988. Foissner's protargol impregnation method). **g:** Infraciliature of ventral side, 196 µm (nuclear apparatus of this specimen, see Fig. 138i). Broken lines connect cirri very likely originating from same anlage; in this specimen, the cirrus (III/2; arrowhead) behind the right frontal cirrus is very likely dislocated posteriorly to near the buccal cirrus. **h:** Infraciliature of dorsal side, 270 µm. Asterisk marks a row with bristles and cirri. CC = caudal cirri, FC = right frontal cirrus, LMR = inner and outer left marginal row, MP = posteriormost midventral pair, MV = midventral row behind cirral pairs (= leftmost ventral row), RMR = outer right marginal row, 1, 6 = dorsal kineties. Page 711.

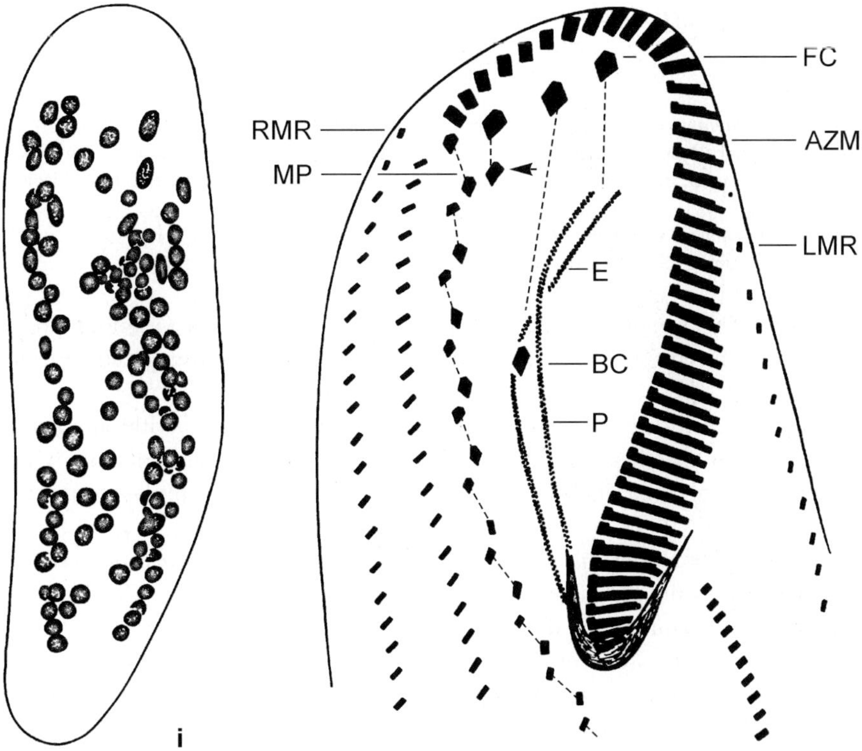

Fig. 138i, j *Australothrix alwinae* (from Blatterer & Foissner 1988. i, Foissner's protargol impregnation method; j, Wilbert's protargol impregnation method). **i:** Nuclear apparatus, 196 µm. Some of the smaller nodules are micronuclei (cp. Fig. 138b). **j:** Infraciliature of oral region, length of adoral zone about 77 µm. Arrow marks cirrus behind right frontal cirrus which is more anteriorly than in the specimen shown in Figure 138g. Broken lines connect cirri very likely originating from same anlage. AZM = adoral zone of membranelles, BC = buccal cirrus, E = endoral, FC = left frontal cirrus, LMR = outer left marginal row, MP = anteriormost midventral pair, P = paroral, RMR = outer right marginal row. Page 711.

Dorsal cilia about 5 µm long in life, arranged in five or six almost bipolar kineties. Rarely an additional row composed of few bristles and cirri. Several caudal cirri at end of each kinety (Fig. 138h, Table 31).

Occurrence and ecology: Terrestrial. Type location of *Australothrix alwinae* is the upper soil layer (0–5 cm with many litter and brown sand) of a coastal forest in the Royal National Park, south of Sydney, Australia. This site is about 100 m above sea-level and the sample, which was collected by H. Blatterer on October 18, 1986, had a pH of 4.5 (Blatterer & Foissner 1988). The abundance in the raw culture was moderate. No records published since then. According to Foissner (1998), *Australothrix alwinae* probably strongly prefers soil habitats.

Polyphagous, that is, feeds on bacteria, hypha, heterotrophic flagellates, testate amoebae (*Trinema lineare*), ciliates, and even rotifers (Blatterer & Foissner 1988). Biomass of 10^6 specimens about 1200 mg (Foissner 1998).

Australothrix steineri Foissner, 1995
(Fig. 139a–h, Table 31)

1995 ***Australothrix steineri* nov. spec.**[1] – Foissner, Arch. Protistenk., 145: 66, Figs. 103–109, Tab. 7 (Fig. 139a–h; original description. 1 holotype slide [registration number 1997/90] and 1 paratype slide [1997/91] each with protargol-impregnated specimens are deposited in the Oberösterreichische Landesmuseum in Linz [LI], Austria).

2001 ***Australothrix steineri* Foissner, 1995** – Berger, Catalogue of ciliate names 1. Hypotrichs, p. 11 (nomenclator containing all basionyms, combinations, and higher taxa of hypotrichs).

Nomenclature: This species was named in honour of Erich Steiner (Vienna), president of the "Mikrographische Gesellschaft Wien", a great lover of microscopy and a friend of W. Foissner.

Remarks: *Australothrix steineri* is an almost worm-like species with only few midventral pairs. As in all other congeners, ontogenetic data are needed to show which of the five cirral rows are marginals and which originate from frontal-midventral cirral anlagen.

For separation from other *Australothrix* species, see key. In life, *Australothrix steineri* is best recognised by the soil habitat, the worm-like body (160–320 × 20–30 µm), the many macronuclear nodules, and the 4–6 cirral rows.

Morphology: Body size 160–320 × 20–30 µm in life; length:width ratio about 9–12:1 both in life and in protargol preparations indicating that it is acontractile. Body very slender and flexible, often slightly sigmoidal and contorted along major axis; widest in oral area, postorally gradually narrowed with posterior end pointed or even elongated tail-like; slightly flattened dorsoventrally (Fig. 139a, b, Table 31). About 70–150 macronuclear nodules, exact number difficult to ascertain because of similar sized and stained food vacuoles; individual nodules usually ellipsoidal, sometimes globular or reniform; nucleoli small. Several compact micronuclei 4.0 × 2.5 µm in life, weakly stained by protargol. Contractile vacuole in mid-body at left cell margin, with two long collecting canals extending to near anterior and posterior end of cell. Cortex and cytoplasm colourless, no cortical granules or cytoplasmic crystals; posterior cell portion often containing shiny fat droplets 1–3 µm across. Food vacuoles 5–10 µm in diameter.

Adoral zone occupies only about 13% of body length on average, of usual shape and fine-structure, composed of an average of 25 membranelles, bases of largest membranelles about 5 µm wide in life (Fig. 139a, c, f). Buccal cavity flat, narrow, and short;

Fig. 139a–e *Australothrix steineri* (from Foissner 1995. a, b, from life; c–e, protargol impregnation). **a:** Ventral view of a representative, slightly twisted specimen, 250 µm. **b:** Dorsal view of a broad specimen showing contractile vacuole with collecting canals. **c–e:** Infraciliature of ventral and dorsal side and nuclear apparatus of same specimen, 190 µm. For detailed labelling of frontal ciliature, see Figure 139f. Arrow marks very short, leftmost (mid)ventral row. CC = caudal cirri, LMR = left marginal row, RMR = right marginal row, 1, 4 = dorsal kineties. Page 716. →

[1] The diagnosis by Foissner (1995) is as follows: Size in vivo 160–320 × 20–30 µm. Many macronuclear nodules. 25 adoral membranelles, 5 ventral cirral rows and 4 dorsal kineties on average. Midventral row slightly shorter than adoral zone of membranelles, consists of 2–3 cirral pairs, continues as ventral row.

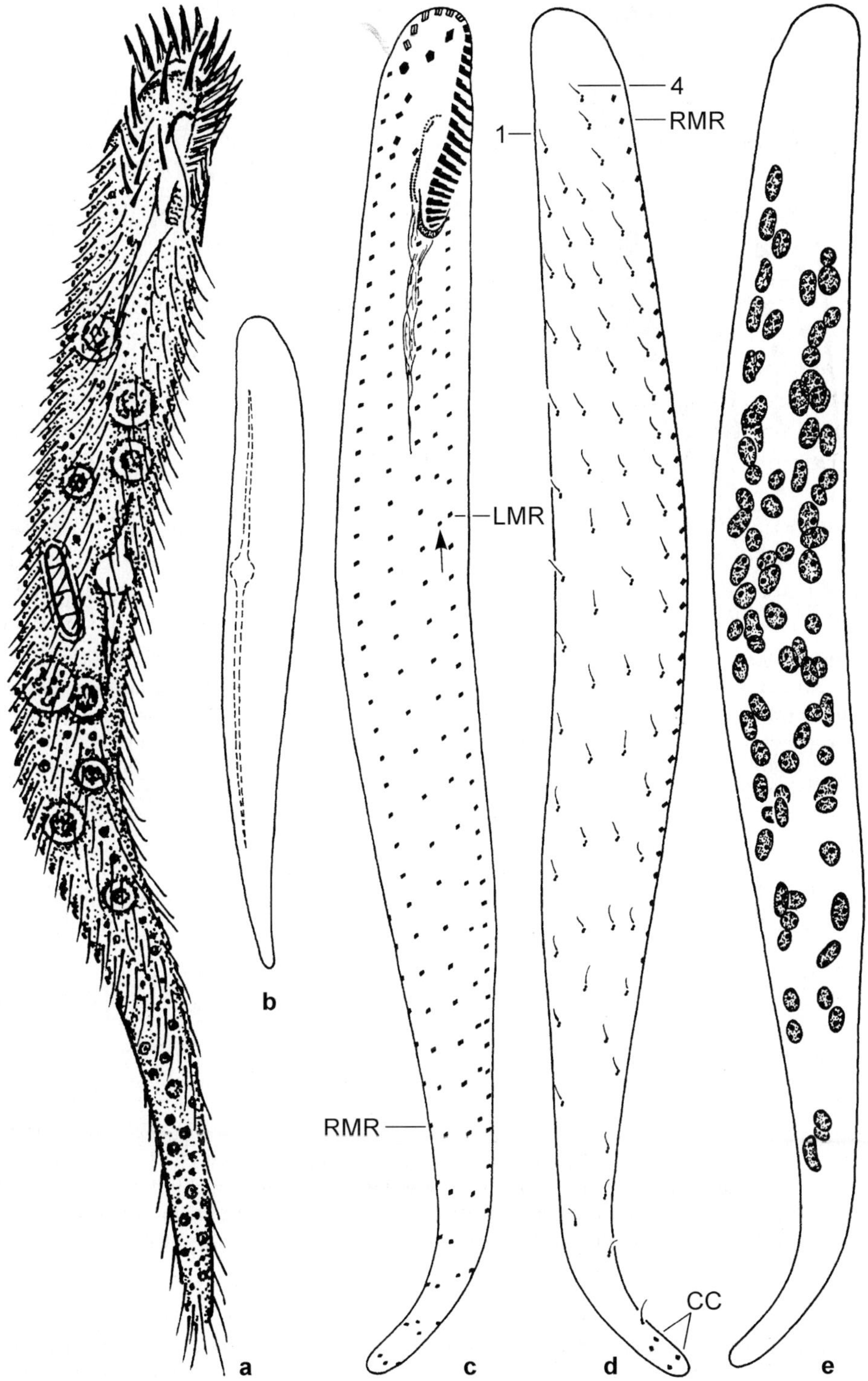
4
RMR
1
LMR
RMR
CC
a
b
c
d
e

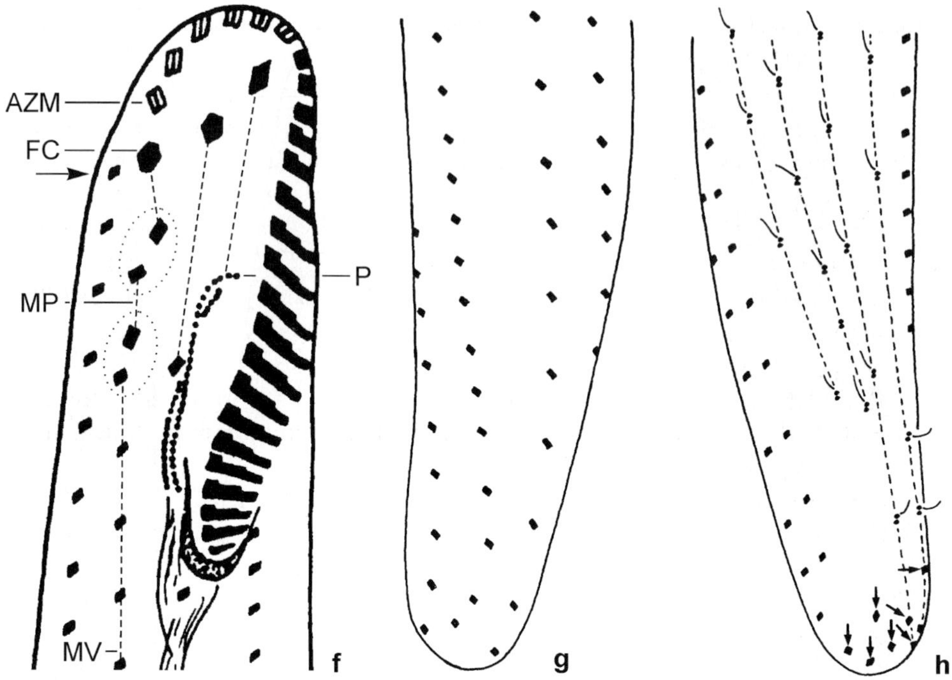

Fig. 139f–h *Australothrix steineri* after protargol impregnation (from Foissner 1995). **f:** Infraciliature of oral region (total view, see Figure 139c). Broken lines connect cirri very likely originating from same anlage; ellipses mark cirral pairs (that is, pseudopairs) as counted by Foissner (1995); arrow denotes first cirrus of rightmost ventral row. **g, h:** Infraciliature of ventral and dorsal side of the posterior body portion (arrows mark caudal cirri). AZM = distal end of adoral zone of membranelles, FC = right frontal cirrus, MP = midventral pair (this is likely the sole one in this specimen), MV = midventral row, P = paroral. Page 716.

buccal lip inconspicuous. Undulating membranes of about same length and distinctly curved near level of buccal cirrus, one upon the other. Pharyngeal fibres distinct after protargol impregnation.

Cirral pattern and number of cirri of usual variability, except for rather strongly varying number of marginal cirri (Fig. 139a, c, f, g, Table 31). All cirri only 8–10 μm long, which is rather short for such a large species. Invariably three enlarged frontal cirri in an oblique row. One cirrus (cirrus III/2) behind right frontal cirrus; likely, cirrus III/2 was considered a part of the anterior midventral pair in the original description. Buccal cirrus about at summit of paroral. Frontoterminal cirri likely lacking; however, this has to be checked by ontogenetic data. Midventral complex composed of two or three cirral pairs[1] only; zigzag-pattern thus very short and inconspicuous, arranged in area between right frontal cirrus and third cirral row from right; cirri of midventral pair(s) of about same size and slightly larger than (mid)ventral and marginal cirri. On

[1] Foissner (1995) likely counted pseudopairs as marked by dotted ellipses in Figure 139f. However, in fact only one (or two) midventral pair(s) is (are) present (Fig. 139f, MP).

average five (mid)ventral and marginal cirral rows on dorsolateral and ventral side; rightmost ventral row (possibly a marginal row; Fig. 139f, arrow) as long as right marginal row; next row commences underneath last midventral pair and is therefore likely a midventral row (Fig. 139f, MV); other ventral rows begin postorally; leftmost ventral row (Fig. 139c, arrow) distinctly shortened, does not extend into posterior half of cell. Distance between individual cirri of (mid)ventral rows increases from anterior to posterior. Transverse cirri lacking. Right marginal row commences near level of right frontal cirrus, extends, like left row, to pointed posterior end of cell. As in congeners, ontogenetic data are needed for a correct interpretation (midventrals, ventrals, marginals?) of the cirral rows.

Dorsal cilia about 3 μm long, arranged in four kineties, which are slightly shortened anteriorly and posteriorly. About 4–8 caudal cirri, exact number difficult to recognise because located at pointed posterior body end; it is unknown from which dorsal kinety(ies) they originate (Fig. 139d, h, Table 31).

Occurrence and ecology: Terrestrial. Type location of *Australothrix steineri* is the upper soil layer of the bank of Rio Corobici at the hacienda "La Pacifica" (Centre Ecologia La Pacifica) near the town of Canas in Costa Rica (about 10°27'N 85°08'W). This site is in the flood area. Foissner (1995) found it also in the upper (0–3 μm) litter and soil layer of a tropical dry forest in the Santa Rosa National Park in Costa Rica. Foissner (1997) recorded it about 40 km west of Manaus (Brazil) on the Anavilhanas archipelago (vicinity of Ariau lodge) in the Rio Negro. This site is a blackwater inundation primary(?) rain forest flooded during high-water periods. The sample was composed of litter, soil, and roots from 0–8 cm with the litter layer up to 5 cm thick, followed by a 3–5 cm thick root-carpet mixed with brown, humic soil (pH 5.1). The records from mainly terrestrial habitats and the very slender body indicate that *Australothrix steineri* is a true soil inhabitant (Foissner 1995, 1998), possibly confined to the neotropical region.

Feeds on fungal spores and, possibly, bacteria and heterotrophic flagellates (Foissner 1995). Biomass of 10^6 specimens about 81 mg (Foissner 1998).

Australothrix simplex Liu, Shen, Song & Gu, 1992 (Fig. 140a, b)

1992 ***Australothrix simplex* n. sp. Liu et al.** – Shen, Liu, Song & Gu, Protozoa, p. 150, Fig. 2–23Ba, Bb (Fig. 140a, b; original description in Chinese. The type slides are likely deposited in the Institute of Hydrobiology of the Academia Sinica in Wuhan, China).

2001 ***Australothrix simplex* Liu et al., 1992** – Berger, Catalogue of ciliate names 1. Hypotrichs, p. 11 (nomenclator containing all basionyms, combinations, and higher taxa of hypotrichs; authorship not totally correct).

Nomenclature: The name *simplex* (Latin adjective; simple) probably refers to the single long (mid)ventral cirral row making the cirral pattern rather simple as compared to, for example, *Australothrix australis* and *A. alwinae*.

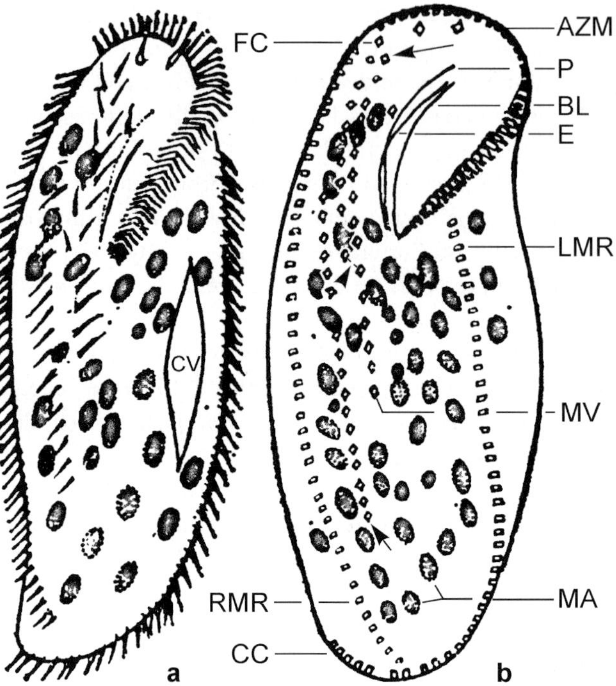

Fig. 140a, b *Australothrix simplex* (from Shen et al. 1992. a, from life?; b, protargol impregnation). **a:** Ventral view, size not indicated. **b:** Infraciliature of ventral side and nuclear apparatus, size not indicated. Long arrow marks cirrus (III/2) behind right frontal cirrus, short arrow denotes rear end of ventral row right of midventral complex. Arrowhead likely marks posteriormost midventral pair. AZM = adoral zone of membranelles, BL = buccal lip, CC = caudal cirri, CV = contractile vacuole, E = endoral, FC = right frontal cirrus, LMR = left marginal row, MA = macronuclear nodules, MV = midventral row, P = paroral, RMR = right marginal row. Page 719.

The headline in the original description indicates that Liu is the senior author of this species. Unfortunately, the order of the other authors (et al.) was not determined. However, it seems justified to arrange them as for the full paper, namely Shen, (Liu), Song, and Gu. Thus, the species should be cited in future as follows: *Australothrix simplex* Liu, Shen, Song & Gu, 1992. By contrast, Foissner (1998, p. 199) ignored the senior position of Liu and assigned the species to Shen, Liu, Song and Gu. In the catalogue of ciliate names, I forgot to give the full authorship (Berger 2001).

Remarks: The original description is in Chinese and lacks a formal diagnosis; however, one of the authors (Shen Yunfen) translated the description (see below) for Wilhelm Foissner. The original description is likely only based on data from protargol-impregnated specimens. Further, the morphometric data are sparse and the illustrations are not detailed. Thus, a comprehensive redescription – containing live data, a morphometric analysis, and clear line drawings of the ventral and dorsal infraciliature – is needed. Ontogenetic data are necessary to know whether the second cirral row from right (Fig. 140b, short arrow) is a marginal row or a midventral. In life, *Australothrix simplex* is best recognised by the soil habitat, the scattered macronuclear nodules, the four cirral rows, and the lack of transverse cirri.

Morphology: Body size 116–187 × 49–77 μm (likely from protargol preparations). Body length:width ratio about 3:1 in life (estimated from Fig. 140a). Body flexible, leaf-shaped, widest near mid-body, anterior end rounded, posterior slightly pointed. Macronuclear nodules scattered, usually ellipsoidal, according to Figure 140b about

40–50 nodules are present. Presence/absence of cortical granules not described. Adoral zone occupies one third or slightly more of body length, probably of usual shape and structure, composed of about 38 membranelles. Buccal field large, at least in protargol preparations. Undulating membranes obviously long and curved with one buccal cirrus near end of first third of membranes. Left and middle frontal cirrus distinctly enlarged. Right frontal cirrus and the one behind (= cirrus III/2) arranged more or less as in congeners. Frontoterminal cirri not shown and thus likely lacking as in congeners. Midventral complex composed of about 11 "right" and about 16 "left" cirri, that is, the zigzag-pattern is formed by about 11 midventral pairs and terminates at level of buccal vertex; a short midventral row extends behind left cirri of zigzagging pattern to near mid-body (Fig. 140b). Right of midventral complex a further cirral row with about 19 cirri extends from slightly ahead of buccal vertex to posterior quarter of cell (Fig. 140b, arrow). Transverse cirri absent. Right marginal row composed of about 47 cirri, commences near distal end of adoral zone and terminates ahead of rear end of left row, which is composed of about 49 cirri and extends in J-shape along body margin. Five dorsal kineties and 4–5 caudal cirri.

Occurrence: Type locality of *Australothrix simplex* is a forest soil (sample collected in November) in the Huang San Mountains (about 30°10'N 114°40'E) in Hunan Province, China (Shen et al. 1992). No records published since then. Likely a true soil inhabitant (Foissner 1998). Biomass of 10^6 specimens about 270 mg (Foissner 1998).

Australothrix zignis (Entz, 1884) Blatterer & Foissner, 1988 (Fig. 141a–d)

1884 ***Uroleptus zignis* n. sp.** – Entz, Mitt. zool. Stn Neapel, 5: 373, Tafel 23 Fig. 14–16 (Fig. 141a–c; original description. No type material available and no formal diagnosis provided).

1884 ***Urostyla zignis*** – Entz, Mitt. zool. Stn Neapel, 5: 443, Legend to Tafel 23, Fig. 14–16 (likely unintentional assignment to *Urostyla*; see nomenclature).

1932 ***Uroleptus zignis* Entz sen., 1884** – Kahl, Tierwelt Dtl., 25: 547, Fig. 101_{13} (Fig. 141d [redrawn from Entz 1884]; first reviser).

1972 ***Kahliella zignis* (Entz, 1884) n. comb.** – Borror, J. Protozool., 19: 9 (combination with *Kahliella* Corliss, 1960).

1988 ***Australothrix zignis* (Entz, 1884) nov. comb.** – Blatterer & Foissner, Stapfia, 17: 39 (combination with *Australothrix*).

2001 ***Australothrix zignis* (Entz, 1884) Blatterer and Foissner, 1988** – Berger, Catalogue of ciliate names 1. Hypotrichs, p. 99 (nomenclator containing all basionyms, combinations, and higher taxa of hypotrichs).

Nomenclature: The name *ζιγνις* (Greek) refers to a lizard-like animal in Aristotle (Entz 1884, p. 373, footnote 1). "*Urostyla zignis*" in the figure legend of the original description is very likely an unintentional combination which is not listed in my nomenclator (Berger 2001); the same name ("*Urostyla zignis* Entz, 1884") was used by Hemberger (1982, p. 278), who did not treat this species in detail. *Uroleptus zignus* in Kopylov (1979, p. 586) is an incorrect subsequent spelling.

Remarks: The original description and illustrations are of rather high quality and are the only available data about this long-known species because Kahl (1932) did not provide own observations. However, as discussed below there exists an uncertainty about the nuclear apparatus. Thus, a redescription, including detailed life observations and morphometric analysis of protargol-impregnated specimens, is strongly recommended. Such an analysis will show above all whether or not the present species is an urostyloid, that is, a species with a midventral complex.

Gruber (1888, p. 62) found a strong population in the Mediterranean Sea in Genoa (Italy) which matched the population by Entz (1884) completely, except for the macronuclear apparatus, which was multi-nodular in Gruber's population. However, no name and illustration were given by Gruber. In spite of this, Gruber's observation is very interesting because from *Pseudokeronopsis rubra* and *P. flava* we know that Entz (1884) misobserved the nuclear apparatus. Interestingly, Entz (incorrectly) described only two nodules for these multimacronuclear species. Thus, it cannot be excluded that *Australothrix zignis* really has not two but many macronucleus nodules.

The illustration by Biernacka (1963, p. 48) is insufficient and thus the record from the Baltic is not listed in the ecology section (Fig. 143a, b). Borror (1972) transferred *Uroleptus zignis* to *Kahliella* Corliss, 1960. He characterised *Kahliella*, in the absence of ontogenetic data, as Urostylidae with "Cirri in 7–10 ventral rows" although *U. zignis* has only six cirral rows. Later, he did not include the present species in the Urostylidae (Borror & Wicklow 1983).

Australothrix zignis is best recognised by the marine habitat, the size (300 × 60 µm), the two macronuclear nodules (however, see previous paragraph), the six cirral rows, and especially by the reddish colour which is due to the cortical granules. *Pseudokeronopsis rubra*, which also lives in the sea and has a similar size and general appearance (including the reddish colour), has two curved rows of frontal cirri (against three distinct frontal cirri), only one right marginal row, and one left marginal row, distinct transverse cirri, and many (about 100–200) macronuclear nodules which do not fuse during ontogenesis. *Metaurostylopsis rubra* has much more cirral rows and very many macronuclear nodules (Fig. 133).

Morphology: Body size around 300 × 60 µm in life when moderately extended; body length:width ratio thus around 5:1. Body elongate spindle-shaped, that is, widest near mid-body, narrowed anteriorly, and converging posteriorly into a rather long, slender, pointed or truncated tail; usually sigmoidal, very contractile and outline therefore rather variable, especially the tail, which varies from short to long; ventral side plane, dorsal vaulted. Two large macronuclear nodules one after the other left of midline (see remarks). Contractile vacuole near mid-body at left cell margin, with an anterior and posterior collecting canal originating from small vesicles. Cytoplasm slightly dirty-brownish, often contains scattered, irregular-shaped, blood-red or brown patches of different size; anterior end of cell almost invariably with a red patch surrounded by highly refractive granules which appear dark at bright-field illumination. Same granules (likely cortical granules) form meridional, regularly arranged strips alternating with bright (non-pigmented) stripes; granules stripes converging on tail (likely because they run along or between the dorsal kineties).

Adoral zone occupies slightly more than 25% of body length in moderately extended specimens, extends far onto right body margin, composed of about 62 membranelles

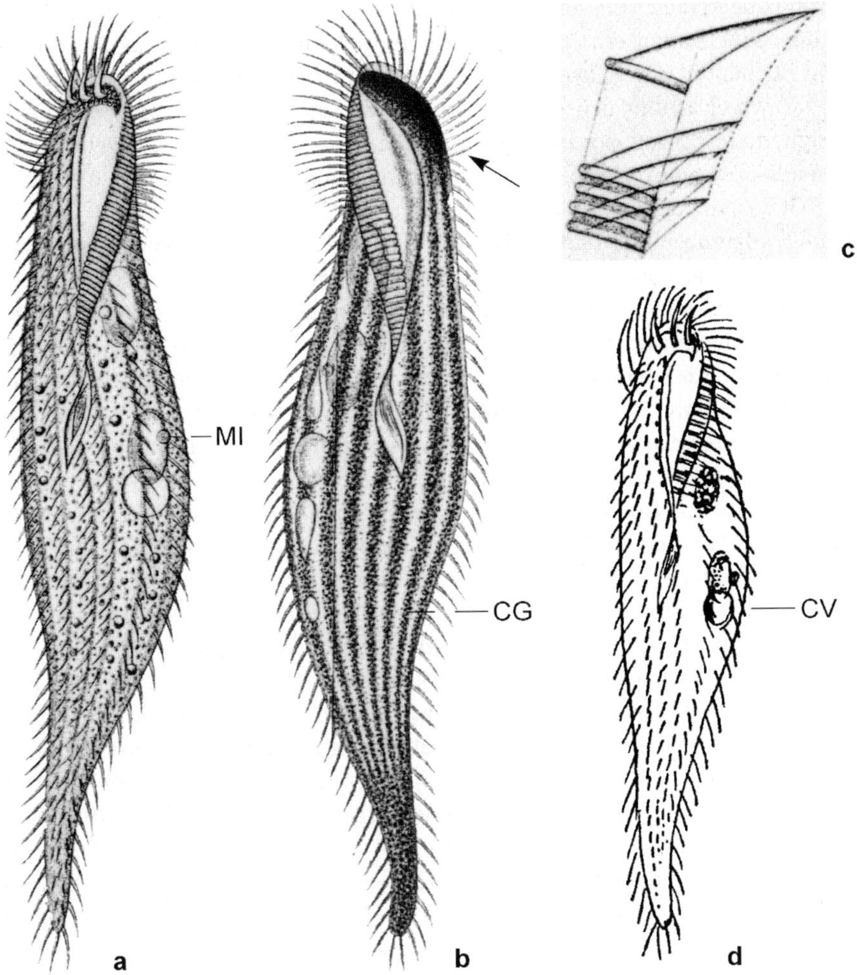

Fig. 141a–d *Australothrix zignis* (a–c, from Entz 1884; d, after Entz 1884 from Kahl 1932. a–d, from life). **a, d:** Ventral view of a representative specimen, size not indicated, species around 300 × 60 µm. **b:** Dorsal view showing contractile vacuole with lacunar collecting canals, adoral zone with cytopharynx, red patch at anterior cell end (arrow), and cortical granules strips (along or between?) dorsal bristle rows. **c:** Part of adoral zone. CG – cortical granules, CV = contractile vacuole, MI = micronucleus attached to rear macronuclear nodule. Page 721.

(number not mentioned by Entz 1884, but counted by myself from Figures 143a, b and thus only a rough estimation which must not be over-interpreted). Frontal scutum well developed. Buccal lip anteriorly curved, bears paroral, which is obviously composed of rather long cilia. Pharyngeal fibres curved, with a bundle of long cilia (likely from the endoral).

Cirral pattern rather constant, variability of cirral number within rows not described. Invariably three frontal cirri and six cirral rows. Median two rows likely form the zigzagging midventral pairs. Two left and two right marginal rows. All rows terminate at end of tail which is thus densely cirrated. All cirri rather strong, those of outer left and outer right marginal row become longer from anterior to posterior. Transverse cirri lacking. Details of ciliature, for example presence/absence of frontoterminal cirri and buccal cirrus, not described.

Dorsal ciliature not described by Entz (1884). Figure 143b shows seven bright and seven dark (due to cortical granules) stripes, indicating that around seven bristle rows are present. However, this is only a presumption which has to be checked by the redescription of protargol-impregnated specimens.

Occurrence and ecology: Type locality of *Australothrix zignis* is the littoral of the Gulf of Naples, Italy, where Entz (1884) discovered it together with *Pseudokeronopsis rubra* (as "*Holotricha flavorubra*"). Records not substantiated by illustrations and/or morphological data: coastal area of Black Sea, USSR (Kopylov 1979); benthal of Gulf of Riga, Baltic Sea (Boikova 1984); aufwuchs from Caspian Sea (Agamaliev 1974, p. 57). Held (1937) found *Australothrix zignis* during April very sparsely in the pelagial of the Heustadelwasser, a groundwater pond influenced by the Danube River in Vienna, Austria; however, very likely, Held did not see the marine *Australothrix zignis*, but, for example, the limnetic *Diaxonella pseudorubra*, which is also red.

According to Kopylov (1979), the carbon content of *Australothrix zignis* is 19% of wet mass and thus distinctly higher than that of other ciliates (for example, 10% in a marine *Euplotes*). The reason for the increased C-content is that *A. zignis* fed on diatoms which had accumulated oil and volutin or polyphosphate droplets (Kopylov 1979; Entz 1884 also described diatoms as main food). Kopylov (1979) estimated a biomass of 837 mg for 10^6 specimens and a caloric value of 2.04 cal mg^{-1} wet mass which, however, is a distinct overestimation as stated by Kopylov (1979) himself; the average for the other three species investigated was about 0.9 cal mg^{-1}.

Australothrix gibba (Claparède & Lachmann, 1858) Blatterer & Foissner, 1988 (Fig. 142a, b)

1838 ***Oxytricha gibba*** – Ehrenberg, Infusionsthierchen, p. 365, Tafel XLI, Fig. II (Fig. 142c–f; see remarks).

1858 ***Oxytricha gibba***[1] – Claparède & Lachmann, Mém. Inst. natn. génev., 5: 144, Planche V, Fig. 8 (Fig. 142a; original description. No type material available).

1859 ***Oxytricha gibba* von Claparede und Lachmann** – Stein, Organismus der Infusionsthiere I, p. 178 (short comment; see remarks).

1866 ***Oxytricha ehrenbergiana*** – Diesing, Sber. Akad. Wiss. Wien, 53: 95 (new species for *Oxytricha gibba* sensu Ehrenberg 1838; see remarks).

1882 ***Uroleptus gibba*, C. & L. sp.** – Kent, Manual Infusoria II, p. 780 (combination with *Uroleptus*, see nomenclature; first reviser).

[1] The diagnosis by Claparède & Lachmann (1858) is as follows: Seulement cinq rangées de pieds-cirrhes sur la face ventrale; pas de queue.

1932 ***Uroleptus (Oxytricha) gibbus*** (Clap. u. L., 1858) – Kahl, Tierwelt Dtl., 25: 547, Fig. 101_{14} (Fig. 142b, redrawn from Claparède & Lachmann 1858).

1972 ***Paraurostyla gibba*** **(Müller, 1786) n. comb.** – Borror, J. Protozool., 19: 10 (pro parte, see remarks).

1988 ***Australothrix gibba*** **(Claparéde & Lachmann, 1859) nov. comb.** – Blatterer & Foissner, Stapfia 17: 39 (combination with *Australothrix*; 1859 is an incorrect year).

1992 ***Paraurostyla gibba*** **(Kahl, 1930–5) Borror, 1972** – Carey, Marine interstitial ciliates, p. 177, Fig. 698 (redrawing of Fig. 142b and thus not shown; guide; incorrect author).

2001 ***Australothrix gibba*** **(Claparède and Lachmann, 1858) Blatterer and Foissner, 1988** – Berger, Catalogue of ciliate names 1. Hypotrichs, p. 55, 92 (nomenclator containing all basionyms, combinations, and higher taxa of hypotrichs).

Nomenclature: Claparède & Lachmann (1858) conserved Ehrenberg's (1838, p. 365) name *Oxytricha gibba*, which is a combination of *Trichoda gibba* Müller, 1786 (p. 179). Müller (1786) wrote in the Latin diagnosis "..., dorso gibbera, ...". Thus, the name *gibb·us -a -um* (Latin adjective; vaulted, humpbacked, convex) likely refers to the vaulted dorsal side.

Oxytricha and *Australothrix* are feminine (Aescht 2001). By contrast, *Uroleptus* is masculine so that Kent's (1882) spelling "*Uroleptus gibba*" is incorrect. Kahl (1932) emended this error, and in the headline he correctly changed the species-group name to *gibbus*. In the last sentence of the description Kahl wrote "... mit *Oxytricha gibbus* Ehrenb.", which is also incorrect. The misleading spelling *Uroleptus (Oxytricha) gibbus* in Kahl (1932) does not mean that he considered *Oxytricha* as subgenus of *Uroleptus*, but should indicate that the species was classified in *Oxytricha* previously.

Remarks: The systematics of this species is very complicated. Müller (1786, p. 179) described a *Trichoda gibba* from the sea[1]. Ehrenberg (1838, p. 365), who supposed synonymy with the marine[1] *Trichoda foetida* Müller, 1786, redescribed this species, under the name *Oxytricha gibba*, from freshwater in the surroundings of Berlin, Germany. Claparède & Lachmann (1858) described a species resembling Ehrenberg's (1838) population and thus conserved the name used by Ehrenberg. Hence, it is uncertain whether their description is a redescription of *Trichoda gibba* Müller or an original description with *Oxytricha gibba* Claparède & Lachmann, 1858 as basionym (Berger 2001, p. 55, 92). Now I consider, for the sake of simplicity, Claparède & Lachmann's paper as original description as indicated (i) by the lack of an author in the headline ("*Oxytricha gibba*" against, for example, "*Oxytricha pellionella*. Ehr." on page 145) and (ii) by the presence of a formal diagnosis as in new species established by Claparède & Lachmann (1858). Stein (1859, p. 178, 184), who redescribed *Oxytricha gibba* from the harbour of Travemünde (Baltic Sea), doubted the identifications by Ehrenberg (1838) and Claparède & Lachmann (1858).

Both the original description of *Trichoda gibba* by Müller (1786) and the "redescription" by Ehrenberg (1838) are insufficient so that an identification will always be arbitrarily. Thus, it seems proper to consider Stein's (1859, p. 184) data as redescription

[1] Müller (1786, p. 179) found *Trichoda gibba* "In aqua littorali passim" and *T. foeta* (p. 180) "In aqua marina sub medium ...". Ehrenberg (1838, p. 366) wrote that Müller found his little animals in brackish coastal water from the Baltic, and a similar in fresh water. Stein (1859, p. 184) also stated that Müller found his *Trichoda gibba* common in sea water, and Kahl (1932, p. 583) mentioned only marine records at *Holosticha gibba* (Müller, 1786).

of the marine species which is now the type species of *Holosticha* (see there). By contrast, the freshwater species described by Ehrenberg (1838) and Claparède & Lachmann (1858) from near Berlin is assigned to Claparède & Lachmann because they provided a rather good illustration and a detailed description, which can also be interpreted as an original description. Furthermore, the marine and limnetic species were not established in the same genus so that they are no primary homonyms.

Stein (1859) stated that the *Oxytricha gibba* of Claparède & Lachmann shows great resemblance with extended, short-tailed forms of *Uroleptus musculus* (Müller, 1773) as described by him. He was uncertain regarding the six cirral rows characterising Claparède & Lachmann's population and proposed two explanations: (i) Claparède & Lachmann overestimated the number of cirral rows, that is, the present species has not six but only four rows, namely two marginal rows and two ventral rows (= one row of midventral pairs). Thus, their population would belong to *Uroleptus*, respectively, would be a synonym of *Uroleptus musculus* sensu Stein. (ii) If in fact six cirral rows are present, then Claparède & Lachmann must have overlooked transverse cirri which would assign the species to *Urostyla*.

Diesing (1866b) proposed a new name for *Oxytricha gibba* sensu Ehrenberg (1838), namely *Oxytricha ehrenbergiana* (see also Diesing 1850, p. 157, 644 and Berger 1999, p. 247). However, because of the solution proposed above, this name is not necessary now.

Kent (1882) transferred *Oxytricha gibba* to *Uroleptus* – an act overlooked by Berger (2001) – because Stein (1859, p. 178) referred to Claparède & Lachmann's species as possibly being identical with *Uroleptus musculus* (see above). However, Kent did not doubt Claparède & Lachmann's observation and considered the different number of cirral rows as sufficient difference.

Kahl (1932) also accepted Claparède & Lachmann's species and separated it from "*Uroleptus zignis*" (now *Australothrix zignis*) by the body size (100–130 μm vs. 300 μm) and contractility (lacking vs. present), and the habitat (freshwater vs. sea).

Borror (1972, p. 10) synonymised *Trichoda gibba* Müller, 1786 (marine; now type species of *Holosticha*) and *Oxytricha gibba* Claparède & Lachmann, 1858 (limnetic; present species), and simultaneously transferred it to *Paraurostyla* Borror, 1972. A comparison of Claparède & Lachmann's (1858) and Stein's (1859) descriptions shows that synonymy is very unlikely. Furthermore, if we accept Stein's report as redescription of Müller's marine species, it seems more reliable to classify it in *Holosticha* than in *Paraurostyla* which is an oxytrichid (Berger 1999, p. 841).

The original description of *Australothrix gibba* by Claparède & Lachmann (1858) is the sole illustrated and thus serious record of this long known species. This could be due to several reasons: (i) it is a very rare species; (ii) it has a well-known synonym, but nobody recognised the identity; (iii) the original description is, as suggested by Stein (1859; see above), misleading (for example, number of cirral rows not correctly counted or transverse cirri overlooked), making an identification impossible. I do not know which point applies. By contrast, there is a considerable number of records not substantiated by morphological data and/or illustrations. Surprisingly, the most recent

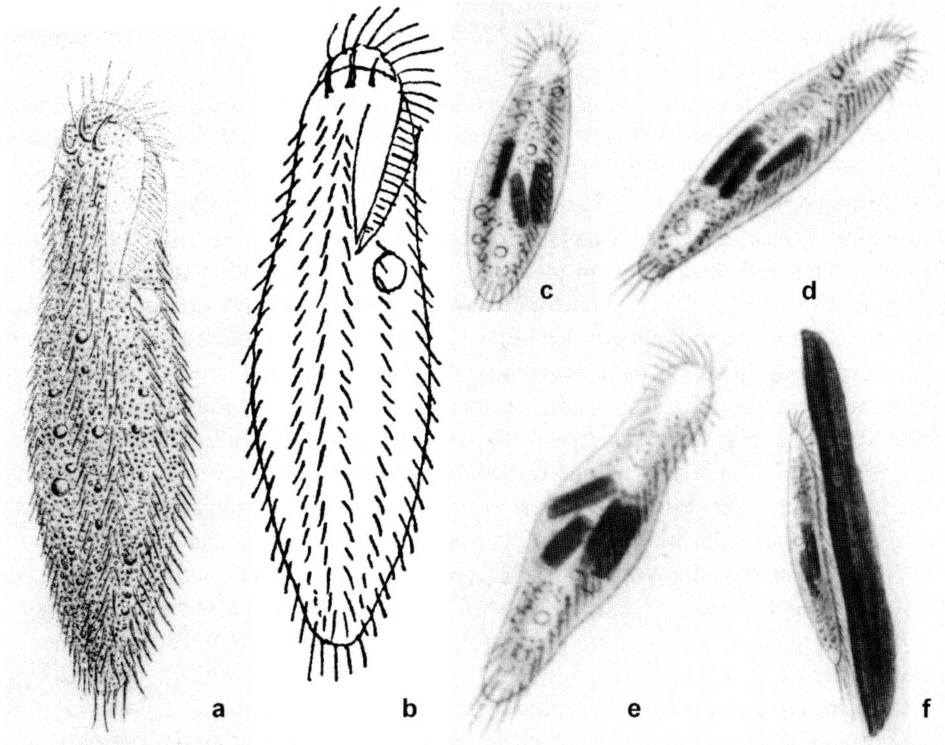

Fig. 142a–f *Australothrix gibba* (a, from Claparède & Lachmann 1858; b, after Claparède & Lachmann 1858 from Kahl 1932) and *Oxytricha gibba* sensu Ehrenberg (1838; c–f) from life. **a, b:** Ventral view, 100–130 μm, according to Kahl this specimen is 120 μm long. Note that this species has six cirral rows, the middle two likely forming a midventral complex. However, this has to be confirmed by a detailed redescription. **c–f:** Shape variants in dorsal/ventral views and lateral view of Ehrenberg's *Oxytricha gibba* population, size around 105 μm. Synonymy with Claparède & Lachmann's population, of course, somewhat arbitrarily (see remarks). Page 724.

record is about 30 years ago, where Haneda (1971) found it in limnetic habitats from the Antarctica. I suppose that many records are misidentifications.

Australothrix gibba is characterised by the limnetic habitat, the small size (100–130 μm vs. more than 200 μm in the congeners, except for the terrestrial *A. simplex* which is also below 150 μm), the six cirral rows, the three frontal cirri, and the lack of transverse cirri. The marine *Holosticha gibba* is, inter alia, larger (around 170 μm) and has prominent transverse cirri and only two marginal rows and a row of midventral pairs (that is, in total four narrowly spaced cirral rows).

Morphology: Body length about 100–130 μm, length:width ratio about 4:1 (from Fig. 142a). Body outline elongate elliptical. Two macronuclear nodules mentioned in text of original description, but no shown in Fig. 142a; according to Kahl (1932), number of macronuclear nodules not known, whereas Blatterer & Foissner (1988,

p. 44) also mention two nodules in their species comparison. Contractile vacuole about at level of buccal vertex at left cell margin. No cortical granules or coloration of cytoplasm mentioned.

Adoral zone of membranelles occupies about 33% of body length (estimated from Fig. 142a). Buccal field obviously of usual shape and size. No further details on oral apparatus known. Three distinctly enlarged frontal cirri. Buccal cirrus neither mentioned nor illustrated, but probably present. In total six cirral rows on ventral side and not five as mentioned in original diagnosis (see footnote at list of synonyms); in the description, however, Claparède & Lachmann (1858, p. 145) correctly mention six (two marginal, four ventral) cirral rows. But as in congeners, ontogenetic data are needed for a correct interpretation of the cirral pattern. Rightmost cirral row extends close to right body margin, commences, like next three rows, near distal end of adoral zone of membranelles and terminates, like all rows, at/near rear body end. Third and fourth row from right rather close together, indicating that it is a row of midventral pairs. Leftmost row commences rather far anteriorly as in *Australothrix australis* and *A. alwinae*; second row from left begins about at level of buccal vertex. Transverse cirri very likely lacking; the "w" in the original illustration marks the cytopyge (anus; see Claparède & Lachmann 1859, p. 477) and not transverse cirri. No details about the dorsal ciliature known, but very likely the bristles are short because they did not illustrate them as they did in *Tachysoma pellionellum* (Müller, 1773) Borror, 1972, which has 8–15 μm long dorsal cilia (Berger 1999, p. 433).

Occurrence and ecology: All limnetic records of *Trichoda gibba*, *Oxytricha gibba*, and *Uroleptus gibbus* which are not substantiated by illustrations and/or morphological data are listed here at *Australothrix gibba* and not under the marine *Holosticha gibba*, irrespective of the authorship (Müller, Ehrenberg, Stein, Claparède & Lachmann) mentioned in the individual papers. However, records from inland salt waters are assigned to *Holosticha gibba*. For explanation of problems, see remarks.

Type locality of *Australothrix gibba* are limnetic habitats in the surroundings of Berlin, Germany (Claparède & Lachmann 1858). Ehrenberg (1838) found his population also in freshwater near Berlin among cyanobacteria and diatoms.

Records from limnetic habitats not substantiated by morphological data and/or illustrations: in old water from the Prater, an amusement park in Vienna, Austria, during February (Schmarda 1846, p. 45); ditch in Tallinn (Reval), Estonia (Eichwald 1849, p. 523; see also Jacobson 1928, p. 103); rare in the Rhone river in the port of Lyon, France, during February (Ormancy 1852, p. 279); from November to August often dominant in a chalk stream in Cambridgeshire (about 80 km north of London), Great Britain (Gray 1952, p. 110); pelagial (20–30 m) of Lake Comer, Italy (Forel 1885, p. 140; Cattaneo 1882, p. 118); in the surface water of Lake Garda (Italy) numerous specimens which were green due to chlorophyll globule (ingested algae? symbiotic algae?) and had two nuclei (Cattaneo 1888, p. 97; 1889, p. 117); rare in an alpine thermal spring near Vinadio (about 44°18'N 07°10'E), a village west of Cuneo, Italy (Issel 1901, p. 69); stagnant waters near Lecce, Italy (Gargiulo 1907, p. 26); a 100 μm long specimen in a draw-well near Rome, Italy during December (Grispini 1938, p. 152, as *Uroleptus gibbus* Clap e L. 1858); alpine limnetic habitats in Italy (Calloni 1890, p. 282, 418); in clean, stagnant water with

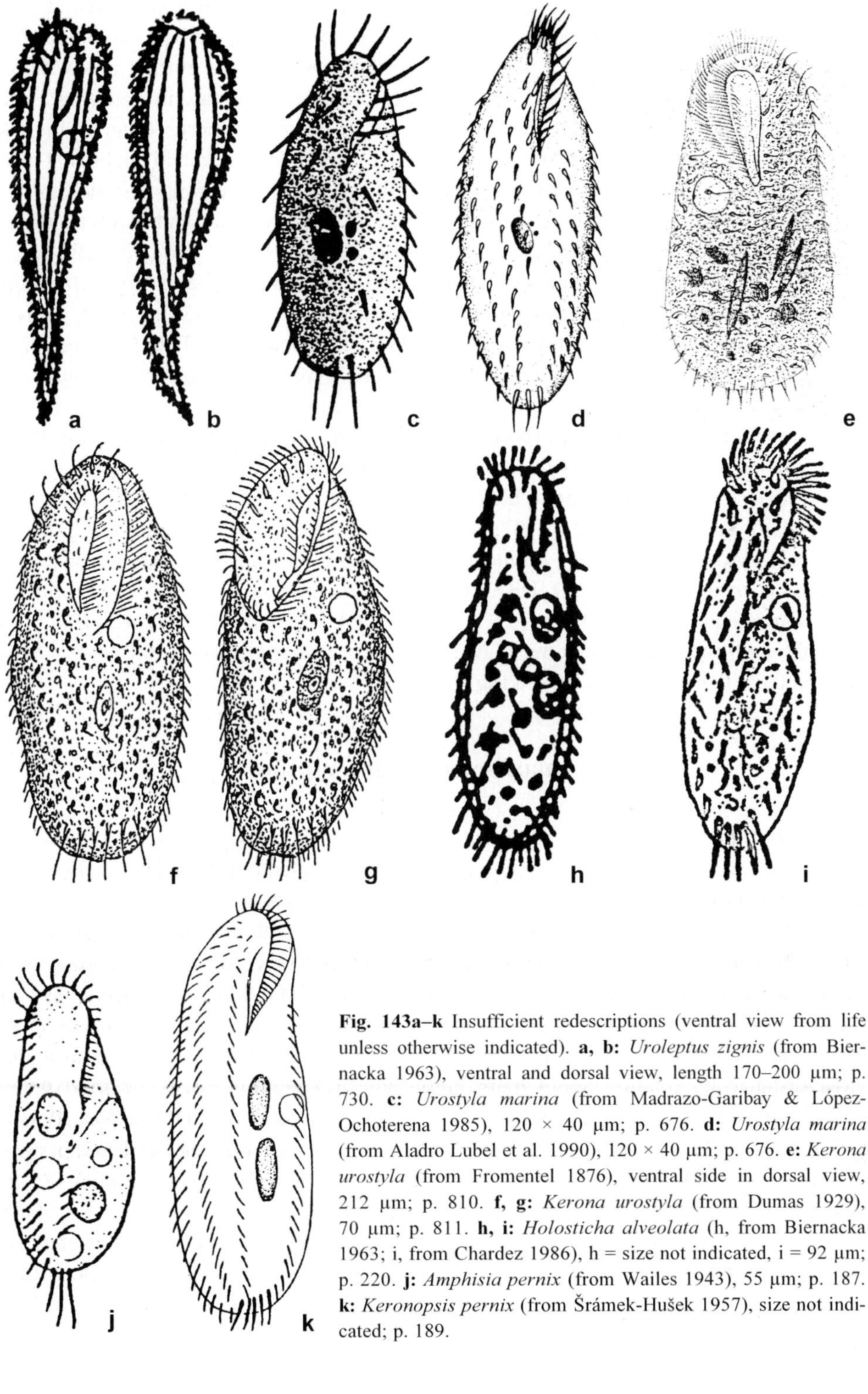

Fig. 143a–k Insufficient redescriptions (ventral view from life unless otherwise indicated). **a, b:** *Uroleptus zignis* (from Biernacka 1963), ventral and dorsal view, length 170–200 μm; p. 730. **c:** *Urostyla marina* (from Madrazo-Garibay & López-Ochoterena 1985), 120 × 40 μm; p. 676. **d:** *Urostyla marina* (from Aladro Lubel et al. 1990), 120 × 40 μm; p. 676. **e:** *Kerona urostyla* (from Fromentel 1876), ventral side in dorsal view, 212 μm; p. 810. **f, g:** *Kerona urostyla* (from Dumas 1929), 70 μm; p. 811. **h, i:** *Holosticha alveolata* (h, from Biernacka 1963; i, from Chardez 1986), h = size not indicated, i = 92 μm; p. 220. **j:** *Amphisia pernix* (from Wailes 1943), 55 μm; p. 187. **k:** *Keronopsis pernix* (from Šrámek-Hušek 1957), size not indicated; p. 189.

Sphagnum near Warsaw, Poland (Wrzesniowskiego 1866, p. 18); in bogs, cold and warm springs, and under ice in Switzerland (Perty 1852, p. 154); very rare between *Scorpidium scorpioides* (moss) in a patch of peat near the Pfäffikersee, a lake in the Canton Zurich, Switzerland (Messikommer 1954, p. 643); rare in limnetic habitats in and near Basle, Switzerland during April (Riggenbach 1922, p. 55); Hinterstockhornsee, an alpine lake south-west of Thun, Switzerland (Baumann 1910, p. 660) and in a limnetic habitat at about 1700 m altitude in the same area (Zschokke 1900, p. 71; for a review of some other records from Switzerland and Italy, see André 1912, p. 134); common in the littoral of lake Baikal, Russia (Gajevskaja 1927, p. 314 and Gajewskaja 1933, p. 146); limnetic habitat near St. Petersburg, Russia (Eichwald 1844, p. 583); soil from the bank of the Umsindusi river near Pretermaritzburg, South Africa (Fantham & Paterson 1926, p. 676); limnetic habitat near the Syowa Station (about 69°47'S 38°17'E), Antarctica (Hada 1966, p. 212, as *Uroleptus gibbus* (Cl. & L.)); Antarctica (Haneda 1971, p. 43; as *Uroleptus gibbus* (Clap. & Lach.)).

The record of one specimen (body length 105 μm) in the surface layer of an Italian soil fertilised with cow dung is likely a misidentification (Luzzatti 1938, p. 101, as *Uroleptus (Oxytricha) gibbus* (Clap, e L., 1858)).

Food not known. Biomass of 10^6 specimen about 16 mg (own calculation). Patrick et al. (1967, p. 324; as *Uroleptus gibbus* C. & L.) found it in the Savannah River (border of South Carolina and Georgia), USA at various sites from January to June at following conditions: 14.5–29.0°C, pH 6.5–7.0, 1–<7 mg l^{-1} Cl, >3–10 mg l^{-1} CO_2, 6.3–10.7 mg l^{-1} O_2, 0.02–0.60 mg l^{-1} Fe, <3–3 mg l^{-1} Mg, 0.001–0.170 mg l^{-1} NH_4-N, 0.07–<0.7 mg l^{-1} NO_3-N, 0.05–0.1 mg l^{-1} PO_4, >1–10 mg l^{-1} SO_4.

Insufficient redescription

***Uroleptus zignis* Entz 1884** – Biernacka, 1963, Polskie Archwm Hydrobiol., 11: 48, Abb. 91 (Fig. 143a, b). Remarks: The description and illustration by Biernacka (1963) are rather superficial. Furthermore, the Baltic specimens are distinctly smaller than that from the type locality, namely 170–200 μm against about 300 μm. Summary of the relevant data of the Baltic population: Length 170–200 μm (according to Biernacka 1963, the small length is due to the low salt content in the habitat where the specimens were found); body sigmoidal and very contractile; yellow-brown; usually swims with the broadened front end anteriorly, however, often shrinks back and swims backwards for a large distance by rotation about main body axis; often gliding between sand grains. Found in coastal pools at the "Westerplatte" in the Bay of Danzig (Baltic Sea, Poland) at 4.0–5.5‰ salt content and 12°C water temperature (see also Biernacka 1962, p. 86).

Urostylidae Bütschli, 1889

1889 **Urostylinae Bütschli** – Bütschli, Protozoa, p. 1741 (original description; diagnosis see Urostyloidea). Type genus: *Urostyla* Ehrenberg, 1830.
1926 **Urostylidae** – Calkins, Protozoa, p. 390 (brief review).
1981 **Urostylidae Butschli, 1889**[1] – Wicklow, Protistologica, 17: 331 (revision).
1983 **Urostylidae Bütschli, 1889** – Borror & Wicklow, Acta Protozool., 22: 120 (revision).
1994 **Urostylidae Bütschli, 1889** – Tuffrau & Fleury, Traite de Zoologie, 2: 128 (revision).
1999 **Urostylidae Bütschli, 1889** – Shi, Song & Shi, Progress in Protozoology, p. 110 (revision).
2001 **Urostylidae Bütschli, 1889** – Berger, Catalogue of ciliate names 1. Hypotrichs, p. 114 (nomenclator containing all basionyms, combinations, and higher taxa of hypotrichs).
2002 **Urostylidae Bütschli, 1889** – Lynn & Small, Phylum Protozoa, p. 441 (guide to representative genera).

Nomenclature: For derivation of the name see this chapter at the Urostyloidea. The list of synonyms contains only the original description (Urostylinae; introduced as subfamily), the first change to the name Urostylidae (family; Calkins 1926), and papers where a urostylid group contains a subgroup based on the same genus group name (Tables 6, 8–11).

Characterisation (Fig. 144a, apomorphy 1): Urostyloidea with many frontal cirri arranged in a bicorona (A).[2] Midventral complex composed of midventral pairs.

Remarks: For the content of this subgroup in other revisions see Tables 6, 8–11. The characterisation above contains only the character combination frontal ciliature and midventral complex as likely present in the last common ancestor of the Urostylidae. The bicorona is very likely an apomorphy of this group. By contrast the midventral complex composed of cirral pairs only is certainly a plesiomorphy in the stem-lineage of the Urostylidae. Of course, within this group the pattern changed. For example, *Tricoronella* formed a tricorona, or *Urostyla grandis* has a multicorona and, in addition, midventral rows.

Fig. 144a shows a hypothesis about the possible relationships within the Urostylidae. In this tree two major groups occur. The Urostylinae are characterised by midventral rows which, however, occur in the Bakuellidae and in *Uroleptopsis ignea* too. A midventral row is not a very complex feature because it originates simply by production of more than two cirri (= the ordinary midventral pair) per anlage. Thus their convergent evolution within the Urostyloidea and other groups of hypotrichs is not a great surprise.

The Retroextendia – comprising the pseudokeronopsids, the pseudourostylids, and *Tricoronella* – are characterised by a high DE-value, that is, the distal end of the adoral zone of membranelles extends far posteriorly (explanation see Fig. 1c and chapter 1.8 in the general section). This feature was also used by Wicklow (1981) to characterise the Keronopsidae (now Pseudokeronopsidae), and Wiackowski (1988) also estimated a close relationship of *Pseudokeronopsis* and *Pseudourostyla*. I checked relevant Figures and found that *Pseudokeronopsis*, *Thigmokeronopsis*, *Tricoronella*, and *Bicoronella*

[1] Wicklow (1981) provided the following diagnosis: Anterior frontal cirri are differentiated from other midventral cirri becoming markedly hypertrophied; the distal end of the adoral zone of paramembranelles extends only to the anterior end of the cell.

[2] Note that *Urostyla* includes some little known species, which have obviously only three frontal cirri. Their classification in *Urostyla* is only provisionally.

species have high DE-values (usually distinctly above 0.28, maximum 0.75, minimum 0.23; n = 30). *Trichototaxis stagnatilis* is little known (Fig. 165a, b). However, the cirral pattern (bicorona, midventral complex likely composed of cirral pairs only, DE-factor about 0.42) indicates that it belongs to the Retroextendia. Since a redescription is lacking it is not considered in the tree (Fig. 144a). By contrast the DE-value of urostylines (*Urostyla grandis*, *Keronella*, *Metabakuella*) is distinctly lower (less than 0.10 to 0.25), indicating that the DE-value is indeed a usable feature.

Key to the subgroups of the Urostylidae

Since the taxa below are mainly distinguished by details of the cirral pattern, that is, certain cirral groups present or not, protargol preparations, or at least very detailed live observations (interference contrast) are needed for successful identification. The bakuellids (e.g., *Bakuella, Holostichides*) also have midventral rows; however, they do not have many frontal cirri arranged in a coronal form, but only three more or less distinct frontal cirri.

1 Midventral complex composed of midventral pairs only; distal end of adoral zone extends far posteriorly (DE-value above 0.28; e.g., Fig. 145a–e) .... Retroextendia (p. 732)

\- Midventral complex composed of midventral pairs (anterior portion) and midventral rows (posterior portion); distal end of adoral zone of membranelles does not extend far posteriorly (DE-value distinctly below 0.28; Fig. 200a–c) Urostylinae (p. 1017)

Retroextendia tax. nov.

Nomenclature: The name of this new taxon is a composite of the Latin words *retro* (back, behind [spatial]) and *extendo* (extend), and refers to the far posteriorly extending distal end of the adoral zone of membranelles.

Characterisation (Fig. 144a, apomorphy 2): Urostylidae with far posteriorly extending (DE-value usually distinctly >0.28) distal end of the adoral zone of membranelles (A).

Remarks: The Retroextendia comprise all Urostylidae which have the plesiomorphic midventral complex composed of cirral pairs only.[1] However, this large group shows an interesting feature which can be interpreted as apomorphy, namely, the far posteriorly extending distal end of the adoral zone of membranelles. For detailed discussion and explanation of this feature (DE-value), see Fig. 1c, chapter 1.8 in the general section, and remarks of the Urostylidae. In some Retroextendia species the DE-value is 0.75, that is, the distal end of the adoral zone is only slightly more anteriorly

[1] Only one species (*Uroleptopsis ignea*) has midventral rows, which must be interpreted as autapomorphy of this species.

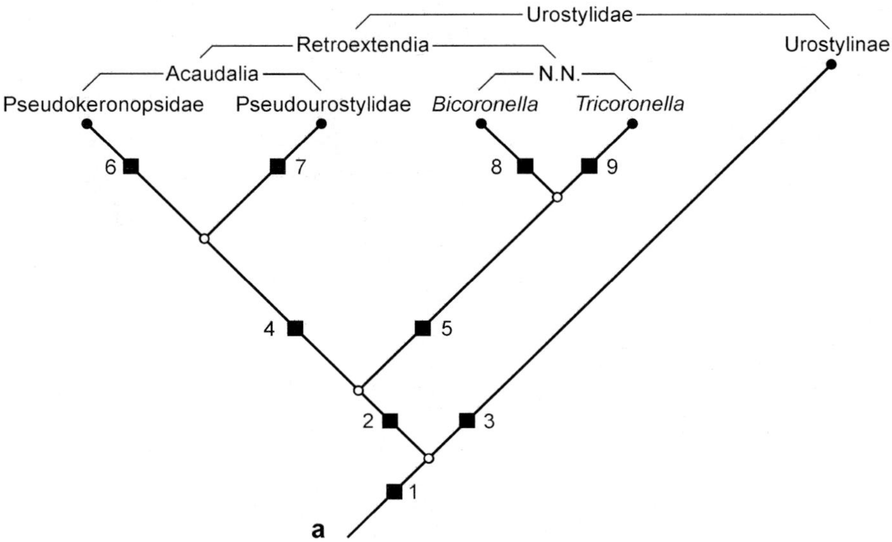

Fig. 144a Diagram of phylogenetic relationships within the Urostylidae (original). Autapomorphies (black squares 1–9): **1** – frontal cirri arranged in bicorona. **2** – distal end of adoral zone of membranelles extends far posteriorly (DE-value >0.28; convergence to, for example, *Pseudoamphisiella* and some taxa in the remaining hypotrichs). **3** – midventral rows present (convergence to Bakuellidae). **4** – caudal cirri lacking (convergence to several other groups in the Urostyloidea and the remaining hypotrichs). **5** – more than 4 dorsal kineties (convergence to *Pseudourostyla*); terrestrial (convergence to *Pseudourostyla franzi* and *P. muscorum*). **6** – the many macronuclear parts fuse to some parts during cell division; parental adoral zone of membranelles totally replaced? (it is uncertain for which group this feature is an autapomorphy because data are lacking for *Bicoronella*, *Tricoronella*, and all Pseudourostylidae (except *Pseudourostyla cristata*)). **7** – more than 2 marginal rows (convergence to some taxa in the Urostylinae, Bakuellidae, and remaining hypotrichs); marginal rows of one side originate from single anlage (convergence to *Diaxonella*). **8** – cirral row ahead of undulating membranes. **9** – frontal cirri arranged in tricorona. N.N. = nomen nominandum (Ax 1999, p. 18; a group, which is still to name; in the present case it would be possible to name this group *Tricoronella* with *Tricoronella* and *Bicoronella* as subgenera). Note that the present tree is only one of several hypothesis and that the "defined endings" -idae and -inae have no meaning in the present book (see chapter 7.2 of the general section).

than the proximal end. Rarely it is beyond 0.28, for example in *Uroleptopsis*. However, this group is certainly the sister group of *Pseudokeronopsis* because they have the same curious mode of macronuclear division. Thus, we have to assume that the DE-value has lowered in *Uroleptopsis*.

Fig. 144a shows a hypothesis about the possible relationships within the Retroextendia. The first group that splits off from the major lineage is *Tricoronella* + *Bicoronella*, which has increased the number of dorsal kineties from three to five or more. Moreover, the two species included are true terrestrial inhabitants whereas all other Retroextendia are limnetic or marine (except for *Pseudourostyla franzi* and *P. muscorum* which are also soil species). This indicates that the we have a change from the limnetic to the terrestrial habitat in the *Tricoronella* + *Bicoronella* lineage. If Fig. 144a is true then *Tricoronella* + *Bicoronella* are the sistergroup to the remaining Retroextendia,

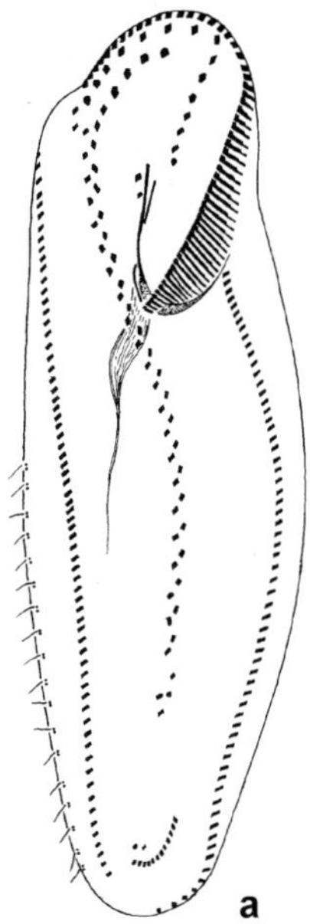

a

Bicoronella
AZM, BI, BC+, 2FT, MC(MP), TC+, CC+, 1LMR, 1RMR, >3DK

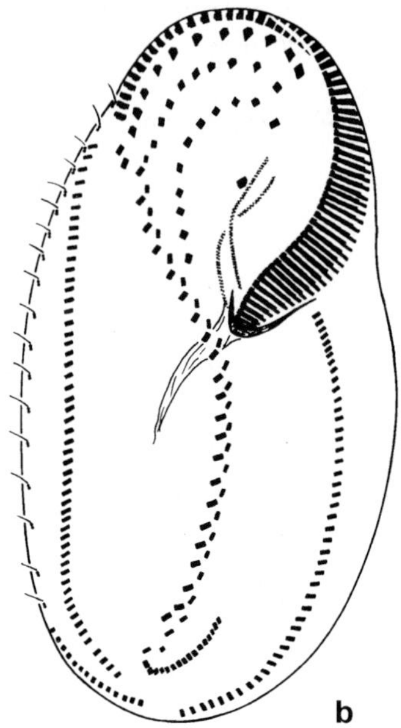

b

Tricoronella
AZM, TI, BC+, 2FT, MC(MP), TC+, CC+, 1LMR, 1RMR, >3DK

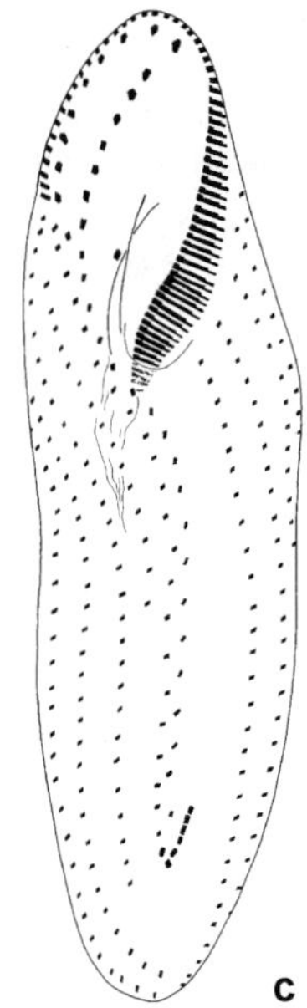

c

Pseudourostyla
AZM, BI, BC+, 2FT, MC(MP), TC+, CC-, >1LMR, >1RMR, >3DK

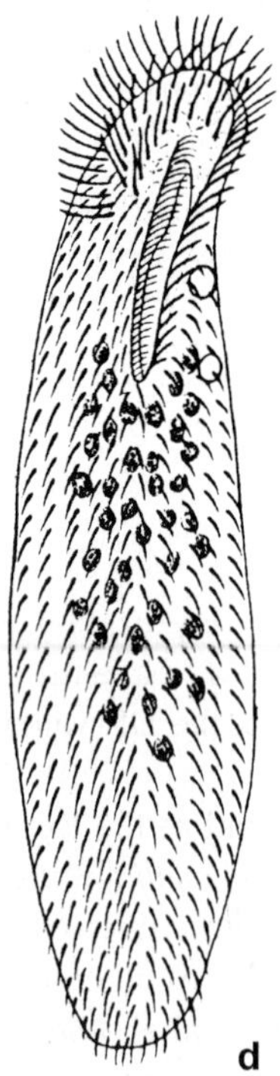

d

Hemicycliostyla
AZM, BI, BC?, FT?, MC(MP?), TC-, CC?, >1LMR, >1RMR, ?DK

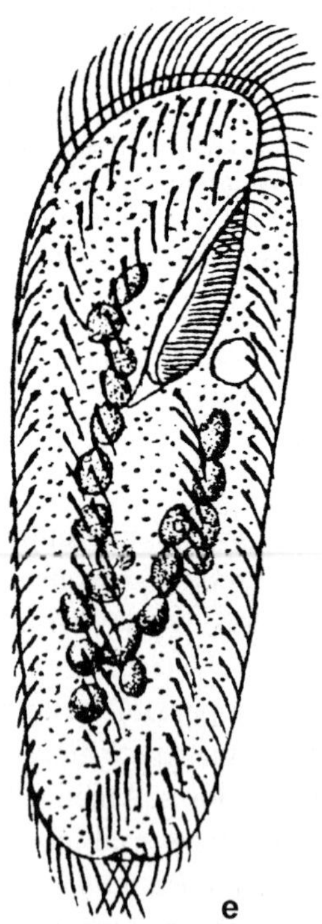

e

Trichototaxis
AZM, BI, BC?, ?FT, MC(MP?), TC+, CC?, >1LMR, 1RMR, ?DK

Fig. 145a–e Ventral cirral pattern in members of the Retroextendia. **a:** *Bicoronella costaricana.* **b:** *Tricoronella pulchra.* **c:** *Pseudourostyla levis.* **d:** *Hemicycliostyla sphagni.* **e:** *Trichototaxis stagnatilis.* Sources of illustrations see individual descriptions. Abbreviations used in short characterisations of infraciliature (explanation of supplemental signs and numbers see legend to Fig. 20a–c): AZM = adoral zone of membranelles, BC = buccal cirrus, BI = bicorona, CC = caudal cirri, DK = dorsal kineties, FT = frontoterminal cirri, LMR = left marginal row, MC(MP) = midventral complex composed of cirral pairs only, RMR = right marginal row, TC = transverse cirri.

which are characterised by the lack of caudal cirri. Admittedly, the loss of the caudal cirri is not a very spectacular feature because it is simple and occurs in almost all groups of hypotrichs. However, in the present case a relatively large number of species, distributed in at least four well-defined groups (*Pseudokeronopsis*, *Uroleptopsis*, *Thigmokeronopsis*, *Pseudourostyla*; and possibly *Hemicycliostyla*), shows this feature. Moreover, Wiackowski (1988) found that *Pseudourostyla* is closely related to *Pseudokeronopsis*. Indeed, if you omit the additional marginal rows of *Pseudourostyla cristata* you will have problems distinguishing it from *Pseudokeronopsis* species, at least as concerns the cirral pattern. The same phenomenon occurred in the Oxytrichidae. *Oxytricha granulifera* and *Onychodromopsis flexilis* (= *Allotricha antarctica* in Berger 1999) differ only in the number of marginal rows. Their sister group relationship proposed by morphologic and ontogenetic data (Berger & Foissner 1997, Berger 1999) was later confirmed by molecular data (Fig. 15a).

Hemicycliostyla, a taxon usually synonymised with *Urostyla*, lacks transverse cirri. The far posteriorly extending distal end of the adoral zone of membranelles indicates that it is closely related to the Retroextendia. The structure of the midventral complex is not known in detail because a redescription is lacking. Since *Hemicycliostyla* has many marginal rows it is preliminarily considered as sister group of *Pseudourostyla*. The lack of transverse cirri, sometimes dismissed as inadequate feature, was confirmed by Berger (2004b) for *Uroleptopsis*, which was synonymised with *Pseudokeronopsis* (e.g., Eigner 2001). Thus we can assume that *Hemicycliostyla* (without transverse cirri) also exists. *Pseudourostyla* and *Hemicycliostyla* are preliminarily united in the Pseudourostylidae. For the group Pseudourostylidae + Pseudokeronopsidae the name Acaudalia is introduced (see below).

Key to the subgroups of the Retroextendia

The features used in the key below can be recognised by detailed live observations (differential interference contrast). The key is guide to the genera (*Pseudourostyla*, *Hemicycliostyla*, *Trichototaxis*) of the Pseudourostylidae.

1 One left and 1 right marginal row (e.g., Fig. 145a, b) **2**
\- More than 1 left marginal row (Fig. 145c–e) **4**
2 Caudal cirri present (Fig. 145a, b) **3**
\- Caudal cirri lacking (Fig. 168a–f) . . . . Pseudokeronopsidae (p. 832)
3 Cirral row ahead of undulating membranes present; frontal cirri arranged in a bicorona (Fig. 145a) . . . . *Bicoronella* (p. 736)
\- Cirral row ahead of undulating membranes lacking; frontal cirri arranged in a tricorona (Fig. 145b) . . . . *Tricoronella* (p. 741)
4 (1) One right marginal row (Fig. 145e) . . . . *Trichototaxis* (p. 824)
\- More than 1 right marginal row (Fig. 145c, d) **5**
5 Transverse cirri present (Fig. 145c) . . . . *Pseudourostyla* (p. 750)
\- Transverse cirri absent (Fig. 145d) . . . . *Hemicycliostyla* (p. 811)

Bicoronella Foissner, 1995

1995 ***Bicoronella* nov. gen.**[1] – Foissner, Arch. Protistenk., 145: 63 (original description). Type species (by original designation on p. 63): *Bicoronella costaricana* Foissner, 1995.

2001 ***Bicoronella* Foissner 1995** – Aescht, Denisia, 1: 31 (catalogue of generic names of ciliates).

2001 ***Bicoronella* Foissner, 1995** – Berger, Catalogue of ciliate names 1. Hypotrichs, p. 13 (nomenclator containing all basionyms, combinations, and higher taxa of hypotrichs).

2002 ***Bicoronella* Foissner, 1995** – Lynn & Small, Phylum Ciliophora, p. 446 (guide to ciliate genera).

Nomenclature: The name *Bicoronella* is a composite of the Latin numeral *bi-* (two), the Greek noun *coron-* (arch, bow), and the diminutive suffix *-ella*, referring to the two arched rows of frontal cirri, a main feature of the genus. Feminine gender because ending with *-ella* (ICZN 1999, Article 30.1.3).

Characterisation (Fig. 144a, autapomorphy 8): Adoral zone of membranelles continuous. Frontal cirri arranged in bicorona. Buccal cirri present. Cirral row ahead of undulating membranes (A). 2 frontoterminal cirri. Midventral complex composed of midventral pairs only. Transverse, pretransverse ventral, and caudal cirri present. 1 left and 1 right marginal row. More than 4 dorsal kineties.

Remarks: See this chapter at single species.

Species included in *Bicoronella*: (1) *Bicoronella costaricana* Foissner, 1995.

Single species

Bicoronella costaricana Foissner, 1995 (Fig. 146a–e, Table 32)

1995 ***Bicoronella costaricana* nov. spec.**[2] – Foissner, Arch. Protistenk., 145: 64, Fig. 98–102, Tab. 6 (Fig. 146a–e; original description. The holotype slide [registration number 1997/92] and one paratype slide [1997/93] are deposited in the Oberösterreichische Landesmuseum in Linz [LI], Austria).

2001 ***Bicoronella costaricana* Foissner, 1995** – Berger, Catalogue of ciliate names 1. Hypotrichs, p. 13 (nomenclator containing all basionyms, combinations, and higher taxa of hypotrichs).

Nomenclature: This species was named after the country (Costa Rica) where it was discovered by Foissner (1995). *Bicoronella costaricana* was fixed as type species of *Bicoronella* by original designation.

Remarks: Foissner (1995) assigned *Bicoronella* to the Pseudokeronopsidae obviously because of the bicorona which is characteristic for the members of this group. However, the assignment was only provisional because morphogenetic data are lacking. The autapomorphy of the pseudokeronopsids is that not all macronuclear nodules fuse

[1] The diagnosis by Foissner (1995) is as follows: Pseudokeronopsidae (?) with transverse and caudal cirri and 2 arched rows of frontal cirri along anterior body margin. 2 frontoterminal cirri.

[2] The diagnosis by Foissner (1995) is as follows: Size in vivo about 150–200 × 50–60 μm. Cortical granules in rows, yellowish, 0.5–1 μm in diameter. On average 65 macronuclear nodules, 53 adoral membranelles, 7 cirri in upper and 5 cirri in lower frontal row, 6 cirri in row extending from first frontal cirrus to buccal cavity, 19 midventral pairs terminating above transverse cirri, 10 transverse cirri, 4 caudal cirri, and 5 dorsal kineties.

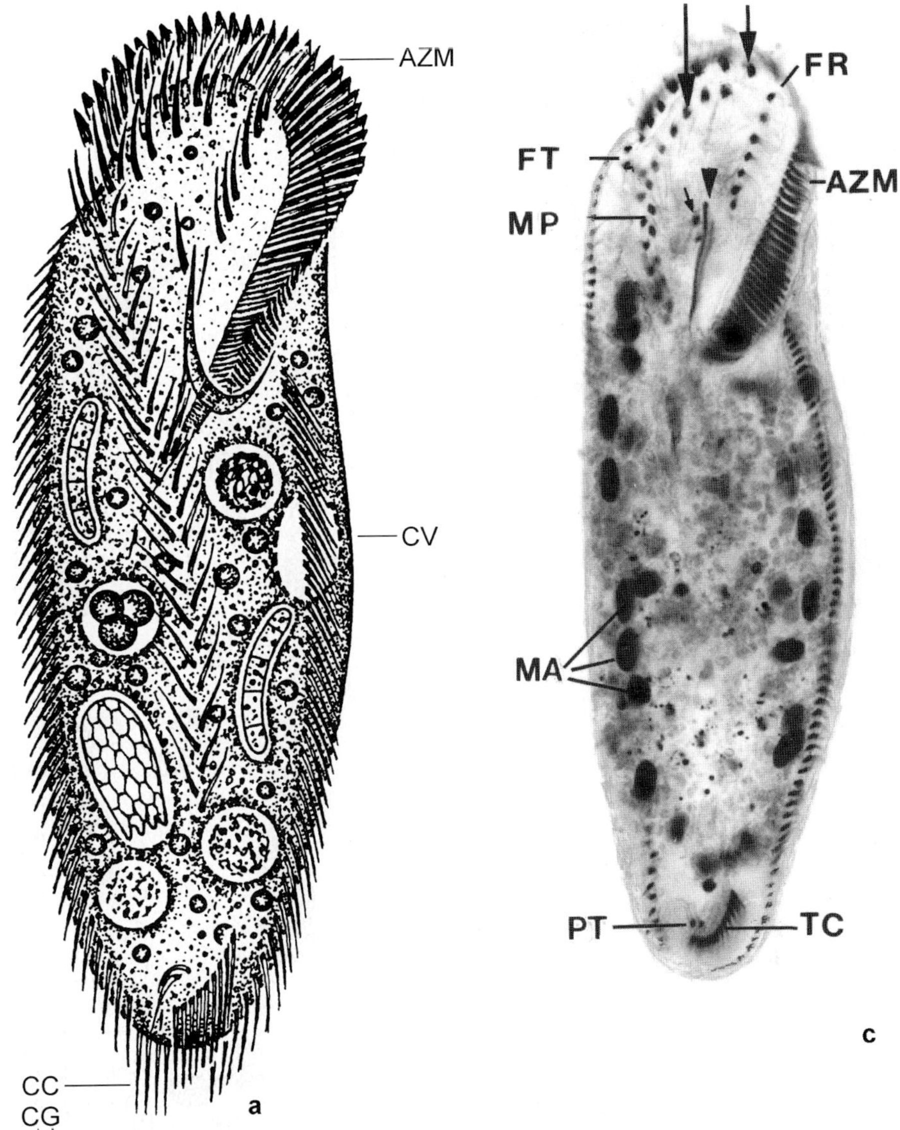

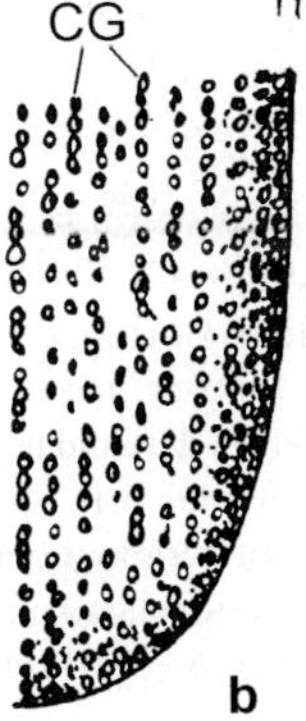

Fig. 146a–c *Bicoronella costaricana* (from Foissner 1995. a, b, from life; c, protargol impregnation). **a:** Ventral view of a representative specimen, 167 µm. This specimen has ingested, inter alia, cyanobacteria, fungal spores, and a *Euglypha rotunda*, a small testate amoebae. **b:** The yellowish cortical granules are about 0.5–1.0 µm across and arranged in meridional rows. **c:** Infraciliature of ventral side and nuclear apparatus, 187 µm (specimen illustrated in Fig. 146d). Short large arrow marks anterior (= front) row of frontal cirri, long large arrow denotes rear bow. Small arrow points to buccal cirri; arrowhead denotes undulating membranes. AZM = adoral zone of membranelles, CC = caudal cirri, CG = cortical granules, CV = contractile vacuole, FR = cirral row extending between buccal cavity and anterior bow of frontal cirri, FT = frontoterminal cirri, MA = macronuclear nodules, MP = midventral pairs, PT = pretransverse ventral cirri, TC = transverse cirri. Page 736.

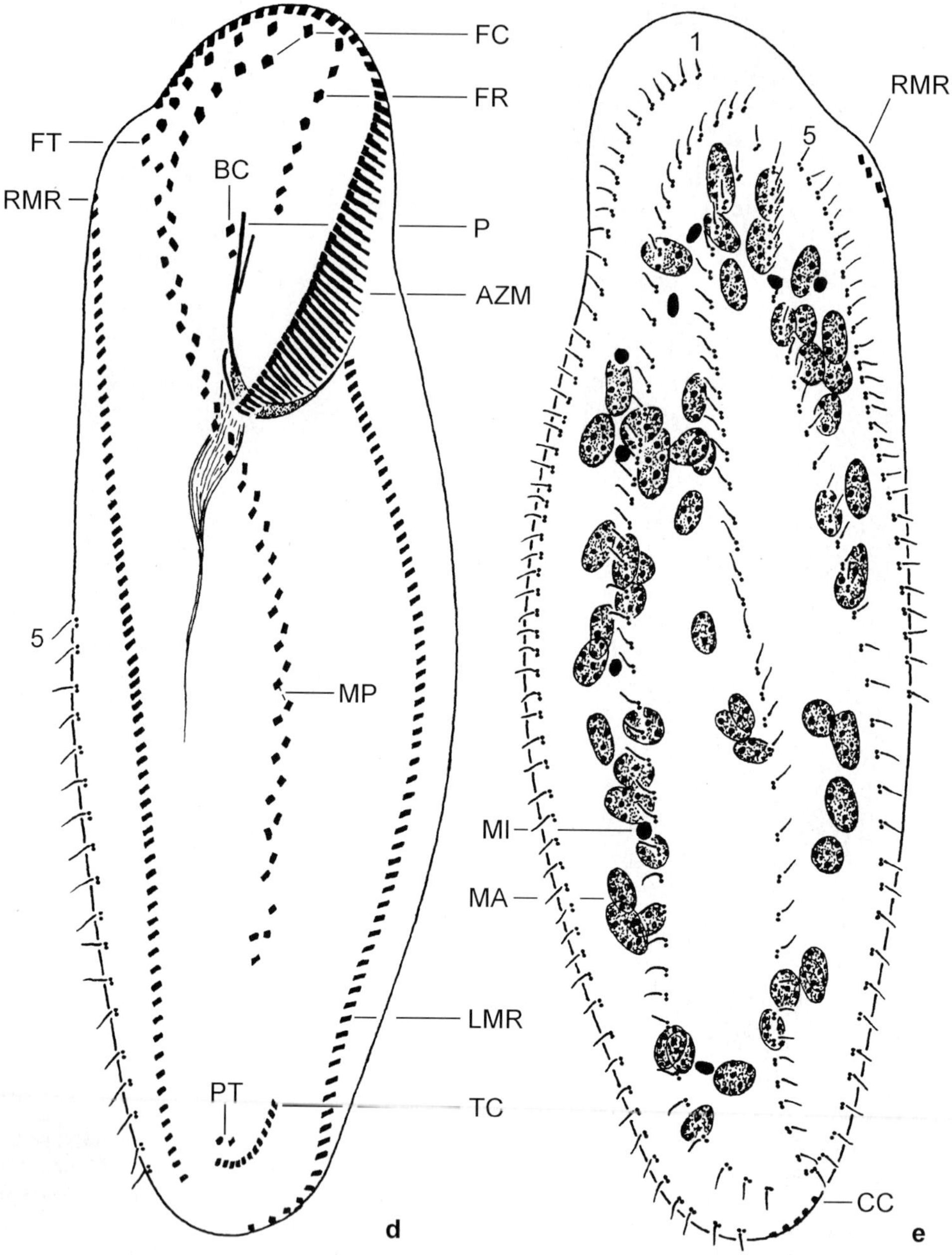

Fig. 146d, e *Bicoronella costaricana* after protargol impregnation (from Foissner 1995). Infraciliature of ventral and dorsal side and nuclear apparatus, 187 µm (micrograph, see Fig. 146c). AZM = adoral zone of membranelles, BC = buccal cirrus, CC = caudal cirri, FC = anterior and posterior bow of frontal cirri, FR = cirral row extending from left end of anterior frontal cirri bow to buccal cavity, FT = frontoterminal cirri, LMR = left marginal row, MA = macronuclear nodules, MI = micronucleus, MP = midventral pair, P = paroral, PT = pretransverse ventral cirri, RMR = right marginal row, TC = transverse cirri, 1, 5 = dorsal kineties. Page 736.

to a single mass during division (Fig. 144a, autapomorphy 6). Since we do not know the fate of the macronucleus in *B. costaricana*, I do not include it in the Pseudokeronopsidae as defined by Berger (2004b; see also same chapter at *Tricoronella pulchra*).

Shi (1999, 1999a) and Shi et al. (1999, p.116) synonymised *Bicoronella* with *Tricoronella* because they do not consider the different number of frontal-cirri bows (two [bicorona] against three [tricorona]) as generic feature. Thus, they "transferred" *Bicoronella costaricana* to *Tricoronella*, however, not formally; hence, I do not mention this binomen. Furthermore, they classified *Tricoronella* in the Holostichidae because they ignored the pseudokeronopsids.

Several taxa with many frontal cirri arranged in two or three arched rows are known, for example, *Pseudokeronopsis*, *Uroleptopsis*, *Tricoronella*, *Keronella*, *Pattersoniella* Foissner, 1987, *Keronopsis* Penard, 1922, and *Paraholosticha* Wenzel, 1953 (= replacement name for *Paraholosticha* Kahl, 1932). *Pattersoniella* is a stylonychine because it has a rigid body and a fragmenting dorsal kinety 3 (Fig. 16f; Berger 1999, p. 766). *Keronopsis* (with transverse cirri) and *Paraholosticha* (without transverse cirri) lack a midventral pattern and are therefore not closely related to the present species. *Keronella* has, inter alia, not only midventral pairs, but also midventral rows and is therefore assigned to the Urostylinae.

Pseudokeronopsis lacks caudal cirri whereas *Uroleptopsis* lacks transverse and caudal cirri. *Caudiholosticha sylvatica*, which has, like *Bicoronella costaricana*, a frontal row ahead of the undulating membranes, has only three frontal cirri (Fig. 45b, h). *Eschaneustyla* lacks midventral pairs. *Tricoronella*, with the single species *T. pulchra*, has three arched rows of frontal cirri. The cirral row ahead of the undulating membranes is, at the present state of knowledge, the sole apomorphy of *B. costaricana*.

In life, *Bicoronella costaricana* is best recognised by the soil habitat, the size (150–200 × 50–60 μm), the yellowish cortical granules, the two arched rows of frontal cirri, and the cirral row extending from the leftmost frontal cirrus to the buccal cavity.

Morphology: Body size about 150–200 × 50–60 μm in life; body length:width ratio 3.2:1 on average in protargol preparations. Body slender, slightly sigmoidal and narrowed to posterior end, left margin distinctly convex, right slightly convex, straight or even slightly concave, both ends broadly rounded; very flexible and dorsoventrally flattened up to 2:1 (Fig. 146a, Table 32). Macronuclear nodules located mainly along body margin, ellipsoidal (length:width ratio about 2:1), contain small nucleoli. Micronuclei almost globular. Contractile vacuole near mid-body at left cell margin, with two longitudinal collecting canals. Cortical granules in longitudinal rows, 0.5–1.0 μm across, yellowish (Fig. 146b); cells appear yellowish to yellow-brown at low magnification due to cortical granulation and food inclusions. Cytoplasm colourless. Movement obviously without peculiarities because not mentioned in original description.

Adoral zone occupies 35% of body length on average, of usual shape and structure, composed of an average of 53 membranelles, bases of largest membranelles 15 μm wide in life (Fig. 146a, c, d). Buccal cavity short and flat, but rather wide. Endoral and paroral short as compared with length of adoral zone of membranelles, almost straight, intersect optically in posterior third; paroral likely composed of dikinetids, endoral of

Table 32 Morphometric data on *Bicoronella costaricana* (from Foissner 1995)

Characteristics [a]	mean	M	SD	SE	CV	Min	Max	n
Body, length	165.8	172.0	17.6	5.9	10.6	140.0	187.0	9
Body, width	52.6	55.0	4.4	1.5	8.5	43.0	58.0	9
Anterior body end to rear end of adoral zone, distance	59.1	62.0	6.0	2.0	10.1	50.0	65.0	9
Anterior body end to end of midventral row, distance	124.9	133.0	17.0	5.7	13.6	98.0	147.0	9
Rear body end to transverse cirri, distance	10.7	11.0	1.6	0.5	14.8	8.0	12.0	9
Macronuclear nodule, length	8.6	8.0	1.5	0.5	17.6	6.0	11.0	9
Macronuclear nodule, width	4.0	4.0	0.7	0.2	16.5	3.0	5.0	9
Macronuclear nodules, number	75.4	65.0	20.4	6.8	27.1	52.0	115.0	9
Micronucleus, length	3.0	3.0	–	–	–	2.2	3.5	9
Micronucleus, width	2.6	2.5	–	–	–	2.2	3.0	9
Micronuclei, number	8.2	8.0	2.4	0.8	29.7	5.0	11.0	9
Adoral membranelles, number	51.6	53.0	5.3	1.8	10.3	43.0	57.0	9
Front row of frontal cirri, number of cirri	6.6	7.0	0.9	0.3	13.4	5.0	8.0	9
Rear row of frontal cirri, number of cirri	4.9	5.0	0.9	0.3	18.9	4.0	6.0	9
Cirral row ahead of buccal cavity, number of cirri	6.2	6.0	0.8	0.3	13.4	5.0	7.0	9
Buccal cirri, number	1.3	1.0	–	–	–	1.0	2.0	9
Frontoterminal cirri, number	2.0	2.0	0.0	0.0	0.0	2.0	2.0	9
Midventral complex, number of cirral pairs	19.0	19.0	3.3	1.1	17.5	15.0	24.0	9
Pretransverse ventral cirri, number	2.4	2.0	–	–	–	2.0	3.0	9
Transverse cirri, number	10.0	10.0	1.8	0.6	18.0	8.0	12.0	9
Right marginal cirri, number	57.0	54.0	7.6	2.5	13.4	49.0	71.0	9
Left marginal cirri, number	52.7	51.0	8.4	2.8	16.0	42.0	64.0	9
Dorsal kineties, number	5.1	5.0	–	–	–	5.0	6.0	9
Caudal cirri, number	4.1	4.0	1.1	0.4	25.7	3.0	6.0	9

[a] All measurements in µm. All data are based on mounted, protargol-impregnated (Foissner's method), and randomly selected specimens. CV = coefficient of variation in %, M = median, Max = maximum value, mean = arithmetic mean, Min = minimum value, n = number of individuals investigated, SD = standard deviation, SE = standard error of arithmetic mean.

monokinetids. Pharyngeal fibres conspicuous in protargol preparations, extend backwards beyond mid-body.

Cirral pattern of usual variability, number of cirri, however, rather strongly varying, that is, coefficient of variability for most features above 10% (Fig. 146a, c–e, Table 32). All cirri 15–20 µm long. Invariably two arched rows (bicorona) with seven frontal cirri on average in front and five in rear; coronal cirri slightly larger than other cirri, with rhomboid or pentagonal bases. Extraordinary cirral row extending from left cirrus of anterior frontal row to near buccal cavity; very likely, it originates from anlage I, but it cannot be excluded that this is the rear (third) row of a tricorona slightly shifted leftwards. Usually one, sometimes two buccal cirri right of anterior end of paroral. Invariably two inconspicuous frontoterminal cirri immediately behind distal end of adoral zone of membranelles. Midventral complex composed of 19 closely spaced midventral pairs on average, prominent because slightly sigmoidally extending in midline of cell from

Table 34 Continued

Characteristics[a]	Species	mean	M	SD	SE	CV	Min	Max	n
Micronucleus, length	cr1	6.2	6.0	1.3	0.3	21.1	3.0	9.0	19
	cr2	5.4	6.0	1.3	–	24.8	4.0	8.0	18
	fra	2.9	3.0	0.4	–	13.6	2.0	4.0	10
Micronucleus, width	cr1	3.7	4.0	0.9	0.2	23.1	2.0	5.0	19
Micronuclei, number	cr1	4.8	5.0	1.5	0.3	30.9	3.0	7.0	19
	cr3	–	–	–	–	–	6.0	8.0	?
	cr4	–	–	–	–	–	6.0	8.0	?
	fra			more than 10					
	le1	6.0	–	–	–	–	3.0	9.0	?
	le3	–	–	–	–	–	3.0	12.0	?
	nov	6.4	6.0	0.9	–	13.6	5.0	8.0	17
Adoral membranelles, number	cr1	87.8	86.0	6.9	1.4	7.9	74.0	100.0	24
	cr2	98.9	100.0	12.4	–	12.5	75.0	115.0	18
	cr4	–	–	–	–	–	90.0	130.0	?
	cr5	62.0	–	–	–	–	–	–	?
	fra	67.1	68.0	4.1	–	6.2	59.0	72.0	10
	le2	82.6	–	3.9	–	–	–	–	49
	nov	50.3	50.5	3.0	–	6.0	45.0	57.0	30
	uro	47.0	47.0	3.5	1.1	7.5	42.0	54.0	10
Buccal cirri, number	cr1	1.3	1.0	0.4	0.1	34.6	1.0	2.0	24
	cr2	1.0	1.0	0.2	–	21.3	1.0	2.0	22
	fra	2.1	2.0	0.6	–	27.0	1.0	3.0	10
	uro	4.2	4.0	0.8	0.3	18.8	3.0	5.0	10
Frontal cirri (anterior bow), number	cr1	12.6	12.0	1.2	0.2	9.4	10.0	14.0	24
	le2	12.0	–	1.3	–	–	–	–	38
	uro	6.2	6.5	0.9	0.3	14.8	5.0	7.0	10
Frontal cirri (posterior bow), number	cr1	9.9	10.0	1.3	0.3	13.3	8.0	12.0	24
	le2	9.9	–	1.3	–	–	–	–	38
	uro	2.9	3.0	0.7	0.2	25.4	2.0	4.0	10
Frontoterminal cirri, number	cr1	2.0	2.0	0.0	0.0	0.0	2.0	2.0	23
	cr2	2.0	2.0	0.0	–	0.0	2.0	2.0	8
	fra	2.0	2.0	0.0	–	0.0	2.0	2.0	10
Frontal cirri (anterior bow) plus right cirri of midventral pairs, number	cr1	34.0	35.0	2.7	0.6	8.0	28.0	38.0	24
Frontal cirri (posterior bow) plus left cirri of midventral pairs, number	cr1	31.4	32.0	2.8	0.6	9.1	25.0	35.0	24
Midventral pairs, number	cr1	21.5	22.0	2.4	0.5	11.0	17.0	26.0	24
Midventral complex, cirral number 1[b]	cr2	34.5	35.0	5.1	–	14.9	27.0	44.0	8
	fra	19.0	19.0	2.0	–	10.8	15.0	23.0	10
	le2	20.0	–	2.3	–	–	–	–	37
	uro	18.8	19.0	2.4	0.8	12.9	16.0	22.0	9
Midventral complex, cirral number 2[b]	cr2	31.6	31.0	4.9	–	15.6	25.0	41.0	9
	fra	18.2	18.0	2.3	–	12.4	13.0	22.0	10
	le2	18.1	–	2.3	–	–	–	–	37
	uro	16.1	17.0	2.2	0.7	13.3	13.0	19.0	9
Ventral cirri, number[h]	cr1	3.0	3.0	0.6	0.1	19.2	2.0	4.0	24
Transverse cirri, number	cr1	7.6	7.5	1.4	0.3	18.9	5.0	10.0	24
	cr2	9.7	10.0	1.6	–	16.1	6.0	12.0	21
	cr3	–	–	–	–	–	8.0	12.0	?
	cr4	–	–	–	–	–	12.0	16.0	?
	fra	7.7	7.0	1.8	–	23.8	5.0	11.0	10

Table 34 Continued

Characteristics[a]	Species	mean	M	SD	SE	CV	Min	Max	n
Transverse cirri, number	le1	8.0	–	–	–	–	–	–	?
	le2	8.0	–	0.9		–	6.0	10.0	49
	nov	7.7	8.0	0.7	–	8.9	7.0	9.0	30
	uro	7.4	7.0	1.3	0.5	17.1	6.0	10.0	7
Left marginal rows, number	cr1	5.1	5.0	0.5	0.1	10.4	4.0	6.0	23
	cr3	7.0	–	–	–	–	–	–	?
	cr4	7.0	–	–	–	–	–	–	?
	cr5	–	–	–	–	–	4.0	6.0	?
	fra	8.4	8.0	0.5	–	6.1	8.0	9.0	10
	le1	5.0	–	–	–	–	4.0	5.0	?
	le2	5.5	–	0.5	–	–	–	–	49
	nov	2.0	2.0	0.0	–	0.0	2.0	2.0	30
Outermost left marginal row, number of cirri	cr1	8.3	6.0	5.2	1.1	63.2	2.0	20.0	23
Innermost left marginal row, number of cirri	cr1	34.3	34.0	2.9	0.6	8.4	29.0	40.0	23
Right marginal rows, number	cr1	4.4	4.0	0.5	0.1	11.2	4.0	5.0	23
	cr3	7.0	–	–	–	–	–	–	?
	cr4	7.0	–	–	–	–	–	–	?
	cr5	4.0	–	–	–	–	–	–	?
	fra	7.8	8.0	0.8	–	10.1	7.0	9.0	10
	le1	4.0	–	–	–	–	3.0	4.0	?
	le2	4.1	–	0.4	–	–	–	–	49
	nov	2.0	2.0	0.0	–	0.0	2.0	2.0	30
Outermost right marginal row, number of cirri	cr1	28.0	28.0[f]	8.5	1.8	30.5	15.0	42.0	22
Innermost right marginal row, number of cirri	cr1	26.9[g]	30.0	6.8	1.4	25.3	10.0	36.0	23
Dorsal kineties, number	cr1				8.0 (see text)				
	fra	4.0	4.0	0.0	–	0.0	4.0	4.0	10
	le2	7.8	–	0.5	–	–	7.0	9.0	45
	nov	7.0	7.0	0.0	0	0.0	7.0	7.0	30
	uro	7.5	8.0	0.8	0.3	10.1	6.0	8.0	8
Dorsal kinety 1, number of bristles[e]	le2	35.0	–	6.5	–	–	–	–	33
Dorsal kinety 2, number of bristles[e]	le2	26.5	–	4.5	–	–	–	–	33
Dorsal kinety 3, number of bristles[e]	le2	20.1	–	5.2	–	–	–	–	33
Dorsal kinety 4, number of bristles[e]	le2	21.3	–	3.6	–	–	–	–	33
Dorsal kinety 5, number of bristles[e]	le2	21.8	–	3.3	–	–	–	–	33
Dorsal kinety 6, number of bristles[e]	le2	22.0	–	3.4	–	–	–	–	33
Dorsal kinety 7, number of bristles[e]	le2	22.4	–	3.3	–	–	–	–	33
Dorsal kinety 8, number of bristles[e]	le2	22.1	–	3.5	–	–	–	–	33

[a] All measurements in μm. Unless otherwise indicated data are from protargol-impregnated specimens. CV = coefficient of variation in %, M = median, Max = maximum value, mean = arithmetic mean, Min = minimum value, n = number of individuals investigated (? = number not given. If only one value is available, then it is listed under the column mean; if two values are given, then they are listed under Min and Max), SD = standard deviation, SE = standard error of arithmetic mean.

[b] Cirral number 1 = number of right cirri of midventral pairs (frontal cirri not included); cirral number 2 = number of left cirri of midventral pairs.

[c] Distance between anterior end of paroral and cytostome.

Table 34 Continued

[d] Bouin-fixed normal specimens. For sizes of amicronucleates and renucleates, see Takahashi (1988).

[e] Values from specimens with eight dorsal kineties.

[f] Oberschmidleitner & Aescht (1996) erroneously wrote 280.

[g] Oberschmidleitner & Aescht (1996) erroneously wrote 269.

[h] The "ventral cirri" are not described or designated in the illustrations. Likely, these are the cirri ahead the transverse cirri which do no form distinct pairs.

[i] From life.

[j] Note that the scale bars in Figures 1, 2 of Grim & Manganaro (1985) are much too short; accordingly the specimen in Fig. 1 would be only 73 μm long (against 200–300 μm mentioned in the text!).

The pseudourostylids with *Pseudourostyla* as type genus were accepted by Small & Lynn (1985), Eigner & Foissner (1992), Tuffrau & Fleury (1994), and Oberschmidleitner & Aescht (1996). Tuffrau (1979, p. 526) classified *Pseudourostyla* in the Urostylidae, while Corliss (1977, p. 137; 1979) considered it as Holostichidae. Shi (1999, p. 245) and Shi et al. (1999) synonymised the pseudourostylids with the urostylids, whereas Shi (1993) classified it in the Kinetodesmophorida Shi, 1993. Tuffrau (1987) obviously overlooked *Pseudourostyla*.

Pseudourostyla, at least *P. cristata* and *P. nova*, has the same type of marginal row formation as *Diaxonella*. All rows of one side originate from a common anlage per filial product. This feature indicates a sister-group relationship between *Pseudourostyla* and *Diaxonella*, which, however, has only three frontal cirri. By contrast, *Trichototaxis* has, like *Pseudourostyla*, a distinct bicorona and at least two left marginal rows, but – like *Diaxonella* – only one right marginal row. Detailed redescription, including morphogenetic data, of *T. stagnatilis* and molecular studies on all taxa are needed to show whether the proposed classification (*Diaxonella* in the Holostichidae; *Pseudourostyla* and *Trichototaxis* in the Urostylidae) is realistic.

Shi (1999a) and Shi et al. (1999) consider *Pseudourostyla* Borror, 1972 as valid, with *Hemicycliostyla* Stokes, 1886 as synonym. However, this is incorrect because of the principle of priority (ICZN 1999, Article 23). Furthermore, the type species of Stokes' genus, *Hemicycliostyla sphagni*, lacks transverse cirri, while such cirri are present in *Pseudourostyla cristata*, type of *Pseudourostyla*. Consequently, synonymy is very unlikely.

Hemberger (1982, p. 75, 77) synonymised *Pseudourostyla* with *Urostyla* because he regarded the different types of marginal row formation as not essential. However, *Pseudourostyla* differs from *Urostyla* not only by the mode of marginal row formation, but also by the presence of frontoterminal cirri. The frontoterminal cirri of *P. cristata* were first recorded by Eigner & Foissner (1992, 1993), who reinvestigated the population studied by Jerka-Dziadosz (1972). The lack of frontoterminal cirri in *Urostyla grandis* was proved by Ganner (1991, p. 118). The different mode of marginal row formation in *P. cristata* and *U. grandis* was discussed by Jerka-Dziadosz (1972), although the specific mode in *P. cristata* was already described by Jerka-Dziadosz (1964).

Marginal row formation in *Ponturostyla enigmatica* (Dragesco- & Dragesco-Kernéis, 1986) Jankowski, 1989 – an 18-cirri oxytrichid with many marginal rows – proceeds very similarly to in *Pseudourostyla cristata* (Song 2001). The mode of marginal row formation in *Pseudourostyla* (several rows originate from a common anlage) is also reminiscent of the formation of dorsomarginal kineties, which are obviously a novelty for most non-urostyloid hypotrichs (Fig. 14a). These dorsomarginal kineties originate from/very near the right side of the anterior end of the right marginal row anlage. Possibly these processes are homologous.

The extrusomes of *P. cristata* and *P. franzi* can explode like trichocysts. Possibly the other species have the same organelles, which would be then an apomorphy of *Pseudourostyla*.

As mentioned above, Borror (1972) transferred three species (*Urostyla cristata, Urostyla muscorum, Oxytricha urostyla*) to *Pseudourostyla*. Borror & Wicklow (1983) synonymised *Urostyla muscorum* with *Urostyla grandis* and *Oxytricha urostyla* with *Urostyla multipes* so that *Pseudourostyla* was monotypic in their revision. In 1991, we (Foissner et al. 1991, p. 260) considered *Oxytricha urostyla* and *Oxytricha multipes* Claparède & Lachmann, 1858 as supposed synonyms of *Paraurostyla weissei* (Stein, 1859) Borror, 1972. Later, I excluded *Oxytricha urostyla* from the synonymy of *P. weissei* because it has a bicorona and not, like *P. weissei* and *O. multipes*, three distinct frontal cirri (Berger 1999, p. 873).

Urostyla pseudomuscorum lacks midventral rows so that it is obviously misclassified in *Urostyla*. The body size, the cirral pattern, the habitat, and especially the two macronuclear nodules agree rather well *P. urostyla*, with which *U. pseudomuscorum* is therefore very likely synonymous.

Metaurostyla polonica Jankowski, 1979, type of *Metaurostyla*, is a junior synonym of *Urostyla grandis*, type of *Urostyla*. Consequently, Jankowski's genus is eliminated. Two other *Metaurostyla* species should not be classified in *Urostyla* because their midventral complex is composed of midventral pairs only (vs. midventral pairs and rows in *U. grandis*). I preliminary assign them to *Pseudourostyla*. Due to the two macronuclear nodules they closely resemble *P. urostyla*.

Species included in *Pseudourostyla* (alphabetically arranged according to basionym): (1) *Pseudourostyla franzi* Foissner, 1987; (2) *Pseudourostyla levis* Takahashi, 1973; (3) *Pseudourostyla nova* Wiackowski, 1988; (4) *Urostyla cristata* Jerka-Dziadosz, 1972; (5) *Urostyla muscorum* Kahl, 1932. Incertae sedis: (6) *Metaurostyla magna* Alekperov, 1984; (7) *Metaurostyla raikovi* Alekperov, 1984; (8) *Oxytricha urostyla* Claparède & Lachmann, 1858; (9) *Urostyla* sp. sensu Shin (1994).

Key to *Pseudourostyla* species

Identification of *Pseudourostyla* species requires the standard set of features (habitat, macronuclear apparatus, cirral pattern). For *P. muscorum*, the number of macronuclear nodules is not reliably known; very likely it has several to many dispersed throughout the cell. Since all *Pseudourostyla* species are rather large, details of the cirral pattern (e.g., one or many buccal cirri) can be recognised even with a well adjusted interference contrast microscope. However, if you are uncertain you have to make protargol preparations.

1 Two macronuclear nodules (Fig. 155a, b) **6**
\- 14 or more macronuclear nodules (Fig. 151f, 153a) **2**
2 Several (around 9) buccal cirri (Fig. 152a) .... *Pseudourostyla muscorum* (p. 791)
\- 1–5 buccal cirri **3**
3 1–2, rarely 3 buccal cirri **4**
\- 3–5, usually 4 buccal cirri (Fig. 154b) .... *Urostyla* sp. (p. 798)
4 On average 18 macronuclear nodules; 2 left and 2 right marginal rows (Fig. 153a, b) .... *Pseudourostyla nova* (p. 793)
\- On average 50 or more (range 27 to about 300) macronuclear nodules; more than 2 left and more than 2 right marginal rows **5**
5 On average 50 macronuclear nodules; midventral complex terminates near transverse cirri; limnetic (Fig. 148a, 150h) .... *Pseudourostyla cristata* group with *P. cristata* (p. 757) and *P. levis* (p. 778)
\- On average about 200 macronuclear nodules; midventral complex terminates near mid-body; terrestrial (Fig. 151a, d) .... *Pseudourostyla franzi* (p. 787)
6 (1) Ten or less cirral rows (the 2 pseudorows formed by the midventral pairs included; Fig. 155a, b, 156a, 157a) .... *Pseudourostyla urostyla* (p. 800)
\- About 15–16 cirral rows (the 2 pseudorows formed by the midventral pairs included; Fig. 158a, 159a) **7**
7 Body length about 180 µm in life (Fig. 158a) ..... *Pseudourostyla raikovi* (p. 807)
\- Body length about 520 µm in life (Fig. 159a) ..... *Pseudourostyla magna* (p. 809)

Pseudourostyla cristata (Jerka-Dziadosz, 1964) Borror, 1972
(Fig. 148a–x, 149a–z, Tables 34, 35)

1956 ***Urostyla*** – Pigon, Acta biochim. pol., 3: 613, Fig. 1–6 (Fig. 148w, x; illustrated record).

1964 ***Urostyla cristata* sp. n.** – Jerka-Dziadosz, Acta Protozool., 2: 123, Fig. 1, 2 and Plates I, II (Fig. 148a, b; original description; no type material available and no formal diagnosis provided).

1965 ***Urostyla cristata*** – Jerka-Dziadosz, Acta Protozool., 3: 133, Fig. 1, 2, Plate I (Fig. 149k–r; physiological and traumatic regeneration).

1972 ***Urostyla cristata*** – Jerka-Dziadosz, Acta Protozool., 10: 74, Fig. 1, 2 in text and Fig. 1–26 on Plates I–VI (Fig. 149a–j; redescription after protargol impregnation and morphogenesis).

1972 ***Pseudourostyla cristata* (Jerka-Dziadosz, 1964) n. comb.** – Borror, J. Protozool., 19: 11 (combination with *Pseudourostyla* Borror, 1972; revision of hypotrichs).

1979 ***Pseudourostyla cristata*** – Borror, J. Protozool., 26: 547, Fig. 4 (Fig. 148y; illustration, redefinition of genus).

1982 ***Urostyla cristata* Jerka-Dziadosz, 1964** – Hemberger, Dissertation, p. 79 (revision of non-euplotid hypotrichs).

1982 ***Urostyla cristata*** – Tchang, Pang & Zou, Zool. Res., 3: 1, Fig. 1–7, Plates I, II (Fig. 148r–u; morphogenesis of Chinese population).

1983 ***Pseudourostyla cristata* (Jerka-Dziadosz, 1964) Borror, 1972** – Borror & Wicklow, Acta Protozool., 22: 107, 115, 124, Fig. 8 (Fig. 148d; revision of urostylids).

1985 ***Pseudourostyla cristata*** – Small & Lynn, Phylum Ciliophora, p. 451, Fig. 5 (Fig. 148d; guide to ciliate genera).

1985 ***Pseudourostyla cristata*** – Grim & Manganaro, Trans. Am. microsc. Soc., 104: 350, Fig. 1–17 (detailed examination of extrusomes).

1992 ***Pseudourostyla cristata*** – Eigner & Foissner, Europ. J. Protistol., 28: 468, Fig. 20, 21 (Fig. 148f, g; brief reinvestigation of Polish population).

1996 ***Pseudourostyla cristata* (Jerka-Dziadosz 1964) Borror 1972**[1] – Oberschmidleitner & Aescht, Beitr. Naturk. Oberösterreichs, 4: 14, Abb. 7–18, Tabelle 5 (Fig. 148h–q; redescription and deposition of a neotype slide in the Oberösterreichische Landesmuseum in Linz [LI], Austria; slide not mentioned by Aescht 2003, p. 384; see remarks).

1999 ***Pseudourostyla cristata*** – Shi, Song & Shi, Progress in Protozoology, p. 196, Fig. 6A–I (morphogenesis).

2001 ***Pseudourostyla cristata* (Jerka-Dziadosz, 1964) Borror, 1972** – Berger, Catalogue of ciliate names 1. Hypotrichs, p. 101 (nomenclator containing all basionyms, combinations, and higher taxa of hypotrichs).

Nomenclature: The species-group name *cristat·us -a -um* (Latin adjective; comb-bearing) refers to the "comb-shaped disposition of the fronto-transversal cirri" (Jerka-Dziadosz 1964, p. 127). Likely, Jerka-Dziadosz meant the comb-shaped arrangement of the frontal-midventral-transversal cirri primordia. The present species was fixed as type of *Pseudourostyla* by original designation.

Remarks: The present species was very likely first described by Pigon (1956) as unidentified *Urostyla*, as indicated by the cirral pattern and nuclear apparatus (Fig. 148w, x). Jerka-Dziadosz (1964) described *P. cristata* in some detail, including several ontogenetic stages which show that the marginal rows of each side arise from a common primordium. In 1972, she compared the morphogenesis of *Urostyla cristata* with that of *U. grandis* and discussed the distinct difference in marginal row formation (Jerka-Dziadosz 1972). Simultaneously, Borror (1972) established *Pseudourostyla* with *Urostyla cristata* as type species because he erroneously assumed that *U. grandis* lacks a midventral pattern, whereas such a zigzagging pattern is present in *U. cristata*. In 1979, when Borror redefined the urostylids, he included the specific mode of marginal row formation into the diagnosis of *Pseudourostyla*.

Eigner & Foissner (1992) reinvestigated a Polish *Pseudourostyla cristata* population supplied by Jerka-Dziadosz. They found that it has, like *P. franzi* and *P. nova*, frontoterminal cirri (Fig. 148f, g).[2] This additional difference to *Urostyla grandis* (see genus section) was confirmed by Oberschmidleitner & Aescht (1996) in an Austrian *P. cristata* population (Fig. 148p) and by Shi et al. (1999a, their Fig. 6I) in a Chinese population. Borror & Wicklow (1983) stated that they have documented the presence of frontoterminal cirri (= migratory cirri in their terminology) in all urostyline genera they had examined, inter alia, *Pseudourostyla cristata* (Fig. 148d). This general statement

[1] The improved diagnosis by Oberschmidleitner & Aescht (1996; according to literature data and their own observations) is as follows: In vivo 220–450 × 60–180 μm. Zahlreiche 3–5 μm lange, sehr dünne pfeilförmige Trichocysten. 1–2 Buccalcirren. Zwei Frontalcirrenreihen in Form einer Bicorona, obere Reihe etwa 12, untere Reihe etwa 10 Cirren. Midventralreihen aus 17–26 Paaren, beiderseits davon 4–7 Marginalreihen. Etwa 50 Makronucleus-Teile und 8 Dorsalkineten.

[2] Wiackowski (1988), who also studied the preparations made by Jerka-Dziadosz (Wiackowski 1988, p. 3), did not see, like Jerka-Dziadosz, migratory cirri (state 0 for feature 15; p. 6, 7) in *P. cristata*, but observed them in a *Pseudourostyla* sp.

of these organelles is comparatively low (Jerka-Dziadosz 1970). Some specimens have further regions of various sizes, often in the posterior half, that are barren of extrusomes (Grim & Manganaro 1985). These authors found, like Jerka-Dziadosz (1970), a narrow, longitudinal band on the left side of the dorsal surface that was barren of extrusomes. This region corresponds with the location of the underlying contractile vacuole system (Jerka-Dziadosz 1970). According to Grim & Manganaro (1985), different specimens of *P. cristata* clearly have extrusomes of different average lengths, although this has not been analysed statistically.

Oberschmidleitner & Aescht (1996, p. 17) made detailed observations on the ejection of the extrusomes (Fig. 148k–n). They explode after treatment with methyl green-pyronin or haematoxylin, after fixation with sublimate-formalin, and after sudden, high coverglass pressure. In the course of the explosion the stalk elongates and broadens. Distally a drop-shaped, little head forms which penetrates the pellicle. Incompletely ejected extrusomes have the head still underneath the pellicle and the stalks anchored in the cytoplasm. This stage is readily apparent on protargol-impregnated cells fixed with sublimate-formalin. When treated with methyl green-pyronin or haematoxylin, the extrusomes explode completely.

The ultrastructure of the extrusomes was studied by Suganuma (1973). The 5 µm long organelles, which look like little umbrellas in life, consist of four parts, namely, a shaft, a tip, a body, and a cap. Tesarova & Foissner (1999) found further details: (i) The cap is bell-shaped and extruded with the organelle, as revealed by the spiral appearance of the head in the SEM; (ii) The cap microtubuli have the same diameter as ordinary cortical microtubuli; (iii) The incus is connected via fine fibres to the unit membrane enveloping the extrusome; (iv) Negative staining reveals a dense reticulum composed of many fibres around the head of extruded extrusomes; (v) The shaft has a narrow transverse striation possibly caused by spirally arranged extrusome material, as indicated by ultrathin sections and negative staining; (vi) At the posterior end of the shaft is a globule recognisable after SEM and negative staining; (vii) The fluffy material around the shaft is more distinct in cells treated with acetic acid; (viii) The shaft ends in the proximal fifth of the extrusome; (ix) There is ramified, fluffy material on the pellicle around the extrusome attachment site.

Jerka-Dziadosz (1964) designated the extrusomes as protrichocysts. Later, she used the term trichocysts (Jerka-Dziadosz 1972). Suganuma (1972, p. 347) stated that "the trichocyst of *Urostyla cristata* closely resembles the trichocyst of *Paramecium*". However, Grim & Manganaro (1985) describe several significant differences between these two types of extrusomes, such as a complex microtubular cap in *Pseudourostyla* (not present in *Paramecium*); a small, but prominent electron opaque shaft or core in *Pseudourostyla* (not present in *Paramecium*); and a nail-shaped cap in *Paramecium* (not present in *Pseudourostyla*). They therefore did not believe that there are sufficient similarities to warrant calling the extrusomes of *Pseudourostyla* trichocysts. Consequently, they used the generic term extrusome and suggested from their results that these organelles release cyst wall material. Thus, their high number in trophozoites with a good food supply is unexpected. They speculated that these organelles perhaps evolved in a milieu in which rapid encystment in a rapidly changing environment had survival value.

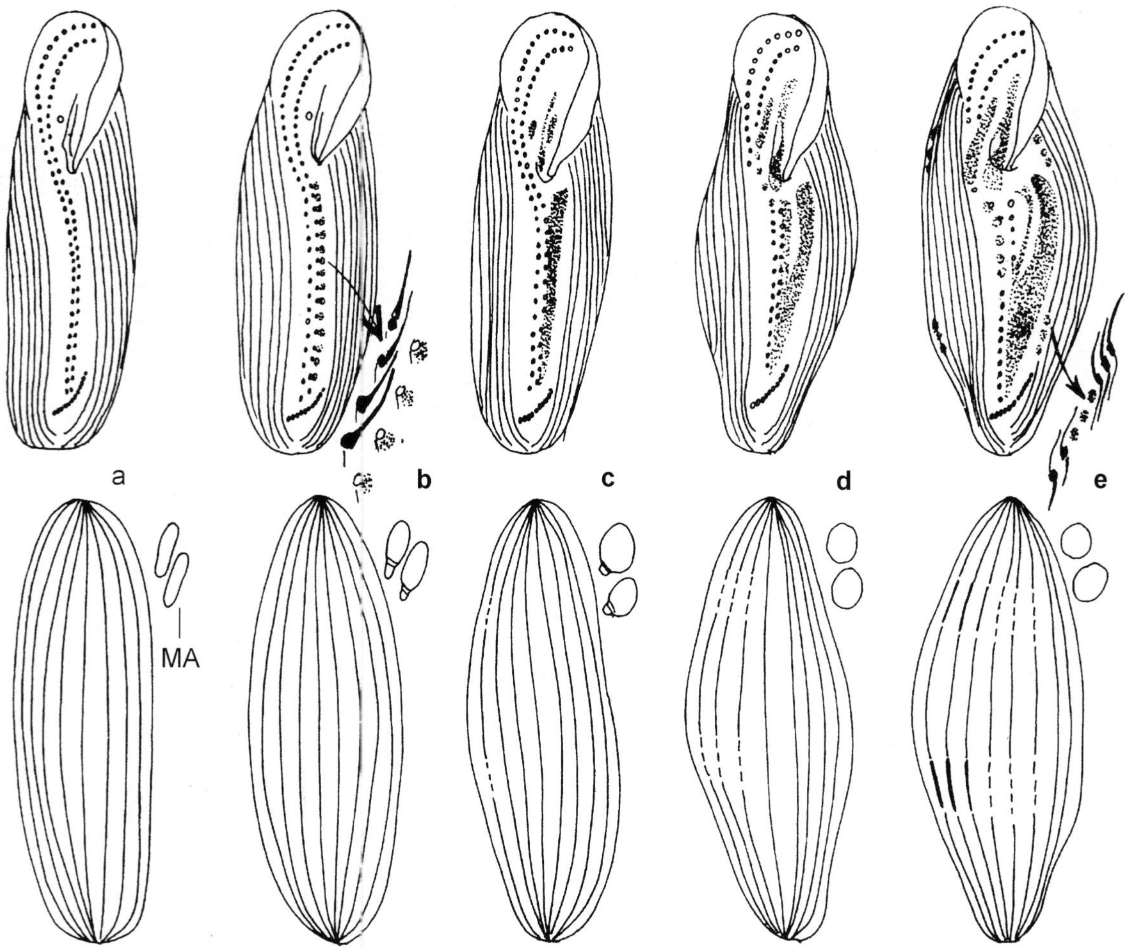

Fig. 149a–e *Pseudourostyla cristata* (from Jerka-Dziadosz 1972. Protargol impregnation). Schematic illustrations of the infraciliature of the ventral (upper line) and dorsal (lower line) side of an interphasic specimen (a) and early and middle dividers (b–e). For a brief description of the main events during cell division, see text. MA = macronuclear nodules. Page 757.

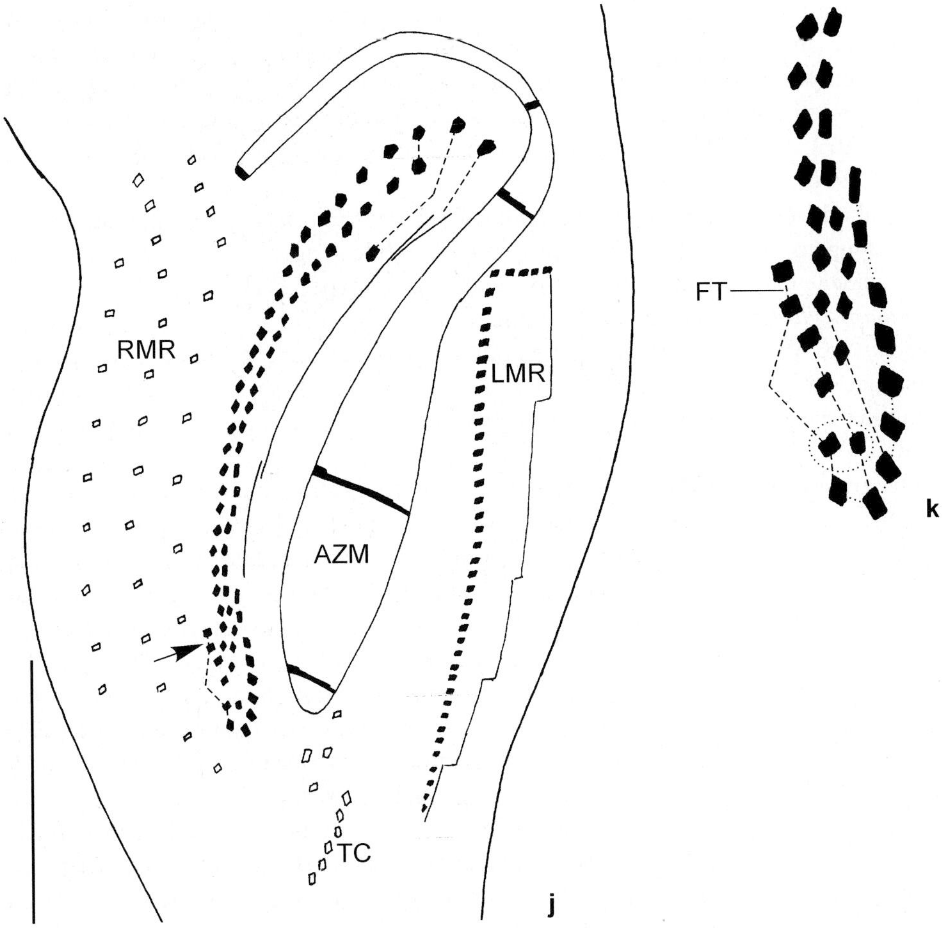

Fig. 150j, k *Pseudourostyla levis* (original from neotype population). Infraciliature of an opisthe of a late divider (scale bar = 30 μm). Note that not all structures are shown because some of them are stained to weakly to be recognised in detail. Cirri which originate from the same anlage are connected by a broken line (only the cirri of some selected anlagen are connected). Arrow in (j) marks frontoterminal cirri (see k). In (k) the new transverse cirri are connected by a dotted line and the pretransverse ventral cirri are circled. Parental cirri white, new black. AZM = adoral zone of membranelles, FT = frontoterminal cirri, LMR = new left marginal rows which originate from same anlage, RMR = parental right marginal rows, TC = parental transverse cirri. Page 778.

(36.3 h) or with normal micronuclei (23.3 h; Takahashi 1983c). Takahashi & Suhama (1989, 1991a) studied the regeneration of amicronucleate fragments (Fig. 150d–g).

Takahashi investigated several aspects of conjugation of *Pseudourostyla levis*, namely, the nuclear behaviour in total and temporary conjugation (Fig. 8a; Takahashi 1974, 1979), the pre-mating behaviour and pairing process in two types of total conjugation (Takahashi 1983a), and the properties of synconjugant clones from total conjugation (Takahashi 1983d). Furthermore, he made a comparison of nuclear changes between slow and rapid types of total conjugation (Takahashi 1983b). Paraconjugation (also called pseudoconjugation) occurs in some combinations of *Pseudourostyla levis* strains, while in other combinations there is true (meiotic) temporary conjugation (Takahashi 1973, 1974, 1983b), suggesting that paraconjugation is genetically determined (Raikov 1996).

Takahashi (1973) found that *Pseudourostyla levis* consists of two mating groups, that is, syngens. Syngen 1 had eight mating types while syngen 2 had three. Later, it became evident that syngen 1 consisted of 28 mating types (Takahashi 1977), that is, *Pseudourostyla levis* has a multiple mating type system (Dini & Nyberg 1993, p. 104). Suganuma & Yamamoto (1980) studied mitochondrial crystals in both *P. cristata* from a pond in Nara City (Japan) and *P. levis* provided by Takahashi, the author of *P. levis*. Unfortunately, they did not comment on differences between these two Japanese populations.

Occurrence and ecology: Due to the neotype designation, the type locality of *P. levis* is the site where the neotype material was collected (ICZN 1999). Clone Hb-25b, to which the neotype specimen belongs, has a rather complicated history. Hb-25b is a descendant of a cross-breeding of stocks HG-3 and HS-3. HG-3 was collected at Kumano-cho, Aki-gun, Hiroshima Prefecture on 02.11.1981, whereas HS-3 is a descendant of a cross-breeding of stock HB-12 and SB-3. Stock HB-12 is from Hiwa-cho, Hiba-gun, Hiroshima Prefecture (collected on 29.04.1970) and SB-3 is from Seburi-son, Kanzaki-gun, Saga Prefecture (collected on 02.05.1972). For simplicity, I fix Kumano-cho (34°19'49"N 132°34'43"E), Aki-gun in the Hiroshima prefecture as type locality of the neotype of *P. levis*. Takahashi collected all these samples from paddy fields.

Takahashi (1973) found *Pseudourostyla levis* in paddy fields from the Japanese cities Hiroshima, Yamaguchi, Fukuoka, and Saga. Unfortunately, he did not designate one of these towns, which are up to 240 km apart, as type locality. This is one reason for the neotype designation (see above). There are some further records of *P. levis* from Japan by Takahashi, namely, Yuda in Yamaguchi City, several localities (Aki, Hongo, Koyo, Yagi, Oshiba, Gokurakuji, Shichikenjaya, Takehara) in the Hiroshima Prefecture (Takahashi 1977, Takahashi & Suhama 1991a, 1991b, p. 108), and Kochi City (Takahashi 1983c).

Takahashi (1973) collected dry rice-stubble to which the cysts were attached. They excysted a week or less after inoculation into a wheat-lettuce infusion with several pieces of rice-stubble. *Pseudourostyla levis* reproduced normally in this medium containing the heterotrophic flagellate *Bodo* sp. and the bacterium *Bacilis subtilis* as food. Later, *Chlorogonium elongatum* was used as supplementary food organism (Takahashi 1983b). Under these culture conditions the generation time was about 24 h at 23° C (see also Takahashi 1972).

Pseudourostyla franzi Foissner, 1987 (Fig. 151a–i, Table 34)

1987 ***Pseudourostyla franzi* nov. spec.**[1] – Foissner, Zool. Beitr., 31: 194, Abb. 3a–f, 37, Tabelle 2 (Fig. 151a–f; original description. The holotype slide [1988/135] and 2 paratype slides [1988/136, 137] are deposited in the Oberösterreichische Landesmuseum in Linz [LI], Austria).

1992 ***Pseudourostyla franzei* Foissner, 1987** – Shen, Liu, Song & Gu, Protozoa, p. 150, Fig. Aa, Ab (Fig. 151h, i; incorrect subsequent spelling; redescription of a Chinese population).

2001 ***Pseudourostyla franzi* Foissner, 1987** – Berger, Catalogue of ciliate names 1. Hypotrichs, p. 77 (nomenclator containing all basionyms, combinations, and higher taxa of hypotrichs).

Nomenclature: Foissner dedicated this species to the collector of the sample, Herbert Franz, a famous Austrian soil zoologist.

Remarks: Differs from the other *Pseudourostyla* species, inter alia, by the higher number of marginal rows and macronuclear nodules (see key for details). *Pseudourostyla muscorum*, which lives in terrestrial mosses, has only two marginal rows per side, but many more buccal cirri. Foissner (pers. comm.) found *P. franzi* several times. These populations agree in all features with the type population. The description below is based solely on Foissner's (1987b) data.

The dorsal ciliature of *P. franzi* is highly interesting. Morphogenetic data are needed to decide whether the cirri behind the dorsal kinety area are true caudal cirri or a kind of marginal cirri (Fig. 151e).

Morphology: Body size 200–300 × 50–70 μm, body length:width ratio about 3.5–3.8:1 in life (Fig. 151a, b), and 4.6:1 on average in protargol preparations (Table 34). Body outline elongate elliptical and often somewhat sigmoidal, lateral margins slightly converging posteriorly; body somewhat flattened dorsoventrally, very flexible, but not contractile. Macronuclear nodules scattered throughout cytoplasm except for the first and last tenth, usually ellipsoidal, rarely globular (Fig. 151a, f). More than 10 micronuclei scattered between macronuclear nodules, of same shape as macronuclear nodules, but slightly smaller (Fig. 151f, Table 34). Contractile vacuole somewhat ahead of mid-body near left cell margin, during diastole with two long collecting canals (Fig. 151b). Cortical granules along cirral rows and dorsal kineties, large (1.0–2.0 × 1.5 μm) and colourless, make cells brownish at low magnification, stain red when methyl green-pyronin is added, and often impregnate strongly with protargol so that the cirral pattern is masked. According to Foissner (pers. comm.) the cortical granules explode similarly to trichocysts, indicating homology with the extrusomes of the *P. cristata* group. Moves moderately rapidly.

Adoral zone occupies 30% of body length on average, extends far onto right body side, composed of an average of 67 membranelles of ordinary fine structure; bases of largest membranelles about 13 μm wide in life. Buccal cavity wide and deep. Undulating membranes long and slightly curved, intersect optically about at level of buccal cirri.

[1] The diagnosis by Foissner (1987b) is as follows: In vivo etwa 200–300 × 50–70 μm große *Pseudourostyla* mit ellipsoiden farblosen subpelliculären Granula und durchschnittlich insgesamt 18 Cirrenreihen, von denen die äußeren auf die Dorsalseite verlagert sind. Etwa 200 Makronucleus-Teile. 4 Dorsalkineten.

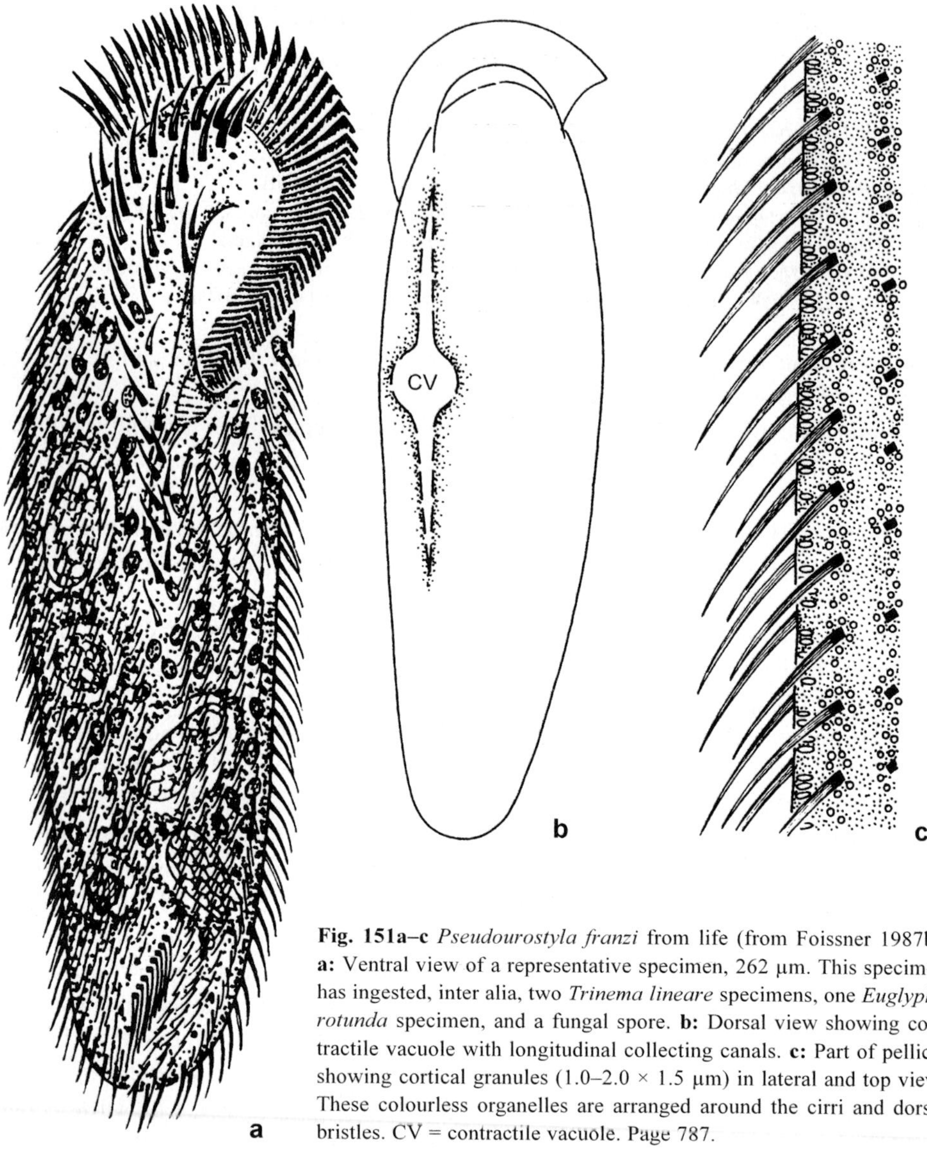

Fig. 151a–c *Pseudourostyla franzi* from life (from Foissner 1987b). **a:** Ventral view of a representative specimen, 262 µm. This specimen has ingested, inter alia, two *Trinema lineare* specimens, one *Euglypha rotunda* specimen, and a fungal spore. **b:** Dorsal view showing contractile vacuole with longitudinal collecting canals. **c:** Part of pellicle showing cortical granules (1.0–2.0 × 1.5 µm) in lateral and top view. These colourless organelles are arranged around the cirri and dorsal bristles. CV = contractile vacuole. Page 787.

Pharynx extends longitudinally backwards, wall after protargol impregnation with many about 1 µm long rods (Fig. 151d).

Cirral pattern and number of cirri of usual variability (Table 34), except for the number of buccal cirri (CV = 27%) and transverse cirri (CV = 24%). All cirri fine and about 20 µm long. Frontal cirri distinctly enlarged, form distinct bicorona along distal portion of adoral zone; specimen shown in Fig. 151d with eight cirri in anterior and seven cirri in posterior bow (however, note that it is difficult to decide where the frontal

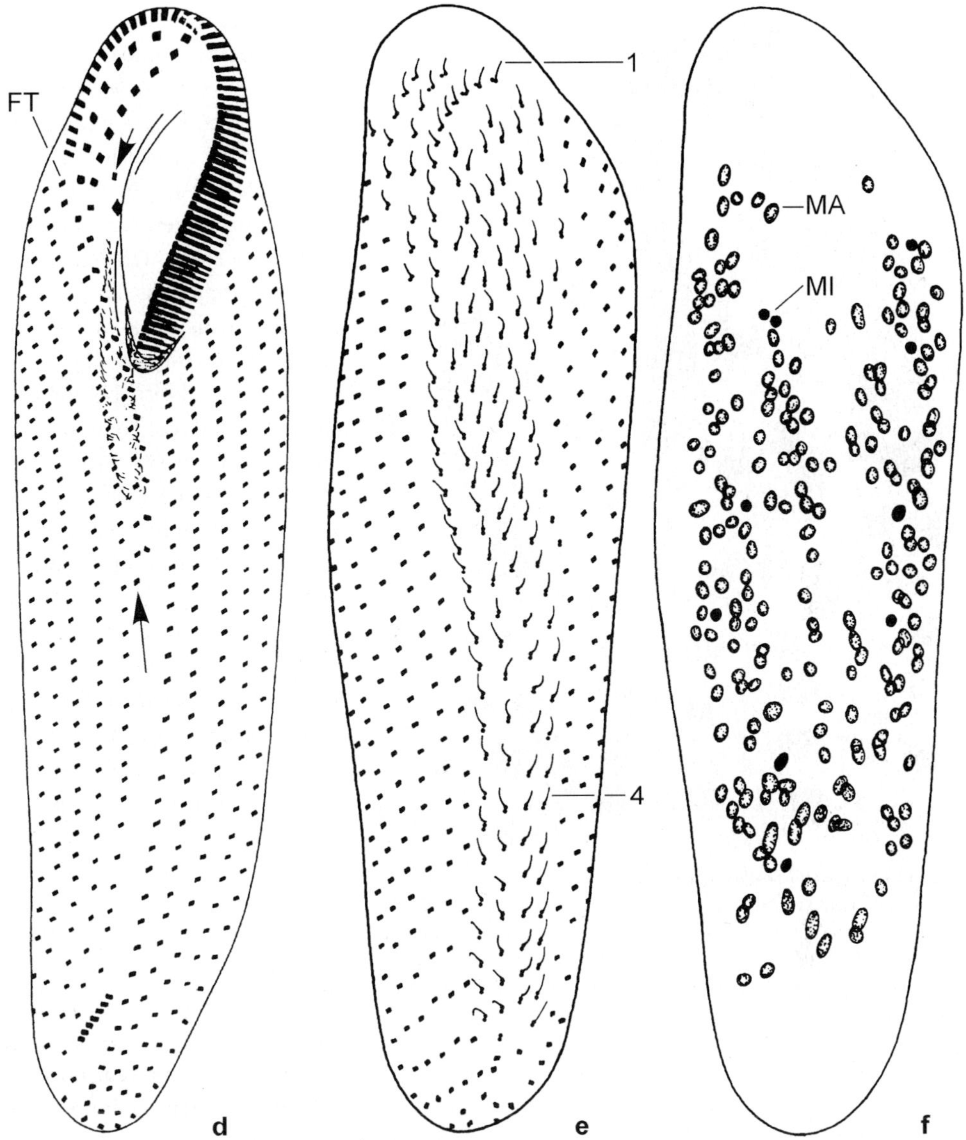

Fig. 151d–f *Pseudourostyla franzi* (from Foissner 1987b. Protargol impregnation). **d:** Infraciliature of ventral side of a representative specimen, 257 µm. The cortical granules (see Fig. 151c) stained very heavily with protargol and thus masked the dorsal infraciliature so that another specimen had to be used to show the dorsal infraciliature (e). Short arrow marks the anterior buccal cirrus which is always smaller than the posterior; possibly they are not from the same anlage. Long arrow denotes rear end of midventral complex. **e, f:** Infraciliature of dorsal side and nuclear apparatus of same specimen, 214 µm. *Pseudourostyla franzi* is easily distinguished from its congeners by the high number of macronuclear nodules and cirral rows; both features are easily recognisable in life. Note the suture formed by dorsal kinety 1, dorsal kineties 2–4, and the rightmost right marginal rows which commence with a more or less high number of dorsal bristles. FT = frontoterminal cirri, MA = macronuclear nodule, MI = micronuclei, 1, 4 = dorsal kineties. Page 787.

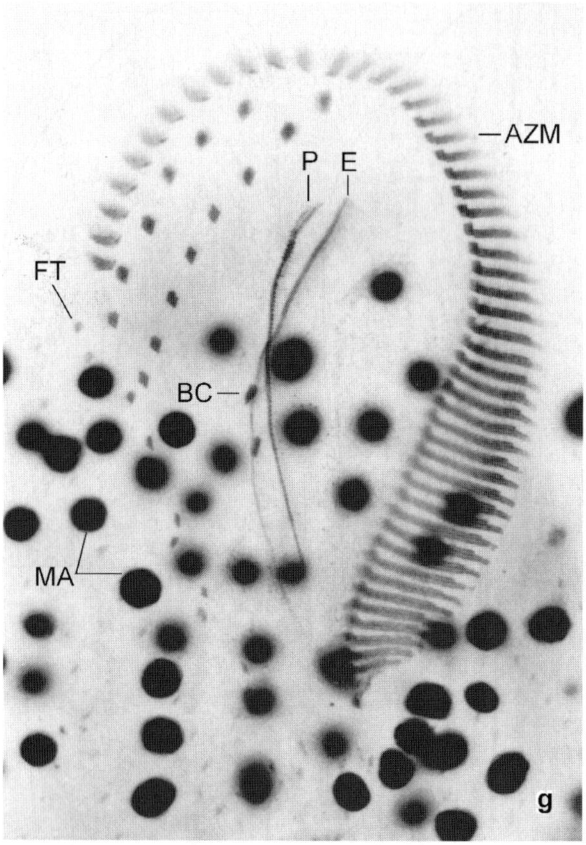

Fig. 151g *Pseudourostyla franzi* (original kindly supplied by W. Foissner, Salzburg. Protargol-impregnated specimen from South Africa). Infraciliature of oral region. AZM = adoral zone of membranelles, BC = anterior buccal cirrus, E = endoral, FT = frontoterminal cirri, MA = macronuclear nodules, P = paroral. Page 787.

ciliature ends and the midventral complex begins). Usually two buccal cirri at level of optical intersection of undulating membranes; posterior buccal cirrus always larger than anterior (morphogenetic data are needed to show whether or not the anterior cirrus is in fact a buccal cirrus, that is, originates from anlage II). Two frontoterminal cirri behind distal end of adoral zone at level of buccal cirri. Midventral complex likely composed of cirral pairs only (whether or not a short midventral row is present can only be recognised in late dividers); all midventral cirri of about same size, but cirri of each pair differently orientated; first of the on average 18 pairs/pseudopairs at about level of large buccal cirrus, last pair at 47% of body length on average. Transverse cirri narrowly spaced, form straight, oblique row terminating 23 μm ahead of rear body end on average in protargol-impregnated specimens; thus, in life only the rearmost (= rightmost) cirri reach posterior body end. On average eight left and eight right marginal rows; all rows end terminal (Fig. 151d); outermost rows arranged on dorsal side and composed of dorsal bristles anteriorly.

Dorsal cilia about 3 μm long in life, arranged in a rather complicated pattern (Fig. 151e): (i) One almost bipolar row (= dorsal kinety 1) in midline of cell with anterior end distinctly curved rightwards. (ii) Three anteriorly distinctly shortened kineties (= dorsal kineties 2–4). (iii) Outermost seven right marginal rows with the anterior portion composed of dorsal bristles and commencing more posteriorly than kinety 1 with which, together with kineties 2–4, they form a conspicuous suture. (iv) Anterior portion of outermost three left marginal rows also composed of dorsal bristles. All dorsal kineties end subterminally with some cirri behind. Ontogenetic data are needed for a correct interpretation of this pattern.

Specimens of the Chinese population in life(?) 210–300 × 50–65 μm (Fig. 151h, i). Fig. 151i is likely a simplified redrawing of Foissner's illustration (Fig. 151d).

Occurrence and ecology: Very likely *P. franzi* is confined to terrestrial habitats (Foissner 1987a, 1998). The type locality of *Pseudourostyla franzi* is the Mount Verde on the Cap Verde Island Santo Vicente, where it occurred in a brown-black soil sample (altitude about 720 m, pH 8.4) collected by Herbert Franz (Vienna) on October 17, 1985 (Foissner 1987b, p. 187, 194). Shen et al. (1992) found this species in Chinese subtropical soils. Foissner (1995, p. 39) recorded *P. franzi* in a soil sample (0–3 cm litter and soil layer) from the Santo Rosa National Park (Costa Rica), about 5 km east of the ranch house "La Casona", near a small path to the Pacific Ocean. Foissner (1998, p. 208; 1999, p. 324; pers. comm.) found *P. franzi* also in Australia (cortical granules 3.0 × 1.5 μm) and Africa (Kenya; Republic of South Africa, Fig. 151g). In Namibia it occurred in three out of 73 soil samples (Foissner et al. 2002, p. 62).

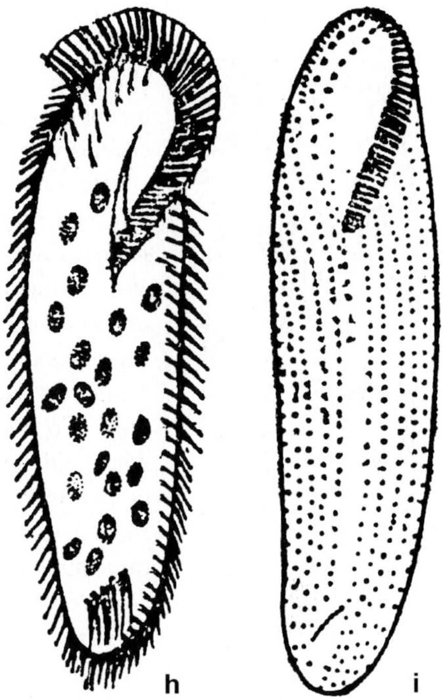

Fig. 151h, i *Pseudourostyla franzi* (from Shen et al. 1992; Fig. 151i is likely a redrawing of Fig. 151d. h, from life?; i, protargol impregnation). **h:** Ventral view, size not indicated. **i:** Infraciliature of ventral side, size not indicated. Page 787.

Feeds on naked and testate amoeba (*Trinema lineare*, *Euglypha rotunda*), ciliates, and fungi (Foissner 1987b). Other populations ingested heterotrophic flagellates (*Polytoma*) and cysts of naked amoeba and ciliates, for example, *Leptopharynx costatus* (Foissner, pers. comm.). Biomass of 10^6 individuals about 375 mg (Foissner 1998, p. 208).

Pseudourostyla muscorum (Kahl, 1932) Borror, 1972 (Fig. 152a, Table 34)

1932 ***Urostyla muscorum* spec. n.** – Kahl, Tierwelt Dtl., 25: 566, Fig. 110_9 (Fig. 152a; original description. No type material available and no formal diagnosis provided).

1972 ***Pseudourostyla muscorum* (Kahl, 1932) n. comb.** – Borror, J. Protozool., 19: 11 (combination with *Pseudourostyla*; revision of hypotrichs).

2001 ***Pseudourostyla muscorum* (Kahl, 1932) Borror, 1972** – Berger, Catalogue of ciliate names 1. Hypotrichs, p. 102 (nomenclator containing all basionyms, combinations, and higher taxa of hypotrichs).

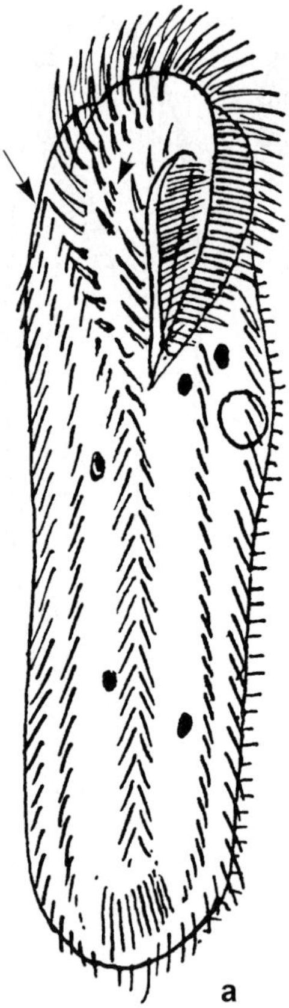

Fig. 152a *Pseudourostyla muscorum* (from Kahl 1932). Ventral view from life, 300 µm. The long arrow marks two small cirri which could be the frontoterminal cirri. The short arrow denotes three cirri which are possibly the rear end of the inner (= posterior) frontal cirral bow. Page 791.

Nomenclature: No derivation of the name was given in the original description. The species-group name *muscorum* (Latin, genitive plural; Hentschel & Wagner 1996, p. 410) means "living in moss" and refers to the habitat where the species was discovered.

Remarks: Kahl (1932) provided a rather detailed description and illustration. Borror (1972) transferred it to *Pseudourostyla* because of the distinct zigzagging midventral pattern and the increased number of marginal rows. Later, he synonymised this moss-inhabiting species with the limnetic *Urostyla grandis* (Borror & Wicklow 1983, p. 120). However, this synonymy is certainly unjustified because *P. muscorum* has only two marginal rows per side and therefore a much less dense ciliature than *Urostyla grandis*, which usually has four or five marginal rows per side and some midventral rows. At superficial observation *P. muscorum* is easily confused with the syntopic *Anteholosticha antecirrata*, which has a similar size, cortical granulation, and nuclear apparatus (Fig. 73a–h). However, it has only three frontal cirri and one left and right marginal row. Detailed redescription necessary, including morphogenesis to show the mode of marginal row formation. Surprisingly, several records not substantiated by morphological data are available; likely, some of them are misidentifications.

Metabakuella bimarginata (Fig. 204a–d) is very similar as concerns size, shape, cortical granulation, and marginal rows. However, this species has distinct midventral rows so that synonymy with *P. muscorum* is unlikely.

Kahl (1932) did not describe the nuclear apparatus in detail. According to question 6 of his *Urostyla* key, the present species has (very likely) many macronuclear nodules. Further, he separated it from the multimacronucleate *Urostyla grandis* only by the habitat (terrestrial mosses vs. limnetic) and the total number of cirral rows (his question 7b), also indicating that *P. muscorum* and *U. grandis* have the same nuclear pattern. Consequently, *Pseudourostyla muscorum* differs from *P. urostyla* and its synonym *U. pseudomuscorum* not only in the number of cirral rows (six vs. eight), but also in the nuclear apparatus (many scattered macronuclear nodules vs. two).

Morphology: Body length 250–350 µm; length:width ratio according to Fig. 152a about 3.7:1, that is, body width about 65–95 µm. Body outline elongate elliptical; right margin according to Fig. 152a straight to slightly concave, left distinctly vaulted at level of contractile vacuole; both ends broadly rounded. Macronuclear nodules scattered (number not estimated by Kahl). Contractile vacuole at about 40% of body length near left cell margin. Cortical granules arranged in longitudinal rows, possibly slightly coloured because cells are brownish due to these organelles. Adoral zone occupies about 31% of body length (Fig. 1152a), distal end extends far onto right side. Buccal cavity moderately wide, buccal lip distinctly curved leftwards anteriorly. Frontal cirri arranged in two bows (coronas), cirri of rear bow possibly smaller than those of anterior bow (Fig. 152a). Several (9 in Fig. 152a) buccal cirri along full length of paroral. A short row of three cirri (Fig. 152a, short arrow) between buccal row and anterior end of midventral complex; possibly these cirri are the rear end of the inner frontal cirral bow. Frontoterminal cirri not mentioned, but possibly illustrated by Kahl (1932; Fig. 152a, long arrow). Midventral complex likely composed of midventral pairs only, extends in cell midline close to transverse cirri. 8–15 transverse cirri subterminal in roughly J-shaped, slightly oblique row, do not protrude beyond rear body end. Two left and two right marginal rows, the outermost rows almost joining posteriorly. Dorsal bristles likely of ordinary length, that is, 3–4 μm (Fig. 152a); number of kineties and presence/absence of caudal cirri not known.

Occurrence and ecology: Kahl (1932) found this species in terrestrial mosses from three sites (North Germany; Bavarian Alps, South Germany; Wisconsin, USA) without fixing one of them as type locality. According to Kahl common, but not very abundant. Omnivorous (even ingests rotifers!), but feeds mainly on ciliates.

Records not substantiated by illustrations and/or morphological data: moderately abundant in two dry mosses from Bavaria (Wenzel 1953, p. 108; body length only 110–200 μm indicating misidentification); dystrophic pond and damped *Sphagnum* between the villages of Ibbenbüren and Hopsten (52°21'N 7°39'E) about 15 km west of the town of Osnabrück, Germany (Mücke 1979, p. 271); in the litter horizon and the matted and fibrous layer (largely litter interwoven with many fine roots) from Lodge Wood, a beech wood on the Chiltern Hills (51°40'N 0°56'W), about 20 km west of London, Great Britain (Stout 1963, p. 285); in submerged mosses (frequency 1.9%), wet mosses (frequency 16.7%), moist mosses (frequency 12.1%), and dry mosses (frequency 6.7%) from the Slovenský raj in the Stratenská hornatin highlands and in dry mosses from localities near Bratislava, Slovakia (Tirjaková & Matis 1987a, p. 11; 1987b, p. 23; Matis et al. 1996, p. 20); moss and soil from the Antarctic region (Sudzuki 1979, p. 124).

Pseudourostyla nova Wiackowski, 1988 (Fig. 153a–d, Table 34)

1988 ***Pseudourostyla nova* sp. nov.**[1] – Wiackowski, J. nat. Hist. (London), 22: 1085, 1092, Fig. 1–18, Table 1 (Fig. 153a–d; original description. The type slide is deposited in the Department of Hydrobiology, Institute of Environmental Biology, Jagiellonian University, Krakow, Poland).

[1] The diagnosis by Wiackowski (1988, p. 1092) is as follows: *Pseudourostyla* with two marginal rows both on left and right side of the cell. The right row of midventral cirri is shorter than the left one. It ends close to the right end of the adoral zone of membranelles. The left midventral row approaches the anterior end of the cell. This pattern results from a particular morphogenetic process occurring in the last phase of the cortical morphogenesis. Number of dorsal kineties: 7; about 15 fragments of macronucleus; several (5–8) micronuclei; dense coat of extrusomes all over the cell surface.

2001 ***Pseudourostyla nova* Wiackowski, 1988** – Berger, Catalogue of ciliate names 1. Hypotrichs, p. 77 (nomenclator containing all basionyms, combinations, and higher taxa of hypotrichs).

Nomenclature: No derivation of the name was given in the original description. The species-group name *nova* (Latin adjective; new) possibly refers to the specific (= new) mode of arrangement of frontal cirri (see morphogenesis).

Remarks: Wiackowski (1988) assigned the present species to *Pseudourostyla* because of the lack of large frontal cirri, the adoral zone which extends far onto the right cell side, and the presence of more than one marginal row per side. Moreover, the presence of frontoterminal cirri, the specific ontogenesis of the marginal rows (both rows of each side originate from a single anlage), and the lack of distinct midventral rows indicate a classification in *Pseudourostyla*.

Wiackowski overlooked that Borror (1972) not only transferred *Urostyla cristata* to *Pseudourostyla*, but also *Urostyla muscorum* and *Oxytricha urostyla*. Thus, he did not compare his species with *Pseudourostyla muscorum*, which also has only two marginal rows per side. However, Wiackowski's species has a lower number of buccal cirri (one vs. about nine) and probably also fewer macronuclear nodules (no detailed data available for *P. muscorum*). Moreover, only one frontal cirral bow is present during interphase. This feature, which is due to unequal migration of the left and right cirri of the cirral pairs, is considered as autapomorphy of *P. nova* by Wiackowski (1988).

Morphology: Body size in life not given; specimen shown in Fig. 153a about 235 × 70 µm, which is only slightly larger than the average values of protargol-impregnated cells (232 × 65 µm; Table 34); length:width ratio about 3.6:1 after protargol impregnation. Body outline elongate elliptical with lateral margins more or less parallel; anterior and posterior end broadly rounded; body distinctly flattened dorsoventrally. Macronuclear nodules arranged in indistinct, wide row underneath midventral complex; individual nodules usually ellipsoidal (length:width ratio about 2:1), in life(?) 15–18 µm long, contain several nucleoli of ordinary size. Micronuclei attached to macronuclear nodules at various positions, ellipsoidal, in life(?) 5–7 µm long. Contractile vacuole distinctly ahead of mid-body close to left body margin. Cortical granules form dense seam throughout cortex, after protargol impregnation (Wilbert's method) about 3.5 µm long. Cytoplasm of live specimens grey (possibly due to the cortical granules). Movement not mentioned, likely without peculiarities.

Adoral zone occupies 34% of body length on average, composed of about 50 membranelles (Table 34), extends around anterior cell margin. Largest membranelles in proximal quarter of adoral zone (Fig. 153b). Undulating membranes begin at 18% of body length in specimen shown in Fig. 153b; both membranes slightly curved and optically intersecting at level of buccal cirrus, paroral shorter than endoral, which terminates near proximal end of adoral zone. Paroral composed of double row of basal bodies, endoral consists of single row.

Cirral pattern and number of cirri of usual variability (Fig. 153b, Table 34). During interphase only the posterior bow of slightly enlarged frontal cirri extends along the frontal cell margin (see morphogenesis, for details). Buccal cirrus distinctly behind anterior end of undulating membranes. Two frontoterminal cirri in ordinary position, that is, left of anterior end of inner right marginal row. Midventral complex composed of cirral pairs only; individual cirral pairs only in posterior body half of ordinary arrangement, that is, right cirrus obliquely ahead of left cirrus; in anterior portion right cirrus at same level or behind left cirrus because cirri of anterior frontal bow do not extend along anterior cell margin, that is, frontal cirrus formed by anlage III is close to distal end of adoral zone (Fig. 153b; details, see morphogenesis). Possibly two pretransverse ventral cirri present. Transverse cirri subterminal, arranged in hook-shaped pattern, according to Fig. 153a about 34 μm long, bases distinctly larger than those of marginal cirri or most midventral cirri (Fig. 153b). Invariably two left and two right marginal rows; inner right marginal row begins at level of frontoterminal cirri, ends subterminally, outer right row commences likely somewhat behind anterior end of inner row, extends along rear body margin to cell midline; inner left marginal row begins about at level of cytostome, terminates near end of outer right marginal row, leaving only a small gap; outer left marginal row commences distinctly more anteriorly than inner row, ends subterminally (Fig. 153b).

Dorsal cilia invariably arranged in seven roughly bipolar kineties; length neither mentioned nor deducible from illustration, likely of ordinary length, that is, 2–4 μm. Caudal cirri lacking (Fig. 153c).

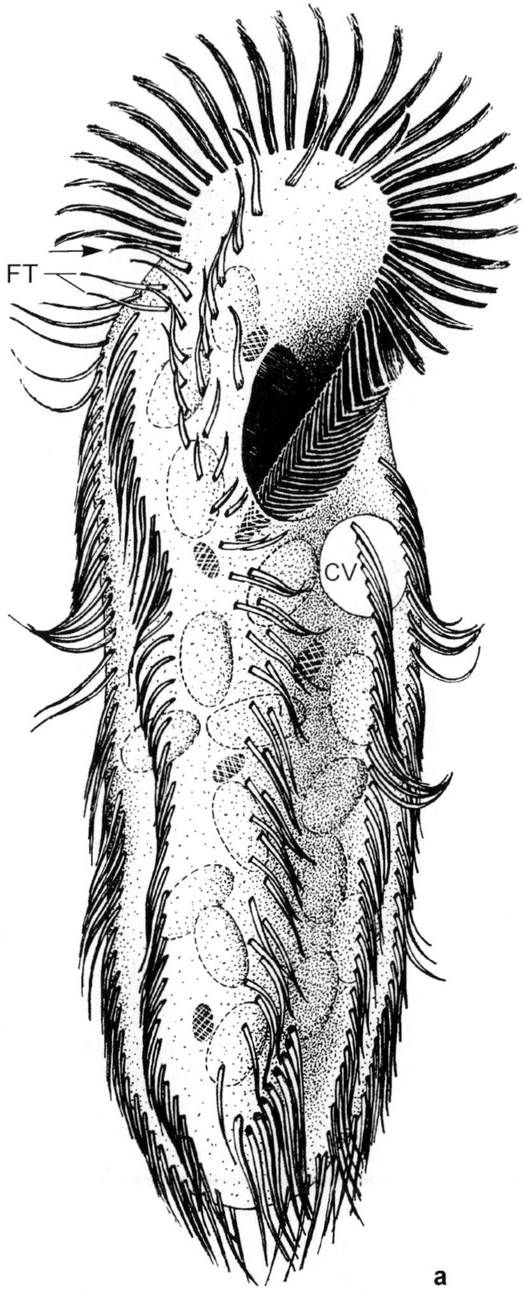

a

Fig. 153a *Pseudourostyla nova* (from Wiackowski 1988a). Ventral view of a representative specimens from life, 235 μm. Arrow marks cirrus III/1 (for details on the rather complicated cirral pattern, see Fig. 153d). CV = contractile vacuole, FT = frontoterminal cirri. Page 793.

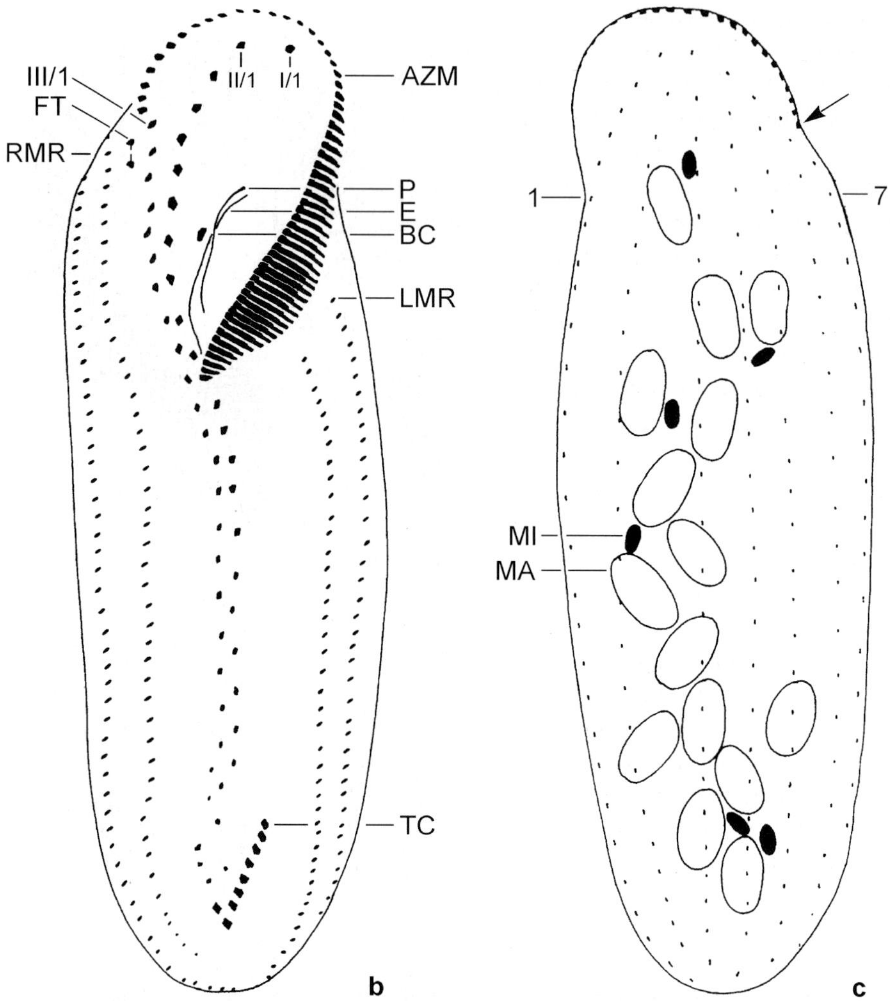

Fig. 153b, c *Pseudourostyla nova* (from Wiackowski 1988a. Protargol impregnation, Wilbert's method). Infraciliature of ventral and dorsal side and nuclear apparatus of representative specimen, 239 µm. Arrow in (c) denotes distal end of adoral zone of membranelles. AZM = adoral zone of membranelles, BC = buccal cirrus, E = endoral, FT = frontoterminal cirri, LMR = outer left marginal row, MA = macronuclear nodule, MI = micronucleus, P = paroral, RMR = inner right marginal row, TC = leftmost transverse cirrus, I, II, III = anteriormost cirrus of anlagen I–III, 1, 7 = dorsal kineties. Page 793.

Cell division: The ontogenesis is described in the original description (Wiackowski 1988; Fig. 153d). Ten stages are documented by micrographs. No line drawings are given so that the reader has to be referred to the original paper. The description of the process is detailed and divided in three parts.

The first basal bodies of the oral primordium appear in patches close to the left cirri of the midventral pairs about in the middle of the cell. Somewhat later, these patches

1932 ***Urostyla (Oxytricha) urostyla* (Clap. u. L., 1858)** – Kahl, Tierwelt Dtl., 25: 568, Fig. 97_{20} (Fig. 155b; combination with *Urostyla*; revision of hypotrichous ciliates; redrawing from Claparède & Lachmann 1858).

1972 ***Oxytricha urostyla* Clap. & Lach., 1858** – Borror, J. Protozool., 19: 9 (synonymy with *Urostyla multipes*; see remarks; revision of hypotrichs).

1972 ***Pseudourostyla urostyla* (Claparède & Lachmann, 1858) n. comb.** – Borror, J. Protozool., 19: 11 (combination with *Pseudourostyla*; revision of hypotrichs).

2001 ***Pseudourostyla urostyla* (Claparède and Lachmann, 1858) Borror, 1972** – Berger, Catalogue of ciliate names 1. Hypotrichs, p. 65 (nomenclator containing all basionyms, combinations, and higher taxa of hypotrichs).

Nomenclature: Claparède & Lachmann (1958) likely used the species-group name *urostyla* (for derivation, see genus *Urostyla*) because the present species resembles *Urostyla grandis*. In my catalogue of ciliate names (Berger 2001) I forgot the combination with *Urostyla* proposed by Kahl (1932).

Remarks: Claparède & Lachmann (1858) provided an extensive description including a detailed illustration so that the existence of this species should be beyond reasonable doubt (Fig. 155a). Surprisingly, it was never reliable recorded since its original description. Claparède & Lachmann unequivocally observed two macronuclear nodules so that identity with *Urostyla grandis*, which has many nodules, can be excluded. Stein (1859, p. 194) mentioned *Oxytricha urostyla* at the end of the description of his *Urostyla weissei* – now *Paraurostyla weissei* (Stein, 1859) Borror, 1972 (for review, see Berger 1999, p. 844) – where he suggested that the considerable differences between these two species could be due to misobservations by Claparède & Lachmann. However, Claparède & Lachmann (1858) described a second similar species, *Oxytricha multipes*, which has a lower number of frontal cirri and the anteriormost cirri distinctly enlarged. Consequently, Kent (1882, p. 765) synonymised it with *Paraurostyla weissei*. Interestingly, Stein, Kent, and later workers ignored the principle of priority. To avoid the suppression of the now well established name *Paraurostyla weissei*, we classified *O. multipes* as supposed synonym of *P. weissei* (Foissner et al. 1991, Berger 1999). By contrast, the present species has distinctly more frontal cirri, which form a distinct bicorona, indicating that *O. urostyla* has a midventral complex. Therefore synonymy with *Paraurostyla weissei* is very unlikely (Berger 1999, p. 873). Only the description of a population matching the original description of *O. urostyla* (two macronuclear nodules, bicorona, midventral complex, two or more marginal rows per side including special mode of row formation, transverse cirri present, with cortical granules?) will finally show that the classification of *O. urostyla* in *Pseudourostyla* is correct and synonymy with *Paraurostyla weissei* unjustified.

Fromentel (1876) provided a brief redescription and one illustration (Fig. 143e). However, the data are too scant to accept the identification. Fromentel's contribution is therefore classified as insufficient redescription. Dumas (1929) provided two poor illustrations, which do not show important features (Fig. 143f, g). His text is a repetition of Fromentel's data. Further details, see insufficient redescriptions at the end of the present genus.

Kahl (1932), who did not provide own data, transferred the present species from *Oxytricha* to *Urostyla*, which he characterised as follows: several (4–10) ventral rows with short cirri, two indistinctly set off marginal rows, and a distinct row of transverse cirri.

Borror (1972) mistakenly classified *Oxytricha urostyla* twice: (i) On page 9 as a synonym of *Urostyla multipes* (Claparède & Lachmann, 1858) Kahl, 1932, which is a supposed synonym of *Paraurostyla weissei* (details, see above). (ii) On page 11, Borror transferred it to *Pseudourostyla* because the cirral pattern resembles that of *Pseudourostyla cristata*, type of the genus. By contrast, Borror & Wicklow (1983, p. 124) considered *Pseudourostyla* as monotypic and again classified *Oxytricha urostyla* as synonym of *Urostyla multipes*. I basically agree with Borror's (1972) classification of *Oxytricha urostyla* in *Pseudourostyla* because the cirral pattern matches the characterisation of the genus rather well.

As mentioned above, *Pseudourostyla urostyla* was not reliable recorded since its original description in 1858 (the redescriptions by Fromentel and Dumas are insufficient, see below). It differs from most other *Pseudourostyla* species, inter alia, by the two macronuclear nodules. *Urostyla grandis* is larger (usually >250 µm) and has many macronuclear nodules and more (usually >10) cirral rows. *Pseudourostyla raikovi* and *P. magna*, which also have only two macronuclear nodules, have more cirral rows (Fig. 158a, 159a). *Pseudourostyla urostyla* can be easily confused with *Paraurostyla weissei*, which has a similar size, shape, cirral pattern, and nuclear apparatus (see above). However, the different frontal ciliature (bicorona vs. few distinctly enlarged frontal cirri) and the midventral complex (very likely present vs. lacking) should allow a clear separation even in life.

Urostyla limboonkengi is also rather similar, but has three enlarged frontal cirri (vs. bicorona), 8–13 cirral rows (vs. about eight in present species), and was discovered in a marine habitat (vs. limnetic). The marine *U. dispar*, which is also bimacronucleate, has a bipartite adoral zone of membranelles and only one left marginal row. *Urostyla agamalievi* also has only two macronuclear nodules, but is larger (300–320 µm vs. 190 µm) and has more cirral rows (in total 15 vs. up to 10). However, both observations clearly show that such binucleate *Urostyla*-like species exist. The overall appearance (size, shape, nuclear apparatus) of *U. algivora* is somewhat reminiscent of *Paraurostyla weissei*, which, however, has only one left marginal row and lacks a bicorona (for review, see Berger 1999).

Urostyla pseudomuscorum (Fig. 156a) and *Urostyla algivora* (Fig. 157a) are classified as supposed synonyms of *P. urostyla* because the differences are rather inconspicuous (see below for details). Considering all three descriptions, the total ranges of some morphometrics are: body length 190–300 µm; 6–15 transverse cirri (this range is unusually high for a single species); 7–10 cirral rows (values must not be over-interpreted because cirral pattern difficult to analyse without silver impregnation).

Morphology: The following description is based solely on Claparède & Lachmann's data. Body size about 220 µm in life, body length:width ratio ca. 2.8:1. Body roughly elongate elliptical with both ends broadly rounded, likely very flexible. Two ellipsoidal macronuclear nodules slightly left of midline. Contractile vacuole in ordinary position, that is, near left body margin slightly behind buccal vertex. Cells intense brown due to

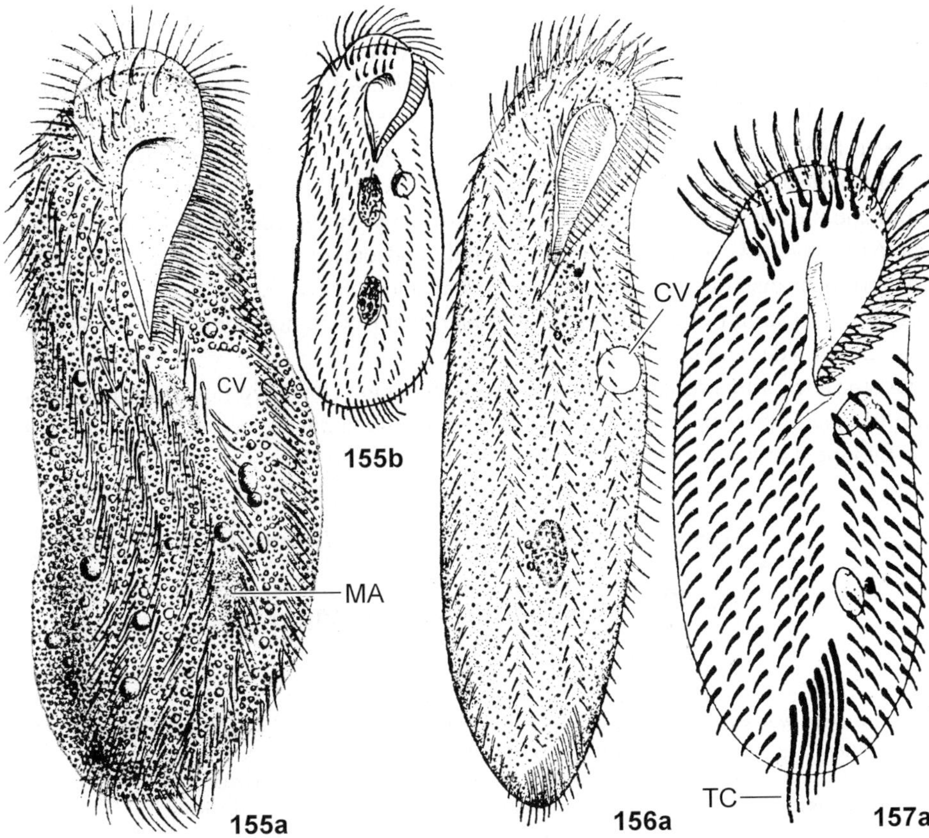

Fig. 155a, b *Pseudourostyla urostyla* from life (a, from Claparède & Lachmann 1858; b, after Claparède & Lachmann 1858 from Kahl 1932). Ventral view, 220 µm. Arrow in (a) marks shortened innermost right marginal. CV = contractile vacuole, MA = posterior macronuclear nodule. Page 800.
Fig. 156a *Urostyla pseudomuscorum*, a supposed synonym of *Pseudourostyla urostyla*, from life (from Wang 1940). Ventral view (140–300 µm) showing, inter alia, cirral pattern, contractile vacuole, nuclear apparatus, and longitudinally arranged brilliant corpuscles (cortical granules of the trichocyst-type?; see remarks for details). CV = contractile vacuole. Page 804.
Fig. 157a *Urostyla algivora*, a supposed synonym of *Pseudourostyla urostyla* (from Gellért & Tamás 1958. Opalblue staining after Bresslau). Infraciliature of ventral side and nuclear apparatus, 190 µm. TC = transverse cirri. Page 806.

pigmented granules in the cytoplasm (possibly cortical granules are present). Food and movement not mentioned.

Adoral zone occupies 40% of body length in specimen shown in Fig. 155a. Buccal cavity deep and wide, buccal lip prominent. Undulating membranes likely without peculiarities, paroral anteriorly distinctly curved. Cirral pattern composed of seven rows including the midventral complex (which forms two rows). Frontal ciliature composed of a distinct bicorona, indicating that a midventral complex is present although this is not evident from the illustration. Midventral complex, if present at all, likely composed of cirral pairs only; cirri of each pair either distinctly separated, so that a distinct zigzag pattern is not formed, but two rather distant rows, or cirri of midventral complex almost arranged in line feigning a single row. No buccal cirrus illustrated, possibly overlooked, that is, not distinguished from the midventral cirri. Eight transverse cirri, not distinctly enlarged, form short row arranged at almost right angles to cell midline near body end and thus project distinctly beyond rear body end. Two left marginal rows. Right of midventral complex three cirral rows; innermost distinctly shortened anteriorly, that is, commencing at level of contractile vacuole; middle and outermost right marginal rows extend from near distal end of adoral zone to rear body end (note that Kahl 1932 did not show the shortening of the innermost right marginal row in his redrawing). Ciliary pattern of dorsal side not known.

Occurrence and ecology: Claparède & Lachmann (1858) did not mention the type locality. Likely, they discovered it in France. Only one faunistic record, namely from the river Lielupe, Latvia (Liepa 1973, p. 33).

Supposed synonyms of *Pseudourostyla urostyla*

Urostyla pseudomuscorum Wang, 1940
(Fig. 156a)

1940 ***Urostyla pseudomuscorum* sp. nov.** – Wang, Sinensia, 11: 28, Fig. 7 (Fig. 156a; original description; no formal diagnosis provided and no type material available).

2001 ***Urostyla pseudomuscorum* Wang, 1940** – Berger, Catalogue of ciliate names 1. Hypotrichs, p. 102 (nomenclator containing all basionyms, combinations, and higher taxa of hypotrichs).

Nomenclature: No derivation of the name is given in the original description. The species-group name *pseudomuscorum* is a composite of *pseudo-* (Greek; wrong) and the species-group name *muscorum* (see previous species for derivation) and likely alludes to the similar cirral pattern of *P. muscorum* and the present species.

Remarks: The description and the illustration provided by Wang (1940) are rather thorough, although some details of the cirral pattern are likely not quite correctly shown. Borror (1972, p. 9) and Borror & Wicklow (1983, p. 120) synonymised *U. pseudomuscorum* with *U. gracilis* Entz, 1884. However, Entz's species is marine (vs. limnetic), smaller (120–200 μm vs. 240–300 μm), and has a different cirral pattern (cp. Fig. 156a and Fig. 217a–c). Hemberger (1982) obviously overlooked the present species. *Pseudourostyla muscorum* has many macronuclear nodules and only two marginal rows per side (Fig. 152a).

Wang mentioned that the middle six of a total of eight cirral rows form three pairs. I suppose that the middle pair of rows is the midventral complex, whereas the other rows are very likely marginal rows. According to the illustration the row formed by the left cirri of the midventral pairs is in line with the buccal cirral row. This is very uncommon and therefore one must assume that Wang (1940) did not analyse this part of the cirral pattern quite correctly so that this detail should not be over-interpreted.

The present species obviously lacks midventral rows, which are present in *Urostyla*, strongly indicating that it does not belong to this group. The size, the cirral pattern (bicorona, number of cirral rows, transverse cirri), and especially the two macronuclear nodules are strongly reminiscent of *Pseudourostyla urostyla*. Consequently, I put *U. pseudomuscorum* into the synonymy of Claparède & Lachmann's species. However, since Wang (1940) described three distinct paired cirral rows, I avoid a final synonymy. Descriptions of further populations are needed for a correct interpretation of the data.

Wang (1940) did not describe cortical granules. However, in the last two sentences he wrote that "Under oil immersion, the endoplasm is palish semitransparent. It appears to contain numerous very minute brilliant corpuscles which are more or less lineally arranged." Although Wang wrote that the corpuscles appear in the endoplasm I suppose that these structures are cortical granules (possibly of the trichocyst-type as in other *Pseudourostyla*-species) because they are brilliant and, more importantly, more or less linearly arranged. Under oil immersion, when the specimens are strongly squeezed, it is often difficult to decide whether a structure is in the cytoplasm or in the cortex.

Morphology: Body size of *U. pseudomuscorum* in life 240–300 × 60–80 µm, body length:width ratio 3–4:1 (Fig. 156a). Body outline elongate oval to slender elliptical, that is, widest at middle portion and slightly narrowing towards both ends. Body slightly flexible and extensible. Two distinctly separated macronuclear nodules, ovoid to ellipsoidal, 26–33 × 18–25 µm; the girdle or constriction, which was usually present in Wang's specimens (Fig. 156a, rear nodule), was the reorganisation band. Usually two, rarely up to four micronuclei closely attached to each macronuclear nodule; individual micronuclei 4.4–5.9 µm across. Contractile vacuole in ordinary position, that is, near left body margin slightly ahead of mid-body. Cortical granules possibly present (see last paragraph of remarks).

Adoral zone of membranelles occupies about one third of body length. Buccal field triangular, of ordinary width, anterior margin strongly curved. Paroral long. Cirral pattern *Pseudourostyla*-like (see remarks). Frontal cirri arranged in a bicorona; anterior bow composed of stronger and more cirri (specimen illustrated with about 11 cirri) than rear bow, which usually consists of 3–6 cirri. According to Wang (1940), the present species has three rows (bows) of frontal cirri. However, the illustration shows that the third (rearmost) row is the leftwards curved anterior portion of the buccal cirral row. Midventral complex obviously composed of cirral pairs only, extends from bicorona to anterior end of transverse cirral row. 9–15, on average 11 (slightly enlarged?) transverse cirri, rightmost 4–5 cirri project beyond rear body end. In total invariably (n >20) eight cirral rows; assuming that the middle, narrowly spaced two rows are the midventral complex, *Urostyla pseudomuscorum* has three marginal rows per side; innermost two rows per side narrowly spaced so that in total three pairs of rows are present (Fig. 156a; see also remarks). Dorsal ciliature (length of dorsal bristles, number and arrangement of kineties, absence/presence of caudal cirri) not described.

Occurrence and ecology: Limnetic. Type locality is a brook near the Bible Institute in Nanyoh, Hunan, China. Wang (1940) found it in great abundance in a glass vessel among filamentous algae one week after sampling (October 1937). No further records published.

Pseudourostyla pseudomuscorum voraciously feeds on diatoms, *Chilomonas*, *Colpidium*, and other species of flagellates and ciliates. It is able to ingest more than 10 *Dexiostoma campylum* specimens consecutively so that it is packed with prey.

Urostyla algivora Gellért & Tamás, 1958 (Fig. 157a)

1958 ***Urostyla algivora* n. sp.** – Gellért & Tamás, Annls Inst. biol. hung. Tihany, 25: 228, 240, Fig. 5 (Fig. 157a; original description; likely no type slides available and no formal diagnosis provided).

2001 ***Urostyla algivora* Gellért and Tamás, 1958** – Berger, Catalogue of ciliate names 1. Hypotrichs, p. 101 (nomenclator containing all basionyms, combinations, and higher taxa of hypotrichs).

Nomenclature: No derivation of the name is given in the original description. The species-group name *algivora* (algae-feeding) refers to the food (small, globular, green-algae).

Remarks: *Urostyla algivora* was described after opalblue staining. The description was published in a local Hungarian journal and likely this was the reason why it was overlooked by Borror (1972), Stiller (1974b), and Borror & Wicklow (1983). Hemberger (1982, p. 33) synonymised it with *U. viridis* Stein, 1859, which, however, has three distinctly enlarged frontal cirri. The distinct bicorona of *U. algivora* indicates that a midventral complex is present, that is, the classification in the urostylids is very likely correct. However, some details of the infraciliature (e. g., presence/absence of buccal cirri, frontoterminal cirri, and midventral rows) and the morphogenesis (e. g., type of marginal row formation) are unknown so that the generic assignment is uncertain.

I did not find a distinct difference between *U. algivora* and *P. urostyla* and therefore consider it, like *U. pseudomuscorum*, as supposed synonym. Of course we will never know whether or not these two populations and *U. pseudomuscorum* (see above) belonged to the same species. Descriptions of further populations from various regions are needed for a more proper solution. However, the (supposed) synonymy proposed seems the best solution with the data available. The differences among the cirral patterns of the three species must not be over-interpreted because none is described after protargol impregnation, which is indispensable to count the number of the rows correctly and to interpret the arrangement of the cirri exactly.

Morphology: Body length of *U. algivora* 190 μm in life(?). Body outline elliptical. Two relatively small macronuclei, each with one micronucleus. Cytoplasm brownish. Presence/absence of cortical granules unknown. Likely slowly moving. Adoral zone occupies about 38% of body length in specimen shown in Fig. 157a, composed of 35 membranelles. Peristomial lip slightly curved and pointed. 6–8 enlarged cirri in frontal corona; specimen illustrated with eight such cirri and each one cirrus behind except for the two leftmost cirri. Bicorona distinctly set off from the many (10 in specimen illustrated) cirral rows densely covering ventral side; according to original description eight ventral rows present, namely five right and three left of median (of course, without ontogenetic data it is impossible to designate the various rows correctly). Six transverse cirri, which, according to the description, do not project beyond rear body end; however, this disagrees with the illustration where the cirri project slightly (left) to distinctly (right). Dorsal infraciliature (length of dorsal bristles, number of dorsal kineties, presence/absence of caudal cirri) not described.

Occurrence and ecology: Type locality of *U. algivora* is the eastern shore of the Tihany peninsula of lake Balaton, Hungary, where Gellért & Tamás (1958) discovered it in detritus drifts permanently sprinkled by spray. No further records published. Feeds on small green algae.

Pseudourostyla raikovi (Alekperov, 1984) comb. nov. (Fig. 158a, b, Addenda)

1979 ***Metaurostyla raikovi*, comb. n.** – Jankowski, Trudy zool. Inst., Leningr., 86: 70 (see nomenclature).

1984 ***Metaurostyla raikovi* Alekperov, sp. n.** – Alekperov, Zool. Zh., 63: 1458, Fig. 1a, b (Fig. 158a, b; original description; type slides [No: 95, 98?] are likely deposited in the Institute of Zoology, Academy of Sciences of Azerbaijan, Baku; no formal diagnosis provided).

2001 ***Metaurostyla raikovi* Alekperov, 1984** – Berger, Catalogue of ciliate names 1. Hypotrichs, p. 47 (nomenclator containing all basionyms, combinations, and higher taxa of hypotrichs).

Nomenclature: Alekperov (1984) dedicated this species to Igor Raikov (†), a famous Russian protozoologist. Jankowski (1979, p. 70) mentioned a *Paraurostyla raikovi* Alekperov, 1977 which he transferred to *Metaurostyla* (see first entry in list of synonyms). I could not find the original description of *P. raikovi*. This and the fact that Alekperov (1984), who mentioned Jankowski's paper, described a *Metaurostyla raikovi* indicate that "*Paraurostyla raikovi* Alekperov, 1977" is a nomen nudum.

Remarks: *Metaurostyla raikovi* was described in Russian, mainly after wet silver nitrate impregnation. Consequently, some data (e.g., presence or absence of cortical granules) are lacking. Moreover, I suppose that the cirral pattern is not quite correctly illustrated due to the inappropriate preparation method. However, the illustration shows that *M. raikovi* has (i) a midventral complex very likely composed of cirral pairs only; (ii) a frontal ciliature of the bicorona type; (iii) frontoterminal cirri (obviously more than two); (iv) more than one marginal row per side; (v) transverse cirri. This combination of features indicates that *M. raikovi* belongs to *Pseudourostyla*, although we do not know the mode of marginal row formation (all rows originate individually vs. from a common anlage per side as in the type species *P. cristata*). Thus, I transfer *M. raikovi* to *Pseudourostyla*.

Pseudourostyla raikovi and *P. magna* have, like *P. urostyla*, the plesiomorphic number of only two macronuclear nodules. Preliminarily I avoid synonymy of these three species because Alekperov's species have distinctly more cirral rows than *P. urostyla* (about 15 or 16 vs. about 10 or less). Moreover, *Pseudourostyla magna* is distinctly larger than the other two species (around 520 μm vs. around 180–220 μm).

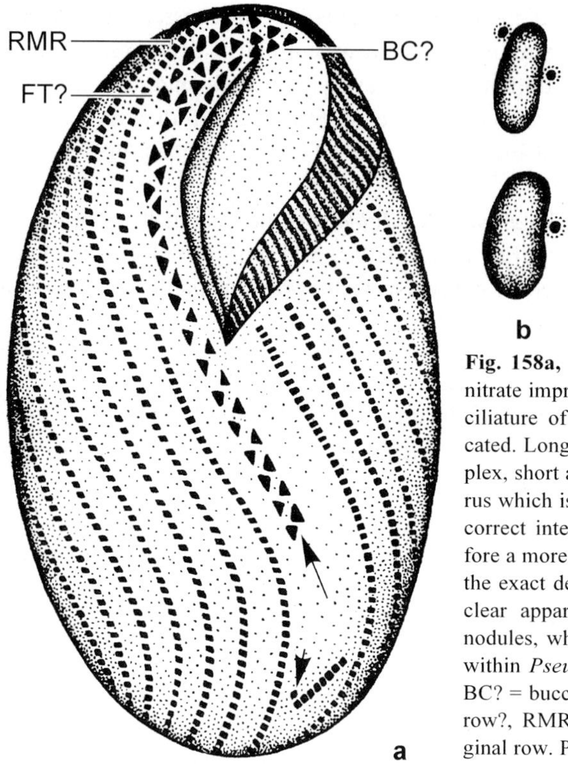

Fig. 158a, b *Pseudourostyla raikovi* after wet silver nitrate impregnation (from Alekperov 1984). **a:** Infraciliature of ventral side, size of specimen not indicated. Long arrow marks rear end of midventral complex, short arrow denotes the rightmost transverse cirrus which is possibly a pretransverse ventral cirrus. A correct interpretation of the cirral pattern and therefore a more proper classification is only possibly after the exact description of the cirral pattern. **b:** The nuclear apparatus is composed of two macronuclear nodules, which is very likely the plesiomorphic state within *Pseudourostyla*, and about three micronuclei. BC? = buccal cirral row?, FT? = frontoterminal cirral row?, RMR = anterior end of innermost right marginal row. Page 807.

Pseudourostyla raikovi has to be redescribed in detail, including cell division, to show whether or not the present classification is correct.

Morphology: The following description is based mainly on the specimen shown in Fig. 158a because I did not translate the original description in detail. Body length about 180 µm in life, about 130–150 µm after wet silver nitrate impregnation. Several features, for example, body outline in life, presence/absence of cortical granules, contractile vacuole, not known. Two roughly bean-shaped macronuclear nodules, each with one or two globular micronuclei attached (Fig. 158b). Adoral zone occupies 45% of body length in specimen illustrated, composed of about 50–60 membranelles. Buccal field wide (Fig. 158a; preparation artefact?). Details of undulating membranes not shown. Cirral pattern likely not correctly illustrated (Fig. 158a). Very likely *P. raikovi* has a bicorona and an indistinctly set off midventral complex composed of cirral pairs only; specimen illustrated with about 37 cirri in bicorona and midventral complex which terminates at 71% of body length. Likely one row each of buccal (about 7) and frontoterminal (about 6) cirri. Nine transverse cirri arranged in hook-shaped, slightly subterminal row. Seven right and six left marginal rows (Fig. 158a). Dorsal ciliature (length of dorsal bristles; number and arrangement of kineties; presence/absence of caudal cirri) not known.

Borror (1972, p. 9) synonymised *H. trichota* with *H. sphagni* and transferred the latter species to *Urostyla*. Likely for that reason *Hemicycliostyla* is lacking in some reviews, for example, Small & Lynn (1985). Hemberger (1982, p. 31) synonymised both *H. sphagni* and *H. trichota* as well as *Urostyla gigas* Stokes, 1886 with *Urostyla caudata* Stokes, 1886 because he did not consider the differences in the cirral pattern and contractile vacuole system described by Stokes as sufficient. Simultaneously he transferred *Urostyla caudata* to *Paraurostyla* because he assumed that it lacks a midventral pattern (further details, see *U. caudata*).

Shi et al. (1999, p. 112) and Shi (1999a, p. 363) put *Hemicycliostyla* Stokes, 1886 into the synonymy of *Pseudourostyla* Borror, 1972 which is an offence against the principle of priority because *Hemicycliostyla* is not a forgotten name (see list of synonyms).

Nowadays *Hemicycliostyla* is classified, if accepted at all, in the urostyloids (e.g., Fauré-Fremiet 1961; Corliss 1977, p. 137, 1979; Stiller 1974a) because it very likely has a midventral complex as indicated by presence of a bicorona. By contrast, Tuffrau (1979, 1987) and Tuffrau & Fleury (1994, p. 137) transferred it to the Kahliellidae, however, without detailed explanations.

All species classified in *Hemicycliostyla* lack modern redescriptions, that is, we do not know details about the cirral pattern. Thus, the characterisation above and the key below are not very precise.

Some authors mentioned undetermined *Hemicycliostyla* species from terrestrial habitats, however, without providing details (e.g., Stout 1958, p. 977; 1961, p. 745; Sudzuki 1979, p. 227; Szabó 2000, p. 14). Possibly they observed *Australothrix* species, which also lack transverse cirri. However, they do not have many cirri arranged in a bicorona, but only three enlarged frontal cirri.

Species included in *Hemicycliostyla* (alphabetically arranged according to basionym): (1) *Hemicycliostyla lacustris* Gellért & Tamás, 1958; (2) *Hemicycliostyla marina* Kahl, 1932; (3) *Hemicycliostyla sphagni* Stokes, 1886; (4) *Urostyla concha* Entz, 1884.

Key to *Hemicycliostyla* species

At least two *Urostyla* species (*U. dispar*, *U. gracilis*) have indistinct transverse cirri; possibly they are even absent in these species. Consequently you should check the *Urostyla* key too, if you have problems identifying your population with one of the species in the key below. *Australothrix* species, also lacking transverse cirri, have three enlarged frontal cirri.

1 Limnetic (Fig. 160a, 161a) **2**
\- Marine (Fig. 162a, 163a) **3**
2 Body length about 400–500 μm (Fig. 160a) .... *Hemicycliostyla sphagni* (p. 814)
\- Body length about 200 μm (Fig. 161a) .... *Hemicycliostyla lacustris* (p. 817)
3 (1) Two macronuclear nodules (Fig. 162a) ...... *Hemicycliostyla concha* (p. 818)
\- Many macronuclear nodules (Fig. 163a) .... *Hemicycliostyla marina* (p. 820)

Hemicycliostyla sphagni Stokes, 1886 (Fig. 160a–h)

1886 ***Hemicycliostyla sphagni*, sp. nov.** – Stokes, Proc. Am. phil. Soc., 23: 22, Fig. 1 (Fig. 160a; original description; no type material available and no formal diagnosis provided).
1886 ***Hemicycliostyla trichota*, sp. nov.** – Stokes, Proc. Am. phil. Soc., 23: 22, Fig. 2 (Fig. 160e; original description of synonym; no type material available and no formal diagnosis provided).
1888 ***Hemicycliostyla sphagni*, Stokes** – Stokes, J. Trenton nat. Hist. Soc., 1: 276, Plate X, fig. 9 (Fig. 160b; review of freshwater ciliates from the USA).
1888 ***Hemicycliostyla trichota*, Stokes** – Stokes, J. Trenton nat. Hist. Soc., 1: 276, Plate X, fig. 10 (Fig. 160f; review of freshwater ciliates from the USA).
1931 ***Hemicycliostyla trichota* Stokes** – Tai, Sci. Rep. natn. Tsing Hua Univ., Ser. B., 1: 51, Plate XIV, Fig. 7 (Fig. 160h; redescription of synonym from life)
1932 ***Hemicycliostyla sphagni* Stokes, 1886** – Kahl, Tierwelt Dtl., 25: 544, Fig. 97 $_1$ (Fig. 160d; revision of hypotrichs).
1932 ***Hemicycliostyla trichota* Stokes, 1886** – Kahl, Tierwelt Dtl., 25: 544, Fig. 97 $_2$ (Fig. 160g; revision of hypotrichs).
1953 ***Hemicycliostyla sphagni* Stokes (1886)** – Jirovec et al., Protozoologie, p. 509, Fig. 234F (redrawing of Fig. 160a; textbook).
1972 ***Urostyla sphagni* (Stokes, 1886) n. comb.** – Borror, J. Protozool., 19: 9 (revision; combination with *Urostyla*).
1974 ***Hemicycliostyla sphagni* Stokes** – Stiller, Fauna Hung., 115: 31, Ábra 19A (redrawing from Stokes).
1974 ***Hemicycliostyla sphagni* Stokes var. *trichota* Stokes** – Stiller, Fauna Hung., 115: 31, Ábra 19B (redrawing from Stokes).
1989 ***Hemicycliostyla sphagni* Stokes, 1886**[1] – Gong & Shen, The aquatic Fauna, p. 105, Plate IV, Fig. 22 (Fig. 160c; redescription).
2001 ***Hemicycliostyla sphagni* Stokes, 1886** – Berger, Catalogue of ciliate names 1. Hypotrichs, p. 30 (nomenclator containing all basionyms, combinations, and higher taxa of hypotrichs).

Nomenclature: The species-group name *sphagni* obviously refers to the habitat (marsh water with *Sphagnum*) where the species was discovered. *Hemicycliostyla sphagni* was fixed as type species of *Hemicycliostyla* Stokes, 1886 by Jankowski (1979) by subsequent designation. The species-group name *trichot·us -a -um* (having hairs; from *trichós*, Greek, the hair) obviously alludes to the high number of cirri.

Remarks: Stokes (1886) described four large *Urostyla*-like species, two in *Urostyla* and two in his new genus *Hemicycliostyla*, which he separated from *Urostyla* by the lack of transverse cirri. Stokes usually made very good live observations so that it is unlikely that he overlooked the transverse cirri (anal styles in his terminology) in the two *Hemicycliostyla* populations. Three years later, Bütschli (1889, p. 1741) synonymised

Fig. 160a–h *Hemicycliostyla sphagni* from life (a, e, from Stokes 1886; b, f, from Stokes 1888; c, from Gong & Shen 1989; d, g, after Stokes 1888 from Kahl 1932; h, from Tai 1931). **a–d:** Ventral views (a, b, d = 420–510 µm; c = 380–450 µm) showing, inter alia, basic cirral pattern, oral apparatus, contractile vacuoles, and nuclear apparatus. **e–h:** Ventral views of the synonym *H. trichota*, e–g = 423 µm, h = 345 µm. For differences between *H. sphagni* and the synonym *H. trichota*, see remarks. CV = contractile vacuoles. Page 814. →

[1] According to Gong & Shen (1989, p. 105), the illustration of *H. sphagni* is Figure 23 on Plate V. However, the correct figure number is Figure 22 on Plate IV.

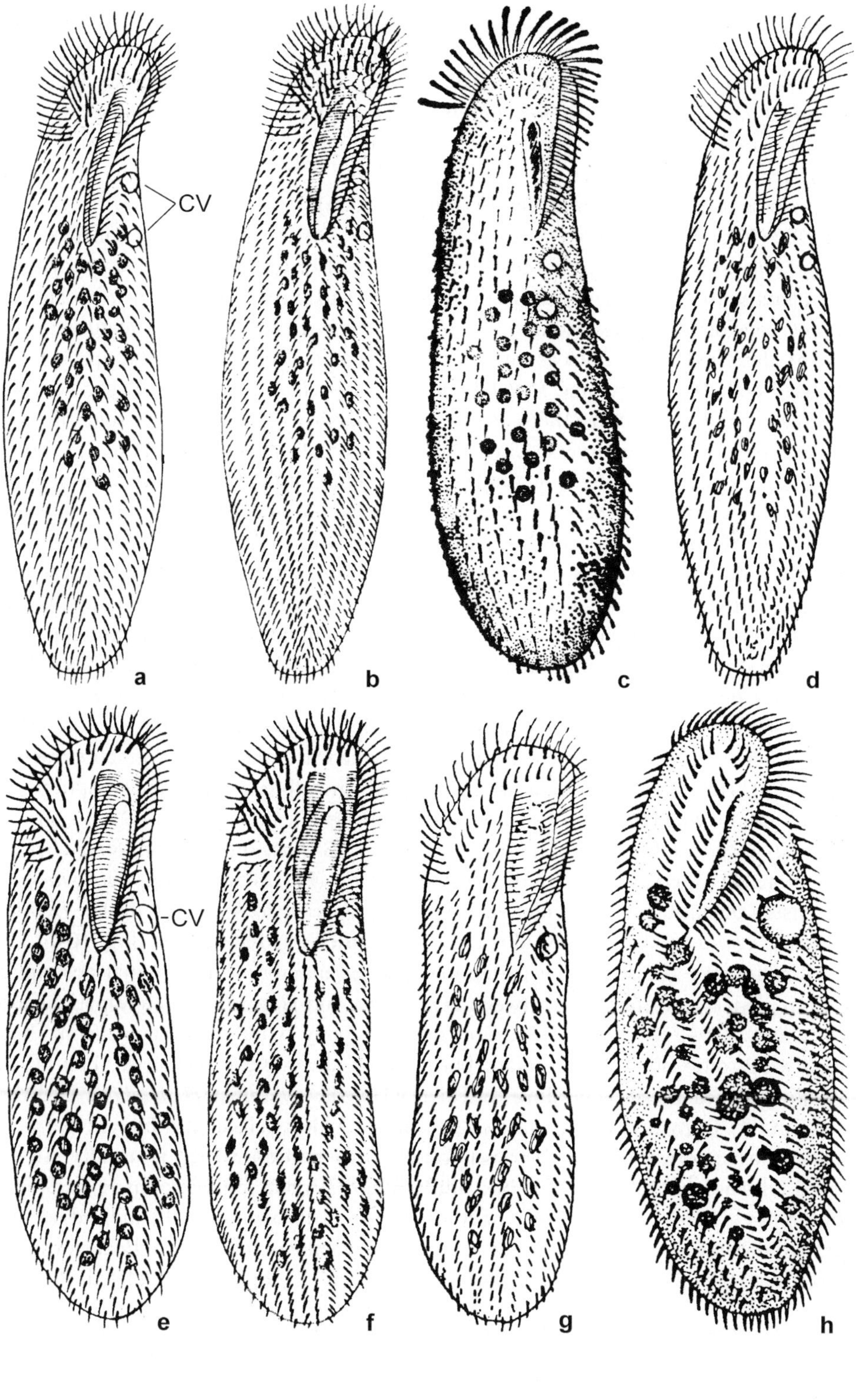
CV
a
b
c
d
CV
e
f
g
h

H. sphagni and the second species described by Stokes (1886) with *Urostyla grandis*, type of *Urostyla*, a proposal later followed by Borror & Wicklow (1983, p. 120). However, *Urostyla grandis* has distinct transverse cirri, whereas this cirral group is, as just mentioned, lacking in *Hemicycliostyla*. Only recently I could confirm the lack of transverse cirri in *Uroleptopsis citrina*, which was synonymised with *Pseudokeronopsis rubra* (transverse cirri present) by Borror & Wicklow (1983), indicating that this cirral group can be used as systematic feature (Berger 2004b).

Kahl (1932) accepted Stokes' genus and *Hemicycliostyla sphagni*, but considered *H. trichota* only as a modification of the former species. Borror (1972) finally synonymised *H. trichota* with *H. sphagni*, and simultaneously transferred it to *Urostyla* because he did not consider the presence/absence of transverse cirri as generic feature. Stiller (1974b) classified *H. trichota* as variety of *H. sphagni*. Hemberger (1982, p. 31) synonymised both *H. sphagni* and *H. trichota* with *Urostyla caudata* and *U. gigas*, which were discovered by Stokes (1886) in the same marsh water. However, since the two *Urostyla* species have transverse cirri, I disagree with Hemberger's proposal. I follow Kahl (1932) and Borror (1972) in that I put *H. trichota* into the synonymy of *H. sphagni* because the differences between the two species are indeed rather inconspicuous (423 μm vs. 423–508 μm; ratio body length:width 3:1 vs. 4:1; body less extensile vs. extensile; one contractile vacuole vs. two; cytoplasm not vacuolised vs. vacuolised). Stokes mentioned further "differences", namely, *H. trichota* has more macronuclear nodules than *H. sphagni*, has "par-oral" cilia on the right margin of the proximal portion of the adoral zone (vs. lacking), and is less active. Interestingly, both species have been redescribed once by Chinese workers. Gong & Shen (1989) rediscovered *H. sphagni* and described, like Stokes in the original description, two contractile vacuoles (Fig. 160c), whereas Tai (1931) found *H. trichota*, which has only one contractile vacuole (Fig. 160h). However, exact data are lacking and therefore detailed redescriptions are needed to show whether or not *H. trichota* is a valid species.

Urostyla trichota[1] sensu Conn (1905, p. 58, Fig. 237) has distinct transverse cirri and is therefore classified as synonym of *Urostyla grandis* (Fig. 205l). *Hemicycliostyla sphagni* sensu Takahashi (1972) is identical with *Pseudourostyla levis* (see there). *Hemicycliostyla trichota* Stokes sensu Naidu (1965) is insufficiently redescribed (Fig. 164h).

Morphology: The following paragraph is based mainly on the original description of *H. sphagni*. For differences between *H. sphagni* and *H. trichota*, see remarks. Body size about 420–510 × 100–130 μm in life (body width estimated from body length:width ratio of 4:1; Fig. 160a); body length according to Gong & Shen (1989) 380–450 μm (in life?), according to Tai (1931), the synonym *H. trichota* is 345 × 64 μm in life. Body outline elongate-ovate, that is, widest behind mid-body, tapering to rounded posterior end and to narrower frontal body portion, which is usually slightly curved leftwards. Body soft, flexible, and (slightly?) contractile (extensile in Stokes' terminology). Many macronuclear nodules dispersed throughout cell, individual nodules ovoid or ellipsoidal, small. Two contractile vacuoles left of proximal portion of adoral zone. Cytopyge dorsal, near posterior body end. Cytoplasm vacuolised. Presence

[1] Conn (1905) made, although not formally, a new combination which was overlooked by Berger (2001).

or absence of cortical granules unknown. Adoral zone occupies about one third of body length, distal end extends far posteriorly; likely composed of a rather high number of membranelles. Buccal field rather more narrow than wide, details about undulating membranes unknown. Circa 20 frontal cirri (possibly slightly enlarged) arranged in a bicorona. About 11 longitudinal cirral rows, that is, ventral side densely covered with midventral and marginal cirri. Details about cirral pattern (e.g., presence/absence of buccal cirri, frontoterminal cirri, and midventral rows) not known. Transverse cirri lacking (it is unlikely that Stokes overlooked them). Dorsal cilia short, details of kinety pattern (number and arrangement of kineties; caudal cirri present/absent) not known.

Occurrence and ecology: Likely confined to freshwater; not reliably recorded from Europe! Type locality of both *H. sphagni* and its synonym *H. trichota* is a marsh water with *Sphagnum* near Stokes' place of residence, that is, Trenton, USA (Stokes 1886). Gong & Shen (1989) collected their population in freshwater habitats from the Suoxiyu Nature Preserve Area, Hunan Province, China. Tai (1931) found the synonym *H. trichota* in a small stream on east campus of Tsing Hua University, China. Records not substantiated by morphological data: small muddy, turbid pool (25°C; 80°31'04"W 37°22'03"E) with little vegetation at piped spring in the Mountain Lake area, Giles County, Virginia, USA (Bovee 1960, p. 357); *Sphagnum fimbriatum* from Lake Memphremagog south of Montreal, Canada (Fantham & Porter 1946, p. 129); Slovakia (Matis et al. 1996, p. 11).

a

Fig. 161a *Hemicycliostyla lacustris* (from Gellért & Tamás 1958. Opalblue-staining). Infraciliature of ventral side and nuclear apparatus. Arrow marks a cirral row commencing behind proximal end of adoral zone of membranelles. Page 817.

Hemicycliostyla lacustris Gellért & Tamás, 1958 (Fig. 161a)

1958 ***Hemicycliostyla lacustris* n. sp.**[1] – Gellért & Tamás, Annls Inst. bio. Tihany, 25: 227, Fig. 4 (Fig. 161a; original description. Likely no type material available).

[1] The diagnosis provided by Gellért & Tamás (1958) is as follows: Grösse 190 μ. Gestalt oval. Mächtiges Wirbelorgan mit 50–60 niedrigen Membranellen. Peristomlippe kurz. Frontalcirren in 2 Reihen (1 kurze und 1 längere), die sich in 2 Ventralreihen fortsetzen. Ausserdem noch 6 Ventralreihen. Makronuclei in 50 Bröckel zerteilt. Nahrung kleine, grüne Kugelalgen und Diatomeen.

2001 ***Hemicycliostyla lacustris* Gellért and Tamas, 1958** – Berger, Catalogue of ciliate names 1. Hypotrichs, p. 30 (nomenclator containing all basionyms, combinations, and higher taxa of hypotrichs).

Nomenclature: No derivation of the name is given in the original description. The species-group name *lacustr·is -is -e* (Latin; occurring in the lake or pond) refers to the type of habitat (Lake Balaton) where the species was discovered.

Remarks: Gellért & Tamás (1958) described this species after opalblue-staining. Consequently the species requires detailed redescription, including live data (presence or absence of cortical granules) and exact illustrations of the infraciliature of the ventral and dorsal side. The bicorona and the lack of transverse cirri is characteristic for *Hemicycliostyla* so that I retain the original classification. Obviously Gellért & Tamás' species was overlooked by most workers (Borror 1972, Stiller 1974b, Hemberger 1982, Borror & Wicklow 1983). The type species of *Hemicycliostyla* is about twice as long, indicating that they are not synonymous.

Morphology: Body length about 190 µm (from life?); ratio of body length:width of opalblue-prepared specimen illustrated about 2.8:1 (Fig. 161a). Outline oval, specimen illustrated broad-elliptical, roughly as in *Urostyla grandis*. About 50 macronuclear nodules dispersed throughout cell. Contractile vacuole not mentioned. Presence/absence of cortical granules unknown. Adoral zone occupies about 40% of body length in specimen illustrated, composed of 50–60 low adoral membranelles. Buccal lip short, undulating membranes not described. Frontal cirri slightly enlarged, arranged in a bicorona, strongly indicating that this species has a midventral complex. In total eight cirral rows, four right of midline, one (innermost left marginal row?) behind proximal end of adoral zone, and three left of midline. Buccal cirri possibly lacking; presence/absence of frontoterminal cirri unknown. Transverse cirri absent (Fig. 161a), as also indicated by the classification in *Hemicycliostyla*. Dorsal ciliature (length of dorsal bristles; number and arrangement of dorsal kineties; presence/absence of caudal cirri) not known.

Occurrence and ecology: Likely limnetic. Type locality is Lake Balaton (Hungary), where Gellért & Tamás (1958) discovered *H. lacustris* among debris washed ashore on the Tihany peninsula. No further records published. Feeds on diatoms and small, green globular algae.

Hemicycliostyla concha (Entz, 1884) Kahl, 1932
(Fig. 162a–c)

1884 ***Urostyla concha* n. sp.** – Entz, Mitt. zool. Stn Neapel, 5: 379, Tafel 23, Fig. 13 (Fig. 162a; original description; no formal diagnosis provided and no type material available).

1932 ***Urostyla concha* Entz, 1884** – Kahl, Tierwelt Dtl., 25: 564, Fig. 97 $_{13}$ (Fig. 162b; revision of hypotrichs; see also next entry).

1932 ***Hemicycliostyla concha*** – Kahl, Tierwelt Dtl., 25: 564 (combination with *Hemicycliostyla*, see nomenclature).

1933 ***Urostyla concha* Entz sen. 1884** – Kahl, Tierwelt N.- u. Ostsee, 23: 108, Fig. 16.21 (Fig. 162c; guide to marine ciliates).

1972 ***Urostyla concha* Entz, 1884** – Borror, J. Protozool., 19: 9 (revision of hypotrichs).

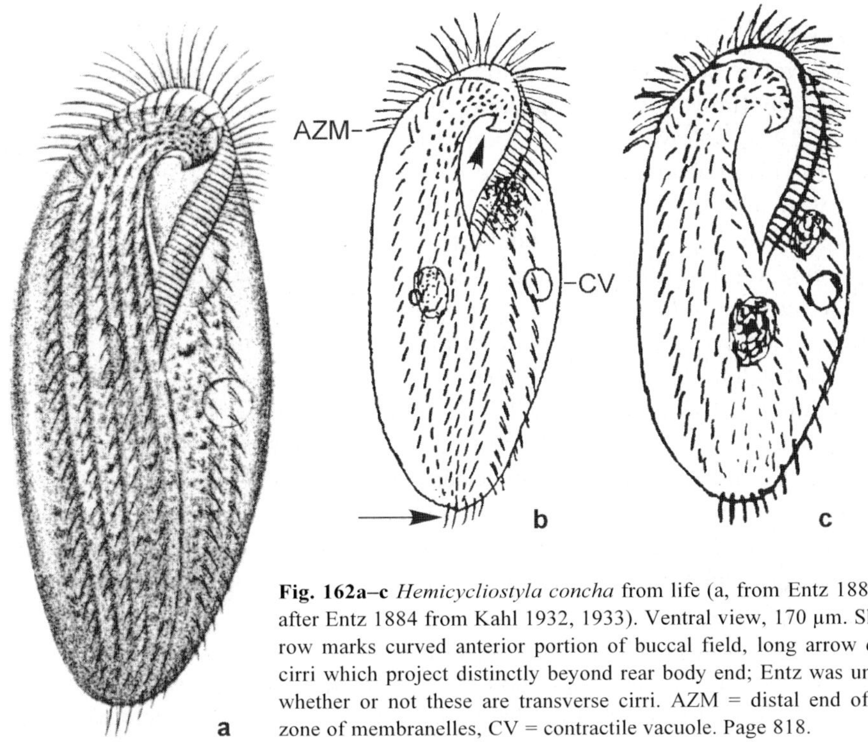

Fig. 162a–c *Hemicycliostyla concha* from life (a, from Entz 1884; b, c, after Entz 1884 from Kahl 1932, 1933). Ventral view, 170 µm. Short arrow marks curved anterior portion of buccal field, long arrow denotes cirri which project distinctly beyond rear body end; Entz was uncertain whether or not these are transverse cirri. AZM = distal end of adoral zone of membranelles, CV = contractile vacuole. Page 818.

2001 ***Urostyla concha* Entz, 1884** – Berger, Catalogue of ciliate names 1. Hypotrichs, p. 101 (nomenclator containing all basionyms, combinations, and higher taxa of hypotrichs).

Nomenclature: The species-group name *concha* (feminine; latinized; Greek, *he konche*, the mussel) refers to the concave and therefore mussel-shaped ventral side of the cell (Entz 1884). Kahl (1932), who classified the present species in *Urostyla*, wrote that it has to be designated as *Hemicycliostyla concha* when the lack of transverse cirri is confirmed. Although he did not transfer it formally to *Hemicycliostyla*, he can be considered as the combining author for the classification in *Hemicycliostyla*. This is one reason why I preliminarily assign Entz's species to *Hemicycliostyla*.

Remarks: Entz (1884) could not clarify whether the cirri at the rear body end are transverse cirri, which form a distinct pseudorow, or if these are only the rearmost cirri which are not arranged in a distinct transverse pseudorow. *Urostyla concha* has, like all other hypotrichs described by Entz (1884), only two macronuclear nodules, a feature especially doubted for his *Holosticha* (now *Pseudokeronopsis*) populations, which usually have many macronuclear nodules dispersed throughout the cell. It would be risky to assume that Entz generally misobserved the nuclear apparatus and therefore I use this feature to separate *H. concha* from *H. marina* Kahl, 1932, which has a very similar general appearance, but many macronuclear nodules. Time will show whether both types (two vs. many macronuclear nodules) or only one is confirmed. Kahl (1932) formally classified Entz's species in *Urostyla*, but doubted the presence of true transverse cirri and therefore supposed that it belongs to *Hemicycliostyla* when the lack of transverse cirri is confirmed (see nomenclature).

Borror (1972) followed Entz and Kahl and classified it in *Urostyla*, whereas Borror & Wicklow (1983) obviously overlooked it. Hemberger (1982, p. 32) considered it, together with *H. marina*, as synonym of *U. gracilis* Entz, 1884. However, *Urostyla gracilis* is red so that synonymy is very unlikely.

According to Entz (1884) the present species has an almost rigid body, whereas most (all?) other urostylids are very flexible. A detailed redescription that either confirms or disproves this observation should be awaited for a serious discussion (e.g., a relationship with the rigid stylonychines; Berger 1999).

Morphology: Body size about 170 × 70 μm. Body outline broadly elliptical. Body almost rigid, dorsal side, as is usual, vaulted, ventral side (slightly) mussel-shaped excavated, right margin bulgy. Two macronuclear nodules, one covered by proximal portion of adoral zone, the other in right body portion about at level of proximal end of adoral zone. Contractile vacuole about in mid-body near left cell margin. Presence/absence of cortical granules unknown; however, cytoplasm granular, straw-coloured (it cannot be excluded that the colour is due to cortical granules).

Adoral zone occupies about one third of body length (in specimen illustrated about 42%!), extends rather far onto right body margin. Buccal field, respectively, buccal lip anteriorly strongly curved leftwards, that is, crescent-shaped. Cirral pattern not known in every detail, for example, presence/absence of buccal cirri, frontoterminal cirri, and midventral rows. In total eight cirral rows, six of them extend onto frontal area where most of them curve leftwards, that is, frontal ciliature possibly of a multicorona type; distinct, that is, enlarged frontal cirri lacking; however, this feature is difficult to recognise in life and therefore must not be over-interpreted. Outermost right (marginal) cirral row displaced inwards due to the bulged right body margin. Two cirral rows left of midline distinctly separated from innermost right row. Rear body end with 7–8 transverse cirri distinctly projecting beyond rear cell margin; Entz himself supposed that these cirri are not transverse cirri sensu stricto, but the rearmost cirri of the cirral rows not arranged in a distinct transverse pseudorow. Dorsal infraciliature unknown.

Occurrence and ecology: Marine. Type locality of *Hemicycliostyla concha* is the Gulf of Naples (Italy), Mediterranean Sea where Entz (1884) found few specimens among dense layers of diatoms and always together with *Onychodactylus acrobates* (*Onychodactylus* is a urodele according to Remane et al. 1976) and the urostyloid *Pseudokeronopsis flava*.

Hemicycliostyla marina Kahl, 1932
(Fig. 163a, b)

1932 ***Hemicycliostyla marina* spec. n.** – Kahl, Tierwelt Dtl., 25: 545, Fig. 94 ${}_{1}$ (Fig. 163a; original description; no type material available and no formal diagnosis provided).

1933 ***Hemicycliostyla marina*** **Kahl 1932** – Kahl, Tierwelt N.- u. Ostsee, 23: 107, Fig. 16.13 (Fig. 163b; guide to marine ciliates).

1972 ***Urostyla marina*** **(Kahl, 1932) n. comb.** – Borror, J. Protozool., 19: 9 (revision; combination with *Urostyla*; see nomenclature).

2001 ***Hemicycliostyla marina*** **Kahl, 1932** – Berger, Catalogue of ciliate names 1. Hypotrichs, p. 30 (nomenclator containing all basionyms, combinations, and higher taxa of hypotrichs).

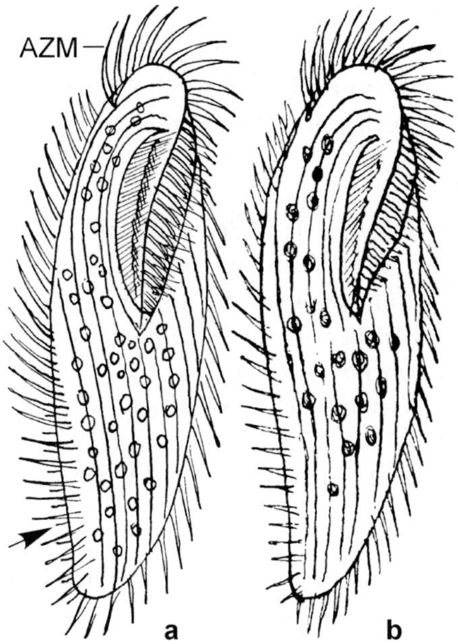

Fig. 163a, b *Hemicycliostyla marina* (a, from Kahl 1932; b, after Kahl 1932 from Kahl 1933. After fixation). Ventral view showing basic cirral pattern, oral apparatus, and nuclear apparatus, 215 μm. Arrow marks an about 5 μm long dorsal bristle. AZM = distal end of adoral zone of membranelles. Page 820.

Nomenclature: No derivation of the name is given in the original description. The species-group name *marin·us -a -um* (Latin adjective; living in the sea, belonging to the sea) refers to the habitat (Atlantic Ocean) where the species was discovered. Borror (1972) transferred the present species to *Urostyla*, and simultaneously *Urostyla marina* Kahl, 1932 to *Paraurostyla* Borror, 1972. Consequently the two species were not secondary homonyms in Borror's paper. *Urostyla marina* Kahl, 1932 is now the type species of *Metaurostylopsis* and can be easily distinguished from the present species, inter alia, by the three enlarged frontal cirri.

Remarks: Kahl (1932) described this species from just one well fixed specimen from a plankton sample and therefore *H. marina* needs detailed reinvestigation. The specimen had many frontal cirri and lacked transverse cirri and therefore Kahl assigned it to *Hemicycliostyla*. Likely, because of the spiral arrangement of the frontal cirri he supposed a close relationship with *Urostyla gracilis* (see next paragraph). Borror (1972) accepted the species, but transferred it to *Urostyla* because he did not consider the presence or absence of transverse cirri as an important taxonomic feature. Borror & Wicklow (1983, p. 117) suggested that it should be of questionable taxonomic placement pending rediscovery.

Hemberger (1982, p. 32) synonymised *H. marina* with *Urostyla gracilis* (see also previous paragraph) and *Urostyla concha*. However, *Urostyla gracilis* is red and has, like *U. concha*, only two macronuclear nodules, whereas *H. marina* has many nodules (Fig. 163a), indicating that they are not synonymous. Since *U. concha* very likely lacks transverse cirri it is classified in *Hemicycliostyla* in the present review.

Morphology: As already mentioned the description of *H. marina* is based on a single, fixed specimen! Consequently, details of the cirral pattern should not be overinterpreted. Body length after fixation (method not indicated) about 215 μm; according to Kahl in life therefore about 300 μm long. Body roughly spindle-shaped. Kahl did not mention the nuclear apparatus in the text; however, from the key (his question 1) and the illustration one has to conclude that *H. marina* has "many" macronuclear nodules dispersed throughout the cell. Presence/absence of contractile vacuole and cortical granules not mentioned. Adoral zone occupies almost 50% of body length (this high value indicates that the specimen was an early postdivider, with high lip and prominent membranelles and undulating membranes (Fig. 163a). Sigmoidal arrangement of cirral rows possibly partially due to fixation. About 10 narrowly spaced rows of closely arranged cirri; in addition two (marginal?) rows of stronger and wider spaced cirri per side. Three ventral rows extend onto the frontal area, where the cirri are not enlarged. Dorsal bristles about 5 µm long; number and arrangement of kineties and presence/absence of caudal cirri not known.

Occurrence and ecology: Marine. Kahl (1932) discovered *Hemicycliostyla marina* in a fixed plankton sample (collected by E. Hentschel) from the Atlantic Ocean between Iceland and Greenland. No further records published.

Insufficient redescriptions

***Hemicycliostyla* sp.** – Lepsi, 1957, Buletin sti. Acad. Repub. pop rom., 9: 234. Remarks. Lepsi provided a brief description in Romanian, but no illustration of a hypotrich which obviously belonged to *Hemicycliostyla*. I do not think that it will ever be possibly to know which species Lepsi observed. Body length about 225 µm, body length:width ratio 4 to 5:1. Likely many macronuclear nodules. Contractile vacuole in ordinary position. Adoral zone about one third of body length. Numerous rows bearing short cirri. Frontal, transverse, and caudal cirri lacking. Found in river Dorna, Romania.

***Hemicycliostyla trichota* Stokes** – Naidu, 1965, Hydrobiologia, 25: 559, Fig. 44 (Fig. 164h). Remarks: The illustration is too primitive to accept the identification; moreover, Naidu illustrated a single macronucleus which does not match the nuclear pattern of *H. sphagni* and its synonym *H. trichota*, which have many macronucleus-nodules. Body size 360–410 × 80–96 µm. Naidu examined a few specimens from gutter water at Vijayawada (India) in mid-May.

Fig. 164a–j Insufficient redescriptions and species indeterminata (ventral views from life unless otherwise indicated). **a:** *Holosticha* sp. (from Lepsi 1932), size not indicated; p. 189. **b:** *Holosticha* sp. (from Chardez 1981), 115 µm (?); p. 189. **c, d:** *Holosticha obliqua* (c, from Aladro Lubel et al. 1986; d, from Aladro Lubel et al. 1990), 130 µm, 125 µm; p. 188. **e:** *Urostyla* sp. (from Lepsi 1957), possibly a reorganiser, 90 µm; p. 1112. **f:** *Urostyla* sp. (from Nikoljuk & Geltzer 1972), 100 µm; p. 1112 **g:** *Urostyla grandis* (from Edmondson 1906), 250–400 µm; p. 1112. **h:** *Hemicycliostyla trichota* (from Naidu 1965), 360–410 µm according to text (only about 116 µm according to data in figure legend); p. 822. **i, j:** *Holosticha aquarumdulcium* (i, from Bürger 1905; j, from Bürger 1905 after Kahl 1932), 320 µm; p. 177. →

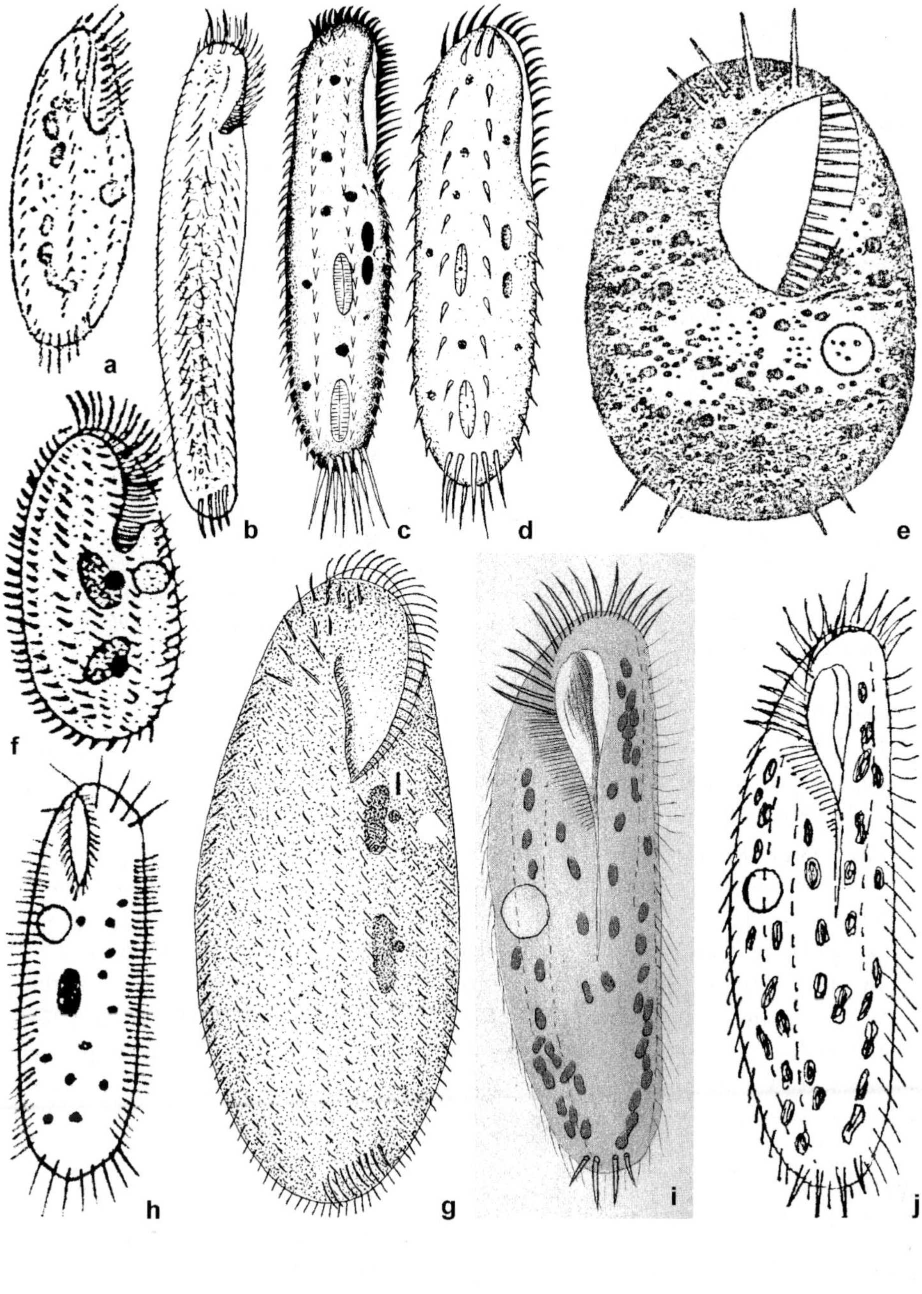
a
b
c
d
e
f
I
h
g
i
j

Trichototaxis Stokes, 1891

1891 ***Trichototaxis* g. n.**[1] – Stokes, Jl R. microsc. Soc., year 1891: 701 (original description). Type species (by monotypy): *Trichototaxis stagnatilis* Stokes, 1891.
1932 ***Trichotaxis* Stokes, 1891** – Kahl, Tierwelt Dtl., 25: 588 (revision; incorrect subsequent spelling, see nomenclature).
1933 ***Trichotaxis* Stokes 1891** – Kahl, Tierwelt N.- u. Ostsee, 23: 111 (guide to marine ciliates).
1950 ***Trichotaxis* Stokes** – Kudo, Protozoology, p. 674 (textbook).
1961 ***Trichotaxis* Stokes** – Fauré-Fremiet, C. r. hebd. Séanc. Acad. Sci., Paris, 252: 3517 (systematics of hypotrichs).
1961 ***Trichotaxis* Stokes** – Corliss, Ciliated protozoa, p. 170 (revision of ciliates).
1965 ***Trichotaxis*** – Lepsi, Protozoologie, p. 972 (textbook).
1972 ***Trichotaxis* Stokes, 1891**[2] – Borror, J. Protozool., 19: 11 (revision of hypotrichs).
1974 ***Trichotaxis* Stokes** – Stiller, Fauna Hung., 115: 52 (guide to hypotrichs).
1979 ***Trichototaxis* Stokes, 1891** – Jankowski, Trudy zool. Inst., 86: 68 (catalogue of generic names of hypotrichs).
1979 ***Trichotaxis* Stokes, 1891** – Corliss, Ciliated protozoa, p. 309 (revision of ciliates).
1979 ***Trichotaxis* Stokes, 1891** – Tuffrau, Trans. Am. microsc. Soc., 98: 526 (brief revision).
1982 ***Trichototaxis* Stokes, 1891**[3] – Hemberger, Dissertation, p. 117 (revision of hypotrichs).
1983 ***Trichototaxis* Stokes, 1891** – Curds, Gates & Roberts, Synopses of the British Fauna, 23: 418 (guide to ciliate genera).
1987 ***Trichotaxis* Stokes, 1891** – Tuffrau, Annls Sci. nat., 8: 115 (classification of hypotrichs).
1992 ***Trichotaxis* Stokes, 1891** – Carey, Marine interstitial ciliates, p. 186 (guide).
1994 ***Trichotaxis* Stokes 1891** – Tuffrau & Fleury, Traite de Zoologie, 2: 128 (revision of hypotrichs).
2001 ***Trichototaxis* Stokes 1891** – Aescht, Denisia, 1: 167 (catalogue of generic names of ciliates).
2001 ***Trichototaxis* Stokes, 1891** – Berger, Catalogue of ciliate names 1. Hypotrichs, p. 96 (nomenclator containing all basionyms, combinations, and higher taxa of hypotrichs).

Nomenclature: *Trichototaxis* is a composite of *trichodes* (Greek adjective; hairy) and *he taxis* (Greek noun; position, row according to Stokes) and obviously refers to the (increased number of) cirral rows. Feminine gender (Aescht 2001, p. 303). Kahl (1932, 1933) classified *Trichototaxis* (Kahl 1932, p. 588) as subgenus of *Holosticha* (Kahl 1932, p. 570, 571) which was frequently overlooked. Thus the correct designation in his papers is *Holosticha (Trichototaxis)*, respectively, *Holosticha (Trichotaxis)* (see next paragraph).

Trichotaxis in Kahl's (1932) revision is an incorrect subsequent spelling which was used by many workers. Since this misspelling occurs very frequently it is not marked as incorrect at the individual entries of the list of synonyms. Foissner (1987d, p. 226) and Berger (2001, p. 95) transferred the species originally established in *Trichotaxis* to *Trichototaxis*. Now I am no longer convinced that this transfer was necessary because *Trichotaxis* is not a synonym, but an incorrect subsequent spelling.

[1] The diagnosis by Stokes (1891) is as follows: Animal free-swimming, hypotrichous, soft and flexible, depressed; frontal styles numerous, in two curved, sub-parallel series; ventral styles forming three longitudinal rows; marginal setae uninterrupted; anal styles well developed; inhabiting fresh-water infusions.

[2] The improved diagnosis by Borror (1972) is as follows: Two rows of left marginal cirri. One row of right marginal cirri.

[3] The diagnosis by Hemberger (1982) is as follows: 1 rechte Marginalreihe; mehr als 1 linke Marginalreihe; familientypische Midventral-Reihen, die aus schrägen Anlagen entstehen; Transversalcirren vorhanden.

Characterisation (A = supposed apomorphy): Adoral zone of membranelles likely continuous. Many frontal cirri arranged in a bicorona. Midventral complex likely composed of cirral pairs only. More than two marginal rows (A?). Transverse cirri present.

Remarks: Stokes (1891) established *Trichototaxis* for a species with a distinct bicorona, a midventral complex composed of cirral pairs, transverse cirri, and one right and two left marginal rows. Of course, Stokes could not recognise the origin of the individual rows, that is, he summarised the inner left marginal row and the two pseudorows formed by the midventral pairs as "three ventral rows".

Kahl (1932) assigned all *Holosticha* species with "three ventral rows" to the subgenus *Trichototaxis*, namely, (i) *Holosticha (Trichototaxis) crassa* Claparède & Lachmann, 1858 (= *Thigmokeronopsis crassa* in present book); (ii) *H. (Trichototaxis) stagnatilis* (Stokes, 1891) Kahl, 1932 (type species); (iii) *H. (Trichototaxis) aquarumdulcium* Bürger, 1905 (species indeterminata); *H. (Trichototaxis) velox* (Quennerstedt, 1869) Kahl, 1932 (= junior synonym of *H. gibba* in present book); *H. (Trichototaxis) fossicola* Kahl, 1932 (likely a *Paraurostyla* species; see species misplaced). Kahl himself recognised that this assemblage is heterogeneous, but avoided a splitting because some species were superficially described.

The improved diagnosis by Borror (1972) is non-specific because he ignored the presence of a bicorona (and therefore of a midventral complex) in the type species. Consequently, he included eight species in *Trichototaxis* (see also species misplaced below). Later he submerged it in *Keronopsis* (now *Pseudokeronopsis*) because in his opinion the number of left marginal rows is variable within a clone (Borror 1979, p. 546, 547).

According to Corliss (1979) and Curds et al. (1983) *Trichototaxis* comprises several species. As an example they show the illustration of *Balladyna euplotes* Dragesco, 1960 (p. 314, Fig. 169), which was transferred to *Trichototaxis* by Borror (1972). However, this species obviously lacks a bicorona and a midventral complex, that is, this species is certainly not closely related to *T. stagnatilis*, type of the present genus.

According to Borror & Wicklow (1983, p. 119) *Trichototaxis* was separated from *Holosticha* formerly (likely they mainly meant Borror 1972) on the number of left marginal rows (two vs. one). Since they found that this feature is so variable within a species, they did not use it any longer as diagnostic feature and therefore synonymised *Trichototaxis* with *Pseudokeronopsis* because they put *T. stagnatilis*, type of *Trichototaxis*, into the synonymy of *Pseudokeronopsis similis*. However, they overlooked (basically they disregarded the rule of priority) that due to this synonymy their own genus *Pseudokeronopsis* would have become superfluous (see also *Uroleptopsis* for same situation). For Borror & Wicklow's distribution of the other species classified in *Trichototaxis* by Borror (1972), see the section on species misplaced in *Trichototaxis*, or the individual descriptions of the species.

Wirnsberger et al. (1987) supposed synonymy of *Trichototaxis* and *Diaxonella*. Later, Wirnsberger (= Aescht) again considered both taxa as valid (Oberschmidleitner & Aescht 1996; further details, see *T. stagnatilis*). *Diaxonella* has three frontal cirri and is therefore assigned to the Holostichidae in the present monograph.

The cirral pattern of *T. stagnatilis* is not known in detail. That is, one cannot exclude, that the inner left marginal row is the anterior portion of the transverse cirral row. Such a cirral pattern is known from *Thigmokeronopsis crassa* (Fig. 176a–p). However, this species is marine and has much more macronuclear nodules so that synonymy can be excluded.

Fauré-Fremiet (1961), Stiller (1974a, p. 130), Corliss (1977, p. 137), Curds et al. (1983), Tuffrau (1979, 1987), and Tuffrau & Fleury (1994) classified *Trichototaxis* in the Holostichidae, that is, the urostyloid character of this group was never doubted. Some authors did not identify their populations to species level (e.g., Gracia et al. 1987, p. 28).

Species included in *Trichototaxis*: (1) *Trichototaxis stagnatilis* Stokes, 1891.

Species misplaced in *Trichototaxis*: Some species, either originally described in *Trichototaxis* and *Holosticha (Trichototaxis)* or transferred to it, are misplaced in this genus/subgenus because they deviate distinctly in one or more important features from the type species.

Holosticha (Trichototaxis) fossicola Kahl, 1932 (basionym: *Trichotaxis fossicola*). Remarks: This species is possibly the senior synonym of *Paraurostyla granulifera* Berger & Foissner, 1989a. For review, see Berger (1999, p. 874).

Trichotaxis aeruginosa Foissner, 1980 (incertae sedis in *Diaxonella*).

Trichotaxis multinucleatus Burkovsky, 1970 (species indeterminata).

Trichotaxis pulchra Borror, 1972a (now *Diaxonella pseudorubra pulchra*).

Trichotaxis rubentis Sarmiento & Guerra, 1960 (species indeterminata).

Trichotaxis villaensis Sarmiento & Guerra, 1960 (species indeterminata).

Trichototaxis aquarumdulcium (Bürger, 1905) Stiller, 1974b (basionym: *Holosticha aquarumdulcium*). Remarks: A species indeterminata (see *Holosticha*).

Trichototaxis caudata (Ehrenberg, 1833) Borror, 1972 (incorrectly spelled *Trichotaxis caudata* in Borror 1972). Remarks: This species is certainly an 18-cirri oxytrichid and was transferred to *Urosoma* Kowalewskiego, 1882 by Berger (1999, p. 398), that is, the recent name is *Urosoma caudata* (Ehrenberg, 1833) Berger, 1999. Borror (1972, p. 11) mentioned *Uroleptus caudatus* (Ehrenberg, 1838) Kahl, 1932 in his list of synonyms of the present species. However, I could not find this combination in Kahl (1932, pp. 547–550). Borror & Wicklow (1983, p. 119), who eliminated *Trichototaxis*, put it into the synonymy of *Uroleptus musculus* (Müller, 1773) Ehrenberg, 1831.

Trichototaxis euplotes (Dragesco, 1960) Borror, 1972 (incorrectly spelled *Trichotaxis euplotes* in Borror 1972). Remarks: This species lacks frontal cirri and a midventral complex and is therefore misplaced in *Trichototaxis*. The sole agreement is in the increased number of cirral rows. Borror & Wicklow (1983, p. 119) recognised the misclassification and supposed that it belongs to *Balladyna,* where Dragesco (1960, p. 314, Fig. 169) originally classified it.

Trichototaxis hembergeri Shao, Li, Hu & Song, 2005. Remarks: This name is disclaimed for nomenclatural purposes (ICZN 1999, Article 8.3; details see *Diaxonella*).

Trichototaxis rubra Plückebaum, Winkelhaus & Hauser, 1997 (see *Diaxonella pseudorubra*).

Single species

Trichototaxis stagnatilis Stokes, 1891
(Fig. 165a, b)

1891 ***Trichototaxis stagnatilis*** – Stokes, Jl R. microsc. Soc., year 1891: 701, Fig. 9 (Fig. 165a; original description; no formal diagnosis provided and no type material available).
1932 ***Trichotaxis stagnatilis* Stokes, 1891** – Kahl, Tierwelt Dtl., 25: 588, Fig. 101 $_{18}$ (Fig. 165b; revision; combination with *Holosticha*, see nomenclature).
1950 ***Trichotaxis stagnatilis* S.** – Kudo, Protozoology, p. 674, Fig. 316h (redrawing of Fig. 165a; textbook).
1972 ***Trichotaxis stagnatilis* Stokes, 1891** – Borror, J. Protozool., 19: 11 (revision of hypotrichs).
1974 ***Trichotaxis stagnatilis* Stokes** – Stiller, Fauna Hung., 115: 52, Fig. 31A (redrawing of Fig. 165a; guide).
2001 ***Trichototaxis stagnatilis* Stokes, 1891** – Berger, Catalogue of ciliate names 1. Hypotrichs, p. 111 (nomenclator containing all basionyms, combinations, and higher taxa of hypotrichs).

Nomenclature: No derivation of the name is given in the original description. The species-group name *stagnatil·is -is -e* (Latin) obviously alludes to the fact this species was discovered in a stagnant water body. *Trichototaxis stagnatilis* is the type species of *Trichototaxis* by monotypy. Kahl (1932) classified *Trichototaxis* (subsequently incorrectly spelled *Trichotaxis*; see genus section) as subgenus of *Holosticha*. Thus, the correct name of the present species in his revision is *Holosticha (Trichototaxis) stagnatilis* (Stokes, 1891) Kahl, 1932.

Remarks: *Trichototaxis stagnatilis*, type of the genus, was never reliably redescribed, indicating that it is either a very rare species or so insufficiently described that identification (synonymy) with a known species is impossible. The general appearance of *T. stagnatilis* resembles *Diaxonella pseudorubra*, the new, senior synonym of *D. trimarginata*, because this species also has a slightly increased number of marginal rows. Wirnsberger et al. (1987, p. 86) therefore suggested synonymy of these two genera. However, Oberschmidleitner & Aescht (1996, p. 24) distanced themselves from this proposal because of the different frontal ciliature (bicorona vs. three frontal cirri). Stokes (1891) wrote that the endoplasm is finely granular and brownish, whereas *Diaxonella pseudorubra* is red (of course, this difference could be explained in that we assume that Stokes did not have a perfectly adjusted microscope so that he did not recognise the true colour). I avoid synonymy, that is, accept both species and genera and suggest trying to find *T. stagnatilis* in the type locality area, that is, in/near Trenton, New Jersey, USA. If the bicorona described by Stokes can be confirmed, synonymy of *Trichototaxis* and *Diaxonella* can be excluded. If *D. pseudorubra*, which has three frontal cirri, is found in the Trenton area, then Stokes very likely did not observe the frontal ciliature correctly, although one cannot exclude that both *T. stagnatilis* and *D. pseudorubra* occur in the Trenton area.

The validity of *T. stagnatilis* was rarely doubted (see genus section). Only Borror (1979, p. 547) and Borror & Wicklow (1983, p. 119) eliminated *Trichototaxis* in that they synonymised the type species *T. stagnatilis* with *Pseudokeronopsis similis*. However, this species lacks a second left marginal row and has a more pronounced moniliform macronucleus, strongly indicating that this synonymy is incorrect.

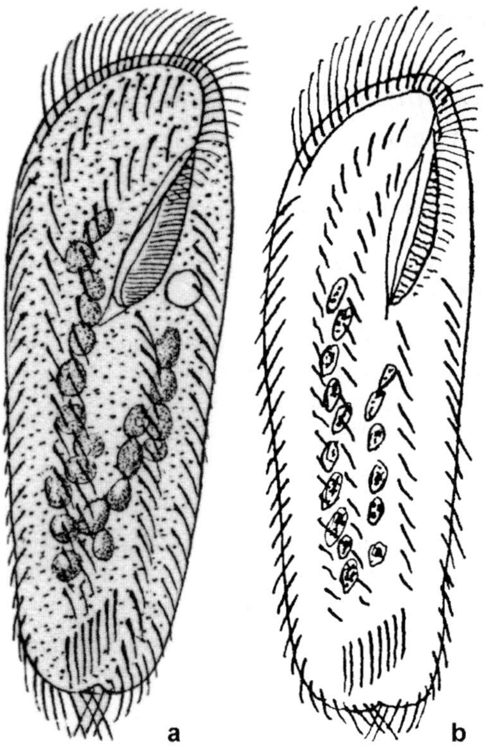

Fig. 165a, b *Trichototaxis stagnatilis* from life (a, from Stokes 1891; b, after Stokes 1891 from Kahl 1932). Ventral view showing, inter alia, the cirral pattern, the nuclear apparatus, and the contractile vacuole (not illustrated by Kahl), 170 μm. Note that Kahl's redrawing is not very exact (e.g., inner left marginal row extends to near body end in [a] against ahead of transverse cirri in [b]). The main difference to *Diaxonella* is the presence of a bicorona (vs. three frontal cirri in *Diaxonella*). Page 827.

Morphology: Body length about 170 × 55 μm (body width estimated from body length:width ratio of about 3:1 given by Stokes). Body flexible, distinctly flattened dorso-ventrally. Body outline obovate, that is, widest in anterior portion; anterior body end obliquely rounded, posterior also rounded and often centrally emarginate; right margin more or less convex, left margin roughly straight. Many macronuclear nodules scattered throughout cell or long (U-shaped) moniliform (Fig. 165a); individual nodules sub-spherical to broadly ovate. Contractile vacuole left of proximal portion of adoral zone, empties via dorsal surface. Cytoplasm finely granular, brownish. Presence/absence of cortical granules unknown. Seldom rapidly moving.

Adoral zone occupies about 38% of body length in specimen illustrated, proximal end extends to near cell-midline. Details of undulating membranes not known. Two cirral rows extend from left anterior cell corner along anterior margin to transverse cirri, indicating that *T. stagnatilis* has a bicorona and a midventral complex composed of cirral pairs. Presence/absence of buccal cirrus and frontoterminal cirri not known. Six or more transverse cirri arranged in oblique, subterminal row; cirri therefore not projecting beyond rear body end. Right marginal row extends from near distal end of adoral zone to rear body end. Two left marginal rows extending to posterior body end, that is, marginal rows confluent posteriorly (Fig. 165a). See remarks of genus section for different interpretation of inner left marginal row. Dorsal infraciliature (length of dorsal bristles, number and arrangement of dorsal kineties, presence/absence of caudal cirri) not known.

Occurrence and ecology: Limnetic. Type locality of *T. stagnatilis* is a freshwater habitat (pond, pool) in/near Trenton, New Jersey, USA (Stokes 1891, p. 697). He found it in an infusion containing decaying *Sphagnum*. No further records published.

Species indeterminata

Trichotaxis multinucleatus Burkovsky, 1970 (Fig. 166a)

1970 ***Trichotaxis multinucleatus* n. sp.** – Burkovsky, Acta Protozool., 8: 59, 64, Fig. 15 (Fig. 166a; original description; no formal diagnosis provided; type material possibly in Burkovsky's private collection).

1972 ***Trichotaxis multinucleatus* Burkovsky, 1970** – Borror, J. Protozool., 19: 11 (revision of hypotrichs).

2001 ***Trichototaxis multinucleatus* (Burkovsky, 1970) comb. nov.** – Berger, Catalogue of ciliate names 1. Hypotrichs, p. 95, 96 (combination with *Trichototaxis*; nomenclator containing all basionyms, combinations, and higher taxa of hypotrichs).

Remarks: This species was established in *Trichotaxis*, which is an incorrect subsequent spelling of *Trichototaxis*. *Trichototaxis* is feminine; thus, the correct species-group name is *multinucleata* (Latin) which refers to the many (macro)nuclei.

Burkovsky (1970d) provided the following brief characterisation: body elongated, 120–200 × 40–50 µm. Two marginal and three ventral rows; five frontal and eight transverse (anal cirri in his terminology) cirri. Nuclear apparatus composed of 25–30 oval nodules. Contractile vacuole behind proximal end of adoral zone. The cirral pattern illustrated is difficult to interpret because it is composed of each one left and right marginal row, three distinctly separated ventral rows (very likely no midventral complex is present), three frontal cirri, a buccal cirrus, and likely a cirrus III/2. Burkovsky's species basically lacks all *Trichototaxis*-features, for example, the bicorona, the midventral complex, and more than two marginal rows. Consequently, his classification is incorrect and I strongly doubt that the description allows a certain identification. Borror (1972, p. 11) and Carey (1992, p. 186, Fig. 742) considered it as valid *Trichototaxis* species, whereas Borror & Wicklow (1983, p. 119) supposed a relationship to *Paraurostyla*. Hemberger (1982, p. 117, 278 [incorrectly spelled *Trichotaxis multinucleatum*]) classified it as incertae sedis in the hypotrichs. If a midventral complex is present, then it could be a *Diaxonella* species. Type locality is the Kandalaksha Gulf of the White Sea (see also Burkovsky 1970a, p. 190). Also recorded from Brazomar Beach (Castro Urdiales), Bay of Biscay (Fernandez-Leborans et al. 1999, p. 742).

Trichotaxis rubentis Sarmiento & Guerra, 1960 (Fig. 166b)

1960 ***Trichotaxis rubentis* n. sp.** – Sarmiento & Guerra, Publnes Mus. Hist. nat., Lima, 19: 18, Lamina III, Fig. 27 (Fig. 166b; original description; no formal diagnosis provided and likely no type material available).

1972 ***Trichotaxis rubentis* Sarmiento & Guerra, 1960** – Borror, J. Protozool., 19: 11 (revision of hypotrichs).

1987 ***Trichototaxis rubentis* (Sarmiento und Guerra, 1960) nov. comb.** – Foissner, Arch. Protistenk., 133: 226 (combination with *Trichototaxis*).

2001 ***Trichototaxis rubentis* (Sarmiento & Guerra, 1960) Foissner, 1987** – Berger, Catalogue of ciliate names 1. Hypotrichs, p. 96 (nomenclator containing all basionyms, combinations, and higher taxa of hypotrichs).

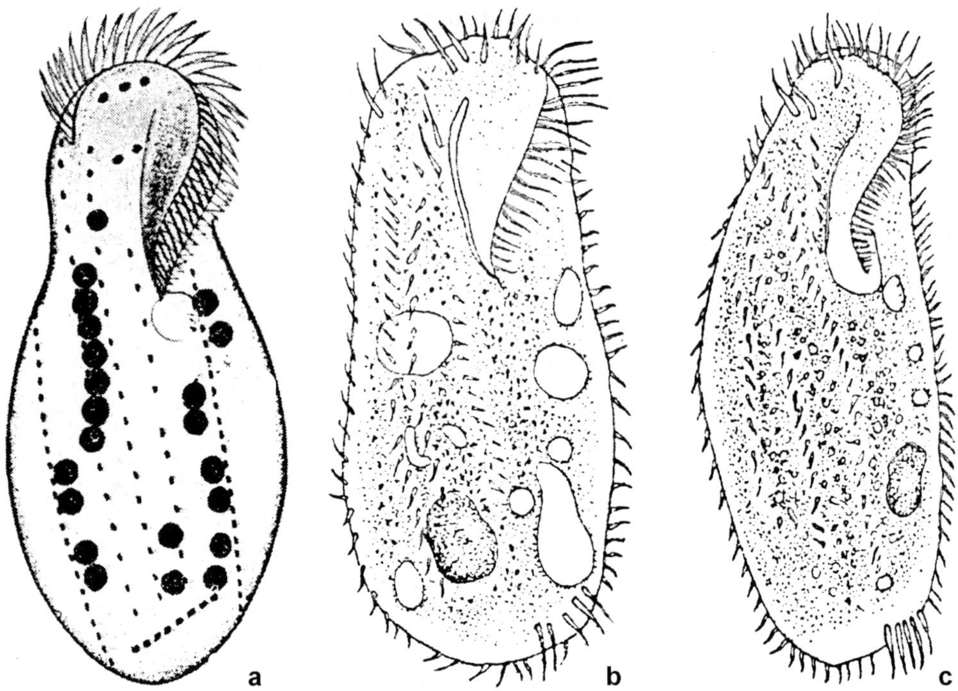

Fig. 166a–c Species indeterminata. **a:** *Trichototaxis multinucleata* (from Burkovsky 1970d. Wet silver nitrate impregnation), infraciliature of ventral side and nuclear apparatus, 195 µm. Page 829. **b:** *Trichototaxis rubentis* (from Sarmiento & Guerra 1960. Method not indicated), ventral view, 134–187 µm. Page 829. **c:** *Trichototaxis villaensis* (from Sarmiento & Guerra 1960. Method not indicated), ventral view, 139–206 µm. Page 831.

Remarks: No derivation of the name is given in the original description. The species-group name *rubentis* likely refers to the red colour of the cell. This species was established in *Trichotaxis*, which is an incorrect subsequent spelling of *Trichototaxis* (see also nomenclature in genus section). The illustration obviously shows a fixed and squeezed specimen. Thus, the cirral pattern is more or less distinctly destroyed. The three enlarged frontal cirri prevent a classification in *Trichototaxis*, whose type species has a distinct bicorona. According to the cirral pattern and the red colour the present species could be a synonym of *Diaxonella pseudorubra*. However, this species has many macronuclear nodules, whereas *T. rubentis* has two nodules (the single nodule also mentioned is likely a misobservation). Likely for that reason, Hemberger (1982, p. 117) wrote that it is definitely an insufficiently observed *Paraurostyla* species. Because of the considerable uncertainties, I classify it as species indeterminata.

Body size about 134–187 × 33–53 µm in life(?). Body flexible, ventral side plane, dorsal side vaulted. One or two macronuclear nodules in posterior body portion. Moves in serpentine curves, when touching an obstacle it contracts. Contractile vacuole present. Cytoplasm granulated, opaque red. Adoral zone occupies about 38% of body length. Three frontal cirri, 4–6 transverse cirri. One left and one right marginal row and three ventral cirral rows. Found in clear, stagnant waters during a survey of Villa lagoon (Peru) and surrounding canals (Sarmiento & Guerra 1960; see also Guillén et al. 2003; p. 180).

Trichotaxis villaensis Sarmiento & Guerra, 1960 (Fig. 166c)

1960 ***Trichotaxis villaensis* n. sp.** – Sarmiento & Guerra, Publnes Mus. Hist. nat., Lima, 19: 18, Lamina III, Fig. 28 (Fig. 166c; original description; no formal diagnosis provided and likely no type material available).

1972 ***Trichotaxis villaensis* Sarmiento & Guerra, 1960** – Borror, J. Protozool., 19: 11 (revision of hypotrichs).

1987 ***Trichototaxis villaensis* (Sarmiento und Guerra, 1960) nov. comb.** – Foissner, Arch. Protistenk., 133: 226 (combination with *Trichototaxis*).

2001 ***Trichototaxis villaensis* (Sarmiento and Guerra, 1960) Foissner, 1987** – Berger, Catalogue of ciliate names 1. Hypotrichs, p. 96 (nomenclator containing all basionyms, combinations, and higher taxa of hypotrichs).

Remarks: No derivation of the name is given in the original description. The species-group name *villaensis* obviously alludes to the locality (Villa lagoon, Peru) were the species was discovered. *Trichotaxis* is a frequently used incorrect subsequent spelling of *Trichototaxis* (see nomenclature at genus section). Therefore Foissner (1987d) made a new combination.

The original description is somewhat superficial and the illustration not very detailed. The three frontal cirri prevent a classification in *Trichototaxis* because the type species (*T. stagnatilis*) has a distinct bicorona. By contrast, *Diaxonella pseudorubra* has basically the same cirral pattern, indicating a relationship of *T. villaensis* to *Diaxonella*. Hemberger (1982, p. 117) wrote that both *Trichototaxis* species described by Sarmiento & Guerra (1960) are definitely insufficiently redescribed *Paraurostyla* species. Very likely he concluded this from the low number of macronuclear nodules (one or two), which is indeed more reminiscent of this oxytrichid genus than of *Diaxonella*, whose type species has, like many other urostyloids, many nodules dispersed throughout the cell. According to Borror & Wicklow (1983, p. 119), who eliminated *Trichototaxis*, the present species possibly belongs to *Holosticha*. Because of the considerable uncertainties I classify it, like *T. rubentis*, as species indeterminata.

Body size about 139–206 × 43–62 µm in life(?). Body outline oval, dorsal side convex, ventral side slightly concave. Two ovoid macronuclear nodules. Contractile vacuole near proximal end of adoral zone. Cytoplasm granulated, dark-grey. Adoral zone of membranelles occupies about 40% of body length. Cirral pattern basically as in the second *Trichototaxis* species describedby Sarmiento & Guerra (1960), *Trichototaxis rubentis*, that is, three frontal cirri, 4–6 transverse cirri, one left and one right marginal row, and three ventral cirral rows. Type locality is Villa lagoon (Peru) and surrounding canals where Sarmiento & Guerra (1960) discovered it in clear waters with algae and diatoms (see also Guillén et al. 2003, p. 180).

Pseudokeronopsidae Borror & Wicklow, 1983

1983 **Pseudokeronopsidae fam. nov.** – Borror & Wicklow, Acta Protozool., 22: 123 (original description; no formal diagnosis provided). Type genus: *Pseudokeronopsis* Borror & Wicklow, 1983.
1985 **Pseudokeronopsidae** – Small & Lynn, Phylum Ciliophora, p. 451 (guide).
1994 **Pseudokeronopsidae Borror et Wicklow, 1983** – Tuffrau & Fleury, Traite de Zoologie, 2: 130 (revision).
2001 **Pseudokeronopsidae Borror and Wicklow, 1983** – Berger, Catalogue of ciliate names 1. Hypotrichs, p. 111 (nomenclator containing all basionyms, combinations, and higher taxa of hypotrichs).
2002 **Pseudokeronopsidae Borror and Wicklow, 1983**[1] – Lynn & Small, Phylum Ciliophora, p. 445 (guide).
2004 **Pseudokeronopsidae** – Berger, Acta Protozool., 43: 116 (phylogenetic analysis).

Nomenclature: The names Pseudokeronopsidae and Pseudokeronopsinae are based on the genus-group name *Pseudokeronopsis*. Originally established as family and subfamily. I do not categorise the supraspecific taxa (see chapter 7.2 in the general section).

Characterisation (Fig. 144a, apomorphy 6; Fig. 167a, apomorphy 1): Acaudalia where the many macronuclear nodules fuse to a strongly branched mass or to some parts during cell division (A). Parental adoral zone of membranelles totally replaced during cell division (A?).

Remarks: Borror & Wicklow (1983) – using basically the system proposed by Wicklow (1981) – united *Pseudokeronopsis* (including *Uroleptopsis* as synonym) and *Thigmokeronopsis* in the Pseudokeronopsidae. As unifying features they mentioned the presence of a bicorona (in the present paper the apomorphy of the Urostylidae) and the far posteriorly extending distal end of the adoral zone of membranelles (in the present paper the apomorphy of the Retroextendia). Eigner & Foissner (1992) also considered *Thigmokeronopsis* and *Pseudokeronopsis* as sister groups. However, they could not provide a synapomorphy because of the lack of appropriate data on *Thigmokeronopsis*.

I basically agree with this grouping, but provide a different and more detailed foundation of the relationships (Fig. 167a; Berger 2004b). The tree proposed is the most parsimonious one of three possible arrangements of *Pseudokeronopsis*, *Thigmokeronopsis*, and *Uroleptopsis*.

The first feature mentioned in the characterisation above is realised in *Thigmokeronopsis*, at least in *T. crystallis* and *T. antarctica*, where the many nodules fuse to several parts during ontogenesis (Petz 1995). Unfortunately, nothing is known about the fate of the macronucleus during this part of the life cycle in the type species *Thigmokeronopsis jahodai* (Wicklow 1981). The plesiomorphic state of this feature is that all nodules (two to many) fuse to a single compact mass during cell division. This state is present in all hypotrichs except the pseudokeronopsids. In *T. rubra* the macronuclear nodules fuse to a single mass which, however, is strongly branched (Fig. 171s), similar

[1] The definition by Lynn & Small (2002) is as follows: Frontal cirri form a conspicuous, arc-like file that parallels the anterior left serial oral polykinetids, which may be doubled by an arc-like extension of the frontoventral zig-zag file; left and right marginal cirri as 1 (rarely 2) rows.

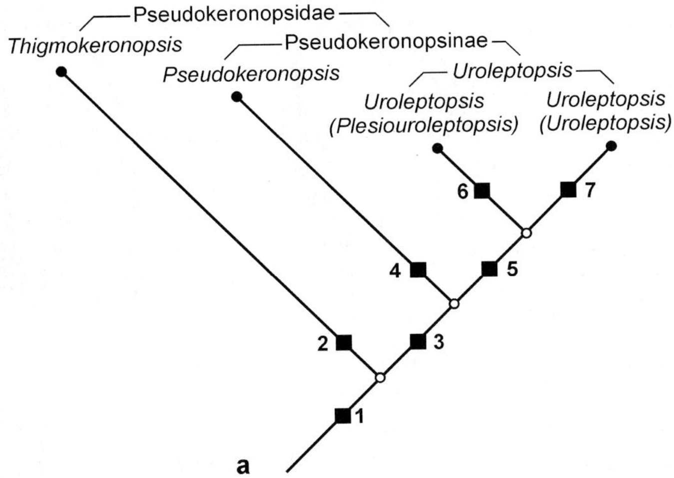

Fig. 167a Diagram of phylogenetic relationships within the Pseudokeronopsidae (from Berger 2004b, slightly modified). Autapomorphies (black squares 1–7): **1** – the many macronuclear nodules fuse to a strongly branched mass or to some parts during cell division; parental adoral zone of membranelles totally replaced during cell division (?; possibly an apomorphy of the Acaudalia or even the Retroextendia, Fig. 144a). **2** – thigmotactic field of cirri present; anlagen of marginal rows and dorsal kineties originate de novo. **3** – each of the many macronuclear nodules divides individually. **4** – four or more dorsal kineties. **5** – cirral anlage I forms two cirri; transverse cirri lacking; gap in adoral zone (convergence to some holostichids). **6** – some midventral cirral anlagen produce midventral rows (convergence to Urostylinae and Bakuellidae). **7** – buccal cirrus not in ordinary position; some midventral cirral anlagen finally produce only one cirrus. Note that the present tree is only one of several hypothesis and that the "defined endings" -idae and -inae have no meaning in the present book (see chapter 7.2 in the general section).

to in *Metaurostylopsis marina* (Song et al. 2001). Interestingly, in *M marina* species the formation of the proter's adoral zone proceeds very similarly (identically?) to in the pseudokeronopsids (see next paragraph). Thus, I supposed that *Metaurostylopsis* could be the sister group of the Pseudokeronopsidae because the macronucleus already shows the tendency not to fuse to a compact, globular mass (Berger 2004b).

Whether the formation of a totally new adoral zone of membranelles for the proter (usually de novo in a pouch in the buccal cavity) – as described for *Thigmokeronopsis*, *Uroleptopsis*, and *Pseudokeronopsis* – is a further apomorphy for the Pseudokeronopsidae or for the Acaudalia or even for the Retroextendia (Fig. 144a) is uncertain because too few data are available. Thus, cell division data on a second *Pseudourostyla* species and, for example, *Tricoronella* are needed for a more serious discussion of this feature. "Unfortunately" a more or less identical mode of adoral zone formation is described for *Metaurostylopsis marina* (Song et al. 2001), *Metaurostylopsis rubra* (Wilbert & Song 2003), a *Diaxonella* population (Shao et al. 2005; misidentified as *Trichototaxis*), and *Anteholosticha multistilata* (Hemberger 1982). In *A. multistilata* and *Diaxonella* the macronuclear nodules fuse to a single, compact mass so that they cannot be included in the Pseudokeronopsidae. By contrast, in *Metaurostylopsis marina* the macronuclear nodules behave like those of *Thigmokeronopsis rubra*, that is, forms a

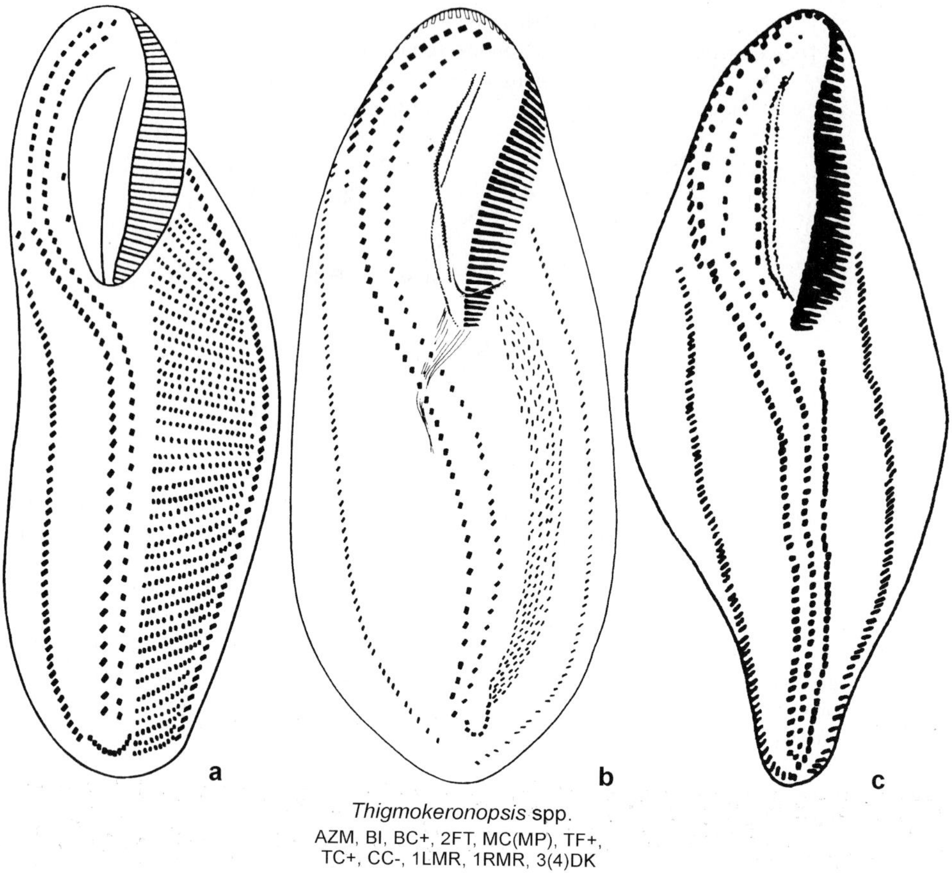

Fig. 168a–c Ventral cirral pattern in members of the Pseudokeronopsidae (part 1). **a:** *Thigmokeronopsis jahodai*. **b:** *Thigmokeronopsis crystallis*. **c:** *Thigmokeronopsis crassa*. Sources of illustrations see individual descriptions. Abbreviations used in short characterisations of infraciliature (explanation of supplemental signs and number see legend to Fig. 20a–c): AZM = adoral zone of membranelles, BC = buccal cirrus, BI = bicorona, CC = caudal cirri, DK = dorsal kineties, FT = frontoterminal cirri, LMR = left marginal row, MC(MP) = midventral complex composed of cirral pairs only, RMR = right marginal row, TC = transverse cirri, TF = thigmotactic field.

branched pattern (see previous paragraph). However, both *Metaurostylopsis* and *A. multistilata* have three enlarged frontal cirri, that is, another type of frontal ciliature, and are therefore not included in the Urostylidae, but in the Bakuellidae (with midventral rows), respectively, Holostichidae (without midventral rows). Thus, I have to interpret the new formation of the adoral zone in the pseudokeronopsids, in *Metaurostylopsis*, and in *A. multistilata* as convergence. Song et al. (1997) described a total new formation of the proter's adoral zone for *Pseudoamphisiella lacazei*. However, in this species the corresponding oral primordium originates behind the parental adoral zone and not in the buccal cavity as in the pseudokeronopsids. Further, the macronuclear

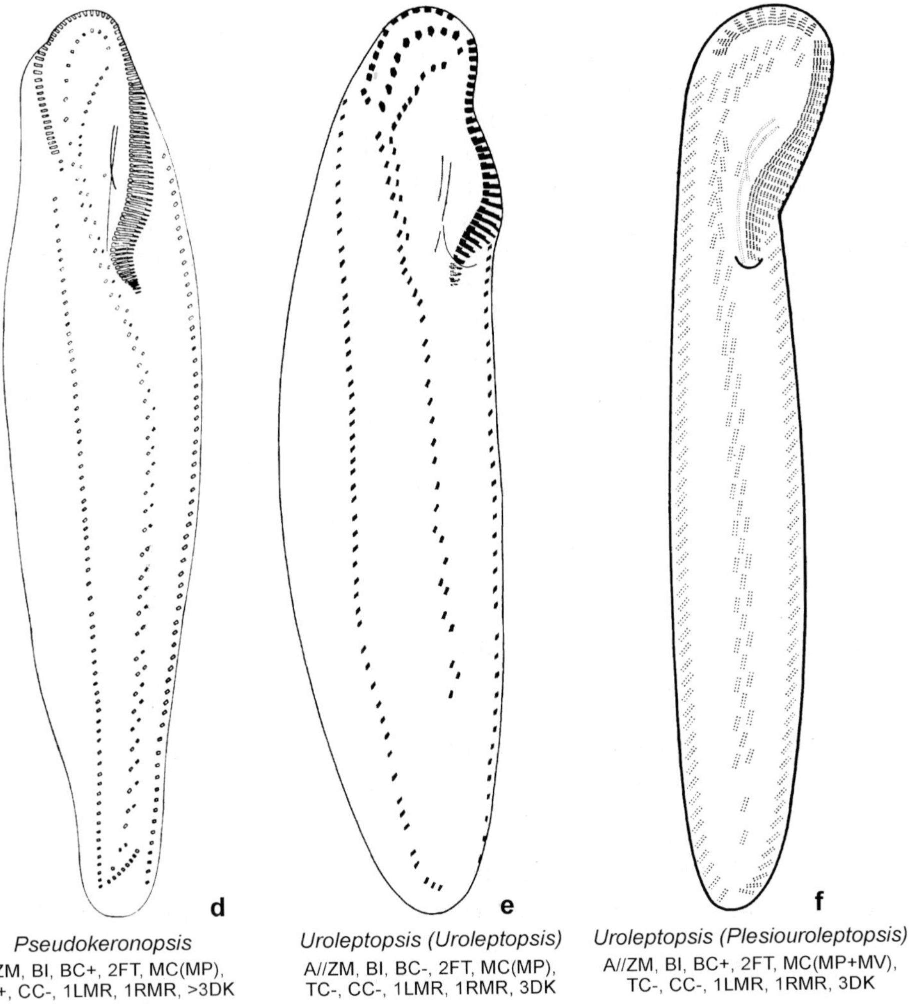

Fig. 168d–f Ventral cirral pattern in members of the Pseudokeronopsidae (part 2). **d:** *Pseudokeronopsis rubra*. **e:** *Uroleptopsis (Uroleptopsis) citrina*. **f:** *Uroleptopsis (Plesiouroleptopsis) ignea*. Sources of illustrations see individual descriptions. Abbreviations used in short characterisations of infraciliature (explanation of supplemental signs and number see legend to Fig. 20a–c): AZM, A//ZM – adoral zone of membranelles (without and with gap), BC = buccal cirrus, BI = bicorona, CC = caudal cirri, DK = dorsal kineties, FT = frontoterminal cirri, LMR = left marginal row, MC(MP), MC(MP+MV) = midventral complex composed of cirral pairs only, respectively, of cirral pairs and midventral rows, RMR = right marginal row, TC = transverse cirri.

nodules fuse to a single mass, so that it cannot be included in the Pseudokeronopsidae. In the other urostyloids, few parental adoral membranelles (e.g., *Bakuella*; Song et al. 1992, Eigner & Foissner 1992) to many (e.g., *Urostyla grandis*; Ganner 1991) are reorganised. This type of reorganisation is obviously distinctly different from the total new formation discussed above.

Borror & Wicklow (1983) characterised the Pseudokeronopsidae, inter alia, by the far posteriorly extending distal end of the adoral zone. In fact, this feature applies to *Pseudokeronopsis* and *Thigmokeronopsis*, and to a somewhat smaller extent also to *Uroleptopsis*, that is, these taxa have a high DE-value (see Fig. 1c and chapter 1.8 of general section). High DE-values are also known from *Pseudourostyla*, *Tricoronella*, and *Bicoronella*, which have, like the pseudokeronopsids, a bi- or tricorona and a midventral complex composed of cirral pairs only. Thus, this feature was used not for the Pseudokeronopsidae, but to characterise the Retroextendia (Fig. 144a, apomorphy 2). For discussion of convergencies in this feature see Retroextendia.

Lynn & Small (2002) included *Thigmokeronopsis*, *Keronella*, *Tricoronella*, *Bicoronella*, and *Pseudokeronopsis* in the Pseudokeronopsidae (Table 11). In *Tricoronella* and *Keronella* the macronuclear nodules fuse to a single mass during ontogenesis, strongly indicating that they do not belong to the Pseudokeronopsidae. For *Bicoronella* no data about the nuclear apparatus during division are available. However, there is some evidence that it is related to *Tricoronella* (Fig. 144a). *Keronella* has, besides midventral pairs, midventral rows, so that it is assigned to the Urostylinae in the present paper (Fig. 144a).

Key to the genera of the Pseudokeronopsidae

Since the taxa below are mainly distinguished by details of the cirral pattern, that is, certain cirral groups present or not, protargol preparations, or at least very detailed live observations (interference contrast) are needed for successful identification.

1 Transverse cirri absent (Fig. 168e, f) .... *Uroleptopsis* (p. 980)
\- Transverse cirri present (Fig. 168a–d) **2**
2 A cirral row or a more or less large field of cirri/cilia between left marginal row and midventral complex (Fig. 168a–c) .... *Thigmokeronopsis* (p. 836)
\- No cirral row or field between left marginal row and midventral complex (Fig. 168d) .... *Pseudokeronopsis* (p. 886)

Thigmokeronopsis Wicklow, 1981

1981 ***Thigmokeronopsis* n. gen.**[1] – Wicklow, Protistologica, 17: 331, 348 (original description). Type species (by original designation on p. 348): *Thigmokeronopsis jahodai* Wicklow, 1981.

1983 ***Thigmokeronopsis jahodai* (n. gen., n. sp.)** – Wicklow, Diss. Abstr. Int., 43B: 2135 (see nomenclature).

1983 ***Thigmokeronopsis* Wicklow, 1981** – Borror & Wicklow, Acta Protozool., 22: 124 (revision of urostylids; see nomenclature).

1985 ***Thigmokeronopsis*** – Small & Lynn, Phylum Ciliophora, p. 451 (guide to ciliate genera).

[1] The diagnosis by Wicklow (1981, p. 348) is as follows: Somatic ciliature includes dorsal bristle rows, one left and one right marginal cirral row; frontal ciliature includes migratory, midventral, transverse, malar, and thigmotactic cirri. Thigmotactic cirri form a left, post-oral ciliary field used for adhesion to substrate. Multimacronucleata.

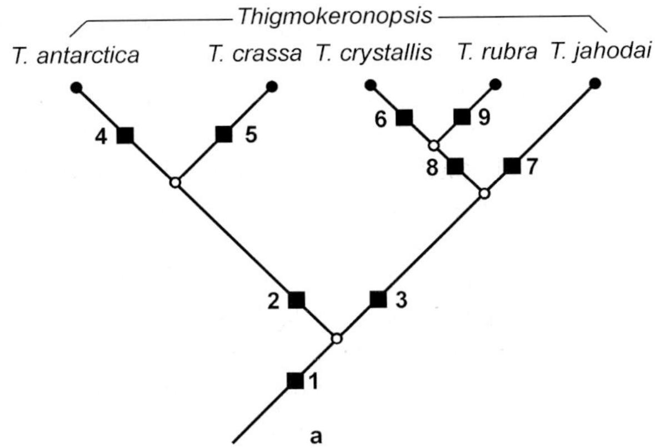

Fig. 169a Diagram of phylogenetic relationships within *Thigmokeronopsis* (original). For relationship of *T. jahodai* and *T. magna*, see *T. magna*. Autapomorphies (black squares 1–9): **1** – very many transverse cirri form long row extending to near proximal end of adoral zone; cirri of midventral pairs distinctly separated, that is, distinct zigzag-pattern lacking; marginal rows and dorsal kineties originate de novo. **2** – number of buccal cirri slightly increased (about 4). **3** – anterior and middle portion of transverse cirral row modified to moderately large thigmotactic field. **4** – midventral complex shortened posteriorly; transverse cirral row shortened anteriorly; cortical granules lacking (convergence to 6). **5** – number of buccal cirri high (about 9); cortical granules red (convergence to 9). **6** – cortical granules lacking (convergence to 4). 7 – thigmotactic field very large; 4 dorsal kineties; 2 buccal cirri. **8** – no autapomorphy known. **9** – cortical granules (one type) red (convergence to 5). Alternate hypothesis for the *crystallis-rubra-jahodai* branch: **3** – thigmotactic field very large. **6** – cortical granules lacking (convergence to 4). 7 – 4 dorsal kineties; 2 buccal cirri. **8** – thigmotactic field reduced to moderate size. **9** – cortical granules (one type) red (convergence to 5). For detailed description and discussion of autapomorphies, see text. Note that this is only a proposal which can change significantly when new taxa and/or new data become available.

(1985, p. 451), Tuffrau (1987, p. 115), and Tuffrau & Fleury (1994). By contrast, Shi (1999a) and Shi et al. (1999) listed *Thigmokeronopsis* in the Urostylidae, whereas Carey (1992) placed it in the Holostichidae. Petz (1995) stated that *T. antarctica* and *T. crystallis* would need a genus and probably a new family of their own if *T. jahodai* differs significantly in the morphogenetic features (especially nuclear division) discussed above.

At present it seems unwise to subdivide *Thigmokeronopsis* (Fig. 169a), for example, into (i) *T. crassa* and *T. antarctica* (Fig. 169a; autapomorphy 2, number of buccal cirri slightly increased), and (ii) *T. crystallis*, *T. rubra* and *T. jahodai* (Fig. 169a, autapomorphy 3, anterior portion of long row of transverse cirri modified to thigmotactic field) because some important data are lacking (see above).

Species included in *Thigmokeronopsis* (alphabetically arranged according to basionyms): (1) *Oxytricha crassa* Claparède & Lachmann, 1858; (2) *Thigmokeronopsis antarctica* Petz, 1995; (3) *Thigmokeronopsis crystallis* Petz, 1995; (4) *Thigmokeronopsis jahodai* Borror & Wicklow, 1983; (5) *Thigmokeronopsis magna* Wilbert & Song, 2005; (6) *Thigmokeronopsis rubra* Hu, Warren & Song, 2004.

Key to *Thigmokeronopsis* species

Identification of *Thigmokeronopsis* species can be done easily by the rather different size of the thigmotactic field, which is a more or less conspicuous patch/stripe of (transverse) cirri between the left marginal row and the midventral complex. However, the field must not be confused with the oral primordium of early dividers of other species.

1 Thigmotactic field very large (Fig. 170a, b, 170.1k) **5**

\- Thigmotactic field forms stripe or single file of cirri **2**

2 Thigmotactic field forms a small to moderate-wide stripe of fine cirri (Fig. 168b; do not confuse the thigmotactic field with the oral primordium of an early divider of another species) **4**

\- Thigmotactic field forms single file of transverse cirri (Fig. 168c) **3**

3 2–5, usually 4 buccal cirri; midventral complex extends to near posterior quarter of cell; cortical granules lacking; distinct gap between proximal end of adoral zone and thigmotactic field ("transverse cirri"), thigmotactic cirri very fine (Fig. 173a, b) . . . . *Thigmokeronopsis antarctica* (p. 862)

\- 6–10 buccal cirri; midventral complex extends close to rear cell end; cortical granules red; thigmotactic field ("transverse cirri") extends close to proximal end of adoral zone, cirri of about same size as other cirri (Fig. 176d) . . . . *Thigmokeronopsis crassa* (p. 873)

4 (2) Cells more or less colourless (Fig. 172a, b) *Thigmokeronopsis crystallis* (p. 855)

\- Cells reddish (Fig. 171a, f) . . . . *Thigmokeronopsis rubra* (p. 852)

5 (1) One buccal cirrus; 3 dorsal kineties (Fig. 170.1g, k) . . . . *Thigmokeronopsis magna* (p. 848)

\- Two buccal cirri; 4 dorsal kineties (Fig. 170a, c) *Thigmokeronopsis jahodai* (p. 842)

Thigmokeronopsis jahodai Wicklow, 1981
(Fig. 170a–j, Table 36)

1981 ***Thigmokeronopsis jahodai* n. g. n. sp.** – Wicklow, Protistologica, 17: 331, Fig. 1–31 (Fig. 170a, c–j; original description; no formal diagnosis provided; site were type material is deposited not mentioned).

1983 ***Thigmokeronopsis jahodai* (n. gen., n. sp.)** – Wicklow, Diss. Abstr. Int., 43B: 2135 (see nomenclature).

1983 ***Thigmokeronopsis jahodai* Wicklow, 1981** – Borror & Wicklow, Acta Protozool., 22: 115, 124, Fig. 9 (Fig. 170b; revision of urostylids).

1985 ***Thigmokeronopsis jahodai*** – Small & Lynn, Phylum Ciliophora, p. 451, Fig. 6 (Fig. 170b; guide to ciliate genera).

1992 ***Thigmokeronopsis jahodai* Wicklow, 1981** – Carey, Marine interstitial ciliates, p. 187, Fig. 741 (redrawing from Wicklow 1981).

2001 ***Thigmokeronopsis jahodai* Wicklow, 1981** – Berger, Catalogue of ciliate names 1. Hypotrichs, p. 89 (nomenclator containing all basionyms, combinations, and higher taxa of hypotrichs).

2001 ***Thigmokeronopsis jahodai*** – Eigner, J. Euk. Microbiol., 48: 77, Fig. 31 (Fig. 170c modified; brief review of urostylids).

2002 ***Thigmokeronopsis jahodai*** – Lynn & Small, Phylum Ciliophora, p. 445, Fig. 13A (Fig. 170b; guide to ciliate genera).

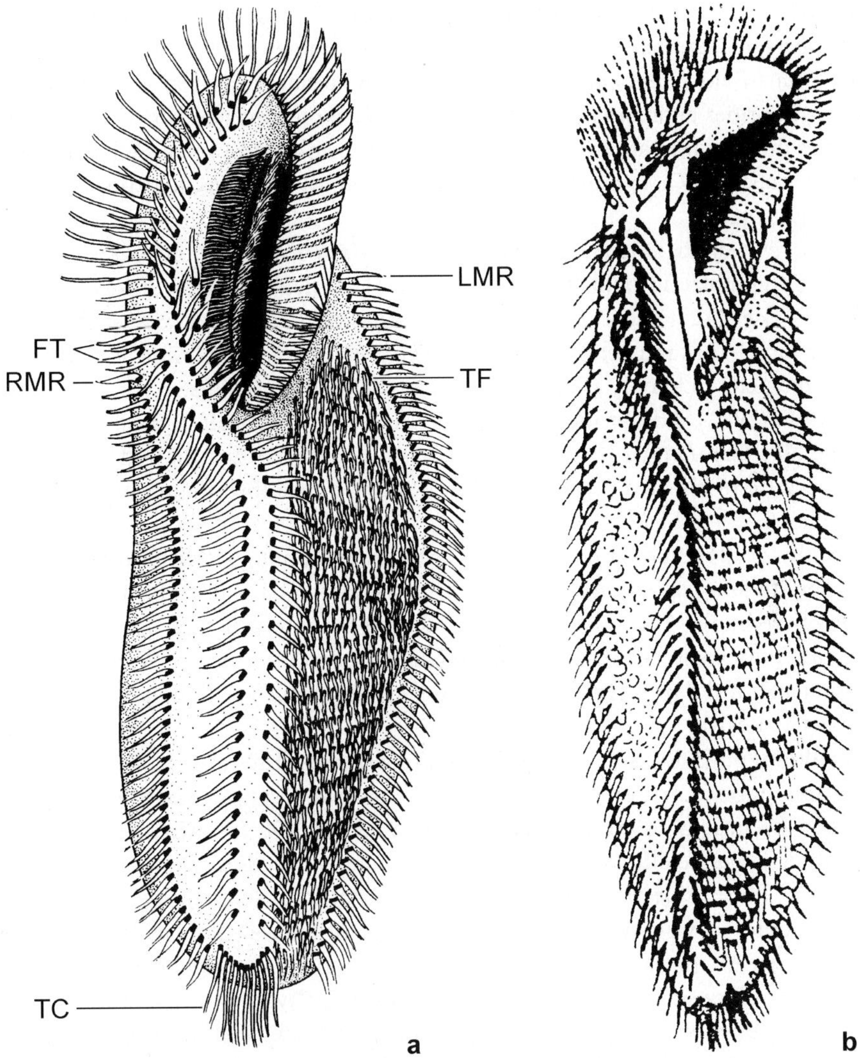

Fig. 170a, b *Thigmokeronopsis jahodai* (a, from Wicklow 1981; b, from Borror & Wicklow 1983. a, protargol impregnation; b, method not indicated). Ventral views, a = 233 µm, b = 178 µm. FT = frontoterminal cirri, LMR = left marginal row, RMR = anteriormost cirrus of right marginal row, TC = transverse cirri, TF = thigmotactic field. Page 842.

Nomenclature: This species is named in honour of William J. Jahodai (Wicklow 1981, p. 349). *Thigmokeronopsis jahohai* in Borror & Wicklow (1983, p. 124) is an incorrect subsequent spelling. *Thigmokeronopsis jahodai* was fixed as type species of *Thigmokeronopsis* by original designation. For a discussion of Wicklow's (1983) dissertation abstract, see the genus section.

Remarks: The most important autapomorphy of *T. jahodai* is the huge cirral field between the left marginal row and the midventral complex (Fig. 169a, autapomorphies 7). Due to this field *T. jahodai* has one of the most outstanding cirral patterns within the hypotrichs. Further autapomorphies are the slightly increased number of dorsal kineties (4 vs. 3 in the ground pattern of the pseudokeronopsids) and two buccal cirri; the supposed sister species *T. crystallis* and *T. rubra* have only one buccal cirrus, which is the plesiomorphic state.

Thigmokeronopsis jahodai differs from *T. crystallis* and *T. antarctica*, besides the size of the thigmotactic field, by the presence of cortical granules. *Thigmokeronopsis crassa* – whose thigmotactic field is composed, as in *T. antarctica*, of a single file of transverse cirri – has red cortical granules. *Thigmokeronopsis rubra* is also reddish, but has a thigmotactic field only slightly wider than that of *T. crystallis*.

Wicklow (1981) did not mention caudal cirri, indicating that *T. jahodai* lacks such cirri, like all other *Thigmokeronopsis* species. It is unlikely that he did not check this feature because in his next paper (Wicklow 1982) he described the presence or absence of caudal cirri.

Wicklow (1981) studied not only the ordinary morphology, but also the ultrastructure of non-dividing specimens and the cell division, which is documented by scanning electron micrographs. For a detailed description of the ultrastructure, see Wicklow's paper because only some details are presented here. I contacted Bary J. Wicklow about the deposition site of the type slides and the mode of macronuclear fission. Unfortunately, I did not get an answer.

Morphology: Body size in protargol preparations 180–240 × 55–85 µm, body length:width ratio 3.1:1 on average (Table 36). Body supple and obviously highly contractile[1] because a 200 µm individual can stretch to 400 µm when feeding actively. Body outline elongate elliptical with both ends broadly rounded; ventral body side more or less plane, dorsal side vaulted. Macronuclear nodules scattered throughout cytoplasm; shape of nodules as well as number, size and shape of micronuclei not mentioned. Presence or absence of contractile vacuole also not mentioned in original description. Irregular groups of yellow-green granules scattered subcortically; granules more numerous on dorsal surface than on ventral side (the description indicates that these are ordinary cortical granules); size of granules not mentioned. Food vacuoles with numerous and sundry diatoms. Ordinary movement, however, when subject to a current – such as water forced from a pipette – cells adhere firmly to substrate by the thigmotactic field.

Adoral zone occupies about 33–37% of body length (Fig. 170a, b), extends far onto right body side, composed of around 75 membranelles of ordinary shape and fine-structure. Membranelles of proximal half of zone composed, as is usual, of four rows of basal bodies: posteriormost two rows each composed of about 40 basal bodies, next row consists of about 25 basal bodies, and anteriormost row composed of only 3–6 basal bodies. Buccal cavity obviously rather long, moderately wide, and likely of ordinary depth. Endoral, as is usual, in dorsal wall of buccal cavity, almost straight, long because

[1] According to Wicklow (1981, p. 334) the body is "slightly contractile". However, this is likely an understatement for a specimen which can stretch from 200 µm to 400 µm when feeding actively.

continuing toward cytostome, well beyond rear end of paroral and proximal membranelles; composed of single row of basal bodies. Anterior end of endoral is oriented with the distal end of basal bodies nearest to ventral surface of cell; as the row descends into the buccal cavity, it rotates 180° along its longitudinal axis until the proximal end of each basal body is nearest the cell's ventral surface. Paroral distinctly curved, composed of a series of short (1–6 basal bodies) oblique rows that lie within a pellicular groove on the right side of the buccal opening; rows longer in the middle and shorter toward anterior and posterior end of paroral. Undulating membranes likely optically not distinctly crossing (Fig. 170c). Pharynx probably of ordinary length and structure.

Cirral pattern very conspicuous due to the thigmotactic field (Fig. 170a, b). Frontal cirri form distinct bicorona, not or only indistinctly set off from midventral complex; on average 54 cirral pairs form bicorona **and** midventral complex which extends roughly S-shaped to near transverse cirri. Two buccal cirri right of proximal half of paroral. Two frontoterminal cirri close to distal end of adoral zone. Zigzagging midventral pattern indistinct. Right cirrus of each midventral pair composed of three rows of 6–7 basal bodies oriented at a 60° angle to cell's main axis; left cirrus composed of three rows of 5–6 basal bodies lying at 75° angle (angles become less acute in cirri ahead of level of cytostome). Transverse cirri form bow near rear body end, cirri thus distinctly projecting beyond body margin (Fig. 170a–c). Area bordered by left marginal row, proximal portion of adoral zone, and midventral complex almost completely covered by a thigmotactic field composed of about 42 transverse rows each consisting of 6–17 cirri; thigmotactic cirri smaller than midventral and frontal cirri, that is, composed of two rows of 2–6 parallelogram-arranged basal bodies (however, groups of only three basal bodies also occur). Right marginal row commences slightly behind level of frontoterminal cirri, terminates close to right transverse cirrus; left marginal row begins distinctly ahead of level of buccal vertex, terminates about at level of transverse cirri; marginal cirri comprise two rows of 5–7 basal bodies arranged at 60° to main body axis.

Dorsal cilia according to scanning electron micrograph about 3 µm long, arranged in four (likely bipolar) kineties. Caudal cirri probably lacking because neither mentioned in the description of the interphase morphology nor in the ontogenesis.

Cell division: Fortunately, Wicklow (1981) studied the division of *Thigmokeronopsis jahodai* so that the origin of the thigmotactic cirri is known. Several stages are documented by scanning electron micrographs, which are, however, not shown in the present book. The reader is therefore referred to Wicklow's paper. The main events are summarised in some schematic illustrations (Fig. 170d–j).

Ontogenesis commences with the formation of oral primordia by proliferation of basal bodies from the left cirri of the midventral complex in the opisthe and from the dedifferentiated endoral membrane in the proter. Development occurs on the cell surface (Fig. 170d).

Next, the anlagen for the frontal-midventral-transverse cirri primordia separate from the oral primordium. Right of the right parental marginal row and left of the left parental marginal row the primordia for the marginal rows and the dorsal kinety 4 and 1 are formed (Fig. 170e). At this stage, membranelles begin to differentiate in a posteriad direction (membranellar organisation in the opisthe proceeds that of the proter).

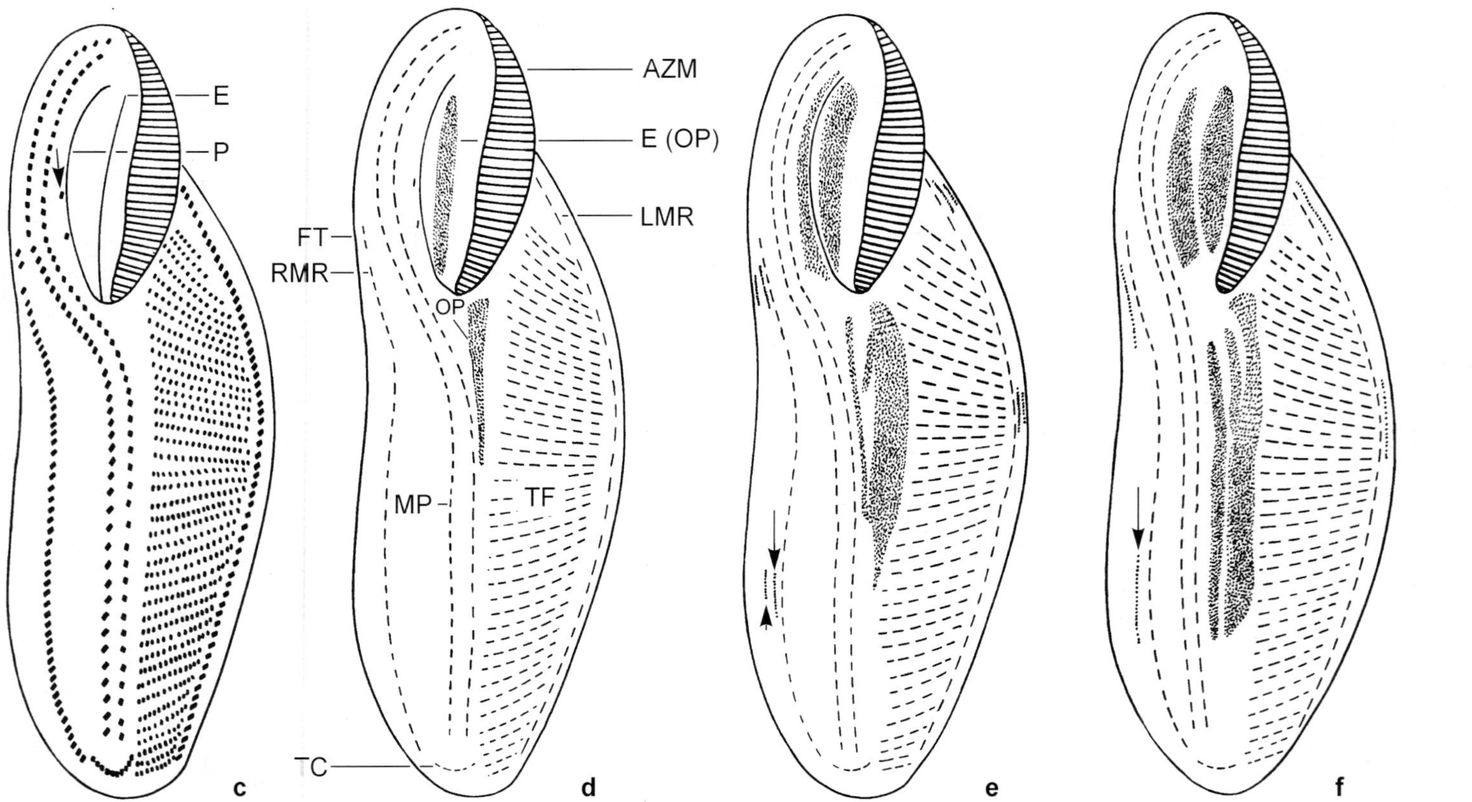

Fig. 170c–f *Thigmokeronopsis jahodai* (from Wicklow 1981. Method not indicated, likely after protargol impregnation. Sizes not indicated). **c:** Infraciliature of ventral side of non-dividing specimen. Arrow marks buccal cirri. **d–f:** Schematic representation of very early and early dividers. Parental structures are depicted as broken lines. Short arrow in (e) denotes opisthe's primordium of dorsal kinety 4; long arrow in (e) and (f) marks opisthe's primordium of right marginal row. Note that all marginal primordia and dorsal anlagen likely originate de novo. AZM = parental adoral zone of membranelles, E = endoral, FT = frontoterminal cirri, LMR = left marginal row, MP = midventral pairs, OP = oral primordium, P = paroral, RMR = right marginal row, TC = transverse cirri, TF = thigmotactic field. Page 842.

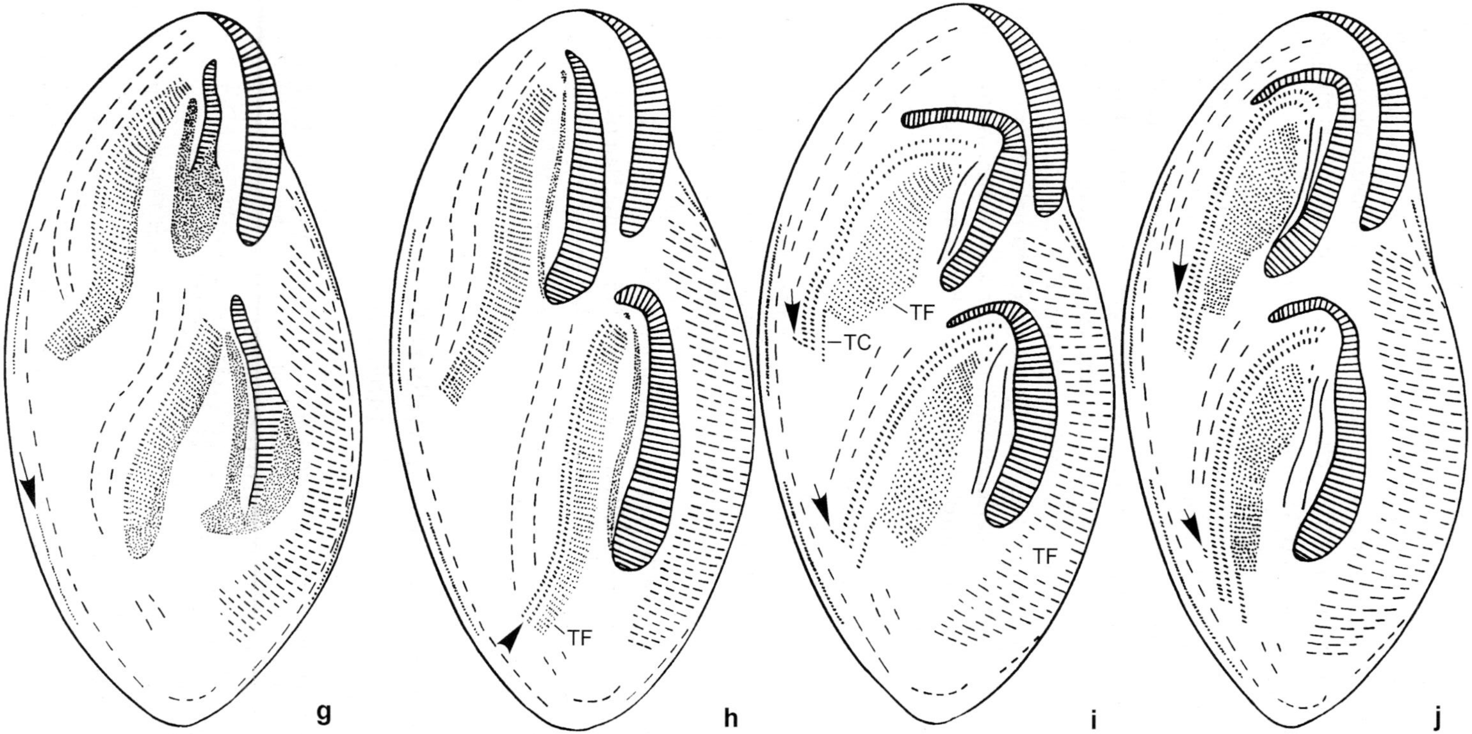

Fig. 170g–j *Thigmokeronopsis jahodai* (from Wicklow 1981. Method not indicated, likely after protargol impregnation. Sizes not indicated). Schematic representation of middle and late dividers. Parental structures are depicted by broken lines. Arrow in (g) marks primordium of opisthe's right marginal row. Arrowhead in (h) denotes new midventral complex including bicorona. Arrows in (i) and (j) mark new frontoterminal cirri. As in other pseudokeronopsids, the parental adoral zone is completely replaced. TC = transverse cirri, TF = new and parental thigmotactic fields. Page 842.

Somewhat later, the oral primordia divide into the anlagen for the undulating membranes and the adoral zone (Fig. 170f, g). Next, the left frontal cirrus (= cirrus I/1) originates, as is usual, from the anterior end of the undulating membrane primordium (Fig. 170h). Furthermore, each oblique frontal-midventral-transverse cirri streak has split into three parts, namely, into (i) a right and (ii) a left cirrus to form the pairs of the midventral complex (Fig. 170h, arrowhead), and (iii) a short row which later is part of the thigmotactic field (Fig. 170h). This new thigmotactic field is formed in a cortical invagination.

Ontogenesis continues with the longitudinal splitting of the undulating membranes anlagen and the shaping of the new adoral zones (Fig. 170i, j). From the frontal-midventral-transverse cirral anlage II, two buccal cirri originate. Furthermore, the short streak of each rearmost anlagen fuse to a single cirrus, a transverse cirrus. The last anlage finally forms five cirri, namely, the two frontoterminal cirri (= migratory cirri sensu Wicklow 1981) which migrate anteriorly, a midventral pair, and the right transverse cirrus (Fig. 170i). This is uncommon because usually the rightmost anlage also produces only four cirri, namely the two frontoterminal cirri, the rightmost pretransverse ventral cirrus, and the rightmost transverse cirrus. Thus, checking this feature is recommended. Only after the new ciliature has differentiated do the remaining parental structures begin to be disassembled and resorbed.

Wicklow did not describe the division of the nuclear apparatus. Thus, we are unable to say whether or not it has the same incompletely fusing macronucleus as *T. antarctica* and *T. crystallis*, or a completely fusing, but distinctly branched structure like *T. rubra*. Studies on *T. jahodai* and *T. crassa* will show whether or not this is a common feature for all *Thigmokeronopsis* species. If yes, it has to be interpreted as autapomorphy of the pseudokeronopsids.

Occurrence and ecology: Marine. The type locality of *Thigmokeronopsis jahodai* is the Great Bay near Jackson Estuarine Laboratory, Adams Point, New Hampshire, USA, where Wicklow (1981) discovered it in the surface sediments mostly composed of gravel and crushed shell. He cultivated *T. jahodai* in 30‰ sea-water at 16° C with the diatom *Phaeodactylum* sp. as food. No further records published since then.

Thigmokeronopsis magna Wilbert & Song, 2005 (Fig. 170.1a–k, Table 36)

2005 ***Thigmokeronopsis magna* nov. spec.**[1] – Wilbert & Song, J. nat. Hist. (London), 39: 956, Fig. 9A–K, 15J–M, O, Table VII (Fig. 170.1a–k; original description; the type material is deposited in the Oberösterreichische Landesmuseum in Linz [LI], Upper Austria).

[1] The diagnosis by Wilbert & Song (2005) is as follows: Large flexible marine *Thigmokeronopsis* about 150–300 × 50–80 µm in vivo; ca 65 adoral membranelles extending to about one-third of cell length; 12–16 left postoral cirral rows forming thigmotactic field; 60 left and 70 right marginal cirri, about 43 and 15 pairs of cirri in midventral and frontal rows, respectively; one buccal cirrus, ten transverse and two frontoterminal cirri; three dorsal kineties; no caudal cirri; more than 150 macronuclear nodules; one contractile vacuole positioned in about mid-body.

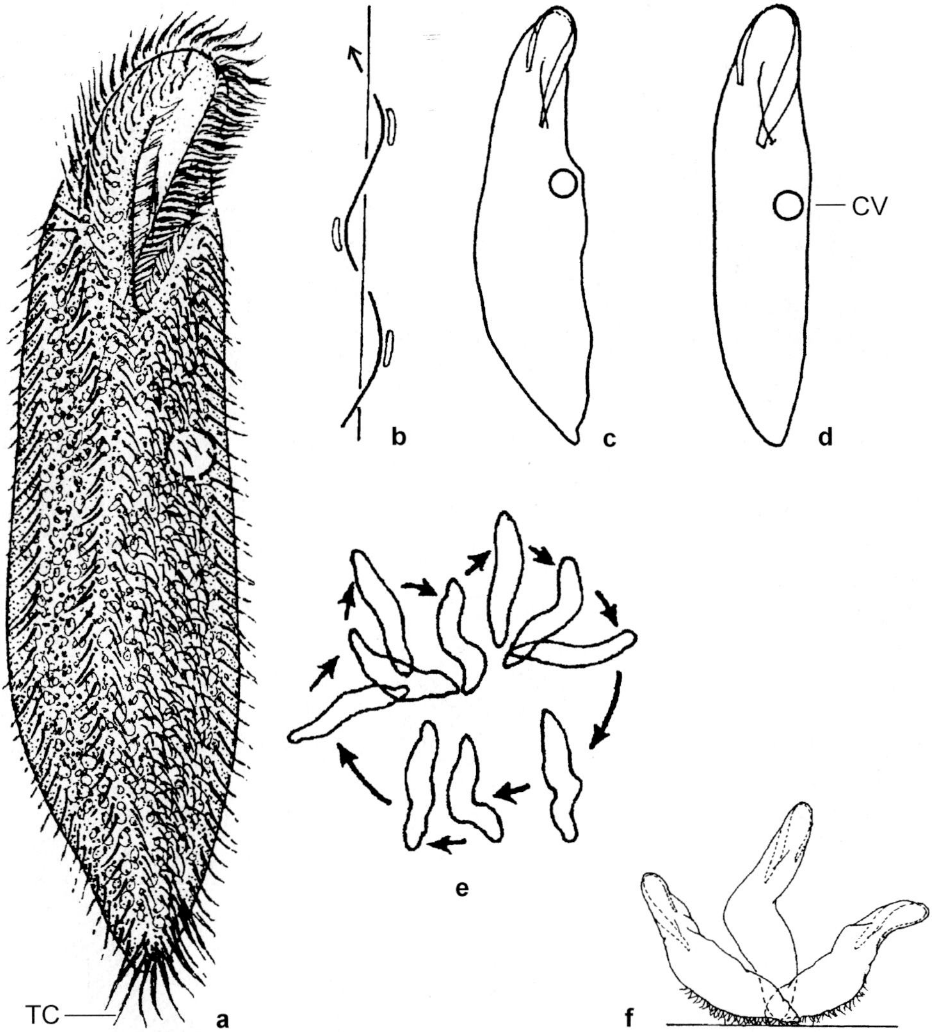

Fig. 170.1a–f *Thigmokeronopsis magna* (from Wilbert & Song 2005. From life). **a:** Ventral view of a representative specimen, 270 µm. **b:** Swimming movement. **c, d:** Body shape variants in ventral view, 270 µm. **e, f:** Type (i) and type (ii) movements; details see text. CV = contractile vacuole, TC = transverse cirri. Page 848.

Nomenclature: No derivation of the name is given in the original description. The species-group name *magn·us -a -um* (Latin adjective; large, strong) likely refers to the large body size.

Remarks: *Thigmokeronopsis magna* is likely the next relative of *T. jahodai* because they differ only in the number of buccal cirri (1 vs. 2) and dorsal kineties (3 vs. 4). Wilbert & Song (2005) mentioned the number of midventral cirri (about 100 vs. 50) as

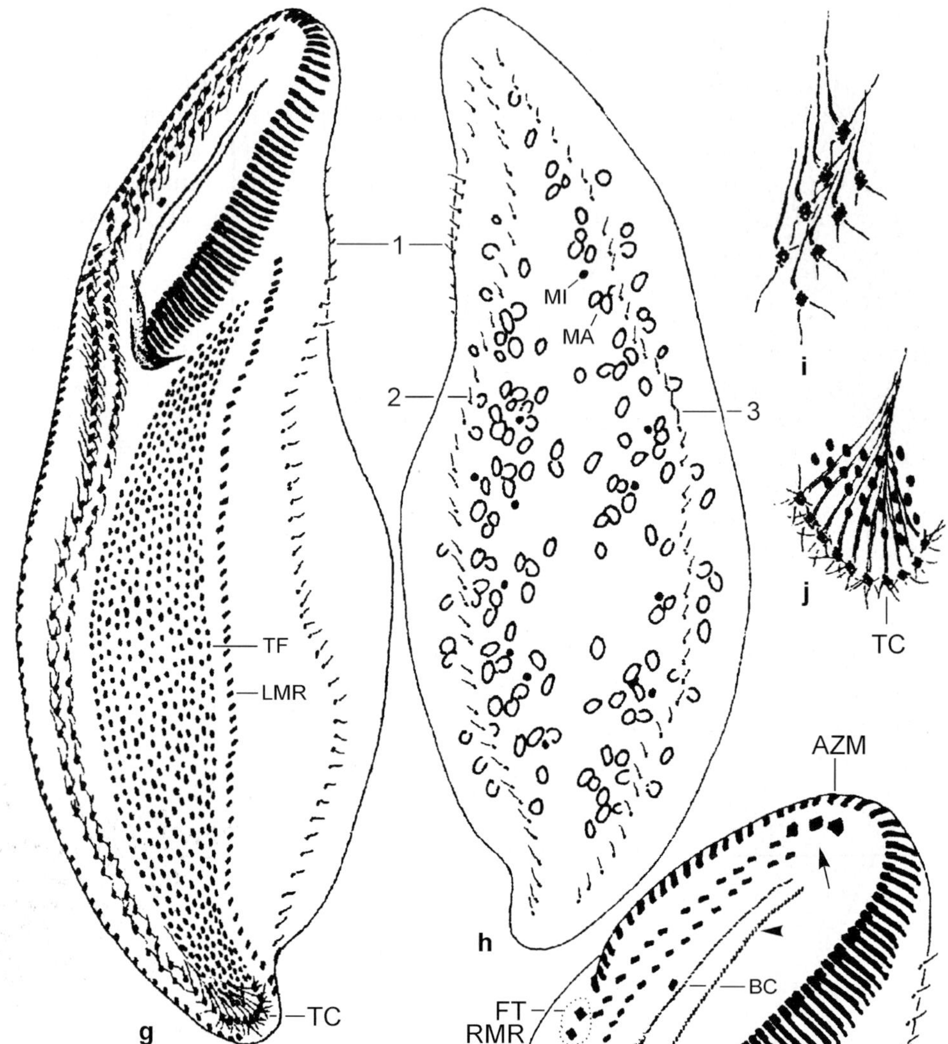

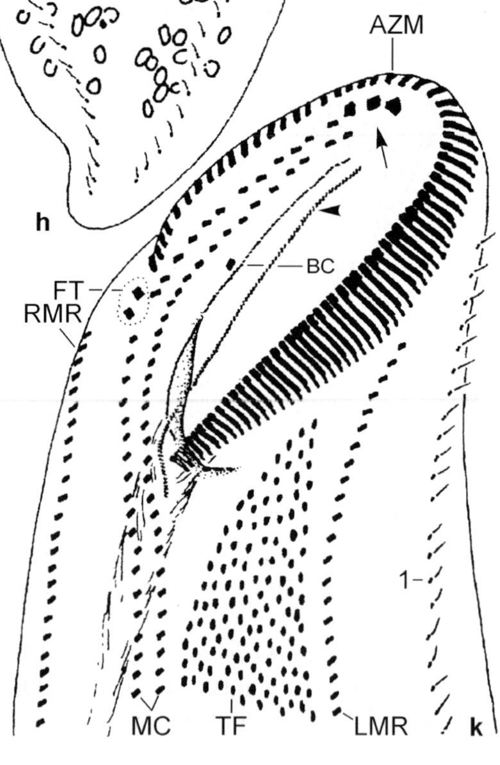

Fig. 170.1g–k *Thigmokeronopsis magna* (from Wilbert & Song 2005. Protargol impregnation). **g, h:** Infraciliature of ventral and dorsal side and nuclear apparatus of same specimen, 116 μm (see footnote k of Table 36). **i, j:** Fibre system associated with midventral cirri and transverse cirri. **k:** Infraciliature of anterior body portion. Arrowhead marks paroral according to Wilbert & Song's opinion; I suppose that this is the endoral. Arrow marks enlarged frontal cirri (ontogenetic data are needed to show whether or not these three cirri are homologous to the common frontal cirri). AZM = adoral zone of membranelles, BC = buccal cirrus, FT = fronto-

further difference. However, this is due to a misinterpretation because *T. jahodai* does not have 50 midventral cirri but 50 cirral pairs, that is, both species have about the same number (around 100) of midventral cirri. Possibly, there exist other differences in some not yet studied features (life data in *T. jahodai*; ontogenetic data in *T. magna*).

Morphology: Body size in life 150–300 × 50–80 µm, on average 200 × 60 µm. Body outline variable, but usually slender elongate with anterior portion slightly narrowed and posterior tapered. Body very flexible, pellicle thin. More than 150 macronuclear nodules scattered throughout cell; individual nodules spherical to ellipsoid, usually with one to several large nucleoli. Contractile vacuole, as is usual, near left cell margin about in mid-body. No cortical granules observed although cytoplasm usually brownish to dark brown (due to food?) under low magnification. Cytoplasm with many tiny lipid droplets and food vacuoles. Movement slow; usually three different modes observed: (i) crawling on bottom of Petri dish or debris, making small circular movement (Fig. 170.1e); (ii) attached by thigmotactic field to bottom while raised anterior portion of cell makes slow left-right movements (Fig. 170.1f); (iii) when disturbed, relatively faster, crawling around irregularly. When swimming in water, *T. magna* moves slowly without specific features (Fig. 170.1b).

Adoral zone occupies about 30% of body length in life (Fig. 170.1a) and 39% on average (Table 36) in protargol preparations, composed of an average of 65 membranelles; extends far onto right body margin (DE-values of prepared specimens illustrated about 0.63 and 0.65). Buccal field large and deep. Undulating membranes long, and almost in parallel, slightly bent and optically intersecting in posterior portion (Fig. 170.1a, g, k).

Cirral pattern and number of cirri of usual variability (Table 36). About 15 cirral pairs (that is, about 30 cirri) form bicorona (Wilbert & Song incorrectly wrote "about 15 frontal cirri forming bicorona"). Usually, leftmost three cirri of anterior corona conspicuously enlarged (Fig. 170.1a, g, k). Buccal cirrus about at mid-level of paroral. Frontoterminal cirri behind distal end of adoral zone. Midventral complex only indistinctly set off from bicorona, terminates near rightmost transverse cirri, composed of an average of 43 cirral pairs; cirri of individual pairs distinctly separate, that is, zigzag-pattern, as is usual for *Thigmokeronopsis*, indistinct. Transverse cirri arranged in bow-shaped figure, slightly enlarged, project distinctly beyond rear body end. Area bordered by left marginal row, proximal portion of adoral zone, and midventral complex almost completely covered by thigmotactic field, which is roughly banana-shaped in specimen shown in Fig. 170.1g; field composed of up to 13 longitudinal rows in midportion (for formation of this field, see ontogenesis at type species; note that these longitudinal rows are in fact pseudorows). Bases of thigmotactic cirri usually smaller than those of other cirri and bases more or less ovoid. Right marginal row commences near level of frontoterminal cirri, left row begins distinctly ahead of proximal end of adoral zone; marginal rows distinctly separated posteriorly.

← terminal cirri, LMR = left marginal row, MA = macronuclear nodules, MC = midventral complex, MI = micronucleus, RMR = right marginal row, TC = transverse cirri, TF = thigmotactic field, 1–3 = dorsal kineties. Page 848.

Dorsal cilia about 3 µm long, arranged in three bipolar kineties. Caudal cirri lacking (Fig. 170.1g, h).

Occurrence and ecology: Marine. Type locality of *T. magna* is the littoral zone of Potter Cove, King George Island (62°14'S 58°40'W), where Wilbert & Song (2005) discovered it in the periphyton on rocks and in tidal pools during February to April 2002.

Thigmokeronopsis magna feeds on small and large pennate diatoms and small protozoans (Wilbert & Song 2005).

Thigmokeronopsis rubra Hu, Warren & Song, 2004 (Fig. 171a–y, Table 36, Addenda)

2004 ***Thigmokeronopsis rubra* sp. nov.**[1] – Hu, Warren & Song, J. nat. Hist., 38: 1060, Fig. 1–35 (Fig. 171a–y; original description; the holotype slide [TPJ-96100201] and a paratype slide [TPJ-96100202] are deposited in the Laboratory of Protozoology, Ocean University, Qingdao; one paratype slide [2002:5:22:1] is deposited in the Department of Zoology, Natural History Museum, London).

Nomenclature: The species-group name *rub·er -ra -rum* (Latin adjective; red) alludes to the reddish appearance of this organism due to its numerous red cortical granules (Hut et al. 2004).

Remarks: This is a "typical" *Thigmokeronopsis*, likely most closely related to *T. crystallis* as indicated by the similar size of the thigmotactic field. It differs mainly by the presence of cortical granules and the smaller size, including some further, size-related morphometrics, e.g., number of adoral membranelles, number of marginal cirri. The morphogenesis shows that the many macronuclear nodules fuse to a single, branched mass, which is a distinct difference to *T. crystallis* and *T. antarctica*, where the nodules fuse only partially. For a more detailed discussion, see genus section.

Morphology: Hu et al. (2004) characterised two populations morphometrically, but without providing further details, indicating that the description and the illustrations are a mixture of the two populations, which are likely from the same locality. Body length 140–200 µm, body length:width ratio of the live specimen illustrated 3.6:1 (Fig. 171a). More than 100 (n = 7) macronuclear nodules dispersed throughout cell; micronuclei likely of ordinary size and number. Contractile vacuole neither mentioned nor illustrated, indicating that it is lacking. Pellicle flexible. Two types of cortical granules: type I pale yellow-green, about 1 µm across, distributed sparsely and located peripherally; type II medium or dark red, about 0.5 µm across, arranged in longitudinal lines, located deeper in cortex than type I granules and making cells reddish at low magnification. Cytoplasm transparent, containing several food vacuoles. Movement relatively slow, crawling on substratum, occasionally stationary with thigmotactic cilia attached to substratum.

---

[1] The diagnosis by Hu et al. (2004) is as follows: Reddish marine *Thigmokeronopsis*, about 140–200 µm long in vivo. Two kinds of cortical granules: (1) pale yellow-green, distributed sparsely, (2) red, arranged in lines. Usually 7–11 left postoral ventral files. Two mid-ventral rows comprising a total of ca 51 cirri, extending almost to posterior end of cell. One buccal, seven transverse and two frontoterminal cirri. Bicorona with 11–16 frontal cirri. Thirty to 49 left and 28–45 right marginal cirri. Three dorsal kineties.

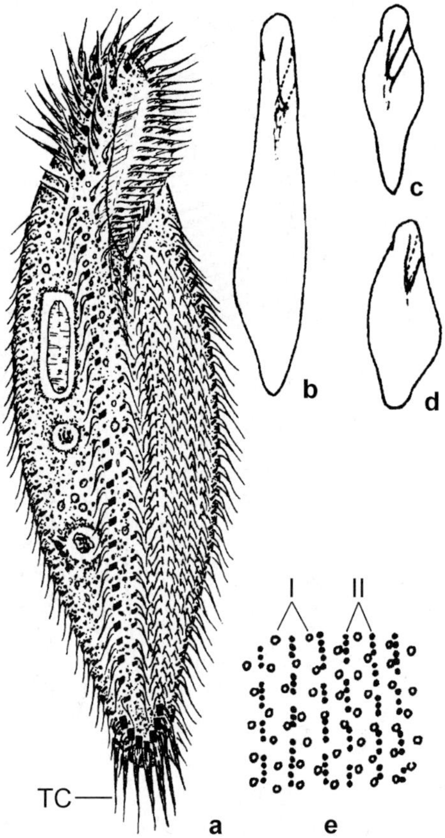

Fig. 171a–e *Thigmokeronopsis rubra* (from Hu et al. 2004. From life). **a:** Ventral view of a representative specimen, 142 µm. **b–d:** Shape variants (d contracted?). **e:** This species has pale yellow-green cortical granules (I) and red granules (II). TC = transverse cirri, I, II = types of cortical granules. Page 852.

Adoral zone occupies about 25% of body length in life, 37–41% on average in protargol preparations (Table 36), indicating a strong shrinkage of the postoral region; composed of an average of about 38 membranelles of ordinary fine structure, extends far onto right body margin; cilia of membranelles about 15 µm long. Undulating membranes long and curved, intersecting optically slightly behind mid-region; commence at same level, paroral slightly shorter than endoral. Buccal field moderately wide in life (Fig. 171a–d).

Cirral pattern and number of cirri of usual variability (Table 36). Bicorona composed of 11–16 cirri, bases of cirri of anterior bow somewhat enlarged, not distinctly set off from midventral complex. Buccal cirrus slightly ahead of level of optical intersection of undulating membranes. Two frontoterminal cirri near distal end of adoral zone of membranelles. Midventral complex extends slightly subterminally (about at 87% of body length in specimen shown in Fig. 171f); cirri of each pair, as in congeners, distinctly separated so that zigzagging pattern is not recognisable; right cirri slightly larger than left cirri. Two small pretransverse ventral cirri close to transverse cirri. 6–8 more or less distinctly enlarged transverse cirri arranged in terminal, U-shaped row; about 18 µm long and therefore distinctly projecting beyond rear body end. Other cirri about 10–12 µm long. Between left marginal row and midventral complex, the thigmotactic field – composed of 9–11 (mean = 10.6, SD = 0.7, SE = 0.2, CV = 6.5 %, n = 11) inconspicuous, slightly irregular longitudinal pseudofiles of fine cirri – extends from proximal end of adoral zone to transverse cirri. Marginal rows commence somewhat behind level of distal end of adoral zone, end near transverse cirri.

Dorsal cilia about 5 µm long, arranged in three roughly bipolar kineties. Caudal cirri absent (Fig. 171g, j).

Cell division (Fig. 171k–y): Hu et al. (2004) studied the morphogenesis in great detail. As expected, the formation of the oral apparatus, cirral pattern, and dorsal kineties proceeds as in the congeners, so that the reader is mainly referred to the original description

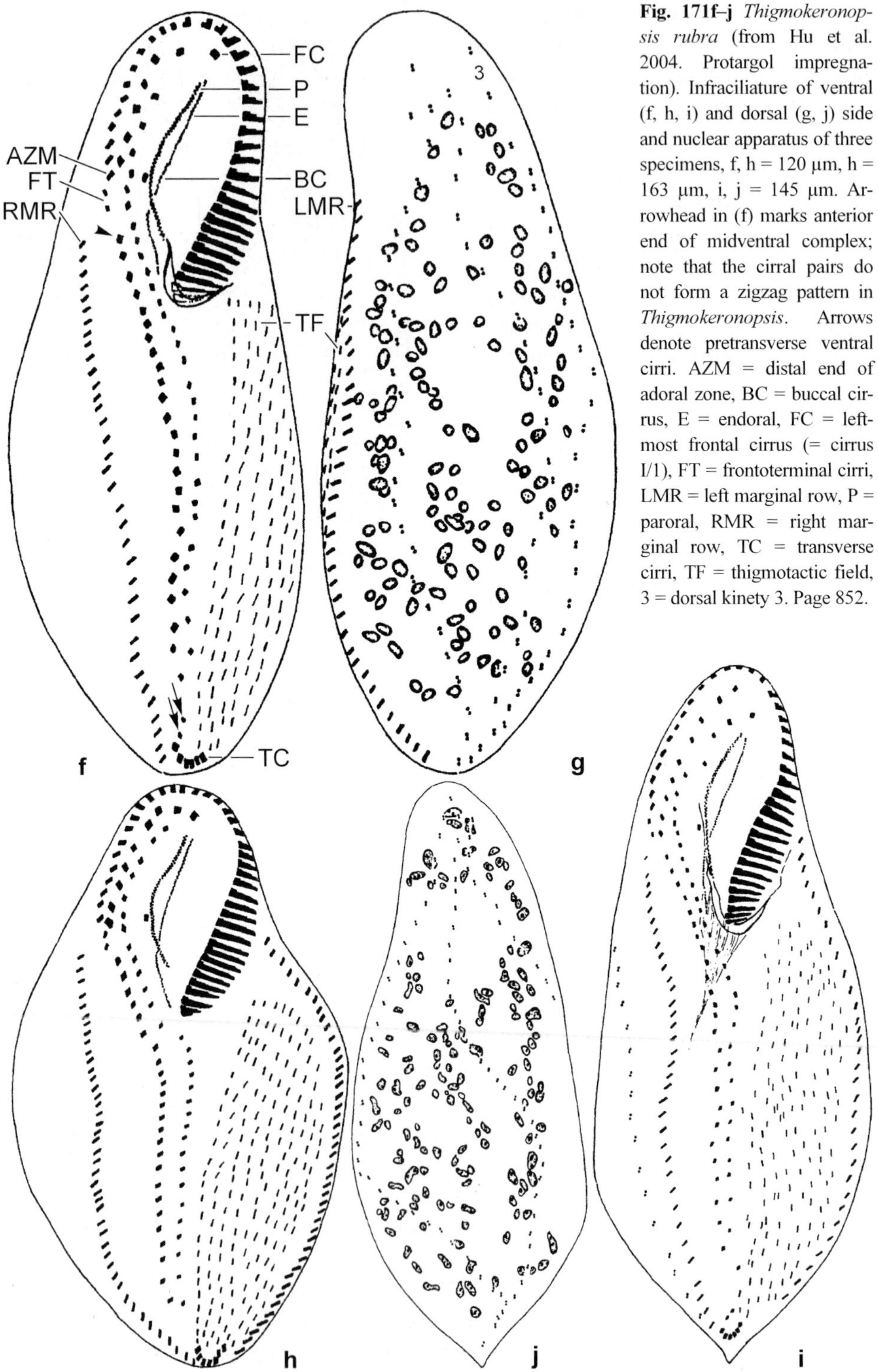

Fig. 171f–j *Thigmokeronopsis rubra* (from Hu et al. 2004. Protargol impregnation). Infraciliature of ventral (f, h, i) and dorsal (g, j) side and nuclear apparatus of three specimens, f, h = 120 µm, h = 163 µm, i, j = 145 µm. Arrowhead in (f) marks anterior end of midventral complex; note that the cirral pairs do not form a zigzag pattern in *Thigmokeronopsis*. Arrows denote pretransverse ventral cirri. AZM = distal end of adoral zone, BC = buccal cirrus, E = endoral, FC = leftmost frontal cirrus (= cirrus I/1), FT = frontoterminal cirri, LMR = left marginal row, P = paroral, RMR = right marginal row, TC = transverse cirri, TF = thigmotactic field, 3 = dorsal kinety 3. Page 852.

and the illustrations and legends in the present book. The main difference to *T. crystallis*, which has a similar-sized thigmotactic field, is in the division of the macronucleus. In the present species all nodules fuse to a single mass (Fig. 171q, s), which is, however, branched as in *Metaurostylopsis*. By contrast, Petz (1995) did not observe a fused macronucleus. The two pretransverse ventral cirri are formed – as in other hypotrichs, for example, oxytrichids (for review see Berger 1999) or amphisiellids (Berger 2004a) – from the two rightmost anlagen.

Occurrence and ecology: Marine. Type locality of *T. rubra* are mollusc-culturing waters (36°08'N 120°43'E) off the coast of Qingdao, China. Hu et al. (2004) studied two populations, without providing details, indicating that they are sampled from the same locality, but at different time. Further record: Jiaouzhou Bay near Qingdao, China (Gong et al. 2005, p. 163).

Thigmokeronopsis crystallis Petz, 1995
(Fig. 172a–c, 174b, d–h, l, m, Table 36)

1995 ***Thigmokeronopsis crystallis* nov. spec.**[1] – Petz, Europ. J. Protistol., 31: 138, Fig. 5–7, 10, 11–15, 19, 20, Table 1 (Fig. 172a–c, 174b, d–h, l, m; original description. The holotype slide [registration number 2001/143] and a paratype slide [2001/144] are deposited in the Oberösterreichische Landesmuseum in Linz [LI], Austria).

2001 ***Thigmokeronopsis crystallis* Petz, 1995** – Berger, Catalogue of ciliate names 1. Hypotrichs, p. 89 (nomenclator containing all basionyms, combinations, and higher taxa of hypotrichs).

2001 ***Thigmokeronopsis crystallis*** – Eigner, J. Euk. Microbiol., 48: 77, Fig. 25 (Fig. 172b modified; brief review of urostylids).

2002 ***Thigmokeronopsis crystallis*** – Lynn & Small, Phylum Ciliophora, p. 445, Fig. 13B (Fig. 172a; guide to ciliate genera).

2005 ***Thigmokeronopsis crystallis* Petz (1995)** – Petz, Ciliates, p. 399, Fig. 14.90a–c, 14.177 (Fig. 172a–c; guide to Antarctic marine ciliates).

Nomenclature: The species-group name *crystallis* (from the Greek noun, *to krystallos*; ice) refers to the habitat where the species was discovered, namely the sea ice of the Weddell Sea, Antarctica (Petz 1995).

Remarks: The thigmotactic field assigns the present species to *Thigmokeronopsis*. According to the size of the field, *Thigmokeronopsis crystallis* (Fig. 172a, b) is, like *T. rubra* (Fig. 171f), between *T. jahodai* (very large field; Fig. 170a–c) and the *T. antarctica/crassa*-group (very inconspicuous field, that is, only long row of transverse cirri present; Fig. 172a, b, 176a). Further, it differs from *T. jahodai* by the lower number of dorsal kineties (3 vs. 4), and from *T. rubra* and *T. jahodai* by the lack of cortical granules, which is likely the sole autapomorphy of the species (Fig. 169a, autapomorphy 6). *Thigmokeronopsis antarctica* and *T. crassa* have more buccal cirri than *T. crystallis* (2–5, respectively, 6–10 vs. 1–2). The thigmotactic field of *T. crystallis* must not be confused with the well developed oral primordium of other species.

---

[1] The diagnosis by Petz (1995) is as follows: In vivo about 170 × 60 μm. Usually 5 left postoral cirral files, midventral rows extending to transverse cirri, 1 buccal, 8 transverse and 2 frontoterminal cirri. Bicorona with 26–41 cirri. 3 dorsal kineties. More than 100 macronuclear nodules. No caudal cirri.

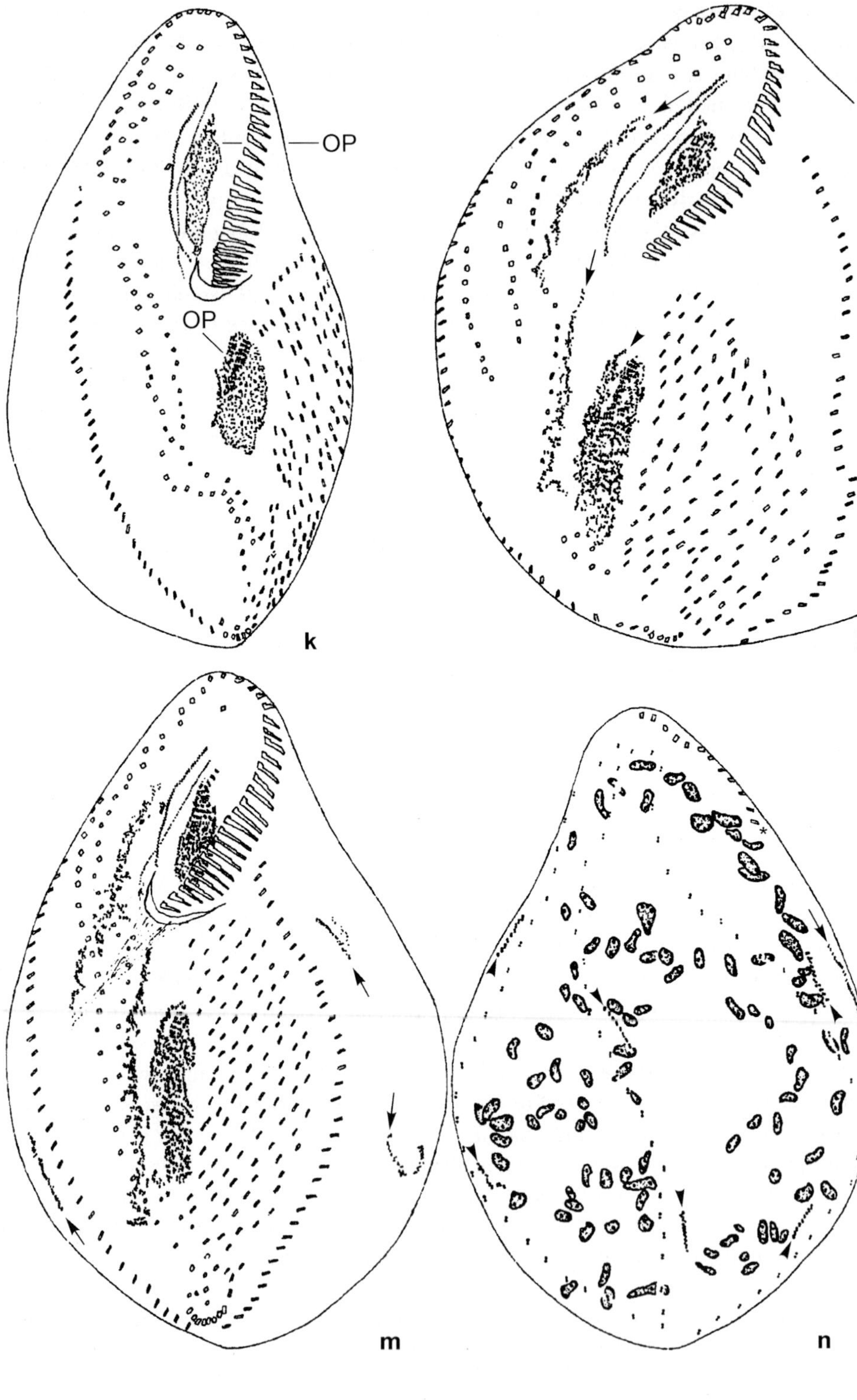
OP
OP
k
l
m
n

Morphology: Body size about 170 × 60 μm in life, smallest measured protargol-impregnated specimen, however, 180 μm long (Table 36), indicating that life data are underestimated; on the other hand it is known that specimens impregnated with Wilbert's (1975) method are often distinctly inflated. Length:width ratio around 3:1 in life (Fig. 172a); by contrast, protargol-impregnated specimens distinctly stouter with an average length:width ratio of 2.2:1 (Table 36). Body outline long-elliptical, right margin more or less straight, left margin slightly convex, anterior and posterior end evenly rounded. Body slightly flattened dorsoventrally. Many macronuclear nodules scattered throughout cell, individual nodules ellipsoidal, usually with one large nucleolus. Micronuclei also ellipsoidal, but often larger than macronuclear nodules; usually arranged as shown in Fig. 172c, that is, in longitudinal row somewhat left of midline. Contractile vacuole left of proximal end of adoral zone, with inconspicuous collecting canals. Cytopyge dorsally in left posterior fourth of cell (Fig. 172a). No cortical granules. Cytoplasm with food vacuoles, some lipid globules up to 7 μm across, and slightly larger green inclusions; individuals therefore appear dark at low magnification and bright field illumination. Slowly crawling on bottom of Petri dish.

Adoral zone occupies about 40% of body length, distal end extends far onto right side, composed of an average of 64 membranelles of ordinary fine structure. Bases of largest membranelles about 15 μm wide, cilia ca. 20 μm long. Buccal cavity large and deep. Paroral slightly shorter than endoral, curved and optically intersecting with endoral about in middle portion; usually composed of two basal body rows, in middle and posterior portion sometimes of three or four rows. Endoral, as is usual, in buccal cavity, commences at about same level as paroral, terminates near proximal adoral membranelle; likely composed of single row of basal bodies bearing about 20 μm long cilia. Pharyngeal fibres 50–70 μm long.

Cirral pattern very conspicuous due to the thigmotactic field (details, see below), which looks like a well developed oral primordium at first glance (Fig. 172a, b). Frontal cirri about 20 μm long, form distinct bicorona; anterior corona extends slightly backwards on left side, cirri composed of four rows of basal bodies, that is, slightly larger than those of posterior row, which usually consist of three rows only. Frontal cirri distinguished from cirri of midventral complex by conspicuous longitudinally extending, fibrillar associates. Buccal cirrus/cirri slightly ahead of optical intersection of undulating membranes, that is, about 20 μm behind anterior end of undulating membranes in specimen shown in Fig. 172b. Usually two frontoterminal cirri behind distal end of adoral zone, bases double-rowed, cilia about 20 μm long. Midventral complex not set off from frontal cirri, extends sigmoidally to transverse cirri; characteristic zigzagging midventral pattern very indistinct; right cirri about 20 μm long and slightly larger than left cirri, which are only 15 μm long. Transverse cirri in ordinary position, that is, about

← **Fig. 171k–n** *Thigmokeronopsis rubra* (from Hu et al. 2004. Protargol impregnation). **k, l:** Infraciliature of ventral side of early dividers, k = 140 μm, l = size not indicated. Arrows in (l) mark frontal-midventral-transverse cirral anlagen; arrowhead denotes undulating membrane anlage of opisthe. **m, n:** Infraciliature of ventral and dorsal side and nuclear apparatus of middle divider, size not indicated. Note that the primordia for the marginal rows (arrows) and dorsal kineties (arrowheads) originate de novo in *Thigmokeronopsis*, that is, not within the parental structures. OP = oral primordium. Page 852.

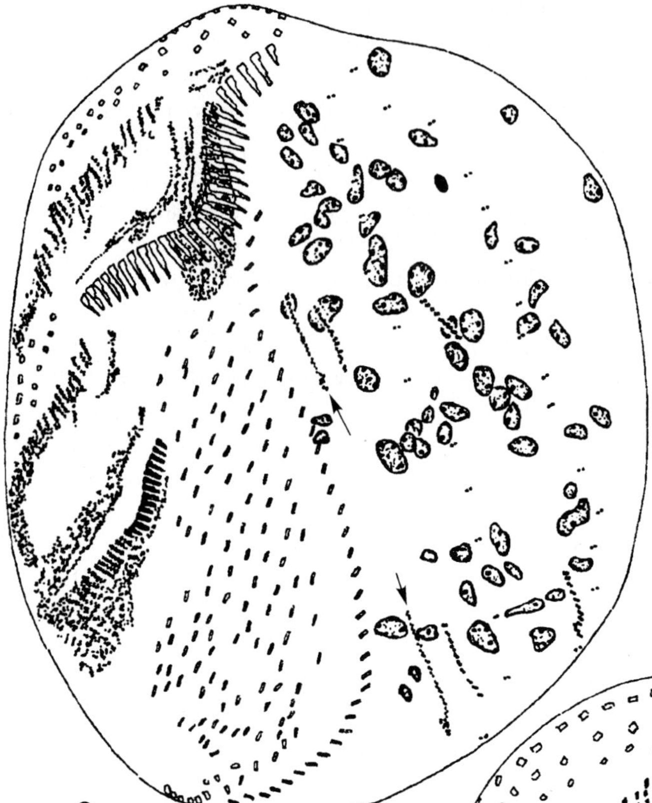

o

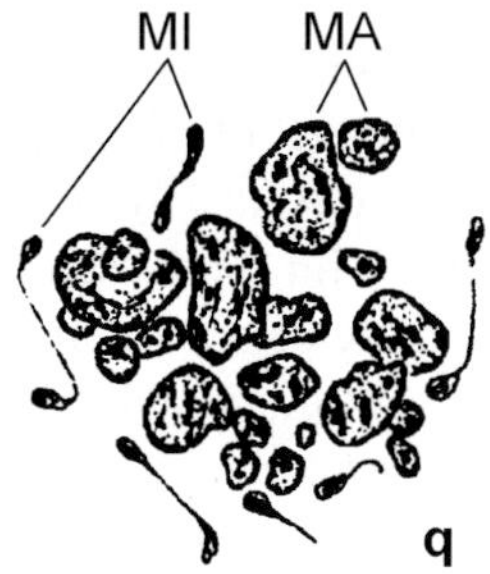

q

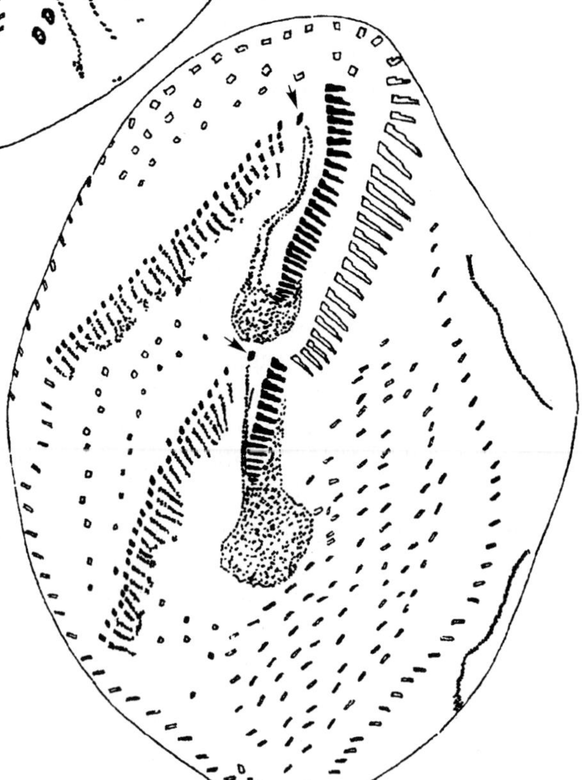

p

Fig. 171o–q *Thigmokeronopsis rubra* (from Hu et al. 2004. Protargol impregnation). **o:** Infraciliature and nuclear apparatus of a middle divider in ventro-lateral view, size likely not indicated. Arrows mark primordia for new left marginal rows, which originate left of the parental left marginal row. **p, q:** Infraciliature of ventral side and nuclear apparatus of a middle divider, 110 µm. Arrows mark new left frontal cirrus (= cirrus I/1). The macronuclear nodules are already fused to some nodules. MA = macronuclear nodules, MI = micronuclei. Page 852.

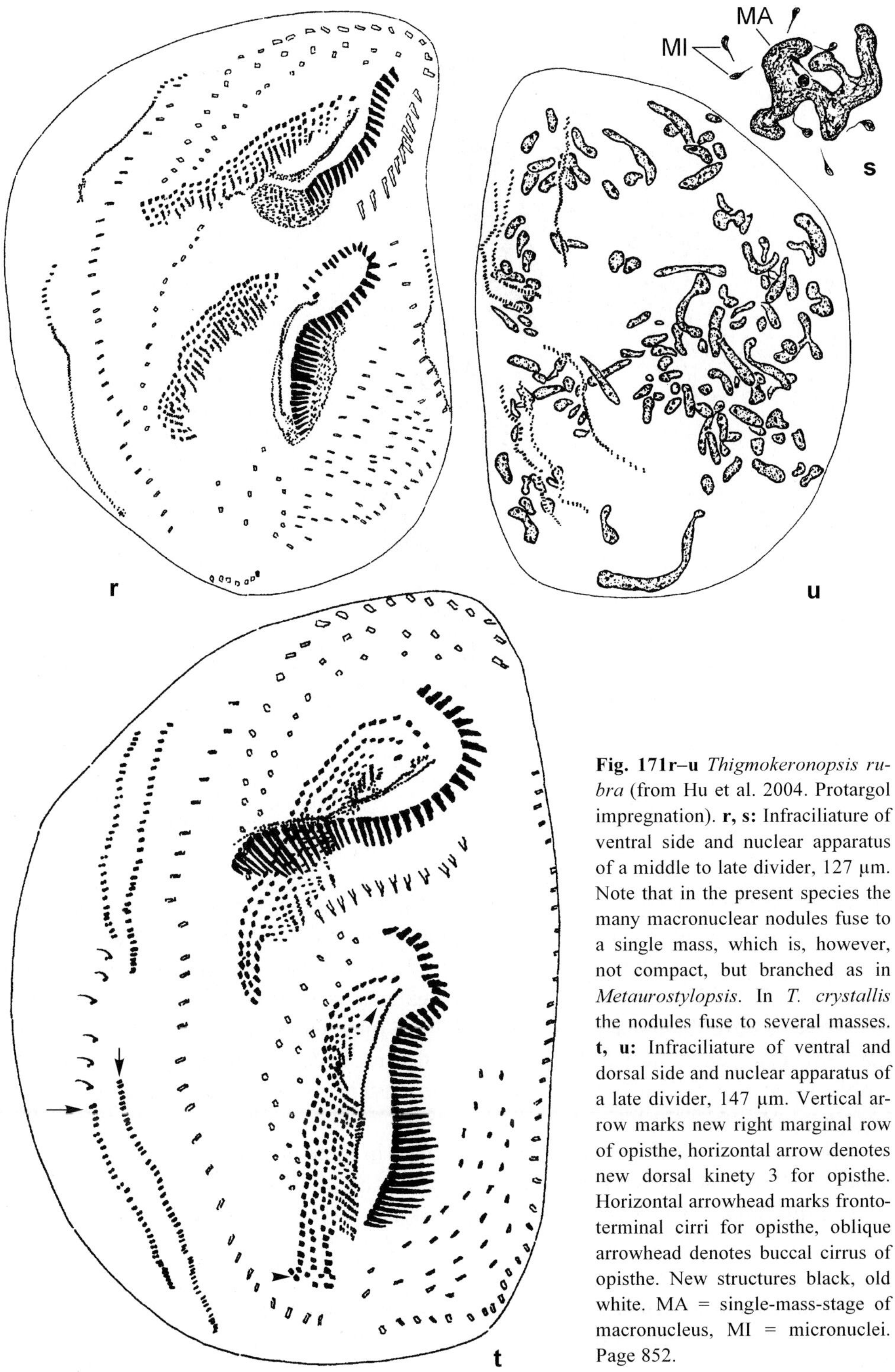

Fig. 171r–u *Thigmokeronopsis rubra* (from Hu et al. 2004. Protargol impregnation). **r, s:** Infraciliature of ventral side and nuclear apparatus of a middle to late divider, 127 µm. Note that in the present species the many macronuclear nodules fuse to a single mass, which is, however, not compact, but branched as in *Metaurostylopsis*. In *T. crystallis* the nodules fuse to several masses. **t, u:** Infraciliature of ventral and dorsal side and nuclear apparatus of a late divider, 147 µm. Vertical arrow marks new right marginal row of opisthe, horizontal arrow denotes new dorsal kinety 3 for opisthe. Horizontal arrowhead marks frontoterminal cirri for opisthe, oblique arrowhead denotes buccal cirrus of opisthe. New structures black, old white. MA = single-mass-stage of macronucleus, MI = micronuclei. Page 852.

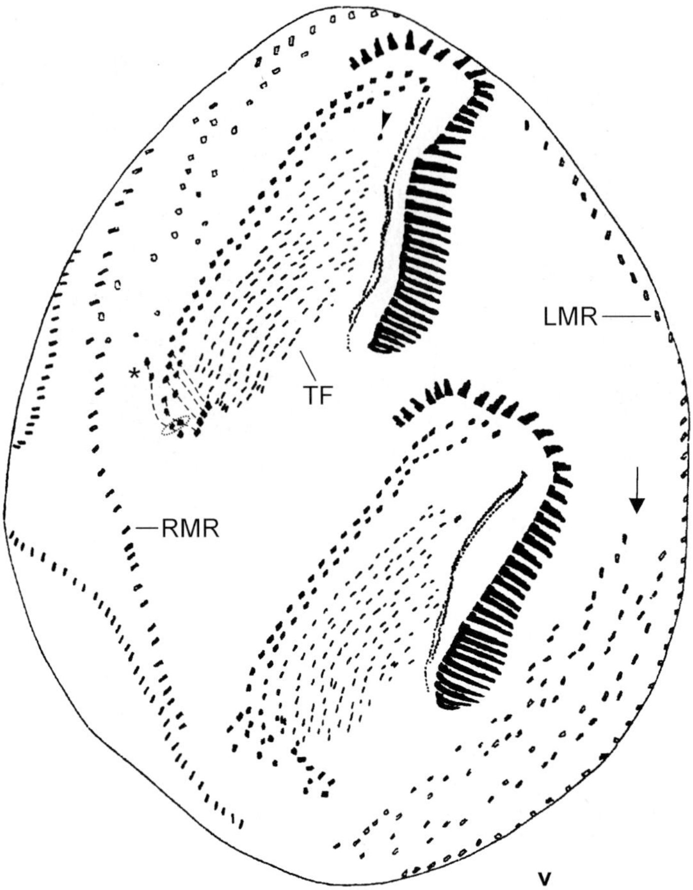

Fig. 171v *Thigmokeronopsis rubra* (from Hu et al. 2004. Protargol impregnation). Infraciliature of ventral side of a very late divider, size not indicated. Arrowhead denotes buccal cirrus of proter, arrow marks parental thigmotactic field; asterisk marks frontoterminal cirri of proter. Cirri which originate from the five rightmost (= posteriormost) anlagen of the proter are connected by broken lines. The two pretransverse ventral cirri of the proter are encircled. Note that the new marginal rows originate right (right side), respectively, left (left side) of the parental structures. LMR = parental left marginal row, RMR = parental right marginal row, TF = new thigmotactic filed of proter. Page 852.

10 μm ahead of rear body end, form J-shaped pattern, slightly larger than marginal cirri. Between left marginal row and midventral complex the thigmotactic field – composed of 4–7 (mean = 5.0, SD = 0.9, SE = 0.3, CV = 17.1%, n = 12) inconspicuous, slightly irregular longitudinal pseudofiles of fine cirri – extends from proximal end of adoral zone to transverse cirri; thigmotactic cirri only 6–7 μm long, bases double-rowed. Right marginal row commences near distal end of adoral zone, ends subterminally while left row terminates at rear end so that they do not join posteriorly; marginal cirri about 22 μm long, bases double-rowed.

Dorsal cilia 4–6 μm long, invariably arranged in three bipolar kineties composed of about 35–48 basal body pairs. Caudal cirri lacking (Fig. 172c).

Cell division: This part of the life cycle proceeds very similarly in *T. antarctica* and *T. crystallis*. Thus, Petz (1995) summarised the descriptions which are presented at *T. antarctica* (Fig. 174a–m).

Occurrence and ecology: Marine; less frequent than *Thigmokeronopsis antarctica* (Petz 1995). Type locality is the Weddell Sea, Antarctica (71°00'S 11°80'W; core number

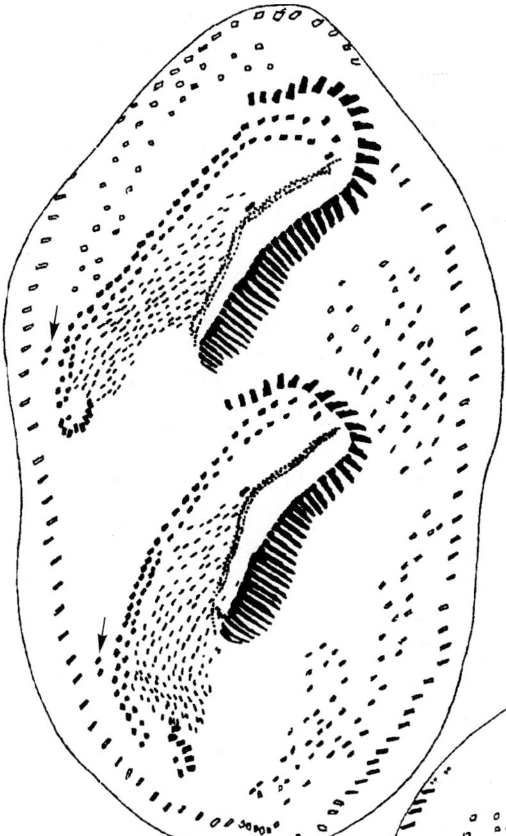

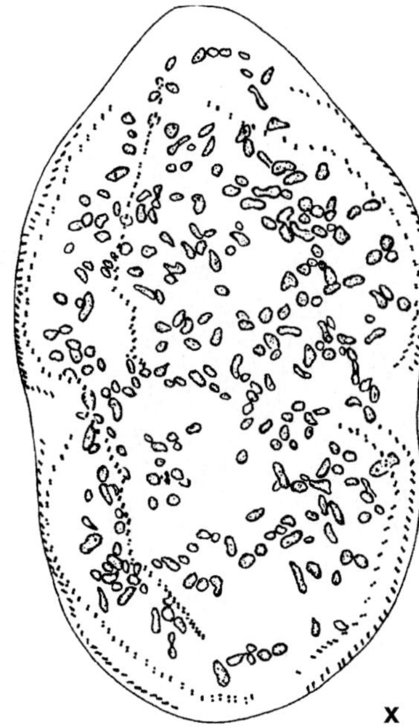

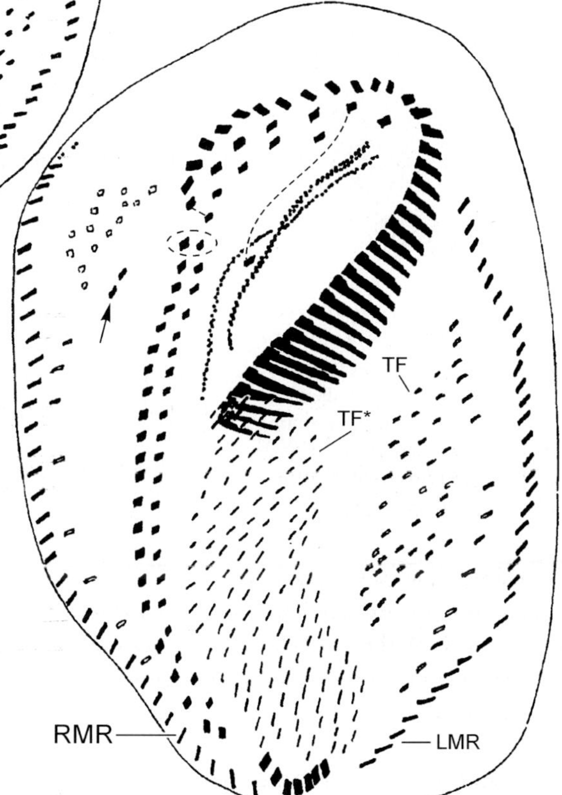

Fig. 171w–y *Thigmokeronopsis rubra* (from Hu et al. 2004. Protargol impregnation). **w, x:** Infraciliature of ventral and dorsal side and nuclear apparatus of a very late divider, 137 µm. The frontoterminal cirri (arrows) migrate anteriad to their final position near the distal end of the adoral zone of membranelles. **y:** Infraciliature of ventral side of a late reorganiser, size not indicated. Arrow marks new frontoterminal cirri. Cirri originating from frontal-midventral-transverse cirral anlagen II and VI are connected by broken lines; anteriormost midventral pair of midventral complex encircled. Note that the right cirrus of each pair is slightly larger than the left cirrus. LMR = new left marginal row, RMR = new right marginal row, TF, TF* = old and new thigmotactic field. Page 852.

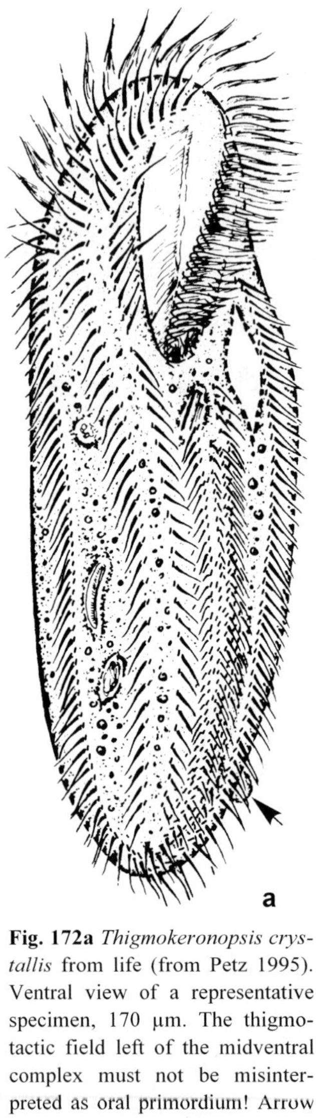

Fig. 172a *Thigmokeronopsis crystallis* from life (from Petz 1995). Ventral view of a representative specimen, 170 μm. The thigmotactic field left of the midventral complex must not be misinterpreted as oral primordium! Arrow marks the cytopyge. Page 855.

AN 103113), where Petz (1995) found it regularly in the brine-filled pore system of newly formed and multi-year sea ice. According to Petz et al. (1995, p. 14), *Thigmokeronopsis crystallis* was found in succeeding stages in sea ice formation, namely in pancake ice (young, roundish pieces), multi-year ice, and multi-year land-fast ice (sea ice connected to land). Also recorded from marine habitats in China (Song & Wang 1999, p. 73).

Thigmokeronopsis crystallis feeds on flagellates and pennate diatoms (up to 80 μm long), rarely centric diatoms are also ingested (Petz 1995). Biomass of 10^6 specimens about 214 mg (Petz 1995). Environmental parameters in brine (Petz 1995): -4.5 to -2.1°C; salinity 37.6–77.0‰; PO_4 1.7–2.3 μmol l^{-1} melted ice; NO_2 0.1 to 0.3 μmol l^{-1}; NO_3 2.6–45.8 μmol l^{-1}; NH_4 3.5 to 15.5 μmol l^{-1}; Si 7.7–39.8 μmol l^{-1}; chlorophyll *a* 13.2 to 120.2 μg l^{-1}. In raw cultures also found at +1°C and only 15.6‰ salinity.

Thigmokeronopsis antarctica Petz, 1995 (Fig. 173a–d, 174a, c, i–k, 175a–d, Table 36)

1995 ***Thigmokeronopsis antarctica* nov. spec.**[1] – Petz, Europ. J. Protistol., 31: 138, Fig. 1–4, 8, 9, 16–18, 21–24, Table 1 (Fig. 173a–d, 174a, c, i–k, 175a–d; original description. The holotype slide [registration number 2001/140] and a paratype slide [2001/152] are deposited in the Oberösterreichische Landesmuseum in Linz [LI], Austria).

2001 ***Thigmokeronopsis antarctica* Petz, 1995** – Berger, Catalogue of ciliate names 1. Hypotrichs, p. 89 (nomenclator containing all basionyms, combinations, and higher taxa of hypotrichs).

2001 ***Thigmokeronopsis antarctica*** – Eigner, J. Euk. Microbiol., 48: 77, Fig. 33 (Fig. 173b modified; brief review of urostylids).

2005 ***Thigmokeronopsis antarctica* Petz (1995)** – Petz, Ciliates, p. 396, Fig. 14.89a–c (Fig. 173a, b, d; guide to Antarctic marine ciliates).

Nomenclature: No derivation of the name was given in the original description. The species-group name *antarctic·us -a -um* (Latin adjective; living in the Antarctic Ocean; Hentschel & Wagner 1996, p. 89) refers to the region (Antarctica) where the species was discovered. *Thigmokeronopsis antarcitica* in Hu et al. (2004, p. 1067) is an incorrect subsequent spelling.

---

[1] The diagnosis by Petz (1995) is as follows: In vivo about 200 × 60 μm. Usually 1 left postoral cirral file, midventral rows shortened, 3–4 buccal and 2 frontoterminal cirri. Bicorona with 20–42 cirri. 3 dorsal kineties. More than 100 macronuclear nodules. No caudal cirri.

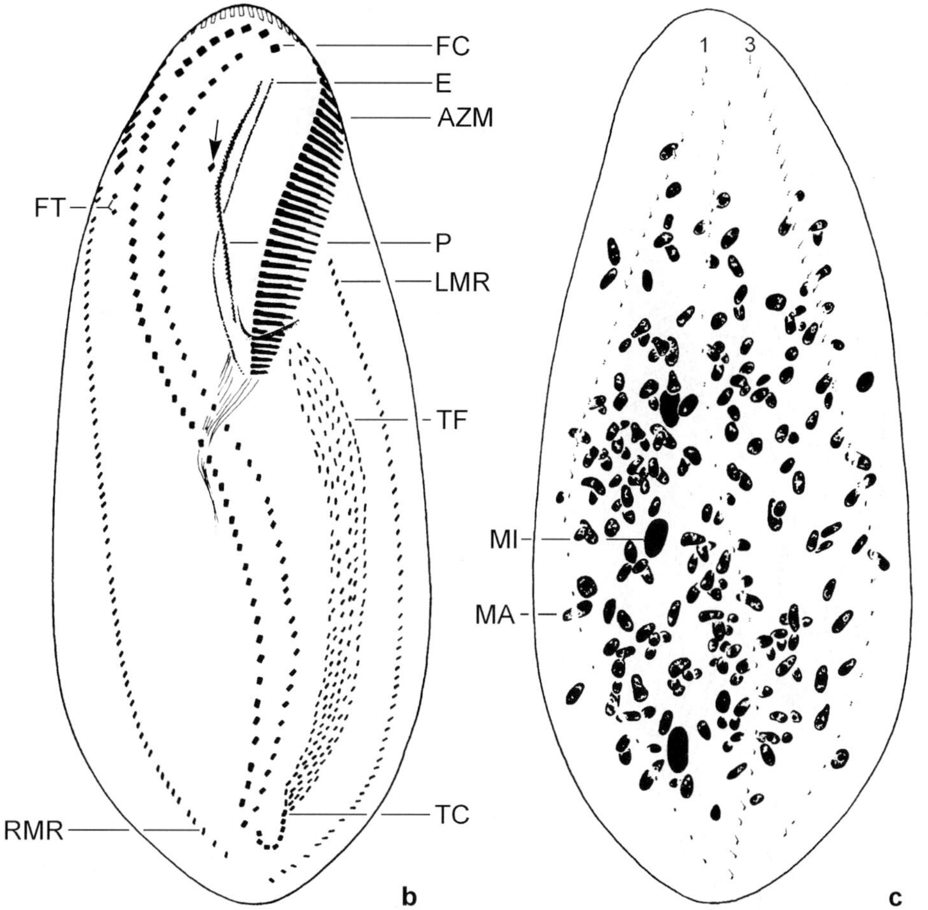

Fig. 172b, c *Thigmokeronopsis crystallis* after protargol impregnation (from Petz 1995). Infraciliature of ventral and dorsal side and nuclear apparatus, 193 µm. Arrow marks buccal cirrus. AZM = adoral zone of membranelles, E = endoral, FC = leftmost frontal cirrus (cirrus I/1), FT = frontoterminal cirri, LMR = left marginal row, MA = macronuclear nodule, MI = micronucleus, P = paroral, RMR = right marginal row, TC = transverse cirri, TF = thigmotactic field, 1, 3 = dorsal kineties. Page 855.

Remarks: The very indistinctly zigzagging midventral cirri and some further morphological and morphogenetic features assign this species to *Thigmokeronopsis*. The thigmotactic field, or rather the long row of transverse cirri of *T. antarctica* (Fig. 173a, b) is inconspicuous against the huge field of *T. jahodai* (Fig. 170a–c) and the moderate sized field of *T. rubra* (Fig. 171a, f, h) and *T. crystallis* (Fig. 172a, b). Furthermore, *Thigmokeronopsis antarctica* differs from the type species by the lower number of dorsal kineties (3 vs. 4) and the lack of cortical granules. *Thigmokeronopsis crystallis* has, inter alia, less buccal cirri than *T. antarctica* (1–2 vs. 2–5), and a longer midventral complex (cp. Fig. 172a, b, 173a, b). For a separation from *T. crassa*, see key. Autapomorphies of *T. antarctica* are likely the lack of cortical granules, the posteriorly shortened

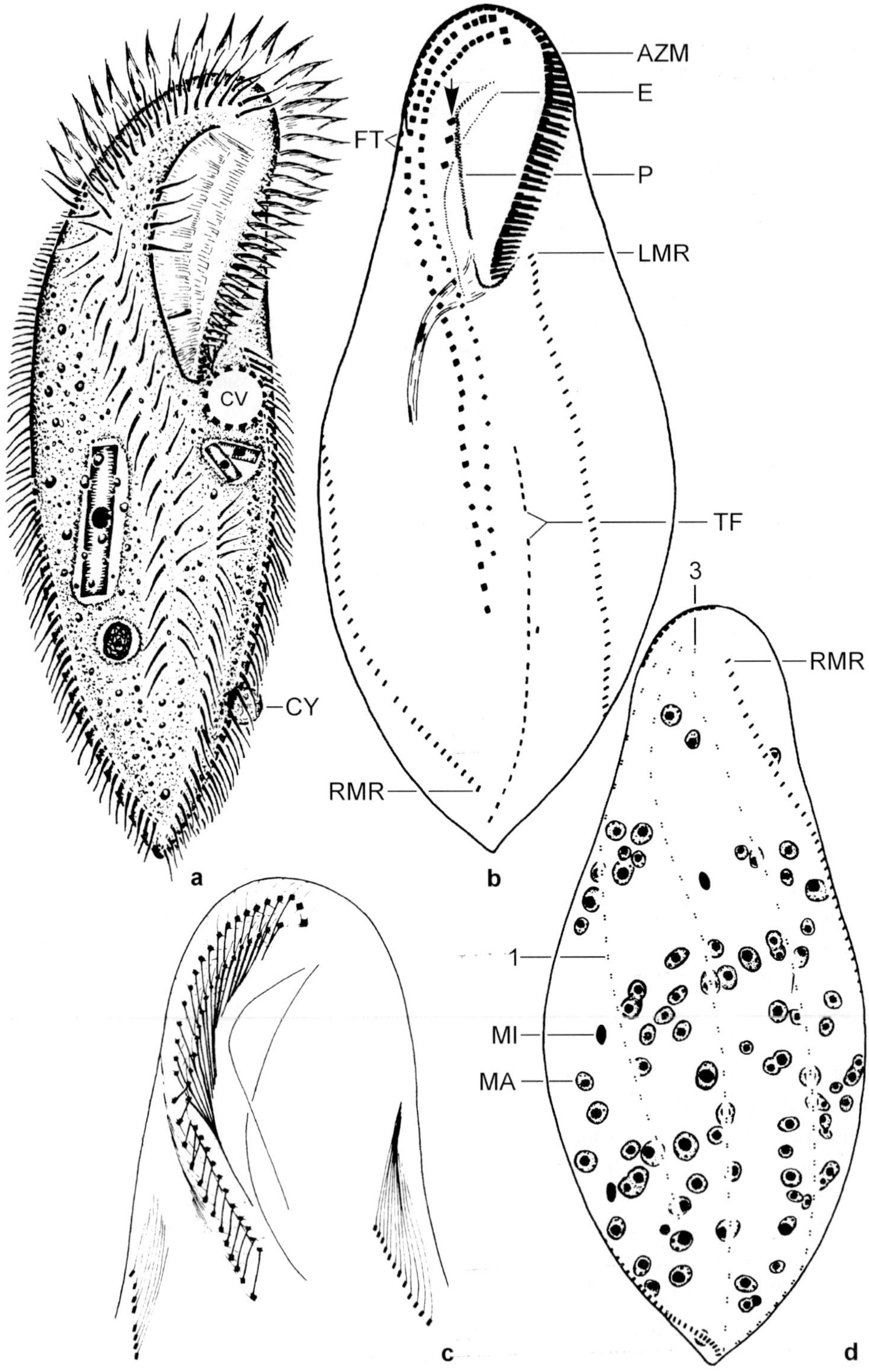
AZM
E
FT
P
LMR
CV
TF
CY
RMR
3
RMR
1
MI
MA
a
b
c
d

midventral complex, and the rather long distance between the proximal end of the adoral zone and the anterior end of the transverse cirral row (Fig. 169a, autapomorphies 4).

Morphology: Body size in life about 200 × 60 μm; body length:width ratio about 3.3:1 in life (Fig. 173a), but only 2.0:1 on average after protargol impregnation (Table 36) because cells distinctly inflated (Fig. 173b). Body outline roughly spindle-shaped; right margin convex, left one straight to slightly S-shaped, anterior end broadly rounded, posterior end tapered (Fig. 173a). Slightly flattened dorsoventrally. Body very likely, as in all other urostylids, rather flexible. Macronuclear nodules scattered throughout cytoplasm, individual nodules spherical to ellipsoidal, usually with one large nucleolus. Several (often three) lenticular to ellipsoidal micronuclei, one always close to proximal end of adoral zone of membranelles, others distributed fairly regularly along left body half (Fig. 173d). Contractile vacuole left of proximal end of adoral zone. Cytopyge dorsally in left posterior quarter of cell (Fig. 173a). No cortical granules, however, cytoplasm brownish-red, possibly due to food, rendering individuals dark at low magnification. Cytoplasm with many greasily shining droplets and food vacuoles. Crawls slowly on bottom of Petri dish.

Adoral zone occupies 35–47% of body length, extends far onto right side, composed of an average of 61 membranelles of usual shape and structure, bases of largest membranelles about 14 μm wide, cilia circa 20 μm long in life. Buccal cavity deep and therefore bright at bright-field illumination. Paroral curved leftwards anteriorly, likely usually composed of two (1–3) rows of basal bodies, distinctly shorter than endoral (Table 36). Endoral in buccal cavity, arched to almost straight, extends to proximal end of adoral zone, optically crossing paroral about at level of buccal cirri; apparently composed of a single row of basal bodies, which bear about 25 μm long cilia. Pharyngeal fibres without peculiarities, about 50–80 μm long.

Cirral pattern and number of cirri of usual variability (Table 36). Cirral pattern not as conspicuous as in *T. jahodai* and *T. crystallis* due to the single file of transverse cirri (Fig. 173b). Frontal cirri about 20 μm long, form distinct bicorona, anterior corona sometimes curved backwards on left side, bases of cirri slightly enlarged, that is, four-rowed, as opposed to usually three-rowed posterior cirri row; frontal cirri differ from midventral cirri by conspicuous, longitudinally extending fibrillar associates (Fig. 173c). Buccal cirri right of optical intersection of undulating membranes (Fig. 173b). Usually two frontoterminal cirri behind distal end of adoral zone, bases double-rowed, cilia about 25 μm long. Midventral complex not set off from frontal cirri, extends slightly sigmoidally to near posterior quarter of cell; urostylid zigzagging midventral pattern very indistinct; right cirri about 25 μm long and slightly larger than left cirri, which are only about 20 μm long; right portion of midventral complex by about one

---

← **Fig. 173a–d** *Thigmokeronopsis antarctica* (from Petz 1995. a, from life; b–d, protargol impregnation). **a:** Ventral view of a representative specimen, 250 μm. **b, d:** Infraciliature of ventral and dorsal side and nuclear apparatus, 325 μm. Arrow marks row of buccal cirri. **c:** Fibrillar associates in anterior region. AZM = adoral zone of membranelles, CV = contractile vacuole, CY = cytopyge, E = endoral, FT = frontoterminal cirri, LMR = left marginal row, MA = macronuclear nodule, MI = micronucleus, P = paroral, RMR = right marginal row, TF = thigmotactic field, 1, 3 = dorsal kineties. Page 862.

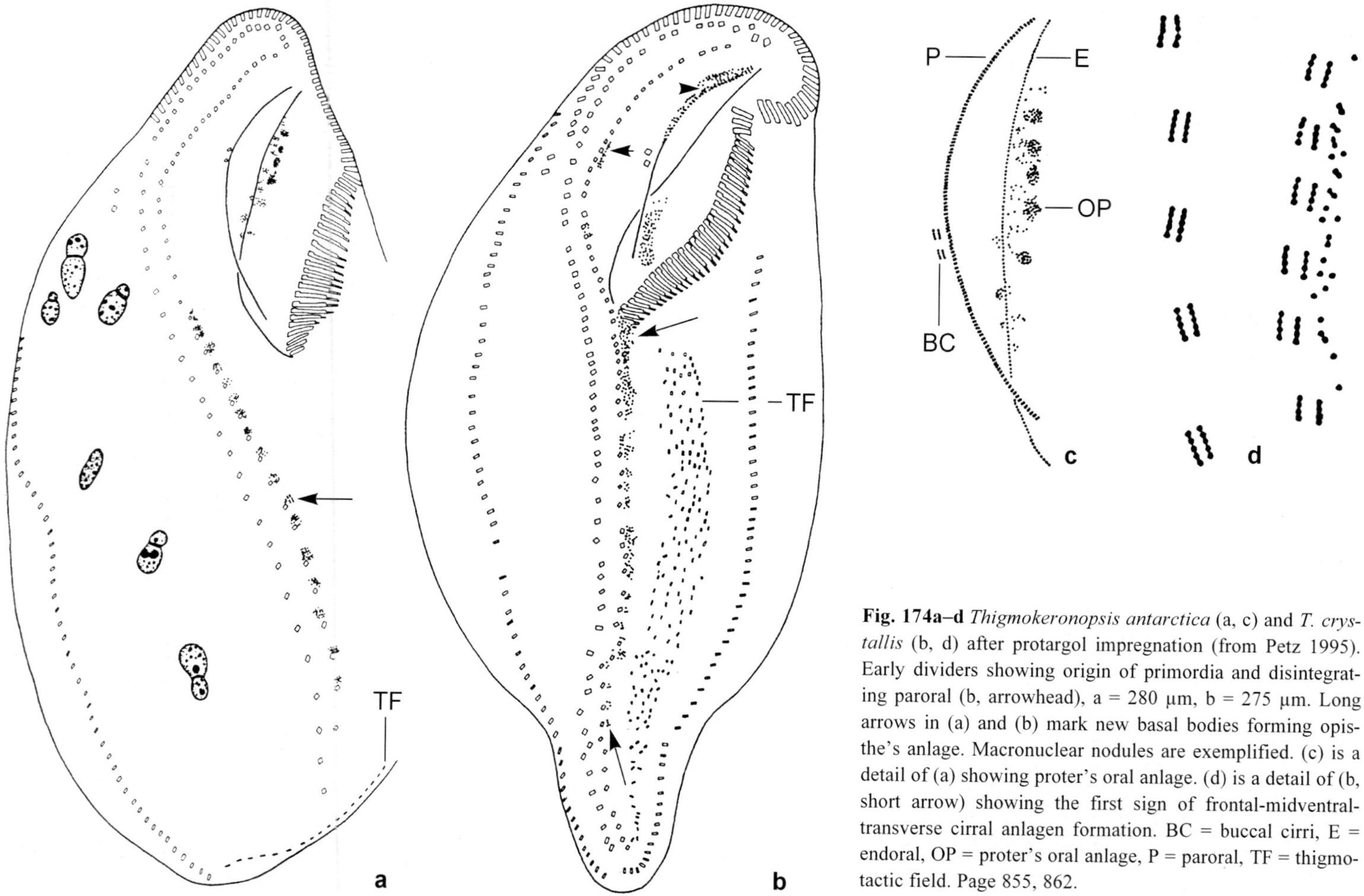

Fig. 174a–d *Thigmokeronopsis antarctica* (a, c) and *T. crystallis* (b, d) after protargol impregnation (from Petz 1995). Early dividers showing origin of primordia and disintegrating paroral (b, arrowhead), a = 280 μm, b = 275 μm. Long arrows in (a) and (b) mark new basal bodies forming opisthe's anlage. Macronuclear nodules are exemplified. (c) is a detail of (a) showing proter's oral anlage. (d) is a detail of (b, short arrow) showing the first sign of frontal-midventral-transverse cirral anlagen formation. BC = buccal cirri, E = endoral, OP = proter's oral anlage, P = paroral, TF = thigmotactic field. Page 855, 862.

cirrus longer than left portion (Fig. 173b); each two midventral cirri linked ladder-like by a fibre (Fig. 173c). Thigmotactic field very inconspicuous, consists of usually one (mean = 1.3, median = 1.0, SD = 0.5, SE = 0.1, CV = 35.5%, Min = 1, Max = 2, n = 30) longitudinal file of cirri extending from about cell centre to near posterior body end; composed of 24–42 (n = 19) fine, double-rowed, 13 µm long cirri. Transverse cirri sensu stricto not recognisable (however, see remarks in genus section for the supposed homonomy of the transverse cirri and the thigmotactic field). Marginal cirri 25–30 µm long, bases two-rowed; anterior third to half of right marginal row extends dorsolaterally, row ends slightly subterminally ahead of rear end of thigmotactic field; left row usually terminates on dorsal side so that marginal rows do not confluent posteriorly; rarely, one or two short additional cirral rows near left marginal row (Fig. 173a).

Dorsal cilia about 6 µm long, arranged in three kineties; kinety 1 slightly shortened posteriorly, rows 2 and 3 bipolar. Caudal cirri lacking (Fig. 173d).

Cell division: The division proceeds very similarly in *Thigmokeronopsis antarctica* and *T. crystallis*. Thus, Petz (1995) described this part of the life cycle in common for both species (Fig. 174a–m).

Stomatogenesis in the opisthe commences with the appearance of small groups of basal bodies adjacent to numerous postoral left midventral cirri (Fig. 174a, b). Proliferation was observed near 16–27 cirral bases, but never at the rearmost midventral and transverse cirri. The anlagen grow by further proliferation and finally form a longitudinal anarchic field. All midventral cirri, even those associated with primordia, are obviously still intact, that is, ciliature and fibres are present. This indicates that parental basal bodies are not incorporated in the anlage.

The proter's oral primordium originates apokinetally, usually to the left of the posterior portion of the paroral approximately in the same focal plane. Paroral and endoral are apparently still intact. However, the paroral starts to dedifferentiate at the anterior end now (Fig. 174b). The endoral disassembles slightly later. A few basal bodies also proliferate close to some anterior left midventral cirri, generating the proter's anlage for the frontal-midventral-transverse cirri (Fig. 174b, d). The anterior midventral and the buccal cirri are still present, indicating that they are not involved in anlagen formation. This is a strong evidence for a de novo origin of the anlagen for the adoral zone, the undulating membranes, and the frontal-midventral-transverse cirri. Marginal row and dorsal kinety primordia are not yet present. The parental left postoral ventral files are fairly disordered, indicating that they commence disintegration (Fig. 174b). A replication band occurs in each macronuclear nodule (Fig. 174a).

In intermediate dividers, the opisthe's anarchic field separates into anlagen for adoral membranelles, undulating membranes, and frontal-midventral-transverse cirri. The frontal-midventral-transverse cirral anlage detaches, whereas the anlagen for adoral membranelles and undulating membranes remain connected posteriorly (Fig. 174e).

The proter's oral anarchic field develops on the dorsal and right wall of the buccal cavity and splits into primordia for adoral membranelles and undulating membranes which are connected posteriorly by many basal bodies. The new adoral membranelles and the undulating membranes differentiate in a posteriad direction. A single cirrus

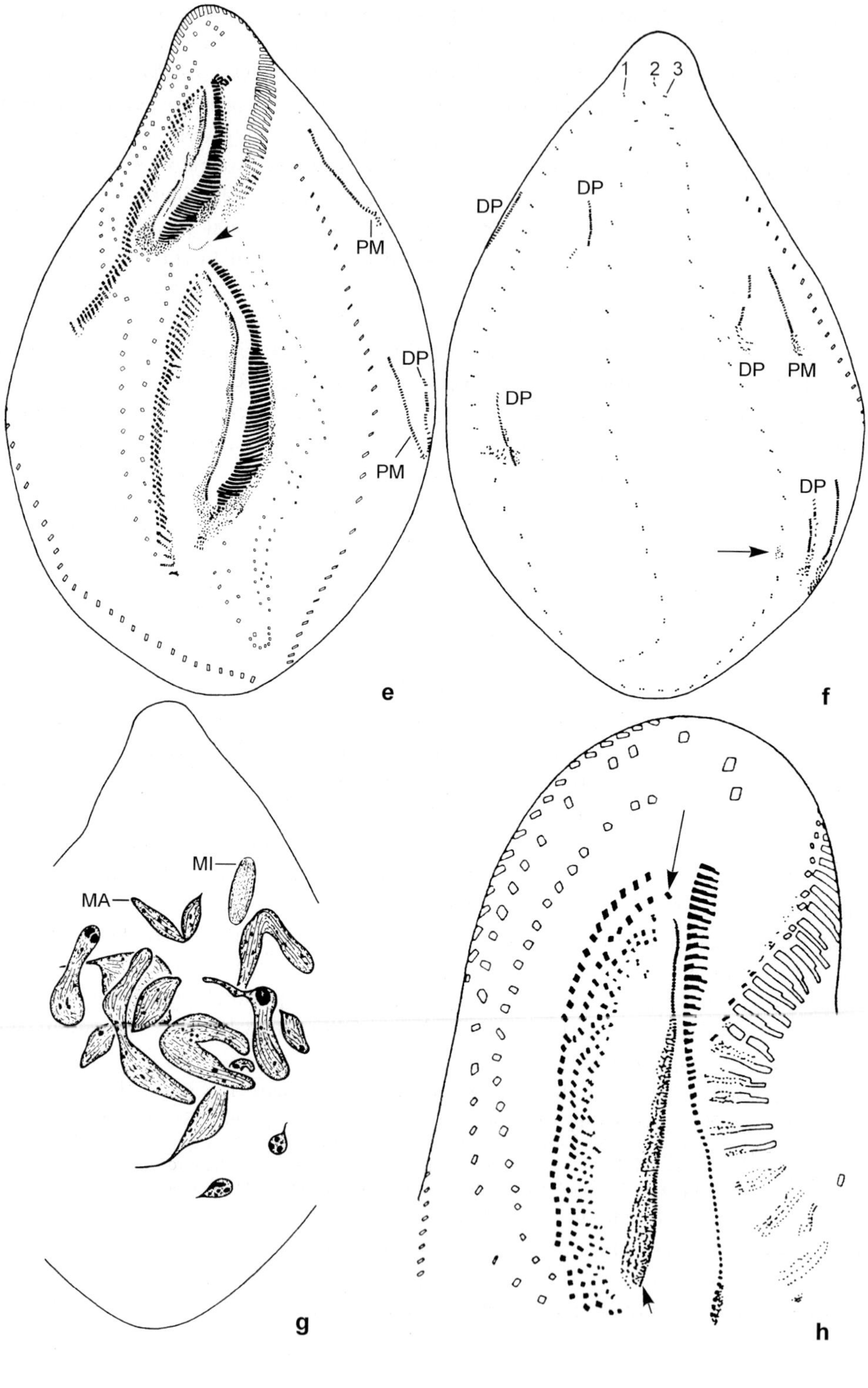

PM
DP
PM
e
1
2
3
DP
DP
DP
PM
DP
DP
f
MI
MA
g
h

(= cirrus I/1) develops from the anterior end of the undulating membrane anlage (Fig. 174h). Subsequently, the adoral zone evaginates.

The proter's frontal-midventral-transverse cirri primordium produces about 50 oblique streaks of basal bodies. Cirri develop from right to left, that is, anterior frontal (= anterior corona) and right midventral cirri first, subsequently posterior bicorona and left midventral cirri, and thigmotactic and transverse cirri last (Fig. 174e, h, i). Both frontal-midventral-transverse cirral anlagen cross the parental midventral rows but resorption of cirri could not be observed in these regions.

The marginal rows apparently originate de novo dorsally to the old ones. Likewise, dorsal kineties form de novo: one anlage each is located near a marginal primordium, whereas the future kinety 2 originates between the old rows 1 and 2 (Fig. 174f). The parental marginal and dorsal rows are evidently fully intact at this stage. Differentiation of new marginal and dorsal structures occurs in a posteriad direction as indicated by a gradual shortening of the cilia. Additional basal bodies proliferate at the posterior ends of the primordia. Rarely, the ends of marginal and dorsal anlagen seem to be connected, probably due to some distortion during preparation (Fig. 174e). Several large, elongated macronuclear nodules containing long chromatin filaments occur (Fig. 174g).

Starting from posterior, the parental adoral membranelles are resorbed. The old thigmotactic cirri are disintegrating and the parental buccal cirri have vanished (Fig. 174e).

Late dividers have differentiated the new endoral and paroral and still proliferate new basal bodies in marginal and dorsal primordia (Fig. 174i, j). Many macronuclear nodules have tapering ends, apparently indicating that they are dividing (Fig. 174k, m).

The following observations relate only to *Thigmokeronopsis crystallis* because very late dividers of *T. antarctica* have not been found. The late stage probably differs, however, only in the development of more thigmotactic cirri, of slightly enlarged transverse cirri, and fewer buccal cirri.

The new buccal cirri are likely derived, as is usual, from anlage II. The frontoterminal cirri originate as rightmost cirri in the last two frontal-midventral-transverse streaks and subsequently move anteriad[1]; the cirri to their left become transverse cirri. The other transverse cirri originate in the preceding 5–8 streaks next to the midventral cirri. All other cirri left of the new frontal and midventral cirri migrate posteriad and leftwards, forming the postoral thigmotactic field (Fig. 174l).

← **Fig. 174e–h** *Thigmokeronopsis crystallis* after protargol impregnation (from Petz 1995). **e–g:** Infraciliature in ventral and dorsal view and nuclear apparatus of a middle divider, 240 µm. Short arrow in (e) marks remnant of, very likely, endoral (with short cilia). Some basal bodies proliferate apparently close to a pair of dorsal basal bodies which is seen only in this specimen (long arrow in f). **h:** Detail of an intermediate divider (slightly later than that in (e)). Long arrow marks frontal cirrus I/1, short arrow denotes undulating membrane primordium. DP = dorsal kinety anlage, MA = macronucleus, MI = micronucleus, PM = marginal row primordium, 1–3 = dorsal kineties. Page 855.

[1] The frontoterminal cirri formation from two anlagen is unusual and was also observed in *Holosticha pullaster* (Fig. 28e) and *Uroleptus caudatus* (Eigner 2001). In *Bakuella edaphoni*, the frontoterminal cirri originate from the penultimate anlage which is also uncommon (Fig. 117l, m). These details should be reinvestigated since in all cases the evidences for these unusual types are not very strong.

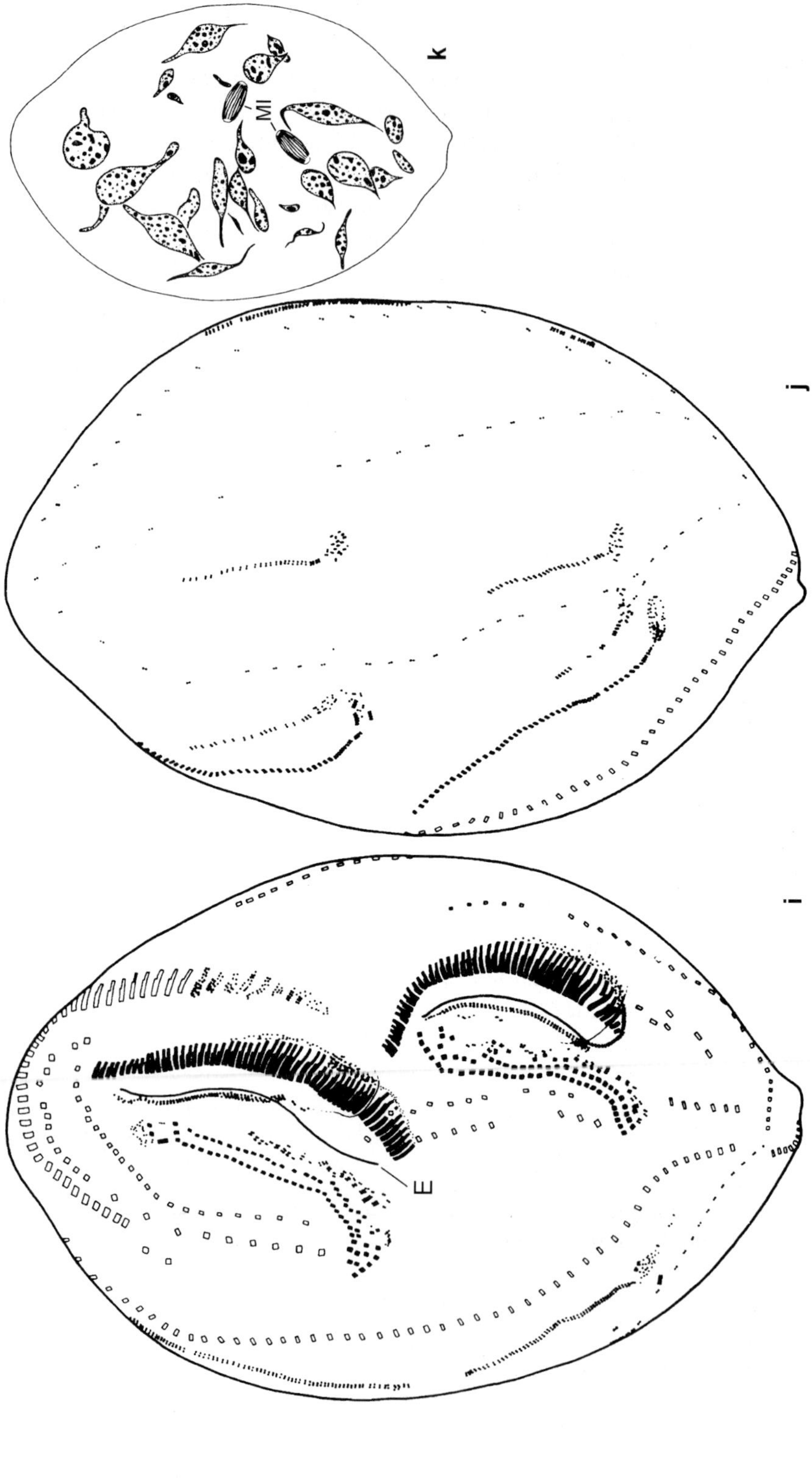

Fig. 174i–k *Thigmokeronopsis antarctica* after protargol impregnation (from Petz 1995). Infraciliature of ventral and dorsal side and nuclear apparatus of middle divider (195 µm) showing differentiation of midventral cirri, proliferation of basal bodies in marginal rows and dorsal kineties, and division of macro- and micronuclei. Parental structures white, new structures black. E = new endoral, MI = micronuclei. Page 862.

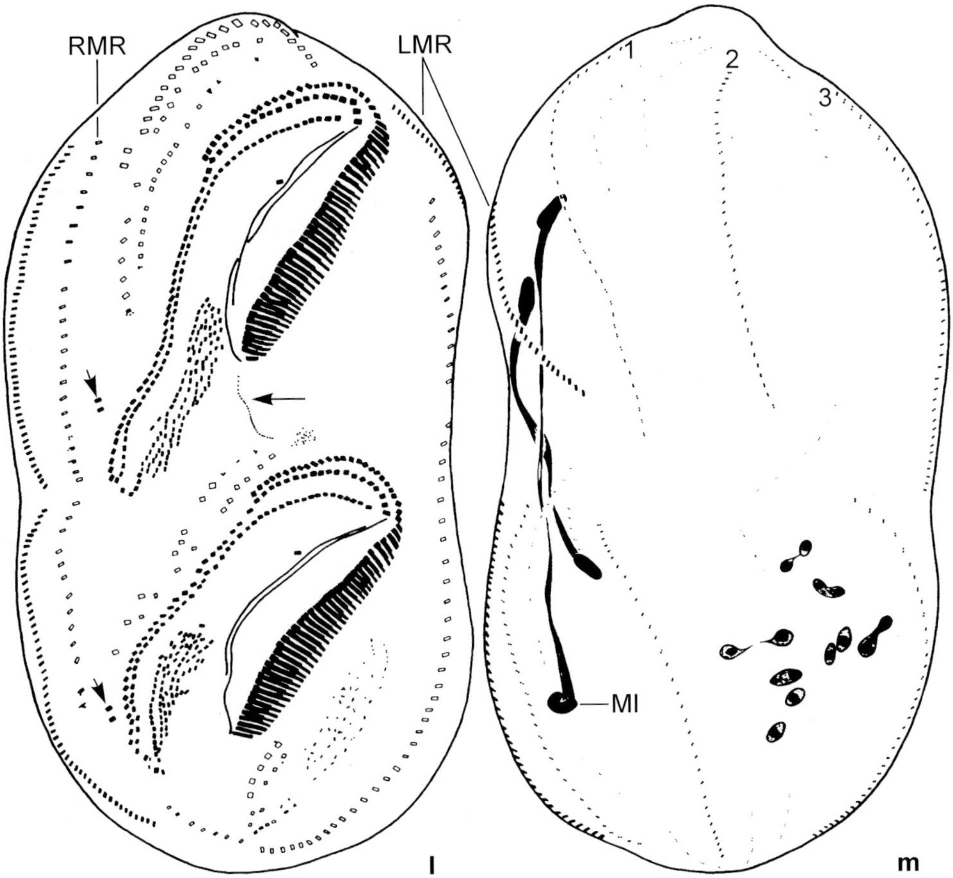

Fig. 174l, m *Thigmokeronopsis crystallis* after protargol impregnation (from Petz 1995). Infraciliature of ventral and dorsal side and nuclear apparatus of a late divider, 241 µm. Short arrows mark new frontoterminal cirri, long arrow points to remnant of, very likely, endoral (apparently some basal bodies still ciliated). Parental structures white, new structures black. LMR = new left marginal row of proter, MI = micronucleus, RMR = parental right marginal row, 1–3 = dorsal kineties. Page 855.

The anterior portions of the new adoral zones have distinctly curved. Marginal and dorsal rows are obviously fully developed (Fig. 174l). Numerous ellipsoidal macronuclear nodules occur, some of which are apparently dividing (Fig. 174m). The parental adoral zone is almost completely resorbed. Disintegration is also evident within old midventral, marginal, and dorsal rows (Fig. 174l, m).

Reorganisation: Physiological regeneration occurred in the field material studied by Petz (1995). As in many other hypotrichs, reorganisation resembles the development of the proter during cell division. Thus, only the illustrations are shown in the present book (Fig. 175a–d). For a detailed description, see Petz (1995, p. 143).

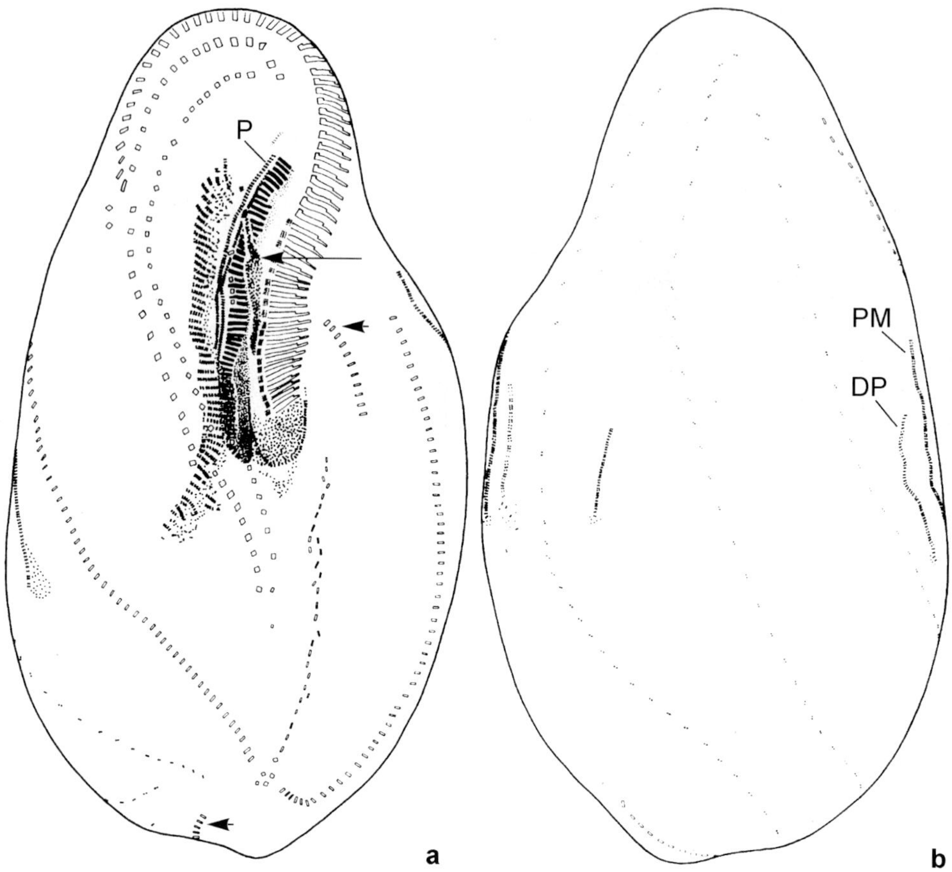

Fig. 175a, b *Thigmokeronopsis antarctica* after protargol impregnation (from Petz 1995). Infraciliature of ventral and dorsal side of an early reorganiser, 198 µm. Long arrow marks dedifferentiated parental paroral, short arrows denote supernumerary left marginal cirri. P = new paroral, PM = anlage of right marginal row, DP = anlage of dorsal kinety 3. Page 862.

Occurrence and ecology: Marine. Type locality of *T. antarctica* is the sea ice of the Weddell Sea, Antarctica (70°17'S 08°53'W; core number AN 103107b), where Petz (1995) found it regularly in the brine-filled pore system of newly formed and multi-year sea ice. It occurred in moderate abundance within the distinctly brownish-coloured, densely populated layer together with diatoms, autotrophic and heterotrophic flagellates, and other ciliates, for example, once also with *T. crystallis*. According to Petz et al. (1995, p. 14), *Thigmokeronopsis antarctica* was found in succeeding stages in sea ice formation, namely in nilas ice (young, sheet ice), pancake ice (young, roundish pieces), multi-year ice, and multi-year land-fast ice (sea ice connected to land). The environmental parameters in brine were: temperature -1.8 to 3.4°C; salinity 32.3–59.1‰; PO_4 <1.0–5.9 µmol l^{-1} melted ice; NO_2 <0.1–0.3 µmol l^{-1}; NO_3 1.0–41.9 µmol l^{-1}; NH_4 1.0 to 18.2 µmol l^{-1}; Si 3.0–56.0 µmol l^{-1}; chlorophyll *a* 0.7–49.3 µg l^{-1}. In raw cultures also

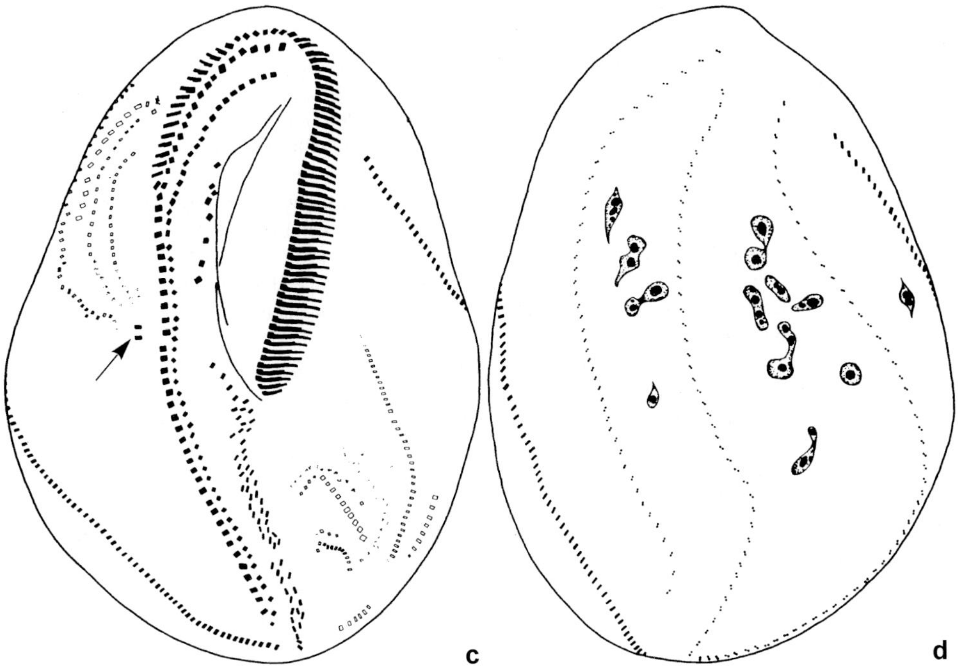

Fig. 175c, d *Thigmokeronopsis antarctica* after protargol impregnation (from Petz 1995). Infraciliature of ventral and dorsal side and some macronuclear nodules of a late reorganiser, 180 µm. Arrow marks new frontoterminal cirri. Page 862.

found at +1°C and a salinity of about 15‰. Biomass of 10^6 individuals about 245 mg (Petz 1995).

Feeds mainly on small and large pennate diatoms (about 10–50 µm long), rarely on ciliates (*Gymnozoum* sp.) and flagellates (Petz 1995).

Thigmokeronopsis crassa (Claparède & Lachmann, 1858) comb. nov. (Fig. 176a–p, Table 36, Addenda)

1858 ***Oxytricha crassa***[1] – Claparède & Lachmann, Mém. Inst. natn. génev., 5: 147, Planche VI, Fig. 7, 7a (Fig. 176f, g; original description; no type material available).

1877 ***Oxytricha crassa*, Cl. u. L.** – Wrześniowski, Z. wiss. Zool., 29: 278 (combination with *Holosticha*, see nomenclature).

1882 ***Amphisia gibba* var. *crassa*** – Kent, Manual Infusoria II, p. 768 (revision; change in rank).

1922 ***Amphista crassa* Cl. et Lach., 1858** – Faria, Cunha & Pinto, Mems Inst. Oswaldo Cruz, 15: 113, 197 (incorrect subsequent spelling of *Amphisia*; combination with *Amphisia*).

1929 ***Amphisia crassa* Cl. u. L.** – Hamburger & Buddenbrock, Nord. Plankt., 7: 89, Fig. 107 (redrawing of Fig. 176f; guide).

[1] The diagnosis by Claparède & Lachmann (1858) is as follows: Oxytrique bossue, sans queue, mais rétrécie soit en avant, soit en arrière. Pieds-cirrhes beaucoup plus courts que chez toutes les espèces précédentes.

1932 ***Trichotaxis (Oxytricha) crassa* (Clap. u. L., 1858)** – Kahl, Tierwelt Dtl., 25: 588, Fig. 101 $_{27}$ (Fig. 176h; revision of hypotrichs).

1933 ***Trichotaxis crassa* (Claparède & Lachmann 1858)** – Kahl, Tierwelt N. u. Ostsee, 23: 111, Fig. 17.14 (redrawing of Fig. 176h; guide).

1972 ***Trichotaxis crassa* (Claparède & Lachmann, 1858) Kahl, 1932** – Borror, J. Protozool., 19: 11, Fig. 19 (redrawing of Fig. 176h; revision of hypotrichs; combination with *Trichototaxis*; incorrect combining author).

1992 ***Trichotaxis crassa* (Claparède and Lachmann, 1858) Kahl, 1935** – Carey, Marine interstitial ciliates, p. 186, Fig. 740 (redrawing of Fig. 176f; guide; incorrect combining author and year).

1999 ***Trichototaxis crassa* (Claparède and Lachmann, 1858) Kahl, 1932** – Petz, Ciliophora, p. 299, Fig. 8.69A B (Fig. 176h and figure from Borror 1972; review; incorrect combining author).

1999 ***Oxytricha crassa* Claparède & Lachmann, 1858** – Berger, Monographiae biol., 78: 243 (revision of oxytrichids; brief note).

1999 ***Pseudokeronopsis qingdaoensis* Hu & Song, in press** – Song & Wang, Progress in Protozoology, p. 73 (nomen nudum, see nomenclature).

2000 ***Pseudokeronopsis qingdaoensis* sp. nov.** – Hu & Song, Acta Zootax. sinica, 25: 364, Fig. 1–5 (Fig. 176a–e; original description of new synonym; see nomenclature for incorrect original spelling; one holotype slide and several paratype slides with protargol-impregnated specimens have been deposited in the Laboratory of Protozoology, Ocean University Qingdao, China).

2001 ***Oxytricha crassa* Claparède and Lachmann, 1858** – Berger, Catalogue of ciliate names 1. Hypotrichs, p. 53 (nomenclator containing all basionyms, combinations, and higher taxa of hypotrichs).

2001 ***Pseudokeronopsis qingdaoensis* Hu and Song, 2000** – Berger, Catalogue of ciliate names 1. Hypotrichs, p. 76 (nomenclator containing all basionyms, combinations, and higher taxa of hypotrichs).

2004 ***Pseudokeronopsis qingdaoensis* Hu & Song, 2000**[1] – Song, Wilbert & Hu, Cah. Biol. mar., 45: 335, Fig. 1A–G, 2A–D, 3–13, Tables 1, 2 (Fig. 176i–m, o, p; detailed redescription; voucher slides are deposited in the Laboratory of Protozoology, Ocean University of China).

Nomenclature: No derivation of the name is given in the original description. The species-group name *crass·us -a -um* (Latin adjective; thick, fat) likely alludes to the wide body outline. Kahl (1932, 1933) classified *Trichototaxis* as subgenus of *Holosticha*. Consequently, the correct name in his reviews is *Holosticha (Trichototaxis) crassa* (Claparède & Lachmann, 1858) Wrześniowski, 1877. This was overlooked by several workers (e.g., Borror 1972, Petz 1999) who assumed that Kahl (1932) transferred the present species to *Trichototaxis*. Unfortunately, I overlooked some of the difficult-to-find combinations in the catalogue of ciliate names (Berger 2001).

Wrześniowski (1877) established *Holosticha* and wrote that, inter alia, *Oxytricha crassa* has to be assigned to this genus. Although he did not transfer it to *Holosticha* formally, he has to be considered as the author for this combination.

Fromentel (1878, p. 161) wrote that *Oxytricha crassa* characterises *Oxytricha* well, a statement which can be considered as a proposal of *O. crassa* as type species. In my

[1] Song et al. (2004a) provided the following new diagnosis: Marine *Pseudokeronopsis*; 150–300 × 40–70 µm in vivo; body colourless, flexible, and contractile. Bicorona comprising about 15 pairs of frontal cirri; 6–13 buccal and 2–4 frontoterminal cirri; transverse row with 25–45 cirri extending to anterior 2/5 of cell; midventral complex consisting of two conspicuously separated rows with c. 40 pairs of cirri; distal end of adoral zone curved posteriorly to about cytostome level; constantly 3 dorsal kineties. Two kinds of cortical granules: one tiny and colorless; another one large and about 2 µm across, brownish to brown-reddish, blood-cell-shaped, distributed sparsely. Contractile vacuole positioned in anterior third of body. More than 100 macronuclear segments.

review on oxytrichids I mistakenly wrote that *O. crassa* was fixed as type species of *Oxytricha* by Fromentel (Berger 1999, p. 243).

The species-group name *qingdaoensis* (adjective) is a composite of the geographical name Qingdao (coast-town in eastern China) and the suffix *-ensis* and refers to the locality where the synonym was discovered. "*Psedokeronopsis*" in "*Psedokeronopsis qingdaoensis* sp. nov." of the original description (p. 361, 364) is an incorrect subsequent spelling of *Pseudokeronopsis*. The name *P. qingdaoensis* was first published by Song & Wang (1999). However, it does not satisfy the provisions of Article 13.1 of the ICZN (1999) and is thus a nomen nudum.

Remarks: The classification of *O. crassa* is difficult because we do not know exactly whether or not the cirral pattern is correctly shown in the original description (Claparède & Lachmann 1858; Fig. 176f). Since other species in Claparède & Lachmann are well described, one has to assume that the cirral pattern illustrated is basically correct, although details must not be over-interpreted because the authors did not have the advantage of modern microscopes and silver impregnation.

Before I present my reflections on *O. crassa*, the history of this somewhat curious species should be recapitulated. Claparède & Lachmann (1858) emphasised the vaulted dorsal side, so that the species-group name *gibba* would be better for this species than for *Trichoda gibba* sensu Ehrenberg. Moreover, they mentioned the five cirral rows, which cannot be considered as misobservation. Wrześniowski (1877) mentioned a resemblance of *O. crassa* and *O. velox* Quennerstedt, a junior synonym of *Holosticha gibba*. The spindle-shaped body is indeed reminiscent of *H. gibba*, which has many transverse cirri longitudinally arranged between the rear portion of the midventral complex and the left marginal row, and there is no doubt that the cirral row behind the proximal end of the adoral zone of *O. crassa* can be transverse cirri although they never reach the proximal end of the adoral zone in *Holosticha*. However, synonymy of *O. crassa* and *H. gibba* is prevented by the lack of enlarged frontal cirri in *O. crassa*. In spite of the different frontal ciliature, Kent (1882) considered *O. crassa* as variety of *Amphisia gibba* (= *Holosticha gibba* in present book). Faria et al. (1922) and Hamburger & Buddenbrock (1929) again raised *O. crassa* to species rank in the genus *Amphisia*, a junior synonym of *Holosticha*.

By contrast, Kahl (1932) proposed a classification of *O. crassa* in (the subgenus) *Trichototaxis* because it has – like *T. stagnatilis*, type of *Trichototaxis* – two cirral rows near the left cell margin. Borror (1972) again raised *Trichototaxis* to genus rank and therefore transferred the present species from *Holosticha (Trichototaxis)* in Kahl (1932) to *Trichototaxis*, a classification also accepted by Carey (1992).

Hemberger (1982, p. 117) supposed synonymy of *O. crassa* with *Holosticha kessleri*, a further junior synonym of *H. gibba*. By contrast, Borror & Wicklow (1983, p. 119) synonymised it with *Pseudokeronopsis rubra*. However, this species is much more slender and has a different cirral pattern, making identity very unlikely.

The discussion above shows that two alternate hypothesis about *O. crassa* exist. Early authors and Hemberger (1982) proposed a classification in *Holosticha*, respectively, a synonymy with *H. gibba*, type of this genus. The body shape, the size, and the

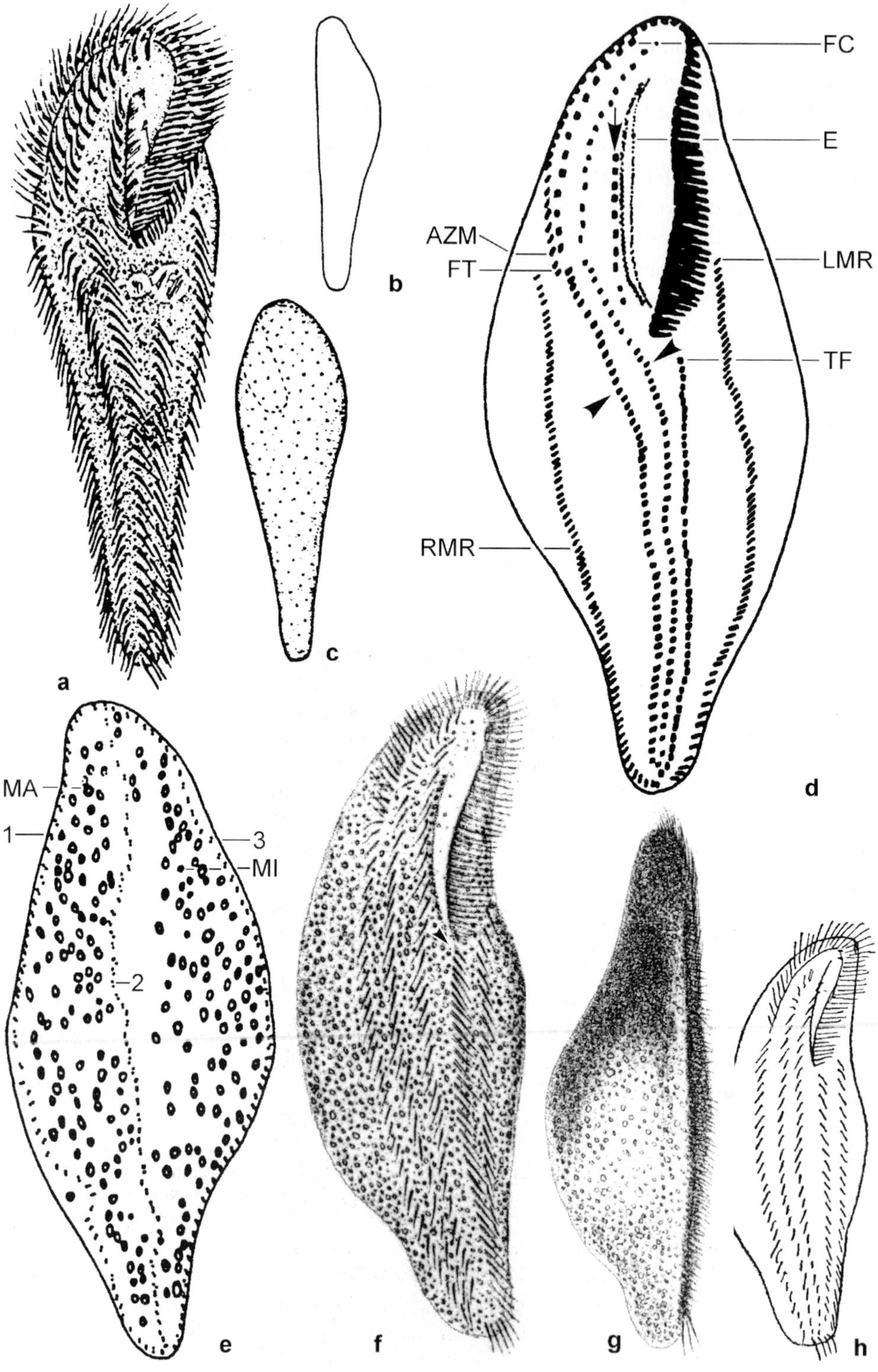
FC
E
AZM
FT
LMR
TF
RMR
MA
1
3
MI
2
a
b
c
d
e
f
g
h

marine habitat indicate that this classification could be correct. However, the lack of three enlarged frontal cirri and the very long row of transverse cirri (in *Holosticha* species the transverse cirral row never extends to near the proximal end of the adoral zone) contradicts this proposal. Claparède & Lachmann did not recognise the nuclear apparatus, indicating that *O. crassa* has many macronuclear nodules dispersed throughout the cell. If this assumption is correct, synonymy with *H. gibba* (two macronuclear nodules) is impossible. Kahl (1932), Borror (1972), and Carey (1992) assumed a relationship of *O. crassa* with *Trichototaxis stagnatilis* mainly because both species have a bicorona and two cirral rows near the left cell margin. Most features support a classification of *O. crassa* in *Trichototaxis*, except for the transverse cirri, which form a distinct compact row in *T. stagnatilis*, whereas in *O. crassa* transverse cirri are either lacking or very inconspicuous if the second cirral row on the left side is considered as inner left marginal row. Claparède & Lachmann (1858) mentioned that the rearmost cirri are slightly longer. However, the illustration shows that a short compact transverse cirral row, as in *T. stagnatilis*, is lacking in *O. crassa* (Fig. 176f). Moreover, the type species of *Trichototaxis* is limnetic whereas *Oxytricha crassa* was discovered in the sea. Although both features must not be over-interpreted, they indicate that the assignment of the present species to *Trichototaxis* is incorrect.

Hu & Song (2000) and Song et al. (2004a) described *Pseudokeronopsis qingdaoensis* from the Yellow Sea. This species has a bicorona, a very long row of transverse cirri extending to the proximal end of the adoral zone, many macronuclear nodules, a spindle-shaped body outline, and a body length of 150–300 µm. Moreover, the distal end of the adoral zone extends very far posteriorly on the right body margin. All these features fit the description of *O. crassa*, making a synonymy very likely. The sole significant difference between the two descriptions concerns the buccal ciliature. *Pseudokeronopsis qingdaoensis* has about nine buccal cirri, whereas Claparède & Lachmann (1858) neither mentioned nor illustrated such cirri. However, this must not be over-interpreted because the buccal cirri are very difficult to recognise in life, especially in species with a bicorona and a midventral complex. Since the last hypothesis, that is, synonymy of *Oxytricha crassa* and *Pseudokeronopsis qingdaoensis*, is the most parsimonious of the three presented I use it in the present book.

← **Fig. 176a–h** *Thigmokeronopsis crassa* (a–e, from Hu & Song 2000; f, g, from Claparède & Lachmann 1858; h, after Claparède & Lachmann 1858 from Kahl 1932. a–c, f–h, from life; d, e, protargol impregnation). **a:** Ventral view of a representative specimen, 229 µm. **b:** Left lateral view showing dorsoventral flattening, 183 µm. **c:** Dorsal view (body length 172 µm) showing arrangement of cortical granules, which are red and about 1.5 µm across. **d, e:** Infraciliature of ventral and dorsal side and nuclear apparatus of same specimen, 154 µm. Arrow marks buccal cirral row. Arrowheads denote the two longitudinal pseudorows formed by the midventral pairs; the lack of a zigzag-pattern is one of the features, indicating that this species is a *Thigmokeronopsis* and not a *Pseudokeronopsis*, as suggested by Hu & Song (2000) and Song et al. (2004a). **f–h:** Ventral (f, h) and lateral (g) view showing mainly cirral pattern, 150 µm. The small arrowhead in (f) marks the anterior end of the extremely long transverse cirral row. Note that Claparède & Lachmann recognised the cirral pattern rather perfectly (cp. Fig. 176d and f). AZM = distal end of adoral zone of membranelles, E = endoral, FC = frontal cirri forming bicorona, FT = frontoterminal cirri, LMR = anterior end of left marginal row, MA = macronuclear nodule, MI = micronucleus, RMR = right marginal row, TF = thigmotactic field (= transverse cirri), 1–3 = dorsal kineties. Page 873.

Oxytricha crassa, respectively its synonym *P. qingdaoensis*, has a midventral complex with a very indistinct zigzagging pattern, that is, the midventral pairs form two longitudinal pseudorows. Furthermore, the numerous transverse cirri are arranged parallel left of the midventral complex, a rather high number of buccal cirri is present, and the undulating membranes are of about equal length (Fig. 176d). These features indicate a close relationship with *Thigmokeronopsis,* especially *T. antarctica* (Fig. 173b). By contrast, the midventral complex of *Pseudokeronopsis* spp. shows a distinct zigzag-pattern, the transverse cirri form a short, oblique row, and the paroral is distinctly shorter than the endoral (Fig. 168d). Thus, I transfer *O. crassa*, the senior synonym of *P. qingdaoensis*, to *Thigmokeronopsis*. The distinctly increased number of buccal cirri is, at the present state of knowledge, the sole autapomorphy of *T. crassa* (Fig. 169a, autapomorphy 5).

Unfortunately, the original description of the junior synonym is in Chinese. However, a morphometric characterisation and an English summary, containing the most important data, are provided. Recently, Song et al. (2004a) provided a detailed redescription of *P. qingdaoensis*. For a separation from the other *Thigmokeronopsis* species see key. *Holosticha* species, which also have an increased number of transverse cirri, do not have a bicorona, but three distinctly enlarged frontal cirri. *Pseudoamphisiella* species are also rather similar, but have – like *Holosticha* spp. – only three frontal cirri. The limnetic *Trichototaxis stagnatilis* has a very similar cirral pattern, but the second cirral row from left is likely a marginal row, and not a conspicuously elongated row of transverse cirri.

Oxytricha crassa sensu Fromentel (1878) and sensu Dumas (1929; Fig. 177a) are insufficiently described.

Morphology: Although the synonymy proposed seems rather solid, I do not mix the descriptions so that the revision can also be used by workers who do not agree with it.

Description according to Claparède & Lachmann (1858; Fig. 176f, g): body length in life about 150 μm, body length:width ratio about 2.9:1. Body outline broadly spindle-shaped, that is, widest in mid-body and distinctly narrowed anteriorly and posteriorly. Ventral side plane, dorsal distinctly vaulted. Nuclear apparatus and contractile vacuole neither mentioned nor illustrated. Presence/absence of cortical granules not known. Yellowish to brownish due to ingested food. Adoral zone occupies about 38% of body length in specimen illustrated. Buccal field narrow. Right marginal row extends from near distal end of adoral zone of membranelles to rear body end; two cirral rows extend from anterior body end to near rear body end, indicating that this is the midventral complex composed of cirral pairs; no enlarged frontal cirri illustrated or mentioned, which means that a bicorona is present. One cirral row commencing at proximal end of adoral zone (Fig. 176f, arrowhead); the synonymy with *P. qingdaoensis* implies that this is the extremely elongated row of transverse cirri. Left marginal row extends to body end, where the cirri are long, indicating that these are the rearmost transverse cirri. No details about the dorsal infraciliature (length of dorsal bristles, number and arrangement of dorsal kineties, presence/absence of caudal cirri) known.

Description of the synonym *P. qingdaoensis* (based on the original description and the redescription by Song et al. 2004a; Fig. 176a–e, i–m, o, p; some features are documented

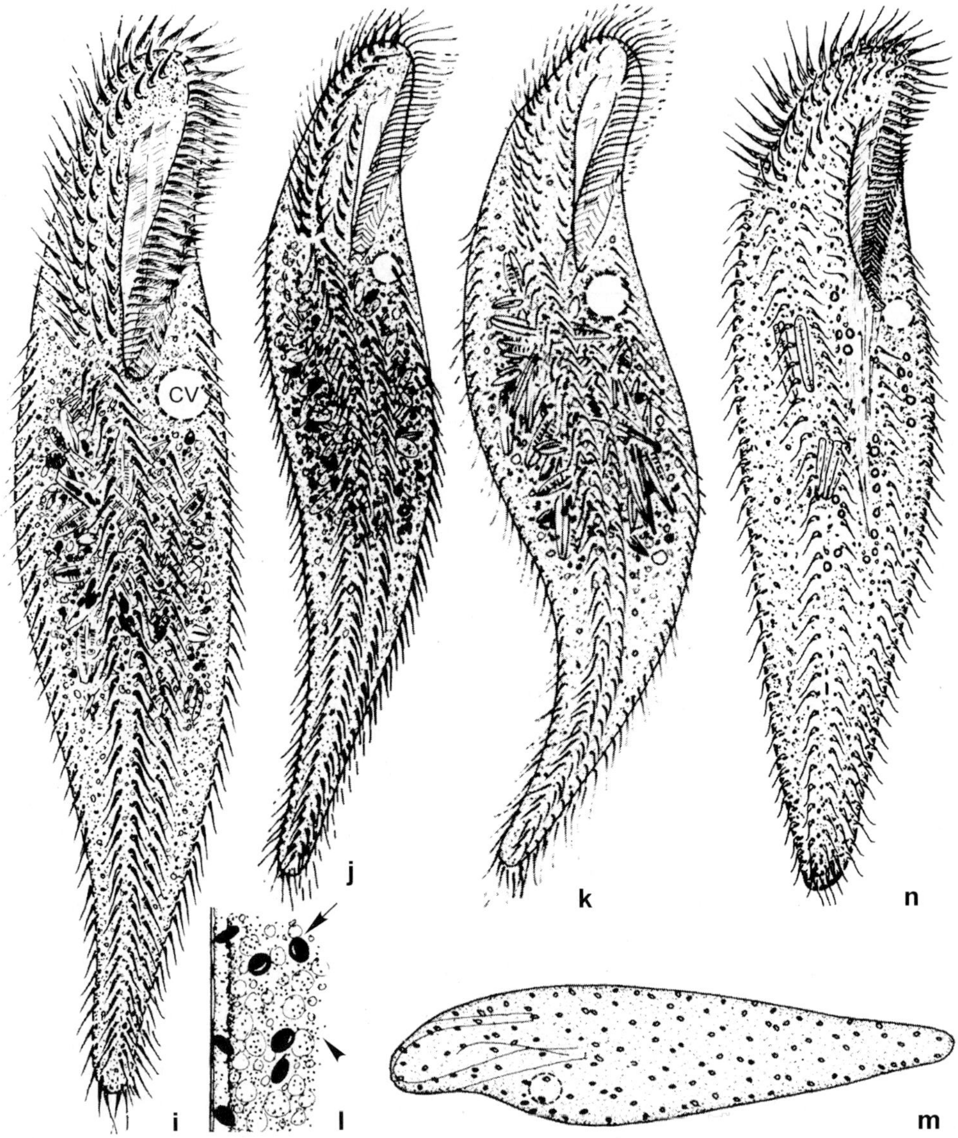

Fig. 176i–n *Thigmokeronopsis crassa* (i–m, from Song et al. 2004a; n, from Hu et al. 2003a. i–n, from life). **i, n:** Ventral view of a representative specimens, i, n = 300 μm. Note that the contractile vacuole pulsates rather slowly, that is, only each 5 minutes. **j, k:** Slender, sigmoidally curved specimens with narrowed posterior body portion, j = 209 μm, k = 184 μm. **l:** Detail of cortex showing tiny (about 0.2 μm across) pigment-like type I granules (arrowhead) and blood-cell-shaped (about 2 μm across) type II granules, which are brown. **m:** Dorsal view showing distribution of cortical granules, 268 μm. CV = contractile vacuole. Page 873.

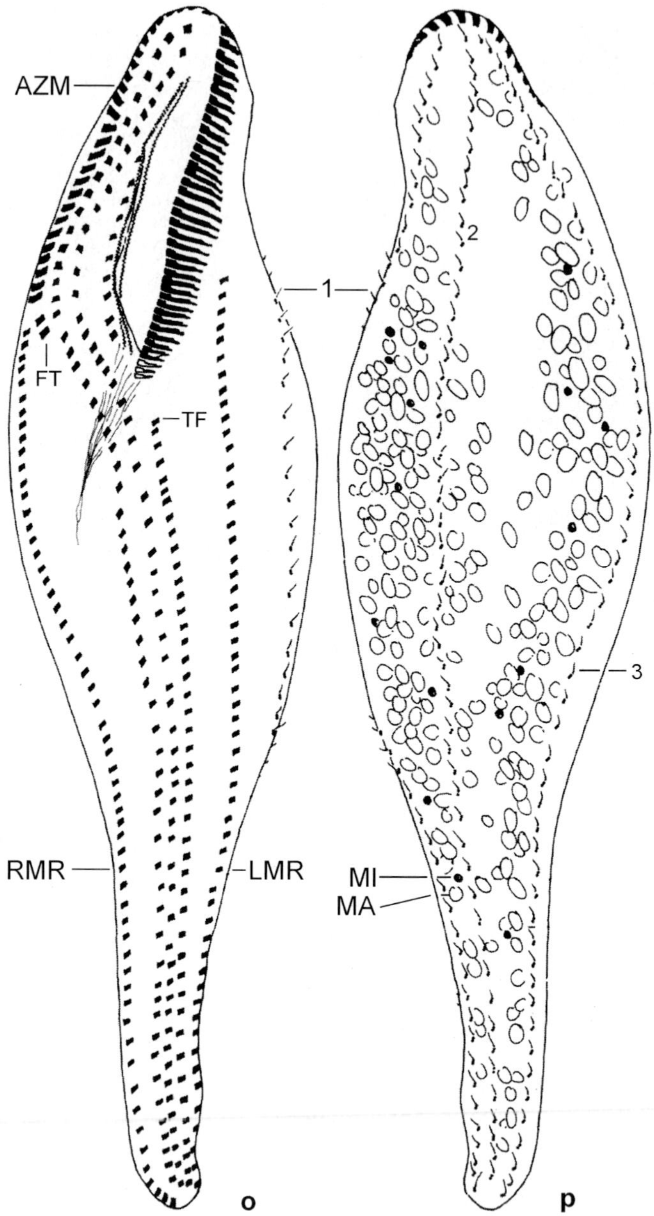

Fig. 176o, p *Thigmokeronopsis crassa* (from Song et al. 2004a. Protargol impregnation). Infraciliature of ventral and dorsal side and nuclear apparatus of same specimen, 254 µm. Note that the bicorona (two bows of frontal cirri) is almost not set off from the midventral complex. The zigzag pattern is very indistinct in *Thigmokeronopsis* so that the midventral cirri and the many transverse cirri feign three ventral rows. Note, however, that all these rows are not true rows (as, for example, the marginal rows) but pseudorows, that is, the individual cirri of each row do not originate from one anlage but from many. AZM = adoral zone of membranelles (extends far posteriorly onto right body margin), FT = frontoterminal cirri, LMR = left marginal row, MA = macronuclear nodule, MI = micronucleus, RMR = right marginal row, TF = thigmotactic field (= transverse cirri), 1–3 = dorsal kineties. Page 873.

by micrographs not shown in present book): body size 130–240 × 50–70 µm in life (Hu & Song 2000); freshly collected specimens of population studied by Song et al. (2004a) usually about 250 µm long in life; length of cells cultured for several days range from 150 µm (posterior portion often short and narrow) to 300 µm (with long, tail-like posterior portion; Fig. 176i–k, m). Body length:width ratio about 4–5:1

Table 36 Morphometric data on *Thigmokeronopsis antarctica* (ant, from Petz 1995), *Thigmokeronopsis crystallis* (cry, from Petz 1995), *Thigmokeronopsis crassa* (qi1, synonym *Pseudokeronopsis qingdaoensis* from Hu & Song 2000; qi2, synonym *P. qingdaoensis* from Song et al. 2004a), *Thigmokeronopsis jahodai* (jah, from Wicklow 1981), *Thigmokeronopsis magna* (mag, from Wilbert & Song 2005), and *Thigmokeronopsis rubra* (ru1, ru2, two populations from Hu et al. 2004)

Characteristics [a]	Species	mean	M	SD	SE	CV	Min	Max	n
Body, length	ant	250.1	251.0	46.0	8.3	18.4	146.0	353.0	30
	cry	226.3	227.0	27.6	8.0	12.2	180.0	279.0	12
	jah	208.0	–	–	–	–	180.0	240.0	20
	mag	202.2	–	34.4	7.7	17.0	135.0[k]	255.0	20
	qi1	194.5	–	29.1	6.1	15.0	150.0	260.0	23
	qi2	178.9	–	11.5	3.1	6.4	158.0	196.0	14
	ru1	133.6	–	8.4	2.5	6.3	120.0	145.0	11
	ru2	126.1	–	15.1	3.8	12.0	90.0	152.0	16
Body, width	ant	124.7	123.0	21.1	3.8	17.0	74.0	161.0	30
	cry	101.3	93.0	17.7	5.3	17.5	84.0	136.0	12
	jah	67.0	–	–	–	–	55.0	85.0	20
	mag	81.7	–	6.3	1.9	7.8	71.0	91.0	11
	qi1	76.8	–	10.6	2.2	13.9	55.0	100.0	23
	qi2	78.5	–	8.0	2.1	10.1	68.0	98.0	14
	ru1	59.6	–	3.3	1.2	5.6	55.0	65.0	8
	ru2	72.7	–	10.1	2.5	13.9	58.0	90.0	16
Anterior body end to proximal end of adoral zone, distance	ant	97.9	98.0	10.8	2.0	11.0	68.0	121.0	30
	cry	90.4	87.0	10.9	2.9	12.0	76.0	112.0	12
	jah [d]	87.0	–	–	–	–	–	–	1
	mag	78.7	–	7.5	1.9	9.6	68.0	90.0	16
	qi1	75.6	–	6.4	1.3	8.5	63.0	83.0	23
	qi2	69.5	–	5.1	1.4	7.3	62.0	79.0	14
	ru1	50.7	–	4.2	1.3	8.3	46.0	62.0	11
	ru2	51.3	–	5.1	1.3	9.9	39.0	59.0	16
Paroral, length	ant	57.7	57.0	8.2	1.5	14.3	41.0	71.0	30
	cry	61.1	58.0	11.6	3.3	19.0	45.0	82.0	12
Endoral, length	ant	80.0	81.0	9.6	1.8	12.0	60.0	97.0	30
	cry	77.5	73.0	13.5	3.9	17.5	64.0	105.0	12
Macronuclear nodule, length	ant	10.7	10.5	4.2	0.8	39.8	4.0	22.0	30
	cry	9.2	8.5	2.9	0.7	31.9	4.5	14.0	12
	ru2	–	–	–	–	–	3.0	7.0	?
Macronuclear nodule, width	ant	6.3	6.0	1.9	0.3	29.8	4.0	11.0	30
	cry	5.1	5.0	1.1	0.3	21.4	3.0	7.0	12
	ru2	–	–	–	–	–	1.5	3.5	?
Macronuclear nodules, number	ant			about 120–170					
	cry			about 110–180					
	qi1	–	–	–	–	–	195.0	253.0	7
Micronucleus, length	ant	7.8	7.0	1.6	0.3	20.3	5.0	10.0	30
	cry	11.1	11.0	0.9	0.3	8.5	10.0	13.0	12
Micronucleus, width	ant	4.1	4.0	0.6	0.1	15.2	3.0	5.5	30
	cry	4.9	5.0	0.6	0.2	13.1	4.0	6.0	12
Micronucleus to left body margin, distance	ant	15.1	13.0	9.0	1.4	59.7	3.0	43.0	30
	cry	28.2	26.0	11.1	2.3	39.5	13.0	52.0	12
Micronuclei, number	ant	3.3	3.0	0.8	0.2	24.1	2.0	5.0	30
	cry	2.9	2.0	1.6	0.5	53.6	1.0	6.0	12
	qi1	–	–	–	–	–	3.0	8.0	9

Table 36 Continued

Characteristics [a]	Species	mean	M	SD	SE	CV	Min	Max	n
Adoral membranelles, number	ant	60.7	61.0	5.6	1.0	9.2	45.0	71.0	30
	cry	64.3	64.5	6.3	1.6	9.7	55.0	72.0	12
	jah	75.0	–	–	–	–	–	–	?
	mag	64.9	–	4.8	1.6	7.4	59.0	71.0	9
	qi1	55.6	–	4.1	1.1	7.4	50.0	65.0	14
	qi2	52.5	–	4.5	1.4	8.5	46.0	60.0	10
	ru1	38.3	–	2.9	0.8	7.5	33.0	43.0	12
	ru2	37.7	–	2.4	0.6	6.3	31.0	42.0	16
Frontal cirri, number	ant [e]	32.9	33.0	5.1	1.1	15.4	20.0	42.0	30
	cry [e]	32.9	33.0	4.8	1.2	14.6	26.0	41.0	12
	mag [i]	14.0	–	1.5	0.4	10.6	12.0	16.0	12
	qi1 [e]	26.6	–	3.8	0.9	14.1	21.0	35.0	19
	qi2 [e]	–	–	–	–	–	27.0	32.0	4
	ru1 [e]	13.8	–	1.8	0.5	13.0	12.0	16.0	12
	ru2 [e]	12.9	–	1.6	0.5	12.1	11.0	15.0	12
Frontoterminal cirri, number	ant	2.1	2.0	0.4	0.1	19.5	1.0	3.0	30
	cry	1.9	2.0	0.3	0.1	16.6	1.0	2.0	12
	jah	2.0	–	–	–	–	–	–	?
	mag	2.0	–	0.0	0.0	0.0	2.0	2.0	13
	qi1	2.0	–	0.0	0.0	0.0	2.0	2.0	19
	qi2	2.3	–	0.8	0.3	33.1	2.0	4.0	7
	ru1	2.0	–	0.0	0.0	0.0	2.0	2.0	12
	ru2	2.0	–	0.0	0.0	0.0	2.0	2.0	16
Buccal cirri, number	ant	3.5	4.0	0.9	0.1	24.1	2.0	5.0	30
	cry	1.3	1.0	0.5	0.1	36.2	1.0	2.0	12
	jah	2.0	–	–	–	–	–	–	?
	mag	1.0	–	0.0	0.0	0.0	1.0	1.0	>20
	qi1	9.3	–	1.2	0.3	12.8	6.0	10.0	16
	qi2	9.4	–	1.9	0.7	20.2	8.0	13.0	7
	ru1	1.0	–	0.0	0.0	0.0	1.0	1.0	12
	ru2	1.0	–	0.0	0.0	0.0	1.0	1.0	16
Midventral complex, number of right cirri	ant [b]	47.0	45.5	6.0	1.2	12.8	41.0	64.0	30
	cry [b]	52.5	50.0	7.7	2.1	14.7	42.0	66.0	12
	jah [b]	54.0	–	–	–	–	–	–	?
	mag [j]	43.3	–	3.8	1.1	7.7	39.0	46.0	12
	qi1 [f]	40.8	–	6.0	1.4	14.8	33.0	56.0	19
	qi2 [f]	–	–	–	–	–	44.0	63.0	5
	ru1	24.6	–	1.8	0.5	7.1	22.0	28.0	11
Midventral complex, number of left cirri	ant [c]	46.0	44.0	5.1	1.0	11.2	40.0	63.0	30
	cry [c]	50.4	51.0	7.7	2.1	15.3	41.0	64.0	12
	jah [c]	54.0	–	–	–	–	–	–	?
	qi1 [g]	40.6	–	5.2	1.2	12.8	33.0	53.0	19
	qi2 [g]	–	–	–	–	–	41.0	59.0	5
	ru1	24.1	–	2.3	0.7	9.7	22.0	30.0	11
	ru2 [h]	50.9	–	4.8	1.2	9.4	40.0	59.0	15
Transverse cirri, number	ant	0.0	0.0	0.0	0.0	0.0	0.0	0.0	30
	cry	8.3	8.0	0.8	0.2	9.3	7.0	10.0	12
	jah [d]	10.0	–	–	–	–	–	–	1
	mag	10.2	–	1.3	0.4	13.2	9.0	13.0	12
	qi1	34.3	–	5.1	1.2	14.9	27.0	45.0	19
	qi2	30.4	–	3.5	1.1	22.4	25.0	36.0	10

Table 36 Continued

Characteristics[a]	Species	mean	M	SD	SE	CV	Min	Max	n
Transverse cirri, number	ru1	7.1	–	0.3	0.1	4.1	7.0	8.0	12
	ru2	7.1	–	0.8	0.2	10.7	6.0	8.0	13
Right marginal cirri, number	ant	62.9	64.0	7.4	1.6	11.7	51.0	77.0	30
	cry	53.3	53.0	6.9	2.3	12.9	41.0	63.0	12
	jah[d]	51.0	–	–	–	–	–	–	1
	mag	70.5	–	3.8	1.2	5.3	64.0	76.0	10
	qi1	49.3	–	12.7	2.9	25.7	43.0	74.0	19
	qi2	52.5	–	5.5	1.5	10.5	43.0	62.0	14
	ru1	36.8	–	4.7	1.4	12.9	29.0	45.0	12
	ru2	34.9	–	3.5	0.9	10.1	28.0	41.0	16
Left marginal cirri, number	ant	62.8	63.0	9.3	2.0	14.8	45.0	80.0	30
	cry	51.1	51.0	5.0	1.9	9.7	43.0	58.0	12
	jah[d]	61.0	–	–	–	–	–	–	1
	mag	59.5	–	3.0	0.9	5.0	55.0	63.0	10
	qi1	53.2	–	8.5	1.9	15.9	38.0	70.0	19
	qi2	50.6	–	5.5	1.5	10.1	42.0	61.0	14
	ru1	39.5	–	5.1	1.5	12.9	31.0	49.0	12
	ru2	36.4	–	3.4	0.8	9.2	30.0	43.0	16
Dorsal kineties, number	ant	3.0	3.0	0.0	0.0	0.0	3.0	3.0	30
	cry	3.0	3.0	0.0	0.0	0.0	3.0	3.0	12
	jah	4.0	–	–	–	–	–	–	?
	mag	3.0	–	0.0	0.0	0.0	3.0	3.0	13
	qi1	3.0	–	0.0	0.0	0.0	3.0	3.0	16
	qi2	3.0	–	0.0	0.0	0.0	3.0	3.0	11
	ru1	3.0	–	0.0	0.0	0.0	3.0	3.0	16
	ru2	3.0	–	0.0	0.0	0.0	3.0	3.0	16
Dorsal kinety 2, number of basal body pairs	ant	32.9	33.0	5.1	1.1	15.4	20.0	42.0	30
	cry	39.3	39.5	2.8	1.0	7.2	36.0	43.0	12

[a] All measurements in µm. All data are from protargol-impregnated specimens. CV = coefficient of variation in %, M = median, Max = maximum value, mean = arithmetic mean, Min = minimum value, n = number of individuals investigated (? = number not given. If only one value is available, then it is listed under the column mean; if two values are given, then they are listed under Min and Max), SD = standard deviation, SE = standard error of arithmetic mean.

[b] Frontal cirri (anterior corona) included.

[c] Frontal cirri (posterior corona) included.

[d] Single value from Fig. 170a.

[e] Cirri forming the bicorona.

[f] Frontal cirri (anterior corona) not included.

[g] Frontal cirri (posterior corona) not included.

[h] Right cirri of midventral complex included (frontal cirri not included).

[i] Cirral pairs forming bicorona.

[j] Cirral pairs forming midventral complex.

[k] Note that the specimen shown in Fig. 170.1g, h is, according to the scale bar, only 116 µm long.

(Fig. 176i). Body outline roughly club-shaped, that is, anterior body half broad and posterior half distinctly narrowed (Fig. 176a, c); prepared specimens spindle-shaped (Fig. 176d, e). Specimens studied by Song et al. (2004a) widest about at mid-body with anterior portion considerably narrowed. Body about flattened 2:1 dorsoventrally, rather flexible, and highly contractile (Fig. 176b). More than 100 (sometimes more than 200) macronuclear nodules scattered throughout cytoplasm, individual nodules 3–7 × 2–4 µm, difficult to recognise in life; several micronuclei, ovoid to globular, about 2.5–3.0 µm across. No contractile vacuole illustrated by Hu & Song (2000; Fig. 176a–c); according to Song et al. (2004a), vacuole near left cell margin slightly behind proximal end of adoral zone, pulsates infrequently, that is, less than one time per 5 minutes (Fig. 176i–k, m). Body brick-red, likely due to the sparsely scattered, red cortical granules about 1.5 µm across (Fig. 176c). Note that Claparède & Lachmann also illustrated many such structures (Fig. 176f). According to the detailed studies by Song et al. (2004a) two types of cortical granules are present (Fig. 176l, m): type I granules sparsely but evenly distributed on dorsal side, often densely and irregularly distributed on ventral side forming pattern of dark patches near buccal area or along cirral rows; individual granules pigment-like and only about 0.2 µm across in life, colourless to greyish, slightly ellipsoidal after fixation. Type II granules sparsely distributed, lower in number than type I granules, not grouped and therefore never rendering cell any colour; individual type II granules blood-cell-shaped (like erythrocytes of mammals), about 3:1 flattened and 2 µm across, reddish-brown to dark-brown when observed under high magnification. Cytoplasm opaque and colourless, often dark grey or even black in central body portion due to food vacuoles and other granules; often with numerous refractive globules 2–5 µm across. Movement slow, without peculiarities, that is, crawling without pause on debris or bottom of Petri dish showing great flexibility.

Adoral zone occupies about 39% of body length on average (Table 36), distal end extends far onto right side (32% in specimen shown in Fig. 176d), composed of an average of 56 membranelles (likely) of usual fine structure; bases of membranelles 6–10 µm wide. Buccal field of ordinary width. Undulating membranes of equal length, indistinctly curved and arranged in parallel. Paroral according to Song et al. (2004a) distinctly wider than endoral (Fig. 176o).

Cirral pattern very conspicuous due to long row of transverse cirri; number of cirri of usual variability, except for cirri in right marginal row, which shows a rather high coefficient of variation (Table 36). Most cirri relatively fine and 10–15 µm long. Note that the cirral pattern almost perfectly matches Claparède & Lachmann's illustration (cp. Fig. 176d, f, o). Frontal cirri form distinct bicorona, more or less distinctly set off from midventral complex. Buccal cirri form row along right side of paroral. Usually two frontoterminal cirri close to distal end of adoral zone, often difficult to recognise. Midventral complex extends to near rear body end, composed of cirral pairs only, which, however, do not form a zigzag-pattern. Transverse cirri of about same size as other cirri, form long row extending from close to proximal end of adoral zone to rear body end (Fig. 176d, o). Right marginal row commences near distal end of adoral zone, extends to rear body end, where it is separated by an inconspicuous gap from rear end of left marginal row (Fig. 176d). Dorsal bristles 2–3 µm (Song et al. 2004a; according to original description 5 μm) long, invariably arranged in three bipolar rows with about 50 basal body pairs each. Caudal cirri lacking (Fig. 176e, p).

Occurrence and ecology: Marine. Type locality of *Thigmokeronopsis crassa* is a fjord near Bergen, Norway (Claparède & Lachmann 1858). The type locality of the synonym *P. qingdaoensis* is a mollusc farm (36°08'N 120°43'E) in the Yellow Sea off Qingdao, China; the population studied by Song et al. (2004a) is from the same locality. Hu & Song (2000) found the present species at about 24 °C, pH 7.9, and 8.0–8.7 mg l^{-1} O_2; Song et al. (2004a) recorded it at a salinity of 33‰ and a water temperature of 28 °C.

Records of *T. crassa* not substantiated by morphological data: Krasnovodsk Bay, Caspian Sea (Agamaliev 1973, p. 1598; 1983, p. 37); Black Sea (Pavlovskaya 1969); Rapallo, Italy (Cuneo 1891, p. 146). Records from Italian freshwater habitats are likely based on misidentifications (Longhi 1892, p. 155; 1894, p. 26; 1895, p. 82). Pavlovskaya (1969) made experiments on algal nutrition, however, without providing details. The synonym *Pseudokeronopsis qingdaoensis* feeds on diatoms (Fig. 176i, h; Song et al. 2004a).

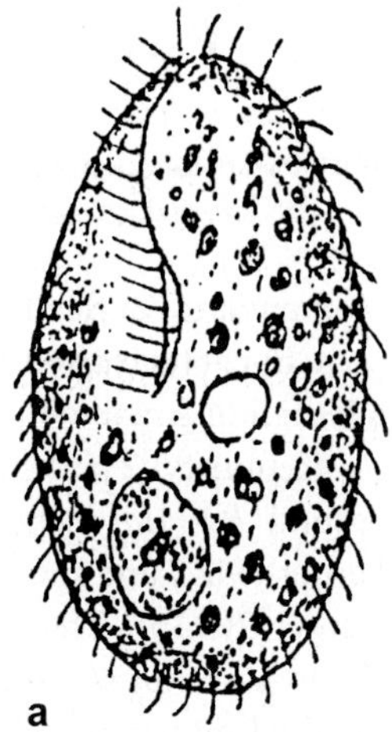

a

Fig. 177a Insufficient redescription of *Oxytricha crassa* from life (from Dumas 1929). Ventral side as seen from dorsal, size not indicated. Page 885.

Insufficient redescriptions

Oxytricha crassa – Dumas, 1929, Microzoaires, p. 72, Planche XXVIII, Fig. 14 (Fig. 177a). Remarks: The illustration and the description are too primitive to allow an identification. France.

Oxytricha crassa – Fromentel, 1878, Microzoaires; p. 265, Planche XII, Fig. 7, 7a. Remarks: The superficial illustrations and the description will never allow a reliable identification. Likely found in France.

Pseudokeronopsinae Borror & Wicklow, 1983

1983 Pseudokeronopsinae subfam. nov. – Borror & Wicklow, Acta Protozool., 22: 123 (original description; no formal diagnosis provided). Type genus: *Pseudokeronopsis* Borror & Wicklow, 1983.

Nomenclature: See Pseudokeronopsidae.

Characterisation (Fig. 167a, autapomorphy 3): Pseudokeronopsidae with individually dividing macronuclear nodules (A).

Remarks: The feature mentioned in the characterisation above is described only for several *Pseudokeronopsis* populations (e.g., Gruber 1884, Rühmekorf 1935, Wirnsberger 1987) and for *Uroleptopsis ignea* and *U. citrina* (Mihailowitsch & Wilbert 1990, Berger 2004b). The plesiomorphic state is the partial fusion which is still present in *Thigmokeronopsis* (see Pseudokeronopsidae). It is more parsimonious to assume that individual division evolved via the intermediate state still realised in *Thigmokeronopsis*, rather than directly from the common mode of fusing to a single, compact mass. In *Protocruzia*, the adelphotaxon to the remaining spirotrichs (Hammerschmidt et al. 1996, Wright et al. 1997, Lynn & Small 1997, van Hoek et al. 1998, Lynn 2003), the macronuclear nodules also divide individually (Ammermann 1968). This has to be interpreted as convergence to the mode found in the Pseudokeronopsinae.

Here, the ring-shaped structures present in *U. citrina* (Fig. 192e, f, i, j, 193b, c) and some (all?) *Pseudokeronopsis* populations (e.g., Prowazek 1900, Hu & Song 2001, Song et al. 2002) have to be discussed. Such organelles are not described for *Thigmokeronopsis* (see there). Thus, one could suggest that this feature is an autapomorphy for the Pseudokeronopsinae. However, these structures are also described for two Holostichidae species (Song & Wilbert 1997, Gong et al. 2001) which are certainly not members of the Pseudokeronopsinae. Thus, this feature cannot be used as phylogenetic marker at the present state of knowledge.

For a key of the genera of the Pseudokeronopsinae, see Pseudokeronopsidae (p. 836).

Pseudokeronopsis Borror & Wicklow, 1983

1983 ***Pseudokeronopsis* gen. nov.**[1] – Borror & Wicklow, Acta Protozool., 22: 99, 123 (original description). Type species (by original designation on p. 100, 123): *Oxytricha rubra* Ehrenberg, 1836.

1984 ***Pseudokeronopsis* Borror und Wicklow, 1983** – Foissner, Stapfia, 12: 111 (brief discussion of *Pseudokeronopsis*, *Trichototaxis*, and *Uroleptopsis*).

1985 ***Pseudokeronopsis*** – Small & Lynn, Phylum Ciliophora, p. 452 (guide to ciliate genera).

1999 ***Pseudokeronopsis* Borror & Wicklow, 1983** – Shi, Acta Zootax. sinica, 24: 365 (revision of hypotrichs).

1999 ***Pseudokeronopsis* Borror & Wicklow, 1983** – Shi, Song & Shi, Progress in Protozoology, p. 114 (revision of hypotrichs).

2001 ***Pseudokeronopsis* Borror & Wicklow 1983** – Aescht, Denisia, 1: 136 (catalogue of generic names of ciliates).

2001 ***Pseudokeronopsis* Borror and Wicklow, 1983** – Berger, Catalogue of ciliate names 1. Hypotrichs, p. 76 (nomenclator containing all basionyms, combinations, and higher taxa of hypotrichs).

---

[1] Borror & Wicklow (1983) did not provide a formal diagnosis.

2002 ***Pseudokeronopsis* Borror and Wicklow, 1983** – Lynn & Small, Phylum Ciliophora, p. 446 (guide to ciliate genera).

Nomenclature: No derivation of the name is given in the original description. *Pseudokeronopsis* is a composite of the Greek adjective *pseudo-* (wrong, lying) and the name of the hypotrich genus *Keronopsis* Penard, 1922. *Keronopsis* is a composite of *Kerona* (an oxytrichid genus whose name is likely derived from the Greek noun *he keronea*, the fruit of the carob tree, *Ceratonia siliqua*; Hentschel & Wagner 1996, p. 341) and the Greek suffix *-opsis* (appearance, looking like). Probably, the name should indicate the resemblance between *Pseudokeronopsis* and *Keronopsis* species, which have a similar (but not homologous) frontal ciliature, but lack a midventral complex. Feminine gender because ending with *-opsis* (ICZN 1999, Article 30.1.2). *Pseudokeronopnopsis* in Shi (1993) is an incorrect subsequent spelling.

Characterisation (Fig. 167a, autapomorphy 4): Adoral zone of membranelles continuous. Frontal cirri arranged in a bicorona. Buccal cirrus present. 2 or more frontoterminal cirri. Midventral complex composed of midventral pairs only. Transverse cirri present. 1 left and 1 right marginal row. 4 or more dorsal kineties on average (A). Caudal cirri lacking. Many macronuclear nodules. Parental adoral zone completely replaced during division. Macronuclear nodules divide individually. Marine and limnetic.

Remarks: Even Entz (1884, p. 360, footnote) recognised that *Holosticha* is heterogeneous because on the one hand it comprised species with three distinctly enlarged frontal cirri and on the other hand species with many normal-sized frontal cirri forming bows. In 1922, Penard established *Keronopsis* with the sole species *K. helluo* Penard, 1922. This species also has many frontal cirri whose exact arrangement is, however, not yet known in detail; basically they are arranged in two curved rows. Kahl (1932) ignored some features of *K. helluo* – for example, the lack of a midventral pattern and the division in a cyst – and classified the holostichid species with many frontal cirri in *Keronopsis*, which he lowered to subgeneric rank within *Holosticha*. Under Borror (1972), *Keronopsis* again attained full generic status, but contained basically the same species as in Kahl (1932). Hemberger & Wilbert (1982) recognised that *Keronopsis helluo* is a non-urostyloid hypotrich and considered it as senior synonym of *Paraholosticha* Kahl, 1932. Moreover, they proposed that *Keronopsis* species with a midventral pattern can be shifted to *Holosticha* without problems, irrespective of whether three or many frontal cirri are present. However, Borror & Wicklow (1983) in their revision on urostylines recognised that these two groups are different and therefore established *Pseudokeronopsis* to include all species with a bicorona.

Pseudokeronopsis is mainly characterised by plesiomorphies so far (see characterisation above). The sole (?) known autapomorphy (Fig. 167a, autapomorphy 4) of this group is likely the increased number of dorsal kineties (4 or more vs. 3). I do not know if this feature is in fact a good autapomorphy for *Pseudokeronopsis*. Anyhow, it is interesting that all *Thigmokeronopsis* species (except *T. jahodai*) and both *Uroleptopsis* species investigated in detail invariably have three dorsal kineties, which is the plesiomorphic state in the Urostyloidae (see ground pattern of this group). In contrast, *Pseudokeronopsis* species have four or more bristle rows (Foissner 1984, Wirnsberger et al.

1987, Hu & Song 2001, Song et al. 2002, Hu et al. 2004a). Only *P. flava* very rarely has only three kineties (Wirnsberger et al. 1987), which, however, must not be over-interpreted.

Eigner & Foissner (1992) used the feature "Parental basal bodies not involved in formation of ciliary structures of daughters". Likely they meant the frontal-midventral-transverse-cirri because the formation of the marginal rows and dorsal kineties proceeds in ordinary manner in *Pseudokeronopsis*, that is, within the parental structures (Wirnsberger 1987). In contrast, the parental midventral complex is obviously not involved in primordia formation. However, the data on *Thigmokeronopsis* (Petz 1995) and *Uroleptopsis* (Mihailowitsch & Wilbert 1990, Berger 2004b) indicate that this feature also applies to these taxa. Thus, it could be an autapomorphy of the Pseudokeronopsidae. However, it is very difficult to decide whether or not parental cirri are definitely involved in primordia formation or not. Thus, this feature is not used further. But it cannot be excluded that it will be a useful marker in future when more ontogenetic data on urostyloids became available.

The phylogenetic relationships between the species included in *Pseudokeronopsis* are not reliably resolvable at present because several species are not described in detail. The differences among some species (*P. rubra*, *P. flava*, *P. carnea*, *P. flavicans*) are not very pronounced so that it is even difficult to identify them properly; if in doubt I recommend writing "species of the *Pseudokeronopsis rubra* group".

Maula et al. (1993) found prokaryotic endosymbionts in the macronuclear nodules, but not in the micronuclei of *Pseudokeronopsis* sp.

Species included in *Pseudokeronopsis* (alphabetically arranged according to basionym): (1) *Holosticha decolor* Wallengren, 1900; (2) *Holosticha (Keronopsis) flavicans* Kahl, 1932; (3) *Holosticha (Keronopsis) ovalis* Kahl, 1932; (4) *Holosticha multinucleata* Maupas, 1883; (5) *Holosticha similis* Stokes, 1886; (6) *Oxytricha flava* Cohn, 1866; (7) *Oxytricha flava carnea* Cohn, 1866; (8) *Oxytricha rubra* Ehrenberg, 1835; (9) *Pseudokeronopsis pararubra* Hu, Warren & Suzuki, 2004; (10) *Pseudokeronopsis sepetibensis* Wanick & Silva-Neto, 2004.

Species misplaced in *Pseudokeronopsis*: Several species transferred to *Pseudokeronopsis* by Borror & Wicklow (1983) and some species originally assigned to this genus belong to other taxa. *Paraholosticha ovata* Horváth, 1933 (for nomenclature, see Berger 2001, p. 68) was transferred to *Pseudokeronopsis* by Borror & Wicklow (1983, p. 120, 124) and previously to *Uroleptopsis* by Borror (1972, p. 11). It has only two macronuclear nodules and lives in freshwater, indicating that it is not a pseudokeronopsid. Furthermore, it lacks a midventral complex so that it is not even an urostyloid. Since *P. ovata* has no transverse cirri, the classification in *Paraholosticha* is correct. By contrast, *Keronopsis* Penard is defined by the presence of such cirri (Dieckmann 1988).

Holosticha pulchra Kahl, 1932, a further species transferred to *Pseudokeronopsis* by Borror & Wicklow (1983), obviously has three enlarged frontal cirri, indicating that it belongs to *Anteholosticha* and not to *Pseudokeronopsis* (Fig. 90a). The specimens illustrated by Borror (1972; Fig. 90c, 180y) and Borror & Wicklow (1983; Fig. 90d) have a distinct bicorona. According to Wirnsberger (1987, p. 154), Borror's (1972) population likely belongs to *P. carnea*. *Holosticha retrovacuolata* Tucolesco has three distinct frontal cirri and thus cannot be a *Pseudokeronopsis*. It is a junior synonym of *Holosticha pullaster*.

A further species transferred to *Pseudokeronopsis* by Borror & Wicklow (1983) is the huge freshwater species *Holosticha (Keronopsis) spectabilis* Kahl, 1932. It has, inter alia, caudal cirri, a dorsal fragmentation, and dorsomarginal kineties (Fig. 242–244) and therefore differs significantly from the other *Pseudokeronopsis* species. Recently, Warren et al. (2002) established *Neokeronopsis* for this highly interesting species. Because of the dorsal kinety fragmentation it belongs to the Oxytrichidae.

Pseudokeronopsis pernix (Wrześniowski, 1877) Borror & Wicklow, 1983 very likely does not belong to the present genus. I preliminarily classify it a supposed synonym of *Holosticha pullaster*.

Pseudokeronopsis trisenestra Dragesco & Dragesco-Kernéis, 1991 is a synonym of *Diaxonella pseudorubra* (Fig. 102c–f). *Pseudokeronopsis ignea* Mihailowitsch & Wilbert, 1990 lacks transverse cirri and thus very likely belongs to *Uroleptopsis* (Fig. 198a–j). The marine *Pseudokeronopsis qingdaoensis* Hu & Song, 2000 is the junior synonym of *Thigmokeronopsis crassa* (Fig. 176a–p).

Pseudokeronopsis pseudorubra in Humpesch & Moog (1994, p. 91) is likely an unintended combination of *Keronopsis pseudorubra* Kaltenbach (= *Diaxonella pseudorubra* in present book). I did not find a formal combination of this species with *Pseudokeronopsis*, which would be incorrect because this species lacks a distinct bicorona.

Holosticha begoniensis Fernandez-Leborans, 1990 is very likely a *Pseudokeronopsis* species. However, since live data are lacking it is classified as species indeterminata (Fig. 99e).

Key to *Pseudokeronopsis* species

Pseudokeronopsis species are rather difficult to identify because the differences are often inconspicuous and the variability within many cirri-pattern features is obviously high. Furthermore, the colour is very important in this group; however, experience shows that this feature, although rather useful, is sometimes difficult to use even for specialists. Note that some *Uroleptopsis* species resemble, for example, *Pseudokeronopsis flava*. Because of these difficulties, each identification of a *Pseudokeronopsis* species should be confirmed by protargol impregnation. Note that three *P. rubra* populations with only two macronuclear nodules have been described so far (e.g., Fig. 178z). Thus check the nuclear apparatus very carefully.

1 Freshwater; macronucleus moniliform **2**
- Marine (salt water); macronucleus composed of many nodules scattered throughout cytoplasm **3**

2 Body length around 200 μm in life; symbiotic algae lacking (Fig. 190a, d, e, j) .... *Pseudokeronopsis similis* (p. 972)
- Body length around 100 μm in life; symbiotic algae present (Fig. 190l) .... *Keronopsis clavata*, likely a junior synonym of *P. similis* (p. 978)

3 (1) Cells more or less colourless **4**
- Cells yellow, orange, or red (at least the cortical granules) **5**

4 Cortical granules likely lacking; adoral zone about 33% of body length (Fig. 188a) . ... *Pseudokeronopsis decolor* (p. 963)
- Cortical granules colourless; adoral zone about 40% of body length (Fig. 189a) (difficult to separate from *P. decolor*, that is, possibly synonymous) .... *Pseudokeronopsis ovalis* (p. 968)

5 (3) 12–13 transverse cirri (Fig. 187a) ... *Pseudokeronopsis multinucleata* (p. 960)
- 10 or less transverse cirri, on average 2–7 (e.g., Fig. 179b) **6**

6 Cells brick-red (e.g., Fig. 179a) **10**
- Cells orange-red or more or less intensely yellow **7**

7 Buccal cirrus lacking; 3 contractile vacuoles (Fig. 186a–e) .... *Pseudokeronopsis sepetibensis* (p. 957)
- Buccal cirrus present; 0–1 contractile vacuole (contraction interval often rather long!) **8**

8 Cells orange-red; on average 5 or more dorsal kineties (e.g., Fig. 181l) **11**
- Cells more or less distinctly yellow; on average 3–4 dorsal kineties **9**

9 Contractile vacuole ahead of mid-body (Fig. 185a, g); bicorona composed on average of 14 cirri; 3–6, on average 4 transverse cirri (Fig. 185a–q) .... *Pseudokeronopsis flavicans* (p. 951)
- Contractile vacuole distinctly behind mid-body (Fig. 184c, p); bicorona composed on average of 9–12 cirri; 1–4, on average 2–4 transverse cirri (Fig. 184h) .... *Pseudokeronopsis flava* (p. 940)

10 (6) Cortical granules brick-red .... *Pseudokeronopsis rubra* (p. 890)
- Cortical granules orange-red .... *Pseudokeronopsis pararubra* (p. 927)

11 (8) Midventral complex extends to transverse cirri (Fig. 180.1h, j, l); on average 6.5 dorsal kineties (Table 37) .... *Pseudokeronopsis pararubra* (p. 927)
- Midventral complex terminates more or less distinctly ahead of transverse cirri (Fig. 181l, 182e, h); on average 5 dorsal kineties (Table 37) .... *Pseudokeronopsis carnea* (p. 930)

Pseudokeronopsis rubra (Ehrenberg, 1836) Borror & Wicklow, 1983
(Fig. 178a–z, 179a–q, 180a–x, Tables 37, 38, Addenda)

1835 ***Oxytricha rubra*** – Ehrenberg, Abh. preuss. Akad. Wiss., Phys.-math. Kl., year 1835: 164 (nomen nudum; see nomenclature).

1836 ***Oxytricha rubra* n. sp.**[1] – Ehrenberg, Mitth. Ges. naturf. Freunde Berlin, 1: 3, 5 (original description with Latin diagnosis; no type material available).

1838 ***Oxytricha rubra*** – Ehrenberg, Infusionsthierchen, p. 364, Tafel XL, Fig. IX_{1-4} (Fig. 178a, b; description and first illustration).

---

[1] The diagnosis by Ehrenberg (1836) is as follows: Expansa 1/10''' longa vermicularis elongata utroque fine rotundata, lateritio-rubra, oris longi rima tertiam fere corporis partem aequante, rependo procedens rarius natans.

1841 ***Oxytricha rubra*** – Dujardin, Zoophytes, p. 419, Planche XI, fig. 13 (Fig. 178e; illustrated record from Mediterranean Sea).
1850 ***Oxytricha rubra* Ehrenberg**[1] – Diesing, Systema Helminthum, p. 156 (revision; no original data).
1865 ***Oxytricha rubra*** – Fresenius, Zool. Gart., Frankf., 6: 127, Fig. 34, 35 (Fig. 178f, g; illustrated record).
1866 ***Oxytricha rubra*. Ehrnbg. char. em.** – Cohn, Z. wiss. Zool., 16: 282, 291, Tafel XV, Fig. 41, 42 (Fig. 178c, d; redescription).
1882 ***Holosticha rubra*, Ehr. sp.** – Kent, Manual Infusoria II, p. 770 (combination with *Holosticha*; revision).
1884 ***Holosticha flavorubra* mihi.** – Entz, Mitt. zool. Stn Neapel, 5: 359 (see nomenclature and remarks).
1884 ***Holosticha flavorubra* var. *rubra*** – Entz, Mitt. zool. Stn Neapel, 5: 359, 363, Tafel 22, Fig. 16, 17 (Fig. 178h, i; see nomenclature and remarks).
1900 ***Holosticha rubra* Ehrbg.** – Wallengren, Acta Univ. lund., 36: 8, Pl. I, Fig. 5–10 (Fig. 178j–n; detailed redescription from life).
1902 ***Holosticha rubra*** – Wallengren, Zool. Jb., 15: 46, Fig. P–U, CC (Fig. 178o, p; description of cell division).
1929 ***Holosticha rubra* Ehrbg.** – Hamburger & Buddenbrock, Nord. Plankt., 7: 86, Fig. 104a (redrawing of Fig. 178i; pro parte; guide to marine ciliates).
1932 ***Keronopsis (Oxytricha, Holosticha) rubra* (Ehrb., 1838)** – Kahl, Tierwelt Dtl., 25: 571, Fig. $101_{16, 17}$ (Fig. 178q, r; revision and redescription).
1932 ***Keronopsis rubra* forma *pentasticha*** – Kahl, Tierwelt Dtl., 25: 573 (see remarks).
1932 ***Keronopsis rubra* forma *heptasticha*** – Kahl, Tierwelt Dtl., 25: 573 (see remarks).
1933 ***Keronopsis rubra* (Ehrenberg 1838)** – Kahl, Tierwelt N.- u. Ostsee, 23: 108, Fig. 16.25 (redrawing of Fig. 178q; revision of marine ciliates).
1933 ***Keronopsis rubra* f. *pentasticha*** – Kahl, Tierwelt N.- u. Ostsee, 23: 109 (guide to marine ciliates).
1933 ***Keronopsis rubra* f. *heptasticha*** – Kahl, Tierwelt N.- u. Ostsee, 23: 109, Fig. 16.26 (Fig. 178s; guide to marine ciliates).
1936 ***Keronopsis (Oxytricha, Holosticha) rubra* (Ehrenberg 1838)** – Kiesselbach, Thalassia, 2: 19, Abb. 42 (Fig. 178y; illustrate record).
1940 ***Holosticha (Oxytricha) rubra* (Ehrbg. 1838) Kent 1881–1882** – Bullington, Pap. Tortugas Lab., 32: 210, Fig. 20A, B, 21A, B, Plate I, Fig. 2 (Fig. 178t–w; detailed redescription from life, contains a rather long diagnosis not repeated here).
1972 ***Keronopsis rubra* (Ehrenberg, 1838) Kahl, 1932** – Borror, J. Protozool., 19: 11 (combination with *Keronopsis*, see nomenclature; revision of hypotrichs).
1972 ***Keronopsis pentasticha* Kahl, 1932** – Borror, J. Protozool., 19: 11 (raised to species rank and combination with *Keronopsis*; revision of hypotrichs).
1972 ***Keronopsis heptasticha* Kahl, 1932** – Borror, J. Protozool., 19: 11 (raised to species rank and combination with *Keronopsis*; revision of hypotrichs).
1981 ***Keronopsis rubra* var. *albino*** – Uhlig, Jber. biol. Anst. Helgoland, 1980: 34, Abb. 18 (see nomenclature and remarks for status of the pigment mutant).
1983 ***Pseudokeronopsis rubra* (Ehrenberg, 1838) nov. comb.** – Borror & Wicklow, Acta Protozool., 22: 99, 123, Fig. 21 (Fig. 178x; combination with *Pseudokeronopsis*; revision of urostylids).
1985 ***Pseudokeronopsis rubra*** – Small & Lynn, Phylum Ciliophora, p. 452, Fig. 7B (Fig. 178x; guide to ciliate genera).
1987 ***Pseudokeronopsis rubra***[2] – Wirnsberger, Larsen & Uhlig, Europ. J. Protistol., 23: 77, Fig. 1–5, Tables 1–3 (Fig. 179a–e; redescription and fixation of neotype [slides 1986/42, 1986/43 in LI]; see nomenclature).
1987 ***Pseudokeronopsis rubra*** – Wirnsberger, Arch. Protistenk., 134: 150, Fig. 1, 2, 5, 6, 8, 9 (Fig. 179i, j, m, n, p, q; cell division).

---

[1] The characterisation by Diesing (1850) is as follows: Corpus lineare utrinque aequaliter rotundatum, setarum serie duplici longitudinali mediana, oris limbo ciliato, lateritio-rubrum. Longit. 1/12–1/10'''.

[2] The rediagnosis provided by Wirnsberger et al. (1987) is as follows: Size in vivo about 170–290 × 30–70 µm. Granules brick-red. Midventral rows extend to the ventral cirri close to the about 7 transverse cirri. Bicorona comprised 14–16 frontal cirri. On the average 6 dorsal kineties.

1992 ***Keronopsis rubra* (Ehrenberg, 1838) Kahl, 1930–5** – Carey, Marine interstitial ciliates, p. 185, Fig. 732 (redrawing of Fig. 178q; guide).

2001 ***Pseudokeronopsis rubra* (Ehrenberg, 1835) Borror and Wicklow, 1983** – Berger, Catalogue of ciliate names 1. Hypotrichs, p. 62 (nomenclator containing all basionyms, combinations, and higher taxa of hypotrichs).

2001 ***Pseudokeronopsis rubra*** – Eigner, J. Euk. Microbiol., 48: 77, Fig. 26, 27 (Fig. 179i; brief review of Urostylidae).

2001 ***Pseudokeronopsis rubra*** – Hu & Song, Acta Protozool., 40: 107, Fig. 1a–l, 2a–h, 3a–d, 4–16, Table 1 (Fig. 180a–x; redescription and ontogenesis; identification uncertain, see remarks).

2002 ***Pseudokeronopsis rubra*** – Lynn & Small, Phylum Ciliophora, p. 446, Fig. 17A, B (Fig. 179b, c; guide to ciliate genera).

Nomenclature: The Latin adjective *rub·us -ra -rum* (red) refers to the dark red colour of the species. *Oxytricha rubra* Ehrenberg was fixed as type species of *Pseudokeronopsis* by original designation. Entz (1884) considered *P. rubra* and *P. flava* as synonyms. To express this assumption he unnecessarily introduced the species-group name *flavorubra* with the two varieties *flava* and *rubra*. The name *flavorubra* is a composite of the Latin adjectives *flav·us -a -um* (yellow, sulphurous yellow) and *rub·us -ra -rum* (red) and should likely indicate that this species is, in Entz' sense, either yellow or red. *Holosticha flaborubra* Entz 1884 in Bullington (1940, p. 210) is an incorrect subsequent spelling.

The species-group names *pentasticha* and *heptasticha* – created by Kahl (1932) to name forms (populations) with five, respectively, seven dorsal kineties – are composites of the Greek numerals *pent-* (five), respectively, *hept-* (seven) and the Greek noun *stich-* (row). These two forms are, according to ICZN (1999, Article 45.6.4), subspecies which were raised to species rank by Borror (1972).

Kahl (1932, p. 570, 571; 1933) classified *Keronopsis* as subgenus of *Holosticha*; thus, the correct names in his papers are *Holosticha (Keronopsis) rubra* (Ehrenberg, 1836) Kent, 1882, *Holosticha (Keronopsis) rubra pentasticha* Kahl, 1932, and *Holosticha (Keronopsis) rubra heptasticha* Kahl, 1932 (see Berger 2001). The misleading spelling "*Keronopsis (Oxytricha, Holosticha) rubra* (Ehrb., 1838)" in Kahl (1932) should simply indicate that this species or its synonyms have already been classified in the genera *Oxytricha* and *Holosticha*. Ganapati & Rao (1958) used Kahl's subgeneric classification of *Keronopsis* and therefore they are not the combining authors for *Keronopsis*. Borror (1972) incorrectly assumed that Kahl (1932) transferred the present species to the genus *Keronopsis*; however, as already mentioned above, Kahl classified *Keronopsis* only as subgenus and thus Kahl is not the author for this combination because subgenera are not considered in this respect (ICZN 1999, Article 51.3.2). Consequently, Borror (1972) himself is the combing author with *Keronopsis*, as also indicated by Borror & Wicklow (1983, p. 123).

Ehrenberg (1835) published *Oxytricha rubra* without description, definition, or indication; thus, the name is a nomen nudum (ICZN 1999, Article 12 and p. 111). Consequently, Ehrenberg's next paper (Ehrenberg 1836), which contains a Latin diagnosis, is the original description. In my catalogue (Berger 2001) I was less rigorous, respectively, correct and therefore considered Ehrenberg (1835) as author of *O. rubra*.

Oxytricha rubra Buitkamp, 1977a is a junior primary homonym of *O. rubra* Ehrenberg, 1836. Foissner (1987d) introduced the replacement name *Oxytricha germanica* for Buitkamp's species, which is a junior synonym of *Cyrtohymena muscorum* (for details see Berger 1999, 2001).

Uhlig (1981a, b) described the albino strain as variety, which is thus of infrasubspecific rank (ICZN 1964, Articles 45d (iii), 45e (ii)). Since infrasubspecific taxa are excluded from the species group, the provisions of the Code do not apply (ICZN 1964, Articles 1, 45b). Wirnsberger et al. (1987, p. 86) proposed the following designation for this unpigmented descendant of the neotype material (see occurrence): *Pseudokeronopsis rubra*, "albino" strain.

To solve the great taxonomic problems (see remarks) of the *Pseudokeronopsis rubra* group, Wirnsberger et al. (1987) fixed a neotype each for *P. rubra*, *P. carnea*, and *P. flava*. They studied three populations of *P. rubra*, including the albino strain. Unfortunately, they failed to declare only one of these three populations as neotype, which would have been important, inasmuch as they are from two different locations. Since the place of origin of the neotype becomes the type locality of the nominal species-group (ICZN 1999, Article 76.3), the new type locality of *P. rubra* was not fixed so far. I fix here, in agreement with E. Aescht (= E. Wirnsberger), the population from South Africa as neotype because the slides deposited in the museum and the specimen shown in Figures 2 and 3 by Wirnsberger et al. (1987) (= Fig. 179b, c of the present paper) are from this site.

Remarks: The systematics of the *Pseudokeronopsis rubra* group (*P. rubra*, *P. pararubra*, *P. carnea*, *P. flava*, *P. flavicans*) is rather difficult and not yet sufficiently solved. For simplicity, I follow Wirnsberger et al. (1987), who recognised all taxa as valid species (*P. flavicans* with doubt; for taxonomy of the recently described *P. pararubra* see below) although the differences are not very distinct. Further populations have to be studied to show whether or not the morphometric differences are stable.

The description provided by Ehrenberg (1836, 1838) is understandably rather incomplete. However, the main characteristics (translated: "band-like, often twisted, ventrally flattened, a conspicuous furrow along the middle of the cell, more or less brick-red, shadowing the water reddish, dying cells yellow-red") match the observations by Wirnsberger et al. (1987). Ehrenberg mentioned two large "spherical glands". Usually these glands are the macronuclear nodules in his papers. This would mean that the multimacronucleata populations identified subsequently as *O. rubra* belong to a different species. However, I agree with Wirnsberger et al. (1987) that this problem must not be over-interpreted. Considering the other well matching characteristics, it may be justified to assume that Ehrenberg misinterpreted some food vacuoles as macronuclear nodules. Ehrenberg (1838) briefly mentioned that *Trichoda patens* Müller, 1786 is possibly identical with his *Oxytricha rubra*. However, Müller (1786, p. 181) did not describe the highly characteristic red colour of Ehrenberg's species, indicating the *T. patens* and *O. rubra* are not the same species.

Dujardin (1841) provided a brief description (with original data?) and one illustration which basically fits Ehrenberg's data. Diesing (1850), who provided no own data, doubted Dujardin's identification, likely because his population was obviously from freshwater. However, according to Stein (1859, p. 182) and Fresenius (1865), Dujardin's population was from the Mediterranean Sea. In spite of this, Stein and Fresenius

Table 37 Morphometric data on *Pseudokeronopsis carnea* (ca1, Heligoland, from Wirnsberger et al. 1987; ca2, Denmark, neotype population from Wirnsberger et al. 1987; ca3, Mediterranean, from Foissner 1984), *P. flava* (fl1, neotype population from Wirnsberger et al. 1987; fl2, Chinese population from Song et al. 2004b), *P. flavicans* (fc1, fc2, two populations from Song et al. 2002; one of them is the neotype population), *P. pararubra* (pa1, population from Japan; pa2, type population; from Hu et al. 2004a), *P. rubra* (ru1, South Africa, neotype population from Wirnsberger et al. 1987; ru2, Indopacific region, from Wirnsberger et al. 1987; ru3, from Hu & Song 2001), *P. rubra* albino strain (rua, from Wirnsberger et al. 1987), *P. sepetibensis* (sep, from Wanick & Silva-Neto 2004)

Characteristics [a]	Species	mean	M	SD	SE	CV	Min	Max	n
Body, length	ca1	152.3	151.0	12.9	3.3	8.5	128.0	177.0	15
	ca2	163.8	167.0	19.3	4.9	11.8	126.0	191.0	15
	ca3	171.0	170.0	24.0	6.2	14.0	140.0	225.0	15
	fc1	204.7	–	14.8	6.1	7.3	180.0	220.0	9
	fc2	209.1	–	21.2	8.0	8.1	193.0	238.0	14
	fl1	173.1	174.0	21.9	5.6	12.2	121.0	209.0	15
	fl2	175.5	–	27.1	6.9	15.5	144.0	246.0	16
	pa1	215.6	–	24.7	4.9	11.4	174.0	260.0	25
	pa2	288.4	–	28.7	7.4	9.9	238.0	336.0	15
	ru1	171.8	173.0	12.5	3.2	7.3	140.0	189.0	15
	ru2	189.5	185.0	17.1	4.4	9.4	159.0	220.0	15
	ru3	180.7	–	37.5	8.2	20.8	124.0	264.0	21
	rua	177.7	172.0	25.7	6.6	14.5	141.0	231.0	15
	sep	120.5	120.0	11.9	2.6	9.9	200.0	140.0	20
Body, width	ca1	38.1	38.0	5.0	1.3	13.2	28.0	46.0	15
	ca2	28.3	28.0	5.3	1.3	18.8	22.0	38.0	15
	ca3	28.6	28.0	3.4	0.9	11.8	22.0	35.0	15
	fc1	44.0	–	2.8	1.2	6.4	40.0	48.0	9
	fc2	60.1	–	4.7	1.8	6.2	56.0	65.0	14
	fl1	34.3	33.0	5.9	1.5	17.3	25.0	44.0	15
	fl2	51.8	–	8.7	2.2	16.8	48.0	72.0	16
	pa1	55.6	–	7.1	1.4	12.8	41.0	66.0	25
	pa2	113.6	–	20.8	5.4	18.3	78.0	152.0	15
	ru1	30.5	29.0	4.4	1.1	14.4	24.0	41.0	15
	ru2	50.4	51.0	8.4	2.2	16.7	37.0	63.0	15
	ru3	50.8	–	8.2	1.9	16.0	32.0	64.0	19
	rua	50.5	51.0	6.2	1.6	12.3	41.0	59.0	15
	sep	20.9	20.0	1.6	0.3	7.6	20.0	26.0	20
Anterior body end to proximal end of adoral zone, distance	ca1	48.9	49.0	4.5	1.1	9.1	42.0	59.0	15
	ca2	39.0	40.0	5.7	1.5	14.5	28.0	50.0	15
	ca3	51.1	50.0	4.3	1.1	8.4	43.0	59.0	15
	fc1	64.7	–	4.5	1.8	7.0	60.0	67.0	9
	fc2	65.1	–	7.4	2.8	9.0	55.0	72.0	14
	fl1	47.0	47.0	5.0	1.3	10.5	38.0	55.0	15
	fl2	55.8	–	5.3	1.3	9.6	46.0	65.0	16
	pa1	82.6	–	5.6	1.1	6.8	70.0	91.0	25
	pa2	95.8	–	4.7	1.2	4.9	88.0	104.0	15
	ru1	52.0	52.0	2.9	0.7	5.6	48.0	57.0	15
	ru2	62.7	61.0	6.2	1.6	9.9	57.0	80.0	15
	ru3 [e]	62.2	–	7.2	1.6	11.6	51.0	76.0	20
	rua	58.3	59.0	7.6	2.0	13.0	47.0	77.0	15
Adoral membranelles, width of largest membranelle	ca1	8.3	8.0	1.2	0.3	14.1	7.0	10.0	15
	ca2	8.1	8.0	1.3	0.3	16.0	7.0	10.0	15
	ca3	6.7	7.0	0.6	0.1	8.8	6.0	8.0	15

Table 37 Continued

Characteristics [a]	Species	mean	M	SD	SE	CV	Min	Max	n
Adoral membranelles, width of largest membranelle	fl1	7.8	8.0	1.0	0.3	12.9	6.0	10.0	15
	ru1	6.5	6.0	0.9	0.2	14.1	5.0	8.0	15
	ru2	9.1	9.0	0.8	0.2	9.1	7.0	10.0	15
	rua	7.0	7.0	1.0	0.3	14.3	5.0	9.0	15
Paroral, length	ca1	16.4	16.0	2.3	0.6	13.8	14.0	21.0	15
	ca2	15.1	15.0	3.3	0.8	21.6	10.0	21.0	15
	ca3	24.0	24.0	3.2	0.8	13.3	18.8	29.0	15
	fl1	15.3	15.0	2.2	0.6	14.5	11.0	19.0	15
	ru1	17.7	18.0	2.4	0.6	13.5	13.0	22.0	15
	ru2	21.2	21.0	4.0	1.0	18.7	14.0	29.0	15
	rua	17.6	17.0	2.2	0.6	12.7	14.0	21.0	15
Anterior body end to first midventral cirrus, distance	ca1	13.4	14.0	2.6	0.7	19.2	9.0	17.0	15
	ca2	12.9	13.0	2.7	0.7	20.8	9.0	20.0	15
	ca3	17.9	18.0	2.4	0.6	13.4	14.0	21.0	15
	fl1	15.3	15.0	2.1	0.5	13.4	11.0	19.0	15
	ru1	21.7	22.0	5.5	1.4	25.5	16.0	40.0	15
	ru2	24.4	22.0	4.4	1.1	18.2	18.0	30.0	15
	rua	18.0	18.0	3.9	1.0	21.6	10.0	24.0	15
Last midventral cirrus to anterior-most transverse cirrus, distance	ca1	8.9	9.0	2.3	0.6	25.5	5.0	13.0	15
	ca2	17.5	15.0	4.6	1.2	26.1	11.0	26.0	15
	ca3	26.3	25.0	5.1	1.3	19.5	18.0	39.0	15
	fl1	30.3	30.0	6.2	1.6	20.5	22.0	42.0	15
	ru1	1.5	1.0	1.3	0.3	85.0	0.0	4.0	15
	ru2	2.5	2.0	1.8	0.5	74.9	0.0	7.0	15
	rua	1.2	1.0	1.5	0.4	122.5	0.0	4.0	15
Macronuclear nodule, length	ca3	3.3	3.0	0.9	0.2	27.9	1.8	5.6	15
Macronuclear nodule, width	ca3	2.1	2.0	0.4	0.1	17.3	1.5	2.8	15
Macronuclear nodules, number	fc2	–	–	–	–	–	72.0	97.0	2
	fl2	–	–	–	–	–	60.0	90.0	3
Adoral membranelles, number	ca1	51.7	51.0	3.4	0.9	6.5	46.0	58.0	15
	ca2	48.7	49.0	4.8	1.2	9.9	39.0	56.0	15
	ca3	66.4	67.0	6.3	1.6	9.5	58.0	80.0	15
	fc1	51.3	–	2.7	1.1	5.3	46.0	54.0	9
	fc2	56.1	–	5.5	2.1	9.8	50.0	66.0	7
	fl1	52.0	53.0	5.4	1.4	10.3	43.0	59.0	15
	fl2	45.4	–	2.6	0.6	5.6	43.0	51.0	16
	pa1	71.8	–	3.9	1.0	5.4	64.0	82.0	25
	pa2	77.9	–	6.5	1.7	8.4	68.0	92.0	14
	ru1	58.0	58.0	3.9	1.0	6.7	50.0	65.0	15
	ru2	78.1	77.0	7.1	1.8	9.1	66.0	92.0	15
	ru3	51.6	–	3.7	0.8	7.2	46.0	60.0	20
	rua	59.2	57.0	6.6	1.7	11.1	50.0	72.0	15
	sep	44.0	43.5	6.1	1.9	13.9	38.0	55.0	10
Bicorona, number of cirral pairs	fc1	–	–	–	–	–	5.0	7.0	5
	fc2	7.4	–	1.0	0.3	13.1	6.0	9.0	10
	fl2	5.6	–	0.7	0.2	13.2	5.0	6.0	13
Anterior corona, number of cirri	ca1	5.1	5.0	0.6	0.2	12.1	4.0	6.0	15
	ca2	6.3	6.0	0.6	0.1	9.8	6.0	8.0	15
	ca3	7.2	7.0	0.9	0.2	11.3	6.0	9.0	15
	fl1	5.3	5.0	0.6	0.2	11.7	4.0	6.0	15
	pa1 [c]	18.4	–	2.1	0.4	11.3	15.0	23.0	25

Table 37 Continued

Characteristics [a]	Species	mean	M	SD	SE	CV	Min	Max	n
Anterior corona, number of cirri	pa2 [c]	21.7	–	2.7	0.7	12.6	18.0	26.0	16
	ru1	8.1	8.0	1.0	0.3	12.8	6.0	10.0	15
	ru2	8.5	8.0	2.6	0.7	31.2	5.0	12.0	15
	ru3 [c]	12.6	–	1.1	0.2	8.4	11.0	14.0	19
	rua	10.2	10.0	1.6	0.4	15.4	7.0	12.0	15
	sep	4.1	4.0	0.3	0.1	7.7	4.0	5.0	10
Posterior corona, number of cirri	ca1	4.4	4.0	0.6	0.2	13.9	3.0	5.0	15
	ca2	5.2	5.0	0.6	0.1	10.8	4.0	6.0	15
	ca3	7.2	7.0	0.9	0.2	12.0	6.0	9.0	15
	fl1	4.3	4.0	0.6	0.2	14.2	3.0	5.0	15
	ru1	5.6	6.0	1.0	0.2	17.5	4.0	7.0	15
	ru2	7.7	8.0	2.1	0.6	27.9	5.0	13.0	15
	rua	8.6	8.0	1.4	0.4	16.3	6.0	12.0	15
	sep	4.4	4.0	0.7	0.2	15.9	4.0	6.0	10
Buccal cirri, number	fc1	1.0	–	0.0	0.0	0.0	1.0	1.0	12
	fc2	1.1	–	0.3	0.1	30.9	1.0	2.0	9
	fl2	1.0	–	0.0	0.0	0.0	1.0	1.0	7
	pa1	1.0	–	0.0	0.0	0.0	1.0	1.0	25
	pa2	1.0	–	0.0	0.0	0.0	1.0	1.0	16
	ru3	1.0	–	0.0	0.0	0.0	1.0	1.0	21
Frontoterminal cirri, number	fc1	2.0	–	0.0	0.0	0.0	2.0	2.0	9
	fc2	2.0	–	0.0	0.0	0.0	2.0	2.0	11
	fl2	2.3	–	0.5	0.1	20.1	2.0	3.0	12
	pa1	2.0	–	0.0	0.0	0.0	2.0	2.0	25
	pa2	2.0	–	0.0	0.0	0.0	2.0	2.0	16
	ru3	2.0	–	0.0	0.0	0.0	2.0	2.0	20
	sep	4.3	4.0	0.4	0.1	11.2	4.0	5.0	10
Transverse cirri, number	ca1 [b]	7.1	7.0	0.7	0.2	9.6	6.0	9.0	15
	ca2 [b]	6.6	7.0	0.7	0.2	11.2	5.0	8.0	15
	ca3 [b]	6.1	6.0	0.7	0.2	11.6	5.0	7.0	15
	fc1	4.1	–	0.5	0.2	12.6	3.0	5.0	10
	fc2	3.6	–	1.0	0.3	26.8	3.0	6.0	9
	fl1 [b]	2.0	2.0	0.4	0.1	19.0	1.0	3.0	15
	fl2	3.7	–	0.5	0.1	12.6	3.0	4.0	14
	pa1	7.6	–	0.7	0.1	8.6	7.0	9.0	24
	pa2	9.4	–	1.5	0.3	12.2	7.0	11.0	16
	ru1 [b]	7.2	7.0	0.9	0.2	13.0	6.0	9.0	15
	ru2 [b]	7.1	8.0	1.5	0.4	21.7	4.0	9.0	15
	ru3	3.1	–	0.6	0.2	19.8	2.0	4.0	16
	rua [b]	6.1	6.0	2.2	0.6	35.4	3.0	10.0	15
	sep	3.0	3.0	0.0	0.0	0.0	3.0	3.0	13
Midventral pairs, number	ca1	31.9	32.0	2.9	0.8	9.2	27.0	37.0	15
	ca2	30.3	30.0	3.7	0.9	12.1	22.0	35.0	15
	ca3	35.0	35.0	4.2	1.1	12.1	29.0	44.0	15
	fc1	28.3	–	2.4	0.8	8.3	25.0	32.0	7
	fc2	33.4	–	4.4	1.5	13.1	29.0	40.0	9
	fl1	32.3	33.0	3.0	0.8	9.4	28.0	38.0	15
	fl2	29.6	–	3.7	0.9	12.3	24.0	36.0	16
	pa1 [c]	74.1	–	7.0	1.4	9.5	62.0	91.0	24
	pa2	79.0	–	9.4	2.4	11.9	65.0	93.0	16
	ru1	38.0	38.0	3.5	0.9	9.2	30.0	43.0	15

Table 37 Continued

Characteristics [a]	Species	mean	M	SD	SE	CV	Min	Max	n
Midventral pairs, number	ru2	36.7	37.0	5.2	1.3	14.2	28.0	47.0	15
	ru3 [c]	61.6	–	8.3	2.1	13.4	49.0	77.0	16
	rua	37.5	36.0	5.7	1.5	15.3	28.0	48.0	15
	sep	43.8	41.5	7.9	2.5	18.1	35.0	57.0	10
Pretransverse ventral cirri, number	ca1	3.0	3.0	0.5	0.1	17.3	2.0	4.0	15
	ca2	3.1	3.0	0.3	0.1	11.3	3.0	4.0	15
	ca3	1.9	2.0	0.3	0.1	18.8	1.0	2.0	15
	fl1	1.7	2.0	0.7	0.2	41.2	1.0	3.0	15
	ru1	3.5	3.0	1.5	0.4	43.2	1.0	7.0	15
	ru2	3.1	3.0	0.7	0.2	21.9	2.0	4.0	15
	rua	6.6	6.0	3.8	1.0	57.3	2.0	16.0	15
Right marginal cirri, number	ca1	59.3	59.0	5.9	1.5	10.0	52.0	71.0	15
	ca2	58.8	58.0	6.3	1.6	10.7	43.0	70.0	15
	ca3	59.1	57.0	6.2	1.6	10.4	53.0	74.0	15
	fc1	–	–	–	–	–	44.0	57.0	5
	fc2	60.2	–	3.2	1.3	5.3	56.0	65.0	8
	fl1	65.0	63.0	9.0	2.3	13.8	49.0	76.0	15
	fl2	52.6	–	5.6	1.4	10.7	43.0	60.0	16
	pa1	59.5	–	6.0	1.2	10.1	46.0	73.0	24
	pa2	69.6	–	6.7	1.8	9.6	59.0	80.0	14
	ru1	79.0	76.0	9.1	2.3	11.5	59.0	95.0	15
	ru2	63.6	67.0	7.8	2.0	12.2	45.0	73.0	15
	ru3	57.9	–	4.6	1.1	8.0	48.0	67.0	17
	rua	72.8	75.0	11.2	2.9	15.4	51.0	88.0	15
	sep	49.3	50.0	9.3	3.0	19.0	29.0	65.0	10
Left marginal cirri, number	ca1	56.4	56.0	3.9	1.0	6.9	49.0	62.0	15
	ca2	63.9	62.0	8.4	2.2	13.1	47.0	82.0	15
	ca3	61.5	61.0	6.7	1.7	10.9	53.0	68.0	15
	fc1	–	–	–	–	–	40.0	51.0	5
	fc2	52.4	–	3.1	1.4	5.8	50.0	57.0	8
	fl1	56.1	55.0	7.2	1.8	12.8	41.0	72.0	15
	fl2	47.8	–	5.0	1.3	10.4	41.0	57.0	16
	pa1	56.9	–	5.4	1.1	9.5	48.0	68.0	24
	pa2	68.5	–	6.7	1.7	9.7	57.0	79.0	15
	ru1	72.0	72.0	7.1	1.8	9.8	57.0	82.0	15
	ru2	62.2	61.0	6.4	1.6	10.2	48.0	71.0	15
	ru3	52.7	–	5.2	1.3	9.9	45.0	62.0	17
	rua	75.9	74.0	10.9	2.8	14.4	57.0	99.0	15
	sep	48.1	48.0	9.4	3.0	19.6	27.0	66.0	10
Dorsal kineties, number	ca1	6.1	6.0	–	–	–	5.0	7.0	15
	ca2	5.2	5.0	–	–	–	5.0	6.0	15
	ca3	5.0	5.0	–	–	–	5.0	5.0	15
	fc1	4.1	–	0.3	0.1	7.7	4.0	5.0	10
	fc2	4.1	–	0.4	0.1	8.8	4.0	5.0	14
	fl1	3.8	4.0	–	–	–	3.0	5.0	15
	fl2	3.0	–	0.0	0.0	0.0	3.0	3.0	16
	pa1	6.6	–	0.7	0.1	9.9	6.0	8.0	24
	pa2	6.4	–	0.6	0.2	9.7	5.0	7.0	16
	ru1	6.3	6.0	–	–	–	5.0	8.0	15
	ru2	6.4	6.0	–	–	–	5.0	8.0	15
	ru3	5.1	–	0.7	0.2	14.0	4.0	7.0	16

Table 37 Continued

Characteristics [a]	Species	mean	M	SD	SE	CV	Min	Max	n
Dorsal kineties, number	rua	7.4	7.0	–	–	–	6.0	9.0	15
	sep	5.0	5.0	0.7	0.2	14.5	4.0	6.0	10

[a] All measurements in µm. Data provided by Wirnsberger et al. (1987) are based on mounted, protargol-impregnated (Foissner's method), and randomly selected specimens. Data provided by Song and co-workers are based on specimens impregnated with Wilbert's protargol method. CV = coefficient of variation in %, M = median, Max = maximum value, mean = arithmetic mean, Min = minimum value, n = number of individuals investigated, SD = standard deviation, SE = standard error of arithmetic mean. Wirnsberger et al. (1987) provided three further characteristics, namely, "sum of marginal cirri", "sum of frontomidventral-transverse cirri", and "number of total cirri".

[b] Accessory cirrus included (see corresponding illustrations which show ventral infraciliature).

[c] Cirri of posterior corona included.

[d] Total number of midventral cirri. The mean of 61.6 in ru3 indicates that about 32 midventral pairs are present on average.

[e] Designated as buccal field length by Hu & Song (2001). I assume that they meant length of adoral zone of membranelles because the average value is about 34% of body length, a value much too high for the buccal field (buccal cavity).

were not convinced that Dujardin observed an *Oxytricha rubra*, mainly because his specimen was narrowed slightly tail-like.

Cohn (1866) mentioned four or six rows composed of scarlet-red granules on the ventral side. Two of these are arranged left of the midline (Fig. 178c, d). Wirnsberger et al. (1987) therefore supposed that Cohn had observed a *Trichototaxis* species. I suspect that the outermost of the two rows left of the midline is a dorsal kinety. Thus, I accept Cohn's identification.

Claparède & Lachmann (1858, p. 150) suggested that the red colour of *O. rubra* stems from red algae and thus synonymised it with *O. caudata* Ehrenberg, 1833. However, this limnetic species is very likely an 18-cirri oxytrichid and was transferred to *Urosoma* by Berger (1999, p. 398).

Kent (1882) transferred *Oxytricha rubra* to *Holosticha* (overlooked by Berger 2001) because he did not agree with the proposal by Claparède & Lachmann (1858) and Stein (1859) to assign it to *Uroleptus*. Kent used the illustration of *P. rubra* provided by Cohn (1866), whose identification was considered as incorrect by Wirnsberger et al. (1987; see above). By contrast, I accept Kent's description of *P. rubra* and therefore mention his revision in the list of synonyms.

Entz (1884) synonymised Cohn's taxa *flava* and *carnea* with *P. rubra* because he did not consider the different colours and the other differences (e.g., contractility) described by the early workers as sufficient for species separation. Unfortunately, he unnecessarily introduced the new species-group name *flavorubra* for this collective species instead of retaining the name *rubra* already introduced by Ehrenberg (see nomenclature). I mention the species *Holosticha flavorubra* in the list of synonyms of *P. rubra* simply because it is the oldest of the species concerned. Entz (1884) provided own data which basically agree with the observations by previous authors. He found two large macronuclear nodules in *H. flavorubra flava* and *H. flavorubra rubra*. Simultaneously, Gruber (1884a, p. 143) reliably reported a nuclear pattern composed of many nodules scattered throughout the cytoplasm for *Oxytricha flava* and its relatives *carnea* and *rubra*. Entz (1884, p. 362) assumed that Gruber studied postconjugants and he felt his observation of two macronuclear nodules confirmed by Ehrenberg's (1838) data: "2 grosse runde Drüsen" (two large globular glands), which was the term for the macronuclear nodules in Ehrenberg's time. It is highly interesting that Wang & Nie (1932, p. 353) – a paper overlooked by Wirnsberger et al. (1987) – also described a "*Holosticha rubra* (Ehrenberg) Kent 1880–1882" with only two macronuclear nodules (Fig. 178z). This record especially indicates that a bimacronucleate *P. rubra* exists. If in fact such a "binucleate *P. rubra*" can be proven by a record substantiated by data on life and protargol preparations, a new species has to be established for such a population because *Pseudokeronopsis rubra* is now defined as multimacronucleate due to the neotypification by Wirnsberger et al. (1987). Entz (1884) recorded both *flava* and *rubra* in the Bay of Naples, indicating species status because sympatry generally excludes subspecies status (Sudhaus & Rehfeld 1992).

Wallengren (1900) provided a rather detailed, eight pages redescription from life, "unfortunately" in Swedish. I did not translate the paper and thus refer to his illustrations. I agree with Wirnsberger et al. (1987) that Wallengren observed a true *P. rubra*. In 1902, Wallengren studied cell division and reorganisation.

Kahl (1932) summarised all colour variants among *P. rubra*. Since previous workers mentioned only three rows of pigment rosettes (that is, bristles rows), he proposed the names forma *pentasticha* and *heptasticha* for populations with five, respectively, seven dorsal kineties. However, the reinvestigation by Wirnsberger et al. (1987) yielded that *P. rubra* has on average six dorsal kineties and shows a rather high variability (min = 5, max = 8) for this feature. Thus, Kahl's separation according to the number of dorsal rows does not seem justified (Wirnsberger et al. 1987), although Borror (1972) raised both forms to species rank. Kahl (1932) obviously did not study the frontal ciliature in detail because he described and illustrated three slightly enlarged frontal cirri, which is not in accordance with the neotype population (Fig. 178q, 179b).

Bullington (1940) provided the first reliable record from North America. His description is not discussed by Wirnsberger et al. (1987). The main features (red colour, 5–6 dorsal rows) indicate that his identification is correct.

Borror (1972, p. 5) studied a late divider of *Keronopsis pulchra*. According to Wirnsberger (1987, p. 154) this is likely a *Pseudokeronopsis carnea*. I show this stage in Fig. 180y, but do not enter it in a list of synonyms because any assignment would be arbitrarily.

Borror (1979, p. 547) mentioned a high variability in the number of transverse cirri in *P. rubra*. He even observed clones lacking caudal cirri. Unfortunately, no details or illustrations are given. Wirnsberger et al. (1987) therefore assumed that Borror (1979) likely observed a *Uroleptopsis*, which is defined, inter alia, by the absence of transverse cirri (Berger 2004b).

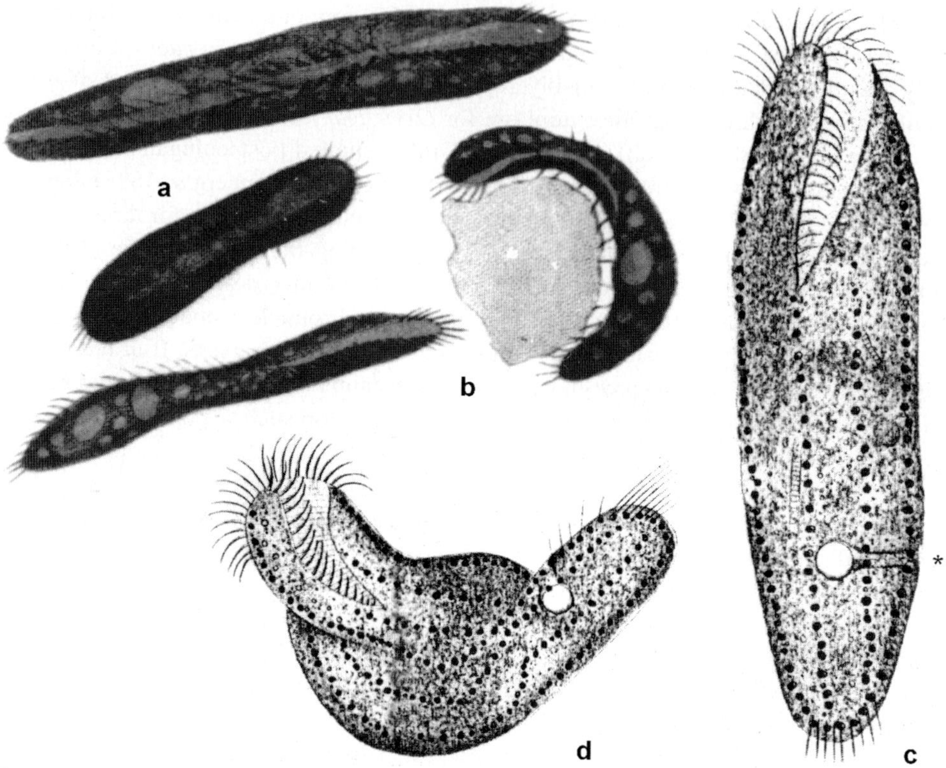

Fig. 178a–d *Pseudokeronopsis rubra* (a, b, from Ehrenberg 1838; c, d, from Cohn 1866. a–d, from life). **a, b:** Various views, 173–208 μm. **c, d:** Almost extended and contracted specimen, 200 μm. Asterisk likely marks the cytopyge. Page 890.

In 1981, Uhlig established the variety *Keronopsis rubra* var. *albino* (for nomenclature see previous chapter). It occurred in the clone of the neotype material from South Africa. For a brief description, see below (Fig. 179f–h, Table 37).

Borror & Wicklow (1983) illustrated one specimen whose cirral pattern does not fit the neotype population of *P. rubra* very well (Fig. 178x). Possibly it is a *P. carnea* as indicated by the low number of frontal cirri (9), but rather high number (5) of transverse cirri. However, since no details about the colour are provided I mention it under *P. rubra*.

Several authors did not consider *P. carnea* and *P. flava* as valid species, but as variants of *P. rubra* (e.g., Kahl 1932, Ganapati & Rao 1958, Borror & Wicklow 1983). By contrast, I follow Wirnsberger et al. (1987), who studied all three species and could find

---

Fig. 178e–i *Pseudokeronopsis rubra* (e, from Dujardin 1841; f, g, from Fresenius 1865; h, i, from Entz 1884. e–i, from life). Ventral and dorsal views showing, inter alia, body shape, cirral pattern (details must not be over-interpreted!), and cortical granulation, e = 200 μm, f, g = 140–200 μm, h, i = 300–400 μm. AZM = adoral zone of membranelles, CV = contractile vacuole, LMR = left marginal row, RMR = right marginal row. Page 890. →

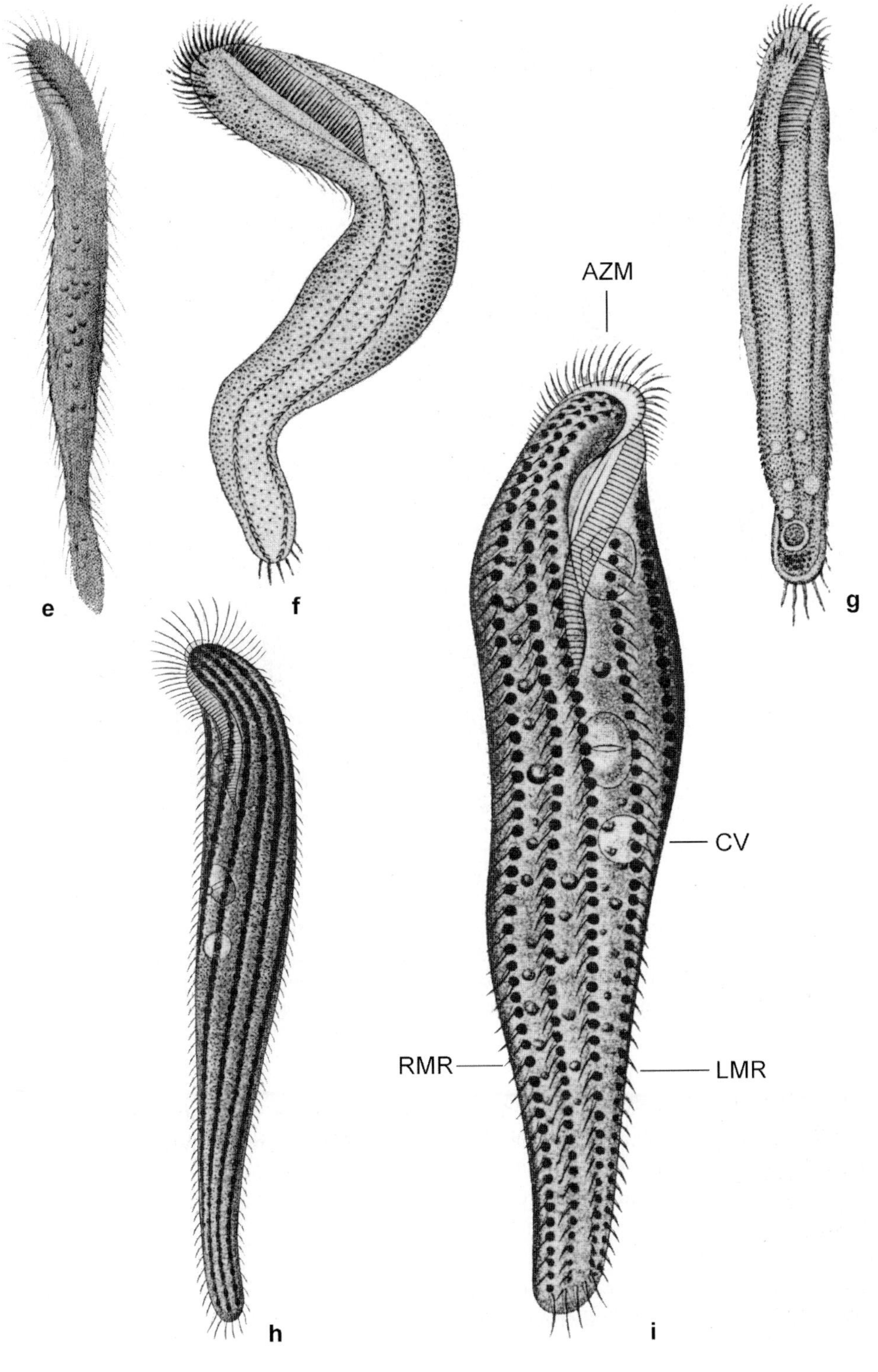
AZM
CV
RMR
LMR
e
f
g
h
i

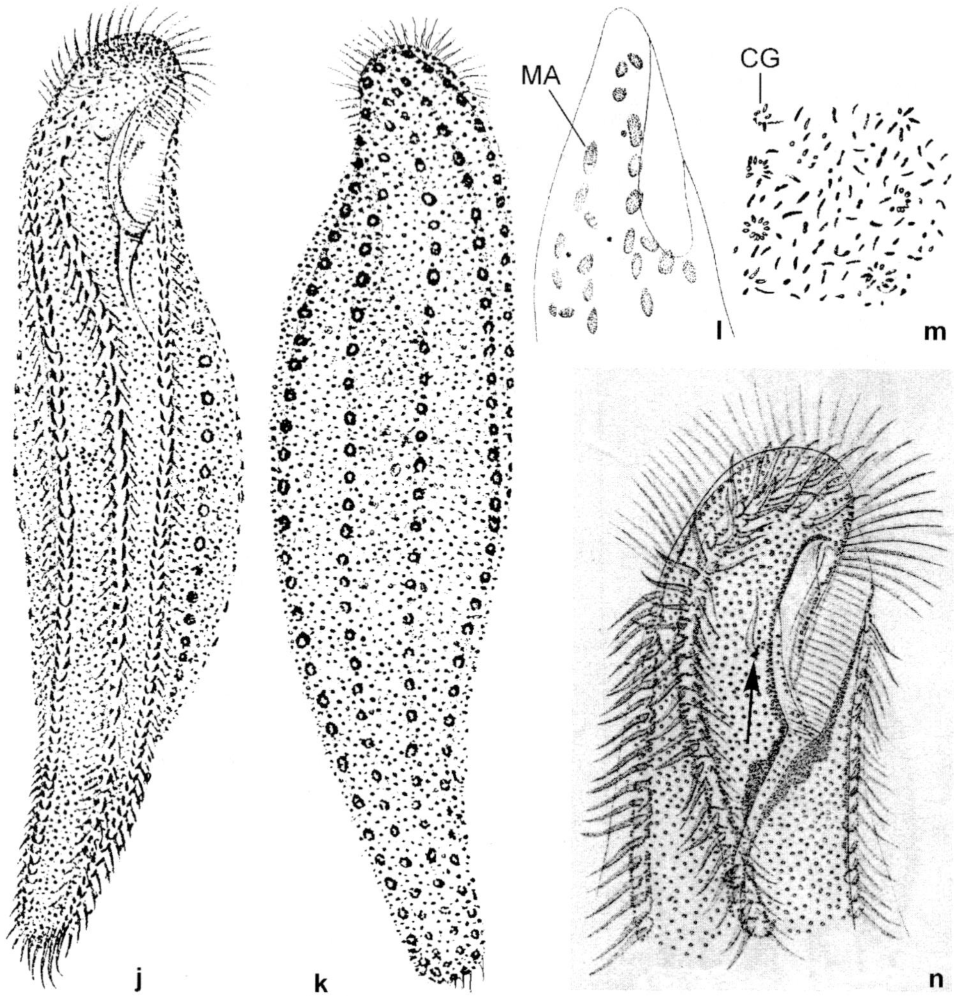

Fig. 178j–n *Pseudokeronopsis rubra* (from Wallengren 1900. j, k, m, n, from life; l, haemalaun stain). **j, k:** Ventral and dorsal view showing cirral pattern and cortical granulation, size obviously not indicated. **l:** Anterior body portion with part of nuclear apparatus. **m:** Cortical granules around dorsal bristles. **n:** Ciliature of anterior body portion. CG = cortical granules, MA = macronuclear nodule. Page 890.

some differences. Possibly, a more detailed morphometric analysis than that provided by Wirnsberger et al. (1987) will separate them still more clearly. Perhaps the location of the buccal cirrus (more anteriorly in *P. rubra* than in *P. carnea* and *P. flava*) and the beginning of the right marginal row (distinctly behind distal end of adoral zone in *P. rubra* and *P. carnea* against very close to distal end of adoral zone in *P. flava*) are such key characteristics. However, there is no doubt that some redescriptions cannot be unequivocally assigned to one of these taxa because in almost all cases some features are insufficiently known. Wirnsberger et al. (1987) fixed, as already mentioned in the nomenclature section, a neotype for *P. rubra*. Thus, their description is authoritative and given entirely.

Hu & Song (2001) studied the morphology and morphogenesis of a Chinese population. Their specimens have, like *P. flavicans* (Fig. 185l) and *Uroleptopsis citrina* (Fig. 192e, f, i, j 193b, c), blood cell-shaped organelles (mitochondria?) in the cortex (Fig. 180c). Previously, such conspicuous structures were described for *P. rubra* only by Prowazek (1900, p. 288, 289); for details on Prowazek's data, see additional observations. Not even Wirnsberger & Hausmann (1988b), in their fine-structural study on *P. carnea*, mentioned organelles of such a curious shape. I know these blood-cell-shaped organelles from *Uroleptopsis citrina*, which strongly indicates that *Uroleptopsis* is the sister group of *Pseudokeronopsis*. However, in the current case, the question is whether or not Hu & Song's population is conspecific with *P. rubra* sensu Wirnsberger et al. (1987). There are several other features (number of transverse cirri and dorsal kineties; contractility) which do not agree very well with the data of the neotype material (3.1 vs. 7.2 transverse cirri on average; 5.1 vs. 6.3 dorsal kineties; distinctly contractile vs. only slightly). For the sake of simplicity I accept Hu & Song's identification, but keep the data separate; possibly, their population belongs to *P. flavicans*.

According to the review by Wirnsberger et al. (1987), some of the *Pseudokeronopsis* populations described so far have to be assigned to other species (for detailed foundations see the remarks in the context of the relevant species). *Oxytricha rubra* sensu Möbius (1888; Fig. 181c–e) is very likely *P. carnea*. The same is true for *P. rubra* sensu Foissner (1984, Fig. 182a–h), as indicated by the colour, the location of the buccal cirrus, the number of dorsal kineties, and the anteriorly slightly shortened right marginal row. However, according to Wirnsberger et al. (1987, p. 86), the identity with *P. carnea* is not quite certain; possibly, Foissner's population is a kind of intermediate taxon whose definite position cannot be completely clarified. *Keronopsis rubra* sensu Ganapati & Rao (1958) is usually dusky yellow and frequently brownish, strongly indicating that it belongs to *P. carnea* too (Fig. 183a).

Holosticha rubra sensu Wang & Nie (1932) needs special consideration because it has only two macronuclear nodules (Fig. 178z); for discussion of this feature, see paragraph on Entz (1884) in the present chapter. This significant difference to the neotype, which has many macronuclear nodules, does not allow a synonymisation with *P. rubra*. For a brief characterisation of Wang & Nie's population, see the end of the *P. rubra* description.

Holosticha rubra sensu Borror (1963), which was not assigned to a valid species by Wirnsberger et al. (1987, p. 86), fits the original description of *Anteholosticha pulchra* Kahl rather well, where it is therefore listed (Fig. 90e–g). *Keronopsis rubra* sensu Jerka-Dziadosz & Janus (1972) from a freshwater habitat is synonymous with *Diaxonella pseudorubra* (Fig. 102a). The data on *K. rubra* sensu Jerka-Dziadosz & Kapuścińska-Czerska (1972) thus likely also refer to *Diaxonella*. Zhang et al. (1985a) and Pang et al. (1986) identified their *Keronopsis rubra* according to Jerka-Dziadosz & Janus (1972); consequently, their populations also belong to *Diaxonella pseudorubra* as indicated, inter alia, by only three distinct frontal cirri and three dorsal kineties.

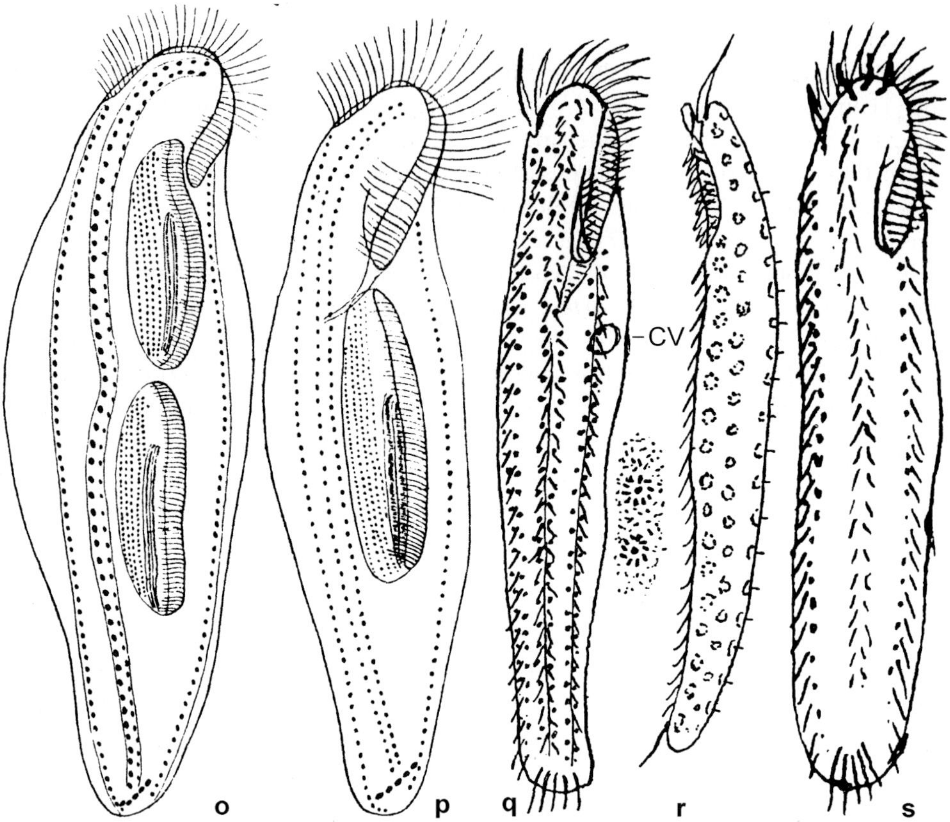

Fig. 178o–s *Pseudokeronopsis rubra* (o, p, from Wallengren 1902; q, r, from Kahl 1932; s, from Kahl 1933. From life). **o:** Late divider showing, inter alia, intact parental midventral complex and bicorona and anlagen complex of both proter and opisthe. **p:** Late reorganiser. **q:** Ventral view, 300 µm. **r:** Left lateral view. Inset between (q) and (r) shows cortical granulation around dorsal bristles. **s:** Ventral view of "*Keronopsis rubra* f. *heptasticha*", size not indicated. CV = contractile vacuole. Page 890.

Several nominal species have been synonymised with *P. rubra* so far. *Uroleptus roscovianus* Maupas, 1883 (Fig. 196a) was synonymised with *P. rubra*, inter alia, by De Morgan (1926) and Ganapati & Rao (1958). However, this species lacks transverse cirri and is therefore very likely not identical with *P. rubra* or its relatives, *P. carnea* and *P. flava*. I follow Kahl (1932), who transferred it to *Uroleptopsis*.

Oxytricha protensa Perty, 1852 (p. 153, Tafel VI, Fig. 20A–E) was synonymised with *P. rubra* by Borror (1972) and Borror & Wicklow (1983). However, Perty's species is from limnetic habitats in Switzerland and is not red, but grey or greenish; according to Berger (1999, p. 252), it is a species indeterminata.

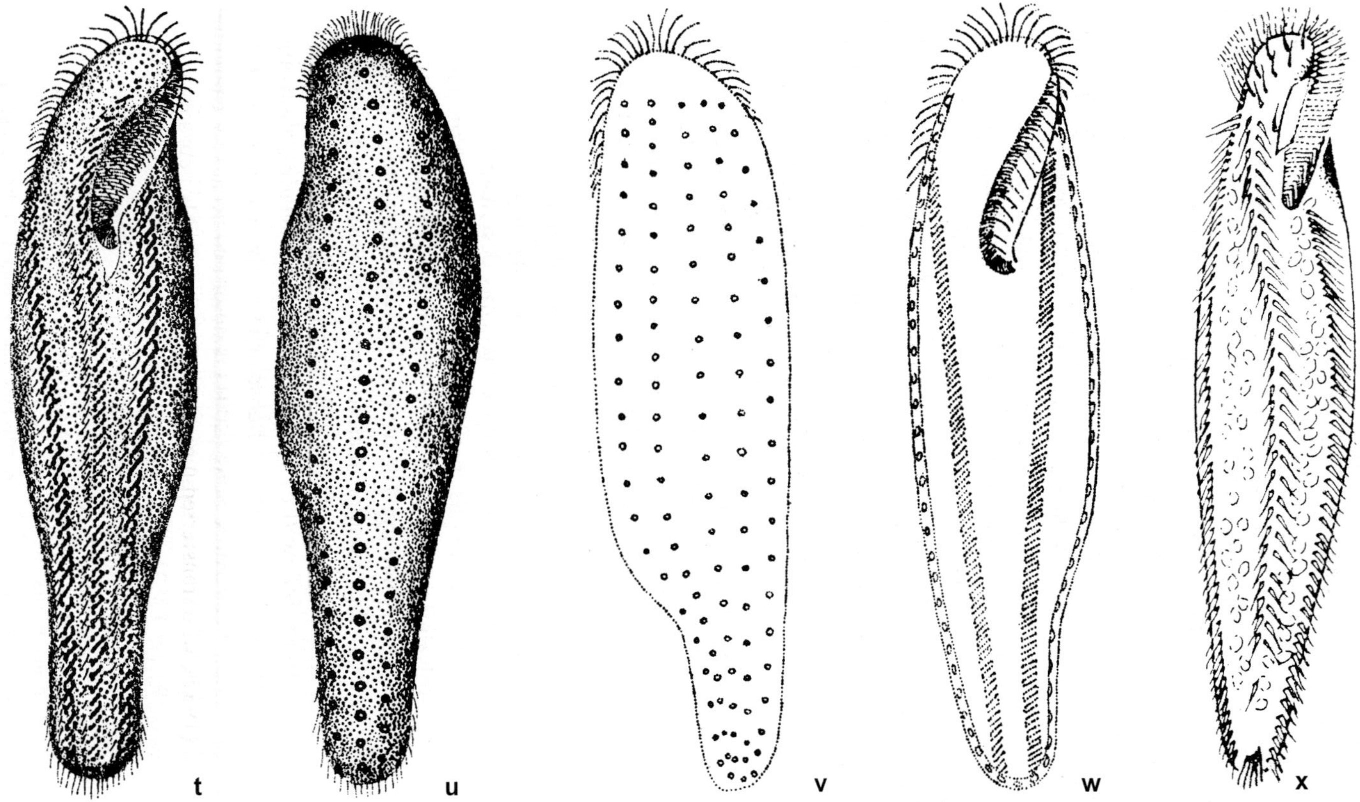

Fig. 178t–x *Pseudokeronopsis rubra* (t–w, from Bullington 1940; x, from Borror & Wicklow 1983. t–w, from life; x, protargol impregnation?). **t–w:** Ventral and dorsal views showing, inter alia, cirral pattern and cortical granulation, t, u = 142 µm, w = 146 µm. **x:** Infraciliature of ventral side and nuclear apparatus, 200 µm. The distance between the rear end of the midventral complex and the low number of cirri forming the bicorona are distinct differences to the neotype population (Fig. 179b); thus, this population possibly belongs to a different species. Page 890.

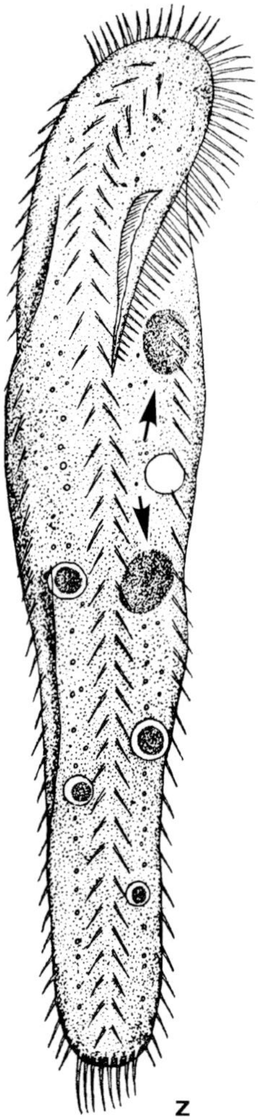

z

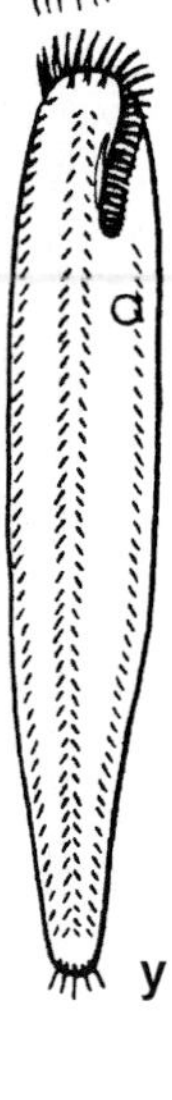

y

Oxytricha crassa Claparède & Lachmann, 1858 (Fig. 176a–h) was synonymised with *P. rubra* by Borror & Wicklow (1983). Kahl (1932) classified it in *Holosticha (Trichototaxis)*, whereas in the present book it is assigned to *Thigmokeronopsis*.

Uroleptopsis citrina Kahl, 1932 (Fig. 192a, c), type species of *Uroleptopsis*, was synonymised with *P. rubra* by Borror & Wicklow (1983). It is broadly band-shaped, yellow, has only three dorsal kineties, and lacks transverse cirri. I consider *U. citrina* as valid species.

The following *P. rubra* populations are superficially described or deviate significantly in some important features (e.g., size, habitat) from the relevant descriptions. Thus, they are classified as insufficient redescriptions although it cannot be excluded that the authors in fact observed the present species. *Keronopsis rubra* sensu Agamaliev (1974, 1983; Fig. 191e, f) from the Caspian Sea has three distinct frontal cirri and therefore cannot be identical with *P. rubra* which has a bicorona. According to Wirnsberger et al. (1987) it could be a new *Holosticha* species; likely it is identical with *Bakuella agamalievi*. *Holosticha rubra* sensu Alzamora (1929, Fig. 191c) is only 100 μm long, which is only about 50% of the minimum value usually reported. *Keronopsis rubra* sensu Chardez (1986, Fig. 191d) is from a freshwater habitat near Lüttich, Belgium, and thus cannot be identical with *P. rubra*.

Beside the brick-red (dark-red) colour, some morphometric features (number of dorsal kineties and transverse cirri) separate this species from the most similar species, *P. carnea* and *P. flava*. For separation from the recently described *P. pararubra*, see remarks of this species. If you are uncertain about the identification, write *Pseudokeronopsis rubra* group.

Morphology: The list of synonyms shows that a lot of data about this common marine hypotrich exists. At first the two populations described by Wirnsberger et al. (1987) are characterised, followed by some additional or deviating observations by previous workers. Subsequently, the Chinese population studied by Hu & Song (2001) is described because it cannot be excluded that it belongs to a different, possibly new species.

Fig. 178y, z *Pseudokeronopsis rubra* (y, from Kiesselbach 1936a; z, from Wang & Nie 1932. From life). **y:** Ventral view, 240 × 40 μm. The frontal ciliature is likely not correctly observed. **z:** Note that this specimen/population (body length 255 μm) from the Bay of Amoy (China) has only two macronuclear nodules (arrows). For discussion see remarks at *P. rubra*. Page 890.

Populations described by Wirnsberger et al. (1987): Body size 170–290 × 30 to 70 µm in life, body length:width ratio 3.8:1 (Danish population) to 5.6:1 (neotype population) on average after protargol impregnation (Table 37). Body slender, right margin straight to slightly concave, left one weakly convex (Fig. 179a). Posterior portion often spoon-like. Up to 2:1 flattened dorso-ventrally. 100–200 macronuclear nodules scattered throughout cell; individual nodules spherical to ovoid, about 4 × 3 µm in life. Several micronuclei, slightly argyrophilic, spherical, about 4 µm across in life. No contractile vacuole. Pellicle soft, flexible, crenellated along cirral rows. Cortex colourless, with underlying brick-red, 0.5–2.0 µm-sized granules scattered throughout endoplasm and grouped (only visible at high magnification) along cirral rows and dorsal cilia; about 6–10 granules form one rosette surrounding a dorsal bristle. Cytoplasm contains yellow lipid droplets about 7 µm across. Movement rather slow, dying cells flow out, granules turn orange-yellow. Conjugation was infrequently observed in Danish population, but never (1979–1986) in neotype population and albino strain.

Adoral zone of membranelles occupies 30% (neotype population) to 33% (Danish population) of body length (Table 37), basically formed like a question mark; distal end extends far onto right body margin (Fig. 179b); largest membranelles in life about 15 µm long. Buccal field moderately wide and rather deep. Paroral and endoral commence at about same level, narrowly spaced and slightly arched, paroral distinctly shorter than endoral. Bicorona composed of about eight slightly enlarged anterior cirri and six to eight posterior cirri, that is, bicorona in total composed of 14–16 cirri; associated fibrillar system conspicuous (Fig. 179d); anterior corona strongly curved leftwards. Invariably one buccal cirrus slightly behind anterior end of undulating membranes. Two frontoterminal cirri at ordinary position, that is, immediately behind distal end of adoral zone. Midventral complex visible as bright furrow at low magnification, only inconspicuously set off from bicorona, composed of distinctly zigzagging cirral pairs, extends very close to the about seven transverse cirri; midventral cirri about 10 µm long. Transverse cirri 11–18 µm long in life and distinctly projecting beyond rear body end, arranged in oblique row with some (two?) pretransverse ventral cirri ahead. Marginal cirri 10–12 µm long, all bases of nearly same size, never confluent posteriorly; right row commences distinctly behind distal end of adoral zone, at least in specimen illustrated (Fig. 179b); left marginal row commences rather far anteriorly (as compared to many other hypotrichs). Dorsal cilia about 4.5 µm long in life, distance between individual bristles of a row 8–14 µm; usually arranged in six bipolar kineties each composed of 15–22 basal body pairs. No caudal cirri (Fig. 179c).

The albino strain (mutant of neotype population) – described by Wirnsberger et al. (1987) – was found by Uhlig (1981a, b) in a culture. He thought the loss of the red component of the granules to be the result of gamma-radiation which had happened a year before. This interpretation, however, must not be considered to be the only possibility because in about 10-month cultures of the Danish population Wirnsberger et al. also found a strong, but not complete loss of the red colour. However, this is a reversible process, which may therefore be influenced by the culture conditions or food uptake, while the mutant has been colourless since 1980. Uhlig (1981a, b, 1982, 1983, 1984) noted the following differences between the red and the white strain: extensive

Table 38 Pigments of neotype populations of *Pseudokeronopsis rubra*, *Pseudokeronopsis carnea*, and *Pseudokeronopsis flava*. Absorption maxima in nm of crude extracts and fractions from preparative thin-layer chromatography (TLC) plates. The fraction numbers 1 to 4 (top and bottom of the plates respectively) indicate "main fractions" with practically identical Rf values. Numbers in parentheses indicate weak maxima. From Larsen (1987) and Wirnsberger et al. (1987)

	Pseudokeronopsis rubra	*Pseudokeronopsis carnea*	*Pseudokeronopsis flava*
Crude extract	446 (339)	453 (321)	409
Fraction 1	432 (338)	435	388
Fraction 2	444 (330)	444 (312)	414
Fraction 3	325 (440)	466 (370)	406
Fraction 4	441	456	–

loss of phototaxy, lower dry weight, volume, and surface area as well as inhibition of the biosynthesis of suspected carotinoids in the albino mutant. Concerning the ciliary pattern, the albino strain has a highly variable appearance (Wirnsberger et al. 1987). They did not find a single normal specimen. The most frequent abnormalities were: supernumerary frontal and ventral cirri appearing randomly arranged, segmentation and overlapping of marginal rows, and no or incomplete displacement of the frontoterminal cirri and the buccal cirrus (Fig. 179f, h). Disorder also occurred on the dorsal side (Fig. 179g). The macronuclear nodules often appeared dumbbell-shaped and the chromatin bodies swollen.

The extracts of the cortical granules of *P. rubra*, *P. carnea*, and *P. flava* were investigated by Wirnsberger et al. (1987; Table 38). Uhlig (1989a) provided some further details about the substance produced by *P. rubra*. Dying cultures liberate considerable amounts of a yellow-red pigment (absorption maximum at 418 nm) which is water soluble, composing about 50% of the total cellular pigmentation. This so-called "Keronopsin" induces strong effects on a wide variety of marine and freshwater flagellates and ciliates (see Lueken et al. 1989a, b), but even on other microfaunal organisms; non-flagellated and non-ciliated protists are obviously not affected. Lyophilised Keronopsin (the yield of about 10^6 cells comes to about 50 mg) is light sensitive, but heat resistant and effective for more than four years (Uhlig 1989a). Höfle et al. (1994) found usually four main components (Keronopsin A1, A2, B1, B2) in about equal parts in aqueous and aqueous-methanolic extracts.

Additional observations from other sources (note that only "important" data – e.g., size, colour – of the old papers are included): body length/size 173–208 μm (Ehrenberg 1838), 180–220 μm (Dujardin 1841), 140–200 μm (Fresenius 1865), 200 × 50 μm (Cohn 1866), 300–400 μm (Entz 1884), 200–300 μm (Kahl 1932), 240 × 40 μm (Kieselbach 1936a); 172–264 μm (mean 206 μm; n = 76) × 36–66 μm (mean = 49 μm, n = 55; Bullington 1940). Body slender, anteriorly slightly widened, posterior quarter narrowed to 66% to 50% of body width (Kahl 1932); contractile (Cohn) to very contractile, flexible, and soft (Bullington 1940). Many macronuclear nodules, usually with two micronuclei (Kahl 1932). Contractile vacuole not recognised by Ehrenberg (1838) and Fresenius (1865). Ehrenberg (1838) and Entz (1884) described the colour of

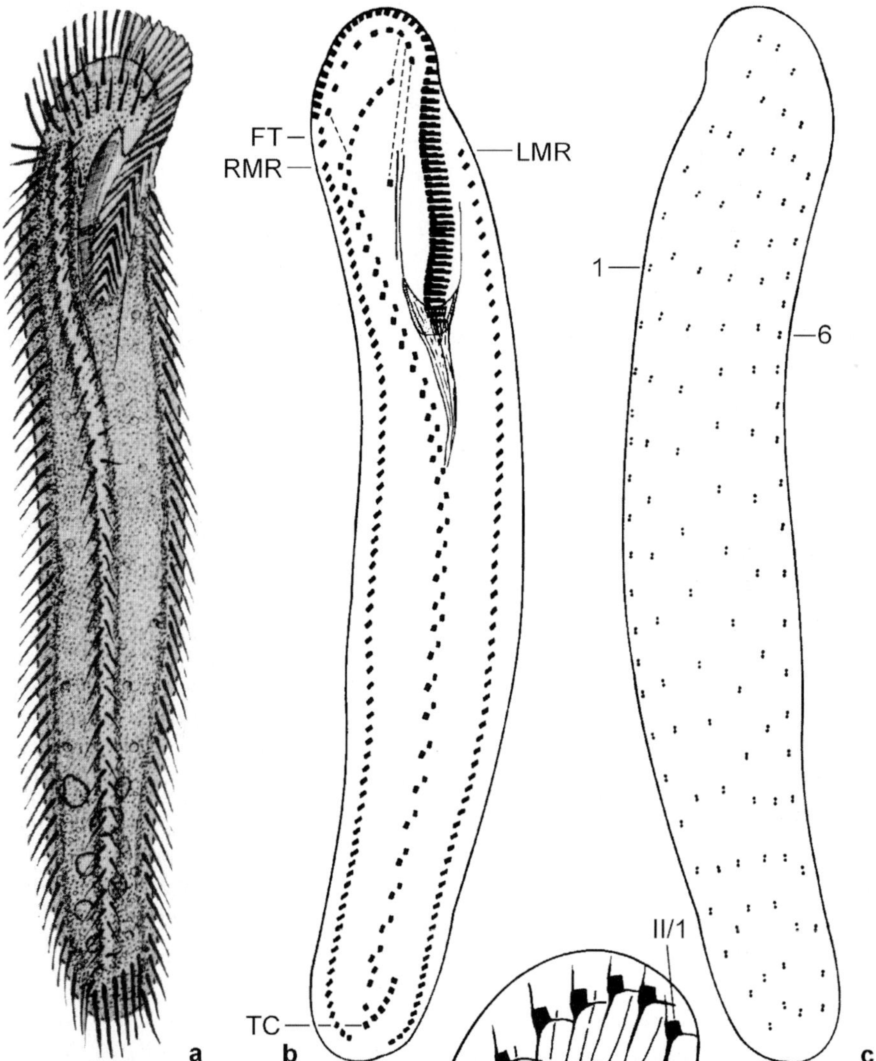

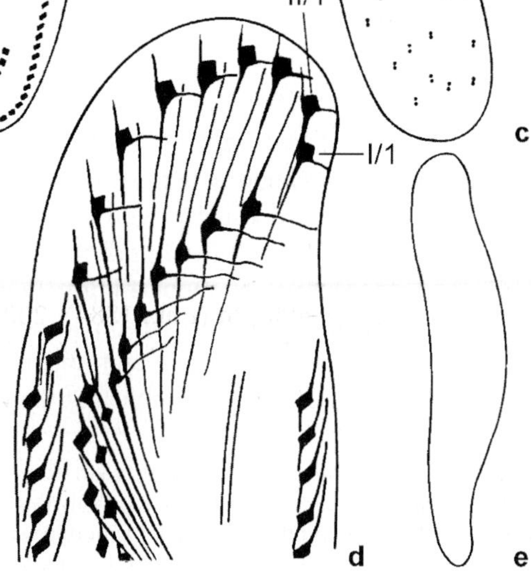

Fig. 179a–e *Pseudokeronopsis rubra* (from Wirnsberger et al. 1987. a, e, from life; b–d, protargol impregnation). **a:** Ventral view of representative specimen, 170 µm. **b, c:** Infraciliature of ventral and dorsal side of a specimen from the neotype population, 181 µm. Broken lines connect cirri, which originate from the same anlage. **d:** Fibrillar associates in the anterior region. **e:** Shape variant of neotype population. FT = frontoterminal cirri, LMR = left marginal row, RMR = right marginal row, TC = transverse cirri, I/1, II/1 = first and second frontal cirrus from left, 1, 6 = dorsal kineties. Page 890.

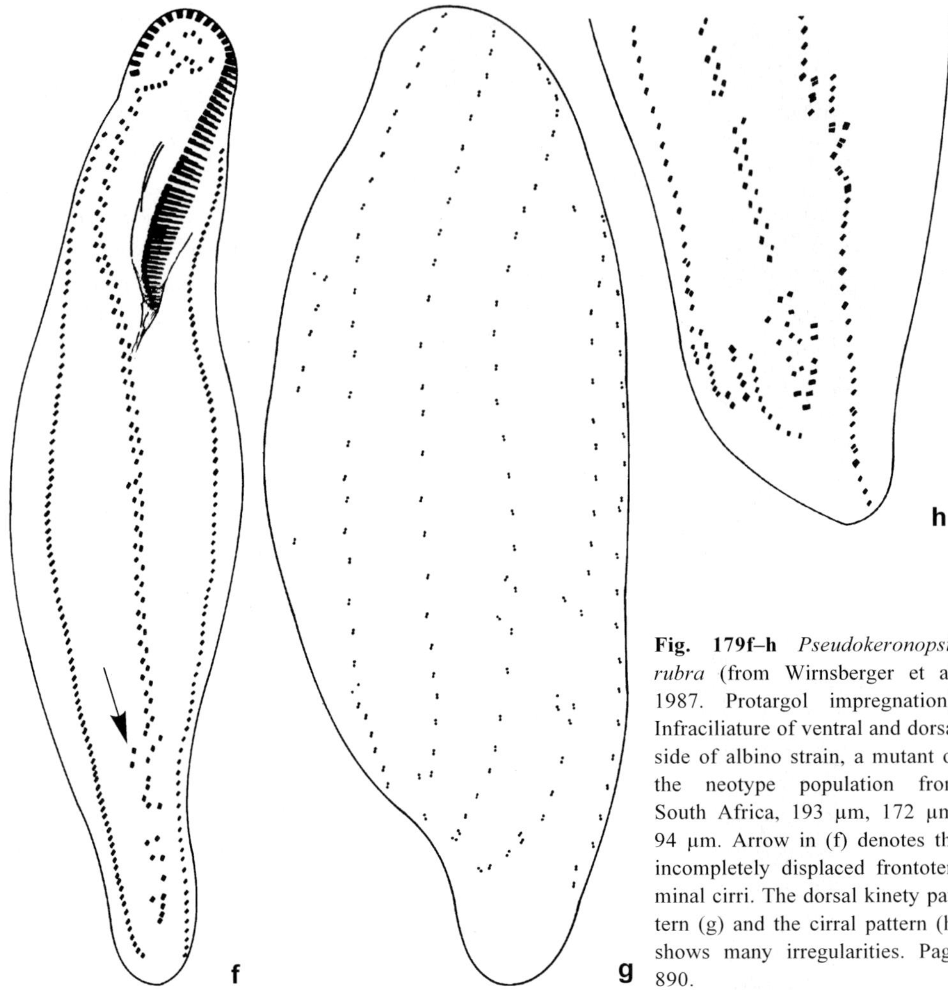

f g h

Fig. 179f–h *Pseudokeronopsis rubra* (from Wirnsberger et al. 1987. Protargol impregnation). Infraciliature of ventral and dorsal side of albino strain, a mutant of the neotype population from South Africa, 193 µm, 172 µm, 94 µm. Arrow in (f) denotes the incompletely displaced frontoterminal cirri. The dorsal kinety pattern (g) and the cirral pattern (h) shows many irregularities. Page 890.

P. rubra as brick-red[1], according to Fresenius (1865) the cytoplasm is red-brown; Bullington (1940) found a reddish-brown colour with a yellow cast when cells were squeezed; dorsal side usually with four rows of pigment patches, chromium-yellow specimens with eight well-developed rows of red patches (Entz 1884); distance between individual dorsal bristles about 14 µm (Kahl 1932). Bullington (1940) counted 5–6 rows of prominent, deep-reddish-brown spots (9.0–11.5 µm apart) usually composed of 10 granules and observed that the entire body was filled with small reddish granules, smaller than those composing the rings; a massing of these granules in the an-

[1] Under the dissecting microscope I observed few specimens of a wine-red hypotrich in a sea water sample from Atatore near Mali Lošinj, Croatia. I did not check these specimens in the microscope because I assumed that it would develop in the raw culture. Unfortunately, it disappeared very rapidly so that I am not certain about the identity of these specimens. However, in any case I was impressed from the very dark-red colour.

terior tip of the body looks like a pigment spot. Prowazek (1900) described the cortical granules as red-yellow; however, he also mentioned round or moderately oval, disc-shaped vesicles which he also described as greenish, double-outlined, fine rings in the ectoplasmatic layer; there is no doubt that this is the first report of the "blood-cell-shaped" structures later reported for *Uroleptopsis citrina* by Kahl (1932) and myself, for *Pseudokeronopsis rubra* by Hu & Song (2001, Fig. 180c), and for *P. flavicans* by Song et al. (2002; Fig. 185l). Perilemma covers, as in other species, cilia of membranelles and cirri (Bardele 1981, p. 415). Details on the oral apparatus, see Foissner & AL-Rasheid (2006). Moves, when freely swimming, in left spirals (Bullington 1940). Child (1949) reported on the intracellular indophenol reaction and intracellular oxidation and reduction of dye indicators.

Chinese population described by Hu & Song (2001, Fig. 180a–i, Table 37): Body size 160–200 × 24–40 μm in life. Body slender, distinctly contractile and very flexible; right and left margin slightly convex, anterior end bluntly rounded, rear end often spoon-like; a conspicuous furrow along midline of ventral surface (not recognisable in cross-section illustrated, Fig. 180b); dorsal side almost flat, ventral side vaulted. Numerous macronuclear nodules scattered throughout cytoplasm; individual nodules spherical to ovoid. Cortex reddish under low magnification, with underlying brick-red pigment granules (<0.5 μm in diameter) grouped along cirral rows (Fig. 180d) and dorsal kineties (Fig. 180c); about 6–10 pigment granules form a rosette around each bristle (Fig. 180c). Blood-cell-shaped structures (designated as cortical granules by Hu & Song) about 1–2 μm across obviously scattered throughout cortex (Fig. 180c). Cytoplasm relatively transparent, contains several food vacuoles. Movement slow. Conjugation was frequently observed under culture conditions. Adoral zone of membranelles occupies 34% of body length on average (Table 37), extends, almost *Gonostomum*-like, along left body margin for a considerable distance (Fig. 180e). Apical membranelles about 15 μm long. Buccal field narrow, but strongly deepened. Pharyngeal fibres conspicuous. Paroral and endoral commence about at same level, paroral, however, distinctly shorter than endoral. Bicorona composed of about seven anterior, slightly enlarged cirri and five or six posterior ones. Buccal cirrus right of optical intersection of undulating membranes. Frontoterminal cirri, as is usual, close to distal end of adoral zone. Midventral complex composed of 49–77, about 7–8 μm long cirri, that is, about 25–39 pairs; extend in inconspicuous furrow (Fig. 180e, g). Transverse cirri very near body end, about 15 μm long and thus distinctly projecting (Fig. 180a). Marginal cirri about 8 μm long, composed of two ciliary rows only; right marginal row commences slightly ahead of level of buccal cirrus, left one distinctly below this level, which is a difference to the neotype population, where both rows commence at the same level. Marginal rows distinctly separated posteriorly. Dorsal kineties bipolar, cilia about 3 μm long.

Cell division (Fig. 178o, p, 179i–q, 180j–x): This process was studied in detail by Wirnsberger (1987; Fig. 179i–p) and Hu & Song (2001; Fig. 180j–v). Individual stages from life were described by Wallengren (1902; for example, Fig. 178o).

Here, I provide the detailed description by Wirnsberger (1987), who found no differences in the overall pattern of cell division in *P. rubra*, *P. carnea*, and *P. flava*. Thus, some illustrations shown in the sequence (Fig. 178i–p) are from *P. carnea* (Fig. 178k, l, o) and all statements apply to all these species.

Stomatogenesis in the opisthe commences with the formation of small groups of basal bodies very close to several left midventral cirri (Fig. 179i). Proliferation was observed near 10–15 left cirri of midventral pairs in *P. rubra*, near 5–11 in *P. carnea*, and near 4–14 in *P. flava*. Because of the overlapping, this is no discriminating feature. The incorporation of basal bodies derived from parental left midventral cirri into the anarchic field can be completely excluded because they remain intact until nearly all primordia are built. In addition, the morphometric data on intermediate dividers clearly show that the mean number of the cirral pairs is statistically not different from those of the interphase cells (*P. rubra*, Indopacific population, mean interphase = 36.7; mean division = 32.1. *Pseudokeronopsis carnea*, Heligoland population, mean interphase =

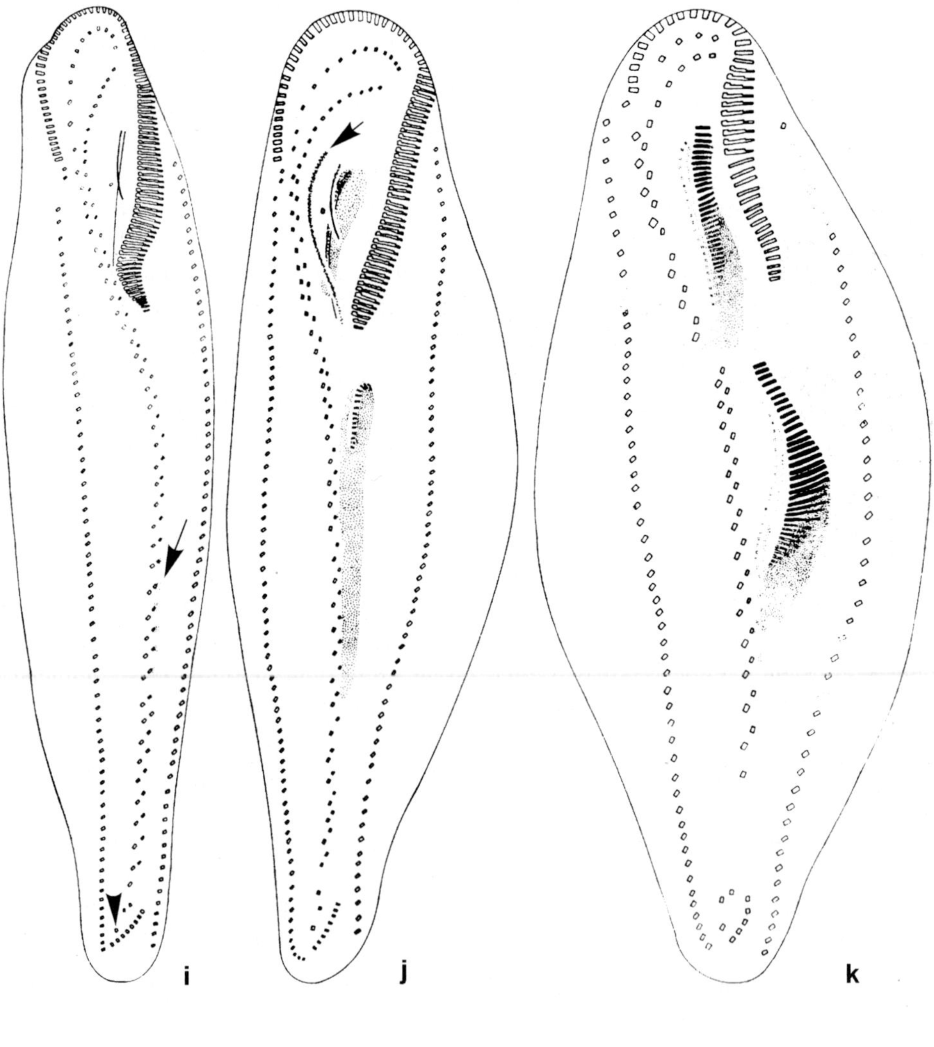

i j k

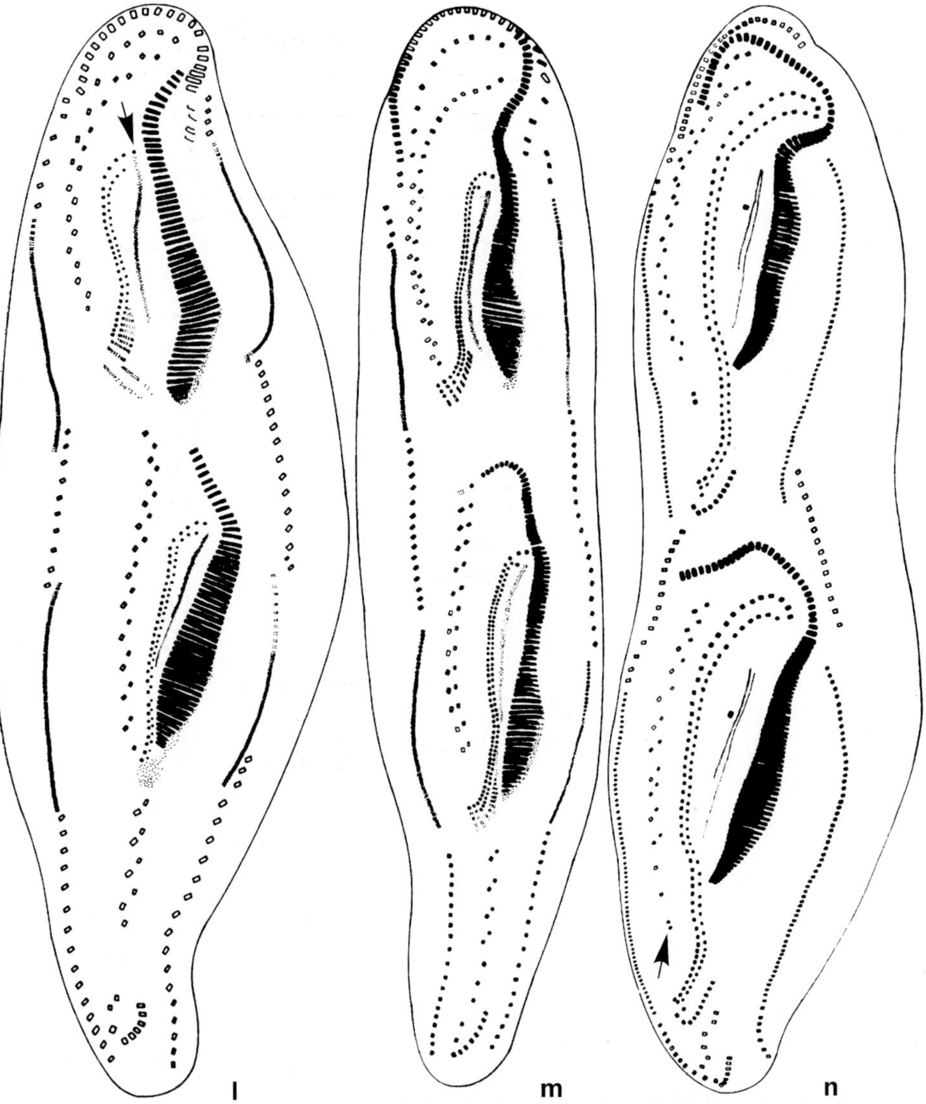

Fig. 179l–n *Pseudokeronopsis* spp. (from Wirnsberger 1987. Infraciliature of ventral side of dividers of *P. carnea* [l] and *P. rubra* [m, n] after protargol impregnation). **l:** Middle to late divider, 159 μm. Arrow marks frontal cirrus I/1 which originates from the undulating membrane anlage. **m:** Late divider, 242 μm. Note the differentiation of the cirri. **n:** Very late divider, 214 μm. Arrow marks the anteriorly migrating frontoterminal cirri. The parental adoral zone is almost completely resorbed. Page 890, 930.

← **Fig. 179i–k** *Pseudokeronopsis* spp. (from Wirnsberger 1987. Infraciliature of ventral side of dividers of *P. rubra* [i, j] and *P. carnea* [k] after protargol impregnation). **i:** Very early divider, 240 μm. Arrow marks one of several regions along the left side of the midventral complex where new basal bodies occur. Arrowhead denotes the so-called ventral cirri, that is, pretransverse ventral cirri. **j:** Early to intermediate stage, 216 μm. Arrow marks the frontal-midventral-cirral anlage of the proter. Note the splitting of the anarchic field in the opisthe. **k:** Middle divider, 147 μm. Page 890, 930.

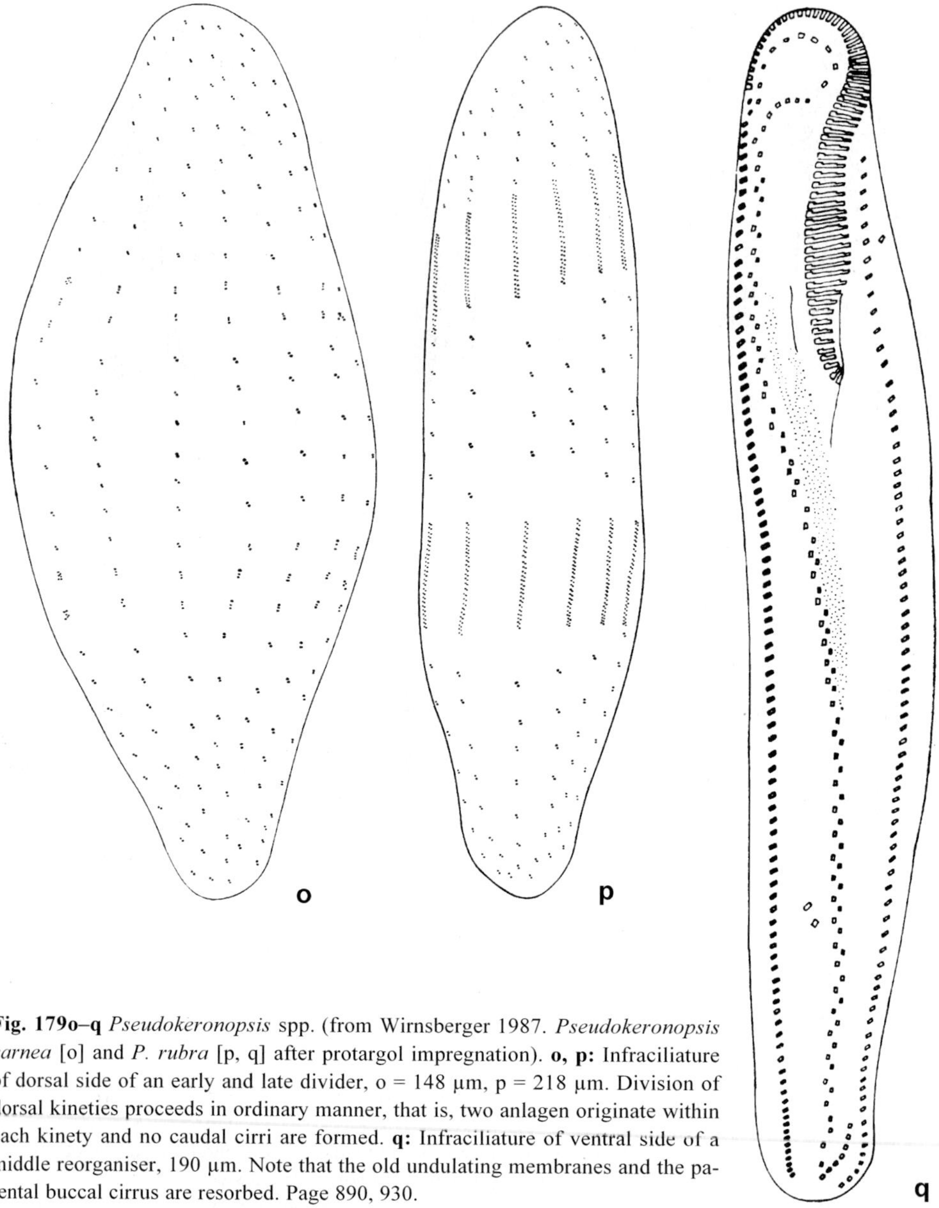

Fig. 179o–q *Pseudokeronopsis* spp. (from Wirnsberger 1987. *Pseudokeronopsis carnea* [o] and *P. rubra* [p, q] after protargol impregnation). **o, p:** Infraciliature of dorsal side of an early and late divider, o = 148 μm, p = 218 μm. Division of dorsal kineties proceeds in ordinary manner, that is, two anlagen originate within each kinety and no caudal cirri are formed. **q:** Infraciliature of ventral side of a middle reorganiser, 190 μm. Note that the old undulating membranes and the parental buccal cirrus are resorbed. Page 890, 930.

31.9; mean division = 29.2; t-test). Later, a small bare patch develops between the cirral pairs of midventral cirri and enlarges during cytokinesis (Fig. 179k, l). The basal bodies increase in number to form a longish field. The length of this primordium is about one third of the parental body length. Soon, the adoral membranelles of the opisthe organise in a posteriad direction, while the primordium for the undulating membranes separates to the right of the anarchic field (Fig. 179j).

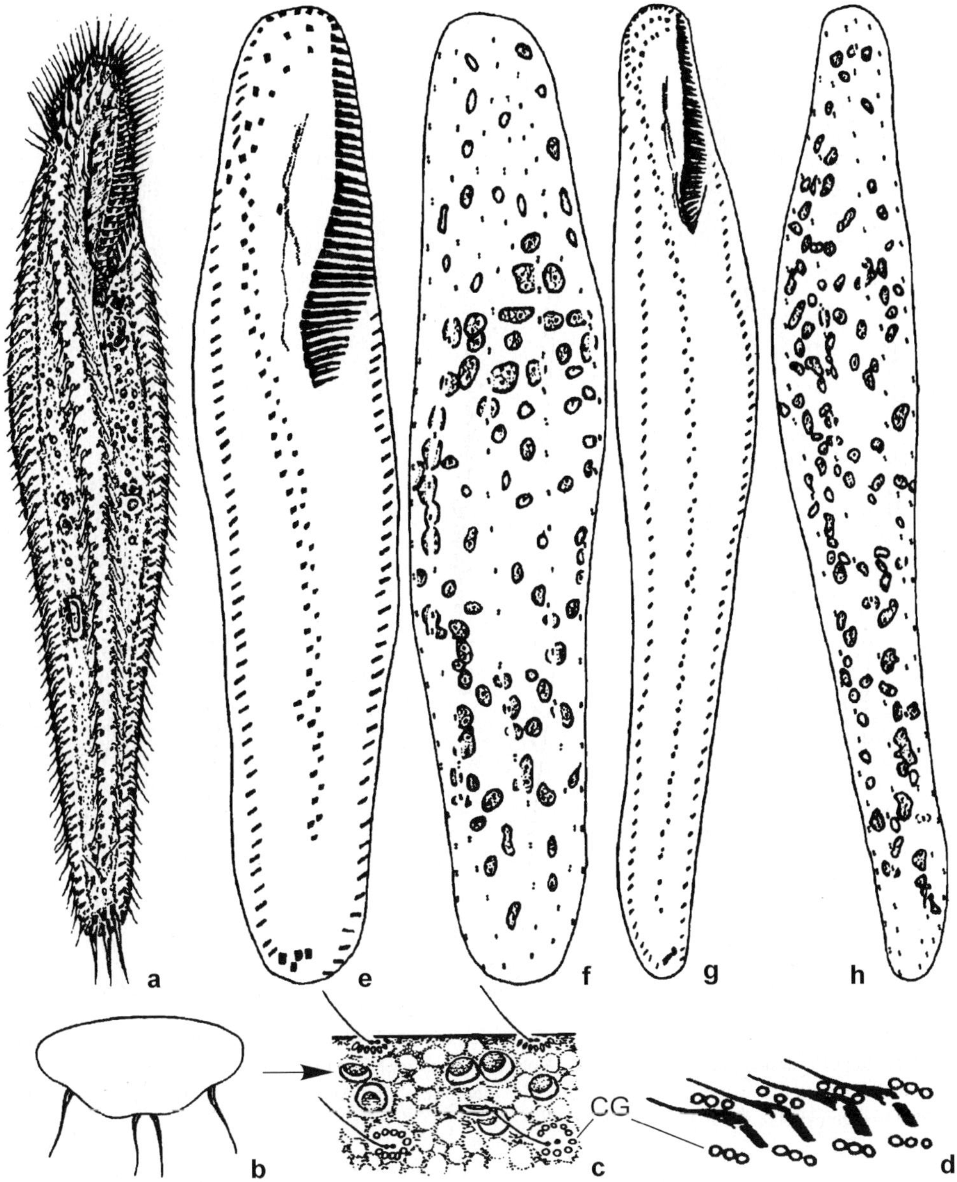

Fig. 180a–h *Pseudokeronopsis rubra* (from Hu & Song 2001. a–d, from life; e–h, protargol impregnation). **a:** Ventral view of a representative specimen, 178 µm. I am not quite certain that this population from China is conspecific with the neotype population of *P. rubra* because there are some distinct differences, for example, large distance between midventral complex and transverse cirri (against small distance) and low number of transverse cirri. **b:** Transverse section showing the vaulted ventral side and the almost plane dorsal side. **c, d:** Cortical granules around dorsal kineties (c) and cirri (d). The arrow marks the blood cell-shaped organelles close underneath the cell surface. **e–h:** Infraciliature of ventral and dorsal side and nuclear apparatus of two specimens, e, f = 180 µm, g, h = 182 µm. CG = cortical granules. Page 890.

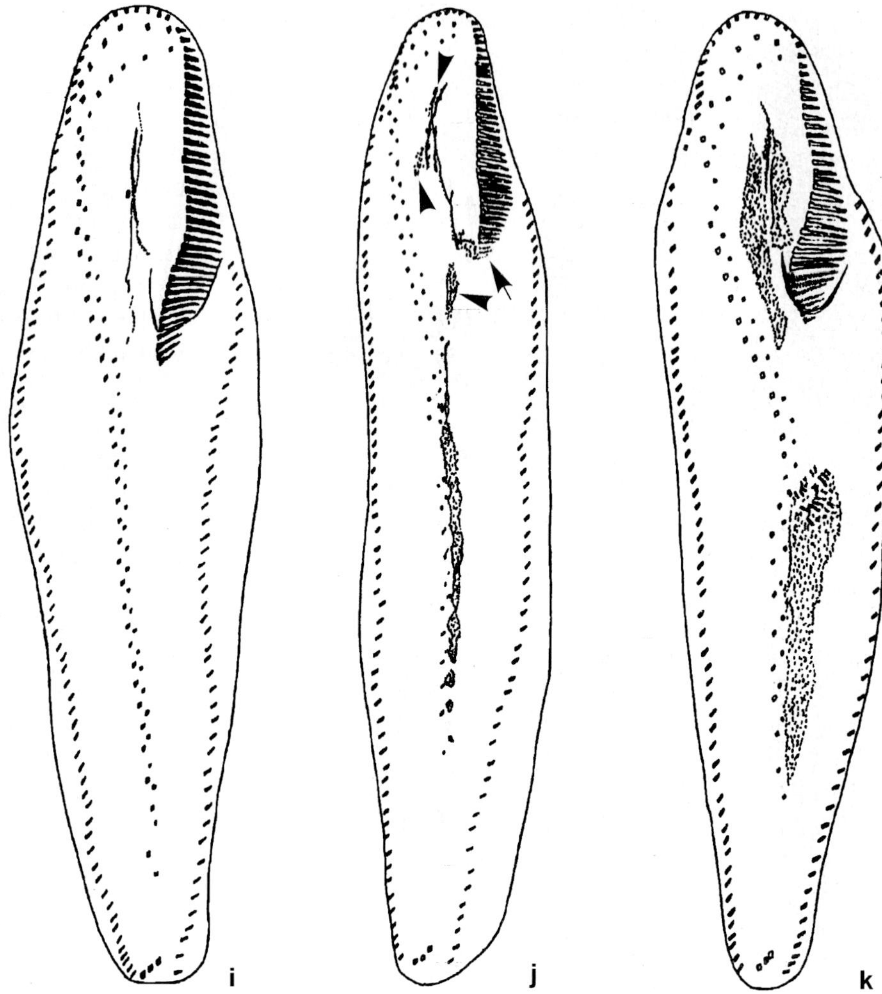

Fig. 180i–k *Pseudokeronopsis rubra* (from Hu & Song 2001. Protargol impregnation). **i:** Infraciliature of ventral side of a non-divider, 230 μm. **j, k:** Very early and early divider, sizes not indicated. Arrow marks the dedifferentiating parental adoral zone of membranelles. Arrowheads denote the early frontal-midventral-transverse cirral anlagen. Page 890.

Stomatogenesis in the proter begins with the dedifferentiation of the endoral starting from its anterior end. However, focus series indicate that the anarchic field has developed independently on the surface of the buccal cavity (Fig. 179j). Very probably the first new basal bodies appear de novo beside the endoral. All parental adoral membranelles are gradually resorbed in an anteriad direction (Fig. 179k–n). This total replacement of the parental adoral zone was already reported by Wallengren (1902, Fig. 178o). As in the opisthe, the anarchic field is soon far advanced in the formation of membranelles and the undulating membranes primordium detaches (Fig. 179k). The parental buccal cirrus is resorbed.

The development of the cirral primordia proceeds as follows: in the opisthe, the frontal-midventral-transverse cirral anlagen are a further result of the splitting anarchic field (Fig. 179k). In the proter, they originate apokinetally alongside the parental paroral (Fig. 179j). In both filial products these three sets of primordia (adoral membranelles, undulating membranes, frontal-midventral-transverse cirri) are joined at their posterior end for a while. The right midventral cirri soon differentiate, while nearby the proliferation of new basal bodies seems to continue within the frontal-midventral-primordium as well as within the undulating membrane anlage (Fig. 179k, l). Finally, a longitudinal series of numerous (30–50) oblique cirral streaks is generated in a posteriad direction.

New marginal rows and dorsal kineties are formed within the parental structures at two levels which correspond to the location of the frontal-midventral-transverse primordia in the proter and the opisthe (Fig. 179k–m). The anlagen arise directly from the parental marginal cirri and dorsal basal bodies, respectively. These basal bodies are the only parental structures participating in the formation of primordia, the rest are resorbed.

The undulating membranes anlage differentiates the first frontal cirrus (Fig. 179k). The buccal cirrus is produced in anlage II, the anterior cirrus formed there becomes the second anterior frontal cirrus (Fig. 179b, l). Anlage III–XXXI (until XXXIX) form two cirri each: anterior and posterior frontal cirri and cirral pairs forming the midventral complex. Four to seven of the more posterior streaks differentiate three cirri each: each a pair of midventral cirri and a transverse cirrus. The posteriormost two anlagen build four cirri each: the left anlage produces a transverse cirrus, one pretransverse ventral cirrus, and the rearmost midventral pair; the right anlage forms the rightmost transverse cirrus, the right pretransverse ventral cirrus, and the frontoterminal cirri (Fig. 179n).

Pseudokeronopsis flava, which is characterised by two transverse cirri, shows similar late divisional stages. In rare cases, additional cirri develop even from several anteriormost anlagen. All these supernumerary cirri are later resorbed in this species. By the time the segregation of new cirri is finished, the adoral zone of membranelles of each filial product starts extension and migration to form the mature cortical pattern (Fig. 179m, n).

Fig. 180y shows a late divider of "*Keronopsis pulchra*". However, according to Wirnsberger (1987, p. 154) it is *P. carnea*. I do not assign it finally but show it because it agrees basically with the data by Wirnsberger (1987) and Hu & Song (2001).

Nuclear division proceeds in the characteristic *Pseudokeronopsis* pattern, that is, a replication band traverses each macronuclear nodule, which divides individually, that is, without prior fusion to a single mass. The micronuclei behave like those of other hypotrichs.

Reorganisation was described by Wallengren (1902; Fig. 178p), Wirnsberger (1987; Fig. 179q), and Hu & Song (2001; Fig. 180w, x). According to Wirnsberger (1987), physiological regeneration was generally observed in normal, well-fed cultures. Only one set of primordia originates, resembling the process in the opisthe. However, prolif-

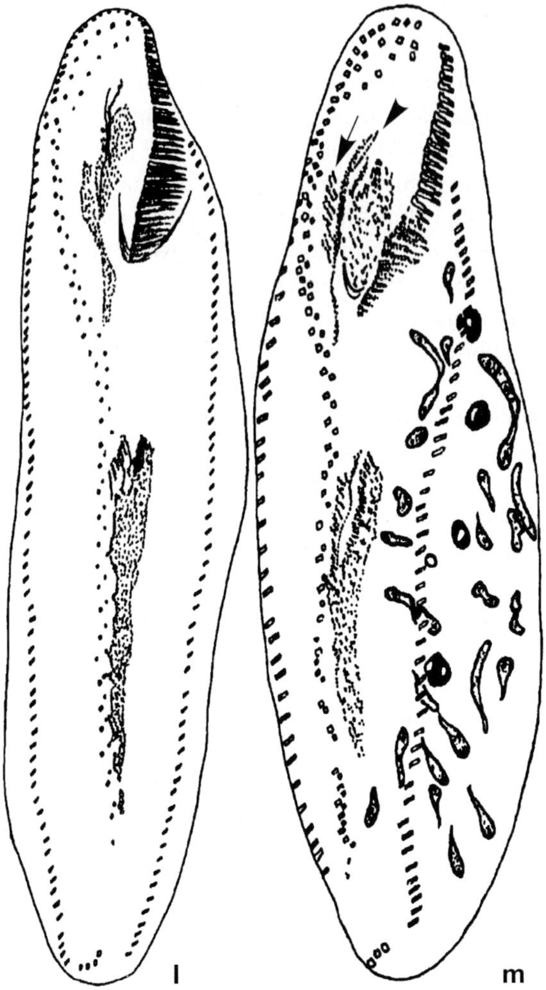

Fig. 180l, m *Pseudokeronopsis rubra* (from Hu & Song 2001. Protargol impregnation). **l:** Ventral view of an early to middle divider, 257 µm. **m:** Ventral view of middle divider showing ciliature and nuclear apparatus, size not indicated. Arrow marks frontal-midventral-transverse cirral anlage of proter, arrowhead denotes undulating membranes anlage. Page 890.

eration occurs close to fewer left cirral pairs, the anarchic field is slightly shorter and forms a small streak of new basal bodies extending onto the frontal area (Fig. 179q). Consequently, the primordia are situated more anteriorly than those of the posterior set in dividers. The old undulating membranes are resorbed beginning at their anterior end (Fig. 179q). In no case could their involvement and that of the buccal cirrus be observed in the formation of the anarchic field. The new adoral membranelles do not join with the remaining part of the old adoral zone, but replace them completely. There are only slight deformations of the macronuclear nodule during reorganisation. All other events correspond to those of cell division.

Molecular data: Ruthmann (1972, 1973) studied division and formation of the macronuclei and the fine structure of the macronuclear anlage. DNA synthesis proceeds synchronously in all macronuclei in the second half of the cell cycle, which takes about 24 h at room temperature. A G2 phase is virtually absent, each nucleus divides as soon the replication band has passed over it. The micronuclear S phase falls within macronuclear G1 and is followed by immediate division. In exconjugants a large, highly polyploid macronuclear anlage is formed, from which condensed chromatin bodies are passed into the cytoplasm, where they are thought to give rise to numerous small macronuclei of the vegetative cell. Electron microscopy revealed that the chromatin bodies within the macronuclear anlage are separated from each other by sheets of low contrast lamellar material. The anlage therefore appears as a composite nucleus containing prepacked units which are extruded into the cytoplasm following

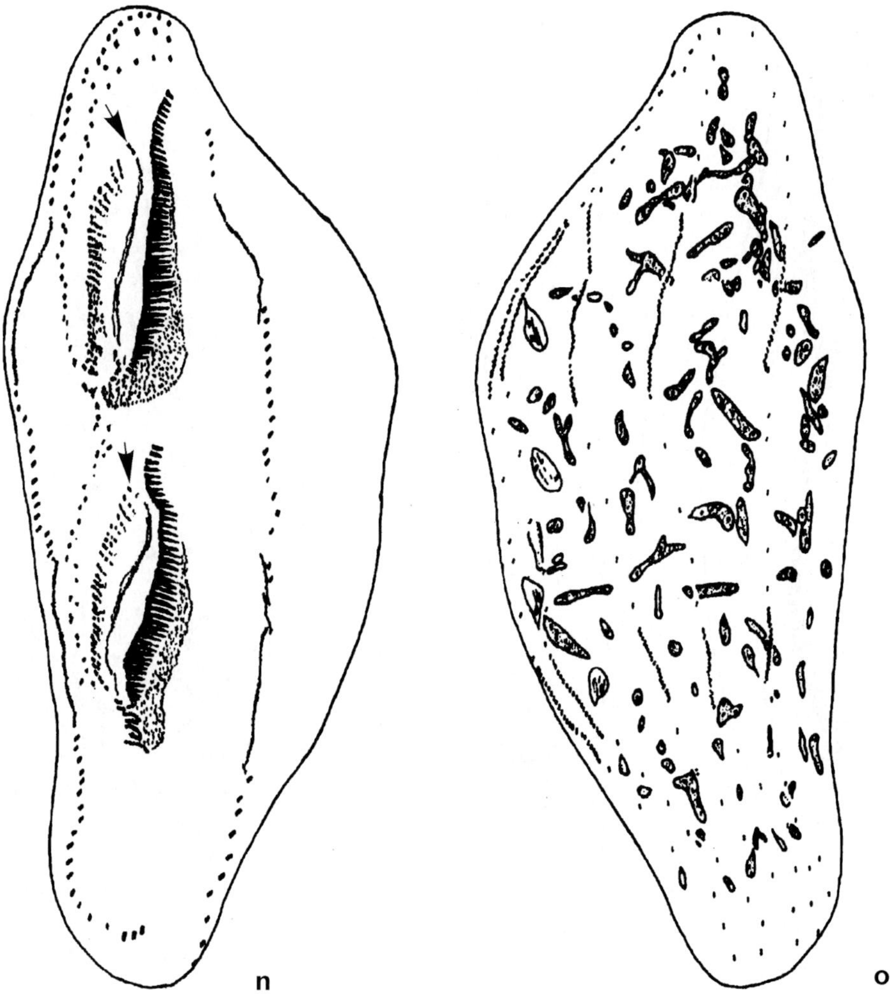

Fig. 180n, o *Pseudokeronopsis rubra* (from Hu & Song 2001. Protargol impregnation). Infraciliature of ventral and dorsal side and nuclear apparatus of middle divider, 248 μm. Arrows in (n) marks new frontal cirrus I/1 which originates, as is usual, from the left-most anlage, which also produces the undulating membranes. Page 890.

condensation. Furthermore, Ruthmann found that the proportion of DNA contents of individual macronuclear nodules to that of a micronucleus varies from 0.25:1 to 3.75:1, that is, the maximum differences between individual macronuclear nodules are about 15-fold while the mean DNA content of a macronuclear nodule is close to that of a micronucleus (see also Raikov 1982, p. 297). As in other hypotrichs, size of DNA is gene-sized; however, giant chromosomes are lacking in *P. rubra* (Ammermann 1987, p. 62f).

Pseudokeronopsis rubra has the same telomere repeat unit (CCCCAAAA) as other hypotrichs (Schlegel & Steinbrück 1986, Steinbrück 1990). Li & Gu (2005a) sequenced

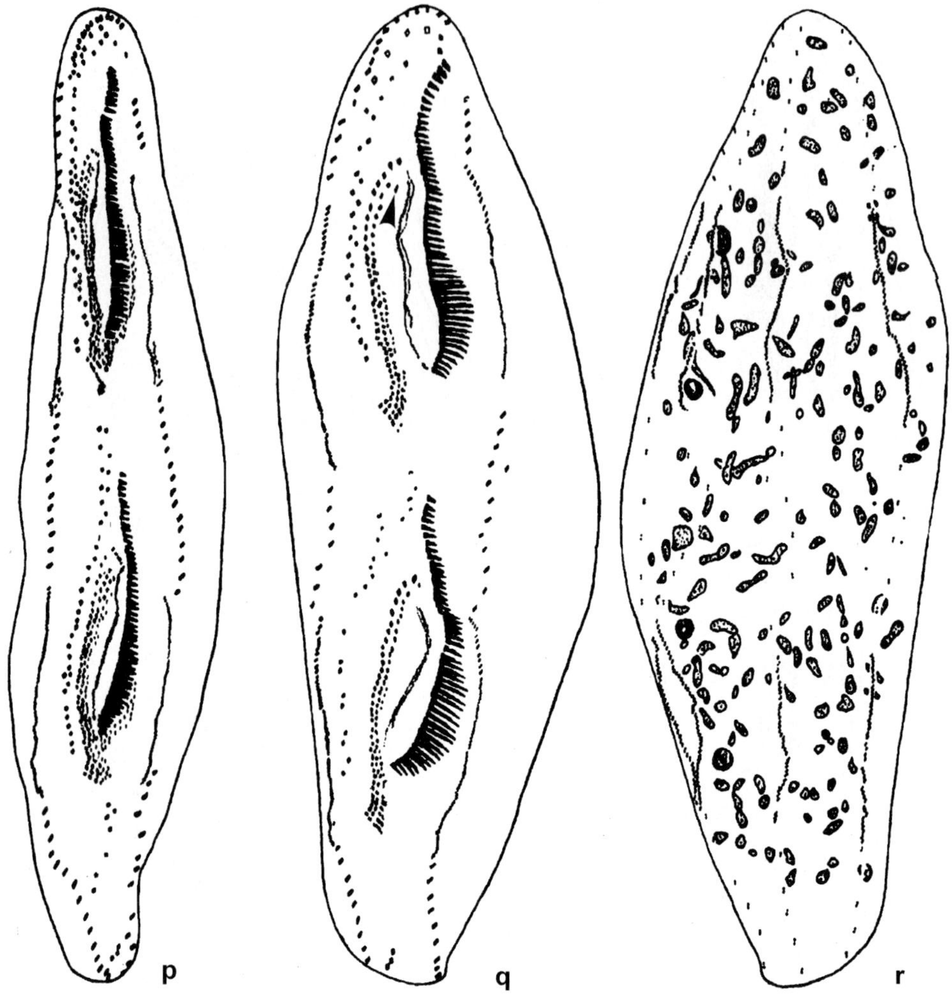

Fig. 180p–r *Pseudokeronopsis rubra* (from Hu & Song 2001. Infraciliature of dividers after protargol impregnation; sizes not indicated). **p:** Ventral side of middle divider. **q, r:** Ventral and dorsal side of late divider. Note that the macronuclear nodules divide individually. Arrowhead in (q) denotes the new buccal cirrus of the proter. Page 890.

the 16s-like SSrRNA of *P. rubra*. Unfortunately, the position in the phylogenetic tree is not described in detail.

Occurrence and ecology: Common in marine sediments and aufwuchs (Patterson et al. 1989, p. 210). Ehrenberg (1836, 1838) discovered *P. rubra* in a sample from the North Sea near Gothenburg, Sweden. Later, he found it also in the Baltic Sea from near Copenhagen, Denmark. The abundance was so high that the water became reddish. Due to the neotype fixation by Wirnsberger et al. (1987), the type locality of *P. rubra* is the sampling site of the neotype population. Unfortunately, they did not declare one of the

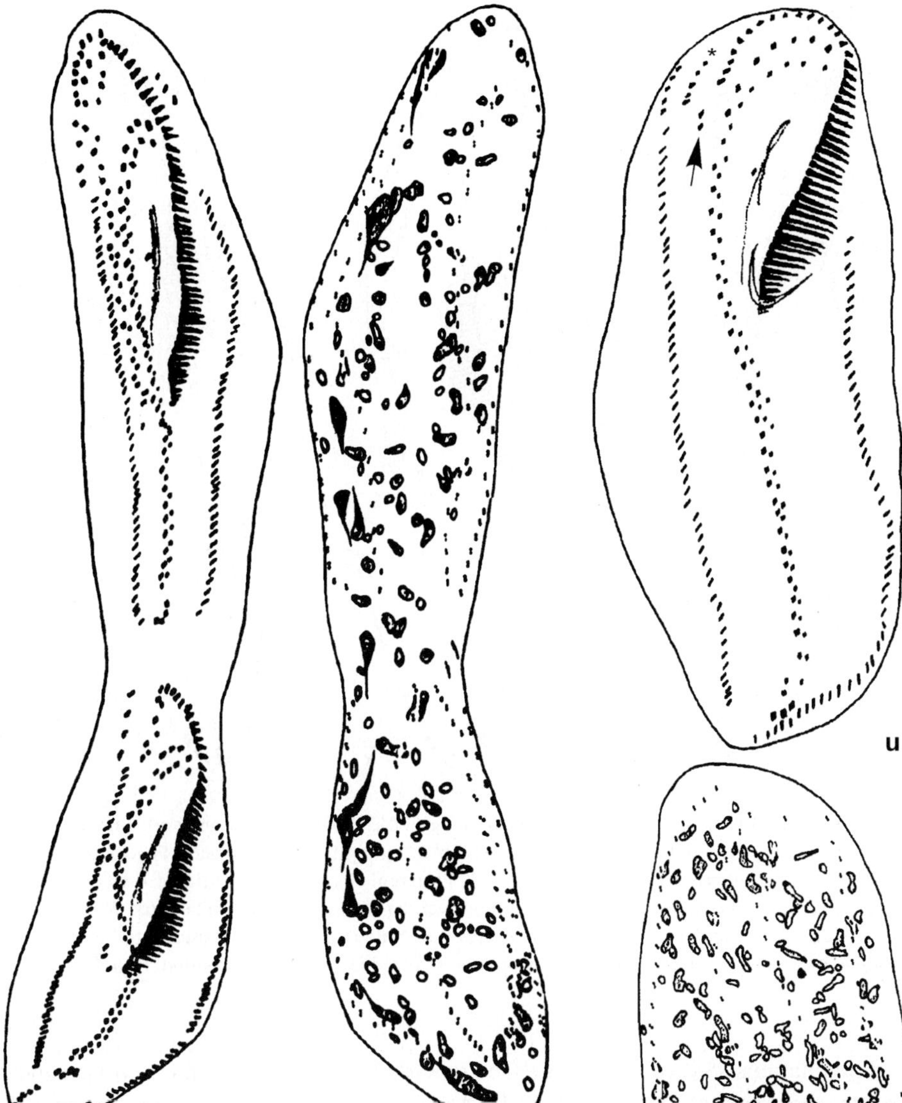

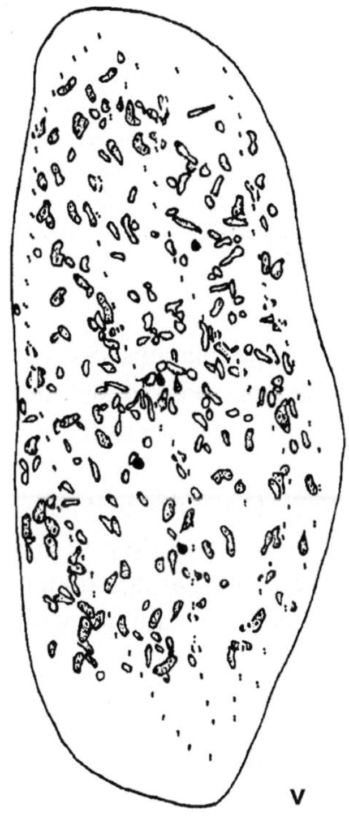

Fig. 180s–v *Pseudokeronopsis rubra* (from Hu & Song 2001. Infraciliature and nuclear apparatus of dividers after protargol impregnation). **s, t:** Ventral and dorsal side of very late stage, 242 μm. It is difficult to define the new frontoterminal cirri in this divider. Unfortunately, the illustrations by Hu & Song are much too small for such a complicated pattern. **u, v:** Ventral and dorsal side of a postdivider. Arrow marks new frontoterminal cirri. The asterisk likely denotes the distal portion of the parental adoral zone which will be replaced completely in *Pseudokeronopsis* and some related taxa. Page 890.

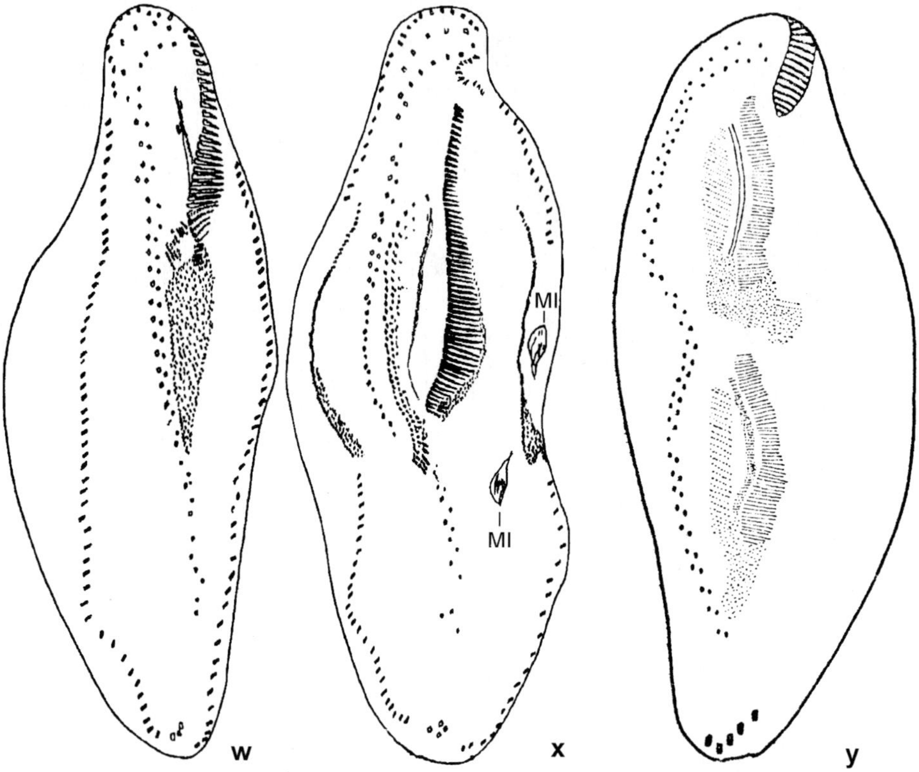

Fig. 180w, x *Pseudokeronopsis rubra* (from Hu & Song 2001. Protargol impregnation). Infraciliature of ventral side of an early and middle reorganiser, w, x (?) = 212 µm. MI = micronucleus. Page 890.
Fig. 180y *Pseudokeronopsis* sp. (from Borror 1972. Protargol impregnation). Middle divider, 145 µm. Borror identified it as *Keronopsis pulchra* which, however, has three enlarged frontal cirri. Wirnsberger et al. (1987) assumed identity with *Pseudokeronopsis carnea*. Note that the parental midventral complex and the adjoining bicorona are retained, that is, they are not involved in primordia formation. Page 899.

two red populations studied as neotype (see remarks). Consequently, the new type locality was not fixed so far. According to Erna Aescht (= E. Wirnsberger; pers. comm.), the neotype slides deposited in the Museum are from the South African population. Thus, the new type locality[1] of *P. rubra* is the Atlantic Ocean from the Cape Town area, South Africa. The albino mutant was isolated by Uhlig (1980) from the neotype population of *P. rubra*. According to Hering, 1000–2000 specimens of the clone population were treated with gamma radiation about one year previously. Possibly, the albino strain is a radiation mutant, which manifested, however, only after 120 generations. The

[1] Unfortunately, no details about the sample site are provided neither by Wirnsberger et al. (1987) nor by Uhlig (1981b) who got the neotype material from Dr. Hering, Groote Schuur Hospital in Cape Town, South Africa. Very likely (but not one hundred per cent), Dr. Hering collected the material from the Atlantic Ocean in the Cape Town area.

third *P. rubra* population studied by Wirnsberger et al. (1987) was from the Indopacific region and was supplied by Dr. Lange, Berlin.

Further records substantiated by morphological data and/or illustrations: France (Dujardin 1841[1]); North Sea at Oostende, Belgium (Fresenius 1865); aquarium with material mainly from the North Sea near Heligoland (Cohn 1866); various sources, including salt water remitted from the Aston Aquarium, Birmingham and also from among decaying organisms collected at St. Helier (Jersey) and Bangor (N. Wales), United Kingdom (Kent 1882); Mediterranean Sea in Bay of Naples, Italy (Entz 1884); Baltic Sea (Wallengren 1900); common (in the Baltic Sea?) near Kiel, Germany (Kahl 1932); scattered in a four-week-old, sometimes aerated, foul-smelling sample from the Val di Lone, Northern Adriatic Sea (Kiesselbach 1936a); Gulf of Mexico at Dry Tortugas and Atlantic Ocean near Beaufort, USA (Bullington 1940); Indian Ocean near Visakhapatnam, India (Ganapati & Rao 1958). Ruthmann (1972, 1973) obtained *P. rubra* from F. Ott, who isolated it from Woods Hole, USA. Walker (1975a, b) studied a population from aquaria with material from coral reefs of the Makatumbe Islands at the entrance of Dares-Salaam harbour (Tanzania), Indian Ocean (species identified by J. Dragesco and M. Tuffrau). The population studied by Hu & Song (2001) was found in a shrimp farm in Qingdao (36°08 N, 120°43E) at 12‰ salinity, 20°C water temperature, and pH 8.0. Specimens were cultured in boiled sea water to which squeezed rice grains were added.

Records from marine habitats not substantiated by meaningful morphological data: harbour of Oostende, Belgium (Pauw 1969, p. 200; other record from Belgium: Chardez 1987, p. 13); Arcachon (France), Gulf of Biscay, Atlantic Ocean (Delphy 1938, p. 68, with two insignificant sketches); Northern Sea in Sylt, Germany (Küsters 1974, p. 174); Loche Eil, west coast of Scotland (Wyatt & Pearson 1982, p. 287, 301); rock pool at the eastern side of Langland Bay, Gower, Swansea, England (Fowell 1944, p. 209); Warberg (= Varberg?), Kattegat, North Sea/Baltic Sea, Sweden (Quennerstedt 1867, p. 39; assigned to *P. rubra* with reservation by Wirnsberger et al. 1987); littoral and sparsely in the psammon of the Caspian Sea (Agamaliev 1967, p. 369; 1967a, p. 1426; 1970, p. 1279; 1971, p. 383; as "*K. rubra* f. *heptasticha*"; 1972, p. 7; 1973, p. 1599; 1974a, p. 20; 1976, p. 92; 1990, p. 62; Agamaliyev 1974, p. 21); mesopsammon of the Black Sea (Bulgaria) at following conditions: 21–25°C, 1.0–18% salinity (in Detcheva 1980 "‰" is given), 18% frequency (Detcheva 1980, p. 34; 1982, p. 249); Gulf of Varna and littoral of Bulgarian Coast and other sites of the Black Sea (Jeliaskowa-Paspalewa 1933, p. 22, 23; Kovaleva & Golemansky 1979, p. 270; Kovaleva 1966, p. 1603; Tucolesco 1962a, p. 813; Detcheva 1983, p. 72; 1992, p. 103); 10 specimens cm^{-2} in 0–2 cm sandy substrate at 24 °C in Bay of Odessa, Ukraine, Black Sea (Dzhurtubayev 1978, p. 65); marine sand near Marseille, Mediterranean Sea, France (Vacelet 1961, p. 15); interstitial of Gulf of Naples, Mediterranean Sea, Italy (Nobili 1957, Tabella I; Dini et al. 1995, p. 69); artificial substrates in Lagoon of Venice, Adriatic Sea, Italy (Coppellotti & Matarazzo 2000, p. 426); Gulf of Biscay (Spain) at Castro Urdiales (Fernandez-Leborans & Novillo 1993, p. 216); frequent in coarse

[1] I am uncertain about the sample site, that is, marine or freshwater as mentioned by Diesing (1850). According to Fresenius (1865, p. 128), Dujardin's population is from the Mediterranean Sea.

(above 500 μm) and medium (200–500 μm) sand and rare in fine (below 200 μm) and medium sand near the cities of Vishakhapatnam and Konarka, Bay of Bengal, India (Rao & Ganapati 1968, p. 90; Rao 1969, p. 92); White Sea, inter alia, on algae and stones and in the interstitial (0.1–0.5 mm) of the Kandalaksha Gulf (Raikov 1962, p. 331; Burkovsky 1970a, p. 189; 1970b, p. 11; 1970c, p. 56); Gulf of Kola, Russia (Gassovsky 1916, p. 142); Barents Sea (Kovaljeva 1967, p. 83); in sand of middle(?) grain size, Japan Sea (Raikov 1963, p. 1757; Raikov & Kovaleva 1968, p. 331); Bay of Amoy, Chinese Sea (Wang & Nie 1934, p. 4210); sand from islands (Al-Batinah, Abu Ali, Tarut) of Arabian Gulf (Al-Rasheid 1996, p. 758; 1997, p. 756, one colour micrograph); from or near oysters near Karachi, Arabic Sea, Pakistan (Laird 1961, p. 457); Gulf of Mexico (Smith 1904; Borror 1962, p. 342); rare from or near oysters in Passamaquoddy Bay, New Brunswick, Canada (Laird 1961, p. 457); from floating material, debris, and bottom samples from the Atlantic Ocean in Woods Hole, USA (Lackey 1936, p. 269; as *Holosticha rubrum*).

Records from inland salt-waters and estuaries: salt marshes of the Dee estuary at Parkgate, Cheshire, Great Britain (Webb 1956, p. 152; Carey & Maeda 1985, p. 568); estuary in Italy (Cuneo 1891, p. 146); Lake Techirghiol, Romania (Tucolesco 1965, p. 160); saline lake near Sewastopol, Krim Peninsula, Ukraine (Dagajeva 1930, p. 35, 41); Dnjepr estuary, Black Sea (Kovalchuk 1989; incorrectly spelled *Keronopsi rubra*); polysaprobic to isosaprobic (only 1.15–2.00 mg l^{-1} O_2) estuary of the Almendares River at northern coast of Havana City, Cuba (Diaz Pérez & Montoto Lima 1989, p. 92).

There are several records from limnetic habitats not substantiated by morphological data, indicating misidentification because *P. rubra* was never reliably recorded from freshwater (possibly confused with *Diaxonella pseudorubra*): with low abundance in Vienna, Austria, in late May (Schmarda 1846, p. 5, 45); brook in the city of Salzburg, Austria (Haslauer et al. 1979, p. 40, see also Foissner & Foissner 1988, p. 86); Belgium (Bervoets 1940, p. 124); Genkeltalsperre, a reservoir in Germany (Nusch 1970, p. 300); polluted stream near Fidenza, Italy (Madoni & Bassanini 1999, p. 395); Slovakia (Tirjaková 1992, p. 293, 1992a, p. 77; Matis et al. 1996, p. 17); reservoir and stream near Madrid, Spain (Fernandez-Leborans et al. 1985, p. 88; Sola et al. 1996, p. 241); Armenia (Zharikov 1982, p. 912; identification marked as uncertain); planktonic in Azerbaijanian reservoirs (Alekperov 1981, p. 57; 1982, p. 45); planktonic in Ukrainian ponds (Oleksiv 1985, p. 91); *Sphagnum* mosses from near Lake Itaska, USA (Bovee 1979, p. 619; as *Kerona rubra*).

Feeds mainly on diatoms and autotrophic flagellates, but also on unicellular cyanobacteria (Fenchel 1969, p. 25) and *Phaeodactylum* (Fenchel 1968, p. 116). Uhlig (1979) fed it with *Dunaliella tertiolecta*. *Pseudokeronopsis rubra* was cultured by several workers. For example, Ruthmann (1972) maintained the cultures in sea water at room temperature with a few wheat grains. Wirnsberger et al. (1987) used artificial sea water (Wimex Seasalt. H. Wiegandt, Krefeld, Germany; 3.0–3.4%) at room temperature with *Dunaliella* sp., rice grains, boiled egg yolk, or yeast as food. Further culture methods, see Walker (1975a).

Intrinsic rate of natural increase (r_m) is 0.55 day^{-1} and body weight is 6.5×10^{-6} g (Fenchel 1968a, p. 129; 1974, p. 320; as *Keronopsis rubrum*; Taylor & Shuter 1981, p. 163).

According to Zaika (1970, p. 98), generation time is 16–17 h at 22–26°C. Walker (1975a, b) found a fission rate of close to 1 d^{-1} at 23°C; the specimens usually divided during the night or in the early morning, indicating a circadian rhythm. Uhlig (1981a, b) cultured *P. rubra* and the *albino* strain at permanent light, permanent dark, and at 12 h day-night rhythm at 24° C. In both populations, per trial reproduction was three times higher in the permanent dark situation as in the permanent light cultures; the light-dark culture had an intermediate rate. Maximum growth-rates were determined at 27°C at continuous darkness. The red *P. rubra* is negatively phototactic, while the *albino* strain does not react phototacticly (Uhlig 1981b). Uhlig (1981b) also found that the red strain has about double the biomass of the albino strain. N-content is about 3.7% in both strains, whereas C-content is slightly higher (24.0%) in the red strain than in the albino strain (22.3%), so that the C:N ratio is 6.5 against 5.9.

The red-orange pigment of *P. rubra* is extractable with ethanol. In both populations (red and albino), pigment extracts show absorption maxima at 200 nm, 420 nm, and 675 nm. However, in the albino strain the 420 nm maximum is very weak, whereas it clearly dominates in the red *P. rubra*. Obviously, this pigment is responsible for the light sensitive behaviour of the intensively pigmented form. The red pigment of *P. rubra*, which is resistant to heat (100°C), but light sensitive, immobilises other protists (Uhlig & Moschny 1988). When offered to the unpigmented, granular-free albino mutant, the liquid pigment is absorbed by the albino-cells and is granularly incorporated into the cell cortex in a *P. rubra*-like pattern. When liquid pigment extract of *P. rubra* is offered to heterotrophic flagellates or ciliates from marine or freshwater habitats, the cells are reversibly immobilised, narcotised, vacuolised (mainly in freshwater species), or irreversibly damaged (Uhlig 1989). Lueken et al. (1989a, b) used *Euplotes vannus* to test whether this pigment (keronopsin), the water-soluble component of the toxic pigment complex, narcotises cells by affecting ion conductance. Keronopsin inhibited the generation of normal action potentials upon constant current injection. Effects on locomotory behaviour, spontaneous membrane depolarisation, action potentials, inward currents and tail outward currents all provide evidence that keronopsin acts on cells by obstructing Ca^{2+} influx. Both forms react to blue light (473 nm); red and green light do not induce significant reactions in both the red and albino strains (Uhlig 1983).

Uhlig (1981b) tried to adapt *P. rubra* and the albino strain to hypo- and hyper-saline conditions. However, reproduction was distinctly reduced at 20‰, respectively, 45‰ salinity.

Machemer (1966) studied the practicability of conditioning with the stimuli "rough surface" and "light" in mass experiments. In *P. rubra*, he found a moderate preference for rough in tests of two and four hours. Conditioning over two and four hours (867 specimens) carried out on chessboard-pattern in both combinations of the stimuli rough surface and light (rough/light-smooth/dark) and rough/dark-smooth/light led to negative results.

According to Fernandez-Leborans & Novillo (1994, p. 203), the occurrence of *Pseudokeronopsis rubra* is not affected by the presence of 1 mg^{-1} lead over a period of 240 hours. Mauch (1976, p 422) classified *P. rubra* as indicator of mesosaprobic conditions because Kahl (1932) stated that it is common in mesosaprobic detritus. By contrast,

Foissner (1979) mentioned it, likely basing on Sládeček (1973, p. 195), as oligosaprobic indicator (o = 4, b = 6, SI = 1.6, weight = 3). However, it was later omitted from the list of saprobic indicators because it does not occur in freshwater (Foissner 1988a, Foissner et al. 1991, Berger & Foissner 2003).

Holosticha rubra sensu Wang & Nie (1932)
(Fig. 178z)

1932 ***Holosticha rubra*** **(Ehrenberg) Kent 1880-1882** – Wang & Nie, Contr. biol. Lab. Sci. Soc. China, 8: 353, Fig. 63 (Fig. 178z).

Remarks: This population has only two macronuclear nodules so that it cannot be synonymised with a species of the *Pseudokeronopsis rubra* group, although the other features correspond rather well. Ehrenberg (1838) and Entz (1884) also described their *P. rubra* populations with only two macronuclear nodules whereas all other authors found many scattered nodules. Thus, the existence of a bimacronucleate species of the *P. rubra* group cannot be excluded (for a more detailed discussion of Ehrenberg's and Entz's paper, see remarks under *P. rubra*). However, the data about binucleate populations are still too scant, so that it would be unwise to describe a new species without original data. Here, I give the most important data on Wang & Nie's (1932) population (see also Fig. 178z). Body size around 255 × 45 μm in life. Body rather flexible and contractile and thus variable in shape, about five times as long as broad. **Two** macronuclear nodules in left body portion, individual nodules elliptical or ovoidal (Fig. 178z); of course it cannot be excluded that Wang & Nie misinterpreted large cytoplasmic inclusions as macronuclear nodules because they very likely did not stain the cells. Contractile vacuole near left margin between the two macronuclear nodules. Cytoplasm (endoplasm) in most specimens deep yellowish, but brick-red specimens also occur. In most specimens, the anterior body end turns slightly leftwards and, therefore, it revolves more rapidly from right to left when the organism swims or swerves through the water; creeping movement rare. Adoral zone occupies about 33% of body length. Cirral pattern pseudokeronopsid. Midventral complex not set off from bicorona, extends to near the eight or nine transverse cirri, which project distinctly beyond rear body end. Buccal cirrus neither mentioned nor illustrated (must not be over-interpreted because the presence/lack of this cirrus is difficult to observe in life). Found in the Bay of Amoy, China. It usually appeared in old cultures of sea water, especially among decaying "zoophytes" collected days or weeks ago.

Pseudokeronopsis pararubra Hu, Warren & Suzuki, 2004 (Fig. 180.1a–k, Table 37, Addenda)

2004 ***Pseudokeronopsis pararubra*** **sp. n.**[1] – Hu, Warren & Suzuki, Acta Protozool., 43: 352, Fig. 1A–L, 2A–F, 3A–E, 6A–K, 7A–M (Fig. 180.1a–k; original description; the holotype slide [registration number 2004:6:2:1] is deposited in the Natural History Museum in London; one paratype slide [HD-2000102001] is deposited in the Laboratory of Protozoology, Ocean University of China).

Nomenclature: No derivation of the name is given in the original description. The species-group name *pararubra* is a composite of the prefix *para-* (beside or deviating) and the species-group name *rubra* (see type species for derivation) and likely indicates that the present species is closely related to *P. rubra*.

Remarks: *Pseudokeronopsis pararubra* is a further species of the *P. rubra* group. The differences to *P. rubra* and *P. carnea*, the most similar species, are sophisticated. According to Hu et al. (2004a) the following features separate *P. pararubra* from *P. rubra*: body outline in life long elliptical against band-like (often a useful feature); buccal cirrus behind mid-portion of paroral against ahead of mid-portion (detailed morphometric data for this feature are lacking); cortical granules orange-red against brick-red (difficult feature when colour is similar); 18–22 cirri in bicorona against 14–16; 8–9 transverse cirri on average against 6–7; erythrocyte-like structures present against likely lacking (a common feature in the pseudokeronopsids and so far possibly overlooked in *P. rubra*). *Pseudokeronopsis pararubra* differs from *P. carnea* (according to Hu et al. 2004a) in the following features: midventral complex terminates very close to transverse cirri against more or less distinctly set off; body broad without narrowed posterior portion against long elliptical with a conspicuously narrowed posterior portion; 15–26 frontal cirri in bicorona against 10–14; 8–9 (average) transverse cirri against 6–7.

I preliminary accept *P. pararubra*; however, detailed genetic and molecular studies are needed for more detailed analyses. So far I recommend the designation "species of the *Pseudokeronopsis rubra* group", if the identification is uncertain.

Hu et al. (2004a) studied the morphology of two populations and the ontogenesis, which proceeds basically as in the other *Pseudokeronopsis* species. Thus, the corresponding illustrations and the micrographs documenting the species are omitted.

Morphology: The description is based on the type population (China) and a population from Japan.

Body size 180–350 × 50–90 µm in life. Body outline usually long elliptical with anterior end broadly rounded and margins of posterior portion slightly converging; left margin convex, right distinctly sigmoidal (Fig. 180.1a). Ratio of body length:width about 4–5:1; ratio of body width:height about 2:1 (Fig. 180.1b). More than 100 macro-

[1] The diagnosis by Hu et al. (2004a) is as follows: Marine *Pseudokeronopsis*, long elliptical in outline, 180–350 × 50–90 µm in vivo and dark reddish in colour. Ciliature comprising: 64–92 adoral membranelles; bicorona of 15–26 frontal cirri; 1 buccal cirrus and 2 frontoterminal cirri; 7–11 transverse cirri; two midventral rows consisting of 62–93 cirri which extend to transverse cirri; 48–79 left and 46–80 right marginal cirri; 5–8 dorsal kineties. Numerous (>100) macronuclear segments. Two types of cortical granules: one orange-red pigment, mainly grouped around cirri and dorsal bristles; the other, colourless and blood-cell-shaped, lying just beneath the former and densely distributed.

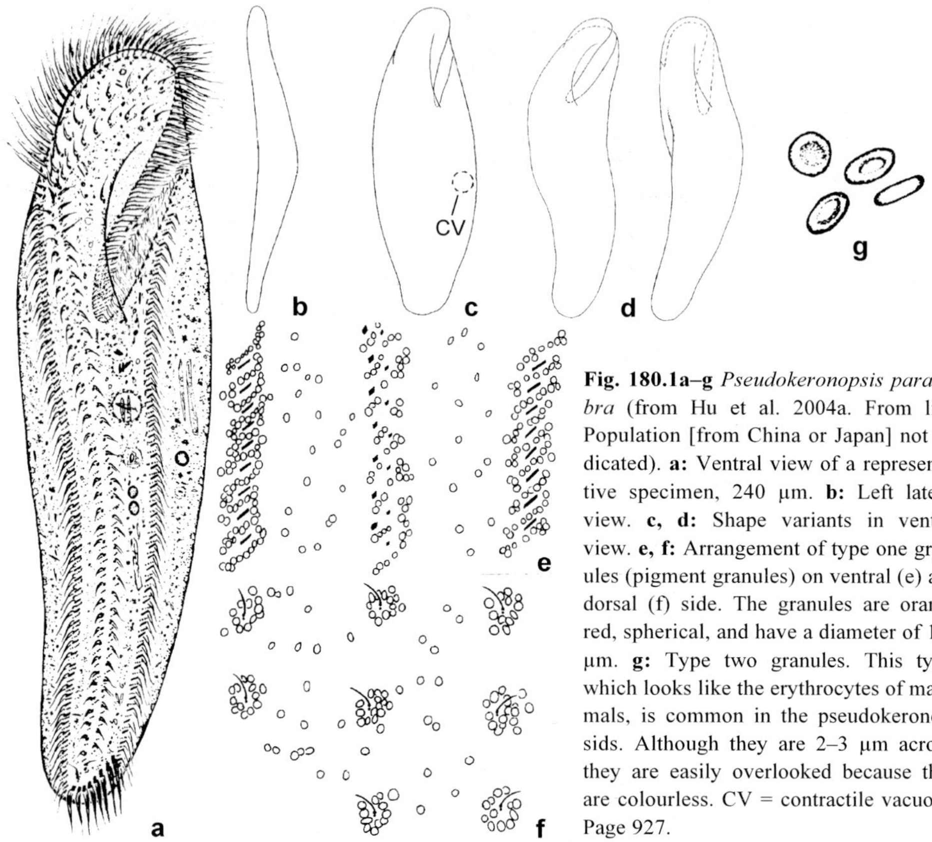

Fig. 180.1a–g *Pseudokeronopsis pararubra* (from Hu et al. 2004a. From life. Population [from China or Japan] not indicated). **a:** Ventral view of a representative specimen, 240 µm. **b:** Left lateral view. **c, d:** Shape variants in ventral view. **e, f:** Arrangement of type one granules (pigment granules) on ventral (e) and dorsal (f) side. The granules are orange red, spherical, and have a diameter of 1–2 µm. **g:** Type two granules. This type, which looks like the erythrocytes of mammals, is common in the pseudokeronopsids. Although they are 2–3 µm across, they are easily overlooked because they are colourless. CV = contractile vacuole? Page 927.

nuclear nodules scattered through cell; individual nodules ellipsoidal, 3–6 µm long. Micronuclei difficult to recognise both in life and in protargol preparations (Fig. 180.1k). In type population sometimes a contractile (?) vacuole was observed near left body margin slightly behind mid-body (Fig. 180.1c). Pellicle comparatively thick but flexible. Two types of cortical granules: type one is orange-red, spherical, 1–2 µm across, mainly grouped around cirri and dorsal bristles; few granules are randomly dispersed (Fig. 180.1e, f); make cells dark reddish at low magnification. Type two granules look like erythrocytes of mammals (Fig. 180.1g), colourless, 2–3 µm in diameter, and packed closely underneath type one granules. Usually several food vacuoles 6–12 µm across. Slowly crawling on substrate showing great flexibility or rotating about main body axis when swimming.

Fig. 180.1h–k *Pseudokeronopsis pararubra* (from Hu et al. 2004a. Protargol impregnation. Populations [from China or Japan] not indicated). Infraciliature of ventral and dorsal side and nuclear apparatus of two specimens, h, i = 300 µm, j, k = 275 µm. Long arrow in (h) marks frontoterminal cirri, short arrow denotes buccal cirrus, which is behind the middle of the paroral. Pretransverse ventral cirri circled by dotted line. Note that the midventral complex terminates very close the transverse cirri, that is, there is no gap between these two cirral groups, which is an important difference to *P. carnea*. 1 = dorsal kinety 1 (= leftmost kinety). Page 927. →

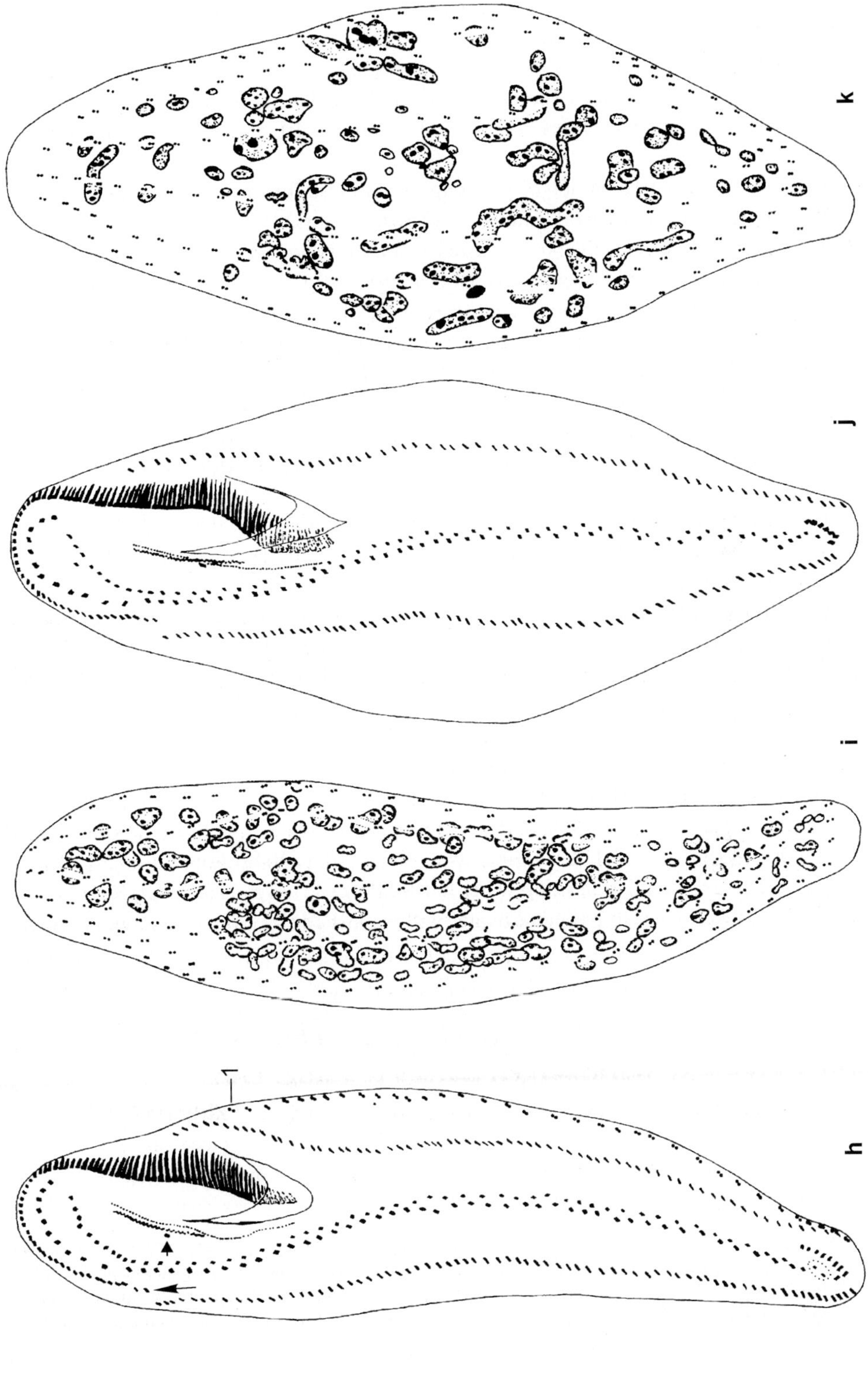
k
j
i
h
1

Adoral zone occupies 25–33% of body length, distal end extends far onto right body margin (13% of body length in specimen shown in Fig. 180.1h), composed of 72–78 membranelles of common fine structure on average (Table 37). Buccal field narrow and deep. Undulating membranes *Pseudokeronopsis*-like, that is, paroral about half as long as endoral. Pharyngeal fibres conspicuous, about 30–50 µm long in Japanese population.

Most cirri rather fine, about 10 µm long, motionless for most of the time; transverse cirri about 15 µm long. Cirral pattern as shown in Figs. 180.1h–k, that is, as in other *Pseudokeronopsis* species (Table 37). Bicorona almost not set off from midventral complex, which extends to transverse cirri. Buccal cirrus slightly behind mid-paroral. Invariable two frontoterminal cirri in ordinary position, that is, between distal end of adoral zone and anterior end of right marginal row. Two pretransverse ventral cirri between midventral complex and transverse cirri. 7–11 transverse cirri arranged in hook-shaped pattern, bases slightly enlarged, project distinctly beyond rear body margin (Fig. 180.1a, h, j). Right marginal row commences about at level of distal end of adoral zone; left row commences distinctly ahead of proximal end of adoral zone; marginal rows widely separated posteriorly.

Dorsal cilia about 5 µm long, arranged in 5–8 roughly bipolar kineties, which are easily recognisable in life due to orange-red cortical granules arranged around bristles. Caudal cirri lacking.

Cell division: Hu et al. (2004a) studied the ontogenesis of *P. pararubra*. It proceeds as in congeners and thus the reader is referred to the original paper. The only difference to the other species is in the connection of the oral primordium and the frontal-ventral-transverse cirral anlagen in the proter. According to Hu et al. (2004a) this is an artefact, that is, these two parts are never connected in the proter. However, I suppose that electron microscopic studies are needed for a final decision.

Occurrence and ecology: Marine. Type locality of *P. pararubra* are molluscs culturing waters (salinity 3.2–3.7%, water temperature 16–24°C, pH 8.0–8.2) off the coast of Qingdao (36°08'N 120°43'E), China, where Hu et al. (2004a) found it on 20.10.2000. They found it also in a fish farming water in Mie Port (32°48'N 129°46'E), Nagasaki, Japan. Cultures were kept in boiled seawater with squashed rice grains to support bacterial growth.

Pseudokeronopsis carnea (Cohn, 1866) Wirnsberger, Larsen & Uhlig, 1987

(Fig. 5a, 6a, 179k, l, o, 181a–l, 182a–h, 183a, Tables 37, 38)

1866 ***Oxytricha flava*, var. *carnea*, nov. var.**[1] – Cohn, Z. wiss. Zool., 16: 288, 300 (original description without illustration; no type material available).

---

[1] The diagnosis, respectively, short characterisation provided by Cohn (1866, p. 300) is as follows: Fleischfarben, mit zwei rothen punctirten Linien auf der Bauchseite und am Seitenrande; Köper um das Doppelte grösser und breiter, äusserst flexil, nicht retractil, nach hinten nicht verdünnt, quer abgestumpft; beim Absterben zerfliessend. Bewegung wie bei *O. flava*, theils kriechend, theils freischwimmend, zwischen faulen Stoffen.

1884 ***Holosticha flava*, Cohn sp.** – Rees, Tijdschr. ned. dierk. Vereen, Supplement Deel I: 644, 645, Plaat XVI, Fig. 19, 20 (Fig. 181a, b; misidentification, redescription from life).

1888 ***Oxytricha rubra* Ehbg.** – Möbius, Arch. Naturgesch., 54: 86, Tafel VI, Fig. 1–3 (Fig. 181c–e; misidentification; see remarks).

1926 ***Holosticha rubra* (Ehrenberg) var. *flava*** – De Morgan, J. mar. biol. Ass. U. K., 14: 40, Fig. 18–20, 22 (Fig. 181f–k; misidentification, see remarks; detailed redescription from life and after nucleus staining).

1932 ***Keronopsis rubra* var. *carnea* Cohn** – Kahl, Tierwelt Dtl., 25: 573 (see remarks; revision).

1958 ***Keronopsis rubra* Ehrenberg, 1838** – Ganapati & Rao, Andra Univ. Mem. Oceanogr., 2: 83, Plate I, Fig. 8 (Fig. 183a; misidentification).

1984 ***Pseudokeronopsis rubra* (Ehrenberg, 1838)** – Foissner, Stapfia, 12: 111, Abb. 58a–h, Tabelle 26 (Fig. 182a–h; misidentification; at least one slide of protargol-impregnated specimens is deposited in the Oberösterreichische Landesmuseum in Linz [LI], Austria).

1987 ***Pseudokeronopsis carnea* (Cohn, 1866) nov. comb.**[1] – Wirnsberger, Larsen & Uhlig, Europ. J. Protistol., 23: 79, Fig. 9, Tables 1–3 (Fig. 181l; combination with *Pseudokeronopsis*, redescription and fixation of neotype [slides 1986/40 and 1986/41 in LI]; see remarks).

1987 ***Pseudokeronopsis carnea*** – Wirnsberger, Arch. Protistenk., 134: 150, Fig. 3, 4, 7 (Fig. 179k, l, o; cell division).

1988 ***Pseudokeronopsis carnea* (Cohn, 1866)** – Wirnsberger & Hausmann, J. Protozool., 35: 182, Fig. 1–17 (Fig. 5a, 6a; detailed analysis of fine structure).

2001 ***Pseudokeronopsis carnea* (Cohn, 1866) Wirnsberger, Larsen and Uhlig, 1987** – Berger, Catalogue of ciliate names 1. Hypotrichs, p. 55 (nomenclator containing all basionyms, combinations, and higher taxa of hypotrichs).

Nomenclature: No derivation of the name is given in the original description. The species-group name *carne·us -a -um* (Latin adjective; fleshy) refers to the flesh-colour of this taxon (vs. the yellow colour in *P. flava* and the dark red colour in *P. rubra*). Kahl (1932) classified *Keronopsis* as subgenus of *Holosticha*; thus, the correct name of the taxon *carnea* in Kahl is *Holosticha (Keronopsis) rubra* var. *carnea* (Cohn, 1866) Kent, 1882 because the taxon *carnea* was erected by Cohn and *Oxytricha rubra* was transferred to *Holosticha* by Kent. *Oxytricha flava* var. *canea* in Ganapati & Rao (1958, p. 83) is an incorrect subsequent spelling.

Remarks: Cohn (1866) found, like Wirnsberger et al. (1987), *Pseudokeronopsis carnea* sympatric with *P. flava* indicating that they are species. He gave no illustration because the general morphology is almost identical to that of *P. flava*. The main differences he found were in size, colour, and consistency. However, he doubted species status and therefore classified it as variety of his *Oxytricha flava*.

Rees' (1884; Fig. 181a, b) population of *Holosticha flava* had orange cortical granules and five dorsal kineties, indicating that he observed not the yellow *P. flava* with usually four kineties (Table 37), but the present species (Wirnsberger et al. 1987). Rees illustrated not a bicorona, but a tricorona, which I interpret as misobservation. Entz (1884, p. 364) considered *P. carnea* as a transitional form between *P. flava* and *P. rubra*.

Oxytricha rubra sensu Möbius (1888) had, like Rees' population, five rows of large, yellow patches – each composed of globular, yellow-red granules – and five transverse

[1] The rediagnosis provided by Wirnsberger et al. (1987) is as follows: Size in vivo about 140–200 × 28–40 μm. Granules orange-red, especially at low magnification, around cirral bases and dorsal cilia. Midventral rows shortened, end about 9–25 μm before the 6–7 transverse cirri. 10–14 bicoronal frontal cirri. 5–6 dorsal kineties.

cirri, strongly indicating that he observed *P. carnea* too (Wirnsberger et al. 1987). The single large, globular macronucleus is likely a (food)vacuole and must not be over-interpreted (Fig. 181c).

De Morgan's (1926) *Holosticha rubra flava* was "dusky yellow", and sometimes he observed dark-brown pigment masses near the base of the cirri. According to Wirnsberger et al. (1987, p. 85) this population likely belongs either to *P. carnea* (with question mark) or *P. rubra*. The midventral complex, which extends to near the transverse cirri, suggests identity with *P. rubra*; however, in the specimen of the neotype population of *P. carnea* (Fig. 181l), the gap between the midventral complex and the transverse cirri is also not very distinct, so the affinity of De Morgan's population remains uncertain.

Kahl (1932) considered Cohn's *Oxytricha flava carnea* as variety of *P. rubra* and the yellow *Oxytricha flava* obviously only as a form of *carnea*. Kahl (1933), in his review on the ciliates from the Northern and Baltic Sea, did not mention *P. carnea*. Kahl (1932) was obviously the last worker who mentioned the present species, until Wirnsberger et al. (1987) investigated the *Pseudokeronopsis rubra* group in detail. They studied not only the fine structure and the ontogenesis, but also fixed a neotype (Wirnsberger et al. 1987, Wirnsberger 1987, Wirnsberger & Hausmann 1988b). Wirnsberger et al. (1987) suggested that Foissner's (1984) *Pseudokeronopsis rubra* belongs to *P. carnea* mainly because of the orange-yellow colour. However, they correctly write that the morphometry confirms Foissner's identification. These somewhat contradictory data indicate that further species exist. For the sake of simplicity I follow Wirnsberger et al. (1987), but keep Foissner's data separate so that later workers are not confronted with a description based on at least two species.

Keronopsis rubra sensu Ganapati & Rao (1958) is dark-yellow and frequently brownish, indicating that they observed *P. carnea*.

Wirnsberger et al. (1987) fixed a neotype for *P. carnea* to solve the taxonomic problem of the *Pseudokeronopsis rubra* group. They studied three populations, including that described by Foissner (1984; see list of synonyms). Unfortunately, they failed to declare only one of these three populations as neotype, which would have been important, inasmuch as all of them are from different locations. Consequently, not only the old (original) type locality is not known (see occurrence and ecology), but also the new one is not fixed. I fix, in accordance with Erna Aescht (= E. Wirnsberger), the population from the Lillebaelt (Denmark) as neotype population because the slide deposited in the museum and the specimen illustrated in Fig. 9 by Wirnsberger et al. (= Fig. 181l in the present paper) are from this site.

For a comparison with other *Pseudokeronopsis* species, see key at the end of the genus section.

Morphology: First, the neotype population studied by Wirnsberger et al. (1987) is described. Then supplementary and/or deviating data from the other descriptions mentioned in the list of synonyms are provided. Foissner's data are kept separate for the reasons mentioned in the remarks.

The following light-microscopic characterisation of the neotype population (Fig. 181l, Tables 37, 38) contains basically only peculiarities not found in *P. rubra* and *P.*

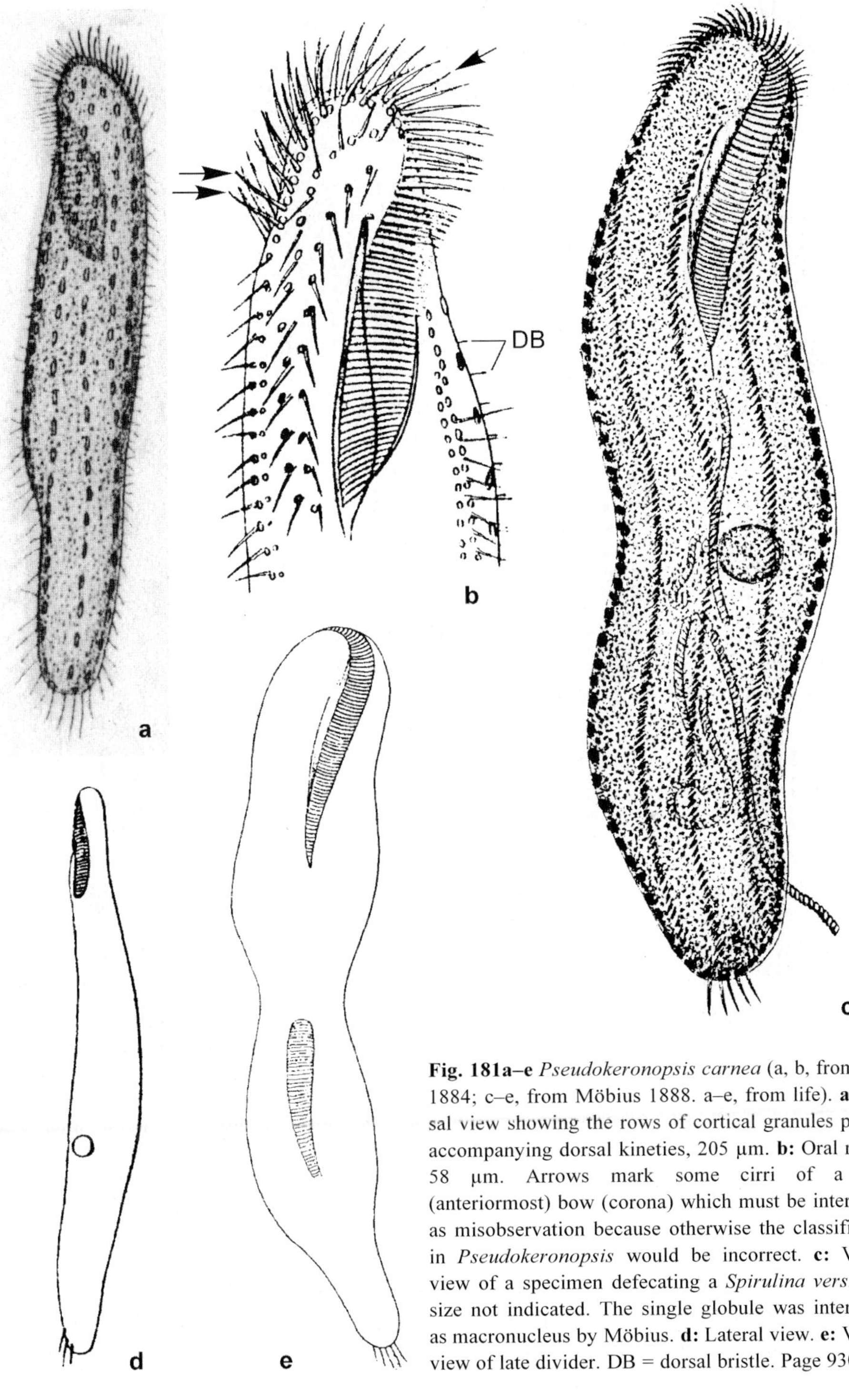

Fig. 181a–e *Pseudokeronopsis carnea* (a, b, from Rees 1884; c–e, from Möbius 1888. a–e, from life). **a:** Dorsal view showing the rows of cortical granules patches accompanying dorsal kineties, 205 µm. **b:** Oral region, 58 µm. Arrows mark some cirri of a third (anteriormost) bow (corona) which must be interpreted as misobservation because otherwise the classification in *Pseudokeronopsis* would be incorrect. **c:** Ventral view of a specimen defecating a *Spirulina versicolor*, size not indicated. The single globule was interpreted as macronucleus by Möbius. **d:** Lateral view. **e:** Ventral view of late divider. DB = dorsal bristle. Page 930.

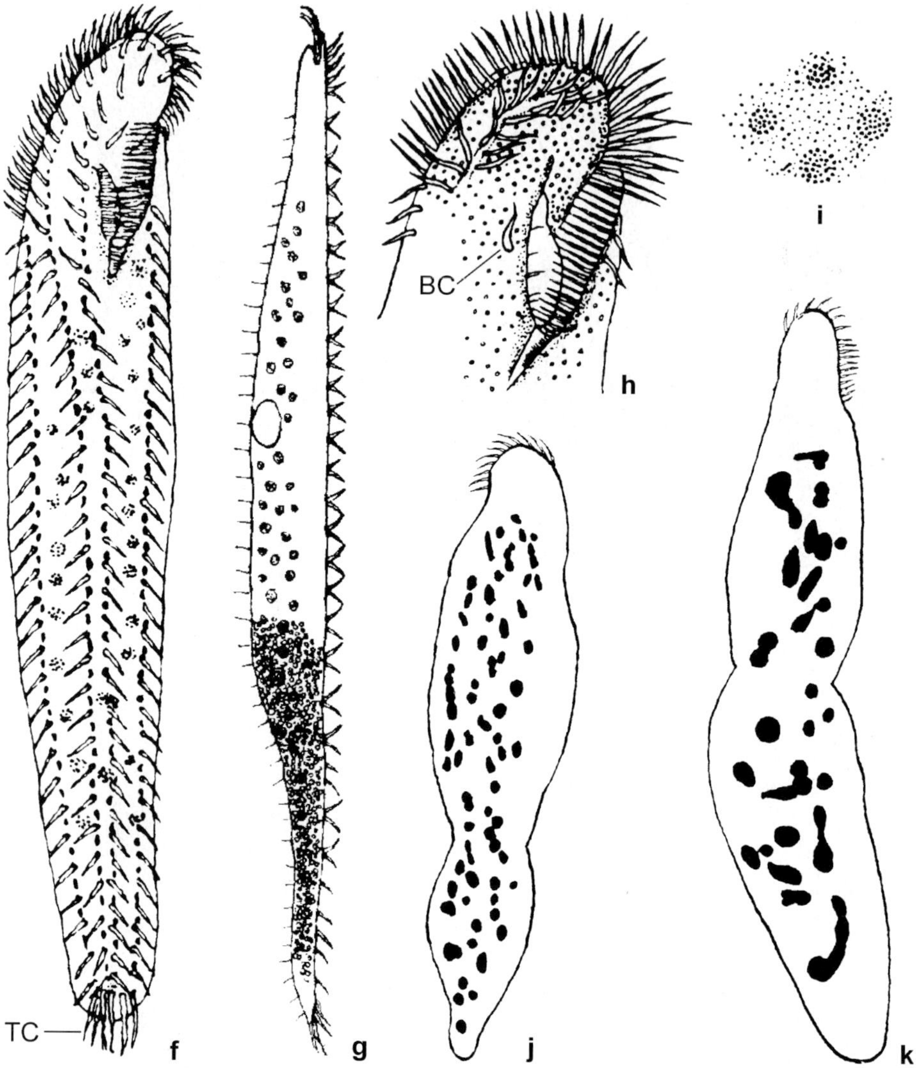

Fig. 181f–k *Pseudokeronopsis carnea* (from De Morgan 1926. f–h, from life; i, from life?; j, k, stain with congo-red or methyl-green). **f, g:** Ventral and right lateral view (180–370 μm) showing, among others, cirral pattern, dorsal bristles, contractile vacuole, macronuclear nodules, and "closely pressed corpuscles" (= blood corpuscle-shaped organelles?). **h:** Oral region showing, inter alia, the buccal cirrus and the paroral. **i:** Four macronuclear nodules. **j, k:** According to De Morgan, each two macronuclear nodules fuse prior to division. BC = buccal cirrus, TC = transverse cirri. Page 930.

flava; thus, see the neotype population of *P. rubra* for a general morphology. Body size in life 140–200 × 28–40 μm. Cortical granulation, see next paragraph. Moves more rapidly than *P. rubra*. Bicorona composed of 9.5–11.5 cirri on average. First midventral pair only about 13 μm behind anterior cell end (against 22–24 μm in *P. rubra*, Table 37).

On average around 85 frontal-midventral-transverse cirri (against around 103 in *P. rubra*). Midventral complex terminates distinctly (9–25 µm on average; Table 37) ahead of transverse cirri. No striking bright furrow along midventral complex at low magnification. 6–7 transverse cirri and 5–6 dorsal kineties.

According to the fine-structural analysis by Wirnsberger & Hausmann (1988b), the striking orange-red colour of *Pseudokeronopsis carnea* is caused by two types of pigment structures (cortical granules), namely the pigment vacuoles and the pigmentocysts (see their paper for a detailed documentation with electron micrographs). The pigment vacuoles are not extrusive and are confined to a characteristic ectoplasmic zone, about 1.5–3.0 µm thick, where they form 2–5, usually three layers. Only a few mitochondria can be found in this ectoplasmic region. The pigment vacuoles show a loose, fluffy periphery and a more intensely stained centre, which is elliptic with a sometimes lamellar appearance. The surrounding membrane of the (perhaps artificially) irregularly shaped and swollen vacuoles is often destroyed, but even in such cases the pigmented central part persists. Possibly, these pigment vacuoles are identical with the blood-cell-shaped organelles described for *P. rubra* (Fig. 180c) by Hu & Song (2001), for *P. flavicans* (Fig. 185l) by Song et al. (2002), and for *Uroleptopsis citrina* (Fig. 192e, 193b, c) by myself. However, it is somewhat surprising that Wirnsberger never mentioned these conspicuous, erythrocyte-shaped structures. The second type of granule, the pigmentocysts, are narrowly arranged around the cirri and dorsal bristles, and recognisable even at low magnification. In smaller numbers, they occur in the endoplasm and between the ciliary organelles. Under the light microscope, the pigmentocysts appear darker red than the pigment vacuoles. They are spherical to egg-shaped, measuring 0.5–1.0 µm in length. Their muciferous content is more or less dense; in Wirnsberger & Hausmann's material, none appeared paracrystalline. A short electron-dense channel is oriented to and connected with the pellicular membranes. Although many empty pigmentocysts with a persisting channel were found, the discharge of their contents

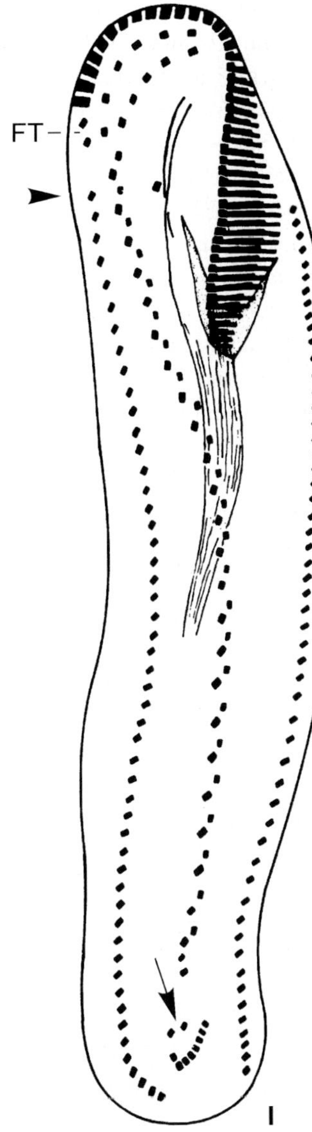

Fig. 181l *Pseudokeronopsis carnea* (from Wirnsberger et al. 1987. Protargol impregnation). Infraciliature of ventral side of a specimen from the neotype population, 202 µm. Arrow marks two pretransverse ventral cirri. Note short distance between rear end of midventral complex and transverse cirri. Arrowhead denotes beginning of right marginal row (see text for details). FT = frontoterminal cirri. Page 930.

has never been observed. For a pigment analysis, see Table 38. Larsen (1989) found that the pigments most likely consist of long-chained aliphatic compounds which may differ in structure and be combined to inorganic chromatophores.

The basic fine structural features of the infraciliature and the cytoplasmic organelles of *P. carnea* are similar to those found in other hypotrichs (Wirnsberger & Hausmann 1988a, b). However, a special kind of linear microtubular array borders the longer sides of the cirral bases and the margins of the adoral membranelles and those of the membranes in the right buccal area. To the left of the endoral, these microtubular arrays result in a highly ordered structure reminiscent of oral ribs. This peculiar arrangement of microtubules in cirri and membranelles has also been found in *Thigmokeronopsis jahodai*, indicating a homogeneity of the fine structure of urostylid hypotrichs (Wirnsberger & Hausmann 1988a). A typical plasma membrane covers flat alveoli that are frequently insignificant or invisible in main parts of the somatic region, but very distinct in the buccal area. Below the somatic pellicle is a single layer of longitudinal subpellicular microtubuli. An epiplasm could not be found (Wirnsberger & Hausmann 1988a).

The extracts of the cortical granules of *P. rubra*, *P. carnea*, and *P. flava* have been investigated by Wirnsberger et al. (1987; Table 38). Solutions of the crude extracts of *P. carnea* and *P. flava* revealed a difference in colour. The extract of *P. carnea* had an orange-brown colour, while that of *P. flava* appeared as clear yellow. The spectrums of the crude extracts and of the main fractions showed different maxima in all three species (Table 38). When examined in daylight, the PSC plates showed almost the same difference in colour as did the solutions of the crude extracts. *Pseudokeronopsis carnea* revealed four distinct fractions, wile *P. flava* showed only three bands. In UV-light at 365 nm the difference between the two preparations was even more striking. The *P. flava* bands showed a strong yellow fluorescence, while that of *P. carnea* had only a faint brownish colour (Wirnsberger et al. 1987, p. 84).

Additional and/or deviating observations by other authors (unless otherwise indicated, the data are from Cohn 1866): Body size 258–330 × 39–42 µm, on average 306 × 40 µm; 175 µm long on average (Rees); 180–370 µm long and 4–5times a long as wide (De Morgan); 100–300 µm (Ganapati & Rao). Body margins more or less in parallel and transversely truncated posteriorly. Dying specimens burst. Many macronuclear nodules about 3–4 µm across scattered throughout cytoplasm (De Morgan, Fig. 181g). Contractile vacuole difficult to recognise. According to De Morgan's illustration the vacuole is at about 40% of body length near the dorsal surface (Fig. 181g); Ganapati & Rao observed it in the middle of the posterior quarter, pulsation was moderate. Cortex with many fine orange-red granules, body therefore opaque; larger, likewise orange-red granules string together mainly along mid-body and lateral margins or scattered. Rees observed two small orange cortical granules per marginal cirrus and a large granule near each frontal and midventral cirrus (Fig. 181b). Möbius counted five dorsal rows of larger, yellow patches with smaller granules in between; he was the first who recognised that the dorsal patches are composed of several granules. Colour dusky-yellow, sometimes dark-brown pigment masses near cirral base (De Morgan). De Morgan found "in the ectoplasm a number of very small spheres of equal size which lie in direct contact with one another. These do not stain *intra vitam*, and require high magnification

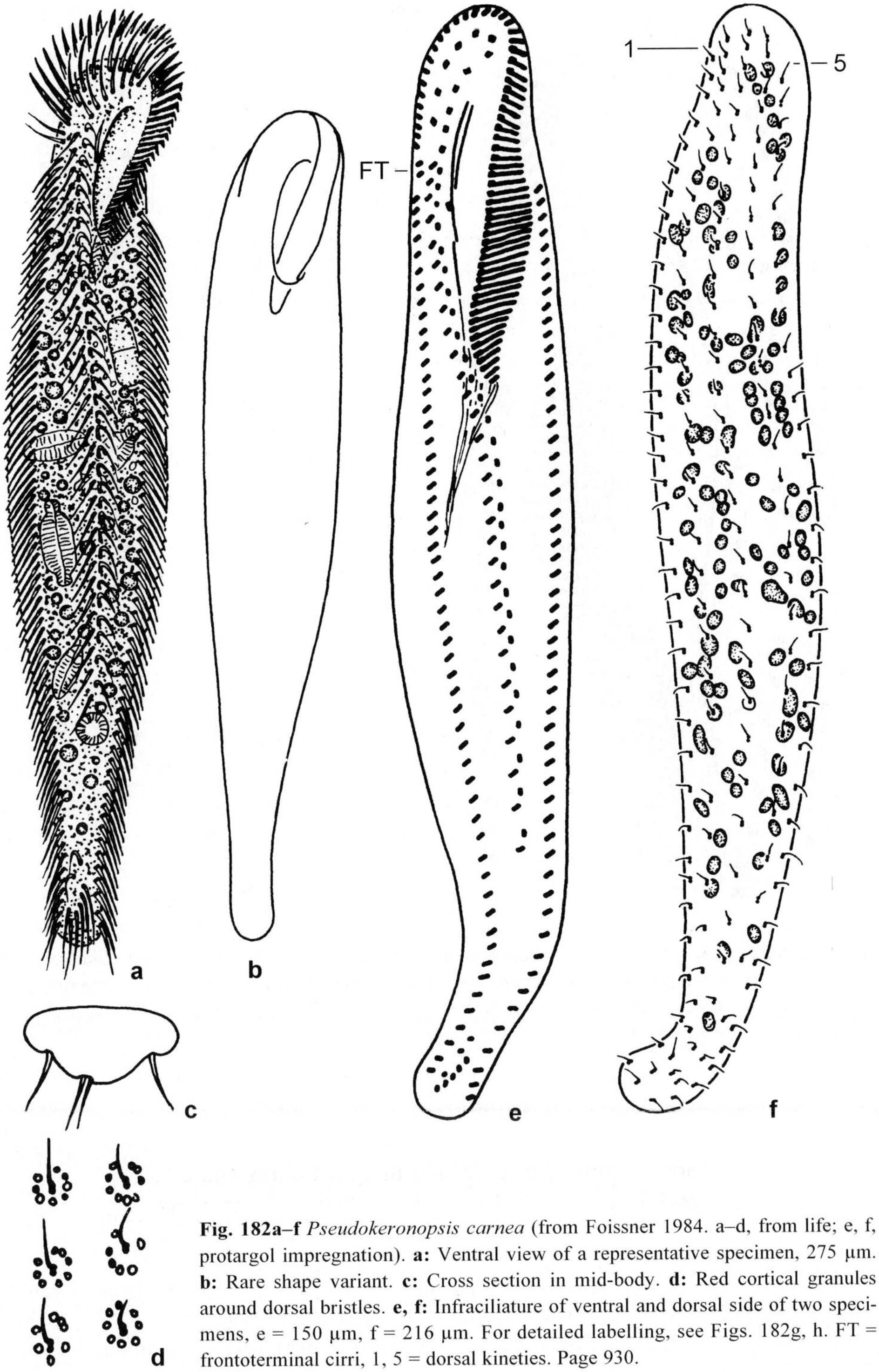

Fig. 182a–f *Pseudokeronopsis carnea* (from Foissner 1984. a–d, from life; e, f, protargol impregnation). **a:** Ventral view of a representative specimen, 275 µm. **b:** Rare shape variant. **c:** Cross section in mid-body. **d:** Red cortical granules around dorsal bristles. **e, f:** Infraciliature of ventral and dorsal side of two specimens, e = 150 µm, f = 216 µm. For detailed labelling, see Figs. 182g, h. FT = frontoterminal cirri, 1, 5 = dorsal kineties. Page 930.

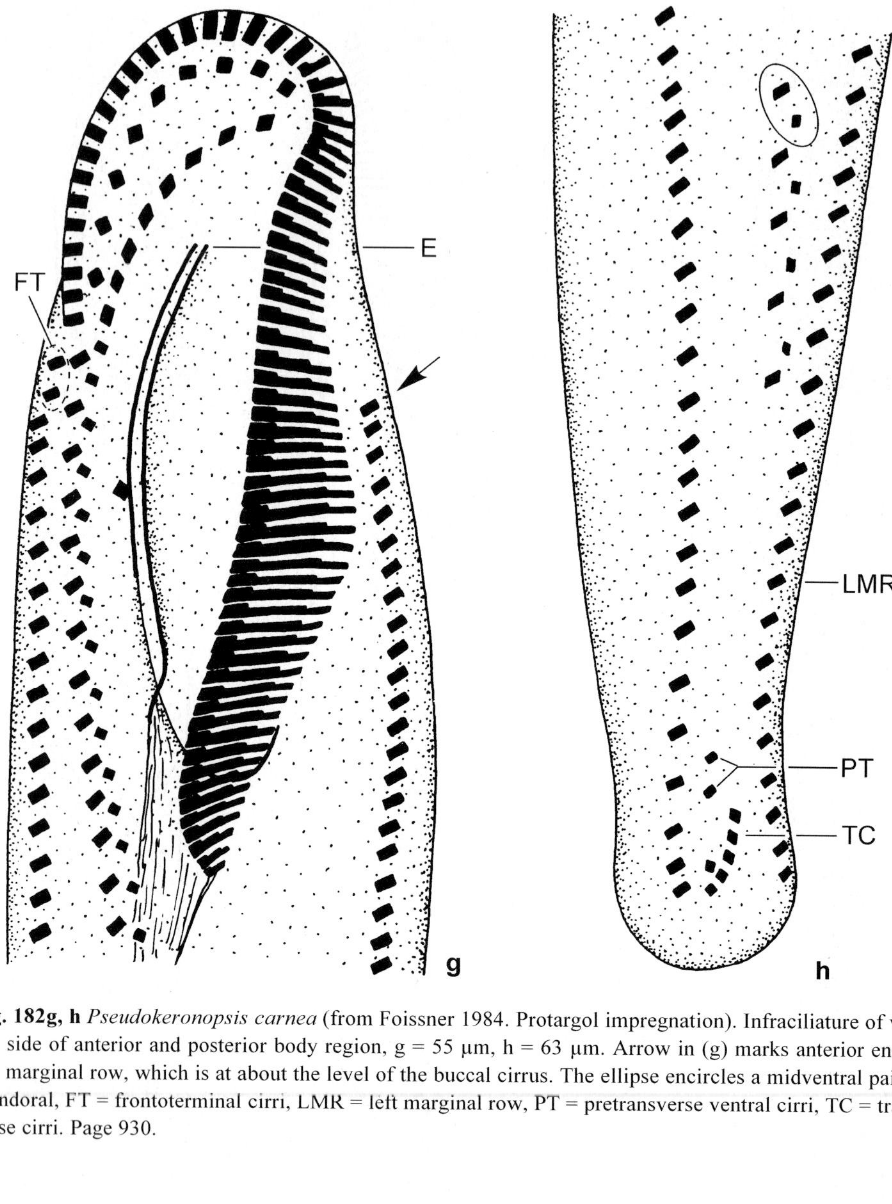

Fig. 182g, h *Pseudokeronopsis carnea* (from Foissner 1984. Protargol impregnation). Infraciliature of ventral side of anterior and posterior body region, g = 55 µm, h = 63 µm. Arrow in (g) marks anterior end of left marginal row, which is at about the level of the buccal cirrus. The ellipse encircles a midventral pair. E = endoral, FT = frontoterminal cirri, LMR = left marginal row, PT = pretransverse ventral cirri, TC = transverse cirri. Page 930.

to make out". Possibly these are the blood-cell-shaped organelles mentioned above. Body very flexible, movement thus worm-like. 40–50 adoral membranelles (Möbius). Cytopharynx short, ciliated (De Morgan; likely he observed the endoral cilia). No buccal cirrus illustrated by Rees (Fig. 181b), so it cannot be excluded that he observed another species (note that *Uroleptopsis* spp. also lack a buccal cirrus). By contrast, De Morgan and Ganapati & Rao described and illustrated such a cirrus (Fig. 181f, g). Midventral complex extends close to the 6–8 transverse cirri projecting about half their length beyond rear body end (De Morgan; see remarks); Ganapati & Rao counted 5–8 transverse cirri. Dorsal cilia short, arranged in five kineties accompanied by large cortical granules (Rees; likely he interpreted the rosette as single granule, as already suggest by Möbius 1888).

Population described by Foissner (1984; Fig. 182a–h, Table 37) 200–320 × 30–50 μm in life. Body contractile by about one third of length; very flexible and fragile, so mutilated specimens often occur; slender, extended individuals slenderly fusiform to pisciform; rear body end always slightly widened in spoon shape. According to Foissner, body 0.5–2.0:1 flattened dorso-ventrally (0.5:1 would mean that a specimen is twice as high as wide, which is unlikely); dorsal side slightly convex, mid-region of ventral side distinctly vaulted (Fig. 182c). 100–200 very small macronuclear nodules scattered throughout cytoplasm (Fig. 182f, Table 37); individual nodules globular to ellipsoidal. Likely several micronuclei. Contractile vacuole not detectable. Specimens pale to dark orange-red (never yellow) due to numerous endoplasmic granules about 0.1 μm across and many cortical granules 0.5–1.0 μm across arranged in rosettes around cirri and dorsal bristles. Endoplasm usually with many 1–5 μm-sized, greasy-shining, colourless globules and ingested diatoms. Movement sluggishly, worm-like. Adoral zone widened in proximal portion, largest membranelles about 20 μm long, extends far onto right body margin. Buccal field narrow, flat, on right side bordered by the long endoral and the slightly shorter paroral. Cirral pattern genus-specific, cirri about 15 μm long (Fig. 182e). Cirri forming bicorona slightly enlarged; midventral complex continuous with bicorona, terminates distinctly ahead of transverse cirri; left cirrus of each midventral pair smaller than right. Invariably (n = 15) two frontoterminal cirri at ordinary position (Fig. 182e, g). Buccal cirrus distinctly ahead of anterior end of undulating membranes. One or two pretransverse ventral cirri ahead of very fine transverse cirri, which are arranged in J-shape and project slightly beyond rear body end (Fig. 182e, h). Marginal rows distinctly separated posteriorly, both rows commence, as in *P. rubra*, about at level of buccal cirrus. Dorsal bristles about 4 μm long in life, arranged in five bipolar kineties. Caudal cirri lacking (Fig. 182f).

Cell division (Fig. 179k, l, o, 181e, j, k): Divisional morphogenesis of *P. carnea* was studied in detail by Wirnsberger (1987). She found a very similar overall pattern of the morphogenetic events in *P. rubra*, *P. carnea*, and *P. flava*. Consequently, the reader is referred to Figs. 179i–p. Wirnsberger et al. (1987) did not observe conjugation over a one-year period. De Morgan (1926) studied the behaviour of the nuclear apparatus during division. Accordingly, two nodules aggregated and then divided, that is, no single large macronuclear mass was formed prior to fission (Fig. 179j, k).

a

Fig. 183a *Pseudokeronopsis carnea* (from Ganapati & Rao 1958. From life). Ventral view, 176 μm. Page 730.

Occurrence and ecology: Marine, likely moderately common. Cohn (1866) discovered *P. carnea* in an aquarium in Breslau, Poland, filled with sea water from Heligoland, Germany. Furthermore, this 25-litre-bottle, which was not aerated artificially, contained material from Dorsetshire (England?). Thus, the original type locality of *O. flava carnea* was not known. Cohn found the present taxon together with *P. flava* among decaying algae and animals, indicating that it tolerates organic pollution.

Due to the neotype fixation by Wirnsberger et al. (1987), the type locality of *P. carnea* is the sampling site of the neotype population (for problems, see remarks). The (new) type location of *P. carnea* is the Lillebælt, Denmark (Wirnsberger et al. 1987), likely in the Nakkebølle Fjord at the south of the island Funen (Larsen 1987). In addition, Wirnsberger et al. (1987) found it in a sample from Heligoland, North Sea.

Further records documented by morphological data and/or illustrations: Oosterschelde, a part of the North Sea near the coast of the Netherlands (Rees 1884); Bay of Kiel, Baltic Sea, Germany (Möbius 1888); throughout the year in "Drake's Island Tank", a large shallow tank standing in front of a south window in the laboratory at Plymouth, United Kingdom (De Morgan 1926); algal layer of a sea water aquarium in the Zoological Institute of the University of Salzburg, filled with water and fish from the coast near Portorose (Slovene), Adriatic Sea (Foissner 1984).

Only one record not substantiated by illustrations and/or morphological data: Gulf of Biscay (Spain), at Castro Urdiales, a beach facing north towards the open sea and with continuous strong wave movement during January (Fernandez-Leborans & Novillo 1993, p. 216).

Wirnsberger et al. (1987) cultured *P. carnea* in artificial sea water (Wimex Seasalt, H. Wiegandt, Krefeld, Germany; 30–34‰ salinity) at room temperature with *Dunaliella*, rice grains, boild egg yolk, or yeast as food supply. Möbius' (1888) specimens ingested *Spirulina versicolor* (Fig. 181c) whereas Foissner's (1984) population devoured, inter alia, diatoms (Fig. 182a).

Pseudokeronopsis flava (Cohn, 1866) Wirnsberger, Larsen & Uhlig, 1987
(Fig. 184a–v, Tables 37, 38)

1866 ***Oxytricha flava* nov. spec.**[1] – Cohn, Z. wiss. Zool., 16: 288, 299, Tafel XV, Fig. 27–29 (Fig. 184a–c; original description; no type material available).

---

[1] The diagnosis, respectively, short characterisation provided by Cohn (1866, p. 299) is as follows: Körper gelb, flexil, nicht retractil, bandförmig platt, linear, nach hinten etwas verdünnt, der Vorderrand der Stirn stärker zugerundet, der mit zahlreichen, gleichlangen, wirbelnden Wimpern (Borsten) besetzte Vorderrand des bis zum ersten Drittel reichenden Peristoms abgerundet, schief nach hinten in den rechten Seitenrand des Peristoms übergehend, welcher mit langen, gekrümmten, der linke dagegen mit geraden, Querstreifen ähnlichen Wimpern besetzt ist. Eine gerade Furche (?) durchzieht die Bauchseite vom hintern Peristomwinkel bis zum After, und ist äusserlich mit einer doppelten Wimpernreihe besetzt; zwei andere Wimpernreihen begleiten sie näher den Seitenrändern. Ein Büschel von Afterwimpern ist unter dem geradabgestutzten hinteren Rand eingefügt. Contractile Blase eine, im ersten Drittel. Nucleus? Cuticula feinkörnig, beim Absterben nicht zerfliessend. Bewegung schwimmend, kriechend, mit steter Flexion des Körpers.

1882 ***Holosticha flava*, Cohn sp.** – Kent, Manual infusoria, II, p. 769, Plate XLIII, Fig. 19, 20 (redrawings from Cohn; revision; combination with *Holosticha*).
1884 ***Oxytricha flava* Cohn** – Gruber, Z. wiss. Zool., 40: 142, 143, Tafel IX, Fig. 30 (brief description of nuclear apparatus).
1884 ***Holosticha flavorubra* mihi. var. *flava*** – Entz, Mitt. zool. Stn Neapel, 5: 359, 360, Tafel 22, Fig. 14, 15 (Fig. 184g; see nomenclature and remarks).
1888 ***Holesticha flava*** – Gruber, Ber. naturf. Ges. Freiburg i. B., 3: 61, Tafel VII, Fig. 3, 4 (Fig. 184d, e; details on nuclear apparatus; incorrect subsequent spelling of *Holosticha*).
1932 ***Keronopsis rubra* var. *carnea* forma *flava*** – Kahl, Tierwelt Dtl., 25: 573, Fig. 104_6 (Fig. 184f, redrawing from Cohn; see nomenclature and remarks).
1933 ***Keronopsis rubra* forma *flava* Cohn** – Kahl, Tierwelt N.- u. Ostsee, 23: 108 (guide to marine ciliates).
1935 ***Keronopsis flava*** – Rühmekorf, Arch. Protistenk., 85: 260, Fig. 1, 3–5, 8–10, 13–15 (Fig. 184i–m; morphology, nuclear division, and starving forms; see nomenclature).
1987 ***Pseudokeronopsis flava* (Cohn, 1866) nov. comb.**[1] – Wirnsberger, Larsen & Uhlig, Europ. J. Protistol., 23: 79, Fig. 10, Tables 1–3 (Fig. 184h; combination with *Pseudokeronopsis*, redescription and fixation of neotype. Two neotype slides [registration numbers 1986/38, 1986/39] are deposited in the Oberösterreichische Landesmuseum in Linz [LI], Upper Austria).
1987 ***Pseudokeronopsis flava*** – Wirnsberger, Arch. Protistenk., 134: 150 (cell division).
2001 ***Pseudokeronopsis flava* (Cohn, 1866) Wirnsberger, Larsen and Uhlig, 1987** – Berger, Catalogue of ciliate names 1. Hypotrichs, p. 55 (nomenclator containing all basionyms, combinations, and higher taxa of hypotrichs).
2004 ***Pseudokeronopsis flava* (Cohn, 1866) Wirnsberger, Larsen & Uhlig, 1987**[2] – Song, Sun & Ji, J. mar. biol. Assoc. U.K., 84: 1137, Fig. 1A–G, 2–22, Tables 1, 2 (Fig. 184p–u; detailed description of Chinese population; voucher slides are deposited in the Laboratory of Protozoology, Ocean University of China).
2005 ***Pseudokeronopsis flava* (Cohn, 1866) Wirnsberger et al., 1987** – Sun & Song, Acta zool. sin., 51: 81, Fig. 1a–e, 2a–e, 3a–d, Plate Ia–m (Fig. 184.1a–l; description of cell division).

Nomenclature: No derivation of the name is given in the original description. The species-group name *flav·us -a -um* (Latin adjective; yellow, sulphurous yellow) refers to the yellow colour of this species. For derivation of the name *flavorubra*, superfluously introduced by Entz (1884), see *P. rubra*.

Kahl (1932) classified *Keronopsis* as subgenus of *Holosticha*; thus, the correct name of the taxon *flava* in Kahl is *Holosticha (Keronopsis) rubra* var. *carnea* f. *flava* (Cohn, 1866) Kent, 1882 because Cohn erected the taxon *flava* and Kent transferred *Oxytricha rubra* to *Holosticha*. Kahl (1933) did not mention the taxon *carnea*, so the correct name of the taxon *flava* in Kahl (1933) is *Holosticha (Keronopsis) rubra flava*. Rühmekorf (1935, p. 257) wrote that she found hypotrichs of the genus *Holosticha*; thus, the correct name of the present species in her paper is *Holosticha (Keronopsis) flava* and not

[1] The rediagnosis provided by Wirnsberger et al. (1987) is as follows: In vivo about 140–260 × 40–57 µm, very flexible. Granules yellow. Midventral rows end on the average 30 µm before the transverse cirri which are nearly always 2 in number. Bicorona comprised of 9 frontal cirri. On the average 4 dorsal kineties.

[2] Song et al. (2004b) provided the following new diagnosis based on previous and original data: Yellow to yellow-brown coloured marine *Pseudokeronopsis*; body ribbon-shaped and highly flexible, about 140–350 × 30–50 µm in vivo; bicorona comprising about five pairs of frontal cirri; one buccal, 2–3 frontoterminal and 2–4 transverse cirri; midventral rows consisting of ~ 30 pairs of cirri, which terminate subcaudally some distance above the inconspicuous transverse cirri; about 50 adoral membranelles; left and right marginal rows separated at posterior end; 3–4 dorsal kineties. Colourless, blood-cell-shaped cortical granules ellipsoid in shape, distributed over whole cell, while yellow to yellow-brownish pigments positioned in a *P. rubra*-pattern; contractile vacuole positioned in posterior 1/4–1/5 of body. Numerous macronuclear segments.

Keronopsis flava, which would be a "new combination" with *Keronopsis*. *Holotricha flavorubra* in Entz (1884, p. 373) is an incorrect subsequent spelling of *Holosticha*.

Remarks: Cohn (1866) provided a rather good description of this yellow representative of the *Pseudokeronopsis rubra* group. He could not recognise the nuclear apparatus, indicating that his population had many macronuclear nodules. Before this separation from *P. rubra*, *Pseudokeronopsis flava* caused a great variability in the colour of *P. rubra*, ranging from dark red to yellowish. Kent (1882), who did not provide own data, transferred *Oxytricha flava* (and *Oxytricha rubra*) to *Holosticha* because of the uninterrupted median cirral rows (against the "sporadically" arranged cirri of the 18-cirri oxytrichids).

Gruber (1884b, p. 482) found four varieties of "*Holosticha (Oxytricha) flava* Cohn" in the Harbour of Genoa, Italy, however, without giving details. In two other papers, he provided some data on the nuclear apparatus of *H. flava* and mentioned a brown (Fig. 184o) and colourless (Fig. 184n) form (Gruber 1884a, 1888). Possibly, the colourless population is *P. decolor*, as indicated by the lack of a distinct colour and the lower number of macronuclear nodules. But the information available seem to be too scant to assign these populations definitely.

Entz (1884) studied the *P. rubra* group in detail. Unfortunately, he introduced the superfluous species-group name *flavorubra* instead of using the name of the oldest synonym, *Oxytricha rubra*. He distinguished two varieties, namely *Holosticha flavorubra flava* and *H. flavorubra rubra*. Entz doubted some observations by Cohn and Gruber, especially that the nuclear apparatus is composed of many macronuclear nodules. Entz himself observed only two nodules in both varieties (Fig. 184g); unfortunately, he could not check this observation when he wrote the paper (see also remarks on Entz 1884 in *P. rubra* and the bimacronucleate *P. rubra* population described by Wang & Nie 1932; Fig. 178z). Furthermore, he found that the right marginal row extends, like the bicorona, around the anterior body margin. Now we know that these observations are likely incorrect because so far no binucleate "*Pseudokeronopsis flava*" or "*P. rubra*" have been reliably recorded. And the right marginal row always commences near the distal end of the adoral zone. However, since we cannot exclude that a binucleate *Pseudokeronopsis* exists, I keep Entz's data separate.

Kahl (1932) classified *Oxytricha flava* only as a form of *Holosticha (Keronopsis) rubra*. His text is somewhat confusing and superficial as concerns the ranking of the taxa. I think that he considered *flava* as a form of "*Holosticha rubra* var. *carnea*". Like Entz he assumed that – according to his own observations – it is only a yellow modification of *Pseudokeronopsis rubra*.

Rühmekorf's (1935) identification is likely correct as already suggested by Wirnsberger et al. (1987) although the midventral complex extends to near the rear body end (Fig. 184i), whereas it terminates distinctly ahead of the transverse cirri in the neotype population (Fig. 184h). Obviously, Rühmekorf forgot to illustrate the transverse cirri she mentioned in the description.

Hamburger & Buddenbrock (1929, p. 86), Borror (1972, p. 11), and Borror & Wicklow (1983, p. 123) synonymised *Oxytricha flava* with *P. rubra*, and Carey (1992) even ignored it in his review on marine interstitial ciliates. By contrast, Wirnsberger et

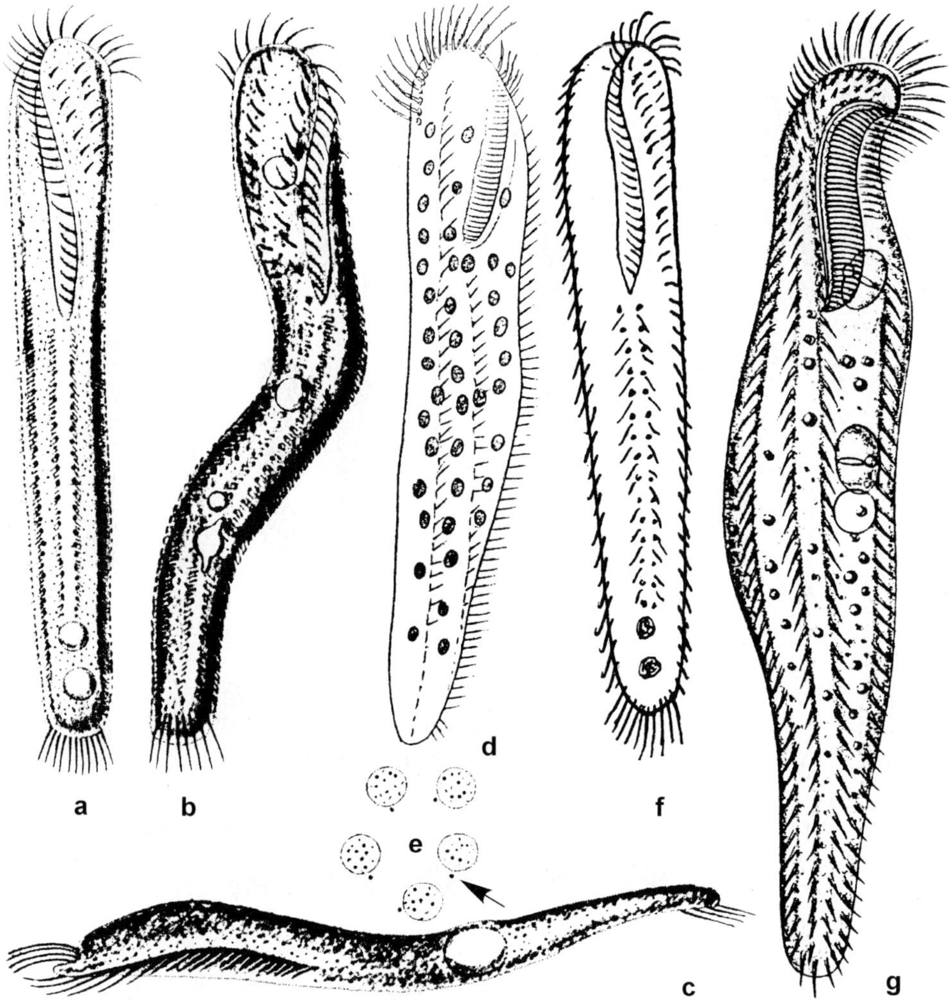

Fig. 184a–g *Pseudokeronopsis flava* (a–c, from Cohn 1866; d, e, from Gruber 1888; f, after Cohn 1866 from Kahl 1932; g, from Entz 1884. a–c, f, g, from life; d, e, nuclear stain). **a–c, f:** Dorsal view showing ventral cirral pattern (a, f), ventral view (b), and left lateral view (c), 160–200 μm. The bundle of cirri at the rear body end are transverse cirri according to Cohn; however, it cannot be excluded that they are the rearmost marginal cirri because, according to the redescription by Wirnsberger et al. (1987), *Pseudokeronopsis flava* has only 1–3 transverse cirri (see Fig. 184h). Note the contractile vacuole in the posterior body half (especially Fig. 184c), a feature confirmed by Song et al. (2004b; Fig. 184p) and Sun & Song (2005a; Fig. 184.1a). **d, e:** Ventral cirral pattern and part of nuclear apparatus, size not indicated. Arrow marks a tiny globule interpreted as micronucleus by Gruber. **g:** Ventral view showing cirral pattern, contractile vacuole, and nuclear apparatus. The two macronuclear nodules do not match the data by other authors so that they must be interpreted as misobservation (see text); on the other hand it cannot be excluded that such a species with only two nodules exists. Page 940.

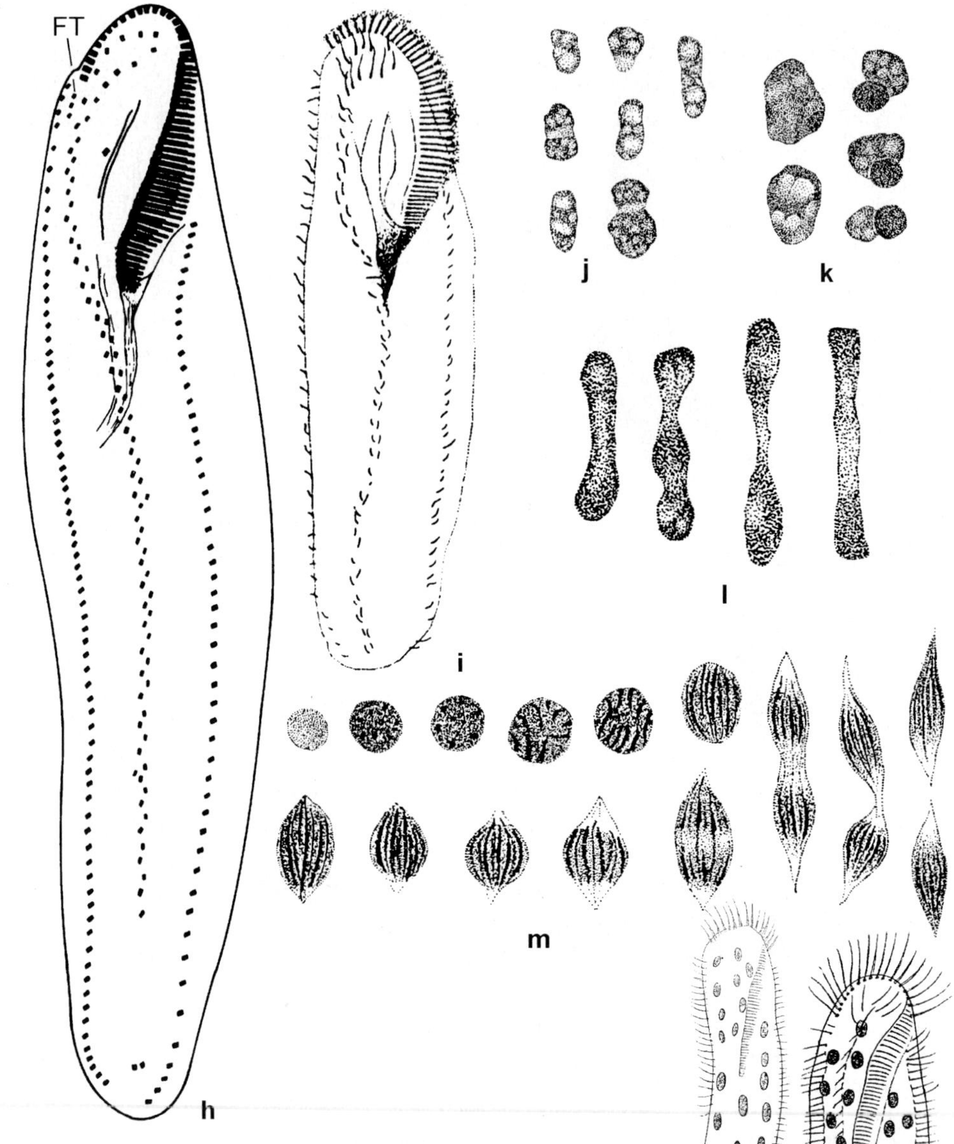

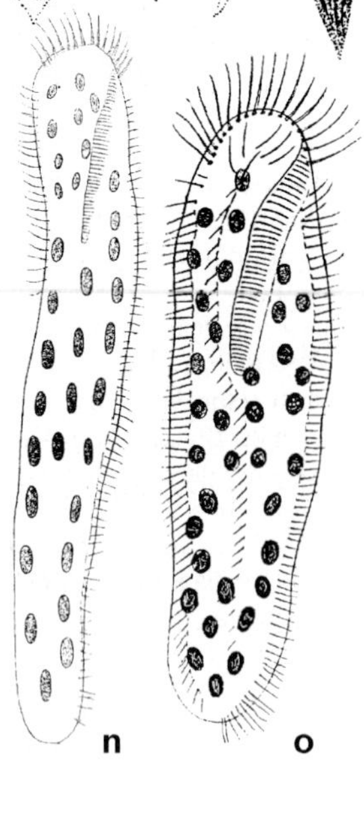

Fig. 184h–o *Pseudokeronopsis flava* (h, from Wirnsberger et al. 1987; i–m, from Rühmekorf 1935; n, o, from Gruber 1888. h, protargol impregnation; i, silver-staining according to Gelei; j–m, Feulgen stain; n, o, nuclear stain [method not specified]). **h:** Infraciliature of ventral side of a specimen from the neotype population, 215 μm. Main features of *P. flava* are the low number (usually 2) of transverse cirri and the distinct gap (on average 30 μm) between these cirri and the midventral complex. **i:** Infraciliature of ventral side, 230–250 μm (size of individual specimen according to data provided [1:800] only 86 μm). **j, k:** Normal and clumsy macronuclear nodules. **l, m:** Dividing macronuclear nodules (l) and micronuclei (m). **n, o:** Colourless and brown specimen. FT = frontoterminal cirri. Page 940.

al. (1987) reactivated Cohn's *Oxytricha flava* and provided a detailed description, including chromatographic analysis of pigments (Table 38). To clear up the complex problem around *Pseudokeronopsis rubra* they designated a neotype not only for *P. rubra*, but also for *P. flava* and *P. carnea*. Thus, the description below commences with the characterisation of the neotype population.

Recently, Song et al. (2004b) described a Chinese population of *P. flava*. It differs from the neotype population in the number of transverse cirri (3–4 vs. 1–3) and dorsal kineties (invariable 3 vs. 3–5, on average 3.8). Song et al. (2004b) described a posteriorly dislocated contractile vacuole, which was already mentioned in the original description (Fig. 184c, p). Unfortunately, this organelle is not described for the neotype population. Sun & Song (2005a, b) describe the morphogenesis of the Chinese population (Fig. 184.1a–l).

The identifications by Rees (1884) and De Morgan (1926) are likely incorrect. Rees' (1884; Fig. 181a, b) population has orange cortical granules and five dorsal kineties, indicating that he observed *Pseudokeronopsis carnea*, as already suggested by Wirnsberger et al. (1987). *Holosticha rubra* var. *flava* sensu De Morgan (1926, Fig. 181f–k) is, according to Wirnsberger et al. (1987, p. 85), likely a *P. carnea* or *P. rubra*. I mention it under *P. carnea* because De Morgan described the colour as dusky yellow with sometimes dark-brown pigments near the cirral bases. Identity with the neotype of *P. flava* is very unlikely because De Morgan's population has 6–8 transverse cirri (against usually 2) and a midventral complex which terminates at the transverse cirri (against distinctly ahead of in *P. flava*).

Possibly, the unidentified population studied by Anigstein (1949) also belongs to *P. flava* because it was 200–250 µm long, yellowish, and rather resistant to lower osmotic pressure. *Holosticha flavorubra flava* sensu Gourret & Roeser (1888, Fig. 191a, b) is insufficiently described and thus their identification not comprehensible.

Due to the yellow colour, *Pseudokeronopsis flava* is easily confused with *Uroleptopsis citrina* at superficial live observation. However, they differ in several features, partially recognisable even by detailed live observation: (i) adoral zone continuous against with break; (ii) buccal cirrus present against lacking; (iii) transverse cirri present against lacking; (iv) paroral of normal length (mean = 15 µm) against short (mean = 9 µm); (v) usually four dorsal kineties against three (however, *P. flava* sensu Song et al. 200b also has only three kineties!). For a separation from the other *Pseudokeronopsis* species, see key.

Morphology: The neotype population described by Wirnsberger et al. (1987) is characterised first. This is followed by the description of the Chinese population (Song et al. 2004b) and additional and/or deviating data from the other sources mentioned in the list of synonyms. Finally, the data by Entz for his binucleate population are provided.

The following characterisation of the neotype population (Fig. 184h, Tables 37, 38) basically contains only peculiarities not found in *P. rubra* and *P. carnea*; thus, see the neotype population of *P. rubra* for a general morphology. Body size 140–260 × 40–57 µm in life; body length:width ratio 5.1:1 on average in protargol preparations (Table 37). Body very flexible and frequently distorted. No contractile vacuole observed.

Cortical granules yellow, but their grouping around cirral and cilia bases is hardly visible (more distinct with the UV-microscope). For the absorption maxima of crude extracts and fractions found by Larsen (1987) and Wirnsberger et al. (1987), see Table 38. Movement still faster than *P. carnea*, nestling worm-like close to particles of mud. Dead cells can be found dried on slides, whereas cells of *P. rubra* and *P. carnea* burst. Conjugation was never observed. Bicorona usually composed of nine cirri. Midventral complex ends on the average 30 µm ahead of the transverse cirri. Supernumerary cirri occur, perhaps due to incomplete resorption during ontogenesis. Frequently three frontoterminal cirri. Transverse cirri, usually two in number, project conspicuously beyond posterior margin. Right marginal row commences near distal end of adoral zone of membranelles, left one begins distinctly behind level of buccal cirrus (Fig. 184h; by contrast, in both *P. rubra* and *P. carnea* both marginal rows commence about at level of buccal cirrus, Fig. 179b, 181l). Usually four dorsal kineties (Table 37).

Chinese population described by Song et al. (2004b; Fig. 184p–u; most features are documented by micrographs not shown in present monograph): freshly collected specimens in life 200–250 µm long, ratio of body length:width about 4–5:1; after several days of culture, both size and shape are rather variable, that is, length 150–300 µm, body length:width ratio 6–9:1. Body slender to long-belt-shaped, acontractile; posterior portion often more or less narrowed and somewhat distorted, that is, folded or twisted in middle portion when crawling on substrate (Fig. 184p, r, s); body margins parallel in slender specimens, sometimes distinctly vaulted in mid-body in small individuals. Dorsal side (sometimes?) irregularly bulged (Fig. 184r). Body very flexible, dorsoventrally flattened by about 2:1. About 100 macronuclear nodules mainly scattered throughout central body portion (Fig. 184u); individual nodules ovoid to ellipsoidal, about 5 × 3 µm, difficult to recognise in life. Several micronuclei, ovoid and about 3 µm long. Contractile vacuole about 15 µm across, pulsating only infrequently (at about 5 min intervals), near left cell margin in posterior body portion (at 72–76% in specimens shown in Fig. 184p, s); in some specimens two contractile vacuoles of different size, which fuse before pulsation. Pellicle thin. Two types of cortical granules: type one pigment-like and only about 0.5 µm across, yellowish to bright-yellow when observed under high magnification (×1000). Brightness of granule colour possibly dependent on life stage(?) or nutrition, that is, freshly collected cells usually more brightly yellow than those after several weeks of culture. Most granules densely arranged around cirri on ventral side forming dark belts along cirral rows (Fig. 184s); some granules scattered throughout cortex rendering cells slightly yellow-brownish under low magnification. Generally, colour is stronger in anterior and posterior portion than in central region. Type two granules formed like erythrocytes of mammals, 1.5–2.0 µm across, colourless, narrowly spaced underneath cell surface (and underneath type one granules). Cytoplasm often with numerous refractive, colourless or greyish globules 3–6 µm across. Food vacuoles not observed. Movement without peculiarities. Adoral zone occupies 20–30% of body length in protargol preparations, bases of membranelles up to 6–10 µm wide, cilia about 15 µm long in life. Distal end of zone extends, as is usual in *Pseudokeronopsis*, distinctly onto right body margin. Buccal field narrow in life. Paroral and endoral about equal in length (Fig. 184t), which is a distinct difference to the neotype population and

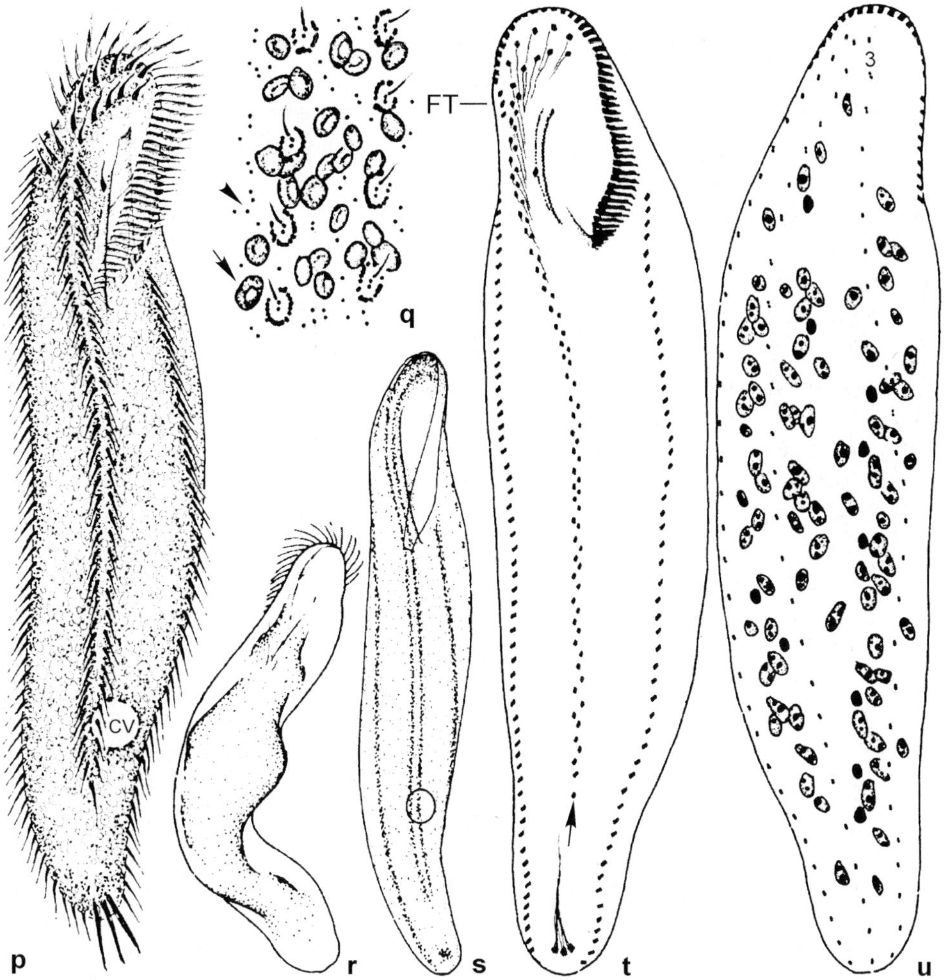

Fig. 184p–u *Pseudokeronopsis flava* (from Song et al. 2004b. p–s, from life; t, u, protargol impregnation). **p:** Ventral view of a representative specimen, 218 μm. **q:** Cortical granulation. Arrow marks an erythrocyte-like colourless granule 1.5–2.0 μm across, arrowhead denotes yellow-brownish to brightly-yellow coloured granules about 0.5 μm across. **r:** A small (160 μm) and curved specimen showing humped dorsal side. **s:** Ventral view showing distribution of small cortical granules and contractile vacuole near left cell body margin in posterior portion. **t, u:** Infraciliature of ventral and dorsal side and nuclear apparatus of same specimen, 200 μm. Arrow in (t) marks posterior end of midventral complex. CV = contractile vacuole, FT = frontoterminal cirri, 3 = dorsal kinety 3 (= rightmost kinety in this population). Page 940.

other *Pseudokeronopsis* species where the paroral is distinctly shorter than the endoral. Cirral pattern basically without peculiarities; thus, the reader is mainly referred to Figs. 184t, u and Table 37. Most cirri relatively fine, 12–15 μm long; midventral complex terminates about 30 μm ahead of transverse cirri. Chinese population invariable

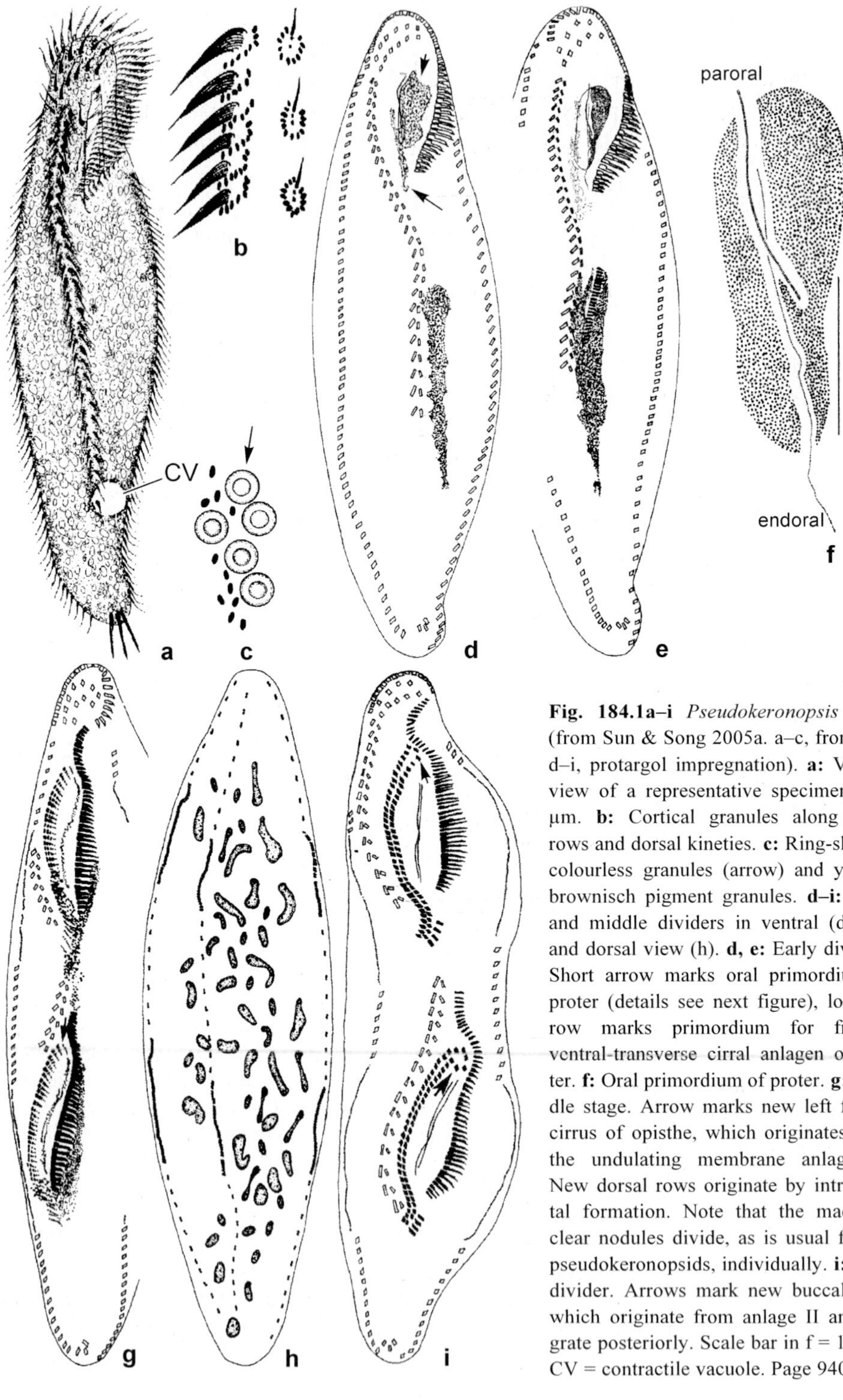

Fig. 184.1a–i *Pseudokeronopsis flava* (from Sun & Song 2005a. a–c, from life; d–i, protargol impregnation). **a:** Ventral view of a representative specimen, 180 µm. **b:** Cortical granules along cirral rows and dorsal kineties. **c:** Ring-shaped, colourless granules (arrow) and yellow-brownisch pigment granules. **d–i:** Early and middle dividers in ventral (d–g, i) and dorsal view (h). **d, e:** Early dividers. Short arrow marks oral primordium of proter (details see next figure), long arrow marks primordium for frontal-ventral-transverse cirral anlagen of proter. **f:** Oral primordium of proter. **g:** Middle stage. Arrow marks new left frontal cirrus of opisthe, which originates from the undulating membrane anlage. **h:** New dorsal rows originate by intrakinetal formation. Note that the macronuclear nodules divide, as is usual for the pseudokeronopsids, individually. **i:** Later divider. Arrows mark new buccal cirri, which originate from anlage II and migrate posteriorly. Scale bar in f = 10 µm. CV = contractile vacuole. Page 940.

with three dorsal kineties, each composed of about 30 basal body pairs bearing 2–3 µm long bristles (Fig. 184u).

Additional observations by Cohn (1866), Gruber (1884a, 1888), and Rühmekorf (1935): Body size according to Cohn 160–200 × 16–23 µm, occasionally much smaller specimens occur, length:width ratio around 10:1; Rühmekorf's specimens 230–250 × 40–50 µm, only around 20 µm high. Rühmekorf (1935) studied starving specimens: at first they became very narrow; later, they reduced the body size in general, namely down to 20–30 × 10–15 µm. 63% of hunger specimens with a length of 90–110 µm regenerate to normal individuals. Body flexible but acontractile, does not burst when dying; slightly but continuously narrowed posteriorly; anterior end broadly rounded, rear end bluntly truncated; distinctly flattened dorso-ventrally (Cohn); body yellow and opaque (Cohn); colour due to yellowish cortical granules, rarely some larger, reddish granules in lateral region of ventral side (Cohn). Cortex finely granulated (Cohn). Nuclear apparatus not recognised by Cohn, which is easily explained by the high number of rather small macronuclear nodules; first recognised by Gruber (1884a, p. 143; 1888, Fig. 184d, e). Rühmekorf found 50–80 macronuclear nodules and 5–8 micronuclei; each micronucleus with eight long, thin chromosomes. A vacuole which Cohn thought was contractile in rear third of cell near left body margin; sometimes Cohn observed a row of vacuoles with one right of the cytostome. Movement restless, bends horseshoe-like or swims straight ahead, often turning back (Cohn); according to Bullington (1925, p. 271; body 226 × 69 µm), *Pseudokeronopsis flava* rotates right spiralling. Adoral zone about one third of body length (Cohn). Bicorona composed of around eight cirri (Cohn). Numerous, slightly enlarged transverse cirri form a bundle; likely Cohn misinterpreted some marginal cirri as transverse cirri because *P. flava* usually has a rather low number according to Wirnsberger et al. (1987).

Population from Bay of Naples described by Entz (1884) 200–270 × 40–60 µm, that is, 5–6 times as long as wide (Fig. 184g). Body elongate-elliptical, rear portion almost tail-like, anterior portion also narrowed and usually curved leftwards; dorsal side vaulted, ventral flat or even hollowed; distinctly contractile. Two (!) macronuclear nodules with obviously one micronucleus each (Fig. 184g; see remarks). Contractile vacuole near left cell margin about in mid-body. Cytoplasm coarsely granulated and always yellow (amber, straw-coloured, brownish-yellow); often both body ends almost colourless, however, even in the palest specimens the anterior tip is golden yellow; both ventral and dorsal side each with four wide, coarsely granulated stripes. Adoral zone about 27% of body length (according to text only about 20%); buccal area narrow, right and anterior margin distinctly bulged. Cirral pattern as shown in Fig. 184g, that is, a bicorona continuing in a midventral complex which extends to near rear body end; however, the anteriorly curved right marginal row is very likely a misobservation. Five enlarged transverse cirri, which are, however, not clearly recognisable in the illustration (Fig. 184g).

Cell division: Cohn (1866) did not see dividers although the abundance was rather high. Entz (1884) often found dividers, but did not provide any data. Gruber (1888) first recognised that the individual macronuclear nodules do not fuse to a single mass prior to division (Fig. 184l, m). Ontogenesis was studied in detail by Wirnsberger

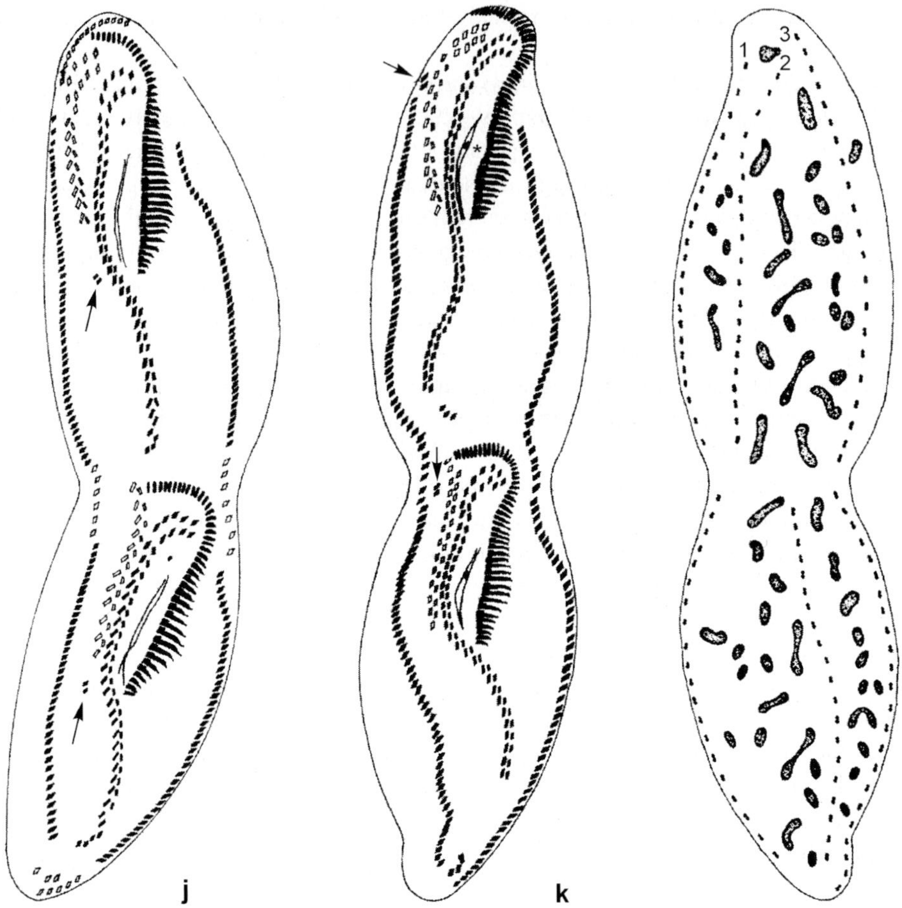

Fig. 184.1j–l *Pseudokeronopsis flava* (from Sun & Song 2005a. Protargol impregnation). Infraciliature of ventral (j, k) and dorsal side (l) of late and very late dividers (j–l = 173 µm). Arrows in (j, k) mark anteriorly migrating frontoterminal cirri. 1–3 = dorsal kineties. Page 940.

(1987), Wirnsberger & Hausmann (1988a), and Sun & Song (2005a, b). Wirnsberger (1987) and Sun & Song (2005a) found a very similar overall pattern of the morphogenetic events in *P. rubra*, *P. carnea*, and *P. flava*. Thus, the reader is referred to Figs. 179i–p and 184.1a–l. The sole peculiarity in *P. flava* is that the posteriormost and even the anteriormost frontal-midventral-transverse cirral anlagen form supernumerary cirri (if compared with the interphase cells), which are resorbed later.

Occurrence and ecology: Marine, likely moderately common. Cohn (1866) found *P. flava* in an aquarium in Breslau, Poland, filled with sea water from Heligoland, Germany. Furthermore, this 25-litre-bottle, which was not aerated artificially, contained material from Dorsetshire (England?). Thus, the original type locality of *O. flava* was not known. Cohn observed most specimens among decaying algae and animals, indicating that it tolerates organic pollution. Now, the type locality is determined by the neotype which was designated by Wirnsberger et al. (1987). They found *P. flava*, sympatric with *P. carnea*, in a sample collected by Hans F. Larsen in the Lillebælt, Denmark, likely in the Nakkebølle Fjord at the south of the island Funen (Larsen 1987). Kahl (1932, 1933) did not provide detailed faunistic data about *P. flava*; accordingly, *Pseudokeronopsis rubra*, with which he synonymised *P. flava*, is rather common in mesosaprobic debris.

Records substantiated by morphological data and or illustrations: Ligurian Sea in Genoa, Italy (Gruber 1884b, 1888); abundant on freshly collected *Ulva* sp. and among algae from the littoral, but also on algae-covered crustaceans from the Bay of Naples, Italy (Entz 1884); marine sludge sample from Plymouth, England (Rühmekorf 1935); late July in littoral water (33‰; water temperature about 28°C) near Zhanjiang, Guangdong Province, South China (Song et al. 2004b, Sun & Song 2005a, b).

Unsubstantiated records: abundant among purple bacteria in a brackish water habitat near Laboe, a village about 10 km north of the city of Kiel, Germany (Sick 1933, p. 66); Bay of Concarneau, Bretagne, France (Fabre-Domergue 1885, p. 568); brackish water habitat in St. John's Bayou, Louisiana, USA (Smith 1904, p. 46).

Pseudokeronopsis flava is polyphagous (Wirnsberger et al. 1987), according to Entz (1884) it feeds mainly on diatoms. By contrast, Rühmekorf (1935) stated that diatoms, heterotrophic flagellates, and other small organisms present in the raw culture were not ingested. She cultivated her populations (*P. flava* and *P. flavicans*) in Erdschreiber-solution with *Dunaliella* sp. as food. Wirnsberger et al. (1987) used artificial sea water (Wimex Seasalt. H. Wiegandt, Krefeld, Germany; 30–34‰) at room temperature with *Dunaliella* sp., rice grains, boiled egg yolk, or yeast as food.

Division rate at optimum conditions (cultivation in Nordsee-Erdschreiber-solution at 20–21° C) about 0.49 per day; at 15° C division rate decreases to 0.23, at 10° C to 0.12 per day (Rühmekorf 1935). Division rate (0.49 d^{-1}) did not change when salinity was lowered to about 16‰; at 8‰ it decreased to 0.35 d^{-1}. At 64‰ specimens divided maximally six times during 57 days.

Statkewitsch (1905, p. 300) studied the influence of electric current on 27 ciliate species, inter alia, *Pseudokeronopsis flava*; however, no details were provided for this species.

Pseudokeronospsis flavicans (Kahl, 1932) Borror & Wicklow, 1983 (Fig. 185a–q, Table 37, Addenda)

1932 ***Keronopsis flavicans* spec. n.**[1] – Kahl, Tierwelt Dtl., 25: 574, Fig. 104_4 (Fig. 185a; original description; not type material available; see nomenclature).

1933 ***Keronopsis flavicans* Kahl 1932** – Kahl, Tierwelt N.- u. Ostsee, 23: 109, Fig. 16.28 (Fig. 185b; guide to marine ciliates).

1935 ***Keronopsis flavicans*** – Rühmekorf, Arch. Protistenk., 85: 257, Fig. 6, 7, 11, 12 (Fig. 185c–f; see nomenclature and remarks).

[1] Kahl did not provide a formal diagnosis.

1972 ***Keronopsis flavicans* Kahl, 1932** – Borror, J. Protozool., 19: 11 (revision of hypotrichs; combination with *Keronopsis*).

1983 ***Pseudokeronopsis flavicans* (Kahl, 1932) nov. comb.** – Borror & Wicklow, Acta Protozool., 22: 123 (revision of urostylids; combination with *Pseudokeronopsis*).

2001 ***Pseudokeronopsis flavicans* (Kahl, 1932) Borror and Wicklow, 1983** – Berger, Catalogue of ciliate names 1. Hypotrichs, p. 36 (nomenclator containing all basionyms, combinations, and higher taxa of hypotrichs).

2002 ***Pseudokeronopsis flavicans* (Kahl, 1932) Borror and Wicklow, 1983**[1] – Song, Wilbert & Warren, Acta Protozool., 41: 151, Fig. 9–17, 38–45, Tables 2, 3 (Fig. 185g–q; detailed redescription from life and after protargol impregnation and designation of a neotype. The neotype slide is deposited in the Natural History Museum in London, U.K., registration number 2001:1z:z8:03. A paraneotype slide is deposited in the Laboratory of Protozoology, Ocean University of Qingdao, China).

Nomenclature: No derivation of the name is given in the original description. The species-group name *flavicans* is likely a composite of the Latin adjectives *flavus* (yellow) and *canus* (white, grey) and probably refers to the (inconspicuous) colour of this species.

Kahl (1932, 1933) classified *Keronopsis* as subgenus of *Holosticha*; thus, the correct name in his papers is "*Holosticha (Keronopsis) flavicans* Kahl, 1932". Rühmekorf (1935, p. 257) wrote that she found hypotrichs of the genus *Holosticha*; thus, the correct name of the present species in her paper is *Holosticha (Keronopsis) flavicans* and not *Keronopsis flavicans* (see list of synonyms), which would be a "new combination" with *Keronopsis*. This was done, although not formally, by Borror (1972), who classified it definitely in the genus *Keronopsis*.

Remarks: Kahl (1932) described this species in more or less detail from life. He focused on the cortical granules, which are arranged, as in other marine *Pseudokeronopsis* species, mainly around the dorsal bristles and cirri. The granules were very delicate and in some specimens almost not recognisable. Unfortunately, Kahl did not clearly state which colour the present species had. According to question 2 of his key, "*Keronopsis flavicans*" is obviously yellow or red. The species-group name *flavicans* indicates that it is yellowish. There is no doubt that the validity of this species is not very well covered. However, since Kahl knew this group rather well and because of the redescription by Song et al. (2002; see below) which fits Kahl's data well, I accept Kahl's decision to consider such populations as distinct species. Wirnsberger et al. (1987) were uncertain about the status of *P. flavicans* and assumed that it is either a synonym of *P. flava* or, as suggested by Kahl, a distinct species.

Rühmekorf's populations were identified by Kahl, that is, the author of *Pseudokeronopsis flavicans* (Rühmekorf 1935, p. 257). Thus, I assume that the identification is

[1] The improved diagnosis by Song et al. (2002) is as follows: Marine *Pseudokeronopsis*, body slender, ribbon-shape with narrow caudal portion, 200–300 × 40–55 μm in vivo and yellow in colour. Ciliature comprising bicorona of 5–9 pairs of frontal cirri; one buccal, 2 frontoterminal and 3–6 transverse cirri; two midventral rows consisting of about 25–40 pairs of cirri which terminate caudally about 15 μm above the inconspicuous transverse cirri; 46–66 adoral membranelles; 40–57 (mean 52) left and 44–65 (mean 60) right marginal cirri; 4–5 dorsal kineties. Numerous macronuclear segments. One contractile vacuole anterior 2/5 of body. Bright yellow-brownish cortical granules grouped around cilia and cirri; colourless blood-cell-shaped granules lying just beneath pellicle and distributed throughout whole cell.

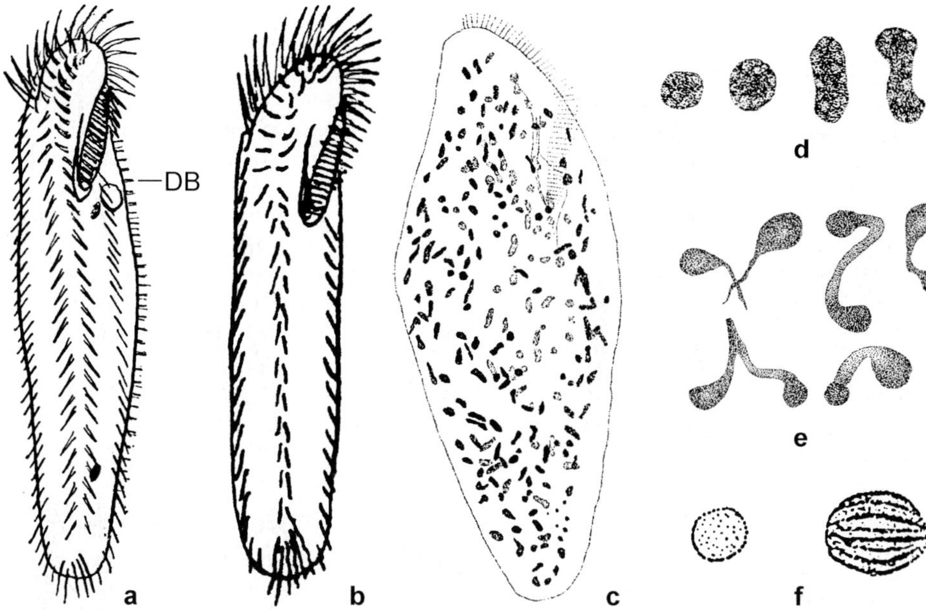

Fig. 185a–f *Pseudokeronopsis flavicans* (a, from Kahl 1932; b, from Kahl 1933; c–f, from Rühmekorf 1935. a, b, from life; c–f, hematoxylin stain). **a, b:** Ventral views, a = 250 µm. The two black dots in (a) are not explained, but likely they are micronuclei. **c:** Nuclear apparatus. **d:** Shape variability of macronuclear nodules. **e:** Dividing macronuclear nodules. **f:** Dividing micronuclei. DB = dorsal bristle. Page 951.

correct. By contrast, Wirnsberger et al. (1987, p. 85) considered Rühmekorf's *P. flavicans* as synonym of *P. carnea*.

Recently, Song et al. (2002) provided a detailed redescription of *P. flavicans* from the Chinese Sea. To clear up the somewhat confusing systematics of this species, they designated a neotype[1]. Their description indeed fits Kahl's data well. They found more or less distinct differences to *P. flava*, the most similar species. For a detailed comparison, see key and next paragraph.

According to Song et al. (2002), *Pseudokeronopsis flavicans* is a distinct species which differs from the most similar congener *P. flava* (as defined by Wirnsberger et al. 1987) in the following features: blood-cell-shaped structures present against absent (however, possibly the blood corpuscle-shaped structures are present in all marine *Pseudokeronopsis* species); colour of pigments (bright yellow-brownish vs. yellow); body size (200–300 µm vs. 140–250 µm long); number of cirri in bicorona (on average 14.8 vs. 9); number of transverse cirri (3–6 vs. 1–3); body length:width ratio (5–6:1 vs. 3–5:1); gap between rear end of midventral complex and transverse cirri (<20 µm vs. about 30 µm). *Pseudokeronopsis carnea* has an orange-red colour (vs. yellow-brown),

[1] They failed to mention which of the two populations studied (fc1 or fc2 in Table 37) is in the neotype slide deposited in the Natural History Museum.

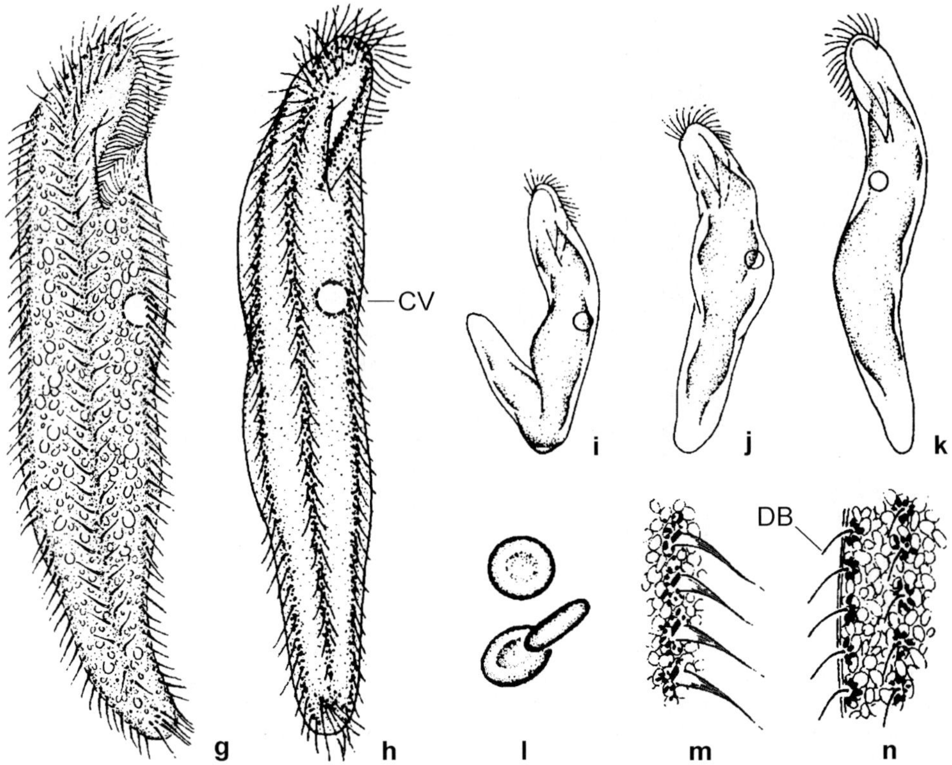

Fig. 185g–n *Pseudokeronopsis flavicans* (from Song et al. 2002. g–n, from life). **g:** Ventral view of a representative specimen, 295 μm. **h:** Ventral view of a slender specimen showing that the cortical granules are arranged mainly along the cirral rows. **i–k:** Dorsal views showing flexibility and contractility. **l:** Blood corpuscle-like granules. **m, n:** Details of cortex showing yellowish cortical granules around cirri and dorsal bristles. CV = contractile vacuole, DB = dorsal bristle. Page 951.

a plumper body shape without narrowed posterior end (vs. slender and narrowed) and a higher number of dorsal kineties (5–8 vs. usually 4).

Morphology: First Kahl's population is described, supplemented with some other data. The description of the neotype population from China follows.

Body size 150–250 × 33–55 μm; width estimated via the body length:width ratio (4.5:1) of the specimen shown in Fig. 185a. Body length according to Münch (1956) 175 μm. Body elongate elliptical (not as slender as in *P. rubra*), posterior portion slightly narrowed. Distal end of adoral zone of membranelles extends distinctly onto right body margin (about 13% of body length in Fig. 185a). Adoral zone "wider and stronger" than in *P. rubra*. Frontal scutum does not form notch on left side. Frontal cirri (that is, cirri of anterior corona) not distinctly enlarged. Kahl did not describe the morphology, especially the ciliature, in detail; however, the following features should be mentioned briefly (Fig. 185a–d): many macronuclear nodules of various shapes and sizes. Rühmekorf (1935) counted around 200 macronuclear nodules and 8–10 micronuclei. Cortical granules rosettes around dorsal bristles very delicate, in some specimens almost not recognisable; in anterior portion of cell bristles arranged very closely so that granule wreathes touch or are only slightly separated; even behind mid-body the wreathes around the bristles are only 4 µm away. Contractile vacuole left of proximal end of adoral zone of membranelles (in neotype population distinctly behind); midventral complex not set off from bicorona, extends close to transverse cirri. Buccal cirrus neither illustrated (Fig. 185a, b) nor mentioned in description; by contrast, this cirrus is present in the neotype population (Fig. 185o), indicating that Kahl overlooked it, like the frontoterminal cirri, because he did not have the advantage of silver staining. Specimens shown in Fig. 185a, b each with five transverse cirri distinctly projecting beyond rear body end. Dorsal bristles about 4 µm long (Fig. 185a).

Description of neotype population by Song et al. (2002; Fig. 185g–q, Table 37): The data are documented by eight micrographs not shown in the present book (Fig. 38–45 in Song et al. 2002). Body size 200–300 × 40–55 µm in life, ratio of body length:width about 5–6:1 (Fig. 185g, h, k). Body slender with margins of rear portion distinctly converging posteriorly; margins usually parallel, sometimes vaulted in middle portion; dorsal side slightly uneven and irregularly bulged; slightly contractile, pellicle soft and thin and cells thus flexible and often somewhat distorted in middle portion, that is, folded or twisted when crawling or gliding on bottom of Petri dish (Fig. 185i–k); dorsoventrally flattened about 2:1. About 100 macronuclear nodules scattered throughout body except in anteriormost portion; individual nodules oval to elliptical, about 5 × 3 µm in size, difficult to observe in life. Several ovoid micronuclei (Fig. 185p). Contractile vacuole small, near left body margin at 33–40% of body length (Fig. 185g–k). Two types of cortical granules. Type 1 spindle-shaped, about 1.0–1.5 µm long, bright yellow-brownish, regularly grouped in dense rosettes around both cirri and dorsal cilia. Type 1 granules thus distributed in belts or lines along cirral rows and dorsal kineties, making whole cell bright yellow (Fig. 185h, m, n). *Pseudokeronopsis flavicans* also appears to contain some dissolved pigments because, in addition to these belts of pigment-like granules, other regions are also yellowish when observed at high magnification. Granules of type 2 (possibly mitochondria) colourless, blood corpuscle-shaped, about 1.5 µm across and 0.5 µm thick (pers. comm. by Song Weibo because values forgotten in description by Song et al. 2002), densely packed and positioned somewhat deeper beneath the cell surface than type 1 granules (Fig. 185l, m). Cytoplasm often containing numerous light-reflecting, colourless or slightly yellowish globules 3–6 µm across. Food vacuoles not observed. Movement relatively slow, permanently crawling on debris or bottom of Petri dish.

Adoral zone occupies 28% (population fc2 in Table 37) to 33% (population fc1) of body length on average (according to the description, the adoral zone extends to about one fourth of body length in fixed specimens); bases of largest membranelles up to about 12 µm wide, cilia about 15 µm long in life. Distal end of adoral zone extends considerably posteriad on right body margin. Buccal area narrow and inconspicuous. Paroral and endoral slightly curved, commence about at same level; endoral almost twice as long as paroral, which optically intersects with endoral in rear portion. Pharyn-

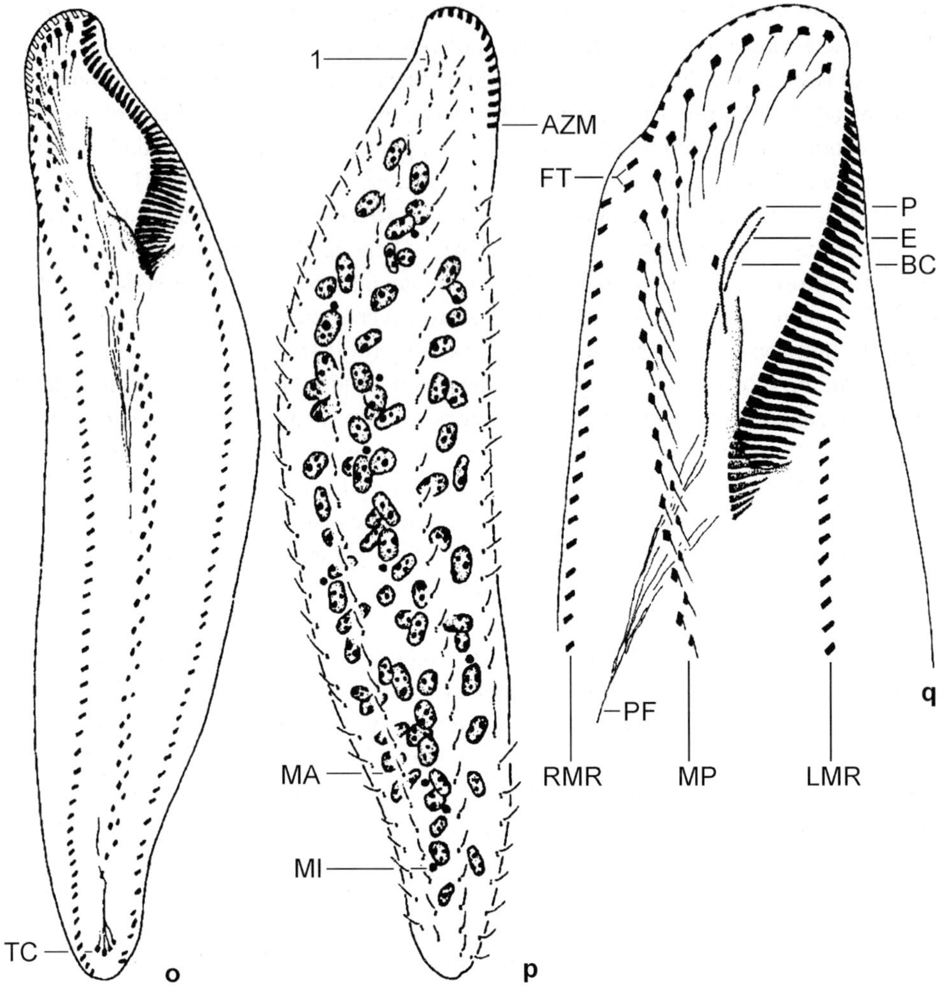

Fig. 185o–q *Pseudokeronopsis flavicans* (from Song et al. 2002. Protargol impregnation). **o, p:** Infraciliature of ventral and dorsal side and nuclear apparatus of same specimen, 215 μm. **q:** Infraciliature of anterior body portion showing, inter alia, adoral zone, undulating membranes, frontoterminal cirri, and midventral complex which is not set off from the bicorona. AZM = adoral zone of membranelles, BC = buccal cirrus, E = endoral, FT = frontoterminal cirri, LMR = left marginal row, MA = macronuclear nodule, MI = micronucleus, MP = midventral pairs, P = paroral, PF = pharyngeal fibres, RMR = right marginal row, TC = transverse cirri. 1 = dorsal kinety 1. Page 951.

geal fibres conspicuous after protargol impregnation, about 50 μm long (Fig. 185g, o, q).

Cirral pattern and number of cirri of usual variability (Fig. 185o–q), except for buccal and transverse cirri in population fc2 (Table 37). Most cirri relatively fine. Bicorona composed of about seven pairs of slightly enlarged cirri, not set off from midventral complex. Usually one, rarely two small buccal cirri (thus likely overlooked by Kahl) right of mid-point of paroral. Two small frontoterminal cirri close behind distal end of adoral zone of membranelles, often difficult to recognise. Midventral complex giving the impression of a narrow longitudinal furrow at low magnification; composed of cirral pairs only, left cirrus of each pair obviously somewhat smaller than right one (Fig. 185q); terminates about 20 μm ahead of transverse cirri. Usually four rather fine and thus inconspicuous transverse cirri distinctly projecting beyond body end (Fig. 185g, h, o). Marginal cirri narrowly spaced, right row commences near distal end of adoral zone; rear end of marginal rows clearly separated (Fig. 185o). Dorsal bristles 3–8 μm long in life, usually arranged in four, rarely five, bipolar kineties each comprising about 30 dikinetids (Fig. 185p). Caudal cirri lacking.

Occurrence and ecology: Marine. Kahl (1932) discovered *P. flavicans* in the Baltic Sea in the Kiel area (Germany), where it was common. He found it mainly in the debris of old cultures. Due to the neotypification by Song et al. (2002), the type locality of *P. flavicans* is now near the coast of Qingdao (36°08'N 120°43'E), China. Song et al. isolated *P. flavicans* from a semi-closed pond used for mollusc culture (salinity 1.0–1.5%) on May 6 and 12, 1997. Rühmekorf (1935) found it in a marine sludge sample from Plymouth, England.

Records from marine habitats not substantiated by morphological data and/or illustrations: rare in the sandy sublittoral of the Stoller Grund, a region in the Bay of Kiel, Germany (Bock 1952, p. 83); coastal groundwater (5.9‰ salinity) from the island Hiddensee, Germany (Münch 1956, p. 434); in the sediment of the Plymouth area, England (Lackey & Lackey 1963, p. 802; see also Carey & Maeda 1985, p. 568); with low abundance in Loch Eil, west coast of Scotland (Wyatt & Pearson 1982, p. 301); littoral of Black Sea in Romania (Tucolesco 1962a, p. 813); Narragansett area, Rhode Island, USA, in summer 1960 (Lackey 1961, p. 276; as *Amphisia (Keronopsis flavicans)*). Tuclesco (1965, p. 160) recorded *P. flavicans* from the saline lake Techirghiol, Romania.

Records from limnetic habitats (must be interpreted as misidentifications because *P. flavicans* was never reliable recorded from freshwater): wet moss from the Slovensky raj, Slovakia (Tirjaková & Matis 1987a, p. 9); dead arm of Danube River in Slovakia (Matis & Tirjaková 1994, p. 52; see also Matis et al. 1996, p. 13); Savannah River, USA (Patrick et al. 1967, p. 321). According to Kahl (1932), *Pseudokeronopsis flavicans* feeds on oscillatorians.

Pseudokeronopsis sepetibensis Wanick & Silva-Neto, 2004 (Fig. 186a–e, Table 37)

2004 ***Pseudokeronopsis sepetibensis* n. sp.**[1] – Wanick & Silva-Neto, Zootaxa, 587: 6, Fig. 2a–e, 3–7, Table 1 (Fig. 186a–e; original description; slides containing the holotypes and paratypes have been deposited in

[1] Wanick & Silva-Neto (2004) provided the following diagnosis: Body size 100–140 μm long × 20–26 μm wide, elongated and dorsoventrally flattened, with round ends. Strangled right above the posterior end. Cytoplasm fully pigmented in light yellow-greenish coloration. Three contractile vacuoles present, arranged in a longitudinal row by the left margin of body, near its median region. Ventral ciliature showing inconspicuous bicorona and mid-ventral row.

the collection of the Laboratório de Protistologia, Instituto de Biologia, Universidade Federal do Rio de Janeiro, Brazil).

Nomenclature: The species-group name *sepetibensis* refers to the locality (Sepetiba Bay) where the species was discovered (Wanick & Silva-Neto 2004).

Remarks: *Pseudokeronopsis sepetibensis* has four frontoterminal cirri and lacks a buccal cirrus. These are distinct differences to other well known species which have only two frontoterminal cirri, but one buccal cirrus. Moreover, the bicorona and the zigzag-pattern of the midventral complex of *P. sepetibensis* are very inconspicuous so that it cannot be excluded that it does not belong to *Pseudokeronopsis*. The break in the adoral zone and the inconspicuous midventral pattern are somewhat reminiscent of *Uroleptopsis*. However, cell division data (transverse cirri really present? buccal cirrus really lacking or moved anteriorly?) should be awaited for a re-assessment of the generic assignment.

Morphology: Body size 100–140 × 20–26 µm in protargol preparations, body length:width ratio of live specimen shown in Fig. 4 of the original paper about 11:1, that is, body outline very slender, posterior portion slightly narrowed, both ends rounded. Body dorsoventrally flattened. 22–25 ovoid macronuclear nodules scattered throughout the cell (Fig. 186e); individual nodules 2–5 µm across (in protargol preparations?). Micronuclei not recognisable, respectively, distinguishable from macronuclear nodules. Three contractile vacuoles longitudinally arranged along left cell margin in a region of 44% to 70% of body length in the specimen shown in their Fig. 4. "Cytoplasm fully pigmented in light yellow-greenish coloration"; whether cortical granules are present or not, not clearly stated.

Adoral zone occupies about 29% of body length in specimen shown in Fig. 186a, composed of an average of 44 membranelles; divided in an anterior and posterior portion by a small gap about at level of leftmost frontal cirrus. Undulating membranes rather short, commence distinctly behind of anterior of end of cell; paroral (= "external oral membrane" in their terminology) about half as long as endoral.

Cirral pattern of usual variability (Table 37). Buccal cirrus lacking. Bicorona inconspicuous, composed on average of four cirral pairs; the specimen illustrated and photographed, however, shows more or less no bicorona (ontogenetic data are needed to elucidate the situation). Usually four frontoterminal cirri in ordinary position, that is, near distal end of adoral zone. Midventral complex composed of cirral pairs only, which, however, form a very inconspicuous zigzag-pattern, especially in the region behind the level of the cytostome (this is reminiscent on *Uroleptopsis citrina*, where one cirrus of each pair is resorbed!). Invariably three inconspicuous transverse cirri; I suppose that the anterior one is a pretransverse ventral cirrus. Marginal rows without peculiarities, that is, commence right of frontoterminal cirri (right side), respectively, left of proximal portion of adoral zone and terminate slightly ahead of rear body end.

4–6, on average five bipolar dorsal kineties; length of dorsal cilia and presence/absence of caudal cirri not mentioned; however, the classification in *Pseudokeronopsis* indicates that caudal cirri are lacking (Fig. 186d).

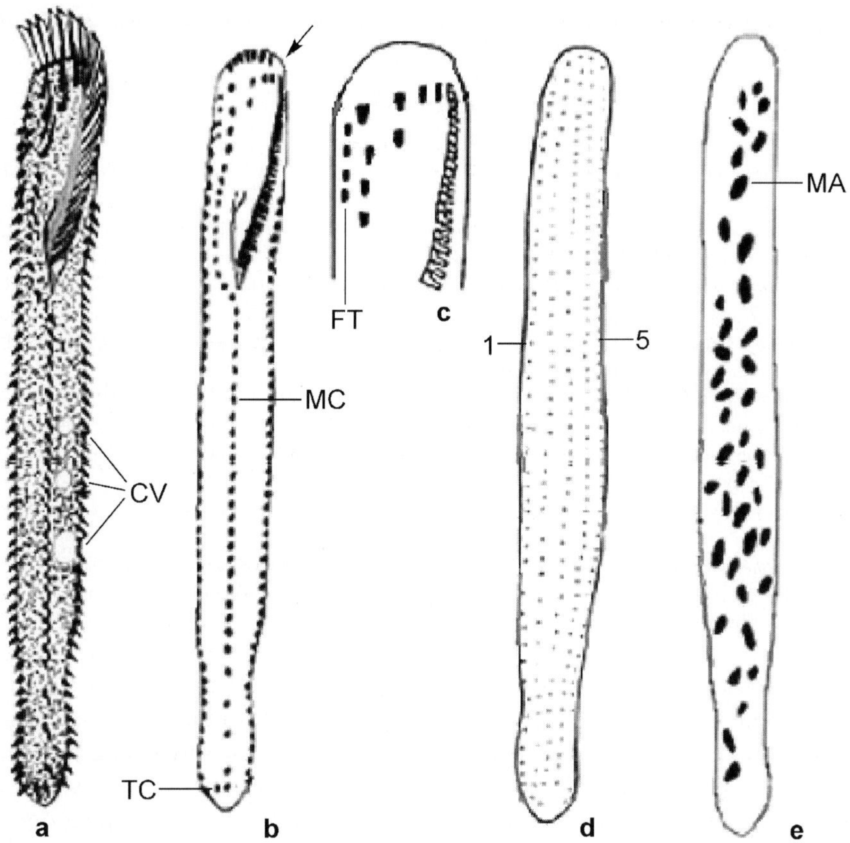

Fig. 186a–e *Pseudokeronopsis sepetibensis* (from Wanick & Silva-Neto 2004. From life, b–e, protargol impregnation). **a:** Ventral view, size not indicated. **b–e:** Infraciliature of ventral and dorsal side and nuclear apparatus, 100–140 µm. Arrow in (b) marks gap in adoral zone. Note that the cirral pattern of the frontal region does not fit the *Pseudokeronopsis* pattern very well because a distinct bicorona is lacking. CV = contractile vacuoles, FT = frontoterminal cirri, MA = macronuclear nodule, MC = midventral complex, TC = transverse cirri, 1, 5 = dorsal kineties. Page 957.

Occurrence and ecology: Marine. *Pseudokeronopsis sepetibensis* is so far recorded only from the type locality, that is, the Sepetiba Bay (about 23°05'S 44°03'W)[1] in the south of Rio de Janeiro state (Brazil) where Wanick & Silva-Neto (2004) found it during May and July; the salinity ranged from 23.4‰ to 31.3‰ and pH from 7.6 to 8.2. They used crushed wheat and rice grains to support bacterial growth in the cultures.

[1] The geographic coordinates provided in the original description are somewhat confusing.

Pseudokeronopsis multinucleata (Maupas, 1883) Borror & Wicklow, 1983
(Fig. 187a–e)

1883 ***Holosticha multinucleata* (nov. sp.)** – Maupas, Archs Zool. exp. gén., 1: 562, Planche XXIII, fig. 1–4 (Fig. 187a–d; original description; no type material available and no formal diagnosis provided).
1929 ***Holosticha multinucleata* Maupas** – Hamburger & Buddenbrock, Nord. Plankt., 7: 87, Fig. 105 (redrawing of Fig. 187a, b; guide to marine ciliates).
1932 ***Keronopsis (Holosticha) multinucleata* (Maupas, 1883)** – Kahl, Tierwelt Dtl., 25: 573, Fig. 101_{19} (Fig. 187e; revision of hypotrichs; see nomenclature).
1933 ***Keronopsis multinucleata* (Maupas 1883)** – Kahl, Tierwelt N.- u. Ostsee, 23: 109, Fig. 16.29 (redrawing of Fig. 187a; guide to marine ciliates; see nomenclature).
1972 ***Keronopsis multinucleata* (Maupas, 1883) Kahl, 1932** – Borror, J. Protozool., 19: 11 (revision of hypotrichs; combination with *Keronopsis*, see nomenclature).
1983 ***Pseudokeronopsis multinucleata* (Maupas, 1883) nov. comb.** – Borror & Wicklow, Acta Protozool. 22: 123 (revision of urostylids; combination with *Pseudokeronopsis*).
1992 ***Keronopsis multinucleatum* Maupas, 1883** – Carey, Marine interstitial ciliates, p. 184, Fig. 727 (redrawing of Fig. 187a; guide).
2001 ***Pseudokeronopsis multinucleata* (Maupas, 1883) Borror and Wicklow, 1983** – Berger, Catalogue of ciliate names 1. Hypotrichs, p. 37 (nomenclator containing all basionyms, combinations, and higher taxa of hypotrichs).

Nomenclature: No derivation of the name is given in the original description. The species-group name *multinucleat·us -a -um* is a composite of the Latin *mult-* (many), the thematic vowel *·i*, the Latin noun *nucle·us* (nucleus), and the suffix *-at·a* (to have something), and refers to the high number of macronuclear nodules. *Keronopsis* and *Pseudokeronopsis* are feminine (ICZN 1999; Article 30.1.2); thus, the ending *·um* (neuter; for example, in Carey 1992) is incorrect.

Kahl (1932, 1933) classified *Keronopsis* as subgenus of *Holosticha*; consequently, the correct name in Kahl's papers is *Holosticha (Keronopsis) multinucleata* Maupas, 1883. His misleading arrangement of names in *Keronopsis (Holosticha) multinucleata* should indicate that this species was originally classified in *Holosticha*. Borror (1972) incorrectly assumed that Kahl (1932) transferred the present species to *Keronopsis*; later, he recognised that he himself did this (see list of synonyms in Borror & Wicklow 1983). Borror & Wicklow (1983) also mention the following name in the list of synonyms: "*Holosticha (Keronopsis) multinucleata* (Maupas, 1883) Kahl, 1932". However, this authorship is incorrect because Kahl did not change the generic classification and therefore must not be cited as combining author because the generic assignment is the same in Maupas and Kahl and the presence of subgeneric names does not affect the authorship (see ICZN 1999, Article 51.3.2).

Remarks: Maupas (1883) described *Holosticha multinucleata* in detail from life. Unfortunately, no modern description of this species is available, making the characterisation of this and similar species (*H. velox*, *H. globulifera*) rather difficult. Thus, a detailed redescription is urgently needed to clarify the complex situation. De Morgan (1926, p. 45; Fig. 188h–k) redescribed this species from Plymouth, Great Britain. However, the lack of the red cortical granules suggests that his identification is incorrect. The curved bicorona indicates that it is a *P. decolor*.

Kahl (1932) considered all three taxa (*multinucleata, decolor, globulifera*) as valid. However, he already stated that *H. globulifera* is possibly a synonym of *P. decolor*.

Borror synonymised several species with *H. multinucleata*. In 1972 he put *H. decolor* into the synonymy of *H. multinucleata*. Eleven years later, Borror & Wicklow (1983) considered both taxa as valid and distinguished them, like Kahl, mainly by the cortical granulation, which seems justified. *Holosticha globulifera* was synonymised with *H. multinucleata* by Borror (1972). By contrast, Borror & Wicklow (1983) classified *H. globulifera* as synonym of *P. decolor*, a proposal also used in the present book. *Keronopsis tannaensis* Shigematsu, 1953 was also classified as synonym of *H. multinucleata* by Borror (1972); however, this species lacks transverse cirri and was thus transferred to *Uroleptopsis* by Berger (2004b). *Uroleptus roscovianus* Maupas was synonymised with *P. multinucleata* by Borror & Wicklow (1983), but – like *Keronopsis tannaensis* – it lacks transverse cirri and therefore belongs to *Uroleptopsis*.

According to Hamburger & Buddenbrock (1929), *Oxytricha rubra* sensu Möbius (1888) is possibly a synonym of *P. multinucleata*. In the present book Möbius' population is listed under *P. carnea* (Fig. 181c–e).

The illustrations by Hamburger & Buddenbrock (1929), Kahl (1932, 1933), and Carey (1992) are redrawings of Figs. 187a, b. Thus, only Kahl's (1932) illustration is shown in the present book.

The sister species of *P. multinucleata* is, at the present state of knowledge, *Pseudokeronopsis decolor*. They differ mainly in the cortical granulation, which is composed of reddish, obliquely arranged, short stripes in *P. multinucleata* and very indistinct colourless granules in *P. decolor*, respectively, its synonym, *Holosticha globulifera*. The species of the *P. rubra* group (*rubra, carnea, flava, flavicans*) are more slender and usually have fewer transverse cirri. In *P. rubra* the red cortical granules form rosettes around the dorsal bristles, while *P. multinucleata* obviously has many short, oblique rows forming a distinct band along dorsal kineties.

Morphology: Due to the lack of a redescription, the following paragraphs are based solely on Maupas' detailed live data (Fig. 187a–d). Body size rather variable, that is, body length 120–270 µm; body width from one fourth of length in small specimens to one third in large. Large cells widest at about mid-body with left margin distinctly vaulted and right side only slightly curved; anterior body end broadly, rear broadly to narrowly rounded. Small specimens often sigmoidal with both ends broadly rounded. Ventral side plane, dorsal side vaulted; cells very flat in anterior and posterior portion (Fig. 187c). Body supple, elastic, and likely slightly contractile. Nuclear apparatus composed of numerous small, globular macronuclear nodules scattered throughout cell; no nucleoli observed. Maupas found that the macronuclear nodules had the same shape and distribution in a conjugating pair. Contractile vacuole dorsally about in mid-body near left body margin. Cortical granules brick red, scattered on ventral surface, but arranged in five longitudinal bands, likely along bristle rows on dorsal side; individual bands composed of fine, obliquely arranged rows of granules (Fig. 187b). Cytoplasm

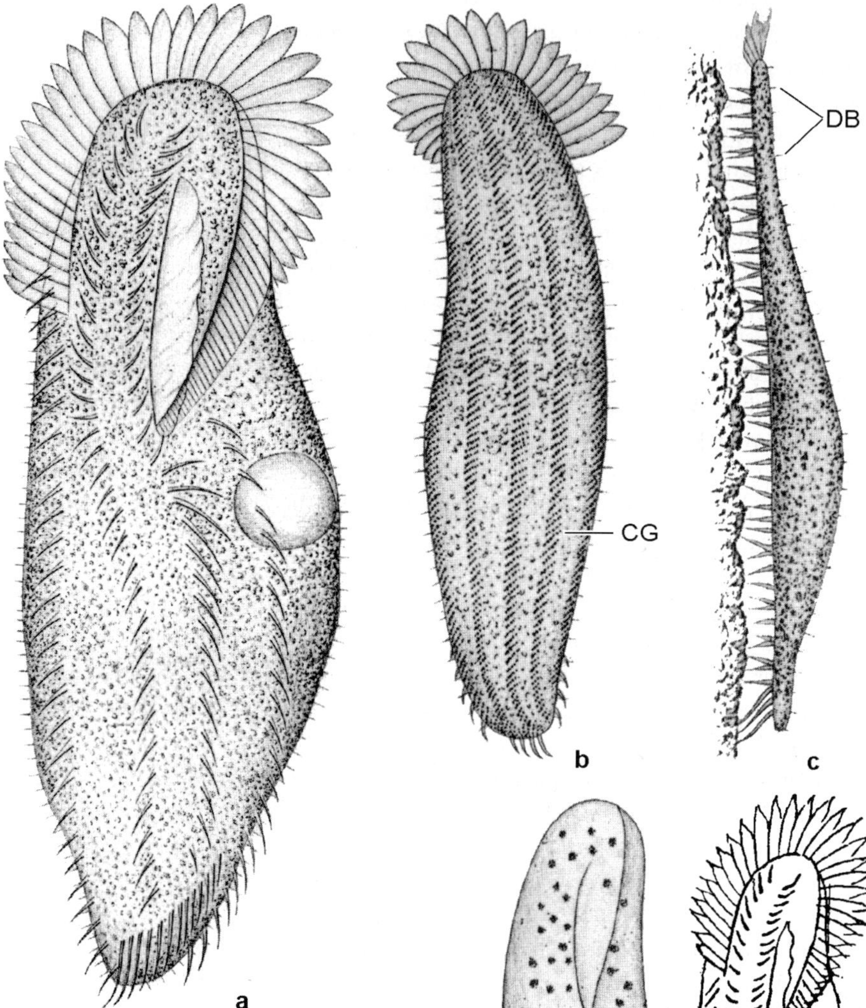

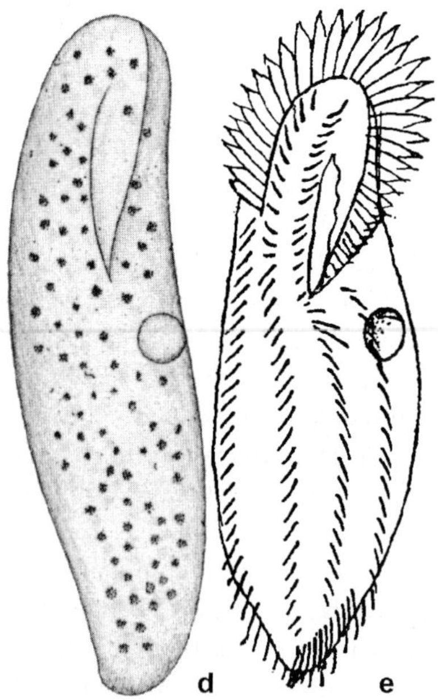

Fig. 187a–e *Pseudokeronopsis multinucleata* (a–d, from Maupas 1883; e, after Maupas 1883 from Kahl 1932. a–c, e, from life; d, nucleus stain). **a–c, e:** Ventral, dorsal, and left lateral view showing, inter alia, cirral pattern, contractile vacuole, oral apparatus, and cortical granulation; a = 270 µm, e = 200 µm (likely incorrect value). Specimen shown in (a) has each 32 cirri in right and left row formed by midventral complex, 23 left and 29 right marginal cirri, 13 transverse cirri, and ca. 63 adoral membranelles (since this illustration is based on live observations, the data must not be over-interpreted). **d:** Ventral view showing nuclear apparatus. CG = brick red cortical granules arranged in many short oblique rows forming distinct bands along dorsal kineties, DB = dorsal bristles. Page 960.

yellow due to scattered, irregularly shaped granules. Movement agile, running along algae when searching for food.

Adoral zone occupies about 38% of body length (Fig. 187a, d), according to text obviously up to 43%, extends very far (25% of body length; Fig. 187a) onto right body margin. Buccal area long triangular with distinct paroral on buccal lip (Fig. 187a). Buccal cirrus neither mentioned nor illustrated by Maupas (must not be over-interpreted because such a cirrus is difficult to recognise in life). Cirral pattern composed of two distinctly separated ventral rows extending onto frontal area, where they form the bicorona, which is not distinctly curved and set off from the midventral complex. Midventral complex extends to near transverse cirri (for some morphometric data of the specimen shown in Fig. 187a, see figure legend). 12–13 slightly to distinctly enlarged transverse cirri arranged in oblique row near posterior cell end and thus rearmost cirri distinctly projecting beyond rear body margin. Right marginal cirral row extends from distal end of adoral zone to rear cell end, left one terminates slightly ahead. Dorsal bristles obviously short, that is, around 3 μm, likely accompanied by the above mentioned bands of red cortical granules (Fig. 187b, c). Caudal cirri probably lacking (Maupas 1883).

Occurrence and ecology: Marine. The type locality of *P. multinucleata* is the harbour of Algiers, Algeria, where Maupas (1883) recorded it several times in pure water, however, always with low abundance.

Records not substantiated by morphological data and/or illustrations: Bay of Concarneau, Bretagne, France (Fabre-Domergue 1885, p. 568); with low abundance in the sandy littoral of the Kandalaksha Gulf, White Sea (Burkovsky 1970a, p. 189; 1970b, p. 11; 1970c, p. 56; as *Keronopsis multinucleatum*); attached to debris in the Woods Hole area, USA (Lackey 1936, p. 269). The record from a draw-well in Italy by Grispini (1938, p. 152) is likely a misidentification.

Pseudokeronopsis decolor (Wallengren, 1900) Borror & Wicklow, 1983 (Fig. 188a–k)

1900 ***Holosticha decolor* n. sp.** – Wallengren, Acta Univ. lund., 36: 16, Pl. I, Fig. 11, 12a–e (Fig. 188a, d, e; original description; no type material available and no formal diagnosis provided).

1926 ***Holosticha multinucleata* Maupas** – De Morgan, J. mar. biol. Ass. U. K., 14: 45, Fig. 23–25 (Fig. 188h–k; misidentification, see remarks).

1929 ***Holosticha multinucleata* var. *decolor* Wallengren** – Hamburger & Buddenbrock, Nord. Plankt., 7: 88 (change in rank).

1932 ***Keronopsis (Holosticha) decolor* Wallengren, 1900** – Kahl, Tierwelt Dtl., 25: 574, Fig. 101_{20} (Fig. 188b; revision of hypotrichs).

1932 ***Keronopsis globulifera* spec. n.** – Kahl, Tierwelt Dtl., 25: 574, Fig. 102 (Fig. 188f; original description of synonym; no type material available).

1933 ***Keronopsis decolor* Wallengren 1900** – Kahl, Tierwelt N.- u. Ostsee, 23: 109, Fig. 16.30 (Fig. 188c; guide to marine ciliates).

1933 ***Keronopsis globulifera* Kahl 1932** – Kahl, Tierwelt N.- u. Ostsee, 23: 109 (guide to marine ciliates).

1983 ***Pseudokeronopsis decolor* (Wallengren, 1900) nov. comb.** – Borror & Wicklow, Acta Protozool., 22: 123, Fig. 22 (Fig. 188g; combination with *Pseudokeronopsis*).

2001 ***Pseudokeronopsis decolor* (Wallengren, 1900) Borror and Wicklow, 1983** – Berger, Catalogue of ciliate names 1. Hypotrichs, p. 34 (nomenclator containing all basionyms, combinations, and higher taxa of hypotrichs).

2001 ***Holosticha (Keronopsis) globulifera* Kahl, 1932** – Berger, Catalogue of ciliate names 1. Hypotrichs, p. 36 (nomenclator containing all basionyms, combinations, and higher taxa of hypotrichs).

Nomenclature: No derivation of the names was given in the original descriptions. The species-group name *decolor* is a composite of the Latin prefix *de-* (away from) and the Latin noun *color* (colour) and likely refers to the colourless cortex (against the red or yellow one in, for example, *P. multinucleata*, *P. rubra*, and *P. flava*). The species-group name *globulifera* is a composite of the Latin noun *globulus -i* (small globule -s) and the Latin verb *fero* (carry, to have) and refers to the globules present mainly in the anterior and posterior portion of the cell.

Kahl (1932, 1933) classified *Keronopsis* as subgenus of *Holosticha*; thus, the correct names in his papers are *Holosticha (Keronopsis) decolor* Wallengren, 1900 and *Holosticha (Keronopsis) globulifera* Kahl, 1932. His misleading arrangement of the genus names in "*Keronopsis (Holosticha) decolor*" (see list of synonyms) should only indicate that this species was originally classified in *Holosticha*.

Borror & Wicklow (1983, p. 123) mentioned the following name in the list of synonyms of *P. decolor*: "*Keronopsis decolor* (Wall., 1900) Borror, 1972". This is incorrect because Borror (1972, p. 11) did not combine the present species with *Keronopsis*, but synonymised it with "*Keronopsis multinucleata*". Thus, Borror (1972) cannot be considered as combing author with *Keronopsis*. A further name mentioned by Borror & Wicklow (1983) in their list of synonyms is "*Holosticha (Keronopsis) decolor* (Wall., 1900) Kahl, 1932". However, Kahl is not the combining author because the generic assignment is the same in Wallengren and Kahl, and subgenera are irrelevant in this respect (ICZN 1999, Article 51.3.2).

Remarks: See *Pseudokeronopsis multinucleata* for a general discussion of the "*P. multinucleata* group" to which the present species belongs. *Pseudokeronopsis decolor* was described extensively from life, however, in Swedish. I did not translate the paper in detail and thus the description below is much shorter than the original one. Wallengren knew about the importance of the cortical granules in hypotrich systematics, as indicated by the redescription of *P. rubra*. He compared his species with *P. multinucleata* and wrote that *P. decolor* does not have a granulation like Maupas' species. Hamburger & Buddenbrock (1929) therefore classified Wallengren's species only as variety of *P. multinucleata*. By contrast I preliminarily consider the differences in the cortical granulation and in the number of transverse cirri (8–9 in *P. decolor* vs. 12–13 in *P. multinucleata*) as sufficient to accept both species. However, there is no doubt that detailed redescriptions of *P. multinucleata* and *P. decolor* are needed to clarify the situation.

Holosticha multinucleata sensu De Morgan (1926, Fig. 188h–k) lacks the red cortical granules of *P. multinucleata* (Fig. 187b). Furthermore, the cirral pattern, especially the distinctly curved bicorona, resembles more *P. decolor* than *P. multinucleata*. Thus, I consider De Morgan's population as conspecific with *P. decolor*.

Holosticha globulifera is a poorly known species never recorded since Kahl (1932, 1933). Kahl (1933) does not show an illustration of this species, but refers to *Ante-*

holosticha multistilata for body shape, and to *P. decolor* (Fig. 188c) for ciliature. Synonymy of *Holosticha globulifera* and *H. decolor* was already suggested by Kahl (1932) himself. Borror (1972) mentioned both species as synonyms of *H. multinucleata*, which, however, has a distinctly different cortical granulation. Later, Borror & Wicklow (1983) considered *H. globulifera* as synonym of *H. decolor*. They provided an illustration of *P. decolor* (Fig. 188g) which strongly resembles Kahl's description of *H. globulifera* (Fig. 188f), except for the number of transverse cirri, which is distinctly lower (about 3) than in Wallengren's and Kahl's populations (about 8). Thus, I am not quite sure that Borror & Wicklow's identification is correct. Both species are more or less colourless. According to Wallengren, his species lacks cortical granules. By contrast, Kahl found such organelles in *H. globulifera*; however, they were very fine and colourless so that it cannot be excluded that Wallengren overlooked them in *P. decolor*. I agree with the synonymisation proposed by Borror & Wicklow because the differences are too indistinct. However, I keep the data separate so that a re-activation of *H. globulifera* is easily possible. Detailed redescription necessary.

Borror & Wicklow (1983) classified – beside *H. globulifera* – *Keronopsis tannaensis* Shigematsu, 1953, and *Oxytricha viridis* Pereyaslawzewa, 1886[1] as synonyms of *P. decolor*. However, both species lack transverse cirri and thus belong to *Uroleptopsis* (Berger 2004b).

Holosticha decolor sensu Peschkowsky (1927, p. 41) is from a freshwater habitat near Moscow, Russia. It has about seven buccal cirri, indicating that it is possibly a *Anteholosticha intermedia*. Peschkowsky studied the fine structure, however, without providing a general view of the species so that a re-identification is not possible.

For a separation from the other *Pseudokeronopsis* species see key and this chapter under *P. multinucleata*.

Morphology: First a brief description of *P. decolor* is provided. The, synonym *Holosticha globulifera* is characterised separately.

Body size of *P. decolor* not given by Wallengren (1900); however, judging by the rather high number of cirri in the marginal rows and the midventral complex, it is at least 150 µm long (Fig. 188a); specimen illustrated by Borror & Wicklow (1983) 176 µm long (in life?). Kahl (1932) wrote that the size is absent; however, in the legend to Fig. 188b he mentioned 270 µm. Body outline of specimen illustrated broad elliptical with a body length:width ratio of 2.6:1; however, specimen possibly slightly squeezed due to preparation. Numerous macronuclear nodules, possibly confined to left body portion (Fig. 188a); individual nodules globular to ellipsoidal, about 4–6 µm long; divide individually (Fig. 188d, e), indicating that it is in fact a *Pseudokeronopsis*. Contractile vacuole in mid-body near left cell margin (Fig. 188a). Cortex colourless (Borror & Wicklow 1983), that is, distinct cortical granules likely lacking, or at least colourless (species name). Adoral zone occupies 33% of body length in specimen shown in Fig. 188a, according to Borror & Wicklow (1983) only about 25%; extends far onto right body margin. Buccal area narrow, undulating membrane long. Cirral pattern as in other *Pseudokeronopsis* species, that is, bicorona not distinctly set off from midventral com-

[1] In my review on oxytrichids (Berger 1999, p. 244) I wrote, par lapsus, that Borror & Wicklow (1983) synonymised it with *Pseudokeronopsis rubra*.

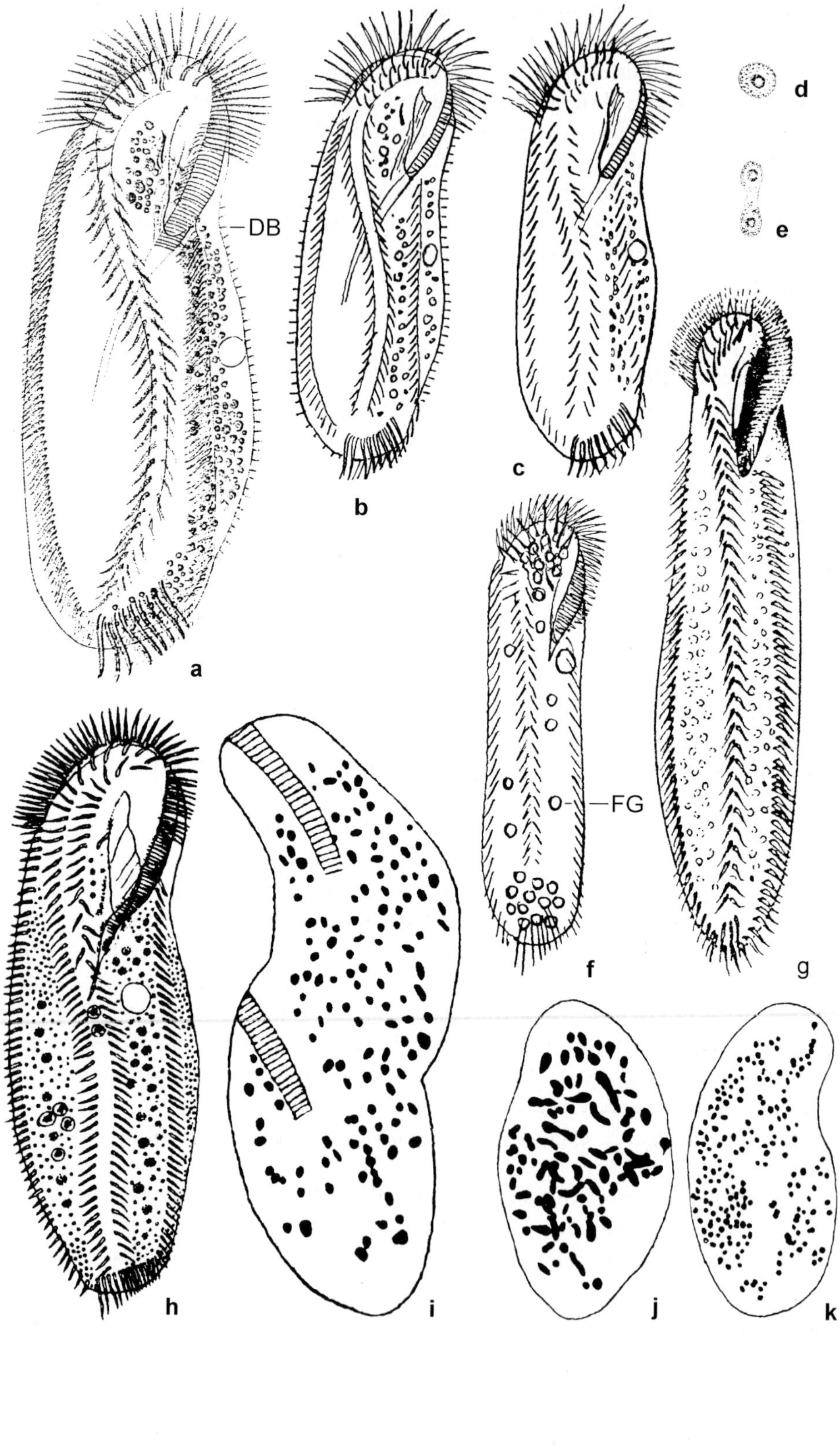
DB
FG
a
b
c
d
e
f
g
h
i
j
k

plex, which is composed of cirral pairs only and extends to near transverse cirri. One buccal cirrus at ordinary position (Fig. 188a). About nine prominent transverse cirri arranged in oblique row near rear cell end; protrude by about half of their length beyond body end; Borror & Wicklow illustrated only three cirri, making their identification uncertain. Marginal cirri narrowly spaced, rows obviously almost confluent posteriorly. Dorsal cilia short, that is, around 3 µm (Fig. 188a).

Population described by De Morgan occasionally up to 300 µm long (Fig. 188h–k); body length:width ratio 3:1, sometimes only 2:1. Slightly contractile and very flexible. Frontal scutum not transparent. Nuclear apparatus as shown in Fig. 188g, h; macronuclear nodules do not fuse prior to fission (Fig. 188i), just as in other pseudokeronopsids; no micronuclei recorded. Cytoplasm greyish-yellow, possibly due to many scattered, small globules; body margins often pure yellow. De Morgan repeated Maupas' data on the brick red cortical granules of *P. multinucleata*, but obviously he himself did not find these organelles. Cirral pattern as shown in Fig. 188h, that is, bicorona distinctly curved as in *P. decolor* and not as straight as in *P. multinucleata*. Adoral zone usually one third, often even one half of body length (likely early postdividers). Single buccal cirrus in ordinary position (Fig. 188h). Cytopharynx long, narrow, and ciliated (likely De Morgan observed the cilia of the endoral).

Synonym *Holosticha globulifera* 200–300 µm long, body length:width ratio 4.6:1 (Fig. 188f). Body outline almost rectangular with ends broadly rounded. Many macronuclear nodules scattered throughout cytoplasm. Contractile vacuole slightly behind buccal vertex. Cortical granules colourless and tiny (irregular dust-like), arranged in rather wide stripes. Cytoplasm, especially in anterior and posterior portion, packed with (hollow) fatty globules (Fig. 188f). Adoral zone about 30% of body length, wide. Buccal area narrow. Cirral pattern as described above, however, anteriormost cirri slightly enlarged, but not distinctly separated from other cirri. 6–8 transverse cirri distinctly projecting beyond rear cell end (Fig. 188f). Dorsal cilia neither mentioned nor illustrated, indicating that they are short.

Occurrence and ecology: Marine. The type locality of *Pseudokeronopsis decolor* is the Baltic Sea near Malmö, Sweden, where Wallengren (1900) discovered it in November. De Morgan (1926) found his population throughout the year in Drake's Island Tank, a large shallow tank standing in front of a south window in the Laboratory at Plymouth, United Kingdom. The locus classicus of the synonym *Holosticha globulifera* is the Nord-Ost-Kanal (a canal connecting the North and the Baltic Sea) near the bridge in the city of Kiel, Germany, where Kahl (1932) found it with high abundance in the detritus. Borror & Wicklow (1983) did not specify the location were they collected *P.*

← **Fig. 188a–k** *Pseudokeronopsis decolor* (a, d, e, from Wallengren 1900; b, c, after Wallengren 1900 from Kahl 1932, 1933; f, from Kahl 1932; g, from Borror & Wicklow 1983; h–k, from De Morgan 1926. a–c, f, h, from life; d, e, i–k, nucleus stain?; g, protargol impregnation?). **a–c:** Ventral view, a = size not indicated; according to Kahl 1932, the length is 270 µm. **d, e:** Non-diving (4–6 µm) and dividing macronucleus-nodule. **f:** Ventral view, 200–300 µm. **g:** Infraciliature of ventral side, 176 µm. This population has a distinctly lower number of transverse cirri than the type material (indicating that the identification is incorrect). **h:** Ventral view showing cirral pattern, size not indicated. **i–k:** Nuclear apparatus of non-diving (j, k) specimens and a middle to late divider (i). DB = dorsal bristle, FG = fat globule. Page 963.

decolor; but very likely, the sample is from a marine habitat in the USA. So far, I know of only one further (unsubstantiated) record of *P. decolor*, namely from the sandy littoral near Chalupy, Bay of Puck, a part of the Danzig Bay, Poland (Czapik & Fyda 1992, p. 110). The synonym *H. globulifera* ingested diatoms (Kahl 1932).

Pseudokeronopsis ovalis (Kahl, 1932) Borror & Wicklow, 1983 (Fig. 189a–g, Table 37)

1932 ***Keronopsis ovalis* spec. n.** – Kahl, Tierwelt Dtl., 25: 575, Fig. 110_2 (Fig. 189a; original description; no type material available and no formal diagnosis provided).

1932 ***Keronopsis ovalis* forma *arenivora* f. n.** – Kahl, Tierwelt Dtl., 25: 575, Fig. 103_1 (Fig. 189b; original description; no type material available).

1933 ***Keronopsis ovalis* Kahl 1932** – Kahl, Tierwelt N.- u. Ostsee, 23: 109, Fig. 16.33 (Fig. 189g; guide to marine ciliates).

1933 ***Keronopsis ovalis* forma *arenivora* Kahl** – Kahl, Tierwelt N.- u. Ostsee, 23: 109 (guide to marine ciliates).

1954 ***Keronopsis arenivorus*, n. sp.** – Dragesco, Bull. Soc. zool. Fr., 79: 69, fig. 3e (Fig. 189f; original description of synonym; likely no type material available).

1960 ***Keronopsis arenivorus* Dragesco** – Dragesco, Trav. Stn biol. Roscoff, 12: 311, Fig. 165 (Fig. 189c–e; detailed description).

1967 ***Keronopsis arenivorus* Drag.** – Petran, Ecol. Mar., 2: 188, Plansa V, C (Fig. 189h; illustrated record).

1972 ***Keronopsis ovalis* Kahl, 1932** – Borror, J. Protozool., 19: 11 (revision of hypotrichs; combination with *Keronopsis*).

1979 ***Holosticha arenivorus* (Dragesco, 1954)** – Jankowski, Trudy zool. Inst., Leningr., 86: 57 (combination with *Holosticha*).

1979 ***Paraholosticha arenivora* (Dragesco, 1954)** – Jankowski, Trudy zool. Inst., Leningr., 86: 60 (combination with *Paraholosticha* Kahl; see nomenclature).

1983 ***Pseudokeronopsis ovalis* (Kahl, 1932) nov. comb.** – Borror & Wicklow, Acta Protozool., 22: 124 (revision of urostylids; combination with *Pseudokeronopsis*).

1992 ***Keronopsis arenivorus* Dragesco, 1954** – Carey, Marine interstitial ciliates, p. 184, Fig. 728 (guide).

1992 ***Keronopsis ovalis* Kahl, 1930-5** – Carey, Marine interstitial ciliates, p. 184, Fig. 734 (guide).

2001 ***Pseudokeronopsis ovalis* (Kahl, 1932) Borror and Wicklow, 1983** – Berger, Catalogue of ciliate names 1. Hypotrichs, p. 36 (nomenclator containing all basionyms, combinations, and higher taxa of hypotrichs).

Nomenclature: No derivation of the names are given in the original descriptions. The species-group name *oval·is -is -e* (Latin adjective; egg-shaped, oval) likely refers to the oval outline of the body. The name *arenivora* is a composite of the Latin noun *aren·a* (sand, sandy location, coast), the thematic vowel *·i-* (because the second word begins with the consonant v), and the Latin verb *vorare* (devour, feed) and refers to the fact that this form ingested sand grains.

Kahl (1932, 1933) classified *Keronopsis* as subgenus of *Holosticha*; thus, the correct names in his papers are *Holosticha (Keronopsis) ovalis* and *H. (Keronopsis) ovalis* f. *arenivora*; in the legend to Fig 103_1 it was designated as *Holosticha ovalis* f *arenivora*. Obviously this was overlooked by Borror (1972), who assumed that Kahl (1932) established the present species in the genus *Keronopsis*. Borror & Wicklow (1983) recognised this mistake and correctly put Kahl in parenthesis and mentioned Borror (1972) as combining author in the corresponding entry in their list of synonyms.

Dragesco (1954) established his species explicitly for Kahl's "*Keronopsis ovalis* fa. *arenicola*" (obviously Dragesco meant "... fa. *arenivora*"). He created a new species although he simply wanted to raise Kahl's forma *arenivora* to species rank. If one considers this form as distinct species, its correct name is *Holosticha arenivora* Kahl, 1932 because it has subspecific rank (ICZN 1999, Article 45.6.4). *Holosticha* and *Keronopsis* are of feminine gender (Aescht 2001); thus, the species-group name *arenivora* (feminine) is correct, and not *arenivorus* (masculine) as supposed by Dragesco.

Jankowski (1979, p. 60) transferred *Keronopsis arenivora* Dragesco to *Paraholosticha* Kahl, 1932, which he divided into three subgenera, namely, *Paraholosticha (Paraholosticha)* Kahl, 1932, *Paraholosticha (Estuaricola)* Jankowski, 1979, and *Paraholosticha (Desicaryon)* Jankowski, 1979. He fixed *K. arenivora* as type species of *P. (Desicaryon)* so that the correct name in Jankowski (1979) is *Paraholosticha (Desicaryon) arenivora* (Kahl, 1932) Jankowski, 1979. However, since Kahl (1932) did not fix a type species for *Paraholosticha*, this genus is invalid (ICZN 1964, Article 13b; ICZN 1999, Article 13.3). Consequently, Jankowski's subgenera established within this invalid genus are likely also invalid.

Remarks: Kahl (1932) mentioned differences between the nominal form of *Holosticha ovalis* (Fig. 189a) and the sand-ingesting form which he therefore named *H. ovalis* forma *arenivora* (Fig. 189b). While the nominal form is not redescribed so far, the forma *arenivora* was redescribed and raised to species rank by Dragesco (1954, 1960). Later, Borror (1972) and Borror & Wicklow (1983) again proposed identity of *K. arenivora* and *Holosticha ovalis*. According to Kahl (1932) the cortical granules of both *P. ovalis* and the *arenivora*-form are rather loosely arranged (Fig. 189a, b, g), whereas Dragesco (1960) illustrated a very dense cortical granulation (Fig. 189d). Furthermore Dragesco's specimen does not really show a bicorona, but a single corona, a feature which, however, must not be over-interpreted because he made only live observations where this characteristic is sometimes difficult to recognise. In addition, he described five (vs. 1) "buccal" cirri, indicating that several similar taxa exist. On the other hand, the body size and the number of transverse cirri are rather similar in Kahl's and Dragesco's *arenivora* population, indicating that this is in fact a distinct species as suggested by Dragesco. However, at the present state of knowledge it is impossible to decide whether *P. ovalis* is a single, rather variable species or a complex of two or more species. Thus, I summarise both taxa (*P. ovalis* and its form *arenivora*), respectively, populations under the name *Pseudokeronopsis ovalis*, but keep the sparse morphological and faunistic data separate. Detailed redescription necessary.

The illustration provided by Petran (1967, Fig. 189h) looks almost like a redrawing of Fig. 189c. However, Petran counted 10–11 transverse cirri, whereas Dragesco (1960) illustrated 12 cirri. In addition, Petran drew the two elongated cirri at the right rear body end in line with the right marginal row and not distinctly set off as in Dragesco's figure.

Jankowski (1979) considered *Keronopsis arenivora* not as urostyloid, but as paraholostichid species, likely because of the frontal ciliature and the widely separated cirral rows. Widely separated ventral cirral rows are also known from the urostyloid *Pseudoamphisiella*, which does not show the characteristic zigzagging pattern although the ciliature originates, as in other urostyloids, from many anlagen. I do not think that *K. arenivora* is a paraholostichid because Kahl knew both the paraholostichids and the urostyloids very well so that a misidentification by Kahl is very unlikely. Furthermore, *Keronopsis arenivora* has a highly fragmented macronucleus and lives in marine habitats, whereas paraholostichids usually have only few macronuclear nodules and live in freshwater.

Pseudokeronopsis ovalis is very similar to *P. decolor*, which possibly lacks cortical granules and has a shorter adoral zone (*P. ovalis* 40% against 33% in *P. decolor*).

Morphology: At first the sparse original data provided by Kahl (1932) on *Pseudokeronopsis ovalis* are provided. Then the data by Kahl (1932) and Dragesco (1954, 1960) on the forma *arenivora* are presented.

Body size of *P. ovalis* 120–140 × 45–52 µm (width estimated from Fig. 189a), body length:width ratio about 2.7:1. Length according to Münch (1956) about 90 µm. Body outline oval. Body soft and slightly contractile, cloudy granulated; ectoplasm (cortex) thick. Many macronuclear nodules, difficult to recognise in life. Contractile vacuole in ordinary position, that is, near left body margin slightly behind level of proximal end of adoral zone of membranelles. Cortical granules large, loosely arranged (Fig. 189a, b). Adoral zone about 40% of body length, extends far onto right body margin. Buccal area without distinct lip, paroral large, endoral "membranous". Bicorona (if present at all!) indistinct (Fig. 189a), more or less continuous with midventral complex, which is obviously composed of cirral pairs only. In contrast to the *arenivora* form, no buccal cirrus is mentioned and illustrated by Kahl (1932; Fig. 189a); in his 1933-paper, however, a buccal cirrus is present (Fig. 189g, arrow). 8–10 transverse cirri, right ones distinctly projecting beyond rear body end. Dorsal cilia likely of ordinary length, that is, 3–5 µm (Fig. 189a).

Body size of *arenivora* form 180–200 × 70–80 µm (width estimated from Fig. 189b), that is, distinctly larger than nominal form. General morphology, nuclear apparatus, and cortical granulation as in nominal form. Contractile vacuole not illustrated by Kahl (1932, Fig. 189b). Cytoplasm always containing many small sand grains. Adoral zone and cirral pattern basically as in the nominal form, except that a distinct bicorona and a buccal cirrus are present. Bicorona continuous with midventral complex, which is composed of two widely separated rows. 12–15 transverse cirri.

Specimens of Dragesco's population 220–300 × 80–110 µm (width estimated from Fig. 189c, f); according to Petran (1967) about 250 µm long. Outline basically ellipsoidal with margins slightly converging posteriorly. Nuclear apparatus composed of several globular macronuclear nodules and large micronuclei. Contractile vacuole left behind proximal end of adoral zone. Cortical granules very densely arranged (Fig. 189d), which is a distinct difference to Kahl's populations (see above). Cytoplasm, as in Kahl's *arenivora* population, with many sand grains. Midventral complex composed of two distinctly separated cirral rows; however, only the left one extends along anterior body margin, that is, only one corona present (see remarks). A row of five (buccal?) cirri ahead of short undulating membrane. 12 transverse cirri distinctly projecting beyond rear body end. Two elongated cirri (caudal cirri?) subterminally on right body

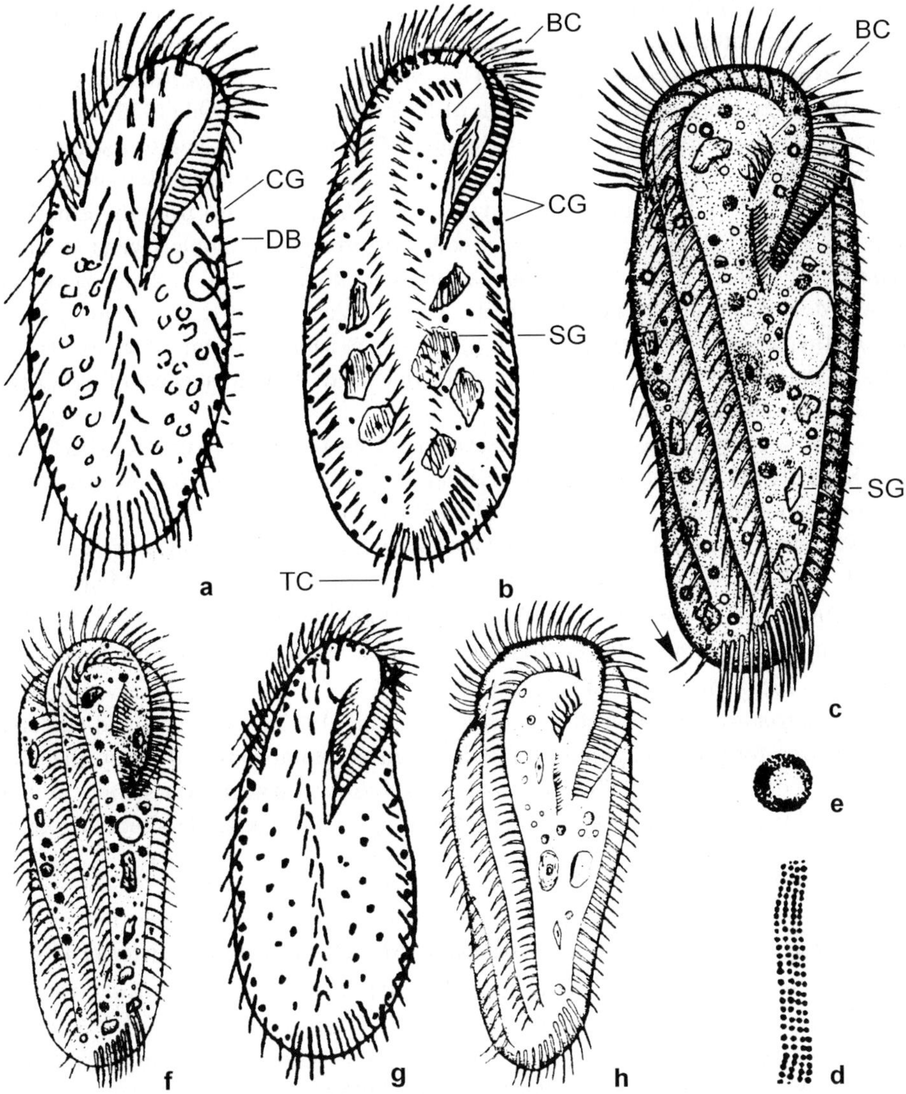

Fig. 189a–h *Pseudokeronopsis ovalis* (a, b, from Kahl 1932; c–e, from Dragesco 1960; f, from Dragesco 1954; g, from Kahl 1933; h, from Petran 1967. a–d, f–h, from life; e, nuclear stain). For the uncertain systematics of this species (group?), see text. **a, g:** Ventral view of *Pseudokeronopsis ovalis*, 130 μm. Note that Kahl (1933) modified the illustration in that he added a buccal cirrus (arrow). **b, c, f, h:** Ventral views of the *arenivora* form, b = 180–200 μm, c = 243 μm, f = 225 μm, h = 140 μm (according to Petran's text it is 250 μm long!). Arrow in (c) marks two cirri distinctly set off from all other cirral groups; possibly, these are caudal cirri. Note that the French populations (c, f) differ in some features (frontal ciliature, buccal ciliature) from the type population (b); for details, see text. **d:** Arrangement of cortical granules in French population. **e:** Macronuclear nodule. BC = buccal cirrus/cirri, CG = cortical granules, DB = dorsal bristle, SG = sand grain, TC = transverse cirri. Page 968.

side; Petran (1967) illustrated these two cirri in line with the right marginal row (Fig. 189h). Dorsal ciliature not known.

Occurrence and ecology: Marine. The type locality of *Pseudokeronopsis ovalis* (Fig. 189a) is the North Sea in the harbour of Helgoland, where Kahl (1932) found it behind the protecting wall. Records not substantiated by illustrations and/or morphological data: rare in the sandy sublittoral of the Stoller Grund, a region in the Bay of Kiel, Germany (Bock 1952, p. 83); coastal groundwater (1.6–7.9‰ salinity) from the island Hiddensee, Germany (Münch 1956, p. 434); sediment from the Atlantic Ocean at Castro-Urdiales near Bilbao, Spain (Fernandez-Leborans & Novillo 1994, p. 203); sandy sediment of Mediterranean Sea near Marseille, France (Vacelet 1961, p. 4; 1961a, p. 15); with low frequency in the mesopsammon (23.5° C, 3% salinity) of the Black Sea coast in Bulgaria (Detcheva 1980, p. 34; 1982, p. 249); fine sand of the west coast of the Caspian Sea (Agamaliev 1967, p. 369).

The type locality of the form *arenivora* (Fig. 189b) is the Bay of Kiel (Baltic Sea) where Kahl (1932) discovered it in the psammon of Bülk, Germany (co-ordinates of the light-house in Bülk 54°27'21.2"N 10°11'54,5"E). Dragesco (1954, 1960) found the *arenivora* form in fine sand from the Atlantic Ocean at Roscoff (France, for review see Bocquet 1971, p. 388) and at Banyuls (likely Banyuls-sur-Mer at the Mediterranean Coast is meant). Records not substantiated by morphological data: psammal of various sites on Romanian coast of Black Sea (Băcescu et al. 1967, p. 7; Kovaleva & Golemansky 1979, p. 275; Petran 1967; 1971, p. 154).

Pseudokeronopsis ovalis feeds, according to Kahl (1932), on ciliates and euglenids. The *arenivora* form by Dragesco (1960) ingested cryptomonads and likely flagellates digested in reddish vacuoles. *Pseudokeronopsis ovalis* was present in a sample to which 1 mg l^{-1} lead was added at periodic intervals during 240 h (Fernandez-Leborans & Novillo 1994).

Pseudokeronopsis similis (Stokes, 1886) Borror & Wicklow, 1983 (Fig. 190a–k)

1886 ***Holosticha similis*, sp. nov.** – Stokes, Proc. Am. phil. Soc., 23: 26, Fig. 7 (Fig. 190a; original description; no type material available and no formal diagnosis provided).

1888 ***Holosticha similis*, Stokes** – Stokes, J. Trenton nat. Hist. Soc., 1: 281, Plate X, fig. 13 (Fig. 190b; review).

1932 ***Keronopsis (Holosticha) similis* (Stokes, 1886)** – Kahl, Tierwelt Dtl., 25: 577, Fig. 106 $_{18}$ (Fig. 190c; revision; see remarks).

1932 ***Keronopsis (Holosticha) monilata* (Kahl, 1928)** – Kahl, Tierwelt Dtl., 25: 577, Fig. 104 $_3$, 110 $_4$; not Fig. 104 $_1$ (Fig. 190d, e; see remarks).

1945 ***Keronopsis monilata* Kahl** – Šrámek-Hušek, Veda prír., 23: 247, Fig. 1, not Fig. 2 (Fig. 190f; misidentification).

1953 ***Keronopsis monilata* (Kahl 1928)** – Jírovec, Wenig, Fott, Bartoš, Weiser & Šrámek-Hušek, Protozoologie, p. 512, Fig. 236A (Fig. 190f; misidentification).

1963 ***Keronopsis monilata* (Kahl, 1928)** – Vuxanovici, Studii Cerc. Biol., 15: 203, Plansa II, Fig. 9 (Fig. 190g; misidentification).

1972 ***Keronopsis similis* (Stokes, 1886) Kahl, 1932** – Borror, J. Protozool., 19: 11 (revision of hypotrichs; combination with *Keronopsis*; see nomenclature).

1974 ***Keronopsis monilata* (Kahl (1928) Kahl, 1932** – Jones, Univ. South Alabama Monogr., 1: 40, Fig. XXVIII-4,5 (Fig. 190h, i; misidentification).
1982 ***Holosticha similis* Stokes, 1886** – Hemberger, Dissertation, p. 105, pro parte; not his Fig. 16a–i (revision of non-euplotid hypotrichs; see remarks).
1983 ***Pseudokeronopsis similis* (Stokes, 1886) nov. comb.** – Borror & Wicklow, Acta Protozool., 22: 124, Fig. 23 (Fig. 190j; pro parte; revision of urostylids; combination with *Pseudokeronopsis*).
1986 ***Keronopsis monilata* Kahl 1928** – Chardez, Revue verviét. Hist. nat., 43: 21, Fig. 25 (Fig. 190k; misidentification).
2001 ***Pseudokeronopsis similis* (Stokes, 1886) Borror and Wicklow, 1983** – Berger, Catalogue of ciliate names 1. Hypotrichs, p. 39 (nomenclator containing all basionyms, combinations, and higher taxa of hypotrichs).
2002 ***Pseudokeronopsis monilata*** – Liu & Jin, Zool. Res., 23: 258, Fig. 1–7 (misidentification).

Nomenclature: No derivation of the name is given in the original description. I do not know to which feature the species-group name *simil·is -is -e* (Latin adjective; similar, identical) refers. Possibly it should indicate the similar size of the numerous frontal cirri and the midventral cirri (against the presence of three enlarged frontal cirri in *Holosticha* species).

Kahl (1932) named the present species "*Keronopsis (Holosticha) similis* (Stokes 1886)". However, he classified *Keronopsis* as subgenus of *Holosticha* (see his page 571) so that the correct name in his paper is *Holosticha (Keronopsis) similis* Stokes, 1886 and not *Keronopsis similis* as assumed, for example, by Borror (1972) who presumed that Kahl had transferred it to *Keronopsis*. Later, Borror & Wicklow (1983) recognised that Borror (1972) himself had transferred *Holosticha similis* Stokes to *Keronopsis*. Simultaneously, Borror & Wicklow (1983) wrote in their list of synonyms "*Holosticha (Keronopsis) similis* (Stokes, 1886) Kahl, 1932" which is also incorrect because a (new) combination is only made when a species is transferred from one genus to another genus; it is not affected by the shift from one subgenus to another subgenus within the same genus (ICZN 1964, Article 51d(i); ICZN 1999, Article 51.3.2).

The name *Holosticha similis* (Kahl, 1932) Borror, 1972 (basionym *Balladyna similis* Kahl, 1932) belongs to an oxytrichid now designated as *Oxytricha balladyna* Song & Wilbert, 1989 (for review see Berger 1999, p. 126). It was not a (junior) secondary homonym of the present species in Borror's (1972) revision because he classified the present species in *Keronopsis* (see list of synonyms). Now both taxa are assigned to quite different genera, namely *Oxytricha* and *Pseudokeronopsis*. In spite of this, the epithet of the oxytrichid had to be replaced due to the classification in *Oxytricha*, where it became the junior homonym of *Oxytricha similis* Engelmann, 1862.

Liu & Jin (2002) transferred *Holosticha monilata* – obviously not formally – to *Pseudokeronopsis*.

Remarks: The systematics of this species has to be explained in detail because *P. similis* was often synonymised with *Holosticha monilata* Kahl (now *Anteholosticha monilata*). Stokes (1886), a meticulous worker, provided a detailed description and an unambiguous illustration showing that *P. similis* has two anteriorly curved cirral rows, that is, a bicorona (Fig. 190a, b) and not three enlarged frontal cirri as *Anteholosticha* species. It is unlikely that Stokes (1886) made a misobservation in this respect because in the same paper he described several hypotrichs, some with three and some with many frontal cirri.

Kahl (1932) supposed identity of *Anteholosticha monilata* and *P. similis*, but did not synonymise these two species definitely. Unfortunately, his redescription of *A. monilata* contains specimens with both a bicorona (Fig. 190d, e) and three frontal cirri (Fig. 55b). Likely this mistake caused great confusion for the following 60 years because many later workers (Foissner & Didier 1981, Hemberger 1982, Borror & Wicklow 1983, Foissner 1988a) did not distinguish between *P. similis* and *A. monilata* although they knew about the difference in the frontal ciliature. Thus, several identifications are incorrect (Šrámek-Hušek 1945[1], Fig. 190f; Jirovec et al. 1953, Fig. 190f; Vuxanovic 1963, Fig. 190g; Chardez 1986, Fig. 190k; Liu & Jin 2002). For a more detailed discussion of Kahl's data on "*Holosticha monilata*" see *Anteholosticha monilata*. Probably, synonymy between *P. similis* and *A. monilata* was mainly assumed by the more or less identical nuclear apparatus, that is, the moniliformly arranged macronuclear nodules. In 1983, Borror & Wicklow credibly confirmed Stokes' observations. Surprisingly, they kept the synonymy of *P. similis* and *A. monilata*, although both taxa have been confirmed by modern methods (Fig. 55o, p, 190j). Only Foissner et al. (1991, p. 231) cleared this confusion in that they recognised both *Pseudokeronopsis similis* and *Anteholosticha monilata*.

Sixty years of confusion make it difficult to assign the faunistic and ecological literature of both species correctly because the identification literature is often not mentioned. The following identifications of "*H. similis*" are certainly incorrect and assigned to *A. monilata*: Foissner & Didier (1981, p. 260) and Song & Wilbert (1989, p. 159). Jones (1974) redescribed both *P. similis* and *A. monilata*. However, there is no doubt that he confused them as indicated by body size and cirral pattern. Thus, I assign his "*Keronopsis monilata*" to the present species and his "*K. similis*" to *Anteholosticha monilata*. *Pseudokeronopsis similis* sensu Wiackowski (1988) cannot be reindentified because he did not illustrate it and, surprisingly, did not use the frontal ciliature as feature in his study on urostylids. It is thus mentioned under *P. similis*.

Beside *Anteholosticha monilata*, three further species have been synonymised with *P. similis* by Borror & Wicklow (1983): (i) *Trichototaxis stagnatilis* Stokes (incorrectly spelled *Trichotaxis stagnatilis*) has two left marginal rows and thus should not be synonymised with *P. similis*, although the remaining ciliature and the nuclear apparatus are very similar (Fig. 165a, b). (ii) *Holosticha aquarumdulcium* Bürger, which is somewhat superficially described, is very large (320 × 80 μm) and has many scattered macronuclear nodules (against moniliform in left body portion) and fewer transverse cirri (Fig. 164i, j). (iii) *Keronopsis clavata* Vuxanovici, which is superficially described, is only about 100 μm long and possibly has some symbiotic algae (Fig. 190l). Since no redescription of this species is available, I preliminarily treat it as supposed synonym of *P. similis*.

---

[1] Šrámek-Hušek (1945) described two populations of *Holosticha monilata*. Fig. 190f certainly represents *Pseudokeronospsis similis* as indicated by the distinct bicorona. According to the author it is very similar to Kahl's specimen shown in Fig. 190e. The second specimen (Fig. 55c), which is, according to Šrámek-Hušek, similar to Kahl's Fig. 55b, has a distinctly lower number of frontal cirri and a rather small size (100–150 μm) so that it cannot be excluded that it is in fact a *Holosticha monilata* (although I have some doubt). Consequently, I identify the specimen shown in Fig. 190f as *P. similis* and that shown in Fig. 55c as *Anteholosticha monilata*.

Although *P. similis* is recorded several times, a detailed redescription is still lacking. And since no type material is available and the type locality is actually not known (see below), such a newly described population should be used for neotypification to finally clear up the systematics.

For a separation from other *Pseudokeronopsis* species, see key. *Anteholosticha monilata*, with which it was synonymised for many years (see above), has basically the same nuclear apparatus, but three distinctly enlarged frontal cirri and is thus an *Anteholosticha*. Furthermore, it has a dense seam of cortical granules which are colourless and therefore difficult to recognise; by contrast, *Pseudokeronopsis similis* obviously lacks a cortical granulation.

Morphology: The following description is based mainly on the original description by Stokes (1886). Supplementary and deviating data from other sources (e.g., Kahl, Jones) mentioned in the list of synonyms are indicated. The data by Vuxanovici (Fig. 190g) are somewhat superficial and must not be over-interpreted.

Body size 195 × 50 µm; width estimated via the body length:width ratio (4.8:1) of the specimen shown in Fig. 190a. Kahl (1932, p. 577), obviously par lapsus, wrote that the length of *P. similis* sensu Stokes is only 100 µm, whereas in the legend to the illustration on page 581 the correct value of about 200 µm (exactly 195 µm) is given. Under the headline of *K. monilata*, he provided a length of 200–300 µm for *P. similis* with individual values of 250 µm (Fig. 190d) and 300 µm (Fig. 190e). Jones' specimens 140–200 µm long, individual illustrated by Chardez about 240 µm (Fig. 190k). Body outline elongate-elliptical (length more than four times the width) with posterior portion slightly more broadly rounded than anterior, which is somewhat curved leftwards. Other populations almost rectangular (Fig. 190d) to slightly converging posteriorly (Fig. 190e). Body flexible and somewhat contractile (respectively extensible as formulated by Stokes). Nuclear apparatus conspicuous because moniliform, that is, macronuclear nodules arranged in one or two rows in left body portion; individual nodules ovate to ellipsoidal. Kahl (1932) counted eight (Fig. 190e) to 16 or more (Fig. 190d) macronuclear nodules; Vuxanovici (1963) found 10–12 nodules arranged almost near midline (Fig. 190g; misobservation?), and Jones gave a range of 25–35 nodules 3–5 µm long and arranged in two indistinct rows in left body portion (Fig. 190h). Micronuclei obviously distinctly recognisable (Fig. 190d), in specimens with only eight nodules one micronucleus between each two nodules (Fig. 190e); Jones counted ten micronuclei (each 3–5 µm across) within the macronuclear complex (Fig. 190h, i). Contractile vacuole close to left body margin near proximal end of adoral zone of membranelles (Fig. 190a), according to Kahl slightly behind this level and obviously with distinct collecting canals (Fig. 190d, e); by contrast, Vuxanovici and Jones observed it near mid-body (Fig. 190g, h). Cortical granules neither mentioned nor illustrated by Stokes or other workers. Cytopyge on dorsal side near posterior body end. Pellicle fragile, cytoplasm cloudy (Vuxanovici, Jones).

Adoral zone occupies about 24% in specimen shown in Fig. 190a; according to Stokes' text the peristomial field is confined to anterior third, which corresponds with, for example, Jones' data (Fig. 190h). Buccal field oblique (likely because the anterior body portion is often bent leftwards), narrow ovate, and bearing the undulating membranes. According to Kahl, buccal lip short (Fig. 190e) to long (Fig. 190d).

Cirral pattern and number of cirri as in other *Pseudokeronopsis* species (Fig. 190a, j). Bicorona not distinctly separated from the midventral complex; composed of about 14 cirri, that is, about seven pairs. No buccal cirrus illustrated or mentioned by Stokes; this must not be over-interpreted because this cirrus is often rather difficult to recognise in life. Kahl (Fig. 190d, not Fig. 190e) and Borror & Wicklow (Fig. 190j) found one buccal cirrus. Midventral complex composed of two straight median rows formed by the cirral pairs; terminates right of rear end of transverse cirral row. About 12–14 transverse cirri arranged in rather oblique row; only most posterior ones protrude slightly behind rear cell end (Fig. 190a). In Kahl's specimens, transverse cirri (about 10–13; Fig. 190e) sometimes not projecting (Fig. 190d) as in Jones' specimen, which has 11 transverse cirri (Fig. 190h); Vuxanovici counted only eight transverse cirri (Fig. 190g). Marginal cirri posteriorly more closely spaced and longer than anteriorly; rows likely not distinctly separated posteriorly. Dorsal bristles short (likely around 3 µm) and immobile.

Liu & Jin (2002) studied the effect of the micronucleus on the somatic function. The data suggest that the micronuclei play an important role for maintaining the stable structure of the cell and the macronucleus.

Occurrence and ecology: Freshwater and possibly also in saline waters. According to my experience, in central Europe certainly not as common as *Anteholosticha monilata*. The type locality of *Pseudokeronopsis similis* is not designated in detail by Stokes (1886). He discovered it in marsh water with *Sphagnum*. I presume that this marsh, neighboured, among others, by elder and azalea, is located near Trenton in New Jersey, USA, were Stokes lived and worked.

Kahl (1932) found the specimen shown in Fig. 190d in a pond in the municipal park of the city of Hamburg, Germany. For the second specimen no locality was given (Fig. 190e); possibly he found it, like *Anteholosticha monilata*, in saline waters near Oldesloe, Germany. Šrámek-Hušek (1945) and Jírovec et al. (1953) found *P. similis* in freshwater habitats in Czechoslovakia, for example, the Jizery River (Fig. 190f). Vuxanovici (1963) recorded many specimens in cultures with half-decayed plants from Lake Fundeni near Bucharest, Romania. Jones (1974) found *P. similis* in the Sea Cliff light area, a region in the Mobile Bay estuary, Alabama, USA in April. Unfortunately, I did not find detailed data about the salinity of the water in this area. The range is from 3% at the mouth to 0.2% at the upper end of the bay (Jones 1974, p. 1). This record strongly indicates that *P. similis* can withstand considerable salt content. Borror & Wicklow (1983) did not specify the location they found it. Very likely, the sample is from the USA. Chardez (1986) isolated the present species from a freshwater habitat in Verviers, a small town near Liège, Belgium.

Records not substantiated by sufficient morphological data and/or illustrations: Danube river in Austria (Humpesch & Moog 1994, p. 91); from April to July in alkaline water bodies (pH 7.0–8.4; 4.2–19.6° C; 6.5–10.5 mg l^{-1} O_2; 374–794 µS cm^{-1} conductivity) in the Hortobágy National Park, Hungary (Szabó 1999a, p. 229; 2000a, p. 8); riverside of the Danube River and its dead arms in Slovakia (Tirjaková 1992, p. 77; Matis et

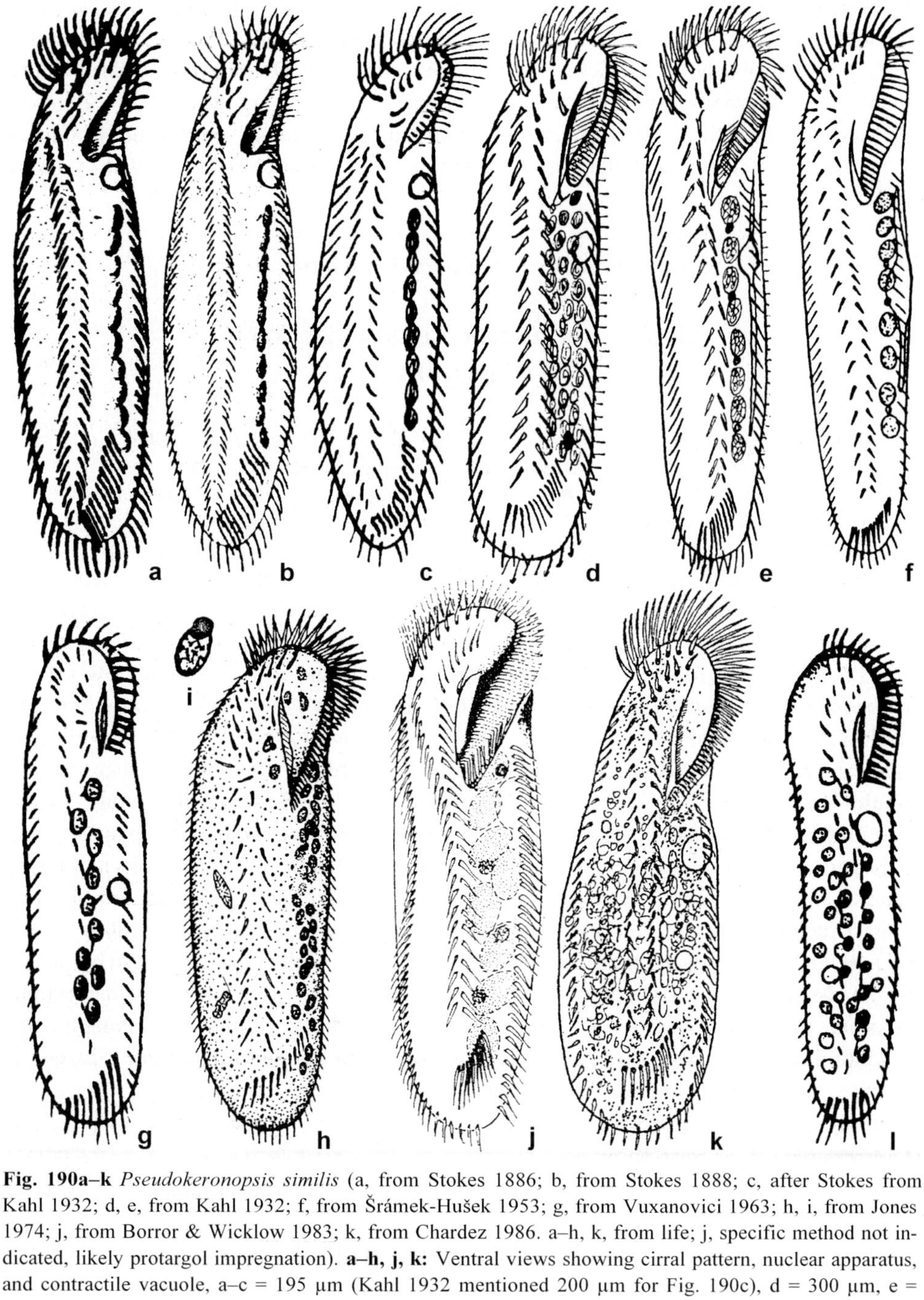

Fig. 190a–k *Pseudokeronopsis similis* (a, from Stokes 1886; b, from Stokes 1888; c, after Stokes from Kahl 1932; d, e, from Kahl 1932; f, from Šrámek-Hušek 1953; g, from Vuxanovici 1963; h, i, from Jones 1974; j, from Borror & Wicklow 1983; k, from Chardez 1986. a–h, k, from life; j, specific method not indicated, likely protargol impregnation). **a–h, j, k:** Ventral views showing cirral pattern, nuclear apparatus, and contractile vacuole, a–c = 195 μm (Kahl 1932 mentioned 200 μm for Fig. 190c), d = 300 μm, e = 250 μm, f = 250 μm, g = 230 μm, h = 155 μm, j = 152 μm, k = 244 μm. **i:** Macronuclear nodule with attached micronucleus (total length about 5 μm). Page 972.

Fig. 190l *Keronopsis clavata* (100 μm) from life (from Vuxanovici 1963), a supposed synonym of *P. similis*. Page 978.

al. 1996, p. 12); aufwuchs in two Eifel maar lakes, Germany (Packroff 1992, p. 211; Packroff & Wilbert 1991, p. 123); moss on a pond bank, likely in Poland (Wiackowski 1988, p. 4); aufwuchs of artificial substrates exposed in a Donor water reservoir and other sites in Azerbaijan (Alekperov 1989a, p. 16).

According to Kahl (1932), who did not distinguish between *P. similis* and *Anteholosticha monilata*, ciliates and diatoms are ingested; the latter are also mentioned by Jones (1974).

Supposed synonym of *Pseudokeronopsis similis*

Keronopsis clavata Vuxanovici, 1963 (Fig. 190l)

1963 ***Keronopsis (?) clavata* n. sp.** – Vuxanovici, Studii Cerc. Biol., 15: 203; Plansa II, Fig. 8 (Fig. 190l; original description; no formal diagnosis provided and no type material available).

2001 ***Keronopsis clavata* Vuxanovici, 1963** – Berger, Catalogue of ciliate names 1. Hypotrichs, p. 44 (nomenclator containing all basionyms, combinations, and higher taxa of hypotrichs).

Remarks: The species-group name *clavat·us -a -um* (Latin adjective; club-shaped) refers to the outline of the body which resembles a club. The description of *K. clavata* is not very detailed. Vuxanovici himself mentioned similarity with *Anteholosticha monilata*. Likely for that reason, Hemberger (1982, p. 105) and Borror & Wicklow (1983) synonymised it with *Pseudokeronopsis similis* (the senior synonym of *H. monilata* according to their assessment), however, without explanation. Indeed, the anteriorly distinctly curved cirral rows and the nuclear apparatus composed of several macronuclear nodules indicate that it is a *Pseudokeronopsis similis*. However, the body length (100 µm) and the possible presence of symbiotic algae ("zooclorele") are against a final synonymisation because *P. similis* is usually more than 200 µm long and lacks symbiotic algae. It therefore cannot be excluded that *Keronopsis clavata* is a distinct species. Thus, I include it in the key and present the illustration and the essential data of the original description. Body length about 100 µm; roughly club-shaped, flexible but acontractile. Many (about 20) scattered macronuclear nodules. Contractile vacuole behind proximal end of adoral zone. Cytoplasm with dark granules (globules) and zoochlorellae. Movement slow, never resting. Bicorona not separated from midventral complex, which terminates slightly ahead of five transverse cirri. Marginal rows likely distinctly separated. Numerous specimens in February in clean water from lake Herăstrău near Bucharest, Romania.

Insufficient redescriptions

***Holosticha flavorubra* var. *flava*, G. Entz.** – Gourret & Roeser, 1888, Archs Biol., Paris, 8: 185, Planche XIV, Fig. 7, 8 (Fig. 191a, b). Remarks: This population from the harbour in Bastia, Corsica, has two macronuclear nodules and in total four cirral rows.

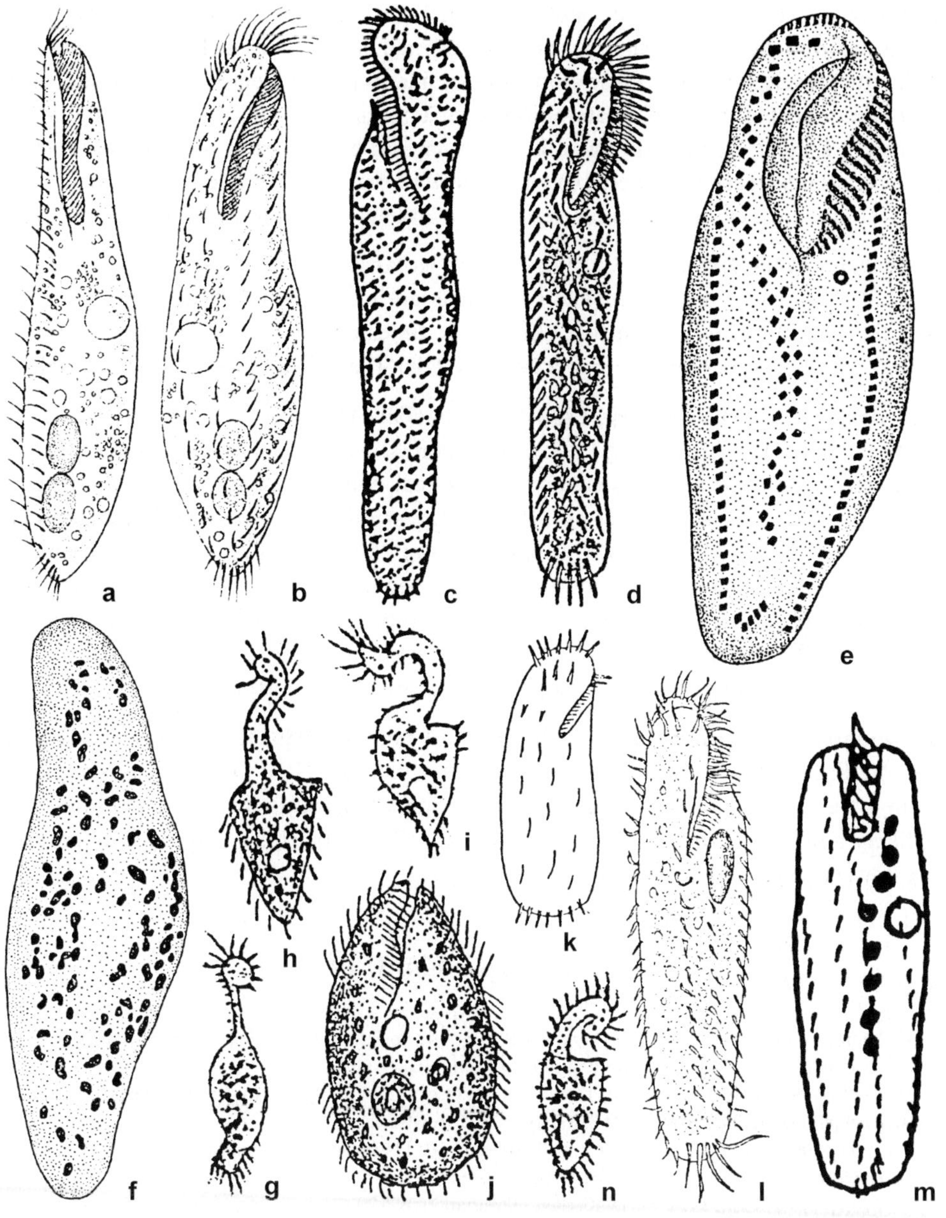

Fig. 191a–n Insufficient redescriptions and species indeterminata (ventral view from life unless otherwise indicated). **a, b:** *Holosticha flavorubra* var. *flava* (from Gourret and Roeser 1888), left lateral and ventral view, size not indicated; p. 978. **c:** *Holosticha rubra* (from Alzamora 1929), ventral cirral pattern from dorsal, 100 µm; p. 980. **d:** *Keronopsis rubra* (from Chardez 1986), 170 µm; p. 980. **e, f:** *Keronopsis rubra* (from Agamaliev 1974. e, Chatton-Lwoff silver impregnation; f, haemalaun stain), infraciliature of ventral side and nuclear apparatus, e = 127 µm, f = 147 µm?; p. 980. **g, i, n:** *Oxytricha pullaster* (from Fromentel 1876), size not indicated; p. 189. **h:** *Oxytricha pullaster* (from Dumas 1929), size not indicated; p. 189. **j:** *Oxytricha alba* (from Dumas 1929), size not indicated; p. 189. **k:** *Amphisia diademata* (from Alzamora 1929), 80 µm; p. 187. **l:** *Holosticha kessleri* (from Sarmiento and Guerra 1960), 77–132 µm; p. 188. **m:** *Holosticha fontinalis* (from Lepsi 1926b), 126 µm; p. 182.

Obviously, two cirral rows run along the left body margin (Fig. 191a, b). The nuclear apparatus and the cirral pattern strongly indicate that the identification is incorrect. Since the description and illustrations are too inexact, a re-identification is impossible. Size not indicated; body elongate-elliptical; macronuclear nodules in rear body portion, narrowly spaced; contractile vacuole about in mid-body, likely incorrectly illustrated near right body margin (Fig. 191b); young specimens rosy, old ones russet due to ochre cortical granules; cytoplasm colourless. Found among filamentous green algae.

***Holosticha rubra* Ehrbg.** – Alzamora, 1929, Notas Resúm. Inst. esp. Oceanogr., 2: 13, Fig. 28 (Fig. 191c). Remarks: The specimens are only about 100 μm long, indicating that the identification is incorrect. About six times as long as broad, posteriorly narrowed. Parduzca(?) coloured. Five transverse cirri. Bay of Palma de Mallorca, Spain.

***Keronopsis rubra* (Ehrenberg, 1838)**[1] – Agamaliev, 1974, Acta Protozool., 13: 72, Fig. 10A, B, Plate III, Fig. 15 (Fig. 191e, f). Remarks: This population has three distinctly enlarged frontal cirri and thus cannot be identical with *Pseudokeronopsis rubra*, which has a bicorona. Wirnsberger et al. (1987) suggest that it is a new *Holosticha* species. Body length 140–170 μm in life? (possibly up to 250 μm long?). 80–120 macronuclear nodules. Adoral zone 37% of body length in specimen illustrated, composed of 36–40 membranelles. Three frontal cirri; midventral complex composed of about 22 cirral pairs in specimen illustrated, rear end of complex possibly composed of short rows. Five transverse cirri; 38 left and 31 or more right marginal cirri. Caspian Sea.

***Keronopsis rubra* (Ehrenberg) Kahl 1922** – Chardez, 1986, Revue verviét. Hist. nat., 43: 21, Fig. 6 (Fig. 191d). Remarks: Incorrect date for Kahl (1932). The sample was from a freshwater habitat, indicating that the identification is incorrect. The illustration is likely a redrawing from Kahl (1932; cp. Fig. 178q with Fig. 191d). No data provided. Freshwater near Lüttich, Belgium.

Uroleptopsis Kahl, 1932

1932 ***Uroleptopsis* gen. n.**[2] – Kahl, Tierwelt Dtl., 25: 543 (original description). Type species (by original designation on p. 543): *Uroleptopsis citrina* Kahl, 1932.
1933 ***Uroleptopsis* Kahl 1932** – Kahl, Tierwelt N.- u. Ostsee, 23: 107 (guide to marine ciliates).
1950 ***Uroleptopsis* Kahl** – Kudo, Protozoology, p. 672 (textbook).
1972 ***Uroleptopsis* Kahl, 1932**[3] – Borror, J. Protozool., 19: 11 (revision of hypotrichs).
1974 ***Uroleptopsis* Kahl** – Stiller, Fauna Hung., 115: 62 (revision of Hungarian ciliates).
1982 ***Uroleptopsis* Kahl, 1932**[4] – Hemberger, Dissertation, p. 120 (revision of non-euplotid hypotrichs).

[1] Note after layout was finished: This paragraph is invalid (see footnote 1 on page 541).
[2] The diagnosis by Kahl (1932) is as follows: Oxytrichidae ohne Transversalcirren; auf dem Frontalfeld 2 gebogene Reihen in ununterbrochener Fortsetzung der beiden Ventralreihen, nach vorn zu schwach verstärkt.
[3] Borror (1972) provided the following definition: One row each of right and left marginal cirri. Transverse cirri absent. No frontal cirri. Macronucleus finely divided.
[4] Hemberger (1982) provided the following diagnosis: Je 1 linke und rechte Marginalreihe; keine Transversalcirren vorhanden; ohne differenzierte Frontalcirren; 2 familientypische Midventral-Reihen; Makronuclei zahlreich.

1992 ***Uroleptopsis* Kahl, 1930–5** – Carey, Marine interstitial ciliates, p. 187 (guide)[1].
2001 ***Uroleptopsis* Kahl 1932** – Aescht, Denisia, 1: 171 (catalogue of generic names of ciliates).
2001 ***Uroleptopsis* Kahl, 1932** – Berger, Catalogue of ciliate names 1. Hypotrichs, p. 97 (nomenclator containing all basionyms, combinations, and higher taxa of hypotrichs).
2004 ***Uroleptopsis* Kahl, 1932** – Berger, Acta Protozool., 43: 110 (redefinition, key, phylogeny; see characterisation below).

Nomenclature: No derivation of the name is given in the original description. *Uroleptopsis* is likely a composite of the genus-group name *Uroleptus* (a composite of the Greek noun *he ura* [tail] and the Greek adjective *lept-* [slender, small]) and the Greek suffix *-opsis* (appearance, looking like). Probably, the name should indicate the (superficial) resemblance of *Uroleptus* and *Uroleptopsis* species. Feminine gender because ending with *-opsis* (ICZN 1999, Article 30.1.2). Incorrect subsequent spellings: *Uroleptapsis multiseta* (Dragesco 1970, p. 98); *Uroletapsis multiseta* Dragesco, 1970 (Dragesco 1980, p. 181, 182); *Uroleptosis citrina* Kahl, 1932 (Borror & Wicklow 1983, p. 123).

Characterisation (Fig. 167a, autapomorphy 5): Gap in adoral zone (A). Frontal cirri arranged in bicorona. Buccal cirrus in ordinary position, that is, right of paroral. 2 or more frontoterminal cirri. Midventral complex basically composed of midventral pairs. Transverse cirri lacking (A). 1 left and 1 right marginal row. Caudal cirri absent. Many macronuclear nodules which divide individually. Parental adoral zone completely replaced during cell division, that is, proter gets totally new adoral zone. Frontal-midventral cirral anlage I forms two cirri (A). Living in saline waters.

Remarks: The characterisation above is based mainly on the well studied species *U. citrina* (type) and *U. ignea*. Thus, one cannot exclude that the characterisation has to be modified when new data on the other species became available. Further symplesiomorphies are likely three dorsal kineties (described for *U. citrina*, *U. ignea*, *U. tannaensis*) and a distinctly posteriorly extending distal end of the adoral zone of membranelles (see Retroextendia).

The history of *Uroleptopsis* – although not very long – is rather eventfully. Kahl (1932) established it because the lack of transverse cirri in some species precluded their classification in *Holosticha (Keronopsis)* (nowadays most species of this subgenus are classified in *Pseudokeronopsis*). Kahl described one new species, the type *U. citrina*, and transferred two species to *Uroleptopsis*, namely *Oxytricha viridis* Perejaslawzewa and *Uroleptus roscovianus* Maupas. *Uroleptopsis* was accepted – besides the workers listed in the synonymy – by Stiller (1974a, p. 130), Corliss (1975, p. 257; 1977, p. 137), Tuffrau (1979, p. 525; 1987, p. 115)[2], and Tuffrau & Fleury (1994, p. 130). By contrast, Borror (1979, p. 546, 547) put it – together with *Trichototaxis* – into the synonymy of *Keronopsis* (sensu lato). Later, he synonymised it with *Pseudokeronopsis* because he considered the type species *U. citrina* as junior synonym of *P. rubra*, the type of *Pseudokeronopsis* (Borror & Wicklow 1983, p. 123). Borror argued – however, without convincing evidence – that some *P. rubra* populations also lack transverse cirri so that

[1] Carey (1992) accepted *Uroleptopsis*, but ignored the type species *U. citrina*.

[2] Tuffrau (1979, 1987) classified *Uroleptopsis* in the Kahliellidae, which is, however, incorrect because this group does not have a midventral complex.

the presence or absence of these cirri cannot be used to define groups. I suggest that he mixed, certainly erroneously, species with and without transverse cirri. In addition, Borror & Wicklow (1983) obviously overlooked that due to this synonymy *Pseudokeronopsis* Borror & Wicklow, 1983 would become invalid, or better its establishment would have been superfluous because it would have been the junior synonym of *Uroleptopsis* Kahl, 1932.

In 1990, Mihailowitsch & Wilbert described *Pseudokeronopsis ignea* which lacks transverse cirri. Thus, Foissner (1995) transferred this species to *Uroleptopsis*. However, the resurrection of Kahl's genus by Foissner was not accepted by Eigner (2001, p. 73), who distinguished two patterns of transverse cirri formation. According to Eigner, *Pseudokeronopsis ignea* has transverse cirri. However, his interpretation of the ontogenetic data of *P. ignea* (Fig. 198c–j) is not comprehensible for me. I agree with Mihailowitsch (1989), Mihailowitsch & Wilbert (1990), and Foissner (1995) that this species does not have transverse cirri as they are usually defined (Fig. 1a). Thus, I accept Foissner's decision to transfer Mihailowitsch & Wilbert's species from *Pseudokeronopsis* to *Uroleptopsis*.

The presence or absence of certain cirral groups is widely used to define genera or subgenera. Examples are the oxytrichid *Tachysoma* Stokes, 1887 (caudal cirri absent; for review see Berger 1999) and the urostyloid *Australothrix* (transverse cirri lacking). In some cases, not the presence or absence, but even the different number of cirri within a cirral group is used, usually of course besides other features, to characterise supraspecific taxa, for example, the Bakuellidae (midventral complex not only composed of cirral pairs, but also of midventral rows against midventral complex composed of cirral pairs only, as, for example, in the Holostichidae). Consequently, it seems logical to accept *Uroleptopsis*, inasmuch as it shows two further apomorphies besides the lack of transverse cirri, namely two frontal cirri originating from the frontal-midventral anlage I against a single cirrus in almost all other hypotrichs, and a gap in the adoral zone of membranelles. *Uroleptopsis* is therefore as well defined as many other genera of hypotrichs and it should not be synonymised with *Pseudokeronopsis* as suggested by Borror & Wicklow (1983) and Eigner (2001).

In the following paragraphs the apomorphies of *Uroleptopsis* are discussed in detail. (i) *Cirral anlage I forms two cirri.* Within the pseudokeronopsids this feature only occurs in *Uroleptopsis*. In *Pseudokeronopsis*, *Thigmokeronopsis*, and most other hypotrichs anlage I – which produces the undulating membranes – forms only a single cirrus; this state must thus be considered as plesiomorphic. *Bicoronella costaricana* and *Caudiholosticha sylvatica* have 1–7 cirri behind the left frontal cirrus, indicating that in these two species anlage I also produces more than one cirrus (Berger & Foissner 1989, Foissner 1982, 1995). *Caudiholosticha sylvatica* has three frontal cirri, indicating that it is not closely related to the pseudokeronopsids. In contrast, *Bicoronella costaricana* has a bicorona and is therefore likely a near relative of the pseudokeronopsids (Fig. 144a). However, ontogenetic data are needed to show the fate of the macronuclear nodules and the origin of the supernumerary cirri behind the left frontal cirrus.

(ii) *Transverse cirri lacking.* As already discussed above this feature was the main reason why Kahl (1932) separated *U. citrina* from *Holosticha (Keronopsis)* species (now *Pseudokeronopsis*). The loss of the transverse cirri, or another cirral group, undoubtedly occurred several times independently within the hypotrichs. Other urostyloid taxa lacking transverse cirri are, for example, *Australothrix*, *Eschaneustyla*, and *Holostichides*. However, in *Eschaneustyla* the macronuclear nodules fuse to a single mass during cell division and *Australothrix* and *Holostichides* lack, in addition, a bicorona, strongly indicating that none of them are members of the Pseudokeronopsidae.

(iii) *Gap in adoral zone.* Within the pseudokeronopsids, *Uroleptopsis citrina* and *U. ignea* are the sole species which have a distinct break in the adoral zone. The gap is not very distinct in life and thus I assume that it has been overlooked in the three other *Uroleptopsis* species. Obviously, this break is homologous with the interruption separating the anterior (= outer, = collar) membranelles from the ventral (= inner, = buccal) membranelles of the oligotrichs (Petz & Foissner 1992, Foissner et al. 1999). In most hypotrichs, this transition site is inconspicuous because not marked by a gap. Very likely such a distinct division evolved several times independently, for instance, in *Holosticha*, *Afrothrix*, *Erniella*, or *Etoschothrix* (Foissner et al. 2002).

Uroleptopsis citrina and *U. ignea* are the sole *Uroleptopsis* species whose morphology and ontogenesis are described by modern methods (Berger 2004b, Mihailowitsch & Wilbert 1990). Thus, they can be thoroughly compared. They share the synapomorphies indicated in the characterisation above and discussed in the previous paragraphs. The following two conspicuous differences exist: (i) the type species lacks a buccal cirrus in the ordinary position, that is, right of the paroral (present in *U. ignea*), and (ii) the pattern of the midventral complex. In *U. citrina* the middle portion of the midventral complex is composed of the right cirri of the cirral pairs only (Fig. 192k, n–p, v, w); the anterior and posterior portion consist of ordinary midventral pairs although, very rarely short midventral rows occur at the end of the complex. In contrast, the anterior portion of the midventral complex of *U. ignea* is composed of midventral pairs, while the middle and posterior portion consist of midventral rows (Fig. 198a, h). A more detailed analysis of the first difference shows that the buccal cirrus is not lacking in *U. citrina*, but it does not migrate posteriorly into the ordinary position; it is – in contrast to *U. ignea* – part of the bicorona even in non-dividers. This difference is of course very conspicuous and therefore I divided *Uroleptopsis* into two subgenera (Berger 2004b). As already mentioned, all other species assigned to *Uroleptopsis* are not described by modern methods (Fig. 194, 195a–c, 196a–f, 197a–d). However, the data available indicate that they lack a buccal cirrus in the ordinary position, that is, they are obviously more similar to *U. citrina* than to *U. ignea*. Consequently, the subgenus for *U. ignea* is monotypic, a condition rejected in phylogenetic systematics. However, a subgenus has the advantage against a genus that the binomen – including the authorship – of the species does not change (see below).

Interestingly, the species assigned to *U. (Uroleptopsis)* are from marine habitats, whereas *U. ignea* – the sole species belonging to *U. (Plesiouroleptopsis)* – was discovered in an inland saltwater. Further studies will show whether or not this ecological separation is confirmed.

Species included in *Uroleptopsis* (alphabetically arranged according to basionym): (1) *Keronopsis tannaensis* Shigematsu, 1953; (2) *Oxytricha viridis* Pereyaslawzewa,

Table 39 Morphometric data on *Uroleptopsis citrina* (cit, from Berger 2004b) and *Uroleptopsis ignea* (ign, from Mihailowitsch & Wilbert 1990)

Characteristics [a]	Species	mean	M	SD	SE	CV	Min	Max	n
Body, length	cit	155.0	154.0	19.9	3.6	12.9	99.0	188.0	31
	ign	300.0	304.9	31.1	9.8	10.4	262.0	353.0	10
Body, width	cit	41.0	42.0	7.2	1.3	17.5	26.0	57.0	31
	ign	33.6	37.1	8.7	2.7	25.8	23.0	46.0	10
Body length:width, ratio	cit	3.8	3.8	0.6	0.1	14.6	3.0	5.2	31
Adoral zone of membranelles, length	cit	46.2	46.0	5.0	0.9	10.9	32.0	54.0	31
Adoral zone , relative length (%)	cit	29.9	29.3	1.9	0.3	6.5	27.0	35.0	31
Anterior body end to distal end of adoral zone, distance	cit	11.3	11.0	2.6	0.5	22.8	5.0	18.0	31
Distance 1 [b]	cit	4.0	4.0	1.9	0.3	47.3	1.0	9.0	31
Gap in adoral zone, width	cit	3.1	3.0	1.1	0.2	34.1	0.8	6.0	31
Anterior body end to paroral, distance	cit	22.3	22.0	3.3	0.6	14.8	15.0	29.0	31
Paroral, length	cit	8.9	9.0	1.6	0.3	17.7	6.0	12.0	31
Anterior body end to first midventral cirral pair, distance	cit	18.4	18.5	3.3	0.6	17.9	11.0	24.0	30
Distance 2 [b]	cit	41.5	40.0	7.6	1.4	18.4	24.0	68.0	31
Distance 3 [b]	cit	72.8	72.0	17.2	3.1	23.6	46.0	112.0	31
End of midventral complex to rear body end, distance	cit	25.6	24.0	6.3	1.1	24.6	14.0	38.0	31
Anterior body end to right marginal row, distance	cit	14.7	15.0	3.5	0.6	23.6	8.0	21.0	29
End of right marginal row to rear body end, distance	cit	4.0	4.0	1.6	0.3	40.4	1.0	6.0	31
Front macronuclear nodule, length	cit	4.7	4.0	1.3	0.2	27.3	3.0	8.0	31
Front macronuclear nodule, width	cit	2.4	2.5	0.5	0.1	21.7	1.5	3.0	31
Front micronucleus, diameter	cit	2.5	2.5	0.3	0.1	13.1	1.5	3.0	28
Adoral membranelles, total number	cit	39.4	39.0	4.0	0.7	10.2	29.0	47.0	31
	ign	41.8	41.5	5.9	1.9	14.1	33.0	55.0	10
Proximal adoral membranelles, number	cit	29.5	29.0	2.8	0.5	9.5	22.0	35.0	31
Distal adoral membranelles, number	cit	9.9	10.0	1.4	0.2	13.8	7.0	12.0	31
Bicorona, number of cirri	cit	15.1	14.0	2.3	0.4	15.2	12.0	20.0	31
	ign	13.0	13.0	0.0	0.0	0.0	13.0	13.0	10
Anterior corona, number of cirri	cit	7.5	7.0	1.2	0.2	15.2	6.0	10.0	31
Posterior corona, number of cirri	cit	7.5	7.0	1.2	0.2	15.3	6.0	10.0	31
Frontoterminal cirri, number	cit	2.3	2.0	0.5	0.1	20.5	2.0	3.0	31
	ign	2.0	2.0	0.0	0.0	0.0	2.0	2.0	10
Buccal cirri, number	ign	1.0	1.0	0.0	0.0	0.0	1.0	1.0	10
Midventral pairs, number	cit [c]	7.1	7.0	2.0	0.4	27.6	4.0	15.0	31
	ign	5.6	5.0	0.8	0.3	14.3	5.0	7.0	10
Single midventral cirri, number	cit	10.3	11.0	4.5	0.8	43.5	4.0	18.0	31
Midventral pairs in posterior portion of midventral complex, number	cit	6.8	5.5	3.3	0.6	48.2	2.5	14.0	31
Midventral rows, number	ign	17.3	17.1	2.7	0.9	15.7	14.0	21.0	10
Midventral cirri, total number	cit	39.2	38.0	6.4	1.2	16.4	26.0	53.0	31
Left marginal cirri, number	cit	38.9	38.0	6.1	1.1	15.7	28.0	49.0	31
	ign	49.3	49.0	5.9	1.9	12.0	41.0	62.0	10
Right marginal cirri, number	cit	47.7	48.0	6.3	1.1	13.3	34.0	63.0	31
	ign	60.0	62.0	5.8	1.8	9.7	51.9	67.0	10

Table 39 Continued

Characteristics [a]	Species	mean	M	SD	SE	CV	Min	Max	n
Dorsal kineties, number	cit	3.0	3.0	0.0	0.0	0.0	3.0	3.0	31
	ign	3.0	3.0	0.0	0.0	0.0	3.0	3.0	10
Dorsal kinety 1, number of bristles	cit	25.2	25.0	4.3	0.8	16.9	14.0	32.0	30

[a] All measurements in µm. CV = coefficient of variation in %, M = median, Max = maximum value, mean = arithmetic mean, Min = minimum value, n = number of individuals investigated, SD = standard deviation, SE = standard error of arithmetic mean. Data are based on protargol-impregnated specimens (cit, Foissner's method; ign, Wilbert's method).

[b] See Fig. 192n.

[c] Midventral pairs in anterior portion of midventral complex.

1886; (3) *Pseudokeronopsis ignea* Mihailowitsch & Wilbert, 1990; (4) *Uroleptopsis citrina* Kahl, 1932; (5) *Uroleptus roscovianus* Maupas, 1883.

Some papers refer to unidentified *Uroleptopsis* species. Walker (1968) studied the response of *Uroleptopsis* sp. from a coral reef near Dar-es-Salaam, Tanzania on changes in the salinity. A transfer from 33‰ (normal concentration) to 66‰ was survived in some experiments for up to seven days, whereas a reduction to about 7‰ (20% of normal concentration) killed most specimens immediately. Cooley & Keltner (1970, p. 17) found a *Uroleptopsis* species in the Santa Rosa Sound adjacent to Sabine Island, Canada(?) Mironov & Avdeeva (1978) recorded a *Uroleptopsis* species in a study dealing with the development of protozoa biocenosis on oil in sea water. Platt-Rohloff et al. (1997) found small, rod-shaped bacteria in the cytoplasm of *Uroleptopsis*. The bacteria were well visible in life, but difficult to stain with DNA-specific dyes. Electron microscopical images revealed 2 × 1 µm-sized, gram-negative bacterial rods.

Species misplaced in *Uroleptopsis*: Three further species have been assigned to *Uroleptopsis*. However, as discussed below, they do not fit into the groundplan of *Uroleptopsis* and thus belong elsewhere. Dragesco (1970, p. 97) described *Uroleptopsis multiseta* which has around seven cirral rows. They are longitudinally arranged and widely spaced and therefore do not form a midventral pattern. Dragesco (1970) himself suggested that this species possibly belongs to a new genus, *Plesiotricha* Dragesco, 1970; however, he did not formally transfer it to his own genus. Later, he classified it in *Kahliella* Corliss, 1960 (Dragesco & Dragesco-Kernéis 1986, p. 431). Due to this act it became a secondary homonym of *Kahliella multiseta* Dragesco, 1970 (p. 105) and thus the species-group name of *U. multiseta* had to be replaced: *Kahliella microstoma* Dragesco & Dragesco-Kernéis, 1986. For a more detailed discussion of this nomenclatural problem, see Foissner (1987d, p. 230).

Borror (1972, p. 11) transferred *Paraholosticha ovata* Horváth, 1933 to *Uroleptopsis*. Stiller (1974a, p. 132) obviously overlooked this act because she transferred it to *Uroleptopsis* too. This species has, inter alia, only two macronuclear nodules, lacks a distinct midventral complex, and lives in freshwater. All these features strongly indicate that it does not belong to *Uroleptopsis*. Likely, it is a junior synonym of *Paraholosticha muscicola* Kahl, 1932 (see Berger 2001).

A further species transferred to *Uroleptopsis* is *Uroleptus kahli* Grolière, 1975 because Jankowski (1979, p. 61) mentioned "*Uroleptopsis kahli*, ibidem" under *Perisincirra* Yankowskij, 1978 for which Grolière's species is the type. I do not understand the word *ibidem* ("the same reference" or "in the same place") in this context because neither Grolière (1975) nor Yankowskij (1978) mentioned a combination with *Uroleptopsis*. I therefore assume that Jankowski (1979) made the combination with *Uroleptopsis*, possibly par lapsus. However, the classification of Grolière's species in *Uroleptopsis* is very likely incorrect because recently we found that *Perisincirra* is a valid group (Foissner et al. 2002).

Key to *Uroleptopsis* species

The key is from Berger (2004b). Note that only two of the five species listed below are described after protargol impregnation. Thus, for the remaining three species the features concerning the cirral pattern are rather uncertain. Further, size and shape as well as the nuclear apparatus do not allow a separation. Consequently, I used the colour of the cells as key character, which is admittedly a difficult feature, especially for beginners. Please be sure that the colour of your specimens is a real colour and not due to badly adjusted optics of your microscope! However, before you can use the key, you must be certain that your specimens belong to a *Uroleptopsis* population. Thus, identification needs both protargol impregnation (cirral pattern) and live observation (colour). Only experienced workers have a chance to identify these species correctly exclusively after live observation.

1 Buccal cirrus present, that is, cirrus arranged right of paroral (Fig. 198a) .... *Uroleptopsis (Plesiouroleptopsis) ignea* (p. 1012)
\- Buccal cirrus lacking, that is, no cirrus immediately right of paroral (e.g., Fig. 192c, k) .... *Uroleptopsis (Uroleptopsis)* (p. 986) **2**
2 Cells more or less colourless (Fig. 194a) .... *Uroleptopsis tannaensis* (p. 1002)
\- Cells light green, rose-carmine, or yellow **3**
3 Cells rose-carmine (Fig. 196a) .... *Uroleptopsis roscoviana* (p. 1006)
\- Cells yellow or light green **4**
4 Cells yellow (Fig. 192a, c) .... *Uroleptopsis citrina* (p. 987)
\- Cells light green (Fig. 195a) .... *Uroleptopsis viridis* (p. 1004)

Uroleptopsis (Uroleptopsis) Kahl, 1932

1932 ***Uroleptopsis* gen. n.** – Kahl, Tierwelt Dtl., 25: 543 (original description). Type species (by original designation on p. 543): *Uroleptopsis citrina* Kahl, 1932.

2004 ***Uroleptopsis (Uroleptopsis)* Kahl, 1932 stat. nov.** – Berger, Acta Protozool., 43: 114 (division of *Uroleptopsis* into two subgenera).

Nomenclature: For the genus-group names the principle of coordination applies. Thus, a name established for a taxon at either rank in the genus-group is deemed to have been simultaneously established by the same author for a nominal taxon at the other rank in the group; both nominal taxa have the same type species (ICZN 1999, Article 43.1). Consequently, Kahl (1932) is the author and *U. citrina* Kahl, 1932 the type species of *Uroleptopsis (Uroleptopsis)*, which is also termed the nominotypical subgenus (ICZN 1999, Article 44.1).

Characterisation (Fig. 167a, autapomorphies 7): *Uroleptopsis* with cirrus II/2 (= buccal cirrus) not in ordinary position right of paroral, but in line with cirri of posterior corona (A). Some midventral cirral anlagen produce only one cirrus (A?).

Remarks: At least in the type species, cirrus II/2 is formed during cell division and present in interphasic specimens; however, it does not migrate posteriorly in the ordinary position right of the paroral. For the other species assigned to *Uroleptopsis (Uroleptopsis)* it is unclear whether or not cirrus II/2 is present at all.

Uroleptopsis citrina has a further highly interesting feature, namely the lack of the left cirri in some cirral pairs of the middle region of the midventral complex. The plesiomorphic state is that a frontal-midventral anlage finally produces two cirri, a so-called midventral pair (the rearmost anlagen of course produce three cirri if they also form a transverse cirrus. At first, *Uroleptopsis citrina* also produces only midventral pairs (Fig. 192v). However, somewhat later the left cirrus of the corresponding pairs is obviously resorbed so that only the right cirrus remains in the interphasic specimen (Fig. 192w). Interestingly, this resorption is confined to the middle portion of the midventral complex. Thus, this section of the complex looks almost like a midventral row. Unfortunately, it is unknown whether or not this character is also present in the other species assigned to *U. (Uroleptopsis)*. Thus, it could be that this feature is only an apomorphy of *U. citrina*. A very similar resorption of cirri is described for *Psammocephalus faurei* by Wicklow (1982) indicating a convergence.

Species included in *Uroleptopsis (Uroleptopsis)* (alphabetically arranged according to basionym): (1) *Keronopsis tannaensis* Shigematsu, 1953; (2) *Oxytricha viridis* Pereyaslawzewa, 1886; (3) *Uroleptopsis citrina* Kahl, 1932; (4) *Uroleptus roscovianus* Maupas, 1883.

Uroleptopsis citrina Kahl, 1932
(Fig. 192a–y, 193a–h, Table 39)

1932 ***Uroleptopsis citrina* spec. n.** – Kahl, Tierwelt Dtl., 25: 543, Fig. 87 (Fig. 192a; original description; no formal diagnosis provided and no type material available).

1933 ***Uroleptopsis citrina* Kahl 1932** – Kahl, Tierwelt N.- u. Ostsee, 23: 107, Fig. 16.12 (Fig. 192b; guide to marine ciliates).

1950 ***Uroleptopsis citrina* K.** – Kudo, Protozoology, p. 672, Fig. 315g (redrawing of Fig. 192a; textbook).

1972 ***Uroleptopsis citrina* Kahl, 1932** – Borror, J. Protozool., 19: 11 (revision of hypotrichs).

2001 ***Uroleptopsis citrina* Kahl, 1932** – Berger, Catalogue of ciliate names 1. Hypotrichs, p. 97 (nomenclator containing all basionyms, combinations, and higher taxa of hypotrichs).

2004 ***Uroleptopsis citrina* Kahl, 1932** – Berger, Acta Protozool., 43: 102, Fig. 5–28, 35–42 (Fig. 192c–y, 193a–h; detailed redescription from life and after protargol preparations; cell division; 5 neotype slides [accession numbers 2004/301–305] have been deposited in the Oberösterreichische Landesmuseum in Linz [LI], Upper Austria).

Nomenclature: No derivation of the name is given in the original description. The species-group name *citrin·us -a -um* (Latin adjective; lemon yellow) obviously refers to the yellow colour of this species (Berger 2004b). If the subgenus name is included, then the name has to be written as follows: *Uroleptopsis (Uroleptopsis) citrina* Kahl, 1932. *Uroleptopsis citrina* was fixed as type species of *Uroleptopsis* by original designation.

Remarks: Kahl (1932) described the type species in some detail from life (Fig. 192a). Subsequently it was listed in several reviews, but no additional data have been provided. Borror & Wicklow (1983) classified it as one of several junior synonyms of *Pseudokeronopsis rubra*, which is hardly comprehensible because *U. citrina* is yellow, has no buccal cirrus right of the paroral, lacks transverse cirri, and invariably has three dorsal kineties; in contrast, *Pseudokeronopsis rubra* is red, has an ordinary buccal cirrus and distinct transverse cirri, and usually six dorsal kineties (Fig. 179b, c, Table 37; Wirnsberger et al. 1987).

The neotype population from the northern Adriatic Sea described by Berger (2004b) very closely resembles the type population from the Baltic. They agree in the following features: (i) marine habitat; (ii) size; (iii) nuclear apparatus; (iv) contractile vacuoles; (v) yellow cortical granules and ring-shaped structures underneath cell surface; (vi) narrow buccal field; (vii) bicorona and long midventral complex basically composed of cirral pairs (the lack of some left cirri in the middle region of the midventral complex is very difficult to recognise without silver impregnation and was therefore possibly overlooked by Kahl although he even wrote that the two ventral rows nearly make the impression of a single row); (viii) lack of a buccal cirrus in ordinary position; (ix) lack of transverse cirri; (x) three dorsal kineties.

Differences concern (i) details of the oral apparatus; (ii) a so-called "Mycetom" (a plasma region containing many bacteria) described by Kahl; (iii) the distance between the rear end of the marginal rows; (iv) the body shape; and (v) the colour. The gap in the adoral zone of the Adriatic population is clearly recognisable only in protargol preparations; thus, it is unknown whether or not this feature was present in Kahl's population as he did not have the advantage of silver impregnation. Further, Kahl did not see a buccal field and a paroral. The buccal field is obviously very narrow in Kahl's population and thus one can also say that it is lacking. The lack of the paroral in his specimens is much more difficult to explain because he described this structure in many other hypotrichs; possibly he overlooked it because it is indeed rather inconspicuous. The mycetom was obviously present in all specimens studied by Kahl; however, he wrote that the constancy of this feature has to be checked. Possibly it was a kind of parasitism. In the specimens of Kahl's population the marginal rows are distinctly separated posteriorly (Fig. 192a). In contrast, they optically almost overlap in the specimens from the Adriatic Sea (Fig. 192c). Kahl's specimens are more or less ribbon-shaped whereas the specimens from the Adriatic Sea are usually elongate elliptical in outline. Possibly, Kahl's specimens did not have their natural outline due to the mycetom whose effect on the cell is not known (malformation?, loss of various structures?, ...).

According to Kahl, the yellow colour of *Uroleptopsis citrina* is due to the yellow cortical granules. However, in my population the number of these granules is usually too low to cause such a distinct colour. In contrast, the yellow colour of the Adriatic population is mainly caused by a diffuse colour of the cytoplasm which is often very distinct in the marginal areas.

Since no type or voucher slides were available for *U. citrina*, and because of the great taxonomic problems discussed above, I fixed the population from the Adriatic Sea as neotype (details see Berger 2004b).

For a separation from the other *Uroleptopsis* species, see the key above. According to Kahl (1932) and my experience, *Uroleptopsis citrina* is very easily confused with *Pseudokeronopsis flava* which is also yellow. However, this species usually has two transverse cirri, four dorsal kineties, one buccal cirrus in ordinary position, and lacks a break in the adoral zone (Fig. 184h; Wirnsberger et al. 1987). Protargol impregnation is thus recommended to check these features.

Morphology (Fig. 192a–p, 193a–g, Table 39): First the neotype population from the Adriatic Sea is described, followed by some data provided by Kahl (1932).

Neotype population described by Berger (2004b): Body size usually 160–220 × 30–50 µm, body length:width ratio about 5:1 in life, 3.8:1 on average in protargol preparations (Table 39). Body outline elongate-elliptical to almost ribbon-shaped; at left anterior corner a minute process likely causing break in adoral zone (Fig. 192c). Body about 1.5–2.0:1 flattened dorsoventrally, very flexible, and rather resistant to coverglass pressure, not distinctly contractile. Pellicle slightly to distinctly crenelated along cirral rows. Nuclear apparatus masked by cytoplasmic inclusions, thus very difficult to recognize in life without staining (Fig. 192c, l). Macronuclear nodules scattered throughout cytoplasm, usually ellipsoidal (length:width ratio 2:1 on average in protargol preparations; Table 39), sometimes globular or dumbbell-shaped, with one or few nucleoli (Fig. 192g, h, l); specimen shown in Figs. 192k–m with about 100 macronuclear nodules. Micronuclei globular, difficult to distinguish from globular macronuclear nodules in protargol preparations and thus difficult to count; number likely around five. Contractile vacuole difficult to recognise in freely motile specimens due to cytoplasmic inclusions (Fig. 192c, 193a); slightly squeezed cells show distinct lacunar systems with several dilatations near left body margin (Fig. 192d). Cortical granules difficult to recognise, although of ordinary size (0.8–1.2 µm across) and yellow colour; variable number of granules arranged in rings around dorsal bristles, but also scattered over whole cell (Fig. 192d, f, 193b, d). Yellow colour of specimens basically not caused by cortical granules but due to diffuse colour. Anterior and posterior body end often more distinctly yellow than remaining body portions. Underneath cell surface a distinct, about 2–3 µm wide seam formed by numerous ring-shaped structures of unknown function (mitochondria?); individual structures colourless, 1.5–2.0 µm across (Fig. 192e, f, i, j, 193b, c). Cytoplasm usually packed with greasily shining globules 2–4 µm across (Fig. 192c, f, 193a, c). Food vacuoles 3–10 µm across, contain bacteria. Movement without peculiarities, that is, moderately rapidly gliding showing great flexibility.

Adoral zone of membranelles occupies 30% of body length on average (Table 39), bipartite by inconspicuous break (gap) about where zone turns from ventral body surface

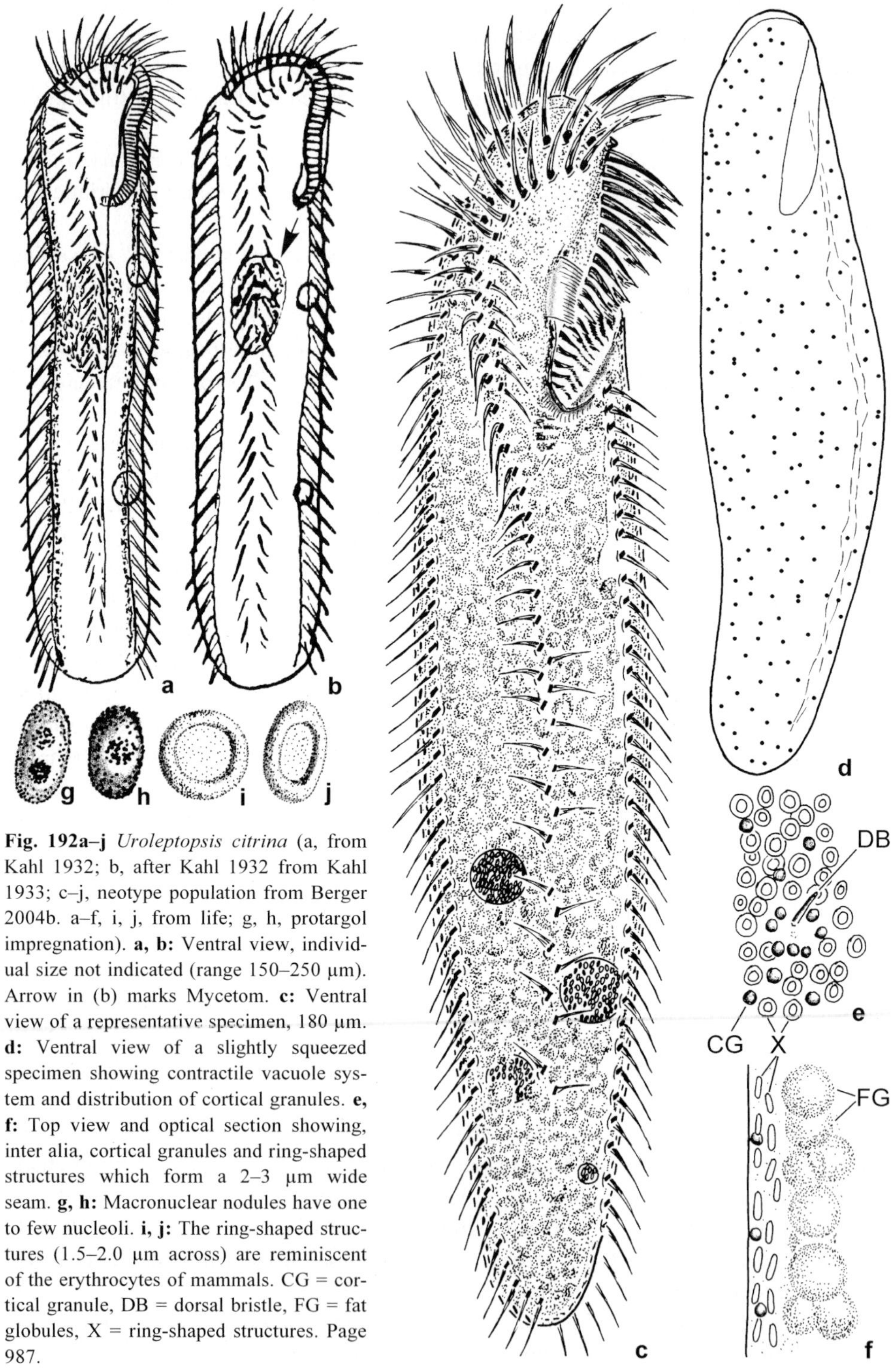

Fig. 192a–j *Uroleptopsis citrina* (a, from Kahl 1932; b, after Kahl 1932 from Kahl 1933; c–j, neotype population from Berger 2004b. a–f, i, j, from life; g, h, protargol impregnation). **a, b:** Ventral view, individual size not indicated (range 150–250 µm). Arrow in (b) marks Mycetom. **c:** Ventral view of a representative specimen, 180 µm. **d:** Ventral view of a slightly squeezed specimen showing contractile vacuole system and distribution of cortical granules. **e, f:** Top view and optical section showing, inter alia, cortical granules and ring-shaped structures which form a 2–3 µm wide seam. **g, h:** Macronuclear nodules have one to few nucleoli. **i, j:** The ring-shaped structures (1.5–2.0 µm across) are reminiscent of the erythrocytes of mammals. CG = cortical granule, DB = dorsal bristle, FG = fat globules, X = ring-shaped structures. Page 987.

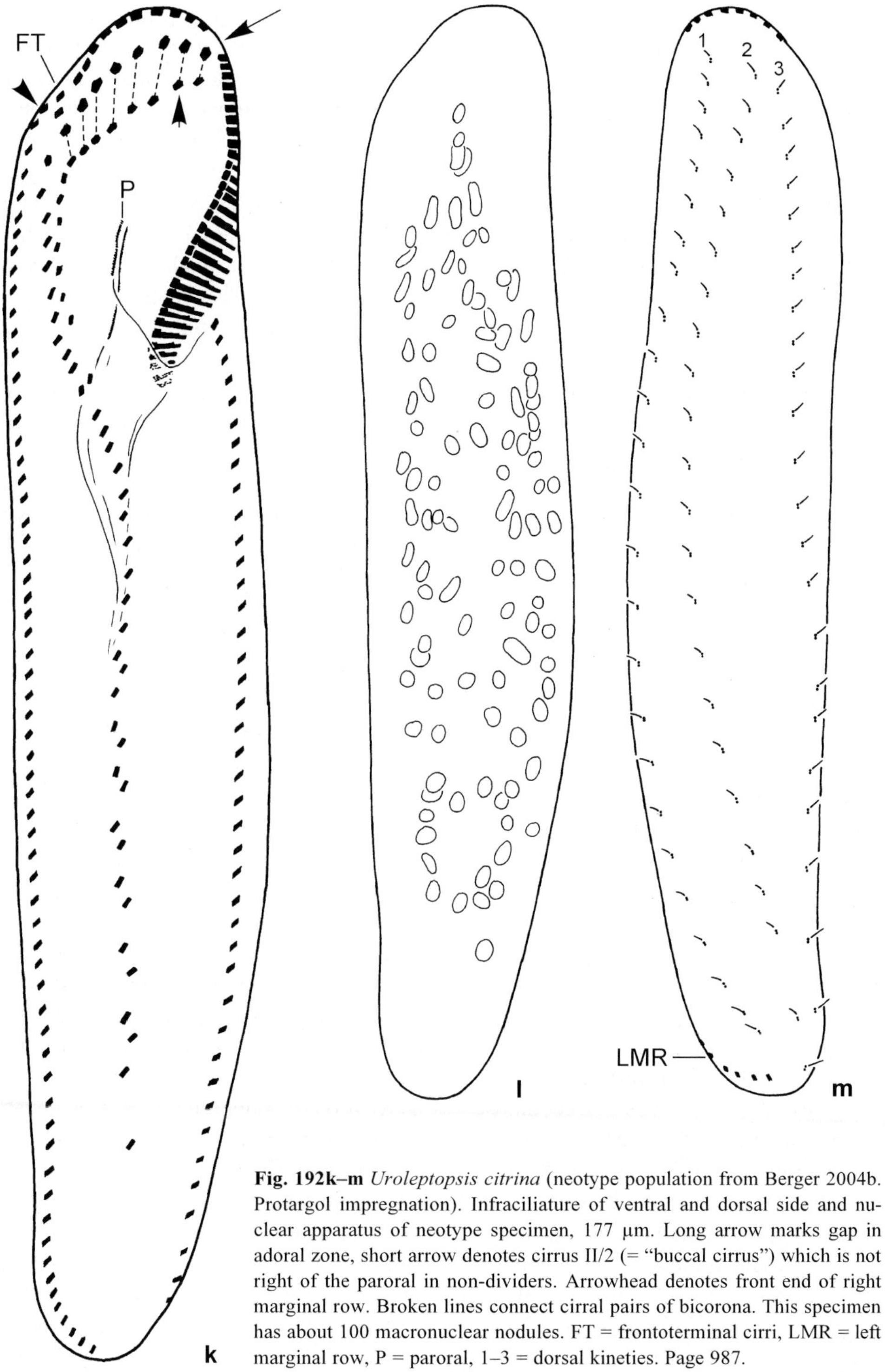

Fig. 192k–m *Uroleptopsis citrina* (neotype population from Berger 2004b. Protargol impregnation). Infraciliature of ventral and dorsal side and nuclear apparatus of neotype specimen, 177 µm. Long arrow marks gap in adoral zone, short arrow denotes cirrus II/2 (= "buccal cirrus") which is not right of the paroral in non-dividers. Arrowhead denotes front end of right marginal row. Broken lines connect cirral pairs of bicorona. This specimen has about 100 macronuclear nodules. FT = frontoterminal cirri, LMR = left marginal row, P = paroral, 1–3 = dorsal kineties. Page 987.

to dorsal side of frontal scutum (Fig. 192c). Gap usually distinct in protargol preparations, on average 3 µm wide, separates zone into about 10 distal and about 30 proximal membranelles. Proximal portion of adoral zone in most specimens roughly in *Gonostomum* pattern, that is, extends along left body margin, performs more or less abrupt right bend and slight clockwise rotation to plunge into the buccal cavity; four-rowed portion of middle membranelles often slightly set off which is sometimes even recognisable in life in that this right portion (or only some cilia) forms a separate ciliary bundle; width of membranelles increases rapidly up to 10 µm from proximal end to level where left marginal row commences. Distal portion of adoral zone extends onto right body margin to 7% of body length on average (Table 39). In several specimens proximal portion of adoral zone sigmoidally curved (Fig. 192p), almost as illustrated by Kahl (1932; Fig. 192a). Buccal area very narrow in life, of ordinary size in protargol preparations possibly due to inflation of buccal cavity (Fig. 192c, 193g). Buccal lip distinctly curved and thickened at vertex (Fig. 192k, 193g). Paroral and endoral begin about at same level, that is, at about 14% of body length (Table 39, Fig. 192k). Paroral short (9 µm on average in protargol preparations; Table 39), straight, composed of 6–8 µm long, likely zigzagging cilia, on anterior portion of buccal lip, optically not intersecting with endoral, which is also more or less straight, but about twice as long as paroral. Pharyngeal fibres inconspicuous in life, clearly recognisable after protargol impregnation, of ordinary length and structure, extend obliquely backwards, with long, fine structures (cilia of endoral?) beating inside.

Cirral pattern and number of cirri of usual variability, except for number of cirral pairs in anterior and posterior portion and number of single midventral cirri in middle portion of midventral complex which vary rather strongly (Fig. 192k, n–p, Table 39). Cirri of bicorona 12–15 µm long, remaining cirri about 12 µm. Frontal ciliature conspicuous because of the bicorona type; anterior and posterior corona composed of seven cirri each on average due to two peculiarities, namely (i) the formation of two cirri from anlage I, and (ii) the buccal cirrus (= cirrus II/2) is formed, but does not migrate posteriorly into the ordinary position (for details, see ontogenesis). Only one out of at least 31 specimens with eight cirri in anterior and only seven cirri in posterior corona. Cirri of anterior corona slightly larger than those of posterior, base of most (all?) coronal cirri of polygonal outline. Bicorona not distinctly set off from anteriormost cirral pair of midventral complex (Fig. 192k, 193e). No cirrus immediately right of paroral, that is, "buccal cirrus" lacking (but see description of bicorona above). Usually two, rarely three frontoterminal cirri in ordinary position, namely near distal end of adoral zone (Fig. 192k, n). Midventral complex composed of 39 cirri on average; due to morphogenetic peculiarity separated in three more or less clearly recognisable portions (Fig. 192k, n): (i) anterior portion composed of seven midventral pairs on average; (ii) middle portion composed of about 10 single cirri forming more or less continuous (not zigzagging) line; and (iii) posterior portion composed of around seven cirral pairs whose cirri are usually more widely separated than those of the anterior portion; length of these three portions highly variable. Midventral complex terminates at 84% of body length on average; rearmost cirri must not be misinterpreted as transverse cirri, which are lacking (checked in many hundred specimens). Right cirri of midventral pairs usually composed of 2 × 4 basal bodies, left cirri often made of 2 × 3 basal bodies. Right marginal row commences close to frontoterminal cirri, curves leftwards to about cell-midline at posterior end; usually terminates about 4 µm ahead of rear cell end. Left marginal row begins in ordinary position, that is, slightly ahead of proximal end of adoral zone, extends onto dorsolateral surface posteriorly, where it ends about in midline; marginal rows thus optically more or less confluent (Fig. 192c, k, m). Dorsa cilia 2–3 µm long in life, arranged in three bipolar kineties (Fig. 192m, 193f, Table 39). Caudal cirri lacking.

Kahl's specimens 150–250 µm long, body width 1/6–1/5 of body length (Fig. 192a, b). Body anteriorly flattened, very flexible, and more or less acontractile. Many macronuclear nodules. Invariably a single large globular to ellipsoidal body in cell centre, stains with methyl green; according to Kahl this is a kind of mycetom, that is, cytoplasm packed with bacteria (Fig. 192a, b). Contractile vacuole near left margin at about 33% of body length, rarely visible; at about 66% obviously a second vacuole, which, however, is not contractile. Cortex packed with pale yellow oval ring-shaped protrichocysts about 1.5 µm across; smaller, solid, lemmon-yellow protrichocysts (= cortical granules) near the cirral and bristle rows which form difficult-to-recognise rings around dorsal bristles; these lemmon-yellow granules cause the colour of the cell. Oral apparatus 1/6–1/5 of body length; adoral zone curiously shaped (Fig. 192a), narrow, turns rightwards proximally and ends in a short cytopharynx. Distinct buccal lip as well as the buccal field and the undulating membrane (paroral) lacking; however, sometimes the oral seam extends slightly rightwards. Marginal rows separated posteriorly, left row extends to near middle portion of proximal part of adoral zone. Cirri of ventral rows (= midventral complex) finer than those of bicorona, thus making the impression of a single row.

Molecular data: Specimens of the neotype population have been sent to M. Schlegel (Leipzig University, Germany). Data will be published elsewhere.

Cell division (Fig. 192q–y, 193h): This part of the life cycle proceeds basically as in *Pseudokeronopsis* and in some details also as in *Thigmokeronopsis*. Thus, the reader is referred mainly to the illustrations and the legends. Here, only some relevant deviations from *Pseudokeronopsis*, which is very likely the sister group of *Uroleptopsis*, are mentioned. A very early stage and one or two stages between those shown in Figs. 192q, s are lacking in the sequence presented.

Stomatogenesis: As in *Thigmokeronopsis* and *Pseudokeronopsis*, the anlage for the new adoral zone of the proter is likely formed on the dorsal wall of the buccal cavity (Fig. 192q). Distinctly later, the newly formed adoral zone with many differentiated membranelles, the undulating membranes anlage, and the many oblique frontal-midventral cirral anlagen are recognisable (Fig. 192s). In this and some later stages the new adoral zone is roughly longitudinally arranged about in midline of cell (Fig. 192t). Meanwhile, the parental adoral zone is successively resorbed. In the stages shown in Figs. 192s, t the disintegrating proximal portion of the parental adoral zone is divided with one part near the anterior cell end and the second about in mid-body. The break in the new adoral zone occurs obviously rather late, that is, shortly before or after the separation of the proter and the opisthe (Fig. 192w, x).

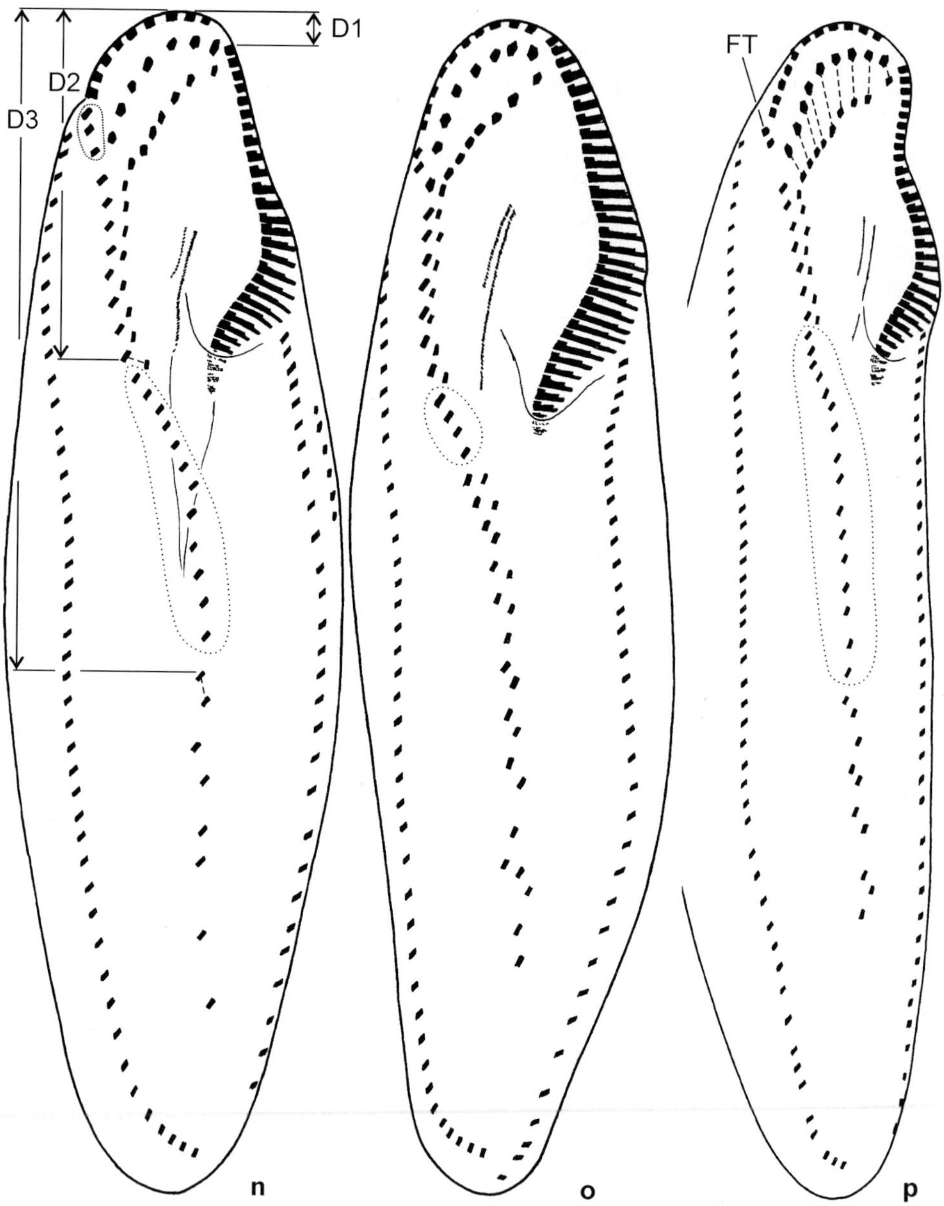

Fig. 192n–p *Uroleptopsis citrina* (neotype population from Berger 2004b. Protargol impregnation). Infraciliature of ventral side. **n:** Specimen with 3 frontoterminal cirri (encircled by very narrowly spaced dots), 146 µm. D1–D3 designate parameters listed in Table 39. **o, p:** Specimen with a short portion of single midventral cirri (circled by dotted line in o; 142 µm) and specimen with a long portion (p; 172 µm). Broken lines in (p) connect cirral pairs of bicorona. D1 = distance between anterior body end and rear end of gap in adoral zone, D2 = distance between anterior body end and end of anterior portion of midventral pairs (last pair of this portion connected by broken line), D3 = distance between anterior body end and anterior end of rear portion of midventral pairs (first pair of this portion connected by broken line), FT = frontoterminal cirri. Page 987.

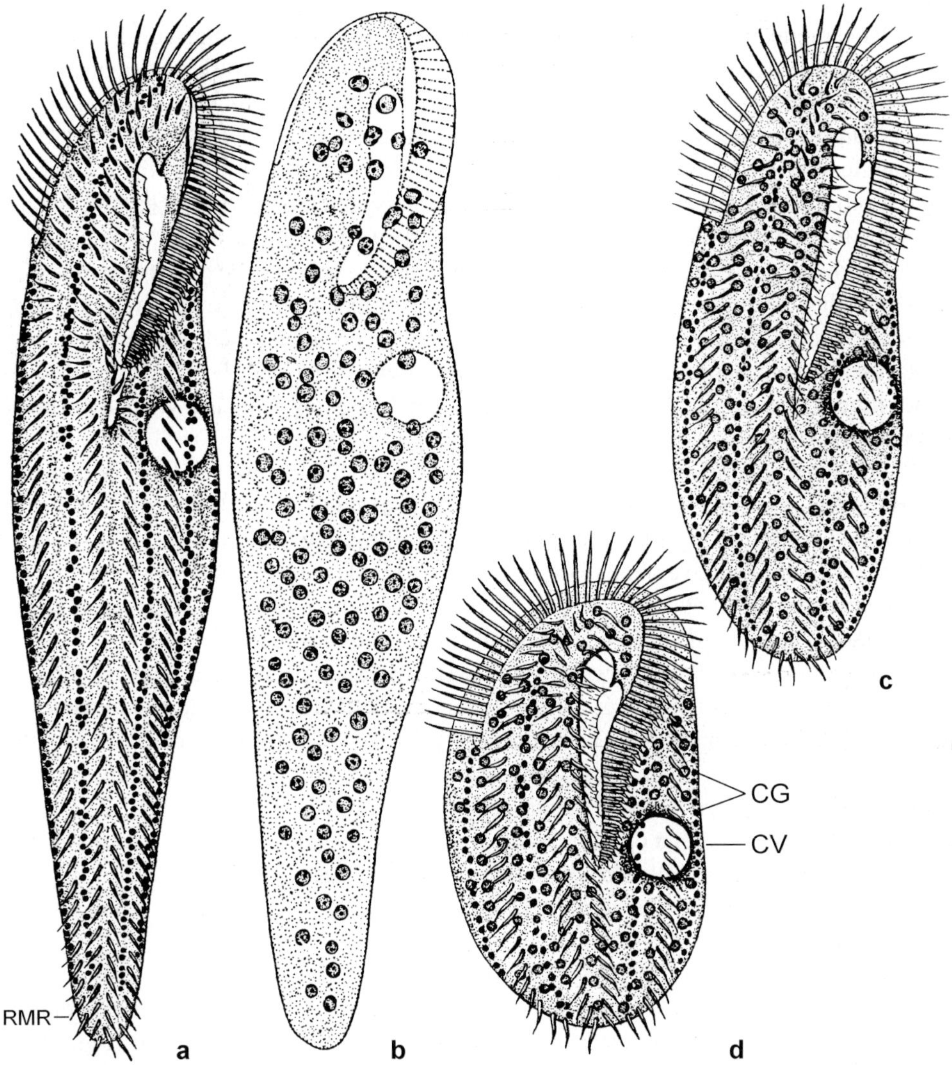

Fig. 197a–d *Uroleptopsis roscoviana* (from Ozaki & Yagiu 1943. From life and after nucleus-stain?). **a, b:** Ventral views of representative (interphasic) specimen (163 µm; likely both figures show the same specimen) showing, inter alia, cirral pattern, contractile vacuole, cortical granules (dark dots between cirral rows), and nuclear apparatus. Note that the right marginal row extends – between adoral zone and anterior bow of bicorona – to the anterior body end (misobservation?). **c, d:** Ventral views of postdividers?, c = 109 µm, d = 76 µm. CG = cortical granules, CV = contractile vacuole, RMR = right marginal row. Page 1006.

Morphology: The descriptions of the two synonyms are kept separate. Body size of type population 190–220 × 47–55 µm, length:width ratio about 4:1. Body widest at level of contractile vacuole, that is, about in mid-body (Fig. 196a); margins moderately converging anteriorly, but strongly converging posteriorly so that rear end rather narrowly rounded; ventral side plane, dorsal portion distinctly vaulted in mid-body and thus about 2:1 flattened dorso-ventrally; anterior and posterior portion rather thin (Fig. 196b, c); very flexible and soft and likely somewhat contractile. Up to 100 macronuclear nodules scattered throughout cytoplasm in large specimens[1], individual nodules 3–4 μm across; central region (nucleolus) homogenous, outer layer granulated (Fig. 196d, e). Contractile vacuole in ordinary position, that is, near left cell margin slightly behind level of proximal end of adoral zone, empties, as is usual, on dorsal surface; during diastole with two spindle-shaped canals (Fig. 196a–d). Cytopyge on dorsal side at about 75% of body length. Body rose-carmine coloured due to fine and irregularly-shaped pigment granules; granules scattered on dorsal side, arranged along cirral rows and body margin on ventral side. In Roscoff, Maupas found a specimen which lacked these pigment granules, but had ones like an unnamed *Styloplotes* (= *Diophrys*) species. Movement without peculiarities, that is, glides moderately rapidly showing great flexibility; never resting.

Adoral zone occupies about 36% (Fig. 196a) to 44% (Fig. 196d) of body length; according to Maupas' text the zone is about one third of body length. The values from the illustrations are rather high for a urostyloid, indicating that at least Fig. 196d shows a post divider. Adoral zone extends far onto right body margin, in specimen shown in Fig. 196a nearly to 20% of body length. Frontal scutum distinct. Buccal area large, shaped like an acute-angled triangle, paroral(?) long and thus prominent.

Cirral pattern composed of two closely spaced median cirral rows, strongly indicating that they form a (although not distinctly curved; Fig. 196a) bicorona and an attached midventral complex composed of cirral pairs; last cirri of midventral complex slightly longer but not stronger than other cirri. Maupas stated that enlarged frontal cirri are lacking. Transverse cirri neither mentioned nor illustrated. Right marginal row commences at level of distal end of adoral zone, not distinctly separated from left row posteriorly, which begins slightly ahead of buccal vertex; rearmost marginal cirri slightly longer and stronger than anterior ones. Dorsal cilia fine and short, that is, around 3 μm. Caudal cirri lacking.

Synonym *H. multiplex* in life(?) 70–160 μm long, specimen shown in Fig. 197a 163 × 35 μm, that is, body length:width ratio about 4.6:1. Body outline elongate elliptical with distinctly narrowed posterior portion (Fig. 197a) to wide elliptical (Fig. 197c, d); specimens shown in Figs. 197c, d are likely postdividers which do not yet have the final shape. Numerous macronuclear nodules scattered throughout cell, individual nodules 2–3 μm across on average. Contractile vacuole near left body margin at about 40% of body length (Fig. 197a). Body reddish brown due to cortical granules arranged in longitudinal rows in parallel with cirral rows and likely also with dorsal kineties (Fig. 197a, c, d). Adoral zone of normal (interphasic) specimens about one third of body length, composed of about 35 proximal membranelles and 28 distal, that is, in total about 63 membranelles. Adoral zone extends very far onto right body margin. Paroral rather long. Buccal cirrus lacking. Bicorona not very distinctly curved and more or less

[1] Maupas fixed the cells with osmium-tetroxide and stained the nuclear apparatus with picro-carmine. He already recognised the high number of macronuclear nodules in this and related species in 1879, however, without mentioning a binomen (Maupas 1879, p. 252).

not set off from midventral complex, which extends to near rear body end. Left marginal row without peculiarities, right row commences at anterior body end, that is, runs in parallel between adoral zone and anterior bow of bicorona (this feature is very unusual and has to be checked carefully). Average length of cirri only about 6 µm. Transverse cirri lacking.

Occurrence and ecology: Marine. Type locality of *U. roscoviana* is likely the French Atlantic Coast near Roscoff because Maupas (1883) found it for the first time in the marine biological laboratory of Roscoff. Later he found it several times among algae in the Mediterranean Sea at the coast and in the harbour of Algiers (Algeria) and finally in the newly built zoological station in Banyuls-sur-Mer, France. Abundance was always low; never occurred among decomposing algae. Only one further record which is, however, not substantiated by illustrations and/or morphological data: Gulf of Mexico (Borror 1962; p. 342).

The type locality of the junior synonym *H. multiplex* is near the Onomiti (= Onomichi?) coast at the Setonaikai, the Inland Sea of Japan. Ozaki & Yagiu (1943) found it there throughout the year in tidal pools and at the sea-bottom where it crept on roots of seaweed (*Sargassum enerve*, *S. thunbergii*).

Uroleptopsis (Plesiouroleptopsis) Berger, 2004

2004 ***Uroleptopsis (Plesiouroleptopsis)* nov. subgen.** – Berger, Acta Protozool., 43: 114 (original description). Type species (by original designation on p. 114): *Pseudokeronopsis ignea* Mihailowitsch & Wilbert, 1990).

Nomenclature: *Plesiouroleptopsis* is a composite of the Greek *plesió* (near, neighbouring; Hentschel & Wagner 1996, p. 480) and the existing genus-group name *Uroleptopsis* (see there for derivation). *Plesiouroleptopsis* alludes to the fact that the type species, *Pseudokeronopsis ignea*, has a buccal cirrus in ordinary position right of the paroral which is the plesiomorphic character state (Berger 2004b). By contrast, the type species of *U. (Uroleptopsis)*, *Uroleptopsis citrina*, has no cirrus immediately right of the paroral which is interpreted as apomorphy. *Plesiouroleptopsis* has, like *Uroleptopsis*, feminine gender because ending with *-opsis* (ICZN 1999, Article 30.1.2).

Characterisation (Fig. 167a, autapomorphy 6): *Uroleptopsis* with midventral complex composed of midventral pairs and midventral rows (A).

Remarks: As already mentioned, *Pseudokeronopsis ignea* was transferred to *Uroleptopsis* by Foissner (1995). It is very likely the sole species in *Uroleptopsis* with the buccal cirrus in the ordinary position, which is of course the plesiomorphic state. However, the midventral rows are a novelty within the Pseudokeronopsidae, respectively, Retroextendia and thus the autapomorphy for the species/subgenus. Midventral rows are not a very complex feature. They originate simply in that more than two cirri per streak are formed (see ground pattern of the Urostyloidea). Thus, their convergent evolution in other urostyloids – for example, the Urostylinae and Bakuellidae – is not a great surprise. However, in none of these taxa do the macronuclear nodules divide individually as in *U. ignea* (and the remaining pseudokeronopsines) and therefore we have to postulate the convergent evolution of midventral rows.

Species included in *Uroleptopsis (Plesiouroleptopsis)* (basionym is given): (1) *Pseudokeronopsis ignea* Mihailowitsch & Wilbert, 1990.

Single species

Uroleptopsis ignea (Mihailowitsch & Wilbert, 1990) Foissner, 1995 (Fig. 198a–j, Table 39)

1989 ***Keronella rubra* n. spec.**[1] – Mihailowitsch, Dissertation, p. 81, Fig. 9a–j, A10a–o (Fig. 198a–j; unpublished thesis).

1990 ***Pseudokeronopsis ignea* nov. spec.**[2] – Mihailowitsch & Wilbert, Arch. Protistenk., 138: 214, Fig. 13–22, Tabellen 5, 6 (Fig. 198a–j; original description; type slides are deposited in the Institut für landwirtschaftliche Zoologie und Bienenkunde, University of Bonn, Germany).

1995 ***Uroleptopsis ignea* (Mihailowitsch & Wilbert, 1990) nov. comb.** – Foissner, Arch. Protistenk., 145: 64 (combination with *Uroleptopsis*).

2001 ***Uroleptopsis ignea* (Mihailowitsch and Wilbert, 1990) Foissner, 1995** – Berger, Catalogue of ciliate names 1. Hypotrichs, p. 76 (nomenclator containing all basionyms, combinations, and higher taxa of hypotrichs).

2004 ***Uroleptopsis (Plesiouroleptopsis) ignea* (Mihailowitsch and Wilbert, 1990) Foissner, 1995** – Berger, Acta Protozool., 43: 115, Fig. 29, 30 (Fig. 198a, h; see nomenclature and remarks).

Nomenclature: No derivation of the name is given in the original description. The species-group name *igne·us -a -um* (Latin adjective; fiery red, flaming) obviously alludes to the reddish cortical granules of this species. If the name of this species is cited with the subgenus name, then this has to be done as follows: *Uroleptopsis (Plesiouroleptopsis) ignea* (Mihailowitsch & Wilbert, 1990) Foissner, 1995. If somebody raises the subgenus *U. (Plesiouroleptopsis)* to genus rank, which is only recommended when a second species is known, then the present species has to be transferred to this genus and the "somebody" will be the combining author (ICZN 1999, Article 51.3.2).

Mihailowitsch (1989) described this species under the name mentioned in the list of synonyms (see corresponding footnote). Since this paper does not constitute a nomenclaturally valid work, Mihailowitsch & Wilbert (1990) provided an original description fulfilling the requirements of the Code. The line drawings are identical in both papers, but of much better quality and larger size in the thesis, from where I thus took them. In addition, several micrographs in the dissertation document the observations.

Remarks: Mihailowitsch & Wilbert (1990) recognised the main features of the present species which separate it from other *Pseudokeronopsis* species, namely the presence of short midventral rows and the lack of transverse cirri. They argued that these

[1] The dissertation by Mihailowitsch (1989) is not a valid publication in the sense of the Code (ICZN 1985, Article 9 (11)). However, to complete the picture I mention this name in the present revision. But to avoid nomenclatural problems, this binomen is disclaimed for nomenclatural purposes (ICZN 1999, Article 8.3).

[2] The diagnosis by Mihailowitsch & Wilbert (1990) is as follows: Länge in vivo 270–355 µm, Körper langgestreckt und sehr metabol. Entlang der Midventral-Cirrenreihe sind dicke rötliche Granula zu sehen. Fortbewegungsweise schlängelnd.

characteristics could be used to establish a genus. However, they refrained from this act because Borror & Wicklow (1983) stated a high variability of the transverse cirri feature in many urostyloids. By contrast, in my experience most cirral groups are rather stable within species, that is, a species either has a cirral group or it does not.

Foissner (1995) discussed Borror & Wicklow's (1983) decision to reject *Uroleptopsis* (actually, already Borror 1979 rejected *Uroleptopsis*; see genus section). He resurrected Kahl's genus and simultaneously transferred *P. ignea* to *Uroleptopsis*.

Mihailowitsch (1989) and Mihailowitsch & Wilbert (1990) studied this species in life, but likely not in much detail. They recognised the presence of reddish cortical granules, however, no further life data are provided. Thus, it is unclear whether or not ring-shaped structures, which form a distinct seam in *U. citrina*, are present.

Morphology: Body length in life 270–355 μm; ratio of body length:width 8.9:1 in protargol preparations (Table 39); body length after protargol impregnation 262 to 353 μm (Table 39), specimen shown in Fig. 198a, however, 405 μm long (incorrect bar?). Body elongate, very flexible. Macronucleus composed of many (more than 100) nodules scattered throughout cell. 3–4 micronuclei. Contractile vacuole near left margin, however, at beginning of last fifth of cell, which is very unusual. Large, reddish cortical granules along midventral complex. Movement winding.

Adoral zone occupies about 30% of body length in specimen shown in Fig. 198a, bipartite in invariably 11 distal membranelles, which are stronger than the 33–55 proximal membranelles; gap between distal and proximal portion approximately of the width of one membranelle; proximal portion almost in *Gonostomum* pattern, that is, extends along body margin with rearmost portion at first curved rightwards and finally backwards. Undulating membranes commence at about same level, but endoral about twice as long as paroral, which optically intersects endoral at level of buccal cirrus. Anterior bow of bicorona composed of seven cirri, rear bow, however, made of six cirri only because cirrus II/2 (= buccal cirrus) not included in rear corona as in *U. citrina,* but in ordinary position, that is, distinctly behind anterior end of paroral right of optical intersection of membranes. Two frontoterminal cirri in ordinary position. Midventral complex composed of 5–7 cirral pairs and 14–21 midventral rows each composed of usually three, sometimes four cirri. Transverse cirri lacking. Marginal rows slightly separated posteriorly. Sizes of individual cirri, see Fig. 198a. Invariably three dorsal kineties, each slightly shortened anteriorly and posteriorly. Caudal cirri absent.

Cell division: Fortunately, Mihailowitsch (1989) and Mihailowitsch & Wilbert (1990) studied the ontogenesis of this species more or less completely (Fig. 198c–j). Cell division commences with the formation of some basal body patches left of the midventral complex (Fig. 198c). Immediately right of the undulating membranes a longish anlage, which also includes the buccal cirrus occurs (Fig. 198c). The patches left of the midventral complex enlarge and fuse to a single field and the anterior anlage already differentiates into membranelles (Fig. 198d). The anlage for the midventral complex likely originates de novo in that one (or more) longitudinal primordium forms many streaks. Between the anlagen for the adoral zone and the midventral complex disordered basal bodies form the undulating membranes anlagen of the proter. The marginal rows are unchanged (Fig. 198d). Fig. 198e shows a middle divider with newly

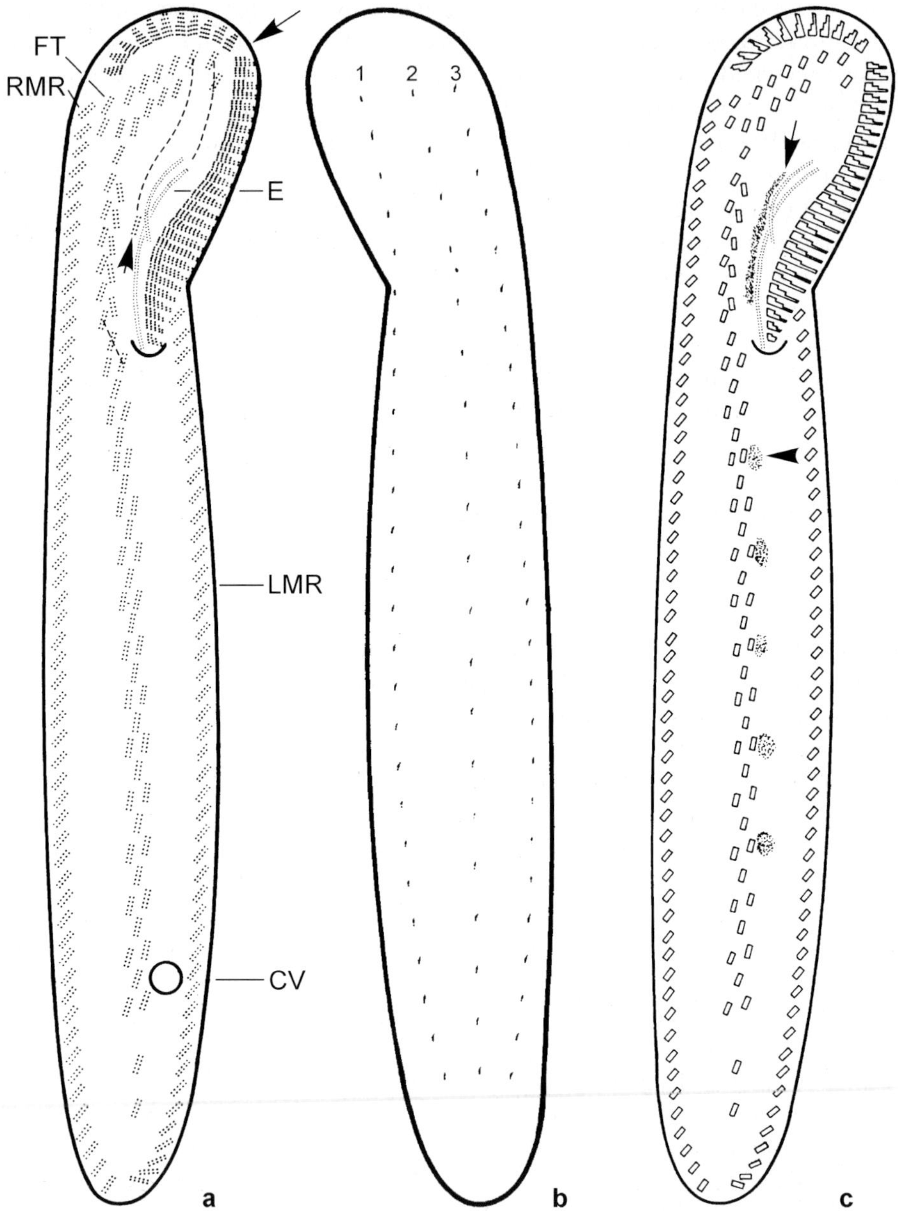

Fig. 198a–c *Uroleptopsis ignea* (from Mihailowitsch 1989. Protargol impregnation). **a, b:** Infraciliature of ventral and dorsal side of a representative specimen, 403 µm (according to the morphometry, see Table 39, the greatest length is 353 µm; possibly the scale bar has a wrong length or this specimen is not included in the data set. Long arrow marks gap in adoral zone, short arrow denotes buccal cirrus (= cirrus II/2) which is in ordinary position, that is, right of paroral. Broken lines connect cirri of anlagen I and II and the anteriormost anlage which produces a midventral row. **c:** Infraciliature of ventral side of a very early divider, 409 µm. Arrow marks anlagen field for proter, arrowhead denotes front of five oral primordium anlagen of opisthe. CV = contractile vacuole, E = endoral, LMR = left marginal row, FT = frontoterminal cirri, RMR = right marginal row, 1–3 = dorsal kineties. Page 1012.

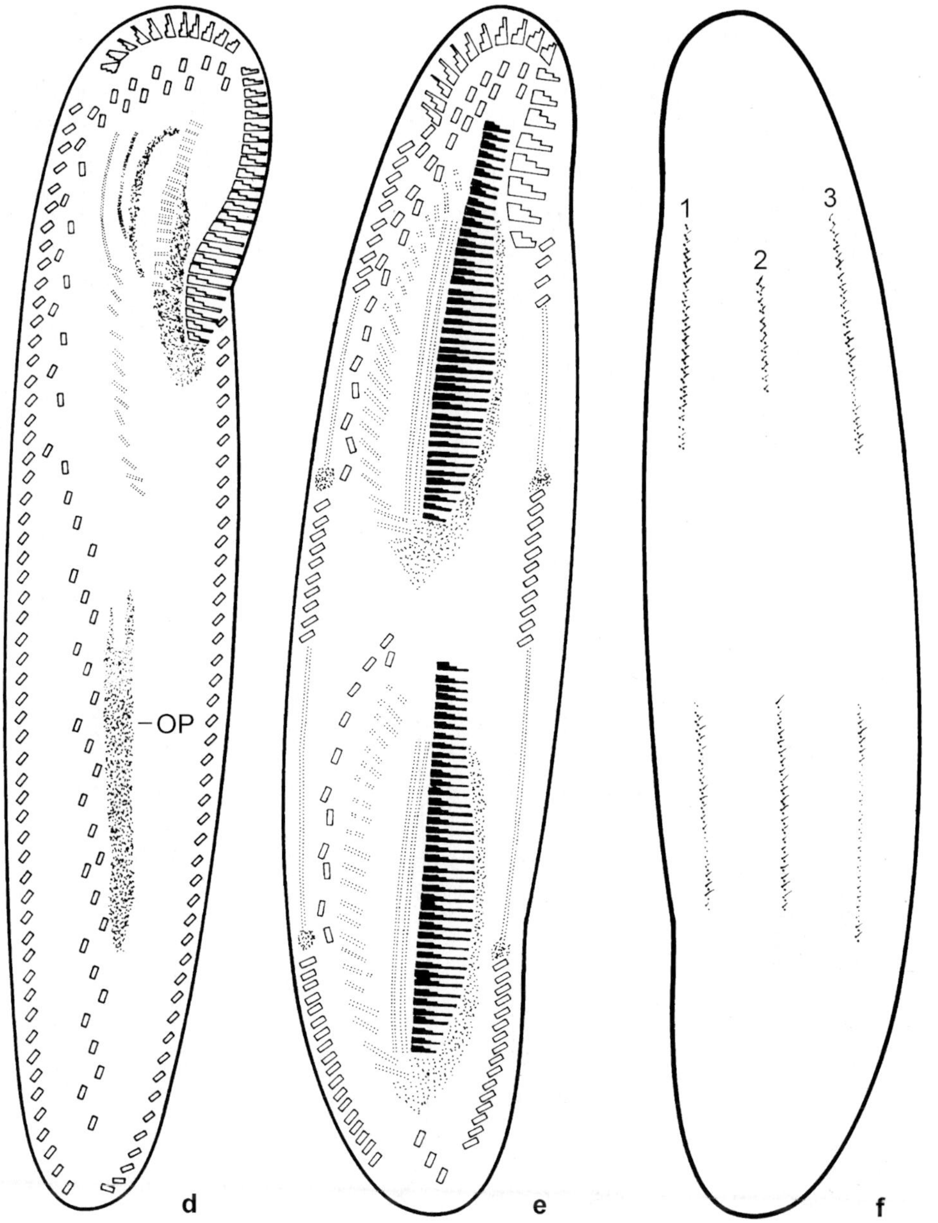

Fig. 198d–f *Uroleptopsis ignea* (from Mihailowitsch 1989. Protargol impregnation). Old structures white, new black. **d:** Infraciliature of ventral side of an early divider, 312 μm. Note that the cirri of the parental midventral complex are not involved in primordia formation. **e, f:** Infraciliature of ventral and dorsal side of a middle divider, 375 μm (I suppose that these two figures show the same specimen although for (f) a different length [350 μm] is given). Note that from the anteriormost frontal-midventral anlage two cirri are formed which is one important autapomorphy of *Uroleptopsis*. OP = oral primordium for opisthe, 1–3 = new dorsal kineties of proter. Page 1012.

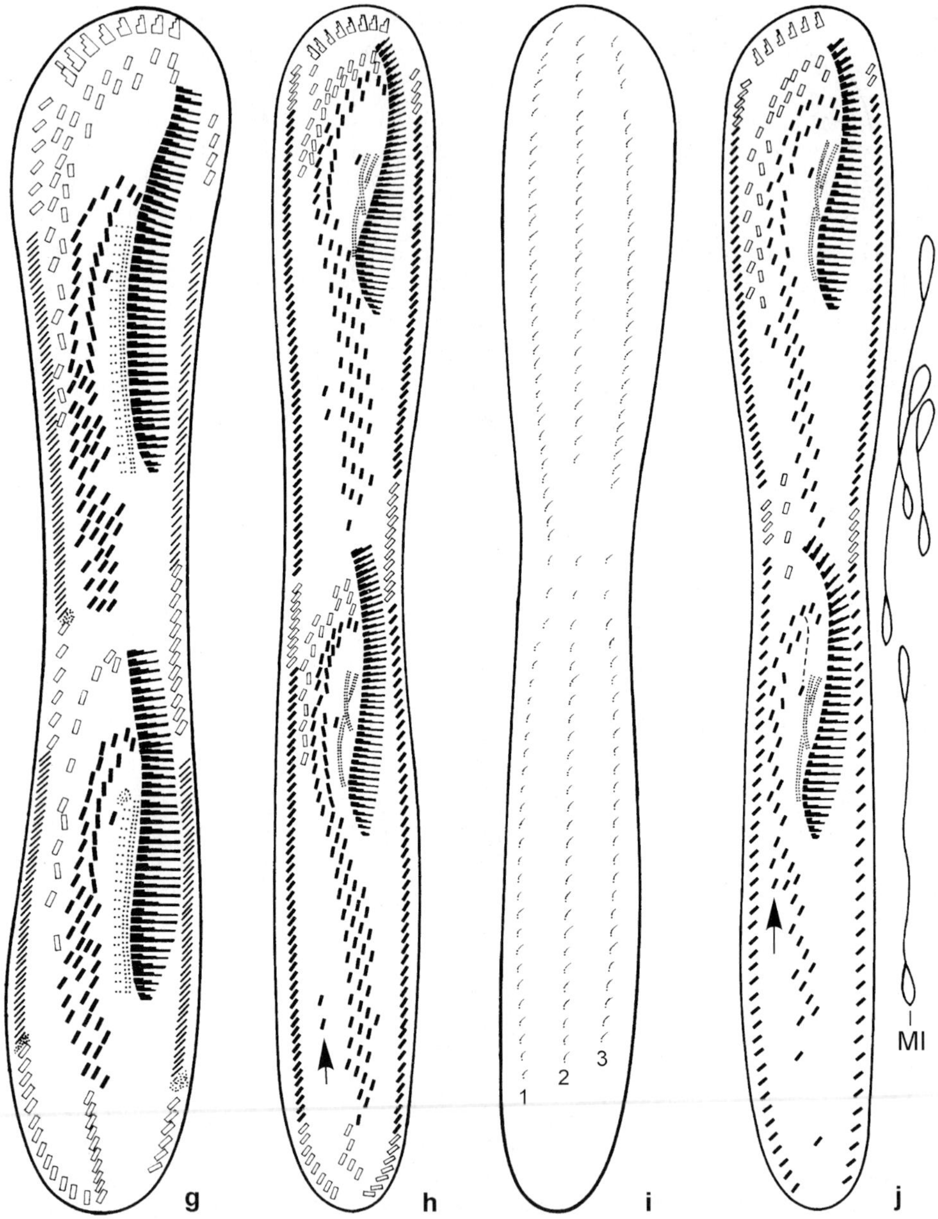

Fig. 198g–j *Uroleptopsis ignea* (from Mihailowitsch 1989. Protargol impregnation). Old structures white, new black. **g:** Infraciliature of ventral side of a late divider, 362 µm. **h, i:** Infraciliature of ventral and dorsal side of a very late divider, 397 µm. Arrow marks anteriorly migrating frontoterminal cirri of opisthe. In the middle region of (i) some parental bristles are still recognisable. **j:** Infraciliature of ventral side and division of micronuclei of a very late divider, 348 µm. Arrow marks anteriorly migrating frontoterminal cirri of opisthe. The broken line connects the buccal cirrus (= cirrus II/2) with the corresponding cirrus of the anterior bow of the bicorona. MI = dividing micronucleus, 1–3 = new dorsal kineties of opisthe. Page 1012.

Mihailowitsch (1989, p. 81, thesis) mentioned a *Keronella rubra*[1]; however, according to the ICZN (1985, Article 9, (11)), a thesis does not constitute a publication in the sense of nomenclature and thus the name is not available. Moreover, it is not included in the Zoological Record. In 1990, Mihailowitsch & Wilbert described it under the name *Pseudokeronopsis ignea* (now *Uroleptopsis ignea*) because its macronuclear nodules do not fuse during division. Hence, *Keronella* is still monotypic.

Wiackowski (1985) established *Keronella* mainly because the anterior corona of frontal cirri is distinctly set off from the zigzagging midventral pattern. However, such an offset also occurs in *Pseudokeronopsis* spp. (Wirnsberger et al. 1987; Fig. 179b) and is thus not included in the characterisation above. However, it cannot be excluded that this feature becomes useful above the species level in future. The formation of the *Keronella* bicorona proceeds as, for example, in *Pseudokeronopsis*, *Thigmokeronopsis*, *Pseudourostyla*, and likely in *Bicoronella*, that is, the frontal ciliature is simply the anterior, leftwards-curved portion of the midventral complex, although the zigzag pattern, so characteristic for the midventral complex, is no longer recognisable in this region. By contrast, in the oxytrichid *Kerona* Müller, 1786, the corona (or bicorona if the cirral row 2 is considered as rear corona; for review and terminology see Berger 1999, p. 825ff) originates from the (two) leftmost cirral anlagen. In *Keronopsis* Penard, 1922 and *Paraholosticha* Wenzel, 1953, the anterior portion of the three leftmost (of a total of five) cirral anlagen form the corona. These different ontogenetic patterns show that the frontal ciliature (corona) of *Keronella* (and of some other urostyloids mentioned above), *Kerona*, and *Keronopsis* and *Paraholosticha* evolved convergently.

Wiackowski (1985, p. 89) assigned *Keronella* to the Pseudokeronopsidae because the numerous frontal cirri form a bicorona as in *Pseudokeronopsis*, the name-bearing type of this group. This classification was kept by Tuffrau (1987), Tuffrau & Fleury (1994), and Lynn & Small (2002). By contrast, Wirnsberger (1987, p. 155) proposed a classification in the bakuellids because macronuclear division is different, that is, the individual nodules fuse prior to division in *Keronella*, whereas they divide individually in *Pseudokeronopsis*. Alekperov (1988, p. 779, 780; 1992) also assigned *Keronella* to the bakuellids because morphology and ontogenesis are as in related genera. Eigner (1994), who obviously overlooked Alekperov's papers, classified it in the bakuellids too and provided, like Franco et al. (1996), a key to the genera of this group. Recently, Shi (1999, 1999a), who omitted the Bakuellinae, included it in the Holostichidae. In the present book *Keronella* is included in the Urostylinae because it has a bicorona (preventing a classification in the Holostichidae and Bakuellidae) and the midventral complex is composed of cirral pairs and midventral rows. In the Pseudokeronopsidae the whole parental adoral zone of membranelles is replaced during cell division (vs. only proximal membranelles in *Keronella*) and the macronuclear nodules divide individually or form at least a branched mass (vs. globular mass in *Keronella*).

Species included in *Keronella*: (1) *Keronella gracilis* Wiackowski, 1985.

Species misplaced in *Keronella*: *Keronella perbella* Alekperov & Musayev, 1988 (now *Metabakuella perbella*).

---

[1] This name is disclaimed for nomenclatural purposes (ICZN 1999, Article 8.3).

Table 40 Morphometric data on *Keronella gracilis* (aus, original data of an Austrian population; pol, type population from Poland; from Wiackowski 1985)

Characteristics [a]	Population	mean	M	SD	SE	CV	Min	Max	n
Body, length	aus	130.6	129.0	15.5	2.3	11.9	98.0	170.0	44
	pol	157.5	–	–	1.6	6.3	133.0	175.0	40
Body, width	aus	50.8	50.0	6.3	1.0	12.4	37.0	66.0	43
	pol	80.9	–	–	1.6	12.6	60.0	101.0	40
Anterior body end to proximal end of adoral zone, distance	aus	44.7	44.0	4.4	0.7	9.8	37.0	58.0	44
	pol	57.9	–	–	0.6	6.1	50.0	65.0	40
Distance 1 [b]	aus	40.1	39.0	7.0	1.1	17.4	26.0	60.0	37
Distance 2 [c]	aus	42.9	43.5	9.2	1.5	19.2	24.0	60.0	30
Largest adoral membranelle, width	pol	12.2	–	–	0.1	6.8	10.0	14.0	40
Marginal rows, maximal distance in between	pol	68.6	–	–	1.3	11.9	50.0	85.0	40
Posterior macronuclear nodule, length	aus	5.7	5.0	1.9	0.3	33.8	4.0	15.0	40
Posterior macronuclear nodule, width	aus	4.1	4.0	0.7	0.1	16.6	3.0	5.0	40
Largest macronuclear nodule, length	pol	14.2	–	–	0.3	12.6	11.0	19.0	39
Largest macronuclear nodule, width	pol	6.3	–	–	0.2	14.9	4.0	8.0	39
Macronuclear nodules, number	aus	55.0	57.0	8.2	1.7	14.8	39.0	68.0	23
	pol	59.4	59.0	–	0.8	5.7	53.0	65.0	17
Micronucleus, diameter	pol	4.4	–	–	0.1	6.6	4.0	5.0	37
Micronuclei, number	pol	6.5	6.0	–	0.5	26.8	4.0	10.0	14
Adoral membranelles, number	aus	46.9	47.0	4.1	0.7	8.7	39.0	54.0	31
	pol	44.2	45.0	–	0.4	5.7	39.0	49.0	40
Anterior corona, number of cirri	aus	8.9	9.0	1.0	0.1	10.7	7.0	11.0	41
	pol	8.3	8.0	–	0.1	7.6	7.0	9.0	40
Posterior corona, number of cirri	aus	7.5	7.0	0.7	0.1	9.6	6.0	9.0	39
Frontoterminal cirri, number	aus	11.9	12.0	1.9	0.3	15.9	5.0	16.0	34
	pol	12.0	12.0	–	0.3	14.7	7.0	15.0	40
Buccal cirri, number	aus	1.0	1.0	0.0	0.0	0.0	1.0	1.0	36
	pol			invariably 1					
Midventral pairs, number	aus	11.9	12.0	2.0	0.5	16.8	9.0	15.0	14
Midventral rows, number	aus	5.0	5.0	1.9	0.5	38.5	2.0	9.0	15
FVT-primordia, number 1 [e]	pol	23.1	23.0	–	0.3	7.3	20.0	27.0	39
FVT-primordia, number 2 [e]	pol	5.4	5.0	–	0.2	18.2	3.0	8.0	37
Rightmost midventral row, number of cirri [d]	aus	9.8	10.0	1.1	0.2	10.8	8.0	12.0	25
Transverse cirri, number	aus	9.4	9.0	1.5	0.2	16.3	6.0	12.0	38
	pol	7.4	8.0	–	0.2	15.9	5.0	9.0	40
Left marginal cirri, number	aus	40.9	41.0	4.8	0.8	11.8	33.0	51.0	35
	pol	41.8	42.0	–	0.6	9.6	33.0	47.0	40
Right marginal cirri, number	aus	46.4	45.5	5.2	0.9	11.1	39.0	59.0	36
	pol	42.8	44.0	–	0.8	11.9	31.0	50.0	40
Dorsal kineties, number	aus	5.0	5.0	0.0	0.0	0.0	5.0	5.0	29
	pol			invariably 5					
Dorsal kinety 1, number of bristles	aus	27.8	28.0	3.9	0.8	14.1	17.0	33.0	25
Dorsal kinety 2, number of bristles	aus	28.2	28.0	2.7	0.6	9.7	23.0	33.0	21
Dorsal kinety 3, number of bristles	aus	23.4	23.5	1.8	0.5	7.8	21.0	26.0	12
Dorsal kinety 4, number of bristles	aus	24.6	25.0	2.7	0.7	11.0	19.0	30.0	15
Dorsal kinety 5, number of bristles	aus	29.5	30.0	3.7	0.8	12.6	22.0	37.0	24
Dorsal bristles ahead of right marginal row, number	aus	1.1	1.0	–	–	–	0.0	2.0	38
	pol			usually 2					

Table 40 Continued

Characteristics[a]	Population	mean	M	SD	SE	CV	Min	Max	n
Caudal cirri, number	aus	3.2	3.0	0.8	0.1	26.1	2.0	5.0	31
	pol	2.4	2.0	–	0.1	22.8	1.0	3.0	39

[a] All measurements in µm. Data of Austrian population are based on mounted, protargol-impregnated (Foissner's method), and randomly selected specimens. Data of Polish population are based on specimens impregnated with Wilbert's protargol method. CV = coefficient of variation in %, M = median, Max = maximum value, mean = arithmetic mean, Min = minimum value, n = number of individuals investigated, SD = standard deviation, SE = standard error of arithmetic mean.

[b] Distance between anterior body end and posterior end of frontoterminal row.

[c] Distance between rear body end and anterior end (Fig. 202a, vertical arrow) of rightmost midventral cirral row.

[d] Cirral row marked with a vertical arrow in Fig. 202a.

[e] FVT = frontal-midventral-transverse. Number 1 is the total number of frontal-midventral-transverse primordia; number 2 is the number of frontal-midventral-transverse primordia producing more than two cirri, that is, it is the number of primordia which produce transverse cirri.

Single species

Keronella gracilis Wiackowski, 1985 (Fig. 201a–s, 202a–d, Table 40)

1985 ***Keronella gracilis* n. gen., n. spec.**[1] – Wiackowski, Protistologica, 21: 81, 91, Fig. 1, 2a, b, 3–8, 9a, b, 10a–h, 11–21, Tables 1, 2 (Fig. 201a–s; original description. The type slide is deposited in the Department of Hydrobiology, Institute of Environmental Biology, Jagiellonian University, Poland).

1988 ***Keronella gracilis* Wiackowski, 1985** – Wiackowski, Acta Protozool., 27: 4 (phenetic classification of urostylids).

1994 ***Keronella gracilis* Wiackowski, 1985** – Eigner, Europ. J. Protistol., 30: 474, Fig. 26 (Fig. 201b; brief revision of Bakuellinae).

1996 ***Keronella gracilis*** – Franco, Esteban & Téllez, Acta Protozool., 35: 329, Fig. 21 (Fig. 201b, c; key to species of the Bakuellinae).

2001 ***Keronella gracilis* Wiackowski, 1985** – Berger, Catalogue of ciliate names 1. Hypotrichs, p. 43 (nomenclator containing all basionyms, combinations, and higher taxa of hypotrichs).

Nomenclature: No derivation of the name is given in the original description. The Latin adjective *grácil·is -is -e* means small, thin, slender, delicate. I do not know to which feature the species-group name refers because the present species is neither small nor slender. *Keronella gracilis* was fixed as type species of *Keronella* by original designation. Voucher slides of the Austrian population are deposited in the Oberösterreichische Landesmuseum in Linz [LI], Austria.

[1] The diagnosis by Wiackowski (1985) is as follows: Corona of frontal cirri consists of 7–9 cirri; frontoterminal (migrating) cirri: 7–15; dorsal kineties: 5; caudal cirri: 2–3; Ma: about 60 oval fragments.

Remarks: A summary of the description of the present species, fortunately without reference to a name, is given by Wiackowski (1984). My Austrian population (Fig. 202a–d) agrees very well with the type population (Fig. 201a–s) from Poland, except for the cortical granules, which are not described by Wiackowski (1985, 1988). However, according to Wiackowski (1988, p. 7, character 26), *Urostyla grandis* also lacks such organelles, indicating a misobservation by Wiackowski in that feature because *Urostyla grandis* always has distinct cortical granules. Wiackowski (1985, p. 82) described the cytoplasm as having a "greenish-yellowish shadow" which is possibly due to the yellowish cortical granulation. Thus, I simple assume that Wiackowski (1985, 1988) overlooked the cortical granules because they are small (>1.0 µm); however, it cannot be excluded that a (sub)species without cortical granules exists.

In life, *Keronella gracilis* is best recognised by the bicorona, the cortical granules, and the midventral rows in the posterior body portion. *Bakuella* species have three distinct frontal cirri.

Morphology: The morphology section below is based mainly on the data from the original description, followed by some additional and deviating observations from the Austrian population. Body size in life not indicated in original description; in protargol preparations (Wilbert's method) 133–175 × 60–101 µm, length:width ratio 1.9:1 on average (Table 40). Body outline according to Fig. 201a wide elliptical with anterior end slightly narrower rounded than posterior; however, I am uncertain whether or not the figure shows a freely motile, that is, unsqueezed specimen. Body distinctly flattened dorsoventrally. Macronuclear nodules scattered, ellipsoidal, with several small nucleoli. Micronuclei also scattered, globular. Contractile vacuole slightly ahead of mid-body near left body margin. Cytoplasm clear with a greenish-yellow shade (I assume, that this is likely due to the cortical granules, which are not described by Wiackowski; see remarks!).

Adoral zone occupies about 37% of body length, of usual shape and structure, composed of an average of 44 membranelles, bases of largest membranelles 12 µm wide on average in protargol preparations. Individual membranelles of ordinary fine structure. Buccal cavity moderately wide. Undulating membranes curved, intersect optically slightly behind level of buccal cirrus. Paroral composed of zigzagging basal bodies, begins distinctly more anteriorly than endoral, which consists of oblique pairs of basal bodies (Fig. 201b, e, j, k).

Cirral pattern and number of cirri of usual variability (Fig. 201b, d, e, h–k). Frontal cirri slightly larger than other cirri, which all have a rather similar size. Invariably two parallel rows, that is, a bicorona of frontal cirri. Anterior corona almost semicircular, distinctly set off from first midventral pair; posterior corona less curved than anterior, left portion hook-shaped, right end very close to first midventral pair. Buccal cirrus distinctly behind anterior end of paroral. Row of frontoterminal cirri commences close to distal end of adoral zone of membranelles, slightly curved leftwards and therefore terminating close to proximal end of adoral zone. Midventral complex composed of 11 midventral pairs and five midventral rows in specimen illustrated (Fig. 201b); usually 5–6 midventral rows, each with three (in anteriormost rows) to more than 12 (in posteriormost row) cirri; short rows usually slightly oblique, long rows usually longitudinally

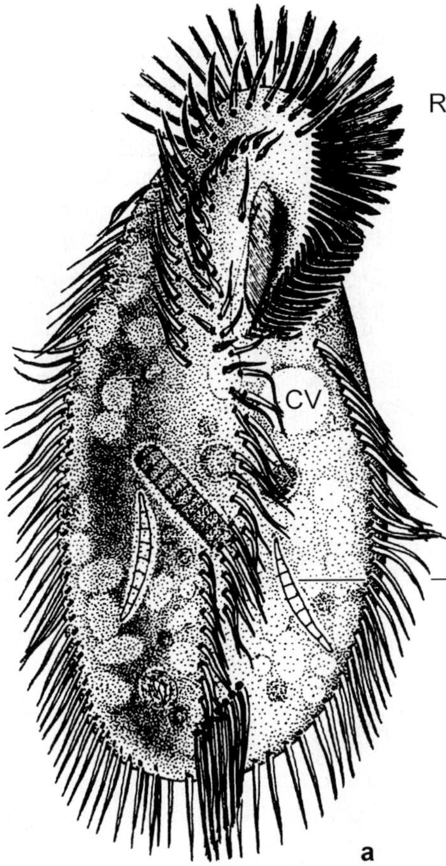

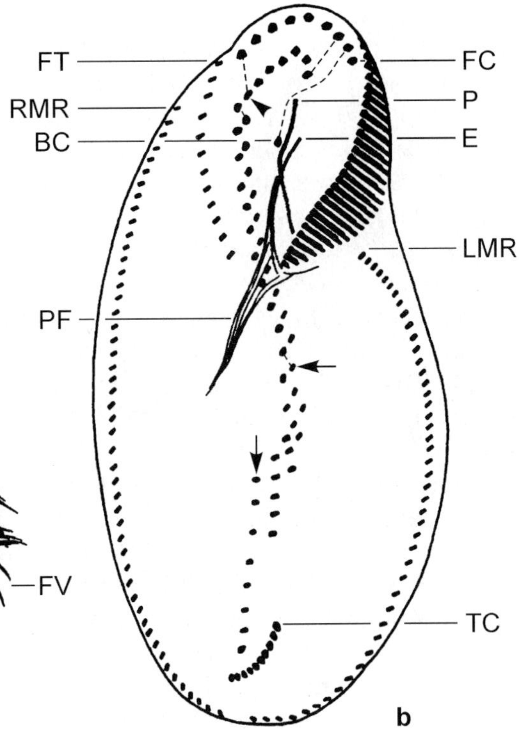

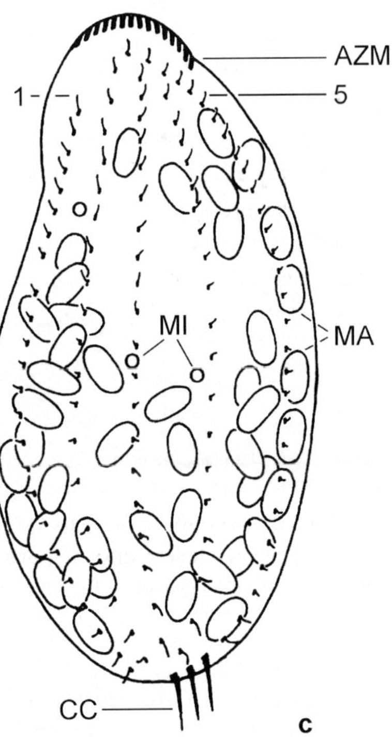

Fig. 201a–c *Keronella gracilis* (from Wiackowski 1985. a, from life?; b, c, protargol impregnation). **a:** Ventral view, 155 μm. I do not know whether or not the body outline is from a freely motile specimen. **b, c:** Infraciliature of ventral and dorsal side and nuclear apparatus of same specimen, 176 μm. Arrowhead in (b) marks rightmost cirrus of posterior corona. Horizontal arrow denotes last midventral pair; vertical arrow marks midventral row ahead of penultimate transverse cirrus from right. Broken lines connect cirri originating from same anlage. Note that the caudal cirri are associated to dorsal kinety 5 (c). AZM = distal end of adoral zone of membranelles, BC = buccal cirrus, CC = caudal cirri, CV = contractile vacuole, E = endoral, FC = leftmost cirrus of anterior corona, FT = anteriormost cirrus of frontoterminal row, FV = food vacuole containing fungal spore, LMR = left marginal row, MA = macronuclear nodule, MI = micronuclei, P = paroral, PF = pharyngeal fibres, RMR = anteriormost cirrus of right marginal row, TC = leftmost transverse cirrus, 1, 5 = dorsal kineties. Page 1023.

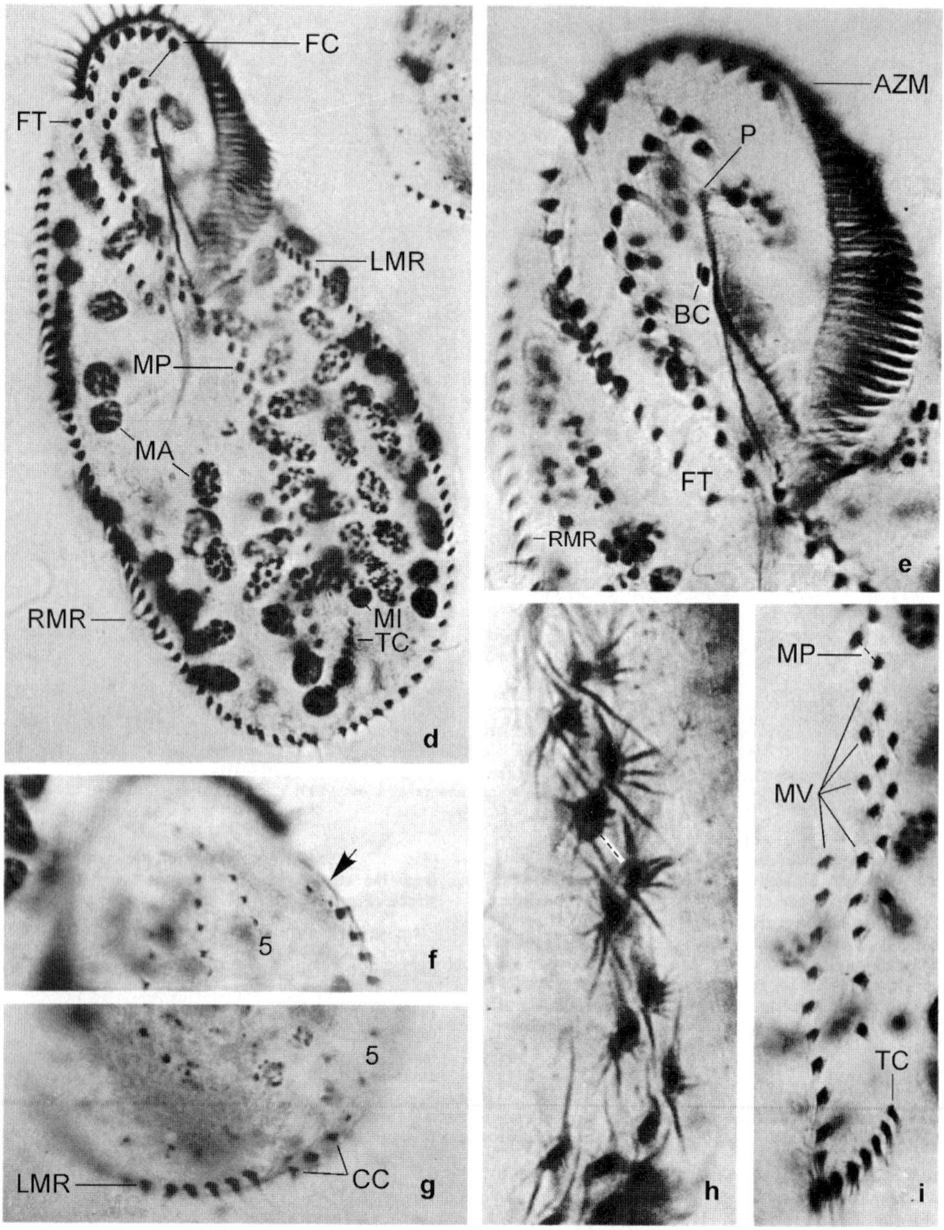

Fig. 201d–i *Keronella gracilis* after protargol impregnation (from Wiackowski 1985). **d, e:** Infraciliature of ventral side and nuclear apparatus and detail of oral region. **f:** Anterior end of dorsal kineties; arrow marks basal body pairs ahead of right marginal row. **g:** Caudal cirri at end of dorsal kinety 5. **h:** Some midventral pairs (and their fibre system) showing characteristic zigzag pattern. **i:** Posterior portion of midventral complex and transverse cirri. AZM = adoral zone of membranelles, BC = buccal cirrus, CC = caudal cirri, FC = left cirrus of anterior and posterior corona, FT = frontoterminal row, LMR = left marginal row, MA = macronuclear nodules, MI = micronucleus, MP = midventral pair, MV = midventral rows, P = paroral, RMR = right marginal row, TC = transverse cirri, 5 = rightmost dorsal kinety. Page 1023.

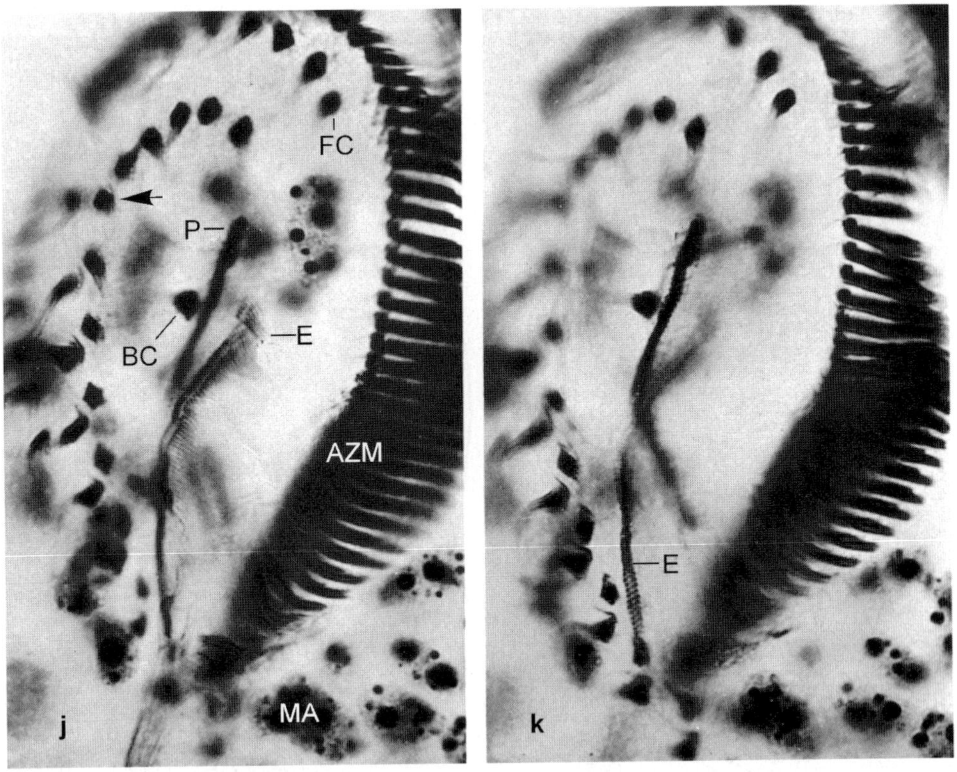

Fig. 201j, k *Keronella gracilis* after protargol impregnation (from Wiackowski 1985). Infraciliature of oral region at two focus levels showing, inter alia, the two curved rows of frontal cirri forming the bicorona, the anterior midventral pairs, and the structure and shape of the undulating membranes. Arrow marks rightmost cirrus of rear corona. AZM = adoral zone of membranelles, BC = buccal cirrus (= cirrus II/2), E = endoral, FC = leftmost cirrus (= cirrus I/1) of anterior corona, MA = macronuclear nodule, P = paroral. Page 1023.

arranged. No distinct pretransverse ventral cirri. Transverse cirri narrowly spaced in slightly curved, oblique row, project by about half of their length beyond rear body end. Right marginal row begins about at level of first midventral pair, ends subterminally slightly right of midline; left row, which commences left of proximal portion of adoral zone, terminates at rear end of cell in midline so that marginal rows do not confluent posteriorly; marginal cirri rather narrowly spaced.

Dorsal cilia short and stiff, invariably arranged in five rows of almost body length; usually two bristles ahead of anterior end of right marginal row. 2–3 caudal cirri at end of kinety 5; sometimes up to three further, very small caudal cirri occur at end of other rows (Fig. 201c, f, g).

Observations on Austrian population (Fig. 202a–d, Table 40): Body size 150–200 × 60–80 μm in life (n = 2), length:width ratio about 2.5:1 on average in life and in protargol preparations. Body outline almost parallel-sided with both ends broadly rounded, left margin at level of adoral zone with slight indentation (Fig. 202c); body very flexi-

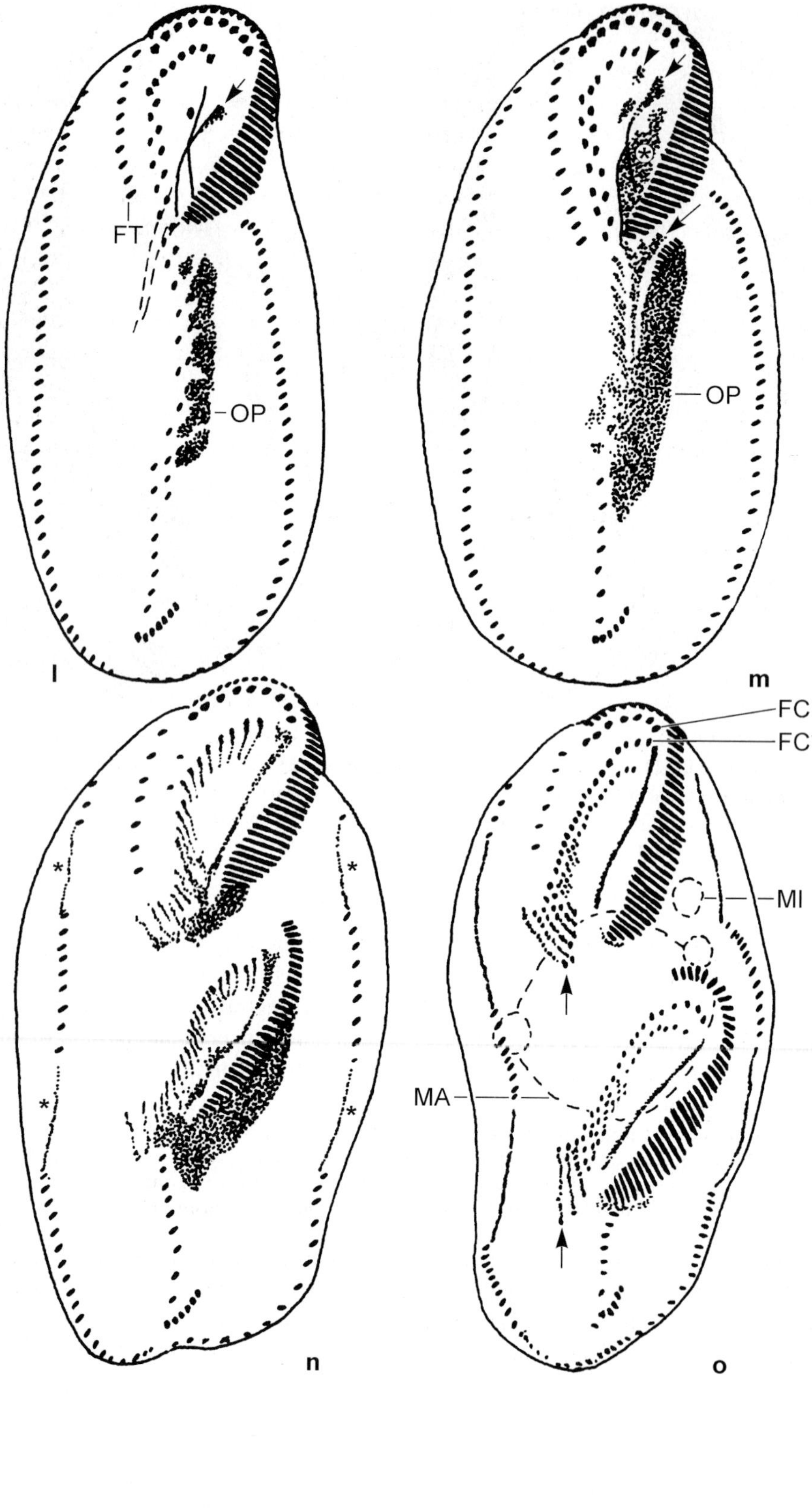
FT
OP
l
OP
m
n
FC
FC*
MI
MA
o

ble and about 2:1 flattened dorsoventrally. Macronuclear nodules scattered, up to 10 × 6 µm in life, nucleoli 3–4 µm across. Micronuclei about 3 µm across in life. Contractile vacuole with distinct, longitudinal collecting canals during diastole. Cytoplasm colourless, contains many greenish (fat?) globules 5–15 µm across. Cortical granules mainly around cirri and dorsal bristles; individual granules about 0.5 µm across, yellowish (Fig. 202b), making cells yellowish or greenish at low magnification, do not stain with Foissner's protargol method used. Movement inconspicuous, glides (jerks finely) moderately rapidly on microscope slide. Adoral zone as in type population, largest membranelles about 12 µm wide in life. Buccal cavity small; pattern and structure of undulating membranes not exactly recognisable in my preparations, but basically as described by Wiackowski (1985). Cirral pattern as in type population (Fig. 202a, d, Table 40). Marginal cirri 12–15 µm long in life, caudal cirri about 15 µm, and dorsal bristles about 3 µm.

Cell division (Fig. 201l–s): Division of *Keronella gracilis* is described in detail by Wiackowski (1985); most stages are documented by micrographs not shown in the present book. The formation/reorganisation of the oral and frontal-midventral-transverse ciliature proceeds spatially independently in proter and opisthe. The processes in the opisthe are described first.

Ontogenesis commences with formation of small patches of basal bodies very close left of the middle portion of the midventral complex (Fig. 201l). While the number of basal bodies increases, the midventral cirri of this region disintegrate and a long oral primordium occurs (Fig. 201l, m). The formation of the adoral membranelles begins at the right anterior corner of the oral primordium. Simultaneously, the primordium for the undulating membranes is formed right of the oral primordium (Fig. 201m, n). The ladder-like structure right of the undulating membrane primordium is, as is usual, the primordium for the frontal-midventral-transverse cirri (Fig. 201n). The formation of the adoral membranelles proceeds from anterior to posterior and from right to left. The left frontal cirrus (= cirrus I/1) splits off, as is usual, from the anterior end of the undulating membrane primordium, while the remaining basal bodies form the paroral and endoral during the next stages (Fig. 201o–s). Differentiation of the frontal-midventral-transverse cirri proceeds from anterior (left anlagen) to posterior (right anlagen). All anlagen, except the 5–9 rightmost, form two cirri (Fig. 201o–q). The second (rear) cirrus of the leftmost anlage (= anlage II) migrates, as is usual, backwards to form the buccal

---

← **Fig. 201l–o** *Keronella gracilis* after protargol impregnation (from Wiackowski 1985). Ventral infraciliature of dividing specimens (for details, see text; sizes not indicated). **l:** Very early divider. Arrow marks disintegrating anterior end of parental endoral. **m:** Early divider with leftmost cirrus (arrowhead) of posterior corona, parental buccal cirrus, and anterior end (short arrow) of parental endoral modified to primordia. Asterisk marks patch of basal bodies on dorsal wall of buccal cavity. Long arrow denotes undulating membrane anlage of opisthe. **n:** Middle divider showing anlagen for marginal rows (asterisks), formation of new adoral membranelles in opisthe, and frontal-midventral-transverse cirral anlagen for both proter and opisthe. **o:** Middle divider showing, inter alia, formation of frontal-midventral-transverse cirri and fused macronucleus. Arrows mark posterior end of rightmost frontal-midventral-transverse cirral anlage which forms the rightmost transverse cirrus and the frontoterminal row. FC = leftmost cirrus of parental anterior corona, FC* = leftmost cirrus of new anterior corona of proter, FT = parental frontoterminal row, MA = fused macronucleus, Mi = micronucleus, OP = oral primordium. Page 1023.

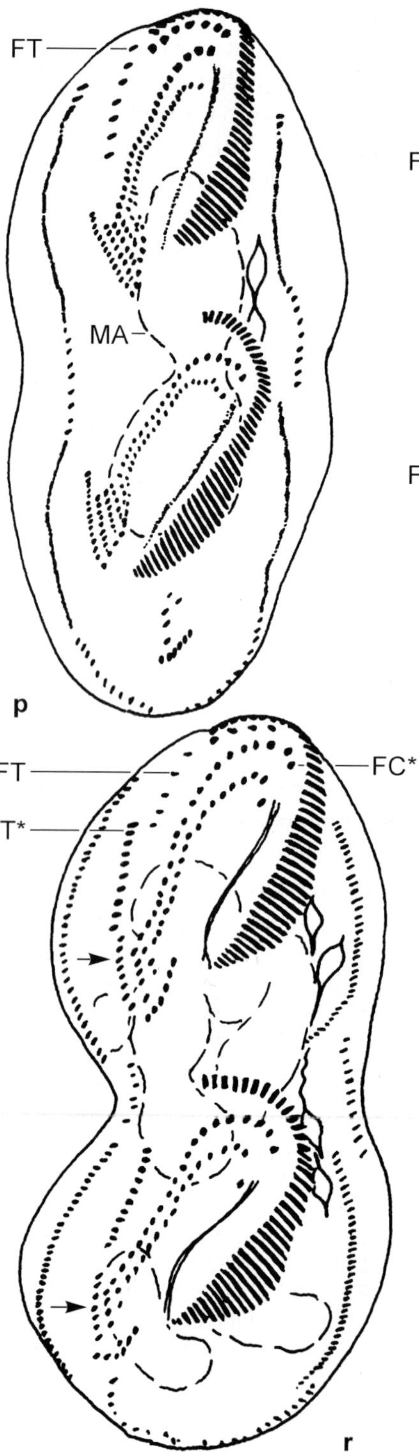
FT
MA
p
FT
FT*
FT*
q
FT
FT*
FC*
r
BC
BC
s

cirrus (Fig. 201q–s). The 5–9 rightmost anlagen produce more than two cirri. Those which form three cirri produce the rearmost midventral pairs and the leftmost transverse cirri. Those primordia which produce four or more cirri – except the rightmost – form the midventral rows and most transverse cirri (Fig. 201p–s). The rightmost anlage segregates the rightmost transverse cirrus and the frontoterminal cirri; they form a rather long row migrating to near the distal end of the adoral zone of membranelles (Fig. 201q–s). The anterior (leftmost) 6–8 anlagen form the frontal cirri, that is, the bicorona; the offset of the anterior corona occurs in very late dividers. The other cirral pairs form the midventral pairs of the midventral complex (Fig. 201s).

In the proter, the first sign of ontogenesis occurs with a slight delay relative to the opisthe. At first, new basal bodies occur at the anterior end of the endoral (Fig. 201l). Later, the endoral and the paroral disintegrate from anterior to posterior; in addition, the buccal cirrus and the leftmost cirrus of the posterior parental corona are transformed to primordia and the right wall of the buccal cavity (peristome according to original description) and the region of the cytostome become packed with basal bodies. Simultaneously, the anterior parental midventral pairs become disorganised and are likely involved in the formation of the frontal-midventral-transverse primordia of the porter (Fig. 201m, n). Some proximal parental adoral membranelles disintegrate and form a basal body field together with those basal bodies which previously occurred near the cytostome (Fig. 201n). None of the dividing specimens found showed evidence that the remaining adoral membranelles are also modified. This strongly suggests that the proter receives the parental adoral zone, in which only few membranelles close to the cytostome are renewed.[1] The development of the undulating membranes and the frontal-midventral-transverse ciliature proceeds as in the opisthe (Fig. 201o–s). The parental frontoterminal cirri do not participate in primordia formation and are resorbed during division (Fig. 201l–r).

Marginal primordia are formed, as is usual, within the parental marginal rows at two levels (Fig. 201n). Old cirri gradually disintegrate and long primordia are formed. Differentiation of new marginal cirri proceeds from anterior to posterior (Fig. 201q). Formation of new dorsal kineties is according to the *Gonostomum* pattern, that is, all kineties form an anlage each in the proter and the opisthe by intrakinetal proliferation of basal

---

← **Fig. 201p–s** *Keronella gracilis* after protargol impregnation (from Wiackowski 1985). Ventral infraciliature of dividing specimens (for details, see text; sizes not indicated). **p:** Late divider showing segregation of frontal-midventral-transverse cirri and division of macronucleus and micronuclei. **q:** Late divider showing, inter alia, migration of frontoterminal rows, segregation of marginal cirri, formation of undulating membranes, and division of macronucleus. Arrow marks new buccal cirrus of proter. **r:** Very late divider. Arrows mark the rightmost midventral row (marked with a vertical arrow in Fig. 201b). **s:** Very late divider with very pronounced division furrow. Arrows mark separation of anterior corona from midventral complex. Both in the proter and the opisthe each three midventral rows are formed in this specimen. BC = new buccal cirrus, FC* = leftmost cirrus of anterior corona of proter, FT = parental frontoterminal row, FT* = new frontoterminal row of proter and opisthe. Page 1023.

---

[1] Wiackowski (1985, p. 86) also discussed the possibility that reorganisation (new formation) of the adoral zone proceeds very rapidly so that he could have overlooked this process. However, this is very unlikely.

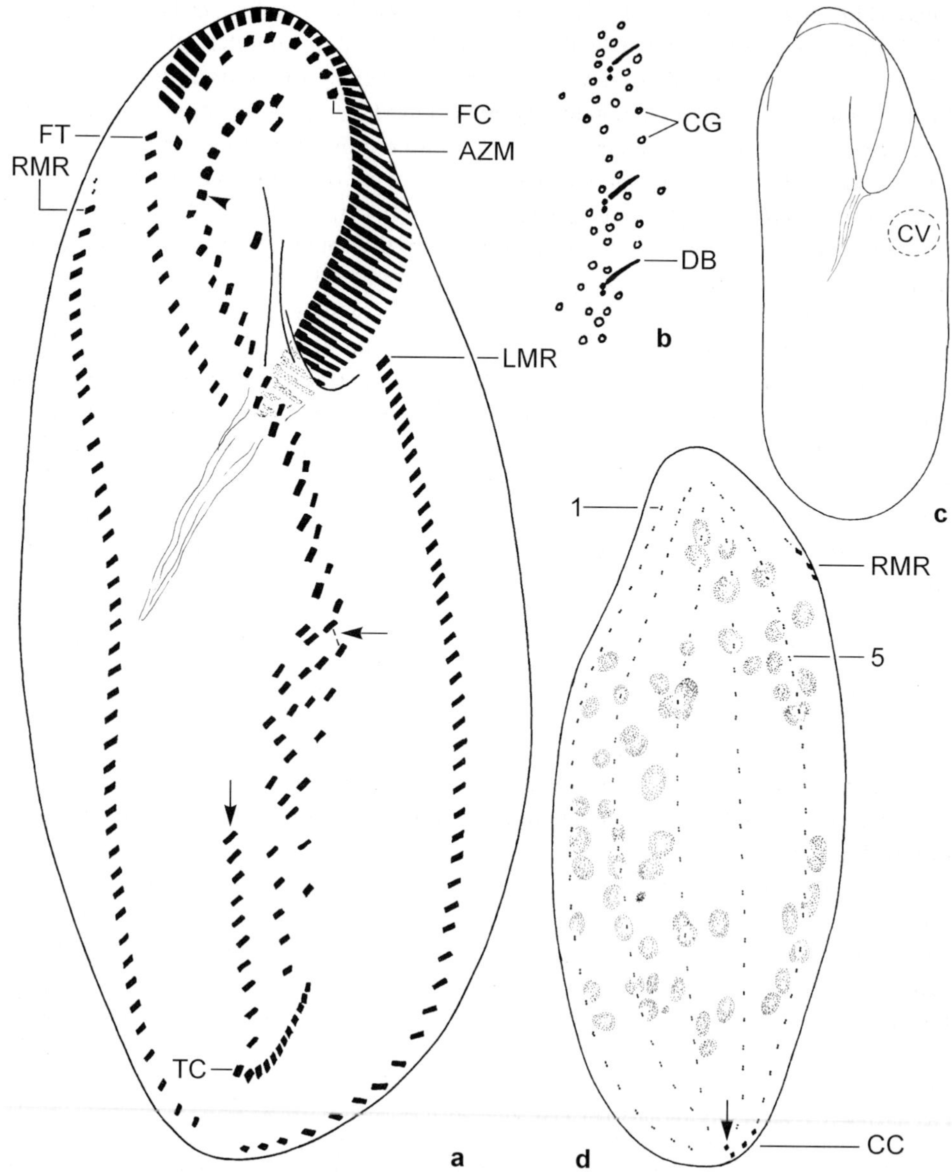

Fig. 202a–d *Keronella gracilis* (originals of specimens from the Salzburg population. a, d, protargol impregnation; b, c, from life). **a:** Infraciliature of ventral side, 127 μm. Arrowhead marks the rightmost cirrus of the posterior corona. Horizontal arrow denotes last midventral pair; vertical arrow marks rightmost midventral row. **b:** Cortical granules are about 0.5 μm across and on dorsal side arranged mainly around bristles. **c:** Ventral view of a freely motile specimen. **d:** Infraciliature of dorsal side (164 μm) with three caudal cirri associated with kinety 5 and one cirrus (arrow) associated with kinety 4. AZM = adoral zone of membranelles, CC = caudal cirri, CG = cortical granules, CV = contractile vacuole, DB = dorsal bristle, FC = leftmost cirrus (cirrus I/1) of anterior corona, FT = anteriormost cirrus of frontoterminal row, LMR = left marginal row, RMR = anterior end of right marginal row, TC = rightmost transverse cirrus, 1, 5 = dorsal kineties. Page 1023.

bodies. No fragmentation of kineties and no dorsomarginal kineties occur. Usually three caudal cirri originate at the rear end of the rightmost kinety (= kinety 5). Rarely single, very small caudal cirri at the end of the other four kineties occur, for example, on kinety 4 in a specimen from my Austrian population (Fig. 201d).

During first stages of ontogenesis, a replication band occurs in all macronuclear nodules. Later, each nodule rounds up. When the new midventral cirri arise, the macronuclear nodules fuse to a single mass (Fig. 201o). The first division of this mass occurs when all new midventral cirri are formed (Fig. 201p, q). Division is finished after the separation of the proter and the opisthe. The mitosis of the micronuclei occurs during the first division of the macronucleus. The proter and the opisthe each contain one offspring of each micronucleus.

Occurrence and ecology: Likely a true soil and moss inhabitant (Foissner 1998) which obviously strongly prefers beech litter and therefore often associated with *Territricha stramenticola*, an oxytrichid also preferring this habitat (Berger 1999, p. 884). The type locality of *Keronella gracilis* is in the region of the "Krakow Gate" in the Ojcow National Park (about 50°12'N 19°37'E) in southern Poland, where Wiackowski (1985) discovered it in mosses growing on calcareous rock. He put the moss into a petri dish with distilled water, and a heterotrophic flagellate as main food was added.

I found this species, inter alia, in the beech litter beside the path (47°48'39"N 13°05'24"E; altitude about 700 m) from Gnigl, a district of the city of Salzburg, to the Gaisberg, a mountain nearby (Fig. 202a–d). Foissner et al. (2005) found it in a beech forest in Salzburg City, Austria. Bonkowski (1996, p. 35) recorded *Keronella gracilis* from beech litter of two German sites, in the Göttinger Wald, a forest east of the city of Göttingen and on the "Kleinen Guldenberg" near Zierenberg, a small town about 30 km north-west of Kassel. Foissner (2000) recorded it from a beech forest in the surroundings of Munich, Germany.

In the moss samples collected by Wiackowski (1985), *Keronella gracilis* ingested fungal spores (obviously *Fusarium*-like taxa; Fig. 201a), fragments of cyanobacteria, small ciliates like *Chilodonella uncinata*, and other small protists. He cultivated it in Pringsheim's medium with *Chlorogonium* sp. as food. Later he fed it with baker's yeast (Wiackowski 1988). The food vacuoles of my population mainly contained crescent-shaped fungal spores (*Fusarium*) about 12–25 × 4–5 μm in size. Biomass of 10^6 specimens about 300 mg (Foissner 1998).

Metabakuella Alekperov, 1989

1989 ***Metabakuella* gen. nov.** – Alekperov, Revision of *Bakuella* and *Keronella*, p. 7 (original description). Type species (by original designation): *Keronella perbella* Alekperov & Musayev, 1988.

1992 ***Metabakuella* Alekperov, 1989** – Alekperov, Zool. Zh., 71: 9 (revision of Bakuellidae).

1996 ***Metabakuella*** – Franco, Esteban & Téllez, Acta Protozool., 35: 329, 330 (key to genera and species of the Bakuellinae).

1999 ***Metabakuella* Alekperov, 1989** – Shi, Acta Zootax. sinica, 24: 245 (revision of hypotrichous ciliates).

1999 ***Metabakuella* Alekperov, 1989** – Shi, Song & Shi, Progress in Protozoology, p. 111 (revision on hypotrichous ciliates).

2001 ***Metabakuella* Alekperov 1989** – Aescht, Denisia, 1: 98 (catalogue of generic names of ciliates).

2001 ***Metabakuella* Alekperov, 1989** – Berger, Catalogue of ciliate names 1. Hypotrichs, p. 47 (nomenclator containing all basionyms, combinations, and higher taxa of hypotrichs).

Nomenclature: No derivation of the name is given in the original description. *Metabakuella* is a composite of *metá* (Greek; after, behind, in between) and the genus-group name *Bakuella* (for derivation see there), indicating an intermediate position (between *Bakuella* and *Keronella*?). Feminine gender because ending with the Latin suffix *-ella* (ICZN 1999, Article 30.1.3).

Characterisation (Fig. 199a, no autapomorphy known): Adoral zone of membranelles continuous. Frontal cirri arranged in a bicorona. Buccal cirrus(i) present. 2 or more frontoterminal cirri. Midventral complex composed of cirral pairs and midventral rows. Transverse cirri present. 2 or more right and 2 or more left marginal cirral rows. Caudal cirri lacking.

Remarks: The two species assigned have, besides the characteristics mentioned above, some further plesiomorphic features in common, namely, body elongate elliptical and likely flexible (?*M. perbella*); many macronuclear nodules; adoral zone occupies about 33% of body length; undulating membranes long and curved (and likely optically intersecting); many buccal cirri; three bipolar dorsal kineties.

Alekperov (1989) established *Metabakuella* for *Keronella perbella* because it does not have, like *K. gracilis* (type of *Keronella*), only one marginal row per side, but more. He also transferred *Bakuella variabilis* to *Metabakuella*, a classification which is very likely incorrect because this species has three frontal cirri, whereas *Metabakuella perbella* has a bicorona. In the present book *Bakuella variabilis* is preliminarily assigned to *Urostyla*, which is, at present, a melting pot for large urostylids with many cirral rows.

Alekperov (1989, 1992) classified *Metabakuella*, together with *Bakuella*, *Keronella*, *Pseudobakuella*, and *Parabakuella* (junior synonym of *Holostichides*), in the Bakuellidae. Eigner (1994) overlooked these two papers and therefore did not consider *Metabakuella* in his review. Franco et al. (1996) described *Metabakuella bimarginata* which fits the definition of *Metabakuella* rather well. Shi (1999) and Shi et al. (1999) classified *Metabakuella* in the Urostylidae.

Franco et al. (1996, their Table 3) characterised *Metabakuella*, inter alia, by the presence of two frontoterminal cirri and two right and two left marginal cirral rows. However, the type species has about 12 frontoterminal cirri and up to three left marginal rows (Fig. 203a). Franco et al. (1996) themselves mentioned invariably three right marginal rows for *M. perbella* in their Table 2. Possibly, their characterisation contains only the features of the last common ancestor.

According to Franco et al. (1996), *Metabakuella perbella* and *M. bimarginata* differ from each other in the number of frontal and frontoterminal cirri, the number of midventral pairs, and the number of right marginal rows. However, especially the data on *M. perbella* (see below), are not very detailed and unambiguous so that the differences in these features are not very convincing. The same is true for the body size, which is very likely distinctly underestimated in *M. perbella*, and the feature cortical granulation cannot be used because it is not known for the type species. And even the increased number of frontoterminal cirri in *M. perbella* is not absolutely certain because confusion with the anterior portion of the inner right marginal row cannot be excluded. In spite of these uncertainties I consider both species mentioned in the next chapter as valid.

In *Bakuella*, the number of frontoterminal cirri was used for the characterisation of the subgenera *Bakuella (Pseudobakuella)* (2 cirri) and *Bakuella (Bakuella)* (>2 cirri). The same distinction would be possible in *Metabakuella* (see key below). However, since this would result in two monotypic taxa, I refrain from such an act.

Species included in *Metabakuella* (alphabetically arranged according to basionyms): (1) *Keronella perbella* Alekperov & Musayev, 1988; (2) *Metabakuella bimarginata* Franco, Esteban & Téllez, 1996.

Key to *Metabakuella* species

See remarks of genus section for some explanations. Protargol impregnation is needed because the frontoterminal cirri are usually not clearly recognisable in live preparations.

1 About 12 frontoterminal cirri (Fig. 203a) .... *Metabakuella perbella* (p. 1035)
\- Two frontoterminal cirri (Fig. 204b) .... *Metabakuella bimarginata* (p. 1038)

Metabakuella perbella (Alekperov & Musayev, 1988) Alekperov, 1989 (Fig. 203a, b, Table 41, Addenda)

1988 ***Keronella perbella* Alekperov et Musaev, sp. n.** – Alekperov & Musayev, Zool. Zh., 67: 1907, Fig. 2a, b (Fig. 203a, b; original description, likely nor formal diagnosis provided. Type slides are probably deposited in the Zoological Institute of the Azerbaijan Academy of Sciences, Baku).

1989 ***Metabakuella perbella* (Al. et M.) comb. nov.** – Alekperov, Revision of *Bakuella* and *Keronella*, p. 7 (combination with *Metabakuella*).

1992 ***Metabakuella perbella* (Alekperov et Musaev, 1988) Alekperov, 19891** – Alekperov, Zool. Zh., 71: 9 (revision; incorrect year).

1996 ***Metabakuella perbella* (Alekperov and Musayev, 1988) Alekperov, 1989** – Franco, Esteban & Téllez, Acta Protozool., 35: 330 (brief revision of Bakuellinae).

2001 ***Metabakuella perbella* (Alekperov and Musayev, 1988) Alekperov, 1989** – Berger, Catalogue of ciliate names 1. Hypotrichs, p. 43 (nomenclator containing all basionyms and combinations of hypotrichs).

Nomenclature: No derivation of the name is given in the original description. The species-group name *perbell·us -a -um* is a composite of *per-* (Latin; very, extra) and the Latin adjective *bellus* (delightful, beautiful, etc.), and obviously alludes to the (very beautiful) general appearance of this species. In the original description, the name of the junior author is differently spelled, namely Musaev in the headline of the description (p. 1907) and Musayev in the English title (p. 1909). I use the latter because this is the relevant name for citing the whole paper in English. "*Bakuella perbella* Alekperov & Mamajeva 1988" in Aescht (2001, p. 98) is an incorrect combination and an incorrect authorship. The present species was fixed as type of *Metabakuella* by original designation.

Table 41 Morphometric data on *Metabakuella bimarginata* (bim, from Franco et al. 1996) and *Metabakuella perbella* (per, from Alekperov & Musayev 1988)

Characteristics [a]	Population	mean	M	SD	SE	CV	Min	Max	n
Body, length	bim	227.0	–	38.2	4.8	–	150.0	300.0	63
	bim	215.8	–	31.9	5.2	–	150.0	280.0	37
Body, width	bim	82.9	–	30.5	4.5	–	50.0	115.0	46
	bim	70.3	–	11.1	1.8	–	50.0	100.0	37
Macronuclear nodule, length	bim	11.3	–	1.8	0.4	–	8.3	14.4	19
Macronuclear nodule, width	bim	4.1	–	0.9	0.2	–	2.7	6.0	19
Macronuclear nodules, number	bim	184.9	–	39.3	7.4	–	110.0	263.0	28
Micronuclei, length	bim	10.3	–	1.4	0.5	–	8.9	13.3	12
Micronuclei, width	bim	4.9	–	0.9	0.3	–	4.0	7.0	12
Micronuclei, number	bim	8.8	–	2.2	0.5	–	4.0	12.0	20
Adoral membranelles, number	bim	46.0	–	3.6	1.3	–	40.0	50.0	7
	per	–	–	–	–	–	35.0	40.0	?
Frontal cirri, number	bim	4.8	–	1.1	0.3	–	4.0	6.0	10
Buccal cirri, number	bim	7.0	–	1.1	0.3	–	5.0	8.0	13
	per [b]	7.0	–	–	–	–	–	–	1
Cirral pairs in midventral complex, number	bim	16.2	–	2.2	0.8	–	13.0	18.0	7
Midventral rows, number	bim	5.3	–	1.2	0.4	–	4.0	7.0	9
	per [b]	8.0	–	–	–	–	–	–	1
Transverse cirri, number	bim	9.8	–	0.9	0.3	–	8.0	11.0	12
	per	10.0	–	–	–	–	–	–	?
Inner left marginal row, number of cirri	bim	52.2	–	7.7	2.0	–	37.0	64.0	15
Outer left marginal row, number of cirri	bim	49.4	–	10.5	2.9	–	32.0	62.0	13
Inner right marginal row, number of cirri	bim	47.4	–	6.0	2.0	–	40.0	55.0	9
Outer right marginal row, number of cirri	bim	53.6	–	8.3	2.8	–	40.0	68.0	9
Dorsal kineties, number	bim	3.0	–	0.0	0.0	–	3.0	3.0	10
	per	3.0	–	–	–	–	–	–	?

[a] All measurements in µm. Data by Franco et al. (1996) are based on cells after Fernández-Galiano silver impregnation, except for body length and body width (upper line from life) and some other features (number of membranelles, buccal cirri, frontal cirri, midventral pairs, dorsal kineties) which are from protargol preparations. ? = sample size not known; if only one value is known it is listed in column mean, if two values are available they are listed as Min and Max. CV = coefficient of variation in %, M = median, Max = maximum value, mean = arithmetic mean, Min = minimum value, n = number of individuals investigated, SD = standard deviation, SE = standard error of arithmetic mean.

[b] From Fig. 203a.

Remarks: Alekperov & Musayev (1988) originally described this species in *Keronella*, likely because of the presence of a bicorona. However, in 1989, Alekperov established *Metabakuella* with *K. perbella* as type because it has more than the ordinary two (one per side) marginal rows. It differs from *M. bimarginata* basically only in the number of frontoterminal cirri (about 12 vs. invariably two). According to the original description the body length is about 90 µm (in life?); according to the scale bar (25 µm) the silver-impregnated specimen illustrated is only 67 µm long. I strongly doubt these values because according to the cirral pattern (number of cirri per marginal row, midventral complex composed of about 10 pairs and 8 rows!) this species must be rather

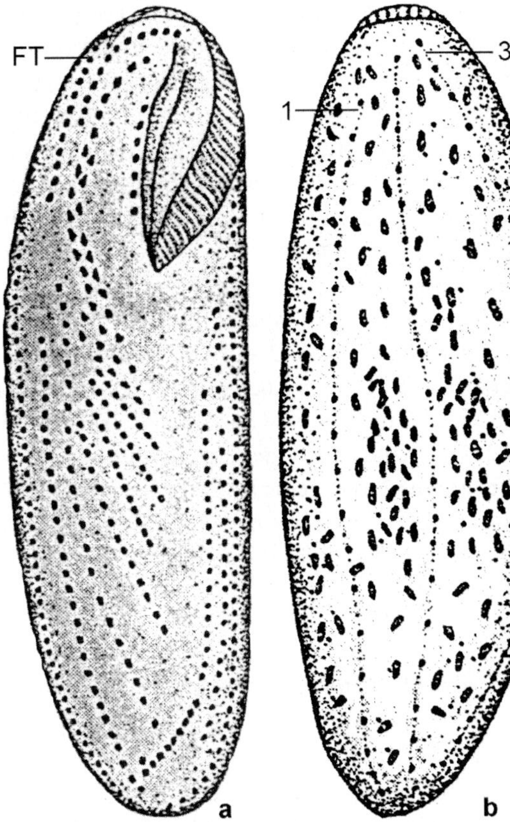

Fig. 203a, b *Metabakuella perbella* (from Alekperov & Musayev 1988. Wet silver nitrate impregnation). Infraciliature of ventral and dorsal side and nuclear apparatus, 67 µm (according to bar! See text, for discussion of body size). Some details of the cirral pattern (e.g., arrangement of frontal cirri and transition to midventral complex possibly not quite correctly illustrated). FT = frontoterminal cirri (this could also be the anterior portion of the inner right marginal row), 1, 3 = dorsal kineties. Page 1035.

large, at least 150–200 µm. In addition, some details of the cirral pattern (bicorona, midventral cirral pairs) are possibly not quite correctly illustrated, respectively, recognised, likely due to the wet silver impregnation procedure. Consequently, detailed redescription, including thorough live observation (presence/absence of cortical granules) is necessary.

Morphology: Body length (in life?) about 90 µm; silver-impregnated specimen illustrated only 67 µm long (Fig. 203a; see remarks). According to the text of the original description about 40 macronuclear nodules, which is obviously an underestimation; specimen illustrated likely with about 123 nodules. 7–12 micronuclei. Contractile vacuole, cortical granulation, and other features recognisable in life likely not described.

The following description of the infraciliature is based mainly on the specimen illustrated. Adoral zone occupies 33% of body length, composed of 35–40 membranelles. Buccal field moderately wide; undulating membranes likely roughly as in *M. bimarginata* and *Bakuella*, that is, long, curved, and optically intersecting. Frontal cirri arranged in a (indistinct) bicorona, transition to midventral complex possibly not clearly recognised. Buccal cirri along paroral. Very likely about 12 frontoterminal cirri form row extending from distal end of adoral zone to about level of rearmost buccal cirrus; however, it cannot be excluded that this row is the anterior portion of the inner right marginal row, which possibly has a distinct break; perhaps, *Metabakuella perbella* has, like *M. bimarginata*, only two frontoterminal cirri not clearly recognised in the original description (further data, including ontogenetic ones, are needed for a final decision). Midventral complex composed of cirral pairs (difficult to count, about 10) and ca. eight midventral rows (distinction of rightmost row from inner right marginal row possibly difficult). Transverse cirri hook-shaped arranged in oblique, slightly subterminal row; bases of cirri not larger than that of remaining cirri. Specimen illustrated with one more or less bipolar and one anteriorly distinctly shortened right marginal row; according to Table 2 in Franco et al. (1996) invariably three right marginal rows present. Three left marginal rows (Fig. 203a), inner row with 20 cirri, middle with about 25, and outer row with circa 47. Likely invariably three bipolar dorsal kineties (Fig. 203b). Caudal cirri lacking.

Occurrence and ecology: Type locality of *Metabakuella perpella* is the Apsheron Peninsula, Azerbaijan, where Alekperov & Musayev (1988) discovered it in soil. Possibly they collected the sample in/near Baku where they lived and worked. No further records published.

Metabakuella bimarginata Franco, Esteban & Téllez, 1996 (Fig. 204a–d, Table 41)

1996 ***Metabakuella bimarginata* sp. n.**[1] – Franco, Esteban & Téllez, Acta Protozool., 35: 322, Fig. 1–12, Table 1 (Fig. 294a–d; original description; site where type slides deposited not mentioned).

2001 ***Metabakuella bimarginata* Franco, Esteban and Téllez, 1996** – Berger, Catalogue of ciliate names 1. Hypotrichs, p. 47 (nomenclator containing all basionyms, combinations, and higher taxa of hypotrichs).

Nomenclature: The species-group name *bimarginata* is a composite of the Latin numeral *bi-* (two) and the Latin adjective *marginát·us -a -um* (to line, having a margin) and refers to the two marginal cirral rows on each side.

Remarks: This species has an inconspicuous bicorona and two marginal rows per side. Consequently, the classification in *Metabakuella* seems correct. Differs from the type species mainly in the number of frontoterminal cirri (two vs. about 12). All other features overlap more or less distinctly (Table 41). The cortical granulation of *M. bimarginata* cannot be used for the distinction because this feature was not investigated in the type species. *Pseudourostyla muscorum* (Fig. 152a) is very similar (see there). However, this species obviously lacks midventral rows so that synonymy is unlikely.

---

[1] The diagnosis by Franco et al. (1992) is as follows: In vivo 225 × 85 μm. Two rows of cirri running along each margin of the ventral surface of the cell. More than 100 ellipsoid macronuclear fragments. One row of buccal cirri. Four to six frontal cirri, 2 frontoterminal cirri, 13–18 pairs of midventral cirri in the anterior part of the cell. Midventral cirri in the posterior half of the cell as rows with a gradually increasing number of cirri towards the rear of the cell. Cytoplasm with distinct rows of greenish cortical granules. Transverse cirri present. Caudal cirri absent.

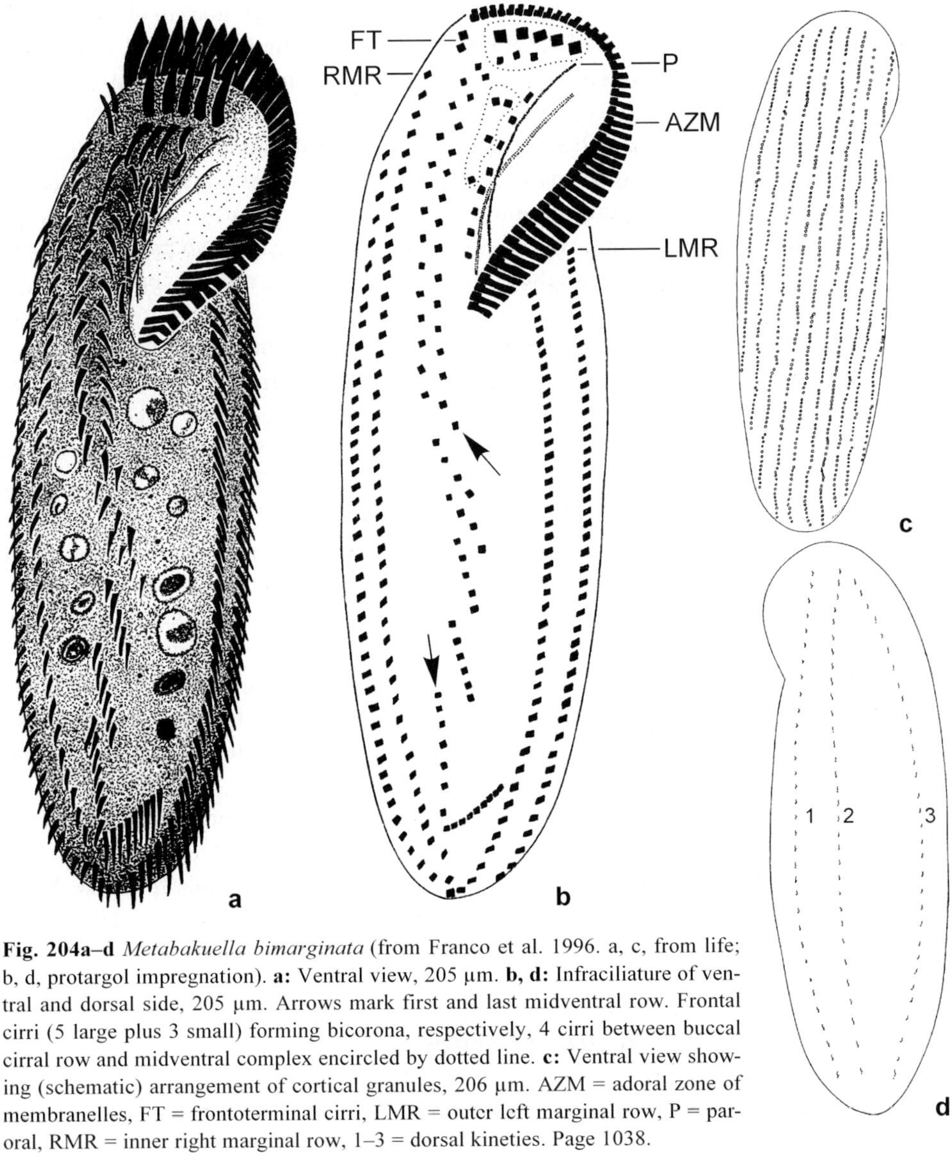

Fig. 204a–d *Metabakuella bimarginata* (from Franco et al. 1996. a, c, from life; b, d, protargol impregnation). **a:** Ventral view, 205 μm. **b, d:** Infraciliature of ventral and dorsal side, 205 μm. Arrows mark first and last midventral row. Frontal cirri (5 large plus 3 small) forming bicorona, respectively, 4 cirri between buccal cirral row and midventral complex encircled by dotted line. **c:** Ventral view showing (schematic) arrangement of cortical granules, 206 μm. AZM = adoral zone of membranelles, FT = frontoterminal cirri, LMR = outer left marginal row, P = paroral, RMR = inner right marginal row, 1–3 = dorsal kineties. Page 1038.

Morphology: Body size about 225 × 85 μm on average in life (Table 41). Body outline elongate elliptical to almost parallel-sided with both ends broadly rounded. Body flexible, dorsoventrally flattened with a slightly concave(?) dorsal surface. 185 macronuclear nodules on average scattered throughout cytoplasm; individual nodules about 11 × 4 μm. On average nine large (10 × 4 μm) micronuclei. Sometimes bacteria present in macronuclear nodules; usually one bacterium per nodule, individual bacteria rod-shaped, with rounded ends, 3–25 μm long and about 1 μm wide (on average 10.9 ×

1 µm); 7–52 bacteria per cell. Contractile vacuole close to left cell margin about at level of proximal end of adoral zone of membranelles. Cortical granules arranged in longitudinal rows (Fig. 204c), individual granules spherical, greenish, and "small", that is, likely 1 µm or less in diameter.

Adoral zone occupies 34% of body length in specimen illustrated (Fig. 204b), composed of 46 membranelles on average (Table 41). Undulating membranes long, optically intersecting about in mid-portion, paroral composed of "irregularly" arranged (likely zigzagging?) basal bodies forming two rows; endoral composed of two parallel rows of basal bodies; paroral commences slightly more anteriorly than endoral, both terminate near proximal end of adoral zone.

4–6 slightly enlarged frontal cirri forming anterior row of bicorona. Specimen illustrated with three smaller cirri behind the three rightmost enlarged frontal cirri, that is, these smaller cirri form the posterior bow of the bicorona; consequently, bicorona indistinct. Buccal cirri almost along whole length of paroral. Between buccal cirral row and anterior portion of midventral complex 2–3 cirri (however, in specimen shown in Fig. 204b four such cirri are present!). Invariably two frontoterminal cirri near distal end of adoral zone. Midventral complex composed of 13–18 cirral pairs with last pair at about 50% of body length (40% in specimen illustrated Fig. 204b) and on average five midventral rows with 3–7 cirri per row, and with rightmost row terminating immediately ahead of rightmost transverse cirrus. On average about 10 transverse cirri arranged in oblique, slightly subterminal row; bases of transverse cirri not larger than that of other cirri. Inner right marginal row commences near frontoterminal cirri, ends slightly subterminally; outer right marginal row distinctly shortened anteriorly, terminates at rear end of cell and therefore almost continuous with inner left marginal row (Fig. 204b).

Dorsal bristles (length not indicated; likely around 3 µm long) arranged in three bipolar kineties (Fig. 204d). Caudal cirri lacking.

Occurrence and ecology: Limnetic. Type locality of *M. bimarginata* is the Manzanares stream, Guadarrama Mountains (about 40°45'N 3°54'W), Madrid, Spain, where Franco et al. (1996) found it in the benthal. No further records published.

Urostyla Ehrenberg, 1830

1830 ***Urostyla grandis*. nov. gen.**[1] – Ehrenberg, Abh. preuss. Akad. Wiss., year 1830: 43 (original description). Type species (by original designation and monotypy; see nomenclature): *Urostyla grandis* Ehrenberg, 1830.

1831 ***Vrostyla* E.** – Ehrenberg, Abh. preuss. Akad. Wiss., year 1831: 119 (review; see nomenclature).

1838 ***Urostyla***[2] – Ehrenberg, Infusionsthierchen, p. 369 (revision).

1859 ***Urostyla*. Ehrbg.**[3] – Stein, Infusionthierchen, p. 191 (revision).

1882 ***Urostyla*, Ehrenberg** – Kent, Manual Infusoria II, p. 764 (revision).

---

[1] Ehrenberg (1830) provided the following brief diagnosis: styli; uncini nulli.

[2] Ehrenberg (1838) provided the following improved diagnosis: U. corpore albo semcylindrico subclavato, utrinque rotundato, antica parte levius incrassata, stylis brevibus.

[3] Stein (1859) provided the following characterisation: Körper sehr metabolisch, langgestreckt, elliptisch, oblong oder eiförmig, vorn und hinten abgerundet; 3 oder mehrere griffelförmige Stirnwimpern; 5–12 dünne griffelförmige Afterwimpern; 5 oder mehrere Längsreihen von borstenförmigen Bauchwimpern.

1886 ***Urostyla* Ehrbg.** – Blochmann, Mikroskopische Thierwelt, p. 75 (review).
1889 ***Urostyla* Ehrbg 1830** – Bütschli, Protozoa, p. 1741 (revision).
1895 ***Urostyla* Ehrbg.** – Blochmann, Mikroskopische Thierwelt, p. 111 (review).
1932 ***Urostyla* Ehrenberg, 1838** – Kahl, Tierwelt Dtl., 25: 564 (revision).
1933 ***Urostyla* Ehrenberg 1838** – Kahl, Tierwelt V.- u. Ostsee, 23: 108 (guide to marine ciliates).
1936 ***Urostyla* Ehrenberg, 1830** – Bhatia, Ciliophora, p. 366 (revision of Indian ciliates).
1950 ***Urostyla* Ehrenberg** – Kudo, Protozoology, p. 672 (textbook).
1961 ***Urostyla* Ehr.** – Corliss, Ciliated Protozoa, p. 170 (revision of ciliates).
1972 ***Urostyla* Ehrenberg, 1830**[1] – Borror, J. Protozool., 19: 8 (revision of hypotrichs).
1974 ***Urostyla* Ehrenberg** – Stiller, Fauna Hung., 115: 37 (guide to hypotrichs).
1979 ***Urostyla* Ehrenberg, 1830** – Jankowski, Trudy zool. Inst., 86: 70 (guide to generic names of hypotrichs).
1979 ***Metaurostyla polonica* gen. et sp. n.** – Jankowski, Trudy zool. Inst., 86: 70 (original description of synonym). Type species (by original designation): *Metaurostyla polonica* Jankowski, 1979.
1979 ***Urostyla* Ehrenberg, 1830** – Corliss, Ciliated Protozoa, p. 309 (revision of ciliates).
1979 ***Urostyla* Ehrenberg, 1830**[2] – Borror, J. Protozool., 26: 549 (redefinition of the Urostylidae).
1982 ***Urostyla* Ehrenberg, 1838**[3] – Hemberger, Dissertation, p. 77 (revision of non-euplotid hypotrichs).
1983 ***Urostyla* Ehrenberg, 1830** – Curds, Gates & Roberts, Synopses of the British Fauna, 23: 397 (guide to ciliate genera).
1983 ***Urostyla* Ehrenberg, 1838** – Borror & Wicklow, Acta Protozool., 22: 120 (revision of urostylids).
1985 ***Urostyla*** – Small & Lynn, Phylum Ciliophora, p. 450 (guide to ciliate genera).
1994 ***Urostyla* Ehrenberg, 1830** – Tuffrau & Fleury, Traite de Zoologie, 2: 128 (textbook).
1999 ***Urostyla* Ehrenberg, 1838** – Shi, Acta Zootax. sinica, 24: 362 (revision of Urostylina; incorrect year).
1999 ***Urostyla* Ehrenberg, 1838** – Shi, Song & Shi, Progress in Protozoology, p. 110 (revision of hypotrichs; incorrect year).
2001 ***Urostyla* Ehrenberg 1830** – Aescht, Denisia, 1: 172 (catalogue of generic names of ciliates).
2001 ***Urostyla* Ehrenberg, 1830** – Berger, Catalogue of ciliate names 1. Hypotrichs, p. 100 (nomenclator containing all basionyms, combinations, and higher taxa of hypotrichs).
2002 ***Urostyla* Ehrenberg, 1830** – Lynn & Small, Phylum Ciliophora, p. 442 (guide to ciliate genera).

Nomenclature: No derivation of the name is given in the original description. *Urostyla* is a composite of the Greek nouns *he urá* (the tail) and *ho stylos* (style, pillar, cirrus) and possibly alludes to the transverse cirri. Ehrenberg (1830, 1838) misinterpreted the transverse cirri region as gap (like the oral apparatus) surrounded by immobile styles. Feminine gender (Aescht 2001, p. 304). The spelling *Vrostyla* in Ehrenberg (1831) is due to the fact that the printer did not distinguish the letters U and V in the italics script. *Metaurostyla* is a composite of the Greek prefix *meta+* (in the middle of, later, together with) and the genus-group name *Urostyla* and likely refers to the systematic position of the type species. *Urostylia* in Winterbourn & Brown (1967, p. 46) is an incorrect spelling.

There is obviously confusion about the kind of type fixation. Borror (1972) and Aescht (2001) assumed that *U. grandis* was fixed as type species of *Urostyla* by monotypy. In the catalogue on ciliate names (Berger 2001) I wrote "type by original designa-

---

[1] The diagnosis by Borror (1972) is as follows: Cirri arise from one primordial field during division and differentiate into 4–12 ventral rows. Transverse cirri present or absent. Frontal cirri either undifferentiated from ventral rows, or occurring in many rows of fine cirri. Usually many macronuclei.

[2] The diagnosis by Borror (1979) is as follows: Several rows of right marginal cirri; midventral cirri in typical zigzag series, 2 cirri per original oblique ciliary streak, except for the first several streaks that differentiate into several isolated frontal cirri each.

[3] The diagnosis by Hemberger (1982) is as follows: Mindestens je 2 rechte und linke Marginalreihen, meist aber mehr; dazwischen familientypische Midventral-Reihen; frontal Cirrendifferenzierung häufig schwer erkennbar; Transversalcirren vorhanden; Morphogenesse familientypisch; meist viele Makronuclei.

tion and monotypy". Another check of the literature showed that it is not unequivocally possible to decide how Ehrenberg (1830) fixed the type species because he included two species, namely "*Urostyla grandis*. nov. Gen." and "*Trichoda patens* M.?". Moreover, in the line below he wrote that *Urostyla* comprises two species. Applying Article 68.2.1 of the ICZN (1999), type fixation is by original designation in the present case because the expression "nov. gen.", which is an equivalent to "gen. n., sp. n.", was applied to *U. grandis* and not to *T. patens*. On the other hand, one can also apply Article 67.2.5 which says that a nominal species is deemed not to be originally included in a genus if it was doubtfully or conditionally included. And in the present case *T. patens* was obviously doubtfully included as indicated by the question mark. Applying this article, *Urostyla grandis* is the type of *Urostyla* by monotypy. Consequently I confirm my earlier decision (Berger 2001) and state that the type fixation was by original designation and monotypy.

Metaurostyla Jankowski, 1979 was established with *M. polonica* Jankowski, 1979 as type species, which is the *Urostyla grandis* described by Jerka-Dziadosz (1972). Very likely *Metaurostyla* is a nomen nudum because Jankowski (1979) published it without description or definition (Aescht 2001, p. 100).

Characterisation (Fig. 199a, autapomorphies 5): Adoral zone of membranelles continuous. Many frontal cirri basically arranged in a bicorona, forming, together with many parabuccal cirri, a multicorona (A). One or more buccal cirri right of paroral. Midventral complex composed of cirral pairs and midventral rows. Frontoterminal cirri lacking (A). Transverse cirri present. Two or more right and 2 or more left marginal rows. Caudal cirri lacking. Proximal portion of parental adoral zone reorganised during division. Marginal rows divide individually. Macronuclear nodules fuse to a single mass prior to division.

Remarks: Ehrenberg (1830) established *Urostyla* with two species, namely *U. grandis* (type species; for fixation, see nomenclature) and *Trichoda patens* Müller, 1786. Just one year later, he (Ehrenberg 1831) included only the type species, and in his 1838-monograph (p. 369, his genus section) he explained that the two species originally included were fused ("verschmolzen") in this 1831-paper. Ehrenberg (1833, p. 278) obviously recognised this incorrect synonymy and transferred *Trichoda patens* to *Uroleptus*, an act overlooked by Berger (2001), who mistakenly ascribed *U. patens* to Ehrenberg (1833). Interestingly, Ehrenberg (1838, p. 365) did not consider *U. patens* further because he was unable to classify it correctly. Anyhow, *Trichoda patens* Müller, 1786 (p. 181, Tab. XXVI, Fig. 1, 2) indeed looks like a *Uroleptus* and, since it was originally recorded from a marine habitat, synonymy with *U. grandis* can be excluded.

Perty (1852, p. 154) doubted the validity of *Urostyla* and suggested synonymising it with *Oxytricha*, which was, however, less narrowly defined as today.

Stein (1859) redescribed *U. grandis* in detail and assigned two further species to the present genus, namely, *Urostyla weissei* (now *Paraurostyla weissei*; for review, see Berger 1999, p. 844) and *Urostyla viridis*. The latter species is not yet redescribed in detail, so the exact position is uncertain. I therefore retain the original generic assignment. Moreover, Stein transferred *Oxytricha multipes* Claparède & Lachmann, 1858 to *Urostyla* (although not formally), an act overlooked by Berger (2001). Berger (1999,

p. 873) mentioned it as supposed synonym of *Paraurostyla weissei*. Stein (1859) himself considered *Oxytricha urostyla* Claparède & Lachmann, 1858 as supposed synonym of *P. weissei*. However, this species has a bicorona and is assigned to *Pseudourostyla* in the present book. Stein also recognised the resemblance of Perty's *Oxytricha fusca* and *U. grandis*, that is, he was the first revisor who synonymised these two species (Kent 1882, p. 765).

Trichogaster Sterki, 1878, respectively, *Prooxytricha* Poche, 1913, which is a replacement name, was classified as junior synonym of *Urostyla* by some authors. However, I follow Kahl (1932, p. 538), who eliminated this genus because the sole species is insufficiently known. For details, see the chapter on taxa not considered.

Kent (1882) included the same species as Stein (1859) in *Urostyla*. Moreover, he included *Urostyla flavicans*, which is, however, a junior synonym of *Paraurostyla weissei* (for review see Berger 1999, p. 844).

Bütschli (1889) synonymised the two *Hemicycliostyla* species described by Stokes with *Urostyla grandis*, that is, he eliminated Stokes' genus.

Kahl (1932) included 18 species in *Urostyla*. He correctly stated that it comprises various groups which could possibly be split off. Indeed, several species have been transferred to other genera since then (for examples, see next paragraph).

Borror (1972) rediagnosed *Urostyla* and included nine species, inter alia, *Eschaneustyla brachytone* (type of *Eschaneustyla*) and *Hemicycliostyla sphagni* (type of *Hemicycliostyla*). Consequently, Borror considered *Eschaneustyla* and *Hemicycliostyla* as junior synonyms of *Urostyla*. However, *Eschaneustyla* has a rather different cirral pattern because midventral pairs and transverse cirri are lacking. *Hemicycliostyla*, which was already synonymised with *Urostyla* by Bütschli (1889; see above) has a cirral pattern which closely resembles that of *U. grandis* except for the transverse cirri, which are lacking in *Hemicycliostyla*. Basically it depends on the phylogenetic relationships (which we do not know) whether or not *Hemicycliostyla* can be retained. In the present monograph both *Eschaneustyla* (see Epiclintidae) and *Hemicycliostyla* (see Pseudourostylidae) are considered as valid. On the other hand, Borror (1972) established *Pseudourostyla* for *Urostyla cristata* and *Paraurostyla* for *Urostyla weissei*. Both taxa are accept by most workers.

Borror (1979) redefined the Urostylidae, and his diagnosis was the first to contain the zigzagging midventral pattern.

Jankowski (1979) established a new genus and species for the *Urostyla grandis* sensu Jerka-Dziados, namely *Metaurostyla polonica*. However, the identification of the population studied by Jerka-Dziadosz as *Urostyla grandis* is beyond reasonable doubt. Consequently, *Metaurostyla* – which is a nomen nudum according to Aescht (2001) – is a junior synonym of *Urostyla*. The other three species originally described in *Metaurostyla* are assigned as follows: (i) *Metaurostyla thompsoni* Jankowski, 1979 is classified as supposed synonym of *Metaurostylopsis marina*; (ii) *Metaurostyla raikovi* Alekperov, 1984, and (iii) *M. magna* Alekperov, 1984 obviously lack midventral rows and are therefore preliminarily transferred to *Pseudourostyla*.

Hemberger (1982, p. 75, 77) synonymised *Pseudourostyla* Borror, 1972 with *Urostyla* because he did not consider the different mode of marginal row formation as sufficient

Table 42 Morphometric data on *Urostyla grandis* (gr1, population 1 from Ganner 1991; gr2, population 2 from Ganner 1991; gr3, from Song & Wilbert 1989; gr4, from Shin 1994[d])

Characteristics [a]	Species	mean	M	SD	SE	CV	Min	Max	n
Body, length	gr1	282.9	285.0	23.0	5.1	8.1	242.5	330.0	20
	gr2	187.5	188.8	18.7	5.9	10.0	166.4	217.0	10
	gr3	209.0	–	23.9	8.5	11.4	162.0	238.0	8
	gr4	222.8	225.0	37.4	11.8	16.7	170.0	298.0	10
Body, width	gr1	118.3	120.0	11.7	2.6	9.9	92.5	132.5	20
	gr2	75.9	74.8	7.3	2.3	9.6	63.7	87.1	10
	gr3	97.6	–	1.3	4.0	11.6	72.0	108.0	8
	gr4	103.8	101.5	12.6	4.0	12.2	87.0	125.0	10
Body length:width, ratio	gr4	21.2	2.2	0.3	0.1	15.9	1.7	2.7	10
Anterior body end to proximal end of adoral zone, distance	gr1	110.6	110.0	7.1	1.6	6.4	95.0	125.0	20
	gr2	71.8	72.2	9.4	3.0	13.1	58.5	85.8	10
	gr3	85.2	–	5.4	2.2	6.4	80.0	93.0	6
	gr4	92.9	90.0	10.5	3.5	11.3	82.0	117.0	9
Largest adoral membranelle, width	gr1	18.3	19.0	3.0	0.7	16.3	15.0	21.0	20
	gr2	13.3	13.0	0.6	0.2	4.4	12.5	14.0	10
Paroral, length	gr1	76.4	76.3	6.5	1.5	8.5	62.5	90.0	20
	gr2	47.6	45.5	6.5	2.9	13.7	39.9	55.9	5
	gr4	76.0	70.0	10.6	3.5	13.9	65.0	100.0	9
Endoral, length	gr1	96.0	95.0	9.0	2.0	9.4	77.5	120.0	20
	gr2	55.4	57.4	8.2	3.7	14.7	42.9	65.0	5
Rear transverse cirrus to rear body end, distance	gr1	31.0	30.0	4.8	1.1	15.4	25.0	40.0	20
	gr2	20.2	20.2	3.4	1.1	16.7	11.7	23.4	10
Macronuclear nodule, length	gr1	12.4	11.0	5.1	1.1	41.4	6.0	27.0	20
	gr2	6.0	5.7	1.1	0.4	18.2	4.7	7.8	10
	gr4	9.1	10.0	1.1	0.4	11.6	8.0	10.0	9
Macronuclear nodule, width	gr1	2.8	2.5	1.0	0.2	34.0	2.0	5.0	20
	gr2	2.7	2.6	0.3	0.1	9.4	2.5	3.2	10
	gr4	2.9	3.0	0.2	0.1	5.7	2.5	3.0	9
Micronucleus, length	gr1	5.7	6.0	0.6	0.1	11.1	5.0	7.0	20
	gr2	4.2	4.0	1.1	0.3	25.1	2.8	5.5	10
Micronucleus, width	gr1	5.4	5.0	0.6	0.1	10.8	5.0	7.0	20
	gr2	3.3	3.3	0.4	0.1	11.3	2.7	3.9	10
Macronuclear nodules, number	gr3	109.3	–	26.8	9.5	24.5	72.0	142.0	8
	gr4	175.5	179.5	35.5	–	20.3	125.0	227	8
Micronuclei, number	gr1	4.4	5.0	1.6	0.3	36.4	2.0	8.0	20
	gr2	8.6	8.5	1.0	0.3	11.2	7.0	10.0	10
Adoral membranelles, number	gr1	59.6	58.5	4.6	1.1	8.0	52.0	69.0	20
	gr2	47.2	48.0	5.0	1.6	10.5	40.0	55.0	10
	gr3	56.8	–	2.8	1.1	4.9	53.0	60.0	6
	gr4	49.9	47.0	6.2	2.1	12.5	43.0	64,0	9
Buccal cirri, number	gr1	8.2	8.0	1.1	0.3	14.0	7.0	11.0	20
	gr2	5.8	6.0	1.0	0.3	16.8	4.0	7.0	9
	gr3	7.2	–	2.2	0.8	30.5	4.0	10.0	8
	gr4	7.1	7.0	0.7	0.3	9.7	6.0	8.0	7
Parabuccal cirri, number	gr1	7.9	8.0	1.4	0.3	17.7	5.0	10.0	20
	gr2	4.7	5.0	1.3	0.4	28.5	2.0	7.0	10
Midventral pairs, number	gr3	14.5	–	2.4	1.0	16.3	12.0	18.0	8

Table 42 Continued

Characteristics [a]	Species	mean	M	SD	SE	CV	Min	Max	n
Transverse cirri, number	gr1	11.0	10.5	2.0	0.4	17.9	9.0	15.0	20
	gr2	7.8	7.0	1.6	0.5	20.8	6.0	10.0	10
	gr3	10.5	–	1.8	0.6	16.9	8.0	13.0	8
	gr4	10.0	10.0	2.3	0.8	22.7	7.0	13.0	8
Left marginal rows, number	gr1	5.7	6.0	0.9	0.2	15.2	4.0	7.0	20
	gr2	5.3	5.0	0.8	0.3	15.5	4.0	7.0	10
	gr3	7.0	–	0.7	0.2	9.5	6.0	8.0	10
Right marginal rows, number	gr1 [b]	6.2	6.0	0.6	0.1	9.6	5.0	7.0	20
	gr2 [b]	8.5	8.0	0.7	0.2	8.3	8.0	10.0	10
	gr3 [b]	6.3	–	0.7	0.2	10.7	6.0	8.0	10
Outermost right marginal row, number of cirri	gr1	66.3	67.0	6.2	1.4	9.4	56.0	80.0	20
	gr2	49.8	48.5	5.6	1.8	11.2	41.0	58.0	10
	gr3 [c]	58.3	–	8.4	3.4	14.3	48.0	67.0	6
Outermost left marginal row, number of cirri	gr1	36.3	35.0	6.8	1.5	18.8	24.0	52.0	20
	gr2	33.5	33.0	2.5	1.3	7.5	31.0	37.0	4
	gr3 [c]	31.7	–	6.7	2.7	21.1	24.0	35.0	6
Innermost left marginal row, number of cirri	gr1	43.1	44.0	7.6	1.7	17.6	21.0	53.0	20
	gr2	33.2	33.5	4.6	1.5	14.0	24.0	40.0	10
Dorsal kineties, number	gr1	3.3	3.0	0.5	0.1	14.3	3.0	4.0	20
	gr2	3.0	3.0	0.0	0.0	0.0	3.0	3.0	10
	gr3	3.2	–	0.4	0.1	12.7	3.0	4.0	16
	gr4	3.0	3.0	0.0	0.0	0.0	3.0	3.0	10
Dorsal kinety 1, number of bristles	gr1	63.7	62.0	10.4	2.4	16.3	48.0	86.0	19
Dorsal kinety 2, number of bristles	gr1	51.5	51.0	5.4	1.2	10.9	44.0	62.0	20
Dorsal kinety 3, number of bristles	gr1	56.1	55.0	6.8	1.5	12.2	45.0	71.0	20

[a] All measurements in µm. CV = coefficient of variation in %, M = median, Max = maximum value, mean = arithmetic mean, Min = minimum value, n = number of individuals investigated, SD = standard deviation, SE = standard error of arithmetic mean.

[b] Midventral rows included.

[c] Row (innermost? outermost?) not specified.

[d] Designation of several structures unclear; morphometric data therefore not included in present table).

difference. However, the two taxa differ not only in this ontogenetic feature, but also in the composition of the midventral complex (only midventral pairs vs. midventral pairs and rows) and the frontoterminal cirri (present vs. absent).

Borror & Wicklow (1983) classified six species in *Urostyla*, namely *U. grandis* (with seven synonyms!), *U. concha* (now *Hemicyclistyla concha*), *U. gracilis* (with two synonyms), *U. marina* (with two synonyms; now *Metaurostylopsis marina*), *U. multipes* (with one synonym; now a supposed synonym of *Paraurostyla weissei*), and *U. dispar*. However, they provided no details, especially as concerns the synonymies.

In the present revision 10 species are assigned to *Urostyla*, seven of which as incertae sedis because they are either not described in every detail so that important data are unknown, or they deviate in at least one important feature from the type species. For example, *Urostyla viridis* has three enlarged frontal cirri (vs. more or less distinct bicorona/multicorona in *U. grandis*) and it is unknown whether or not a midventral complex is present. A classification of these species in another genus would make these genera inhomogenous, and the establishment of new genera seems unwise as long as new data are lacking.

The type species *Urostyla grandis* is the sole species assigned to *Urostyla* which is described in detail. Although it shows a highly characteristic combination of features, it is difficult to recognise the autapomorphies of *Urostyla*. Probably the increased number of parabuccal cirri is one candidate. They originate from the anteriormost anlagen, which produce not only two cirri each to form the bicorona, but distinctly more cirri which cause a somewhat irregular pattern between the anterior end of the midventral complex and the buccal cirral row. Basically one could also designate this pattern as multicorona. Another autapomorphic feature is possibly the lack of frontoterminal cirri. However, several other taxa which also lack this cirral group have been described, for example, *Australothrix* (three enlarged frontal cirri, transverse cirri absent; caudal cirri present) or *Pseudoamphisiella* (three enlarged frontal cirri; transverse cirri form long row; caudal cirri present; zigzag pattern very indistinct; etc.). I suppose that the frontoterminal cirri have been lost, as well as some other cirral groups (e.g., caudal cirri), two or several times independently.

Species included in *Urostyla* (alphabetically arranged according to basionyms): (1) *Urostyla caudata* Stokes, 1886; (2) *Urostyla gigas* Stokes, 1886; (3) *Urostyla grandis* Ehrenberg, 1830. Incertea sedis: (4) *Bakuella variabilis* Borror & Wicklow, 1983; (5) *Urostyla agamalievi* Alekperov, 1984; (6) *Urostyla dispar* Kahl, 1932; (7) *Urostyla gracilis* Entz, 1884; (8) *Urostyla limboonkengi* Wang & Nie, 1932; (9) *Urostyla naumanni* Lepsi, 1935; (10) *Urostyla viridis* Stein, 1859.

Species misplaced in *Urostyla*: The following species – largely originally classified in *Urostyla* – are now assigned to other genera within the urostyloids, or they do not belong to the urostyloids at all. Synonyms of "true" *Urostyla* species and species classified as incertae sedis in *Urostyla* are not mentioned in the following list. If you do not find a certain name in the list below, see the index.

Urostyla agilis Stokes in Sládecek et al. (1981, p. 127) and Wegl (1983, p. 124). Remarks: This is an unintended combination of *Balanitozoon agile* Stokes, 1886a. Correct, present combination: *Urotricha agilis* (Stokes, 1886a) Kahl, 1930; see Foissner (1988a, p. 33) and Foissner et al. (1994, p. 349).

Urostyla algivora Gellért & Tamás, 1958. Remarks: A supposed synonym of *Pseudourostyla urostyla.*

Urostyla coei Turner, 1939. Remarks: A junior synonym of *Paraurostyla weissei* Stein, 1859 (for review see Berger 1999, p. 844).

Urostyla concha Entz, 1884 (now *Hemicycliostyla concha*).

Urostyla cristata Jerka-Dziadosz, 1964 (now *Pseudourostyla cristata*).

Urostyla flavicans Wrzesniowskiego, 1867. Remarks: A junior synonym of *Paraurostyla weissei* Stein, 1859 (for review see Berger 1999, p. 844).

Urostyla hologama Heckmann, 1965. Remarks: A junior synonym of *Paraurostyla weissei* Stein, 1859 (for review see Berger 1999, p. 844).

Urostyla intermedia Bergh, 1889 (now *Anteholosticha intermedia*).

Urostyla latissima Dragesco, 1970. Remarks: Likely not a urostyloid (see remarks at *Urostyla viridis*).

Urostyla lynchi Horváth, 1939. Remarks: A junior synonym of *Paraurostyla weissei* (Stein, 1859) (for review see Berger 1999, p. 844).

Urostyla marina Kahl, 1932 (now *Metaurostylopsis marina*).

Urostyla multipes (Claparède & Lachmann, 1858) Kahl, 1932. Remarks: A supposed synonym of *Paraurostyla weissei* (see Berger 1999, p. 873).

Urostyla muscorum Kahl, 1932 (now *Pseudourostyla muscorum*).

Urostyla paragrandis Wang, 1930. Remarks: A junior synonym of *Paraurostyla weissei* Stein, 1859 (for review see Berger 1999, p. 844).

Urostyla polymicronucleata Merriman, 1937. Remarks: A species of the *Paraurostyla weissei* complex (for review see Berger 1999, p. 844).

Urostyla pseudomuscorum Wang, 1940. Remarks: A supposed synonym of *Pseudourostyla urostyla*.

Urostyla rubra Andrussowa, 1886. Remarks: A species indeterminata.

Urostyla sp. in Shin (1994). Remarks: The specimens of this population have frontoterminal cirri and distinctly more than three dorsal kineties, indicating that it is a *Pseudourostyla* (see there).

Urostyla sp. in Thompson (1972). Remarks: Jankowski (1979) established *Metaurostyla thompsoni* for this population which is classified as supposed synonym of *Metaurostylopsis marina* in the present monograph.

Urostyla vernalis Stokes, 1894. Remarks: A junior synonym of *Paraurostyla weissei* Stein, 1859 (for review see Berger 1999, p. 844).

Urostyla weissei Stein, 1859. Remarks: Type species of *Paraurostyla* Borror, 1972. For review see Berger (1999, p. 844).

Key to *Urostyla* species

Since only the type species is described in detail, that is, both after live and silver preparations, details of the infraciliature cannot be used in the key below. If the transverse cirri are very indistinct, that is, possibly lacking, see *Hemicycliostyla*.

1 Two macronuclear nodules (e.g., Fig. 219a) **2**
\- More than 2 macronuclear nodules (e.g., Fig. 208a) **6**
2 Limnetic **3**
\- Marine **4**
3 Body length below 200 µm; cells green due to symbiotic green algae; 3 enlarged frontal cirri (Fig. 222a) .... *Urostyla viridis* (p. 1106)
\- Body length distinctly above 200 µm; green symbiotic algae lacking; several frontal cirri arranged in (sometimes indistinct) bicorona (Fig. 215a) .... *Urostyla agamalievi* (p. 1093)
4 (2) Cells pale copper-red, brownish-pink, or splendid dark crimson (Fig. 217a, 218a) .... *Urostyla gracilis* (p. 1097)

- Cells more or less colourless **5**

5 Three frontal cirri; more than 1 left marginal row; adoral zone of membranelles continuous (Fig. 219a) .... *Urostyla limboonkengi* (p. 1101)

- Many frontal cirri; 1 left marginal row; adoral zone bipartite (Fig. 216a) .... *Urostyla dispar* (p. 1094)

6 (1) About 6–8 macronuclear nodules arranged in line (Fig. 220a) .... *Urostyla naumanni* (p. 1103)

- Many (small) macronuclear nodules dispersed in cell (e.g., Fig. 208h) **7**

7 Three (or at least very few) frontal cirri (Fig. 221a) .. *Urostyla variabilis* (p. 1104)

- Several frontal cirri form more or less distinct bicorona (e.g., Fig. 208e) **8**

8 Body outline broad elliptical (Fig. 208s, t); rear body end without distinctly elongated cirri (Fig. 208a) .... *Urostyla grandis* (p. 1048)

- Body outline elongate elliptical to roughly spindle-shaped (Fig. 213a, 214a); rear body end with bundle(s) of distinctly elongated cirri (Fig. 213a, 214a) **9**

9 Body length ca. 600 µm; several contractile vacuoles (Fig. 213a, arrowheads) .... *Urostyla caudata* (p. 1088)

- Body length ca. 800 µm; 1 contractile vacuole (Fig. 214a) *Urostyla gigas* (p. 1091)

Urostyla grandis Ehrenberg, 1830
(Fig. 2a–g, 3a–l, 4a–h, 7a–t, 9a–f, 10a–d, 11a, 17a–y, 205a–z, 206a–d, 207a–w, 208a–t, 209a–e, 210a–f, 211a–g, Tables 12, 42, Addenda)

1830 ***Urostyla grandis*. nov. Gen.** – Ehrenberg, Abh. preuss. Akad. Wiss., year 1830: 43 (original description; no formal diagnosis and no illustration provided and no type material available).

1831 ***Bursaria vorax* E.** – Ehrenberg, Abh. preuss. Akad. Wiss., year 1831: 110 (brief original description of synonym without illustration; no type material available; see nomenclature).

1831 ***Vrostyla grandis* E.**[1] – Ehrenberg, Abh. preuss. Akad. Wiss., year 1831: 119 (brief description without illustration; see nomenclature).

1838 ***Bursaria vorax*** – Ehrenberg, Infusionsthierchen, p. 327, Tafel XXXV, Fig. I (Fig. 205x; first illustration and review).

1838 ***Urostyla grandis*** – Ehrenberg, Infusionsthierchen, p. 369, Tafel XLI, Fig. VIII (Fig. 205a–e; first illustration and review).

1849 ***Oxytricha fusca***[2] – Perty, Mitt. naturf. Ges. Bern, year 1849: 169 (original description of synonym; no formal diagnosis provided and no type material available).

1850 ***Urostyla grandis* Ehrenberg** – Diesing, Systema Helminthum I, p. 161 (brief review).

1852 ***Oxytricha fusca*** – Perty, Kenntniss kleinster Lebensformen, p. 154, Tafel VI, Fig. 19A, B (Fig. 205f; first illustration of synonym).

1859 ***Urostyla grandis*. Ehrbg.** – Stein, Organismus der Infusionsthiere I, p. 195, Tafel XIII, Fig. 5–12, Tafel XIV, Fig. 1–6 (Fig. 17a–y, 206a–d, 209a–d; very detailed redescription from life).

1865 ***Urostyla grandis*** – Quennerstedt, Acta Univ. lund., 2: 59, Pl. II, Fig. 16 (Fig. 205g; illustrated record from Sweden).

---

[1] The diagnosis provided by Ehrenberg (1831) is as follows: Körperdurchmesser 1/9 Linie. Körper oben gewölbt, unten flach, sehr groß, drei- bis viermal so lang als breit, hinten und vorn kleine Borsten und überdies hinten Griffel, sonst überall gewimpert.

[2] The diagnosis by Perty (1849) is as follows: Gestreckt elliptisch, oben flach gewölbt, unten flach concav, Mundspalte weit, Leib gewöhnlich durch Nahrung gelbbraun bis schwärzlich. L 1/14–1/7'''.

1882 ***Urostyla grandis*, Ehr.** – Kent, Manual Infusoria II, p. 765, Plate XLIII, Fig. 6–8 (Fig. 205y, redrawing of Fig. 17c, 206c; revision).

1885 ***Urostyla trichogaster*. sp. nov.** – Stokes, Ann. Mag. nat. Hist., 15: 444, Plate XV, fig. 3 (Fig. 205h; original description of synonym; no formal diagnosis provided and no type material available).

1888 ***Urostyla trichogaster*, Stokes** – Stokes, J. Trenton nat. Hist. Soc., 1: 278, Plate X, fig. 12 (Fig. 205z; review).

1889 ***Urostyla grandis*** – Bergh, Archs Biol., 9: 497, Planche XXXV, Fig. 1–10 (Fig. 209e; division of nuclear apparatus).

1891 ***Urostyla elongata*** – Stokes, Jl R. microsc. Soc., 1891: 700, Fig. 7 (Fig. 205i; original description of synonym; no formal diagnosis provided and no type material available).

1891 ***Urostyla fulva*** – Stokes, Jl R. microsc. Soc., 1891: 700, Fig. 8 (Fig. 205j; original description of synonym; no formal diagnosis provided and no type material available).

1901 ***Urostyla grandis* Ehrbg.** – Roux, Mém. Inst. natn. génev., 19: 95, Planche V, fig. 17 (Fig. 205k; redescription).

1905 ***Urostyla trichota* (*Hemiclostyla* of Stokes)** – Conn, Bull. Conn. St. geol. nat. Hist. Surv., 2: 58, Fig. 237 (Fig. 205l; illustrated record; see nomenclature and remarks).

1905 ***Urostyla vernalis* (?) Stokes** – Conn, Bull. Conn. St. geol. nat. Hist. Surv., 2: 58, Fig. 239 (Fig. 205m; illustrated record).

1905 ***Urostyla trichogastra* Stokes** – Conn, Bull. Conn. St. geol. nat. Hist. Surv., 2: 58, Fig. 241 (Fig. 205n; illustrated record; see nomenclature).

1908 ***Urostyla grandis*** – Fauré-Fremiet, Bull. Inst. gén. psychol., 7: 441, Fig. 1, 2 (Fig. 205o–r; redescription).

1912 ***Urostyla grandis* Stein** – André, Catalogue des invertébrés de la Suisse, 6: 123 (review of Swiss ciliates).

1914 ***Urostyla grandis* Ehr.** – Smith, Kans. Univ. Sci. Bull., 9: 164, Plate XLIV, Fig. 48 (Fig. 205s; redescription from life; see remarks).

1925 ***Urostyla grandis* Ehrenberg** – Wang, Contr. biol. Lab. Sci. Soc. China, 1: 49, Fig. 124 (Fig. 205u; redescription).

1932 ***Urostyla grandis* Ehrb., 1838** – Kahl, Tierwelt Dtl., 35: 565, Fig. 97_3, 99 (Fig. 207a, b; revision).

1932 ***Urostyla trichogaster* Stokes, 1885** – Kahl, Tierwelt Dtl., 25: 566, 97_8 (Fig. 207c; revision).

1932 ***Urostyla elongata* Stokes, 1891** – Kahl, Tierwelt Dtl., 25: 566, Fig. 97_7 (Fig. 207d; revision).

1932 ***Urostyla fulva* Stokes, 1891** – Kahl, Tierwelt Dtl., 25: 566, Fig. 97_{10} (Fig. 207e; revision).

1935 ***Urostyla grandis* Ehrenberg** – Tittler, Cellule, 44: 191, Fig. 1–12 (Fig. 2a–g, 3a–l, 4a–h, 205t; description, division, encystment, and endomixis).

1945 ***Urostyla trichogaster* Stokes** – Fauré-Fremiet, Bull. biol. Fr. Belg., 79: 119, Fig. 10–16 (Fig. 205v; homopolar doublets).

1950 ***Urostyla*** – Gelei, Acta biol. hung., 1: 112, Abb. 23e (Fig. 205w; illustration).

1950 ***Urostyla grandis* E.** – Kudo, Protozoology, p. 672 (redrawing of Fig. 206a; textbook).

1952 ***Urostyla grandis* Ehrenberg 1838** – Šrámek-Hušek, Chekh. Biol., 1: 368, 376, Fig. 1–7 (Fig. 17e, 206a, b, 207a, f–h; redescription; see nomenclature and remarks).

1952 ***Urostyla grandis typica* var. n.** – Šrámek-Hušek, Chekh. Biol., 1: 370, 376, Fig. 1–3 (Fig. 17e, 206a, b; erection of variety; see nomenclature and remarks).

1952 ***Urostyla grandis*, var. *kahli* var. n.** – Šrámek-Hušek, Chekh. Biol., 1: 370, 376, Fig. 4–7 (Fig. 207a, f–h; erection of variety; see nomenclature and remarks).

1952 ***Urostyla grandis* Ehrenberg 1838** – Šrámek-Hušek, Cslká Biol., 1: 176, Fig. 1–7 (Fig. 17e, 206a, b, 207a, f–h; redescription).

1952 ***Urostyla grandis typica* var. n.** – Šrámek-Hušek, Cslká Biol., 1: 177, Fig. 1–3 (Fig. 17e, 206a, b; erection of variety; see nomenclature and remarks).

1952 ***Urostyla grandis kahli* var. n.** – Šrámek-Hušek, Cslká Biol., 1: 176, Fig. 4–7 (Fig. 207a, f–h; erection of variety; see nomenclature and remarks).

1961 ***Urostyla grandis* Ehrenberg, 1838** – Reuter, Acta zool. fenn., 99: 17, Abb. 22 (Fig. 207i; illustrated record).

1963 ***Urostyla grandis* Ehrbg.** – Jerka-Dziadosz, Acta Protozool., 1: 43, Fig. 1–3, Plates I–IV (Fig. 9a–f, 10a–d, 210a–f; description of cell division and regeneration).

1963 ***Urostyla grandis*** **Ehrenberg** – Lundin & West, Free-living protozoa, p. 67, Plate 27, Fig. 9 (Fig. 207j; illustrated record).

1963 ***Urostyla trichogaster*** **Stokes** – Lundin & West, Free-living protozoa, p. 67, Plate 27, Fig. 10 (Fig. 207k; illustrated record).

1965 ***Urostyla*** – Lepsi, Protozoologie, p. 980 (textbook).

1968 ***Urostyla grandis*** **Ehrenberg, 1838** – Chorik, Free-living ciliates, p. 126, Fig. 115 (Fig. 207l; redescription in Russian).

1972 ***Urostyla grandis*** **Ehrenberg, 1830** – Borror, J. Protozool., 19: 8, Fig. 11 (redrawing of Fig. 9a; revision of hypotrichs).

1972 ***Urostyla grandis*** – Jerka-Dziadosz, Acta Protozool., 10: 83, Fig. 3, 4, Plate VI, Fig. 27, Plates VII–X (Fig. 7a–t; morphogenesis).

1974 ***Urostyla grandis*** **Ehrenberg, 1838** – Jones, Univ. South Alabama Monogr., 1: 39, Plate XXVIII, Fig. 1 (Fig. 207m; redescription).

1974 ***Urostyla grandis*** **Ehrenberg** – Pätsch, Arb. Inst. landw. Zool. Bienenkd., 1: 55, Abb. 42 (Fig. 207n; redescription after protargol impregnation).

1974 ***Urostyla trichogaster*** **Stokes 1885 var. *elongata* (Stokes, 1891) comb. n.** – Stiller, Annls hist.-nat. Mus. natn. hung., 66: 132 (change in rank, see nomenclature).

1974 ***Urostyla trichogaster*** **Stokes 1885 var. *fulva* (Stokes, 1891) comb. n.** – Stiller, Annls hist.-nat. Mus. natn. hung., 66: 132 (change in rank, see nomenclature).

1974 ***Urostyla grandis*** **Ehrenberg** – Stiller, Fauna Hung., 115: 39, Fig. 23A (redrawing of Fig. 207b; revision).

1974 ***Urostyla trichogaster*** **Stokes** – Stiller, Fauna Hung., 115: 39, Fig. 24A (redrawing of Fig. 205h; revision).

1974 ***Urostyla trichogaster*** **var. *elongata* Stokes** – Stiller, Fauna Hung., 115: 40, Fig. 24B (redrawing of Fig. 205i; revision).

1974 ***Urostyla trichogaster*** **var. *fulva* Stokes** – Stiller, Fauna Hung., 115: 40, Fig. 24C (redrawing of Fig. 205j; revision).

1979 ***Urostyla grandis*** – Borror, J. Protozool., 26: 547, Fig. 4 (Fig. 207h1; redefinition of the urostylids).

1979 ***Metaurostyla polonica*** **gen. et sp. n.** – Jankowski, Trudy zool. Inst., 86: 70 (junior synonym; see remarks).

1980 ***Urostyla chlorelligera*** **nov. spec.** – Foissner, Ber. Nat.-Med. Ver. Salzburg, 5: 103, Abb. 23a–c (Fig. 212a–c; original description of new [supposed] synonym).

1982 ***Urostyla grandis*** **Ehrenberg, 1838** – Hemberger, Dissertation, p. 78 (revision of non-euplotid hypotrichs).

1983 ***Urostyla grandis*** **Ehrenberg, 1830** – Borror & Wicklow, Acta Protozool., 22: 120, Fig. 7 (Fig. 207o; revision of urostylids).

1985 ***Urostyla grandis*** – Small & Lynn, Phylum Ciliophora, p. 450, Fig. 1 (Fig. 207o; guide to ciliate genera).

1987 ***Urostyla grandis*** **Ehrenberg, 1838** – Tuffrau, Annls Sci. nat. (Zool.), 8: 13, Fig. 3 (Fig. 207p; in-annotated illustration).

1989 ***Urostyla grandis*** **Ehrenberg, 1830** – Song & Wilbert, Lauterbornia, 3: 157, Abb. 88, Tabelle 30 (Fig. 207q–t; redescription after protargol impregnation and morphometric characterisation).

1991 ***Urostyla grandis*** **Ehrenberg, 1830** – Foissner, Blatterer, Berger & Kohmann, Informationsberichte des Bayer. Landesamtes für Wasserwirtschaft, 1/91: 222, Fig. 1–21 (Fig. 208h–q; review of ciliates of the saprobic system).

1991 ***Urostyla grandis*** – Foissner, Europ. J. Protistol., 27: 323, Fig. 30, 31 (Fig. 208k; paper on taxonomic methods).

1991 ***Urostyla grandis*** **Ehrenberg, 1830** – Ganner, Dissertation, p. 101, Abb. 108–121 (Fig. 208a–g, 211a–g; detailed redescription including morphogenesis).

1994 ***Urostyla grandis*** **Ehrenberg, 1830** – Shin, Dissertation, p. 94, Fig. 13A–C (Fig. 207u–w; redescription).

1994 ***Urostyla grandis*** **Ehrenberg, 1838** – Tuffrau & Fleury, Traite de Zoologie, 2: 128, Fig. 46a (Fig. 207p; textbook).

2001 ***Urostyla grandis*** **Ehrenberg, 1830** – Berger, Catalogue of ciliate names 1. Hypotrichs, p. 101 (nomenclator containing all basionyms, combinations, and higher taxa of hypotrichs).

2001 ***Urostyla grandis* Ehrenberg, 1830** – Eigner, J. Euk. Microbiol., 48: 76, Fig. 15–23 (Fig. 208a, e, 211a–g; brief review of Urostylidae).
2002 ***Urostyla grandis*** – Lynn & Small, Phylum Ciliophora, p. 443, Fig. 5 (Fig. 207o; guide to ciliate genera).
2004 **Urostyliden** – Gruber, Universum Magazin, Juni 2004: 101, 1 micrograph (Fig. 208s; brief report about present book).

Nomenclature: No derivation of the names are given in the original descriptions. The species-group name *grand·is -is -e* (Latin adjective; large, important) obviously alludes to the large body size. For discussion of the spelling *Vrostyla* in Ehrenberg (1831), see the genus section. *Urostyla grandis* was fixed as type species of *Urostyla* (see nomenclature of genus section).

The species-group name *trichogaster* ("belly-hair") is a composite of the Greek nouns *he thrix* (hair-) and *he gaster* (belly) and obviously alludes to the high number of cirri on the ventral side. The species-group names *fusc·us -a -um* (Latin adjective; dark-brown, dark, grey, black) and *fulv·us -a -um* (Latin adjective; bronze-coloured, fire-red, red-yellow, brownish; Hentschel & Wagner 1996) likely refer to the colour of the cell. The species-group name *elongat·us -a -um* (Latin adjective; elongated, stretched) probably alludes to the "elongated or sub-elliptical" body outline. The species group-name *polonica* refers to the country (Poland) where the type population (*Urostyla grandis* sensu Jerka-Dziadosz 1972) was collected. *Metaurostyla polonica* was designated as type species of *Metaurostyla* by original designation.

Šrámek-Hušek (1952a, b) distinguished two varieties within *U. grandis*; the variety name *typic·us -a -um* (Greek adjective; typical, normal, genuine, prototypical) was used for the ordinary form as described, for example, by Stein (1859); the variety *U. grandis kahli* was dedicated to Alfred Kahl who described a population which closely resembled that described by Šrámek-Hušek himself.

Stiller (1974a) considered the synonyms *Urostyla elongata* Stokes and *Urostyla fulva* Stokes as varieties of another synonym (*Urostyla trichogaster*) described by Stokes.

Urostyla tricogaster Stokes 1885 in Grispini (1938, p. 152) and *Urostyla trichogastra* in Conn (1905, p. 58) are incorrect subsequent spellings. *Hemiclostyla* in Conn (1905, p. 58; see list of synonyms) is an incorrect subsequent spelling of *Hemicycliostyla*.

Remarks: The history of *Urostyla grandis* is very comprehensive, and therefore not every paper mentioned in the list of synonyms is discussed in detail.

Ehrenberg (1830) provided neither a description nor an illustration. Nevertheless, this paper is accepted as original description of *U. grandis* (e.g., Borror 1972, Foissner et al. 1991). The first illustration and more or less detailed description appeared in Ehrenberg (1838), where he discussed that he had synonymised *U. grandis* with *Trichoda patens* Müller, 1786 (p. 181, Tab. XXVI, Fig. 1, 2) in one of his earlier papers (see genus section). Ehrenberg (1838) misinterpreted the transverse cirri region as gap (like the oral apparatus) bearing 5–8 small cirri (= transverse cirri) on its left margin. Moreover, he wrote that the *Trichoda uva* of Müller could be a postdivider of *U. grandis*. I checked Müller's papers (Müller 1773, 1776, 1779, 1786), but did not find a

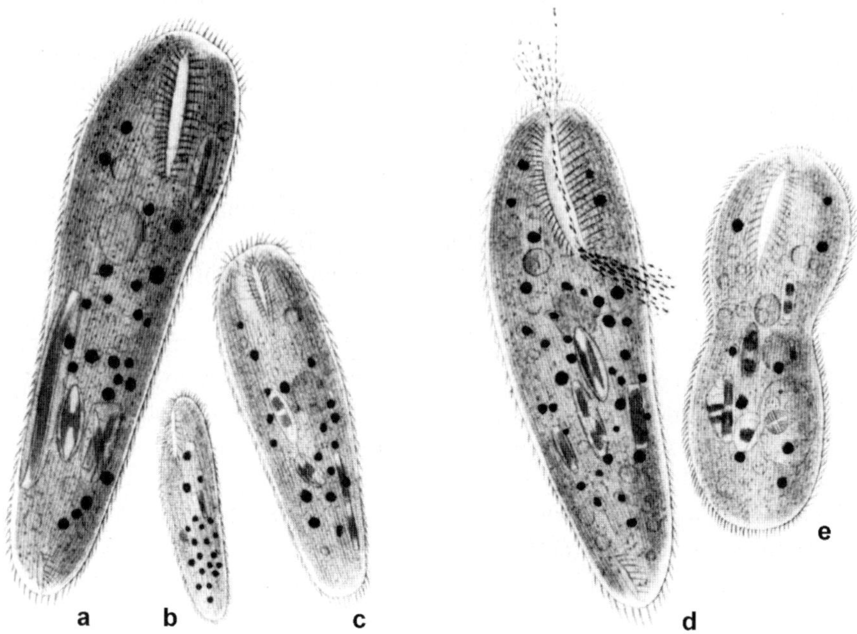

Fig. 205a–e *Urostyla grandis* from life (from Ehrenberg 1838). Various views, showing, inter alia, oral apparatus, cirral rows, and ingested food. The specimen shown in (e) is a divider. According to Ehrenberg, body length is 173–260 μm. Page 1048.

T. uva. I suppose that Ehrenberg meant *T. uvula* Müller, 1773 (p. 94), which could indeed be a hypotrich according to the illustrations by Müller (1786; Tafel XXVI, Fig. 11, 12). However, an identification with the present species would be arbitrary and destabilise nomenclature.

Stein (1867, p. 327) discussed that *Leucophrys sanguinea* Ehrenberg, 1833 (p. 253, Tafel III, Fig. 5) is possibly a junior synonym of *U. grandis* (see also Entz 1884, p. 378 for discussion). However, *Leucophrys sanguinea* is red, so that synonymy can be excluded. Possibly it is identical with *Diaxonella pseudorubra*, which is also limnetic and red. However, an identification with one of these species would be very arbitrary and should therefore be avoided.

Dujardin (1841, p. 422) did not provide own observations, but briefly discussed Ehrenberg's results. Somewhat later, Perty (1852, p. 154) doubted the validity of *Urostyla* (see genus section) and even *U. grandis*. He assumed that Ehrenberg's species is possibly only a "higher evolutionary stage" of *Oxytricha eurystoma* Ehrenberg, 1838, which is a junior objective synonym of *Steinia platystoma* (Ehrenberg, 1831) Diesing, 1866 (see Berger 1999, p. 626, for description). Perty even took into account that *U. grandis* is only a modification of his *Oxytricha fusca*, that is, Perty himself supposed identity of these two species. However, Perty (1852) did not realise that Ehrenberg's species had priority.

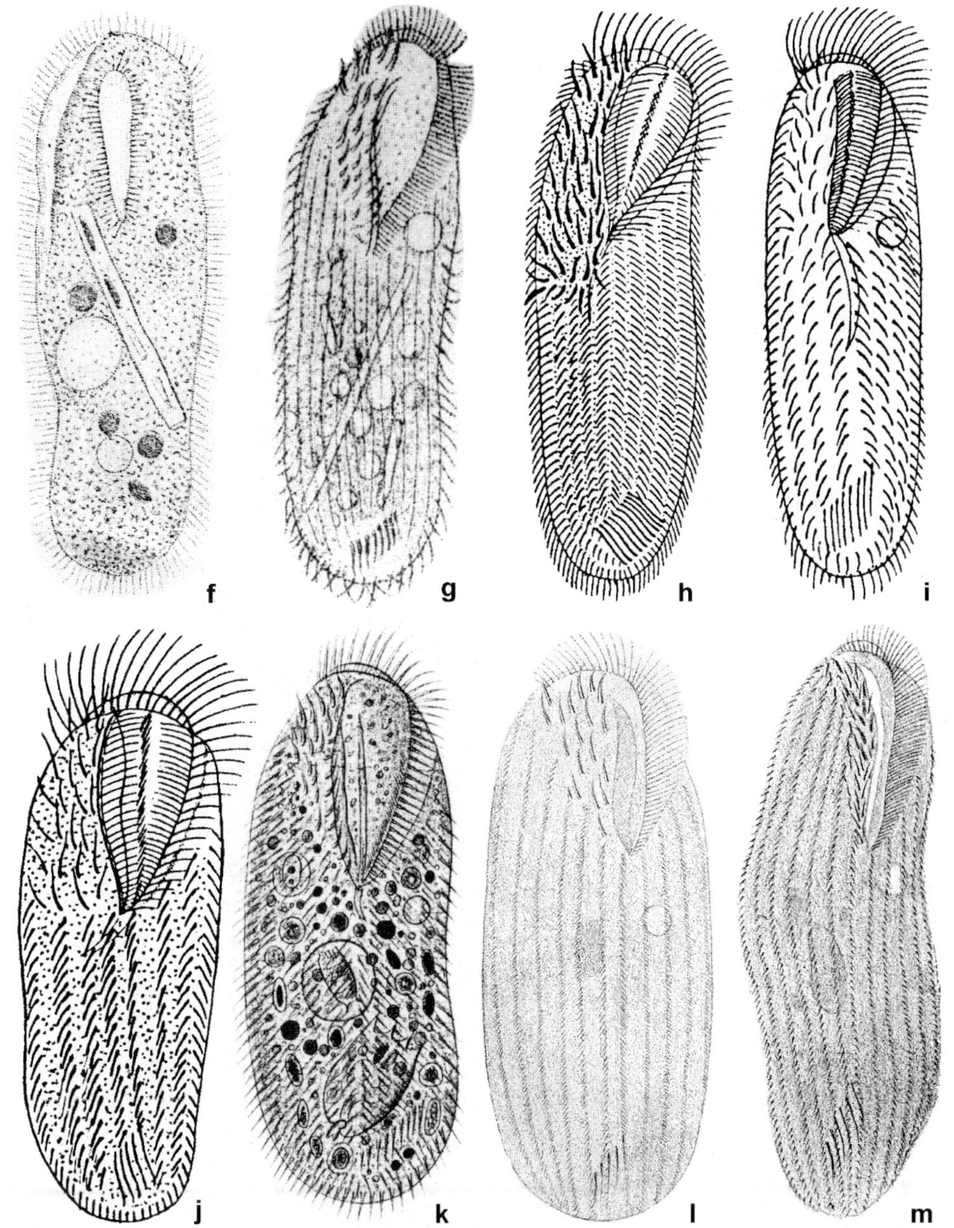

Fig. 205f–m *Urostyla grandis* from life (f, from Perty 1852; g, from Quennerstedt 1869; h, from Stokes 1885; i, j, from Stokes 1891; k, from Roux 1901; l, m, from Conn 1905). Note that some illustrations are very faint in the original papers and therefore rather difficult to reproduce. However, it is always better to show the original illustration than a redrawing. **f:** Ventral view of the synonym *Oxytricha fusca*, 150–300 µm. **g, k–m:** Ventral views, g = size not indicated, k = 300–500 µm, l = 296 µm, m = 194 µm. **h:** Ventral view of the synonym *Urostyla trichogaster*, 254–339 µm. **i:** Ventral view of the synonym *Urostyla elongata*, 300 µm. **j:** Ventral view of the synonym *Urostyla fulva*, 254 µm. Page 1048.

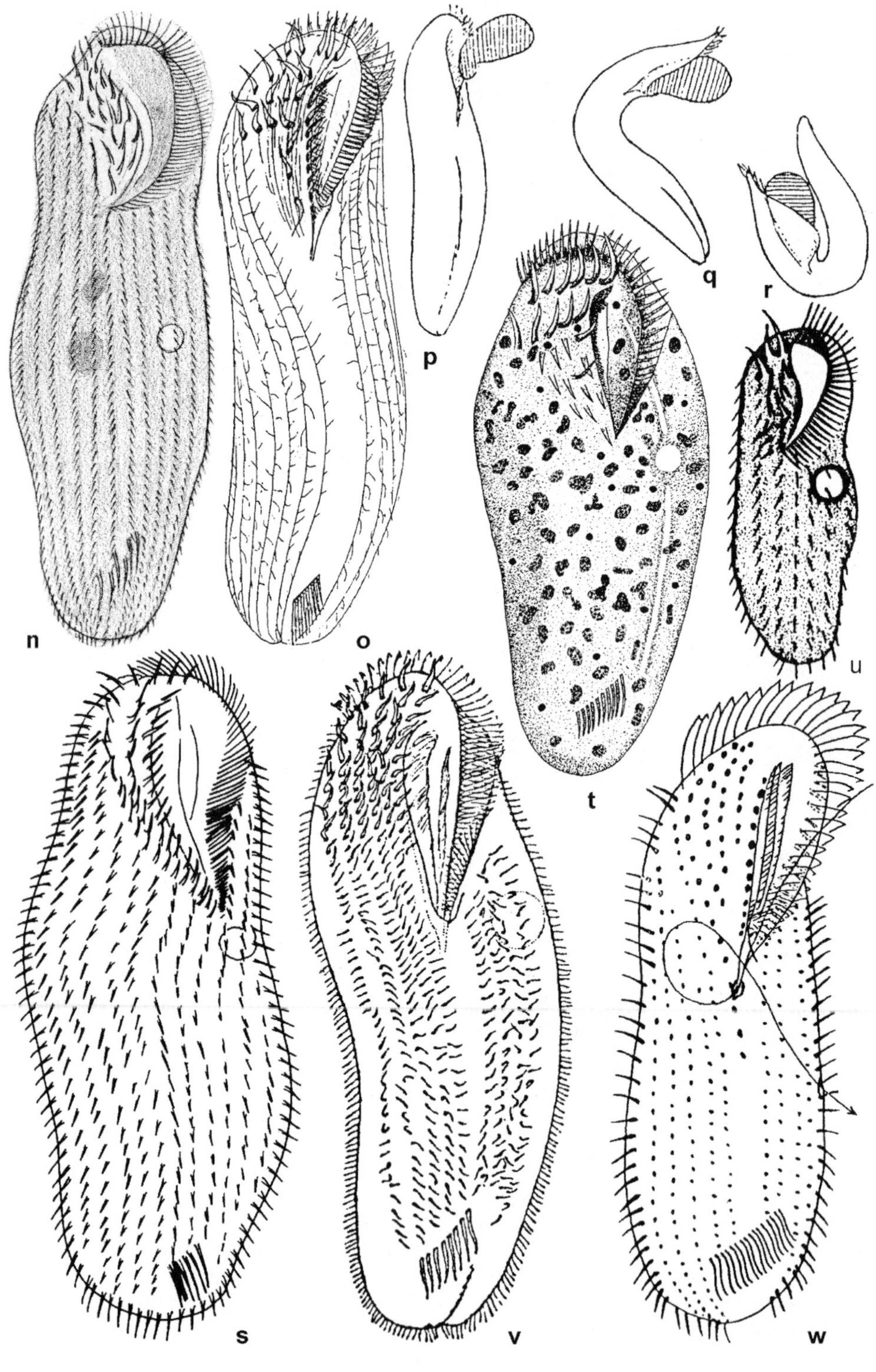
n
o
p
q
r
t
u
s
v
w

Claparède & Lachmann (1858, p. 142) found specimens which resembled *Oxytricha fusca*. However, they correctly stated that Perty described this species rather imperfectly.

Stein (1859) provided very detailed life observations of *U. grandis* concerning (i) the interphasic morphology, (ii) the resting cyst, (iii) the cell division, and (iv) the infestation by the endoparasite *Podophrya urostylae*. He discussed the shortcomings of Ehrenberg's original description and recognised that *U. grandis* has several synonyms. Stein (1859, p. 195, 204) considered *Trichoda patula* Müller, 1786 (p. 181, Tab. XXVI, Fig. 3–5) as supposed (because marked with a question mark) synonym of *U. grandis*, and he criticised that Ehrenberg had transferred *T. patula* to the hymenostome genus *Leucophrys* (for early history, see Ehrenberg 1838, p. 311). Later, Ehrenberg's description of this species was considered as original description of *Tetrahymena patula* (Ehrenberg, 1830) Corliss, 1951 (for review, see Corliss & Dougherty 1967 and Corliss 1973). According to Stein, *Bursaria vorax* Ehrenberg is a junior synonym of *U. grandis*. Ehrenberg (1838) himself mentioned great similarity of *B. vorax* with *Urostyla grandis* and *Stylonychia lanceolata* (now *Pleurotricha lanceolata*; for review see Berger 1999, p. 699). Of course, the identification of *B. vorax* with *U. grandis* is somewhat arbitrary because several details of the cirral pattern are not known for *B. vorax*. Foissner (1993, p. 423), in his review on *Bursaria*, supposed that *B. vorax* is either a *Condylostoma* or a *Urostyla* species. Stein (1859) also synonymised *Oxytricha fusca* Perty with *U. grandis* and explained the differences between these two species (e.g., transverse cirri lacking in *O. fusca* vs. present) by the superficial observations made by Perty (however, note that Perty himself supposed identity of these two species; see above). Stein (1859, p. 205) considered two further species described by Perty as junior synonyms of the present species, namely, *Oxytricha decumana* Perty, 1852a (see also Perty 1852, p. 154, not illustrated) and *O. protensa* Perty, 1852a (see Perty 1852, p. 153, Tafel VI, Fig. 20A–E for description). In the review on oxytrichids I classified both species as indeterminable (Berger 1999, p. 247, 252).

Stein (1859, p. 205) mentioned that *Oxytricha multipes* Claparède & Lachmann, 1858 (p. 143, Planche 5, Fig. 1) has to be classified between *U. grandis* and *U. weissei* (now *Paraurostyla weissei*). Now *Oxytricha multipes* is considered as supposed synonym of *P. weissei* (for review, see Berger 1999, p. 844, 873).

Quennerstedt's (1865) paper is written in Swedish and therefore unreadable for me. However, the illustration shows that the identification is beyond reasonable doubt (Fig. 205g). Kent (1882) provided three illustrations, two of which are certainly redrawings of Stein's figures. One illustration deviates distinctly from the pattern, indicating that it is an original (Fig. 205y).

Stokes (1885, 1891) described three new *Urostyla* species, namely, *U. trichogaster*,

---

← **Fig. 205n–w** *Urostyla grandis* (n, from Conn 1905; o–r, from Fauré-Fremiet 1908; s, after Smith 1914; t, from Tittler 1935; u, from Wang 1925; v, from Fauré-Fremiet 1945; w, from Gelei 1950. n, o–r, s?, u?, from life; t, v, w, after fixation and staining? [methods not clearly indicated]). **n, o, s–w:** Ventral views, o = size not indicated, s = 250 μm, t = 232 μm, u = 176 μm, v = 305 μm, w = 400 μm. The arrow in (w) obviously denotes the food passage. **p–r:** Ingestion of a *Glaucoma*. Page 1048.

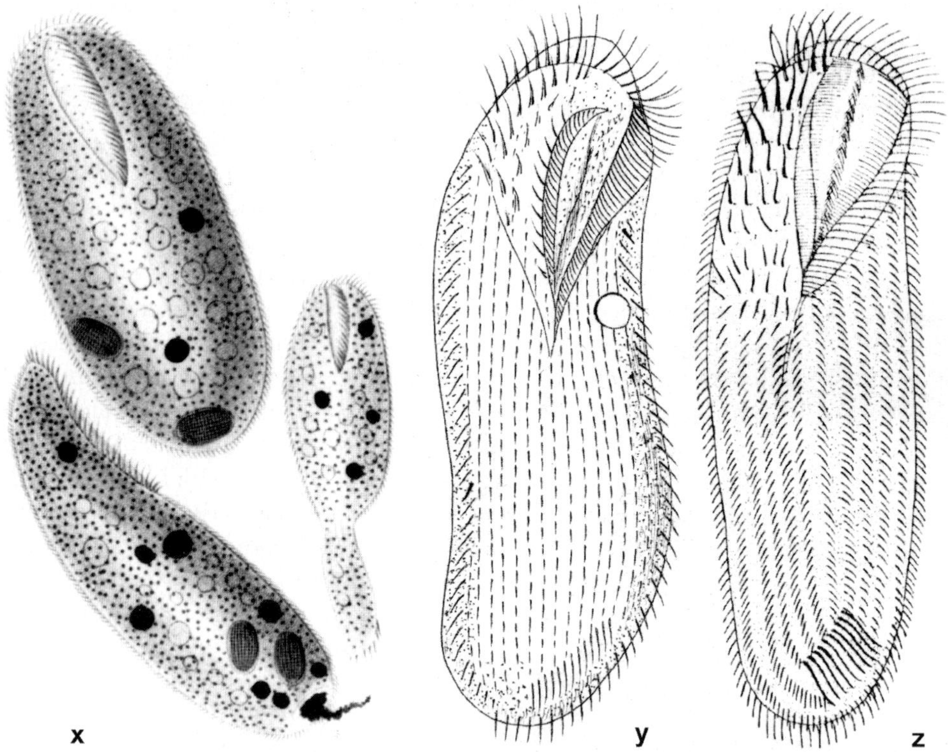

Fig. 205x–z *Urostyla grandis* from life (x, from Ehrenberg 1838; y, from Kent 1882; z, from Stokes 1888). **x:** Various views of the synonym *Bursaria vorax*, 173–231 μm. **y, z:** Ventral views showing cirral pattern and contractile vacuole (y). Page 1048.

U. elongata, and *U. fulva*. Unfortunately, Stokes – who made very good live observations – did not compare his data carefully with the literature although the type species is listed in his review on ciliates from the USA (Stokes 1888, p. 277). I checked the descriptions and could not find features which justify the recognition of these populations as distinct species. Thus, the synonymy proposed by Kahl (1932) and formalised by Borror (1972) has to be accepted and therefore I mention only body length and number of transverse cirri in the morphology section below. The faunistic records on Stokes' species are kept separate so that workers who do not agree with this synonymy can also use these data.

Bütschli (1889, p. 1741) synonymised both *Hemicycliostyla sphagni* and *H. trichota* with *U. grandis*, which is, however, incorrect because *Hemicycliostyla* species lack transverse cirri, whereas this cirral group is present in *U. grandis*.

Conn (1905, p. 58) recorded and illustrated three species originally described by Stokes. I am certain that all specimens/populations described by Conn belong to *U. grandis*. Conn himself wrote that "perhaps all of these are only varieties of *U. grandis* Ehrbg.", a statement which is only correct for *U. trichogaster*, which is generally ac-

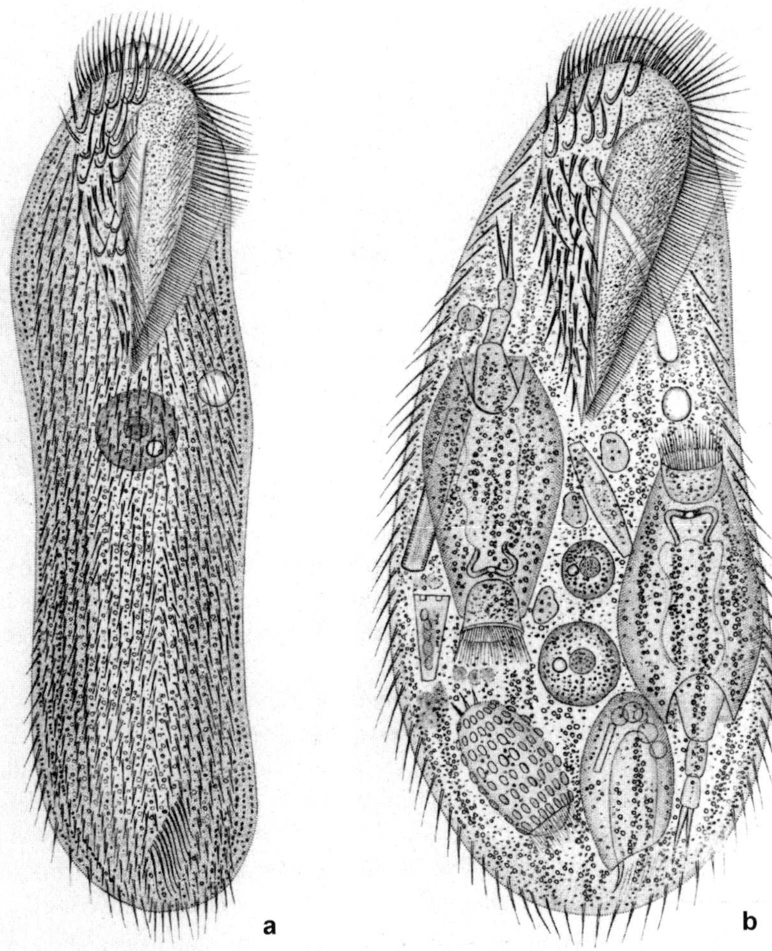

Fig. 206a, b *Urostyla grandis* from life (from Stein 1859). Ventral views showing shape variability, size range see text. Note that Stein likely slightly overestimated the number of cirral rows. The specimen shown in (b) is packed with food (rotifers, ciliates, diatoms, etc.). Page 1048.

cepted as synonym of *U. grandis*. The other two identifications made by Conn are incorrect. His *Urostyla trichota* population has transverse cirri and therefore he obviously transferred this species from *Hemicycliostyla* (without transverse cirri) to *Urostyla* (with transverse cirri), an incorrect act which was overlooked by Berger (2001). *Urostyla vernalis* Stokes, 1894 – at present classified as synonym of *Paraurostyla weissei* – has only two macronuclear nodules (see Berger 1999, p. 844).

Smith (1914) mentioned two macronuclear nodules, but obviously did not illustrate them (my Xerox copy of this paper is not quite perfect so that I cannot exclude that the macronuclear nodules are illustrated quite faintly). I suppose that he used some data from Edmondson (1906), whose *U. grandis* is very likely a *Paraurostyla weissei* (Fig.

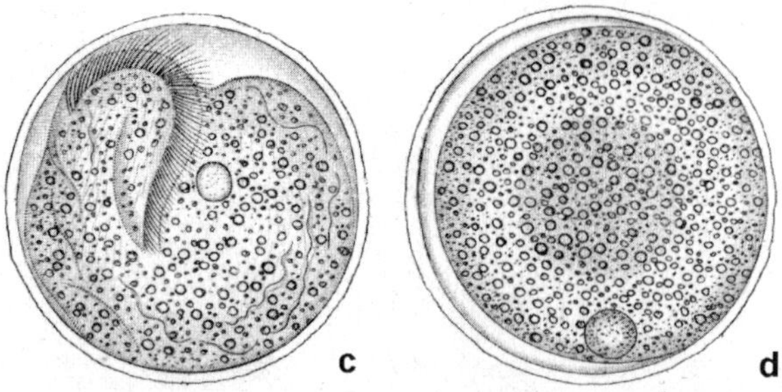

Fig. 206c, d *Urostyla grandis* from life (from Stein 1859). Resting cysts, diameter ranges from about 80 to 120 µm. Page 1048.

164g). However, the cirral pattern of Smith's population is indeed *U. grandis*-like, so his identification can be accepted (Fig. 205s). In his text he wrote "Length of type specimen, 250 microns". I suppose that he meant "type specimen" to be the specimen he illustrated, and not a type specimen in the nomenclatural sense. In the legend to Plate XLIV he mistakenly wrote that the description of the species is on page 167.

Wang (1925) did not illustrate and describe *U. grandis* in detail. His figure does not show transverse cirri, whereas in the description a number of 10–12 is mentioned. Either he forgot to illustrate the transverse cirri, or his specimens indeed had no such cirri and he took the value from the literature. In spite of these uncertainties, I accept Wang's identification.

Kahl (1932) redescribed *Urostyla grandis* and found some differences to Stein's data without discussing them in detail. In spite of these differences he did not doubt conspecificity of his and Stein's populations, and also accepted the synonymy with *U. fusca* proposed by Stein. Kahl (1932) did not synonymise *U. trichogaster* with *U. grandis* formally, but doubted that Stokes' species is valid. Moreover, he discussed that *U. elongata* and *U. fulva* very closely resemble *U. trichogaster,* respectively, *Urostyla grandis*.

Šrámek-Hušek (1952a, b) distinguished two varieties mainly differing in the cirral pattern. For Stein's population he proposed the name *typica* and for populations resembling Kahl's and his description he suggested the name *kahli*. However, the differences in the cirral pattern are very likely mainly due to minor misobservations, as already suggested by Kahl (1932), and therefore should not be over-interpreted. Probably, Stein (1859; Fig. 206a) distinctly overestimated the number of cirral rows (18 vs. about 12), causing a rather different cirral pattern and indicating that the distinction is unjustified. However, if one accepts Šrámek-Hušek's separation, the names *typica* and *kahli* are of subspecific rank according to ICZN (1999, Article 45.6.4). Probably, the name *typica* is

---

Fig. 207a–h1 *Urostyla grandis* (a, b, from Kahl 1932; c, after Stokes 1885 from Kahl 1932; d, e, after Stokes 1891 from Kahl 1932; f–h, from Šrámek-Hušek 1952a, b; h1, from Borror 1979. a–h, from life; h1, protargol impregnation). **a–f:** Ventral views, a = 350 µm, b = size not indicated, c, d = 300 µm, e = 250 µm, f = 300–500 µm. **g, h:** Contracted specimen and cortical granules along cirral row. **h1:** Infraciliature of ventral side, 360 µm. Page 1048. →

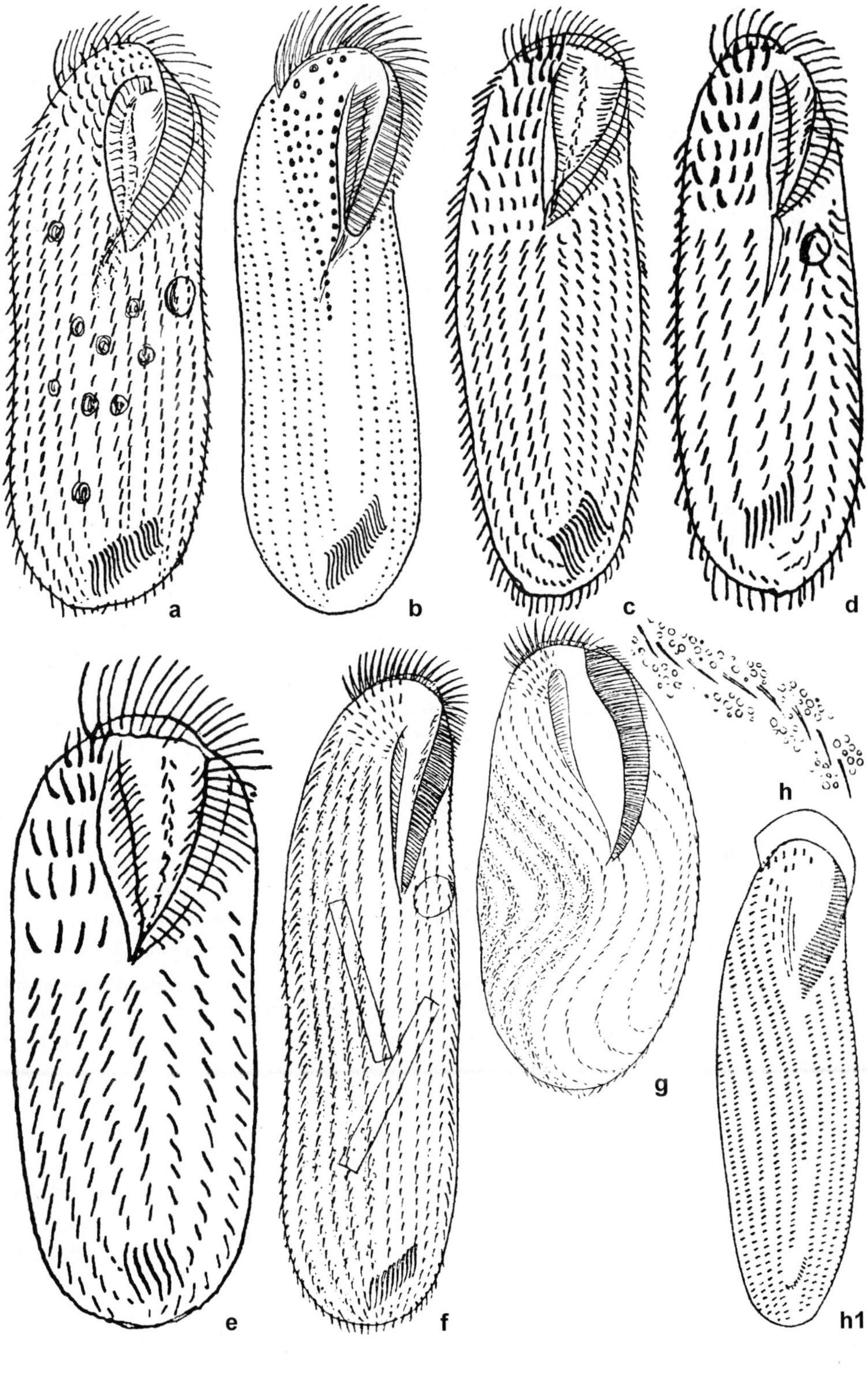
a
b
c
d
e
f
g
h
h1

superfluous because for the nominotypical subspecies the subspecies name is the same as for the species.

The identifications by Reuter (1961), Jerka-Dziadosz (1963), Lundin & West (1963), and Chorik (1968) are beyond reasonable doubt, although the descriptions are partially not very detailed, especially those by Lundin & West (1963).

Borror (1972) put five species into the synonymy of *U. grandis*. Beside *Oxytricha fusca*, which was already synonymised by Stein (1859), *Urostyla trichogaster* Stokes, *U. elongata* Stokes, *U. fulva* Stokes, and *U. limboonkengi* Wang & Nie are listed. The synonymy of Stokes' species is beyond reasonable doubt. By contrast, *Urostyla limboonkengi* should not be identified with *U. grandis* because it has only two macronuclear nodules (vs. many) and only three frontal cirri (vs. many). In the present book it is considered as distinct species although its systematic position is not yet clear. Consequently, *Urostyla limboonkengi* is preliminarily classified as incertae sedis in *Urostyla*.

Jerka-Dziadosz provided the first protargol preparations of *Urostyla grandis* clearly showing that *U. grandis* has midventral cirri (Jerka-Dziadosz 1972, Plates VII–X). Jankowski (1979) established the new genus and species *Metaurostyla polonica* (see also Curds et al. 1983, p. 397) for this population. However, the identification by Jerka-Dziasdosz's was never doubted by other authors, so that Jankowski's species is a junior synonym of *U. grandis*. By contrast, the illustration by Pätsch (1974) – also based on protargol preparations – does not show the midventral complex, and moreover, she described three fine caudal cirri which are difficult to recognise. Both observations by Pätsch do not fit authoritative redescriptions (Song & Wilbert 1989, Ganner 1991, Foissner et al. 1991), indicating that Pätsch interpreted the slides inaccurately.

The specimens of the population from a slightly saline (2‰) bay, described by Jones (1974, Fig. 207m), are only 140–200 μm long. However, the cirral pattern indicates that the identification is correct.

Hemberger (1982) correctly put Stokes' species *U. trichogaster*, *U. elongata*, and *U. fulva* into the synonymy of *U. grandis*. Moreover, he synonymised *U. naumanni* Lepsi with the present species. However, Lepsi's species is marine and has a moniliform macronucleus composed of 6–8, nodules whereas *U. grandis* lives in limnetic habitats and has many small nodules dispersed throughout the cell. In the present book, *Urostyla naumanni* is classified as incertae sedis in *Urostyla*. Hemberger provisionally assigned *Hemicycliostyla lacustris* Gellért & Tamás, 1958 to *U. grandis*. However, this species lacks transverse cirri and therefore should not be synonymised with *U. grandis*, which has a distinct row of such cirri. Consequently, I accept the original generic assignment by Gellért & Tamás.

Borror & Wicklow (1983) synonymised seven species (*Oxytricha fusca*, *Urostyla trichogaster*, *Hemicycliostyla sphagni*, *H. trichota*, *Urostyla caudata*, *U. gigas*, *U. muscorum*) with *U. grandis*. I accept only *O. fusca* and *U. trichogaster* as synonyms, and consider the other species as valid (for generic assignment of these species in the present book, see the index). Borror & Wicklow obviously ignored the presence/absence of transverse cirri not only for the characterisation of higher taxa, but also as species character.

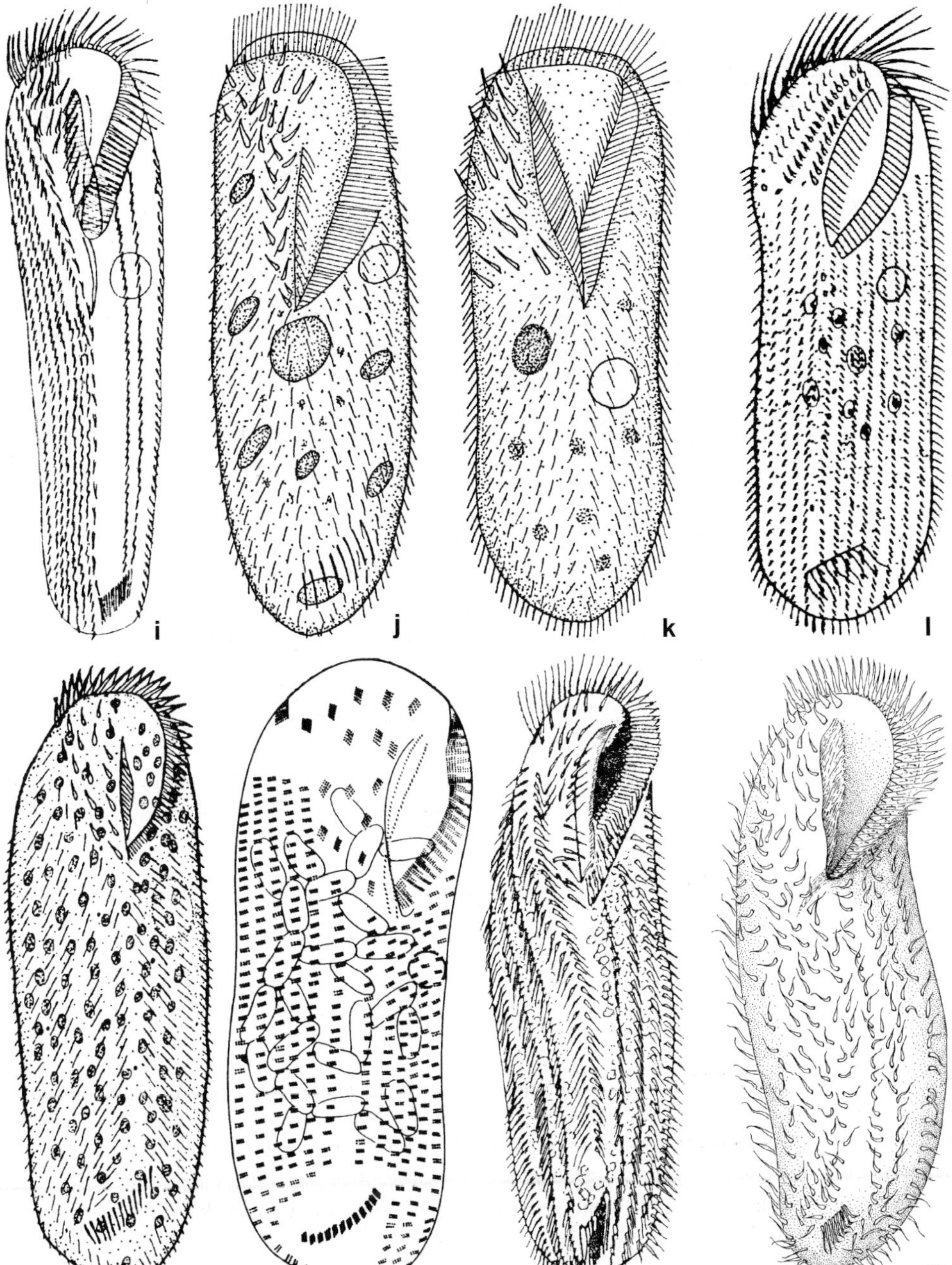

Fig. 207i–p *Urostyla grandis* (i, from Reuter 1961; j, k, from Lundin & West 1963; l, from Chorik 1968; m, from Jones 1974; n, from Pätsch 1974; o, from Borror & Wicklow 1983; p, from Tuffrau 1987. i–m, from life; n, o?, p?, protargol impregnation). Ventral views, i = about 250 μm, j, k, p = size not indicated, l = 415 μm, m = 165 μm, n = 356 μm, o = 200 μm. Page 1048.

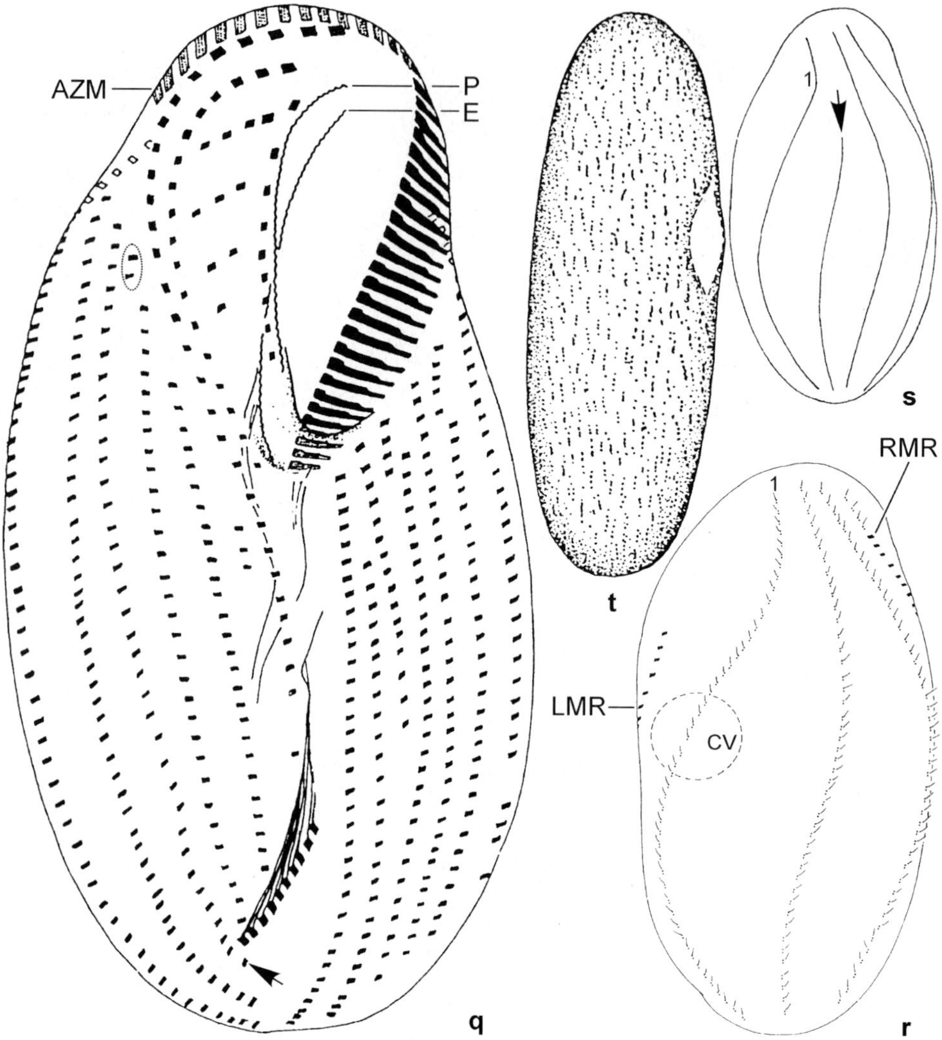

Fig. 207q–t *Urostyla grandis* (from Song & Wilbert 1989. q–s, protargol impregnation; t, from life). **q:** Infraciliature of ventral side of a representative specimen, 292 µm. Arrow in (q) denotes innermost right marginal row. The two circled cirri look like frontoterminal cirri. However, ontogenetic data show that *Urostyla grandis* does not have such a cirri-group, indicating that these two cirri are an irregularity of this specimen. **r, s:** Infraciliature of dorsal side of a specimen with three dorsal kineties (ordinary pattern) and a specimen with four kineties (rare; note that kinety 2 [arrow] is distinctly shortened anteriorly). **t:** Dorsal view showing cortical granulation and contractile vacuole, 434 µm. AZM = distal end of adoral zone of membranelles, CV = contractile vacuole, E = endoral, LMR = outermost left marginal row, P = paroral, RMR = outermost right marginal row, 1 = dorsal kinety 1. Page 1048.

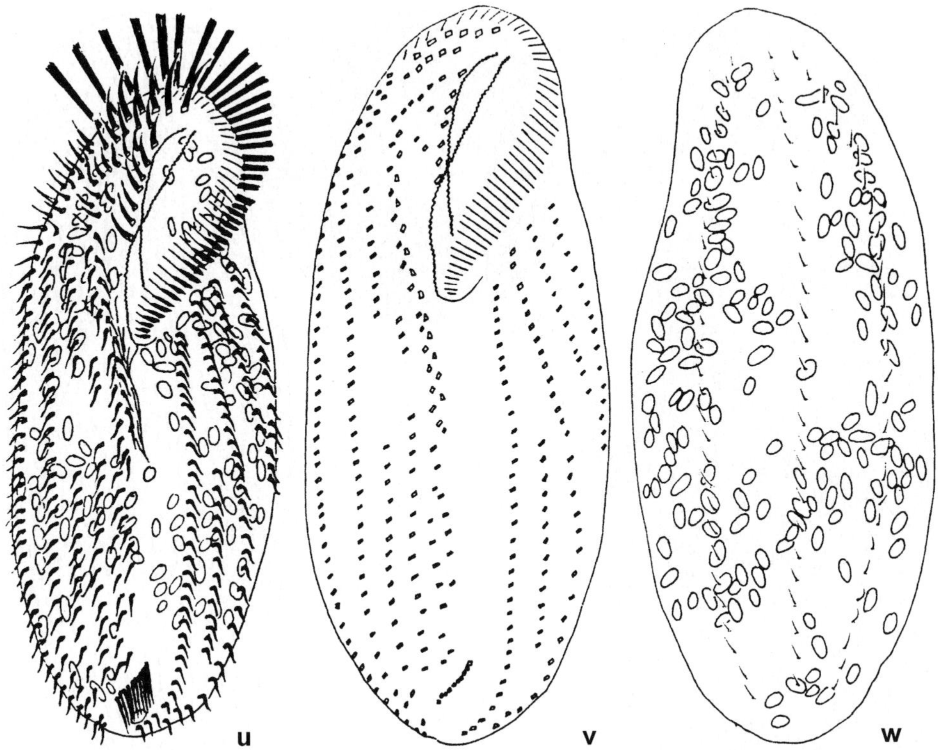

Fig. 207u–w *Urostyla grandis* (from Shin 1994. u, from life; v, w, protargol impregnation). Ventral view and infraciliature of ventral and dorsal side, u–w = 187 μm. Page 1048.

Song & Wilbert (1989) provided the first detailed morphometric characterisation of *Urostyla grandis* and an exact illustration of the ventral cirral pattern and the dorsal kinety arrangement. Their data were used by Foissner et al. (1991) in their review on saprobic ciliates to define *Urostyla grandis* morphologically.

Ganner (1991) made detailed studies on the morphology and cell division. One main result of this thesis is the proof that *U. grandis* lacks frontoterminal cirri, a feature which has to be interpreted as apomorphy in Fig. 199a. However, since the other species classified in *Urostyla* are not yet described after silver preparations we do not know at which level this feature is an apomorphy.

Shin (1994) redescribed a Korean population, mainly after protargol impregnation. His data agree rather well with that of other populations, so that the identification is beyond reasonable doubt although he did not mention the conspicuous cortical granules.

The *Urostyla grandis* population used by Croft et al. (2003) for molecular analyses was neither described morphologically nor identified by a systematist. However, its placement in the molecular trees together with some other urostyloids indicates that the identification is correct.

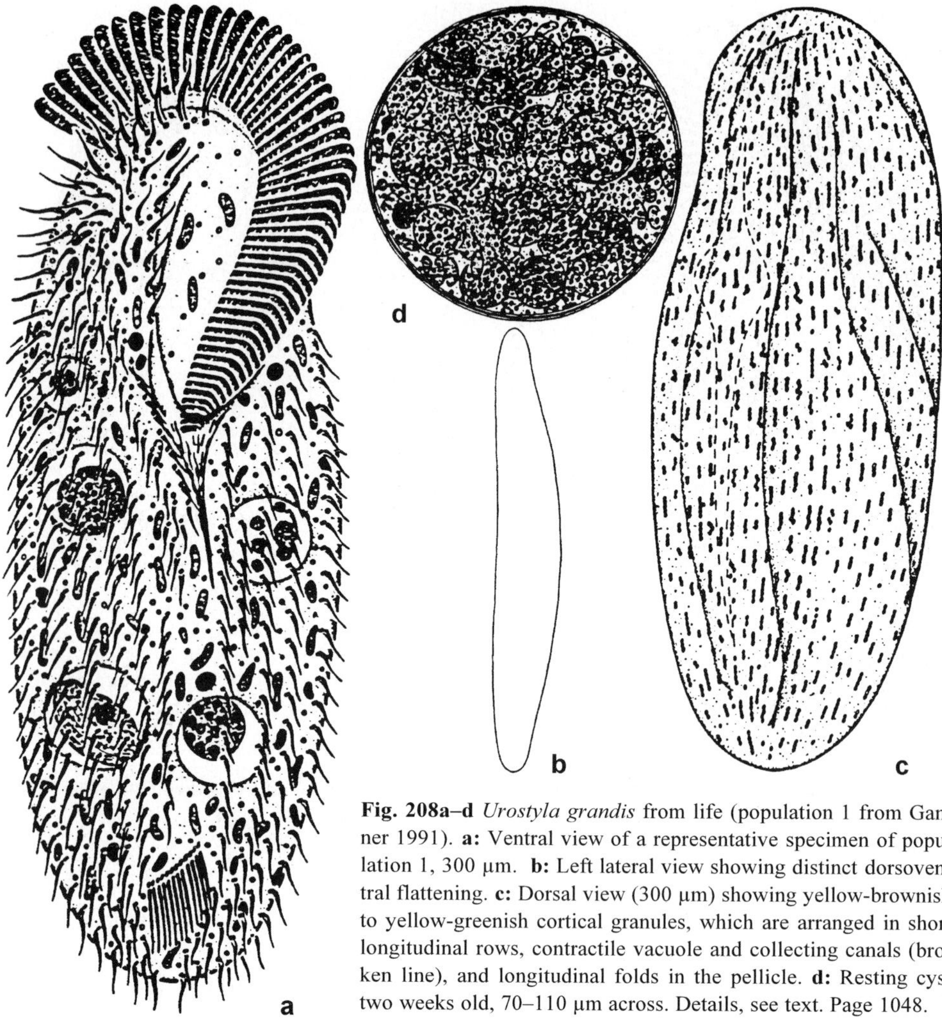

Fig. 208a–d *Urostyla grandis* from life (population 1 from Ganner 1991). **a:** Ventral view of a representative specimen of population 1, 300 µm. **b:** Left lateral view showing distinct dorsoventral flattening. **c:** Dorsal view (300 µm) showing yellow-brownish to yellow-greenish cortical granules, which are arranged in short longitudinal rows, contractile vacuole and collecting canals (broken line), and longitudinal folds in the pellicle. **d:** Resting cyst two weeks old, 70–110 µm across. Details, see text. Page 1048.

A new synonym of *U. grandis* is very likely *U. chlorelligera* Foissner, 1980 (see below). *Urostyla grandis* sensu Edmondson (1906) is possibly a *Paraurostyla weissei* (see insufficient redescriptions; Fig. 164g). *Urostyla grandis* sensu Woodruff (1921, her Fig. 3, 4) is certainly a misidentification because this population has two large macronuclear nodules. *Urostyla grandis* sensu Wiackowski (1988, p. 6, 7) lacks cortical granules (mucocysts; his feature 26), which makes the identification very uncertain because the cortical granules are mentioned in all detailed descriptions of *U. grandis*. Consequently, Wiackowski's data are not considered further.

Morphology: There are many morphological data available on *U. grandis*. The most detailed, modern description was provided by Ganner (1991, Fig. 208a–g, Table 42). Thus, his populations are described first, followed by supplementary and deviating

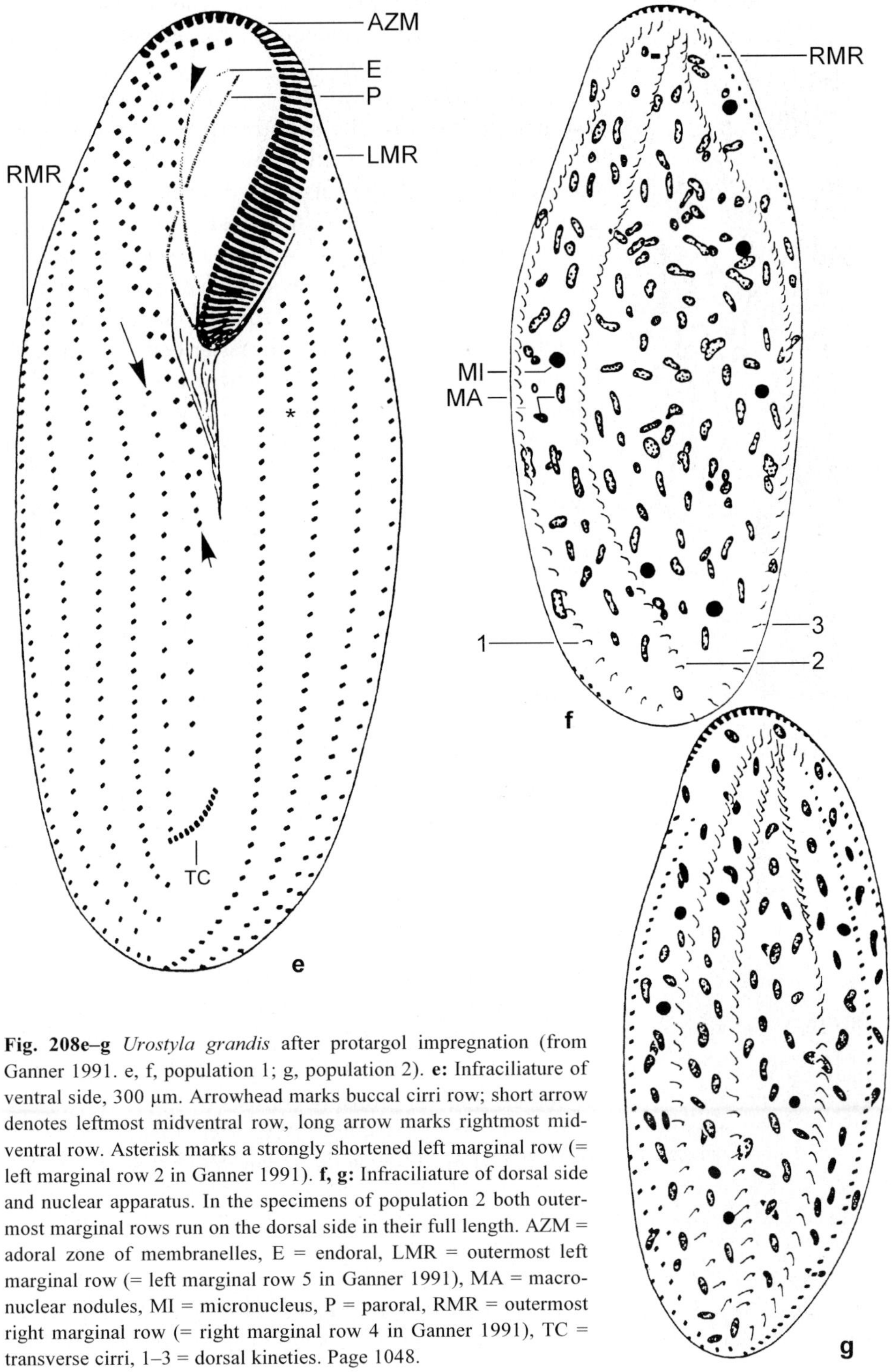

Fig. 208e–g *Urostyla grandis* after protargol impregnation (from Ganner 1991. e, f, population 1; g, population 2). **e:** Infraciliature of ventral side, 300 μm. Arrowhead marks buccal cirri row; short arrow denotes leftmost midventral row, long arrow marks rightmost midventral row. Asterisk marks a strongly shortened left marginal row (= left marginal row 2 in Ganner 1991). **f, g:** Infraciliature of dorsal side and nuclear apparatus. In the specimens of population 2 both outermost marginal rows run on the dorsal side in their full length. AZM = adoral zone of membranelles, E = endoral, LMR = outermost left marginal row (= left marginal row 5 in Ganner 1991), MA = macronuclear nodules, MI = micronucleus, P = paroral, RMR = outermost right marginal row (= right marginal row 4 in Ganner 1991), TC = transverse cirri, 1–3 = dorsal kineties. Page 1048.

data from other sources. Since the synonymy in the list above is beyond reasonable doubt, the individual original descriptions are not given.

Ganner (1991) studied two populations, one from the river Oichten near the city of Salzburg, and the other from the Salzach River in the city of Salzburg, Austria. Unless otherwise indicated, the data refer to the Oichten population. Body size 246–370 × 86–123 μm in life, on average 300 (n = 5) × 103 (n = 4) μm; Salzach specimens 150–300 μm long in life. Body outline elliptical, left margin convex, right convex to straight, sometimes even slightly concave. Anterior and posterior body end broadly rounded. Body dorsoventrally flattened about 3:1 (Fig. 208a–c). Pellicle soft, very flexible; body slightly contractile. About 100–150 macronuclear nodules scattered throughout cytoplasm; nodules 6.5–10.5 × 2.6–3.9 μm, ellipsoidal, bean-shaped, or pear-shaped. Micronuclei globular to slightly ellipsoidal, about 5 μm across in life. Contractile vacuole near left body margin, distinctly ahead of mid-body, with anterior and posterior collecting canal extending to near front and rear cell end; vacuole up to 25 μm across. Cortical granules yellow-brownish to yellow-greenish, 1.0–1.5 μm across, globular to slightly ellipsoidal, arranged in about 25 longitudinal rows on dorsal and ventral side. Cytoplasm colourless, usually packed with colourless inclusions about 0.5–2.5 μm across. Food vacuoles about 15–30 μm across. Cytopyge subterminal. Movement slowly gliding; late dividers (division furrow clearly recognisable) do not move, although cirri beat heavily.

Adoral zone occupies about one third of body length, on average composed of 47–60 membranelles of ordinary fine structure. Buccal field wide and deep. Undulating membranes long and curved, optically intersecting about in mid-region; paroral likely composed of zigzagging basal bodies. Endoral probably consists of single row of basal bodies, each basal body likely with a parasomal sack attached. Cytopharynx about 50 μm long in life.

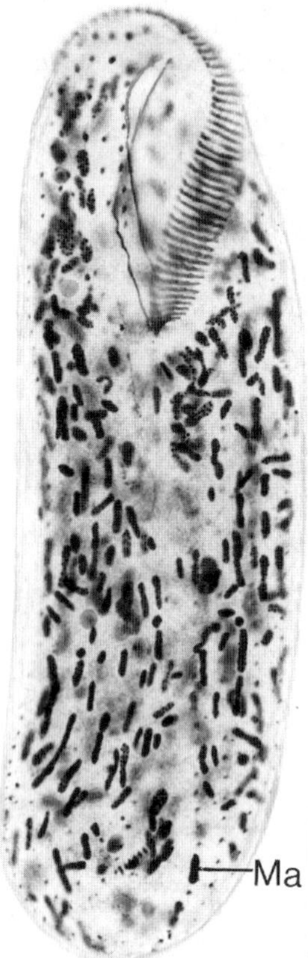

h

i

Fig. 208h, i *Urostyla grandis* (from Foissner et al. 1991. Protargol impregnation). **h:** Ventral view showing nuclear apparatus and infraciliature of anterior body portion. **i:** Arrangement of cortical granules which sometimes impregnate with protargol. Ma = macronuclear nodule. Page 1048.

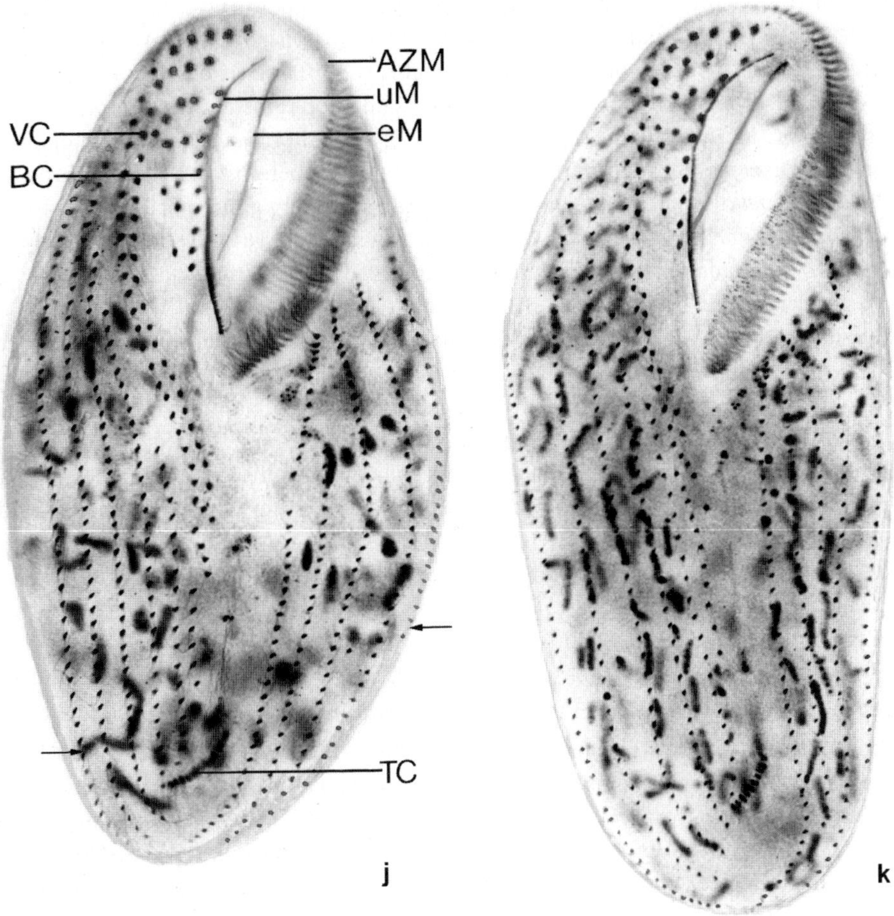

Fig. 208j, k *Urostyla grandis* (from Foissner et al. 1991. Protargol impregnation). Infraciliature of ventral side. Arrows in (j) mark the outermost marginal rows. Explanation of original labelling: AZM = adoral zone of membranelles, BC = buccal cirri, eM = endoral, TC = transverse cirri, uM = paroral, VC = anterior, zigzagging portion of midventral complex. Page 1048.

Cirral pattern and number of cirri of usual variability (Fig. 208e, Table 42). Frontal cirri arranged as shown in Fig. 208e. Pattern basically composed of a bicorona with slightly enlarged cirri and some smaller parabuccal cirri in area between buccal cirral row and anterior portion of midventral complex, producing a somewhat irregular multicorona (for detailed explanation of origin, see cell division, especially Fig. 211g). On average six, respectively, eight buccal cirri. Frontoterminal cirri lacking. Midventral complex not set off from frontal ciliature, composed of about 16[1] cirral pairs and 3–4

[1] The value is from the proter in Fig. 211g in that anlage VII is considered as first midventral pair because this is the anteriormost anlage which produces only two cirri (the remnant at the rear end is likely

midventral rows with number of cirri increasing from left to right. On average about eight (Salzach river population), respectively, 11 (Oichten population) transverse cirri arranged in slightly curved oblique, subterminal row so that cirri do not project beyond rear body end. Specimen shown in Fig. 208e with four right marginal rows. 4–7 left marginal rows; usually the second or third row is very short and can also be displaced posteriorly. Distance between midventral complex and innermost left marginal row distinctly wider than distance between other cirral rows forming a more or less conspicuous postoral field (Fig. 208l). Outermost marginal rows either partially or completely displaced on dorsal side (Fig. 208f, g). Cirrus right of leftmost frontal cirrus slightly displaced posteriad.

Dorsal bristles about 4 µm long in life, arranged in three bipolar kineties; rarely four kineties occur. Caudal cirri lacking (Fig. 208f).

As already mentioned, Ganner (1991) studied two populations which differ rather distinctly in body size after protargol impregnation (Table 42). This difference is only partly due to different fixation procedures, as indicated by the smaller values of many morphometric data. The values from population 1 agree rather well with those provided by Song & Wilbert (1989).

Important additional and deviating data from other sources (see also Figures of individual redescriptions and Table 42): Body length/size 170–260 µm (Ehrenberg 1838); 170–230 µm (*Bursaria vorax*; Ehrenberg 1838); 150–300 µm (Perty 1852); 300–500 × 120–180 µm (Roux 1901); 176 µm (Wang 1925); 300–400 µm (Kahl 1932; Jerka-Dziadosz 1963; Pätsch 1974); 140–300 µm (Tittler 1935); 300–500 µm (Šrámek-Hušek 1952a, b); about 250 µm (Reuter 1961); 200–300 × 100–120 µm (original data from a population collected from a brook in Großgmain, Salzburg), 200–250 × 80–100 µm (Salzach river, Salzburg; own observations). Synonym *U. trichogaster* 250–340 µm long, *U. elongata* about 300 µm, and *U. fulva* about 254 µm (Stokes 1885, 1891). More than 100 macronuclear nodules and 6–8 micronuclei (Kahl 1932); macronuclear nodules up to 15 × 5 µm in life (own observations, population from the Garstnerbach, Upper Austria); the many macronuclear nodules are all structurally and functionally the same (Prescott 1989, p. 18); for some further data on the nuclear apparatus, including historic information, see cell division. A brief, unimportant note about the systole/diastole of contractile vacuole is given by Lieberkühn (1856, p. 33). Cortical granules very conspicuous because yellow-green or yellow-brown, about 1 µm long, slightly ellipsoidal, and arranged in many short, longitudinal rows so that cells are yellowish (Foissner et al. 1991), which was already mentioned by Ehrenberg (1838); cortical granules sometimes citrine (Foissner, pers. comm.), clearly recognisable even at a magnification of ×200 (Fig. 208q, r). According to Stein (1859) the presence or absence of the cortical granules (Oelbläschen or Oeltröpfchen in his terminology) depends on the kind of food. According to my experience, *Urostyla grandis* never lacks the cortical granules. Cytoplasm packed with mitochondria (Fauré-Fremiet 1910a, p. 515). Moves left spiralling, like many other ciliates (Bullington 1925, p. 271; for review, see Seravin 1970).

---

resorbed in later dividers), and the anlage XXII because this is the rearmost anlage which produces only three cirri (one midventral pair and a transverse cirrus).

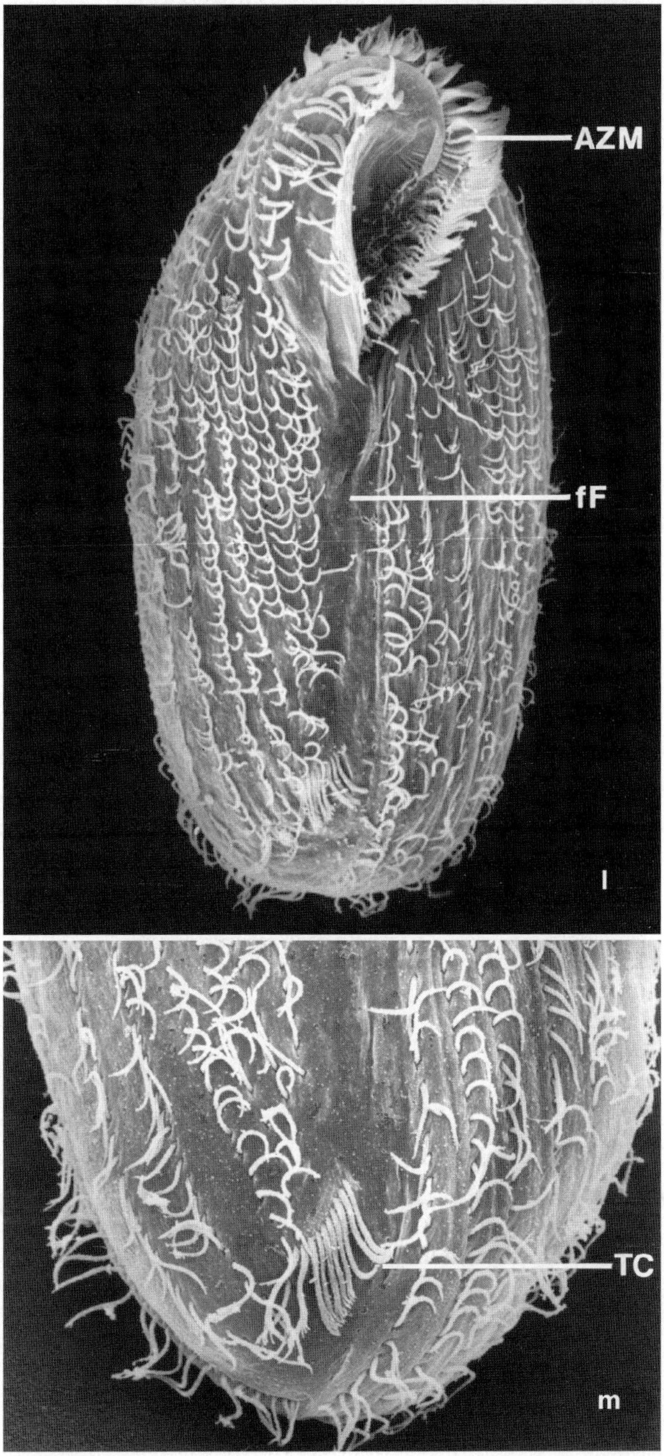

Fig. 208l, m *Urostyla grandis* (from Foissner et al. 1991. Scanning electron micrographs). **l:** Ventral view showing cirral pattern and oral apparatus. **m:** Ventral view of posterior body portion showing, inter alia, transverse cirri which are distinctly displaced anteriad and therefore do not project beyond the rear body end. Explanation of original labelling: AZM = adoral zone of membranelles, fF = postoral area which is free of cirri, TC = transverse cirri. Page 1048.

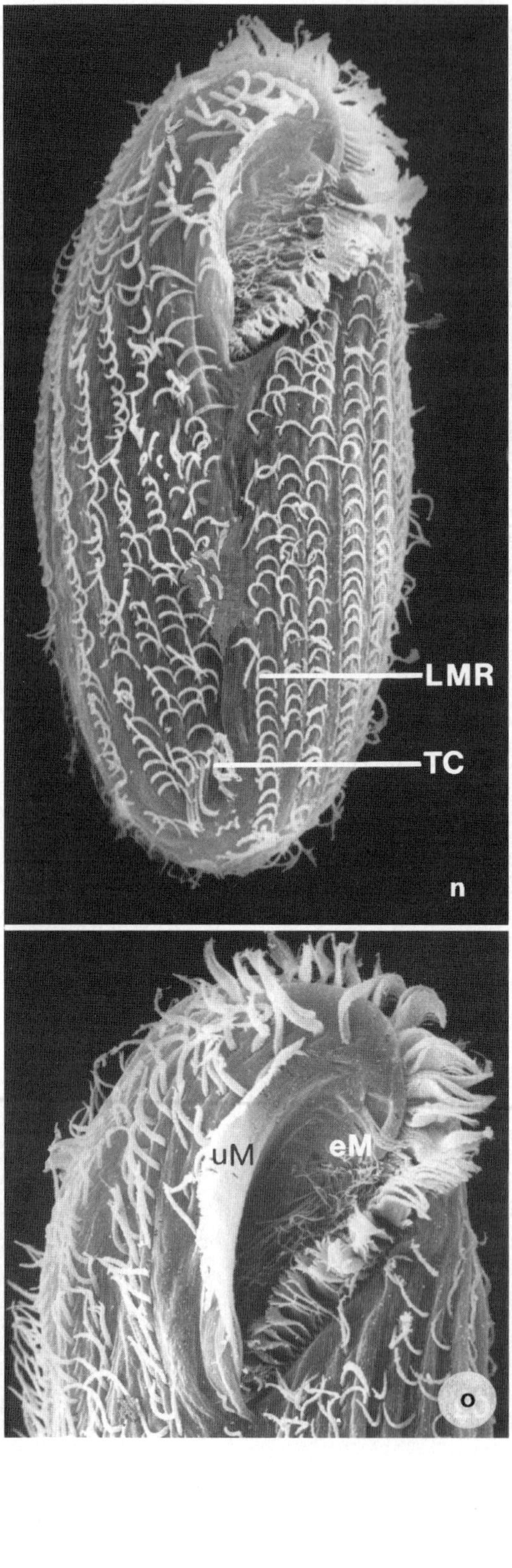

Kowalewskiego (1882, Planche XXIX, Fig. 1, 1a) studied some details of the oral apparatus. Total number of cirral rows (including marginal rows; individual values from life observations must not be over-interpreted because it is rather difficult to count the rows exactly): 8 (*U. elongata* and *U. fulva*; Stokes 1891); 11–12 (Kahl 1932, Šrámek-Hušek 1952a, b; 4–5 rows on frontal area not included); 12 (Jerka-Dziadosz 1963; Pätsch 1974); 10 to 14 (Jerka-Dziadosz 1972). Number of transverse cirri 8–10 (*U. elongata*; Stokes 1891); 5–6 (*U. fulva*; Stokes 1891); 10–12 (*U. trichogaster*; Stokes 1885); 10–12 (Roux 1901; Wang 1925, see remarks; Tittler 1935); 5–12 (Conn 1905), 10–20 (Kahl 1932), 12–16 (Jerka-Dziadosz 1963, 1972); 12 (Pätsch 1974; from Fig. 207n); transverse cirri about 20 μm long in life (own observations).

Cell division: This part of the life cycle was studied several times in more or less detail, namely, by Stein (1859; Fig. 209a–d), Jerka-Dziadosz (1963, Fig. 210a–f; 1972, 7a–t), and by Ganner (1991, Fig. 211a–g). For review and minor comments, see Doroszewski & Raabe (1966, p. 131) and Wiackowski (1984a). The following description is based mainly on

Fig. 208n, o *Urostyla grandis* (from Foissner et al. 1991. Scanning electron micrographs). Ventral views of total cell and anterior body portion. Explanation of original labelling: eM = endoral, LMR = innermost left marginal row, TC = transverse cirri, uM = paroral. Page 1048.

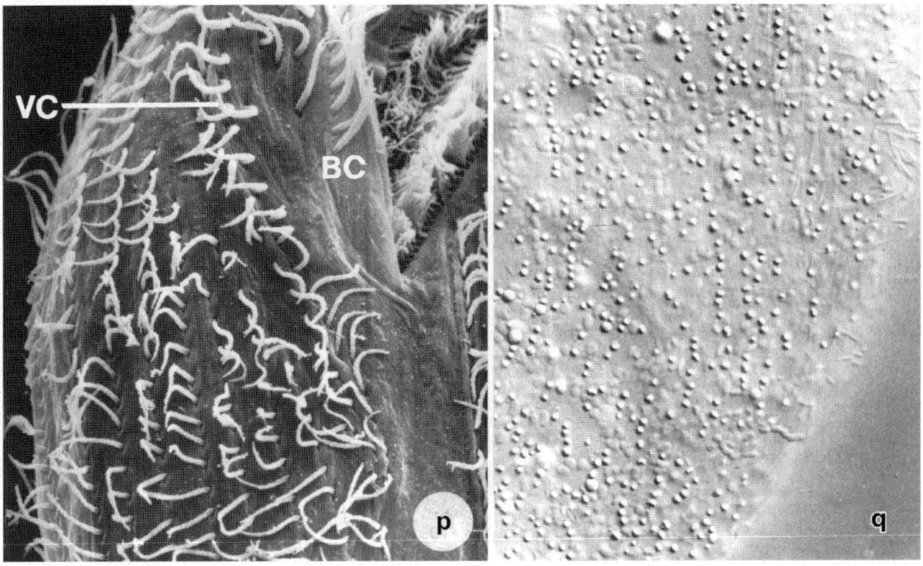

Fig. 208p, q *Urostyla grandis* (from Foissner et al. 1991. p, scanning electron micrograph; q, interference contrast micrograph). **p:** Ventral view showing mainly the midventral complex which is composed of midventral pairs and midventral rows in *U. grandis*. **q:** The cortical granules are about 1 μm in size, yellowish, and arranged in longitudinal rows (see also Fig. 208r). Explanation of original labelling: BC = buccal cirral row, VC = zigzagging (= midventral pair) portion of midventral complex. Page 1048.

Ganner's data, who studied and illustrated division in great detail and simultaneously discussed some differences to the results by Jerka-Dziadosz (1972).

Stomatogenesis of proter: It commences with a successive resorption of basal bodies of the endoral from posterior to anterior and of the posterior adoral membranelles from right to left (Fig. 211a). The resorption of basal bodies of the endoral and the rear adoral membranelles continues. The anterior portion of the endoral modifies to a small basal body field which enlarges by proliferation (Fig. 211b, c). By contrast, Jerka-Dziadosz (1972) described a basal body field near the proximal end of the endoral (Fig. 211b). Next, the basal bodies of the paroral are disorganised, forming a narrow, longitudinal basal body field. Later this field fuses with the anlage formed by the endoral to form the oral primordium of the proter (Fig. 211c, d). According to Jerka-Dziadosz (1972) the primordium originating from the parental buccal cirri also contributes to the oral primordium of the proter (Fig. 7c, d). At the left margin of the oral primordium the new endoral is formed (Fig. 211d–f). The formation of new adoral membranelles is confined to the proximal and middle third of the parental adoral zone; the membranelles of the anterior third are not reorganised (Fig. 211a–g). The membranelles of the rear third are reorganised under contribution of the oral primordium, while those of the middle third are reorganised without contribution of the oral primordium.

Stomatogenesis of opisthe: The formation of the oral primordium of the opisthe commences at three sites, namely (i) left of the rear half of parental midventral pairs

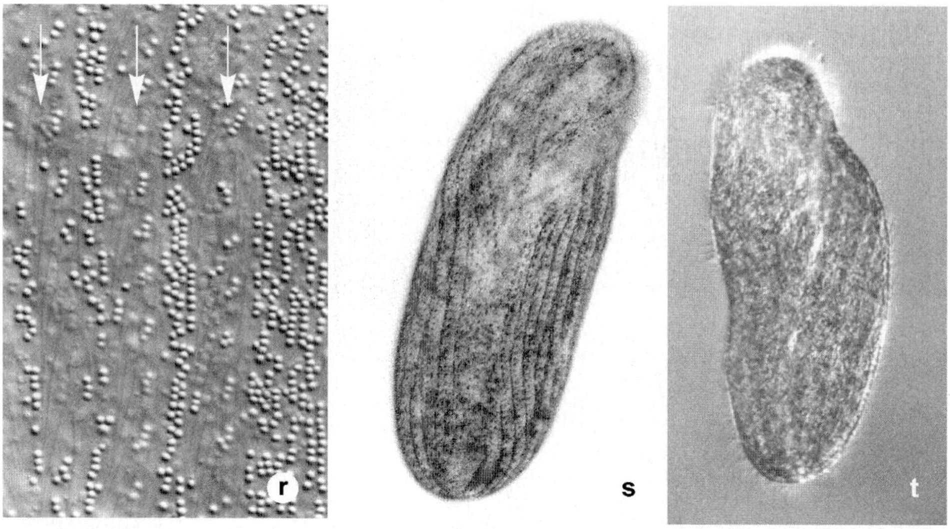

Fig. 208r–t *Urostyla grandis* (originals of a population from a brook [Hainbach] in Upper Austria. Interference contrast micrographs). **r:** The cortical granules (about 1 μm across, yellowish) are arranged mainly between the cirral rows (arrows). **s, t:** Freely motile specimen when moving forwards and backwards. Page 1048.

(Fig. 211a, short arrows); (ii) left of the leftmost midventral row; and (iii) immediately ahead of the leftmost transverse cirrus (Fig. 211b). The cirri close to these primordia are not changed in these early stages; perhaps one or two transverse cirri are modified (Fig. 211c). According to Jerka-Dziadosz (1972) the oral primordium of the opisthe originates close behind the parental adoral zone, and parental cirri are only incorporated later (Fig. 7b–e).

Origin of cirral primordia: The frontal-midventral-transverse cirri primordia of the proter originate from the buccal cirri and some parabuccal cirri (Fig. 211c–f). The rear primordia originate from midventral pairs. In the opisthe the primordia are formed right of the oral primordium. The anterior primordia originate from basal bodies of the oral primordium, the middle one originate from the left cirri of the midventral pairs, and the rear primordia are formed from cirri of the left midventral rows (Fig. 211d–f).

Differentiation of new cirri: The following description refers to the specimen shown in Fig. 211g. Anlage I produces the left frontal cirrus (= cirrus I/1); anlagen II–VI form the frontal cirri (two bows of slightly enlarged cirri), the buccal cirri (from anlage II), and the parabuccal cirri (from anlagen III–VI). The anlagen VII–XIII each produce a midventral pair. Anlagen XIV–XXII form each three cirri, that is, a midventral pair and a transverse cirrus (anlage XV does not form a transverse cirrus in this specimen).

Fig. 209a–e *Urostyla grandis* (a–d, from Stein 1859; e, from Berg 1889. a–d, from life; e, nucleus stain). → Late and very late dividers and a postdivider (d) in ventral and dorsal (c) view. The many macronuclear nodules fuse to a single mass which is the plesiomorphic type of macronuclear division. Page 1048.

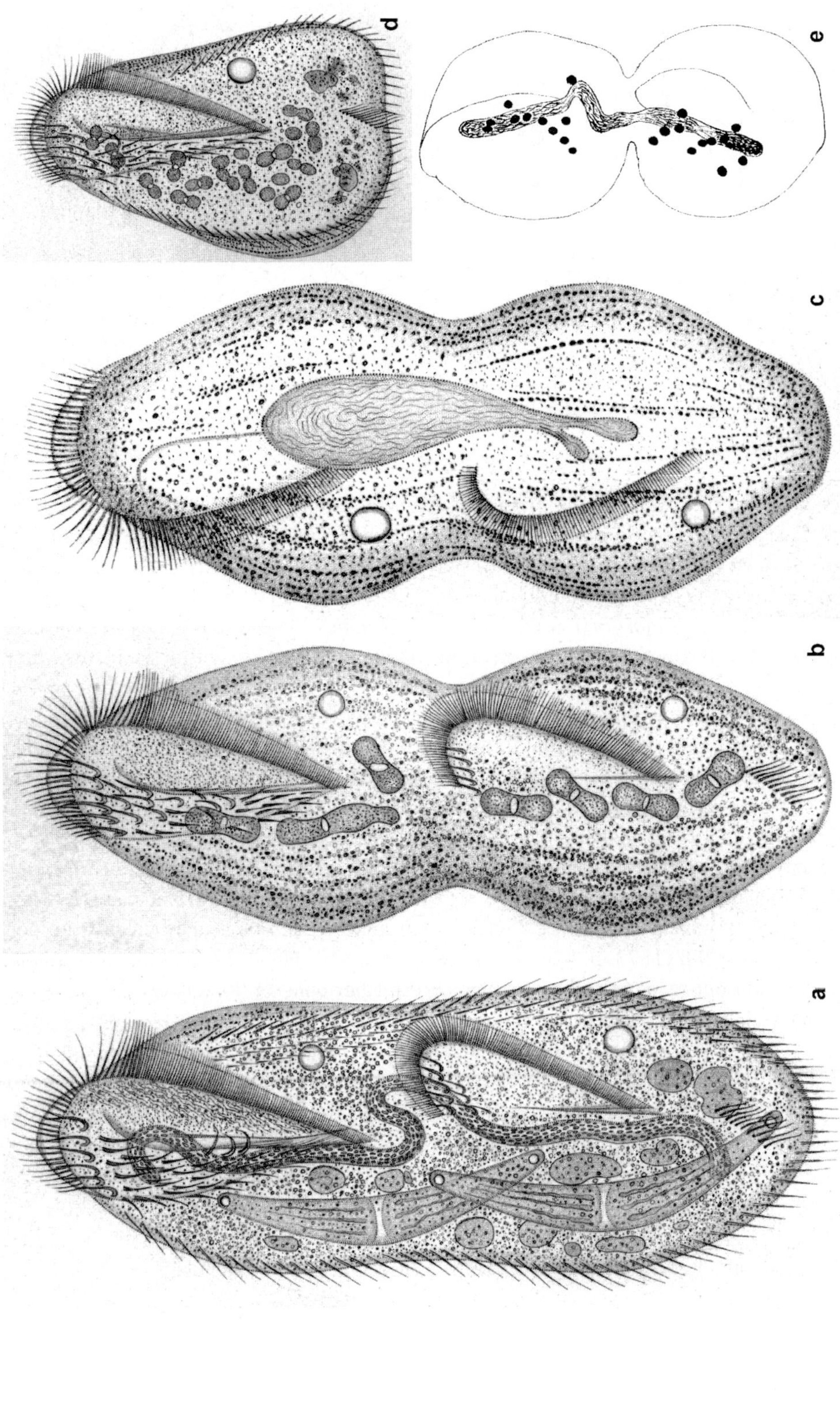

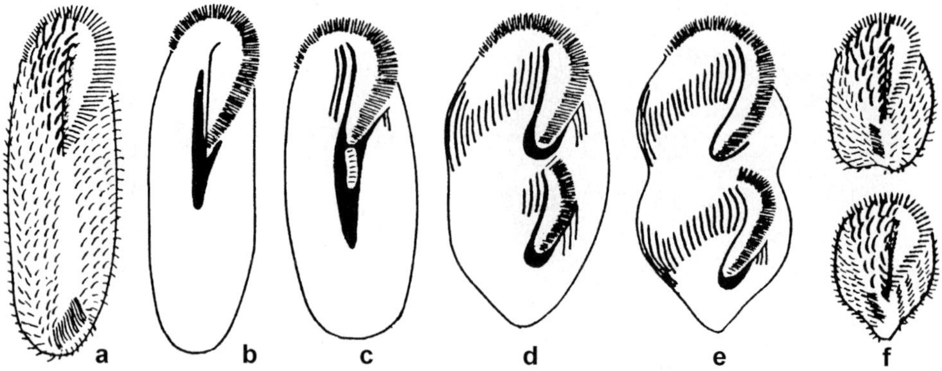

Fig. 210a–f *Urostyla grandis* (from Jerka-Dziadosz 1963. Iron haematoxylin staining). Schematic representation of cell division. Note that these figures do not show the process as exactly as those by Ganner (1991). Details on cell division see text. Page 1048.

The midventral rows and the rightmost transverse cirri originate from the anlagen XXIII–XVI. The anteriormost cirri of the rightmost anlage do not set off anteriorly, showing that *Urostyla grandis* lacks frontoterminal cirri. The short left marginal rows originate from the innermost marginal cirral anlage (Fig. 211f).

Development of marginal rows and dorsal kineties: The formation of the marginal rows and dorsal kineties proceeds in ordinary manner, that is, within each row/kinety two primordia occur (Fig. 211e–g).

Division of nuclear apparatus: Urostyla grandis has many macronuclear nodules dispersed through the cell. At first a replication band passes through each macronuclear nodule (Tittler 1935, Fig. 2a–g). Prior to division the nodules fuse to a single mass, and later divide into the high number of nodules (Fig. 3a–l, 209e). The micronuclei divide in ordinary manner (Fig. 4a–h).

Stein (1859) redescribed the non-dividers of *U. grandis* without nucleus. Balbiani (1861, Planche VIII, Fig. 17A–E; see also Balbiani 1862) found that it is dispersed in many nodules which fuse during division. The many macronuclear nodules were confirmed by Bütschli (1873, p. 670, Tafel XXVI, Fig. 15). He also found the micronuclei, however, did not recognise them as such. For further data on the nuclear apparatus, including division, see general section and Bergh (1889; Fig. 209e; for review, see Belař 1926, p. 134), Raabe (1946, 1947; for review, see Sonneborn 1949, p. 57), Inaba & Suganuma (1966), Sugnanuma & Inaba (1966, 1967), and Prescott (1994).

Cyst: *Urostyla grandis* forms resting cysts. This stage of the life cycle was first described by Stein (1859, Fig. 206c, d). According to Ganner (1991), cysts are globular and about 70–112 μm across in life (mean = 85.4 μm; n = 20). The wall is composed of a highly refractive middle layer and a less refractive inner and outer layer. About two weeks after encystation, the cyst does not show a special surface structure, that is, the wall is smooth. Yellowish cortical granules exclusively arranged close to cyst wall (Fig.

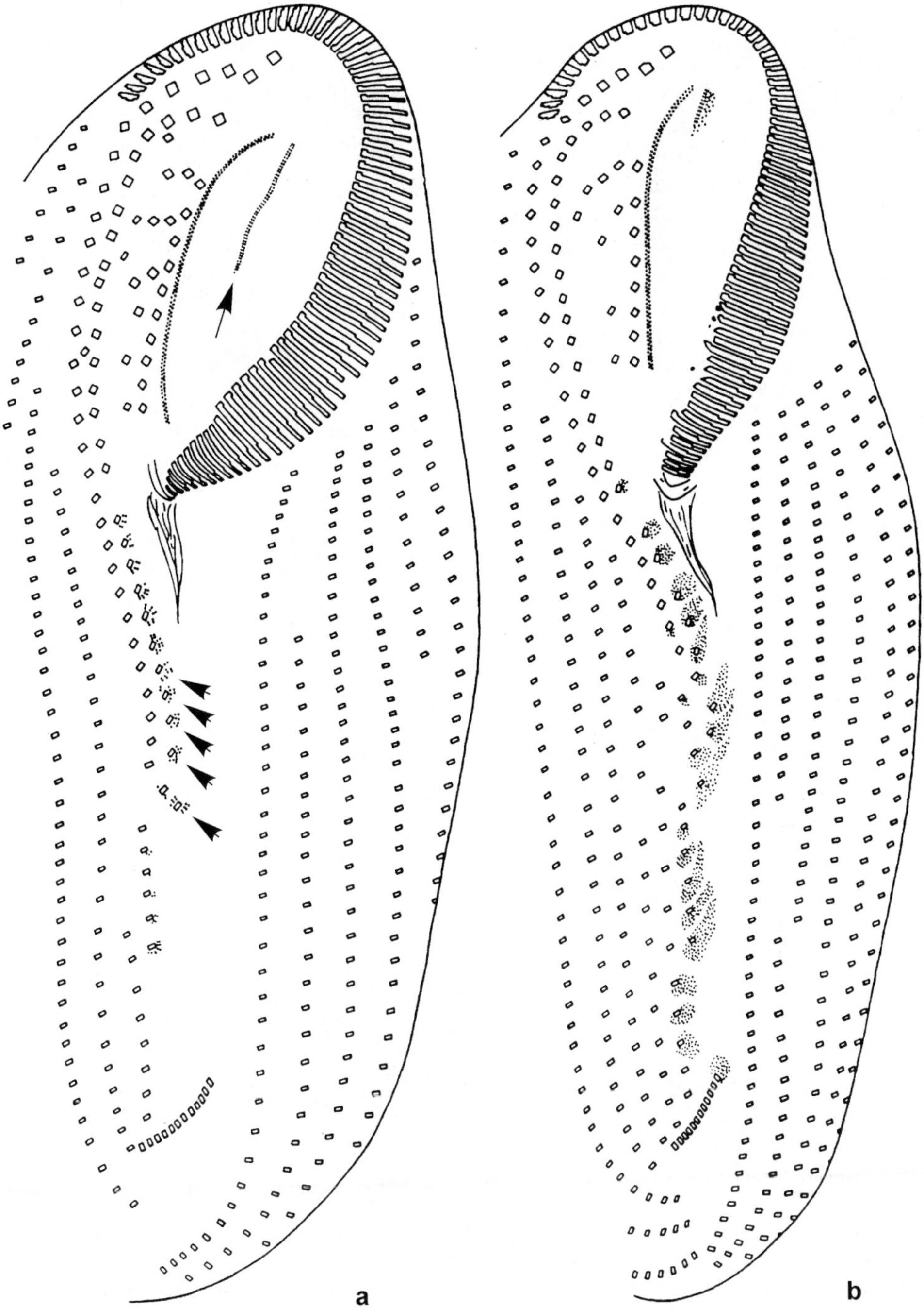

a b

Fig. 211a, b *Urostyla grandis* (population 1 from Ganner 1991. Protargol impregnation). Infraciliature of ventral side of very early dividers. The parental endoral is resorbed from rear to front (long arrow). The oral primordium of the opisthe is formed, inter alia, from basal bodies proliferating left of the midventral complex (short arrows). Page 1048.

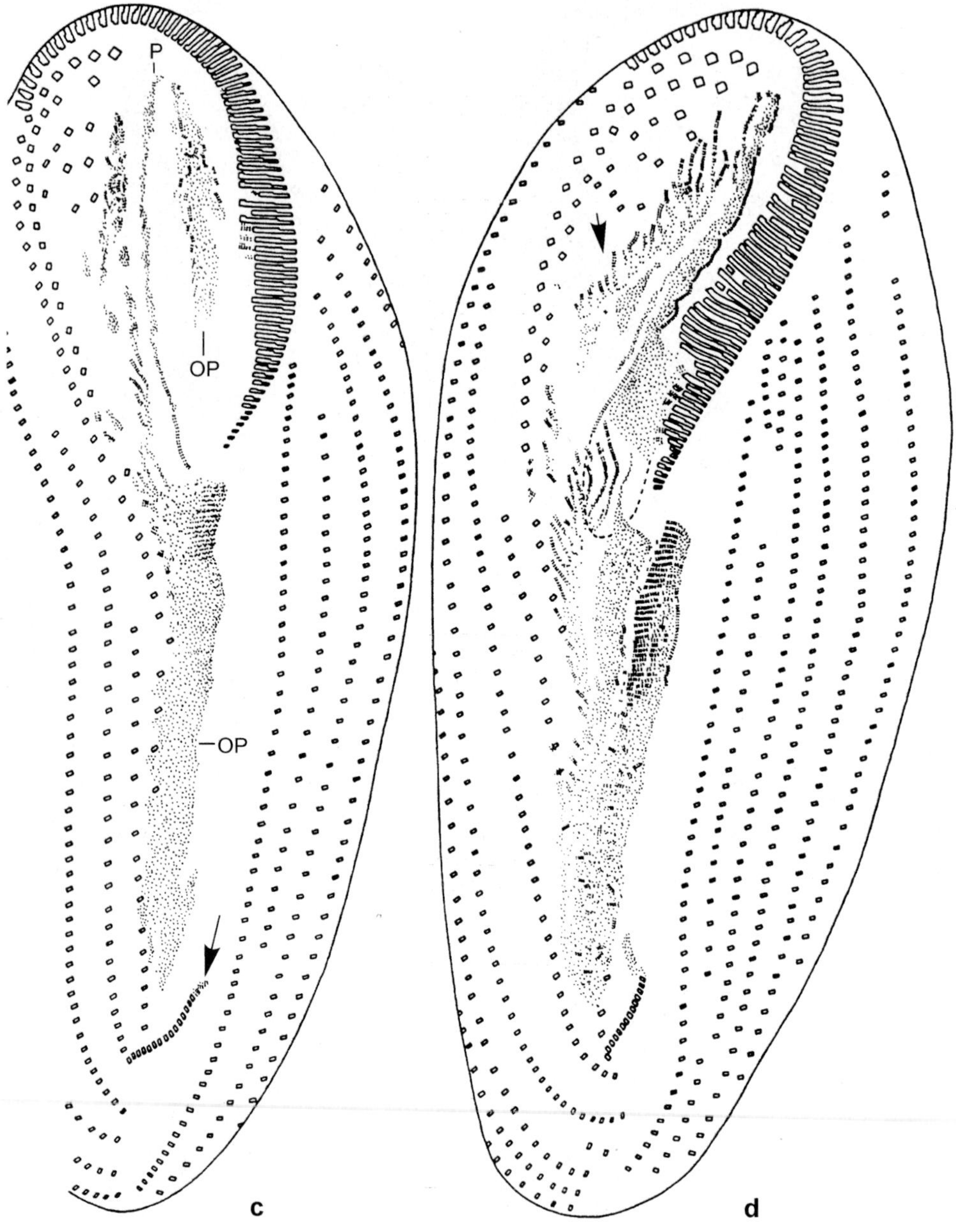

Fig. 211c, d *Urostyla grandis* (population 1 from Ganner 1991. Protargol impregnation). Infraciliature of ventral side of an early and middle divider. Arrow in (c) marks proliferating basal bodies close to the left end of the transverse cirral row. Arrow in (d) denotes the frontal-midventral-transverse cirri anlage field of the proter. OP = oral primordium of proter and opisthe, P = disorganised parental paroral which contributes to the formation of the oral primordium of the proter. Page 1048.

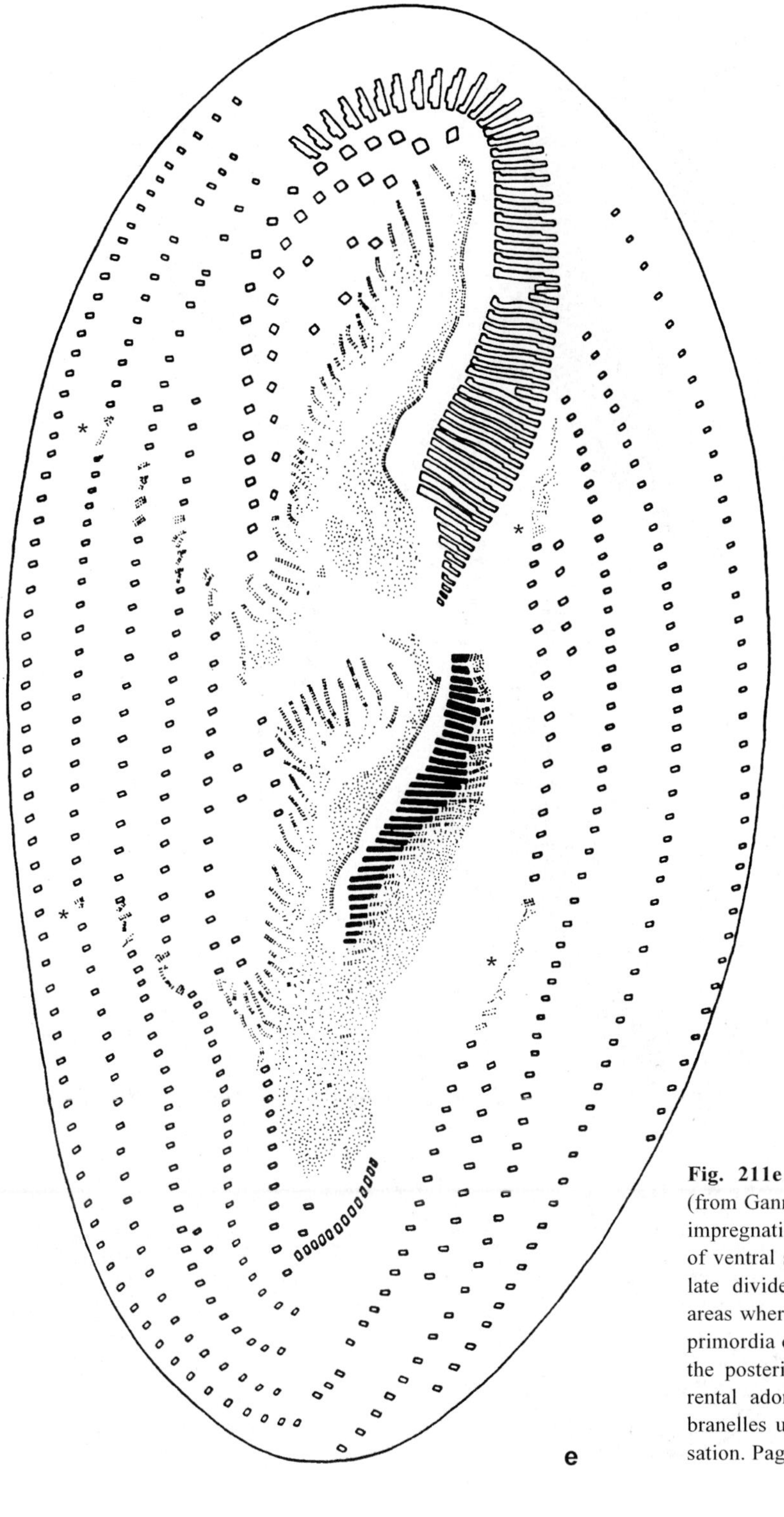

e

Fig. 211e *Urostyla grandis* (from Ganner 1991. Protargol impregnation). Infraciliature of ventral side of a middle to late divider. Asterisks mark areas where the marginal row primordia originate. Note that the posterior half of the parental adoral zone of membranelles undergoes reorganisation. Page 1048.

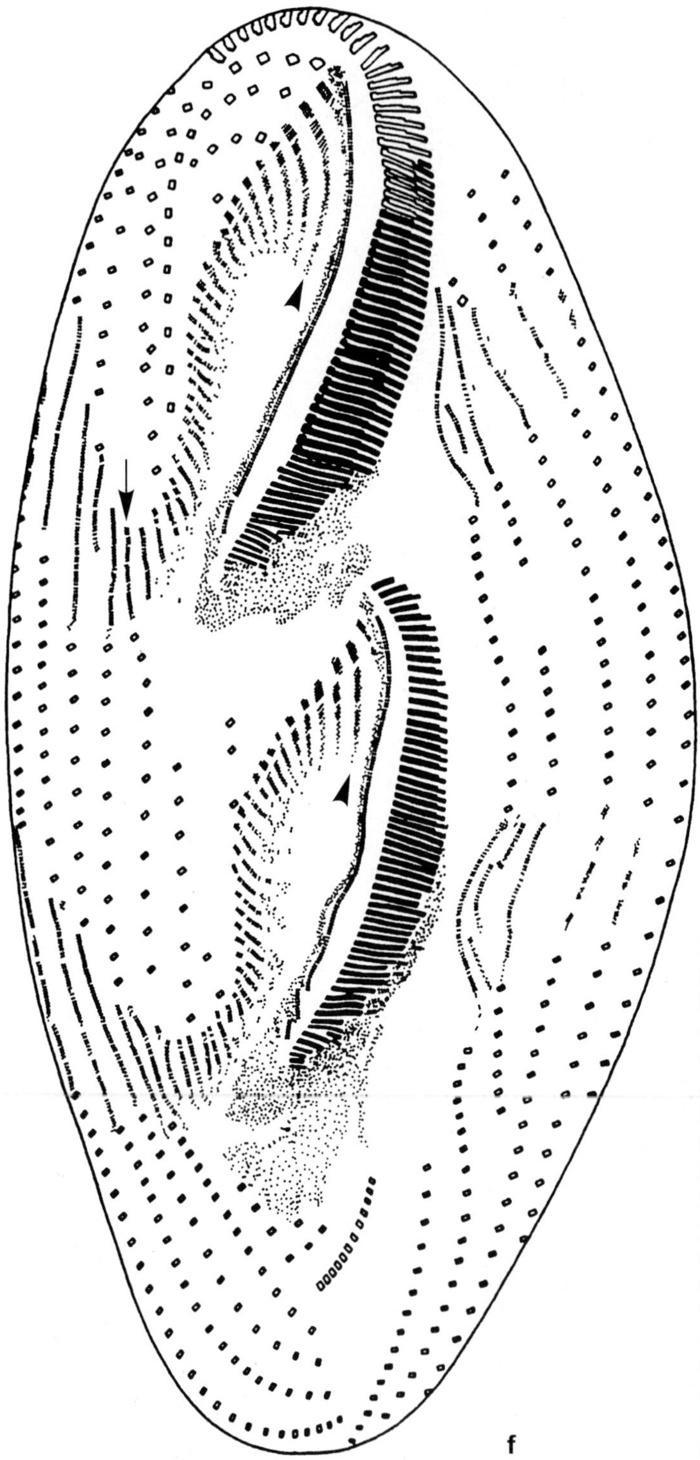

Fig. 211f *Urostyla grandis* (from Ganner 1991. Protargol impregnation). Infraciliature of ventral side of a late divider. Arrowheads denote the anlage II which produces, as is usual, the buccal cirri and a frontal cirrus. Arrow marks the primordium of the innermost right marginal of the proter; the anlage left of this streak is (very likely) the rightmost frontal-midventral-transverse cirral anlage. Note that each marginal row divides individually, whereas in *Pseudourostyla* all marginal rows per side originate from a common anlage. Page 1048.

208d). Cyst described by Fauré-Fremiet (1910a, p. 515) about 60 μm across after staining.

Rios et al. (1985, 1988) studied the ultrastructure and the chemical composition of *U. grandis* cysts which are, however, only 22–24 μm across (according to their Figure 1 about 37 μm), strongly indicating a misidentification because it is very unlikely that such a large species can make such small cysts. Unfortunately, Rios et al. (1985, 1988) did not provide data on the interphasic specimen, so that a re-identification is impossible. Consequently, the data should be used with great caution and therefore cannot serve as phylogenetic marker. However, their cysts contained many macronuclear nodules, indicating that they studied a urostyloid

According to Bussers & Jeuniaux (1974), the cyst of the synonym *U. trichogaster* does not contain chitin. Tittler (1935) found that several macronuclear nodules degenerate during encystment, so that the resting cyst has a lower number of macronuclear nodules than the vegetative cell. Further literature: Gutiérrez et al. (1998); Li & Gu (2005).

Ultrastructure: There are only few ultrastructural data available. Cilia of membranelles and cirri covered by a perilemma (Bardele 1981, p. 415; Preisig et al. 1994, p. 19). Wetzel (1925, p. 281; his Fig. F_1b) studied the fine structure of the buccal cavity. He described three undulating membranes using stained thin-sections, namely the praeoral, the paroral, and the endoral. His "praeoral" membrane is what we call paroral and his paroral is possibly the buccal seal, described recently by Foissner & Al-Rasheid (2006).

Ruthmann & Noll-Altmann (1980; see also Lee et al. 1985a, p. 382 for review) found that the cytoplasm of *U. grandis* contains numerous bacteria. From experiments with antibiotics they concluded that the bacteria are symbionts which are necessary for the completion of normal cell division. If the bacteria are lacking then double monsters are formed.

Suganuma & Inaba (1966, 1967) and Inaba & Suganauma (1966) studied the nuclear apparatus and the macronuclear division with electron microscopy. For details, see nuclear apparatus in the general section. The fine structure of the undulating membrane in *U. grandis* is very similar to that found in other species, for example, *Paraurstyla weissei*, *Pseudourostyla cristata*, *Pseudokeronopsis rubra* (Bakowska & Jerka-Dziadosz 1978, p. 297). Further literature: Fauré-Fremiet & André (1968).

Molecular data: Sapra et al. (1985) electrophoresed the DNA of seven species, including *U. grandis* provided by Klaus Heckmann (Germany). All species showed at least two common features, namely, (i) the DNA banding pattern in the gel consisted of up to 100 bands representing DNA molecules ranging in size from about 0.4 to 20 kb, and (ii) one common band was distinctly visible for all species at about 7.6 kb range. This band corresponds to the ribosomal genes which code for 25 S and 19 S rRNA. Sapra et al. (1985) therefore concluded that the size of rDNA is about the same for all the hypotrichs investigated. Schlegel & Steinbrück (1986) studied the alpha and beta tubulin genes.

Croft et al. (2003; GenBank accession number AF508054) analysed the length (in base pairs) of the actin-encoding macronuclear molecule (1482) and of its 5' leader (120), ORF (1131), 3' trailer (231), and encoded amino acid chain (376). Simultaneously,

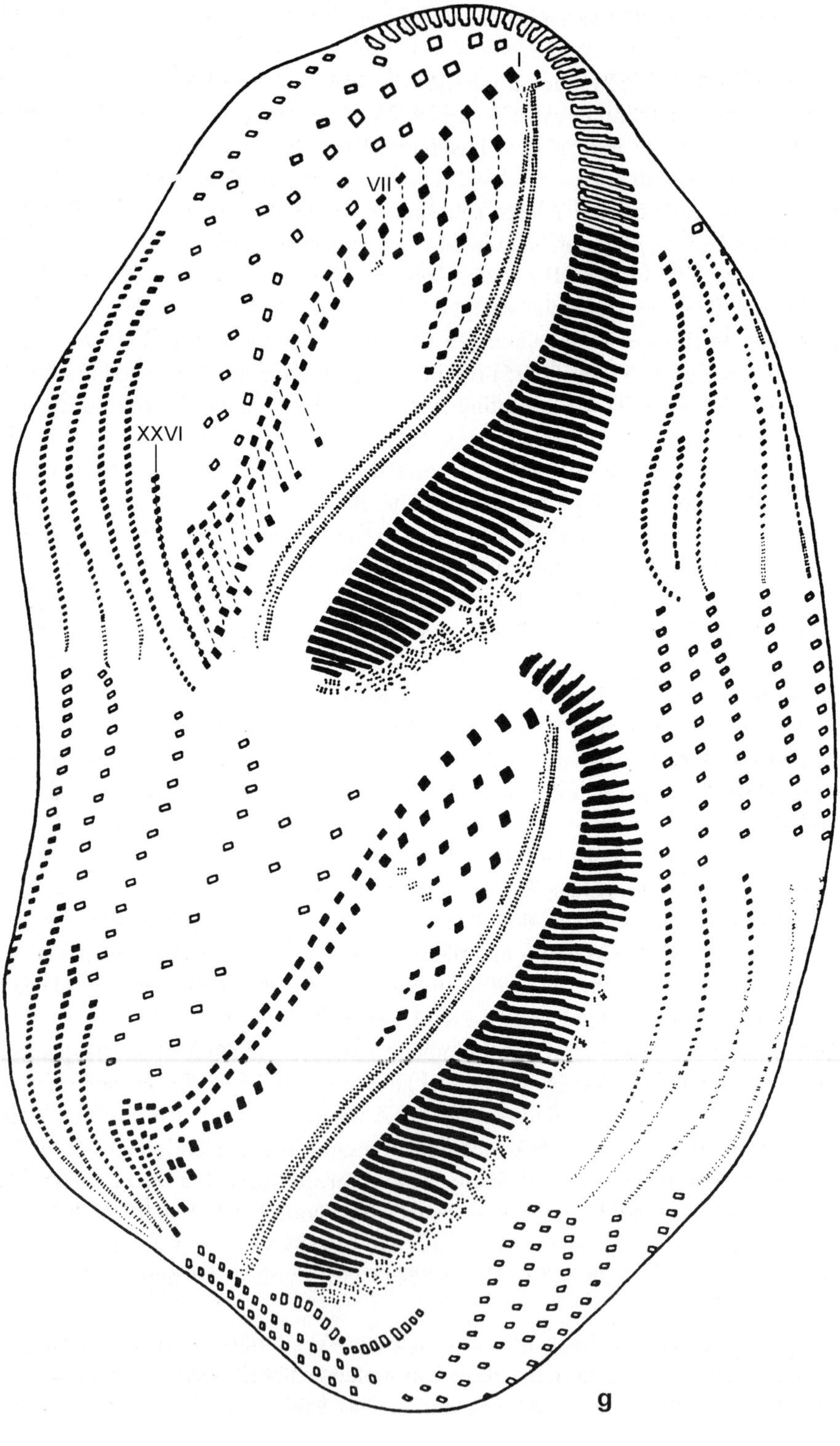
I
VII
XXVI
g

Hewitt et al. (2003; GenBank accession number AF508781) provided the lengths (in base pairs) of the SSU rDNA (1768), ITS 1 (130), 5.8S (153), ITS 2 (198), and LSU rDNA (1366). In all trees published in these two papers and in the studies by Kim et al. (2004), Dalby & Prescott (2004), and Coleman (2005), *Urostyla grandis* invariably clustered with *Holosticha polystylata*, a junior synonym of *Diaxonella pseudorubra*. In Foissner et al. (2004a), *Urostyla grandis* clustered with *Holosticha multistilata* (= *Anteholosticha multistilata* in present book) showing that the urostylids are a monophyletic group. Chen & Song (2001) mentioned AF164129 as GenBank number for the SSU rRNA. Hoffman & Prescott (1997) sequenced DNA polymerase alpha of *U. grandis* and submitted the complete amino acid sequence to the GenBank under the accession number U89706 (see also Curtis & Landweber 1999). Hogan et al. (2001) obtained the macronuclear actin I sequence from the GenBank database (accession no. AF188160). They analysed the micronuclear version of the actin I gene, which is not scrambled (for review see John & Klobutcher 2002, p. 504). Two IESs (internal eliminated segment; short, AT-rich, non-coding segments) are present and divide the micronuclear precursor into three MDSs (macronuclear destined segments).

In Snoeyenbos-West et al. (2002), *Urostyla grandis* clustered with *Engelmanniella mobilis*, respectively, *Paraurostyla weissei*, both of which do not belong to the urostyloids. Likely this is due to undersampling.

Parasitism: *Urostyla grandis* is sometimes infected by the suctor *Podophrya urostylae* (Maupas, 1881) Jankowski, 1963 (Matthes 1988, p. 165; Tirjaková 2004, p. 4). For details see general section (Fig. 17a–y, 18a–c).

Further data: For data on regeneration and doublet formation, see general section.

Occurrence and ecology: *Urostyla grandis* is common, but usually not abundant in limnetic habitats. The type locality is Berlin (Germany), where Ehrenberg (1830) discovered it on slimy, dead leaves of *Phragmites* in slowly running waters (Ehrenberg 1838). He found it mainly in spring, sometimes in rather high numbers. The synonym *Bursaria vorax* was discovered in muddy water in Berlin, Germany (Ehrenberg 1831, 1838). Perty (1849b, 1852) mentioned at least three localities (Bern, throughout the year; Lake Neuenburger, September; Lugano, August) in Switzerland, where he found the synonym *Oxytricha fusca* in brooks, ditches, ponds, and bog water throughout the year. Unfortunately, he did not fix one of these sites as type locality. The type locality of the synonym *U. trichogaster* is not unequivocally fixed because Stokes (1885) mentioned two sites (shallow ponds in central New Jersey; more or less concentrated infusion of fallen leaves with water from the Delaware river) in his material section. On page 445 he wrote that *U. trichogaster* occurred in a vegetable infusion, indicating that he discovered it in the Delaware river. The type locality of the synonyms *Urostyla elongata* and *U. fulva* is a pond in Trenton, New Jersey, USA (Stokes 1891, p. 697).

← **Fig. 211g** *Urostyla grandis* (from Ganner 1991. Protargol impregnation). Infraciliature of ventral side of a very late divider with 26 frontal-midventral-transverse cirral anlagen in the proter. Anlage VII produces the anteriormost midventral pair. None of the cirri of anlage XXVI splits of anteriorly, strongly indicating, together with interphasic data, that *U. grandis* lacks frontoterminal cirri. Cirri which originate from the same anlage are connected by a broken line. Parental cirri white, new black. I, VII, XXVI = some selected frontal-midventral-transverse cirral anlagen of the proter. Page 1048.

Records of *U. grandis* substantiated by morphological data: river Oichten (population 1) near the city of Salzburg and the Salzach River (population 2) in the city of Salzburg, Austria (Ganner 1991); Hainbach, a small brook in Upper Austria and other sites in Austria (own observations; Fig. 208r–t); Austrian and German running waters, for example, the clean river Illach and some mesosaprobic rivers (Amper, Windach, Schwebelbach) in Bavaria (Foissner 1997a, p. 185; Foissner et al. 1991; 1992, p. 51; 1992a, p. 102); often associated with *Paraurostyla weissei* in stagnant and slowly running water bodies from/near the villages of Tharand (Saxony) and Niemegk (Brandenburg) in Germany, and in Prague, Czech Republic (Stein 1859; see also Šrámek-Hušek 1953, p. 79); mesosaprobic pond near the Bohemian village Písek (Šrámek-Hušek 1952a, b); in floated dry grass from rock pools in Finland (Reuter 1961); limnetic habitats near Geneva (Switzerland) in February and October (Roux 1901); various limnetic habitats (mainly among macrophytes) in Germany (Kahl 1932); ponds and brooks in Rhineland (Naturlehrpark Wilderath) in Germany (Pätsch 1974); abundant during winter and spring in a eutrophic pond (Poppelsdorfer Weiher) in Bonn, Germany (Song & Wilbert 1989); Moldavian water basins (Chorik 1968); river Jeziorka in the village Jeziorna near Warsaw, Poland (Jerka-Dziadosz 1963); pond of Sadyba in Warsaw, Poland (Jerka-Dziadosz 1972); limnetic habitat in Sweden (Quennerstedt 1865; 1869, p. 31); limnetic habitat in Nanking, China (Wang 1925); limnetic habitats in western South Korea (Shin 1994); abundant in a culture from Van Cortlandt Park near New York City, USA, during autumn 1930 (Tittler 1935); rare in December at 0.2% salinity in Choctaw Point, Mobile Bay, Alabama, USA (Jones 1974); infusions of hay and leaves (source of water [pond water, tap water] not mentioned) from Kansas, USA (Smith 1914); Baraga, Upper Peninsula of Michigan, USA (Lundin & West 1963).

Many records from freshwater not substantiated by morphological data. Eurasia: Vienna, Austria during June (Riess 1840, p. 38); common among plants in a lake (Ambrassersee) and in a stoup near Innsbruck, Austria (Dalla Torre 1891, p. 205); pond at Salzburg University, Austria (Blatterer 1989, p. 10); mesosaprobic sites in the river Traun, Upper Austria (Foissner & Moog 1992, p. 103); muddy rain-puddle and groundwater pool in Austria (Spandl 1926a, p. 91); various habitats (spring, ditch, muddy rain puddles, dead arm) near Vienna, Austria (Kühn 1940, p. 187); river Enns, Austria (Meisriemler & Riedl 1985, p. 174); Belgium (Chardez 1987, p. 14); Bulgaria (Spandl 1926b, p. 534); Detschewa 1972, p. 77); sporadically in a pond in Bohemia, Czechoslovakia (Švec 1897, p. 33); river Vltava in Prague, Czechoslovakia (Kalmus 1928, p. 395); chalk streams in Dorset, England (Baldock et al. 1983, p. 241); mesosaprobic running waters and other sites near the cities of Krefeld and Bonn, Germany (Heuss 1976, p. 155; Jutrczenki 1982, p. 109; Schmidt 1913, p. 80; 1916, p. 93); freshwater habitats near the village Hinsbeck, Rhine region, Germany (Schneider 1930, p. 526); freshwater section of the Hamburg Harbour, Germany (Tent 1981, p. 12; Bartsch & Hartwig 1984, p. 557); river Schussen near its mouth into the Lake Bodensee, Germany (Wetzel 1928, p. 258); mesotrophic lake (Erdfallsee) and ponds near the city of Münster, Germany (Mücke 1979, p. 271; Kuhlmann & Heckmann 1994, p. 220); pelagic(?) in the Danube river near Esztergom and Budapest, Hungary (Bereczky 1972, p. 215; 1977, p. 64); farmyard manure, Sopron, Hungary (Varga 1953, p. 377); brook in Hungary (Vörösváry 1950, p. 376); alkaline ponds in the Hortobágy National Park, Hungary (Szabó 1999a, p. 229); during December in a draw-well near Rome, Italy (Grispini 1938, p. 152); common in mesosaprobic running waters in the northern Apeninnes, Italy (Madoni & Bassanini 1999, p. 395); Ferrara, Italy (Canella 1954, p. 111, Tavola XXVII, fig. 175, the micrograph does not allow a re-identification); ditch with brackish water near Brondolo, Northern Italy during August (Schmarda 1846, p. 45; 1847, p. 16; further records from Italy, see Dini et al. 1995, p. 70); sometimes dominant in Latvian rivers (Liepa 1973, p. 33; 1983, p. 137; 1984, p. 63; 1986, p. 226; 1990, p. 67; Veylande & Liyepa 1985, p. 83); pelagic among *Scirpus* and *Phragmites* in the oligotrophic lake Vaydavas, Latvia (Liyepa 1984, p. 114); Pradnik stream near Cracow, Poland (Czapik 1975, p. 28); pond(?) near Warsaw, Poland (Wrzesniowski 1861, p. 331; Wrzesniowskiego 1866, p. 19); various dystrophic lakes in the Wigry National Park in Poland (Czapik & Fyda 1995, p. 68); Biała Przemsza River, Poland (Czapik 1982, p. 31); bottom mud of unfertilised and unpolluted fishponds in Poland (Czapik 1959, p. 191; Siemiñska & Siemiñska 1967, p. 60 [review paper]; Grabacka 1971, p. 13; 1977, p. 380); Slovakia, inter alia, with a constancy of 7.7% in the Danube river system near Bratislava (Tirjaková 1992, p. 294; 1992a, p. 80; Szentivány & Tirjaková 1994, p. 94; Matis et al. 1996, p. 20); dead arm of Danube river in Čičov, Slovakia (Matis & Tirjaková 1994, p. 53); Turiec River in Slovakia (Tirjaková 1993, p. 135); River Henares (Spain) at two polysaprobic sites near mouth into River Jarama (Sola et al. 1996, p. 241); alphamesosaprobic sites of the river Llobregat, Spain (Gracia & Igual 1987, p. 3; 1987a, p. 120; Gracia et al. 1989, p. 28); peat bogs in Sainte-Croix and near the lake Pfäffikersee, Switzerland (Mermod 1914, p. 102; Messikommer 1954, p. 642); Lake Geneva and other sites in Switzerland (Perty 1849a, p. 22; Roux 1900, p. 464; Forel 1904, p. 132; André 1916, p. 622; Bourquin-Lindt 1919, p. 73; Riggenbach 1922, p. 52; for a review on old Swiss records, see André 1912, p. 123); rivers, estuaries, reservoirs, and other sites in Azerbaijan (Adil 1934, p. 5; Agamaliyev & Aliyev 1983, p. 21; Alekperov 1982a, p. 87; 1983, p. 22; 1984a, p. 19; 1984b, p. 18; Aliev 1982, p. 807); cooling plant of a Moldavian power station (Chorik & Vikol 1973, p. 69); during spring, summer, and autumn benthic and pelagic in the River Tisa, Ukraine (Kovalchuk 1997, p. 98; 1997a, p. 445); Volga river (Mamaeva 1979a, p. 409; 1979b, p. 71); limnetic habitats in the Golf of Kola region, USSR (Gassovsky 1916, p. 145); littoral of Lake Baikal, USSR (Rossolimo 1923, p. 76); littoral of Rybinsk reservoir, USSR (Mylnikova 1981, p. 25); Saint Petersburg, USSR (Weisse 1848a, 46; 1848c, p. 362); Mozhaisk reservoir near Moscow, USSR (Belova 1988, p. 66); Salem City of Hakodate and Oonuma Park, Japan (Muramatsu 1957, p. 468); small pond in the botanical garden of the Nara Women's University, Nara, Japan (Inaba & Suganuma 1966); freshwater in the city of Sapporo, Japan (Hayashi 1959).

America: Lake Cromwell, Terrebonne, Canada (Puytorac et al. 1972, p. 435); pond on the University of Colorado campus, Boulder, USA (Hogan et al. 2001, p. 15101; Croft et al. 2003, p. 342); Iowa (Shawhan et al. 1947, p. 365); Louisiana, USA (Bamforth 1963, p. 134); Massachusetts, USA (Cole 1853, p. 48); Douglas Lake, Michigan, USA (Cairns & Plafkin 1975, p. 51); Mirror Lake, Ohio State University Campus,

Ohio, USA (Stehle 1920, p. 120); Crystal Lake, Norman, Oklahoma, USA (Bragg 1960); from September to January rare in a pond in the botanical gardens of the University of Pennsylvania, USA (Wang 1928, p. 441); Moutain Lake Region, Giles County, Virginia, USA (Bovee 1960, p. 357); pond in Montgomery County, Virginia, USA (Henebry & Cairns 1980, p. 112); campus of the University of Costa Rica (Ruiz 1961, p. 213); Argentina (Seckt 1924, p. 94); Amazon river near Iquitos, Peru (Cairns 1966, p. 61); limnetic(?) mosses from Venezuela (Scorza & Núñez Montiel 1954, p. 222).

Africa: Madagascar and Mozambique (Sondheim 1929, p. 20).

Terrestrial records are likely mainly based on misidentifications, because *U. grandis* was never reliably recorded from a true terrestrial habitat (Fantham & Porter 1946, p. 130 [moss *Polytrichum juniperum*]; Fantham et al. 1927, p. 354; Geptner 1973, p. 153 [the illustration is from Kahl 1932]; Goodey 1911, p. 169; Grandori & Grandori 1934, p. 285 [the illustrations are from Stein 1859 and Kahl 1932]; Luzzatti 1938, p. 101; Nikoljuk & Geltzer 1972, p. 130 [the illustration is from Kahl 1932]; Sztrantowicz 1984, p. 74; Yakimoff & Zérèn 1924, p. 42; Sandon 1927, p. 191). However, Reuter (1961) found it in flooded, dry grass; unfortunately, he did not provide details, but from the method section one can conclude that the grass was in a (dry) rock pool, so that the occurrence of the present species in such a sample is not impossible, inasmuch as it forms resting cysts.

Records from inland salt waters or brackish and marine habitats (Boutchinsky 1895, p. 144; Butschinsky 1897, p. 195; Smith 1904, p. 45 [reviewed by Borror 1962, p. 342]) are likely based on misidentifications because Bick (1967, p. 202; 1968, p. 265) found it only in freshwater, but not in experiments with artificial marine brackish water of 3.5‰ and 7.0‰ salinity. Agamaliev (1986, p. 207) recorded *U. grandis* from a low salinity (2.0–3.5 ‰) lagoon of the Caspian Sea.

Records of the synonym *Oxytricha fusca*: Perty (1852, p. 154; 1852a, p. 64), Zschokke (1900, p. 71); Baumann (1910, p. 660).

Records (mainly without morphological data) of the synonym *Urostyla trichogaster*: lake (Loch Leven) near Edinburgh, Scotland (Bryant & Laybourn 1974, p. 268); Gif-sur-Yvette, France (Fauré-Fremiet 1945b; Bussers & Jeuniaux 1974); brook (Kalanos) in Hungary (Vörösváry 1950, p. 376); draw-well near Rome, Italy (Grispini 1938, p. 152); Mount Niwot, Colorado, USA (Hamilton 1943, p. 51); fresh waters in Connecticut, USA (Conn 1905; including two further synonyms); various sites in the Upper Peninsula of Michigan, USA (Lundin & West 1963; West 1953, p. 278); polyurethane foam in Douglas Lake, Michigan, USA (Cairns et al. 1973, p. 653; Yongue & Cairns 1976, p. 752); Cape Fear River in the vicinity of Fayetteville, North Carolina, USA (Cairns & Yongue 1973, p. 32); Reelfoot Lake, Tennessee, USA, during summer (Bevel 1938, p. 145); New River, Virginia, USA (Yongue & Cairns 1979, p. 76).

At first Jerka-Dziadosz (1963) used tap water as culture medium for *Urostyla grandis* and *Colpidium* as food; later, she used Pringsheim medium and *Tetrahymena pyriformis*. Altmann & Ruthmann (1979) cultured *U. grandis* at a constant temperature of 18° C in a medium of freshwater and soil extract in 10 ml Boveri dishes and fed every

24 h with *Chlorogonium* sp. Tittler (1935) maintained *U. grandis* in hay infusion with *Colpidium*, *Halteria*, and *Chilomonas* as food.

Voracious omnivore (Kahl 1932, Ganner 1991), that is, feeds on diatoms, alga (*Closterium*; Fig. 209a), protozoan cysts, heterotrophic flagellates (*Polytoma* sp.; Ganner 1991), ciliates (*Colpidium colpoda*, *Dexiostoma campylum*; Ganner 1991) and detritus (Ehrenberg 1838, Perty 1852, Stein 1859, Kalmus 1928, Jones 1974), but also on rotifers (Kahl 1932) like *Rotifer vulgaris* (Ehrenberg 1838), *Lepadella* and *Squamella oblonga* (Stein 1859, Fig. 206b; Quennerstedt 1865), *Colurella* (Ganner 1991), and nematodes like *Anguillula* (Stokes 1885). Stein found specimens which had ingested up to five rotifers. Ehrenberg (1838) reported that the synonym *Bursaria vorax* had ingested several *Coleps hirtus* specimens, Stein observed *Coleps hirtus* and *Lembadion bullinum* in the food vacuoles. Czapik (1975) recorded cyanobacteria, diatoms, and small ciliates as food. Fauré-Fremiet (1945b, 1961a) fed the synonym *U. trichogaster* with *Dexiostoma camplyum*, *Cyclidium*, and *Chilomonas*. Stokes (1885) observed *Trinema enchelys* and a small *Anguillula* in the food vacuoles. Feeds also on euplotids (*Euplotes octocarinatus*, *E. patella*, *E. aediculatus*), which, however, develop lateral "wings" making engulfment by predators more difficult (Kuhlmann & Heckmann 1985, 1994, Heckmann 1995, Kuhlmann et al. 1999). Wicklow (1997) discovered *Aspidisca turrita* (Ehrenberg, 1831) Claparède & Lachmann, 1858 coexisting with *U. grandis* and hypothesised that the dorsal horn of *A. turrita* was an antipredator structure induced by a cue released by *U. grandis*. The *Urostyla* factor (U-factor) induces defensive morphological changes in species of *Euplotes* and *Sterkiella* (Wicklow 1997). In addition, he reported the formation of keel-like dorsal projections in *Sterkiella* sp.

Biomass of 10^6 specimens about 473 mg when 180 µm long (Nesterenko & Kovalchuk 1991, p. 28), according to Foissner et al. (1991) around 500 mg. However, because of the rather variable size, the biomass of *U. grandis* should be calculated for each population. Michiels (1974, p. 135) estimated only 80 mg per 10^6 specimens which is certainly too low. According to Dillon & Hobbs (1973), the synonym *U. trichogaster* has a volume of 10×10^7 µm^3. Respiration rate of a cyst is about 0.15 nl O_2 cell^{-1} h^{-1}, of a starved specimen about 1.7 nl O_2 cell^{-1} h^{-1} (Pigon 1953, 1954; reviewed by Fenchel & Finlay 1983, p. 111 and Khlebovich 1987, p. 219).

Bragg (1960, p. 55) found *U. grandis* in an American lake at following conditions: 6–24° C, pH 6.0–8.4, 3.6–10.0 mg l^{-1} O_2. Bick & Drews (1973, p. 401) found it even at pH 4.7. In the river Llobregat (Spain), *Urostyla grandis* occurred at following average values: 11.3° C, pH 8.43, alcalinity 170 mg l^{-1} $CaCO_3$, 213 mg l^{-1} Cl^-, 12.8 mg O_2 l^{-1}, 1.1 mg l^{-1} NH_4^+, 0.19 mg l^{-1} NO_2^-, 5.2 mg l^{-1} BOD (Igual 1990, p. 8). Schmerenbeck (1975, p. 57) found it in the aufwuchs community of experiments with up to 40 cm s^{-1} current velocity. Münch (1970, p. 570) studied the influence of the temperature on the decomposition of peptone under laboratory conditions (see also Wilbert 1977). *Urostyla grandis* occurred only at 20°C with low abundance. Fernandez-Leborans & Antonio-Garcia (1986, p. 209) found that *U. grandis* does not tolerate the presence of 20 µg l^{-1} zinc and 10 µg l^{-1} lead. By contrast it is rather resistant against free chlorine (Cairns & Plafkin 1975). Pütter (1900, p. 284), based on observations made by Verworn, provided some data on the thigmotaxis and galvanotaxis or *Urostyla grandis*.

According to Wetzel (1928a, p. 314) *Urostyla grandis* avoids anaerobic regions. However, Hempstead & Jahn (1940, p. 415) found it even at 9.2 mg l^{-1} hydrogen sulfide in the Silver Lake Bog, Dickinson County, Iowa (USA). Eurythermic, that is, occurs throughout the year (Hayashi 1959).

Urostyla grandis has been widely used as indicator of water quality for a long time (e.g., Mez 1898, p. 240; Mauch 1976, p. 459). Usually it is classified as alphamesosaprobic indicator of water quality (Kolkwitz 1950, p. 39), according to Kahl (1932), however, *Urostyla grandis* is katharobic. Sládeček (1988) proposed b = 3, a = 7, SI = 2.7, I = 4 (Table 12), whereas Sládeček & Sládečková (1997, p. 138) proposed a slightly modified classification: b = 4, a = 6, I = 3, SI = 2.6. Foissner et al. (1991), in their review on saprobic ciliates, however, used Sládeček's (1988) classification, which is also listed by Berger & Foissner (2003, p. 147).

Supposed synonym of *Urostyla grandis*

Urostyla chlorelligera Foissner, 1980
(Fig. 212a–c)

1980 ***Urostyla chlorelligera* nov. spec.**[1] – Foissner, Ber. Nat.-Med. Ver. Salzburg, 5: 103, Abb. 23a–c (Fig. 212a–c; original description. No type slides available).

2001 ***Urostyla chlorelligera* Foissner, 1980** – Berger, Catalogue of ciliate names 1. Hypotrichs, p. 101 (nomenclator containing all basionyms, combinations, and higher taxa of hypotrichs).

Nomenclature: No derivation of the name is given in the original description. The species-group name *chlorelligera* refers to the symbiotic algae (cortical granules?; see remarks) present in this species.

Remarks: This species, which is described only after live observations, was obviously overlooked by Borror & Wicklow (1983) in their review on urostylids. The symbiotic algae of *U. chlorelligera* are very small (about 1.5 µm across vs. usually about 4–6 µm; Foissner et al. 1999, p. 75), pale green, and arranged in rows in the ectoplasm. This combination of features is strongly reminiscent of the conspicuous cortical granules of *U. grandis* which are about 1 µm across. W. Foissner (pers. comm.) and I suppose that Foissner (1980a) misinterpreted the cortical granules of *U. grandis* as small zoochlorellae and therefore consider *U. chlorelligera* as junior synonym ot the type species. However, since one cannot definitely exclude that a species with such tiny symbiotic algae (possibly cyanobacteria) exists, I keep the data separate.

The distinguishing features discussed by Foissner (1980a), namely body length, size of macronuclear nodules, position of transverse cirri (projecting in *U. chlorelligera* vs.

[1] The diagnosis by Foissner (1980a) is as follows: 200 bis 270 µm große, im Umriß orthogonale *Urostyla*, deren Transversalcirren den Körperrand überragen. Dicht unter der Pellicula liegen in einer gelartigen Ektoplasmazone viele in Reihen angeordnete, 1,5 bis 2,0 µm große, blaßgrüne Zoochlorellen. Etwa 100, 10 bis 15 µm große, bohnenförmige bis ellipsoide Makronuclei und ca. 8 kugelförmige, etwa 4 µm große Mikronuclei.

non-projecting in *U. grandis*), are overlapping and therefore not usable for the separation.

Morphology: Body length 200–270 µm in life; length:width ratio of specimen illustrated about 3.3:1 (Fig. 212a). Body outline rectangular, that is, lateral margins in parallel, anterior and posterior body end broadly rounded; anterior end distinctly oblique on right side. Body flexible, ventral side plane, dorsal side vaulted. About 100 macronuclear nodules scattered throughout cytoplasm; individual nodules 10–15 µm long, bean-shaped to ellipsoidal (Fig. 212b). About eight globular micronuclei c. 4 µm across. Contractile vacuole left of proximal end of adoral zone of membranelles. Close underneath pellicle in a gel-like ectoplasmic layer many pale green symbiotic algae 1.5–2.0 µm across; 2–3 rows of symbiotic algae between each two cirral rows (Fig. 212c; see remarks for different interpretation of these globules). Cytoplasm strongly cloudy due to many tiny, colourless inclusions.

Adoral zone occupies about 40% of body length (Fig. 212a). Buccal field deep. Paroral commences near frontal cirri, bears long, fine cilia. Pharyngeal fibres distinct. Cirral pattern obviously very similar to that of *Urostyla grandis* and therefore not described in every detail (Fig. 212a). About five enlarged frontal cirri along anterior body end; likely a row of about 10 buccal cirri and 2–3 rows of parabuccal cirri (Fig. 212a); according to the text about eight cirral rows on the frontal field (long right marginal rows included?). Presence/absence of frontoterminal cirri not known. No (zigzagging) midventral pattern described, however, very likely present, as indicated by illustration (Fig. 212a). About 15, circa 20 µm long transverse cirri arranged in oblique row left of midline, protrude beyond rear body end, bases of about same size as those of other cirri; distinct bundle of fibres extends anteriorly from transverse cirri. Right side with six cirral rows, the innermost four more narrowly spaced than the outer two rows (this indicates that these are midventral cirri). Four cirral rows left of midline, innermost two terminate ahead of transverse cirral row, outermost rows end subterminally.

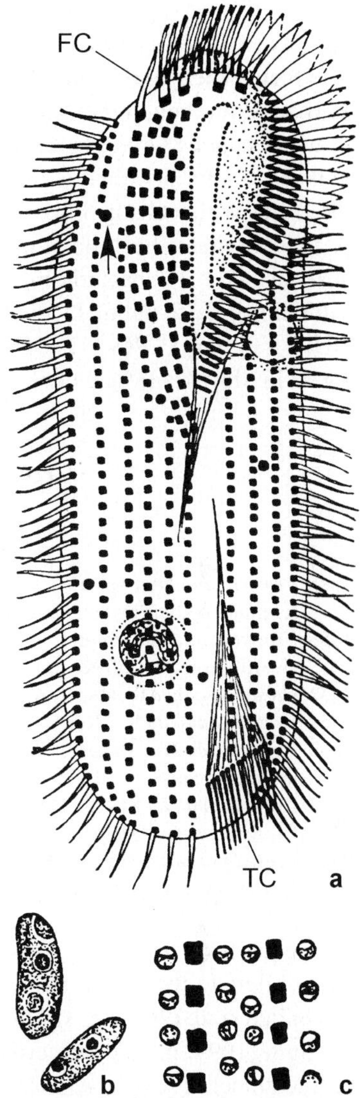

Fig. 212a–c *Urostyla chlorelligera* (from Foissner 1980a. From life). **a:** Ventral view showing cirral pattern, 260 µm. Details must not be overinterpreted. Arrow marks a micronucleus. **b:** Macronuclear nodules are bean-shaped or ellipsoidal and 10 to 15 µm long in life. **c:** The small (1.5 to 2.0 µm) pale green symbiotic algae are arranged in longitudinal rows between the cirral rows (see remarks). FC = rightmost frontal cirrus, TC = transverse cirri. Page 1086.

Dorsal ciliature (number of dorsal kineties, length of dorsal bristles, presence/absence of caudal cirri) not described.

Occurrence and ecology: Type locality of the synonym *U. chlorelligera* is the Piffkar-Alm (about 47°08'42" N 12°49'00"E), an alpine pasture (altitude about 1620 m) near the famous Grossglockner-Hochalpenstrasse where Foissner (1980a) discovered it in mossy footprints filled with rainwater. He found it also in a pasture pond (pond 1 near the Hotel Wallackhaus; Foissner 1980b, p. 111) in the same area. Foissner et al. (1982, p. 97) provided the following autecological data, which are, however, from a single analysis: 12×10^6 bacteria ml^{-1}; 9°C water temperature; pH 4.9; 10.9 mg l^{-1} O_2 (128% saturation); 0.35 mmol l^{-1} total hardness; 54 mg l^{-1} $KMnO_4$-consumption; 0.4 mg l^{-1} NH_4^+; 0.07 mg l^{-1} PO_4^{3-}. Feeds on *Vorticella*-species (Foissner 1980a). Obviously also found in Slovakia (Matis et al. 1996, p. 20).

Urostyla caudata Stokes, 1886
(Fig. 213a–d)

1886 ***Urostyla caudata*, sp. nov.** – Stokes, Proc. Am. phil. Soc., 23: 24, Fig. 4 (Fig. 213a; original description; no type material available and no formal diagnosis provided).

1888 ***Urostyla caudata*, Stokes** – Stokes, J. Trenton nat. Hist. Soc., 1: 279, Plate X, Fig. 15 (Fig. 213b; review of freshwater ciliates from the USA).

1932 ***Urostyla caudata* Stokes, 1886** – Kahl, Tierwelt Dtl., 25: 566, Fig. 97_{11} (Fig. 213c; revision of hypotrichs).

1950 ***Urostyla caudata* S.** – Kudo, Protozoology, p. 672, Fig. 316a (redrawing of Fig. 213a; textbook).

1963 ***Urostyla caudata* Stokes** – Lundin & West, Free-living protozoa, p. 67, Plate 27, Fig. 8 (Fig. 213d; illustrated record).

1972 ***Urostyla caudata* Stokes, 1886** – Borror, J. Protozool., 19: 9 (revision of hypotrichs).

1974 ***Urostyla caudata* Stokes** – Stiller, Fauna Hung., 115: 40, Fig. 23C (Fig. 213c; review).

2001 ***Urostyla caudata* Stokes, 1886** – Berger, Catalogue of ciliate names 1. Hypotrichs, p. 101 (nomenclator containing all basionyms, combinations, and higher taxa of hypotrichs).

Nomenclature: The species-group name *caudát·us -a -um* (Latin adjective; tailed) obviously refers to the broad, tail-like prolongation of the body. "*Urostyla caudate* Stokes" in Guillén et al. (2003, p. 180) is an incorrect subsequent spelling

Shin (1994, p. 85) redescribed *Paruroleptus caudatus* (Stokes, 1886). As basionym he mentioned *Urostyla caudata* Stokes, 1886 which is incorrect because the basionym of *P. caudatus* is *Holosticha caudata* Stokes, 1886.

Remarks: Stokes (1886) described this species in some detail. Of course, particulars of the cirral pattern (presence/absence of buccal cirri, frontoterminal cirri, midventral rows, and caudal cirri) remain unknown. Consequently, the classification in *Urostyla* is only preliminary. Interestingly, Stokes (1886) illustrated 10 cirral rows, whereas two years later he drew 11 rows, indicating that Stokes (1888) had additional observations. Kahl (1932), Borror (1972), and Stiller (1974b) accepted this huge species without providing new data. The illustration in Lundin & West (1963) is rather simple, but shows the significant features indicating that the identification is correct. Borror & Wicklow (1983, p. 120) synonymised *U. caudata* with *U. grandis*. However, these two species differ, inter alia, in body length (around 630 µm vs. 200–400 µm), body shape

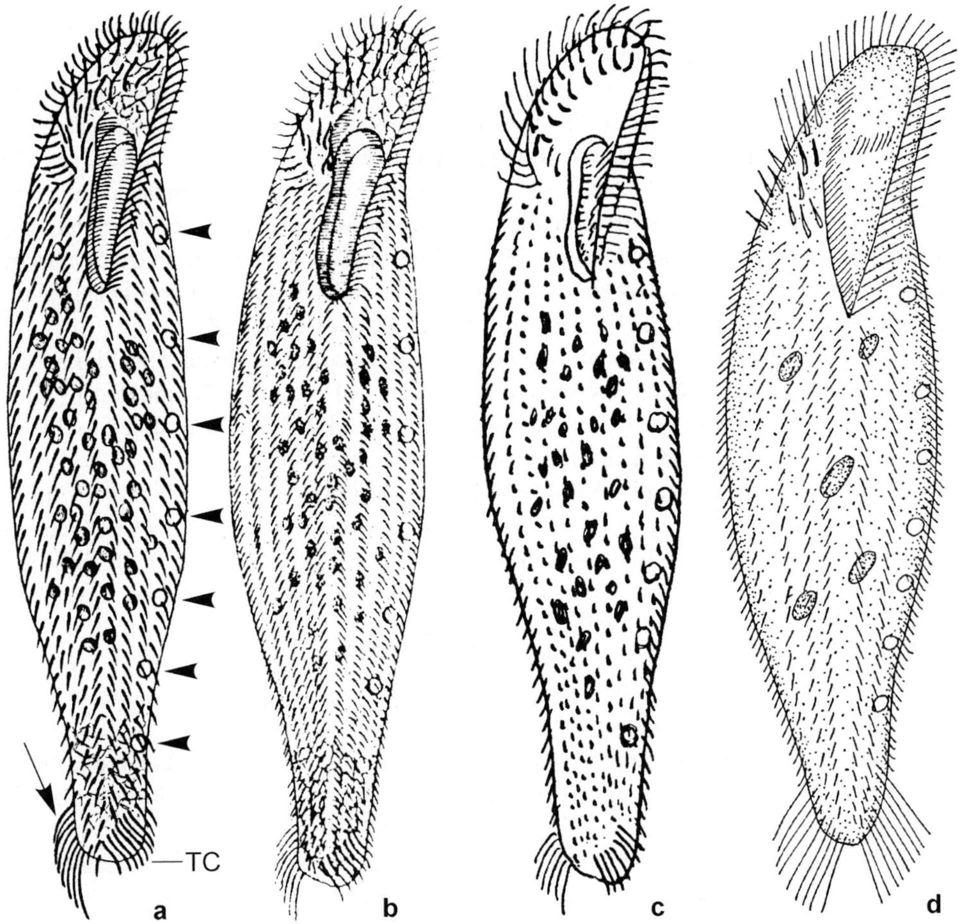

Fig. 213a–d *Urostyla caudata* from life (a, from Stokes 1886; b, from Stokes 1888; c, after Stokes 1888 from Kahl 1932; d, from Lundin & West 1963). Ventral views showing, inter alia, contractile vacuoles (arrowheads in a) which form a row near the left cell margin, and a conspicuous bundle of elongated right marginal cirri (arrow in a) near rear body end, a–c = about 635 µm, d = size not indicated. Note that Stokes (1886) illustrated in total 10 cirral rows, whereas the specimen shown in (b) has 11 rows. TC = transverse cirri. Page 1088.

(narrowed posteriorly vs. not narrowed), contractile vacuole (multiple vs. single), and right marginal cirri at the rear body end (elongated vs. not elongated), indicating that the synonymy is unjustified. Especially the increased number of contractile vacuoles, the conspicuous bundle of marginal cirri (Fig. 213a, arrow), and of course the huge size should make *Urostyla caudata* easily recognisable. For comparison with *U. gigas*, see key and remarks at *U. gigas*.

Hemberger (1982, p. 31) transferred *U. caudata* Stokes, 1886 to *Paraurostyla* Borror, 1972 because he obviously assumed that it lacks a midventral complex. Moreover, he considered *U. gigas*, *Hemicycliostyla sphagni*, and *H. trichota* – all species also described by Stokes (1886) from the same site – as synonyms of *U. caudata* because it is unlikely that four such large and rather similar species exist in the same freshwater marsh. He ignored the differences in the cirral pattern and contractile vacuole system described by Stokes (1886). By contrast, I consider both *Urostyla* species and *H. sphagni* as valid because I and many other workers assume that the presence/absence of a certain cirral group, including the transverse cirri, can be successfully used for the systematics of hypotrichs. However, Hemberger (1982, p. 32) correctly stated that an exact classification of these species is impossible without detailed redescriptions of freshwater populations from the Trenton area (USA), where Stokes lived and worked.

As already stated, *Urostyla caudata* is a very conspicuous freshwater species so that it is unlikely that it has been overlooked in Europe. All records substantiated by morphological data are from North America, strongly indicating that it is confined to this region. Detailed redescription, including neotypification, necessary.

Morphology: Body length 635 µm in life; body length:width ratio about 5:1 (Stokes 1886). Body outline roughly spindle-shape, anterior portion rounded and slightly curved leftwards, posterior portion narrowed into a straight, broad tail. Body soft, flexible, and extensile (respectively contractile). Many macronuclear nodules scattered throughout cytoplasm. About 10–12 contractile vacuoles (specimen illustrated with 7, that is, some are very likely contracted) in single series along left body margin. Presence/absence of cortical granules unknown. Cytoplasm vacuolised. Cytopyge on dorsal side near posterior end. Movement not described.

Adoral zone occupies about 33% of body length, distal end extends far onto right body margin (Fig. 213a). Buccal field of ordinary relative size; undulating membranes likely without peculiarities.

Cirral pattern not known in detail, that is, presence/absence of buccal cirri and frontoterminal cirri as well as midventral rows not described and not recognisable from the illustration (Fig. 213a). About 20 frontal cirri form a rather distinct bicorona; the presence of this bicorona strongly indicates that a midventral complex is present. Specimens illustrated with 10 (Fig. 213a; five left marginal rows, two rows formed by the midventral pairs, three? right marginal rows [possibly one or two of these three rows is/are a midventral row(s)]) and 11 (Fig. 213b, c) cirral rows. About 8–10 long, slender, that is, not enlarged transverse cirri in oblique row; usually projecting beyond rear body end. Posterior end of right marginal row composed of rather long, curved cirri forming a conspicuous bundle (Fig. 213a).

Occurrence and ecology: Possibly confined to limnetic habitats of (North) America. Stokes (1886) discovered *U. caudata* in marsh water with *Sphagnum*, likely somewhere near Trenton, New Jersey (USA), where he lived and worked. West (1953, p. 282) and Lundin & West (1963) found *U. caudata* in natural waters of the Upper Peninsula of Michigan, USA.

Records not substantiated by morphological data: *Sphagnum girgensohnii* (collected on 24.10.1936) from a roadside ditch on the main road from Rawdon, a town near Montreal, Canada (Fantham & Porter 1946, p. 119); freshwater habitats from the

Hiroshima region, Japan (Matsuoka et al. 1983, p. 15); ponds in Pantanos de Villa, Chorrillos, Lima, Peru (Guillén et al. 2003, p. 180).

Urostyla gigas Stokes, 1886
(Fig. 214a–c)

1886 ***Urostyla gigas*, sp. nov.** – Stokes, Proc. Am. phil. Soc., 23: 23, Fig. 3 (Fig. 214a; original description; no formal diagnosis provided and no type material available).
1888 ***Urostyla gigas*, Stokes** – Stokes, J. Trenton nat. Hist. Soc., 1: 277, Plate X, Fig. 11 (Fig. 214b; review of freshwater ciliates from the USA).
1932 ***Urostyla gigas* Stokes, 1886** – Kahl, Tierwelt Dtl., 25: 566, Fig. 97_9 (Fig. 214b; revision of hypotrichs).
1972 ***Urostyla gigas* Stokes, 1886** – Borror, J. Protozool., 19: 9 (revision of hypotrichs).
2001 ***Urostyla gigas* Stokes, 1886** – Berger, Catalogue of ciliate names 1. Hypotrichs, p. 101 (nomenclator containing all basionyms, combinations, and higher taxa of hypotrichs).

Nomenclature: The species-group name *gigas* (Greek adjective; gigantic, huge, enormous) refers to the huge body size (about 850 μm).

Remarks: This is one of the largest hypotrichs ever described. Stokes (1886) provided a rather detailed description and illustration, so the validity of *U. gigas* is beyond reasonable doubt although it was never redescribed since then, indicating that it is very rare and possibly confined to (North) America. It was accepted by Kahl (1932) and Borror (1972). By contrast, Borror & Wicklow (1983, p. 120) synonymised *U. gigas* with *U. grandis*, type of *Urostyla*. However, these species differ, inter alia, in body size (847 μm vs. 250–400 μm long), body outline (elongate elliptical vs. wide with parallel margins), frontal ciliature (5–6 frontal cirri, that is, distinct bicorona very likely lacking vs. distinct bicorona present), and marginal ciliature (elongated marginal cirri at posterior body end present vs. lacking). Hemberger (1982) obviously did not mention *U. gigas*.

Urostyla caudata, described by Stokes in the same paper, is somewhat smaller (about 600 μm) and has more frontal cirri and contractile vacuoles (Fig. 213a). Both species have elongated marginal cirri near the rear body end, indicating a close relationship (of course one cannot exclude that they are synonyms).

The distal end of the adoral zone of membranelles extends far posteriorly (DE-value about 0.41). Thus, it cannot be excluded that the present generic assignment is incorrect; possibly it belongs to the Retroextendia, which also have high DE-values.

The classification of the present species in *Urostyla* is uncertain because the exact cirral pattern (e.g., midventral rows present or not) is unknown. Consequently, for a final assignment a detailed redescription, including ontogenetic data, is needed.

Morphology: Body length about 850 μm, length:width ratio of extended specimens about 5:1. Body outline elongate elliptical, that is, widest in mid-body and tapering towards both ends, posterior end slightly curved leftwards, anterior slightly narrower rounded than posterior and slightly curved rightwards. Body soft, flexible, contractile (extensile according to Stokes' terminology). Many macronuclear nodules scattered throughout cytoplasm; Stokes estimated 40 to 60 nodules (this number must not be over-interpreted because it is very difficult to estimate the number in life in such a huge

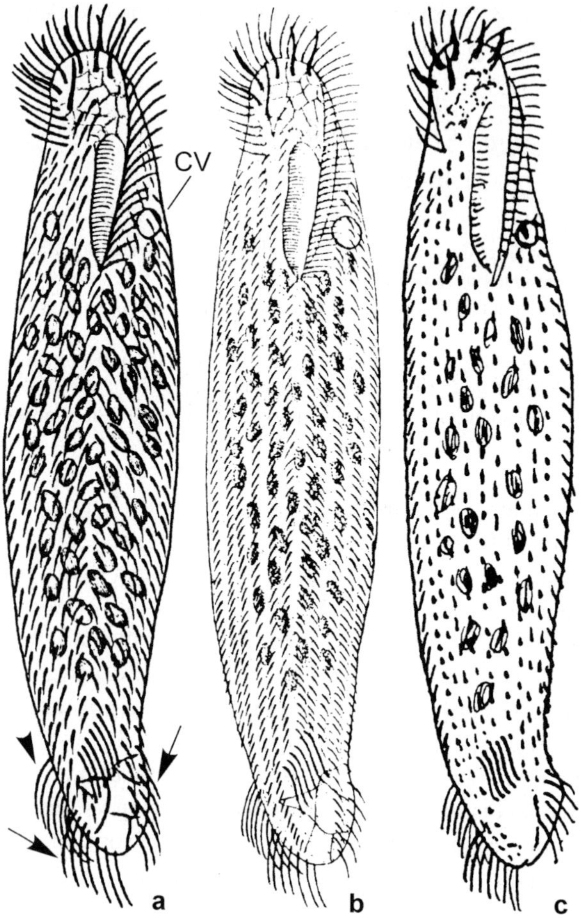

Fig. 214a–c *Urostyla gigas* from life (a, from Stokes 1886; b, after Stokes 1886 from Stokes 1888; c, after Stokes from Kahl 1932). Ventral view showing, inter alia, contractile vacuole, nuclear apparatus, and cirral pattern. Arrows mark elongated marginal cirri; arrowhead denotes a row of elongated cirri arranged on the dorsal side, indicating that these are caudal cirri. Note also the anteriorly dislocated transverse cirri. Moreover the present species has only one contractile vacuole in ordinary position (near the buccal vertex), whereas *U. caudata*, which is similar in size (about 600 μm) and body shape, has several contractile vacuoles along the left body margin. CV = contractile vacuole. Page 1091.

species!). Contractile vacuole left of proximal portion of adoral zone of membranelles (Fig. 214a). Cytopyge subterminally on dorsal side. Cytoplasm heavily vacuolised, especially in the extremities (similar to in *Loxodes rostrum*; Foissner et al. 1995, p. 378). Movement slow, often twisting.

Adoral zone occupies about 25% of body length, that is, zone more than 200 μm long! Distal end of zone extends far posteriorly onto right body margin (see remarks). Buccal field of ordinary relative size, undulating membranes (paroral) distinctly shortened anteriorly (Fig. 214a). 5–6 enlarged frontal cirri, according to Fig. 214a five of them form curved row, one cirrus behind right end of anterior corona. Whole ventral side covered by longitudinal cirral rows; specimen shown in Fig. 214a with 11 rows; interpretation of cirral pattern (midventral complex, marginal rows) needs detailed redescription and likely ontogenetic data. Six transverse cirri arranged in rather oblique, distinctly subterminal row; individual cirri small, fimbriated, and not projecting beyond rear body end. Rear body portion with short rows of conspicuously long arcuate cirri;

right side with two such rows, one of which originates on dorsal side (possibly caudal cirri); left side with one row (Fig. 214a). Dorsal bristles short and immobile (number and arrangement of dorsal kineties not known).

Occurrence and ecology: Possibly confined to limnetic habitats of (North) America; very rare. Stokes (1886) discovered *U. gigas* in marsh water with *Sphagnum*, likely somewhere near Trenton, New Jersey (USA), where he lived and worked. No further records published. Omnivorous, occasionally it also ingests angular sand grains.

Incertae sedis in *Urostyla*

The following species do not fir the characterisation of *Urostyla* very well because they have three frontal cirri or possibly lack a midventral complex. However, since details are lacking I classify them in *Urostyla*, which is a melting pot for little known "urostyloid" hypotrichs. A classification of these species in other genera would make these taxa unnecessarily inhomogenous.

Urostyla agamalievi Alekperov, 1984 (Fig. 215a, b, Addenda)

1984 ***Urostyla agamalievi*** **Alekperov, sp. n.** – Alekperov, Zool. Zh., 63: 1460, Fig. 3a, b (Fig. 215a, b; no formal diagnosis provided; type slides are likely deposited in the Institute of Zoology, Academy of Sciences of Azerbaijan, Baku).

2001 ***Urostyla agamalievi*** **Alekperov, 1984** – Berger, Catalogue of ciliate names 1. Hypotrichs, p. 101 (nomenclator containing all basionyms, combinations, and higher taxa of hypotrichs).

Nomenclature: Alekperov (1984) dedicated this species to his colleague F. G. Agamaliev, Baku, Azerbaijan.

Remarks: Alekperov (1984) described this species mainly after wet silver nitrate preparations. The arrangement of the frontal cirri and the midventral cirri is very likely not quite correctly shown, that is, it remains unknown whether or not buccal cirri, frontoterminal cirri, and midventral row(s) are present. Consequently, the generic assignment is uncertain. However, since new data are lacking I preliminarily accept the classification in *Urostyla*. Detailed redescription – including live data (presence/absence of cortical granules) and morphogenetic features (e. g., origin of marginal rows, presence or absence of midventral rows and frontoterminal cirri) – necessary.

Morphology: Body length in life 300–320 μm, in wet silver nitrate preparations about 250–270 μm. Two ellipsoidal macronuclear nodules, each with one micronucleus (Fig. 215b). Adoral zone occupies about 40% of body length in specimen illustrated; composed of about 65–70 membranelles. Undulating membranes obviously long and curved (no details recognisable). Note that details about the cirral pattern must not be over-interpreted (see remarks). Frontal cirri slightly enlarged and likely arranged in more or less distinct bicorona; no distinct buccal cirrus illustrated. Presence/absence of frontoterminal cirri not known (see right marginal rows). Midventral complex terminates at

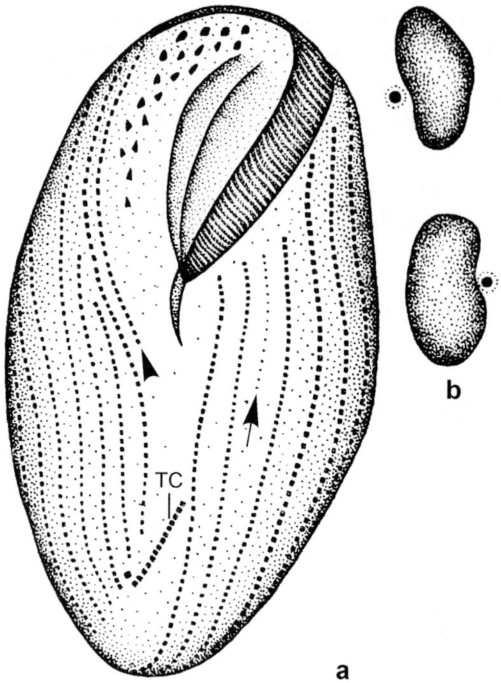

Fig. 215a, b *Urostyla agamalievi* (from Alekperov 1984. Wet silver nitrate impregnation). **a:** Infraciliature of ventral side, 250–270 µm. Arrowhead marks innermost right cirral row; arrow denotes a posteriorly shortened left marginal row. **b:** Nuclear apparatus. TC = transverse cirri. Page 1093.

30% of body length in specimen illustrated (Fig. 215a), possibly composed of cirral pairs only. 15 transverse cirri form oblique, subterminal, hook-shaped row; bases of cirri of about same size as marginal cirri. Seven left marginal rows. Eight cirral rows right of median, innermost row commences near distal end of adoral zone, terminates, about in mid-body (possibly this is a frontoterminal row; Fig. 215a, arrowhead), next row extends from about same level as previous row to near right portion of transverse cirral row; third and fourth row begin about at level of buccal vertex, terminate near right transverse cirri; remaining four rows likely more or less of body length (Fig. 215a). Dorsal infraciliature (length of dorsal bristles, number and arrangement of dorsal kineties, presence/absence of caudal cirri) not known.

Occurrence and ecology: Limnetic. Alekperov (1984) found *Urostyla agamalievi* pelagic(?) in freshwater of Azerbaijan, likely near Baku. No further records published.

Urostyla dispar Kahl, 1932
(Fig. 216a–c)

1932 ***Urostyla dispar* spec. n.** – Kahl, Tierwelt Dtl., 25: 565, Fig. 98, 110 19 (Fig. 216a, b; original description, no formal diagnosis provided and no type material available).

1933 ***Urostyla dispar* Kahl 1932** – Kahl, Tierwelt N.- u. Ostsee, 23: 108, Fig. 16.16 (Fig. 216c; guide to marine ciliates).

1972 ***Paraurostyla dispar* (Kahl, 1932) n. comb.** – Borror, J. Protozool., 19: 10 (revision of hypotrichs; combination with *Paraurostyla*).

1983 ***Urostyla dispar* Kahl, 1932** – Borror & Wicklow, Acta Protozool., 22: 121 (revision of urostylids).

1992 ***Paraurostyla dispar* (Kahl, 1930–5) Borror, 1972** – Carey, Marine interstitial ciliates, p. 177, Fig. 695 (the illustration is a redrawing of Fig. 216a; guide).

2001 ***Urostyla dispar* Kahl, 1932** – Berger, Catalogue of ciliate names 1. Hypotrichs, p. 101 (nomenclator containing all basionyms, combinations, and higher taxa of hypotrichs).

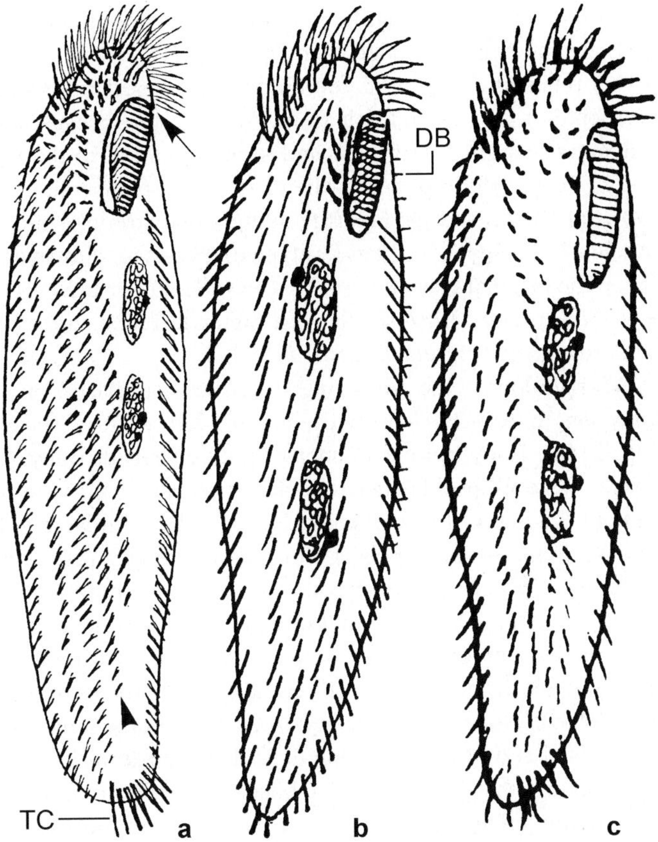

Fig. 216a–c *Urostyla dispar* (a, b, from Kahl 1932; c, from Kahl 1933. a–c, from life). **a:** Ventral view of a representative specimen of the type population from the Baltic Sea, 200–250 µm. Arrow marks hook dividing the adoral zone in two portions. Arrowhead denotes the row formed by the right cirri of the midventral pairs. Note that a detailed redescription and ontogenetic data are needed to show whether or not the present species is a urostyloid at all. However, the bicorona strongly indicates that it indeed has a midventral complex. **b:** Specimen from the North Sea population, 200 µm. Note that this population had no transverse cirri (see text for remarks). **c:** This figure is possibly a combination of Fig.s 216a, b. DB = dorsal bristles, TC = transverse cirri. Page 1094.

Nomenclature: No derivation of the name is given in the original description. The species-group name *dispar* (Latin adjective; dissimilar, different) likely alludes to the interrupted adoral zone which is different from the continuous zone of, for example, *Urostyla grandis*, type of *Urostyla*.

Remarks: Kahl (1932) described this species in some detail. The bipartite adoral zone, the cirral pattern, and the two macronuclear nodules are a very conspicuous combination of features so that *U. dispar* should be rather easily recognisable. Kahl (1932) studied two populations which differ mainly in the presence/absence of transverse cirri. Kahl therefore concluded that this cirral group is sometimes absent in *U. dispar*. According to my experience this cirral group is either invariably present or invariably absent within a species. Consequently, I assume that he either overlooked the transverse cirri in the Sylt population (Fig. 216b), or the two populations are not conspecific. Detailed redescription(s) is(are) needed to clear up the situation.

Borror (1972) transferred *U. dispar* to *Paraurostyla*, obviously mainly because the cirri are arranged in many longitudinal rows. Other diagnostic features of *Paraurostyla*, for example, frontal cirri clearly differentiated, reduced in number, and occurring in obvious groups or short rows do not apply for the present species or are unknown (e.g., locomotory cirri arise from longitudinal streaks of cilia during division; Borror 1972, p. 4). Later, he again included it in *Urostyla* (Borror & Wicklow 1983), whereas Carey (1992) retained the generic combination proposed by Borror (1972). In the review on oxytrichids I also excluded it from *Paraurostyla* because the bicorona indicates that a midventral complex is present and a classification in the urostyloids is therefore more likely (Berger 1999, p. 842). The exact cirral pattern (e.g., midventral rows present or not) is unknown and the bipartite adoral zone strongly indicates that *Urostyla dispar* is not closely related to *U. grandis*, type of *Urostyla*. However, since new data are lacking I keep the original classification in *Urostyla*. Other urostyloids with a bipartite bicorona (e.g., *Holosticha*, *Afrothrix*) have a rather different cirral pattern, indicating that these taxa are not very closely related.

Morphology: The following description is based on Kahl's text and Fig. 216a. Body length 200–250 μm, body length:width ratio of specimen illustrated about 4.2:1 (Fig. 216a). Body outline slenderly spindle-shaped. Body soft, slightly twitching, contractile. Two ellipsoidal macronuclear nodules, each with a micronucleus. Contractile vacuole lacking. Cytoplasm colourless, usually with large yellowish food inclusions. No cortical granules mentioned or illustrated, strongly indicating that they are lacking because in *U. grandis* and many other species Kahl described such organelles. Burrows in debris, sometimes lies there motionless.

Adoral zone conspicuous because bipartite by a hook-shaped structure (Fig. 216a, arrow) of the frontal field in a ventral (proximal) portion and a dorsal (distal) portion (Fig. 216a). Adoral zone occupies 23% of body length in specimen shown in Fig. 216a; distal end of distal portion extends far posteriorly onto right body margin. Buccal field lacking (that is, very small), buccal lip and undulating membranes present. Frontal cirri slightly enlarged, arranged in a bicorona; leftmost two cirri of anterior corona distinctly larger than remaining cirri. Between bicorona and anterior portion of paroral 6–7 slightly enlarged cirri in two short rows (possibly one [or more] of these cirri is a buccal cirrus; Fig. 216a). Bicorona not distinctly set off from the two median cirral rows, strongly indicating that this is a midventral complex composed of cirral pairs; row formed by left cirri of pairs ends more anteriorly than row formed by right cirri, indicating that the rear portion of the complex is not composed of pairs (Fig. 216a). Right of midventral complex in total three cirral rows; possibly the inner two rows are midventral rows, however, a final decision can only be made by means of ontogenetic data. Right marginal row (= rightmost cirral row) extends to near rear body end. Only one left marginal row. Type population (Bülk, Baltic Sea) invariably with 6–8 fine transverse cirri which project distinctly beyond rear body end. Dorsal ciliature (length of dorsal kineties, number and arrangement of kineties, presence/absence of caudal cirri) not known.

North Sea population (Fig. 216b) without transverse cirri (see remarks). Dorsal cilia obviously of ordinary length, that is, about 3 μm (Fig. 216b).

Occurrence and ecology: Marine, rare. Kahl (1932) described two populations. From the text one has to conclude that the population from Bülk near the German city of Kiel (Baltic Sea) is the type population (Fig. 216a). Kahl found it there frequently ("not rare") on coarse, dirty sand from a depth of about 5 m. Fig. 216b is the product of two earlier observations on material from the North Sea (depth near the German Island Sylt). Records not substantiated by morphological data: North Sea at Sylt (Küsters 1974, p. 175); White Sea estuary (Mazei & Burkovsky 2002, p. 186).

Urostyla gracilis Entz, 1884
(Fig. 217a–c)

1884 ***Urostyla gracilis* n. sp.** – Entz, Mitt. zool. Stn Neapel, 5: 376, Tafel 23, Fig. 8–10, not Fig. 11, 12 (Fig. 217a–c; original description; see nomenclature and remarks; no formal diagnosis provided and no type material available).

1884 ***Urostyla gracilis* n. sp. var. *pallida*** – Entz, Mitt. zool. Stn Neapel, 5: 376, Tafel 23, Fig. 8–10 (Fig. 217a–c; original description; no formal diagnosis provided and no type material available).

1932 ***Urostyla gracilis* Entz, 1884** – Kahl, Tierwelt Dtl., 25: 564 (revision of hypotrichs; see remarks).

1933 ***Urostyla gracilis* Entz sen. 1884** – Kahl, Tierwelt N.- u. Ostsee, 23: 108 (guide to marine ciliates; see remarks).

1972 ***Urostyla gracilis* Entz, 1884** – Borror, J. Protozool., 19: 9 (revision of hypotrichs).

1983 ***Urostyla gracilis* Entz, 1884** – Borror & Wicklow, Acta Protozool., 22: 120 (revision of urostylids).

2001 ***Urostyla gracilis* Entz, 1884** – Berger, Catalogue of ciliate names 1. Hypotrichs, p. 101 (nomenclator containing all basionyms, combinations, and higher taxa of hypotrichs).

Nomenclature: No derivation of the name is given in the original description. The species-group name *grácil·is -is -e* (Latin adjective; thin, fine, slender, delicate) likely refers to the relatively slender body shape (as compared to, for example, *Urostyla grandis*). The variety names *pallid·us -a -um* (Latin adjective; pale) and *sanguine·us -a -um* (Latin adjective; bloody, blood-red; see supposed synonym) refer to the colour of the cell.

Entz (1884) established the present species and simultaneously distinguished two varieties for which he proposed the names *pallida*, respectively, *sanguinea*. According to Article 45.6.4 of the ICZN(1999) these two species-group names are subspecific because (i) they were first published before 1961; (ii) the author expressly used the term variety; and (iii) the content of the work does not unambiguously reveal that the names were proposed for infrasubspecific entities. However, Entz found the varieties simultaneously in the same vessel, respectively, sample site, so that they cannot be subspecies. Consequently, two possibilities exist: (i) *pallida* and *sanguinea* are phenotypes of the same species; or (ii) *pallida* and *sanguinea* are different species. At present it is impossible to falsify one of the two hypotheses because new data are lacking. Thus, I provide the following pragmatic solution: I first give a description of *U. gracilis*, which contains the data of the dominant variety *pallida*[1]. The variety, respectively, subspecies *sanguinea* (see above for determination of rank) is raised to species rank with Entz (1884) as author (Article 50.3.1 of ICZN 1999) and briefly described as supposed synonym of *U. gracilis*.

Remarks: Entz (1884; see also Entz 1904, p. 126, for a brief note) established two varieties which differ mainly in size and colour (details, see previous paragraph). Kahl

[1] The name *pallida* becomes superfluous because, in the case of subspecies, the subspecific name of the nominotypical subspecies must be the same as the specific name.

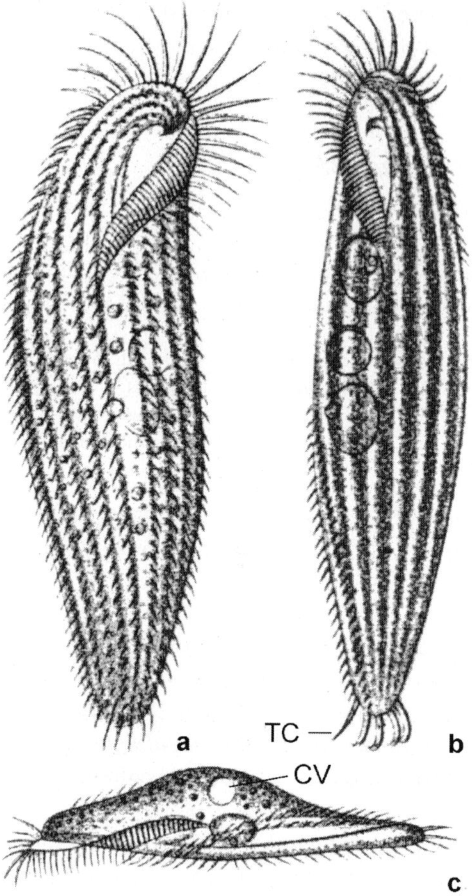

Fig. 217a–c *Urostyla gracilis* from life (from Entz 1884). Ventral, dorsal, and left lateral view (individual sizes not indicated). The dark stripes on the dorsal surface are very likely rows of cortical granules. CV = contractile vacuole, TC = transverse cirri. Page 1097.

(1932) supposed – although somewhat cryptically because he referred to a note at *Urostyla concha* – that it is a *Hemicycliostyla* species because transverse cirri are possibly lacking. However, Entz (1884) clearly described and illustrated five transverse cirri – at least for the dominant variety *pallida* (Fig. 217b) – so that a transfer to *Hemicycliostyla* is not justified. Kahl (1932, 1933) provided only Entz's illustration of the variety *sanguinea* (Fig. 218c, d). In addition, in his 1933 review he doubted the two macronuclear nodules of *U. gracilis* and suspected that Entz illustrated them schematically, that is, Kahl assumed that *U. gracilis* has many macronuclear nodules.

Borror (1972) considered *Urostyla pseudomuscorum*, which also has only two macronuclear nodules, as junior synonym. However, this species is limnetic (vs. marine) and larger (240 to 300 μm vs. 120–200 μm), has 9–15 (vs. 5) transverse cirri, and lacks a distinct colour (vs. reddish to violet). In addition, *U. pseudomuscorum* invariably has eight cirral rows with the middle six rows arranged in three pairs, whereas Entz's species has 9–10 cirral rows. Although this difference must not be over-interpreted because it is solely based on live observations, it indicates – together with the other features – that *U. gracilis* and *U. pseudomuscorum* are not conspecific. Since the cirral pattern of *U. pseudomuscorum* is reminiscent of *Pseudourostyla* species I transferred it to this genus.

Hemberger (1982, p. 278) mentioned *Urostyla gracilis* var. *pallida* and var. *sanguinea* in a list of species which are either indeterminable or of questionable position. However, on page 32 he transferred it to *Paraurostyla* and simultaneously synonymised it with *Urostyla concha* Entz, 1884 (= *Hemicycliostyla concha* in present book) and *Hemicycliostyla marina* Kahl, 1932. But neither *U. concha* nor *H. marina* have a red colour, indicating that the synonymy is incorrect, although the general appearance is indeed rather similar.

Borror & Wicklow (1983) considered not only *U. pseudomuscorum* as junior synonym of the present species, but also *U. limboonkengi* Wang & Nie, 1932 (Fig. 219a). Indeed, they agree in size, habitat, and nuclear apparatus. However, Wang & Nie did not mention any colour (although they knew about the significance of this feature) and described three distinctly enlarged frontal cirri, indicating that *U. gracilis* and *U. limboonkengi* are not identical.

Without doubt the classification of *U. gracilis* is uncertain because details of the cirral pattern (composition of midventral complex; frontal ciliature) are not yet known. Consequently, I classify it as incertae sedis in *Urostyla*. Detailed redescription needed.

According to Entz (1884, p. 379), *Leucophrys sanguinea* Ehrenberg sensu Eichwald (1852, p. 514, Tab. VI, fig. 13) is identical with *U. sanguinea*. However, the description of *L. sanguinea* by Eichwald is much too superficial to allow an identification and it is therefore not considered further.

The marine *Metaurostylopsis rubra*, which has a red cytoplasm, has many macronuclear nodules, about 15 cirral rows, and three distinctly enlarged frontal cirri. The common *Pseudokeronopsis rubra* has, besides the midventral complex which is made of cirral pairs only, one left and one right marginal row and many macronuclear nodules.

Morphology: *Urostyla gracilis* is about 120–200 × 30–40 µm in life (this is the range for both *pallida* and *sanguinea*, but *sanguinea* was usually the larger one). Body flexible and contractile. Body outline elongate elliptical or almost spindle-shaped, that is, widest about in mid-body and margins posteriorly more converging than anteriorly. Ventral side plane, dorsal side vaulted. Two macronuclear nodules and micronuclei in left body portion (Fig. 217a, b). Contractile vacuole near left body margin about in mid-body. Cytoplasm glutinous, coarse (likely because containing many greasily shining globules), cells usually do not dissolve. Cortical granules obviously present because Entz mentioned that the ectoplasm is composed of narrow, closely-spaced, fine-grained rows; size and colour of cortical granules not mentioned, but the colour of the cell (diffuse pale copper-red or brownish pink) could be due to the cortical granules.

Adoral zone of membranelles occupies about 25% of body length, distal end extends slightly (about 12% of body length) posteriorly on right body side (details must not be over-interpreted). Buccal field of moderate size, buccal lip conspicuously curved leftwards and inwards. Frontal scutum distinct.

Cirral pattern not known in detail and classification therefore uncertain. All cirri fine, except transverse cirri. In total 9–10 cirral rows with 5–6 rows on right side; cirri-free stripe behind buccal vertex usually somewhat wider than stripes among other cirral rows. Cirral rows of right body half extend onto frontal field where they obviously form a whirl, that is, *Urostyla gracilis* does not have distinctly enlarged frontal cirri, but a "multicorona", indicating that it is indeed a *Urostyla*. Five enlarged transverse cirri, which project distinctly beyond rear body end (Fig. 217b); distal end often hook-shaped. In fusiform specimens, right marginal (= outermost right row) row extends onto dorsolateral surface anteriorly, left marginal row (and often next row too) extends dorsolaterally posteriorly. Dorsal ciliature (length of dorsal kineties, number and arrangement of kineties, presence/absence of caudal cirri) not known.

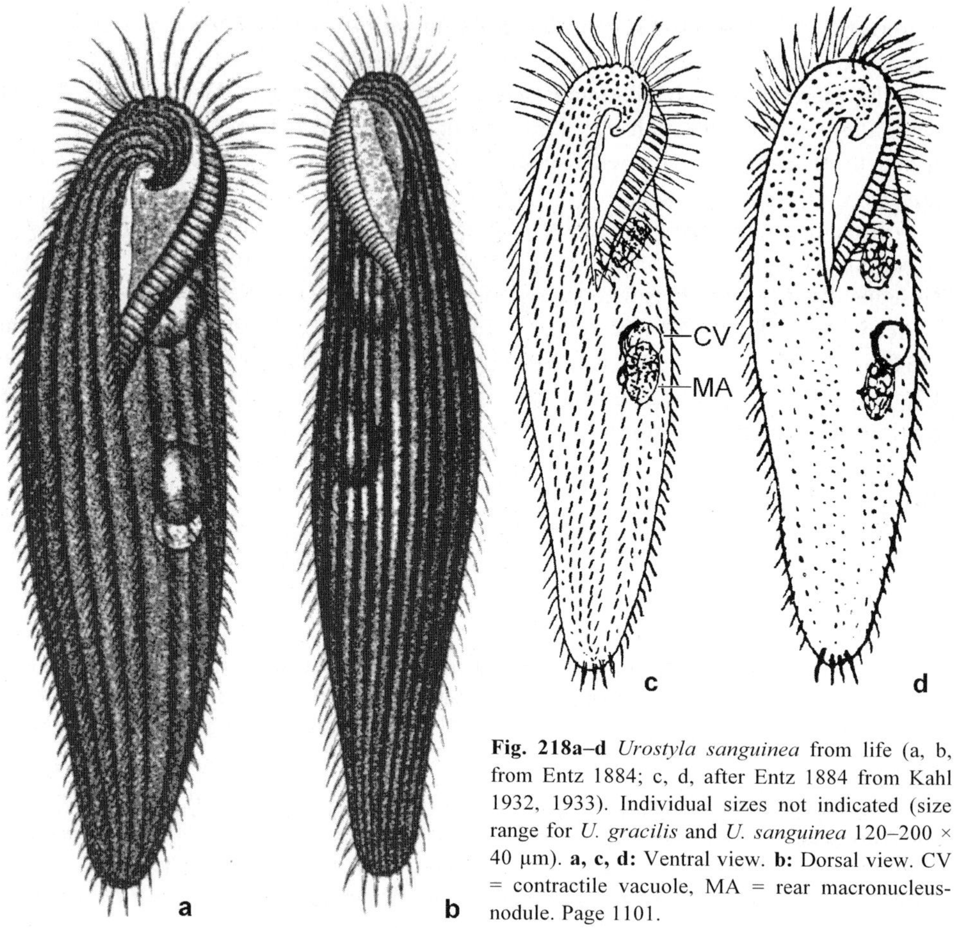

Fig. 218a–d *Urostyla sanguinea* from life (a, b, from Entz 1884; c, d, after Entz 1884 from Kahl 1932, 1933). Individual sizes not indicated (size range for *U. gracilis* and *U. sanguinea* 120–200 × 40 μm). **a, c, d:** Ventral view. **b:** Dorsal view. CV = contractile vacuole, MA = rear macronucleus-nodule. Page 1101.

Occurrence and ecology: Marine. Type locality of *Urostyla gracilis* is the Gulf of Naples, Italy (Entz 1884). Entz found the variety *pallida* (now *U. gracilis*) highly abundantly everywhere in the vessel which contained decomposing algae. The variety *sanguinea* (now *U. sanguinea*) occurred in the same vessel, but only sparsely and among reddish algae (Ceramien). They occurred together with, for example, *Pseudokeronopsis rubra*, *Oxytricha saltans*, and *Australothrix zignis*. The records of *U. gracilis* from oysters from the Passamaquoddy Bay and Malpeque Bay (New Brunswick, Canada) and from Conway (North Wales) by Laird (1961, p. 457) are not substantiated by morphological data.

Records from freshwater are very likely misidentifications: Budapest, Hungary (Krepuska 1917, p. 176); Antarctic lakes (Hawthorn & Ellis-Evans 1984, p. 74).

Supposed synonym of *Urostyla gracilis* Entz, 1884

Urostyla sanguinea Entz, 1884
(Fig. 218a–d)

1884 ***Urostyla gracilis* n. sp. var. *sanguinea*** – Entz, Mitt. zool. Stn Neapel, 5: 376, Tafel 23, Fig. 11, 12, not Fig. 8–10 (Fig. 218a, b; original description; no formal diagnosis provided and no type material available).

1932 ***Urostyla gracilis* Entz, 1884** – Kahl, Tierwelt Dtl., 25: 564, Fig. 97 $_{12}$ (Fig. 218c; revision of hypotrichs).

1933 ***Urostyla gracilis* Entz sen. 1884** – Kahl, Tierwelt N.- u. Ostsee, 23: 108, Fig. 16.20 (Fig. 218d; guide to marine ciliates).

Remarks: See nomenclature and remarks of *U. gracilis* for details. *Urostyla sanguinea* closely resembles *U. gracilis* (= *U. gracilis* var. *pallida* of Entz 1884). Consequently, only additional or deviating data on *U. sanguinea* are provided: body size in life 120–200 × 30–40 µm, usually close to 200 × 40 µm. More flexible and contractile than *U. gracilis*. Colour of cell splendid dark crimson, like stained with haematoxylin. Occurrence see *U. gracilis*.

Urostyla limboonkengi Wang & Nie, 1932
(Fig. 219a, b)

1932 ***Urostyla limboonkengi* sp. nov.** – Wang & Nie, Contr. biol. Lab. Sci. Soc. China, 8: 358, Fig. 66 (Fig. 219a; original description; no formal diagnosis provided and no type material available).

1934 ***Urostyla limboonkengi*** – Wang & Nie, Proc. Pacific. Sci. Congr., 5: 4210 (brief review; see nomenclature).

1935 ***Urostyla limboonkengi* Wang u. Nie, 1932** – Kahl, Tierwelt Dtl., 30: 841, Fig. 155 $_{19}$ (Fig. 219b; revision).

2001 ***Urostyla limboonkengi* Wang and Nie, 1932** – Berger, Catalogue of ciliate names 1. Hypotrichs, p. 101 (nomenclator containing all basionyms, combinations, and higher taxa of hypotrichs).

Nomenclature: Wang & Nie (1932) dedicated this species to Lim Boon Keng, President of the University of Amoy. Wang & Nie (1934) provided a brief characterisation of the present species. On page 4211 they wrote that a more complete presentation will be published in the "Contributions from the Biological Laboratory of the Science Society of China, Zool. Ser. vol. 8, no. 9". However, this detailed paper appeared already in 1932, indicating that the 1934 congress paper was in print for several years.

Remarks: The description and illustration are relatively detailed although some data, for example, midventral complex present or absent, are not known. The three enlarged frontal cirri show that it is misclassified in *Urostyla*, which is characterised, via the type species *Urostyla grandis*, by the presence of a coronal frontal ciliature. In spite of this obvious misclassification, I do not change the generic assignment because new data are needed for a transfer. The body shape, the oral apparatus, the nuclear apparatus, and in some respects also the cirral pattern are reminiscent of *Paraurostyla*

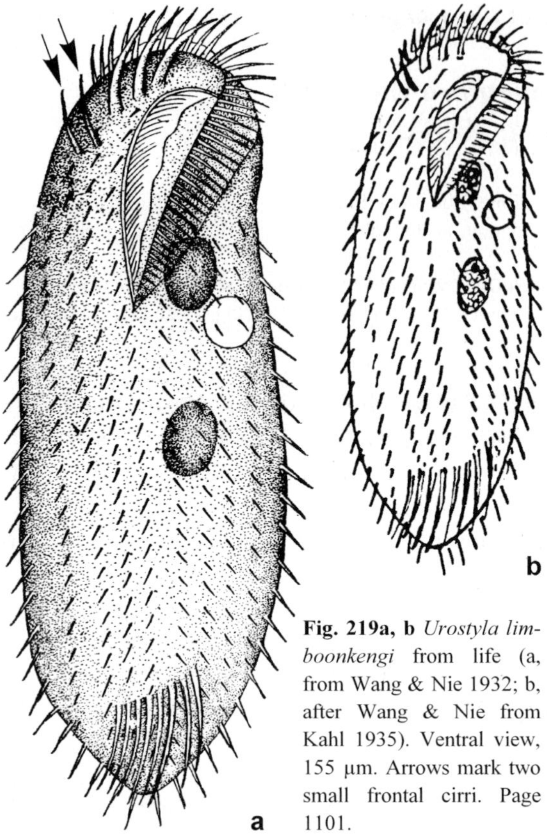

Fig. 219a, b *Urostyla limboonkengi* from life (a, from Wang & Nie 1932; b, after Wang & Nie from Kahl 1935). Ventral view, 155 µm. Arrows mark two small frontal cirri. Page 1101.

weissei. However, this oxytrichid is limnetic and has, inter alia, only one left marginal row (for review see Berger 1999, p. 844).

Borror (1972, p. 9) synonymised the present species with the limnetic *Urostyla grandis*, which, however, has, inter alia, many macronuclear nodules and frontal cirri. Hemberger (1982, p. 33) put *U. limboonkengi* into the synonymy of *U. vernalis* Stokes, a species synonymised with *Paraurostyla weissei* by Borror (1972) and Berger (1999). By contrast, Borror & Wicklow (1983, p. 120) synonymised it with *U. gracilis*. Indeed, they agree in size, habitat, and nuclear apparatus. However, Wang & Nie (1932) did not mention any colour although they knew about the significance of this feature and described three distinctly enlarged frontal cirri, indicating that *U. limboonkengi* and *U. gracilis*, which is reddish or crimson and has many frontal cirri, are not synonymous.

Morphology: In life about 155 × 55 µm, body length:width ratio about 3:1. Body outline roughly elongate elliptical, widest slightly behind mid-body, rounded anteriorly, bluntly tapering posteriorly. Body somewhat flexible and contractile. Two macronuclear nodules left of midline in middle body region. Contractile vacuole in ordinary position, that is, at left cell margin at level of buccal vertex. Presence/absence of cortical granules neither mentioned nor illustrated. Adoral zone occupies about one third of body length, buccal field roughly triangular. Paroral prominent because *Cyrtohymena*-like. Three distinctly enlarged frontal cirri, right of them two smaller cirri (possibly frontoterminal cirri). In total 8–13 rows of fine cirri; midventral pattern neither described nor clearly illustrated, that is, midventral complex possibly lacking. Eight somewhat enlarged transverse cirri arranged in slightly oblique row, almost not projecting beyond rear body end. Marginal rows (outermost left and right cirral row) confluent posteriorly. Dorsal ciliature (length of dorsal kineties, number and arrangement of kineties, presence/absence of caudal cirri) not known.

Occurrence and ecology: Marine. Type locality of *Urostyla limboonkengi* is the Bay of Amoy (now Xiamen), South-China Sea. Wang & Nie (1932) found several specimens among the decaying algae of a sea-water sample collected during summer. No further records published.

Urostyla naumanni Lepsi, 1935 (Fig. 220a)

1935 ***Urostyla naumanni* n. sp.** – Lepsi, Bul. Muz. natn. Ist. nat. Chisinău, 6: 16, 1 figure (Fig. 220a; original description; no formal diagnosis provided and no type material available).

1972 ***Paraurostyla naumanni* (Lepsi, 1935) n. comb.** – Borror, J. Protozool., 19: 10 (combination with *Paraurostyla*; revision of hypotrichs).

2001 ***Urostyla naumanni* Lepsi, 1935** – Berger, Catalogue of ciliate names 1. Hypotrichs, p. 102 (nomenclator containing all basionyms, combinations, and higher taxa of hypotrichs).

Nomenclature: Lepsi (1935) dedicated this species to Einar Naumann, founder of the International Society of Limnology.

Remarks: The illustration provided by Lepsi (1935) is not very detailed, but the pearl necklace-shaped macronucleus – together with some other features – is rather conspicuous so that an identification should be possible. The very long adoral zone of membranelles (47% of body length) and the low body length:width ratio (about 2:1) indicate that the specimen illustrated is a postdivider not yet fully developed.

Since it is not known whether or not *U. naumanni* has a midventral complex, its correct systematic position is not known. Consequently the transfer to *Paraurostyla* by Borror (1972) was unfounded as already briefly discussed by Berger (1999, p. 842). By contrast, Hemberger (1982, p. 78) synonymised *U. naumanni* with *U. grandis* because their cirral pattern is similar. However, the nuclear apparatus of the two species is quite different (moniliform vs. scattered) and *U. grandis* has distinctly more than eight frontal cirri and more than 2–3 transverse cirri. In addition *U. grandis* is confined to limnetic habitats, whereas *U. naumanni* was discovered in the Black Sea. Consequently, I retain Lepsi's species in *Urostyla*, but classify it as incertae sedis. Detailed redescription needed.

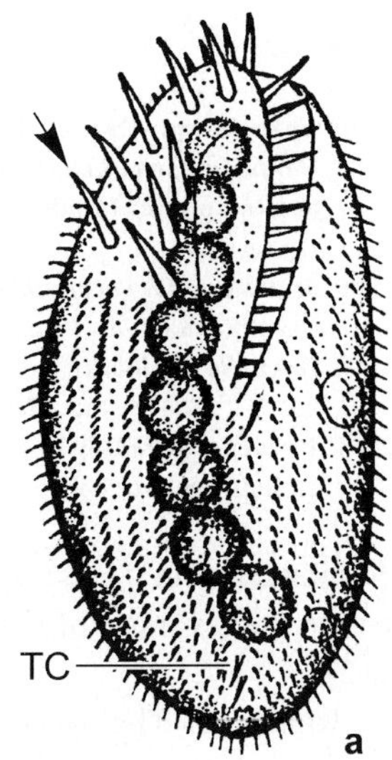

Fig. 220a *Urostyla naumanni* from life (from Lepsi 1935). Ventral view (170 µm) showing, inter alia, cirral pattern, nuclear apparatus, contractile vacuoles (left margin), and longitudinal rows of dots (cortical granules?) between cirral rows. Arrow marks rearmost frontal cirrus of the anterior row. TC = transverse cirri. Page 1103.

Morphology: Body about 170 × 85 µm in life, that is, body length:width ratio 2:1. Body outline broad elliptical with left margin less vaulted than right. Body almost acontractile. Macronucleus pearl necklace-shaped, composed of 6–8 globules; apparatus rather large (size possibly overestimated by Lepsi). Presence/absence of contractile vacuole not unequivocally observed; Lepsi observed two vacuoles near the left cell margin, but did not see the contraction (in marine species the period between two cycles is often very long and contraction therefore difficult to observe). Cells yellowish. Probably *U. naumanni* has cortical granules because Lepsi described rows of dots among the cirral rows (Fig. 220a); he interpreted them as vestigial cirri. Adoral zone of membranelles occupies about 47% of body length in specimen illustrated (Fig. 220a; see remarks). Buccal field of ordinary size, undulating membrane (paroral) likely long and curved. Eight distinctly enlarged frontal cirri arranged in two rows with five cirri along right anterior cell margin and three cirri right of paroral. In total about 15 cirral rows more or less equally spread over ventral side. One slightly enlarged cirrus behind buccal vertex, indicating that the present species is not a urostyloid. Subterminally 2–3 slightly enlarged transverse cirri. Dorsal ciliature (length of dorsal kineties, number and arrangement of kineties, presence/absence of caudal cirri) not known.

Occurrence and ecology: Marine. Type locality is the Black Sea at Constanta (Romania), where Lepsi (1935) discovered it in August 1931. No further records published. Likely feeds on bacteria.

Urostyla variabilis (Borror & Wicklow, 1983) comb. nov. (Fig. 221a–c)

1979 ***Bakuella* sp.** – Borror, J. Protozool., 26: 547, Fig. 4 (Fig. 221b, c; illustration and brief note).

1983 ***Bakuella variabilis* sp. n.** – Borror & Wicklow, Acta Protozool., 22: 111, Fig. 2 (Fig. 221a; original description; no formal diagnosis provided. Type slides are deposited in the slide collection of A. C. Borror).

1989 ***Metabakuella variabilis* (B. & W.) comb. nov.** – Alekperov, Revision of *Bakuella* and *Keronella*, p. 7 (combination with *Metabakuella*).

1992 ***Bakuella variabilis* Borror & Wicklow, 1983** – Song, Wilbert & Berger, Bull. Br. Mus. nat. Hist. (Zool.), 58: 146, Fig. 56, 57 (Fig. 221a–c; revision of *Bakuella*).

2001 ***Bakuella variabilis* Borror and Wicklow, 1983** – Berger, Catalogue of ciliate names 1. Hypotrichs, p. 12 (nomenclator containing all basionyms, combinations, and higher taxa of hypotrichs).

Nomenclature: The species-group name *variabil·is -is -e* (Latin adjective; variable) refers to the diversity of the cirral pattern within a population (Borror & Wicklow 1983, p. 112).

Remarks: The synonymy of *Bakuella* sp. in Borror (1979) and *Bakuella variabilis* was already supposed by Song et al. (1992). Interestingly, Borror & Wicklow (1983) did not mention Borror's (1979) description, possibly because they overlooked it.

Bakuella variabilis has 2–5 left marginal rows and the midventral rows are rather long and mainly longitudinally arranged right of the zigzagging midventral portion. Thus, the classification in *Bakuella* is likely incorrect because this group is characterised by one left and one right marginal row and midventral rows which are rather short, oblique, and arranged behind the zigzagging cirral pairs. A classification of *B. variabilis* in *Metabakuella*, as suggested by Alekperov (1989), is also uncertain, because the type species, *Metabakuella perbella*, has a bicorona (vs. 3[?] frontal cirri in *U. variabilis*).

The arrangement of the cirral rows of *B. variabilis* is reminiscent of *Urostyla grandis*, type of the genus. Although *U. grandis* has many frontal cirri (vs. only three in *U. variabilis*) I preliminary transfer *Bakuella variabilis* to *Urostyla* (as incertae sedis) because this genus already contains several species which do not fit the *U. grandis* pattern very well. Synonymy of *U. variabilis* and *U. grandis*, as supposed by Ganner (1991, p. 124), is very unlikely because *Urostyla variabilis* has, inter alia, frontoterminal cirri (vs. lacking), only three frontal cirri (vs. many), and only one right marginal row (vs. several). As in many other "*Urostyla*"-species, a reinvestigation – including morphogenesis and molecular data – are needed for a more proper classification. Song et al. (1992) supposed that it likely needs a genus of its own. However, since the establishment of monotypic taxa should be avoided, the discovery of a further species showing this combination of features should be awaited. Possibly it is related to *Metaurostylopsis*. However, since dorsal ciliature data (number of dorsal kineties, presence/absence of caudal cirri) are lacking I avoid a classification in this genus.

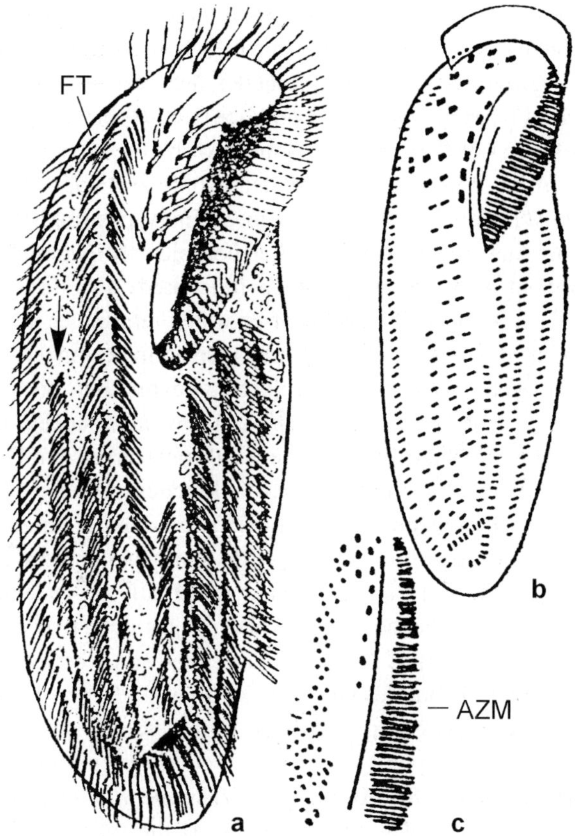

Fig. 221a–c *Urostyla variabilis* (a, from Borror & Wicklow 1983; b, c, from Borror 1979. a, protargol impregnation?; b, c, protargol impregnation). **a:** Ventral view (175 µm) showing, inter alia, cirral pattern and nuclear apparatus. Arrow marks rightmost midventral row. **b:** Infraciliature of ventral side, 270 µm. **c:** Frontal-midventral-transverse cirral anlagen of a middle divider. FT = frontoterminal cirri. Page 1104.

Morphology: Body length 225–240 µm (in life?). Body length:width ratio of specimen illustrated 2.9:1 (Fig. 221a), that is, outline broad elliptical. Body flexible and opaque. More than 100 small macronuclear nodules scattered throughout cytoplasm. Contractile vacuole near left cell margin about in mid-body, empties dorsally. Cortical granules in groups of 2–4 granules near each cirrus and additional dense groups between cirral rows; on dorsal surface, granules in oblique rows of 3–10 granules per row; size and colour of granules not mentioned. Feeds while crawling forward very slowly (about 1 body length per 4 sec.) either straight or in a slight left-hand curve.

Swims only when strongly stimulated, or mechanically lifted from substratum; then it moves slowly in a wide counterclockwise helix.

Adoral zone of membranelles occupies about 33% of body length. Buccal field narrow. Three (enlarged?) frontal cirri. 8–10 buccal cirri along anterior and middle portion of paroral. Specimen illustrated with five cirri between buccal cirral row and anterior portion of midventral complex. Frontoterminal cirri (migratory cirri according to Borror & Wicklow's terminology) in ordinary position, that is, between anterior end of right marginal row and midventral complex. Midventral complex composed of cirral pairs (about 43 pairs in specimen illustrated) and three longitudinal midventral rows (Fig. 221a); cirral pair portion of complex terminates at about 66% of body length in specimen illustrated. About 12 transverse cirri arranged in oblique, subterminal row, that is, transverse cirri just not projecting beyond rear body end. 2–5 left marginal rows of different length. One right marginal row commencing distinctly behind anterior body end, terminates slightly behind level of transverse cirri. Dorsal ciliature (length of dorsal bristles, number and arrangement of kineties, presence/absence of caudal cirri) not known.

Cell division: Borror (1979) provided a detail of a middle divider showing that many anlagen form a midventral complex composed of cirral pairs and midventral rows (Fig. 221c). The anlagen which eventually produce midventral pairs also form more than two cirri, indicating that in later stages only two cirri remain. The origin (from one or many anlagen) of the paramalar cirri (= cirri between buccal row and anterior portion of midventral complex) remains unknown.

Occurrence and ecology: Limnetic and semiterrestrial. Type locality of *U. variabilis* is a temporary pool in a flooded agricultural field in Lee (43°08'N 70°58'W), New Hampshire, USA. Borror (1979) found *Bakuella* sp. in freshwater in New Hampshire (possibly this was the same site as the type locality). Feeds on flagellates (Borror & Wicklow 1983).

Urostyla viridis Stein, 1859
(Fig. 222a–c, Table 12)

1859 ***Urostyla viridis*. Stein**[1] – Stein, Organismus der Infusionsthiere I, p. 206, Tafel XIII, Fig. 13, 14 (Fig. 222a, b; original description. No type material available).

1901 ***Urostyla viridis* Stein** – Roux, Mém. Inst. natn. génev., 19: 96, Planche V, fig. 18 (Fig. 222c; redescription).

1912 ***Urostyla viridis* Stein** – André, Catalogue des invertébrés de la Suisse, 6: 124 (review of Swiss ciliates).

1972 ***Paraurostyla viridis* (Stein, 1859) n. comb.** – Borror, J. Protozool., 19: 10 (combination with *Paraurostyla*; see remarks).

1979 ***Onychodromopsis viridis* comb. n.** – Jankowski, Trudy zool. Inst., 86: 84 (combination with *Onychodromopsis*; see remarks).

1991 ***Paraurostyla viridis* (Stein, 1859) Borror, 1972** – Foissner, Blatterer, Berger & Kohmann, Informationsberichte des Bayer. Landesamtes für Wasserwirtschaft, 1/91: 258, Fig. 1–3, not Fig. 4, 5 (Fig. 222a–c; review of ciliates of the saprobic system).

---

[1] The diagnosis provided by Stein (1859) is as follows: Körper lanzettförmig, mit 3 Stirnwimpern, 5 Afterwimpern und vielen über die ganze Bauchfläche vertheilten Bauchwimperreihen; Nucleus doppelt.

2001 ***Urostyla viridis* Stein, 1859** – Berger, Catalogue of ciliate names 1. Hypotrichs, p. 102 (nomenclator containing all basionyms, combinations, and higher taxa of hypotrichs).

Nomenclature: No derivation of the name is given in the original description. The species-group name *virid·is -is -e* (Latin; green) obviously refers to the symbiotic algae making the cells bright green.

Remarks: Stein (1859) described and illustrated this species in some detail. Nevertheless, a final generic assignment is impossible because it is unknown whether or not a midventral complex is present. The transfer from *Urostyla* to *Paraurostyla* by Borror (1972) was therefore unfounded. Consequently, I did not deal with it in the review on oxytrichids (Berger 1999, p. 843).

Kahl (1932, p. 567) provided own data for *U. viridis*. Unfortunately, he did not include Stein's figures in his revision so that his misidentification was not recognised for a long time. Foissner et al. (1991) in their review on saprobic ciliates discussed that the redescriptions by Kahl (1932) and Pätsch (1974, p. 56) differ from the original description and the redescription by Roux (1901) in the oral apparatus and the cirral pattern. They correctly stated that Kahl's and Pätsch's populations are reminiscent of *Onychodromopsis* Stokes. A relationship to *Onychodromopsis* was already recognised by Jankowski (1979), who even transferred *U. viridis* to Stokes' genus, but obviously without distinguishing between the earlier (Stein, Roux) and later (Kahl, Pätsch) descriptions. In the present monograph I omit Kahl's and Pätsch's redescriptions. Consequently, the morphology section below contains only data from the original description by Stein (1859), supplemented by some observations by Roux. The ecology section also comprises only data where the identification is based on Stein or Roux, that is, post-1932 identifications, which are very likely based on Kahl's (1932) review, are only briefly mentioned in a separate paragraph.

As just mentioned, *Urostyla viridis* sensu Kahl (1932) and Pätsch (1974) are reminiscent of *Onychodromopsis flexilis* Stokes, 1887 (Fig. 143a–d in Berger 1999). All these populations have very short/inconspicuous undulating membranes, three long right and two or more long left marginal rows, and lack the characteristic 18-cirri pattern of the oxytrichids. In my review (Berger 1999) I overlooked that Petz & Foissner (1996, p. 258, 270) fixed their Antarctic population as neotype of *Onychodromopsis flexilis*. However, this population – for which I (Berger 1999, p. 268) mistakenly[1] established *Allotricha antarctica* – has undulating membranes of ordinary length (although they are straight as in the other populations mentioned above) and usually only two right marginal rows, and the outer left marginal row consists of five cirri only on average. Further, it has the characteristic pattern of the 18-cirri oxytrichids. Because of these differences I doubt that the neotype of *O. flexilis* is identical with Stokes' (1887) population and the populations mentioned above. In spite of this, the neotypification has to be accepted because it cannot be undone. Pätsch (1974), who identified her population according to Kahl (1932), described symbiotic algae for her *Urostyla viridis*. However, one can imagine that the symbionts were ingested autotrophic flagellates

[1] It is nomenclatorically impossible/incorrect to establish a new species for a population which is the neotype of a known species.

which were present in the eutrophic pond where she found her population (Pätsch 1974, p. 74ff). Kahl (1932) mentioned zoochlorellae only in the key, but it is unclear whether or not his population indeed had symbiotic algae.

According to Hemberger (1982, p. 33), *Urostyla viridis* has two junior synonyms, namely, *U. algivora* and *Urostyla latissima* Dragesco, 1970. *Urostyla algivora* is classified as supposed synonym of *Pseudourostyla urostyla* (Fig. 157a). *Urostyla latissima* (for review on this species, see Dragesco & Dragesco-Kernéis 1986, p. 439) was described after protargol preparations which rather clearly show that a midventral complex is lacking. Consequently, it is not treated in the present review. Unfortunately, the data are not very exact so that a serious generic assignment is impossible. Previously I supposed, like Hemberger (1982), that it is a synonym of "*Paraurostyla viridis*" (Berger 1999, p. 842). The general appearance of the infraciliature is reminiscent of *U. viridis* sensu Pätsch, but I do not believe that they are conspecific.

Entz (1884, p. 378) assumed a very close relationship of his *U. gracilis* (Fig. 217a–c) and the present species. However, there are several distinct differences (e.g., marine vs. limnetic; many frontal cirri vs. 3 enlarged frontal cirri; reddish or crimson vs. green due to symbiotic algae), indicating that they are rather different.

The illustrations in Nikoljuk & Geltzer (1972, his Fig. 238) and Sládeček (1963, Plate 24, Fig. 16) are redrawings from Kahl (1932) and therefore not considered further.

Hoffman & Prescott (1997), Kelminson et al. (2002), Hewitt et al. (2003), and Croft et al. (2003) investigated some molecular markers (see below). Unfortunately, no morphological data of this population, which was not checked by a systematist, are provided. The positions of *U. viridis* within the trees obtained from these molecular data are rather different. In Hoffman & Prescott's paper it is, inter alia, the sister of *Uroleptus gallina*, whereas it is the sister of *Oxytricha granulifera* in Hewitt et al. (2003). In the actin tree calculated by Kim et al. (2004) it clustered with *Engelmanniella mobilis*, a species of uncertain position according to morphological data. Especially the close relationship with *O. granulifera* strongly indicates a misidentification, that is, molecular data/trees based on insufficiently determined species can produce great confusion.

The discussion above shows that *U. viridis* is involved in complex problems, that is, (i) it is not defined objectively because type specimens are lacking; (ii) there exist some uncertain redescriptions and synonyms; and (iii) the type locality is not fixed (see below). Consequently, a detailed redescription should involve neotypification.

Morphology: As discussed above, the data are largely from the original description unless otherwise indicated because Roux's population (Fig. 222c) agrees very well with Stein's observations. Body length 115–175 µm, body length:width ratio about 3:1. Roux's specimens 100–120 × 40 µm (André 1912 mentioned 100–110 × 40–45 µm). Body outline elongate elliptical, anteriorly rounded, posteriorly lanceolate tapered; anterior portion slightly curved rightwards. Strongly flattened dorso-ventrally, that is, ventral side plane to slightly excavated, dorsal side only slightly vaulted. Cells not very flexible. Invariably two ellipsoidal macronuclear nodules narrowly spaced behind proximal portion of adoral zone; each nodule with a micronucleus attached at left side; the presence of a reorganisation band ("spaltförmige Höhle" in Stein's terminology) indicates

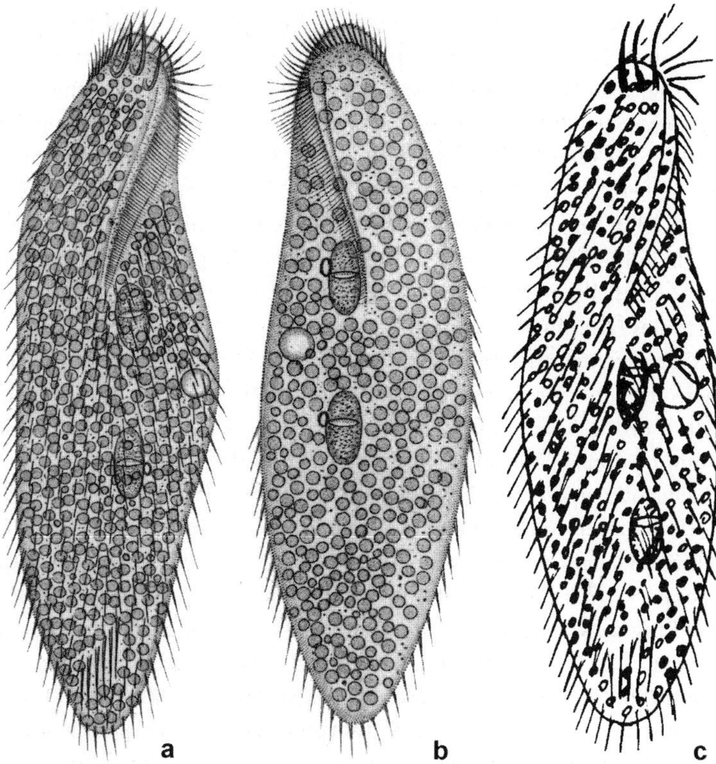

Fig. 222a–c *Urostyla viridis* from life (a, b, from Stein 1859; c, after Roux 1901). Ventral (a, c) and dorsal views (b) showing, inter alia, cirral pattern, nuclear apparatus, contractile vacuole, and symbiotic green algae. Individual sizes of Stein's specimens (a, b) not indicated (size range of population 115 to 175 µm), c = 100 µm. The generic assignment of the present species is uncertain, that is, only provisional because the cirral pattern is not known in detail (e.g., midventral complex present or not). Page 1106.

that Stein studied a thriving population. Contractile vacuole near left margin about in mid-body. Cytoplasm packed with symbiotic green algae making cells bright green. Cortical granules lacking because Stein (1859), who knew about this feature, neither mentioned nor illustrated them. Movement not described.

Adoral zone occupies about one third of body length, proximal portion extends obliquely to near midline of cell. Buccal field very narrow, frontal scutum distinct. Three enlarged frontal cirri triangularly arranged. Many rows of fine cirri narrowly and equidistantly arranged over whole ventral side, number difficult to determine because of underlying symbiotic algae; specimen shown in Fig. 222a with about 15 rows including marginal ones. Rows right of midline extend to near frontal cirri, cirri of frontal region not larger than remaining ventral cirri. Presence/absence of midventral complex not known, assignment to urostyloids therefore uncertain (see remarks). Five indistinctly enlarged transverse cirri arranged in oblique, subterminal row; do not project beyond rear body end. Marginal cirri (cirri of outermost cirral rows) become longer posteriad. Dorsal ciliature (length of dorsal bristles, number and arrangement of kineties, presence/absence of caudal cirri) not known.

Molecular data: As discussed in the remarks, the population isolated by Hoffman & Prescott (1997) is neither described morphologically, nor checked by a specialist. Indeed the position of "*Paraurostyla viridis*" as sister-group of *Oxytricha granulifera* in the molecular trees provided by Hewitt et al. (2003) indicates that the identification is incorrect. In the tree shown by Dalby & Prescott (2004) it clustered with *Engelmanniella mobilis*, which is very likely not a urostyloid either. Thus, I only list the markers investigated: (i) macronuclear DNA polymerase alpha gene (Hoffman & Prescott 1997, GenBank accession number U89701); (ii) 17S ribosomal RNA gene, internal transcribed spacer 1, 5.8S ribosomal RNA gene and internal transcribed spacer 2, complete sequence; and 26S ribosomal RNA gene, partial sequence (Hewitt et al. 2003, GenBank accession number AF508766); (iii) macronuclear actine I gene (Hogan & Prescott; direct submission GenBank accession number AY044840; Croft et al. 2003); (iv) macronuclear histone H4 molecules (Kelminson et al. 2002).

Occurrence and ecology: As mentioned above, this chapter contains only records where the identification is based on the original description, the redescription by Roux (1901), or the review by Foissner et al. (1991), who clearly stated that the identification should be done according to the original description.

Urostyla viridis is a limnetic species. Stein (1859) found *U. viridis* at two sites without designating one of them as type locality. At first (September 1857) he found it with high abundance in a plot of peat near the village of Niemegk (Brandenburg, Germany). In January 1858 he found it with low abundance in a marshy ditch in the "Baumgarten" near Prague, Czech Republic. Roux (1901) recorded *U. viridis* in various freshwater habitats near Geneva (Pinchat, Lalubin, Châtelaine, Petit-Saconnex, Bel-Air, Lignon, Rouelbeau, Plan-les-Quates, Bessinge), Switzerland, throughout the year.

Records not substantiated by morphological data: sporadically in a puddle on a decaying tree stump near the village of Ober-Hollabrunn, Lower Austria during September (Spandl 1926a, p. 91); ditch in Estonia in July (Jacobson 1928, p. 103); pond in the Zoological Garden of Freiburg in Breisgau, Germany (Henderson 1905, p. 17); during March, April, November, and December in alkaline water bodies (at pH 6.2–8.4, 1.8 to 7.8°C; 6.0–12.5 mg O_2 l^{-1}, 286–794 $\mu S\ cm^{-1}$) in the Hortobágy National Park, Hungary (Szabó 1999a, p. 229; 2000a, p. 8); mesosaprobic region of Stirone River, northern Apennines, Italy (Madoni & Bassanini 1999, p. 395); sludge cultures from the Danube river in Romania (Spandl 1926b, p. 534); ponds in Switzerland almost throughout the year (Bourquin-Lindt 1919, p. 73); clean, oxygen-rich waters near Basle, Switzerland (Riggenbach 1922, p. 52); Rhone plancton in Geneva, Switzerland (André 1926, p. 262); bog in Switzerland (Mermod 1914, p. 102; refers to Schlenker's paper which I do not have); saline lakes near Odessa, Ukraine (Boutchinsky 1895, p. 145; Butschinsky 1897, p. 196, reviewed by Hammer 1986, p. 371); Chaohu Lake, China (Xu et al. 2005, p. 188); Brazil (Cunha 1913, p. 107).

The population used for molecular analyses was isolated by Hoffman & Prescott (1997) without giving an origin. According to Kelminson et al. (2002), Hewitt et al. (2003, p. 259), and Croft et al. (2003, p. 342) it was isolated from the Misty Creek Pond, Sarasota, Florida, USA. Bouvier (1893, p. 127) mentioned *U. viridis* in a review on animals with symbiotic algae.

The following papers very likely deal with *U. viridis* sensu Kahl (1932), which is certainly not identical with Stein's material and thus only briefly mentioned: Bervoets (1940, p. 136), Gel'cer & Geptner (1976, p. 177), Grimm (1968, p. 365), Heinis (1937,

p. 63), Ma (1994, p. 95), Messikommer (1954, p. 642), Michiels (1974, p. 135), Patrick (1961, p. 244), Patrick et al. (1967, p. 325), Xu & Wood (1999, p. 105).

Species indeterminata

Urostyla rubra Andrussowa, 1886 (Fig. 223a)

1886 ***Urostyla rubra.* (nova. sp.)** – Andrussowa, Trudy imp. S-peterb. Obshch. Estest., 17: 246, Table II, Fig. 10 (original description; no type material available).

1932 ***Urostyla rubra* Andrussowa, 1886** – Kahl, Tierwelt Dtl., 25: 568, Fig. 86 $_{25}$ (Fig. 223a; revision of hypotrichs).

1972 ***Paraurostyla rubra* (Andrussowa, 1886) n. comb.** – Borror, J. Protozool., 19: 10 (combination with *Paraurostyla*; revision of hypotrichs).

2001 ***Urostyla rubra* Andrussowa, 1886** – Berger, Catalogue of ciliate names 1. Hypotrichs, p. 102 (nomenclator containing all basionyms, combinations, and higher taxa of hypotrichs).

Remarks: The species-group name *rub·er -ra -rum* (Latin; red) obviously refers to the colour of the cell. The original description is in Russian and the illustration not very detailed. I did not translate the paper, but the illustration shows that an identification will hardly be possible. This basically agrees with Hemberger (1982, p. 34), who classified this species as incertae sedis in *Urostyla*, but doubted that it can be identified. Possibly *U. rubra* is identical with *U. gracilis* which is also reddish. *Urostyla naumanni* (Fig. 220a), which was discovered – like the present species – in the Black Sea, is yellowish and has a pearl necklace-shaped macronucleus.

Borror (1972) transferred the present species to *Paraurostyla* Borror, 1972. Borror & Wicklow (1983, p. 117) mistakenly wrote that Borror (1972) considered it as member of *Urostyla*. Moreover Borror & Wicklow stated that it should be of questionable taxonomic placement pending rediscovery.

My photocopy of this article has not a very good quality and therefore I only show Kahl's redrawing. Size not indicated. Body outline broadly oval. Nuclear apparatus likely not described. Cells obviously red (species name). Several enlarged frontal cirri. Specimen illustrated with about 10 cirral rows. Transverse cirri obviously lacking, indicating that the generic assignment was incorrect. Type locality is the Bay of Kertsch, Black Sea.

Fig. 223a *Urostyla rubra*, a species indeterminata, from life (after Andrussowa 1886 from Kahl 1932). Ventral view, size not indicated. Page 1111.

Insufficient redescriptions

***Urostyla grandis* Ehr.** – Edmondson, 1906, Proc. Davenport Acad. Sci., 11: 97, Plate XXII, Fig. 159 (Fig. 164g). Remarks: The two macronuclear nodules prove that the identification is incorrect. Very likely it is a species of the *Paraurostyla weissei* complex (for review of this oxytrichid see Berger 1999). Body length 250–400 μm in life. Both body ends rounded, slightly narrowed anteriorly. Two ellipsoidal macronuclear nodules each with an attached micronucleus. Contractile vacuole slightly ahead of mid-body. Cytoplasm yellowish. Adoral zone occupies about one third of body length. Several frontal cirri. Many cirral rows (specimen illustrated with 10). 10 or 12 transverse cirri. Feeds on diatoms or other unicellular algae. Freshwater in Iowa, USA.

***Urostyla* sp.** – Lepsi, 1957, Buletin sti. Acad. Repub. pop rom., 9: 234, Fig. 6 (Fig. 164e). Remarks: An identification of this population is certainly impossible because the illustration is much too superficial. Body size 90–105 μm. Found in the ombrogenic bog of Poiana Stampei (Eastern Carpathians, Romania).

***Urostyla* sp.** – Nikoljuk & Geltzer, 1972, Pocvennye prostejsie SSSR, p. 131, Plate XII, Fig. 239 (Fig. 164f). Remarks: Possibly this is a *Kahliella* species as indicated by the cirral pattern and the habitat (soil from USSR). About 100 μm long. In total six cirral rows.

Epiclintidae Wicklow & Borror, 1990

1983 **Epiclintina (n. subord.).** – Wicklow, Diss. Abstr. Int., 43B: 2135 (original description). Type genus: *Epiclintes* Stein, 1863.

1990 **Epiclintidae (n. fam.)**[1] – Wicklow & Borror, Europ. J. Protistol., 26: 192 (original description). Type genus: *Epiclintes* Stein, 1863.

1994 **Epiclintidae Wicklow et Borror, 1990** – Tuffrau & Fleury, Traite de Zoologie, 2: 140 (revision).

2001 **Epiclintidae Wicklow & Borror, 1990** – Berger, Catalogue of ciliate names 1. Hypotrichs, p. 106 (nomenclator containing all basionyms, combinations, and higher taxa of hypotrichs).

2002 **Epiclintidae Wicklow and Borror, 1990** – Lynn & Small, Phylum Ciliophora, p. 450 (guide to representative genera).

Nomenclature: The name Epiclintidae, established as family, is based on the genus-group name *Epiclintes*. I use the name without category (see chapter 7.2 of the general section). Wicklow (1983) did not provide a diagnosis or characterisation, strongly indicating that the taxon Epiclintina is a nomen nudum.

Characterisation (Fig. 224a, autapomorphies 1): Urostyloidea with a multicorona, that is, frontal ciliature composed of several cirral bows (A?). Midventral complex consists of midventral rows only (A).

Remarks: Wicklow & Borror (1990) concluded from their ultrastructural and morphogenetic studies on *Epiclintes ambiguus* (= *E. auricularis* in present book) that *Epiclintes* is a specialised descendent from *Kahliella*-like stichotrichines. Also likely to have radiated from this group are, according to Wicklow & Borror (1990), *Engelmanniella* in soils, *Psilotricha* and *Eschaneustyla* in mosses, *Stichotricha* in the periphyton of both freshwater and marine habitats, and freshwater, planktonic forms such as *Hypotrichidium* and *Pseudokahliella*. Divergent features of the cortex, morphogenetic pattern, and degree of contractility, however, caused them to establish the monotypic, stichotrichine family Epiclintidae.

Tuffrau & Fleury (1994) classified the Epiclintidae in the order Oxytrichida, suborder Stichotrichina. Lynn & Small (2002) assigned it to the order Stichotrichida. In the papers of all workers the Epiclintidae are monotypic, that is, contain only the type genus. Eigner (2001) ignored the Epiclintidae and assigned *Epiclintes* to the urostyloids, without, however, providing details.

The classification of the Epiclintidae in the Urostyloidea is indicated by the many oblique frontal-(mid)ventral-transverse cirral anlagen, the many macronuclear nodules, and the simple dorsal kinety pattern composed of bipolar kineties only. I agree with Wicklow & Borror (1990) that *Eschaneustyla* is likely related with *Epiclintes* (Fig. 224a). Both taxa have a frontal ciliature composed of cirral bows and the midventral complex consists of midventral rows only, that is, even the anterior portion of the com-

[1] Wicklow & Borror (1990) provided the following diagnosis: Ventral ciliature is comprised of numerous oblique rows of cirri, the majority of which differentiate during morphogenesis from parental cirral rows and ventral primordia; frontal cirral rows are minimally represented and differentiate from primordia medial to the oral primordium. A longitudinal row of transverse cirri is located medial to the left margin cirri in the posterior half of the cell. Transverse cirri differentiate from the posterior of frontal streaks, ventral (within-rows) streaks, and streaks that develop within ventral primordia. Dorsal cilia project from cylindrical papillae. A system of multiple, membrane-like material is present within the cortex.

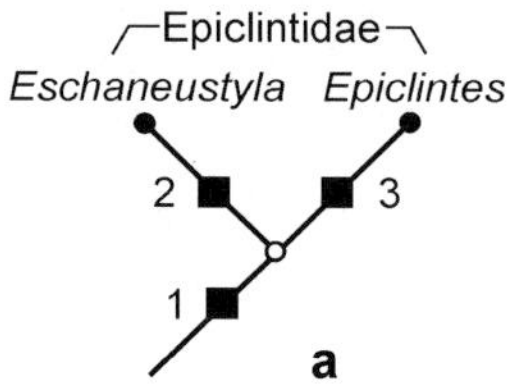

Fig. 224a Diagram of phylogenetic relationships within the Epiclintidae (original). Autapomorphies (black squares 1–3): **1** – frontal ciliature and midventral complex composed of midventral rows only. **2** – many frontoterminal cirri form distinct row; transverse cirri lacking; 4 dorsal kineties; more than one caudal cirrus per dorsal kinety. **3** – body tripartite in head, trunk, and tail; number of transverse cirri distinctly higher than number of frontal and midventral rows; frontoterminal cirri lacking; dorsal papillae present; caudal cirri lacking; a system of multi-layered, membrane-like materials lie within cortex. The features of *Epiclintes* refer only to the well-described type species! Note that there are several convergencies (e.g., transverse cirri lacking, frontoterminal cirri lacking).

plex forms rows instead of pairs (Fig. 225a–c). Unfortunately, there are no further convincing features uniting *Epiclintes* and *Eschaneustyla*. Only one species of *Eschaneustyla*, *Eschaneustyla lugeri* shows a slight cephalisation, a feature rather distinct in *Epiclintes*. Both genera, but especially *Epiclintes*, are characterised by some good apomorphies. For example, *Epiclintes* has dorsal papillae, which occur, interestingly, also in *Paramitrella*, a (little known) marine species of similar shape, but with a midventral complex composed of cirral pairs only (Fig. 240a, b). I do not know whether or not the papillae are a synapomorphy or convergence. Detailed studies on *Paramitrella* are needed to show whether or not it is closely related with *Epiclintes*. A further interesting feature of *Epiclintes* is the presence of a multiple, membrane-like material in the cortex. A similar (homologous?) structure is present in *Engelmanniella mobilis* (Wirnsberger-Aescht et al. 1989), a species of uncertain position; according to molecular data it is often related to oxytrichids (Fig. 15a).

The last common ancestor of *Eschaneustyla* obviously had no transverse cirri and possibly a slightly increased number of dorsal kineties, namely four against three in the ground pattern of the urostyloids. The plesiomorphic value of three kineties is still present in *Epiclintes*.

Epiclintes has very high DE-values (0.56–0.73; Fig. 228f, w). In *Eschaneustyla* these values are distinctly lower (around 0.25). The high values are reminiscent of the Retroextendia, which, however, lack midventral rows.

Further data (e.g., fine structure of *Eschaneustyla*, molecular features) are likely needed to show whether or not *Epiclintes* and *Eschaneustyla* are sister groups (Fig. 224a).

Possibly the Epiclintidae are closely related to *Urostyla* because *U. grandis*, type of the genus, also has a multicorona and the rear portion of the midventral complex is already composed of midventral rows.

Key to the genera of the Epiclintidae

The two taxa assigned to the Epiclintidae are rather easily distinguished by the presence/absence of transverse cirri. Moreover, they do not overlap in habitat. There are some further species with a similar body outline as *Epiclintes auricularis*, namely

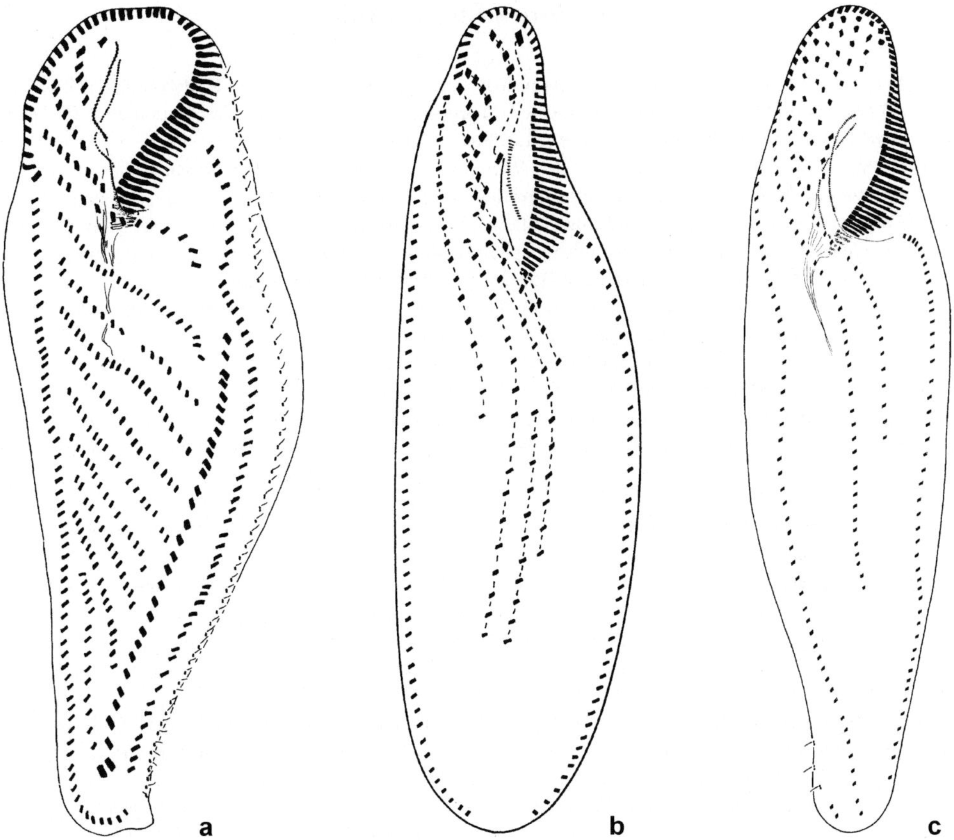

Fig. 225a–c Ventral cirral pattern in members of the Epiclintidae. **a:** *Epiclintes auricularis*. **b, c:** *Eschaneustyla terricola* and *E. lugeri*. Sources of illustrations see individual descriptions. Abbreviations used in short characterisations of infraciliature (explanation of supplementary signs and numbers see Fig. 20a–c): AZM = adoral zone of membranelles, BC = buccal cirrus, CC = caudal cirri, DK = dorsal kineties, FT = frontoterminal cirri, LMR = left marginal row, MC(MV) = midventral complex composed of midventral rows only, MU = multicorona, RMR = right marginal row, TC = transverse cirri.

Psammomitra (Fig. 43a) and *Paramitrella* (Fig. 240a). Both are marine, but have, inter alia, a rather different cirral pattern so that confusion is almost impossible.

1 Transverse cirri present; marine (Fig. 225a, 228t) .... *Epiclintes* (p. 1116)
- Transverse cirri absent; limnetic or terrestrial (Fig. 225b, c) .... *Eschaneustyla* (p. 1146)

Epiclintes Stein, 1863

1863 ***Epiclintes*** – Stein, Amtliche Berichte Deutscher Naturforscher und Ærzte in Karlsbad, 37: 162 (original description; no formal diagnosis provided). Type species (by subsequent designation by Stein 1864, p. 44): *Oxytricha auricularis* Claparède & Lachmann, 1858.
1864 ***Epiclintes* St.** – Stein, Sber. K. böhm. Ges. Wiss., year 1864: 44 (fixation of type species).
1867 ***Epiclintes*** – Stein, Organismus der Infusionsthiere II, p. 150 (review).
1882 ***Epiclintes*, Stein** – Kent, Manual Infusoria II, p. 773 (pro parte; revision).
1889 ***Epiclintes* Stein 1862** – Bütschli, Protozoa, p. 1742 (revision; incorrect year).
1932 ***Epiclintes* Stein, 1859** – Kahl, Tierwelt Dtl., 25: 569 (revision; incorrect year).
1933 ***Epiclintes* Stein 1862** – Kahl, Tierwelt N.- u. Ostsee, 23: 108 (guide to marine ciliates; incorrect year).
1950 ***Epiclintes* Stein** – Kudo, Protozoology, p. 673 (textbook).
1961 ***Epiclintes* Stein** – Fauré-Fremiet, C. r. hebd. Séanc. Acad. Sci., Paris, 252: 3517 (revision).
1961 ***Epiclintes* St.** – Corliss, Ciliated Protozoa, p. 170 (revision).
1972 ***Epiclintes* Stein, 1862**[1] – Borror, J. Protozool., 19: 9 (revision; incorrect year).
1979 ***Epiclintes* Stein, 1863** – Jankowski, Trudy zool. Inst., Leningr., 86: 53 (revision/catalogue of hypotrichs).
1979 ***Epiclintes* Stein, 1862** – Corliss, Ciliated Protozoa, p. 310 (revision; incorrect year).
1982 ***Epiclintes* Stein, 1862**[2] – Hemberger, Dissertation, p. 26 (revision of non-euplotine hypotrichs; incorrect year).
1983 ***Epiclintes* Stein, 1862**[3] – Carey & Tatchell, Bull. Br. Mus. nat. Hist. (Zool.), 45: 48 (revision of *Epiclintes*).
1987 ***Epiclintes* Stein, 1859** – Tuffrau, Annls Sci. nat. (Zool.), 8: 116 (revision; incorrect year).
1992 ***Epiclintes* (Stein, 1862) Carey and Tatchell, 1983** – Carey, Marine interstitial ciliates, p. 189 (guide).
1994 ***Epiclintes* Müller, 1786** – Tuffrau & Fleury, Traite de Zoologie, 2: 141 (review; incorrect author and year).
1995 ***Epiclintes ambiguus* Müller, 1786**[4] – Wilbert, Acta Protozool., 34: 278 (improved diagnosis of *Epiclintes*).
1999 ***Epiclintes* Stein, 1859** – Shi, Song & Shi, Progress in Protozoology, p. 96 (revision; incorrect year).
2001 ***Epiclintes* Stein, 1863** – Aescht, Denisia, 1: 68 (catalogue of generic names of ciliates).
2001 ***Epiclintes* Stein, 1863** – Berger, Catalogue of ciliate names 1. Hypotrichs, p. 20 (nomenclator containing all basionyms, combinations, and higher taxa of hypotrichs).
2002 ***Epiclintes* Müller, 1786** – Lynn & Small, Phylum Ciliophora, p. 450 (guide to ciliate genera; incorrect author and year).

Nomenclature: *Epiclintes* (masculine; Aescht 2001, p. 282) refers to the ability to shoot (Stein 1864, p. 46).

Jankowski (1979) mentioned "*Epiclintes auricularis* Stein, 1863" as type species; however, Stein (1863) is neither the author of the species, nor the combining author. Bütschli (1889) and many other workers considered 1862 as year of the original description of *Epiclintes*. The meeting where Stein presented his data was in September 1862; but the resulting congress paper was published only in 1863.

---

[1] The diagnosis by Borror (1972) is as follows: Cirri in several oblique ventral rows. Transverse cirri present. Elongate, strongly contractile, free-living (often psammobiotic), with ciliated "tail".

[2] The diagnosis by Hemberger (1982) is as follows: Cirren in mehreren schrägen Ventralreihen; Besitz von Frontal- und Transversalcirren nicht eindeutig geklärt; Körper langgestreckt, in "Kopf" und "Schwanz" gegliedert; sehr kontraktil.

[3] Carey & Tatchell (1983) provided a very long diagnosis for *Epiclintes* which is therefore not repeated in the present book.

[4] The diagnosis by Wilbert (1995) is as follows: ventrally 1 row of left marginal cirri and 1 row of right marginal cirri, as well as many diagonal rows of cirri. No differentiated frontal cirri. Transverse cirri present. No caudal cirri. Body elongated, subdivided in "head" and "tail", extremely contractile.

There is some uncertainty about the type species of *Epiclintes*, although the situation is rather simple (see remarks). According to Borror (1972), *Epiclintes ambiguus* is the type species of *Epiclintes* by subsequent designation, whereas Aescht (2001) mentioned *Trichoda felis* as type by monotypy. However, Stein (1864) explicitly fixed *Oxytricha auricularis* as type, which is, admittedly, considered as junior subjective synonym of *Trichoda felis* and/or *T. ambigua* by some workers.

Carey & Tatchell (1983, p. 42) wrote that "Stein (1862)" (= Stein 1863 in present book) was in error in attributing the specific name "*auricularis*" to his new genus *Epiclintes* since the name *T. felis* – discussed as supposed synonym of *O. auricularis* by Stein (1863, 1864) – was published about 70 years before *O. auricularis*. However, they overlooked that Stein did not finally synonymise *O. auricularis* and *T. felis* and therefore Stein's decision to combine *O. auricularis*, and not *T. felis* with *Epiclintes*, was correct. Carey & Tatchell (1983) also wrote that Wallengren (1900) made a taxonomic error which led to great confusion in that he called the present species *Epiclintes ambiguus*. Obviously they overlooked that Wallengren simply followed Bütschli (1889), who made the new combination (see remarks at *E. auricularis*).

Incorrect subsequent spellings of *Epiclintes*: *Epiclinetes pluvialis* Smith (Webb 1961, p. 140); *Epiclinites ambiguus* O.F.M. (Petran 1971, p. 154); *Epiclinthes auricularis* Clap. Lachm., Stein (Mereschkowsky 1879, p. 164).

Characterisation (Fig. 224a, autapomorphies 3): Body tripartite in head, trunk, and tail (A); highly flexible and contractile. Adoral zone of membranelles continuous. Frontal and midventral ciliature composed of several oblique rows. Frontoterminal cirri lacking. Transverse cirri present. 1 right and 1 left marginal row. Dorsal cilia emerge from distinct papillae (A). Caudal cirri lacking (A). Posteriormost frontal-midventral-transverse cirral anlagen produce only transverse cirri (A). A system of multilayered, membrane-like materials lie within cortex (A).

Remarks: For discussion of the history of *Epiclintes*, see this chapter at the type species *E. auricularis*. Note that the characterisation above refers mainly to the well-described type species. The other two preliminarily included species are only little known.

Epiclintes is a rather curious hypotrich because of the high number of transverse cirri and oblique cirral rows. Due to this somewhat unusual cirral pattern, it was classified in various higher taxa. According to Lepsi (1929, p. 297), *Epiclintes* and *Stichotricha* are very primitive hypotrichs. Kahl (1932), who summarised all non-euplotid hypotrichs in the Oxytrichidae, arranged it in the sequence ... *Urostyla*, *Kerona*, *Epiclintes*, *Holosticha* ..., that is, he assumed a urostyloid relationship. Such a classification was also supposed by Calkins (1926, p. 410) and Borror (1972), who, however, definitely assigned it to the Urostylidae. Others classified *Epiclintes* in the Oxytrichidae (Corliss 1961; Bamforth 1962), the Amphisiellidae (Hemberger 1982), the Keronidae (Fauré-Fremiet 1961; Corliss 1977, p. 138; 1979, as incertae sedis; Tuffrau 1987, Carey 1992), the Spirofilidae (Shi 1999, Shi et al. 1999), and the Stichotrichida as incertae sedis (Dini et al. 1995, p. 71). Borror (1979, p. 548) was uncertain about the systematic position of *Epiclintes*, and Borror & Wicklow (1983) did not consider it in their revision on urostylids. Wicklow & Borror (1990) supposed from ultrastructural and morphogenetic data that *Epiclintes* is a specialised descendent from *Kahliella*-like stichotrichines. Interestingly enough, Carey & Tatchell (1983) and Song & Warren (1996) did not say any more on this topic.

According to Wicklow (1979, 1983) and Wicklow & Borror (1990a), including *Epiclintes* in any of the stichotrichid suborders (e.g., Urostylidae) is artificial. Thus, Wicklow (1983) established the suborder Epiclintina, whereas Wicklow & Borror (1990, p. 192) established the family Epiclintidae containing only *Epiclintes* with the two species *E. ambiguus* and *E. caudatus*. The Epiclintidae were accepted by Tuffrau & Fleury (1994) and Lynn & Small (2002). Recently, Eigner (2001) classified *Epiclintes* in the Urostylidae again, without, however, providing an explanation.

As already mentioned, Fauré-Fremiet (1961) put *Epiclintes*, together with *Eschaneustyla* and *Kerona*, into the Keronidae. Although I do not agree that *Epiclintes* is a keronid, I consider the idea of a close relationship with *Eschaneustyla* as highly interesting. A relationship between these two taxa (including *Psilotricha*, *Engelmanniella*, *Stichotricha*) is also supposed by Wicklow & Borror (1990, p. 192).

However, before I discuss this topic, the features indicating a relationship of *Epiclintes* to the urostyloids should be mentioned: (i) Adoral zone of membranelles distinctly reorganised during cell division (Fig. 228m–s). A conspicuous reorganisation of the adoral zone of membranelles is characteristic for many urostyloids, but lacking in all(?) other groups. (ii) *Epiclintes auricularis* has between 40 and 120 macronuclear nodules (Table 43; Fig. 226j, 227m, 228x). The urostyloids are possibly the sole group which includes species with a very high number of macronuclear nodules. (iii) The cirral pattern originates from rather many oblique frontal-(mid)ventral-transverse cirri anlagen (Fig. 228m–s). The urostyloids are the sole group producing a high number of oblique cirral anlagen. In most other groups (e.g., oxytrichids, amphisiellids, kahliellids) the number of anlagen is around six and the resulting rows are more or less longitudinally arranged; rarely 10 or more anlagen occur, for example, in *Paraurostyla weissei* and *Onychodromus quadricornutus* (for review see Berger 1999; now *Styxophrya quadricornuta*). (iv) Three dorsal kineties (e.g., Fig. 226n, 228g). Many urostyloids invariably have three bipolar dorsal kineties. (v) The cirral pattern, especially the fact that only more or less oblique cirral rows are present, is reminiscent of *Eschaneustyla*, which is rather certainly a urostyloid. Admittedly, some of the features are not very convincing, for example, the plesiomorphic number of dorsal kineties. But it is likely the most parsimonious solution to assume that *Epiclintes* is a urostyloid. Wicklow & Borror (1990) provided the lack of midventral cirri as single feature for an exclusion from the urostyloids. However, the zigzagging midventral pairs are a feature of the ground-pattern, which is obviously strongly modified in *Epiclintes* because the anlagen do not form midventral pairs, but midventral rows like, for example, in *Eschaneustyla*. However, as in other cases molecular data should be considered to show which of the proposed relationships is most likely.

According to Bütschli (1889), *Diplagiotricha* Bory, 1824 (see Lamouroux et al. 1824, p. 530) is a senior synonym of *Epiclintes*. However, Bory de Saint-Vincent in Lamouroux et al. (1824, p. 530) did not mention *Diplagiotricha*, but "Diplagiotriques", which was not intended as genus name (Aescht 2001, p. 60). Moreover, Bory de Saint-

Vincent mentioned *Oxytricha ambigua* (for *Trichoda ambigua*) as an example of this group, a species which is not considered as synonym of *E. auricularis* in the present book. Corliss (1979, p. 208) listed *Diplagiotricha* as nomen oblitum, that is, a forgotten name.

Bütschli (1889) provided two illustrations which are, according to the figure legend, from "Lieberkühn's Tafeln von 1855". However, in Bütschli's reference section no paper by Lieberkühn (1855) is mentioned, indicating that these plates were never published.

Unidentified *Epiclintes* species where recorded, inter alia, from the following sites: in the 25 foot deep, 1 million gallon ocean of the Biosphere II project (Spoon & Alling 1993); shallow waters from the Terra Nova Bay, Antarctic region (Petz & Valbonesi 1998); soil samples from the sub-Antarctic Ile de la Posseession, Iles Crozet (Smith 1975, p. 525; 1978, p. 27).

Species included in *Epiclintes* (alphabetically arranged according to basionyms): (1) *Oxytricha auricularis* Claparède & Lachmann, 1858. Incertae sedis: (2) *Epiclintes pluvialis* Smith, 1900; (3) *Epiclintes vermis* Gruber, 1884.

Species misplaced in *Epiclintes*: *Epiclintes caudatus* Bullington, 1940 (now *Paramitrella caudatus*).

Epiclintes caudatus Lackey, 1961 (a nomen nudum; see species indeterminata).

Epiclintes tortuosus Lackey, 1961 (a nomen nudum; see species indeterminata).

Key to *Epiclintes* species

Only the type species, *Epiclintes auricularis*, is well described. The cirral pattern and several other details of the other two species provisionally included are not known. Consequently, only the body shape and the habitat can be used for separation.

1 Limnetic (Fig. 230a) .... *Epiclintes pluvialis* (p. 1142)
- Marine (Fig. 228t, 231a) **2**

2 Body tripartite in head, trunk, and tail (Fig. 228t) .. *Epiclintes auricularis* (p. 1119)
- Body elongate, reminiscent of a microturbellarian (Fig. 231a) ....
.... *Epiclintes vermis* (p. 1143)

Epiclintes auricularis (Claparède & Lachmann, 1858) Stein, 1864 (Fig. 226a–s, 227a–z, 228a–z, 229a, Table 43)

1858 ***Oxytricha auricularis***[1] – Claparède & Lachmann, Mém. Inst. natn. génev., 5: 148, Planche V, Fig. 5, 6 (Fig. 226a, d; original description; no type material available).

1864 ***Epiclintes auricularis*** – Stein, Sber. K. böhm. Ges. Wiss., year 1864: 44, 46 (detailed redescription without illustration; combination with *Epiclintes*).

[1] The diagnosis by Claparède & Lachmann (1858) is as follows: Partie antérieure élargie. Pieds-cirrhes tout à fait rudimentaires. Une queue non rétractile.

1866 ***Claparedia? auricularis* Diesing** – Diesing, Sber. Akad. Wiss. Wien, 53: 99 (transfer to *Claparedia*).

1867 ***Epiclintes auricularis*** – Stein, Organismus der Infusionsthiere II, p. 150 (review).

1877 ***Epiclintes auricularis* Clap. Lachm., Stein** – Mereschkowsky, Trudy imp. S-petersb. Obshch. Estest., 8: 237, Plate I, Fig. 21 (Fig. 226e; redescription).

1879 ***Epiclinthes auricularis* Clap. Lachm., Stein** – Mereschkowsky, Arch. mikrosk. Anat. EntwMech., 16: 164, Tafel X, Fig. 16 (redrawing of Fig. 226e; redescription; incorrect subsequent spelling of *Epiclintes*).

1882 ***Epiclintes auricularis*, C. & L. sp.** – Kent, Manual Infusoria II, p. 773, Plate XLIII, Fig. 28–30 (redrawings of Fig. 226a, d, e; revision).

1884 ***Epiclintes auricularis*, Cl. L. sp.** – Rees, Tijdschr. ned. dierk. Vereen, Suppl. I: 642, 643, Planche XVI, Fig. 17 (Fig. 226g; redescription).

1886 ***Oxytricha auricularis* Clap.** – Pereyaslawzewa, Zap. novoross Obshch. Estest., 10: 92, second plate, Fig. 17 (Fig. 228y, z; redescription).

1888 ***Epiclinites auricularis* Clap und Lachm.** – Gruber, Ber. naturf. Ges. Freiburg i. B., 3: 62, Fig. 8 (Fig. 226j; description of nuclear apparatus; incorrect subsequent spelling of *Epiclintes*).

1889 ***Epiclintes ambiguus* O. F. M. sp.** – Bütschli, Protozoa, Legend to Tafel LXX, Fig. 12a, b (Fig. 226b, c; combination of *Trichoda ambigua* with *Epiclintes*).

1900 ***Epiclintes ambiguus* O. F. Müller**[1] – Wallengren, Acta Univ. lund., 36: 1, Platta I, Fig. 1–4 (Fig. 226m–o; detailed redescription).

1929 ***Epiclintes ambiguus* O. F. M.** – Alzamora, Botas Resúm. Inst. esp. Oceanogr., 2: 12, Fig. 27 (Fig. 226f; illustrated review).

1929 ***Epiclintes ambiguus* O. F. M.** – Hamburger & Buddenbrock, Nord. Plankt., 7: 83, Fig. 99a, b (Fig. 226b, c; guide to marine plankton).

1932 ***Epiclintes (Trichoda) ambiguus* (Müller, 1786) Bütschli, 1889** – Kahl, Tierwelt Dtl., 25: 569, Fig. 97_{17}, 110 $_{16}$ (Fig. 226i, s; revision of hypotrichs).

1933 ***Epiclintes ambiguus* (O. F. Müller 1786)** – Kahl, Tierwelt N.- u. Ostsee, 23: 108, Fig. 16.24 (Fig. 226h; guide to marine ciliates).

1935 ***Epiclintes (Trichoda) ambiguus* (O. F. Müller 1786) Bütschli 1889** – Kiesselbach, Note Ist. italogerm. Biol. mar. Rovigno, 18: 1, Abb. 1–5 (Fig. 227a–h; redescription from life).

1936 ***Epiclintes (Trichoda) ambiguus* (O. F. Müller 1786) Bütschli 1889** – Kiesselbach, Thalassia, 2: 19, Abb. 38–40 (Fig. 227i–k; redescription).

1936 ***Epiclintes ambiguus* (O. F. Müller 1786)** – Kiesselbach, Ciliati della Laguna Veneta, p. 10, Fig. 1 (Fig. 227l; illustrated record).

1943 ***Epiclintes (Trichoda) ambiquus* (Müller, 1786) Butschli, 1889** – Ozaki & Yagiu, J. Sci. Hiroshima Univ., 10: 31, Fig. 11A, B, 12 (Fig. 227m–o; redescription; incorrect subsequent spelling).

1960 ***Epiclintes ambiguus* (Müller) Bütschli** – Dragesco, Trav. Stn. biol. Roscoff, 12: 310, Fig. 166 (Fig. 226k, l; description of nuclear apparatus).

1963 ***Epiclintes ambiguus* (Müller, 1786)** – Borror, Arch. Protistenk., 106: 509, Fig. 113–115 (Fig. 227p–r; redescription).

1972 ***Epiclintes ambiguus* (Müller, 1786) Bütschli, 1889** – Borror, J. Protozool., 19: 9, Fig. 62 (Fig. 227s; revision of hypotrichs).

1973 ***Epiclintes ambiguus*** – Hartwig, Mikrokosmos, 62: 335, Bild 6A (micrograph).

1976 ***Epiclintes ambiguus* (O. F. Müller, 1787)** – Czapik & Jordan, Acta Protozool., 15: 442, Fig. 16B (Fig. 226p; illustrated record; incorrect year).

1979 ***Epiclintes ambiguus* (O. F. Müller, 1786)** – Borror, J. Protozool., 26: 548, Fig. 6 (Fig. 227t; brief review of urostylids).

1980 ***Epiclintes ambiguus* (Müller, 1786)** – Borror, J. Protozool., 27: 11, Fig. 3 (Fig. 227u; brief review on marine ciliates).

1982 ***Epiclintes ambiguus* (Müller, 1786) Bütschli, 1889** – Hemberger, Dissertation, p. 26 (revision).

1983 ***Epiclintes felis* (Muller, 1786) comb. n.** – Carey & Tatchell, Bull. Br. Mus. nat. Hist. (Zool.), 45: 43, 50, Fig. 1–9 (Fig. 228a–e; combination with *Epiclintes*; detailed redescription including ultrastructure; a voucher slide [accession number 1982:2:12] has been deposited in the British Museum in London).

---

[1] The first page of Wallengren (1900), where the description of *E. ambiguus* commences, is lacking in my xerox copy; the entry in the list is from the legend to the figures.

1985 ***Epiclintes ambiguus*** **(O. F. Müller, 1786)** – Aladro Lubel, An. Inst. Biol. Univ. Méx., 55: 27, Lamina 13, Fig. 3 (Fig. 226q; illustrated record).

1990 ***Epiclintes ambiguus*** **(Müller, 1786) Bütschli, 1889** – Wicklow & Borror, Europ. J. Protistol., 26: 184, Fig. 1–26, Tables 1, 2 (Fig. 228f–s; detailed redescription including morphogenesis and ultrastructure; Figure 1, 4 also shown by Small & Lynn 1985, p. 461 and Lynn & Corliss 1991, p. 355).

1990 ***Epiclintes ambiguus*** **(O. F. Müller, 1786)** – Aladro Lubel, Martínez Murillo & Mayén Estrada, Manual de Ciliados, p. 136, one figure on same page (Fig. 226r; review).

1992 ***Epiclintes felis*** **(Müller, 1786) Carey and Tatchell, 1983** – Carey, Marine interstitial ciliates, p. 189, Fig. 750 (redrawing of Fig. 228a; guide).

1994 ***Epiclintes ambiguus*** **Stein, 1859** – Tuffrau & Fleury, Traite de Zoologie, 2: 140, Fig. 52a–e (Fig. 227v; an unpublished illustrations by Fauré-Fremiet; review; incorrect author and year).

1995 ***Epiclintes ambiguus*** **Müller, 1786** – Wilbert, Acta Protozool., 34: 278, Fig. 11, Table 11 (Fig. 227w–z; redescription of saline lake population).

1996 ***Epiclintes ambiguus*** **(Müller, 1786) Bütschli, 1889** – Song & Warren, Acta Protozool., 35: 233, Fig. 12–16, Table 1 (Fig. 228t–x; detailed redescription; voucher slides are deposited in the Laboratory of Protozoology, College Fisheries, Ocean University of Qingdao, China).

1997 ***Epiclintes felis*** **(Müller 1786) Carey and Tatchell 1983** – Al-Rasheid, Arab. Gulf J. Scient. Res., 15: 756, Fig. 7g (record substantiated by micrograph).

2001 ***Epiclintes ambiguus*** – Eigner, J. Euk. Microbiol., 48: 77, Fig. 24 (modified Fig. 228f; brief review of urostylids).

2001 ***Epiclintes auricularis*** **(Claparède & Lachmann, 1858) Stein, 1864** – Berger, Catalogue of ciliate names 1. Hypotrichs, p. 52 (nomenclator containing all basionyms, combinations, and higher taxa of hypotrichs).

2002 ***Epiclintes ambiguus*** – Lynn & Small, Phylum Ciliophora, p. 450, Fig. 30 (Fig. 228l; guide to ciliate genera).

Nomenclature: Claparède & Lachmann (1858) selected the species-group name *auricular·is -is -e* (Latin adjective; concerning the ear, auricular) because the present species has the shape of an ear-cleaner. The species-group name *ambigu·us -a -um* (Latin adjective; bending to both [two] sides; on both sides; Hentschel & Wagner 1996) possibly refers to the movement (see below). The species-group name *felis* (Latin noun) means cat.

Oxytricha auricularis was fixed as type species of *Epiclintes* by subsequent designation by Stein (1864). Kahl (1932) and some later workers wrote *Trichoda* between the genus-group name and the species-group name. However, they did not consider *Trichoda* as subgenus of *Epiclintes*, but wanted to indicate *E. ambiguus* was originally classified in *Trichoda*. Shi (1999, p. 250) and Shi et al. (1999, p. 96) assigned *E. ambiguus* to Stein (1859); however, this is incorrect because *Trichoda ambigua* was described by Müller (1786).

Incorrect subsequent spellings: *Epiclintes ambigum* (Chardez 1956, p. 92); *Epiclintes ambiguua* and *Trichoda ambiguua* (Carey & Tatchell 1983, p. 42); *Epiclintes ambiquus* (Burkovsky 1970b, p. 11; Azovsky 1996, p. 6; Azovsky et al. 1996, p. 30).

Remarks: The synonymy of the present species is muddled and therefore its history has to be explained in detail. Stein (1859, p. 183) discussed Claparède & Lachmann's (1858) paper and found that their *Oxytricha auricularis* and *O. retractilis* belong to a new genus, without, however proposing a new name. Few years later, Stein (1863) found a contractile hypotrich with oblique ventral cirral rows in the Baltic Sea. He recognised that this species was one of the two curious *Oxytricha* species described by Claparède & Lachmann (1858), but he did not remember the species-group name when

Table 43 Morphometric data on *Epiclintes auricularis* (au1, Great Bay, New Hampshire population from Wicklow & Borror 1990; au2, Lake Qarun population from Wilbert 1995; au3, from Song & Warren 1996; au4, from Carey & Tatchell 1983)

Characteristics [a]	Population	mean	M	SD	SE	CV	Min	Max	n
Body, length [c]	au1	–	–	–	–	–	170.0	265.0	8
	au2	77.7	76.0	11.8	–	–	66.0	103.0	10
	au3	142.7	–	16.7	5.6	11.7	120.0	162.0	9
Body, width [c]	au1	–	–	–	–	–	35.0	54.0	8
	au2	37.8	38.0	2.7	–	–	34.0	41.0	10
	au3	61.0	–	5.7	1.9	9.3	54.0	72.0	9
Anterior body end to proximal end of adoral zone, distance [c]	au3	41.8	–	3.9	1.3	9.4	35.0	48.0	9
Macronuclear nodules, length	au3	–	–	–	–	–	5.0	7.0	?
Macronuclear nodules, width	au3	–	–	–	–	–	3.0	5.0	?
Macronuclear nodules, number	au2	56.0	50.0	18.0	–	–	40.0	100.0	11
	au3	–	–	–	–	–	109.0	123.0	2
Adoral membranelles, number	au1	63.0	–	5.0	1.0	8.0	51.0	71.0	25
	au2	44.8	45.0	4.3	–	–	41.0	56.0	11
	au3	54.4	–	4.3	1.5	8.0	48.0	61.0	8
	au4	–	–	–	–	–	25.0	35.0	?
Cirri anlage I, number of cirri formed	au1	1.5	–	0.5	0.1	33.5	1.0	2.0	25
	au2	0.5	1.0	0.5	–	–	0.0	1.0	8
	au3	1.0	–	0.0	0.0	0.0	1.0	1.0	9
Frontal plus midventral cirral rows, number [b]	au1	13.8	–	0.8	0.2	5.9	13.0	15.0	25
	au2	11.4	12.0	1.7	–	–	8.0	13.0	9
	au3	12.4	–	0.5	0.2	4.2	12.0	13.0	8
Frontal plus midventral cirri, number	au1	217.1	–	25.2	5.0	11.3	18.0	34.0	25
Transverse cirri, number	au1	30.4	–	3.4	0.7	11.3	18.0	34.0	25
	au2	25.6	24.0	3.4	–	–	20.0	29.0	8
	au3	26.1	–	2.0	0.7	7.5	24.0	29.0	7
Left marginal cirri, number	au1	77.0	–	7.7	1.5	10.0	65.0	93.0	25
	au2	52.0	51.0	6.7	–	–	45.0	63.0	8
	au3	53.1	–	2.7	1.0	5.1	49.0	56.0	7
Right marginal cirri, number	au1	82.3	–	7.2	1.4	8.8	63.0	94.0	25
	au2	61.3	60.0	7.6	–	–	52.0	72.0	7
	au3	66.1	–	4.4	1.7	6.7	63.0	75.0	7
Cirri, total number	au1	408.4	–	39.0	7.8	9.5	345.0	460.0	25
Dorsal kineties, number	au1	3.0	–	–	–	–	3.0	3.0	25
	au2	4.0	4.0	0.0	–	–	4.0	4.0	7
	au3	3.0	–	0.0	0.0	0.0	3.0	3.0	9

[a] All measurements in µm. Data are based on protargol-impregnated specimens. CV = coefficient of variation in %, M = median, Max = maximum value, mean = arithmetic mean, Min = minimum value, n = number of individuals investigated (? = number not given. If only one value is available, then it is listed under the column mean; if two values are given, then they are listed under Min and Max), SD = standard deviation, SE = standard error of arithmetic mean.

[b] Cirri of anlage I not included.

[c] Data based on contracted cells.

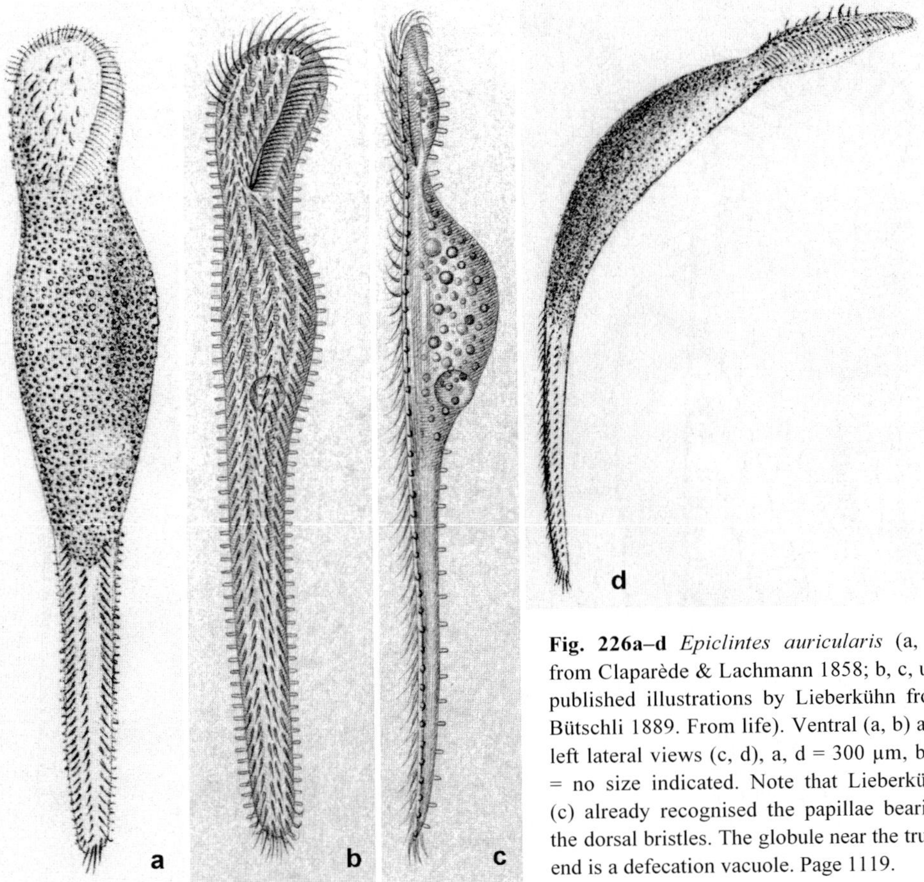

Fig. 226a–d *Epiclintes auricularis* (a, d, from Claparède & Lachmann 1858; b, c, unpublished illustrations by Lieberkühn from Bütschli 1889. From life). Ventral (a, b) and left lateral views (c, d), a, d = 300 µm, b, c = no size indicated. Note that Lieberkühn (c) already recognised the papillae bearing the dorsal bristles. The globule near the trunk end is a defecation vacuole. Page 1119.

he presented his results in a meeting in 1862 (corresponding paper = Stein 1863). For this species he established *Epiclintes*. Stein (1863) also wrote that it is probably identical with *Trichoda felis* Müller, 1786 (p. 213, Tab. XXX, fig. 15). In 1864, Stein again referred to *Epiclintes* and wrote that the species which he could not remember in his 1863 congress paper was *Oxytricha auricularis* Claparède & Lachmann, 1858. Consequently, this is a type fixation by subsequent designation. Stein (1864) again discussed *T. felis* and wrote that this species could be, according to the illustration provided by Müller (1786), a senior synonym of *O. auricularis*. However, he also mentioned that this assumption remains uncertain, because Müller (1786) forgot to describe the sample site, that is, it is unknown whether *T. felis* was discovered in a marine or limnetic habitat. Consequently, Stein (1863, 1864, 1867) did not finally synonymise these two species and therefore transferred only *O. auricularis* to *Epiclintes*. *Oxytricha retractilis* Claparède & Lachmann, 1858, also transferred to *Epiclintes* by Stein (1864), belongs to *Psammomitra*.

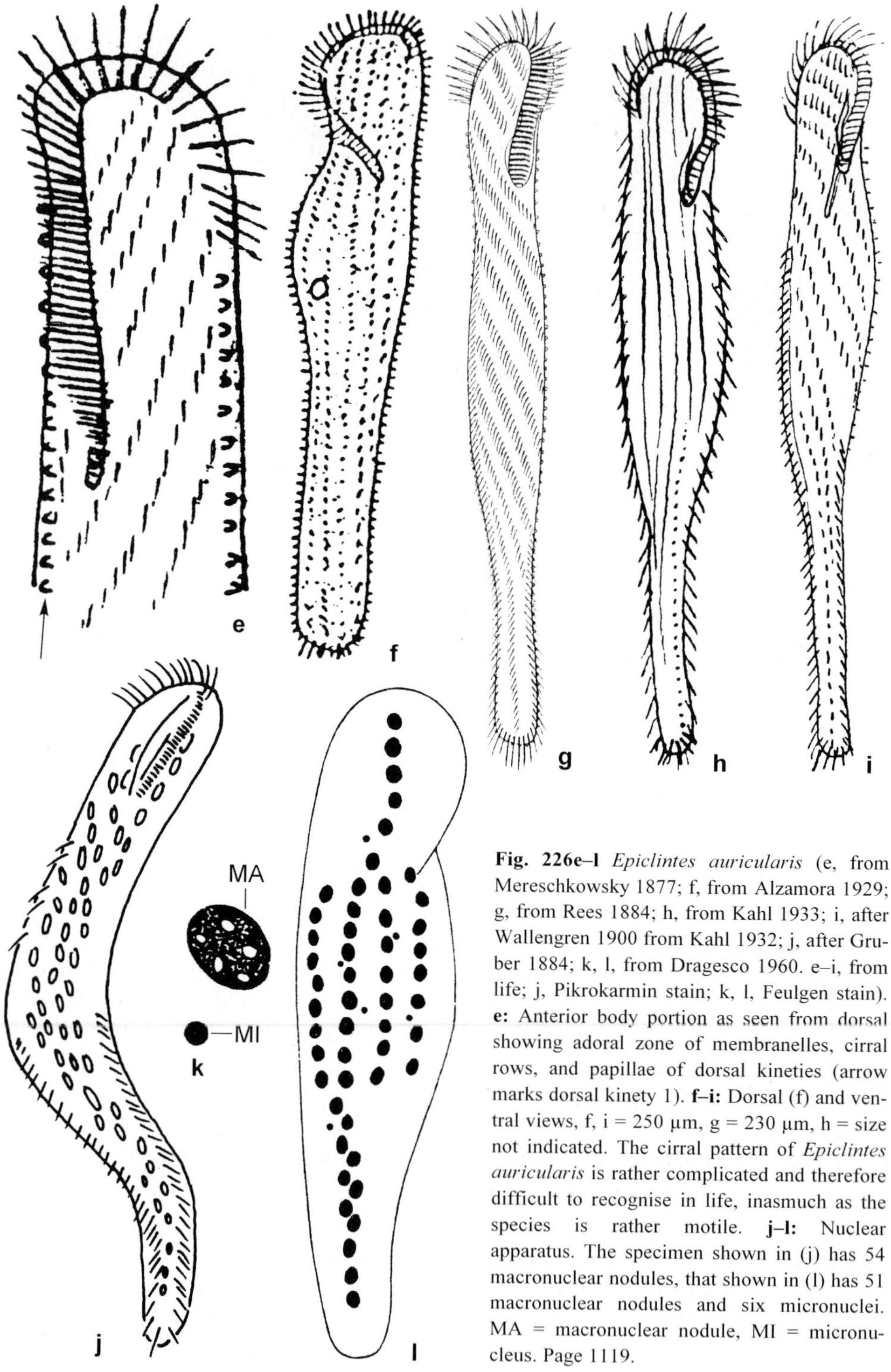

Fig. 226e–l *Epiclintes auricularis* (e, from Mereschkowsky 1877; f, from Alzamora 1929; g, from Rees 1884; h, from Kahl 1933; i, after Wallengren 1900 from Kahl 1932; j, after Gruber 1884; k, l, from Dragesco 1960. e–i, from life; j, Pikrokarmin stain; k, l, Feulgen stain). **e:** Anterior body portion as seen from dorsal showing adoral zone of membranelles, cirral rows, and papillae of dorsal kineties (arrow marks dorsal kinety 1). **f–i:** Dorsal (f) and ventral views, f, i = 250 μm, g = 230 μm, h = size not indicated. The cirral pattern of *Epiclintes auricularis* is rather complicated and therefore difficult to recognise in life, inasmuch as the species is rather motile. **j–l:** Nuclear apparatus. The specimen shown in (j) has 54 macronuclear nodules, that shown in (l) has 51 macronuclear nodules and six micronuclei. MA = macronuclear nodule, MI = micronucleus. Page 1119.

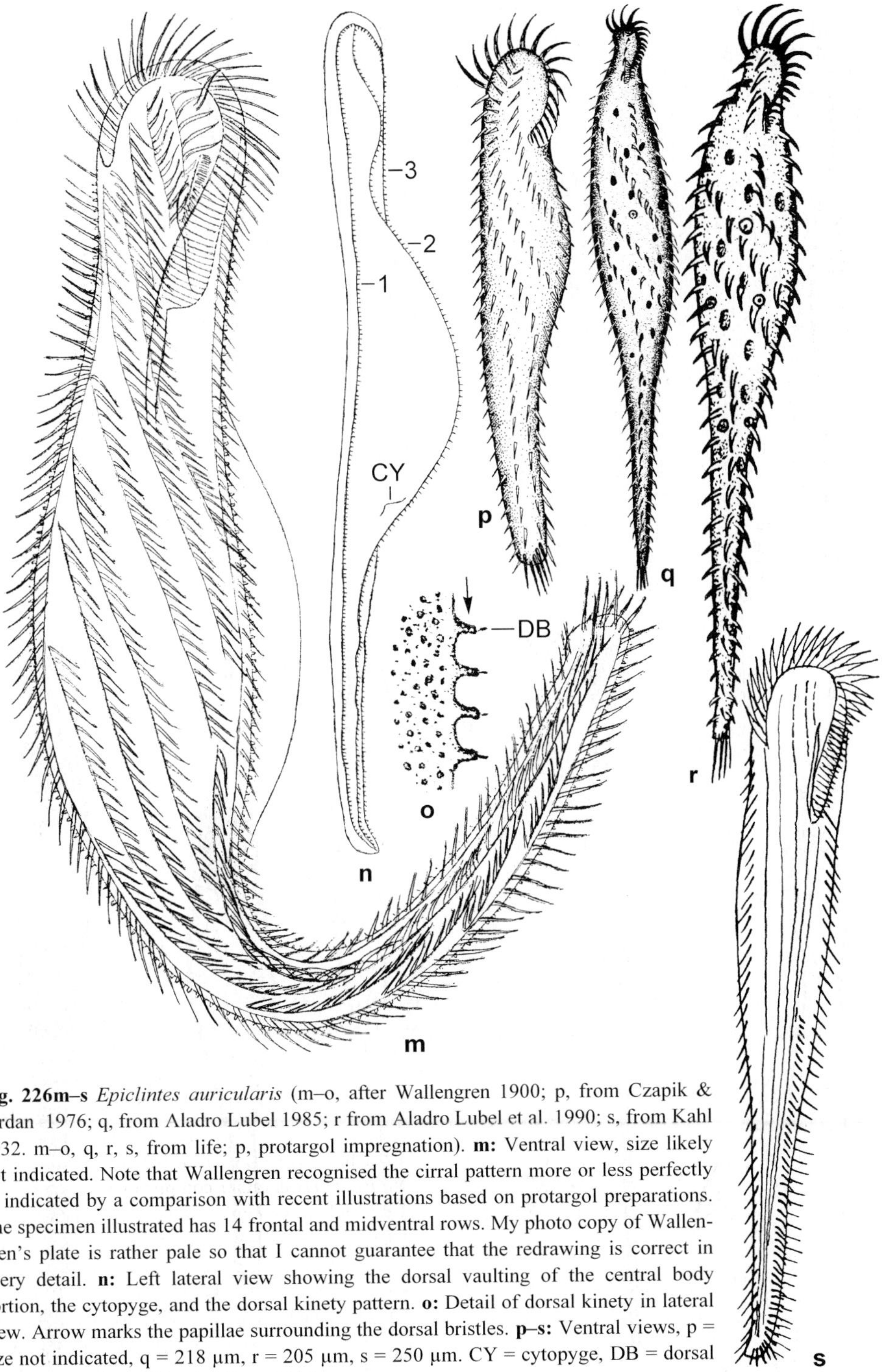

Fig. 226m–s *Epiclintes auricularis* (m–o, after Wallengren 1900; p, from Czapik & Jordan 1976; q, from Aladro Lubel 1985; r from Aladro Lubel et al. 1990; s, from Kahl 1932. m–o, q, r, s, from life; p, protargol impregnation). **m:** Ventral view, size likely not indicated. Note that Wallengren recognised the cirral pattern more or less perfectly as indicated by a comparison with recent illustrations based on protargol preparations. The specimen illustrated has 14 frontal and midventral rows. My photo copy of Wallengren's plate is rather pale so that I cannot guarantee that the redrawing is correct in every detail. **n:** Left lateral view showing the dorsal vaulting of the central body portion, the cytopyge, and the dorsal kinety pattern. **o:** Detail of dorsal kinety in lateral view. Arrow marks the papillae surrounding the dorsal bristles. **p–s:** Ventral views, p = size not indicated, q = 218 µm, r = 205 µm, s = 250 µm. CY = cytopyge, DB = dorsal bristle, 1–3 = dorsal kineties. Page 1119.

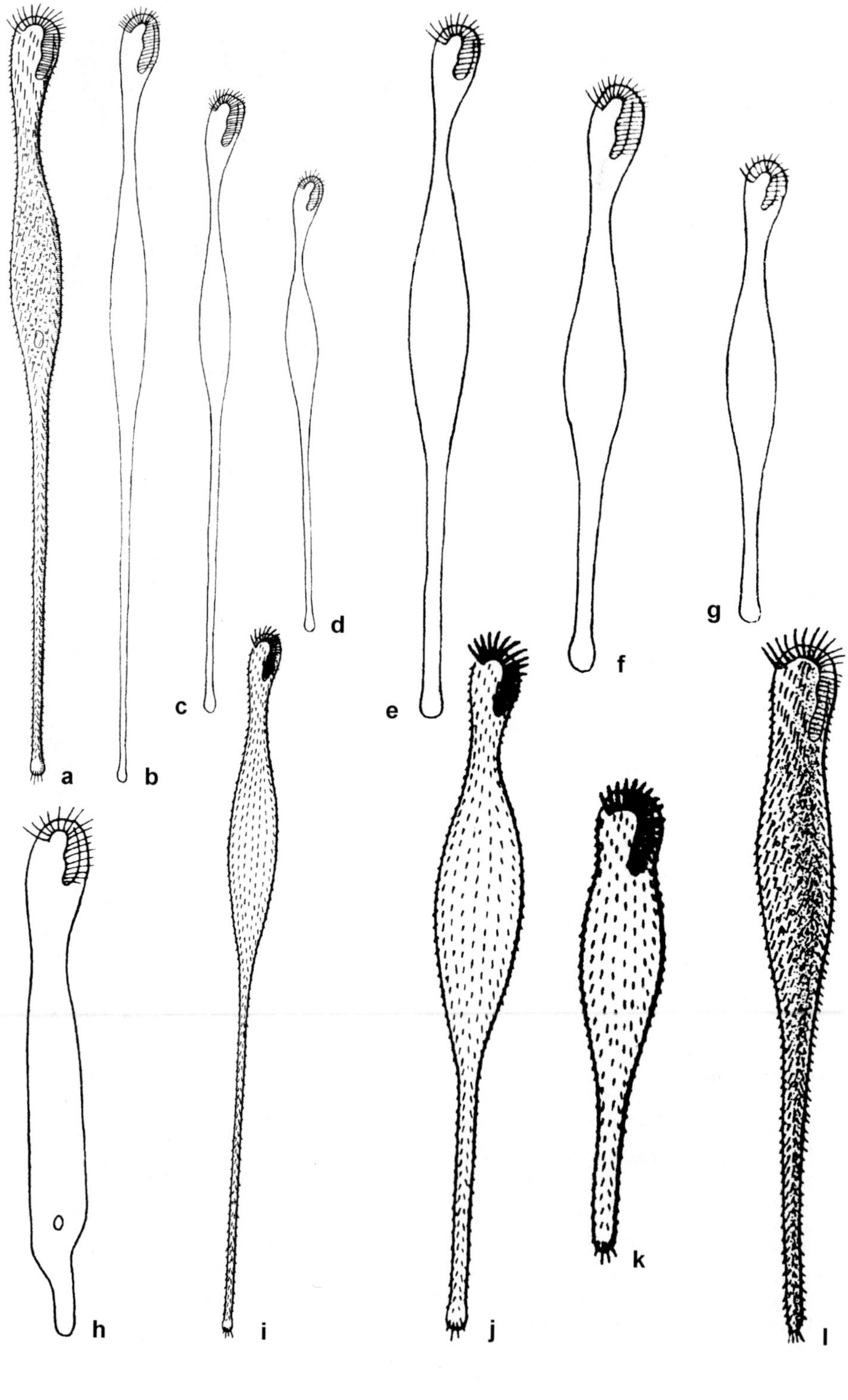
a
b
c
d
e
f
g
h
i
j
k
l

Subsequently, the present species was mainly[1] designated as *Epiclintes auricularis* (see list of synonyms). In 1889, Bütschli synonymised it with the marine *Trichoda ambigua* Müller, 1786 (p. 200, Tab. XXVIII, fig. 11–16) and with *T. felis* Müller, 1786 (with doubt because habitat unknown) and simultaneously transferred *T. ambigua* to *Epiclintes*. The synonymy of *E. auricularis* and *T. ambigua* – proposed, but not founded by Bütschli (1889) – is, as in many other cases with Müller's species, a matter of taste because the description and the illustrations provided by Müller do not allow a reliable identification. However, this somewhat arbitrary synonymy is not the main problem in the present case. The principal problem of "*Epiclintes ambiguus*" is that *Trichoda ambigua* Müller, 1786 is also the basionym of the heterotrich ciliate *Spirostomum ambiguum* (Müller, 1786) Ehrenberg, 1835 (p. 165). For detailed lists of synonyms of this species, see Ehrenberg (1838, p. 332) and Foissner et al. (1992b, p. 317). This species was fixed as type species of *Spirostomum* Ehrenberg by Fromentel (1875, p. 175; see Aescht 2001, p. 151). Interestingly, Bütschli (1889, p. 1724) accepted *S. ambiguum*, but obviously he overlooked that he had used the same basionym for the marine hypotrich "*Epiclintes ambiguus*". This proves that Bütschli (1889) made a serious mistake because he did not consider the authoritative literature (Ehrenberg 1838 and his earlier papers; Stein 1863, 1864, 1867; Fromentel 1875) properly. Although there is a problem with the habitat of *S. ambiguum* – Müller (1786) discovered it in the sea whereas most (all?) later records of this species are from freshwater (Kahl 1932, Foissner et al. 1992b) – the name *Trichoda ambigua* Müller, 1786 should be the basionym of *Spirostomum ambiguum*. As a consequence, "*Epiclintes ambiguus*" cannot be the correct name for the present hypotrich inasmuch as there is no hint that *Trichoda ambigua* sensu Müller (1786) is a mixture of two or more species.

As already discussed, Müller (1786) did not describe the type locality of *Trichoda felis*, which was mentioned as supposed synonym of *E. auricularis* by Stein (1863, 1864, 1867) and some later authors. Because of this uncertainty and because the identification of *Oxytricha auricularis* with *Trichoda felis* would be – as in the case of *T. ambigua* – more or less arbitrary, it seems wise to follow Stein who transferred *O. auricularis* to *Epiclintes*. As a result of this analysis I strongly recommend designating the present species as *Epiclintes auricularis* (Claparède & Lachmann, 1858) Stein, 1864.

From Bütschli's review in 1889 until Carey & Tatchell's (1983) paper, the present species was designated as *E. ambiguus*. Carey & Tatchell studied the present species in

← **Fig. 227a–l** *Epiclintes auricularis* (a–h, from Kiesselbach 1935; i–k, from Kiesselbach 1936a; l, from Kiesselbach 1936. From life and after fixation with osmium tetroxide). **a:** Ventral view of a large, slightly contracted specimen (breed A), 550 µm. **b–d, i:** Largest (b; 680 µm), average (c; i; 540 × 35 µm), and smallest (d; 400 µm) specimen of breed A. Width of tail about 7 µm. **e–g, j:** Largest (e; 350 µm), average (f, j; 300 × 35 µm), and smallest (g; 250 µm) specimen of breed B. Width of tail about 10 µm. **h:** Postdivider (proter) of breed A beginning with formation of a new tail, 345 µm. **k:** Interstitial form from Cuvi, 150 × 26 µm. **l:** Specimen from Venice, 210 µm. Page 1119.

[1] Diesing (1866b) transferred *O. auricularis*, together with *Oxytricha retractilis* and *O. longicaudata*, to *Claparedia* although he noticed that Stein had established *Epiclintes* for the present species. For a more detailed discussion of *Claparedia*, see *Psammomitra*.

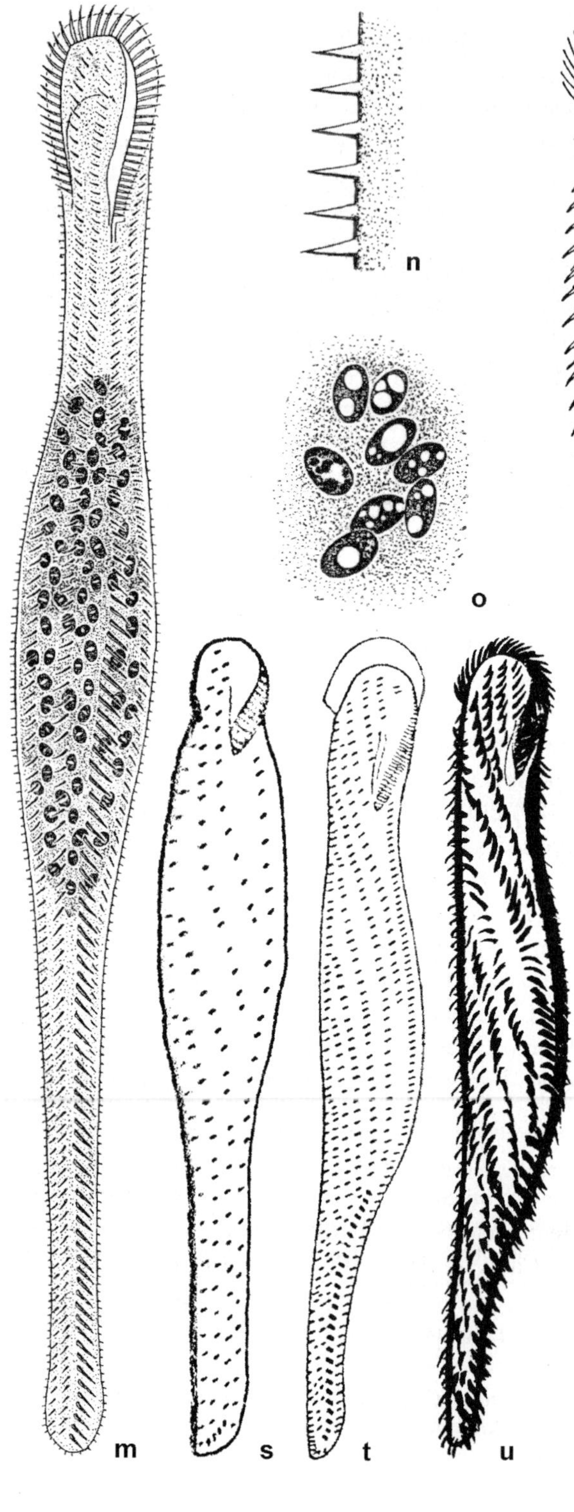
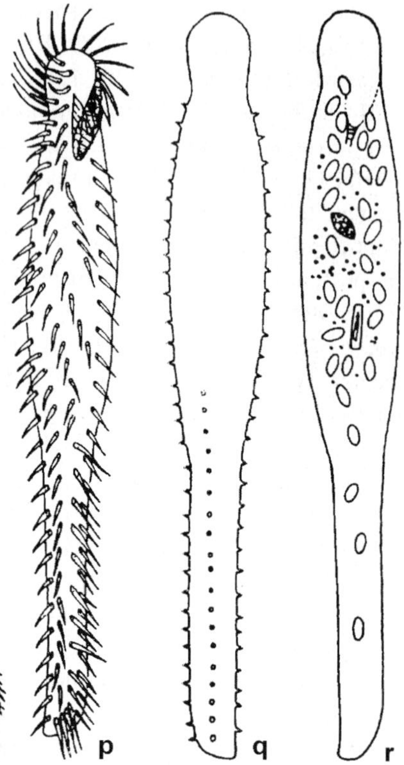

Fig. 227m–u *Epiclintes auricularis* (m–o, from Ozaki & Yagiu 1943; p–r, from Borror 1963; s, from Borror 1972; t, from Borror 1979; u, from Borror 1980. m–o, from life; p, q, wet silver nitrate impregnation?; r, Heidenhain's iron haematoxylin stain; s–u, method not indicated). **m:** Ventral view, 200 μm. Note that Ozaki & Yagiu did not recognise the cirral pattern correctly. **n:** Dorsal bristle complexes. **o:** Detail of nuclear apparatus. **p, s–u:** Infraciliature of ventral side, p = 200 μm, s = 212 μm, t = 324 μm, u = 300 μm. **r:** Nuclear apparatus and ingested diatoms. Page 1119.

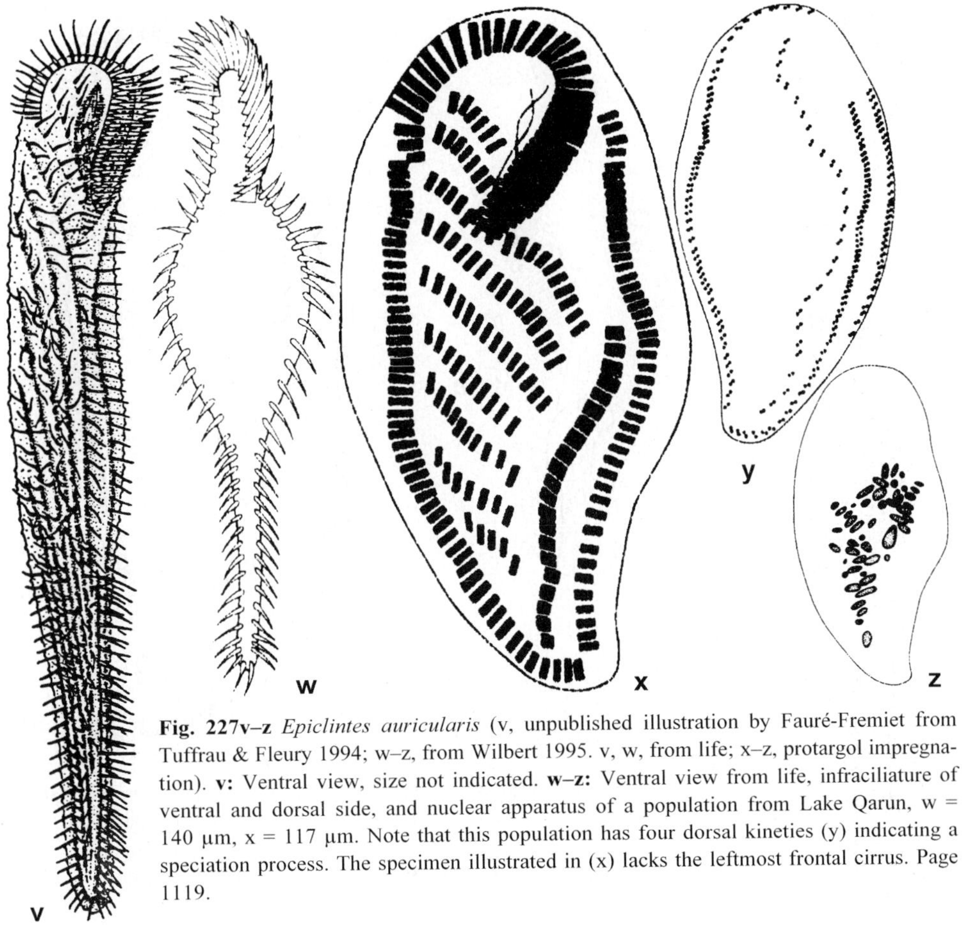

Fig. 227v–z *Epiclintes auricularis* (v, unpublished illustration by Fauré-Fremiet from Tuffrau & Fleury 1994; w–z, from Wilbert 1995. v, w, from life; x–z, protargol impregnation). **v:** Ventral view, size not indicated. **w–z:** Ventral view from life, infraciliature of ventral and dorsal side, and nuclear apparatus of a population from Lake Qarun, w = 140 μm, x = 117 μm. Note that this population has four dorsal kineties (y) indicating a speciation process. The specimen illustrated in (x) lacks the leftmost frontal cirrus. Page 1119.

detail and discussed the nomenclature and synonymy. According to their analysis, *Epiclintes felis* (Müller, 1786) would be the correct name. They argued that Stein had made an error in that he transferred *O. auricularis* and not *T. felis* to *Epiclintes*. However, Carey & Tatchell obviously overlooked that Stein did not synonymise *T. felis* and *O. auricularis* because Müller (1786) did not know the sample site (marine or limnetic) of *T. felis* (see above). Moreover, Carey & Tatchell assumed that Wallengren (1900) had caused great confusion in that he synonymised *E. auricularis* with *Trichoda ambigua*. However, this assumption is incorrect because synonymy of these two species was already proposed by Bütschli (1889). Carey & Tatchell's proposal to name the present species *Epiclintes felis* basically failed because most later authors retained the name *E. ambiguus* (see list of synonyms). However, as already discussed above, this combination is highly problematical and should be replaced by the most proper name, *E. auricularis*. However, very likely the plenary power of the International Commission of Zoological Nomenclature is needed for a final decision.

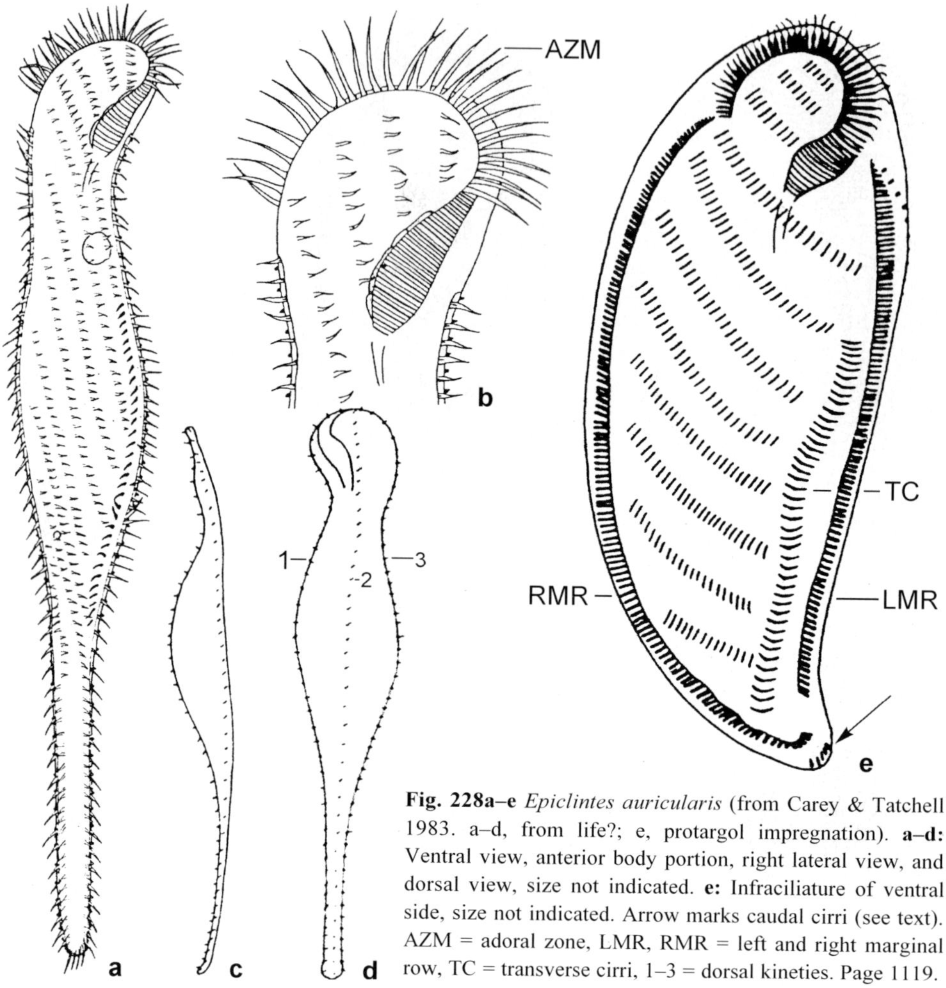

Fig. 228a–e *Epiclintes auricularis* (from Carey & Tatchell 1983. a–d, from life?; e, protargol impregnation). **a–d:** Ventral view, anterior body portion, right lateral view, and dorsal view, size not indicated. **e:** Infraciliature of ventral side, size not indicated. Arrow marks caudal cirri (see text). AZM = adoral zone, LMR, RMR = left and right marginal row, TC = transverse cirri, 1–3 = dorsal kineties. Page 1119.

Epiclintes auricularis is a rather common hypotrich, which has therefore been redescribed several times in more or less detail. The conspecificity of the populations listed above is beyond reasonable doubt, except for the Lake Qarun population described by Wilbert (1995), which has four dorsal kineties against three in marine populations. Lake Qarun has existed for 9000 years (Wilbert 1995, p. 273) so that a speciation process is likely.

Kahl (1932) provided two drawings (Fig. 226i, s). Fig. 226i should be a redrawing of an illustration by Wallengren (Fig. 226m), and Fig. 226s is an original based on a single specimen from the coast of the German island of Sylt. Later he found *E. auricularis* several times in the Bay of Kiel, but could not include his new, correct observations in his review.

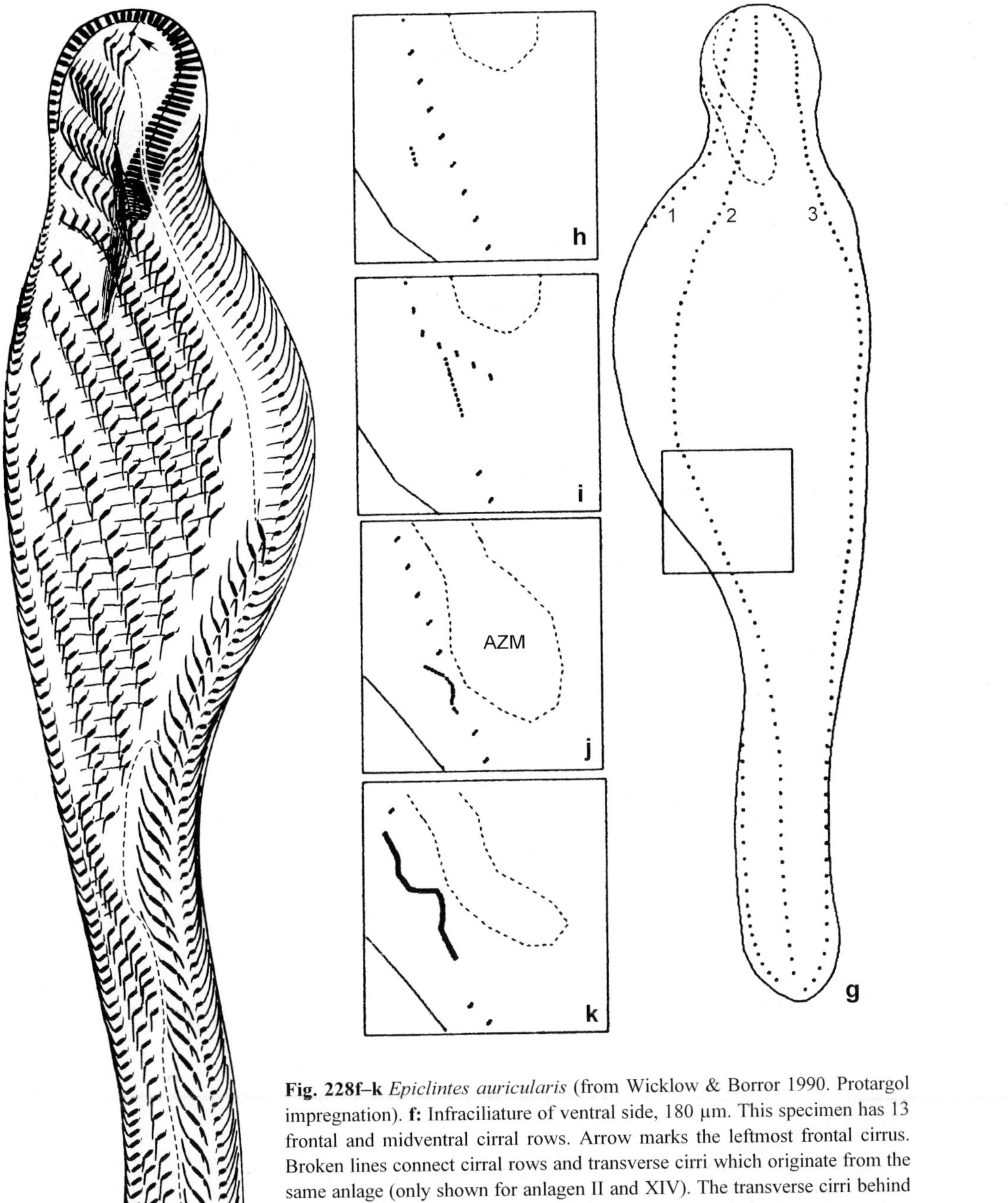

Fig. 228f–k *Epiclintes auricularis* (from Wicklow & Borror 1990. Protargol impregnation). **f:** Infraciliature of ventral side, 180 µm. This specimen has 13 frontal and midventral cirral rows. Arrow marks the leftmost frontal cirrus. Broken lines connect cirral rows and transverse cirri which originate from the same anlage (only shown for anlagen II and XIV). The transverse cirri behind anlage XIV are formed by anlagen which produce only a transverse cirrus (see text and Fig. 228l–s). **g–k:** Infraciliature of dorsal side (g) and successive stages (h–k) of early morphogenesis of the median kinety in the opisthe (area enclosed on rectangle in g), showing cell periphery (solid line), opisthe buccal cavity (dash line), and parental and opisthe basal bodies. AZM = adoral zone of membranelles of opisthe, 1–3 = dorsal kineties. Page 1119.

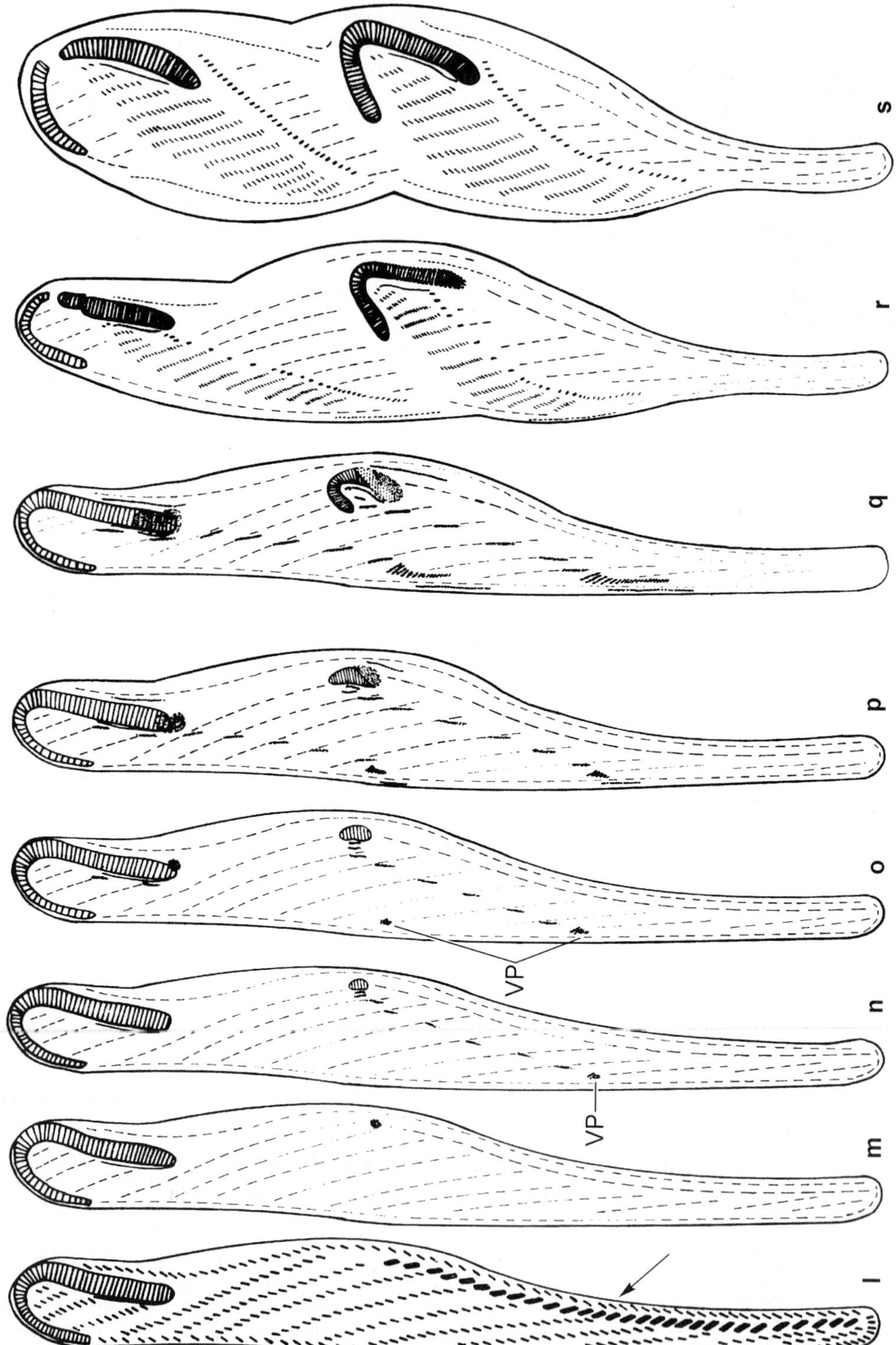
s
r
q
p
o
VP
n
VP
m
l

Borror (1972) considered *O. auricularis*, *Epiclintes vermis* Gruber, 1888, *E. pluvialis* Smith, 1900, and *E. caudatus* Bullington, 1940 as synonyms of *Epiclintes ambiguus*. Hemberger (1982) accepted this synonymy, except for *E. caudatus*. By contrast, I consider all of them as distinct species (*E. auricularis*; *E. vermis*; *E. pluvialis*; *Paramitrella caudata*; *Spirostomum ambiguum*).

Small & Lynn (1985, p. 461) presented two scanning electron micrographs of Wicklow's population, but did not mention *Epiclintes* in the text. Patterson et al. (1989, p. 209) mentioned both *E. ambiguus* and *E. felis* in the list of benthic marine species; without, however explaining the differences.

Morphology: This chapter is based mainly on the three most recent descriptions by Carey & Tatchell (1983), Wicklow & Borror (1990), and Song & Warren (1996). It is supplemented by additional and/or deviating data from the other sources mentioned in the list of synonyms. Old, obviously incorrect data, for example, on the cirral pattern are omitted. The saline lake population described by Wilbert (1995) is kept separate (see below).

Body size in life of more or less extended specimens 210 × 26 µm (Kiesselbach 1936); 182–220 × 22–25 µm (Borror 1963a; head 25 × 17 µm, tail 75–85 µm long); 218–230 × 25 µm (Aladro Lubel 1985, Aladro Lubel et al. 1990); 250–400 × 45–60 µm (Song & Warren 1996); body length on average 300 µm (Claparède & Lachmann 1858, Kent 1882, Bütschli 1889, Kahl 1932); 250 µm (Mereschkowsky 1877, 1879); 200 to 240 µm (Rees 1884); 150 µm (Alzamora 1929); 80–350 µm (Kahl 1933; contracted and extended?); 220–300 µm (Dragesco 1960); 236–400 µm, usually over 300 µm (n = 7; Wicklow & Borror 1990); 140–300 µm (Al-Rasheid 1997). Body length according to Carey & Tatchell (1983) 100–300 µm, but it may shorten to at least 25% of this initial length (this would mean a minimum length of only 25 µm which is likely incorrect). Kiesselbach (1935, 1936a) found two breeds in the Adriatic Sea (near Rovigno): breed A was 400–680 µm long (ratio of head-trunk length:tail length about 1:1; Fig. 227a–d, i); breed B was 250–350 µm long (ratio of head-trunk length:tail length 1:0.6–0.7; Fig. 227e–g, j); there were no intermediate stages. Especially in breed A the tail was distinctly longer and more slender than in all other populations. Body, especially tail region, highly flexible and contractile; according to Kiesselbach (1935) the head-trunk region is more contractile than the tail; cells can contract to about one third of the maximum length.

Body elongate and cephalised, respectively, tripartite in head, trunk, and tail (e.g., Fig. 226a–c, 227a, 228t); tripartition less clearly recognisable on right margin than on left margin. Head auriform, that is, anteriorly broadly rounded and distinctly dorsoventrally flattened. Head of extended specimens about 1.0–1.5 times as wide as trunk. Trunk at least twice as long as head region in extended specimen; when contracted it may equal the head region; semicircular in cross-section (Borror 1963a), that is, ventral

← **Fig. 228l–s** *Epiclintes auricularis* (from Wicklow & Borror 1990. Protargol impregnation). Interphasic specimen (l) and cell division. For details see text. Dashed lines are parental cirral rows. VP = ventral primordium of proter, respectively, opisthe. Arrow in (l) marks the transverse cirrus which is formed by the same anlage as the rearmost midventral row. Page 1119.

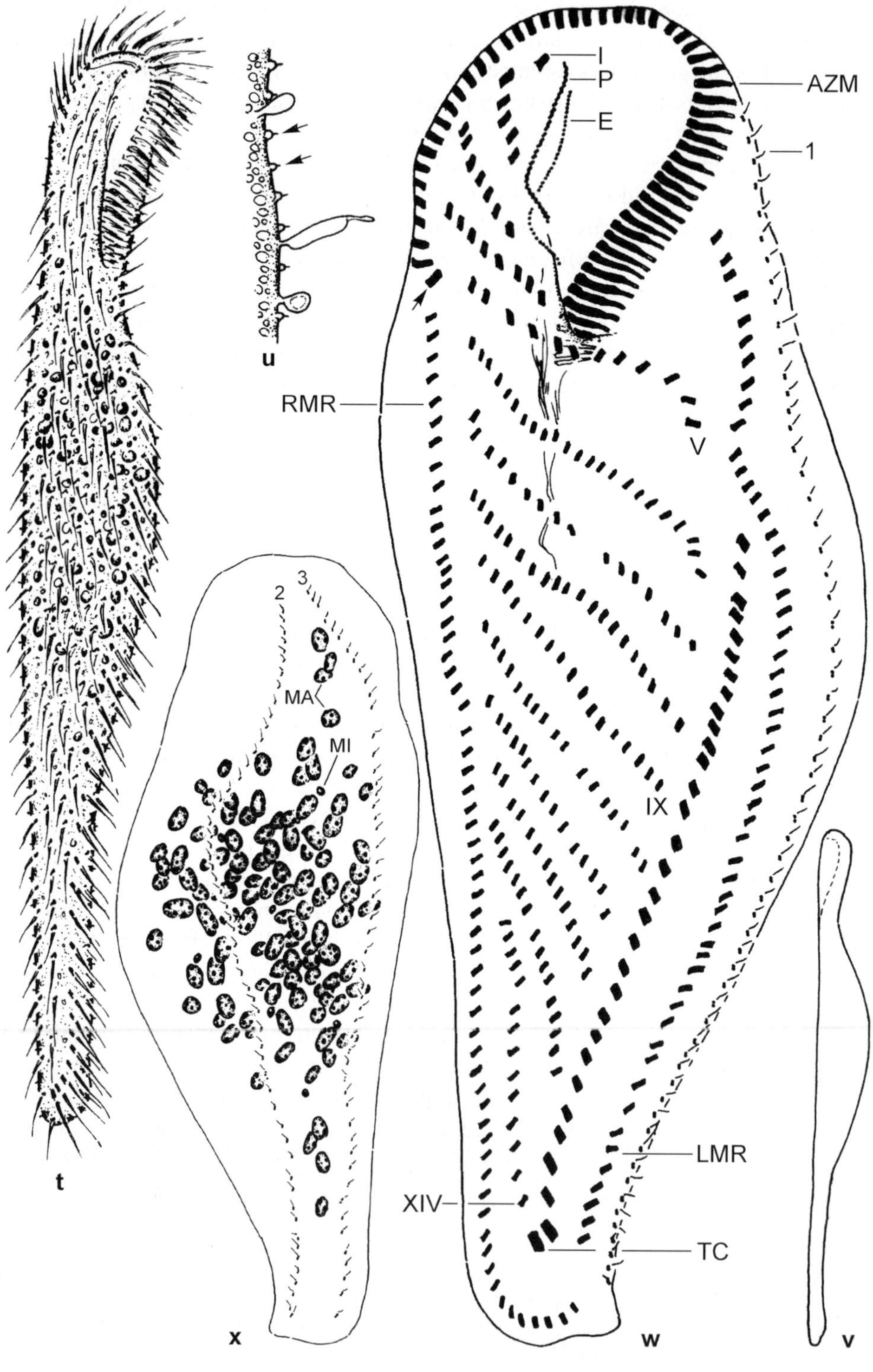

u
I
P
E
AZM
1
RMR
V
IX
LMR
XIV
TC
2
3
MA
MI
t
x
w
v

side plane, dorsal side slightly vaulted in extended specimens to distinctly vaulted in contracted cells. Tail slender, parallel-sided, about as long as head and trunk combined, rear end rounded; distinctly flattened dorsoventrally, that is, almost ribbon-like. Head region and especially tail usually bright, respectively, translucent. Trunk usually opaque due to food and cytoplasmic inclusions (e.g., 226a, 227m, 228t).

Nuclear apparatus difficult to recognise in life and therefore not clearly recognised by earlier authors; composed of about 40–120, usually globular or ellipsoidal macronuclear nodules arranged mainly in trunk (Fig. 227m, o, 228x); according to Wicklow & Borror (1990) nodules are elongate and irregularly shaped. According to Gruber (1888) and Dragesco (1960), macronuclear nodules arranged roughly in longitudinal rows (Fig. 226j, l). Kiesselbach (1935) counted both in breed A and B (see above) about 40 macronuclear nodules. Several micronuclei distributed among macronucleus-nodules (Song & Warren 1996).

Presence/absence of contractile vacuole not clearly known. According to Stein (1864) in ordinary position, that is, near left cell margin about at level of proximal end of adoral zone. Kiesselbach (1935), who studied specimens in some detail and over a long period could not observe a contractile vacuole. Cytopyge clearly visible at dorsal side, just at junction of trunk and tail region (Wallengren 1900, Fig. 226n; Carey & Tatchell 1983); obviously rather often marked by a distinct defecation vacuole in this region (e.g., Fig. 226a–c, f) which was misinterpreted as contractile vacuole by Claparède & Lachmann (1858), Kent (1882), and Carey & Tatchell (1983). Kiesselbach (1936a) also observed a vacuole in this region, but stated that it is acontractile. Ectoplasm thin with numerous ovoid granules 1 μm (Borror 1963a) to about 2–3 μm across (Fig. 228u). Cortical granules lacking, however, under cover glass, specimens frequently with spherical, pyriform, or bar-shaped "extrusome"-like structures emerging from between dorsal bristles (Fig. 228u; Song & Warren 1996).

Epiclintes auricularis is thigmotactic. Cells may remain motionless for long periods (30 s according to Wicklow & Borror 1990) except for the activity of the membranelles and brief retractions of the cell's anterior. When moving rapidly forward, the head may bend from side to side in a similar fashion to *Psammomitra retractilis*. Stimulated cells may reverse rapidly for a distance of about one body length (Song & Warren 1996). The tail serves to periodically jerk the cell backwards during normal locomotion, that is, tail highly motile and flexible (Carey & Tatchell 1983). Cells may also back up to change direction; under severe stimuli such as a change in osmolarity, cells can back up

---

← **Fig. 228t–x** *Epiclintes auricularis* (from Song & Warren 1996. t–v, from life; w, x, protargol impregnation). **t:** Ventral view of an extended specimen, 323 μm. **u:** Part of cortex showing dorsal cilia (arrows), "extrusomes", and ectoplasmic granules. **v:** Left lateral view showing vaulting of dorsal side. **w, x:** Infraciliature of ventral and dorsal side and nuclear apparatus of same specimen, size not indicated. Arrow marks distal end of adoral zone of membranelles. Morphometric data of this specimen: 58 adoral membranelles; 60 right marginal cirri; 54 left marginal cirri; 1 leftmost frontal cirrus; frontal-midventral cirral row II composed of 5 cirri; III 7 cirri; IV 8 cirri; V 12 cirri; VI 23 cirri; VII 16 cirri; VIII 20 cirri; IX 17 cirri; X 17 cirri; XI 15 cirri; XII 14 cirri; XIII 16 cirri; XIV 14 cirri; 31 transverse cirri. AZM = adoral zone of membranelles, E = endoral, LMR = left marginal row, MA = macronuclear nodules, MI = micronucleus, P = paroral, RMR = right marginal row, TC = transverse cirri, I = leftmost frontal cirrus, V = midventral row originating from anlage V, XIV = rearmost midventral row, 1–3 = dorsal kineties. Page 1119.

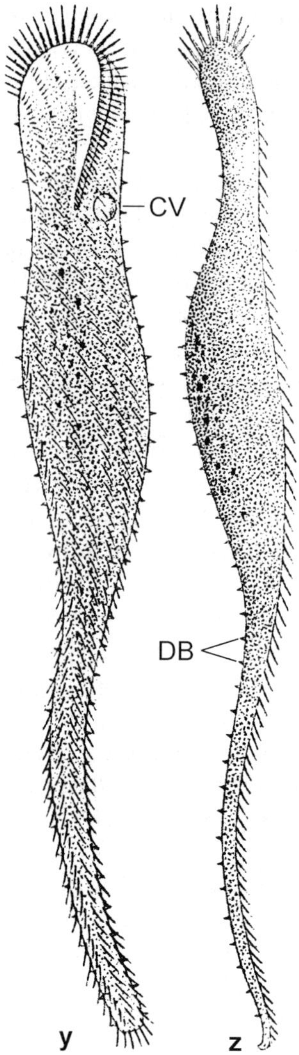

Fig. 228y, z *Epiclintes auricularis* (from Pereyaslawzewa 1886. From life). Ventral view and right lateral view, size not indicated. CV = contractile vacuole, DB = dorsal bristles. Page 1119.

rapidly for a distance of several millimetres (Wicklow & Borror 1990). *Epiclintes auricularis* also shoots up by to stretching the right-angularly bent tail (Stein 1864).

Adoral zone conspicuous because surrounding head region almost completely, that is, extends far onto right body margin; composed of 25–71, on average around 50 membranelles (Table 43); Carey & Tatchell (1983) counted only 25–35 membranelles, which is likely an underestimation. Membranelles of ordinary fine-structure. Paroral and endoral distinctly curved, optically intersecting in posterior portion, paroral composed of dikinetids and arranged in cortical furrow, commences somewhat ahead of endoral, which is made of a row of single basal bodies and lies, as is usual, within the buccal cavity. Buccal field rather narrow in life (e.g., Fig. 226m), distinctly inflated after protargol preparation (Fig. 228w). Cytopharynx clearly recognisable in life.

Cirral pattern rather unusual because composed of about 12–15 oblique frontal-midventral cirral rows and a rather long row of transverse cirri covering ventral side more or less densely. Exact arrangement as shown in Figs. 228f, w. Note that already Wallengren (1900) recognised the pattern more or less perfectly (Fig. 226m). Usually one, sometimes two cirri ahead/right of anterior end of undulating membranes (cirri of anlage I); leftmost frontal cirrus obviously sometimes lacking (Fig. 228e), although it cannot be excluded that it was overlooked. Wicklow & Borror (1990) designated this cirrus as paroral on their page 184, but as buccal cirrus in their Table 2. Right of undulating membranes three relatively short rows of cirri originating from anlagen II–IV (Fig. 226m, 228e, f, w). Anteriormost cirri of these rows not enlarged and no isolated buccal cirrus/cirri present, that is, a clear distinction between various cirral groups (frontal cirri, buccal cirri, midventral cirri) is not possible. Frontoterminal cirri lacking. Row V is the first row which extends obliquely behind proximal end of adoral zone. Transverse cirri enlarged (see ultrastructure), form long and therefore conspicuous row between left marginal row and left (= rear) end of midventral rows; transverse cirral row commences at about 40% of body length in specimen shown in Fig. 228f. Cirri of Borror's (1963a) population about 12 μm long and 8 μm apart; posteriormost four cirri more closely set, at a diagonal (Fig. 227p). Left marginal row commences distinctly ahead of proximal end of adoral zone, ends subterminally; right row begins near distal end of adoral zone, extends beyond rear body end, and terminates near rear end of left marginal row. Marginal cirri about 10 µm long and about 6 µm apart (Borror 1963a); cirri of right row appear slightly longer than that of left row (Carey & Tatchell 1983).

Dorsal ciliature rather conspicuous because non-motile bristles emerge from characteristic, easy to recognise papillae (e.g., Fig. 226a–c, e, o), similar (homologous?) to that in *Paramitrella caudata* (Fig. 240a, b). Bristles emerge about 1 µm from papillae (Song & Warren 1996); bristles according to Borror (1963a) about 2 µm long and 4 µm apart. According to Stein (1864), bristles of left kinety longer and stronger than those of the other kineties. Bristles invariably arranged in three bipolar kineties clearly recognisable even in life due to the papillae (Fig. 226a–c, e, n, o, 228c, d, g, u, w, x; Table 43); each kinety composed of about 70 dikinetids (Wicklow & Borror 1990). Note that Wilbert 1995 found four kineties in a saline lake population described in the next paragraph (Fig. 227y). Carey & Tatchell (1983, p. 45) wrote about "dorsal cirri", which is an incorrect term. Caudal cirri according to most authors lacking; except for Carey & Tatchell (1983; see also Carey 1992), who found a short row of 4–5 such cirri which are finer and slightly longer than the other cirri (Fig. 228e); likely they misinterpreted their preparations. Interestingly, neither Wicklow & Borror (1990) nor Song & Warren (1996) made a comment about this feature. Wicklow & Borror (1990) studied the cell division and found that no caudal cirri are formed at the end of the dorsal kineties, indicating that Carey & Tatchell (1983) made a misobservation.

Lake Qarun population described by Wilbert (1995; Fig. 227w–z, Table 43): body length 80 µm (contracted) to 200 µm (extended). Leftmost frontal cirrus sometimes lacking (Fig. 227x, Table 43). Transverse cirri composed of three kineties. Marginal cirri composed 2 × c. 8 basal bodies. At the slightest disturbance it responds by contracting and adhering to the substrate like a suction cup so that it can no longer be observed. Obviously par lapsus, Wilbert wrote that Wicklow & Borror (1990) did not investigate their population biometrically; likely, he meant Carey & Tatchell (1983), who did not provide a morphometry.

Ultrastructure: The ultrastructure was studied by Carey & Tatchell (1983) and in great detail by Wicklow & Borror (1990; a summary was provided by Wicklow 1979). Adoral membranelles of ordinary fine structure, that is, composed of three or four rows of basal bodies. Longest two rows composed of about 11 basal bodies, next row of about nine, and fourth of about two.

Each cirrus of the frontal-midventral rows composed of 2 × 6–8 basal bodies, lies in a cortical indentation at a 60° angle to the long axis of the cell. Transverse cirri consist of 5 × 8–10 basal bodies. Sets of microtubules form a herring-bone pattern along the cell margin (Wicklow & Borror 1990), a feature already recognised by Kahl (1932). For further details on the fibrils, see Wicklow & Borror (1990).

Ciliary units of dorsal side composed of dikinetids with the anterior basal body ciliated and the posterior unciliated. Transverse microtubules arise at the anterior basal body, whereas postciliary microtubules and a kinetodesmal fibre originate from the posterior basal body (Wicklow & Borror 1990). Microtubule-associated electron dense material occurs on the posterior and right side of the dikinetid. Nematodesmal microtubules descend into the cytoplasm from fibrillar material at the base of the bristle basal bodies. Electron dense granules (diameter 55 nm) are arranged in a linear array at 30 nm intervals along the nematodesmal microtubules; adjacent granules are linked by fine connections (Wicklow & Borror 1990).

The dorsal side of *E. auricularis* is covered by a multilamellate pellicle of about 16 layers (Carey & Tatchell 1983, Wicklow & Borror 1990). This thick pellicle extends down either side of the cell, but is modified on the ventral surface. Near the cirri the pellicle reverts to the more normal ciliate structure, but in other regions the multilamellate appearance is seen; however, the number of layers is reduced to 5–6. This multilamellate pellicle does not extend up onto the dorsal bristles, but terminates in a funnel-shaped structure (papillae). The repeat distance of this thick pellicle is about 6 nm, thus giving a total width of about 100 nm. A similar (homologous?) multilayered membrane system has been found in *Engelmanniella mobilis* (Wirnsberger et al. 1987a, 1989), a species of unknown position according to morphological data. Molecular markers assign *E. mobilis* at various positions (e.g., Hogan et al. 2001, Croft et al. 2003, Foissner et al. 2004a).

Cell division: This process of the life cycle was studied in detail by Wicklow & Borror (1990). The main events are summarised in the paragraphs below and in Figs. 228h–s.

Stomatogenesis commences with the formation of the opisthe oral primordium near the cell surface as an elaboration of basal bodies associated with the dedifferentiation of the anterior-most transverse cirrus (Fig. 228m). This primordium grows into an oval, anteriorly truncated field within which the basal bodies begin with the formation of membranelles. As usual, the differentiation proceeds from anterior to posterior (Fig. 228n–s). The oral primordium of the proter develops later as a field of basal bodies just posterior to the parental proximal membranelles (Fig. 228o, p). Subsequent differentiation of proter promembranelles proceeds dorsally to the parental membranelles. The parental proximal membranelles and the paroral then dedifferentiate and resorb in a posterior to anterior direction. Proter promembranelles emerge onto the cell surface anteriorly as parental membranelles are resorbed. Anlage of paroral and endoral differentiate along the medial edge of the oral primordium. Eventually one or two frontal cirri originate from the anterior end of the paroral anlage and form, as in other hypotrichs, the leftmost frontal cirrus/cirri (Fig. 228q–s).

As opisthe membranelles begin to differentiate, a dikinetid appears right to the oral primordium. The dikinetid may arise de novo or by dedifferentiation of a parental cirrus; its distance from the oral primordium make an origin from this structure unlikely. Initially only the rear basal body of the pair is ciliated. The basal bodies then proliferate to form one or two frontal streaks that give rise to the anteriormost cirral rows (anlagen II and III) of the opisthe (Fig. 228n–q). The proter frontal ciliature develops later but in a similar way. Basal bodies appear right of the developing paroral, proliferate to form first one then two frontal streaks, from which two anterior cirral rows differentiate.

The formation of the remaining frontal-midventral-transverse ciliature proceeds highly interestingly. The process commences with the formation of one anlage each in the posterior portion of some cirral rows (rows VI–XI in specimen shown in Fig. 228n). From the anteriormost cirrus of the parental cirral row XII (= 8th postcytostomal ventral row according to Wicklow & Borror 1990) the so-called ventral primordium develops. Somewhat later at the anterior end of row X the ventral primordium of the proter occurs (Fig. 228o). The right marginal row is not involved in the formation of the two ventral primordia. At first, the ventral primordium forms 2–3 ventral cirral streaks plus 2–3 transverse cirral streaks. Subsequently, the ventral primordium expands anteriorly to form 5–6 ventral cirral streaks and posteriorly to form about 16 transverse cirral streaks (Fig. 228p–r). Whereas each of the anterior 5–6 ventral cirral streaks gives rise to entire ventral cirral row and a transverse cirrus, each of the posterior 16 transverse cirral streaks gives rise to a transverse cirrus only. Additional ventral cirral rows develop by the formation of one streak each within the parental cirral rows (rows VI–XI for opisthe and rows III or IV–IX in specimen shown in Fig. 228p).

Marginal row and dorsal kinety formation proceeds basically as in other hypotrichs, that is, by within-row proliferation (Fig. 228h–k, p–s). However, at least during the first stages no parental structures are obviously involved, which is reminiscent of *Thigmokeronopsis*, where marginal rows and dorsal kineties originate de novo.

Macronuclear nodules show replication bands as in other hypotrichs (Wicklow & Borror 1990). Unfortunately, Wicklow & Borror forgot to describe the behaviour of the macronucleus during cell division, that is, we do not know whether the nodules fuse to a single mass as in the majority of hypotrichs, or whether they divide individually as in the pseudokeronopsids.

According to Eigner (2001, his Fig. 24), anlage II, which usually forms the buccal cirrus, does not produce a transverse cirrus. However, there is no evidence for such an assumption. Quite the reverse, Fig. 228r clearly shows that anlage II forms, as usual, the first (= anteriormost) transverse cirrus (Fig. 228f).

Occurrence and ecology: *Epiclintes auricularis* likely occurs world-wide in marine habitats, mainly on or adjacent to the sand surface. According to Borror (1963) it is numerous only in fresh cultures. Type locality is the North Sea near Bergen, Norway (Claparède & Lachmann 1858). Lieberkühn (unpublished; Fig. 226b, c) and Stein (1863, 1864, 1867) found it in the Baltic Sea near the German city of Wismar.

Further records substantiated by more or less detailed morphological data: Rovigno, Cuvi, and lagoon of Venice, Adriatic Sea (Kiesselbach 1935, 1936, 1936a); Bay of Palma de Mallorca, Mediterranean Sea (Alzamora 1929); Gulf of Genoa, Mediterranean Sea, Italy (Gruber 1884b, p. 482; 1888; see also Petz & Leitner 2003, p. 119); sediment in Chichester Harbour area, Atlantic Ocean, England (Carey & Maeda 1985, p. 565); small, shallow (5–10 cm) lagoons just below the high tide line at the south shore (Oare creek and Faversham creek) of the Thames estuary (salinity 70% of sea water) near Faversham north Kent and on the north shore of the Thames at Shoeburyness, Essex (Carey & Tatchell 1983); coast of the German island of Sylt (North Sea; Fig. 226s) and Bay of Kiel, Baltic Sea (Kahl 1932); Roscoff and Concarneau area (France), Atlantic Ocean (Dragesco 1960); Osterschelde, Netherlands (Rees 1884);

North Sea, Sweden (Wallengren 1900); brackish water of the Gdansk Bay and other sites of the Baltic Sea (Czapik & Jordan 1976); Black Sea (Pereyaslawzewa 1886); abundant among algae at the Solowetzky islands and in the Kloster Bay, White Sea (Mereschkowsky 1877, 1879); Inland Sea of Japan (Ozaki & Yagiu 1943); eutrophic pond (salinity 32‰, water temperature 12–15° C, pH about 8.3) used for storing marine shellfish in Taipingjiao, Qingdao, China (Song & Warren 1996); in sediment samples from the coastline of the Saudi Arabian Gulf Island Tarut (Al-Rasheid 1997); Gulf of Mexico (Aladro Lubel 1985, Aladro Lubel et al. 1990, Borror 1963a); Chondrus detritus and the surface of intertidal mud of Great Bay Estuary, Adams Point, Durham, New Hampshire and from intertidal sand at Sea Point Beach, Kittery, Maine (USA) between 1976 and 1989 (Wicklow & Borror 1990); Lake Qarun, a warm polymictic tropical saline (24.8‰) lake in the Fayum Oasis, Egypt (Wilbert 1995; see remarks).

Records mainly not substantiated by morphological data (however, *Epiclintes auricularis* is easy to identify and therefore all records are reliable): French Coast at Concarneau, Atlantic Ocean (Faurè-Fremiet 1950, p. 50); German island Sylt, North Sea (Hartwig 1973, p. 65; 1973b, 123; 1974, p. 17; Küsters 1974, p. 174); Hiddensee, Baltic Sea (Münch 1956, p. 434); Schlei, a brackish water habitat near the Baltic Sea (Bock 1960, p. 63; Jaeckel 1962, p. 13); Bay of Kiel, Baltic Sea (Möbius 1887; 1888, p. 88; Bock 1952, p. 81); Jadebusen, German Bay, North Sea (Hartwig 1984, p. 127); Mediterranean Sea near Marseille and Banyuls-sur-mer, France (Vacelet 1961, p. 4; 1961a, p. 15; Dragesco 1953, p. 629); Gulf of Naples, Mediterranean Sea (Entz 1884, p. 294; Nobili 1957); Lagoon of Venice, Italy (Coppellotti & Matarazzo 2000, p. 426); upper sublittoral sands of Bulgarian coast of Black Sea (Kovaleva 1966, p. 1603; Kovaleva & Golemansky 1979, p. 275); Romanian coast of Black Sea (Petran 1968, p. 445; 1971, p. 154); Ukrainian coast of Black Sea (Pavloskaya 1969); psammon of western coast of Caspian Sea (Agamaliev 1971, p. 383; 1983, p. 36; Agamaliyev 1974, p. 21); Plymouth area, England (Lackey & Lackey 1963, p. 802); English Channel in the Ambleteuse area, France (Chardez 1956, p. 92); sandy beaches in North Yorkshire, North Sea (Hartwig & Parker 1977, p. 751); Kandalaksha Gulf and other sites of the White Sea (Azovsky et al. 1996, p. 30; Burkovsky 1970a, p. 187; 1970b, p. 11; 1970c, p. 56; 1971, p. 1774; Raikov 1962, p. 331); Barents Sea (Azovsky 1996, p. 6); sediment samples from the Peter the Great Bay, Sea of Japan (Myskova 1976, p. 85); mesopsammon of the Ussuri Gulf and Posjet Gulf, Sea of Japan (Raikov 1963, p. 1757; Raikov & Kovaleva 1968, p. 331); China (Song & Wang 1999, p. 73); Beach sand of Waltair Coast, India (Rao & Ganapati 1968, p. 89); Orissa coast, Bay of Bengal (Rao 1969, p. 92; 1974, p. 170); Coast of Cape Cod and Woods Hole area (USA), Atlantic Ocean (Lackey 1936, p. 269; Fauré-Fremiet 1951, p. 62); Gulf of Mexico (Borror 1962, p. 342; Aladro Lubel et al. 1986, p. 240; Aladro-Lubel et al. 1988, p. 438); Port-Etienne, Mauritania (Dragesco 1965, p. 397); Tuckers Town Beach, Bermuda (Hartwig 1980, p. 427); beach of Embarè, Sao Paulo, Brazil (Kattar 1970, p. 145).

Hartwig (1973b) studied the ecology of *E. auricularis* in the mesopsammon of the German island of Sylt in some detail. It occurred throughout the year, however, with a distinct peak during summer (Fig. 229a). The maximum abundance was 65 specimens

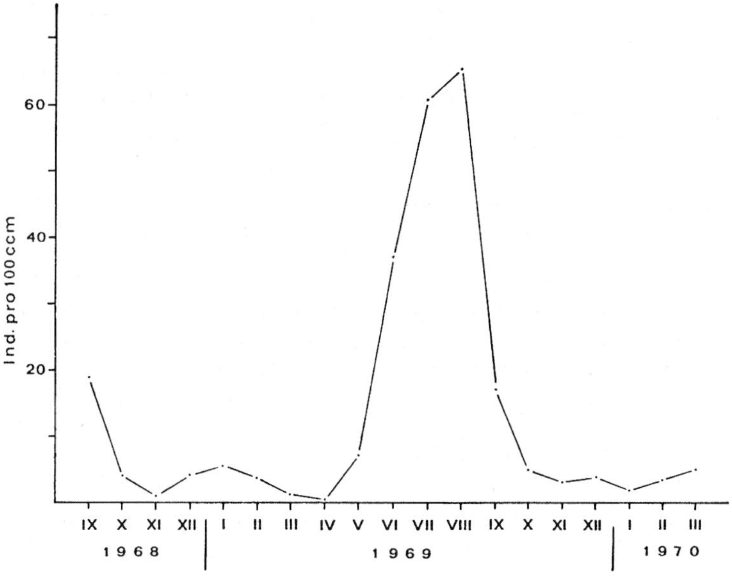

Fig. 229a *Epiclintes auricularis* (from Hartwig 1973b). Population dynamics in the mesopsammon of the island Sylt, North Sea (abundance in specimens per 100 ml). Page 1119.

per 100 cm^3 in August. During the warm period most specimens occurred in the surface layer (0–5 cm), during winter most specimens could be found at 5–10 cm. The maximum depth was 30 cm, that is, *Epiclintes auricularis* occurred in oxidative and reductive milieu. It was absent in the highly lotic beaches of the west coast of the island.

Patterson & Hedley (1992, p. 127) found an *Epiclintes* species (possibly *E. auricularis*) in the sediments of a very low salinity estuarine site (Australia?). Records from freshwater habitats are very likely based on misidentifications: Kiev reservoir, Ukraine (Kovalchuk 1984).

Feeds mainly on diatoms (e.g., Kiesselbach 1935, 1936a, Borror 1963a), for example, *Amphiprora*, *Navicula* sp. (Pavlovskaya 1970; Fenchel 1968, p. 116), sometimes also on bacteria (Carey & Tatchell 1983). Occasionally very long diatoms are engulfed, specimens then often malformed. Generation time at 17–19°C 24–36 h; formation of tail in postdivider needs about 12–18 h (Kiesselbach 1935). Further reproduction data (in Russian), see Zaika (1970; see also Burkovsky 1984, p. 40). Ratio of weight increment to the ingested food is 32% (Pavlovskaya 1969, 1970).

Kiesselbach (1935) found *E. auricularis* in an algal culture with Erd-Schreiber medium. Carey & Tatchell (1983) cultured it in filtered, seventy per cent sea water (pH 7.8, 18°C) under constant illumination from an 8 W fluorescent lamp. Wicklow & Borror (1990) cultured *E. auricularis* on mixed populations of diatoms or on single species diatom populations of *Bellerochea polymorpha* or *Phaeodactylum tricornutum* in 18 to 25 ‰ sea water at 15–16°C.

Incertae sedis in *Epiclintes*

Epiclintes pluvialis Smith, 1900
(Fig. 230a–c)

1900 ***Epiclintes pluvialis* sp. n.** – Smith, Trans. Am. microsc. Soc., 21: 91, Plate VI, Fig. 5 (Fig. 230a; original description; no formal diagnosis provided and no type material available).
1932 ***Epiclintes pluvialis* Smith, 1899** – Kahl, Tierwelt Dtl., 25: 570, Fig. 97 $_{19}$ (Fig. 230b; revision of hypotrichs; incorrect year).
1963 ***Epiclintes pluvialis* Smith** – Lundin & West, Free-living Protozoa, p. 68, Plate 28, Fig. 1 (Fig. 230c; illustrated record).
2001 ***Epiclintes pluvialis* Smith, 1900** – Berger, Catalogue of ciliate names 1. Hypotrichs, p. 20 (nomenclator containing all basionyms, combinations, and higher taxa of hypotrichs).

Nomenclature: No derivation of the name is given in the original description. The species-group name *pluvial·is -is -e* (Latin adjective; belonging to the rain; bringing rain) possibly refers to the fact that it was discovered in a freshwater habitat.

Remarks: Smith (1900) assigned the present species to *Epiclintes* likely because of the body shape, which is, as in the type species, divided into three distinct regions. The cirral pattern is not known so that a final assignment is not yet possible. Consequently, I classify it as incertae sedis in the present genus. Possibly it is a misinterpreted *Uroleptus*-species, as indicated by the tailed body and the limnetic habitat. *Ancystropodium maupasi* Fauré-Fremiet has a similar habitus, but the tail is much thinner (for review see Berger 1999, p. 778). Detailed redescription needed.

Borror (1972, p. 9) synonymised the limnetic *E. pluvialis* with the marine *Epiclintes ambiguus* (= *E. auricularis* in present book). By contrast, Carey & Tatchell (1983, p. 50) excluded it from *Epiclintes* because it has a symmetrical peristome, lacks ventral cirri, and has long and "hispid" dorsal cilia. However, they made no proposal for a more proper classification.

Morphology: Body length 357 µm (no details about variability given); body length:width ratio 5–7:1. Body very elastic, elongate, divided into three distinct regions; central portion widest, convex on dorsal side, flat on ventral, about twice as long as anterior portion, which is much compressed and rounded at anterior border. Rear portion elongate, attenuated tail-like, subcylindrical, and very variable in length. Nuclear apparatus not known. Contractile vacuole in ordinary position, that is, dorsally near left cell margin slightly behind adoral zone of membranelles (Fig. 230a). Cytopyge located at the lower ventral extremity of the central portion. Movements eccentric. Oral apparatus, respectively, adoral zone of membranelles unusual because inverted U-shaped, that is, distal end of adoral zone at same level as proximal end. Further details (e.g., shape and arrangement of endoral and paroral) not known. Cirral pattern not recognised in detail (Fig. 230a); each side with a marginal row (specimen illustrated with about 25 right and 26 left marginal cirri; values must not be over-interpreted), central body portion with about 24 ventral cirri of unknown arrangement (longitudinal or oblique rows?). Dorsal cilia rather long and stiff; number and arrangement of kineties and presence/absence of caudal cirri not known. Reproduction by transverse fission.

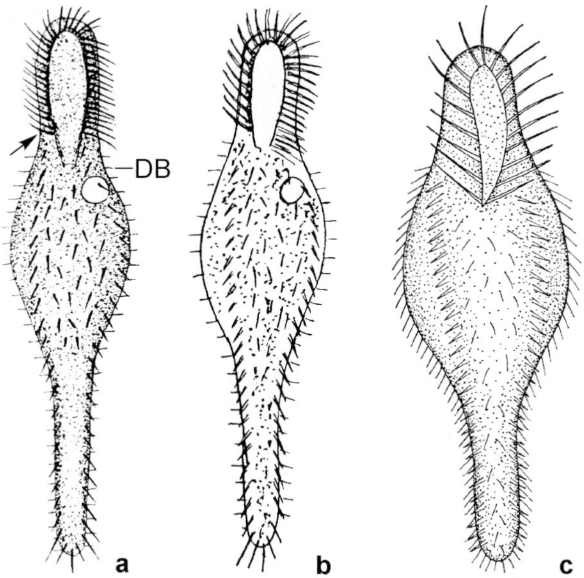

Fig. 230a–c *Epiclintes pluvialis* (a, from Smith 1900; b, after Smith 1900 from Kahl 1932; c, from Lundin & West 1963. From life). Ventral views, a, b = 357 µm (Kahl wrote, likely par lapsus, 375 µm), c = size not indicated. Arrow marks distal end of adoral zone of membranelles. The cirral pattern of *E. pluvialis* is basically unknown and its classification therefore uncertain. The contractile vacuole is in ordinary position, that is, near left cell margin slightly behind proximal end of adoral zone of membranelles. DB = long and stiff dorsal bristles. Page 1142.

Occurrence and ecology: Limnetic and possibly confined to America. Type locality of *E. pluvialis* is a small pond with *Myriophyllum* at Slidell, Louisiana, USA, where it occurred in large quantities in company with a three-horned variety of *Ceratium hirundinella* (Smith 1900). It is a ravenous feeder packed with food so that Smith could not recognise the nuclear apparatus. In one instance the present species was seen to swallow eight specimens of *Trachelomonas armata*. *Epiclintes pluvialis* has the peculiar habit of resting alongside of some debris or algal filament, and collecting around its body a quantity of debris, from which it protrudes most of its body when feeding, and into which it withdraws itself when disturbed. This feature is exactly similar to that of *Stichotricha*.

Records largely not substantiated by morphological data: in aerobic bottom samples (5–10 m depth) from Esthwaite, one of the most eutrophic lakes in English Lake District (Webb 1961, p. 140; sole record from outside America); freshwater habitats in Upper Peninsula of Michigan, USA (West & Lundin 1963, p. 105; Lundin & West 1963, Fig. 230c); limnetic habitats from the USA (Pratt & Cairns 1985, p. 422); Río Bella, Amazonas drainage basin (Cairns 1966, p. 61).

Epiclintes vermis Gruber, 1888
(Fig. 231a, b)

1888 ***Epiclinites vermis*** – Gruber, Ber. naturf. Ges. Freiburg i. B., 3: 62, Tafel VI, Fig. 9, 10 (Fig. 231a, b; original description; no formal diagnosis provided and no type material available; see nomenclature).

2001 ***Epiclintes vermis* Gruber, 1888** – Berger, Catalogue of ciliate names 1. Hypotrichs, p. 20 (nomenclator containing all basionyms, combinations, and higher taxa of hypotrichs).

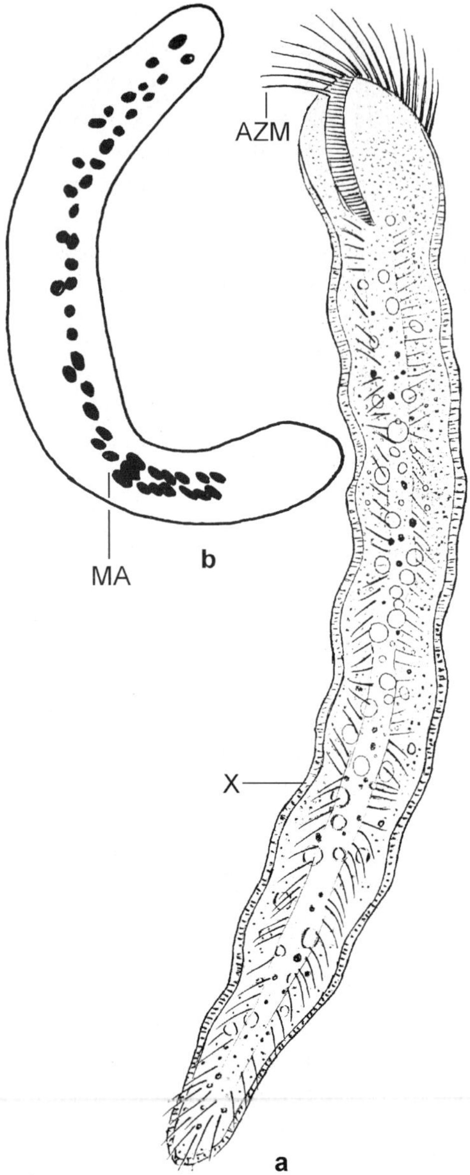

Fig. 231a, b *Epiclintes vermis* (after Gruber 1888. a, from life; b, Pikrokarmin-staining). **a:** Ventral side as seen from dorsal, about 500 µm. **b:** Pattern formed by macronucleus-nodules. AZM = adoral zone of membranelles, MA = macronuclear nodules, X = seam. Page 1143.

Nomenclature: The species-group name *vermis* (Latin noun; the worm; masculine) refers to the vermiform movement. *Epiclinites* in Gruber (1888) is an incorrect spelling. Carey & Tatchell (1983, p. 50, 54) incorrectly assumed that Gruber's paper appeared in 1884.

Remarks: The description of this species is based on two specimens only. Gruber (1888) assigned it to *Epiclintes* without explanation. Kahl (1932, 1933) obviously overlooked it, whereas Borror (1972, p. 9) synonymised *E. vermis* with *E. ambiguus* (= *E. auricularis* in present monograph), however, without providing an explanation. By contrast, Carey & Tatchell (1983, p. 50) excluded it from *Epiclintes* because it has a vermiform body, indicating, in their opinion, a relationship with *Holosticha*. Since the cirral pattern of *E. vermis* is not known in detail it cannot be classified properly. Thus, I retain the original assignment. Detailed redescription needed.

Epiclintes vermis has a vermiform body (against distinctly tripartite in *E. auricularis*) and therefore must not be confused with small annelids (Gruber 1888), respectively, microturbellarians.

Morphology: Body length of extended specimens about 500 µm; body height only 4 µm (incorrect measurement?), that is, body very strongly flattened dorsoventrally. Body outline elongate, roughly like a microturbellarian; anterior portion slightly broadened. Nuclear apparatus difficult to recognise in life; specimen illustrated in Fig. 231b with about 45 globular to ellipsoidal macronuclear nodules arranged mainly in one (likely left) half of body. Cell with conspicuous seam containing (formed by?) granules or rods; according to Gruber these organelles are not trichocysts. Cytoplasm with many granules and rods so that cells are opaque. Winds like a worm. Adoral zone of membranelles occupies only about 13% of body length (Fig. 231a), bears strong membranelles. Further details (e.g., shape and arrangement of endoral and paroral) on oral apparatus lacking. Cirral pattern not described in detail. Postoral area with two long rows of cirri (marginal cirri? midventral pairs?). Posterior portion with many bristles. Dorsal ciliature (length of bristles, number and arrangement of kineties, presence/absence of caudal cirri) not known.

Occurrence and ecology: Marine. Type locality of *E. vermis* is the harbour of Genoa, Italy (Ligurian Sea). No further records published. Food not mentioned.

Species indeterminata

Epiclintes caudatus Lackey, 1961

1961 ***Epiclintes caudatus* p. n.** – Lackey, Limnol. Oceanogr., 6: 276 (see remarks).

Remarks: Lackey mentioned this name in a list of species from the Narragansett Marine Laboratory, Rhode Island. Unfortunately, he did not explain the abbreviation "p. n." in the Table. This abbreviation is obviously not very common because I did not find it in relevant textbooks (e.g., Lincoln et al. 1985, CBE 1996, Winston 1999). However, according to Lackey's text (p. 274), a binomen marked with p. n. is a provisional name. Such a name is not available, that is, it is a nomen nudum because it is neither accompanied by a description or definition, nor by a bibliographic reference to a published statement (ICZN 1999, Article 13.1). Previously I thought that *E. caudatus* sensu Lackey (1961) refers to *Epiclintes caudatus* Bullington, 1940 (now *Paramitrella caudata*) and therefore did not mention Lackey's name in the catalogue (Berger 2001).

Epiclintes tortuosus Lackey, 1961

1961 ***Epiclintes tortuosus* p. n.** – Lackey, Limnol. Oceanogr., 6: 276 (see remarks).
2001 ***Epiclintes tortuosus* Lackey, 1961** – Berger, Catalogue of ciliate names 1. Hypotrichs, p. 20 (nomenclator containing all basionyms, combinations, and higher taxa of hypotrichs).

Remarks: According to Berger (2001) a provisional name mentioned in a species list on ciliates encountered at sampling locations at the Narragansett Marine Laboratory, Rhode Island, in summer 1960. The name is not available, that is, it is a nomen nudum because it is neither accompanied by a description or definition, nor by a bibliographic reference to a published statement (ICZN 1999, Article 13.1).

Eschaneustyla Stokes, 1886

1886 ***Eschaneustyla*, gen. nov.**[1] – Stokes, Proc. Am. phil. Soc., 23: 28 (original description). Type species (by monotypy): *Eschaneustyla brachytona* Stokes, 1886.
1888 ***Eschaneustyla*, Stokes** – Stokes, J. Trenton nat. Hist. Soc., 1: 283 (review of ciliates from the USA).
1932 ***Eschaneustyla* Stokes, 1886** – Kahl, Tierwelt Dtl., 25: 541 (revision).
1974 ***Eschaneustyla* Stokes** – Stiller, Fauna Hung., 115: 44 (guide to hypotrichs).
1979 ***Eschaneustyla* Stokes, 1886** – Jankowski, Trudy zool. Inst., 86: 53 (catalogue of generic names of hypotrichs).
1979 ***Eschaneustyla* Stokes, 1886** – Corliss, Ciliated protozoa, p. 310 (revision).
1982 ***Eschaneustyla* Stokes, 1886** – Foissner, Arch. Protistenk., 126: 37 (discussion of synonymy).
1983 ***Eschaneustyla* Stokes, 1886** – Curds, Gates & Roberts, Synopses of the British Fauna, 23: 426 (guide to ciliate genera).
1985 ***Eschaneustyla*** – Small & Lynn, Phylum Ciliophora, p. 458 (guide to ciliate genera).
1994 ***Eschaneustyla* Stokes, 1886**[2] – Eigner, Europ. J. Protistol., 30: 474 (improved diagnosis and inclusion in subfamily Bakuellinae).
1999 ***Eschaneustyla* Stokes, 1886** – Shi, Song & Shi, Progress in Protozoology, p. 98 (revision of hypotrichs).
1999 ***Eschaneustyla* Stokes, 1886** – Shi, Acta Zootax. sinica, 24: 251 (revision of hypotrichs).
2001 ***Eschaneustyla* Stokes 1886** – Aescht, Denisia, 1: 70 (catalogue of generic names of ciliates).
2001 ***Eschaneustyla* Stokes, 1886** – Berger, Catalogue of ciliate names 1. Hypotrichs, p. 20 (nomenclator containing all basionyms, combinations, and higher taxa of hypotrichs).
2002 ***Eschaneustyla* Stokes, 1886** – Lynn & Small, Phylum Ciliophora, p. 442 (guide to ciliate genera).

Nomenclature: *Eschaneustyla* (Greek) is, according to Stokes (1886), a composite of *εσχατια* (the furthest part), *ανευ* (without), and *στυλοζ* (a style), which obviously alludes to the lack of transverse cirri (anal styles in Stokes' terminology). Feminine gender (Aescht 2001, p. 282).

Characterisation (Fig. 224a, autapomorphies 2): Adoral zone of membranelles continuous. Many frontal cirri in two or more short rows. One or more buccal cirri right of paroral. Midventral complex composed of short midventral rows in anterior portion and long midventral rows in posterior portion. More than two frontoterminal cirri form distinct row (A). Transverse cirri absent (A). 1 right and 1 left marginal row. 4 dorsal kineties (A). Number of caudal cirri distinctly increased (A).

Remarks: The three species assigned to *Eschaneustyla* have, besides the characteristics mentioned above, several other (plesiomorphic) features in common: body moderately large to large (120–260 µm); body very flexible; many macronuclear nodules scattered throughout cytoplasm (nuclear apparatus not described for type species indicating that many macronuclear nodules are present); contractile vacuole slightly ahead of mid-body, during diastole with distinct collecting canals; cortical granules (very likely) present (feature not described for type species); buccal field narrow and flat in live specimens; paroral broadened anteriorly (possibly a further apomorphy of *Es-*

[1] The diagnosis by Stokes (1886) is as follows: Animalcules free-swimming, elliptical or ovate, not encuirassed; frontal styles numerous, more or less uncinate; ventral setae in three unequal longitudinal lines; anal styles none; marginal setae uninterrupted; contractile vesicle canal-like, near the left-hand border. Inhabiting fresh water.

[2] The improved diagnosis by Eigner (1994) is as follows: One buccal cirrus. Several short oblique midventral rows in anterior ventral surface. Transverse cirri absent. Caudal cirri present.

chaneustyla); anterior end of left marginal row slightly curved rightwards (this is somewhat reminiscent of *Holosticha*, where, however, the anterior end is behind the proximal end of the adoral zone versus left in *Eschaneustyla*); usually 2–3 caudal cirri per dorsal kinety.

The cirral pattern of *Eschaneustyla* was – mainly due to the lack of midventral pairs – difficult to interpret before the description of *E. terricola* by Foissner (1982) and the morphogenetic studies by Eigner (1994). This resulted in some misclassifications. Kahl (1932) mentioned *Eschaneustyla* as third genus within the hypotrichs, indicating that he considered it as rather "primitive" (basal) group. Borror (1972) synonymised it with *Urostyla* in that he transferred the type species to *Urostyla*, and Hemberger (1982) put it into the synonymy of *Paraurostyla* Borror, 1972 (details see same chapter at type species). In contrast, the following authors accepted *Eschaneustyla*, but placed it in different higher taxa: Fauré-Fremiet (1961, p. 3517), Stiller (1974a, p. 130), and Corliss (1977, p. 138; 1979, p. 310, with doubt) classified it in the Keronidae Dujardin; Tuffrau (1979, p. 525; 1987, p. 15), Tuffrau & Fleury (1994, p. 137), Curds et al. (1983), and Foissner & Foissner (1988, p. 82) in the Kahliellidae Tuffrau; and Foissner (1982) and Small & Lynn (1985) in the Amphisiellidae Jankowski, respectively, Amphisiellidae Small & Lynn. Eigner (1994) assigned *Eschaneustyla* to the Bakuellinae because of homologous morphogenetic processes, especially the formation of midventral rows. More recently, Shi (1999) and Shi et al. (1999) transferred it to the Spirofilidae Gelei, 1929, likely because of the twisted (curved) cirral rows (I did not translate Shi's papers). Lynn & Small (2002) considered *Eschaneustyla* a urostylid. I suppose that the present genus is closely related with *Epiclintes* (Fig. 224a) because they have a very similar (identical?) frontal ciliature and midventral complex (details see Epiclintidae).

The three species now included in *Eschaneustyla* differ in the arrangement of the frontal cirri. *Eschaneustyla brachytona* and *E. terricola* have a single left frontal cirrus followed by two short cirral rows whose anteriormost cirri can be homologised without any difficulty with the middle and right frontal cirrus of other hypotrichs. *Eschaneustyla lugeri* has a somewhat more complex frontal ciliature which is difficult to interpret without morphogenetic data. Possibly the formation of this pattern proceeds basically as in species with a bicorona, that is, the anteriormost anlagen (e.g., I–IX) form short cirral rows, which consist of two cirri, for example, in *Pseudokeronopsis*, but of three or more cirri in *E. lugeri*. Since pseudokeronopsids form only two cirri per anlage they have a bicorona. By contrast, *Eschaneustyla* species have, like *Urostyla grandis* (type of *Urostyla*), a multicorona.

The characterisation above is not very precise because the type species is not described after silver impregnation. Possibly some features mentioned in the first paragraph of the remarks (e.g., paroral broadened anteriorly) are diagnostic too.

Species included in *Eschaneustyla* (alphabetically arranged according to basionyms): (1) *Eschaneustyla brachytona* Stokes, 1886; (2) *Eschaneustyla lugeri* Foissner, Agatha & Berger, 2002; (3) *Eschaneustyla terricola* Foissner, 1982.

Key to *Eschaneustyla* species

Eschaneustyla brachytona and *E. terricola* are very similar and thus have been synonised by Eigner (1994). By contrast, Foissner et al. (2002) accept both species and separate them by the habitat (limnetic/terrestrial) and the colour of the cell (colourless/yellowish). *Eschaneustyla lugeri* was so far recorded only from the Fiji Islands, indicating that it is lacking (or at least extremely rare) in Europe/Holarctis.

1 Limnetic (Fig. 232a) .... *Eschaneustyla brachytona* (p. 1148)
\- Terrestrial (Fig. 233a, 236a, o) **2**
2 Cortical granules yellowish; usually 1 buccal cirrus; 1 distinctly isolated frontal cirrus in left anterior corner of frontal field (Fig. 233a, b) .... *Eschaneustyla terricola* (p. 1150)
\- Cortical granules colourless; usually 4 buccal cirri; no isolated frontal cirrus in left anterior corner of frontal field (Fig. 236a, i) ...... *Eschaneustyla lugeri* (p. 1161)

Eschaneustyla brachytona Stokes, 1886
(Fig. 232a–c)

1886 ***Eschaneustyla brachytona*, sp. nov.** – Stokes, Proc. Am. phil. Soc., 23: 28, Fig. 11 (Fig. 232a; original description; no type material available and no formal diagnosis provided).
1888 ***Eschaneustyla brachytona*, Stokes** – Stokes, J. Trenton nat. Hist. Soc., 1: 283, Plate XI, fig. 2 (Fig. 232b; review of freshwater ciliates from the USA).
1932 ***Eschaneustyla brachytona* Stokes, 1886** – Kahl, Tierwelt Dtl., 25: 541, Fig. 97 $_{14}$ (Fig. 232c; revision of hypotrichs).
1972 ***Urostyla brachytona* (Stokes, 1886) n. comb.** – Borror, J. Protozool., 19: 9 (revision; combination with *Urostyla*).
1974 ***Eschaneustyla brachytona* Stokes** – Stiller, Fauna Hung., 115: 45, Ábra 26B (redrawing from Stokes).
2001 ***Eschaneustyla brachytona* Stokes, 1886** – Berger, Catalogue of ciliate names 1. Hypotrichs, p. 20 (nomenclator containing all basionyms, combinations, and higher taxa of hypotrichs).

Nomenclature: No derivation of the name is given in the original description. The species-group name *brachytona* is a composite of the Greek adjective *brachýs* (short, small, little), the Greek word *ton-* (stretch, to tense), and the inflectional ending *·a*, and possibly refers to the fact that this species is "somewhat extensile" (Stokes 1886). The present species is the type species of *Eschaneustyla* by monotypy.

"*Exhaneustyla brachytona* Stokes 1886" in Grispini (1938, p. 153) is an incorrect subsequent spelling.

Aescht (2003, p. 382) mistakenly assumed that Eigner's voucher slides of *E. brachytona* (note that Eigner's population is classified as *E. terricola* in the present paper) are the holotype and paratype slide of *E. brachytona* (see *E. terricola* for details).

Remarks: Stokes (1886) described the present species from a freshwater habitat in the USA. The original description is rather detailed, but lacks two somewhat difficult features: (i) Stokes wrote "nucleus not observed" implying that the type population had many small, scattered (and therefore difficult to recognise) macronuclear nodules; and

(ii) Stokes did not mention a colour. Since it is generally known that he was a careful observer, this indicates that his population was more or less colourless, that is, cortical granules where either lacking or colourless. Kahl (1932) accepted Stokes' species without adding new data. Borror (1972) transferred it to *Urostyla*, but without detailed explanation. However, *Urostyla* has, inter alia, distinct transverse cirri (vs. lacking in *Eschaneustyla*) and lacks caudal cirri (vs. present). Later, he observed this species in interphase and morphogenesis (no details provided), which were similar to those of *Kahliella* (Borror 1979, p. 549). Thus, Borror & Wicklow (1983, p. 117) excluded it from the urostyloids. Hemberger (1982, p. 31) transferred it to *Paraurostyla* because of the long ventral rows. However, *Paraurostyla* has a fragmenting dorsal kinety and is thus classified in the oxytrichids by Berger (1999).

Eigner (1994) described a population from a disused coconut doormat on lawn, that is, from a terrestrial habitat. He identified his population as *E. brachytona* and simultaneously synonymised *E. terricola* with the type species. This synonymy was accepted by Franco et al. (1996, p. 329) and also by Foissner (1998, p. 203). However, Foissner et al. (2002) discussed this act again and proposed that Eigner's synonymy should not be followed until a limnetic *Eschaneustyla* population has been investigated in detail. Thus, in the present book all three *Eschaneustyla*-species described so far are accepted. Consequently, the terrestrial population described by Eigner – who did not neotypify *E. brachytona* – is classified as *E. terricola*.

As already discussed, detailed redescription of a limnetic *E. brachytona* population is needed.

Morphology: This chapter contains only data from the original description (Stokes 1886). Body length 170–210 μm; body length:width ratio 3.5–4.0:1 (that is, body width ranges from about 40–60 μm). Body outline elliptical, rear end slightly broader rounded than anterior, which is somewhat curved leftwards and has an inconspicuous constriction slightly behind the anterior end. Body soft, flexible, and somewhat extensible (that is, slightly contractile!). Nuclear apparatus not observed, indicating that the macronucleus is composed of many scattered nodules. Contractile vacuole left of proximal end of adoral zone, during diastole with long, distinct collecting canals (Fig. 232a; according to Stokes' text a second spherical or subfusiform dilatation is present about in mid-body). Cytopyge "postero-terminal" (that is, likely on dorsal side in rear body portion). Presence/absence of cortical granules, respectively, colour of cell not mentioned, indicating that coloured granules are lacking (see remarks).

Adoral zone occupies about one third of body length, distal end extending not very far posteriorly. Buccal field obviously rather narrow. About 25 frontal cirri arranged in oblique rows (specimen illustrated has four such rows); two or three supplementary cirri form anteriormost row (this indicates that the anteriormost cirri [left, middle, and right frontal cirrus] are slightly set off, as in *E. terricola*). Three long midventral cirral rows; right row shortest, middle row longest (terminates at 85% of body length in specimen illustrated), left row possibly commences at level of buccal vertex. Transverse cirri lacking. Marginal rows/cirri without peculiarities, except that they are uninterrupted and longest posteriorly, strongly indicating that Stokes misinterpreted caudal

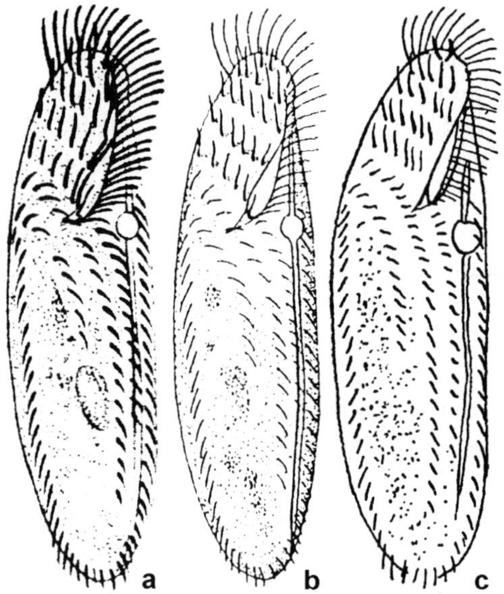

Fig. 232a–c *Eschaneustyla brachytona* (a, from Stokes 1886; b, from Stokes 1888; c, after Stokes 1888 from Kahl 1932. From life). Ventral view, 170–211 μm (individual size not indicated). Very likely Stokes misinterpreted the caudal cirri as marginal cirri. Page 1148.

cirri (not described by Stokes) as marginal cirri. Dorsal ciliary pattern, that is, number of kineties, length of bristles, and presence/absence of caudal cirri, not known.

Morphogenesis: Borror (1979, p. 549) and Borror & Wicklow (1982; 1983, p. 117) obviously studied cell division, however, without providing details.

Occurrence and ecology: Possibly confined to freshwater. Type locality of *Eschaneustyla brachytona* not known; likely near Trenton, New Jersey, USA, where Stokes lived and worked. Stokes (1886) found it in standing freshwater with dead leaves. Records not substantiated by morphological data: draw-well in Italy (Grispini 1938, p. 153); rivers (Ottawa, Rock Creek) in the USA (Patrick 1961, p. 243); Mount Niwot, Boulder County, Colorado, USA (Hamilton 1943, p. 46; identified by Barber R. in a thesis in 1935); limnetic habitats in southeastern Louisiana, USA (Bamforth 1963, p. 133). Sudzuki (1992, p. 65) reported a spirotrich-genus looking like an *Eschaneustyla* in freshwater habitats in south-western islands of Japan.

Eschaneustyla terricola Foissner, 1982
(Fig. 233a–e, 234a–o, 235a, b, Table 44)

1982 ***Eschaneustyla terricola* nov. spec.**[1] – Foissner, Arch. Protistenk., 126: 37, Abb. 3a–e, 45, 47, Tabelle 6 (Fig. 233a–e; original description; type slides are likely deposited in the Oberösterreichische Landesmuseum in Linz [LI], Austria; slides not mentioned by Aescht 2003, p. 398).

1994 ***Eschaneustyla brachytona* Stokes, 1886**[2] – Eigner, Europ. J. Protistol., 30: 464, Fig. 1–21, Table 1 (Fig. 234a–o; redescription, morphogenesis, and reorganisation; see remarks; voucher slides are depos-

[1] The diagnosis by Foissner (1982) is as follows: In vivo etwa 160 × 45 μm große, sehr biegsame, durch zahlreiche subpelliculäre Granula bei kleiner Vergrößerung bräunlich gefärbte *Eschaneustyla* mit aufallend großer adoraler Membranellenzone. Marginalreihen hinten nicht geschlossen. 8–9 unterschiedlich lange Ventralreihen. 6–12 dorsal inserierte Caudalcirren. 4 Dorsalkineten.

[2] The improved diagnosis by Eigner (1994) is as follows: Long-elliptical, in vivo 120–220 × 35–65 μm. One frontal cirrus and 2 distinct cirral rows anterior of undulating membranes. Yellowish cortical granules in longitudinal rows. One buccal cirrus and 1 long frontoterminal row. On average, 5 short and 3 long midventral rows, 46 adoral membranelles and 50 macronuclear segments. Four to six dorsal kineties and 8–25 caudal cirri.

ited in the Oberösterreichische Landesmuseum in Linz [LI], Austria, accession numbers 1993/14 and 1993/15; see nomenclature).

2001 ***Eschaneustyla terricola* Foissner, 1982** – Berger, Catalogue of ciliate names 1. Hypotrichs, p. 20 (nomenclator containing all basionyms, combinations, and higher taxa of hypotrichs).

2001 ***Eschaneustyla brachytona*** – Eigner, J. Euk. Microbiol., 48: 77, Fig. 29 (Fig. 234d modified; brief review of urostylids).

2002 ***Eschaneustyla brachytona*** – Lynn & Small, Phylum Ciliophora, p. 442, Fig. 3 (Fig. 234d, e; guide to ciliate genera).

2002 ***Eschaneustyla terricola* Foissner, 1982** – Foissner, Agatha & Berger, Denisia, 5: 578, Fig. 381f, g (Fig. 235a, b; additional observations from a Namibian population; removal from synonymy with *E. brachytona*).

Nomenclature: No derivation of the name is given in the original description. The species-group name *terricola* (Latin; living in soil) refers to the habitat where the species was discovered.

Eigner (1994, p. 464) wrote that "two slides of protargol impregnated cells have been deposited ... ... Accession number: 93/15, 93/15". This shows that he "simply" deposited voucher slides. By contrast, Aescht (2003, p. 382) designated these two slides as holotype and paratype slide of *E. brachytona*, which is incorrect because these two kinds of types can only stem from the type series which was studied by Stokes (1886), who did not make permanent preparations.

Remarks: Foissner (1982) separated the type species from *E. terricola* by the following features: marginal rows continuous posteriorly (vs. interrupted; however, Foissner explained that Stokes could have misinterpreted the caudal cirri as marginal cirri); adoral zone narrow and inconspicuous (vs. conspicuously large); ventral cirral rows differently arranged (even in view of minor misobservations by Stokes); coloured cortical granules lacking (vs. yellowish granules present). Admittedly these are not very pronounced differences, so one can understand Eigner (1994), who synonymised *E. terricola* with the type species. However, recently we suggested considering Stokes' and Foissner's population as distinct species (limnetic and colourless vs. terrestrial and yellowish) until a limnetic *Eschaneustyla* population is described in detail (Foissner et al. 2002).

According to Eigner (1994, p. 472), his population differs from Foissner's *E. terricola* in three ways, namely frontal cirri, midventral cirri, and variability of ciliature. However, the first two differences are not real differences, but due to different terminology, respectively, interpretation. The small variation in Foissner's population was explained by the use of field material, whereas Eigner studied cultures, where variability is usually higher (Eigner 1994). The most important difference between these two populations exists, in my opinion, in the arrangement of the cortical granules, namely in longitudinal rows in Eigner's population (no illustration provided to show arrangement) against irregularly arranged in groups in Foissner's population (Fig. 233d). The Namibian population differs from both populations in this feature (see below), indicating high variability of cortical granulation or cryptic speciation (Foissner et al. 2002).

Morphology: Because the populations described so far differ distinctly in some features, the descriptions are kept separate.

Table 44 Morphometric data on *Eschaneustyla terricola* (te1, from Foissner 1982; te2, from Eigner 1994)

Characteristics[a]	Species	mean	M	SD	SE	CV	Min	Max	n
Body, length	te1	129.9	128.0	12.0	3.8	9.2	112.0	155.0	10
	te2	154.4	–	35.1	6.9	22.7	112.0	200.0	26
Body, width	te1	39.9	38.0	7.3	2.3	18.2	29.0	53.0	10
	te2	46.6	–	7.2	1.4	15.5	32.0	58.0	25
Macronuclear nodules, number	te1	43.9	52.0	8.3	2.6	15.5	45.0	75.0	10
	te2	49.6	–	13.6	2.7	27.4	27.0	70.0	25
Macronuclear nodule, length	te1	6.9	6.6	1.3	0.4	19.1	5.3	9.3	10
	te2	7.3	–	2.9	0.6	39.7	4.0	16.0	25
Macronuclear nodule, width	te1	3.4	3.3	1.0	0.3	28.3	2.0	5.3	10
	te2	3.9	–	1.3	0.3	33.3	2.0	6.0	25
Micronuclei, number	te1	2.1	2.0	0.3	0.1	14.3	2.0	3.0	10
	te2	3.2	–	1.2	0.4	37.5	2.0	6.0	10
Micronucleus, length	te1	2.9	2.9	0.2	0.1	8.0	2.6	3.2	10
Micronucleus, width	te1	2.6	2.6	0.2	0.1	8.4	2.0	2.9	10
Adoral membranelles, number	te1	39.4	39.0	2.6	0.8	6.5	36.0	43.0	10
	te2	46.2	–	6.5	1.3	14.1	35.0	62.0	25
Adoral zone of membranelles, length	te1	40.1	40.0	2.5	0.8	6.2	37.0	44.0	10
	te2	48.1	–	6.6	1.3	13.7	35.0	60.0	25
Frontal cirri, number	te1	3.0	3.0	0.0	0.0	0.0	3.0	3.0	10
Buccal cirri, number	te1	1.0	1.0	0.0	0.0	0.0	1.0	1.0	10
	te2	–	–	–	–	–	1.0	2.0	25
Cirral rows[c], number	te1	8.1	8.0	0.3	0.1	3.7	8.0	9.0	10
	te2	11.0	–	1.6	0.3	14.5	8.0	14.0	25
Midventral rows with 3–4 cirri, number	te2	4.7	–	1.3	0.3	27.7	2.0	7.0	25
Midventral rows with 5 or more cirri, number	te2	3.3	–	0.8	0.2	24.2	2.0	5.0	25
Anterior body end to rear end of midventral complex, distance	te2	118.8	–	17.4	3.5	14.6	86.0	148.0	25
Rearmost cirral row[b], number of cirri	te1	15.5	15.5	1.6	0.5	10.	14.0	19.0	10
Left marginal row, number of cirri	te1	34.7	34.5	2.6	0.8	7.4	31.0	39.0	10
	te2	47.4	–	7.8	1.6	16.4	36.0	63.0	23
Right marginal row, number of cirri	te1	42.3	41.5	3.8	1.2	9.1	36.0	48.0	10
	te2	54.5	–	9.4	2.0	17.2	42.0	69.0	23
Caudal cirri, number	te1	8.3	8.5	1.9	0.6	23.5	6.0	12.0	10
	te2	14.2	–	4.8	1.2	33.8	8.0	25.0	16
Dorsal kineties, number	te1	4.0	4.0	0.0	0.0	0.0	4.0	4.0	10
	te2	4.4	–	0.6	0.1	13.6	4.0	6.0	19
Cirral anlage I, number of cirri formed	te2	–	–	–	–	–	1.0	3.0	25
Cirral anlage II, number of cirri formed	te2	7.2	–	1.5	0.3	20.8	5.0	10.0	25
Cirral anlage III, number of cirri formed	te2	4.8	–	1.3	0.3	27.1	2.0	8.0	25
Cirral anlage IV, number of cirri formed	te2	3.9	–	0.9	0.2	23.1	3.0	6.0	25
Cirral anlage V, number of cirri formed	te2	3.4	–	0.6	0.1	17.6	3.0	5.0	25
Cirral anlage VI, number of cirri formed	te2	3.2	–	0.6	0.1	18.8	2.0	5.0	25
Cirral anlage VII, number of cirri formed	te2	5.0	–	4.2	0.8	84.0	2.0	16.0	25
Cirral anlage VIII, number of cirri formed	te2	6.8	–	6.0	1.2	88.2	2.0	22.0	25
Cirral anlage IX, number of cirri formed	te2	9.0	–	6.6	1.4	73.3	3.0	24.0	23
Cirral anlage X, number of cirri formed	te2	12.0	–	6.5	1.4	54.2	3.0	24.0	21
Cirral anlage XI, number of cirri formed	te2	15.3	–	5.4	1.4	35.3	6.0	25.0	15
Cirral anlage XII, number of cirri formed	te2	18.5	–	4.7	1.5	25.4	11.0	27.0	10
Cirral anlage XIII, number of cirri formed	te2	22.7	–	4.8	2.0	21.1	15.0	29.0	6

Table 44 Continued

Characteristics[a]	Species	mean	M	SD	SE	CV	Min	Max	n
Cirral anlage XIV, number of cirri formed	te2	24.0	–	–	–	–	–	–	1
Frontoterminal row, number of cirri formed	te2	20.6	–	4.3	0.9	20.9	13.0	27.0	25

[a] All measurements in µm. CV = coefficient of variation in %, M = median, Max = maximum value, mean = arithmetic mean, Min = minimum value, n = number of individuals investigated, SD = standard deviation, SE = standard error of arithmetic mean.

[b] Marked with an asterisk in Fig. 233b.

[c] Rows behind middle and right frontal cirrus (anlagen II and III), midventral rows, and frontoterminal row (rightmost row) included.

Body size of type population (Foissner 1982) about 160 × 45 µm in life. Body outline long oval; posterior portion often distinctly converging or sometimes slightly pointed; left margin usually distinctly convex (Fig. 233a, d). Body very flexible, about 2:1 flattened dorsoventrally (Fig. 233e). Macronuclear nodules mainly arranged along body margins; individual nodules about 10 × 4 µm in life, usually contain one, rarely two large nucleoli surrounded by several very small nucleoli. Micronuclei inconspicuous in life, about 4 µm across, usually arranged as shown in Fig. 233c. Contractile vacuole near left body margin slightly ahead of mid-body, during diastole with long collecting canals (Fig. 233d). Cortical granules (protrichocysts?) arranged in groups (Fig. 233d); individual granules about 0.4 µm across, yellowish at high magnification, make cells brownish at low magnification, are ejected and about 1 µm large when methyl-green pyronin is added. Cytoplasm colourless, with few about 3 µm-sized, colourless to yellowish, shining globules and many food vacuoles up to 10 µm across. Movement slow, gliding.

Adoral zone occupies 31% of body length on average in protargol preparations (Table 44), composed of about 40 membranelles of ordinary fine structure. Bases of largest membranelles about 8 µm wide making zone rather conspicuous. Buccal field narrow and flat. Paroral (erroneously designated as adoral by Foissner) and endoral short (about 10–15 µm in specimen illustrated; Fig. 233b), slightly curved and arranged more or less in parallel. Cytopharynx without peculiarities.

Cirral pattern and number of cirri of usual variability (Table 44), basically as shown in Fig. 233b. For general comment on the rather unusual pattern, see genus section. Left frontal cirrus (= cirrus I/1) isolated in left anterior corner of frontal field, slightly enlarged. About five short, oblique rows in frontal field, each composed of about 3–5 cirri. Anteriormost cirri of anteriormost two rows slightly enlarged (these two cirri are very likely homologous with the middle and right frontal cirrus of the other hypotrichs with three frontal cirri). Buccal cirrus right of anterior portion of paroral. Frontoterminal cirri (13 cirri in specimen shown in Fig. 233b) form long row extending to about level of proximal end of adoral zone. At least two slightly curved cirral rows in postoral region. Cirri of midventral rows very fine, about 10 µm long, anteriorly slightly enlarged;

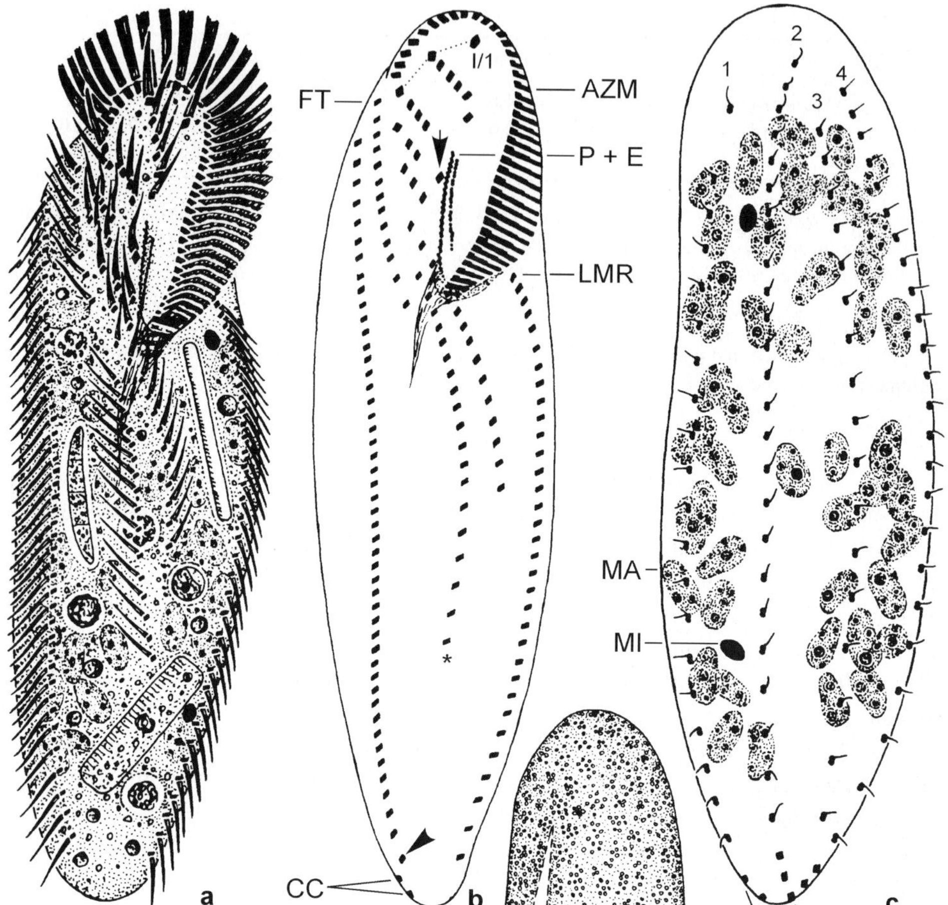

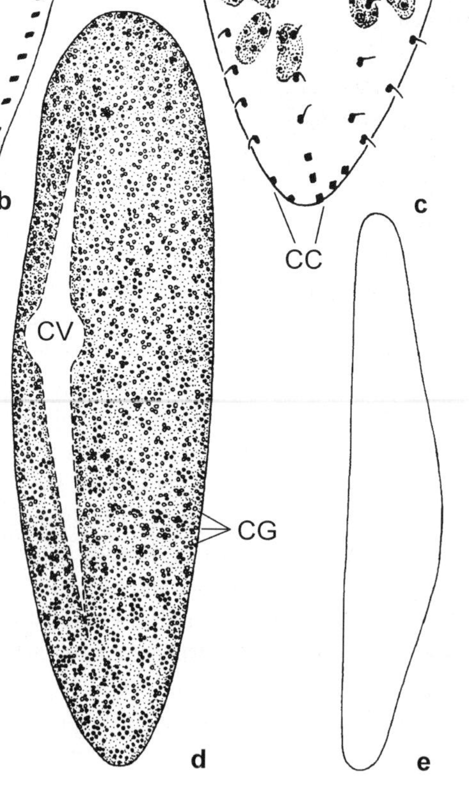

Fig. 233a–e *Eschaneustyla terricola* (from Foissner 1982. a, d, e, from life; b, c, protargol impregnation). **a:** Ventral view, 160 µm. Note the ingested diatoms and fungal spores. **b:** Infraciliature of ventral side, 142 µm. Arrow marks buccal cirrus, arrowhead denotes rear end of right marginal row. The isolated left frontal cirrus and the anteriormost cirrus of the two anteriormost cirral rows are connected by a dotted line; likely these three cirri are homologous to the ordinary three frontal cirri of many hypotrichs. **c:** Infraciliature of dorsal side and nuclear apparatus, 120 µm. **d:** Dorsal view showing contractile vacuole and cortical granulation, 140 µm. **e:** Left lateral view showing dorsoventral flattening. AZM = adoral zone of membranelles, CC = caudal cirri, CG = yellowish cortical granules, CV = contractile vacuole, FT = anterior end of frontoterminal row, LMR = anterior end of left marginal row, MA = macronuclear nodule, MI = micronucleus, P + E = undulating membranes (paroral and endoral), 1–4 = dorsal kineties. Page 1150.

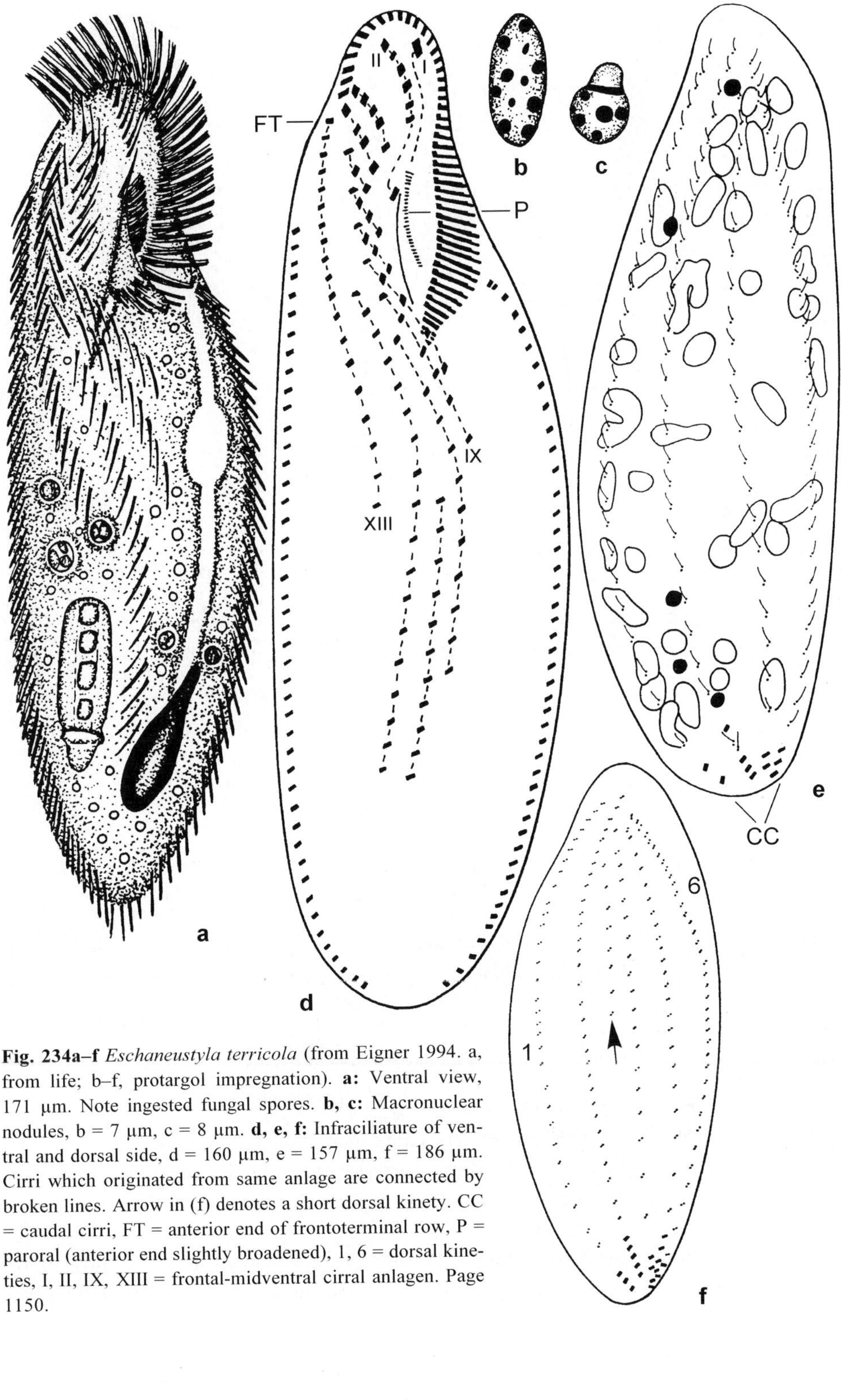

Fig. 234a–f *Eschaneustyla terricola* (from Eigner 1994. a, from life; b–f, protargol impregnation). **a:** Ventral view, 171 μm. Note ingested fungal spores. **b, c:** Macronuclear nodules, b = 7 μm, c = 8 μm. **d, e, f:** Infraciliature of ventral and dorsal side, d = 160 μm, e = 157 μm, f = 186 μm. Cirri which originated from same anlage are connected by broken lines. Arrow in (f) denotes a short dorsal kinety. CC = caudal cirri, FT = anterior end of frontoterminal row, P = paroral (anterior end slightly broadened), 1, 6 = dorsal kineties, I, II, IX, XIII = frontal-midventral cirral anlagen. Page 1150.

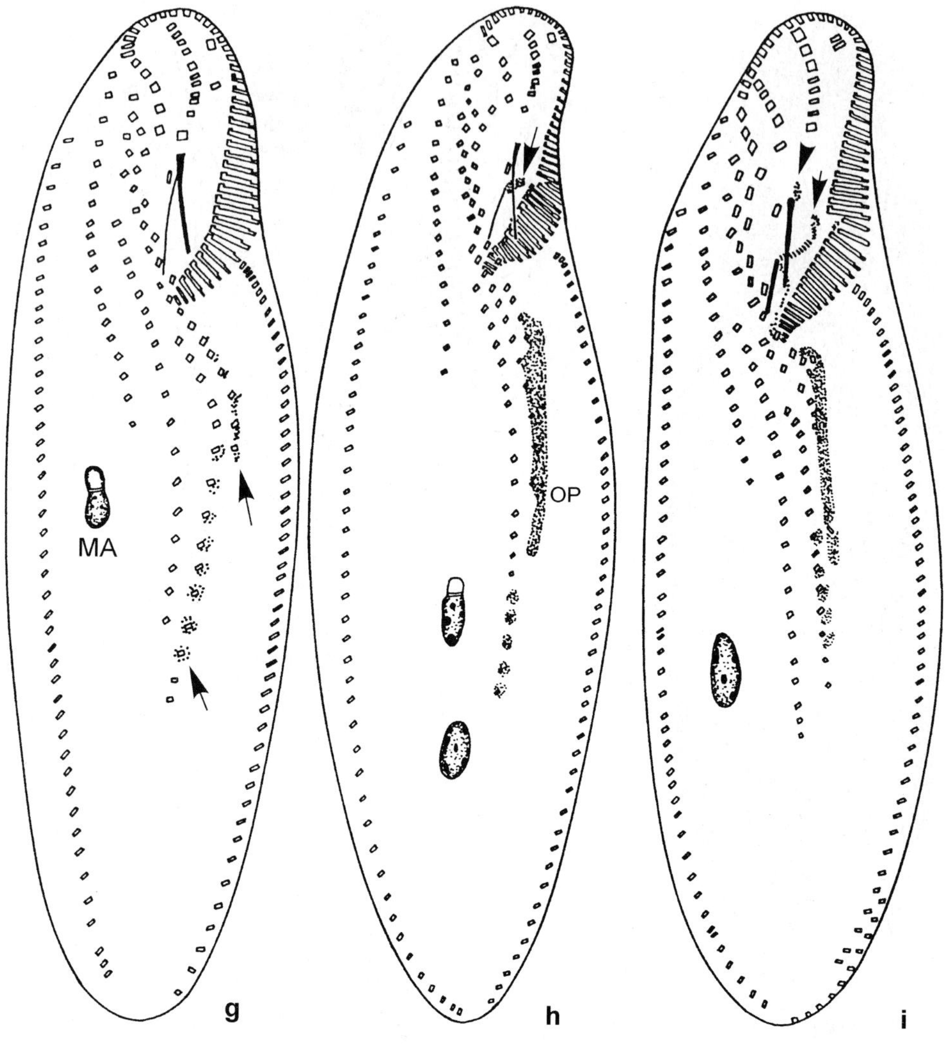

Fig. 234g–i *Eschaneustyla terricola* (from Eigner 1994. Protargol impregnation). Infraciliature of ventral side of very early and early dividers, g = 146 µm, h = 190 µm, i = 166 µm (only one or two of the many macronuclear nodules are shown). Arrows in (g) mark proliferation of basal bodies close to parental cirri. Arrow in (h, i) denotes disaggregating endoral, arrowhead in (i) marks a small field of basal bodies ahead of the paroral. MA = macronuclear nodule, OP = oral primordium. Page 1150.

distance between individual cirri posteriorly wider than anteriorly. Transverse cirri lacking. Marginal rows without peculiarities, posteriorly distinctly separated; left row slightly curved rightwards anteriorly, commences at level where proximal adoral membranelles become smaller (Fig. 233b).

Dorsal cilia about 2 µm long, arranged in four kineties; rows 1 and 3 slightly shortened anteriorly. Bristles within kinety 1 about equidistant, bristles within other rows

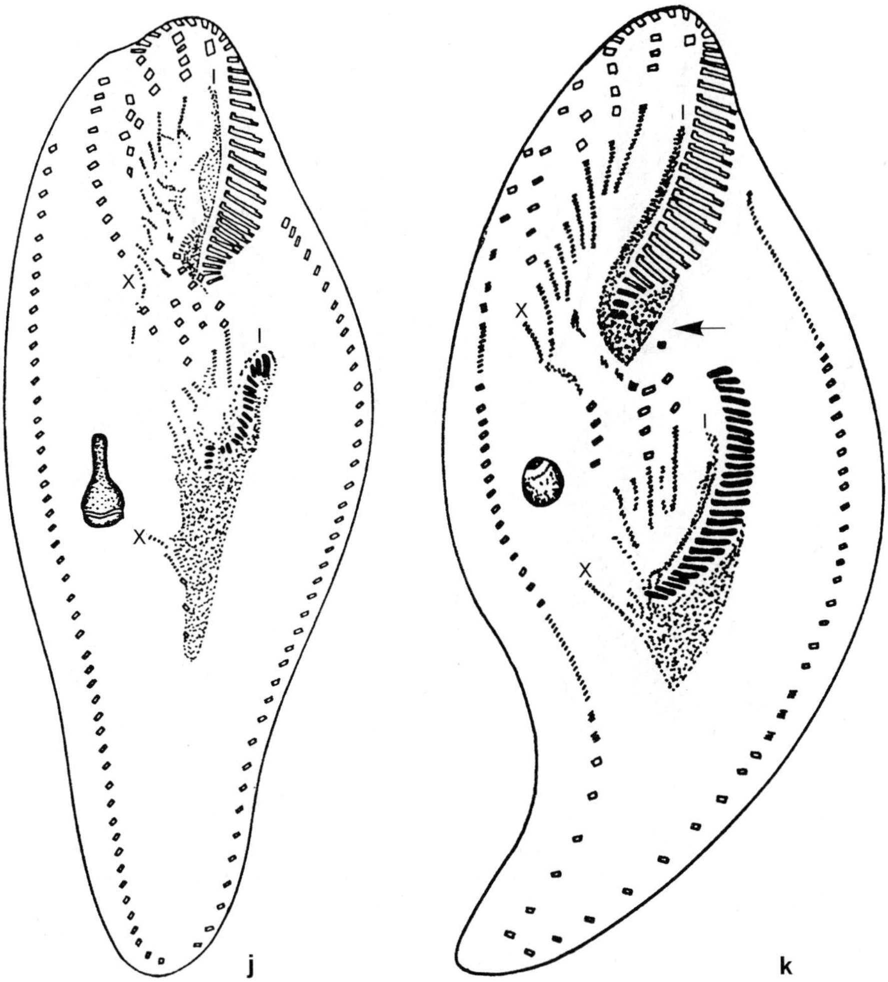

Fig. 234j, k *Eschaneustyla terricola* (from Eigner 1994. Protargol impregnation). Infraciliature of ventral side of middle dividers, j = 130 μm, k = 105 μm (only one of the many macronuclear nodules is shown). Arrow in (k) denotes dissolving proximal parental membranelles. I, X = frontal-midventral cirral anlagen. Page 1150.

posteriorly more widely spaced than anteriorly (Fig. 233c). Each kinety usually with two or three caudal cirri (Fig. 233b, c, Table 44).

Description of Eigner's (1994) population (only deviating or supplementary data provided; see also Fig. 234a–f and Table 44): Body size 120–220 × 35–65 μm. Anterior body end narrowed. Macronuclear nodules about 8 × 5 μm in life, highly variable in size, shape, and number (Fig. 234b, c, e). Micronuclei usually faintly stained in protargol preparations. Cortical granules yellowish, 1 μm across, arranged in longitudinal rows. Adoral zone curved rightwards in life, (slightly?) interrupted in posterior portion in about 15% of specimens (early reorganisers?). Cilia of adoral membranelles about

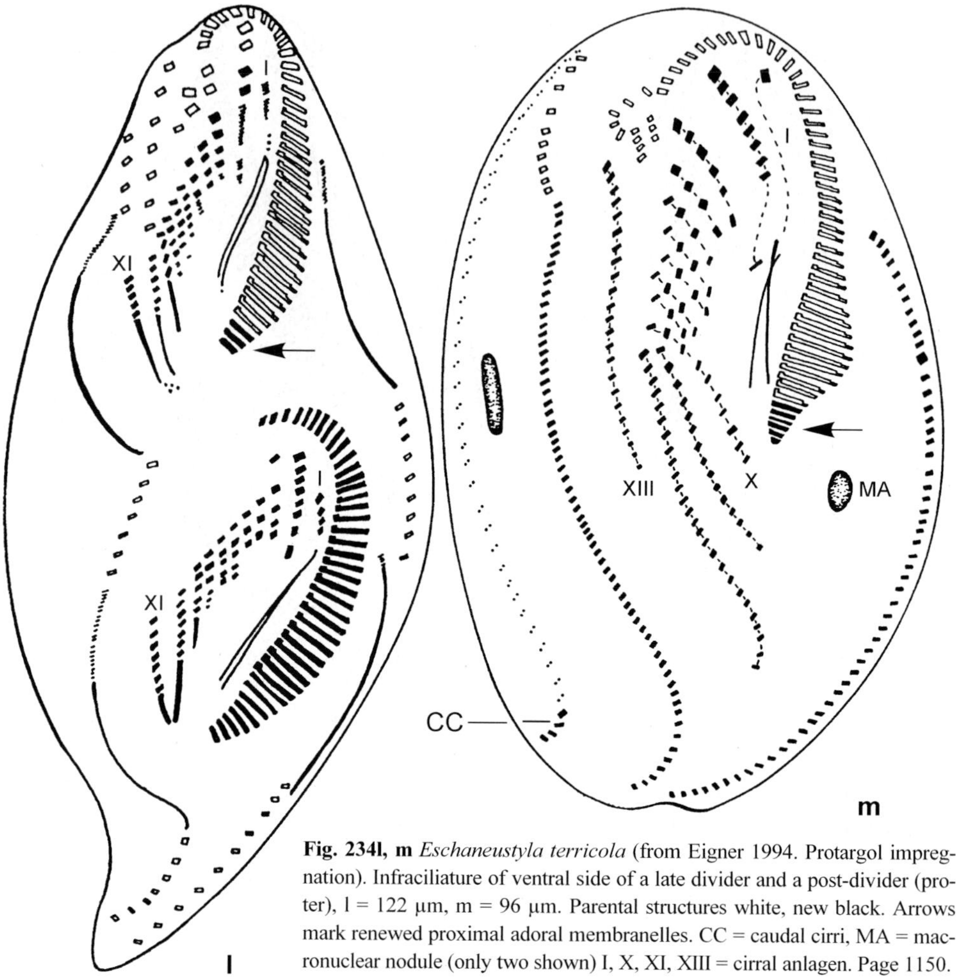

Fig. 234l, m *Eschaneustyla terricola* (from Eigner 1994. Protargol impregnation). Infraciliature of ventral side of a late divider and a post-divider (proter), l = 122 µm, m = 96 µm. Parental structures white, new black. Arrows mark renewed proximal adoral membranelles. CC = caudal cirri, MA = macronuclear nodule (only two shown) I, X, XI, XIII = cirral anlagen. Page 1150.

15 µm long. Paroral thickened at anterior end (Fig. 234d). Cirral row originating from anlage II highly variable in number and size of cirri. Next row usually commences close to distal end of adoral zone, often has all cirri enlarged. Behind this row 2–7 short midventral rows with usually 3–4 cirri, of which the anteriormost usually show zigzag pattern. Behind these short rows 2–5 curved, long midventral rows extending to middle or posterior third of cell. Frontoterminal row commences near distal end of adoral zone, extends to near cell centre; sometimes two frontoterminal rows present (no details provided). Right marginal row commences distinctly behind anterior end of cell (at 20% of body length in specimen shown in Fig. 234d). Dorsal bristles 3–4 µm long, arranged in 4–6 kineties; in specimens with more than four rows, one is usually distinctly shortened (Fig. 234e, f).

Specimens of Namibian population about 200 × 40 µm. Cortical granules brilliantly yellowish, form conspicuous clusters around dorsal bristles, while they are loosely arranged in the unciliated body parts; furthermore, the clusters consist of two size and shape types of granules: ellipsoidal to ovate and about 1.0 × 0.5 µm and globular 0.4–0.6 µm across (Fig. 235a, b). Distinct midventral pairs in anterior portion of midventral complex present (slides available, but not yet analysed; possibly a different species).

Cell division: Morphogenesis of cell division is described in detail by Eigner (1994; Fig. 234g–o). Here only the most important features are described (for a much more detailed description, see original paper).

Stomatogenesis commences with the proliferation of basal bodies around and next to the posterior cirri of two long midventral rows, which, however, appear unchanged (Fig. 234g). A narrow field of basal bodies develops (most likely dissolving cirri of long midventral rows contribute). A small field of basal bodies develops at the anterior end of the endoral (Fig. 234h). The oral primordium lengthens slightly at the anterior end. The endoral disorganises successively in a posteriad direction (Fig. 234i). A streak of basal bodies extends from the proximal membranelles to the anterior end of the posterior part of the endoral. At the anterior, thickened end of the paroral a small field of basal bodies develops (Fig. 234i).

Somewhat later, membranelles differentiate at the left anterior end of the enlarged oral primordium (Fig. 234j). Most posterior cirri of the long midventral rows have dissolved. A streak of basal bodies from the oral primordium develops to the right of the new membranelles (opisthe's anlage I). Four streaks develop from disaggregating parental cirri (opisthe's anlagen II–V). The posteriormost five streaks develop from the oral primordium (opisthe's anlagen VI–XI). The parental undulating membranes dissolve completely and form a large field of basal bodies to the right of the parental adoral membranelles (the anterior portion of this field forms proter's anlage I, the posterior portion contributes to the renewal of proximal membranelles; dissolving cirri from the long midventral rows possibly also contribute to the renewal of proximal membranelles; Fig. 234j). Distinct streaks develop from the posterior cirri of the frontal rows (proter's anlagen II and III). Three streaks (proter's anlagen IV–VI) develop from disaggregating posterior cirri of the short midventral rows (the dissolved buccal cirrus possibly contributes to these anlagen). The posterior streaks are formed by disaggregated cirri of the long midventral rows (proter's anlagen VII–X). Anlage XI is formed by disaggregating cirri of the frontoterminal row.

In the next stage the opisthe's membranelles differentiate posteriad (Fig. 234k). Cirral anlagen align and lengthen. The field of basal bodies right of the parental membranelles has narrowed its anterior part and enlarged its posterior part, forming a pipe-shaped pattern. Two posteriormost membranelles dissolve.

In the next stages the cirri are formed, the midventral cirral rows migrate to their final positions, and about three proximal membranelles are renewed (Fig. 234l, m). Briefly, anlage I splits longitudinally to form the paroral and endoral and cuts off the anterior end to form, as is usual, the left frontal cirrus (= cirrus I/1); anlage II forms the buccal cirrus and the anteriormost short cirral row (anteriormost cirrus obviously homologous with middle frontal cirrus); anlage III forms a short cirral row (right frontal

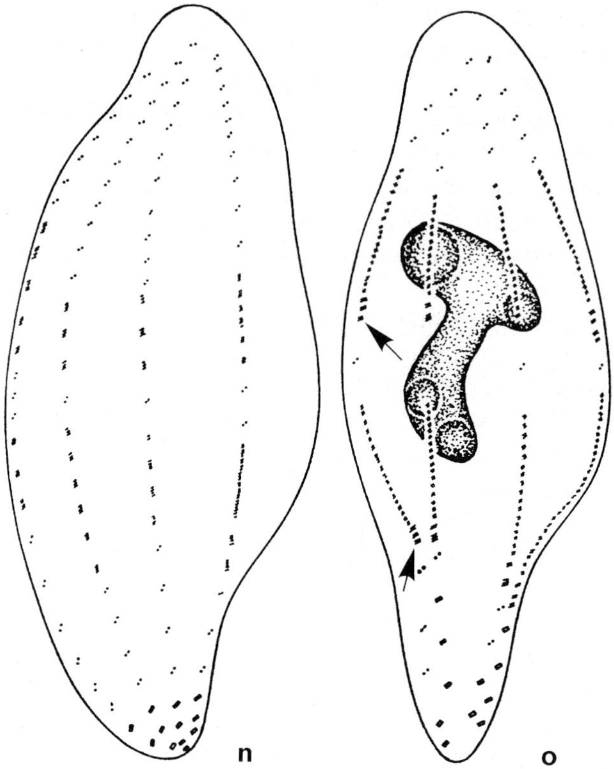

Fig. 234n, o *Eschaneustyla terricola* (from Eigner 1994. Protargol impregnation). Infraciliature of dorsal side of an early (n) and middle (o) divider, n = 134 μm, o = 130 μm. Arrows in (o) denote the caudal cirri of the new dorsal kineties 1. Page 1150.

row according to Eigner 1994; anteriormost cirrus obviously homologous with right frontal cirrus); anlagen IV to n–1 form short and long midventral rows (number of short and long rows variable); rightmost anlage produces, as is usual, the frontoterminal cirri which form a rather long row (Fig. 234m).

Division of marginal rows and dorsal kineties shows no peculiarities. Caudal cirri are formed at the end of each dorsal kinety (Fig. 234n, o). The macronuclear nodules fuse to a single mass and later divide again (Fig. 234o).

Eigner (1994) also studied reorganisation in great detail (see this paper for details). Briefly, the reorganisation of the ciliature proceeds basically as in dividers. The number of macronuclear nodules is not lower in any stage of reorganisation than the average of interphasic specimens. By contrast, in *Stylonychia mytilus* the two macronuclear nodules fuse (Zou & Ng 1991).

Occurrence and ecology: Likely confined to terrestrial habitats. Type locality of *E. terricola* is a xerothermic site without trees (Heißlände Althann; 181 m above sea-level; 48°21'46"N 15°55'54"E) near the village of Zwentendorf, Lower Austria (Foissner 1982). For a detailed description of this site, see Foissner et al. (1985, p. 85, Profil 1). Eigner (1994) found it in a disused coconut doormat which had been lying on a lawn in the village of Schrötten, Deutsch Goritz, Austria, for several years. It was dried for three weeks in December, cut into pieces, and put into petri dishes which were then filled with distilled water. After three (!) hours a fully excysted specimen was observed. Squeezed wheat grains together with baker's yeast were added to support microbial growth.

Further record of *E. terricola*: soil of a tropical dry forest, about 5 km east of the ranch house "La Casona" in the Santa Rosa National Park, Costa Rica (Foissner 1995, p. 39).

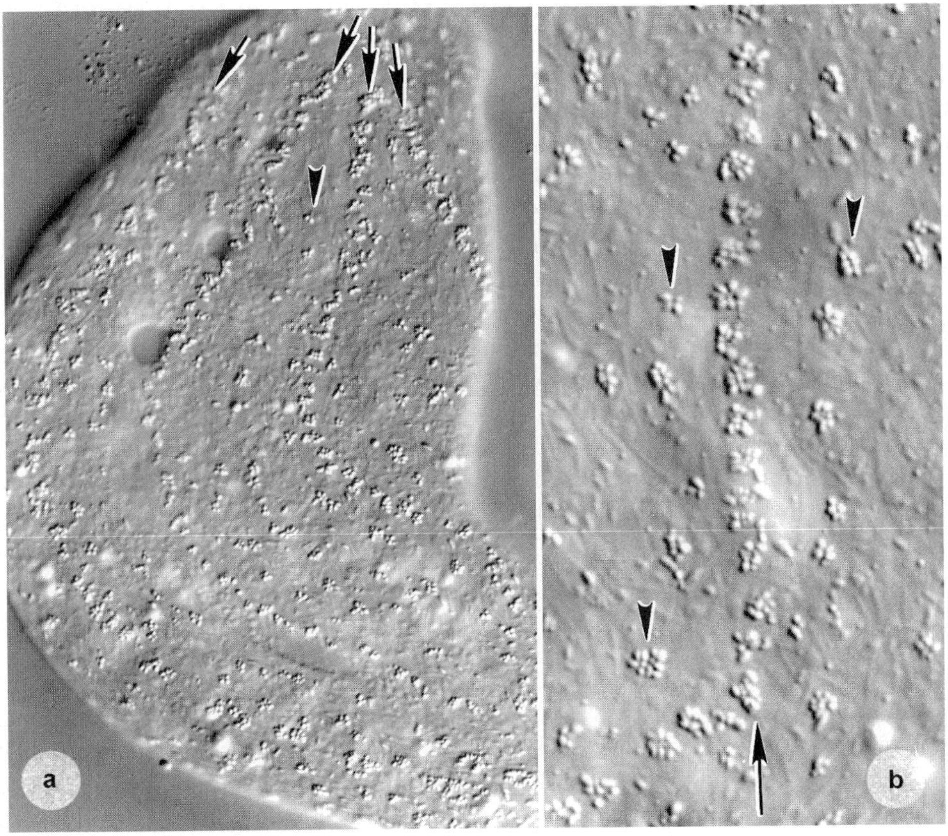

Fig. 235a, b *Eschaneustyla terricola* (from Foissner et al. 2002. Interference contrast). The cortical granules of the Namibian site specimens form conspicuous clusters around the bristles of the dorsal kineties (arrows) and are loosely arranged (arrowheads) in the unciliated areas. Page 1150.

Eschaneustyla terricola feeds on diatoms, fungal spores, green algae, and ciliates (Foissner 1982); food vacuoles of Eigner's specimen often contained large fungal spores, rarely testate amoebae and small ciliates. Biomass of 10^6 specimens about 146 mg (Foissner 1987a, p. 123; 1998, p. 203).

Eschaneustyla lugeri Foissner, Agatha & Berger, 2002 (Fig. 236a–q, Table 45)

2002 ***Eschaneustyla lugeri* nov. spec.**[1] – Foissner, Agatha & Berger, Denisia, 5: 572, Fig. 131a–n, 382a–c, Table 113 (Fig. 236a–q; original description; the holotype slide [2002/738] and 4 paratype slides [2002/739–742] are deposited in the Oberösterreichische Landesmuseum in Linz [LI], Austria).

[1] The diagnosis by Foissner et al. (2002) is as follows: Size about 220 × 55 µm in vivo, slightly contractile. Outline elongate elliptical to slightly sigmoidal. On average 60 macronuclear nodules. Cortical granules

Nomenclature: This species was dedicated to Gerhard Luger (Salzburg, Austria), who collected the sample.

Remarks: At first glance, this species looks like the representative of a new genus because of the numerous frontal cirri forming distinct coronas (Fig. 236a, i). However, the following features indicate that it is closely related to *E. brachytona* and *E. terricola*: (i) ventral cirral pattern in short and long rows; (ii) buccal cavity very flat and narrow compared to size of cell; (iii) slightly sigmoidal adoral zone extending onto right side of cell; (iv) broadened anterior end of paroral; (v) four dorsal kineties each associated with more than one caudal cirrus; (vi) lack of transverse cirri. Furthermore, both terrestrial species now assigned to *Eschaneustyla* have cortical granules, many small macronuclear nodules, and an anteriorly shortened right marginal row. A broadened anterior end of the paroral is also described in the oxytrichid *Notohymena* Blatterer & Foissner, 1988 (see Berger 1999 for review). However, *Notohymena* is an 18-cirri oxytrichid so that convergent evolution of this feature has to be postulated. Morphogenetic data are needed for the correct interpretation of the frontal cirri-pattern, which is more complex than that of *E. terricola*. However, I am certain that the present species is very closely related to the other two *Eschaneustyla* species as indicated by the other matching features listed above.

Eschaneustyla lugeri differs from the congeners by the numerous and therefore much more conspicuous frontal cirri evenly distributed on the frontal area; no isolated (left) frontal cirrus, so conspicuous in *E. terricola*, is recognisable. Furthermore, the left ventral cirral row (Fig. 236i, arrowhead) of *E. lugeri* borders the cirri on the frontal field, while it commences farther subapically in *E. terricola*, where the oblique frontal rows thus abut onto the frontoterminal cirral row. There are also several distinct quantitative differences: four buccal cirri vs. one; one long cirral row (= long midventral row; rightmost row [= frontoterminal row] not included) vs. 2–5; on average 56 vs. 39–46 adoral membranelles (Tables 44, 45). In vivo, *Eschaneustyla lugeri* can be easily recognised by the following combination of features: length 180–260 µm; many macronuclear nodules; many cirri on frontal field forming several coronas, that is, a multicorona; four buccal cirri; two long cirral rows (midventral and frontoterminal).

Morphology: Body size 180–260 × 45–65 µm in life, usually near 220 × 55 µm; body length:width ratio 3.5–5:1 in life, on average 4:1 in life and protargol preparations (Table 45). Body outline elongate elliptical and often slightly sigmoidal, frequently somewhat irregular, that is, with small convexities and concavities, possibly due to slight contractions, as indicated by specimens under mild cover glass pressure which contract by up to 30%. Anterior body portion frequently slightly set off from body proper, that is, cephalised. Flattened about 2:1 dorsoventrally with anterior and posterior portion rather thin (Fig. 236a–d, o). Macronuclear nodules moderately variable in number and shape (Table 45), usually arranged as shown in Fig. 236k; individual nodules about 10 × 5 µm in life, ellipsoidal to elongate ellipsoidal, globular, or dumbbell-shaped, each

colourless, around cirri and dorsal bristles and scattered throughout cortex. 4 buccal cirri, 1 long ventral row consisting of an average of 28 cirri, and a conspicuously long row of frontoterminal cirri ending in rear third of body. Frontal area densely ciliated by an average of 32 cirri forming distinct curved rows.

Table 45 Morphometric data on *Eschaneustyla lugeri* (from Foissner et al. 2002)

Characteristics [a]	mean	M	SD	SE	CV	Min	Max	n
Body, length	202.7	213.0	28.7	8.6	14.1	162.0	246.0	11
Body, width	50.8	50.0	4.6	1.4	9.0	45.0	58.0	11
Body length:width, ratio	4.0	3.9	0.6	0.2	14.2	3.2	5.4	11
AE to proximal end of adoral zone, distance	61.4	61.0	4.8	1.4	7.8	52.0	68.0	11
AE to distal end of adoral zone, distance	18.9	18.0	2.3	0.7	12.4	16.0	22.0	11
Body length:length of adoral zone, ratio	3.3	3.3	0.5	0.1	13.6	2.7	4.1	11
AE to paroral, distance	27.8	28.0	3.0	1.0	10.9	24.0	32.0	9
Paroral, length	26.0	24.0	3.5	1.3	13.6	22.0	32.0	8
AE to first buccal cirrus, distance	33.1	33.0	3.1	0.9	9.5	27.0	38.0	11
AE to last buccal cirrus, distance	43.6	44.0	3.7	1.1	8.5	37.0	50.0	11
AE to RMV, distance	27.5	26.0	2.7	0.8	10.0	24.0	32.0	11
AE to end of RMV, distance	99.3	104.0	11.1	3.3	11.2	78.0	111.0	11
AE to frontoterminal row, distance	23.1	24.0	2.5	0.8	10.9	20.0	28.0	11
AE to end of frontoterminal row, distance	140.2	149.0	16.2	4.9	11.5	114.0	156.0	11
AE to right marginal row, distance	37.5	40.0	5.3	1.6	14.2	28.0	46.0	11
PE to right marginal row, distance	6.0	6.0	0.7	0.2	11.1	5.0	7.0	10
PE to left marginal row, distance	18.0	18.0	7.7	2.4	43.0	9.0	33.0	10
AE to first macronuclear nodule, distance	18.5	20.0	4.0	1.2	21.6	10.0	24.0	11
Anteriormost macronuclear nodule, length	9.0	10.0	2.1	0.6	23.3	6.0	12.0	11
Anteriormost macronuclear nodule, width	4.5	5.0	1.3	0.4	29.0	3.0	6.0	11
PE to rearmost macronuclear nodule, distance	31.2	35.0	10.3	3.1	32.9	10.0	46.0	11
Posteriormost macronuclear nodule, length	7.6	6.0	2.8	0.8	36.2	5.0	13.0	11
Posteriormost macronuclear nodule, width	4.7	5.0	1.1	0.3	23.3	3.0	6.0	11
Macronuclear nodules, number	60.3	60.0	9.0	2.7	14.9	49.0	76.0	11
Anteriormost micronucleus, length	3.8	4.0	0.8	0.2	20.0	3.0	5.0	11
Anteriormost micronucleus, width	3.2	3.0	0.5	0.1	14.5	2.5	4.0	11
Micronuclei, number	8.7	8.0	3.2	1.0	36.6	5.0	14.0	11
Adoral membranelles, number	56.1	57.0	5.1	1.5	9.1	48.0	64.0	11
Frontal cirri, number	32.4	31.0	3.8	1.3	11.7	27.0	39.0	9
Buccal cirri, number	4.0	4.0	–	–	–	3.0	5.0	11
RMV, number of cirri	27.9	28.0	2.5	0.8	9.0	23.0	31.0	11
Frontoterminal row, number of cirri	38.7	38.0	2.3	0.7	5.9	35.0	42.0	11
Left marginal row, number of cirri	45.3	45.0	3.0	0.9	6.6	42.0	52.0	11
Right marginal row, number of cirri	47.6	45.0	3.0	0.9	6.6	42.0	52.0	11
Caudal cirri, total number	9.9	11.0	1.4	0.5	13.8	8.0	11.0	9
Dorsal kineties, number	4.0	4.0	0.0	0.0	0.0	4.0	4.0	11

[a] Data based on all protargol-impregnated (Foissner's protocol) and mounted specimens from a non-flooded Petri dish culture. Measurements in µm. AE = anterior end of cell, CV = coefficient of variation in %, M = median, Max = maximum, mean = arithmetic mean, Min = minimum, n = number of individuals investigated, PE = posterior end of cell, RMV = rightmost midventral row (= left ventral row in Foissner et al. 2002), SD = standard deviation, SE = standard error of arithmetic mean.

with some small nucleoli. Micronuclei scattered, sometimes clumped, about 4 µm across, compact and thus easy to recognise in vivo and protargol preparations. Contractile vacuole with two conspicuous collecting canals at left body margin slightly above mid-body. Cortex very flexible, contains two size types of colourless granules and rather conspicuous crystals (Fig. 236f–h): type I granules about 1.0 × 0.5 µm, around

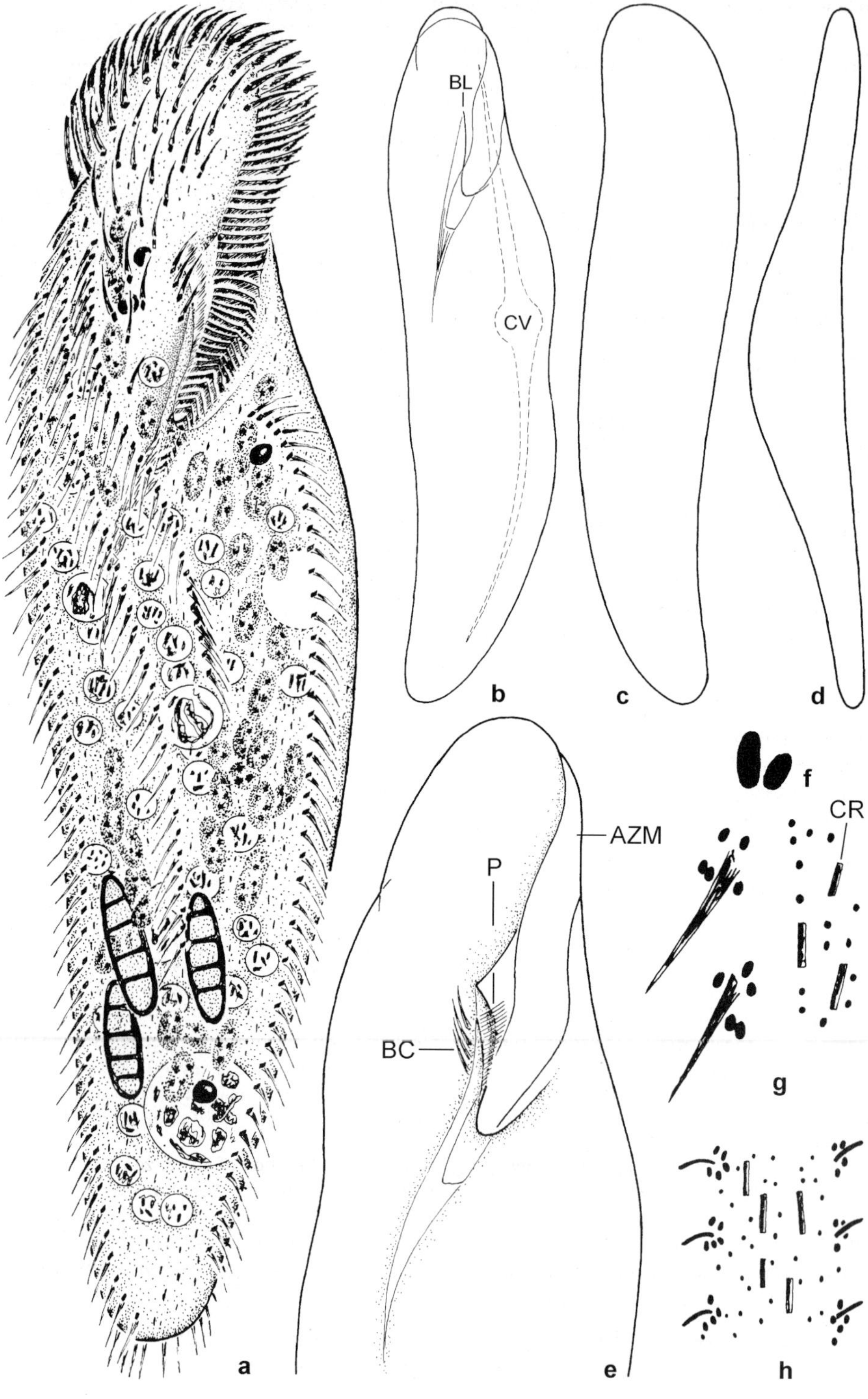
BL
CV
b
c
d
AZM
P
BC
e
a
f
CR
g
h

cirri and dorsal bristles; type II granules <0.5 μm, scattered, highly refractive and thus rather conspicuous despite their minuteness; both granule types do not impregnate with protargol, but stain pink and swell to 2–3 μm long rods forming a thin cover after addition of methyl green-pyronin. Cortical or subcortical crystals rod-shaped, 2–4 μm long, and often orientated in main body axis, do not occur in deeper cytoplasmic areas. Cytoplasm colourless, without conspicuous inclusions. Food vacuoles with bacteria only about 5 μm across, those with heterotrophic flagellates and organic debris up to 20 μm across; also feeds on up to 100 μm long hyphae, 20–30 × 5–10 μm-sized conidia, and small *Euglypha* species (Fig. 236a, o). Movement inconspicuous, glides rather rapidly on microscope slide and soil particles showing great flexibility.

Adoral zone occupies 25–38%, on average 30% of body length, composed of an average of 56 membranelles, bases of largest membranelles about 13 μm wide in life; distal portion extends far on right body side causing the slight cephalisation mentioned above; proximal portion slightly sigmoidal, a conspicuous (compared to many other hypotrichs) feature not described in the congeners studied by Foissner (1982) and Eigner (1994), but recognisable in their figures; such peculiarities are usually lost in protargol preparations, as also evident from the following observations. Buccal apparatus in vivo as in congeners, that is, extremely narrow and flat compared to size of cell and many other hypotrichs (Fig. 236a, b, e, o–q); much more prominent and thus of ordinary appearance in protargol preparations, possibly due to slight inflation of cavity and/or some contraction of the oral region (Fig. 236i). Likewise, paroral and endoral pattern very different in vivo and after protargol impregnation: paroral in life at base of buccal lip, almost straight and slightly curved rightwards, while distinctly curved leftwards and optically intersecting with endoral in silver preparations. Both membranes likely composed of dikinetids, paroral slightly thickened (broadened) in anterior 5 μm, cilia up to 6 μm long (Fig. 236i, n, o–q). Pharyngeal fibres clearly recognisable in life and after protargol impregnation, of ordinary length and structure, extend obliquely backwards.

Cirral pattern and number of cirri of usual variability (Fig. 236i, n; Table 45). Most cirri only 10 μm long in vivo. Frontal area very conspicuous because densely covered by an average of 32 slightly enlarged cirri forming rather distinct rows[1]; prominent frontal cirri and midventral pairs absent. Usually four buccal cirri right of mid-portion of paroral, last cirrus often slightly smaller. Invariably two long, slightly oblique cirral

---

← **Fig. 236a–h** *Eschaneustyla lugeri* (from Foissner et al. 2002. From life). **a:** Ventral view of a representative specimen, 220 μm. **b, c:** Shape variants in ventral and dorsal view. **d:** Right lateral view showing dorsoventral flattening. **e:** Oral apparatus showing sigmoidal proximal portion of adoral zone and the minute buccal cavity, which becomes heavily inflated in protargol preparations (Fig. 236n). **f:** Colourless cortical granules, about 1.0 × 0.5 μm in size, occur around cirri and dorsal bristles. **g, h:** Arrangement of large (about 1.0 × 0.5 μm) and small (about 0.2 μm across) cortical granules and subcortical crystals in ventral and dorsal cortex. AZM = adoral zone of membranelles, BC = buccal cirri, BL = buccal lip, CR = crystals, CV = contractile vacuole with collecting canals, P = paroral. Page 1161.

---

[1] Foissner et al. (2002) did not find dividers and thus could not define the pattern; a proposal is given in Fig. 236n.

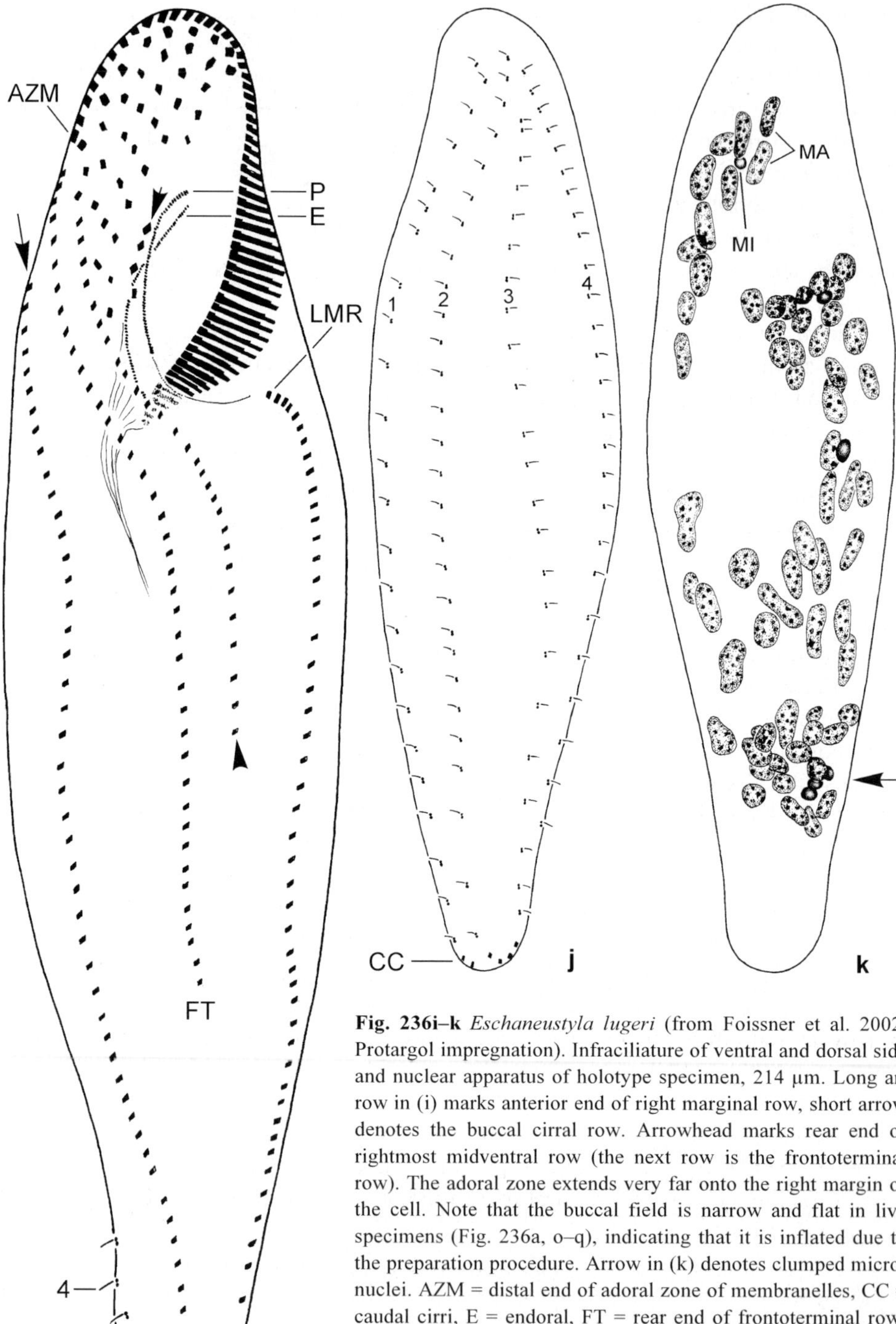

Fig. 236i–k *Eschaneustyla lugeri* (from Foissner et al. 2002. Protargol impregnation). Infraciliature of ventral and dorsal side and nuclear apparatus of holotype specimen, 214 µm. Long arrow in (i) marks anterior end of right marginal row, short arrow denotes the buccal cirral row. Arrowhead marks rear end of rightmost midventral row (the next row is the frontoterminal row). The adoral zone extends very far onto the right margin of the cell. Note that the buccal field is narrow and flat in live specimens (Fig. 236a, o–q), indicating that it is inflated due to the preparation procedure. Arrow in (k) denotes clumped micronuclei. AZM = distal end of adoral zone of membranelles, CC = caudal cirri, E = endoral, FT = rear end of frontoterminal row, LMR = anterior end of left marginal row, MA = macronuclear nodules, MI = micronucleus, P = paroral (anterior end widened), 1–4 = dorsal kineties. Page 1161.

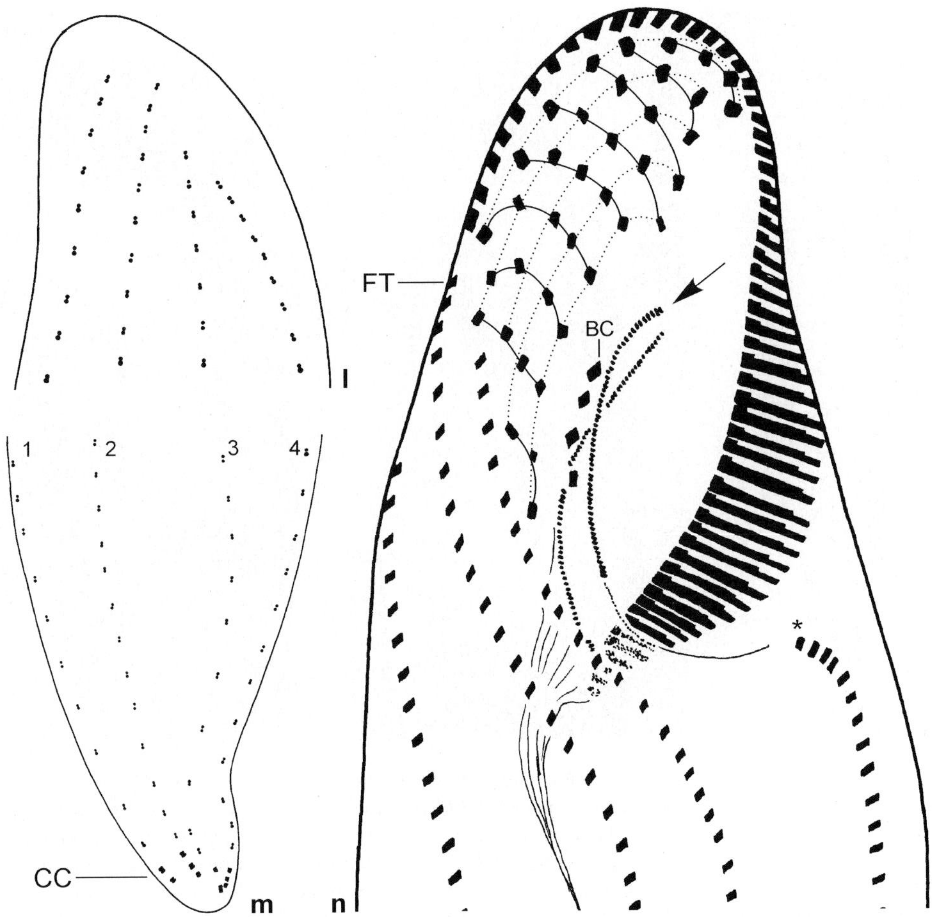

Fig. 236l–n *Eschaneustyla lugeri* (from Foissner et al. 2002. Protargol impregnation). **l, m:** Anterior and posterior portion of dorsal side of an excellently impregnated specimen. **n:** Anterior portion of holotype specimen (Fig. 236i). The many cirri in the frontal area of *E. lugeri* can be interpreted as belonging to about nine oblique cirral rows (solid lines) or about five coronal cirral rows (dotted lines); morphogenetic data are needed for correct interpretation. Arrow marks broadened (thickened) anterior portion of paroral. Asterisk marks curved anterior end of left marginal row. BC = anteriormost buccal cirrus, CC = caudal cirri, FT = anterior end of frontoterminal row, 1–4 = dorsal kineties. Page 1161.

rows, left begins at 13% of body length and terminates at 49% on average (I suppose that this is the rightmost midventral row). Right row commences at 11% of body length and terminates at 69% (its rightmost position indicates that this is the frontoterminal row). Transverse cirri lacking. Right marginal row distinctly shortened anteriorly, ends subterminal; left row distinctly curved rightwards anteriorly, terminates on average 18 μm ahead of rear body end.

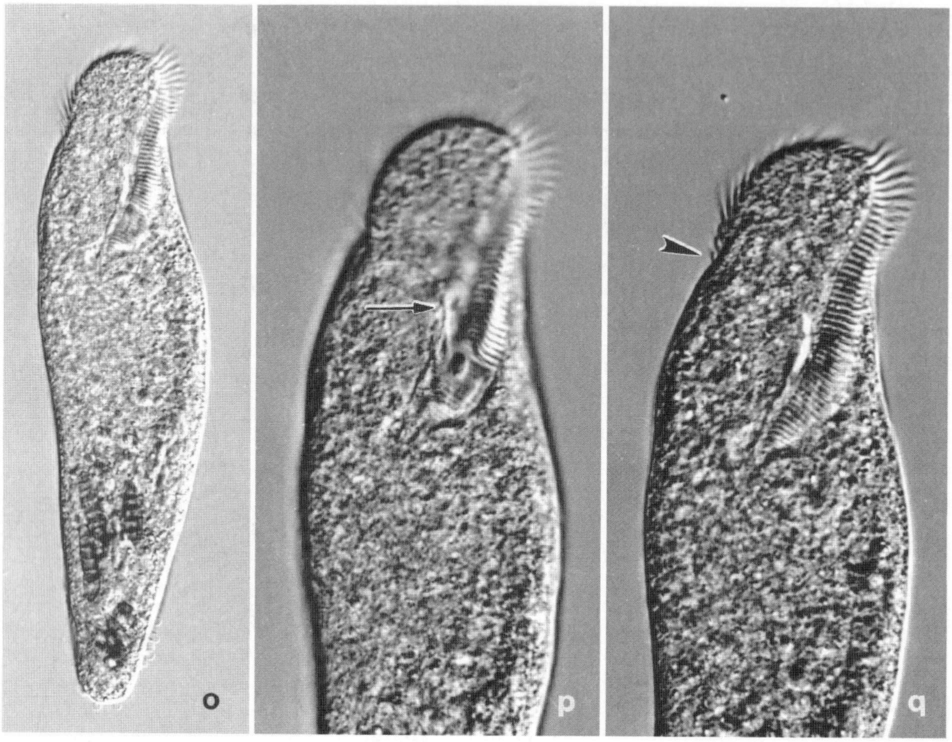

Fig. 236o–q *Eschaneustyla lugeri* (from Foissner et al. 2002. Interference contrast). **o:** The outline of this 180–260 µm long species is frequently somewhat irregular, that is, has convexities and concavities. It feeds, inter alia, on hyphae and conidia (dark inclusions in posterior body portion). **p, q:** The adoral zone, on average 30% of body length and composed of 56 membranelles, extends far onto the right body margin (arrowhead in p) and is slightly sigmoidal in the proximal portion. In life, the buccal cavity (arrow in p) is extremely narrow and flat, while of ordinary size in protargol-impregnated specimens (Fig. 236i). Such artefacts emphasise the need for live observations. Page 1161.

Dorsal cilia about 3 µm long in vivo, arranged in four longitudinal rows; rows 3 and 4 usually slightly shortened anteriorly. Each row with two or three small and thus inconspicuous caudal cirri (Fig. 236i, j, m, n).

Occurrence and ecology: *Eschaneustyla lugeri* is a rare soil species likely not occurring in European/Holarctic soils. As yet found only at the type locality which is a forest soil from Taveuni island, Fiji Islands (16°52'S 180°00'W).

Taxa of Unknown Position within the Urostyloidea

Notocephalus, *Biholosticha*, and *Paramitrella* have a ventral ciliature composed of cirral pairs. This strongly indicates that they belong to the Urostyloidea. However, since they have neither three enlarged frontal cirri nor a bicorona, they are not assigned to one of the four major subgroups (Holostichidae, Bakuellidae, Urostylidae, Epiclintidae) of the urostyloids. *Biholosticha* species obviously have only two enlarged frontal cirri and lack a buccal cirrus. *Notocephalus parvulus* has two large cirri and one slightly smaller cirrus near the anterior end of the undulating membranes (Fig. 237e). Possibly, the two large ones are, as in *Biholosticha*, frontal cirri, and the smaller one is a buccal cirrus. However, detailed redescriptions (*Biholosticha*) and cell division data (*Biholosticha*, *Notocephalus*) are needed to get a better idea of the systematic position of these marine species. *Paramitrella caudata* is, due to its curious shape (Fig. 240a), an easily recognisable hypotrich. Its frontal ciliature is unknown (strongly reduced? overlooked?).

Uncinata gigantea is a huge (up to 1100 µm long!) ciliate which is preliminarily assigned to the urostyloids (Fig. 241a, b). Some workers doubt that it belongs to the Hypotricha at all. Redescription is needed for a more proper classification.

Notocephalus Petz, Song & Wilbert, 1995

1995 ***Notocephalus* nov. gen.**[1] – Petz, Song & Wilbert, Stapfia, 40: 169 (original description). Type species (by original designation on p. 169): *Tachysoma parvulum* Corliss & Snyder, 1986.

1999 ***Notocephalus* Petz, Song & Wilbert, 1995** – Shi, Acta Zootax. sinica, 24: 366 (revision of the Hypotrichida; list of genera and higher taxa).

1999 ***Notocephalus* Petz, Song & Wilbert, 1995** – Shi, Song & Shi, Progress in protozoology, p. 117 (revision of hypotrichous ciliates).

2001 ***Notocephalus* Petz, Song & Wilbert 1995** – Aescht, Denisia, 1: 106 (catalogue of generic names of ciliates).

2001 ***Notocephalus* Petz, Song and Wilbert, 1995** – Berger, Catalogue of ciliate names 1. Hypotrichs, p. 48 (nomenclator containing all basionyms, combinations, and higher taxa of hypotrichs).

2002 ***Notocephalus* Petz, Song & Wilbert, 1995** – Lynn & Small, Phylum Ciliophora, p. 442 (guide to ciliate genera).

Nomenclature: The name *Notocephalus* is a composite of the Greek nouns *ho notos* (the south) and *he kephale* (the head); the former refers to the geographical region where the type species occurs (southern hemisphere, Antarctica), the latter refers to the head-shaped anterior cell portion (Petz et al. 1995). Masculine gender.

Characterisation: Body elongate and cephalised. Adoral zone of membranelles continuous, extends far onto right body margin; proximal half distinctly spoon-shaped, with bases longest in posterior portion. Three more or less distinctly enlarged cirri (3 frontal cirri? 2 frontal cirri plus 1 buccal cirrus?) triangularly arranged on frontal area.

[1] The diagnosis by Petz et al. (1995) is as follows: Body elongate, cephalised. Adoral zone extends far onto right side; proximal half spoon-shaped, with bases longest in posterior portion. 1 right and 1 left marginal row. Midventral and transverse cirri present. Without caudal cirri.

Buccal cirrus lacking (see previous feature). Frontoterminal cirri lacking. Midventral complex composed of midventral pairs only. Transverse cirri present. 1 left and 1 right marginal cirral row. Caudal cirri absent.

Remarks: See this chapter at *N. parvulus*.

Species included in *Notocephalus* (basionym is given): (1) *Tachysoma parvulum* Corliss & Snyder, 1986.

Single species

Notocephalus parvulus (Corliss & Snyder, 1986) Petz, Song & Wilbert, 1995 (Fig. 237a–f, Table 46)

1986 ***Tachysoma parvulum* n. sp.** – Corliss & Snyder, Protistologica, 22: 44, Fig. 6A, B (Fig. 237a, b; original description; no formal diagnosis provided. Holotype material has been deposited in the International Collection of Ciliate Type Specimens housed in the United States National Museum of the Smithsonian Institution, Washington, D.C.; Corliss & Snyder 1986, p. 40)[1].

1995 ***Notocephalus parvulus* (Corliss & Snyder, 1986) nov. comb.**[2] – Petz, Song & Wilbert, Stapfia, 40: 170, Fig. 50a–d, Table 24 (Fig. 237c–f; detailed redescription and combination with *Notocephalus*. One voucher slide [accession number 2001/137; see nomenclature] of protargol-impregnated specimens has been deposited in the Oberösterreichische Landesmuseum in Linz [LI], Austria).

2001 ***Notocephalus parvulus* (Corliss and Snyder, 1986) Petz, Song and Wilbert, 1995** – Berger, Catalogue of ciliate names 1. Hypotrichs, p. 88 (nomenclator containing all basionyms, combinations, and higher taxa of hypotrichs).

2005 ***Notocephalus parvulus* (Corliss & Snyder) Petz, Song & Wilbert (1995)** – Petz, Ciliates, p. 396, Fig. 14.83a–c (Fig. 237c, e, f; guide to Antarctic marine ciliates).

Nomenclature: The species-group name *parvul·us -a -um* (Latin adjective; diminutive of *parvum*; very small) refers to the small length (60 µm) given in the original description; unfortunately, this value is likely much to low (misobservation? see below) and the name therefore somewhat misleading. *Tachysoma* is neuter gender. Thus, mandatory change of ending was necessary (ICZN 1985), because the latinised "*cephalus*" from Greek "*kephale*" is masculine (for details, see Aescht 2001). *Notocephalus parvulus* was fixed as type species of *Notocephalus* by original designation.

Petz et al. (1995, p. 7) did not mention *N. parvulus* in the paragraph where they listed the species for which they have designated a neotype. Consequently, Aescht (2003, p. 393) obviously erroneously designated the voucher slide 2001/137 deposited by Petz et al. (1995) as neotype.

---

[1] Petz et al. (1995, p. 7, 172) write that "types of the latter species [*Tachysoma parvulum*] have not yet been deposited".

[2] The improved diagnosis by Petz et al. (1995) is as follows: In vivo about 155–160 × 40 µm. Body fusiform. Adoral zone composed of about 55–58 membranelles. Midventral row shortened, composed of 10–18 cirri. 3 frontal cirri, usually 5 transverse cirri. 3 dorsal kineties. 2 macro-, 2 micronuclei. No buccal cirri. Marine.

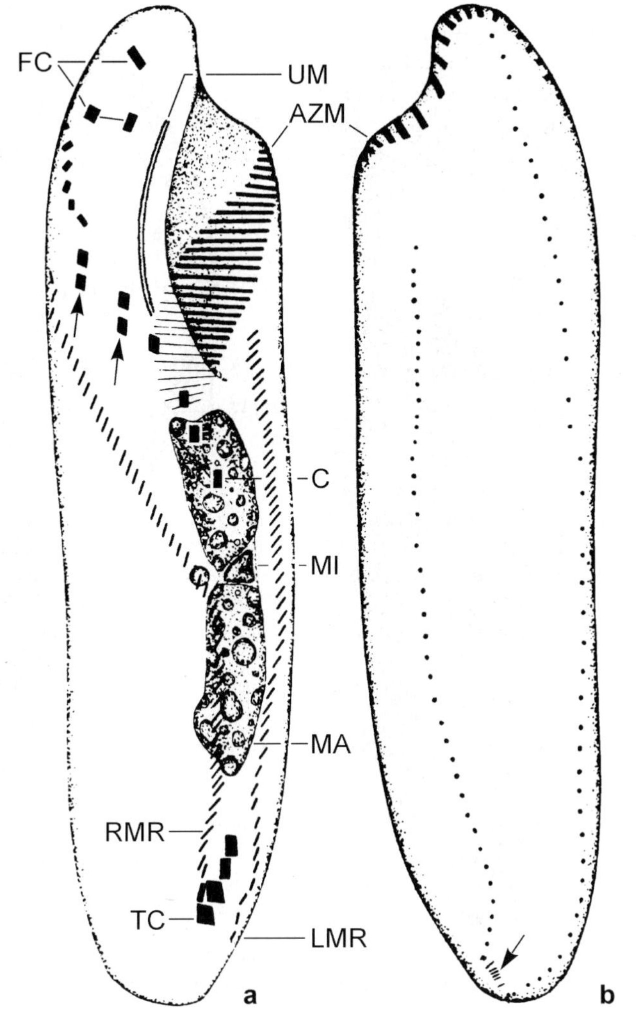

Fig. 237a, b *Notocephalus parvulus* after protargol impregnation (from Corliss & Snyder 1986). Infraciliature of ventral and dorsal side and nuclear apparatus of type specimen, 95 µm. Note that the length given in the original description, namely 60 µm, is obviously incorrect (see text). Arrows in (a) mark likely pseudopairs, arrow in (b) denotes some basal body files (?) between rear end of dorsal kineties (note that *Notocephalus parvulus* has three dorsal kineties and not two as drawn by Corliss & Snyder 1986; see text and Fig. 237f); this feature was not observed by Petz et al. (1995; see Fig. 237f). In protargol preparations, *Notocephalus* is obviously easily recognisable by the notched anterior end (also recognisable in the second population; Fig. 237e, f), the lack of a buccal cirrus, and the two macronuclear nodules. AZM = adoral zone of membranelles, C = last cirrus of midventral row, FC = frontal cirri, LMR = last cirrus of left marginal row, MA = posterior macronuclear nodule, MI = micronucleus in between the macronuclear nodules, RMR = right marginal row, TC = transverse cirri, UM = undulating membranes. Page 1170.

Remarks: The original description is based on preserved material only and thus not very detailed. Petz et al. (1995) rediscovered this curious species in the type location (Weddell Sea, Antarctica). They provided an improved diagnosis (see chapter morphology) and a redescription, including live observations and a morphometry which is, however, based on three specimens only. Furthermore, they reinvestigated one specimen (likely a paratype and not the holotype) of the type material. Accordingly, the original description contains some misobservations, namely (i) body length is not 60 µm, but 140 µm; (ii) the number of dorsal kineties is not two, but three; and (iii) the number of midventral cirri was underestimated. In addition, micronuclei were not discernible in the specimen reinvestigated. The check of the type material by Petz et al. (1995) confirmed the conspecificity with their population.

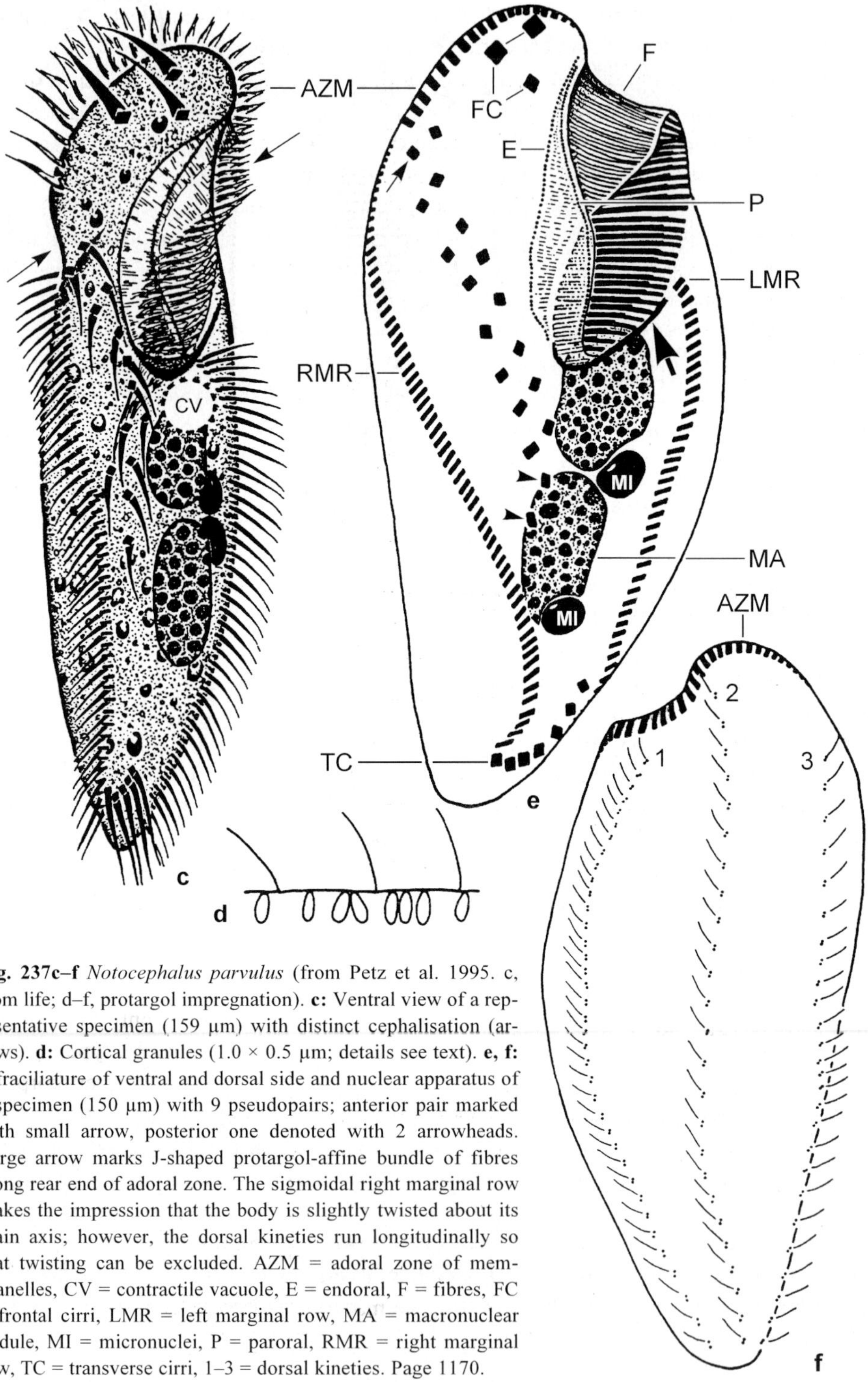

Fig. 237c–f *Notocephalus parvulus* (from Petz et al. 1995. c, from life; d–f, protargol impregnation). **c:** Ventral view of a representative specimen (159 µm) with distinct cephalisation (arrows). **d:** Cortical granules (1.0 × 0.5 µm; details see text). **e, f:** Infraciliature of ventral and dorsal side and nuclear apparatus of a specimen (150 µm) with 9 pseudopairs; anterior pair marked with small arrow, posterior one denoted with 2 arrowheads. Large arrow marks J-shaped protargol-affine bundle of fibres along rear end of adoral zone. The sigmoidal right marginal row makes the impression that the body is slightly twisted about its main axis; however, the dorsal kineties run longitudinally so that twisting can be excluded. AZM = adoral zone of membranelles, CV = contractile vacuole, E = endoral, F = fibres, FC = frontal cirri, LMR = left marginal row, MA = macronuclear nodule, MI = micronuclei, P = paroral, RMR = right marginal row, TC = transverse cirri, 1–3 = dorsal kineties. Page 1170.

Corliss & Snyder (1986) assigned the species to *Tachysoma* Stokes, 1887, a group of flexible 18-oxytrichids without caudal cirri (for review, see Berger 1999, p. 431). Petz et al. (1995) found that the midventral pattern does not suggest inclusion in *Tachysoma*. They recognised a relationship with the holostichids and established *Notocephalus* because the peculiar shape of the adoral zone and the cephalisation do not match any genus in the midventral hypotrichs. In spite of the redescription by Petz et al. (1995), it remains unclear whether the species has three curiously arranged frontal cirri and lacks a buccal cirrus, or has only two enlarged frontal cirri near the anterior body end and a buccal cirrus. However, this question can be answered and the final classification within the urostyloids found only by ontogenetic studies.

Shi (1999, 1999a) and Shi et al. (1999) classified *Notocephalus*, like Petz et al. (1995), in the Holostichidae, Lynn & Small (2002) listed it in the Urostylidae. Shi et al. (1999, p. 118, Fig. 51A, B) provided two illustrations of *Notocephalus parvulus*. According to the legend of these figures and the literature section they are from Foissner (1996, Acta Protozool., 35, 95–123), which is, however, incorrect. According to a personal communication by Weibo Song, the figures are composites of the illustrations by Corliss & Snyders (1986) and Petz et al. (1995), and not originals. Thus, I do not include them in the present book.

According to Corliss & Snyder (1985), *Tachysoma rigescens* (Kahl, 1932) Borror, 1972 appears most closely related to the present species. However, this species is a true *Tachysoma* with an ordinary adoral zone, that is, does not show a cephalisation (for review, see Berger 1999, p. 461). *Biholosticha discocephalus* (Fig. 238a–c) shows some resemblance with *Notocephalus parvulus* (Petz et al. 1995). However, Kahl's species is larger (180–280 µm vs. below 170 µm) and has (i) a globular head (against an oblique one); (ii) a longer and thus more prominent midventral complex composed of many cirral pairs (against few); (iii) 8–10 transverse cirri (against 5–7); (iv) about 15 µm long dorsal bristles (against 4 µm); and, most importantly, (v) many scattered macronuclear nodules (against two). In life, *Notocephalus parvulus* is thus best recognised by the cephalised body, the oblique head, the two macronuclear nodules, the lack of a buccal cirrus, and the marine habitat.

Morphology: The following description is based solely on the redescription by Petz et al. (1995), which fits the type material rather well (see remarks). A summary of the main data of the original description is given below and in Table 46.

Body size 155–160 × 40 µm in life, length:width ratio 2.6:1 on average in protargol preparations. Body fusiform, posterior portion tapering with end narrowly rounded, frontal portion set off head-like from body proper with oblique end broadly rounded; in protargol preparations cephalisation is lost but distinct notch at left anterior body corner present, possibly due to slight inflation of anterior body portion (Fig. 237c, e, f, Table 46). Dorsoventrally slightly flattened, protargol-impregnated specimens twisted. Consistency (flexible or rigid) of body not mentioned in redescription[1]. Invariably two ellipsoidal

[1] According to Wolfgang Petz (pers. comm.), the species is probably very flexible (as all non-stylonychine and non-euplotid hypotrichs; for details, see Berger 1999) and changes body shape distinctly by the fixation. Furthermore it is likely not or only slightly contractile because the length in life (155–160 µm) is almost the same as in protargol preparations (134–166 µm). In addition, twisting about main body axis is

(length:width ratio about 2:1) macronuclear nodules near mid-body slightly left of midline, contain globular nucleoli about 2 μm across. Micronuclei lenticular to slightly ellipsoidal (length:width ratio 1.7:1 on average), each one adjacent to a macronuclear nodule. Contractile vacuole not observed in life; according to protargol-impregnated specimens probably, as is usual, behind buccal vertex. Cortical granules not observed in life; found only in one protargol-impregnated specimen with few granules (1.0 × 0.5 μm) in groups along dorsal kineties and near marginal cirri (Fig. 237d). Cytoplasm with many lipid droplets 3–4 μm across, somewhat greenish and cells thus moderately dark at low magnification. Crawls slowly on substrate.

Oral apparatus prominent because continuous adoral zone occupies about 41% of body length and composed of 57 membranelles on average (Fig. 237c, e). Distal portion of adoral zone extends far onto right side, membranelles rather small, appear cirri-like in life, bases about 5 μm wide; proximal portion on ventral side, conspicuously spoon-shaped because membranelles of rather different size with largest ones about 24 μm wide; cilia of membranelles about 10 μm long. Buccal cavity large and deep, bottom with tightly spaced, protargol-affine fibres transversely extending from adoral membranelles to undulating membranes. Buccal lip conspicuous because covering right and posterior area of proximal portion of adoral zone. Refractive and thus in life conspicuous structure (bundle of fibres?) along posterior margin of adoral zone; bundle 23–27 μm long in protargol preparations (Fig. 237e, large arrow). Undulating membranes almost equally long, paroral runs along margin of buccal lip, likely composed of monokinetids bearing about 7 μm long cilia; endoral deep in buccal cavity, also likely composed of single file of basal bodies with cilia about 20 μm long. Pharyngeal fibres of ordinary length in life, extend obliquely backwards.

Cirral pattern conspicuous because frontal cirri unusually arranged and buccal cirrus and frontoterminal cirri lacking. Three frontal cirri, two of which distinctly enlarged and near anterior body end, third distinctly smaller, that is, of same size as midventral cirri and behind left enlarged frontal cirrus at level of anterior end of undulating membranes (possibly this is the buccal cirrus and one of the three ordinary frontal cirri is lacking). Midventral complex commences at distal end of adoral zone, extends obliquely backwards and terminates behind mid-body; midventral cirri larger than marginal cirri, form distinct, but slightly irregularly arranged pairs (that is, it is difficult [impossible?] to decide which cirri form a midventral pair and which a pseudopair); 1–2 cirri behind last cirral pair. Transverse cirri of about same size as midventral cirri, about 20 μm long, arranged in slightly curved, rather oblique row, extend by about half of their length beyond rear body end. Marginal cirri narrowly spaced in left and right row; individual cirri composed of two basal body rows. Right marginal row roughly sigmoidal, commences near distal end of adoral zone, that is, slightly dorsolaterally and terminates immediately ahead of right transverse cirrus. Left row commences, as is usual, left of proximal portion of adoral zone, terminates left of transverse cirri row, and thus does not overlap with right row. Dorsal cilia about 4 μm long in life, arranged in three kineties: rows 1 and 2 bipolar, row 3 slightly shortened anteriorly. Caudal cirri lacking.

---

likely inconspicuous as indicated by the longitudinally (not spirally) arranged dorsal kineties (Fig. 237f).

Table 46 Morphometric data on *Notocephalus parvulus* (pa1, from Petz et al. 1995; pa2, from Corliss & Snyder 1986)

Characteristics [a]	Population	mean	M	SD	SE	CV	Min	Max	n
Body, length	pa1	151.0	153.0	16.1	9.3	10.7	134.0	166.0	3
	pa2 [f]	95.0	–	–	–	–	–	–	1
Body, width	pa1	58.0	58.0	15.6	11.0	26.8	47.0	69.0	3
	pa2 [e, f]	20.0	–	–	–	–	–	–	1
Anterior body end to rear end of adoral zone, distance	pa1	66.0	66.0	2.8	2.0	4.3	64.0	68.0	3
	pa2 [f]	39.0	–	–	–	–	–	–	1
Undulating membranes, length	pa1	51.3	49.0	4.0	2.3	7.9	49.0	56.0	3
Anterior macronuclear nodule, length	pa1	25.7	27.0	2.3	1.3	9.0	23.0	27.0	3
	pa2 [d, e]	32.0	–	–	–	–	–	–	1
Anterior macronuclear nodule, width	pa1	15.0	14.0	2.7	1.5	17.6	13.0	18.0	3
	pa2 [d, e]	5.0	–	–	–	–	–	–	1
Posterior macronuclear nodule, length	pa1	26.7	26.0	2.1	1.2	7.8	25.0	29.0	3
	pa2 [f]	16.7	–	–	–	–	–	–	1
Posterior macronuclear nodule, width	pa1	12.7	13.0	2.5	1.5	19.9	10.0	15.0	3
	pa2 [f]	6.0	–	–	–	–	–	–	1
Macronuclear nodules, number	pa1	2.0	2.0	0.0	0.0	0.0	2.0	2.0	3
	pa2 [e]	2.0	–	–	–	–	–	–	1
Micronucleus, length	pa1	8.8	9.0	0.4	0.2	4.8	8.0	9.0	3
	pa2 [e]	2.0	–	–	–	–	–	–	1
Micronucleus, width	pa1	5.3	5.0	1.2	0.5	21.9	4.0	7.0	3
	pa2 [e]	2.0	–	–	–	–	–	–	1
Micronuclei, number	pa1	2.0	2.0	0.0	0.0	0.0	2.0	2.0	3
	pa2 [e]	1.0	–	–	–	–	–	–	1
Adoral membranelles, number	pa1 [b]	56.5	56.5	2.1	1.5	3.8	55.0	58.0	2
	pa2 [e]	56.0	–	–	–	–	–	–	1
Pseudopairs, number [h]	pa1	6.7	6.0	2.1	1.2	31.2	5.0	9.0	3
	pa2 [f]	2.0	–	–	–	–	–	–	1
Transverse cirri, number	pa1	5.7	5.0	1.2	0.7	20.4	5.0	7.0	3
	pa2 [e]	5.0	–	–	–	–	–	–	1
Right marginal cirri, number	pa1 [c]	63.0	–	–	–	–	–	–	1
	pa2 [f]	49.0	–	–	–	–	–	–	1
Left marginal cirri, number	pa1 [c]	41.0	–	–	–	–	–	–	1
	pa2 [f]	48.0	–	–	–	–	–	–	1
Dorsal kineties, number	pa1	3.0	3.0	0.0	0.0	0.0	3.0	3.0	3
	pa2 [g]	3.0	–	–	–	–	–	–	1

[a] All measurements in µm. All data are based on protargol-impregnated specimens; Petz et al. (1995) used Wilbert's method, Corliss & Snyder (1986) followed the directions in Lee et al. (1985). Note, that Petz et al. (1995) analysed only three specimens (probably because of the low abundance); thus, the calculation of most statistics is likely somewhat overdone. The data from the type population are single values from the text of the original description and the illustrations of the holotype specimen. CV = coefficient of variation in %, M = median, Max = maximum value, mean = arithmetic mean, Min = minimum value, n = number of individuals investigated, SD = standard deviation, SE = standard error of arithmetic mean.

[b] Likely only two individuals investigated because the median is a decimal number.

[c] Not analysed. The values are from Fig. 237e.

[d] Nodule (anterior or posterior) not specified.

[e] From text of original description.

[f] From Fig. 237a, b.

[g] From Petz et al. (1995, p. 172); Corliss & Snyder (1986) counted only two (see Fig. 237b).

[h] In *N. parvulus* it is difficult to decide what is a midventral pair and what is a pseudopair (see text).

Observations on type population (from Corliss & Snyder 1986; Fig. 237a, b, Table 46): This paragraph contains only some supplementary or deviating data to the redescription above (see remarks, for explanation). Body size according to text 60 × 20 µm, according to the illustration, however, about 95 × 22 µm; Petz et al. (1995) found a body length of 140 µm in a paratype specimen (see remarks). Length:width ratio about 4.3:1 (Fig. 237a, b). Most specimens contractile to some degree. Single micronucleus between the two macronuclear nodules. Eleven frontoventral cirri arranged basically as redescribed by Petz et al. (1995), except that only two (vs. eight) pairs are present. Two dorsal kineties (however, see remarks) with a set of single rows of basal bodies between posterior ends (Fig. 237b, arrow).

Occurrence and ecology: Salt water. The type locality of *Notocephalus parvulus* is the pack ice of the Weddell Sea, Antarctica, where Corliss & Snyder (1986) commonly found it in Bouin-fixed samples of both sea-ice slush and sea-ice pore water collected by D.L. Garrison (University of California at Santa Cruz, USA). For a review on the Antarctic sea ice biota, see Garrison (1991). Petz et al. (1995) found it very rarely in the endopagial of mainly multiyear sea ice in the same region between 69°02'–70°21'S and 08°02'–08°53'W, where it occurred together with diatoms, autotrophic and heterotrophic flagellates, and other ciliates. Petz (2005) found *N. parvulus* in nilas ice of the Ross Sea area. *Notocephalus parvulus* feeds on diatoms (Corliss & Snyder 1986) and likely also on flagellates (Petz et al. 1995).

Petz et al. (1995) recorded the following parameters in brine: temperature -3.4 to -2.4°C; salinity 5.95‰; PO_4 1.5 µmol l^{-1}; NO_2 0.1 µmol l^{-1}; NO_3 2.9 µmol l^{-1}; NH_4 2.7 µmol l^{-1}; Si 7.8 µmol l^{-1}; chlorophyll *a* 49.2 µg l^{-1}. Biomass of 10^6 individuals 111 mg (Petz et al. 1995). *Notocephalus parvulus* is possibly confined to cold, marine habitats, perhaps even to sea ice biota.

Biholosticha Berger, 2003

2003 ***Biholosticha* nov. gen.** – Berger, Europ. J. Protistol., 39: 378 (original description). Type species (by original designation on p. 378): *Holosticha (Holosticha) discocephalus* Kahl, 1932.

Nomenclature: The name *Biholosticha* is a composite of the Latin numeral *bi-* (two) and the genus-group name *Holosticha* (see there for derivation) and alludes to the two frontal cirri and the fact that both species included where previously classified in *Holosticha* (Berger 2003). Like *Holosticha*, feminine gender.

Characterisation (A = supposed apomorphies): Adoral zone of membranelles continuous. Buccal field very narrow (A?). Two enlarged frontal cirri (A?). Buccal cirrus lacking (A?). Midventral complex composed of cirral pairs only. Transverse cirri arranged in slightly curved, transverse (not oblique) row (A). 1 left and 1 right marginal cirral row.

Remarks: Although I did not have new data, I transferred the species mentioned below to a separate group because they show some good agreements (synapomorphies) which clearly distinguish them from *Holosticha*, *Anteholosticha*, *Caudiholosticha*, or any other group (Berger 2003). Inclusion in one of these taxa would make them unnecessarily inhomogenous.

Unfortunately, both *Biholosticha discocephalus* and *B. arenciola* are not described in all details. Thus, the characterisation above is not perfectly founded and certainly has to be modified when new data become available. However, Kahl made good live observations so that we can assume that the cirral pattern shown in Fig. 238a is basically correct. Dragesco described his species after protargol preparations essentially confirming Kahl's observations as concerns the cirral pattern (Fig. 239a).

The supposed apomorphies of *Biholosticha* are: (i) *Two enlarged frontal cirri*. Both species have only two distinctly enlarged frontal cirri, which are arranged more or less ahead of the midventral complex. It is unlikely that both authors made a misobservation. I do not know how this curious pattern originates and with which of the three ordinary frontal cirri the two are homologous. Ontogenetic data are needed to answer this interesting question. Anyhow, I am rather certain that this is a derived state against the three cirri present in many other groups.

(ii) *Buccal cirrus lacking*. In both species no buccal cirrus is described or illustrated. Kahl knew about this cirrus and he illustrated and described it in many other species which are distinctly smaller and thus much more difficult to investigate, that is, it is unlikely that he overlooked it. Dragesco made protargol preparations and therefore it is also unlikely that he did not see it. The buccal cirrus is also absent in a few other urostyloid species or groups, for example, in *Paragastrostyla* or in *Periholosticha*. The loss of such a single cirrus is indeed not a very complex feature and thus can be explained easily by convergent evolution.

(iii) *Transverse cirri arranged in slightly curved, transverse row*. It is conspicuous that in both species the transverse cirri are rather similarly arranged, namely in a distinctly subterminal, slightly curved, transverse (pseudo)row. Usually this cirral group forms a roughly J-shaped, more obliquely arranged pattern. Admittedly this is a vague feature which has to be confirmed or rejected.

(iv) *Buccal field very narrow*. Interestingly, in both species the buccal field is obviously rather small, that is, the buccal lip, which bears the paroral, is very near the right margin of the proximal portion of the adoral zone of membranelles (Fig. 238a, 239a). The adoral zone itself appears rather different in both species. However, both Kahl and Dragesco wrote that the zone is wide, although Kahl's illustration does not give this impression (Fig. 238a). The adoral membranelles of the distal portion of the zone are very widely spaced in the type species (Fig. 238a). If this specific feature also applies to *B. arenicola* (re-investigation needed), then it would be a further apomorphy of *Biholosticha*.

Further interesting agreements are the rather narrowly spaced marginal cirri, the lack of cortical granules (absence mentioned in both descriptions), and the occurrence in marine sands. But there is also a distinct difference, besides the nucleus apparatus, separating the two species unequivocally, namely the shape of the anterior body portion (Fig. 238a, 239a). However, we must consider that Dragesco did not provide data about the body outline of live specimens and Fig. 239a certainly does not show the correct body shape, but a (inflated?) specimen after protargol impregnation. Thus, we cannot exclude that Dragesco's species is also slightly cephalised.

Notocephalus parvulus (Fig. 237) is also slightly cephalised and has only two enlarged frontal cirri (plus a third cirrus behind). Whether or not it is related to *Biholosticha* is not known.

I designated *Holosticha (Holosticha) discocephalus* as type species because in total this species seems better described than *Keronopsis arenciola* (Berger 2003).

Species included in *Biholosticha* (alphabetically arranged basionyms are given): (1) *Holosticha (Holosticha) discocephalus* Kahl, 1932; (2) *Keronopsis arenciola* Dragesco, 1963.

Key to *Biholosticha* species

If you know that your specimen/population is a *Biholosticha* species, identification is rather simple. The main feature for determination is the nucleus apparatus.

1 Two macronuclear nodules (Fig. 239a) .... *Biholosticha arenicola* (p. 1181)
\- Many macronuclear nodules (Fig. 238a) .... *Biholosticha discocephalus* (p. 1178)

Biholosticha discocephalus (Kahl, 1932) Berger, 2003 (Fig. 238a–c)

1932 ***Holosticha discocephalus* spec. n.** – Kahl, Tierwelt Dtl., 25: 579, Fig. 122 24, 25 (Fig. 238a, b; original description; no type material available and no formal diagnosis provided; see nomenclature).

1933 ***Holosticha discocephalus* Kahl 1932** – Kahl, Tierwelt N.- u. Ostsee, 23: 109, Fig. 17.1 (Fig. 238c; guide to marine ciliates).

1972 ***Holosticha discocephalus* Kahl, 1932** – Borror, J. Protozool., 19: 11 (revision of hypotrichs).

1973 ***Holosticha discocephalus* Kahl, 1932** – Hartwig, Abh. math.-naturw. Kl. Akad. Wiss. Mainz, Mikrofauna des Meeresbodens, 18: 452 (record substantiated by some morphological data).

1973 ***Holosticha discocephalus*** – Hartwig, Mikrokosmos, 62: 336, Bild 8 (micrograph).

1980 ***Holosticha discocephalus* Kahl, 1932** – Hartwig, Cah. Biol. mar., 21: 426 (record substantiated by morphological data).

1983 ***Holosticha discocephalus* Kahl, 1932** – Borror & Wicklow, Acta Protozool., 22: 121 (revision of urostylids).

1992 ***Holosticha discocephalus* Kahl, 1930–5** – Carey, Marine interstitial ciliates, p. 182, Fig. 713 (guide).

2001 ***Holosticha (Holosticha) discocephalus* Kahl, 1932** – Berger, Catalogue of ciliate names 1. Hypotrichs, p. 35 (nomenclator containing all basionyms, combinations, and higher taxa of hypotrichs).

2003 ***Biholosticha discocephalus* (Kahl, 1932) n. comb.** – Berger, Europ. J. Protistol., 39: 378 (combination with *Biholosticha*).

Nomenclature: The species-group name *discocephalus* is a composite of the Greek nouns *ho diskos* (disc) and *he kephalé* (head) and alludes to the disc-shaped anterior end (head, oral region) of this species. Kahl (1932, 1933) divided *Holosticha* into several subgenera. Thus, the correct name in his reviews is *Holosticha (Holosticha) disco-*

cephalus Kahl, 1932. This species was fixed as type species of *Biholosticha* by original designation.

Remarks: This is a rather curious species, especially as concerns the body shape. For a detailed discussion of the present classification, see genus section. Detailed redescription necessary.

Hartwig (1973a) provided a micrograph of a slightly squeezed specimen. Some main features, for example, cephalisation and transverse cirri, are clearly recognisable, indicating that the identification is correct.

The present species is easy to separate from *B. arenicola*, which has only two macronuclear nodules (Fig. 239a). Due to the curious body outline easy to distinguish from *Anteholosticha*-species, which have a similar cirral pattern (main difference: three frontal cirri!). As concerns body shape, it is reminiscent of *Oxytricha discifera* Kahl, 1932, which, however, has an 18-cirri pattern (for review see Berger 1999, p. 227). Discocephaline hypotrichs, for example, *Prodiscocephalus* spp., lack a distinct zigzagging cirral pattern (Wicklow 1982, Lin et al. 2004).

Morphology: Body length 180–280 μm, body length:width ratio according to Figs. 238a, b between 3.2:1 and 4.1:1. Body outline highly characteristic because anterior body portion (oral region) conspicuously disc-shaped, that is, cephalised and thus distinctly set off from body proper, which is roughly elongate elliptical (Fig. 238a) to elongate oval in well fed specimens (Fig. 238b). Body very fragile (Kahl), indicating that the consistency is soft. Macronucleus scattered in many globular nodules; micronuclei not found. Contractile vacuole likely lacking. Cortical granules absent (see question 6 on p. 578 in Kahl 1932). Cytoplasm usually darkly granulated. Movement not described.

Adoral zone of membranelles conspicuous, although occupying only about 21% of body length, because almost completely encircling disc-shaped anterior body portion; five (Fig. 238a, c) to six (Fig. 238b) large and widely spaced membranelles on dorsolateral side of anterior body margin; proximal portion of adoral zone very short and wide. Buccal(?) lip thick. Two large frontal cirri immediately ahead of anterior end of midventral complex. Buccal cirrus obviously lacking. Midventral complex long, extends from frontal cirri to near transverse cirri; specimen shown in Fig. 238a with about 31 cirral pairs (value must not be over-interpreted); further, protargol preparations are needed to show that these two cirral rows are in fact caused by midventral pairs. 8–10 distinctly enlarged transverse cirri form slightly curved, almost transversely arranged row distinctly ahead of rear body end so that cirri do not project beyond body margin. Both marginal rows commence in constriction (neck) between "head" and "body", right row extends on ventral side to midline of cell; left row J-shaped, extends onto dorsolateral body side posteriorly; marginal cirri very narrowly spaced. Dorsal cilia about 17 μm long! (length estimated from Fig. 238b) and very fine; number of kineties not known.

Specimens of Hartwig's (1973, 1973a) population only 115–196 μm long; very sensitive against handling; body proper strongly granulated. 70 to more than 90 macronuclear nodules. Specimens of Bermuda population about 200 μm long; distinctly cephal-

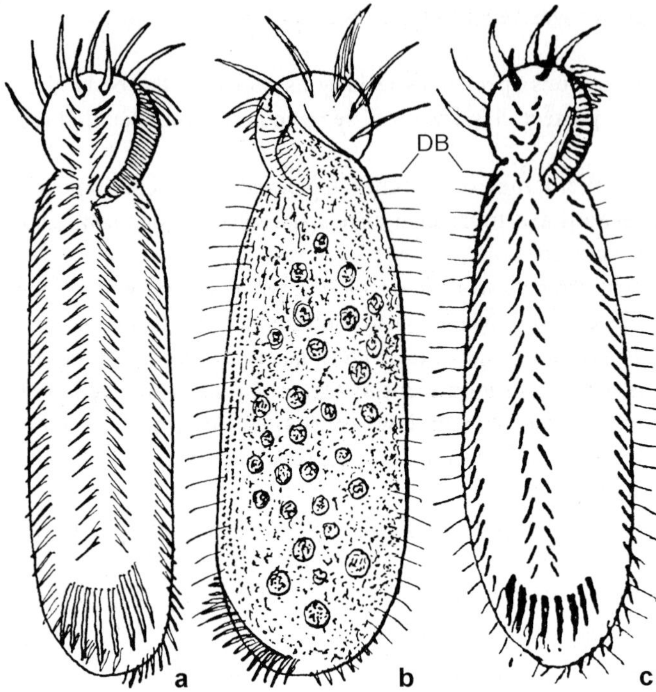

Fig. 238a–c *Biholosticha discocephalus* (a, b, from Kahl 1932; c, from Kahl 1933. From life). Ventral and dorsal view showing cirral pattern, dorsal cilia, and nucleus apparatus, 200 µm. Note the disc-shaped head (oral-region) which makes this species, in combination with the prominent midventral complex, unmistakable. Note the large and widely spaced distal adoral membranelles and the transversely arranged row of strong transverse cirri which do not project beyond rear body end. DB = dorsal bristles (about 17 µm long). Page 1178.

ised; numerous oval macronucleus nodules; two distinct frontal cirri; two ventral rows; nine transverse cirri not projecting beyond rear body margin (Hartwig 1980).

Occurrence and ecology: Saltwater. Type locality *Biholosticha discocephalus* is the Bay of Kiel, Baltic Sea, where Kahl (1932) discovered it on the sandy sediment in a water depth of 5–6 m. In Heligoland he found it very rarely, and only very small specimens (Kahl 1933). Hartwig (1973, 1973a) recorded it mainly in the mesopsammal of different beaches of the islands of Sylt, Heligoland (Germany), and Jordsand (Denmark). Hartwig (1980) found it in marine sands of the Bermuda Islands.

Records not substantiated by morphological data and or illustrations from marine habitats: Bay of Kiel, Baltic Sea (Bock 1952, p. 81; 1953, p. 255); tidal sand flat of Crildumersiel, Jade Bay (Hartwig 1984, p. 127); waters in Plymouth area, England (Lackey & Lackey 1963, p. 802); Roscoff area, Atlantic Ocean (Dragesco 1960, p. 314); Mediterranean Sea near Marseilles (Vacelet 1961, p. 3; 1961a, p. 15); Kandalakshskij Bay, White Sea (Burkovsky 1970a, p. 190; 1970b, p. 11; 1970c, p. 56); Gulf of Mexico (Borror 1962, p. 342). Aliev (1982, p. 87) found *B. discocephalus* in saline lakes (salinity 29‰) in Azerbaijan. *Biholosticha discocephalus* feeds on diatoms (Hartwig 1973).

Biholosticha arenicola (Dragesco, 1963) Berger, 2003 (Fig. 239a)

1963 ***Keronopsis arenicola* n. sp.** – Dragesco, Cah. Biol. mar., 4: 263, Fig. 10 (Fig. 239a; original description; location of type material not indicated, likely in private collection of J. Dragesco; no formal diagnosis provided).

1983 ***Holosticha dragescoi* nom. nov.** – Borror & Wicklow, Acta Protozool., 22: 115, 120 (replacement name, see nomenclature).

1985 ***Erionella* (formerly *Keronopsis*) *arenicola*** – Small & Lynn, Phylum Ciliophora, p. 517, Fig. 30B (Fig. 239a; combination with *Erionella*; see remarks).

2001 ***Holosticha dragescoi* Borror and Wicklow, 1983** – Berger, Catalogue of ciliate names 1. Hypotrichs, p. 34 (nomenclator containing all basionyms, combinations, and higher taxa of hypotrichs).

2001 ***Keronopsis arenicola* Dragesco, 1963** – Berger, Catalogue of ciliate names 1. Hypotrichs, p. 43 (nomenclator containing all basionyms, combinations, and higher taxa of hypotrichs).

2003 ***Biholosticha arenicola* (Dragesco, 1963) n. comb.** – Berger, Europ. J. Protistol., 39: 378 (combination with *Biholosticha*).

Nomenclature: The species-group name *arenicola* (living in sand) is a composite of the Latin noun *aréna* (sand), the thematic vowel ·*i*- (at the end of the first word stem when the second begins with a consonant; Werner 1972, p. 37), and the Latin verb *cólere* (to live in, inhabit) and refers to the habitat (marine sand) where the species was discovered.

Borror & Wicklow (1983) transferred the species from *Keronopsis* to *Holosticha*. Thus, it became a junior secondary homonym of *Holosticha (Holosticha) arenicola* Kahl, 1932. Consequently, they introduced the replacement name *dragescoi* for Dragesco's species. Now, Kahl's species is classified in *Anteholosticha* so that the two species (*arenicola* Kahl and *arenicola* Dragesco) are no longer congeneric. Consequently, the replacement name *dragescoi* Borror & Wicklow is no longer valid (ICZN 1999, Article 59.4).

Remarks: The species was originally classified in *Keronopsis*. Borror & Wicklow (1983) recognised this misplacement and transferred it to *Holosticha* because distinctly enlarged frontal cirri are present. This decision was basically correct, but since it lacks all apomorphies of the true *Holosticha* species this assignment is outdated. It shows several agreements (synapomorphies) with *Holosticha discocephalus* (see above), strongly indicating a close relationship of these two species. Thus, Berger (2003) merged them in *Biholosticha* (see genus section for details).

Biholosticha arenicola is basically described after protargol preparations, but contains some uncertainties (only two frontal cirri? [apomorphy of *Biholosticha*?]; frontoterminal cirri present or absent?; dorsal infraciliature?; caudal cirri absent or present?). Dragesco mentioned that protrichocysts (cortical granules) are lacking, indicating that he made some live observations. However, he certainly did not study the body outline of living, freely motile specimens. Thus, the species should be redescribed in detail.

Borror (1972, p. 10) synonymised *B. arenicola* with *Pseudoamphisiella alveolata*. However, *Pseudoamphisiella* species have a distinct alveolar seam and the cirri of the midventral pairs are distinctly separated, so that the zigzag-pattern is very inconspicuous. In contrast, the present species has distinctly zigzagging cirri. Small & Lynn (1985)

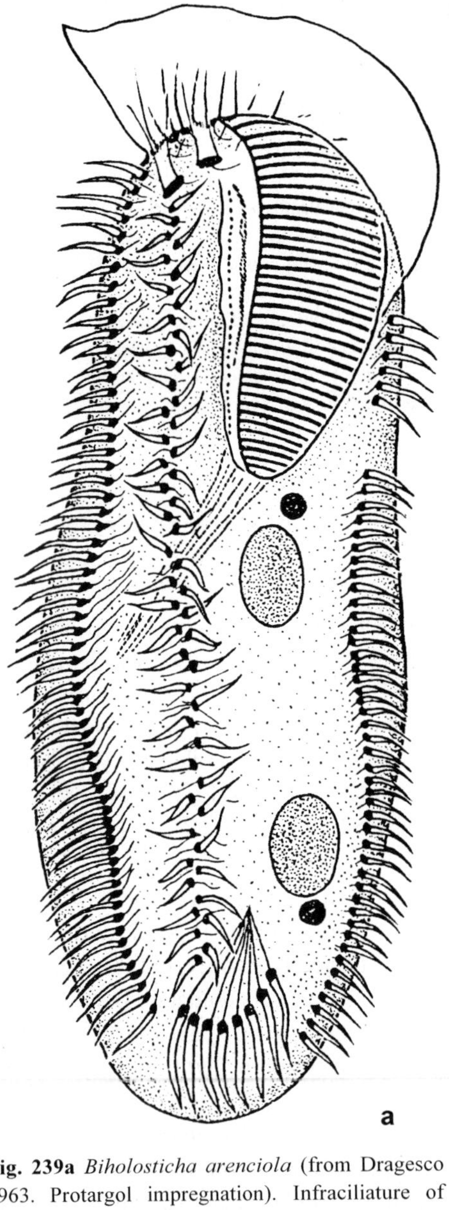

Fig. 239a *Biholosticha arenciola* (from Dragesco 1963. Protargol impregnation). Infraciliature of ventral side and nuclear apparatus, 200 µm. Important features of this species are the two frontal cirri, the lack of a buccal cirrus, the subterminally and transversely arranged transverse cirri, and the two macronucleus-nodules. Page 1181.

transferred it to *Erionella*, however, without explanation. *Erionella* is possibly a synonym of *Pseudoamphisiella* (see there for details).

Biholosticha arenicola is easy to distinguish from *B. discocephalus* by the number of macronuclear nodules (two vs. many). *Pseudoamphisiella* species have a distinct alveolar seam and widely spaced midventral cirri which are not arranged in a conspicuous zigzag-pattern. *Anteholosticha* species have, inter alia, three (vs. two in *Biholosticha*) distinctly enlarged frontal cirri.

Morphology: Body length 200 µm (method [from life; protargol preparations] not indicated). Two ellipsoidal macronucleus nodules left of midline; one slightly ahead of mid-body, second in the rear body third; each nodule associated with a large, globular micronucleus. Contractile vacuole neither mentioned nor illustrated, indicating that it is lacking. Cortical granules seemingly lacking (Dragesco 1963). Movement not described.

Adoral zone prominent because occupying about 37% of body length, number of membranelles not given, specimen shown in Fig. 239a with slightly more than 40 membranelles (value must not be over-interpreted); anteriormost 10 membranelles strong, membranelles of central portion rather wide. Buccal field narrow, right margin bordered by paroral. Only two distinctly enlarged frontal cirri. Buccal cirrus and frontoterminal cirri obviously lacking. Midventral complex composed of narrowly spaced cirral pairs, giving the impression of a single row, extends from near frontal cirri to near transverse cirri; no value given, specimen shown in Fig. 239a with 53 midventral cirri. Eight enlarged transverse cirri in curved, subterminally and transversely arranged row, cirri thus not projecting beyond rear body end. Right marginal row commences near distal end of adoral zone, extends to near right end of transverse cirral row, composed of about 60 cirri. Left marginal row commences distinctly ahead of level of proximal end of adoral zone, ends subterminally, composed of about 40 cirri; specimen shown in Fig. 239a with distinct break in anterior portion (if I understand Dragesco's text correctly, this gap occurs often). Most cirri with distinct fibres. Dorsal infraciliature (length of bristles, number and arrangement of kineties) not known; caudal cirri likely lacking because neither mentioned nor illustrated.

Occurrence and ecology: *Biholosticha arenicola* was as yet found only at the type locality, which is in Roscoff (France) where Dragesco (1963) discovered it in the sand of the Atlantic Ocean (for review, see Bocquet 1971, p. 388).

Paramitrella Jankowski, 1978

1978 ***Paramitrella* gen. n.** – Jankowski, Tezisy Dokl. zool. Inst. Akad. Nauk SSSR, year 1978: 40 (original description; Russian diagnosis not translated). Type species (by original designation on p. 40): *Epiclintes caudatus* Bullington, 1940.

1979 ***Paramitrella* Jk., 1978** – Jankowski, Trudy zool. Inst., Leningr., 86: 60 (catalogue of generic names of hypotrichs).

2001 ***Paramitrella* Jankowski 1978** – Aescht, Denisia, 1: 117 (catalogue of generic names of ciliates).

2001 ***Paramitrella* Jankowski, 1978** – Berger, Catalogue of ciliate names 1. Hypotrichs, p. 69 (nomenclator containing all basionyms, combinations, and higher taxa of hypotrichs).

Nomenclature: No derivation of the name is given in the original description. *Paramitrella* is a composite of the Greek prefix *para+* (closely related), the Latin noun *mitra* (genus group name; mitre), and the Latin diminutive suffix *~ella*. Very likely it alludes to the resemblance with *Psammomitra radiosa*. *Paramitrella* is feminine because ending with *-ella* (ICZN 1999, Article 30.1.3).

Characterisation (A = supposed apomorphies): Body quadripartite in head, neck, trunk, and tail (A). Adoral zone of membranelles continuous. Frontal ciliature unknown (overlooked? lacking?). Midventral complex composed of midventral pairs only. 1 right marginal row and 1 left marginal row. Dorsal cilia emerge from cortical papillae (A?).

Remarks: See same chapter at single species.

Species included in *Paramitrella*: (1) *Epiclintes caudatus* Bullington, 1940.

Single species

Paramitrella caudata (Bullington, 1940) Jankowski, 1978 (Fig. 240a, b)

1940 ***Epiclintes caudatus* n. sp.**[1] – Bullington, Pap. Tortugas Lab., 32: 206, Fig. 18 (Fig. 240a; original description; no type material available).

---

[1] The diagnosis by Bullington (1940) is as follows: *Epiclintes caudatus* n. sp. is an extremely long, narrow

1978 ***Paramitrella* gen. n. for *Epiclintes caudatus* Bullington** – Jankowski, Tezisy Dokl. zool. Inst. Akad. Nauk SSSR, year 1978: 40 (combination with *Paramitrella*; see nomenclature).

1983 ***Epiclintes caudatus* Bullington, 1940** – Carey & Tatchell, Bull. Br. Mus. nat. Hist. (Zool.), 45: 51, Fig. 10 (redrawing of Fig. 240a; revision of *Epiclintes*).

1990 ***Epiclintes caudatus* Bullington, 1940** – Wicklow & Borror, Europ. J. Protistol., 26: 192 (review of Epiclintidae).

2001 ***Paramitrella caudatus* (Bullington, 1940) Jankowski, 1978** – Berger, Catalogue of ciliate names 1. Hypotrichs, p. 20 (nomenclator containing all basionyms, combinations, and higher taxa of hypotrichs).

Nomenclature: No derivation of the name is given in the original description. The species-group name *caudat·us -a -um* (Latin adjective; tailed, having a tail) refers to the long tail-like process of the cell. *Epiclintes* is masculine, *Paramitrella* feminine; consequently, the ending of the species-group name has to be changed from *-us* to *-a*. *Epiclintes caudatus* is the type species of *Paramitrella*. Jankowski (1978) did not write the binomen *Paramitrella caudata*; however, in that he established *Paramitrella* for *E. caudatus* he is the combining author. For *Epiclintes caudatus* sensu Lackey (1961), see *Epiclintes*.

Remarks: Bullington (1940) assigned this species to *Epiclintes* because of similarities in the body shape, the oral apparatus, the spines of the dorsal side, and the ventral cirri. The first three features are indeed reminiscent of *E. auricularis*. However, the cirral pattern of *E. auricularis* and *P. caudata* are rather different. *Epiclintes* has many oblique cirral rows, that is, the midventral complex is obviously composed of midventral rows only, whereas in *P. caudata* the midventral complex consists of distinctly zigzagging cirral pairs so that Jankowski's establishment of *Paramitrella* is understandable.

Interestingly enough, both species have distinct papillae associated with the dorsal bristles as already recognised by Bullington (1940). This feature is rather curious and possibly a synapomorphy of *Epiclintes* and *Paramitrella*. *Psammomitra* has a similar body shape, but lacks the dorsal papillae (Fig. 42, 43).

I did not find *E. caudatus* in Hemberger's (1982) review. Borror (1972, p. 9) synonymised it with *Epiclintes auricularis*, However, this synonymy presupposes that Bullington (1940) observed the cirral pattern completely incorrectly which cannot be concluded from the descriptions and redescription of the other species in his paper. I consider Bullington's description of *P. caudata* as basically correct, although some details, for example, presence/absence of transverse cirri, frontoterminal cirri, and/or buccal cirri, are not known. Later, Borror rejected this synonymy and classified it as incertae sedis in the Epiclintidae (Wicklow & Borror 1990). Detailed redescription needed.

Morphology: Body length 260–440 μm (average 354 μm; n = 12), body width 22–44 μm (average 28.6 μm; n = 6); body length:width ratio 10–14:1; body extremely extensile and contractile, distinctly flattened dorsoventrally. Body outline as shown in Fig. 240a, that is, quadripartite in a broadly rounded head, a relatively long, slender neck, a fusiform trunk, and a long, slender tail slightly widened posteriorly. Nuclear ap-

---

ciliate, with extreme power of extension and contraction, a very long, distinct tail-like process, a distinct necklike process, and a widened anterior extremity; middle of body is widest part, but anterior peristomial area is only slightly less wide. Lateral spines; 2 lateral (marginal) and 2 ventral rows of cirri running lengthwise of the body; peristome has less resemblance to an ear than in *E. ambiguus*; average size about 354 × 28.6 microns; swims in left spirals.

paratus not described, indicating that many macronuclear nodules are dispersed throughout the cell. Contractile vacuole neither mentioned nor illustrated. Cell yellowish (reason not mentioned). Freely motile specimens swim in left spirals, but when running around over the bottom of the dish, or on the underside of the surface film, they move without spiralling, in circles, sometimes to the right, sometimes to the left.

Adoral zone of membranelles occupies almost full length of head region, obviously extends far onto right head margin. Buccal field large and bright. Cirral pattern not known in all details. Presence/absence of frontal cirri, buccal cirri, frontoterminal cirri, and transverse cirri unknown; possibly a detailed study of the cirral pattern will provide an autapomorphy for *Paramitrella caudata*. Midventral complex likely composed of distinctly zigzagging cirral pairs only; commences, according to illustration, at proximal end of adoral zone, possibly extends to near cell end (details difficult to recognise in life, data should not be over-interpreted). One left and one right marginal row, possibly commence about at same level somewhat behind anterior cell end, terminate obviously at rear end.

Dorsal ciliature (presence/absence of caudal cirri, number and arrangement of kineties, length of dorsal bristles) not mentioned. However, short, blunt, almost transparent papillae line the margins of the cell; located at regular intervals, about 65 per side. Whether there is an additional (third) row of such papillae (spines in Bullington's terminology) is unknown; very likely they are homologous to the dorsal bristle complex of *Epiclintes auricularis* indicating a sister-group relationship of these two taxa (however, note that the cirral pattern is likely rather different).

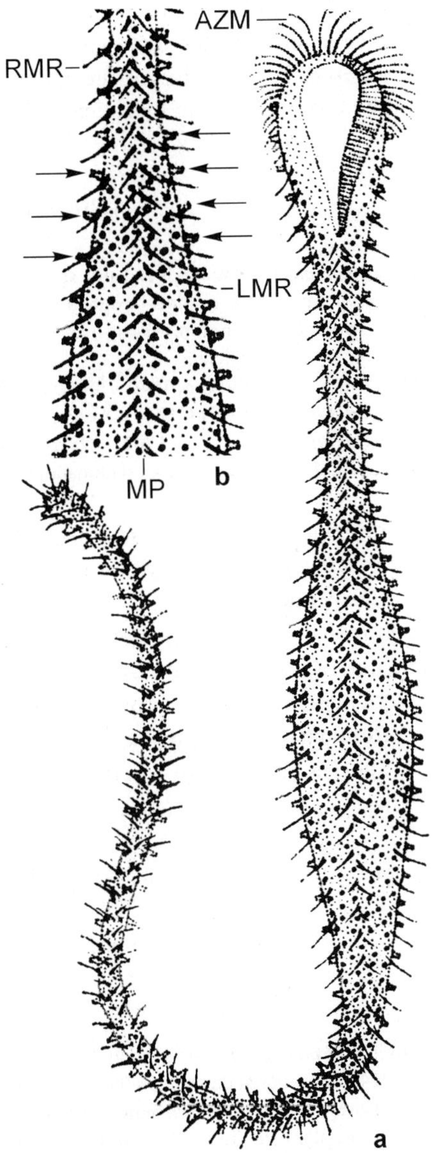

Fig. 240a, b *Paramitrella caudata* (from Bullington 1940. From life). Ventral view of a representative specimen, average body length = 354 µm. (b) is an enlargement of the transition zone of neck and trunk. Arrows mark papillae, that is, dorsal bristle complexes, on dorsal margin. The zigzagging midventral cirri prevent a synonymy with *Epiclintes auricularis*. AZM = adoral zone of membranelles, LMR = left marginal row, MP = midventral pairs, RMR = right marginal row. Page 1183.

Occurrence and ecology: Marine. Type locality is the Gulf of Mexico at Tortugas, Florida, USA (Bullington 1940; see also Borror 1962, p. 342). However, Bullington could not say from which habitat (interstitial, surface layer, etc.) the material came. The slender, vermiform body shape indicates that *P. caudata* is a member of the psammon. Neither mentioned by Patterson et al. (1989) nor by Carey (1992) in their lists/guides to marine benthic/interstitial ciliates.

Uncinata Bullington, 1940

1940 ***Uncinata gigantea* n. gen. and sp.** – Bullington, Pap. Tortugas Lab., 32: 207 (original description). Type species (by original designation and monotypy): *Uncinata gigantea* Bullington, 1940.

1979 ***Uncinata* Bullington, 1940** – Jankowski, Trudy zool. Inst., Leningr., 86: 68 (catalogue of generic names of hypotrichs).

1979 ***Uncinata* Bullington, 1940** – Corliss, Ciliated Protozoa, p. 310 (revision).

2001 ***Uncinata* Bullington 1940** – Aescht, Denisia, 1: 169 (catalogue of generic names of ciliates).

2001 ***Uncinata* Bullington, 1940** – Berger, Catalogue of ciliate names 1. Hypotrichs, p. 96 (nomenclator containing all basionyms, combinations, and higher taxa of hypotrichs).

Nomenclature: *Uncinata* (Latin; that, which has a hook) refers to the prominent beak-like extension of the anterior end of the cell (Bullington 1940, p. 209). Feminine gender.

Characterisation (A = supposed apomorphies): Body elongate, anterior portion ends in a beak-like projection (uncinus; A). Adoral zone of membranelles bipartite by uncinus (A?). Cirral pattern composed of 5 rows; two leftmost rows begin at proximal end of adoral zone, other rows commence on frontal field. Marine.

Remarks: See same chapter at single species.

Species included in *Uncinata*: (1) *Uncinata gigantea* Bullington, 1940.

Single species

Uncinata gigantea Bullington, 1940
(Fig. 241a–c)

1940 ***Uncinata gigantea* n. gen. and sp.**[1] – Bullington, Pap. Tortugas Lab., 32: 206, Fig. 19A, B (Fig. 241a–c; original description; no type material available).

2001 ***Uncinata gigantea* Bullington, 1940** – Berger, Catalogue of ciliate names 1. Hypotrichs, p. 96 (nomenclator containing all basionyms, combinations, and higher taxa of hypotrichs).

Nomenclature: No derivation of the name is given in the original description. The species-group name *gigante·us -a -um* (Latin adjective; huge, extremely large) obviously

[1] The diagnosis by Bullington (1940) is as follows: *Uncinata gigantea* n. gen. and sp. is an extremely long ciliate with two expansions in its body – one at the mouth, the other a short distance back of the mouth – and with a very prominent beaklike projection to the left at the anterior end, characteristic of the genus; body highly extensile and contractile; mouth midway between right and left sides, about 104 microns from point of break; color yellowish; 5 rows of cirri on ventral surface, 2 marginal and 3 ventral; posterior half to two-thirds reduced somewhat in size but not constituting a definite tail-like process; average size approximately 618 × 47 microns.

refers to the huge body size of this species. *Uncinata gigantea* is the type species of *Uncinata*.

In an abstract, Bullington (1935) reported about the morphology and taxonomy of ciliates from the Tortugas region. In the last paragraph he wrote "About 20 have now been studied and drawn to scale. One at last seems to fit in no previously described genus; the other belong to the well-known genera *Coleps*, *Ophryoglena*, *Strombidium*, *Trachelocerca*, *Oxytricha*, *Condylostoma*, *Uncinata*, *Peritromus* and *Stentor*." Obviously par lapsus, Bullington designated *Uncinata* as a well-known genus. Anyhow, since the name is not accompanied by a description or illustration, it is a nomen nudum and was therefore available only when he provided the description of this species in his 1940 paper.

Remarks: *Uncinata gigantea* is only known from the original description. A confusion with a large species of a non-hypotrich genus – for example, *Loxodes*, *Remanella*, or *Geleia*, which have a similar beak-like anterior body end – cannot be excluded, but is very unlikely. Borror (1972) obviously did not mention this species in his review on hypotrichs, and Hemberger (1982, p. 276) at least doubted that *Uncinata* is a hypotrich. However, since Corliss (1977, p. 137; 1979) assigned it to the Holostichidae and since the infraciliature is indeed reminiscent of a urostyloid, it is treated in the present book. Tuffrau (1979, p. 525; 1987, p. 115) and Tuffrau & Fleury (1994, p. 137) classified it in the Kahliellidae.

Although Bullington (1940) described this huge species in some detail, its classification is uncertain because details of the cirral (ciliary?) pattern, for example, presence or absence of a midventral complex, are not known. Possibly, *U. gigantea* has a rather simple cirral pattern, that is, two left marginal rows commencing near the proximal end of the adoral zone, a midventral complex composed of pairs only and forming a bicorona, and one right marginal row. If *Uncinata* is indeed a hypotrich then at least the body shape is an apomorphy. However, as in other cases, a detailed redescription is needed for a more proper statement.

Bullington (1940) observed a similar, but smaller species (about one third of body length). However, the specimens were very fragile preventing detailed studies. *Epiclintes auricularis* has a different cirral pattern and anterior body end and therefore cannot be confused with the present species.

Morphology: Body length 264–1100 µm (average 618 µm; n = 20), body width 25–75 µm (average 47 µm; n = 12), that is, both parameters strongly varying depending on contraction. Body elongate, indistinct tripartite in head, trunk, and tail. Head composed of very prominent characteristic beak-like, leftwards curved projection at anterior end and an expanded cytostome region; cell narrowed posteriorly of this point for a short distance and followed by trunk. Tail region narrowed to about 2/3 the width of trunk; extreme posterior end rounded, slightly enlarged. Body very soft and flexible, yellowish. Nuclear apparatus not described, indicating that many macronuclear nodules are present. Presence/absence of contractile vacuole and cortical granules not described. Ordinary movement not described, swims in left spirals.

Adoral zone of membranelles very conspicuous because obviously bipartite by beak-like projection. Distance from beak to cytostome, which is roughly in cell midline,

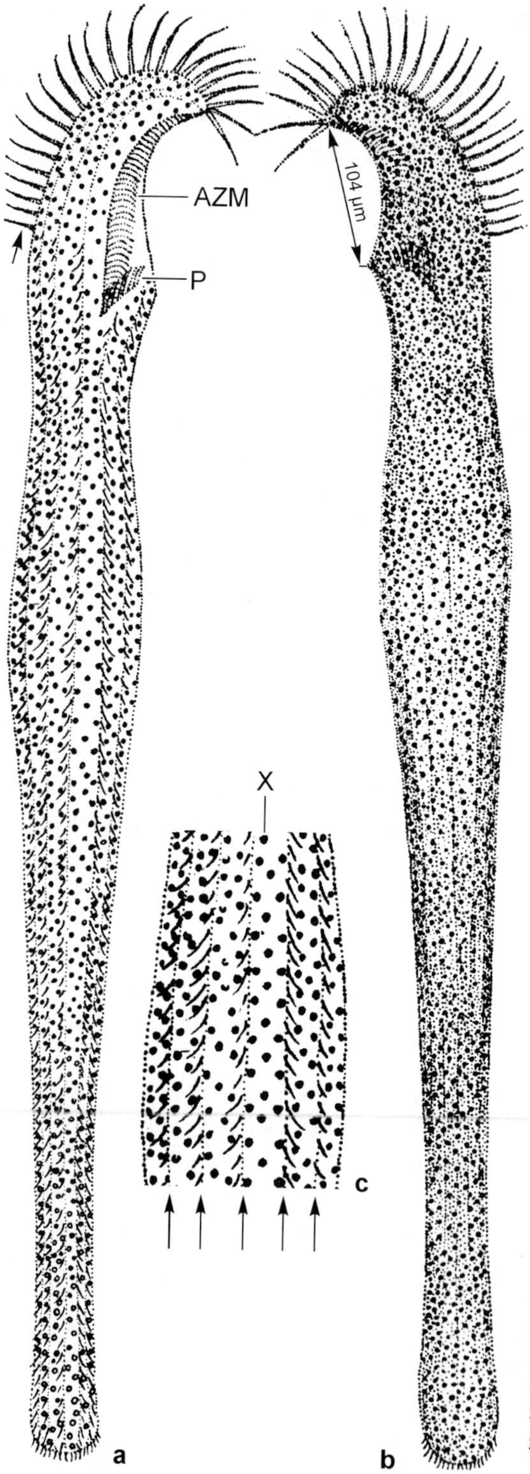

about 104 µm. Distal portion of zone extending from beak far backwards onto right body margin (DE-value about 0.66; see Fig. 1c and chapter 1.8 for explanation), composed of long membranelles (Fig. 241a, b). Proximal portion of adoral zone anteriorly curved leftwards. Details of oral apparatus (number of membranelles [likely very high], shape and arrangement of undulating membranes) not known; the short, leftwards extending (from posterior to anterior) membrane at the base of the adoral zone is not very characteristic for hypotrichs, although such details must not be over-interpreted. Five rows of cirri; according to Bullington two marginals and three ventrals. Leftmost two rows commence near level of cytostome indicating that these are left marginal cirri. Remaining three rows extend from frontal field (beak?; see Fig. 241a) to posterior body end. Whether or not *U. gigantea* has a midventral complex is not known. No frontal cirri, transverse cirri, caudal cirri, or other special cirri have been observed. Dorsal ciliature (length and arrangement of dorsal kineties) not known.

Occurrence and ecology: Very rare. Type locality is the Gulf of

Fig. 241a–c *Uncinata gigantea* (from Bullington 1940. From life). Ventral (a) and dorsal (b) view of representative specimens and detail of ventral side of trunk (c; enlargement of a), average body length 618 µm. Arrow in (a) denotes distal end of adoral zone of membranelles; arrows in (c) mark the five cirral rows. AZM = proximal portion of adoral zone of membranelles, P = paroral?, X = granules not explained by Bullington. Page 1186.

Mexico at Tortugas, Florida, USA, where Bullington (1940; see also Borror 1962, p. 343) found only few specimens, inter alia, during summer 1939. Bullington was uncertain about the exact sample site, possibly it was Long Key. The slender body shape indicates that *U. gigantea* is a member of the psammon, that is, marine sediments (Patterson et al. 1987, p. 211). However, it is not mentioned by Carey (1992) in his guide to marine interstitial ciliates.

Supplement to the Oxytrichidae

Neokeronopsis and *Urostyloides* have, like urostyloids, zigzagging ventral cirri. However, the dorsal ciliature is highly complex due to fragmentation, which proves that they belong to the Oxytrichidae. Since I did not include them in my revision on this group (Berger 1999), I review them in the present book. Other oxytrichids feigning a urostyloid origin are *Territricha stramenticola* and *Pattersoniella vitiphila* (Berger 1999). Molecular data on *Pattersoniella* (Bernhard et al. 2001) confirmed my "morphological" classification of this species in the Stylonychinae (Berger 1999).

Neokeronopsis Warren, Fyda & Song, 2002

2002 ***Neokeronopsis* nov. gen.**[1] – Warren, Fyda & Song, Europ. J. Protistol., 38: 196 (original description). Type species (by original designation on p. 196): *Holosticha (Keronopsis) spectabilis* Kahl, 1932.

Nomenclature: *Neokeronopsis* is a composite of the Greek adjective *ne-* (young, new, unusual), the thematic vowel *·o-* (used at the end of the first root of a word when the second one begins with a consonant; Werner 1972, p. 37), and the name of the hypotrichous genus *Keronopsis* Penard, 1922. For a derivation of the name *Keronopsis*, see *Pseudokeronopsis*. Likely, *Neokeronopsis* should indicate that it is "related to *Keronopsis*" sensu Kahl, that is, *Pseudokeronopsis*. It has, like *Keronopsis*, feminine gender because it ends with *-opsis* (ICZN 1999, Article 30.1.2).

Characterisation (A = supposed apomorphies): Adoral zone of membranelles continuous. Two arched rows of frontal cirri (A). Buccal cirrus present. 2 frontoterminal cirri. Several cirral pairs feigning a midventral pattern (A). Transverse and caudal cirri present. 1 left and 1 right marginal row. Parental adoral zone retained for proter. Dorsal kinety 3 with multiple fragmentation (A). Dorsomarginal kineties present. Several frontal-ventral-transverse cirri anlagen, each producing a cirral pair plus a transverse cirrus, inserted between ordinary six cirral anlagen (A). Macronuclear nodules fuse to single mass during division.

Remarks: Warren et al. (2002) established *Neokeronopsis* because *Holosticha spectabilis* shows a unique combination of features: frontal cirri arranged in a bicorona; many cirral pairs; one marginal row per side; frontoterminal, transverse, and caudal cirri present; fragmentation of dorsal kineties and dorsomarginal kineties present. They compared their new taxon with (mainly urostyloid) genera also having a bicorona. It differs from all of them, except for *Pattersoniella* Foissner, 1987b, by the presence of a fragmenting dorsal kinety and dorsomarginal kineties (for review on these features see Berger 1999 and chapters 1.7 and 2 of general section and ground pattern of Urostyloidea in present book). *Pattersoniella* and *Neokeronopsis* indeed have a very similar ventral and dorsal ciliature and ontogenesis (for review of *Pattersoniella* see Berger 1999,

[1] The diagnosis by Warren et al. (2002) is as follows: Urostylidae with two arched rows of frontal cirri; long midventral row of cirri arranged in paris; one marginal row on each side of the cell; frontoterminal, transverse and caudal cirri present; formation of dorsal kineties being of an *Oxytricha*-pattern.

p. 766). Warren et al. (2002, p. 196) therefore correctly mentioned the rigid body (against flexible) and the lack of cortical granules (against present) as main differences (the species status of both *P. vitiphila* Foissner, 1987b and *N. spectabilis*, however, is beyond reasonable doubt; see below). Because of this astonishing similarity in the infraciliature, Warren et al. (2002, p. 203) doubted my classification of *Pattersoniella* in the oxytrichids (Berger 1999). By contrast, Warren et al. assumed that *Pattersoniella* and *Neokeronopsis* are urostyloids because of the origin of the ventral ciliature from many oblique cirral streaks. My classification of *Pattersoniella* and *Neokeronopsis* in the Dorsomarginalia, respectively, Oxytrichidae (Fig. 14a) is based on the presence of dorsomarginal kineties, respectively, a fragmentation in dorsal kinety 3. The rigid cortex and the lack of cortical granules in *Pattersoniella* assigned it to the Stylonychinae (Berger 1999, p. 499, 766). Most species in the stylonychines have the plesiomorphic number of 18 frontal, ventral, and transverse cirri which originate from the characteristic six more or less longitudinal anlagen. Consequently, I proposed an increase in the anlagen number for *Pattersoniella* (Berger 1999, p. 769). A few years later, my classification of *Pattersoniella* in the stylonychines was confirmed by molecular data (Bernhard et al. 2001). In contrast, *Neokeronopsis spectabilis* has a flexible body and distinct cortical granules. Thus, this species is certainly not a stylonychine and a close relationship of *Pattersoniella* and *Neokeronopsis* can be excluded. According to the cyrtohymenid undulating membrane pattern and the heavy cortical granulation it might be a relative of *Cyrtohymena* species. However, molecular data should be awaited for a more detailed discussion.

Species included in *Neokeronopsis* (basionym given): (1) *Holosticha (Keronopsis) spectabilis* Kahl, 1932.

Single species

Neokeronopsis spectabilis (Kahl, 1932) Warren, Fyda & Song, 2002 (Fig. 242a–h, 243a–m, 244a–j, Table 47)

1932 ***Keronopsis spectabilis* spec. n.** – Kahl, Tierwelt Dtl., 25: 578, Fig. 97_{16} (Fig. 242a; original description; no type material available and no formal diagnosis provided; see nomenclature).

1972 ***Keronopsis spectabilis* Kahl, 1932** – Borror, J. Protozool., 19: 11 (combination with *Keronopsis*, see nomenclature; revision of hypotrichs).

1979 ***Holosticha spectabilis* comb. n.** – Jankowski, Trudy zool. Inst., Leningr., 86: 57 (combination with *Holosticha*; see nomenclature).

1983 ***Pseudokeronopsis spectabilis* (Kahl, 1932) nov. comb.** – Borror & Wicklow, Acta Protozool., 22: 116, 124 (combination with *Pseudokeronopsis*; revision of urostylids).

2001 ***Pseudokeronopsis spectabilis* (Kahl, 1932) Borror and Wicklow, 1983** – Berger, Catalogue of ciliate names 1. Hypotrichs, p. 37 (nomenclator containing all basionyms, combinations, and higher taxa of hypotrichs).

2002 ***Neokeronopsis spectabilis* (Kahl, 1932) nov. comb.**[1] – Warren, Fyda & Song, Europ. J. Protistol., 38: 197, Fig. 1–25, Table 1 (Fig. 243a–m; detailed redescription after protargol impregnation including cell

[1] The improved diagnosis by Warren et al. (2002) is as follows: Large, freshwater *Neokeronopsis*, 368–500 × 152–212 µm wide following protargol fixation, with 2 macronuclei. On average, 81 adoral membranelles;

division. 5 voucher slides of protargol-impregnated specimens have been deposited in the Natural History Museum, London, UK, registration numbers 2000:10:10:1–5).

2005 ***Apoamphisiella* sp.** – Wang, Qiu, Qin & Chen, Int. Congr. Protozool., 12: 9 (description of a Chinese population).

Nomenclature: No derivation of the name is given in the original description. The species-group name *spectabil·is -is -e* (Latin adjective; worth seeing; considerable) likely refers to the huge size (up to 500 μm long!) of this species. It was fixed as type of *Neokeronopsis* by original designation.

Kahl (1932) classified *Keronopsis* as subgenus of *Holosticha*; thus, the correct name in his revision is *Holosticha (Keronopsis) spectabilis* Kahl, 1932. This was obviously overlooked by Borror (1972), who assumed that Kahl had established it in the genus *Keronopsis* (see list of synonyms). Consequently, Borror (1972) should be considered as combining author when it is assigned to *Keronopsis* although he did not transfer it formally; however, at the present state of knowledge this classification is irrelevant anyway. Jankowski (1979) and Hemberger (1982) also did not recognise the subgeneric status of *Keronopsis* in Kahl (1932) because they transferred it unnecessarily to *Holosticha*. The classification in *Holosticha* is incorrect because this group has, inter alia, only three frontal cirri (against a bicorona in *N. spectabilis*).

Remarks: The description of the type population by Kahl (1932) is rather detailed and contains all main features recognisable in life. Kahl established this species in the subgenus *Holosticha (Keronopsis)* because it has many (and not only three), not distinctly differentiated frontal cirri arranged in two rows. Borror (1972) basically kept this classification, but raised *Keronopsis* to genus rank (see nomenclature). When Borror & Wicklow (1983) established *Pseudokeronopsis* with *P. rubra* as type species, they transferred *H. spectabilis* to this genus mainly because of the bicorona. Neither Borror (1972) nor Borror &Wicklow (1983) provided new data. Warren et al. (2002) obviously overlooked that *H. spectabilis* was classified in *Pseudokeronopsis* because they did not mention this combination in their list of synonyms. Their detailed redescription, which includes three morphogenetic stages, contains many new data clearly showing distinct differences to all genera in which the present species was previously classified. Thus, they established *Neokeronopsis*, which is monotypic, at least so far. Unfortunately, Warren et al. (2002) could not make detailed live observations and therefore could not confirm the presence of cortical granules described by Kahl (1932). Consequently, it was reasonable that they did not designate a neotype because the question will always remain whether or not their population had cortical granules.

I found the present species twice, however, on both occasions with very low abundance. Identification of this huge species was rather simple and I can confirm Kahl's data, including the cortical granulation (Fig. 244a–h). The second time I found it I impregnated the sample. As expected, there were only few specimens in the slides and just one individual could be used to make a more or less complete illustration of the ventral infraciliature (Fig. 242h). It fits the data provided by Warren et al. very well so that the

2 frontoterminal cirri; 19–26 pairs of frontoventral cirri; 15–22 transverse cirri in a single row extending about 25% of cell length; 43–65 and 43–54 cirri in left and right marginal rows respectively; 6–9 caudal cirri; 9–12 dorsal kineties.

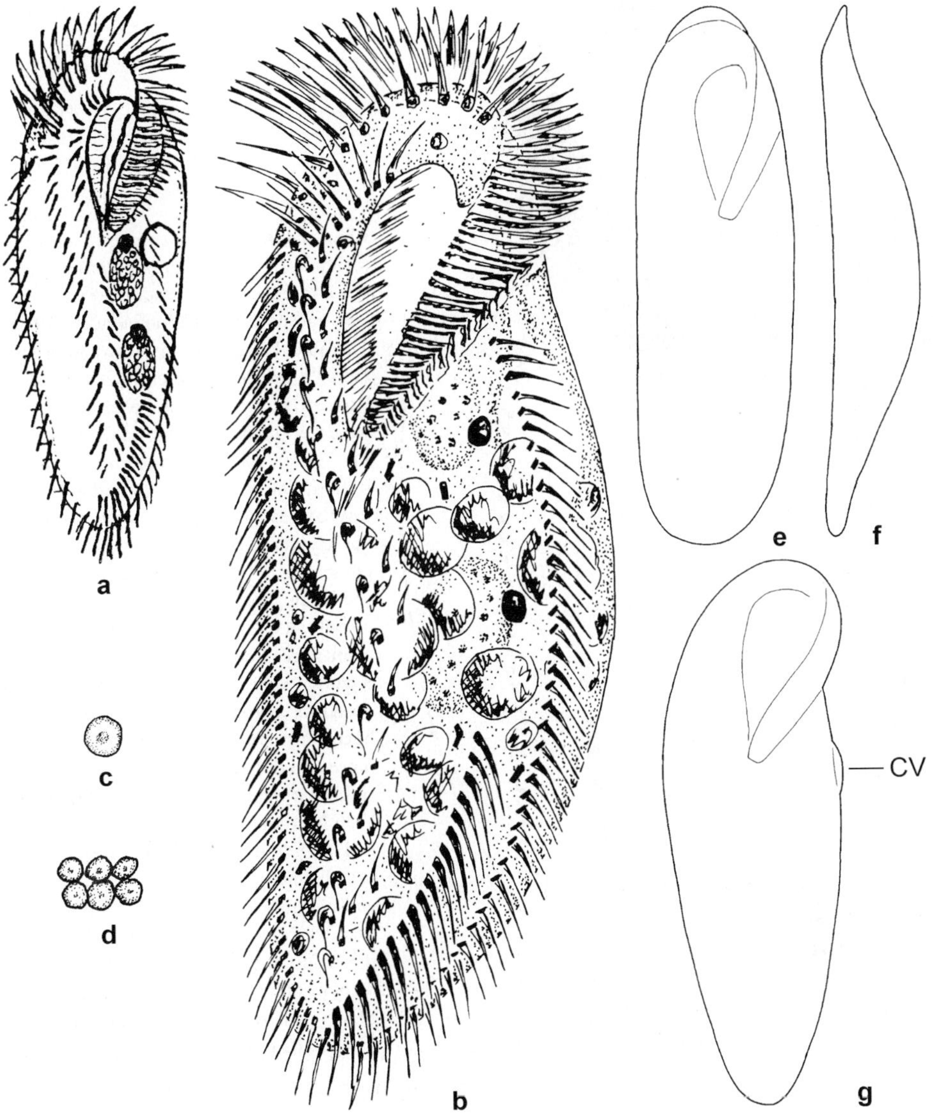

Fig. 242a–g *Neokeronopsis spectabilis* (a, from Kahl 1932; b–g, originals of Carinthian population. a–g, from life). **a:** Ventral view of a representative specimen, 300 μm. **b, e, g:** Ventral views showing variability of outline, b = 360 μm. Outline of specimen shown in Fig. 242b is from micrograph 244b. **f:** Left lateral view showing dorsoventral flattening. **c, d:** The individual cortical granules are 1–2 μm across, often slightly polygonal, and usually have a small central dent. CV = contractile vacuole. Page 1191.

conspecificity of all these populations is beyond reasonable doubt. I also do not fix the specimen shown in Fig. 242h as neotype because the impregnation is of low quality. The Polish specimens are larger than the Austrian individuals. This is also indicated by

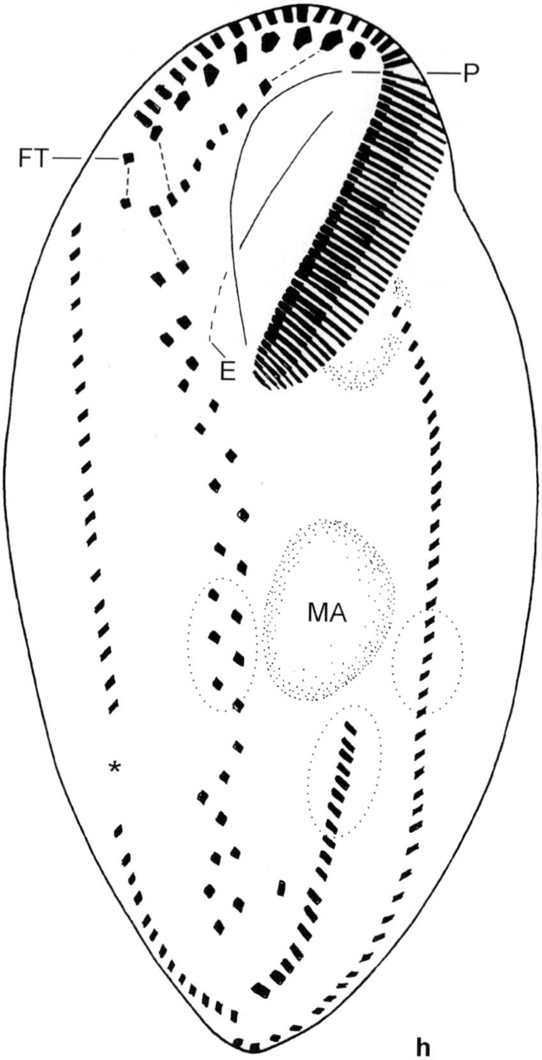

Fig. 242h *Neokeronopsis spectabilis* (original of Carinthian population. Protargol impregnation). Infraciliature of ventral side and macronuclear apparatus, 236 μm. The abundance was very low and this was the sole usable specimen in several slides. Broken lines connect, exemplary, cirri which originate from the same anlage (cp. Fig. 243c). Asterisk marks a non-species-specific interruption in the right marginal row. Areas encircled by dotted lines are not clearly recognisable because the impregnation is of mediocre quality and the specimen mounted upside down (that is, dorsal side above). Brief morphometric characterisation: 61 adoral membranelles; 8 frontal cirri in anterior corona; 7 cirri in posterior corona (leftmost cirrus is the buccal cirrus); 2 frontoterminal cirri; 12 or more cirral pairs; 18 transverse cirri; 32 right marginal cirri; 43 left marginal cirri; each 2 caudal cirri on dorsal kineties 1, 2, and "3" (not illustrated). This specimen has distinctly fewer adoral membranelles and marginal cirri than the Polish population described by Warren et al. (Table 47). E = endoral (posterior portion not clearly recognisable), FT = frontoterminal cirri, MA = posterior macronuclear nodule, P = paroral. Page 1191.

the higher number of adoral membranelles (81 on average vs. 61 in the specimen shown in Fig. 242h).

Neokeronopsis spectabilis is also illustrated in the guide by Streble & Krauter (1982, p. 256f, Fig. 6). However, this illustration is likely a modified redrawing from Kahl and thus not included in the present book. Hemberger (1982, p. 91) synonymised *N. spectabilis* with *Holosticha contractilis*, a junior synonym of *Uroleptus musculus*.

Apomorphies of *N. spectabilis* are likely the huge size and the multiple fragmentation of dorsal kinety 3 anlage, a feature which, however, evolved convergently in some stylonychines (for review, see Berger 1999).

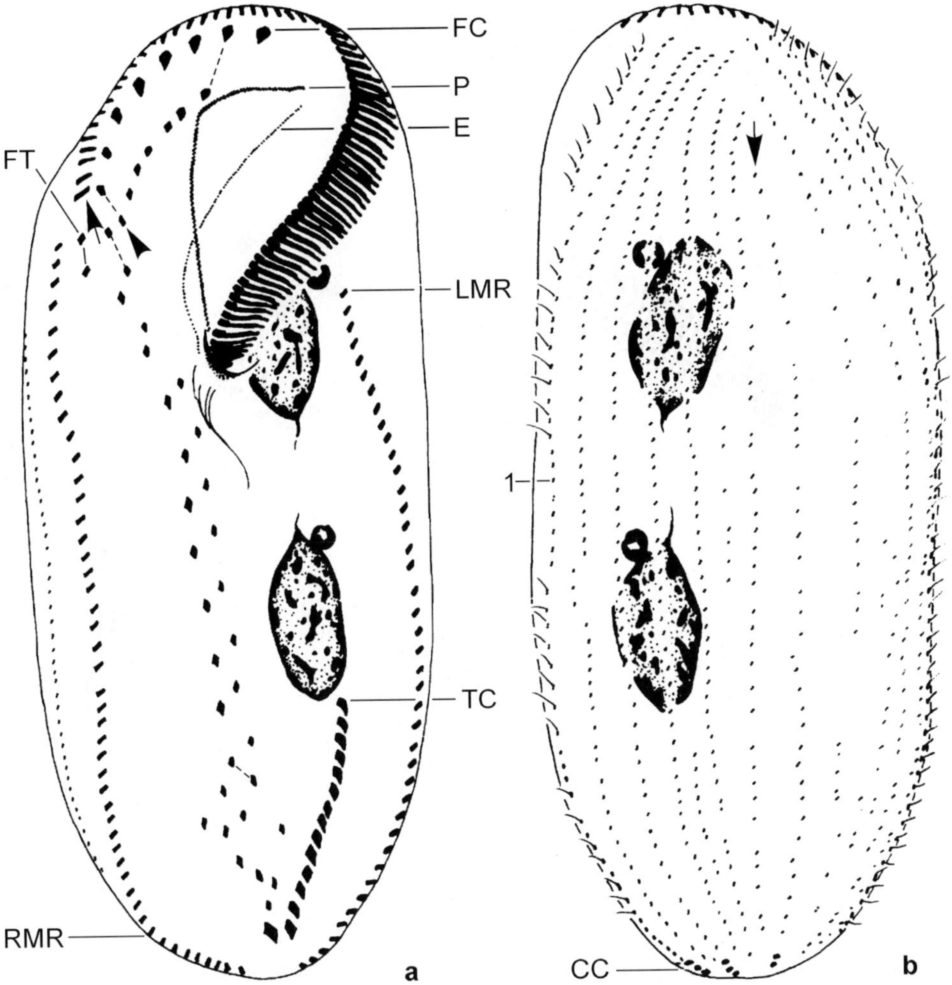

Fig. 243a, b *Neokeronopsis spectabilis* (from Warren et al. 2002. Protargol impregnation). Infraciliature of ventral and dorsal side and nuclear apparatus of same specimen, 425 µm. Arrow in (a) marks distal end of adoral zone of membranelles, arrowhead denotes rightmost cirral pair of bicorona. Broken lines connect, exemplary, cirri originating from the same anlage. Arrow in (b) marks rear fragment of dorsal kinety 3 as indicated by the presence of caudal cirri. CC = caudal cirri, E = endoral, FC = leftmost frontal cirrus (= cirrus I/1), FT = frontoterminal cirri, LMR = left marginal row, P = paroral, RMR = right marginal row, TC = transverse cirri, 1 = dorsal kinety 1 (= leftmost kinety). Page 1191.

Neokeronopsis spectabilis is a very conspicuous freshwater hypotrich which can be easily identified, inter alia, by its enormous size. However, at superficial live observation it can be confused with the sympatric *Urostyla grandis* which has a similar size. In addition, both species are yellowish because they have the same cortical granulation. But there are many differences between *N. spectabilis* and *U. grandis* easily recognisable even at middle magnification: long and prominent transverse cirral row against

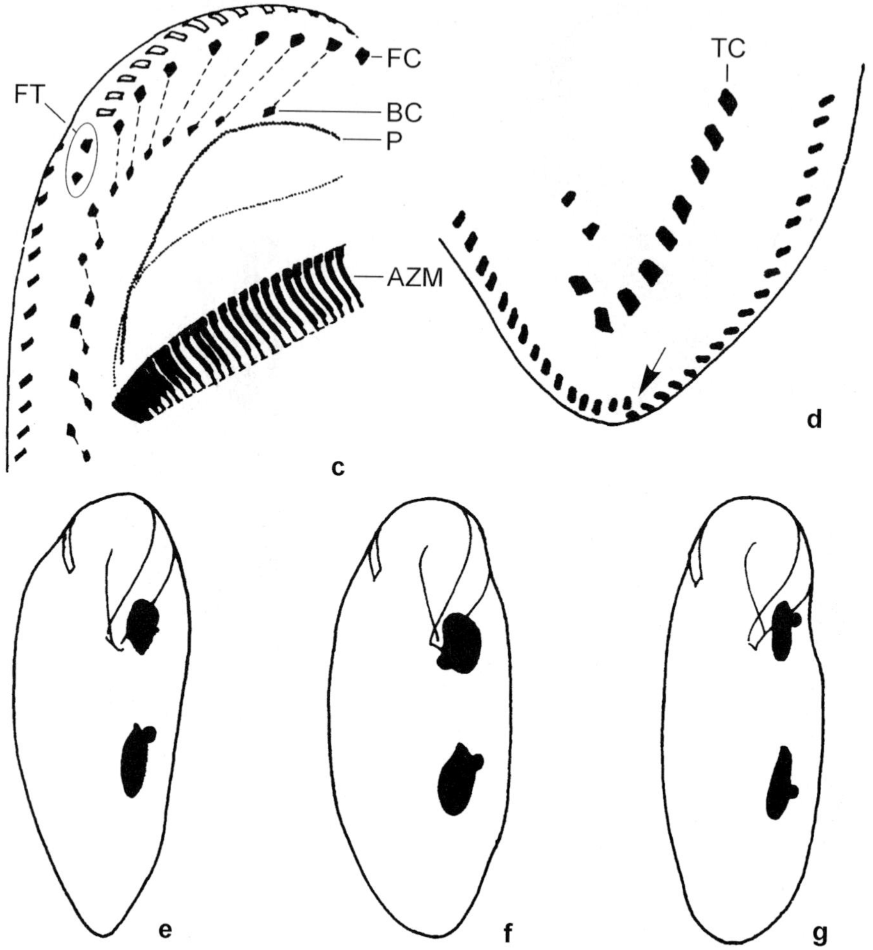

Fig. 243c–g *Neokeronopsis spectabilis* (from Warren et al. 2002. Protargol impregnation). **c:** Anterior portion of infraciliature of ventral side showing, inter alia, bicorona (cirri originating from same anlage connected by a broken line), frontoterminal cirri, and anterior cirral pairs. **d:** Rear cell end showing posterior transverse cirri and end of marginal rows which are slightly overlapping (arrow) in this specimen. **e–g:** Shape variants after protargol impregnation and variability of nuclear apparatus. AZM = adoral zone of membranelles, BC = buccal cirrus, FC = leftmost frontal cirrus (= cirrus I/1), FT = frontoterminal cirri, P = paroral, TC = transverse cirri. Page 1191.

inconspicuous; one left and one right marginal row against many rows on both sides; two macronuclear nodules against very many. Furthermore, *Neokeronopsis spectabilis* has about 10 dorsal kineties, whereas *Urostyla grandis* has only three. *Allotricha mollis* Sterki, 1878 and *Paraurostyla weissei* (Stein, 1859) Borror, 1972 also live in freshwater and have, like *Neokeronopsis spectabilis*, two macronuclear nodules and similar cortical granules. However, the transverse cirral row of both species is much less distinct and they lack the characteristic cirral pairs (for details, see Berger 1999, p. 262, 844).

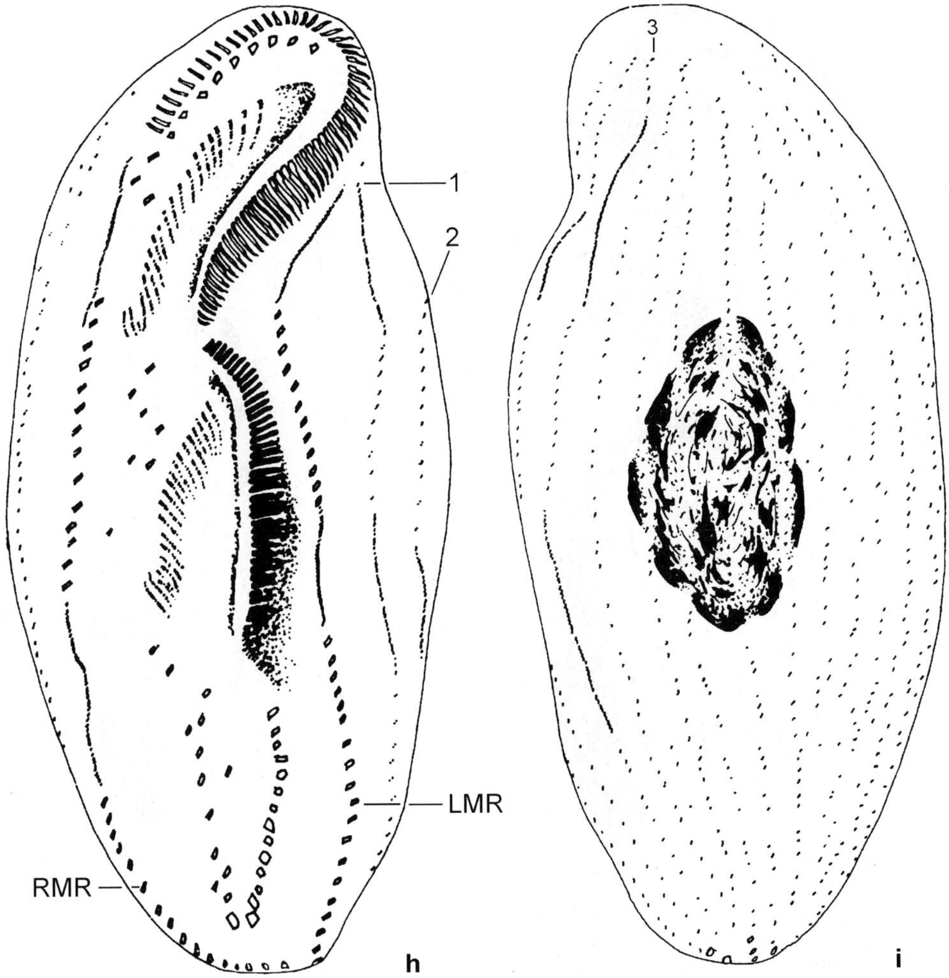

Fig. 243h, i *Neokeronopsis spectabilis* (from Warren et al. 2002. Protargol impregnation). Infraciliature of ventral and dorsal side and nuclear apparatus of an early to middle divider, 425 μm. LMR, RMR = left and right marginal row, each with anlage for proter and opisthe, 1–3 = dorsal kineties each with two anlagen. Page 1191.

Pattersoniella vitiphila Foissner, 1987b (for review see Berger 1999, p. 766) lives in terrestrial habitats, has a firm body (vs. flexible), few cirral pairs (vs. many), 13–18 macronuclear nodules (vs. two), several contractile vacuoles (vs. one), and lacks cortical granules and caudal cirri (vs. both present).

Recently, Wang et al. (2005) described an *Apoamphisiella* sp. from a freshwater pond in Harbin, China (I do not know what the expression "(by Helmut Berger)" means because I neither established *Apoamphisiella* nor was I involved in the identification process). The data indicate that this population is identical with *Neokeronopsis spectabilis*.

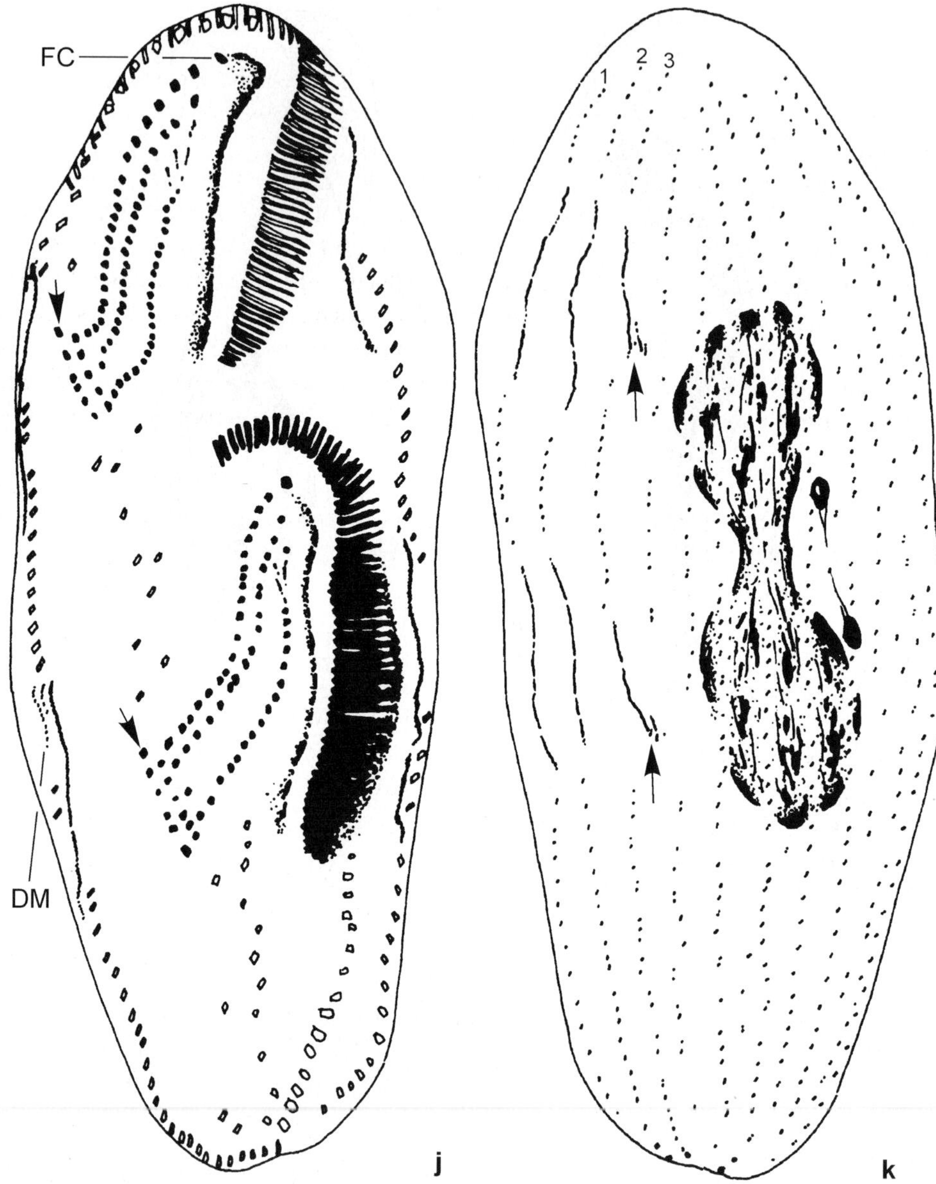

Fig. 243j, k *Neokeronopsis spectabilis* (from Warren et al. 2002. Protargol impregnation). Infraciliature of ventral and dorsal side and nuclear apparatus of middle divider, 410 μm. Arrows in (j) mark new frontoterminal cirri originating from the rightmost frontal-ventral-transverse cirral anlage. Arrows in (k) denote fragmenting dorsal kinety 3. Parental structures white, new black. DM = dorsomarginal kineties anlagen, FC = leftmost frontal cirrus originating, as is usual, from the undulating membrane anlage. Page 1191.

Morphology: First the data of the original description (Kahl 1932) are presented, followed by original observations from an Austrian population. Subsequently, the detailed data by Warren et al. (2002) on protargol-impregnated specimens are provided, supplemented with some own observations from protargol preparations.

Body size of type population (Fig. 242a) 250–350 × 85–125 µm in life (body width estimated from body length:width ratio which is, according to Kahl, 3.0:1). Body obovoidal, distinctly contractile. Two macronuclear nodules. Cortical granules yellow, globular, form rows of small groups on dorsal side and along marginal rows. Adoral zone of membranelles one third of body length. Buccal lip anteriorly distinctly curved backwards extending to adoral zone. Cirral pairs form two distinct (pseudo)rows extending from transverse cirri to anterior body end with anteriormost cirri of right row distinctly enlarged. About 20 transverse cirri in rather oblique row, only posterior cirri somewhat projecting beyond rear body end. Dorsal cilia short and fine. 6–8 caudal cirri continuous with left marginal row.

Observations on Austrian population from the Glan stream (Fig. 242b–h, 244a–j): Body size of two specimens about 400 × 160 µm, respectively, 320 × 120 µm, that is, body length:width ratio 2.5:1, respectively, 2.7:1; length:width ratio of specimen shown in Fig. 242b about 2.6:1. Outline usually as shown by Kahl (cp. Fig. 242a, 244a). Distinctly flattened dorso-ventrally (Fig. 242f). Body very flexible, but only slightly (by about 10%) contractile. In life very resistant against cover-glass pressure; however, occasionally it bursts. Macronuclear nodules although rather large (50 × 33 µm, 54 × 32 µm, 48 × 32 µm) difficult to recognise in life because cells often packed with large food vacuoles masking the nuclear apparatus (Fig. 244a–c); nucleoli up to 3 µm across. In two specimens investigated in life, one micronucleus attached to anterior macronuclear nodule and two micronuclei associated with posterior; one specimen with two micronuclei per nodule; individual micronuclei 8–10 µm across in life (Fig. 242b, h, 244i, j). Contractile vacuole left of proximal end of adoral zone of membranelles, sometimes vaulting left body margin (Fig. 242b, g); with longitudinal collecting canals, the anterior one sometimes with a vesicle. Cortical granules easily recognisable, that is, even at a magnification of × 200 because lemon-yellow and 1–2 µm across; form groups composed of up to 10 granules; groups on dorsal side basically arranged along dorsal kineties (Fig. 244d–h), on ventral side few granules close to most cirri; frontal scutum and area right of rear end of right marginal row with many cortical granules. Individual granules usually globular, sometimes roughly polygonal in outline; occasionally with a small central dent (Fig. 242c, d). Cytoplasm colourless, with many ordinary crystals up to 4 µm long and many food vacuoles up to 40 µm across (242b, 244a–c). Movement without peculiarities, that is, moderately rapid gliding and bending very flexibly around various obstacles. Largest adoral membranelles up to 28 µm wide and 25 µm long. Buccal area huge and bright because deep, *Cyrtohymena*-like. Infraciliature, including frontoterminal cirri, well recognisable in life due to huge size of cells. Frontal cirri about 25 µm long in life; about 17 transverse cirri up to 40 µm long, rearmost cirri (basis about 4 × 4 µm) only slightly projecting beyond posterior body end; most other cirri 12–15 µm long. Individual dorsal cilia difficult to recognise in life because only about 3 µm long and flexible; arranged in many rows recognisable even at low magnification

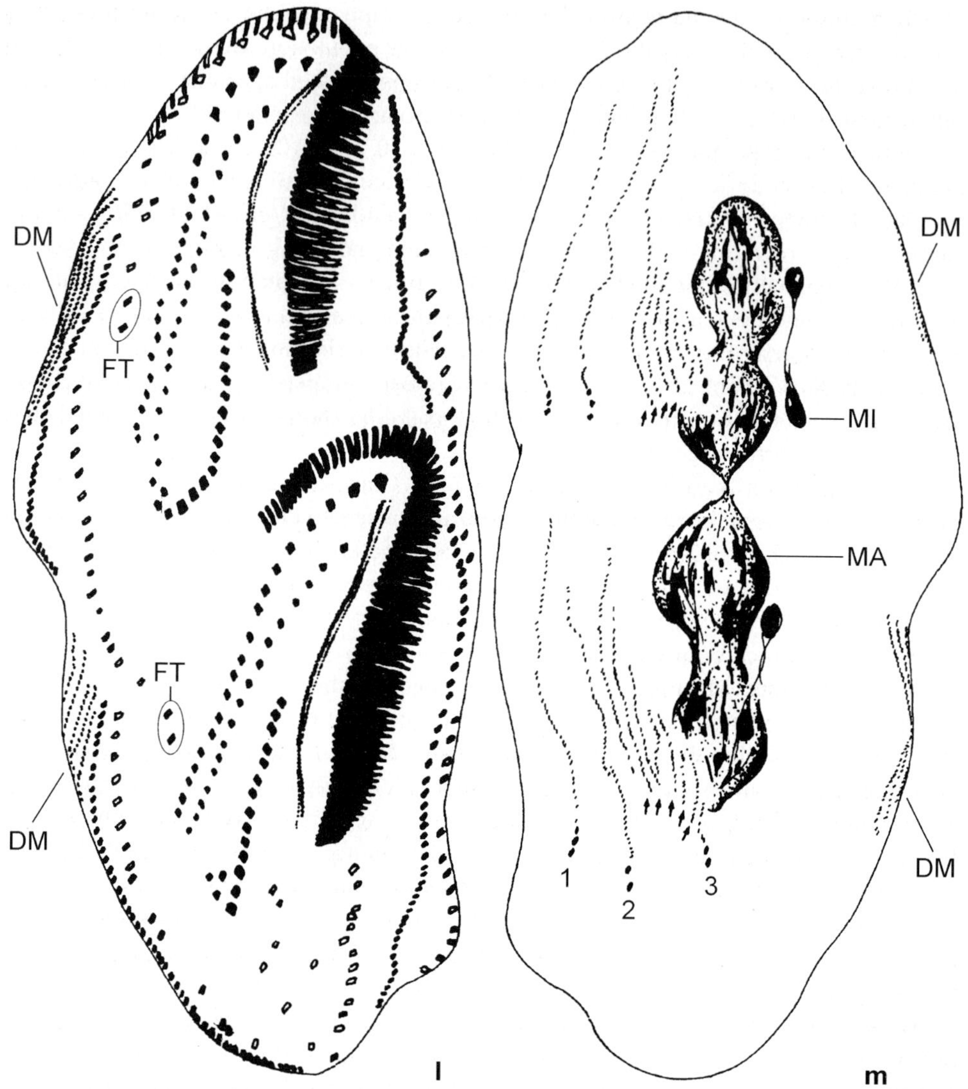

Fig. 243l, m *Neokeronopsis spectabilis* (from Warren et al. 2002. Protargol impregnation). Infraciliature of ventral and dorsal side and nuclear apparatus of a late divider, 454 µm. Arrows in (m) mark dorsal kineties originating by multiple fragmentation of dorsal kinety 3. Parental structures white, new black. DM = dorsomarginal kineties, FT = frontoterminal cirri, MI = micronuclei, MA = macronucleus, 1–3 = dorsal kineties 1–3 with caudal cirri (of course, only the leftmost [= rearmost] fragment of dorsal kinety 3 ends with caudal cirri). Page 1191.

due to the accompanying cortical granules (Fig. 244d, e). For a brief characterisation of the infraciliature, see legend to Fig. 242h.

Observations on Polish population by Warren et al. (2002; Fig. 243a–g, Table 47): As mentioned above, Warren et al. did not provide live data. Body length:width ratio

2.5:1 on average after protargol impregnation. Macronuclear nodules ellipsoidal with a length:width ratio of 2.4:1 on average after protargol impregnation; arranged left of main body axis in ordinary position (Fig. 243a, e–g). Usually a single micronucleus associated with each macronuclear nodule.

Adoral zone occupies 38% of body length in specimen shown in Fig. 243a, extends to 18% of body length on right side; composed of an average of 82 membranelles of ordinary fine structure. Bases of largest membranelles 20–25 µm wide. Buccal area large with undulating membranes more or less in *Cyrtohymena*-pattern, that is, anterior portion of paroral perpendicularly (Fig. 243a) to semicircular (Fig. 243c) curved leftwards while posterior portion almost extending longitudinally; endoral extends diagonally across dorsal wall of buccal cavity forming with anterior half of paroral a more or less bow-shaped structure. Endoral terminating slightly posterior to proximal end of paroral (Fig. 243a, c). Endoral composed of a single row of basal bodies, paroral consists of two rows.

Ciliary pattern of usual variability (Table 47), conspicuous due to prominent transverse cirral row (Fig. 243a). Frontal cirri (usually eight) of anterior corona slightly to distinctly enlarged, corresponding cirri of posterior bow of same size as ventral cirri; leftmost (= anteriormost) cirrus of rear bow homologous to buccal cirrus of other hypotrichs (see cell division). Invariably two frontoterminal cirri left of anterior end of right marginal row; sometimes frontoterminal cirri very close to distal end of adoral zone. Ventral cirral pattern basically composed of cirral pairs "feigning" urostyloid midventral pattern, except for the rearmost portion which likely consists of about two short rows (see cell division); ventral cirri only indistinctly set off from bicorona, rearmost cirri near rear end of transverse cirral row. Transverse cirri arranged in hook-shape, rather prominent because distinctly enlarged, narrowly spaced, and comparatively high in number; row commences at 71% of body length in specimen shown in Fig. 243a. Marginal cirri narrowly spaced, right row commences slightly behind level of distal end of adoral zone, left row begins, as is usual, slightly ahead of level of proximal end of adoral zone; rear end of marginal rows distinctly separated (Fig. 243a) or slightly overlapping (Fig. 243d).

Dorsal cilia arranged in 10 kineties on average (Table 47); variability of number rather high because multiple fragmentation of dorsal kinety 3 does not always result in same number of fragments; furthermore, number of dorsomarginal kineties is variable; most kineties are more or less bipolar, some are rather short. In total seven caudal cirri on average (Table 47); in specimen shown in Fig. 243b three cirri associated with dorsal kinety 1 (leftmost one), three with kinety 2, and two with the posterior fragment of dorsal kinety 3 (see cell division, for details).

Chinese population described by Wang et al. (2005; Table 47): body length in life about 180–250 µm, length:width ratio 3:1; adoral zone occupies about one third of body length; one buccal cirrus; bicorona composed of 15–22 cirri; three rows of caudal cirri.

Cell division (Fig. 243h–m): Ontogenesis of *Neokeronopsis spectabilis* was described by Warren et al. (2002). However, only middle and late dividers were observed so that the beginning of the division process is not known. Recently, Wang et al. (2005)

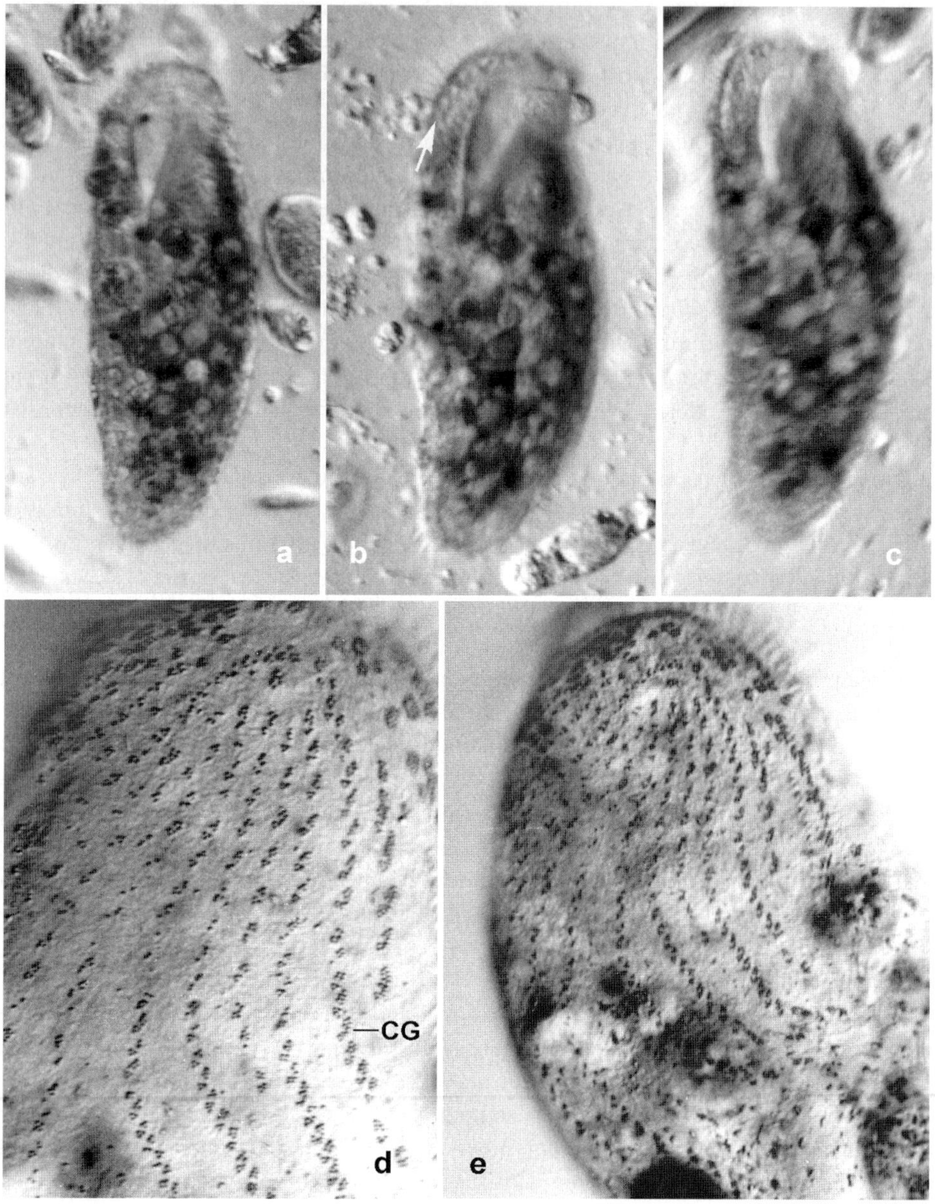

Fig. 244a–e *Neokeronopsis spectabilis* (originals of Carinthian population. Normarski differential interference contrast micrographs). **a–c:** Ventral views of a freely motile specimen. Arrow in (b) marks anterior cirri of bicorona. Note the deep and therefore translucent buccal cavity. **d, e:** Dorsal views of a slightly squeezed specimen showing cortical granules which are yellow, 1–2 μm across, and arranged in groups mainly along dorsal kineties. At superficial live observation, *Neokeronopsis spectabilis* is easily confused with *Urostyla grandis* which has, however, much more cirral rows and macronuclear nodules. CG = cortical granules. Page 1191.

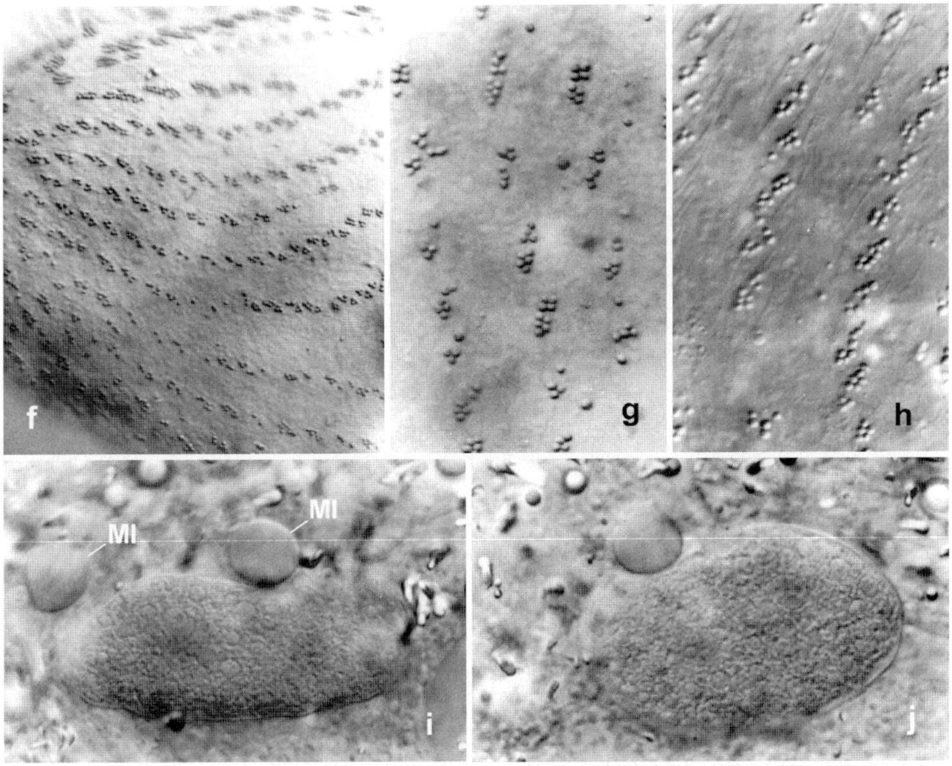

Fig. 244f–j *Neokeronopsis spectabilis* (originals of Carinthian population. Normarski differential interference contrast micrographs). **f–h:** Cortical granules along dorsal kineties. **i, j:** Elongate and ellipsoidal macronuclear nodule with attached micronucleus. MI = micronucleus 8–10 μm across. Page 1191.

noted that the oral primordium is formed de novo between the midventral complex and the left marginal row.

A middle divider shows a well developed adoral zone in the opisthe, right of which is the undulating membrane anlage and the oblique series of 21 frontal-ventral-transversal cirral anlagen (Fig. 243h; according to Wang et al. 2005, 19–25 anlagen are formed). The parental adoral zone is retained. The undulating membrane of the proter originates from the parental undulating membrane, and the cirral anlagen show a similar arrangement as in the opisthe. The parental cirri of the anterior corona, the old frontoterminal cirri, some old midventral cirri, and the old transverse cirri are obviously not involved in anlagen formation. The new marginal rows originate, as in most hypotrichs, within the parental rows (Fig. 243h). Dorsal ontogenesis is basically in *Oxytricha*-pattern as described by Berger & Foissner (1997) and Berger (1999). It commences with the formation of each two anlagen within the parental kineties 1–3, that is the left-most bristle rows (Fig. 243h, i). New dorsomarginal kineties are not yet visible. In this

Table 47 Morphometric data on *Neokeronopsis spectabilis* (sp1, from Warren et al. 2002; sp2, from Wang et al. 2005)

Characteristics[a]	population	mean	M	SD	SE	CV	Min	Max	n
Body, length	sp1	441.8	–	33.0	–	7.0	368.0	500.0	22
Body, width	sp1	176.8	–	13.6	–	8.0	152.0	212.0	21
Macronuclear nodule, length	sp1	78.4	–	9.9	–	13.0	60.0	108.0	22
Macronuclear nodule, width	sp1	33.3	–	7.8	–	23.0	24.0	48.0	22
Macronuclear nodules, number	sp1	2.0	–	0.0	–	0.0	2.0	2.0	22
	sp2	2.0	–	–	–	–	–	–	?
Micronucleus, diameter	sp1	10.5	–	1.7	–	16.0	8.0	13.0	15
Micronuclei, number	sp1	2.2	–	0.9	–	43.0	1.0	4.0	15
	sp2	–	–	–	–	–	2.0	9.0	?
Adoral membranelles, number	sp1	81.5	–	4.3	–	5.0	72.0	90.0	20
	sp2	–	–	–	–	–	70.0	93.0	?
Ventral cirral pairs, number[b]	sp1	22.3	–	1.7	–	8.0	19.0	26.0	22
Frontoterminal cirri, number	sp1	2.0	–	0.0	–	0.0	2.0	2.0	15
	sp2	2.0	–	–	–	–	–	–	?
Transverse cirri, number	sp1	18.4	–	1.8	–	10.0	15.0	22.0	22
	sp2	–	–	–	–	–	15.0	18.0	?
Left marginal cirri, number	sp1	50.2	–	5.7	–	11.0	43.0	65.0	21
	sp2	–	–	–	–	–	49.0	59.0	?
Right marginal cirri, number	sp1	47.5	–	3.0	–	6.0	42.0	54.0	21
	sp2	–	–	–	–	–	42.0	53.0	?
Dorsal kineties, number	sp1	10.3	–	1.3	–	13.0	9.0	12.0	12
	sp2	–	–	–	–	–	8.0	14.0	?
Caudal cirri, number	sp1	7.2	–	1.1	–	15.0	6.0	9.0	18

[a] All measurements µm. Data are based on specimens impregnated with Wilbert's protargol method. CV = coefficient of variation in %, M = median, Max = maximum value, mean = arithmetic mean, Min = minimum value, n = number of individuals investigated, SD = standard deviation, SE = standard error of arithmetic mean, ? = sample size not known.

[b] Pairs forming the bicorona likely included.

stage of division the two macronuclear nodules have fused to a single mass; micronuclei were not observed (Fig. 243i).

In the next stage, the adoral zone of the opisthe is almost completed (Fig. 243j). In both the proter and the opisthe the leftmost frontal cirrus has, as is usual, formed from the undulating membrane anlage. The new frontal-ventral-transverse cirri of both filial products are already recognisable. The anteriormost two cirri of the rightmost anlage begin to migrate anteriorly to form the frontoterminal cirri (Fig. 243j, arrow). The dorsal kinety 3 anlagen begin to fragment (Fig. 243k, arrows). Right of the anterior end of the right marginal row anlagen the new dorsomarginal kineties (= rightmost dorsal kineties) are formed. The macronucleus and the micronuclei divide.

The last stage studied by Warren et al. shows an almost finished adoral zone for the opisthe (Fig. 243l). The undulating membranes anlagen divide longitudinally to form the paroral and endoral. The frontal-ventral-transverse cirri are more or less completely developed and migrate to their final positions. Dorsal kinety 3 anlage has split several times (multiple fragmentation; Fig. 243m, small arrows). On the right body side, dorsomarginal kineties have developed. Caudal cirri are formed at the rear end of the new kineties 1 and 2 and at the rear end of the rightmost fragment of kinety anlage 3. As the division of the single nuclear mass is completed, the macro- and micronucleus within each daughter undergo a further division (Fig. 243m).

Occurrence and ecology: Rare, limnetic species so far only recorded from Europe and China. Type location not mentioned by Kahl, possibly near the German city of Hamburg, where he lived and worked. According to him, *Neokeronopsis spectabilis* is common among *Lemna* and *Utricularia* above sapropelic ground. I found this species twice, unfortunately, on both occasions with low abundance. The first time I recorded it in January 1997 in the Upper Austrian Schwemmbach stream in the village of Teichstätt (48°01'39"N 13°09'09"E). The second time I found it in a sample collected in the Glan stream (Carinthia, Austria) near the St. Martin-Sittich railway station (46°44'00"N 14°09'16"E) in January 2002 (Fig. 242b–h, 244a–j). The ciliate community of this sample was composed of 81 species providing a saprobic index according to Blatterer (1995) of 2.2, indicating betamesosaprobic water quality. Both sampling dates were in January, when the water temperature of Austrian running waters is only slightly above 0° C, indicating that *N. spectabilis* is psychrophilic. I was unsuccessful in the cultivation of this curious species although I put one of the cultures into the refrigerator. Unfortunately, Warren et al. (2002) did not provide details about the sampling date. They found it in 1987 in sediment samples from the upper part of the Ratanica, a small stream which flows into the Dobczyce Reservoir in southern Poland. Again, the abundance was low. Wang et al. (2005) found their population in a freshwater pond near the city of Harbin, China.

Records not substantiated by illustrations and/or morphological data: Belgium (Chardez 1987, p. 13); Gidra River Basin, Slovakia (Tirjaková 2003, p. 36); Tisa river in Carpathian mountains, Ukraine (Kovalchuk 1997, p. 97).

Very voracious (Kahl 1932). In my experience, it feeds on algae and ciliates like the hymenostome *Deltopylum rhabdoides* Fauré-Fremiet & Mugard sensu Song & Wilbert (1989, p. 101). Warren et al. (2002) started short-term cultures in Petri dishes containing filtered stream water plus prey organisms such as a *Colpidium* and *Chlorogonium* species. They failed to establish long-term cultures.

Urostyloides Shi & He, 1989

1989 ***Urostyloides sinensis* gen. and sp. nov.** – Shi & He, Int. Congr. Protozool., 8: 99 (original description). Type species (by monotypy): *Urostyloides sinensis* Shi & He, 1989.

1993 ***Urostyloides* gen. nov.** – Shi, Int. Congr. Protozool., 9: 116 (see nomenclature).

2001 ***Urostyloides* Shi and He, 1989** – Berger, Catalogue of ciliate names 1. Hypotrichs, p. 102 (nomenclator containing all basionyms, combinations, and higher taxa of hypotrichs).

Nomenclature: No derivation of the name is given in the original description. *Urostyloides* is a composite of *Urostyla* (genus group name, see there for derivation) and the suffix *~id·es* (Greek; similarity, especially in shape) and obviously alludes to the similarity of *U. sinensis* to *Urostyla* species. Masculine gender because a compound genus-group name ending with the suffix *-ides* is to be treated as masculine unless its author, when establishing the name, stated that it had another gender or treated it as such by combining it with an adjectival species-group name in another gender form (ICZN 1999, Article 30.1.4.4).

Shi & He (1989) established *Urostyloides* in an abstract without illustration, but with a more or less detailed description. The type fixation is obviously by monotypy because *U. sinensis* was the only included species (ICZN 1985, Article 68d; Article 68b (i) does not apply, because the species was established after 1931). Berger (2001) considered *Urostyloides* and *U. sinensis* as nomina nuda because they were established in an abstract which is not a published work in the sense of the Code (ICZN 1999, Article 9.9). However, since the taxa were published in 1989, the ICZN (1985), which was less rigorous in this respect, has to be applied.

Urostyloides was established a second time as new genus by Shi (1993). Since no type species was fixed in this abstract it is invalid. Aescht (2001) overlooked *Urostyloides*.

Characterisation: Body large. Midventral complex (?) short. Many left and right marginal rows. Transverse and caudal cirri present. Dorsal kineties in disordered arrangement likely originating by fragmentation from 3 dorsal anlagen and one dorsomarginal primordium.

Remarks: See *Urostyloides sinensis*.

Species included in *Urostyloides*: (1) *Urostyloides sinensis* Shi & He, 1989.

Single species

Urostyloides sinensis Shi & He, 1989

1989 ***Urostyloides sinensis* gen. and sp. nov.** – Shi & He, Int. Congr. Protozool., 8: 99 (original description; no formal diagnosis provided; without illustration; site where type material is deposited not mentioned).

2001 ***Urostyloides sinensis* Shi and He, 1989** – Berger, Catalogue of ciliate names 1. Hypotrichs, p. 102 (nomenclator containing all basionyms, combinations, and higher taxa of hypotrichs).

Nomenclature: No derivation of the name is given in the original description. The species-group name *sinens·is -is -e* (Latin adjective; Chinese) obviously refers to the fact that the species was discovered in China. *Urostyloides sinensis* is the type species of *Urostyloides* (see genus section).

Remarks: As already mentioned above, *Urostyloides sinensis* is described in more or less detail in an abstract without illustration. Thus, the characterisation above is rather incomplete. According to the authors (Shi Xinbai & He Wei), it has to be placed between *Urostyla* and *Pseudourostyla*. Interestingly, Shi et al. (1999; Shi Xinbai is the third author of this paper) did not mention this genus and species in their systematic revision on hypotrichous ciliates. Two possibilities exist: (1) They overlooked Shi's own genus and species; or (2) they considered it as invalid and therefore omitted it from the revision.

I do not know which point applies and therefore mention it in the present book to complete the picture.

Shi & He (1989) mentioned a zigzag pattern indicating that it is a urostyloid. On the other hand, they described 15–20 dorsal kineties in disordered arrangement which originate from three dorsal anlagen and one "lateral" primordium, that is, fragmentation and/or retention of parental kineties must occur to produce such a high number of kineties from only four anlagen. However, the presence of fragmentation strongly indicates that this species is not a urostyloid, but an oxytrichid possibly closely related to *Gigantothrix herzogi* Foissner, 1999. This terrestrial species also has a high number of cirral rows and dorsal kineties, but has more macronuclear nodules (33 on average vs. 2 in *U. sinensis*), lacks transverse cirri (vs. 9–17 present) and caudal cirri (vs. present). Thus, synonymy of these two species can be excluded. More detailed morphological (e.g., consistency [rigid or flexible] of body, presence/absence of cortical granules) and cell division data are needed for a proper classification.

Morphology: Body large, however length not given. Two macronuclear nodules, 6–13 micronuclei. "Two rows of frontoventral cirri as zigzag files only about half cell length" (whether or not this is a true zigzagging pattern or two cirral rows roughly arranged in such a pattern is not known). Frontal ciliature (three frontal cirri, bicorona, ...) and presence/absence of buccal and frontoterminal cirri not described. 9–17 transverse cirri arranged in J-shaped pattern. 11–14 right and 8–14 left marginal rows; every marginal cirral base with a long rootlet extending to the left. About 15–20 dorsal kineties in irregular arrangement; originate from three dorsal and one lateral (obviously dorsomarginal) anlagen (see remarks). Three groups of caudal cirri, difficult to recognise because of the many marginal rows. About the formation of the new adoral zone of membranelles of the proter, the authors write: "The parental AZM in the proter is neither partially reorganized like *Urostyla* and *Pseudourostyla* nor left intact like *Paraurostyla*, but remaining trace of AZM primordium vanishes as soon as it appears".

Occurrence and ecology: Type locality is Morshan, a small mountain town 100 km from Harbin, China. Unfortunately, Shi & He (1989) did not mention the habitat (running/standing water, soil) where they found the species twice.

Taxa not Considered

The following taxa, included in the urostyloids or holostichids by some modern (post 1970) workers (e.g., Tables 2–11), are either based on insufficiently described species or they lack a more or less distinct midventral complex and are therefore not considered in the present monograph. Moreover, *Uroleptus* and its supposed synonym *Paruroleptus* are briefly discussed in this chapter.

***Amphisiella* Gourret & Roeser, 1888**, Archs Biol., 8: 180. Type species (by monotypy): *Amphisiella marioni* Gourret & Roeser, 1888. Remarks: Classified in the Urostylidae by Borror (1972, p. 9) and in the Holostichidae by Stiller (1974b, p. 94), Corliss (1979, p. 309), Tuffrau (1979, p. 526; 1987, p. 115), and Carey (1992, p. 178). *Amphisiella* species lack a midventral complex and are therefore not closely related to the urostyloids (e.g., Wicklow 1982, Eigner & Foissner 1994, Petz & Foissner 1996, Berger 2004a). *Amphisiella* will be treated in the third volume of the hypotrich monograph.

***Balladyna* Kowalewskiego, 1882**, Pam. fizyogr.; 2: 408. Type species (by original designation and monotypy): *Balladyna parvula* Kowalewskiego, 1882. Remarks: Classified in the Urostylidae by Borror (1972, p. 9) and in the Holostichidae by Stiller (1974b, p. 96), Corliss (1979, p. 309), Tuffrau (1979, p. 526; 1987, p. 115), and Carey (1992, p. 180). *Balladyna parvula* is a very small hypotrich with about seven frontoventral cirri, five prominent transverse cirri, and relatively long marginal cirri and dorsal bristles. The habitus closely resembles that of species of the *Oxytricha setigera* group, a relationship not discussed by Berger (1999). Since there is no evidence for a midventral complex, it is not treated in the present book.

***Balladinopsis* Ghosh, 1921**, J. R. microsc. Soc., year 1921: 248. Type species (by original designation and monotypy): *Balladinopsis nuda* Ghosh, 1921. Remarks: Classified in the Holostichidae by Stiller (1974b, p. 99). The type species has no frontoventral cirri, that is, lacks a midventral complex. Thus, it is not considered in the present review. Very likely *B. nuda* is a misobserved oxytrichid, that is, a species indeterminata.

***Balladynella* Stiller, 1974**, Annls hist.-nat. Mus. natn. hung., 66: 129, 130. Remarks: Stiller (1974a) overlooked to fix a type species. Consequently, *Balladynella* Stiller, 1974 is not available (ICZN 1964, Article 13b). According to Aescht (2001, p. 30), now available as *Balladynella* Jankowski, 1979 (p. 51) with *Balladyna fusiformis* Kahl, 1932 as type species by original designation. Classified in the Holostichidae by Stiller (1974b, p. 98), Corliss (1979, p. 309), and Tuffrau (1979, p. 526; 1987, p. 115). Since there is no evidence for a midventral complex, it is not treated in the present book.

***Banyulsella* Dragesco, 1954**, Vie Milieu, 4: 637. Type species (by monotypy): *Banyulsella viridis* Dragesco, 1954. Remarks: First mentioned, as nomen nudum, in a species list by Dragesco (1953, p. 629), and described in detail by Dragesco (1960, p. 316).

Classified in the Urostylidae, inter alia, by Borror (1972, p. 9), Corliss (1979, p. 309), and Carey (1992, p. 177). The single species, *B. viridis*, is very small (about 50 µm), has a very large (more than 50% of body length), U-shaped adoral zone, three enlarged frontal cirri, two short and two long "fronto-ventral" cirral rows, and a long row of transverse(?) cirri. Marginal cirri obviously lacking. One row of fine cirri on rear portion of dorsal side. Six small macronuclear nodules. Mesopsammon of Banyuls-sur-Mer, Mediterranean Sea, France. The two long ventral rows are distinctly separated, strongly indicating that *B. viridis* does not have a midventral complex. Thus, it is not treated in the present monograph. Jankowski (1975, p. 27) established the Banyulsellidae (incorrectly spelled Banylsellidae) for this species. It will be treated in one of the next volumes of the hypotrich monograph.

***Coniculostomum* Njine, 1979**, Protistologica, 15: 353. Type species (by monotypy): *Laurentia monilata* Dragesco & Njine, 1971. Remarks: Classified in the Holostichidae by Tuffrau (1979, p. 526; 1987, p. 115). A rigid oxytrichid and therefore assigned to the Stylonychinae (for review, see Berger 1999, p. 606).

***Discocephalus* Ehrenberg, 1829**, Abh. preuss. Akad. Wiss., year 1929: 9, 16. Type species (by monotypy): *Discocephalus rotatorius* Ehrenberg, 1829. Remarks: Classified in the Holostichidae by Tuffrau (1979, p. 526). Name-bearing type of the discocephalids (see Berger 2001, p. 106). This species lacks an urostyloid midventral pattern and is therefore not treated in the present book. However, in some taxa – for example, *Marginotricha faurei* (Dragesco, 1963) Lin, Song & Warren, 2004 – of the discocephalids oblique cirral anlagen, which form cirral pairs, are produced (Wicklow 1982, his Fig. 43). Later, one cirrus of each pair is resorbed so that pairs (and a zigzagging cirral pattern) are lacking during interphase. Thus, I would not be surprised if molecular data indicate a close relationship of urostyloids and discocephalids. On the other hand, a relationship with the stylonychines cannot be excluded because both groups have a rigid body and lack cortical granules. An increase in the number of cirral anlagen from six to more occurred several times independently (see general section).

***Gonostomum* Sterki, 1878**, Z. wiss. Zool., 31: 57. Type species (by original designation): *Oxytricha affinis* Stein, 1859. Remarks: Classified in the Holostichidae by Stiller (1974b, p. 88). Morphological and cell division data indicate that *Gonostomum* is a modified 18-cirri oxytrichid (for review see Berger 1999, p. 367). Molecular studies support (Modeo et al. 2003), respectively, contradict (Affa'a et al. 2004) this view.

***Isosticha* Kiesselbach, 1936**, Thalassia, 2: 18. Type species (by monotypy): *Isosticha contractilis* Kiesselbach, 1936. Remarks: Kiesselbach (1936a) established this monotypic genus in the family Oxytrichidae. Later, it was assigned to the Urostylidae by Corliss (1977, p. 137; 1979, p. 309), Tuffrau (1979, p. 526; 1987, p. 115), and Tuffrau & Fleury (1994, p. 128). By contrast, Borror (1972) and Jankowski (1979) did not mention it in their papers indicating that they did not consider it as hypotrich. Hemberger (1982, p. 275) also assumed that Kiesselbach's species does not belong to the hypotrichs. Aescht (2001) overlooked *Isosticha*. The habitus of *I. contractilis* is reminiscent of a condylostomatid, as already stated by Kiesselbach (1936a). I also assume that it belongs to this group and therefore do not include it in the present review.

***Kahliella* Corliss, 1960**, J. Protozool., 7: 275. Type species (by original designation): *Kahlia acrobates* Horváth, 1932. Remarks: Classified in the Urostylidae by Borror (1972, p. 9) and Stiller (1974b, p. 35). *Kahliella* species lack a midventral complex (e.g., Fleury et al. 1985, Berger & Foissner 1987) and are therefore not treated in the present review. Tuffrau (1979) established a new family for *Kahliella* which will be reviewed in one of the next volumes of the monograph of hypotrichs.

***Kerona* Müller, 1786**, Animalcula Infusoria, p. 233. Type species (by subsequent designation by Ehrenberg 1838): *Cyclidium pediculus* Müller, 1773. Remarks: Classified in the Urostylidae by Borror (1972, p. 9). This ectocommensal on hydras and bryozoans lacks a midventral pattern. It is likely a highly specialised oxytrichid with multiple fragmentation of dorsal kineties reviewed in detail by Berger (1999, p. 825).

***Keronopsis* Penard, 1922**, Infusoires, p. 238. Type species (by original designation): *Keronopsis helluo* Penard, 1922. Remarks: Classified in the Holostichidae by Borror (1972, p. 11), Stiller (1974b, p. 63), Corliss (1979, p. 309), and Carey (1992, p. 183) and in the Urostylidae by Borror (1979, p. 546) and Tuffrau (1979, p. 526). Borror & Wicklow (1983) transferred most species to *Pseudokeronopsis*, a characteristic urostyloid. Although the type species *K. helluo* is not yet redescribed in detail, a sistergroup relationship of *Keronopsis* and *Paraholosticha* Wenzel, 1953 (see below) is very likely. It will be treated in one of the next volumes of the monograph of hypotrichs.

***Klonostricha* Jankowski, 1979**, Trudy zool. Inst., 86: 57. Type species (by original designation): *Klonostricha horrida* Vuxanovici, 1963. Remarks: Classified in the Holostichidae by Stiller (1974b, p. 101). *Klonostricha* was actually established by Vuxanovici (1962, p. 200) who, however, did not fix a type species so that *Klonostricha* Vuxanovici is not available (ICZN 1999, Article 13.3; Aescht 2001). In my catalogue I was of different opinion (Berger 2001, p. 45). Vuxanovici illustrated some cirral pairs which very likely caused Stiller to assume a urostyloid relationship. However, I doubt that an identification will ever be possible. It will be reviewed in one of the next volumes of the monograph of hypotrichs.

***Lacazea* Dragesco, 1960**, Trav. Stn. biol. Roscoff, 122: 330. Type species (by original designation): *Lacazea ovalis* Dragesco, 1960. Remarks: Classified in the Urostylidae by Borror (1972, p. 9). *Lacazea ovalis* is almost globular, has a single macronucleus, and a difficult-to-interpret cirral pattern which does not show zigzag cirri. It will be reviewed in one of the next volumes of the monograph of hypotrichs.

***Lamtostyla* Buitkamp, 1977**, Acta Protozool., 16: 270. Type species (by original designation and monotypy): *Lamtostyla lamottei* Buitkamp, 1977. Remarks: Classified in the

Holostichidae by Corliss (1979, p. 309). *Lamtostyla* lacks a midventral complex and is now assigned to the amphisiellids (e.g., Berger & Foissner 1988, Petz & Foissner 1996). It will be treated in the third volume of the monograph of hypotrichs.

***Laurentiella* Dragesco & Njine, 1971**, Annls Fac. Sci. Univ. féd. Cameroun, 7–8: 125. Type species (by original designation): *Laurentia macrostoma* Dragesco, 1966. Remarks: Classified in the Holostichidae by Corliss (1979, p. 309). Classified in the Stylonychinae (rigid oxytrichids) by Berger (1999, p. 752), inter alia, because of the rigid body and the lack of cortical granules. This assignment was later confirmed by molecular data (Bernhard et al. 2001, Modeo et al. 2003).

***Onychodromopsis* Stokes, 1887**, Ann. Mag. nat. Hist., 20: 107. Type species (by monotypy): *Onychodromopsis flexilis* Stokes, 1887. Remarks: Classified in the Holostichidae by Tuffrau (1987, p. 115). *Onychodromopsis* is an oxytrichid reviewed by Berger (1999, p. 475). Note the problems with *Allotricha* (see also Petz & Foissner 1996).

***Paraholosticha* Kahl, 1932**, Tierwelt Dtl., 25: 545. Remarks: Kahl (1932) did not fix a type species; thus, the genus is invalid (ICZN 1999, Article 13.3). Replaced by *Paraholosticha* Wenzel, 1953 (Arch. Protistenk., 99: 104; type by original designation: *Paraholosticha muscicola* Kahl, 1932). Classified in the Holostichidae by Borror (1972, p. 11) and the Urostylidae by Stiller (1974b, p. 31), Corliss (1977, p. 137; 1979, p. 309), and Carey (1992, p. 177). *Paraholosticha* and the adelphotaxon *Keronopsis* Penard, 1922 lack a midventral complex and are therefore not closely related to the urostyloids (Berger & Foissner 1987, Dieckmann 1989). It will be treated in one of the next volumes of the hypotrich monograph.

***Paraurostyla* Borror, 1972**, J. Protozool., 19: 9. Type species (by original designation): *Urostyla weissei* Stein, 1859. Remarks: Classified in the Urostylidae by Borror (1972), Corliss (1979, p. 309), and Carey (1992, p. 177). Species of this group lack a midventral complex. *Paraurostyla* was assigned to the oxytrichids because the formation of the ventral and dorsal infraciliature, strongly indicated a close relationship with the 18-cirri oxytrichids (e.g., Borror 1979, Wirnsberger et al. 1985; for review, see Berger 1999, p. 841). Later, this relationship was confirmed by molecular data (e.g., Bernhard et al. 2001, Hewitt et al. 2003).

***Paruroleptus* Kahl, 1932**, Tierwelt Dtl., 25: 586. Remarks: Kahl (1932) classified this group as subgenus of *Holosticha*. Thus, the correct basionym is *Holosticha (Paruroleptus)* Kahl, 1932. Unfortunately, Kahl (1932) did not fix a type species so that his subgenus is not available (ICZN 1964, Article 13). Replaced by *Holosticha (Paruroleptus)* Wenzel, 1953 with *Holosticha caudata* Stokes, 1886 as type species by original designation. Likely a junior synonym of *Uroleptus* (further details see below).

***Parurosoma* Gelei, 1954**, Acta biol. hung., 5: 332. Type species (by monotypy): *Holosticha (Parurosoma) dubium* Gelei, 1954. Remarks: Classified in the Holostichidae by

Corliss (1979, p. 309) and Tuffrau (1979, p. 526; 1987, p. 115). The ventral cirral pattern is strongly reminiscent of 18-cirri oxytrichids. Thus, Berger & Foissner (1997) and Berger (1999, p. 491) assigned it to the oxytrichids.

***Pattersoniella* Foissner, 1987**, Zool. Beitr., 31: 207. Type species (by original designation): *Pattersoniella vitiphila* Foissner, 1987. Remarks: Classified in the urostyloids by Shi et al. (1999, p. 119), who even established the urostyloid subgroup Pattersoniellidae. However, *Pattersoniella* is a rigid oxytrichid and was therefore included in the stylonychines by Berger (1999, p. 766), a classification later confirmed by molecular data (Bernhard et al. 2001).

***Prooxytricha* Poche, 1913**, Arch. Protistenk., 30: 261. Type species (same as for *Trichogaster* Sterki, 1878): *Trichogaster pilosus* Sterki, 1878. Remarks: *Prooxytricha* is the replacement name for *Trichogaster* Sterki, 1878 (further details, see *Trichogaster*).

***Psilotricha* Stein, 1859**, Lotos, 9: 5. Type species (by original designation): *Psilotricha acuminata* Stein, 1859. Remarks: Classified in the Holostichidae by Stiller (1974b, p. 47). *Psilotricha* lacks a midventral pattern (Esteban et al. 2001a) and is therefore not treated in the present book. It will be reviewed in one of the next volumes of the hypotrich monograph.

***Stichochaeta* Claparède & Lachmann, 1858**, Mém. Inst. natn. génev., 5: 152. Type species (by monotypy): *Stichochaeta cornuta* Claparède & Lachmann, 1858. Remarks: Classified in the Holostichidae by Carey (1992, p. 185). *Stichochaeta* is a junior synonym of *Stichotricha* Perty, 1849b, and *Stichochaeta cornuta* a junior synonym of *Stichotricha secunda* Perty, 1849b (for review see Foissner et al. 1991, p. 210). It will be reviewed in one of the next volumes of the hypotrich monograph.

***Strongylidium* Sterki, 1878**, Z. wiss. Zool., 31: 58. Type (by monotypy): *Strongylidium crassum* Sterki, 1878. Remarks: A little known group which is, according to Kahl (1932, p. 551), closely related to *Uroleptus*. It will be treated in one of the next volumes of the hypotrich monograph.

***Territricha* Berger & Foissner, 1988**, Zool. Anz., 220: 127. Type species (by original designation): *Territricha stramenticola* Berger & Foissner, 1988. Remarks: Classified in the urostylids by Shi et al. (1999, p. 119) and Lynn & Small (2002, p. 444). When we described this species we assumed that it is closely related to the urostyloids. Later I classified it in the oxytrichids mainly because it has a fragmenting dorsal kinety 3 (Berger 1999, p. 884).

***Trachelochaeta* Šrámek-Hušek, 1954**, Arch. Protistenk., 100: 265. Type species (by original designation): *Trachelochaeta bryophila* Šrámek-Hušek, 1954. Remarks: Classified in the Holostichidae by Stiller (1974b, p. 87), Corliss (1979, p. 309), and Tuffrau (1987, p. 115). Since there is no evidence that this species has a midventral complex it is not treated in the present book. It will be included in one of the next volumes of the hypotrich monograph.

***Trachelostyla* Kahl, 1932**, Tierwelt Dtl., 25: 596. Remarks: Classified in the Holostichidae by Corliss (1979, p. 310), Tuffrau (1979, p. 526; 1987, p. 115), and Carey (1992, p. 185). Kahl (1932) did not fix a type species so that *Trachelostyla* Kahl, 1932 is not available according to ICZN (1999, Article 13.3). For a brief explanation of the complicated nomenclature, see Berger (1999, p. 894; 2001, p. 89) and Aescht (2001, p. 165). There is no evidence that *Stichochaeta pediculiformis* Cohn, 1866, type of the genus *Trachelostyla* Borror, 1972, has a midventral complex. It will be treated in one of the next volumes of the hypotrich monograph.

***Trichogaster* Sterki, 1878**, Z. wiss. Zool., 31: 38, 58. Type species (by monotypy): *Trichogaster pilosus* Sterki, 1878. Remarks: *Trichogaster* Sterki, 1878 is a junior homonym of a fish genus and was therefore replaced by *Prooxytricha* by Poche (1913; see above). Sterki (1878) found only one specimen and wrote that he could not study it in detail. According to Sterki, who did not provide an illustration, *Trichogaster* is closely related to the "Urostylen" (I do not know whether he meant *Urostyla* or urostylids). Likely for this reason, Corliss (1960, p. 273; 1961, p. 170) classified it as supposed synonym of *Urostyla*. I follow Kahl (1932, p. 538), who considered *Trichogaster*, respectively, *Prooxytricha* as unidentifiable taxon, that is, it will very likely never be possible to identify a specimen as *T. pilosus*, respectively, to synonymise it with a better known species. In the following reviews, *Trichogaster*, respectively, *Prooxytricha* is mentioned: Kent (1882, p. 764; valid); Blochmann (1886, p. 75; 1895, p. 111; valid genus and species); Bütschli (1889, p. 1741; described as insufficiently known); Conn (1905, p. 56; valid genus); Calkins (1909, p. 55; 1926, p. 410; valid); Lepsi (1926b, p. 83; valid); Escomel (1929, p. 31; record from Peru?); Corliss (1977, p. 138; 1979, p. 310) and Tuffrau (1979, p. 527; incertae sedis in Stichotrichina); Jankowski (1979, p. 68; catalogue); Tuffrau (1987, p. 116) and Tuffrau & Fleury (1994, p. 144; incertae sedis in Hypotrichida).

Sterki (1878) provided the following description of *T. pilosus*: body length about 230 µm; elongate oval; cytoplasm bright; contractile vacuole and four nuclei (macronuclear nodules?) distinct; food vacuoles mainly with diatoms; adoral zone about one third of body length, wide, shape similar to that of the Urostylen; almost completely covered with very fine and short cilia (much more delicate and more closely spaced as in the Urostylen), likely arranged in longitudinal rows; some slightly enlarged cirri on the frontal area, behind the proximal end of the adoral zone, as well as 4–5 transverse cirri; adoral membranelles short, thin, and narrowly spaced. Peristome (buccal field?) covered with very fine, short cilia. Found in a freshwater habitat in Switzerland.

***Uroleptoides* Wenzel, 1953**, Arch. Protistenk., 99: 107. Type species (by original designation): *Uroleptoides kihni* Wenzel, 1953. Remarks: Classified in the Holostichidae by Stiller (1974b, p. 61) and Corliss (1979, p. 310). Likely an amphisiellid. It will be treated in the third volume of the hypotrich monograph.

***Uroleptus* Ehrenberg, 1831**, Abh. preuss. Akad. Wiss., year 1831: 116. Type species (by subsequent designation by Borror 1972, p. 12): *Trichoda musculus* Müller, 1773. Remarks: For problems with the fixation of the type species, see Foissner et al. (1991), Aescht (2001), and Berger (2001). *Uroleptus* species and species of its supposed synonym *Paruroleptus* have zigzagging cirri and have therefore been assigned to the urostylids/holostichids by most authorities (e.g., Borror 1972, Corliss 1979, Hemberger 1982, Wiackowski 1988, Foissner & Foissner 1988, Borror & Wicklow 1983, Shi et al. 1999, Eigner 2001, Lynn & Small 2002). Only Tuffrau (1979, 1987) and Tuffrau & Fleury (1994) assigned it to the Kahliellidae.

Molecular studies suggested that the inclusion of *Uroleptus* in the urostyloids is incorrect (Hewitt et al. 2003, Croft et al. 2003, Foissner et al. 2004a). Thus, we put forward a hypothesis which tries to explain the convergent evolution of the zigzag pattern of the ventral cirri in *Uroleptus* and the urostyloids (Foissner et al. 2004a). For a more detailed discussion of the supposed convergent evolution of urostyloids and *Uroleptus*, see chapter 2.4 in the general section.

***Wallackia* Foissner, 1976**, Acta Protozool., 15: 387, 390. Type species (by original designation and monotypy): *Wallackia schiffmanni* Foissner, 1976. Remarks: Classified in the Holostichidae by Corliss (1979, p. 310) and Tuffrau (1979, p. 526; 1987, p. 115). *Wallackia* species lack a midventral complex and are therefore not treated in the present book (Foissner et al. 2002, p. 635ff). They will be reviewed in one of the next volumes of the hypotrich monograph.

Addenda

The morphological data by Song (2003), Hu et al. (2003a), Lei et al. (2005a), and Alekperov (2005) and some further papers became available too late for inclusion in the main text and thus are added here at the end of the book.

Song (2003) recorded the following urostyloids in shrimp-farming waters of the Yellow Sea:

(i) *Pseudokeronopsis flavicans* (Kahl, 1932) Borror & Wicklow, 1983; p. 102, Fig. 3-11, A–C (see p. 951, Fig. 185g, o, p).

(ii) *Holosticha manca* Kahl, 1932; p. 102, Fig. 3-11, D–F, M (see *Anteholosticha manca*, p. 422, Fig. 87c, e–g).

(iii) *Uroleptus retractilis* (Claparède & Lachmann, 1858) Song & Warren, 1996; p. 104, Fig. 3-11, G–L (see *Psammomitra retractilis*, p. 222, Fig. 43a–e, i, j).

Hu et al. (2003a) recorded the following urostyloids in scallop-farming waters of the Yellow Sea:

(i) *Pseudokeronopsis rubra* (Ehrenberg, 1838); p. 163, Fig. 5-5, D–F (see p. 890, Fig. 180a, g, h).

(ii) *Pseudokeronopsis pulchra* (Kahl, 1932) Borror & Wicklow, 1983; p. 165, Fig. 5-5, A–C (see *Pseudokeronopsis pararubra*, p. 927, Fig. 180.1h, i, 245a). Remarks: Hu et al. (2004a) recognised that the identification by Hu et al. (2003a) is incorrect and described the species *Pseudokeronopsis pararubra*. The illustration of the live specimen is slightly different in Hu et al. (2003a) and Hu et al. (2004a).

(iii) *Pseudokeronopsis qingdaoensis* Hu & Song, 2000; p. 165, Fig. 5-5, G–J (see *Thigmokeronopsis crassa*, p. 873, Fig. 176n, 245b–d).

(iv) *Holosticha heterofoissneri* Hu & Song, 2001; p. 166, Fig. 5-6, F–H, K–M, Plate IX, Fig. C (see p. 152, Fig. 32g, i, j, l, m, o, p).

(v) *Holosticha bradburyae* Gong et al., 2001; p. 166, Fig. 5-6, A–E, I, J, Plate X, Fig. C, D, Plate XI, Fig. E (see p. 167, Fig. 35a–d, h–j).

(vi) *Holosticha warreni* Song & Wilbert, 1997; p. 168 (see *Anteholosticha warreni*, p. 412).

(vii) *Parabirojimia similis* Hu et al., 2002; p. 168, Fig. 5-7, A–D, J, K (see p. 691, Fig. 136a–e, j, k).

(viii) *Thigmokeronopsis rubra* Hu et al., 2003; p. 169, Fig. 5-7, E–I, L–N (see p. 852, Fig. 171a–j).

(ix) *Bakuella agamalievi* Borror & Wicklow, 1983; p. 171, Fig. 5-8, J–L (see p. 541, Fig. 115a, f, g).

Alekperov (2005) provided a review about free-living ciliates, likely mainly from the Caspian Sea. The following urostyloid species are included:

(i) *Pseudokeronopsis rubra* (Ehrenberg, 1838); p. 221, Fig. 69.5, Plate 22, Fig. 3 (see p. 890, Fig. 246a). Remarks: I did not translate the paper and thus do not know whether or not live data, which are very important for the identification of pseudokeronopsids, are provided. According to the illustration the identification could be correct.

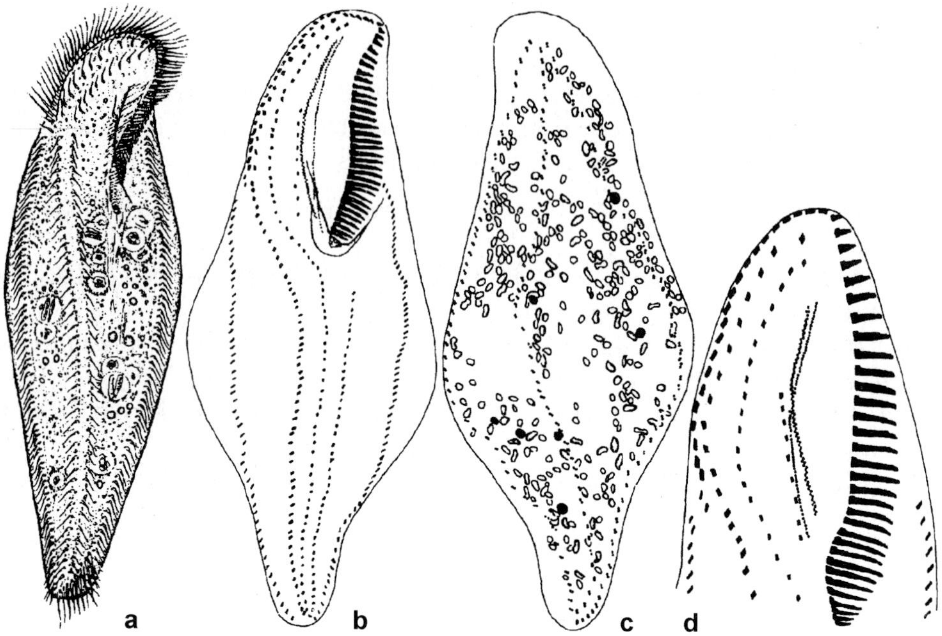

Fig. 245a *Pseudokeronopsis pararubra* (from Hu et al. 2003a). Ventral view of live specimen, 320 µm. Page 1215.

Fig. 245b–d *Thigmokeronopsis crassa* (from Hu et al. 2003a). Infraciliature of ventral and dorsal side and nuclear apparatus, size not indicated. Page 1215.

(ii) *Urostyla grandis* Ehrenberg, 1838; p. 222, Fig. 70.1-2 (see p. 1048, Fig. 246c, d). Remarks: I did not translate the description and thus do not know whether or not live data, which are important for the identification of urostyloids, are provided. According to the illustration the identification could be correct.

(iii) *Urostyla agamalievi* Alekperov, 1984; p. 223, Fig. 70.3-4 (see p. 1093, Fig. 215a, b).

(iv) *Urostyla raikovi* (Alekperov, 1984) comb. nov.; p. 225, Fig. 70.5-6 (see *Pseudourostyla raikovi*, p. 807, Fig. 158a, b). Remarks: For a foundation of the transfer from *Metaurostyla* to *Pseudourostyla* see page 807. A classification in *Urostyla*, as suggested by Alekperov (2005), is likely incorrect because *U. grandis*, type of *Urostyla*, has a midventral complex composed of cirral pairs and rows, while it is composed of cirral pairs only in *P. raikovi*.

(v) *Urostyla magna* (Alekperov, 1984) comb. nov.; p. 225, Fig. 70.7-8 (see *Pseudourostyla magna*, p. 809, Fig. 159a, b). Remarks: For a foundation of the transfer from *Metaurostyla* to *Pseudourostyla* see page 809. A classification in *Urostyla*, as suggested by Alekperov (2005), is likely incorrect because *U. grandis*, type of *Urostyla*, has a midventral complex composed of cirral pairs and rows, while it is composed of cirral pairs only in *P. magna*.

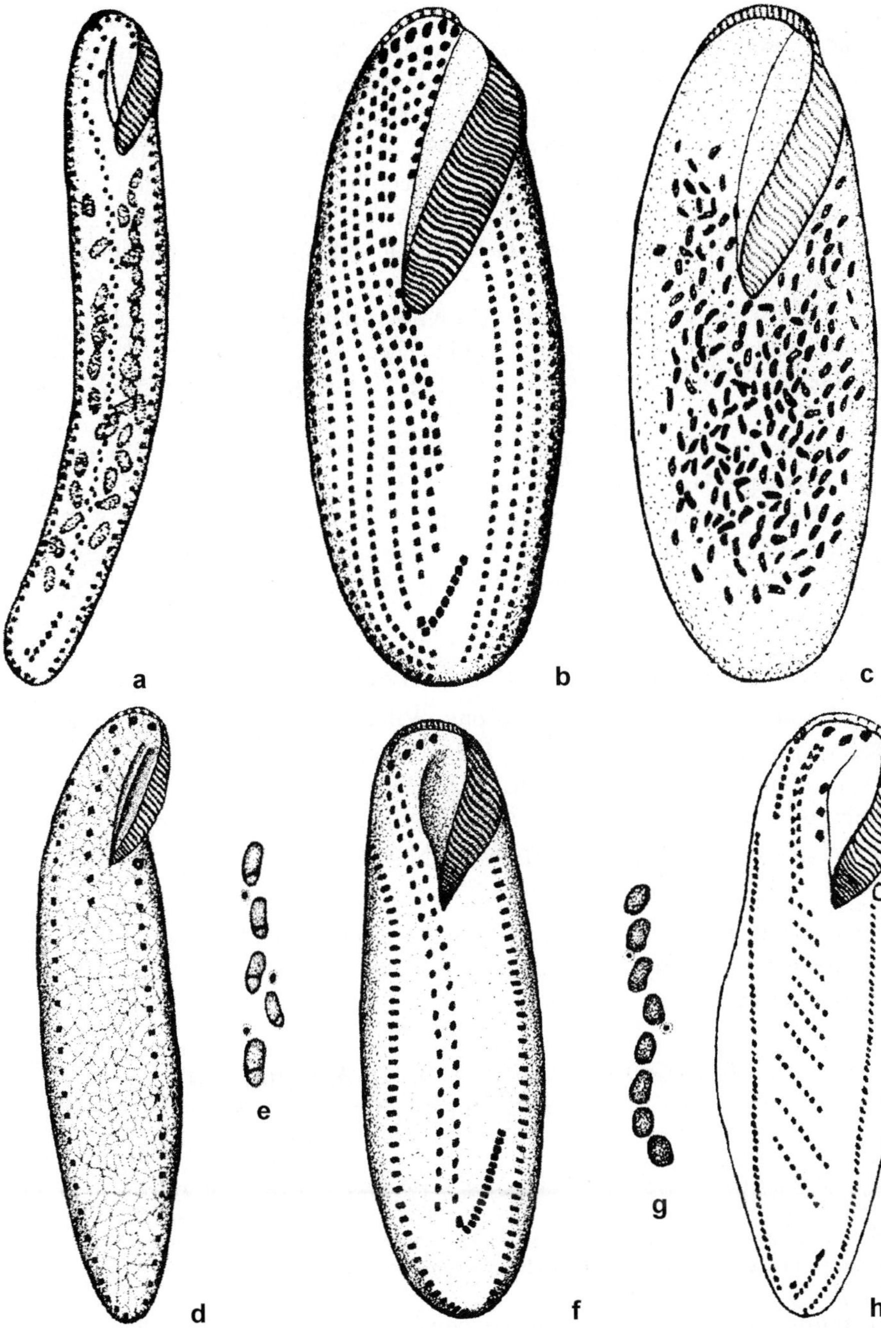

Fig. 246a–h Urostyloids from Alekperov (2005; silver-impregnated specimens). **a:** *Pseudokeronopsis rubra*, 190 µm; page 1215. **b, c:** *Urostyla grandis*, 407 µm; page 1216. **d, e:** *Periholosticha lanceolata*, 145 µm; page 1218. **f, g:** *Anteholosticha monilata*, 120 µm; page 1218. **h:** *Bakuella marina*, 120 µm; page 1218.

(vi) *Periholosticha lanceolata* Hemberger, 1982; p. 226, Fig. 70.9-10 (see p. 502, Fig. 246d, e). Remarks: I did not translate the paper. According to the cirral pattern, the identification could be correct. However, Alekperov's population has a comparatively low number of macronuclear nodules (5–7 vs. 10–27; see Table 24).

(vii) *Holosticha intermedia* (Kahl, 1935); p. 228, Fig. 71.4-5 (Fig. 246f, g). Remarks: I did not translate the description. According to the cirral pattern and the nuclear apparatus the population could be an *Anteholosticha monilata* (see p. 297). *Anteholosticha intermedia* has about 100 macronuclear nodules scattered throughout the cytoplasm (see p. 317).

(viii) *Holosticha azerbaijanica* Alekperov, Asadullayeva, 1999; p. 228, Fig. 71.6 (see *Anteholosticha azerbaijanica*, p. 454, Fig. 97a).

(ix) *Bakuella polycirrata*, 1988; p. 232, Fig. 72.1-3 (see *Bakuella crenata*, p. 548, Fig. 116e, f).

(x) *Bakuella kreuzkampii* Song, Wilbert, Berger, 1992; p. 233, Fig. 72.4 (see *Bakuella agamalievi*, p. 541, Fig. 115l).

(xi) *Bakuella crenata* Agamaliev et Alekperov, 1976; p. 234, Fig. 72.5-6, Plate 22, Fig. 4 (see p. 548, Fig. 116b).

(xii) *Bakuella imbricata* Alekperov, 1982; p. 236, Fig. 72.7, Plate 22, Fig. 5 (see *Bakuella marina*, p. 536, Fig. 114a).

(xiii) *Bakuella marina* Agamaliev et Alekperov, 1976; p. 236, Fig. 73.1 (see p. 536, Fig. 246h).

(xiv) *Pseudobakuella walibonensis* (Song, Wilbert, Berger, 1992); p. 237, Fig. 73.2 (see *Bakuella walibonensis*, p. 581, Fig. 122a). Remarks: In the present monograph this species is classified in the subgenus *Bakuella (Pseudobakuella)*. *Pseudobakuella walibonensis* is a new combination made, but not indicated, by Alekperov (2005).

(xv) *Metabakuella perbella* (Alekperov et Musaev, 1988) Alekperov, 1992; p. 239, Fig. 73.3-4 (see p. 1035, Fig. 203a, b).

Lei et al. (2005a) described a new *Holosticha* species and a Korean population of *H. heterofoissneri* which matches the Chinese populations very well. The new species is briefly reviewed below.

Holosticha hamulata Lei, Xu & Choi, 2005 (Fig. 247a–e)

2005 ***Holosticha hamulata* n. sp.** – Lei, Xu & Choi, J. Euk. Microbiol., 52: 310, 317, Fig. 1–14, 26–35, Table 1 (Fig. 247a–e; original description. The holotype slide [accession number KH-010519-02] and one paratype slide [KH-010324-01] have been deposited in the Marine Biological Specimen Depository of the Chinese Academy of Sciences, Qingdao).

Nomenclature: The species-group name *hamulat·us -a -um* (Latin adjective; armed with small hook) refers to the hook-like structure at the rear body end, a main feature of this species (Lei et al. 2005a).

Remarks: *Holosticha hamulata* differs from the other *Holosticha* species, inter alia, by the long, narrowed posterior body portion. Thus, it is very distinctly tripartite and

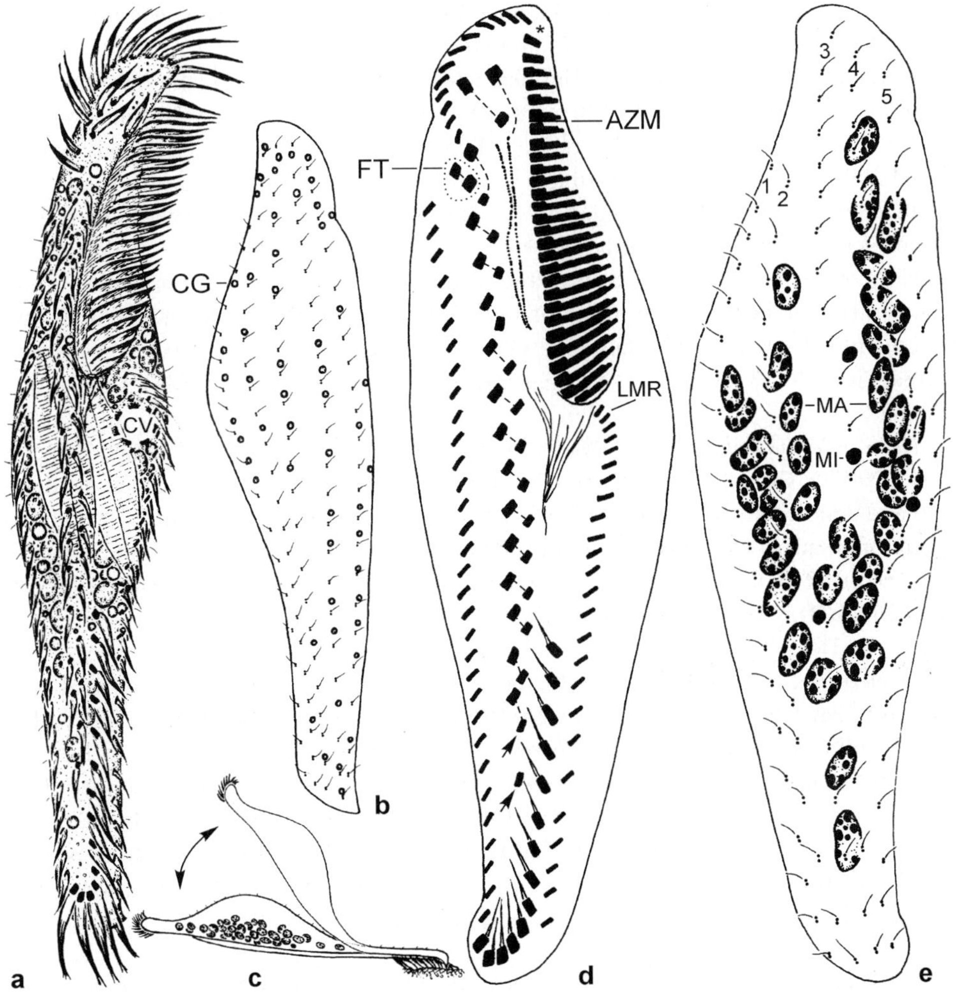

Fig. 247a–e *Holosticha hamulata* (from Lei et al. 2005a. a, c, from life; b, d, e, protargol impregnation). **a:** Ventral view of a representative specimen, 138 µm. **b:** Dorsal view showing dorsal kineties and arrangement of cortical granules (mucocysts?), 103 µm. **c:** Cells are very thigmotactic and moving up and down while attached to the substrate with the posterior body end. **d, e:** Infraciliature of ventral and dorsal side and nuclear apparatus of holotype specimen, 129 µm. Asterisk marks gap in adoral zone, a feature of *Holosticha*. Broken lines connect cirri originating from same anlage (transverse cirri not included). Arrows mark pretransverse ventral cirri. AZM = adoral zone of membranelles, CG = granules (mucocysts?; not observed in life, 0.5–1.0 µm across), CV = contractile vacuole, FT = frontoterminal cirri, LMR = rightwards curved anterior end of left marginal row, MA = macronuclear nodules, MI = micronucleus, 1–5 = dorsal kineties. Page 1218.

easily recognisable. In addition, it is the sole *Holosticha* species with the contractile vacuole ahead of mid-body. Very likely this is due to the tail-like elongation of the posterior body portion. The description below is confined to the comprehensive diagnosis and some relevant figures. For a detailed description, including a morphometric charaterisation of two populations and some micrographs, and another key to *Holosticha* species see the original description of *H. hamulata* (Lei et al. 2005a).

Morphology: The following brief characterisation contains basically only the diagnostic features. Body size 110–180 × 18–30 μm, usually about 150 × 25 μm. Body tripartite with narrow head, inflated trunk, and narrowed, elongated posterior body portion that distally projects ventrally, forming hook-like structure (species name). On average, 33 macronuclear nodules arranged in Y-shaped pattern. Contractile vacuole slightly ahead of mid-body. Cortical granules (mucocysts?) likely present (Fig. 247b). Adoral zone, as is usual for *Holosticha*, bipartite, composed of about 40 membranelles. Three frontal cirri, one buccal cirrus ahead of straight and parallel undulating membranes, two frontoterminal cirri, 12–17 transverse cirri, 26–30 midventral cirri, two pretransverse ventral cirri, 17–28 left and 25–41 right marginal cirri, five dorsal kineties. Caudal cirri lacking.

Holosticha hamulata has a rather conspicuous movement described in detail by Lei et al. (2005a). Crawls very fast on substrate. When disturbed, specimens can suddenly stop and adhere firmly to the substrate, that is, they are highly thigmotactic. Cells may remain motionless for a short period or move up and down by attaching to the substrate with the distal hook and posterior cirri, especially the five posteriormost transverse cirri (Fig. 247c). This behaviour is somewhat reminiscent of *Ancystropodium maupasi* (for review see Berger 1999, p. 778).

Occurrence and ecology: Marine. Type locality of *H. hamulata* are intertidal sediments of Ganghwa Island (37°37'N 126°20'E), Inchon, Korea. This area of the Yellow Sea is nutrient-rich due to domestic sewage, industrial effluents, and waste loadings from an estuary. Sediment temperature ranged from 15–33°C and salinity from 18–33‰; pH was 8.1 (Lei et al. 2005a). *Holosticha hamulata* was highly abundant during diatom blooms in March and May 2001. Lei et al. (2005a) cultured the May population (type) using sediment percolate supplied with squeezed wheat grains to support growth of bacteria and small protozoa. Feeds on diatoms up to 50 μm long (Fig. 247a).

Supplement to the occurrence and ecology section of *Urostyla grandis* (see p. 1048): ponds north of Deuchny Wood, Earn, England (Craigie 1921, p. 119); surroundings of the city of Budapest, Hungary (Krepuska 1917, p. 176); geothermal sulphur spring in Northern Italy (Madoni & Uluhogian 1997, p. 165); alphamesosaprobic site of Parma River, Northern Italy (Madoni 1993, p. 132); Lago Maggiore, Italy (André 1915, p. 108); surroundings of the city of Bologna, Italy (Enriques 1912, p. 106); well near the village of Metkovic, Yugoslavia (Spandl 1926c, p. 26; Karaman 1935, p. 49); plankton of River Plate (rivers Paraná and Uruguay; Carbonell 1935, p. 516). Feeds on testate amoebae like *Euglypha rotunda* and *Trinema lineare* (Chardez 1985, p. 193).

Foissner & AL-Rasheid (2006) found a *Holosticha* sp. in a brackish pond on the coast of the Saudi Arabian Gulf. The low number of transverse cirri and the straight anterior end of the left marginal row indicate that it is not a *Holosticha*.

Hu & Suzuki (2006) described the morphology and cell division of *Pseudoamphisiella alveolata* (see p. 211) in detail. As expected, the morphogenesis proceeds very similar to that of the type species. The population is from the New Fishing Port of Nagasaki, Japan.

Aguirre & Katia (2002) found the following urostyloids in the Wetland Villa, District of Chorrillos, Department of Lima, Peru: *Urostyla caudata*; *Holostricha fasciola* (incorrect subsequent spelling of *Holosticha*); *Holostricha kessleri* (incorrect subsequent spelling); *Trichotaxis rubentis* (species indeterminata); *Trichotaxis villaensis* (species indeterminata).

Azovsky & Mazei (2005, p. 89) found the following urostyloids in the Caucasian coast of the Black Sea: *Holostycha diademata* (incorrect subsequent spelling of *Holosticha*); *Holosticha fasciola*.

Mazei & Burkovsky (2005, p. 112) found *Holosticha extensa* (now *Anteholosticha extensa*) and *H. diademata* in the Chernaya River estuary (Kandalaksha Bay, White Sea). In addition, they provide a total species list of the benthic ciliates of the White Sea.

Chang et al. (2005) made a phylogenetic analysis using the DNA polymerase alpha gene. *Urostyla grandis* and *Holosticha* sp. (similar to *H. kessleri*) do not form a monophyletic group.

Foissner & Stoeck (2006) briefly characterised a highly interesting species from the Niger floodplain in the tropical Africa: body length about 230 μm in life; body outline *Stylonychia*-like, but tailed; body rigid; ventral cirri form distinct zigzag pattern; oral apparatus roughly stylonychine; dorsomarginal kineties present; dorsal kinety fragmentation lacking. The zigzag pattern of the ventral cirri indicates a urostyloid or uroleptid origin, although oxytrichid taxa with such a pattern are known (e.g., *Neokeronopsis*, see present book; *Pattersoniella*, see Berger 1999). However, the presence of dorsomarginal kineties strongly suggests that it is not an urostyloid, but – like the uroleptids – a member of the Dorsomarginalia (see p. 38). According to the rigid body and the stylonychine oral apparatus it *could* even belong to the stylonychines (absence [characteristic for the stylonychines]/presence of cortical granules not mentioned). However, the lack of a dorsal kinety fragmentation would be unusual for a stylonychine, although it cannot be excluded that it was lost during the evolution. The 18S rRNA gene sequence indicates a close relationship with *Oxytricha granulifera*, a position which contradicts all morphological data.

References

Adam H., Czihak G. (1964): Arbeitsmethoden der makroskopischen und mikroskopischen Anatomie. Ein Laboratoriumshandbuch für Biologen, Mediziner und technische Hilfskräfte. Fischer Verlag, Stuttgart. I–XIV, 583 pp.

Adil A.S. (1934): Beitrag zur Hydrofauna der Apscheron-Halbinsel. – Trudy azerb. Otd. zakavk. Fil. Akad. Nauk. SSSR **7**: 3–13 (in Russian with German summary).

Adl S., Simpson A.G.B., Farmer M.A., Andersen R.A., Anderson O.R., Barta J.R., Bowser S.S., Brugerolle G., Fensome R.A., Fredericq S., James T.Y., Karpov S., Kugrens P., Krug J., Lane C.E., Lewis L.A., Lodge J., Lynn D.H., Mann D.G., McCourt R.M., Mendoza L., Moestrup Ø., Mozley-Standridge S.E., Nerad T.A., Shearer C.A., Smirnov A.V., Spiegel F.W., Taylor M.F.J.R. (2005): The new higher level classification of eukaryotes with emphasis on the taxonomy of protists. – J. Euk. Microbiol., 52: 399–451.

Aescht E. (2001): Catalogue of the generic names of ciliates (Protozoa, Ciliophora). – Denisia **1**: 1–350.

Aescht E. (2003): Typen-Liste der Sammlung "Wirbellose Tiere" (ohne Insekten) am Biologiezentrum Linz. – Beitr. Naturk. Oberösterreichs **12**: 377–406.

Aescht E., Foissner W. (1993): Effects of organically enriched magnesite fertilizers on the soil ciliates of a spruce forest. – Pedobiologia **37**: 321–335.

Affa'a F.-M., Hickey D.A., Strüder-Kypke M., Lynn D.H. (2004): Phylogenetic position of species in the genera *Anoplophrya*, *Plagiotoma*, and *Nyctotheroides* (Phylum Ciliophora), endosymbiotic ciliates of annelids and anurans. – J. Euk. Microbiol. **51**: 301–306.

Agamaliev F.G. (1967): Faune des ciliés mésopsammiques de la côte ouest de la mer Caspienne. – Cah. Biol. mar. **8**: 359–402.

Agamaliev F.G. (1967a): Ecology of psammophilous ciliates. – Dokl. Akad. Nauk SSSR **176**: 1425–1427 (in Russian).

Agamaliev F.G. (1970): Vertical distribution of psammophilous ciliates in the Caspian Sea. – Zool. Zh. **49**: 1277–1284 (in Russian with English summary).

Agamaliev F.G. (1971): Complements to the fauna of psammophilic ciliates of the Western coast of the Caspian Sea. – Acta Protozool. **8**: 379–404, Plates I, II (in Russian with English summary).

Agamaliev F.G. (1972): Ciliates from microbenthos of the islands of Apseronskij and Bakinskij archipelagos of the Caspian Sea. – Acta Protozool. **10**: 1–27 (in Russian with English summary).

Agamaliev F.G. (1973): Ciliates of microbenthos in the Krasnovodsk Bay of the Caspian Sea. – Zool. Zh. **52**: 1597–1601 (in Russian with English summary).

Agamaliev F.G. (1973a): Infusoria from Agrakhan Bay, Caspian Sea. – Izv. Akad. Nauk azerb. SSR (Biologicueskie) **year 1973**: 58–63 (in Russian).

Agamaliev F.G. (1974): Ciliates of the solid surface overgrowth of the Caspian Sea. – Acta Protozool. **13**: 53–83, Plates I–IV (in Russian with English summary).

Agamaliev F.G. (1974a): Ciliates of the Turkmen Bay (Caspian Sea). – Zool. Zh. **53**: 19–24 (in Russian with English summary).

Agamaliev F.G. (1978): Morphology of some free-living ciliates of the Caspian Sea. – Acta Protozool. **17**: 419-444, Plates I–IV.

Agamaliev F.G. (1983): Infuzorii Kaspijskogo morja. Sistematika, ekologija, zoogeografija (Ciliates of the Caspian Sea. Systematics, ecology, zoogeography). Nauka, Leningrad (in Russian). 232 pp, Plates I–XX.

Agamaliev F.G. (1986): Ciliates of the low-salinity lagoons of the Caspian Sea. – Arch. Protistenk. **131**: 201–214.

Agamaliev F.G. (1990): The fauna and distribution of ciliates in supralittoral zone of the Apsheron Coast of the Caspian Sea. – Protozoology **13**: 58–63 (in Russian with English summary).

Agamaliev F.G., Alekperov I.K. (1976): A new genus *Bakuella* (Hypotrichida) from the Caspian Sea and the Djeiranbatansky water reservoir. – Zool. Zh. **55**: 128–131 (in Russian with English summary).

Agamaliyev F.G. (1974): Benthic infusoria in the western Caspian Sea. – Hydrobiol. J. **10**: 18–23.

Agamaliyev F.G. (1976): Planctonic and periphytonic ciliophora in the eastern part of the Middle Caspian. – Izv. Akad. Nauk azerb. SSR (Biologicueskie) **year 1976**: 90–94 (in Russian).

Agamaliyev F.G., Aliyev A.R. (1983): Benthic infusoria from the Divichinskiy estuary of the Caspian Sea. – Hydrobiol. J. **18** (years 1982/1983): 20–24.

Agatha S. (2003): Morphology and ontogenesis of *Novistrombidium apsheronicum* nov. comb. and *Strombidium arenicola* (Protozoa, Ciliophora): a comparative light microscopical and SEM study. – Europ. J. Protistol. **39**: 245–266.

Agatha S. (2004): A cladistic approach for the classification of oligotrichid ciliates (Ciliophora: Spirotricha). – Acta Protozool. **43**: 201–217.

Agatha S., Strüder-Kypke M.C., Beran A. (2004): Morphologic and genetic variability in the marine planktonic ciliate *Laboea strobila* Lohmann, 1908 (Ciliophora, Oligotrichia), with notes on its ontogenesis. – J. Euk. Microbiol. **51**: 267–281.

Aguirre G., Katia G. (2002): Diversidad Protozoológica de los Pantanos de Villa, Chorrillos – Lima - Perú. – Dissertation Universidad Nacional Mayor de San Marcos, Lima, Perú. 143 pp.

Aladro Lubel M.A. (1985): Algunos ciliados intersticiales de Isla de Enmedio, Veracruz, México. – An. Inst. Biol. Univ. Méx., Ser. Zoologia (1) **55 (year 1984)**: 1–59, Lamina 1–17.

Aladro-Lubel M.A., Martín-Murilo M.E., Mayén-Estrado R., Hernández- Anaya M., Sánchez-Calderón G. (1986): Cuarenta y nueve especies de ciliados intersticiales identificados en Boca del Río, Veracruz, México. – Revta lat.-am. Microbiol. **28**: 231–242.

Aladro-Lubel M.A., Martínez-Murillo M.E., Mayén-Estrada R. (1988): Lista de los ciliados bentonicos salobres y marinos registrados en Mexico. – An. Inst. Biol. Univ. Méx., Ser. Zoologia (1) **58** (year 1987): 403-447.

Aladro Lubel M.A., Martínez Murillo M.E., Mayén Estrada R. (1990): Manual de ciliados psamofilos marinos y salobres de Mexico. Cuadernos del Instituto de Biologia 9. Universidad Nacional Autonoma de Mexico. 174 pp.

Albrecht J. (1983): Salzbelastung und Ciliatenbesiedlung (Protozoa: Ciliophora) im Weser-Flußsystem (Fulda, Werra, Weser, Leine, Innerste). Dissertation Rheinischen Friedrich- Wilhelms-Universität, Bonn. 195 pp.

Albrecht J. (1984): Zur Autökologie ausgewählter Aufwuchsciliaten des Weser-Flußsystems (Protozoa: Ciliophora). – Decheniana **137**: 132–167.

Albrecht J. (1986): Periphyton (Aufwuchs) communities of ciliated protozoa in salt-polluted running waters of the Weser River Basin - Their structure and indicator value (including model ecosystems). – Int. Revue ges. Hydrobiol. Hydrogr. **71**: 187–224.

Alekperov I.K. (1981): Planktonic infusoria as indicators of the extent of organic pollution in Azerbaidzhanian reservoirs. – Gidrobiol. Zh. **17**: 54–60 (in Russian with English summary. Translation of the title from Zoological Record 1981; I suppose that Alekperov I.K., Alekperov I.H., and Alekperov I.C. is the same author).

Alekperov I.K. (1982): Planktonic infusoria as indicators of the extent of organic pollution in Azerbaidzhanian reservoirs. – Hydrobiol. J. **17** (years 1981/1982): 43–49.

Alekperov I.K. (1982a): Infusoria of the very deep Terterchay Reservoir. – Gidrobiol. Zh. **18**: 83–89 (in Russian with English summery. Translation of the title from Zoological Record 1982).

Alekperov I.H. (1982b): *Bakuella imbricata* sp. n. (Ciliophora, Hypotrichida) from the Djeiranbatansky water reservoir. – Zool. Zh. **61**: 1253–1255 (in Russian with English summary).

Alekperov I.K. (1983): Infusoria of the very deep Terterchay Reservoir. – Hydrobiol. J. **18** (years 1982/1983): 20–25.

Alekperov I.K. (1984): New freshwater ciliates (Hypotrichida) from water bodies of Azerbaijan. – Zool. Zh. **63**: 1458–1463 (in Russian with English summary).

Alekperov I.K. (1984a): Free-living infusoria of Khachinchay Reservoir. – Gidrobiol. Zh. **20**: 17–22 (in Russian with English summary. Translation of the title from Zool. Rec. 1984).

Alekperov I.K. (1984b): Free-living infusoria of Khachinchay Reservoir. – Hydrobiol. J. **20**: 17–22.

Alekperov I.K. (1986): Comparative analysis of the faunal communities of freshwater infusoria from Azerbaidjan with other regions of the world. – Izv. Akad. Nauk azerb. SSR (Biologicueskie) **year 1986**: 53–59 (in Russian).

Alekperov I.K. (1988): Two new species of infusoria (Ciliophora, Hypotrichida) from fresh waters of Azerbaijan. – Zool. Zh. **67**: 777–780 (in Russian with English summary).

Alekperov I.K. (1989): The revision of the *Bakuella* Agamaliev et Alekperov 1976 and *Keronella* Wiackowski 1985 genera (Hypotrichida, Ciliophora). In: Poljansky G.I., Zhukov B.F., Raikov I.B. (eds.). Ecology of

Bergh R.S. (1889): Recherches sur les noyaux de l'*Urostyla grandis* et de l'*Urostyla intermedia* n. sp. – Archs Biol. **9**: 497–514, Planche XXXV.

Bernard C., Fenchel T. (1995): Mats of colourless sulphur bacteria. II. Structure, composition of biota and successional patterns. – Mar. Ecol. Prog. Ser. **128**: 171–179.

Bernerth H. (1982): Ökologische Untersuchungen im Kühlwassersystem eines konventionellen Großkraftwerks am Untermain unter besonderer Berücksichtigung der Ciliaten (Protozoa). – Cour. Forsch.-Inst. Senckenberg **57**: 1–246.

Bernhard D., Schlegel M. (1998): Evolution of histone H4 and H3 genes in different ciliate lineages. – J. molec. Evolut. **46**: 344–354.

Bernhard D., Stechmann A., Foissner W., Ammermann D., Hehn M., Schlegel M. (2001): Phylogeneti relationships within the class Spirotricha (Ciliophora) inferred from small subunit rRNA genes sequences. – Molecular Phylogenetics and Evolution **21**: 86–92.

Bervoets W. (1940): Bijdrage tot de studie der ciliaten van Belgie. – Biol. Jaarb. **7**: 117–137.

Bevel N. (1938): Some notes on the protozoa & of Reelfoot Lake. – J. Tenn. Acad. Sci. **13**: 137–159, Plates I–III.

Bhatia B.L. (1936): Protozoa: Ciliophora. In: Sewell R.B.S. (ed.). The fauna of British India, including Ceylon and Burma, Taylor and Francis, London. XXII & 493 pp & Plates I–XI, 1 map.

Bick H. (1967): Vergleichende Untersuchung der Ciliatensukzession beim Abbau von Zellstoff in marinem Brackwasser und athalassogenem Brackwasser vom Typ der Kaliwerksabwässer. – Hydrobiologia **29**: 185–204.

Bick H. (1968): Untersuchungen zur Verträglichkeit von Meer- und Brackwasser für Ciliaten des Saprobiensystems der Wassergütebeurteilung. – Helgoländer wiss. Meeresunters. **17**: 257–268.

Bick H. (1972): Ciliated protozoa. An illustrated guide to the species used as biological indicators in freshwater biology. World Health Organization, Geneva. 198 pp.

Bick H., Buitkamp U. (1976): Ciliated protozoa from Canadian grassland soils. – Trans. Am. microsc. Soc. **95**: 490–491.

Bick H., Drews E.F. (1973): Selbstreinigung und Ciliatenbesiedlung in saurem Milieu (Modellversuche). – Hydrobiologia **42**: 393–402.

Biernacka I. (1962): Die Protozoenfauna in der Danziger Bucht I. Die Protozoen in einigen Biotopen der Seeküste. – Polskie Archwm Hydrobiol. **10**: 39–109.

Biernacka I. (1963): Die Protozoenfauna in Danziger Bucht II. Die Charakteristik der Protozoen in untersuchten Biotopen der Seekuste. – Polskie Archwm Hydrobiol. **11**: 17–75.

Biernacka I. (1967): Einige Protozoa der Uferzone der Insel Hiddensee. – Wiss. Z. Ernst Moritz Arndt-Univ. Greifswald, Mathematisch-naturwissenschaftliche Reihe **16**: 241–248.

Blatterer H. (1989): Uni-Teich. Weitere Faunistik: Ciliaten (Wimpertiere). – BUFUS-Info **Nr. 5**: 7–10 (ISSN 0256-4173; obviously published just 1990 since the last collection was made on 09.01.1990).

Blatterer H. (1994): Die Ciliaten oberösterreichischer Fliessgewässer mit besonderer Berücksichtigung der südlichen Inn-Zubringer. – Kataloge des O. Ö. Landesmuseums Linz, Neue Folge **71**: 149–163.

Blatterer H. (1995): Verbessertes Verfahren zur Berechnung des Saprobienindex mittels Ciliaten (Ciliophora, Protozoa). – Lauterbornia **20**: 23–36.

Blatterer H., Foissner W. (1988): Beitrag zur terricolen Ciliatenfauna (Protozoa: Ciliophora) Australiens. – Stapfia **17**: 1–84.

Blochmann F. (1886): Die mikroskopische Thierwelt des Süsswassers. Haering, Braunschweig. 122 pp & Tafeln I–VII.

Blochmann F. (1895): Die mikroskopische Thierwelt des Süsswassers. Abtheilung I: Protozoa. Lucas Gräfe & Sillem, Hamburg. 184 pp & Tafeln I–VIII.

Bock K.J. (1952): Zur Ökologie der Ciliaten des marinen Sandgrundes der Kieler Bucht I. – Kieler Meeresforsch. **9**: 77–89.

Bock K.J. (1953): Zur Oekologie der Ciliaten des marinen Sandgrundes der Kieler Bucht II. – Kieler Meeresforsch. **9**: 252–256.

Bock K.J. (1960): Biologische Untersuchungen, insbesondere der Ciliatenfauna, in der durch Abwässer belasteten Schlei (westliche Ostsee). – Kieler Meeresforsch. **16**: 57–68.

Bocquet C. (1971): Espèces nouvelles décrites de la region de Roscoff entre 1945 et 1970. – Cah. Biol. mar. **12**: 381–404.

Boikova E.E. (1984): Benthos infusoria in the coastal area of the Gulf of Riga. Hydrobiology of the Gulf of Riga, pp. 120–127. Zinátne, Riga (in Russian with English summary on p. 8).

Bonkowski M. (1996): Protozoen und Lumbriciden in einem Kalkbuchenwald: Untersuchungen zur Interaktion von Populationen der Bodenfauna und ihrer Wirkung auf Stoffdynamik und Pflanzenwachstum. – Berichte des Forschungszentrums Waldökosysteme, Reihe A **134**: I–VI, 1–133.

Borror A.C. (1962): Ciliate protozoa of the Gulf of Mexico. – Bull. mar. Sci. Gulf Caribb. **12**: 333–349.

Borror A.C. (1963): Morphology and ecology of some uncommon ciliates from Alligator Harbor, Florida. – Trans. Am. microsc. Soc. **82**: 125–131.

Borror A.C. (1963a): Morphology and ecology of the benthic ciliated protozoa of Alligator Harbor, Florida. – Arch. Protistenk. **106**: 465–534.

Borror A.C. (1972): Revision of the order Hypotrichida (Ciliophora, Protozoa). – J. Protozool. **19**: 1–23.

Borror A. (1972a): Tidal marsh ciliates (Protozoa): morphology, ecology, systematics. – Acta Protozool. **10**: 29–71, Plate I.

Borror A.C. (1978): Morphogenesis of *Urostyla marina* Kahl, 1932; redefinition of the family Urostylidae (Ciliophora, Hypotrichida). – J. Protozool. **25**: 10A, Abstract 13.

Borror A.C. (1979): Redefinition of the Urostylidae (Ciliophora, Hypotrichida) on the basis of morphogenetic characters. – J. Protozool. **26**: 544–550.

Borror A.C. (1980): Spatial distribution of marine ciliates: micro- ecologic and biogeographic aspects of protozoan ecology. – J. Protozool. **27**: 10–13.

Borror A.C., Evans F.R. (1979): Cladotricha and phylogeny in the suborder Stichotrichina (Ciliophora, Hypotrichida). – J. Protozool. **26**: 51–55.

Borror A.C., Hill B.F. (1995): The order Euplotida (Ciliophora): Taxonomy, with division of *Euplotes* into several genera. – J. Euk. Microbiol. **42**: 457–466.

Borror A.C., Wicklow B.J. (1982): Non-homology of median rows in hypotrichs with only three longitudinal rows of cirri. – J. Protozool. Suppl. **29**: 285A, Abstract.

Borror A.C., Wicklow B.J. (1983): The suborder Urostylina Jankowski (Ciliophora, Hypotrichida): morphology, systematics and identification of species. – Acta Protozool. **22**: 97–126, Plate I.

Body de Saint-Vincent J.B. (1824): Essai d'une classification des animaux microscopiques. Agasse, Paris. 104 pp.

Bourquin-Lindt E. (1919): Contribution à l'étude des protozoaires de la vallée de La Chaux-de-Fonds. – Bull. Soc. neuchâtel. Sci. nat. **43** (years 1917–1918): 39–95.

Boutchinsky P. (1895): Protozoaires trouvés dans les lacs salés (limans) de Khadgibei et de Kouialnik. – Zap. novoross. Obshch. Estest. **20**: 137–148 (in Russian).

Bouvier E.-L. (1893): La chlorophylle animale et les phénomènes de symbiose entre les algues vertes unicellulaires et les animaux. – Bull. Soc. philomat. Paris, 8. Série **5** (years 1892–1893): 72–149.

Bovee E.C. (1960): Protozoa of the Mountain Lake Region, Giles County, Virginia. – J. Protozool. **7**: 352–361.

Bovee E.C. (1979): Protozoa from acid-bog mosses and forest mosses of the Lake Itasca Region (Minnesota, USA). – Kans. Univ. Sci. Bull. **51**: 615–629, Plates I–V.

Bragg A.N. (1960): An ecological study of the protozoa of Crystal Lake, Norman, Oklahoma. – Wasmann J. Biol. **18**: 37–85.

Brunberg Nielsen L. (1968): Investigations on the microfauna of leaf litter in a Danish beech forest. – Natura jutl. **14**: 79–87.

Bryant V.M.T., Laybourn J.E.M. (1974): The vertical distribution of Ciliophora and Nematoda in the sediments of Loch Leven, Kinross. – Proc. R. Soc. Edinb., Section B (Biology) **74**: 265–273.

Buchar J. (1957): Fauna nálevniku dolního úseku potoka botice. – Cas. národ. Mus. **126**: 137–143 (in Czech with German summary).

Buck H. (1961): Zur Verbreitung der Ciliaten in den Fließgewässern Nordwürttembergs. – Jh. Ver. vaterl. Naturk. Württ. **116**: 195–217.

Buitkamp U. (1977): Die Ciliatenfauna der Savanne von Lamto (Elfenbeinküste). – Acta Protozool. **16**: 249–276.

Buitkamp U. (1977a): Über die Ciliatenfauna zweier mitteleuropäischer Bodenstandorte (Protozoa; Ciliata). – Decheniana **130**: 114–126.

Buitkamp U. (1979): Vergleichende Untersuchungen zur Temperaturadaptation von Bodenciliaten aus klimatisch verschiedenen Regionen. – Pedobiologia **19**: 221–236.

Buitkamp U., Wilbert N. (1974): Morphologie und Taxonomie einiger Ciliaten eines kanadischen Präriebodens. – Acta Protozool. **13**: 201–210.

Bullington W.E. (1925): A study of spiral movement in the ciliate infusoria. – Arch. Protistenk. **50**: 219–274.

Bullington W.E. (1935): Morphology and taxonomy of ciliata. – Yb. Carnegie Instn Wash. **34**: 77–78, Abstract.

Bullington W.E. (1940): Some ciliates from Tortugas. – Pap. Tortugas Lab **32**: 179–221, Plate I.

Bürger O. (1905): Estudios sobre protozoos chilenos del agua dulce. – An. Univ. Chile **117**: 403–449, Lámina I–X.

Bürger O. (1908): Nuevos estudios sobre protozoos chilenos del agua dulce. – An. Univ. Chile **122**: 137–204, Lámina 1–15.

Burkovsky I.V. (1970a): Ciliates of the sand littoral and sublittoral of Kandalaksha Gulf (White Sea) and the analysis on the fauna of benthic ciliates of other seas. – Acta Protozool. **8**: 183–201 (in Russian with English summary).

Burkovsky I.V. (1970b): Psammophilous infusoria of the Kanadalakshskiy Bay (White Sea). Dissertation, Moscow University Publisher. 21 pp. (in Russian).

Burkovsky I.V. (1970c): Ciliates of the mesopsammon of the littoral and sublittoral areas in the Bay of Kandalaksha. – Trudy Belomorskoi Biologicheskoi Stantsii. M. G. U. **3**: 51–59 (in Russian).

Burkovsky I.V. (1970d): The ciliates of the mesopsammon of the Kandalaksha Gulf (White Sea). II. – Acta Protozool. **8**: 47–65 (in Russian with English summary).

Burkovsky I.V. (1971): Ciliates of the rocky supralittoral zone in the White Sea. – Zool. Zh. **50**: 1569–1572 (in Russian with English summary).

Burkovsky I.V. (1971a): A comparative study of the eology of free-living ciliates in the Rugozersky inlet (Kandalaksha Bay, White Sea). – Zool. Zh. **50**: 1773–1779 (in Russian with English summary).

Burkovsky I.V. (1976): Ciliates of the freshened sites of the White Sea. – Zool. Zh. **55**: 287–289 (in Russian with English summary).

Burkovsky I.V. (1984): Ekologiya svobodnozivushchikh infuzorij (Ecology of free-living ciliates). Moscow University Press, Moscow (in Russian. Translation of title from the book review of Seravin L.N., Raikov I.B., 1985, Arch. Protistenk. 129: 293–294). 208 pp.

Bussers J.C., Jeuniaux C. (1974): Recherche de la chitine dans les productions métaplasmatiques de quelques ciliés. – Protistologica **10**: 43–46.

Butschinsky P. (1897): Die Protozoen-Fauna der Salzsee-Limane bei Odessa. – Zool. Anz. **20**: 194–197.

Bütschli O. (1873): Einiges über Infusorien. – Arch. mikrosk. Anat. EntwMech. **9**: 657–678, Tafeln XXV, XXVI.

Bütschli O. (1876): Studien über die ersten Entwicklungsvorgänge der Eizelle, die Zelltheilung und die Conjugation der Infusorien. – Abh. senckenb. naturforsch. Ges. **10**: 213–457, Tafeln I–XV.

Bütschli O. (1889): Protozoa. III. Abtheilung: Infusoria und System der Radiolaria. In: Bronn H.G. (ed.). Klassen und Ordnungen des Thier-Reichs, wissenschaftlich dargestellt in Wort und Bild, Vol. **I**, pp. 1585–2035. Winter, Leipzig. Tafeln LVI–LXXIX.

Cairns J.Jr. (1966): The Catherwood Foundation Peruvian-Amazon Expedition III - Protozoa. – Monogr. Acad. nat. Sci. Philad. **14**: 53–61.

Cairns J.Jr., Henebry M.S. (1982): Interactive and noninteractive protozoan colonization processes. In: Cairns J.Jr. (ed.). Artificial substrates, pp. 23–70. Ann Arbor Science Publishers Inc., Ann Arbor.

Cairns J.Jr., Plafkin J.L. (1975): Response of protozoan communities exposed to chlorine stress. – Arch. Protistenk. **117**: 47–53.

Cairns J.Jr., Yongue W.H.Jr. (1973): A comparison of the protozoan communities in a coastal plain river through space and time. – Revta Biol., Lisb. **9**: 15–34.

Cairns J.Jr., Yongue W.H.Jr., Boatin H.Jr. (1973): The protozoan colonization of polyurethane foam units anchored in the benthic area of Douglas Lake, Michigan. – Trans. Am. microsc. Soc. **92**: 648–656.

Calkins G.N. (1902): Marine protozoa from Woods Hole. – Bull. U. S. Fish Commn **21** (year 1901): 415–468.

Calkins G.N. (1909): Protozoölogy. Lea & Febiger, New York, Philadelphia. 349 pp.

Calkins G.N. (1926): The biology of the Protozoa. Lea & Febiger, Philadelphia, New York. 623 pp.

Calloni S. (1890): La fauna nivale con particolare riguardo ai viventi delle alte alpi. Pavia. 478 pp, I–XX.

Calvo P., Torres A., Fedriani C., Rios R.M., Perez Silva J. (1986): Ultrastructure chez *Histriculus similis* (Cilié hypotriche). – Acta Protozool. **25**: 23–32, Planches I–IV.

Canella M.F. (1954): Ricerche sulla microfauna delle acque interne Ferraresi. Introduzione allo studio dei ciliati e dei rotiferi. – Pubbl. civ. Mus. Stor. nat. Ferrara **4**: 1–154, Tavola I–XLV.

Carbonell J.J. (1935): Some micrographic observations of the waters of the River Plate. – Verh. Int. Ver. theor. angew. Limnol. **7**: 513–516.

Carey, P.G. (1992): Marine interstitial ciliates. An illustrated key. Chapman and Hall, London, New York, Tokyo, Melbourne, Madras. 351 pp.

Carey P.G., Maeda M. (1985): Horizontal distribution of psammophilic ciliates in fine sediments of the Chichester harbour area. – J. nat. Hist. (London) **19**: 555–574.

Carey P.G., Tatchell E.C. (1983): A revision of the genus *Epiclintes* (Ciliophora: Hypotrichida) including a redescription of *Epiclintes felis* comb. n. – Bull. Br. Mus. nat. Hist. (Zool.) **45**: 41–54.

Cattaneo G. (1882): Sui protisti del Lago di Como. – Boll. scient., Pavia **3**: 111–121.

Cattaneo G. (1888): Note sui protozoi lacustri. – Boll. scient., Pavia **10**: 90–100.

Cattaneo G. (1889): Notes sur les protozoaires lacustres. – Journal de Micrographie **13**: 88–93, 115–120.

[CBE] Council of Biology Editors (1996): Scientific Style and format. The CBE manual for authors, editors, and publishers. 6th edition. Cambridge University Press, Cambridge. 825 pp.

Chang W.-J., Bryson P., Landweber L. (2003): Searching for the origin of ciliate scrambled genes: hints from several DNA polymerase alpha genes. – International Conference on the Molecular Biology and Biotechnology of Ciliates and Anaerobic Protozoa in connection with 22nd Annual meeting of the German Society for Protozoology, Nijmegen (March 4–8, 2003).

Chang W.-J., Stover N.A., Addis V.M., Landweber L.F. (2004): A micronuclear locus containing three protein-coding genes remains linked during macronuclear development in the spirotrichous ciliate *Holosticha*. – Protist **155**: 245–255.

Chang W.-J., Bryson P.D., H. Liang, M. K. Shin, Landweber L.F. (2005): The evolutionary origin of a complex scrambled gene. – Proc. nat. Acad. Sci. USA, **102**: 15149–15154.

Chaouite J., Groliere C.A., Pepin D. (1990): Etude du peuplement infusorien des eaux minérales et thermominérales en Auvergne. – J. Protozool. Suppl. **37**: 31A, Abstract 149.

Chardez D. (1956): Quelques infusoires ciliés observés dans la région d'Ambleteuse (Pas-de-Calais). – Revue verviét. Hist. nat. **13**: 91–92.

Chardez D. (1981): Les protozoaires cilies des fagnes. – Revue verviét. Hist. nat. **38**: 51–58.

Chardez D. (1984): Protozoaires des eaux de la Vesdre a Verviers (Sarcomastigophora et Ciliophora). – Revue verviét. Hist. nat. **41**: 18–22.

Chardez D. (1985): Protozoaires prédateurs de thécamoebiens. – Protistologica **21**: 187–194.

Chardez D. (1986): Protozoaires cilies de la Vesdre (Protozoa Ciliophora). – Revue verviét. Hist. nat. **43**: 19–24.

Chardez D. (1987): Catalogue des protozoaires ciliés de Belgique (Protozoa Ciliophora). – Notes Fauniques de Gembloux **14**: 1–16.

Chatton E., Séguéla J. (1940): La continuité génétique des formations ciliaires chez les ciliés hypotriches. Le cinétome et l'argyrome au cours de la division. – Bull. biol. Fr. Belg. **74**: 349–442.

Chen Z., Song W. (2001): Phylogenetic positions of *Uronychia transfuga* and *Diophrys appendiculata* (Euplotida, Hypotrichia, Ciliophora) within hypotrichous ciliates inferred from the small subunit ribosomal RNA genes sequences. – Europ. J. Protistol. **37**: 291–301.

Chen Z., Song W. (2002): Phylogenetic positions of *Aspidisca steini* and *Euplotes vannus* within the order Euplotida (Hypotrichia: Ciliophora) inferred from complete small subunit ribosomal RNA gene sequences. – Acta Protozool. **41**: 1–9.

Child C.M. (1949): A further study of indicator patterns in ciliate protozoa. – J. exp. Zool. **111**: 315–347.

Chorik F.P. (1968): Free-living ciliates in Moldavian water basins. Akad. Nauk MSSR, Kischinev (in Russian). 251 pp.

Chorik F.P., Vikol M.M. (1973): Bottom-dwelling, free-living infusoria of the Kuchurgansk cooling plant of the Moldavian F. R. E. S.. – Biologicheskie Resursy Vodoenoy Moldavii **11**: 56–72 (in Russian).

Claparède É., Lachmann J. (1858): Études sur les infusoires et les rhizopodes. – Mém. Inst. natn. génev. **5** (year 1857): 1–260, Planches 1–13.

Claparède É., Lachmann J. (1859): Études sur les infusoires et les rhizopodes. – Mém. Inst. natn. génev. **6** (year 1858): 261–482, Planches 14–24.

Cleven E.-J. (2004): Seasonal and spatial distribution of ciliates in the sandy hyporheic zone of a lowland stream. – Europ. J. Protistol. **40**: 71–84.

Cohn F. (1851): Beiträge zur Entwickelungsgeschichte der Infusorien. – Z. wiss. Zool. **3**: 257–279, Tafel VII.

Cohn F. (1866): Neue Infusorien im Seeaquarium. – Z. wiss. Zool. **16**: 253–302, Tafeln XIV, XV.

Cole T. (1853): List of infusorial objects, found chiefly in the neighborhood of Salem, Mass., with a sketch of the progress in this branch of natural history. – Proceedings of the Essex Institute **1**: 33–48.

Coleman A.W. (2005): *Paramecium aurelia* revisited. – J. Euk. Microbiol. **52**: 68–77.

Collins K. (1999): Ciliate telomerase biochemistry. – A. Rev. Biochem. **68**: 187–218.

Conn H.W. (1905): A preliminary report on the protozoa of the fresh waters of Connecticut. – Bull. Conn. St. geol. nat. Hist. Surv. **No. 2**: 5–69, Plates I–XXXIV.

Cooley N.R., Keltner J.M.Jr. (1970): Effects of pesticides on estuarine ciliates. – Circ. Fish Wildl. Serv., Wash. **No. 335**: 16–18.

Coppellotti O., Matarazzo P. (2000): Ciliate colonization of artificial substrates in the Lagoon of Venice. – J. mar. biol. Ass. U. K. **80**: 419–427.

Coppellotti O., Guidolin L. (1999): "Neglected" microscopic organisms (Protozoa): possible bioindicators for karst groundwaters, pp. 73–78. Atti Tavola Rotonda Un importante sistema carsico dei Monti Lessini (VR): i Covoli di Velo, Verona-Composilvano.

Coppellotti O., Matarazzo P. (2000): Ciliate colonization of artificial substrates in the Lagoon of Venice. – J. mar. biol. Ass. U. K. **80**: 419–427.

Corliss J.O. (1951): On nomenclature for congeneric ciliates in the *Colpidium-Glaucoma-Leucophrys-Tetrahymena* group. – Soc. Syst. Zool. NewsLetterno. 5: 9–11 (abstract).

Corliss J.O. (1960): The problem of homonyms among generic names of ciliated protozoa, with proposal of several new names. – J. Protozool. **7**: 269–278.

Corliss J.O. (1961): The ciliated protozoa: characterization, classification, and guide to the literature. In: Kerkut G.A. (ed.). International series of monographs on pure and applied biology, Zoology division, Vol. 7, Pergamon Press, Oxford, London, New York, Paris. 310 pp.

Corliss J.O. (1973): History, taxonomy, ecology, and evolution of species of *Tetrahymena*. In: Elliott A.M. (ed.). Biology of *Tetrahymena*, Vol. 1, pp. 1–55. Dowden Hutchinson & Ross, Stroudsburg.

Corliss J.O. (1975): Taxonomic characterization of the suprafamilial groups in a revision of recently proposed schemes of classification for the phylum Ciliophora. – Trans. Am. microsc. Soc. **94**: 224–267.

Corliss J.O. (1977): Annotated assignment of families and genera to the orders and classes currently comprising the Corlissian scheme of higher classification for the phylum Ciliophora. – Trans. Am. microsc. Soc. **96**: 104–140.

Corliss J.O. (1979). The ciliated protozoa. Characterization, classification and guide to the literature. Pergamon Press, Oxford, New York, Toronto, Sydney, Paris, Frakfurt. I–XVI, 455 pp.

Corliss J.O. (1994): An interim utilitarian ("user-friendly") hierarchical classification and characterization of the protists. – Acta Protozool. **33**: 1–51.

Corliss J.O., Dougherty E.C. (1967): An appeal for stabilization of certain names in the protozoan family Tetrahymenidae (Subphylum Ciliophora, Order Hymenostomatida), with special reference to the generic name *Tetrahymena* Furgason, 1940. – Bull. zool. Nom. **24**: 155–185.

Corliss J.O., Esser S.C. (1974): Comments on the role of the cyst in the life cycle and survival of free-living protozoa. – Trans. Am. microsc. Soc. **93**: 578–593.

Corliss J.O., Lom J. (1985): An annotated glossary of protozoological terms. In: Lee J.J., Hutner S.H., Bovee E.C. (eds.). An illustrated guide to the protozoa, pp. 576–602. Society of Protozoologists, Lawrence.

Corliss J.O., Lom J. (2002): An annotated glossary of protozoological terms. In: Lee J.J., Leedale G.F., Bradbury P. (eds.). An illustrated guide to the protozoa, second edition, pp. 1346–1385. Society of Protozoologists, Lawrence.

Corliss J.O., Snyder R.A. (1986): A preliminary description of several new ciliates from the Antarctica, including *Cohnilembus grassei* n. sp. – Protistologica **22**: 39–46.

Craigie J. (1921): The ciliata. – Trans. Proc. Perthsh. Soc. nat. Sci. **7**: 107–121.

Croft K.E., Dalby A.B., Hogan D.J., Orr K.E., Hewitt E.A., Africa R.J., DuBois M.L., Prescott D.M. (2003): Macronuclear molecules encoding actins in spirotrichs. – J. molec. Evolut. **56**: 341–350.

Cuneo G. (1891): Ricerche sui protisti delle acque di Rapallo. – Boll. scient., Pavia **12**: 140–149.

Cunha A.M.da (1913): Contribuiçao para o conhecimento da fauna de protozoarios do Brazil (Beitraege zur Kenntnis der Protozoenfauna Brasiliens). – Mems Inst. Oswaldo Cruz **5**: 101–122, Estampas 9, 10 (in Portuguese and German).

Curds, C.R., Gates, M.A., Roberts, D.M. (1983): British and other freshwater ciliated protozoa Part II Ciliophora: Oligohymenophora and Polyhymenophora. Keys and notes for the identification of the free-living genera. In: Kermack, D.M., Barnes, R.S.K. (eds.). Synopses of the British fauna, No. 23, pp. 1–474. Cambridge University Press, Cambridge, London, New York, New Rochelle, Melbourne, Sydney.

Curtis E.A., Landweber L.F. (1999): Evolution of gene scrambling in ciliate micronuclear genes. – Ann. N. Y. Acad. Sci. **870**: 349–350.

Czapik A. (1959): Microbenthos of fish-ponds in Mydlniki. – Acta hydrobiol., Kraków **1**: 187–196.

Czapik A. (1975): Les associations des ciliés (Ciliata) dans le ruisseau Pradnik pollué par les eaux residuelles d'une laiterie. – Acta hydrobiol., Kraków **17**: 21–34.

Czapik A. (1982): The effect of waste water on ciliate communities in the Biala Przemsza River. – Acta hydrobiol., Kraków **24**: 29–37.

Czapik A., Fyda J. (1992): Contribution à la connaissance des ciliés psammophiles de la Baltique. – Acta Protozool. **31**: 109–114.

Czapik A., Fyda J. (1995): Wsteppe badania nad mikro- i meiofauna sucharów Wigierskiego Parku Narodowego (Micro- and meiofauna of dystrophic lakes in the Wigry National Park). – Przegl. zool. **39**: 65–72.

Czapik A., Jordan A. (1976): Les ciliés psammophiles de la mer Baltique aux environs de Gdansk. – Acta Protozool. **15**: 423–445, Planches I–IV.

Dagajeva V. (1930): Infusorien des Salzigen Sees der Bucht Kruglaja in der Nähe von Sebastopol (Krim). – Trudy sevastopol'. biol. Sta. **2**: 31–46 (in Russian with German summary).

Dalby A.B., Prescott D.M. (2004): The scrambled actin I gene in *Uroleptus pisces*. – Chromosoma **112**: 247–254.

Dallas W.S. (1851): List of the specimens of hemipterous insects in the collection of the British Museum, part 1. British Museum Trustees, London. 368 pp, 11 plates.

Dalla Torre K.W.v. (1891): Studien über die mikroskopische Thierwelt Tirols. III. Theil: Infusoria Ciliata und Tentaculifera Huxley. – Z. Ferdinand. Tirol, 3. Folge **35. Heft**: 191–209.

Darbyshire J.F. (1994): Soil protozoa. CAB International, Wallingford. I–X, 209 pp.

Delphy J. (1938): Études de morphologie et de physiologie sur la faune d'Arcachon. – Bull. Stn. biol. Arcachon **35**: 49–75.

Dembowska W.S. (1926): Studies on the regeneration of protozoa II. Regeneration of the ciliary apparatus in some marine hypotricha. – J. exp. Zool. **43**: 485–504.

De Morgan W. (1926): Further observations on marine ciliates living in the laboratory tanks at Plymouth. – J. mar. biol. Ass. U. K. **14**: 23–53.

Detcheva R.B. (1986): Composition and distribution of the ciliates (Protozoa, Ciliata) from the Tundza River. – Hydrobiology **28**: 61–65 (in Russian with English summary).

Detcheva R. (1977): Nouveaux ciliés de la mer Noire près de la côte bulgare. – Acta Zool. Bulgarica **8**: 3–6.

Detcheva R. (1979): Übersicht über die Ciliatenfauna der bulgarischen Donauzuflüsse. – Arbeitstagung Internacionale Arbeitsgemeinschaft Donauforschung der Societas Internationalis Limnologiae **21**: 357–364.

Detcheva R. (1979a): Repartition des cilies de la riviere Strouma par rapport a certains facteurs abiotiques. – Annls Stn limnol. Besse **13**: 236–246, Tableau 1–3.

Detcheva R.B. (1980): Composition et répartition des ciliés mésopsammiques des plages bulgares. – Hydrobiology **11**: 28–38.

Detcheva R.B. (1982): Recherches, en Bulgarie, sur les cilies de certaines plages de la Mer Noire et des rivieres y aboutissant. – Annls Stn limnol. Besse **15** (years 1981/1982): 231–253.

Detcheva R.B. (1983): Contribution sur la faune des ciliés mesopsammiques de quelques plages bulgares et celles des rivières qui se jettent dans la mer. – Hydrobiology **18**: 64–76.

Detcheva R.B. (1992): Catalogi Faunae Bulgaricae 1 Protozoa, Ciliophora. Academia Scientiarum Bulgaricae, Sofia. 134 pp.

Detcheva R.B. (1993): Les données ßecologiques pour les ciliés (Protozoa – Ciliophora) du bassin de la rivièra Iskar. – Hydrobiology **38**: 33–44 (in Russian with French summary).

Detschewa R.B. (1972): Beitrag zur Kenntnis der Infusorienfauna (Protozoa-Ciliata) in den Binnengewässern Bulgariens. – Izv. zool. Inst., Sof. **36**: 61–79.

Diaz Pérez P.A., Montoto Lima A. (1989): Clasificación saprobiótica de un estuario cubano altamente contaminado. – Revista de Investigaciones Marinas **10**: 89–98.

Dieckmann J. (1988): Infraciliature and morphogenesis of *Paraholosticha sterkii* (Garnjobst, 1934) n. comb. (Ciliophora, Hypotrichida). – Europ. J. Protistol. **23**: 218–228.

Dieckmann J. (1989): Neubeschreibung der Morphogenese von *Paraholosticha muscicola* Kahl, 1932 (Ciliophora, Hypotrichida). – Arch. Protistenk. **137**: 143–156.

Diesing C.M. (1850): Systema Helminthum Vol. I. W. Braumüller, Vindobonae. I–XVI, 680 pp.

Diesing K.M. (1866a): Revision der Prothelminthen. Abtheilung: Amastigen. I. Amastigen ohne Peristom. – Sber. Akad. Wiss. Wien, Mathematisch-naturwissenschaftliche Klasse, Abt. I **52** (year 1865): 505–579.

Diesing K.M. (1866b): Revision der Prothelminthen. Abtheilung: Amastigen. II. Amastigen mit Peristom. – Sber. Akad. Wiss. Wien, Mathematisch-naturwissenschaftliche Klasse, Abt. I **53**: 49–144.

Dillon R.D., Hobbs J.T. (1973): Estimating quantity and quality of the biomass of benthic protozoa. – Proc. S. Dak. Acad. Sci. **52**: 47–49.

Dingfelder J.H. (1962): Die Ciliaten vorübergehender Gewässer. – Arch. Protistenk. **105**: 509–658.

Dini F., Nyberg D. (1993): Sex in ciliates. – Adv. microb. Ecol. **13**: 85–153.

Dini F., Lucchesi P., Macchioni G. (1995): Protozoa. In: Minelli A., Ruffo S., La Posta S. (eds.). Checklist delle specie della fauna italiana, Calderini, Bologna. 92 pp.

Doroszewski M., Raabe Z. (1966): Wzorce morfogenetyczne w podziale i regeneracji orzesków. – Kosmos, Warsz., Seria A. Biologia **15**: 125–137 (in Polish).

Dovgal I.V. (2002): Evolution, phylogeny and classification of Suctorea (Ciliophora). – Protistology **2**: 194–270.

Dragesco J. (1953): Sur l'écologie des ciliés psammophiles littoraux de la région de Banyuls-sur-mer (Pyr.-Or.) (Note préliminaire) (I). – Vie Milieu **4**: 627–632.

Dragesco J. (1954): Diagnoses préliminaires de quelques ciliés nouveaux des sables. – Bull. Soc. zool. Fr. **79**: 62–70.

Dragesco J. (1954a): Diagnoses préliminaires de quelques ciliés nouveaux des sables de Banyuls-sur-mer (I). – Vie Milieu **4**: 633– 637.

Dragesco J. (1960): Ciliés mésopsammiques littoraux. Systématique, morphologie, écologie. – Trav. Stn biol. Roscoff **122**: 1–356.

Dragesco J. (1962): L'orientation actuelle de la systématique des ciliés et la technique d'impregnation au protéinate d'argent. – Bull. Micr. appl. **11:** 49–58.

Dragesco J. (1963): Compléments a la connaissance des ciliés mésopsammiques de Roscoff. – Cah. Biol. mar. **4**: 251–275.

Dragesco J. (1965): Ciliés mésopsammiques d'Afrique Noire. – Cah. Biol. mar. **6**: 357–399.

Dragesco J. (1966): Ciliés libres de Thonon et ses environs. – Protistologica **2**: 59–95.

Dragesco J. (1970): Ciliés libres du Cameroun. – Annls Fac. Sci. Univ. féd. Cameroun **(Numéro hors-série)**: 1–141.

Dragesco J. (1972): Ciliés libres de la cuvette tchadienne. – Annls Fac. Sci. Univ. féd. Cameroun **11**: 71–91.

Dragesco J. (1980): Les protozoaires. – Office de la Recherche Scientifique et Technique Outre-Mer Paris, Initiations, documentations, techniques **44**: 153–192.

Dragesco J. (2003): Infraciliature et morphometrie de vingt espèces de ciliés hypotriches recoltés au Rwanda et Burundi, comprenant *Kahliella quadrinucleata* n. sp., *Pleurotricha multinucleata* n. sp. et *Laurentiella bergeri* n. sp. – Trav. Mus. Hist. nat. Gr. Antipa **45**: 7–59.

Dragesco J., Njine T. (1971): Compléments à la connaissance des ciliés libres du Cameroun. – Annls Fac. Sci. Univ. féd. Cameroun **7–8**: 97–140, Planches I, II.

Dragesco J., Dragesco-Kernéis A. (1986): Ciliés libres de l'Afrique intertropicale. Introduction à la connaissance et à l'étude des ciliés. – Faune tropicale (Éditions de l'ORSTOM) **26**: 1–559.

Dragesco J., Dragesco-Kernéis A. (1991): Free-living ciliates from the coastal area of Lake Tanganyika (Africa). – Europ. J. Protistol. **26**: 216–235.

Dujardin F. (1841): Histoire naturelle des zoophytes. Infusoires, comprenant la physiologie et la classification de ces animaux, et la maniére de les étudier a l'aide du microscope. Librairie Encyclopédique de Roret, Paris. I–XII, 684 pp, 14 pp Explication des Planches, Planches 1–22.

Dumas, E. (1929): Faune du centre. Les microzoaires ou infusoires proprement dits. "Les imprimeries réunies", Moulins. I–X, 216 pp, Planches I–XXXXIV.

Dunger W. (1991): Zur Primärsukzession humiphager Tiergruppen auf Bergbauflächen. – Zool. Jb. Syst., **118**: 423–447.

Dzhurtubayev M.M. (1978): Species composition and vertical distribution of psammophilic infusoria in Odessa Bay. – Hydrobiol. J. **13**: 64–67.

Eddy S. (1925): The distribution of marine protozoa in the Friday harbor waters (San Juan Channel, Washington Sound). – Trans. Am. microsc. Soc. **44**: 97–108.

Edmondson C.H. (1906): The protozoa of Iowa. A study of species known to occur in the waters of this state. – Proc. Davenport Acad. Sci. **11**: 1–123, Plates I–XXX.

Ehrenberg C.G. (1829): Die geographische Verbreitung der Infusionsthierchen in Nord-Afrika und West-Asien, beobachtet auf Hemprich und Ehrenbergs Reisen. – Abh. preuss. Akad. Wiss., Phys.-math. Kl. **year 1829**: 1–20.

Ehrenberg C.G. (1830): Beiträge zur Kenntniß der Organisation der Infusorien und ihrer geographischen Verbreitung, besonders in Sibirien. – Abh. preuss. Akad. Wiss., Phys.–math. Kl. **year 1830**: 1–88, Tafeln I–VIII (Corliss 1979, p. 371, gives following comment to this reference: "Although this particular publication did not appear under this exact title, etc. until 1832, its identical content had, in effect, been available in distributable form since 1830, so I am taking the liberty of so dating the work for historical and nomenclatural purposes and usage". The title of the World List deviates somewhat from the original title which is "Physikalische Abhandlungen der Königlichen Akademie der Wissenschaften zu Berlin").

Ehrenberg C.G. (1831): Über die Entwickelung und Lebensdauer der Infusionsthiere; nebst ferneren Beiträgen zu einer Vergleichung ihrer organischen Systeme. – Abh. preuss. Akad. Wiss., Phys.- math. Kl. **year 1831**: 1–154, Tafeln I–IV (see note following entry on Ehrenberg 1830; similar situation, with "final" appearance again in 1832, but with initialy availability in 1831 in this case).

Ehrenberg C.G. (1833): Dritter Beitrag zur Erkenntniß großer Organisation in der Richtung des kleinsten Raumes. – Abh. preuss. Akad. Wiss., Phys.-math. Kl. **year 1833**: 145–336, Tafeln I–XI (See note following entry on Ehrenberg 1830; similar situation, except "final" appearance in 1835, in this case, but with initial availability in 1833).

Ehrenberg C.G. (1835): Zusätze zur Erkenntniß großer organischer Ausbildung in den kleinsten thierischen Organismen. – Abh. preuss. Akad. Wiss., Phys.-math. Kl. **year 1835**: 151–180, Tafel I (See note following entry on Ehrenberg 1830; similar situation, except "final" appearance in 1837, in this case, but with initial availability in 1835).

Ehrenberg C.G. (1836): Ueber das Leuchten des Meeres etc. – Mittheilungen der Gesellschaft naturforschender Freunde zu Berlin **year 1835**: 1–5.

Ehrenberg C.G. (1838): Die Infusionsthierchen als vollkommene Organismen. Ein Blick in das tiefere organische Leben der Natur. L. Voss, Leipzig. 548 pp, Tafeln I–LXIV.

Ehrenberg C.G. (1849): Über weitere atmosphärische mikroskopische Verhältnisse während der berliner Cholerazeit von 1848 und über schalenlose Infusorien der Atmosphäre. – Bericht über die zur Bekanntmachung geeigneten Verhandlungen der Königl. Preuss. Akademie der Wissenschaften zu Berlin **year 1849**: 91–97.

Eichwald E. (1844): Beitrag zur Infusorienkunde Russland's. – Bull. Soc. impér. nat. Moscou **17**: 480–587.

Eichwald E. (1847): Erster Nachtrag zur Infusorienkunde Russlands. – Bull. Soc. impér. nat. Moscou **20**: 285–366, Tafeln VIII, IX.

Eichwald E.v. (1849): Zweiter Nachtrag zur Infusorienkunde Russlands. – Bull. Soc. impér. nat. Moscou **22**: 400–548.

Eichwald E. (1852): Dritter Nachtrag zur Infusorienkunde Russlands. – Bull. Soc. impér. nat. Moscou **25**: 388–536.

Eigner P. (1994): Divisional morphogenesis and reorganization in *Eschaneustyla brachytona* Stokes, 1886 and revision of the Bakuellinae (Ciliophora, Hypotrichida). – Europ. J. Protistol. **30**: 462–475.

Eigner P. (1995): Divisional morphogenesis in *Deviata abbrevescens* nov. gen., nov. spec., *Neogeneia hortualis* nov. gen., nov. spec., and *Kahliella simplex* (Horvath) Corliss and redefinition of the Kahliellidae (Ciliophora, Hypotrichida). – Europ. J. Protistol. **31**: 341–366.

Eigner P. (1997): Evolution of morphogenetic processes in the Orthoamphisiellidae n. fam., Oxytrichidae, and Parakahliellidae n. fam., and their depiction using a computer method (Ciliophora, Hypotrichida). – J. Euk. Microbiol. **44**: 553–573.

Eigner P. (1999): Comparison of divisional morphogenesis in four morphologically different clones of the genus *Gonostomum* and update of the natural hypotrich system (Ciliophora, Hypotrichida). – Europ. J. Protistol. **35**: 34–48.

Eigner P. (2001): Divisional morphogenesis in *Uroleptus caudatus* (Stokes, 1886), and the relationship between the Urostylidae and the Parakahliellidae, Oxytrichidae, and Orthoamphisiellidae on the basis of morphogenetic processes (Ciliophora, Hypotrichida). – J. Euk. Microbiol. **48**: 70–79.

Eigner P., Foissner W. (1992): Divisional morphogenesis in *Bakuella pampinaria* nov. spec. and reevaluation of the classification of the urostylids (Ciliophora, Hypotrichida). – Europ. J. Protistol. **28**: 460–470.

Eigner P., Foissner W. (1993): Divisional morphogenesis in *Bakuella* sp. and re-evaluation of the classification of urostylid hypotrichs (Ciliophora). – J. Euk. Microbiol. Suppl. **40**: 17A, Abstract 102.

Eigner P., Foissner W. (1994): Divisional morphogenesis in *Amphisiellides illuvialis* n. sp., *Paramphisiella caudata* (Hemberger) and *Hemiamphisiella terricola* Foissner, and redefinition of the Amphisiellidae (Ciliophora, Hypotrichida). – J. Euk. Microbiol. **41**: 243–261.

Eisler K., Fleury A., Foissner W., Grain J. (1995): Morphogenesis and evolution in ciliates. – Protistological Actualities (Proceedings of the Second European Congress of Prostistology and Eight European Conference on Ciliate Biology in Clermont-Ferrand) **year 1995**: 102–127.

Elliott P.B., Bamforth S.S. (1975): Interstitial protozoa and algae of Louisiana salt marshes. – J. Protozool. **22**: 514–519.

El-Serehy H.A. (1993): Species composition and distribution of sand-dwelling ciliates in the great Bitter lake of Suez Canal. – J. Egypt. Ger. Soc. Zool. **10(D)**: 127–141.

Enãceanu V., Brezeanu G. (1970): Repartitia si componenta florei si faunei dunarii de la izvoare la varsare. I. Fauna (Die Verteilung und der Bestand der Flora und Fauna der Donau von der Quelle bis zur Mündung). – Hidrobiologia **11**: 227–264 (in Rumanian with German summary).

Engelmann T.W. (1862): Zur Naturgeschichte der Infusionsthiere. – Z. wiss. Zool. **11**: 347–393, Tafeln XXVIII–XXXI.

Engelmann T.W. (1876): Ueber Entwickelung und Fortpflanzung von Infusorien. – Morph. Jb. **1**: 573–634, Tafeln XXI, XXII.

Enriques P. (1912): Ricerche biologiche sugli infusori dei dintorni di Bologna. Nota seconda. – Rc. Sess. Accad. Sci. Ist. Bologna **17** (year 1912/1913): 103–117.

Entz G. (1884): Über Infusorien des Golfes von Neapel. – Mitt. zool. Stn Neapel **5**: 289–444, Tafeln 20–25.

Entz G. (1900): A sósvizek faunája. - Pótfüzetek a Természettudományi Közlönyhöz. Allattani Közlemények **55**: 99–119 (in Hungarian).

Entz G. (1904): Einiges über das Variieren der Infusorien. – Math. naturw. Ber. Ung. **19** (year 1901): 125–144.

Entz G. (1904a): Die Fauna der kontinentalen Kochsalzwässer. – Math. naturw. Ber. Ung. **19** (year 1901): 89–124.

Escomel E. (1929): Protozoarios de las aguas de Arequipa. – Fauna de Arequipa **year 1929**: 16–32.

Esteban G.F., Finlay B.J., Olmo J.L., Tyler P.A. (2000): Ciliated protozoa from a volcanic crater-lake in Victoria, Australia, – J. nat. Hist. **34**: 159–189.

Esteban G.F., Finlay B.J., Charubhun N., Charubhun B. (2001): On the geographic distribution of *Loxodes rex* (Protozoa, Ciliophora) and other alleged endemic species of ciliates. – J. Zool., London **255**: 139–143.

Esteban G.F., Olmo J.L., Finlay B.J. (2001a): Redescription of *Psilotricha acuminata* Stein, 1859 and revisions of the genera *Psilotricha* and *Urospinula* (Ciliophora, Hypotrichida). – J. Euk. Microbiol. **48**: 280–292.

Fabre-Domergue P.L. (1885): Note sur les infusoires ciliés de la baie de Concarneau. – J. Anat. Physiol., Paris **21**: 554–568, Planches XXVIII–XXIX.

Fantham H.B., Paterson N.F. (1926): Some protozoa found in certain South African soils. – VI. – S. Afr. J. Sci. **23**: 667–705.

Fantham H.B., Porter A. (1946): The microfauna, especially the protozoa, found in some Canadian mosses. – Proc. zool. Soc. Lond. **115** (years 1945–1946): 97–174.

Fantham H.B., Schuurman J.F.M., Robertson K.G. (1927): Some protozoa found in certain South African soils. – VII. – S. Afr. J. Sci. **24**: 346–376.

Faria J.G.de, Cunha A.M.Da, Pinto C. (1922): Estudos sobre protozoarios do mar (Studien ueber Meeresprotozoen). – Mems Inst. Oswaldo Cruz **15**: 186–208, 101–115 (German text), Estampas 23–25.

Fauré-Fremiet E. (1908): Les donditions organiques du comportement chez les cytozoaires l'*Urostyla grandis*. – Bull. Inst. gén. psychol. **7**: 441-446.

Fauré-Fremiet E. (1910a): Études sur les mitochondries des protozoaires et des cellules sexuelles. – Archs Anat. microsc. **2**: 457–648, Planches XIX–XXII.

Fauré-Fremiet E. (1910b): La division de l'*Urostyla grandis*. – Bull. scient. Fr. Belg. **44**: 215–219.

Fauré-Fremiet E. (1945a): Duplicité homopolaire et symétrie chez les *Urostyla*. – C. r. Séanc. Soc. Biol. **139**: 637.

Fauré-Fremiet E. (1945b): Symétrie et polarité chez les ciliés bi- ou multicomposites. – Bull. biol. Fr. Belg. **79**: 106–150.

Fauré-Fremiet E. (1948): Les mecanismes de la morphogénèse chez les cilies. – Folia biotheor. **3**: 25–58.

Fauré-Fremiet E. (1950): Écologie des ciliés psammophiles littoraux. – Bull. biol. Fr. Belg. **84**: 35–75.

Fauré-Fremiet E. (1951): The marine sand-dwelling ciliates of Cape Cod. – Biol. Bull. mar. biol. Lab., Woods Hole **100**: 59–70.

Fauré-Fremiet E. (1961): Remarques sur la morphologie comparée et la systématique des ciliata Hypotrichida. – C. r. hebd. Séanc. Acad. Sci., Paris **252**: 3515–3519.

Fauré-Fremiet E. (1961a): Documents et observations écologiques et pratiques sur la culture des infusoires ciliés. – Hydrobiologia **18**: 300–320.

Fauré-Fremiet E. (1967): La régénération chez les protozoaires. – Bull. Soc. zool. Fr. **92**: 249–272.

Fauré-Fremiet E., André J. (1968): L'organisation corticale des ciliata. – C. r. hebd. Séanc. Acad. Sci., Paris **266**: 487–490.

Fenchel T. (1967): The ecology of marine microbenthos I. The quantitative importance of ciliates as compared with metazoans in various types of sediments. – Ophelia **4**: 121–137.

Fenchel T. (1968): The ecology of marine microbenthos II. The food of marine benthic ciliates. – Ophelia **5**: 73–121.

Fenchel T. (1968a): The ecology of marine microbenthos III. The reproductive potential of ciliates. – Ophelia **5**: 123–136.

Fenchel T. (1969): The ecology of marine microbenthos IV. Structure and function of the benthic ecosystem, its chemical and physical factors and the microfauna communities with special reference to the ciliated protozoa. – Ophelia **6**: 1–182.

Fenchel T. (1974): Intrinsic rate of natural increase: the relationship with body size. – Oecologia (Berlin) **14**: 317–326.

Fenchel T., Finlay B.J. (1983): Respiration rates in heterotrophic, free-living protozoa. – Microb. Ecol. **9**: 99–122.

Fenchel T., Finlay B.J. (1991): The biology of free-living anaerobic ciliates. – Europ. J. Protistol. **26**: 201–215.

Fenchel T., Finlay B.J. (1995): Ecology and evolution in anoxic worlds. Oxford University Press, Oxford, New York, Tokyo. I–XII & 276 pp.

Fenchel T., Lee C.C. (1972): Studies on ciliates associated with sea ice from Antarctica I. The nature of the fauna. - Arch. Protistenk. **114**: 231-236.

Fernandez-Galiano D., Calvo P. (1992): A comparative analysis of the morphology of *Holosticha corlissi* n. sp. (Ciliophora, Hypotrichida) and related species. – J. Protozool. **39**: 600–604.

Fernandez-Leborans G. (1985): The paroral formation of various hypotrichous ciliates. – Arch. Protistenk. **130**: 367–380.

Fernandez-Leborans G. (1990): Morphology and taxonomic position of a ciliate of the subclass Stichotrichia (Protozoa), *Holosticha begoniensis* n. sp. – Arch. Protistenk. **138**: 51–64.

Fernandez-Leborans G. (2001): Relative importance of protozoan functional groups in three marine sublittoral areas. – J. mar. biol. Ass. U. K. **81**: 735–750.

Fernandez-Leborans G., Antonio-Garcia M.T. (1986): Interaction lead-zinc in a natural community of protozoans. – Acta Protozool. **25**: 195–211.

Fernandez-Leborans G., Antonio-García M.T. (1988): Effects of lead and cadmium in a community of protozoans. – Acta Protozool. **27**: 141–159.

Fernandez-Leborans G., Novillo A. (1993): Sublittoral protistan communities of the shores of the sea of Cantabria (Bay of Biscay). – Int. Revue ges. Hydrobiol. Hydrogr. **78**: 201–218.

Fernandez-Leborans G., Novillo A. (1993a): Morphology of a stichotrichid marine ciliate: considerations about some species of the suborder Urostylina (Protozoa). – Zool. Jb. Anat. **123**: 197–210.

Fernandez-Leborans G., Novillo A. (1994): Effects of periodic addition of lead on a marine protistan community. – Aquatic Sciences **56**: 191–205.

Fernandez-Leborans G., Novillo A. (1996): Protozoan communities and contamination of several fluvial systems. – Water Environment Research **68**: 311–319.

Fernandez-Leborans G., Antonio-Garcia M.T., Corpas-Vazquez I. (1985): The effects of lead on some protozoan communities. Ciliates of the Santillana Reservoir and Ciudad Universitaria (Madrid, Spain). – Acta Protozool. **24**: 85–109.

Fernandez-Leborans G., Castro de Zaldumbide M., Perales C. (1990): Biomass and functional distribution according to the feeding modes of two freshwater protozoan communities. – Int. Revue ges. Hydrobiol. Hydrogr. **75**: 507–532.

Fernandez-Leborans G., Valganon B., Castro de Zaldumbide M. (1999): Characterization of a marine sublittoral area facing the open sea, using epibenthic protists. – Bull. mar. Sci. **65**: 725–743.

Finlay B.J., Rogerson A., Cowling A.J. (1988): A beginner's guide to the collection, isolation, cultivation and identification of freshwater protozoa. Culture Collection of Algae and Protozoa, Freshwater Biological Association, The Ferry House, Ambleside, Cumbria, United Kingdom. Titus Wilson and Son, Kendal, United Kingdom. 78 p.

Finlay B.J., Esteban G.F., Olmo J.L., Tyler P.A. (1999): Global distribution of free-living microbial species. – Ecography **22**: 138–144.

Finlay B.J., Esteban G.F., Clarke K.J., Olmo J.L. (2001): Biodiversity of terrestrial protozoa appears homogeneous across local and global spatial scales. – Protist **152**: 355–366.

Fleury A., Iftode F., Deroux G., Fryd-Versavel G. (1985): Unité et diversité chez les hypotriches (Protozoaires ciliés): II – Éléments d'ultrastructure comparée chez divers représentants du sous-ordre des Euhypotrichina. – Protistologica **21**: 505–524.

Fleury A., Philippe H., Baroin A. (1995): Cortical organization and molecular phylogeny. – Protistological Actualities (Proceedings of the Second European Congress of Protistology and Eight European Conference on Ciliate Biology in Clermont-Ferrand) **year 1995**: 103–107, 123–127.

Florentin R. (1899): Études sur la faune des mares salées de lorraine. – Annls Sci. nat., Zoologie **10**: 209–349, Planches 8–10.

Foissner W. (1976): *Wallackia schiffmanni* nov. gen., nov. spec. (Ciliophora, Hypotrichida) ein alpiner hypotricher Ciliat. – Acta Protozool. **15**: 387–392.

Foissner W. (1979): Wimpertiere als Bioindikatoren. – ÖKO-L, Zeitschrift für Ökologie, Natur- und Umweltschutz **4**: 3–7.

Foissner W. (1979a): Methylgrün-Pyronin: Seine Eignung zur supravitalen Übersichtsfärbung von Protozoen, besonders ihrer Protrichocysten. – Mikroskopie **35**: 108–115.

Foissner W. (1980a): Taxonomische Studien über die Ciliaten des Grossglocknergebietes (Hohe Tauern, Österreich). IX. Ordnungen Heterotrichida und Hypotrichida. – Ber. Nat.-Med. Ver. Salzburg **5**: 71–117.

Foissner W. (1980b): Artenbestand und Struktur der Ciliatenzönose in alpinen Kleingewässern (Hohe Tauern, Österreich). – Arch. Protistenk. **123**: 99–126.

Foissner W. (1981): Die Gemeinschaftsstruktur der Ciliatenzönose in alpinen Böden (Hohe Tauern, Österreich) und Grundlagen für eine Synökologie der terricolen Ciliaten (Protozoa, Ciliophora). – Veröff. Österr. MaB-Programms **4**: 7–52.

Foissner W. (1982): Ökologie und Taxonomie der Hypotrichida (Protozoa: Ciliophora) einiger österreichischer Böden. – Arch. Protistenk. **126**: 19–143.

Foissner W. (1983): Die Morphogenese von *Urosoma macrostyla* (Wrzesniowski, 1870) (Ciliophora: Oxytrichidae). – Arch. Protistenk. **127**: 413–428.

Foissner W. (1984): Infraciliatur, Silberliniensystem und Biometrie einiger neuer und wenig bekannter terrestrischer, limnischer und mariner Ciliaten (Protozoa: Ciliophora) aus den Klassen Kinetofragminophora, Colpodea und Polyhymenophora. – Stapfia **12**: 1–165.

Foissner W. (1985): Protozoologische Untersuchungen an Almböden im Gasteiner Tal (Zentralalpen, Österreich). III. Struktur und Dynamik der Testaceen- und Ciliatentaxozönose. – Veröff. Österr. MaB-Programms **9**: 65–95.

Foissner W. (1986): Wimpertiere (Protozoa: Ciliophora) in Flechten. – Katalog des Oberösterreichischen Landesmuseums Linz, Neue Folge **5**: 43–46.

Foissner W. (1987a): Soil protozoa: fundamental problems, ecological significance, adaptations in ciliates and testaceans, bioindicators, and guide to the literature. – Progr. Protistol. **2**: 69–212.

Foissner W. (1987b): Neue und wenig bekannte hypotriche und colpodide Ciliaten (Protozoa: Ciliophora) aus Böden und Moosen. – Zool. Beitr. (N. F.) **31**: 187–282.

Foissner W. (1987c): Neue terrestrische und limnische Ciliaten (Protozoa, Ciliophora) aus Österreich und Deutschland. – Sber. öst. Akad. Wiss., Mathematisch-naturwissenschaftliche Klasse, Abt. I **195 (year 1986)**: 217–268.

Foissner W. (1987d): Miscellanea nomenclatorica ciliatea (Protozoa: Ciliophora). – Arch. Protistenk. **133**: 219–235.

Foissner W. (1987e): Faunistische und taxonomische Notizen über die Protozoen des Fuscher Tales (Salzburg, Österreich). – Jber. Haus Nat. Salzburg **10**: 56–68.

Foissner W. (1988): Gemeinsame Arten in der terricolen Ciliatenfauna (Protozoa: Ciliophora) von Australien und Afrika. – Stapfia **17**: 85–133.

Foissner W. (1988a): Taxonomic and nomenclatural revision of Sládecek's list of ciliates (Protozoa: Ciliophora) as indicators of water quality. – Hydrobiologia **166**: 1–64.

Foissner W. (1989): Morphologie und Infraciliatur einiger neuer und wenig bekannter terrestrischer und limnischer Ciliaten (Protozoa, Ciliophora). – Sber. öst. Akad. Wiss., Mathematisch-naturwissenschaftliche Klasse, Abt. I **196** (year 1987): 173–247.

Foissner W. (1991): Basic light and scanning electron microscopic methods for taxonomic studies of ciliated protozoa. - Europ. J. Protistol. **27**: 313–330.

Foissner W. (1993): Colpodea (Ciliophora). – Protozoenfauna **4/1**: I– X + 1–798 pp.

Foissner W. (1994): *Pentahymena corticicola* nov. gen., nov. spec., a new colpodid ciliate (Protozoa, Ciliophora) from bark of Acacia trees in Costa Rica. – Arch. Protistenk. **144**: 289–295.

Foissner W. (1994a): Morphology and morphogenesis of *Circinella arenicola* nov. gen., nov. spec., a cephalized hypotrich (Ciliophora, Hypotrichida) from sand dunes in Utah, USA. – Europ. J. Protistol. **30**: 156–170.

Foissner W. (1995): Tropical protozoan diversity: 80 ciliate species (Protozoa, Ciliophora) in a soil sample from a tropical dry forest of Costa Rica, with descriptions of four new genera and seven new species. – Arch. Protistenk. **145**: 37–79.

Foissner W. (1996): Terrestrial ciliates (Protozoa, Ciliophora) from two islands (Gough, Marion) in the southern oceans, with description of two new species, *Arcuospathidium cooperi* and *Oxytricha ottowi*. – Biol. Fertil. Soils **23**: 282–291.

Foissner W. (1996a): Faunistics, taxonomy and ecology of moss and soil ciliates (Protozoa, Ciliophora) from Antarctica, with description of new species, including *Pleuroplitoides smithi* gen. n., sp. n.. – Acta Protozool. **35**: 95–123.

Foissner W. (1996b): Ciliates in rapid gravity filters of waterworks exploiting deep groundwaters. – Microscopy Research and Technique **33:** 12–22.

Foissner W. (1996c): Ontogenesis in ciliated protozoa, with emphasis on stomatogenesis. In: Hausmann K., Bradbury P.C. (eds.): Ciliates: Cells as organisms, pp. 95–177. Gustav Fischer, Stuttgart, Jena, Lübeck, Ulm.

Foissner W. (1997): Soil ciliates (Protozoa: Ciliophora) from evergreen rain forests of Australia, South America and Costa Rica: diversity and description of new species. – Biol. Fertil. Soils **25**: 317–339.

Foissner W. (1997a): Faunistic and taxonomic studies on ciliates (Protozoa, Ciliophora) from clean rivers in Bavaria (Germany), with descriptions of new species and ecological notes. – Limnologica (Berlin) **27**: 179–238.

Foissner W. (1998): An updated compilation of world soil ciliates (Protozoa, Ciliophora), with ecological notes, new records, and descriptions of new species. – Europ. J. Protistol. **34**: 195–235.

Foissner W. (1999): Notes on the soil ciliate biota (Protozoa, Ciliophora) from the Shimba Hills in Kenya (Africa): diversity and description of three new genera and ten new species. – Biodiversity and Conservation **8**: 319–389.

Foissner W. (2000): A compilation of soil and moss ciliates (Protozoa, Ciliophora) from Germany, with new records and descriptions of new and insufficiently known species. – Europ. J. Protistol. **36**: 253–283.

Foissner W. (2004): Some new ciliates (Protozoa, Ciliophora) from an Austrian floodplain soil, including a giant; red "flagship", *Cyrtohymena (Cyrtohymenides) aspoecki* nov. subgen., nov. spec. – Denisia **13**: 369–382.

Foissner W., Adam H. (1983): Morphologie und Morphogenese des Bodenciliaten *Oxytricha granulifera* sp. n. (Ciliophora, Oxytrichidae). – Zool. Scr. **12**: 1–11.

Foissner W., AL-Rasheid K. (2006): A unified organization of the stichotrichine oral apparatus, including a description of the buccal seal (Ciliophora: Spirotrichea). – Acta Protozool. **45**: 1–16.

Foissner W., Berger H. (1996): A user-friendly guide to the ciliates (Protozoa, Ciliophora) commonly used by hydrobiologists as bioindicators in rivers, lakes, and waste waters, with notes on their ecology. – Freshwat. Biol. **35**: 375–482.

Foissner W., Berger H. (1999): Identification and ontogenesis of the nomen nudum hypotrichs (Protozoa: Ciliophora) *Oxytricha nova* (= *Sterkiella nova* sp. n.) and *O. trifallax* (= *S. histriomuscorum*). – Acta Protozool. **38**: 215–248.

Foissner W., Blatterer H. (1990): New ciliates (Protozoa, Ciliophora) from Australia and Africa. – J. Protozool. Suppl. **37**: 9A, Abstract 53.

Foissner W., Didier P. (1981): Morphologie und Infraciliatur einiger kinetofragminophorer und hypotricher Ciliaten aus den Fließgewässern von Besse-en-Chandesse (Frankreich). – Annls Stn limnol. Besse **15**: 254–275.

Foissner W., Foissner I. (1988): Stamm: Ciliophora. – Catalogus Faunae Austriae **I c**: 1–147.

Foissner W., Moog O. (1992): Die Gewässergüte der Unteren Traun im Spiegel ihrer Wimpertier-Gesellschaften. – Kataloge des O. Ö. Landesmuseums Linz, Neue Folge **54**: 99–107.

Foissner W., Peer T. (1985): Protozoologische Untersuchungen an Almböden im Gasteiner Tal (Zentralalpen, Österreich). I. Charakteristik der Taxotope, Faunistik und Autökologie der Testacea und Ciliophora. – Veröff. Österr. MaB-Programms **9**: 27–50.

Foissner W., Stoeck T. (2006): A new "flagship" ciliate from the Niger floodplain breaks flexibility-dogma in the classification of the stichotrichine spirotrichs. – Wissenschaftliche Tagung der Deutschen Gesellschaft für Protozoologie (Liebenwalde bei Berlin), **25**: 28, Abstract P1.

Foissner W., Adam H., Foissner I. (1982): Daten zur Autökologie der Ciliaten stagnierender Kleingewässer im Grossglocknergebiet (Hohe Tauern, Österreich). – Ber. Nat.-Med. Ver. Salzburg **6**: 81–101.

Foissner W., Peer T., Adam H. (1985): Pedologische und protozoologische Untersuchung einiger Böden des Tullnerfeldes (Niederösterreich). – Mitt. öst. bodenk. Ges. **30**: 77–117.

Foissner W., Buchgraber K., Berger H. (1990): Bodenfauna, Vegetation und Ertrag bei ökologisch und konventionell bewirtschaftetem Grünland: Eine Feldstudie mit randomisierten Blöcken. – Mitt. öst. bodenk. Ges. **41**: 5–33.

Foissner W., Oleksiv I., Müller H. (1990a): Morphologie und Infraciliatur einiger Ciliaten (Protozoa: Ciliophora) aus stagnierenden Gewässern. – Arch. Protistenk. **138**: 191–206.

Foissner W., Blatterer H., Berger H., Kohmann F. (1991): Taxonomische und ökologische Revision der Ciliaten des Saprobiensystems – Band I: Cyrtophorida, Oligotrichida, Hypotrichia, Colpodea. – Informationsberichte des Bayer. Landesamtes für Wasserwirtschaft **1/91**: 1–478.

Foissner W., Unterweger A., Henschel T. (1992): Beitrag zur Ciliatenfauna (Protozoa: Ciliophora) einiger Seitenbäche der Amper (Oberbayern, Deutschland). – Lauterbornia **Heft 9**: 45–57.

Foissner W., Unterweger A., Henschel T. (1992a): Comparison of direct stream bed and artificial substrate sampling of ciliates (Protozoa, Ciliophora) in a mesosaprobic river. – Limnologica (Berlin) **22**: 97–104.

Foissner W., Berger H., Kohmann F. (1992b): Taxonomische und ökologische Revision der Ciliaten des Saprobiensystems – Band II: Peritrichia, Heterotrichida, Odontostomatida. – Informationsberichte des Bayer. Landesamtes für Wasserwirtschaft **5/92**: 1–502.

Foissner W., Berger H., Kohmann F. (1994): Taxonomische und ökologische Revision der Ciliaten des Saprobiensystems – Band III: Hymenostomata, Prostomatida, Nassulida. – Informationsberichte des Bayer. Landesamtes für Wasserwirtschaft **1/94**: 1–548.

Foissner W., Berger H., Blatterer H., Kohmann F. (1995): Taxonomische und ökologische Revision der Ciliaten des Saprobiensystems – Band IV: Gymnostomatea, *Loxodes*, Suctoria. – Informationsberichte des Bayer. Landesamtes für Wasserwirtschaft **1/95**: 1–540.

Foissner W., Berger H., Blatterer H., Kohmann F. (1995a): Teil II - Protozoa II. 1 Ciliophora. In: Moog O. (ed.). Fauna Aquatica Austriaca, Lieferung Mai/95, Wasserwirtschaftskataster Bundesministerium für Land- und Forstwirtschaft, Wien. 45 pp.

Foissner W., Berger H., Schaumburg J. (1999): Identification and ecology of limnetic plankton ciliates. – Informationsberichte des Bayer. Landesamtes für Wasserwirtschaft **3/99**: 1–793.

Foissner W., Agatha S., Berger H. (2002): Soil ciliates (Protozoa, Ciliophora) from Namibia (Southwest Africa), with emphasis on two contrasting environments, the Etosha region and the Namib desert. Part I: Text and line drawings. Part II: Photographs. – Denisia **5**: 1–1459.

Foissner W., Moon-van der Staay S.Y., van der Staay G.W.M, Hackstein J.H.P., Krautgartner W.-D., Berger H. (2004a): Reconciling classical and molecular phylogenies in the stichotrichines (Ciliophora, Spirotrichea), including new sequences from some rare species. – Europ. J. Protistol. **40**: 265–281.

Foissner W., Berger H., Xu K., Zechmeister-Boltenstern S. (2005): A huge, undescribed soil ciliate (Protozoa: Ciliophora) diversity in natural forest stands of Central Europe. – Biodiversity and Conservation **14**: 617–701.

Forel F.-A. (1885): La faune profonde des lacs suisses. – Neue Denkschr. allg. schweiz. Ges. ges. Naturw. **29**: 1–229.

Forel F.-A. (1904): Le Léman. Monographie limnologique. Vol. **3.** F. Rouge & Cie, Lausanne. 715 pp.

Fowell R.R. (1944): The ecology of a rock pool. – Proc. Swansea scient. Fld Nat. Soc. **2**: 192–212.

Franco C., Esteban G.F., Téllez C. (1996): Description of *Metabakuella bimarginata* sp. n., and key to the ciliate subfamily Bakuellinae Jankowski, 1979. – Acta Protozool. **35**: 321–330.

Frankel J. (1974): Positional information in unicellular organisms. – J. theor. Biol. **47**: 439–481.

Frankel J. (1989): Pattern formation. Ciliate studies and models. Oxford University Press, New York, Oxford. I–XVII & 314 pp.

Fresenius G. (1865): Die Infusorien des Seewasseraquariums. – Zool. Gart., Frankf. **6**: 81–89, 121–129, 1 Tafel.

Fromentel E.de (1875): Études sur les microzoaires ou infusoires proprement dits comprenant de nouvelles recherches sur leur organisation, leur classification et la description des espèces nouvelles ou peu connus. pp. 89–192. G. Masson, Paris.

Fromentel E.de (1876): Études sur les microzoaires ou infusoires proprement dits comprenant de nouvelles recherches sur leur organisation, leur classification et la description des espèces nouvelles ou peu connus. pp. 193–364, Planches I–XXX. G. Masson, Paris.

Funke W. (1986): Tiergesellschaften im Ökosystem "Fichtenforst" (Protozoa, Metazoa-Invertebrata) – Indikatoren von Veränderungen in Waldökosystemen. – Kernforschungszentrum Karlsruhe GmbH, KfK-PEF **9**: 1–150.

Gajevskaja N. (1927): Zur Kenntnis der Infusorien des Baikalsees. – Dokl. Akad. Nauk SSSR **19**: 313–318.

Gajewskaja N. (1933): Zur Oekologie, Morphologie und Systematik der Infusorien des Baikalsees. – Zoologica, Stuttg. **32**: I–VIII, 1–298, Anhang I–III, Tafeln I–XXV.

Galtsoff P.S., Lutz F.E., Welch P.S., Needham J.G. (1959): Culture methods for invertebrate animals. Dover Publications, New York (on the cover the authors are arranged as follows: Lutz, Welch, Galtsoff, Needham. On the spine only Needham is cited). I–XXXII, 590 pp.

Gan H. & Xu R. (2005): Studies on the ecology of benthic ciliates community in intertidal mangrove. – Int. Congr. Protozool., 12: 56, Abstract S2-32.

Ganapati P.N., Rao M.V.N. (1958): Systematic survey of marine ciliates from Visakhapatnam. – Andhra Univ. Mem. Oceanogr. **2**: 75–90, Plate I.

Ganner B. (1991): Zur Taxonomie einiger Ciliaten (Protozoa) aus Fliessgewässern. Dissertation Universität Salzburg. 159 pp.

Ganner B., Unterweger A., Jäger P. (2002): Die Biologie der Salzburger Kläranlagen im Zeitraum von 1991 bis 2000. Zur Evaluation der Beurteilungskriterien bei der mikroskopischen Belebtschlammuntersuchung kommunaler Kläranlagen mit Stickstoffelimination. – Gewässerschutz **6**: 115–170.

Gargiulo A. (1907): I protisti nelle acque stagnanti dei dintorni di lecce. – Riv. ital. Sci. nat. **27**: 71–76, 24–30.

Garrison D.L. (1991): Antarctic sea ice biota. – Amer. Zool. **31**: 17–33.

Gassovsky G.N. (1916): Sur la faune des infusoires du golf de Kola et de ses environs. – Trudy leningr. Obshch. Estest. **45**: 139–215 (in Russian with French summary).

Gayewskaya N.S. (1924): On the question of the function of pulsating and non-pulsating vacuoles of salt-water infusoria. – Russk. gidrobiol. Zh. **3**: 239–252 (in Russian with English summary).

Gel'cer J.G., Geptner V.A. (1976): O protistocidnom dejstuii gerbicidov (The protisticide effect of herbicides). – Pedobiologia **16**: 171–184 (in Russian with English summary).

Gelei J.v. (1937): Der schraubige Körperbau in der Ciliatenwelt im Vergleich zu den Symmetrieverhältnissen der vielzelligen Tiere. – Arch. Protistenk. **88**: 314–338.

Gelei J. (1950): Die Morphogenese der Einzeller mit Rücksicht auf die morphogenetischen Prinzipien von Sewertzoff. – Acta biol. hung. **1**: 69–134.

Gelei J. (1954): Über die Lebensgemeinschaft einiger temporärer Tümpel auf einer Bergwiese im Börzsönygebirge (Oberungarn) III. Ciliaten. Acta biol. hung. **5:** 259–343.

Gelei J., Szabados M. (1950): Tömegprodukció városi esövízpocsolyában (Massenproduktion in einer städtischen Regenwasserpfütze). – Annls biol. Univ. szeged. **1**: 249–294 (in Hungarian with German translation and Russian summary).

Gelei J., Megyeri J., Szabados M., Varga L. (1954): Über die Lebensgemeinschaft einiger temporärer Tümpel auf einer Bergwiese im Börzsönygebirge (Oberungarn) VIII. Allgemeine Betrachtungen. – Acta biol. hung. **5**: 363–384.

Gellért J. (1942): Életegyüttes a fakéreg zöldporos bevonatában (Lebensgemeinschaft in dem grünpulverigen Überzug der Baumrinde). – Acta Scientiarum Mathematicarum et Naturalium, Universitas Francisco-Josephina Kolozsvár **year 1942 (No. 8)**: 1– 36 (in Hungarian; German translation of the tittle from Gellért 1956).

Gellért J. (1956): Die Ciliaten des sich unter der Flechte *Parmelia saxatilis* Mass. gebildeten Humus. – Acta biol. hung. **6**: 77–111.

Gellért J. (1956a): Ciliaten des sich unter dem Moosrasen auf Felsen gebildeten Humus. – Acta biol. hung. **6**: 337–359.

Gellért J. (1957): Néhány hazai lomblevelü és tülevelü erdö talajának ciliáta-faunája Ciliatenfauna im Humus einiger ungarischen Laub- und Nadelholzwälder. – Annls Inst. biol. Tihany **24**: 11–34 (in Hungarian with German summary).

Gellért J., Tamás G. (1958): Detritusz-turzások kovamoszatainak és csillósainak ökológiai vizsgálata a Tihanyi-félsziget keleti partján (Ökologische Untersuchungen an Diatomeen und Ciliaten der Detritus-Drifte am Ostufer der Halbinsel Tihany). – Annls Inst. biol. Tihany **25**: 217–240 (in Hungarian with German comprehensive summary).

Gellért J., Tamás G. (1959): Ecological studies on the diatoms and ciliate infusorians in the detritus-drifts along the shores of the Tihany-Peninsula. – Acta biol. hung. **10**: 117–125.

Gellért J., Tamás G. (1960): Detritusz-turzások kovamoszatainak és csillósainak ökológiai vizsgálata a Balaton déli partján (Ökologische Untersuchung der Kieselalgen und Ciliaten von Detritusdriften an der südlichen Uferseite des Balaton). – Annls Inst. biol. Tihany **27**: 55–64 (in Hungarian with German summary).

Geptner V.A. (1973): Ciliata (Infusoria). In: Lepinis A.K., Geltzer Y.G., Tschibisova D.I., Geptner V.A. (eds.). Key to soil protozoa of European part of the USSR, pp. 124–171. Mintis, Vilnius (in Russian).

Ghosh E. (1921): New hypotrichous infusoria from Calcutta. – J. R. microsc. Soc. **year 1921**: 248–250.

Gierig M. (1993): Biologische Aufbereitung von Tiefengrundwässern. – Informationsberichte des Bayer. Landesamtes für Wasserwirtschaft **2/93**: 1–106.

Gittleson S.M., Hoover R.L. (1969): Cavernicolous protozoa review of the literature and new studies in Mammoth Cave, Kentucky. – Annls Spéléol. **24**: 737–776.

Gmelin K.C. (editor; 1790): Caroli a Linné Systema Naturae. 13th edition (1788–1793). Tom. I., Pars VI. Leipzig.

Goodey T. (1911): A contribution to our knowledge of the protozoa of the soil. – Proc. R. Soc., Series B **84**: 165–180, Plate 4.

Golińska K., Jerka-Dziadosz M. (1973): The relationship between cell size and capacity for division in *Dileptus anser* and *Urostyla cristata*. – Acta Protozool. **12**: 1–21.

Gong J., Song W., Hu X., Ma H., Zhu M. (2001): Morphology and infraciliature of *Holosticha bradburyae* nov. spec. (Ciliophora, Hypotrichida) from the Yellow Sea, China. – Hydrobiologia **464**: 63–69.

Gong J., Song W., Warren A. (2005): Periphytic ciliate colonization: annual cycle and responses to environmental conditions. – Aquat. Microbiol. Ecol., **39**: 159–170.

Gong X., Shen Y. (1989): The new species of protozoa from the Suoxiyu nature preserve area, Hunan Province. In: Shanghao L. (ed.). The aquatic fauna of Suoxiyu nature preserve area, Hunan, China, pp. 96–114. Academia Sinica Press, Peking (in Chinese with English summary).

Goniakowska L., Pigon A. (1968): Effect of tricarboxylic acid cycle intermediates and related compounds on the respiratory rate of *Urostyla cristata* and *Colpoda cucullus*. – J. Protozool. **15**: 716–719.

Gourret P., Roeser P. (1888): Contribution à l'étude des protozoaires de la Corse. – Archs Biol. **8**: 139–204, Planches XIII–XV.

Grabacka E. (1971): Ciliata of the bottom of rearing fishponds in the Golysz complex. – Acta hydrobiol., Kraków **13**: 5–28.

Grabacka E. (1977): The influence of beet sugar factory wastes on the bottom microfauna in fish ponds. – Acta hydrobiol., Kraków **19**: 373–387.

Gracia M.P., Igual J. (1987): Los ciliados como organismos saprobios de las aguas. – Misc. Zool. **11**: 1–11.

Gracia M.P., Igual J. (1987a): Estudio faunístico ecológico de los protozoos ciliados del río Llobregat (España). – Boln R. Soc. esp. Hist. nat. **83**: 113–127.

Gracia P., Castellón C., Igual J., Sunyer R. (1985): Study on the communities of ciliata protozoa of the Llobregat River (Barcelona, Spain). – Int. Congr. Protozool. 7: 54, Abstract 8.

Gracia M.P., Castellon C., Igual J. (1987): Protozoos ciliados del rio Llobregat. – Publnes Dept. Zool. Barcelona **13**: 23–34.

Gracia M.P., Castellon C., Igual J., Sunyer R. (1989): Ciliated protozoan communities in a fluvial ecosystem. – Hydrobiologia **183**: 11–31.

Grandori R., Grandori L. (1934): Studî sui protozoi del terreno. – Boll. Lab. Zool. agr. Bachic. R. Ist. sup. agr. Milano **5**: 1–341, Tavole I–XIV.

Gray E. (1952): The ecology of the ciliate fauna of Hobson's Brook, a Cambridgeshire chalk stream. – J. gen. Microbiol. **6**: 108–122.

Grell K.G. (1968): Protozoologie. Springer, Berlin, Heidelberg, New York. I–VIII, 511 pp.

Griebler C., Sonntag B., Mindl B., Posch T., Klammer S., Psenner R. (2002): Assessment of the ecological integrity of Traunsee (Austria) via analysis of sediments and benthic microbial communities. – Water, Air and Soil Pollution: Focus **2**: 33–62.

Grim J.N. (1989): Changes in body surface features and the densities of some organelles during encystment of the ciliated protozoan *Pseudourostyla cristata*. – J. Protozool. Suppl. **36**: 1A, Abstract 2.

Grim J.N., Manganaro C.A. (1985): Form of the extrusomes and secreted material of the ciliated protozoon *Pseudourostyla cristata*, with some phylogenetic interpretations: a light, scanning electron, and transmission electron microscopic study. – Trans. Am. microsc. Soc. **104**: 350–359.

Grimes G.W. (1972): Cortical structure in nondividing and cortical morphogenesis in dividing *Oxytricha fallax*. – J. Protozool. **19**: 428–445.

Grimes G.W., Adler J.A. (1978): Regeneration of ciliary pattern in longitudinal fragments of the hypotrichous ciliate, *Stylonychia*. – J. exp. Zool. **204**: 57–79, Plates 1–6.

Grimm R. (1968): Biologie der gestauten Elbe. Die Auswirkungen der Staustufe Geesthacht auf die benthale Fauna im oberen Grenzbereich des Elbe-Aestuars. – Arch. Hydrobiol. Suppl. **31**: 281–378.

Grispini M. (1938): I ciliati di una pozza di «macchia» della campagna romana. – Boll. Zool. **9**: 147–158.

Grolière C.-A. (1975): Descriptions de quelques ciliés hypotriches des tourbières a sphaignes et des étendues d'eau acides. – Protistologica **11**: 481–498.

Grolière C.-A. (1977): Contribution a l'étude des ciliés des sphaignes: II. Dynamique des populations. – Protistologica **13**: 335–352.

Grolière C.-A. (1978): Contribution a l'étude des ciliés des sphaignes III. Étude mathématique des résultats. – Protistologica **14**: 295–311.

Jerka-Dziadosz M. (1968): Study on resorption of ciliature in *Urostyla* (Hypotricha). – Acta Protozool. **5**: 359–367, Plates I–IV.

Jerka-Dziadosz M. (1968a): Device for mass fragmentation of large ciliates. – Acta Protozool. **6**: 383–385, Plate I.

Jerka-Dziadosz M. (1970): Studies on the distribution of trichocysts in the normal life cycle and during regeneration of *Urostyla cristata* Jerka-Dziadosz, 1964 (Hypotircha). – Acta Protozool. **7**: 505–512, Plates I–V.

Jerka-Dziadosz M. (1972): Cortical development in *Urostyla*. I. Comparative study on morphogenesis in *U. cristata* and *U. grandis*. – Acta Protozool. **10**: 73–100, Plates I–X.

Jerka-Dziadosz M. (1974): Cortical development in *Urostyla*. II. The role of positional information and preformed structures in formation of cortical pattern. – Acta Protozool. **12**: 239–274, Plates I–XVI.

Jerka-Dziadosz M., Janus I. (1972): Localization of primordia during cortical development in *Keronopsis rubra* (Ehrbg., 1838) (Hypotrichida). – Acta Protozool. **10**: 249–262, Plates I–VI.

Jerka-Dziadosz M., Kapuścińska-Czerska W. (1972): Destruction of cellular components in pigmented protozoa by laser microbeam. – Acta Protozool. **9**: 373–386, Plates I–III.

Jin L. (1991): Cytological pattern of the parachute stage and development of the macronuclear anlage in *Pseudourostyla cristata*. – Acta Scientiarum Naturalium Universitatis Sunyatseni **30**: 81–88 (in Chinese with English summary).

Jin L. (1993): Obtainment and identification of the amicronucleates of *Pseudourostyla cristata*. – Acta Scientiarum Naturalium Universitatis Sunyatseni **32**: 106–110 (in Chinese with English summary).

Jin L.-P. (1995): Four rounds of cortical development during sexual reproduction in *Pseudourostyla cristata*. – Acta Scientiarum Naturalium Universitatis Sunyatsen **34**: 68–78 (in Chinese with English summary).

Jin L.-P. (1996): Regulation of the membranelle number of AZM in proter during the binary division of *Pseudourostyla cristata*. – Acta Scientiarum Naturalium Universitatis Sunyatsen **35**: 86–89 (in Chinese with English summary).

Jin L.-P., Gu F.-K. (1997): Stomatogenesis and remake of adoral zone of membranelles during physiological reorganization in *Pseudourostyla cristata*. – Acta zool. sin. **43**: 420–425 (in Chinese with English summary).

Jin L.-P., Jin H.-Z. (2002): Micronuclear effect on structural stability of the oral apparatus in hypotrichous ciliate. – Acta zool. sin. **48**: 258–263 (in Chinese with English summary).

Jin L., Liu Z. (1996): Synthesis of DNA, RNA and protein during conjugation in *Pseudourostyla cristata*. – Acta Scientiarum naturalium Universitatis Sunyatsen I **35**: 84–89 (in Chinese with English summary).

Jin L.-P., Ng S.F. (1989a): The somatic function of the germ nucleus in *Pseudourostyla cristata*: asexual reproduction and stomatogenesis. – J. Protozool. **36**: 315–326.

Jin L.-P., Ng S.F. (1989b): Somatic function of the micronucleus in *Pseudourostyla*. – Int. Congr. Protozool. **8**: 118, Abstract 2–F–12.

Jin L., Zhang Z. (1997): Cortical development and pattern determination during sexual reproduction in *Pseudourostyla cristata*. – Acta Scientiarum naturalium Universitatis Sunyatsen I **36**: 60–63 (in Chinese with English summary).

Jin L., Zhuang H., Tchang T. (1985): Studies on conjugation in *Pseudourostyla cristata* I. Nucleogenesis. – Zool. Res. **6**: 159–167 (in Chinese with English summary).

Jin L.-P., Liu X.-Y., Jin H.-Z. (2001a): Nuclear control of cortical development during conjugation in *Pseudourostyla cristata*. – Zool. Res. **22**: 89–92 (in Chinese with English summary).

Jin L.-P., Liu X.-Y., Jin H.-Z. (2001b): Nuclear control of cortical development durin conjugation in *Pseudourostyla cristata*. – Zool. Res. **22**: 99–104 (in Chinese with English summary).

Jin L.-P., Liu X.-Y., Jin H.-Z. (2002): Morphogenetic role of the micronucleus during conjugation in the hypotrichous ciliate (*Pseudourostyla cristata*). – Acta zool. sin. **48**: 408–414 (in Chinese with English summary).

Jírovec O., Wenig K., Fott B., Bartoš E., Weiser J., Šramek-Hušek R. (1953): Protozoologie. Nakladatelství Československé Akademie Věd, Praha (in Czech). 643 pp.

Johnson M.D., Tengs T., Oldach D.W., Delwiche C.F., Stoecker D.K. (2004): Highly divergent SSU rRNA genes found in the marine ciliates *Myrionecta rubra* and *Mesodinium pulex*. – Protist **155**: 347–359.

Jones E.E. (1974): The protozoa of Mobile Bay, Alabama. – Univ. South Alabama Monogr. **1**: 1–113.

Jones R.C., Cairns J.Jr., Yonguc W.II.Jr. (1976): Vertical gradients in artificial substrate-associated protozoan community structure in a stratified freshwater lake. – J. Elisha Mitchell scient. Soc. **92**: 1–8.

Jósa Z. (1974): Effect of town waste-water on the development of ciliata plankton in the Szolnok section of the Tisza. – Tiszia. Dissertationes Biologicae a Collegio Exploratorum Fluminis Tisciae Editae **9**: 53–63.

Jutrczenki J.v. (1982): Ökologische Untersuchung der Ciliatenfauna zweier Bonner Waldbäche. – Decheniana **135**: 104–116.

Kahl A. (1928): Die Infusorien (Ciliata) der Oldesloer Salzwasserstellen. – Arch. Hydrobiol. **19**: 189–246.

Kahl A. (1928a): Die Infusorien (Ciliata) der Oldesloer Salzwasserstellen. – Arch. Hydrobiol. **19**: 50–123.

Kahl A. (1930): Urtiere oder Protozoa I: Wimpertiere oder Ciliata (Infusoria) 1. Allgemeiner Teil und Prostomata. – Tierwelt Dtl. **18**: 1–180.

Kahl A. (1932): Urtiere oder Protozoa I: Wimpertiere oder Ciliata (Infusoria) 3. Spirotricha. – Tierwelt Dtl. **25**: 399–650.

Kahl A. (1933): Ciliata libera et ectocommensalia. – Tierwelt N.- u. Ostsee **Lieferung 23 (Teil II. c_3)**: 29–146.

Kahl A. (1934): II. c_5 Suctoria. – Tierwelt N.- u. Ostsee **26 (Teil II, c_5)**: 184–226.

Kahl A. (1935): Urtiere oder Protozoa I: Wimpertiere oder Ciliata (Infusoria) 4. Peritricha und Chonotricha. – Tierwelt Dtl. **30**: 651–886, I–V.

Kalmus H. (1928): Über die Bodenfauna der Moldau im Gebiete von Prag. Ein Jahreszyklus. II. Protozoa, Hydrozoa, Crustacea, Tardigrada, Hydracarina. Mit einem Anhang: Ökologische Beobachtungen und Versuche. – Int. Revue ges. Hydrobiol. Hydrogr. **19**: 349–429.

Kaltenbach A. (1960): Ökologische Untersuchungen an Donauciliaten. – Wass. Abwass. Wien **1960**: 151–174.

Kamra K., Sapra G.R. (1990): Partial retention of parental ciliature during morphogenesis of the ciliate *Coniculostomum monilata* (Dragesco & Njiné, 1971) Njiné, 1978 (Oxytrichidae, Hypotrichida). – Europ. J. Protistol. **25**: 264–278.

Karaman S. (1935): Die Fauna der unterirdischen Gewässer Jugoslaviens. – Verh. int. Ver. theor. angew. Limnol. **7**: 46–73.

Kattar M.R. (1970): Estudo dos protozoários ciliados psamófilos do litoral brasileiro. – Zoologia e biologia marinha N. S. **Nº 27**: 123–206.

Kelminson K., Hewitt E., Arkind C., Kneeland P., DuBois M.L., Prescott D.M. (2002): Histone H4 genes in some stichotrichous ciliates. – Europ. J. Protistol. **38**: 235–246.

Kent W.S. (1880–1882): A manual of the infusoria: including a description of all known flagellate, ciliate, and tentaculiferous protozoa, British and foreign, and an account of the organization and affinities of the sponges. Volumes I–III. David Bogue, London (Vol. I 1880: 1–432; Vol. II 1881: 433–720; Vol. II 1882: 721–913; Vol. III 1882: Plates I–LI). 913 pp.

Khlebovich T.V. (1987): The rate of oxygen consumption in protozoans. – Trudy zool. Inst., Leningr. **165**: 212–223 (in Russian).

Kiesselbach A. (1935): Einige morphologische und biologische Beobachtungen an *Epiclintes ambiguus* (Ciliata Hypotricha). – Note Ist. italo-germ. Biol. mar. Rovigno **18**: 1–9.

Kiesselbach A. (1936): Contributo alla conoscenza dei ciliati della Laguna Veneta. In: Monografia della Laguna veneta, pp. 5–15. Venice (year uncertain; editor unknown).

Kiesselbach A. (1936a): Zur Ciliatenfauna der nördlichen Adria. – Thalassia **2**: 1–53, 8 Tafeln.

Kim O.T.P., Yura K., Go N., Harumoto T. (2004): Highly divergent actins from karyorelictean, heterotrich, and litostome ciliates. – J. Euk. Microbiol. **51**: 227–233.

Kirby H. (1945): The structure of the common intestinal trichomonad of man. – J. Parasitol. **31**: 163–174.

Knüpling J. (1979): Ökologische Untersuchungen an Aufwuchsciliaten im Brackwasserbereich des Elbe-Aestuars. – Arch. Hydrobiol. Suppl. **43**: 273–288.

Kolkwitz R. (1950): Oekologie der Saprobien. Über die Beziehungen der Wasserorganismen zur Umwelt. – SchrReihe Ver. Wass.- Boden- u. Lufthyg. **Nr. 4**: 1–64.

Kopylov A.I. (1979): Chemical composition and caloric value of infusoria. – Oceanology **19**: 586–589.

Kovalchuk A.A. (1984): Structurally functional features of the basic ciliocenoses groups in Kiev reservoir. In: Academy of Sciences of the USSR. The All-Union Society of Protozoologists & Academy of Sciences of the Latvian SSR (ed.). Ecology of marine and fresh water protozoans (Proceedings of the first symposium), pp. 46–48. Academy of Sciences of the USSR. The All- Union Society of Protozoologists & Academy of Sciences of the Latvian SSR, Salaspils (in Russian).

Kovalchuck A.A. (1989): The free living ciliates of the Dnieper-Rug estuary and their role in production-destructional processes. In: Poljansky G.I., Zhukov B.F., Raikov I.B. (eds.). Ecology of marine and fresh=water protozoans. Proceedings of the II Symposium, p. 30. Academy of Sciences of the USSR. The All-Union Society of Protozoologists, Yaroslavl. Abstract (in Russian).

Kovalchuk A.A.(1997): Cilioperiphyton of the River Tisa (in Ukraine). In: Carpathian Biosphere Reserve. Carpathian Agency for Regional Development, Carpathian Association of National Parks and Protected Areas: International aspects of study and consveration of the Carpathians biodiversity, pp. 94–98. Proceedings of the International Scinetic-Practical Conference, Rakhiv.

Kovalchuk A.A. (1997a): Cilioplankton of the River Tisa (ín Ukraine). – Proceedings of the International Regional Seminar Environment Protection: Modern Studies in Ecology and Microbiology **1**: 441–446.

Kovaleva V.G. (1966): Infusoria of the mesopsammon in sand bays of the Black Sea. – Zool. Zh. **45**: 1600–1611 (in Russian with English summary).

Kovaleva V.G., Golemansky V.G. (1979): Psammobiotic ciliates of the Bulgarian coast of the Black Sea. – Acta Protozool. **18**: 265–284 (in Russian with English summary).

Kovaljeva V.G. (1967): New data on the infusorian fauna of the mesopsammon of the Barentz Sea. – Acta Protozool. **5**: 81–88 (in Russian with English summary).

Kowalewski M. (1883): Beiträge zur Naturgeschichte der Oxytrichinen. – Biol. Zbl. **3**: 235–243 (German translation of Kowalewskiego 1882 reported by A. Wrzesniowski; without figures).

Kowalewskiego M. (1882): Przyczynek do historyi naturalnéj oxytrichów. – Pam. fizyogr. **2**: 395–413, Tablica XXIX, XXX (in Polish with French explanation of plates).

Kravchenko V.M. (1969): On fauna of infusoria in the reservoirs of the Seversky Donets Basin. – Vést. zoologii. Institut Zoologii Akademia Nauk Ukrainskoi SSR **year 1969**: 69–75 (in Russian with English summary).

Krepuska G. (1917): Budapest véglényei (Die Protisten von Budapest). – Allatt. Közl. **16**: 86–116, 154–184, 222 (in Hungarian with German summary).

Kudo R.R. (1950): Protozoology. Thomas Publisher, Springfield. I–XIV, 778 pp.

Kuhlmann H.-W., Heckmann K. (1985): Interspecific morphogens regulating prey-predator relationships in protozoa. – Science **227**: 1347–1349.

Kuhlmann H.-W., Heckmann K. (1994): Predation risk of typical ovoid and 'winged' morphs of *Euplotes* (Protozoa, Ciliophora). – Hydrobiologia **284**: 219–227.

Kuhlmann H.-W., Kusch J., Heckmann K. (1999): Predator-induced defenses in ciliated protozoa. In: Tollrian R., Harvell C.D. (eds.). The ecology and evolution of inducible defenses, pp. 142–159. Princeton Press, Princeton.

Kühn G. (1940): Zur Ökologie und Biologie der Gewässer (Quellen und Abflüsse) des Wassersprengs bei Wien. – Arch. Hydrobiol. **36**: 157–262.

Küsters E. (1974): Ökologische und systematische Untersuchungen der Aufwuchsciliaten im Königshafen bei List/Sylt. – Arch. Hydrobiol. Suppl. **45**: 121–211.

Lachmann C.F.J. (1856): Ueber dic Organisation der Infusorien, besonders der Vorticellen. – Archiv für Anatomie, Physiologie und wissenschaftliche Medicin **year 1856**: 340–398, Tafeln XIII, XIV.

Lackey J.B. (1936): Occurrence and distribution of the marine protozoan species in the Woods Hole area. – Biol. Bull. mar. biol. Lab., Woods Hole **70**: 264–278.

Lackey J.B. (1938): A study of some ecologic factors affecting the distribution of protozoa. – Ecol. Monogr. **8**: 501–527.

Lackey J.B. (1961): Bottom sampling and environmental niches. – Limnol. Oceanogr. **6**: 271–279.

Lackey J.B., Lackey E.W. (1963): Microscopic algae and protozoa in the waters near Plymouth in August 1962. – J. mar. biol. Ass. U. K. **43**: 797–805.

Laird M. (1961): Microecological factors in oyster epizootics. – Can. J. Zool. **39**: 449–485.

Lamouroux J.-V.-F., Bory de Saint-Vincent J.B, Deslongchamps E. (1824): Encyclopédie méthodique. Histoire naturelle des zoophytes, ou animaux rayonnés, faisant suite a l'histoire naturelle des vers de bruguière. Vol. **II**, Agasse, Paris (according to the "Catalogue général des livres imprimés de la bibliothèque nationale, Paris" 1824 is the correct year). I–VIII, 819 pp.

Larsen H.F. (1987): A preliminary study on the pigments of three *Pseudokeronopsis* species collected in Denmark. – European Conference on Ciliate Biology **6**: 19, Abstract.

Larsen H.F. (1989): Further studies on the pigments of the marine ciliate *Pseudokeronopsis carnea* Cohn, 1866. – J. Protozool. Suppl. **36**: 29A, Abstract 171.

Laval M. (1971): Mise en évidence par la microscopie électronique d'un organite d'un type nouveau chez les ciliés tintinnides. - C. r. hebd. Séanc. Acad. Sci., Paris **273**: 1383–1386, Planches I, II.

Laval-Peuto M. (1975): Cortex, périlemme et réticulum vésiculeux de *Cyttarocylis brandti* (Cilié tintinnide). Les ciliés a périlemme. – Protistologica **11**: 83–98.

Lee J.J., Small E.B., Lynn D.H., Bovee E.C. (1985): Some techniques for collecting, cultivating and observing protozoa. In: Lee, J.J., Hutner, S.H., Bovee, E.C. (eds.). An illustrated guide to the protozoa, pp. 1–7. Society of Protozoologists, Lawrence.

Lee J.J., Lee M.J., Weis D.S. (1985a): Possible adaptive value of endosymbionts to their protozoan hosts. – J. Protozool. **32**: 380–382.

Lehle E. (1989): Beiträge zur Fauna der Ulmer Region II. Ciliaten (Protozoa:Ciliophora) Bioindikatoren in Waldböden. – Mitt. Ver. Naturwiss. Math. Ulm (Donau) **35**: 131–156.

Lehle E. (1992): Wimpertiere und andere Einzeller im Boden eines Fichtenbestandes im Schwarzwald. – Mikrokosmos **81**: 193–198.

Lehle E. (1993): Bodenciliaten (Protozoa: Ciliophora) als sensitive Indikatoren tiefgreifender Veränderungen in Waldökosystemen. – Kernforschungszentrum Karlsruhe GmbH, KfK-PEF **113**: 1–91.

Lehle E. (1994): Die Auswirkungen von Düngung und Kalkung auf die Bodenciliaten (Protozoa: Ciliophora) eines Fichtenbestandes im Schwarzwald (Süddeutschland). – Arch. Protistenk. **144**: 113–125.

Lehle E., Jans W., Funke W. (1992): Reaktionen von Bodenciliaten auf Düngemaßnahmen in Fichtenforsten des Schwarzwaldes. – Kernforschungszentrum Karlsruhe GmbH, KfK-PEF **94**: 273–287.

Lei Y., Choi J.K., Xu K., Petz W. (2005): Morphology and infraciliature of three species of *Metaurostylopsis* (Ciliophora, Stichotrichia): *M. songi* n. sp., *M. salina* n. sp., and *M. marina* (Kahl 1932) from sediments, saline ponds, and coastal waters. – J. Euk. Microbiol. 52: 1–10.

Lei Y., Xu K., Choi J.K. (2005a): *Holosticha hamulata* n. sp. and *Holosticha heterofoissneri* Hu and Song, 2001, two urostylid ciliates (Protozoa, Ciliophora) from intertidal sediments of the Yellow Sea. – J. Euk. Microbiol. **52**: 310–318.

Lemullois M., Fryd-Versavel G., Fleury-Aubusson A. (2004): Localization of centrins in the hypotrich ciliate *Paraurostyla weissei*. – Protist **155**: 331–346.

Lepinis A.K., Geltzer Y.G., Tschibisova D.I., Geptner V.A. (1973): Key to soil protozoa of European part of the USSR. Mintis, Vilnius (in Russian). 172 pp.

Lepsi J. (1926a): Über die Protozoenfauna einiger Quellen der Dobrudscha. – Arch. Hydrobiol. **17**: 751–770.

Lepsi J. (1926b): Die Infusorien des Süsswassers und Meeres. Bermühler, Berlin-Lichterfelde. 100 pp, Tafeln I–XIV.

Lepsi I. (1927): Studii faunistice, morfologice si fiziologice asupra infuzoriilor din România. – Studii Cerc. Acad. RPR **12**: 1–176 (in Rumanian).

Lepsi J. (1929): Zur Phylogenie der Ciliaten. – Bul. Fac. Sti. Cernauti **3**: 258–303.

Lepsi I. (1932): Recherches préliminaires sur les protozoaires des torrents de Sinaïa. – Publnile Soc. Nat. Rom. **Nr. 10**: 5–53, 1 plate on p. 57.

Lepsi I. (1935): Un nouvel infusoire de la Mer Noire: *Urostyla naumanni* n. sp. – Bul. Muz. natn. Ist. nat. Chisinau **6**: 16–17.

Lepsi I. (1951): Modificarea faunei de protozoare tericole, prin irigatii agricole (La modification de la faune des protozoaires terricoles par des irrigations agricoles). – Buletin sti. Acad. Repub. pop rom., Seria: Geologie, geografie, biologie, stiinte tehnice si agricole. Sectiunea de stiinte biologice, agronomice, geologice si geografice **3**: 513–523 (in Rumanian with Russian and French summary).

Lepsi I. (1957): Infuzori spirotrichi si peritrichi din tinoavele de la poiana stampei, raionul vatra dornei (Spirotriche und peritriche Infusorien aus den Hochmooren von Poiana Stampei, Bezirk Vatra Dornei). – Buletin sti. Acad. Repub. pop rom., Sectia de biologie si stiinte agricole, Seria zoologie **9**: 231–240 (in Rumanian with Russian and German summary).

Lepsi I. (1965): Protozoologie. Academia Republicii Socialiste România, Bucarest (in Romanian). 999 pp.

Levander K.M. (1894): Liste über im Finnischen Meerbusen in der Umgebung von Helsingfors beobachtete Protozoen. – Zool. Anz. **17**: 209–212.

Levander K.M. (1894a): Materialien zur Kenntniss der Wasserfauna in der Umgebung von Helsingfors, mit besonderer Berücksichtigung der Meeresfauna. I. Protozoa. – Acta Soc. Fauna Flora fenn. **12**: 1–115, Tafeln I–III.

Levander K.M. (1901a): Zur Kenntnis des Planktons und der Bodenfauna einiger seichten Brackwasserbuchten. – Acta Soc. Fauna Flora fenn. **20**: 1–34.

Levander K.M. (1901b): Übersicht der in der Umgebung von Esbo-Löfö im Meereswasser vorkommenden Thiere. – Acta Soc. Fauna Flora fenn. **20**: 1–20.

Lewin K.R. (1913): The division of *Holosticha scutellum*. – Proc. Camb. phil. Soc. biol. Sci. **17:** 241, Abstract.

Li Y. & Gu F. (2005): Comparative study on genetic character in resting cells and vegetative cells of two species of ciliate. – Int. Congr. Protozool., 12: 78, Abstract S3-34.

Li Y., Gu F. (2005a): Phylogenetic study on four species of ciliate inferred from 16s-like small subunit rRNA gene sequence. – Int. Congr. Protozool., 12: 114, Abstract S4.1-42.

Lieberkühn N. (1856): Beiträge zur Anatomie der Infusorien. – Archiv für Anatomie, Physiologie und wissenschaftliche Medicin **year 1856**: 20–36.

Liebmann H. (1962): Handbuch der Frischwasser- und Abwasser-Biologie. Biologie des Trinkwassers, Badewassers, Fischwassers, Vorfluters und Abwassers. Band I. R. Oldenbourg, München. I–XV, 588 pp, Tafeln I–XXII.

Liepa R. (1973): The seasonal development of ciliata in the Lielupe. – Latv. PSR Zinát. Akad. Vest. **year 1973**: 31–37 (in Russian with English summary).

Liepa R.A. (1983): Ecologo-faunistic characteristics of ciliates in water bodies of higher saprobity. – Protozoology **8**: 134–141 (in Russian with English summary).

Liepa R.A. (1984): Characteristic structure of protozoan benthos in the small rivers of Latvia. In: Academy of Sciences of the USSR. The All-Union Society of Protozoologists & Academy of Sciences of the Latvian SSR (ed.). Ecology of marine and fresh water protozoans (Proceedings of the first symposium), pp. 62–64. Academy of Sciences of the USSR. The All-Union Society of Protozoologists & Academy of Sciences of the Latvian SSR, Salaspils (in Russian).

Liepa R. (1986): Infusoria-substrate relationships in the small streams of Latvia. – Symposia Biologica Hungarica **33**: 225–232.

Liepa R.A. (1990): Organization and structure of benthic associations of ciliates in rivers. – Protozoology **13**: 64–69 (in Russian with English summary).

Lin M., Prescott D.M. (1986): Fine structure of the macronuclei and micronuclei in five hypotrichs. – Arch. Protistenk. **131**: 33–43.

Lin X., Song W., Warren A. (2004): Redescription of the rare marine ciliate, *Prodiscocephalus borrori* (Wicklow, 1982) from shrimp-culturing waters near Qingdao, China, with redefinitions of the genera *Discocephalus*, *Prodiscocephalus* and *Marginotricha* (Ciliophora, Hypotrichida, Discocephalidae). – Europ. J. Protistol. **40**: 137–146.

Lincoln R.J., Boxshall G.A., Clark P.F. (1985): A dictionary of ecology, evolution and systematics. Cambridge University Press, Cambridge, London, New York, New Rochelle, Melbourne, Sydney. 298 pp.

Liu X.-Y., Jin L.-P. (2000): Studies on the morphology and morphogenesis of *Pseudokeronopsis monilata*. – Acta Scientiarum naturalium Universitatis Sunyatsen I **39 (Suppl. 2)**: 82–86 (in Chinese with English summary).

Liu X.Y., Jin L.-P. (2002): A preliminary study on somatic function of micronucleus in *Pseudokeronopsis monilata*. – Zool. Res. **23**: 258–260 (in Chinese with English summary).

Liu X., Mao Y., Liu T., Chen J., Lee K.-L. D. Lee, Lin L. (2005a): Differential gene expression between amicronucleate and micronucleate of the ciliate *Pseudourostyla cristata* during stationary phase. – Int. Congr. Protozool., 12: 83, Abstract S3-42.

Liu X., Liu T., Chen J., Mao Y., K.-L. D. Lee, L. Jin (2005b): The heat shock protein Hsp70 prevents disorganization of the oral apparatus AZM in the amicronucleate cells of *Pseudourostyla cristata*. – Int. Congr. Protozool., 12: 84, Abstract S3-43.

Liu X, Mao Y., Lee K.-L.D., Jin L. (2005c): Cloning of mitochondrial cytochrome b gene from *Pseudourostyla cristata* by heat shock. – J. Euk. Microbiol. **52**: 19S, Abstract 50A.

Liu X., Mao Y., K.-L.D. Lee, Jin L. (2005d): Identification of mRNAs expressed differentially between micronucleates and amicronucleates of *Pseudourostyla cristata* during starvation. – J. Euk. Microbiol. **52**: 19S, Abstract 51A.

Liyepa R.A. (1984): Classification of Latvian lakes based on species composition of infusoria. – Hydrobiol. J. **20**: 12–16.

Longhi P. (1892): Protisti delle acque dolci di Genova e dintorni. – Atti Soc. ligust. Sci. nat. geogr. **3**: 137–161.

Longhi P. (1894): Prime ricerche intorno ai protisti del distretto di Belluno e suoi dintorni. – Atti Soc. ligust. Sci. nat. geogr. **5**: 13–27.

Longhi P. (1895): Seconda serie di ricerche intorno ai protisti delle acque dolci del distretto di Belluno e dintorni. – Atti Soc. ligust. Sci. nat. geogr. **6**: 67–83, Tavola III.

Lopez-Ochoterena E., Madrazo-Garibay M., Calderon-Aragon L.del C., Coronado-Gutierrez R. (1976): Protozoarios ciliados de Mexico XXI Algunos aspectos biologicos de doce especies recolectadas en la Costa del Golfo de Mexico. – Revta Soc. mex. Hist. nat. **37**: 205–219.

Lozupone C.A., Knight R.D., Landweber L.F. (2001): The molecular basis of nuclear genetic code change in ciliates. – Current Biology **11**: 65–74.

Lueken W., Uhlig G., Krüppel T. (1989a): Soluble poison of *Pseudokeronopsis rubra* affects inward current in *Euplotes vannus*. – J. Protozool. Suppl. **36**: 20A, Abstract 118.

Lueken W., Uhlig G., Krüppel T. (1989b): Soluble poison of *Pseudokeronopsis rubra* affects inward current in *Euplotes vannus*. – J. Protozool. **36**: 350–353.

Lüftenegger G., Foissner W., Berger H., Adam H. (1987): Ciliatostasis and its disruption by human influences. – Society of Protozoologists **1986 Abstracts**: Abstract 139.

Lüftenegger G., Petz W., Foissner W., Adam H. (1988): The efficiency of a direct counting method in estimating the numbers of microscopic soil organisms. – Pedobiologia **31**: 95–101.

Lundin F.C., West L.S. (1963): The free-living protozoa of the Upper Peninsula of Michigan. Northern Michigan College Press, Marquette. 175 pp.

Luzzatti E. (1938): I ciliati di un terreno di «macchia» della campagna romana. – Boll. Zool. **9**: 91–113.

Lynn D.H. (1991): The implications of recent descriptions of kinetid structure to the systematics of the ciliated protists. – Protoplasma **164**: 123–142.

Lynn D.H. (2003): Morphology or molecules: how do we identify the major lineages of ciliates (Phylum Ciliophora)? – Europ. J. Protistol. **39**: 356–364.

Lynn D.H., Corliss J.O. (1991): Ciliophora. In: Harrison F.W. (ed.). Microscopic anatomy of invertebrates Volume 1: Protozoa, pp. 333–467. Wiley-Liss, New York, Chichester.

Lynn D.H., Small E.B. (1997): A revised classification of the phylum Ciliophora Doflein, 1901. – Revta Soc. mex. Hist. nat. **47**: 65–78.

Lynn D.H., Small E.B. (2002): Phylum Ciliophora. In: Lee J.J., Leedale G.F., Bradbury P. (eds.). An illustrated guide to the Protozoa. Second edition. Organisms traditionally referred to as Protozoa, or newly discovered groups. Volume I, pp. 371–656. Society of Protozoologists, Lawrence. (This edition of the illustrated guide is dated with 2000, however, it was available, that is, published just in spring of 2002).

Lynn D.H., Sogin M.L. (1988): Assessment of phylogenetic relationships among ciliated protists using partial ribosomal RNA sequences derived from reverse transcripts. – BioSystems **21**: 249–254.

Ma S., Dang J. (1994): Monitoring and assessment of waters of Fen-River Taiyuan section and Jinyang Lake with protozoa. – Journal of Shanxi University (Nat. Sci. Ed.) **17**: 449–456 (in Chinese with English summary).

Ma Z. (1994): Investigation on protozoan in Yellow River, Lanzhou, Gansu. – Journal of the Northwest Normal University (Natural Science) **30**: 93–96 (in Chinese with English summary).

Machemer H. (1966): Versuche zur Frage nach der Dressierbarkeit hypotricher Ciliaten unter Einsatz hoher Individuenzahlen. – Z. Tierpsychol. **23**: 641–654.

Madoni P. (1979): The ciliated protozoa in Torrente Parma, Torrente Stirone, and Fiume Po. In: Ghetti P.F. (ed.). 3rd technical seminar. Biological water assessment methods Torrente Parma, Torrente Stirone, Fiume Po, Parma, October 1978, **Volume 1**, pp. 297–312. Commission of the European Communities, Bruxelles.

Madoni P. (1980): Zonazioni longitudinali dei corsi d'acqua e distribuzione dei protozoi ciliati. – Atti V Conv. Gr. "G. Gadio" **year 1980**: 39–58.

Madoni P. (1983): Relationship between ciliated protozoa and ecological zones in some water courses of the River Po Basin. – Int. J. Ecol. Environm. Sci. **9**: 87–97.

Madoni P. (1993): Ciliated protozoa and water quality in the Parma River (Northern Italy): long-term changes in the community structure. – Hydrobiologia **264**: 129–135.

Madoni P. (1994): A sludge biotic index (SBI) for the evaluation of the biological performance of activated sludge plants based on the microfauna analysis. – Wat. Res. **28**: 67–75.

Madoni P. (2000a): The acute toxicity of nickel to freshwater ciliates. – Environm. Pollut. (Series A) **109**: 53–59.

Madoni P. (2000b): The acute toxicity of nicekl to twelve species of freshwater ciliates. – J. Euk. Microbiol. Suppl. **47**: 15A, Abstract 113.

Madoni P. (2005): Ciliated protozoan communities and saprobic evaluation of water quality in the hilly zone of some tributaries of the Po River (northern Italy). – Hydrobiologia, **541**: 55–69.

Madoni P., Bassanini N. (1999): Longitudinal changes in the ciliated protozoa communities along a fluvial system polluted by organic matter. – Europ. J. Protistol. **35**: 391–402.

Madoni P., Ghetti P.F. (1977): Indagine preliminare sulla distribuzione dei ciliati e altri Protozoi nei corsi d'acqua della Val Parma. – Rivista di Idrobiologia **16**: 35–53.

Madoni P., Ghetti P.F. (1981): Ciliated protozoa and water quality in the Torrente Stirone (Northern Italy). – Acta hydrobiol., Kraków **23**: 143–154.

Madoni P., Uluhogian F. (1997): Ciliated protozoa of a geothermal sulphur spring. – Hydrobiologia **353**: 161–170.

Madoni P., Zangrossi S. (2005): Ciliated protozoa and saprobical evaluation of water quality in the Taro River (northern Italy). – Ital. J. Zool., **72**: 21–25.

Madoni P., Davoli D., Fontani N., Cucchi A., Rossi F. (2001): Spatial distribution of microorganisms and measurements of oxygen uptake rate and ammonia uptake rate activity in a drinking water biofilter. – Environm. Technol. **22**: 455–462.

Madrazo-Garibay M., López-Ochoterena E. (1985): Protozoarios ciliados de México. XXVI. Análisis morfológico y taxonómico de treinta y cinco especies de la Laguna de Términos, Campeche. – Anales Instituto de Ciencias del Mar y Limnologia Universidad Nacional Autónomia de México **12**: 199–212, Laminas 1, 2.

Mahachanchawalit P., Maeda M., Simidu U. (1986): Succession of ciliated protozoa on the solid surfaces of substrata in seawater. – Nihon Biseibutsu Seitai Gakkaiho **1**: 9–17.

Mamaeva N.V. (1979a): Class Ciliata. – Monographiae biol. **33**: 408–411.

Mamaeva N.V. (1979b): Infusoria of the Volga basin. Ecological survey. Nauka, Leningrad (in Russian). 151 pp.

Mansfeld K. (1923): 16 neue oder wenig bekannte marine Infusorien. – Arch. Protistenk. **46**: 97–140.

Margalef López R. (1945): Infusorios de las aguas continentales de la provincia de Barcelona. – Boln R. Soc. esp. Hist. nat. **43**: 369–381 (possible the surname is Lopes).

Martin E.C., Sharp R.E. (1983): Soil protozoa from a cedar glade in Rutherford County, Tennessee. – J. Tenn. Acad. Sci. **58**: 31–36.

Martin J., Fedriani C., Nieto J. (1981): Étude comparée des processus morphogénétiques d'*Uroleptus* sp. (Kahl, 1932) et de *Holosticha (Paruroleptus) musculus* (Kahl, 1932) (Ciliés hypotriches). – Protistologica **17**: 215–224.

Martin-Cereceda M., Pérez-Uz B., Serrano S., Guinea A. (2002): An integrated apporach to analyse biofilms of a full scale wastewater treatment plant. – Wat. Sci. Tech. **46**: 199–206.

Martin-Gonzalez A., Benitez L., Gutierrez J.C. (1991): Cortical and nuclear events during cell division and resting cyst formation in *Colpoda inflata*. – J. Protozool. **38**: 338–344.

Martin-Gonzalez A., Benitez L., Gutierrez J.C. (1992): Ultrastructural analysis of resting cysts and encystment in *Colpoda inflata* 2 Encystment process and a review of ciliate resting cyst classification. – Cytobios **72**: 93–106.

Matis D. (1977): Príspevok k poznaniu nálevníkov (Ciliata) Zemplínskej síravy (Contribution to the knowledge of infusoria of Zemplínska Sírava). – Acta Fac. Rerum nat. Univ. comen., Bratisl., Series Zoologia **22**: 21–43 (in Czech with English and Russian summary).

Matis D., Straková-Striešková M. (1991): Nálevníky (Ciliophora) teplého potoka a termálnych jazierok bojnickych kúpel'ov (Ciliates (Ciliophora) of the Teply brook and thermal lakes in Bojnice spa (Czecho-Slovakia). – Biológia, Bratisl. **46**: 113–118 (in Czech with English and Russian summary).

Matis D., Tirjaková E. (1994): Nálevníky (Ciliophora) mrtveho ramena Dunaja v Cicove (Ciliated protozoa (Ciliophora) in dead arm of Danube in Cicov (Slovakia)). – Acta Fac. Rerum nat. Univ. comen., Bratisl., Series Zoologia **16** (year 1992): 49–56 (in Slovakian with English summary).

Matis D., Tirjaková E. (1994a): Contribution to the knowledge of Ciliophora of the river Morava. – Ekologia (CSSR) Supplement **1/1994**: 55–61.

Matis D., Tirjaková E., Stloukal E. (1996): Nálevníky (Ciliophora) v databanke fauny Slovenska (Ciliophora in the database of Slovak fauna). – Folia Faunistica Slovaca **1**: 3–38.

Matsuoka T., Matsuo N., Maesako J., Shigenaka Y. (1983): Distribution of fresh-water protozoa I. Municipal and suburban districts of Hiroshima and Higashi-Hiroshima. – Bull. biol. Soc. Hiroshima Univ. **49**: 13–18 (in Japanese with English summary).

Matsusaka T., Noguchi O., Yonezawa F. (1989): Cortical morphogenesis during encystment in a ciliate, *Euplotes encysticus* Yonezawa, 1985. – Europ. J. Protistol. **24**: 133–137.

Matthes D. (1988): Suctoria (Sauginfusorien). – Protozoenfauna **7/1**: I–XIII, 1–226, 307–309.

Mauch E. (1976): Leitformen der Saprobität für die biologische Gewässeranalyse Teil 4. – Cour. Forsch.-Inst. Senckenberg **21**: 339–563.

Maula M.M., Small E.B., Maugel T.K. (1993): A transmission electron microscopic study of a *Pseudokeronopsis* sp. – J. Euk. Microbiol. Suppl. **40**: 42A, Abstract 247.

Maupas E. (1879): Sur quelques protorganismes animaux et végétaux multinucléés. – C. r. hebd. Séanc. Acad. Sci., Paris **89**: 250–253.

Maupas E. (1881): Contribution a l'étude des acinétiens. – Archs Zool. exp. gén. **9**: 299–368.

Maupas E. (1883): Contribution a l'étude morphologique et anatomique des infusoires ciliés. – Archs Zool. exp. gén., 2e Série **1**: 427–664, Planches XIX–XXIV.

Mayén Estrada R., Aladro Lubel M.A. (1987): Treinta especies de protozoarios ciliados bentonicos de Laguna de la Mancha, Veracruz. – Universidad y Ciencia **4**: 69–80.

Mayer M. (1981): Kultur und Präparation der Protozoen. Franckh'sche Verlagshandlung, Stuttgart. 83 pp.

Mazei Y.A., Burkovsky I.V. (2002): Spatial and temporal changes of psammophilous ciliate community in the White Sea estuary. – Uspekhi sovremennoj biologii **122**: 183–189 (ISSN: 0042-1324; in Russian with English summary).

Mazei Y.A., Burkovsky I.V. (2005): Species composition of benthic ciliate community in the Chernaya River estuary (Kandalaksha Bay, White Sea) with a total checklist of the White Sea benthic ciliate fauna. – Protistology **4**: 107–120.

Meisriemler P., Riedl H.E. (1985): Die Limnologie der Enns. In: Bundesministerium für Land- und Forstwirtschaft (ed.). Limnologie der Österreichischen Donau-Nebengewässer. Eine monographische Darstellung Teil I, pp. 159–187.

Mereschkowsky C.von (1877): Studien über Protozoen des nördlichen Russland. – Trudy imp. S-peterb. Obshch. Estest. **8**: 203–385, Tafeln I–III (in Russian; translation of authors name and title from Mereschkowsky 1879).

Mereschkowsky C.von (1879): Studien über Protozoen des nördlichen Russland. – Arch. mikrosk. Anat. EntwMech. **16**: 153–248, Tafeln X, XI.

Mereschkovsky K.C. (1880): Materiale dla faune infusorii Chernova Morya. – Trudy imp. S-peterb. Obshch. Estest. **11**: 23–35 (in Russian).

Mermod G. (1914): Recherches sur la faune infusorienne des tourbières et des eaux voisines de Sainte-Croix (Jura vaudois). – Revue suisse Zool. **22**: 31–114, Planches 2, 3.

Merriman D. (1937): Description of *Urostyla polymicronucleata*, sp. nov. – Arch. Protistenk. **88**: 427–430.

Messikommer E. (1954): Zur Kenntnis der niederen Sumpf- und Wasserfauna der Gegend des Pfäffikersees (Kt. Zürich). – Revue suisse Zool. **61**: 635–656.

Mez, C. (1898): Mikroskopische Wasseranalyse. Anleitung zur Untersuchung des Wassers mit besonderer Berücksichtigung von Trink- und Abwasser. Springer, Berlin. 631 pp.

Michiels M. (1974): Biomass determination of some freshwater ciliates. – Biol. Jaarb. **42**: 132–136.

Mihailowitsch B. (1989): Taxonomische und ökologische Untersuchungen an Ciliaten (Protozoa, Ciliophora) in solebelasteten Fliessgewässern. Dissertation Universität Bonn. 291 pp, Anhang 33 pp.

Mihailowitsch B., Wilbert N. (1990): *Bakuella salinarum* nov. spec. und *Pseudokeronopsis ignea* nov. spec. (Ciliata, Hypotrichida) aus einem solebelasteten Fließgewässer des östlichen Münsterlandes, BRD. – Arch. Protistenk. **138**: 207–219.

Milne W. (1886): On a new tentaculiferous protozoon and other infusoria, with notes on reproduction and the function of the contractile vesicle. – Proc. Manchr Fld Nat. Archaeol. Soc. **18**: 48–56, Plate I (usually cited as "Proceedings. Philosophical Society of Glasgow". However, according to the World List the other name is correct).

Mironov O.G., Avdeeva S.U. (1978): The development of protozoa biocenosis on oil in sea water. – Nauch. Dokl. vyssh. Shk. **1978**: 48–50 (in Russian).

Möbius K. (1887): Über den Bau der adoralen Wimperorgane heterotricher und hypotricher Infusorien der Kieler Bucht und über die Fortpflanzung von *Freia ampulla*. – Biol. Zbl. **6** (year 1886–1887): 539–540.

Möbius K. (1888): Bruchstücke einer Infusorienfauna der Kieler Bucht. – Arch. Naturgesch. **54**: 81–116, Tafeln IV–X.

Modeo L., Petroni G., Rosati G., Montagnes D.J.S. (2003): A multidisciplinary approach to describe protists: redescriptions of *Novistrombidium testaceum* Anigstein 1914 and *Strombidium inclinatum* Montagnes, Taylor, and Lynn 1990 (Ciliophora, Oligotrichia). – J. Euk. Microbiol. **50**: 175–189.

Moog O., Chovanec A., Hinteregger J., Römer A. (1999): Richtlinie zur Bestimmung der saprobiologischen Gewässergüte von Fliessgewässern. Bundesministerium für Land- u. Forstwirtschaft, Wasserwirtschaftskataster, Wien. 144 pp.

Mori G., Mattioli M., Madoni P., Ricci N. (1998): The ciliate communities of different habitats of Lake Massaciuccoli (Tuscany): species composition and distribution. – Ital. J. Zool. **65**: 191–202.

Moskowitz N. (1950): The use of protein silver for staining protozoa. – Stain Technol. **25**: 17–20.

Mücke G. (1979): Ökologische Untersuchungen der Ciliaten in Gewässern des Naturschutzgebietes "Heiliges Meer" unter besonderer Berücksichtigung zönologischer Gesichtspunkte. – Arb. Inst. landw. Zool. Bienenkd. **5**: 1–275.

Müller O.F. (1773): Vermium Terrestrium et Fluviatilium, seu Animalium Infusoriorum, Helminthicorum et Testaceorum, non Marinorum, Succincta Historia. Heineck & Faber, Havniae & Lipsiae. 135 pp.

Müller O.F. (1776): Zoologiae danicae prodromus, seu animalium daniae et norvegiae indigenarum characteres, nomina, et synonyma imprimis popularium. Havniae. 282 pp.

Müller O.F. (1779): Zoologia Danica sev Animalivm Daniae et Norvegiae rariorvm ac Minvs notorvm Descriptiones et Historia. Havniae & Lipsiae. 103 pp.

Müller O.F. (1786): Animalcula Infusoria Fluviatilia et Marina, quae Detexit, Sytematice Descripsit et ad Vivum Delineari Curavit. Mölleri, Hauniae. 367 pp, Tabula I–L.

Münch H.-D. (1956): Spirotriche und peritriche Ciliaten, Rotatorien und Archianneliden aus dem Küstengrundwasser. (Anhang: Oligochaeten und Copepoden). – Wiss. Z. Ernst Moritz Arndt-Univ. Greifswald, Mathematisch-naturwissenschaftliche Reihe **5**: 433–436.

Münch F. (1970): Der Einfluß der Temperatur auf den Peptonabbau und die damit verknüpfte Organismensukzession unter besonderer Berücksichtigung der Populationsdynamik der Ciliaten. – Int. Revue ges. Hydrobiol. Hydrogr. **55**: 559–594.

Muramatsu S. (1957): A preliminary survey of protozoa in the City of Hakodate and Oonuma Park. – Misc. Rep. Yamashina Inst. Orn. Zool. **32**: 466–471 (in Japanese with English summary).

Mylnikova S.M. (1981): The study of perifyton ciliates of the inshore zone of the Rybinsk Reservoir. – Informacionnij Byulleten. Biologija Vnutrennich Vod. Akademija Nauk SSSR **No. 52**: 23–26 (in Russian).

Myskova O.M. (1976): Supplement to the fauna of psammophilious infusoria from Peter the Great Bay (Sea of Japan). – Sb. Rab. Inst. biol. Morya **5**: 83–86 (in Russian with English summary).

Naidu K.V. (1965): Studies on freshwater protozoa of South India II: Ciliophora. – Hydrobiologia **25**: 545–570.

Nauwerck A. (1996): Trophische Strukturen im Pelagial des meromiktischen Höllerersees (Oberösterreich). – Ber. Nat.-Med. Ver. Salzburg **11**: 147–178.

Nesterenko G.V., Kovalchuk A.A. (1991): Determination of the ciliates' individual mass by the improved "volumes ratio" method. – Acta hydrochim. hydrobiol. **19**: 23–28.

Ng S.F. (1986): The somatic function of the micronucleus of ciliated protozoa. – Progr. Protistol. **1**: 215–286.

Ng S.F. (1990): Developmental heterochrony in ciliated protozoa: overlap of asexual and sexual cycles during conjugation. – Biol. Rev. **65**: 19–101.

Niebuhr J.W.H. (1989): Ökologische Studien an Bodenprotozoenpopulationen (Protozoa: Ciliophora) auf dem Gelände eines ehemaligen Textilindustriebetriebes in Nordhorn in Bezug auf ihre Verwendung als Bio-Indikatororganismen für anthropogen belastete Böden. – unter Berücksichtigung einiger Verfahren aus der Spurenanalytik. Diplomarbeit Universität Oldenburg. 181 pp.

Nikoljuk V.F., Geltzer J.G. (1972): Pocvennye prostejsie SSSR (Soil protozoa of the USSR). Soviet Academy of Sciences, Tashkent (in Russian). 311 pp.

Ning Y., Wang S., Ma Z., et al (1993): Investigations on the freshwater protozoa in Lanchou. – Chin. J. Zool. **28**: 1–3 (in Chinese).

Nixon K.C., Wheeler Q.-D. (1990): An amplification on the phylogenetic species concept. – Cladistics **6**: 211–223.

Njine T. (1977): Contribution a la connaissance des cilies du Cameroun: ecologie – cytologie I. Etude ecologique. – Annls Stn limnol. Besse **11**: 1–55.

Njine T. (1979): Compléments a l'étude des ciliés libres du Cameroun. – Protistologica **15**: 343–354.

Nobili R. (1957): Contributo all'ecologia dei ciliati psammofili del Golfo di Napoli. – Boll. Zool. **24**: 211–225.

Nosek J.N., Bereczky M.Cs. (1981): Untersuchungen der Bestandsstruktur des Ciliatenplanktons im Haupt- und in einem Nebenarm der Donau mit Hilfe der Produkt-Moment-Korrelations- und der Pfadanalyse. – Arch. Protistenk. **124**: 173–192.

Nusch E.A. (1970): Ökologische und systematische Untersuchungen der Peritricha (Protozoa, Ciliata) im Aufwuchs von Talsperren und Flußstauen mit verschiedenem Saprobitätsgrad (mit Modellversuchen). – Arch. Hydrobiol. Suppl. **37**: 243–386.

Oberschmidleitner R., Aescht E. (1996): Taxonomische Untersuchungen über einige Ciliaten (Ciliophora, Protozoa) aus Belebtschlämmen oberösterreichischer Kläranlagen. – Beitr. Naturk. Oberösterreichs **4**: 3–30.

Oleksiv I. (1985): Species composition and abundance of planktonic infusoria in ponds. – Gidrobiol. Zh. **21**: 89–93 (in Russian with English summary).

Olins D.E., Olins A.L. (1994): The replication band of ciliated protozoa. – Int. Rev. Cytol. **153**: 137–170.

ÖNORM M6118 (2005): Richtlinien für die ökologische Untersuchung und Bewertung von Fließgewässern – Ciliaten (Protozoa). Österreichisches Normungsinstitut, Wien (in press).

Ormancey P. (1852): Catalogue des animaux infusoires observés dans les environs de Lyon. – Annls Soc. linn. Lyon **years 1850–1852**: 277–282.

Ozaki Y., Yagiu R. (1941a): Studies on the marine ciliates of Japan, mainly from the Setonaikai (the Inland Sea of Japan). I. – J. Sci. Hiroshima Univ., Series B. Division I. Zoology **8**: 165–184.

Ozaki Y., Yagiu R. (1941b): Studies on the marine ciliates of Japan, mainly from the Setonaikai (the Inland Sea of Japan). II. – J. Sci. Hiroshima Univ., Series B. Division I. Zoology **9**: 159–180, Plates I, II.

Ozaki Y., Yagiu R. (1943): Studies on the marine ciliates of Japan, mainly from the Setonaikai (the inland sea of Japan). III. – J. Sci. Hiroshima Univ., Series B. Division I. Zoology **10**: 21–52 (in Japanese with English descriptions of new species).

Packroff G. (1992): Faunistic studies on ciliates of three Eifel maar lakes. – Arch. Hydrobiol. Beih. Ergebn. Limnol. **38**: 209–221.

Packroff G., Wilbert N. (1991): Taxonomische Studien über die Ciliatenfauna (Protozoa, Ciliophora) der Eifelmaare. – Arch. Protistenk. **140**: 121–139.

Packroff G., Zwick P. (1996): The ciliate fauna of an unpolluted foothill stream, the Breitenbach, 1: Qualitative aspects. – Limnologica (Berlin) **26**: 255–262.

Palka L. (1991): Étude expérimentale de la recolonisation par les protozoaires d'un sol stérilisé I. – Diversification du peuplement de ciliés. – Revue Écol. Biol. Sol **28**: 125–131.

Pang Y.-B., Zou S.-F., Shen X.-Q., Hou L.-S. (1981): A study on morphogenesis and experimental regeneration of *Urostyla cristata*. – Int. Congr. Protozool. **6**: 286, Abstract.

Pang Y., Gu F., Zhang Z. (1986): Study of the physiological reorganization or *Keronopsis rubra*. – Zool. Res. 7: 263–271 (in Chinese with English summary).

Patrick R. (1961): A study of the numbers and kinds of species found in rivers in Eastern United States. – Proc. Acad. nat. Sci. Philad. **113**: 215–258.

Patrick R., Cairns J.Jr., Roback S.S. (1967): An ecosystematic study of the fauna and flora of the Savannah River. – Proc. Acad. nat. Sci. Philad. **118** (year 1966): 109–407.

Pätsch B. (1974): Die Aufwuchsciliaten des Naturlehrparks Haus Wildenrath. Monographische Bearbeitung der Morphologie und Ökologie. – Arb. Inst. landw. Zool. Bienenkd. **1**: 1–82.

Patterson D.J., Hedley S. (1992): Free-living freshwater protozoa. A colour guide. Wolfe Publishing Ltd., London. 223 pp.

Patterson D.J., Larsen J., Corliss J.O. (1989): The ecology of heterotrophic flagellates and ciliates living in marine sediments. – Progr. Protistol. **3**: 185–277.

Pauw N.De (1969): Contribution a l'etude du plancton dans le port d'Ostende. – Biol. Jaarb. **37**: 186–261.

Pavlovskaya T.W. (1969): Experiments on algal nutrition of some Black Sea ciliates. – Int. Congr. Protozool. **3**: 174, Abstract.

Pavlovskaya T.V. (1970): Feeding of *Uroletopsis viridis* (Infusoria) on diatoms. – Zool. Zh. **49**: 1775–1779 (in Russian with English summary).

Penard, E. (1922): Études sur les infusoires d'eau douce. Georg & Cie, Genève. 331 pp.

Pereyaslawzewa S. (1886): Protozoaires de la mer Noire. – Zap. novoross. Obshch. Estest. **10**: 79–114, 3 plates (in Russian).

Persoone G. (1968): Ecologie des infusoires dans les salissures de substrats immergés dans un port de mer I. - Le film primaire et le recouvrement primaire. – Protistologica **4**: 187–194.

Perty M. (1849a): Über verticale Verbreitung mikroskopischer Lebensformen. – Mitt. naturf. Ges. Bern **year 1849**: 17–45.

Perty M. (1849b): Mikroskopische Organismen der Alpen und der italienischen Schweiz. – Mitt. naturf. Ges. Bern **year 1849**: 153–176.

Perty M. (1852): Zur Kenntniss kleinster Lebensformen nach Bau, Funktionen, Systematik, mit Specialverzeichniss der in der Schweiz beobachteten. Jent & Reinert, Bern. 228 pp, Tafeln I–XVII.

Perty M. (1852a): System der Infusorien. – Mitt. naturf. Ges. Bern **year 1852**: 57–68.

Peschkowsky L. (1927): Skelettgebilde bei Infusorien. – Arch. Protistenk. **57**: 31–57.

Petran A. (1963): Contributii la cunoasterea microfaunei de ciliate psamofile din marea neagra - litoralul Rominesc (Contribution à la connaissance de la microfaune de ciliés psammophiles. Mer Noire - littoral Roumain). – Studii Cerc. Biol., Seria "biologie animala" **15**: 187–197 (in Rumanian with Russian and French summary).

Petran A. (1967): Certecari asupra faunei de ciliate psamobionte la plajele dn sudul litoralulu romanesc al marii negre. – Ecol. Mar. **2**: 169–191.

Petran A. (1968): Sur l'écologie des ciliés psammobiontes de la Mer Noire (littoral Roumain). – Revue roum. Biol., Zoologie **13**: 441–446.

Petran A. (1971): Sur la faune des cilies des sediments sabloneux du littoral Roumain de la Mer Noir. – Cercetari marine **1**: 149–166.

Petroni G., Dini F., Verni F., Rosati G. (2002): A molecular approach to the tangled intrageneric relationships underlying phylogeny in *Euplotes* (Ciliophora, Spirotrichea). – Molecular Phylogenetics and Evolution **22**: 118–130.

Petz W. (1995): Morphology and morphogenesis of *Thigmokeronopsis antarctica* nov. spec. and *T. crystallis* nov spec. (Ciliophora, Hypotrichida) from Antarctic sea ice. – Europ. J. Protistol. **31**: 137–147.

Petz W. (1999): Ciliophora. In: Boltovskoy D. (ed.). South Atlantic Zooplankton, pp. 265–319. Backhuys Publishers, Leiden.

Petz W. (2005): Ciliates. In. Scott F.J. & Marchant H.J. (eds). Antarctic marine protists, pp. 347–448. Australian Biological Resources Study, Canberra and Australian Antarctic Division, Hobart.

Petz W., Foissner W. (1992): Morphology and morphogenesis of *Strobilidium caudatum* (Fromentel), *Meseres corlissi* n. sp., *Halteria grandinella* (Müller), and *Strombidium rehwaldi* n. sp., and a proposed phylogenetic system for oligotrich ciliates (Protozoa, Ciliophora). – J. Protozool. **39**: 159–176.

Petz W., Foissner W. (1993): Morphogenesis in some freshwater tintinnids (Ciliophora, Oligotrichida). – Europ. J. Protistol. **29**: 106–120.

Petz W., Foissner W. (1996): Morphology and morphogenesis of *Lamtostyla edaphoni* Berger and Foissner and *Onychodromopsis flexilis* Stokes, two hypotrichs (Protozoa: Ciliophora) from Antarctic soils. – Acta Protozool. **35**: 257–280.

Petz W., Foissner W. (1997): Morphology and infraciliature of some soil ciliates (Protozoa, Ciliophora) from continental Antarctica, with notes on the morphogenesis of *Sterkiella histriomuscorum*. – Polar Record **33 (187)**: 307–326.

Petz W., Leitner A.R. (2003): Protozoen. In: Hofrichter R. (ed.). Das Mittelmeer. Fauna, Flora, Ökologie. Band II/1: Bestimmungsführer. Prokaryota, Protista, Fungi, Algae, Plantae, Animalia (bis Nemertea), pp. 102–147. Spektrum Akademischer Verlag, Heidelberg Berlin.

Petz W., Valbonesi A. (1998): Bioloogi antartica: biodiversita', strategie evolutive e colonizzazione. Attività a Baia Terra Nova. – Programma Nazionale di Ricerche in Antartide. Rapporto sulla Camapgna Antartica Estata Australe 1997–98 **ANT 98/02**: 97–99.

Petz W., Wirnsberger E., Foissner W. (1988): Die Kleintiere in den Fichtenwaldböden des Oberen Mühlviertels. Leicht zu übersehen, aber sehr wichtig. In: Amt der oberösterreichischen Landesregierung, Abteilung Kultur (ed.). Das Mühlviertel. Natur, Kultur, Leben, pp. 77–88. Wimmer-Druck, Linz.

Petz W., Song W., Wilbert N. (1995): Taxonomy and ecology of the ciliate fauna (Protozoa, Ciliophora) in the endopagial and pelagial of the Weddell Sea, Antarctica. – Stapfia **40**: 1–223.

Pfannkuche O., Jelinek H., Hartwig E. (1975): Zur Fauna eines Süßwasserwattes im Elbe-Aestuar. – Arch. Hydrobiol. **76**: 475–498.

Pigon A. (1953): Oddychanie i zawartosc niektórych fermentów oddechowych u wymoczków I. *Urostyla grandis* Ehrenb. – Folia biol., Kraków **1**: 225–248 (in Polish with Russian and English summary).

Pigon A. (1954): Respiration and cytochrome oxidase content in certain infusoria (*Urostyla grandis* Ehrb., *Spirostomum minus* Roux, *Spirostomum intermedium* Kahl). – Bull. Acad. pol. Sci. Cl. II Sér. Sci. biol. **2**: 131–134.

Pigon A. (1956): Wielocukrowce w organizmie *Urostyla* w czasie zycia aktywnego i encystacji (Polysaccharides in a ciliate *Urostyla* during active life and encystment). – Acta biochim. pol. **3**: 613–625 (in Polish with English summary).

Pigon A. (1960): Changes of enzyme activity during starvation and encystment of a ciliate (*Urostyla*). Amylase, catalase. – Acta biol. cracov., Série zoologique **3**: 59–70.

Pigon A. (1961a): Changes in the respiratory activity during starvation and encystment of a ciliate (*Urostyla*). – Acta biol. cracov., Série zoologique **4**: 33–45.

Pigon A. (1961b): Changes of enzyme activity during starvation and encystment of a ciliate (*Urostyla*). Dipeptidase, amino acid oxidase. – Acta biol. cracov., Série zoologique **4**: 123–142.

Pigon A. (1962): Changes of enzyme activity during starvation and encystment of a ciliate (*Urostyla*). Cathepsin, acid phosphatase. – Acta biol. cracov., Série zoologique **5**: 175–197.

Pigon A. (1963): Observations on the physiology of encystment in a ciliate (*Urostyla*). – Int. Congr. Protozool. **1** (year 1961): 246–249.

Pigon A. (1965): Activation of catalase in homogenates of active and encysted forms of a ciliate (*Urostyla*). – Acta biol. cracov., Série zoologique **8**: 17–29.

Pigon A., Edström J.-E. (1959): Nucleic acid changes during starvation and encystment in a ciliate (*Urostyla*). – Expl Cell Res. **16**: 648–656.

Platt-Rohloff L., Baier E., Görtz H-D., Hausmann K. (1997): New discoveries of endosymbiotic bacteria in ciliates. – J. Euk. Microbiol. **44**: 36A, Abstract 159.

Plückebaum H., Winkelhaus S., Hauser M. (1997): Taxonomic classification and morphological analysis of a red hypotrichous freshwater ciliate. – J. Euk. Microbiol. Suppl. **44**: 27A, Abstract 103.

Poche F. (1913): Das System der Protozoa. – Arch. Protistenk. **30**: 125–321.

Pomp R., Wilbert N. (1988): Taxonomic and ecological studies of ciliates from Australian saline soils: colpodids and hymenostomate ciliates. – Austr. J. mar. Freshwat. Res. **39**: 479–495.

Pratt J.R., Cairns J.Jr. (1985): Functional groups in the protozoa: roles in differing ecosystems. – J. Protozool. **32**: 415–423.

Pratt J.R., Kepner R.L.Jr. (1992): Collecting Aufwuchs on artificial substrata. In: Lee J.J., Soldo A.T. (eds.): Protocols in Protozoology, B-9.1–B-9.7. Society of Protozoologists, Lawrence.

Preisig H.R., Anderson O.R., Corliss J.O., Moestrup Ø., Powell M.J., Roberson R.W., Wetherbee R. (1994): Terminology and nomenclature of protist cell surface structures. – Protoplasma **181**: 1–28.

Prescott D.M. (1989): DNA gains, losses, and rearrangements in eukaryotes. In: DiBerardino M.A., Etkin L.D. (eds.). Developmental Biology, Volume 6, pp. 13–29. Plenum Publishing Corporation.

Prescott D.M. (1994): The DNA of ciliated protozoa. – Microbiol. Rev. **58**: 233–267.

Prescott D.M. (1998): Invention and mystery in hypotricha DNA. – J. Euk. Microbiol. **45**: 575–581.

Primc-Habdija B., Spoljar M., Matonickin R. (2000): Influence of current velocity on ciliate assemblages on travertine barriers in karstic biotopes. – Internat. Assoc. Danube Res. **33**: 279–284.

Primc-Habdija B., Habdija I., Spoljar M. (2000a): Trophic structure of ciliate associations in periphytic communities in karstic waters. – Verh. int. Verein. theor. angew. Limnol. **27**: 2600–2604.

Primc-Habdija B., Habdija I., Plenkovic-Moraj A. (2001): Tufa deposition and periphyton overgrowth as factors affecting the ciliate community on travertine barriers in different current velocity conditions. – Hydrobiologia **457**: 87–96.

Provasoli L., Balamuth W., Becker E.R., Corliss J.O., Hall R.P., Holz G.G.Jr., Levine N.D., Manwell R.D., Nigrelli R.F., Starr R.C., Trager W. (= The Committee on Cultures of the Society of Protozoology) (1958): A catalogue of laboratory strains of free-living and parasitic Protozoa (with sources from which they may be obtained and directions for their maintenance). – J. Protozool. **5**: 1–38.

Prowazek S. (1900): Protozoenstudien II. – Arb. zool. Inst. Univ. Wien **12**: 243–300, Tafeln I, II.

Pütter A. (1900): Studien über Thigmotaxis bei Protisten. – Arch. Anat. Physiol. Suppl. **year 1900**: 243–302.

Puytorac P.de, Mignot J.P., Grain J., Grolière C.A., Bonnet L., Couillard P. (1972): Premier relevé de certains groupes de protozoaires libres sur le territoire de la station de biologie de l'université de Montréal (Saint-Hippolyte, Comté de Terrebonne, Québec). – Naturaliste can. **99**: 417–440.

Puytorac P.de, Grain J., Rodrigues de Santa Rosa M. (1976): A propos de l'ultrastructure corticale du cilie hypotriche *Stylonychia mytilus* Ehrbg., 1838: les caracteristiques du cortex buccal adoral et paroral des Polyhymenophora Jankowski, 1967. – Trans. Am. microsc. Soc. **95**: 327–345.

Quennerstedt A. (1865): Bidrag till Sveriges infusorie-fauna jemte en kort framställning af infusionsdjurens organisation. I. – Acta Univ. lund. **2**: 1–64, Plates I, II (in Swedish).

Quennerstedt A. (1867): Bidrag till sveriges infusorie-fauna. II. – Acta Univ. lund. **4**: 1–48, Plates I, II (in Swedish).

Quennerstedt A. (1869): Bidrag till sveriges infusorie-fauna. III. – Acta Univ. lund. **6**: 1–35, 1 plate (in Swedish).

Raabe H. (1946): Aparat jadrowy *Urostyla grandis* Ehrbg. Czesc I. Aparat jadra malego (L'appareil nucléaire d '*Urostyla grandis* Ehrbg. Partie I. Appareil micronucléaire). – Annls Univ. Mariae Curie-Sklodowska, Sect. C. Biologia **1**: 1–34, Tablice I–III (in Polish and French).

Raabe H. (1947): Appareil nucléaire d'*Urostyla grandis* Ehrbg. Partie II. Appareil macronucléaire. – Annls Univ. Mariae Curie- Sklodowska, Sect. C. Biologia **1**: 133–170, Table V–VII.

Raikov I.B. (1962): Les ciliés mésopsammiques du littoral de la Mer Blanche (U.R.S.S.) avec une description de quelques espèces nouvelles ou peu connues. – Cah. Biol. mar. **3**: 325–361.

Raikov I.B. (1963): Ciliates of the mesopsammon of the Ussuri Gulf (Japan Sea). – Zool. Zh. **42**: 1753–1767 (in Russian with English summary).

Raikov I.B. (1969): The macronucleus of ciliates. In: Chen T.-T. (ed.). Research in Protozoology Volume 3, pp. 1–128. Pergamon Press, Oxford, London, Edinburgh, New York, Toronto, Sydney, Paris, Braunschweig.

Raikov I.B. (1982): The protozoan nucleus. Morphology and evolution. In: Alfert M., Beermann W., Goldstein L., Porter K.R., Sitte P. (eds.). Cell biology monographs, Volume 9, Springer, Wien, New York. I–XV, 474 pp.

Raikov I. (1996): Advances in the study of nuclear phenomena during sexual processes in ciliates. – Europ. J. Protistol. **32**, Suppl. I: 128–134.

Raikov I.B., Kovaleva V.G. (1968): Complements to the fauna of psammobiotic ciliates of the Japan Sea (Posjet gulf). – Acta Protozool. **6**: 309–333.

Rao G.C. (1969): The marine interstitial fauna inhabiting the beach sands of Orissa Coast. – J. zool. Soc. India **21**: 89–104.

Rao G.C. (1974): On the geographical distribution of interstitial fauna of marine beach sand. – Proc. Indian Acad. Sci., Section B **38**: 164–178.

Rao G.C., Ganapati P.N. (1968): The interstitial fauna inhabiting the beach sands of Waltair Coast. – Proc. natn. Inst. Sci. India, Ser. B (Biological sciences) **34**: 82–125.

Rees J.van (1884): Protozën der Oosterschelde (Protozoaires de l'escaut de l'est). – Tijdschr. ned. dierk. Vereen. Supplement Deel I (years 1883–1884): 592–673, Plaat XVI (in Dutch with French translation).

Remane A., Storch V., Welsch U. (1976): Systematische Zoologie. Stämme des Tierreichs. G. Fischer, Stuttgart. I–XV + 678 pp.

Reuter J. (1961): Einige faunistische und ökologische Beobachtungen über Felsentümpel-Ziliaten. – Acta zool. fenn. **99**: 1–42.

Reuter J. (1963): Some ciliates from Western Norway. – Univ. Bergen Sarsia **10**: 1–8.

Ricci N., Santangelo G., Luporini P. (1982): Researches on the coast of Somalia. Sand-dwelling ciliates (Pupplicazioni del centro di studio per la faunistica ed ecologia tropicali del C. N. R.: CCXL). – Monitore zool. ital., (N. S.) **Supplemento 17**: 115–148.

Riedel-Lorjé J.C. (1981): Untersuchungen über den Indikationswert von Aufwuchs in Süß- und Brackwasserzonen des Elbe-Aestuars unter Berücksichtigung industrieller Einleitungen. – Arch. Hydrobiol. **Suppl. 61**: 153–226.

Riess F. (1840): Beiträge zur Fauna der Infusorien, mit dem beigefügten Ehrenberg'schen Systeme. – Dissertation Universität Wien, 40 pp.

Riggenbach E. (1922): Beiträge zur Faunistik, Biologie und Oekologie der Heliozoen u. Ciliaten von Basel und Umgebung. Dissertation Universität Basel. 82 pp.

Rios R.M., Pérez-Silva J., Fedriani C. (1988): Cytochemical and enzymatic studies on the cyst wall of *Urostyla grandis* (Hypotrichida, Urostylidae). – Cytobios **56**: 163–169.

Rios R.M., Torres A., Calvo P., Fedriani C. (1985): The cyst of *Urostyla grandis* (Hypotrichida: Urostylidae): ultrastructure and evolutionary implications. – Protistologica **21**: 481–485.

Robinson B.S., Bamforth S.S., Dobson P.J. (2002): Density and diversity of protozoa in some arid Australian soils. – J. Euk. Microbiol. **49**: 449–453.

Rossolimo L.L. (1923): Zur Protozoenfauna des Baikalsees. – Russk. gidrobiol. Zh. **2**: 74–82 (in Russian with German summary).

Roux J. (1900): Note sur les infusoires ciliés du Lac Léman. – Revue suisse Zool. **8**: 459–465.

Roux J. (1901): Faune infusorienne des eaux stagnantes des environs de Genève. – Mém. Inst. natn. génev. **19**: 1–149, Planches I–VIII.

Rühmekorf T. (1935): Morphologie, Teilung und Hungerformen von *Keronopsis*. – Arch. Protistenk. **85**: 255–288.

Ruinen J. (1938): Notizen über Ciliaten aus konzentrierten Salzgewässern. – Zoöl. Meded., Leiden **20**: 243–256.

Ruiz A. (1961): Ciliados de vida libre de la ciudad Universitaria, Costa Rica. – Revta Biol. trop. **9**: 209–214.

Ruthmann A. (1972): Division and formation of the macronuclei of *Keronopsis rubra*. – J. Protozool. **19**: 661–666.

Ruthmann A. (1973): Fine structure of the macronuclear anlage of the hypotrichous ciliate *Keronopsis rubra*. – J. Protozool. **20**: 648–651.

Ruthmann A., Noll-Altmann G. (1980): Effect of bacterial symbiont on cell division in a ciliate. In: Schwemmler W., Schenk H.E.A. (eds.). Endocytobiology endosymbiosis and cell biology Volume 1, pp. 361–370. Walter de Gruyter & Co, Berlin, New York.

Ryan P.G., Watkins B.P., Lewis Smith R.I., Dastych H., Eicker A., Foissner W., Heatwole H., Miller W.R., Thompson G. (1989): Biological survey of Robertskollen, western Dronning Maud Land: area description and preliminary species lists. – S. Afr. T. Antarkt. **19**: 10–20.

Sandon H. (1927): The composition and distribution of the protozoan fauna of the soil. In: Crew F.A.E., Cutler D.W. (eds.). Biological monographs and manuals, Volume 7, Oliver & Boyd, Edinburgh London. I–XV + 237 pp + Charts I–III.

Santangelo G., Lucchesi P. (1995): Spatial distribution pattern of ciliated protozoa in a Mediterranean interstitial environment. – Adv. microb. Ecol. **9**: 47–54.

Sapra G.R., Steinbrueck G., Ammermann D. (1985): Characterization of gene sized DNA molecules in several species of hypotrichous ciliates. – Indian J. exp. Biol. **23**: 625–628.

Sarmiento L., Guerra H. (1960): Protozoarios de las aguas de Villa, con la descripcion de tres nuevas especies. – Publnes Mus. Hist. nat., Lima, Ser. A. Zoologia **19**: 1–25, Laminas I–III.

Schewiakoff W. (1893): Über die geographische Verbreitung der Süsswasser-Protozoën. – Zap. imp. Akad. Nauk, 7e Série **41**: 1–201, Tafeln I–IV.

Schlegel M., Steinbrück G. (1986): Interspezifische Variabilität bei hypotrichen Ciliaten. – Z. zool. Syst. Evolut.-forsch. **24**: 247–266.

Schmarda L.K. (1846): Kleine Beiträge zur Naturgeschichte der Infusorien. Haas, Wien. 62 pp, Tafeln I, II.

Schmarda L.C. (1847): Zur Kenntniss der adriatischen Infusorien- Fauna. – Medicinische Jahrbücher des kaiserl. königl. österreichischen Staates **59**: 1–20.

Schmerenbeck W. (1975): Experimentelle Untersuchungen an strömenden Modellgewässern zur Frage der Beziehung zwischen dem Abbau organischer Substanz und der Ciliatenbesiedlung. – Arb. Inst. landw. Zool. Bienenkd. **2**: 1–95.

Schmidt H. (1913): Beitrag zur Kenntnis der Mastigophoren und Ciliaten der Umgegend von Bonn. – Sber. naturh. Ver. preuss. Rheinl. Westf. **year 1913**: 73–83.

Schmidt H. (1916): Beitrag zur Protozoenfauna der Rheinprovinz und Westfalens. – Verh. naturh. Ver. preuss. Rheinl. **72** (year 1915): 59–95.

Schmidt S.L., Bernhard D., Foissner W., Berger H., Schlegel M. (2004a): Phylogenetische Analyse der Spirotrichea (Ciliophora) mittels Gensequenzen der small subunit (ssu) rDNA. – 23. Jahrestagung der Deutschen Gesellschaft für Protozoologie, Innsbruck Abstract.

Schmidt S.L., Bernhard D., Foissner W., Berger H., Schlegel M. (2004b): Phylogenetic analysis within the class Spirotrichea (Ciliophora) inferred from small subunit (ssu) rDNA. – Jahresversammlung der Deutschen Zoologischen Gesellschaft **97**: 174, Abstract V–S1.3.

Schmitz M. (1983): Ökologische und systematische Untersuchungen an Ciliaten (Protozoa, Ciliophora) des Rheines bei Bonn, Bundesrepublik Deutschland. – Preprint S. I. L. **1983**: 1–5.

Schmitz M. (1985): Ökologische und systematische Untersuchungen an Ciliaten (Protozoa, Ciliophora) des Rheines bei Bonn, Bundesrepublik Deutschland. – Verh. int. Verein. theor. angew. Limnol. **22**: 2292–2296.

Schmitz, M.M. (1986): Ökologische und systematische Untersuchungen an Ciliaten (Protozoa, Ciliophora) im oberen Niederrhein. Dissertation Universität Bonn. 246 pp, 149 pp Anhang.

Schneider W. (1930): Die Verbreitung des Tektins bei den Ciliaten. – Arch. Protistenk. **72**: 482–537.

Schwarz M.V.J., Frenzel P. (2003): Population dynamics and ecology of ciliates (Protozoa, Ciliophora) in an anoxic rice field soil. – Biol. Fertil. Soils **38**: 245–252.

Scorza J.V., Núñez Montiel B.O. (1954): Estudio experimental sobre la sucesion de protozoarios que se desarrolla en las infusiones de musgo y de las variaciones de pH. qe la acompañan. – Acta biol. venez. **1**: 213–230.

Seckt H. (1924): Estudios hidrobiologicos en la Argentina. Observaciones sobre la «autopurificación» de los ríos. – Revta Univ. nac. Córdoba **11**: 182–190.

Senler N.G., Yildiz I. (1998): Van Gölü'ne Dökülen Bes Akarsudai Siliyat Türleri (Ciliophora: Protozoa) Üzerine Bir Arastirma (An investigation on ciliates species on five rivers flowing into lake Van). – Yüzüncü Yil Üniversitesi Fen Bilimerli Enstitüsü Dergisi **5**: 1–19 (in Turkish with English summary).

Senler N.G., Yildiz I. (2004): Faunistic and morphological studies on ciliates (Protozoa, Ciliophora) from a small pond, with responses of ciliate populations to changing environmental conditions. – Turkish Journal of Zoology **28**: 245–265.

Senler N.G., Biyik H., Ögün E. (1996): The pollution parameters and protozoological investigations on some rivers flowing into lake Van. Vol. **13**, pp. 180–190. Ulusal Biyoloji Kongresi, Istanbul (in Turkish with English summary).

Senler N.G., Biyik H., Ögün E., Yildiz I. (1998): The pollution parameters and protozoological investigations on three rivers flowing into Lake Van. – Bulletin of Pure and Applied Sciences **17A**: 35–50.

Seravin L.N. (1970): Left and right spiralling round the long body axis in ciliate protozoa. – Acta Protozool. **7**: 313–323.

Shao C., Li L., Hu X., Song W. (2005): Morphology and morphogenesis of the limnetic ciliate *Trichototaxis hembergeri* nov. spec. (Ciliophora, Hypotrichida) from China. – Int. Congr. Protozool., 12: 9, Abstract S1.1-02 (this name is disclaimed for nomenclatural purposes; ICZN 1999, Article 8.3).

Shawhan F.M., Johnson L.P., Jahn T.L. (1947): Protozoa of Iowa. – Proc. Iowa Acad. Sci. **54**: 353–367.

Shen Y.-F. (1983): Protozoa of the Tibetan Plateau. In: Chiang S.-C., Shen Y.-F., Gong X.-J. (eds.): Aquatic invertebrates of the Tibetan Plateau, pp. 39–334. Academia Sinica Press, Peking (in Chinese). 492 pp.

Shen Y., Gong X. (1989): Analysis of protozoan fauna in Suoxiyu nature preserve area. In: Li S. (ed.). The aquatic fauna of Suoxiyu nature preserve area, Hunan, China, pp. 78–87. Academia Sinica Press, Peking (in Chinese with English summary).

Shen Y., Jiang X. (1979): Evaluation of the self-purification effects of water bodies through zooplankton. – Oceanologia Limnol. sin. **10**: 161–173 (in Chinese with English summary).

Shen Y., Liu J., Song B., Gu M. (1992): Protozoa. In: Yin W. et al. (eds.). Subtropical soil animals of China, pp. 97–156. Science Press, Beijing (in Chinese).

Shen Y.F., Feng W.S., Gu M.R., Wang S.D., Wu J.Z., Tan Y.Y. (1994): Monitoring of river pollution. China Architecture & Building press. 308 pp.

Shi X. (1993): How to distinguish the natural groups in hypotrichous ciliates. – Int. Congr. Protozool. **9**: 116, Abstract 454.

Shi X.-L. (1999): Systematic revision of the order Hypotrichida I. Protohypotrichina and Stichotrichina (Ciliophora). – Acta Zootax. sinica **24**: 241–264 (in Chinese with English summary).

Shi X.-L. (1999a): Systematic revision of the order Hypotrichida II. Urostylina (Ciliophora). – Acta Zootax. sinica **24**: 361–371 (in Chinese with English summary).

Shi X., Song W., Shi, X. (1999): Systematic revision of the hypotrichous ciliates. In: Song W. (ed.). Progress in Protozoology, pp. 77–154. Qingdao Ocean University Press, Qingdao (in Chinese).

Shi X., Song W., Shi X. (1999a): Morphogenetic modes of hypotrichous ciliates. In: Song W. (ed.). Progress in Protozoology, pp. 189–210. Qingdao Ocean University Press, Qingdao (in Chinese).

Shi X. (1993): How to distinguish the natural groups in hypotrichous ciliates. – Int. Congr. Protozool. **9**: 116, Abstract 454.

Shi X., He W. (1989): Morphology and morphogenesis in *Urostyloides sinensis* gen. and sp. nov., a new hypotrich ciliate from China. – Int. Congr. Protozool. **8**: 99, Abstract 17D0945.

Shigematsu A. (1953): Two new marine ciliates. – J. Sci. Hiroshima Univ., Series B. Division I. Zoology **14**: 47–50.

Shin M.K. (1994): Systematics of Korean hypotrichs (Ciliophora, Polyhymenophora, Hypotrichida) and molecular evolution of hypotrichs. Dissertation Seoul National University, Seoul. 270 pp.

Shin M.K., Kim W. (1993): Redescription of two holostichid species of genus *Holosticha* Wrzesniowski 1877 (Ciliophora, Hypotrichida, Holostichidae) from Seoul, Korea. – Korean J. syst. Zool. **1**: 251–259.

Shin M.K., Kim W. (1995): Hypotrichs (Ciliophora, Hypotrichida) from Ullung Island, Korea. – Korean J. Zool. **38**: 160–166.

Shin M.K., Hwang U.W., Kim W., Wright A.-D., Krawczyk C., Lynn D.H. (2000): Phylogenetic position of the ciliates *Phacodinium* (order Phacodiniida) and *Protocruzia* (subclass Protocruziidia) and systematics of the spirotrich ciliates examined by small subunit ribosomal RNA gene sequences. – Europ. J. Protistol. **36**: 293–302.

Sick F. (1933): Die Fauna der Meeresstrandtümpel des Bottsandes (Kieler Bucht). Ein Beitrag zur Ökologie und Faunistik von Brackwassergebieten. – Z. wiss. Zool. **2 (B)**: 54–96.

Siemińska A., Siemińska J. (1967): Flora i fauna w rejonie Zespolu Gospodarstw Doswiadczalnych PAN i Zbiornika Goczalkowickiego na Slasku (Flora and fauna in the region of the experimental farms of the Polish Academy of Sciences and of Goczalkowice Reservoir, Silesia). – Acta hydrobiol., Kraków **9**: 1–109 (in Polish with English summary).

Sládeček V. (1963): A guide to limnosaprobical organisms. – Sb. vys. Sk. chem.-technol. Praze **7 (2)**: 543–612.

Sládeček V. (1973): System of water quality from the biological point of view. – Arch. Hydrobiol. Beih. Ergebn. Limnol. **7**: I– IV, 1–218.

Sládeček V. (1988): Spodobrví nálevníci v mikrobentosu vod. – Metodicky pokyn MLVH CSR **61**: 9–16 (in Czech).

Sládeček V., Sládecková A. (1997): Atlas vodních organismu se zretelem na vodárenství, povrchové vody a cistírny odpadních vod 1. díl: Konzumenti. Ceská vedeckotechnická vodohospodárská spolecnost, Praha. 358 pp.

Sládeček V., Zelinka M., Rothschein J., Moravcova V. (1981): Biologicky rozbor povrchové vody. Komentár k CSN 83 0532 - cásti 6: Stanoveni saprobniho indexu. Vydalo Vydavatelstvi Uradu pro normalizaci a mereni, Praha (in Czech). 186 pp.

Small E.B., Lynn D.H. (1985): Phylum Ciliophora Doflein, 1901. In: Lee J.J., Hutner S.H., Bovee E.C. (eds.). An illustrated guide to the protozoa, pp. 393–575. Society of Protozoologists, Lawrence.

Smith H.G. (1975): Protozoaires terricoles de l'Ile de la Possession. – Revue Écol. Biol. Sol **12**: 523–530.

Smith H.G. (1978): The distribution and ecology of terrestrial protozoa of sub-Antarctic and maritime Antarctic islands. – Br. Antarct. Surv. Sci. Rep. **No. 95**: 1–104.

Smith I.F. (1914): A preliminary report on the infusoria of Kansas. – Kans. Univ. Sci. Bull. **9**: 147–174, Plates XXXIII–XLIX.

Smith J.C. (1900): Notices of some undescribed infusoria, from the infusorial fauna of Louisiana. – Trans. Am. microsc. Soc. **21**: 87–94, Plate VI.

Smith J.C. (1904): A preliminary contribution to the protozoan fauna of the Gulf Biologic Station with notes on some rare species. – Rep. Lousiana Gulf Biologic Station for 1903 **2** (year 1904): 43–55.

Snoeyenbos-West O.L., Salcedo T., McManus G.B., Katz L.A. (2002): Insights into the diversity of choreotrich and oligotrich ciliates (Class: Spirotrichea) based on genealogical analyses of multiple loci. – Int. J. syst. evolut. Microbiol. **52**: 1901–1913.

Snoeyenbos-West O.L.O., Cole J., Campbell A., Coats D.W., Katz L.M. (2004): Molecular phylogeny of phyllopharyngean ciliates and their group I introns. – J. Euk. Microbiol. **51**: 441–450.

Sola A., Longás J.F., Serrano S., Guinea A. (1996): Influence of environmental characteristics on the distribution of ciliates in the River Henares (Central Spain). – Hydrobiologia **324**: 237–252.

Sondheim M. (1929): Protozoen aus der Ausbeute der Voeltzkowschen Reisen in Madagaskar und Ostafrika. – Abh. senckenb. naturforsch. Ges. **41**: 283–313, Tafeln 1, 2.

Song W. (1990): Morphologie und Morphogenese des Bodenciliaten *Periholosticha wilberti* nov. spec. (Ciliophora, Hypotrichida). – Arch. Protistenk. **138**: 221–231.

Song W. (1996): Description of the marine ciliate *Pseudoamphisiella lacazei*, (Maupas, 1883) nov. gen., nov. comb. (Protozoa, Ciliophora, Hypotrichida). – Oceanologia Limnol. sin. **27**: 18–22 (in Chinese with English summary).

Song W. (2001): Morphology and morphogenesis of the marine ciliate *Ponturostyla enigmatica* (Dragesco & Dragesco-Kernéis, 1986) Jankowski, 1989 (Ciliophora, Hypotrichida, Oxytrichidae). – Europ. J. Protistol. **37**: 181–197.

Song W. (2003): Free-living pathogenic ciliates in shrimp-farming waters. In: Song W., Y. Zhao, Xu K., Hu X., Gong J. (eds): Pathogenic protozoa in mariculture, pp. 71–112. Science Press, Beijing (in Chinese).

Song W., Chen Z. (1999): Ecological studies on aufwuchs ciliates from a eutrophic freshwater pond. In: Song W. (ed.). Progress in Protozoology, pp. 325–362. Qingdao Ocean University Press, Qingdao (in Chinese).

Song W., Packroff G. (1993): Beitrag zur Morphogenese des marinen Ciliaten *Diophrys scutum* (Dujardin, 1841) (Ciliophora, Hypotrichida). – Zool. Jb. Anat. **123**: 85–95.

Song W., Wang M. (1993): A brief revision of disease-causing ciliates (Protozoa: Ciliophora) from marine culture water bodies. – Marine Sciences **No. 4**: 41–47 (in Chinese with English summary).

Song, W., Wang, M. (1999): New name list of marine ciliates in China. In: Song W. (ed.). Progress in protozoology, pp. 65–76. Qingdao Ocean University Press, Qingdao (in Chinese).

Song W., Warren A. (1996): A redescription of the marine ciliates *Uroleptus retractilis* (Claparède and Lachmann, 1858) comb. n. and *Epiclintes ambiguus* (Müller, 1786) Bütschli, 1889 (Ciliophora, Hypotrichida). – Acta Protozool. **35**: 227–234.

Song W., Warren A. (2000): *Pseudoamphisiella alveolata* (Kahl, 1932) nov. comb., a large marine hypotrichous ciliate from China (Protozoa, Ciliophora, Hypotrichida). – Europ. J. Protistol. **36**: 451–457.

Song W., Wilbert N. (1988): *Parabakuella typica* nov. gen., nov. spec. (Ciliata, Hypotrichida) aus dem Edaphon eines Standortes in Qingdao, China. – Arch. Protistenk. **135**: 319–325.

Song W., Wilbert N. (1989): Taxonomische Untersuchungen an Aufwuchsciliaten (Protozoa, Ciliophora) im Poppelsdorfer Weiher, Bonn. – Lauterbornia **Heft 3**: 2–221.

Song W., Wilbert N. (1994): Morphogenesis of the marine ciliate *Diophrys oligothrix* Borror, 1965 during the cell division (Protozoa, Ciliophora, Hypotrichida). – Europ. J. Protistol. **30**: 38–44.

Song W., Wilbert N. (1997): Morphological studies on some free living ciliates (Ciliophora: Heterotrichida, Hypotrichida) from marine biotopes in Qingdao, China, with descriptions of three new species: *Holosticha warreni* nov. spec., *Tachysoma ovata* nov. spec. and *T. dragescoi* nov. spec. – Europ. J. Protistol. **33**: 48–62.

Song W., Wilbert N. (1997a): Morphological investigations on some free living ciliates (Protozoa, Ciliophora) from China Sea with description of a new hypotrichous genus, *Hemigastrostyla* nov. gen. – Arch. Protistenk. **148**: 413–444.

Song W., Wilbert N. (2002): Faunistic studies on marine ciliates from the Antarctic benthic area, including descriptions of one epizoic form, 6 new species and, 2 new genera (Protozoa: Ciliophora). – Acta Protozool. **41**: 23–61.

Song W., Wilbert N., Berger H. (1992): Morphology and morphogenesis of the soil ciliate *Bakuella edaphoni* nov. spec. and revision of the genus *Bakuella* Agamaliev & Alekperov, 1976 (Ciliophora, Hypotrichida). – Bull. Br. Mus. nat. Hist. (Zool.) **58**: 133–148.

Song W., Cheng X., Liu G., Wu L. (1993): Ecological studies on the periphytonic protozoans in Xiaoxihu Lake, Qingdao. – J. Ocean Univ. Qingdao **23**: 99–106 (in Chinese with English summary).

Song W., Warren A., Hu X. (1996/1997): Morphology and morphogenesis of *Pseudoamphisiella lacazei* (Maupas, 1883) Song, 1996 with suggestion of establishment of a new family Pseudoamphisiellidae nov. fam. (Ciliophora, Hypotrichida). – Arch. Protistenk. **147**: 265–276.

Song W., Petz W., Warren A. (2001): Morphology and morphogenesis of the poorly-known marine urostylid ciliate, *Metaurostylopsis marina* (Kahl, 1932) nov. gen., nov. comb. (Protozoa, Ciliophora, Hypotrichida). – Europ. J. Protistol. **37**: 63–76.

Song W., Wilbert N., Warren A. (2002): New contribution to the morphology and taxonomy of four marine hypotrichous ciliates from Qingdao, China (Protozoa: Ciliophora). – Acta Protozool. **41**: 145–162.

Song W., Wilbert N., Chen Z., Shi X. (2004): Considerations on the systematic position of *Uronychia* and related euplotids based on the data of ontogeny and 18S rRNA gene sequence analyses, with morphogenetic redescription of *Uronychia setigera* Calkins, 1902 (Ciliophora: Euplotida). – Acta Protozool. **43**: 313–328.

Song W., Wilbert N., Hu X. (2004a): New contributions to the marine hypotrichous ciliate, *Pseudokeronopsis qingdaoensis* Hu & Song, 2000 (Protozoa: Ciliophora: Stichotrichida). – Cah. Biol. mar. **45**: 335–342.

Song W., Sun P., Ji D. (2004b): Redefinition of the yellow hypotrichous ciliate, *Pseudokeronopsis flava* (Hypotrichida: Ciliophora). – J. mar. biol. Assoc. U.K. **84**: 1137–1142.

Sonneborn T.M. (1949): Ciliated protozoa: cytogenetics, genetics, and evolution. – A. Rev. Microbiol. **3**: 55–80.

Spandl H. (1926a): Die Tierwelt vorübergehender Gewässer Mitteleuropas. – Arch. Hydrobiol. **16**: 74–132.

Spandl H. (1926b): Wissenschaftliche Forschungsergebnisse aus dem Gebiete der unteren Donau und des Schwarzen Meeres. II. Die Süßwasser-Mikrofauna. (Protozoa, Rotatoria, Gastrotricha, Oligochaeta, Hirudinea, Euphyllopoda, Cladocera, Ostracoda, Copepoda, Mysidacea et Amphipoda.). – Arch. Hydrobiol. **16**: 528–604.

Spandl H. (1926c): Die Tierwelt der unterirdischen Gewässer. – Speläologische Monographien **11**: I–XII & 235 pp.

Sparagano O., Grolière C.-A. (1991): Evaluation de la qualité des eaux d'une rivière avec les protozoaires ciliés comme bioindicateurs de pollution. Comparaison avec la physico-chimie. – Hydroecologie Appliquee **1**: 43–62.

Spoon D.M., Alling A.K. (1993): Preclosure survey of aquatic microbiota of biosphere II. – J. Euk. Microbiol. Suppl. **40**: 41A, Abstract 245.

Šrámek-Hušek R. (1945): Nové a málo známé vodní organismy. 4. *Keronopsis monilata* Kahl (Ciliata-Hypotricha). – Veda prír. **23**: 247–248 (in Czech).

Šrámek-Hušek R. (1952a): O nekolika zajímavych nálevnících z ceskych mechu. – Cslká Biol. **1**: 175–184, Tabule I (in Czech; identical with Srámek-Hušek 1952b).

Šrámek-Hušek R. (1952b): Über einige bemerkenswerte Ciliaten aus böhmischen Moosen. – Chekh. Biol. **1**: 367–376 (in Russian with German summary; identical with Srámek-Hušek 1952a).

Šrámek-Hušek R. (1953): Predbezny seznam nálevníku (Ciliata) z cech, moravy a slezska II. Spirotricha, Peritricha a Chonotricha (Vorläufiges Verzeichnis der in Böhmen, Mähren und Schlesien bis zum Jahre 1950 gefundenen Ciliatenarten). – Cas. národ. Mus. **122**: 76–88 (in Czech with Russian and German summary).

Šrámek-Hušek R. (1954): Neue und wenig bekannte Ciliaten aus der Tschechoslowakei und ihre Stellung im Saprobiensystem. – Arch. Protistenk. **100**: 246–267.

Šrámek-Hušek R. (1957): K poznání nálevniku ostravského kraje (Zur Kenntnis der Ciliaten des Ostrauer-Gebietes [Tschechoslovakei]). – Vest. csl. zool. Spol. **21**: 1–24 (in Czech with German summary).

Statkewitsch P. (1905): Galvanotropismus und Galvanotaxis der Ciliata. Erste Mitteilung. – Z. allg. Physiol. **4**: 296–332.

Stehle M.E. (1920): A preliminary survey of the protozoa of Mirror Lake, on the Ohio State University campus. – Ohio J. Sci. **20**: 89–127, Plates I–III.

Stein, F. (1859): Der Organismus der Infusionsthiere nach eigenen Forschungen in systematischer Reihenfolge bearbeitet. I. Abtheilung. Allgemeiner Theil und Naturgeschichte der hypotrichen Infusionsthiere. Engelmann, Leipzig. I–XII, 206 pp, Tafeln I–XIV.

Stein F. (1859a): Characteristik neuer Infusorien-Gattungen. – Lotos **9**: 2–5, 57–60.

Stein F. (1863): Neue Infusorienformen in der Ostsee. – Amtliche Berichte Deutscher Naturforscher und Ærzte in Karlsbad **37. Versammlung im September 1862**: 161–162, 165–166.

Stein F. (1864): Über den Proteus tenax von O. F. Müller und über die Infusoriengattungen *Distigma* Ehrbg. und *Epiclintes* Stein. – Sber. K. böhm. Ges. Wiss. **year 1864**: 41–46.

Stein F. (1867): Der Organismus der Infusionsthiere nach eigenen Forschungen in systematischer Reihenfolge bearbeitet. II. Abtheilung. 1) Darstellung der neuesten Forschungsergebnisse über Bau, Fortpflanzung und Entwickelung der Infusionsthiere. 2) Naturgeschichte der heterotrichen Infusorien. Engelmann, Leipzig. I–VIII, 355 pp + Tafeln I–XVI.

Steinbrück G. (1990): Recent advances in the study of ciliate genes. – Europ. J. Protistol. **26**: 2–14.

Stella E. (1947): Il ciclo planctonico annuale di un fontanile: la fonte di giuturna al Foro Romano. – Boll. Pesca Piscic. Idrobiol. **23**: 21–35.

Stella E., Campea R. (1948): La fauna protozoaria del Tevere. – Boll. Pesca Piscic. Idrobiol. **24**: 151–169.

Stephanides T. (1948): A survey of the freshwater biology of Corfu and of certain other regions of Greece. – Praktika ell. 'udrobiol. Inst. **2**: 1–263.

Sterki V. (1878): Beiträge zur Morphologie der Oxytrichinen. – Z. wiss. Zool. **31**: 29–58, Tafel IV.

Stiller J. (1974a): Ergänzungen der von Fauré-Fremiet vorgenommenen Neuordnung der hypotrichen Ciliaten. – Annls hist.-nat. Mus. natn. hung. **66**: 129–133.

Stiller J. (1974b): Járólábacskás csillósok – Hypotrichida. – Fauna Hung. **115**: 1–187 (in Hungarian).

Stokes A.C. (1885): Some new infusoria from American fresh waters. – Ann. Mag. nat. Hist., Serie 5 **15**: 437–449, Plate XV.

Stokes A.C. (1886): Some new hypotrichous infusoria. – Proc. Am. phil. Soc. **23**: 21–30, 1 plate.

Stokes A.C. (1886a): Some new infusoria from American fresh waters. – No. 2. – Ann. Mag. nat. Hist., Serie 5 **17**: 98–112, Plate I.

Stokes A.C. (1887): Some new hypotrichous infusoria from American fresh waters. – Ann. Mag. nat. Hist., Serie 5 **20**: 104–114, Plate III.

Stokes A.C. (1888): A preliminary contribution toward a history of the fresh-water infusoria of the United States. – J. Trenton nat. Hist. Soc. **1**: 71–319, Plates I–XIII.

Stokes A.C. (1891): Notes of new infusoria from the fresh waters of the United States. – Jl R. microsc. Soc. **year 1891**: 697–704, Plate X.

Stokes A.C. (1894): Notices of presumably undescribed infusoria. – Proc. Am. phil. Soc. **33**: 338–345, Plate XXI.

Stout J.D. (1958): Biological studies of some tussock-grassland soils VII. Protozoa. – N. Z. Jl agric. Res. **1**: 974–984.

Stout J.D. (1960): Biological studies of some tussock-grassland soils XVIII. Protozoa of two cultivated soils. – N. Z. Jl agric. Res. **3**: 237–243.

Stout J.D. (1961): Biological and chemical changes following scrub burning on a New Zealand hill soil 4. Microbiological changes. – N. Z. Jl Sci. **4**: 740–752.

Stout J.D. (1962): An estimation of microfaunal populations in soils and forest litter. – J. Soil Sci. **13**: 314–320.

Stout J.D. (1963): Some observations on the protozoa of some beechwood soils on the Chiltern Hills. – J. Anim. Ecol. **32**: 281–287.

Stout J.D. (1968): The significance of the protozoan fauna in distinguishing mull and mor of beech (*Fagus silvatica* L.). – Pedobiologia **8**: 387–400.

Stout J.D. (1970): The bacteria and protozoa of some soil samples from Scoresby Land, East Greenland. – Meddr Grønland **184**: 3–24.

Stout J.D. (1984): The protozoan fauna of a seasonally inundated soil under grassland. – Soil Biol. Biochem. **16**: 121–125.

Streble H., Krauter D. (1982): Das Leben im Wassertropfen: Mikroflora und Mikrofauna des Süßwassers; ein Bestimmungsbuch. Franckh'sche Verlagshandlung, Stuttgart. 367 pp.

Strethill Wright T. (1862): Observations on British Protozoa. – Q. Jl microsc. Sci., New Series **2**: 217–221, Plates VIII, IX.

Strüder-Kypke M.C., Lynn D.H. (2003): Sequence analyses of the small subunit rRNA gene confirm the paraphyly of oligotrich ciliate sensu lato and support the monophyly of the subclasses Oligotrichia and Choreotrichia (Ciliophora, Spirotrichea). – J. Zool., London **260**: 87–97.

Su L. and others (1988): A preliminary investigation on the protozoa of Chongqing. – Sichuan Journal of Zoology **7**: 1–4.

Sud G.C. (1969): Observations on zoochlorellae in fresh-water ciliates. – Res. Bull. Panjab Univ. Sci. **20**: 429–441.

Sudhaus W., Rehfeld K. (1992): Einführung in die Phylogenetik und Systematik. Gustav Fischer Verlag, Stuttgart, Jena, New York. I–XII, 241 pp.

Sudzuki M. (1964a): Zur biologischen Analyse der mikroskopischen Süsswassertierwelt geringster Wassermengen I. Die in ärophytischen Moospolstern lebende Mikrofauna und ihre Veränderung. – Zool. Mag., Tokyo **73**: 163–174 (in Japanese with German summary).

Sudzuki M. (1964b): Zur biologischen Analyse der mikroskopischen Süsswassertierwelt geringster Wassermengen. II. Ein Entwurf zum theoretischen Faunenbild auf Grund der Umwandlung der Moosfaunenzusammensetzung. – Zool. Mag., Tokyo **73**: 245–250 (in Japanese with German summary).

Sudzuki M. (1965): Zur biologischen Analyse der mikroskopischen Süßwassertierwelt geringster Wassermengen III. Die in ärophytischen größeren Moospolstern lebende Mikrofauna und ihre Veränderung (im Frühjahr). – Zool. Mag., Tokyo **74**: 135–148.

Sudzuki M. (1979): On the microfauna of the Antarctic region III. Microbiota of the terrestrial interstices. – Mem. nat. Inst. Polar Res., Tokyo **Special Issue 11**: 104–126, Plates 1–10.

Sudzuki M. (1979a): Some approaches to the estimation of the biomass for microfauna communities III. Individual density and presumable biomass of the microbiota inhabiting terrestrial ecosystems in a subtropical island, named Iriomote-Jima. Biomass of Animals of terrestrial ecosystems in Japan, pp. 213–242. Report from the result of 1978. Environmental Agency, Tokyo (in Japanese with English summary).

Sudzuki M. (1992): Seasonal and local occurrences of the rotifera in Southwestern Islands of Japan with lists and notes on other microscopic leading animals. – Proc. Jap. Soc. syst. Zool. **46**: 29–70, Plates I–XXXII.

Suganuma Y. (1973): Electron microscopy of the trichocyst in *Urostyla cristata*, a hypotrichous ciliate. – J. Electron Microsc., Chiba Cy **22**: 347–352.

Suganuma Y., Inaba F. (1966): Electron microscopy of the macronucleus in interphase of *Urostyla grandis*. – Zool. Mag., Tokyo **75**: 172–177 (in Japanese with English summary).

Suganuma Y., Inaba F. (1967): Electron microscopic study of the macronuclei of *Urostyla grandis* during binary fission. – Zool. Mag., Tokyo **76**: 110–118 (in Japanese with English summary).

Suganuma Y., Yamamoto H. (1980): Occurrence, composition, and structure of mitochondrial crystals in a hypotrichous ciliate. – J. Ultrastruct. Res. **70**: 21–36.

Sun P., Song W. (2005a): Morphogenesis of the marine ciliate *Pseudokeronopsis flava* (Cohn, 1866) Wirnsberger et al., 1987 (Protozoa: Ciliophora: Hypotrichida). – Acta zool. sin., 51: 81–88 (in Chinese with English summary and figure explanations).

Sun P., Song W. (2005b): Morphogenesis of the marine ciliate *Pseudokeronopsis flava* (Cohn, 1866) Wirnsberger et al., 1987 (Protozoa: Ciliophora: Hypotrichida). – Int. Congr. Protozool., 12: 14, Abstract S1.1-13.

Sveč F. (1897): Beiträge zur Kenntnis der Infusorien Böhmens. I. Die ciliaten Infusorien des Unterpocernitzer Teiches. – Bull. int. Acad. tchéque Sci. **4**: 29–47, Tafeln I, II.

Szabó A. (1993): Protozoological (Ciliata) studies in the different subtypes of solonetz soils. – Int. Congr. Protozool. **9**: 125, Abstract 489.

Szabó A. (1995): Microfaunistical (protozoa, ciliata) research in the salinized biotopes in the Hortobagy National Park (HNP) I. Protozoological studies in salinized soils. – Debreceni Agrartudomanyi Egyetem Tudomanyos Közle **31**: 13–22 (in Hungarian with English summary).

Szabó A. (1999a): Protozoological research of alkaline biotopes in Hortobágy National Park. I. Ciliata fauna of small alkaline ponds. – Acta Biol. Debr. Oecol. Hung. **9**: 219–241 (in Hungarian with English abstract).

Szabó A. (1999b): Protozoological research of alkaline biotopes in Hortobágy National Park (HNP). II. Ciliata fauna of salinized soils. – Acta Biol. Debr. Oecol. Hung. **9**: 243–256.

Szabó A. (2000): A contribution to the ciliata (Protozoa) fauna of chernozem soils. – Miscellanea Zoologica Hungarica **13**: 11–19.

Szabó A. (2000a): Ciliaa (Protozoa, Ciliophora) species in alkaline biotopes of the Hortobágy National Park new to Hungary. – Miscellanea Zoologica Hungarica **13**: 5–10.

Szentivány M., Tirjaková E. (1994): The structure and dynamics of the community of Ciliophora in the benthos of the Karlova Ves (Bratislava) branch of the Danube. – Acta Fac. Rerum nat. Univ. comen., Bratisl., Series Zoologia **38**: 83–102.

Sztrantowicz H. (1984): Soil protozoa as indicators of environment degradation by industry. – Polish ecological Studies **10**: 67–91.

Tai L.-S. (1931): Notes on fresh water protozoa of Peiping. – Sci. Rep. natn. Tsing Hua Univ., Ser. B. Biological and Psychological Sciences **1**: 1–61, Plate I–XVI.

Takahashi T. (1972): Behavior of multiple macronuclei during the cell cycle of a hypotricha ciliate, *Hemicycliostyla sphagni*. – Zool. Mag., Tokyo **81**: 32–40 (in Japanese with English summary).

Takahashi T. (1973): Mating types and two conjugation types of *Pseudourostyla levis* sp. n. (Ciliata). – J. Sci. Hiroshima Univ., Series B. Division I. Zoology **24**: 145–163, Plates I, II.

Takahashi T. (1974): Nuclear behavior in total and temporary conjugation of *Pseudourostyla levis* (Hypotrichida, Ciliata). – J. Sci. Hiroshima Univ., Series B. Division I. Zoology **25**: 157–171.

Takahashi T. (1977): New mating types of syngen 1, with reference to stocks that induce total or temporary conjugation in *Pseudourostyla levis* (Ciliata). – J. Sci. Hiroshima Univ., Series B. Division I. Zoology **27**: 165–173.

Takahashi T. (1979): Nuclear changes of total conjugation in *Pseudourostyla* (Ciliata). – Zool. Mag., Tokyo **88**: 623, Abstract (in Japanese).

Takahashi T. (1983a): Pre-mating behavior and pairing processes in two types of total conjugation in *Pseudourostyla levis* (Ciliophora). – J. Sci. Hiroshima Univ., Series B. Division I. Zoology **31**: 1–15.

Takahashi T. (1983b): Comparison of nuclear changes between slow and rapid types of total conjugation in *Pseudourostyla levis* (Ciliophora). – J. Sci. Hiroshima Univ., Series B. Division I. Zoology **31**: 27–37.

Takahashi T. (1983c): Several features of minute micronuclei in *Pseudourostyla levis* (Ciliophora). – J. Sci. Hiroshima Univ., Series B. Division I. Zoology **31**: 39–48.

Takahashi T. (1983d): Properties of synconjugant clones from total conjugation in *Pseudourostyla levis* (Ciliophora). – J. Sci. Hiroshima Univ., Series B. Division I. Zoology **31**: 17–25.

Takahashi T. (1988): Reorganization in amicronucleates with defective mouth of the ciliate *Pseudourostyla levis*. – J. Protozool. **35**: 142–150.

Takahashi T., Suhama M. (1989): Regeneration in fragments of amicronucleates of the ciliate *Pseudourostyla levis*. – Int. Congr. Protozool. **8**: 85, Abstract 12E1630.

Takahashi T., Suhama M. (1991a): Regeneration of amicronucleate fragments in the hypotrichous ciliate *Pseudourostyla levis*. – Europ. J. Protistol. **26**: 308–318.

Takahashi T., Suhama M. (1991b): Ciliated protozoan fauna in paddy fields. – Bull. Jap. Soc. Microb. Ecol. **6**: 103–115 (in Japanese with English summary).

Tamás G., Gellért J. (1958): Parti kövek bevonatának kovamoszatai és csillósai a Tihanyi-féslsziget keleti részén (Über Diatomeen und Ciliaten aus dem Aufwuchs der Ufersteine am Ostufer der Halbinsel Tihany). – Annls Inst. biol. Tihany **25**: 241–250 (in Hungarian with German summary).

Tamás G., Gellért J. (1959): Parti kövek bevonatának kovamoszatai és csillósai a Tihanyi-félsziget déli részén (Kieselalgen und Ciliaten im Aufwuchs von Ufersteinen an dem Südufer der Halbinsel Tihany). – Annls Inst. biol. Tihany **26**: 237–245 (in Hungarian with German summary).

Tamás G., Gellért J. (1960): Adatok a Balatoni hidropszammon élovilágának ismeretéhez (Daten zur Erkennung der Lebewelt des Hydropsammons im Balaton). – Acta biol. hung. **27**: 65–73 (in Hungarian with German summary).

Taylor W.D., Shuter B.J. (1981): Body size, genome size, and intrinsic rate of increase in ciliated protozoa. – Am. Nat. **118**: 160–172.

Tchang T., Pang. Y., Zou S. (1982): A study on morphogenesis of *Urostyla cristata*. – Zool. Res. **3**: 1–9 (in Chinese with English summary).

Tent L. (1981): Der Aufwuchs im Hamburger Hafen. Struktur einer Biocoenose in einem Belastungszentrum des Elbe-Aestuars. – Arch. Hydrobiol. **Suppl. 61**: 1–58.

Tesarova P., Foissner W. (1999): New observations on the extrusomes of *Pseudourostyla cristata* (Ciliophora, Hypotrichia). – J. Euk. Microbiol. **46**: 13A, Abstract 98.

Thompson J.C.Jr. (1972): Ciliated protozoa of the Antarctic Peninsula. In: Llano G.A. (ed.). Antarctic terrestrial biology, Antarctic Res. Ser., **20**, pp. 261–288. Amercian Geophysical Union, Washington.

Tirjaková E. (1988): Structures and dynamics of communities of ciliated Protozoa (Ciliophora) in field communities. I. Species composition, group dominance, communities. – Biológia, Bratisl. **43**: 497–503.

Tirjaková E. (1992): Nálevníky (Ciliophora) v spr rojkovské raselinisko. – Ochrana prírody **year 1992**: 291–295 (in Czech).

Tirjaková E. (1992a): Ciliophora of the riverside zone of the Danube river system. – Acta Fac. Rerum nat. Univ. comen., Bratisl., Series Zoologia **36**: 71–80.

Tirjaková E. (1993): Štruktúra a dynamika populácií nálevníkov (Ciliophora) rieky Turiec. – Biológia, Bratisl. **48**: 131–139.

Tirjaková E. (2003): Micro- and meiozoobenthos with focus on ciliates (Ciliophora) of the Gidra River Basin. – Acta Zool. Univ. comen. **45**: 29–40.

Tirjaková E. (2004): Protozoans parasiting on ciliates (Protozoa: Ciliophora) in Slovakia. – Acta Zool. Univ. comen. **46**: 1–5.

Tirjaková E., Matis D. (1987a): Ciliated Protozoa (Ciliophora) from submerged, wet, moist and dry mosses in selected localities of Slovensky raj. – Acta Fac. Rerum nat. Univ. comen., Bratisl., Series Zoologia **29**: 1–16.

Tirjaková E., Matis D. (1987b): Ciliates of dry mosses in Bratislava in relation to air pollution. – Acta Fac. Rerum nat. Univ. comen., Bratisl., Series Zoologia **29**: 17–31.

Tittler I.A. (1935): Division, encystment and endomixis in *Urostyla grandis* with an account of an amicronucleate race. – Cellule **44**: 189–218.

Torch R. (1960): Nuclear division in *Holosticha lacazei* Maupas. – Biol. Bull. mar. biol. Lab., Woods Hole **119**: 343–344, Abstract.

Tucolesco J. (1962a): Écodynamique des infusoires du littoral Roumain de la mer Noire et des bassins salés para-marins. – Annls Sci. nat. (Zool.), 12e série **3**: 785–845.

Tucolesco J. (1962b): Protozoaires des eaux souterraines. I. 33 espèces nouvelles d'infusoires des eaux cavernicoles roumaines. – Annls Spéléol. **17**: 89–105.

Tucolesco J. (1962c): Études protozoologiques sur les eaux Roumaines I. Espèces nouvelles d'infusoires de la mer Noire et des bassins salés paramarins. – Arch. Protistenk. **106**: 1–36.

Tuculescu I. (1965): Biodinamica lacului Techirghiol. Biocenozele si geneza namolului. Academiie Republicii Socialiste România, Bucarest. 525 pp.

Tuffrau M. (1964): Quelques variants techniques del'imprégnation des ciliés par le protéinate d'argent. – Archs Zool. exp. gén. **104**: 186–190.

Tuffrau M. (1967): Perfectionnements et pratique de la technique d'imprégnation au protargol des infusoires ciliés. – Protistologica **3**: 91–98.

Tuffrau M. (1979): Une nouvelle famille d'hypotriches, Kahliellidae n. fam., et ses consequences dans la repartition des Stichotrichina. – Trans. Am. microsc. Soc. **98**: 521–528.

Tuffrau M. (1987): Proposition d'une classification nouvelle de l'ordre Hypotrichida (Protozoa, Ciliophora), fondée sur quelques données récentes. – Annls Sci. nat. (Zool.), 13e Série **8** (years 1986–1987): 111–117.

Tuffrau M., Fleury A. (1994): Classe des Hypotrichea Stein, 1859. – Traite de Zoologie **2**: 83–151.

Tümpling W.v., Friedrich G. (1999): Methoden der biologischen Wasseruntersuchung. Band 2 Biologische Gewässeruntersuchung. Gustav Fischer Verlag, Jena Stuttgart Lübeck Ulm. 545 pp.

Turner H.J.Jr. (1939): A new species of hypotrichous ciliate, *Urostyla coei*. – Trans. Am. microsc. Soc. **58**: 395–397.

Uhlig G. (1979): Als potentielle Futterorganismen wurden im Berichtsjahr zwei weitere Ciliatenspecies in Kultur genommen: *Euplotes* spec. und *Keronopsis rubra*. – Jber. Biol. Anst. Helgoland **1979**: 61.

Uhlig G. (1981a): Isolation and analysis of a pigment mutant of *Keronopsis rubra*, var. *albino*. – Int. Congr. Protozool. **6**: 375, Abstract.

Uhlig G. (1981b): Isolation und Analyse einer Pigment-Mutante von *Keronopsis rubra* var. *albino*. – Jber. Biol. Anst. Helgoland **1980**: 34–36.

Uhlig G. (1982): Vergleichende Untersuchungen an *Keronopsis rubra* und *K. rubra* strain "*albino*". – Jber. Biol. Anst. Helgoland **1981**: 39–44.

Uhlig G. (1983): Vergleichende Untersuchungen an *Keronopsis rubra* und einer weißen Pigmentmutante "*albino*". – Jber. Biol. Anst. Helgoland **1983**: 48–50.

Uhlig G. (1984): Isolation and comparative analysis of a pigment mutant of *Keronopsis rubra*. – J. Protozool. **31:** 57A, Abstract 208.

Uhlig G. (1989): Keronopsin, an antibiotic component of the pigment complex of *Pseudokeronopsis rubra*. – J. Protozool. Suppl. **36**: 22A, Abstract 127.

Uhlig G. (1989a): Bioassay study on the antibiotic "Keronotopsin" produced by *Pseudokeronopsis rubra*. – Int. Congr. Protozool. **8**: 81, Abstract 12D1515.

Uhlig G., Moschny S. (1988): "Keronotoxin", ein Antibiotikum aus Ciliaten-Extrakt Entwicklung eines Biotest-Verfahrens. – Jber. Biol. Anst. Helgoland **1988**: 56–57.

Vacelet E. (1960): Note preliminaire sur la faune infusorienne des "sables a amphioxus" de la baie de Marseille. – Recl Trav. Stn mar. Endoume **20**: 53–57.

Vacelet E. (1961): La faune infusorienne des "sables à amphioxus" des environs de Marseille. – Bull. Inst. océanogr. Monaco **58 (N° 1202)**: 1–12.

Vacelet E. (1961a): Les cilies de la microfaune des "sables mal cibres" des environs de Marseille. – Recl Trav. Stn mar. Endoume **22**: 13–19.

van Hoek A.H.A.M., van Alen T.A., Sprakel V.S.I., Hackstein J.H.P., Vogels G.D. (1998): Evolution of anaerobic ciliates from the gastrointestinal tract: phylogenetic analysis of the ribosomal repeat from *Nyctotherus ovalis* and its relatives. – Mol. Biol. Evol. **15**: 1195–1206.

Varga L. (1953): Die Wirkung der verschiedenen Düngervergärungsmethoden auf die Mikrofauna des Düngers. – Acta agron. hung. **3**: 343–384.

Varga L. (1959): Untersuchungen über die Mikrofauna der Waldstreu einiger Waldtypen im Bükkgebirge (Ungarn). – Acta zool. hung. **4**: 443–478.

Varga L. (1960): Über die Mikrofauna der Waldstreu einiger auf Szikböden angelegter Waldtypen. – Acta zool. hung. **6**: 211–225.

Varga L. (1962): Über die Mikrofauna der Waldstreu von zwei Waldtypen des Durmitor-gebirges (Jugoslawien). – Glasn. Muz. Beogr. Ser. B. **18**: 145–158.

Verhoeven R. (1999): Ciliaten (Protozoa) in Primärdünen. – Faun.-Ökol. Mitt. **Suppl. 26**: 61–67.

Verhoeven R. (2001): Response of soil microfauna to organic fertilisation in sandy virgin soils of coastal dunes. – Biol. Fertil. Soils **34**: 390–396.

Verhoeven R. (2001a): Küstendünen als Lebensraum für Ciliaten. Besiedlungsstruktur, Habitateinfluß und Beziehungen im Nahrungsnetz. – Forschen und Wissen - Umweltwissenschaft (Dissertation Universität Bremen, D 46), GCA-Verlag, Herdecke. 94 pp.

Verhoeven R. (2002): Ciliates in coastal dune soils of different stages of development. – Europ. J. Soil. Biol. **38**: 187–191.

Verni F., Rosati G. (2000): Protozoi ciliati del muschio come bioindicatori uno studio nell'area urbana di Pisa. – Acqua & Aria **year 2000**: 67–70.

Verschaffelt F. (1930): Bijdrage tot de kennis der nederlandsche zoet- en brakwaterprotozoën. – Bot. Jaarb. **21**: 1–199, 1 plate (in Flemish; also published as dissertation in 1929 by Erasmus, Ledeberg/Gent).

Veylande G.K., Liyepa R.A. (1985): Bacterial and protozoan benthos of small rivers of Latvia. – Hydrobiol. J. **21**: 79–86.

Vörösváry B. (1950): A "Kalános patak" csillós véglényei (Die Ciliaten des "Kalános"-Baches). – Annls biol. Univ. szeged. **1**: 343–387 (in Hungarian with Russian summary and German summary and description of the new taxa).

Voss H.J. (1989): Vergleichende morphogenetische Untersuchungen von 13 Arten der Gattung *Euplotes* (Ciliophora, Hypotrichida). – Arch. Protistenk. **137**: 331–344.

Vuxanovici A. (1961): Cercetari asupra unor infuzori dulcicoli din lacurile regiunii Bucuresti (Recherches sur les infusoires d'eau douce de la région de Bucarest). – Studii Cerc. Biol., Seria "biologie animala" **13**: 431–443 (in Rumanian with Russian and French summary).

Vuxanovici A. (1963): Contributii la studiul speciilor din subordinul Hypotricha (Ciliata) (Nota I) (Contribution à l'étude des espèces appartenant au sous-ordre Hypotricha (Ciliata) (Note I)). – Studii Cerc. Biol., Seria "biologie animala" **15**: 199–222 (in Rumanian with Russian and French summary).

Wägele J.W. (2001): Grundlagen der phylogenetischen Systematik. Pfeil, München. 320 pp.

Wailes G.H. (1943): 1. Protozoa 1f. Ciliata 1g. Suctoria. – Can. Pacif. Fauna **1**: 1–46.

Walker G.K., Maugel T.K. (1980): Encystment and excystment in hypotrich ciliates II. *Diophrys scutum* and remarks on comparative features. – Protistologica **16**: 525–531.

Walker G.K., Goode D., Maugel T.K. (1978): Critical-point dried macronuclear and micronuclear chromatin fibers of hypotrich ciliates. – Trans. Am. microsc. Soc. **97**: 340–350.

Walker I. (1968): The problem of somatic pasticity in the theory of evolution. – Arch. Julius Klaus-Stift. Vererb-Forsch. **43**: 40–51.

Walker I. (1975a): The effect of saline concentration on the dynamics of the fission rate in the marine ciliate *Keronopsis rubra* Ehrenberg (Hypotricha, Oxytrichidae). – Revue suisse Zool. **82**: 349–367.

Walker I. (1975b): Density-dependent control of fission rates in the marine ciliate *Keronopsis rubra* Ehrenberg (Hypotricha: Oxytrichidae). – J. Anim. Ecol. **44**: 707–717.

Wallengren H. (1900): Studier öfver ciliata infusorier. IV. – Acta Univ. lund. **36**: 1–54, Platta I, II.

Wallengren H. (1900a): Zur Kenntnis der vergleichenden Morphologie der hypotrichen Infusorien. – Bih. K. svensk VetenskAkad. Handl. **26**: 1–31.

Wallengren H. (1902): Zur Kenntniss des Neubildungs- und Resorptionsprocesses bei der Theilung der hypotrichen Infusorien. – Zool. Jb. Anat. **15**: 1–58, Tafel 1.

Wang C.C. (1925): Study of the protozoa of Nanking Part I. – Contr. biol. Lab. Sci. Soc. China **1**: 1–60, Plates I–XII.

Wang C.C. (1928): Ecological studies of the seasonal distribution of protozoa in a fresh-water pond. – J. Morph. **46**: 431–478.

Wang C.C. (1930): Notes on some new and rare species of hypotrichous infusoria. – Contr. biol. Lab. Sci. Soc. China, Zoological Series **6**: 9–18.

Wang C.C. (1940): Notes on some new fresh-water infusoria. – Sinensia, Shanghai **11**: 11–32.

Wang C.C., Nie D. (1932): A survey of the marine protozoa of Amoy. – Contr. biol. Lab. Sci. Soc. China, Zoological Series **8**: 285–385.

Wang C.C., Nie D. (1934): On the marine protozoa of Amoy. – Proc. Pacif. Sci. Congr. **5**: 4207–4211.

Wang J. (1977): Protozoa from some districes of Tibetan plateau. – Acta zool. sin. **23**: 131–160 (in Chinese with English summary).

Wang R., Qiu Z., Qin Y., Chen J. (2005): Observations on the morphology and morphogenesis of a new hypotrich ciliate (Ciliophora, Hypotrichida) from North China. – Int. Congr. Protozool., 12: 9, Abstract S1.1-03.

Wanick R.C., Silva-Neto I.D. Da. (2004): Benthic ciliates from Sepetiba Bay (Rio de Janeiro, Brazil) with the description of *Pseudokeronopsis sepetibensis* n. sp. (Spirotrichea: Urostylida). – Zootaxa **587**: 1–11.

Warren A., Fyda J., Song W. (2002): The morphology of the poorly-known freshwater urostylid ciliate *Neokeronopsis spectabilis* (Kahl, 1932) nov. gen. nov. comb., (Ciliophora: Urostylidae), with notes on its morphogenesis. – Europ. J. Protistol. **38**: 195–206.

Wasik A., Mikolajczyk E. (1992): The morphology and ultrastructure of the Antarctic ciliate, *Cymatocylis convallaria* (Tintinnina). – Acta Protozool. **31**: 233–239.

Webb M.G. (1956): An ecological study of brackish water ciliates. – J. Anim. Ecol. **25**: 148–175.

Webb M.G. (1961): The effects of thermal stratification on the distribution of benthic protozoa in Esthwaite Water. – J. Anim. Ecol. **30**: 137–151.

Wegl R. (1983): Index für die Limnosaprobität. – Wass. Abwass. Wien **26**: 1–175.

Weisse F. (1845): Verzeichniss von 155 in St-Petersburg beobachteten Infusorienarten, nebst Bemerkungen über dieselben. – Bull. Acad. Saint-Petersbourg **3**: 19–26.

Weisse J.F. (1848a): Drittes Verzeichniss St. Petersburgischer Infusorien. – Bull. Acad. Saint-Petersbourg **5–6**: 39–47.

Weisse J.F. (1848b): Viertes Verzeichniss St. Petersburgischer Infusorien, nebst Beschreibung zweier neuer Arten. – Bull. Acad. Saint-Pétersbourg **6**: 106–112, 1 plate.

Weisse J.F. (1848c): Fünftes Verzeichniss St. Petersburgischer Infusorien, nebst Beschreibung einer neuen Limnias. – Bull. Acad. Saint-Pétersbourg **6**: 353–364.

Weninger G. (1967): Impfungsversuche an synthetischem Abwasser. – Gesundheitsingenieur **10**: 317–320.

Wenzel F. (1953): Die Ciliaten der Moosrasen trockner Standorte. – Arch. Protistenk. **99**: 70–141.

Werner F.C. (1972): Wortelemente lateinisch-griechischer Fachausdrücke in den biologischen Wissenschaften. Suhrkamp-Taschenbuch-Verlag, Frankfurt am Main. 475 pp.

West L.S. (1953): Protozoa of the upper peninsula of Michigan. – Pap. Mich. Acad. Sci. **38**: 269–284.

West L.S., Lundin F.C. (1963): Further progress in the study of Michigan protozoa. – Pap. Mich. Acad. Sci. **48**: 87–111, Plates I, II.

Westheide W., Rieger R. (1996): Spezielle Zoologie. Erster Teil: Einzeller und wirbellose Tiere. G. Fischer Verlag, Stuttgart Jena New York. 909 pp.

Wetzel A. (1925): Vergleichend cytologische Untersuchungen an Ciliaten. – Arch. Protistenk. **51**: 209–304.

Wetzel A. (1928): Zur Frage der Verunreinigung des Bodenseewassers durch den Zufluß der Schussen und der Möglichkeit ihrer Beurteilung durch die Beobachtung der Protozoenbiocönosen. – Int. Revue ges. Hydrobiol. Hydrogr. **19**: 217–260.

Wetzel A. (1928a): Der Faulschlamm und seine ziliaten Leitformen. – Z. Morph. Ökol. Tiere **13**: 179–328.

Wiackowski K. (1984): Un nouvel hypotriche dans le s. o. Urostylina. – J. Protozool. **31**: 77A, Abstract 279.

Wiackowski K. (1984a): Nouvelles données sur la morphogenèse d'*Urostyla*. – J. Protozool. **31**: 77A, Abstract 278.

Wiackowski K. (1985): The morphology and morphogenesis of *Keronella gracilis* n. gen., n. spec. (Hypotrichida, Ciliophora). – Protistologica **21**: 81–91.

Wiackowski K. (1988): Phenetic and cladistic numerical estimates of phylogenetic relationships in Urostylina (Ciliophora: Hypotrichida). – Acta Protozool. **27**: 1–20.

Wiackowski K. (1988a): Morphology and morphogenesis of a new species in the genus *Pseudourostyla* (Hypotrichida, Ciliophora). – J. nat. Hist. (London) **22**: 1085–1094.

Wiackowski K. (1991): *Urostyla thompsoni* Jankowski, 1979 a little known marine ciliate (Ciliophora, Hypotrichida). – Acta Protozool. **30**: 55–60.

Wickham S., Gieseke A., Berninger U.-G. (2000): Benthic ciliate identification and enumeration: an improved methodology and its application. – Aquatic Microbial Ecology **22**: 79–91.

Wicklow B.J. (1979): *Epiclintes ambiguus* and *Kerona polyporum*: comparative ultrastructure, cortical morphogenesis and systematics of two hypotrich ciliates. – J. Protozool. Suppl. **26**: 16A, Abstract.

Wicklow B.J. (1981): Evolution within the order Hypotrichida (Ciliophora, Protozoa): ultrastructure and morphogenesis of *Thigmokeronopsis johadai* (n. gen., n. sp.); phylogeny in the Urostylina (Jankowski, 1979). – Protistologica **17**: 331–351.

Wicklow B.J. (1982): The Discocephalina (n. subord.): ultrastructure, morphogenesis and evolutionary implications of a group of endemic marine interstitial hypotrichs (Ciliophora, Protozoa). – Protistologica **18**: 299–330.

Wicklow B.J. (1983): Morphogenetic and ultrastructural investigation in the Hypotrichida (Ciliophora, Protozoa): implications in ciliate evolution. – Diss. Abstr. Int. **43B**: 2135, Abstract.

Wicklow B.J. (1997): Signal-induced defensive phenotypic changes in ciliated protists: morphological and ecological implications for predator and prey. – J. Euk. Microbiol. **44**: 176–188.

Wicklow B.J., Borror A.C. (1990): Ultrastructure and morphogenesis of the marine epibenthic ciliate *Epiclintes ambiguus* (Epiclintidae, n. fam.; Ciliophora). – Europ. J. Protistol. **26**: 182–194.

Wicklow B.J., Borror A.C. (1990a): *Epiclintes* (Ciliophora): phylogeny uncertain based on ultrastructure and morphogenesis. – J. Protozool. Suppl. **37**: 7A, Abstract 42.

Wilbert N. (1969): Ökologische Untersuchung der Aufwuchs- und Planktonciliaten eines eutrophen Weihers. – Arch. Hydrobiol. Suppl. **35**: 411–518.

Wilbert N. (1975): Eine verbesserte Technik der Protargolimprägnation für Ciliaten. – Mikrokosmos **64**: 171–179.

Wilbert N. (1977): Une méthode simple d'étude expérimentale de l'écologie des ciliés libres. – Année biol. **16**: 545–562.

Wilbert N. (1986): Ciliates from saline lakes in Saskatchewan. – Symposia Biologica Hungarica **33**: 249–256.

Wilbert N. (1995): Benthic ciliates of salt lakes. – Acta Protozool. **34**: 271–288.

Wilbert N., Kahan D. (1981): Ciliates of Solar Lake on the Red Sea Shore. – Arch. Protistenk. **124**: 70–95.

Wilbert N., Song W. (2005): New contributions to the marine benthic ciliates from the Antarctic area, including description of seven new species (Protozoa, Ciliophora). – J. nat. Hist. (London) **39**: 935–973.

Winston J.E. (1999): Describing species. Practical taxonomic procedure for biologists. Columbia University Press, New York.

Winterbourn M.J., Brown T.J. (1967): Observations on the faunas of two warm streams in the Taupo thermal region. – N.Z. J. mar. Freshwater Res. **1**: 38–50.

Wirnsberger E. (1987): Division and reorganization in the genus *Pseudokeronopsis* and relationships between urostylids and oxytrichids (Ciliophora, Hypotrichida). – Arch. Protistenk. **134**: 149–160.

Wirnsberger E., Foissner W. (1987): Morphologie von *Holosticha xanthichroma* sp. n. und die Variabilität der Infraciliatur in der Gattung *Holosticha* (Ciliophora, Hypotrichida). – Acta Protozool. **26**: 1–8.

Wirnsberger E., Hausmann K. (1988a): Morphological, morphometric, and fine structural investigation of some species of the genus *Pseudokeronopsis* (Ciliophora, Hypotrichida). – Society of Protozoologists **1987 Abstracts**: Abstract 128.

Wirnsberger E., Hausmann K. (1988b): Fine structure of *Pseudokeronopsis carnea* (Ciliophora, Hypotrichida). – J. Protozool. **35**: 182–189.

Wirnsberger E., Foissner W., Adam H. (1985): Cortical pattern in non-dividers, dividers and reorganizers of an Austrian population of *Paraurostyla weissei* (Ciliophora, Hypotrichida): a comparative morphological and biometrical study. – Zool. Scr. **14**: 1–10.

Wirnsberger E., Larsen H.F., Uhlig G. (1987): Rediagnosis of closely related pigmented marine species of the genus *Pseudokeronopsis* (Ciliophora, Hypotrichida). – Europ. J. Protistol. **23**: 76–88.

Wirnsberger E., Foissner W., Foissner I. (1987a): Morphogenesis and fine structure of *Engelmanniella mobilis* (Ciliophora, Hypotrichida). – European Conference on Ciliate Biology **6**: 58, Abstract.

Wirnsberger-Aescht E., Foissner W., Foissner I. (1989): Morphogenesis and ultrastructure of the soil ciliate *Engelmanniella mobilis* (Ciliophora, Hypotrichida). – Europ. J. Protistol. **24**: 354–368.

Wise B.N. (1965): The morphogenetic cycle in *Euplotes eurystomus* and its bearing on problems of ciliate morphogenesis. – J. Protozool. **12**: 626–648.

Woodruff L.L. (1921): Micronucleate and amicronucleata races of infusoria. – J. exp. Zool. **34**: 329–337.

Wright A.-D.G., Dehority B.A., Lynn D.H. (1997): Phylogeny of the rumen ciliates *Entodinium*, *Epidinium* and *Polyplastron* (Litostomatea: Entodiniomorphida) inferred from small subunit ribosomal RNA sequences. – J. Euk. Microbiol. **44**: 61–67.

Wrzesniowski A.-O. (1861): Observations sur quelques infusoires. – Annls Sci. nat. (Zool.), 4ᵉ Série **16**: 327–336, Planche 8, 9.

Wrześniowski A. (1877): Beiträge zur Naturgeschichte der Infusorien. – Z. wiss. Zool. **29**: 267–323, Tafeln XIX–XXI.

Wrzesniowskiego A. (1866): Spis wymoczków spostrzeganych w Warszawie i jéj okolicach w latach 1861 do 1865 (Verzeichnis der Infusorien, welche in Warschau und seiner Umgebungen von 1861 – 65 gesammelt wurden). – Wykaz Szkoty Glownej Warszawskiej **No. 5**: 15–28 (in Polish with Latin diagnosis of the new species; German translation of the title from Bütschli 1889).

Wrzesniowskiego A. (1867): Przyczynek do historyi naturalnéj wymoczków. Rozprawa napisana przez Augusta Wrzesniowskiego w celu uzyskania stopnia Doktora Nauk Przyrodzonych w Warszawskiéj Szkole Glównéj, Kraków (Dissertation in Polish; text and figures identical with Wrzesniowskiego 1867). 116 pp + Tablica I–VII.

Wyatt C.E., Pearson T.H. (1982): The Loch Eil project: population characteristics of ciliate protozoans from organically enriched sea-loch sediments. – J. exp. mar. Biol. Ecol. **56**: 279–303.

Xu M., Wood B. (1999): Preliminary study of protozoa of Lough Neagh, Northern Ireland. – Proc. R. Ir. Acad., Series B. Biological, Geological and Chemical Sciences **99B**: 103–108.

Xu M., Cao H., Xie P., Deng D., Feng W., Xu J. (2005): The temporal and spatial distribution, composition and abundance of Protozoa in Chaohu Lake, China: relationship with eutrophication. – Europ. J. Protistol., 41: 183–192.

Yakimoff W.L., Zérèn S. (1924): Contribution à l'étude des protozoaires des sols de russie. 1-re communication. Les protozoaires du sol de Pétrograde et du gouvernement Pétrograde. – Zentbl. Bakt. Parasitkde, Abt. III. **63**: 33–57.

Yang H. (1989): A winter sarvey of the protozoa from fresh waters of the Yuelushan area. – Acta Sci. nat. Univ. Normalis Hunanensis **12**: 151–157 (in Chinese with English summary).

Yankovskij A.V. (1978): The revision of the system of Polyhymenophora class (Spirotrichia). – Tezisy Dokl. zool. Inst. Akad. Nauk SSSR **year 1978**: 39–40 (in Russian; translation of autor, title, and source after Zoological Record, 116, year 1979, p. 144; = Jankowski 1978).

Yasuzumi G., Suganuma Y. (1970): Electron microscope studies on the mitosis of binary fission of micronucleus of the ciliata, *Urostyla*. – J. Electron Microsc., Chiba Cy **19**: 308, Abstract.

Yasuzumi G., Suganuma Y., Onishi T., Kamino T., Matsumoto K. (1972): Electron microscope studies of the cilia of *Urostyla cristata* Jerka-Dziadosz. – Monitore zool. ital., (N. S.) **6**: 1–9.

Yongue W.H.Jr., Cairns J.Jr. (1976): Comparison between numbers and kinds of freshwater protozoans colonizing autoclaved and unautoclaved polyurethane foam substrates. – Appl. environm. Microbiol. **31**: 750–753.

Yongue W.H.Jr., Cairns J.Jr. (1979): Effects of Radford Army ammunition plant wastes on protozoan communities in the New River, Virginia. - Riv. Biol. **11**: 65 78.

Zaika V.E. (1970): Rate of reproduction in infusoria. In: Vodyanitzkii V.A. (ed.). Biology of the sea. 19. Production and food chains in plankton communities, pp. 96–105. Akademiya Nauk Ukrainskori SSR. Naukova Dumka, Kiev (in Russian).

Zaika V.E. (1988): Duration of cytokinesis related to cell weight in ciliates. – Tsitologiya **30**: 757–761 (in Russian with English summary).

Zhang Z. (1984): Macronucleata development of *Urostyla cristata* during the asexual reproduction. – Chin. J. Zool. **year 1984**: 1–2 (in Chinese).

Zhang Z.R., Zhuang H., Jin L.P. (1985): Studies on conjugation in *Pseudourostyla cristata*. II. Morphogenesis of cortical structures. – Zool. Res. **6**: 79–92 (in Chinese with English summary).

Zhang Z., Pang Y., Gu F. (1985a): Studies on the morphology and morphogenesis of *Keronopsis rubra*. – Acta zool. sin. **31**: 59–64, Plates I, II (in Chinese with English summary).

Zhang Z., Zhuang H., Jin L. (1985b): Studies on conjugation in *Pseudourostyla cristata* II. Morphogenesis of cortical structures. – Zool. Res. **6**: 79–92 (in Chinese with English summary).

Zharikov V.V. (1982): Study of the free-living infusoria in some waterbodies of Armenia. – Biol. Zh. Armenii **35**: 910–913, Plates I–III (in Russian with Armenian and English summary).

Zou S.F., Ng S.F. (1991): Physiological reorganization and post-traumatic regeneration in *Stylonychia mytilus*: reflections on developmental constraints and the evolutionary origin of alternative modes of asexual morphogenesis. – J. Protozool. **38**: 182–192.

Zschokke F. (1900): Die Tierwelt der Hochgebirgsseen. – Neue Denkschr. allg. schweiz. Ges. ges. Naturw. **37**: 1–400, 8 Tafeln, 4 Karten.

Systematic Index

The index contains all names of hypotrichous ciliates mentioned in the book, including vernacular names, for example, amphisiellids; food organisms and other ciliates, except for those mentioned in the general section, are not included. The index is two-sided, that is, species appear both with the genus-group name first (for example, *Holosticha pullaster*) and with the species-group name first (*pullaster, Holosticha*). Valid (in my judgement) urostyloid species and genera are in ***boldface italics*** print. Invalid taxa, junior homonyms, synonyms, outdated combinations, incorrect spellings, nomina nuda, and non-urostyloid taxa are given in *italics*. Homonyms are not included separately. The scientific name of a subgenus, when used with a binomen or trinomen, must be interpolated in parentheses between the genus-group name and species-group name (ICZN 1999, Article 6.1). In the following index, these parentheses are omitted to simplify electronic sorting. Thus, the name *Holosticha (Keronopsis) pernix* is listed as *Holosticha Keronopsis pernix*. Note that this name is also listed under "*Keronopsis pernix, Holosticha*" and "*pernix, Holosticha Keronopsis*". Suprageneric taxa are represented in normal type, valid urostyloid ones in **boldface. Boldface page number** indicates the location of a valid taxon in the systematic section. "T" indicates the location of the table with the morphometric characterisation; "K" indicates the location where the name is mentioned in a key.

Table Index

Monographiae Biologicae

Series Editor: **Prof. Henri J. Dumont**
State University of Ghent, Institute of Animal Ecology, Ghent, Belgium,
henri.dumont@rug.ac.be

Published titles within the series:

Volume 85 **Monograph of the Urostyloidae (Ciliophora, Hypotricha)**
Berger H.
2006, 1303 p., Hardcover, ISBN 1-4020-5272-3

Volume 84 **Diapause in Aquatic Invertebrates**
Alekseev, Victor A.
Forthcoming, Hardcover

Volume 83 **Bridging Divides - Maritime Canals as Invasion Corridors**
Gollasch, Stephan, Galil, Bella S., and Andrew N. Cohen
2006, 315 p., Hardcover, ISBN: 1-4020-5046-1

Volume 82 **Biogeography and Ecology of Bulgaria**
Fet, V.
Forthcoming, Hardcover, ISBN: 1-4020-4417-8

Volume 81 **Monograph of the Spathidiidae**
Foissner, W., and Xu, K.
2006, 486 p., Hardcover, ISBN: 1-4020-4210-8

Volume 80 **Nouragues**
Dynamics and Plant-Animal Interactions in a Neotropical Rainforest
Bongers, F.; Charles-Dominique, P.; Forget, P.-M.; Théry, M. (Eds.)
2002, 456 p., Hardcover, ISBN: 1-4020-0123-1

Volume 79 **Environmental Change and Response in East African Lakes**
Lehman, J.T. (Ed.),
1998, 250 p., Hardcover, ISBN: 0-7923-5118-5

Volume 78 **Monograph of the Oxytrichidae (Ciliophora, Hypotrichla)**
Berger, H.
1999, 1092 p., Hardcover, ISBN: 0-7923-5795-7

Volume 77 **The Pantanal of Poconé**
Biota and Ecology in the Northern Section of the World's Largest
Pristine Wetland
Heckman, C.W.
1998, 592 p., Hardcover, ISBN: 0-7923-4863-X

Volume 76 **An Introduction to Saline Lakes on the Qinghai-Tibet Plateau**
Zheng Mianping
1997, 328 p., Hardcover, ISBN: 0-7923-4098-1